1996
WINTER SIMULATION CONFERENCE PROCEEDINGS

1996
WINTER SIMULATION CONFERENCE PROCEEDINGS

Edited By

JOHN M. CHARNES
University of Kansas

DOUGLAS J. MORRICE
University of Texas

DANIEL T. BRUNNER
Systemflow Simulations, Inc.

JAMES J. SWAIN
University of Alabama in Huntsville

Hotel Del Coronado
Coronado, California

8–11 December 1996

Additional copies can be obtained from:

Association for Computing Machinery
Order Department
P.O. Box 12114
Church Street Station
New York, NY 10257
Inside U.S.A. and Canada: (800) 342-6626
Outside U.S.A. and Canada: (212) 626-0500
FAX: (212) 944-1318

The Institute of Electrical and Electronics Engineers
Customer Service Department
445 Hoes Lane
P. O. Box 1331
Piscataway, NJ 08855-1331
U.S. and Canada: (800) 678-IEEE
International: (908) 981-0060
FAX: (908) 981-9667

The Society for Computer Simulation, International
P. O. Box 17900
San Diego, CA 92117-7900
(619) 277-3888
FAX: (619) 277-3888

ISBN	0-7803-3383-7	casebound
	0-7803-3384-5	microfiche
IEEE Catalog Number	96CB35957	
Library of Congress Number	87-654182	

Contents

Advanced Tutorials

State-of-the-Art Reviews

Analysis Methodology

Military Applications

Science of Military Simulation

After Action Reviews

Military Medical Simulation

Operations and Effectiveness

Distributed Interactive Simulation

Military Application of Parallel and Object-Oriented Simulation

Manufacturing Applications

Preface

From the Editors

It seems to me the tide of destiny is moving steadily in our favor,
though the voyage will be long and rough.

—Winston S. Churchill

The 1996 Winter Simulation Conference continues a tradition established nearly three decades ago to bring together individuals interested in simulation to share developments in the field and to provide illustrations of its application and effectiveness. Originally called the "Conference on the Applications of Simulation Using GPSS," the Conference grew quickly to cover all aspects of discrete-event and combined continuous and discrete-event simulation.

Although the idea of simulation was hardly new even in 1967, the development of programs such as GPSS and SIMSCRIPT specifically to facilitate the building of models had made simulation practical in a wide variety of settings. From its beginning, the conference has been about both the methodology and the applications of simulation. Along the way, product tutorials and exhibits were added to the conference as commercial enterprises arose to provide an increasing array of simulation products and services.

The 1996 Conference has the theme of "Surfing the Future," and is being held in Coronado, California at the historic del Coronado Hotel. The locale is suggestive of sun and surf of course, but the theme is also meant to evoke the juxtaposition of "surfing the web" and the inherent ability of simulation to help decision makers select among the myriad ways the future might evolve. Regardless of whether the alternative realities do exist that physicists and science fiction writers love to consider, simulation always makes it possible to create worlds that do not yet exist, in order to bring about worlds that we would like to exist. Whether that reality is a more-robust business process or a more-efficient telecommunication system, it is upon these realities that the rest of culture is made possible.

This year's conference is divided into the traditional groupings of tutorial, application, and methodology tracks. The tutorial tracks offer introductory tutorials for the newcomer to simulation, advanced tutorials for the experienced user, and software/modelware tutorials as a venue for showcasing commercial products and services. The latter tutorials allow prospective users to view the tools of simulation for the first time, and experienced users to learn about new tools and enhancements to existing tools. The application tracks provide examples of simulation in all aspects of manufacturing, military applications, and general applications such as transportation, medical and other service systems, civil and construction, and the environment. As an extension of these applications we offer the "Simulation Pipeline" track containing minitracks in transportation and education. Featured in the Simulation Pipeline is a special session entitled "Simulation Works: Simulation Success Stories," included expressly to demonstrate the effectiveness and importance of simulation in real world applications. Finally, methodology tracks in analysis, state-of-the-art reviews, and modeling methodology round out the program with the theory for the analysis of systems, the statistical methods for building and making use of simulation results, and recent developments that may in time lead to faster, more effective simulation models and modeling interfaces.

Of course, as central as the program is to the Conference, it is worth noting that this conference is above all a gathering of diverse people whose interaction is the *raison d'être* for the meeting. So, while you are here, take advantage of what the program offers in the *Proceedings* and in the sessions themselves, in the exhibits area, and in the other offerings of the conference. But remember that the conference is more than just the program—be sure also to take advantage of the chance to network with others in the field, to compare notes, see what's in development for next year, and what's new both in application and in commercial products. The combination of practitioners, researchers, and commercial vendors is relatively rare in this or any other field and is one of the particular strengths of the conference. The field is ultimately only as strong as these three pillars.

A conference of this size and scope takes the work of many dedicated volunteers whose efforts span several years of planning and preparation. The WSC'96 Conference Committee includes ourselves, Kerim Tumay, Catherine Drury, Deborah Kotlarek, Randall Gibson, and David Graehl. Serving on the Program Committee were Diane Bischak, Barry Nelson, Christos Alexopoulos, Laurel Travis, Mike Taaffe, Russell Barton, John Miller, Gerald (Jerry) Evans, Phillip Farrington, Dean Hartley, Harriet Nembhard, Mani Manivannan, and Chuck White. Providing invaluable support to the program committee was the work of dozens of referees, whose names are listed on page xxviii of this volume. We also acknowledge the invaluable support of the WSC Board, particularly our Board Liaison, Jim Palmer, who was always there to listen to our questions and to provide a "reality check" when needed.

The General Chair must oversee everything from site selection and hotel arrangements to marketing, finance, publishing, and registration. Words of experience from recent past General Chairs David Kelton, Bob Crain, Deb Sadowski, and Bill Lilegdon were a big help. Thanks also go to the many other people in the simulation community who stepped up at various times to offer advice or assistance to the General Chair, most notably Jerry Banks, Jim Henriksen, Averill Law, Van Norman, Tom Schriber, Mike Thompson, and Jim Wilson. Co-workers at Systemflow Simulations, particularly Joe Brill, had to put up with the demands of WSC on the General Chair's time over an extended period and their tolerance is appreciated. Finally, without the patience and understanding of Holly Brunner, who spent four years listening to tales from the front and offered encouragement at every turn plus occasional assistance with various WSC tasks, the General Chair's job would have been much more difficult.

The Program Chair oversees the construction and organization of the program, and is heavily involved in the publishing of the programs and the *Proceedings*. He is grateful for the assistance and advice from Dave Goldsman, Barry Nelson, and James R. Wilson. Efforts above and beyond his duties of Track Coordinator by Phil Farrington, and help with the initial data entry by Tim Blanton and Brad Messer, are also appreciated. We are indebted to the understanding and support of Avi Madan-Swain who endured not only conference preparations, but a change of residence and a new job during this period. Finally, we acknowledge support from the Department of Industrial and Systems Engineering at UAH, and the understanding of the faculty and students during the worst "crunch periods" of the summer and fall.

The *Proceedings* Co-Editors would like to thank Suzanne Charnes and Deanna Morrice for their precious assistance on this project and their extreme patience and understanding during those times when the project seemed to take over our lives. We would also like to thank Dorothy Embers, Anne Madden, and Karla Wallace of The University of Kansas School of Business Faculty Services, and Stan Walker for assistance with the compilation of the information contained in this *Proceedings*. Schlumberger Austin Product Center provided the superb assistance of LaVerne Schuler as well as the use of their services and facilities, all of which are greatly appreciated.

We are particularly indebted to James R. Wilson for constant support and encouragement, sagacious advice, and a seemingly inexhaustible source of experience and expertise from his own years of service to the conference. More than once our conversations took a detour to find out what Jim did in a similar situation or to get the background on some matter long ago adopted by the conference. Most of all, we are grateful to Jim for putting together a series of articles about the conference in *OR/MS Today* (which appears on our web page as well), *Simulation*, *IIE Solutions* and *Simulation Digest*. Jim drew upon his broad knowledge of the history of the conference and the themes he developed as Program Chair for the 1992 (Twenty Fifth Anniversary) Conference.

A successful Winter Simulation Conference depends on the quality work of many paid suppliers. The list is too long to name them all but we would like to acknowledge the consistently upbeat attitude and unflagging efforts of the entire staff of conference management company Talley Management Group and in particular Todd Tate and Maggie Devine-Pennock in the meetings department, who worked closely and effectively with the WSC '96 conference committee all year.

We are also grateful for the can-do spirit and attention to detail of webmasters Peter Welch and Josephine van der Hoeven. On the subject of the World Wide Web, 1996 brought the adoption of a permanent Winter Simulation Conference domain name. Anyone interested in WSC who wants to pay a mid-year visit to the conference web site to check the latest call for papers, program or registration information, or other WSC-related news need only remember the URL:

www.wintersim.org

We thank all attendees for their participation. We hope you enjoy a productive 1996 Winter Simulation Conference.

John M. Charnes	Douglas J. Morrice	Daniel T. Brunner	James J. Swain
Proceedings Co-Editor	*Proceedings* Co-Editor	General Chair	Program Chair

About the Editors

JOHN M. CHARNES is Associate Professor in the School of Business at The University of Kansas in Lawrence. He received a B. Civil Eng., MBA, and Ph.D. in Business Administration from the University of Minnesota. His research interests are in the areas of applied statistics, time series, and simulation of stochastic systems. He is a member of INFORMS and the INFORMS College on Simulation, for which he is currently Secretary/Treasurer. He is Co-Editor of the *Proceedings of the 1996 Winter Simulation Conference.*

DOUGLAS J. MORRICE is Associate Professor in the Department of Management Science and Information Systems at The University of Texas at Austin. He is currently on sabbatical leave in the End to End Simulation Department at Schlumberger Austin Research Center in Austin, Texas. Dr. Morrice received his undergraduate degree in Operations Research at Carleton University in Ottawa, Canada. He holds an M.S. and a Ph.D. in Operations Research and Industrial Engineering from Cornell University. His research interests include discrete event and qualitative simulation modeling and the statistical design and analysis of large scale simulation experiments. Dr. Morrice is a member of the Institute for Operations Research and Management Science (INFORMS) and the Council of Logistics Management. He served as the Secretary for the INFORMS College on Simulation (1994-1996) and is Co-Editor of the *Proceedings of the 1996 Winter Simulation Conference.*

DANIEL T. BRUNNER is President of Systemflow Simulations, Inc., a services firm active in manufacturing, material handling, distribution, transportation, health care, computer systems, and mining. He received a B.S.E.E. from Purdue University and an MBA from The University of Michigan. He has served as Winter Simulation Conference Publicity Chair (1988) and Business Chair (1992), and is the General Chair for WSC'96. He is a member of the Institute of Industrial Engineers and the Society for Computer Simulation.

JAMES J. SWAIN is Associate Professor of Industrial and Systems Engineering at the University of Alabama in Huntsville. He received a B.A. in Liberal Arts, B.S. in Engineering Science and M.S. in Mechanical Engineering from the University of Notre Dame, and a Ph.D. in Industrial Engineering from Purdue University. His research interests include applied statistical and simulation modeling. He has served the INFORMS College on Simulation as *Newsletter* Editor and Secretary/Treasurer, and has been active in the Winter Simulation Conference, including service as *Proceedings* Editor for WSC'92, Program Chair for WSC'96, and the representative of the American Statistical Association to Board of the WSC.

About the Conference

The year 1996 is the twenty-ninth anniversary of the Winter Simulation Conference, a conference that brings together academics, practitioners, and vendors in simulation and related fields. Actually, it is the twenty-eighth conference because the WSC skipped a couple of years in the mid-1970's. The scope of the WSC encompasses all aspects of discrete-event simulation, combined discrete-continuous simulation, and gaming. In what follows, we outline the organizational structure of the WSC, and give some historical data.

The conference itself is sponsored by a group of nine professional organizations. A member from each of the sponsoring organizations serves on the WSC Board of Directors, whose function is to assure that the conference runs smoothly from year to year. The Board selects members from the simulation community to serve on a Conference Committee for each WSC. A Conference Committee consists of a General Chair (who runs that particular conference), a Program Chair (who is responsible for the program of presentations), *Proceedings* Co-Editors, as well as other important functionaries. The Program Chair forms a Program Committee consisting of Track Coordinators, each of whom is responsible for the organization of a number of presentation sessions during the conference. Each Track Coordinator arranges for referees to review the individual papers that are submitted (or invited) for inclusion in the WSC.

Sponsoring Organizations

The Winter Simulation Conference is sponsored by a wide and varied group of professional organizations. Each of the following societies contributes in important ways to the ongoing success of the WSC.

American Statistical Association (ASA)

Association for Computing Machinery/Special Interest Group on Simulation (ACM/SIGSIM)

Institute for Operations Research and the Management Sciences (INFORMS)

Institute for Operations Research and the Management Sciences/College on Simulation (INFORMS/CS)

Institute of Electrical and Electronics Engineers/Computer Society (IEEE/CS)

Institute of Electrical and Electronics Engineers/Systems, Man, and Cybernetics Society (IEEE/SMCS)

Institute of Industrial Engineers (IIE)

National Institute of Standards and Technology (NIST)

The Society for Computer Simulation, International (SCS)

WSC Board of Directors

The Board of Directors is responsible for the year-to-year continuity and organization of the Winter Simulation Conference. It also selects the members of the Conference Committee for each WSC. The Board consists of one member from each of the nine sponsoring organizations, and has four rotating officers: Chair, Vice Chair, Secretary, and Liaison to each year's conference.

ACM/SIGSIM
Herbert D. Schwetman
Mesquite Software, Inc.
3925 West Baker Lane
Austin, TX 78759–5321
(512) 305–0080

ASA
James J. Swain
Department of ISE
University of Alabama in Huntsville
Huntsville, AL 35899
(205) 895–6749

IEEE/CS
Sallie V. Sheppard
Office of Associate Provost
104 Academic Bldg.
Texas A&M University
College Station, TX 77843
(409) 845–3210

IEEE/SMCS
Board Chair
WSC'96 Board Liaison
James D. Palmer
860 Cashew Way
Fremont, CA 94536
(510) 794–0367

IIE
Robert E. Shannon
Department of Industrial Engineering
Texas A&M University
College Station, TX 77843
(409) 845–0501

INFORMS/CS
Board Secretary
W. David Kelton
Dept. of Quant. Analysis & Operations Management
College of Business Administration
University of Cincinnati
Cincinnatti, OH 45221–0130
(513) 556–0404

Stephen D. Roberts
Department of Industrial Engineering
Box 7906
North Carolina State University
Raleigh, NC 27695-7906
(919) 515–2362

NIST
Albert Jones
National Institute of Standards and Technology
Manufacturing Engineering Laboratory
Building 220
Room A127
Gaithersburg, MD 20899
(301) 975–3354

SCS
Board Vice Chair
Brian W. Unger
Department of Computer Science
University of Calgary
Calgary, AB T2N 1N4
CANADA
(403) 220–6316 (6778)

WSC'96 Conference Committee

A Conference Committee organizes and runs each WSC. It is headed by the General Chair, who is the chief overseer of the conference. The Program Chair is responsible for the program of presentations, including the selection of Track Coordinators. The *Proceedings* Co-Editors are charged with producing the document containing the papers submitted by the WSC's presenters. Other Conference Committee positions include the Associate General Chair, the Associate Program Chair, as well as Business, Publicity, Exhibits, Local Arrangements, and Audio/Visual Chairs. The WSC Board Liaison serves as a link between the Conference Committee and the WSC Board of Directors.

General Chair
Daniel Brunner
Systemflow Simulations, Inc.
6366 Guilford Avenue, Suite 310
Indianapolis, IN 46220
(317) 466–0664
brunner@systemflow.com

Associate General Chair (WSC97 General Chair)
David Withers
LEXIS-NEXIS
P. O. Box 933
Dayton, OH 45401
(513) 865–1912
david.withers@lexis-nexis.com

Program Chair
James J. Swain
Lakeside Drive and South Loop Road
Dept of Industrial and Systems Engineering, EB 236
The University of Alabama in Huntsville
Huntsville, AL 35899–0001 U.S.A.
(205) 895–6749
jswain@ebs330.eb.uah.edu

Associate Program Chair
Barry L. Nelson
Department of Ind. Eng. and Management Sciences
2225 North Campus Drive
Northwestern University
Evanston, IL 60208–3119 U.S.A.
(708) 491–3747
nelsonb@random.iems.nwu.edu

Business Chair
Kerim Tumay
CACI Products Company
3333 North Torrey Pines Court
La Jolla, CA 92037
(619) 457–9681
kerim@caciasl.com

***Proceedings* Co-Editors**
John M. Charnes
U. of Kansas School of Business
345 Summerfield Hall
Lawrence, KS 66045–2003 U.S.A.
(913) 864–7591
john.charnes@ukans.edu

Douglas J. Morrice
MSIS Department, CBA 5.202
The University of Texas at Austin
Austin, TX 78712–1175
(512) 471–7857
morrice@mail.utexas.edu

Publicity Chair
Catherine Drury
Micro Analysis & Design, Inc.
4900 Pearl East Circle, Suite 201E
Boulder, CO 80301
(303) 442–6947
cdrury@madboulder.com

Exhibits Chair
Deborah Kotlarek
HK Systems, Inc.
2855 S. James Drive
New Berlin, WI 53151
(414) 860–6462
debbie.kotlarek@hksystems.com

Local Arrangements Chair
Randall Gibson
Automation Associates, Inc.
512 Via de la Valle, Suite 209
Solana Beach, CA 92075
(619) 481–6050
rgibson@aaism.cts.com

Audio/Visual Chair
David Graehl
Ann Arbor Computer
1201 E. Ellswsorth
Ann Arbor, MI 48108
(313) 973–7878
jbwebb!aac!dwg@jwebbus.attmail.com

WSC '96 Program Structure and Track Coordinators

The Program Committee is headed by the Program Chair. The *Proceedings* Co-Editors perform all duties related to the timely publication of the *Proceedings*. The Program Chair divides the WSC program into a number of tracks, and then selects Track Coordinators to head the tracks. Each track consists of a number of sessions; sessions typically have from one to three presentations, some of which are invited. The Track Coordinators solicit referees to review the submitted papers and are instrumental in maintaining the high quality of the presentations.

Introductory Tutorials
Professor Diane P. Bischak
Department of Business Administration
University of Alaska Fairbanks
Fairbanks, AK 99775-6080
(907) 474-6539
ffdpb@aurora.alaska.edu

Advanced Tutorials
Professor Barry L. Nelson
Department of Industrial Engineering and Management Sciences
McCormick School of Engineering and Applied Science
2225 N. Campus Drive
Northwestern University
Evanston, IL 60208-3119 U.S.A.
(708) 491-3747
nelsonb@primal.iems.nwu.edu

State of the Art Reviews
Professor Christos Alexopoulos
School of Industrial and Systems Engineering
Georgia Institute of Engineering
Atlanta, GA 30332-0205 U.S.A.
(404) 894-2361
christos@isye.gatech.edu

Software/Modelware Tutorials (2 Tracks)
Professor Laurel Travis
Metropolitan State University
730 Hinnepin Ave.
Minneapolis, MN 55403-1897 U.S.A.
(612) 341-7462
travis@iems.nmu.edu

Professor Michael Taaffe
Department of Operations and Management Science
Carlson School of Business
University of Minnesota
271 19th Ave., S
Minneapolis, MN 55455 U.S.A.
(612) 626-7724
taaffe@ux.acs.umn.edu

Analysis Methodology
Professor Russell R. Barton
Department of Industrial and Management Systems Engineering
The Pennsylvania State University
207 Hammond Building
University Park, PA 16802 U.S.A.
(814) 863-7289
barton@simplex.psu.edu

Modeling Methodology
Professor John A. Miller
Department of Computer Science
415 Graduate Studies Research Center
University of Georgia
Athens, GA 30602-7404 U.S.A.
(706) 542-3440
jam@cs.uga.edu

Distributed and Parallel Simulation Minitrack
Rajive Bagrodia
Department of Computer Science
University of California, Los Angeles
(310) 825-0956
rajive@cs.ucla.edu

Simulation Environments Minitrack
Ray J. Paul
Department of Computer Science
Brunel Unversity
Uxbridge, Middlesex UB8 3PH, UK
(0895) 274-000
paulr@vax.lse.ac.uk

General Applications
Professor Gerald W. Evans
Department of Industrial Engineering
University of Louisville
Louisville, KY 40292 U.S.A.
(502) 852-6342
gwevan01@ulkyvm.louisville.edu

Transportation and Logistics Minitrack

Chuck White
E.I. Dupont Nemours Co.
Engineering Services
P.O. Box 6090
Wilmington, DE 19714-6090
(302) 774-2342
white@pccvax.dnet.dupont.com

Mani Mannivannen
Consolidated Freightways, Inc.
Dept. of Intelligent Systems Engr.
1717 NW 21st Avenue
Portland, OR 97209 USA
(503) 499-3947
mani@eniac.cnf.com

Education Minitrack

Harriett Black Nembhard
Department Industrial and Systems Engineering
Auburn University
310 Dunstan Hall
(334) 844-1426
nembhard@eng.auburn.edu

Manufacturing Applications

Professor Phillip A. Farrington
Dept of Industrial and Systems Engineering
The University of Alabama in Huntsville
Huntsville, AL 35899-0001 U.S.A.
(205) 895-6568
paf@ebs330.eb.uah.edu

Referees

The high quality of the contributed papers appearing in this volume is a direct result of the joint efforts of the authors, the Program Committee, and the following referees.

Suraj Alexander	University of Louisville
Mary Bailey	University of Arizona
Rajive Bagrodia	UCLA
Terry Beaumariage	Oregon State
Bill Biles	University of Louisville
Bob Bulfin	Auburn University
Yuan Chen	UCLA
Russell Cheng	University of Canterbury
Samir Das	University of Texas
Michael Drevna	Systems Modeling Corp.
Luc Devroye	McGill University
Adel Elmaghraby	University of Louisville
Jerry Evans	University of Louisville
Phillip A. Farrington	UAH
Richard Fujimoto	Georgia Tech
Tesham Gor	University of Louisville
Mahesh Gupta	University of Louisville
Jorge Haddock	RPI
Dean S. Hartley	Center for Modeling, Simulation, and Gaming
Sheldon Jacobson	VPI
Vikas Jha	UCLA
Voratas Kachitvichyanukul	Motorola
Felipe Knop	Purdue University
Anup Kumar	University of Louisville
Pierre L'ecuyer	Universite de Montreal
S. Manivannen	Consolidated Freightways
Russell Meller	Auburn University
Loretta Moore	Auburn University
David Nembhard	Auburn University
David Nicol	William & Mary College
Sundeep Prakash	UCLA
David Pratt	Oklahoma State University
Stephen Robinson	University of Wisconsin—Madison
Rammohan Ragade	University of Louisville
Ralph Rogers	University of Central Florida
Susan M. Sanchez	University of Missouri—St. Louis
Robert Sargent	Syracuse University
Thomas Schriber	University of Michigan
Chien-Chung Shen	Bellcore
Darrell Starks	University of Louisville
James R. Wilson	North Carolina State University
Chuck White	Dupont
Edward Williams	Ford Motor Corp
Philip Wisely	University of Cincinnati
Enver Yücasen	INSEAD

The Winter Simulation Conferences

Below we give information concerning the history of the Winter Simulation Conference. For the periods 1967–1968 and 1974–1995, the attendance figures shown below represent total paid attendance including students, but excluding exhibits-only registrants. For the period 1969–1973, no surviving records of WSC attendance are held by any of the sponsoring societies; consequently, attendance estimates were obtained by discussions with the Program Chairs and General Chairs for this period—namely, Harold Hixson, Julian Reitman, Arnold Ockene, Philip Kiviat, Michael Araten, Joseph Sussman, Austin Hoggatt, and Michael Morris. Although the figures for the years 1969–1973 are subject to uncertainty, there is little doubt that in the period 1970–1971, attendance at the WSC was substantially higher than in all other years. In the following table, the abbreviations GC, PC, APC, PE, PCE, and APE denote General Chair, Program Chair, Associate Program Chair, *Proceedings* Editor, *Proceedings* Co-Editor, and Associate *Proceedings* Editor, respectively.

	Conference	Conference Leaders	Attendance
1	Conference on the Applications of Simulation Using GPSS 13–14 November 1967 Hilton Hotel, New York, NY	Harold G. Hixson (GC) Julian Reitman (PC)	401
2	Second Conference on the Applications of Simulation 2–4 December 1968 Hotel Roosevelt, New York, NY	Julian Reitman (GC) Arnold Ockene (PC)	856
3	Third Conference on the Applications of Simulation 8–10 December 1969 International Hotel, Los Angeles, CA	Arnold Ockene (GC) Philip J. Kiviat (PC)	400
4	Fourth Conference on the Applications of Simulation 9–11 December 1970 Waldorf-Astoria Hotel, New York, NY	Philip J. Kiviat (GC) Michael Araten (PC)	1100
5	The 1971 Winter Simulation Conference 8–10 December 1971 Waldorf-Astoria Hotel, New York, NY	Michael Araten (GC) Joseph Sussman (PC)	1200
6	The 1973 Winter Simulation Conference 17–19 January 1973 St. Francis Hotel, San Francisco, CA	Joseph Sussman (GC) Austin C. Hoggatt (PC)	600
7	The 1974 Winter Simulation Conference 14–16 January 1974 Washington Hilton Hotel, Washington, DC	Michael F. Morris (GC) Harold Steinberg (PC) Harold J. Highland (PE)	463
8	The 1976 Bicentennial Winter Simulation Conference 6–8 December 1976 National Bureau of Standards, Gaithersburg, MD	Harold J. Highland (GC) Thomas J. Schriber (PC) Robert G. Sargent (APC)	306

	Conference	Conference Leaders	Attendance
9	The 1977 Winter Simulation Conference 5–7 December 1977 National Bureau of Standards, Gaithersburg, MD	Robert G. Sargent (GC) J. William Schmidt (PC) Harold J. Highland (PE)	465
10	The 1978 Winter Simulation Conference 4–6 December 1978 The Deauville Hotel, Miami Beach, FL	Larry G. Hull (GC) Norman R. Nielsen (PC) Harold J. Highland (PE)	388
11	The 1979 Winter Simulation Conference 3–5 December 1979 Holiday Inn Embarcadero, San Diego, CA	Mitchell G. Spiegel (GC) Robert Shannon (PC) Harold J. Highland (PE)	375
12	The 1980 Winter Simulation Conference 3–5 December 1980 Orlando Marriott, Orlando, FL	Paul Roth (GC) Tuncer I. Ören (PC,PE) Charles M. Shub (APC)	205
13	The 1981 Winter Simulation Conference 9–11 December 1981 Peachtree Plaza Hotel, Atlanta, GA	Claude M. Delfosse (GC) Charles M. Shub (PC) Tuncer I. Ören (PE)	267
14	The 1982 Winter Simulation Conference 6–8 December 1982 Holiday Inn Embarcadero, San Diego, CA	Yen W. Chao (GC) Orlando Madrigal (PC) Harold J. Highland (PE)	274
15	The 1983 Winter Simulation Conference 12–14 December 1983 Crystal Gateway Marriott Hotel, Arlington, VA	Jerry Banks (GC) Bruce W. Schmeiser (PC) Stephen D. Roberts (PE)	416
16	The 1984 Winter Simulation Conference 28–30 November 1984 Sheraton Dallas Hotel, Dallas, TX	Udo W. Pooch (GC) C. Dennis Pegden (PC) Sallie Sheppard (PE)	350
17	The 1985 Winter Simulation Conference 11–13 December 1985 San Francisco Hilton and Tower, San Francisco, CA	Gerard C. Blais (GC) Susan L. Solomon (PC) Donald T. Gantz (PE)	369
18	The 1986 Winter Simulation Conference 8–10 December 1986 Radisson Mark Plaza Hotel, Washington, DC	James O. Henriksen (GC) Stephen D. Roberts (PC) James R. Wilson (PE)	531

	Conference	Conference Leaders	Attendance
19	The 1987 Winter Simulation Conference 14–16 December 1987 The Ritz Carlton, Buckhead, Atlanta, GA	Hank Grant (GC) W. David Kelton (PC) Arne Thesen (PE)	475
20	The 1988 Winter Simulation Conference 12–14 December 1988 San Diego Marriott, San Diego, CA	Peter L. Haigh (GC) John C. Comfort (PC) Michael A. Abrams (PE)	535
21	The 1989 Winter Simulation Conference 4–6 December 1989 Capital Hilton Hotel, Washington, DC	Kenneth J. Musselman (GC) Philip Heidelberger (PC) Edward A. MacNair (PE)	619
22	The 1990 Winter Simulation Conference 9–12 December 1990 The Fairmont Hotel, New Orleans, LA	Randall P. Sadowski (GC) Richard E. Nance (PC) Osman Balci (PE)	517
23	The 1991 Winter Simulation Conference 8–11 December 1991 The Arizona Biltmore, Phoenix, AZ	W. David Kelton (GC) Gordon M. Clark (PC) Barry L. Nelson (PE)	540
24	The 1992 Winter Simulation Conference 13–16 December 1992 Crystal Gateway Marriott Hotel, Arlington, VA	Robert C. Crain (GC) James R. Wilson (PC) James J. Swain (PE) David Goldsman (APE)	724
25	The 1993 Winter Simulation Conference 12–15 December 1993 The Biltmore Hotel, Los Angeles, CA	Edward C. Russell (GC) William E. Biles (PC) Gerald W. Evans (PE) Mansooreh Mollaghasemi (APE)	565
26	The 1994 Winter Simulation Conference 11–14 December 1994 Walt Disney World Swan Hotel, Orlando, FL	Deborah A. Sadowski (GC) Andrew F. Seila (PC) Jeffrey D. Tew (PCE) S. Manivannan (PCE)	667
27	The 1995 Winter Simulation Conference 3–6 December 1995 Hyatt Regency Crystal City, Arlington, VA	William R. Lilegdon (GC) David Goldsman (PC) Christos Alexopoulos (PCE) Keebom Kang (PCE)	652
28	The 1996 Winter Simulation Conference 8–11 December 1996 Hotel Del Coronado, Coronado, CA	Daniel T. Brunner (GC) James J. Swain (PC) John M. Charnes (PCE) Douglas J. Morrice (PCE)	

Keynote Address

Keynote Speaker Karl G. Kempf, Principal Scientist for Manufacturing Systems, Intel Corporation

Proceedings of the 1996 Winter Simulation Conference
ed. J. M. Charnes, D. J. Morrice, D. T. Brunner, and J. J. Swain

SIMULATING SEMICONDUCTOR MANUFACTURING SYSTEMS : SUCCESSES, FAILURES, AND DEEP QUESTIONS

Karl G. Kempf

Manufacturing Systems
Intel Corporation
5000 W. Chandler Blvd.
Chandler, Arizona 85226, U.S.A.

ABSTRACT

Over the past several years, engineers at Intel Corporation have used modeling and simulation techniques to solve various high volume manufacturing problems. Our current goal is to use a persistent model over the entire life cycle of a factory to promote integration and continuous improvement of all of the components of the manufacturing system. The uses of this model are detailed as well as our progress towards realization of this goal. A summary of problems that we have encountered along the way is included as both a warning to those who have a similar goal and as a work list for vendors of simulation packages. Two disturbing questions about the basis of manufacturing simulation are asked that should be of concern to practicing simulation engineers as well as university researchers.

1 INTRODUCTION

Whether the domain is governmental or commercial or educational, whether the focus is business or technical or organizational, whether the formulation is spreadsheet or linear programming or discrete event, simulation is a powerful tool for understanding and enhancing the performance of complex man/machine systems. In countries where the Gross National Product is based largely on the production of goods for sale, understanding and enhancing the performance of complex manufacturing systems is particularly important. In this application, simulation in its many forms can have an impact measured in many percents of corporate productivity, market share, and profitability. In the extreme, global competitiveness can be positively influenced.

At Intel Corporation, simulation plays an indispensable role. Critical business questions are studied from various perspectives with a host of simulation tools. Product designers, seeking to realize ever more useful and reliable chips, employ simulation to study the behavior of everything from individual transistors to full computer architectures. Process designers, striving for ways to build ever smaller and lower power devices, use simulation to study everything from the behavior of plasmas to the interaction of individual steps in long manufacturing processes. But the focus of this paper is manufacturing systems where the ephemeral business plans, product designs, and process specifications come together and are realized in actual products. In manufacturing, Intel betters the ancient philosophers dream of turning lead into gold. We turn sand into computers.

For manufacturing engineers, simulation in its various forms has application in each life cycle phase of a semiconductor factory including 1) design, 2) production ramp, 3) early high volume production (when the market could absorb more units than can be made with the new capacity), and 4) late commodity production (when old capacity can supply more units than the market desires). This is true whether the factory under discussion performs device fabrication, component packaging, or circuit board assembly. In the next section on Integration, the goal of our simulation efforts will be described along with a report on our progress toward that goal. The following section on Practical Difficulties will make available for others with similar goals the pitfalls to avoid and problems that vendors of commercial simulation packages need to help solve. The final section on Theoretical Difficulties will briefly describe issues that we have encountered in the march towards our goal that question the foundation assumptions of simulation and should stimulate the imaginations of university researchers.

2 INTEGRATION THROUGH SIMULATION

The cost of a fully outfitted semiconductor facility has reached two and a half billion US dollars, with a potential revenue stream of more than twice that per year. At this level of investment and return, nothing can be left to chance. The decisions needed to design, ramp, and operate the manufacturing system must be data as well as experience driven. All of the components of the system - from layout to lot size, from decision policies to automation systems - must be well integrated during every stage of the factory life cycle. Continuous improvement of every component must be supported across the life cycle. Our vision is to achieve integration and continuous improvement by using a persistent simulation model throughout the life of the factory. This model would be used by a variety of simulation tools for solving a variety of problems. We have been focused for a number of years on the process of realizing this vision.

2.1 Simulation for Manufacturing System Design

The first step in the process is to design the high volume manufacturing (HVM) system for "manufacturability". Even before excavation for the building begins, the factory exists as a fully functioning entity in simulation. To build the basic simulation, a number of things must be known, including :

- the business goals of the factory,
- the flow of operations in the manufacturing process,
- the types of processing equipment to be used, and
- the level and types of automation to be included.

Given the basic simulation model, verified as well as possible in the absence of the working factory, a long sequence of iterative studies begin. The sequence is long because there are a variety of stakeholders who must agree and there are numerous factors that must be considered. The sequence is iterative since few if any of the factors are independent.

The number of each type of equipment required to satisfy the business goals is determined (capacity), complicated by the integer nature of equipment and the uncertainty in run rates and availabilities. The floor execution policies for material release, management of work in progress (WIP), equipment maintenance, cross training, and engineering improvements are designed and tested for their positive impact on business goals. Transfer sizes and equipment layouts are evaluated and used to begin to scope the automation system. Consideration of safety, ergonomics, and productivity are factored into the automation design and the staffing levels. Construction constraints including local building codes and federal environmental rules are applied. Since execution policies modify capacity, construction constraints complicate layout optimization, automation impacts policies, and staffing impacts business goals - to mention a very few of the interactions - the iteration proceeds through many cycles. But the hard won result, the HVM system design with tradeoffs quantified, is useful over the life of the factory.

2.2 Simulation for the Production Ramp

With the high volume manufacturing system design as a goal, and a completed building available at some estimated point in the future, the second step in the process is to choreograph the ramp from zero production to full production in the shortest time and most organized manner possible. This generally means getting one instance of each equipment type running so that initial production material can traverse the whole process, then adding further pieces of each equipment type to smoothly ramp capacity up to the designed maximum. At least three factors come into play, and all can be studied with the help of the HVM model developed during the design phase.

One factor is the realistic delivery schedule that can be achieved by the various suppliers of equipment. Early delivery ties up capital unnecessarily. Late delivery delays the critical production ramp. Meeting a tight delivery schedule may strain the supplier beyond what is physically possible. Compromises are required to get the right equipment delivered at the right time.

Another factor is the rate at which skilled personnel can be available to install the equipment on the production floor and to qualify it to accurately execute the process specification. Since each equipment type involved requires a different skill mix, having the right skills available at the right time over the whole ramp is a challenge.

The third factor is the rate at which competent personnel can be made available to operate the installed and qualified production equipment. This rate can be constrained by the availability of local people to join the company, the ability of these people to acquire the operational competencies, or the availability of trainers to facilitate the training.

Since each of these three rates has multiple components, and there can be uncertainty and fluctuation in each component, simulation is a useful approach to determining what the constraint to ramping will be under a variety of scenarios before the ramp actually starts. And of course, as surprises occur during the actual ramp, the same simulation can be used to evaluate their impact, plot work-arounds where possible, and to re-plan when necessary.

2.3 Simulation for Early Production

The high volume manufacturing design may be completed many months before the ramp begins, and the production ramp may take many more months to realize. This is more than enough time for conditions in the market place to change causing modification of the business goals of the factory. And of course, any number of difficulties may have precluded the precise execution of the plan for the ramp yielding a factory that is different in some ways from the one that was initially designed. Each of these changes and deviations must be identified and addressed.

An excellent starting point for optimizing early production, and one that can be used well before the ramp is completed, is the model built during the HVM design phase. Since the model is maintained and used throughout the ramp, it is available to evaluate HVM solutions to changed business conditions and modified ramp execution. Before the end of the ramp, it may be possible to order and install more or less equipment than originally planed. After the end of the ramp, policies can be modified and personnel can be re-trained. The impact of each of these changes on early production can be quantified through simulation using the HVM model.

In fact, even if everything is just as expected at the end of the ramp, and the product issuing from the factory is successful in the market place, the HVM model will continue to be used in at least two ways through the early production phase.

On one hand, as production matures over the first few years of factory operations, execution policies will need to be continuously improved to satisfy the dynamics of the market and the pressure that manufacturing is always under to become more efficient. Again, the continuously updated and validated HVM model is the tool of choice to support evaluation of policy changes and interactions between policy changes prior to implementation on the production floor.

On the other hand, automation tools will have been built into the shop floor control system to support the execution of the material release, WIP management, equipment maintenance, and engineering improvement policies. The HVM simulation model will be used at the core of these tools for automated policy execution since it will have been available early in the development cycle of the automation system and will always be the most complete, correct, and consistent model obtainable. Policies developed by off-line simulation using this model will be communicated to on-line automation systems using this model to direct policy execution.

2.4 Simulation for Late Production

The ramp and early production scenario just described assumed that the market could absorb all of the product that the factory could produce, and so a major part of the business goal was to maximize the output. Later in the factory life cycle when the product for which the factory was initially built becomes a commodity, the business goals change considerably. The capacity is mostly amortized. A few more processes have probably been added to the factory along with a number of other products. The factory is now selling throughput time (TPT) - the ability to offer and deliver short lead times for all of its products. And the factory is very cost sensitive.

Running multiple processes and multiple products over one equipment set can be very complicated. All of the executional policies that were vital to maximizing output previously now become vital to minimizing TPT and cost. The material release and WIP management policies must evolve to handle low volumes of a wide diversity of products. Engineering improvement policies focus more on time saving and cost cutting projects as do the equipment maintenance policies. Staffing and cross training policies are now focused on headcount reduction and productivity improvement even more than previously.

Once again the HVM model that has been maintained and updated since the initial design of the factory can be used in a variety of ways. The model can continue to support the modification and improvement of executional policies to satisfy the new business goals. The model along with the improved policies can continue to support floor execution through the automation system. And, of course, the model is still the best capacity estimator for the factory, providing sanity checks for market-driven changes in product mixes and volumes experienced frequently in a commodity factory.

Finally, the HVM model is invaluable for the last ramp of the factory - ramp down and closure. Ramping down in an orderly fashion is at least as hard as ramping up, although the problem set to be considered is different. There is much choreography to be done to de-install machines, and to re-train and re-deploy personnel. And it must all be done without the adrenaline of the ramp up. But it may be just as economically important to the company to have a clean ramp down as it was to have an aggressive ramp up, and it is probably even more important to the psyche of the manufacturing organization.

2.5 Intel Today and Tomorrow

The vision described here is based on a single continuously updated simulation model that is persistent over the life cycle of the factory. It is key to designing and ramping the high volume manufacturing system since it is one of the only sources of reliable data to drive critical decision making processes. It is vital to early and late production phases since it contains most of the details required for continuous improvement and automated execution of the operational policies that drive floor execution. In all these uses, it is superior to any documentation since it is a computer executable specification not open to different interpretation by different personnel. Given the difference between this vision and conventional manufacturing practice, and given the scope of this vision, it is being implemented at Intel in piecewise fashion with an opportunistic approach.

It has been common practice for many years among industrial and manufacturing engineers in our fabrication and assembly/test facilities to build simulation models to answer capacity and staffing questions, with spreadsheets giving way to discrete event tools. These models have been built earlier and earlier in the factory life cycle, resulting in the current practice of developing the model well before groundbreaking to support the earliest capital purchase and personnel hiring forecasts. Unfortunately, once these early questions have been answered, the models tend not to be maintained. We are striving to use this experience to launch the concept of the design phase model that persists (Hilton et al. 1996, Sohn et al. 1996).

Over the past few years in operating factories, we have begun to simulate operational policies, especially material release and WIP management policies, before sending them to the shop floor for manual execution. It has been the case that new models are built and validated to pursue this work. Since continuous improvement of the policies is required, a few of these models have been updated and maintained their usefulness over several years. We are striving to modify this current practice to promote maintaining the design phase model instead of building a new model.

Although our manufacturing automation group has used simulation to design material handling systems for some time, only recently has it begun to develop and deploy model-based manufacturing planning and scheduling systems. A few assembly/test lines now use a simulation-driven planning system that regularly passes scheduling data to the shop floor control system, with proliferation through all assembly/test factories underway. Similar systems are in development for our fabrication factories. These systems support off-line policy development linked directly to on-line policy execution, and force models to be maintained by appropriate factory personnel. This new direction in our automation group fully supports our vision of the persistent model for design and execution of HVM.

Our most recent advance has been to work toward deciding important HVM questions while fabrication and assembly/test processes are still under development, and of course models and simulations are the tools that have been used. This promises to much improve both the result and the efficiency of the HVM design process that previously was strongly driven by experience and opinion rather than data. It also facilitates the use of existing simulation tools to control ramping since the detailed HVM model will exist at the appropriate time in the hands of the appropriate people.

Although we have followed a jagged trajectory, we will be in position to realize the full vision within the next few years in both fabrication and assembly/test factories. There will be HVM design processes in place that require a model to be built well before building the factory, and require the model be maintained for ramping. There will be automation tools in place that will use the model to continuously improve and automatically execute shop floor policies through all production phases. We expect the results of our integrated efforts to be a step forward in the theory and practice of manufacturing system science.

3 PRACTICAL DIFFICULTIES

Neither our progress so far nor our efforts over the next few years are without practical difficulties. Almost all of these difficulties will be encountered by anyone interested in the modeling and simulation of manufacturing systems. The solution of most of the difficulties will come from advances in modeling and simulation software tools. Other difficulties will be solved by improved training of simulation and manufacturing engineers. The process that we have developed over the years for modeling and simulation has only a few phases, but it can be used to motivate the description of the practical difficulties we have suffered.

3.1 Designing the Model

At Intel, the model design phase centers around the stakeholders, the HVM personnel who are the customers of the modeling and simulation effort. It is first necessary to collect all of the questions that the stakeholders desire to have answered. These questions drive the "what" and the "how" of the modeling effort. We are striving to design the simplest model that can answer all of the questions. The "what" is deciding the entities and behaviors that need to be included. The "how" has to do with the consistent level of abstraction at which to represent. Too many inclusions with too much detail adds to development and run time. Too little of either leads to poor or wrong answers. This initial model design effort will clarify assumptions, including ones about the availability and accuracy of data, all of which need to be checked with the stakeholders.

In our experience, it is vitally important to include the plan for model validation and the planned set of experiments (along with metrics) in the design discussions with the stakeholders. This calibrates the stakeholders who do not always know much about modeling and simulation, and calibrates the modeling and simulation personnel who are not always experienced in manufacturing. It also builds teamwork between the stakeholders and the modelers for the hard work that lies ahead.

Only at the end of the design phase can one decide whether the modeling and simulation project should go ahead, should be delayed, or should terminate. In hindsight, it is clear that we have had projects that should have been terminated, or at least delayed at this point because the questions of the stakeholders could not be answered by modeling and simulation, or the appropriate data was not available to the required accuracy, or some other related overwhelming difficulty. This is indeed a tough call, but pressing ahead in these circumstance only delays and increases the pain. It is better to take the pain early. It is unfortunate that this very important phase of design and negotiation is rarely included in the formal education of modeling and simulation personnel in either university or vendor training.

3.2 Building the Model

Another decision that should be made after the model design has been completed is tool selection. One is trying to select the simplest tool that can be used to answer all of the questions. Unfortunately, we have often made mistakes here. We have certainly taken problems that could have been solved adequately with a spreadsheet and applied a discrete event simulator, with much wasted time and effort. And we have used a spreadsheet on problems that should have been answered with a discrete event simulator, with poor answers as a result. The error has always been deciding the tool before designing the model, usually based on the convenience or preference of personnel on the modeling team. Better to design the model, then select the tool, even if it means changing team personnel.

There is an obvious practical problem here. Ideally, the team would have a suite of tools available to it, and be trained in the entire suite. Unfortunately, given the cost and complexity of many commercial tools, the best that might be achieved practically is the availability of two or three tools, with only one or two team members trained in each. This is origin of many poor model-to-tool mappings, and subsequent difficulties.

Even having selected the most appropriate tool, modeling building has frequently been difficult for us. With the best intentions, manufacturing organizations do not always collect all of the data required for the model. They frequently have great difficulty describing floor procedures to the model engineer in adequate detail. And facilities for automated data transfer from shop floor data systems to modeling and simulation tools are rare indeed. For these and other reasons, model building is a laborious manual process that almost always over-runs its schedule. Again, the formal education of simulation engineers is usually lacking in overcoming any of these difficulties. In addition, tools from simulation vendors are too often weak in this area.

Selecting the best tool and building the model gets us to the most difficult problem in this phase, that is model validation where the model is tested to see if it gives "correct" answers. When validating against an existing factory, historical scenarios can be used. Since all of the details of everything that happens on the factory floor have not been (and in fact, never could have been) deposited in the model, deciding what is correct is not a trivial matter. When validating a model for a non-existent facility, a wide variety of simple tests must be done. In either case, thinking through the validation approach and seeking stakeholder agreement to the approach during the previous design phase is critical because stakeholders have intuition. Four results are possible, and we have experienced all of them.

a) The answer produced by the tool is in agreement with the intuition of the stakeholder, and the answer is correct. This is a happy circumstance. We have more than our intuition to justify our actions.

b) The answer produced by the tool is in agreement with the intuition of the stakeholder, and the answer is wrong. This is the worst possibility. We will charge off in the wrong direction using the results from the tool as a shield against all adversity. It is very difficult to dissuade someone whose intuition is in agreement with a (wrong) simulation result. The mistake here is usually that we built the model including all of the elements that went into the intuition, and so the tool produces results in line with the intuition (as you might expect). The problem is detecting that the answer produced by the tool is wrong.

c) The answer produced by the tool is not in agreement with the intuition of the stakeholder, and the answer is wrong. This is what we see in testing the tool as we develop and debug it. We sometimes use a suite of increasingly difficult test problems about which we have sound intuition. The worst that can happen here is that the stakeholder sees the tool before we have removed all the bugs, and looses confidence in the tool, perhaps undeservedly.

d) The answer produced by the tool is not in agreement with the intuition of the stakeholder, and the answer is correct. This is the most important case. If the tool disagrees with the intuition of the stakeholder, the tool is wrong from the stakeholder's viewpoint. However, if the tool is in fact correct, the unexpected answer is very important. The only way to make the correct answer stick in the face of disagreement with the stakeholder is to have the tool very carefully validated, and to have had the stakeholder buy into the validation process before the disagreement surfaces.

Better analysis tools would make validation much easier, and would also be very useful in the last step in this phase, executing the experiments planned during the design phase. The task of the simulation engineer, from experiment to experiment (and validation to validation) is to track down every performance change in the simulated system. Doing this using the analysis tools available in most commercial simulation packages is error prone and time consuming at best. And, of course, once a change has been detected, explaining its origin is the reason for doing the experiments.

Missing changes and mis-explaining noticed changes are the major errors made during validation and experimentation. The reason that automated analysis tools would be so useful is that experience shows our intuition to be rather lacking when dealing with small changes in complex manufacturing systems. This holds for both the stakeholders who know the real manufacturing system and the simulation engineers who know the model and the tool.

3.3 Maintaining the Model

Given our goal of using the model through all of the life cycle of the factory, the first two phases described here are only the beginning. The design, construction, validation, and initial experiments to define HVM are important, but are really only the infancy of the model. Maintaining the model through the initial production ramp then through early and late volume production is where the majority of effort will be spent.

Not many new problems arise during maintenance of the model. The model must be updated with all of the same problems described above for initially building the model. Our experience indicates that a significant number of things change in a working factory that are a concern on a daily or weekly basis. The model must be re-validated, if not at every update, then at least at every few updates or after a defined time period has passed. All of the validation problems listed above are still encountered here, and must be addressed to keep the model from loosing the confidence of the users, both stakeholders and modelers. Since the model is there to supply data upon which to base operational decisions, for both off-line continuous improvement and on-line execution, all of the problems associated with doing experiments recur.

The difficulty that we have seen in this phase is discipline. The engineers chartered to keep the model vital frequently have a variety of other assignments and responsibilities. Maintaining a complex model requires a lot of time, and underestimation of that time, or de-prioritization of that time to instead address urgent tactical problems can lead to situations where the model is required, but is stale. Then a big effort must be launched, and yet another mechanism of loss of confidence in the model can be accidentally activated. The best answer is simply to develop the discipline to give the model regular maintenance, even when its next use is not clearly in sight.

4 THEORETICAL DIFFICULTIES

In all of the work described so far, at least two high level assumptions have been made implicitly. One is that we are basically modeling the appropriate entities to answer the questions that our stakeholders need answered. The other is that the manufacturing systems that we are modeling are basically well behaved. From our work, we have strong indications that both of these assumptions need to be very closely examined. The deep questions that we have been asking ourselves are outlined here along with pointers to our published work on these topics for those who wish to delve further.

4.1 What Should be Modeled

On the manufacturing floor of a typical factory, one comes into contact with two distinct types of entities. On one hand are entities that have no inherent intelligence. They are dumb. This includes machines, jobs, tools, buffers, and transporters as examples. These entities have status that is tracked since decisions must be made about how they should be used on a minute to minute basis to achieve the performance goals of the factory. But these dumb entities can not make allocation decisions themselves. On the other hand are entities that have high inherent intelligence. They are smart. This includes machine operators, maintenance technicians, area supervisors, and shift managers as examples. These entities can make decisions about the use of the dumb entities, and employ status information to do so, including decisions about WIP movement, capacity allocation, scheduled and unscheduled equipment maintenance, batching, setups, and other related activities. In our factories, one goal of the automation system is to supply status and support decision making.

The point is that both the dumb entities and the smart entities are important to the performance of a manufacturing system. It is clear that the capacity and speed of a factory depend heavily on the capabilities of the equipment set and that variability in factory performance is strongly tied to pseudo-random equipment breakdowns. However, it is equally clear that speed and capacity is directly impacted by the quality of decision making by the smart entities, and that random behavior in the smart agents also shows up as factory variability. In fact, in factories with which we have been associated, the "equipment problems" usually pale by comparison to the "people problems" in terms of achieving and sustaining uniformly high system performance.

Furthermore, to understand and improve manufacturing systems, we need tools to study the dumb entities, the smart entities, and the interactions between the two. Unfortunately, most modeling and simulation tools that are available for the study of manufacturing systems are equipment-centered. Factories can be simulated in great detail without any consideration of the human element. The quantities, properties, and arrangement of the equipment can be altered to play out a wide variety of "what-if" scenarios. The equipment is given decision making capability though dispatching rules that are executed instantaneously and uniformly throughout the simulation run. When operators and technicians are included, the equipment retains the capacity to make decisions while the actual decision makers are treated as tools. A machine decides what to run next, then checks the current simulation data base to see if an appropriately qualified operator is available to load the chosen job.

We propose that new and important insights would be gained if dumb entities and smart entities were modeled and simulated with equal fidelity. We desire the ability to play a wide variety of "what-if" scenarios around the smart entities too, and to include the interactions between the different types of entities in the simulation. We speculate that a significant fraction of the complex global behavior observed in a semiconductor manufacturing system emerges from the interactions between the smart entities, and between smart and dumb entities.

A model has been constructed and a new simulator implemented to test these ideas. The full results have been published elsewhere (Spier and Kempf 1995), and only one experiment is described here to pique the reader's interest. Suffice it to say that the factory is a simple one containing a manufacturing process with six steps, an equipment set with five machines, and a staff of three floor personnel (two operators, one maintenance technician). The experimental data presented in Table 1 was generated by varying the characteristics of the smart objects, an impossibility in most other systems.

The rows marked "NON-COOP" and "COOPERATIVE" test the factory performance of two methods by which the three smart objects interact. In "NON-COOP" mode, the tactic for breaks is simply to take them whenever there is no machine requesting service and it has been long enough since the last break. The advanced decision procedure considers when preventive maintenance is due and whether other smart objects are off the manufacturing floor. For example, operators will not go on breaks together, will service the machines of the operator who is on break in addition to their own machines, and will try to take breaks when machines are being maintained. This is called "COOPERATIVE" mode.

The rows marked "LAZY", "REACTIVE", and "PROACTIVE" test the factory performance of three methods by which the smart objects function individually. The default mode for executing WIP management policies is called "REACTIVE" mode. The smart object simply waits for a machine request, services that request as dictated by the current decision policy, then waits for the next request. There is also "PROACTIVE" mode in which the smart object, upon finishing a request, uses the information available to determine the next machine which will make a request and moves into position before the request is made. Finally, there is "LAZY" mode, where, with multiple requests for service, the smart object selects to service the machine that is closest to its current position.

Table 1 - The Impact of Including Personnel

PRIMARY CHARACTERISTIC	secondary characteristic	TPT	OUT
NON-COOP	reactive	3837	780
COOPERATIVE	reactive	3605	840
LAZY	non-coop	4303	700
REACTIVE	non-coop	3837	780
PROACTIVE	non-coop	3818	793
COOP/PROACTIVE	-	3477	862

The row marked "COOP/PROACTIVE" shows the performance of the best of both approaches. Notice that the performance of the manufacturing system changes dramatically with a constant process and equipment model, but varying character and interaction of the smart objects that are normally excluded from consideration. Notice that the differences between the best (lowest) and worst throughput times and the best (highest) and worst factory output approach 20%. We propose that this might be the magnitude of error in answering questions about a real factory using an equipment-centered simulation model.

4.2 What can be Simulated

In an informal manner, the word "chaos" frequently comes to the lips of engineers trying to operate manufacturing facilities. From a formal perspective, the discovery and application of the theory of deterministic chaos to natural systems has revolutionized work in many branches of physics, chemistry, and biology. We have been interested in applying the formal theory of chaotic behavior to begin to quantify the informal intuition of manufacturing engineers. The goal of this work has been to provide answers to three questions. 1) Can chaotic behavior be found in simulations of simple manufacturing systems? 2) If so, can chaotic behavior be observed in actual manufacturing systems? 3) If so, how can this behavior be utilized, minimized, or avoided? The experiments we have reported previously begin to investigate the first question.

There have been various attempts to define chaotic behavior in the literature, but no general agreement has been reached in this relatively young science. A component common to every definition reflects the chaotic system characteristic that small changes lead to large effects. This is in direct conflict with the doctrine of Newtonian determinism and its small perturbation assumption about natural systems. Such determinism concerns prediction of the future course of a system given a set of natural laws, the structure of the system, and a set of initial conditions. The Newtonian doctrine states that small approximations in the natural laws, small inaccuracies in the structural description, or small errors in the initial conditions could have only small effects on the resulting predictions.

Chaotic systems behave quite differently. Very small perturbations in structures, laws, or conditions produce very large changes in performance predictions as an inescapable consequence of the intertwining of small scale and large scale phenomena. Simple deterministic systems can exhibit chaos as orderly disorder. Their behavior is bounded, but in some important time scales unpredictable.

The manufacturing system that we have studied by deterministic simulation has only four processing steps and four pieces of equipment. But it includes one of the most distinctive characteristics of semiconductor fabrication, re-entrant flow (where every job comes back again and again to the same equipment type in completing the full process flow). Two metrics have been used to evaluate the performance of the modeled system. One is the distribution of TPT over all jobs, where TPT is defined as the time elapsed between a job starting execution at the first step in the process to finishing execution of the last step in the process (in units of minutes). The distribution is summarized as a minimum, an average, a maximum, and a standard deviation. The other metric is the interdeparture time of finished jobs leaving the last step in the process flow. Exit times are recorded (in units of minutes), the sequential deltas between them computed, and basic patterns of interdeparture times identified. The patterns are named with an integer stating their length, and a character indicating the details of their sequence.

Small changes in this system fall into two categories. We have explored changes to policies for the release of raw materials and for the withdrawal of jobs from queues of work in progress, and have investigated changes to the contents and ordering of initial and dynamic queues. Large changes to TPT distributions and temporal patterns of finished jobs flowing out of the system have resulted.

Having shown that small policy or queue changes induce large performance changes, we altered the structure of the manufacturing system to find the origins of the chaotic behavior. This includes the mapping of processing steps onto production machines, the volume of work being added to the system relative to its capacity, and the size of batches being placed into batchable machines. It has been shown that each of these factors contributes to the complexity required for the onset of chaotic behavior.

Although the full results have been published elsewhere (Beaumariage and Kempf 1994), one experiment is described here to stimulate the interest of the reader. The experiment revolves around process Step-1 and process Step-3 in the model. These two steps run on the same machine type, and so there is one queue for work in process at this machine type which holds jobs waiting to run Step-1 and jobs waiting to run Step-3. The machine type involved batches three jobs at a time that must all be waiting for the same processing step (1 or 3). Batches are made up by finding the first three like jobs starting at the head of the queue. The experimental data presented in Table 2 was generated to demonstrate that very small changes in the dynamic contents of a queue precipitates very large changes in the overall performance of this simple manufacturing system.

The row marked "STANDARD" represents the simulation run where the queue was not dynamically altered. The row marked "SUB" shows data from an experiment in which a job waiting for Step 3 was removed from the tail of the queue at an arbitrary time during the run. In the row marked "ADD", data is presented from an experiment in which a job waiting for

Step 3 is added to the tail of the queue at the same arbitrary time as in "SUB". The data in the row marked "SUB:ADD" is from an experiment in which a job waiting for Step 3 is removed from the head of the queue one second before a batch of Step 3s would have been loaded, held for two seconds, then placed back into the tail of the queue one second after a batch of Step 1s was loaded. This is equivalent in the simulation to swapping the order in which two batches were done at one particular point in a long simulation. In the base case, a batch of Step 3s were processed first followed by a batch of Step 1s. In the "SUB:ADD" case, the reverse sequence was forced.

Table 2 : The Impact of Small Queue Changes

	TPT min	TPT ave	TPT max	TPT std dev	Pattern
STD	567	739.3	922	61.8	33-A
SUB	552	817.8	1109	157.9	33-C
ADD	532	788.1	1195	153.1	165-A
SUB:ADD	645	858.0	1146	123.6	132-B

Three very minor queue perturbations produce very large differences in performance results. If unperturbed, all four runs would have shown the same results. But after minor perturbations midway through runs of a year of simulated time, the resulting TPT distributions are each very different than the base system, and are different than each other. Furthermore, all three resulting patterns are different from the base system and very different from each other. We propose that this indicates that manufacturing systems may not be as well-behaved as we assume. The possibility of chaotic behavior should be considered each time a simulation produces a surprise result.

5 CONCLUSIONS

Our work over the past several years is moving us closer and closer to applying modeling and simulation techniques during every life cycle phase of all of our manufacturing systems. We believe that the value of a model that persists over the life of a factory is great in the context of supporting both integration of all components of the systems and continuous improvement of the components and the system.

Unfortunately, there are many hurdles in the path of achieving this goal. They include issues in designing and gaining consensus for the model, building and validating the model, and using and maintaining the model over the life of the factory. Better education and improved tools are required to overcome these difficulties.

But even then, two questions remain that must be addressed further. One deals with what we include in our models, specifically how we include the human decision processes that drive the manufacturing system. The other deals with what we can expect from any of our various simulation techniques given the possibility of chaotic behavior in manufacturing systems. These questions require much more research to settle.

ACKNOWLEDGMENTS

This work has been supported by too broad a set of Intel manufacturing personnel to exhaustively acknowledge here. However, special thanks go to Ken Thompson, Don Rose, and Gene Meieran for management support and to Stephen P. Smith, Naiping Keng, and Vats Srivatsan for technical collaboration.

REFERENCES

Beaumariage, T., and K. Kempf. 1994. The Nature and Origin of Chaos in Manufacturing Systems. In *Proc. SEMI/IEEE Advanced Semiconductor Manufacturing Conference*, 169-174. Institute of Electrical and Electronics Engineers, Piscataway, New Jersey.

Hilton, C. M., G. Mazenko, L. Solomon, and K. Kempf. 1996. Assembly Floor Layout and Operation. In *Proc. Inter. Symp. on Semiconductor Manufacturing*, to appear. Institute of Electrical and Electronics Engineers, Piscataway, New Jersey.

Sohn, B., A. L. Romero, and S. Mouck. 1996. Designing Fabs for High Performance. In *Proc. Inter. Symp. on Semiconductor Manufacturing*, to appear. Institute of Electrical and Electronics Engineers, Piscataway, New Jersey.

Spier, J., and K. Kempf. 1995. Simulation of Emergent Behavior in Manufacturing Systems. In *Proc. SEMI/IEEE Advanced Semiconductor Manufacturing Conference*, 90-94. Institute of Electrical and Electronics Engineers, Piscataway, New Jersey.

AUTHOR BIOGRAPHY

KARL G. KEMPF is Principal Scientist for Manufacturing Systems at Intel Corporation in Chandler, Arizona. He is interested in applying simulation to understand and enhance the performance of complex man/machine systems. He has practiced various forms of modeling and simulation in Gran Prix motor racing at Ferrari S.p.A., in cinematic special effects at Pinewood Movie Studios, at McDonnell Douglas in Space Station design, and at Intel Corporation in manufacturing automation.

Introductory Tutorials

Proceedings of the 1996 Winter Simulation Conference
ed. J. M. Charnes, D. J. Morrice, D. T. Brunner, and J. J. Swain

AN INTRODUCTION TO SIMULATION MODELING

Martha A. Centeno

Department of Industrial & Systems Engineering
College of Engineering and Design
Florida International University
Miami, Florida 33199, U.S.A.

ABSTRACT

This paper offers an introductory overview of discrete simulation, with emphasis on the simulation modeling process rather than on any specific simulation package. Examples are used to demonstrate the application of the methodology at each step of the process. An extensive list of additional readings is given for the reader who wishes to deepen his/her knowledge in a specific area.

1 SIMULATION MODELING

The word simulation means different things depending on the domain in which it is being applied. In this tutorial, simulation should be understood as the process of designing a model of a real system and conducting experiments with this model for the purpose of understanding the behavior of the system or of evaluating various strategies for the operation of the system (Shannon, 1975). In other words, simulation modeling is an *experimental* technique and *applied* methodology which seeks

- ✓ to describe the behavior of systems,
- ✓ to construct theories or hypotheses that account for the observed behavior, and
- ✓ to use these theories to predict future behavior or the effect produced by changes in the operational input set.

1.1 Types of Simulation

Simulation is classified based on the type of system under study. Thus, simulation can be either continuous or discrete. Our emphasis in this tutorial will be on discrete simulation. There are two approaches for discrete simulation: *event-driven* and *process driven.*

Under event driven discrete simulation, the modeler has to think in terms of the events that may change the status of the system to describe the model. The status of the systems is defined by the set of variables (measures of performance) being observed (e.g. number in queue, status of server, number of servers). On the other hand, under the process driven approach, the modeler thinks in terms of the processes that the dynamic entity will experience as it moves through the system.

1.2 Some Terminology

Every simulation model represents a *system* of some form. The system may be a manufacturing cell, an assembly line, a computer network, a service facility, or a health care facility, just to name a few. Although these systems seems to be completely different, they are not really that different if we expressed them in terms of their components. In general the components of a system are dynamic entity(ies) and resource(s).

Dynamic entities are the objects moving through the system that request one of the services provided by the system resources. Entities have *attributes* which describe a characteristic of it. Entities may experience *events* which is an instantaneous occurrence that (may) change(s) the state of the system. An *endogenous event* is an event that occurs within the system, whereas an *exogenous event* is an event that occurs outside the system.

Resources are the static entities that provide a service to an entity. Resources are further classified based on the type of service they provide; e.g. servers (machines or persons), free path transporters, guided transporters, conveyors, and so forth. Resources engage in *activities* which is a time period of specified length.

The *state* of the system is a collection of variables needed to describe the system's performance.

Simulation will model the *random* behavior of a system, over time, by utilizing an internal *simulation clock* and by sampling from a *stream of random numbers*. A *random number generator* is a computational algorithm that enables the creation of a new number every time the function is invoked. The sequence of these numbers is known as a *random number stream*. A true *random number* cannot be generated by a computer. Computers can only generate *pseudo-random numbers*. However, these numbers are statistically random; thus, their usage in simulation modeling is valid.

1.3 The Simulation Modeling Process

The simulation modeling methodology has many stages including: definition of the project and its goals, abstraction of a model, digital representation of the model, experimenting with the model, and producing complete documentation of the project. It tends to be an iterative process in which a model is designed, a scenario defined, the experiment run, results analyzed, another scenario chosen, another experiment was run, and so forth. However, advances in computer hardware, software engineering, artificial intelligence (AI), and databases are exerting a major influence on the simulation modeling process (SMP) and the development philosophy of support tools. If we look at the SMP (Figure 1), it is clear that simulation modeling is a knowledge and data intensive process. Notice that the flow through the SMP is not a sequential one; rather, it is a highly iterative one as discussed in the following paragraphs.

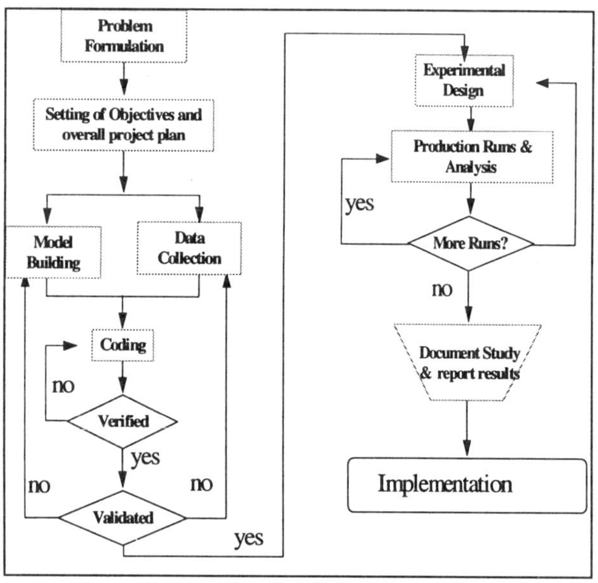

Figure 1: The Simulation Modeling Process

A study of a system starts when there is a problem with an existing system, or when it is not possible to experiment with the real system, or when the system is under design, i.e. the system does not exist yet. After clearly understanding and assessing management needs and expectations, the modeler must determine whether or not simulation is indeed an adequate tool for the analysis of the system under scrutiny. In some instance, an analytical technique may suffice to obtain adequate solutions. Analytical tools should be preferred over simulation whenever possible. The modeler must establish a set of assumptions under which a technique, or a combination of techniques, is applicable, feasible and sufficient.

If simulation modeling is the technique, or it is part of a combination of techniques, the modeler proceeds to gather data and performs proper statistical analysis to support models of the system. The needed data may be available in many different places and/or in different formats. For example, data may reside on paper, or it may reside within a database of a computerized information system. In either case, statistical tests must be performed without altering or destroying the original raw data. When there is no data available, the modeler must define the inputs to the model about the system using rules of thumb and/or personal experience.

A conceptual model of a system must be devised and converted into a digital model. The modeler resorts to tools that will make the transition less difficult and less time consuming such as model generator front ends, simulation packages and so on. However, converting a conceptual model into a digital one is by itself meaningless, unless such a digital model is thoroughly verified and validated. The reliability of the digital model is directly affected by the quality of the verification and validation processes. *Verification* entails assuring that the digital code of the model performs as expected and intended, and *validation* seeks to show that the model behavior represents that of the real-world system being modeled.

With a reliable and accurate digital model, the modeler proceeds to the experimentation phase. Statistical experiments are designed to meet the objectives of the study. Observing a model under one set of experimental conditions usually provides little or incomplete information. Thus, a set of experimental conditions must be analyzed within the framework of multiple experimental conditions.

The capstone of the SMP is the preparation of a set of recommendations into a formal management report. This report includes implementation and operations guidelines.

1.4 Advantages/Disadvantages of Using Simulation

Simulation modeling is a highly flexible technique because its models do not require the many simplifying assumptions needed by most analytical techniques.

Furthermore, simulation tends to be easier to apply (once the programming side is overcome) than analytic methods. In addition, simulation data is usually less costly than data collected using a real system. However, despite its flexibility, simulation modeling is not the magic wand that solves all the problems. Constructing simulation models may be costly, particularly because they need to be thoroughly verified and validated. Additionally, the cost of the experiment may increase as the computational time increases. The statistical nature of simulation requires that many runs of the same model be done to achieve reliable and accurate results.

Despite its disadvantages, simulation remains a valuable tool to address a variety of engineering problems, at the design, planning, and operations levels. In each of this areas, simulation can be used to provide information for decision making, or it may actually provide a solution (Pritsker, 1992).

2 STATISTICAL ASPECTS

2.1 Statistical Tools

It is not uncommon for a new comer to the simulation modeling methodology to feel hesitant about it because of its intense use of statistics. Fortunately, there are a variety of support tools to overcome this hesitation.

Simulation packages such as SIMAN/ARENA (Banks, et al., 1994), GPSS/H (Henriksen, 1989), and SLAMSYSTEM (O'Reilly, 1994) provide some support for both the analysis of data for inputs as well as for the analysis of output. In addition, there are several packages that has been designed to support the statistical analysis needed in simulation. For example, ExpertFit (Law and Vincent, 1995) enables the input data modeling for up to twenty seven probabilistic models; a variety of statistical tests and measures of accuracy are implemented in this package. Another example is SIMSTAT (Blaisdell and Haddock, 1993) which is an interactive program that allows the analysis of both input and output data. Both packages are Windows-Based, and they both allow for the user to save their outputs in various formats.

In the following paragraphs, a discussion of some of the issues and techniques needed are discussed.

2.2 Analysis of Inputs

A simulation model requires data and/or data models for to come alive. Without input data models, the simulation model itself is incapable of generating any data about the behavior of the system it represents. A simulation practitioner may be confronted with one of three possible scenarios at this stage of the SMP:

- ☑ *There seems to be an immense amount of data*
- ☑ *There is no data at all*
- ☑ *There is a few amount of data*

On the onset, the first situation may appear to be ideal: the more data, the merrier. But, caution must be exercise as the data available may be irrelevant for the purposes of the study. Even if the data presents itself as relevant to the study, traditional common sense practices must be done before using the data to develop models for simulation. Is the data reliable? How was it collected? Are there any circumstances that occurred during the data collection process that may lead to skewed or bias data models? Once sufficient information about the data and the data collection process has been compiled, classical statistical techniques should be used to develop data models. Some of these techniques are correlation analysis, point and interval estimation, linear and non-linear regression, and curve fitting in general.

Assume that there is a sample in the available data set that represent the number of customers that arrived at a bank in a 15-minute time interval.. Further, assume that these data was collected over the course of two weeks, between the hours of 9:00 AM to 2:00 PM and 5:00 PM to 6:00 PM. Correlation analysis will help answer questions such as *is the customer arrival rate constant through the day? throughout the week? If it is not, how does it vary? Could the arrival rate be modeled differently for each type of service the bank provide? If yes, do the data lend themselves for proper breakdown?* The use of graphical techniques such as line graphs and scatter plots will also help in establishing how to cluster the data set. Figure, for instance, reveals that there are changes in the rate of arrival throughout the day. Further, it seems that there is a variation of the rate of arrival depending not only on the time, but also on the day of the week. Correlation analysis should be used to confirm the visual hypothesis.

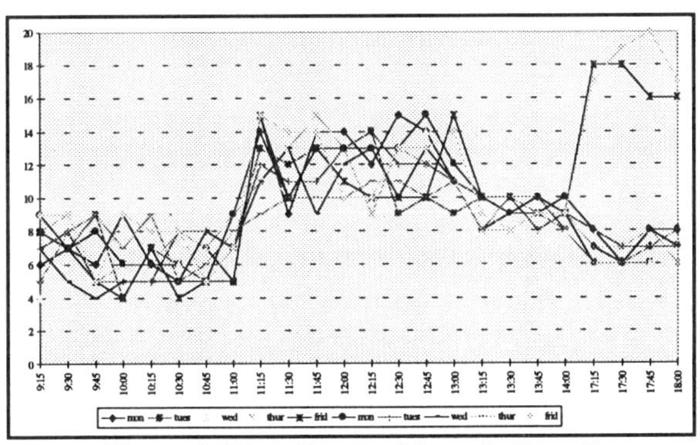

Figure 2: Line Graph of Sample Data Set

It is important that these previous questions be answer by analyzing the data because if it is assume that the entire sample of 1000 observations comes from the same *population*, a single data model may be built for it which will not reflect the true arrival rate behavior.

In our example, it is possible to construct three different probability data models to account for the change in the arrival rate. Goodness of fit tests, such as the χ^2 test and the Kolmogorov-Smirnov test need to be done to test the hypothesized distribution. In our example, we could formulate that

$\mathbf{H_0}$ = *arrival rate follows a uniform distribution between 9:00 AM and 11:00 AM*

After using point estimators for the lower and the upper bounds of the uniform distribution, we can test $\mathbf{H_0}$ using the χ^2 test if the number of observations is adequate and if the expected number of observations in each and every interval is at least five.

The second situation is clearly the worst case scenario because the simulation practitioner has to rely on his/her experience and knowledge about the system to establish appropriate input data models. In this case, the modeler has to quantify his/her educated guesses based on any estimated values available. Such estimates may come from the people who operate the system. In some cases the estimates given by the operators may just be an "average" or mean value. Depending on the actual system, the modeler may use such value as a constant or as the mean value of a distribution. To decide on the distribution, the modeler should use knowledge about the areas of application of the distributions. Table 1 give a summary of some classical distributions and what they are capable of modeling.

The third situation requires some of the efforts of the first one (assessing quality of data, data collection process and so forth). Usually, there is a small sample from which one has to develop data input models. Statistical methods based on the *t*-student distribution are utilized. If the sample size is "large enough", then techniques based on the Normal distribution may be used.

2.3 Analysis of Outputs

The actual analysis of the data provided by the simulation model will depend on the initial objectives of the study and on the type of system being modeled (Sadowski, 1993). There are two basic types of systems: terminating and non-terminating.

Distribution Name:	***Binomial***	***Poisson***
Definition of Variable:	X = Number of successes among *n* Bernoulli trials	X = Number of events during *one unit* of measure (time, volume, length,..)
Some Assumptions:	It can be used in cases where experiment consists of a sequence of n trials and n is fixed. Trials are dichotomous & identical (Bernoulli trials). Trials are independent and have same probability of success. Commonly used to model quality control and reliability data models.	Binomial converges to a Poisson ($n \rightarrow \infty$ and $\pi \rightarrow 0$). It provides probabilities of how many events will occur over a period of time. Events occur at some average rate. Rate of events remains constant over the duration of a unit of time. No two events can occur simultaneously. Commonly used in queueing analysis and to model rate of arrivals / rate of service.
Distribution Name:	***Normal***	***Exponential***
Definition of Variable:	X = *depends on the system under study*	X = time or distance between two consecutive events
Some Assumptions:	It can be applied to model physical, chemical properties. It is the best know and most widely used distribution. It can be used to model both discrete and continuous variables, but with caution because its domain is real and expands from $-\infty$ to $+\infty$	It is a special case of the Gamma model (r=1). It is *memory-less* when modeling time In a Poisson process, the rate of arrival / service is Poisson and the time between arrivals / services is Exponential.

Table 1: Summary of Some Distributions

Terminating systems are systems that have a clear point in time when they *start* operations and a clear point in time when the *end* operations. For these type of systems, it is required to established the sample size and the simulation length. The latter is usually established by the context of the problem. For a bank facility, it may be the entire day, or just the rush hours during Friday afternoon. The sample size is established, based on the accuracy, reliability, and variation desired for the study, using the equation

$$n = \frac{z_{\alpha/2}^2 \sigma^2}{d^2}$$

where d is the accuracy expressed in the same units as those of the measure of performance (e.g. within 5 minutes), z is the critical value from the standard normal table at a given reliability level, 1-α, (e.g. 95% reliability leads to an $\alpha = 0.05$), and σ is the standard deviation desired.

The value of n is the minimum number of *replications* (not runs) needed to have statistically valid results. It is very common for the novice to confuse a replication worth a simulation run. A run is what happens from the moment the user clicks on the run option of the main menu, to the moment in which the software finishes outputting data and comes back to the main menu. A replication, on the other hand, is what happens from the simulated *start* time to the *closing* simulation time. A simulation run of this type of systems has n replications. The summary report from replication to replication is different, but the summary reports from run to run are exactly the same, unless the random number stream is changed between runs.

Non terminating systems are systems that have no clear beginning time when they start operations, nor do they have a clear point in time when they stop operations. For this kind of systems, it is necessary first to bring the simulation model to steady state before the model generated data is used in analysis. All the data generated before the model reaches steady state should be thrown away. Upon reaching steady state, it is necessary to decide the length of the single replication that is to be run. Usually, the method of batching is used to analyze this data. Batching refers to the breaking the highly correlated, sequential observations generated by the model, into groups or batches. These batches would then be treated in the same fashion as the replication in the case of terminating systems.

A batch is formed based on the level of correlation between observations separated by a given lag k. Statistical theory tells us that the farther apart two correlated observations are, the smaller the correlation. So what one needs to do is to compute the correlation factor for various values of the lag k (20-500), graph the correlation factor, and identify the lag value at which the correlation actor ρ is approximately 0. From that process, the number of observations in the batch is established taking into account that this is a random process, so a safety factor needs to be added.

Once the batches are formed, each batch is treated as an independent replication, and similar analysis to that for a terminating system should be done. Regardless of the type of system, the modeler must keep in mind that reporting point estimates of the measure of performance are not that helpful. Confidence intervals on critical measures of performance should be built to add credibility and robustness to the study.

3 CODING THE MODEL

This the stage that, at first, seems to be the most time consuming. However, without proper planning and proper analysis, the effort at this stage is useless. It does not matter how "beautiful" the model looks like (animation), or how fancy (full of details) the model is. If it does not addressed the original problem, it is useless.

The modeler must be knowledgeable about programming and/or a simulation package. Although simulation models can be built using general purpose languages such as C/C++ and FORTRAN, nowadays there are a significant number of packages that have proven themselves as versatile and robust. Some of these packages are domain specific, while other are more general purpose. These packages include GPSS/H™, GPSS/World™, SIMAN/ARENA®, SLAMSYSTEM®, WITNESS, SIMFACTORY, ProModel, AutoMod, AIM, and Taylor (Banks, 1995).

4 VERIFICATION & VALIDATION

Verification is the process of determining that the model operates as ***intended***. In other words, it is the process of debugging the model. All the verification activities are closely related to the model. An empirical way to check if the model is behaving as the modeler intended is to put constant values for all the random processes in the model, and to calculate the outputs the model is supposed to give. This process is by no means sufficient or enough to declare a model as verified, but it helps to identify subtle errors. In fact, determining that a model is **absolutely** verified may be impossible to attain.

During the verification stage, the modeler must test

each major component of the model. These can be done by doing *static* testing (correctness proofs, structured walk through) or by doing *dynamic* testing (execute the model under various conditions) (Sargent, 1984 and 1994)

Validation is the process of ascertaining that a model is an acceptable representation of the "real world" system. The validation process is concerned with establishing the confidence that the end user will have on the model. Some critical questions to ask during this stage are *does the model adequately represents the relationships of the "real" system?, is the model generated output "typical" of the real system?*. Validation requires that the model be tested for reasonableness (continuity, consistency, and degeneracy). A point to start the validation process empirically is to question the structural assumptions of the model as well as the data assumptions made early in the SMP. A more sophisticated way to validate a model may require the use of some analytical tools such as Queueing Theory, but for it additional data may be needed (Sargent, 1994).

In summary, verification and validation can be informal (audits, face validations, Turing test, walk through), Static (consistency checking, data flow analysis), Dynamic (Black-box testing, regression testing), Symbolic (cause-effect diagrams, path analysis), Constraint (assertion checking, inductive assertions), and Formal (Logical deduction, Predicate calculus, proof of correctness). For more details in any of these test, see Balci (1995 and 1996).

5 ENSURING SUCCESS

Simulation modeling is very susceptible to the ability of the project to leader to organize the study. It is paramount to clearly define the objectives of the study; otherwise, there will be no clear guidelines as to what type of experimentation should be done. Further, establishing which assumptions are acceptable depends on what the objectives are.

A modeling strategy needs to be developed early in the process. The complexity of the system may be too much to include in the model. In fact, such level of complexity may not be desirable. A model can be simplified by *omission*, *aggregation*, or *substitution*. Omission refers to leaving out details that are negligible; for example, sojourn time over a two-foot distance by human being. Aggregation refers to the lumping of details / activities; for example, it may not be possible to model the time it takes to move a box from one table to another, when the tables are next to each other, but it may be possible to measure the time it takes from the moment the server seizes the box, opens it, performs whatever activity is supposed to perform, and places it on a cart. Substitution refers to replacing an object in the system for another one with similar, not equal, characteristics; for example, assuming that two cashiers at a fast food place work at the same speed, when indeed they do not.

It is very typical for novice simulation practitioners to resist omission. They want to build an "exact" replica of the real world system. This situation may lead to costly and delayed simulation studies. Similarly, a very experienced modeler may fall I in the trap of simplifying a bit too much. The impact of the simplifications made on the model generated data must be taken into consideration before implementing the simplifications. Again, the impacts are assessed in light of the objectives of the study.

It is very important that the modeler, or someone in the team, be properly trained in both simulation modeling and statistical analysis. A common pitfall of simulation studies is poor analysis of both the input data (to formulate data models) and the model generated data.

A simulation study is not a simple task. Instead of setting an absolute delivery date, plan on having intermediate progress reports. This approach will help you and the end user of the model to verify, validate and accept the recommendations of the study. It is very likely that as you make a progress report, the owner of the system points out that you have not quite understood a portion of the operations of the system.

In summary, keep in mind the maxims outlined by Musselman (1994), and transcribed below:

- ✓ Define the problem to be studied, including a written statement of the problem-solving objective
- ✓ Work on the right problem; fuzzy objectives lead to unclear success.
- ✓ Listen to the customer; do not look for a solution without firsts listening to the problem.
- ✓ Communicate; keep people informed, for the journey is more valuable than the solution.
- ✓ Only by knowing where you started can you judge how far you have come.
- ✓ Direct the model; advance the model by formulating it backwards.
- ✓ Manage expectations; it is easier to correct an expectation now, than to change a belief later.
- ✓ Do not take data for granted.
- ✓ Focus on the problem more than the model.
- ✓ Add complexity; do not start with it.
- ✓ Do not let the model become so sophisticated that it compensates for a bad design, or so complex that it goes beyond your ability to implement.
- ✓ Verbal agreements are not worth the paper they are printed on.

- ✓ Customer perceptions require attention.
- ✓ Report successes early and often.
- ✓ If it does not make sense, check it out.
- ✓ People decide; models do not.
- ✓ Know when to stop; ultimate truth is not affordable.
- ✓ Present a choice; people do not resist their own discoveries.
- ✓ Document, Document, Document!

6 CONCLUDING REMARKS

This tutorial has been intended as an introduction to this powerful technique. In-depth study of other sources is strongly recommended before engaging in a simulation study. Sources for simulation modeling include Nelson (1995), Banks and Carson (1984), Law and Kelton (1991), Sadowski (1993), Pidd 91994). For statistical analysis, recommended source include Nelson (1992), Law and Kelton (1991), and Alexopoulos (1995). For software selection, Banks (1995) is a good starting point. For the methodology in general, excellent sources include Pegden, Shannon, and Sadowski 1995), Shannon (1975), Banks and Carson (1991), and Law and Kelton (1991).

REFERENCES

Alexopoulos, C. 1995. Advanced Methods for Simulation Output Analysis. In *Proceedings of the 1995 Winter Simulation Conference*, Eds. C. Alexopoulos, K. Kang, W. R. Lilegdon, and D. Goldsman. 101-109.

Balci, O. 1995. Principles and Techniques of Simulation Validation, Verification, and Testing. In *Proceedings of the 1995 Winter Simulation Conference*, Eds. C. Alexopoulos, K. Kang, W. R. Lilegdon, and D. Goldsman.

Balci, O. 1996. Principles of Simulation Model Validation, Verification, and Testing. *International Journal in Computer Simulation*, to appear.

Banks, J. 1995. Software Simulation. In *Proceedings of the 1995 Winter Simulation Conference*, Eds. C . Alexopoulos, K. Kang, W. R. Lilegdon, and D. Goldsman. 32-38

Banks, J., B. Burnette, H. Kozloski, and J. Rose. 1994. *Introduction to SIMAN V and CINEMA*. New York: John Wiley & Sons USA.

Banks, J. and J. S. Carson. 1984. *Discrete-Event System* Simulation. Englewood Cliffs, New Jersey, Prentice Hall.

Blaisdell, W. E. and J. Haddock. 1993. Simulation Analysis Using SIMSTAT 2.0. In *Proceedings of the 1993 Winter Simulation Conference*, Eds. G. W. Evans, M. Mollaghasemi, E. C. Russell, W. E. Biles. 213-217.

Henriksen, J. O. and R. C. Crain. 1989. *GPSS/H Reference Manual*. Annandale, Va. Wolverine Software Corporation.

Law, A. and W. D. Kelton. 1991. *Simulation Modeling & Analysis*. 2nd edition, New York, McGraw-Hill, Inc. USA.

Law, A. M. and S. Vincent. 1995. *ExpertFit™ User's Guide*, Averill M. Law & Associates, P.O. Box 40996, Tucson, Arizona 85717.

Musselman, K. J. 1994. Guidelines for Simulation Project Success. In *Proceedings of the 1994 Winter Simulation Conference*, Eds. J. D. Tew, S. Manivannan, D. A. Sadowski, and A. F. Seila. 88-95

Nelson, B. L. 1992. Designing Efficient Experiments. In *Proceedings of the 1992 Winter Simulation* Conference, ed. J. J. Swain, D. Goldsman, R. C. Crain, and J. R. Wilson, 126-132.

Nelson, B. L. 1995. *Stochastic Modeling: Analysis & Simulation*. McGraw-Hill, Inc. USA.

O'Reilly, J. J. 1994. Introduction to SLAM II and SLAMSYSTEM. In *Proceedings of the 1994 Winter Simulation Conference*, Eds. J. D. Tew, S. Manivannan, D. A. Sadowski, and A. F. Seila. 415-419.

Pegden, C. D., R. E. Shannon, and R. P. Sadowski. 1995. *Introduction to Simulation Using SIMAN*. 2nd Edition. New York, McGraw Hill USA.

Pidd, M. 1994. An Introduction to Computer Simulation. In *Proceedings of the 1994 Winter Simulation Conference*, Eds. J. D. Tew, S. Manivannan, D. A. Sadowski, and A. F. Seila. 7-14.

Pritsker, A.A. B. 1992. Simulation: The Premier Technique of Industrial Engineering. *Industrial Engineering*, 24(7) : 25-26.

Sadowski, R. 1993. Selling Simulation and Simulation Results. In *Proceedings of the 1993 Winter Simulation Conference*. ed. G. W. Evans, M. Mollaghasemi, E. C. Russell, W. E. Biles, 65-68.

Sargent, R. G. 1984. Simulation Model Validation. *Simulation and Model-Based Methodologies: An Integrative View*, Eds. Oren, et al. Springer-Verlag.

Sargent, R. 1994. Verification and Validation of Simulation Models. In *Proceedings of the 1994 Winter Simulation Conference*, Eds. J. D. Tew, S. Manivannan, D. A. Sadowski, and A. F. Seila. 77-87.

Shannon, R. E. 1975. *Systems Simulation: The Art and the Science*, New Jersey: Prentice-Hall, Inc.

Systems Modeling Corporation. 1994. *SIMAN V Reference Guide.* Sewickley, Pennsylvania.

AUTHOR BIOGRAPHY

MARTHA A. CENTENO is an assistant professor in the Department of Industrial and Systems Engineering at Florida International University. She received a B.S. in Chemical Engineering from ITESO University (Guadalajara, Jalisco, Mexico), a M.S. in Industrial Engineering from Louisiana State University, and a Ph. D. in Industrial Engineering from Texas A&M University. Dr. Centeno's current research interest include the utilization of artificial intelligence and database technologies to develop comprehensive and smart simulation modeling environments. Dr. Centeno is a member of ASA, Alpha Pi Mu, Tau Beta Pi, IIE, INFORMS, and SCS.

Proceedings of the 1996 Winter Simulation Conference
ed. J. M. Charnes, D. J. Morrice, D. T. Brunner, and J. J. Swain

INSIDE SIMULATION SOFTWARE: HOW IT WORKS AND WHY IT MATTERS

Thomas J. Schriber

Computer and Information Systems
The University of Michigan
Ann Arbor, Michigan 48109-1234, U.S.A.

Daniel T. Brunner

Systemflow Simulations, Inc.
6366 Guilford Avenue, Suite 310
Indianapolis, Indiana 46220-1750, U.S.A.

ABSTRACT

This paper provides simulation practitioners and interested simulation consumers with a grounding in how discrete-event simulation software works. Topics include discrete-event systems and modeling; entities, resources and operations; simulation runs; entity states; entity lists; and entity-list management. The implementation of these generic ideas in SIMAN, ProModel, and GPSS/H is described. The paper concludes with several examples of "why it matters" for modelers to know in fine detail how their simulation software works. (This paper is an updated version of an identically named paper appearing in the *Proceedings of the 1995 Winter Simulation Conference,* pp. 110-117.)

1 INTRODUCTION

1.1 Background

A "black box" approach is often taken in teaching and learning discrete-event simulation software. The external characteristics of the software are studied, but the foundation on which the software is based is ignored or the foundation is touched on only briefly. Choices made in implementation of the foundation might not be studied at all and related to step-by-step model execution. The modeler therefore might not be able to reason things through when faced with such needs as developing good approaches for modeling complex situations, using interactive tools to come to a rapid understanding of error conditions arising during model development, and using interactive tools to verify that complex system logic has been accurately captured in a model. The objective of this paper, then, is to present the logical underpinnings of discrete-event simulation and illustrate this material in terms of three implementations of discrete-event simulation software.

1.2 Structure of the Paper

In Sections 2, 3 and 4 we comment on the nature of discrete-event simulation; entities, resources and operations; and simulation experiments. Sections 5 and 6 deal with entity states and entity management structures. Section 7 discusses the implementation of the preceding generic material in terms of SIMAN, ProModel, and GPSS/H. Section 8 explores "why it matters."

1.3 Terminology and Conventions

Throughout this paper we use terms that we define as well as terms reserved by the developers of particular simulation tools. Terms we define are **boldfaced** on first use. Tool-specific terms are Capitalized or, where appropriate, are spelled out in ALL CAPS.

2 ABOUT DISCRETE-EVENT SIMULATION

2.1 The Transaction-Flow World View

The "transaction-flow world view" often provides the basis for discrete-event simulation. In this view, a system consists of discrete units of traffic that move ("flow") from point to point in the system while competing with each other for the use of scarce resources. The units of traffic are sometimes called "transactions," giving rise to the phrase "transaction flow."

Numerous systems fit the preceding description. Included are many manufacturing, health care, transportation, civil, communication, defense and information processing systems, and queuing systems in general.

2.2 The Nature of Discrete-Event Simulation

A discrete-event simulation is one in which the state of a model changes at only a discrete, but possibly random, set of time points, known as event times. Two or more traffic units often have to be manipulated at one and the same time point. Such "simultaneous" movement of traffic at a time point is achieved by manipulating units of traffic *serially* at that time point. This often leads to logical complexities in discrete-event simulation.

2.3 Discrete-Event Modeling Languages

The challenges faced by a *modeler* escalate for the *designer* of a modeling language. The designer must take the logical requirements of discrete-event simulation into account in a generalized way. Choices and tradeoffs exist. As a result, although discrete-event simulation languages are similar in broad terms, they can and typically do differ in subtle but important particulars.

3 ENTITIES, RESOURCES AND OPERATIONS

The generic term **entity** is used here to designate a unit of traffic (a "transaction"). Entities instigate **events**. An event is a happening that changes the state of a model (or system). In a model of an order-filling system, for example, the arrival of an order, which is an event, might be simulated by bringing an entity into the model.

It is useful to distinguish between two possible types of entities, here referred to as **external entities** and **internal entities**. External entities are those whose creation and movement is explicitly arranged for by the modeler. In contrast, internal entities are created and manipulated implicitly by the simulation software itself. For example, internal entities might be used in some languages to simulate machine *failures*, whereas external entities might be used to simulate the *use* of machines by work-in-process.

The generic term **resource** is used to designate constructs that model system elements (such as servers) that can cause unwanted delays for units of traffic. All tools have "resource" constructs for the *direct* representation of servers, and other constructs (e.g., switches, counters, variables) whose state or value can be the basis for entity delay. The view here is that any of these constructs can at least sometimes appropriately be called a "resource."

An **operation** is a step carried out by or on an entity while it moves through a system. The operations applicable to a ship at a harbor might be these: arrive; capture a berth; capture a tugboat; get pulled into the berth; free the tugboat; load cargo; etc. "Capture" is an example of a compound operation, e.g., wait (if necessary) for a tugboat, then take control of the tugboat. Some operations require time to pass, e.g., the loading of cargo.

4 OVERVIEW OF MODEL EXECUTION

4.1 Experiments, Replications, and Runs

A simulation project is composed of **experiments**. Experiments are differentiated from each other by the use of alternatives in a model's logic and/or data. An alternate part sequencing rule might be tried, for example, or the quantity of various machines might be varied.

Each experiment consists of one or more **replications** (trials). A replication is a simulation that uses the experiment's model logic and data but a different set of random numbers, and so produces different statistical results that can then be analyzed across a set of replications.

A replication involves initializing the model, running it until a run-ending condition is met, and reporting results. This "running it" phase is called a **run**.

4.2 Inside a Run

During a run the simulation **clock** (an internally managed, stored data value) tracks the passage of *simulated* time (as distinct from *wall-clock* time). The clock advances in discrete steps (usually of unequal size) during the run. After all possible actions have been taken at a given simulated time, the simulation clock is advanced to the time of the next earliest scheduled event. Then the appropriate actions are carried out at this new simulated time, etc.

The execution of a run thus takes the form of a two-phase loop: "carry out all possible actions at the current simulated time," followed by "advance the simulated clock," repeated over and over again until a run-ending condition comes about. The two phases are here respectively called the **Entity Movement Phase** (EMP) and the **Clock Update Phase** (CUP).

5 ENTITY STATES

Entities migrate from state to state while they work their way through a model. An entity is always in exactly one of five alternative states, as detailed below.

5.1 The Active State

The **Active State** is the state of the currently moving entity. Only one entity moves at any instant of wall-clock time. This entity progresses through its operations nonstop until it encounters a delay. It then migrates to an alternative state (or leaves the system). Some other entity then becomes the next active entity. And so on.

5.2 The Ready State

During an Entity Movement Phase there may be *more than one* entity ready to move, and yet entities can only move (be in the Active State) one-by-one while wall-clock times goes by. The **Ready State** is the state of those entities waiting their turn to enter the Active State during the current Entity Movement Phase.

5.3 The Time-Delayed State

The **Time-Delayed State** is the state of entities waiting for a *known* future simulated time to be reached so that they can then (re)enter the Ready State. A "part" entity is in a Time-Delayed State, for example, while waiting for the future simulated time at which an operation currently being performed on it by a machine will come to an end.

5.4 The Condition-Delayed State

The **Condition-Delayed State** is the state of entities delayed until some specified condition comes about, e.g., a "part" entity might wait in the Condition-Delayed State until its turn comes to use a machine. Condition-Delayed entities are removed *automatically* from the Condition-Delayed state when conditions permit.

5.5 The Dormant State

Sometimes it is desirable to put entities into a state from which no escape will be triggered automatically by changes in model conditions. We call this state the **Dormant State.** Dormant-State entities rely on modeler-specified logic to transfer them from the Dormant State back to the Ready State. Job-ticket entities might be put into a Dormant State, for example, until an operator entity decides which job-ticket to pull next.

6 ENTITY MANAGEMENT STRUCTURES

Simulation software uses the following lists to organize and track entities in the five entity states.

6.1 The Active Entity

The active entity forms a single list of length one. This "list" is not given a name here. The Active-State entity moves nonstop until encountering an operation that puts it into another state (transfers it to another list) or removes it from the model. A Ready-State entity then becomes the next Active-State entity. Eventually there is no possibility of further action at the current time. The EMP then ends and a Clock Update Phase begins.

6.2 The Current Events List

Entities in the Ready State are kept in a single list here called the **current events list** (CEL). Entities migrate to the current events list from the future events list, delay lists, and user-managed lists. (Each of these latter lists is described below). In addition, any entities cloned from the Active-State entity usually start their existence on the current events list.

6.3 The Future Events List

Entities in the Time-Delayed State belong to a single list into which they are inserted at the beginning of their time-based delay. This list, called the **future events list** (FEL) here, is usually ranked by increasing entity **move time.** (Move time is the simulated time at which an entity is scheduled to try to move again.) At the time of entity insertion into the FEL, the entity's move time is calculated by adding the value of the simulation clock to the known (sampled) duration of the time-based delay.

After an Entity Movement Phase is over, the Clock Update Phase sets the clock's value to the move time of the FEL's highest ranked (smallest move time) entity. This entity is then transferred from the FEL to the CEL, migrating from the Time-Delayed State to the Ready State and setting the stage for the next EMP to begin.

The preceding statement assumes there are not *other* entities on the FEL whose move time matches the clock's updated value. In the case of move-time ties, some tools will transfer all the time-tied entities from the FEL to the CEL during a single CUP, whereas other tools take a "one entity transfer per CUP" approach.

Languages that work with internal entities usually use the FEL to support the timing requirements of these entities. The FEL is typically composed both of external and internal entities in such languages.

6.4 Delay Lists

Delay lists are lists of entities in the Condition-Delayed State. These entities are waiting for a condition to come about (e.g., waiting their turn to use a machine) so they can be transferred automatically into Ready State on the current events list. Delay lists, which are created automatically by the simulation software, are managed by using **related waiting** or **polled waiting**. (The approach taken may depend on the particular software being used.)

If delay can be related easily to events in the model that eliminate the delay, then related waiting can be used to manage the delay list. For example, suppose a machine's status changes from busy to idle. In response, the software can then automatically remove the next machine-using entity from the appropriate delay list and put it in Ready State on the current events list. Related waiting is the prevalent approach used to manage delay.

If the delay condition is too complex to be related easily to delay-resolving events in the model, polled waiting can be used to manage the delay list. With polled waiting the software checks regularly and routinely to see if entities can now be transferred automatically from delay lists to the Ready State.

Complex delay conditions for which polled waiting can be useful can include Boolean (*and/or*) combinations of state changes, e.g., a part supply runs low *or* an output bin needs to be emptied.

6.5 User-Managed Lists

User-managed lists are lists of entities in the Dormant State. The modeler must take steps to establish such lists and provide the logic needed to transfer entities to and from the lists. (The underlying software has no way to know why entities are put into user-managed lists and so has no basis for removing entities from such lists automatically.)

7 IMPLEMENTATION IN THREE TOOLS

The three tools chosen here for commentary on implementation particulars are Systems Modeling Corporation's SIMAN V; ProModel Corporation's ProModel, Version 3; and Wolverine Software Corporation's GPSS/H, Releases 2.x and 3. (See the References.) These three are among more than *fifty* tools reported in 1995 for discrete-event simulation (Swain 1995). Some other tools might be better suited than any of these three for particular modeling activities, but we think that these three tools are representative.

7.1 SIMAN

SIMAN V equivalents of the generic terms in earlier sections are given in Table 1. For example, SIMAN uses *Blocks* to specify operations for *Entities*.

Generic Term	SIMAN Equivalent
Entity	Entity
Resource	Resource; Blockage; Conveyor; Transporter
Operation	Block
Current Events List	Current Events Chain
Future Events List	Future Events Heap
Delay List	Attached or Internal Queue
User-Managed List	Detached Queue

Table 1: SIMAN Terminology

7.1.1 The Current Events Chain

The current events list is called the Current Events Chain (CEC) in SIMAN. The SIMAN CEC is composed of all Ready-State Entities. The first step in SIMAN's Entity Movement Phase is to remove the Entity from the head of the CEC and make it active.

If Entities are placed on the CEC while the Active-State Entity is moving, they are inserted in *last-in, first-out* order. For example, if an Entity produces a **clone** (a copy of itself placed immediately on the CEC), then after the clone-producing Entity is no longer active, its clone will be the next Active-State Entity. If the Active-State Entity produces two or more clones simultaneously, they will be inserted at the front of the CEC (that is, LIFO in terms of other Ready-State Entities) *but* they will be FIFO among themselves.

When the active Entity leaves the Active State and there are no more Ready-State Entities, the EMP checks all polled wait conditions, transfers any qualifying Condition-Delayed Entities to the Ready State on the CEC, etc., until no more action can be taken at the current simulated time. Then the next CUP takes place.

7.1.2 The Future Events Heap

Time-Delayed Entities in SIMAN reside in a structure named the Future Events Heap (FEH). This structure behaves like a list ranked on increasing move time. The Entity with the earliest move time is the next one off the Future Events Heap when a Clock Update Phase occurs.

If there are ties for earliest move time, SIMAN will remove all the tied Entities from the Future Events Heap during one and the same Clock Update Phase.

The FEH can contain internal Entities resulting from elements specified by the modeler in the SIMAN experiment file. An example is beginning-of-downtime and end-of-downtime Entities. When an internal Entity is encountered on the FEH during a CUP, it is processed immediately. (In contrast, external Entities are simply put into Ready State on the CEC.) Because of internal Entities, the CEC might be empty after a CUP takes place. An EMP nevertheless takes place, insuring that a timely check of polled wait conditions will be made.

7.1.3 Attached and Internal Queues

Attached and Internal Queues are the two types of SIMAN lists containing Entities engaged in related waiting. Related waiting results from the use of Hold Blocks. For example, SEIZE, the Block used by an Entity (e.g., a part) to capture a Resource (e.g., a machine), is a Hold Block. Other related-waiting Hold Blocks are ACCESS, ALLOCATE, PREEMPT, PROCEED, REQUEST, and WAIT.

Associated with each Hold Block is an Attached or Internal Queue in which Entities wait for the Hold condition to be satisfied. If a QUEUE Block immediately precedes a Hold Block, an Attached Queue results. An Attached Queue is a *named* list of Entities waiting to execute the associated Hold Block.

Sometimes it is convenient to use identical Hold Blocks at multiple points in a model, e.g., to use a "SEIZE DRILL" Block two or more places in a model. The modeler can choose to associate with each Hold Block its own Attached Queue. These are *unshared* Attached Queues, because *two or more* lists of Entities then wait for the *same* Resource. Hold-Block Priority is used to determine the next Entity to get the Resource.

Alternatively, the modeler can put Entities delayed at two or more identical Hold Blocks into *one and the same* Attached Queue, a *shared* Attached Queue. Shared Attached Queues require use of the keyword SHARED in the QUEUES element in the SIMAN experiment file.

Entities are put into Attached Queues FIFO or LIFO, or are inserted into the Queues based on the value of a modeler-supplied expression.

If no QUEUE Block precedes a Hold Block, SIMAN provides an Internal Queue for that Hold. An Internal Queue is an *unnamed, non-sharable FIFO* Queue.

The SCAN Hold Block is an exception to the rule that Hold Blocks implement related waiting. A SCAN Block delays Entities until a user-supplied expression (involving system-state information and/or data values) is true. Queues (delay lists) that form at SCAN Blocks are polled at the end of each Entity Movement Phase.

7.1.4 Detached Queues

Detached Queues are Entity lists used by SIMAN to implement the Dormant State. Entities are put into Detached Queues when they execute QUEUE Blocks at which a DETACHED modifier is specified. Such Entities are later transferred from their Dormant State to the Ready State by other Entities that use SEARCH and REMOVE or QPICK and MATCH Blocks for this purpose.

7.2 ProModel

ProModel equivalents of the earlier generic terms are given in Table 2. In ProModel, *Entities* compete for *Locations* and *Resources* and engage in operations on the basis of *Operation* and *Routing Statements*.

Generic Term	ProModel Equivalent
Entity	Entity
Resource	Location; Resource; Variable; Node
Operation	Operation Statement; Routing Statement
Current Events List	Action List
Future Events List	Future Events List
Delay List	Waiting List
User-Managed List	None (but see 7.2.3)

Table 2: ProModel Terminology

A Location models the physical space an Entity occupies. An Entity can only occupy one Location at a time. ProModel Resources are used to model resources other than Locations, e.g., forklift trucks; humans. Entities can control multiple Resources simultaneously.

Resources themselves can compete for *Nodes*, moving independently through a Node network in search of something to pick up or a place to be idle. In this sense Resources can behave like Entities. They can migrate among Entity states, and they are tracked in lists.

A ProModel *Variable* is a general-purpose data element whose value can be the object of a WAIT UNTIL and for which ProModel collects statistics.

7.2.1 The Action List

The ProModel Action List contains Entities (and Resources) in the Ready State. The list is ranked LIFO and is empty by the end of each EMP. Deactivation of the active Entity or Resource causes the first Entity or Resource on the Action List to become active.

7.2.2 The Future Events List

Entities undergoing WAIT operations, and Resources (while moving), along with certain internally generated Entities and events, can wait on the Future Events List (FEL). Processing is "first out based on earliest move time." ProModel will remove only one Entity or event per CUP. In case of time ties there will be successive EMPs that occur at the same instant of simulated time.

Many ProModel model-definition constructs have optional user-defined *Logic* fields, e.g., Downtime Logic and Location Exit Logic. Logic is a collection of Operation Statements automatically executed when appropriate. An Entity can launch *Independent Logic* which is like a subroutine call that is to be executed by a clone. We mention Logic here because Downtime Logic and Independent Logic can produce non-Entity-related events on the FEL. When processed, these events may go into another Future-Events-List wait or into some type of delay list. They may cause Entities (or Resources) to materialize on the Action List.

7.2.3 Waiting Lists

ProModel's *Waiting Lists* function as delay lists to implement a variety of related waiting conditions. There is no polled waiting (for reasons given below) and there are no user-managed lists (see the last 7.2.3 paragraph).

To understand the interaction among Locations, Resources, and Variables, we need to consider the model definition framework of ProModel. The entity-flow part of ProModel is specified by the modeler via an ordered collection of *Process Steps* that make up a *Process Table*. Every Process Step includes the name of an Entity Type (or *All*) and the name of a Location (or *All*). An Entity "flows" from one Process Step to the next by jumping to the next Process Step that matches its Type and Location (starting over again at the top of the Table if necessary). This determines "what this Entity Type is supposed to do at this Location."

A Process Step is composed of *Operation Logic* and/or *Routing Logic* components. Competition among Entities for non-transportation Resources is spelled out in the Operation Logic. Competition among Entities for Locations and transportation Resources is spelled out in the Routing Logic. Routing Logic is acted upon after Operation Logic has been acted upon.

A Waiting List (for Entities) is attached to each Location, to each Resource, and to each Variable. A Waiting List (for Resources) is attached to each Node. (Competition among Resources for Nodes takes place automatically based on Path Networks, Work/Park Lists, and Node/Location associations defined outside the table of Process Steps.)

A single Entity (or Resource) can be represented in two or more delay lists of the same type simultaneously. As a result, ProModel does not require a polling mechanism for modeling certain Boolean conditions. (The Entity representations are deleted from all relevant delay lists when a delay-ending condition comes about.)

There are various Routing Rule options for specifying next-Location alternatives. And it is possible to define a Location in such a way that it can override the ranking of its delay list when it is ready to accept another occupant.

ProModel has no user-managed lists as such. However, JOIN, LOAD, and SEND are all Routing Logic options that place Entities on special Location-specific lists where they await a JOIN, LOAD, or SEND Operation Statement, respectively, to be executed by another Entity at the destination Location. This explicit triggering makes these special lists *resemble* user-managed lists. But because the lists and conditions are just for Locations and are not custom-managed, we consider the waiting to be related waiting and the lists to be delay lists, not user-managed lists.

7.3 GPSS/H

GPSS/H equivalents of the generic terms in the preceding sections are given in Table 3. For example, GPSS/H uses *Blocks* to specify operations for *Transactions*.

Generic Term	GPSS/H Equivalent
Entity	Transaction
Resource	Facility; Storage; Logic Switch
Operation	Block
Current Events List	Current Events Chain
Future Events List	Future Events Chain
Delay List	Current Events Chain
User-Managed List	User Chain

Table 3: GPSS/H Terminology

7.3.1 The Current Events Chain

As in SIMAN, the current events list is named the Current Events Chain in GPSS/H. A striking difference between GPSS/H and SIMAN is that by default in GPSS/H, Condition-Delayed Transactions (Xacts) are commingled with Ready-State Transactions on the CEC. For such Xacts, the CEC itself can be thought of as a single global GPSS/H delay list.

Other than the CEC and some internal delay lists, there are no delay lists in GPSS/H. (GPSS/H has a Queue construct and a QUEUE Block that do not perform list management functions; they are for statistics gathering purposes only.)

A characteristic of GPSS/H is that Transactions on the CEC are ranked FIFO within Priority Class. (Priority Class is a Transaction attribute.) This reflects the CEC's global-delay-list function.

Like other types of delay lists and unlike other types of current events lists, the GPSS/H CEC is frequently *not* empty whenever an EMP ends.

7.3.2 The Scan Phase

The EMP in GPSS/H is called the *Scan Phase*. The GPSS/H Scan Phase is more involved than the EMP in SIMAN and ProModel. (See Schriber 1991.)

GPSS/H starts a Scan Phase with the Transaction (Xact) at the head of the CEC and tries to move that candidate-Xact into its next Block. If the Block is one that can deny entry (SEIZE, ENTER, GATE, TEST or PREEMPT) and entry *is* denied, then the Xact is in a Condition-Delayed State and GPSS/H *leaves the candidate on the CEC* and examines the sequential CEC Xact. If entry is not denied, then the candidate becomes the active Xact (without being removed from the CEC) and begins executing Blocks.

If the active Transaction tries to execute a Block and entry is denied, the Transaction shifts to the Condition-Delayed State and remains on the CEC. GPSS/H then resumes scanning the CEC for the next active Transaction. However, because of possible state changes precipitated by the previously active Transaction's Block execution(s), the scan will either continue sequentially or *restart* (see below).

The GPSS/H mechanism of keeping certain Condition-Delayed Transactions on the CEC and examining them one or more times during the Scan Phase to see if they are in the Ready State at the instant of examination implies that all of these Transactions are fundamentally in a polled wait condition.

7.3.3 Restarting the Scan

GPSS/H has an internal status change flag (SCF) that is set to *true* whenever a *unique blocking condition* (see Section 7.3.4) is resolved. If the SCF is *true* when the active Transaction ceases to be active, the SCF is set back to *false* and the scan restarts at the head of the CEC as if the EMP had just begun; otherwise the scan of the CEC continues with the sequential CEC Xact.

Scan restarts occur because there may be Transactions at or near the top of the CEC that should be given first crack at moving in response to the resolution of a unique blocking condition. The net effect of scan restarts and CEC Xact ranking within Priority Class is to provide FIFO-within-Priority-Class queuing in GPSS/H for condition-delayed Xacts resident on the CEC.

The active Xact can execute a YIELD (synonym: BUFFER) Block in GPSS/H to return itself temporarily to the Ready State and force an *immediate* scan restart. The restarted scan will eventually re-encounter the yielding Xact at the same simulated time, which will then again become active. The ability of an Xact to yield control deliberately but only temporarily to one or more other Xacts is quite useful in discrete-event modeling.

7.3.4 Related Waiting on the CEC

State changes involving unique blocking include the transition of a Facility (server) into or out of use; the transition of a Storage (a GPSS/H counter with a capacity) to a smaller count, or out of the empty or into the full state; and a change in the setting of an on-or-off Logic Switch. Transactions waiting to SEIZE a Facility or ENTER a Storage or waiting at a GATE for a Storage to become non-empty or full or for a Logic Switch to change are in a unique blocking condition. (Other types of unique blocking are possible as well.)

Scan restarts imply extra processing demands while GPSS/H re-encounters and re-evaluates CEC Transactions. To offset this each Transaction has a Scan Skip Indicator (SSI) that flags those Transactions waiting for unique blocking conditions to be resolved. The SSI flag is checked before an attempt is made to move a candidate-for-active Transaction into its next Block, allowing the scan to quickly skip over such Condition-Delayed Transactions.

An Xact's SSI is cleared automatically at the moment of resolution of the unique blocking condition which has been forcing the Xact to wait. Internal delay lists are used to track which Xacts' SSIs need to be cleared in response to a given state change. These lists *are* related to an underlying condition, so the polled-waiting nature of the GPSS/H CEC scan is in fact a hybrid polled/related approach in the case of unique blocking.

7.3.5 The Future Events Chain

The GPSS/H Future Events Chain (FEC) is like future events lists in other tools. The GPSS/H CUP will remove multiple Transactions from the FEC if they are tied for the earliest move time, inserting them one by one into their appropriate place on the CEC.

GPSS/H does not use internal entities to model downtimes. GPSS/H models downtimes (and some other control conditions as well) with actual Xacts. These are ordinary Xacts (external entities) that go through the ordinary Time-Delayed State to simulate time-between-failures and time-to-repair.

7.3.6 User Chains

GPSS/H implements the Dormant State with User Chains, which are user-managed lists of Xacts. After a Transaction puts itself onto a User Chain (by executing a LINK Block), it can only be removed by another Transaction (which triggers the removal by executing an UNLINK Block). If UNLINK execution transfers one or more Dormant-State Transactions to Ready State, the SCF will be made *true* (to trigger a scan restart) so these CEC newcomers will have their turn to become active before the next CUP. User Chains can achieve performance improvements over CEC-based queuing because User Chains (like delay lists in other tools) need never be scanned except when an UNLINK is executed.

8 WHY IT MATTERS

8.1 Overview

We now describe three situations that reveal some of the practical differences in implementation particulars among SIMAN, ProModel and GPSS/H. (Space restrictions limit us to three situations.) We then conclude with comments on how knowledge of software internals is needed to make effective use of software checkout tools.

8.2 Trying to Re-capture a Resource Immediately

Suppose a part releases a machine, then immediately re-competes for the machine (e.g., RELEASE followed by SEIZE in SIMAN or GPSS/H; FREE or USE followed by GET or USE in ProModel). The objective is to let a more highly qualified waiting part be the next to capture the machine; in the absence of such a part, the releasing part itself is to re-capture the machine.

Of interest here is the order of events following the giving up of a server. There are at least three alternatives: (1) Coupled with the giving up of the server is the immediate choosing of the next user of the server, without the releasing entity having yet become a contender for the server. (2) The choosing of the next user of the server is deferred until the releasing entity has become a contender. (3) "Neither of the above"; that is, without paying heed to other contenders, the releasing entity recaptures the server immediately.

SIMAN, ProModel, and GPSS/H respectively implement the first, second and third alternatives by default. And so one or another alternative is in effect in the tools considered here, reflecting differing implementation choices made by the software designers.

Note that these alternatives are not intrinsically either "right" or "wrong." The modeler must be aware of the alternative in effect and work with it to produce the desired outcome. (If a modeler is unaware of the alternative in effect in the simulation software being used, it is possible to model a given situation with an unintended effect and perhaps not even become aware of this fact.)

8.3 The First in Line is Still Delayed

Suppose two Condition-Delayed entities are waiting in a list because no units of a particular resource are idle. Suppose the first entity needs *two* units of the resource, whereas the second entity only needs *one* unit. Now assume that one unit of the resource becomes idle. The needs of the first list entity cannot yet be satisfied, but the needs of the second entity can. What will happen?

There are at least three possible alternatives: (1) Neither entity claims the idle resource unit. (2) The first entity claims the one idle resource unit and waits for a second unit. (3) The second entity claims the idle resource unit and goes immediately on its way.

As in Section 8.2, each of these alternatives comes into play in the tools considered here. SIMAN (SEIZE), ProModel (GET or USE), and GPSS/H (ENTER or TEST) respectively implement the first, second and third alternatives by default.

6.4 Yielding Control

Suppose the active entity wants to give control to one or more Ready-State entities, but then needs to become the active entity again before the simulated clock has been advanced. This might be useful, for example, if the active entity has opened a switch permitting a set of other entities to move past a point in the model, and then needs to re-close the switch after the forward movement has been accomplished. (Perhaps a group of identically-flavored cartons of ice cream is to be transferred from an accumulation point to a conveyor leading to a one-flavor-per-box packing operation.)

In SIMAN, the effect can be accomplished approximately with a DELAY that puts the active Entity into a Time-Delayed State for an arbitrarily short but non-zero simulated time.

In ProModel, "WAIT 0" can be used to put the active Entity back on the FEL. It will be returned later (at the same simulated time) by the CUP to the Active State.

In GPSS/H, the active entity can execute a YIELD (BUFFER) Block to move from the Active State to the Ready State and restart the CEC scan. Higher-priority Xacts on the CEC can then become active, one by one, before the control-yielding Xact itself again becomes active at the same simulated time. (If all relevant Xacts have the same priority, a "PRIORITY PR,YIELD" Block can be used to reposition the active Xact behind equal-priority Xacts on the CEC, shift the active Xact to the Ready State, and restart the scan of the CEC.)

6.5 Interactive Model Verification

We now comment briefly on why a detailed understanding of "how simulation software works" supports interactive probing of simulation-model behavior.

In general, simulation models can be run interactively or in batch mode. Interactive runs are of use in checking out (verifying) model logic during model-building and in troubleshooting a model when execution errors occur. Batch mode is then used to make production runs.

Interactive runs put a magnifying glass on a simulation model while it executes. The modeler can follow the active entity step by step and display the current and future events lists and the delay and user-managed lists as well as other aspects of the model. These activities yield valuable insights into model behavior for the modeler who knows the underlying concepts. Without such knowledge, the modeler might not take full advantage of the interactive tools provided by the software or, worse yet, might even avoid using the tools.

ACKNOWLEDGMENTS

Much of the information in the original version of this paper was derived from conversations with software-vendor personnel. The authors gratefully acknowledge the support provided by David T. Sturrock, Deborah A. Sadowski, C. Dennis Pegden and Vivek Bapat, all of Systems Modeling Corporation; Charles Harrell and Eric Du, of ProModel Corporation; Kerim Tumay (now of CACI Products Company); and Robert C. Crain and James O. Henriksen, of Wolverine Software Corporation.

REFERENCES

Baird, S. P., and J. J. Leavy. 1994. Simulation Modeling Using ProModel for Windows. In *Proceedings of the 1994 Winter Simulation Conference*, 527-532. LaJolla, California: Society for Computer Simulation.

Banks, J., J. S. Carson, and J. N. Sy. 1995. *Getting Started with GPSS/H*, Second Edition. Annandale, Virginia: Wolverine Software Corporation.

Banks, J., B. Burnette, H. Kozloski, and J. Rose. 1995. *Introduction to SIMAN V and Cinema V*. New York, New York: John Wiley & Sons.

Crain, R. C. 1995. GPSS/H Release 3. In *CHECKPOINT,* Vol. 11, No. 1. Annandale, Virginia: Wolverine Software Corporation.

Crain, R. C., and D. S. Smith. 1995. Industrial Strength Simulation Using GPSS/H. In *Proceedings of the 1995 Winter Simulation Conference,* 487-493. LaJolla, California: Society for Computer Simulation.

Henriksen, J. O. 1995. An Introduction to SLX. In *Proceedings of the 1995 Winter Simulation Conference,* 502-509. LaJolla, California: Society for Computer Simulation.

Pegden, C. D., R. E. Shannon, and R. P. Sadowski. 1995. *Introduction to Simulation Using SIMAN,* Second Edition. New York, New York: McGraw-Hill.

Profozich, D. M., and D. T. Sturrock. 1995. Introduction to SIMAN/Cinema. In *Proceedings of the 1995 Winter Simulation Conference,* 515-518. LaJolla, California: Society for Computer Simulation.

ProModel Corporation. 1995. *ProModel Version 3 User's Guide.* Orem, Utah: ProModel Corporation.

Schriber, T. J. 1991. *An Introduction to Simulation Using GPSS/H.* New York, New York: John Wiley.

Swain, J. J. 1995. Simulation Survey: Tools for Process Understanding and Improvement. *OR/MS Today*, August '95, 64-79, Baltimore, Maryland: INFORMS.

AUTHOR BIOGRAPHIES

DANIEL T. BRUNNER is President of Systemflow Simulations, Inc., a services firm active in manufacturing, material handling, distribution, transportation, health care, computer systems, and mining. He received a B.S.E.E. from Purdue University and an MBA from The University of Michigan. He has served as Winter Simulation Conference Publicity Chair (1988) and Business Chair (1992), and is General Chair for the 1996 WSC. He is a member of IIE and SCS.

THOMAS J. SCHRIBER is a Professor of Computer and Information Systems at The University of Michigan. He teaches a number of subjects while doing research and consulting in discrete-event simulation. He is a member of ASIM (the German-language simulation society), DSI, IIE, and INFORMS.

Proceedings of the 1996 Winter Simulation Conference
ed. J. M. Charnes, D. J. Morrice, D. T. Brunner, and J. J. Swain

SOFTWARE FOR SIMULATION

Jerry Banks

School of Industrial and Systems Engineering
Georgia Institute of Technology
Atlanta, Georgia 30332-0205

ABSTRACT

This tutorial describes software for conducting computer simulation other software that supports simulation.

1 INTRODUCTION

The next two sections describe general purpose and manufacturing oriented software, respectively. The fourth section describes simulation software for business process reengineering. Next, simulation based scheduling software is introduced. Then, an animator for simulation is discussed. Finally, some simulation support software is described.

2 GENERAL PURPOSE SOFTWARE

Simulation applications are usually accomplished with the use of specially developed software. This tutorial describes the software in two categories. The first of these is software for general purposes. This type of software can solve almost any discrete simulation problem. In this section, seven products, GPSS/H™, GPSS/World™, SIMAN®/Cinema®, SIMSCRIPT II.5®, AweSim®, SIMPLE++™, and Extend™ are discussed.

GPSS/H is a product of Wolverine Software Corporation (Crain and Smith 1995). It is a flexible, yet powerful tool for simulation. It provides improvements over GPSS V released many years earlier by IBM. These enhancements overcome the need to use external routines in FORTRAN to accomplish complex modeling tasks, significantly faster execution, an interactive debugging environment, a floating-point clock, built-in file I/O, use of named entity parameters, extended simulation control statements, built-in math and trig functions, and ampervariables that allow complex arithmetic combinations to be used in the simulation. Version 3.0 adds 23 new random variate distributions, generalized data assignment statements and simplified syntax, enhancements to manipulating entities on user-defined chains, and several new system attributes for extracting data from the simulation. Options available include 32-bit GPSS/H Professional for unlimited model size, Personal GPSS/H for size-limited models, and Student GPSS/H.

GPSS World, from Minuteman Software, is a complete redesign of GPSS/PC™ (Cox 1991). It is designed as a high power environment for simulation professionals. It includes both discrete and continuous simulation. Its features include interactivity, visualizability, and configuration flexibility. It utilizes 32-bit computing, virtual memory, preemptive multitasking, symmetric multiprocessing, and distributed simulation.

Highlights include drag-and-drop model building, 512 megabytes of virtual memory for models, point-and-shoot debugging, an embedded programming language, built-in probability distributions, multiple data types, and many other improvements to GPSS/PC.

There is an enhanced memory version of GPSS/PC that is also available. It allows access of up to 32 mb of memory.

SIMSCRIPT II.5, from CACI Products Company, is a language that allows models to be constructed that are either process oriented or event oriented (Russell 1993). The microcomputer and workstation versions include the SIMGRAPHICS animation and graphics package. SIMSCRIPT can be used to produce both dynamic and static presentation quality graphics such as histograms, pie charts, bar charts, levels of meters and dials and time plots of variables. Animation of the simulation output is also constructed using SIMGRAPHICS. SIMGRAPHICS can be used also to produce interactive graphical front-ends or forms for entering model input data. An input form may include such graphical elements as menu bars with pull-down menus, text or data boxes, and buttons that are clicked on with a mouse to select an alternative. The graphical model front-end allows for a certain set of modifications to the model to be made without programming, facilitating model use by those that are not programmers.

SIMAN/Cinema, from Systems Modeling Corporation, is a general-purpose simulation language and animation system designed to model discrete event, continuous, and combined discrete/ continuous systems (Profozich and Sturrock 1995; Banks, Burnette, Jones, and Kozloski 1995; Pegden, Shannon, and Sadowski 1995).

SIMAN models generally are constructed in a graphical form. In this form, the model is built in the Arena environment and the appropriate data for each SIMAN Block and Element are entered through one or more interactive dialog boxes. Models also can be constructed in a statement form, in which the Block and Element statements are entered into a file using any editor that can create an ASCII file.

Some important aspects of SIMAN V are as follows:

1. Special features useful in modeling manufacturing systems include the ability to describe environments as workcenters (stations) and the ability to define a sequence for moving entities through the system.

2. Constructs that enable the modeling of material handling systems including accumulating and non-accumulating conveyors, transporters, and guided vehicles.

3. An interactive run controller permits breakpoints, watches, and other execution control procedures.

4. The ARENA environment includes menu-driven point-and-click procedures for constructing and animating models.

5. The Input Analyzer assists in fitting distributions to data.

6. The Output Analyzer can be used to obtain confidence intervals, histograms, correlograms, and so on.

7. Models created on any platform are compatible on all other platforms.

AweSim, from Pritsker Corporation, is an integrated simulation system for PC's based on the Microsoft® Windows™ interface. All features are accessible through pull-down menus and dialog boxes, and are selected from the AweSim Executive Window. An AweSim project consists of one or more scenarios, each of which represents an alternative system configuration. A project maintainer examines the components of the current scenario to determine if any of them have been modified, indicates whether tasks such as model translation should be performed, and allows the user to accomplish these tasks before the next function is requested. AweSim allows multiple tasks to be performed in parallel while the simulation is operating in the background.

Some of the features of AweSim are as follows:

1. Models, i.e., networks, are built graphically. Most numeric fields accept expressions. The number of resources to allocate, or the activity number to select, can be determined as an entity flows through the network.

2. Output analysis includes a "report browser" that allows alternative text outputs to be compared side-by-side. Output may be viewed in the form of bar charts, histograms, pie charts, and plots. Output from multiple scenarios can be displayed at the same time in bar chart form. Using the Windows environment, multiple output windows can be opened simultaneously. Built on a relational database, output data can be accessed with commercial database software. Thus, custom reports can be developed.

3. Animations are defined via point and click operations. Animations may be previewed without running the simulation. Multiple animations can provide different concurrent views of the system.

4. AweSim was designed to be used in an integrated manner. For example, historic data may be read to drive the simulation. CAD drawings may be loaded. Output charts and plots created by AweSim may be exported via the clipboard to other applications.

5. Interactive simulation is attained via a control panel that enables stepping through a model, setting breakpoints, examining queues and system variables, and saving model status for a later restart.

SIMPLE++, from AESOP Corporation, is a fully object-oriented simulation system with an integrated graphical user interface (Geuder, 1995). The user creates models by making a library of objects. These library objects represent classes (or parents), whose instances (or children) can be inserted into the models. Simple++ takes advantage of the features of object orientation including modularity, class structure, inheritance, hierarchy, and polymorphism.

Extend, from Imagine That, Inc., is a visual, interactive simulation tool (Krahl, 1995) that contains a built-in development system that allows the user to construct components and build custom user interfaces. Models are constructed graphically by dragging and dropping blocks (high-level model components) from library windows onto the model worksheet. Data can be entered directly into block dialogs, interactively using controls, or read from files as the simulation runs. Output is in the form of plots, histograms, tables, and customizable reports. The block development environment allows simulation modelers to add custom functionality, including hooks to external languages such as C or Fortran. Additional features include multiple scenario analysis, unlimited hierarchical decomposition, animation, client/server interprocess communication, custom icons, work areas for centralized model control and reporting, over 300 built-in functions, 32-bit computing, and compatibility across platforms.

3 MANUFACTURING ORIENTED SOFTWARE

The software discussed in this section is limited to only seven within the category including ProModel®, AutoMod™, Taylor II®, WITNESS®, AIM™, Arena® and Extend+Manufacturing™. For the most part, these software are used for the simulation of manufacturing and material handling systems. References for these software packages include the following: Dewsnup and Bollenbach (1995) for ProModel, Rohrer (1994) for AutoMod, Thompson (1995) for WITNESS, and Nordgren (1995) for Taylor II. For AIM, the reference is Lilegdon, Martin, and Pritsker (1994). For Arena, the reference is Hammann and Markovitch (1995). The reference for Extend+Manufacturing is Krahl (1995).

ProModel, from PROMODEL Corporation, has programming features within the environment, and the capability to add C or Pascal type subroutines to a program. Some of the features of ProModel, Release 3, are as follows:

1. Models are created using a point and click approach. Intuitive interfaces, interactive dialog and online help are provided. An auto-build feature guides the user through the model building process. An online trainer is available.
2. The software operates in the Windows environment, as a 32-bit application, taking advantage of memory management techniques, synchronized windowing and data exchange. Windows fonts, printer drivers, cooperative multitasking, and the Dynamic Link Library are available.
3. Virtually unlimited model size is offered.
4. The simulator offers a 2-D graphics editor with scaling, rotating, and so on. Icons can be defined using either vector based or pixel graphics. These icons are saved as bitmaps at runtime for fast animation during the simulation.
5. CAD drawings as clipart can be imported as well as process information and schedules. Customized output reports and spreadsheet files can be produced. If the data is another Windows application, cutting and pasting can be accomplished.
6. The static and dynamic elements of the animation are developed while defining the model. That is, the simulation model and animation are integrated.
7. Business output graphics are automatically provided and may be printed in color.
8. Only standard hardware is required (IBM or compatible with VGA graphics). No special graphics cards, monitors, or math coprocessor chip is needed.
9. Preprogrammed constructs are provided. This allows for fast modeling of multi-unit and multi-capacity locations, shared and mobile resources, downtime, shifts, and so on.
10. Automatic statistics are available.
11. Submodels allow the creation of a library of templates of work steps, activities, or sub-processes that can be reused. This allows for model construction to be accomplished by a team with later merger of submodels into one model.
12. A free runtime, multiple scenario, capability is provided. Statistics from multiple scenarios can be viewed simultaneously.
13. Multiple views can be saved for easy navigation and graphic layout.
14. Cranes can be added with multiple bridges.
15. Models can be encrypted for protection of data.
16. Interactive subroutine can be added to allow changes to model parameters during model execution.

A model is constructed by defining a route for a part or parts, defining the capacities of each of the locations along the route, defining additional resources such as operators or fixtures, defining the transporters, scheduling the part arrivals, and specifying the simulation parameters. The software then prompts the user to define the layout and the dynamic elements in the simulation.

AutoMod, from AutoSimulations, Incorporated, has general programming features including the specification of processes, resources, loads, queues, and variables. Processes are specified in terms of traffic limits, input and output connections, and itineraries. Resources are specified in terms of their capacity, processing time, MTBF, and MTTR. Loads are defined by their shape and size, their attributes, generation rates, generation limits, and start times, as well as their priority.

The simulator is very powerful in its description of material handling systems. AGVs, conveyors, bridge cranes, AS/RS's, and power and free devices can be defined. The range of definition is extensive. For example, an AGV can be defined in terms of the following: multiple vehicle types, multiple capacity vehicles, path options (unidirectional or bidirectional), variable speed paths, control points, flexible control and scheduling rules, arbitrary blocking geometries, automatic shortest-distance routing, and vehicle procedures.

Numerous control statements are available. For example, process control statements include If-Then-Else, While-Do, Do-Until, Wait-Until, and Wait For. Load control, resource control and other statements are also available. C functions may be defined by the user. Attributes and variables may be specified.

The animation capabilities include true 3-D graphics, rotation, and tilting, to mention a few. A CAD-like drawing utility is used to construct the model. Business graphics can be generated.

In 1993, AutoSimulations added the Simulator to AutoMod. Features of the Simulator include its spreadsheet interface. This eliminates the need for programming in building models.

Another option as a separate utility is AutoStat. It provides simulation warmup capability, scenario management, confidence interval generation, and design of experiments capability.

Taylor II is a product developed by F&H Simulations. It runs under Windows. Working with Taylor II starts with building a model. All model building is menu driven. A model in Taylor II consists of four fundamental entities: elements, jobs, routings, and products. The element types are inout, machine, buffer, conveyor, transport, path, aid, warehouse, and reservoir. One or more operations can take place at an element. The three basic operations are processing, transport, and storage. Defining a layout is the first step when building a model. Layouts consist of element types. By selecting the elements in sequence, the product path or routing is defined. Routing descriptions may be provided from external files.

The next step is detailing the model. In this step the parameters are provided. In addition to a number of default values, Taylor II uses a macro language called TLI for Taylor Language Interface. TLI is a programming language that permits modifications of model behavior in combination with simulation-specific predefined and user-definable variables. TLI can also be used interactively during a simulation run to make queries and updates. Interface to C, Basic, and Pascal is also possible. Local and global attributes are available.

During simulation, zoom, pan, rotate, and pause are options. Modifications can be made on-the-fly. The time representation is fully user-definable (hours, days, seconds, and so on can be mixed). Output analysis possibilities include predefined graphics, user-defined graphics, predefined tabular reports, and user-defined reports. Examples of predefined graphics are queue histograms and utilization pies. User-defined outputs include bar graphs, stacked bars, and other business graphics. Predefined tables include job, element and cost reports.

Animation capabilities include both 2-D and 3-D. The 3-D animation can be shaded. Standard indicators can be shown for elements. Icon libraries for both 2-D and 3-D animation are provided. Each of these libraries contains more than 50 icons.

Additional features include 650 pages of online, context sensitive help with index and page-search capability. Educational support materials are available.

WITNESS, a Windows application from AT&T Istel, contains many elements for discrete-part manufacturing For example, machines can be single, batch, production, assembly, multi-station, or multi-cycle. Conveyors can be accumulating or non-accumulating. Options exist for labor, vehicles, tracks, and shifts. WITNESS also contains elements for continuous processing including processors, tanks, and pipes.

Variables and attributes may be specified. Parts can be smart (having their routing attached) or dumb (elements of the process decide the appropriate routing). Distributions and functions can be used for specifying operation times and for other purposes. Machine downtime can be scheduled on the basis of operations, busy time, or available time. Labor is a resource that can be preempted, use a priority system, and be scheduled based on current model conditions.

Track and vehicle logic allow requests for certain types of jobs, vehicle acceleration and deceleration, park when idle, and change destinations dynamically. Many types of routing logic are possible in addition to the standard push and pull. For example, If-Then-Else conditions may be specified.

Simulation actions, performed at the beginning and end of simulation events, may employ programming constructs such as For-Next, While-End, and GoTo-Label. The user can look at an element at any time and determine the status of a part.

Reporting capabilities include dynamic on-screen information about all modeled elements. Reports may be exported to spreadsheet software.

C-LINKS allows detailed programming and subroutines to be attached to WITNESS.

Data inputs can be numeric, variables, distributions, or a user-defined equation.

Debugging or brainstorming can be accomplished by stopping the model, changing desired parameters, and continuing with the model from the same point in simulation time.

An animation is constructed along with the model definition. This animation and statistical feedback can be turned on or off during any run. Many changes to the model may be made at any time.

Built-in experimentation capabilities are available from the menu bar. The results of the experiments are output to a CSV file by default, or other file types by user choice. The CSV file is in a format that allows the internal statistics package to create confidence intervals.

The capabilities of Witness, version 7.3, include the following:

1. Unlimited number of variables or attributes. Variables can be arrayed.
2. Up to 1000 distinct random number streams.
3. Bitmap import/export with icon sizes increased to 256x256 pixels.
4. 'Module' element for hierarchical modeling. One icon represents the detail existing in another submodel.
5. Double click on an object to open a menu for filling fields with rules, distributions, built-in functions, and so on, for defining elements 'on the fly.'
6. Utilizes WIN32S to run in 32-bit mode.
7. OLE2 compatibility.

AIM (Analyzer for Improving Manufacturing), from

Pritsker Corporation, is one of three applications of **FACTOR**. The other applications are Factor Production Manager and FI-2. Factor Production Manager performs detailed planning/scheduling of operations, order promising, order release, and supply-chain management. FI-2 is an interactive graphical scheduling board. All of these applications use the same data base.

AIM models are OS/2 based and built graphically with icons that represent machines, operators, conveyors, and so on, placed directly on the screen. The animations are created in a virtual window. The current release of AIM is version 6.0. During a simulation, the model can be stopped to check its status or add other components, then continue the simulation. Performance data is dynamically updated and displayed while the simulation is running. A dynamic Gantt chart is provided for tracking machine and operator status. Inventory levels and material handling utilization can also be graphed dynamically. Outputs include bar charts, pie charts, and plots of inventory levels. Alternately, information can be transferred to other software for development of presentation graphics.

Features of AIM include the following:

1. Manufacturing representation. Manufacturing specific modeling components can represent a variety of discrete manufacturing processes. Standard rules provide choices that are interpreted with processes. Custom rules may be written to extend the logic.
2. Integrated with scheduling applications. Models written with AIM can be used with other FACTOR applications providing support for capacity, logistics, production scheduling, supply-chain analysis, and schedule management.
3. Manufacturing data. AIM is built around a relational database that stores the manufacturing operation and simulation output. Part descriptions, process plans, order release schedules, machine locations and schedules, shift schedules, and so on can be transferred from other data sources to the AIM database.
4. Animation support. AIM models are built graphically and are animated automatically during model construction.
5. Interactive model building and simulation. Components are located on a scaled facility background. Intelligent defaults are provided for all components. Components are customized by completing forms. During execution, the modeler can change the status of a component and observe the simulated impact on the manufacturing system.
6. Comparison of alternatives. The AIM project framework organizes all aspects of a manufacturing simulation project. Alternative models of the manufacturing process are stored. Comparison reports show model performance data to identify differences between alternatives.
7. Gantt charts. AIM supports the creation of Gantt charts for the improved verification and validation of models. Model performance can be reviewed in the Gantt chart to follow a single load or the decisions of a resource.
8. Cost modeling methodology. The latest version of the software includes a detailed cost modeling capability. AIM models can be used to represent alternative costing philosophies such as standard cost, ABC, and so on.

ARENA, a product of Systems Modeling Corporation, is an extendible simulation and animation software package. It provides a complete simulation environment that supports all steps in a simulation study. Arena® combines the modeling power and flexibility of the SIMAN simulation language and the Cinema animation system, while offering the ease-of-use of Application Solution Templates.

Arena is a graphical modeling/animation system that is based on hierarchial modeling concepts. It allows users to create new modeling objects called modules, which are the building blocks of model creation. All aspects of a particular process-logic, data, animation, and statistics collection-can be modules to represent the process through which entities flow.

Modules are organized into groups contained in templates. These templates can be used to tailor the software to a specific animation. Application Solution Templates in areas such as business process reengineering, health care, high speed manufacturing, semiconductor wafer fabrication, transportation, and others currently exist or are under development.

The Arena system also includes the Input Analyzer, designed to give users the ability to read raw input data, and the Output Analyzer, for simulation data viewing and analysis.

EXTEND+MANUFACTURING is a product of Imagine That, Inc. It has all of the features of Extend, mentioned previously, plus the following:

1. Blocks that represent machines, stations. Labor buffers, bins, fixtures, and so on.
2. Reneging, preemption, and interruptible processes.
3. Ability to specify sequences for merging and routing streams of entities.
4. Scheduled and unscheduled down times.
5. Material handling constructs.
6. Automatic statistical reporting.

4 BUSINESS PROCESS REENGINEERING

Several vendors have developed software to facilitate the reengineering process. These include BPSimulator™, ProcessModel™, SIMPROCESS®,

Time Machine™, and Extend+BPR™.

BPSimulator, a product of Systems Modeling Corporation and Technology Economics International, as an Application Solution Template under Arena. Model building is focused on the activities that comprise a business process. Models are created by placing activity modules in the Arena work space and providing information about each activity. The software can be used for activity based costing by defining busy and idle costs for any resource.

BPSimulator also provides constructs for modeling human and technological resources. Costs, schedules, downtimes, efficiencies, and other resource attributes can be specified.

An interface to Lotus and Excel is provided. The software is capable of reading pertinent process data from existing IDEF0 models.

ProcessModel, a product of PROMODEL Corporation, interfaces with the ABC Graphics Suite™. Flowcharts are developed using ABC FlowCharter®. Modeling capabilities include the following:

1. Staff scheduling and shift planning.
2. Task allocation prioritization and interruption.
3. Use of multiple or alternative resources.
4. Individual or group processing.
5. Matching of orders to correct customers.
6. Document splitting and tracking.
7. Customers dropping out of line.
8. Activity-based costing.
9. Resource scheduling for breaks and downtimes.
10. Ability to change graphics during simulation.

SIMPROCESS, a product of CACI Products Company, integrates icon-based process flowcharting, hierarchical event-driven simulation, activity-based costing, and data analysis capabilities into a tool for business process reengineering. Other features include reusable templates and advanced modeling constructs, such as If-Then-Else logic. SIMPROCESS is based on CACI's MODSIM®, an object-oriented architecture for simulation. It runs under both Windows and UNIX operating systems.

Time Machine is a product of F&H Simulations. It is a Windows based dynamic simulation worksheet used for flowcharting, process modeling, what-if-analysis, concept validation and the dommunication of ideas. It is built on Taylor II, described previously.

Extend+BPR, a product of Imagine That, Inc., is based on the Extend software discussed previously. It has the following features:

1. Blocks that represent operations, transactions, workflow stacks, labor resources, decisions, and so on.
2. Ability to have entities renege.
3. Specification of rules for the flow of work throughout the operation.
4. Activity-based costing and resource planning.
5. Automatic statistical reporting.

5 SIMULATION-BASED SCHEDULING

Several of the many simulation-based scheduling software currently available are Preactor®, AutoSched™, and FACTOR®. These are discussed briefly in the following paragraphs.

PREACTOR, from Systems Modeling Corporation, is a visually interactive finite capacity scheduling system. Preactor features include automatic sequencing using forward, backward, and bi-directional rules, job selection by priority, due date, or first come-first served. A variety of output reports are available. Comprehensive data import and export capabilities for integration with other manufacturing technologies (e.g., MRP, MEP) are included.

AutoSched, from AutoSimulations, Incorporated, is based on AutoMod, a simulation language that was described previously. The following data/information is handled by AutoSched for capacity analysis and scheduling:

1. Production resources consisting of workstations, storage locations, operators and tools.
2. Products consisting of parts to be manufactured along with their routings.
3. Production requirements consisting of orders and lots.
4. Operating rules including how tasks are selected and calendars that specify when workstations are unavailable. Much of the information required by AutoSched may already exist in a database such as an MRP. If it does exist, AutoSched can import it.

AutoSched allows task selection rules for each resource. More than one rule (logic filter) is allowed. Significant scheduling improvements can be achieved through the application of these sophisticated rules.

AutoSched performs a simulation of the shop floor according to the task selection rules provided by the user. Two outputs are provided, graphical and statistical. Business graphs can be created to track any statistic(s). These graphs are updated dynamically. AutoSched also contains an interactive Gantt chart from which an event can be selected for detailing to include the quantity of orders in a workstation's queue when an order was selected, the quantity of orders in the next workstation's queue, etc. Statistical or historical reports are also available. These include the master schedule file, performance report, workstation report, and other user-defined reports.

FACTOR, from Pritsker Corporation, is an integrated software system that provides capacity management applications to help manufacturers meet customer demand (Lilegdon, 1993). FACTOR includes finite capacity scheduling, operations planning and loading, and order promising capabilities. Since its introduction, the software has been enhanced to include relational database capabilities, Client-Server installations, interactive schedule adjustment, and automatic scheduling capability.

To use FACTOR, a model is built by combining components consisting of order characteristics, shop floor status, production calendar, shift schedules, resources, functional resource groupings, preventive maintenance plans, parts definitions with the associated routings, material availability, tooling requirements, and operator capability. Any of these model components may be defined directly or taken from pre-existing manufacturing system files. A simulation of selected alternatives is conducted for various scheduling strategies with results stored in a database for analysis and eventual shop distribution.

In practice, FACTOR is used to schedule operations at regular intervals (e.g., shifts) or upon demand (when a significant event occurs). In either case, up-to-date status information is transferred into the FACTOR database and the FACTOR scheduling engine generates a recommended schedule based on the objective (e.g., maximize throughput, minimize order lateness). The user reviews a summary of the performance anticipated from the schedule and may then accept it, revise the objective and generate another schedule, or interactively make manual adjustments to the schedule.

6 ANIMATORS

Most simulation animators are integrated with the software. However, this is not always the case, and the introduction of general purpose animation packages allows the use of custom made environments.

Proof Animation is a product of Wolverine Software Corporation (Earle and Henriksen 1995). Any software that can write ASCII data to a file can drive Proof Animation™. Thus, BASIC, C, FORTRAN, GPSS/H, GPSS World, SIMAN V, and SIMSCRIPT II.5, among others, can serve as drivers. Animation is accomplished by using a static background, the layout file, and a trace file that contains dynamic events. Some of the features of the software are as follows:

1. Graphics are vector based, similar to CAD programs.
2. AutoCad-compatible layouts can be imported and exported.
3. Zoom in, zoom out and multi-window animations are supported. Maximum resolution is assured at any scale.
4. Drawing takes place on a coordinate grid using mouse-driven primitives.
5. Moving objects are defined internally by their geometry.
6. Statistics, graphs, and plots can be displayed dynamically.
7. Animation occurs in a postprocessing mode.
8. Motion is smooth on VGA PC's.
9. There is a steady ratio of animated (simulation) time to viewing (wall clock) time. This ratio may be varied while the animation is running.
10. Top view can be changed to isometric and back to top view instantly.
11. An option allows the construction of a demo disk.

7 SIMULATION SUPPORT SOFTWARE

Two products, among many that are available, are discussed in this section. The first is ExpertFit™ from Averill M. Law and Associates, for input data modeling (Law and Vincent 1995). The second is SIMSTAT from MC² Analysis Systems for input and output data analysis (Blaisdell and Haddock 1993).

ExpertFit is used to model input data distributions. It is the Windows-based successor to UniFit II. It puts a distribution into the proper format for thirty simulation software products. The software augments the built-in distribution capability of most of the software described previously. For example, GPSS/H has four built-in distributions, but ExpertFit increases that number to 40. Selection of the best probability distribution for a data set is accomplished automatically. The software determines whether the selected distribution is good in an absolute sense or whether an empirical distribution should be used. It provides more than thirty 2-D and 3-D graphical plots of the data. Lastly, ExperFit assists in picking a distribution in the absence of data, including models for random machine downtime.

SIMSTAT 2.0 is an interactive graphical software tool that performs statistical analysis on simulation input and output data. The software uses pull down menus and is integrated into the Windows environment. Data is maintained in a spreadsheet format for editing, examination and analysis. SIMSTAT takes advantage of the Windows Clipboard. Some of the many graphical capabilities of SIMSTAT include the fitting of input distributions to data, the determination of initialization bias, and autocorrelation plots.

8 SUMMARY

This tutorial describes software for simulation organized in six categories. The first of these is general purpose software. The second is manufacturing oriented software. Then, business process reengineering software is presented. Next, simulation-based scheduling software is described. Then, an animator is described. Finally, some simulation support software is discussed.

REFERENCES

Banks, J., B. Burnette, J.D. Rose, and H. Kozloski. 1995. *Introduction to SIMAN V and Cinema V.* New York: John Wiley.

Blaisdell, W.E., and J. Haddock. 1993. Simulation analysis using SIMSTAT 2.0. In *1993 Winter Simulation Conference Proceedings*, ed. G.W. Evans, M. Mollaghasemi, E.C. Russell, W.E. Biles, 213-217. Association for Computing Machinery, Baltimore, MD.

Crain, R.C. and D.S. Smith. 1995. Industrial strength simulation using GPSS/H. In *1995 Winter Simulation Conference Proceedings*, eds. C. Alexopoulos, K. Kang, W.R. Lilegdon, D. Goldsman 487-593. Association for Computing Machinery, Baltimore, MD.

Cox, S.W. 1991. GPSS World: A brief preview. In *1991 Winter Simulation Conference Proceedings*, ed. B.L. Nelson, G.M. Clark, 59-61. Association for Computing Machinery, Baltimore, MD.

Dewsnup, M.C., and E. Bollenbach. 1995. How to model automated guided vehicle systems using ProModel for Windows. In *1995 Winter Simulation Conference Proceedings*, eds. C. Alexopoulos, K. Kang, W.R. Lilegdon, D. Goldsman 482-486. Association for Computing Machinery, Baltimore, MD.

Earle, N.J. and J.O. Henriksen. 1995. The power and performance of PROOF Animation. In *1995 Winter Simulation Conference Proceedings*, eds. C. Alexopoulos, K. Kang, W.R. Lilegdon, D. Goldsman 494-501. Association for Computing Machinery, Baltimore, MD.

Geuder, D. 1995. Object-oriented modeling with Simple++. In *1995 Winter Simulation Conference Proceedings*, eds. C. Alexopoulos, K. Kang, W.R. Lilegdon, D. Goldsman 529-533. Association for Computing Machinery, Baltimore, MD.

Hammann, J.E., and N.A. Markovitch. 1995. Introduction to Arena. In *1995 Winter Simulation Conference Proceedings*, eds. C. Alexopoulos, K. Kang, W.R. Lilegdon, D. Goldsman 515-518. Association for Computing Machinery, Baltimore, MD.

Krahl, D. 1995. Building end-user applications with EXTEND. In *1995 Winter Simulation Conference Proceedings*, eds. C. Alexopoulos, K. Kang, W.R. Lilegdon, D. Goldsman 413-419. Association for Computing Machinery, Baltimore, MD.

Law, A.M., and S. Vincent. 1995. EXPERTFIT: Total support for simulation input modeling. In *1995 Winter Simulation Conference Proceedings*, eds. C. Alexopoulos, K. Kang, W.R. Lilegdon, D. Goldsman 395-401. Association for Computing Machinery, Baltimore, MD.

Lilegdon, W.R. 1993. Manufacturing decision making with FACTOR. In *1993 Winter Simulation Conference Proceedings*, ed. G.W. Evans, M. Mollaghasemi, E.C. Russell, W.E. Biles, 159-164. Association for Computing Machinery, Baltimore, MD.

Lilegdon, W.R., D.L. Martin, and A.A.B. Pritsker. 1994. FACTOR/AIM: A manufacturing simulation system. *Simulation* 62:367-372.

Nordgren, B. 1995. Taylor II manufacturing simulation software. In *1995 Winter Simulation Conference Proceedings*, eds. C. Alexopoulos, K. Kang, W.R. Lilegdon, D. Goldsman 401-404. Association for Computing Machinery, Baltimore, MD.

Pegden, C.D., R.E. Shannon, and R.P. Sadowski. 1995. *Introduction to simulation using SIMAN,* 2nd ed. New York: McGraw-Hill.

Profozich, D.M., and D.T. Sturrock. 1995. Introduction to SIMAN/Cinema. In *1995 Winter Simulation Conference Proceedings*, eds. C. Alexopoulos, K. Kang, W.R. Lilegdon, D. Goldsman 515-518. Association for Computing Machinery, Baltimore, MD.

Rohrer, M. 1994. AutoMod. In *1994 Winter Simulation Conference Proceedings*, eds. J.D. Tew, M.S. Manivannan, D.A. Sadowski, and A.F. Seila, 487-492, Association for Computing Machinery, Baltimore, MD.

Russell, E.C. 1993. SIMSCRIPT II.5 and SIMGRAPHICS tutorial. In *1993 Winter Simulation Conference Proceedings*, ed. G.W. Evans, M. Mollaghasemi, E.C. Russell, W.E. Biles, 223-227. Association for Computing Machinery, Baltimore, MD.

Thompson, W.B. 1995. Introduction to WITNESS and linking to process mapping tools. In *1995 Winter Simulation Conference Proceedings*, eds. C. Alexopoulos, K. Kang, W.R. Lilegdon, D. Goldsman 462-466. Association for Computing Machinery, Baltimore, MD.

AUTHOR BIOGRAPHY

JERRY BANKS is a Professor in the School of Industrial and Systems Engineering at Georgia Tech. He is a regular attendee and supporter of the Winter Simulation Conference having served on the Board for eight years including two years as Chair.

This tutorial is an updated (as of June 26, 1996) and expanded version of a tutorial with the same title appearing in the *1995 Winter Simulation Conference Proceedings*.

Proceedings of the 1996 Winter Simulation Conference
ed. J. M. Charnes, D. J. Morrice, D. T. Brunner, and J. J. Swain

DISCRETE-EVENT SIMULATION INPUT PROCESS MODELING

Lawrence M. Leemis

Department of Mathematics
College of William & Mary
Williamsburg, VA 23187-8795, U.S.A.

ABSTRACT

General guidelines for selecting probabilistic input models as part of a discrete-event simulation study are presented. Two short examples illustrating input modeling decisions are also presented, as opposed to a complete treatment of the subject.

1 INTRODUCTION

Discrete-event simulation models typically have stochastic components that mimic the probabilistic nature of the system under consideration. Successful input modeling requires a close match between the input model and the true underlying probabilistic mechanism associated with the system. The general question considered here is how to model an element (e.g., arrival process, service times) in a discrete-event simulation given a data set collected on the element of interest.

Since time and space for this tutorial is limited, the following simplifying assumptions have been made.

- The modeler has access to a reliable source of random numbers. Most introductory simulation textbooks (e.g., Law and Kelton 1991) consider random number generation algorithms.
- An algorithm is available for converting these random numbers to random variates associated with the input model to drive the simulation (Devroye 1986).
- Data is available on the aspect of the simulation of interest. For examples of input modeling in the absence of data, see Schmeiser and Deutsch (1977) or Law, McComas, and Vincent (1994).

With these assumptions limiting the scope of this tutorial, the focus turns to selecting the appropriate probabilistic models for the random components in a simulation model. Many simulation textbooks have a much broader treatment of input modeling than presented here (e.g., Law and Kelton 1991). These texts include more specific information on statistical tests for independence, graphical methods for model selection, parameter estimation techniques, and goodness-of-fit tests. Advanced input modeling is considered by Nelson et al. (1995).

An input model can be specified in a variety of ways, such as a cumulative distribution function, hazard function, intensity function, or a variate-generation algorithm. An input model characterizes each of the stochastic elements of a discrete-event simulation.

Figure 1 contains a taxonomy whose purpose is to illustrate the scope of potential input models that are available to simulation analysts. There is certainly no uniqueness in the branching structure of the taxonomy. The branches under *stochastic processes*, for example, could have been *state* followed by *time*, rather than *time* followed by *state*, as presented.

Examples of specific models that could be placed on the branches of the taxonomy appear at the far right of the diagram. Mixed, univariate, time-independent input models have empirical/trace-driven given as an possible model. All of the branches include this particular model. A *trace-driven* input model simply generates a process that is identical to the collected data values so as not to rely on a parametric model. A simple example is a sequence of arrival times collected over a 24-hour time period. The trace-driven input model for the arrival process is generated by having arrivals occur at the same times as the observed values.

The upper half of the taxonomy contains models that are independent of time. These models could have been called *Monte Carlo* models. Models are classified by whether there is one or several variables of interest, and whether the distribution of these random variables is discrete, continuous, or contains both continuous and discrete elements. Examples of univariate discrete models include the binomial distribution and a degenerate distribution with all of its

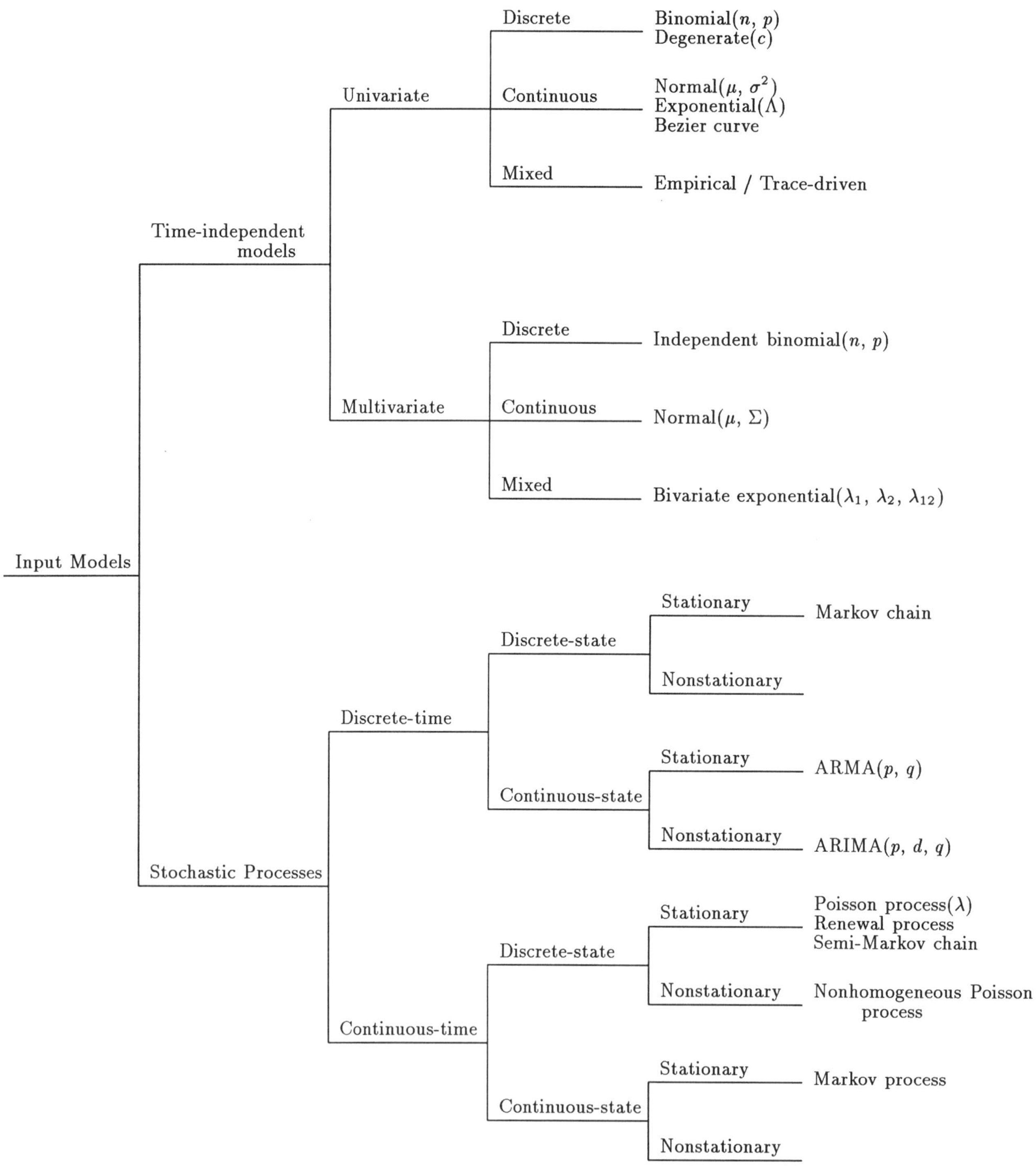

Figure 1: A Taxonomy for Input Models

mass at one value. Examples of continuous distributions include the normal distribution, an exponential distribution with a random parameter Λ (see, for example, Martz and Waller 1982), and Bézier curves (Flanigan–Wagner and Wilson 1993). Bézier curves offer a unique combination of the parametric and nonparametric approaches. An initial distribution is fitted to the data set, then the modeler decides whether differences between the empirical and fitted models represent sampling variability (chance variation) or an aspect of the distribution that should be included in the input model.

Examples of k-variable multivariate input models (see Johnson 1987) include a sequence of k independent binomial random variables, a multivariate normal distribution with mean μ and variance-covariance matrix Σ and a bivariate exponential distribution (Barlow and Proschan 1981).

The lower half of the taxonomy contains stochastic process models. These models are often used to solve problems at the system level, in addition to serving as input models for simulations with stochastic elements. Models are classified by how time is measured (discrete/continuous), the state space (discrete/continuous) and whether the model is stationary in time. For Markov models, the discrete-state/continuous-state branch typically determines whether the model will be called a "chain" or a "process", and the stationary/nonstationary branch typically determines whether the model will be preceded with the term "homogeneous" or "nonhomogeneous". Examples of discrete-time stochastic processes include homogeneous, discrete-time Markov chains (Ross 1993) and ARIMA time series models (Box and Jenkins 1976). Since point processes are counting processes, they have been placed on the continuous-time, discrete-space branch. Although the Poisson, renewal and nonhomogeneous Poisson processes are all pure birth processes, more general point processes, such as one to model the number of customers in a queue, can be placed on one of the continuous time, discrete-space branches.

2 EXAMPLES

Two simple examples illustrate the types of decisions that often arise in input modeling. The first example determines an input model for service times and the second example determines an input model for an arrival process.

2.1 Service Time Model

Consider a data set of $n = 23$ service times collected to determine an input model in a discrete-event simulation of a queuing system. The service times in seconds are

105.84 28.92 98.64 55.56 128.04 45.60
67.80 105.12 48.48 51.84 173.40 51.96
54.12 68.64 93.12 68.88 84.12 68.64
41.52 127.92 42.12 17.88 33.00.

[Although these service times come from the life testing literature (Lieblein and Zelen 1956), the same principles apply to both input modeling and survival analysis.]

The first step is to assess whether the observations are independent and identically distributed (iid). The data must be given in the order collected for independence to be assessed. Situations where the iid assumption would *not* be valid include:

- A new teller has been hired at a bank and the 23 service times represent a task that has a steep learning curve. The expected service time is likely to decrease as the new teller learns how to perform the task more efficiently.
- The service times represent 23 completion times of a physically demanding task during an 8-hour shift. If fatigue is a significant factor, the expected time to complete the task is likely to increase with time.

If a simple linear regression of the observation numbers regressed against the service times shows a significant nonzero slope, then the iid assumption is probably not appropriate.

Assume that there is a suspicion that a learning curve is present. An appropriate hypothesis test is

$$H_0 : \beta_1 = 0$$

$$H_1 : \beta_1 < 0$$

associated with the linear model (Neter, Wasserman, and Kutner 1989)

$$Y = \beta_0 + \beta_1 X + \epsilon,$$

where X is the observation number, Y is the service time, β_0 is the intercept, β_1 is the slope, and ϵ is an error term. Figure 2 shows a plot of the (x_i, y_i) pairs for $i = 1, 2, \ldots, 23$, along with the estimated regression line. The p-value associated with the hypothesis test is 0.14, which is not enough evidence to conclude that there is a statistically significant learning curve present.

There are a number of other graphical and statistical methods for assessing independence. These include analysis of the sample autocorrelation function associated with the observations and a scatterplot of

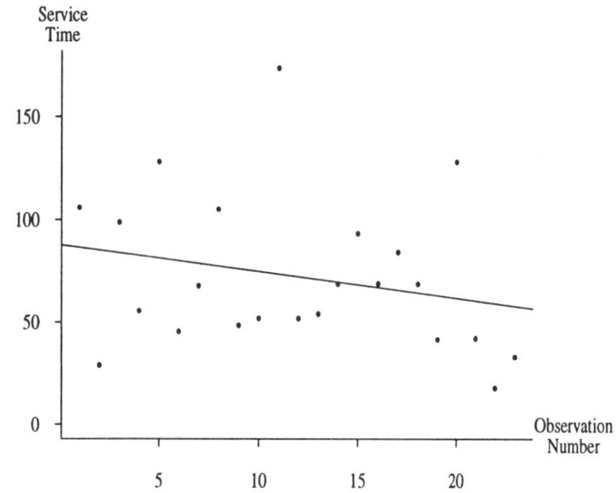

Figure 2: Service Time Vs. Observation Number

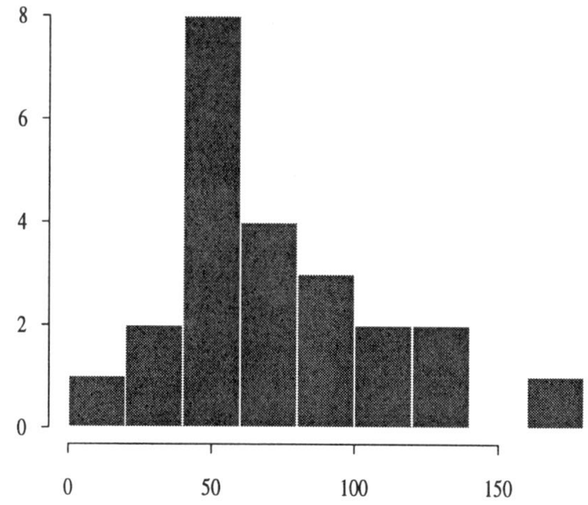

Figure 3: Histogram of Service Times

adjacent observations. For this particular example, assume that we are satisfied that the observations are truly iid in order to perform a classical statistical analysis.

The next step in the analysis of this data set includes plotting a histogram and calculating the values of some sample statistics. A histogram of the observations is shown in Figure 3. Although the data set is small, a skewed bell-shaped pattern is apparent. The largest observation lies in the far right-hand tail of the distribution, so care must be taken to assure that it is representative of the population. The sample mean, standard deviation, coefficient of variation, and skewness are

$$\bar{x} = 72.22 \qquad s = 37.49 \qquad \frac{s}{\bar{x}} = 0.52$$

$$\frac{1}{n}\sum_{i=1}^{n}\left(\frac{x_i - \bar{x}}{s}\right)^3 = 0.88.$$

Examples of the interpretations of these sample statistics are:

- A coefficient of variation $s/\bar{x}$ close to 1, along with the appropriate histogram shape, indicates that the exponential distribution is a potential input model.

- A sample skewness close to 0 indicates that a symmetric distribution (e.g., a normal distribution) is a potential input model.

The next decision that needs to be made is whether a parametric or nonparametric input model should be used. One simple nonparametric model would repeatedly select one of the service times with probability 1/23. The small size of the data set, the tied value, 68.64 seconds, and the observation in the far right-hand tail of the distribution, 173.40 seconds, tend to indicate that a parametric analysis is more appropriate. Since the input model is for service times, the accurate modeling of the right-hand tail of the distribution is critical. These long service times significantly impact queuing statistics. For this particular data set, a parametric approach is chosen.

There are dozens of choices for a univariate parametric model for the service times. These include general families of scalar distributions, modified scalar distributions and commonly-used parametric distributions (see Schmeiser 1990). Since the data is drawn from a continuous population and the support of the distribution is positive, a time-independent, univariate, continuous input model is chosen. The shape of the histogram indicates that the gamma, inverse Gaussian, log logistic, log normal, and Weibull distributions (Lawless 1982) are good candidates. The Weibull distribution is analyzed in detail here. Similar approaches apply to the other distributions.

Parameter estimates for the Weibull distribution can be found by least squares, the method of moments, and maximum likelihood. Due to desirable statistical properties, maximum likelihood is emphasized here. The Weibull distribution has probability density function

$$f(x) = \lambda^\kappa \kappa x^{\kappa-1} e^{-(\lambda x)^\kappa} \qquad x \geq 0,$$

where λ is a positive scale parameter and κ is a positive shape parameter. Let $x_1, x_2, \ldots, x_n$ be the data values. The likelihood function is

$$L(\lambda, \kappa) = \prod_{i=1}^{n} f(x_i) = \lambda^{n\kappa}\kappa^n \left[\prod_{i=1}^{n} x_i\right]^{\kappa-1} e^{-\sum_{i=1}^{n}(\lambda x_i)^\kappa}.$$

The log likelihood function is

$$\log L(\lambda, \kappa) = n \log \kappa + \kappa n \log \lambda + (\kappa - 1) \sum_{i=1}^{n} \log x_i - \lambda^{\kappa} \sum_{i=1}^{n} x_i^{\kappa}.$$

The 2×1 score vector has elements

$$\frac{\partial \log L(\lambda, \kappa)}{\partial \lambda} = \frac{\kappa n}{\lambda} - \kappa \lambda^{\kappa-1} \sum_{i=1}^{n} x_i^{\kappa}$$

and

$$\frac{\partial \log L(\lambda, \kappa)}{\partial \kappa} = \frac{n}{\kappa} + n \log \lambda + \sum_{i=1}^{n} \log x_i - \sum_{i=1}^{n} (\lambda x_i)^{\kappa} \log \lambda x_i.$$

When these equations are equated to zero, the simultaneous equations have no closed-form solution for $\hat{\lambda}$ and $\hat{\kappa}$:

$$\frac{\kappa n}{\lambda} - \kappa \lambda^{\kappa-1} \sum_{i=1}^{n} x_i^{\kappa} = 0$$

$$\frac{n}{\kappa} + n \log \lambda + \sum_{i=1}^{n} \log x_i - \sum_{i=1}^{n} (\lambda x_i)^{\kappa} \log \lambda x_i = 0.$$

To reduce the problem to a single unknown, the first equation can be solved for λ in terms of κ yielding

$$\lambda = \left(\frac{n}{\sum_{i=1}^{n} x_i^{\kappa}} \right)^{1/\kappa}.$$

Law and Kelton (1991, p. 334) give an initial estimate for κ that can be used in Newton's method to numerically solve for the maximum likelihood estimators. Qiao and Tsokos (1994) consider numerical problems with Newton's method and give an alternative algorithm for calculating the maximum likelihood estimators.

The score vector has a mean of $\mathbf{0}$ and a variance-covariance matrix $I(\lambda, \kappa)$ given by the 2×2 Fisher information matrix

$$I(\lambda, \kappa) = \begin{bmatrix} E\left[\frac{-\partial^2 \log L(\lambda,\kappa)}{\partial \lambda^2}\right] & E\left[\frac{-\partial^2 \log L(\lambda,\kappa)}{\partial \lambda \partial \kappa}\right] \\ E\left[\frac{-\partial^2 \log L(\lambda,\kappa)}{\partial \kappa \partial \lambda}\right] & E\left[\frac{-\partial^2 \log L(\lambda,\kappa)}{\partial \kappa^2}\right] \end{bmatrix}.$$

The observed information matrix

$$O(\hat{\lambda}, \hat{\kappa}) = \begin{bmatrix} \frac{-\partial^2 \log L(\hat{\lambda},\hat{\kappa})}{\partial \lambda^2} & \frac{-\partial^2 \log L(\hat{\lambda},\hat{\kappa})}{\partial \lambda \partial \kappa} \\ \frac{-\partial^2 \log L(\hat{\lambda},\hat{\kappa})}{\partial \kappa \partial \lambda} & \frac{-\partial^2 \log L(\hat{\lambda},\hat{\kappa})}{\partial \kappa^2} \end{bmatrix},$$

can be used to estimate $I(\lambda, \kappa)$.

For the 23 service times, the fitted Weibull distribution has maximum likelihood estimators $\hat{\lambda} = 0.0122$ and $\hat{\kappa} = 2.10$. The log likelihood function

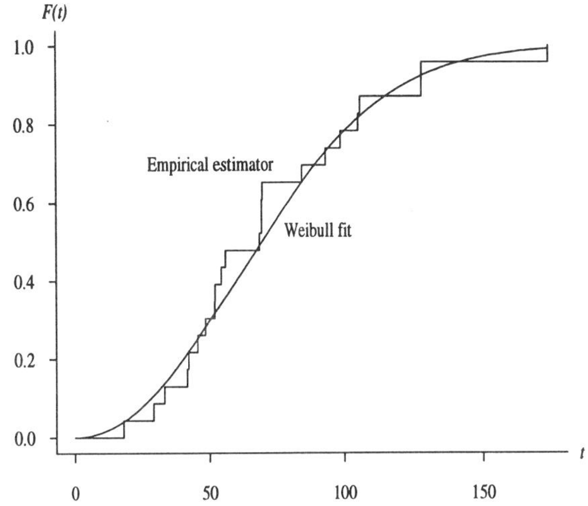

Figure 4: Empirical and Fitted Cumulative Distribution Functions for the Service Times

evaluated at the maximum likelihood estimators is $\log L(\hat{\lambda}, \hat{\kappa}) = -113.691$. Figure 4 shows the empirical cumulative distribution function (a step function with each step height $1/n$) along with the Weibull fit to the data.

The observed information matrix is

$$O(\hat{\lambda}, \hat{\kappa}) = \begin{bmatrix} 681{,}000 & 875 \\ 875 & 10.4 \end{bmatrix},$$

revealing a positive correlation between the elements of the score vector. Using the fact that the likelihood ratio statistic, $2[\log L(\hat{\lambda}, \hat{\kappa}) - \log L(\lambda, \kappa)]$, is asymptotically χ^2 with 2 degrees of freedom and that $\chi^2_{2,0.05} = 5.99$, a 95% confidence region for the parameters is all λ and κ satisfying

$$2[-113.691 - \log L(\lambda, \kappa)] < 5.99.$$

The 95% confidence region is shown in Figure 5. The line $\kappa = 1$ is not interior to the region, indicating that the exponential distribution is not an appropriate model for this particular data set.

As further proof that κ is significantly different from 1, the standard errors of the distribution of the parameter estimators can be computed by using the inverse of the observed information matrix

$$O^{-1}(\hat{\lambda}, \hat{\kappa}) = \begin{bmatrix} 0.00000165 & -0.000139 \\ -0.000139 & 0.108 \end{bmatrix}.$$

This is the asymptotic variance-covariance matrix for the parameter estimators $\hat{\lambda}$ and $\hat{\kappa}$. The standard errors of the parameter estimators are the square roots of the diagonal elements

$$\hat{\sigma}_{\hat{\lambda}} = 0.00128 \qquad \hat{\sigma}_{\hat{\kappa}} = 0.329.$$

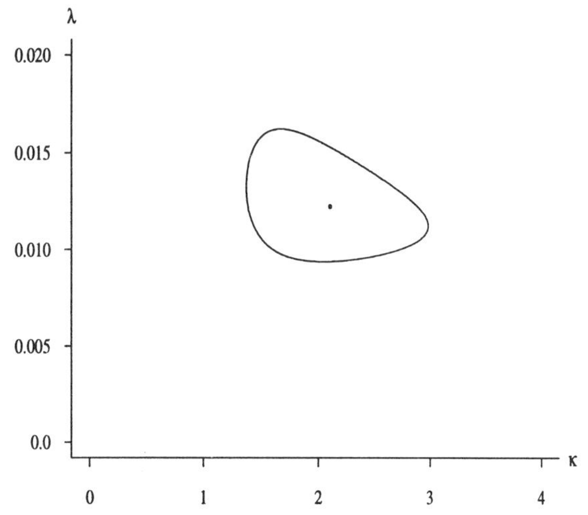

Figure 5: 95% Confidence Region Based on the Likelihood Ratio Statistic

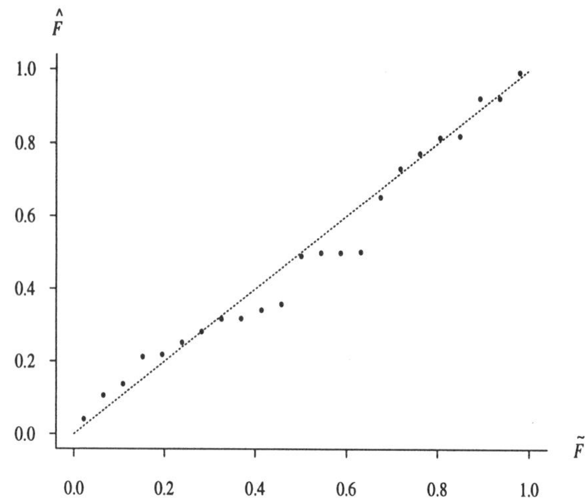

Figure 6: A P–P Plot for the Service Times

Thus an asymptotic 95% confidence interval for κ is

$$2.10 - (1.96)(0.329) < \kappa < 2.10 + (1.96)(0.329)$$

or

$$1.46 < \kappa < 2.74,$$

since $z_{0.025} = 1.96$. Since this confidence interval does not contain 1, the inclusion of the Weibull shape parameter κ is justified.

At this point, model adequacy should be assessed. Since the chi-square goodness-of-fit test suffers from arbitrary interval limits and should not be applied to small data sets, the Kolmogorov–Smirnov, Cramer–von Mises, or Anderson–Darling goodness-of-fit tests (Lawless 1982) are appropriate here. The Kolmogorov-Smirnov test statistic, for example, for this data set with a Weibull fit is 0.152, which measures the maximum difference between the empirical and fitted cumulative distribution functions. This test statistic corresponds to a p-value of approximately 0.15 (Law and Kelton 1991, page 391), so the Weibull distribution provides a reasonable model for these service times. The Kolmogorov–Smirnov test statistic values for several potential models are shown below.

Model	Test statistic
Exponential	0.301
Weibull	0.152
Gamma	0.123
Inverse Gaussian	0.099
Log normal	0.090

Many of the discrete-event simulation packages exhibited at the *Winter Simulation Conference* have the capability of determining maximum likelihood estimators for several parametric distributions. If the package also performs a goodness-of-fit test such as the Kolmogorov–Smirnov or chi-square test, the distribution that best fits the data set can quickly be determined.

P–P and Q–Q plots can also be used to assess model adequacy. A P–P plot, for example, is a plot of the fitted cumulative distribution function at the ith order statistic $x_{(i)}$, i.e., $\hat{F}(x_{(i)})$, versus the adjusted empirical cumulative distribution function, i.e. $\tilde{F}(x_{(i)}) = \frac{i-0.5}{n}$, for $i = 1, 2, \ldots, n$. A plot where the points fall close to a line indicates a good fit. For the 23 service times, a P–P plot for the Weibull fit is shown in Figure 6, along with a line connecting (0, 0) and (1, 1). P–P plots should be constructed for all competing models.

2.2 Arrival Process Model

Arrival times to a lunch wagon between 10:00 AM and 2:30 PM are collected on three days. The realizations were generated from a hypothetical arrival process given by Klein and Roberts (1984). A total of $n =$ 150 arrival times were observed, including $n_1 = 56$, $n_2 = 42$ and $n_3 = 52$ on the $k = 3$ days. Defining $(0, 4.5]$ be the time interval of interest (in hours) the three realizations are

$$0.2152 \quad 0.3494 \quad 0.3943 \quad \ldots \quad 4.175 \quad 4.248,$$

$$0.3927 \quad 0.6211 \quad 0.7504 \quad \ldots \quad 4.044 \quad 4.374,$$

and

$$0.4499 \quad 0.5495 \quad 0.6921 \quad \ldots \quad 3.643 \quad 4.357.$$

One preliminary statistical issue concerning this data is whether the three days represent processes drawn from the same population. External factors such as the weather, day of the week, advertisement, and workload should be kept fixed. For this particular example, these factors have been fixed and the three processes are representative of the population of arrival processes to the lunch wagon.

The input model for the process comes from the lower branch (stochastic processes) of the taxonomy in Figure 1. Furthermore, the arrival times constitute realizations of a continuous-time, discrete-state stochastic process, so the remaining question concerns whether or not the process is stationary.

If the process proves to be stationary, the techniques from the previous example, such as drawing a histogram, and choosing a parametric or nonparametric model for the *interarrival* times are appropriate. This results in a Poisson or renewal process. On the other hand, if the process is nonstationary, a nonhomogeneous Poisson process might be an input appropriate model. A nonhomogeneous Poisson process is governed by an intensity function $\lambda(t)$ which gives an arrival rate [e.g., $\lambda(2) = 10$ means that the arrival rate is 10 customers per hour at time 2] that can vary with time.

Figure 7 contains a plot of the empirical cumulative intensity function estimator suggested by Leemis (1991) for the three realizations. The solid line denotes the point estimator for the cumulative intensity function $\Lambda(t) = \int_0^t \lambda(\tau)d\tau$ and the dashed lines denote 95% confidence intervals. The cumulative intensity function estimator at time 4.5 is $150/3 = 50$, the point estimator for the expected number of arriving customers per day. If $\hat{\Lambda}(t)$ is linear, a stationary model is appropriate. Since people are more likely to arrive to the lunch wagon between 12:00 ($t = 2$) and 1:00 ($t = 3$) than at other times and the cumulative intensity function estimator has an S-shape, a nonstationary model is indicated. More specifically, a nonhomogeneous Poisson process will be used to model the arrival process.

The next question to be determined is whether a parametric or nonparametric model should be chosen for the process. Figure 7 indicates that the intensity function increases initially, remains fairly constant during the noon hour, then decreases. This may be difficult to model parametrically, so a nonparametric approach, possibly using $\hat{\Lambda}(t)$ in Figure 7 might be appropriate.

There are many potential parametric models for nonstationary arrival processes. The power law, or Weibull process has intensity function

$$\lambda(t) = \lambda^{\kappa}\kappa t^{\kappa-1} \qquad t > 0,$$

where λ and κ are positive parameters. This popular model would *not* be appropriate for this data

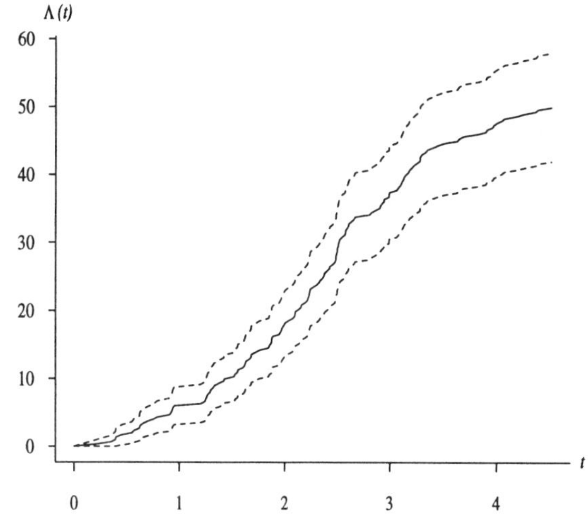

Figure 7: Point and 95% Confidence Interval Estimators for the Cumulative Intensity Function

set since the intensity function can only increase, decrease or remain constant, and can not model an intensity function that increases, then decreases. Since the intensity function is analogous to the hazard function for time-independent models, an appropriate 2-parameter distribution to consider would be one with a hazard function that increases initially, then decreases. A log-logistic process, for example, with intensity function

$$\lambda(t) = \frac{\lambda\kappa(\lambda t)^{\kappa-1}}{1+(\lambda t)^{\kappa}} \qquad t > 0,$$

for $\lambda > 0$ and $\kappa > 0$, would certainly be an improved choice. A more general EPTF (exponential-polynomial-trigonometric function) model is given by Lee, Wilson and Crawford (1991) with intensity function

$$\lambda(t) = \exp\left[\sum_{i=0}^{m} \alpha_i t^i + \gamma\sin(\omega t + \phi)\right] \qquad t > 0.$$

The trigonometric function is capable of modeling the intensity function that increases, then decreases.

In all of the parametric models, the likelihood function for the vector of unknown parameters $\theta = (\theta_1, \theta_2, \ldots, \theta_p)$ from a single realization on $(0, c]$ is

$$L(\theta) = \left[\prod_{i=1}^{n} \lambda(t_i)\right] \exp\left[-\int_0^c \lambda(t)dt\right].$$

Maximum likelihood estimators can be determined by maximizing $L(\theta)$ or its logarithm with respect to all unknown parameters. Confidence intervals for the unknown parameters can be found in a similar manner to the service time example.

ACKNOWLEDGMENTS

The author thanks Steve Tretheway for his help in developing Figure 1 while at The University of Oklahoma. Thanks also goes to Steve Park for reading a draft of this tutorial. The figures have been done using the Splus and xfig programs.

REFERENCES

Barlow, R. E., and F. Proschan. 1981. *Statistical theory of reliability and life testing: probability models.* Silver Springs, Maryland: To begin with.

Box, G., and G. Jenkins. 1976. *Time series analysis: forecasting and control.* Oakland, California: Holden–Day.

Devroye, L. 1986. *Non-uniform random variate generation.* New York: Springer–Verlag.

Flanigan–Wagner, M., and J. R. Wilson. 1993. Using univariate Bézier distributions to model simulation input processes. In *Proceedings of the 1993 Winter Simulation Conference*, ed. G. W. Evans, M. Mollaghasemi, E. C. Russell, and W. E. Biles, 365–373. Institute of Electrical and Electronics Engineers, Piscataway, New Jersey.

Johnson, M. E. 1987. *Multivariate statistical simulation.* New York: John Wiley & Sons.

Klein, R. W., and S. D. Roberts. 1984. A time-varying Poisson arrival process generator. *Simulation* 43:193–195.

Law, A. M., and W. D. Kelton. 1991. *Simulation modeling and analysis.* 2d ed. New York: McGraw–Hill.

Law, A. M., M. G. McComas, and S. G. Vincent. 1994. The crucial role of input modeling in successful simulation studies. *Industrial Engineering* 26:55–59.

Lawless, J. F. 1982. *Statistical models & methods for lifetime data.* New York: John Wiley & Sons.

Lee, S., J. R. Wilson, and M. M. Crawford. 1991. Modeling and simulation of a nonhomogeneous Poisson process having cyclic behavior. *Communications in Statistics — Simulation and Computation* 20:777–809.

Leemis, L. M. 1991. Nonparametric estimation of the intensity function for a nonhomogeneous Poisson process. *Management Science* 37:886–900.

Lieblein, J., and M. Zelen. 1956. Statistical investigation of the fatigue life of deep-groove ball bearings. *Journal of Research of the National Bureau of Standards* 57:273–316.

Martz, H. F., and R. A. Waller. 1982. *Bayesian reliability analysis.* New York: John Wiley & Sons.

Nelson, B. L., P. Ware, M. C. Cario, C. A. Harris, S. A. Jamison, J. O. Miller, J. Steinbugl, J. Yang. 1995. Input modeling when simple models fail. In *Proceedings of the 1995 Winter Simulation Conference*, ed. C. Alexopoulos, K. Kang, W. R. Lilegdon, and D. Goldsman, 93–100. Institute of Electrical and Electronics Engineers, Piscataway, New Jersey.

Neter, J., W. Wasserman, and M. H. Kutner. 1989. *Applied Linear Regression Models.* 2d ed. Boston: Irwin.

Ross, S. M. 1993. *Introduction to probability models.* 5th ed. Boston: Academic Press.

Schmeiser, B. 1990. Simulation experiments. In *Handbooks in OR & MS*, ed. D. P. Heyman and M. J. Sobel, 296–330. New York: Elsevier Science Publishers.

Schmeiser, B., and S. J. Deutsch. 1977. A versatile four-parameter family of probability distributions, suitable for simulation. *AIIE Transactions* 2:176–182.

Qiao, H., and C. P. Tsokos. 1994. Parameter estimation of the Weibull probability distribution. *Mathematics and Computers in Simulation* 37:47–55.

AUTHOR BIOGRAPHY

LAWRENCE M. LEEMIS is a professor in the Mathematics Department at the College of William & Mary. He received his BS and MS degrees in Mathematics and his PhD in Industrial Engineering from Purdue University. He has also taught courses at Baylor University, The University of Oklahoma, and Purdue University. His consulting, short course, and research contract work includes contracts with AT&T, NASA/Langley Research Center, Delco Electronics, Department of Defense, Air Logistic Command, ICASE, Komag, Federal Aviation Administration, Tinker Air Force Base, Magnetic Peripherals, and Argonne National Laboratory. His research and teaching interests are in reliability and simulation. He is a member of ASA, IIE, and INFORMS.

Proceedings of the 1996 Winter Simulation Conference
ed. J. M. Charnes, D. J. Morrice, D. T. Brunner, and J. J. Swain

STATISTICAL ISSUES IN SIMULATION

W. David Kelton

Department of Quantitative Analysis and Operations Management
College of Business Administration
University of Cincinnati
Cincinnati, Ohio 45221–0130, U.S.A.

ABSTRACT

This paper describes, in general terms, methods to help design the runs for simulation models and interpreting their output. Statistical methods are described for several different purposes, and related problems like comparison, variance reduction, sensitivity estimation, metamodeling, and optimization are mentioned. The main point is to call attention to the challenges and opportunities in using simulation models carefully and effectively.

1 INTRODUCTION AND SCOPE

Building a good simulation model can be a lot of work. You have to figure out how to model the system, express the model in whatever software you're using, collect data on the corresponding real system (if any) to set the simulation model's input parameters, verify that the simulation model, as expressed in the software, is working properly, and validate the simulation's output against the corresponding output from the real system (if any). After all that, you should feel pretty good.

But not *too* good. If you stop there you've wasted a lot of effort, since now it's the simulation model's turn to go to work for you. And I don't mean just running it (once) on your computer and looking at the results (which you doubtless did anyway just to get it to run). What you really have now is far more than just "a" simulation model—you have a great vehicle to test out a lot of different ideas without a lot more work on your part (although your computer will now get very busy, but that's good—and cheap, unlike your time invested to build the model), and to learn a lot about your model and the system it's simulating in terms of performance and possible improvement.

To do all this effectively, though, you have to think carefully about just how you're going to exercise your model. And, perhaps unfortunately, the most common kinds of simulation models can fool you (although not intentionally) if you're not circumspect about how you interpret their output.

My purpose in this tutorial is to call your attention to these issues and indicate in general terms how you can deal with them. I won't be going into great depth on a lot of technical details, but will refer you instead along the way to any of several texts on simulation that do, to introductory tutorials on this subject in the last few WSCs, and to more advanced and specialized WSC reviews.

Section 2 takes up the issue of randomness in simulation, Section 3 considers planning your runs, and Section 4 looks at the role of time in simulation. Analysis of a single variant is described in Section 5, and of alternative variants in Section 6. Sections 7–10 touch on variance reduction, sensitivity estimation, metamodels, and simulation optimization. This paper is an update of Kelton (1994, 1995).

2 DIDO VS. RIRO

Some simulations take as input only fixed, nonrandom values, typically representing parameters that describe the model and the particular variant of it you're evaluating. If the system you're simulating is really like this, then you can get by with such a *deterministic* simulation model. The nicest thing about this is, since there's no randomness in the input, there's no randomness in the output either—if you repeat the simulation you'll get the same thing over again. Thus, your answers are exact, at least up to roundoff. Figure 1 illustrates the idea in a manufacturing example, where the inputs are the machine cycle times, the interarrival times between successively arriving batches of parts, and the sizes of the these batches; the outputs are the hourly production and the machine utilization. The big dots for the inputs represent their (deterministic) values, and the big dots for the outputs represent the (determinis-

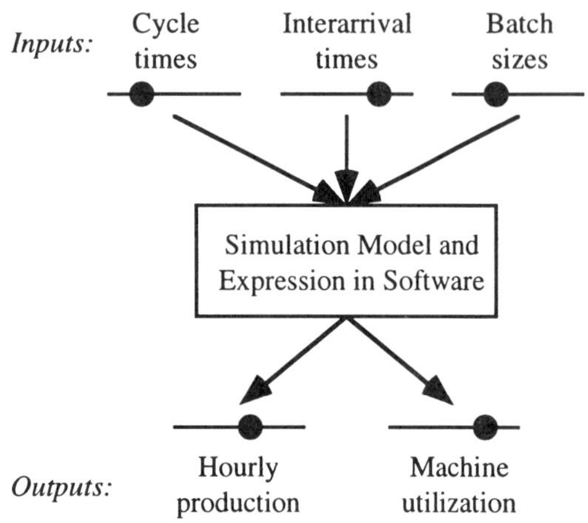

Figure 1: DIDO Simulation

tic) output performance measures obtained by transforming the input via the simulation's logic into the output. To abuse the computer-science anti-maxim of GIGO (Garbage In, Garbage Out), this situation might be called DIDO (Deterministic In, Deterministic Out). You might still have to make a lot of different runs, but for the purpose of evaluating a lot of different input-parameter combinations rather than to deal with uncertainty in the output.

But many (maybe most) systems involve some kind of uncertain, random input, so realistic simulation models ought to provide for such variable input as well; these are called *stochastic* simulation models. In fact, ignoring randomness in the input can make for dangerous errors in the simulation output. For instance, even in simple queueing models, which form the basic building blocks for a lot of simulations, the averages (expected values) of output performance measures like queue length and waiting time depend directly on the *variance* (as well as other things) of the service-time distributions. So ignoring randomness actually gets you the wrong answer, rather than just complicates your life. Besides, you might be interested in the output's randomness itself—like variability in hourly production. Of course, if you put random stuff into the simulation logic it's going to give you random stuff out—RIRO. Figure 2 illustrates the idea in the same manufacturing example, except now the inputs are probability distributions for the three quantities, regarded as random variables. The simulation proceeds by "drawing" realizations from the input probability distributions (indicated by the multiplicity of big dots from the input distributions) and transforms them into *an* observation on each of the (unknown) output distributions (indicated by the *single* big dot from each of the output distributions).

The whole point of this tutorial can be pretty much summed up by the fact that there's only a single big dot from the output distributions in Figure 2. The purpose of such a simulation is to learn (infer) something about these unknown output distributions, like maybe their expected values, variances, or probabilities on one side of some fixed tolerances. But all you get from one run of a stochastic simulation is a single observation on each of the output distributions, from which you obviously can't tell much about the governing output distribution (especially if, by unluck of the draw, you got an unusual value of the output under its distribution).

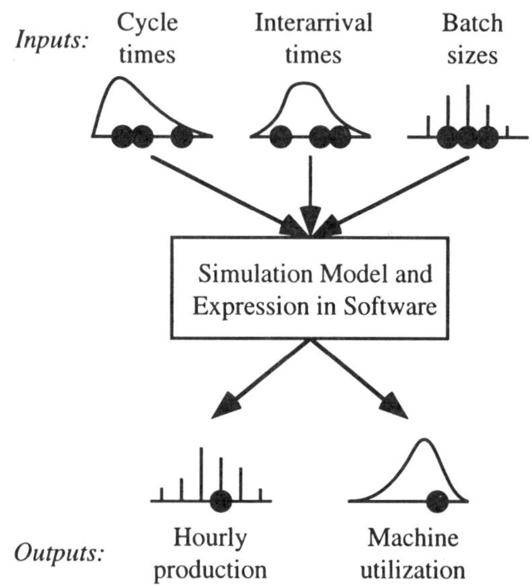

Figure 2: RIRO Simulation

So you have to think of *a* simulation run's output on some performance measure as *a* single observation on the output distribution; something very different could just as well have happened, just as something very different could just as well have happened on some other day in the actual manufacturing facility. Thus, you have to take care to perform the right kinds of simulation runs (design the simulation experiments) and do the right kinds of statistical analyses on the output data generated from the simulation. The rest of this tutorial will indicate some of the issues involved in these statistical questions as they apply to output data from a stochastic simulation.

3 EXPERIMENTAL DESIGN, OR THE SIMULATION AS TOMATO

Regardless of the type of simulation you have, you need to think ahead of time about exactly what scenarios you'll be asking your model to evaluate. Sometimes this is easy, having been specified by executive fiat or being just plain obvious. But most of the time it's not so clear, and you'll find yourself wondering what values the input parameters should take, and in what combinations with each other. You might also want to know what the effect is of changing some input parameter, and perhaps whether the effect of such a change might depend on (interact with) where the other parameters are set.

In these situations, formal experimental-design procedures can be of great help; you just need to think differently about them in the context of a simulation model. Traditionally, the "experiment" is some kind of physical situation; an agricultural experiment might be aimed at evaluating the effect of factors like different seed hybrids, fertilizers, and watering schedules on the yield of tomatoes, so a tomato-growing experiment with different factor-level combinations would be carried out and the results analyzed. The only things different about a simulation experiment are that you have a computer program rather than a tomato plant, the responses are the output measures like hourly production and machine utilization rather than yield of tomatoes, and the input factors are parameters like mean cycle times, variance of interarrival times, and maximum batch sizes, rather than seed hybrid, fertilizer, and watering (I guess another difference is that sometimes simulations run faster than tomatoes grow). So you can design the simulation experiment in the same way, and analyze it similarly as well in terms of measuring effects of factors and interactions among them.

Big simulations usually involve a lot of input factors, and you'll have to do some paring down of their numbers before you can do a workable analysis. For this purpose there are several factor-screening designs to help separate which factors matter (and should thus be retained as factors) and which ones don't (which should be frozen at some reasonable values and eliminated as factors).

For more on experimental design in the simulation context, see chapter 12 of Banks and Carson (1984), Cook (1992), Hood and Welch (1992), Chapter 12 of Law and Kelton (1991), Ramberg *et al.* (1991), Swain and Farrington (1994), and Sanchez (1994).

4 DOES TIME GO BY?

An issue that has as great an impact on what you do with your model as the deterministic/stochastic issue is whether time plays a role in the system and your model of it. Some simulations don't involve the passage of time, and are called *static*; examples include Monte Carlo evaluation of integrals and prediction with a cross-sectional regression model. The design-and-analysis approach is conceptually simple here (although may still be computationally scary): just repeat, or *replicate*, the model as many times as necessary to get the precision you need. Methods from classical statistical analysis can usually be used directly. For instance, in estimating an integral via a static Monte Carlo simulation, just get many independent estimates and then take the average, standard deviation, and maybe form a confidence interval in the elementary way.

But most simulations of industrial interest involve the passage of time as an important element; these are *dynamic* simulations, and the design-and-analysis approach can be a lot harder (as discussed in Section 5). From here on I'll assume that a dynamic simulation is what you've got.

5 EVALUATING A SINGLE CONFIGURATION

As a first step, you might want to evaluate the output from just a single configuration of the model. This section discusses issues involved in doing so, which will then be components of more ambitious goals like comparing alternative configurations or optimizing.

5.1 What to Watch?

In a stochastic simulation you'd really like to know all about the output distributions in Figure 2, but that's asking way too much in terms of the number and maybe length of the replications. So you usually have to settle for various summary measures of the output distributions. Traditionally, people have focused on estimating the expected value (mean) of the output distribution, and this can be of great interest. For instance, knowing something about the *average* hourly production is obviously important.

But things other than means might be interesting as well, like the *standard deviation* of hourly production, or the *probability* that the machine utilization for the period of the simulation will be above 0.80. In another example you might observe the *maximum* length of the queue of parts in a buffer somewhere to plan the floor space; in this connection it might be more reasonable to seek a value (called a *quantile*) below which the maximum queue length will fall with probability 0.95.

So think beforehand about what you'd like to get out of your simulation; it's easier to ignore things you have than go back and get things you forgot.

5.2 Multivariate Output

You'll probably want to get several different things out of your simulation; the stylized simulation of Figure 2 indicates two outputs, but dozens or scores might be more like it. Since these are all coming out of the same simulation runs, they're likely to be related in some way. For instance, high hourly production is probably associated with high utilization of the machine. So what you really have is a *vector* of output measures, and so multivariate statistical analyses can sometimes help you with estimating all the output parameters simultaneously, as well as with understanding how they might be related to each other.

For details in the simulation context and further references see Charnes (1991), section 9.7 of Law and Kelton (1991), and Seila (1991, 1992).

5.3 How Long?

A fundamental issue in your planning is whether you want performance measures over the long run (technically infinite, sometimes called *steady state*) or for a specific (finite, sometimes called *terminating*) fixed time period. The answer to this question is not a simulation issue, but rather one concerning the goals of your study. The answer also has obvious impact on how long you run your simulations; it also, perhaps less obviously, affects the kind of statistical analyses you can do on your output.

If a terminating simulation is appropriate for your goals, things are easy (at least in concept). Just run your model for whatever time period is called for, and get your output measures. That's one (replication, that is). Then repeat (replicate) this until you're happy with your results (described in more detail in Section 5.4 below). A complete run of the simulation constitutes a sample of size *one* (so isn't worth much), but standard statistical methods can be applied to the results *across* independent replications. The important thing is that you *do* multiple replications, *not* just one run, and use the output values from the entire replications as the fundamental "data points" for the basic ingredients in statistical analysis.

On the other hand, if you *really* want steady-state measures, the statistical-analysis problems become a lot harder (and, of course, your simulation runs become a lot longer). There are some things you can do, though, which are described in Section 5.5 below.

5.4 How to Express Things?

Traditionally, people have expressed statistical analyses of simulation output data in the form of confidence intervals (or confidence regions in the case of multivariate output). Compared to hypothesis tests, many people feel that confidence intervals are more informative and useful.

Increasingly, though, clever graphical displays are being used, which may not even involve formal inferential statistical analysis. For instance, histograms or dot plots of the output can indicate clear patterns that might not otherwise emerge from numerical measures. For more on graphical tools for describing simulation output, see Grier (1992).

Of course, animation has become very popular and, in some important ways, effective. But it's essential not to let yourself get swept along in the obvious visual appeal of animation to the exclusion of a proper statistical evaluation. For the fifteen simulated minutes that you had the patience to watch the animation, how do you know that the model was not in some weird state that's not representative of conditions as a whole?

5.5 Difficulties and Cautions

Alas, there are some pretty bad things you can do to yourself if you're not pretty careful about how your statistical analysis goes. Maybe the biggest mistake is to take as the basic "data" points for statistical analysis the *individual* observations coming out of a simulation over time. For instance, you dare not use the sequence of times in queue of successive parts in their raw form in standard statistical calculations (like "sample" variances). The problem is that they're not independent—if one part has a big delay in queue, the next one probably will too—which renders most of classical statistical theory invalid, sometimes with disastrous consequences. For instance, the "sample" variance of the individual part delays in queue will be biased low, perhaps severely, causing you to underestimate the variance and place more confidence in your results' precision than you ought to.

On the bright side, though, people have worked out some fairly simple and practical methods for dealing with simulation-generated data that usually work out pretty well. If you have a terminating simulation, for instance, you just make multiple replications and treat the summary statistics from each replication (averages, proportions, extremes, etc.) as the basic "data" points, which *can* be plugged into standard statistical formulas since the replications *are* identically distributed and independent of each other.

With steady-state simulations, though, things aren't quite so easy. Here are some ideas that people have come up with and tested out:

Replication. Even though you want (simulated) time to go to infinity, you can't. So just make runs as long as you can and then replicate them, pretending that the results give you a picture of being "in" steady state. You then have independent replications, just like in the terminating case, that you can plug into classical statistics. The problem with this is that the initial conditions you use to start the simulation (like everything's empty) are probably pretty atypical of steady state, which biases the run's output, at least for a while. You can make some plots and try to see where things stabilize, deleting the output data prior to that point, or maybe try to find better initial conditions that are more representative of steady state.

Batch means. Since you want to get as close as you can to steady state, just make one enormously long run. But then you really have only one replication (data point), so you can't estimate variances or do statistical analysis. To "manufacture" more observations out of this, split the run up into "batches" of observations, and treat the means of these batches as being independent unbiased observations of what's going on in steady state. While the initial-condition bias is less severe than with the Replication method, the batch means are not really independent; the key is to have big batches, and people have developed ways to help you decide how big the batches need to be for your situation.

Time-series models. The correlated, nonstationary simulation output series can be thought of as a time series, just like economic data such as stock prices or housing starts over time. Then a time-series model (like AR or ARMA) is fit to the data, and the fitted model is used for inference.

Standardized time series. A process version of the central limit theorem is applied to "standardize" the output series, and methods for statistical analysis have been worked out based on this.

Regeneration cycles. Some simulations return now and then to a state from which they "start over" probabilistically. For instance, if a queue empties out at some point it looks just like it did at the beginning (assuming it started empty). This creates independent *cycles* that are manipulated for statistical analysis.

Spectral analysis. Estimates of the correlation structure of the process are used to form a variance estimate for statistical analysis.

You get the idea that this is a hard problem, and that there is no completely satisfactory solution.

There's a very large literature on this subject, and the above list is a pretty thin tour of these methods. But they're all explained in detail elsewhere; see Alexopoulos (1993, 1994), chapter 11 of Banks and Carson (1984), chapter 3 of Bratley, Fox, and Schrage (1987), Charnes (1993), chapters 2, 3, and 5 of Fishman (1978), Goldsman (1992), chapter 7 of Khoshnevis (1994), Kleijnen (1987), chapter 9 of Law and Kelton (1991), Lewis and Orav (1989), chapter 6 of Ripley (1987), Seila (1991, 1992), and chapter 6 of Thesen and Travis (1992).

6 COMPARING ALTERNATIVES

Most of the time you'll be considering several different configurations of a simulation model, perhaps distinguished from each other by input-parameter values or by logical and structural differences. On the basis of some output performance measure, you might like to estimate the difference between various pairings of the configurations, perhaps expressed as a confidence interval for a difference or maybe a test of the null hypothesis that there is no difference. Most of the methods described in Section 5.5 can be adapted for these kinds of goals. For instance, in a terminating simulation you can use paired-sample confidence intervals for the difference, discussed in any elementary statistics book. The same difficulties and cautions apply, though, if you're interested in steady state.

Simulation is an ideal setting in which to apply any of several *selection and ranking* techniques. For instance, you can invoke statistical methods (basically telling you how much data you need to collect) that allow you to declare one of your alternatives as being the best on some criterion, and be highly confident that you're right about your choice. What makes simulation an attractive setting for this is that these techniques often require two-stage or sequential sampling (deciding on the sample size on the fly), which is a lot easier to do in simulation than in growing tomatoes.

For more depth on these subjects, see chapter 12 of Banks and Carson (1984), Goldsman, Nelson, and Schmeiser (1991), chapter 10 of Law and Kelton (1991), Goldsman and Nelson (1994), or chapter 7 of Thesen and Travis (1992).

7 VARIANCE REDUCTION

Section 5.5 dwelt on some of the difficulties and dangers in dealing with simulation data, but on the positive side there are some important opportunities not available when experimenting with tomatoes. Ease of sequential sampling, as mentioned in Section 6, was one example.

But a more important example is to reduce the variance of the output without doing any (well, hardly any) extra work. Such *variance-reduction* techniques often proceed by exploiting your ability to control the random-number generator driving the simulation, and re-use random numbers to induce helpful correlations that reduce noise in the output.

For instance, when comparing several alternative configurations of a manufacturing facility you could use the *same* random numbers, properly synchronized, to drive all configurations. This would result in the same jobs' arriving to the facilities at the same times, and with the same processing requirements. Whatever differences in performance you observe are thus attributable to configuration differences rather than to "environmental" differences in the arriving jobs (since there weren't any such differences). While intuitively appealing, there is actually firm statistical foundation for this, and the variance of the difference is usually reduced. This strategy, known as *common random numbers*, is often used unconsciously by just starting the runs for all alternatives with the same random-number streams and seeds.

There are many other sophisticated variance-reduction ideas; for details see chapter 2 of Bratley, Fox, and Schrage (1987), chapter 3 of Fishman (1978), chapter 11 of Law and Kelton (1991), Kleijnen (1987), L'Ecuyer (1994), Lewis and Orav (1989), chapter 7 of Morgan (1984), Nelson (1992), and chapter 5 of Ripley (1987).

8 WHAT IF YOU WANT SENSITIVITIES?

Related to the question of comparing alternatives is the more micro-level question of measuring the effect on the output due to a change in one or several of the inputs. For example, how much would hourly production increase if the mean cycle time on a critical machine were reduced by a small amount? Viewing the output as a (complicated) function of the input, this is a question about a partial derivative of the output with respect to one of the inputs.

A direct way to address this is to make two sets of runs—one at the original value and another at the changed value of the input parameter—and then look at the difference. There are other ways of doing this, though, that are more clever (and maybe more complicated), yet are also more economical from the point of view of the amount of simulating you have to do.

Details on these methods can be found in chapter 12 of Law and Kelton (1991), Glasserman and Glynn (1992), Ho (1992), L'Ecuyer (1991), and Strickland (1993).

9 METAMODELS

Thinking of the simulation logic and action as being a transformation of inputs into outputs, the notion arises that a simulation is just a function, albeit a pretty complicated one that you can't write down as some little formula. But it might be possible to *approximate* what the simulation does with some little formula, which could be particularly useful if a large number of input-factor combinations are of interest and it takes a long time to run the simulation.

So people sometimes fit a regression model to the simulation model, with the dependent variable's being the simulation output and the independent variables' being the input parameters to the simulation. All the usual techniques for building regression models come into play, like selecting important subsets of the independent variables, modeling nonlinearities, and considering interactions. Since this is a (regression) model of a (simulation) model, it's sometimes called a *metamodel*.

For more on metamodeling, see Barton (1992), Hood and Welch (1993), Kleijnen (1987), and chapter 12 of Law and Kelton (1991).

10 DOING THE BEST YOU CAN TO FIND OPTIMAL CONFIGURATIONS

The ultimate, maybe, in using a simulation model is to find input-factor settings that optimize some performance measure. This could involve several of the above issues, including gradient estimation, metamodeling, and comparing alternatives. Now optimization of nonlinear functions is a hard enough problem in itself, but in a stochastic simulation you have uncertainty in terms of measuring the response, as well as the statistical difficulties described in Section 5.5. So this is truly a tall order.

People have made important advances in this, though. One idea is to estimate the partial derivatives at a point (the gradient), then move in the direction of steepest descent (if you're minimizing) or steepest ascent (if you're maximizing). You could also fit a regression metamodel as in Section 9 and then use simple calculus to optimize it in lieu of the simulation itself. There are, to be sure, many more techniques (like adaptation of stochastic-programming methods) that have been developed or are under investigation; for more details see Azadivar (1992), Fu (1994), or chapter 12 of Law and Kelton (1991).

11 CONCLUSIONS

While all the details, methods, and cautions of doing a good job at output analysis may seem bewildering, you really owe it to yourself to try to get as much honest, precise information out of your hard-built simulation model as you can. While there are dangers and difficulties at times, there are also reliable and robust methods available. Moreover, some simulation-software products now have output-analysis capabilities built in to facilitate things.

ACKNOWLEDGMENT

I'm grateful for support from the Minnesota Supercomputer Institute.

REFERENCES

Alexopoulos, C. 1993. Advanced simulation output analysis for a single system. In *Proceedings of the 1993 Winter Simulation Conference*, ed. G.W. Evans, M. Mollaghasemi, E.C. Russell, and W.E. Biles, 89-96. WSC Board of Directors.

Alexopoulos, C. 1994. A review of advanced methods for simulation output analysis. In *Proceedings of the 1994 Winter Simulation Conference*, ed. J.D. Tew, M.S. Manivannan, D.A. Sadowski, and A.F. Seila, 133–140. WSC Board of Directors.

Azadivar, F. 1992. A tutorial on simulation optimization. In *Proceedings of the 1992 Winter Simulation Conference*, ed. J.J. Swain, D. Goldsman, R.C. Crain, and J.R. Wilson, 198–204. WSC Board of Directors.

Banks, J. and J.S. Carson, II. 1984. *Discrete-event system simulation.* Englewood Cliffs, New Jersey: Prentice-Hall.

Barton, R.R. 1992. Metamodels for Simulation Input-Output Relations. In *Proceedings of the 1992 Winter Simulation Conference*, ed. J.J. Swain, D. Goldsman, R.C. Crain, and J.R. Wilson, 289–299. WSC Board of Directors.

Bratley, P., B.L. Fox, and L.E. Schrage. 1987. *A guide to simulation.* 2nd ed. New York: Springer-Verlag.

Charnes, J.M. 1991. Multivariate simulation output analysis. In *Proceedings of the 1991 Winter Simulation Conference*, ed. B.L. Nelson, W.D. Kelton, and G.M. Clark, 187–193. WSC Board of Directors.

Charnes, J.M. 1993. Statistical analysis of output processes. In *Proceedings of the 1993 Winter Simulation Conference*, ed. G.W. Evans, M. Mollaghasemi, E.C. Russell, and W.E. Biles, 41–49. WSC Board of Directors.

Cook, L.S. 1992. Factor screening of multiple responses. In *Proceedings of the 1992 Winter Simulation Conference*, ed. J.J. Swain, D. Goldsman, R.C. Crain, and J.R. Wilson, 174–180. WSC Board of Directors.

Fishman, G.S. 1978. *Principles of discrete event simulation.* New York: John Wiley & Sons.

Fu, M.C. 1994. A tutorial review of techniques for simulation optimization. In *Proceedings of the 1994 Winter Simulation Conference*, ed. J.D. Tew, M.S. Manivannan, D.A. Sadowski, and A.F. Seila, 149–156. WSC Board of Directors.

Glasserman, P. and P.W. Glynn. 1992. Gradient estimation for regenerative processes. In *Proceedings of the 1992 Winter Simulation Conference*, ed. J.J. Swain, D. Goldsman, R.C. Crain, and J.R. Wilson, 280–288. WSC Board of Directors.

Goldsman, D. 1992. Simulation output analysis. In *Proceedings of the 1992 Winter Simulation Conference*, ed. J.J. Swain, D. Goldsman, R.C. Crain, and J.R. Wilson, 97–103. WSC Board of Directors.

Goldsman, D. and B.L. Nelson. 1994. Ranking, selection and multiple comparisons in computer simulation. In *Proceedings of the 1994 Winter Simulation Conference*, ed. J.D. Tew, M.S. Manivannan, D.A. Sadowski, and A.F. Seila, 192–199. WSC Board of Directors.

Goldsman, D., B.L Nelson, and B. Schmeiser. 1991. Methods for selecting the best system. In *Proceedings of the 1991 Winter Simulation Conference*, ed. B.L. Nelson, W.D. Kelton, and G.M. Clark, 177–186. WSC Board of Directors.

Grier, D.A. 1992. Graphical techniques for output analysis. In *Proceedings of the 1992 Winter Simulation Conference*, ed. J.J. Swain, D. Goldsman, R.C. Crain, and J.R. Wilson, 314–319. WSC Board of Directors.

Ho, Y.-C. 1992. Perturbation analysis: concepts and algorithms. In *Proceedings of the 1992 Winter Simulation Conference*, ed. J.J. Swain, D. Goldsman, R.C. Crain, and J.R. Wilson, 231–240. WSC Board of Directors.

Hood, S.J. and P.D. Welch. 1992. Experimental design issues in simulation with examples from semiconductor manufacturing. In *Proceedings of the 1992 Winter Simulation Conference*, ed. J.J. Swain, D. Goldsman, R.C. Crain, and J.R. Wilson, 255–263. WSC Board of Directors.

Hood, S.J. and P.D. Welch. 1993. Response surface methodology and its application in simulation. In *Proceedings of the 1993 Winter Simulation Conference*, ed. G.W. Evans, M. Mollaghasemi, E.C. Russell, and W.E. Biles, 115–122. WSC Board of

Directors.

Kelton, W.D. 1994. Analysis of output data. In *Proceedings of the 1994 Winter Simulation Conference*, ed. J.D. Tew, M.S. Manivannan, D.A. Sadowski, and A.F. Seila, 62–68. WSC Board of Directors.

Kelton, W.D. 1995. A Tutorial on design and analysis of simulation experiments. In *Proceedings of the 1994 Winter Simulation Conference*, ed. C. Alexopoulos, K. Kang, W.R. Lilegdon, and D. Goldsman, 24–31. WSC Board of Directors.

Khoshnevis, B. 1994. *Discrete systems simulation.* New York: McGraw-Hill.

Kleijnen, J.P.C. 1987. *Statistical tools for simulation practitioners.* New York: Marcel Dekker, Inc.

Law, A.M. and W.D. Kelton. 1991. *Simulation modeling and analysis.* 2nd ed. New York: McGraw-Hill.

L'Ecuyer, P. 1991. An overview of derivative estimation. In *Proceedings of the 1991 Winter Simulation Conference*, ed. B.L. Nelson, W.D. Kelton, and G.M. Clark, 28–36. WSC Board of Directors.

L'Ecuyer, P. 1994. Efficiency improvement and variance reduction. In *Proceedings of the 1994 Winter Simulation Conference*, ed. J.D. Tew, M.S. Manivannan, D.A. Sadowski, and A.F. Seila, 122–132. WSC Board of Directors.

Lewis, P.A.W. and E.J. Orav. 1989. *Simulation methodology for statisticians, operations analysts, and engineers, volume I.* Belmont, California: Wadsworth, Inc.

Morgan, B.J.T. 1984. *Elements of simulation.* London: Chapman and Hall.

Nelson, B.L. 1992. Designing efficient simulation experiments. In *Proceedings of the 1992 Winter Simulation Conference*, ed. J.J. Swain, D. Goldsman, R.C. Crain, and J.R. Wilson, 126–132. WSC Board of Directors.

Ramberg, J.S., S.M. Sanchez, P.J. Sanchez, and L.J. Hollick. 1991. Designing simulation experiments: Taguchi methods and response surface metamodels. In *Proceedings of the 1991 Winter Simulation Conference*, ed. B.L. Nelson, W.D. Kelton, and G.M. Clark, 167–176. WSC Board of Directors.

Ripley, B.D. 1987. *Stochastic simulation.* New York: John Wiley & Sons.

Sanchez, S.M. 1994. A robust design tutorial. In *Proceedings of the 1994 Winter Simulation Conference*, ed. J.D. Tew, M.S. Manivannan, D.A. Sadowski, and A.F. Seila, 106–113. WSC Board of Directors.

Seila, A.F. 1991. Output analysis for simulation. In *Proceedings of the 1991 Winter Simulation Conference*, ed. B.L. Nelson, W.D. Kelton, and G.M. Clark, 28–36. WSC Board of Directors.

Seila, A.F. 1992. Advanced output analysis for simulation. In *Proceedings of the 1992 Winter Simulation Conference*, ed. J.J. Swain, D. Goldsman, R.C. Crain, and J.R. Wilson, 190–197. WSC Board of Directors.

Strickland, S.G. 1993. Gradient/sensitivity estimation in discrete-event simulation. In *Proceedings of the 1993 Winter Simulation Conference*, ed. G.W. Evans, M. Mollaghasemi, E.C. Russell, and W.E. Biles, 97–105. WSC Board of Directors.

Swain, J.J. and P.A. Farrington. 1994. Designing simulation experiments for evaluating manufacturing systems. In *Proceedings of the 1994 Winter Simulation Conference*, ed. J.D. Tew, M.S. Manivannan, D.A. Sadowski, and A.F. Seila, 69–76. WSC Board of Directors.

Thesen, A. and L.E. Travis. 1992. *Simulation for decision making.* St. Paul, Minnesota: West Publishing Company.

AUTHOR BIOGRAPHY

W. DAVID KELTON is a Professor in the Department of Quantitative Analysis and Operations Management in the College of Business Administration at the University of Cincinnati, as well as a Fellow of the campus-wide Institute for Data Sciences. He received a B.S. in Mathematics from the University of Wisconsin–Madison, an M.S. in Mathematics from Ohio University, and M.S. and Ph.D. degrees in Industrial Engineering from Wisconsin. In 1987 he was WSC Program Chair and in 1991 was WSC General Chair. He is currently the co-representative of the INFORMS College on Simulation to the WSC Board of Directors.

Proceedings of the 1996 Winter Simulation Conference
ed. J. M. Charnes, D. J. Morrice, D. T. Brunner, and J. J. Swain

VERIFYING AND VALIDATING SIMULATION MODELS*

Robert G. Sargent

Simulation Research Group
College of Engineering and Computer Science
439 Link Hall
Syracuse University
Syracuse, New York 13244

ABSTRACT

This paper discusses verification and validation of simulation models. The different approaches to deciding model validity are presented; how model verification and validation relate to the model development process are discussed; various validation techniques are defined; conceptual model validity, model verification, operational validity, and data validity are described; ways to document results are given; and a recommended procedure is presented.

1. INTRODUCTION

Simulation models are increasingly being used in problem-solving and in decision-making. The developers and users of these models, the decision-makers using information derived from the results of the models, and people effected by decisions based on such models are all rightly concerned with whether a model and its results are "correct". This concern is addressed through model verification and validation. Model validation is usually defined to mean "substantiation that a computerized model within its domain of applicability possesses a satisfactory range of accuracy consistent with the intended application of the model" (Schlesinger, et al. 1979) and is the definition used here. Model verification is often defined as "ensuring that the computer program of the computerized model and its implementation is correct", and is the definition adopted here. A model sometimes becomes certified or accredited which is an "official" determination that a model is acceptable with respect to a set of explicit standards for a specific purpose. A related topic is model acceptability (or credibility), which is developing in the (potential) users (e.g., decision-makers) of information (e.g., results) from the models sufficient confidence in the information that they are willing to use it.

* This paper is a modified version of "Verification and Validation of Simulation Models", *Proceedings of the 1994 Winter Simulation Conference*, pp. 77-87.

A model should be developed for a specific purpose (or application) and its validity determined with respect to that purpose. If the purpose of a model is to answer a variety of questions, the validity of the model needs to be determined with respect to each question. Several sets of experimental conditions are usually required to define the domain of a model's intended applicability. A model may be valid for one set of experimental conditions and be invalid in another. A model is considered valid for a set of experimental conditions if the model's accuracy is within its acceptable range of accuracy, which is the amount of accuracy required for the model's intended purpose. This generally *requires* that the model's output variables of interest (i.e., the model variables used to be used in answering the questions that the model is being developed to answer) be identified and that their required amount of accuracy be specified. The amount of accuracy required should be specified prior to starting the development of the model or very early in the model development process. If the variables of interest are random variables, then properties and functions of the random variables such as their means and variances are usually what is of primary interest and are what are used in determining model validity. Several versions of a model are usually developed prior to obtaining a satisfactory valid model. The substantiation that a model is valid, i.e., model verification and validation, is generally considered to be a process and is usually part of the model development process.

It is often too costly and time consuming to determine that a model is *absolutely* valid over the complete domain of its intended applicability. Instead, tests and evaluations are conducted until sufficient confidence is obtained that a model can be considered valid for its intended application (Sargent 1982, 1984 and Shannon 1975, 1981). The relationships of cost (and a similar relationship holds for the amount of time) of performing model validation and the value of the model to the user as a function of model confidence are illustrated in Figure 1. The cost of model validation is usually quite significant; in particular when extremely

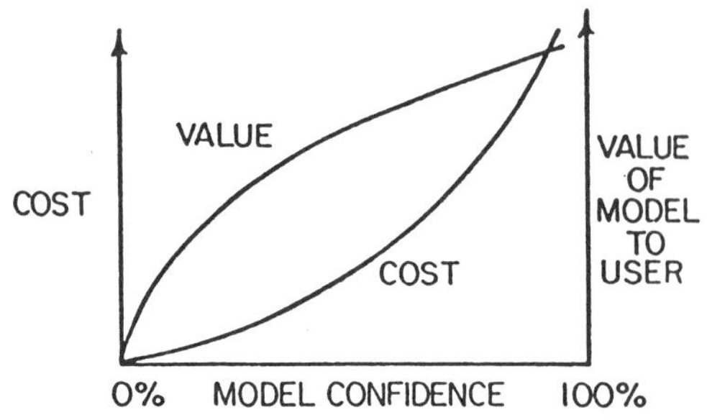

Figure 1: Model Confidence

high confidence is required because of the consequence of an invalid model.

The remainder of this paper is organized as follows: Section 2 discusses the three basic approaches used in deciding model validity; Section 3 defines the validation techniques used; Sections 4, 5, 6, and 7 contain descriptions of data validity, conceptual model validity, computerized model verification, and operational validity, respectively; Section 8 describes ways of presenting results; Section 9 contains a recommended validation procedure; and Section 10 gives the conclusions.

2. VALIDATION PROCESS

There are three basic approaches used in deciding whether a simulation model is valid or invalid. Each of these approaches require the model development team to conduct verification and validation as part of the model development process and this is discussed below in some detail. The most common approach is for the model development team to make the decision whether the model is valid. This decision is a subjective decision based on the results of the various tests and evaluations conducted as part of the model development process.

Another approach, often called independent verification and validation (IV&V), uses a third (independent) party to decide whether the model is valid. The third party is independent of both the model development team and the model sponsor/user(s). After the model has been developed, the third party conducts an evaluation to determine whether the model is valid. Based upon this validation, the third party makes a subjective decision on the validity of the model. This approach is usually used when there is a large cost associated with the problem the simulation model is being used for and/or to help in model acceptability. (A third party is also usually used for model certification or accreditation.)

The evaluation performed in the IV&V approach ranges from simply reviewing the verification and validation conducted by the model development team to a complete verification and validation effort. Wood (1986) describes experiences over this range of evaluation by a third party on energy models. One conclusion that Wood (1986) makes is that a complete IV&V evaluation is extremely costly and time consuming for what is obtained. This author's view is that if a third party is to be used, they should be involved *during* the model development process. If the model has already been developed, this author believes that a third party should usually only evaluate the verification and validation that has already been performed. (Also see Davis (1986) for an approach that simultaneously specifies and validates a model).

The last approach to determine whether a model is valid is to use a scoring model (see, e.g., Balci 1989, Gass 1979, and Gass and Joel 1987). Scores (or weights) are determined subjectively when conducting various aspects of the validation process. Then these scores are combined to determine category scores and an overall score for the simulation model. A simulation model is considered valid if its overall and category scores are greater than some passing score(s). This approach is infrequently used in practice.

This author does not believe in the use of a scoring model for determine validity. One reason is that the subjectiveness of this approach tends to be hidden and thus it appears to be objective. A second reason is "how are passing scores" to be determined. A third reason is that a model may receive a passing score and yet have a defect that needs correction. A fourth reason is that the score(s) may cause over confidence in a model or be used to argue one model is better than another.

We now discuss how model verification and validation relate to the model development process. There are two common ways to view this relationship. One way uses a detail model development process and the other uses a simple model development process. Banks, Gerstein, and Searles (1988) reviewed work in both of these ways and concluded that the simple way more clearly illuminates model verification and validation. This author recommends the use of the simple way (see e.g., Sargent 1982) and is the way presented here.

Consider the simplified version of the modelling process in Figure 2. The *problem entity* is the system (real or proposed), idea, situation, policy, or phenomena to be modelled; the *conceptual model* is the mathematical/logical/verbal representation (mimic) of the problem entity developed for a particular study; and the *computerized model* is the conceptual model implemented on a computer. The conceptual model is developed through an *analysis and modelling phase*, the computerized model is developed through a *computer programming and implementation phase*, and inferences about the problem entity are obtained by conducting computer experiments on the computerized model in the *experimentation phase*.

We will now relate model validation and verification to this simplified version of the modelling process (See Figure 2). *Conceptual model validity* is defined as determining that the theories and assumptions underlying the conceptual model are correct and that the model

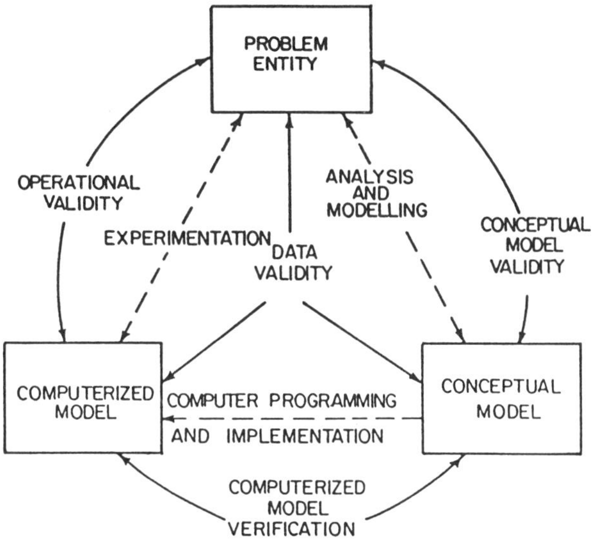

Figure 2: Simplified Version of the Modelling Process

representation of the problem entity is "reasonable" for the intended purpose of the model. *Computerized model verification* is defined as ensuring that the computer programming and implementation of the conceptual model is correct. *Operational validity* is defined as determining that the model's output behavior has sufficient accuracy for the model's intended purpose over the domain of the model's intended applicability. *Data validity* is defined as ensuring that the data necessary for model building, model evaluation and testing, and conducting the model experiments to solve the problem are adequate and correct.

Several versions of a model are usually developed in the modelling process prior to obtaining a satisfactory valid model. During each model iteration, model verification and validation are performed (Sargent 1984). A variety of (validation) techniques are used, which are described below. No algorithm or procedure exists to select which techniques to use. Some attributes that effect which techniques to use are discussed in Sargent (1984).

3. VALIDATION TECHNIQUES

This section describes various validation techniques (and tests) used in model verification and validation. Most of the techniques described here are found in the literature (see Balci and Sargent (1984a) for a detailed bibliography), although some may be described slightly different. They can be used either subjectively or objectively. By objectively, we mean using some type of statistical test or mathematical procedure, e.g., hypothesis tests and confidence intervals. A combination of techniques is generally used. These techniques are used for validating and verifying the submodels and overall model.

Animation: The model's operational behavior is displayed graphically as the model moves through time. For example, the movements of parts through a factory during a simulation are shown graphically.

Comparison to Other Models: Various results (e.g., outputs) of the simulation model being validated are compared to results of other (valid) models. Examples are (i) simple cases of a simulation model may be compared to known results of analytic models, and (ii) the simulation model may be compared to other simulation models that have been validated. (Sometimes the simulation model being validated requires modification to allow comparisons to be made.)

Degenerate Tests: The degeneracy of the model's behavior is tested by removing portions of the model or by appropriate selection of values of the input and internal parameters. For example, does the average number in the queue of a single server continue to increase with respect to time when the arrival rate is larger than the service rate.

Event Validity: The "events" of occurrences of the simulation model are compared to those of the real system to determine if they are the same. An example of events are deaths in a fire department simulation.

Extreme-Condition Tests: The model structure and output should be plausible for any extreme and unlikely combination of levels of factors in the system; e.g., if in-process inventories are zero, production output should be zero.

Face Validity: Face validity is asking people knowledgeable about the system whether the model and/or its behavior are reasonable. This technique can be used in determining if the logic in the conceptual model is correct and if a model's input-output relationships are reasonable.

Fixed Values: Fixed values (e.g., constants) are used for all model input and internal variables. This should allow checking the model results against values easily calculated.

Historical Data Validation: If historical data exist (or if data is collected on a system for building or testing the model), part of the data is used to build the model and the remaining data is used to determine (test) if the model behaves as the system does. (This testing is conducted by driving the simulation model with either Distributions or Traces (Balci and Sargent 1982a, 1982b, 1984b).)

Historical Methods: The three historical methods of validation are *Rationalism, Empiricism, and Positive Economics.* Rationalism assumes that everyone knows whether the underlying assumptions of a model are true. Then logic deductions are used from these assumptions to develop the correct (valid) model. Empiricism requires every assumption and outcome to be empirically validated. Positive Economics requires only that the model be able to predict the future and is not concerned with a model's assumptions or structure (causal relationships or mechanisms).

Internal Validity: Several replications (runs) of a stochastic model are made to determine the amount of (internal) stochastic variability in the model. A high amount of variability (lack of consistency) may cause the model's results to be questionable, and, if typical of the problem entity, may question the appropriateness of the policy or system being investigated.

Multistage Validation: Naylor and Finger (1967) proposed combining the three historical methods of Rationalism, Empiricism, and Positive Economics into a multistage process of validation. This validation method consists of (1) developing the model's assumptions on theory, observations, general knowledge, and function, (2) validating the model's assumptions where possible by empirically testing them, and (3) comparing (testing) the input-output relationships of the model to the real system.

Operational Graphics: Various performance measures, e.g., number in queue and percentage of servers busy, are shown graphically as the model moves through time; i.e., the dynamic behavior of performance indicators are visually displayed as the simulation model moves through time.

Parameter Variability - Sensitivity Analysis: This technique consists of changing the values of the input and internal parameters of a model to determine the effect upon the model behavior and its output. The same relationships should occur in the model as in the real system. Those parameters that are sensitive, i.e., cause significant changes in the model's behavior or output, should be made sufficiently accurate prior to using the model. (This may require iterations in model development.)

Predictive Validation: The model is used to predict (forecast) the system behavior and then comparisons are made between the system behavior and the model's forecast to determine if they are the same. The system data may come from an operational system or from experiments performed on the system, e.g., field tests.

Traces: The behavior of different types of specific entities in the model are traced (followed) through the model to determine if the model's logic is correct and if the necessary accuracy is obtained.

Turing Tests: People who are knowledgeable about the operations of a system are asked if they can discriminate between system and model outputs. (Schruben (1980) contains statistical tests for use with Turing tests.)

4. DATA VALIDITY

Even though data validity is usually not considered part of model validation, we discuss it because it is usually difficult, time consuming, and costly to obtain sufficient, accurate and appropriate data, and is frequently the reason that attempts to validate a model fail. Data are needed for three purposes: for building the conceptual model, for validating the model, and for performing experiments with the validated model. In model validation, we are concerned only with the first two types of data.

To build a conceptual model, we must have sufficient data on the problem entity to develop theories that can be used in building the model, to develop the mathematical and logical relationships in the model for it to adequately represent the problem entity for its intended purpose, and to test the model's underlying assumptions. In addition, behavior data is needed on the problem entity to be used in the operational validity step of comparing the problem entity's behavior with the model's behavior. (Usually, these data are system input/output data.) If these data are not available, high model confidence usually cannot be obtained because sufficient operational validity cannot be achieved.

The concern with data is that appropriate, accurate, and sufficient data are available, and if any data transformations are made, such as disaggregation, they are correctly performed. Unfortunately, there is not much that can be done to ensure that the data are correct. The best that can be done is to develop good procedures for collecting and maintaining data, test the collected data using techniques such as internal consistency checks, and screen for outliers and determine if they are correct. If the amount of data is large, a data base should be developed and maintained.

5. CONCEPTUAL MODEL VALIDATION

Conceptual model validity is determining that (i) the theories and assumptions underlying the conceptual model are correct and (ii) the model representation of the problem entity and the model's structure, logic, and mathematical and causal relationships are "reasonable" for the intended purpose of the model. The theories and assumptions underlying the model should be tested using mathematical analysis and statistical methods on problem entity data. Examples of theories and assumptions are linearity, independence, stationary, and Poisson arrivals. Examples of applicable statistical methods are fitting distributions to data, estimating parameter values from the data, and plotting the data to determine if they are stationary. In addition, all theories used should be reviewed to ensure they were applied correctly; for example, if a Markov chain is used, does the system have the Markov property and are the states and transition probabilities correct.

Next, each submodel and the overall model must be evaluated to determine if they are reasonable and correct for the intended purpose of the model. This should include determining if the appropriate detail and aggregate relationships have been used for the model's intended purpose, and if the appropriate structure, logic, and mathematical and causal relationships have been used. The primary validation techniques used for these evaluations are face validation and traces. Face validation is having experts on the problem entity evaluate the conceptual model to determine if they believe it is correct and reasonable for its purpose. This usually requires examining the flowchart or graphical model, or the set of model equations. The use of traces is the tracking of entities through each submodel and the overall model to determine if the logic is correct and the necessary accuracy is maintained. If any errors are found in the conceptual model, it must be revised and conceptual model validation performed again.

6. COMPUTERIZED MODEL VERIFICATION

Computerized model verification is ensuring that the computer programming and implementation of the conceptual model is correct. To help ensure that a correct computer program is obtained, program design and development procedures found in the field of Software Engineering should be used in developing and implementing the computer program. These include such techniques as top-down design, structured programming, and program modularity. A separate program module should be used for each submodel, the overall model, and for each simulation function (e.g., time-flow mechanism, random number and random variate generators, and integration routines) when using general purpose higher order languages, e.g., FORTRAN, PASCAL or C, and where possible when using simulation languages (Chattergy and Pooch 1977). (See Whitner and Balci (1989) for a more detailed discussion on model verification.)

One should be aware that the type of computer language used effects the probability of having a correct program. The use of a special purpose simulation language, if appropriate, generally will result in having less errors than if a general purpose simulation language is used, and using a general purpose simulation language will generally result in having less errors than if a general purpose higher order language is used. Not only does the use of simulation languages increase the probability of having a correct program, they usually reduce programming time significantly. (However, flexibility is usually reduced.)

After the computer program has been developed, implemented, and hopefully most of the programming "bugs" removed, the program must be tested for correctness and accuracy. First, the simulation functions should be tested to see if they are correct. Usually straightforward tests can be used here to determine if they are working properly. Next, each submodel and the overall model should be tested to see if they are correct. Here the testing is more difficult. There are two basic approaches to testing: static and dynamic testing (analysis) (Fairley 1976). In static testing the computer program of the computerized model is analyzed to determine if it is correct by using such techniques as correctness proofs, structured walk-through, and examining the structure properties of the program. The commonly used structured walk-through technique consists of each program developer explaining their computer program code statement by statement to other members of the modelling team until all are convinced it is correct (or incorrect).

In dynamic testing, the computerized model is executed under different conditions, and the resulting values obtained are used to determine if the computer program and its implementations are correct. This includes both the values obtained during the program execution and the final values obtained. There are three different strategies used in dynamic testing: bottom-up testing which means, e.g., testing the submodels first and then the overall model; top-down testing which means, e.g., testing the overall model first using programming stubs (sets of data) for each of the submodels and then testing the submodels; and mixed testing, which is using a combination of bottom-up and top-down testing (Fairley 1976). The techniques commonly used in dynamic testing are traces, investigations of input-output relations using different validation techniques, internal consistency checks, and reprogramming critical components to determine if the same results are obtained. If there are a large number of variables, one might aggregate to reduce the number of tests needed or use certain types of design of experiments (Kleijnen 1987), e.g., use factor screening experiments (Smith and Mauro 1982) to identify the key variables, in order to reduce the number of experimental conditions that need to be tested.

One must continuously be aware while checking the correctness of the computer program and its implementation that errors may be caused by the data, the conceptual model, the computer program, or the computer implementation.

7. OPERATIONAL VALIDITY

Operational validity is primarily concerned with determining that the model's output behavior has the accuracy required for the model's intended purpose over the domain of its intended applicability. This is where most of the validation testing and evaluation takes place. The computerized model is used in operational validity and thus any deficiencies found may be due to an inadequate conceptual model, an improperly programmed or implemented conceptual model (e.g., due to programming errors or insufficient numerical accuracy), or due to invalid data.

All of the validation techniques discussed in Section 3 are applicable to operational validity. Which techniques and whether to use them objectively or subjectively must be decided by the model development team and other interested parties. The major attribute effecting operational validity is whether the problem entity (or system) is observable or not, where observable means it is possible to collect data on the operational behavior of the program entity. Figure 3 gives a classification of the validation approaches for operational

	OBSERVABLE SYSTEM	NON-OBSERVABLE SYSTEM
SUBJECTIVE APPROACH	* COMPARISON USING GRAPHICAL DISPLAYS * EXPLORE MODEL BEHAVIOR	* EXPLORE MODEL BEHAVIOR * COMPARISON TO OTHER MODELS
OBJECTIVE APPROACH	* COMPARISON USING STATISTICAL TESTS AND PROCEDURES	* COMPARISON TO OTHER MODELS USING STATISTICAL TESTS AND PROCEDURES

Figure 3: Operational Validity Classification

validity. "Comparison" means comparing/testing the model and system input-output behaviors, and "explore model behavior" means to examine the output behavior of the model using appropriate validation techniques and usually includes parameter variability-sensitivity analysis. Various sets of experimental conditions from the domain of the model's intended applicability should be used for both comparison and exploring model behavior.

To obtain a *high* degree of confidence in a model and its results, comparison of the model's and system's input output behaviors for at *least* two different sets of experimental conditions is usually required. There are three basic comparison approaches used: (i) graphs of the model and system behavior data, (ii) confidence intervals, and (iii) hypothesis tests. Graphs are the most commonly used approach and confidence intervals are next.

7.1 Graphical Comparison of Data

The model's and system's behavior data are graphed for various sets of experimental conditions to determine if the model's output behavior has sufficient accuracy for its intended purpose. Three types of graphs are used: Histograms, Box (and Whisker) Plots (Johnson 1994), and Behavior Graphs. (See Sargent (1996) for a thorough discussion on the use of these for model validation.) An example of a box plot is given in Figure

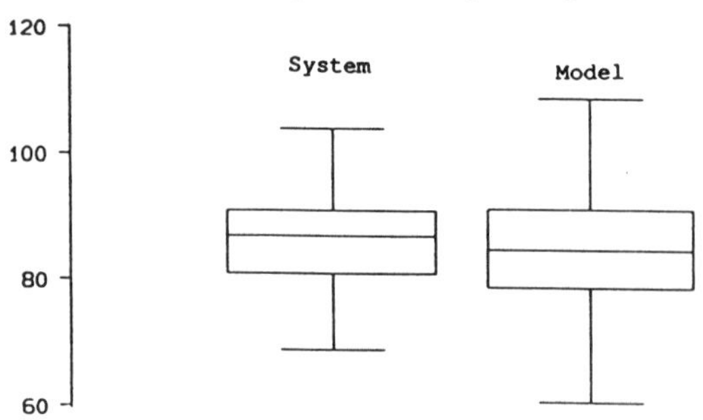

Figure 4: Box Plot

4 and examples of behavior graphs are shown in Figures 5 and 6. A variety of graphs using different types of measures and relationships are required such as (i) means, variances, maximums, distributions, and time series of each output variable, (ii) relationships between parameters of each output variable, e.g., means and variances, and (iii) relationships between different output variables. It is important that appropriate measures and relationships be used in validating a model and that they be determined with respect to the model's intended purpose. See Anderson and Sargent (1974) for an example of a set of graphs used in the validation of a simulation model.

These graphs can be used in model validation in three ways. First, the model development team can use the graphs in the model development process to make a subjective judgement on whether the model does or does not possess sufficient accuracy for its intended purpose. Secondly, they can be used in the face validity technique where experts are asked to make subjective judgements on whether a model does or does not possess sufficient accuracy for its intended purpose. The third way the graphs can be used is in Turing Tests.

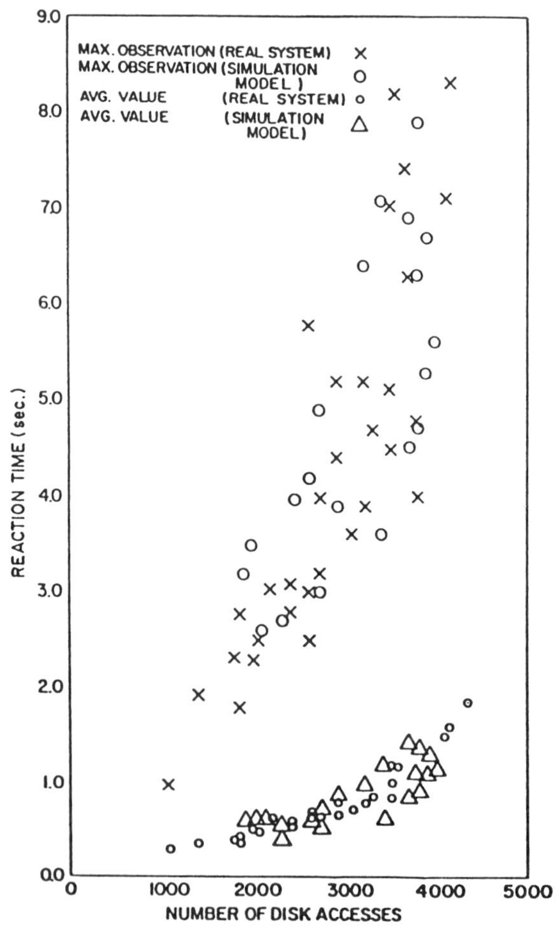

Figure 5: Disk Access

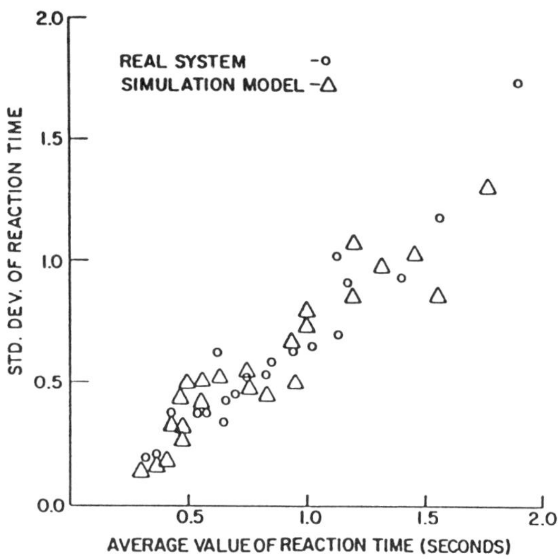

Figure 6: Reaction Time

7.2 Confidence Intervals

Confidence intervals (c.i.), simultaneous confidence intervals (s.c.i.), and joint confidence regions (j.c.r.) can be obtained for the differences between the population parameters (e.g., means and variances) and distributions of the model and system output variables for each set of experimental conditions. These c.i., s.c.i., and j.c.r. can be used as the model range of accuracy for model validation.

To construct the model range of accuracy, a statistical procedure containing a statistical technique and a method of data collection must be developed for each set of experimental conditions and for each parameter of interest. The statistical techniques used can be divided into two groups: (A) univariate statistical techniques and (B) multivariate statistical techniques. The univariate techniques can be used to develop c.i. and with the use of the Bonferroni inequality (Law and Kelton 1991) s.c.i. The multivariate techniques can be used to develop s.c.i. and j.c.r. Both parametric and nonparametric techniques can be used.

The method of data collection must satisfy the underlying assumptions of the statistical technique being used. The standard statistical techniques and data collection methods used in simulation output analysis (Banks, Carson and Nelson 1996, Law and Kelton 1991) can be used for developing the model range of accuracy; namely (1) replication, (2) batch means, (3) regenerative, (4) spectral, (5) time series, (6) standardized time series, and (7) overlapping batch means.

It is usually desirable to construct the model range of accuracy with the lengths of the c.i. and s.c.i. and the sizes of the j.c.r. as small as possible. The shorter the lengths or the smaller the sizes, the more useful and meaningful the model range of accuracy will usually be. The lengths and the sizes (i) are affected by the values of confidence levels, variances of the model and system response variables, and sample sizes and (ii) can be made smaller by decreasing the confidence levels or increasing the sample sizes. (Variance reduction techniques (Law and Kelton 1991) can be used when collecting data from a simulation model to decrease the variability of an estimate and thus obtain a smaller c.i., s.c.i., or j.c.r.) A tradeoff needs to be made among the sample sizes, confidence levels, and estimates of the length or sizes of the model range of accuracy, i.e., c.i., s.c.i., or j.c.r. Tradeoff curves can be constructed to aid in the tradeoff analysis.

Details on the use of c.i., s.c.i. and j.c.r. for operational validity, including a general methodology, are contained in Balci and Sargent (1984b). A brief discussion on the use of c.i. for model validation is also contained in Law and Kelton (1991).

7.3. Hypothesis Tests

Hypothesis tests can be used in the comparison of parameters, distributions, and time series of the output data of a model and a system for each set of experimental conditions to determine if the model's output behavior has an acceptable range of accuracy. An acceptable range of accuracy is the amount of accuracy that is required of a model to be valid for its intended purpose.

The first step in hypothesis testing is to state the hypotheses to be tested:

H_0: Model is valid for the acceptable range of accuracy under the set of experimental conditions. (1)

H_1: Model is invalid for the acceptable range of accuracy under the set of experimental conditions.

Two types of errors are possible in testing the hypotheses in (1). The first or type I error is rejecting the validity of a valid model; the second or type II error is accepting the validity of an invalid model. The probability of a type error I is called *model builder's risk*, α, and the probability of type II error is called *model user's risk*, β (Balci and Sargent 1981). In model validation, model user's risk is extremely important and must be kept small. Thus *both* type I and type II errors must be carefully considered in using hypothesis testing for model validation.

The amount of agreement between a model and a system can be measured by a validity measure. The validity measure is chosen such that the model accuracy or the amount of agreement between the model and the system decreases as the value of the validity measure increases. The acceptable range of accuracy can be used to determine an acceptable validity range, $0 \leq \lambda \leq \lambda^*$.

The probability of acceptance of a model being valid, P_a, can be examined as a function of the validity measure by using an Operating Characteristic Curve (Johnson 1994). Figure 7 contains three different operating characteristic curves to illustrate how the sample size of observations affect P_a as a function of λ.

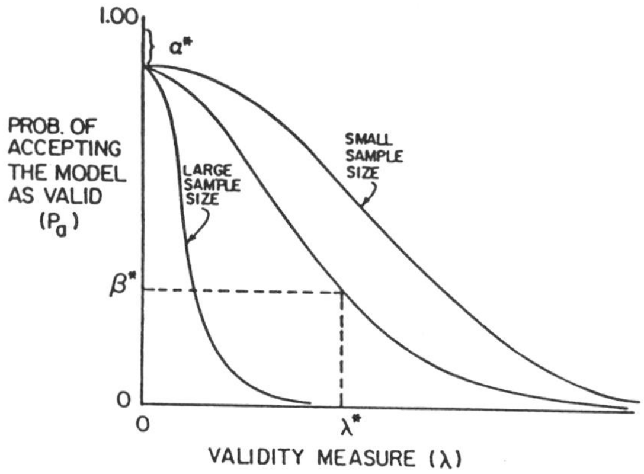

Figure 7: Operating Characteristic Curves

As can be seen, an inaccurate model has a high probability of being accepted if a small sample size of observations are used and an accurate model has a low probability of being accepted if a large sample size of observations are used. The location and shape of the operating characteristic curves is a function of the statistical technique being used, the value of α chosen for $\lambda = 0$, i.e., α^*, and the sample size of observations. Once the operating characteristic curves are constructed, the intervals for the model user's risk $\beta(\lambda)$ and the model builder's risk α can be determined for a given λ as follows:

$$\alpha \leq \text{model builder's risk } \alpha \leq (1-\beta^*) \qquad (2)$$

$$0 \leq \text{model user's risk } \beta(\lambda) \leq \beta^*.$$

Thus, there is a direct relationship among builder's risk, model user's risk, acceptable validity range, and sample size of observations. A tradeoff among these must be made in using hypothesis tests in model validation.

Details of the methodology of using Hypothesis Tests in comparing model's and system's output data for model validations are given in Balci and Sargent (1981). Examples of the application of this methodology in the testing of output means for model validation are given in Balci and Sargent (1982a, 1982b, 1983). Also see Banks, Carson, and Nelson (1996).

8. DOCUMENTATION

Documentation on model verification and validation is usually critical in convincing users of the "correctness" of a model and its results, and should be included in the simulation model documentation (For a general discussion on documentation of computer-based models, see Gass (1984).) Both detailed and summary documentation is desired. The detailed documentation should include specifies on the tests, evaluations made, data, results, etc. The summary documentation should contain a separate evaluation table for data validity, conceptual model validity, computer model verification, operational validity, and an overall summary. See Figure 8 for the evaluation table of computer model verification as an example of the first four tables. (See Sargent (1994) for examples of evaluation tables of conceptual model validity and overall summary.) The columns of the table are self explanatory except for the last column which refers to the confidence the evaluators have in the results or conclusions and this is often expressed as low, medium, or high.

9. RECOMMENDED PROCEDURE

This author recommends that, as a minimum, the following steps be performed in model validation:

(1) An agreement be made between (i) the model development team and (ii) the model sponsors and users (if possible) on the basic validation approach and on a minimum set of specific validation techniques to be used in the validation process ***prior*** to developing the model.

(2) The amount of accuracy required of the model's output variables of interest for the model's intended application be specified prior to starting the development of the model or very early in the model development process.

(3) The assumptions and theories underlying the model be tested, when possible.

(4) In each model iteration, at least face validity be performed on the conceptual model.

(5) In each model iteration, explore the model's behavior using the computerized model.

(6) In at least the last model iteration, comparisons be made between the model and system behavior (output) data for at ***least*** two sets of experimental conditions, when possible.

(7) Validation documentation be included in the model documentation.

(8) If the model is to be used over a period of time, a schedule for periodic review (and possible revalidation) of it be made and followed.

Models occasionally are developed to be used more than once. A procedure for reviewing the validity of these models over their life cycles needs to be developed as specified by step (8). No general procedure can be given as each situation is different. For example, if no data were available on the problem entity (e.g., a system) when the model was initially developed and validated, then revalidations of it should take place prior to each time the model is used if additional data or system understanding has occurred since its last validation.

10. SUMMARY

Model verification and validation are critical in the development of a simulation model. Unfortunately, there are no set of specific tests that can be easily applied to determine the "correctness" of the model. Furthermore, no algorithm exists to determine what techniques or procedures to use. Every new simulation project presents a new and unique challenge.

There is considerable literature on verification and validation. (See the following UHL on the WWW for a

Category/Item	Technique(s) Used	Justification for technique used	Reference to supporting report	Result/ conclusion	Confidence in result
• *Correctness of computer programs* • *Correctness of implementation*	• *Static testing* • *Dynamic testing*				

Strengths	
Weaknesses	

Overall evaluation for Computer Model Verification	*Overall conclusion*	*Justification for conclusion*	*Confidence in conclusion*

Figure 8: Evaluation Table for Computer Model Verification

recent bibliography: http://manta.cs.vt.edu/biblio/.) Articles given in the limited bibliography can be used as a starting point for furthering your knowledge on model verification and validation.

LIMITED BIBLIOGRAPHY

Anderson, H.A. and R.G. Sargent. 1974. An Investigation into Scheduling for an Interactive Computer System, *IBM Journal of Research and Development*, 18, 2, pp. 125-137.

Balci, O. 1989. How to Assess the acceptability and Credibility of Simulation Results, *Proceedings of the 1989 Winter Simulation Conference*, Washington, D.C. pp. 62-71.

Balci, O. and R.G. Sargent. 1981. A Methodology for Cost-Risk Analysis in the Statistical Validation of Simulation Models, *Comm. of the ACM*, 24, 4, pp. 190-197.

Balci, O. and R.G. Sargent. 1982a. Validation of Multivariate Response Simulation Models by Using Hotelling's Two-Sample T^2 Test, *Simulation*, Vol. 39, No. 6, pp. 185-192.

Balci, O. and R.G. Sargent. 1982b. Some Examples of Simulation Model Validation Using Hypothesis Testing, *Proceedings of the 1982 Winter Simulation Conference*, pp. 620-629.

Balci, O. and R.G. Sargent. 1983. Validation of Multivariate Response Trace-Driven Simulation Models, *Performance* 83, edited by Agrawada and Tripathi, North Holland, pp. 309-323.

Balci, O. and R.G. Sargent. 1984a. A Bibliography on the Credibility Assessment and Validation of Simulation and Mathematical Models, *Simuletter*, 15, 3, pp. 15-27.

Balci, O. and R.G. Sargent. 1984b. Validation of Simulation Models via Simultaneous Confidence Intervals, *American Journal of Mathematical and Management Science*, pp. 375-406.

Banks, J., J.S. Carson II, and B.L. Nelson. 1996. *Discrete-Event System Simulation*, 2nd Edition Prentice-Hall, Englewood Cliffs, N.J.

Banks, J., D. Gerstein, and S.P. Searles. 1988. Modeling Processes, Validation, and Verification of Complex Simulations: A Survey, *Methodology and Validation*, Simulation Series, Vol. 19, No. 1, The Society for Computer Simulation, pp. 13-18.

Chattergy, R. and V.W. Pooch. 1977. Integrated Design and Verification of Simulation Programs, *Computer*, Vol. 10, No. 4, pp. 40-46.

Davis, E.A. 1986. Use of Seminar Gaming to Specify and Validate Simulation Models, *Proceedings of the 1986 Winter Simulation Conference*, Washington, D.C., pp. 242-247.

DOD Simulations: Improved Assessment Procedures Would Increase the Credibility of Results 1987. United States General Accounting Office, PEMD-88-3.

Fairley, R.E. 1976. Dynamic Testing of Simulation Software, *Proceedings of the 1976 Summer Computer Simulation Conference*, Washington, D.C., pp. 40-46.

Gass, S.I. 1977. A Procedure for Evaluation of Complex Models, *Proceedings of the First International Conference on Mathematical Modeling*, University of Missouri, pp. 247-257.

Gass, S.I. 1983. Decision-Aiding Models: Validation, Assessment, and Related Issues for Policy Analysis, *Operations Research*, 31, 4, pp. 601-663.

Gass, S.I. 1984. Documenting a Computer-Based Model, *Interfaces*, Vol. 14, No. 3, pp. 84-93.

Gass, S.I. and L. Joel. 1987. Concepts of Model Confidence, *Computers and Operations Research*, 8,4, pp. 341-346.

Gass, S.I. and B.W. Thompson. 1980. Guidelines for Model Evaluation: An Abridged Version of the U.S. General Accounting Office Exposure Draft, *Operations Research*, 28, 2, pp. 431-479.

Johnson, R.A. 1994. *Miller and Freund's Probability and Statistics for Engineers*, 5th edition, Prentice-Hall, Englewood Cliffs, N.J.

Kleijnen, J.P.C. 1987. *Statistical Tools for Simulation Practitioners*, Marcel Dekker, New York.

Kleindorfer, G.B. and R. Ganeshan. 1993. The Philosophy of Science and Validation in Simulation, *Proceedings of 1993 Winter Simulation Conference*, pp. 50-57.

Law, A.M. and W.D. Kelton. 1991. *Simulation Modeling and Analysis*, 2nd Edition, McGraw-Hill Book Company.

Naylor, T.H. and J.M. Finger. 1967. Verification of Computer Simulation Models, *Management Science*, 14, 2, pp. B92-B101.

Oren, T. 1981. Concepts and Criteria to Assess Acceptability of Simulation Studies: A Frame of Reference, *Communications of the ACM*, 24, 4, pp. 180-189.

Rao, M.J. and R.G. Sargent. 1988. An Advisory System for Operational Validity, *Artificial Intelligence and Simulation: The Diversity of Applications*, edited by T. Hensen, Society for Computer Simulation, San Diego, CA, pp. 245-250.

Sargent, R.G. 1979. Validation of Simulation Models, *Proceedings of the 1979 Winter Simulation Conference*, edited by Highland, H.J., et al., San Diego, California, pp. 497-503.

Sargent, R.G. 1981. An Assessment Procedure and a Set of Criteria for Use in the Evaluation of Computerized Models and Computer-Based Modelling Tools, Final Technical Report RADC-TR-80-409.

Sargent, R.G. 1982. Verification and Validation of Simulation Models, Chapter IX in *Progress in Modelling and Simulation*, edited by F.E. Cellier, Academic Press, London, pp. 159-169.

Sargent, R.G. 1984. Simulation Model Validation, *Simulation and Model-Based Methodologies: An Integrative View*, edited by Oren, et al., Springer-Verlag.

Sargent, R.G. 1985. An Expository on Verification and Validation of Simulation Models, *Proceedings of the 1985 Winter Simulation Conference*, Dallas, Texas, pp. 15-22.

Sargent, R.G. 1986. The Use of Graphic Models in Model Validation, *Proceedings of the 1986 Winter Simulation Conference*, Washington, D.C., pp. 237-241.

Sargent, R.G. 1988. A Tutorial on Validation and Verification of Simulation Models, *Proceedings of 1988 Winter Simulation Conference*, San Diego, CA, pp. 33-39.

Sargent, R.G. 1990. Validation of Mathematical Models, *Proceedings of GEOVAL-90: Symposium on Validation of Geosphere Flow and Transport Models*, Stockholm, Sweden, pp. 571-579.

Sargent, R.G. 1991. Simulation Model Verification and Validation, *Proceedings of 1991 Winter Simulation Conference*, Phoenix, AZ, pp. 37-47.

Sargent, R.G. 1992. Validation and Verification of Simulation Models, *Proceedings of 1992 Winter Simulation Conference*, Arlington, VA, pp. 104-114.

Sargent, R.G. 1994. Verification and Validation of Simulation Models, *Proceedings of 1994 Winter Simulation Conference*, Lake Buena Vista, FL, pp. 77-87.

Sargent, R.G. 1996. Some Subjective Validation Methods Using Graphical Displays of Data, *Proceedings of 1996 Winter Simulation Conference*.

Schlesinger, et al. 1979. Terminology for Model Credibility, *Simulation*, 32, 3, pp. 103-104.

Schruben, L.W. 1980. Establishing the Credibility of Simulations, *Simulation*, 34, 3, pp. 101-105.

Shannon, R.E. 1975. *Systems Simulation: The Art and the Science*, Prentice-Hall.

Shannon, R.E. 1981. Tests for the Verification and Validation of Computer Simulation Models, *Proceedings of the Winter Simulation Conference*, pp. 573-577.

Smith, D.E. and C.A. Mauro. 1982. Factor Screening in Computer Simulation, *Simulation*, 38, 2, pp. 49-54.

Whitner, R.B. and O. Balci. 1989. Guidelines for Selecting and Using Simulation Model Verification Techniques, *Proceedings of 1989 Winter Simulation Conference*, Washington, D.C., pp. 559-568.

Wood, D.O. 1986. MIT Model Analysis Program: What We Have Learned About Policy Model Review, *Proceedings of the 1986 Winter Simulation Conference*, Washington, D.C., pp. 248-252.

Zeigler, B.P. 1976. *Theory of Modelling and Simulation*, John Wiley and Sons, Inc., New York.

AUTHOR BIOGRAPHY

ROBERT G. SARGENT is a Professor at Syracuse University. He received his education at the University of Michigan. Dr. Sargent has served his profession in numerous ways and has been awarded the TIMS College on Simulation Distinguished Service Award for long-standing exceptional service to the Simulation Community. His research interests include the methodology areas of modeling and discrete event simulation, model validation, and performance evaluation. Professor Sargent is listed in *Who's Who in America*.

Proceedings of the 1996 Winter Simulation Conference
ed. J. M. Charnes, D. J. Morrice, D. T. Brunner, and J. J. Swain

DESIGN OF OBJECT-ORIENTED SIMULATIONS IN C++

Jeffrey A. Joines
Stephen D. Roberts

Department of Industrial Engineering
Campus Box 7906
North Carolina State University
Raleigh, NC 27695-7906, U.S.A.

ABSTRACT

An object-oriented simulation (OOS) consisting of a set of object classes written in C++ can be used to create simulation models and packages. The simulations built with these tools possess the benefits of an object-oriented design, including the use of encapsulation, inheritance, polymorphism, run-time binding, and parameterized typing. These concepts are illustrated by creating a set of object frames which encapsulate simulation requirements. Simulation modeling is contained within a set of modeling frameworks. A network queuing simulation language is developed which has several notable features not available in other non-OOS languages. OOS provides full accessibility to the base language, faster executions, portable models and executables, a multi-vendor programming language, and a growing variety of complementary development tools.

1 INTRODUCTION

Previous WSC tutorial papers (see Joines, Powell, and Roberts 1992, 1993, 1994, 1995) described the benefits of object-oriented simulation (OOS). This paper will focus on the fundamental class structure (Section 2) and the design of the simulation system (Section 3). The network simulation language, YANSL, is presented as an illustration within the context of this design (Section 4). Everything is implemented in C++, which impacts the implementation. C++ is an object-oriented extension to the C programming language (Ellis and Stroustrup 1990). The source code in this paper is available from anonymous ftp: `ftp.eos.ncsu.edu/pub/simul`.

1.1 The Object-Oriented Context

The general conceptual design of the object-oriented context for simulation, as conceived in this approach, is illustrated in the following figure.

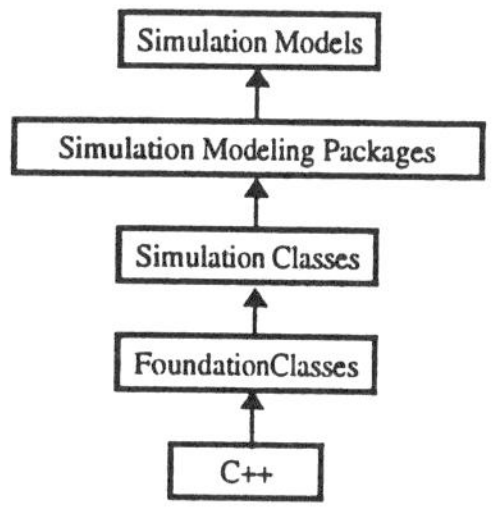

The lowest level construct is the C++ general programming language. Specific simulation models are at the highest level. The common notion of a simulation language falls somewhere in the middle of this design.

"Users" may "relate" to this design at any of the design levels. Persons interested only in the results may simply execute the models, while very knowledgeable persons may employ "raw" C++. The concepts at each level are "encapsulated" so that a simulation model user, for example, needs not be concerned about the concepts at a lower level. The more interested user, however, can delve deeper into the design, ultimately reaching the C++ level. Implicit in this design is a "hierarchy" of information, ranging from the behavior of specific models to general program behavior.

One perspective on the contribution of object-oriented simulation is that "simulation software engineering" is now being added to "simulation applications engineering." This addition will provide simulation users with not only a full array of simulation tools, but also the means to add new or modify existing tools.

1.2 OOS Appeal to Simulation Application Users

The idea of an "object-oriented" simulation has great intuitive appeal in applications because it is very easy to view the real world as being composed of objects. In a manufacturing cell, objects that should come to mind include the machines, the workers, the parts, the tools, and the conveyors. Also, the part routings, the schedule, as well as the work plan could be viewed as objects.

It is also easy to describe existing simulation languages using object terminology (see Joines, Powell, and Roberts 1992). A simulation language provides a user with a set of pre-defined object classes from which the simulation modeler can create needed objects. The modeler declares objects and specifies their behavior through the parameters available. The integration of all the objects into a single bundle provides the simulation model.

1.2.1 Problems with Extensibility

Because many simulation languages offer pre-specified functionality produced in another language (assembly language, C, FORTRAN, etc.), the user cannot access the internal function of the language. Instead, only the vendor can modify the internal functionality.

Also, users have only limited opportunity to extend an existing language feature. Some simulation languages allow for certain programming-like expressions or statements, which are inherently limited. Most languages allow the insertion of procedural routines written in other general-purpose programming languages.

None of this is fully satisfactory because, at best, any procedure written cannot use and change the behavior of a pre-existing object class. Also, any new object classes defined by a user in general programming language do not co-exist directly with vendor code.

The Arena software (Hammann and Markovitch 1995) provides some upward extensibility by a template approach to representing blocks of Siman statements (which may include the graphical representation). SLX (Henriksen 1995) will provide both higher and lower extensibility. Full object-oriented simulation systems include Smalltalk (Goldberg and Robson 1989), Modsim III (Belanger and Mullarney 1990), and Simple++ (Geuder 1995). C++ based simulation packages include C++Sim (Little and McCue 1994), C++/CSIM17 (Schwetman 1995) and Simpack (Fishwick 1995).

1.2.2 An Object-Oriented Simulation Approach

OOS deals directly with the limitation of extensibility by permitting full data abstraction as well as procedural abstraction. Data abstraction means that new data types with their own behavior can be added arbitrarily to the programming language. When a new data type is added, it can assume just as important a role as any implicit data types. For example, in the simulation language context, a new user-defined robot class can be added to a language that contains standard resources without compromising any aspect of the existing simulation language, and the robot may be used as a more complex resource.

2 CLASSES in C++

The class concept is fundamental to object-oriented software. The class provides a "pattern" for creating objects and defines the "type." An example (as it appears in C++) is the following `Exponential` class, which is used to obtain exponential random variates:

```
#include "random.h"
/*Class Exponential describes inverse transfor-
mation generator for exponential variates. */

class Exponential: public Random {
 private:
   double mu;
 public:
   Exponential(double, unsigned int=0, long=0);
   Exponential( int, unsigned int=0, long=0 );
   virtual double sample();
   void setMu(double initMu) {mu = initMu;}
   double getMu() { return mu; }
};
```

2.1 Class Properties

The class definition specifies the object's properties, namely the data and functions, and are generally grouped into "public" and "private" sections (C++ also permits another grouping called "protected"). When the object is created, the public properties can be accessed from outside the object. The private properties are information kept strictly locked within the object and are available only to object's functions. For example, the object `mu` (exponential mean) is declared as a private data member of type double and cannot be directly accessed. However, a public function called `getMu()` does return the value of `mu`, while `setMu()` allows the user to change the value of `mu`. Making a property private restricts use outside the class and encapsulates the object's properties.

2.2 Inheritance

The `Exponential` class was not defined "from scratch." It doesn't say anything explicitly about how it obtains random numbers. Because the random number generator establishes the source of randomness for all random processes, it is defined in its own class. Hence, the `Exponential` class is derived from the `Random` class so it has access to all the public properties of the `Random` class without having to re-code them.

This use of prior classes is called "inheritance." In fact, inheritance makes the `Exponential` class a "kind of" `Random` class. In object-oriented terminology this is considered an "is-a" relationship. The other major kind of relationships between two classes is the "has-a." In the case of the Exponential, the exponential has a double object called `mu`. A "has-a" relationship is not the result of inheritance, but an act of composition.

2.3 Construction and Initialization of Objects

When a class object is needed, the creation and initialization of it is provided by a function called a "constructor." C++ will provide one if it isn't included in the class definition. In the case of the `Exponential` class, there are two constructors. One takes a double and the other takes an integer. Notice that some of the arguments have specified defaults, so the user doesn't have to specify all the potential features of an exponential object (these additional arguments pertain to the control of the random number stream). Within the constructors (details not shown), space is allocated for the object and parameters are assigned. Although, not used in `Exponential`, C++ permits user specified destructors. A destructor will clean-up any object responsibilities (like collecting statistics) and deallocate the space.

2.4 Run-time Binding

The `sample()` function is specified as a "virtual" function in both the `Random` (not shown) and `Exponential` class. Therefore, sampling can be specified as a `Random` object and at run-time, the program will decide from which random variate to sample. This approach of tying the variate to the sample at run-time is also called "delayed" or "run-time" binding. Run-time binding may extract a small run-time penalty, but makes this entire specification of sampling from variates much easier to write, maintain, and use. With run-time binding, new variate types can be added through inheritance without altering existing simulation code.

2.5 Polymorphism

The `Exponential` class has two constructors, so users may specify either floating point or integer arguments for the mean interarrival time. Although it is not necessary in this case (C++ will make the right conversions), it does illustrate the use of polymorphism--where the same property applies to different objects. Under other circumstances, polymorphism allows users to produce the same behavior with different object types. For instance, requesting a material handling does not depend on the the device being sought. For example, one message "request" can be used for AGV or Trucks (e.g. request("AGV") or request("Truck"))rather than a message for each type (e.g., requstAGV and requestTruck).

3 CLASS HIERARCHY AND FRAMES FOR OOS

A key to the creation of a fully integrated simulation package is the use of a *class inheritance hierarchy* (introduced in Section 2.2). With C++ being the most abstract form (lowest level) of a simulation package, the more concrete elements are added so that, at the highest level, the final product may be a specific simulation model. A specific simulation model is also a kind of simulation model, which is a kind of simulation, which is a kind of C++ program.

An inheritance hierarchy can be viewed as a tree. The base of the tree is the most abstract class and the leaves present the most specific class. In order to collect classes into levels of abstraction, we introduce object-based "frames." A *frame* is a set of classes that provide a level of abstraction in the simulation and modeling platform. A frame is a convenient means for describing various "levels" within the simulation class hierarchy and is a conceptual term (not formalized within C++).

3.1 Foundation Frame

The foundation classes provide a base structure from which more simulation-specific classes may be created. These foundation classes are not simulation specific.

3.1.1 Abstract Objects

The `AbstractObject` forms the *fundamental base class* for the entire design and all other classes are derived from this base class. The `AbstractObject` defines all the essential properties and gives uniform character to the design. It provides for the following common properties and is called a "nice" class (see Carroll and Ellis 1995): a *default constructor* which construct objects without user-specified parameters; a *copy constructor* which establishes a mechanism for creating a new object as a copy of another within this class; an *assignment operator* which allows objects to be the target of an assignment using this operator; an *equality operator* which tests the "equality" of one object with another using this operator; a *less than operator* which test whether one object is "less than" another using this operator; and a *destructor* which provides for the orderly destruction of an object.

3.1.2 Foundation Support Classes

The foundation support classes provide general manipulation of objects important in the creation of simulation languages/packages including classes for strings, arrays, and linked lists. These are now available in the new standard template library(STL) (Plauger 1995).

3.2 Simulation Frame

This frame provides basic simulation functionality including random number and random variate generation, statistics collection, and base simulation elements as seen below.

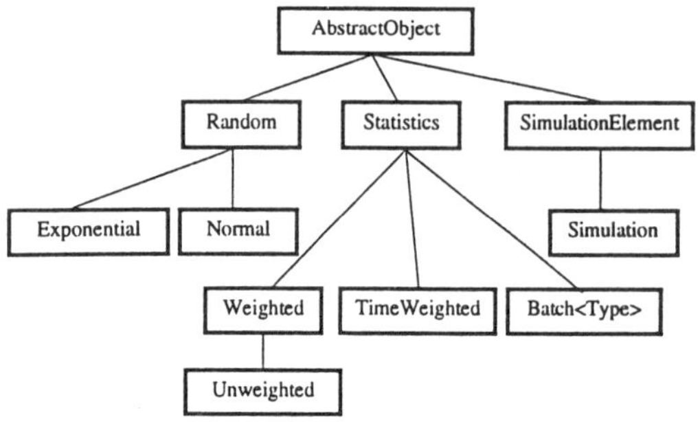

As can be seen, all the classes are derived from `AbstractObject` in order to maintain a common class design throughout any simulation project.

3.2.1 Random Numbers and Random Variates

Random number generation is obtained from the `Random` class. Random variate generators are derived from the `Random` class so that each source of variate generation has its own random number generator(or generators). This design has two benefits by facilitating the use of inverse transform method of random variate generation and by associating each variate generator with its own random number stream, variance reduction through correlated sampling is possible. Random number and variate generation properties include: (1) setting/getting generator parameters, (2) obtaining random numbers/variates, and (3) creating antithetic sampling.

3.2.2 Statistics Collection

Basic statistics can be collected on `Weighted`, `Unweighted`, and `TimeWeighted` variables. Also, statistics may be "batched" from any of the basic statistic types. Tables, plots, and histograms may be displayed for basic or batched statistics. Statistics collection properties include: (1) stopping and starting statistics collection, (2) clearing the statistics, and (3) reporting statistics. Basic statistics are collected during the simulation and provide: (1) observation base of (weighted) observations or time, (2) mean and standard deviation, and (3) minimum and maximum observations. Batched statistics are also collected during the simulation and provide both over-batch and current batch results. Batches can be based on time intervals or numbers of observations.

3.2.3 Simulation Component Classes

`SimulationElement` contains the simulation time and manages the event calendar. It provides for event and time management by being capable of: (1) scheduling events, (2) getting the next event, and (3) getting and setting the current time. This class provides an important base class from which modeling classes are derived.

The `Simulation` class has the run control properties which manage the complete simulation and include: (1) getting the current replication number, (2) setting the number of replications, (3) setting the length of the run, (4) stopping the simulation or current replication, and (5) printing summary and individual output reports.

3.3 Simulation Modeling Frame

To aid in the construction of simulation languages and packages, several simulation modeling classes have been designed and implemented. The components of the modeling frame are events, entities, processes, nodes, and choices. These components are derived from both the `SimulationElement` and the `Link` classes. In the following figure and subsequent figures, the solid lines are inherited relationships ("is-a") while the dashed lines are composition relationships ("has-a").

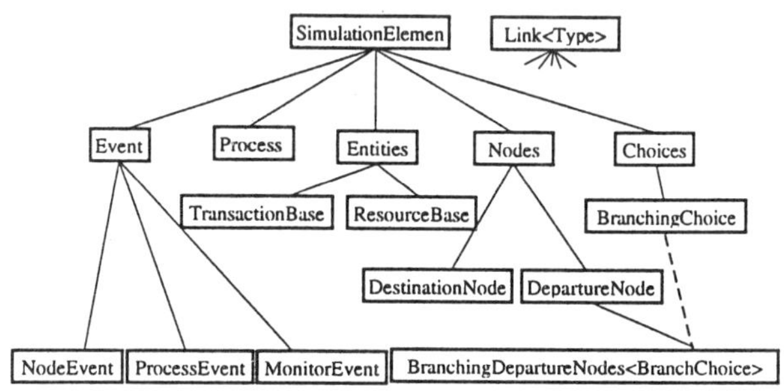

`Events` contain the properties related to simulation event management and provide: (1) the means for setting and getting the event time, (2) setting and getting other event information, and (3) processing the event. `Node`, `Process`, and `Monitor` events provide specialized properties that are needed when events occur within a node or a process, or are independently specified.

`Entities` provide active elements for the simulation, whether permanent or temporary. The properties of entities include: (1) getting the entity's creation time, (2) obtaining its status, (3) getting its current location, (4) obtaining the entry time of the entity's current state, and (5) getting the entity's time in the system. `TransactionBase` and `ResourceBase` classes are derived from `Entities` and extend the entities for use in general networks. The `TransactionBase` class provides entities that may need service and has properties for: (1) getting/setting the node entry time, (2) getting the creation node, and (3) getting/setting the identification number. The `ResourceBase` class provides entities that can provide service and has properties for: (1) getting/setting the resource name, (2) getting/setting resource states, and (3) defining the resource states.

`Processes` provide an encapsulated means for describing simulation processes (not computer tasks) such as seizing and releasing resources and reneging at queues. The process class is generally used to provide a means of decomposing a complex simulation activity, like preempting a resource, and is a form of a "helper" class for the simulation.

`Nodes` are used for network modeling and contain the properties which include: (1) getting/setting the node count, (2) getting node identification number, (3) obtaining the node type, (4) accessing a list of all nodes in the network, and (5) finding the entities at the current node. Derived from `Node` are the `Destination` and `Departure` nodes. A destination node can be entered while a departure node may be exited. Often departure nodes have branches connected to them and therefore need a "BranchingChoice" and are called `Branching-DepatureNodes` The properties of the destination node include the "entering process" while the departure nodes provides the "exiting process", the branching choice, and related branching specifications.

`Choices` are used to give the simulation model "intelligence." Routes, rules, and policies may be modeled through the various choices.

3.3.1 Frames and Frameworks

While frames provide a convenient means to describe the levels of abstraction within the entire object-oriented simulation platform, another means of encapsulation is needed to deal with the broad simulation modeling concepts and features contained within the design. In a sense, the frames are quite similar to class libraries which can be called upon in the development of an actual simulation modeling language or package. However, for the higher level modeling classes, these library-like collections of classes are too complexly interrelated to be represented simply as a single level of abstraction. A better approach to the description of these higher level complex interactions is the notion of "frameworks." For our purposes, *frameworks* are used to describe those collections of classes that provide a set of specific modeling facilities. The frameworks may consist of one or more class hierarchies. These collections make the use and reuse of simulation modeling features more intuitive and provide for greater extensibility.

There are several frameworks that compose the modeling frame and include: (1) ***Transaction Framework*** which establishes the basic properties of the transaction and provides a means to create transactions, to bring them into the network, to branch them from node to node, to cause them to exit the network, and to destroy them; (2) ***Resource Framework*** which establishes the basic properties of resources, provides for resource teams and resource groups, provides for preemption of resources, seize resources, and release resources; (3) ***Queuing Framework*** which stores transactions awaiting resources, ranks transactions in the queue, provides conditional and unconditional reneging from queues, and gates transactions in queues until conditions are appropriate for their future movement; and (4) ***Activity Framework*** which delays transactions for some specified time, determines among resource alternatives and requires resources, and abort transactions from the activity.

4 CREATING A SPECIFIC OOS

Special simulation languages and packages may be created from these object classes. In this section we present the YANSL network queuing simulation language that has been presented in prior WSC papers. YANSL is an acronym for "Yet Another Network Simulation Language." YANSL is just one *instance* of the kind of simulation capability that can be developed within an object-oriented simulation environment.

4.1 Basic Concepts and Objects in YANSL

YANSL was developed to illustrate the importance of object-oriented simulation. YANSL is a network queuing simulation package of roughly the power of a GPSS (Schriber 1991), SLAM (Pritsker 1995), SIMAN (Pegden, Shannon, and Sadowski 1995), or INSIGHT (Roberts 1983), but without the "bells and whistles." Users familiar with any of these languages should recognize, however, that it is a very powerful alternative.

4.1.1 Classes Specific to YANSL

Several classes are chosen from the modeling frameworks to create the YANSL modeling package. These classes are collected together to form a "simple" modeling/simulation language which can be extended to create more complicated features. The general simulation support classes, such as variate generation, statistics collection, and time management, are used indirectly throughout the modeling frameworks. The network concepts are somewhat enhanced, but are taken from the modeling framework. The node hierarchy for YANSL is:

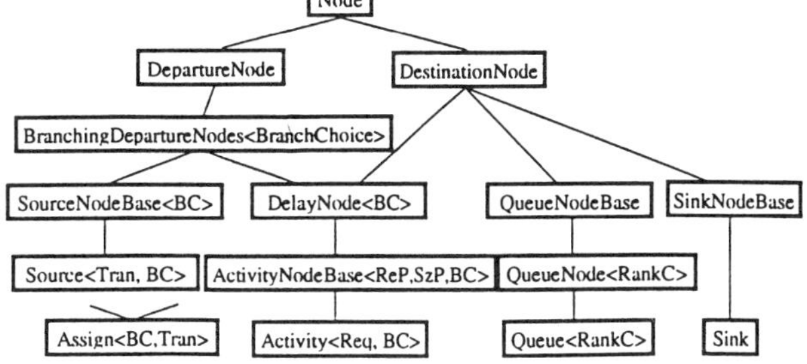

The higher level nodes (`Assign`, `Activity`, `Queue`, `Source`, and `Sink`) are used directly by the YANSL modeler. Lower level nodes provide abstractions which are less specific thus allowing specialization for other simulation constructs (e.g., the QueueNodeBase class excludes ranking and statistics). Sink and queue nodes can have transactions branched to them and are therefore destination nodes, while the source node is a departure node. The delay and assign nodes are both departure and destination nodes, so they inherit from both the departure and destination node classes. This inheritance from multiple parents is called "multiple inheritance." An activity is a "kind of" delay but includes resource requirements. The properties of the YANSL nodes allow transactions to be created at source nodes, wait at queue nodes, receive attribute assignment at assign nodes, be delayed at activity nodes, and exit the network at sink nodes.

Resources may service transactions at activity nodes. The resource framework for YANSL allows resources to be identified as individuals, as member of alternative groupings, or as members of teams.

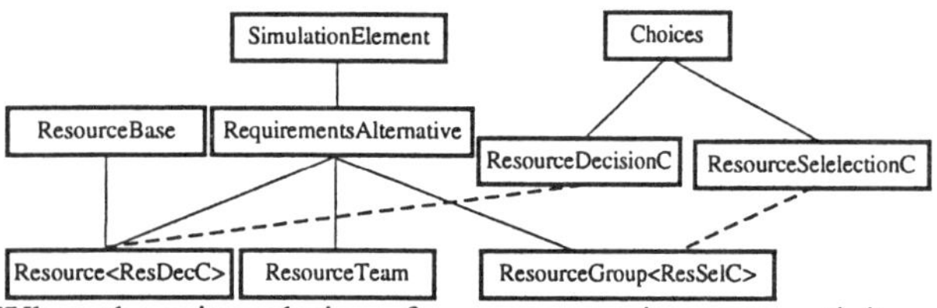

When there is a choice of resource service at an activity, then a resource selection method is used. The ability to request a resource service at run-time without specifying it explicitly is another example of polymorphism.

`Choices` available in YANSL extend those in the modeling frameworks (see the following class hierarchy):

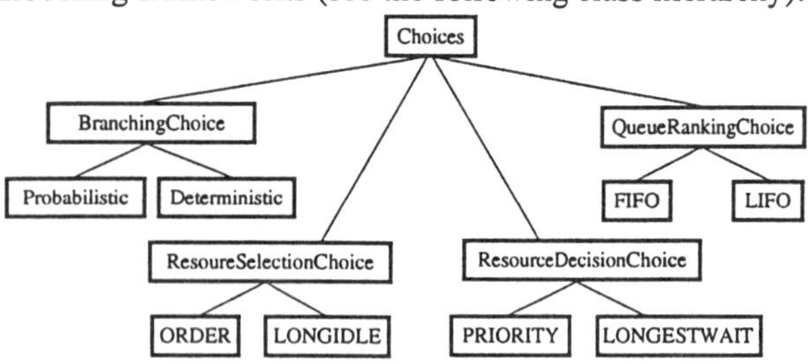

The choices available add broad flexibility to the decision-making functions in the simulation, without needing different classes for each different function. Instead, classes are parameterized with these choice classes and the choices consist of several methods. Specifically in YANSL, they allow for the selection of alternative branches from a departure node, selection among alternative resources in requirements at an `Activity`, as well as provide the decision making ability for resources to choose what to do next, and ranking choices among transactions at an `Queue`. The choices are used to represent the time-dependent and changing decisions that need to be modeled

4.1.2 Modeling with YANSL

When modeling with YANSL, the modeler views the model as a network of elemental queuing processes (graphical symbols could be used). Building the simulation model requires the modeler to select from the predefined set of node types and integrate these into a network. Transactions flow through the network and have the same interpretation they have in the other simulation languages. Transactions may require resources to serve them at activities and thus may need to queue to await resource availability. Resources may be fixed or mobile in YANSL, and one or more resources may be required at an activity. Unlike some network languages, resources in YANSL are active entities, like transactions, and may be used to model a wide variety of real-world items.

4.2 The TV Inspection and Repair Problem

As a portion of their production process, TV sets are sent to a final inspection station (refer to Joines, Powell, and Roberts, 1992 and to the harbor problem in Joines, Powell, and Roberts, 1993). Some TVs fail inspection and are sent for repair. After repair, the TVs are returned for re-inspection. Transactions are used to represent the TVs. The resources needed are the inspector and the repairman. The network is composed of a source node which describes how the TVs arrive, a queue for possible wait at the inspect activity, the inspect activity and its requirement for the inspector, a sink where good TVs leave, a queue for possible wait at the repair activity, and the repair activity. Transactions branch from the source to the inspect queue, are served at the inspect activity, branch to either the sink or to the repair queue, are served at the repair activity and return to the inspect queue. The data used in the simulation is that the interarrival time of TVs, the inspect service time, and repair service time are exponentially distributed with a mean of 5, 3.5, and 8 minutes respectively, and the probability that a TV is good after being inspected is .85.

4.3 A YANSL Model

The YANSL network has all the graphical and intuitive appeal of any network based simulation language. A graphical user interface could be built to provide "convenient" modeling with error checking and help offered to the user. Whatever the modeling system used, the ultimate computer readable representation of the model would appear as follows:

```
#include "simulation.h"

main(){

// SIMULATION INFORMATION
```

```
Simulation      tvSimulation(1); //1 replication

// DISTRIBUTIONS
Exponential     interarrival( 5 ),
                inspectTime( 3.5 ),
                repairTime( 8.0 );

// RESOURCES
Resource< PRIORITY >  inspector, repairman;

// NETWORK NODES

   /** Transactions Arrive **/
Source< Transaction, DETERMINISTIC >
       tvSource( interarrival, 0.0, 480 );
     // Begin at 0.0 and quit at 480.0

   /** Inspection **/
Queue< FIFO >  inspectQueue("Inspect Queue");
     inspector.addQueue( inspectQueue );
Activity<RequirementSet,PROBABILITY>
                       inspect( inspectTime );
     inspect.addRequirement( inspector );
     inspectQueue.addActivity( inspection );

   /** Repair **/
Queue< FIFO >  repairQueue("Repair Queue");
     repairman.addQueue( repairQueue );
Activity<RequirementSet,DETERMINISTIC>
                       repair( repairTime );
     repair.addRequirement( repair );
     repairQueue.addActivity( repair );

   /** Transactions Leave **/
Sink finish;

//NETWORK BRANCHES

tvSource.addBranch( inspectQueue );
inspect.addBranch( finish, .85 );
               // 85% are good and leave
inspect.addBranch( repairQueue, .15 );
               // 15% need repair
repair.addBranch( inspectQueue );

//RUN the Simulation
tvSimulation.run();
}
```

The previous model has an almost one-to-one correspondence to the problem entities. The statements are highly readable and follow a simple format. The pre-defined object classes give the user wide flexibility.

While the "statements" in YANSL are very similar to those in SIMAN, SLAM, or INSIGHT, it is all legitimate C++ code. Also this model runs in less than half the time a SIMAN model runs on the same machine! But the real advantage of YANSL is its extensibility.

4.4 The Objects and their Specification

The YANSL "statement" model is enclosed in the recognizable C/C++ framework, namely having a `#include` statement that includes all the simulation requires, a `main()` function header, and `{}` which enclose the block of code (YANSL statements). This framework is left only to reveal it is C++ code since even this elements could be camouflaged using the C preprocessor.

There are two types of YANSL "statements." The first is the declaration of objects in the model. These statements describe the elements in the simulation. The second type of statement is member function calls or messages to structure the model. The same division of statements occurs in existing simulation languages. The only order requirement for statements is that an object must be declared before it is used.

The objects in YANSL are declared in a form consistent with C++ . The object class is specified first, then the objects are named. Initialization of specific objects are done in parentheses. Some object declarations appear more complex because the object class is also parameterized by information in <>. In object-oriented terminology, these are called "parameterized types" and provide "has-a" relationships. Parameterized types are created by class *templates* so that the ultimate specification of a class is not known until that class is declared in the model to create the object (both the class and the object are created). Templates make it easy for a user to specify a kind of class rather than having a whole bunch of classes whose similarities are greater than their differences. As an example, consider

```
Queue< FIFO > inspectQueue( "Inspect Queue" );
```

where the `Queue` class needs some ranking choice class called `FIFO`, while the object `inspectQueue` is initialized with a string, the name of the queue. Notice that a class will be parameterized with another class, while an object is parameterized with another object.

4.5 Running the Simulation

The prior model is compiled under a C++ compiler(a compiler should be AT&T version 3.0 compatible), linked with the YANSL simulation library, and executed. Currently, the YANSL simulation library has been compiled under Borland C++ 5.0 (Borland 1995). C++ is strongly typed, so error checking is very good. Also, the simulation is easily linked into other C++ libraries.

4.6 Embellishments

The lack of distinction between the base features of YANSL and any extensions illustrate the "seamless" nature of user additions. Many embellishments are possible. For example, the embellishments shown in the earlier papers (Joines and Roberts 1992, 1993) could be applied here. Such embellishments can be added for a single use or they can be made a permanent part of YANSL, say YANSL II. In fact, a different kind of simulation language, say for modeling and simulating logistics systems, might be created and called LOG-YANSL for those special users. Perhaps the logistics users would get together and share extensions and create a more general LOG-YANSL II. And so it goes! For those familiar with some existing network simulation language, consider the difficulty of doing the same.

5 CONCLUSIONS

Modeling and simulation in an object-oriented language possesses many advantages. As shown, internal functionality of a language now becomes available to a user (at the discretion of the class designer). Such access means that existing behavior can be altered and new objects with new behavior introduced. The O-O approach provides a consistent means of handling these problems.

The user of a simulation in C++ is granted lots of speed in compilation and execution. The C language has been a language of choice by many computer users and now C++ is beginning to supplant it. With the new ANSI standard, all C++ compilers are expected to accept the same C++ language. To take full advantage of object-oriented simulation will require more skill from the user. However, that same skill would be required of any powerful simulation modeling package, but with greater limitations.

REFERENCES

Borland. 1995. *Borland C++ version 5.0.* Borland International, Inc. 100 Borland Way, Scotts Valley, CA.

CACI. 1995. *A quick look at Modsim III.* CACI Products Company, La Jolla, CA.

Ellis, M, and B Stroustrup. 1990. *The annotated C++ reference manual.* Reading, Massachusetts: Addison-Wesley.

Fishwick, P. A. 1995. *Simulation Model Design and Execution*, Englewood Cliffs, N.J. Prentice-Hall.

Geuder, D.. 1995. Object-Oriented Simulation Modeling with Simple++ In *Proceedings of the 1995 Winter Simulation Conference*, ed., C. Alexopoulos, K. Kang, W. R. Lilegdon, and D. Goldsman, 519-523. Institute of Electrical and Electronics Engineers, Washington, D.C.

Goldberg, A., and D. Robson. 1989. *Smalltalk-80: the language.* Reading, Massachusetts: Addison-Wesley.

Hammann, J. E. and N. A. Markovitch. 1995. Introduction to Arena. In *Proceedings of the 1995 Winter Simulation Conference*, op cit.

Henriksen, J. O. 1995. Introduction to SLX. In *Proceedings of the 1995 Winter Simulation Conference.* 502-509, op.cit.

Joines, J. A., K. A. Powell, Jr., and S. D. Roberts. 1992. Object-oriented modeling and simulation with C++. In *Proceedings of the 1992 Winter Simulation Conference*, ed., J. J. Swain, D. Goldsman, R. C. Crain, and J. R. Wilson, 154-162. Institute of Electrical and Electronics Engineers, Washington, D.C.

Joines, J. A., K. A. Powell, Jr., and S. D. Roberts. 1993. Building object-oriented simulations with C++. In *Proceedings of the 1993 Winter Simulation Conference*, ed., G. W. Evans, M. Mollagasemi, E. C. Russell, and W. E. Biles, 205-212. Institute of Electrical and Electronics Engineers, Los Angeles, CA.

Joines, J.A. and S. D. Roberts. 1994. Design of Object-Oriented Simulations in C++. In *Proceedings of the 1994 Winter Simulation Conference*, ed., Jeffrey Tew, S. Manivannan, Deborah Sadowski, and Andrew Seila, 157-165. Institute of Electrical and Electronics Engineers, Orlando FL.

Joines, J.A. and S. D. Roberts. 1995. Design of Object-Oriented Simulations in C++. In *Proceedings of the 1995 Winter Simulation Conference*, 82-92, op.cit.

Little, M.C. and McCue, D.L., 1994, "Construction and Use of a Simulation Package in C++,"*C User's Journal*, 12(3).

Pegden, C. D., R. E. Shannon, and R. P. Sadowski. 1995. *Introduction to simulation using SIMAN*, Second Edition, New York: McGraw-Hill.

Plauger, P. 1995. *The draft C++ library*, Englewood Cliffs, N.J. Prentice-Hall.

Pritsker, A. A. B. 1995. *Introduction to simulation and SLAM II*, Fourth Edition. New York: Halsted Press.

Roberts, S. D. 1983. *Modeling and simulation with INSIGHT.* Indianapolis, Indiana: Regenstrief Institute.

Schriber, T. J. 1991. *An introduction to simulation using GPSS/H.* New York: John Wiley and Sons.

Schwetman, H. 1995. Object-Oriented Simulation Modeling with C++/CSIM17. In *Proceedings of the 1995 Winter Simulation Conference*, 529-533, op.cit.

AUTHOR BIOGRAPHIES

JEFFERY A. JOINES is a Ph.D Candidate in the Department of Industrial Engineering and a Research Associate in Furniture Manufacturing and Management Center at North Carolina State University. He received his B.S.I.E, B.S.E.E, and M.S.I.E from NCSU. He is a member of INFORMS, IIE, and IEEE. His interests include object-oriented simulation, cellular manufacturing, and genetic algorithms.

STEPHEN D. ROBERTS is Professor and Head of the Department of Industrial Engineering at NCSU. He received his B.S.I.E., M.S.I.E., and Ph.D. from Purdue University. He was the recipient of the 1994 Distinguished Service Award. He has served as Proceedings Editor and Program Chair for the Winter Simulation Conference. He is an INFORMS/CS representative to and past Chair of the Board of Directors of WSC.

Proceedings of the 1996 Winter Simulation Conference
ed. J. M. Charnes, D. J. Morrice, D. T. Brunner, and J. J. Swain

SIMULATION OF COMMUNICATIONS NETWORKS

Averill M. Law
Michael G. McComas

Averill M. Law and Associates
P.O. Box 40996
Tucson, Arizona 85717

ABSTRACT

This paper discusses how simulation is used to design and analyze communications networks. Topics discussed include: network issues addressed by simulation, simulation software for network modeling, techniques for building valid and credible models, and statistical considerations. A comprehensive example will also be given in the conference presentation.

1 INTRODUCTION

In this paper we present an overview of the use of simulation in the design and analysis of communications networks. A detailed discussion of simulation, in general, may be found in Law and Kelton (1991). A practical discussion of the steps in a sound simulation study is given in Law and McComas (1990). General references on communications are Halsall (1996), Stallings (1994), and Tannenbaum (1989).

It is often of interest to study a proposed or existing communications network to improve its performance. However, it is generally necessary to use a model for this purpose, since experimentation with the network itself is either disruptive, not cost effective, or simply impossible (e.g., the network has not yet been built).

If the relationships that compose the model are simple enough, it may be possible to use mathematical methods (such as algebra, calculus, or probability theory) to obtain *exact* answers to the questions of interest; this is called an *analytic solution*. As a matter of fact, analytic queueing models have been used for years to study performance issues for communications networks and computer systems (see, for example, Kleinrock 1976). However, as network topologies and protocols have become more complex, analytic methods have become increasingly inadequate. Additional shortcomings of analytic queueing models are as follows:

- Only steady-state results are typically possible
- It is difficult to obtain performance measures other than mean values (e.g., the 95th percentile of end-to-end delay)
- Original analytic solutions require considerable mathematical sophistication on the part of the analyst

Because of the drawbacks of analytic methods, there has been a considerable increase in the use of simulation for network analyses during the past five to ten years. This has in turn resulted in the introduction of a number of new simulation products specifically for communications networks. In a *simulation* a mathematical/logical model is numerically evaluated over a time period of interest, and performance measures are *estimated* from model-generated data. Simulation analyses are applicable to systems of almost any level of complexity. Perhaps the only impediment to the use of simulation is the potentially large amount of computer execution time required to process messages for high-traffic-rate networks.

2 OBJECTIVES OF SIMULATION IN COMMUNICATIONS

The following are some of the benefits of using simulation to design and analyze communications networks:

- Determination of the system-wide impact of making "local" changes to the network
- Improved system performance (delays, throughput, etc.)
- Reduced expenditures
- Insurance that performance objectives are met before equipment is bought or leased
- Identification of bottlenecks before system implementation

- Reduced system development time

Simulation has addressed a number of specific communications issues, including:

- How will my network perform when the traffic load increases?
- What are the requirements for number/speed of links, number/speed of switches, and buffer sizes for my wide area network?
- What will be the impact of a link failure?
- What protocols will provide the best network performance?
- What is the best design for my new communications network?
- What is the desirability of ATM or frame relay?
- What will happen when additional PCs or workstations are added to my local area network (LAN)?
- What impact will adding a new application such as image processing have on my LAN?
- Will my client-server strategy provide the required performance?
- How many satellites are needed to provide a certain level of service between two earth stations?

The following is a list of performance measures that are commonly used in simulation studies of networks:

- Throughput (e.g., in kilobits per second)
- End-to-end delay
- Delay from point A to point B in a network
- Number of "data units" in a queue or a buffer
- Utilizations of nodes or links
- Probability of a blocked call
- Probability of a lost call (mobile system)
- Number of collisions and deferrals (LAN)

3 SIMULATION SOFTWARE FOR NETWORKS

One of the major tasks in building a simulation model of a communications network is that of converting a system description into a computer program. An analyst may use either a general-purpose programming language (e.g. C or C++) or simulation software for this purpose. Some advantages of a programming language are as follows:

- Most modelers already know a programming language, but this is often not the case with simulation software.
- C or C++ are available on virtually every computer, but a particular simulation software product may not be available for the analyst's computer.
- Software cost will generally be considerably lower (but not necessarily project cost).

The major advantage of using simulation software is that they automatically provide most of the features needed in programming a simulation model, resulting in a significant decrease in programming time (and usually project cost). Simulation software also provide a more natural framework for system modeling. In general, we believe that an analyst would be prudent to use simulation software to model a communications network.

There are three types of software for simulating communications networks (see Law and McComas 1994). A *general-purpose simulation language* is a simulation package that is general in nature (e.g., it could also be used for modeling manufacturing systems or for combat modeling), but may have special features for communications such as explicit modules for Ethernet or token ring. Examples of simulation languages are Arena, AweSim, BONeS DESIGNER, GPSS/H, MODSIM III, SES/*workbench*, and SIMSCRIPT II.5 (only BONeS DESIGNER and SES/*workbench* have communications modules).

A model is developed in a simulation language by writing a program using the language's modeling constructs, which include entities (messages), attributes (message type or destination), resources (nodes or links), and queues (buffers). The major advantage of most languages is their ability to model almost any kind of communications network, regardless of its complexity or uniqueness. Possible drawbacks of languages, as compared to some simulators (see below), are the need for programming expertise and possibly the long time spent coding and debugging that is associated with modeling complex networks.

A *communications-oriented simulation language* is a simulation language that is specifically oriented toward communications networks--OPNET Modeler is such a product. Advantages are possibly reduced programming time and modeling constructs oriented toward communications systems.

A *communications-oriented simulator*, in its most basic form, is a simulation package that allows one to simulate a network in a specific class of communications networks with no programming. Examples of basic simulators are COMNET III, NETWORK II.5, OPNET Planner, and SES/*strategizer*. The particular network of interest (in the domain of the package) is selected for simulation by choosing items from menus (typically using a point-and-click approach), by filling in dialogue boxes (forms), and by the use of graphics. Typical modeling constructs for a LAN simulator are LAN types (Ethernet, token ring, etc.), stations on a LAN (PCs or work stations), LAN interconnection devices (bridges and routers), and traffic (message) generators. The major advantage of a simulator is that "program" development time may be considerably less than for a simulation language. This may be very important given the tight time constraints in many business environments.

Another advantage is that simulators have modeling constructs closely related to the components of a communications network, a very desirable feature for someone like a network manager. Also, people without programming backgrounds and those who use simulation only occasionally often prefer simulators because of their ease of use.

The major drawback of *basic* simulators is that they are limited to modeling only those network configurations allowed by the package's standard building blocks. Thus, if a communications system has some unique features, they might have to be modeled in an approximate manner when using certain simulators. This difficulty can be largely overcome in the Developmental Version of COMNET III, which allows existing modeling constructs to be modified and new constructs to be added. Also, OPNET Modeler can be used to create/modify modules for OPNET Planner.

Note that an important feature for simulation software to be used for network modeling is fast model execution speed, because in some networks a very large number of messages will need to be simulated.

4 DEVELOPING VALID AND CREDIBLE SIMULATION MODELS

A simulation model is a surrogate for actually being able to experiment with a communications system. Thus, an idealized goal in building a simulation model is for it to be *valid* enough so that any conclusions drawn from the model would be similar to those derived from physically experimenting with the system (if this were possible). It is also important for a model to be *credible*; otherwise, its results may never be used in the decision-making process, even if the model is valid.

The following are some important ideas/techniques for deciding the appropriate level of model detail, for validating a simulation model, and for developing a model with high credibility:

- State definitively the issues to be addressed and the performance measures for evaluation at the beginning of the study.
- Collect information on the network topology and protocols based on conversations with all important people associated with the system.
- Delineate all information and data summaries in an "assumptions document."
- Interact with the manager on a regular basis throughout the study.
- Perform a structured walk-through (before programming) of the conceptual simulation model as embodied in the assumptions document before an audience of all key project personnel.
- Use sensitivity analyses (see Law and Kelton 1991) to determine important model factors.
- Compare performance measures (e.g., utilization) for the existing network (if there is one) to comparable performance measures for a simulation model of the existing network.

5 STATISTICAL ISSUES IN NETWORK SIMULATION

Since random samples from input probability distributions (e.g., the distribution for interdeparture times of messages) are used to "drive" a simulation model through time, basic simulation output data (e.g., end-to-end delays of messages) or an estimated performance measure computed from them (e.g., average end-to-end delay from the entire run) are also random. Therefore, it is important to model the random inputs to a simulation model correctly and also to design and analyze simulation experiments in a proper manner. These topics are briefly discussed in this section.

5.1 Modeling System Randomness

The most important source of randomness for network simulations is usually that associated with message traffic. In general, one should model *messages* (or *transactions*) not packets. The messages are fragmented into packets by the network protocols employed in the simulation. Note also that messages may not be independent of each other (e.g., there are often acknowledgement messages). The following methods of generating traffic are often used, with the first approach generally being the most statistically valid:

- Message departure times and message sizes for a particular node are application based (file transfer, word processing, E-mail, etc.), and may depend on the receipt of an acknowledgement message
- Message interdeparture times and message sizes for a particular node are each independent samples from respective probability distributions (usually exponential for interarrival times)
- Traffic data are read into the simulation model from a network analyzer

Because network components are generally quite reliable, equipment breakdowns are not typically modeled in a simulation. An exception is where one is interested in the transient response of the network, e.g., the ability of the network to reconfigure itself after a link failure. In such cases, the operational status of a component can be modeled as an "up" period of random duration followed by a "down" (or repair) period of random duration.

5.2 Design and Analysis of Simulation Experiments

Because of the random nature of simulation input, a simulation model produces a statistical estimate of the (true) performance measure not the measure itself. In order for a simulation estimate to be statistically precise (have a small variance) and free of bias, the analyst must specify for each network configuration appropriate choices for the following:

- Length of each simulation run
- Number of independent simulation runs
- Length of the warmup period, if one is appropriate

We recommend always making at least three to five independent runs for each configuration, and using the average of the estimated performance measures from the individual runs as the overall estimate of the performance measure. (Independent runs mean using different random numbers for each run, starting each run in the same initial state, and resetting the model's statistical counters back to "zero" at the beginning of each run.) This overall estimate should be more statistically precise than the estimated performance measure from one run. Note that independent runs (as compared to one very long run) are required to obtain legitimate and simple variance estimates and confidence intervals.

When simulating certain types of communications systems, we are often interested in the long-run (or steady-state) behavior of the system, i.e., its behavior when operating in a "normal" manner. On the other hand, simulations of these kinds of systems generally begin with the system in an empty and idle state. This results in the output data from the beginning of the simulation not being representative of the desired "normal" behavior of the system. Therefore, simulations are often run for a certain amount of time, the *warmup period*, before the output data are actually used to estimate the desired measure of performance. Use of these warmup-period data would bias the estimated performance measure.

A comprehensive treatment of simulation output-data analysis can be found in Law and Kelton (1991).

6 SIMULATION ANALYSIS OF A COMMUNICATIONS SYSTEM

In the actual conference presentation, we will give a detailed simulation analysis of a wide area communications network. The following performance issues will be addressed:

- Is the performance of the existing network satisfactory?
- What impact will a link failure have on system performance?
- How much can the traffic rates be increased before the system "blows up"?
- What impact will changing the message-size distribution and the form of the traffic have on system performance?
- What will be the effect of faster links or processors?

REFERENCES

Halsall, F. 1996. *Data communications, computer networks and open systems.* 4th ed. New York: Addison-Wesley.

Kleinrock, L. 1976. *Queueing systems.* Volume 2: Computer applications. New York: Wiley.

Law, A. M., and W. D. Kelton. 1991. *Simulation modeling and analysis.* 2d ed. New York: McGraw-Hill.

Law, A. M., and M. G. McComas. 1990. Secrets of successful simulation studies. *Industrial Engineering* 22:47-48,51-53,72.

Law, A. M., and M. G. McComas. 1994. Simulation software for communications networks: The state of the art. *IEEE Communications Magazine* 32:44-50.

Stallings, W. 1994. *Data and computer communications.* 4th ed. New York: Macmillan.

Tannenbaum, A. S. 1989. *Computer networks.* 2d ed. Englewood Cliffs, New Jersey: Prentice-Hall.

AUTHOR BIOGRAPHIES

AVERILL M. LAW is President of Averill M. Law & Associates (Tucson, Arizona), a company specializing in simulation model building, training, and software. He has been a simulation consultant to more than 100 organizations, including General Motors, IBM, AT&T, General Electric, Nabisco, Xerox, NASA, the Air Force, the Army, and the Navy. He has presented more than 250 simulation short courses in 16 countries, and delivered more than 100 talks on simulation modeling at technical conferences.

He is the author (or coauthor) of three books and more than 35 papers on simulation, manufacturing, communications, operations research, and statistics, including the textbook *Simulation Modeling and Analysis* that is used by more than 47,000 people worldwide. His series of papers on the simulation of manufacturing systems won the 1988 Institute of Industrial Engineers' best publication award. He is the codeveloper of the ExpertFit software package for selecting simulation input probability distributions, and he has developed several simulation videotapes. Dr. Law wrote a regular column on simulation for *Industrial Engineering* magazine from 1990 through 1991.

He has been a tenured faculty member and has taught simulation at the University of Wisconsin and the University of Arizona. Dr. Law has a Ph.D. in industrial engineering and operations research from the University of California at Berkeley.

MICHAEL G. MCCOMAS is Vice President for Consulting Services of Averill M. Law & Associates. He has considerable simulation modeling experience in a wide variety of application areas, and is the coauthor of seven published papers on simulation. His educational background includes an M.S. in Systems and Industrial Engineering from the University of Arizona.

Proceedings of the 1996 Winter Simulation Conference
ed. J. M. Charnes, D. J. Morrice, D. T. Brunner, and J. J. Swain

INTRODUCTION TO SIMULATION IN HEALTH CARE

Julie C. Lowery

VA Health Services Research and Development Field Program
P.O. Box 130170
Ann Arbor, MI 48113-0170

ABSTRACT

The purpose of this article is to discuss some of the more challenging issues associated with conducting simulation in healthcare. The current healthcare environment is ripe for the use of simulation. The pressure to control costs is higher than ever, so, there is a critical need for powerful tools which can help clinicians and administrators (our clients) make good decisions on how to achieve objectives of reducing costs while maintaining high quality care. In addition, the highly stochastic nature of disease processes, as well as the complexity of subsystem interactions, makes simulation the decision-support tool of choice for analyzing the organization and delivery of healthcare services.

However, for simulation to reach its potential as a major weapon in the fight against spiraling healthcare costs, pragmatic approaches to several challenging technical questions must be offered and discussed. Therefore, this article will present approaches to dealing with the following, frequently encountered tactical issues in simulating healthcare services--degree of model complexity, definitions of input distributions, model validation, and interpretation of findings. The last issue to be discussed is less of a technical concern, and instead addresses the promotion of simulation in healthcare.

1 INTRODUCTION

The purpose of this article is to discuss some of the more challenging issues associated with conducting simulation in healthcare. Healthcare is a somewhat difficult arena in which to conduct simulation studies for a number of reasons. A panel discussion at the 1994 Winter Simulation Conference (Hakes et al., 1994) offered several interesting insights into the barriers associated with implementing simulation in healthcare, including the following:

- Historical disincentives to control costs. (This is changing!)
- Healthcare managers' traditional reliance on simpler, deterministic analytic techniques for decision-making.
- Administrators' and providers' resistance to the unfamiliar and dehumanizing nature of simulation.
- Highly technical nature of simulation.
- Number and variety of customers with competing priorities for solutions suggested by simulation.

Despite the existence of these barriers, the current healthcare environment is ripe for the use of simulation. The pressure to control costs is higher than ever, so, there is a critical need for powerful tools which can help clinicians and administrators (our clients) make good decisions on how to achieve objectives of reducing costs while maintaining high quality care. In addition, the highly stochastic nature of disease processes, as well as the complexity of subsystem interactions, makes simulation the decision-support tool of choice for analyzing the organization and delivery of healthcare services.

However, for simulation to reach its potential as a major weapon in the fight against spiraling healthcare costs, not only must the barriers discussed in 1994 be overcome, but pragmatic approaches to several challenging technical questions must be offered and discussed. Shuman and Wolfe (1992), in their assessment of the status of simulation in healthcare, write:

> Building a simulation model may be the easiest part of the process; addressing the tactical concerns of simulation, designing a valid simulation experiment and conducting a rigorous analysis of the results remain sophisticated endeavors. We have a fear that these factors are being increasingly overlooked by would-be modelers.

Therefore, this article will present approaches to dealing with the following, frequently encountered tactical issues in simulating healthcare services--degree of model complexity, definitions of input distributions, model validation, and interpretation of findings. The last issue to be discussed is less of a technical concern, and instead addresses the promotion of simulation in healthcare.

Finally, it should be noted that this article is not a review of the applications of simulation in healthcare. Others have offered extensive bibliographies of such applications (Klein et al., 1993; Valinksy, 1975); and it is not my intent to repeat or improve upon those discussions here. However, a review of these bibliographies is strongly recommended to those just embarking on a healthcare simulation project. While previous simulation applications are frequently not available for widespread use, one can always learn from the techniques used and the problems encountered by others.

2 MODEL COMPLEXITY

One of the barriers to implementing simulation in healthcare discussed at WSC '94 was the highly technical nature of simulation. This issue includes the perception that the time, effort, and skill involved in developing, validating, and then experimenting with a simulation model are not worth the expected benefits. The time and skill level required for model building is certainly decreasing as the power and flexibility of available simulation software increases. Nevertheless, the time and effort required to construct a valid model and conduct carefully designed experiments with the model can still be significant, even for analysis of a relatively simple system. Therefore, it is important that the effort not be lengthened with the development of a needlessly complex model. Especially today where the healthcare environment is in a state of such rapid change, we cannot afford to spend too much time developing models of systems that are going to be outdated as soon as they are completed.

In his keynote address to the 1993 Winter Simulation Conference, John Salt provided an excellent discussion of the benefits of keeping simulation models "constructively simple" (Salt, 1993). The rule of thumb is to develop as simple a model as possible that you think will meet the project's immediate objectives. You can always add to the complexity of the model later if necessary, generally without having wasted any time by first developing a simple model.

One example of an area for which the appropriate level of complexity must be decided is that of patient case-mix. Obviously the case-mix of patients moving through a healthcare system can be modeled at a tremendous level of complexity--e.g., down to the level of individual diagnosis, if necessary. Of course, the effort associated with modeling at this level of complexity is significant, as are the data requirements for defining the arrival time and input distributions associated with such a large number of patient categories.

The key to deciding on the level of detail required in case-mix categories is the objective(s) of the model. If one of the objectives is to enable investigation of the effects of changes in case-mix on system performance, then case-mix needs to be defined at the level at which current planning activities are performed. For example, if planning presently takes place by projecting workload by department (e.g., number of surgery patients, number of medicine patients), then case-mix need only be defined at that level. Or, if planning takes place by DRGs, then patients should be modeled at this level of specificity. But there is no reason to further subdivide patient categories beyond what your facility currently uses or plans to use in the near future.

Before embarking on a complex simulation study, one should always consider the option of using a simpler, analytic model. While simulation offers a number of important advantages over analytic models, often these simpler models will suffice. Especially if the objective of the project is to plan for average performance, and variability around the average is not of significant consequence, then an analytic model may be fine. For example, if the objective is to construct a model for determining the annual staffing requirements of a department or facility, then an analytic model comprised of average time requirements should be sufficient for calculating the number of full-time equivalent positions to budget for the year. However, if you are interested in determining how to schedule those positions on any given day or week, you may then want to consider use of a simulation model, because workload variability will play more of a role in scheduling decisions.

Similarly, the design of an operating room block scheduling system, which requires allocation of blocks of time to individual surgeons or groups of surgeons, can easily be accomplished using an analytic model and average operating room time requirements. Even variability can be investigated using an analytic model, if the objective is to look at the effect of only one or two parameters on that variability. For example, to determine how much operating room time should be set aside for emergency cases, by time of day or day of week, an analytic model which considers variability in emergency arrivals can be designed. Historical data on emergency arrivals can be collected, arrival time distributions graphed, and a decision made regarding the desired probability of having a room available when one is needed. In contrast, if the effect of the interaction of multiple variables (e.g., emergency arrivals, variability in surgical procedure times, and a patient scheduling system) on resource requirements (e.g., operating room nurse staffing) is to be investigated, a simulation model is desirable.

3 DEFINITIONS OF INPUT DISTRIBUTIONS

Common input distributions in healthcare models include the arrival times of unscheduled patients and the service times of resources (e.g., providers, rooms, beds, equipment, etc.). Two choices are available for defining these input distributions: theoretical or empirical. Both kinds can be defined using distribution fitting software such as UniFit II (Law and Vincent, 1993). The advantage of empirical distributions is that they fit the data, which might not be adequately represented by a theoretical distribution. However, the advantage of theoretical distributions is that they are easier to use for planning purposes.

For example, if you want to determine the effect of a change in demand (i.e., arrival rates) or practice patterns (i.e., service times) on performance, you will need to define the change in the corresponding input distribution. If the input distribution is a theoretical one, the change is fairly straightforward (if one assumes that the shape of the distribution is likely to remain the same)--the distribution's parameters (e.g., mean and standard deviation) simply need to be revised. In contrast, if the distribution is empirical, the entire distribution must be redefined. When making workload projections, it is generally easier to think in terms of revising individual parameters than entire distributions.

In addition, for those facilities or projects with little available historical data for defining empirical distributions, it is possible to use theoretical distributions that have been found to adequately represent healthcare processes in previous research. For example, the exponential distribution is routinely used to model the arrival of emergency patients to healthcare facilities (Lowery 1991; Esogbue and Singh, 1976; Cooper and Corcoran, 1974; Kao, 1974; Clipson and Wehrer, 1973; Shonick and Jackson, 1973). The exponential process works well for interarrival times when there is little or no control over the arrival process (Schriber, 1991, p. 254). Length of stay in hospitals has been modeled using a variety of theoretical distributions; but it appears that the lognormal distribution has the most documented support (Lowery, 1991; Magerlein, 1978; Storer and Hancock, 1976; Whitmore, 1975; Thomas, 1968; Balintfy, 1962; Flagle, 1960). Both of these distributions, as well as many others, can be easily modeled in commercially available simulation software.

When using distribution fitting software to test the degree of fit between a set of actual data and a theoretical distribution, one must consider the tradeoff between obtaining a possibly better fit using an empirial distribution and the flexibility of using a theoretical one. Therefore, even though statistical tests may indicate a poor to fair fit of the data with any theoretical model, look carefully at the more subjective, graphical comparisons of the actual data with the best theoretical model--the fit may look very acceptable. Consider the accuracy of the data set as well--is it possible that system problems or deficiencies might be skewing or distorting the data and, in turn, preventing the theoretical model from matching the data more precisely? If you do not want to include all the system distortions as part of your input distributions (which is generally the case), the theoretical model may be more appropriate.

The bottom line for deciding whether a selected input distribution, theoretical or empirical, is adequate for a modeling effort is whether the model eventually validates (see "Model Validation" below). If a model is not deemed sufficiently valid for its intended purposes, one may need to return to the assumptions underlying the definition of the input distributions. In contrast, a valid model provides some evidence that the selected input distributions are adequate, even if they do not precisely match the actual data from which they were derived. There is a need for more published research on the results of fitting actual healthcare data to theoretical distributions, and on the contribution of these distributions to model validity, to help other modelers who do not have access to historical data for performing their own distribution fitting.

4 MODEL VALIDATION

Many clinicians and administrators doubt the capability of computer models to capture the complexity and unpredictable nature of healthcare. Yet it is precisely these characteristics which lend themselves to modeling through simulation. One possible explanation for the lack of credibility of simulation models in healthcare applications is the limited attention modelers often pay to the model validation process. Demonstration of a simulation model's validity--i.e., its ability to accurately represent the system under investigation--is key to the acceptance of simulation as a technique. However, if published studies are any indication of the level of effort that goes into the validation component of modeling efforts, it is little wonder that simulation as a technique still meets with some skepticism.

Few of the published results of healthcare simulation efforts offer much insight into the results of validation, let alone the validation process itself. This is unfortunate, because many of the textbook recommendations for validating simulation models do not lend themselves easily to healthcare applications. (See tutorial on "Model Validation" elsewhere in these *Proceedings* for more detailed information on validation techniques.) For example, recommended techniques of confidence intervals and hypothesis testing require independent observations. Many of the measures of interest in healthcare applications--e.g., average daily census of inpatient units, utilization of inpatient resources by day of week, patient waiting times for outpatient services by time of day--are strongly autocorrelated.

While it is possible to generate independent observations of these measures from simulation model output using such methods as batched means, it is often very difficult to use the same techniques for obtaining the real-world data for comparison, due to limited availability of these data. Performance data on healthcare systems are becoming increasingly available with the widespread implementation of computer-based information systems; but even with the implementation of these systems, the rapidly changing nature of healthcare today often precludes the availability of data for an extended period of time under the same set of input conditions. (An objective of validation is to compare a sample of model observations with a sample of actual observations under the same set of input conditions.)

Another problem with use of hypothesis testing as a method of validation is that because a simulation model is only an approximation of an actual system, the null hypothesis that model behavior and system behavior are the same will almost certainly be false (Law and Kelton, 1991, p. 312), especially when modeling such complex systems as healthcare applications. While formal statistical tests may lead to the conclusion that a model is not an accurate representation of the real world system under investigation, the model may still be valid for the purpose for which it is intended. This is especially true for models that are designed primarily for comparing alternatives than for predicting absolute answers.

However, the fact that it may be difficult or even impossible to develop a model that passes traditional statistical tests of accuracy does not at all preclude the need to conduct as thorough a validation as possible and justify a conclusion of model validity. Subjective validation techniques can be used and are discussed elsewhere in these *Proceedings* by Sargent (see "Some Subjective Validation Methods Using Graphical Displays") and Lowery (see "Design of Hospital Admissions Scheduling System Using Simulation"). Publication of validation methodologies and conclusions would go a long way in helping others design studies and interpret findings for healthcare applications.

5 INTERPRETATION OF FINDINGS

This section discusses two concerns that frequently arise when clinicians and healthcare managers review the results of simulation models. Unfortunately, there are no solutions to these concerns. I can only present the warning that they may occur and offer suggestions for words of reassurance. The two concerns are that (1) simulation does not provide the single, best answer to the problem at hand; and (2) simulation models do not predict the future. These statements may seem obvious and simplistic to researchers, management engineers, and statisticians who frequently work with simulation models. But they are not apparent to those most interested in the findings from these models--especially if simulation has been touted as the be-all and end-all solution to the problem at hand. Thus, while we certainly want to promote simulation as an effective decision-support tool, we must be careful to also explain its limitations.

Unlike analytic models, simulation models do not automatically provide the single, optimal solution to the problem under investigation. Instead simulation provides answers to "what if" questions via a series of trial and error experiments; or the results of simulation experiments are analyzed using such statistical techniques as analysis of variance, to determine the relationships between independent and dependent variables of interest. These approaches may disappoint some clients, who want a model that can provide *the* answer to their question, and quickly. The idea of having to review lots of output, only to decide that values of the input variables should be revised and the model rerun to see if performance can be improved further may not be very appealing.

With some explanation and encouragement, clients can be shown the advantages of the iterative and experimental approach of simulation over the use of analytic models which provide a single, "optimal" solution. The biggest advantage is that simulation models allow for the consideration of multiple, often competing performance objectives. Efforts to reduce healthcare costs cannot be implemented at the expense of compromising quality. Thus, it is critical that an investigation of the effect of any change in resources or procedures include the examination of multiple performance measures. In contrast, analytic models generally accommodate single objectives subject to constraints on other objectives. Unfortunately, managers may not know the values at which the constraints should be set, because they may not understand the nature of the tradeoff between competing objectives. Simulation allows the explicit review of these tradeoffs.

For example, the decision to allocate beds among clinical services requires the consideration of the tradeoff between maximizing the utilization of beds (and associated costly resources such as staff and equipment) and minimizing adverse occurrences such as turning patients away because of lack of available beds. If managers could set one of these objectives at a desired maximum or minimum level to serve as a constraint—e.g., no more than 2 turnaways per month—then it would be a relatively straightforward process to determine the number of beds required to maximize utilization at that level.

However, managers are often unable or unwilling to set constraints, until they can see the relationship between the performance measures—e.g., by accepting an additional turnaway each month, how much of an increase in utilization can I obtain? Only after this relationship has been identified can decision-makers then determine what is acceptable. Simulation allows the examination of these tradeoffs, at the expense of some additional time and effort on the part of the decision-makers to experiment with the model. Once clients understand the importance and benefits of the experimentation process, hopefully they will embrace simulation as the powerful tool it is.

The other major "limitation" of simulation is its inability to predict the future. However, this characteristic can only be considered a limitation if predictive powers are expected. Therefore, it is important that managers understand up front that simulation is actually a "what if" tool. *What* would happen to performance *if* we changed values of the input variables. The values of the input variables must be specified. Unfortunately the values of the input variables must often be predicted, and simulation usually cannot help with this task.

In the previous example of bed allocation among clinical services, inpatient workload is a model input. If bed requirements for the next five years need to be determined, inpatient workload must first be predicted, and then the effect of that workload on bed utilization at various bed levels can be investigated (Lowery, 1992). On more than one occasion my clients have been disappointed to find out that the model does not predict inpatient workload.

While there is no actual solution to this problem, clients' disappointment and disillusionment can be mitigated by (1) explaining as soon as possible in the design process what the model will and will not do; and (2) offering sensitivity analysis as a way of diffusing the issue. Sensitivity analysis allows one to determine which input variables are the most important in terms of their ability to affect a change in the model's performance measures. Variables for which small changes in their values result in large changes in model performance should receive significant care and attention in measuring their true magnitude. Conversely, less time and effort should be spent collecting data on variables to which the model is not sensitive.

Thus, if a client expresses concern that a simulation model will not provide much useful information because future values of the input variables cannot be predicted, the answer is to encourage the use of sensitivity analysis. Results of the analysis may indicate that the input variables most difficult to predict have little effect on performance, thus leaving a wide margin for error in predictions. On the other hand, if results show that model performance is sensitive to a particular variable, then sensitivity analysis will at least provide the client with information on the magnitude and nature of the effect of the variable. Such information can provide an incentive to expend resources on input data collection or estimation, and can be helpful in the decision-making process even if accurate predictions are not possible.

6 PROMOTION OF SIMULATION

The promotion of simulation should not be limited to extolling its virtues as a "what if" experimentation tool. It can also make extremely valuable contributions to decisions by serving as a means of documenting assumptions, organizing the decision-making process, and identifying potential problem areas—regardless of whether or not the model is actually ever run. It is amazing how much time is spent in the planning process (especially in meetings) arguing over differences of opinion, where these differences are due to disagreements over assumptions never actually stated or acknowledged. Without a structure for identifying and documenting the numerous assumptions underlying decisions, it is impossible to conduct a meaningful discussion of the alternative solutions. The process of designing a simulation model provides such a structure. Documenting assumptions in this manner enables all participants to start discussions with the same set of information, which in turn, significantly improves the efficiency of the decision-making process. While disagreements may still ensue over the content of the assumptions, the arguments become very clear and focused when the assumptions are in black and white in front of all participants.

Similarly, designing a model forces the decision-making process into a very defined format, which can assist in organizing the process. Finally, one of the frequently cited benefits of simulation is that the knowledge gained during the actual design process may suggest improvements in the system under investigation (Banks and Carson, 1984). Designing a simulation model requires a thorough investigation and documentation of current processes, which can identify unexpected problems unrelated to the original objectives of the study. Solutions to these problems may be readily apparent and relatively easy to implement, without the assistance of the simulation model.

Thus, it is not necessary to commit to a full-blown, large-scale simulation project to be able to realize benefits from the modeling effort. If clients are concerned with the time and effort involved in simulation, the intermediate benefits should be emphasized. Mahachek (1992) explains that one of the major barriers to the implementation of simulation in healthcare is the perception that "simulation is an *additional* layer of effort rather than an *organizer* of all your current efforts." He goes on to explain that, "Simulation is not a chasm which can only be crossed with one all-consuming leap. It can be nibbled at." And each of those nibbles can yield some benefit. I hope the suggestions provided in this article will serve to ensure the achievement of those benefits.

REFERENCES

Balintfy JL. *Mathematical Models and Analysis of Certain Stochastic Processes in General Hospitals.* PhD dissertation, Johns Hopkins University, 1962.

Banks J, Carson JS. *Discrete Event System Simulation.* Englewood Cliffs, NJ: Prentice-Hall, Inc., 1984.

Clipson CW, Wehrer JJ. *Planning for Cardiac Care.* Ann Arbor, MI: Health Administration Press, 1973.

Cooper, JK and TM Corcoran. Estimating bed needs by means of queueing theory. *New England Journal of Medicine* 1974;291(8):404-405.

Esogbue AO, Singh AJ. A stochastic model for an optimal priority bed distribution problem in a hospital ward. *Operations Research* 1976;24(5):884-898.

Flagle, CD. The problem of organization for hospital inpatient care. In *Management Sciences: Models and Techniques.* Edited by CW Churchman and M Verhulst. New York: Pergamon Press, 1960.

Hakes B, Keller L, Lilegdon WR, Mabrouk K, McGuire F, Lowery JC. Barriers to implementing simulation in health care. In *Proceedings of the 1994 Winter Simulation Conference.* Edited by JD Tew and S Manivannan. Baltimore: Association of Computing Machinery, 1994, pp. 868-875.

Kao EPC. Modeling the movement of coronary patients within a hospital by semi-Markov process. *Operations Research* 1974;22(4):683-699.

Klein RW, Dittus RS, Roberts SD, Wilson JR. Simulation modeling and health-care decision making. *Medical Decision Making* 1993;13(4)347-354.

Law AM, Kelton WD. *Simulation Modeling & Analysis.* New York: McGraw-Hill, Inc., 1991.

Law AM, Vincent SG. *UniFit II User's Guide.* Tucson: Averill M. Law & Associates, 1993.

Lowery JC. *Design and Validation of Methodologies for Determining Critical Care Bed Requirements.* PhD dissertation, Health Services Organization and Policy, University of Michigan, 1991.

Lowery JC, Martin JB. Design and validation of a critical care simulation model. *Journal of the Society for Health Systems* 1992;3(3):15-36.

Magerlein JM. *Hospital Operating Room and Admissions Interactions; A Scheduling Methodology for Surgical Patients.* PhD dissertation, Department of Industrial and Operations Engineering, University of Michigan, 1978.

Mahachek AR. An introduction to patient flow simulation for health-care managers. *Journal of the Society for Health Systems* 1992;3(3):73-81.

Salt JD. Keynote address: Simulation should be easy and fun. In *Proceedings of the 1993 Winter Simulation Conference.* Edited by GW Evans, M Mollaghasemi, EC Russell, WE Biles. New York: Association of Computing Machinery, 1993, pp. 1-5.

Schriber TJ. *An Introduction to Simulation Using GPSS/H.* New York: John Wiley & Sons, 1991.

Shonick W, Jackson JR. An improved stochastic model for occupancy-related random variables in general acute hospitals. *Operations Research* 1973;21(4):952-965.

Shuman LJ, Wolfe H. Introduction. *Journal of the Society for Health Systems* (issue focus: simulation) 1992;3(3):2-4.

Storer RH, Hancock WM. *A Simulation Study of Parameters Affecting Hospital Unit Occupancy and Implications for Unit Sizing*. Report 76-4, Bureau of Hospital Administration, University of Michigan, December 1976.

Thomas WH. A model for predicting the recovery progress of coronary patients. *Health Services Research* 1968;3:185-213.

Valinsky D. Simulation. In *Operations Research in Health Care*. Edited by LJ Shuman, RD Speas, JP Young. Baltimore: The Johns Hopkins University Press, 1975.

Whitmore GA. The inverse Gaussian distribution as a model of hospital stay. *Health Services Research* 1975;10:297-302.

AUTHOR BIOGRAPHY

JULIE C. LOWERY is Associate Director of the Center for Practice Management and Outcomes Research, VA Health Services Research and Development Field Program, Ann Arbor, MI. She has a BS in Microbiology and a Master's in Health Services Administration, both from the University of Michigan. She received her PhD in Health Services Organization and Policy from the School of Public Health, University of Michigan, where she specialized in operations research and information systems in health care.

Proceedings of the 1996 Winter Simulation Conference
ed. J. M. Charnes, D. J. Morrice, D. T. Brunner, and J. J. Swain

INTRODUCTION TO MANUFACTURING APPLICATIONS

Gordon M. Clark

Systems Solutions Inc.
400 Longfellow Ave.
Worthington, Ohio 43085-3074, U.S.A.

ABSTRACT

This tutorial introduces manufacturing applications of simulation through four illustrative example applications. These examples illustrate the additional understanding of system behavior gained by the use of simulation models. Individuals using simulation should use a structured process in applying simulation. The second example illustrates this structured process. The examples also illustrate the use of both stochastic and deterministic variables in modeling manufacturing systems.

1 INTRODUCTION

Manufacturing is one of the earliest simulation application areas (Naylor et al. 1966), and the attendance at the manufacturing application track of the Winter Simulation conferences indicates that manufacturing remains as one of the most popular application areas. We use simulation to improve the performance of manufacturing systems because:

- Many manufacturing systems are too complex to be analyzed and improved by simply thinking and talking about possible approaches.
- Simulation can predict system performance resulting from interactions among system components.

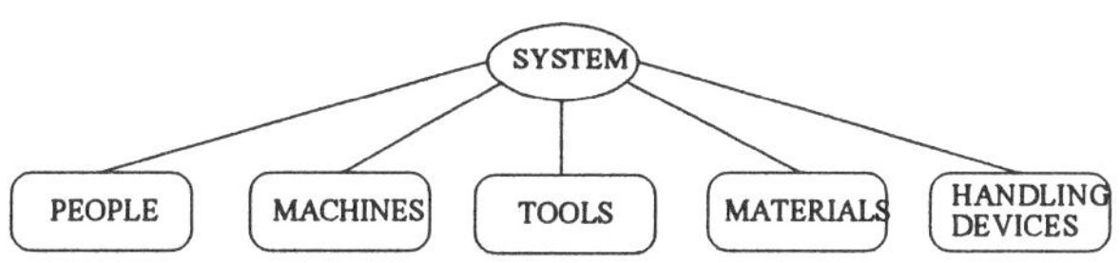

Figure 1 System Components

System components can be people, machines, tools, material handling devices, and materials as depicted in Figure 1. The result of interactions among these components may be very difficult to predict without the use of a model, and a simulation model is frequently the easiest model to use.

This tutorial introduces simulation applications to manufacturing systems by illustrating:

- Diverse uses of simulation.
- Use of random and deterministic variables in simulation models.
- A structured process for applying simulation.

This tutorial uses three example applications to illustrate the above points. The first example is a simulation model of mold production cell which uses a robot. The model illustrates the use of simulation to support the design of the cell. The second example is a study to determine effective operating policies for a cell. This example illustrates the steps and process that one ought to follow in a simulation study. The third example is an on-line simulation used to schedule a manufacturing system. The simulation model is a completely deterministic model in this application.

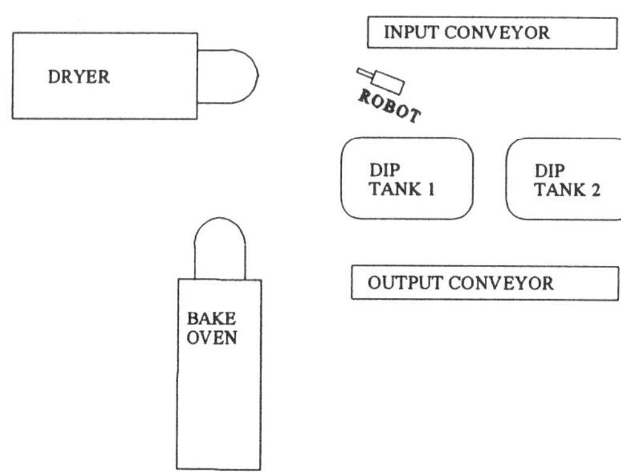

Figure 2 Mold Cell

2 MOLD PRODUCTION CELL MODEL

Bradken at Marion, Ohio produces molds for castings, and their current production process moves parts from station to station manually. Bradken wanted to increase the capacity of their mold operations so they designed a new cell using a robot for this purpose. They used a simulation to insure that the cell met their design objectives and to assist in specifying equipment performance capabilities.

Figure 2 depicts a simplified layout of the mold cell. The material handling operations are:

- Move part from the input conveyor to dip tank 1.
- Move part from dip tank 1 to the dryer conveyor.
- Move part from the dryer conveyor to dip tank 2.
- Move part from dip tank 2 to the dryer conveyor.
- Move part from the dryer conveyor to the bake oven.
- Move part from the bake oven to the output conveyor.

All parts enter dip tank 1 exactly once, but parts enter dip tank 2 repeatedly, and the number of dips in tank 2 depends on the type of part. That is, each part undergoes two different cycles.

- A dip in tank 1 and then a drying operation.
- A dip in tank 2 and then a drying operation.

A part repeats the second cycle several times.

The cell design team had a number of questions that had to be answered prior to installing equipment. Robots vary with respect to their speeds and the possibility existed of having two robots. The team had a preference for a single robot from a particular manufacturer. The length of the dryer conveyor is important from a part quality viewpoint. A longer conveyor permitted more drying time and more time for parts to cool before being handled by the robot. Another question concerns the loading of the dryer conveyor. Should the dryer conveyor have parts that completed the same number of tank dips or should the conveyor have parts with a mixture of dips completed? The major questions motivating the use of simulation were:

- What will the system capacity be given the operating characteristics of the robot?
- How many positions should the dryer conveyor contain?
- What should the loading pattern be for the dryer conveyor?

To answer the above questions, we constructed a simulation model. The model was deterministic since the robot and conveyor moves times were predicted to have little variation. However, some uncertainty existed as to the values of the mean robot handling times and mean times to dip a part in a tank. To explore the implications of this uncertainty, we ran three sets of simulations, i.e., one with optimistic robot times, expected robot times, and pessimistic robot times. Another source of uncertainty is the possibility of equipment malfunctions causing a loss of production. This effect was assessed by lowering the capacity estimates.

The simulation model predicted the following performance measures:

- The time to initialize the system with parts. This time is the system operation time required to load conveyors prior to producing a finished mold.
- The parts produced in a shift after initialization.
- The average work-in-process.
- The throughput time for a part.

The results showed the following:

- System capacity is very sensitive to robot handling and dip times.
- A preferred dryer conveyor length. We identified this length by taking the shortest conveyor that met quality constraints. Shorter conveyors result in less work-in-process.
- A desirable loading pattern for the dryer conveyor. We simulated a number of different loading patterns and selected the loading pattern that simplified cell operation.

Figure 3 Gear Production Flow

3 GEAR MANUFACTURING THROUGHPUT TIME

Clark and Cash (1993) used simulation in a study to identify preferred operating policies for a rough steel cell used in the manufacture of precision gears. Figure 3 depicts the three cells used to produce gears. The manufacturer produces gears to order, rather than making gears for stock. Customer orders may specify a gear that the manufacturer has produced in the past, but the elapsed time between repeat orders is so long that making gears for stock is not economical. Each customer order specifies a quantity of gears that varies over a wide range. The flow allowance is the lead time quoted to the customer specifying the promised delivery date. Currently, the type of gear determines the flow allowance, but no allowance is made for the number of gears in the customer order. The manufacturer currently releases work to production as soon as raw materials are available to produce the customer order; thus, the number of gears in a job has a considerable range of variation. The manufacturer uses manual procedures for tracking and scheduling work in the plant. This study illustrates the process of using simulation to generate recommendations for management action. This process includes the following steps:

- Specify study objectives.
- Specify performance measures.
- Determine alternatives to investigate.
- Describe systems to be simulated.
- Specify system experimental conditions.
- Create simulation model.
- Prepare input data.
- Formulate experimental design.
- Conduct simulation experiments.
- Analyze results.
- Make recommendations.

3.1 Study Objectives

The manager of manufacturing engineering and the director of engineering requested a study to determine policies for scheduling work in the rough steel cell. These scheduling policies consist of policies for controlling the release of work to the cell and sequencing work in the cell. The objectives for these policies are to:

- Reduce throughput time through the cell.
- Reduce WIP.
- Reduce quoted lead time.
- Reduce tardiness.
- Reduce cost.

The tardiness objective requires establishing flow allowances and due dates specifically for the rough steel cell. This study emphasized simplified procedures for scheduling because of:

- The objective of reducing cost, and
- The lack of a computerized procedure for tracking work in the plant.

3.2 Performance Measures

The primary performance measures are:

- Average WIP.
- Average system time.
- Average number of tardy jobs per year.
- Average time a tardy job is late.
- Quoted lead times for each type of job.

3.3 Alternatives Investigated

The alternatives investigated included fixed capacity buffers, a modified due-date procedure, an upper limit on job size, and a sequencing rule. The following paragraphs describe the alternatives.

Fixed Capacity Buffers: The use of fixed capacity buffers at each work station is a simplified means for reducing WIP. A station is blocked, becoming inactive, when it completes work on a job and the next station in a job's route has a full buffer. Reducing WIP also simplifies the scheduling problem for work in the cell. If the buffers do not significantly reduce capacity, by the occurrence of blocking, the reduced WIP will reduce throughput time. The use of buffers forces incoming orders to wait in a backlog when the first work station in the processing plan has a full buffer. Thus, the use of buffers introduces a control on the timing of production release. A similar alternative is to define an upper limit on the number of jobs in the entire cell. This alternative is known as the CONWIP alternative (Spearman et al. 1990).

Modified Due-Date Procedure: We defined a modified due-date procedure that incorporated the number of gears in an order to determine the flow allowance. The modified flow allowance has two components. One for the aggregate setup time, and one for the aggregate run time per gear. The run time component is proportional to the order quantity. For most customer orders, the modified procedure has a shorter flow allowance than the current flow allowances.

Job Size: An upper limit on job size or the number of gears in a job will reduce the large variation in the number of gears in a job. Large jobs tend to create floating bottlenecks. A customer order for more gears than the job size limit will result in multiple jobs to fill an order.

Sequencing Rules: We investigated two sequencing rules for work at a work station. They were first-in-first-out (FIFO) and earliest due date (EDD).

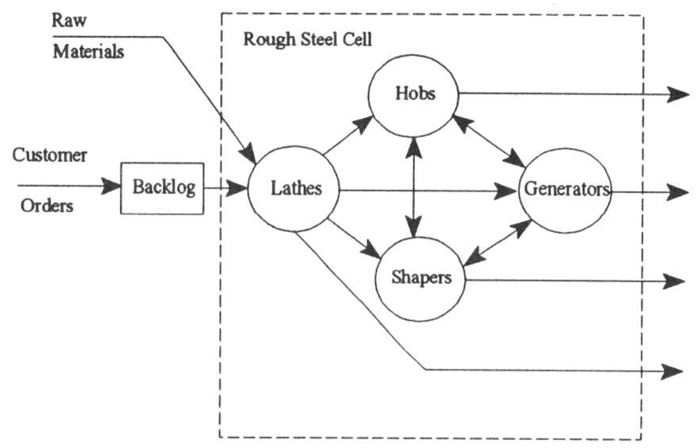

Figure 4 System Studied

3.4 System Description

Figure 4 depicts the system studied. The rough steel cell has the following work stations: lathes, hobs, shapers, and generators. Each work station may have a buffer and multiple machines. The buffer sizes and number of machines in each work station are inputs. The service time for a job at a machine has a setup time and a run time component. The setup times and single part run times are lognormal random variables. The total service time for a job is the sum of the setup time and lotSize independent run times, where lotSize is the number of gears in the job. The system represents numGears different gear types, where each gear type has its own processing plan. A processing plan gives the route for a gear type through the cell and the standard setup and run times. The arrival times of customer orders are exogenous, deterministic inputs.

3.5 Experimental Conditions

The director of engineering and the manager of manufacturing engineering selected 50 gear types for analysis. That is, numGears = 50. The gears selected are representative of future business. They supplied the process plans for each gear type.

The manager of information systems supplied historical job release times over the previous four years. These data became the basis for the exogenous customer order times. The study used three different customer order patterns, known as release schedules, i.e., RS1, RS2, and RS3. Each release schedule gives specifies the time materials are available for production for each customer order over a one-year period. The intent is to represent more than a single scenario to increase the robustness of study conclusions. These release schedules present the lathe work station with average utilization levels of 65%, 85%, and 95% for RS1, RS2, and RS3, respectively. These averages apply over a one year period.

3.6 Study Requirements

The five previous steps; i.e., specify study objectives, performance measures, alternatives to investigate, system to simulate, and experimental conditions; place requirements on the study. They dictate the detail in the simulation model and the data to be collected. All concerned parties should review the results of these steps prior to making simulation runs and recommendations.

3.7 Simulation Model

The simulation model was programmed in WITNESS which permitted animation of the simulations. The animated display was effective in showing company management the nature of the simulation. Two additional programs, written in C++, simplified the use of WITNESS considerably. These programs prepared inputs for WITNESS and analyzed the WITNESS output data. The extensive inputs required to represent the large number of different gear processing plans, i.e., 50, and their flow allowances motivated the input program.

3.8 Prepare Input Data

An analysis of shop labor records, supplied by the manager of information systems, provided historical data on actual times to implement the process plans for the fifty gears. The study assumed that the coefficient of variation for setup and run times at a machine group is the same for all 50 gears. That is, the ratio between the standard deviation and mean of a setup (run) time is a constant for a machine group. The estimation of these coefficients of variation used historical data.

3.9 Experimental Design

The primary objective of the first set of simulation experiments was to determine the effectiveness of buffers in limiting WIP without significantly reducing capacity. This set of experiments imposed no limit on job size. These experiments had four factors, i.e., buffer configuration, flow allowance procedure, release schedule, and sequencing rule. The following table shows the levels of each factor. Each possible combination of the levels for each factor was simulated in the first set of experiments for a total of 72 simulation runs.

Factor	Levels
Buffer Configuration	Buffer size of 1 at each station Buffer size of 2 at each station Buffer size of 3 at each station CONWIP with WIP limited to x CONWIP with WIP limited to y No WIP Control (unlimited buffer size)
Flow Allowance	Current procedure Modified
Release Schedule	RS1 RS2 RS3
Sequencing Rule	FIFO EDD

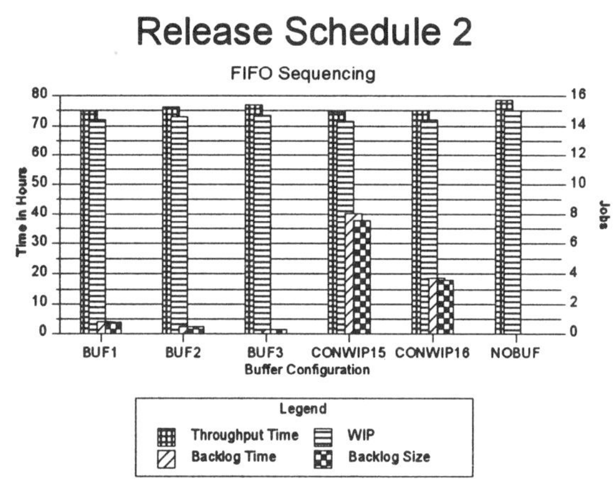

Figure 5 Throughput Time and Wip

Stochastic simulations of this type present two experimental problems (Law and Kelton 1991). That is, the initial condition effect and run length so that confidence intervals are sufficiently narrow. The release schedules apply over a one-year period. Each simulation run consisted of eleven consecutive years by repeating the appropriate release schedule eleven times. Thus, the final simulation state at the end of December became the initial condition for the next January. The C++ post-processor program deleted the first year to reduce the initial condition effect. The analysis assumed that statistics for each subsequent year are independent and identically distributed, which is the batch means procedure. The post-processor program employed these assumptions in calculating 90% confidence intervals which were sufficiently narrow.

Based on results from the first set of experiments, the analysis identified a preferred buffer configuration, sequencing rule, and flow allowance procedure. Further simulation experiments investigated the upper limit on job size.

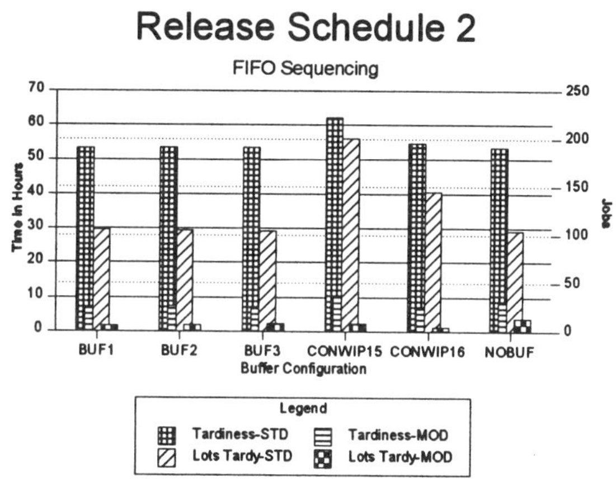

Figure 6 Tardiness

3.10 Simulation Results

Figures 5 and 6 illustrate the results from the simulation experiments. In both figures, the use of buffers at each station dominates the CONWIP results. The system and WIP performance measures apply to the shop after leaving the backlog. Throughput time and WIP are less with a buffer size of 1 at each station. However, the total of backlog time and throughput time are slightly larger than the results with no buffers. Figure 6 clearly show the superiority of the modified due-date procedure.

Also, for the modified procedure, the tardiness results for a buffer size of 1 are slightly less than tardiness with no buffers. The manufacturer prefers a buffer size of 1 since:

- WIP is less reducing costs, improving quality, and simplifying scheduling.
- Tardiness is lower.

3.12 Major Points Illustrated

The gear manufacturing throughput time example illustrates the overall steps required to apply simulation and influence management decisions. An important milestone is to review the first five steps with all concerned parties before collecting data and programming the model. Then, affected individuals will feel they are a part of the study. Also, the simulation experimental results can address the study objectives and provide the proper outputs. The effort in programming the simulation may not be the major factor in the overall study effort. Data collection in this simulation study was the major part of the study effort.

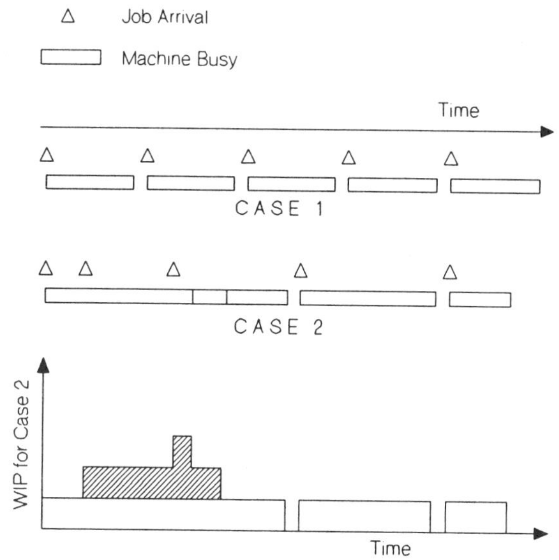

Figure 7 Effect of Statistical Fluctuations on WIP

The operation times in this example are stochastic rather than deterministic. Variability in operation times is very important in estimating WIP and throughput time. We may model that variability by representing operation times as stochastic variables. Any utilization close to one will result in excessive work-in-process (WIP) if there is any variation in service times or times between arrival of lots to the respective machines. Figure 7 illustrates the effect of fluctuations in job inter-arrival times and job operation times on WIP for a single machine. The top sequence, called case 1, of job arrivals and service times for each arrival follows a perfect uniformly spaced pattern that has no variation. That is, the times between each arrival are all equal and the service times are also all equal. The proportion of time the machine is busy represents the machine utilization which is close to one. Because of this lack of variation, case 1 gives no queueing and no instances of WIP greater than 1. The lower sequence, called case 2, of service and inter-arrival times has precisely the same mean and gives the same overall machine utilization which is the proportion of time the machine is busy. However, this statistical fluctuation increases the WIP which becomes as large as three. The shaded area in plot at the bottom of the figure shows the jobs waiting in the machine queue.

Potential sources of variations in job service times are:

- Tooling failures.
- Machine cycle length changes due to different types of jobs, i.e., a machine performs operations on non-identical parts. For example, a machine processes an XYZ123 job and then an ABC123.
- Machine failures and adjustments.
- Variations in human paced task times.

Variations in inter-arrival times could result from:

- Any variation in the times between release to production due to the company planning system or customer order times, e.g., job release times that vary with the hour of the day or the day of the week.
- Variations in the times materials arrive from vendors.
- Variations in initiation of production caused by tooling not being available.
- Variations in the times jobs depart from upstream work stations in the job's route.

See Law and Kelton (1991) for another list of potential sources of statistical variations in manufacturing simulations.

4 SIMULATION-BASED SCHEDULER

FACTOR (Pritsker Corporation 1989) is an example of a simulation-based scheduling system. A scheduler will use FACTOR in an on-line mode. That is, FACTOR will take inputs from an existing data base and then generate schedules after a short time delay such as a half hour. The data base will specify the status of all jobs in the system, process plans for these jobs, standard setup and run times, and the status of resources such as machines. For many applications, the principal output for the simulation is a schedule giving the processing times of jobs by resources. Shop personnel can use this schedule to insure that other resources such as tools are available when the schedule requires them. The schedule also identifies which jobs will be probably be late. The simulation can do "what if" comparisons. As an example, the simulation may compare sequencing rules such as earliest due date and shortest processing time. The schedules are realistic in the sense that the simulation represents the finite capacity of resources in a detailed manner.

FACTOR has a completely deterministic simulation. That is, FACTOR does not sample from probability distributions in generating a schedule. Since the scheduler must generate a single schedule, a deterministic representation simplifies this task. Also, by accessing a data base specifying the process plans and standard times for all jobs, the nature of each simulated task is known in more detail than simulating in a planning mode. For example, when simulating to identify preferred designs for a production line, the precise sequence of each job type may not be known.

Cheselka (1992) describes the use of FACTOR to schedule Timken's Gambrinus Thermal Treatment Facility. Scheduling that facility is challenging because the scheduler must balance three conflicting objectives.

- Complete orders by their due date.
- Maximize furnace utilization.
- Minimize energy costs.

These objectives can conflict because maximizing utilization and minimizing energy costs would sequence jobs to avoid changes in furnace temperature and speed of material handling devices transporting jobs through the furnace. FACTOR uses a scheduling logic that first identifies the highest priority jobs using critical slack.

Critical_Slack = Firm_Plan_Date - Current_Time - Estimated_Processing_Time

If the critical slack for an job is less than 30 hours then the job is considered critical. The system assigns a higher priority to critical jobs, and they are scheduled first. Within the same priority level, FACTOR will maximize furnace utilization by searching for a job that matches the current furnace setup after completing a job. The setup includes furnace temperature and speed of the material handling device.

The timeliness of schedules depends on the ability to quickly obtain inputs from an existing data base. At Timken, the data inputs to FACTOR include data from the following data bases.

- The VAPP data base supplies job due dates and current job work center locations.
- The RODS data base supplies detailed order information such as product size and special processing data.
- The Heat Chemistry data base supplies a heat chemistry analysis for each job .

5 LOT SIZE

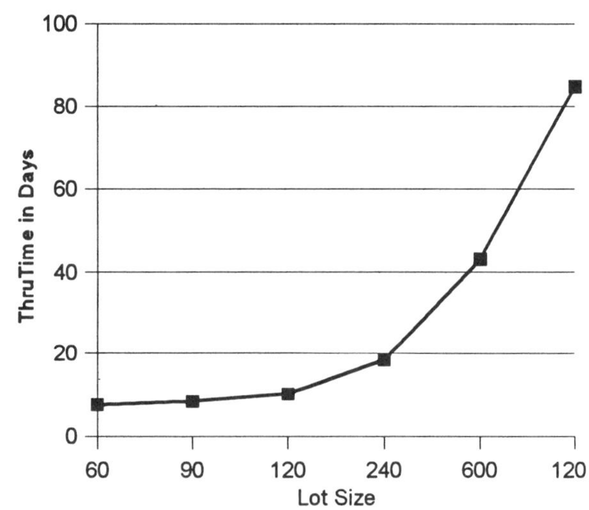

Figure 8 Effect of Lot Size on Throughput Time

Manufacturers may consider a number of different lot sizes when they produce a variety of products and deliver these products at different times. Some manufacturers use economic order quantity models (Elsayed 1994) in choosing production lot sizes. These models consider such factors as the demand forecast, setup costs and inventory carrying costs. They omit the queueing effects resulting from varying the lot size. Simulation can represent these effects and be useful in selecting a preferred lot size. The author bases this application of simulation to selecting lot sizes on his experience as a member of a team performing manufacturing assessments for the Total Quality Joining program managed by the Edison Welding Institute.

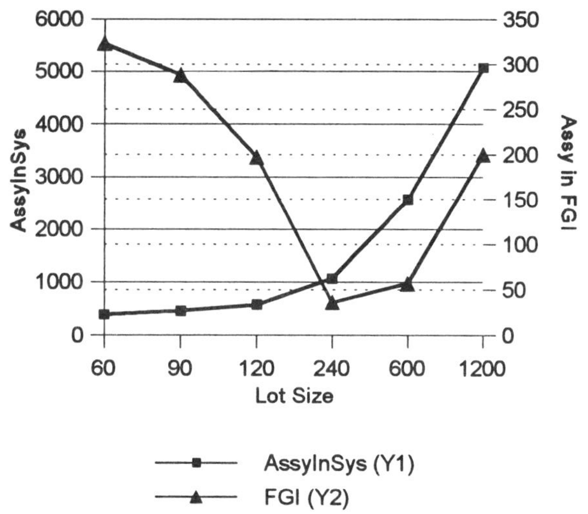

Figure 9 Effect of Lot Size on WIP and FGI

Consider a supplier producing an assembly of sheet metal parts joined by welding. The customer is an Original Equipment Manufacturer (OEM) who receives a daily shipment of the assemblies. The OEM gives the supplier one day's advance notice specifying the amount the supplier must deliver the next day. This amount is a random variable with mean 60. Each assembly has eight different parts, and the process for making parts requires two steps, i.e., sheering and then forming. The setup times for the sheering and forming machines have means of .6 hours. Once setup, each machine produces parts at a mean rate of 200 per hour. The setup times and part production times are random variables. Once the machines make all eight parts in a lot, the parts go to a welding machinewhere the mean time to fabricate an assembly is .12 hours. Once welded, an assembly has the following production steps:

- Cleaning.
- Inspection.
- Painting.
- Inspection.

Fork trucks move parts or assemblies between production steps; however, inspection occurs at the current work center.

The supplier uses a reorder point rule to determine times for starting production for the eight constituent parts. In this case the reorder point is 600 assemblies. Let the inventory position be the amount of work-in-process and finished goods inventory that is not committed to satisfying a delivery requirement to the OEM. When the supplier receives advance notice of the amount to be shipped, the inventory position is decreased by this amount. When the manufacturer initiates production for a new lot by releasing materials for the eight constituent parts, he increases the inventory position by the lot size. When the inventory position is not larger than 600 assemblies, the manufacturer starts new lots until the inventory position is not less than the inventory objective which is the reorder point plus the lot size.

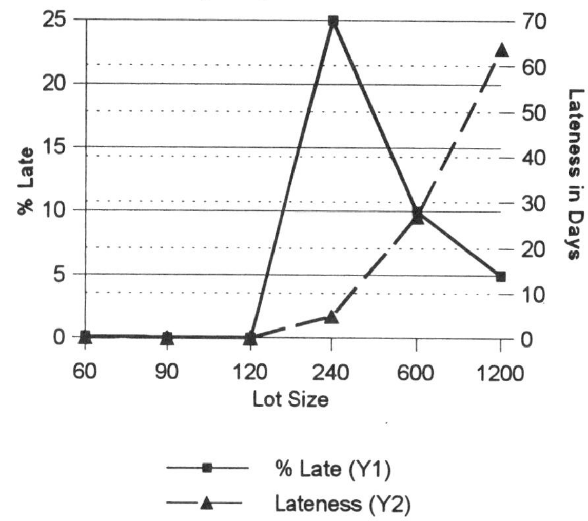

Figure 10 Effect of Lot Size on Delivery Performance

Using the equation on page 73 of Elsayed (1994), the economic order quantity is 1200 assemblies when:

- The inventory carrying charge is 15% per year.
- The material cost is $5 per part.
- The variable machine cost is $20 per hour.

Figures 8, 9 and 10 display results from simulating this system for lot sizes of 60, 90, 120, 240, 600 and 1200. Figure 8 shows the dramatic effect of lot size on throughput time. The average throughput time increases from an average of 7.5 days with a lot size of 60 to 85 days with a lot size of 1200. Figure 9 shows the effect on Finished Goods Inventory (FGI) and WIP. FGI is highest with a lot size of 60. That lot size gives an average FGI of 322 assemblies. Work-in-process (WIP) is lowest, i.e., 388 assemblies, for a lot size of 60, and WIP increases to over 5000 assemblies with a lot size of 1200. Figure 10 shows the effect on delivery performance. Delivery performance is excellent for lot sizes no larger than 120. Delivery performance drops sharply for a lot size of 240 because of the associated decline in FGI. The figures suggest that the supplier would prefer a lot size of 60; however, this performance has a cost of added setup labor. A lot size of 60 requires an average of 9.6 hours of setup labor per day. The lot size of 1200 decreases this daily setup time to .48 hours.

This application of simulation to predicting the effect of lot size on manufacturing system performance clearly shows the dramatic effects of lot size and the capability of simulation to represent those effects.

5 CONCLUSIONS

The four examples summarized in this paper illustrate important applications of simulation in manufacturing. Simulation is a powerful approach to modeling manufacturing systems in that many complex and diverse systems can be represented. Simulation can predict system performance measures that are difficult to assess without a model. However, simulation requires data that characterizes the behavior of system components. Also, individuals contemplating the use of simulation should use a structured process such as the one described in Section 3.

REFERENCES

Cheselka, Michael C. 1992. The Timken Company's Gambrinus Thermal Treatment Facility Scheduling System. In *Proceedings of the 1992 Winter Simulation Conference*, ed. J.J. Swain, D. Goldsman, R.C. Crain, and J.R. Wilson, 833-841, Institute of Electrical and Electronics Engineers, San Francisco, California.

Clark, Gordon M. and Charles R. Cash. 1993. Reduction of Gear Manufacturing Throughput Time: Rough Steel Cell Operating Policies. Technical Report, Department of Industrial and Systems Engineering, The Ohio State University, Columbus, Ohio.

Elsayed, Elsayed A. and Thomas O. Boucher. 1994. *Analysis and Control of Production Systems*. 2d Ed. Englewood Cliffs: Prentice Hall.

Law, Averill M. and W. David Kelton. 1991. *Simulation Modeling & Analysis*. New York: McGraw-Hill, Inc.

Naylor, T.H., J.L. Balintfy, D.S. Burdick, and K. Chu. 1966. *Computer Simulation Techniques*. New York: John Wiley & Sons, Inc.

Pritsker Corporation. 1989. *FACTOR Version 4.0 Implementation Guide.* Pritsker Corporation, Indianapolis, Indiana.

Spearman, M. L., D. L. Woodruff, and W. J. Hopp. 1990. CONWIP: A Pull Alternative to Kanbans. *International Journal of Production Research* 22:879-894.

AUTHOR BIOGRAPHY

GORDON M. CLARK is president of System Solutions Inc. and a professor emeritus in the department of Industrial, Welding and Systems Engineering at The Ohio State University. He received the B.I.E. degree from The Ohio State University in 1957, the M.Sc. in Industrial Engineering from The University of Southern California in 1965, and the Ph.D. degree from The Ohio State University in 1969. He is actively involved as an industrial engineer performing manufacturing assessments. His current technical interests include the development of finite capacity scheduling systems. He is a member of the editorial board for the *Journal of Operations Management*. In 1991, he served as program chair for the WSC.

Proceedings of the 1996 Winter Simulation Conference
ed. J. M. Charnes, D. J. Morrice, D. T. Brunner, and J. J. Swain

BUSINESS PROCESS SIMULATION

Kerim Tumay

CACI Products Company
3333 North Torrey Pines Court
La Jolla, CA 92037

ABSTRACT

This introductory tutorial provides an overview of business process simulation and how it works. Descriptions of modeling elements and model performance measures are presented. Classification of business processes is followed by unique modeling, simulation procedure and analysis considerations. Types of business process simulation tools are discussed. Finally, a purchasing process simulation exercise is presented to illustrate the power and suitability of simulation for analyzing a business process.

1 INTRODUCTION

Industrial and service enterprises are faced with increasing pressures to minimize the time it takes to service customers, develop products and fulfill demand. A business manager's ability to minimize risks while maximizing profits depends on his ability to quickly evaluate alternatives. Whether the alternatives involve automation or outsourcing of a process, expansion or downsizing of the workforce, businesses must be able to evaluate the impact of change with accuracy and speed.

Understanding and estimating the time and cost to complete a product development process is a key business challenge. Typically, managers have relied on project management tools to planning purposes. However, highly variable activity times and interdependencies between resources make it very difficult analyze activity costs and resource requirements using traditional project management tools.

Today, customers are much more demanding and cost-conscious. Such performance metrics as waiting time and activity costs are critical to providing quality service at competitive prices. Typically, staffing and pricing decisions for a customer service process have been made by analytical tools which fail to take into account the randomness and system dynamics that result in queuing and variable costs.

For an industrial enterprise, the key problem in supply chain management is how to balance inventory. Variability in demand and process times, complexity of the supply chain, and system dynamics create uncertainty that can not accurately be modeled and analyzed with spreadsheets and flowcharts.

The power of computer simulation combined with simplicity of flowcharting and spreadsheets provides the most cost effective, accurate, and rapid strategic weapon for businesses to evaluate alternatives prior to committing expensive resources and time. The ability to visualize how a process would behave, measure its performance, and try "what ifs" in a computer model makes computer simulation an invaluable tool for decision making.

Like activity-based costing (ABC), business process simulation (BPS) embodies the concept that a business is a series of inter-related processes, and that these processes consist of activities that convert inputs to outputs. A BPS model realistically captures the resource constraints, decision rules, and stochastic behavior of real-world situations. A process model, when simulated, mimics the operations of the business. This is accomplished by stepping through the events in compressed time while displaying an animated picture of the workflow. Because simulation software keeps track of statistics about model elements, performance of a process can be evaluated by analyzing the model output data.

2 HOW BPS WORKS

Creating a process model starts with mapping the business processes. Most process modeling tools provide a drag and drop type of a graphical interface with process documentation facilities. Then, you drill-down into the processes where sub-processes and activities are defined. Describing the entities that flow through the processes and linking the processes using connectors facilitates the workflow definition. Finally, you define the resources and assign them to the activities where they are used.

Before simulating a model, you need to select the performance measures of interest. For example, you may be interested in throughput and cycle time reports for entities, activity costs for processes, and utilization reports for resources. When you run a simulation, the process simulation tool automatically verifies your model and begins advancing the simulation clock. During the simulation, you see an animated picture of the flow that helps you visualize the process in motion. You can also have BPS generate real-time graphs, letting you view key performance measures, during the simulation.

When the simulation is over, you can bring up the model results and analyze the performance measures of interest. To draw useful and correct conclusions from simulation results requires statistical input and output data analysis. Questions such as which distribution to use for representing activity times or how many data points to collect are important is developing a statistically valid model. And, questions such as how long to simulate a process or how many replications to run become significant for producing valid and accurate results.

The strength of process simulation is in its ability to incorporate stochastic situations in the model. It cannot offer an optimum solution. In order to find the best solution, one has to determine various scenarios and simulate them. This is where Design of Experiments can aid in searching and finding the best solution. With the aid of Design of Experiments, one can also compare the performance measures from Current State with measures from Future State alternatives.

3 MODELING ELEMENTS AND MODEL PERFORMANCE MEASURES

Even though model building constructs may have different names or characteristics in different products, most business process simulation models contain some basic building blocks, activity modeling constructs, and advanced modeling elements. In this section, an overview of these modeling elements are presented.

3.1 Basic Model Building Blocks

The four basic building blocks of a process model are entities, resources, activities, and connectors.

Entities: Otherwise referred to as flow objects, tokens or transactions, these are the objects that are processed by resources. Examples of entities are customers, products, documents, orders, and calls. Entities may have attributes such as order quantity, priority, and due date.

Resources: These are the agents that are used for adding value to Entities. Examples of resources are service representatives, automated process equipment, and transportation equipment. Human resources can be defined as members of departments and workgroups. Resources are allocated to activities and may have attributes such as cost and expertise level.

Activities: Activities are linked by connectors to represent the flow of entities through the simulation model. Activities may be value added or non-value added. Examples of activities are branching, assembly, batching.

Connectors: Connectors are used for linking processes and activities. Entities follow the connectors as they are processed by the model. Connectors are helpful for defining parallel flows and rework situations based on deterministic, probabilistic, or conditional decision rules.

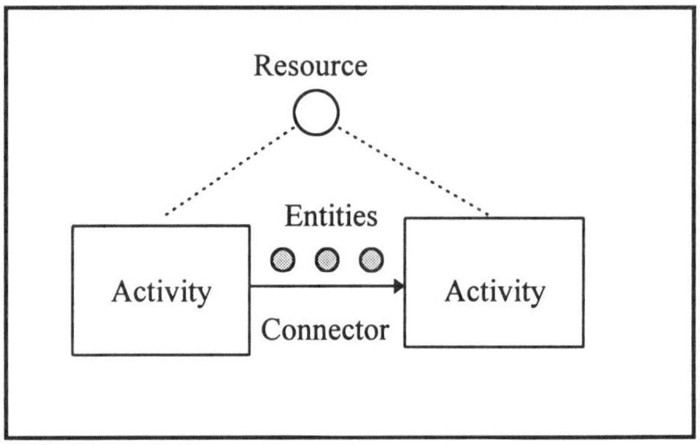

Figure 1: Basic Model Building Blocks

3.2 Activity Modeling Constructs

A minimum set of activity modeling constructs are needed for modeling the dynamic behavior of business processes. Examples of activity modeling constructs include generate, branch, assemble, batch, gate, split, and join.

Generate: A GENERATE activity generates the arrival of entities into the model. Arrivals may be random, deterministic or conditional. An example of a GENERATE activity is the arrival of patients in a clinic. A GENERATE activity may have values for arrival time, quantity, frequency and occurrences.

Branch: A BRANCH activity allows for defining alternative routings for Entities. Branching may be based on a probability or a condition. For example, the outcome of an inspection process may be modeled using probabilistic branching.

Assemble: An ASSEMBLE activity assembles multiple entities coming from multiple sources to create a single entity. For example, the development of a business proposal may contain three documents that are merged using an assembly activity.

Batch: A BATCH activity combines a given quantity of entities into a single batch. A batch may be defined as a permanent or a temporary batch. An example of a batching activity is the accumulation of mail for delivery.

Gate: A GATE activity holds entities in a queue, until a signal is received. For example, a GATE activity would be used to model orders held in inventory until a signal is received is from the distributor to fulfill the demand.

Split: A SPLIT activity takes an incoming entity and creates children of that entity as well as providing an output of the parent entity. For example, copies of a purchase order may be created with a SPLIT activity and sent to accounts payable and shipping.

Join: A JOIN activity takes the children and parent entities that were split up, and matches them to produce the original one. For example, a JOIN activity may be used for matching the paperwork with the shipment.

3.3 Advanced Modeling Functions

In addition to basic building blocks and activity modeling constructs, users may need to have access to more advanced modeling functions that provide the power and flexibility to deal with the complexity of real world business processes. Some of these functions are attributes, expressions, resources schedules, interruptions, and user defined distributions.

Attributes: Attributes are logical values that may be used for decision making or performance tracking. For example, the capacity or trigger level for an activity may be an attribute that is used for decision making in an IF-THEN statement. Entities may have distinct characteristics such as time or cost. For example, you may define orders as entities that arrive in the system where the order quantity would be an attribute that is different from one order to another.

Expressions: In order to define complex decisions or resource assignment rules, programming-like statements such as IF THEN ELSE, INCREMENT, and DECREMENT are needed. These statements provide the flexibility to model unique, real-world situations.

Resource Schedules: Since the activities are constrained by the availability of resources, assuming that all resources are available throughout a simulation is unrealistic. Resource schedules allow a realistic way to define the times when resources are available to work and when they are off shift.

Interruptions: Most business processes experience interruptions that may be planned or unplanned. Because of the non-linear impact of interruptions on processing, representation of these situations provide a more accurate picture of the performance measures of a business.

User defined distributions and functions: User distributions or functions may be used for representing the variability associated with arrival times, activity times, or move times. User distributions may be discrete or continuous.

3.4 Model Performance Measures

Basic model performance measures are cycle time, entity count, resource utilization, and activity cost.

Cycle Time: Cycle time is the total time an entity spends traversing a process. This means cycle time includes value-added process time, waiting time, movement time, etc. One of the most valuable outputs of BPS is the automatic tracking and calculation of cycle time for each entity. Furthermore, BPS tools provide average and maximum values based on the total count of entities that traversed the process.

Entity Count: Entity counts include the total number of entities that traverse a process or that are still being processes. This means that the total entity count is the sum of entities processed plus the entities that are in-process. Once again, BPS tools automatically track and calculate entity counts for each type of entity in a model.

Resource Utilization: During a simulation, resources change states from busy to idle, from unavailable to reserved. Resource utilization defines the percentage of time that a resource spends in each state. The availability and assignment of resources dictate the allocation of resources to activities in a model. So, the resource utilization results provide useful statistics in measuring and analyzing under-utilization or over-utilization of resources.

Activity Cost: When a resource is defined in a process model, it is defined by the number of available units, usage costs, setup costs, and fixed costs. When an activity is defined, it is defined by the resources required to perform it, the duration for the activity, and the entities that it processes. During a simulation, the BPS tool automatically keeps track of the time each entity spends in an activity and the time each resource is assigned to that activity. Activity cost calculations provide a realistic way for measuring and analyzing the costs of activities. Object-oriented BPS tools allow detail breakdown of activity costs by resource or entity type as well as aggregated process costs.

Advanced model performance measures are measures that need to be customized using advanced modeling elements such as attributes and expressions. Examples are service level, on-time performance level, inventory level, and order fill rate.

4 MODELING, PROCEDURE, AND ANALYSIS CONSIDERATIONS

Simulation models can provide the most accurate and insightful means to analyze and predict the performance measures of business processes. However, one must be aware of the dangers of using incorrect modeling and analysis procedures which can result in erroneous results. In this section, business processes are classified into 4 major categories to discuss the unique modeling, simulation procedure and analysis considerations. They are:

1. Project-based processes,
2. Production-based processes,
3. Distribution-based processes,
4. Customer service based processes.

This classification is not intended to imply that all business processes fall clearly into one of these four types. In fact, they do not. For example, a customer service process may have both front-room and back-room activities where the front-room activity is a customer service based process while the back-room activity is a production-based process. Nevertheless, the considerations presented in this section should prepare the user for planning a business process simulation study. They might also be helpful for assessing the suitability of a particular simulation software for that application.

4.1 Project-based Processes

These processes are usually provided by a single person or a group of people. Typical examples are product development processes and administrative processes. Usually, project-based processes are analyzed using project management tools. However, process cycle times and resource requirements analyzed by simulation techniques are more accurate because activity times are highly variable and shared resources create multiple interdependencies.

Modeling priorities, preemption, shifts, downtime, overtime, learning curves for resources are some of the important considerations in building a valid simulation of a project.

Model Elements	Examples
Entities	proposal, report
Resources	consultants, workers
Activities	design, testing, review
Workflow	parallel flow, feedback loops

Table 1: Examples For Project-based Processes

One of the important procedure and analysis considerations regarding the simulation of a project is multiple replications. Since activity times are highly variable, a single simulation run will produce only a single observation. Multiple replications will produce several observations which will provide a more accurate estimate and confidence interval for the performance measures.

4.2 Production-based Processes

In production-based processes, outputs are produced in a batch or continuous flow mode in relatively high volumes. Typical examples are order fulfillment, accounts payable, and claims processing processes.

Activities such as batching, unbatching, assembly, disassembly, setup, inspection, and rework are typical activities in production-based processes. In order to accurately model these activities, a model must allow for keeping track of individual Entities and their attributes. Other important modeling consideration with production based processes are queuing rules and downtime modeling.

Model Elements	Examples
Entities	orders, electronic forms
Resources	equipment, staff
Activities	batching, assembly, inspection
Workflow	sequential flow, feedback loops

Table 2: Examples For Production-based Processes

Production-based process models are usually simulated to obtain steady-state behavior since a repeating sequence of products is always being produced. An important procedural concept in analyzing performance is determination of warm-up period and elimination of bias associated with statistics collected during warm-up.

4.3 Distribution-based Processes

Distribution-based processes include transportation and delivery processes where products or people are carried between locations via a distribution network. A fundamental difference between transportation and delivery is that the Entities being transported are people rather than goods. Typical transportation processes are found in mass transit systems. Typical delivery processes are found in manufacturing distribution, mail delivery and moving services.

When modeling distribution processes, it is important to define attributes for Entities in order to keep track of unique characteristics such as destination, size, or cost. When modeling movement, it may sometimes be more appropriate to represent transportation resources as entities.

Model Elements	Examples
Entities	people, loads
Resources	trucks, rail cars, planes
Activities	load, move, unload
Workflow	alternative routes, looping

Table 3: Examples For Distribution-based processes

Most distribution processes are transient in behavior. Therefore, simulation period should be long enough to include a cycle period and multiple replications should be made to analyze the performance measures.

4. 4 Customer Service-based Processes

The customer service-based processes present a major area of application for simulation because total waiting time may be as high as ninety five percent of the total processing time in a typical service process. Typical customer service processes are telephonic services (call centers), service factories (restaurants, copy centers), service shops (hospitals, repair shops), and retail stores.

Simulation of customer service processes present a unique challenge because both the entities and resources are humans. Humans have much more complex and unpredictable behavior than products, documents, equipment, or vehicles. For example, customers waiting in a line may balk, jockey, or renege. Modeling these types of situations require programming logic to realistically represent the behavior. Resource interactions with entities also create situations where resources may change their behavior based on states of the entities. Once again, programming flexibility may be required to model such situations. Usually, service activities times are highly variable and customer arrivals are random. Therefore, the use of probability distributions are required for accurate representations.

Because arrivals are cyclical and random, customer service systems rarely reach steady state. Therefore, it is appropriate to view the operations of a service system in terms of time windows (periods) and define the model elements accordingly. For example, it is most appropriate to define a 24 hour day in terms of three periods (morning, afternoon, and evening) for an around-the-clock service system and define the arrivals of customers and shifts for resources in terms of those windows. Because activity times and costs are highly variable, it is important to run multiple replications of a model for statistically valid analysis of model results.

Model Elements	Examples
Entities	customers, patients
Resources	service representative, nurse
Activities	take order, service, assist
Workflow	based on customer type or state

Table 4: Examples For Customer Service Processes

5 SIMULATION TOOLS

In the past few years, several new software tools have been developed specifically for modeling business processes. Most of these tools define business processes using graphical symbols or objects, with individual process activities depicted as a series of boxes and arrows. Special characteristics of each process or activity may then be attached as attributes of the process. Many of these tools also allow for some type of analysis depending on the sophistication of the underlying methodology of the tool.

Business Process Simulation software tools can be broken into four major categories:

1) Flow Diagramming Based BPS Tools: At the most basic level are flow diagramming tools that help define activities and routings. Flowchart based models are methodology independent therefore they are the easiest to learn. Unfortunately, the ease-of-use results in limited modeling capability and simulation analysis capability. Examples of flow charting based simulation tools are Process Charter and ProcessModel.

2) System Dynamics Based BPS Tools: At the next level are continuous simulation software products that utilize the systems dynamics methodology. Models built with these tools consist of methodology specific constructs such as levels, stacks, flows, converters and connectors. Examples of these tools are Ithink and Powersim.

3) Discrete-event Based BPS Tools: Discrete-event based simulation tools are more powerful that flow charting tools. They are also better suited for modeling business processes as compared to system dynamics based tools. These tools provide modeling of entity flows with animation capabilities that allow the user to see how entities are routed through the system.

Examples are Extend BPR, BPSimulator, and ServiceModel.

4) Object-Oriented, Discrete-event Based BPS Tools: The most capable and powerful tools for business process simulation are the object-oriented, discrete-event driven simulation products. These tools provide object-oriented and hierarchical modeling which facilitates the development of enterprise process models through the use of reusable templates. Examples of object-oriented, discrete-event based simulation tools are SimProcess and Rethink.

6 A PURCHASING PROCESS SIMULATION

During the presentation of this paper at the Winter Simulation Conference '96, a presentation of a typical business simulation exercise will be shown. In this section, a brief description of the exercise is provided.

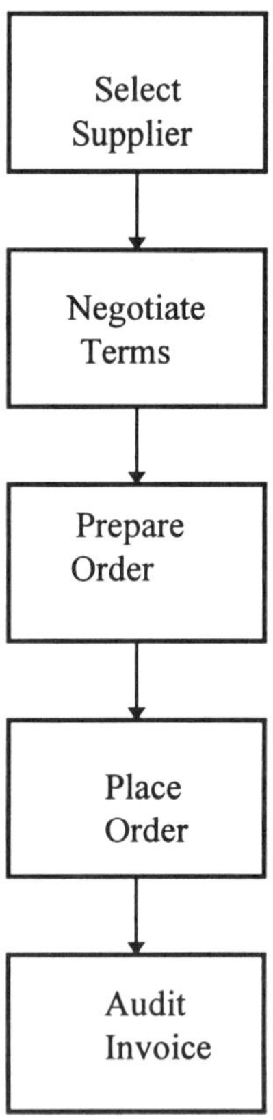

Figure 2: Purchasing Process Flow Diagram

This purchasing process simulation example presents three alternatives ways for organizing resources. The purpose of the exercise is to determine the best organization.

Alternative 1: The purchasing process is performed by centralized, functional organization where the five functions are performed by staff dedicated to each function.

Alternative 2: The purchasing process is performed by three decentralized, product organizations where the each product organization performs all 5 functions for its product line.

Alternative 3: The purchasing process is performed by a hybrid organization where the supplier selection and terms and pricing functions are performed by a centralized organization; and the other 3 functions are performed by decentralized, product organizations.

7 CONCLUSIONS

Business processes are too complex and dynamic to be analyzed with flowcharting and spreadsheet analysis. Discrete event simulation is the most powerful and realistic tool for analyzing the performance of business processes. Simulation takes into account the variability of activity times, interdependencies of resources, and other complexities that affect performance over time. Discrete event based simulation tools provide statistical input and output capabilities and advanced modeling elements that are necessary to accurately simulate business processes.

REFERENCES

Tumay, K. 1995. Business Process Simulation. In *Proceedings of the 1995 Winter Simulation Conference,* ed. C. Alexopoulos, K. Kang, W.R. Lilegdon, and D. Goldsman, 55-60. Institute of Electrical and Electronics Engineers, Washington D.C.

Gladwin B., and K. Tumay. 1994. Modeling Business Processes With Simulation Tools. In *Proceedings of the 1994 Winter Simulation Conference,* ed. J.D. Tew, S. Manivannan, D.A. Sadowski, and A.F. Seila, 114-121. Institute of Electrical and Electronics Engineers, Orlando, Florida.

Harrell, C.R., and K. Tumay. 1995. *Simulation Made Easy: A Manager's Guide.* Norcross, Georgia: IE Management Press.

Scheer, A.W. 1994. *Business Process Engineering: Reference Models For Industrial Enterprises.* Berlin, Germany: Springer-Verlag.

AUTHOR BIOGRAPHY

KERIM TUMAY is the Director of Process Simulation for CACI Products Company. He received his B.S. and M.S. degrees in Industrial Engineering from Arizona State University. He has published over 30 papers and is the co-author of the simulation book titled "Simulation Made Easy" by IE Management Press.

Advanced Tutorials

Proceedings of the 1996 Winter Simulation Conference
ed. J. M. Charnes, D. J. Morrice, D. T. Brunner, and J. J. Swain

VIRTUAL REALITY AND SIMULATION

Martin Barnes

Deneb Robotics, Inc.
3285 Lapeer Road West
P.O. Box 214687
Auburn Hills, MI 48321, USA

ABSTRACT

Virtual Reality is a leading edge technology which has only very recently become available on platforms and at prices accessible to the majority of simulation engineers. The media representation of Virtual Reality has been focused on the entertainment aspects of the technology. As a result many simulation engineers are unaware of or confused by the concepts embodied in applying Virtual Reality to simulations.

The objective of this state-of-the-art review is to provide an awareness of Virtual Reality with respect to simulation.

A brief tutorial will be presented to clarify the terminology employed in the field. Examples of applications in a range of fields will be given, using Virtual Reality software combined with a variety of simulation engines. Virtual Reality devices will be available affording attendees to experience different modes of involvement with the simulated environment.

1.0 INTRODUCTION

Virtual Reality (VR) is a leading edge technology that has only recently become available to simulation professionals. The newer workstations and PCs provide performance that supports VR at an affordable price. VR software is now available to run on these platforms. Perhaps most important for this audience, some of this software comes with simulation engines that support simulation modelling and analysis.

The evolution of modern simulation modelling tools has encompassed prototypes, physical models, programming languages, simulation languages, animations of simulation results and visual interactive simulations. At each step in this progress there was confusion and misunderstanding concerning both the new technology and its value to the simulation process. The purpose of this paper is to describe the tools and methods that are particular to VR and to illustrate how these are being applied to simulation.

2.0 WHAT IS VIRTUAL REALITY

VR is the term used to describe advanced methods of involvement and interaction for humans with a computer-generated graphical (usually 3D) environment. Normally referred to as a VR "world," this environment is experienced by a human participant through the use of special VR equipment. In simulation terms, VR can be thought of as an advanced user interface to 3D simulation models.

VR equipment provides the human with sensory information about the VR world. This can include 3D viewing, sound, touch and smell which have all been employed to stimulate the senses so as to increase the illusion that the human is a participant in this VR world. The objective of the technology is to obtain a level of human involvement (sometimes referred to as immersion) with the VR world that is suitable to the task at hand. Different tasks require different levels of involvement. For example, airplane flight crisis simulator VR worlds are most effective when they persuade the human to generate an emotional involvement with the situation. Scientific VR worlds designed to promote insight and analysis do not usually require this level of emotional involvement.

As important as reception of information by the human senses is, it becomes much more valuable when the human can interact with the VR world. VR equipment allows the human to change position and orientation in the VR world. This is the equivalent to "moving" through the VR world. There is also the capability to interact with the VR world, obtaining information and making instantaneous changes to the VR world.

3.0 VR IN THE MEDIA

The popular media representation of VR has tended to emphasize the entertainment aspects of the technology. The depiction of VR use by characters in science fiction movies and television programs has a limited impact. This is because the real world viewer is watching a 2D animation of actors who are supposed to be experiencing

VR. There is no VR experience for the real world viewer, no 3D viewing and no interaction. In an effort to overcome this perception gap, graphics special effects are often used to ensure a high impact on the real world viewer. For example 'morphing' and graphics panning, zooming and flying around graphics provide impressive special effects. This representation has promoted an awareness about VR but has caused some confusion about what VR technology is. Whilst these graphical tricks can be applied within VR they do not by themselves constitute VR.

True VR can be experienced with some computer games. These VR games allow one or more participants using VR equipment to be able to see and "move" within a 3D VR world. The human(s) can also perform specific actions in this VR world. In this type of game, the human is truly involved and interacting with a VR world. These games have tended to be fairly simple so far, however they have a high impact on participants because they utilize all the essential elements of VR.

Much of the technical literature has concentrated on the capability to move around a VR world that is unchanging or "static." A "walk through" of a planned building or around a conceptual automobile or airplane is an excellent communication exercise. It can even provide problem identification and resolution capabilities. These VR worlds are an extension of the 3D CAD tools that allow for the creation of a 3D representation, or model, of some static object or objects. In these examples the VR aspect is mainly the manner in which the human views the model. In more advanced VR worlds of this static type, the human can interact with the VR world. For example, lifting the hood on a car or selecting a change of color for the car body.

4.0 VR AND ANIMATION

Animation of some or all of the objects in the VR world provides the human with an experience that can seem quite realistic. Because humans have an acute awareness of time passing, animation which represents change through time increases the human sense of involvement dramatically. Many VR worlds are based purely on fixed animations and therefore human interaction is limited to changing such factors as position and orientation of view. This is particularly so when the VR world is simply replaying a 3D animation, without any underlying simulation engine. When the human can interact with a VR world that is being animated, there is a strong sense of involvement and participation. Where the level of interaction enables changes to be made to the VR world that are immediately reflected in the behavior of the animated VR world, human interactions are closely parallel to human interactions with the real world.

5.0 VR AND THE REAL WORLD

In this paper VR worlds will be described that are closely related to existing or proposed real environments. These VR worlds provide animation that is based on rules of behavior and data in the model. In these worlds, the human can interact to change rules or data and the animation will change accordingly. As a result, these VR worlds provide human participant(s) with experiences and feedback data that enable valid and valuable inferences to be made about the real environments.

6.0 VR AND SIMULATION

If this sounds familiar to an audience of simulation professionals, this is to be expected. Similar terms could be used to describe simulation models. However, there are aspects of VR that differentiate it from simulation tools to date. These aspects include the manner in which the human experiences the simulation model, the degree to which the experience appears true to life and the manner in which the human can interact with the model. To reiterate an opening remark, VR in simulation can be regarded as an advanced user interface to a 3D simulation model.

7.0 VR EQUIPMENT

In the context of this paper, the order of importance for VR devices is sight, view control, and interaction (including touch). In order to place these in perspective, a description is given of the status of devices for sound, smell, taste and even (yes really!) mind control.

7.1 Sight

The most common VR equipment handles how the human sees the VR world. The stereo glasses employed by 3D movies are a commonly understood example of this type of equipment. Put simply, two coordinated and simultaneous images are produced and each image is directed to a separate eye. In this way the human viewer sees a 3D image.

A computer monitor can produce the two images, which can then be interpreted by a set of special VR glasses that will show a 3D view to the human. In a similar way, higher quality devices can be used. For example, head sets and other VR devices employ liquid crystal displays to provide each high quality image.

7.2 View Control and Tracking

The control of the human's view of the VR world is an important interactive capability of VR. The use of keyboard and mouse becomes difficult quickly when the human cannot see the devices. Hand held devices such as a remote mouse allows much better interactive view control. This works well with glasses or head-mounted devices.

A Binocular Omni-Orientation Monitor (BOOM) allows the user to move the viewing device itself with periscope-like movements. The computer registers the movements and changes the view position and orientation accordingly. There are now desk top devices that combine 3D viewing and view control features.

The most attractive view control VR device may well be a tracker. A tracker allows the computer to monitor the position of the human and to alter the VR world view according to the human's movements. There are a variety of tracker devices.

The ability of the VR world to react quickly to a view control command is very important to the human's sense of the realism of the VR world. If the VR world view change is too slow to react, it is difficult for the human to maintain involvement. As an example, if you turn your head in the real world you will instantly see a new view. That is because the real world is "on display" constantly. However, in the VR world a fast turn of the head may get ahead of the computer's ability to generate the new views and a "lag" effect will be experienced. This is a significant issue in the design and selection of VR equipment and software.

7.3 Interaction

Tracker devices by their nature provide input to the computer on the location of the tracker device. It is a natural extension to link similar devices to the VR world as interactive input devices that do more than register location. A glove that contains flexible fiber optic cabling and 20 or more sensors can be used to register complex combinations of locational information in the VR world. One use of such devices is to monitor complex manual activity as input to control equivalent manual operations in the VR world. An extension of this would be to provide complex commands to the VR world such as by the medium of International Sign Language. Another extension would be to provide interactions with the VR world such as grabbing or manipulating VR objects.

Tracker technology extends to full body suits which can be used to control an animated figure within a VR world.

7.4 Voice Control

Voice control is now available on standard PCs. However, the quality of general tools for this type of interactive control is not at a point that its application to any VR world is simple and cheap.

7.5 Miscellaneous Devices

Laser pointer devices that display within the VR world provide the equivalent of point and click control on a screen.

In summary, there are many alternative interactive control mechanisms. The choice is likely to depend on the application.

7.6 Touch

With the use of a set of trackers, the various parts of the human's extremities can be tracked. In this way, the human extremities can be tracked to see if they are in contact with some object in the VR world.

This mechanism can be enhanced by the use of force feedback, which exerts pressure on the human body to simulate touching an object. Simulating the texture of touch is beyond current technology.

7.7 Sound

Humans are by now very used to the various methods of sound reproduction. Most typically with a computer device, digital sound recordings can be replayed and fed to the user through speakers or headphones. Stereo or even surround sound is already established technology. The integration of tracker devices with the sound reproduction mechanism will allow the human to experience omnidirectional and distance-based sounds. This technology has a clear attraction outside the VR world.

7.8 Smell

The application of smell in VR is still at the level of basic research.

Over 30 different basic smells have been identified. By employing combinations of these, a wide range of aromas can be generated. Omnidirectional and distance-based smell production requires some development but there is a reasonably clear path towards it.

7.9 Taste

This technology is further away than smell. Artificial flavors provide some capability. However the application of this to Virtual Reality is at the research level now. Problems of the association between taste, texture and smell are complex problems which have yet to be tackled for Virtual Reality applications.

7.10 Biocontrol

This technology exists only as research projects. Eye movements and muscle movements can be monitored and used to interact with the VR world. As with several other aspects of VR, the technology has applications outside of VR. For example, eye movement has been used in camcorders to assist the focusing mechanism.

Successful experiments have even been performed where the subject was able to learn to control their own brain waves by using EEG brain wave monitoring device feedback. The EEG devices were then hooked up to a computer and control rules established so that certain brain wave combinations would be interpreted as commands to the computer. In this way subjects were able to mentally manipulate a cursor.

8.0 AREAS OF SIGNIFICANT INTEREST

8.1 Shared VR worlds

VR worlds can be shared by more than one human at a time and, through networks, humans in different physical locations can participate simultaneously in a single VR world. Bandwidth limitations make this difficult to implement, however there are some applications of this that are commercially available.

8.2 VR for remote control

The control of remote robotics devices by an operator in a VR world is available. One application of this is for hazardous environments.

9.0 APPLICATIONS

In the applications described below, VR worlds have been created that are 3D and that realistically represent existing or planned environments. The VR world is animated with behavior controlled by a simulation engine which uses simulated behavior rules and model data.

9.1 Primary Elements of Virtual Manufacturing

Visualization allows the user to assemble the components in a Virtual Manufacturing cell. This could be by utilizing the 3D models of the part designs using popular CAD packages, such as Unigraphics, Intergraph, CATIA or Pro/ENGINEER to visualize how the components relate to each other. Sikorsky used Deneb's IGRIP simulation software to recreate the actual manufacturing process. In the model, the user can manipulate and view each of the components to develop a manufacturing approach (reference Figure 1).

Figure 1. 3D Visualization

Virtual Manufacturing modelling allows the user to introduce the engineering design to the processes that will be used to create the actual part, assembly, or installation. The tool design is imported to the model, combined with the part and dynamic representations of the machines that will produce the part. Machines such as 5-axis milling centers or processes such as aircraft joining are programmed to operate identically in the virtual environment as they would in the real environment. Design problems such as collisions, clearances, missing manufacturing features, fit, and manufacturing sequence issues are quickly identified. Manufacturing and design concepts are easily developed with minimal cost to the program. The ability to insert people in the environment and analyze their activities minimizes ergonomic problems.

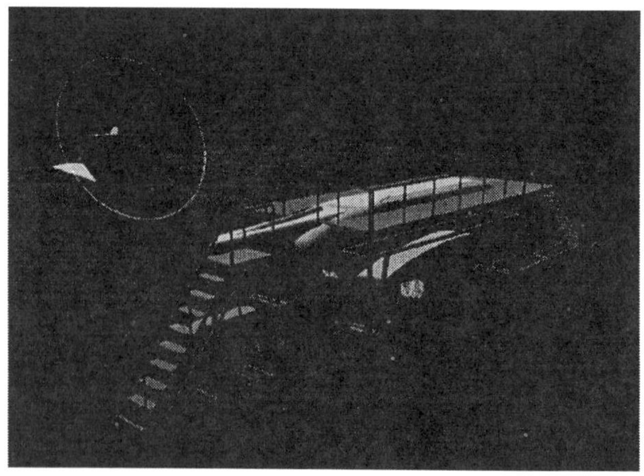

Figure 2. Workcell Modelling

Factory modelling allows engineers to predict the cost and schedule impact of potential design or process changes. The discrete process modelling tool, QUEST, is used by Sikorsky to model the impact of machine run times, set up times, resource constraints, part lot quantity, and many other factors to develop a simulation that mirrors the entire manufacturing process under review. The information provided by the simulation enables "what if" analysis. The model provides support personnel with a tool to optimize the manufacturing process, try new concepts without disrupting manufacturing and predict changes due to load variations. Figure 3 is an example of a model of a sheet metal manufacturing center. The graphical interface allows the manufacturing experts to actively participate in validating the model.

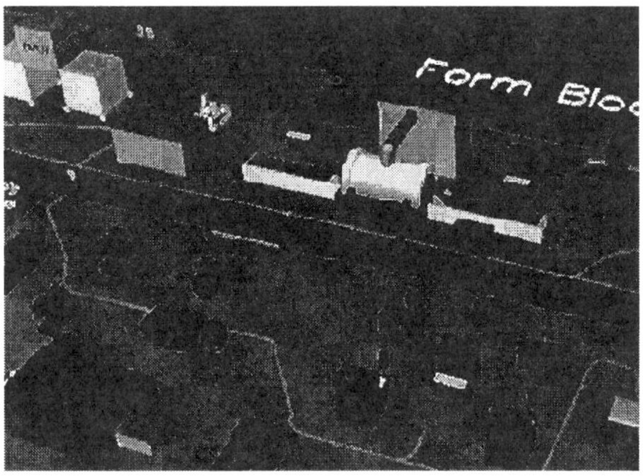

Figure 3. Discrete Process Modelling

Virtual Manufacturing has enabled Sikorsky manufacturing personnel to directly influence the design process. They improved transfer of data between partners and suppliers, allowed for in-depth product visualization and enabled advanced graphic database management — all in an effort to rapidly respond to customer demands.

The three common elements in the evolution of a product are cost, schedule, and information. Virtual Manufacturing allows engineers, designers, suppliers, and others to understand the cost and schedule impact of decisions and to consolidate processes.

Overhead costs are addressed by improving schedule performance. The amount of time it takes to design, plan, tool and manufacture a product directly impacts overhead costs. As illustrated in Figure 4, a reduction in the time it takes to get a product to market equates to a return on investment (ROI) improvement.

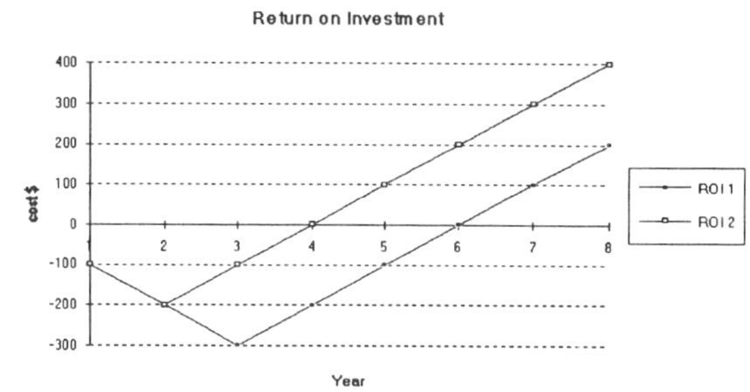

Figure 4. Courtesy of Sikorsky Aircraft

9.1.1 Scenario

The following discussion represents a typical manufacturing and design scenario at Sikorsky where they have integrated Virtual Manufacturing into their IPPD process.

In Phase 1, they derive a sound manufacturing concept to optimize cost, schedule, and account for all design issues such as weight and stress. This is the original conceptualization phase where the IPPD team determines the cost and schedule impact of decisions.

In Phase 2, they validate the manufacturing and design plan by simulating the actual manufacturing processes and equipment that are used either in-house, at suppliers, or at partners, to ensure that the design intent is being met. At this time, the IPPD team predicts the tolerance the process can yield and validates the manufacturing processes (reference Figure 5).

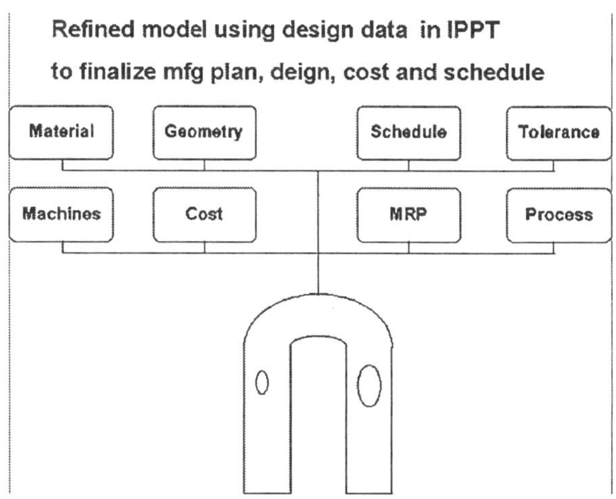

Figure 5. Manufacturing Validation

In the Phase 3, they use Virtual Manufacturing to

manage internally and externally generated changes by predicting what effects the changes will have. Results provide instant access to cost and schedule impact and provide visibility as to manufacturing capability.

9.1.2 Conclusion

At Sikorsky Aircraft, the ability to remain competitive will be closely tied to reducing lead times, improving schedules, and eliminating unwarranted costs. Virtual Manufacturing will be one of the key enabling technologies used to realize these goals by bringing all the decision drivers up front, dramatically reducing the overhead costs commonly associated with product/process development activities.

9.2 Airfoil Manufacturing Cell

British Aerospace, United Kingdom, is using QUEST, Virtual NC®, and evaluating IGRIP and Deneb/ERGO™ to conduct a number of aircraft manufacturing studies.

The site is familiar with the use of simulation and already have a capability. They are however, particularly interested in the use of QUEST for high value new start projects, where the superior visualization is a valuable tool within their concurrent engineering environment. It also provides a good insight to the maturity of facility definition for their customers.

One QUEST-based model was created to determine how increased build rates, spare parts, and repairs would impact production. The issues they wished to investigate included facility sizing, shared facility analysis, floor space management, and estimation of tooling requirements.

To create the QUEST model, legacy CAD data of the production facility was imported from CATIA, Intergraph, AutoCAD and ANVIL CAD systems. This highlighted an additional benefit of allowing CAD data from multiple courses to be viewed on a single screen. BAe modelled proposed facility enhancements to accommodate increased production whilst maintaining a robust and flexible manufacturing facility. The impact of enhancements are difficult to visualize in 2D simulators. Issues for BAe included space and labor allocation, material flow, bottlenecks, throughput, and accurate modelling of manufacturing cells.

The machining and treatments facility at BAe, Samlesbury is using Virtual NC for detailed process analysis. They have modelled 4 and 5 axis milling machines tooling and jigs to prove-out a large percentage of complex part programs before beginning actual machining. This particular approach has contributed to a reduction in part program prove out time on line machines.

One particular problem BAe had experienced in the past was "shanking." Shanking is largely due to a product/tooling design flaw, and occurs when the shank of the tool collides with the workpiece or its fixture. Virtual NC's near-miss and collision detection facilities allowed BAe to prove-out the complex NC programs and eliminates all such collisions.

Today, Virtual NC is also being used to evaluate new machine tools, fixtures and jigs, verifying the reachability of a component to the machine and programs before implementation in the production facility.

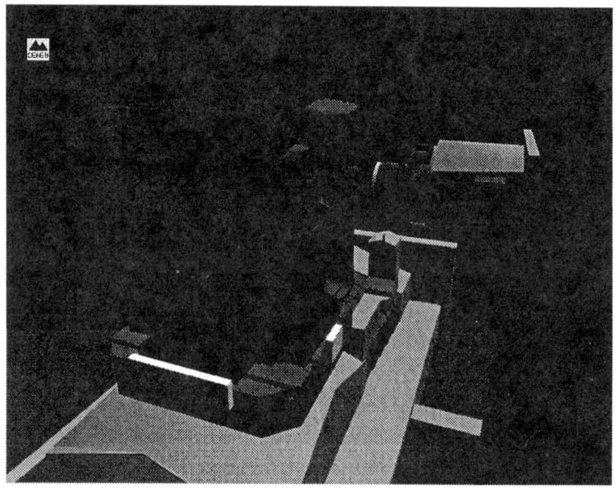

Figure 6. Courtesy of British Aerospace

The benefits from Virtual Manufacturing the BAe engineers realized include:

- Integration of the CAD systems allow data from multiple sources to be viewed on a single screen. These floor layouts can be populated with tooling geometry. This allows fast reviews to be carried out, and checks to be made on whether the tooling fits within the layout as expected and if there is any interference between buildings, tooling, and piece parts.
- The ability to replay existing Virtual NC, Deneb/ERGO, IGRIP, and QUEST models means there is no overhead in reusing existing data.
- 3D simulation allows better communication of ideas/concepts to both management and shop floor operators.
- More accurate modelling of material handling systems making it simpler to view the progress of production through the facility.
- Checking NC programs for collisions and workpiece damage prior to production. This reduces downtime, scrap, and tape prove-out time on the machine.

9.3 Sacramento Air Logistics Command

When stringent air regulation amendments to the 1990 Clean Air Act threatened U.S. foundries with significant capital costs for retrofitting or forced closure, the government and private sector formed the Casting Emission Reduction Program (CERP) to address the environmental issues. QUEST interactive 3D graphics discrete event simulation software supported the CERP project team at U.S. Air Force Advanced Manufacturing Center of the Sacramento Air Logistic Command at McClellan Air Force Base, Sacramento, CA, to demonstrate the viability of a pilot foundry prior to construction. The foundry will be built at McClellan AFB in 1996 to conduct environmental and process experimentation.

CERP participants include McClellan Air Force Base, SALC, and the Environmental Research Consortium, a committee of the U.S. Council for Automotive Research (USCAR) which is a U.S. auto company coalition. The consortium includes Chrysler Corp., Ford Motor Co., GM, and Navistar International. The Environmental Protection Agency (EPA), the California Air Resources Board, and the American Foundrymen's Society are also participants.

Over 90 percent of all manufactured items used by the DoD either are castings, contain castings, or are derived from castings. The weapon systems used by the military make large scale use of metal castings. Castings are also critical in the private sector. Automobiles and transportation equipment use 50 to 60 percent of all castings produced.

To revitalize the U.S. foundry industry, CERP's initial goal is to develop new materials, processes or equipment for metal casting manufacturing that will achieve a near-zero effect on the environment.

CERP's objectives included building a pilot facility for testing and research, collecting data on actual emissions, developing low level emission testing equipment, and modelling foundry processes.

QUEST and IGRIP were used by SALC manufacturing engineers to develop the foundry simulation model.

The initial model was conceptual. Even so, the content, like the object library that was created, could be used to prepare the technical model. During design review, this initial model served as a baseline from which to conduct analyses to optimize the layout. Very specific data from the design group was incorporated — feeds, flows, and equipment.

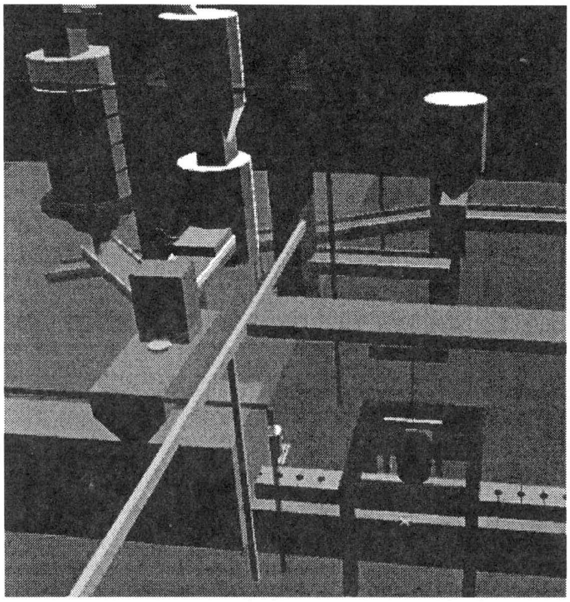

Figure 7. The Virtual Prototype of the Proposed Foundry

Eventually, the software will be used for time and motion studies, and to analyze human factors and do time study measurement in lieu of doing engineered time standards.

As the simulation matured and was shown to molders who actually work in the foundry, their input on how specific setups would work were added to the database. The simulation also became a valuable tool for training employees.

Using virtual reality technology, the government and private sector CERP team has demonstrated the usefulness of virtual reality modelling in design. QUEST is helping in the design of a flexible and relevant pilot foundry that will help the industry find new, low environmental impact ways of doing business. The simulations have proven to be a valuable design tool, as well as a communication and visualization tool to help make the project "come to life" prior to construction.

9.4 Other DoD Applications: Next Generation Submarine being Developed with Virtual Technologies

Prototypes of a nuclear attack submarine being developed by General Dynamics Electric Boat Division for the Navy are being designed, evaluated, and optimized in a virtual environment. Operating under a contract from the Advanced Research Project Agency's (ARPA)

Marine Systems Division, Electric Boat demonstrated the feasibility of state-of-the-art simulation-based design (SBD). The objective was to implement an accurate, efficient, and dynamic environment for design, rapid prototyping, concurrent product and process development, mission planning, operation maintenance, and training. Traditional prototyping methods can take several months to physically build and test, and at a great expense. Subsequent to the ARPA demonstration, Electric Boat has adapted the SBD approach to design and now utilizes 21 seats of ENVISION software.

By simulating the kinematic, dynamic, mechanical, and other characteristics of the submarine, its components and subsystems, engineers can create a multi-disciplinary environment in which to evaluate a wide range of parameters and optimize the design based on the results.

The first phase of the SBD project was to demonstrate the feasibility of designing and evaluating the next generation submarine in a virtual environment. Electric Boat selected ENVISION to visualize design concepts and integrate components and subsystems. Electric Boat also uses ENVISION to visualize the results of the model's maneuverability tests in a virtual undersea environment. From the data provided by an external hydrodynamic analysis program being visualized (Figure 8) in the simulation, engineers can optimize the conceptual designs, including consideration of both the submarine's performance characteristics and the cost ramifications of a proposed design change.

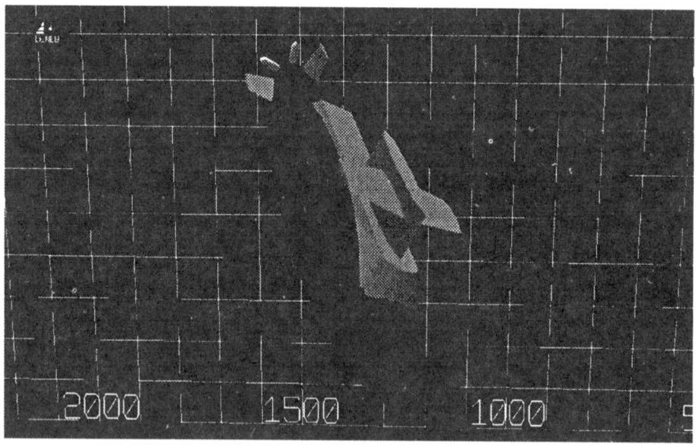

Figure 8. ENVISION's 3D graphics allow engineers to evaluate the impact of design modification by conducting maneuverability tests in a virtual, undersea environment

In a proof of concept demonstration to the ARPA, Electric Boat featured its SBD capabilities and concurrent engineering methods enabled by the virtual technologies. Developmental submarine models were created by importing CAD data from Parametric Technology, Computervision, CATIA, and Intergraph software packages. The submarine hull was simulated operating at various speeds and directions in a "virtual sea" to evaluate its performance. Using Pro/ENGINEER CAD software, the hull design was modified, re-imported into ENVISION, and simulated again. The comparison simulations enabled the engineers to evaluate the impact of the design modifications on maneuverability. Because of the seamless migration from CAD systems to ENVISION, Electric Boat engineers were able to generate and simulate each design iteration in about 10 minutes.

The driving force behind Electric Boat's pioneering work with virtual technologies is the critical need to reduce the time and cost required for designing, engineering, and building a submarine. The development window has been as long as 14 years for previous submarines. Electric Boat feels they can reduce this time to six or seven years with virtual technologies.

Virtual engineering allows the parties involved with design, manufacturing, operations, and maintenance to jointly contribute to the design process early in the cycle. Electric Boat engineers often experience communication difficulties when trying to describe innovative ideas to people not trained in their particular discipline. The physically accurate 3D models simulated in ENVISION help engineers and technicians from various disciplines to comprehend new ideas and concepts. In addition, virtual reality technologies such as immersion enable engineers and their naval customers to "walkthrough" the model with visual feedback. Immersion provides for evaluation of other design criteria, such as ergonomics, before a design is built.

Figure 9.

Deneb, IBM Federal Systems (acquired by Loral Corp.), Intergraph, the University of Iowa, Silicon Graphics, and Parametric Technologies are working with Electric Boat to prove out SBD concepts and the applicability of virtual technologies to engineering.

Already the benefits of applying virtual technologies to engineering, manufacturing, mission planning, operations, maintenance, and training are evident. At Electric Boat, the ability to optimize systems through concurrent engineering, lowered costs and reduced times to product delivery are just a few of the more significant benefits of SBD and virtual engineering. The virtual engineering systems that evolve from this project will be applicable to most complex product design endeavors with similar benefits. Electric Boat has adopted the SBD paradigm and the ENVISION software as key elements of the design process.

9.5 Virtual Collaborative Engineering

The most recent development in the use of Virtual Manufacturing software is the use of the Virtual Collaborative Engineering (VCE) environment that links multiple users at multiple locations to discuss, analyze, and review simulations over a wide area network. VCE users interactively evaluate design concepts, manufacturing tooling, processes, and factory layouts — even at geographically remote locations. Any VCE user can assume control of a simulation, make changes, or view changes made by others on the VCE network. Engineers, manufacturing personnel, system operators, or other users interact within the same simulation, creating a "virtual conference room" for design discussion, evaluation, and approval. VCE participants can use a mixture of UNIX and Windows NT platforms.

By using any one of the Deneb simulations in a VCE environment, users increase productivity, speed decision making, reduce costs, and improve time to market. Also, improved communications between engineers, clients, suppliers, and/or product teams prior to product launch reduces the need for multiple revisions and prototypes.

Some current uses for VCE include:

- Interactively evaluate virtual prototypes of designs
- Link remote product teams for Integrated Product and Process Development
- Integrate suppliers, users and clients to gain early design feedback
- Evaluate alternative scenarios
- Conduct daily design/process review sessions
- Train workers at multiple locations
- Link integrated product teams early in acquisition programs

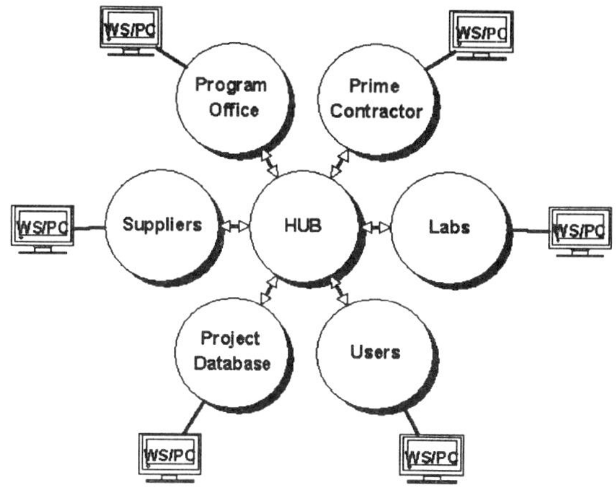

Figure 10. Virtual Collaborative Environment

10 VIRTUAL MANUFACTURING BENEFITS

Besides improved communications across multi-disciplinary product teams, specific benefits demonstrated and projected for key industries include:

- 50 percent reduction in time to market based on the increased speed of decision making and the parallel development of product and process designs
- $1-5 billion life cycle cost savings for new Navy/Air Force Aircraft development and production
- Achieve learning curve of third ship set on first ship set for new shipbuilding programs
- Integration of global customers, subcontractors, suppliers and users early in the design and development cycle
- 25 percent reduction in the cost of new products due to the utilization of the Design for Assembly and Design for Manufacturing capabilities
- 70-80 percent reduction in rework costs for tooling and fixtures since they were proven out in the Virtual Manufacturing environment before initial fabrication

REFERENCES

Armed Forces Journal, "New and Newer Submarines," July 1995

Beaudreaux, J., General Dynamics — Electric Boat, "ARPA SBD Contract."

CIME, "Full Speed Ahead," Spring 1996

McDonald, M, Sandia National Laboratories, "Virtual Collaborative Engineering," 1996

NASA Tech Briefs, "Resource Report," May, 1995

R&D Magazine, "New Horizons for Aerospace — Simulate Your System Before you Build it," January 1996

Scotton, T.W., United Technologies Research Center, "Virtual Manufacturing," Deneb User Group 1995 Proceedings, October 1995

Ulowetz, K., McClellan AFB, "CERP Virtual Foundry," 1995

Varanay, S., Sikorsky Aircraft Co., "Implementing Virtual Manufacturing at Sikorsky Aircraft Co.," Agility Forum Annual Meeting, January 1996

AUTHOR BIOGRAPHY

MARTIN BARNES was educated in England, receiving his honors degree in Maths and Statistics from the University of Leeds and his Masters in Operational Research from the University of Lancaster. After working in Operational Research and Production management in British Leyland, Martin emigrated to the USA to set up the support activities for ISTEL, where he was Customer Support manager for SEE WHY and WITNESS simulation software packages. Martin was Customer Services manager for Waterloo Manufacturing Software where he was involved in the design and implementation of TACTIC, a finite capacity scheduling software package. Currently, Martin is the Support Manager for Deneb Robotic's simulation software package QUEST.

Proceedings of the 1996 Winter Simulation Conference
ed. J. M. Charnes, D. J. Morrice, D. T. Brunner, and J. J. Swain

SO YOU WANT TO BE A SIMULATION CONSULTANT

John S. Carson II

AutoSimulations
1827 Powers Ferry Road
Bldg 17, Suite 100
Atlanta, Georgia 30339, U.S.A.

ABSTRACT

This presentation is for those who want to be a simulation consultant, whether in an independent consulting company or within a consulting group in a larger company. Both practical project management and good technical fundamentals are critical for success. Does a simulation consultant need to be an expert in all disciplines? Does he or she need to be an engineer, a modeler, a statistician, and a computer scientist? The ideas presented in this tutorial are based on many years of practical simulation experience. Examples from real-world projects will be used to illustrate various points. The author's woeful experience will be used to illustrate pitfalls to avoid.

1 INTRODUCTION

Practical project management and sound technical fundamentals both play key roles in being a successful simulation consultant. Section 2 provides a brief overview of the key management issues; section 3 covers some of the technical issues that have arisen in the author's experience over 20 years in consulting, model development, and management of simulation analysts. During the first part of the presentation, the author will reveal his biases and discuss his past and present employment and simulation experience.

2 MANAGEMENT ISSUES

A well managed simulation project proceeds by a set of steps that provide general guidelines for success (Banks, Carson and Nelson 1996, Law and Kelton 1991). Some steps involve mostly the simulation analyst, but all involve the customer or model user to a more or lesser extent. These steps include:

- Problem formulation:
  - Statement of objectives
  - Specification document
- Model design
- Data collection
- Model development
- Verification and validation
- Experimentation and analysis
- Documentation

Although called "steps", in reality they are activities that the consultant will do repeatedly and iteratively, often not exactly in the order outlined here. An excellent set of guidelines for implementing these steps in a project are presented in Musselman (1994) in an entertaining and instructive fashion. In the following subsections, we discuss some of the key ideas.

2.1 Problem Formulation

Good communication with the customer is the single most important key to success. Communication includes determining the customer's true problem, setting objectives, setting project scope, coming to agreement on model assumptions, and developing understanding of model input requirements and model outputs. These items should be formalized in a Specification Document and a Project Plan.

2.2 Model Design

Model design involves abstracting the system and determining its elements and their interactions. The objectives and questions being asked determine the level of detail of each of the model's components.

2.3 Data Collection

To assist the customer in collecting the necessary data, the consultant must identify and clearly define the data needed to drive the model. While the model itself determines what data is needed, practical considerations and data availability may affect a model's structure and level of detail. The consultant should question the data:

How was it collected? How was it measured? Are their outliers that do not make sense? Has the process fundamentally changed since the data was collected?

2.4 Model Development

Keep the model as simple as possible. Any complexity introduced should be related to project objectives; without it, it would be impossible to address one of the questions being asked. Always keep in mind that the project goal is not to get the model running, but to provide possible solutions and insights into the customer's problem.

2.5 Verification and Validation

To verify the model, the analyst attempts to confirm that it works as intended. A good test for model robustness is to make as many runs as computer time allows, as soon as the model is running, over different random number streams and scenarios; a quick check for reasonableness of key outputs and re-running the scenarios with errors to make fixes is a powerful way to move in the direction of model accuracy. If the model runs perfectly, you haven't tested it enough.

To validate the model, the analyst involves the customer to assure, within reasonable accuracy, that the model corresponds to the real world system, or for new systems, meets customer expectations. Before validation, the model must be updated with the customer's most recent and accurate data.

Keys to success include regular model reviews with all members of the customer's team, controlling (and minimizing) changes in scope and assumptions, and most importantly, quick fixes for any perceived problems. If the customer thinks it's a mistake, it is.

2.6 Experimentation and Analysis

Conduct experimentation with the Base Model as early as possible, even before the model is complete. Early feedback keeps the customer's attention. Capture numerous outputs beyond the primary outputs that address customer's questions; model verification continues throughout model experimentaiton. Question and thoroughly examine all outputs.

Keep in mind that the purpose of modeling is to provide insight and offer possible solutions. It does not replace careful analysis and human thought. It can help in avoiding costly errors and can suggest solutions, but it does not provide *the answer*.

2.7 Documentation

With updating, the Functional Specification Document provides the basis for documentation of model assumptions, model inputs and outputs. Monthly progress reports for longer projects, and written confirmation of assumptions discussed over the telephone, are also important. Internal documentation of model components (resources, variables, etc) is invaluable when a model is modified at a later date.

3 TECHNICAL ISSUES

Simulation modeling and analysis requires a range of skills and fundamental technical competencies.

3.1 Simulation Software: Power versus Ease of Use

Many articles have been written on simulation software, including the overview by Banks (1995) and articles written by vendors in each year's Proceedings of the Winter Simulation Conference. One point to consider for the budding simulation consultant: The package that is easy for an end user who develops the occasional simulation model, may or may not be the right tool for the consultant who does models of larger and more complex systems, especially data-driven generic or template models that allow a user to select from a wide range of scenarios.

If a package requires no programming, then somebody somewhere who never saw your customer's system wrote the model for it. Are you buying a house, or the hammer and the saw? It may be easy to use off-the-shelf for direct models, but will it help you build a more generic model that will cover all desired scenarios and be easy to use by your customer? Do you like pulling teeth (your own)?

Talk to a number of independent simulation consultants who use a range of products and find the one (or ones) most suitable for the types of systems you simulate. (Forget not the author's biases.)

3.2 Knowing the Insides

A detailed knowledge of the simulation package's internals is essential for knowing exactly what your model is doing, and for debugging those complex situations that often arise in a consultant's life between midnight and 3 am. For details and examples of how seemingly similar models can do quite different things, see Schriber and Brunner (1995).

3.3 Statistical Issues

Papers on statistical guidelines and pitfalls in input modeling include Vincent and Law (1995); background material is provided by Leemis (1995). Kelton (1995) provides an overview of experimental design and output analysis.

3.3.1 Do Downtimes Matter?

Low frequency, high impact downtimes make huge differences in throughput; for example, for one downtime per day of duration 6 hours on average, the statistical distribution assumed makes a large difference in many model outputs. High frequency, short downtimes (many downtimes per production cycle, each of short duration) can often be accounted for with a deterministic adjustment to production rates, without causing undue changes to model output.

3.3.2 Animation

Statistical analysis is no substitute for a good animation. The most technically competent statistical analysis will do no good without model credibility. The single tool that has increased modeling credibility the most in recent years is animation.

A good animation requires a good story line and immediate recognition by the customer. A good animation provides insight into poor system performance and directs attention toward problem areas. However, without a good explanation of why something is happening, its value decreases and credibility suffers. A poor animation is like a cartoon with the sound turned off.

3.4 Other Topics

When do statistical replications matter? What can you compute ahead of time to verify models? What types of systems are prone to model lock-up and can have too simple a model? What type of data do customers most often provide incorrectly? It's difficult to tell when steady state is reached, but how do you tell if it's ever reached? How do you model rush hour or seasonal arrivals?

4 SUMMARY

Both practical project management and technical competency play key roles in successful simulation consulting.

REFERENCES

Banks, J. 1995. Software for Simulation. In *Proceedings of the 1995 Winter Simulation Conference*, ed. C. Alexopoulos, K. Kang, W.R. Lilegdon, and D. Goldsman, 32-38. Arlington, VA.

Banks, J., J.S. Carson, and B.L. Nelson. 1996. *Discrete-Event System Simulation*. Second Edition. Upper Saddle River, NJ: Prentice-Hall.

Kelton, W.D. 1995. A Tutorial on Design and Analysis of Simulation Experiments. In *Proceedings of the 1995 Winter Simulation Conference*, ed. C. Alexopoulos, K. Kang, W.R. Lilegdon, and D. Goldsman, 24-31. Arlington, VA.

Law, A.M. and W.D. Kelton. 1991. *Simulation Modeling and Analysis.* Second Edition. New York: McGraw-Hill.

Leemis, L.M. 1995. Input Modeling for Discrete-Event Simulation. In *Proceedings of the 1995 Winter Simulation Conference*, ed. C. Alexopoulos, K. Kang, W.R. Lilegdon, and D. Goldsman, 16-23. Arlington, VA.

Musselman, Kenneth J. 1994. Guidelines for Simulation Project Success. In *Proceedings of the 1994 Winter Simulation Conference*, ed. J.D. Tew, S. Manivannan, D.A. Sadowski, and A.F. Seila, 88-95. Lake Buena Vista, Florida.

Schriber, T.J. and D.T. Brunner. 1995. Inside Simulation Software: How It Works and Why It Matters. In *Proceedings of the 1995 Winter Simulation Conference*, ed. C. Alexopoulos, K. Kang, W.R. Lilegdon, and D. Goldsman, 110-117. Arlington, VA.

Vincent, S. and Law, A.M. 1995. ExpertFit: Total Support for Simulation Input Modeling. In *Proceedings of the 1995 Winter Simulation Conference*, ed. C. Alexopoulos, K. Kang, W.R. Lilegdon, and D. Goldsman, 395-400. Arlington, VA.

AUTHOR BIOGRAPHY

JOHN S. CARSON, Senior Simulation Analyst, has been with AutoSimulations since 1994. He has over 20 years experience in simulation modeling for manufacturing, distribution, warehousing and material handling, transportation, port operations, and medical/health care systems. He has been a simulation consultant, and has taught at Georgia Tech, the University of Florida, and the University of Wisconsin-Madison. He is the co-author of two textbooks, including Discrete-Event System Simulation, Second Edition (1996). He holds a Ph.D. in Industrial Engineering and Operations Research from the University of Wisconsin-Madison, and is a member of IIE and INFORMS.

Proceedings of the 1996 Winter Simulation Conference
ed. J. M. Charnes, D. J. Morrice, D. T. Brunner, and J. J. Swain

SUPPORTING MANUFACTURING WITH SIMULATION: MODEL DESIGN, DEVELOPMENT, AND DEPLOYMENT

Frank Chance

Chance Industrial Solutions
6900G Lake Drive
Dublin, California 94568
U.S.A.

Jennifer Robinson

University of Massachusetts
581 VFW Parkway
Chestnut Hill, Massachusetts 02167
U.S.A.

John Fowler

Arizona State University
Dept. of I.M.S.E., Box 875906
Tempe, Arizona 85287-5906
U.S.A.

ABSTRACT

In this paper, we identify and discuss the features we believe are key to the successful use of simulation as a manufacturing support tool. The discussion begins with three sample projects drawn from the authors' industrial and consulting experiences. Using these projects as motivation, we discuss the ideal project lifecycle — model design, development, and deployment. For model design, we emphasize the importance of a clear and consistent specification, articulated in a written document. This specification should identify project customers, goals, and deliverables. We next review a range of model development options, stressing the existence of many non-simulation alternatives. We also discuss methods for model verification and validation. Finally, we consider the difficulties of model deployment, including simulation output analysis, data maintenance, and model integration. We close with several suggestions on how best to present simulation results to a management audience.

1 THREE SAMPLE PROJECTS

To motivate the remainder of this paper, we begin by describing three sample manufacturing projects. These projects are drawn from the authors' experiences.

1.1 Planning Toolset and Staffing Levels

In this project, the client is planning for a new factory, and wishes to compare toolset and staffing levels required to support a variety of different product mix and volume scenarios. The results of the project will be used for three purposes: 1) to determine how many of each type of tool to order; 2) to decide how many operators and maintenance engineers will be needed; and 3) to estimate the relative costs of producing the different products. Because of the high likelihood of product mix changes, the model should allow for quick analysis of multiple scenarios, and should provide sensitivity information. The model will likely be used throughout facility planning and construction.

1.2 Optimizing Cycle Time with Equipment

In this project, the client has already begun building a new single-product factory, and wishes to optimize cycle time by making additional capital equipment purchases. The factory will produce the company's latest product, which is expected to have high demand. A pilot line in an existing facility is currently producing the new product at very low volumes. However, the actual process parameters are in a constant state of flux as engineers make refinements. The client has developed an in-house Excel capacity model. This model is periodically updated with the latest process changes. From this project, the client expects to get a list of additional tools that should be ordered beyond the minimum cost toolset, and an estimate of the expected cycle time once the factory is operational.

1.3 Optimizing Cycle Time with Scheduling

In this project, the client wishes to implement a scheduling system that will balance the competing demands of many different customers for specialized orders. The total production volume is high, and between-product setup times are significant. The client has a sophisticated in-house shop-floor tracking system that is continuously updated with the location of all work in process. The system also has product routing data and equipment status information. The client needs to be able to test out different dispatching strategies, to minimize overall cycle time. Also, the client wishes to be able to track the status of individual orders, so that expected delivery dates can be given to customers based on the latest available information.

2 PROJECT LIFECYCLES

The sample projects outlined above pose questions that can be answered a variety of ways. One such method is physical experimentation — making changes in the real factory and analyzing the results. An example might be a new configuration of a workcell, which is tested out in a small area before being implemented in the whole factory. Physical experimentation is not always possible, however, especially when planning for a new factory. Even in an existing factory, it is often prohibitively expensive to experiment with the real facility. Increasing equipment quantities, for example, can require many months of lead time, not to mention sizable capital investment. Therefore, it is usually necessary to build a model of the real factory, and perform experiments in this virtual environment.

In an ideal world, modeling projects would progress in an orderly fashion through three distinct phases — model design, development, and deployment. In reality, however, these phases often overlap, or are repeated as later phases shed light on errors or omissions in earlier work. Changes in project scope can also cause iterations through these phases. Often, the later work cannot even be clearly defined until the earlier phases are completed. Interaction is usually needed between the client and the analyst to clear up details and discuss new possibilities. However, by following as closely as possible the model design, development, and deployment phases, we believe that it is possible to minimize false starts. In the remaining sections of this paper, we outline some key features that we have found contribute to successful manufacturing simulation projects.

3 MODEL DESIGN

Building and using a factory model can be a daunting task. The sheer size and complexity of most factories makes it difficult to completely view and understand all of the many elements found there. These elements include products, processes, tools, material handling systems, inventory, and operators. We believe that the goal of any modeling project is to develop a model that reflects enough, but not too much, of this complexity. Models with excessive detail take longer to build, debug, and understand, are harder to maintain, and have longer run times than less detailed models. The guiding light in determining how much complexity is necessary should be the questions posed of the model. Clarification and documentation of these questions, then, is of paramount importance, and is the primary goal of the design phase.

3.1 Identifying Project Customers

The first step in the design phase is identifying who will use the results the project is meant to create. Potential customers might be from the shop floor, or from a corporate planning department. In many cases, a project will have several customers, each of whom has his or her own expectations. Prioritizing among the needs of these different customers is the client's internal responsibility. However, the analyst must sometimes consolidate various customer objectives, and identify which of these, if any, conflict.

For example, the customers for the scheduling project outlined above probably include the computer integrated manufacturing (CIM) group and the factory operations manager. The CIM organization might want to completely automate lot selection, while the operations manager might wish for more operator empowerment. Resolving these issues in the product design phase can be much more cost effective than dealing with them after the model is built.

3.2 Identifying Project Goals

As indicated above, a project may have multiple customers, and/or multiple goals. Trade-offs may be necessary. The goals ultimately selected help determine the planning horizon, and should be stated in terms of quantifiable performance measures. The specification of project goals should also include a project timeline with deliverables highlighted. The three goals in our first example are clearly stated in Section 1.1. The goal in the second example is to optimize the cycle time of the new factory by adding equipment. However, more specific information such as how much the cycle time needs to be reduced, or how much money can be spent to reduce it, is also needed. The third example, the shop floor scheduling model, is a potentially huge project that requires the definition of smaller, tangible goals.

3.2.1 Identifying the Planning Horizon

The planning horizon to which the model will be applied is important in determining the level of detail to include. A common classification of planning horizons is strategic, tactical, and operational. Although no formal definition of these horizons exists, strategic models are usually concerned with long-term questions like "should I build a new factory, and if so, what should I produce in it?" The first two examples in this paper are primarily strategic questions. Tactical models address questions like "how many units should I start this quarter?" Operational models are concerned with production over the next day or week. The third example in this paper, the scheduling system, is mainly an operational model, though questions such as "what is the best dispatch rule to use every day?" might be considered tactical.

The way in which a model adds value is highly influenced by the project's planning horizon. For example, at the strategic level the analyst has the opportunity to make recommendations that save (or cost) the company millions of dollars in equipment purchases. In this case, there is a large value to answering these few key questions correctly. At the operational level, the analyst has the opportunity to set in place a system that will help with many day-to-day or hour-to-hour decisions ("what job should I run next on this machine?" or "when should I promise this order will be shipped?"). Here, there is a large cumulative value to making these many small decisions correctly.

3.2.2 Identifying Performance Measures

Performance measures are necessary to quantify project goals. Performance measures should be tracked in the real world, not just in models. The performance measures in the first example will likely be expressed primarily in terms of cost, both toolset cost and product cost. The primary performance measure in the second example is cycle time, though toolset cost is also important. Performance measures in a scheduling system might include percentage of on-time delivery, product cycle time, or average work-in-process (WIP).

3.3 Writing a Project Specification

Ideally, every project should have a document that provides a black-and-white synopsis of project design. This document should identify the customer or customers, the planning horizon, the project goals (stated in terms of quantifiable performance measures), and the expected deliverables. Where possible, the expected deliverables should be expressed in the form of mock results charts. The analyst is then responsible for filling in the charts with actual model output. This helps to ensure that the results provided from the project are in the format that the customer needs.

For example, Figure 1 displays a mock results chart for the first sample project (planning toolset and staffing levels with mix/volume uncertainty). This chart displays two performance measures, toolset cost and annual profit, for four candidate scenarios. Figure 2 displays a mock results chart for the second sample project (cycle time optimization via capital purchases). The third sample project would have a mock results chart similar to Figure 1, except the performance measures would likely be cycle time or on-time delivery percentage.

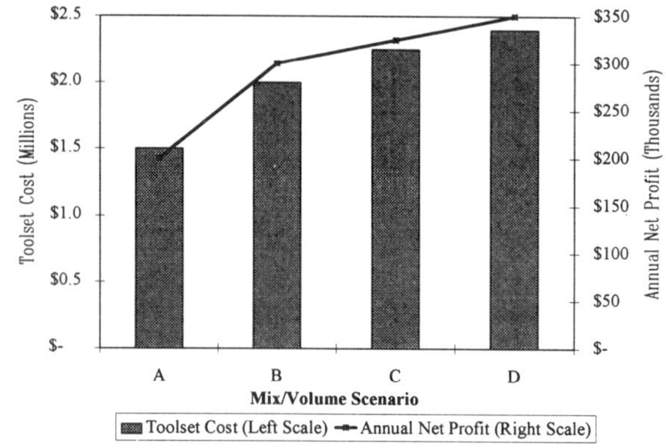

Figure 1: Toolset Cost and Annual Net Profit for Four Mix/Volume Scenarios

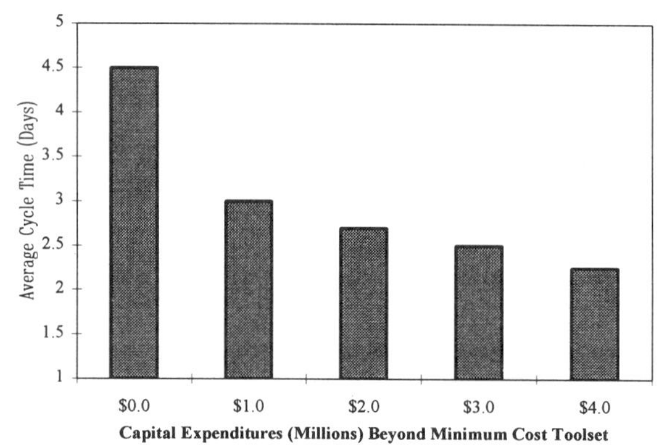

Figure 2: Average Cycle Time vs. Additional Capital Expenditures

In addition to summarizing what the analyst will deliver, the specification should also include client responsibilities. In particular, the specification should document what data the client will provide, and in what basic format (electronically, on paper, in spreadsheets, etc.). In our experience, misunderstandings about the quality of the data available, and who is responsible for improving it, can result in considerable project delays.

The specification should be a living document. As the project scope or deliverables evolve, the specification should be updated to reflect these changes and their impacts. These changes, however, require that both the client and the analyst are in agreement, and the specification often serves as a good negotiation tool. While the specification should not be changed radically in the midst of the project, adding specific details and refinements as they become apparent is beneficial.

4 MODEL DEVELOPMENT

Once the project specification has been written, model development can begin. This requires selecting a methodology and tool, building the model, populating the model with the necessary data, and verifying and validating both the model and the data. Decisions in this phase of the project depend in particular on whether the model is intended for single use or for on-going support. For the latter, the question of who will actually own and maintain the model is important. Also of obvious importance is the type of question being answered by the model. Certain modeling methodologies are more appropriate for certain types of questions. In some cases, what types of tools are available can drive the questions that are asked in model design. In other cases, the client may specify the use of a particular tool or methodology. In general, we feel that the client should identify questions to be answered, while the analyst should be primarily responsible for actual model selection and development.

4.1 Selecting a Methodology and Tool

Many different methodologies and tools are available to support manufacturing projects, including spreadsheets, analytic models, data-driven simulation models, simulation languages, and general-purpose programming languages with simulation libraries. Some projects may require the use of more than one tool to answer different types of questions. Some strengths and weaknesses of these tools for different applications are discussed below.

Spreadsheet models are appealing for projects where a large number of scenarios must be evaluated quickly. They are fast, easy to understand and use, and easy to modify. Managers are generally accustomed to reviewing results presented in spreadsheet-based formats. However, there are limits to the types of questions that can be addressed with spreadsheet models. Spreadsheet models are static, generally grouping time into large buckets. They are also usually deterministic. Though spreadsheets typically include the ability to sample from distributions, this capability is not usually exploited in manufacturing applications. Because of these limitations, spreadsheet models are often not capable of estimating cycle time or WIP.

Analytic models include capacity analysis, queueing, linear programming, and other math programming models. For an example of the successful application of linear programming-based methods to manufacturing production planning, see Leachman et al. (1996). Like spreadsheet models, analytic models are useful for evaluating multiple scenarios, because they are typically very fast. Analytic models can also include detail beyond that of spreadsheets. For example, capacity analysis models can accurately capture complexity such as rework, batch processing, and re-entrant flows. Queueing models can improve upon spreadsheet models by including variability with arrival and service time distributions. Most of the results available in the area of queueing models, however, are only applicable to long-term, steady state behavior. Queueing models can thus be used to estimate long-run average cycle times and WIP, but not short-term behavior (although there is active research in this area, we have not seen any commercial applications of the methodology to date).

The results from analytical models are relatively easy to interpret, because they consist of a single number for each performance measure. However, the simplifying assumptions necessary to get closed form results are not always appropriate, especially in queueing models. Analytic models are best suited to strategic and tactical questions, where the emphasis for dynamic behavior is on relative performance.

Discrete event simulation can capture virtually any level of manufacturing detail, and is potentially very accurate. It captures both dynamic (time-dependent) and stochastic (random) behavior. Simulation is sometimes used in the early stages of a project to help understand how the system works. It has intuitive appeal for managers, especially when animation is used, because they can 'see' what is going on in the model. However, there are several disadvantages to using simulation models. They typically take much longer to run than analytic models, and the results from a simulation model can be difficult to interpret. Statistical analysis of the output is necessary, because each simulation run represents a single possible sample path. For these reasons, simulation can be an expensive option.

When simulation is chosen for a project, it is also necessary to decide between data-driven simulation models (simulators), and user-developed models written in a simulation or general-purpose programming language. In our opinion, languages are more appropriate for building small, detailed models, while simulators tend to be more applicable for modeling large-scale manufacturing systems. The decision also depends on the modeler's proficiency with different tools, and on whether or not the model will be reused.

Overall, tool choice for a project should be most heavily influenced by the needs of the client, particularly if the model is to be used on an on-going basis after the project is complete. Attention must be paid to what the client will be able to learn, use, and modify as needed. The planning horizon and the project goals will also influence which tool is appropriate. Clearly, if accurate cycle time estimation is critical, a spreadsheet model is probably not the best choice. If many different scenarios need to be evaluated quickly, with comparisons based on toolset capacity and cost rather than on cycle time, simulation may not be the best choice. We think that it is important to remember that simulation is not the only tool available. In many cases, simulation may not be required at all. The goal of the project is to answer the questions posed by the client, using the simplest model.

For the first project in this paper, we would recommend a spreadsheet or capacity analysis model. The second project is better suited to a hybrid capacity analysis and simulation model. The third probably requires a simulation for evaluating potential dispatch rules. Analytic models might also be embedded in short-term dispatch support systems.

4.2 Developing the Model and Collecting Information

Once the tool and methodology have been selected, the basic structure of the model should be constructed. This requires gathering information about the real system, including things like the total number of products, the number of tools, the general sequence of steps followed by operators at the different machines, and the structure of the material handling system. The analyst should rely on the project specification to resolve disputes about what the model should be able to do and what it will not be expected to do. Tools that grow haphazardly over time, as different functionalities are included in response to input from various sources, are extremely difficult to use. At this stage, it is important to keep in mind the overall goals of the model. The actual detailed steps followed in model development depend on the tool being used.

4.3 Populating Model Data

Perhaps the biggest roadblock to building models for real manufacturing applications lies in the difficulty of obtaining the proper data. In our experience, there are three classes of data: 1) the data you want; 2) the data you need; and 3) the data you get. These sets are typically decreasing in size, but are not necessarily overlapping. That is, data may be provided that is not needed for the model, even as data critical to building the model is unavailable. One suggestion is to find out what types of data will be available before spending much effort designing and developing the model. In many cases, the accuracy level of the data does not justify a complex model. If some data is simply not available, it may be possible to perform sensitivity analysis to see how important the missing data is. Only if the data will likely have a significant impact on the model are new data collection efforts necessary. Carrying this approach to its extreme, it is sometimes advantageous to build a model with a bare minimum of actual data. At that point, sensitivity analysis can be used to determine the priority areas for data collection.

Of our three sample projects, only the third is likely to have very detailed, accurate data. Of course, even with a detailed shop floor control system, some information may not be entered into the system at all, or may be entered inaccurately. In some manufacturing facilities, detailed information is only kept as it relates to the bottleneck pieces of equipment. In such cases, it may make more sense to only build a detailed model of the top few bottlenecks. For the other two projects, it is likely that any existing data will be stored in spreadsheet form. Often, data for the same factory will be stored in a variety of spreadsheets, and these spreadsheets will be maintained by several different employees.

4.4 Verifying and Validating the Model

We distinguish between model verification and model validation as follows: verification is the process of ensuring that the model produces a correct output given a specified input, while validation is the process of ensuring the model accurately represents reality to the necessary degree (100% accurate representation is usually not required for a successful project).

For verification purposes, we recommend beta-testing the model early and often. These tests should attempt to produce solid, repeatable results with the smallest possible set of input data. One way to assist this process over time is to build up a library of verification tests, the expected results of which are known in advance. Whenever possible, such verification suites should be automated. This reduces any temptation to cut corners. Each model should also be stress-tested with a subset of actual data from the facility being modeled (if possible). The analyst should make changes to the input data and examine the results to ensure that outputs move in the expected direction, and that no counter-intuitive situations arise. Counter-intuitive results can be significant, however, since they sometimes provide important new insights into system behavior.

4.5 Verifying and Validating the Data

As with model beta-testing, we recommend starting data verification and validation efforts as early as possible in the modeling process. An example of data verification might involve having end-users check data forms or data summary charts to see that data has been entered correctly. An example of validation would be reviewing outputs with end-users to ensure that results look reasonable. Often, it will take numerous passes before bugs in the data are completely worked out. For existing factories, another method of validation is a variant on the Turing test (see Schruben 1980). This involves taking model output data and transforming it into the same format as typical manufacturing reports. If manufacturing personnel cannot distinguish between the model output and actual factory output, then the model is probably a reasonable representation of reality. Tool utilization, cycle time, and production volumes are possible candidates for this method, as they are likely to be included on existing production reports. In some cases, there may be other models that can be used for validation. For example, in our second project, a new capacity and simulation model could be validated against the existing spreadsheet-based capacity model.

5 PROJECT DEPLOYMENT

Frequently, model development and debugging consume the lion's share of the project schedule. This is in spite of the fact that the model itself is not the project's end goal. The goal is to use the model to answer the questions outlined in the project specification. For most simulation projects, this requires analyzing the simulation output, presenting the results to management, and setting up the model for on-going data collection and analysis (if needed). These topics are outlined below, followed by a brief discussion of common simulation project pitfalls.

5.1 Analyzing Steady-State Simulation Output

Often, simulation projects do not require steady-state analysis. When required, however, steady-state analysis presents a variety of difficulties not found in finite-horizon analysis. Simulation texts usually treat this topic in depth (see, for example, Law and Kelton 1991). Nelson (1992) also provides a helpful overview of steady-state analysis. Two features often found in manufacturing simulations cause special difficulties — highly correlated output data and initialization bias.

Highly correlated data makes it difficult to produce within-run confidence intervals using traditional statistical methods. If the correlation is not taken into account, the width of confidence intervals is usually biased on the low side. This bias can cause modelers to predict significant differences between alternatives where none exist. It may be possible to batch the output data in such a way as to reduce the serial correlation to a negligible level. Otherwise, the best way to avoid the problem of correlated data is to make independent replications and use cross-replication results to generate confidence intervals. Using multiple replications, however, can aggravate initialization bias problems.

Initialization bias is present in any simulation where the initial simulation state is not drawn randomly from its steady-state distribution. Since this steady-state distribution is never known exactly (otherwise, the problem would be trivial), and may not exist at all, initialization bias is present to some extent in nearly all steady-state manufacturing simulations. Since this bias usually dies away over time, one popular method is to run the simulation for a very long time and then truncate the first portion of the output. Another option is to initialize the factory with a certain amount of work-in-process inventory. With the former method, the difficulty lies in deciding how much to throw away; with the latter method, it lies in deciding initial WIP levels.

Neither method is entirely satisfactory, but two qualitative observations are in order. First, the best way to determine if initialization bias is a problem is to look at time series output. For example, Figure 3 displays time series output for cycle time in a manufacturing simulation. From this chart, it appears there is negligible initialization bias. Figure 4, displays cycle time output that appears to contain significant initialization bias.

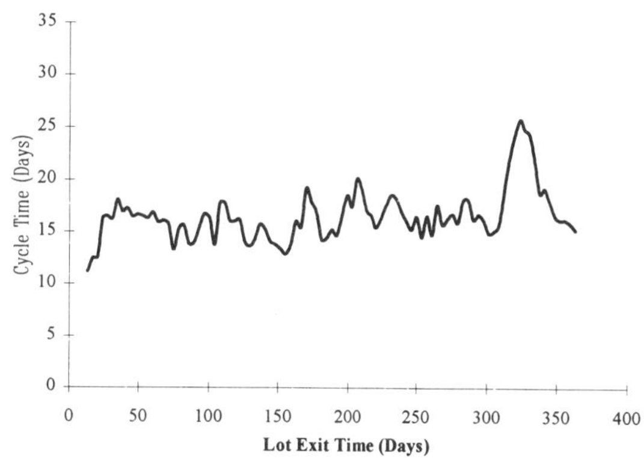

Figure 3: Cycle Time vs. Lot Exit Time for Run with Negligible Initialization Bias

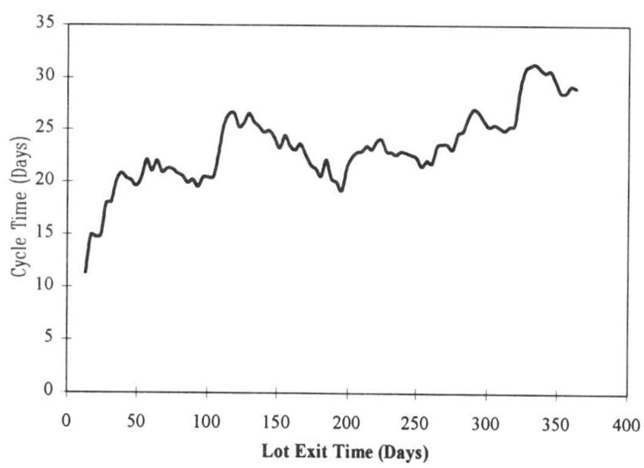

Figure 4: Cycle Time vs. Lot Exit Time for Run with Significant Initialization Bias

Our second qualitative observation is that the effect of initialization bias in manufacturing simulations generally increases as utilization rises. Thus, a truncation point that successfully eliminates most initialization bias for a highly loaded system will generally work quite well for more lightly loaded systems. In fact, the two series shown above are from simulations of the same model. Figure 3 displays output from a 65% utilization run, while Figure 4 displays output from a 95% utilization run. Not only does the second run have a higher cycle time, as we would expect, it also appears to have a longer transient effect.

To set a truncation point, visually inspect output charts of data averaged across multiple replications. This averaging will smooth out random fluctuations and will make the trend easier to pick out. For example Figure 5 displays the cycle time output series for 95% utilization averaged across ten replications. From this chart, it appears that a truncation point of six months or more is required. At this point, we would recommend making multiple replications of somewhat longer runs to confirm that six months is in fact a reasonable truncation point.

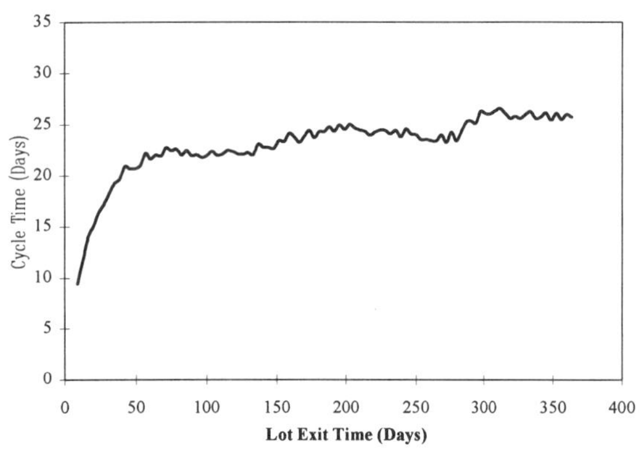

Figure 5: Cycle Time vs. Lot Exit Time Averaged Across 10 Independent Replications

When it is not possible to visually examine all relevant output, one can employ a statistical test, such as the one described by Schruben, Singh, and Tierney (1983).

5.2 Presenting Simulation Results to Management

When presenting results to management, there are several things to keep in mind that can increase the likelihood of model acceptance. First, display results graphically whenever possible. This makes it easier to understand trade-offs in a short amount of time. The graphs should follow the formats identified in the project specification. Second, present only a few vital pieces of output. Do not waste managers' time by forcing them to wade through a myriad of details. In particular, make sure that the results presented answer the questions asked in the original project specification. Finally, whenever possible present results in terms of dollars. Managers have a ready grasp of the bottom line. And, ultimately, project success will be judged in terms of tangible financial results.

5.3 Maintaining an On-Going Model

Considerably more work is required to build a model that will be used on an on-going basis. Such a model must be robust enough to be used by different people, and flexible enough to be changed over time. Systems that will be used on a daily basis will probably have to be integrated directly into existing systems. For example, a scheduling system, to be effective, must receive at least periodic updates from the shop floor control system. A strategic planning model, such as the one identified in our first example, is more likely to exist as a stand-alone model. A model like the one in the second example, with a single process flow from an existing product, could be integrated with the Excel capacity planning model. This would have the advantage of being able to automatically update the simulation model when changes were made to the Excel model. If the model is really only going to be used once, such integration is probably not worth the effort. However, models intended for a single use often end up being used on an on-going basis. This can lead to much more work than integrating the models in the first place, and can be the source of many data maintenance nightmares.

5.4 Common Simulation Project Pitfalls

Simulation projects can fail for a variety of reasons. Perhaps the most common problem is lack of a good project design. Models are built without regard to the questions that they will answer, and then they are not able to adequately provide the answers that are needed. Models built for single-use often end up being maintained and used over time. If they have not been designed to be used by someone other than the model developer, inaccurate results can be generated. Lack of a clear design can also result in insufficient buy-in and participation from the client.

Another common problem is data. Sometimes the necessary data does not exist, or is prohibitively expensive to obtain. Other times inaccurate data is used through oversight, or lack of clear lines of responsibility for collecting the data. A prevalent problem with simulation models is that they are too detailed. As a result, they are not maintained, because data collection is such an effort, or they are not run at all because the runs take such a long time.

A more subtle reason why simulation projects sometimes fail is that the performance measures optimized in the model are different from the performance measures rewarded in real life. For example, few simulation models will be able to motivate workers to reduce cycle time when pay standards are based upon throughput.

In general, we believe that more of these problems stem from project management / environmental issues than from lack of simulation expertise on the part of the analyst.

6 SUMMARY

Despite the cautionary tone we have taken in this paper, we believe it is possible to successfully use simulation in a manufacturing environment. Based on our experience, however, doing so requires more than just expertise in simulation. First and foremost, it requires that the analyst work with the client to prepare a written project specification. This document should clearly spell out project customers, goals, and deliverables. Once goals are established (along with suitable performance measures), the analyst should consider the entire range of available analysis tools. Depending on the project goals, a non-simulation methodology may be preferable. Even within the class of simulation tools, there is usually a spectrum of choices ranging from data-driven simulators to full-blown simulation languages. No matter the choice of tool, data-related problems should never be ignored or underestimated, as they often cause significant delays in model development. Model completion, however, does not guarantee project success. That comes only when the model is used to answer the questions it was built to satisfy. Then, and only then, will simulation be viewed by management as an effective decision-support tool.

REFERENCES

Law, A. M., and W.D. Kelton. 1991. *Simulation Modeling and Analysis.* McGraw-Hill.

Leachman, R., R. F. Benson, C. Liu, and D. J. Rarr. 1996. IMPReSS: An Automated Production Planning and Delivery Quotation System at Harris Corporation - Semiconductor Sector. *Interfaces* 26#1 (Jan/Feb): 6-37.

Nelson, B. L. 1992. Statistical Analysis of Simulation Results. In *Handbook of Industrial Engineering*, 2nd Edition: 2567-2593.

Schruben, L. W. 1980. Establishing the Credibility of Simulation. *Simulation* 34: 101-105.

Schruben, L. W., H. Singh, and L. Tierney. 1983. Optimal Tests for Initialization Bias in Simulation Output. *Operations Research* 31: 1167-1178.

AUTHOR BIOGRAPHIES

FRANK CHANCE is a specialist in the modeling and simulation of complex manufacturing facilities. He is the author of the Delphi and Factory Explorer™, integrated capacity, cost, and simulation analysis tools. He is the founder and principal of Chance Industrial Solutions (CIS), which distributes products and consulting services through Wright Williams and Kelly. During 1994-1995, he was a member of the SEMATECH MIMAC (Measurement and Improvement of Manufacturing Capacity) team. Dr. Chance holds M.S. (1991) and Ph.D. (1993) degrees in Operations Research from Cornell University. He has taught at Cornell University and was a visiting assistant professor at the University of California, Berkeley, prior to founding CIS.

JENNIFER ROBINSON is an independent consultant with experience in capacity analysis, cost modeling, and simulation of manufacturing facilities. During 1993 and 1994 she worked for the Operational Modeling department at SEMATECH, and was a member of the MIMAC (Measurement and Improvement of Manufacturing Capacity) project team. Her consulting clients have also included Digital Equipment Corporation, Chance Industrial Solutions, and Wright Williams and Kelly. She received an M.S. degree in Operations Research from the University of Texas at Austin in 1992, and is currently working towards a Ph.D. degree in Industrial Engineering at the University of Massachusetts at Amherst. Her research is in the area of performance analysis of manufacturing systems, with emphasis on queueing network models.

JOHN W. FOWLER is an Assistant Professor in the Industrial and Management Systems Engineering Department at Arizona State University. Prior to his current position, he was a Senior Member of Technical Staff in the Modeling, CAD, and Statistical Methods Division of SEMATECH and an Adjunct Assistant Professor in the Graduate Program in Operations of the Mechanical Engineering Department at the University of Texas at Austin. He received his Ph.D. in Industrial Engineering from Texas A&M University and spent the last 1.5 years of his doctoral studies as an intern at Advanced Micro Devices. His research interests include modeling, analysis, and control of manufacturing (especially semiconductor) systems. He is a member of IIE, INFORMS, and SCS.

Proceedings of the 1996 Winter Simulation Conference
ed. J. M. Charnes, D. J. Morrice, D. T. Brunner, and J. J. Swain

BATCHING METHODS IN SIMULATION OUTPUT ANALYSIS: WHAT WE KNOW AND WHAT WE DON'T

Bruce W. Schmeiser

School of Industrial Engineering
Purdue University
West Lafayette, Indiana 47907-1287, U.S.A.

Wheyming Tina Song

Department of Industrial Engineering
National Tsing Hua University
Hsinchu, Taiwan 300, REPUBLIC OF CHINA

ABSTRACT

As an advanced tutorial, we discuss batching methods for determining point-estimator precision for steady-state simulation experiments. We emphasize batching methods in which each batch provides a point estimator analogous to that of the experiment, but we mention other methods that use batches, especially the more-general idea of standardized time series. Despite the preponderance of literature on confidence-interval estimation for the mean using adjacent nonoverlapping batches, we focus on estimating the point estimator's standard error and consider both general point estimators and general batching relationships. Literature on multivariate batching exists, but we focus on the univariate problem. We consider the initial-transient problem only in passing. Specific issues include form of the point estimator, definition of the batch statistics, form of the batch-statistics estimator, optimal batch size (including various definitions of *optimal*), and determining batch size. This paper is a short summary of the issues, with a fairly complete bibliography.

1 INTRODUCTION

Consider a simulation experiment that produces steady-state data $\{Y_i; i = 1, 2, ...\}$. The purpose of the experiment is to estimate θ, a property of the steady-state distribution function F_Y. The property θ is a performance measure of the model being simulated, often a mean or variance or quantile.

2 POINT ESTIMATION

Point estimation is typically straightforward. Early observations whose distributions might differ substantially from F_Y are weighted lightly, often discarded entirely. The point estimator, $\widehat{\theta}$, is chosen to mimic θ, typically the sample mean, sample variance, or a simple function of the relevant order statistics. Although sometimes point estimators are defined to be the average of batch statistics, the grand estimator (obtained by using all of the data at once) is often simpler to compute and has better statistical properties, in particular smaller bias. (The average of batch means is the grand mean, so in that special important case the point estimators are identical.) The purpose of batching is to estimate the precision of the point estimator, not to improve the point estimator.

3 BATCH STATISTICS

The ith *batch* is composed of the observations $Y_i, ..., Y_{i+m-1}$ for $i = 1, 2, ..., n-m+1$, where m is the batch size and n is the run length. The ith batch statistic $\widehat{\theta}_i$ is a miniature version of $\widehat{\theta}$, again typically a sample mean, sample variance, or simple function of the relevant order statistics. When appropriate, a grand estimator can be used in the batch statistic; for example, centering a batch variance on the grand mean is better than centering it on the batch mean.

4 STANDARD-ERROR ESTIMATION

Batch statistics are combined to estimate the variance of $\widehat{\theta}$, or its square root the standard error. The batch-statistics estimator for the variance is

$$\widehat{\text{Var}}_A[\widehat{\theta}] = \frac{\sum_A (\widehat{\theta}_i - \widehat{\theta})^2}{d|A|},$$

where d is a function of n and m, chosen, for example, to yield $\text{E}\widehat{\text{Var}}_A[\widehat{\theta}] = \text{Var}_A[\widehat{\theta}]$ for independent data. A is a subset of $\{1, 2, ..., n-m+1\}$ and $|A|$ is the cardinality of A. Not surprisingly, the choice of A has little effect on bias, and variance is minimized if $A = \{1, 2, ..., n-m+1\}$, the case of overlapping batch statistics. Maybe surprisingly, computation of batch means and batch variances is $O(n)$; batch quantiles require $O(n \ln n)$ computation.

Rather than thinking of $\widehat{\mathrm{Var}}_A[\widehat{\theta}]$ as an estimator of $\mathrm{Var}_A[\widehat{\theta}]$, one can think of estimating the asymptotic constant $n\,\mathrm{lim}\mathrm{Var}_A[\widehat{\theta}]$. The change in the batch-statistics estimator is simple: multiply by n. But this relationship is true only asymptotically, and the difference can be noticeable for surprisingly long runs.

5 OPTIMAL BATCH SIZE

Optimal batch size is usually discussed in one of two contexts: confidence-interval estimation or standard-error estimation. Criteria for confidence-interval procedures include probability of covering θ, expected interval length, variance of interval length, and (more generally) the probability of covering an arbitrary point θ_0. Because confidence-interval estimation is inherently multi-criterian, seemingly contradictory advice abounds in the literature. Choosing ten to thirty batches is fine for fixed n if only coverage probability and expected length are considered, but does not yield the consistency needed for the asymptotic results underlying sequential methods. And more batches are useful when using control variates or obtaining smaller variance for the confidence-interval length.

Standard-error estimation is simpler because it allows a single criterion. The usual criterion is the mean squared error (mse) of $\widehat{\mathrm{Var}}_A[\widehat{\theta}]$. If $\widehat{\theta}$ is the sample mean, then asymptotic arguments lead to an mse-optimal batch size

$$\vec{m}^* = \left(2n\frac{c_{\mathrm{b}}^2}{c_{\mathrm{v}}}\frac{\gamma_1}{\gamma_0}\right)^{1/3} + 1,$$

where c_{b}^2 and c_{v} are the known bias constant and variance constant of the batch-statistics estimator, $\gamma_j = \sum_{h=-\infty}^{\infty} |h|^j \,\mathrm{Corr}(Y_i, Y_{i+h})$, and the additive constant is chosen to yield batch size of one for independent data. Much less is known when $\widehat{\theta}$ is not a sample mean, but that $\vec{m}^*$ grows with the cube root of n seems quite general.

6 DETERMINING BATCH SIZE

Many papers over the last few decades have suggested methods for choosing the number of adjacent nonoverlapping batches to obtain reasonable confidence intervals for the mean, often based on tests of independence. In the last few years papers have appeared on estimating $\vec{m}^*$, the central problem being how to efficiently estimate γ_1/γ_0, which can be interpreted as the autocorrelation's center of gravity. Very recently some papers have considered mse-optimal batch sizes for specific point estimators that are not means. Given n saved observations, $O(n)$ batch-means algorithms exist that can be used without tuning; no such algorithms exist for general point estimators.

7 WHAT WE DON'T KNOW

Despite knowing quite a lot about batch means, no method is available for estimating the variance of $\bar{Y}$ when the data come one at a time and are not saved. Such a method should use finite memory and somehow increase batch size as the run length increases. It should not need to know the run length *a priori.*

Little is known about batch sizes when the point estimator is not $\bar{Y}$. Creating a literature for each possible point estimator seems daunting. General methods will probably be created from two general results: that $\mathrm{Var}_A[\widehat{\theta}]$ is $O(n^{-1})$ and that $\vec{m}^*$ is $O(n^{1/3})$.

How to transfer the technology is also an open question. In practice, there are often tens of performance measures θ, all needing a point estimator and associated measure of precision. The practitioner has not the time, and often not the knowledge, to use any method that is not completely automated. Batch means has an advantage compared to other methods of estimating point-estimator precision because it generalizes so well to non-mean point estimators, which might allow the same method to be used for all point estimators.

More fundamentally, the research community does not agree on a set of criteria for good algorithms, and the research community and the practitioner community too seldom discuss the criteria. For example, the authors think that a good beginning would be the creation of an automated method to allow statistically meaningless point-estimator digits to be deleted from simulation output reports. Even such a minor suggestion for indicating point-estimator precision in practice leads, however, to interesting, non-trivial issues.

ACKNOWLEDGMENTS

This work was supported in part by NSF Grant DDM–93–00058 to Purdue University and by NSC Grant 84–0415–E–007–012 to the National Tsing Hua University.

BIBLIOGRAPHY

Adam, N. R. 1983. Achieving a confidence interval for parameters estimated by simulation. *Management Science* 29: 856–866.

Bischak, D. P. 1988. *Weighted batch means for improved confidence intervals for steady-state processes.* Ph.D. Thesis, Department of Industrial and Operations Engineering, The University of Michigan.

Bischak, D. P., W. D. Kelton and S. M Pollock. 1993. Weighted batch means for confidence intervals in steady-state simulations. *Management Science* 39: 1002–1019.

Blackman, R. B. and J. W. Tukey. 1958. *The Measurement of Power Spectra.* New York: Dover.

Brillinger, D. R. 1973. Estimation of the mean of a stationary time series by sampling. *Journal of Applied Probability* 10: 419–431.

Carlstein, E. 1986. The use of subseries for estimating the variance of a general statistic from a stationary sequence. *Annals of Statistics* 14:1171–1179.

Ceylan, D. 1995. *Variance and quantiles in dynamic-system performance: point estimation and standard errors.* Ph.D. Thesis, Department of Industrial Engineering, Purdue University.

Ceylan, D. and B. Schmeiser. 1993. Interlaced variance estimators. In *Proceedings of the Winter Simulation Conference*, ed. G. W. Evans, M. Mollaghasemi, E. C. Russell and W. E. Biles, 1382–1383. Piscataway, New Jersey: IEEE.

Charnes, J. M. 1989. *Statistical analysis of multivariate discrete-event simulation output.* Ph.D. Thesis, Department of Operations and Management Science, University of Minnesota.

Charnes, J. M. 1990. Power comparisons for the multivariate batch-means method. In *Proceedings of the Winter Simulation Conference*, ed. O. Balci, R. P. Sadowski and R. E. Nance, 281–287. Piscataway, New Jersey: IEEE.

Charnes, J. M. and W. D. Kelton 1988. A comparison of confidence region estimators for multivariate simulation output. In *Proceedings of the Winter Simulation Conference*, ed. M. A. Abrams, P. L. Haigh and J. C. Comfort, 458–465. Piscataway, New Jersey: IEEE.

Chen, R. D. and A. F. Seila. 1987. Multivariate inference in stationary simulation using batch means. In *Proceedings of the Winter Simulation Conference*, ed. A. Thesen, H. Grant and W. D. Kelton, 302–304. Piscataway, New Jersey: IEEE.

Chien, C. 1988. Small-sample theory for steady state confidence intervals. In *Proceedings of the Winter Simulation Conference*, ed. M. A. Abrams, P. L. Haigh and J. C. Comfort, 408–413. Piscataway, New Jersey: IEEE.

Chien, C. 1989. *Small sample theory for steady-state confidence intervals.* Ph.D. Thesis, Department of Operations Research, Stanford University.

Chien, C. 1994. Batch size selection for the batch means method. In *Proceedings of the Winter Simulation Conference*, ed. J. D. Tew, S. Manivannan, D. A. Sadowski and A. F. Seila, 345–352. Piscataway, New Jersey: IEEE.

Chien, C., D. M. Goldsman and B. Melamed. 1996. Large-sample results for batch means. *Management Science*, forthcoming.

Chun, Youngsoo. 1989. *Time-series models of batch-means processes in simulation analysis.* Ph.D. Thesis, Department of Operations and Management Science, University of Minnesota.

Conway, R. W. 1963. Some tactical problems in digital simulation. *Management Science* 10:47–61.

Damerdji, H. 1987. *Topics in discrete-event stochastic systems.* Ph.D. Thesis, Department of Industrial Engineering, University of Wisconsin—Madison.

Damerdji, H. 1991. Strong consistency and other properties of the spectral variance estimator. *Management Science* 37: 1424–1440.

Damerdji, H. 1993. Mean-square consistency of the variance estimator in steady-state simulation output analysis. *Operations Research* 43:282–291.

Damerdji, H. 1994a. Strong consistency of the variance estimator in steady-state simulation output analysis. *Mathematics of Operations Research* 19:494–512.

Damerdji, H. 1994b. On the batch means and area variance estimators. In *Proceedings of the Winter Simulation Conference*, ed. J. D. Tew, S. Manivannan, D. A. Sadowski and A. F. Seila, 340–344. Piscataway, New Jersey: IEEE.

Damerdji, H. and M. K. Nakayama. 1996. Two-stage procedures for multiple comparisons with a control in steady-state simulations. In *Proceedings of the Winter Simulation Conference*, ed. J. Charnes, D. Morrice, D. Brunner and J. Swain. Piscataway, New Jersey: IEEE.

Fishman, G. S. 1967. Problems in the statistical analysis of simulation experiments: the comparison of means and the length of sample records. *Communications of the ACM* 10:94–99.

Fishman, G. S. 1978. Grouping observations in digital simulation. *Management Science* 24: 510–521.

Fishman, G. S. and L. S. Yarberry. 1993. An implementation of the batch means method. Technical-Report UNC/OR/TR/93-1, Department of Operations Research, University of North Carolina, Chapel Hill, North Carolina.

Fox, B. L., D. M. Goldsman and J. J. Swain. 1991. Spaced batch means. *Operations Research Letters* 10: 255-266.

Glynn, P. W. and D. L. Iglehart. 1990. Simulation output analysis using standardized time series. *Mathematics of Operations Research* 15:1–16.

Glynn, P. W. and W. Whitt. 1991. Estimating the asymptotic variance with batch means. *Operations Research Letters* 10:431–435.

Goldsman, D. M. 1984. *On using standardized time series to analyze stochastic processes.* Ph.D. Thesis, Department of Operations Research and Industrial Engineering, Cornell University.

Goldsman, D. M. and M. S. Meketon. 1986. A comparison of several variance estimators for stationary increment stochastic processes. Technical Report, School of Industrial and Systems Engineering, Georgia Institute of Technology.

Goldsman, D., M. Meketon, and L. W. Schruben. 1990. Properties of standardized time series weighted area variance estimators. *Management Science* 36:602-612.

Goldsman, D. M. and L. W. Schruben. 1984. Asymptotic properties of some confidence interval estimators for simulation output. *Management Science* 30:1217–1225.

Goldsman, D. M., B. L. Nelson and B. Schmeiser. 1991. Methods for selecting the best system. In *Proceedings of the Winter Simulation Conference*, ed. B. L. Nelson, W. D. Kelton and G. M. Clark, 177–186. Piscataway, New Jersey: IEEE.

Kang, K. 1984. *Confidence Interval Estimation via Batch Means and Time Series Modeling.* Ph.D. Thesis, Department of Industrial Engineering, Purdue University.

Kang, K. and D. M. Goldsman. 1990. The correlation between mean and variance estimators in computer simulation. *IIE Transactions* 22:15–23.

Kang, K. and B. Schmeiser. 1987. Properties of batch means from stationary ARMA time series. *Operations Research Letters* 6:19–24.

Kang K. and B. Schmeiser. 1990. Graphical methods for evaluating and comparing confidence-interval procedures. *Operations Research* 38:546–553.

Kim, Y. B. 1992. *Output analysis of single replication methods in simulation experiments.* Ph.D. Thesis, Department of Engineering Science, Rensselaer Polytechnic Institute.

Law, A. M. 1977. Confidence intervals in discrete event simulation: a comparison of replication and batch means. *Naval Research Logistics Quarterly* 24: 667-678.

Law, A. M. 1983. Statistical analysis of simulation output data. *Operations Research* 31: 983–1029.

Law, A. M. and J. S. Carson. 1979. A sequential procedure for determining the length of a steady-state simulation. *Operations Research* 27: 1011–1025.

Law, A. M. and W. D. Kelton. 1982. Confidence intervals for steady-state simulations, II: a survey of sequential procedures. *Management Science* 28, 550-562.

Law, A. M. and W. D. Kelton. 1984. Confidence intervals for steady-state simulations, I: a survey of fixed sample size procedures. *Operations Research* 32: 1221–1239.

Loh, W. W. 1994. *On the method of control variates*, Ph.D. Thesis, Department of Operations Research, Stanford University.

Mechanic, H. and W. McKay. 1966. Confidence intervals for averages of dependent data in simulation II. Technical report ASDD 17-202, IBM Corporation, Yorktown Heights, NY.

Meketon, M. S. and B. Schmeiser. 1984. Overlapping batch means: something for nothing? In *Proceedings of the Winter Simulation Conference*, ed. S. Sheppard, U. Pooch and C. D. Pegden, 227–230. Piscataway, New Jersey: IEEE.

Moran, P. A. P. 1975. The estimation of standard errors in Monte Carlo simulation experiments. *Biometrika* 62:1–4.

Muñoz, D. F. 1991. *Cancellation methods in the analysis of simulation output.* Ph.D. Thesis, Department of Operations Research, Stanford University.

Muñoz, D. F. and P. W. Glynn. 1991. Multivariate standardized time series for steady-state simulation output analysis. Technical report, Department of Operations Research, Stanford University.

Muñoz, D. F. and P. W. Glynn. 1997. A batch means methodology for estimation of a nonlinear function of a steady-state mean. *Management Science* 43, forthcoming.

Nelson, B. L. 1989. Batch size effects on the efficiency of control variates in simulation. *European Journal of Operational Research* 43: 184–196.

Nelson, B. L. 1990. Control variate remedies. *Operations Research* 43: 184–196.

Nelson, B. L. 1992. Statistical analysis of simulation results. In *Handbook of Industrial Engineering*, ed., G. Salvendy. New York: John Wiley.

Pedrosa, A. 1994. *Automatic batching in simulation output analysis.* Ph.D. Thesis, Department of Industrial Engineering, Purdue University.

Pedrosa, A. and B. Schmeiser. 1993. Asymptotic and finite-sample correlations between obm estimators. In *Proceedings of the Winter Simulation Conference*, ed. G. W. Evans, M. Mollaghasemi, E. C. Russell and W. E. Biles, 481–488. Piscataway, New Jersey: IEEE.

Pedrosa, A. and B. Schmeiser. 1994. Estimating the variance of the sample mean: Optimal Batch-Size Estimation and 1-2-1 Overlapping Batch Means. Technical Report SMS94-3, School of Industrial Engineering, Purdue University.

Percival, D. B. 1993. Three curious properties of the sample variance and autocovariance for stationary processes with unknown mean. *The American Statistician* 47:274–276.

Politis, D. N. and J. P. Romano. 1992a. A general resampling scheme for triangular arrays of α-mixing random variables with application to the problem of spectral density estimation. *Annals of Statistics* 20:1985–2007.

Politis, D. N. and J. P. Romano. 1992b. A circular block-resampling procedure for stationary data. In *Exploring the Limits of Bootstrap*, ed., R. LePage and L. Billard, 263-270. New York: John Wiley

Politis, D. N. and J. P. Romano. 1993. On the sample variance of linear statistics derived from mixing sequences. *Stochastic Processes and Their Applications* 45:155–167.

Politis, D. N. and J. P. Romano. 1994a. Large sample confidence regions based on subsamples under minimal assumptions. *Annals of Statistics* 22:2031–2050.

Politis, D. N. and J. P. Romano. 1994b. The stationary bootstrap. *Journal of the American Statistical Association* 89:1303–1313.

Politis, D. N. and J. P. Romano. 1995. Bias-corrected nonparametric spectral estimation. *Journal of Time Series Analysis* 16:67–104.

Sargent, R. G., K. Kang and D. Goldsman. 1992. An investigation of finite-sample behavior of confidence interval estimates. *Operations Research* 40: 898–913.

Schafer, R. E. 1974. On assessing the precision of simulations. *Journal of Statistical Computation and Simulation* 3: 67–69.

Schmeiser, B. 1982. Batch size effects in the analysis of simulation output. *Operations Research* 30:556–568.

Schmeiser, B. 1990. Chapter 7: Simulation experiments. In *Handbooks in Operations Research and Management Science, Volume 2: Stochastic Models*, ed. D. P. Heyman and M. J. Sobel, 295–330. Amsterdam: North-Holland.

Schmeiser, B. 1992. Modern simulation environments: Statistical issues. In *Proceedings of the First IE Research Conference*, ed. G.-A. Klutke, D. Mitta, B. Nnaji and L. Seiford, 139–144. Norcross, Georgia: Institute of Industrial Engineers.

Schmeiser, B., T. Avramidis and S. Hashem. 1990. Overlapping batch statistics. In *Proceedings of the Winter Simulation Conference*, ed. O. Balci, R. P. Sadowski and R. E. Nance, 395–398. Piscataway, New Jersey: IEEE.

Schmeiser, B. and M. D. Scott. 1991. SERVO: Simulation experiments with random-vector output. *Proceedings of the Winter Simulation Conference*, ed. B. Nelson, W. D. Kelton and G. Clark, 927–936. Piscataway, New Jersey: IEEE.

Schmeiser, B. and W. T. Song. 1987. Correlation among estimators of the variance of the sample mean. In *Proceedings of the Winter Simulation Conference*, ed. A. Thesen, H. Grant and W. D. Kelton, 309–317. Piscataway, New Jersey: IEEE.

Schriber, T. J. and R. W. Andrews. 1979. Interactive analysis of simulation output by the method of batch means. In *Proceedings of the Winter Simulation Conference*, ed. Highland, H. J., M. G. Spiegel and R. Shannon, 513–525. Piscataway, New Jersey: IEEE.

Schruben, L. W. 1980. A coverage function for interval estimators of simulation response. *Management Science* 26: 18–27.

Schruben, L. W. 1983. Confidence interval estimation using standardized time series. *Operations Research* 31: 1090–1108.

Seila, A. F. 1982. A batching approach to quantile estimation in regenerative simulations. *Management Science* 28:573–581.

Sherman, M. 1995a. Variance estimation for statistics computed from spatial lattice data. *Journal of the Royal Statistical Society B* 58: 509-523 (1996).

Sherman, M. 1995b. On batch means in the simulation and statistics communities. In *Proceedings of the Winter Simulation Conference*, ed. C. Alexopoulos, K. Kang, W. Lilegdon, and D. Goldsman, 297–303. Piscataway, New Jersey: IEEE.

Sherman, M. 1996. On Databased Choice of Batch Size for Simulation Output Analysis. Technical Report 250, Department of Statistics, Texas A&M University.

Sherman, M. and E. Carlstein. 1996. Replicate Histograms. *Journal of the American Statistical Association* 91:566-57.

Simonoff, J. S. 1993. The relative importance of bias and variability in the estimation of the variance of a statistic. *The Statistician* 42:3–7.

Song, W. T. 1988. *Estimators of the variance of the sample mean: quadratic forms, optimal batch sizes, and linear combinations.* Ph.D. Thesis, Department of Industrial Engineering, Purdue University.

Song, W. T. 1996. On the estimation of optimal batch sizes in the analysis of simulation output analysis. *European Journal of Operational Research* 88:304–309.

Song, W. T. and B. Schmeiser. 1988a. On the dispersion matrix of estimators of the variance of the sample mean in the analysis of simulation output. *Operations Research Letters* 7:259–266.

Song, W. T. and B. Schmeiser. 1988b. Minimal-mse linear combinations of variance estimators of the sample mean. In *Proceedings of the Winter Simulation Conference*, ed. M. A. Abrams, P. L. Haigh and J. C. Comfort, 414–421. Piscataway, New Jersey: IEEE.

Song, W. T. and B. Schmeiser. 1988c. Estimating standard errors: empirical behavior of asymptotic mse-optimal batch sizes. In *Computing Science and Statistics: Proceedings of the* 20th *Symposium on the Interface*, ed. E.J. Wegman, D. T. Gantz, and J. J. Miller, 575–580. Alexandria, Virginia: American Statistical Association.

Song, W. T. and B. Schmeiser. 1993. Variance of the sample mean: Properties and graphs of quadratic-form estimators. *Operations Research* 41:501–517.

Song, W. T. and B. Schmeiser. 1994. Reporting the precision of simulation experiments. In *New Directions in Simulation for Manufacturing and Communications*, ed. S. Morito, H. Sakasegawa, K. Yoneda, M. Fushimi, and K. Nakano, 402–407. Tokyo: Operations Research Society of Japan.

Song, W. T. and B. Schmeiser. 1995. Optimal mean-squared-error batch sizes. *Management Science* 41:110–123.

Walsh, J. E. 1956. Questionable usefulness of variance for measuring estimate accuracy in Monte Carlo importance sampling problems. In *Symposium on Monte Carlo Methods*, ed. H. A. Meyer, 141–144. New York: John Wiley.

Welch, P. D. 1987. On the relationship between batch means, overlapping batch means and spectral estimation. In *Proceedings of the Winter Simulation Conference*, ed. A. Thesen, H. Grant and W. D. Kelton, 309–317. Piscataway, New Jersey: IEEE.

Wood, D. Ceylan and B. Schmeiser. 1994. Consistency of overlapping batch variances. In *Proceedings of the Winter Simulation Conference*, ed. J. Tew, S. Manivannan, D. Sadowski, and A. Seila, 316–319. Piscataway, New Jersey: IEEE.

Wood, D. Ceylan and B. Schmeiser. 1995. Overlapping batch quantiles. In *Proceedings of the Winter Simulation Conference*, ed. C. Alexopoulos, K. Kang, D. Goldsman, and W. Lilegdon, 303–308. Piscataway, New Jersey: IEEE.

Yang, W. N. and B. L. Nelson. 1988. Multivariate estimation and variance reduction in terminating and steady-state simulation. In *Proceedings of the Winter Simulation Conference*, ed. M. A. Abrams, P. L. Haigh and J. C. Comfort, 466-472. Piscataway, New Jersey: IEEE.

Yarberry, L. S. 1993. *Incorporating a dynamic batch size selection mechanism in a fixed-sample-size batch means procedure.* Ph. D. thesis, Department of Operations Research, University of North Carolina, Chapel Hill.

AUTHOR BIOGRAPHIES

BRUCE W. SCHMEISER is a professor in the School of Industrial Engineering at Purdue University. He received his Ph.D. from the School of Industrial and Systems Engineering at the Georgia Institute of Technology in 1975. His primary research interests are the probabilistic and statistical aspects of digital-computer stochastic simulation. He is an active participant in the Winter Simulation Conference, including being Program Chair in 1983 and chairing the Board of Directors during 1988–1990. He is also active in the American Statistical Association, IIE, INFORMS, and Omega Rho, The Operations Research Honor Society.

WHEYMING TINA SONG is a professor in the Department of Industrial Engineering at the National Tsing Hua University in Taiwan. She received a B. S. in statistics and an M. S. in industrial engineering from Cheng-Kung University in Taiwan, M. S. degrees in applied mathematics and industrial engineering from the University of Pittsburgh, and a Ph. D. in industrial engineering from Purdue University. Her research interests are the statistical aspects of simulation.

Proceedings of the 1996 Winter Simulation Conference
ed. J. M. Charnes, D. J. Morrice, D. T. Brunner, and J. J. Swain

PRINCIPLES OF CONSERVATIVE PARALLEL SIMULATION

David M. Nicol

Department of Computer Science
Dartmouth College
Hanover, NH 03755

ABSTRACT

This tutorial describes considerations in writing parallelized discrete-event simulations. We identify key principles behind various synchronization methods tailored to simulate complex systems.

1 INTRODUCTION

The use of parallel computers to execute discrete-event simulations has been a topic of active research interest for the last fifteen years. The seminal work by Chandy and Misra was done well before parallel computers were in common use, and (perhaps surprisingly), was principally motivated by concerns other than performance. What has become known as the "null-message" algorithm (Chandy and Misra, 1979) was developed as an example of a distributed algorithm to be validated; Chandy and Misra's later work on detecting and breaking deadlock also retained a strong flavor of algorithm validation. Jefferson's seminal Time Warp paper (Jefferson, 1985) had interests in proving the correctness of the method, but importantly was proposed as a way to avoid "blocking" inherent in the Chandy and Misra work, so as to improve upon available parallelism and possibly improve performance.

Since these early times, a great deal of research has gone into new algorithms for synchronization and optimizations for the old ones. So-called "conservative" algorithms have their philosophical base in the Chandy and Misra approach. A defining characteristic of a conservative approach is that no computation is ever done that might possibly be incorrect. A defining characteristic of so-called "optimistic" approaches is that some incorrect computations may be done, but the synchronization system is capable of automatically detecting and correcting such errors. Reynolds made a number of important refinements to this classification (Reynolds, 1988), for our purposes two stand out. Optimism has the components of *aggressiveness*—doing computations that may turn out to be incorrect—and *risk*—exporting the results of an incorrect computation on one processor to other processors. As we will see, it is possible, even desirable, to use a protocol where there is aggressiveness but not risk.

As medium-scale parallel processors become common, it is natural for a simulationist to ask "Should I use parallel simulation?", and if the answer is yes, to ask just how that simulation ought to be constructed. One option is to use a parallel simulation package, the other is to handle the synchronization issues yourself, in your own code. One principle disadvantage of using a package is that it may not work on the architectures available to you. Another is that these (few) packages are products of universities, so you cannot expect much support as you encounter difficulties or discover that the package does not allow you to to model quite the way you had intended. The principle advantage of parallel simulation packages is that the details of synchronization are largely hidden from you. The advantage of writing your own simulator from scratch is that you are fully aware of all the details, and can optimize the synchronization to the problem you are simulating. The disadvantage is that it is definitely more work to build your own synchronization mechanisms, and, depending on the type of synchronization you use, as your model evolves you made need to alter the synchronization strategy. Nevertheless, home-grown parallel simulators are viable for target applications where the semantics are well understood and unlikely to change, e.g., queueing networks, Petri nets.

There is an extremely important application area for roll-your-own parallel simulations emerging, the "High Level Architecture" (HLA) specifications that will determine how to construct a federated distributed simulation from cooperating simulators. The HLA specs principally call for conservative synchronization between individual simulators, and explicitly provide a mechanism for a simulator to report its "lookahead" to the controlling agent. Consequently, understanding the principles of conservative synchronization for that context will be quite important.

The purpose of this tutorial is to provide those who would build their own simulators with the concepts necessary to approach such a task.

2 DISCRETE EVENT SIMULATION

2.1 The DES Paradigm

Discrete-event simulation is a algorithmic paradigm that is suitable for emulating complex systems whose "state" is discrete, and which changes at discrete instants in time. Queueing networks or Petri nets are examples—their states are vectors of queue lengths or place markings, the state-vector changes when a job departs or a transition fires. It may happen that other information is also needed in the state, e.g., if a queue does not have exponential service then some record of residual service time must be part of the state information.

The key idea behind discrete-event simulation is to focus the computational effort of the program on only those instants where the state changes, and only at those portions of the state that change. This offers large computational savings over "time-stepping" methods that advance the entire state-vector at each of many small time-increments. Some sort of time-stepping is generally required to simulate so-called "continuous systems" whose state evolution behavior is described using differential equations. Whether one uses discrete-event or continuous simulation techniques depends entirely on the nature of the model. It is worth noting in this discussion that continuous simulation models frequently admit to parallelization in a much more straightforward fashion than discrete-event simulations, because the synchronization structure is much simpler.

2.2 World Views

Simulation modelers frequently describe models as having an "event" world view, or a "process" world view. In the event world view the behavior of the model is described in terms of how the simulation model changes when an event is processed. For instance, the code for an event-oriented queueing simulator will identify events such as *DepartQueue*, *ArriveQueue* and with each event will proscribe code that affects the simulation state as a result of the event occurring.

A process-oriented view expresses the logic of the simulation at a higher level of abstraction. In the case of the queueing network it might describe the simulation from the point of view of a job—join a queue, wait for service, join another queue, and so on. The critical difference is that the job's behavior can generally be expressed without reference to the state of the queue, something that is not possible in an event-oriented simulator. Details concerning contention resolution are buried—from the modeler's point-of-view—in what ever process-oriented simulation package is being used.

The choice of world view does have a significant impact on parallelization. By its nature, implementing process orientation is a delicate business. Process-oriented simulations are almost exclusively developed using software packages; to go parallel with a process-oriented view means either to parallelize these packages or write one's own parallelized process-oriented system. The later task can be quite daunting if one is to use optimistic synchronization. Such simulators need to be able to save and restore the state of the *simulator*, as well as the simulation model state. We have had success using conservative techniques with existing process oriented simulators, but in ways limited by the need to not change the base process-oriented simulator.

Most simulations written from scratch are event-oriented. This makes the simulator easier to write, but makes the development of simulation models somewhat more difficult. If one is tailoring the synchronization protocol to the simulation problem, then an event oriented view will almost certainly be the view of choice.

3 SYNCHRONIZATION

Fundamentally, the heart of a discrete-event simulation is an event-list (even process-oriented simulators use these in their internal structure). Causality is maintained in a serial simulator because the next event executed is always the one with least time-stamp in the entire simulation model. This ensures that any event that might alter state information upon which the next event depends has already been executed. A useful way to view a parallel simulation is as a set of cooperating serial simulators, each with its own event list. Now, however, an event executed by one simulator may change model state upon which events in another simulator depend, which is tantamount to one simulator changing the event list of another. One way to ensure causality is to prohibit a simulator from executing its next event until it is sure that no other simulator will cause an earlier event to be inserted, or the next event to be removed. Such an approach is said to be *conservative*. An *optimistic* method will not block a simulator; instead it goes ahead and executes the next event, but first saves enough state information so that it can return ("rollback") to that point again if it needs to. An optimistic system detects when a "straggler" event (one with a time-stamp that is smaller than the current simulation clock) is received, does a rollback to that time and undoes the effects of wrong computations it might have performed at times larger than that of the straggler message. Another important distinction is that it is possible to have one simulator execute optimistically but temporarily withhold affecting other simulators with its results until it is certain that its inter-processor interactions are correct. Such an approach is said to be "aggressive" but "risk-free".

The main question for a conservative method is how a simulator can determine if its next event is safe to process. There is no single answer, it depends on the simulation model. The general label for one's ability to extract such information from a model is called *lookahead*. The advantage to an aggressive but risk free approach is that it avoids the necessity for explicitly computing lookahead—it allows the computation to *observe* lookahead (in a way we'll make clear later)—but avoids disseminating incorrect information.

4 LOOKAHEAD

Lookahead is the key ingredient for all conservative synchronization methods. A loose definition is that it is the ability of a simulator to predict future behavior with respect to modifying the event lists of other simulators. More precisely, a process p has lookahead with respect to process q if p's simulation clock is at time s, and yet p can determine that under no circumstances will it insert or delete an event from process q's event list with time-stamp $t > s$. Even so, this definition does not capture the many aspects of lookahead, knowledge of which is important when developing a synchronization strategy for a given simulation problem. We now explore these aspects.

4.1 Subtleties of Lookahead

To help motivate a classification of lookahead, we first look at some situations that illustrate different facets of lookahead.

Models of just about any kind of network involve movement of objects (e.g., packets, jobs, parts, patients, vehicles, etc.) and contention among these objects for network resources. The way one process typically affects another is by sending it an object, after that object has acquired and used some resource. A concrete example of this is a queueing network. The network is partitioned among processors; a job contends for service, and after receiving it is routed to another another queue, possibly to one on a different processor.

Consider a resource mapped to process p, such that after an object uses that resource it migrates to contend for a resource that is mapped to a different process q. The nature of the resource allocation scheme affects p's lookahead. If allocation is non-preemptive (that is, once a resource is allocated to an object it is not released until that object has completed its service), then p knows that an object in service will remain in service without the possibility of another object acquiring that same resource, and possibly releasing it and migrating to q sooner than would the original process. Another factor is the mechanism by which the resource is released and the object migrates. In a queueing network, a job frequently receives a known amount of service, and its departure time is known at the point the job enters service. In other models it is possible that the object holds the resource until told to release it. p's ability to look ahead relies then on being able to predict when such release directives are issued.

There are further complications. It might be that an object does not know where it is migrating until its point of departure. This happens, for instance, if a job leaves one queue and joins the shortest queue from among a set, at the point of departure. It might be that at the point the object begins its service p knows what the state of the object will be when it migrates, it might be that it does not (for instance, if the object carries with it information relating to the simulation state at the time of departure).

Another type of lookahead arises when prior to simulating the timing of activities, certain measurements are made that yield lower bounds on delays until those activities occur. For example, in the trace-driven simulation of a shared memory multiprocessor, one can analyze an individual processor's trace file to determine the numbers of local references between local references. By assuming that each local reference is resolved as quickly as possible one can compute a lower bound on inter-global-reference delays. As a process simulates the memory behavior described by a trace, it can look ahead over the remaining local references until the next global one (which presumably affects another simulation process).

As a final example of the subtleties of lookahead, we observe that lookahead may be "sampled", in the sense of generating random numbers in order to obtain lookahead, and "observed". An example of the former type arises in stochastic simulations, where random variables may be sampled before they are actually needed, for the purpose of computing lookahead. A common instance of this is to sample the service time of the next job to receive service at a queue, before the job actually arrives. If the queue is non-preemptive, we know that at least the pre-sampled service time must elapse before the next job departs the queue. A case of observed lookahead occurs in the parallelized direct-execution simulation of computer programs. The program is executed between points where it interacts with other simulators, and the number of instructions executed in such an interval are measured. To exploit this we can interleave execution of the computer program to get the measurements, and the timing simulator.

4.2 Dimensions of Lookahead

We now classify dimensions of lookahead, knowledge of which will prove useful when we develop synchronization protocols.

4.2.1 Time / Content Lookahead

Our informal definition of lookahead focuses on the temporal aspect, and has nothing to say about the content of the next event p may send to q. Consider: p may know simply that it will not affect q at a time less than t, or, it may know that it *will* affect q at time t, and it knows how it will do so. The latter case is what we call "content" lookahead; its presence clearly indicates that p has a better ability to predict future behavior. Intuitively, process q may be able to use content lookahead in a more advanced way than purely time lookahead.

Another possibility is that p knows that it will affect q at time t, but will not know exactly how it will do so until later—perhaps only at time t. From q's perspective, knowledge that *something* specific will occur at time t may be handled differently from knowledge that nothing will happen before time t. The difference is subtle, but may have an impact on how q is implemented.

4.2.2 Bounded / Exact Time Lookahead

There is a difference between p knowing that it will not affect q before time t, and p knowing that it will not affect q before time t and that it *will* affect q at time t. The former case is "bounded-time lookahead" and the latter case "exact-time lookahead". The distinction has an effect on the way that q deals with the information. In fact, if p has content lookahead and exact-time lookahead, then the lookahead information suffices to actually deliver the forecast event to q.

4.2.3 Directed / Semi-directed / Undirected Lookahead

Our loose definition of lookahead was given in terms of p knowing how it might affect q in the future. This is an example of what we'll call "directed" lookahead. It might be the case that p knows that it will not affect some set of processes before time t. For instance, if a server routes a job to the queue with shortest length among a subset of queues at the instant of departure, then its lookahead is limited to that set of queues. This we call "semi-directed" lookahead. Finally, if p knows only that it will not affect *any* process before time t, we say that it has "undirected" lookahead.

4.2.4 Conditional / Unconditional Lookahead

An important class of methods rely on exploiting something called "conditional lookahead". The idea is that at time s, process p knows that it will not affect process q before time t, *provided* that the state of p does not change before time t. An example is a job in a preemptive queue, nominally scheduled to depart at time t. If no other higher priority job enters before time t and interrupts it, the lookahead is to t. The lookahead in this case is unconditional if the queue is non-preemptive.

5 USING LOOKAHEAD

Synchronization protocols vary in how they use the lookahead that applications provide. One distinction is whether a protocol is synchronous or asynchronous (or both). This distinction is drawn based on how the lookahead is used and distributed. In an asynchronous protocol, when process q is specifically told that p might affect it at time t, process q then blocks at t until told otherwise. A synchronous protocol involves some sort of global reduction synchronization. Processes that participate in the reduction provide lookahead values to it; the result of the reduction governs how far a process may advance before blocking. Synchronous protocols find a way to combine lookahead information from multiple sources.

Some synchronous protocols use the notion of conditional lookahead. For instance, the YAWNS (Nicol et al. 1989, Nicol 1993) protocol works using two principle ideas. First, that every message sent by one process to another is pre-sent in time and content. For instance, if a job enters service at a non-preemptive queue and we know its service time and routing destination, then at the point it enters service its arrival at the next queue may be reported. The second idea is that a process be able to examine its state and determine a lower bound on the time of the next message it sends, *conditioned on the assumption that no further messages are received.* Each process is thus computing conditional lookahead. Each offers that conditional lookahead to a global min-reduction. The value produced by the reduction is a simulation time up to which all processes may advance asynchronously of all others, without concern for receiving a message in its past.

Another way lookahead is combined is through "lookahead propagation networks". This can be effective when the underlying simulation model is of a network where one can view the nodes as adding delay to received objects, then propagating the objects. The fundamental idea is best described with an example. Say we have a queueing server that is presently idle but has pre-sampled the service time of the next arrival. If the server has no idea of when next an arrival might occur, it must assume the worst, that an arrival will come immediately. That estimate can obviously be improved if a way is found to bound the arrival time of the next arrival. The lookahead propagation network does this. The idea is to compute minimal length paths through a network that has the topology of the simulated network; nodes in the lookahead propagation network are weighted with the value of the service time to be given to the next arrival. Receiving a lower bound on the arrival time of a job from any source, an idle node adds to this bound its presampled service delay and offers to its successors the sum as a bound on when next it will propagate a job to them.

6 PROTOCOLS

As a concrete application of these ideas, we now look at some synchronization protocols and observe how they differ in their use and requirements of lookahead.

6.1 Null Message Protocol

The original protocol (Chandy and Misra, 1979) (discovered independently by Byrant, 1977) describes a distributed simulation in terms of "logical processes" (LPs), and "channels" between those processes. It is assumed that the channels are static; one might imagine a directed graph where LPs are nodes and channels are edges. One LP affects another by sending it a message over a channel. The message has a time-stamp on it, the time at which it affects the receiver. All such communication is point-to-point. It is assumed that messages from one LP to another appear at the receiver in the order they were sent. A distinction is made between having the message show up at the receiver, and having that LP "accept" the message, e.g., pull it out of the channel and incorporate it into its own state. An LP can detect when an input channel is empty.

An LP can be certain that it does not execute an event out of order so long as it never accepts a message with a time-stamp less than that of the next internal event the LP has to perform. Consequently, if any of its input channels are empty, the LP must block. Without further structure, this blocking rule makes it quite easy to deadlock; consider—a channel may be declared between two LPs, but through the vagrancies of random sampling no message is ever sent through that channel. To avoid this specific problem one may use so-called "null-messages". The rule becomes that when an LP accepts a message of any kind with time-stamp t, it posts a null-message on each of its output ports for which the next message time is not known. One might naively put time-stamp t on such null-messages, so that a null-message is interpreted as a declaration from sender to receiver that the sender has advanced to time t and so implicitly will send no subsequent message with time-stamp less than t. This is still not enough to avoid deadlock, lookahead needs to be incorporated. The net effect is that the null-messages sent in response to an accepted message have time-stamps greater than the accepted message. The increase in time-stamps, say v, reflects how good the lookahead is at the LP. The value v depends, of course, on the model under simulation. Intuitively, the increase in time-stamp value reflects a priori knowledge that *any* response by the LP to *any* accepted message will require at least v units of simulation time to "propagate" through the LP. As we have discussed earlier, that value may be derived from pre-sampling random variables or by knowledge that any such delay (even stochastic) exceeds some fixed value v.

Now consider our earlier classifications of lookahead. Null-message type protocols are seen to use time-lookahead as only the message time-stamps are involved in the sequencing. These protocols also use bounded-time-lookahead in that the values placed on null-messages are only bounds. The protocols use semi-directed lookahead, provided that the LP-channel graph is not a complete graph—a null message is sent to all receivers of output channels from an LP, but it is sent only to those receivers. Finally, null-message type protocols use unconditional lookahead.

6.2 Appointments

Imagine a simulation with the same type of static LP-channel structure as is assumed by null-message protocols, but with additional assumed intelligence on the part of LPs. We now require an LP to maintain on each of its output channels an "appointment" time (Nicol and Reynolds, 1984; Nicol 1988). One can think of the transferral of an appointment from sender to receiver as a null-message, but with some critical differences. First, the semantics of message acceptance becomes different from null-messages. In the null-message approach a message is not accepted from a channel before the LP is prepared to process that null-message, at its posted time-stamp. In an appointment-based protocol the receiver understands the appointment as a promise by the sender not to transmit any message with larger time-stamp. For its part the receiver tacitly promises not to advance its simulation clock beyond any appointment. Consequently the receiver may accept the appointment at any time, and bases its blocking decisions on accepted appointment values rather than unaccepted null-message times. The practical import of this is that at the point it accepts an appointment, the information there contained may allow the receiver to improve upon an appointment it provides to yet another LP. For example, suppose process p gives process q an appointment with time-stamp t. Process q is a stochastic queueing server, and knows that it routes all jobs accepted from p to some other process r. p's knowledge of a lower bound on the arrival time of the next job from p may allow it to immediately improve its appointment with r, even if the appointment time lies far into the future.

Whereas (Nicol and Reynolds, 1984) introduced the concept of appointments, it was only later in (Nicol, 1988) where they were actually used. The proposed appointments calculation procedure also introduced the idea of using a "shadow network", or as we have termed it here, a lookahead propagation network. Shortest-path computations on the the shadow network were used to push lookahead forward across LPs.

According to our classification, appointment-based protocols differ from null-message-based protocols in their ability to use content-lookahead (if available), and directed lookahead. A very important difference is that the null-message protocol can be implemented using very little model-specific information. An aggressive appointment protocol may involve some complex calculations and logic on the part of the LPs. The tradeoff is increased protocol complexity in exchange for sometimes markedly increased performance.

6.3 PUCS

The PUCS protocols (Parallel Uniformized Continuous time Simulator) were developed specifically for simulating continuous-time Markov chains (CTMC) (Heidelberger and Nicol, 1993). These protocols are essentially appointment-based; their appointments exploit the mathematically structure of CTMCs.

The simplest form of PUCS requires that each sending process identify for each process to which it may send messages a maximum "rate" at which those messages may be dispatched.

A typical example is a queue p with n servers, each with service rate μ. Jobs leaving p are routed to queue q with probability α. The fastest possible rate at which p routes jobs to q is when all of its servers are busy; that rate is $n \times \alpha \times \mu$.

Using uniformization, the mathematical structure of CTMCs allow the simulation to be performed (at least conceptually) in two phases. In the first phase, each process generates a synchronization schedule for all other processes it may affect, and conveys that schedule to them. If the maximum rate at which p affects q is λ_{pq}, then the spacing of synchronization points from p to q is random, with an exponential distribution having rate λ_{pq}. These schedules govern synchronization in the second phase, for the synchronization points are appointments. When process p reaches an appointment time t that it sent to process q, p randomly decides to actually affect q, or not. It does so by measuring the actual transition rate at which it dispatches messages to q, given p's state at t, call this rate a_{pq}. Then, with probability a_{pq}/λ_{pq} it chooses to undergo a state transition that affects q; with complimentary probability it does not, and merely sends q a message reporting a "pseudo" event instead. In the example of a multi-server given earlier, if k out of n servers are busy at time t, then an actual state transition occurs at t with probability k/n.

From the point of view of the lookahead classification scheme, PUCS has bounded-time-lookahead, and directed lookahead. It is properly seen as a specific way to generate and manage appointments, for a specific problem class.

6.4 TNE

The *Time-of-Next-Event-Algorithm* protocols (Groselj and Tropper, 1988) compute "link times" between LPs, a link time being a lower bound on the time of the next message sent from one LP to another. A link time is clearly an appointment. It computes link times using (i) knowledge of the time of the next event in each LP, (ii) knowledge of a minimum delay added by an LP to any message that it processes, (iii) knowledge of existing link times, (iv) a shortest path algorithm. Variants of TNE have focused on how to propagate shortest path information across processors. The TNE algorithms are thus seen to computing appointments based on the topology of the LP network and the lookahead of minimum delays across LPs. With respect to our classifications, TNE algorithms have bounded-time lookahead, directed lookahead, and do not have content lookahead. The lookahead computed is unconditional.

6.5 Bounded Lag

Like the other conservative mechanisms described so far, the Bounded Lag protocol (Lubachevsky,1989) assumes a static LP interconnection topology. Like some of the other conservative mechanisms, it relies upon a lookahead propagation network. Two things distinguish the Bounded Lag algorithm from others. First, it is synchronous. It does a global lookahead calculation, involving all processors, to identify for each LP a point in simulation time up to which it may safely execute events. That point is identified through analysis of the lookahead propagation network to identify for each LP the earliest time at which it might be affected in the future by another LP, OR, a lower bound on that time, the "bounded-lag". The key idea here is that for any LP, given some lag B, an analysis of the LP interconnection network and lookahead at each of the LPs can determine a sphere-of-influence, a set of LPs that might possibly affect the LP within B simulation time units. If none can, then B units from the present time serves as the LPs upper limit. What is critical here is that the lag allows one to bound the effects of all LPs outside of the sphere-of-influence, without explicitly considering them.

The lookahead discussed in the Bounded Lag papers focuses on minimum propagation delays between LPs (equivalent to minimum service times at queues), and so-called opaque periods which are localized periods during which the time of an LP's next action is insensitive to the receipt of any further messages. This is like a non-preemptive server.

This protocol has bounded-time lookahead, both directed and undirected lookahead (the lag is undirected). Like the appointments protocol it does not specify precisely how the opaque periods or through-LP lookaheads are defined, so content-lookahead might be exploited (although no examples of that have have appeared in the bounded lag literature). Applications of the bounded-lag algorithm have always used unconditional lookahead, but the method is robust enough that conditional lookahead can be transformed into unconditional lookahead in the course of the lookahead phase.

6.6 Conditional Events

The "Conditional Events" paper (Chandy and Sherman, 1989) recognized that in ordinary discrete-event simulation, without knowledge of model-specific information, all events on the event-list are "conditional" in the sense that any one of the *except the one with least time-stamp* might be removed as a consequence of executing an event with smaller time-stamp. Of course, with more model specific information we might identify events on the event list that are unconditional, and we might identify events that do not affect existing events on the event list.

In a parallel simulation context a process has a set of events; some are known to be unconditional, some are known not to affect existing events on this processor's event list, some are conditional. However, the one with least time-stamp is not unconditional, at least not before we can establish that the process will not be affected by another at an earlier time.

Chandy and Sherman showed how to transform unconditional events into conditional events. Each process offers to a global min-reduction the least time-stamp among all its conditional events. The reduction delivers the least such to all processes, and this defines a simulation time up to which all processes may simulate concurrently.

Details of how one identifies conditional or unconditional events are model dependent. The next protocol we examine, YAWNS, gives some refinement to the concept.

The conditional events approach is synchronous; its lookahead is bounded, conditional and undirected. Content-lookahead might be employed in identifying unconditional events on a model specific basis.

6.7 YAWNS

The YAWNS (Yet Another Windowing Network Simulator) protocol (Nicol et al., 1989; Nicol, 1993) is an application of conditional event approach. Its principle contribution is to show by mathematics and implementation the utility of a simple conservative tightly synchronized approach in situations where there is ample parallel workload, and lookahead of a specific type.

The key ideas behind YAWNS were described earlier. The specific resolution YAWNS provides to the conditional event proposal is identification of unconditional events (service completions), and conditional events (next messages out from a process).

6.8 The Event Horizon

As a final topic we examine an approach that resembles the YAWNS approach, except that the method is not completely conservative. The concept of "event horizon" (Steinman, 1991) codifies the logic behavior conditional events: given all processes are synchronized at t, the event horizon is the simulation time stamp of the next message to pass between any two processes. Strictly conservative applications of conditional event approaches compute lower bounds on the event horizon. Yet, as pointed out in (Steinman, 1991) that isn't necessary. So long as one saves state and withholds sending messages until it is certain that they should be sent, one can *compute* up to the event-horizon. From a synchronization point, each process executes events (saving state) up to the point where its time of next event exceeds the time-stamp on any message the process generated but hasn't yet delivered. That time-stamp is the processes local event horizon. A min-reduction on all local event horizons yields the global event horizon. All messages generated by events with time-stamps no greater than the horizon are delivered; as a result, a processor may receive a message with a time-stamp smaller than its local event horizon; that process will roll back. But, importantly, such rollbacks are entirely local to the process. They do not propagate. Construction of a simulation using this technique is considerablely simpler than one that goes fully optimistic.

While the event horizon idea is not strictly conservative, it is a natural and useful extension to ideas explored in the conservative context. It should be considered in cases where predicted lookahead is difficult to acquire.

7 SUMMARY AND CONCLUSIONS

This tutorial has highlighted key ideas useful for writing one's own parallel discrete-event simulation.

REFERENCES

Bryant, R.E. 1977. Simulation of packet communication architecture computer systems. *MIT-LCS-TR-188, Massachusetts Institute of Technology.*

Chandy, K.M. and J. Misra. 1979. Distributed simulation:a case study in design and verification of distributed programs. *IEEE Trans. on Software Engineering*, 5, 440-452.

Chandy, K.M. and Sherman, R. 1989. The conditional event approach to distributed simulation. In: *Proceedings of the 1989 Conference on Parallel and Distributed Simulation*, 93-99, SCS Simulation Series.

Groselj, B. and Tropper, C. 1988. The time of next event algorithm. In: *Proceedings of the 1988 Conference on Parallel and Distributed Simulation*, 25-29, SCS Simulation Series.

Heidelberger, P. and D.M. Nicol. 1993. Conservative parallel simulation of continuous time Markov chains using uniformization. *IEEE Trans. on Parallel and Distributed Systems*,4.

Jefferson, D.R. 1985. Virtual time. *ACM Transactions on Programming Languages and Systems*, 3, 404-425.

Lubachevsky, B.D. 1989. Efficient distributed event-driven simulations of multiple-loop networks. *Communications of the ACM*, 32, 111-123.

Nicol, D.M. and Reynolds, P.F. Jr. 1984. Problem oriented protocol design. In: *Proceedings of the the 1984 Winter Simulation Conference*, 471-474.

Nicol, D.M. 1988. Parallel discrete-event simulation of FCFS stochastic queueing networks. In: *Proceedings of the ACM/SIGPLAN PPEALS 1988. Parallel Programming: Experiences with Applications, Languages and Systems*, 124-137. ACM Press.

Nicol, D.M., Micheal C., and Inouye, P. 1989. Efficient aggregaton of multiple LP's in distributed memory parallel simulations, In: *Proceedings of the 1989 Winter Simulation Conference*, 680-685.

Nicol, D.M. 1993. The cost of conservative synchronization in parallel discrete-event simulations. *Journal of the ACM*, 40, 304-333.

Reynolds, P.F. Jr. 1988. A spectrum of options for parallel simulation. In: *Proceedings of the 1988 Winter Simulation Conference*, 325-332.

Steinman, J.S. 1991. SPEEDES: synchronous parallel environment for emulation and discrete event simulation, In: *Proceedings of the 1991 Conference on Parallel and Distributed Simulation*, 95-103, SCS Simulation Series.

AUTHOR BIOGRAPHY

DAVID M. NICOL received the Ph.D. in Computer Science from the University of Virginia in 1985 and is presently an Associate Professor of Computer Science at Dartmouth College. He is an area editor for the ACM's *Transactions on Modeling and Computer Simulation* and an associate editor for the *INFORMS Journal on Computing.* He has served as the 1989 Program Chairman and the 1990 General Chairman of the Workshop on Parallel and Distributed Simulation (PADS), has served on the PADS Steering Committee, and in 1996 was the Program Chairman for the ACM Sigmetrics Conference. His interests are in parallel simulation, performance analysis, and algorithms for mapping parallel workload.

Proceedings of the 1996 Winter Simulation Conference
ed. J. M. Charnes, D. J. Morrice, D. T. Brunner, and J. J. Swain

PERILS AND PITFALLS OF PARALLEL DISCRETE-EVENT SIMULATION

Rajive L. Bagrodia

Computer Science Department
University of California, Los Angeles
Los Angeles, California 90095, U.S.A.

ABSTRACT

The design of efficient parallel discrete-event simulation (PDES) models often appears to be a mysterious art practiced primarily by academic researchers who have been rigorously ordained in this task. This tutorial attempts to unravel some of the mysteries. It describes the process of generating an efficient parallel implementation of a discrete-event simulation (DES) model. Common pitfalls in the parallel execution of the models are described together with suggestions on their avoidance.

1 INTRODUCTION

Parallel (or distributed) discrete-event simulation refers to the execution of a discrete-event simulation program on a parallel (or distributed) architecture (Fujimoto 1990). In recent years, interest in exploiting parallelism in the execution of discrete-event simulations in a number of domains including network design and configuration, personal communication systems, parallel programs, digital battlefields, and digital circuits has been growing. This demand has been fueled both by the increasing availability of parallel computers (PCs with multiple processing units have become widely available!) and by the increasing complexity and scalability of systems that is making sequential model execution computationally intractable.

The focus of many PDES studies remains on the design of a *parallel* simulation model rather than on the design of a discrete-event simulation (DES) model for which parallelization can be explored as *one* execution option. Part of the reason for this is that it is harder to use parallel model execution as an 'afterthought.' Unless a modeler pays careful attention to some of the complex issues that must be addressed to support parallel execution from the initial stages of model *design*, subsequent parallelization efforts may prove to be overwhelming. In this regard, there is a close analogy between the design of general purpose parallel programs and PDES: it is much harder to port a 'dusty deck' FORTRAN or C program to a parallel machine than it is to port a program that was designed for eventual migration to a parallel architecture.

This tutorial describes the process of generating an efficient parallel implementation of a DES model. It outlines some of the common pitfalls in the design of the initial DES model that can make subsequent parallelization considerably harder, if not impossible. A companion paper in this volume (Liu et al. 1996) describes a case study in porting a model for wireless network simulation to a distributed memory architecture. The next section is an overview of parallel synchronization protocols. Section 3 discusses the role of simulation languages in the design of a PDES. Section 4 addresses the range of issues that must be addressed in preparing a DES model for parallel execution. Section 5 is the conclusion.

2 PARALLEL SIMULATION PROTOCOLS

Three primary types of synchronization protocols have been described in the literature: conservative (Misra 1986), optimistic (Jefferson 1985), and mixed (Jha and Bagrodia 1994), where the latter may include sub-models that execute in either conservative or optimistic modes. This section gives an overview of the synchronization problem in PDES and presents an algorithm for its solution.

2.1 Model

A typical simulation is assumed to consist of a collection of logical processes (or LPs), where each LP models some physical process in the system. Many papers in the PDES literature use an LP to represent the sequential unit of computation in a model. However the decomposition of a model into LPs may be driven by issues of modularity and software design rather than by concerns of parallel performance. In this paper, we view a PDES as a collection of Sequential Discrete Event Simulation (SDES) models. Each SDES models a subsystem of the physical system and executes on a unique processor; the processor may represent a machine in a network of workstations or a single node of a shared or distributed memory parallel architecture. Each SDES consists of one or more LPs and uses a sequential synchronization algorithm (for example, the Global Event List algorithm) to schedule local events in their correct timestamp order.

Some mechanism must be defined to allow the SDES models to communicate with each other. These constructs may be provided in the form of explicit messages, remote procedure calls, or other mechanisms that depend on the conceptual framework that is used by the language. In this paper, we do not make any assumptions about the language or notation used by an analyst to describe the SDES. For simplicity and uniformity of exposition in this paper, we will assume that SDES models communicate with each other using time-stamped messages, and any simulation algorithm must execute the model as if all events were executed in their timestamp order.

We assume that the following set of variables are defined for each SDES (Jha and Bagrodia 1994):

- Earliest Output Time (EOT): For a given SDES *s*, EOT_s refers to the (lower bound) on the future time at which *s* will cause an event to be scheduled at another SDES. For some s, if EOT_s is infinity, the remaining SDES can be executed independently of *s*. A sink process is an example of such an SDES.
- Earliest Input Time (EIT): EIT_s of an SDES *s* refers to the earliest (future) event that may be scheduled on *s* by another SDES. (An SDES may contain unprocessed messages with timestamp smaller than its EIT). The EIT of an SDES is infinity if all events scheduled on that SDES are generated locally; a *source* process is an example.
- Lookahead: At any simulation time t, the lookahead of an SDES is a lower bound on the duration after which it will schedule an event at another SDES. The lookahead is used by an SDES to compute its EOT.

We use a simple closed queuing network model to illustrate the preceding concepts. The network consists of three FIFO servers. A job arriving at a server, waits for service, and after completion of the service proceeds with equal probability to any of the other servers in the network (Figure 1). We are interested in measuring the average and maximum time that is spent by a job at any server. Each server may be viewed as an SDES; interesting events in the system are the arrival of a job at a server (henceforth called the arrival event) and departure of a job from a server. The jobs in the system are abstracted via the events and need not be modeled as explicit logical processes. The arrival (and departure) events are modeled by sending messages among the corresponding servers.

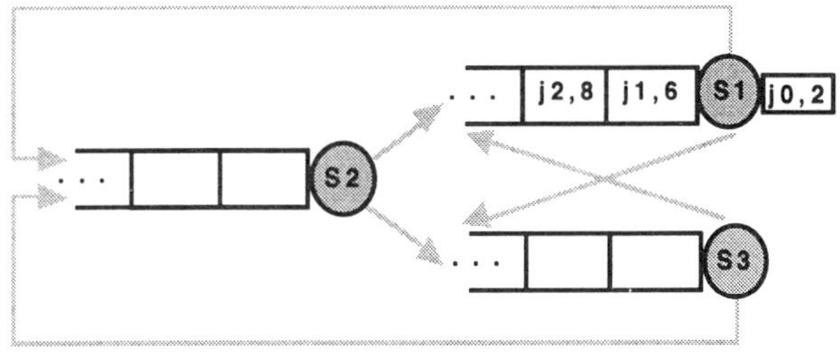

Figure 1: Closed Queuing Network

Assume that the service time at each server is 6 time units. Consider a server that is busy servicing a job: its EOT is the departure time of the current job. For install server, in Figure 1, *s1* is busy servicing job j0. EOT_{s1} is $2 + 6 = 8$ (we assume that the job began service as soon as it arrived at the server). If the server is idle, its EOT will be its EIT plus its lookahead.

2.2 Synchronization

Traditionally, a PDES model was either optimistic (all SDES executed in the optimistic mode) or conservative (all SDES executed in a conservative mode). A conservative SDES cannot tolerate causality errors; hence it will only process events with timestamps less than its EIT. An optimistic SDES may additionally process events with timestamps greater than its EIT; however, the underlying synchronization protocol must detect and correct violations of the causality constraint. The simplest mechanism for this is to have the SDES (or each LP) periodically save (or checkpoint) its state. Subsequently, if it is discovered that the LP processed messages in an incorrect order, it can be rolled back to an appropriate checkpointed state, following which the events are processed in their correct order. An optimistic algorithm is also required to periodically compute a lower bound on the timestamp of the earliest global event, also called the Global Virtual Time or GVT. All checkpoints timestamped earlier than GVT can be reclaimed. Using our model, it is sufficient for an optimistic SDES to preserve at least one checkpointed state with a timestamp smaller than its EIT. (The minimum of the EIT of all optimistic SDES is a reasonable lower bound on the GVT of the model). Given appropriate mechanisms to advance the EIT and EOT of the conservative or optimistic SDES, it is possible to implement a PDES model which contains sub-models that are executed using *optimistic, conservative, or sequential synchronization* mechanisms (Jha and Bagrodia 1994).

For a given SDES *s*, we define two sets IN_s and OUT_s respectively as the set of SDES from which s can either receive or send event messages (in the case when the communication topology is not known precisely, the sets can be constructed conservatively; in the default case, each set may contain every SDES). SDES s computes EIT_s as the minimum of the EOT of all SDES in IN_s. Given its EIT, an SDES can compute its EOT locally. The frequency and method used by an SDES to compute its EIT and the frequency with which it communicates its EOT to other SDES can have a significant impact on the performance of the PDES model.

In general, any of the GVT computation algorithms, conservative algorithms, or even a combination of the preceding algorithms can be used by a PDES to compute the EIT of each SDES (regardless of the execution mode of the individual SDES in the model). The choice of a specific algorithm for a given scenario is an efficiency rather than a correctness issue. We outline an aggressive null message based scheme: whenever the EOT of an SDES, say s, changes, EOT_s is sent using a null message to all SDES in OUT_s; the null message may of course be piggy backed on a regular message when feasible. On receipt of a new EOT from an SDES in IN_s, the SDES s recomputes its EIT (and perhaps its EOT) and propagates changes to OUT_s. It is easy to show that given a model with no zero delay cycles, such an algorithm will eventually advance the EIT of every SDES.

Consider the queuing network example. For a server process s, OUT_s and IN_s both consist of all server processes in the system, other than s. Consider the computation of EIT for a server using the null message scheme described earlier. The EOT of a server changes every time a job departs from a busy server or a job arrives at a server. When a server say *s1* forwards a job to *s2* it also sends a *null message* to *s3* (the null message to *s2* can be piggy backed with the *job* message) possibly allowing it to advance its EIT as explained earlier. Other techniques to improve the computation of EIT and lookahead are discussed subsequently.

3 SIMULATION MODEL DESIGN

A number of factors govern the choice of the specific programming language that is used by an analyst to describe the model. Among many others, these include the *conceptual framework* or *world view* adopted by the language (Evans 1988), the availability of, or familiarity of the analyst with specific languages, and perhaps the simulation libraries and support facilities provided by the language. From the perspective of this paper, an important concern in choosing a simulation language is its support for parallel execution of the model. Henceforth we refer to a language that supports parallel execution of the model as a Parallel Simulation Language (or PSL). (Bagrodia 1994) is a recent survey of languages and software to develop PDES models.

3.1 Parallel Simulation Languages

Most extant PSLs provide a set of DES primitives (scheduling events, advancing simulation time, etc.) together with a set of parallel programming primitives for thread (or object) definition, creation and interprocess (or inter-object) communication and synchronization. We use *thread* or *process* to refer to the basic unit of parallelism that is provided by the language.

A parallel language (and by extension a PSL) can use either a shared-nothing or a shared-everything model. In the former case, the processes do not have access to any shared variables; communication is entirely via messages or procedure calls. In the latter paradigm, every data item declared in the program is assumed to be accessible by every process. Note that a shared-everything programming model may be implemented on a distributed memory machine and similarly a shared-nothing model may be implemented on a shared memory architecture. Some parallel languages may also use a hybrid model, with restricted forms of data sharing. A majority of existing PSLs use the shared nothing programming model; henceforth we will use the term MP-PSL to refer to such a language. Shared-everything and hybrid models are active areas of research in PSL design. In particular, some PSLs distinguish between processes on the same and different processors allowing the former to access shared variables.

Some PSLs additionally provide constructs to modify the attributes of a model that can affect the synchronization overheads and hence its parallel performance. For instance, the Apostle (Wonnacott and Bruce 1996) simulation language provides support for granularity control of a simulation object to reduce overheads for event handling, SPEEDES (Steinman 1991) requires user-directed checkpointing as a way to provide incremental state saving, Maisie (Bagrodia and Liao 1994) and U.P.S. (Nicol and Heidelberger 1996) support user-specified lookahead specification, and Maisie also provides a variety of other constructs to monitor and optimize parallel performance including dynamic topology information, a set of constructs to reduce rollback costs, and modifying the state saving interval or time window.

PSLs also differ in their support for specific synchronization protocols. Many existing PSLs support only optimistic protocols — Sim++ (Baezner, Lomow, and Unger 1990), ModsimII (West and Mullarney 1988), Apostle, and SPEEDES to name a few. Relatively few support only conservative protocols; and even fewer, like Maisie, support conservative, optimistic, and mixed protocols. Even though conservative implementations are considerably easier to develop and have relatively few operating parameters to tune, optimistic protocols are generally perceived to be more widely applicable. However, with aggressive exploitation of lookahead in models, it has been possible to extract significant speedups from a variety of applications using conservative protocols.

3.2 Parallel Simulation Overheads

There are three primary sources of overheads in the parallel execution of a simulation model:

- Partitioning related overheads: Specific decompositions of a model into its component SDES modules sets an upper bound on the speedup that can be obtained from parallel execution of the model. The allocation of LPs to SDES determines the contribution to the overhead by load (im)balance, message communications, and related factors.
- Synchronization protocol overheads: These are the overheads that are contributed by the specific protocol that is used to synchronize the execution of the SDES. For instance, a conservative protocol introduces null message or blocking overheads whereas an optimistic protocol may introduce rollback or checkpointing overheads.
- Target architecture overheads: These costs include architecture-specific costs like message latency and context switching overheads that, in many cases, are beyond the direct control of the modeler.

As described in the next section, the analyst must recognize and reduce these overheads for successful parallelization of the DES model.

4 PARALLEL MODEL EXECUTION

The use of a PSL is a necessary, though hardly sufficient, condition for the corresponding model to run efficiently on a parallel architecture. In this section, we outline common pitfalls that must be overcome by an analyst to design a efficient parallel implementation. As with any parallel program, there is no guarantee that parallel execution of a model will yield performance benefits; however following the guidelines in this section will certainly increase those chances.

Pitfall 1 – Shared Variables: Even though a MP-PSL uses a shared-nothing programming model, it may not be possible (or efficient) for the compiler to detect uses of shared variables. Shared variables may not pose a problem for sequential implementations, but can have a serious consequence when porting the model to a parallel architecture. Consider the implementation of global variables: as the language assumes a shared-nothing model, no attempt is made to maintain consistency among the copies of the global variables stored on multiple processors.

In general, shared variables may be eliminated from a program by transforming it such that the variables are stored in one or more SDES, and all read and write operations to the corresponding data are implemented via messages or other mechanisms provided by the language for interprocess communication. However such a transformation might create a significant bottleneck in the program. In many cases, efficient alternatives are available. For instance, a common use of global variables is as write-once variables that are used to store input data like connectivity matrices or boundary values that are input to the model and are not modified subsequently. In this case, global variables can be eliminated by replicating the initialization code on each processor or by suspending all SDES until the initial values have been broadcast.

Some languages like Maisie allow the programmer to use write-once globals by providing an explicit library routine for synchronous broadcasts of initialized global variables. Other languages like Apostle allow variables to be shared among processes on the same processor. However, indiscriminate use of shared or global variables in a DES model is likely to present serious obstacles in its migration to a parallel architecture.

Pitfall 2 – Pointer Data Structures: The use of pointer data structures is similar to the issue of shared variables. An MP-PSL must not pass pointers among SDES (or logical processes) as there is no logical sharing of data between them. For PDES models that are synchronized using optimistic algorithms, even local use of pointer data structures may complicate checkpointing. Many optimistic systems use copy-state checkpointing where the entire state of a process is copied and stored in a time-stamped queue. Dynamic data structures like queues and trees make it hard to use copy-state checkpointing.

Elimination of pointers in interprocess communication typically requires that all data be passed by value. For large structured data types, this might significantly increase the communication overhead and hence program execution time. A common solution is to allow pointer passing among processes mapped to a common processor, although this adversely affects maintainability.

Pitfall 3 – Zero Delay Cycles: A model is said to contain a cycle if there is a sequence of SDES e_i, e_{i+1}, e_{i+2}, ..., e_{i+n}, e_i, n >= 0, such that each SDES may send a message to the next SDES in the sequence. The model is said to contain a zero-delay cycle if the sequence of messages exchanged in the cycle have the same timestamp. A simple example of a model with a cycle is shown in Figure 2. The merge process simply outputs messages received from its inputs in increasing timestamp order. If a path exists in the model from the merge process to a server process, and the server process has a lookahead of zero, the model could include a zero-delay cycle. A PDES model cannot contain zero-delay cycles as they can cause deadlocks in a conservative simulation (Misra 1986) or instabilities in an optimistic system.

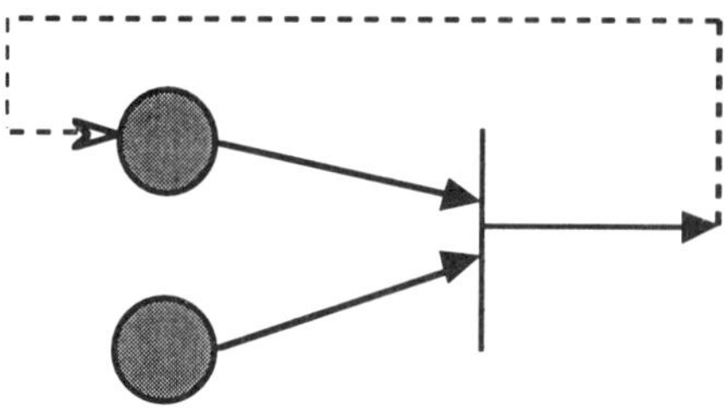

Figure 2: Merging Processes

Some parallel simulation environments (e.g. TWOS) avoid zero-delay cycles by requiring that the receive timestamp of a message be strictly greater than the (simulation) time at which it is sent (Jefferson, Beckman and Wieland 1987). This might be a significant restriction on the modeler particularly because sequential languages do not impose such constraints. A somewhat less restrictive approach is to ensure that at least one process in every cycle in the model has a non-zero lookahead (Misra, 1986).

Pitfall 4 – Poor Lookahead: Assume that a message with timestamp t_s arrives at an SDES at physical time t_r and is processed by the SDES at physical time t_p, $t_p >= t_r$ (for optimistic systems, t_p refers to the time at which the corresponding message was processed and not canceled subsequently). The duration $[t_r, t_p]$ represents simulation overhead if the corresponding SDES was conservative and its processor was idle in the duration, or if it was optimistic and the computation performed during the duration was subsequently canceled. It follows that to reduce synchronization overheads, the EIT of an SDES should be as high as possible. The EIT of an SDES is determined by the number and EOT of its predecessors. Given a model with a specific connectivity, we first examine techniques that improve the lookahead (Fujimoto 1988) for each component of the model. Subsequently we look at techniques to improve the performance of a model by reducing its connectivity. Traditionally, superior lookahead is assumed to impact the performance of only conservative SDES; however if an optimistic SDES processes a message with a timestamp smaller than its EIT, it can assert that corresponding messages with a lower timestamp can never be canceled and hence reduce its checkpointing overheads.

A commonly used technique to improve lookahead for stochastic models is to presample random distributions that are used to model various temporal intervals (Nicol 1988). For instance, consider a FIFO server that is idle at simulation time t_l. In the absence of presampling, its EOT will only be $t_l + \varepsilon$, where ε is some minimum value that can be generated by the corresponding random distribution. However, if the idle server presamples the next random value from the specified distribution, it can use that to advance its EOT. This technique is also used to improve the performance of a priority-preemptible server which would otherwise have a poor lookahead even when it was busy serving a low priority job. Language level facilities have been developed to allow the runtime system to extract the lookahead from the model or to allow a programmer to directly express the lookahead in the model (Jha and Bagrodia 1993, Nicol and Heidelberger 1996, Cota and Sargent 1989).

Compile and run time analyses of a model have also been shown to be useful in improving the application lookahead. Consider the simulation of asynchronous parallel programs where the tasks or threads communicate with each other using explicit send and receive commands. The lookahead for these applications is typically the time between two consecutive communication commands, referred to as the local block time (LBT). A common approach to estimating the LBT is by direct execution where the physical time needed to execute the code is measured by the processor clock and used to advance the local simulation clock. Techniques like predictive barrier scheduling (Legedza and Weihl 1996) use compile and run-time analysis of the program to estimate a lower bound on the time between successive communication statements in a program. Further, semantic information can be used to identify the source of an incoming message in the receive statement allowing a thread to proceed as soon as the corresponding message is available locally rather than waiting for all messages to be synchronized (Prakash and Bagrodia 1995).

Pitfall 5 – High Connectivity: Improving the connectivity information available for each SDES can improve the performance of a conservative SDES. By default, the runtime system assumes that the model is fully connected, requiring each SDES to send information about its EOT to all other SDES in the system, considerably increasing the synchronization overhead. By providing dynamic connectivity information to the model, this overhead can be reduced significantly (Jha and Bagrodia 1993).

It may also be possible to tailor the decomposition of the entities in a model to improve its connectivity. For instance, it is often possible to use clustering in network and VLSI simulations to collapse strongly connected sub-models into a single SDES and reduce the connectivity among the SDES (Gerasoulis and Yang 1993).

Pitfall 6 – Load Imbalance: Traditional techniques to decompose parallel programs (and hence parallel models) typically try to balance the computation among the partitions while minimizing the message communications among them (Sarkar and Hennessy 1988). The assignment of processes to processors may be accomplished using either *static* or *dynamic* algorithms. The primary advantage of static scheduling algorithms is the absence of any run time overheads in distributing the computation. The primary drawback of static algorithms is the difficulty in accurately estimating the computation and communication costs for each LP at compile time. For instance, in a gate-level circuit simulation, the computation load may be approximated by the number of gates in each partition and the communication load by the number of 'nets' that are cut between two partitions. A static partitioning algorithm may use these metrics to generate a statically-balanced decomposition. Clearly the level of dynamic circuit activity (which is a function of the input vectors that are fed to the circuit) among the different partitions may not be correlated well with the number of gates.

In the case of dynamic partitioning algorithms, the load at each processor is monitored at runtime. When the load is determined to be unbalanced based on the criteria specified by the system (e.g. CPU utilization, number of LPs in the scheduled queue, etc.) LPs are selected for transfer based on a specific *transfer* policy. The selected LPs are subsequently moved from one SDES to another at runtime based on specific *placement* policies that may be specified for the system. The primary advantage of dynamic techniques is that they can be much more responsive and accurate in identifying and rectifying load imbalance in a computation. However the techniques may impose a significant overhead for monitoring the load and transferring LPs. Because of the significant runtime overhead of migrating objects, extant PDES systems generally use static model decompositions to assign LPs to an SDES.

Although a significant amount of research has been undertaken in the area of load balancing for parallel and distributed computing (Shirazi, Hurson, and Kavi 1996), there has been significantly less exploration of this issue in the context of PDES (Wilson and Nicol 1996). A majority of the load balancing research in PDES has been dedicated to VLSI simulations (Bailey, Briner, and Chamberlain 1994). A parallel program is typically partitioned in a manner that minimizes message communications among the components. However, for a parallel simulation other factors may be more important: for instance, the communication topology has been found to have a significant impact on performance where models with an acyclic communication topology perform significantly better than ones that contain cycles (Cong, Li, and Bagrodia 1994). In addition, although it may not be possible to eliminate cycles completely from a model, reductions in the number of cycles can also have a significant impact on the performance. Thus from a PDES perspective, it may be preferable to use a partitioning that has a higher total communication cost that another provided that the former also has a smaller number of cycles among the partition.

Pitfall 7 – High Message Traffic: This factor is related closely to the load imbalance issue in that the decomposition of a model should be aimed at reducing the message traffic among the partitions. In addition it is often possible to aggregate messages from one SDES to another, particularly when multiple processes may be mapped to an SDES. As a large number of small messages typically contribute greater communication overhead than a small number of large messages, appropriate use of message aggregation or piggy-backing can reduce this overhead.

Pitfall 8 – Low Event or Computation Granularity: Context-switching costs can be a significant contributor to simulation overheads when a large number of processes are mapped to a single SDES. A number of factors including cache behavior, task switching times, and processes scheduling costs contribute to this overhead. Some of these costs, e.g. the scheduling policy, are determined by the design choices that have been made by the language designer and may be beyond the control of the programmer. For instance, the traditional Global Event List (GEL) algorithm schedules events across different LPs mapped to an SDES in the order of their timestamps. It is typically possible to reduce the scheduling overheads by using a parallel simulation algorithm: compute the EIT for each LP; select an LP that has the largest number of pending safe events (i.e. events with timestamps smaller than its EIT) and process all *safe* events for that LP before switching to another LP. Although this policy will cause some events to be scheduled out of their timestamp order, it is easy to show that causality will not be violated. For queuing network and network simulation benchmarks (Jha and Bagrodia 1993), this form of local scheduling has been shown to significantly improve overall execution times for a model.

In general, the overhead costs tend to increase with the number of LPs. However it is often possible to *hide* the communication latency of a parallel program by multiprocessing the LPs mapped to an SDES. This behavior has been observed for parallel programs in many applications and has led to the design of numerous thread packages. The support for multiple processes at an SDES can similarly improve the performance of a PDES model.

Pitfall 9 – Low Inherent Parallelism: It may sometimes be the case that the parallel implementation of an application fails to yield significant performance improvements because the application itself or a specific decomposition contains a low degree of inherent parallelism. As an example of such a scenario consider the simulation of a network containing a large number of nodes but a very small number of jobs. As the parallel activity is determined by the processing associated with each job, there will be insufficient parallelism in the model. In general, it is a non-trivial problem for an analyst to identify the extent of parallelism that is present in a PDES model.

Theoretical metrics like *critical path* have been used to provide a loose lower bound on the parallel execution time of a model (Lin 1990). The critical path has limited practical utility because it ignores many common and often unavoidable sources of overhead including message latency, load imbalance etc. Recently, the notion of an *Ideal Simulation Protocol* or *ISP* (Jha and Bagrodia 1996) has been suggested to experimentally compute a realistic lower bound on parallel execution time. This protocol includes all overheads that arise in the parallel execution of a model *other than those that are directly attributed to the synchronization protocol*. In other words ISP allows an analyst to compute the speedup that could be obtained from the parallel execution of a PDES model, if the synchronization overheads were zero. ISP could thus be used by an analyst to determine the parallelism potential of a model (or a specific decomposition) before expending significant effort in trying to tune other simulation-specific parameters.

Pitfall 10 – High Checkpointing Overheads: Checkpointing is used in optimistic protocols to support rollbacks to cancel incorrect computation. The simplest technique is copy-state checkpointing which copies the entire state of an SDES or LP before processing each event. It is often more efficient to use *interval checkpointing* where the state of an SDES is saved after processing multiple events. Periodic or interval checkpointing reduces the total state saving time; however it increases the computation time because some correct events must now be recomputed. The checkpointing frequency that will yield optimal performance is a tradeoff between the preceding factors and is typically application-dependent. An analytical formulation to select an optimal checkpointing interval has been described in (Lin et al. 1993). The preceding paper also describes an algorithm that can be used to select an 'optimal' checkpoint interval during the execution of a model. Subsequent work has extended these results to use adaptive checkpointing intervals.

For many applications including circuit and battlefield simulations, the state of an SDES or even an LP can be very large. Further it is often the case that only a small fraction of its total state space is modified when an event is processed. For such applications the use of copy-state checkpointing can increase the synchronization overheads sufficiently to offset any performance benefits of parallel execution. An alternative is to use *incremental* state saving, where only the portion of the object's state that is modified by an event is saved. Incremental state saving can either be programmer-directed or system-directed. In the latter case, two possibilities exist: the run-time system can explicitly compare the old and new states of an object and only save the modified portions, or save a history of all modifications as they are made to an object. Incremental state saving can reduce checkpointing costs but considerably increase rollback costs because the previous state of the object must now be reconstructed using the modification history or the incrementally saved states. Comparison of the two methods for checkpointing is an ongoing research area (Palaniswamy and Wilsey 1993).

5 CONCLUSIONS

The increasing complexity of many DES models has led to an increased the demand for PDES. However, the process of developing an efficient parallel model, even in languages that support PDES, remains a challenging task. This tutorial described some of the more common pitfalls that analysts need to be aware of when embarking on the journey to parallelize a DES model.

ACKNOWLEDGMENTS

This work was supported by the U.S. Advanced Research Projects Agency ARPA-CSTO, Air Force Contract F-30602-94-C-0273, 'Scalable Systems Software Measurement and Evaluation' and Dept. of Defense Contract DABT-63-94-C-0080, 'Transparent Virtual Mobile Environment.' The author would like to thank Vikas Jha, Yu-an Chen, Jay Martin, Rich Meyer, Brian Park and others in the UCLA Parallel Computing Laboratory, for their role in the many useful discussions that led to this paper.

REFERENCES

Bagrodia, R. 1994. Language support for parallel discrete-event simulations. In *Proc. 1994 Winter Simulation Conference*, ed. J. D. Tew, S. Manivannan, D. Sadowski, and A. Seila, 1324-1331.

Bagrodia, R. and W.-T. Liao. 1994. MAISIE: A language for the design of efficient discrete-event simulations, *IEEE Trans. Software Eng.*, 20(4), 225-238.

Baezner, D., G. Lomow, and B. W. Unger. 1990. Sim++: The transition to distributed simulation. In *Proc. 1990 SCS Multiconference on Distributed Simulation*, 211-218.

Bailey, M. L., J. V. Briner, Jr. and R. D. Chamberlain. 1994. Parallel logic simulation of VLSI systems.

ACM Computing Surveys (26) 3:255-294.

Cong, J., Z. Li, and R. Bagrodia. 1994. Acyclic multi-way partitioning of boolean networks. In *Proc. ACM/IEEE Design Automation Conf.*, 670-675.

Cota, B. A. and R. G. Sargent. 1989. Automatic lookahead computation for conservative distributed simulation. Technical Report CASE Center 8916, Simulation Research Group and CASE Center, Syracuse University, Syracuse, NY.

Evans, J. B. 1988. *Structures of discrete event simulation: an introduction to the engagement strategy*. Elis Horwood Ltd.

Fujimoto, R. 1988. Lookahead in parallel discrete event simulation. *Int'l Conference on Parallel Processing*.

Fujimoto, R. 1990. Parallel discrete event simulation. *CACM*, 33(10):30-53.

Gerasoulis, A. and T. Yang. 1993. On the granularity and clustering of directed acyclic task graphs. *IEEE Trans. on Parallel and Distributed Systems*, 686-701.

Jefferson, D. 1985. Virtual Time, *ACM TOPLAS*, 7(3):404-425.

Jefferson, D., B. Beckman, and F. Wieland. 1987. Distributed simulation and the time warp operating system. In *Symposium on Operating Systems Principles*.

Jha, V. and R. Bagrodia. 1993. Transparent implementation of conservative algorithms in parallel simulation languages. In *Proc. 1993 Winter Simulation Conference*, ed. G. W. Evans, M. Mollaghasemi, E. C. Russell, and W. E. Biles, 677-686.

Jha, V. and R. Bagrodia. 1994. A unified framework for conservative and optimistic distributed simulation. In *Proc. 8th Workshop on Parallel and Distributed Simulation*, ed. D. K. Arvind, R. Bagrodia, J. Y. Lin, 12-19, IEEE Computer Society Press.

Jha, V. and R. L. Bagrodia. 1996. A performance evaluation methodology for parallel simulation protocols. In *Proc. 10th Workshop on Parallel and Distributed Simulations*, 180-183, IEEE Computer Society Press.

Legedza, U. and W. E. Weihl. 1996. Reducing synchronization overhead in parallel simulation. In *Proc. Tenth Workshop on Parallel and Distributed Simulation*, 86-95, IEEE Computer Society Press.

Lin, Y. 1990. Understanding the limits of optimistic and conservative parallel simulation. PhD thesis, University of Washington, Seattle.

Lin, Y., B. R. Preiss, W. M. Loucks, and E. D. Lazowska. 1993. Selecting the checkpoint interval in time warp simulation. In *Proc. Seventh Workshop on Parallel and Distributed Simulation,* ed. R. Bagrodia and D. Jefferson, 3-10. IEEE Computer Society Press.

Liu, W., C. Chiang, H. Wu, V. Jha, M. Gerla, and R. Bagrodia. 1996. Parallel simulation environment for mobile wireless networks. In *Proc. 1996 Winter Simulation Conference*, ed. J. Charnes and D. Morice.

Misra, J. 1986. Distributed discrete-event simulation, *ACM Computing Surveys* 18 (1): 39-65.

Nicol, D. M. 1988. Parallel discrete event simulation of FCFS stochastic queuing networks. In *Parallel programming: experience with applications, languages and systems*, 124-137. ACM SIGPLAN.

Nicol, D. M. and P. Heidelberger. 1996. On extending more parallelism to serial simulators. In *Proc. Tenth Workshop on Parallel and Distributed Simulation*, 202-205, IEEE Computer Society Press.

Palaniswamy A. C. and P. A. Wilsey. 1993. An analytical comparison of periodic checkpointing and incremental state saving. In *Proc. Seventh Workshop on Parallel and Distributed Sim.*, ed. R. Bagrodia and D. Jefferson, 127-134. IEEE Computer Society Press.

Prakash, S. and R. Bagrodia. 1995. Parallel simulation of data parallel programs. In *Proc. Eighth Workshop on Languages and Compilers for Parallel Computing*.

Sarkar V. and J. Hennessy. 1988. Compile-time partitioning and scheduling scheme of parallel programs. In *Proc. SIGPLAN '88 Symposium on Compiler Contruction*, 17-26.

Shirazi, B., A. R. Hurson, K. Kavi. 1996. *Scheduling and Load Balancing in Parallel and Distributed Systems*. IEEE Computer Society Press.

Steinman, J. 1991. SPEEDES: synchronous parallel environment for emulation and discrete event simulation, In *Advances in Parallel and Distributed Simulation*, 95-103. SCS Multiconference, Anaheim, CA.

West, J. and A. Mullarney. 1988. ModSim: a language for distributed simulation. In *Proceedings of 1988 SCS Multiconference on Distributed Simulation*, San Diego, CA, 155-159.

Wilson, L. F. and D. M. Nicol. 1996. Experiments in automated load balancing. In *Proc. Tenth Workshop on Parallel and Distributed Simulation*, 4-11, IEEE Computer Society Press.

Wonnacott, P. and D. Bruce. 1996. The APOSTLE simulation language: granularity control and performance data. In *Proc. Tenth Workshop on Parallel and Distributed Simulation*, 114-123, IEEE Computer Society Press.

AUTHOR BIOGRAPHY

RAJIVE L. BAGRODIA is an Associate Professor of Computer Science at UCLA. He obtained a Bachelor of Technology in Electrical Engineering from the Indian Institute of Technology, Bombay, in 1981, and the M.A. and Ph.D. degrees in Computer Science from the University of Texas at Austin, in 1983 and 1987 respectively. His research interests include distributed algorithms, parallel languages, programming methodology and performance evaluation.

Proceedings of the 1996 Winter Simulation Conference
ed. J. M. Charnes, D. J. Morrice, D. T. Brunner, and J. J. Swain

NEW ADVANCES AND APPLICATIONS OF COMBINING SIMULATION AND OPTIMIZATION

Fred Glover
James P. Kelly
Manuel Laguna

Graduate School of Business
University of Colorado
Boulder, Colorado 80309-0419, U.S.A.

ABSTRACT

The area of integrating simulation and optimization has recently undergone remarkable changes. New advances are making available applications of simulation that previously had been considered infeasible or beyond the scope of current technology to handle.

This paper describes recently developed computer software that effectively integrates simulation and optimization. The software employs graphics to show the performance of the search mechanisms that control the integration. We also demonstrate the ability of the system to find optimal and near optimal solutions in minutes for applications where an exhaustive examination of relevant alternatives requires days or months. The scaling of reduced time applies equally to settings where simulation runs require greater time. As a result, for a selected practical time limit, the new approach provides the opportunity to obtain decisions and scenarios whose quality greatly exceeds the quality available in the past.

1 INTRODUCTION

The ability to guide a series of simulations in the most effective way, instead of blindly itemizing scenarios (with the hope that at least one of those itemized will be one that is most desirable to implement) has been a long standing goal. The integration of simulation and optimization is putting this goal within practical reach.

Now, practical software exists that is capable of interfacing simulation and special search processes, to guide a series of simulations to uncover optimal or near optimal scenarios. Applications include the goals of finding:

- the best configuration of machines for production scheduling;
- the best investment portfolio for financial planning;
- the best utilization of employees for workforce planning;
- the best location of facilities for commercial distribution;
- the best operating schedule for electrical power planning;
- the best assignment of medical personnel in hospital administration;
- the best setting of tolerances in manufacturing design;
- the best set of treatment policies in waste management;
- and many other practical objectives.

Current advances not only open new doors for simulation, but also extend the areas to which optimization can be applied. The great advantage of simulation, which has been lacking in traditional optimization, is the ability to handle uncertainties and complex interactions (at a level that can scarcely be formulated in standard optimization models). The marriage of simulation and optimization offers a way to overcome this limitation.

Earlier attempts to create methods for optimizing simulations have largely been based on ad hoc approaches, or have relied on the user to run through a cumbersome "seat of the pants" analysis. Alternatively, they have been based on stochastic approximation designs whose main focus involves the analysis of convergence behavior in an infinite time frame. Not surprisingly, the results of such efforts have left a great deal to be desired from a practical standpoint. As a step in the direction of greater rigor within a finite time horizon, a systematic catalog of all possible alternatives may be examined by complete enumeration algorithms. Although this approach guarantees optimal solutions, it has very limited application. As an example of an exceedingly simple setting where enumeration may be practicable, suppose that a simulation model depends on only two input factors, as for example, a job shop with parallel machines of 2 types. If a feasible shop design would allow from 1 to 10 machines of each type, then 100 simulation runs are needed to enumerate all possibilities. If each simulation is relatively short (e.g., 3 seconds), then the entire process could be done in 5 minutes of computer time. However, if instead of 2 machine types we allow a very modest increase to consider 5 types, then enumerating all alternatives to find an optimal one would require 10^5 = 100,000 simulations, or approximately 3.5 days of computer time. Of course, most simulation settings are not really so simple as the one described. It is easily possible for complete enumeration to take weeks or even months to carry out.

Recent developments in the area of optimization have led to the creation of intelligent search procedures capable of finding optimal or near optimal solutions to complex problems with large solution spaces, by exploring only a small fraction of the possible alternatives. In particular, the system we describe in this paper is the result of implementing a search technology known as *scatter search,* by customizing its operation to the context of optimizing simulations. The system searches for the best possible solution to an optimization problem, defined on a set of input factors to a simulation model.

2 CLASSICAL METHODS AND META-HEURISTICS

To give a background for understanding the rationale underlying our approach, we provide a brief review of classical optimization perspectives and the emerging role of meta-heuristics.

Formally, optimization deals with finding the best (optimal) solution to problems that in general can be expressed in the form of an objective function (to be optimized) and a set of constraints (which restrict the values of the decision variables). The best known optimization tool is linear programming, which model assumes that the objective function and constraints can be expressed using linear functions. Linear programming techniques are able to find optimal solutions to problems without the need to evaluate all possible alternatives. Models with thousands and even millions of variables can be solved with reasonable amounts of computer time.

Evidently, however, not all business and industrial problems can be expressed by means of a linear objective and linear equalities or inequalities. Many complex systems may not even have a convenient mathematical representation, linear or non-linear. Techniques such as linear programming and its cousins (non-linear programming and integer programming) generally require a number of simplifying assumptions about the real system to be able to properly frame the problem. One usual simplifying assumption is to disregard the so called "statistical fluctuations" of the system. For instance, a production process may be modeled by assuming that production times do not vary, permitting a single time estimate to be used for modeling purposes. The advantage, of course, is that once the problem has been formulated, well-established techniques are often able to find an optimal solution, provided the formulation is "congenial." However, there are problems such as production scheduling where even the deterministic versions remain hard to solve. This is due to the combinatorial nature of these problems, as illustrated by the situation where the goal is to determine an optimal order in which to process a set of jobs. Even for fewer that a hundred jobs to be ordered, the number of alternative configurations is astronomical. (Seventy jobs would require longer than the age of the universe to enumerate using all of today's computers working at once!)

Due to the realization that practical problems are not going to become any easier to solve, and that practitioners are seeking solutions to increasingly more complex problems, researchers have actively developed solution procedures that are referred to as *meta-heuristics.* Heuristics have been used for many years to provide approximate solutions to complex problems. For example, a production heuristic may be to give priority to jobs with the shortest estimated processing time. Depending on the context, this heuristic (or processing rule) may actually work fairly well. However, in some other situations the results may be disastrous, with dire consequences for equipment utilization, production lead times, and work-in-process inventory.

The alternatives seem at first to be less than encouraging: either to seek optimal solutions to simplified problems or to seek sub-optimal and possibly very poor solutions to complex systems. The area of meta-heuristics arose with the goal of providing something better, based on integrating high-level intelligent procedures and fast computer implementations. Numerous successful applications emerged, but with an accompanying limitation: typically they required highly problem-specific designs, customizing the solution procedures to each particular case. Every time a new problem surfaced, a new procedure needed to be developed. Meta-heuristic approaches are based on general principles, but they also owe their efficiency to the knowledge of characteristics particular to each situation. In this sense, there is no separation between the model and the solution procedure. In fact, the solution procedure may be seen as a way of modeling the problem.

Of course, it is preferable to separate the solution procedure from the system we are trying to optimize if such a separation can be achieved successfully. The disadvantage of this "black box" approach (see Figure 1), is that the optimization procedure is generic and does not know anything about what goes on inside of the box.

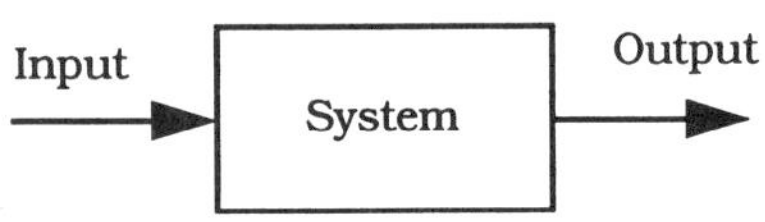

Figure 1: System as a black box.

The clear advantage, on the other hand, is that the same optimizer can be used for many systems. Our approach is an implementation of a generic optimizer that successfully embodies the principle of separating the method from the model. The optimization problem is defined outside the system, which is represented in this case by a simulation model. Therefore, the simulation model can change and evolve to incorporate additional elements, while the optimization routines remain the same. Hence, there is a complete separation between the model that represents the system and the procedure that is used to solve optimization problems defined within this model.

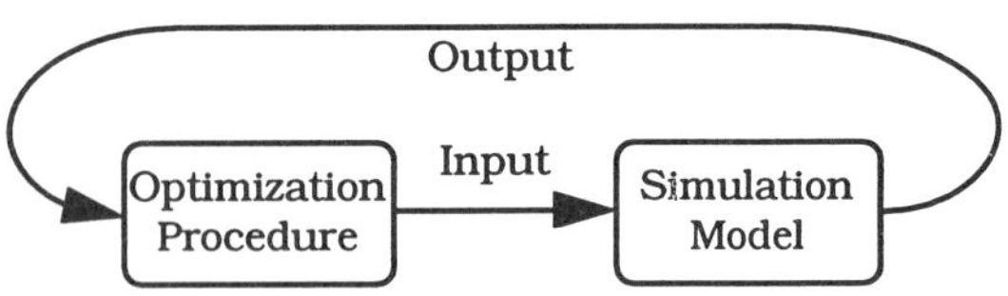

Figure 2: Coordination between optimization and simulation.

The optimization procedure uses the outputs from the simulation model which evaluate the outcomes of the inputs that were fed into the model. On the basis of this evaluation, and on the basis of the past evaluations which are integrated and analyzed with the present simulation outputs, the optimization procedure decides upon a new set of input values (see Figure 2). The optimization procedure is designed to carry out a special "non-monotonic search," where the successively generated inputs produce varying evaluations, not all of them improving, but which over time provide a highly efficient trajectory to the best solutions. The process continues until some termination criterion is satisfied (usually given by a limit expressing the user's preference for the amount of time to be devoted to the search). The underlying components of our method, scatter search and tabu search, are briefly sketched in the next two sections.

3 SCATTER SEARCH

Two of the best-known meta-heuristics are genetic algorithms and tabu search. Genetic Algorithm (GA) procedures were developed by John Holland in the early 1970s, at the University of Michigan (Holland 1975). Parallel to the development of GAs, Fred Glover, at the University of Colorado, established the principles and operational rules for tabu search (TS) and a related methodology know as *scatter search* (Glover 1977).

Scatter search has some interesting commonalties with GA ideas, although it also has a number of quite distinct features. Several of these features have come to be incorporated into GA approaches after an intervening period of approximately a decade, while others remain largely unexplored in the GA context.

Scatter search (Glover 1994) is designed to operate on a set of points, called *reference points,* that constitute good solutions obtained from previous solution efforts. The approach systematically generates linear combinations of the reference points to create new points, each of which is mapped into an associated feasible point. Tabu search is then superimposed to control the composition of reference points at each stage. Tabu Search (see e.g., Glover and Laguna 1993) has its roots in the field of artificial intelligence as well as in the field of optimization. The heart of tabu search lies in its use of adaptive memory, which provides the ability to take advantage of the search history in order to guide the solution process. In its simplest manifestations, adaptive memory is exploited to prohibit the search from reinvestigating solutions that have already been evaluated. However, the use of memory in our implementation is much more complex and calls upon memory functions that encourage search diversification and intensification. These memory components allow the search to escape from locally optimal solutions and in many cases find a globally optimal solution.

Similarities are immediately evident between scatter search and the original GA proposals. Both are instances of what are sometimes called "population based" approaches. Both incorporate the idea that a key aspect of producing new elements is to generate some form of combination of existing elements. On the other hand, several contrasts between these methods may be noted. The early GA approaches were predicated on the idea of choosing parents randomly to produce offspring, and further on introducing randomization to determine which components of the parents should be combined. By contrast, the scatter search approach does not correspondingly make recourse to randomization, in the sense of being indifferent to choices among alternatives. However, the approach is designed to incorporate strategic probabilistic biases, taking account of evaluations and history. Scatter search focuses on generating relevant outcomes without losing the ability to produce diverse solutions, due to the way the generation process is implemented. For example, the approach includes the generation of new points that are not convex combinations of the original points. The new points then may contain information that is not contained in the original reference points.

Scatter search is an *information driven* approach, exploiting knowledge derived from the search space, high-quality solutions found within the space, and trajectories through the space over time. The combination of these factors creates a highly effective solution process. The incorporation of such designs is responsible for endowing our system with the ability to solve complex simulation-based problems with unprecedented efficiency.

4 TABU SEARCH BASICS

One way of intelligently guiding a search process is to forbid (or discourage) certain solutions from being chosen based on information that suggests these solutions may duplicate, or significantly resemble, solutions encountered in the past. In tabu search, this is often done by defining suitable attributes of moves or solutions, and imposing restrictions on a set of the attributes, depending on the search history. Two prominent ways for exploiting search history in TS are through *recency* and *frequency* memories. Recency memory is typically (though not invariably) a short-term memory that is managed by structures or arrays called "tabu lists," while frequency memory more usually fulfills a long term search function. A standard form of recency memory discourages moves that lead to solutions with attributes shared by other solutions recently visited. A standard form of frequency memory discourages moves leading to solutions whose attributes have often been shared by solutions visited during the search, or alternately encourages moves leading to solutions whose attributes have rarely been seen before. Another standard form of frequency memory is defined over subsets of elite solutions to fulfill an intensification function.

Short and long term components based on recency and frequency memory can be used separately or together in complementary TS search strategies. Note that this approach operates by implicitly modifying the neighborhood of the current solution. Tabu search in general includes many enhancements to the scheme sketched here, and we refer the interested reader to Glover and Laguna (1993) or Glover (1996). The details of the short-term and long-term adaptive memories, and a recovery strategy for both intensifying and diversifying the search are discussed in the following section.

5 AN OVERVIEW OF THE ALGORITHM

We assume that a solution to the optimization problem can be represented by a n-dimensional vector $\mathbf{x}$, where x_i may be a real or an integer bounded variable (for $i = 1, ..., n$). In addition, we assume that the objective function value $f(\mathbf{x})$ can be obtained by running a related simulation model that uses $\mathbf{x}$ as the value of its input factors. Finally, a set of linear constraints (equality or inequality) may be imposed on $\mathbf{x}$.

The algorithm starts by generating an initial population of reference points. The initial population may include points suggested by the user, and it always includes the following midpoint:

$$x_i = l_i + (u_i - l_i)/2,$$

where u_i and l_i are the upper and lower bounds on x_i, respectively. Additional points are generated with the goal of creating a diverse population. A population is considered diverse if its elements are "significantly" different from one another. We use a distance measure to determine how "close" a potential new point is from the points already in the population, in order to decide whether the point is included or discarded.

Every reference point $\mathbf{x}$ is subjected to a feasibility test before it is evaluated (i.e., before the simulation model is run to determine the value of $f(\mathbf{x})$). The feasibility test consists of checking (one by one) whether the linear constraints imposed by the user are satisfied. An infeasible point $\mathbf{x}$ is made feasible by formulating and solving a linear programming (LP) problem. The LP (or mixed-integer program, when $\mathbf{x}$ contains integer variables) has the goal of finding a feasible $\mathbf{x}^*$ that minimizes the absolute deviation between $\mathbf{x}$ and $\mathbf{x}^*$.

The population size is automatically adjusted by the system considering the time that is required to complete one evaluation of $f(\mathbf{x})$ and the time limit the user has allowed the system to search. Once the population is generated, the procedure iterates in search of improved outcomes. At each iteration two reference points are selected to create four offspring. Let the parent-reference points be $\mathbf{x}_1$ and $\mathbf{x}_2$, then the offspring $\mathbf{x}_3$ to $\mathbf{x}_6$ are found as follows:

$$x_3 = x_1 + d$$
$$x_4 = x_1 - d$$
$$x_5 = x_2 + d$$
$$x_6 = x_2 - d$$

where $d = (\mathbf{x}_1 - \mathbf{x}_2)/3$. The selection of $\mathbf{x}_1$ and $\mathbf{x}_2$ is biased by the values $f(\mathbf{x}_1)$ and $f(\mathbf{x}_2)$ as well as the tabu search memory functions. An iteration ends by replacing the worst parent with the best offspring, and giving the surviving parent a tabu-active status for given number of iterations. In subsequent iterations, the use of two tabu-active parents is forbidden.

5.1 Restarting Strategy

In the course of searching for a global optimum, the population may contain many reference points with similar characteristics. That is, in the process of generating offspring from a mixture of high-quality reference points and ordinary reference points member of the current population, the diversity of the population may tend to decrease. A strategy that remedies this situation considers the creation of new population.

Our implementation of a restarting mechanism has the goal of creating a population that is a blend of high-quality points found in earlier explorations (we call these the *elite* points) complemented with points generated in the same way as during the initialization phase. The restarting procedure, therefore, injects diversity through newly generated points and preserves quality through the inclusion of elite points.

5.2 Adaptive Memory and the Age Strategy

Some of the points in the initial population may have poor objective function values. Therefore, they may never be chosen to play the role of a parent and would remain in the population until restarting. To additionally diversify the search, we increase the "attractiveness" of these unused points over time. This is done by using a form of long-term memory that is different from the conventional frequency-based implementation.

In particular, we introduce the notion of "age" and define a measure of "attractiveness" based on the age and the objective function value of a particular point. The idea is to use search history to make reference points not used as parents "attractive," by modifying their objective function values according to their age.

At the start of the search process, all the reference points $\mathbf{x}$ in a population of size p have zero age. At the end of the first iteration, there will be p-1 reference points from the original population and one new offspring. The ages of the p-1 reference points are made one and that of the new offspring zero. The process then repeats for the subsequent iterations, and the age of every reference point increases by one in each iteration except for the age of the new population member whose age is initialized to zero. (A variant of the above procedure sets the surviving parent's age also to 0.)

Each reference point in the population has an associated age an objective function value. These two values are used to define a function of attractiveness that makes an old high-quality point the most attractive. Low-quality points become more attractive as their age increases.

5.3 Neural Network Accelerator

This strategy is designed to increase the power of the system's search engine. The concept behind embedding a neural network is to "screen out" values $\mathbf{x}$ that are likely to result in a very poor value of $f(\mathbf{x})$. The neural network is a prediction model that helps the system accelerate the search by avoiding simulation runs whose results can be predicted as inferior. Engaging the neural network accelerator is an option to the user. When the neural network is used, information is collected about the objective function values obtained by different optimization variable settings. This information is then used to train the neural network during the search. The system automatically determines how much data is needed and how much training should be done, based once again on both the time to perform a simulation and the optimization time limit provided by the user.

The neural network is trained on the historical data collected during the search and an *error* value is calculated during each training round. This error refers to the accuracy of the network as a prediction model. That is, if the network is used to predict $f(\mathbf{x})$ for $\mathbf{x}$-values found during the search, then the error indicates how good those predictions are. The error term can be calculated by computing the differences between the known $f(\mathbf{x})$ and the predicted $\hat{f}(\mathbf{x})$ objective function values. The training continues until the error reaches a minimum prespecified value.

The neural network accelerator can be used at several risk levels. The risk is associated with the probability of discarding $\mathbf{x}$ when $f(\mathbf{x})$ is better than $f(\mathbf{x}_{best})$, where $\mathbf{x}_{best}$ is the best solution found so far. The risk level is defined by the number of standard deviations used to determine how close a predicted value $\hat{f}(\mathbf{x})$ is of the best value $f(\mathbf{x}_{best})$. A risk-averse user would, for instance, would only discard $\mathbf{x}$ if $\hat{f}(\mathbf{x})$ is at least three standard deviations larger than $f(\mathbf{x}_{best})$, in a minimization problem.

6 THE OPTQUEST SYSTEM

A commercial implementation of the system described above has been recently released under the name of OptQuest (1996). In its current version, OptQuest has been specifically customized to help users to find optimal input parameter settings to simulation models built with Micro Saint 2.0. (Micro Saint is registered trademark of Micro Analysis and Design, Inc.) In order to use OptQuest the user first creates a Micro Saint model. Once the simulation model has been created, an "Optimize" option can be selected within Micro Saint to access OptQuest. The interaction between OptQuest and Micro Saint can be summarized as follows:

1) Micro Saint invokes OptQuest and the most current version of the simulation model is stored in the corresponding model file.

2) OptQuest reads the model file from disk to allow the user to select optimization variables from a "Variable Catalog."

3) The user describes the optimization problem (which may include constraints) in terms of the selected variables and the system variable named *objective*, which calculates $f(\mathbf{x})$ for a given simulation run with input factors $\mathbf{x}$.

4) OptQuest repeatedly calls Micro Saint to perform simulation runs during the search for the optimal solution.

5) Micro Saint returns simulation results (i.e., the value of *objective*) to OptQuest.

6) Upon exiting OptQuest, the model may be updated with the values found during the search.

The OptQuest design is quite flexible, since with limited programming effort the simulator can be replaced to customize the system for different applications.

6.1 A Job Shop Simulation

We illustrate the operation an some of the features of OptQuest with a simulation of a Job Shop. The job shop is simulated as network of queues. The job shop has several different processing centers with one or more identical machines in each center. Several types of jobs have different routings through the job shop. We consider a job shop with five different machine centers (*drills*, *grinders*, *lathes*, *punches*, and *saws*). We also consider that three different types of jobs (A, B, and C) arrive to the job shop for processing. The Micro Saint network representation of this model is shown in Figure 3.

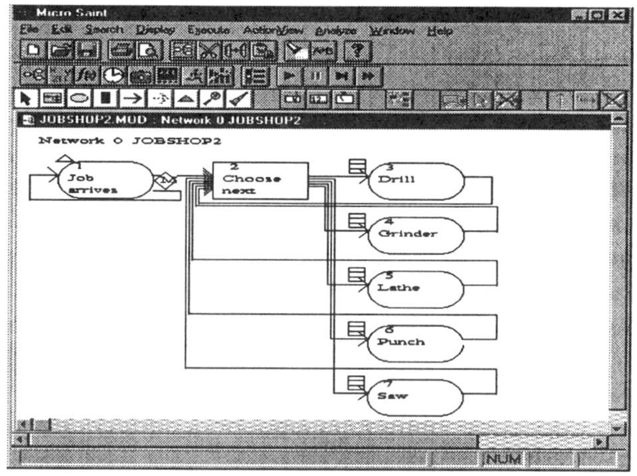

Figure 3: Micro Saint task network.

The objective function in this problem creates a tradeoff between minimizing the makespan and the cost associated with the number of machines used in each work center. Let the lower and the upper bounds be 1 and 10, respectively, for the number of machines in each work center. Then the solution $\mathbf{x} = \{1, ..., 1\}$ minimizes the capital cost of equipment, while the solution $\mathbf{x} = \{10, ..., 10\}$ minimizes the makespan. We consider that $\mathbf{x} = \{drills, grinders, lathes, punches, saws\}$. The OptQuest screen shown in Figure 4 is used to select the variables for optimization:

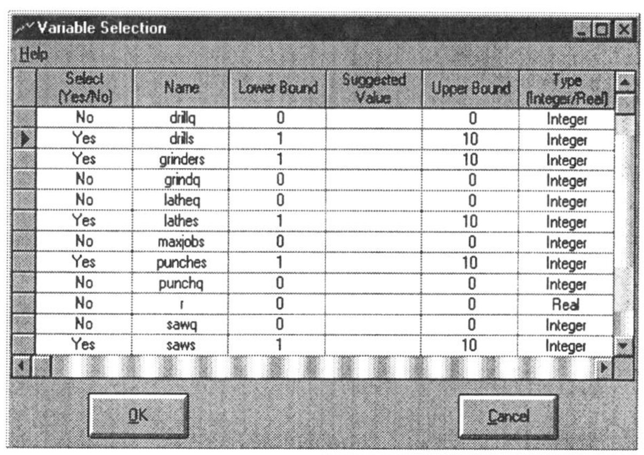

Variable Selection

Help

Select (Yes/No)	Name	Lower Bound	Suggested Value	Upper Bound	Type (Integer/Real)
No	drillq	0		0	Integer
Yes	drills	1		10	Integer
Yes	grinders	1		10	Integer
No	grindq	0		0	Integer
No	latheq	0		0	Integer
Yes	lathes	1		10	Integer
No	maxjobs	0		0	Integer
Yes	punches	1		10	Integer
No	punchq	0		0	Integer
No	r	0		0	Real
No	sawq	0		0	Integer
Yes	saws	1		10	Integer

OK Cancel

Figure 4: Variable selection screen.

The column labeled "Suggested Value" allows the user to suggest a reference point $\mathbf{x}$ to be included in the initial population. The "Variable Selection" screen shows all the variables defined in the simulation model, however, only a few of them are appropriate to become part of the optimization model. For instance, setting the value of *drillq* as part of the optimization process is meaningless, since this variable is used in the model to keep track of the number of jobs waiting to be processed by an available drill. The main OptQuest screen looks as depicted in Figure 5 after a 10-minute run using the "Aggressive Neural Network" option.

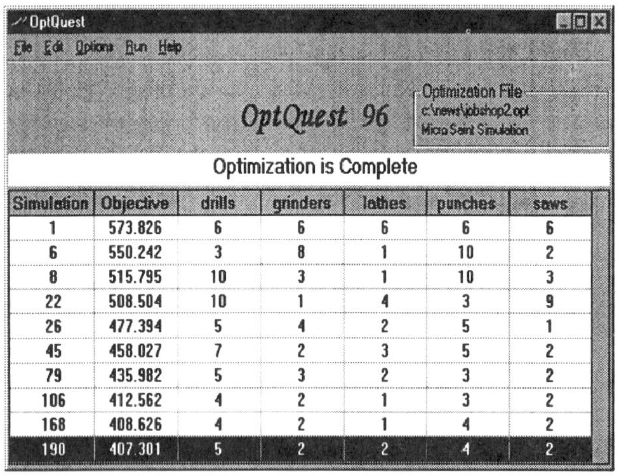

Figure 5: Main OptQuest screen.

6.2 Adding Constraints

We have mentioned that an important feature of OptQuest is the flexibility to impose constraints on the variables selected for optimization. In many situations it is desirable to search for solutions that satisfy restrictions that must be met by the system being simulated. OptQuest allows the user to impose any number of constraints, as long as they can be expressed as linear combinations of the optimization variables.

To illustrate the use of constraints, let's turn our attention again to the job shop problem. Suppose that we would like to search for an optimal job shop configuration that meets the following requirements:

1) The ratio of *lathes* to *drills* should be at least 2 to 1.

2) The number of *drills* and *punches* should not represent more than 40% of the total number of machines in the shop.

3) Given that machine costs (in thousands) are $100 for a drill, $150 for a grinder, $80 for a lathe, $120 for a punch, and $200 for a saw, the total cost of the equipment should not exceed $2,000K.

These restrictions can be represented in terms of linear constraints through simple algebraic transformations. Once they have been transformed, the constraints can be entered using the "Constraint Editor" in OptQuest. The screen in Figure 6 shows the result of adding the linear constraints that represent the requirements imposed on the problem.

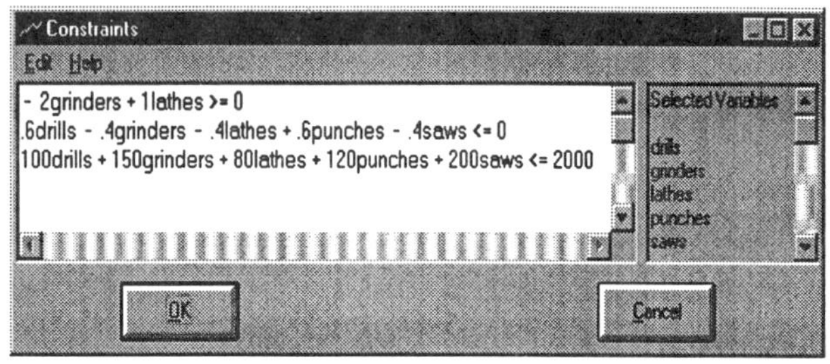

Figure 6: Constraint editor in OptQuest.

After running the optimization process for 10 minutes, the solution $\mathbf{x} = \{2, 2, 5, 2, 2\}$ is reported as the best. It can be easily verified that this solution satisfies all the constraints with $f(\mathbf{x}) = 405.206$.

It is interesting to note that the final objective function value (i.e., the best found in 10 minutes of search) is better than the one found for the unconstrained model (i.e., 407.301). This might seem surprising at first glance. However, restricting the search to regions where good solutions lie allows OptQuest to find high quality solutions faster. Therefore, if the user knows that good shop configurations meet the specifications set by the constraints, then by creating a constrained optimization model, better configurations can be found in a smaller amount of time.

6.3 Optimizing an Average Objective Function Value

We devote this subsection to address the optimization of the average response (or average objective function value). Instead of seeking a minimum or a maximum value of the objective function in a single run, it may sometimes be desirable to optimize its average value. This can be done by changing the number of runs that the simulator is asked to perform. One thing that it must remembered is that an increase in the number of runs causes an increase in the simulation time. So, the number of search iterations that can be performed in a fixed amount of time will decrease.

Suppose we change the number of runs to be 5 in our example. Now, OptQuest attempts to find the best values for *drills*, *grinders*, *lathes*, *punches*, and *saws* in order to minimize the average objective function value obtained after 5 runs of the simulation. In other words, the system attempts to find the best **x** such that $\bar{f}(\mathbf{x})$ is minimized, where the average value is found by running the simulation a given number of times with the same **x** value and different seeds for the random number generator. Detailed information concerning individual runs is displayed by the system on a separate window. This window is labeled "Run Data" and takes on the form depicted in Figure 7 during the search process in the job shop example.

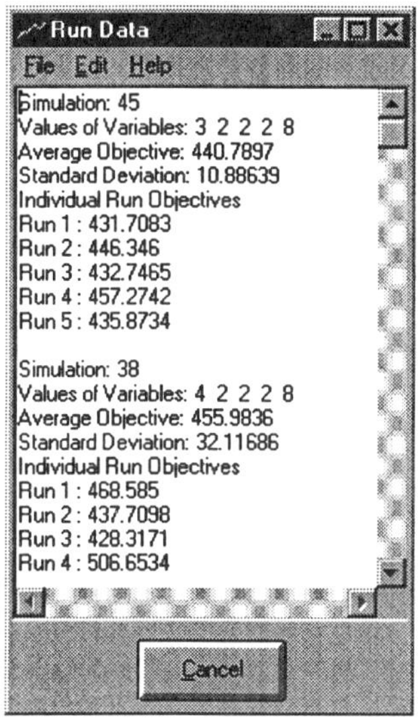

Figure 7: Run data screen.

The 10-minute limit allows the search in this case to only evaluate 85 solutions. This is due to the longer simulation time needed to perform 5 runs instead of one every time a configuration is tested.

7 CONCLUSIONS

In this paper we have described recent advances in optimization and simulation technologies that have made possible the development of a system that can effectively perform the task of optimizing simulations. We showed the benefits of adapting the approach known as scatter search in this context, while complementing this methodology with appropriate tabu search elements.

We have also demonstrated the capabilities of a commercial software package that successfully implements the notion of optimizing a complex system (in this case represented by a simulation model). The software package includes a scatter search / tabu search module, a mixed integer programming solver, and a procedure to configure and train neural networks. All these tools are integrated by a user-friendly interface.

We are currently undertaking additional research in the area of optimizing simulations with the goal of producing systems that are still more effective for dealing with the growing complexity of practical problems. The importance of integrating the complementary realms of optimization and simulation assures that future advances will have a high impact on real world applications

REFERENCES

Glover, F. 1977. Heuristics for integer programming using surrogate constraints. *Decision Sciences* 8:156-166.

Glover, F. 1994. Genetic algorithms and scatter search: unsuspected potentials. *Statistics and Computing* 4:131-140.

Glover, F. 1996. Tabu search and adaptive memory programming—advances, applications and challenges. To appear in *Interfaces in Computer Science and Operations Research,* eds. Barr, Helgason and Kennington, Kluwer Academic Publishers.

Glover, F. and M. Laguna. 1993. Tabu search. In *Modern Heuristic Techniques for Combinatorial Optimization*, ed. C. Reeves, 60-150. Blackwell Publishers: Oxford.

Holland, J. H. *Adaptation in Natural and Artificial Systems.* University of Michigan Press, Ann Arbor.

OptQuest 96 User's Guide—The Micro Saint Simulation Optimizer. Optimization Technologies, Inc. Boulder, Colorado.

AUTHOR BIOGRAPHIES

FRED GLOVER is the U S West Chaired Professor in Systems Science at the University of Colorado, Boulder. He has authored or co-authored more than two hundred published articles in the fields of mathematical optimization, computer science and artificial intelligence, with particular emphasis on practical applications in industry and government. In addition to holding editorial posts for journals in the U.S. and abroad, Dr. Glover has been featured as a National Visiting Lecturer by the Institute for Operations Research and the Management Sciences and has served as a host and lecturer in the U.S. National Academy of Sciences Program of Scientific Exchange.

JAMES P. KELLY is an Assistant Professor of Management Science in the College of Business and Administration and Graduate School of Business Administration at the University of Colorado in Boulder.

He received his bachelors and masters degrees in Chemical Engineering from Bucknell University in 1981. He worked four years as a systems analyst and project manager at Texas Instruments in Dallas, Texas. He returned to graduate school and received his doctoral degree in Applied Mathematics and Operations Research from the University of Maryland in 1990. While at the University of Maryland, his research into methods for providing statistical confidentiality was funded by a series of grants from the United States Bureau of the Census. His current research interests are in the area of heuristic combinatorial optimization and applied artificial intelligence. Dr. Kelly has published several papers on topics such as tabu search and simulated annealing in various journals such as *Operations Research, European Journal of Operational Research,* and the *INFORMS Journal on Computing*. He is a member of INFORMS. He is an associate editor for the *INFORMS Journal on Computing* and an editor for the *Journal of Heuristics.*

MANUEL LAGUNA is an Assistant Professor of Operations Management in the College of Business and Administration and Graduate School of Business Administration of the University of Colorado at Boulder. He received master's and doctoral degrees in Operations Research and Industrial Engineering from the University of Texas at Austin. He was the first U S WEST postdoctoral fellow in the Graduate School of Business at the University of Colorado. He has done extensive research in the interface between artificial intelligence and operations research to develop solution methods for problems in the areas of production planning and inventory control, routing and network design in telecommunications, and automated drawing. Dr. Laguna co-edited the Tabu Search volume of *Annals of Operations Research*, and he is currently editor of the *Journal of Heuristics* and *Combinatorial Optimization: Theory and Practice.* He is a member of the Institute for Operations Research and the Management Sciences, the Institute of Industrial Engineers, and the International Honor Society Omega Rho.

Proceedings of the 1996 Winter Simulation Conference
ed. J. M. Charnes, D. J. Morrice, D. T. Brunner, and J. J. Swain

MODELING WITH EVENT GRAPHS

Arnold H. Buss

Operations Research Department
Naval Postgraduate School
Monterey, CA 93943-5000, U.S.A.

ABSTRACT

Event Graphs are a way of graphically representing discrete-event simulation models. Also known as "Simulation Graphs," they have a minimalist design, with a single type of node and two types of edges with up to three options. Despite this simplicity, Event Graphs are extremely powerful. The Event Graph is the only graphical paradigm that directly models the event list logic. There are no limitations to the ability of Event Graphs to create a simulation model for any circumstance. Their simplicity, together with their extensibility, make them an ideal tool for rapid construction and prototyping of simulation models. In this paper we will demonstrate the ability of Event Graphs to leverage simple models into more complex ones with very few additional features.

1 INTRODUCTION

Several extensions to Event Graph capabilities have been introduced (see Schruben, 1995), including cancelling edges, passing parameters on scheduling edges, and the use of data structures. This paper presents some of the ways these features may be utilized to enhance Event Graph modeling. Of particular interest is the ability to easily leverage simple models into more complex ones.

When considering extensions to any model or methodology, care must be taken to avoid the methodology becoming burdened with too many features, destroying the elegance and utility of the original. With this paper we hope to demonstrate the viability of the extensions to the basic Event Graph paradigm as genuine enhancements to the construction of models. While the formal modeling power is enhanced, the ease of use and the quality of the resulting models are both improved.

In the following section we review the basics of Event Graph methodology and Section 3 we present some examples. Following that, we describe the enhanced features and show examples of their use.

2 BASICS OF EVENT GRAPH MODELS

We assume the reader is familiar with the basic concepts of discrete event simulation (see any introductory text such as Law and Kelton 1991), so we will only briefly review the components.

Two fundamental components of a discrete event simulation model are a set of *state variables*, and a set of *events*. The model emulates the system being studied by producing state trajectories, that is, time plots of the values of the system's state variables. Measures of performance are determined as statistics of these state trajectories. Discrete event models have state trajectories that are piecewise constant. Events are the points in time at which at least one state variable changes value. It is important to note that an event is an *instantaneous* occurrence in the discrete event model. No simulated time passes when an event occurs; simulated time passes only *between* the occurrence of events.

The timing of the occurrence of events is controlled by the *Future Event List* (or simply the *Event List*), which is nothing more than a "to-do" list of scheduled events. Whenever an event is scheduled to occur, an *event notice* is created and stored on the future events list. Every event notice contains two pieces of information: (1) What event is being scheduled; and (2) The (simulated) time at which the event is to occur. The future event list keeps the event notices in order by ranking them based on the lowest scheduled time. Events occurring simultaneously in simulated time must be prioritized according to some secondary rule.

The future events list is managed by a "Timemaster" who controls the flow of time in the simulated world of the model. The Timemaster examines the event list to see if there are any scheduled events.

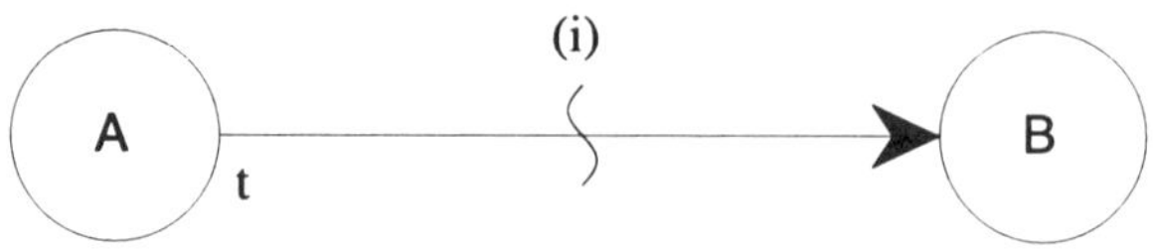

Figure 1: Fundamental Event Graph Construct

An empty list means there is nothing to do, so the Timemaster terminates (i.e. the simulation run ends). If the event list is not empty, the Timemaster updates the simulated clock to the time of the first event notice and executes the associated event — that is, the state transitions associated with that event are invoked. Note that the terminating condition (empty event list) means the simulation must be initiated with at least one scheduled event for any event to actually occur. We will follow Schruben's (1995) convention of a single distinguished event (Run) that is always on the event list initially.

When an event occurs, all state changes are made. Next, all further events are scheduled, and finally the event notice is removed from the Event List. The events scheduled are specified by the occurring event itself and may be conditional on certain values of the current state. The order of execution for these three steps could be altered, but the resulting models would be different. Although it is possible to mix up the actions (e.g. First change some states, then schedule some events, then change some more states, etc.), the resulting model would be confusing and prone to errors. There is considerable benefit from adapting a convention such as the one above.

Event Graphs are a way of representing the Future Event List logic for a discrete-event model. An Event Graph consists of *nodes* and *edges*. Each node corresponds to an event, or state transition, and each edge corresponds to the scheduling of other events. Each edge can optionally have an associated Boolean condition and/or a time delay. Figure 1 shows the fundamental construct for Event Graphs and is interpreted as follows: the occurrence of Event A causes Event B to be scheduled after a time delay of t, providing condition (i) is true (after the state transitions for Event A have been made). By convention, the time delay t is indicated toward the tail of the scheduling edge and the edge condition is shown just above the wavy line through the middle of the edge. If there is no time delay, then t is omitted. Similarly, if Event B is *always* scheduled following the occurrence of Event B, then the edge condition is omitted, and the edge is called an *unconditional* edge.

Thus, the basic Event Graph paradigm contains only two elements (event node and scheduling edge)

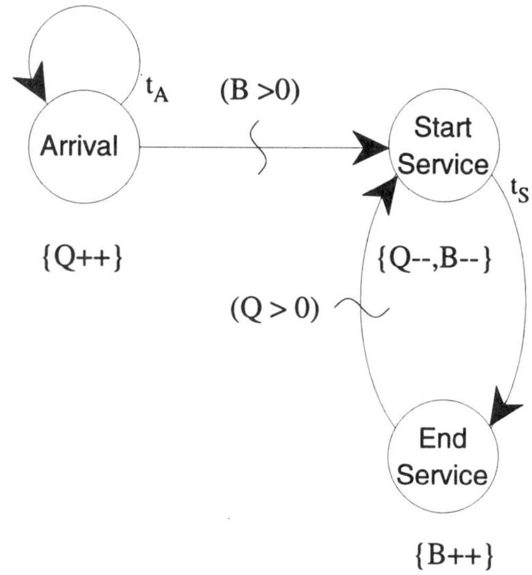

Figure 2: Discrete Event Model for Multiple Server Queue

with two options on the edges (time delay and edge condition). The simplicity of the Event Graph paradigm is evident from the fact that we can represent *any* discrete event model using *only* these constructs (Schruben 1992, 1995; Schruben and Yücesan 1993). A major advantage of the minimalist approach of Event Graphs is that the modeler can spend more time on model formulation and less on learning the constructs of the paradigm.

There is a price to the simplicity of Event Graphs, however. Since Event Graphs represent the event scheduling relationship, rather than the physical movement of, say, customers through a queueing system, Event Graphs require a higher degree of abstraction on the part of the user than other graphical systems. The author's experience using Event Graphs in an introductory simulation course indicates that this higher abstraction is easy to master and provides rich payoffs for understanding and creating discrete event simulations. Indeed, the use of Event Graphs tends to accelerate the understanding of the Discrete Event paradigm.

3 EXAMPLES

3.1 The $G/G/k$ Queue

An Event Graph model for the standard $G/G/k$ queue is shown in Figure 2. The state transitions for each event are shown in curly braces beside the corresponding node. The model utilizes the C notation for incrementing ('++') and decrementing ('−−') variables).

3.2 Tandem Queue Model

The $G/G/k$ queueing model of the previous section may be extended to a tandem queueing model by connecting copies of the model together, as shown in Figure 3. The two stations have their respective events and state variables indexed by the station, so that Q_i is the number of parts in the queue for workstation $i = 1, 2$, for example.

This stringing together of models superficially resembles the corresponding models in simulation languages with process interaction world views (GPSS, SIMAN, etc.) in which transaction block diagrams are connected. However, in the process languages, connecting blocks indicates transactions that get passed, whereas for Events Graphs it is the scheduling of events that is added. Since neither transactions nor entities are being passed, Event Graph models may be connected in more intricate ways, leading to much greater flexibility and more potential for modularity.

While very straightforward, this approach to the tandem queue model does not leave enough flexibility in the number of stations that can be modeled. The number of workcenters is hard-wired into the structure of the model, so that a different model is needed for each size shop. This is needless repetition of the model's logic, since the fundamental dynamics of an N-workstation tandem queueing model are exactly the same regardless of the value of N. A more robust approach is to have a *single* model for the tandem queueing structure and have the capability to specify N at runtime. That is, the number of stations in the queue is considered *data* for the model rather than a fundamental structural part of the model. Such a model is presented below when we discuss advanced features of Event Graphs. First, however, we will give another example of connecting two models in a useful way.

3.3 Worker Interference Model

A workcenter has K identical machines with a single worker operating them. Arriving parts must be loaded on a machine (if available) by the worker. If all machines are busy, the parts wait in a queue. Even if a machine is available, the parts must wait until the worker is free to load them on a machine, a situation called worker interference. More generally, interference can occur whenever multiple resources are required to perform a task but not all are available when the part is ready. Once loaded, machines automatically process a part with no further input needed from the worker until finished. However, when completed, the part must be unloaded by the worker.

We will construct an Event Graph model for this scenario by combining a $G/G/k$ queueing model (for the loading part) with a piece that resembles the downstream workstations for the transfer line model. First, define the following variables:

Q = # parts in queue awaiting processing
B = Worker status (1 if idle, 0 if busy)
M = # available machines ($0 \leq M \leq K$)
U = # parts waiting to be unloaded from machines ($0 \leq U \leq M$)
P = Total number of parts processed
t_A = Time between part arrivals
t_L = Loading times
t_U = Unloading times
t_S = Part processing times

The loading process is identical to the $G/G/k$ queue described above, as shown in Figure 4, with two exceptions: (1) The condition for starting to load a part requires *both* a machine and a worker available, reflected in the edge condition on the Arrival–Start Loading edge. (2) The loading activity requires both a machine and a worker, reflected in the state transition for Start Loading.

The unloading piece is shown in Figure 5 Note that the unloading portion only differs structurally from a $G/G/k$ model in the arrival process: unloading is triggered by the completion of processing by a machine, not by outside arrivals. The only other difference is the fact that a machine is freed by the last event (Finish Unloading) in addition to the worker.

Now all we need to do is connect the two pieces of the model. Since a machine starts processing as soon as it is loaded, we can schedule a Finish Processing after Finish Loading with a delay of t_S, the service time. After the worker is finished loading a part, another loading/unloading task may be performed, if necessary. At this point there is some ambiguity in the problem description, since it is not specified what the worker is to do if there are parts waiting to be both loaded and unloaded. Assume that the worker's priority is unloading parts over loading parts. Then after the Finish Loading event, a Start Unloading is scheduled, providing there is a part waiting to be unloaded (i.e. if ($U > 0$)). On the other hand, if there are no parts waiting to be unloaded ($U = 0$) *and* there is at least one part waiting to be loaded ($Q > 0$) *and* there is a machine available to load the part onto ($M > 0$), then a Start Loading may be scheduled after a Finish Loading. Similarly, after a Finish Unloading event, if no parts are waiting to be unloaded ($U = 0$)

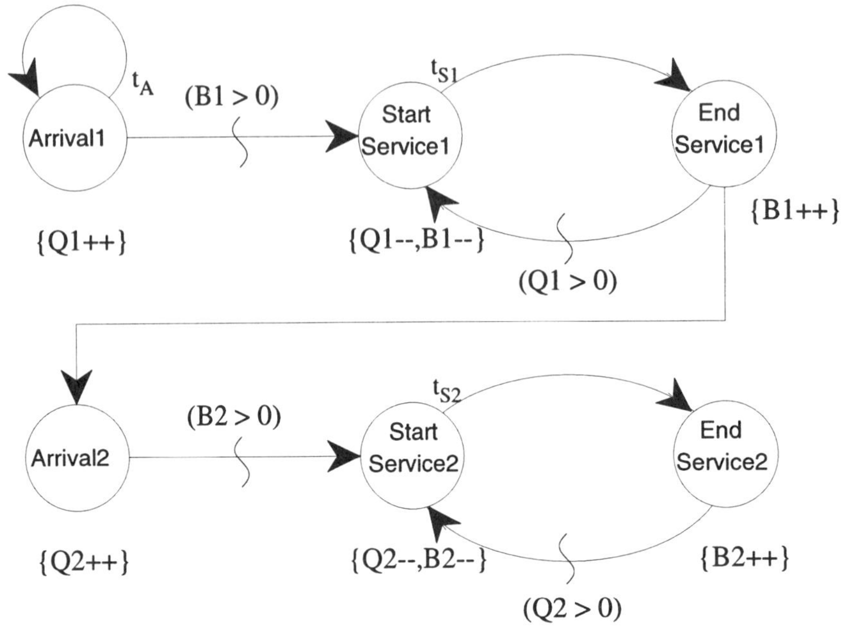

Figure 3: A Two Station Tandem Queue

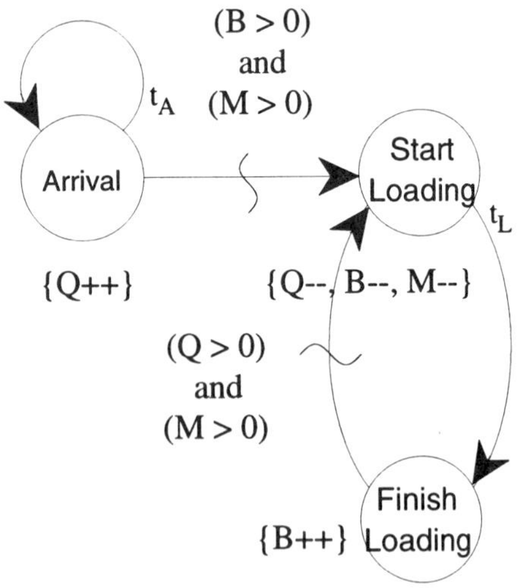

Figure 4: Event Graph for Loading Portion of Model

and there are parts waiting to be loaded, then a Start Loading event may also be scheduled. Note that following a Finish Unloading event there is no need to check the condition $(M > 0)$ since M has just been incremented. The final model is shown in Figure 6.

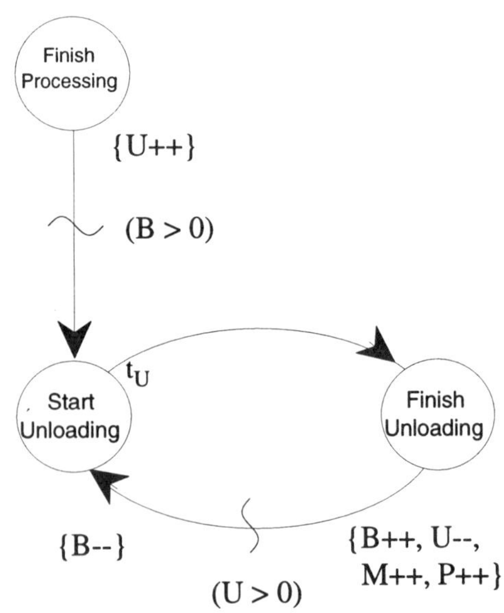

Figure 5: Event Graph for Unloading Portion of Model

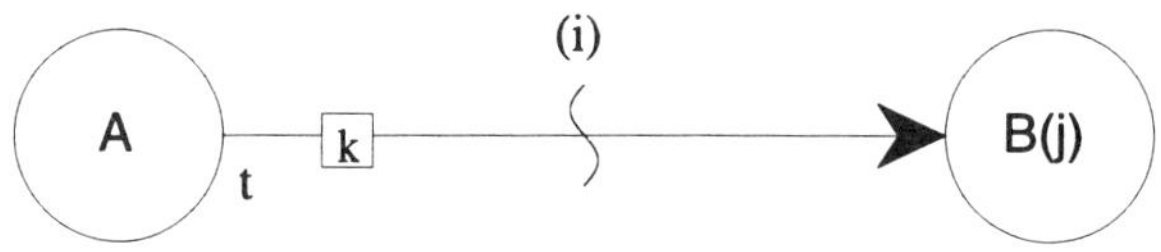

Figure 7: Passing Attributes on Edges

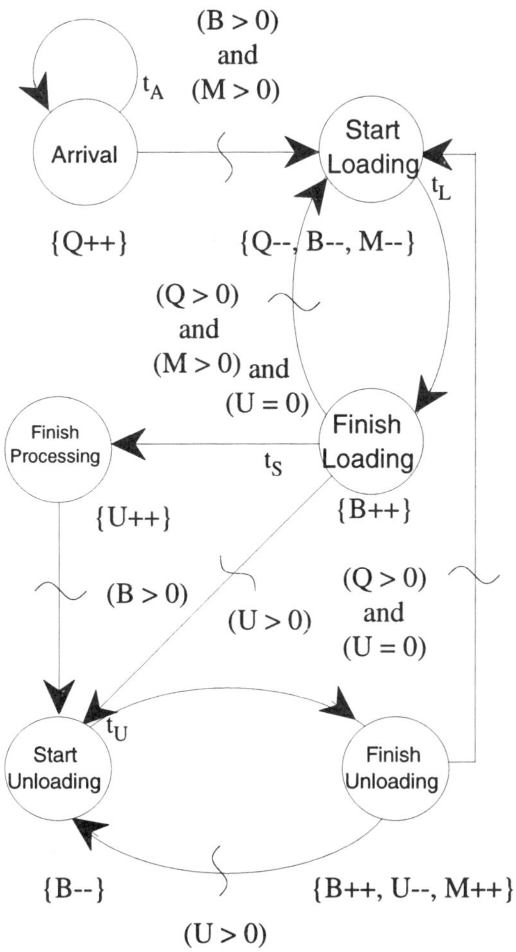

Figure 6: Final Event Graph Model for Worker Interference Problem

4 ADVANCED FEATURES OF EVENT GRAPHS

The Event Graph paradigm described above is a simple and elegant way to represent discrete event logic. Without any further enhancements it has sufficient flexibility and power to represent *any* discrete event model. We will discuss three such enhancements of the basic Event Graph paradigm: passing attributes to events on scheduling edges, event-canceling edges, and the use of data structures (instead of just simple data types). As noted previously, these enhancements do not increase the formal power of Event Graphs, only their readability, ease of construction, and in some cases the quality of the model itself.

4.1 Passing Attributes on Edges

The first enhancement provides the event node with the capability to pass attributes on an event scheduling edge to the scheduled event. Figure 7 illustrates the basic construction and is interpreted as follows: When event A occurs, A's state transitions are made and expression k and condition (i) evaluated. If condition (i) is true, then event B is scheduled to occur after a delay of t time units with parameter j set equal to the computed value of k. Note that k could be a parameter list, as with a procedure call with arguments.

This simple enhancement allows complex models to be built up from simpler components in a relatively straightforward manner. To illustrate we will extend the queueing model of the previous section to the tandem queueing model discussed earlier. A production facility consists of N machine groups, each group having a single waiting line. Jobs enter at machine group 1 and upon leaving go to machine group 2, etc. For simplicity, we will assume the queues (buffers) all have infinite capacity.

Modeling this system is made much simpler by the observation that each machine group operates like the multiple-server queueing system with two exceptions: the departure from a machine group schedules the arrival of a job to the next machine group, and the only arrival of jobs from outside the facility are to machine group 1. The state space must also be expanded to

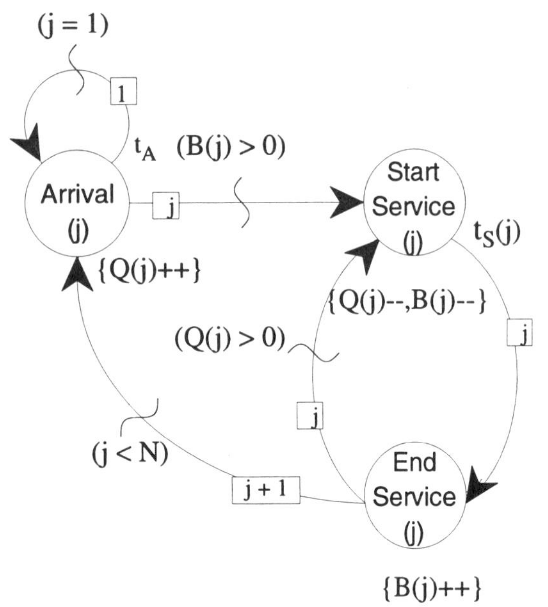

Figure 8: Event Graph for Transfer Line Model

identify the number of jobs in queue as well as the number of available machines at *each* workcenter. It is convenient to simply make Q and B arrays, with $Q(j)$ the number in queue and $B(j)$ the number of available machines at machine group j. Similarly, the parameters of the system are now an array, with $k(j)$ the number of machines in machine group j.

Figure 8 shows the Event Graph model for the transfer line. The similarity of this model to the queueing model in Figure 2 is self-evident. The self-scheduling edge for the Arrival(j) event adds the condition that $j = 1$ to generate the arrival of jobs from outside the shop. The other Arrival(j) events are scheduled from the previous machine group. However, an End Service(j) event with $j = N$ results in a job leaving the system. Therefore, there is the condition $j < N$. All other edges in the model are the same as the corresponding ones in Figure 2, with parameter j representing the current machine group being passed. The state transitions for the events are similarly indexed by the corresponding machine group number.

The transfer line model could have been modeled using just the basic constructs in Section 2 (without passing attributes) by simply stringing together copies of the model in Figure 2 and making the appropriate adjustments in edges. However, that approach would "hard-wire" the number of machine groups N into the model. To simulate facilities having different numbers of machine groups a new model would have

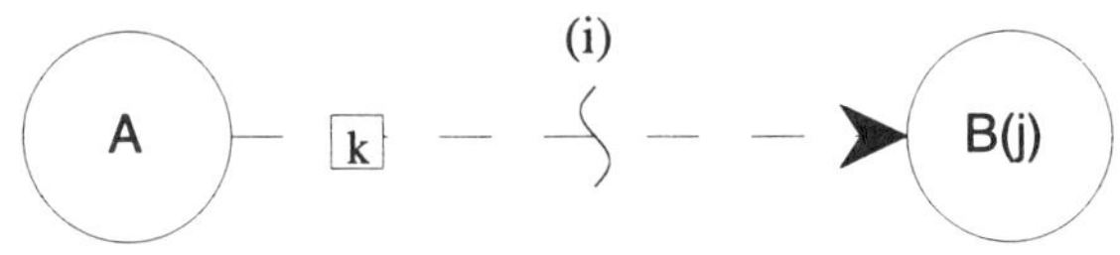

Figure 9: A Canceling Edge

to be constructed. In contrast, the Event Graph in Figure 8 can be used to model transfer lines of *any* size by simply setting the appropriate value of N and of $k(j)$ for $j = 1, \ldots, N$.

4.2 Canceling Edges

The second enhancement covers situations in which the modeler wishes to have an event notice removed from the event list. That is, a scheduled event needs to be canceled. This is accomplished in a Event Graph by the addition of *canceling edges* denoted by dashed arrows; Figure 9 shows the basic construction of a canceling edge. The interpretation of Figure 9 is: When event A occurs, then (after the appropriate state transitions are made), if Condition (i) is true, the next occurrence of event B with parameter j equal to k is removed from the Event List. If there is no such event notice on the event list, nothing happens. The parameter is optional and, if omitted, the next occurrence of B is cancelled.

4.3 Use of Data Structures

The modeling power of Event Graphs is considerably increased with the judicious use of data structures, such as lists and priority queues. For example, in the $G/G/k$ queue of Figure 2, suppose statistics on individual customers' times in queue are needed. These can be determined by creating a queue containing the times of arrival for each customer. Assume the current value of simulated time is available in a global variable called "Clk" (after Schruben, 1995). Upon arrival, the value of Clk is stored in a fifo queue called ArrivalTimes. When service is started, the time of arrival for that customer is removed, and the difference between it and the current value of Clk is that customer's time in queue. We will utilize a stylized syntax for a fifo queue (the data structure, not the system being modeled). Add(<*Queue*>, <*Value*>) puts <*Value*> at the end of the list <*Queue*>, and Remove(<*Queue*>) removes and returns the first element of <*Queue*>. The only changes that need to be made to Figure 2 are in the state transitions for Arrival and Start Service events as follows:

Arrival
$Q{++}$
Add(ArrivalTimes, Clk)

Start Service
$Q--, B--$
TimeOfArrival = Remove(ArrivalTimes)
TimeInQueue = Clk − TimeOfArrival

End Service
$B{++}$

Another use for lists in queueing simulations is when service disciplines other than first-come first-served are employed. One such rule that is often used is shortest processing time (SPT). The state changes for the Event Graph of Figure 2 can be modified by generating each customer's service time t_S upon arrival to the system. The service times are then put on a priority queue called ServiceTimes, ranked according to the smallest value. The new state changes are:

Arrival
$Q{++}$
Generate t_S
Add(ServiceTimes, t_S)

Start Service
$Q--, B--$
t_S = Remove(ServiceTimes)

End Service
$B{++}$

The use of data structures such as fifo and priority queues for Event Graph modeling is generic in that no specifications are made with regard to their actual structure or implementation. Any implementation of Event Graph methodology should utilize the most efficient data structure for the task.

4.4 A Queue with Reneging

In a $G/G/k$ model, suppose customers who wait in the queue for more than t_R time units *renege*, that is leave the system without receiving service. The reneging behavior can be modeled in a straightforward manner using a canceling edge, as shown in Figure 10. Each customer's renege event is 'scheduled' upon entry to the system. Since each customer's renege time is different, the order of reneges is not necessarily the same as the order of arrival to the system. Therefore, we add the state variable A which represents the cumulative number of customer arrivals.

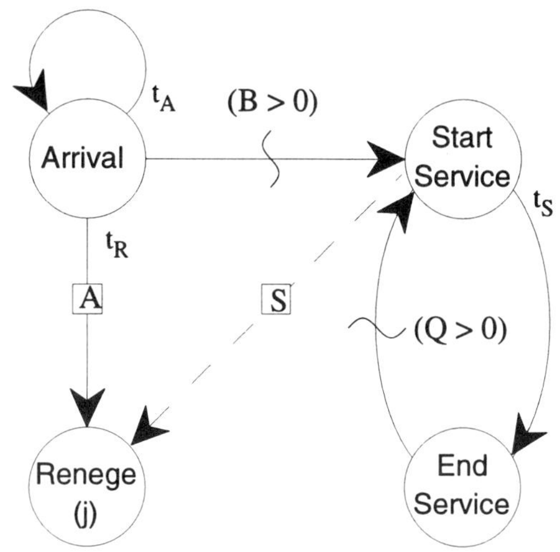

Figure 10: Modeling Customers' Reneging

Thus, A gives each customer a unique number in the Arrival event which is then passed to the Renege event via its scheduling edge. If the Renege event occurs before Start Service, the corresponding customer is removed from the queue and the number in queue is decremented. On the other hand, if the Start Service event occurs first, then the Renege event corresponding to that customer is canceled. The state changes are summarized as follows:

Arrival
$Q{++}, A{++}$
Generate t_S
Add(Queue, A)

Start Service
$Q--, B--$
S = Remove(Queue)

Renege(j)
$Q--$
Remove(Queue, j)

End Service
$B{++}$

Figure 10 utilizes passing a parameter on a canceling edge. When Start Service occurs, the number of the customer about to receive service (S) is removed from Queue and the event list is scanned for a Renege event with parameter value S. When that event is found, it is removed from the Event List, thus canceling the scheduled Renege event. Since a Renege event may have occurred prior to the Start Service, successive values of S are not necessarily sequential.

5 CONCLUSIONS

We have given a brief overview of the use of some advanced features of Event Graphs for discrete-event simulation models. Event Graphs are currently the only graphical tool that directly models the Event List paradigm. The enhancements we have described here allow the modeler to easily leverage simple models into more complex ones. More important, the visual power of the Event Graph gives the modeler a unique perspective on the model and allows the key underlying relationships to be vividly represented. The availability of Event Graph software in the form of Sigma^{TM} (Schruben, 1995) allows simulation modelers to take advantage of the benefits offered by the Event Graph paradigm.

ACKNOWLEDGEMENTS

The author wishes to thank Paul Sanchez and Brian Widdowson for very helpful comments on an earlier draft of this paper. Support for this work from the Naval Postgraduate School is gratefully acknowledged.

REFERENCES

Buss, A. 1995. A Tutorial on Discrete-Event Modeling with Simulation Graphs, *Proceedings of the 1995 Winter Simulation Conference*, C. Alexopoulis, K. Kang, W. Lilegdon, D. Goldsman (eds).

Schruben, L. 1983. Simulation Modeling with Event Graphs, *Communications of the ACM*, **26**, 957–963.

Schruben, L. 1992. *Sigma: A Graphical Simulation Modeling Program*, Boyd and Fraser Publishing Company, Danvers, MA.

Schruben, L. 1995. *Graphical Simulation Modeling and Analysis Using Sigma for Windows*, Boyd and Fraser Publishing Company, Danvers, MA.

Schruben, L and E. Yücesan. 1993. Modeling Paradigms for Discrete Event Simulation, *Operations Research Letters*, **13**, 265–275.

AUTHOR BIOGRAPHY

ARNOLD BUSS is a Visiting Assistant Professor of Operations Research at the Naval Graduate School.

Proceedings of the 1996 Winter Simulation Conference
ed. J. M. Charnes, D. J. Morrice, D. T. Brunner, and J. J. Swain

INTRODUCTION & OVERVIEW OF "ARTIFICIAL LIFE" -- EVOLVING INTELLIGENT AGENTS FOR MODELING & SIMULATION

A. Martin Wildberger

Electric Power Research Institute
3412 Hillview Ave.
Palo Alto, CA 94304-1395 U.S.A.

ABSTRACT

"Artificial Life," despite its biological analogy and the hyperbole that its name implies, is really a collection of methods for building discrete event simulations with evolving multiple agents. It consists mainly of representing parts of systems or natural phenomena as individual active objects that may be both persistent and self-modifiable, operating on them with genetic algorithms or other evolutionary computing techniques and treating their multi-dimensional parameter (state) space discretely, by using cellular automata or similar coupled map lattices. The attractiveness of these methods for general purpose modeling and simulation lies in their ability to produce complex emergent phenomena out of a small set of relatively simple rules, constraints and relationships couched in either quantitative or qualitative terms. This tutorial includes brief introductions to both GA and CA, a description of a tool kit for building multi-agent simulations, and an outline of a current application to electric power systems and to the evolution of the electric power industry itself. References are included to demonstration software and source code available on the internet.

1 BACKGROUND

For at least fifteen years, many simulations, and simulation tools, have employed multiple agents, usually called "actors" or "demons" and represented by separate processes, which operate independently and interact by communicating selected information through messages. Until recently, these agents and their interactions have been specifically modeled to represent relatively unchanging functions, operations or physical entities. For most of these simulations, consistency and reproducibility were essential requirements. In general, artificial intelligence techniques were used only when necessary to model human decision making and were couched in the form of policy rules to be followed, for instance, by processes simulating additional crew members in a team trainer.

More recently, some researchers have been building intelligent agent simulations with some form of self-modifiability in an attempt to model natural or societal processes whose complexity would otherwise lead to exponentially increasing computational demands. The chief difference is that these agents change as the simulation progresses. They "evolve" by "adapting" their behavior in both competitive and cooperative ways to meet general goals which are assigned by the simulation designer but whose details may also change through the same "evolutionary" process. Like Monte Carlo, this simulation technique uses statistical methods to obtain an end result that is, nevertheless, deterministic and reproducible within the bounds of numerical precision, although the route and the CPU time taken may not be identical over all runs of the simulation.

These simulations go by various names such as: artificial life, animats or softbots, all of which may suggest more capability than is actually being delivered, and some of the practitioners of these simulations have been guilty of unrealistic "hype." Nevertheless, this approach has been particularly successful in modeling biological, bio-molecular and ecological phenomena, as might be expected since it is loosely based on a biological analogy. What is more surprising is the success these methods have had in modeling economic and financial processes such as commodity markets and currency exchanges.

The attractiveness of these methods for general purpose modeling and simulation lies in their ability to produce complex emergent phenomena out of a small set of relatively simple rules, constraints and relationships couched in either quantitative or qualitative terms. Inventing the right set of the local rules to achieve the desired global behavior is not always easy -- although it often seems obvious afterward. For instance, in a simulation of freeway traffic flow, simple rules that govern the actions of individual vehicles ("travel at such-and-such a speed," "don't hit anything") result in global behavior that is highly complicated (structured "platoons" of fast-moving autos form, as do convoluted traffic jams). In other examples, flocking arises in a simulation of multiple individual birds, and elasticity in polymers emerges from a simulation of molecular interactions.

A variety of evolutionary computational methods are employed in these simulations, but most are variations on John Holland's genetic algorithms (GA). (Holland 1975) The GA structure, based on individual membership in "species," most easily lends itself to modeling individuals as unique instances of more general (evolving) classes. These simulations typically include displays of the agent behavior over time in a physical space, if appropriate, or, more often, in a state or phase space. Therefore, cellular automata (CA) and similar spatially extended, coupled map lattices are also frequently used as another tool for intelligent agent based modeling.

2 GENETIC ALGORITHMS

Genetic algorithms (GA) provide a technique for the simulation of complex systems by modeling them in the form of multiple, interacting, relatively simplistic agents, represented simply by bit strings of ones and zeros, that evolve and adapt through competition and cooperation in a (usually) stable environment defined by the simulation-builder, modeled as another bit string and called a "fitness function" by analogy with biological evolution. The bits may stand for characteristics, capabilities, or relatively simple strategies which can be selected and recombined to form different individuals by "crossover" (cutting up two strings and patching them together) or by "mutations" (random changes in bit settings). By comparing randomly selected individuals' relative success with respect to the fitness function and then creating a new "generation" of individuals from crossover and mutation biased toward using the most successful individuals, eminently successful individuals are eventually "evolved."

Genetic algorithms are examples of a very broad class of systems that can be described as "adaptive non-linear networks" (Holland 1975) Some artificial neural network paradigms also fit into this class. All these systems consist of a large number of units that interact in a non-linear, competitive manner, and are modified by their own interaction or by external operators so that the overall system adapts to its environment. That environment is conventionally defined in terms of the system's input, its output, and some criterion, either internal or external to the system, by which its success in adapting can be measured. The competitive nature of this process does not necessarily imply a zero-sum game in which individual units within the system only benefit at the expense of other units. Combining and/or cooperating among units can be beneficial to the system, but ultimately the whole system adapts through change, growth, or even loss of individual units.

Genetic algorithms are based loosely on the theory of evolution, genetic diversity, and the "survival of the fittest." To use genetic algorithms to model individual agents, it is first necessary to invent, select, or otherwise define, a number of possible candidate strategies, rules or procedures to make available to each agent. Each agent may also be given some number of possible candidate attributes, qualities or characteristics. If any of these strategies or attributes are relatively complex in itself, it should be described in terms of its constituent parts or aspects, some of which it may have in common with other strategies or attributes. These (whole or partial) strategies and attributes may be considered analogous to biological genes, and an individual agent defined in terms of whether it has or does not have particular "genes".

Conceptually, all the agents evolve in parallel over a series of generations through competition and "selective breeding." Each generation of individuals is tested as to how well they perform on the basis of a specified fitness function, which results in an adjustment of their current "strength", the measure of their success. The next generation of new individuals are produced by genetic recombination and some mutation of the most successful individuals. Most unsuccessful individuals do not reproduce at all. Eventually almost all individuals belong to the same successful set. If we wish to maintain, or to evolve, two or more different classes of agents, we can limit the interbreeding to members of the same class (or species).

2.1 Mathematical Formulation Of GA

Mathematically, genetic algorithms are similar to other forms of "hill climbing" optimization using randomized gradients. However, they allow a convenient representation of a multi-dimensional problem space when different coordinates of that space are measured by continuous, discrete, and symbolic values. Genetic algorithms also provide for direct parallel implementation whereas conventional algorithms require the additional effort to convert them from their original sequential form.

To begin describing the computational implementation of genetic algorithms, it is first necessary to define formal classifiers. First, take as primitive elements, all possible binary strings: $m_1 m_2 ...... m_j ...... m_k$ (of length k, where m_j is taken from the set {1,0}, and m_j is either zero or one). These binary strings are called "messages". Next, define a "condition" to be any subset of messages specified by a string of the form: $c_1 c_2 ...... c_j ...... c_k$ (of length k, where c_j is taken from the set {1,0,#}, c_j is either zero, or one, or #, and the number sign, #, is a pure symbol that can intuitively be taken to mean: "don't care whether it's zero or one"). The subset of messages that meet a particular condition have, for all the positions in them:

(1) $m_j = 1$ when $c_j = 1$

(2) $m_j = 0$ when $c_j = 0$

(3) m_j is either 1 or 0 when $c_j = \#$

For instance, the message: "0001100" meets the condition: "00##100". (In this example, k = 7.)

Conditions can be combined according to all the usual Boolean operations: **and**, **or**, **not**, and any combination of them. Any set of one or more such conditions is defined to be a "classifier" on the space of messages.

The definition of "genetic classifiers" requires another level of abstraction. A "schema" is defined to be any subset of conditions. For length k = 11, an example of a schema is: s = *0##1****** where "*" means "don't care whether it's zero, one, or #". Members of the subset specified by a schema are called "instances" of that schema. Examples of two possible instances of the schema s would be: x = 10##1001010 and y = 00##1010011. A schema defines a subset of the set of all possible conditions, whereas each condition defines a subset of the set of all possible messages. Any set of one or more such schema is defined to be a "genetic classifier" on the space of messages.

As these genetic classifiers evolve through many generations, they must be tested for how well they meet the original fitness function. At time t, test the available instances of s and record their "strengths"; i.e.: their fitness. Then assign a "payoff" value v(s,t) to s at time t by averaging the strengths of all its instances. Let v(t) be the average strength at time t of all instances of all schema under consideration and let m(s,t) be the number of instances available of schema s at time t. Then bias the selective breeding so that, at some future time, t + T, the number of instances of s will increase or decrease according to the ratio of the strength of s to the average strength of all schema: $m(s,t + T) = b[v(s,t)/v(t)]m(s,t)$ where b is an arbitrary (but strictly positive) constant.

The selection of strings for reproduction is performed randomly, but the probability of a given string being selected is made proportional to the normalized value of its strength ratio as described above. The total number of strings selected for each generation can be controlled so as to maintain the size of the population within tractable limits. Since a single condition is an instance of 2^k substring schema, explicit calculation of these ratings would be impractical. Genetic operators accomplish this same effect implicitly by constructing hybrid offspring from combinations of substring schema carried by high strength parents.

The simplest method of string reproduction is to copy without change the strings selected by the random process. This clones the most successful instances, and causes the "fittest" schema to grow exponentially until the population limit is reached. This approach usually converges rapidly but it assumes that one of the trial solutions is, in fact, the best. It does not provide for the "discovery" of any other solution than those candidates included by the algorithm designer. Other "genetic operators" combine parts of more than one string (selected randomly as above) to form wholly new individuals. These approaches are mostly variations on a basic operation called "crossover". Crossover is applied to a pair of strings as follows: select at random a position i, $1 < i < k$, and exchange the segments to the left of position i in the two strings. Crossover is considered to be analogous with sexual reproduction. Clearly, this method can be expanded by breaking the two strings in more than one place and/or by selecting pieces from more than two strings. In general, the simplest technique is as effective as the more complex ones, although the latter may speed convergence in certain cases.

Operators of the crossover type mix aspects of the trial solutions in an attempt to produce a better result than any one of the original set could achieve. However, they add nothing really new to that trial set. Another type of genetic operators, "mutation", can be used to produce random changes in a string, thus potentially inventing a wholly new strategy or characteristic. Experience with this type of operation indicates that the rate of mutation must be kept relatively low for best results.

2.2 Other Evolutionary Computing Methods

Two similar computing approaches have developed in parallel to GA. One, the "evolutionary programming" technique of Lawrence J. Fogel (1966) is based on evolution at the "behavioral" (performance) level--where selection for fitness actually takes place--rather than at the genetic level of GA. In this method, evolutionary change is produced solely through random mutation of the most successful candidates: no explicit connection is made between "behavioral" fitness and "genetic" change. Despite this difference, evolutionary programming generally produces results similar to GA. The second approach involves "evolutionary strategies," a technique introduced by Ingo Rechenberg (1973), and Hans-Paul Schwefel (1975). This method focuses on evolution of machine learning through trial-and-error search procedures with automated reinforcement of successful strategies.

Although evolutionary programming, evolutionary strategies, and GA developed separately and continue to represent areas of independent research, they are beginning to be viewed as different facets of evolutionary computing rather than fundamentally different techniques.

2.3 Cautions on the Use of GA

Genetic algorithms, and other evolutionary computing methods, must be used with care. They cannot be guaranteed to produce an absolutely optimum solution. It is also difficult to predict how many generations it will take to produce a satisfactory result. The inherent combinatoric explosion of which they are capable, is only avoided because most possible strings are never tested. Genetic algorithms are most beneficial when used in a parallel processing environment. Like neural networks, they can easily be simulated on a single sequential processor, but either the time or the computer power must be available to test hundreds of solutions in each generation until a satisfactory set is obtained. Designing the genetic operators, describing candidate solutions as strings of genes, and defining the pay-off function are still more of an art than a science. The process must be tuned to guard against "super" individuals winning too soon and causing the irrevocable loss of genetic material. Furthermore, problem spaces can be constructed (using Walsh functions) whose maxima will never be found by a genetic algorithm.

3 CELLULAR AUTOMATA

Cellular automata (CA) are massively parallel computing engines that provide tools for simulating phenomena characterized by simple local rules and complicated, unpredictable global behavior. CA can produce complex "emergent" behaviors even though they do not appear to have been included in the original CA design. CA in use range from J.H. Conway's classic computer game "Life" (Conway 1982) to the ECHO and SWARM (Santa Fe Institute 1996) series of programs designed by researchers at the Santa Fe Institute (SFI) to simulate some of the emergent phenomena mentioned previously.

CA are built from an array of ordered sites; input data or boundary conditions are represented by the initial site configuration. Values assigned to sites change synchronously in discrete steps over time by application of relatively simple, local rules. This process is somewhat similar to the evolution of successive GA generations except that no external fitness function is used to evaluate computations: all information is endogenous.

Construction of a CA and examination of its operation is relatively simple; "programming" it to produce meaningful results, which requires selection of the correct initial configuration and proper behavioral rules, is not--and the programming considerations are usually difficult to express formally. Despite considerable effort, the mathematical fundamentals of CA design are still not well understood. However, the ability of GA to evolve good solutions to poorly formulated problems may provide an approach to improved CA design. Melanie Mitchell (Mitchell and Forrest 1993) and colleagues at SFI are investigating GA that can evolve an initial CA configuration and a set of local rules appropriate for specific computational tasks.

3.1 Mathematical Formulation of CA

The simplest CA can be built by starting with a linear array of zeros and ones: for example, 01100101, and applying a rule to construct another such array. For instance, the rule: "If both nearest neighbors are equal, change; otherwise stay the same," produces, sequentially:

01100101
11100010
10101001
11010001

In this example, it is assumed that the first and last cells are neighbors: i.e., the array is circular.

Although it is clear that this deterministic, finite machine will eventually repeat itself, it is not immediately obvious how soon repetition will occur, nor is it easy to predict the length and membership of the limit cycle into which any finite CA must eventually fall. Small changes in the starting values and in the rules can have significant effects. By displaying each new linear array beneath (or above) the last and using contrasting colors (or black and white) to represent the ones and zeros, it is possible to watch the sequence of developing patterns and immediately recognize stable states or limit cycles. CA of higher dimensions and/or more complex rules can evolve to chaotic, aperiodic patterns that are the analog of strange attractors in dynamical systems. Since these extremely simple CA produce global effects from strictly local operations, they make it possible to model the emergence of the most complex known phenomena, perhaps life itself, from multiple, simple activities by many individual elements that appear to be affecting only their immediate neighbors.

3.2 CA Notation and Terminology

The simplest form of finite CA is made up of a lattice of L cells, each of which, at time t, can be in one of K states. A single fixed rule is used to update the state of each cell. The rule may be different for each cell, but the rule associated with each cell is not subject to change. The rule for each cell operates on that cell simultaneously with all the rest to convert all their states, (called the configuration or global state of the CA) at time t, to a new configuration at time t+1. The rule is a mapping from the current state of the cell and that of its nearest neighbors to the updated state of the cell in question. The number of neighbors that enter into the rule is known as the cell's radius, r. Note that the terminology in this paper is used by most, but not all researchers in the field.

The rule is usually the same for all cells and is often expressed as a table. The following is a rule table for a binary-state (0,1) nearest neighbor (r=1) CA. All 8 possible neighborhood patterns are shown on the left while the right hand bit shows the rule's output bit, which becomes the center cell's value at the next time step.

000 -> 0
001 -> 0
010 -> 0
011 -> 1
100 -> 0
101 -> 1
110 -> 1
111 -> 1

This rule is known in shorthand as #232 because the output values for sequentially increasing inputs make up the binary version of that number, i.e.:
$11101000_2 = 232_{10}$

CA can be extended to any number of dimensions, the number of cells allowed to become infinite, and their rules made as complicated as desired. The rules can include probabilities, and uncertainty (noise) can also be allowed to affect the accuracy with which values can be read.

3.3 Visualizing CA

There are two common ways to visualize CA. A state space (or phase space) plot is best for showing space-time patterns, but a state transition graph is more effective in illustrating CA behavior as a non-linear dynamical system. Software is available on the Internet (Wuensche 1996) that demonstrates these visualizations and allows the user to explore various rules and initial configurations for CA.

3.3.1 Space-Time Patterns in State Space

A finite sized CA may be bounded in any way that is desired, but the boundary cells will then require special rules to handle their unique situations. The simplest CA architecture is a small, 1-D array of cells of length L, with binary values (K=2) and a small local neighborhood (r<<L). This array is treated as circular, thus creating periodic boundary conditions which do not require any special rules for cells 1 and L since they are considered to be neighbors. Evolution of the CA may be represented an a sequence of configurations wrapped around a cylinder.

However, it is convenient to split the cylinder between cells 1 and L, flatten it out, and represent its evolution as a 2-D space-time pattern, with space running across the screen or page, and time progressing downward. By painting pixels in differing colors depending on the value of each cell (or black and white in the case of a binary valued CA), it is easy to detect patterns, observe static, periodic, or complex (possibly chaotic) behavior, and to compare the effects of different rules and different initial configurations. Although many researchers have developed rule classification schemes on the basis of such space-time patterns, these effects are necessarily phenomenological and not all observers may detect the same patterns.

3.3.2 State Transition Graphs

A finite, deterministic CA must eventually repeat a configuration (or global state) that occurred at some earlier time. Its trajectory will then become trapped in a repeating sequence of the same configurations. This becomes a cyclic attractor, with a period of 1 or more, but no greater than K^L. All configurations are either part of one such attractor or belong to some transient sequence of configurations that lead to an attractor. Configurations at the beginning of such a transient cannot be reached from any other such global state of the CA and can only exist if the creator of the CA initializes it in one of those configurations. (These configurations have become known as "garden-of-Eden" global states.) All the configurations on a cyclic attractor, along with all configurations on transient trajectories leading to that attractor, make up its basin of attraction. Any particular CA may have one or many such basins, but it is impossible to transition from a configuration in one of these basins to a configuration in any other one. All of the basins of attraction for a particular CA make up that CA's basin field.

If we represent each configuration of a CA by drawing a small circle, and connect each of those circles to all the other circles that represent just those configurations that can transition to the first state, we will produce a directed graph that links all of the configurations belonging to each basin of attraction for that CA. This graph of the entire basin of attraction field displays all the possible dynamics of that CA. The set of basins partitions all the configurations of the CA into disjunct classes. The CA cannot evolve from one basin to another.

In any finite CA, it is always theoretically possible to compute the complete basin of attraction field by exhaustively working either forward (as described above) or backward (by computing all the pre-images of each configuration). However, the time to compute pre- or post- images increases exponentially with lattice size so that this approach is impractical for larger arrays. Wuensche and Lesser (1992) have developed a computational shortcut, the "reverse algorithm," which directly computes pre-images and improves the computation time by several orders of magnitude. They have also defined a parameter, Z, as the probability that the next unknown cell in a partial pre-image can be uniquely determined from the values of the cells already known. A relatively high value for Z implies the ability to compute all configurations of a CA without an exhaustive search.

3.4 Using CA as System Simulators

The remarkable computing capability of CA make them useful tools for modeling and simulation, especially for phenomena whose local actions are easily described but exhibit complex behavior of mixed deterministic and random origins.

As a very simple example of pattern recognition or image enhancement that is, nevertheless, actually used in pixel-based image processing, consider this CA that automatically computes whether its initial configuration contained more black or white sites. The CA is represented as a looped string of m ordered, black-or-white sites. It follows the rule:

the color of any site at time = t_{n+1} is the color of the majority of the site and its two neighbors at time = t_n

Eventually the CA becomes all white or all black, reflecting the initial majority.

In an area of more engineering interest, Professor Kai Nagel of the University of Cologne has constructed a CA model for freeway traffic. (Nagel and Rasmussen 1994) The deterministic version produces a high/low density phase transition at the point of maximum throughput with self-organized criticality driven by the speed of the slowest car. Adding random noise causes spontaneous formation of traffic jams. Their distribution responds to a scaling law near the critical point and expansion of the scaling region can be characterized by the percentage of drivers using "cruise control" to reduce their fluctuations at high speed.

Lattice gases (LGA) constitute a distinct subclass of CA that can simulate complex fluid dynamics governed by non-linear equations such as the Navier-Stokes equation. There have been some significant recent developments in programming languages and computer architectures that strongly support CA (and especially LGA) based parallel computations. (Margolus 1995, Toffoli 1995)

3.5 Modeling Wave Equations with CA

In a project sponsored by EPRI at San Jose State University, Prof. Rudy Rucker and his students are developing techniques for using complex cellular automata to model certain aspects of the electric power grid. The long term goal is to perfect a distributed-computation simulation of the global behavior of a circuit (for instance, its stability) based on the local behavior of the individual components, and to apply this approach to power quality (harmonics, load induced transients, etc.) where CA cells might represent individual loads (such as electrical appliances) within a building.

A cellular-automaton implementation of partial differential equations has been developed, including: the heat equation, the wave equation (or telegrapher's equation), the damped driven oscillator, the damped driven oscillator coupled with the wave equation, and the Fermi-Pasta-Ulam non-linear soliton wave. These simulations have never before been implemented as pure cellular automata. The CA implementation makes it possible to rapidly explore alternate parameter settings, to breed and mutate parameter settings, and to view space-time diagrams of the simulation in real time. These simulations can run rapidly on a standard desktop Windows computer. Research is continuing on how best to use these techniques for real world problems, as well as on extending these techniques to branching networks and to higher-dimensional systems. Version 3.0 of CAPOW (Cellular Automata Power Simulator) has been posted on the Internet. (Rucker 1996)

This modeling approach, while it makes use of computations carried out by local rules on a spatially extended lattice, does not include any explicit intelligence or agency in the individual cells. However, the possibility of using genetic algorithm-based agents to guide the evolution of the CA is being explored. The rules developed in these CA-based experiments can also be provided as methods for the agent-objects used to model components in a distributed agent simulation of the electric power grid. (See Section 5 below.)

4 SWARM -- A TOOL FOR MULTI-AGENT SIMULATIONS

The SWARM Simulation System has been developed by the Santa Fe Institute as a tool kit for the study of complex adaptive systems using multi-agent discrete event simulation. The full source code is available free. (SFI 1996) It requires the use of the GNU C Compiler, UNIX, and X-Windows.

The basic unit of a SWARM simulation is the agent, which may be any entity that can generate events that affect itself and other agents. Simulations consist of groups of many interacting agents, called “swarms,” which may, themselves, be grouped into more comprehensive swarms. The logical structure of swarms of agents interacting through discrete events is implemented in Objective C, an object-oriented (OO) language. As in most OO programming, software consists primarily of definitions of various classes of objects. An object is then a combination of instance variables for the object's state and methods (services, procedures) that implement the object's behavior. Each object carries with it its own state variables, but the generic definition of its behavior is provided by the class.

The SWARM system itself is an object framework: a set of class libraries that are designed to work together. These include support libraries with potential use outside of SWARM as well as domain-specific libraries that provide facilities for building two dimensional discrete lattices, genetic algorithms, and neural networks. Agents are created by taking a class from the SWARM libraries, specializing it for the particular modeling domain, and then instantiating it, with each instantiated object an agent. A schedule of discrete events involving the agent-objects defines a process occurring over time and is implemented as a partially ordered series of actions to be performed by objects on objects. In the simplest case, a model consists of one swarm inhabited by a group of agents and a schedule of activity for those agents. In SWARM, the environment is itself modeled by one or more other agents. In the general case, the environment for each agent consists of many or all of the other agents, some of whom might have a larger influence than others.

In addition to all the common OO facilities, SWARM implements a “probe” facility which allows any object's state to be read or set and any method to be called in a generic fashion, without requiring additional user code. Probes are used to make data analysis tools work in a general way and are also the basis of graphical tools to inspect objects in a running system. SWARM also provides data collection tools in the form of observer agents, special objects whose purpose it is to observe other objects via the probe interface. The observer agents themselves are a swarm, a group of agents and a schedule of activity. By combining this swarm with a model swarm running as a subswarm of the observer, a full experimental apparatus is created. By using hierarchical swarms to separate data collection from the model, the model itself remains pure and self-contained, a simulated world under glass. Different observer swarms can be used to implement different data collection and experimental control protocols, but the model itself remains unchanged.

The SWARM software package is currently being used in a research project started in 1996 with the support of the Electric Power Research Institute (EPRI). Professor V. C. Ramesh at the Illinois Institute of Technology is investigating a multiple agent approach to contingency analysis. All electric utility control centers run computer simulations continuously to test whether any single accident or failure will cause any security constraint to be violated for longer than a specified short time period and thus lead to the possibility of a serious blackout. Prof. Ramesh is retaining the conventional model of the problem as constrained non-linear flow, but he is building (five to ten) individual intelligent agent solvers, each of which will attack the solution of the model in a different way. He expects that their cooperation and competition will achieve a faster convergence than could any one alone.

5 MODELING THE ELECTRIC POWER GRID

The North American power network may realistically be considered to be the largest machine in the world since its transmission lines connect all the electric generation and distribution on the continent. A design for distributed control of an electric power system by intelligent agents operating locally with minimal supervisory control is being developed systematically to include modeling, computation, sensing and control. (Wildberger 1994)

The distributed intelligent agent model of the electric power grid is being developed to serve two purposes:

1. Concurrent simulation of the grid for real-time distributed control using new developments in active, high-power, electronic control devices and self-calibrating, self-diagnostic sensors. (Wildberger 1994)
2. "What if" studies and computer experiments intended to provide insight into the evolution of the entire electric power industry under various forms of organization imposed or evolved as the result of varying degrees of deregulation, competition and unbundling of services.

Since intelligent agents have been used successfully to model the traders in commodities and the commodity futures markets, model development began by using artificial agents represent the buyers and sellers of bulk power. Software developed by Prof. Gerald Sheblé at Iowa State University, based on a form of genetic algorithm, permits the user to assign combinations of relatively simple strategies to a small number of agent-traders and to compare their profits and losses as the market clears. Rather than the bilateral trading between utilities which has been the rule in the past, this models the "power pool" arrangements that are now being established both by institutions and private parties. The trading model is being extended to include futures trading and retail as well as wholesale contracts, but, in order to include the effects of each transaction on power flow and stability in the electrical network itself, the network's physical components must also be modeled as intelligent agents.

The basic concept is to represent each component of the network by an agent of limited intelligence which seeks to ensure its own survival while optimizing its performance in the context of all the other agents. For instance, a single bus will strive to stay within its voltage and power flow limits while still carrying the voltages and power flows imposed on it by the combination of other agents representing generators, loads, transformers, etc. Specific classes of components will have additional survival constraints which may not be exceeded or, if exceeded, would require replacement or off-line maintenance. For instance, the chemical state of the oil bath in a large transformer must stay within specified limits for safe operation. It's state may be tested periodically by taking samples, or instrumentation may be installed for continuous monitoring. The cost-benefit of this instrumentation for each transformer is one example of management information this distributed model and simulation is intended to provide.

More complex components, such as a generating plant or a substation, must be modeled as class and object hierarchies of simpler components. An Integrated Knowledge Framework (IKF) has been developed that describes the management and operational functions at a generic, coal-fired, steam power plant as an OO model containing 422 classes, interconnected through three different types of relations. (EPRI 1996) This framework identifies the data, information and knowledge required for the important functions typically performed at a fossil power plant, as well as the flow of that data, information and knowledge between the function that generates it and those that use it. Although the IKF was not developed solely as a design for distributed agent control and it contains many classes which would not be appropriate for instantiation as active objects, this model defines, at an abstract level, all the structure needed for an intelligent agent model of the generation aspects of electric power.

6 SUMMARY

What is generally called "Artificial Life," is essentially a combination of methods for building discrete event simulations with evolving multiple agents As such, it provides another important tool for Simulationists to use in the modeling and analysis of many complex adaptive systems produced by humans either through engineering design or as a (possibly unintended) result of their social and economic behavior. The tools for artificial life simulations include a variety of evolutionary computing techniques operating on a coupled map lattice. Genetic algorithms and cellular automata, respectively, are the most common ones used. These methods are particularly useful in modeling and simulating very large and complex systems for two major reasons:

1. their ability to produce complex emergent phenomena out of a small set of relatively simple rules, constraints and relationships couched in either quantitative or qualitative terms
2. their inherent parallel, distributed structure

REFERENCES

Conway, J.H. 1982. What is Life? In *Winning Ways for Your Mathematical Plays*. eds. E. Berlekamp, J.H. Conway and R. Guy, Vol. 2, Chap. 25. New York, NY: Academic Press

Electric Power Research Institute. 1996. Integrated Knowledge Framework (IKF) for Coal-Fired Power Plants. EPRI Technical Report TR-106211-V1/2/3, (Mar). Pleasant Hill, CA: EPRI Dist. Ctr.

Fogel, L.J., A.J. Owens, and M.J. Walsh. 1966. *Artificial Intelligence Through Simulated Evolution*. New York: Wiley

Holland, J. H. 1975. *Adaptation in Natural and Artificial Systems*. University of Michigan Press, Ann Arbor, MI.

Koza, J.R. 1992. *Genetic Programming: On the Programming of Computers by Means of Natural Selection*. Cambridge. MA: MIT Press.

Koza, J.R. 1994. *Genetic Programming II: Scalable Automatic Programming by Means of Automatically Defined Functions*. Cambridge. MA: MIT Press

Margolus, N. 1995. Ultimate Computers. In *Proceedings of the Seventh SIAM Conference on Parallel Processing for Scientific Computing* (San Francisco, Feb. 15-17) eds. D.H. Bailey et al. 181-186. Society for Industrial and Applied Mathematics, Philadelphia, PA.

Mitchell, M. and S. Forrest. 1993. Genetic Algorithms and Artificial Life. Technical Report SFI 93-11-072, Santa Fe Institute, Santa Fe, NM.

Nagel, K. and S. Rasmussen. 1994. Traffic at the Edge of Chaos. Technical Report SFI 94-06-032. Santa Fe Institute, Santa Fe, NM.

Rechenberg, I. 1973. Evolutionsstrategie: *Optimierung technischer Systeme nach Prinzipien der biologischen Evolution*. Stuttgart: Frommann-Holzboog

Reynolds, C.W. 1987. Flocks, Herds, and Schools: A Distributed Behavioral Model." In Proceedings of SIGGRAPH'87. *Computer Graphics* V 21(4): 25-34.

Rucker, R. (rucker@jupiter.sjsu.edu). 1996. "CAPOW3B?.ZIP," from WEB site: http://www.mathcs.sjsu.edu/capow/.

Santa Fe Institute (sfi@santafe.edu). 1996. ftp://ftp.santafe.edu/pub/swarm/, or from WEB site: http://www.santafe.edu/projects/swarm/.

Schwefel, H-P. 1981. *Numerical Optimization of Computer Models*. Chichester, UK: Wiley

Toffoli, T. 1995. Fine-Grained Models and Massively-Parallel Architectures: The Case for Programmable Matter. In *Proceedings of the Seventh SIAM Conference on Parallel Processing for Scientific Computing* (San Francisco, Feb. 15-17) eds. D.H. Bailey et al., 181-186. Philadelphia, PA: Society for Industrial and Applied Mathematics.

Von Neumann, J. 1949. Theory of Self-Reproducing Automata. In *1949 University of Illinois Lectures on the Theory and Organization of Complicated Automata*. (reprinted 1966) ed. A.W. Burks. Urbana, IL: University of Illinois Press

Wildberger, A.M. 1994. Automated Management for Future Power Networks: A Long-Term Vision. *Public Utilities Fortnightly* **132**, 20 (Nov): 38-41.

Wuensche, A. 1996. http://www.santafe.edu/~wuensche

Wuensche, A. and M.J. Lesser. 1992. *The Global Dynamics of Cellular Automata: an Atlas of Basin of Attraction Fields of One-Dimensional Cellular Automata*. (diskette included). Santa Fe Institute Studies in the Sciences of Complexity. Reference Vol. I. Reading, MA: Addison -Wesley

BIOGRAPHY

Dr. A. Martin Wildberger manages Strategic Research and Development projects in Applied Mathematics and Information Science at the Electric Power Research Institute (EPRI). He also provides institute-wide applied research support in mathematical modeling and computer simulation.

Dr. Wildberger received his B.S. degree (cum laude) from Fordham University, his M.S. degree from the U.S. Naval Postgraduate School Engineering School and his Ph.D. from the Catholic University of America. He has published or presented over sixty technical papers and has been, since 1987, the editor of a monthly column in *Simulation* entitled: "AI and Simulation"

State-of-the-Art Reviews

Proceedings of the 1996 Winter Simulation Conference
ed. J. M. Charnes, D. J. Morrice, D. T. Brunner, and J. J. Swain

DEVELOPING A REAL-TIME EMULATION/SIMULATION CAPABILITY FOR THE CONTROL ARCHITECTURE TO THE RAMP FMS

Wayne J. Davis
Joseph G. Macro
Andrew L. Brook
Michael S. Lee
Guoyan S. Zhou

Department of General Engineering
University of Illinois at Urbana-Champaign
Urbana, IL 61801 U.S.A

ABSTRACT

This paper first discusses an object-oriented, control architecture and then applies the architecture to produce a real-time software emulator for the Rapid Acquisition of Manufactured Parts (RAMP) flexible manufacturing system (FMS). In specifying the control architecture, the coordinated object is first defined as the primary modeling element. These coordinated objects are then integrated into a Recursive, Object-Oriented Coordination Hierarchy. A new simulation methodology, the Hierarchical Object-Oriented Programmable Logic Simulator, is then employed to model the interactions among the coordinated objects. The final step in implementing the emulator is to distribute the models of the coordinated objects over a network of computers and to synchronize their operation to a real-time clock. Specialized displays have also been developed to allow one to monitor the detailed, real-time operation of each coordinated object. The research will also convert the real-time emulation model into a real-time simulation model for the RAMP FMS. Both the real-time emulation and simulation capabilities will be demonstrated at the presentation.

1. INTRODUCTION

Although simulation remains the tool of choice for modeling the behavior of FMSs, the accuracy of current simulation tools in modeling modern manufacturing systems such as Flexible Manufacturing Systems (FMSs) has been questioned. Mize et al. [1991] have concluded that current simulation tools cannot accurately estimate the true performance of an FMS. Rather, these merely permit the modeler to compare the relative performance differences among alternative designs for the FMS.

In our research, we have consistently observed that available simulation tools overestimate the performance for an operational FMS (see Flanders and Davis [1995] for a real-world example). We believe that the inherent inaccuracy of current simulation tools arises from the inability to model *all the constraints* associated with the operation of an FMS. Current simulation tools focus on modeling the primary job entity as it flows through a stochastic queueing network representation of the FMS. Few, if any, simulation tools readily permit the modeler to consider the flow of the supporting resources (e.g. tooling, part kits, fixtures, and processing plans). The coordination of all entity flows in an FMS is crucial, and it is the interactions among the controllers within the FMS that coordinate these flows (see Davis et al. [1993]).

A new simulation methodology has been developed to explicitly model the controller interactions. This methodology also permits the immediate consideration of the detailed processing plans for the parts to be produced in the FMS. In this manner, all entity flows are considered. This paper discusses an application of this simulation methodology to construct a software emulator for the control architecture of a RAMP FMS that is operated by the Naval Air Warfare Center/Aircraft Division/Indianapolis (NAWC AD Indpls). This FMS will produce new and spare electronic circuit boards for defense systems using through-the-hole assembly technologies. The complexity of this system is evidenced by the fact that the cell controller requires nearly a million lines of computer code to implement, excluding the numerous off-the-shelf software packages contained within the controller. Given this complexity, no prior attempt to simulate the entire system has been successful, and production capabilities are still unconfirmed. Furthermore, given the current failure to model the dynamics for this FMS, the ability to schedule this FMS has been compromised. The development of the emulator is the first step toward the development of a new scheduler for this system. This scheduler will be developed within the framework of a comprehensive methodology for the modeling, scheduling and control of FMSs as discussed in Davis et al. [1993].

In this State-of-the-Art tutorial, a detailed discussion of the development of the real-time emulator for the RAMP FMS will be provided. The conversion of the real-time emulator into a real-time simulator for the RAMP FMS will also be addressed. Finally, an on-line demonstration of the real-time emulation and simulation capability will be provided.

2. MODELING APPROACH

2.1 The Coordinated Object

In order to define the set of subsystems contained within a FMS, we sought to find a single modeling template which could be recursively applied to decompose the system into its component subsystems. Working in collaboration with the Government Systems Group at Motorola, we first published the fractal architecture as a mechanism for decomposing the factory into its constituent processing elements (see Tirpak et al. [1992]). During the past few years, the basic fractal unit defined in that architecture has been generalized to become the coordinated object, and we now refer to the fractal architecture as the Recursive Object-Oriented Coordination Hierarchy.

The generalized coordinated object (CO) is depicted in Figure 1. It represents the most fundamental hierarchical element where integrated planning and control are implemented. Each CO contains one or more subordinate task-capable resources or subsystems, P_n (n=1,...,N), which can be allocated to execute the tasks that have been assigned to the CO. In order to execute these tasks, entities, (including both jobs or supporting resources), enter the CO through its input port and eventually exit through its output port. These entities are assumed to be under the control of the CO from the time of their arrival at the input port until they exit through the output port. Consequently, any entity residing in the CO's output queue is controlled by the CO's supervisor since it can be assumed that the CO no longer has any assigned tasks to be performed upon an entity in its output queue.

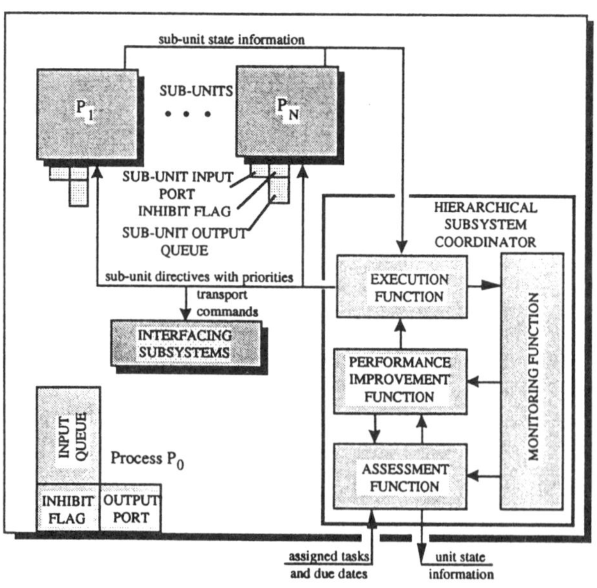

Figure 1. Schematic of the Coordinated Object: A Basic Module for Planning and Control

In general, the CO does not perform the tasks itself. Rather, it decomposes its assigned tasks into subtasks using detailed process plans. These subtasks are then implemented at the subordinate processes or subsystems. When the CO allocates a given entity with an assigned subtask(s) to one of its subordinate subsystems P_n, the physical control of that entity is relegated to that subsystem. Note that in Figure 1 the Input Port and the Output Queues of each subordinate subsystem belong to the CO, while the Input Queue and the Output Port belong to that subordinate subsystem. Therefore, a consistent chain of command for the control of a given entity is defined as the entity flows among the various subsystems contained within the CO.

To move the entities from one subordinate subsystem to another, we assume that each CO has the essential interfacing (material handling) subsystems which contain the transport processes within the CO. These interfacing subsystems are also task-capable resources and must be under the control of the CO. In general, a subordinate subsystem cannot execute its assigned tasks until essential entities and resources are delivered to its control domain. The CO must be able to manage this flow of entities and resources through the control of its subordinate Interfacing Subsystems. Again, the control of the Interfacing Subsystems is manifested via the CO's assignment of tasks to be executed by the interfacing subsystem.

2.2 The Recursive Object-Oriented Coordination Hierarchy (ROOCH)

The recursive nature of the ROOCH arises from the fact that any subordinate subsystem or object within a coordinated object, (including the Interfacing Subsystems), can also be a coordinated object. This recursive approach may be applied to construct the ROOCH with the essential number of hierarchical levels needed to model any FMS.

For an application of the ROOCH, consider the Rapid Access to Manufactured Parts (RAMP) FMS. The ROOCH for the RAMP FMS is pictured in Figure 2.

The recursive nature of the ROOCH is immediately apparent. The RAMP FMS forms the primary CO. Included in the RAMP CO are ten individual subordinate processing centers which perform all the tasks that are required to assemble a circuit board. For the RAMP CO, the primary interfacing subsystem is the central Automated Storage and Retrieval System (AS/RS). The AS/RS is also a CO which includes the primary AS/RS controller and 30 additional controllers to manage the subordinate transport processes within the AS/RS. Another interfacing subsystem is the communication network which downloads the essential processing plans to a given processing center.

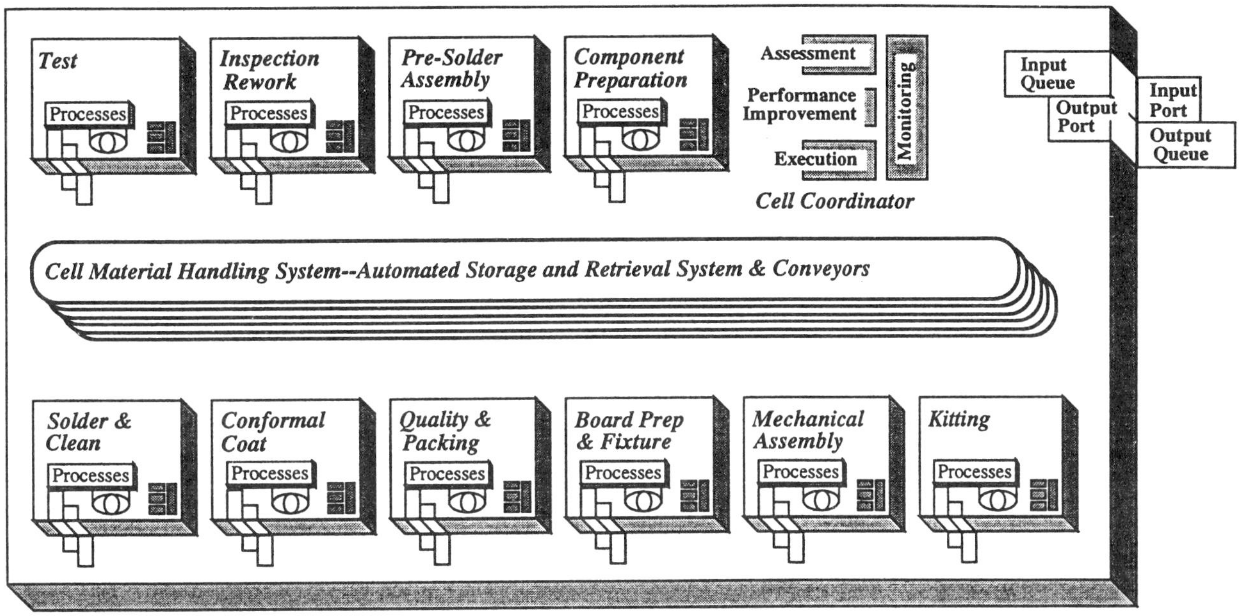

Figure 2. The ROOCH for the RAMP FMS

Each processing center is also a CO containing one or more primary unit processes and a material handler or interfacing subsystem. At most processing centers in the RAMP FMS, the human operator provides the primary material handling. In many cases, the human operator also functions as one of the unit processors. It should be noted, however, that the human operator typically functions only in one capacity at a time. The notion of a unit or transport process is therefore very generic as it can describe a human, a machine or any other type of subsystem that can execute a task.

Despite the apparent complexity of Figure 2, there is a simplicity in the proposed ROOCH. Each included function is a CO except the lowest level process controllers. Each CO is responsible for executing tasks assigned by its supervisor. In most cases, however, the CO cannot execute these tasks itself but rather, defines subtasks to be executed by a subordinate subsystem.

In addition to the assignment of tasks to its subordinate COs, the CO must also direct the flow of jobs and supporting resources to the subordinate COs and processes. This flow of entities is implemented by the CO assigning tasks to its subordinate interfacing subsystems. As is the case for the AS/RS in the RAMP FMS, a given interfacing subsystem may also be a coordinated object. It, too, is responsible for scheduling the execution of its assigned transport tasks. Hence, interfacing subsystems may also possess a dedicated coordination hierarchy which is responsible for both planning and execution of transport tasks.

We usually prefer to distinguish the Processing Subsystem, P_n (n=1,...,N), from the Interfacing Subsystems. The Processing Subsystems are responsible for planning the execution of assigned tasks which physically modify the state of an entity using instructions contained in a process plan. The interfacing subsystems, on other hand, address tasks that support the processing subsystems in generating the desired change in state for a given entity. For example, a material handling system such as the AS/RS may change the location of an entity or store an entity until the next processing task is scheduled. These Interfacing COs do not execute the processing steps contained within the processing plan for the product being manufactured. Rather, the manner in which these execute their assigned tasks is known by the controller and is inherent within the subsystem itself. Therefore, the ensemble of tasks that the Interfacing COs can execute is independent of the product being manufactured. For example, there is a physical layout and control system for most material handling systems which specifies which tasks the material handling system can address and how it will execute each task.

The lowest-level control object in the ROOCH is either a Transport or Unit Process Object. Unit processes are always subordinate to a CO for processing subsystems while transport processes reside within an interfacing subsystem. Neither transport nor unit process objects can have subordinates.

2.3 The Hierarchical Object-Oriented Programmable Logic Simulator (HOOPLS)

As is the case for most large-scale, discrete-event systems, numerous types of entity flows must be considered for the RAMP FMS, including the several types of totes that contain both the production jobs and the supporting resources (tools, fixtures, part kits and so forth).

The flow (location) of all entity types must be modeled by the simulation, and the control of every entity must also be explained. The entities in a modern FMS do not flow of their own volition. Rather, their flows are managed by controllers. Furthermore, whenever there is an interaction among the controllers, there is likely an exchange of control for one or more entities among the interacting controllers. A new simulation approach, the Hierarchical Object-Oriented Programmable Logic Simulator (HOOPLS), has been formulated to explicitly address these controller interaction concerns.

A HOOPLS-based model for a FMS consists of four primary frames. The first frame is the **model frame** which contains the specifications for the ROOCH associated with the modeled FMS. The second frame is the **control frame** which provides for the exhaustive definition of the control messages that will be issued or received by each controller contained within the ROOCH. This frame also defines the state transition mechanisms which occur upon the receipt of a control message and the subsequent control message(s) that are issued.

Given that HOOPLS explicitly models the interaction among the controllers within the ROOCH and considers the flow of all entities to be a consequence of these interactions, *HOOPLS has abandoned the use of a traditional event calendar.* Instead, HOOPLS employs a message relay which stores the control messages that will be passed among the controllers, chronologically-ordered based upon their delivery time. Each control message designates the controller that issued the message, the recipient controller for the message, the message content, and the scheduled delivery time. The message relay is responsible for delivering the control message to the recipient controller at the appropriate simulated time. It is obvious that the prescribed operation of the message relay physically mimics the communication network which links the controllers in an actual FMS (for the RAMP FMS, this is a dedicated Local Area Network).

The third simulation frame in the HOOPLS-based model is the **processing plan frame**. The processing plan details not only which manufacturing processes will be required, and in which order, but also details which supporting resources will be required to complete a processing task.

The fourth simulation frame is the **experimental frame** which specifies the experimental parameters governing the simulation study. The experiment frame also includes extensive capabilities for initializing the simulation to a known system state.

To demonstrate the modeling objectives for HOOPLS, a simulation/emulation model developed for the RAMP FMS will now be summarized. HOOPLS intrinsically employs an object-oriented architecture, and the C++ programming language was selected for implementing of the simulation model. In addition, the object-oriented design and analysis methods detailed by Booch [1991] were used to specify the class hierarchy for the objects.

3. IMPLEMENTING THE EMULATOR

3.1 Prior Modeling Effort

Space limitations do not permit a detailed discussion of all the essential steps that are required to construct the emulator. Rather, we can only provide an overview of our basic approach. In an earlier paper, we discuss an initial effort to construct the basic object classes for the controllers, entities, and control messages that are needed to describe the RAMP FMS (see Davis et al. [1994]). The ROOCH used for this paper is depicted in Figure 2. This same ROOCH is also employed in the definition of the model frame for the HOOPLS-based emulator.

The distinction between an emulation and simulation is simple. Both employ the same model. The primary difference arises with the manner in which the model is executed. In a HOOPLS-based simulation approach, the messages are stored chronologically on a message calendar. After the state transitions arising from the receipt of a given message at a given controller are implemented and the new controlled messages derived from the processing of the current message are chronologically stored in the message calendar, the next message is popped from the message calendar for processing. At this moment, the simulation time is advanced to the delivery time for the next control message to be popped. A HOOPLS-based emulation uses the same approach save one fundamental difference. The emulation is also tied to a real-time clock. When the next message is removed from the event calendar, it will not be executed until the actual time at which that message is to be delivered to the recipient controller. Thus, the difference between an emulation and simulation may simply be summarized by saying that under a simulation, message processing procedures control the advancement of time while under an emulation, the advancement of time is controlled by a real-time clock.

In the development of the control frame, over seventy controllers were described. As shown in Table 1, the AS/RS alone includes 31 controllers. These controllers are differentiated into four classes of objects. The Coordinate Nodes model the cell and station controllers. The Transport Nodes, (a special case of the Coordinate Node), model the AS/RS and station-level material handler controllers. The Unit Process Nodes model the process controllers at each station. Finally, Transport Process Nodes model the dedicated transport process controllers within the AS/RS. Using this class definition schema, Coordinate Nodes can have other Coordinate Nodes, Transport Nodes and Unit Process Nodes as subordinates. A Transport Node can have only Transport Process Nodes as subordinates. Unit and Transport Process Nodes cannot have subordinates.

In the control frame, we establish the communication scheme for the controller interactions. Here, we define two classes of message. An Action Message requests a subordinate to take execute an action. A Status Mes-

RAMP Cell Controller
- Kitting Station
  - Processor : Human Operator
  - Material Handler : Human Operator
- Board Preparation Station
  - Processor : Human Operator
  - Material Handler : Human Operator
- Component Preparation Station
  - Processor : Human Operator
  - Material Handler: Human Operator
  - Part Carousel
  - Tinning Robot
- Pre-Solder Assembly Station
  - Processor : Human Operator
  - Material Handler: Human Operator
  - Part Carousel
  - Part Location Indicator
- Test Facility Station
  - Processor : Human Operator
  - Material Handler: Human Operator
  - Burn In Tester
  - Conductivity Tester
- Inspection/Rework Station
  - Similar to Board Preparation Station
- Mechanical Assembly Station
  - Similar to Board Preparation Station
- Conformal Coat Station
  - Similar to Board Preparation Station
- Quality & Packaging Station
  - Similar to Board Preparation Station
- Wave Solder and Clean Station
  - Similar to Board Preparation Station
- Automated Storage and Retrieval System
  - Storage layer Conveyors (6)
  - Inserter/Extractor (5)
  - Machine Station Input/Output Conveyors (20)

Table 1. Controllers in the RAMP FMS

sage provides feedback to the supervisor at the completion of the requested action.

With the interaction of either a Coordinate Node or Transport Node with a subordinate of any class other than a Transport Node, we have three basic action messages. AcceptItem instructs the subordinate that an entity is in its input queue and that the subordinate now has control of that entity. ReturnItem instructs the subordinate to place the entity in its output queue and return the control to the requesting supervisor. ExecuteTask requests the subordinate to execute a specific task upon a given item. If the subordinate is a Coordinate Node or Unit Process Node, this task will be defined within the process plan database. If the subordinate is a Transport Node, then the instruction is predefined within the process's capabilities.

Kitting: Generate the parts kits for the order
- Input: KIT tote containing all order components
- Output: Bare Board Tote containing up to 10 boards
Pre- and Post-Solder Assembly Tote(s)
Mechanical Tote holding large parts

Board Preparation: Place board into fixture
- Input: Bare Board Tote with bare boards
- Output: One Pallet Tote for each fixtured board

Component Preparation: Bend/tin component leads
- Input: Pre- and Post-Solder Assembly Tote(s)
- Output: Same

Pre-Solder Assembly: Place components on board
- Input: Pre-Solder Assembly Tote(s) and all Pallet Totes for given order
- Output: Same

Pre-Solder Inspect/Rework:
- Input: Pallet Totes and Pre-Solder Assembly Tote(s) if rework is needed
- Output: Same

Wave Solder and Clean:
- Input: All Pallet Totes
- Output: Same

Post-Solder Assembly: Mount small parts
- Input: All Pallet and Post-Solder Assembly Tote
- Output: Same

Post-Solder Inspection
- Input: All Pallet Totes with Pre- and Post-Solder Assembly Totes if reworked
- Output: Same

Mechanical Assembly: Remove board from fixture and mount larger mechanical parts.
- Input: All Pallet and Mechanical Part Tote
- Output: Printed Wiring Assembly Tote for each board and Mechanical Part Tote

Test: Perform bed-of-nails and burn-in test.
- Input: Printed Wiring Assembly Totes for job and various part totes if reworked
- Output: Same

Conformal Coat: Apply protective coating to board.
- Input: Printed Wiring Assembly Totes for order
- Output: Same

Final Quality Control and Inspection/Packaging:
- Input: Post-Wiring Assembly Totes
- Output: Packaged boards

Table 2. Processing Steps for Production in the RAMP FMS

When the subordinate to the Coordinate Node is a Transport Node, there are two action messages. DelieverItem instructs the Transport Node to deliver an entity to a requested station while PickupItem instructs the Transport Node to pickup an entity at a specified station. Each of the above action messages has a corresponding status message which notifies the supervisor when the requested action is completed.

The process planning frame specifies the details for all the processing instructions needed to manufacture a given part. A generic processing plan for any part manufactured in the RAMP FMS is given in Table 2. Note, however, that the actual process plan contains considerably more detail pertaining to the execution of each task. For example, detailed instructions on how to bend the leads for each component and where the component is to be placed upon the board are provided.

Finally, the experiment frame is very similar to that of most simulations detailing the length of the simulation run and other parameters that are needed to specify the simulation experimental trial. The experiment frame under the HOOPLS paradigm also provides extensive provision for initializing the simulation to a known state.

3.2 On-Going Development of RAMP Emulator

The development of the emulator is initiated as a "clean-sheet" design. All that is being retained from the prior development of the HOOPLS-based simulation model is the basic structure of the ROOCH and the control messages that will be employed. However, the list of control messages will probably be expanded. A distributed object architecture is being employed in the development of the emulator such that each controller can be positioned on a different computer if desired. The controller objects are being coded in C++.

In developing the new emulation model, the greatest effort has been devoted to the definition of the state variables for each coordinated object, especially the state variables for the AS/RS. There are several reasons for the focus upon the AS/RS. The AS/RS includes 30 transport processes. The complete AS/RS will have a total of 31 controllers, nearly half the total number of controllers in the entire RAMP FMS.

In defining the state variables for each control object, there are fundamental issues to be addressed. For example, should the AS/RS controller know the contents of the totes which contain jobs and supporting resources while they are being stored and transported by the AS/RS? This single issue has been the subject of hours of debate. Our current design assumes that the AS/RS will know the contents of the totes. This decision extends the general capability of the formulation as the cell controller can request to the AS/RS that a given item be moved without specifying the tote that it is in. By making this design decision, the AS/RS can easily be replaced by another material handling system, one which may not employ totes. In fact, there are presently other versions of the RAMP FMS which do not include an AS/RS.

Distributing the controllers across several computers creates several other problems. First and foremost, a method of communication between the objects on different computers must be developed. The computers, (in this case, Sun workstations), are networked. If a control object wants to send a message to another control object across this network, it needs to know on which computer the object resides. To fulfill the communication requirements, we have written our own mail system which knows the address of every control object. When an object desires to send a message, it must specify the recipient of the message, the time that the message is to be delivered, and the contents of the message. The message is then placed into the local computer's mail box. If the recipient object resides on the same computer, the message-passing is handled by the local mailer. If the recipient controller is located at another computer, then the message is routed to the mailer on the computer where the recipient object resides.

Messages are retained in the recipient object's mailbox until the scheduled time of their delivery. When the messages are delivered, the recipient controller executes the appropriate state transitions and generates its response messages which it then places in its mail box for transmission to other controllers at the desired time.

Using the same message service, we have also included programmed features which will permit the emulation to proceed at a rate that is faster than real time. Specifically, we have included a clock object which monitors a real-time clock and advances the emulated time at a time-scale factor greater than real time. The clock object sends messages to each computer's mail service specifying the current emulated time, which the mail service then uses to determine when messages are to be forwarded to their recipient controllers.

Each controller type also has its own specialized display, which was developed under X-Windows/Motif®. Within each controller display, the state of the controlled subsystem will be constantly updated. Much time has also been devoted to the development of the displays. Our desire is for an individual to be able to view a given controller's window and visualize the operation of the subsystem that is being controlled. These displays also are especially useful when validating the model. These provide a far more detailed depiction of the operation of the system than that which can be achieved using current simulation and animation techniques. Current animation capabilities provide iconic displays which permit only the basic entity flow among the various objects to be depicted. Using our emulation capabilities, the modeler will be able to access the detailed state information for each modeled object within the simulation as it evolves in real-time. The interaction among the primary modeled controller objects can also be visualized by monitoring the control messages that are being transmitted among the controllers.

3.3 The Demonstration

At the State-of-the-Art tutorial, we plan to demonstrate the real-time emulation of the RAMP FMS. The ability to execute this demonstration is derived from another unique capability which we have included within the emulator. As stated above, we chose to employ a distributed-object programming architecture within the emulation which permits each controller to be executed on a different platform. Each platform is connected to the Internet. To provide a future capability to model all types of RAMP FMSs, it was also necessary to implement the display window to each controller as a distinct programmed object from the programmed object representing the controller itself. Thus, a given controller and its associated control window are separate programmed objects which can also run on different computers. This design decision also enhanced the virtual manufacturing capability by permitting the operations of the RAMP FMS to be viewed and eventually controlled from any computer on the Internet.

The computer code for the control displays is situated at the computer where the displays are viewed. When this code is executed, the display object initially logs into the message service for the emulation and provides the message service with its IP address on the Internet. Eventually, the message service will request the viewer's name and password which it will then employ to control access to certain state information and to limit the viewer's capability to interact with the operating RAMP FMS. Once the viewer has gained access, the primary control window for the RAMP FMS cell controller is opened at the remote location. The message service then notifies the cell controller for the RAMP FMS to begin sending state information to the requesting window object at intervals of approximately 1 second duration.

The displaying software is contained at the distal location. Therefore, when the display object receives the state information, it then employs the state information based upon the control window being viewed. It is important to observe that only textual state information is being transmitted across the Internet, not the entire pixel information that is needed to construct the control window. Therefore, the information transmission requirements are minimal.

The remote viewer can interact with the system as various elements of a given controller's window are user-selectable icons. For example, when the viewer clicks the icon of the AS/RS within the cell controller window, the viewing programming immediately opens a window for the AS/RS controller. The program then identifies the message service that the viewer wishes to monitor the operations of the AS/RS. The message service then tells the cell controller to temporarily suspend state updates to the remote viewer and simultaneously tells the AS/RS controller to start sending state information updates. Within the AS/RS controller window, the viewer may choose to have detailed information presented for a given insertor/extractor or level of the storage conveyors.

As stated above, a simulation and an emulation differ only in the manner in which they advance time. When the emulation of the RAMP FMS is complete, the emulation model will immediately be converted to a simulation model for the RAMP FMS. Multiple instances of the resulting RAMP FMS simulation model will then be assigned to distinct processors. Each instance of a simulation model (referred to as a simulation engine) will operate under a different control strategy and each simulation engine will receive current state information updates from the control objects executing the real-time emulation. Operating under the selected control strategy, each simulation will perform a real-time simulation of the future response of the system over a future planning horizon given the current state of the emulated system. Each simulation engine will attempt to generate real-time simulation trials as quickly as possible.

Using the included message service, the output of these real-time simulation trials will be forwarded to the various programmed objects which perform the essential real-time statistical and compromise analyses which are needed to select the best control strategy among the considered strategies for implementation. This capability will be demonstrated at the tutorial.

Given space limitations, we cannot discuss all the components of the real-time simulation analysis in this paper. Our basic approach to real-time simulation has been previously discussed in Davis, Wang and Hsieh [1991] and Tirpak, Deligiannis and Davis [1992]. In a chapter of the forthcoming Handbook on Simulation (see Davis [1997]), we discuss evolving directions in simulation tools and real-time simulation. This chapter will provide an overview of the real-time simulation and emulation technology which will be demonstrated at the State-of-the-Art tutorial. Harmonosky [1995] also provides an abridged summary of ongoing research in real-time simulation and scheduling.

5. FUTURE WORK

The emulator will be completed in time for this paper to be presented. After the emulator is completed, the emulation code will be modified such that it can operate as a second generation, HOOPLS-based simulation model for the RAMP FMS. This simulation model then will be employed to perform real-time simulation (see Davis, Wang and Hsieh [1991]). This is a necessary first step to develop a real-time production scheduler for the RAMP FMS (see Davis [1992] and Davis et al. [1993].

REFERENCES

Booch, G. 1991. *Object Oriented Design with Applications*, Benjamin Cummings, Redwood City, California.

Davis, W. J., H. Wang and C. Hsieh. 1991. Experimental Studies in Real-Time, Monte Carlo Simulation. *IEEE Systems, Man and Cybernetics*, 21(4), 802-81.

Davis, W. J. 1992. "A Concurrent Computing Algorithm for Real-time Decision Making." *Proc. of the ORSA Computer Science and Operations Research: New Developments in their Interfaces Conference*, eds. O. Balci,

R. Sharda and S. Zenios, 247-266, Pergamon Press, London.

Davis, W. J., Setterdahl, D., Macro, J., Izokaitis, V. and Bauman, B. 1993. "Recent Advances in the Modeling, Scheduling and Control of Flexible Automation." *Proc. of the 1993 Winter Simulation Conference*, eds. G. W. Evans, M. Mollaghasemi, E. C. Russell, W. E. Biles, 143-155, The Society for Computer Simulation, San Diego, CA.

Davis, W. J., J. G. Macro and D. L. Setterdahl. 1994. "An Object-Oriented, Coordination-Based Simulation Model for the RAMP Flexible Manufacturing System." *Proc. of the Flexible Automation and Integrated Manufacturing Conf.*, eds. M. M. Ahmad and W. G. Sullivan, 138-147, Begell House, Inc., New York.

Davis, W. J. 1997. "Reat-Time Simulation: The Need and the Evolving Research Requirements." To appear in *The Simulation Handbook*,Wiley, new York.

Flanders, S. W. and W. J. Davis. 1995. "Scheduling a Flexible Manufacturing System with Tooling Constraints: An Actual Case Study." *Interfaces,* 25, 42-55.

Harmonosky, C. M. 1995. Simulation-Based Real-Time Scheduling: Review of Recent Developments. *Proc. of the 1995 Winter Simulation Conf.*,eds. C. Alexopoulos, K. Kang, W. R. Lilegdon and D. Goldsman, 220-225, The Society for Computer Simulation, San Diego, CA.

Mize J. H., H. C. Bhuskute, and M. Kamath. 1992. " Modeling of Integrated Manufacturing Systems." *IIE Transactions* , 24(3):14-26.

Tirpak, T. M., S. M. Daniel, J. D. LaLonde and W. J. Davis. 1992. "A Fractal Architecture for Modeling and Controlling Flexible Manufacturing Systems." *IEEE Trans. on Systems, Man and Cybernetics*, 22(5), 564-567.

Tirpak, T. M., S. J. Deligiannis and W. J. Davis. 1992. Real-Time Scheduling of Flexible Manufacturing. *Manufacturing Review* (ASME), 5(3), 193-212.

ACKNOWLEDGMENT

This research is supported by the US Army Research Laboratory/Battelle Grant TCN 95-300 and the corporate members of the Hyper-Linked Manufacturing Consortium.

BIOGRAPHICAL SKETCHES

WAYNE J. DAVIS is a professor of General Engineering at the University of Illinois at Urbana-Champaign. His research areas include distributed planning and control architectures for large-scale, discrete-event systems; computer-integrated manufacturing; and simulation.

JOSEPH G. MACRO is a Ph.D. candidate in Mechanical and Industrial Engineering at the University of Illinois at Urbana-Champaign. His research interests include advanced simulation methodologies and computer-integrated manufacturing.

ANDREW L. BROOK is an undergraduate student in Computer Science at the University of Illinois at Urbana-Champaign.

MICHAEL S., LEE is an undergraduate student in Computer Science and General Engineering at the University of Illinois at Urbana-Champaign.

GUOYAN ZHOU is a master's degree candidate in Computer Science at the University of Illinois at Urbana-Champaign.

Proceedings of the 1996 Winter Simulation Conference
ed. J. M. Charnes, D. J. Morrice, D. T. Brunner, and J. J. Swain

Distributed Interactive Simulation: It's Past, Present, and Future

Deb Fullford

MäK Technologies
185 Alewife Brook Parkway
Cambridge, MA 01238, U.S.A.

ABSTRACT

In the mid 1980's the U.S. Department of Defense, Defense Advanced Research Projects Agency (DARPA) launched the SIMNET project, which gave birth to a technology era and networking protocol presently known as Distributed Interactive Simulation (DIS). The DIS Protocol is a widely used IEEE Standard. At first it was primarily used by the military for mission rehearsal, training, and weapon evaluations. Today, other areas, such as transportation, medical care, entertainment, Internet commerce, and manufacturing, are using DIS complaint simulations to meet their needs. This paper will provide a detailed overview of DIS. It will focus on the fundamental concepts, such as peer to peer connectivity, protocol data units, and dead reckoning. It will also cover some advanced topics, such as smoothing for visualization and shared virtual environments. Finally, this paper will discuss the future of DIS. In particular, MäK will discuss the next generation DIS, otherwise known as the High Level Architecture (HLA), new protocols, and bandwidth reduction techniques.

1 INTRODUCTION

Over the past 30 years, the military has developed computer-generated synthetic environments to train people in very complex, dangerous, or financially risky tasks. However, since the technology for developing synthetic environments was very expensive, simulators were only developed for systems in which the cost of failed performance outweighed the cost of the training simulator.

The solution of this problem led to the invention of a technology called Distributed Interactive Simulation (DIS), a networking technique for connecting thousands of simulators together on a computer network. Instead of spending $50 Million for a single flight simulator, the demand was created for thousands of simulators to be networked in a synthetic environment for training groups of combatants in collaborative tasks.

The following sections will discuss the history of this cutting edge technology, provide high level technical discussions on how it works and provide some insight on where it is going.

2 HISTORY OF SYNTHETIC ENVIRONMENTS IN THE MILITARY

Until about 1983, the common perception of a training simulator in the Department of Defense was that of a stand-alone, high-fidelity device designed to train a single individual, or single crew, in the skills necessary to operate a weapon system; skills such as flying, shooting, targeting, etc. The goals of the engineers who designed and built these devices was to provide the highest possible degree of realism to the crew, accurately emulating the real-world cues that the crew would experience in an actual combat situation. These lofty goals tended to drive the performance requirements and prices of these systems through the roof.

Depending on performance capabilities, these simulators could cost anywhere from $1 Million to $50 Million each. Very often, the simulator would be more expensive than the vehicle it simulated. The cost per man-hour of training time was staggering because each simulator accommodated only a single crew (often only one person), while the training facility required several operators to run the simulation. If a $50 Million simulator were operated for 8 hours a day, 200 days per year, for 10 years, it would cost $3,125.00 per hour to amortize its purchase price, without considering hourly operating costs or repairs.

Moreover, while these high-fidelity simulations could be used for individual crew training, they could not address the need to train teams of soldiers in collective or collaborative tasks such as practice of tactics, strategies and procedures. To train these groups, it was necessary to transport all participating soldiers and equipment to an exercise site to rehearse battle scenarios. These live training exercises, often held in unpleasant and hard-to-reach locations such as the North Pole or the Sudan Desert, usually concluded with a small percentage of casualties, and even some fatalities.

Scheduling such a live exercise can be as daunting a logistical nightmare as scheduling an actual war, and may take several months to complete. This limits the number of collaborative training exercises to one or two per year.

Beyond their cost and logistics problems, live exercises are a security nightmare. To provide a realistic training environment, the exercises must be carried out in an area which closely resembles the area of conflict. These areas may be close to enemy land where an exercise may be easily observed, or on U.S. soil where the exercise may be observed via foreign satellites.

With the live training exercises' associated loss of life, high costs, poor security, and infrequent staging, the need for a better collaborative training solution was obvious. In 1983, Air Force Lieutenant Colonel Jack Thorpe, of the Defense Advanced Research Projects Agency (DARPA), decided to solve this problem.

The specific mission of DARPA was to fund high risk/high payoff advanced technologies which could be used by our armed forces to maintain an advantage on the battlefield. Almost all exotic technologies in the US arsenal, including Stealth radar avoidance, cruise missiles, laser guided weapons, Virtual Reality, and even the Internet, have all come from DARPA projects.

Jack Thorpe, along with his consultant, Colonel (retired) Gary Bloedorn, initiated one of the most influential technology revolutions ever to have occurred within the DoD. Jack and Gary envisioned a multi-user computer-generated synthetic environment where collaborative training, weapon concept development and tactics development could be performed in a very controlled, cost effective, safe, and secure manner. The outcome was SIMNET (SIMulator NETworking), a major paradigm shift for the training community. Instead of developing a small number of high-fidelity, multimillion dollar simulators for part-task training, DARPA developed inexpensive $250,000 simulators which had less fidelity than their older cousins.

"Selective fidelity" was a buzzword coined in the SIMNET program, meaning that a simulator need only be good enough to accomplish the training goal at hand. Hundreds of these "selective fidelity" simulators were networked together to provide an electronic world for collaborative tactics development and rehearsal. The tank simulators which SIMNET fielded had relatively low resolution (320 X 128 pixels per vision block), a relatively low frame rate (15 Hz), no motion base, an inexpensive fiberglass hull, and a reduced set of controls. For an 8 hour a day, 200 day a year, 10 year service life, the amortized cost of a SIMNET simulator costs only $15.63 per hour.

Proponents of high-fidelity simulators and live training exercises vilified the SIMNET system, claiming that it's cartoonish graphics and reduced fidelity man-machine interface rendered it useless as a training device. However, for its targeted intention, namely collaborative training, the SIMNET system ultimately proved to be an excellent value. Over its 10 year life span (now entering its twilight) SIMNET enjoyed widespread success and acclaim for its reduction of training costs, increased quality and quantity of tactical team training, and usefulness as a testbed for new weapon concepts.

SIMNET's biggest accomplishment, however, was the spawning of a standard networking protocol which enables heterogeneous simulations to interact in a shared synthetic environment. This standard, called the DIS Protocol, has forever changed the way the Pentagon trains our soldiers. DIS compliance is now mandatory for all new training devices procured by the DoD.

There is still decades of research and development to be done in the areas of networked simulation for collaborative training. The most notable present-day need is in the area of immersive virtual reality technology, used in the simulation of individual soldiers. Even with the mighty budget of the DoD, we still cannot field a helmet-mounted display with the field-of-view, resolution, and frame rate necessary to render a useable world. Current offerings leave the user legally blind, and without necessary visual cues to perform required tasks. Force and tactile feedback are also lagging far behind. State-of-the-art systems currently use exercise bicycles, stair climbers, or linear treadmills to simulate human locomotion.

It is clear, however, that the infrastructure and architecture for a global network of synthetic environment portals for training, entertainment and education is in place. This infrastructure is in constant use and is driving development of associated technologies.

3 OVERVIEW OF DIS

3.1 Technical Details

The DIS Protocol consists of some 29 different network packets, known as Protocol Data Units (PDUs) which are passed among simulation nodes. The structure of the PDUs are outlined in a DIS standards document. The most current document being the "IEEE Standard for Distributed Interactive Simulation - Application Protocols" which outlines the IEEE 1278.1 DIS Standard PDUs (IEEE 1995).

The application specific information is encapsulated in UDP/IP Ethernet frames and transported in broadcast mode (one to all). The packets can be sent over any network media, from voice grade phone lines to ATM switches. More bandwidth allows a greater number of entities to be supported. DIS has also been used over the Internet using IP multicasting with excellent results.

A DIS PDU can describe the state information of a dynamic entity (Entity State), a combat event (Fire and Detonate), a resupply interaction (Logistic), or an electromagnetic emission such as light, radar, or energy weapon emissions (Electromagnetic Emission). Because the heritage of the protocol is military in nature, versions up to DIS 1278.1 have retained many military specific fields in the packets. However, current proposals for the next generation protocols remove most of the military content and provide a much more efficient, reconfigurable, and general purpose packet structure.

DIS uses predictive algorithms, known as dead reckoning, that allow entities on the network to greatly reduce the frequency of rebroadcasts of state. Each entity broadcasts its type, location, velocity, acceleration, orientation, and angular velocity. All the receiving simulators can then propagate the sending entity into the future, relieving the sending entity of the responsibility to continually rebroadcast. When the error between the exact position of the entity and the predicted position exceeds a certain threshold, the sending entity will update the network with its new kinematic state.

Each entity sends out information on how it should be dead reckoned. The Dead Reckoning Models (DRM) notation consists of three elements:

DRM (F or R, P or V, W or B)

The first element specifies whether the coordinate system of the entity is fixed (F) or rotating (R). The second element specifies whether rate is constant (P) or varying (V). The third elements specifies the coordinate system to be used, either world (W) or body (B). (IEEE 1995). A common dead reckoning algorithm used is DRM (F,P,W), shown below in Equation 1.

$$P = Vo + V\,dt \quad (1)$$

Networked latency and dropped packets can cause a jittering effect when visualizing the location of the entities. This occurs when there is a discrepancy between the values most recently received from a PDU and the calculated dead reckoned values. To counteract this, several visualization tools use translational and rotational smoothing to interpolate between the dead reckoned location and the location received from the PDU. As a cautionary note, high level smoothing (such as rotational) utilizes precious CPU resources.

DIS enables disparate simulators to interact in a shared virtual world. In order to ensure interoperability between these simulations a common coordinate system is used to transmit the entities' state within the shared synthetic environment. Each simulator needs to be able to transform between their local system and the DIS coordinate system, knows as the Geocentric Cartesian Coordinate System (GCS). It is a right-handed system with the origin at the centroid of the earth. The positive X-axis passes through the Prime Meridian at the equator, the positive Y-axis passes through 90 degrees East Longitude at the equator and the positive Z-axis passes through the North Pole (IEEE 1995).

The DIS Entity State packet, which makes up most of the network traffic in DIS 2.0.4, is about 140 bytes long, and is broadcast anywhere from once every 30 seconds, to four or five times per second. The next generation will further increase efficiency, using a trimmed-down Entity State packet that reduces network traffic by a factor of 3 or better (see Figure 1) (Taylor 1996b).

3.2 Benefits

The DIS architecture provides flexible tradeoffs between computational loading, positional error, and network bandwidth. If highly accurate position is required, the error threshold can be reduced, which will result in more network broadcasts of state. Conversely, if the only network bandwidth available is that provided by a 9600 baud modem, and 25 entities are required on the system, the error thresholds can be increased, and more compute-intensive prediction algorithms can be used on the receiving ends. State-of-the-art DIS systems currently in use by the government are running over 10,000 entities on Ethernets, long haul networked from several locations. Future developments, such as low-cost commercial data services (bi-directional cable TV, ISDN, etc.), next generation protocols, and support for multicast routing, will enable DIS exercises to reach the 100,000 player level.

Another important feature of DIS is its self-healing nature. When a new entity enters the world, he begins to broadcast Entity State packets. If recipients have never heard from this entity before, they simply add him to their remote entity database. If an entity is not heard from within five seconds, recipients will time him out, removing him from their remote entity database. Players can enter and leave at will without disturbing other participants, causing no effect other than their appearance and disappearance. Dropped packets don't cause system failure. In this architecture there is no central server, thus no single point of failure.

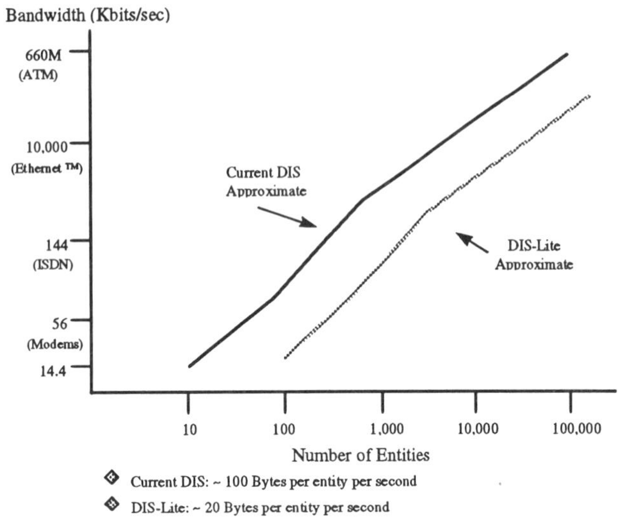

Figure 1: Bandwidth Requirements for Current and Future DIS Versions.

Heterogeneous nodes can interact with each other using DIS. The network protocol provides a standard mechanism for communication between simulators which may have radically different architectures. An entity broadcasting an Entity State packet is simply informing the network of what kind of entity it is, and its kinematic information. Two different recipients of the same Entity State packet may render the remote entity with very different levels of fidelity. For example, a PC may render a remote F-15 with only 10 flat shaded polygons, and may only have an internal simulation frame rate of 5 frames per second. A high-end Silicon Graphics workstation may render the same F-15 with 500 Gouroud shaded, phototextured polygons at a rate of 60 frames per second. In the entertainment world, this would allow a high-fidelity arcade-based system costing tens of thousands of dollars to interact, via the Internet, with a $200 home-based game system.

As compared to the computational power necessary to render out-the-window images in real-time, the DIS prediction algorithms do not represent a significant load. The most expensive prediction algorithm that DIS commonly uses consumes about 100 floating point operations per remote entity per simulation frame. Considering that the average PC can do about 10 million of these operations per second, and a simulation might run at 20 frames per second, a PC can handle several thousand remote entities using the most expensive algorithm. Naturally, this load must be balanced with other computational tasks, but this is a good indication as to the level of hardware necessary to accommodate fairly large DIS entity populations in a virtual world.

Each year, a demonstration of DIS interoperability is conducted at the annual Interservice/Industry Training Systems and Education Conference (I/ITSEC). Booths representing over 50 different organizations are networked together for a simulated exercise. DIS node systems, ranging from PCs to high fidelity flight simulators with multi-million dollar image generators, are connected for a week of interactive exercises. This is the third straight year such a demo was shown.

Though there are several commercial-off-the-shelf DIS developer's toolkits available, most of the industry uses a product called VR-Link™ produced by MäK Technologies of Cambridge, MA (www.mak.com). VR-Link, developed by some of the original SIMNET engineers, currently dominates the networked simulation market worldwide.

4 CURRENT NON-MILITARY USES OF DIS

It is no surprise that DIS technology has been finding its way into other government and commercial industries. For example, Air Traffic Control Simulations and Intelligent Vehicle Highway System research are among the government simulation communities using DIS.

In the commercial realm, DIS has been used for virtual reality systems, networking of games and entertainment systems, and most recently for multi-user virtual worlds on the World Wide Web. This section discusses some of the emerging non-military applications.

4.1 Government Applications

4.1.1 FAA Air Traffic Control Trainers

A DIS-based Air Traffic Control Training System was developed by the MITRE Corporation of Virginia. The system enables students to interact with simulated aircraft, either manned or automated, which could be coming from anywhere around the country. The aircraft are represented as DIS entities. Different situations can be scripted, repeated, and analyzed for performance. A national dial-in network for pilots in training is certainly possible, and has been proposed to the FAA.

4.1.2 Intelligent Transport Systems

The Department of Transportation has been sponsoring work in Intelligent Transportation Systems, formerly known as Intelligent Vehicle Highway Systems. The Analytical Sciences Corporation (TASC) of Massachusetts has developed a DIS-based automotive simulation to be used during their research. One experiment examined the effects of transportation on pollution. Like the FAA's pilot training network, a national network for driver training and city planning is also on the drawing board.

4.2 Commercial Applications

4.2.1 Networked Computer Games

Several game companies have adopted DIS as their networking mechanism for multi-user games on the Internet. Among them are Electronic Arts, Zombie, and Serum, three of the leading game developers for the new generation of 3-D first person interactive games. Several titles will be in the stores by Christmas of 1996.

4.2.2 Interactive Television

Telephone and cable TV companies are gearing up for high-bandwidth, bi-directional data services for the home, a move fueled by the passage of the new Telecommunications bill. The future promises such services as movies-on-demand, interactive video games, home shopping and more. By using DIS, the service provider can bring interactive virtual reality to users in their homes. The DIS bandwidth reduction techniques which allow thousands of entities to exist on Local Area Networks (LANs) will allow people in their living rooms to interact with each other in multi-participant virtual worlds. Each channel can be associated with a specific activity or experience. User's will be able to select from myriad sporting activities, games of chance and skill, and even fantasy adventures.

5 FUTURE DEVELOPMENTS IN DISTRIBUTED INTERACTIVE SIMULATION

DIS, being a relatively new technology, can still be improved. This section will discuss several research projects currently underway which will make DIS the logical choice for future Internet applications. The projects focus on the dynamic transfer of high-order data (VR-Protocol), accurate representation of contact forces (Newtonian Protocol), interoperability (High Level Architecture), and bandwidth reduction (DIS-Lite and Multicast).

5.1 VR-Protocol

In DIS, it is hard to add new behaviors or visual representations dynamically. This prohibits entities with experimental behaviors or visual characteristics from jumping into an existing exercise. A research project examining this problem has developed a VR-Protocol. The VR-Protocol enables the dynamic transfer of both behavioral and visual characteristics between systems (Taylor 1996a).

The VR-Protocol research project has demonstrated the ability to transfer a polygonal file for a new entity (used for visual representations such as avatars) on a query-response basis when needed. New players can now join the game and inform the pre-existing systems of their appearance. Though this capability is similar to Virtual Reality Markup Language (VRML), the transaction is not initiated by a person but is automatic and transparent to the users of the system.

The VR-Protocol also can transfer the behavior of an entity, via scripts. The scripts define the new behaviors based on a predefined set of basic primitive behaviors. For example, if a hundred virtual people are being simulated on a single workstation, a very efficient algorithm for wandering around can be defined and sent to the other players. The complex behavior can then be used by each player to simulate the entities without using any bandwidth. VR-Protocol is incorporating several scripting languages, such as JAVA. The VR-Protocol is really the best of VRML and Java rolled into one.

5.2 Newtonian Protocol

A drawback of DIS is its inability to transmit contact force information between two objects. This leads to low-fidelity emulations of collisions and logistical operations. The Newtonian Protocol is an extension to DIS which allows objects to exchange Newtonian forces. Using the protocol, systems can emulate tactile and force feedback, towing, and combat operations. The Newtonian Protocol uses a novel transfer function exchange paradigm to reduce the amount of network traffic required to maintain physical contact between two entities across a network (Taylor and Katz 1995)

5.3 DIS-Lite

Even with vast technological advancements in the communications arena, we are still far away from infinite bandwidth. In order to reduce the network bandwidth, a next generation DIS protocol, DIS-Lite, is being developed. DIS-Lite reduces bandwidth requirements by eliminating the transmission of redundant information. Figure 1 on page 4 compares the current version of DIS and DIS-Lite with respect to the required bandwidth versus the number of participants. DIS-Lite is approximately 30-70% more efficient than the current generation DIS (Taylor 1996b).

5.4 Multicast Traffic Management

Currently, DIS uses UDP Broadcast messages to send information across the network. This means that each participant receives every message that is sent whether or not they need or want the information. This unnecessary traffic is a waste of bandwidth. To alleviate this problem, a Multicast Traffic Management system is being implemented for the next generation of DIS. This capability will enable large scale DIS exercises to run over the Internet. The network traffic is separated into groups of interest, possibly by exercise ID, location, or even fidelity level. On a packet-switched internetwork, traffic will be routed only to those listeners who subscribe to a particular group of interest. This enables players to selectively send packets to subsets of listeners.

5.5 High Level Architecture

Although the existing DIS protocols work extremely well for certain applications, the DoD needs increased interoperability among simulations for which the DIS protocols may not be well suited, for example, higher order war game models. A fairly new research effort, funded by the Defense Modeling and Simulation Office (DMSO), called High Level Architecture (HLA) is concentrating on creating an interoperable architecture where the different types of systems can interact.

Under HLA all participating simulations are known as federates. A collection of simulations is known as a federation. An Object Model Template (OMT) is used to define simulation object models (SOMs) and federate object models (FOMs). The FOM is used by the participating federates to specify how information is sent. The OMT provides a standard framework for describing a simulation in terms of its objects and the interactions between objects (DoD 1996b).

HLA also encompasses a Runtime Infrastructure (RTI) which binds the federates together. Federates send information to the RTI which in turn sends the information to the appropriate parties. The RTI is comprised of five service groups:

- Federation Management
- Declaration Management
- Object Management
- Ownership Management
- Time Management

Federation management handles the creation, dynamic control, modification, and deletion of a federation execution. Declaration management enables federates to declare their desire to generate (publish) and receive (subscribe/reflect) object state and interaction information. Object management is responsible for the creation and deletion of objects as well as the transmittal of object and interaction data. Ownership management allows federates to transfer ownership of object attributes. Time management provides useful services for setting and modifying simulation clocks. (DoD 1996a)

Even though there has been a lot of hype about HLA, it is still a long way off. The release of a fully functional RTI has been pushed back to May of 1997.

6 CONCLUSION

DIS lets users interact with one another, and their environment, over a network. DIS' predictive algorithms dramatically reduces the amount of network traffic. Its self-healing property enables users to come and go without affecting the other participants. In addition, heterogeneous nodes can interact in the same synthetic environment.

The future is bright for DIS. The VR-Protocol will give DIS the ability to transfer visual representations, as in VRML, and behavioral information, as in Java. The Newtonian Protocol provides a mechanism to represent sustained contact force between entities over a network. DIS-Lite and Multicast Traffic Management Systems are aimed at reducing the required bandwidth making large scale DIS operations across the Internet possible.

REFERENCES

Department of Defense. 1996a. *High Level Architecture for Simulations Version 0.5 Interface Specification*.

Department of Defense. 1996b. *High Level Architecture Object Model Template Version 0.3.*

Institute of Electrical and Electronics Engineers. 1995. *IEEE Standard for Distributed Interactive Simulation - Application Protocols*. IEEE Publishing. NY.

Taylor, Darrin. 1996a. The VR-Protocol. In *Proceedings of the 14th DIS Workshop on Standards for the Interoperability of Distributed Simulations*.

Taylor, Darrin. 1996b. DIS-Lite and Query Protocol: Message Structures. In *Proceedings of the 14th DIS Workshop on Standards for the Interoperability of Distributed Simulations*.

Taylor, Darrin and Katz, Warren. 1995. Aspects of the Newtonian Protocol and its Application to Distributed Simulation. In *Proceedings of the 10th DIS Workshop on Standards for the Interoperability of Distributed Simulations*.

AUTHOR BIOGRAPHIES

DEBORAH A. FULLFORD is a member of the technical staff at MäK Technologies. She has a B.S and M.S in Aeronautical Engineering. She is a member of the Society of Computer Simulation and has served as both Session and Panel Chair for the Distributed Simulation Track of the Summer Computer Simulation Conference and the Multi-Simulation Conference. She is also a member of the DIS Standards Protocol Working Group.

Proceedings of the 1996 Winter Simulation Conference
ed. J. M. Charnes, D. J. Morrice, D. T. Brunner, and J. J. Swain

ADAPTIVE PROTOCOLS FOR PARALLEL DISCRETE EVENT SIMULATION

Samir R. Das

Division of Computer Science
The University of Texas at San Antonio
San Antonio, TX 78213-0667, USA

ABSTRACT

This paper reviews issues concerning the design of adaptive protocols for parallel discrete event simulation (PDES). The need for adaptive protocols are motivated in the background of the synchronization problem that has driven much of the research in this field. Traditional conservative and optimistic protocols and their hybrid variants are also discussed. Adaptive synchronization protocols are reviewed with special reference to their characteristics regarding the aspects of the simulation state that influence the adaptive decisions and the control parameters used. Finally, adaptive load management and scheduling strategies and their relationship to the synchronization protocol are discussed.

1 INTRODUCTION

Parallel discrete event simulation or PDES refers to parallel execution of discrete event simulation programs on a multiprocessor system or on a network of workstations. Over the past decade, there has been a considerable amount of activity in this field. There are several motivating factors. First, with the advent of technology, many application areas (e.g., simulations of computer architecture, VLSI circuits, communication networks) are getting larger and more complex. Thus speeding up simulations of such systems has become more important so that a sufficiently large design parameter space can be explored. The interest is also fueled by the common availability of multiprocessor systems (especially symmetric multiprocessors) and high-speed network based computing platforms (e.g., a cluster of workstations connected by high-speed networking hardware such as ATM or Myrinet switches). Second, it has been observed that many real simulations indeed contain a significant amount of parallelism. Third, the synchronization problem inherent in PDES is traditionally considered a challenging problem. Unlike many other areas of parallel computing (e.g., scientific computing) most discrete event simulation models are asynchronous and possess irregular, random or data dependent behavior. Thus, most of the research in PDES so far is centered around design and evaluation of synchronization protocols. For a review of the current state-of-the-art in PDES research see (Fujimoto 1990; Nicol and Fujimoto 1994; Ferscha 1995a).

There are two major classes of synchronization protocols that evolved: *conservative* and *optimistic*. The *conservative* protocols often rely on certain information on the behavior of the simulation model, such as the communication topology, or *lookahead* (i.e., the model's ability to predict the future course of events) (Fujimoto 1990) to determine which events are "safe" to process. They may also require certain implements such as message delivery in timestamp order, as in (Chandy and Misra 1979), or periodic global synchronizations as in (Lubachevsky 1989; Nicol 1993). All these make it difficult to develop general-purpose parallel simulators that can perform efficiently with little or no knowledge of the behavior of the underlying simulation model. Even if such knowledge is available, conservative mechanisms, being based on blocking, may fail to exploit much of the parallelism in the simulation model (Fujimoto 1989).

On the other hand, optimistic synchronization protocols such as *Time Warp* (Jefferson 1985) take a very different approach to resolve dependence between simulation events. They try to resolve dependence *a posteriori* by letting dependence violations to occur. If such violation is detected, erroneously executed computations are "undone" using a *rollback* mechanism. Time Warp uses *checkpointing* in addition to a *message cancellation* mechanism (based on antimessages) to implement rollback. The principal advantage of Time Warp over blocking-based, conservative protocols is that Time Warp offers the potential for greater exploitation of parallelism and, perhaps more importantly, greater transparency of the synchronization mechanism to the simulation programmer (Fujimoto

1990). The latter is due to the fact that Time Warp is less reliant on *a priori*, application specific information regarding which simulation events depend on which others.

Time Warp, however, is prone to inefficient execution in many situations because of over-optimistic behavior. For example, some LPs may execute at a much higher simulation time than others. This may lead to several performance problems, such as the possibility of long and/or cascaded rollbacks (Lubachevsky, Shwartz, and Weiss 1991) and significant memory management overheads (Das and Fujimoto 1993; Das and Fujimoto 1994). Regardless of the over-optimistic behavior, one key performance problem in Time Warp is the checkpointing related overheads.

In spite of the above problems, both conservative and optimistic protocols had a fair amount of success in parallelizing various simulation models (Fujimoto (1993a) has a review of different PDES application areas studied). However, PDES has a very limited penetration in the general simulation modeling community (Fujimoto 1993b). One often cited reason is the current level of sophistication required to effectively exploit the technology. For an acceptable level of performance justifying the use of parallel resources, careful attention must be paid to the interplay of the synchronization mechanism with the application simulation model and the underlying systems architecture. Without an insight in all the above, it is often unclear which synchronization protocol will be the best or even how certain details of the protocol (e.g., checkpointing interval in Time Warp (Fleischmann and Wilsey 1995)) should be selected. Thus there is a recent interest in the PDES community in investigating "adaptive" or dynamic mechanisms to choose an appropriate synchronization protocol (conservative, optimistic, or some hybrid variant) or parameters (such as the checkpointing interval in Time Warp) of a possibly statically chosen protocol. This paper reviews the state-of-the-art in such adaptive mechanisms for PDES. We start with some background in the traditional conservative vs. optimistic debate in the next section and then introduce the hybrid protocols in Section 3. In Section 4, we discuss the adaptive synchronization protocols, followed by dynamic load balancing/scheduling mechanisms in Section 5. Conclusions are presented in Section 6.

2 CONSERVATIVE VS. OPTIMISTIC DEBATE

The question of relative merits of conservative or optimistic protocols has often been raised. However, it is now a widely accepted belief in the PDES community that a formulation of general rules of superiority of one protocol versus another is hard (Ferscha 1995a). Still, some analytical results tend to favor optimistic mechanisms more than their conservative counterparts. Lipton and Mizell (1990) showed that optimistic mechanisms can arbitrarily outperform conservative mechanisms. But conservative mechanisms can outperform the optimistic mechanisms by at most a constant factor. This issue may be of purely theoretical interest, as it only represents extreme case scenarios. Also, the analysis is based on simple assumptions such as constant rollback cost and zero message communication or state saving/restoration cost, which may be biased towards optimistic mechanisms. In an unrelated work, Lin and Lazowska (1990) showed that Time Warp is "optimal" (i.e., can complete in "critical path" time) under the assumption that no correct computation is rolled back by incorrect computations. This study also ignores all overheads. It is also unclear how to guarantee such an assumption except in specific simulation models so that an optimal performance can be expected.

Without any clear choice between optimistic and conservative protocols, a conventional wisdom in the PDES community has been to study *hybrid* variations that take an intermediate approach between purely conservative (block-resume) and purely optimistic (lookahead-rollback) extremes. Intuitively, these protocols use a form of "reasonable" or "calculated" optimism so that they can exploit the benefits of optimism (e.g., ability to extract more parallelism, less reliance on model-specific information) without some its liabilities (e.g., rollback overhead, large memory usage). Reynolds, Jr. (1988) was one of the first to study a framework for specifying hybrid PDES protocols. He defined the notions of "aggressiveness" and "risk." An aggressive PDES protocol may process events in out-of-timestamp order and they may be required to rollback. A risky protocol, in addition to being aggressive, may pass messages that have been generated based on aggressive processing. Thus a message generated in a risky protocol may need to be canceled. Time Warp exhibits the maximal form of aggressiveness and risk. On the other hand, conservative protocols do not employ any aggressiveness or risk. According to Reynolds, Jr. (1988), aggressiveness and risk may be considered two important design variables that may influence the performance of a PDES protocols.

3 HYBRID PROTOCOLS

Hybrid protocols belong to primarily two categories: those that impart optimism to conservative protocols and those that introduce blocking to optimistic protocols. Indeed some of them is able to seamlessly switch between conservative and optimistic extremes with appropriate tuning of one or more control parameters (Rajaei, Ayani, and Thorelli 1993), and thus may be difficult to classify. The classification followed below is based more on the spirit of the protocol, rather than its ability to move seamlessly between two extremes.

3.1 Adding Optimism to Conservative Protocols

In the *speculative execution* protocol proposed by Mehl (1991) logical processes compute future events in their idle time, however such speculative computation does not modify the local state or the event queues. A scratchpad area in the memory is used as state and the events generated are stored in a private buffer. If it turns out later that the speculative execution is correct, local state can be quickly updated and the events are sent to the appropriate destinations. Thus support for rollback is not required. Dickens and Reynolds, Jr. (1990) earlier presented a similar, albeit more aggressive approach, where local state and local event queues can be modified by such speculative execution. However, events destined to remote processors are not sent out until they are determined to be correct. Thus antimessages are not required. Rollbacks are possible, but they are only "local" in the sense that there cannot be any cascades.

Steinman (1992a) proposed the *Breathing Time Bucket* algorithm, which also has similar, aggressive, but risky-free approach. Here, logical processes process events in cycles. In each cycle the maximum number of independent events are processed in the following way. Each LP processes events in timestamp order until the timestamp (called the local "event horizon") of the earliest new event it generates in the current cycle. New events generated are accumulated in local buffers and are not actually passed to the destination processes. The LPs synchronize (at least implicitly) at the end of the cycle to compute the minimum of all local event horizons, This minimum is the global event horizon. The messages generated by events with timestamp less than or equal to the global event horizon are then sent to their appropriate destinations. This may cause rollbacks, which, however, are only local. Analytical modeling as well as experimental results showed that this approach can be quite efficient if enough events can be processed in each cycle.

Lubachevsky, Shwartz, and Weiss (1989) proposed an extension to the *bounded lag* protocol (Lubachevsky 1989) to incorporate optimism to a purely conservative mechanism. The bounded lag protocol is a conservative protocol that relies on a conservative estimate of the minimum simulation time distance (called *lag*) between a pair of logical processes as a basis for determining events safe to process. In the extension, called *filtered rollback*, the simulation time distance is more aggressively estimated and the protocol runs the risk of overestimation (possibly rarely). Thus dependence violations are possible which are taken care of by rollbacks and event cancellation just like Time Warp. The protocol is thus both aggressive and risky.

3.2 Throttling Optimistic Protocols

Many protocols try to control the optimism in Time Warp. One key technique often used is to limit all event computations within a window of simulated time beyond the *global virtual time* or GVT (GVT in Time Warp denotes a lower bound on the timestamps of all future rollbacks and thus defines a commitment horizon (Jefferson 1985)). This limits length of rollbacks, thus preventing long cascades. This also bounds memory usage, a major concern in Time Warp simulations. The original work exploiting this idea was by Sokol, Briscoe, and Wieland (1988), where they proposed the *moving time window* (MTW) protocol. A key problem here is the determination of the appropriate size of the time window (as observed by Reiher, Wieland, and Jefferson (1989)) Too narrow a window will reduce rollbacks, but will admit only a small amount of parallelism. Too large a window can potentially exploit more parallelism, but rollbacks may increase as well. A similar idea is also explored by Turner and Xu (1992) in the *bounded Time Warp* (BTW) protocol, where no events are processed beyond a bound in simulation time until all processes have reached that bound, when a new bound is established.

In the MIMDIX system Madisetti, Hardaker, and Fujimoto (1993) explored the idea of *probabilistic resynchronization*. All processes in the simulation are rolled back to close to GVT at probabilistically chosen intervals of real time. Special processes, called *genies*, broadcast special SYNC messages with timestamp slightly larger than the current GVT to induce such rollbacks. This scheme was shown to effectively eliminate over-optimistic behavior for a suitable choice of the resynchronization interval.

Steinman (1993) proposed an extension to his breathing time bucket algorithm by allowing it to take risks. The resulting protocol is called *breathing Time Warp* and is essentially a mixture of Time Warp and breathing time bucket. The idea is to release the events generated by the first N (say) events beyond GVT in each cycle. This is similar to Time Warp, as these messages may need to be canceled later. The protocol switches back to the risk-free breathing time bucket from the $N + 1$ event in the cycle.

In Composite ELSA protocol (Arvind and Smart 1992) a node can switch between conservative and optimistic modes by using additional information regarding whether an event is safe. Both local and cascaded rollbacks are allowed. Rajaei, Ayani, and Thorelli (1993) explored a *local Time Warp* approach, where logical processes use Time Warp within clusters and use a synchronous, time window based conservative protocol across clusters. This hierarchical, cluster-based approach limits the progress of erroneous computations within a cluster. All inter-cluster messages are correct messages.

Several protocols have also been suggested to stop the spread of incorrect computation as soon as possible. Examples include "Wolf calls" (Madisetti, Walrand, and Messerschmitt 1988) and "Filter" (Prakash and Subramanian 1991). They broadcast or multicast special correction messages as soon there is a rollback in any LP. These correction messages stop the spread of incorrect computation in remote processes quicker than regular antimessages as they are sent directly instead of an indirect, recursive fashion as antimessages. However, these techniques are reactive, rather than proactive as action is taken only after the primary rollback (an observation made by Srinivasan and Reynolds (1996)).

4 ADAPTIVE PROTOCOLS

Many of the hybrid protocols can be adjusted in the continuum between the conservative and optimistic extremes. For example, the width of the time window in the MTW protocol can be tightened to equal to the *lag* and MTW will behave as the conservative bounded lag protocol. On the other hand if the width is infinite, MTW is equivalent to Time Warp. Similarly, the resynchronization interval in MIMDIX or the value of N in breathing Time Warp can be controlled to impart some form of extreme nature to the protocol. Not unexpectedly, experimental results indicate that overall performance of the protocol is sensitive to the choice of such control parameters. Best performance usually is obtained by setting such parameters somewhere between its extremes.

However, it is difficult for a simulation modeler, who is not intimately familiar with PDES protocols and the underlying parallel architecture, to set these parameters appropriately for an optimal performance. Furthermore, many simulation models are very dynamic in their runtime characteristics. The synchronization mechanism must adapt itself to a change in the model characteristics to optimize performance. This observation triggered a recent emergence of hybrid protocols that automatically "adapt" themselves in the continuum between purely conservative and purely optimistic strategies. The key idea is to monitor the state of the parallel simulation to estimate the appropriate trade-off between the conservatism (blocking or lost opportunity cost) and optimism (rollback and memory management costs) and accordingly adjust the control parameters.

The initial ideas about the usefulness of such adaptive protocols date back to the work by Reynolds, Jr. (1988), where he defined "adaptability" as the ability of logical processes to dynamically change one or more control parameters based on "knowledge of selected aspects of the state of the simulation." The objective is to minimize the execution time of the parallel simulation. The key issues are, of course, what control parameter(s) is (are) appropriate, and which aspects of the simulation state influence the binding of such control parameter(s) and how. Any adaptive protocol must take appropriate design decisions that address the above issues. As predicted by Reynolds, Jr. (1988), there are indeed quite a few very reasonable design choices. In the following we describe the adaptive protocols described in recent literature by broadly classifying them according to whether the adaptive decisions are taken based on purely the local state of the each LP or the global state of all LPs.

4.1 Protocols Based on Local State

One of the earliest attempts to develop adaptive protocols was made by Reiher, Wieland, and Jefferson (1989), where they proposed a *penalty based throttling* mechanism, where the logical processes that experienced rollbacks in the recent past are penalized. The penalty is represented by reduced scheduling quanta. This method was implemented in the JPL Time Warp Operating System (TWOS) but failed to produce any significant performance gain over Time Warp.

In the *Adaptive Time Warp* or ATW (Ball and Hoyt 1990) a logical process waits for an interval of real time (called *blocking window* or BW) between processing successive events. The BW is adjusted such that the sum total of the CPU time spent in blocked state and in faulted state (i.e., undoing incorrect computation) is minimized. The time spent in blocked state in directly proportional to the width the BW. The time spent in faulted state is modeled by a logistic response function. A numerically efficient approximate method is used to compute the minima. Ferscha and Lüthi (1995) proposed a similar, but more sophisticated method based on *probabilistic cost expectation function* or PCEF. An explicit cost model involving the probability and cost of rollback is used instead of a logistic response function. The probability of rollback is assumed to be dependent on both real time and simulation time difference between the next scheduled event and the last message arrival. In order to compute this probability each LP monitors the timestamp and real time instants of arrival of every message as well as the rollback behavior. The cost model is used to evaluate the optimal real time blocking window. This method was shown to outperform Time Warp in experiments involving simulation of stochastic petri nets. Similar cost model has also been proposed for evaluating an optimal simulation time window rather than a real time blocking window (Das 1996). Here, simulation time distance from GVT is used as a parameter for evaluating the probability of rollback. This, however, depends on global state as GVT is needed.

In the conservative-optimistic or CO-OP protocol, (Steinman 1992b) LP scheduling on a processor uses some knowledge of the probability of the next event being correct. He also suggested a local CO-OP protocol where conditions on the input channels of the LPs control a variable that influence the LP scheduling decisions on a processor. (Ferscha and Chiola 1994) proposed a *probabilistic distributed simulation protocol* where Time Warp delays event computation artificially event computations based on the probability of an external message arrival inducing a rollback. Hamnes and Tripathi (1994b) proposed a *local adaptive protocol* or LAP, which uses input channel specific information for each LP to determine a real time blocking window. An LP blocks if it estimates that an event arriving later in one of the empty input channels will cause a rollback. This estimate is based on average inter-arrival times (both in simulation time and real time) of messages in each input channel in recent past. Null messages are used to prevent deadlock and to preempt unnecessary blocking. LAP was shown to outperform both conservative and optimistic methods in a performance study with closed queuing systems simulations (Hamnes and Tripathi 1994a).

Ferscha (1995b) presented a *probabilistic adaptive direct optimism control* or PADOC protocol, which is similar in spirit, but adds a probabilistic component and computes the blocking window incrementally. Each logical process maintains a history record of the simulation time differences of the arriving messages in the recent past. This record is used to forecast the timestamps of forthcoming messages. Several forecasting methods have been explored, such as methods based on central tendencies as well as autoregressive moving average (ARMA) models. Other models, such as neural networks, are also deemed possible. In addition to the forecast about the timestamp of the next arriving message, a confidence (ξ) in the forecast is also computed. The LP is blocked for a small predetermined period (same as the average event granularity) with probability $P(\xi)$ and advances to the next event in the event list with probability $1 - P(\xi)$. Then a new forecast is computed (in case there are new message arrivals in between) and this cycle continues. $P(\xi)$ was assumed to be a sigmoid function dependent on ξ as well as the difference between the local simulation time and the timestamp forecast for the next arriving message. PADOC was shown to outperform Time Warp for stochastic petri net simulations, especially under load imbalance.

Palaniswamy and Wilsey (1993) exploited an *adaptive bounded time window* approach where width of the time window (similar to MTW) is controlled dynamically, based on the concept of *useful work* done by an LP. Useful work is a measure of the productive work done by the LP and is a function of a number of parameters such as the ratio of the number of events committed to the number of events executed, number of rollbacks, average rollback length, number of antimessages sent etc. Each LP has its own value of window which is increased or decreased periodically depending on the change in useful work. Experiments with digital logic simulations demonstrated superiority of this method over ordinary Time Warp.

4.2 Protocols Based on Global State

Some protocols rely primarily on some aspects of the global state of the simulation for making adaptive decisions. Important examples are the *Adaptive Memory Management* protocol by Das and Fujimoto (1994) and the *Near Perfect State Information* (NPSI) protocols by Srinivasan and Reynolds (1995). In the adaptive memory management protocol an indirect approach is used for adaptation. It has been observed that over-optimistic Time Warp not only incurs high rollback costs, but also high memory management costs, because of loss of locality and possibly other virtual memory management overheads. Also, artificial memory limitation automatically restricts Time Warp's optimism. Thus the total amount of memory used by the simulation is used as the control parameter, by exploiting Time Warp's ability to continue simulation using any amount of memory larger than a lower bound. The cancelback protocol (Jefferson 1990) is used to restrict Time Warp within the allocated amount of memory. This scheme reduces both rollback and memory related costs and attempts to run Time Warp with just the sufficient amount of memory required for the best overall performance. Several monitored information about the rollback behavior, frequency of fossil collection and cancelback is used to decide the right amount of memory to be allocated.

NPSI protocols are a class of adaptive protocols relying on the availability of "near-perfect" information on the global state of the simulation. NPSI protocols use a quantity called *error potential* EP_i associated with each LP_i. The value of EP_i is used to control LP_i's optimism by, for example, introducing artificial blocking. Note that an inherent assumption in NPSI protocols is that the NPSI is available with minimal cost. This limits the use of such protocols to shared memory multiprocessors or distributed memory systems with parallel reduction network support. The usefulness of the latter for PDES has been demonstrated in (Reynolds, Jr., Pancerella, and Srinivasan 1993).

Different NPSI protocols can be envisaged for different choices for EP and different mechanisms to control optimism. A specific NPSI protocol called the *Elastic Time Algorithm* or ETA is described in (Srinivasan and Reynolds 1995). In ETA, the difference between the next event time of LP_i and the GVT is used as EP_i. More recently, the definition of EP_i is refined to replace GVT by the minimum of the next event time of all LPs that can send events to LP_i and timestamps of any messages in transit that are destined to LP_i (Srinivasan and Reynolds 1996). Optimism is controlled by blocking for an amount of real time proportional to EP_i between processing of successive events. The proportionality constant s (called the *scale factor*) requires some tuning for the optimal performance, which can also be adapted (Srinivasan and Reynolds 1996).

5 DYNAMIC LOAD BALANCING AND SCHEDULING

It is interesting to note how dynamic load balancing can interact with the synchronization mechanism. In particular, optimistic mechanisms introduce "a new wrinkle" to dynamic load balancing: high processor utilization may not imply good performance as processor may be busy incorrect computation that will be undone later (Nicol and Fujimoto 1994). Several load balancing algorithms have been proposed for Time Warp. (Reiher and Jefferson 1990) used a metric called *effective processor utilization* which is defined as the fraction of time a processor is executing correct computations (which are not later rolled back). Based on this metric, processes are migrated from processors with high effective utilization to processors with low utilization. Glazer and Tropper (1993) proposed an allocating *virtual time slices* to LPs based on their observed rate of progress in simulation time.

It is conceivable that a load distribution strategy that targets equalization of simulation clocks thus imparting some form of temporal (in the sense of simulation time) locality in the simulation system will potentially reduce rollbacks (Jefferson 1985). One (extreme) way of doing it is to migrate LPs to processors in such a way that the n (n being the number of processors) slowest (in simulation time) LPs in the system are all assigned to different processors (Ahmed, Rönngren, and Ayani 1994). If the above condition is strict, then there is a possibility of LP migration after processing each simulation event. Such strict condition on LP scheduling improves temporal locality at the expense of spatial locality, as different events on an LP is processed at different processors. Also, use of a centralized scheduling queue may introduce bottlenecks. A milder technique was explored by Burdorf and Marti (1993) that may improve spatial locality. It is expected for most systems a tradeoff between temporal and spatial locality will yield the best performance. However, in our knowledge no such tradeoff has been evaluated for PDES systems.

6 CONCLUSIONS

Even though the adaptive protocols differ in many details, they all (i) use some aspect of the simulation state, local or global, for making adaptive decisions, (ii) use one or more control parameters to tune the dynamics of the simulation, and (iii) provide a mapping that specifies the binding of the control parameter depending on the state. Past rollback behavior or past message arrival pattern is frequently used as the state information if the protocol relies on local state alone. Availability of global state information (as in NPSI protocols) yields more complete knowledge about the dynamics of the simulation as the relative progress of all LPs or overall memory usage can be known. The control parameters commonly used are scheduling quanta, real time blocking window, simulation time window, or amount of allocated memory. The sole use of the control parameters is to selectively reduce the simulation time progress of some LPs per unit real time by starving it of processor or memory resources. The mapping between the state information and the control parameter is often linear, but may be non-linear (as in the PADOC protocol). Some protocols (such as the ETA) even use tuning of the parameters of the mapping function.

One view of any adaptive protocol is that of a feedback based control system (Palaniswamy and Wilsey 1994), and there is an inherent possibility of instability in such systems, as the feedback is always applied with a lag in real time. Instability is very probable if the characteristics of the simulation model (e.g., model parallelism) changes too fast thus not giving the protocol sufficient time to react to the changes. Study of stability properties is important because of the dynamic nature of many simulation models. With the initial success in developing performance-efficient adaptive PDES protocols, we feel that this is one area that needs to be explored in future.

REFERENCES

Ahmed, H., R. Rönngren, and R. Ayani 1994. Impact of event scheduling on performance of Time Warp parallel simulations. In *Proceedings of the 27th Annual Hawaii Intl. Conf. on System Sciences.*

Arvind, K. and C. Smart 1992. Hierarchical parallel discrete event simulation in composite ELSA. In 6th *Workshop on Parallel and Distributed Simulation*, pp. 147–158.

Ball, D. and S. Hoyt 1990. The adaptive Time-Warp concurrency control algorithm. *Proceedings of the SCS Multiconference on Distributed Simulation 22*(1), 174–177.

Burdorf, C. and J. Marti 1993. Load balancing strategies for Time Warp on multi-user workstaions. *The Computer Journal 36*(2), 168–176.

Chandy, K. M. and J. Misra 1979. Distributed simulation: A case study in design and verification of distributed programs. *IEEE Transactions on Software Engineering SE-5*(5), 440–452.

Das, S. R. 1996. Estimating the cost of throttled execution in Time Warp. In *Proceedings of the 10th Workshop on Parallel and Distributed Simulation*, pp. 186–189.

Das, S. R. and R. M. Fujimoto 1993. A performance study of the cancelback protocol for Time Warp. *Proceedings of the 7th Workshop on Parallel and Distributed Simulation 23*(1), 135–142.

Das, S. R. and R. M. Fujimoto 1994. An adaptive memory management protocol for Time Warp parallel simulation. In *Proceedings of the 1994 ACM SIGMETRICS Conference on Measurement and Modeling of Computer Systems*, pp. 201–210.

Dickens, P. M. and P. F. Reynolds, Jr. 1990. SRADS with local rollback. *Proceedings of the SCS Multiconference on Distributed Simulation 22*(1), 161–164.

Ferscha, A. 1995a. Parallel and distributed simuation of discrete event systems. In A. Y. H. Zomaya (Ed.), *Parallel and Distributed Computing Handbook*, Chapter 35. McGraw-Hill.

Ferscha, A. 1995b. Probabilistic adaptive direct optimism control in Time Warp. In *Proceedings of the 9th Workshop on Parallel and Distributed Simulation*, pp. 120–129.

Ferscha, A. and G. Chiola 1994. Self adaptive logical processes: The probabilistic distributed simulation protocol. In *Proceedings of the 27th Annual Simulation Symposium*, pp. 78–88.

Ferscha, A. and J. Lüthi 1995. Estimating rollback overhead for optimism control in Time Warp. In *Proceedings of the 28th Annual Simulation Symposium.*

Fleischmann, J. and P. A. Wilsey 1995. Comparative analysis of periodic state saving techniques in Time Warp simulators. In *Proceedings of the 9th Workshop on Parallel and Distributed Simulation*, pp. 50–58.

Fujimoto, R. M. 1989. Performance measurements of distributed simulation strategies. *Transactions of the Society for Computer Simulation 6*(2), 89–132.

Fujimoto, R. M. 1990. Parallel discrete event simulation. *Communications of the ACM 33*(10), 30–53.

Fujimoto, R. M. 1993a. Parallel and distributed simulation: Algorithms and applications. In *1993 Winter Simulation Conference Proceedings*, pp. 106–114.

Fujimoto, R. M. 1993b. Parallel discrete event simulation: Will the field survive? *ORSA Journal of Computing 5*(3), 213–230.

Glazer, D. W. and C. Tropper 1993. On process migration and load balancing in Time Warp. *IEEE Transactions on Parallel and Distributed Systems 3*(4), 318–327.

Hamnes, D. O. and A. Tripathi 1994a. Evaluation of a local adaptive protocol for distributed discrete event simulation. In *Proceedings of the 1994 International Conference on Parallel Processing*, pp. III:127–134.

Hamnes, D. O. and A. Tripathi 1994b. Investigations in adaptive distributed simulation. In *Proceedings of the 8th Workshop on Parallel and Distributed Simulation*, pp. 20–23.

Jefferson, D. R. 1985. Virtual time. *ACM Transactions on Programming Languages and Systems 7*(3), 404–425.

Jefferson, D. R. 1990. Virtual time II: The cancelback protocol for storage management in distributed simulation. In *Proc. 9th Annual ACM Symposium on Principles of Distributed Computing*, pp. 75–90.

Lin, Y.-B. and E. D. Lazowska 1990. Optimality considerations of "Time Warp" parallel simulation. *Proceedings of the SCS Multiconference on Distributed Simulation 22*(1), 29–34.

Lipton, R. J. and D. W. Mizell 1990. Time Warp vs. Chandy-Misra: A worst-case comparison. *Proceedings of the SCS Multiconference on Dis-*

tributed Simulation 22(1), 137–143.

Lubachevsky, B. D. 1989. Efficient distributed event-driven simulations of multiple-loop networks. *Communications of the ACM 32*(1), 111–123.

Lubachevsky, B. D., A. Shwartz, and A. Weiss 1989. Rollback sometimes works ... if filtered. In *1989 Winter Simulation Conference Proceedings*, pp. 630–639.

Lubachevsky, B. D., A. Shwartz, and A. Weiss 1991. An analysis of rollback-based simulation. *ACM Transaction on Modeling and Computer Simulation 1*(2), 154–193.

Madisetti, V., J. Walrand, and D. Messerschmitt 1988. Wolf: A rollback algorithm for optimistic distributed simulation systems. In *1988 Winter Simulation Conference Proceedings*, pp. 296–305.

Madisetti, V. K., D. A. Hardaker, and R. M. Fujimoto 1993. The MIMDIX operating system for parallel simulation and supercomputing. *Journal of Parallel and Distributed Computing 18*(4), 473–483.

Mehl, H. 1991. Speedup of conservative distributed discrete-event simulation methods by speculative computing. *Proceedings of the Multiconference on Advances in Parallel and Distributed Simulation 23*(1), 163–166.

Nicol, D. M. 1993. The cost of conservative synchronization in parallel discrete event simulations. *Journal of ACM 40*(2), 304–333.

Nicol, D. M. and R. M. Fujimoto 1994. Parallel simulation today. *Annals of Operations Research 53*, 249–286.

Palaniswamy, A. C. and P. A. Wilsey 1993. Adaptive bounded time windows in an optimistically synchronized simulator. In *Great Lakes VLSI Conference*, pp. 114–118.

Palaniswamy, A. C. and P. A. Wilsey 1994. Scheduling time warp processes using adaptive control techniques. In *Proceedings of the 1994 Winter Simulation Conference*, pp. 731–738.

Prakash, A. and R. Subramanian 1991. Optimistic distributed simulations. In *Proceedings of the 24th Annual Simulation Symposium*, pp. 123–132.

Rajaei, H., R. Ayani, and L.-E. Thorelli 1993. The local Time Warp approach to parallel simulation. In *Proceedings of the 7th Workshop on Parallel and Distributed Simulation*, pp. 119–126.

Reiher, P. L. and D. Jefferson 1990. Dynamic load management in the Time Warp Operating System. *Transactions of the Society for Computer Simulation 7*(2), 91–120.

Reiher, P. L., F. Wieland, and D. R. Jefferson 1989. Limitation of optimism in the Time Warp Operating System. In *Proceedings of the 1989 Winter Simulation Conference*, pp. 765–770.

Reynolds, Jr., P. F. 1988. A spectrum of options for parallel simulation. In *1988 Winter Simulation Conference Proceedings*, pp. 325–332.

Reynolds, Jr., P. F., C. M. Pancerella, and S. Srinivasan 1993. Design and performance analysis of hardware support for parallel simulation. *Journal of Parallel and Distributed Computing 18*(4), 435–453.

Sokol, L. M., D. P. Briscoe, and A. P. Wieland 1988. MTW: a strategy for scheduling discrete simulation events for concurrent execution. *Proceedings of the SCS Multiconference on Distributed Simulation 19*(3), 34–42.

Srinivasan, S. and P. F. Reynolds 1995. NPSI adaptive synchronization algorithms for PDES. In *1995 Winter Simulation Proceedings*, pp. 658–665.

Srinivasan, S. and P. F. Reynolds 1996. Elastic time. *ACM Transactions on Modeling and Computer Simulation*. Submitted.

Steinman, J. 1992a. SPEEDES: A multiple-synchronization environment for parallel discrete event simulation. *International Journal of Computer Simulation 2*(3), 251–286.

Steinman, J. 1992b. SPEEDES: A unified approach to parallel simulation. In *6th Workshop on Parallel and Distributed Simulation*, pp. 75–84.

Steinman, J. S. 1993. Breathing Time Warp. In *Proceedings of the 7th Workshop on Parallel and Distributed Simulation*, pp. 109–118.

Turner, S. J. and M. Q. Xu 1992. Performance evaluation of the bounded Time Warp algorithm. *Proceedings of the SCS Multiconference on Parallel and Distributed Simulation 24*(3), 117–126.

AUTHOR BIOGRAPHY

SAMIR R. DAS is an assistant professor in the Division of Computer Science at The University of Texas at San Antonio. He received his ME in computer science and engineering from the Indian Institute of Science, Bangalore, in 1988 and PhD in computer science from the Georgia Institute of Technology, Atlanta, in 1994. His research interests include parallel and distributed systems, performance evaluation and discrete event simulation. In the recent past he worked on performance aspects of Time Warp based parallel simulation systems.

Proceedings of the 1996 Winter Simulation Conference
ed. J. M. Charnes, D. J. Morrice, D. T. Brunner, and J. J. Swain

ADVANCED TUTORIAL - SIMULATION-BASED SCHEDULING AND CONTROL

Brett A. Peters
Jeffrey S. Smith
James Curry
Cynthia LaJimodiere

Department of Industrial Engineering
Texas A&M University
College Station, Texas 77843, U.S.A.

Glenn R. Drake

Systems Modeling Corporation
504 Beaver Street
Sewickley, Pennsylvania 15143, U.S.A.

ABSTRACT

This paper presents a simulation control system developed by the Texas A&M Computer Aided Manufacturing Laboratory (TAMCAM). This control system is a research tool that TAMCAM uses to explore the advantages and disadvantages of on-line simulation for process control. The simulation control system is developed directly from information about the shop floor stored in a relational database. Processing times and fallout rates are also estimated directly from external data sources. The system is implemented in Arena, Microsoft Access, and Microsoft Visual C++. The advanced tutorial will demonstrate the software tools described in this paper.

1 INTRODUCTION

Simulation is commonly used to gain insight into manufacturing systems. Simulation is based on a logical model of how a series of processes interact combined with statistical and deterministic information about the individual processes. Historically, simulations are developed off-line using custom software packages/languages with limited direct connections to the actual data generated by the production system (Drake and Smith 1996). Traditional simulation generally examines long-term system performance for planning and design purposes. These models are usually "throw away models" because they are seldom used after the initial plans or designs of a project are finalized (Thompson 1994). Primary reasons for this inflexibility are that the input data of the simulation are gathered and analyzed outside the simulation environment and that the simulation system cannot communicate automatically with the Manufacturing Execution System (MES) that is responsible for gathering shop floor status information and executing tasks.

On-line simulation integrates the information system with the simulation model. On-line simulation for process scheduling, real-time "intelligent" shop floor control, performance forecasting, process capability estimation, real-time control systems emulation, real-time displays of system status, and short term decision making is an active area of research (See, for example, Erickson et. al. 1987, Wu and Wysk 1989, Harmonosky and Robohn 1991, Rogers and Flanagan 1991, Moser and Engell, 1992, Smith et al. 1994, Drake and Smith. 1995, Jones et. al. 1995, Tunali 1995, and Grabot et al. 1996). By using the most current system information, accurate predictions about the system and future control alternatives can be developed. The reason for using simulation is that simulation can often capture and describe the complex interactions within a particular FMS where analytical methods fail (Erickson et al. 1987, Wu and Wysk 1989).

An important element to make on-line simulation effective is the ability to do retrospective studies. To model the past, the TAMCAM Simulation Control System (TSCS) writes the transitions between system states to a temporal database for later analysis. Queries on the database provide estimates of process times, rework, and fallout rates. These estimates for the distribution of the time required to do a task drive the simulation model as inputs in look-ahead mode. This paper describes the structure of the TSCS and the associated advanced tutorial demonstrates the prototype software tools and describes our limited implementation experience.

2 ENVIRONMENT

The TAMCAM system is based on a three level control system containing a manufacturing information system, a real-time simulation model operating as a Manufacturing Execution System (MES), and execution level controllers for automated and manual equipment (Figure 1). The highest level is a manufacturing information system usually a MRP system that contains information about inventory level and production schedules. This system drives a simulation controller that handles the actual shop floor control. The semantics of the manufacturing information system are enforced at the control and simulation levels.

The simulation level acting as a MES uses a discrete event simulation for decision making (Smith et al. 1994). In addition to decision making, on-line simulation, as proposed by Drake and Smith (1995, 1996), uses a real time link between a simulation model and the production system as a direct method of process control. The TSCS simulation is generated directly from the manufacturing information system. This direct generation from a relational database structure, using Simon or Arena, is similar to the methodology proposed by Centeno and Standridge (1991). This direct link insures consistency between the two levels of the process.

The simulation model can be run in a real-time control mode or in a look-ahead simulation mode to test proposed schedules. In real-time mode, the simulation prompts the client for a task using the task initiation queue (TIQ). After the task is complete, the client returns the length of time required for the task and the outcome of the task using the task completion queue (TCQ). In look-ahead mode, the real-time delay blocks are replaced with statistical distributions for the time required in the delay block. The commercial versions of the Arena simulation package now include the capability for directly implementing real-time control.

The lowest level contains operational clients. The clients of the shop floor controller include automated shop floor equipment or Human Machine Interfaces (HMI) for processes that are not completely automated. When messages pass through the router, the content of the messages is written to the system state database with a time stamp. The workstations are based on a message-based part state graph (MPSG) controller (Smith and Joshi 1995, Smith 1996).

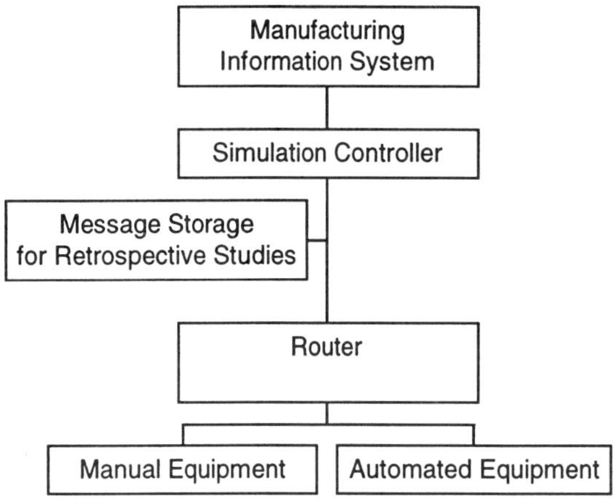

Figure 1: Diagram of the System

3 DEFINITION OF STATES

The key to on-line simulation is determining at what level to model the system. The TSCS model is based on constructing the system state information such that operations only occur on one part where all resources employed in the operation work on the part for the duration of the operation. This restriction avoids estimating process times and fallout rates from processes that are mixed together and facilitates direct estimation of process times from the messages being sent in the system. An example of a mixed process to be avoided is a worker loading a machine and then the worker being released after the completion of the loading task before the completion of the machining process. This process is modeled as two tasks since the number of resources required by the part changes during the operation. If the worker is seized for the entire period of time then the operation would be modeled as one task in the simulation.

All operations that act on multiple parts are modeled in the logic of the simulation. For assembly, the parts are combined in the logic of the simulation before the actual operation occurs. For disassembly, parts in the simulation model are created for the parts disassembled after the operation has occurred. Batching is modeled as an assembly operation followed by the actual operation with a disassembly operation to return the parts to their original status.

The task initiation message contains the operation and the part ID. All tasks are sent only to free resources that have the parts required to complete a task in their part queue or buffers. This restriction avoids including queuing times of parts waiting for resources or resources waiting for parts in the processing time. Queuing times are excluded from the model because queuing time arises primarily from the scheduling choices developed by the decision maker.

The task completion message contains the outcome of the task and the active processing time required for the task. The unique identifier for the task completion message is the part number and workstation. The basic active processing time includes set up time, down time, and transportation time within the workstation. For particular resources where a more detailed description of the system is required, the active processing time is split into subgroups including down time, reaction time, setup time and activity time for particular sub-tasks (Table 1). These additional attributes are only required if needed by the simulation.

Table 1: Messages Sent within the System

	Initiation Sequence	Completion Messages
Required Information	Part Number, Work Station	Part Number , Workstation , Active Processing Time,
Optional Information	Due Date, Special Instructions, Priority	Setup Time, Reaction Time, Sub-Operation, Processing Times, Success of the Operation, Additional Routing Based on the Operation.

4 INFORMATION SYSTEM

The information system is spilt into a description of the manufacturing system and a historic record of the on-line simulation.

The description of the manufacturing system contains a list of parts (i.e., part types), a list of specific parts to be produced with due dates, a list of resources for the production process with feasible messages for each resource, routings for the parts, alternative routings for the parts, and any additional information required to develop the simulation. Many of these inputs including the current production schedule, due dates, part routings, and resource relationships are deterministic and can be directly accessed from a relational database using SQL as suggested by Drake (1996). However, several key inputs to the proposed simulation model, including distribution of the process times, fallout rates, and shift schedules, are stochastic and vary for different part types and equipment stations and cannot be directly accessed from a static database.

This information system can access the simulation to run the simulation in look-ahead mode. The look-ahead simulation tool can predict whether a given production schedule is feasible by using the simulation as a query tool. The feasibility can be expressed as either the probability of completing a set of tasks within a given period of time or the entire estimated cumulative density function for completing the task as a function of time. Using this ability, multiple simulations can be used to improve the scheduling of resources by changing the operating rules (Tautuo and Pierreval 1995). The instruction for on-line simulation tools for manufacturing systems should resemble commands commonly used in database systems. A method for achieving this goal is to have the commands as an extension of SQL (Balasubramanian and Tuzhilin 1996). Even though our model does not follow the command set proposed by Balasubramanian and Tuzhilin, the TSCS model tries where possible to use a command structure that is easily understood to database and industrial/manufacturing engineering professionals.

The results of the real-time simulation execution are stored in a database. The real-time results are the data on how the system actually operates. The combination of a MES system with a database system is extremely common with most vendors using Open Database Connectivity (ODBC) software or a similar methodologies to connect controllers to database systems. These systems warehouse this information for many purposes including statistical quality control.

The TSCS database system stores the results of operations including process times and fallout rate. The user can exclude data points from the estimation of processing times for statistical quality control or look-ahead simulation. The primary reasons for excluding data are that the observations are influenced by a problem where the underlying cause is resolved and will not reoccur or that the data is in error.

From the database system, a data analysis tool estimates the distribution for process times and fallout rates at each workstation for each part type. These estimates are stored and used in the simulation model. The default distribution for process times is the gamma distribution. If the sample is not sufficiently large, then the estimation of the process times is based on a grouping of parts with similar characteristics. The estimation procedure identifies extreme values and allows the user to delete these values from the estimation.

The results of the simulated system in look-ahead mode can also be compared to the actual flow time for the system in real time. A difference between the simulated system in real time and the estimated performance in look-ahead mode could reveal problems in the simulation or the actual system that occur in a series of stations.

5 IMPLEMENTATION ISSUES

The demonstration of the TSCS consists of a simulation-based controller developed in Arena, a message router and client controls developed in Microsoft Visual C++, and an external database system developed in Microsoft Access. The simulation-based controller is built in Arena using the Arena Real-Time template. The Arena model also uses a user-coded dynamic link library (DLL) written in Microsoft Visual C++ to provide the implementation specific communications functions required by the router. All connections within the real-time system are implemented using the TCP/IP protocols. The connections within the forecast system are implemented with Database Access Objects (DAO). The connections between the different software applications are shown in Figure 2 where the solid lines are TCP/IP connections and the dashed lines are DAO connections.

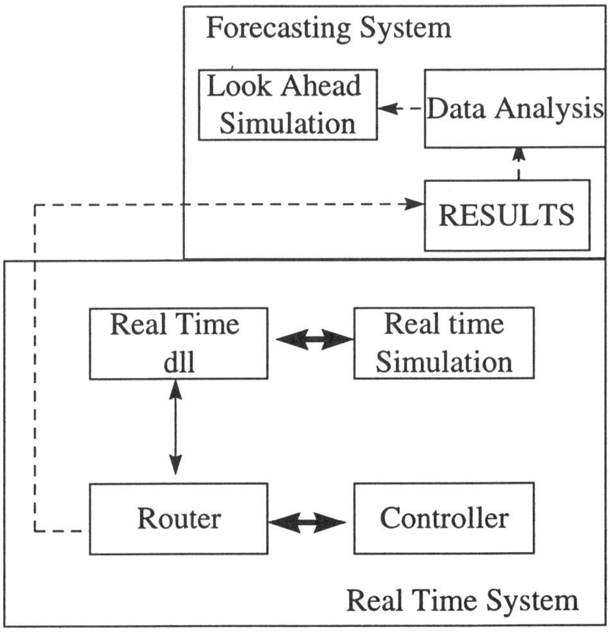

Figure 2: Layout of Software System

The ability to estimate statistical distribution directly from the information system is critical to this model. For implementing real-time simulation in an actual manufacturing environment, the key information challenge is insuring semantic correctness in a distributed database system. For information systems that are loosely federated together, the accessing of data may include changes to the data structure, attribute naming, and data schema. An example of these differences is extracting processing times from a database system that stores the processing time as an attribute in a table and a system that stores processing times as the difference between start time and completion time in a series of records. The major obstacle to this integration is not the ability to communicate between different data sources that is standardized with ODBC, DAO and embedded SQL; the challenge is identifying the ways information is stored and having a clear understanding of what the information implies about the manufacturing process. For data quality in on-line simulation, the information must be accurate and represent what the model thinks it represents.

For the TSCS control system, these information system design problems were solved by making the information system meet the needs of the simulation model. For a commercial implementation of these concepts, substantial commitment for standardization of system states would be required. Based on the amount of resources spent on the simulation of manufacturing processes compared to relational database systems and MRP systems, the majority of the flexibility must come from the simulation package. This flexibility includes the ability to analyze data that arise from a mixture of processes, the ability to access different databases and a clear set of rules for defining the states of the system.

6 CONCLUSIONS

Combining simulation, information systems, and real-time control is a very promising framework for the study and optimization of the dynamic characteristics of a flexible manufacturing system. Look-ahead simulation is also a promising tool for predicting future lead times and calculating due dates. As flexible and agile manufacturing become prevalent, the ability to describe the short term future performance of a manufacturing system becomes critical. This tutorial session will describe the framework developed for simulation-based control. In addition, it will demonstrate the implementation and use of a simulation control system using Systems Modeling's Arena, Microsoft Visual C++, and Microsoft Access as the core demonstration platform.

7 REFERENCES

Balasubramanian, P. and Alexander Tuzhilin. 1996. Using query driven simulation for querying outcomes of business processes, *Decision Support Systems*, Vol. 16, 275-295.

Centeno, M. and C. Standridge. 1991. Modeling manufacturing systems: an information base approach, *Simulation Symposium*.

Drake, Glenn R. and Jeffrey Smith. 1996. Simulation systems for real time planning scheduling and control, *In the Proceedings of the 1996 Winter Simulation Conference.*

Drake, Glenn R. and Jeffrey Smith. 1995. Simulation as a planning and scheduling tool for flexible manufacturing systems, *In the Proceedings of the 1995 Winter Simulation Conference*, December, Washington DC.

Erickson, C., Vandenberge, A. and T. Mills. 1987. Simulation, animation, and shop floor control, *In the Proceedings of the 1987 Winter Simulation Conference*, 649-652.

Grabot, B., Blanc, J., and C. Binda. 1996. A decision support system for production activity control, *Decision Support System*, 87-110.

Godwin, A. N. and D. Sandars. 1995. Forcast standards and existing databases, *Factory 2000*, IEE Conf. Publ. No. 398.

Harmononosky, C. M. and S. F. Robohn. 1991. Real-time scheduling in computer integrated manufacturing: A review of recent literature. *International Journal of Computer Integrated Manufacturing*, 331-340.

Jones, A., Rabelo, L., and Y. Yuehwern. 1995. A hybrid approach for real-time sequencing and scheduling. *International Journal of Computer Integrated Manufacturing*, 145-154.

Moser, M. and S. Engell. 1992. Avoiding scheduling errors by partial simulation of the suture, *In the Proceedings of the 31st Conference on Decision and Control*, Tucson, Az., December.

Prabhu, V. and Neil Duffie. 1995. A distributed simulation approach for enabling cooperation between entities in heterarchical manufacturing systems, *SPIE*, vol. 2596, 234-242.

Rogers, P., and M. Flanagan. 1991. On-line simulation for real-time scheduling of manufacturing systems, *Industrial Engineering*, Vol. 23, 37-40.

Smith, J. S., Wysk, R. A., Sturrock, D., Ramaswamy, S., Smith, G., and S. Joshi. 1994. Discrete event simulation for shop floor control, *In the Proceedings of the 1994 Winter Simulation Conference*, December, Lake Buena Vista, FL, 962-969.

Smith, J. and Joshi, S. 1995. A shop floor controller class for computer-integrated manufacturing, International *Journal of Computer Integrated Manufacturing*, Vol. 8, No. 5, 327-339.

Smith, J. S. 1996. Design and implementation of FMS control software, *In the Proceedings of the 1996 FAIM Conference*, Atlanta, GA, May.

Smith, J. S. and Joshi, S. B. 1993. Message-based part state graphs (MPSG) for shop floor control, *In the Proceedings of the 2nd IIE Research Conference*, Los Angels, CA, May 26-28.

Tautuo, L. and H. Pierreval. 1995. Using evolutionary algorithms and simulation for the optimization of manufacturing systems, *Emerging Technologies and Factory Automation INRIA/ IEEE Symposium*.

Thompson, M. B. 1994. Expanding simulation beyond planning and design, *Industrial Engineering*, Vol. 26, No. 10,65-67.

Tunali, S. 1995. Simulation for evaluating machine and AGV scheduling rules in an FMS environment, *Engineering Management Conference*, 433-438.

Wu, D. S. and R. A. Wysk. 1989. An application of discrete event simulation to on-line control and scheduling in flexible manufacturing, *International Journal of Production Research*, Vol. 27, 1603-1624.

AUTHOR BIOGRAPHIES

JEFFREY S. SMITH is an assistant professor in the Industrial Engineering Department at Texas A&M University. His research interests are shop floor control, manufacturing systems design, analysis, control, and simulation.

BRETT A. PETERS is an assistant professor in the Industrial Engineering Department at Texas A&M University. His research interests are manufacturing systems design, analysis and control, and simulation. He is an active member of IIE, SME, and INFORMS.

JAMES CURRY is a research assistant in the Industrial Engineering Department at Texas A&M University.

GLENN R. DRAKE is an associate applications engineer for Systems Modeling Corporation. He received his B.S.I.E. in 1994 and his M.S.I.E. in 1996 from Texas A&M University.

CYNTHIA LAJIMODIERE is a research assistant in the Industrial Engineering Department at Texas A&M University.

Proceedings of the 1996 Winter Simulation Conference
ed. J. M. Charnes, D. J. Morrice, D. T. Brunner, and J. J. Swain

SELECTION OF INPUT MODELS USING BOOTSTRAP GOODNESS-OF-FIT

Russell C. H. Cheng
Wayne Holland
and
Neil A. Hughes

Institute of Mathematics and Statistics
The University of Kent at Canterbury
Canterbury, Kent CT2 7NF
ENGLAND

ABSTRACT

Bootstrap methods are a natural adjunct of computer simulation experiments; both use resampling techniques to construct the statistical distributions of quantities of interest. In this paper we consider how bootstrap methods can be used in selecting appropriate input models for use in a computer simulation experiment. The proposed method uses a goodness-of-fit statistic to decide on which of several competing input models should be used. We use bootstrapping to find the distribution of the test statistic under different assumptions as to which model is the correct fit. This allows the quality of fit of the different models to be compared.

The bootstrapping process can be extended to the simulation experiment itself, allowing the effect of variability of estimated parameters on the simulation output to be assessed.

The methodology is described and illustrated by application to a queueing example investigating the delays experienced by motorists caused by toll booths at a bridge river crossing.

1 INTRODUCTION

We consider the selection of input models in computer simulation experiments. It is supposed that samples of real data sets exist drawn from input distributions needed in the simulation, but that there is uncertainty concerning the underlying form of these distributions. We suppose however that we are able to draw up a list, for each data set, of candidate distributions that they are likely to have been obtained from. The main task is therefore to fit these possible distributions to the data, and based on the quality of fit, to decide on which distribution is the best fit. In addition to selecting input models, an assessment has to be made concerning the adequacy of the selected models.

Goodness-of-fit procedures are well-known in the statistical literature. A good review of their use in input model selection is discussed by Law and Kelton (1991). See also Banks et al. (1984). A difficulty that arises is that many of the most sensitive tests of fit are based on statistics whose distribution is dependent on both the model being considered and also on whether parameters are estimated or not. These distributions are usually hard to obtain in general. Thus a goodness-of-fit test is limited to only those models for which its own distribution is actually known. In this paper we show that the parametric bootstrap provides a convenient way of applying goodness-of-fit for selecting input models that overcomes this difficulty. We give a method of doing this and illustrate it with an example involving a queueing system.

An important aspect is that the variability of the final simulation output will be influenced by uncertainty in the input model fitting process as well as by uncertainty arising from the (pseudo) random nature of the simulation itself. Both these source of variation need to be taken into account in assessing the overall variability of the simulation output. For a good review of sensitivity analysis see Kleijnen (1995). Cheng and Holland (1995, 1996) have considered this problem using bootstrap techniques to measure the variability of the simulation output arising from estimating unknown parameters when input models are being fitted. The present paper is thus an extension of their technique to include selection as well as fitting of appropriate input models.

For related work see Swain et al. (1988), Barton and Schruben (1993), and Shanker and Kelton (1994).

2 INPUT MODEL SELECTION

We use a framework which highlights the input models used in the simulation. We assume that the simulation uses k univariate input distributions (or models), with distribution functions

$$F_i(x, \theta_i) \quad i = 1, 2, ..., k \tag{1}$$

where

$$\theta_i = (\theta_{1i}, \theta_{2i}, ..., \theta_{p_i i}) \quad i = 1, 2, ..., k$$

is the vector of p_i unknown parameters on which the ith distribution depends. We assume that each model is selected separately from the others, with the same method of selection used for each model. We can therefore, for simplicity, drop the subscript and denote the typical model being considered by $F(x, \theta)$. We suppose that $F(x, \theta)$ is the unknown true distribution, and that we have narrowed down F to one of m possibilities:

$$G_j(x, \theta_j) \quad j = 1, 2, ..., m$$

In addition we assume that there is available a sample of empirical data for each model:

$$\mathbf{x}_i = (x_{i1}, x_{i2}, ..., x_{in_i}) \; i = 1, 2, ..., k.$$

The combined samples will be denoted by

$$\mathbf{X} = (\mathbf{x}_1, \mathbf{x}_2, ..., \mathbf{x}_k).$$

Again for simplicity we suppress the subscript and write

$$\mathbf{x} = (x_1, x_2, ..., x_n).$$

for a sample drawn from a typical $F(x, \theta)$. The initial problem is therefore to use $\mathbf{x}$ to to assess the goodness-of-fit of the $G_j's$ and to select one of the G_j to represent F.

Our proposed method is to fit each G_j to the sample $\mathbf{x}$, then assess the goodness-of-fit by calculating the value, T_j, of a goodness-of-fit test statistic, T, under the assumption that G_j is the correct model. In general the distribution of T will depend on the model being fitted and also on whether the parameters have been estimated or not. There is therefore no guarantee that goodness-of-fit of the models will follow the ranked order of the T_j values. We do not therefore compare the T_j values directly. Instead we calculate the probability value of each T_j; i.e.

$$p_j = \Pr\{T \leq T_j | \, A_j\}$$

where A_j is the assumption that G_j is the correct model, that it has been used to estimate θ_j and that $G_j(., \hat{\theta}_j)$ has been used to compute T_j. The model deemed to be the best fit corresponds to that with the smallest p_j-value.

The fact that the distribution of T is altered if parameters are fitted is a difficulty of some awkwardness, as different significance levels are needed for different distributions (See D'Agostino and Stephens, 1986). However the difficulty is overcome by use of the bootstrap to construct an approximation of the required distribution for each T_j. Basically the method is to draw a number of so-called bootstrap samples from $G_j(., \hat{\theta}_j)$. An estimate of θ_j and a corresponding T_j is calculated for each such sample. Then the empirical distribution function (EDF) of these T_j's estimates the true distribution of T_j. For a discussion of bootstrap sampling methods, see Efron (1979, 1987) and Efron and Tibshirani (1994). We use the following result.

Theorem 1. Suppose that $\mathbf{x}$ is a random sample of size n drawn from the distribution with CDF $G(x, \theta)$ and that

$$T = T(\mathbf{x}, \; G(., \theta))$$

is a goodness-of-fit statistic dependent on $\mathbf{x}$ and on $G(., \theta)$. Suppose that (i) the CDF, $F_T(t, \theta)$, of T is a continuous function of θ for each t, (ii) $\hat{\theta}$ is a consistent estimator of θ, calculated from $\mathbf{x}$. Then the distribution of $T^* = T(\mathbf{x}^*, \; G(., \hat{\theta}))$, where $\mathbf{x}^*$ is a random sample of size n drawn from $G(., \hat{\theta})$, tends to that of T, in probability, as $n \to \infty$.

Proof. The distribution of T^* is simply that of T when calculated at $\hat{\theta}$. By assumption (ii) $\hat{\theta}$ is a consistent estimator, that is $\hat{\theta} \to \theta$ in probability. Hence, by (i), $F_{T^*}(t, \hat{\theta}) \to F_T(t, \theta)$ for each t in probability, i.e. the distribution of T^* tends to that of T in probability. □

Theorem 1 shows that, providing n is sufficiently large, we can approximate the distribution of T by bootstrap sampling of T^* from the fitted distribution $G(., \hat{\theta}_j)$. Our model selection method is thus as follows.

For each of the possible models, $j = 1, 2, ..., m$:

1. Fit the model G_j by estimating θ_j from the sample $\mathbf{x}$. Let the estimator be $\hat{\theta}_j$.
2. Calculate an appropriate goodness-of-fit statistic T_j for the fitted model, $G_j(., \hat{\theta}_j)$.
3. Use bootstrap sampling to estimate the distribution of T_j:
   (a) Generate B bootstrap samples $\mathbf{x}^{(i)}$, $i = 1$, 2, ..., B from the fitted model $G_j(., \hat{\theta}_j)$.
   (b) For each sample, $\mathbf{x}^{(i)}$, fit $G_j(x, \theta_j)$ by estimating θ_j, giving estimates $\hat{\theta}_j^{(i)}$ $i = 1$, 2, ..., B.
   (c) Calculate the goodness-of-fit statistic $T_j^{(i)}$ for the fitted model $G_j(., \hat{\theta}_j^{(i)})$ for $i = 1$, 2, ..., B.

(d) Form the EDF of the $T_j^{(i)}$ and hence find the p-value of T_j :

$$p_j = \frac{\#\ of\ T_j^{(i)} \le T_j}{B}.$$

The model G_j with the smallest p_j value is selected as being the best fit.

3 GOODNESS-OF-FIT STATISTICS

We consider examples of the method of model selection proposed in the previous section. Table 1 gives data of the times in seconds to serve three classes of vehicle: Private Cars (PC), Light Vans (LV) and Heavy Goods Vehicles (HG) at the toll booths of the Severn Bridge River crossing in Britain. This example will be used later to illustrate the complete procedure suggested for carrying out a simulation experiment. Here we consider simply the model selection procedure. There is some evidence that an appropriate distribution should allow for a minimum processing time to handle a vehicle. We thus consider a three parameter versions of the Weibull, gamma and lognormal distributions as possible candidates. An interesting aspect of fitting these models is that they all possess embedded two-parameter special cases, and this possibility must be allowed for; see Cheng and Iles (1990) for a discussion of how this should be done. In the HG data set there is evidence that the service times are a mixture of distributions. We therefore consider, in addition, a five parameter mixture model of two normal models with different means and variances, with an unknown mixing proportion. We use maximum likelihood estimation for the parameters. There are many possible goodness-of-fit test statistics available. For illustration we use the Anderson-Darling test statistic which is known to be particularly sensitive and powerful for detecting differences in tail behaviour. If the ordered sample is $x_1 < x_2 < ... < x_n$, and the model under consideration is $G(., \theta)$ then, writing $Z_i = G(x_i, \hat{\theta})$, the test statistic is

$$A^2 = -n - n^{-1} \sum_{i=1}^{n} (2i-1)\{\ln(Z_i) + \ln(1 - Z_{n-i+1})\} \tag{2}$$

Table 2 gives the values of A^2 and their associated $p - values$ for each of the data sets and each model.

It will be seen that the best fit is the mixture model for the PC and HG data sets, and the lognormal model for the LV data set.

It is of interest to see how sensitive the test is in distinguishing between models. To illustrate this we consider the lognormal model and mixture models

Table 1: Service Times (secs) for PCs, LVs and HGs

PC			LV		HG	
5.7	10.9	4.9	4.3	10.9	8.6	7.2
5.4	5.7	4.3	4.7	3.1	32.0	7.7
5.7	6.3	6.9	6.7	4.5	8.7	6.9
3.8	4.9	8.5	7.2	6.6	12.1	8.7
5.3	7.8	5.2	6.3	4.7	25.5	5.1
5.1	7.1	4.4	6.2	4.2	16.7	7.1
7.0	4.9	7.6	4.8	3.3	7.1	9.3
10.5	5.5	7.6	6.3	4.0	37.1	6.4
7.6	5.1	3.5	3.5	7.8	4.6	8.1
4.3	10.1	12.0	5.7	5.8	8.8	11.8
3.2	6.5	3.0	5.2	8.0	22.9	30.4
15.1	3.2	5.9	4.9	6.1	11.0	8.3
7.4	4.4	3.8	7.7	4.3	8.9	6.9
5.6	6.2	5.7	7.9	3.9	8.1	8.3
4.1	6.9	7.2	4.4	6.7	8.7	4.1
5.1	5.1	5.9	6.4	7.2	7.7	12.5
2.7	4.9	6.1	4.7	5.0	8.2	10.6
4.2	4.4	5.6	5.2	6.4	8.8	5.9
2.9	13.1	4.9	3.6	10.5	11.0	6.1
5.7	2.7	6.8	5.8	8.0	11.3	8.1
6.1	3.0	3.2	8.2	12.5	7.4	10.5
5.0	3.8	4.8	4.1	4.0	8.1	14.3
4.2	5.9	7.1	4.6	3.8	7.2	9.7
3.2	5.9	4.5	3.1		7.1	5.0
5.2	3.4	5.9			6.8	12.5
9.2	3.7	3.4			7.7	11.5
4.9	4.6	3.6			7.4	13.4
5.1	3.3	4.9			12.9	9.7
3.9	5.2	5.9			13.4	10.1
4.1	4.6	6.7			9.4	10.3
5.2	6.5	6.1			7.6	18.0
15.6	5.7	5.7			11.8	21.6
5.7	4.2	3.7			9.4	4.5
8.2	9.0	4.7			12.5	6.8
5.3	8.2	6.9			12.2	11.2
6.6	4.6	5.8			8.2	13.0
3.9	3.6	6.4			6.7	25.8
4.7	4.4	7.4			9.4	5.2
					12.6	

Table 2: Anderson-Darling Statistic and p-Values

Model		PC Data	LV Data	HG Data
Weibull	A^2	1.269	0.277	2.282
	$P(A^2)$	0.999	0.385	> 0.999
Lognormal	A^2	0.544	0.227	1.041
	$P(A^2)$	0.902	0.229	0.998
Gamma	A^2	0.798	0.238	1.833
	$P(A^2)$	0.968	0.237	> 0.999
Mixture	A^2	0.256	0.541	0.423
	$P(A^2)$	0.544	0.897	0.870

fitted to the HG data. Figure 1 shows ten separate EDF's of A^2, where each EDF is constructed as follows. The fitted lognormal model is treated as the stage-1 model and a bootstrap sample of the same size as the original sample, is obtained from it. The lognormal model (stage-2) is fitted to this bootstrap sample. 1000 secondary bootstrap samples, each of size 20, are then drawn from this stage-2 model, and the lognormal model (stage-3) is then fitted to each sample. The corresponding A^2 is calculated as in (2) for this stage-3 fit. The variability between EDF's is due to the bootstrap sampling of the stage-1 model, and gives an indication of the variabiity due to using fitted parameters instead of the unknown "true" parameter values. For comparison Figure 1 also shows ten EDF's each formed from 1000 A^2 values calculated in exactly the same way except that the lognormal model of stages 1 and 2 is replaced by the normal mixture model; however the lognormal is still the fitted model at stage 3, and each A^2 is calculated for this fitted lognormal model. These latter EDF's therefore indicate how the distribution of A^2 when the incorrect (lognormal) model is fitted to data drawn from the mixture model. There is significantly more variability in the EDF when the incorrect model is fitted. This is due to the nature of the mixture distribution with its far greater inherent variability. With data generated from simpler alternative distributions like the gamma, there should be considerably less variation, though this has yet to be investigated. Figure 1 does however give an indication of the reasonable power of the test in rejecting the lognormal model when the alternative mixture model is actually the correct one.

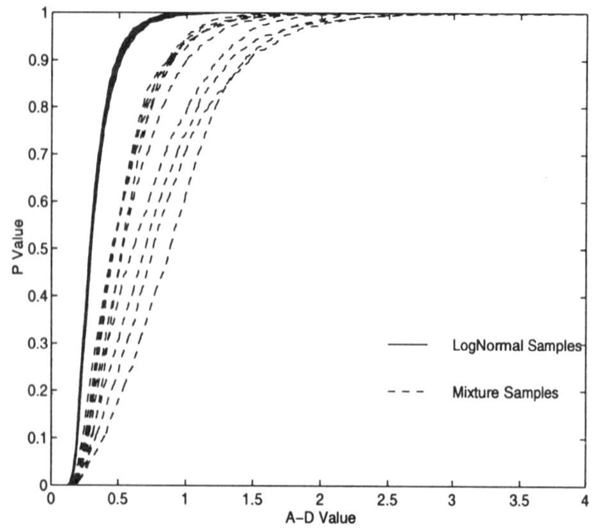

Figure 1: Comparison of A^2 EDFs for Lognormal Model Fitted to Lognormal and Mixture Samples

4 SENSITIVITY ANALYSIS

Cheng and Holland (1995, 1996) show how bootstrap sampling can be applied to the simulation experiment itself. We follow their terminology. The simulation study is assumed to consist of making a number of runs of a computer simulation model and observing the output of interest, y, from each run. Let the length of each run be l (measured in simulation time, say). In each run, the input models (1) are used to generate k streams of random variate. The variates used in one simulation run will be denoted by

$$\xi_i = (\xi_{i1}, \xi_{i2}, ..., \xi_{im_i}) \; i = 1, 2, ..., k.$$

Each ξ_i is assumed to be a random sample with the individual observations, ξ_{ij}. The number of variates, m_i, used is different in each stream. In what follows the m_i can be variable, but to simplify the discussion the run length, l, can be regarded as fixed, with the m_i also fixed. The F_i are assumed to have been selected using the method of the previous section.

Though they may be generated in various ways, it will be convenient to regard the input variates as generated by the inverse transform method (see Law and Kelton, 1991, for example):

$$\xi_{ij} = F_i^{-1}(u_{ij}, \theta), \quad j = 1, 2, ..., m_i$$

where F^{-1} is the inverse of F, and the u_{ij} are independent uniform $U(0, 1)$ variates. We write

$$\mathbf{u}_i = (u_{i1}, u_{i2}, ..., u_{im_i}) \; i = 1, 2, ..., k$$

and

$$\mathbf{U} = (\mathbf{u}_1, \mathbf{u}_2, ..., \mathbf{u}_k).$$

Helton (1993, 1994) distinguishes two types of uncertainty: *parameter uncertainty* and *simulation uncertainty.* The parameter uncertainty is due to the θ_j

being unknown and having to be estimated, whilst the simulation uncertainty occurs because of the randomness of **U**. Thus the output of interest from the simulation run, y, has random variability because it depends on **U**, and θ:

$$y = y(\mathbf{U}, \theta).$$

Let

$$\eta(\theta) = E(y, \theta) = \int y(\mathbf{U}, \theta)\mathbf{dU}$$

be the expected value of y, and let θ^0 denote the true parameter value. We assume that the objective is to estimate $\eta(\theta^0)$ in the case where θ^0 is not known.

We consider the overall simulation experiment as being made up of r runs. Assuming, for the moment, that the runs are made at some given θ, the responses or outputs from these runs can be written as:

$$y_j(\mathbf{U}_j, \theta) = \eta(\theta) + e_j(\mathbf{U}_j, \theta), \ j = 1, 2, ..., r. \quad (3)$$

The 'error' variable e_j is the random difference between the jth simulation run output and $\eta(\theta)$. We shall assume $E(e_j|\theta) = 0$ and $Var(e_j|\theta) = \tau^2(\theta)/l$ for $j = 1, 2, ..., r$. Thus

$$E[y_j(\mathbf{U}_j, \theta)|\theta] = \eta(\theta),$$

and the mean of the outputs

$$\bar{y} = \sum_{j=1}^{r} y_j(\mathbf{U}_j, \theta)/r,$$

is an unbiased estimator of $\eta(\theta)$ with

$$Var(\bar{y}|\theta) = \tau^2(\theta)/rl. \quad (4)$$

This variance measures the simulation uncertainty within the model. It is the total variance of the response only when θ is fixed and known.

When θ is unknown, it has to be estimated. If we have estimates $\hat{\theta}_j$ that are $\sqrt{n}$-consistent (where n is the typical size of real data samples), as would be the case if we use maximum likelihood estimation, then the total variation of $\bar{y}$ is given by

$$\begin{aligned} Var(\bar{y}) &= \underset{\hat{\theta}}{Var}\,(\underset{y}{E}\,(\bar{y}|\hat{\theta})) + \underset{\hat{\theta}}{E}\,(\underset{y}{Var}\,(\bar{y}|\hat{\theta})) \\ &= \sigma^2(\theta^0)/n + \tau^2(\theta^0)/rl \ + \ R \end{aligned} \quad (5)$$

where θ^0 is the unknown true parameter value, $\sigma^2(\theta^0)/n$ is the parameter uncertainty, $\tau^2(\theta^0)/rl$ is the simulation certainty, and the remainder term, R, involves only terms of order $O(n^{-3/2})$ and $O[(nrl)^{-1}]$.

Cheng and Holland (1996) show that the variance for $\bar{y}$ given $\hat{\theta}$ is:

$$Var(\bar{y}|\hat{\theta}) = \tau^2(\hat{\theta})/(rl) = [\tau^2(\theta^0) + O_p(n^{-1/2})]/(rl),$$

This means that we can use

$$t_0^2 = \sum_{j=1}^{r}(y_j - \bar{y})^2/(r-1), \quad (6)$$

to estimate the stochastic uncertainty, $\tau^2(\theta^0)/l$.

Assessing the parameter uncertainty is more difficult. Cheng and Holland give two methods for doing this. We consider here the bootstrap method that they propose.

A sample $\mathbf{X}^1 = (\mathbf{x}_1^1, \mathbf{x}_2^1, ..., \mathbf{x}_k^1)$ is obtained with the $\mathbf{x}_i^1$ drawn from the selected distributions $G_j(x, \hat{\theta})$ $(j = 1, ..., k)$. This sample, $\mathbf{X}^1$, can be used to produce an estimate $\hat{\theta}^1$,in exactly the same way that $\hat{\theta}$ was obtained from **X**. Repeating this B times yields B such estimates: $\hat{\theta}^1$, $\hat{\theta}^2$, ..., $\hat{\theta}^B$. We can then carry out B bootstrap simulation experiments, one for each $\hat{\theta}^i$, with all runs of length l as in the original experiment, but with the bootstrap experiment containing r' runs, where r' may be different from r. (The run lengths can be different from l but there is actually no loss of generality in assuming them to be the same.) This yields B sets of responses, with r' responses in each set:

$$y_1^i, y_2^i, ..., y_{r'}^i, \quad i = 1, 2, ..., B. \quad (7)$$

Let the means of each set be

$$\bar{y}^i, \quad i = 1, 2, ..., B. \quad (8)$$

Cheng and Holland (1996) show how the variability of these $y's$ depends on both the simulation variance τ^2 and on the parameter variance σ^2.

Theorem 2.

$$Var(\bar{y}^i|\hat{\theta}) = \sigma^2(\theta^0)/n + \tau^2(\theta^0)/r'l + R. \quad (9)$$

where the remainder, R, is of order

$$R = O_p(n^{-3/2}) + O_p[(n^{1/2}r'l)^{-1}].$$

Proof: See Cheng and Holland (1996). □

An estimate of this variance, (9), is the sample variance of the $\{\bar{y}^i\}$ (8):

$$s_B^2 = \sum_{i=1}^{B}(\bar{y}^i - \bar{\bar{y}})^2/(B-1). \quad (10)$$

As t_0^2 (6) estimates $\tau^2(\theta^0)/l$, this can then used to adjust s_B^2 to give the following estimate of the total variance of $\bar{y}$ (5):

$$\widehat{Var}(\bar{y}) = s_B^2 + (\frac{1}{r} - \frac{1}{r'})t_0^2. \qquad (11)$$

When $r' = r$, i.e. when each bootstrap is an exact replica of the original experiment, then $r = r'$, and the total variance is estimated by s_B^2 on its own.

5 APPLICATION TO A QUEUEING SYSTEM

To illustrate the applicability of the above, we consider the simulation of a queueing situation, investigating the delays experienced by motorists caused by toll booths at a bridge river crossing. The bridge considered crosses the River Severn in the United Kingdom. The bridge is approached by the M4 motorway, consisting of three carriageways. The approach widens to accommodate eight toll booths.

Data for service times at toll booths for three types of vehicles - private cars (PC), light vans (LV) and heavy goods vehicles (HG) have already been given in Table 1. Preliminary investigation of the three data sets indicated that the service time distribution was different for each type of vehicle. The mean and standard deviation of service time (in seconds) was 5.73 and 2.2 for PCs, 5.80 and 2.03 for LVs and 10.88 and 6.20 for HGs. These were based on 114, 47 and 77 observations respectively. No data was available for inter-arrival times. However, Griffiths and Williams (1984), in their study of the Severn Bridge, conclude that it may be reasonably assumed that inter-arrival times follow a negative exponential distribution for each type of vehicle. Further, they state that past records indicate that the composition of traffic types is 75.4% PCs, 19.4% HGs and 5.4% LVs. It was further predicted that traffic flow by 1995 would reach 51,051 vehicles per day. This leads to estimates for mean arrival rates per hour of 1604 for PCs, 413 for HGs and 111 for LVs.

Within the simulation, vehicles are assumed to choose an approach lane at random, unless there are queues in which case the carriageway with the shortest queue is selected. Driver choice of toll booth is selected by conditional probability distribution (conditional upon the lane of approach). The actual probabilities used are given in Table 3. Thus three probability distributions are constructed for choice of toll booth (out of eight) dependent upon lane of approach (choice of three). These distributions are then modified by dividing by the queue length at each booth, and then normalised so that the probabilities summed

Table 3: Distribution of Driver Choice of Toll Booth

Lane	1	2	3
Booth			
1	0.20	0.05	0.01
2	0.25	0.25	0.09
3	0.25	0.25	0.10
4	0.10	0.15	0.15
5	0.09	0.12	0.15
6	0.07	0.08	0.15
7	0.03	0.06	0.15
8	0.01	0.04	0.20

to unity. This means that each driver has an individual probability distribution for choice of booth dependent on lane of approach and queue length at each toll booth at the moment of selection. The maximum queue that can develop at a toll booth is ten, at which point the traffic commences queueing on the approach road. Once vehicles are queueing it is assumed that PCs and LVs will take two seconds to move up one place in the queue; HGs are assumed to take four seconds.

The simulation procedure may be outlined in the following manner. The best distributions are fitted to the service time data sets and a simulation run of length $l = 36,000$ simulation seconds is repeated $r = 250$ times. The output from the simulation experiment consists of the mean delay for PCs, LVs, HGs and a combined overall mean for each simulation run. Considering just the overall mean for a moment, these may be thought of as $(y_1, y_2, ..., y_{250})$ from (3). These may be used to obtain an estimate of the simulation variance (4). Following this, bootstrap samples are generated from each of the three specified service time distributions. Bootstrapping is not performed on the inter-arrival time distributions because we are assuming that they are known to be negative exponential. As discussed in the previous section, the purpose of bootstrapping here is to obtain estimates of variability caused by fitting distributions where there is uncertainty. Each of the bootstrap samples is used to fit parameter estimates for the appropriate distribution and the simulation then proceeds with these estimates. This time, a single simulation run is performed ($r' = 1$). This is of length 36,000 simulation seconds, as before. The reason for this is that we have our estimate of simulation variability from the first experiment. We are now interested in variability between bootstrap experiments to give us the variability caused by unknown input parameters. Thus this bootstrapping procedure is repeated $B = 250$

Table 4: Results for Toll Booth Experiment

	Mean Delay	**Para Var**	**Sim Var**	**Total Var**
Best Fit	15.66	1.0400	2.5×10^{-4}	1.0402
Worst Fit	15.70	1.0532	2.1×10^{-4}	1.0532
Results for PCs				
	Mean Delay	**Para Var**	**Sim Var**	**Total Var**
Best Fit	15.78	0.9428	3.2×10^{-4}	0.9460
Worst Fit	15.78	0.9581	3.2×10^{-4}	0.9584
Results for LVs				
	Mean Delay	**Para Var**	**Sim Var**	**Total Var**
Best Fit	24.56	2.2739	1.1×10^{-3}	2.2751
Worst Fit	24.74	1.8377	9.2×10^{-4}	1.8386
Results for HGs				
	Mean Delay	**Para Var**	**Sim Var**	**Total Var**
Best Fit	16.14	1.0284	2.6×10^{-4}	1.0287
Worst Fit	16.18	1.0258	2.3×10^{-4}	1.0260
Results for All Vehicles Combined				

times. The output provided by each bootstrap simulation is the mean delay for PCs, LVs, HGs and the overall mean delay. For mean overall delay, say, we may regard the output as that described in (7). The overall variance (5) may be calculated using (10) and (11). Table 4 summarises the results for delays to PCs, LVs, HGs and overall delay respectively, when the best distributions have been used, and for comparison when the worst distributions have been used. It can be seen that the choice of distribution has little effect on the results, except for the the parameter variance for the HGs. This data was unusual and was not particularly well fitted by any of the distributions considered. However, there is clearly some sensitivity to the choice of input distribution. In all other cases, the observed data could be reasonably fitted by a number of the distributions. Therefore, less sensitivity would be expected in such cases. The lack of sensitivity to choice of input model is heightened by two further points. Firstly, we were only able to conduct the exercise for service time distributions. The inter-arrival time distributions were taken as fixed, yet optimal selection of these distributions could contribute considerably to the variability in the context of a queueing analysis. Secondly, we have chosen here to calculate only mean measures. It would be sensible to consider the calculation of some probability measures, for instance the probability of queueing occurring on the approach road to the toll booths or the probability of a toll booth being idle. Once would expect such measures to be much more sensitive to input distribution and this will be the subject of further investigation.

The mean delay, $\bar{y}$, has been calculated here from the simulation runs at the fitted parameters only. The bootstrap results have been used only in the variance estimation. It is possible to incorporate the results for the mean from the bootstrap simulation runs with the main simulation result to produce an estimator with lowest overall variance. Details of the procedure may be found in Cheng and Holland (1995).

REFERENCES

Banks, J., Carson II, J.S. and Nelson, B.L. (1984). *Discrete-Event Simulation, (2nd Edn).* Upper Saddle River, NJ: Prentice Hall.

Barton, R.R. and Schruben, L.W. (1993). Uniform and Bootstrap Resampling of Empirical Distributions. In *Proceedings of the 1993 Winter Simulation Conference* (ed. G.W. Evans, M. Mollaghasemi, E.C. Russell and W.E. Biles), IEEE Piscataway, New Jersey, 503-508.

Cheng, R.C.H. and Holland, W. (1995). The Effect of Input Parameters on the Variability of Simulation Output. *Proceedings of the Second United Kingdom Simulation Society Conference (*Eds R.C.H. Cheng and R.J. Pooley), North Berwick, April 1995, Edinburgh University, pp 29-36.

Cheng, R.C.H. and Holland, W. (1996). Sensitivity of Computer Simulation Experiments to Errors in Input Data. To appear in *J. of Statistical Computation and Simulation.*

Cheng, R.C.H. and Iles, T.C. (1990). Embedded Models in Three-Parameter Distributions and their Estimation. *J. R. Statist. Soc.* B, **52**, 135-149.

D'Agostino, R.B. and Stephens, M.A. (1986). *Goodness of Fit Techniques.* New York: Marcel Dekker.

Efron B. (1979). Bootstrap Methods : Another Look at the Jackknife. *The Annals of Statistics*, **7**, pp 1-26.

Efron, B. (1982). *The Jackknife, the Bootstrap and Other Resampling Plans.* Vol. **38** of CBMS-NSF Regional Conference Series in Applied Mathematics. SIAM.

Efron, B. and Tibshirani, R.J. (1993). *An Introduction to the Bootstrap.* New York and London: Chapman and Hall.

Griffiths, J.D. and Williams J.E. (1984). Traffic Studies on the Severn Bridge. *Traffic Engineering and Control*, **25**, pp 268-71, 274

Helton, J.C. (1993). Uncertainty and Sensitivity Analysis Techniques for Use in Performance Assessment for Radioactive Waste Disposal. *Reliability Engineering and System Safety,* **42**, 327-367.

Helton, J.C. (1994). Treatment of Uncertainty in Performance Assessments for Complex Systems. *Risk Analysis,* **14**, 483-511.

Kendall, M.G. and Stuart, A. (1979). *The Advanced Theory of Statistics. Vol. 2: 4th Edn.* London: Griffin.

Kleijnen J.P.C. (1995). Sensitivity Analysis and Related Analyses : a Survey of Statistical Techniques (submitted for publication).

Law, A.M. and Kelton, W.D. (1991). *Simulation Modeling and Analysis 2nd Edition.* New York: McGraw-Hill.

Shanker, A. and Kelton, W. D. (1994). Measuring Output Error due to Input Error in Simulation: Analysis of Fitted vs. Mixed Empirical Distributions for Queues. To appear.

Swain, J.J., Venkatraman, S. and Wilson, J.R. (1988). Distribution Selection and Validation. *J. of Statist. Comput. and Simul.*, **29**, 271-297.

AUTHOR BIOGRAPHIES

RUSSELL C. H. CHENG is Professor of Operational Research in the Institute of Mathematics and Statistics at the University of Kent at Canterbury. He has an M.A. and the Diploma in Mathematical Statistics from Cambridge University, England. He obtained his Ph.D. from Bath University. He is General Secretary of the U.K. Simulation Society, a Fellow of the Royal Statistical Society, Member of the Operational Research Society. His research interests include: variance reduction methods and parametric estimation methods. He is Joint Editor of the IMA Journal on Mathematics Applied to Business and Industry, and an Associate Editor for *Management Science.*

WAYNE HOLLAND is a lecturer in Operational Research in the Institute of Mathematics and Statistics at the University of Kent at Canterbury. He has a B.Sc. in Mathematics and its Applications from the University of Wales. He also obtained his Ph.D. from the University of Wales. He is a member of the Operational Research Society. His research interests include numerical approximation of transient queueing measures, analysis of queueing networks with particular application to transportation and computer communication systems.

NEIL HUGHES is a Research Assistant in the Institute of Mathematics and Statistics at the University of Kent at Canterbury. He has a B.Sc. in Mathematics and its Applications from the University of Wales, and is preparing his doctoral dissertation on the automatic generation of terrain databases for use in computer generated imagery.

Proceedings of the 1996 Winter Simulation Conference
ed. J. M. Charnes, D. J. Morrice, D. T. Brunner, and J. J. Swain

MODELING AND SIMULATING TIME SERIES INPUT PROCESSES WITH ARTAFACTS AND ARTAGEN

Marne C. Cario

Delphi Packard Electric Systems
P.O. Box 431
Warren, Ohio 44486, U.S.A.

ABSTRACT

We develop an efficient numerical method for fitting ARTA processes for use as simulation input. ARTA processes are stationary time series with arbitrary marginal distributions and autocorrelations specified through finite lag p. We discuss the software package **ARTAFACTS**, which implements the numerical method, and the package **ARTAGEN**, which generates observations from ARTA processes. To demonstrate the use of the software, we present a real-world example.

1 INTRODUCTION

Many systems contain processes that are inherently dependent. For instance, the temperatures in an oven of a steel mill, measured over time, are likely to be autocorrelated. The sizes of demands on an inventory system are often dependent; for instance, a small demand may follow a very large demand. Daily reserve levels in a blood bank are autocorrelated. The prevalence of dependent input processes, as well as the potential error in the simulation output if dependencies are ignored (see, for example, Livny, Melamed, and Tsiolis 1993), motivates the need for simulation input models that represent autocorrelated time series.

ARTA (AutoRegressive To Anything), introduced in Cario and Nelson (1996), is a transformation-oriented approach for modeling and generating a time-series process $\{Y_t; t = 1, 2, \ldots\}$ with an arbitrary marginal distribution and autocorrelation structure specified through lag p. This approach takes a process with a known autocorrelation structure, the base process $\{Z_t\}$, and transforms it to achieve the desired marginal distribution for the input process, $\{Y_t\}$. The target autocorrelation structure of $\{Y_t\}$ is obtained by adjusting the autocorrelation structure of the base process. In our model, the base process is a standardized Gaussian autoregressive process of order p (AR(p)). Thus, the major step in constructing an ARTA process is finding the autocorrelation structure of the AR(p) process that results in the desired autocorrelation structure for the ARTA process. We present an efficient numerical procedure for performing this step.

The paper is organized as follows: Section 2 discusses other methods for generating time-series inputs for simulation. Section 3 gives a brief review of ARTA processes. In Section 4 we present a specialized numerical method for finding the required autocorrelation structure of the AR(p) process. We describe **ARTAFACTS** (ARTA Fitting Algorithm for Constructing Time Series), a Fortran implementation of the numerical procedure, in Section 5. In Section 6 we discuss the software package **ARTAGEN** (AutoRegressive To Anything GENeration), which generates observations from an ARTA process for use as simulation input. Section 7 provides a real-world application of an ARTA process. Conclusions appear in Section 8.

2 BACKGROUND

A variety of methods exist for modeling time-series input processes with specified marginal distributions and autocorrelation structures. These methods differ in their degree of generality and their ease of use.

In Song, Hsiao, and Chen (1996), the authors use a transformation-oriented approach with an AR(1) base process. However, they attempt to match only the lag-1 autocorrelation, and they use simulation to find the lag-1 autocorrelation of the base process that gives the desired lag-1 autocorrelation of the input process. This approach becomes computationally prohibitive if extended to more than two or three autocorrelations. The ARTA approach matches $p \geq 1$ autocorrelations, and is computationally more efficient.

TES (Melamed 1991) also uses a transformation-oriented approach to modeling time-series input processes. TES, which is implemented in the software package **TEStool** (Melamed, Hill, and Goldsman 1992), uses a series of autocorrelated uniform random variates as the base process. To obtain an input process with the desired characteristics using **TEStool**, the user must adjust the distribution of the base-process noise term. Unfortunately, there is no standard procedure for adjusting the noise term, so constructing a TES process is an art as well as a science. Furthermore, TES processes often exhibit unusual sample paths. The sample paths can be improved at the expense of further complicating the modeling process.

A completely different approach, for which there is a large literature, is to construct a time-series process that exploits properties that are specific to the particular marginal distribution of interest for $\{Y_t\}$. An example is Lewis, McKenzie and Hugus (1989) who construct time series with gamma marginals. The primary shortcoming of this type of approach is that it is not general: a different model is required for each marginal distribution of interest. In addition, the sample paths of these processes, while adhering to the desired marginal distribution and autocorrelation structure, sometimes have unexpected features.

3 ARTA PROCESSES

This section reviews the definition of an ARTA process and summarizes the results of Cario and Nelson (1996). Formally, an ARTA process is a model of a stationary time series $\{Y_t\}$ with the following properties:

1. $Y_t \sim F_Y, t = 1, 2, \ldots$, where F_Y is an arbitrary cumulative distribution function (cdf); and
2. $(\rho_1, \rho_2, \ldots, \rho_p) = \boldsymbol{\rho} = (\mathrm{Corr}[Y_t, Y_{t+1}], \mathrm{Corr}[Y_t, Y_{t+2}], \ldots, \mathrm{Corr}[Y_t, Y_{t+p}])$

where F_Y and $\boldsymbol{\rho}$ are given. In an ARTA process, $\{Y_t\}$ is represented as a transformation of a standardized Gaussian AR(p) process, as follows:

ARTA Process

1. Let $\{Z_t; t = 1, 2, \ldots\}$ be a stationary Gaussian AR(p) process

$$Z_t = \alpha_1 Z_{t-1} + \alpha_2 Z_{t-2} + \cdots + \alpha_p Z_{t-p} + \varepsilon_t,$$

where $\{\varepsilon_t\}$ is a series of independent $\mathrm{N}(0, \sigma^2)$ random variables and σ^2 is selected so that the marginal distribution of the $\{Z_t\}$ process is $\mathrm{N}(0, 1)$, where N denotes the normal distribution.

2. Define the ARTA process $Y_t = F_Y^{-1}[\Phi(Z_t)]$, $t = 1, 2, \ldots$, where Φ is the standard normal cdf.

The transformation $F_Y^{-1}[\Phi(\cdot)]$ ensures that $\{Y_t\}$ has the desired marginal distribution F_Y. Therefore, the central problem is to select the autocorrelation structure $\mathbf{r} = (r_1, r_2, \ldots, r_p)$ for the AR(p) process, $\{Z_t\}$, that gives the desired autocorrelation structure $\boldsymbol{\rho}$ for the input process, $\{Y_t\}$.

The autocorrelation structure of the AR(p) base process directly determines the autocorrelation structure of the input process, since $\mathrm{Corr}[Y_t, Y_{t+h}] = \mathrm{Corr}\left\{F_Y^{-1}[\Phi(Z_t)], F_Y^{-1}[\Phi(Z_{t+h})]\right\}$. To adjust this correlation, we can restrict attention to adjusting $\mathrm{E}[Y_t Y_{t+h}]$, since

$$\mathrm{Corr}[Y_t, Y_{t+h}] = \frac{\mathrm{E}[Y_t Y_{t+h}] - (\mathrm{E}[Y_t])^2}{\mathrm{Var}[Y_t]},$$

and $\mathrm{E}[Y_t]$ and $\mathrm{Var}[Y_t]$ are fixed by F_Y. Then, since (Z_t, Z_{t+h}) has a standard bivariate normal distribution with $\mathrm{Corr}[Z_t, Z_{t+h}] = r_h$, we have

$$\begin{aligned} \mathrm{E}[Y_t Y_{t+h}] &= \mathrm{E}\left\{F_Y^{-1}[\Phi(Z_t)] F_Y^{-1}[\Phi(Z_{t+h})]\right\} \\ &= \int_{-\infty}^{\infty}\int_{-\infty}^{\infty} F_Y^{-1}[\Phi(z_t)] F_Y^{-1}[\Phi(z_{t+h})] \\ &\quad \times \varphi_{r_h}(z_t, z_{t+h})\, dz_t\, dz_{t+h}, \qquad (1) \end{aligned}$$

where φ_{r_h} is the standard bivariate normal probability density function (pdf) with correlation r_h. We are only interested in processes for which this expectation exists.

Observe from Equation (1) that the lag-h autocorrelation of $\{Y_t\}$ is a function only of the lag-h autocorrelation of $\{Z_t\}$, which appears in the expression for φ_{r_h}. We denote this function by $\rho(r_h)$. Thus, the problem of determining the autocorrelations for $\{Z_t\}$ that give the desired autocorrelations for $\{Y_t\}$ reduces to p independent problems: For each lag $h = 1, 2, \ldots, p$, find the value r_h for which $\rho(r_h) = \rho_h$.

Cario and Nelson (1996) prove that for each lag h all feasible values of ρ_h are attainable under ARTA processes. By feasible, we mean that $\underline{\rho} \le \rho_h \le 1$, where

$$\underline{\rho} = \frac{\int_0^1 F_Y^{-1}(u) F_Y^{-1}(1-u)\, du - (\mathrm{E}[Y_t])^2}{\mathrm{Var}[Y_t]}$$

is the minimum feasible bivariate correlation for F_Y (see Whitt 1976).

In general, it is not possible to find the r_h-values analytically; therefore, we develop an efficient numerical procedure to find the r_h-values to within any precision. Cario and Nelson (1996) prove that $\rho(r)$ is nondecreasing and, under very mild conditions on F_Y, it is also continuous. These results guarantee that our numerical procedure will converge.

4 NUMERICAL PROCEDURE

There are two main considerations in constructing an efficient numerical procedure for finding the autocorrelation structure of $\{Z_t\}$ which gives the target autocorrelation structure for $\{Y_t\}$. The first consideration is the choice of a method to integrate Equation (1) numerically. The second consideration is the development of a numerical search procedure that selects and updates the values of r_h in an efficient manner.

Several properties of the integrand make it nontrivial to integrate Equation (1) numerically. Foremost is the fact that the integrand is bivariate, and there are few general-purpose numerical-integration methods for multivariate integrals. Our application requires a general-purpose method since we do not make any restrictions on the functional form of the marginal distribution F_Y other than those required for it to be a cdf and for the integral (1) to exist. For instance, we do not require F_Y to be continuous. The doubly infinite ranges of the inner and outer integrals further complicate the integration since many numerical-integration methods for multivariate integrals apply to subspaces with special shapes, such as spheres and cubes. Finally, since Equation (1) must be calculated for many values of r_h, and it is often computationally expensive to calculate F_Y^{-1}, we need a method that efficiently manages the number of function evaluations.

Taking the previously mentioned factors into consideration, our solution is to iteratively apply a univariate numerical-integration method to compute Equation (1). Specifically, we use a bivariate implementation of the Gauss-Kronrod quadrature rule (Piessens et al. 1983, pp. 16–17). Dropping the time subscripts t and $t+h$ from z, the Gauss-Kronrod quadrature rule approximates Equation (1) as:

$$\widehat{\mathrm{E}}[Y_t Y_{t+h}] = \sum_{i=1}^{15}\sum_{j=1}^{15} w(z_i)w(z_j)F_Y^{-1}[\Phi(z_i)] \times F_Y^{-1}[\Phi(z_j)]\varphi_{r_h}(z_i, z_j), \qquad (2)$$

where $w(z)$ is a weight function and the z_is and z_js are abscissae. Piessens et al. (1983) give the abscissae for the interval $(-1,1)$, but they may easily be mapped onto the range of interest. In our implementation, we truncate each doubly infinite range to the finite range $(-10,10)$, and use a linear mapping of the abscissae from $(-1,1)$ to $(-10,10)$. Using the finite range does not decrease the accuracy of the approximation significantly since the value of the bivariate normal pdf is effectively zero outside of this range.

We set absolute-error tolerances for the inner and outer integral using the method given on pp. 112–113 of Piessens et al. (1983). The absolute error is taken to be the absolute error of the difference between two sums: the Kronrod sum, for which all of the weights are non-zero, and the Gauss sum, for which some of the weights are zero. If either the inner or the outer integral exceeds its absolute-error tolerance on the initial integration range $(-10,10)$, then the corresponding integration range is bisected and the integral is computed on each half-interval. This procedure is repeated until all of the inner and outer integrals meet the absolute-error tolerances, giving us a bound on the overall integration error as well.

Given a value of r, we may compute the value of $\rho(r)$ using the Gauss-Kronrod quadrature rule. The remaining step is to construct an efficient numerical search procedure to find the autocorrelation structure $\mathbf{r}^* = (r_1^*, r_2^*, \ldots, r_p^*)$ of $\{Z_t\}$, such that $\boldsymbol{\rho}^* = (\rho(r_1^*), \rho(r_2^*), \ldots, \rho(r_p^*)) \approx \boldsymbol{\rho}$. Many distributions have special properties which enhance the efficiency of such a numerical search procedure.

When F_Y is a continuous uniform distribution, a closed-form solution for $\rho(r)$ is given in Li and Hammond (1975, Equation 7) as:

$$\rho(r) = 2\sin\frac{\pi}{6}r.$$

Furthermore, for some distributions—such as the exponential distribution with rate parameter λ, or any other scale family—$\rho(r)$ is independent of the distribution parameters. To increase the efficiency of the search procedure when $\rho(r)$ is parameter-independent, we access a precomputed table of $(r, \rho(r))$ values to find a good starting point for the search. For symmetric distributions, $\rho(-r) = -\rho(r)$ (Song, Hsiao, and Chen 1996); thus, it is only necessary to compute Equation (1) for positive values of r. Our numerical search procedure makes use of these results.

When Y_t is a discrete random variable with finite support, we use a special-purpose numerical-integration method. In this case, which includes the empirical cdf, it is easy to show that Equation (1) is a finite sum of terms of the form

(constant) $\times$ (area under the bivariate normal pdf).

In general, the form of Equation (2) leads to additional computational savings. Observe that in Equation (2), φ_{r_h} is the only term that is a function of r_h. Since for many distributions calculating F_Y^{-1} is computationally intensive, we compute Equation (2) for many values of r_h simultaneously. Specifically, at each pair of abscissae (z_i, z_j), we calculate $w(z_i)w(z_j)F_Y^{-1}[\Phi(z_i)]F_Y^{-1}[\Phi(z_j)]$ once, followed by $\varphi_{r_h}(z_i, z_j)$ for several values of r_h. Then, after multiplying the two terms together, we keep track of the Gauss and the Kronrod sums for each value of r_h separately. The error for each integration interval is taken to be the maximum absolute error over all values of r_h, and the interval is bisected if the error does not meet the given tolerance.

In brief, our numerical search procedure selects a grid of r_h-values at each iteration. At least initially, these grid points depend upon F_Y and $\boldsymbol{\rho}$. Then, the Gauss-Kronrod quadrature rule is used to integrate Equation (1) simultaneously for all of the grid points. For each autocorrelation lag h, the procedure finds the two grid points r_{j_h} and r_{j_h+1} such that $\rho(r_{j_h}) \leq \rho_h \leq \rho(r_{j_h+1})$. If one of these two grid points meets the relative-error criterion, then the search is complete for lag h. Otherwise, several points between r_{j_h} and $r_{j_{h+1}}$ are added to the grid for the next iteration. The procedure is repeated until the relative-error criteria are met for all p lags.

5 ARTAFACTS—SOFTWARE FOR FITTING ARTA PROCESSES

`ARTAFACTS` is a Fortran implementation of the numerical search procedure outlined in Section 4. User input, which is read from a file, includes the marginal distribution F_Y, the number of autocorrelation lags p to match, the autocorrelations $\boldsymbol{\rho}$ to match, the relative-error criterion for each autocorrelation lag, the parameters of the ARTA-process marginal distribution, and the name of the output file. The following marginal distributions are supported: Normal, Student's t, Continuous Uniform, Exponential, Gamma, Weibull, Lognormal, Johnson Unbounded, Discrete with finite support and Empirical. The output consists of the first p autocorrelation lags of the AR(p) process, along with the resultant autocorrelation values for the ARTA process, which are within the user-specified relative errors. The output also includes the AR parameters $\boldsymbol{\alpha}= (\alpha_1, \alpha_2, \ldots, \alpha_p)$.

If necessary, `ARTAFACTS` can read in time-series data from a file, calculate the sample autocorrelation function of the series and test whether the autocorrelations are significantly different from 0. For this option, the user specifies the number of autocorrelation lags to calculate and the name of the file to which the sample autocorrelation structure will be written.

Prior to invoking the numerical search procedure, `ARTAFACTS` performs several checks on the feasibility of the user input. For instance, `ARTAFACTS` checks that the correlation matrix implied by the autocorrelation structure for the input process is positive definite. Furthermore, `ARTAFACTS` checks that $\underline{\rho} \leq \rho_h \leq 1, h = 1, 2, \ldots, p$, where $\underline{\rho}$, the minimum feasible bivariate correlation for F_Y, is calculated by `ARTAFACTS`. These two conditions are necessary, but not sufficient, for the ARTA process to be stationary. After the numerical procedure terminates, `ARTAFACTS` determines whether or not the underlying AR(p) process is stationary. If the AR(p) process is not stationary, a warning appears in the output file and no ARTA representation is possible.

Figure 1 displays a sample `ARTAFACTS` input file for an application that will be presented in the next section. The first line contains the name of the output file, `weib.out`. The `0` on the second line indicates that we do not want `ARTAFACTS` to read in an empirical time series and estimate its autocorrelation structure. Instead, the third line tells `ARTAFACTS` that we want the marginal distribution of the ARTA process to be a Weibull distribution, which has code number `6`. The `3` on the next line indicates that we want to match three autocorrelation lags of the input process. The following six lines give the desired autocorrelations and their corresponding relative-error criteria for lags 1 through 3. The final two lines of input contain the scale parameter, 0.9432, and shape parameter, 5.14, of the Weibull distribution.

```
weib.out
0
6
3
0.749
0.01
0.409
0.01
0.121
0.01
0.9432
5.14
```

Figure 1: Sample `ARTAFACTS` Input File, `input.dat`

Figure 2 displays the `ARTAFACTS` output file `weib.out` corresponding to the input file given in Figure 1. The output file reiterates that the ARTA distribution is a Weibull(0.9432, 5.14) distribution and that we want to match three autocorrelation lags. The output file also gives the desired autocorrelations and their required relative errors. Next, it displays three columns of program output, the three ARTA-process autocorrelations $\boldsymbol{\rho}^* = (0.751, 0.407, 0.121)$, their underlying AR-process autocorrelations $\mathbf{r}^* = (0.752, 0.408, 0.122)$ such that $\rho(r_h^*) \approx \rho_h$, for $h = 1, 2, 3$, and the actual relative errors that were attained. The final column of output contains the parameters α_1, α_2, and α_3 of the AR(3) base process.

```
**********  USER INPUT  **********

ARTA Distribution:  Weibull
Scale Parameter:      0.94320
Shape Parameter:      5.14000

Number of ARTA autocorrelation lags to match:   3

Lag   Desired Autocorrelation   Desired Relative Error
***   ***********************   **********************
  1      0.7490000000000            0.01000
  2      0.4090000000000            0.01000
  3      0.1210000000000            0.01000

Minimum feasible bivariate correlation is:  -0.98875


 **********************************


 **********  ARTA PROGRAM OUTPUT **********


 Lag        ARTA Correl.          AR Correl.        Rel. Error

   1      0.750955             0.752000          0.002610
   2      0.406725             0.408000          -.005563
   3      0.121049             0.121600          0.000402


 i              Alpha(i)
 1                 1.0017
 2                -0.2978
 3                -0.0632
```

Figure 2: Sample **ARTAFACTS** Output File

6 ARTAGEN—SOFTWARE FOR GENERATING ARTA PROCESSES

After fitting an ARTA process using **ARTAFACTS**, we can use the AR output parameters $\boldsymbol{\alpha}$ to generate simulation inputs (see Cario and Nelson 1996). **ARTAFACTS** automatically writes the input file for **ARTAGEN**, a software package that generates observations from ARTA processes.

ARTAGEN reads input from a file called `input.gen`. A sample input file for **ARTAGEN** appears in Figure 3. The first line of the input file specifies the name of the file, `weib.gen`, to which the generated observations will be written. The second line contains the name of the file, `weib.sum`, to which summary statistics about the generated observations will be written. The next three lines contain the code number (`6`) of the ARTA marginal distribution, the number of autocorrelation lags that have been matched (`3`), and the number of observations to generate (`1000`). The autocorrelations for the AR(p) process followed by the autocorrelations for the ARTA process appear on the next six lines. The final two lines contain the parameters of the marginal distribution.

```
weib.gen
weib.sum
6
3
1000
0.752
0.408
0.1216
0.749
0.409
0.121
0.9432
5.14
```

Figure 3: Sample **ARTAGEN** Input File, `input.gen`

7 APPLICATION

Continuous-flow production lines, such as those used to extrude plastics, are common in the chemical industry. Process variables such as temperatures and pressures are often key parameters on these types of production lines, and understanding their effects on the manufacturing system is critical. Systems simulation can be used to model new and existing continuous-flow lines, as well as to train new operators in proper responses to process changes.

In this section, we present time-series data for a pressure variable from a continuous-flow polymer process at a manufacturing plant of a large chemical company. We fit a marginal distribution using the commercially available ARENA input processor and estimated the autocorrelation structure using **ARTAFACTS**. We then used **ARTAFACTS** to fit an ARTA process to the data set. Finally, we generated 1000 observations from the ARTA process using **ARTAGEN**. To assess the fit of the ARTA process, we compared the properties of the empirical time series with those of the generated time series.

Figure 4 displays the time-series plot of the empirical data. The sample contains 519 observations, with sample average $\hat{\mu} = 45.48$ and sample variance $s^2 = 0.039$. The sample autocorrelation function through lag 3 is $\hat{\rho} = (0.749, 0.409, 0.121)$. No other autocorrelations are statistically different from zero.

ARENA selected 44.6+Weibull(0.9432, 5.14) as the best fit for the marginal distribution of the pressure

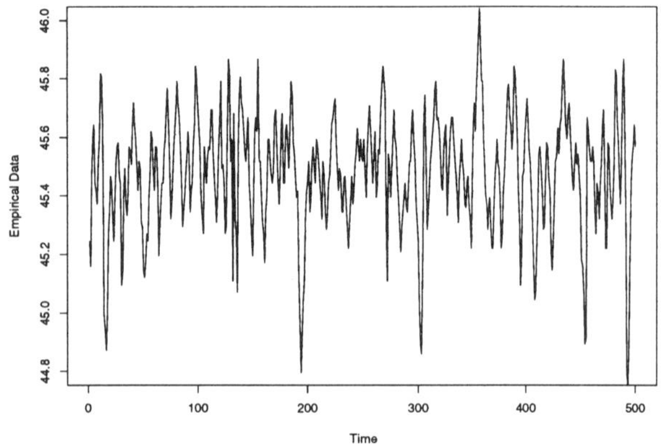

Figure 4: Time-series Plot of the Empirical Data

variable. We used this Weibull distribution (minus the constant) and $\hat{\rho}$ as the input to **ARTAFACTS**, which is displayed in Figure 1. Recall from the output file in Figure 2 that the autocorrelation structure for the underlying AR(3) process was $\mathbf{r}^* = (0.752, 0.408, 0.122)$ and the implied ARTA-process autocorrelation structure was $\boldsymbol{\rho}^* = (0.751, 0.407, 0.121)$.

Given $\mathbf{r}^*$, we used **ARTAGEN** to generate an ARTA process with the specified Weibull distribution and autocorrelation structure $\boldsymbol{\rho}^* \approx \boldsymbol{\rho}$. Recall that the **ARTAGEN** input file for this process appears in Figure 3. Figure 5 contains the **ARTAGEN** summary statistics file.

As indicated in Figure 5, the sample average for this series is $\hat{\mu} = 44.6 + 0.86 = 45.46$ and the sample variance is $s^2 = 0.0387$. The sample autocorrelation function through lag 3 is $\hat{\rho} = (0.739, 0.383, 0.081)$. The differences between the sample statistics and the theoretical values are due to sampling error—they would converge to the theoretical values as the number of generated observations increased.

Figure 6 displays the time-series plot of the observations from the ARTA process that we fit to the pressure variable. Comparing Figures 4 and 6, observe that the sample paths are qualitatively similar, although the height of the spikes is more variable in the empirical time series, and the ARTA process varies more consistently about its mean than does the empirical time series. The differences between the two time series reflect sampling error as well as the fact that the marginal distribution and autocorrelation structure do not capture all of the characteristics of a time-series process.

The scatterplots of (Y_t, Y_{t+1}) for the empirical and ARTA data on the pressure variable appear in Figures 7 and 8, respectively. Observe that the ARTA data

```
**********  USER INPUT  **********

ARTA Distribution:  Weibull
Scale Parameter:     0.94320
Shape Parameter:     5.14000

Lag      ARTA Corr.      AR Corr.
 1        0.7490          0.7520
 2        0.4090          0.4080
 3        0.1210          0.1216
**********************************

***** SUMMARY STATISTICS *****

Average of Output (Y) Process:       0.86158
Sample Variance of Output (Y) Process:       0.03869
Lag       Correlation of Output (Y) Process
 1           0.73937780
 2           0.38340840
 3           0.08111539


Average of Input (Z) Process:             -0.02773
Sample Variance of Input (Z) Process:               1.02594
Lag      Correlation of Input (Z) Process
 1           0.74147892
 2           0.38993248
 3           0.08886040
```

Figure 5: Sample **ARTAGEN** Summary File, **weib.sum**

is more scattered than the empirical data, and that there are a few more points in the bottom, left corner of the scatterplot for the empirical data. Similar differences appear in the scatterplots for lags two and three, which are not displayed. The differences are compatible with the differences between the fitted distribution and the empirical distribution, along with the differences between the time-series plots. Overall, the ARTA process provides a plausible model for the empirical time series.

8 CONCLUSIONS

The example in Section 7 demonstrates the usefulness of ARTA processes for modeling time-series inputs for simulation. Certainly, they provide a more faithful representation of the empirical systems than do series of independent, identically distributed random variables, which are often used in practice. Furthermore, the software packages **ARTAFACTS** and **ARTAGEN** simplify the tasks of building and generating observations from ARTA models. Instructions for downloading and using **ARTAFACTS** and **ARTAGEN** appear on the ARTA web page at **http://primal.iems.nwu.edu/~nelsonb/ARTA/**.

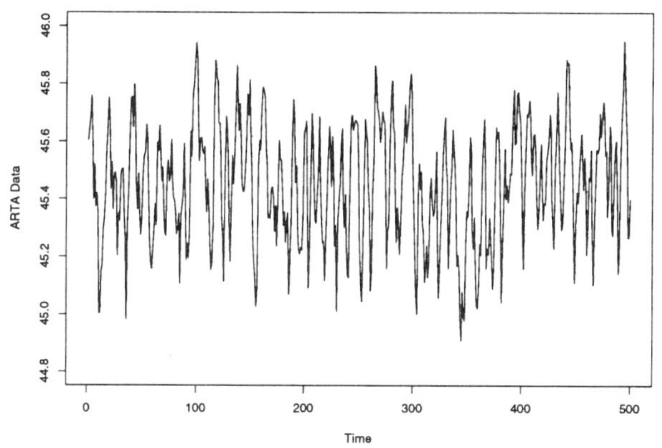

Figure 6: Time-series Plot of Observations from the ARTA-process Model

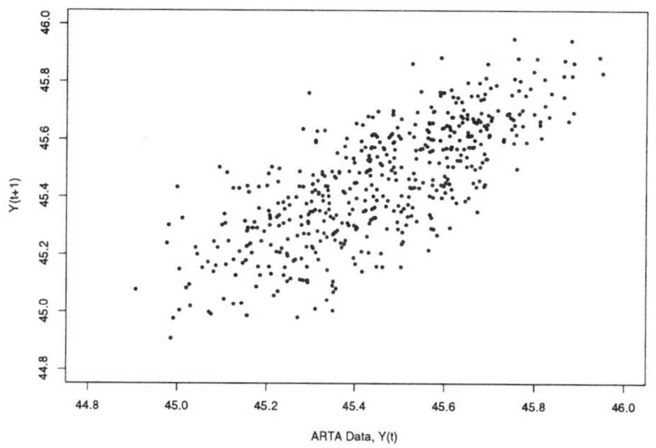

Figure 8: Scatterplot of (Y_t, Y_{t+1}) for the ARTA-process Model

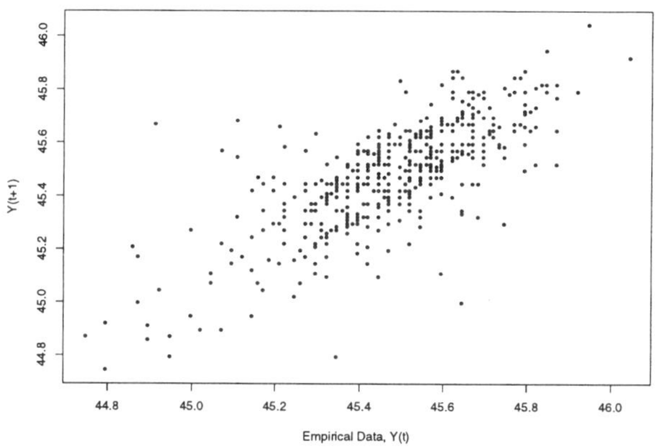

Figure 7: Scatterplot of (Y_t, Y_{t+1}) for the Empirical Data

ACKNOWLEDGMENTS

The author would like to thank Barry L. Nelson of Northwestern University for his active participation in this research and in writing this paper.

REFERENCES

Cario, M. C., and B. L. Nelson. 1996. Autoregressive to anything: Time-series input processes for simulation. *Operations Research Letters*, in press.

Lewis, P. A. W., E. McKenzie, and D. K. Hugus. 1989. Gamma processes. *Communications in Statistics–Stochastic Models* 5:1–30.

Li, S. T., and J. L. Hammond. 1975. Generation of pseudorandom numbers with specified univariate distributions and correlation coefficients. *IEEE Transactions on Systems, Man and Cybernetics* 5:557–561.

Livny, M., B. Melamed, and A. K. Tsiolis. 1993. The impact of autocorrelation on queuing systems. *Management Science* 39:322–339.

Melamed, B. 1991. TES: A class of methods for generating autocorrelated uniform variates. *ORSA Journal on Computing* 3:317–329.

Melamed, B., J. R. Hill, and D. Goldsman. 1992. The TES methodology: Modeling empirical stationary time series. In *Proceedings of the 1992 Winter Simulation Conference*, ed. J. J. Swain, D. Goldsman, R. C. Crain and J. R. Wilson, 135–144. Institute of Electrical and Electronics Engineers, Piscataway, New Jersey.

Piessens, R., E. de Doncker-Kapenga, C. W. Überhuber, and D. K. Kahaner. 1983. *Quadpack: A subroutine package for automatic integration.* New York: Springer-Verlag.

Song, W. T., L. Hsiao, and Y. Chen. 1996. Generation of pseudo-random time series with specified marginal distributions. *European Journal of Operational Research*, forthcoming.

Whitt, W. 1976. Bivariate distributions with given marginals. *The Annals of Statistics* 4:1280–1289.

AUTHOR BIOGRAPHY

MARNE C. CARIO is a systems analyst at Delphi Packard Electric Systems in Warren, Ohio. Her research interests focus on simulation input modeling and the application of simulation to manufacturing systems.

Proceedings of the 1996 Winter Simulation Conference
ed. J. M. Charnes, D. J. Morrice, D. T. Brunner, and J. J. Swain

IMPLEMENTING THE BATCH MEANS METHOD IN SIMULATION EXPERIMENTS

Christos Alexopoulos

School of Industrial and Systems Engineering
Georgia Institute of Technology
Atlanta, Georgia 30332–0205, U.S.A.

Andrew F. Seila

Terry College of Business
University of Georgia
Athens, GA 30602–6255, U.S.A.

ABSTRACT

This paper reviews and evaluates strategies for implementing the batch means method for estimating the mean of a stationary simulation output process.

1 INTRODUCTION

Suppose $\{X_i, i \geq 1\}$ is a discrete-time stochastic process. The method of batch means is frequently used to estimate the steady-state mean μ of $\{X_i\}$ or the $\mathsf{Var}(\overline{X}_n)$ (for finite n) and owes its popularity to its simplicity and effectiveness. The original references on the method are Conway (1963), Fishman (1978a,b), and Law and Carson (1979).

The classical approach divides the output $X_1, \ldots, X_n$ of a long simulation run into a number of contiguous *batches* and uses the sample means of these batches (or *batch means*) to produce point and interval estimators.

To motivate the method, suppose temporarily that the process $\{X_i\}$ is weakly stationary, that is, $\mathsf{E}(X_i) = \mu$, $\mathsf{Var}(X_i) = \sigma^2$, and the $\mathsf{Cov}(X_i, X_j)$ depends only on the *lag* $|j - i|$. Also assume that $\lim_{n\to\infty} n\mathsf{Var}(\overline{X}_n) < \infty$. Then split the data into k batches, each consisting of b observations. (Assume $n = kb$.) The ith batch consists of the observations

$$X_{(i-1)b+1}, X_{(i-1)b+2}, \ldots, X_{ib}$$

for $i = 1, 2, \ldots, k$ and the ith batch mean is given by

$$Y_i(b) = \frac{1}{b}\sum_{j=1}^{b} X_{(i-1)b+j}.$$

For fixed m, let $\sigma_m^2 = \mathsf{Var}(\overline{X}_m)$. Since the batch means process $\{Y_i(b), i \geq 1\}$ is also weakly stationary, some algebra yields

$$\begin{aligned}\sigma_n^2 &= \frac{\sigma_b^2}{k} + \frac{1}{k^2}\sum_{i\neq j}\mathsf{Corr}[Y_i(b), Y_j(b)] \\ &= \frac{\sigma_b^2}{k}\left(1 + \frac{n\sigma_n^2 - b\sigma_b^2}{b\sigma_b^2}\right). \qquad (1)\end{aligned}$$

Since $n \geq b$, $(n\sigma_n^2 - b\sigma_b^2)/(n\sigma_b^2) \to 0$ as first $n \to \infty$ and then $b \to \infty$. As a result, σ_b^2/k approximates σ_n^2 with error that diminishes as b and n approach infinity. Equivalently, the correlation among the batch means diminishes as b and n approach infinity.

To use the last limiting property, one forms the grand batch mean

$$\overline{Y}_k = \overline{X}_n = \frac{1}{k}\sum_{i=1}^{k} Y_i(b),$$

estimates σ_b^2 by

$$\hat{V}_B(n,k) = \frac{1}{k-1}\sum_{i=1}^{k}(Y_i(b) - \overline{Y}_k)^2,$$

and computes the following approximate $1-\alpha$ confidence interval for μ:

$$\overline{Y}_k \pm t_{k-1,1-\alpha/2}\sqrt{\hat{V}_B(n,k)/k}. \qquad (2)$$

The main problem with the application of the batch means method in practice is the choice of the batch size b. If b is too small, the means $Y_i(b)$ can be highly correlated and the resulting confidence interval will frequently have coverage below the user-specified nominal coverage $1-\alpha$. Alternatively, a large batch size will likely result in very few batches and potential problems with the application of the central limit theorem to obtain (2). An extensive study of batch size effects for fixed sample size was conducted by Schmeiser (1982).

Remark 1 For fixed sample size, a plot of the batch means is a very useful tool for checking the effects of initial conditions, non-normality of batch means, and existence of correlation between batch means. For example, consider the M/M/1 queueing system with interarrival rate $\tau = 0.09$ and service rate $\omega = 0.1$. The limiting mean customer delay is $\mu = \tau/[\omega(\omega - \tau)] = 90$. A sample of 100,000 customer

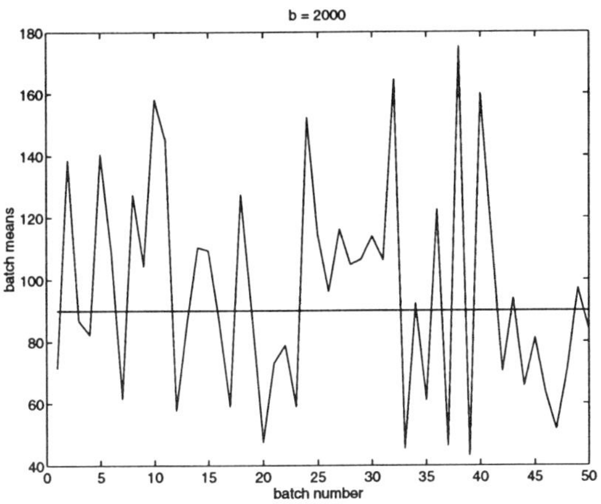

Figure 1: Batch means for delay times in an M/M/1 queue

delays, D_i, was generated by means of Lindley's recursion $D_{i+1} = \max\{D_i+S_i-A_{i+1}, 0\}$, $i \geq 1$, starting with an empty system ($D_1 = 0$), where A_i is the ith interarrival time and S_i is the service time of the ith customer. For moderate-to-heavy traffic intensity $\nu = \tau/\omega$, the autocorrelation function for the process $\{D_i\}$ has a very long tail (see Blomqvist 1967). This property makes the M/M/1 system a good model for evaluating simulation methodologies.

Figure 1 shows the plot of the batch means $Y_1(2000), \ldots, Y_{50}(2000)$ for batch size $b = 2000$. The first batch mean is small but not the smallest, relaxing one's worries about the effect of the initial transient period. This also hints that $l = 2000$ is a reasonable truncation index for removing transient observations. Had the first batch mean been smaller than the other batch means, one can assess the effect of the initial conditions by removing the first batch and comparing the new grand batch mean with the old. Although the plot does not indicate the presence of serious autocorrelation among the batch means, the asymmetric dispersion of the batch means about the actual mean should make the experimenter concerned about the coverage of the confidence interval (2).

Example 1 shows how an asymptotically optimal batch size can be obtained in special cases.

Example 1 Consider the stationary AR(1) process

$$X_i = \mu + \rho(X_{i-1} - \mu) + Z_i, \quad i \geq 1,$$

where $|\rho| < 1$, $X_0 \sim N(\mu, 1)$, and the Z_i's are i.i.d. $N(0, 1-\rho^2)$. Carlstein (1986) showed that

$$\mathsf{Bias}(\hat{V}_B(n,k)) = -\frac{2\rho}{(1-\rho)^3(1+\rho)}\frac{1}{b} + o\left(\frac{1}{b}\right) \quad (3)$$

and

$$\mathsf{Var}(\hat{V}_B(n,k)) = \frac{2}{(1-\rho)^4}\frac{b}{n} + o\left(\frac{b}{n}\right),$$

where $o(h)$ is a function such that $\lim_{h\to 0} o(h)/h = 0$. Then the batch size that minimizes the asymptotic (as $n \to \infty$ and $k \to \infty$) mean squared error $\mathsf{MSE}(\hat{V}_B(n,k)) = \mathsf{Bias}^2(\hat{V}_B(n,k)) + \mathsf{Var}(\hat{V}_B(n,k))$ is

$$b_0 = \left(\frac{2|\rho|}{1-\rho^2}\right)^{2/3} n^{1/3}. \quad (4)$$

Clearly, the optimal batch size increases with the absolute value of the correlation ρ between successive observations. Unfortunately, such an analysis cannot be performed for the majority of output processes. Furthermore, asymptotically optimal batch sizes may differ considerably from optimal batch sizes for finite sample sizes (as Song and Schmeiser (1995) observed for a congested M/M/1 queue).

2 CONSISTENT ESTIMATION BATCH MEANS METHODS

Consistent estimation batch means methods assume the existence of a parameter σ^2 (the time-average variance of the process $\{X_i\}$), such that a central limit theorem holds

$$\sqrt{n}(\overline{X}_n - \mu) \xrightarrow{d} \sigma N(0,1) \quad \text{as } n \to \infty \quad (5)$$

and aim at constructing a consistent estimator for σ^2 and an asymptotically valid confidence interval for μ. [Notice that the X_i's in (5) need not be i.i.d.] Consistent estimation methods are often preferable to methods that "cancel" σ^2 (see Glynn and Iglehart 1990) because: (a) The expectation and variance of the halfwidth of the confidence interval resulting from (5) is asymptotically smaller for consistent estimation methods; and (b) Under reasonable assumptions $n\mathsf{Var}(\overline{X}_n) \to \sigma^2$ as $n \to \infty$.

Chien, Goldsman, and Melamed (1996) considered stationary processes and, under quite general moment and sample path conditions, showed that as both $b, k \to \infty$, $\mathsf{MSE}[b\hat{V}_B(n,k)] \to 0$. Notice that mean squared error consistency differs from consistency.

The limiting result (5) is implied under the following two assumptions, where $\{W(t), t \geq 0\}$ is the standard Brownian motion process (see Resnick 1994, Chapter 6).

Assumption of Weak Approximation (AWA). There exist finite constants μ and $\sigma > 0$ such that

$$\frac{n(\overline{X}_n - \mu)}{\sigma} \xrightarrow{d} W(n) \quad \text{as } n \to \infty.$$

Assumption of Strong Approximation (ASA). There exist finite constants μ, $\sigma > 0$, $\lambda \in (0, 1/2]$, and a finite random variable C such that, with probability one,

$$|n(\overline{X}_n - \mu) - \sigma W(n)| \leq Cn^{1/2-\lambda} \quad \text{as } n \to \infty.$$

Both AWA and ASA state that the process $\{n(\overline{X}_n - \mu)/\sigma\}$ is close to a standard Brownian motion. However the stronger ASA addresses the convergence rate of (5).

The ASA is not restrictive as it holds under relatively weak assumptions for a variety of stochastic processes including Markov chains, regenerative processes and certain queueing systems (see Damerdji 1994 for details). The constant λ is closer to 1/2 for processes having little autocorrelation while it is closer to zero for processes with high autocorrelation. In the former case the "distance" between the processes $\{n(\overline{X}_n - \mu)/\sigma\}$ and $\{W(n)\}$ "does not grow" as n increases.

2.1 Batching Rules

Fishman and Yarberry (1994) and Fishman (1996, Chapter 6) presented a thorough discussion of batching rules. Both references contain detailed instructions for obtaining FORTRAN implementations for various platforms via anonymous ftp.

Equation (1) suggests that fixing the number of batches and letting the batch size grow as $n \to \infty$ ensures that $\sigma_b^2/k \to \sigma_n^2$. This motivates the following rule.

The Fixed Number of Batches (FNB) Rule. Fix the number of batches at k. For sample size n, use batch size $b_n = \lfloor n/k \rfloor$.

The FNB rule along with AWA lead to the following result.

Theorem 1 (Glynn and Iglehart 1990) *If $\{X_i\}$ satisfies AWA, then as $n \to \infty$, $\overline{X}_n \xrightarrow{p} \mu$ and (5) holds. Furthermore, if k is constant and $\{b_n, n \geq 1\}$ is a sequence of batch sizes such that $b_n \to \infty$ as $n \to \infty$, then*

$$\frac{\overline{X}_n - \mu}{\sqrt{\hat{V}_B(n,k)/k}} \xrightarrow{d} t_{k-1} \quad \text{as } n \to \infty.$$

The primary implication of Theorem 1 is that

$$\overline{Y}_k \pm t_{k-1,1-\alpha/2}\sqrt{\hat{V}_B(n,k)/k} \tag{6}$$

is an asymptotically valid confidence interval for μ. Unfortunately, the FNB rule has two major limitations: (a) $b_n\hat{V}_B(n,k)$ is not a consistent estimator of σ^2. Therefore the confidence interval (6) tends to be wider than the interval a consistent estimation method would produce; (b) Statistical fluctuations in the halfwidth of the confidence interval (6) do not diminish relative to statistical fluctuation in the sample mean (see Fishman 1996, pp. 544–545).

The following theorem proposes batching assumptions which along with ASA yield a strongly consistent estimator for σ^2.

Theorem 2 (Damerdji 1994) *If $\{X_i\}$ satisfies ASA, then $\overline{X}_n \xrightarrow{a.s.} \mu$ as $n \to \infty$. Furthermore suppose that $\{(b_n, k_n), n \geq 1\}$ is a batching sequence satisfying*

(1) *$b_n \to \infty$ and $k_n \to \infty$ monotonically as $n \to \infty$*

(2) *$b_n^{-1}n^{1-2\lambda}\ln n \to 0$ as $n \to \infty$*

(3) *there exists a finite positive integer a such that*

$$\sum_{n=1}^{\infty}(b_n/n)^a < \infty.$$

Then

$$b_n\hat{V}_B(n,k_n) \xrightarrow{a.s.} \sigma^2 \tag{7}$$

and

$$Z_{k_n} = \frac{\overline{X}_n - \mu}{\sqrt{\hat{V}_B(n,k_n)/k_n}} \xrightarrow{d} N(0,1). \tag{8}$$

The last display implies that

$$\overline{X}_n \pm t_{k_n-1,1-\alpha/2}\sqrt{\hat{V}_B(n,k_n)/k_n}$$

is an asymptotically valid $1-\alpha$ confidence interval for μ.

Theorem 2 motivates the consideration of batch sizes of the form $b_n = \lfloor n^\theta \rfloor$, $0 < \theta < 1$. In this case one can show that the conditions (1)–(3) are met if $\theta \in (1-2\lambda, 1)$. In particular, the assignment $\theta = 1/2$ and the SQRT rule below are valid if $1/4 < \lambda < 1/2$. Notice that the last inequality is violated by processes having high autocorrelation ($\lambda \approx 0$).

The Square Root (SQRT) Rule. For sample size n, use batch size $b_n = \lfloor\sqrt{n}\rfloor$ and number of batches $k_n = \lfloor\sqrt{n}\rfloor$.

Under some additional moment conditions, Chien (1989) showed that the convergence of Z_{k_n} to the $N(0,1)$ distribution is fastest if both b_n and k_n grow proportionally to $\sqrt{n}$. Unfortunately, in practice the SQRT rule tends to seriously underestimate the $\mathsf{Var}(\overline{X}_n)$ for fixed n.

Example 2 (The M/M/1 queue) Consider an M/M/1 queueing system with interarrival rate $\tau = 0.9$ and service rate $\omega = 1$, and assume that the system starts empty. Table 1 contains performance statistics for 0.95 confidence intervals for the steady-state mean customer delay $\mu = 9$. The confidence intervals resulted from 500 independent replications. The FNB rule used 16 batches and batch sizes 2^m, $m \geq 0$. The SQRT rule started with batch size $b_1 = 1$ and number of batches $k_1 = 8$, and computed confidence intervals with batch sizes

$$b_l = 2^{(l-1)/2} \times \begin{cases} b_1 & \text{if } l \text{ is odd} \\ 3/(2\sqrt{2}) & \text{otherwise} \end{cases}$$

and numbers of batches

$$k_l = 2^{(l-1)/2} \times \begin{cases} k_1 & \text{if } l \text{ is odd} \\ 11/\sqrt{2} & \text{otherwise.} \end{cases}$$

Columns 2 and 4 contain the estimated coverage probabilities of the confidence intervals produced by the FNB rule and the SQRT rule, respectively. Columns 3 and 5 display the respective average interval halfwidths. Specifically, for sample size $n \approx 2^{17} =$ 131,072, roughly 94 percent of the confidence intervals resulting from the FNB rule contained μ whereas only 78 percent of the confidence intervals resulting from the SQRT rule contained μ. However, the latter intervals were 43 percent narrower. Experiments by Fishman and Yarberry showed that the disparity in coverage between the two rules grows with increasing traffic intensity.

Table 1: Performance statistics for the FNB and SQRT rules on 0.95 confidence intervals for the mean customer delay in an M/M/1 queue with utilization $\nu = 0.9$

	FNB Rule		SQRT Rule	
$\log_2 n$	Coverage	Average Halfwidth	Coverage	Average Halfwidth
10	0.544	3.244	0.326	1.694
11	0.640	3.506	0.366	1.665
12	0.746	3.304	0.414	1.437
13	0.798	2.963	0.466	1.271
14	0.838	2.435	0.498	1.063
15	0.880	1.901	0.604	0.904
16	0.912	1.437	0.664	0.738
17	0.944	1.053	0.778	0.599
18	0.934	0.756	0.810	0.471
19	0.950	0.541	0.854	0.369
20	0.940	0.385	0.858	0.283

With the contrasts between the FNB and SQRT rules in mind, Fishman and Yarberry proposed two strategies that dynamically shift between the two rules. Both strategies perform "interim reviews" and compute confidence intervals at times $n_l \approx n_1 2^{l-1}, l = 1, \ldots, L+1$.

The LBATCH Strategy. At time n_l, if an hypothesis test detects autocorrelation between the batch means, the batching for the next review is determined by the FNB rule. If the test fails to detect correlation, all future reviews omit the test and employ the SQRT rule.

The ABATCH Strategy. If at time n_l the hypothesis test detects correlation between the batch means, the next review employs the FNB rule. If the test fails to detect correlation, the next review employs the SQRT rule.

Both strategies LBATCH and ABATCH yield random sequences of batch sizes. Under relatively mild assumptions, these sequences imply convergence results analogous to (7) and (8) (see Fishman and Yarberry 1994 and Fishman 1996).

Test for Correlation

We will briefly review a test for the hypothesis H_0: the batch means $Y_1(b), \ldots, Y_k(b)$ are uncorrelated. A commonly used test is due to von Neumann (1941) and is very effective when the number of batches k is small.

Assume that the process $\{X_i\}$ is weakly stationary and let

$$\rho_l(b) = \mathsf{Corr}[Y_i(b), Y_{i+l}(b)], \quad l = 0, 1, \ldots$$

be the autocorrelation function of the batch means process. The von Neumann test statistic for H_0 is

$$R_k = \sqrt{\frac{k^2-1}{k-2}} \left[\hat{\rho}_1(b) + \frac{(Y_1(b) - \overline{X}_n)^2 + (Y_k(b) - \overline{X}_n)^2}{2\sum_{i=1}^{k}(Y_i(b) - \overline{X}_n)^2} \right],$$

where

$$\hat{\rho}_1(b) = \frac{\sum_{i=1}^{k-1}(Y_i(b) - \overline{X}_n)(Y_{i+1}(b) - \overline{X}_n)}{\sum_{i=1}^{k}(Y_i(b) - \overline{X}_n)^2}$$

is an estimator for the lag-1 autocorrelation $\rho_1(b)$.

Under H_0, $R_k \approx N(0,1)$ for large b (the batch means become approximately normal) or large k (by the central limit theorem). If $\{X_i\}$ has a monotone decreasing autocorrelation function (e.g., the delay process for an M/M/1 queueing system), one rejects H_0 at level β if $R_k > z_{1-\beta}$. Alternatively, if $\{X_i\}$ has an autocorrelation function with damped harmonic behavior around the zero axis (e.g., an AR(1) process with $\rho < 0$), the rejection of H_0 when $R_k > z_{1-\beta}$ can lead to erroneous conclusions. In this case, repeated testing under the ABATCH strategy reduces this possibility.

The p-value $= 1 - \Phi(R_k)$ of the test is the largest value of the type I error $\beta = P(\text{reject } H_0 \mid H_0 \text{ is true})$ given the observed value of R_k. Hence, a p-value close to zero implies low credibility for H_0. The plot of the p-values versus the batch size is a useful graphical device.

2.2 Implementing the LBATCH and ABATCH Strategies

To understand the role of the hypothesis test in the LBATCH and ABATCH algorithms, define the random variables

$$\overline{R}_l \;=\; \text{fraction of rejected tests for } H_0 \text{ on reviews } 1,\ldots,l.$$

A sufficient condition for strong consistency (equation (7)) and asymptotic normality (equation (8)) is $\beta_0 > 1 - 4\lambda$ (or $\lambda > (1-\beta_0)/4$), where $\beta_0 = \lim_{l\to\infty} \overline{R}_l$ is the long-run fraction of rejections. In practice, β_0 differs from but is expected to be close to the type I error β. Clearly, $\lambda > 1/4$ guarantees (7) and (8) regardless of β_0. However, β_0 plays a small role when $\lambda \le 1/4$. Specifically, for β_0 equal to 0.05 or 0.10, the lower bound $(1-\beta_0)/4$ becomes 0.2375 or 0.2225, respectively, a small reduction from 1/4.

On review l, both strategies induces batch size

$$b_l \;=\; 2^{(l-1)(1+\overline{R}_{l-1})/2} \times \begin{cases} b_1 & \text{if } (l-1)(1+\overline{R}_{l-1}) \text{ is even} \\ \tilde{b}_1/\sqrt{2} & \text{otherwise,} \end{cases}$$

where

$$\tilde{b}_1 = \begin{cases} 3/2 & \text{if } b_1 = 1 \\ \lfloor \sqrt{2}\, b_1 + 0.5 \rfloor & \text{if } b_1 > 1, \end{cases}$$

and number of batches

$$k_l \;=\; 2^{(l-1)(1-\overline{R}_{l-1})/2} \times \begin{cases} k_1 & \text{if } (l-1)(1-\overline{R}_{l-1}) \text{ is even} \\ \tilde{k}_1/\sqrt{2} & \text{otherwise,} \end{cases}$$

where $\tilde{k}_1 = \lfloor \sqrt{2}\, k_1 + 0.5 \rfloor$.

The resulting sample sizes are

$$n_l = k_l b_l = \begin{cases} 2^{l-1} k_1 b_1 & \text{if } (l-1)(1+\overline{R}_{l-1}) \text{ is even} \\ 2^{l-2} \tilde{k}_1 \tilde{b}_1 & \text{otherwise} \end{cases}$$

and the definitions for $\tilde{b}_1$ and $\tilde{k}_1$ guarantee that if H_0 is never rejected, then both b_l and k_l grow approximately as $\sqrt{2}$ with l (i.e., they follow the SQRT rule).

The final implementation issue is the relative difference between the potential terminal sample sizes

$$\Delta(b_1,k_1) = \frac{|2^L k_1 b_1 - 2^{L-1}\tilde{k}_1\tilde{b}_1|}{2^L k_1 b_1} = \frac{|2k_1 b_1 - \tilde{k}_1\tilde{b}_1|}{2k_1 b_1}.$$

This quantity is minimized (i.e., the final sample size is deterministic) when $2k_1 b_1 = \tilde{k}_1\tilde{b}_1$. Although this condition excludes several practical choices for b_1 and k_1, such as $b_1 = 1$ (to test the original sample for independence) and $8 \le k_1 \le 10^5$, $\Delta(b_1,k_1)$ remains small for numerous choices of b_1 and k_1.

2.3 Tests for the Batching Rules

The experiments in Examples 3, 4 and 5 compare the LBATCH and ABATCH strategies by means of three queueing systems with traffic intensity $\nu = 0.9$. Each system starts empty and has a first-come, first-served discipline. Each experiment computed 0.95 confidence intervals for the long-run mean customer delay from 500 independent replications. Both strategies started with $k_1 = 8$ batches of size $b_1 = 1$ and used type I error $\beta = 0.1$ for H_0.

Example 3 (Example 2 continued) The entries of Tables 1 and 2 indicate that the ABATCH strategy comes closer to the FNB rule's superior coverage with shorter confidence intervals.

Table 2: Performance statistics for the LBATCH and ABATCH strategies on 0.95 confidence intervals for the mean customer delay in an M/M/1 queue with utilization $\nu = 0.9$

	LBATCH Strategy		ABATCH Strategy	
$\log_2 n$	Coverage	Average Halfwidth	Coverage	Average Halfwidth
10	0.398	2.085	0.562	3.384
11	0.420	1.992	0.632	3.450
12	0.464	1.693	0.712	3.100
13	0.518	1.477	0.760	2.686
14	0.562	1.227	0.816	2.168
15	0.652	1.029	0.850	1.708
16	0.714	0.834	0.902	1.296
17	0.808	0.663	0.932	0.955
18	0.852	0.513	0.938	0.688
19	0.866	0.395	0.930	0.493
20	0.876	0.298	0.936	0.353

Example 4 (An M/G/1 queue) Consider an M/G/1 queueing system with i.i.d. interarrival times from the exponential distribution with parameter $\tau = 0.9$ and i.i.d. service times S_i from the hyperexponential distribution with density function

$$f(x) = 0.9\left(\frac{1}{0.5}e^{-x/0.5}\right) + 0.1\left(\frac{1}{5.5}e^{-x/5.5}\right), \quad x \ge 0.$$

This distribution applies when customers are classified into two types, 1 and 2, with respective probabilities 0.9 and 0.1, type 1 customers have exponential service times with mean 0.5, and type 2 customers have exponential service times with mean 5.5. The service times have mean $\mathsf{E}(S) = 0.9(0.5) + 0.1(5.5) = 1$, second moment $\mathsf{E}(S^2) = 0.9(0.5^2) + 0.1(5.5^2) = 6.5$, and coefficient of variation $\sqrt{\mathsf{Var}(S)}/\mathsf{E}(S) = 2.345$, which is larger than the unit coefficient of variation of the exponential distribution.

The long-run mean delay time in queue is given by the Pollaczec-Khintchine formula (Heyman and Sobel

1982, pp. 250–252)

$$\mu = \lim_{i\to\infty} \mathsf{E}(D_i) = \frac{\tau\mathsf{E}(S^2)}{2[1-\tau\mathsf{E}(S)]} = 29.25. \quad (9)$$

Notice that the M/M/1 system in Example 2 with the same traffic intensity $\nu = 0.9$ has much smaller long-run mean delay time.

Table 3 displays the results of this experiment. As n increases, the conservative ABATCH strategy produces 0.95 confidence intervals for μ that are roughly 50 to 100 percent wider than the respective confidence intervals produced by the LBATCH strategy but have coverage rates that are acceptably close to 0.95 for substantially smaller sample sizes (as small as $2^{17} = 131{,}072$).

Table 3: Performance statistics for the LBATCH and ABATCH strategies on 0.95 confidence intervals for the mean customer delay in an M/G/1 queue with hyperexponential service times and utilization $\nu =$ 0.9

	LBATCH Strategy		ABATCH Strategy	
$\log_2 n$	Coverage	Average Halfwidth	Coverage	Average Halfwidth
10	0.204	5.865	0.356	10.305
11	0.254	5.962	0.436	11.426
12	0.294	5.552	0.566	11.635
13	0.354	5.083	0.652	11.166
14	0.392	4.418	0.746	10.147
15	0.452	3.863	0.794	8.658
16	0.540	3.215	0.856	7.057
17	0.620	2.678	0.898	5.483
18	0.632	2.178	0.896	4.090
19	0.694	1.761	0.900	2.997
20	0.748	1.387	0.924	2.145
21	0.806	1.083	0.926	1.525

Example 5 (An M/D/1 queue) Consider an M/G/1 queueing system with i.i.d. interarrival times from the exponential distribution with parameter $\tau = 0.9$ and fixed unit service times. Then, by (9), the long-run mean delay time in queue is $\mu = 4.5$.

The results of this experiment are contained in Table 4. As in Examples 3 and 4, the performance of the ABATCH strategy makes it an attractive compromise between the "extreme" FNB and SQRT rules.

Example 6 tests the LBATCH and ABATCH methods by means of an AR(1) process.

Example 6 Consider the stationary AR(1) process $X_i = -0.9X_{i-1} + Z_i$ with mean 0 (see Example 1). The autocorrelation function $\rho_j = (-0.9)^j$, $j \geq 0$, of this process oscillates around the zero axis and the time-average process variance is $\sigma^2 = (1-0.9)/(1+0.9) = 0.053$.

Table 4: Performance statistics for the LBATCH and ABATCH strategies on 0.95 confidence intervals for the mean customer delay in an M/D/1 queue with unit service times and utilization $\nu = 0.9$

	LBATCH Strategy		ABATCH Strategy	
$\log_2 n$	Coverage	Average Halfwidth	Coverage	Average Halfwidth
10	0.460	1.062	0.616	1.631
11	0.548	0.962	0.720	1.538
12	0.598	0.842	0.788	1.391
13	0.648	0.686	0.842	1.101
14	0.696	0.556	0.858	0.852
15	0.794	0.445	0.884	0.632
16	0.808	0.351	0.924	0.472
17	0.862	0.271	0.942	0.343

Table 5: Performance statistics for the LBATCH and ABATCH strategies on 0.95 confidence intervals for the mean $\mu = 0$ of the stationary AR(1) process $X_i = -0.9X_{i-1} + Z_i$

	LBATCH Strategy		ABATCH Strategy	
$\log_2 n$	Coverage	Average Halfwidth	Coverage	Average Halfwidth
5	1.000	0.1217	1.000	0.1153
6	0.978	0.0364	0.980	0.0367
7	0.980	0.0244	0.980	0.0246
8	0.982	0.0166	0.980	0.0167
9	0.972	0.0111	0.966	0.0111
10	0.984	0.0076	0.980	0.0075
11	0.976	0.0051	0.978	0.0050
12	0.982	0.0035	0.984	0.0035
13	0.964	0.0024	0.962	0.0023
14	0.960	0.0016	0.962	0.0016

The entries of Table 5 were obtained from 500 independent replications. The type I error for H_0 was $\beta = 0.1$. The 0.95 confidence intervals for μ produced by both methods have roughly equal halfwidths and coverages. In fact, almost all cover age estimates are greater than the nominal coverage 0.95. This behavior is due to the fact that $b\mathsf{Var}(\hat{V}_B(n,k))$ tends to overestimate σ^2 (the coefficient of $1/b$ in equation (3) is $2.624 > 0$).

From equation (4), the batch size that minimizes $\mathsf{MSE}(\hat{V}_B(n,k))$ is $b_0 = 113.71$. 500 independent replications with 144 batches of size 114 (sample size 16416) produced 0.95 confidence intervals with estimated coverage 0.958 and average halfwidth 0.0016 — not a substantial improvement over the statistics in the last row of Table 5 (for sample size roughly equal to $2^{14} = 16384$).

3 OVERLAPPING BATCH MEANS

An interesting variation of the traditional batch means method is the method of *overlapping* batch means (OBM) proposed by Meketon and Schmeiser (1984). For given batch size b, this method uses all $n - b + 1$ overlapping batches to estimate μ and $\mathsf{Var}(\overline{X}_n)$. The first batch consists of observations $X_1, \ldots, X_b$, the second batch consists of $X_2, \ldots, X_{b+1}$, etc. The OBM estimator of μ is

$$\overline{Y}_O = \frac{1}{n-b+1} \sum_{i=1}^{k} Y_i(b)$$

and has sample variance

$$\hat{V}_O = \frac{1}{n-b} \sum_{i=1}^{n-b+1} (Y_i(b) - \overline{Y}_O)^2.$$

The following list contains properties of the estimators $\overline{Y}_O$ and $\hat{V}_O$:

(i) The OBM estimator is a weighted average of non-overlapping batch means estimators.

(ii) Asymptotically (as $n, b \to \infty$ and $b/n \to 0$), the OBM variance estimator $\hat{V}_O$ and the non-overlapping batch means variance estimator $\hat{V}_B \equiv \hat{V}_B(n,k)$ have the same expectation. Furthermore,
$$\frac{\mathsf{Var}(\hat{V}_O)}{\mathsf{Var}(\hat{V}_B)} \to \frac{2}{3}.$$
In words, the asymptotic ratio of the mean squared error of $\mathsf{Var}(\hat{V}_O)$ to the mean squared error of $\mathsf{Var}(\hat{V}_B)$ is equal to 2/3 (Meketon and Schmeiser 1984).

(iii) The behavior of $\mathsf{Var}(\hat{V}_O)$ appears to be less sensitive to the choice of the batch size than the behavior of $\mathsf{Var}(\hat{V}_B)$ (Song and Schmeiser 1995, Table 1).

(iv) If $\{X_i\}$ satisfies ASA and $\{(b_n, k_n), n \geq 1\}$ is a sequence that satisfies the assumptions (A.1)–(A.3) in Theorem 2 and
$$\frac{b_n^2}{n} \to 0 \quad \text{as } n \to \infty,$$
then (Damerdji 1994)
$$b_n \hat{V}_O \overset{a.s.}{\to} \sigma^2.$$

Song and Schmeiser (1995) considered weakly stationary processes with $\gamma_m = \sum_{j=-\infty}^{\infty} j^m C_j < \infty$ for $m = 0, 1$ and studied batch means variance estimators with

$$\mathsf{Bias}(\hat{V}) = -c_b \gamma_1 \frac{1}{b} + o\left(\frac{1}{b}\right)$$

and

$$\mathsf{Var}(\hat{V}) = c_v \gamma_0^2 \frac{b}{n} + o\left(\frac{b}{n}\right).$$

The constants c_b and c_v depend on the amount of overlapping between the batches. In particular, the estimator $\hat{V}_B$ has $c_b = 1$ and $c_v = 2$, while $\hat{V}_O$ has $c_b = 1$ and $c_v = 4/3$. Then the asymptotic batch size that minimizes $\mathsf{MSE}(\hat{V}) = \mathsf{Bias}^2(\hat{V}) + \mathsf{Var}(\hat{V})$ is

$$b^* = \left(\frac{2c_b^2 \gamma_1^2}{c_v \gamma_0^2}\right)^{1/3} n^{1/3}. \tag{10}$$

Song (1996) developed methods for estimating the ratio $(\gamma_1/\gamma_0)^2$ for a variety of processes, including moving average processes and autoregressive processes. Then one can obtain an estimator for b^* by plugging the ratio estimator into equation (10). Sherman (1995) proposed an alternative method that does not rely on the estimation of $(\gamma_1/\gamma_0)^2$.

Remark 2 Welch (1987) noted that both traditional batch means and overlapping batch means are special cases of spectral estimation and, more importantly, suggested that overlapping batch means yield near-optimal variance reduction when one forms sub-batches within each batch and applies the method to the sub-batches. For example, a batch of size 64 is split into 4 sub-batches and the first (overlapping) batch consists of observations $X_1, \ldots, X_{64}$, the second consists of observations $X_{17}, \ldots, X_{80}$, etc.

REFERENCES

Blomqvist, N. 1967. The covariance function of the M/G/1 queueing system. *Skandinavisk Aktuarietidskrift* 50: 157–174.

Carlstein. E. 1986. The use of subseries for estimating the variance of a general statistic from a stationary sequence. *Annals of Mathematical Statistics* 14:1171–1179.

Chien, C.-H. 1989. Small sample theory for steady state confidence intervals. Technical Report No. 37, Department of Operations Research, Stanford University, Palo Alto, California.

Chien, C., Goldsman, D., and B. Melamed. 1996. Large-sample results for batch means. *Management Science*, to appear.

Conway, R. W. 1963. Some tactical problems in digital simulation. *Management Science* 10:47–61.

Damerdji, H. 1994. Strong consistency of the variance estimator in steady-state simulation output analysis. *Mathematics of Operations Research* 19:494–512.

Fishman, G. S. 1978a. Grouping observations in digital simulation. *Management Science* 24:510–521.

Fishman, G. S. 1978b. *Principles of Discrete Event Simulation.* John Wiley, New York, New York.

Fishman, G. S. 1996. *Monte Carlo: Concepts, Algorithms and Applications.* Chapman and Hall, New York, New York.

Fishman, G. S., and L. S. Yarberry. 1994. An implementation of the batch means method. Technical Report UNC/OR/TR/93–1, Department of Operations Research, University of North Carolina, Chapel Hill, North Carolina.

Glynn, P. W., and D. L. Iglehart. 1990. Simulation analysis using standardized time series. *Mathematics of Operations Research* 15:1–16.

Heyman, D. P., and M. J. Sobel. 1982. *Stochastic Models in Operations Research, Volume I.* McGraw-Hill, New York, New York.

Law, A. M., and J. S. Carson. 1979. A sequential procedure for determining the length of a steady-state simulation. *Operations Research* 27:1011–1025.

Meketon, M. S., and B. W. Schmeiser. 1984. Overlapping batch means: Something for nothing? In *Proceedings of the 1984 Winter Simulation Conference*, eds. S. Sheppard, U. W. Pooch, and C. D. Pegden, 227–230. IEEE, Piscataway, New Jersey.

Resnick, S. I. 1994. *Adventures in Stochastic Processes.* Birkhaüser, Boston, Massachusetts.

Schmeiser, B. W. 1982. Batch size effects in the analysis of simulation output. *Operations Research* 30:556–568.

Sherman, M. 1995. On batch means in the simulation and statistical communities. In *Proceedings of the 1995 Winter Simulation Conference*, eds. C. Alexopoulos, K. Kang, W. R. Lilegdon, and D. Goldsman, 297–302. IEEE, Piscataway, New Jersey.

Song, W.-M. T., and B. W. Schmeiser. 1995. Optimal mean-squared-error batch sizes. *Management Science* 41:110–123.

Song, W.-M. T. 1996. On the estimation of optimal batch sizes in the analysis of simulation output. To appear in *European Journal of Operations Research.*

von Neumann, J. 1941. Distribution of the ratio of the mean square successive difference and the variance. *Annals of Mathematical Statistics* 12:367–395.

Welch, P. D. 1987. On the relationship between batch means, overlapping batch means and spectral estimation, In *Proceedings of the 1987 Winter Simulation Conference*, eds. A. Thesen, H. Grant, and W. D. Kelton, 320–323. IEEE, Piscataway, New Jersey.

AUTHOR BIOGRAPHIES

CHRISTOS ALEXOPOULOS is an Associate Professor in the School of Industrial and Systems Engineering at the Georgia Institute of Technology. He received his Ph.D. in Operations Research from the University of North Carolina at Chapel Hill. His research interests are in the areas of applied probability, statistics, and optimization of stochastic systems. His recent work involves problems related to the optimal design of telecommunications and transportation systems. Dr. Alexopoulos is a member of INFORMS and the INFORMS College of Simulation. He served as a Co-Editor for the *Proceedings of the 1995 Winter Simulation Conference*.

ANDREW SEILA is an Associate Professor in the Terry College of Business at the University of Georgia, Athens, Georgia. He received his B.S. in Physics and Ph.D. in Operations Research and Systems Analysis, both from the University of North Carolina at Chapel Hill. Prior to joining the faculty of the University of Georgia, he was a Member of Technical Staff at the Bell Laboratories in Holmdel, New Jersey. His research interests include all areas of simulation modeling and analysis, especially output analysis. He has been actively involved in the *Winter Simulation Conference* since 1977, and served as a Program Chair for the 1994 conference.

Proceedings of the 1996 Winter Simulation Conference
ed. J. M. Charnes, D. J. Morrice, D. T. Brunner, and J. J. Swain

ESTIMATING STEADY STATE MEAN FROM SHORT TRANSIENT SIMULATIONS

Pieter A. Voss
Jorge Haddock
Thomas R. Willemain

Department of Decision Sciences and Engineering Systems
Rensselaer Polytechnic Institute
Troy, New York 12180-3590, USA

ABSTRACT

This paper presents a new approach to estimating and constructing confidence intervals for the steady state mean of a stochastic process from short simulations which may exhibit significant transient response. Specifically, we examine the conditional least squares estimator for the mean of an autoregressive process. If the process is autoregressive with normal innovations, this estimator is the conditional maximum likelihood estimate (MLE) of the mean. We show that the MLE is asymptotically normal, and derive a finite-sample approximation to its distribution. This provides the basis for two asymptotically valid single-replication confidence intervals which do not require choosing a batch size. As a point estimator, the MLE is a generalization of the estimator due to Snell and Schruben (1984) and is related to the weighted batch mean (Bischak, Kelton, and Pollock, 1993).

Empirical results for a queuing network show that the autoregressive process is a reasonable model of transient response. For short series with reasonable initializations (e.g., empty and idle), the MLE yields confidence intervals which are comparable or superior to those of existing procedures, in both single and parallel replication simulations.

1 INTRODUCTION

In this paper we present a new approach to the problem of estimating and constructing confidence intervals for the steady state mean μ of a stochastic process from short simulations which may exhibit significant transient response. We assume that the process may be strongly autocorrelated (positive or otherwise), that sufficient knowledge of steady state conditions is not available to initialize the process in steady state, and that the cost of collecting simulation observations is large so that available data is limited to a finite series of n observations $\{X_1, \ldots, X_n\}$.

The problem of estimating the steady state mean in general, and of controlling the effects of initial transient response in particular have been widely studied; many existing methods are surveyed by Law and Kelton (1991) and Pawlikowski (1990), and progress continues to be made. However, all existing confidence interval procedures explicitly assume that the observed process is in steady state, stationary in both mean and covariance. These include (unweighted) batch means (see Law, 1977), Fishman's autoregressive approach (Fishman, 1978), standardized time series (Schruben, 1983), weighted batch means (Bischak, Kelton, and Pollock, 1993), and intervals based on the ratio estimator due to Glynn and Heidelberger (1992). Since initial transient effects may invalidate the assumption of stationarity, either truncation or initialization methods are employed to minimize bias due to initial transient response. A third approach, weighting initial observations, is due to Snell and Schruben (1984) and is developed further in this paper.

For short transient simulations, there is as yet no completely satisfactory confidence interval procedure. In particular, Sargent, Goldsman, and Swain (1992) show that most confidence interval procedures are invalid for small sample sizes, even when assumptions such as stationarity and normality are satisfied. Constructing valid confidence intervals from short stationary series is the motivation for weighted batch means (Bischak, Kelton and Pollock, 1993). However, their derivation assumes the process is in steady state.

The ability to analyze short series is particularly valuable when simulating for long periods of time is costly. When the cost of simulating is high, parallelization is a simple and direct method of exploiting parallel processors to obtain results in shorter "wall-clock" time. In this context, two ways in which the problem of confidence interval estimation can arise are:

(1) Different simulation models may be independently executed on separate processors; confidence intervals may be periodically computed for each model and particular systems may be then selected or rejected on the basis of the results.

(2) Independent replications of the same simulation model may be executed on parallel processors, and the results of each simulation run are then combined to form a single estimate and confidence interval.

While the above situations lend themselves readily to parallel simulation, similar situations also occur when simulating on single-processor computers.

This paper investigates the conditional maximum likelihood estimator for the mean of autoregressive processes. Based on two expressions for its asymptotic distribution, we derive asymptotically valid confidence intervals procedures for both single-replication and multiple-replication experiments. These confidence intervals are often superior to those obtained by existing methods in short series with significant initialization bias. In long series, the estimator reduces to the sample mean, and the confidence intervals are asymptotically equivalent to those of existing procedures. Thus the new procedure applies to simulations of any length.

2 AUTOREGRESSIVE MODEL

The presence of transient behavior necessarily indicates that the process is autocorrelated, that is, sequential observations are related to previous observations. The principal distinction between transient and steady state observations is that initial observation are typically more extreme than steady state observations, and they tend to be systematically so (e.g., initializing a queuing simulation empty and idle). To study the transient response, we therefore require a model of the process which captures this dependence. While no single model is able to capture every situation, the linear autoregressive process is widely used to approximate the autocorrelation structure of a wide range of discrete-time stochastic processes. Further, autoregressive processes exhibit the sum-of-exponentials transient mean characteristic of many short-memory stochastic processes. Processes for which transient response has been theoretically shown to be bounded by sum-of-exponential functions include ARMA processes (Fuller, 1976), continuous time Markov chains (Glynn, 1984) and M/M/k queuing systems (Odoni and Roth, 1983).

The autoregressive model assumes that each new observation X_t is linearly related to previous observations with the addition of a random "shock" or innovation. An autoregressive process of order p, AR(p), is represented as follows:

$$X_t = \phi_0 + \sum_{i=1}^{p} \phi_i X_{t-i} + \varepsilon_t, \; t = 1, 2, \ldots \qquad (1)$$

where p is finite and $\{\varepsilon_t, t = 1, 2, \ldots\}$ is a sequence of independent and identically distributed random variables with mean 0 and variance $\sigma_\varepsilon^2 < \infty$. The value $\phi = \{\phi_0, \ldots, \phi_p\}$ is a vector of autoregressive coefficients. Provided that the mean system response converges to a unique limiting steady state, the steady state mean response $\mu = \mu(\phi)$ is:

$$\mu(\phi) = \lim_{t \to \infty} E[X_t] = \phi_0 \left(1 - \sum_{i=1}^{p} \phi_i\right)^{-1}. \qquad (2)$$

Although there may exist a value for (2), it is not necessarily true that mean system response will converge to a unique finite value. For example, consider a queue where customers arrive slightly more rapidly than they can be served, in which case the sequence of time-in-system for each customer will diverge. While this will become apparent after many observations, it may not be immediately apparent from the first customers that the server cannot keep up with demand. If it is not known in advance that the server can keep up with demand, it may be necessary to empirically establish the existence of a steady state mean before estimating its value.

Taking the expectation of (1) yields the linear difference equation $EX_t - \sum \phi_i EX_{t-i} = 0$. Solving this equation yields the sequence of expected values, called the transient expectation function:

$$EX_t = \frac{\phi_0}{1 - \sum_{i=1}^{p} \phi_i} + \sum_{i=1}^{p'} \sum_{j=1}^{n_i} c_{ij} t^j r_i^{-t}. \qquad (3)$$

where c_{ij} are constants determined by the initial p observations, and $\{r_1, \ldots, r_p\}$ are the roots to the characteristic equation $1 - \sum_{i=1}^{p} \phi_i x^i = 0$. Since the roots are not necessarily distinct, we let n_i denote the multiplicity of root r_i. From (3), we can see that the transient response decays exponentially to zero for large t if and only if $|r_i| > 1, i = 1, \ldots, p$; otherwise, the transient response grows exponentially. This stationarity condition can also be written as:

$$1 - \sum_{i=1}^{p} \phi_i x^i = 0 \quad \forall |x| \le 1 \qquad (4)$$

3 ESTIMATION OF STEADY STATE MEAN

Since we assume initialization effects may be large, we must estimate the autoregressive model parameters without assuming that the process is stationary. We choose here the method of maximum likelihood, conditioning on the first p observations. For the purpose of deriving the maximum likelihood estimator and its distribution, the order of autoregression p is assumed known. In practice, it is also necessary to identify the order of autoregression; however, to establish consistency of the estimator, it is sufficient to assume that the estimators of the autoregressive order and coefficients are consistent (Voss, Haddock, and Willemain, 1996).

Let $\hat{\phi}$ be an estimate of the vector of autoregressive coefficients ϕ. We can define an estimate for the mean $\hat{\mu} = \mu(\hat{\phi})$ as follows:

$$\mu(\hat{\phi}) = \begin{cases} \hat{\phi}_0\left(1 - \sum_{j=1}^{p} \hat{\phi}_j\right)^{-1}, & if\ 1 - \sum_{i=1}^{p} \hat{\phi}_i x^i \neq 0 \quad \forall |x| \leq 1 \\ undefined, & otherwise \end{cases} \tag{5}$$

Note that this estimator does not require the assumption that a unique limiting steady state response exists. If the estimated autoregressive parameters satisfy the stationarity condition, then the estimate is defined and finite. On the other hand, if the estimated parameters fall outside the stationary region, system response is to all indications nonstationary in mean or covariance, and no estimate of steady state is meaningful. If $\hat{\phi}$ is a consistent estimator of ϕ, and $\{X_t\}$ converges to a covariance stationary process, then $\hat{\mu}$ is defined almost surely and is a consistent estimator of μ (Voss, Haddock, and Willemain, 1996).

We can compute the conditional least squares estimate of ϕ as follows (Fuller, 1976):

$$\hat{\phi} = A_n^{-1} v_n \tag{6}$$

where

$$A_n = \frac{1}{n-p} \sum_{t=p+1}^{n} \underline{X}_t{}' \underline{X}_t$$

$$\underline{X}_t = (1, X_{t-1}, \ldots, X_{t-p})$$

$$v_n = \frac{1}{n-p} \sum_{t=p+1}^{n} \underline{X}_t{}' X_t,$$

Provided that the stationarity condition (4) is satisfied, this estimator is asymptotically normally distributed:

$$n^{1/2}(\hat{\phi} - \phi) \rightarrow N\left(0, A^{-1}\sigma_\varepsilon^2\right), \tag{7}$$

where $A = \underset{n \to \infty}{p\ lim}\ A_n$. The elements of A are given by

$$a_{ij} = \begin{cases} \gamma_X(|i-j|) + \mu^2; & i,j=2,3,\ldots,p+1 \\ 1; & i=j=1 \\ \mu; & otherwise \end{cases}, \tag{8}$$

where $\gamma_X(k)$ is the process autocovariance at lag k. The asymptotic properties for the linear least squares estimators also apply to estimates obtained by any method which is asymptotically equivalent to the method of maximum likelihood, such as solution of the Yule-Walker equations, which differ from the linear least squares estimates by a quantity which approaches zero in probability as n increases (Fuller, 1976).

When model (1) holds and the innovations ε_t are normally distributed, then $\hat{\phi}$ is the conditional maximum likelihood estimate of ϕ. Further, wherever $\mu(\hat{\phi})$ is defined, it is the maximum likelihood estimate of μ. This follows from the invariance principle, since the mapping $\mu: \mathfrak{R}^{p+1} \rightarrow \mathfrak{R}$ defines a continuous interval on $\mathfrak{R}$, and $\mu(\hat{\phi})$ is locally differentiable with respect to ϕ. For this reason, we shall refer to $\hat{\mu}$ as the MLE.

Voss, Haddock and Willemain (1996) show that the estimator $\hat{\mu}$ is related to other estimators in the literature. In particular, $\hat{\mu}$ is equivalent to the weighted batch mean Bischak, Kelton, and Pollock (1993) when all the data are grouped into a single batch, and is the minimum-variance instance of the weighted batch mean. However, Bischak, et al. assume that the process $\{X_t\}$ is in steady state and that any initial transient observations have been deleted, and they use a different consistent estimator of ϕ. $\hat{\mu}$ can readily be computed for autoregressive processes of any order, and reduces to the GLS estimator derived by Snell and Schruben (1984) for the AR(1) process.

For autoregressive processes, we expect $\hat{\mu}$ to be nearly unbiased regardless of the initial observations. If the autocorrelation structure of the process is known, then $\hat{\mu}$ is exactly unbiased and has the same variance as the sample mean. If model (1) holds, and $\phi_1, \ldots, \phi_p$ are known and satisfy the stationarity condition of (4), then the MLE is exactly unbiased for the mean, $E(\hat{\mu}) = \mu$ $\forall \{X_1, \ldots, X_p\}$ and $Var(\hat{\mu}) = \sigma_\varepsilon\left[1 - \sum_{i=1}^{p} \phi_i\right]^{-2}$.

4 CONFIDENCE INTERVALS FOR STEADY STATE MEAN

In this section, we develop confidence intervals for the steady state for both single-replication and multiple-replication experiments. In order to develop a confidence interval procedure based on $\hat{\mu}$, we need to determine its distribution. We first show that the MLE is asymptotically normal, which suggests a single-replication interval similar to Fishman (1978) and a multiple-replication procedure very much like that of the traditional replication-deletion method. Next we show that the MLE is asymptotically distributed as the index of a bivariate normal random variable, which suggests a more complex interval procedure which is likely more accurate in short series.

Single-replication intervals

LEMMA 1:

Let $\bar{X}_{n-p} = (n-p)^{-1}\sum_{t=p+1}^{n} X_t$ denote the p-truncated sample mean. Assume $\{X_t\}$ converges to a covariance stationary process satisfying (1). If $\hat{\mu}_n$ is the conditional least squares estimate (5) where p is fixed and finite, then $\hat{\mu}_n$ converges to the sample mean in probability

$$n^{1/2}(\hat{\mu}_n - \bar{X}_{n-p}) \to 0 \qquad (9)$$

Proof: We can rewrite () as the sample mean plus a bias correction:

$$\hat{\mu}_n = \bar{X}_{n-p} + \frac{\sum_{i=1}^{p} \hat{\phi}_i \left(\sum_{t=p-i+1}^{p} X_t - \sum_{t=n-i+1}^{n} X_t \right)}{(n-p)\left(1 - \sum_{i=1}^{p} \hat{\phi}_i\right)} \qquad (10)$$

The bias correction is a finite sum of less than $2p^2$ terms of X_t. Since X_t is covariance stationary for large t with finite variance, it satisfies the stationarity condition (4), so the $1 - \sum_{i=1}^{p} \hat{\phi}_i$ term in the denominator converges to a nonzero quantity in probability. Further, since the bias correction term is a finite sum of variables with finite mean and variance, and is divided by $n-p$, it converges to zero in probability as $n \to \infty$. Therefore, $n^{1/2}(\hat{\mu}_n - \bar{X}_{n-p}) \to 0$ in probability.

Corollary:

If (1) holds and satisfies the stationarity condition (4), and $\hat{\mu}_n$ is the maximum likelihood estimate of μ, then $\hat{\mu}$ is also asymptotically normal, $n^{1/2}(\hat{\mu}_n - \mu) \to N\left(0, \sigma_\varepsilon^2 \left[1 - \sum_{i=1}^{p} \phi_i\right]^{-2}\right)$. This follows from Equation (1) and from Yuan and Nelson, (1994), who note that $n^{1/2}(\bar{X}_n - \mu) \to N\left(0, \sigma_\varepsilon^2 \left[1 - \sum_{i=1}^{p} \phi_i\right]^{-2}\right)$.

Since the MLE has the same asymptotic variance as the sample mean and is based on an autoregressive model of the output process, a single-replication confidence interval similar to that of Fishman (1978) and Yuan and Nelson (1994):

$$\hat{\mu}_n \pm t_{1-\alpha/2,\hat{f}} \sqrt{\hat{Var}(\hat{\mu}_n)} \qquad (11)$$

where $t_{1-\alpha/2,\hat{f}}$ denotes the corresponding quantile of the student-t distribution, and $\hat{Var}(\hat{\mu}_n) = \hat{\sigma}_\varepsilon^2 \left[1 - \sum_{i=1}^{p} \hat{\phi}_i\right]^{-2} / (n-p)$. An estimate for the degrees of freedom is

$$\hat{f} = \frac{n(1 - \sum \hat{\phi}_i)}{(1+2p)(1 - \sum \hat{\phi}_i) + 4\sum i\hat{\phi}_i}. \qquad (12)$$

The approximate estimator (12) is specifically derived for short series. However, in simulation experiments of AR(1) processes with $\phi_1 = 0.95$ and $n = 128$, this often yielded estimates of less than 1.0, leading to confidence intervals with tails heavier than even the Cauchy distribution. Here, we arbitrarily restrict $\hat{f} \geq 2$.

Since the MLE is the ratio of two random variables, $\hat{\phi}_0$ and $(1 - \sum \hat{\phi}_i)$, and the variability of the denominator may be considerable at small sample sizes, a more accurate solution may be derived as follows. For convenience, let us write $\hat{\mu}_n = \hat{W}_{n,1} / \hat{W}_{n,2}$ where

$$\boldsymbol{W} = \begin{bmatrix} W_1 \\ W_2 \end{bmatrix} = \begin{bmatrix} \phi_0 \\ 1 - \sum_{i=1}^{p} \phi_i \end{bmatrix}$$

and $\hat{\boldsymbol{W}}_n$ is defined similarly in terms of $\hat{\phi}_n$. From (7), we can see that $n^{1/2}(\hat{\boldsymbol{W}}_n - \boldsymbol{W}) \to N(0, \Sigma)$, since finite sums of asymptotically normal variables are themselves asymptotically normal. Therefore $\hat{\mu}_n$ is the index of an asymptotically bivariate normal random variable. Let $\boldsymbol{B} = \boldsymbol{A}^{-1}\sigma_\varepsilon^2$ with elements $\{b_{ij}\}$ be the limiting covariance matrix of $n(\hat{\phi} - \phi)$. We can write the limiting covariance matrix Σ as:

$$\Sigma = \begin{bmatrix} \sigma_1^2 & \rho\sigma_1\sigma_2 \\ \rho\sigma_1\sigma_2 & \sigma_2^2 \end{bmatrix} = \begin{bmatrix} b_{11} & -\sum_{i=2}^{p+1} b_{i1} \\ -\sum_{i=2}^{p+1} b_{i1} & \sum\sum_{i,j=2}^{p+1} b_{ij} \end{bmatrix} \qquad (13)$$

The following result is based on Fieller (1932), who derived the distribution of the ratio of two correlated noncentral normal random variables. For a bivariate normal rv with mean $\sqrt{n}\boldsymbol{W}$ and covariance matrix Σ, the distribution of the index is asymptotically:

$$g_n(y) = \frac{k_1}{2(k_2 a)^{3/2}} \left(2\sqrt{k_2 a}\, exp((-k_2 c)) + k_2 b \sqrt{\pi}\, exp\left(-\frac{k_2}{4a}(4ac - b^2)\right) erf\left(\frac{k_2 b}{2\sqrt{k_2 a}}\right) \right) \qquad (14)$$

where

$$a = n\left[\left(\frac{y}{\sigma_1}\right)^2 - 2\rho\left(\frac{y}{\sigma_1}\right)\left(\frac{1}{\sigma_2}\right) + \left(\frac{1}{\sigma_2}\right)^2\right]$$

$$b = 2\sqrt{n}\left[\left(\frac{y}{\sigma_1}\cdot\frac{W_1}{\sigma_1}\right) - \rho\left(\frac{y}{\sigma_1}\cdot\frac{W_2}{\sigma_2} + \frac{W_1}{\sigma_1}\cdot\frac{1}{\sigma_2}\right) + \left(\frac{W_2}{\sigma_2^2}\right)\right]$$

$$c = \left[\left(\frac{W_1}{\sigma_1}\right)^2 - 2\rho\left(\frac{W_1}{\sigma_1}\cdot\frac{W_2}{\sigma_2}\right) + \left(\frac{W_2}{\sigma_2}\right)^2\right]$$

$$k_1 = n\left(2\pi\sigma_1\sigma_2\sqrt{1-\rho^2}\right)^{-1}$$

$$k_2 = \frac{1}{2(1-\rho^2)}$$

and $G_n(y) = \int_{-\infty}^{y} g_n(t)\,dt$.

Let $f_n(y|\phi)$ denote the true finite-sample probability density function of $\hat{\mu}_n$. We can approximate $f_n(y|\phi)$ by $g_n(y|\phi)$. Similarly, we can approximate the cumulative distribution function, $F_n(y|\phi)$, by $G_n(y|\phi)$. For any fixed y, $g_n(y|\phi) \to f_n(y|\phi)$ and $G_n(y|\phi) \to F_n(y|\phi)$.

If $\{\hat{\phi}, \hat{\sigma}_\varepsilon^2\}$ are the maximum likelihood estimates, and $\hat{\mu}$ is defined, the finite sample distribution of $\hat{\mu}$ may be approximated as follows. An estimate of the covariance matrix, $\boldsymbol{B}$, is $\hat{\boldsymbol{B}} = \hat{A}^{-1}\hat{\sigma}^2$ where the elements of $\hat{A}$ are

$$\hat{a}_{ij} = \begin{cases} \tilde{\gamma}_X(|i-j|) + \hat{\mu}^2; & i,j=2,3,\ldots,p+1 \\ 1; & i=j=1 \\ \hat{\mu}; & \text{otherwise} \end{cases} \quad (15)$$

Estimates of the autocovariances, $\tilde{\gamma}(i)$, may be determined by solving the system of $p+1$ linear equations: $\tilde{\gamma}(i) - \hat{\phi}_1\tilde{\gamma}(|i-1|) - \ldots - \hat{\phi}_p\tilde{\gamma}(|i-p|) = 0$, $i = 1, \ldots, p$ and $\tilde{\gamma}(0) = \hat{\sigma}_\varepsilon^2$. The estimated covariance matrix $\hat{\boldsymbol{B}}$ may be substituted into (13) to obtain the estimated covariance matrix $\hat{\Sigma}$ and the estimated probability density and cumulative distribution functions, $\hat{g}_n(y)$ and $\hat{G}_n(y)$, respectively. This leads to the following $1-\alpha$ asymptotically valid confidence internal for μ:

$$\left[\hat{\mu}_n - \left(\hat{G}_n^{-1}\left(1-\frac{\alpha}{2}\right) - \hat{\mu}_n\right), \hat{\mu}_n + \left(\hat{\mu}_n - \hat{G}_n^{-1}\left(\frac{\alpha}{2}\right)\right)\right] \quad (16)$$

Multiple-replication intervals

In multiple-replication experiments, we can let the data suggest the sampling distribution of $\hat{\mu}$. Consider R independently-seeded replications of the same simulation model running either in parallel on R processors, or in sequence on a single processor. Further assume that for each replication, the same number of observations are collected, avoiding the issue of ratio bias (Glynn and Heidelberger, 1992). For each replication, we can identify and estimate an autoregressive model and obtain an estimate for the mean, $\hat{\mu}(j) \quad j = 1, \ldots, R$. Since $\hat{\mu}$ is asymptotically normal, this suggests the following confidence interval for μ:

$$\bar{\hat{\mu}} \pm t_{1-\alpha/2, R-1}\sqrt{\hat{Var}(\bar{\hat{\mu}})} \quad (17)$$

where $\bar{\hat{\mu}} = R^{-1}\sum_{j=1}^{R}\hat{\mu}(j)$ and $\hat{Var}(\bar{\hat{\mu}}) = R^{-1}(R-1)^{-1}\sum_{j=1}^{R}(\hat{\mu}(j) - \bar{\hat{\mu}})^2$. This is similar to the traditional confidence interval for the method of independent replications (see Law, 1991) except that it is based on the MLE $\hat{\mu}$ instead of the sample mean $\bar{X}$.

5 NUMERICAL EXPERIMENTS

The performance of the proposed estimator and confidence interval procedure is demonstrated empirically in a series of numerical experiments for the queuing network studied by Schruben (1982). The sequence of the number of entities in the system just before each customer arrival is shown in Figure 2 for two independent replications. On this graph is also superimposed the steady state mean, which is 22.40 ± 0.12 with 95% confidence, based on 100,000 observations. We see here that in short series, the initial transient period can be a substantial portion of the available data.

We compare the proposed procedure with the method of unweighted batch means (UBM) (see Law and Kelton, 1991) and weighted batch means (WBM). In single-replication experiments, MLE (F) indicates intervals based on Fieller's distribution (16) and MLE (N) indicates intervals based on the normal distribution (11); in multiple-replication experiments, intervals are computed by (17). We evaluate estimators on the basis of mean error (bias) and root-mean-square-error, and confidence intervals on the basis of coverage and mean interval width

FIGURE 1. Number of customers in system vs. time

In UBM, the series is optimally truncated to minimize the mean square error of the sample mean, as in Snell and Schruben (1984). For the series in Figure 1, the optimal truncation point is 36 when initialized empty and idle, and 0 when initialized in steady state. The entire observed series is always used for MLE and WBM since these are robust to initialization bias. For UBM and WBM, the batch size is 8 when $n = 128$ and 16 when $n \geq 256$. The autoregressive order is estimated by minimizing the finite-sample information criterion (FIC) (Broerson and Wensink,1993). The maximum candidate order p_{max}, is one unit less than the batch size since this is the largest value which can be chosen with WBM; p_{max} is therefore 7 when $n = 128$ and 15 when $n \geq 256$.

The output analysis procedures and process (1) simulation are written in C++ and compiled using the SparcWorks on a Sun workstation. Simulations are conducted under SIMAN IV on an IBM RS6000 workstation, and the results are exported to a text file for analysis.

Single replication Intervals:

In these experiments, series of length $n = \{128, 256, 512\}$ are generated for 1000 super-replications, and series of length $n = \{8192, 16384\}$ are generated for 200 super-replications. Table 1 summarizes performance of the estimators and interval procedures for single-replication experiments. First, in short simulations when the process is initialized empty and idle, MLE and WBM exhibit nearly the same bias, and are much less biased than even the optimally truncated sample mean in UBM, though MLE and WBM also exhibit considerably higher root-mean-square error than UBM. Second, in short samples initialized at steady state, all three estimators are practically unbiased, though MLE and WBM exhibit larger root-mean-square error for short series. Third, for large samples, all three estimators, UBM, WBM and MLE perform nearly identically in both mean error and root mean square error regardless of initial conditions. While optimal truncation can greatly minimize the root-mean-square error of the sample mean in short series, enough residual transient response remains that the truncated sample mean is still somewhat biased.

In addition to an estimate of the steady state mean, we also require an accurate assessment of how accurate the estimator is. Table 1 shows coverage frequencies and mean widths for nominally 90% confidence intervals. In short series initialized empty and idle, MLE(F) and MLE(N) yield coverages equal or very close to nominal coverages for all run lengths studied, even for very short run lengths $n = 128$. Further, MLE yields coverages which are much closer to nominal coverage than those obtained by UBM and WBM. In short series which are initialized from the steady state distribution, none of the three interval procedures attain nominal coverages, but MLE(F) and MLE(N) again yield coverages closer to

TABLE 1. Estimator Performance — single replication experiments

		Mean Error			Root Mean Square Error			Coverage Frequency				Mean Interval Width			
N	Initialized	UBM	WBM	MLE	UBM	WBM	MLE	UBM	WBM	MLE (F)	MLE (N)	UBM	WBM	MLE (F)	MLE(N)
128	Empty	-4.8	0.30	0.26	5.8	15.	15.	25%	71%	90%	91%	2.9	7.2	13.	12.
256	Empty	-2.1	-0.80	-0.80	3.1	2.4	2.4	40%	81%	93%	95%	2.0	4.3	7.1	7.2
512	Empty	-0.88	-0.48	-0.48	1.8	1.9	1.9	62%	81%	89%	89%	1.8	2,8	3.5	3.8
8192	Empty	-0.048	-0.036	-0.035	0.35	0.35	0.35	89%	89%	88%	88%	0.62	0.62	0.54	0.55
16392	Empty	-0.046	-0.039	-0.038	0.24	0.24	0.24	89%	89%	89%	89%	0.43	0.43	0.43	0.43
128	SSD	-0.32	0.28	0.29	2.7	4.0	3.9	56%	73%	77%	78%	2.0	3.9	5.3	5.1
256	SSD	-0.10	0.20	0.23	2.0	2.9	2.6	57%	78%	80%	81%	1.5	3.0	3.8	3.9
512	SSD	0.019	0.14	0.14	1.4	1.5	1.5	67%	81%	82%	84%	1.5	2.2	2.3	2.4
8192	SSD	-0.040	-0.025	-0.025	0.30	0.30	0.30	92%	91%	93%	93%	0.60	0.60	0.54	0.55
16392	SSD	-0.0057	-0.0029	-0.0027	0.24	0.24	0.24	91%	91%	90%	90%	0.43	0.43	0.43	0.43

nominal than those of UBM and WBM. In long series with $n \geq 8192$, all intervals yield coverages nearly equal to nominal. In every case, WBM also yields coverages closer to nominal than UBM, though MLE yields coverages closer to nominal than WBM.

The improved coverage of MLE(F) and MLE(N) come at the expense of wider intervals, just as the improved bias comes at the expense of increased root-mean-square error. While for very long series all the interval widths are nearly equal, for very short series, MLE(F) and MLE(N) yield intervals somewhat wider than WBM and much wider than UBM. However, this appears to be an unavoidable result of the short series. Interval-width is a valid measure of relative performance only among those methods which achieve nominal, or at least superior, coverage. Lastly, for this example, the simpler MLE(N) yields intervals nearly as accurate as MLE(F) on each of the four performance measures.

Multiple-Replication Intervals

Table 2 summarizes the performance of UBM, WBM, and MLE for multiple-replication confidence interval procedures. In these experiments, the total number of observations, N, is held constant, and are obtained by running R independent replications of length $n = N/R$. Intervals are computed as in (17), except for $R = 1$ where intervals are obtained using the corresponding single-replication procedures. Statistics are collected over 200 super-replications. Because of difficulty in generating and storing simulation results, these results are based on simulations of an AR model identified and estimated from an observed series of 100,000 observations of the model in Figure 1.

When all independent replications are initialized from the steady state distribution, all three estimators, UBM, WBM and MLE, result in confidence intervals with nominal coverage and comparable interval widths, as expected. However, initializing all replications from the steady state distribution is usually only possible by truncating excessive numbers of observations.

When each independent replication is initialized empty and idle, the MLE emerges as the superior estimator. As the available observations consist of more replications of shorter series, the optimally-truncated sample mean in UBM gradually becomes more biased, and the coverage of nominally 90% intervals based on the sample mean deteriorates rapidly. However, parallel intervals based on the MLE and WBM are much more robust to initialization bias. For $R = 32$, 90% intervals based on UBM yield coverages of 72%, whereas MLE and WBM yield about 86%. Further, the interval widths for all methods are nearly equal, regardless of whether the series are initialized empty or in steady state, or the number of replications into which available data is distributed. Even more extreme, for $R = 64$, nominally 90% intervals based on UBM yield coverages of 0%, whereas those based on MLE and WBM yield about 80%. Note again that for relatively few replications of long series, all methods yield nearly identical results, regardless of initial conditions.

These results indicate that when running multiple replications, there is little harm in using the MLE as the point estimator, and that in highly replicated series, there can be significant advantage. One would do almost as well using the WBM, but the MLE performs somewhat better for the example considered here.

6 SUMMARY

These numerical experiments demonstrate that the proposed estimator and distribution are robust to transient

TABLE 2. Estimator Performance — multiple replication experiments

		Mean Error			Root Mean Square Error			Coverage Frequency			Mean Interval Width		
N	Initialized	UBM	WBM	MLE	UBM	WBM	MLE	UBM	WBM	MLE	UBM	WBM	MLE
128	Empty	-.046	-.039	-.039	0.25	0.24	0.24	89%	89%	89%	0.43	0.43	0.39
256	Empty	-0.12	-0.025	-0.023	0.27	0.25	0.24	84%	88%	89%	0.41	0.41	0.41
512	Empty	-0.26	-0.074	-0.074	0.36	0.26	0.26	72%	87%	86%	0.41	0.41	0.41
8192	Empty	-1.33	-0.29	-0.27	1.37	0.87	0.76	0%	80%	81%	0.56	0.81	0.75
16392	Empty	-1.3	-0.36	-0.33	1.3	0.50	0.45	0%	64%	65%	0.39	0.52	0.48
128	SSD	-0.0056	-0.0030	-0.0027	0.24	0.24	0.24	90%	91%	90%	0.43	0.44	.039
256	SSD	-0.028	-0.016	-0.019	0.23	0.23	0.23	90%	90%	90%	0.40	0.40	0.40
512	SSD	-0.048	-0.032	-0.035	0.22	0.23	0.23	90%	90%	90%	0.39	0.41	0.41
8192	SSD	-0.032	-0.015	-0.015	0.24	0.24	0.24	89%	89%	89%	0.38	0.40	0.40
16392	SSD	-0.018	-0.018	-0.018	0.22	0.23	0.22	90%	90%	90%	0.43	0.43	0.39

behavior resulting from initial conditions, yielding estimates and confidence intervals which are comparable or superior to those of unweighted batch means and weighted batch means, even for extreme initializations. The confidence interval procedures presented here permit rapid evaluation of complex systems on the basis of steady state mean in short simulations. They can be usefully applied for providing decision support in evaluating many systems in real-time, where the cost of simulating long series of observations is prohibitively expensive. An areas for further investigation is rapid ranking and selection based on the MLE.

ACKNOWEDGMENTS

This research was partially supported by NASA under grant NAG-1-1469, and by the Department of Decision Sciences and Decision Systems, Rensselaer Polytechnic Institute.

REFERENCES

Bischak, D. P., W. D. Kelton, and S. M. Pollock. 1993. Weighted Batch Means For Confidence Intervals In Steady-State Simulations. Management Science, 39, 1002-1019.

Broerson, P. M. T., and H. E. Wensink. 1993. On Finite Sample Theory for Autoregressive Model Order Selection. IEEE Transactions on Signal Processing, 41, 194-204.

Fieller, E. C. 1932. The Distribution of the Index in a Normal Population. Biometrika, 24, 428-440.

Fishman, G. S. 1978 Principles of Discrete Event Simulation. John Wiley and Sons, New York.

Fuller, W. A. 1976. Introduction to Statistical Time Series. John Wiley and Sons, New York.

Glynn, P., and P. Heidelberger. 1992. Experiments with Initial Transient Deletion for Parallel, Replicated Steady State Simulations. Management Science, 38, 400-419.

Kelton, W. D., and A. M. Law. 1985. The Transient Behavior of the M/M/s Queue, with Implications for Steady-State Simulation. Operations Research, 33, 378-396.

Law, A. M. 1977. Confidence Intervals in Discrete Event Simulation: A Comparison of Replications and Batch Means. Naval Logistics Review Quarterly, 24, 667-678.

Law, A. M., and W. D. Kelton. 1991. Simulation Modeling and Analysis, 2nd ed. McGraw-Hill, New York.

Murray, J. R., and W. D. Kelton. 1988. The Transient Behavior of the M/E2/2 Queue and Steady-State Simulation. Computers & Operations Research, 15, 357-367.

Odoni, A. R., and E. Roth. 1983. An Empirical Investigation of the Transient Behavior of Stationary Queueing Systems. Operations Research, 31, 432-455.

Pawlikowski, K. 1990. Steady-State Simulation of Queueing Processes: A Survey of Problems and Solutions. ACM Computing Surveys, 22, 123-171.

Roth, E. 1994. The Relaxation Time Heuristic for the Initial Transient Problem in M/M/k Queueing Systems. European Journal of Operational Research, 72, 376-386.

Sargent, R. G., K. Kang, and D. Goldsman. 1992. An Investigation of Finite-Sample Behavior of Confidence Interval Estimators. Operations Research, 40, 898-913.

Schruben, L. 1982. Detecting Initialization Bias in Simulation Output. Operations Research, 30, 569-590.

Snell, M. and L. Schruben. 1984. Weighting Simulation Data to Reduce Initialization Effects. IIE Transactions, 17, 354-363.

Voss, P. A., J. Haddock, and T. R. Willemain. 1996. Estimating Steady State Mean in Short Simulations with Initial Transient. Technical Report 37-96-443. Dept. of Decision Sciences and Engineering Systems, Rensselaer Polytechnic Institute. Troy, New York.

Yuan, M., and B. Nelson. 1994. Autoregressive-output-analysis methods revisited, Annals of Operations Research, 53, 391-418.

AUTHOR BIOGRAPHY

Pieter A. Voss is a Ph.D. Candidate in Operations Research and Statistics at Rensselaer. Research interests include simulation analysis and engineering pedagogy. As instructor of Engineering Economics, he received the Master Teaching Fellowship and Outstanding TA Award. He is president of the RPI Student Chapter of InfORMS. He has industrial experience with the Cedel Group, General Electric, the US Army, and Ingersoll-Rand.

Jorge Haddock is Professor of Industrial Engineering and Operations Research and Director of Undergraduate Programs in the department of Decision Sciences and Engineering Systems at Rensselaer. He received his Ph.D. in Industrial Engineering from Purdue University. His research interests are modeling of production and service systems, and analysis methodology. He received the Outstanding Young Industrial Engineer Award in 1990.

Thomas R. Willemain is Associate Professor of Decision Sciences and Engineering Systems at Rensselaer. He received the BSE from Princeton University and the SM and Ph.D. from Massachusetts Institute of Technology, all in Electrical Engineering. His research interests include time series analysis and forecasting.

Proceedings of the 1996 Winter Simulation Conference
ed. J. M. Charnes, D. J. Morrice, D. T. Brunner, and J. J. Swain

INDEPENDENT VERIFICATION AND VALIDATION: A MISSING LINK IN SIMULATION METHODOLOGY?

James D. Arthur
Richard E. Nance

Systems Research Center
and
The Department of Computer Science
Virginia Tech
Blacksburg, VA 24061

ABSTRACT

Independent verification and validation (IV&V) is a powerful tool that can be used to mitigate the increasing complexities associated with an ever-expanding set of modeling and simulation problems. In this paper we discuss the use of independent V&V within the modeling and simulation community. Literature reviews and conversations with experienced technical managers serve as a basis for our conjecture that (a) validation is the major focus of most modeling and simulation efforts, (b) verification plays only a secondary role, and (c) independent V&V is, for all practical purposes, being ignored. In an effort to raise the awareness of the benefits and applicability of independent V&V within the modeling and simulation community, we describe in a step-by-step fashion the application of independent V&V to one particular life cycle model of a simulation study.

1 INTRODUCTION

Because this paper discusses IV&V as it relates to both software engineering and modeling and simulation, we present definitions of verification and validation based on those found in (Balci 1994). In general, validation is the process of substantiating that the behavior of a model or software system conforms to stated requirements. In the modeling and simulations domain, those requirements are derived from the objectives of the simulation study. Verification, on the other hand, is the process of substantiating that each successive model or software system representation is transformed from one form into another, as intended, with sufficient accuracy. In effect, validation deals with building the *right* model or system; verification deals with building the model or system *right*.

Within the software engineering community the complexity of today's systems is dictating the need for and expanded use of *independent* verification and validation (IV&V). Although standard "in-house" V&V techniques are used during many software development activities, their effectiveness and potential to achieve maximal benefits are often constrained by:

- organizational objectives and concerns, and
- an inability of developers to perform critical analyses without being influenced by their active participation in the development effort.

The discrete event modeling and simulation (M&S) community shares similar goals, and impediments to achieving those goals, in their quest to build simulation models and related software systems (e.g., retrospective analysis, animation generators, etc.) that accurately reflect real-world interactions. Clearly simulation software exhibits similar complexity and criticality characteristics as found in non-simulation software. *Why then does there appear to be no movement, or desire, to introduce independent V&V into the M&S development process?*

While it is not our intent to answer the above question, we do pose it to express a concern and to promote *independent* V&V as an additional "tool" to help cope with the growing demand of increasingly complex simulation studies. In doing so, this paper focuses on substantiating the perceptions that IV&V is being ignored and outlining the potential use and benefits of IV&V in the M&S community.

2 INDEPENDENT VERIFICATION AND VALIDATION: AN OVERVIEW

2.1 Definitions and Objectives

From the software engineering perspective, Lewis (Lewis 1993, p. 7) defines verification as an iterative process aimed at determining whether the product of each step in the development cycle:

- fulfills the requirements levied on it by previous steps, and
- is internally complete, consistent, and sufficiently correct to support the next phase.

Validation, on the other hand, is defined as the process of

executing the software to exercise the hardware and comparing the test results to the specification requirements.

An examination of the above definitions reveals that both verification and validation strongly correlate with those definitions found in the simulation literature (Caughlin 1995, Balci 1994). In particular, M&S software is validated relative to perceived or actual systems behavior, while verification is the activity that ensures that the simulation software accurately reflects the model.

Independent Verification and Validation is defined as a series of technical and management activities performed by someone other that the developer of a system with the objectives of

- improving the quality of that system, and
- assuring that the delivered product satisfies the user's operational needs.

While not totally absent in the simulation literature, see for example (Sargent 1994), references to independent V&V in the simulation literature are rare indeed.

The advantages of an independent IV&V process are many. In particular, the *independence* in V&V

- provides an objective assessment of the product during its creation,
- adds a new analytical perspective not present in the development environment,
- brings its own set of tools and techniques to bear on ensuring development accuracy and validity,
- introduces "intermediate" users of the system who serve as "beta testers" before the product goes to market, and finally
- significantly enhances testing and the discovery of design flaws and coding errors.

2.2 IV&V in the Software Development Life Cycle

Within the conventional software development life cycle, independent V&V is best illustrated in Figure 1.

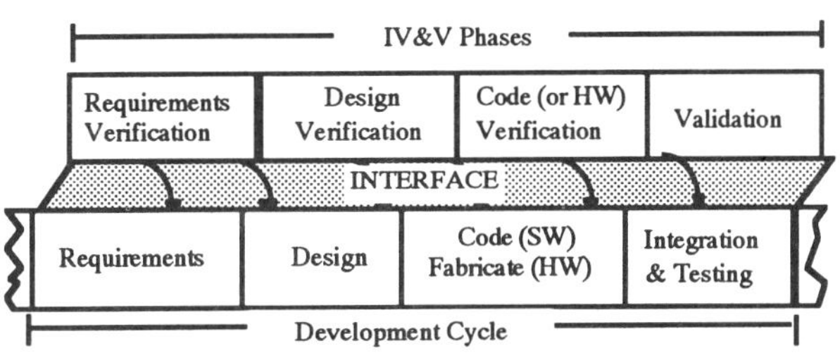

Figure 1: IV&V Interface with Development Cycle

Each of the verification phases and the concluding validation phase begins as soon as possible after initiation of its development counterpart. Development artifacts are examined through phase-specific V&V activities; results of those examinations are then factored back into the development process. In an independent V&V scenario, the organization performing the V&V activities is not part of the development organization. The interface between the development and IV&V organization is defined in a contractual agreement stipulated by the sponsoring organization.

Achieving effectiveness in an independent V&V effort requires that its integration with the software development effort be:

(1) viewed as an overlay process that is *separate* and *distinct* from the development effort,
(2) designed to complement the software development effort and to minimize any adverse impact on the development schedule, and
(3) tailored to fit the application and, as much as possible, the development organization's software development process.

3 INDEPENDENT VERIFICATION AND VALIDATION IN MODELING AND SIMULATION

Clearly we believe that independent V&V is not only desirable, but has much to offer the M&S community. An examination of current literature reveals that V&V is an integral part of most modeling and simulation efforts, but also tends to confirm our suspicions that the use of *independent* V&V methods is severely lacking. We outline the results of that investigation as it relates to the treatment of independent V&V in "theory" and in practice.

3.1 Treatment in Theory

The first part of our investigation focuses on determining the role that *independent* V&V plays in the theoretical domain. That is, we focus on the extent to which IV&V is being touted in simulation texts and research papers emphasizing simulation methodologies. The selected texts focus on topics ranging from simulation modeling and analysis, to those that stress the instruction of simulation methods. The publication dates spanned the time period from 1987 through 1993. Within that group, four of the books lack any treatment of V&V (Cassandras 1993, Karian and Dudewicz 1990, Schruben 1992, Zeigler 1990); three of the remaining four books provide only a cursory introduction to V&V (Bratley, Fox and Schrage 1987, Davies and O'Keefe 1989, Ross 1990), and only one devotes multiple sections to V&V (Law and Kelton 1991). Not surprisingly, none of the eight text books addressed independent verification and validation.

An examination of the last four proceedings from the Winter Simulation Conference (WSC) and several back issues of *Transactions on Modeling and Computer*

Simulation (TOMACS) reveals that many researchers are aware of the necessary role that V&V plays in model development and simulation. Although we found no papers in TOMACS that had V&V as its major focus, several were present in the WSC proceedings, e.g. (Caughlin 1995, Sargent 1994, Balci 1994, Kleindorfer and Ganeshan 1993, Pace 1993, Yucesan and Jacobson 1992, Bailey and Kemple 1992, and Legge and Wyatt 1992). Each paper reiterates the critical need for V&V; several of them describe new and extant methods for implementing V&V practices. Only one author, Sargent (Sargent 1994), advises the use of independent V&V to address the growing complexity of simulation software. More specifically, he states that independent V&V is usually applied to large-cost simulation studies and is most effective when applied during the development process.

3.2 Treatment in Practice

The second aspect of our investigation examines the extent to which independent V&V is being used in practice. Two sources of information are considered: published papers describing practical applications of M&S, and persons actively involved in developing or managing M&S efforts. Papers describing the application of M&S to problems found in industry and government abound in the WSC proceedings. During a cursory examination of several papers, we noted that most of them describe some form of verification and/or validation activities. We then selected eight papers at random to review more closely. We observed that seven of the eight describe some form of verification or validation: two on verification and validation efforts, four on validation only, and one on verification efforts only. Not surprisingly, validation assumes a prominent role in the development of simulation models and software. Verification appears to be relegated to a lesser role. Unfortunately again, we found no papers describing independent V&V being applied to the development of the simulation models and software.

As a final data point we discussed the role of IV&V with a technical manager possessing 25 years of experience working with major DoD software systems, including simulations. His position is that independent V&V should be included as part of the development effort, but in most cases it is not. He further states that one need not necessarily employ an independent organization to perform the V&V. We note that this manager's primary experience with simulation modeling is to produce "scaffolding software" for testing embedded systems software under development. Thus, the developed software is tested by a "simulator/stimulator" that initiates the interfaces needed with the remainder of the system in which the developed system is embedded. Within this type of framework, independence is gained if one group develops the embedded software, one group develops the stimulator/simulator, and a third party provides the real-world data to validate interaction of the two. In reality, however, this approach achieves only independent validation and misses out on those benefits attributed to independent verification, e.g., early error detection and reduced effort to fix those errors.

3.3 In the Final Analysis

Based on the analysis outlined above we conclude the following:

(1) Validation is "alive and well." Most simulation developers are aware of and attempt to validate their models and products relative to real-world expectations.
(2) Verification plays a secondary role in the development process. That is, the prevailing wisdom is that validation is the crucial hurdle and little benefit is recognized in employing verification to assure the desired outcome of the validation process.
(3) *Independent verification and validation is being ignored.*

Why is there such a stratification in the level of use among validation, verification and independent V&V? Clearly, validation is deemed crucial in showing that one's simulation model and software system adequately reflects the behavior of the system under study. We believe that verification plays a reduced role because of the additional costs in terms of time and effort. Moreover, there is some truth to the prevailing wisdom stated in (2) above. Yet, what is the recourse when no real system exists to produce observable behavioral data? We believe that the subjective techniques described in (Balci 1994) must be buttressed by meticulous attention to verification.

Finally, why is the use of independent V&V totally lacking in development efforts encompassing M&S? The circumstances outline below offer some assistance in answering this question.

- The costs and benefits of independent V&V must be judged relative to the size, complexity and critical nature of the product being developed. Because simulation models and systems have exhibited only modest growth in size and complexity over the last years, the demand for independent V&V has not been justifiable.
- Even for those sizable development efforts where independent V&V can produce cost savings, the entrenchment of existing M&S procedures hinder the introduction of new approaches.
- Finally, the real benefits of independent V&V and risk mitigation (i.e., balancing IV&V costs against derived benefits) are only now beginning to surface. This has contributed to the perception that IV&V is cost prohibitive and to a lack of real understanding

as to why, when and how IV&V should be employed.

4 THE BENEFITS OF INDEPENDENT VERIFICATION AND VALIDATION

Underscoring the increasing importance of verification and validation, Caughlin states that "as our ability to model the real-world grows, our ability to verify or validate these models shrinks" (Caughlin 1995, p. 1405). While his proposal to use reduced order metamodels is appealing, we question the extent to which additional detail can be hidden without sacrificing critical information.

4.1 Placing IV&V in Perspective

In "bottom line parlance" we must recognize the necessity and importance of both verification and validation, and be prepared to expend the requisite effort to adequately perform such activities. As noted in Section 3.3 validation is receiving proper attention. Minimally, one should expect validation to be performed because its results confirm the validity of the model and simulation system. On the other hand, verification (i.e., establishing correctness between the model and simulation) appears to be somewhat lacking. As M&S problems grow in size and complexity, the role of verification becomes even more critical. More specifically, the more complex a system is, the higher the probability of introducing an error during development. Verification stresses the early detection of errors, which in turn, translates into reduced development costs and reduced schedule slippage.

In discussing the benefits of independent V&V we first make the following assumptions:

- The problem being addressed is of sufficient size and complexity to warrant the use of independent V&V. In part, this assumption is driven by the fact that the additional costs imposed by IV&V must not be prohibitive. Lewis (Lewis 1993) states that an IV&V effort should be allocated 16%-17% of the development budget. This translates to 3%-4% of the life-cycle costs.
- The model and simulation software is being developed for a customer who is not affiliated with the development organization. In large organizations, however, the customer might be employed by a branch of the organization different from the development branch.
- The IV&V organization possesses expertise in both simulation and the application domain.
- Verification and validation play equally important roles in the M&S development process.

4.2 The Activities and Derived Benefits

To place our discussion in context, we examine the advantages of independent V&V relative to the life-cycle model defined by Balci and Nance (Balci 1994, Nance 1994). That model is illustrated in Figure 2.

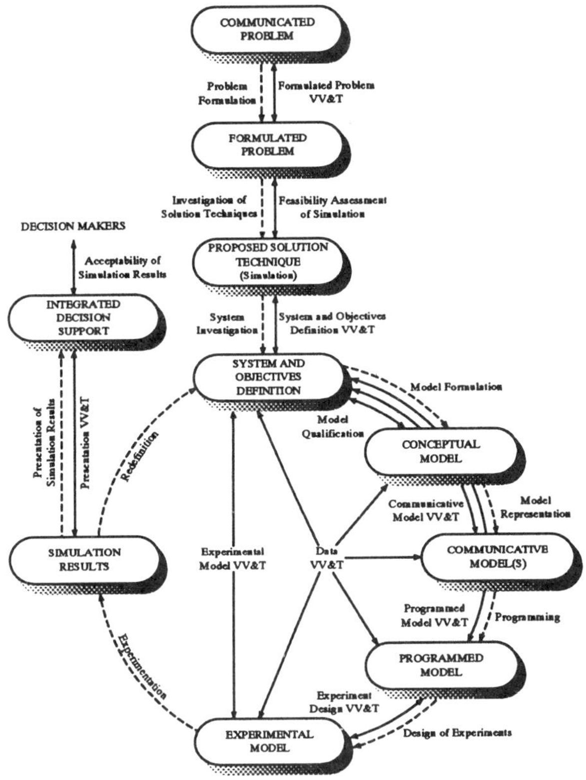

Figure 2: The Life Cycle of a Simulation Study

The first three steps address problem identification, formulation and the feasibility of using simulation in defining a solution. In effect, these activities capture the critical characteristics of the problem and determine to what extent simulation provides a viable solution approach. The independent agent contributes to the process by:

- acting as an "interpreter" between the customer and developer, and thereby minimizing the possibility of mis-communication,
- providing an additional perspective on the problem, and thereby, complementing and augmenting problem identification and formulation,
- recording decisions and details to ensure that they are reflected in the proposed simulation solution,
- contributing to the proposed solution through suggested approaches and alternatives, and
- providing an independent feasibility assessment.

Transitioning from the Proposed Solution Technique (step three) to System and Objectives Definition (step four) is most crucial because it is during this process where the requirements for the model and simulation system are defined. Correspondingly, the responsibilities of the independent agent include:

- verifying that the evolved requirements accurately reflect and address
  - (a) the critical characteristics of the formulated problem, and
  - (b) the proposed solution approach,
- providing feedback to the developer when discrepancies are detected, and
- verifying that the resulting document presents a set of requirements that are complete, consistent and measurable.

Steps five and six address development activities focusing on model conceptualization and physical realization. During each activity, the independent agent is concerned with both validation and verification. In particular, the agent:

- verifies that the conceptual model accurately reflects the requirements derived during System and Objectives Definition,
- verifies that the physical (or communicative) model(s), is an accurate reflection of the conceptual model,
- assesses the validity of the data to be used to validate the conceptual and physical model,
- monitors the validation tests performed by the developer,
- reviews the test results,
- performs additional validation tests as needed, and
- reports verification and validation exceptions to both the customer and to the developer for correction.

The transition from step six to step seven results in a programmed representation of the communicative model. Similar to the activities listed during model development, the independent agent:

- verifies that all elements described by the model are accurately reflected in the programmed representation,
- examines the validity of the data to be used for validating the programmed representation,
- monitors validation tests,
- examines validation test results for inconsistencies and discrepancies,
- when specified, examines the programmed model for adherence to
  - (a) coding standards and practices, and
  - (b) quality criteria, and
- reports all verification and validation exceptions to the customer and to the developer for correction.

Step eight comprises the simulation experiment design. More specifically, it reflects the design of one or experiments that have defined objectives, input data sets, output formats, and an evaluation procedure. The experiments are performed using the programmed model and results are examined to determine how well they correlate with real-world expectations. Primarily, the tasks of the independent agent are focused on verifying and validating the design of the experiments, their execution and result interpretation. Those tasks entail:

- verifying the accuracy of the design relative to specified design objectives,
- verifying that experiment instantiation (or implementation) correctly reflects the design,
- verifying that the parameters of the experiments fall within the boundaries that the programmed model is intended to operate,
- examining the validity of the data to be used to be used in the experiment relative to experiment objectives and expected output,
- monitoring the experiment to ensure that specified procedures are performed,
- assisting in the interpretation and validation of the results, and
- reporting validation results to the customer and developer.

At this point in time both the communicative and experimental models have been validated, i.e., the correct models have been developed. Verification has proceeded in a stepwise fashion to ensure that each intermediate artifact is built correctly. Throughout the development process, the independent agent plays an active role in:

- assessing the validity of each artifact,
- verifying that each artifact is an accurate reflection of its predecessor,
- providing constant feedback to both the customer and developer, and although not explicitly stated,
- ensuring that
  - (a) all deficiencies are corrected, and
  - (b) all changes are appropriately propagated back through previously developed artifacts.

5 SUMMARY AND CONCLUSIONS

Within the modeling and simulation community, verification and validation techniques are being applied during the development effort. Unfortunately, validation appears to play the more prominent role, with verification playing a secondary role, and only cursory attention being given to independent V&V. Independent verification and validation is not intended to replace existing V&V activities. Instead, it is intended to supplement those activities, and to provide an effective way to address the increasing complexities of modeling challenges facing the M&S community. Because of the additional costs, independent V&V is not a viable option for all development efforts. Nonetheless, we ask that the reader consider that as the size and complexity of simulation models increase, so do the benefits of independent V&V. In a brief review of those benefits, we illustrate how independent V&V can assist in building a better product and mitigate costs through early error detection.

REFERENCES

Bailey, M.P. and W.G. Kemple (1992). The Scientific Method of Choosing Model Fidelity. *Proceedings of the 1993 Winter Simulation Conference*, 791-797. Los Angeles, CA.

Balci, O. (1994). Validation Verification, and Testing Techniques Throughout the Cycle of a Simulation Study. *Annals of Operations Research* 53, 121-173.

Bratley, P., B.L. Fox and L.E. Schrage (1987). *A Guide to Simulation.* New York, NY: Springer-Verlag.

Cassandras, C.G. (1993). *Discrete Event Systems - Modeling and Performance Analysis.* Homewood, IL: Aksen Associates Incorporated.

Caughlin, D. (1995). Verification, Validation, and Accreditation (VV&A) of Models and Simulations Through Reduced Order Metamodels. *Proceedings of the 1995 Winter Simulation Conference*, 1405-1412. Arlington, VA.

Davies, R.M. and R.M. O'Keefe (1989). *Simulation Modelling With Pascal.* New York, NY: Prentice Hall.

Karian, A.A. and E.J. Dudewicz (1991). *Modern Statistical, Systems, and GPSS Simulation - The First Course.* New York, NY: Computer Science Press.

Kleindorfer, G.B. and R. Ganeshan (1993). The Philosophy of Science and Validation in Simulation. *Proceedings of the 1993 Winter Simulation Conference*, 50-57. Los Angeles, CA.

Law, A.M. and W. D. Kelton (1991). *Simulation Modeling & Analysis.* New York, NY: McGraw-Hill, Inc.

Legge, G. and D.L. Wyatt (1992). A Software Mechanism To Enhance Simulation Model Validity. *Proceedings of the 1993 Winter Simulation Conference*, 798-806. Los Angeles, CA.

Lewis, R.O. (1993). *Independent Verification and Validation: A Life Cycle Engineering Process for Quality.* New York, NY: John Wiley & Sons, Inc.

Nance, R.E. (1994). The Conical Methodology and the Evolution of Simulation Model Development. *Annals of Operations Research* 53, 1-46.

Pace, D.K. (1993). Naval Modeling and Simulation Verification Validation and Accreditation. *Proceedings of the 1993 Winter Simulation Conference*, 1077-1080. Los Angeles, CA.

Ross, S.M. (1990). *A Course in Simulation.* New York, NY: Macmillan Publishing.

Sargent, R.G. (1994). Verification and Validation of Simulation Models. *Proceedings of the 1994 Winter Simulation Conference*, 77-87. Lake Buena Vista, FL.

Schruben, L.W. (1992). *SIGMA: Graphical Simulation Modeling Program.* San Francisco, CA: The Scientific Press.

Yucesan, E. and S.H. Jacobson (1992). Building Correct Simulation Models is Difficult. *Proceedings of the 1993 Winter Simulation Conference*, 783-790. Los Angeles, CA.

Zeigler, B.P. (1990). *Object-Oriented Simulation with Hierarchical, Modular Models.* - Intelligent Agents and Endomorphic Systems. Boston, MA: Academic Press, Inc.

AUTHOR BIOGRAPHIES

JAMES D. ARTHUR is an Associate Professor of Computer Science at Virginia Tech (VPI&SU). He received B.S and M.A. degrees in Mathematics from the University of North Carolina at Greensboro in 1972 and 1973, M.S. degree in Computer Science from the University of Houston in 1979, and M.S. and Ph.D. degrees in Computer Science from Purdue University in 1981 and 1983. His research interests include Software Engineering (Methods and Methodologies supporting Software Quality Assessment and IV&V Processes), Parallel Computation, and User Support Environments. Dr. Arthur is the author of over 27 papers on software engineering, software quality assessment, IV&V, and user/machine interaction. He has served as: participating member of IEEE Working Group on Reference Models for V&V Methods; an advisor to the U.S. Navy Commonality Working Group; Chair of Education Panel for National Software Council Workshop; Guest Editor for *Annals of Software Engineering* special volume on Process and Product Quality Measurement; and Principal Investigator or Investigator on 14 externally funded research projects totaling in excess of $1.4 million. His instructional expertise includes: (a) Roles, Models and Activities for Software Verification and Validation, (b) Fundamentals of Software Engineering and Product Quality Assessment, and (c) Languages and Language Translation. Dr. Arthur is a member of Pi Mu Epsilon (Math Honor Society), Upsilon Pi Epsilon (Computer Science Honor Society), Golden Key National Honor Society, Sigma Xi (National Research Society), ACM, and the IEEE Computer Society.

RICHARD E. NANCE is the RADM John Adolphus Dahlgren Professor of Computer Science and the Director of the Systems Research Center at Virginia Polytechnic Institute and State University (Virginia Tech). He received B.S. and M.S. degrees from N.C. State University in 1962 and 1966, and the Ph.D. degree from Purdue University in 1968. He has served on the faculties of Southern Methodist University and Virginia Tech, where he was Department Head of Computer

Science, 1973-1979. Dr. Nance has held research appointments at the Naval Surface Weapons Center (NSWC) and at the Imperial College of Science and Technology (UK). Within ACM, he has chaired two special interest groups: Information Retrieval (SIGIR), 1970-71 and Simulation (SIGSIM), 1983-85. He has served as Chair of the External Activities Board and several committees. The author of over 100 papers on software engineering, modeling and simulation, performance evaluation, and computer networks, Dr. Nance has served on the Editorial Panel of the *Communications of the ACM* for research contributions in simulation and statistical computing, 1985-89, as Area Editor for Computational Structures and Techniques of *Operations Research*, 1978-82, and as Department Editor for Simulation, Automation, and Information Systems of *IIE Transactions*, 1976-81. He served as Area Editor for Simulation, 1987-89 and as a member of the Advisory Board, 1989-92, *ORSA Journal on Computing*. He was the founding Editor-in-Chief of the ACM *Transactions on Modeling and Computer Simulation*, 1990-1995. He served as Program Chair for the 1990 Winter Simulation Conference. Dr. Nance received a Distinguished Service Award from the TIMS College on Simulation in 1987. He received an award for Distinguished Service to SIGSIM and the simulation community in 1995. Dr. Nance is a member of Sigma Xi, Alpha Pi Mu, Upsilon Pi Epsilon, ACM, IIE, and INFORMS.

Analysis Methodology

Proceedings of the 1996 Winter Simulation Conference
ed. J. M. Charnes, D. J. Morrice, D. T. Brunner, and J. J. Swain

COMPUTATIONAL STRUCTURE OF A PERFORMANCE ASSESSMENT INVOLVING STOCHASTIC AND SUBJECTIVE UNCERTAINTY

Jon C. Helton

Department of Mathematics
Arizona State University
Tempe, Arizona 85287-1804, U.S.A.

ABSTRACT

A recent performance assessment for the Waste Isolation Pilot Plant (WIPP), which is being developed by the U.S. Department of Energy for the geologic disposal of transuranic waste, is used to illustrate the computational structure of a large analysis that maintains a separation between stochastic (i.e., aleatory) and subjective (i.e., epistemic) uncertainty. In this analysis, stochastic uncertainty arises from the many possible disruptions that could occur over the 10,000 yr regulatory period that applies to the WIPP, and subjective uncertainty arises from the imprecision with which many of the quantities required in the analysis are known. Important parts of the computational structure are (1) the use of Latin hypercube sampling to incorporate the effects of subjective uncertainty, (2) the use of Monte Carlo (i.e., random) sampling to incorporate the effects of stochastic uncertainty, and (3) the efficient use of the necessarily limited number of mechanistic calculations that can be performed to support the analysis.

1 INTRODUCTION

The Waste Isolation Pilot Plant (WIPP) is located in southeastern New Mexico and is being developed by the U.S. Department of Energy (DOE) for the geologic disposal of transuranic (TRU) waste (U.S. DOE 1990). Waste disposal will take place in panels excavated in bedded salt approximately 2000 ft below the land surface (Figure 1).

As part of the development process for the WIPP, a sequence of performance assessments (PAs) has been carried out by Sandia National Laboratories (SNL) to organize knowledge currently available about the WIPP and to provide guidance for future research and development efforts (WIPP PA 1991, 1992; Helton et al. 1993, 1995a, 1996). The structure of these PAs derives from the U.S. Environmental Protection Agency's (EPA's) regulation for the geologic disposal of radioactive waste: 40 CFR 191, Subpart B: *Environmental Radiation Protection Standards for the Management and Disposal of Spent Nuclear Fuel, High-Level and Transuranic Radioactive Wastes* (U.S. EPA 1985, 1993). Assessing compliance with 40 CFR 191 presents an interesting analysis problem due to the need to use detailed numerical models to evaluate potential radionuclide releases from the WIPP and requirements in the regulation for a detailed representation of the effects of different types of uncertainty.

At present (June 1996), the most recent iteration of these PAs is underway and will support an application by the DOE to the EPA for the certification of the WIPP for the disposal of TRU waste. This presentation provides an overview of the computational structure being used in this PA to assess compliance with 40 CFR 191.

2 SUMMARY OF 40 CFR 191, SUBPART B

The following is the central requirement of 40 CFR 191, Subpart B, and the primary focus of this paper:

§ 191.13 Containment requirements.

(a) Disposal systems for spent nuclear fuel or high-level or transuranic radioactive wastes shall be designed to provide a reasonable expectation, based upon performance assessments, that cumulative releases of radionuclides to the accessible environment for 10,000 years after disposal from all significant processes and events that may affect the disposal system shall:
  (1) Have a likelihood of less than one chance in 10 of exceeding the quantities calculated according to Table 1 (Appendix A); and
  (2) Have a likelihood of less than one chance in 1,000 of exceeding ten times the quantities calculated according to Table 1 (Appendix A).

(b) Performance assessments need not provide complete assurance that the requirements of 191.13(a) will be

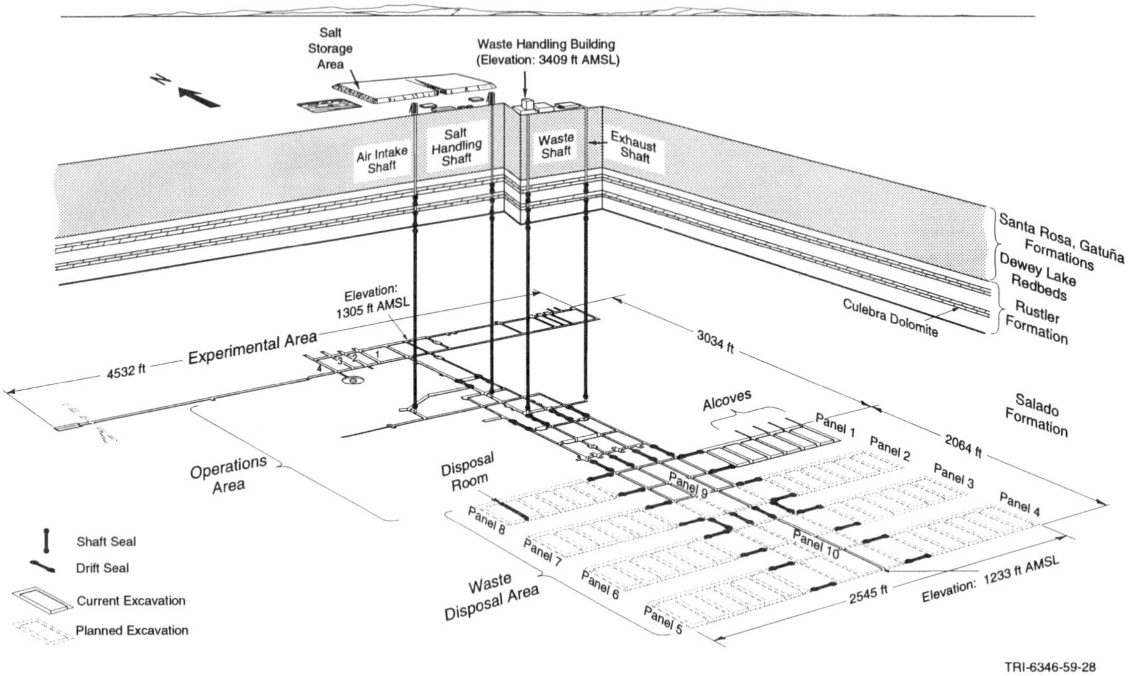

Figure 1: Cross-sectional view of the WIPP (Fig.1-9, Vol. 1, WIPP PA 1991; see Sect. 2.2, Vol. 2, WIPP PA 1992 for detailed stratigraphy)

met. Because of the long time period involved and the nature of the events and processes of interest, there will inevitably be substantial uncertainties in projecting disposal system performance. Proof of the future performance of a disposal system is not to be had in the ordinary sense of the word in situations that deal with much shorter time frames. Instead, what is required is a reasonable expectation, on the basis of the record before the implementing agency, that compliance with 191.13(a) will be achieved.

Containment Requirement 191.13(a) refers to "quantities calculated according to Table 1 (Appendix A)," which means a normalized radionuclide release to the accessible environment based on the type of waste being disposed of, the initial waste inventory, and the release that takes place (App. A, U.S. EPA 1985). Table 1 (Appendix A) of U.S. EPA 1985 specifies allowable releases (i.e., release limits) for individual radionuclides. The WIPP is intended for TRU waste, which is defined to be "waste containing more than 100 nanocuries of alpha-emitting transuranic isotopes, with half-lives greater than twenty years, per gram of waste" (p. 38084, U.S. EPA 1985). Specifically, the normalized release R for transuranic waste is defined by

$$R = \sum_i (Q_i / L_i)\left(1 \times 10^6 \text{ Ci} / C\right), \qquad (1)$$

where Q_i is the cumulative release of radionuclide i to the accessible environment during the 10,000-yr period following closure of the repository (Ci), L_i is the release limit (Ci) for the radionuclide i (Table 1, App. A, U.S. EPA 1985) and C is the amount of TRU waste emplaced in the repository (Ci). For the 1996 WIPP PA, $C = 4.07 \times 10^6$ Ci.

To help clarify the intent of 40 CFR 191, the EPA also published 40 CFR 194, *Criteria for the Certification and Recertification of the Waste Isolation Pilot Plant's Compliance with 40 CFR Part 191 Disposal Regulations; Final Rule* (U.S. EPA 1996). There, the following elaboration of the intent of 40 CFR 191.13 appears (pp. 5242-5243, U.S. EPA 1996):

§ 194.34 Results of performance assessments.

(a) The results of performance assessments shall be assembled into "complementary, cumulative distribution functions" (CCDFs) that represent the probability of exceeding various levels of cumulative release caused by all significant processes and events.

(b) Probability distributions for uncertain disposal system parameter values used in performance assessments shall be developed and documented in any compliance application.

(c) Computational techniques, which draw random samples from across the entire range of the probability distributions developed pursuant to paragraph (b) of this section, shall be used in generating CCDFs and shall be documented in any compliance application.
(d) The number of CCDFs generated shall be large enough such that, at cumulative releases of 1 and 10, the maximum CCDF generated exceeds the 99th percentile of the population of CCDFs with at least a 0.95 probability.
(e) Any compliance application shall display the full range of CCDFs generated.
(f) Any compliance application shall provide information which demonstrates that there is at least a 95 percent level of statistical confidence that the mean of the population of CCDFs meets the containment requirements of § 191.13 of this chapter.

When viewed at a high level, three basic entities underlie the results required in 191.13 and 194.34 and ultimately determine the conceptual and computational structure of the 1996 WIPP PA:

EN1: a probabilistic characterization of the likelihood of different futures occurring at the WIPP site over the next 10,000 yr,
EN2: a procedure for estimating the radionuclide releases to the accessible environment associated with each of the possible futures that could occur at the WIPP site over the next 10,000 yr,
EN3: a probabilistic characterization of the uncertainty in the parameters used in the definition of EN1 and EN2.

Together, EN1 and EN2 give rise to the CCDF specified in 191.13(a) (Figure 2), and EN3 corresponds to the distributions indicated in 194.34(b).

3 PROBABILISTIC CHARACTERIZATION OF DIFFERENT FUTURES

The entity EN1 is the outcome of the scenario development process for the WIPP and provides a probabilistic characterization of the likelihood of different futures that could occur at the WIPP over the next 10,000 yr, with the period of 10,000 yr specified in 191.13(a). When viewed formally, EN1 is defined by a probability space $(\mathcal{S}_{st}, \mathscr{S}_{st}, p_{st})$, with the sample space $\mathcal{S}_{st}$ given by

$$\mathcal{S}_{st} = \{\mathbf{x}_{st}: \mathbf{x}_{st} \text{ is a possible 10,000 yr sequence of occurrences at the WIPP}\}. \qquad (2)$$

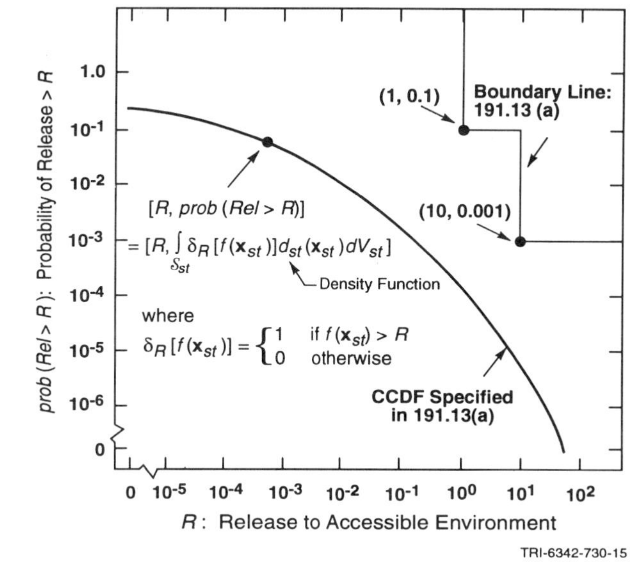

Figure 2: Boundary Line and Associated CCDF Specified in 191.13(a)

The subscript *st* refers to stochastic (i.e., aleatory) uncertainty and is used because $(\mathcal{S}_{st}, \mathscr{S}_{st}, p_{st})$ is providing a probabilistic characterization of occurrences that may take place in the future.

The following guidance (p. 5242, U.S. EPA 1996)

§ 194.32 Scope of performance assessments.

(a) Performance assessments shall consider natural processes and events, mining, deep drilling, and shallow drilling that may affect the disposal system during the regulatory time frame.
(b) Assessments of mining effects may be limited to changes in the hydraulic conductivity of the hydrogeologic units of the disposal system from excavation mining for natural resources. Mining shall be assumed to occur with a one in 100 probability in each century of the regulatory time frame.

and an extensive review of possible disruptions at the WIPP led to drilling intrusions and potash mining being the only occurrences incorporated into the definition of $\mathcal{S}_{st}$. Specifically, the elements $\mathbf{x}_{st}$ of $\mathcal{S}_{st}$ are vectors of the form

$$\mathbf{x}_{st} = [\underbrace{t_1, a_1, b_1, l_1, p_1}_{1^{\text{st}} \text{ intrusion}}, \underbrace{t_2, a_2, b_2, l_2, p_2}_{2^{\text{nd}} \text{ intrusion}}, \ldots, \underbrace{t_n, a_n, b_n, l_n, p_n}_{n^{\text{th}} \text{ intrusion}}, t_{min}] \qquad (3)$$

in the 1996 WIPP PA, where n is the number of drilling intrusions, t_i is the time (yr) of the ith intrusion, a_i designates the type of waste penetrated by the ith intrusion (i.e., no waste, contact-handled (CH) waste, remote-handled (RH) waste), b_i designates whether or not the ith intrusion penetrates pressurized brine in the Castile Formation, l_i designates the location of the ith intrusion, p_i designates the plugging procedure used with the ith intrusion (i.e., continuous plug, two discrete plugs, three discrete plugs), and t_{min} is the time (yr) at which potash mining occurs.

In consistency with the following guidance (p. 5242, U.S. EPA 1996)

§ 194.33 Consideration of drilling events in performance assessments.

(2) In performance assessments, drilling events shall be assumed to occur in the Delaware Basin at random intervals in time and space during the regulatory time frame.
(3) The frequency of deep drilling shall be calculated in the following manner:
  (i) Identify deep drilling that has occurred for each resource in the Delaware Basin over the past 100 years prior to the time at which a compliance application is prepared.
  (ii) The total rate of deep drilling shall be the sum of the rates of deep drilling for each resource.

the drilling rate in the vicinity of the WIPP was determined to be 46.8 intrusions/km^2 10^4 yr, which leads to a rate of

$$\lambda_d = \left(46.8 / \text{km}^2\ 10^4\ \text{yr}\right)\left(0.6285\ \text{km}^2\right)$$
$$= 2.94 \times 10^{-3}\ \text{yr}^{-1} \quad (4)$$

for intrusions into the area (0.6285 km^2) marked by a berm used as part of a passive marker system (Figure 3). Further, 100 yr of active institutional control (§194.41, p. 5243, U.S. EPA 1996) and 600 yr of passive institutional control (§194.43, p. 5243, U.S. EPA 1996) lead to the following time-dependent drilling rate:

$$\begin{aligned}\lambda_d(t) &= 0\ \text{yr}^{-1} \quad 0 \le t \le 100\ \text{yr}\\ &= 2.94 \times 10^{-5}\ \text{yr}^{-1} \quad 100 < t \le 700\ \text{yr}\\ &= 2.94 \times 10^{-3}\ \text{yr}^{-1} \quad 700 < t \le 10000\ \text{yr}. \quad (5)\end{aligned}$$

Drilling intrusions are assumed to be equally likely to occur at each node in Figure 3. Further, the analysis uses specified probabilities for: encountering no waste (0.80), CH waste (0.18), or RH waste (0.02); encountering pressurized brine (0.08); and use of a one (0.02), two

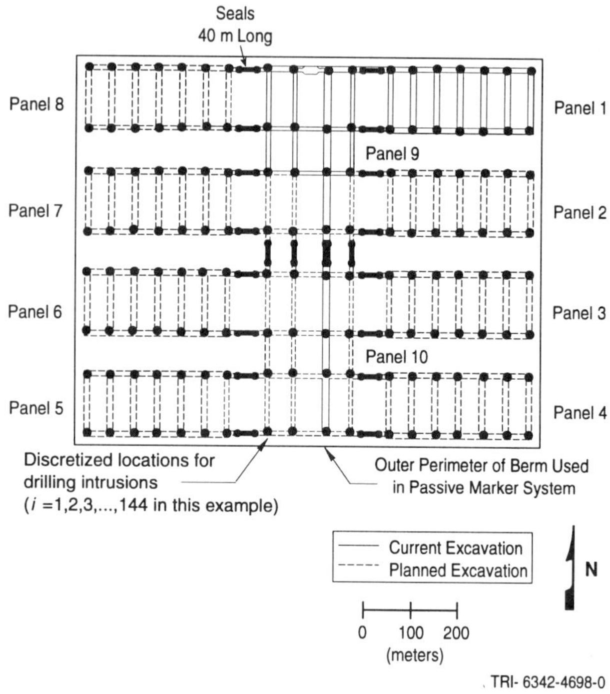

Figure 3: Discretized Locations for Drilling Intrusions Used in 1996 WIPP PA

(0.68) or three plug (0.30) procedure to seal boreholes. The CH waste is emplaced in the repository in 55-gallon drums that come from 569 distinct waste streams, which also have assigned probabilities. As the CH waste is emplaced in the repository in drums stacked three high, each drilling intrusion into CH waste is assumed to intersect three randomly selected waste streams. Finally, the distribution for t_{min} is defined by the assumption that potash mining occurs at a rate of $\lambda_m = 1 \times 10^{-4}$ yr^{-1} (194.32(b)). The preceding assumptions define $(\mathcal{S}_{st}, \mathcal{s}_{st}, p_{st})$.

4 ESTIMATION OF RELEASES

The entity EN2 is the outcome of the model development process for the WIPP and provides a way to estimate radionuclide releases to the accessible environment (i.e., values for Q_i and hence R in Eq. (1)) for the different futures (i.e., elements $\mathbf{x}_{st}$ of $\mathcal{S}_{st}$) that could occur at the WIPP. Estimation of environmental releases corresponds to evaluation of the function f in Figure 2. Release mechanisms associated with f include direct removal to the surface at the time of a drilling intrusion (i.e., cuttings, spallings, brine flow) and release subsequent to a drilling intrusion due to brine flow up a borehole with a degraded plug (i.e., groundwater transport).

The primary computational models in the 1996 WIPP PA are illustrated in Figure 4. Most of these models involve the numerical solution of partial differential equations used to represent material deformation, fluid flow or radionuclide transport. It is the models in Figure 4 that actually define the function f in Figure 2.

As indicated in Figure 2, the CCDF specified in 191.13(a) can be formally defined by an integral of f over $\mathcal{S}_{st}$. In practice, this CCDF is never obtained by direct evaluation of an integral due to the complexity of f and $\mathcal{S}_{st}$. Rather, an approximation procedure based on importance sampling (Helton and Iuzzolino 1993, Helton 1994) or Monte Carlo (random) sampling (Helton and Shiver 1996) is used. The 1996 WIPP PA uses a Monte Carlo procedure. Specifically, elements

$$\mathbf{x}_{st,i},\ i = 1, 2, \ldots, nS, \tag{6}$$

are randomly sampled from $\mathcal{S}_{st}$ in consistency with the definition of $(\mathcal{S}_{st}, \mathscr{S}_{st}, p_{st})$. Then, the integral in Figure 2, and hence the associated CCDF, is approximated by

$$\begin{aligned} prob(Rel > R) &= \int_{\mathcal{S}_{st}} \delta_R\left[f(\mathbf{x}_{st})\right] d_{st}(\mathbf{x}_{st})\, dV_{st} \\ &\doteq \sum_{i=1}^{nS} \delta_R\left[f(\mathbf{x}_{st,i})\right] / \, nS. \end{aligned} \tag{7}$$

The models in Figure 4 are too computationally intensive to permit their evaluation for every element $\mathbf{x}_{st,i}$ of $\mathcal{S}_{st}$ in Eq. (6). Due to this constraint, the models in Figure 4 are evaluated for representative elements of $\mathcal{S}_{st}$ and then the results of these evaluations are used to construct values of f for the large number of $\mathbf{x}_{st,i}$ in Eq. (7).

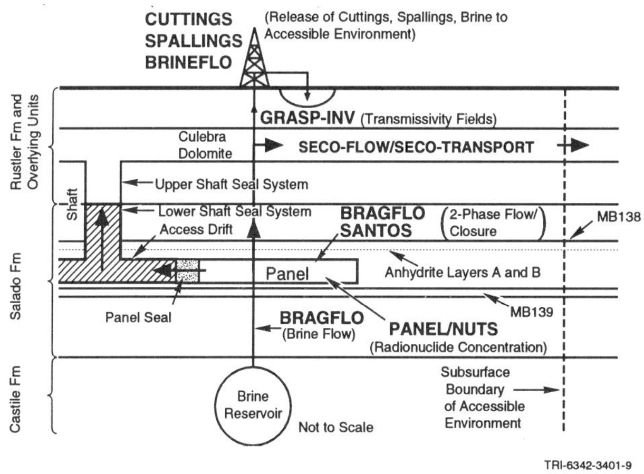

Figure 4: Computer Programs (Models) Used in 1996 WIPP PA

5. PROBABILISTIC CHARACTERIZATION OF PARAMETER UNCERTAINTY

The entity EN3 is the outcome of the data development effort for the WIPP and provides a probabilistic characterization of the uncertainty in the parameters that underlie the WIPP PA. When viewed formally, EN3 is defined by a probability space $(\mathcal{S}_{su}, \mathscr{S}_{su}, p_{su})$, with the sample space $\mathcal{S}_{su}$ given by

$$\mathcal{S}_{su} = \{\mathbf{x}_{su}: \mathbf{x}_{su} \text{ is possibly the correct vector of parameter values to use in the WIPP PA}\}. \tag{8}$$

The subscript *su* refers to subjective (i.e., epistemic) uncertainty and is used because $(\mathcal{S}_{su}, \mathscr{S}_{su}, p_{su})$ is providing a probabilistic characterization of where the appropriate inputs to use in the WIPP PA are believed to be located. In practice, $\mathbf{x}_{su}$ is a vector of the form

$$\mathbf{x}_{su} = \left[x_1, x_2, \ldots, x_{nV}\right], \tag{9}$$

where nV is the number of uncertain variables under consideration, and $(\mathcal{S}_{su}, \mathscr{S}_{su}, p_{su})$ is obtained by specifying a distribution

$$D_j,\ j = 1, 2, \ldots, nV, \tag{10}$$

for each element x_j of $\mathbf{x}_{su}$. The preceding distributions correspond to the distributions in 194.34(b).

In concept, some elements of $\mathbf{x}_{su}$ can affect the definition of $(\mathcal{S}_{st}, \mathscr{S}_{st}, p_{st})$ (e.g., the rate constant λ_d in Eq. (4) used to define the Poisson process for drilling intrusions) and other elements relate to the models in Figure 4 that determine the function f in Figure 2 and Eq. (7) (e.g., radionuclide solubilities in Castile brine or fracture spacing in the Culebra Dolomite). However, all elements of $\mathbf{x}_{su}$ in the 1996 WIPP PA relate to the models in Figure 4 (Table 1).

If the value for $\mathbf{x}_{su}$ was precisely known, then the CCDF in Figure 2 could be determined with certainty and compared with the boundary line specified in 191.13(a). However, given the complexity of the WIPP site and the 10,000 yr period under consideration, $\mathbf{x}_{su}$ can never be known with certainty. Rather, uncertainty in $\mathbf{x}_{su}$ as characterized by $(\mathcal{S}_{su}, \mathscr{S}_{su}, p_{su})$ will lead to a distribution of CCDFs as indicated in 194.34(c) and (e) (Figure 5). The proximity of this distribution to the boundary line in Figure 2 provides an indication of the confidence that 191.13(a) will be met as required in 191.13(b).

Table 1: Four Examples of the $nV = 57$ Uncertain Input Variables (i.e., elements x_j of $\mathbf{x}_{su}$) Considered in the 1996 WIPP PA

WGRMICI: Gas generation rate due to microbial degradation of cellulosics under inundated conditions. Used in BRAGFLO. Range: 3.17×10^{-10} to 9.51×10^{-9} mol/kg-cellulosics. Distribution: Uniform.

BHPERM: Borehole permeability. Used in BRAGFLO. Range: 1×10^{-14} to 1×10^{-11} m^2. Distribution: Loguniform.

WTAUFAIL: Shear strength of waste. Used in CUTTINGS. Range: 0.05 to 10 Pa. Distribution: Uniform.

CKDPU3: Distribution coefficient for plutonium in the +3 oxidation state in the Culebra Dolomite. Used in SECO-TRANSPORT. Range: 0.02 to 0.5 m^3/kg. Distribution: Uniform.

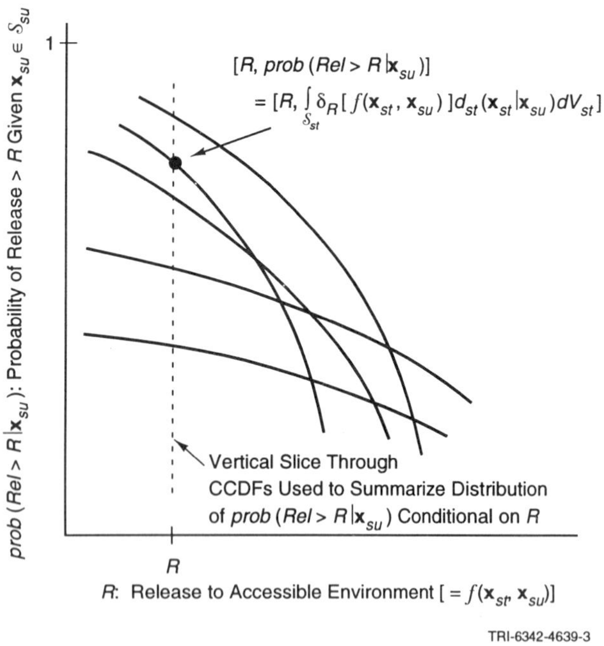

Figure 5. Distribution of CCDFs Resulting from Possible Values for $\mathbf{x}_{su} \in \mathcal{S}_{su}$

The distribution of CCDFs in Figure 5 can be summarized by distributions of exceedance probabilities conditional on individual release values (Figure 6). Specifically, Figure 6 shows the distribution of exceedance probabilities associated with the release of size R indicated in Figure 5. In concept, this distribution is defined

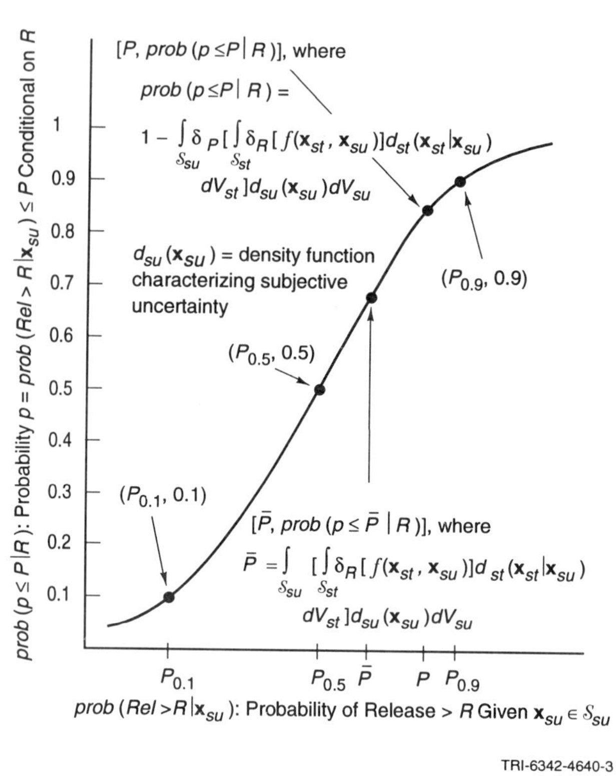

Figure 6: Distribution of Exceedance Probabilities for a Normalized Release of Size R (Figure 5) Due to Subjective Uncertainty

by a double integral over $\mathcal{S}_{su}$ and $\mathcal{S}_{st}$. In practice, this integral is too complex to permit a closed-form evaluation. Instead, the 1996 WIPP PA uses Latin hypercube sampling (McKay et al. 1979) to evaluate the integral over $\mathcal{S}_{su}$ and, as indicated in Eq. (7), simple random sampling to evaluate the integral over $\mathcal{S}_{st}$.

Specifically, a Latin hypercube sample (LHS)

$$\mathbf{x}_{su,k},\ k = 1, 2, \ldots, nLHS, \tag{11}$$

is generated from $\mathcal{S}_{su}$ in consistency with the definition of $(\mathcal{S}_{su}, \mathcal{s}_{su}, p_{su})$ and a random sample as indicated in Eq. (6) is generated from $\mathcal{S}_{st}$ in consistency with the definition of $(\mathcal{S}_{st}, \mathcal{s}_{st}, p_{st})$. The percentile values in Figure 6 (i.e., $P_{0.1}$, $P_{0.5}$, $P_{0.9}$) are then approximated by solving

$$prob\left(p \le P|R\right) \doteq$$

$$1 - \sum_{k=1}^{nLHS} \delta_P\left[\sum_{i=1}^{nS} \delta_R\left[f\left(\mathbf{x}_{st,i}, \mathbf{x}_{su,k}\right)\right] / \, nS\right] / \, nLHS \tag{12}$$

for P with $prob(p \le P|R) = 0.1$, 0.5 and 0.9, respectively. Similarly, the mean exceedance probability $\overline{P}$ is approximated by

$$\overline{P} \doteq \sum_{k=1}^{nLHS} \left[\sum_{i=1}^{nS} \delta_R \left[f\left(\mathbf{x}_{st,i}, \mathbf{x}_{su,k}\right) \right] / \, nS \right] / \, nLHS \,. \quad (13)$$

The results of the preceding calculations are typically displayed by plotting percentile values (e.g., $P_{0.1}$, $P_{0.5}$, $P_{0.9}$) and also the mean values (i.e., $\overline{P}$) for the exceedance probabilities above individual release values (i.e., R) and then connecting these points to form continuous curves (Figure 7). The proximity of these curves to the boundary line provides an indication of the confidence with which 191.13(a) will be met.

6. COMPUTATIONAL DETAILS OF 1996 WIPP PA

The requirements in 194.34(c), (d) and (f) interact in determining the details of the 1996 WIPP PA. Requirements 194.34(c) and (d) can be satisfied with a random sample from $\mathcal{S}_{su}$ of size 298 (i.e., $1 - 0.99^n > 0.95$ yields $n = 298$). However, the WIPP PA decided to use Latin hypercube sampling because of the efficient manner in which it stratifies across the range of each uncertain parameter (Iman and Helton 1988) and the observed stability of uncertainty and sensitivity analysis results produced in past analyses that involved a separation of stochastic and subjective uncertainty (Iman and Helton 1991, Helton et al. 1995b).

Given that Latin hypercube sampling is to be used, the confidence intervals required in 194.34(f) can be obtained with a replicated sampling technique proposed by R.L. Iman (1982). In this technique, the LHS in Eq. (11) is repeatedly generated with different random seeds. These samples lead to a sequence $\overline{P}_r(R)$, $r = 1, 2, \ldots, nR$, of estimated mean exceedance probabilities, where $\overline{P}_r(R)$ defines the mean CCDF obtained for sample r (i.e., $\overline{P}_r(R)$ is the mean probability that a normalized release of size R will be exceeded) and nR is the number of independent LHSs generated with different random seeds. Then,

$$\overline{P}(R) = \sum_{r=1}^{nR} \overline{P}_r(R) \, / \, nR \quad (14)$$

and

$$SE(R) = \left\{ \sum_{r=1}^{nR} \left[\overline{P}_r(R) - \overline{P}(R) \right]^2 / \, nR(nR-1) \right\}^{1/2} \quad (15)$$

provide an additional estimate of the mean CCDF and an estimate of the standard error associated with the mean exceedance probabilities. The t-distribution with $nR-1$ degrees of freedom can be used to place confidence intervals around the mean exceedance probabilities for individual R values (i.e., around $\overline{P}(R)$). Specifically, the $1-\alpha$ confidence interval is given by $\overline{P}(R) \pm t_{1-\alpha/2}$ $SE(R)$, where $t_{1-\alpha/2}$ is the $1-\alpha/2$ quantile of the t-distribution with $nR-1$ degrees of freedom (e.g., $t_{1-\alpha/2} = 4.303$ for $\alpha = 0.05$ and $nR = 3$). The same procedure can also be used to place pointwise confidence intervals around percentile curves.

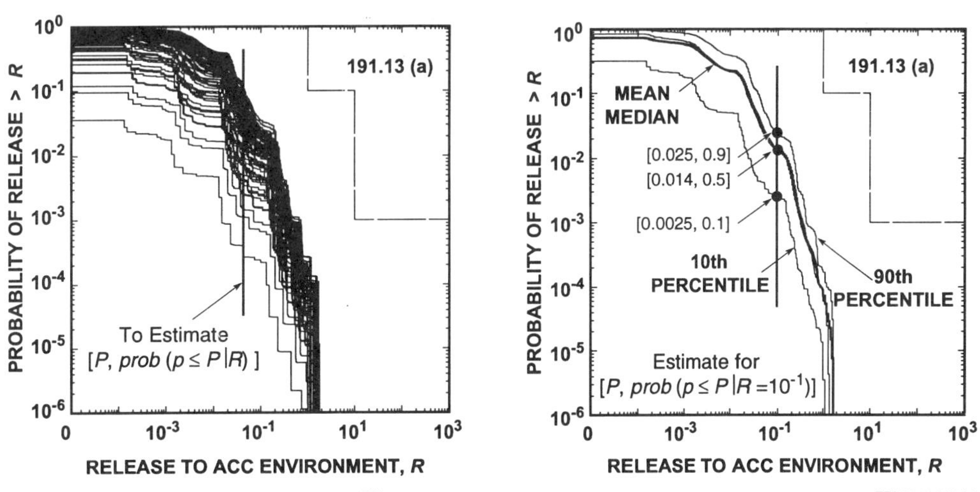

Figure 7: Example CCDF Distribution from 1992 WIPP PA (WIPP PA 1992)

To implement the preceding procedure, the 1996 WIPP PA is using $nR = 3$ replicated LHSs of size $nLHS = 100$ each (see Eq. (11)). This produces a total of 300 observations, which is approximately the same as the sample size of 298 indicated above. Each sample is generated with the restricted pairing technique developed by Iman and Conover (1982) to induce specified rank correlations between correlated variables and also to assure that uncorrelated variables have correlations close to zero.

Once the indicated LHSs are generated, calculations are performed with the models in Figure 4 for the individual sample elements. The number of individual model calculations is too large to describe here. However, the basic strategy is to avoid the unnecessary proliferation of computationally demanding calculations by identifying situations where (1) a single computationally demanding calculation can be used to supply input to several less demanding calculations, (2) mathematical properties of the models can be used to extend the results of a single calculation to many different situations, or (3) a relatively inexpensive screening calculation can be used to determine if a more detailed, and hence more expensive, calculation is needed. As examples, (1) each BRAGFLO calculation, which involves the numerical solution of a system of nonlinear partial differential equations and is quite demanding computationally (i.e., 1 - 2 hrs of CPU time on a Digital VAX Alpha using VMS), is used to supply conditions that are used in a number of different calculations with the SPALLINGS, BRINEFLO, NUTS and PANEL models indicated in Figure 4; (2) the linearity of the system of partial differential equations that underlies SECO-TRANSPORT makes it possible to perform transport calculations for unit releases of individual radionuclides to the Culebra Dolomite and then use the outcome of these calculations to construct transport results for arbitrary time-dependent radionuclide releases into the Culebra; and (3) transport calculations with NUTS are initially performed with a nondecaying tracer and then calculations with radionuclides are only performed for those cases that have a potential to result in a radionuclide release from the repository.

The analysis effort for all three replications was still quite large and involved 1800 BRAGFLO calculations, 15,600 CUTTINGS/SPALLINGS calculations (Note: What is designated as CUTTINGS and SPALLINGS in Figure 4 is actually a single program), 15,600 BRINEFLO calculations (Note: BRINEFLO is actually a special configuration of BRAGFLO used to estimate brine releases, i.e., blowout, at the time of a drilling intrusion), approximately 1500 screening and 500 full calculations with NUTS, 2100 PANEL calculations, 100 GRASP_INV calculations, 600 SECO-FLOW calculations, and 1200 SECO-TRANSPORT calculations. The outcome of these calculations is a set of results for each LHS element.

As discussed in conjunction with Eq. (7), Monte Carlo procedures are then used to construct a CCDF for each LHS element. Specifically, this CCDF is produced from $nS = 10{,}000$ randomly selected futures of the form shown in Eq. (3), where nS is the sample size in Eq. (6). Once each future $\mathbf{x}_{st,i}$ is sampled, the corresponding normalized release $f(\mathbf{x}_{st,i})$ is constructed from releases to the accessible environment calculated with CUTTINGS, SPALLINGS, BRINEFLO, NUTS and SECO-TRANSPORT. In this procedure, extensive algebraic manipulations and interpolations are performed to estimate releases for futures involving multiple intrusions from the results of the previously indicated calculations for one or two intrusions at fixed points in time. Once values for $f(\mathbf{x}_{st,i})$ are determined, which correspond to the normalized release R in Eq. (1), the CCDF specified in 191.13(a) is readily constructed.

Repetition of the preceding procedure for each LHS element yields a distribution of CCDFs of the form in Figure 7 for each of the $nR = 3$ replicates as requested in 194.34(e). Further, the replicated samples and the procedure in Eqs. (14) and (15) provide a basis for the estimation of confidence intervals as requested in 194.34(f).

To this point, the emphasis of this presentation has been on uncertainty analysis. However, the techniques in use also provide a basis for sensitivity analysis. Specifically, Latin hypercube sampling generates a mapping from analysis input (i.e., elements $\mathbf{x}_{su}$ of $\mathcal{S}_{su}$) to analysis results that can be explored with sensitivity analysis techniques based on examination of scatterplots, correlation analysis, regression analysis, and other procedures for investigating multivariate data (Helton 1993). Quantities that can be investigated include both the results from individual models and the final CCDFs. Sensitivity analysis plays three important roles by (1) identifying the effects of individual variables (i.e., elements x_j of $\mathbf{x}_{su}$), (2) providing programmatic direction as to where resources can be invested to reduce the uncertainty in analysis outcomes, and (3) facilitating quality assurance by systematically examining the effects of many combinations of analysis inputs on a variety of analysis results. Sensitivity analysis procedures will be used extensively to study the outcomes of the 1996 WIPP PA.

7 STATUS

The calculations described in this presentation are currently underway at SNL and are anticipated to be completed by the end of July 1996. Extensive documentation of these calculations will be published by SNL and the DOE.

REFERENCES

Helton, J.C. 1993. Uncertainty and Sensitivity Analysis Techniques for Use in Performance Assessment for Radioactive Waste Disposal. *Reliability Engineering and System Safety* 42: 327-367.

Helton, J.C. 1994. Treatment of Uncertainty in Performance Assessments for Complex Systems. *Risk Analysis* 14: 483-511.

Helton, J.C., and H.J. Iuzzolino. 1993. Construction of Complementary Cumulative Distribution Functions for Comparison with the EPA Release Limits for Radioactive Waste Disposal. *Reliability Engineering and System Safety* 40: 277-293.

Helton, J.C., and A.W. Shiver. 1996. A Monte Carlo Procedure for the Construction of Complementary Cumulative Distribution Functions for Comparison with the EPA Release Limits for Radioactive Waste Disposal. *Risk Analysis* 16: 43-55.

Helton, J.C., et al. 1993. Uncertainty and Sensitivity Analysis Results Obtained in a Preliminary Performance Assessment for the Waste Isolation Pilot Plant. *Nuclear Science and Engineering* 114: 286-331.

Helton, J.C., et al. 1995a. Effect of Alternative Conceptual Models in a Preliminary Performance Assessment for the Waste Isolation Pilot Plant. *Nuclear Engineering and Design* 154: 251-344.

Helton, J.C., et al. 1995b. Robustness of an Uncertainty and Sensitivity Analysis of Early Exposure Results with the MACCS Reactor Accident Consequence Model. *Reliability Engineering and System Safety* 48: 129-148.

Helton, J.C., et al. 1996. Uncertainty and Sensitivity Analysis Results Obtained in the 1992 Performance Assessment for the Waste Isolation Pilot Plant. *Reliability Engineering and System Safety* 51: 53-100.

Iman, R.L. 1982. Statistical Methods for Including Uncertainties Associated with Geologic Isolation of Radioactive Waste Which Allow for a Comparison with Licensing Criteria. *Proceedings of the Symposium on Uncertainties Associated with the Regulation of the Geologic Disposal of High Level Radioactive Waste*, Gatlinburg, Tennessee, March 9-13, 1981, ed. D.C. Kocher. NUREG/CP-0022, CONF-810372. Oak Ridge National Laboratory, Oak Ridge, Tennessee, 145-157.

Iman, R.L., and W.J. Conover. 1982. A Distribution-Free Approach to Inducing Rank Correlation Among Input Variables. *Communications in Statistics* B11: 311-334.

Iman, R.L., and J.C. Helton. 1988. An Investigation of Uncertainty and Sensitivity Analysis Techniques for Computer Models. *Risk Analysis* 8: 71-90.

Iman, R.L., and J.C. Helton. 1991. The Repeatability of Uncertainty and Sensitivity Analyses for Complex Probabilistic Risk Assessments. *Risk Analysis* 11: 591-606.

McKay, M.D., W.J. Conover, and R. J. Beckman. 1979. A Comparison of Three Methods for Selecting Values of Input Variables in the Analysis of Output from a Computer Code. *Technometrics* 21: 239-245.

U.S. DOE (Department of Energy). 1990. *Final Supplement Environmental Impact Statement, Waste Isolation Pilot Plant.* DOE/EIS-0026-FS. U.S. Department of Energy, Office of Environmental Restoration and Waste Management, Washington, DC.

U.S. EPA (Environmental Protection Agency). 1985. Environmental Standards for the Management and Disposal of Spent Nuclear Fuel, High-Level and Transuranic Radioactive Waste; Final Rule, 40 CFR Part 191. *Federal Register* 50: 38066-38089.

U.S. EPA (Environmental Protection Agency). 1993. Environmental Radiation Protection Standards for the Management and Disposal of Spent Nuclear Fuel, High-Level and Transuranic Radioactive Waste; Final Rule, 40 CFR Part 191. *Federal Register* 58: 66398-66416.

U.S. EPA (Environmental Protection Agency). 1996. 40 CFR Part 194: Criteria for the Certification and Recertification of the Waste Isolation Pilot Plant's Compliance With the 40 CFR Part 191 Disposal Regulations; Final Rule. *Federal Register* 61: 5224-5245.

WIPP PA (Performance Assessment) Division. 1991. *Preliminary Comparison with 40 CFR Part 191, Subpart B for the Waste Isolation Pilot Plant, December 1991*, Vols. 1-4. SAND91-0893/1-4. Sandia National Laboratories, Albuquerque, New Mexico.

WIPP PA (Performance Assessment) Department. 1992. *Preliminary Performance Assessment for the Waste Isolation Pilot Plant, December 1992*, Vols. 1-5. SAND92-0700/1-5. Sandia National Laboratories, Albuquerque, New Mexico.

AUTHOR BIOGRAPHY

JON C. HELTON is a professor in the Department of Mathematics at Arizona State University. His research interests include environmental modeling, risk assessment, and uncertainty and sensitivity analysis techniques.

Proceedings of the 1996 Winter Simulation Conference
ed. J. M. Charnes, D. J. Morrice, D. T. Brunner, and J. J. Swain

FIVE-STAGE PROCEDURE FOR THE EVALUATION OF SIMULATION MODELS THROUGH STATISTICAL TECHNIQUES

Jack P.C. Kleijnen

Department of Information Systems and Auditing(BIKA)/Center for Economic Research (CentER)
School of Management and Economics (FEW)
Tilburg University (KUB)
5000 LE Tilburg, NETHERLANDS

ABSTRACT

This paper recommends the following sequence for the evaluation of simulation models. 1) Validation: the availability of data on the real system determines the proper type of statistical technique. 2) Screening: in the simulation's pilot phase the important inputs are identified through a novel technique, namely sequential bifurcation, which uses aggregation and sequential experimentation. 3) Sensitivity or what-if analysis: the important inputs are analyzed in more detail, including interactions between inputs; relevant techniques are design of experiments (DOE) and regression analysis. 4) Uncertainty or risk analysis: important environmental inputs may have values not precisely known, so the resulting uncertainties in the model outputs are quantified; techniques are Monte Carlo and Latin hypercube sampling. 5) Optimization: policy variables may be controlled, applying Response Surface Methodology (RSM), which combines DOE, regression analysis, and steepest-ascent hill-climbing. This paper summarizes case studies for each stage.

1 INTRODUCTION

The main message of this paper is that most simulation studies should go through the following *five stages* and apply the following statistical *techniques* (which have already been applied in many practical simulations).

Stage 1: Validation and verification (V & V)
The analysts may use design of experiments or DOE (for example, a fractional factorial design) if there are no data on the input/output (I/O) of the simulation model. However, if there are enough data, then the simulation model may be validated by a special type of regression analysis (see §4).

Stage 2: Screening: sequential bifurcation (SB)
If the *pilot* phase of a simulation study has *many factors*, then straightforward experimentation may require too much computer time. In those situations the analysts may use SB, which is simple, efficient, and effective (see §2).

Stage 3: Sensitivity analysis (SA)
Sensitivity analysis is defined in this paper as the investigation of the reaction of the simulation response to either *extreme values* of the model's quantitative factors (parameters and input variables) or *drastic changes* in the model's qualitative factors (modules). (This is also called global, instead of local SA.) However, if the model's I/O behavior is non-monotonic, it may be dangerous to consider extreme input values only (see Saltelli, Andres, and Homma 1995). Non-monotonicity may be quantified by quadratic effects. Notice that this paper concentrates on a single response per run; also see Helton (1996) and McKay (1995).

The analysts may use *regression analysis* to generalize the results of the simulation experiment. To obtain better regression estimators, the analysts should apply *DOE* (see §3).

Stage 4: Uncertainty analysis (UA)
In UA (sometimes called risk analysis) the model inputs are sampled from prespecified distributions. So the input values range between the extreme values investigated in SA. The goal of UA is to quantify the *probability* of specific output values, whereas SA does not tell how likely a specific result is.

UA uses *Monte Carlo* sampling, possibly including variance reduction techniques such as *Latin hypercube sampling* or LHS; the results may be analyzed through regression analysis (see §5).

Stage 5: Optimization
To optimize the controllable (policy) inputs, the analysts may use *Response Surface Methodology* (RSM), which builds on regression analysis and DOE (see §6).

This five-stage procedure implies that SA should precede UA, which is controversial (see McKay 1995, pp. 7, 33, and also Fűrbringer and Roulet 1995).

Uncertainty about a simulated system's response may have two different causes: (i) The system's process is represented by a *deterministic* model, but its *parameters* are not known exactly. (ii) Some models are *intrinsically stochastic*: without the randomness the problem disappears; examples are queuing problems. Helton (1996) points out that stochastic uncertainty is a property of the system, whereas subjective uncertainty is a property of the analysts.

Uncertainty is the central problem in *mathematical probability theory*. This discipline, however, has two schools: *frequentists* versus *subjectivists*. Many UA studies concern unique events; for example, a specific investment, a nuclear accident. See Cooke (1995, pp. 4-6) and Winkler (1996).

Note that Bayesians try to combine prior subjective data with new factual data; see Draper (1995). Bayesians average the outcomes, using the probabilities of the various input scenarios. It might be argued, however, that in general it is the job of 'managers' to make decisions, pondering facts and opinions; it is the job of scientists to prepare a basis for such decisions.

Zeigler (1976)'s seminal book on the theory of modeling and simulation distinguishes between *input variables* and *parameters*. By definition, a variable can be directly observed; an example is the number of check-out lanes in a supermarket. A parameter, however, can *not* be observed directly, so its value must be inferred from observations; an example is the arrival rate of customers. Hence mathematical statistics may be used to quantify the probability of certain values for a parameter (see §5). This paper will use the term *factor* for parameters, input variables, and modules that are changed from (simulation) run to run, applying DOE. (For modules treated as factors also see Helton et al. 1995, McKay 1995, pp. 51-54.)

This paper combines, updates, and revises Kleijnen (1994, 1995b, 1996). For didactic reasons, the paper is not organized in the order of the five stages; for example, DOE is simpler explained for SA (stage 3, §3) than for V & V (stage 1, §4). Hence §2 discusses screening, especially SB. §3 explains SA using regression analysis and DOE. §4 briefly addresses V & V. §5 discusses UA, distinguishing between deterministic and stochastic simulations. §6 explains optimization based on RSM. §7 summarizes the conclusions.

2 SCREENING: SEQUENTIAL BIFURCATION

Screening is the search for the few (say) k important factors among the many (say) K potentially important factors ($k << K$). In practice, simulated systems often do involve many factors. Andres (1996) gives an example with 3,300 factors, Bettonvil and Kleijnen (1996) 281, De Wit (1995) 82, and McKay (1995) 84 factors. These examples concern deterministic simulations. Other examples would be stochastic simulations of queueing networks; in this field, however, applications of screening are yet unknown.

The problem is that a simulation run may require so much *computer time* that the number of runs (say) n must be much smaller than the number of factors ($n << K$). Therefore the analysts assume that only a few factors are important ($k << K$): *parsimony, Occam's razor*.

The theory on group screening goes back to the 1960s; see Bettonvil and Kleijnen (1996) and Kleijnen (1995b). This paper recommends a *novel* group screening technique, namely SB, which is more efficient than competing group screening techniques. Moreover, SB has been used in practice, namely in the greenhouse simulation in Bettonvil and Kleijnen (1996), and in the building thermal simulation in De Wit (1995). In De Wit's case study, SB gives the 16 most important inputs among the 82 factors after only 50 runs. De Wit checks these results through a different technique, namely Morris (1991)'s randomized one-factor-at-a time design, which takes 328 runs.

All group screening techniques including SB use the following two assumptions.

Assumption 1: Low-order polynomial metamodel
A metamodel implies that the underlying simulation model is treated as a *black box*. The advantages of a low-order polynomial is that it is simple and it may apply to *all* types of random and deterministic simulation. The disadvantage is that it cannot exploit the special structure of the simulation model at hand (for alternative techniques see Helton 1996, Ho and Cao 1991, and Kleijnen and Rubinstein 1996).

Low-order polynomials are often used in DOE with its concomitant regression analysis or Analysis of Variance (ANOVA). Applications, however, are then limited to systems with a small number of factors (also see §3).

It is prudent not to assume a first-order polynomial, but a metamodel with interactions. In SB such a metamodel requires twice as many runs as a first-order metamodel does (foldover principle; see Andres 1996 and Kleijnen 1987).

Assumption 2: Known signs
Known signs are needed to know with certainty that individual effects do *not compensate* each other within a group. In practice, signs may be known indeed

(magnitudes, however, are unknown, so simulation is used). For example, higher speeds of servers are supposed to have non-negative effects on the throughput.

Details on SB are given in Bettonvil and Kleijnen (1996). Different screening techniques are Andres (1996)'s Iterated Fractional Factorial Design (IFFD) and McKay (1995)'s replicated LHS design (also see Saltelli et al. 1995).

3 SENSITIVITY ANALYSIS: REGRESSION/DOE

Once the important factors are identified (screening, §2), further analysis is recommended, using fewer assumptions: account for interactions and curvature, and assume no known signs. (The metamodeling methodology of this section is discussed at length in Kleijnen and Sargent 1996.)

3.1 Regression Analysis

It is good practice to make *scatter plots* (see Eschenbach 1992, Helton 1996). However, superimposing many plots is cumbersome. Moreover, their interpretation is subjective: are the response curves really parallel and straight lines? These shortcomings are removed by regression analysis. See Kleijnen (1995b).

Consider the well-known *second degree polynomial* metamodel with Y_i denoting the simulation response of factor combination i (capitals denote stochastic variables); β_0 the overall mean response or regression intercept; β_h the main or first-order effect of factor h; $x_{i,\,h}$ the value of the standardized factor h in combination i; $\beta_{h,\,h'}$ the interaction between the factors h and h' with $h < h'$; $\beta_{h,\,h}$ the quadratic effect of factor h; E_i the additive fitting error of the regression model for factor combination i; and n the number of simulated factor combinations.

Then the *relative importance* of a factor is obtained by sorting the absolute values of the main effects β_h, provided the factors are *standardized* such that they range between -1 and +1; see Kleijnen (1995b). The original factor h is denoted by z_h, which ranges between a lowest and a highest values ($l_h \leq z_h \leq u_h$). So either the simulation model is not valid outside that range (see §4) or in practice that factor can range over that domain only. DOE addresses the question whether z_h is to be set at the extreme values only or also at intermediate values; see below. (Other measures of factor importance, such as partial correlation coefficients, are discussed in Helton 1996, McKay 1995, Saltelli and Sobol 1995, and Sobol 1996. When the logarithmic scale for Y and z_h is used, the regression parameters denote elasticity coefficients; these coefficients are popular in econometrics.)

To make statistical inferences (for example, about the importance of a factor), a *Gaussian* distribution is normally assumed. To satisfy this assumption, the analysts may apply transformations such as log(Y). An alternative is to replace both Y and x_h by their ranks: rank regression. Moreover, the analysts may investigate whether the various transformations give the same qualitative results. See Andres (1996), Helton (1996), Kleijnen (1987), Kleijnen, Bettonvil, and Van Groenendaal (1996), and Saltelli and Sobol (1995).

Of course, it is necessary to check whether the fitted regression metamodel is an adequate approximation of the underlying simulation model: (cross)validation (see Kleijnen 1995b and Kleijnen and Sargent 1996).

3.2 DOE

Let q denote the number of regression parameters; for example, q equals $k + 1 + k(k - 1)/2$ if there are main effects and two-factor interactions. Which n combinations to simulate can be determined such that the accuracy of the estimated parameters is maximized (variance minimized); this is the goal of DOE.

Consider a *first-order polynomial* metamodel with k main effects, and an overall mean. By definition, a *resolution III* or R-3 design permits the unbiased estimation of such a first-order polynomial. In *practice*, analysts usually first simulate the 'base-case', and next they change *one factor at a time*; hence $n = 1 + k$. DOE, however, gives *orthogonal* designs, which yield unbiased estimators with *smaller variances*. A well-known class of such designs are 2^{k-p} fractional factorials; for example, a simulation with $4 \leq k \leq 7$ factors requires $n = 2^{7-4} = 8$ factor combinations. Many *simulation applications* of these designs can be found in Kleijnen (1987) and Kleijnen and Van Groenendaal (1992).

In practice, however, it is prudent to assume that *two-factor interactions* may be important. A *resolution IV* or R-4 design permits the unbiased estimation of all k main effects, even if two-factor interactions are present. These designs do not give unbiased estimators of individual two-factor interactions; they may give an indication of the importance of (confounded, biased) interactions. Compared with R-3 designs, R-4 designs require that the number of simulated factor combinations be *doubled*; for example, $k = 7$ now requires $n = 2 \times 8 = 16$ runs. Obviously, changing one factor at a time (a less accurate R-3 design) does not enable estimation of interactions! Applications of R-4 designs are mentioned in Kleijnen (1995b).

Suppose the analysts wish to estimate the *individual* two-factor interactions. A *resolution V* or R-5 design gives unbiased estimators for all main effects and all two-factor interactions. Obviously many more simulation runs are now necessary; for example, $k = 8$ factors implies $q = 37$ effects. Therefore practitioners study only small values of k; for example, $k < 5$ requires *full or complete factorial two-level designs*, denoted as 2^k. For higher values of k, however, *fractional* factorials are recommended, for example, 2^{8-2} ($n = 64$) for $k = 8$ (so $q = 37$); see Kleijnen (1987).

If all 2^k combinations are simulated, then *high-order interactions* can be estimated. However, these interactions are hard to interpret. Hence, the analysts may apply either transformations such as the logarithmic transformation or they may restrict the experimental domain. Also see Helton (1996), and Saltelli and Sobol (1995).

If factors are quantitative and their *quadratic* effects are to be estimated, then factors must be simulated for more than two values (taking extreme values minimizes the variance of the estimated main effects; see Cheng and Kleijnen 1996). *Central composite designs* have five values per factor. These designs require relatively many runs (the number of parameters is also high): $n >> q$. See Kleijnen (1987) and Kleijnen and Van Groenendaal (1992). Applications are found in the optimization of simulation models (see §6).

Simulating as many as five values per factor resembles UA, in the sense that the range of factor values is well covered (see §5).

4 VALIDATION

This paper confines the discussion of V & V to the role of SA (§3); other statistical techniques for V & V are discussed in Kleijnen (1995a).

If the simulation is fed with real-life input data (trace driven simulation), then Kleijnen, Bettonvil, and Van Groenendaal (1996) propose a novel test. This test uses regression analysis, but does *not* hypothesize that real and simulated data lie on a straight line through the origin! Instead, the difference between simulated and real data is regressed on their sum.

If no data are available, then the analysts and their clients still have *qualitative* knowledge; that is, they do know in which direction certain factors affect the response (see §2). If the regression metamodel (see §3) gives an estimated effect with the *wrong sign*, this is a strong indication of a wrong simulation model or a wrong computer program. An application is Kleijnen (1995c), concerning the hunt for mines on the bottom of the sea. This case-study further shows that the validity of a simulation model is restricted to a certain domain of factor combinations, called *experimental frame* in Zeigler (1976).

5 UNCERTAINTY ANALYSIS

The regression metamodel (see §3) shows which factors are important; for the important *environmental* inputs the analysts should try to collect data on the values that occur in practice. If they do not succeed, then they may use UA: its goal is to quantify the *probability* of specific output values (whereas SA does not tell how likely a specific result is).

First the analysts derive a *probability function* for the input values, based on sample data if available; otherwise this distribution must use subjective expert opinions (see Draper 1995, p. 92, Helton et al. 1995, p. 288, Kleijnen 1996). Correlated inputs are discussed in Cooke (1995), Helton (1996), and Kleijnen (1996).

Next the analysts use pseudorandom numbers to sample input values: *Monte Carlo*. They often use *Latin hypercube sampling* (LHS), which forces the sample (of size n) to cover the whole experimental area; see Avramidis and Wilson (1996) and Helton (1996).

This paper claims that LHS is a *variance reduction technique*, not a *screening* technique. For screening purposes the inputs should be changed to their extreme values (see §2). Of course, the larger sample in LHS gives more insight than the small sample in screening does; however, for a large number of factors such a large sample may be impossible. Also see Kleijnen (1996) and McKay (1992).

5.1 Deterministic Simulation

The distribution of the simulation response may be characterized by its location (measured by the mean, modus, or median) and its dispersion (quantified by the standard deviation or various percentiles and quantiles). Which quantities sufficiently summarize the distribution, depends on the users' *risk attitude*: risk neutral (the mean is then sufficient), risk averse, or risk seeking. See Helton (1996) and Kleijnen (1996).

Combining UA with *regression analysis* gives estimates of the effects of the various inputs; that is, regression analysis shows which inputs contribute most to the uncertainty in the output. (Technically, in the regression metamodel x_{ik} is replaced by X_{ik}.) Because more values are sampled per factor, more complicated metamodels might now be used; for example, for prediction purposes splines may be used. For explanatory purposes and SA, however, simple metamodels may be preferred; see Kleijnen and Sargent (1996).

Applications in business and economics are, for example, investment studies on the probability of negative Net Present Values; see Kleijnen (1996) and Van Groenendaal and Kleijnen (1996). Applications in the natural sciences are, for example, Sandia investigations on nuclear waste disposal and reactor safety (Helton 1996); Oak Ridge studies of radioactive doses in humans; European Communities investigations on nuclear reactor safety; and Dutch studies on ecology; see Kleijnen (1996).

It is prudent to study the effects of different *specifications* of the input distributions. Sometimes, this type of SA is called *robustness analysis*. Examples are given by Helton et al. (1995), Kleijnen (1987, 1996), and McKay (1995, p. 31). The analysis may also use sophisticated, fast sampling techniques based on *importance sampling* or *likelihood ratios*; see Kleijnen and Rubinstein (1996). (Importance sampling is indispensable whenever rare events are simulated; see Helton 1996, and Kleijnen and Rubinstein 1995.)

5.2 Stochastic Simulation

A well-known example of stochastic simulation is a queueing model with interarrival times sampled from an exponential distribution with parameter λ (Helton 1996 discusses UA of three stochastic models in nuclear science). For this example Kleijnen (1983) discusses UA, assuming that the *central limit theorem* (CLT) applies. For example, suppose λ is estimated from a sample of independent interarrival times, and assume the distribution of the estimated interarrival parameter approximates a normal distribution. Then a value for λ can be sampled from this distribution, and used as input to the simulation. That simulation is run for 'enough' customers. Next the procedure is repeated: sample λ, and so on. Cheng and Holland (1996), however, do not rely on the CLT, but apply *bootstrapping*.

Still, the question remains *which* response to report to the users: the unconditional, ex post variance (see Cheng and Holland 1996); the ex post variance, mean, and various quantiles (see Haverkort and Meeuwissen 1995); or the conditional moments (conditioned on the values of the estimated parameters)? The discussion in Draper (1995, pp. 78, 83) clearly demonstrates how *controversial* this issue is.

6 OPTIMIZATION: RSM

The decision makers should control the *policy variables*. This paper is limited to RSM, which combines regression analysis and DOE (see §3) with a hill-climbing technique called steepest ascent; see Myers and Montgomery (1995).

Four *general characteristics of RSM* are: (i) RSM uses first and second order polynomial regression metamodels, now called *response surfaces* (see §3). (ii) RSM employs the *statistical designs* of DOE (see §3). (iii) RSM augments DOE with the *steepest ascent* technique, to determine in which direction the decision variables should be changed. (iv) RSM uses *canonical analysis* to analyze the shape of the optimal region (unique maximum, saddle point, ridge).

Kleijnen (1995b) mentions several simulation studies that use RSM. Kleijnen (1995d) discusses a system dynamics model for coal transport.

7 CONCLUSIONS

There are several *related types of analysis* in simulation studies. Questions are: *when* should *which* type of analysis be applied; which *statistical techniques* should then be used?

This paper proposes a *five-stage procedure* for the analysis of simulation models: (i) validation and verification (V & V), (ii) screening, (iii) sensitivity analysis (SA) (iv) uncertainty analysis (UA), and (v) optimization.

Each type of analysis may apply its own set of *statistical techniques*, for example, SA may use 2^{k-p} fractional designs, and UA may apply LHS. Some techniques may be applied in both analyses, for example, regression modelling.

Applications of the statistical techniques recommended above, are quite plentiful. Further *research* is needed on UA of stochastic simulation models.

SA and UA remain *controversial* topics; for example, this paper claims that SA should precede UA. *SA* shows which model inputs are really important. From the users' perspective, these important inputs are either controllable or not. The controllable inputs may be *optimized*. The values of the uncontrollable inputs may be well-known, in which case these values can be used for *validation* of the simulation model. If, however, these values are not well known, then the likelihood of their values can be quantified objectively or subjectively, and the probability of specific output values can be quantified by *UA*.

More specifically, SA means that the model is subjected to *extreme value* testing. A model is valid only within its experimental frame. Mathematically that frame might be defined as the hypercube formed by the k standardized inputs of the model. Experimental designs such as 2^{k-p} fractional factorials specify *which* combinations are actually observed or simulated; for example, a 2^{-p} fraction of the 2^k corner points of that hypercube (also see Draper 1995, p. 55). The n observed input combinations and their corresponding responses are analyzed through a regression (meta)model, which is an approximation of the simulation model's I/O behavior. That regression model quantifies the importance of the simulation inputs. DOE gives better estimates of those input effects.

SA does not tell *how likely* a particular combination of inputs (specified by DOE) is, whereas UA does account for the probabilities of input values.

Further development of a methodology for the evaluation of complicated simulation models certainly requires continued communication within and among scientific disciplines!

REFERENCES

Andres, T.H. (1996), Sampling methods and sensitivity analysis for large parameter sets. *Journal Statistical Computation and Simulation* (accepted).

Avramidis, A.N. and J.R. Wilson (1996), Integrated Variance Reduction Strategies for Simulation. *Operations Research* 44: 327-346.

Bettonvil, B. and J.P.C. Kleijnen, (1996), Searching for important factors in simulation models with many factors: sequential bifurcation. *European Journal of Operational Research* (accepted).

Cheng, R.C.H. and W. Holland (1996), The sensitivity of computer simulation experiments to errors in input data. *Journal Statistical Computation and Simulation* (accepted).

Cheng, R.C.H. and J.P.C. Kleijnen (1996), Improved design of queueing simulation experiments with highly heteroscedastic responses. *Operations research* (accepted).

Cooke, R.M. (1995), UNICORN: methods and code for uncertainty analysis. The SRD Association, AEA Technology, Thomson House, Risley, Warrington WA3 6AT, UK.

De Wit, M.S. (1995), Uncertainty analysis in building thermal modelling. *SAMO95*, Office for Official Publications of the European Communities, Luxembourg.

Draper, D. (1995), Assessment and propagation of model uncertainty, *Journal Royal Statistical Society, Series B*, 57: 45-97.

Eschenbach, T.G. (1992), Spider plots versus tornado diagrams for sensitivity analysis. *Interfaces*, 22: 40-46.

Fűrbringer, J.-M. and C.A. Roulet (1996), Comparison and combination of factorial and Monte-Carlo design in sensitivity analysis. *Building and Environment*, 30: 505-519.

Haverkort, B.R. and A.M.H. Meeuwissen (1995), Sensitivity and uncertainty analysis of Markov-reward models. *IEEE Transactions on Reliability*, 44: 147-154.

Helton, J.C. (1996), Uncertainty and sensitivity analysis in the presence of stochastic and subjective uncertainty. *Journal Statistical Computation and Simulation* (accepted).

Helton, J.C. et al. (1995), Effect of alternative conceptual models in a preliminary performance assessment for the waste isolation pilot plant. *Nuclear Engineering and Design*: 251-344.

Ho,Y. and X. Cao. 1991. *Perturbation analysis of discrete event systems*. Dordrecht: Kluwer.

Kleijnen, J.P.C. (1983) Risk analysis and sensitivity analysis: antithesis or synthesis? *Simuletter*, 14: 64-72.

Kleijnen, J.P.C.(1987), *Statistical Tools for Simulation Practitioners*. Marcel Dekker, Inc., New York.

Kleijnen, J.P.C. (1994), Sensitivity analysis versus uncertainty analysis: when to use what? *Predictability and Nonlinear Modelling in Natural Sciences and Economics*, edited by J. Grasman and G. van Straten, Kluwer, Dordrecht, The Netherlands, 1994, pp.322-333. (Preprinted in *Kwantitatieve Methoden*, 15, 1994: 3-15.)

Kleijnen, J.P.C. (1995a), Verification and validation of simulation models. *European Journal of Operational Research* 82: 145-162.

Kleijnen, J.P.C. (1995b), Sensitivity Analysis and Optimization in Simulation: Design of Experiments and Case Studies. *Proceedings of the 1995 Winter Simulation Conference* (edited by C. Alexopoulos, K. Kang, W.R. Lilegdon, D. Goldsman).

Kleijnen, J.P.C. (1995c), Case-study: statistical validation of simulation models. *European Journal of Operational Research*, 87: 21-34.

Kleijnen, J.P.C. (1995d), Sensitivity analysis and optimization of system dynamics models: regression analysis and statistical design of experiments. *System Dynamics Review*, 11: 1-14.

Kleijnen, J.P.C. (1996), Sensitivity analysis and related analyses: a survey of statistical techniques. *Journal Statistical Computation and Simulation* (accepted)

Kleijnen, J.P.C. and R.Y. Rubinstein (1996) Optimization and sensitivity analysis of computer simulation models by the score function method. *European Journal of Operational Research*, 88: 1-15.

Kleijnen, J.P.C., B. Bettonvil, and W. Van Groenendaal (1996), Validation of simulation models: a novel regression test. *Management Science* (accepted).

Kleijnen, J.P.C. and W. Van Groenendaal (1992), *Simulation: a Statistical Perspective*. Wiley, Chichester.

Kleijnen, J.P.C. and R. Sargent (1996), Metamodeling methodology. Tilburg University, Tilburg, Netherlands.

McKay, M.D. (1992), Latin hypercube sampling as a tool in uncertainty analysis of computer models. *1992 Winter Simulation Conference Proceedings* (edited by J.J. Swain, D. Goldsman, R.C. Crain, and J.R. Wilson).

McKay, M.D. (1995), Evaluating prediction uncertainty. Los Alamos National Laboratory, NUREG/CR-6311 (LA-12915-MS).

Myers, R.H. and D.C. Montgomery (1995), *Response Surface Methodology*.

Morris, M.D. (1991), Factorial plans for preliminary computational experiments, *Technometrics*, 33: 161-174.

Saltelli, A., T.H. Andres, and T. Homma (1995), Sensitivity analysis of model output; performance of the iterated fractional factorial design method. *Computational Statistics & Data Analysis*, 20: 387-407.

Saltelli, A. and I.M. Sobol (1995), About the use of rank transformation in sensitivity analysis of model output. *Reliability Engineering and System Safety*, 50: 225-239.

Sobol, I.M. (1996), Sensitivity analysis of nonlinear models using sensitivity indices. *Journal Statistical Computation and Simulation* (accepted).

Winkler, R.L. (1996), Uncertainty in probabilistic risk assessment. *Reliability Engineering and System Safety*, Special Issue, edited by J.Helton and D. Burmaster.

Zeigler, B. (1976), *Theory of Modelling and Simulation*. Wiley Interscience, New York.

AUTHOR BIOGRAPHY

JACK P.C. KLEIJNEN is Professor of Simulation and Information Systems in the Department of Information Systems and Auditing; he is also associated with the Center for Economic Research (CentER). Both the Department and the Center are within the School of Management and Economics of Tilburg University (Katholieke Universiteit Brabant) in Tilburg, Netherlands. He received his Ph.D. in Management Science at Tilburg University. His research interests are in simulation, mathematical statistics, information systems, and logistics. He published six books and more than 130 articles; lectured at numerous conferences throughout Europe, the USA, Turkey, and Israel; he was a consultant for various organizations; and is a member of several editorial boards. He spent some years in the USA, at different universities and companies. He was awarded a number of fellowships, both nationally and internationally.

Proceedings of the 1996 Winter Simulation Conference
ed. J. M. Charnes, D. J. Morrice, D. T. Brunner, and J. J. Swain

UNCERTAINTY AND SENSITIVITY STUDIES OF MODELS OF ENVIRONMENTAL SYSTEMS

E. Marian Scott

Department of Statistics
University of Glasgow
Glasgow G12 8QW
UNITED KINGDOM

ABSTRACT

Traditionally, sensitivity and uncertainty analyses in environmental modelling contribute quantitative descriptions of the relative importances of individual parameters and processes, highlighting areas of significant contributors to the overall uncertainties of model predictions and giving markers for areas requiring substantial improvement (perhaps through directed experimental work). Both sensitivity and uncertainty analysis require extensive re-sampling from input data and simulation of model response and there is a large and growing literature concerning their use in environmental modelling. There is however growing interest in the sensitivity of predictions to model structure, and in evaluating the contribution of model structural uncertainty to the overall uncertainty within the general framework of sensitivity analysis (Draper, 1995, Beven, 1993, Beven and Binley, 1992).

Sensitivity and uncertainty analysis contribute to all stages of model development, testing and assessment and their impact on model reliability and validity will be described. After a general introduction to sensitivity and uncertainty analysis and discussion of their contributions, some applications of sensitivity analysis to several environmental modelling studies will be presented.

2 INTRODUCTION

An environmental model will very often be composed of a number of linked sub-models, representing physical processes understood to varying degrees. They may be dynamic and stochastic, and the model extent may also be spatial. Such models may be used for example, to predict levels of Cs-137 on pasture over several seasonal cycles following the Chernobyl power plant accident; they may be used to model the global carbon cycle in climate studies and to model the dispersal of a radioactive pollutant discharged into the Irish Sea from Sellafield nuclear fuel reprocessing plant. The model, depending on its purpose and field of application will include many different processes, the sub-processes may operate at quite different time and space scales and for the modeller there will be choice of which processes to explicitly model and how to parameterise these processes. Some parameters will be known precisely, for others, there may be conflicting evidence so that the parameter can only be defined in terms of a range of values which may span several orders of magnitude. Some of the parameters may be time- and space-dependent.

Sensitivity analysis encourages the exploration of the interactions between the various modelled processes and helps throw light on the properties of complex computational models by in its simplest form perturbing one parameter at a time and studying its effect on the response. For large computer models, with perhaps hundreds of parameters, sensitivity analysis may involve perturbing all parameters simultaneously, but in such a way that the main effects of individual parameters and their interactions can be estimated. Uncertainty analysis contributes estimates of uncertainties on the final predictions by propagating through the model all quantified sources of uncertainty. Thus, sensitivity analysis identifies the key contributors to uncertainties, while uncertainty analysis quantifies the overall uncertainty, so that together they contribute to a reliability assessment for the model.

3 MODEL RELIABILITY EVALUATION

Modelling the environment may be done with a number of goals in mind: they include description of the system, prediction, impact assessment and to provide a basis for management decisions. It is crucial that the model should be fit for the purpose to which it was designed.

There are a number of properties the modeller wishes the model and the model results to possess, but one of the most important is reliability. What properties might be considered to define model reliability? One fundamental requirement is that the model should reproduce the behaviour of the system under study, within the limits of uncertainty of the data and the model predictions.

As part of this requirement,therefore, an assessment of the uncertainty associated with the model predictions is required. This uncertainty should be acceptable for decision making purposes and it should also be possible to identify the factors contributing to this uncertainty.

3.1 Factors affecting reliability

The factors affecting reliability can be broadly classified corresponding to the main stages in model development: specification, conceptualisation, computation and parameterisation (IAEA, 1989).

In the specification of the problem, the modeller must consider the spatial and temporal resolution required in the predictions, the purpose of the modelling and the scales at which the predictions will be required. He must also consider the processes which operate.

He must formulate an appropriate conceptual model, defining the model structure, processes and the scale at which they operate and any interconnections.

The conceptual model must be translated into a computational model, whose code must be verified.

Finally, any unknown parameters must be defined. The parameters may have associated with them large uncertainties , arising from two sources: natural stochastic variation and lack of knowledge on the modeller's behalf (Hoffman & Hammonds, 1994).

Thus, at each stage of the modelling process uncertainties are introduced, not only in the definition of the parameters. Sensitivity analysis contributes at each stage and only after these stages have been completed, can the critical stage of model testing and validation be reached. The testing and validation must take into account the uncertainties introduced in the preceeding stages (McKay, 1995).

3.2 The role of sensitivity and uncertainty analysis in model development

3.2.1 Parametric sensitivity analysis

Sensitivity analysis involves determination of the amount and kind of change produced in a given system parameter by a change in another parameter. It is a tool in the investigation of factors which are important or influential in determing the system response. To perform a sensitivity analysis, it is necessary to first identify the parameters of interest, define for each a probability density function (p.d.f.) to reflect the belief that the parameter will take on various values within its possible range, account for dependencies amongst the parameters and propagate the uncertainties through the model to generate a p.d.f. of predicted values which can then be analysed.

Definition of the parameter p.d.f. can often prove difficult, occasionally there may be experimental information available which would allow empirical based estimation of the p.d.f., but more commonly there may be little information and the p.d.f. then fundamentally reflects the modeller's belief. The elicitation of belief based p.d.f.s in environmental contexts is a challenging task, but which which if successfully achieved raises the possibility of directly incorporating expert opinion into the formal analysis through Bayesian methodology (Cooke,1994a,b, EC Munvar report, 1995).

Given the p.d.f., there are a number of widely used methods of sampling, they include simple random sampling (SRS), (parameter values chosen at random from the p.d.f.) and Latin hypercube sampling (LHS), (the range of each parameter is partitioned into n intervals and one value is selected from each interval of each parameter). SRS is simple, reliable easy to analyse but inefficient. LHS is basically a stratified sampling procedure and is generally more efficient than SRS (Andres, 1987, Andres, 1997, Iman and Conover, 1980).

A large number of methods for the analysis of the results from the sensitivity analysis exist, and indeed there have been a number of comparisons of the methods (Iman and Helton, 1988, Saltelli and Homma, 1992, Hamby, 1995). Techniques used include response surface replacement, correlation and partial rank correlation coefficients, regression (on ranks) as well asa number of standard test statistics (Smirnov, Vramer-Von Mises and Mann-Whitney), (Saltelli et al, 1992). Whatever the method, its goal is to explore the model response to changing parameters.

However, parametric uncertainty reflects only one source (and perhaps not the dominant source) of uncertainty. The other key contributor is uncertainty about the model structure itself.

3.2.2 Model uncertainty

In the development of the model, the modeller must work with an imperfect and incomplete description of the physical system. He must select features and processes to be parameterised, he must synthesise sometimes conflicting evidence and he must prioritise.

Model or structural uncertainty reflects that an acceptable model prediction may be achieved in many different ways and that the modeller must use judgement to decide on the structure of the model. In an attempt to explicitly consider this source of uncertainty in the predictions (particularly important when the model is being used as a management decision tool), a number of problems must first be resolved. It may be more difficult to quantify the uncertainty in the model structure when compared to a single parameter.

Models may differ in terms of the physical, chemical or biological processes included; they may have different state spaces; they may have been constructed using a different knowledge base; the complexity of the models may thus be different, yet all may be valid but still provide different responses. The issue of how uncertainty about model structure can be quantified and incorporated in a standard sensitivity analysis is important one which is yet to be fully addressed. The issue of structural uncertainty is one which can be addressed using a Bayesian formulation, within which, it is possible to elicit and make use of expert beliefs. Following Draper (1995):

let M denote the model, S the structural assumptions and Θ the model parameters. Suppose we have a discrete set of models, say m. Let x denote the data and y the quantity to predict, then

$$p(y \mid x,S) = \Sigma p(S_i \mid x)p(y \mid x,S_i) \qquad (1)$$

and S_i, i=1,..,m denotes the different structural alternatives within the models. The analysis presented in (1) allows, through the application of Bayes Theorem, the distribution of the predicted quantity to be defined in terms of prior beliefs $p(S_i \mid x)$ for the model structure given the data and the predictive distribution for y given the model structure S_i. However, the prior distributions $p(S_i \mid x)$ must be defined. Draper suggests creating cross-validation or bootstrap samples (Efron and Tibshirani, 1993) from the available data and conducting parallel modelling activities on each sample. However, in many environmental studies, there is little available data so that this may not be a viable alternative. A simple alternative to this approach which makes use of all the data, but rather re-samples from the modellers has often been used in international model inter-comparisons. Modellers are given access to the same information, and are tasked to independently develop models for the same, defined purposes. In this way, a more robust and hopefully reliable set of predictions can be achieved. In addition, the variability in the model predictions provides a direct measure of uncertainty based on structural uncertainty.

4. EXAMPLES

A number of recent modelling studies have had as a stated objective an assessment of the uncertainty in predictions due to differences in model structures. In such work, often described as 'model benchmarking', a number of modellers have been provided with the same information concerning physical environment as well as clear modelling objectives. Model endpoints have been selected and the results compared both across the different models and also, when possible, with observational data. Results from three such studies are presented and discussed.

4.1 IAEA VAMP programme (IAEA, 1995)

The International Atomic Energy Agency programme had as one of its objectives:

'to provide a mechanism for validation of assessment models using Chernobyl related data'. As part of this work, a further objective was defined: 'to test predictive capabilities of models and identify the most important reasons for model misprediction'. The study was designed in the form of a test scenario, describing in some detail, conditions in a location (unknown to modellers) in central Europe. A wide range of model endpoints were identified including Cs-137 concentrations in a number of agricultural products at different points in time and also an estimate of dose to the local population. Modellers provided their best estimate as well as an estimate of the uncertainty on the result. Finally the model results could be compared both internally and against experimental data from the region (Hoffman & Thiessen, 1995, Hoffman et al, 1996).

The results showed variation amongst the predictions, and differences between the model predictions and experimental data. On further analysis, one of the main sources of variation in the results was found to be 'modeller interpretation' (IAEA, 1995). In addition to best estimates, the modellers also provided in some instances, uncertainty estimates, one of the main sources of difference in these estimates was the judgement on how representative the measured data.

4.2 BIOMOVS (1993)

BIOspheric Model Validation Study (BIOMOVS) was an international study for testing models for ecological transfer and bioaccumulation of radionuclides and other trace substances. One of the primary objectives of BIOMOVS was 'to explain differences in model predictions due to structural deficiencies, invalid assumptions...' (BIOMOVS, 1993). Within the programme, one approach adopted involved the formulation of test scenarios based on suitable data and a comparison of model predictions against these independent data sets, while for the other approach, model predictions and associated estimates of uncertainty were compared. As in VAMP, when the model comparisons were made substantial differences were found, related in part to the expertise and judgement of the modeller. More recently, the BIOMOVS programme has developed a number of test cases to further investigate the role of expert judgement in determining uncertainty.

4.3 IAEA Arctic assessment programme (Scott et al, 1997)

The Arctic assessment programme was concerned with evaluating the present and potential impacts of the dumping of nuclear waste within shallow coastal waters in the Arctic (specifically the Kara and Barents Seas). The waste of most concern was reactor compartments from nuclear submarines, some of which still contained spent nuclear fuel. A modelling group was created with the task of providing reliable and realistic estimates of the potential radiological impact on biota and human populations.

The modelling group comprised 9 modellers, and in the first phase of the work they undertook an extensive benchmarking exercise, with a view to providing provisional estimates of the uncertainties in the predictions due to differences in the model structure. The endpoints of the benchmarking were spatially and temporally discriminating, and it was of interest to investigate the variation in the endpoints.

The models used in the work simulate the dispersion of radioactivity due to advection and diffusion within the water column and include interactions with suspended material and sediment. There were two main modelling approaches, one was based on compartmental models while the other used detailed 3-D hydrodynamic models.

In addition to a quantification of predictive uncertainty as a result of model structure, a directed sensitivity analysis was also carried out on sub-processes. Results again emphasised the importance of structural uncertainty in the overall predictive uncertainty.

5 SUMMARY AND CONCLUSIONS

Sensitivity and uncertainty analysis are important tools in the modelling toolkit. They should be performed routinely, as the modeller might assess the goodness of fit or check on outliers.

In model assessment and testing, sensitivity and uncertainty analysis provide the essential information for determining the acceptability of the uncertainties and the influential factors.

In model validation, the agreement of the model to observed or measured data can only be judged in the light of the predictive uncertainties.

In model design, sensitivity analysis encourages the identification of key processes to be further investigated and gives a metric by which to compare different model structures.

Taken together, sensitivity and uncertainty analysis contributes to a more reliable model, and one whose abilities are better understood.

REFERENCES

Andres T H, 1987. Statistical sampling strategies. Procs NEA Workshop on Uncertainty analysis for Performance Assessments of Radioactive Waste Disposal Systems, Seattle 1987.

Andres T H, 1997. Sampling methods and sensitivity analysis for large parameter sets. Journal of Statistical Computation and Simulation, to appear.

Beven K, Binley A, 1992. The future of distributed models: Model calibration and uncertainty prediction. Hydrological Processes, 6, 279-298.

Beven K, 1993. Prophecy, reality and uncertainty in distributed hydrological modelling. Advances in Water Research 16, 41-51.

BIOMOVS, 1993. Final report, BIOMOVS Technical report 15, Swedish Radiation Protection Institute, Stockholm.

Cooke R, 1994a. Parameter fitting for uncertain models: modelling uncertainty in small models. Reliability Engineering and System Safety 44, 89-102.

Cooke R, 1994b. Uncertainty in dispersion and deposition accident consequence modelling assessed with performance-based expert judgement. Reliability Engineering and System Safety, 45 35-46.

Draper D, 1995. Assessment and propagation of model uncertainty. Journal of the Royal Statistical Society, Series B

EC, 1995. Review on development of methodologies for modelling with uncertainty and variability: Munvar project. European Commission report, EUR 16174 EN.

Efron B, Tibshirani R, 1993. An introduction to the Bootstrap. Chapman & Hall, New York.

Hamby D M, 1995. A comparison of sensitivity analysis techniques. Health Physics, 68(2), 195-204.

Hoffman F O, Hammonds J S, 1994. Propagation of uncertainty in risk assessments: the need to distinguish between uncertainty due to lack of knowledge and uncertainty due to variability. Risk Analysis 14, 707-712.

Hoffman F O, Thiessen K M, 1995. Use of Chernobyl data to test predictions and uncertainty estimates from exposure assessment models. Proceedings of 'Environmental impact or radioactive releases', IAEA-SM-339/20, 325-336.

Hoffman F O, Thiessen K M, Watkins B, 1996. Opportunities for the testing of environmental transport models using data obtained following the Chernobyl accident. Health Physics 70, 5-7.

Iman R L, Conover W J, 1980. Small sample sensitivity analysis techniques for computer models, with an application to risk assessment. Communications in Statistical Theory and Methods, A9(17), 1749-1842.

Iman R L, Helton J C, 1988. An investigation of uncertainty and sensitivity analysis techniques for computer models. Risk Analysis 8, 71-90.

International Atomic Energy Agency, 1989. Evaluating the reliability of predictions made using environmental transfer models. Safety Series, 100.

International Atomic Energy Agency, 1995. Validation of models using Chernobyl fallout data from the central Bohemia region of the Czech republic. IAEA-Tecdoc 795.

McKay M D, 1995. Evaluating prediction uncertainty. Los Alamos National Lab Technical report, NUREG/CR-6311.

Saltelli A, Homma T, 1992. Sensitivity analysis for model output. Computational Statistics and Data analysis, 13(1), 73-94.

Saltelli A, Andres T H, Homma T, 1992. Sensitivity analysis of model output: an investigation of new techniques. Computational Statistics and Data Analysis, 14, 1-28.

Scott E M, Gurbutt P, Harms I, Heling R, Nielsen S P, Preller R, Osvath I, Sazykina T, Wada A, Sjoeblom K L, 1997. Benchmarking of numerical models describing the dispersion of radionuclides in the Arctic Seas. Science of the Total Environment, to appear.

AUTHOR BIOGRAPHY

E. MARIAN SCOTT is a senior lecturer in the Department of Statistics, University of Glasgow. She is associate editor of the Journal of Statistical Computation and Simulation, and will be guest editor of a special issue on Sensitivity and Uncertainty analysis. Her research interests include environmental modelling, sensitivity analysis and model uncertainties.

Proceedings of the 1996 Winter Simulation Conference
ed. J. M. Charnes, D. J. Morrice, D. T. Brunner, and J. J. Swain

THE HIT-AND-RUN SAMPLER: A GLOBALLY REACHING MARKOV CHAIN SAMPLER FOR GENERATING ARBITRARY MULTIVARIATE DISTRIBUTIONS

Robert L. Smith

Department of Industrial and Operations Engineering
The University of Michigan
Ann Arbor, Michigan 48109, U.S.A.

ABSTRACT

The problem of efficiently generating general multivariate densities via a Monte Carlo procedure has experienced dramatic progress in recent years through the device of a Markov chain sampler. This procedure produces a sequence of random deviates corresponding to a random walk over the support of the target distribution. Under certain regularity conditions, the corresponding Markov chain converges in distribution to the target distribution. Thus the sample of points so generated can serve as a statistical sample of points drawn from the target distribution. A random walk that can globally reach across the support of the distribution in one step is called a Hit-and-Run sampler. Hit-and-Run Markov chain samplers offer the promise of faster convergence to the target distribution than conventional small step random walks. Applications to optimization are considered.

1 INTRODUCTION

The Monte Carlo problem of efficiently generating univariate distributions can be attacked from a variety of directions including the simple device of evaluating the inverse of the random variables cdf at a uniform [0,1] deviate. However, the multivariate case is more problematic with no corresponding transform method for the general case (Schmeiser (1980)). There are transform methods for special cases, including multivariate normals and uniform distributions over simplices and ellipsoids (Rubinstein (1982)), but no constant Jacobian analytic function is known that maps from the unit cube to the epigraph of a general multivariate density function. Nonetheless, the ability to generate general multivariate deviates would have many important applications. The simulation of complex systems can involve generating multivariate densities that are not only not normal but may not belong to a named family of distributions. An interesting and potentially very significant application from the field of optimization is to generate samples from multivariate distributions that have a high likelihood of being close to a global optimum.

The basic idea is very intuitive. Consider the problem of minimizing a continuous objective function f over a unit cube C in R^n,

$$\max_{x \in C \subseteq R^n} f(x)$$

If we were to generate a point X uniformly distributed in the epigraph of f on C, i.e., in $S = \{(x, f(x)) | x \in C\}$, then points $x \in C$ with higher objective function values $f(x)$ would be more likely to be generated than those with lower values. In fact, the pdf of X would be a scalar multiple of f. We could increase this bias toward higher values of f by selecting X uniform from the epigraph of a transformed version of f that made higher values of f correspondingly more likely. For example, consider the epigraph of the function $e^{f(x)/T}$ where $T > 0$. It is clear that as $T \to \infty$, the distribution of X would converge to the singular distribution of unit mass at the global optimum $x^* = \text{argmax}\{f(x) | x \in C\}$. This observation in fact forms the basis for an extension of simulated annealing to the continuous case (Romeijn and Smith (1994a,1994b)). However since the objective function f is arbitrary apart from being continuous, the epigraph of the function $g(x) = e^{f(x)/T}$ can also be quite general and in particular non-convex.

A straightforward approach to generating a uniform point from such a general bounded region $S \subseteq R^n$ would be to enclose it in a sphere say and then use Von Neumann's rejection method (Hammersley and Handscomb (1964)) by generating a sequence of independent uniform deviates in the sphere until a point falls in S. Unfortunately, the number of such trial points grows exponentially fast in dimension n even for nice regions S such as a cube. For example, the expected number of points that must be generated within a circumscribed sphere around a cube S

grows from 1.5 for $n = 2$ to 10^{30} for dimension $n = 100$. (The n-dimensional content of an n-dimensional sphere of radius r is $2r^n\pi^{n/2}(n\Gamma(n/2))^{-1}$, so that the volume of the sphere enclosing an n-dimensional unit cube is $2(n\pi/4)^{n/2}(n\Gamma(n/2))^{-1}$).

Lacking either an efficient transformation or rejection method of generating a uniform deviate in a general region S, people have turned to the device of a so-called Markov chain sampler. Probably the first instance of such a Monte Carlo procedure was the celebrated Metropolis Method (Metropolis et al (1953)). The basic idea is to begin with any point in the region S and initiate a random walk within S from this starting point. Since the distribution of each successor point depends only upon the current point, the sequence of points generated form a homogeneous continuous state Markov chain. The generator that produces a successor point from the current point is so constructed that the limiting distribution of the Markov chain is the target distribution over S, e.g., the uniform distribution.

2 HIT-AND-RUN SAMPLERS

An important class of Markov chain samplers is the so-called Hit-and-Run sampler. They differ from other samplers, such as the neighborhood walk of discrete simulated annealing, by their ability to execute movements that can span the entire region S in a single step. This ability to globally reach across a region has led researchers to conjecture that their rate of convergence is superior to neighborhood random walks.

When the number of iterations of the walk required to obtain a variational distance from the target distribution of at most ϵ is a polynomial function of the size of the problem (e.g., the dimension of the region), then the sampler is called rapidly mixing. Recently, Lovasz and Simonovits (1993) have demonstrated a neighborhood walk that achieves rapid mixing for a class of "well rounded" convex regions. The walk proceeds by generating successor points within the intersection of the region S with a small ball around the current point. Surprisingly however, the polynomiality of Hit-and-Run remains an open question.

2.1 Uniform Target Distributions

We now turn to describing a series of Hit-and-Run samplers that have in practice experienced excellent rates of convergence to their target distributions. The first is the first instance of a Hit-and-Run sampler which was termed at the time a symmetric mixing algorithm (Smith (1984); see also Smith (1980), Boneh and Golan (1979)). It is only required that the region S be an open subset of R^n (and that the target distribution be the uniform distribution over S). In particular, S may be non-convex and indeed disconnected. However, since S is open, for every point $x \in S$ there must exist a small ball around it that lies totally within S.

Hit-and-Run Sampler (Uniform Target Distribution)

1. Select a starting point $x_0 \in S$ and set $i = 0$.
2. Generate a random direction d_{i+1} uniformly distributed over a direction set $D \subseteq R^n$. Find the line set $L = S \cap \{x | x = x_i + \lambda d_{i+1}, \lambda \text{ a real scalar}\}$ and generate a random point x_{i+1} uniformly distributed over L.
3. If $i = N$, stop. Otherwise, set $i = 1 + 1$ and return to 2.

Since the region S is arbitrary apart from being open, the step of generating x_{i+1} uniform over L can be computationally very demanding. To avoid this, in practice one would execute step 2 by having first enclosed the region S in a hypercube. Then after generating a uniform direction d over D (easily done by simply normalizing a vector of n independent normal [0,1] components), the end points y^1 and y^2 of the bidirectional line emanating from the current iterate x and intersecting the enclosing cube are determined. Then a rejection method is employed by generating one-dimensional uniform points on the line segment from y^1 to y^2 until one falls within L. Since this subproblem is one-dimensional, a rejection method is typically very efficient for this task. The remaining $(n-1)$-dimensional effort lies in choosing d uniform over D, an easy task for D a sphere for example. In this latter case the procedure is called the Hypersphere Directions Hit-and-Run Sampler.

The reason that the Markov chain corresponding to the iterates $X_0, X_1, X_2, \ldots$ converges in distribution to a uniform distribution over S is easily seen from the fact that 1) it is possible to go from any point in S to a neighborhood of any other point in one step, and 2) the uniform distribution is a stationary distribution of the chain. The latter observation is clear if we observe that from the directional symmetry of the direction distribution and next iterate chosen, the probability of transitioning from a point x to a neighborhood of a point y in S is the same as the probability of transitioning from the point y to a neighborhood of the point x, for all x and y in S (see Smith (1984) for a formal proof of this result). Smith (1984) reports on empirical tests that suggest rapid convergence to a uniform distribution for relatively large dimensional regions. An analytic bound on the error from uniform after a fixed number of iterations is also provided. The bound is smallest for the case where the region is a sphere since the bound is decreasing in the ratio of the volume of the region to the volume of the smallest circumscribed sphere. Thus elongated ellipsoids can be expected to require more iterations to achieve a given closeness to uniformity. This is also intuitively clear since most of the directions generated from a current iterate will result in small movements from the current point (thus the conductance is small). Since the uniform distribution remains the limiting distribution regardless of how directions are generated (Smith (1984)), it can prove profitable to alter the distribution employed to better fit the region. Kaufman and Smith (1996) show how to select a direction distribution to accelerate convergence and in particular show how to achieve a sphere-case rate of convergence for any ellipsoid region. Incidentally, if we choose the direction distribution set D to be a hypercube, we obtain the so-called Coordinate Directions Hit-and-Run Sampler, a version of the Gibbs Sampler except that the sequence of coordinate directions chosen is not deterministic. See Chen and Schmeiser (1993) for a computational comparison of the two approaches.

Turning to applications, in addition to their direct use for the Monte Carlo generation of uniform samples, these Hit-and-Run Samplers with uniform target distributions have also been employed as random probes to discover the properties of various regions in R^n. Berbee et al (1987) explored their use in identifying redundant constraints in a linear program. See also Caron, Hlynka and McDonald (1993). The idea here is that the bidirectional line emanating from the current iterate x_i will hit two points on the boundary of the polyhedral feasible region of the LP, thus identifying two non-redundant constraints. By repeating for each iterate, we play a coupon-collectors problem of non-redundant constraint identification until all non-redundant constraints have been identified with a given probability. By then removing the redundant constraints, the resulting smaller LP may be more efficiently solved for an optimal solution (if the resulting solution is feasible for the original LP, it must be optimal as well). The construction of a stopping rule that guarantees that every non-redundant constraint has been identified with a certain probability is complicated by the fact that the hit point on the boundary of the polyhedron will not in general be uniformly distributed on that surface (see Caron and MacDonald (1989) for how to address the dependency of the two hit points). In fact, the distribution will in general be quite irregular and complexly dependent on the geometry even though the interior iterates $X_0, X_1, X_2, \ldots$ will asymptotically be uniform within the polyhedron. The problem is that the probability of hitting a neighborhood of a facet depends on the angle the facet makes with the bidirectional line emanating from the current iterate: the more orthogonal the line meeting the facet, the better the chance. Boender et al (1991) alter the Hit-and-Run Sampler to reject hit points in proportion to the cosine of that angle to remove this bias and thereby produce hit points that are asymptotically uniform on the boundary of the polyhedron. Since the volume of the surface of a polyhedron is zero, rejection methods cannot be used here; nor can transformational methods since the number of simplices in a decomposition of the surface is exponentially large in the dimension of the polyhedron. The resulting Hit-and-Run sampler, somewhat whimsically called the Shake-and-Bake algorithm, is arguably the only practical way to generate uniform samples on the surface of large dimensional polyhedrons.

2.2 General Target Distributions

Thus far we have only discussed Hit-and-Run Samplers for generating *uniform* distributions on regions $S \subseteq R^n$. The problem of devising a Markov chain sampler to generate essentially arbitrary distributions f over $S \subseteq R^n$ is addressed in Belisle, Romeijn, and Smith (1993). They propose the following sampler:

Hit-and-Run Sampler (General Target Distribution)

1. Select a starting point $x_0 \in S$ and set $i = 0$.
2. Generate a direction d_{i+1} in D with distribution ν. Find the line set $L = S \cap \{x | x = x_i + \lambda d_{i+1}, \lambda \text{ a real scalar}\}$ and generate a point $x_{i+1} = x_i + \lambda_i d_{i+1} \in L$ with λ_i having the density
$$f_i(\lambda) = \frac{f(x_i + \lambda d_{i+1})}{\int f(x_i + r d_{i+1}) dr}.$$
3. If $i = N$, stop. Otherwise return to 2.

The sampler for arbitrary densities f over S then proceeds as in uniform Hit-and-Run except that the next iterate x_{i+1} is chosen according to the conditionalized density of f in the direction d_{i+1}. The direction distribution ν needs to span R^n but is otherwise arbitrary. In Belisle, Romeijn, and Smith (1993), it is shown that for S a bounded open set in R^n and f the pdf of the target distribution π, the sequence of iterates $X_0, X_1, X_2, \ldots$ converges in total variation to the target distribution π.

Romeijn and Smith (1994a) extend this sampler to a Metropolis type sampler that also converges to arbitrary target distributions. They utilize the sampler within a simulated annealing type optimal search procedure to generate densities that come from a temperature parametrized family of Boltzman distributions to converge to a global optimal solution. See also Zabinsky et al (1993) for a more direct use of Hit-and-Run within an optimization framework.

3 CONCLUSION

The field of Markov Chain Samplers is currently quite active both from the theoretical and practical points of view. At least some of the interest stems from the recent results that random procedures can achieve polynomial performance for tasks such as approximating the volume of convex bodies that it has been demonstrated no deterministic procedure can ever achieve (Dyer and Frieze (1991), Dyer, Frieze, and Kannan (1991)). However in a larger context the problem of generating general multivariate distributions is a fundamental problem of simulation whose importance hardly needs justification.

REFERENCES

Bélisle, C.J.P., H.E. Romeijn and R.L. Smith, "Hit-and-Run Algorithms for Generating Multivariate Distributions," *Mathematics of Operations Research* 18, pp. 255–266,1993.

Berbee, H.C.P., C.G.E. Boender, A.H.G. Rinnooy Kan, C.L. Scheffer, R.L. Smith, and J. Telgen "Hit-and-Run Algorithms for the Identification of Nonredundant Linear Inequalities," *Mathematical Programming* 37, pp. 184–207,1987.

Boender, C.G.E., R.J. Caron, A.H.G. Rinnooy Kan, J.F. McDonald, H. Edwin Romeijn, R.L. Smith, J. Telgen, and A.C.F. Vorst,"Shake-and-Bake Algorithms for Generating Uniform Points on the Boundary of Bounded Polyhedra," *Operations Research* 39, pp. 935-953, 1991.

Boneh, A., and A. Golan, "Constraints' Redundancy and Feasible Region Boundedness by Random Feasible Point Generator (RFPG)," Third European Congress on Operations Research (EURO III), Amsterdam, 1979.

Caron, R. J. and J. F. McDonald, "A New Approach to the Analysis of Random Methods for Detecting Necessary Linear Inequality Constraints," *Mathematical Programming* 43, pp. 97-102, 1989.

Caron, R. J., Hlynka, M., McDonald, J. F., "On the Best Case Performance of Hit and Run Methods for Detecting Necessary Constraints," *Mathematical Programming* 54, pp. 233–249, 1992.

Chen, M. and B. Schmeiser, "Performance of the Gibbs, Hit-and-Run and Metropolis Samplers," *Journal of Computational and Graphical Statistics* 2, pp. 251–272, 1993.

Dyer, M and Frieze, A., "Computing the Volume of Convex Bodies: A Case Where Randomness Provably Helps." *Probabilistic combinatorics and its applications*, San Francisco, pp. 123–169, 1991.

Dyer, M, Frieze, A., and Kannan, R., "A Random Polynomial-time Algorithm for Approximating the Volume of Convex Bodies," *Journal of the ACM* 38, pp. 1–17. 1991.

Hammersley, J.M. and Handscomb, D.C., *Monte Carlo Methods*, John Wiley and Sons, NY, 1964.

Kaufman, D. and R. L. Smith, "Direction Choice For Accelerated Convergence In Hit-and-run Sampling," *Operations Research*, in press, 1996.

Lovasz, L. and Simonovits, M., "Random Walks in a Convex Body and an Improved Volume Algorithm," *Random Stuctures and Algorithms* 4, pp. 359-412, 1993.

Metropolis, N., A.W. Rosenbluth, M.N. Rosenbluth, A.H. Teller and E. Teller, "Equations of State Calculations by Fast Computing Machines," *The Journal of Chemical Physics* 21, pp. 1087–1092, 1953.

Romeijn, E. and R.L. Smith, "Simulated Annealing for Constrained Global Optimization," *Journal of Global Optimization* 5, pp. 101-126, 1994a.

Romeijn, E. and R.L. Smith, "Simulated Annealing and Adaptive Search in Global Optimization," *Probability in the Engineering and Informational Sciences* 8, pp. 571-590, 1994b.

Rubinstein, R.Y., "Generating Random Vectors Uniformly Distributed Inside and on the Surface of Different Regions," *European Journal of Operations Research* 10, pp. 205-209, 1982.

Schmeiser, B.W., "Random Variate Generation: A Survey". *Simulation with Discrete Models: State of the Art View*, T. I. Oren, C.M. Shub, P. F. Roth (eds), IEEE, pp. 79-104, 1980.

Smith, R.L., "Monte Carlo Procedures for Generating Random Feasible Solutions to Mathematical Programs," ORSA/TIMS Conference, Washington, May, 1980.

Smith, R.L. Efficient Monte Carlo Procedures for Generating Points Uniformly Distributed over Bounded Regions. *Operations Research* 32, pp. 1296–1308,1984.

Zabinsky, Z.B., R.L. Smith, J.F. McDonald, H.E. Romeijn, and D.E. Kaufman, "Improving Hit and

Run for Global Optimization," *Journal of Global Optimization* 3, pp. 171–192, 1993.

AUTHOR BIOGRAPHIES

ROBERT L. SMITH is Professor of Industrial and Operations Engineering at the University of Michigan, Ann Arbor. He received his Ph.D. in Engineering Science from the University of California at Berkeley where he was an NSF Fellow. He holds a bachelors degree in Physics from Harvey Mudd College and an MBA from Berkeley. He has been a visiting scholar at the Statistical Laboratory, University of Cambridge; the Econometric Institute, Erasmus University, Rotterdam, The Netherlands; the Laboratoire d'Automatique et d'Analyse du Systems du CNRS, Toulouse, France; the Faculty of Industrial Engineering and Management, Technion, Haifa, Israel; the Mathematical Sciences Research Institute, Berkeley, California; the Operations Research Group, Institute of Pure and Applied Mathematics, Rio de Janeiro, Brazil; and the Department of Industrial Engineering and Operations Research, University of California, Berkeley. His research is directed toward the modeling, analysis, and optimization of dynamical systems over time. He is a member of the Institute for Operations Research and Management Sciences and The Mathematical Programming Society, an Associate Editor of *Management Science*, and is the author of over sixty peer reviewed articles.

Proceedings of the 1996 Winter Simulation Conference
ed. J. M. Charnes, D. J. Morrice, D. T. Brunner, and J. J. Swain

RANDOM VARIATE GENERATION IN ONE LINE OF CODE

Luc Devroye
School of Computer Science
McGill University
Montreal, Canada H3A 2A7
luc@cs.mcgill.ca

ABSTRACT

A random variate with a given non-uniform distribution can often by generated in one assignment statement if a uniform source and some simple functions are available. We review such one-line methods for most of the key distributions.

1 A MODEL OF COMPUTATION

Random variate generators that are conceptually simple and quick to program become invariably popular, even if they are not as efficient as some more complicated methods. We explore and survey the simplest end of the spectrum—the generators that can be implemented in one line of code. We assume throughout that an unlimited source of i.i.d. uniform $[0,1]$ random variates $U_1, U_2, \ldots$ is available. When discussing one-liners, we must distinguish between two situations: in the ordinary case, each request of a uniform variate is fulfilled by another number from this sequence. In the extended case, we may index our requests by U_1 , U_2, and so forth, so that repetitions of the same uniform variate within the code are possible. This will be called an *extended one-liner.*

The standard operators $+, -, *, /$ are available, as are mod, round, $\lfloor . \rfloor$, sign, $\lceil . \rceil$, $|.|$, sin, cos, exp, log, tan, atan. Many functions may be derived from these using only a constant number of combinations. For example, the indicator function $I_{x>0}$ is simply $I_{x>a} = 2\,\mathrm{sign}(x-a) - 1$, and $I_{a>x>b} = (\mathrm{sign}(x-a) - \mathrm{sign}(x-b))/2$. Furthermore, max is included as

$$\max(a,b) = a + (b-a) I_{b>a} \ .$$

Alternatively,

$$\max(a,b) = \frac{a+b}{2} + \left| \frac{a-b}{2} \right| \ .$$

Some may include more complicated functions such as Γ or ζ, but these will not be required for the discussion below.

One may think of a one-liner as an expression tree in which the leaves are uniform $[0,1]$ random variables or constants, and the internal nodes are the operators or functions in the accepted class of operators, which we shall call $\mathcal{F}$. In a *simple one-liner*, each leaf has a different uniform random variate associated with it. In an extended one-liner, repetitions may occur. We may put this differently. Each expression may be represented as a directed acyclic graph (or "dag"), in which the leaf nodes contain constants or U_i's, but each U_i occurs only once. If multiple U_i nodes are disallowed, Extended one-liners are implementable by dags, while simple one-liners are implementable by trees. We should point out here that some smart compilers may transform expressions with repetitions into dags before machine translation.

The well-known Box-Muller formula for normal random variates,

$$X = \sqrt{-2 \log U_1} \cos(2\pi U_2) \ ,$$

is thus a simple one-liner. However, the equivalent form

$$X = \exp((1/2)\log(-2\log U_1)) \left(\cos^2(\pi U_2) - \sin^2(\pi U_2)\right)$$

is an extended one-liner. The figure below depicts the expression trees for both forms of the Box-Muller formula.

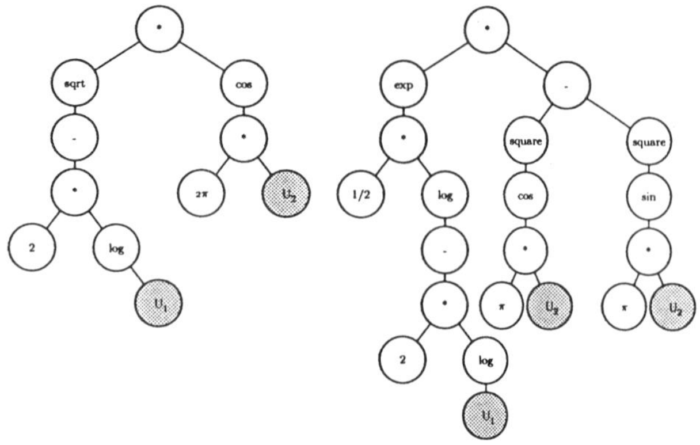

When we have a family of distributions with parameter(s) θ, then it is of interest to have expression trees that have a fixed structure, independent of θ. The value θ appears at best in one or more leaves. For example, if a gamma (k) random variable is generated in one line by summing k independent exponential random variates, then the structure itself of the expression tree changes with k, and indeed, its size grows proportionally with k. For a family of distributions, we define a *fixed one-liner* as one whose expression tree has a given structure, whose internal nodes have fixed operators, and whose leaves have constants, the value θ, or uniform random variates. The operator "take the i-th component of a vector (such as θ)" is in the set of accepted operators.

With the previous set-up, if the family of operators and functions has k members (which is fixed of course), and if each member is unary or binary say, then the number of possible trees on n internal nodes does not exceed

$$\frac{1}{n+1}\binom{2n}{n} \times n^k \approx n^k = \Theta\left(n^{k-3/2}4^n\right) .$$

This limits the numbers of families we may construct using the basic operations. Still, by making n even moderately large, the possibilities are virtually unlimited. We will take a little tour of the popular distributions and exhibit a number of (mostly known) one-liners. Notable exceptions are the gamma and Poisson distributions.

2 NOTATION

We give different symbols for different random variables. For example, U is uniform $[0,1]$, N is standard normal (with density $e^{-x^2/2}/\sqrt{2\pi}$), A is arcsine (with density $1/(\pi\sqrt{1-x^2})$ on $[-1,1]$), $B_{a,b}$ is beta (a,b) (with density $x^{a-1}(1-x)^{b-1}/B(a,b)$ on $[0,1]$ where $a,b>0$), G_a is gamma (a) (with density $x^{a-1}e^{-x}/\Gamma(a)$ on $[0,\infty)$, where $a>0$), E is exponential (with density e^{-x} , $x>0$), L is Laplace (with density $e^{-|x|}/2$), C is Cauchy (with density $1/(\pi(1+x^2))$), T_a is Student t(a) (with density

$$1/(B(a/2,1/2)\sqrt{a}(1+x^2/a)^{\frac{a+1}{2}}) ,$$

where $a>0$).

Some densities are best defined in terms of their characteristic functions φ. A partial list follows below (note that $K(\alpha)=\alpha-2I_{\alpha>1}$): $S_{\alpha,0}$ is symmetric stable with $\varphi(t)=e^{-|t|^\alpha}$, $0<\alpha\le 2$, $S_{\alpha,\beta}$ is stable (α,β) with $\varphi(t)=e^{-|t|^\alpha e^{-i(\pi}}$, $0<\alpha\le 2$, $\alpha\ne 1$, $\beta\in[-1,1]$, $S_{1,\beta}$ is stable $(1,\beta)$ with $\varphi(t)=e^{-|t|(\pi/2+i\beta\,\mathrm{sign}(t)\log|t|)}$, $-1\le\beta\le 1$, $S_{\alpha,1}$ is positive extreme stable with $\varphi(t)=e^{-|t|^\alpha e^{-i(\pi}}$, $0<\alpha<1$, G_a is gamma (a) with $\varphi(t)=1/(1-it)^a$, $a>0$, L is Laplace with $\varphi(t)=1/(1+t^2)$, M_a is Mittag-Leffler with $\varphi(t)=1/(1+(-it)^a)$, $a\in(0,1]$. $P_{a,b}$ is Pillai with $\varphi(t)=1/(1+(-it)^a)^b$, $a\in(0,1], b>0$. L_a is Linnik with $\varphi(t)=1/(1+|t|^a)$, $a\in(0,2]$.

3 THE INVERSION METHOD

The inversion method is based upon the property that $F^{\mathrm{inv}}(U)$ has distribution function F if U is uniformly distributed on $[0,1]$. It leads to one-liners only if F is explicitly invertible in terms of functions that are in $\mathcal{F}$. In the table below, a, b and c are positive constants that serve as parameters.

In this manner, we note that $E \stackrel{\mathcal{L}}{=} -\log U$ as $F(x)=1-e^{-x}$, $B_{1,a}\stackrel{\mathcal{L}}{=}U^{1/a}$, as $F9x)=x^a$ $(0<x<1)$, $B_{a,1}\stackrel{\mathcal{L}}{=}1-U^{1/a}$, $A\stackrel{\mathcal{L}}{=}\cos(\pi U)$, as $F(x)=1-\arccos(x)/\pi$, and $C\stackrel{\mathcal{L}}{=}\tan(\pi U)$ as $F(x)=1/2+\arctan(x)/\pi$.

Other notable examples include the logistic $(F(x)=1/(1+e^{-x}))$ which can be obtained as $-\log(U/(1-U))$. $(U^{-1/a}-1)^{1/c}$ yields a Burr XII random variate $(F(x)=-1/(x^c+1)^a$ $(x>0))$, and $(U^{-1/a}-1)^{-1/c}$ yields a Burr III random variate $(F(x)=1/(x^{-c}+1)^a$ $(x>0))$. A Fréchet or Weibull random variate with $F(x)=1-e^{-x^a}$ $(x>0)$ may be obtained as $\log^{1/a}(1/U)$. For the Gumbel distribution $(F(x)=e^{-ae^{-x}})$, we suggest $-\log((\log(1/U)/a))$. A Pareto or Pearson XI $(F(x)=a/x^{a+1}$ $(x\ge 1))$ may be obtained by $U^{-1/a}$. A tail of the Rayleigh distribution has distribution function $F(x)=1-e^{\frac{a^2-x^2}{2}}$ $(x\ge a>0)$, so that random variates may be obtained as $\sqrt{a^2-2\log U}$. The hyperbolic secant distribution function is $F(x)=1-(2/\pi)\arctan(e^{-\pi x/2})$. Random variates may be obtained as $(2/\pi)\ \log\tan(\pi U/2)$.

4 COMBINATIONS OF TWO OR MORE RANDOM VARIABLES

Mixtures of the form $X = Y$ with probability p and $X = Z$ with probability $1-p$ are easily taken care of in one-liners by setting

$$X = YI_{U<p} + ZI_{U\geq p} = Y + (Z-Y)I_{U\geq p}$$

where U is uniform $[0,1]$ and independent of (Y,Z). However, countably infinite mixtures are not easy to transform into one-liners.

Special distributional properties often lead to elegant one-liners. The triangular density provides a textbook example:

THE TRIANGULAR DENSITY. Assume that we wish to obtain a one-liner for the triangular density with support on $[a,b]$ and mode at m. This could be achieved in a number of ways, but two possibilities are

$$X = m + (a + U_1(b-a) - m)\sqrt{U_2}\ ,$$

and

$$X = m + (a + U_1(b-a) - m)\max(U_2, U_3)\ .$$

Here all the U_i's are i.i.d. uniform $[0,1]$ random variables.

5 THE POLAR METHOD

In the standard polar method, one generates a random pair (X,Y) as $(R\cos\Theta, R\sin\Theta)$, where R and Θ are (random) polar coordinates. Typically, Θ is uniformly distributed on $[0,2\pi]$, and R has a given distribution that is easy to sample from. Often, R is independent of Θ. In the context of one-liners, we thus have

$$X = R\cos(2\pi U)$$

where U is uniform $[0,1]$. Another way of writing this is $X = RA$, where A is a random variable on $[-1,1]$ with the arcsine density (note: our arcsine density is in fact a linear transformation of the standard arcsine density $2/(\pi\sqrt{z(1-z)})$, $0 < z < 1$). A simple exercise in analysis shows that if R and A are independent, and R has density f on $[0,\infty)$, then $X = RA$ has density

$$\int_0^\infty \frac{f(\sqrt{x^2+y^2})}{\pi\sqrt{x^2+y^2}}\,dy\ .$$

EXAMPLE 1: THE NORMAL DENSITY. If

$$f(r) = re^{-r^2/2}\ ,\ r > 0\ ,$$

which is the Rayleigh density (so that $R = \sqrt{-2\log(U)}$ has density f when U is uniform $[0,1]$), then the density of RA is

$$\int_0^\infty \frac{e^{-x^2/2}e^{-y^2/2}}{\pi}\,dy = \frac{1}{\sqrt{2\pi}}e^{-x^2/2}\ .$$

We thus rediscover the Box-Muller method given in the introduction:

$$N \stackrel{\mathcal{L}}{=} \sqrt{-2\log U_1}\cos(2\pi U_2)\ .$$

EXAMPLE 2: THE SYMMETRIC BETA DISTRIBUTION. In the above context, define

$$f(r) = 2cr(1-r^2)^{c-1}\ ,\ 0 < r < 1\ ,$$

where $c > 0$ is a parameter. The distribution function is $F(r) = 1-(1-r^2)^c$, so that, by the inversion method, R is distributed as $\sqrt{1-U^{1/c}}$ when U is uniform $[0,1]$. The density of RA is supported on $[-1,1]$ and is given by

$$\begin{aligned} h(x) &= \int_0^{\sqrt{1-x^2}} \frac{2c(1-x^2-y^2)^{c-1}}{\pi}\,dy \\ &= \frac{2c(1-x^2)^{c-1/2}}{\pi} \times \int_0^1 (1-u^2)^{c-1}\,du \\ &= \frac{2c(1-x^2)^{c-1/2}}{\pi} \times \frac{2^{2c-1}\Gamma^2(c)}{\Gamma(2c)} \\ &= \frac{\Gamma(2c+1)(1-x^2)^{c-1/2}}{2^{2c}\Gamma^2(c+1/2)}, \end{aligned}$$

in which we recognize a shifted version of the symmetric beta density. In the last step, we used a property due to Binet (property 23 on page 261 of Whittaker and Watson, 1927). The last density is that of $2B_{c+1/2,c+1/2} - 1$. Thus, our one-liner for symmetric betas with parameter $a > 1/2$ is based on

$$B_{a,a} \stackrel{\mathcal{L}}{=} \frac{1 + \sqrt{1 - U_1^{\frac{2}{2a-1}}}\cos(2\pi U_2)}{2}\ .$$

This is Ulrich's formula (Ulrich, 1984). For $a = 1/2$, note $B_{1/2,1/2} \stackrel{\mathcal{L}}{=} (1+\cos(2\pi U_2))/2 = \cos^2(\pi U_2)$. Just as for the normal distribution, equivalent respresentations are easy to construct. We note here that the density h plays a central role in nonparametric estimation theory. The case $c = 3/2$ leads to the Epanechnikov or Bartlett density. The case $c = 5/2$ is usually referred to as the quartic kernel. Note that Ulrich's formula is not valid for $a < 1/2$. A fixed one-liner for the entire symmetric beta family is given in the next section.

EXAMPLE 3: THE HALF-BETA DISTRIBUTION. The half-beta density is the density of $B_{1/2,a}$. If X has density h from the previous section, then on $[0,1]$, $Y = X^2$ has density

$$\frac{\Gamma(2c+1)(1-y)^{c-1/2}}{2^{2c}\Gamma^2(c+1/2)\sqrt{y}} = \frac{\Gamma(c+1)(1-y)^{c-1/2}}{\Gamma(1/2)\Gamma(c+1/2)\sqrt{y}} ,$$

which is the density of $B_{1/2,c+1/2}$. Thus, for $a \geq 1/2$,

$$B_{1/2,a} \stackrel{\mathcal{L}}{=} \left(1 - U_1^{\frac{2}{2a-1}}\right) \cos^2(2\pi U_2) .$$

EXAMPLE 4: THE T DISTRIBUTION. If B is beta (a,a), then

$$T_{2a} \stackrel{\mathcal{L}}{=} \frac{\sqrt{2a}(B-1/2)}{2\sqrt{B(1-B)}} .$$

This method yields an extended fixed one-liner for the t distribution. Another fixed one-liner is developed in the next section.

6 THE POLAR METHOD AND NORMAL SCALE MIXTURES

Assume that X is a normal scale mixture random variable, i.e., X can be written as YN, where N is standard normal, and Y is an arbitrary positive random variable, independent of N. If N_1, N_2 are two independent normals independent of Y, then

$$(X_1, X_2) \stackrel{\text{def}}{=} (YN_1, YN_2) \stackrel{\mathcal{L}}{=} (Z_1, Z_2) \times (Y\sqrt{2E}) ,$$

where E is exponential, and (Z_1, Z_2) is uniform on the unit circle and independent of E and Y. This says that (X_1, X_2) has a radially symmetric distribution with random radius distributed as $Y\sqrt{2E}$. Generating X_1 based upon this formula might be called a polar method for X. In some cases, the distribution of $Y\sqrt{2E}$ is very simple. For example, if $Y \stackrel{\mathcal{L}}{=} \sqrt{a/2G_{a/2}}$, then simultaneously

$$Y\sqrt{2E} = \sqrt{\frac{aE}{G_{a/2}}} \; ; \; YN \stackrel{\mathcal{L}}{=} T_a .$$

The last relationship in fact describes the genesis of the t-distribution. The quantity under the square root in the first equation has a beta II (or F) distribution. Routine calculations show that

$$Y\sqrt{2E} \stackrel{\mathcal{L}}{=} \sqrt{a(U^{-2/a} - 1)} .$$

One could thus generate T_a in one line using two independent uniform $[0,1]$ random variates U_1 and U_2 since

$$T_a \stackrel{\mathcal{L}}{=} \sqrt{a(U_1^{-2/a} - 1)} \cos(2\pi U_2) .$$

This method for t-variates was pointed out by Bailey (1994). This could also have been obtained by the method of the previous section, but the method above requires less integration work. We do not make claims that this is the fastest method for generating t variates. There is another interesting observation here: the t density is a mixture of bimodal densities with infinite peaks (arcsine densities)!

ANOTHER ONE-LINER FOR SYMMETRIC BETAS. Best (1978) has proved that

$$B_{a,a} \stackrel{\mathcal{L}}{=} \frac{1}{2}\left(1 + \frac{T_{2a}}{\sqrt{2a + T_{2a}^2}}\right) .$$

Thus, if S denotes a random sign ($S = I_{V \leq 1/2}$ for V uniform $[0,1]$), the following distributional identity yields a one-liner for all symmetric beta distributions:

$$B_{a,a} \stackrel{\mathcal{L}}{=} \frac{1}{2} + \frac{S}{2\sqrt{1 + \frac{1}{\left(U_1^{-1/a} - 1\right)\cos^2(2\pi U_2)}}} .$$

This (new) method is applicable for all values of a—Ulrich's formula required $a > 1/2$.

ANOTHER ONE-LINER FOR HALFBETAS. As $(2B_{a,a} - 1)^2 \stackrel{\mathcal{L}}{=} B_{1/2,a}$, the previous paragraph suggests yet another one-liner for halfbetas:

$$B_{1/2,a} \stackrel{\mathcal{L}}{=} \frac{T_{2a}^2}{2a + T_{2a}^2} = \frac{1}{1 + 2a/T_{2a}^2} .$$

Thus, if S denotes a random sign ($S = I_{V \leq 1/2}$ for V uniform $[0,1]$), the following distributional identity yields a one-liner for all halfbeta distributions:

$$B_{1/2,a} \stackrel{\mathcal{L}}{=} \frac{1}{1 + \frac{1}{\left(U_1^{-1/a} - 1\right)\cos^2(2\pi U_2)}} .$$

This (new) method is applicable for all values of a.

7 DISTRIBUTIONS DEFINED AS ONE-LINERS

Systems of distributions such as Pearson's usually have simple analytic formats. Yet, random variate generation may cause problems. The Pearson IV family, for example, requires quite a bit of work (see Devroye, 1986, p. 480). In modeling, it may be useful to define a distribution by specifying first a one-liner (so that generation is easy), and then worrying about the choice of the parameters and the fine-tuning of the model. There are literally hundreds of such attempts. Tukey (1960) defined a symmetric family by

$$X = \frac{U^\lambda - (1-U)^\lambda}{\lambda},$$

where $\lambda \in \mathbb{R}$ is a parameter and U is uniform $[0,1]$. This was later generalized by Ramberg and Schmeiser (1974) by using different λ's for the exponents.

Omitting location and scale parameters, the **Schmeiser-Deutch family** (Schmeiser and Deutch, 1977) is the family of distributions of the random variables

$$X = \begin{cases} -(\lambda - U)^\mu , & \text{if } U \le \lambda \\ (U-\lambda)^\mu , & \text{if } U > \lambda , \end{cases}$$

where λ and μ are shape parameters.

In hydrology, one uses the Wakeby distribution (Johnson and Kotz, 1988, vol.9, p.513) because it is a versatile five-parameter distribution, gives occasionally outliers, and is easy to simulate. A random variate is obtained as

$$X = a + \frac{b}{c}\left(1-(1-U)^c\right) - \frac{b'}{c'}\left(1-(1-U)^{c'}\right),$$

where U is uniform $[0,1]$, $b' \ge 0$, $b+b' \ge 0$, and either $c+c' > 0$ or $c+b' = c' = 0$.

Burr (1942) (see Tadikamalla, 1980), Johnson (1949) and many others since then have invented their own families of distributions based on this convenient principle. Tadikamalla (1980) reviews many systems. For example, in the **Tadikamalla-Johnson system** Tadikamalla and Johnson (1990), we begin with a logistic random variate

$$Z = \log \frac{U}{1-U},$$

and define three random variables:

$$\begin{aligned} Y_L &= \xi + \lambda e^{(Z-\gamma)/\delta} , \\ Y_B &= \xi + \lambda / \left(1 + e^{-(Z-\gamma)/\delta}\right) , \\ Y_U &= \xi + \lambda \sinh((Z-\gamma)/\delta) . \end{aligned}$$

Just as with the Johnson family, the family covers the entire skewness-kurtosis plane. And the one-liners are fixed as well for the family.

8 REPRESENTATION THEOREMS

Sometimes densities, distribution functions or characteristic functions can be written as integrals, which, upon closer inspection, reveal some method for generating random variates. Two examples follow that lead to useful one-liners.

A NEW ONE-LINER FOR LINNIK'S DISTRIBUTION. Kawata (1972, pp. 396–397) derives the following representation for the density f of the Linnik distribution with parameter $a \in [1,2]$:

$$f(x) = f(-x) = \frac{1}{\pi}\sin\left(\frac{\pi a}{2}\right)\int_0^\infty u e^{-xu} g(u)du, x > 0,$$

where

$$g(u) = \frac{1}{\pi}\sin\left(\frac{\pi a}{2}\right)\frac{u^{a-1}}{1+2\cos(\frac{\pi a}{2})u^a + u^{2a}}, u > 0.$$

Thus, a Linnik random variate can be generated as SE/W where S is a random sign, E is exponentially distributed, and W has density g. Using the transformation, $v = u^a$, it is easy to establish that W is distributed as

$$\left(C \sin\frac{\pi a}{2} - \cos\frac{\pi a}{2}\right)^{1/a},$$

where C is a Cauchy random variable restricted to $C \ge 1/\tan\frac{\pi a}{2}$. Equivalently, C is distributed as $\tan(\frac{\pi}{2}(1-aU))$, where U is uniformly distributed on $[0,1]$. In summary, we have

$$L_a \stackrel{\mathcal{L}}{=} \frac{SE}{\left(\sin\frac{\pi a}{2}\tan(\frac{\pi}{2}(1-aU)) - \cos\frac{\pi a}{2}\right)^{1/a}} .$$

Not in partcular that if E is exponential, S is a random sign, L is Laplace, C is Cauchy, and N_1, N_2 are i.i.d. normal random variates, then

$$L_1 \stackrel{\mathcal{L}}{=} \frac{EN_1}{N_2} \; ; \; L_1 \stackrel{\mathcal{L}}{=} EC \; ; \; L_1 \stackrel{\mathcal{L}}{=} \frac{E}{C} \; ; \; L_2 \stackrel{\mathcal{L}}{=} SE \stackrel{\mathcal{L}}{=} L .$$

THE STABLE LAWS. Ibragimov and Chernin (1952) and Zolotarev (1966, 1986) derived various useful representations for the stable laws. As an example, define $K(a) = a - 2I_{a>1}$, and let $G_{a,b}$ be the distribution function for $S_{a,b}$. Set $\theta = bK(a)/a$. Note that for $x > 0$, $b > 0$,

$$G_{a,b} = \begin{cases} 1 - \frac{1}{2}\int_{-\theta}^1 e^{-x^{\frac{a}{a-1}}U_a(z,\theta)}\,dz & (a > 1) \\ \frac{1-\theta}{2} + \frac{1}{2}\int_{-\theta}^1 e^{-x^{\frac{a}{a-1}}U_a(z,\theta)}\,dz & (a < 1) \\ \frac{1}{2}\int_{-1}^1 e^{-e^{-x/b}U_1(z,b)}\,dz & (a = 1) \end{cases} .$$

The values for $x < 0$ are obtained by noting that $G_{a,b}(x) + G_{a,-b}(-x) \equiv 1$. The values for $b < 0$ are obtained by noting that $S_{a,b} \stackrel{\mathcal{L}}{=} -S_{a,-b}$. The functions U are defined as follows:

$$U_1(z,b) = \frac{\pi(1+bz)}{2\cos\left(\frac{\pi z}{2}\right)}e^{\frac{\pi}{2}(z+1/b)\tan\left(\frac{\pi}{2}\right)} ;$$

$$U_a(z,\theta) = \left(\frac{\sin\left(\frac{\pi a(z+\theta)}{2}\right)}{\cos\left(\frac{\pi z}{2}\right)}\right)^{\frac{a}{1-a}} \frac{\cos\left(\frac{\pi((a-1)z+a\theta)}{2}\right)}{\cos\left(\frac{\pi z}{2}\right)} .$$

All $S_{a,b}$'s are supported on the real line except $S_{a,1}$ for $a < 1$ (which is supported on the positive halfline) and $S_{a,-1}$ for $a < 1$ (which is supported on the negative halfline). Another representation is that the distribution of $S_{a,1}^{a/(a-1)}$ for $a < 1$ is given by

$$\frac{1}{\pi}\int_0^{\pi} e^{-xA(z)}dz ,$$

where

$$A(z) = \frac{\sin((1-a)z)}{\sin(az)} \left(\frac{\sin(az)}{\sin z}\right)^{\frac{1}{1-a}} .$$

In the integrals, we recognize exponential power mixtures, which lead to a variety of one-liners. Kanter (1975) used the last representation to suggest that

$$S_{a,1} \stackrel{\mathcal{L}}{=} \left(\frac{A(\pi U)}{E}\right)^{\frac{1-a}{a}} .$$

For $a < 1$, this would suffice for all values of $b \in [-1,1]$ as

$$S_{a,b} \stackrel{\mathcal{L}}{=} \left(\frac{1+b}{2}\right)^{1/a} S_{a,1} + \left(\frac{1-b}{2}\right)^{1/a} S_{a,-1}$$

(Zolotarev, 1986, p. 61). Defining θ as above and

$$B_a(z) = \frac{\sin\left(\frac{\pi a(z+\theta)}{2}\right)}{\cos\left(\frac{\pi((a-1)z+a\theta)}{2}\right)} \left(\frac{\cos\left(\frac{\pi((a-1)z+a\theta)}{2}\right)}{\cos\left(\frac{\pi z}{2}\right)}\right)^{\frac{1}{a}} ,$$

Chambers, Mallows and Stuck (1976) (see also Zolotarev, 1986) suggest the extended one-liner

$$S_{a,b} \stackrel{\mathcal{L}}{=} B_a(U-1/2)E^{1-1/a}$$

valid for all $a \neq 1$ and $b \in [-1,1]$. For $a = 1$, they obtain

$$S_{1,b} \stackrel{\mathcal{L}}{=} B_1(U-1/2) - \frac{2b}{\pi}\log E ,$$

where

$$B_1(z) = \frac{2b}{\pi}\log\left(\frac{1+bz}{\cos\left(\frac{\pi z}{2}\right)}\right) + (1+bz)\tan\left(\frac{\pi z}{2}\right) .$$

9 SCALE MIXTURES

We say that X is a scale mixture if $X = YZ$ can be decomposed as the product of two independent random variables Y and Z. Such mixtures are convenient ways of trying to discover one-liners. Famous scale mixtures occur when Y is uniform $[0,1]$. In that case, the distribution of X is unimodal with a peak at the origin, and the mixture is called a Khinchine mixture. One should try replacing Y with all random variables for which one-liners are already known. The following are prime candidates: $Y = U^a$, Y is normal, Y is exponential, and Y is Cauchy. At this junction, it is impossible to be exhaustive. We will rather limit ourselves to a few nice examples.

KHINCHINE MIXTURES. The density of $X = U_1U_2$ (with U_1, U_2 i.i.d. uniform $[0,1]$) is $-\log(x)$ on $[0,1]$.

NORMAL SCALE MIXTURES. Let N be standard normal, and let X be a positive random variable with two-sided La-place transform $L(s) = \mathbf{E}e^{-sX}$. Then $Y = N\sqrt{X/2}$ has characteristic function $L(t^2)$. This is easily see by noting that

$$\begin{aligned}\mathbf{E}e^{itY} &= \mathbf{E}e^{itN\sqrt{X/2}}\\ &= \mathbf{E}e^{-t^2X} \quad \text{(condition on } X)\\ &= L(t^2) .\end{aligned}$$

Three main examples come to mind:

A. If X is exponential (thus, $L(s) = 1/(1+s)$), then $N\sqrt{X/2}$ is Laplace, as it has characteristic function $1/(1+t^2)$.

B. Assume that $0 < \alpha < 1$ and set $X = S_{\alpha,1}$. From Zo-lo-ta-rev (1986, p. 112), we know that for the positive stable distribution, $L(s) = e^{-s^\alpha}$ if $s \geq 0$. Therefore, $N\sqrt{S_{\alpha,1}/2}$ has characteristic function $e^{-|t|^{2\alpha}}$. That is, $N\sqrt{S_{\alpha,1}/2} \stackrel{\mathcal{L}}{=} S_{2\alpha,0}$. Symmetric stables can be built up from positive stables and normals. For the latter distributions, one-liners were exhibited aerlier in the paper.

C. If we take in the previous example $\alpha = 1$, then we note that $N\sqrt{S_{1,1}/2}$ has characteristic function

$$e^{-2t^2\log|t|} ,$$

where we used the fact that the two-sided La-place transform of $S_{1,1}$ is $e^{-s\log s}$ for $s > 0$ (Zo-lo-ta-rev, 1986, p. 112).

CAUCHY SCALE MIXTURES. Let C be standard Cauchy and let X be a positive random variable with two-sided La-place transform $L(s)$. Then $Y = CX$ has characteristic function $L(|t|)$. This is easily see by noting that

$$\begin{aligned}\mathbf{E}e^{itY} &= \mathbf{E}e^{itXN}\\ &= \mathbf{E}e^{-|t|X} \quad \text{(condition on } X)\\ &= L(|t|) .\end{aligned}$$

Three examples follow:

A. Since G_a has two-sided La-place transform $L(s) = 1/(1+s)^a$, valid for $\Re(s) \geq 0$, CG_a has characteristic function $1/(1+|t|)^a$.

B. Assume that $0 < \alpha < 1$. Arguing as in the previous subsection, we note that $S_{\alpha,1}C \stackrel{\mathcal{L}}{=} S_{\alpha,0}$. Thus, symmetric stables can also be obtained from Cauchy variables and positive extreme stables.

C. Finally, $S_{1,1}C$ has characteristic function $e^{-|t|\log|t|}$.

SYMMETRIC STABLE MIXTURES. Scale mixtures with stable distributions are best studied via characteristic functions. In particular, if X has density f, then the characteristic function of $X^c S_{a,0}$ is

$$\varphi(t) = \mathbf{E}e^{itX^c S_{a,0}} = \mathbf{E}e^{-|tX^c|^a} = \int f(x)e^{-|t|^a|x|^{ca}}\,dx\,.$$

This is particularly helpful if $f(x)$ has a factor $e^{-|x|^b}$. For example, if $f(x) = e^{-x^b}/\Gamma(1+1/b)$, $x > 0$, then, with $c = b/a$,

$$\varphi(t) = \int_0^\infty \frac{e^{-x^b(1+|t|^a)}}{\Gamma(1+1/b)}\,dx = \frac{1}{(1+|t|^a)^{1/b}}.$$

This is a generalized form of Linnik's distribution (Linnik (1962), Laha (1961), Lukacs (1970), pp. 96–97), which is obtained for $b = 1$ (and thus $c = 1/a$). The one-liner suggested by

$$L_a \stackrel{\mathcal{L}}{=} S_{a,0}E^{\frac{1}{a}}$$

is due to Devroye (1990). If $L_{a,b}$ denotes a generalized Linnik random variable, other consequences of the relationship given above include:

$$\begin{aligned} L_{a,b} &\stackrel{\mathcal{L}}{=} S_{a,0}G_b^{1/a}\;; \\ L_{1,1} &\stackrel{\mathcal{L}}{=} CE\;; \\ L_{2,1} &\stackrel{\mathcal{L}}{=} N\sqrt{E/2} \stackrel{\mathcal{L}}{=} L\;; \\ L_{a,1/2} &\stackrel{\mathcal{L}}{=} S_{a,0}(|N|/\sqrt{2})^{2/a}\;; \\ L &\stackrel{\mathcal{L}}{=} N_1N_2 + N_3N_4\,. \end{aligned}$$

The last statement involves four independent standard normals, and follows from the previous statement.

It is equally simple to verify that if $\{\alpha_j\}$ is a sequence of numbers from $(0,2]$, $\gamma_j \ge 0$, and $S_{\alpha_j,0}$ is a sequence of independent symmetric stable random variables with the given parameters, then $\sum_{j=1}^n S_{\alpha_j}(\gamma_j E)^{1/\alpha_j}$ has characteristic function

$$\frac{1}{1+\sum_{j=1}^n \gamma_j |t|^{\alpha_j}}.$$

POSITIVE EXTREME STABLE MIXTURES: THE MITTAG-LEFFLER DISTRIBUTION. The one-liner

$$X = S_{a,1}G_b^{1/a}$$

yields a random variable X with characteristic function

$$\varphi(t) = \left(\frac{1}{1+(-it)^a}\right)^b\,,\quad a \in (0,1], b > 0\,.$$

This distribution was studied by Pillai (1990). A related distribution was studied by Klebanov, Maniya and Melamed (1984). For $b = 1$, we obtain the Mittag-Leffler distribution with parameter a. Note that the stable $(a,1)$ random variate mentioned here has characteristic function

$$\psi(t) = \begin{cases} \exp(-t^a e^{-i\pi a/2}) & (t > 0) \\ \exp(-(-t)^a e^{i\pi a/2}) & (t < 0)\,. \end{cases}$$

Using Kanter's one-liner for $S_{a,1}$, we see that if E, E^* are i.i.d. exponential random variables and U is uniformly distributed on $[0,1]$, then

$$M_a \stackrel{\mathcal{L}}{=} \left(\frac{E\sin(a\pi U)}{\sin(\pi U)}\right)^{1/a}\left(\frac{E^*\sin(a\pi U)}{\sin((1-a)\pi U)}\right)^{(a-1)/a}$$

REFERENCES

R. W. Bailey, "Polar generation of random variates with the t distribution," *Mathematics of Computation*, vol. 62, pp. 779–781, 1994.

D. J. Best, "A simple algorithm for the computer generation of random samples from a Student's t or symmetric beta distribution," in: *COMPSTAT 1978: Proceedings in Computational Statistics*, ed. L. C. A. Corsten and J. Hermans, pp. 341–347, Physica Verlag, Wien, Austria, 1978.

I. W. Burr, "Cumulative frequency functions," *Annals of Mathematical Statistics*, vol. 13, pp. 215–232, 1942.

J. M. Chambers, C. L. Mallows, and B. W. Stuck, "A method for simulating stable random variables," *Journal of the American Statistical Association*, vol. 71, pp. 340–344, 1976.

L. Devroye, *Non-Uniform Random Variate Generation*, Springer-Verlag, New York, 1986.

L. Devroye, "A note on Linnik's distribution," *Statistics and Probability Letters*, vol. 9, pp. 305–306, 1990.

I. A. Ibragimov and K. E. Chernin, "On the unimodality of stable laws," *Theory of Probability and its Applications*, vol. 4, pp. 417–419, 1959.

M. E. Johnson, "Computer generation of the exponential power distribution," *Journal of Statistical Computation and Simulation*, vol. 9, pp. 239–240, 1979.

N. L. Johnson, "Systems of frequency curves generated by methods of translation," *Biometrika*, vol. 36, pp. 149–176, 1949.

N. L. Johnson, "Systems of frequency curves derived from the first law of Laplace," *Trabajos de Estadistica*, vol. 5, pp. 283–291, 1954.

N. L. Johnson and S. Kotz, "Encyclopedia of the Statistical Sciences," vols. 1-4 (1983), vols. 5-6 (1985), vol. 7 (1986), vols. 8-9 (1988), supplement (1989), John Wiley, New York,

N. L. Johnson and S. Kotz, "Extended and multivariate Tukey lambda distributions," *Biometrika*, vol. 60, pp. 655–661, 1973.

M. Kanter, "Stable densities under change of scale and total variation inequalities," *Annals of Probability*, vol. 3, pp. 697–707, 1975.

T. Kawata, *Fourier Analysis in Probability Theory*, Academic Press, New York, N.Y., 1972.

L. B. Klebanov, G. V. Maniya, and I. A. Melamed, "A problem of Zolotarev and analogs of infinitely divisible and stable distributions in a scheme for summing a random number of random variables," *Theory of Probability and its Applications*, vol. 29, pp. 791–794, 1984.

R. G. Laha, "On a class of unimodal distributions," *Proceedings of the American Mathematical Society*, vol. 12, pp. 181–184, 1961.

R. G. Laha and E. Lukacs, "On a problem connected with quadratic regression," *Biometrika*, vol. 47, pp. 335–343, 1960.

Yu. V. Linnik, "Linear forms and statistical criteria: I, II," *Selected Translations in Mathematical Statistics and Probability*, vol. 3, pp. 1–40, 41-90, 1962.

E. Lukacs, *Characteristic Functions*, Griffin, London, 1970.

R. N. Pillai, "Semi-α Laplace distributions," *Communications in Statistics — Theory and Methods*, vol. 14, pp. 991–1000, 1985.

I. Popescu, "Algorithms for generating the logistic variable," *Economic Computation and Economic Cybernetics Studies and Research*, vol. 10(1), pp. 55–61, 1976.

J. S. Ramberg and B. W. Schmeiser, "An approximate method for generating asymmetric random variables," *Communications of the ACM*, vol. 17, pp. 78–82, 1974.

B. W. Schmeiser and S. J. Deutch, "A versatile four parameter family of probability distributions suitable for simulation," *AIEE Transactions*, vol. 9, pp. 176–182, 1977.

P. R. Tadikamalla, "A look at the Burr and related distributions," *International Statistical Review*, vol. 48, pp. 337–344, 1980.

P. R. Tadikamalla, "Random sampling from the exponential power distribution," *Journal of the American Statistical Association*, vol. 75, pp. 683–686, 1980.

P. R. Tadikamalla, "On simulating non-normal distributions," *Psychometrika*, vol. 45, pp. 273–279, 1980.

P. R. Tadikamalla and N. L. Johnson, "Tables to facilitate fitting Tadikamalla and Johnson's L_B distributions," *Communications in Statistics, Series Simulation*, vol. 19, pp. 1201–1229, 1990.

J. W. Tukey, "The practical relationship between the common transformations of percentages of counts and of amounts," Technical Report 36, Statistical Techniques Research Group, Princeton University, 1960.

G. Ulrich, "Computer generation of distributions on the m-sphere," *Applied Statistics*, vol. 33, pp. 158–163, 1984.

E. T. Whittaker and G. N. Watson, *A Course of Modern Analysis*, Cambridge University Press, Cambridge, 1927 (1980).

V. M. Zolotarev, "On the representation of stable laws by integrals," *Selected Translations in Mathematical Statistics and Probability*, vol. 6, pp. 84–88, 1966.

V. M. Zolotarev, *One-Dimensional Stable Distributions*, American Mathematical Society, Providence, R.I., 1986.

AUTHOR BIOGRAPHY

LUC DEVROYE is professor in the School of Computer Science at McGill University in Montreal, Canada. His research interests include simulation, random number generation, and the probabilistic analysis of algorithms. His 1986 book *Non-Uniform Random Variate Generation* (now out of print) may be consulted as background for the present article. Although he has tried hard, he still does not know how to generate a gamma random variate in one line of code.

Proceedings of the 1996 Winter Simulation Conference
ed. J. M. Charnes, D. J. Morrice, D. T. Brunner, and J. J. Swain

IMPORTANCE SAMPLING FOR MARKOV CHAINS : COMPUTING VARIANCE AND DETERMINING OPTIMAL MEASURES

Indira Kuruganti
Stephen Strickland

Department of Systems Engineering
University of Virginia
Charlottesville, VA 22903

ABSTRACT

In this paper we describe several computational algorithms useful in studying importance sampling (IS) for Markov chains. Our algorithms compute optimal IS measures and evaluate the estimate variance for a given measure. As knowledge of the optimal IS measure implies knowledge of the quantity to be estimated, our algorithms produce this quantity as a by-product. Since effective IS measures must often closely approximate the optimal measure, the use of these algorithms for small problems may produce insights that lead to effective measures for larger problems of actual interest. We consider two classes of problems: hitting times and fixed-horizon costs.

1 INTRODUCTION

Most investigators who experiment with importance sampling quickly discover that the "holy grail" of variance reduction is highly elusive. One seems as apt to make things worse as to improve them. While theoretical results provide guidance in some problem domains (e.g. single station queues, highly reliable Markovian systems - see Heidelberger (1995) for a review), little support exists in other situations.

The few general results in importance sampling indicate that any IS measure with "nice" variance properties (e.g. bounded relative error) must closely approximate the optimal IS measure in many aspects (e.g. support, relative distribution over the sample space). Thus it seems reasonable to compute optimal IS measures for "small" problems in the hope that the insights may be applied to the larger problems of actual interest. In particular, computationally reasonable approximations of the optimal measure may have good variance behavior.

In this paper we describe several computational algorithms useful in studying importance sampling (IS) for Markov chains. Our algorithms compute optimal IS measures and evaluate the estimate variance for a given measure. Since knowledge of the optimal IS measure implies knowledge of the quantity to be estimated, our algorithms produce this quantity as a by-product. Our algorithms employ dynamic-programming-like ideas and compute IS measures in a recursive form amenable to event-by-event simulation.

We consider two classes of problems: hitting times and fixed-horizon costs. Typical examples of hitting time problems include estimating the mean-time-to-overflow in finite capacity Markovian queueing systems, or the mean-time-to-failure in a system with exponential component lifetimes and repair times. Fixed-horizon costs occur when estimating inventory costs or queueing times over a fixed time period.

The paper is organized as follows. In Section 2 we review basic concepts of importance sampling and summarize some earlier results on the properties of the optimal (zero-variance) importance sampling scheme. In Section 3 we discuss the problem of estimating the first passage time to state F starting from state O. We present different approaches of finding the optimal dynamic IS scheme (Section 3.1) and develop a useful tool for assessing the variance of a constructed scheme without using simulation (Section 3.2). Next, in Section 4 we consider the problem of estimating the mean performance measure over a fixed (finite) time horizon. We discuss two different formulations of this problem - a multiplicative formulation in Section 4.1 and an additive formulation in Section 4.2. In each case, we present the optimal IS scheme and develop a tool for variance computation. We conclude in Section 5 with a mention of some heuristics for constructing "good" IS schemes suggested by the results of this paper.

2 BACKGROUND

Let X denote a stochastic process defined on the probability space $(\Omega, \mathcal{F}, \mathbf{P})$, and assume that we wish to estimate the mean $\alpha \equiv \mathsf{E}[g(x)]$ via simulation. The **naive simulation approach** for estimating α involves generating M independent sample paths $x_1, \ldots, x_M$ and forming the estimator

$$\hat{\alpha} = \frac{1}{M}\sum_{i=1}^{M} g(x_i)$$

Since $\mathsf{E}_{\mathsf{P}}[g(x_i)] = \alpha$ the naive estimator is always unbiased.

To motivate the need for importance sampling, let us consider the problem of estimating a rare event probability α. The performance measure in this case is $g(x) = 1_A(x)$ (where $1_A(\cdot)$ is the indicator function of A). Here, the naive estimator has standard deviation $\sigma_{\hat{\alpha}} = \sqrt{\alpha(1-\alpha)/M}$. So, the relative error $\sigma_{\hat{\alpha}}/\alpha$ grows without bound as $\alpha \rightarrow 0$. Equivalently, as the event becomes rarer, the number of samples needed to estimate α to a fixed precision grows without bound. Thus, for many practical situations where simulation is warranted, direct simulation of the system is infeasible.

Importance sampling (IS) is based on the observation that

$$\begin{aligned} \mathsf{E}_{\mathsf{P}}[g(x)] &= \int g(x) d\mathsf{P}(x) \\ &= \int g(x) \frac{d\mathsf{P}(x)}{d\widetilde{\mathsf{P}}(x)} d\widetilde{\mathsf{P}}(x) \\ &= \mathsf{E}_{\widetilde{\mathsf{P}}}[g(x)\mathcal{L}(x)] \end{aligned} \quad (1)$$

where $\widetilde{\mathsf{P}}$ is any measure absolutely continuous w.r.t. $g(\cdot)\mathsf{P}$ and $\mathcal{L}(x) = d\mathsf{P}(x)/d\widetilde{\mathsf{P}}(x)$ is the Radon-Nikodym derivative of P w.r.t. $\widetilde{\mathsf{P}}$ (often called the likelihood ratio) Glynn and Iglehart (1989). Absolute continuity requires that $\widetilde{\mathsf{P}}(x) > 0$ whenever $g(x)\mathsf{P}(x) > 0$.

From (1) it is clear that we can form a new (unbiased) estimator for α, namely,

$$\tilde{\alpha} = \frac{1}{M}\sum_{i=1}^{M} g(x_i)\mathcal{L}(x_i)$$

where the x_i are sampled from the measure $\widetilde{\mathsf{P}}$. The variance of this estimator is given by

$$\sigma^2_{\tilde{\alpha}} = \frac{\mathsf{E}_{\widetilde{\mathsf{P}}}[(g(x)\mathcal{L}(x))^2] - \alpha^2}{M} = \frac{m^2_{\widetilde{\mathsf{P}}} - \alpha^2}{M}. \quad (2)$$

Here $m^2_{\widetilde{\mathsf{P}}}$ represents the second moment of the estimator $g(x_i)\mathcal{L}(x_i)$ corresponding to a single sample. If $\widetilde{\mathsf{P}}$ is chosen properly, then $\tilde{\alpha}$ will have significantly lower variance than $\hat{\alpha}$ - often orders of magnitude lower. However, a poor choice of $\widetilde{\mathsf{P}}$ can lead to increased (even infinite) variance - a fact most recently illustrated in a paper by Andradóttir et. al. (1995). Thus, choosing $\widetilde{\mathsf{P}}$ properly is the key issue in employing importance sampling.

Since the goal is to minimize the variance of $\tilde{\alpha}$, we also make the following

DEFINITION 1 *An* **optimal IS measure** *is one which produces a zero-variance estimate of* α.

Note that if

$$d\widetilde{\mathsf{P}}(x) = g(x) \cdot d\mathsf{P}(x)/\mathsf{E}_{\mathsf{P}}[g(x)] =: d\mathsf{P}^*, \quad (3)$$

then

$$g(x) \cdot \mathcal{L}(x) = \mathsf{E}_{\mathsf{P}}[g(x)]$$

and the variance is zero. Thus, an optimal IS measure always exists. Note that P^* assigns probability proportional to a path's contribution to α, thereby giving rise to the name importance sampling.

3 HITTING TIMES

Consider a continuous-time Markov chain defined on a finite state space S. Given an initial state O, we wish to estimate the hitting time τ_F for a set of states $F \subset S$, $O \notin F$ under the assumption that probability of reaching F from O is very small. Since we are only interested in the *first passage* to F, we will model F as a single absorbing state. Such a model may be used for example to determine the mean time to system failure in a highly reliable system or the the mean time to buffer overflow in a queuing system.

If τ_O is the first passage time to O, and $\alpha = \mathsf{P}[\tau_F < \tau_O]$, then for the regenerative process considered here, $\tau_F = \mathsf{E}_{\mathsf{P}}[\min(\tau_O, \tau_F)]/\alpha$ (see Bratley, Fox, and Schrage (1987) and Heidelberger (1995) for details). This result can be heuristically explained as follows. Since visits to F from O are rare, each sample path which starts in O and eventually ends in F may visit O multiple times in the intermediate epochs. Suppose in an arbitrary sample path, there are M "cycles" (i.e., sections of the sample path beginning in state O and ending with first passage into either state O or F). Let the time duration of the i−th cycle be denoted by T_i. Then, since the system probabilistically regenerates itself at each re-visit to O, $\{T_1, T_2, \ldots, T_M\}$ are i.i.d. variables. Thus, the time until failure is the sum of a random number (M) of i.i.d. random variables and by Wald's lemma we have,

$$\tau_F = \mathsf{E}_{\mathsf{P}}[\sum_{i=1}^{M} T_i] = \mathsf{E}_{\mathsf{P}}[T_i]\mathsf{E}_{\mathsf{P}}[M].$$

The ratio formula for τ_F follows from the fact that $\mathbb{E}_{\mathbb{P}}[M] = 1/\alpha$ and since F is reached very rarely, $\min(\tau_O, \tau_F) \approx \tau_O$ most of the time. Hence, we can write,

$$\mathbb{E}_{\mathbb{P}}[T_i] = \mathbb{E}_{\mathbb{P}}[\min(\tau_O, \tau_F)] \approx \mathbb{E}_{\mathbb{P}}[\tau_O].$$

Now, estimating τ_O involves only a few state transitions and moreover, most cycles end in state O in a typical simulation. So we can easily estimate $\mathbb{E}_{\mathbb{P}}[\min(\tau_O, \tau_F)]$ using the naive simulation approach.

Thus the problem of estimating τ_F essentially reduces to that of estimating the rare event probability $\alpha \equiv \mathbb{P}[A] = \mathbb{E}_{\mathbb{P}}[1_A(x)]$. Here, A is the set of sample paths that begin in O and end in F without visiting O along the way. Since S is finite, A is either finite or at most countably infinite.

It is clear that α does not depend on the time spent in each state; rather it depends on the particular sequence of states visited in a given simulation run. Hence, we can replace the continuous time Markov process by the corresponding embedded discrete time Markov chain (DTMC) $\{X_t, t = 0, 1, 2, \ldots\}$ with state space S and one-step transition probability matrix $P = [p_{s,t}]_{m \times m}$. *In the remaining sections on this problem we will work with this DTMC.*

The optimal IS scheme is given by:

$$\mathbb{P}^*(x) = \begin{cases} \frac{\mathbb{P}(x)}{\alpha} & \text{for } x \in A \\ 0 & \text{otherwise.} \end{cases}$$

A Markov chain construction of $\mathbb{P}^*$ is given in the following theorem:

THEOREM 1 *Suppose the system behaves as a DTMC with state space S and transition matrix P. Then the probability assignment of (3) corresponds to a DTMC whose transition probability matrix P^* has elements:*

$$p^*(s,t) = \frac{p(s,t)\gamma(t)}{\gamma(s)} \tag{4}$$

where $\gamma(s)$ denotes the probability that the system reaches F starting from state s without ever returning to the initial state O.

For proof of this theorem see Kuruganti and Strickland (1995).

Thus if $\gamma(s)$ for each state s in S be known then we can compute the optimal scheme using (4). Since $\gamma(O) = \alpha$, this result is not useful for importance sampling; however, it shows that $\mathbb{P}^*$ can be realized by a Markov chain.

3.1 Recursive Calculation of the Optimal IS Measure

Observe that

$$\gamma(s) = \lim_{N \to \infty} \gamma_{\leq N}(s), s \in S - F$$

where $\gamma_{\leq N}(s)$ denotes the probability that the system first reaches F starting from state s *in N steps or less* without ever returning to the initial state O.

Let $\gamma_N(s)$ denote the corresponding probability of hitting F in exactly N steps without returning to O. Clearly, $\gamma_1(s) \equiv p(s, F)$ and $\gamma_N(F) = 1$ for $N = 1, 2, \ldots$. Set

$$q(s,t) = \begin{cases} 0 & t = O \text{ or } s = F, t \neq F \\ 1 & s = t = F \\ p(s,t) & \text{otherwise} \end{cases}$$

and let Q be the matrix with $q(s,t)$ as its elements.

Now, if the system is in state $s \in S - F$, then:

$$\begin{aligned} \gamma_{\leq N}(s) &= q(s,F) + \sum_{t \notin F} q(s,t)\gamma_{\leq N-1}(t) \\ &= \sum_t q(s,t)\gamma_{\leq N-1}(t). \end{aligned} \tag{5}$$

As in dynamic programming methods (e.g., see Bertsekas (1987)), we can regard $\gamma_{\leq N}(s)$ as the "cost-to-go" at the $N-$th stage. (Note that we are simply accumulating costs without the optimization at each stage that occurs in dynamic programming.)

Suppose $\Gamma_{\leq N} = [\gamma_{\leq N}]_{m \times 1}$ and $\Gamma_0 = [1_F(s)]_{m \times 1}$. Then, equation (5) becomes

$$\Gamma_{\leq N} = Q\Gamma_{\leq N-1} = Q^N \Gamma_0$$

Assuming that the expected hitting time is bounded, we have $\gamma_{\leq N} \to \gamma$ as $N \to \infty$ (see Kemeny and Snell (1976) for a detailed explanation). So if $\Gamma = [\gamma(s)]_{m \times 1}$

$$\Gamma = Q\Gamma$$

which can be solved by a variety of methods. Γ can also be approximated accurately by $Q^N \Gamma_0$ for large N at a computational cost of $O(\log N)$.

3.2 Variance Computation

The recursive approach above can be adapted to compute the estimate variance for any IS scheme characterized by a transition matrix $\tilde{P}$. From (2), this amounts to computing

$$\mathbb{E}_{\tilde{\mathbb{P}}}\left[(1_A(x)\mathcal{L}(x))^2\right] = \mathbb{E}_{\mathbb{P}}\left[(1_A(x)\mathbb{P}(x)/\tilde{\mathbb{P}}(x)\right].$$

Let $\psi_N(s)$ be the contribution to $m^2_{\tilde{\mathbb{P}}}$ of all paths starting from state s in $S-F$ which hit F in exactly N steps. And let $\psi_{\leq N}(s) = \sum_1^N \psi_j(s)$. As before, let $\tilde{Q}$ be a matrix whose elements are given by

$$\tilde{q}(s,t) = \begin{cases} 0 & t = O \text{ or } s = F, t \neq F \\ 1 & s = t = F \\ \tilde{p}(s,t) & \text{otherwise} \end{cases}$$

Then

$$\begin{aligned} \psi_{\leq N}(s) &= \frac{q(s,F)^2}{\tilde{q}(s,F)} + \sum_{t \notin F} \frac{q(s,t)^2}{\tilde{q}(s,t)} \psi_{\leq N-1}(t) \\ &= \sum_t \frac{q(s,t)^2}{\tilde{q}(s,t)} \psi_{\leq N-1}(t). \end{aligned} \quad (6)$$

The desired second moment is given by :

$$m^2_{\tilde{\mathbb{P}}} = \psi(O) = \lim_{N\to\infty} \psi_{\leq N}(O).$$

Letting Λ be the matrix with elements

$$l(s,t) = \begin{cases} \frac{q(s,t)^2}{\tilde{q}(s,t)} & \text{if } q(s,t) > 0 \\ 0 & \text{otherwise} \end{cases} \quad (7)$$

and $\Psi = [\psi]_{m\times 1}$ we can write (6) as

$$\psi_{\leq N} = \Lambda \psi_{\leq N-1} = \Lambda^N \psi_0.$$

where $\psi_0(s) = 1_F(s)$. If the IS scheme has finite variance, so $\psi_{\leq N} \to \psi$ as $N \to \infty$, then $\Psi = [\psi(s)]_{m\times 1}$ is given by

$$\Psi = \Lambda \Psi.$$

Note though that this tool does not address the effect of the IS measure on the computational cost of each sample. We ignore this issue here because in most practical situations of interest, with an appropriate choice for $\tilde{\mathbb{P}}$ the reduction (relative to the naive approach) in number of samples needed to estimate α far exceeds the increased cost of generating each sample under importance sampling (see Sadowsky (1993)).

4 FIXED HORIZON COST FUNCTIONS

Let τ be an arbitrary (but fixed) number of time epochs and $Y = g(X_0, X_1, \ldots, X_\tau)$ some performance measure of interest. Suppose we wish to estimate the mean $\alpha_\tau = E_P[Y]$. Let $\tilde{P} = [\tilde{p}(s,t)]_{m\times m}$ be the one-step transition probability matrix used during importance sampling and x_i the state visited at the i-th epoch. For simplicity, we assume that x_0 is fixed under both P and $\tilde{P}$ i.e., the system always starts in the same initial state. (For a more general treatment we can include the initial probability distribution vectors used under the original and importance sampling measures respectively).

From (3) the optimal (zero-variance) importance sampling scheme for this problem is given by :

$$\mathbb{P}^*(x_0, x_1, \ldots, x_\tau) = \frac{\mathbb{P}(x_0, x_1, \ldots, x_\tau) g(x_0, x_1, \ldots, x_\tau)}{\alpha_\tau} \quad (8)$$

If the function $g(X_0, X_1, \ldots, X_n)$ can be expressed in terms of a recursive equation, then we can exploit some of the techniques developed in the first segment of the paper to find the optimal IS scheme and compute the IS variance. In particular, we show how this may be accomplished for two formulations which cover many applications of practical interest :

Multiplicative:

$$g(X_0, X_1, \ldots, X_\tau) = \prod_{i=0}^{\tau} g(X_i)$$

Additive:

$$g(X_0, X_1, \ldots, X_\tau) = \sum_{i=0}^{\tau} g(X_i).$$

Let $G = [g(s)]_{m\times 1}$. *In both cases we assume that the performance measure of interest is a positive quantity i.e.,* $g(s) > 0$ *for every state* s *in the state space* S. Since the multiplicative form is easier to work with we first examine this in the following section.

4.1 Multiplicative Cost

In this formulation, the mean performance measure α_τ is given by :

$$\alpha_\tau = \sum_{x_1, x_2, \ldots, x_\tau} \prod_{i=0}^{\tau} g(x_i) \prod_{i=0}^{\tau-1} p(x_i, x_{i+1})$$

which can be calculated recursively as explained below. Define

$$\begin{aligned} h_j(x_j) & \\ &= \sum_{x_1, \ldots, x_{j-1}} \prod_{i=0}^{j} g(x_i) \prod_{i=0}^{j-1} p(x_i, x_{i+1}) \\ &= \sum_{x_{j-1}} h_{j-1}(x_{j-1}) p(x_{j-1}, x_j) g(x_j) \end{aligned} \quad (9)$$

with $h_0(x_0) = g(x_0) \equiv \alpha_0$ and

$$\alpha_j = \sum_{x_j} h_j(x_j), \quad j = 1, 2, \ldots, \tau \quad (10)$$

(recall that we assume a fixed starting state x_0). In matrix form, the recursive equation (9) can be written as :

$$H_j = DP^T H_{j-1}, \quad j = 1, 2, \ldots, \tau \tag{11}$$

where D is a $m \times m$ diagonal matrix with elements $g(s)$ along the principal diagonal and P^T is the transpose of P. Here $H_j = [h(s)]_{m\times 1}$ is a vector that is recursively accumulated.

4.1.1 Optimal Scheme

Comparing equations (9) and (10) with (5) we note that in this dynamic programming formulation, α_j behaves as the "cost-to-reach" state x_j, the state visited at the j-th epoch. So, by analogy with (4) we hypothesize the following form for the elements of the Markovian optimal IS transition matrix P^*:

$$p^*(x_j, x_{j+1}) = \frac{\alpha_j p(x_j, x_{j+1}) g(x_{j+1})}{\alpha_{j+1}} \tag{12}$$

where $j = 0, 1, \ldots, \tau-1$. Note that non-negativity of $p^*(x_j, x_{j+1})$ is assured by the positive form assumed for the function $g(.)$. Moreover, equations (9) and (10) ensure that every row of P^* sums to one. Since (12) expresses the optimal IS transition probabilities in terms of the known simulation inputs P and G, the optimal IS scheme can be computed and actually implemented in a simulation.

That this is indeed the optimal IS scheme can be verified as follows. Let $(x_0, \ldots, x_\tau)$ be a particular sample path generated under P^*. Now:

$$\begin{aligned}
&\mathbb{P}^*(x_0, x_1, \ldots, x_\tau) \\
&= \prod_{i=0}^{\tau-1} p^*(x_i, x_{i+1}) \\
&= \frac{\alpha_0 p(x_0, x_1) g(x_1)}{\alpha_1} \frac{\alpha_1 p(x_1, x_2) g(x_2)}{\alpha_2} \\
&\quad \cdots \frac{\alpha_{\tau-1} p(x_{\tau-1}, x_\tau) g(x_\tau)}{\alpha_\tau} \\
&= \frac{\prod_{i=0}^{\tau} g(x_i) \prod_{i=0}^{\tau-1} p(x_i, x_{i+1})}{\alpha_\tau}, \text{ and} \\
&\mathbb{P}(x_0, x_1, \ldots, x_\tau) = \prod_{i=0}^{\tau-1} p(x_i, x_{i+1}).
\end{aligned}$$

So, under $\mathbb{P}^*$ each sample leads to the following I.S. estimate for α_τ :

$$\begin{aligned}
&g(X_0, X_1, \ldots, X_\tau)\mathcal{L}(X_0, X_1, \ldots, X_\tau) \\
&= \left(\prod_{i=0}^{\tau} g(x_i)\right) \frac{\mathbb{P}(x_0, x_1, \ldots, x_\tau)}{\mathbb{P}^*(x_0, x_1, \ldots, x_\tau)} = \alpha_\tau.
\end{aligned}$$

which is the (constant) true value. Hence the variance of the IS estimator is zero i.e., $\mathbb{P}^*$ is the optimal IS scheme.

4.1.2 Variance Computation

As in Section 3.2 we can compute the variance of an arbitrary IS scheme for this problem using dynamic programming and consequently develop a matrix-based formulation.

Equation (11) suggests a matrix-based method for finding $\alpha_\tau \equiv h_0(x_0)$. So, using the same arguments as in Section 4.1 we can derive the recursive matrix equation:

$$V_j = D^2 \Lambda^T V_{j-1}, \; j = 1, 2, \ldots, \tau$$

where $V_0 \equiv G_2 \equiv [g(s)^2]_{m\times 1}$, $m^2_{\tilde{\mathbb{P}}} \equiv \sum_{x_\tau} v_\tau(x_\tau)$ and Λ^T is the transpose of the matrix Λ defined in (7). The vector $V_j = [v_j(x_j)]_{m\times 1}$ can be recursively computed.

Using induction, it can be shown that:

$$\begin{aligned}
V_j &= \left[D^2 \Lambda^T\right]^j V_0, \; j = 1, 2, \ldots, \tau \\
\Rightarrow V_\tau &= \left[D^2 \Lambda^T\right]^\tau G_2.
\end{aligned}$$

So, the variance of an arbitrary (Markovian) IS scheme for this problem can be found using :

$$\tilde{\sigma}^2 = \sum_{x_\tau} v_\tau(x_\tau) - \alpha_\tau^2. \tag{13}$$

4.2 Additive Cost

Now assume that:

$$\alpha_\tau \equiv E_P[Y] = \sum_{x_1, \ldots, x_\tau} \left(\sum_{i=0}^{\tau} g(x_i)\right) \prod_{i=0}^{\tau-1} p(x_i, x_{i+1}).$$

Typical examples include the net payoff from a fixed number of successive gambling trials and the total waiting time experienced by a single customer at different stations in a queueing network. This representation can also be used when the performance measure of interest is the number of times a particular state (or set of states) F is visited in a simulation run of fixed length. In this case $g(x_i) = 1_F(x_i)$. (Note that this last example does not satisfy our earlier assumption of strictly positive values for $g(x_i)$.)

4.2.1 Optimal Scheme

Equation (8) and the fact that

$$\alpha_\tau \equiv E_{\mathbb{P}}\left[\sum_{i=0}^{\tau} g(x_i)\right] = \sum_{i=0}^{\tau} E_{\mathbb{P}}\left[g(x_i)\right]$$

imply that

$$\mathbb{P}^*(x_0, x_1, \ldots, x_{j+1}) = \frac{\left\{\sum_{i=0}^{j+1} g(x_i)\right\}\prod_{i=0}^{j+1} p(x_i, x_{i+1})}{\sum_{i=0}^{j+1} E_{\mathbb{P}}\left[g(x_i)\right]}.$$

But by the Markov property, this probability is given by:

$$\mathbb{P}^*(x_0, x_1, \ldots, x_{j+1}) = \mathbb{P}^*(x_0, x_1, \ldots, x_j)\, p^*(x_j, x_{j+1}).$$

We therefore conclude that

$$\begin{aligned} p^*(x_j, x_{j+1}) &= \frac{\mathbb{P}^*(x_0, x_1, \ldots, x_{j+1})}{\mathbb{P}^*(x_0, x_1, \ldots, x_j)} \\ &= \left(\frac{\sum_{i=0}^{j+1} g(x_i)}{\sum_{i=0}^{j} g(x_i)}\right)\left(\frac{\prod_{i=0}^{j} p(x_i, x_{i+1})}{\prod_{i=0}^{j-1} p(x_i, x_{i+1})}\right) \left(\frac{\sum_{i=0}^{j} E_{\mathbb{P}}\left[g(x_i)\right]}{\sum_{i=0}^{j+1} E_{\mathbb{P}}\left[g(x_i)\right]}\right) \\ &= p(x_j, x_{j+1}) \left(\frac{\sum_{i=0}^{j} E_{\mathbb{P}}\left[g(x_i)\right]}{\sum_{i=0}^{j+1} E_{\mathbb{P}}\left[g(x_i)\right]}\right)\left(\frac{\sum_{i=0}^{j+1} g(x_i)}{\sum_{i=0}^{j} g(x_i)}\right) \\ &= p(x_j, x_{j+1})\left(\frac{\alpha_j}{\alpha_{j+1}}\right)\left(1 + \frac{g(x_{j+1})}{\sum_{i=0}^{j} g(x_i)}\right). \end{aligned}$$

Since all the components of the above expression can be computed off-line or along the sample path, this $P^* = [p^*(s,t)]_{m\times m}$ can be used for simulation. But note that the transition probabilities at epoch i depend on the sample path history up to i. Thus it appears that the optimal IS measure cannot be realized by a Markov chain. Also, observe that as $j \to \infty$,

$$\frac{\alpha_j}{\alpha_{j+1}} \to 1 \text{ and } \frac{g(x_{j+1})}{\sum_{i=0}^{j} g(x_i)} \to 0,$$

so

$$p^*(x_j, x_{j+1}) \to p(x_j, x_{j+1}).$$

Thus, as the sample path length increases, the optimal IS transition matrix P^* approaches the original matrix P. This is consistent with the result derived in Andradóttir et. al (1995) that for fixed time-horizon problems, for best results, the IS transition matrix $\tilde{P}$ must approach the original transition matrix P as the sample path length (i.e., time-horizon) increases.

4.2.2 Variance Computation

We begin by establishing that α_τ can be recursively computed. So, we define

$$\begin{aligned} h_j(x_j) &= \sum_{x_1,\ldots,x_{j-1}} \sum_{i=0}^{j} g(x_i) \prod_{i=0}^{j-1} p(x_i, x_{i+1}) \\ &= \sum_{x_{j-1}} h_{j-1}(x_{j-1}) p(x_{j-1}, x_j) \\ &\quad + P^j(x_0, x_j) g(x_j) \end{aligned}$$

with $h_0(s) = g(s) 1_{x_0}(s)$ and $\alpha_j = \sum_{x_j} h_j(x_j)$. Note that $P^j(x_0, x_j)$, the total probability of going from x_0 to x_j in exactly j steps, is an element of the row for x_0 in the matrix P^j. So, in matrix form the above recursive equation is:

$$H_j = P^T H_{j-1} + D {P^j}^T I_0, j = 1, 2, \ldots, \tau \quad (14)$$

where D is a diagonal $m \times m$ matrix with elements $g(s)$ along the principal diagonal and $I_0 = [1_{x_0}(s)]_{m\times 1}$. Thus the vector $H_j = [h_j(x_j)]_{m\times 1}$ (and hence α_j) can be recursively computed using (14).

Similarly, we can show that the I.S. variance can be recursively computed by defining

$$\begin{aligned} v_j(x_j) &= \sum_{x_1,\ldots,x_{j-1}} \sum_{i=0}^{j} g(x_i)^2 \prod_{i=0}^{j-1} \frac{p(x_i, x_{i+1})^2}{\tilde{p}(x_i, x_{i+1})} \\ &= \sum_{x_{j-1}} v_{j-1}(x_{j-1}) l(x_{j-1}, x_j) \\ &\quad + \Lambda^j(x_0, x_j) g^2(x_j) \\ &\quad + 2g(x_j)\tilde{h}_j(x_j) \end{aligned}$$

where the last term itself has a recursive structure as indicated below:

$$\begin{aligned} \tilde{h}_j(x_j) &= \sum_{x_1,\ldots,x_{j-1}} \left(\sum_{i=0}^{j-1} g(x_i)\right) \prod_{i=0}^{j-1} \frac{p(x_i, x_{i+1})^2}{\tilde{p}(x_i, x_{i+1})} \\ &= \sum_{x_{j-1}} \Lambda^{j-1}(x_0, x_{j-1}) g(x_{j-1}) l(x_{j-1}, x_j) \\ &\quad + \sum_{x_{j-1}} \tilde{h}_{j-1}(x_{j-1}). \end{aligned}$$

Writing these results in matrix form we get:

$$V_j = \Lambda^T V_{j-1} + \Lambda^j G_2 + 2D\tilde{H}_j \quad \text{for } j = 1, 2, \ldots, \tau$$

where

$$\tilde{H}_j = \Lambda^T \left(\tilde{H}_{j-1} + D(\Lambda^{j-1})^T I_0\right)$$

$$\text{So, } \tilde{h}_j = \Lambda^{T^j}\tilde{h}_0 + \sum_{i=1}^{j-1} \Lambda^{T^i} D(\Lambda^{j-i})^T I_0.$$

Here $\tilde{h}_0 = [g(s)^2 1_{x_0}(s)]_{m\times 1}$, $V_0 \equiv [g(s)^2 1_{x_0}(s)]_{m\times 1}$, $G_2 \equiv [g(s)^2]_{m\times 1}$, and $m^2_{\tilde{\mathbb{P}}} \equiv \sum_{x_\tau} v_\tau(x_\tau)$. So, the variance of an arbitrary (Markovian) IS scheme for this problem can be once again computed using Equation (13).

5 CONCLUSIONS

In this paper we have presented several recursive algorithms for constructing optimal IS measures and computing the variance of arbitrary IS measures. (A different non-recursive approach for finding the optimal IS measure for the hitting time problem was presented in Kuruganti and Strickland (1995).) While these techniques are limited to relatively "small" problems, they may prove useful for developing effective schemes for large problems, either through approximation of the optimal scheme or through exploratory evaluation of trial schemes for small prototypes.

For example, when estimating the first passage time to F, using approximations of the methods developed in Section 3.1 we can generate guesstimates for $\gamma(s)$ and substitute these in (4) to construct a sub-optimal yet computationally attractive IS measure that is dynamic (state-dependent). Another possible approach is to replace the original model by a simpler model (with fewer states and/or simpler transition structure) for which $\gamma(s)$ can be computed easily. Then, we could use these results to approximate $\gamma(s)$ for the original more complex model.

Approximations of the recursive method of computing $\gamma(s)$ (and thence the optimal change of measure) may also lead to heuristic strategies for constructing sub-optimal but computationally attractive IS simulation schemes with desirable asymptotic properties. One strategy in this context involves considering at each stage of the infinite-horizon dynamic programming algorithm the contributions from only the dominant most likely paths to failure. In an earlier paper, it was established by Strickland (1993) that the resultant IS scheme behaves like the optimal IS scheme in the limit as the rare event of interest has vanishingly small probability. An alternative technique is to use a myopic approach i.e., restrict attention to only sample paths of length N or less (where N is a small positive integer) in calculating $\gamma(s)$ via equation (5). Similar approximation heuristics may also be investigated for the fixed-horizon problem considered in this paper.

These techniques potentially present a simple and intuitive alternative to the approaches inspired by large deviations theory that have thus far been discussed in the literature - see for example Cottrell et. al (1983), Parekh and Walrand (1989) and Sadowsky (1993).

ACKNOWLEDGMENTS

The work of the authors was supported in part by the National Science Foundation under contract No. ECS-9209902.

REFERENCES

Andradóttir, S., Heyman, D.P., and Teunis, J.Ott, 1995, "On the Choice of Alternative Measures in Importance Sampling with Markov Chains," *Operations Research*, Vol. 43, No. 3, May-June 1995.

Bertsekas, D.P., 1987, *Dynamic Programming : Deterministic and Stochastic Models*, Prentice Hall, New Jersey.

Bratley, P., Fox, B., and Schrage, L., 1987, *A Guide to Simulation*, 2nd Ed., Springer-Verlag, New York.

Cottrell, M., Fort, J. and Malgouyres, G., 1983, " Large Deviations and Rare Events in the Study of Stochastic Algorithms," *IEEE Transactions on Automatic Control*, Vol. AC-28, No. 9, pp. 907-920.

Glynn, P., and Iglehart, D., 1989, "Importance Sampling for Stochastic Simulation," *Management Science* , Vol. 35, No. 11, pp. 1367-1392.

Heidelberger, P., 1995, "Fast Simulation of Rare Events in Queuing and Reliability Models," *ACM Transactions on Modeling and Computer Simulation*, Vol. 5, No. 1, pp 43-85.

Kemeny,J.G., and Snell,J.L., 1976, *Finite Markov Chains*, Springer-Verlag, New York.

Kuruganti, I., and Strickland, S.G., 1995, "Optimal Importance Sampling For Markovian Systems," *Proceedings of the 1995 IEEE Conference on Systems, Man and Cybernetics.*

Parekh, S., and Walrand, J., 1989, "A Quick Simulation Method for Excessive Backlogs in Networks of Queues," *IEEE Transactions on Automatic Control*, Vol. 34, No.1, pp. 54-66.

Sadowsky, J., S., 1993, "On the Optimality and Stability of Exponential Twisting in Monte Carlo Estimation," *IEEE Transactions on Information Theory*, Vol. 39, pp. 119-128.

Strickland, S., 1993, "Optimal Importance Sampling for Quick Simulation of Highly Reliable Markovian Systems," *Proceedings of the 1993 Winter Simulation Conference*, pp 437-444.

AUTHOR BIOGRAPHIES

INDIRA KURUGANTI is a doctoral candidate in the Department of Systems Engineering at the University of Virginia. Her research interests include Markov processes and rare event simulation.

STEPHEN G. STRICKLAND has been an Assistant Professor in the Department of Systems Engineering at the University of Virginia since 1990. His research interests include gradient/sensitivity estimation and rare event simulation.

Proceedings of the 1996 Winter Simulation Conference
ed. J. M. Charnes, D. J. Morrice, D. T. Brunner, and J. J. Swain

SELECTING THE BEST SYSTEM IN TRANSIENT SIMULATIONS WITH VARIANCES KNOWN

Halim Damerdji
Department of Industrial Engineering
North Carolina State University
Raleigh, NC 27695–7906, U.S.A.

Peter W. Glynn
Department of Operations Research
Stanford University
Stanford, CA 94305–4022, U.S.A.

Marvin K. Nakayama
Department of Computer and Information Science
New Jersey Institute of Technology
Newark, NJ 07102, U.S.A.

James R. Wilson
Department of Industrial Engineering
North Carolina State University
Raleigh, NC 27695–7906, U.S.A.

ABSTRACT

Selection of the best system among k different systems is investigated. This selection is based upon the results of finite-horizon simulations. Since the distribution of the output of a transient simulation is typically unknown, it follows that this problem is that of selection of the best population (best according to some measure) among k different populations, where observations within each population are independent, and identically distributed according to some general (unknown) distribution. In this work in progress, it is assumed that the population variances are known. A natural single-stage sampling procedure is presented. Under Bechhofer's indifference zone approach, this procedure is asymptotically valid.

1 INTRODUCTION

Simulation is often used in order to select the best system among a set of k, say, different systems. For example, in the design of an (s, S)-inventory system one may want to try k different settings for the parameters (s, S), the objective being to find the system with the smallest "mean inventory level at end of day." The decision of selecting the best of these systems might be based upon the results of finite-horizon simulations.

In statistical terminology the setting is the following. There are k distinct populations, and a sample of independent and identically distributed variables $X_{i,1}$, $X_{i,2}$, ... can be collected from each population ($i = 1,\ldots,k$). (When simulating the ith system, observation $X_{i,j}$ is the output of the jth replicate.) Let F_i be the cumulative distribution function (cdf) of $X_{i,1}$, whose exact form is unknown, and

$$\mu(F_i) \equiv \mathsf{E}_{F_i} X_{i,1}$$

its mean. (The expectation is indexed to stress that it is with respect to F_i.) The goal is to find an efficient procedure so that the system with the largest mean is selected with probability at least P^*, which is a prespecified parameter (e.g., $P^* = 0.9$). Under the indifference-zone approach, one will want to select the best population with a high likelihood when the second largest mean is at least δ units away from the largest. The procedure's tolerance parameter δ is prespecified.

The parametric Gaussian procedures (e.g., the Dudewicz-Dalal and Rinott procedures) seem to be used in practice, even when the distributions of the samples depart from normality. The small robustness study of Sullivan and Wilson (1989) shows clearly that in populations with even moderate skewness, there can be substantial discrepancy between nominal and actual probabilities of correct selection that are obtained with classical two-stage Dudewicz-Dalal-type procedures. Selection of the best system in the nonparametric setting has both practical and theoretical importance.

It will be assumed here that population variances are known. In practice, it is unlikely that cdf's are not known, yet variances are. However, the problem is difficult. The proofs we have used relied heavily on Berry-Esseen bounds and local limit theorems. We believe that some of these techniques will also be fruitful for tackling the general unknown-variances case. We also believe that solving the general i.i.d. nonparametric case is a step towards solving the problem of selection of the best system in the (stationary) dependent case (which arises in steady-state simulation).

The rest of the paper is organized as follows. Section 2 contains a brief literature review. Some preliminary notations are given in Section 3. The procedure is presented in Section 4 along with the paper's main result, whose proof is sketched in Section 5. A two-stage sampling procedure for the case of unknown variances is presented in Section 6. Section 7 is the conclusion.

2 LITERATURE REVIEW

Selection of the best system, under the indifference-zone approach, is briefly reviewed. Related topics such as subset selection, elimination, and sequential procedures are not discussed. The problem of selection of the best system has been extensively studied in the i.i.d. Gaussian context. The setting considered by Bechhofer (1954), in his seminal paper, was that of known variances. The case of unknown but common variances was investigated by Bechhofer, Dunnett, and Sobel (1954). Dudewicz and Dalal (1975) proposed two two-stage sampling procedures for the case of most interest in practice, where variances are unknown and possibly unequal. Dudewicz and Dalal (1975) then showed the validity of one of the methods. Rinott (1978) considered a modification of the other procedure and showed its validity.

Selection of the best system in the i.i.d. non-Gaussian context has been investigated for certain probability distributions, such as the exponential, gamma, Weibull, Bernoulli (selection of the population with largest probability of success), and multinomial (selection of the cell with largest probability). Nonparametric models have been considered, but mostly in the context of location parameter distributions. (These are distributions with cdf's $F(x;\theta) \equiv F(x-\theta)$.) There are also results on scale distributions. (These are distributions with cdf's $F(x;\theta) \equiv F(x/\theta)$.) Gupta and McDonald (1980) discuss the case where the populations have stochastically increasing (in the parameter) cumulative distribution functions. For relevant discussions on selection of the best system in the nonparametric setting, see Chapter 8 of Gupta and Panchapakesan (1979), Chapter 7 of Gibbons, Olkin, and Sobel (1977), Section 8.6 of Mukhopadhyay and Solanky (1994), and Bechhofer, Santner, and Goldsman (1995, p. 68).

3 PRELIMINARY NOTATIONS

Let Λ be the set of possible distribution functions the various populations can be coming from. In the inventory example, Λ is the set of cdf's that can model the end-of-day inventory level. Let $\tilde{F} = (F_1, \ldots, F_k) \in \Lambda^k$ and $\mu(\tilde{F}) = (\mu(F_1), \ldots, \mu(F_k))$. Denote $\mu(\tilde{F}, [i])$ as the ith largest mean in $\tilde{F}$ (and so $\mu(\tilde{F}, [k])$ is the mean of the best population in $\tilde{F}$). The preference region is given by

$$\Theta(\delta) \equiv \{\tilde{F} \in \Lambda^k : \mu(\tilde{F}, [k]) \geq \mu(\tilde{F}, [k-1]) + \delta\}.$$

The preference region is the set of joint cdf's for which the best and second best means are at least δ units apart from one another. When $\tilde{F} \in \Theta(\delta)$, the procedure should select the best system, with probability at least P^*. The least favorable configuration is the set

$$\begin{aligned}\bar{\Theta}(\delta) \equiv \{\tilde{F} \in \Lambda^k : \mu(\tilde{F}, [1]) &= \cdots = \mu(\tilde{F}, [k-1]) \\ &= \mu(\tilde{F}, [k]) - \delta\},\end{aligned}$$

which is assumed to be nonempty. The procedure will have most difficulty in selecting the best system when $\tilde{F} \in \bar{\Theta}(\delta)$.

Once the systems are chosen, the joint experiment depends upon $\tilde{F}$, which is simply the product of the k marginal cdf's $F_1, \ldots, F_k$. The probability measure associated with this particular experiment will be denoted by $P_{\tilde{F}}$. It is assumed that the variance $\sigma^2(i, \tilde{F}) = \mathsf{E}_{F_i}[(X_{i,1} - \mu(F_i))^2]$ is known for each population ($i = 1, \ldots, k$).

Another parameter used in the selection procedure is the solution h to the equation

$$\int_{-\infty}^{\infty} \Phi(z+h)^{k-1} \Phi(dz) = P^*,$$

where $\Phi(.)$ is the standard-normal cdf. The solution to the equation exists and is unique from classical arguments. Bechhofer, Santner, and Goldsman (1995, pp. 61–63 and pp. 293–294) contains a table of values, and a computer program used to determine h when this parameter cannot be deduced from the table.

4 THE PROCEDURE

Since variances are assumed to be known, the following single-stage procedure for selecting the best system is proposed.

1. Choose P^* ($1/k < P^* < 1$) and $\delta > 0$. Compute h.

2. For $i = 1, \ldots, k$, compute

$$N(i, \tilde{F}, \delta) \equiv \left\lceil \frac{h^2 \sigma^2(i, \tilde{F})}{\delta^2} \right\rceil,$$

where $\lceil x \rceil$ is the smallest integer greater or equal to x.

3. For every population i, sample $X_{i,1}, \ldots, X_{i,N(i,\tilde{F},\delta)}$. Let

$$\bar{X}_i(\delta) \equiv \frac{1}{N(i, \tilde{F}, \delta)} \sum_{j=1}^{N(i,\tilde{F},\delta)} X_{i,j}$$

be the sample mean over the available observations.

4. Select the population with the largest sample mean.

Note that $N(i, \tilde{F}, \delta)$ is not a random variable but a deterministic integer. The procedure uses no distributional information beyond the second moments of the underlying distributions. Since the forms of the cdf's are unknown, it is unrealistic to expect the procedure to be exact for a finite sample; asymptotics have then to be considered in order to show its (asymptotic) validity.

The asymptotics we consider are as the tolerance parameter δ shrinks to zero, which will result in the sample size for each population tending to infinity. The second effect of letting $\delta \to 0$ will be that the preference region gets closer to the whole set Λ^k, making the correct selection increasingly more difficult (because when $\tilde{F}$ belongs to the least favorable configuration, the best mean will be closer and closer to the remaining $(k-1)$ means); nontrivial results are thereby obtained in the limit. Consider the following two assumptions.

Assumption 1 *For all $F \in \Lambda$, F admits a density f, the distribution mean is finite, its variance is finite and uniformly bounded away from zero, and its absolute centered third moment is uniformly bounded (from above).*

Assumption 2 *There exists a finite constant K such that, for all $F \in \Lambda$, $\sup_{\{-\infty<x<\infty\}} f(x) \leq K$.*

Our goal is to show that the above procedure is asymptotically valid. One must show then that the probability of correct selection is at least P^* for any joint distribution $\tilde{F}$ in the preference region. (Correct selection will be denoted by CS(δ).) Since the procedure must be efficient (from a sampling standpoint), it is actually desired that

$$\lim_{\delta\to 0} \inf_{\{\tilde{F}=(F_1,\ldots,F_k)\in\Theta(\delta)\}} P_{\tilde{F}}\Big[\mathrm{CS}(\delta)\Big] = P^*. \qquad (1)$$

Consider the following results.

Proposition 3 *Under Assumptions 1 and 2, we have that for every $\epsilon > 0$, there exists $\delta_0 \equiv \delta_0(\epsilon)$ such that when $0 < \delta \leq \delta_0$,*

$$P_{\tilde{F}}\Big[\mathrm{CS}(\delta)\Big] \;\geq\; P^* - \epsilon \qquad \text{for all } \tilde{F} \in \Theta(\delta).$$

Proposition 4 *Under Assumptions 1 and 2, we have that for all $\epsilon > 0$, there exists $\delta_0 \equiv \delta_0(\epsilon)$ such that when $0 < \delta \leq \delta_0$,*

$$P_{\tilde{F}}\Big[\mathrm{CS}(\delta)\Big] \;\leq\; P^* + \epsilon \qquad \text{for all } \tilde{F} \in \bar{\Theta}(\delta).$$

The main result of the paper follows as a corollary.

Corollary 5 *Under Assumptions 1 and 2, (1) holds, i.e., the procedure is asymptotically valid (as the tolerance parameter $\delta \to 0$).*

The proofs of the propositions are sketched in the next section. The complete proofs are given in Damerdji et al. (1996).

5 SKETCH OF THE PROOFS

As previously mentioned, Berry-Esseen bounds on the rate of convergence in the central limit theorem for i.i.d. variables as well as local limit theorems are pivotal.

Lemma 6 (Petrov, 1975) *Let $Y_1,\ldots,Y_n$ be independent random variables having a common distribution. Suppose $\mathsf{E}Y_1 = \mu$, $\mathsf{E}[(Y_1-\mu)^2] = \sigma^2$, and $\mathsf{E}[|Y_1-\mu|^3] < \infty$. Then*

$$\sup_x \left| P\Big(\frac{1}{\sigma\sqrt{n}}\sum_{j=1}^{n}(Y_j-\mu) < x\Big) - \Phi(x)\right| \leq A\frac{\mathsf{E}[|Y_1-\mu|^3]}{\sigma^3}\frac{1}{\sqrt{n}},$$

where A is a universal constant.

Lemma 7 (Petrov, 1975) *Let $Y_1,\ldots,Y_n$ be independent random variables having a common distribution, with density $p(x)$, such that $\mathsf{E}Y_1 = \mu$, $\mathsf{E}[(Y_1-\mu)^2] = \sigma^2$, $\mathsf{E}[|Y_1-\mu|^3] < \infty$, and $\sup_{-\infty<x<\infty} p(x) \leq K$ for some constant K. Let $p_n(x)$ be the density of the random variable $(1/(\sigma\sqrt{n}))\sum_{j=1}^{n}(Y_j-\mu)$. Then*

$$\sup_{-\infty<x<\infty}\left|p_n(x) - \frac{1}{\sqrt{2\pi}}e^{-x^2/2}\right| \leq A'\frac{\mathsf{E}[|Y_1-\mu|^3]}{\sigma^3\sqrt{n}}\max(1,K^3),$$

and for all x

$$\left|p_n(x) - \frac{1}{\sqrt{2\pi}}e^{-x^2/2}\right| \leq A'\Big(\frac{\mathsf{E}[|Y_1-\mu|^3}{\sigma^3]}\Big)^{2m-1}\frac{\max(1,K^{2m+1})}{\sqrt{n}(1+|x|^m)}$$

for $m = 2$ and $m = 3$, where A' is a universal constant.

Denote $\bar{X}_{(i)}(\delta)$ as the sample mean of the population with ith largest mean. The probability of correct selection is

$$P_{\tilde{F}}\Big[\mathrm{CS}(\delta)\Big] = P_{\tilde{F}}\Big[\bar{X}_{(i)}(\delta) < \bar{X}_{(k)}(\delta) \quad \forall i \neq k\Big].$$

Let

$$a(t; \tilde{F}, (i), \delta) \equiv P_{\tilde{F}}\Big[\frac{\bar{X}_{(i)}(\delta) - \mu(\tilde{F}, [i])}{\delta/h} < t\Big].$$

We now sketch the proof of Proposition 3. For all $\tilde{F} \in \Theta(\delta)$, we have that

$$P_{\tilde{F}}\Big[\mathrm{CS}(\delta)\Big] \geq P_{\tilde{F}}\Big[\frac{\bar{X}_{(i)}(\delta) - \mu(\tilde{F}, [i])}{\delta/h} < \frac{\bar{X}_{(k)}(\delta) - \mu(\tilde{F}, [k])}{\delta/h} + h \quad \forall i \neq k\Big]$$
$$= \int_{-\infty}^{\infty} \Big(\prod_{i \neq k} a(z+h; \tilde{F}, (i), \delta)\Big) a(dz; \tilde{F}, (k), \delta). \tag{2}$$

Some more notation is needed. Let

$$b(t; \tilde{F}, (i), \delta) \equiv P_{\tilde{F}}\Big[\frac{\bar{X}_{(i)}(\delta) - \mu(\tilde{F}, [i])}{\sqrt{\sigma^2((i), \tilde{F})/N((i), \tilde{F}, \delta)}} < t\Big],$$

$$c(t; \tilde{F}, (i), \delta) = a(t; \tilde{F}, (i), \delta) - b(t; \tilde{F}, (i), \delta),$$

and

$$d(t; \tilde{F}, (i), \delta) = b(t; \tilde{F}, (i), \delta) - \Phi(t).$$

We naturally have that

$$a(z+h; \tilde{F}, (i), \delta) = \Phi(z+h) + c(z+h; \tilde{F}, (i), \delta) + d(z+h; \tilde{F}, (i), \delta).$$

The expression $\prod_{i \neq k} a(z+h; \tilde{F}, (i), \delta)$ in (2) can then be expanded into a sum of terms involving $c(z+h; \tilde{F}, (i), \delta)$, $d(z+h; \tilde{F}, (i), \delta)$, and $\Phi(z+h)$. Lemma 6 is used to show that for all $\tilde{F} \in \Theta(\delta)$,

$$P_{\tilde{F}}\Big[\mathrm{CS}(\delta)\Big] \geq \int_{-\infty}^{\infty} \Big(\Phi(z+h)\Big)^{k-1} a(dz; \tilde{F}, (k), \delta) - H(\tilde{F}, (k), \delta),$$

where the exact form of $H(\tilde{F}, (k), \delta)$ is given in Damerdji et al. (1996). Under Assumptions 1 and 2, this term can be made arbitrarily small uniformly in $\tilde{F}$.

We now rewrite

$$a(dz; \tilde{F}, (k), \delta) = \Phi(dz) + \Big(a(dz; \tilde{F}, (k), \delta) - \Phi(dz)\Big),$$

and use Lemma 7 to bound $|a(dz; \tilde{F}, (k), \delta) - \Phi(dz)|$.

As for Proposition 4, we start by considering $\tilde{F}$ in the least favorable configuration. For such $\tilde{F}$, we get that

$$P_{\tilde{F}}\Big[\mathrm{CS}(\delta)\Big] \leq \int_{-\infty}^{\infty} \Big(\Phi(z+h)\Big)^{k-1} a(dz; \tilde{F}, (k), \delta) + H(\tilde{F}, (k), \delta).$$

Using the bound on $|H(\tilde{F}, (k), \delta)|$ developed earlier, we get that $P_{\tilde{F}}\Big[\mathrm{CS}(\delta)\Big] \leq P^* + \epsilon$, where ϵ depends only upon δ. The detailed proof is given in Damerdji, Glynn, Nakayama, and Wilson (1996).

6 THE TWO-STAGE SAMPLING PROCEDURE

The following is a two-stage procedure for selection of the best system for i.i.d. observations with unknown distributions (and variances). The (asymptotic) validity of this procedure is yet unavailable.

1. Choose P^* $(1/k < P^* < 1)$ and $\delta > 0$. Compute h.
2. Let $m(\delta) \equiv \lceil \delta^{-2} \rceil$ be the size of the first subsample for each population.
3. For every population $i = 1, \ldots, k$, let $X_{i,1}$, $X_{i,2}$, $\ldots$, $X_{i,m(\delta)}$ be the corresponding sample. Let
$$\bar{X}_{i,1}(\delta) \equiv \frac{1}{m(\delta)} \sum_{j=1}^{m(\delta)} X_{i,j}$$
and
$$S_i^2(\delta) \equiv \frac{1}{m(\delta)-1} \sum_{j=1}^{m(\delta)} \Big(X_{i,j} - \bar{X}_{i,1}(\delta)\Big)^2.$$
4. For every population i, compute the total sample size:
$$N(i, \tilde{F}, \delta) = \max\Big\{m(\delta), \Big\lceil \frac{h^2 S_i^2(\delta)}{\delta^2} \Big\rceil\Big\}.$$
5. If $N(i, \tilde{F}, \delta) \geq m(\delta) + 1$, sample $X_{i,m(\delta)+1}$, $\ldots$, $X_{i,N(i,\tilde{F},\delta)}$.
6. Let
$$\bar{X}_i(\delta) \equiv \frac{1}{N(i, \tilde{F}, \delta)} \sum_{j=1}^{N(i,\tilde{F},\delta)} X_{i,j}.$$
7. Select the population with the largest sample mean.

Note that in the unknown-variances case, the total sample sizes are not deterministic quantities but, rather, random variables. This is not a major issue in the Gaussian case; in that setting the mean and sample variances are independent (the total sample is function of the sample variance), which allows passage to the t-distribution. See Remark 4.1 of Dudewicz and Dalal (1975). Another great difficulty in showing the validity of this procedure is that we want equality in (1) so the procedure is computationally efficient. This work is still in progress.

7 CONCLUSION

Suppose a number of systems are to be investigated via simulation, where the nature of the problem is such that the simulations are of the finite-horizon type. Simulation of each system will then involve running independent replicates, each replicate providing a single "observation," whose distribution is unknown.

Selection of the best system is undertaken here. In this work in progress, it is assumed that the population variances are known. A single-stage procedure was presented, and its asymptotic validity discussed. The detailed proofs are given in Damerdji, Glynn, Nakayama, and Wilson (1996).

The assumption of known variances is unrealistic in practice. For the unknown-variances case, a two-stage procedure was presented. Asymptotic validity of this procedure, i.e., Equation (1), is still an open problem for the two-stage case. Some of the tools described here might be useful. Another problem under investigation is that of selection of the best system in steady-state simulation, i.e., when the problems are such that the simulations are of the steady-state type.

REFERENCES

Bechhofer, R. E. 1954. A single-sample multiple decision procedure for ranking means of normal populations with known variances. *The Annals of Mathematical Statistics,* **25**, 16–39.

Bechhofer, R. E., Dunnett, C. W., and M. Sobel. 1954. A two-sample multiple-decision procedure for ranking means of normal populations with a common unknown variance. *Biometrika,* **41**, 170–176.

Bechhofer, R. E., T. J. Santner, and D. M. Goldsman. 1995. *Design and Analysis of Experiments for Statistical Selection, Screening, and Multiple Comparisons.* Wiley, New York.

Damerdji, H., P. W. Glynn, M. K. Nakayama, and J. R. Wilson. 1996. Selection of the best system in a finite-horizon simulation context. Working paper.

Dudewicz, E. J., and S. R. Dalal. 1975. Allocation of observations in ranking and selection with unequal variances. *Sankhyā,* **B37**, 28–78.

Gibbons, J. D., I. Olkin, and M. Sobel. 1977. *Selecting and Ordering Populations: A New Statistical Methodology.* Wiley, New York.

Gupta, S. S., and G. C. McDonald. 1980. Nonparametric procedures in multiple decisions (ranking and selection procedures). *Colloquia Mathematica Societatis János Bolyai,* **32**, 361–389.

Gupta, S. S., and S. Panchapakesan. 1979. *Multiple Decision Procedures: Theory and Methodology of Selecting and Ranking Populations.* Wiley, New York.

Mukhopadhyay, N., and T. K. S. Solanky. 1994. *Multistage Selection and Ranking Procedures.* Marcel Dekker, New York.

Petrov, V. V. 1975. *Sums of Independent Random Variables.* Springer-Verlag, New York.

Rinott, Y. 1978. On two-stage selection procedures and related probability-inequalities. *Communication in Statistics: Theory and Methods,* **A7(8)**, 799–811.

Sullivan, D. W., and J. R. Wilson. 1989. Restricted subset selection procedures for simulation. *Operations Research,* **37**, 52–71.

AUTHOR BIOGRAPHIES

HALIM DAMERDJI is an assistant professor in the Department of Industrial Engineering at North Carolina State University. He received a Ph.D. degree in industrial engineering from the University of Wisconsin-Madison. He has held positions at the Ecole Nationale Polytechnique of Algiers and Purdue University. His research interests are in simulation and applied stochastic processes. He is a member of INFORMS.

PETER W. GLYNN received his Ph.D. from Stanford University, after which he joined the faculty of the Department of Industrial Engineering at the University of Wisconsin-Madison. In 1987, he returned to Stanford, where he currently holds the Thomas Ford Faculty Scholar Chair in the Department of Operations Research. He was a co-winner of the 1993 Outstanding Simulation Publication Award sponsored by the TIMS College on Simulation. His research interests include discrete-event simulation, computational probability, queueing, and general theory for stochastic systems.

MARVIN K. NAKAYAMA is an assistant professor in the Department of Computer and Information Science at the New Jersey Institute of Technology. Previously, he has held positions in the Management Science/Computer Information Systems Department of the Graduate School of Management at Rutgers University, and at the IBM Thomas J. Watson Research Center in Yorktown Heights, New York. He received Ph.D. in Operations Research from Stanford University. His research interests include applied probability, simulation output analysis, gradient estimation, rare-event simulation, and random-variate generation.

JAMES R. WILSON is Professor and Director of Graduate Programs in the Department of Industrial Engineering at North Carolina State University. He received a B.A. degree in mathematics from Rice University, and he received M.S. and Ph.D. degrees in industrial engineering from Purdue University. His current research interests are focused on the design and analysis of simulation experiments. He also has an active interest in applications of operations research techniques to all areas of industrial engineering. From 1988 to 1992, he served as Departmental Editor of *Management Science* for Simulation. He was *Proceedings* Editor for WSC '86, Associate Program Chair for WSC '91, and Program Chair for WSC '92. He has also held various offices in TIMS (now INFORMS) College on Simulation. He is a member of ASA, ACM/SIGSIM, IIE, and INFORMS.

Proceedings of the 1996 Winter Simulation Conference
ed. J. M. Charnes, D. J. Morrice, D. T. Brunner, and J. J. Swain

GETTING MORE FROM THE DATA IN A MULTINOMIAL SELECTION PROBLEM

J. O. Miller

Department of Industrial, Welding
& Systems Engineering
The Ohio State University
Columbus, OH 43210, U.S.A.

Barry L. Nelson

Department of Industrial Engineering
& Management Sciences
Northwestern University
Evanston, IL 60208, U.S.A.

Charles H. Reilly

Department of Industrial Engineering & Management Systems
University of Central Florida
Orlando, FL 32816, U.S.A.

ABSTRACT

We consider the problem of determining which of k simulated systems is most likely to be the best performer based on some objective performance measure. The standard experiment is to generate v independent vector observations (replications) across the k systems. A classical multinomial selection procedure, BEM (Bechhofer, Elmaghraby, and Morse), prescribes a minimum number of replications so that the probability of correctly selecting the true best system meets or exceeds a prespecified probability. Assuming that larger is better, BEM selects as best the system having the largest value of the performance measure in more replications than any other. We propose using these same v replications across k systems to form v^k pseudoreplications (no longer independent) that contain one observation from each system, and again select as best the system having the largest value of the performance measure in more pseudoreplications than any other. We expect that this new procedure, AVC (all vector comparisons), dominates BEM in the sense that AVC will never require more independent replications than BEM to meet a prespecified probability of correct selection. We present analytical and simulation results to show how AVC fares versus BEM for different underlying distribution families, different numbers of populations and various values of v. We also present results for the closely related problem of estimating the probability that a specific system is the best.

1 INTRODUCTION AND MOTIVATION

Motivating Example: As tactical war planning analysts, we are directed to provide the Joint Task Force Commander with the best plan to cripple the enemy's command and control. "Best" means achieving the highest level of cumulative damage expectancy (CDE) against a selected set of targets given current intelligence estimates of enemy defense capabilities and available friendly forces. Our team prepares four independent attack plans and we simulate v replications across all four plans. For each replication we compare the CDE between each of the four plans. Since the chosen plan can only be executed a single time, we select as the best plan the one that has the largest CDE in most of the replications.

We consider the general problem of selecting the best of $k \geq 2$ independent populations, $\pi_1, \pi_2, \ldots, \pi_k$, where in our context "populations" is taken to mean simulated systems. This is known as the multinomial selection problem (MSP). Let $\mathbf{X}_i = (X_{1i}, X_{2i}, \ldots, X_{ki})$ represent a vector of independent observations of some common performance measure across all populations on the ith replication. For each i, the best population is the population with the largest X_{ji}. The goal is to find the population that is most likely to be the best performer among the populations, as opposed to identifying the best average performer over the long run. Applications include selecting the best of a set of tactical or strategic military actions, as presented above. In the areas of marketing research or opinion surveys, we might determine the most popular brand, flavor, etc., or the most favored candidate or position on a political issue. An example in the area of structural engineering is finding the design that performs best in a one-time catastrophic event, such as an earthquake. The goal in any MSP is to achieve a prespecified probability of correctly selecting the best population with a minimum amount of data.

Let $Y_{ji} = 1$ if $X_{ji} > X_{\ell i}$, for $\ell = 1, 2, \ldots, k$, but $\ell \neq j$; and let $Y_{ji} = 0$ otherwise. In other words, $Y_{ji} = 1$ if X_{ji} is the largest observation in $\mathbf{X}_i$. In case of a tie for the largest value, we randomly select one of the tied populations as the best. Suppose that there are v independent replications across all populations, and let $Y_j = \sum_{i=1}^{v} Y_{ji}$ represent the number of times population j wins out of these v replications. So $\sum_{j=1}^{k} Y_j = v$ and the k-variate discrete random variable $\mathbf{Y} = (Y_1, Y_2, \ldots, Y_k)$ follows a multinomial distribution with $p_j = \Pr\{\pi_j \text{ wins}\}, j = 1, 2, \ldots, k$, where $0 < p_j < 1, j = 1, 2, \ldots, k$, and $\sum_{j=1}^{k} p_j = 1$. The probability mass function for $\mathbf{Y}$ with parameter v and success probabilities $\mathbf{p} = (p_1, p_2, \ldots, p_k)$ is

$$\Pr\{Y_1 = y_1, Y_2 = y_2, \ldots, Y_k = y_k\} = \frac{v!}{\prod_{j=1}^{k} y_j!} \prod_{j=1}^{k} p_j^{y_j}$$

The goal of MSP is to find the population π_i associated with the largest p_i.

The paper is organized as follows: We first provide a brief review of MSP and the classical approach to solving it. Then we describe our new procedure that uses the same data and increases the probabilitly of correctly selecting the best population. Some analytical results are presented next, along with empirical results for a number of specific distributions for the X_{ji}. Finally, we describe the closely related problem of estimating p_j, the probability that population j will be the best.

2 BACKGROUND

Bechhofer, Elmaghraby and Morse (1959) describe a single-stage procedure for selecting the multinomial event (population) which has the largest success probability. BEM requires the specification of P^* (where $1/k < P^* < 1$), a minimum probability of correctly identifying the population with the largest success probability (i.e., the best population), and θ^* (where $1 < \theta^* < \infty$), the ratio of the largest success probability to the second largest success probability. The procedure consists of the following steps:

Procedure BEM

1. For given k and θ^*, find the minimum value of v that guarantees that the probability of selecting the best population is at least P^*.
2. Generate v independent replications for each population.
3. Compute $Y_j = \sum_{i=1}^{v} Y_{ji}$, for $j = 1, 2, \ldots, k$.
4. Let $y_{[1]} \leq y_{[2]} \leq \cdots \leq y_{[k]}$ be the ranked sample counts from step 3. Select the population associated with the largest count, $y_{[k]}$, as the best population. In case of a tie for the largest count, randomly select one of the tied populations as the best.

To determine the appropriate v in step 1, let $p_{[1]} \leq p_{[2]} \leq \cdots \leq p_{[k]}$ denote the ranked success probabilities for the k populations. Since only values of the ratio $\theta = p_{[k]}/p_{[k-1]}$ greater than or equal to θ^* are of interest, we are indifferent between the best and the next-best population for values of $\theta < \theta^*$. A procedure of this type is referred to as an *indifference-zone approach.* Select v as the minimum number of independent vector observations required to achieve a probability of correct selection (PCS) greater than or equal to P^* whenever $\theta \geq \theta^*$.

If we obtain a PCS $\geq P^*$ with our selected v under the least favorable configuration (LFC) of $\mathbf{p} = (p_{[1]}, p_{[2]}, \ldots, p_{[k]})$, a PCS of at least P^* can be guaranteed for any configuration of $\mathbf{p}$ with $\theta \geq \theta^*$. Keston and Morse (1959) prove that the LFC for BEM is given by:

$$p_{[1]} = p_{[2]} = \cdots = p_{[k-1]} = \frac{1}{(\theta^* + k - 1)}$$

$$p_{[k]} = \frac{\theta^*}{(\theta^* + k - 1)}$$

However, the PCS can be calculated for any configuration of $\mathbf{p}$ with $p_{[k]} > p_{[k-1]}$. Let $\pi_{[j]}$ be the population associated with $p_{[j]}$ and let $Y_{[j]}$ represent the number of wins for $\pi_{[j]}$. So the subscripts for the populations and the associated number of wins are based on the ranking of the p_j s. We refer to the PCS using BEM for a fixed k and v as PCS^{bem}. For any fixed k and v, PCS^{bem}can be expressed as

$$\text{PCS}^{\text{bem}}(\mathbf{p}) = \sum \frac{1}{t} \frac{v!}{\prod_{j=1}^{k} Y_{[j]}\,!} \prod_{j=1}^{k} p_{[j]}^{Y_{[j]}},$$

where the summation is over all vectors $\mathbf{Y} = (Y_{[1]}, \ldots, Y_{[k]})$ such that $Y_{[k]} \geq Y_{[j]}$ ($j = 1, 2, \ldots, k - 1$), and t is the number of populations tied for the most wins.

3 NEW METHOD

We propose a method to provide a PCS greater than or equal to PCS^{bem} using the same replications $\mathbf{X}_i, i = 1, 2, \ldots, v$. In other words, an improvement in PCS for "free." We use the BEM parameters k, P^*, and θ^*, and we execute the first step of BEM to find a value of v. Then a total of v^k pseudoreplications are formed by associating each X_{ji} ($j = 1,2,\ldots,k$; $i = 1,2,\ldots,v$), with all possible combinations of the remaining $X_{\ell h}$ ($\ell = 1,2,\ldots,k$; $\ell \neq j$; $h = 1,2,\ldots,v$). Each such pseudoreplication contains one observation from each population. Note that the v^k pseudoreplications include the v independent replications from which the pseudoreplications are formed.

Define

$$Z_j = \sum_{a_1=1}^{v} \sum_{a_2=1}^{v} \cdots \sum_{a_k=1}^{v} \prod_{\ell=1;\ell\neq j}^{k} \phi(X_{ja_j}, X_{\ell a_\ell}) \quad (1)$$

for $j = 1,2,\ldots,k$, where $\phi(X_{ji}, X_{\ell i})$ is an indicator function that takes the value 1 only if X_{ji} is larger than $X_{\ell i}$. Thus, Z_j represents the number of times out of v^k pseudoreplications that population π_j wins. This new procedure consists of the following steps:

Procedure AVC

1. Given values for k, P^*, and θ^*, use step 1 of procedure BEM to determine a value for v.
2. Generate v independent replications for each population and construct the additional $v^k - v$ pseudoreplications possible with one value from each of the populations.
3. Compute Z_j using equation (1).
4. Let $z_{[1]} \leq z_{[2]} \leq \cdots \leq z_{[k]}$ be the ranked sample counts from step 3. Select the population associated with the largest count, $z_{[k]}$, as the best population. In case of a tie for the largest count, randomly select one of the tied populations as the best.

We conjecture that the PCS with AVC, referred to as PCS^{avc}, is greater than or equal to PCS^{bem}. PCS^{avc} can be expressed

$$\text{PCS}^{\text{avc}}(\mathbf{p}) = \sum \frac{1}{t} \Pr\{Z_{[1]} = z_{[1]}, \ldots, Z_{[k]} = z_{[k]}\},$$

where the summation is over all vectors $\mathbf{Z} = (Z_{[1]}, \ldots, Z_{[k]})$ such that $Z_{[k]} \geq Z_{[j]}$, $j = 1,2,\ldots,k-1$, and t is the number of populations tied for the most wins. Since $\mathbf{Z}$ does not follow a multinomial distribution, we must go back to the distributions of the original observations, X_{ji}, $j = 1,2,\ldots,k$; $i = 1,2,\ldots,v$ to calculate PCS^{avc}. Analytical and simulation results using a number of different population distributions show that PCS^{avc} depends on the underlying distributions of the X_{ji} s.

4 ANALYTICAL RESULTS

Initially, we restrict our attention to continuous distributions for the X_{ji} s which eliminates the possibility of ties among observations. We can calculate PCS^{avc} by conditioning on the joint density of all the order statistics for the v independent replications from $\pi_{[k]}$. Let $\pi_{[k]}$ follow some distribution and let X represent an observation from $\pi_{[k]}$. Let all the remaining populations be identically distributed and let O represent an observation from any of these populations. This set up gives us the LFC for BEM.

Consider permutations of the ranks of the observations from all populations. For any such permutation we can determine the value of $Z_{[k]}$ and calculate the probability of obtaining that arrangement of ranks. We refer to such an arrangement as a *rank order*.

As an example, suppose $k = 3, v = 2$. Then

$$\begin{aligned} \Pr\{Z_{[3]} = 8\} &= \Pr\{O_{(1)} < O_{(2)} < O_{(3)} \\ &\quad < O_{(4)} < X_{(1)} < X_{(2)}\} \quad (2) \\ \Pr\{Z_{[3]} = 6\} &= 4\Pr\{O_{(1)} < O_{(2)} < O_{(3)} \\ &\quad < X_{(1)} < O_{(4)} < X_{(2)}\} \quad (3) \end{aligned}$$

Probability statement (2) uniquely covers all 4! permutations of the O's that are less than both X's. However, in probability statement (3), since all the O's are not adjacent, we need to consider how many of the 4! permutations of the O's result in a unique combination of adjacent O's. This is why the coefficient '4' appears on the right-hand side of equation (3). Similar arguments can be used to derive expressions for possible values of $Z_{[k]}$ for integers $k, v \geq 2$. For this example, there is only a single rank order that results in each value of $Z_{[k]}$. As k or v get even moderately large, there will be multiple rank orders that result in the same value for $Z_{[k]}$ and the calculation of the probability of each value of $Z_{[k]}$ becomes extremely tedious.

If we pick a particular distribution family for X and O, then we can obtain experimental results with simulation as well as derive formulas to compare PCS^{avc} with PCS^{bem}. First, consider $X \sim \exp(\lambda)$ and $O \sim \exp(\mu)$ and let $\lambda < \mu$, where λ and μ are exponential rates. For $k = 2$, $v = 2$, we have $p_{[2]} = \Pr\{X > O\} = \mu/(\lambda + \mu)$ and $p_{[1]} = \Pr\{X < O\} = \lambda/(\lambda + \mu)$. To calculate PCS^{bem}, we need to consider vectors $\mathbf{Y} = (Y_{[1]}, Y_{[2]})$ such that $Y_{[2]} \geq Y_{[1]}$.

With $v = 2$, the only possible $\mathbf{Y}'s$ with $Y_{[2]} \geq Y_{[1]}$ are $(0,2)$ and $(1,1)$. This gives us

$$\begin{aligned} \text{PCS}^{\text{bem}} &= \Pr\{Y_{[2]} = 2\} + \frac{1}{2}\Pr\{Y_{[2]} = 1\} \quad (4) \\ &= p_{[2]}^2 + \frac{1}{2}\, 2\, p_{[1]} p_{[2]} \\ &= \frac{\mu}{\lambda + \mu} \end{aligned}$$

Similarly, to calculate PCS^{avc}, we need to consider vectors $\mathbf{Z} = (Z_{[1]}, Z_{[2]})$ such that $Z_{[2]} \geq Z_{[1]}$. With $v^k = 4$, the only $\mathbf{Z}'s$ with $Z_{[2]} \geq Z_{[1]}$ are: $(0,4)$, $(1,3)$, $(2,2)$. So

$$\begin{aligned} \text{PCS}^{\text{avc}} &= \Pr\{Z_{[2]} = 4\} + \Pr\{Z_{[2]} = 3\} \\ &\quad + \frac{1}{2}\Pr\{Z_{[2]} = 2\}. \quad (5) \end{aligned}$$

When $X \sim \exp(\lambda)$, the joint distribution of $(X_{[1]}, X_{[2]})$ is $f_{12}(a,b) = 2\lambda^2 e^{-\lambda(a+b)}$. The probabilities for the values of $Z_{[2]}$ in Equation (5) and PCS^{avc} can then be found as follows:

$$\begin{aligned} \Pr\{Z_{[2]} = 4\} &= \int_0^\infty \int_0^b (1 - e^{-\mu a})^2 \times \\ &\quad 2\lambda^2 e^{-\lambda(a+b)} da\, db \\ &= \frac{\mu^2}{(2\lambda + \mu)(\lambda + \mu)} \\ \Pr\{Z_{[2]} = 3\} &= 2\int_0^\infty \int_0^b (1 - e^{-\mu a})(e^{-\mu a} - e^{-\mu b}) \times \\ &\quad 2\lambda^2 e^{-\lambda(a+b)} da\, db \\ &= \frac{2\lambda^2}{(2\lambda + \mu)(\lambda + \mu)^2} \\ \Pr\{Z_{[2]} = 2\} &= \int_0^\infty \int_0^b (e^{-\mu a} - e^{-\mu b})^2 \times \\ &\quad 2\lambda^2 e^{-\lambda(a+b)} da\, db\, + \\ &\quad 2\int_0^\infty \int_0^b (1 - e^{-\mu a}) e^{-\mu b} \times \\ &\quad 2\lambda^2 e^{-\lambda(a+b)} da\, db \\ &= \frac{2\lambda\mu(4\lambda\mu + \mu^2 + \lambda^2)}{(2\lambda + \mu)(\lambda + 2\mu)(\lambda + \mu)^2} \\ \text{PCS}^{\text{avc}} &= \frac{\mu(\lambda^2 + 6\lambda\mu + 2\mu^2)}{(2\lambda + \mu)(\lambda + 2\mu)(\lambda + \mu)} \end{aligned}$$

Given expressions for both PCS^{bem} and PCS^{avc}, we can find the improvement in PCS with AVC from

$$\begin{aligned} \Delta\text{PCS} &= \text{PCS}^{\text{avc}} - \text{PCS}^{\text{bem}} \\ &= \frac{\lambda\mu(\mu - \lambda)}{(2\lambda + \mu)(\lambda + 2\mu)(\lambda + \mu)} > 0 \quad (6) \end{aligned}$$

The $(\mu - \lambda) > 0$ term in equation (6) shows that when X is the best population, AVC always shows an improvement in PCS over BEM. Similar analytical results have been obtained for exponential populations with $k = 2, v = 3$ and $k = 3, v = 2$. Additionally, results for continuous uniform populations ($k = 2, v = 2, 3$; $k = 3, v = 2$) show an increase in PCS with AVC, but the increase is different than it is for the exponential populations.

As an illustration of how AVC compares to BEM for discrete distributions, let $X \sim \text{Bern}(p_x)$ and $O \sim \text{Bern}(p_o)$ with $p_x > p_o$. For $k = 2$, $v = 2$, we have $p_{[2]} = \Pr\{X > O\} = 1/2(p_x + 1 - p_o)$ and $p_{[1]} = \Pr\{X < O\} = 1/2(p_o + 1 - p_x)$. From equation (4) we obtain

$$\text{PCS}^{\text{bem}} = \frac{1}{2}(p_x + 1 - p_o)$$

PCS^{avc} can be calculated as we did for our exponential example by finding the probabilities for the following values of $Z_{[2]}$ and using Equation (5).

$$\begin{aligned} \Pr\{Z_{[2]} = 4\} &= \frac{9}{16}p_x^2 - \frac{5}{8}p_x^2 p_o + \frac{1}{8}p_x^2 p_o^2 \\ &\quad + \frac{3}{8}p_x - \frac{3}{4}p_x p_o + \frac{3}{8}p_x p_o^2 \\ &\quad - \frac{1}{8}p_o + \frac{1}{16}p_o^2 + \frac{1}{16} \\ \Pr\{Z_{[2]} = 3\} &= -\frac{3}{4}p_x^2 + \frac{3}{2}p_x^2 p_o - \frac{1}{2}p_x^2 p_o^2 \\ &\quad + \frac{1}{2}p_x - \frac{1}{2}p_x p_o^2 - \frac{1}{2}p_o \\ &\quad + \frac{1}{4}p_o^2 + \frac{1}{4} \\ \Pr\{Z_{[2]} = 2\} &= -\frac{1}{8}p_x^2 - \frac{3}{4}p_x^2 p_o + \frac{3}{4}p_x^2 p_o^2 \\ &\quad - \frac{1}{4}p_x + \frac{3}{2}p_x p_o - \frac{3}{4}p_x p_o^2 \\ &\quad - \frac{1}{4}p_o - \frac{1}{8}p_o^2 + \frac{3}{8} \\ \text{PCS}^{\text{avc}} &= \frac{1}{4}(p_o^2 - p_x^2) + \frac{1}{2}p_x p_o (p_x - p_o) \\ &\quad + \frac{3}{4}(p_x - p_o) \end{aligned}$$

Then calculating the difference in PCS

$$\Delta\text{PCS} = \frac{1}{4}(p_x - p_o)(1 + 2p_x p_o - p_x - p_o) > 0 \quad (7)$$

We again see a term, $(p_x - p_o) > 0$ in Equation (7), which shows an improvement in PCS with AVC when X is the better population. Figure 1 illustrates the difference in the improvement with AVC for exponential and Bernoulli populations. Notice that the

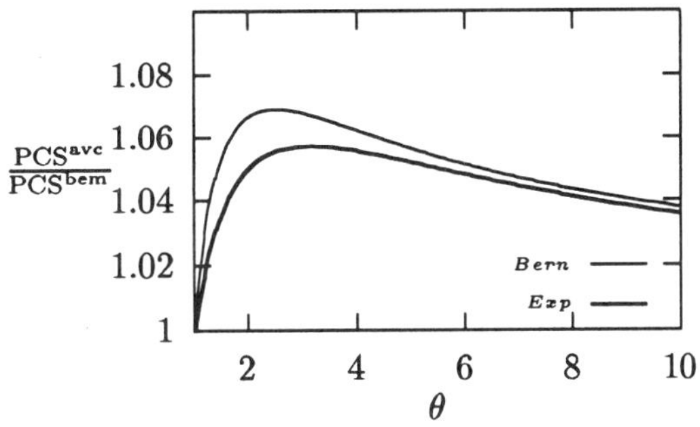

Figure 1: Exponential and Bernoulli Populations: $k = 2$, $v = 2$, $p_x = \frac{7}{8}$

maximum improvement with AVC occurs at a θ value between 2 and 3.

We have determined that the increase in PCS with AVC can be attributed to a subset of the total possible rank orders. Most rank orders result in the same conclusion with either BEM or AVC. For illustration consider $k = 2$. For any rank order with $Z_{[2]} > v^2/2$, AVC always makes the correct selection, however; for values of $Z_{[2]}$ close to $v^2/2$, BEM can make an incorrect selection. We refer to such rank orders as a *gain*. Similarly, for rank orders with $Z_{[2]} < v^2/2$, AVC always makes an incorrect selection; however, for such values of $Z_{[2]}$ close to $v^2/2$, BEM can make a correct selection. We refer to such a rank order as a *loss*.

Given a particular rank order, the probability of a gain (loss) is simply the fraction of the time AVC is correct and BEM is incorrect (BEM is correct and AVC is incorrect). The contribution to the overall PCS is then the probability of obtaining a particular rank order times the probability of the gain or loss with that rank order. Summing this contribution over all possible rank orders with a gain or loss provides an alternative way to quantify the difference in PCS between BEM and AVC. Calculations using this approach with exponential and continous uniform populations ($k = 2, v = 2, 3$; $k = 3, v = 2$) match results for finding the gain in PCS by taking $PCS^{avc} - PCS^{bem}$. Our conjecture is that we can systematically identify all rank orders with a gain or a loss and then group these gains and losses in some fashion to show the sum of the probabilties of the gains exceeds the sum of the probabilties of the losses.

5 EMPIRICAL RESULTS

Due to the difficulty in calculating analytical results for even small k and v, analytical results were only presented for a single k and v over a range of values for θ. The simulation results presented in this section, can easily show the relationship between PCS^{bem} and PCS^{avc} over a range of vectors and number of populations for a fixed θ. Mirroring the analytical results already presented, Figures 2 - 4 illustrate the improvement in PCS with AVC for up to $v = 50$ vectors for exponential and Bernoulli populations. All results are for 100,000 replications using a separate random number stream for each population and common random numbers for each set of distributions. The standard errors for PCS^{bem} and PCS^{avc} are approximately 0.0015. All runs were done using the LFC for BEM with $\theta = 1.2$. This was accomplished by fixing the parameter for one distribution and varying the other with increasing k to maintain a constant θ. For the exponential populations $\lambda = 1$ and for the Bernoulli populations $p_x = 1/2$.

Figure 2 clearly illustrates the improvement in PCS^{avc} over PCS^{bem} for 2 to 5 exponential populations. Likewise, Figure 3 shows similiar results for 2 and 3 Bernoulli populations. Figure 4 demonstrates the dependence of PCS^{avc} on the underlying population distributions.

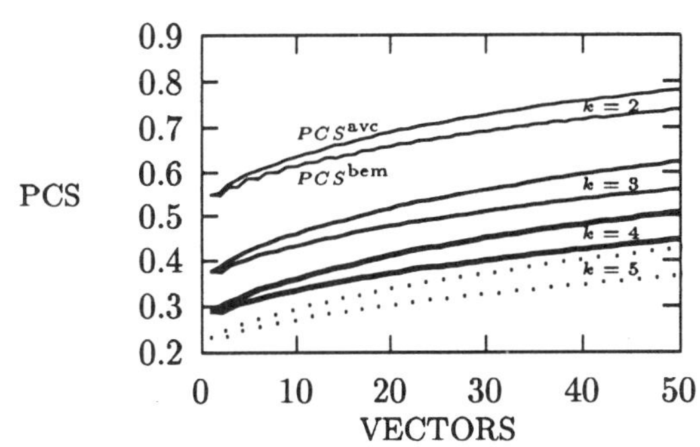

Figure 2: Exponential Populations, $\theta = 1.2$

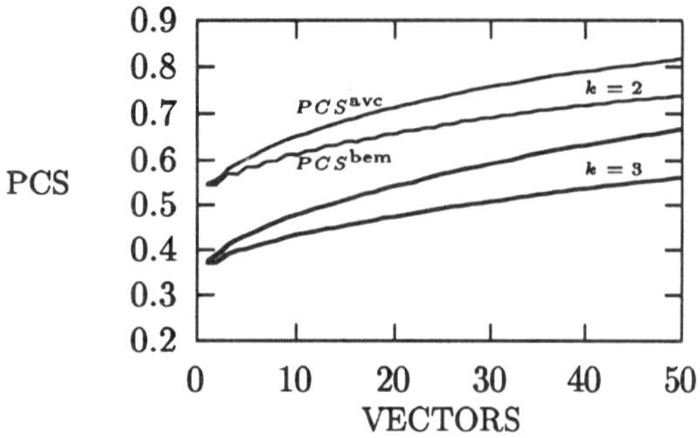

Figure 3: Bernoulli Populations, $\theta = 1.2$

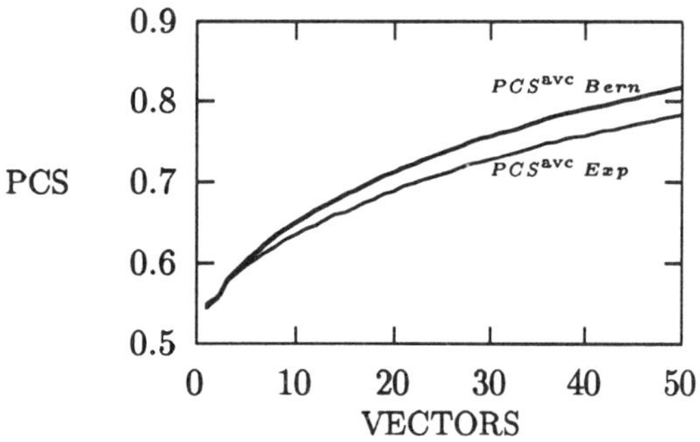

Figure 4: Exponential and Bernoulli Populations: $k = 2$, $\theta = 1.2$

These results clearly show an improvement in PCS with AVC for all values of k and v considered, and also clearly illustrate the dependence of PCS^{avc} on the underlying population distributions.

6 POINT ESTIMATION

Although the focus of this paper is selecting the best population, a closely related problem is *estimation* of $p_j = \Pr\{\pi_j \text{ wins}\}$. If v vector observations are taken, then the natural point estimator associated with the BEM method is

$$\widehat{p}_j = \frac{Y_j}{v}$$

the fraction of wins out of v replications achieved by population j. It is well known that $\text{E}[\widehat{p}_j] = p_j$ and $\text{Var}[\widehat{p}_j] = p_j(1-p_j)/v$.

The natural point estimator associated with the AVC method is

$$\bar{p}_j = \frac{Z_j}{v^k}$$

the fraction of wins out of v^k psuedoreplications achieved by population j, where the pseudoreplications are obtained by taking all combinations of k observations with one from each population. Clearly, $\text{E}[\bar{p}_j] = p_j$. Unfortunately, the variance of $\bar{p}_j$ is quite complex. However, the following analysis shows that we should anticipate a variance reduction relative to $\text{Var}[\widehat{p}_j]$.

To simplify the development from here on, suppose that we are trying to estimate p_1. Let

$$h(X_{1i}, X_{2i}, \ldots, X_{ki}) = \prod_{\ell=2}^{k} \phi(X_{1i}, X_{\ell i})$$

an indicator function that will take the value 1 only if X_{1i} is largest among $(X_{1i}, X_{2i}, \ldots, X_{ki})$. Then equation (1) can be rewritten as

$$Z_1 = \sum_{a_1=1}^{v} \sum_{a_2=1}^{v} \cdots \sum_{a_k=1}^{v} h(X_{1a_1}, X_{2a_2}, \ldots, X_{ka_k}).$$

When written in this way we can verify that $\bar{p}_1 = Z_1/v^k$ is a *k-sample U-statistic* with kernel h of order $(1, 1, \ldots, 1)$ (Randles and Wolfe 1979, pp. 104–109). We can show that

$$\text{Var}[\widehat{p}_1] = \text{Var}[\bar{p}_1] + \text{E}\left[\text{Var}(\widehat{p}_1 | \widehat{F}_1, \widehat{F}_2, \ldots, \widehat{F}_k)\right]$$

where $\widehat{F}_j$ is the empirical cdf of the X_{ji}. This shows that $\text{Var}[\bar{p}_1] \le \text{Var}[\widehat{p}_1]$. But how much variance reduction should we expect?

Since $\bar{p}_1$ is a U-statistic, its asymptotic distribution (as v goes to infinity) is known. Specifically,

$$\sqrt{kv}(\bar{p}_1 - p_1) \Longrightarrow \text{N}(0, \sigma^2_{\text{avc}})$$

where σ^2_{avc} depends on the distribution of the X_{ji}, but is much less complicated than the finite-sample variance.

To compare $\bar{p}_1$ to $\widehat{p}_1$, we note that

$$\sqrt{kv}(\widehat{p}_1 - p_1) \Longrightarrow \text{N}(0, \sigma^2_{\text{bem}})$$

where $\sigma^2_{\text{bem}} = kp_1(1-p_1)$. Therefore, a ratio of $\sigma^2_{\text{avc}}/\sigma^2_{\text{bem}} < 1$ implies that $\bar{p}_1$ is asymptotically superior to $\widehat{p}_1$. Table 1 shows one such calculation for exponential populations with $\lambda_1 = 1.0$, $\lambda_2 = \cdots = \lambda_k = 0.99$ (we have changed the problem slightly in that a smaller performance measure is considered better, allowing us to exploit the fact that the minimum of exponentially distributed random variables is also exponentially distributed).

The table shows a substantial variance reduction for $\bar{p}_1$, increasing to a limit of 50% as the number of populations to compare gets large (the limit of 1/2 holds for exponential distributions with any parameters). Thus, when there are more populations, making the comparison problem harder, the payoff is largest.

Table 1: Ratio of the Asymptotic Variances of Estimators for p_1

k	$\sigma^2_{\text{avc}}/\sigma^2_{\text{bem}}$
2	0.667
3	0.600
4	0.572
5	0.556
10	0.527
20	0.513
100	0.503
∞	0.500

7 CONCLUDING REMARKS

When trying to pick the best system out of k systems, there are many instances when this selection should be based on one-time performance rather than long-run average performance. Multinomial selection procedures provide a framework for defining such a problem, and Procedure BEM is the classical approach for solving it. Procedure AVC is an alternative approach designed to obtain a higher PCS by performing all possible comparisons across all systems for a given set of system performance data. Construction of procedure AVC closely follows that of BEM allowing researchers to easily move from a standard approach to our new approach.

Given fixed values of k, P^*, and θ^*, we conjecture that $\text{PCS}^{\text{avc}} \geq \text{PCS}^{\text{bem}}$. An interesting question is how many fewer replications are needed for an AVC-like procedure to perform just as well as BEM. Table 2 presents some preliminary comparisons of the minimum number of independent replications needed to achieve a given P^* for AVC and BEM. Values for BEM are taken from Bechhofer, Santner, and Goldsman (1995). The AVC values are from simulations (10,000 replications) using exponential populations under the LFC for BEM. As P^* increases and the difference between the best population and the other populations decreases, we see a more dramatic reduction in the number of vector observations needed with AVC to achieve the same P^*.

Table 2: Minimum v with AVC and BEM

		$P^* = .90$		$P^* = .95$	
k	θ^*	avc	bem	avc	bem
2	2.0	12	15	19	23
	1.2	116	199	128	327
3	2.0	21	29	33	42

ACKNOWLEDGMENTS

Dr. Nelson's work was partially supported by National Science Foundation Grant No. DMI-9622065.

REFERENCES

Bechhofer, R. E., S. Elmaghraby, and N. Morse. 1959. A single-sample multiple-decision procedure for selecting the multinomial event which has the highest probability. *Annals of Mathematical Statistics* 30:102–119.

Bechhofer, R. E., T. Santner, and D. Goldsman. 1995. *Design and analysis of experiments for statistical selection, screening and multiple comparisons.* New York: John Wiley & Sons, Inc.

Kesten, H., and N. Morse. 1959. A property of the multinomial distribution. *Annals of Mathematical Statistics* 30:120–127.

Randles, R. H., and D. A. Wolfe. 1979. *Introduction to the theory of nonparametric statistics.* New York: John Wiley.

AUTHOR BIOGRAPHIES

J.O. MILLER is a Major in the United States Air Force and a Ph.D. student in the Department of Industrial, Welding and Systems Engineering at The Ohio State University. His research interests include simulation, stochastic processes, and ranking and selection.

BARRY L. NELSON is an associate professor in the Department of Industrial Engineering and Management Sciences at Northwestern University. He is interested in the design and analysis of computer simulation experiments, particularly statistical efficiency, multivariate output analysis and input modeling. He is the simulation area editor for *Operations Research* and will be Program Chair for the 1997 Winter Simulation Conference.

CHARLES H. REILLY is a Professor and Chair in the Department of Industrial Engineering and Management Systems at the University of Central Florida. His principal research interests are generation methods for synthetic optimization problems and the relationships between problem generation methods and the outcomes of computational experiments. Dr. Reilly is a member of IIE, INFORMS, and ASEE; he is the Immediate Past Director of the Operations Research Division of IIE.

Proceedings of the 1996 Winter Simulation Conference
ed. J. M. Charnes, D. J. Morrice, D. T. Brunner, and J. J. Swain

A COMPARISON OF PERTURBATION ANALYSIS TECHNIQUES

Michael C. Fu

College of Business and Management
University of Maryland at College Park
College Park, Maryland 20742, U.S.A.

Jian-Qiang Hu

Department of Manufacturing Engineering
Boston University
Boston, Massachusetts 02215, U.S.A.

ABSTRACT

Perturbation analysis (PA) is a technique for estimating gradients of performance measures, particularly applicable to the simulation of discrete-event systems. Over the past two decades, various "versions" have been developed. In this paper, we compare and contrast some of these perturbation analysis techniques by applying them to a simple example. This example also serves to highlight the issue of process representation that can play a very crucial role in the application of perturbation analysis.

1 INTRODUCTION

Sensitivity analysis and optimization of discrete-event systems via simulation has become an increasingly important area of research in the past two decades (cf. Fu 1994). One line of this research is perturbation analysis (cf. Ho and Cao 1991, Glasserman 1991). Even in the field of perturbation analysis, different techniques have been developed. For gradient estimation, the "original" technique is infinitesimal perturbation analysis (IPA), which remains the easiest PA technique to apply in practice. However, its ease of implementation is often offset by its relatively limited domain of applicability, which has led to the development of various other generalizations or alternatives, among them the following: smoothed perturbation analysis (SPA), rare perturbation analysis (RPA), discontinuous perturbation analysis (DPA), and augmented perturbation analysis (APA) . The purpose of this paper is to use a very simple example to illustrate these techniques and the role of process representation in their application. Much of this material is taken from Fu and Hu (1996).

The rest of the paper is organized as follows. We present the simple example in Section 2, including two different process representations. Section 3 will summarize briefly the techniques of IPA, SPA, DPA, RPA, and APA, and illustrate their application on the simple example. Obviously, full justice cannot be given due to space limitations, and Fu and Hu (1996) can be consulted for a more detailed exposition of the individual techniques. In Section 4, three specific cases of the examples are carried out to illustrate the specific form of each of the estimators. Some conclusions based on the example and the specific cases are provided in Section 5.

2 THE EXAMPLE

We consider a single random variable X, defined on an appropriate probability space, for which our goal is to estimate

$$\frac{dE[X]}{d\theta},$$

where θ is a scalar parameter of the distribution.

The example is the following:

$$X = \begin{cases} X_+ & \text{w.p. } \theta \\ X_- & \text{w.p. } 1-\theta \end{cases} . \tag{1}$$

Thus, we also allow X_+ and X_- to be random variables themselves with possible dependence on θ, but we impose the following conditions (cf. (4) and (5)):

$$E\left[\frac{dX_+}{d\theta}\right] = \frac{dE[X_+]}{d\theta}, \quad E\left[\frac{dX_-}{d\theta}\right] = \frac{dE[X_-]}{d\theta}.$$

This corresponds to the requirement that the components are sufficiently smooth so that IPA applies.

We consider two representations corresponding to two different methods for generating X from random numbers. Assume that each of X_+ and X_- can be expressed as functions of a single random variable $U \sim U(0,1)$, i.e., besides the possible implicit dependence on θ that we do not display, we write $X_+ = X_+(U)$ and $X_- = X_-(U)$. The first representation of (1) uses two independent random numbers, whereas the second uses a single random number.

Representation 1:

$$X = \begin{cases} X_+(U_2) & \text{if } U_1 \le \theta \\ X_-(U_2) & \text{if } U_1 > \theta \end{cases}. \quad (2)$$

Representation 2:

$$X = \begin{cases} X_+(U_1/\theta) & \text{if } U_1 \le \theta \\ X_-((1-U_1)/(1-\theta)) & \text{if } U_1 > \theta \end{cases}. \quad (3)$$

3 PA TECHNIQUES

In this section, we apply in turn the various PA techniques to our example after a very brief discussion of each. To keep things simple, we will limit the discussion of the techniques to a single random variable X, as in the example. Recall that our goal is to estimate

$$\frac{dE[X]}{d\theta}.$$

For discrete-event systems, θ can be more general. For example, it could represent a vector operating parameters such as the re-order point and order quantity in an inventory control system.

3.1 IPA

We will denote a sample point by ω with corresponding random variable value $X(\omega)$. The IPA estimator is simply the sample path derivative of the quantity of interest, defined by

$$\frac{dX(\theta,\omega)}{d\theta} = \lim_{\Delta\theta\to 0} \frac{X(\theta+\Delta\theta,\omega) - X(\theta,\omega)}{\Delta\theta}. \quad (4)$$

Thus, for the IPA estimator to be an unbiased gradient estimator, we need

$$\frac{dE[X]}{d\theta} = E\left[\frac{dX}{d\theta}\right], \quad (5)$$

i.e., it becomes a question of an interchange of expectation and limit:

$$\lim_{\Delta\theta\to 0} E\left[g_{\Delta\theta}\right] = E\left[\lim_{\Delta\theta\to 0} g_{\Delta\theta}\right],$$

$$\text{where} \quad g_{\Delta\theta} = \frac{X(\theta+\Delta\theta) - X(\theta)}{\Delta\theta}.$$

For our example, the IPA estimator is obtained by simply taking the derivatives in Equations (2) and (3) to obtain the following.

Representation 1:

$$\frac{dX}{d\theta} = \begin{cases} \frac{d}{d\theta}X_+(U_2) & \text{if } U_1 \le \theta \\ \frac{d}{d\theta}X_-(U_2) & \text{if } U_1 > \theta \end{cases}.$$

Representation 2:

$$\frac{dX}{d\theta} = \begin{cases} \frac{d}{d\theta}X_+(U_1/\theta) & \text{if } U_1 \le \theta \\ \frac{d}{d\theta}X_-((1-U_1)/(1-\theta)) & \text{if } U_1 > \theta \end{cases}.$$

In general, these sample derivatives are biased, i.e., they do not satisfy Equation (5). We now establish necessary and sufficient conditions for their unbiasedness.

Rewriting the representations in terms of indicator functions, we have

$$X = X_+(U_2)\mathbf{1}\{U_1 \le \theta\} + X_-(U_2)\mathbf{1}\{U_1 > \theta\},$$

$$X = X_+(U_1/\theta)\mathbf{1}\{U_1 \le \theta\} + X_-((1-U_1)/(1-\theta))\mathbf{1}\{U_1 > \theta\},$$

where $\mathbf{1}\{\cdot\}$ is the set indicator function. In the first representation, we can uncouple to get

$$E[X] = \theta E[X_+(U_2)] + (1-\theta)E[X_-(U_2)],$$

so that differentiation yields

$$\begin{aligned} \frac{dE[X]}{d\theta} &= \theta\frac{dE[X_+]}{d\theta} + E[X_+] \\ &\quad + (1-\theta)\frac{dE[X_-]}{d\theta} - E[X_-] \\ &= E\left[\frac{dX}{d\theta}\right] + E[X_+ - X_-], \end{aligned}$$

the latter following from our assumption on X_+ and X_-. Thus, we have the following result:

Proposition 1 For Representation 1, a necessary and sufficient condition for unbiasedness of IPA is given by:

$$E[X_+ - X_-] = 0.$$

In addition, the unbiasedness of subsequent SPA estimators imply the following corresponding result for Representation 2:

Proposition 2 For Representation 2, a necessary and sufficient condition for unbiasedness of IPA is given by:

$$X_+(1) = X_-(1).$$

i.e., X is a.s. continuous across its breakpoint.

Note that $X_+(1)$ and $X_-(1)$ correspond to setting $U_1 = \theta$ for Representation 2.

The techniques SPA, RPA, DPA, and APA can be applied even when the interchange equation (5) is not satisfied. We now consider each of these in turn.

3.2 SPA

SPA uses conditional expectation to derive alternative estimators for which a modified version of the interchange equation is satisfied. Introduced by Gong and Ho (1987), this technique has been developed into a very general methodology in the generalized semi-Markov process (GSMP) framework, e.g., Fu and Hu (1992) and Dai and Ho (1995). The resulting estimator contains an IPA term (possibly 0) and a conditional contribution that is the product of a probability rate term and a conditional performance difference term. We illustrate the estimator on the example.

Let $Y = \mathbf{1}\{U_1 \leq \theta\}$. Then, Y is a Bernoulli random variable with parameter θ, i.e., $P(Y=1)=\theta$ and $P(Y=0)=1-\theta$. We first consider the right-hand (RH) derivative, i.e., $\Delta\theta > 0$. If $Y=1$, then there is no possibility of a change to $Y=0$ from the perturbation $\Delta\theta$, since

$$P(U_1 > \theta + \Delta\theta | U_1 \leq \theta) = 0, \quad \Delta\theta > 0.$$

However, if $Y = 0$, then we have

$$P(U_1 \leq \theta + \Delta\theta | U_1 > \theta) = \frac{\Delta\theta}{1-\theta},$$

and the probability rate term is given by

$$\lim_{\Delta\theta \to 0^+} \frac{P(U_1 \leq \theta + \Delta\theta | U_1 > \theta)}{\Delta\theta} = \frac{1}{1-\theta}.$$

Similarly, for the left-hand (LH) derivative, $\Delta\theta < 0$, the probability rate term is given by $1/\theta$.

The conditional difference term is given by

$$X(U_1 \to \theta^-) - X(U_1 \to \theta^+),$$

which depends on the representation used.

Thus, for our example, we have the following SPA derivative estimators:

Representation 1 (LH and RH, respectively):

$$\frac{dX}{d\theta} + \frac{1}{\theta}(X_+(U_2) - X_-(U_2))\,\mathbf{1}\{U_1 \leq \theta\}, \quad (6)$$

$$\frac{dX}{d\theta} + \frac{1}{1-\theta}(X_+(U_2) - X_-(U_2))\,\mathbf{1}\{U_1 > \theta\}. \quad (7)$$

Representation 2 (LH and RH, respectively):

$$\frac{dX}{d\theta} + \frac{1}{\theta}(X_+(1) - X_-(1))\,\mathbf{1}\{U_1 \leq \theta\}, \quad (8)$$

$$\frac{dX}{d\theta} + \frac{1}{1-\theta}(X_+(1) - X_-(1))\,\mathbf{1}\{U_1 > \theta\}. \quad (9)$$

3.3 RPA

Intuitively, IPA corresponds to the case where small perturbations in the parameter cause small perturbations in the output random variable. In contrast, RPA corresponds to the case where small perturbations in the parameter cause no change in the output random variable much of the time, and the "rare" times when a change does occur results in a relatively large change in the output random variable. First introduced by Brémaud and Vázquez-Abad (1992), a maximal coupling interpretation was provided in Brémaud (1993), which is presented here.

Letting f_θ represent the p.d.f. of X and defining

$$\begin{aligned} g_\theta(\Delta\theta, x) &= \frac{\min(f_\theta(x), f_{\theta+\Delta\theta}(x))}{1-\delta_\theta(\Delta\theta)}, \\ g_\theta^1(\Delta\theta, x) &= \frac{(f_\theta(x) - f_{\theta+\Delta\theta}(x))^+}{\delta_\theta(\Delta\theta)}, \\ g_\theta^2(\Delta\theta, x) &= \frac{(f_{\theta+\Delta\theta}(x) - f_\theta(x))^+}{\delta_\theta(\Delta\theta)}, \\ \text{where } \delta_\theta(\Delta\theta) &= \frac{1}{2}\int |f_{\theta+\Delta\theta}(x) - f_\theta(x)|dx, \end{aligned}$$

the main result is the following. If

$$\delta_\theta' = \lim_{\Delta\theta \to 0^+} \frac{\delta_\theta(\Delta\theta)}{\Delta\theta} \quad \text{exists,}$$

and the c.d.f.'s corresponding to densities $g_\theta(\Delta\theta,\cdot), g_\theta^1(\Delta\theta,\cdot), g_\theta^2(\Delta\theta,\cdot)$, converge weakly to c.d.f.'s $F_\theta(\cdot), G_\theta^1(\cdot), G_\theta^2(\cdot)$, respectively, as $\Delta\theta \to 0^+$, then

$$\delta_\theta'[V^{(2)} - V^{(1)}] \quad (10)$$

is an unbiased estimator for $dE[X]/d\theta$, where $V^{(1)}$ and $V^{(2)}$ are independent r.v.'s with respective c.d.f.'s G_θ^1, G_θ^2. This is the two-sided RPA estimator. One-sided (left and right) RPA estimators are also available, but will not be presented here. For our particular example, we cannot write out a more explicit estimator without specification of the distributions involved in X_- and X_+ in Equation (1). This will be taken up in Section 4 for the three specific cases.

3.4 DPA

The main idea in DPA is to represent the jumps in the sample performance function by using step functions (Shi 1996). For example, if the random variable X depends on θ through N random numbers as follows:

$$X = X_i^k \text{ if } P_i^{k-1}(\theta) < U_i \leq P_i^k(\theta),$$

$$k = 1, ..., M,\ 0 = P_i^0 < P_i^1 < ... < P_i^k < ... < P_i^M = 1,$$

where the P_i^k are cumulative probabilities, then the DPA estimator is given by

$$\sum_{i=1}^{N}\sum_{k=1}^{M-1}\left(L_i^k - L_i^{k+1}\right)\frac{dP_i^k(\theta)}{d\theta}.$$

For our example, we obtain the following estimators.

Representation 1:

$$\frac{dX}{d\theta} + \left(X_+(U_2) - X_-(U_2)\right). \qquad (11)$$

Representation 2:

$$\frac{dX}{d\theta} + \left(X_+(1) - X_-(1)\right). \qquad (12)$$

3.5 APA

The APA method is an extension of IPA that attempts to partition the sample space into a finite number of sets with the same event sequence (Gaivoronski et al. 1992). Here, for the case of a single random variable, we will simplify the exposition considerably. Assume that the sample space can be partitioned into a finite number of sets $\{E_1, ..., E_M\}$ (events in the probability space) such that for each E_j, there is a corresponding $(P_j^{\min}, P_j^{\max}] \subset (0,1]$ on which X is uniformly differentiable with respect to θ. Then, the APA estimator is given by

$$\frac{dX}{d\theta}+\sum_{j=1}^{M}\frac{X(P_j^{\max})\frac{d}{d\theta}P_j^{\max} - X(P_j^{\min})\frac{d}{d\theta}P_j^{\min}}{P_j^{\max} - P_j^{\min}}\mathbf{1}\{E_j\}.$$

For our example, we obtain

Representation 1:

$$\frac{dX}{d\theta} + \frac{X_+(U_2)}{\theta}\mathbf{1}\{U_1 \le \theta\} - \frac{X_-(U_2)}{1-\theta}\mathbf{1}\{U_1 > \theta\}. \qquad (13)$$

Representation 2:

$$\frac{dX}{d\theta} + \frac{X_+(1)}{\theta}\mathbf{1}\{U_1 \le \theta\} - \frac{X_-(1)}{1-\theta}\mathbf{1}\{U_1 > \theta\}. \qquad (14)$$

4 SPECIFIC CASES

We now consider three specific cases of (1). In the first case, IPA works for neither representation. In the second case, IPA works for Representation 2, but not for 1. In the third case, IPA works for Representation 1 (but not for 2), in spite of the fact that the function is a.s. *discontinuous*. Note that this cannot happen for Representation 2 (across the breakpoint). The corresponding SPA, RPA, DPA, and APA estimators obtained via Equations (6) through (14) are then presented for each of the two representations (RPA is discussed first, because it is independent of the representation). Note that even when the IPA estimator is unbiased, the other estimators may be different (though equal in expectation).

4.1 CASE 1

$$X = \begin{cases} U(0,\theta) & \text{w.p. } \theta \\ 0.5 & \text{w.p. } 1-\theta \end{cases}.$$

Then, $E[X] = 0.5(\theta^2 - \theta + 1)$, $dE[X]/d\theta = \theta - 0.5$. Since $X_+(U) = \theta U$, $X_-(U) = 0.5$, applying Propositions 1 and 2, IPA fails for both representations. For this case, the RPA "estimator" turns out to be the exact answer $(\theta - 0.5)$, since $\delta_\theta(\Delta\theta) = \Delta\theta$, and $g_\theta^1(\Delta\theta,\cdot)$ and $g_\theta^2(\Delta\theta,\cdot)$ converge to masses at θ and 0.5, respectively. The other estimators follow.

Representation 1:

$$X = \begin{cases} \theta U_2 & \text{if } U_1 \le \theta \\ 0.5 & \text{if } U_1 > \theta \end{cases}.$$

$$\frac{dX}{d\theta} = \begin{cases} U_2 & \text{if } U_1 \le \theta \\ 0 & \text{if } U_1 > \theta \end{cases}.$$

The IPA estimator has expectation $E[dX/d\theta] = 0.5\theta \ne dE[X]/d\theta$, and hence is biased.

SPA (LH):

$$\frac{dX}{d\theta} + \frac{1}{\theta}\left(\theta U_2 - 0.5\right)\mathbf{1}\{U_1 \le \theta\}.$$

SPA (RH):

$$\frac{dX}{d\theta} + \frac{1}{1-\theta}\left(\theta U_2 - 0.5\right)\mathbf{1}\{U_1 > \theta\}.$$

DPA:

$$\frac{dX}{d\theta} + \left(\theta U_2 - 0.5\right).$$

APA:

$$\frac{dX}{d\theta} + \frac{\theta U_2}{\theta}\mathbf{1}\{U_1 \le \theta\} - \frac{0.5}{1-\theta}\mathbf{1}\{U_1 > \theta\}.$$

Representation 2:

$$X = \begin{cases} U_1 & \text{if } U_1 \le \theta \\ 0.5 & \text{if } U_1 > \theta \end{cases}.$$

$$\frac{dX}{d\theta} = 0.$$

IPA is clearly biased, since it yields identically zero.

SPA (LH):

$$\frac{1}{\theta}\left(\theta - 0.5\right)\mathbf{1}\{U_1 \le \theta\}.$$

SPA (RH):

$$\frac{1}{1-\theta}(\theta-0.5)\,\mathbf{1}\{U_1>\theta\}.$$

DPA:

$$\theta-0.5.$$

APA:

$$\mathbf{1}\{U_1\le\theta\}-\frac{0.5}{1-\theta}\mathbf{1}\{U_1>\theta\}.$$

From this case, we note the following:

- the DPA estimator can be obtained from the SPA estimators via appropriate unconditioning;
- the DPA estimator for Representation 2 coincides with the RPA estimator; however, this is not true for Representation 1;
- the RPA estimator is independent of the representation.

4.2 CASE 2

$$X=\begin{cases} U(0,\theta) & \text{w.p. } \theta \\ \theta & \text{w.p. } 1-\theta \end{cases}.$$

Then, $E[X]=\theta-0.5\theta^2$, $dE[X]/d\theta=1-\theta$. Since $X_+(U)=\theta U$, $X_-(U)=\theta$, applying Propositions 1 and 2, IPA works for Representation 2, but fails for 1. For this example, RPA does not apply, as the limit defining δ_θ' is infinity. The other estimators follow.

Representation 1:

$$X=\begin{cases} \theta U_2 & \text{if } U_1\le\theta \\ \theta & \text{if } U_1>\theta \end{cases}.$$

$$\frac{dX}{d\theta}=\begin{cases} U_2 & \text{if } U_1\le\theta \\ 1 & \text{if } U_1>\theta \end{cases}.$$

The IPA estimator has expectation $E[dX/d\theta]=1-0.5\theta\neq dE[X]/d\theta$, and hence is biased.

SPA (LH):

$$\frac{dX}{d\theta}+(U_2-1)\,\mathbf{1}\{U_1\le\theta\}.$$

SPA (RH):

$$\frac{dX}{d\theta}+\frac{\theta}{1-\theta}(U_2-1)\,\mathbf{1}\{U_1>\theta\}.$$

DPA:

$$\frac{dX}{d\theta}+\theta\,(U_2-1).$$

APA:

$$\frac{dX}{d\theta}+U_2\mathbf{1}\{U_1\le\theta\}-\frac{\theta}{1-\theta}\mathbf{1}\{U_1>\theta\}.$$

Representation 2:

$$X=\begin{cases} U_1 & \text{if } U_1\le\theta \\ \theta & \text{if } U_1>\theta \end{cases}.$$

$$\frac{dX}{d\theta}=\begin{cases} 0 & \text{if } U_1\le\theta \\ 1 & \text{if } U_1>\theta \end{cases}.$$

Here the IPA estimator is unbiased, since $E[dX/d\theta]=1-\theta$. Furthermore, we have

$$X_+(1)=\theta=X_-(1),$$

so both SPA estimators and the DPA estimator reduce to the IPA estimator, which is unbiased. The additional term in the APA estimator, however, is nonzero, but it clearly must have zero expectation.

APA:

$$\frac{dX}{d\theta}+\mathbf{1}\{U_1\le\theta\}-\frac{\theta}{1-\theta}\mathbf{1}\{U_1>\theta\}.$$

4.3 CASE 3

$$X=\begin{cases} U(0,\theta) & \text{w.p. } \theta \\ 0.5\theta & \text{w.p. } 1-\theta \end{cases}.$$

Then, $E[X]=0.5\theta$, $dE[X]/d\theta=0.5$. Since $X_+(U)=\theta U$, $X_-(U)=0.5\theta$, applying Propositions 1 and 2, IPA works for Representation 1, but fails for 2. Again, for this example, RPA does not apply. The other estimators follow.

Representation 1:

$$X=\begin{cases} \theta U_2 & \text{if } U_1\le\theta \\ 0.5\theta & \text{if } U_1>\theta \end{cases}.$$

$$\frac{dX}{d\theta}=\begin{cases} U_2 & \text{if } U_1\le\theta \\ 0.5 & \text{if } U_1>\theta \end{cases}.$$

We note that IPA works for this case, $E[dX/d\theta]=0.5$, even though it is not a.s. continuous. In fact, it is continuous when $U_2=0.5$, which occurs w.p. 0. Even though IPA works, the other estimators still have additional terms that clearly must have expectation zero.

SPA (LH):

$$\frac{dX}{d\theta}+(U_2-0.5)\mathbf{1}\{U_1\le\theta\}.$$

SPA (RH):

$$\frac{dX}{d\theta}+\frac{\theta}{1-\theta}(U_2-0.5)\mathbf{1}\{U_1>\theta\}.$$

DPA:

$$\frac{dX}{d\theta}+\theta(U_2-0.5).$$

APA:

$$\frac{dX}{d\theta} + U_2\mathbf{1}\{U_1 \le \theta\} - \frac{0.5\theta}{1-\theta}\mathbf{1}\{U_1 > \theta\}.$$

Representation 2:

$$X = \begin{cases} U_1 & \text{if } U_1 \le \theta \\ 0.5\theta & \text{if } U_1 > \theta \end{cases}.$$

$$\frac{dX}{d\theta} = \begin{cases} 0 & \text{if } U_1 \le \theta \\ 0.5 & \text{if } U_1 > \theta \end{cases}.$$

The IPA estimator has expectation $E[dX/d\theta] = 0.5(1-\theta) \neq dE[X]/d\theta$, and hence is biased.

SPA (LH):

$$\frac{dX}{d\theta} + 0.5\mathbf{1}\{U_1 \le \theta\}.$$

SPA (RH):

$$\frac{dX}{d\theta} + \frac{0.5}{1-\theta}\mathbf{1}\{U_1 > \theta\}.$$

DPA:

$$\frac{dX}{d\theta} + 0.5\theta.$$

APA:

$$\frac{dX}{d\theta} + \mathbf{1}\{U_1 \le \theta\} - \frac{0.5\theta}{1-\theta}\mathbf{1}\{U_1 > \theta\}.$$

5 CONCLUSIONS

For this simple example, we have developed necessary and sufficient conditions for unbiasedness of the IPA estimator. The conditions clearly depend on the representation, as Propositions 1 and 2 indicate. When IPA fails, the other techniques (SPA, DPA, RPA, APA) can almost always be applied. All but RPA can be considered extensions or generalizations of IPA, and this is evident in the example where each contains the IPA estimator as a component of the total estimator. Furthermore, RPA is independent of the process representation. In fact, it has been pointed out by a number of researchers that RPA is more closely akin to the so-called weak derivative estimators (Pflug 1992). On the other hand, in certain problems, the RPA estimator coincides with the DPA estimator (Shi 1996), but the latter depends on the representation. The example also illuminates the fact that DPA and SPA can be viewed as conditional forms of one another. The simple example also showed that aside from IPA, the APA estimator is the only one that can always be estimated directly from the original sample. In fact, the APA estimator often corresponds to the likelihood ratio/score function method estimator when the IPA contribution is zero.

ACKNOWLEDGMENTS

Michael Fu was supported in part by a 1996 Summer Research Grant from the Maryland Business School and by the National Science Foundation under grant EEC-9402384. Jian-Qiang Hu was supported in part by the National Science Foundation under grants EEC-9527422 and DDM-9212368.

REFERENCES

Brémaud, P. 1993. Maximal coupling and rare perturbation sensitivity analysis, *Queueing Systems: Theory and Applications* 11: 307-333.

Brémaud, P. and F.J. Vázquez-Abad. 1992. On the pathwise computation of derivatives with respect to the rate of a point process: the phantom RPA method, *Queueing Systems: Theory and Applications* 10: 249-270.

Dai, L., and Y.C. Ho. 1995. Structural infinitesimal perturbation analysis for derivative estimation in discrete event dynamic systems, *IEEE Transactions on Automatic Control* 40: 1154-1166.

Fu, M.C. 1994. Optimization via simulation: a review. *Annals of Operations Research* 53: 199-248.

Fu, M.C. and J. Q. Hu. 1992. Extensions and generalizations of smoothed perturbation analysis in a generalized semi-Markov process framework, *IEEE Transactions on Automatic Control* 37: 1483-1500.

Fu, M.C. and J. Q. Hu. 1996. *Conditional Monte Carlo: Gradient Estimation and Optimization Applications*, Kluwer Academic Publishers, forthcoming.

Gaivoronski, A., L.Y. Shi, and R.S. Sreenivas. 1992. Augmented infinitesimal perturbation analysis: an alternate explanation, *Discrete Event Dynamic Systems: Theory and Applications*, 2: 121-138.

Glasserman, P. 1991. *Gradient Estimation Via Perturbation Analysis*, Kluwer Academic.

Gong, W.B. and Y.C. Ho. 1987. Smoothed perturbation analysis of discrete-event dynamic systems, *IEEE Transactions on Automatic Control* 32: 858-867.

Ho, Y.C. and X.R. Cao. 1991. *Discrete Event Dynamic Systems and Perturbation Analysis*, Kluwer Academic.

Pflug, G. Ch. 1992. Gradient estimates for the performance of Markov chains and discrete event processes, *Annals of Operations Research* 39: 173-194.

Shi, L. Y. 1996. Discontinuous perturbation analysis of discrete event dynamic systems, to appear in *IEEE Transactions on Automatic Control.*

AUTHORS BIOGRAPHIES

MICHAEL C. FU is an Associate Professor of Management Science & Statistics in the College of Business and Management, with a joint appointment in the Institute for Systems Research, at the University of Maryland at College Park. He received a bachelor's degree in mathematics and bachelor's and master's degrees in electrical engineering from MIT in 1985, and M.S and Ph.D. degrees in applied mathematics from Harvard University in 1986 and 1989, respectively. His research interests include simulation optimization and sensitivity analysis, particularly with applications towards manufacturing systems, inventory control, and the pricing of financial derivatives. He teaches courses in applied probability, stochastic processes, discrete-event simulation, and operations management, and in 1995 was awarded the Maryland Business School's annual Allen J. Krowe Award for Teaching Excellence. He is a member of **IEEE** and **INFORMS**, and was on the Program Committee for the Spring 1996 INFORMS National Meeting in Washington, D.C.

JIAN-QIANG HU is an Associate Professor in the Department of Manufacturing Engineering at Boston University. He received a B.S. degree in applied mathematics from Fudan University, China in 1985, and M.S. and Ph.D. degrees in applied mathematics from Harvard University in 1987 and 1990, respectively. His research interests include modeling, sensitivity analysis, simulation, optimization, and control of discrete event stochastic systems and queueing networks with applications to manufacturing systems and communication networks. He is a member of **IEEE** and **INFORMS**, and is on the Organizing Committee for the 1997 INFORMS Applied Probability Section Conference in Boston.

Proceedings of the 1996 Winter Simulation Conference
ed. J. M. Charnes, D. J. Morrice, D. T. Brunner, and J. J. Swain

SPLITTING FOR RARE EVENT SIMULATION: ANALYSIS OF SIMPLE CASES

Paul Glasserman
Columbia University
New York, NY 10027

Philip Heidelberger
IBM T.J. Watson Research Center
Yorktown Heights, NY 10598

Perwez Shahabuddin
Columbia University
New York, NY 10027

Tim Zajic
Columbia University
New York, NY 10027
& IBM T.J. Watson Research Center
Yorktown Heights, NY 10598

ABSTRACT

An approach to rare event simulation uses the technique of splitting. The basic idea is to split sample paths of the stochastic process into multiple copies when they approach closer to the rare set; this increases the overall number of hits to the rare set for a given amount of simulation time. This paper analyzes the bias and efficiency of some simple cases of this method.

1 INTRODUCTION

Estimations of the small probabilities of rare events are required in the design and operation of many engineering systems. Consider the case of a telecommunications network. It is customary to model such systems as a network of queues, with each queue having a buffer of finite capacity. Information packets that arrive to a queue when its buffer is full are lost. The rare event of interest may be the event of a packet being lost. Current standards stipulate that the probability of packet loss should not exceed 10^{-9}. Or in a reliability model of a space craft computer, we may be interested in estimating the probability of the event that the system fails before the mission completion. Naturally, one would want this to be extremely low. The main problem with using standard simulation to estimate such small probabilities is that a large number of events have to be simulated in the model before any samples of the rare event may occur. Hence special simulation techniques are needed to make the events of interest occur more frequently.

Importance sampling is a technique that has been widely used for this purpose. The reader is referred to Heidelberger (1995) and Shahabuddin (1995) for some surveys. In importance sampling, the stochastic model is simulated with a new probability dynamics (called a change of measure), that makes the events of interest occur more frequently. The sample value is then adjusted to make the final estimate unbiased. However, choosing any change of measure that makes the event of interest occur frequently is not enough; *how* it is made to happen more frequently is also very important. For example, an arbitrary change of measure that makes the rare event happen more frequently may give an estimator with an infinite variance. Thus the main problem in importance sampling is to come up with an appropriate change of measure for the rare event simulation problem in hand. This may be difficult or almost impossible for complicated models. Hence, even though importance sampling works very well for a large class of stochastic models, the scope of application of importance sampling is limited to systems with "nice" structure.

This paper deals with an alternate approach to rare event simulation that uses the simulation technique of splitting (see, e.g., Hammersley and Handscomb 1965). In standard simulation, the stochastic process being simulated, spends a lot of time in a region of the state space which is "far away" from the rare set of interest, i.e, from where the chance of it entering the rare set is extremely low. In splitting a region of the state space that is "closer" to the rare set is defined. Each time the process reaches this region, from the "far away" region, many identical copies of this process are generated. This way we get more instances of the stochastic process spending time in a region where the rare event is more likely to occur. The boundary between the far away region and the closer region is called a threshold. The above described a case with one-threshold; one can easily extend it to the case where we have multiple thresholds. This approach to rare event simulation was introduced in Kahn and Harris (1951) and used later in Bayes (1970) and Hopmans and Kleijnen(1979). Recently it was revisited in a significant way by Villén-Altamirano and Villén-Altamirano (1991), Villén-Altamirano et al. (1994) and Villén-Altamirano and Villén-Altamirano (1994), who used it for estimating the probability of rare events in computer and communication systems. They called their version of this approach RESTART. A software package called ASTRO (Villén-Altamirano and Villén-Altamirano

1994) was created that implements their method. They also did some approximate efficiency analysis that gave some insights into threshold selection and number of split paths generated at each threshold. But a formal and thorough analysis was lacking.

Glasserman, Heidelberger, Shahabuddin, Zajic (1996) (henceforth referred to as GHSZ) describe a unifying class of models and implementation conditions under which this type of method is provably effective and even optimal (in an asymptotic sense) for rare event simulation. The theory of branching processes (see, e.g., Harris 1989) was used to derive the unbiasedness and efficiency results. Experimental results supporting the theoretical analysis and exploring the robustness of the splitting method, are also reported in GHSZ. In this paper we introduce and derive some biasedness and efficiency results that supplement those in that paper. We begin with a simple setting, and give conditions under which the splitting method is optimal. We then give reasons why deviations from this simple setting result in difficulties. Some of these have been handled in GHSZ, whereas others are currently being investigated. Some analytical results on the optimal selection of thresholds are introduced next. Finally we give an analysis of the bias introduced in one implementation of this method that truncates sample paths to save simulation effort.

2 A SIMPLE SETTING

Consider the problem of estimating $\gamma = P(A)$, thinking of A as a rare event. Let $A = A_k \supset A_{k-1} \cdots \supset A_1$ be a nested sequence of events which we think of as intermediate thresholds. Let $p_1 = P(A_1)$ and $p_{i+1} = P(A_{i+1}|A_i)$, $i = 1, \ldots, k-1$; then

$$\gamma \equiv \gamma_k = p_1 p_2 \cdots p_k.$$

We think of k increasing to infinity and $\gamma \to 0$ (this would happen, for example, if $p_i = p$ for all i, where p is some fixed constant between 0 and 1).

To motivate the above setting, consider a single server queueing system with a finite buffer B. Define the state of the system to be the number of jobs in the queue. The problem may be to estimate the probability that starting from state 0, the system reaches state B before visiting 0. We can think of this event as the event A. Estimating probabilities of this type are crucial to the simulation based estimation of performance measures like the steady state probability of packet loss (see, e.g., Heidelberger 1995). Clearly, if the overall arrival rate is smaller than the overall service rate (which is a requirement for the stability of the queue), and B is large, then the event A is a rare event. Suppose now that we place $k-1$ intermediate thresholds between 0 and B (with B being the kth threshold). Let A_i be the event that starting from state 0, the number of jobs in the system reaches threshold i before reaching 0. Then clearly $A_{i+1} \subset A_i$ and we have an example of the setting mentioned in the previous paragraph.

Suppose that for each i we can generate n_i Bernoulli random variables with parameter p_i, all independent of each other. These are the building blocks of a splitting estimator in this simple setting. From each successful Bernoulli outcome at stage i, we generate n_{i+1} stage-$(i+1)$ Bernoullis. Thus, at the first stage we have Bernoullis

$$\mathbf{1}_1, \mathbf{1}_2, \ldots, \mathbf{1}_{n_1};$$

the jth of these, if successful, spawns

$$\mathbf{1}_{j1}, \mathbf{1}_{j2}, \ldots, \mathbf{1}_{jn_2},$$

and so on. The estimator is

$$I_k = \frac{1}{n_1 \cdots n_k} \sum_{i_1=1}^{n_1} \cdots \sum_{i_k=1}^{n_k} \mathbf{1}_{i_1} \cdots \mathbf{1}_{i_1 \cdots i_k}.$$

It is easy to show that I_k is an unbiased estimator. By conditioning on $\mathcal{F}_k$ we mean conditioning on the outcomes of all Bernoullis up to stage k. Then

$$\begin{aligned} E(I_k) &= E(E(I_k|\mathcal{F}_{k-1})) \\ &= E\left(\frac{1}{n_1 \cdots n_k} \sum_{i_1=1}^{n_1} \cdots \sum_{i_{k-1}=1}^{n_{k-1}} \mathbf{1}_{i_1} \cdots \mathbf{1}_{i_1 \ldots i_{k-1}} \right. \\ &\qquad \left. \sum_{i_k=1}^{n_k} E(\mathbf{1}_{i_1 \cdots i_k}|\mathcal{F}_{k-1}).\right) \\ &= E(I_{k-1}\frac{1}{n_k} n_k p_k) = p_k E(I_{k-1}). \end{aligned}$$

Doing this iteratively we get that $E(I_k) = p_1 \ldots p_k = \gamma_k$.

Returning to the simple queueing example introduced above, a Bernoulli random variable might be the indicator that a simulated process reaches the next threshold from the current one, without visiting state 0. In particular, the p_i should be considered unknown, so the n_{i+1} Bernoullis from each of the successful outcomes at stage i would typically be generated implicitly by simulating n_{i+1} samples of the underlying process starting from stage i, until it either hits threshold $i+1$, or it hits 0 (see Figure 1). If a sample path hits threshold $i+1$ before hitting 0, then the corresponding Bernoulli random variable is set to 1; else it is set to zero. Of course, since the Bernoullis are all independent, the queuing process must satisfy the assumption that the dynamics of the process after it hits the ith threshold, is independent of the past and depends only on i. A simple example where this is true is the M/M/1/N queue.

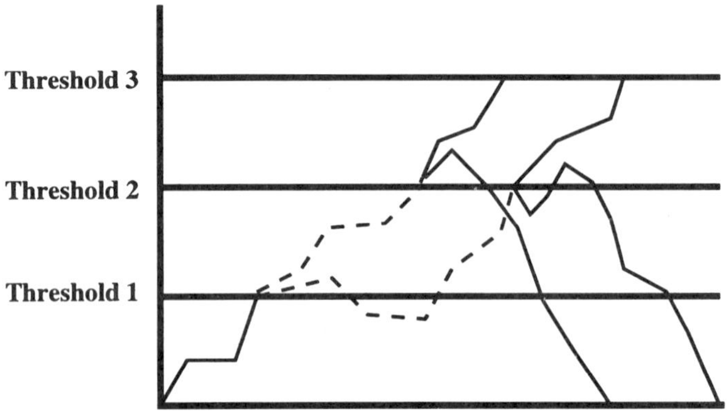

Figure 1: Splitting with Three Thresholds and Two Split Subpaths at Each Intermediate Threshold.

We now calculate the variance of this estimator:

$$\begin{aligned}\sigma_{k+1}^2 &\equiv \mathrm{Var}[I_{k+1}] \\ &= \mathrm{Var}[E[I_{k+1}|\mathcal{F}_k]] + E[Var[I_{k+1}|\mathcal{F}_k]] \\ &= \frac{1}{n_{k+1}^2}\{\mathrm{Var}[I_k n_{k+1} p_{k+1}] \\ &\quad + E[I_k n_{k+1} p_{k+1}(1-p_{k+1})]\} \\ &= p_{k+1}^2 \sigma_k^2 + \frac{p_1 \cdots p_k p_{k+1}(1-p_{k+1})}{n_1 \cdots n_{k+1}}.\end{aligned}$$

This recurrence relation can be solved to get

$$\sigma_k^2 = \sum_{j=1}^{k} \left(\prod_{i=j+1}^{k} p_i^2 \right) \frac{p_1 \cdots p_j (1-p_j)}{n_1 \cdots n_j},$$

which can also be written as

$$\sigma_k^2 = (p_1 \cdots p_k)^2 \sum_{j=1}^{k} \frac{1-p_j}{(p_1 n_1) \cdots (p_j n_j)}. \quad (1)$$

The next result examines the behavior of this variance as k increases.

Lemma 1 *(i) If* $\liminf_{j\to\infty} \frac{1}{j}\sum_{i=1}^{j} \log(n_i p_i) > 0$,

$$\sigma_k^2 = O((p_1 \cdots p_k)^2); \quad (2)$$

(ii) if $-\infty < \liminf_{j\to\infty} \sum_{i=1}^{j} \log(n_i p_i)$,

$$\sigma_k^2 = O(k(p_1 \cdots p_k)^2);$$

(iii) if $\limsup_{k\to\infty} p_k < 1$ *and* $\liminf_{j\to\infty} \frac{1}{j}\sum_{i=1}^{j} \log(n_i p_i) < 0$,

$$(p_1 \cdots p_k)^2 = o(\sigma_k^2).$$

Proof. The three cases are related to the convergence of the sum on the right side of (1). By the root test, the series

$$\sum_{j=1}^{\infty} \frac{1-p_j}{(p_1 n_1) \cdots (p_j n_j)} \quad (3)$$

converges if

$$\limsup_{j\to\infty} \left[\frac{1-p_j}{(p_1 n_1) \cdots (p_j n_j)} \right]^{1/j} < 1;$$

i.e., if

$$\limsup_{j\to\infty} \frac{1}{j} \left[\log(1-p_j) + \sum_{i=1}^{j} \log\left(\frac{1}{n_i p_i}\right) \right] < 0.$$

The condition in (i) is equivalent to

$$\limsup_{j\to\infty} \frac{1}{j} \sum_{i=1}^{j} \log(\frac{1}{n_i p_i}) < 0$$

which ensures convergence of (3) and proves (2). The reverse inequality in (iii) similarly ensures divergence of (3). For (ii), notice that

$$\sum_{j=1}^{k} \frac{1-p_j}{(p_1 n_1) \cdots (p_j n_j)} = O(k)$$

if

$$\limsup_{j\to\infty} (1-p_j) \prod_{i=1}^{j} \left(\frac{1}{p_i n_i}\right) < \infty,$$

which holds under the condition in (ii). □

The conditions in this lemma simplify in the important special case that the p_i approach some limit p and all n_i equal some n for sufficiently large i. In this setting, the three cases in the lemma can be replaced with $np > 1$, $np = 1$, and $np < 1$. The corresponding results for this special cases may be found in GHSZ.

In case (i) of the lemma, the second moment of the estimator is also $O((p_1 \cdots p_k)^2)$, because the first moment is $p_1 \cdots p_k$. Nonnegativity of variance makes this the best possible rate of decrease for the second moment. In contrast, straightforward simulation (corresponding to a single Bernoulli with parameter $p_1 \cdots p_k$) has variance

$$(p_1 \cdots p_k)[1 - (p_1 \cdots p_k)] = O(p_1 \cdots p_k)$$

per replication and a second moment of the same order.

We now supplement results for the variance with an assessment of the computational effort. We assume, for simplicity, that the work per sample is constant across stages. (In many cases this may not be true. For example, in many queueing models the expected cost of simulating a trial from threshold i grows proportionally with i. This is because, with positive probability bounded away from zero, the system never reaches threshold $i+1$ and therefore many trials consist of simulating the queue until it empties again. However, these other cases can be handled similarly and lead to similar conclusions.) Then the expected work is proportional to the expected number of samples, which is

$$n_1 + (n_1p_1)n_2 + \cdots + (n_1p_1 \cdots n_{k-1}p_{k-1})n_k = n_1 \sum_{j=0}^{k-1} \prod_{i=1}^{j} p_i n_{i+1}.$$

For the expected work we have:

Lemma 2 *(i) If* $\limsup_{j\to\infty} \frac{1}{j}\sum_{i=1}^{j} \log(p_i n_{i+1}) > 0$, *the expected work per run grows exponentially in* k.

(ii) If $\sum_{i=1}^{\infty} \log(p_i n_{i+1}) < \infty$, *then the expected work per run is* $O(k)$.

(iii) If $\limsup_{j\to\infty} \frac{1}{j}\sum_{i=1}^{j} \log(p_i n_{i+1}) < 0$, *then the expected work per run is* $O(1)$.

Proof. For case (i), note that

$$\frac{1}{k}\log\left(\sum_{j=0}^{k-1}\prod_{i=1}^{j} p_i n_{i+1}\right) \geq \frac{1}{k}\log\left(\prod_{i=1}^{k-1} p_i n_{i+1}\right) = \frac{1}{k}\sum_{i=1}^{k-1}\log(p_i n_{i+1})$$

so a positive limsup for this expression indicates exponential growth of expected work. The expected work is $O(k)$ if

$$\frac{1}{k}\sum_{j=0}^{k-1}\prod_{i=1}^{j} p_i n_{i+1}$$

converges, and a sufficient condition for this is the condition in case (ii) above. The condition in case (iii) above ensures that the series

$$\sum_{j=0}^{\infty}\prod_{i=1}^{j} p_i n_{i+1}$$

converges, by the root test. □

As in Lemma 1, the conditions here can be replaced with $np >, =,$ or < 1 in the case of $p_i \to p$ and fixed $n_i = n$. The corresponding results for these special cases may also be found in GHSZ.

The work-normalized variance, balancing computational effort and estimator variance, is the product of the variance and the expected work per run; see Glynn and Whitt (1992) for full justification of this criterion. Combining Lemmas 1 and 2 yields a condition for optimal splitting:

Theorem 1 *If*

$$-\infty < \liminf \sum_{i=1}^{j} \log(p_i n_i)$$

and

$$\sum_{i=1}^{\infty} \log(p_i n_{i+1}) < \infty,$$

then I_k *is asymptotically efficient in the sense that*

$$\lim_{k\to\infty} \frac{\log O(k^2(p_1\cdots p_k)^2)}{\log E[I_k]} = 2.$$

We interpret this result to mean that splitting is most effective when $n_i \approx 1/p_i$. GHSZ discuss the use of a random number of splits in order to get the expected number of subpaths equal to $1/p_i$ when p_i is not the reciprocal of an integer.

The analysis above is based on a very simple model of splitting in which the success probabilities p_i are constant at each threshold, regardless of what may have happened at previous thresholds. Consider estimating the probability that a Markov chain reaches some rare set before returning to its initial state. We label the initial state 0 and assume it is recurrent. Imagine introducing intermediate thresholds in the state space of the Markov chain and splitting each path that reaches a threshold into some number of subpaths. In general, the probability that the chain will reach the ith threshold before 0, given that it has reached the $(i-1)$th threshold before 0, will depend on the state of the chain when it reached the $(i-1)$th threshold. The assumption of constant p_i would hold if, say, there were just one state through which the $(i-1)$th threshold could be reached; but, more typically, p_i would vary depending on the entry state.

The case where we have a fixed and finite number of entry states into each threshold, and the probability dynamics of the process is homogeneous (in some limiting sense) with respect to the thresholds, is further analyzed in GHSZ. To get a sense of the possible impact of the variability of the p_i's in a more general setting (i.e., uncountably infinite number of entry states into thresholds), we consider a simple two-threshold problem. Our objective is to estimate $\gamma = p_1p_2$ where, now, $p_2 = E[\tilde{p}_2]$, with $\tilde{p}_2$ stochastic. The mechanism we have in mind is this: each path has probability p_1 of reaching the first threshold; upon reaching that threshold, its offspring are randomly assigned a sample of $\tilde{p}_2$ as their common second-stage success probability. This accurately represents the Markovian setting described above. The estimator is

$$I_2 = \frac{1}{n_1n_2}\sum_{i_1=1}^{n_1}\sum_{i_2=1}^{n_2} \mathbf{1}_{i_1}\mathbf{1}_{i_1i_2},$$

with each $\mathbf{1}_{i_1} \sim$ Bernoulli(p_1) and each $\sum_{i_2=1}^{n_2} \mathbf{1}_{i_1 i_2}$ conditionally Binomial$(n_2, \tilde{p}_2)$, independent for different i_1.

Let $N_2 = \sum_{i_2=1}^{n_2} \mathbf{1}_{i_1 i_2}$ for some i_1. Then

$$\mathrm{Var}[N_2] = n_2 p_2(1-p_2) + (n_2^2 - n_2)\mathrm{Var}[\tilde{p}_2].$$

Proceeding as we did for (1), we get

$$\mathrm{Var}[I_2] = \frac{1}{n_1^2 n_2^2}\left\{n_2^2 p_2^2 n_1 p_1(1-p_1) + n_1 p_1 \mathrm{Var}[N_2]\right\}.$$

This becomes

$$\frac{p_2^2 p_1(1-p_1)}{n_1} + \frac{p_1 p_2(1-p_2)}{n_1 n_2} + \frac{p_1(n_2^2 - n_2)\mathrm{Var}[\tilde{p}_2]}{n_1 n_2^2}.$$

The last term gives the effect of a random $\tilde{p}_2$ compared with a fixed p_2. To get a sense of its impact, divide through by $\gamma^2 = p_1^2 p_2^2$, and suppose that $n_i \approx 1/p_i$, $i = 1, 2$. The contribution of the first two terms is then $O(1)$ whereas the new variability term contributes $O(n_2^2 Var[\tilde{p}_2]) = O(Var[\tilde{p}_2]/p_2^2)$.

This simple observation has important implications for the effectiveness of multithreshold splitting procedures: splitting will be most effective if there is little variability in the success probability at each threshold. This further suggests (at least heuristically) that the thresholds should be chosen in a way that is consistent with the most likely path to a rare set. For then each subpath will draw a success probability close to that for the most likely path, resulting in little variation across subpaths. Understanding the large deviations behavior of a rare event may therefore be useful in designing a splitting procedure.

3 OPTIMAL PARTITIONS

We now return to the simple setting from the start of Section 2. In particular, the p_i are constant at each threshold and we want to estimate $\gamma \equiv \gamma_k = p_1 \cdots p_k$, with $p_1 = P(A_1)$ and $p_i = p(A_i|A_{i-1})$, continuing to think of $k \to \infty$ and $\gamma_k \to 0$. We consider the problem of choosing the intermediate events $A_0, A_1, \ldots, A_{k-1}$ and make two observations: choosing these events so that the p_i converge as $i \to \infty$ has an asymptotic optimality property, and there is a connection between being able to choose the thresholds so that the p_i converge and being able to analyze the large deviations behavior of a rare event.

We begin by examining the optimal choice of $p_1, \ldots, p_k$ for fixed γ. Based on the analysis in Section 2, we restrict attention to the case $n_i = 1/p_i$, ignoring the integrality constraint on the n_i. In this case, the variance becomes

$$\sigma_k^2 = \gamma^2 \sum_{j=1}^{k}(1-p_j),$$

and the expected work per run becomes

$$\sum_{j=1}^{k} \frac{1}{p_j}.$$

Our objective is then to minimize

$$g(p_1, \ldots, p_k) = \sum_{i,j}^{k} \frac{1-p_i}{p_j}$$

subject to $p_1 \cdots p_k = \gamma$. Rewriting the constraint and appending it with a Lagrange multiplier yields

$$\sum_{i,j}^{k} \frac{1-p_i}{p_j} + \lambda\left(\sum_{j=1}^{k} \log p_j - \log\gamma\right).$$

The first-order conditions

$$-\frac{1}{p_i^2}\sum_{j=1}^{k}(1-p_j) - \sum_{j=1}^{k}\frac{1}{p_j} + \frac{\lambda}{p_i} = 0, \quad i = 1, \ldots, k,$$

and $\sum_{j=1}^{k} \log p_j - \log\gamma = 0$, are solved by taking $p_i = p \equiv \gamma^{1/k}$ and $\lambda = k/p$. Moreover, the objective g is convex because it is a sum of terms $(1-p_i)/p_j$, each of which is convex in (p_i, p_j) (or simply p_i in case $j = i$). Thus, it is optimal to make the p_i equal.

It is now a simple matter to conclude that partitioning so that the p_i converge is asymptotically optimal (at least among schemes with $n_i = 1/p_i$). For each k let $q_1^{(k)}, \ldots, q_k^{(k)}$ be any probabilities multiplying to γ_k. We claim that if $p_k \to p$ as $k \to \infty$, then

$$\limsup_{k\to\infty} \frac{g(p_1, \ldots, p_k)}{g(q_1^{(k)}, \ldots, q_k^{(k)})} \le 1.$$

In light of the optimization carried out above,

$$\frac{g(\gamma_k^{1/k}, \ldots, \gamma_k^{1/k})}{g(q_1^{(k)}, \ldots, q_k^{(k)})} \le 1$$

for all k. In addition, we now argue that

$$\frac{g(p_1, \ldots, p_k)}{g(\gamma_k^{1/k}, \ldots, \gamma_k^{1/k})} \to 1, \tag{4}$$

whenever the p_i converge. To see this, notice that

$$\frac{1}{k^2} g(p_1, \ldots, p_k) = \frac{1}{k}\sum_{j=1}^{k}(1-p_j)\frac{1}{k}\sum_{i=1}^{k}\frac{1}{p_i} \to \frac{1-p}{p}.$$

Also, since $\prod_{j=1}^{k} p_j = \gamma_k$, $\gamma_k^{1/k} \to p$, so

$$\frac{1}{k^2} g(\gamma_k^{1/k}, \ldots, \gamma_k^{1/k}) = (1-\gamma_k^{1/k})\frac{1}{\gamma_k^{1/k}} \to \frac{1-p}{p},$$

as $k \to \infty$, which verifies (4). We conclude that choosing the p_i so that they converge to a limit is asymptotically as effective (as $k \to \infty$) as using the optimal partition at each k.

What does choosing the p_k to converge entail for the sets A_k? We now point out that the availability of a convergent p_k sequence is related to the A_k satisfying a limit theorem of the large deviations type. More specifically, if the p_k converge then the $P(A_k)$ have a logarithmic limit; and if the $P(A_k)$ have an asymptotically exponential decay, then the p_k converge. For if $p_k \to p$ then

$$\lim_{k\to\infty} \frac{1}{k} \log P(A_k) = \lim_{k\to\infty} \frac{1}{k} \sum_{j=1}^{k} \log p_j = \log p.$$

And if

$$P(A_k) \sim Ce^{-\alpha k}, \quad \text{for some } \alpha > 0,$$

as $k \to \infty$ through integer values, then

$$p_k = \frac{P(A_k)}{P(A_{k-1})} \to e^{-\alpha}.$$

This gives another sense in which knowing something about the large deviations behavior of a rare event could be useful in designing a splitting procedure. Knowing the large deviations behavior should be useful in setting thresholds for which the resulting p_k converge.

4 TRUNCATION BIAS

As mentioned before, in many queueing models the expected cost of simulating a trial from threshold i grows proportionally with i. This is because, with positive probability bounded away from zero, the system never reaches threshold $i+1$ and therefore many trials consist of simulating the queue until it empties again. As such unsuccessful trials do not contribute positive weight to the estimation of γ_k, it seems wasteful to devote significant computing resources to them. Therefore, it is desirable to "throw away" trials that have dropped many thresholds from the starting threshold and thus are very unlikely to reach the next highest threshold. However, doing so introduces some bias in the estimator. In this section we analyze this "truncation" bias for a simple example, which should nevertheless yield insight into more complex situations.

We assume there is a truncation threshold d. If a trial started at threshold i where $i \leq d$, then we simulate the sample path the same way as in the case without truncation. If a trial started from threshold i, where $i > d$, ever drops to threshold $(i-d)$, that trial is counted as a failure and discarded. We analyze this bias for the simplest possible queueing system, the M/M/1 queue. We let λ and μ denote the arrival and service rates, respectively, and define $\rho = \lambda/\mu < 1$. We assume $\lambda + \mu = 1$. The embedded discrete time Markov chain is a random walk with increments that take on the value +1 with probability λ and -1 with probability μ. In this case we let the thresholds correspond to queue sizes of $1, 2, \ldots k$. To estimate the bias we first need to calculate p_i, which is the probability that the queue length ever reaches $(i+1)$ before emptying, given that the initial queue length is i. Such probabilities are known from analysis of the "gambler's ruin" problem; see pages 344-348 of Feller (1968). More generally, if $r_j = P\{\text{hit } 0 \text{ before } n | \text{ start at } j\}$, then

$$r_j = \frac{1 - \rho^{n-j}}{1 - \rho^n}. \tag{5}$$

Specializing (5) to $j = i$ and $n = (i+1)$, we have $p_i = \rho[1-\rho^i]/[1-\rho^{i+1}]$. Now let p'_i denote the probability of reaching threshold $i+1$ before threshold $i-d$, given an initial queue length of i. Splitting using truncation yields an unbiased estimate of $\gamma'_k = \prod_{i=1}^{k-1} p'_i$. Note that $p'_i = p_i$ for $0 \leq i \leq d$. The formula for p'_i for $i \geq d$ is determined from the right-hand-side of (5) with $j = d$ and $n = (j+1)$: $p'_i = \rho[1-\rho^d]/[1-\rho^{d+1}]$.

We wish to compare γ'_k to γ_k and in particular wish to know how $d = d_k$ should be chosen so that $\gamma'_k/\gamma_k \to 1$ as $k \to \infty$.

$$\frac{\gamma'_k}{\gamma_k} = \prod_{i=d+1}^{k-1} \frac{p'_i}{p_i} = \left[\frac{1-\rho^d}{1-\rho^{d+1}}\right]^{k-d-1} \prod_{i=d+1}^{k-1} \frac{1-\rho^{i+1}}{1-\rho^i}. \tag{6}$$

The product term on the right-hand-side of (6) telescopes to $[1-\rho^k]/[1-\rho^{d+1}]$ which $\to 1$ provided both k and $d \to \infty$. Thus we require $[(1-\rho^d)/(1-\rho^{d+1})]^{k-d} \to 1$. This will be true provided $(1-\rho^d)^{k-d} \to 1$ or, equivalently, $(k-d)\log(1-\rho^d) \to 0$. Using the Taylor series expansion $\log(1-\epsilon) \approx -\epsilon$ for small ϵ, we then require that $k\rho^d \to 0$ (since $d\rho^d \to 0$ as $d \to \infty$). That is, we require that $d \to \infty$ and that k not grow too quickly with respect to d, specifically:

$$k = o(\rho^{-d}). \tag{7}$$

In an asymptotically optimal splitting procedure the expected cost to simulate all of the offspring from a single trial from threshold 0 without truncation is of order $w = k^2$. With truncation, this is reduced to order $w' = d \times k$. Thus $w/w' = k/d$ can grow arbitrarily large and still satisfy (7), i.e., by appropriately choosing the truncation threshold we obtain significant computational savings without introducing significant bias. As a numerical example, when Equation 6 is computed with $\rho = 0.5$, $k = 20$ and $d = 5$, $\gamma'_k/\gamma_k = 0.81$ representing a truncation bias of about 20%, but when d is increased to 10, γ_k'/γ_k increases to 0.996. Even with $k = 50$ (and $d = 10$), $\gamma_k'/\gamma_k = 0.98$, representing only 2% bias.

ACKNOWLEDGEMENT

This work is supported by NSF grants DMI-94-57189 and DMS-9508709.

REFERENCES

Bayes, A.J. 1970. Statistical techniques for simulation models. *The Australian Computer Journal* 2: 180-184.

Feller, W. 1968. *An Introduction to Probability Theory and Its Applications, Volume I*, Third Edition. New York: John Wiley & Sons, Inc.

Glasserman, P., P. Heidelberger, P. Shahabuddin, and T. Zajic. 1996. Multilevel Splitting for Estimating Rare Event Probabilities. IBM Research Report, Yorktown Heights, New York.

Glynn, P.W., and W. Whitt. 1992. The Asymptotic Efficiency of Simulation Estimators. *Operations Research* 40: 505–520.

Hammersley, J., and D. Handscomb. 1965. *Monte Carlo Methods.* Methuen & Co. Ltd., London.

Harris, T. 1989. *The Theory of Branching Processes.* Dover, New York.

Heidelberger, P. 1995. Fast simulation of rare events in queueing and reliability models. *ACM Transactions on Modeling and Computer Simulation* 5, 43-85.

Hopmans, A.C.M., and J.P.C. Kleijnen. 1979. Importance sampling in system simulation: A practical failure? *Mathematics and Computing in Simulation XXI*: 209-220.

Kahn, H., and T.E. Harris. 1951. Estimation of Particle Transmission by Random Sampling. *National Bureau of Standards Applied Mathematics Series* 12, 27-30.

Shahabuddin, P. 1995. Rare Event Simulation in Stochastic Models. In *Proceedings of the 1995 Winter Simulation Conference*, 178-185, IEEE Press.

Villén-Altamirano, M., and J. Villén-Altamirano. 1991. RESTART: A method for accelerating rare events simulation. In *Proceedings of the 13th International Teletraffic Congress, Queuing performance and control in ATM*, 71-76, North Holland Publishing Company.

Villén-Altamarino, M., A. Martinez Marron, J. Gamo and F. Fernandez-Cuesta. 1994. Enhancement of acclerated simulation method RESTART by considering multiple threshholds. In *Proceedings of the 14th International Teletraffic Congress, The fundamental role of teletraffic in the evolution of telecommunication networks*, 787-810, Elsevier.

Villén-Altamarino, M., and J. Villén-Altamirano. 1994. RESTART: A straightforward method for fast simulation of rare events. In *Proceedings of the 1994 Winter Simulation Conference*, 282-289, IEEE Press.

AUTHOR BIOGRAPHIES

PAUL GLASSERMAN is a Professor in the Management Science division of the Columbia University Graduate School of Business. Prior to joining the Columbia faculty he was a Member of Technical Staff in the Operations Research department of AT&T Bell Laboratories in Holmdel, NJ. He holds a Ph.D. from Harvard University and an A.B. from Princeton University.

PHILIP HEIDELBERGER received a B.A. in mathematics from Oberlin College in 1974 and a Ph.D. in Operations Research from Stanford University in 1978. He has been a Research Staff Member at the IBM T.J. Watson Research Center since 1978. While on sabbatical in 1993-1994, he was a visiting scientist at Cambridge University and at ICASE, NASA Langley Research Center. He is the Editor-in-Chief of *ACM TOMACS*, was Program Chairman of the 1989 Winter Simulation Conference, and Program Co-Chairman of the ACM Sigmetrics/Performance '92 Conference. He is a Fellow of both the IEEE and the ACM.

PERWEZ SHAHABUDDIN is an Assistant Professor at the Industrial Engineering and Operations Research Department at Columbia University, New York, NY, since Fall 1995. He is currently on a leave of absence from the IBM T.J. Watson Research Center, Yorktown Heights, NY, where he has been a Research Staff Member since 1990. He received his B.Tech in Mechanical Engineering from the Indian Institute of Technology, Delhi, in 1984, followed by a M.S. in Statistics and a Ph.D in Operations Research from Stanford University in 1987 and 1990, respectively. From 1984 to 1985 he worked as a system analyst at Engineers India Limited, India. Currently he is serving as an Associate Editor for *IEEE Transactions on Reliability* and is on the Editorial Board of *IIE Transactions-Operations Engineering.* He is a recipient of a 1996 National Science Foundation Career Award.

TIM ZAJIC received a Ph.D. in Operations Research from Stanford University in 1993. From 1994 to 1995 he was a visitor at the Centro de Investigación en Matemáticas (CIMAT) in Guanajuato, México. He is currently a postdoctoral fellow at the Department of Industrial Engineering and Operations Research at Columbia University and the IBM T.J. Watson Research Center.

Proceedings of the 1996 Winter Simulation Conference
ed. J. M. Charnes, D. J. Morrice, D. T. Brunner, and J. J. Swain

IMPORTANCE SAMPLING FOR LARGE ATM-TYPE QUEUEING NETWORKS

Pierre L'Ecuyer and Yanick Champoux

Département d'Informatique et de Recherche Opérationnelle
Université de Montréal, C.P. 6128, Succ. Centre-Ville
Montréal, H3C 3J7, CANADA

ABSTRACT

We estimate, by simulation, the cell-loss rate in an ATM switch modeled as a queueing network. Cell losses are rare events, so estimating their frequency by simulation is hard. We experiment with importance sampling as a mean of improving the simulation efficiency in that context.

1. INTRODUCTION

An Asynchronous Transfer Mode (ATM) communication switch can be modeled as a network of queues with finite buffer sizes. Cells (or packets) of information join the network in a stochastic manner and some may be lost due to buffer overflow. The long-term (or steady-state) fraction of cells that are lost at a given node is called the cell-loss rate (CLR) at that node. Typical CLRs are small and the cell losses also tend to occur in bunches. They are therefore rare events, so estimating the CLRs with reasonable precision by straightforward simulation is extremely time-consuming—in some cases practically impossible.

Efficiency improvement methods have been proposed to deal with such a situation. Most of these methods improve the efficiency by reducing the variance of the estimator, and are called *variance reduction techniques*. For rare events, *importance sampling* (IS) seems the method of choice. It changes the probability laws governing the system so that the rare events of interest occur more frequently, and eventually are no longer rare events. The estimator is also changed accordingly (multiplied by a likelihood ratio) so that it remains unbiased.

For general background on efficiency improvement, consult Bratley, Fox, and Schrage (1987), Glynn (1994) and L'Ecuyer (1994). For more on IS, see Glynn and Iglehart (1989), Heidelberger (1995), Shahabuddin (1994) and the several other references given there. Application of IS to the simulation of communication systems is studied by Chang et al. (1994), Chang, Heidelberger, and Shahabuddin (1995), Fleming, Schaeffer, and Simon (1995), among several others. A somewhat related approach, which also concentrates the simulation effort on the more interesting events, is *splitting* (called "RESTART" by some authors); see, e.g., Glasserman et al. (1996).

By far the most difficult problem in the practical application of importance sampling is to find an appropriate change of measure; that is, figure out how to change the probability laws so that the variance gets reduced to an acceptable level. Theoretically, there always exists a change of measure that reduces the variance to an arbitrary small value, but it is usually much too complicated and too difficult to find.

Chang et al. (1994) proposed an approach, based on the theories of *effective bandwidth* and *large deviations*, to derive an "asymptotically optimal" change of measure for estimating the probability p that a queue length exceeds a given level x before returning to empty, given that the queue is started from empty, for a single queue with multiple independent arrival sources. Roughly, *asymptotically optimal* means that the standard error of the IS estimator converges to zero exponentially fast with the same decay rate (exponent) as the quantity to be estimated, as a function of the level x. For a more precise mathematical statement, see Chang et al. (1994) and Heidelberger (1995). An asymptotically optimal change of measure does not (generally) minimize the variance, but can reduce it by several orders of magnitude.

The probability p just described is closely related to the CLR, so this change of measure could be used as well to estimate the CLR in a single queue with finite buffer size. Chang et al. (1994) extended their method to *intree* networks of queues, which are acyclic tree networks where cells flow only towards the root of the tree. The arrival sources can feed any node. For intree networks, they did not prove that their proposed change of measure is asymptotically optimal, but conjectured that it is close and gave an upper bound on the variance. They reported numerical experiments with single-node and two-node queueing systems, and indeed observed spectacular variance reductions.

This paper reports further experimentation with this approach, for larger queueing networks of a specific type. Our findings confirm the large variance reductions observed by Chang et al. (1994) in general. We make an heuristic adaptation to a certain class of non-intree networks and observe large variance reductions as well, under certain conditions. The next section describes the queueing network model. Section 3 recalls how to compute confidence intervals for the CLR via the A-cycle method. Section 4 explains how importance sampling is applied. The numerical results are reported in Section 5.

2. THE MODEL

We consider an acyclic queueing network whose *nodes* are partitioned into four *levels*. Each node is a single-server FIFO queue with finite buffer size. Levels 2 and 3 have m_2 nodes each, while levels 1 and 4 have $m_1 m_2$ nodes each. Each level-2 node is fed by m_1 level-1 nodes, while each level-3 node feds m_1 nodes at level 4. The "customers" in the network are (indentical) *cells* containing bits of information. They arrive at level 1, then move ahead to levels 2, 3, and 4, in succession, before leaving the network. Each level-1 node is fed by m_0 independent arrival *sources*. Each source is assigned to a specific destination at level 4, so all cells from this source have a common trajectory in the network. The assignments of destinations to sources is fixed (deterministically) beforehand. In practice, we may be interested in a random assignment of destinations to sources, but in that case it is probably better to stratify the experiment over the set of possible assignments. Here, we concentrate on what to do after the assignment has been fixed.

For $\ell = 1, \ldots, 4$, all level-ℓ nodes have the same buffer size B_ℓ and the same constant service time $1/c_\ell$ (so c_ℓ is the service rate). Whenever a cell arrives at a node where the buffer is full, it is *lost* and just disappears.

The $m_0 m_1 m_2$ arrival sources are *iid* Markov modulated processes. A source is *off* for a while, then *on* for a while, then *off* for a while, and so on. During a *on* period, cells arrive at a constant rate, one cell per unit of time, whereas during a *off* period, none arrives from that source. The durations of *off* and *on* periods are independent geometric random variables with respective means κ_0 and κ_1. The parameter κ_1 is called the *average burst size*.

We want to estimate the fraction of cells lost (the CLR) at a given level of the network (among those reaching that level), in steady-state. For this, we concentrate on a selected node of the network, say node q^* at level ℓ^*, and trim down from the network all nodes at level 3 or 4 from which node q^* cannot be reached. An alternative would be to take the average CLR for all nodes at a given level as the estimator. With a straightforward simulation approach, this would yield a better estimator than concentrating on a single node. But with IS, it seems better to concentrate on a single node, and increase only the traffic to that node, to control the variance of the likelihood ratio.

Other variants of the model could be considered. For example, each arriving cell could have its destination determined randomly, or the destination could be generated randomly for each *on* period (or "burst") of each source. These models may be more difficult to handle with IS (in general) than the one we consider, because the likelihood ratio would tend to have more terms. Our fixed-assignment model is reasonable because in communication networks, a typical connection between a source and a destination would last for a period of time several orders of magnitude larger than the average time between bursts.

3. CONFIDENCE INTERVALS

To compute a confidence interval on μ, one needs to estimate the variance of $\hat{\mu}$. For this, we apply a generalization of the classical regenerative method, called the *A-cycle* method, introduced and used by Nicola et al. (1993) and Chang et al. (1994). Let A be a subset of the state space of the system. In this paper, A is taken as the set of states for which the queue at node q^* is empty. Let $t_0 = 0$ and let $t_1, t_2, \ldots$ be the successive times at which the system's state enters the set A. The system's state at those entering times t_i has a steady-state distribution π defined by:

$$\pi(\cdot) \stackrel{\text{def}}{=} \lim_{n\to\infty} \frac{1}{n} \sum_{i=1}^{n} P\{\text{state} \in \cdot \text{ at time } t_i\}.$$

The process between times t_{i-1} and t_i is called the ith A-cycle. Let X_i denote the number of arrivals to node q^* during the ith A-cycle and Y_i be the number of lost cells among those X_i arrivals. Let E_π denote the mathematical expectation over an A-cycle when the initial state (at the beginning of the A-cycle) has distribution π. One has:

$$\mu = \frac{\mathrm{E}_\pi[Y_1]}{\mathrm{E}_\pi[X_1]}. \tag{1}$$

In the limit, as the number of A-cycles increases, the *average* distribution of the system states at the times

t_i approaches π. To reduce the initial bias, one may *warm-up* the system by discarding (say) the first n_0 A-cycles from the statistics. Then, take the averages of the Y_i and X_i over the next n A-cycles, which yields the estimator:

$$\hat{\mu} = \frac{\sum_{i=1}^{n} Y_i}{\sum_{i=1}^{n} X_i}.$$

Under mild conditions which are assumed to hold here, π is also a steady-state distribution in the pointwise sense. The A-cycles are then asymptotically identically distributed (their initial state follows π) but *not independent*. To reduce the dependence, and also improve the normality, one can *batch* them, as in the batch means method (Bratley, Fox, and Schrage 1987). One then applies the standard methodology for computing a confidence interval for a ratio of expectations, using the batch means as observations.

Suppose we take b batches of $k = n/b$ successive observations (X_i, Y_i) each. For $1 \le j \le b$, let $(\bar{X}_j, \bar{Y}_j)$ be the mean within batch j, and put

$$\bar{Z}_j = \bar{Y}_j - \mu \bar{X}_j.$$

The $\bar{Z}_j$ have zero expectation and variance

$$\mathrm{Var}[\bar{Z}_j] = \mathrm{Var}[\bar{Y}_j] - 2\mu \mathrm{Cov}[\bar{X}_j, \bar{Y}_j] + \mu^2 \mathrm{Var}[\bar{X}_j],$$

which can be estimated by

$$S_Z^2 = S_Y^2 - 2\hat{\mu} S_{XY} + \hat{\mu}^2 S_X^2$$

where $\bar{X}$, S_X^2, $\bar{Y}$, S_Y^2, S_{XY} denote the sample mean and variance of the $\bar{X}_j$'s, the sample mean and variance of the $\bar{Y}_j$'s, and the sample covariance between the $\bar{X}_j$'s and $\bar{Y}_j$'s, respectively. Assuming that the Z_j are iid normally distributed and denoting $\bar{Z} = \bar{Y} - \mu \bar{X}$, one obtains that

$$\frac{\sqrt{b}\bar{Z}}{S_Z} = \frac{\sqrt{b}\bar{X}(\hat{\mu} - \mu)}{S_Z}$$

has a Student-t distribution with $b - 1$ degrees of freedom, which is approximately standard normal for large enough b. From this, one can compute a confidence interval on μ in a standard way.

4. APPLYING IMPORTANCE SAMPLING

To estimate the CLR μ at node q^*, the straightforward approach is to simulate until n cells reach node q^*, and define $\hat{\mu}$ as the fraction of those lost due to a full buffer. This estimator turns out to have *relative error* (the standard deviation divided by the mean) $\mathrm{RE}[\hat{\mu}] = O((n\mu)^{-1/2})$), which is unbounded as $\mu \to 0$, so one must find a better estimator.

We will use IS, with the methodology developed by Chang et al. (1994) for choosing the change of measure. We now explain how this can be applied to our model. The two quantities to estimate are the numerator and denominator in (1). The denominator $\mathrm{E}_\pi[X_1]$ is easy to estimate just by simulating several A-cycles without IS. The numerator is more difficult because it involves rare events, and IS will be used for it, as follows.

Let 0 and 1 denote the states *off* and *on* of a source. Due to the geometric assumption, each source evolves as a discrete-time Markov chain with state space $\{0, 1\}$ and transition matrix R with elements r_{ij} given by $r_{01} = 1/\kappa_0$, $r_{00} = 1 - r_{01}$, $r_{10} = 1/\kappa_1$, $r_{11} = 1 - r_{10}$. Its arrival rate is $\rho = \kappa_1/(\kappa_1 + \kappa_0)$.

Let S^* denote the set of sources feeding q^*. To increase the traffic at node q^* (and get more cell losses), we change the matrix R for all the sources in S^*, so they spend more time in the *on* state. At the beginning of an A-cycle with IS, the transition matrix R for the sources in S^* is replaced by

$$\tilde{R} = \begin{pmatrix} \tilde{r}_{00} & \tilde{r}_{01} \\ \tilde{r}_{10} & \tilde{r}_{11} \end{pmatrix} = \begin{pmatrix} r_{00}/K_0 & r_{01}e^\theta/K_0 \\ r_{10}/K_1 & r_{11}e^\theta/K_1 \end{pmatrix},$$

where $\theta \ge 0$, and K_0 and K_1 are (positive) normalizing constants ensuring that the lines sum up to 1. We will explain later on how to choose θ. Note that $\theta > 0$ increases the average input rate from the source while $\theta = 0$ leaves it (and R) unchanged. This new matrix $\tilde{R}$ is in effect until buffer q^* fills up or empties again, whichever comes first. When q^* fills up, the IS is turned off ($\tilde{R}$ is replaced by R) until the end of the A-cycle. We call this an *A-cycle with IS*.

For a given initial state, let $\tilde{\mathrm{E}}$ denote the expectation over an A-cycle with IS. Let N_{ij} be the number of times a source in S^* goes from state i to state j *while using the new probabilities* $\tilde{r}_{ij}$, during the A-cycle, for $i = 0, 1$ and $j = 0, 1$, and let $N_T = N_{00} + N_{01} + N_{10} + N_{11}$, the total number of transitions generated from $\tilde{R}$. The *likelihood ratio* associated with this change of measure is then

$$L = \left(\frac{r_{00}}{\tilde{r}_{00}}\right)^{N_{00}} \left(\frac{r_{01}}{\tilde{r}_{01}}\right)^{N_{01}} \left(\frac{r_{10}}{\tilde{r}_{10}}\right)^{N_{10}} \left(\frac{r_{11}}{\tilde{r}_{11}}\right)^{N_{11}}.$$

If V is a random variable computed during an A-cycle, then

$$\mathrm{E}[V] = \tilde{\mathrm{E}}[LV],$$

so computing LV over an A-cycle with IS yields an unbiased estimator of $\mathrm{E}[V]$.

Generally, the variance of L may increase exponentially fast with N_T, so (intuitively) one would like to keep N_T small while still making cell losses frequent enough. This is why the A-cycle methodology, with short A-cycles, is to be preferred over using IS with longer simulations.

One simulates two "versions" of each A-cycle, one with IS and the other without, both starting from the same initial state. Thus, the A-cycles come in pairs. For the ith A-cycle pair, the simulation with IS provides an estimation $W_i = L_i Y_i$ of the numerator, where L_i and Y_i are the value of the likelihood ratio and the number of cell losses for this cycle, while the no-IS one provides an estimator X_i of the denominator. The final state of the no-IS A-cycle, which obeys approximately distribution π, is taken as the initial state for the next pair of A-cycles.

For $\ell^* = 1$ or 2, one has an intree network, for which all sources feed q^*, so θ can be chosen as proposed by Chang et al. (1994). Let $\lambda(\theta)$ be the spectral radius (largest eigenvalue) of the matrix

$$\begin{pmatrix} r_{00} & r_{01}e^{\theta} \\ r_{10} & r_{11}e^{\theta} \end{pmatrix}$$

and $f(\theta)$ be the corresponding eigenvector. Define $\psi_0(\theta) = \theta^{-1}\ln(\lambda(\theta))$. Compute θ and the corresponding $\tilde{r}_{ij}$ as follows:

Algorithm 1 (For $\ell^* = 1$ or 2.)
1. If $\ell^* = 1$, then
   compute θ_1^*, a solution of $m_0\psi_0(\theta) = c_1$
   else
   compute $\tilde{\theta}_1$, solution of $d(m_0 \ln \lambda(\theta))/d\theta = c_1$;
   Let
   $$\psi_1(\theta) = \begin{cases} m_0\psi_0(\theta) & \text{if } \theta \le \tilde{\theta}_1; \\ c_1 - \tilde{\theta}_1(c_1 - m_0\psi_0(\tilde{\theta}_1)) & \text{otherwise;} \end{cases}$$
   Compute θ_2^*, a solution of $m_1\psi_1(\theta) = c_2$;
   Let $\theta_1^* = \min(\theta_2^*, \tilde{\theta}_1)$;
2. For each (i, j), define
   $$\tilde{r}_{ij} = \frac{\exp(j\theta_1^*) f_j(\theta_1^*)}{\lambda(\theta_1^*) f_i(\theta_1^*)} r_{ij}.$$

For $\ell^* = 3$ or 4, the network is no longer intree, in the sense that many cells exit before reaching the root q^*. We nevertheless heuristically adapt the algorithm of Chang et al. (1994) as follows. We change r_{ij} to $\tilde{r}_{ij}$ only for the souces in S^*. When choosing θ, we neglect all the traffic not directed towards node q^*. To simplify, we also neglect the possibility that the effective bandwidth exceeds the service rate at a node other than q^*. This is reasonable because in our setup, S^* typically contains only a small fraction of the sources. This yields the following algorithm, where s^* is the cardinality of S^*, and where λ, f, and ψ_0 are defined as before.

Algorithm 2 (For $\ell^* = 3$ or 4.)
1. Compute θ^*, a solution to $s^*\psi_0(\theta) = c_{\ell^*}$;
2. Define
   $$\tilde{r}_{ij} = \frac{\exp(j\theta^*) f_j(\theta^*) r_{ij}}{\lambda(\theta^*) f_i(\theta^*)}.$$

5. NUMERICAL RESULTS

5.1. The Setup

In the examples that follow, the simulation was run first using the straightforward approach without IS, for b batches of k A-cycles each, then with IS, for b batches of k pairs of A-cycles each. The tables report the value of the CLR estimator $\hat{\mu}$, its variance estimate $\hat{\sigma}^2 = S_Z^2/(b\bar{X}^2)$, the relative half-width $\hat{\Delta} = 2.57 S_Z/(\sqrt{b}\bar{X}\hat{\mu}) = 2.57 S_Z/(\sqrt{b}\bar{Y})$ of a 99% confidence interval on μ (under the normality assumption), the CPU time t (in seconds) required to perform the simulation, and the estimated *relative efficiency*, defined as $\hat{\mu}^2/(t\hat{\sigma}^2)$. These values are noisy but give a rough indication of what happens. For the cases where no cell loss was observed in all the A-cycles simulated, we put $\hat{\mu} = 0$ and the entries for the variance and efficiency estimates are left blank. The IS adds overhead: it takes more CPU time than no-IS for the same total number of simulated cells—up to 15 times more in our implementation. So, for the IS estimator to win (be more efficient) it must have approximately 15 times less variance. To compare IS with no-IS, one must look at the efficiency (eff.) columns. Beware of comparing the CPU times and efficiencies across the tables, because the experiments were run on different machines (SUN SparcStations 4, 5, and 20).

For all the examples b has been fixed to 200.

5.2. CLR Estimation at Level 1

Example 1 Take $B_1 = 512$, $m_0 = 4$, $\rho = 1/10$, $c_1 = 1$ and vary the average burst size κ_1. Let $k = 3\,000$ for IS and $k = 50\,000$ for no-IS. Table 1 gives the results. For large average burst sizes, the CLR μ is high and easy to estimate, with or without IS. But for small burst sizes (the other parameters remaining the same), μ is small and much more difficult to estimate. Then, IS is much better than no-IS. With $\kappa_1 = 50$ the no-IS estimator is no longer trustworthy, and for $\kappa_1 = 10$ and 25 not even a single cell loss was observed in the kb A-cycles of the no-IS sample.

We made several other experiments where we varied the buffer size B_1 or the number of sources m_0 (with $m_0\rho$ fixed). As B_1 increases, μ decreases exponentially fast and the no-IS estimator quickly becomes useless, whereas IS works fine. As a function of m_0, μ increases slowly, so IS produces larger gains for small m_0.

5.3. CLR Estimation at Level 2

Example 2 Let $B_1 = 256$, $B_2 = 1024$, $m_0 = 8$,

$m_1 = 4$, $c_1 = 2$, $c_2 = 4$, $\rho = 1/21$, and vary the average burst size κ_1. The average input rate at the level 2 node is thus approximatively 1.5 cells per unit of time, whereas the service rate is 4. Here, $k = 250$ for IS and $k = 7\,500$ for no-IS. Table 2 gives the results, which are similar to those of Example 1, but with smaller CLR values and more spectacular improvement for the IS over the no-IS. For any κ_1, it appears difficult to estimate μ with significant precision with no-IS.

Example 3 Same as for the previous example, except that κ_1 is now fixed at 50 and we vary the buffer size B_2. We take $k = 250$ for IS and $k = 7\,500$ for no-IS. The results are given in Table 3. As the cell-loss rate is smaller than 10^{-6} (the value around which no-IS estimation passes from difficult to near-impossible) for all buffer sizes, no cell losses was observed in any of the no-IS simulations. The CPU time used by IS increases with the buffer size while it remains constant for no-IS. This is because, when the buffer size is larger, it takes more time to fill it (and ultimately observe a cell loss) under IS. The CPU time for the no-IS is not affected by the buffer size simply because the simulation does not try to cause an overflow, and so does not work more when the buffer is larger.

5.4. CLR Estimation at Level 3

Example 4 Let $B_1 = B_2 = 512$, $B_3 = 256$, $c_1 = 1$, $c_2 = c_3 = 2$, $m_0 = 2$, $m_1 = 3$, $m_2 = 10$, $\rho = 1/21$, and we vary the average burst size κ_1. We take $k = 500$ and $9\,000$ for IS and no-IS, respectively. We assign 6 sources to the node of interest at level 3. Table 4 gives the results. While no-IS has difficulty to observe a cell loss, IS gives reasonable estimations.

Example 5 Same as for the previous example, except that κ_1 is fixed at 50 and we vary the buffer size B_3. We take $k = 500$ and $9\,000$ (for IS and no-IS). Results appears in Table 5. Again, the IS works fine while the no-IS observes no cell loss except at the lowest buffer size.

Example 6 Same as the two previous example, except that κ_1 is fixed at 50 and the buffer size B_3 at 256. We take $k = 500$ and $9\,000$ (for IS and no-IS) and we vary the number of sources directed towards the observed node. Table 6 gives the results. IS again dominates as cell losses become sufficiently rare.

5.5. CLR Estimation at Level 4

Example 7 Let $B_1 = B_2 = B_3 = B_4 = 512$, $c_1 = c_4 = 1$, $c_2 = c_3 = 4$, $m_0 = 5$, $m_1 = 10$, $m_2 = 6$, $\rho = 1/41$ and we vary the average burst size κ_1. We assign 6 sources to the node of interest at level 4. We take $k = 250$ and $4\,000$, for IS and no-IS, respectively. The results are in Table 7. IS still gives far more better performances than no-IS.

Example 8 Same as the previous example, except that κ_1 is fixed at 50 and we vary the buffer size B_4. We take $k = 250$ and $4\,000$, for IS and no-IS. The results appears in Table 8. Again, IS is more effective than no-IS.

5.6. Variants of the Algorithms

The algorithms of Chang et al. (1994), used here, provide good changes of measures in an asymptotic sense, but not the best possible values of θ. Moreover, their aim is to reduce the variance and they do not take into account the differences in computational costs. We made additional experiments where θ was varied around the value given by the algorithm, to see whether the variance and efficiency would improve. For all levels, the optimal θ was generally slightly smaller but very close to the one prescribed by the algorithm.

When estimating the CLR at level 4 with IS, when the target buffer overflows at level 4 and the IS is turned off, there should be a large number of cells already in the network at previous levels, which may produce more cell losses than necessary. So, perhaps IS could be turned off earlier; e.g., when the total number of cells in buffer q^* or at previous nodes but on their way to q^*, reaches some threshold. We experimented with this idea and (for our model) obtained no significant improvement over the basic method which turns off the IS when the buffer q^* overflows.

Another idea is to play with different definitions of the A-cycles. For example, instead of starting a new A-cycle whenever q^* is empty, start it whenever the number of cells in the buffer crosses β upward, where β is a fixed integer. We tried this but did not obtain much success in terms of efficiency improvement. When increasing β, the no-IS A-cycles tend to become excessively long.

One can also impose a lower bound, say, t_0 on the length of the A-cycles, to avoid lots of extremely short A-cycles, which tends to occur under both the IS and no-IS setup. In our experiments, values of t_0 between 50 and 100 (roughly) gave slight efficiency improvements.

ACKNOWLEDGMENTS

This work has been supported by NSERC-Canada grant # OGP0110050 and FCAR-Québec grant # 93-ER-1654 to the first author, as well as a scholarship

Table 1: CLR Estimation at Level 1 for Different Burst Sizes

	no-IS					IS				
κ_1	$\hat{\mu}$	$\hat{\sigma}^2$	$\hat{\Delta}$	cpu	eff.	$\hat{\mu}$	$\hat{\sigma}^2$	$\hat{\Delta}$	cpu	eff.
10	0			1270		1.10E-22	1.7E-49	0.95%	2492	29
25	0			1302		7.27E-10	1.8E-23	1.5%	1425	21
50	4.42E-6	3.0E-12	101%	1314	2.0E-2	8.91E-6	4.8E-15	2.0%	934	18
100	8.71E-4	2.6E-9	15%	1323	2.4E-1	9.09E-4	1.3E-10	3.2%	661	10
150	4.38E-3	1.5E-8	7.2%	1329	9.3E-1	4.32E-3	3.8E-9	3.6%	583	8.5

Table 2: CLR Estimation at Level 2 for Different Burst Sizes

	no-IS					IS				
κ_1	$\hat{\mu}$	$\hat{\sigma}^2$	$\hat{\Delta}$	cpu	eff.	$\hat{\mu}$	$\hat{\sigma}^2$	$\hat{\Delta}$	cpu	eff.
10	0			2208		2.89E-41	8.2E-83	81%	2613	3.9E-3
25	0			2190		4.62E-21	3.9E-43	35%	2453	2.2E-2
50	0			2972		2.60E-11	4.5E-24	21%	2340	6.4E-2
100	1.54E-6	2.4E-12	256%	3161	3.7E-6	1.66E-7	3.3E-16	28%	2230	3.7E-2
150	7.34E-6	1.3E-11	123%	3269	3.7E-4	3.86E-6	2.2E-13	31%	2288	3.0E-2

Table 3: CLR Estimation at Level 2 for Different Buffer Sizes

	no-IS					IS				
B_2	$\hat{\mu}$	$\hat{\sigma}^2$	$\hat{\Delta}$	cpu	eff.	$\hat{\mu}$	$\hat{\sigma}^2$	$\hat{\Delta}$	cpu	eff.
512	0			2184		1.91E-7	1.3E-16	15%	1273	2.3E-1
764	0			2183		2.36E-9	2.8E-20	18%	1496	1.3E-1
1024	0			2184		2.60E-11	4.5E-24	21%	1719	8.7E-2
1280	0			2186		3.43E-13	1.7E-27	30%	1929	3.7E-2
1536	0			2184		5.35E-15	2.5E-30	76%	2113	5.4E-3

Table 4: CLR Estimation at Level 3 for Different Burst Sizes

	no-IS					IS				
κ_1	$\hat{\mu}$	$\hat{\sigma}^2$	$\hat{\Delta}$	cpu	eff.	$\hat{\mu}$	$\hat{\sigma}^2$	$\hat{\Delta}$	cpu	eff.
10	0			1088		5.44E-20	8.6E-42	14%	1454	2.4E-1
25	0			1050		1.29E-10	2.8E-23	11%	1213	4.9E-1
50	0			1125		6.04E-7	7.2E-16	11%	1042	4.8E-1
100	2.09E-5	2.1E-10	178%	1134	6.9E-3	4.06E-5	6.7E-12	16%	881	2.8E-1
150	1.09E-4	1.6E-9	94%	1136	1.5E-2	1.68E-4	9.1E-11	15%	813	3.8E-1

Table 5: CLR Estimation at Level 3 for Different Buffer Sizes

	no-IS					IS				
B_3	$\hat{\mu}$	$\hat{\sigma}^2$	$\hat{\Delta}$	cpu	eff.	$\hat{\mu}$	$\hat{\sigma}^2$	$\hat{\Delta}$	cpu	eff.
128	2.37E-5	1.1E-10	112%	7036	2.2E-3	4.13E-5	5.3E-12	14%	5779	5.6E-2
256	0			7024		6.04E-7	7.2E-16	11%	7316	6.9E-2
512	0			7059		3.40E-10	2.9E-22	13%	10246	4.0E-2
768	0			7037		2.51E-13	1.7E-28	13%	12930	2.9E-2
1024	0			7027		2.08E-16	1.4E-34	14%	15649	2.0E-2

Table 6: CLR Estimation at Level 3 for Different Numbers of Directed Sources

	no-IS					IS				
sources	$\hat{\mu}$	$\hat{\sigma}^2$	$\hat{\Delta}$	cpu	eff.	$\hat{\mu}$	$\hat{\sigma}^2$	$\hat{\Delta}$	cpu	eff.
4	0			2263		1.93E-8	1.2E-18	15%	2899	1.1E-1
5	0			2357		1.43E-7	4.5E-17	12%	2578	1.8E-1
6	0			2447		6.04E-7	7.2E-16	11%	2413	2.1E-1
7	0			2551		2.09E-6	9.1E-15	12%	2357	2.0E-1
8	1.10E-5	4.9E-11	163%	2651	1.8E-4	4.89E-6	2.2E-14	7.8%	2300	4.7E-1

Table 7: CLR Estimation at Level 4 for Different Burst Sizes

	no-IS					IS				
κ_1	$\hat{\mu}$	$\hat{\sigma}^2$	$\hat{\Delta}$	cpu	eff.	$\hat{\mu}$	$\hat{\sigma}^2$	$\hat{\Delta}$	cpu	eff.
10	0			3562		2.13E-28	5.9E-58	29%	9978	7.7E-3
25	0			3575		1.34E-12	1.5E-25	75%	5347	2.2E-3
50	0			3584		1.37E-7	4.4E-16	39%	3496	1.2E-2
100	2.06E-4	2.1E-8	180%	3599	6.1E-5	6.81E-5	1.1E-10	39%	2505	1.7E-2
150	1.08E-4	8.9E-9	223%	3589	6.7E-3	4.63E-4	7.8E-9	49%	2194	1.3E-2

Table 8: CLR Estimation at Level 4 for Different Buffer Sizes

	no-IS					IS				
B_4	$\hat{\mu}$	$\hat{\sigma}^2$	$\hat{\Delta}$	cpu	eff.	$\hat{\mu}$	$\hat{\sigma}^2$	$\hat{\Delta}$	cpu	eff.
128	1.59E-3	7.5E-8	44%	3580	4.3E-3	1.08E-3	4.0E-9	15%	1881	1.5E-1
256	8.27E-6	6.8E-11	255%	3593	1.3E-2	5.54E-5	1.4E-10	55%	2440	8.8E-3
516	0			3586		1.37E-7	4.4E-16	39%	3580	1.2E-2
768	0			3592		3.63E-10	2.7E-21	36%	4488	1.1E-2
1024	0			3595		1.02E-12	1.5E-26	31%	5550	1.2E-2

from *Newbridge Networks Corporation* to the second author.

REFERENCES

Bratley, P., B. L. Fox., and L. E. Schrage. 1987. *A Guide to Simulation.* Second ed. New York: Springer-Verlag.

Chang, C. S., P. Heidelberger., S. Juneja., and P. Shahabuddin. 1994. Effective bandwidth and fast simulation of ATM intree networks. *Performance Evaluation*, 20:45–65.

Chang, C. S., P. Heidelberger., and P. Shahabuddin. 1995. Fast simulation of packet loss rates in a shared buffer communications switch. *ACM Transactions on Modeling and Computer Simulation*, 5(4):306–325.

Fleming, P. J., D. Schaeffer., and B. Simon. 1995. Efficient Monte Carlo simulation of a product-form model for a cellular system with dynamic resource sharing. *ACM Transactions on Modeling and Computer Simulation*, 5(1):3–21.

Glasserman, P., P. Heidelberger., P. Shahabuddin., and T. Zajic. 1996. Splitting for rare event simulation: Analysis of simple cases. In these proceedings.

Glynn, P. W. 1994. Efficiency improvement techniques. *Annals of Operations Research*, 53:175–197.

Glynn, P. W., and D. L. Iglehart. 1989. Importance sampling for stochastic simulations. *Management Science*, 35:1367–1392.

Heidelberger, P. 1995. Fast simulation of rare events in queueing and reliability models. *ACM Transactions on Modeling and Computer Simulation*, 5(1):43–85.

L'Ecuyer, P. 1994. Efficiency improvement via variance reduction. In *Proceedings of the 1994 Winter Simulation Conference*, 122–132. IEEE Press.

Nicola, V. F., P. Shahabuddin., P. Heidelberger., and P. W. Glynn. 1993. Fast simulation of steady-state availability in non-markovian highly dependable systems. In *Proceedings of the 23rd International Symposium on Fault-Tolerant Computing*, 38–47. IEEE Computer Society Press.

Shahabuddin, P. 1994. Importance sampling for the simulation of highly reliable markovian systems. *Management Science*, 40(3):333–352.

AUTHOR BIOGRAPHIES

PIERRE L'ECUYER is a professor in the "Département d'Informatique et de Recherche Opérationnelle", at the University of Montréal. He received a Ph.D. in operations research in 1983, from the University of Montréal. From 1983 to 1990, he was with the computer science department, at Laval University, Québec. He obtained the *E. W. R. Steacie* grant from NSERC-Canada for the period 1995–97. His main research interests are random number generation, sensitivity analysis and optimization of discrete-event stochastic systems, and discrete-event simulation in general. He is the Departmental Editor for the Simulation Department of *Management Science* and an Area Editor for the *ACM Transactions on Modeling and Computer Simulation*. Check his web page at `http://www.iro.umontreal.ca/~lecuyer`.

YANICK CHAMPOUX is currently M.Sc. student in the "Département d'Informatique et de Recherche Opérationnelle", at the University of Montréal. He works on the application of efficiency improvement techniques for the simulation of large queueing networks. He completed his B.Sc. in mathematics (statistics) at the same university in 1996.

Proceedings of the 1996 Winter Simulation Conference
ed. J. M. Charnes, D. J. Morrice, D. T. Brunner, and J. J. Swain

A FRAMEWORK FOR RARE EVENT SIMULATION OF STOCHASTIC PETRI NETS USING "RESTART"

Christian Kelling

Technische Universität Berlin • Institut für Technische Informatik
Franklinstr. 28/29 • 10587 Berlin • Germany
email: ck@cs.tu-berlin.de

ABSTRACT

For the performability evaluation of complex systems simulation remains often the only feasible method. Unfortunately, simulation experiments tend to be very time consuming if rare events have to be considered. This paper describes an algorithmic approach for fast simulation of rare events applied in a Petri net modeling environment. The technique is based on the RESTART method, which is applicable for rare events in a wide range of simulation models, and has the potential to reduce the simulation overhead extremely. The paper presents selection and refinement techniques for the most important input parameters of RESTART. The techniques allow an efficient execution of RESTART simulations especially in a flexible evaluation tool. The results show run length reductions up to six orders of magnitude.

1 INTRODUCTION

Model-based performance and dependability evaluation of real-world systems usually requires large and complex models to be considered. Most analytical evaluation methods need the generation of the complete state space of the model. Despite several approaches to reduce the complexity of this task are known (Buchholz and Kemper 1995, Ciardo and Trivedi 1993, Ziegler and Szczerbicka 1995), state space explosion remains the most serious limitation of these techniques. In this situation simulation methods offer a valuable instrument.

The work presented here considers evaluation methods for models of so-called soft real-time systems. In those systems, rare events as failures, missed deadlines, or frame losses do not lead to a serious failure, but to degrading performance or limited quality of service, which can be tolerated in most cases. However, designers of applications like telephone switches, video transmission links, or data bases are interested in the probability of those rare events. For these reasons, the availability of simulation methods for rare events is of interest especially if they support well established modeling paradigms as Stochastic Petri Nets. Thus, the analyst is able to consider performability evaluation as a substantial part of system design. The Petri net framework here allows for both, the qualitative as well as the quantitative investigation of the models under study.

Traditional simulation techniques require a huge amount of computing time to achieve reliable estimates of performance measures if the probabilities of the related events are very small, say about 10^{-9} to 10^{-12}. The investigation of even small models by crude simulation under these constraints is simply impossible. To decrease the overhead of stochastic simulation, fast simulation techniques are applied. They offer the potential to reduce the simulation run length by some orders of magnitude.

On the other hand, most fast simulation techniques require individually adapted simulation algorithms representing special knowledge about the simulation model under study. This makes its incorporation into flexible, user-friendly evaluation tools difficult. In addition, for the most prominent technique, the importance sampling method, some serious constraints apply for the considered model. Stationary simulation using importance sampling is only possible if the model has regeneration points. Identification of regeneration points may become very costly in complex models. Moreover, a single model parameter which corresponds to the rare event of interest must be identified for the application of the likelihood ratios (Heidelberger 1995). Some promising results have been published applying importance sampling to the transient simulation of Stochastic Activity Networks in the tool UltraSAN (Obal and Sanders 1994).

A technique with higher flexibility is the RESTART method (REpetitive Simulation Trials After Reaching Thresholds (Villén-Altamirano 1991)). This method does not require the model to have regeneration points, moreover it is not necessary to identify a single model parameter for biasing. The basic idea is to repeat the simulation of those phases of the model evolution with higher probability which lead to the rare event under study. The starting points of the repetitions are called thresholds. At these points, the model state is saved and repetitive trials are simulated, each of them beginning at the threshold. Using this approach, the portion of simulation time spent in model phases which do not contribute to the estimate of the rare event probability is reduced significantly.

The developers of the method demonstrate for optimal parameters a speedup up to 10^8 for probabilities of about 10^{-11} in a queueing example (Villén-Altamirano 1994). However, this run length reduction can be achieved only if (nearly) optimal thresholds can be applied. Moreover, the probabilities related to the thresholds must be known to derive the corresponding number of retrials. The authors provide probabilities for optimal thresholds and the optimal number of retrials, but their determination requires to know the model behavior prior to the simulation study.

Hence, the determination of these input parameters remains the most difficult part for the application of RESTART. Only little research has been done concerning the selection of suitable input parameters and the effectiveness of the practical application of the method in a flexible simulation environment. In this paper we enhance a recently proposed selection procedure for thresholds of the RESTART method using results of a crude pre-simulation study and some information derived automatically from the considered model (Kelling 1995b). Furthermore, a new technique is described, which estimates the threshold probability during the simulation run and performs both, the adaptation of the threshold location according to the known optimality criterions and the computation of the optimal number of retrials based on the estimates.

The presented technique is part of TimeNET (Timed Net Evaluation Tool (German et al. 1995)), a modeling environment for Stochastic Petri Nets. SPNs are the most general class of timed Petri nets and allow for generally distributed firing times without structural restrictions. Despite analytical solution techniques are available for a large subclass of SPNs, discrete event simulation remains the sole method to handle complex models or nets with special properties. However, all Petri net properties based on the net structure can be also derived for simulation models. These properties are usually part of a qualitative analysis and are applied to proof an intended model behavior. The analytical components of TimeNET can handle SPN models with up to 10^6 states. The simulation component of the tool (Kelling 1995a) provides a framework for sequential as well as parallel and distributed execution of simulation experiments. The RESTART module enhances the simulator that includes standard simulation for the transient as well as the stationary case and a variance reduction method using control variates. Simulation studies with SPNs containing several hundred transitions and places and thousands of tokens have been performed with TimeNET.

The paper is organized as follows. In the next section, a short review of the RESTART method is given. Section 3 provides some background information about the considered class of Petri nets, introduces an example model, and derives preconditions for the applicability of RESTART. The fast simulation with RESTART in TimeNET is described in section 4 and results in section 5 demonstrate its performance. Finally we conclude and give directions for future work.

2 THE "RESTART" METHOD

RESTART has been proposed by M. and J. Villén-Altamirano and is based on ideas originally published by Bayes 1972, Hopmans 1979, and Wilson 1984. A complete introduction can be found in (Villén-Altamirano 1991) and (Villén-Altamirano 1994). The short review given here is mainly based on these papers.

The technique is developed from the observation that rare events tend to occur more often in a simulation run if a model state is chosen as the starting point which is significantly different from the average and closer to the rare event of interest. These biased starting points are called thresholds. Using the term event, we refer to an event in the probabilistic sense, i.e. a subset of the sample space or a state of the model.

Figure 1 shows a RESTART simulation. Suppose a simulation where $s(\omega)$ is defined for all $\omega \in \Omega$ and a rare event A given by $A = \{\omega \in \Omega : s(\omega) \geq S_A\}$ whose probability is to be estimated. We assume m events C_i with $C_i = \{\omega \in \Omega : s(\omega) \geq S_i\}$. Thus, an event C_i is related to each of the thresholds S_i and $C_1 \subset C_2 \subset \ldots \subset C_m \subset A$.

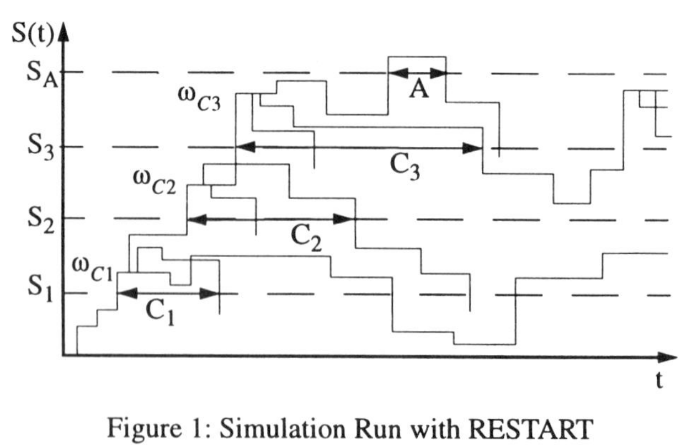

Figure 1: Simulation Run with RESTART

The RESTART algorithm comprises:

- If a transition from $\overline{C_i}$ to C_i occurs, the model state ω_{Ci} is saved.
- If a transition from C_i to $\overline{C_i}$ occurs, i.e. a state $\omega_{\overline{Ci}}$ is observed, the model state ω_{Ci} is restored and the

interval $[\omega_{Ci}, \omega_{\overline{C}i})$ is simulated again using a different stream of random numbers.

- This procedure is repeated R_i times (R_i is the number of retrials according to threshold i). The starting state is always ω_{Ci}, but in general the trials end up with different states $\omega_{\overline{C}i1}, \omega_{\overline{C}i2}, \ldots, \omega_{\overline{C}iR_i}$.
- While a retrial is performed at level i, a transition from C_i to C_{i+1} may occur. In this case, R_{i+1} retrials are performed at level $i+1$ before the run proceeds at level i.
- If the trial R_i is complete, the simulation proceeds without retrials.

Given this scenario, the measure of interest can be computed by

$$\hat{P}(A) = \delta_A \left(t_{\text{first}} \prod_{i=1}^{m} R_i \right)^{-1}$$

where δ_A is the model time during A is observed in all retrials and t_{first} is the model time spent in the first retrials.

As the authors of RESTART show in (Villén-Altamirano 1991), the variance of this estimator can be computed, if the correlation between successive samples is known. The autocorrelation has to be taken into account, since the events A are no longer independent due to the forcing. However, to estimate the covariances requires much more overhead than the evaluation of the rare event itself. Our experiences with RESTART demonstrate that, with well known techniques to reduce the influence of autocorrelation (i.e. independent replications, batching), the method provides a very high accuracy, even if a standard variance estimation is applied. Detailed results are given in section 5 of this paper.

The main input parameter of this simulation technique is the location of the thresholds. Let $P_i = P(C_i | C_{i-1})$ ($1 \le i \le m$) be the conditional probability that a threshold S_i is reached given all states in which threshold S_{i-1} is already reached. To maximize the gain, under some simple assumptions optimal thresholds should be located so that $P_i = e^{-2}$ and the optimal number of retrials after crossing one of the thresholds is $R_i = P_i^{-1}$, which in the case of optimal locations obviously leads to $R_i = e^2$. In discrete models thresholds might be selected only from some discrete values. In this situation, the optimal number of retrials is derived from the thresholds probabilities by $R_i = (P_i \cdot P_{i+1})^{-1/2}$. Thus, the threshold selection criterion does not depend on the model under study. This result makes the incorporation of RESTART into a flexible simulation tool much more easier.

The remaining problem is the determination of suitable thresholds and the computation of the corresponding number of retrials. Before we turn to these new techniques, in the next section the relation between performance measures in Petri nets and the RESTART method is described and some requirements for its applicability are derived.

3 [RELATION TO STOCHASTIC PETRI NETS AND REQUIREMENTS

This part describes the links between measures derived from Petri net models and the RESTART method. A review of the considered class of timed Petri nets we use for designing and evaluating our simulation models is omitted. A detailed definition is given e.g. in (Ciardo et al. 1994).

Performance and reliability measures of timed Petri nets are usually expressed by tokens to be in a particular place. We distinguish between two typical kinds of estimators, the mean number of tokens in a place (E-measure, expectation) and the probability for a logical condition to be true (P-measure), in which this logical condition is also derived from a number of tokens to be in a place. Consider the example model of an ATM switch given in figure 2. A typical application of those ATM systems is the transmission of video data under real-time constraints. The arrival process is described by a Markov-modulated Poisson process. These processes are used for the modeling of superposed data and video traffic consisting of small packets (cells) in ATM systems (Lucantoni 1993). The model can be solved analytically, thus the simulation results can be validated. Simulation becomes necessary, if a transmission link with more than one switch and additional traffic sources is investigated.

The parameters of the SPN model are similar to (Blondia 1992). The transitions T1 and T2 model the modulator which can be in one of two states. A token in place P1 models a phase with a high cell arrival rate. The length of this phase is given by the firing delay of transition T1. The delay of T3 is the interarrival time in this case. When T1 fires, a phase with a lower cell traffic intensity starts. The arrivals are now determined by the delay of transition T4, the length of the period is given by T2. All transition delays in the modulator are exponentially distributed. The service in the queue is modelled by the subnet consisting of place P4 and transition T5. The service time is deterministic. A buffer is provided by place P4. We assume that the buffer capacity is limited to 50 cells. If the buffer is full and a new cell arrives, this cell gets lost (t2 fires, since the inhibitor arc from place P5 is without effect). This event corresponds to a frame loss in the video transmission.

Typical measures are related to the buffer. The mean number of cells in the buffer is expressed by the E-measure E{#P4}. A more important measure is the probability of a frame loss which occurs whenever a new cell arrives and the buffer is full. The loss probability is one of the service qualities an ATM provider has to ensure, thus

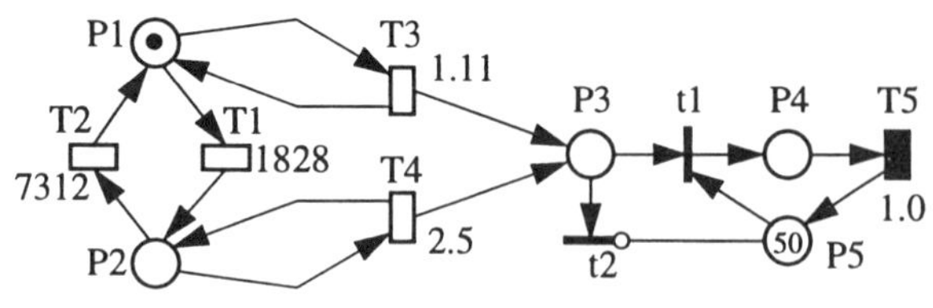

Figure 2: SPN of the MMPP/D/1/50-Queue

video transmission is a typical soft real-time application in this context. The measure is derived from a certain number of cells waiting for service and therefore a typical P-measure. The measure definition P{#P4>=c} describes that the number of waiting cells in the buffer reaches or exceeds the value *c*.

For two special cases of P-measures the RESTART method can be used. Our implementation deals with the probability that the number of tokens in the considered place reaches a certain limit. This type of measure usually corresponds to rare events in the modeled system, e.g. to buffer overflows or to losses as in the example above. From those results, probabilities for deadline violations due to failures can be derived.

The technique presented here determines suitable thresholds for one measure of a Petri net model which needs to be associated to only one place. The measure can either represent an upper limit of a number of tokens in a place (as the loss probability in the example) or a lower limit of that number. Therefore, subject of the evaluation is the rare event that a predefined number of tokens in the place can be observed.

Some requirements for the efficient application of RESTART must be met. First, for the measure monotonous behavior is required, i.e. given the rare event *A*, no events which are a subset of *A* may have a smaller probability than *A*. For measures which represent lower limits, an upper bound of the number of tokens must exist for the place of interest. This is a Petri net property which can be derived from so-called P-invariants. These invariants are one of the structural characteristics that can easily be determined for the underlying Petri net exploiting its structure. In addition, the number of tokens that can be added to or removed from the place of interest at a given time must be known. This can also be determined easily from the Petri net. The last precondition ensures that at a given time only the next threshold can be reached.

4 FAST SIMULATION WITH RESTART IN THE TOOL TimeNET

This section introduces the application of the RESTART technique. Despite fast simulation methods in general offer a huge potential for the acceleration of experiments, their applicability is often limited due to a lack of an user-friendly modeling framework or absent tool support or both. Often for each model an individually adapted simulation software is required.

The approach presented here allows each model to be evaluated automatically in a Petri net modeling tool using the RESTART technique as long as the rare event of interest can be expressed in a certain form (cf. section 3). Therefore the simulation technique is available for the users of the tool without programming overhead or model modification. The modeling framework of SPNs supports an automatic parameter selection. Finally, RESTART is the only fast simulation method, which is applicable to the transient and stationary simulation of non-regenerative models.

As shown above, the main problem with the application of RESTART is the selection of appropriate thresholds as starting points for the retrials and the determination of the number of these retrials. According to (Villén-Altamirano 1994) thresholds should be selected so that $P_i = e^{-2}$. Obviously, the optimal selection of the thresholds is only possible if some results of the model under study are known. Therefore, as a first step a pre-selection of thresholds prior to the simulation run is required. The next section reviews a new technique for the determination of suitable thresholds originally proposed in (Kelling 1995b).

4.1 Determination of Thresholds

The algorithm for the threshold selection is shown in figure 3. Suppose, for a considered place a rare event is defined exceeding a number of tokens *c*. Then, a maximum number of M ($M < c$) P-measures

$$p_i = P\{\#P_x \geq \left\lfloor \frac{ci}{M} \right\rfloor\},\ i \in \{1 \ldots (M-1)\}$$

is defined. A pre-simulation with predefined length is started estimating the probability of the measures p_i. This pilot run yields reliable estimates for the measures of high probability (small *i*). From those measures (nearly) optimal thresholds can be selected according to the optimality criterion (cf. section 2). Assume that $M' < M$ thresholds could be determined, then we have to derive the missing thresholds in the interval $(\lfloor c/M' \rfloor \ldots c)$. This interval is divided in equal sections defining approximate locations for those remaining thresholds.

Note, that this algorithm for upper limits can also be applied to lower limits, but in this case the marking of the considered place needs to be bounded by *c*, what can be derived from the corresponding P-invariant.

Applying this heuristic, at least some optimal thresholds can be found in the pilot run. Depending on the computing time the experimenter is willing to spent with the pre-simulation, the number of optimally selected thresholds can be increased easily. In any case, some approximately chosen thresholds are added to the optimal ones. As stated in

(Villén-Altamirano 1994), RESTART is sufficiently robust to deal with non-optimal thresholds. However, the efficiency can be increased significantly with optimally located thresholds. But, we found in experiments that the gain drops dramatically if a non-optimal number of retrials is used. This leads to a second step, that refines both, the initially selected thresholds and the number of retrials during the simulation run according to the optimality criterion. This algorithm is described in the next section.

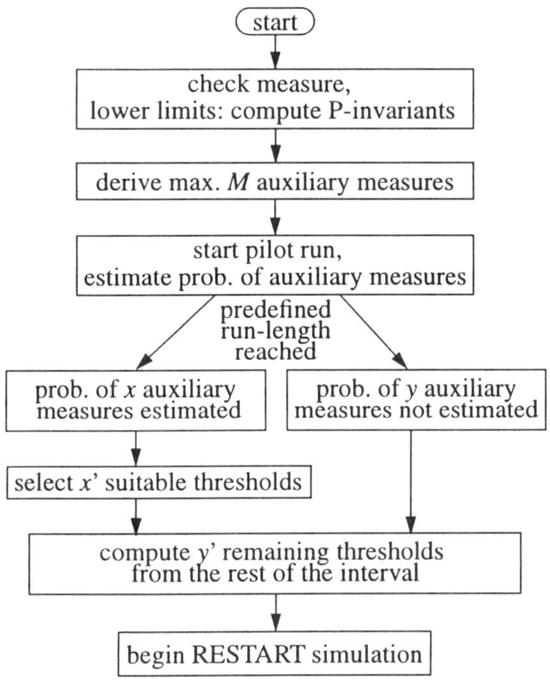

Figure 3: Threshold Determination for RESTART

4.2 Refinement of Threshold Location

As stated in section 2, the simulation proceeds without retrials whenever R_i retrials are complete. In the special case that R_1 retrials are done, the simulation behaves like a standard simulation, since no biased starting point was considered. That enables us to divide the whole RESTART run into K phases, in which the probability of the rare event can be estimated for each of the phases. The refinement method applies two main observations:

First, for each of the phases $1 \le k \le K$ different thresholds $S_i^{(k)}$ and different numbers of retrials $R_i^{(k)}$ can be applied. Each parameter set leads to one sample of the estimated rare event probability, which can be computed by

$$\hat{P}^{(k)}(A) = \delta_A^{(k)} \left(t_{\text{first}}^{(k)} \prod_{i=1}^{m^{(k)}} R_i^{(k)} \right)^{-1}.$$

Note, that the estimated sample of $P(A)$ is always correct, regardless which parameter set is applied (but it might be obtained inefficiently). By incorporating the samples from previous phases, the simulation results are not lost, if a modified parameter set is used.

Second, during each of the phases, not only the probability of the rare event A can be estimated, but additionally an estimate for each of the conditional probabilities P_i can be obtained. These intermediate results can justify the preselection of the threshold locations and the numbers of retrials. This is important for all thresholds for which no estimate could be found in the pilot run, and which therefore were established heuristically. The location will be examined by checking the optimality for the assumed $P_i^{(k)}$. Subsequently, new threshold locations can be used in the phase k+1, if the set of thresholds applied in phase k turns out to be inadequate. Additionally, the number of retrials for each threshold level is optimized using the estimated conditional probabilities P_i. This becomes necessary especially if constraints apply for the threshold values, as it is always the case in discrete models. Finally, the total number of thresholds is adjusted considering the first estimate of $P(A)$.

Since the small probabilities related to heuristically established thresholds describe always a tail of the distribution density function, for the determination of the threshold location linear regression is applied to predict the tail of the function. For this region of the most common pdfs, linearity turns out to be a useful assumption. Each of the decisions derived from the estimates of the various probabilities is only be taken, if the estimates are obtained from a sufficient number of samples. Of course, the accuracy requirements can be assumed to be significantly lower than those for the final estimation of $P(A)$. Additional tests, which compare the probabilities of the first thresholds (obtained in the pre-simulation) with estimates of the same measures obtained in the RESTART simulation ensure further stability.

4.3 The RESTART Component of TimeNET

After preliminary studies, a RESTART algorithm has been implemented in the simulator TimeNET-Sim (Kelling 1995a), which is the simulation component of the tool TimeNET. It allows for complete simulation experiments to estimate rare event probabilities. User have to provide the model with the performance measure specified. As an additional input parameter the number of thresholds might be given. This value can be approximated using the expected order of magnitude of the probability to be estimated. If this parameter is not specified, the adaptive procedures start with default values and determine the optimal number automatically.

The experiment starts with a pre-simulation to obtain the initial location of thresholds as described in section 4.1. With these thresholds the RESTART algorithm will be executed. During the run, the parameters are optimized as described in section 4.2. The simulation may run in parallel, i.e. distributed in a workstation cluster. Each of the distributed replications obtains an estimate for the performance measure and a centralized variance estimation determines a confidence interval. The optimization is done in each of the replications independently in order to omit synchronization. The simulation stops when a predefined accuracy is reached.

5 SIMULATION RESULTS

In this section we demonstrate the performance of the method. The simulation technique is employed to the evaluation of the quality of service in a real-time transmission setting. The example in figure 2 describes an ATM switch, the model together with the measures considered here has been explained in section 3.

The small probabilities for different numbers of ATM cells in the buffer (place P4) are investigated. All measures were estimated with a maximum relative half-width of the confidence interval of 0.1 at 0.95 confidence level. The length of the pre-simulation was set to 10 minutes. Thus, at least three thresholds can be selected optimally. All runs were executed using 10 parallel and independent replications in a workstation cluster. For the first two measures, also a crude simulation was performed to derive the speedup ratios. The overhead of this experiment ranges from 208 to (fictitious) $1.6 \cdot 10^8$ minutes. All other speedups are approximations based on these values.

Table 1 shows the results for a first experiment. The total number of thresholds was initially set to a default value of 10. This is a non-optimal choice for probabilities different from about 10^{-10}, but a realistic situation, since the experimenter does not know the model behavior in advance. Without refinement during the simulation, in some cases the run length was disastrous. It can be shown that the refinement reduces the run length in all cases. Note that the improvement is better for higher run length of the first approach.

A second experiment addresses the improvement caused by the refinement of the threshold location and the number of retrials only. Here the number of thresholds was set to their optimal values. The results in table 2 attest that the heuristic for the selection of the initial thresholds works very well and in some cases no improvement can be achieved by the refinement. On the other hand a further run length reduction by about 60% is possible. A detailed study is needed here to develop conditions for the successful application of the refinement.

Table 1: Speedups for a Default Number of Thresholds

meas.	prob.	no refinement		with refinement	
		sim. time (min.)	speed-up	sim. time (min.)	speed-up
b≥26	$1.27 \cdot 10^{-6}$	> 2700	<<1	140	1.5
b≥31	$1.0 \cdot 10^{-7}$	644	4	156	17
b≥35	$1.34 \cdot 10^{-8}$	187	104	71	275
b≥40	$1.07 \cdot 10^{-9}$	61	4180	24	$1.1 \cdot 10^4$
b≥44	$1.40 \cdot 10^{-10}$	48	$4 \cdot 10^4$	39	$4.8 \cdot 10^4$
b≥48	$1.35 \cdot 10^{-11}$	126	$1.3 \cdot 10^5$	112	$1.5 \cdot 10^5$
b=50	$1.59 \cdot 10^{-12}$	154	$1 \cdot 10^6$	146	$1.1 \cdot 10^6$

Table 2: Effect of the Refinement of Threshold Location (Number Optimal)

meas.	prob.	no refinement		with refinement	
		sim. time (min.)	speed-up	sim. time (min.)	speed-up
b≥26	$1.27 \cdot 10^{-6}$	37	5	30	7
b≥31	$1.0 \cdot 10^{-7}$	40	65	52	50
b≥35	$1.34 \cdot 10^{-8}$	60	325	130	150
b≥40	$1.07 \cdot 10^{-9}$	61	4180	24	$1.1 \cdot 10^4$
b≥44	$1.40 \cdot 10^{-10}$	58	$3.3 \cdot 10^4$	39	$4.8 \cdot 10^4$
b≥48	$1.35 \cdot 10^{-11}$	50	$3.8 \cdot 10^5$	105	$1.6 \cdot 10^5$
b=50	$1.59 \cdot 10^{-12}$	96	$1.7 \cdot 10^6$	206	$7.8 \cdot 10^6$

The last study comprises the sensitivity of the method, if initially a non-optimal number of thresholds is chosen. Table 3 summarizes the values. The RESTART approach without refinement shows drastically decreasing speedup, if the number of thresholds is smaller than the optimal one. Using the refinement, this effect can be compensated in most cases and a further run length reduction of more than one order of magnitude can be achieved.

The results show that the new simulation component offers excellent speedups. With the optimal number of thresholds given, the heuristic establishes thresholds as a set of input parameters for RESTART. Then, the simulation run reaches speedups up to $1.7 \cdot 10^6$ for probabilities of about 10^{-12}. Compared to the accurate numerical results, non of the relative errors was greater than 0.02. The optimization of the number of retrials turned out to have a greater impact than the refinement of the actual threshold location. The improvement by the factor 20 demonstrates that this feature is helpful when the optimal number of

Table 3: Effect of the Refinement if the Optimal Number of Thresholds Is Unknown

initial number of thresholds	run length (min.)	
	no refinement	with refinement
7	6000	305
8	100	177
9	61	69
10 (optimal)	70	24
11	93	73

thresholds is unknown prior to the simulation run. However, in some situation the refinement overhead is remarkable and simulations without this procedure are shorter. Thus, additional studies are required to discover the reasons for the increasing run length due to the refinement and to improve the algorithm.

The execution of the experiments in parallel provides independent samples for the variance estimation and yields an additional gain of about one order of magnitude.

6 SUMMARY

We have introduced an automatic simulation framework for fast simulation of rare events with RESTART. The implemented heuristic to determine the thresholds allows the incorporation of the RESTART acceleration into a Petri net simulation environment, runs transparently to the analyst and does not require detailed knowledge about the applied algorithm. Furthermore, an additional refinement to improve locations of thresholds and the number of retrials with respect to the optimality criterion is presented. The method conserves the simulation overhead done in previous phases by incorporating these samples into the final result estimation. Thus, no simulation time is lost, but the efficiency increases during the simulation run, if non-optimal parameters were selected at the beginning. A remarkable gain can be reached applying the refinement strategies, if the initially chosen parameters turn out to be significantly different from the optimal ones. Examples show that an investigation of probabilities as small as 10^{-12} is possible and that very accurate estimates can be obtained.

The techniques extend the work in (Villén-Altamirano 1994) by providing selection heuristics for thresholds and by proposing an adaptive algorithm for the refinement of the parameter setting during the simulation run. It presents the first integration of the RESTART method into a Petri net modeling tool.

Some additional studies are necessary for the detailed investigation of complex measures. Future work will address the reasons for the increasing run length due to the refinement and transient simulation with RESTART.

ACKNOWLEDGEMENTS

I am thankful to my students Patrick Drescher and Henning Ziegler who implemented the RESTART techniques and did patiently most of the simulation experiments. My thanks go to Manuel and Jose Villén-Altamirano for inviting me to Madrid, for the fruitful discussion, and their interest in my work.

REFERENCES

Bayes, A.J. 1972. A Minimum Variance Sampling Technique for Simulation Models, *Journal of the ACM* 19: 734–741.

Blondia, C. 1992. Performance Evaluation on an M/1-Stage in an ATM Switching Element. *Performance Evaluation* 15(1): 1–20.

Buchholz, P. and P. Kemper 1995. Numerical Analysis of Stochastic Marked Graph Nets. In *Proc of the Sixt Int. Workshop on Petri Nets and Performance Models*, Durham, North Carolina, USA, IEEE Press: 32–41.

Ciardo, G. and K. S. Trivedi 1993. A Decomposition Approach for Stochastic Reward Net Models. *Performance Evaluation* 18(1): 37–59.

Ciardo, G., R. German, and C. Lindemann 1994. A Characterization of the Stochastic Process Underlying a Stochastic Petri Net. *Trans. on Softw. Eng.*, 20(7): 506–515.

German, R., C. Kelling, A. Zimmermann and G. Hommel 1995. TimeNET - A Toolkit for Evaluating Non-Markovian Stochastic Petri Nets. *Performance Evaluation* 24 (1995): 69–87.

Heidelberger, P.1995. Fast Simulation of Rare Events in Queueing and Reliability Models. *ACM Transactions on Modeling and Computer Simulation*, 5(1): 43–85.

Hopmans, A.C.M. and J.P.C. Kleijnen 1979. Importance Sampling in Systems Simulation: A Practical Failure. In *Mathematics and Computing in Simulation XXI*. North-Holland: 209–220.

Kelling, C. 1995(a). TimeNET-Sim - a Parallel Simulator for Stochastic Petri Nets. In *Proc. of the 28th Annual Simulation Symposium*, Phoenix, AZ, USA, IEEE Press: 250–258.

Kelling, C. 1995(b). Rare Event Simulation with RESTART in a Petri Net Modeling Environment. In *Proc of the 1995 European Simulation Symposium*, Erlangen, Germany. SCS: 370–374.

Lucantoni, D. 1993. The BMAP/G/1-Queue: A Tutorial. In *Models and Techniques for the Performance Evaluation of Computer and Communications Systems*. L. Donatiello and R. Nelson. Berlin, Springer Verlag: 330–358.

Obal, W. D. and W. H. Sanders 1994. Importance Sampling Simulation in UltraSAN. *Simulation* 62(2): 98–111.

Villén-Altamirano, M. and J. Villén-Altamirano 1991. RESTART: A Method for Accelerating Rare Event Simulations. In *Queueing Performance and Control in ATM (ITC-13)*. J. W. Cohen and C. D. Pack, Elsevier Science Publishers B. V. (North-Holland): 71–76.

Villén-Altamirano, M. and J. Villén-Altamirano 1994. Enhancement of the Accelerated Simulation Method RESTART by Considering Multiple Thresholds. In *ITC-14*. J. Labetoulle and J. W. Roberts, Elsevier Science Publishers B. V. (North-Holland): 797–810.

Wilson, J.R. 1984. Variance Reduction Techniques for Digital Simulation. *American Journal on Mathematics in Managment Science*, 4(3, 4): 277–312.

Ziegler, P. and H. Szczerbicka 1995. A Structure Based Decomposition Approach for GSPN. In *Proc of the Sixt Int. Workshop on Petri Nets and Performance Models*, Durham, North Carolina, USA, IEEE Press: 261–270.

AUTHOR BIOGRAPHY

CHRISTIAN KELLING is a research assistent in the Computer Science Department of the Berlin University of Techology in Germany. He received his Diploma degree in electrical engineering from the Humboldt-University of Berlin in 1992, and received his Ph. D. degree in 1995 from the Berlin University of Techology. His research interests focus on fast simulation techniques for design and evaluation of highly reliable systems as real-time systems and embedded architectures. He is a member of GI and SCS.

Proceedings of the 1996 Winter Simulation Conference
ed. J. M. Charnes, D. J. Morrice, D. T. Brunner, and J. J. Swain

GRID–BASED SIMULATION AND THE METHOD OF CONDITIONAL LEAST SQUARES

Katherine B. Ensor

Department of Statistics
Rice University
Houston, TX 77251-1892, U.S.A.

Peter W. Glynn

Department of Engineering–Economic
Systems and Operations Research
Stanford University
Stanford, CA 94305-4022, U.S.A.

ABSTRACT

This paper is concerned with the use of simulation to compute the conditional expectations that arise in the method of conditional least squares. Our approach involves performing simulations at each point on a discrete grid imbedded within a statistical parameter space. Our main result concerns the number of grid points and amount of simulation necessary in order to obtain a degree of accuracy comparable to that in the case in the which the conditional expectations are available in closed form.

1 INTRODUCTION

In this paper, we discuss a method known as "conditional least squares" that is widely used for purposes of statistical parameter estimation in the stochastic process setting; see Hall & Heyde (1980) for an introduction to the method. This method requires minimizing a function over the parameter space that involves conditional expectations defined in terms of the stochastic process under consideration. In certain applications, it is natural to compute the conditional expectations via Monte Carlo simulation. In doing so, it is clearly practical only to perform simulations at a finite number of different parameter values. This leads naturally to the concept of "grid–based simulation", in which simulations are performed at various points comprising a grid.

In Section 2, we introduce the method of conditional least squares in the context of parameter estimation for continuous time Markov chains (CTMCs). Section 3 concerns the asymptotic analysis of conditional least squares under the assumption that the relevant conditional expectations can be computed in closed form. Finally, Section 4, we study the use of grid–based simulation in the CTMC context and prove our main result (see Theorems 1 and 2). We show that if the CTMC is observed at n equally spaced time points, then one needs to simulate on the order of n^2 time units of the CTMC so as to ensure that the simulation based minimization gives roughly the same solution quality as that associated with the case in which the conditional expectations are available in closed form. This is accomplished by starting with a coarse grid, and refining it successively as more information on the likely location of the minimizer becomes available.

2 PROBLEM FORMULATION

Suppose that we observe a stationary finite–state continuous–time Markov chain $X = (X(t) : t \geq 0)$, with the intention of using the observed data to make inferences about the generator underlying X. In this paper, we shall adopt a parametric statistical formulation for this inference problem. Specifically, we shall require that the generator underlying X be a member of a parametric family $(A(\theta) : \theta \in \Lambda)$ of generators defined on the state space S associated with X. Our goal, then, is to develop a means of estimating the "true" value of the d-dimensional parameter θ, call it θ^*, underlying X. For example, in the context of the $M/M/1/\infty$ single-server queueing model, this would correspond to attempting to estimate the vector $\theta^* = (\lambda^*, \mu^*)$, where λ^* and μ^* are the arrival and service rates for the queue, respectively.

Without any significant loss of generality, we shall assume that the parameter set Λ is the d-dimensional unit hypercube. We shall further assume that:

A1. i) $|S| < \infty$;

ii) $A(\theta^*)$ is irreducible;

iii) $A(\cdot)$ is three times continuously differentiable on Λ;

iv) $\{ (x, y) : A(\theta, x, y) \neq 0 \}$ is independent of $\theta \in \Lambda$;

v) θ^* lies in the interior of Λ.

If X is observed continuously over some finite interval $[0,t]$, the method of maximum likelihood applies directly here (see, for example, Billingsley 1961). In particular, let $A(\theta,x,y)$ be the (x,y)'th element of the generator $A(\theta)$, let $Y=(Y_n : n\geq 0)$ be the embedded discrete–time Markov chain associated with X, and let $J(t)$ be the number of jumps of X over $[0,t]$. Then, the likelihood function $L_c(\theta,t)$ can be easily written down explicitly:

$$L_c(\theta,t) = \prod_{i=0}^{J(t)-1} A(\theta,Y_i,Y_{i+1}) \qquad (1)$$
$$\cdot \exp(\textstyle\int_0^t A(\theta,X(s),X(s)ds).$$

The maximum likelihood estimator for θ^* is then taken to be the maximizer of $L_c(\cdot,t)$ over Λ.

However, in many applications, X is observed only discretely, see for example Bridges, Ensor and Thompson (1992). In particular, we shall be concerned with the situation in which X is observed only at the integer times $0,1,2,\ldots,n$. Let $P_\theta(\cdot)$ be the probability measure on the path space of X under which X evolves according to a stationary process with generator $A(\theta)$, and set $P(\theta,t,x,y) = P_\theta(X(t)=y \mid X(0)=x)$. If we put $X_i \stackrel{\Delta}{=} X(i)$, then the likelihood function $L_n(\theta)$ associated with the discrete sample $(X_0,\ldots,X_n)$ is given by

$$L_n(\theta) = \prod_{i=0}^{n-1} P(\theta,1,X_i,X_{i+1}).$$

However, in contrast to (1), the likelihood function here is not a simple function of the matrices $(A(\theta) : \theta\in\Lambda)$ that are typically directly specified by the modeler. Rather, in order to compute $L_n(\theta)$, it is necessary to compute the $P(\theta,1,x,y)$'s from $A(\theta)$. Setting $P(\theta,t)=(P(\theta,t,x,y) : x,y\in S)$, this may be accomplished either by taking advantage of the fact that

$$P(\theta,t)=\exp(A(\theta)t)$$

or by noting that the transition semigroup $(P(\theta,t) : t\geq 0)$ is the unique solution of both the backward Kolmogorov differential equations

$$\begin{aligned} P'(\theta,t) &= A(\theta)P(\theta,t) \\ \text{such that } P(\theta,0) &= I \end{aligned}$$

and the forward Kolmogorov differential equations

$$\begin{aligned} P'(\theta,t) &= P(\theta,t)A(\theta) \\ \text{such that } P(\theta,0) &= I. \end{aligned}$$

Clearly, significant numerical effort will generally be required to compute $L_n(\theta)$ for a fixed value of θ. Given that $L_n(\cdot)$ needs to be maximized over Λ in order to compute the maximum likelihood estimator, the numerical challenge is even more daunting.

In an effort to develop a more tractable numerical approach to such inference problems, Klimko and Nelson (1978) proposed the method of conditional least squares. In particular, for a given $f : S\to I\!R$, the idea is to define the estimator θ_n^* as the minimizer of the sum of squares

$$\frac{1}{n}\sum_{i=0}^{n-1}(f(X_{i+1}-E_\theta[f(X_{i+1})|X_i])^2.$$

Setting $g(\theta,x)=E_\theta[f(X_1) \mid X_0=x]$, we note that the method of conditional least squares requires the computation of $g(\theta)=(g(\theta,x) : x\in S)$. This can be accomplished by, for example, solving the linear system of differential equations

$$\begin{aligned} u'(\theta,t) &= A(\theta)u(\theta,t) \qquad (2) \\ \text{subject to } u(\theta,0) &= f, \end{aligned}$$

in which case $g(\theta)=u(\theta,1)$. However, our interest in this paper stems from the fact that $g(\theta)$ can also be computed via Monte Carlo simulation. The simulation alternative is particularly attractive, relative to (2), when $|S|$ is large. The idea that simulation has a useful role to play in the statistical estimation context has received significant attention from the statistics and econometric communities; see, for example, Cook and Stefanski (1994), Diggle & Gratton (1984), Duffie and Singleton (1993), Ensor (1994), Lee (1992), Keane (1994), Maa et al. (1993), McFadden (1989), Pakes and Pollard (1989), Thompson, Brown and Atkinson (1988).

Before concluding this section, it should be noted that conditional least squares typically exacts a cost from a statistical standpoint. While more numerically tractable then maximum likelihood, the asymptotic variance of θ_n^* tends to be larger than that of the maximum likelihood estimator. Thus, θ_n^* does not extract as much of the statistical information present in the sample as does the method of maximum likelihood. This is typical of the trade off between statistical efficiency and computation tractability that is common to the area of statistical inference for stochastic processes.

3 LIMIT THEORY FOR CONDITIONAL LEAST SQUARES IN THE CTMC SETTING

In this section, we fully work out the asymptotic limit theory for conditional least squares in the CTMC setting. (The existing literature tends to focus on discussion of the method in a general framework, under hypotheses that need to be verified on a case–by–case basis).

Our first goal is to verify consistency of θ_n^* as an estimator of θ^*. Our argument requires that we start by establishing smoothness of $g(\cdot)$. Letting e_i be the i'th unit vector in $\mathbb{R}^d$. Assume, without any loss of generality, that e_i is an admissable direction from the point θ_0, in the sense that $\theta_0 + he_i$ belongs to Λ for h sufficiently small. Then we can use A1 iv), (1), and Taylor's theorem to write

$$\begin{aligned} h^{-1}(P(\theta_0 + he_i, t, x, y) - P(\theta_0, t, x, y)) \quad (3) \\ = E_{\theta_0}\left[I(X(t) = y) \frac{\partial_i L_c(\xi, t)}{L_c(\theta_0, t)} \mid X(0) = x \right] \end{aligned}$$

where ξ lies on the line segment connecting θ_0 and $\theta_0 + he_i$ and $\partial_i L_c(\xi, t)$ is the i'th component of the gradient of $L_c(\cdot, t)$ with respect to θ, evaluated at ξ. Now,

$$\begin{aligned} \frac{\partial_i L_c(\theta, t)}{L_c(\theta, t)} \partial_i \log L_c(\theta, t) \\ = \sum_{i=0}^{J(t)-1} \frac{\partial_i A(\theta, Y_i, Y_{i+1})}{A(\theta, Y_i, Y_{i+1})} \\ + \int_0^t \partial_i A(\theta, X(s), X(s) ds \end{aligned}$$

By A1 i) and iii) it follows that $|\partial_i L_c(\xi, t)| \leq (a + bJ(t))L_c(\xi, t)$ for deterministic constants a and b. Furthermore, for c arbitrarily small, positive, and deterministic, we can find h_0 so that for $|h| < h_0$, $L_c(\xi, t)/L_c(\theta_0, t) \leq (1 + c)^{J(t)}d$ with $d > 0$ and deterministic. Since $J(t)$ is stochastically dominated by a Poisson random variable having mean equal to $\sup\{ -A(\theta_0 + he_i, x, x) : |h| < h_0, x \in S \}$ we may conclude that $E_{\theta_0}(a + bJ(t))(1 + c)^{J(t)}d < \infty$ for c sufficiently samll, thereby permitting the application of the dominated convergence theorem in (3). Hence, $P(\cdot, t, x, y)$ is differentiable on Λ. One may easily proceed to show that $P(\cdot, t, x, y)$ is, in fact, three times differentiable under A1 iii).

Let $P(\cdot)$ and $E(\cdot)$ denote the probability and expectation operators on the path–space of X associated with $A(\theta^*)$. Also, let $\pi = (\pi(x) : x \in s)$ be the stationary distribution of X under P, and let $P(x, y) = P(\theta^*, 1, x, y)$. The strong law of large numbers for irreducible CTMC's guarantees that for $x, y \in S$,

$$\begin{aligned} \pi_n(x, y) \triangleq \tfrac{1}{n} \textstyle\sum_{i=0}^{n-1} I(X_i = x, X_{i+1} = y) \quad (4) \\ \to \pi(x)P(x, y) \quad P \text{ a.s.} \end{aligned}$$

as $n \to \infty$. Setting $\alpha(\theta) = E[(f(X_1) - g(\theta, X_0))^2]$, we note that

$$\alpha_n(\theta) = \sum_{x,y} (f(y) - g(\theta, x))^2 \pi_n(x, y)$$

and hence

$$\begin{aligned} \sup_{\theta \in \Lambda} |\alpha_n(\theta) - \alpha(\theta)| \\ \leq \sup\{ (f(y) - g(\theta, x))^2 : x, y \in S, \theta \in \Lambda \} \\ \cdot \max_{x,y} |\pi_n(x, y) - \pi(x)P(x, y)|. \end{aligned}$$

The supremum of $|g(\theta, x)|$ over $\theta \in \Lambda$ and $x \in S$ is finite because of the continuity of $P(\cdot, 1, x, y)$ over Λ (a compact set), and the finiteness of S. In view of (4), we have therefore proved the following result.

Proposition 1 *Under A1, $\alpha_n(\cdot)$ converges uniformly P a.s. to $\alpha(\cdot)$ over Λ.*

Let θ_n^* be any (measurable) selection from the set of global minimizers of $\alpha_n(\cdot)$. In view of Proposition 1, strong consistency of θ_n^* to θ^* follows if we show that θ^* is the unique global minimizer of $\alpha(\cdot)$. For this, we need an identifiability assumption.

A2. If $\theta_1, \theta_2 \in \Lambda$ and $g(\theta_1) = g(\theta_2)$, then $\theta_1 = \theta_2$.

Observe that

$$\begin{aligned} \alpha(\theta) = \alpha(\theta^*) + 2E\Big[(f(X_1) - g(\theta^*, X_0)) \\ \cdot (g(\theta^*, X_0) - g(\theta, X_0))\Big] \\ + E\Big[(g(\theta^*, X_0) - g(\theta, X_0))^2\Big]. \end{aligned}$$

Since $f(X_1) - g(\theta^*, X_0)$ is a martingale difference under P, the second term on the right-hand side vanishes. So, under A2, θ^* is indeed the unique global minimizer of $\alpha(\cdot)$, proving our next result.

Proposition 2 *Under A1-A2, $\theta_n^* \to \theta^*$ P a.s. as $n \to \infty$.*

To deal with the central limit theory for the estimator θ_n^*, note that since both $\alpha_n(\cdot)$ and $\alpha(\cdot)$ are smooth, it is evident that $\nabla\alpha_n(\theta_n^*) = \nabla\alpha(\theta^*) = 0$ (note also conditon A1 v)), so that

$$\nabla\alpha_n(\theta_n^*) - \nabla\alpha_n(\theta^*) = \nabla\alpha(\theta^*) - \nabla\alpha_n(\theta^*).$$

Let $\partial_{ij}^2 \alpha_n(\xi)$ be the second partial derivative of α_n with respect to θ_i and θ_j, evaluated at ξ. Then, by

Taylor's theorem, there exists ξ_{ni} lying on the line segment connecting θ_n^* and θ^* such that

$$(\partial_{ij}^2 \alpha_n(\xi_{ni}) : 1 \le j \le d)(\theta_n^* - \theta^*) = \partial_i \alpha(\theta^*) - \partial_i \alpha_n(\theta^*)$$

holds for $1 \le i \le d$. Set $H_n = (\partial_{ij}^2 \alpha_n(\xi_{ni}) : 1 \le i,\ j \le d)$. Then,

$$H_n(\theta_n^* - \theta^*) = \nabla\alpha(\theta^*) - \nabla\alpha_n(\theta^*).$$

The proof of Proposition 1 also carries over to showing that $\partial_{ij}^2 \alpha_n(\cdot)$ converges uniformly P a.s. to $\partial_{ij}^2 \alpha(\cdot)$ on Λ. Since $\theta_n^* \to \theta^*$ P a.s. as $n \to \infty$, it follows that $H_n \to H$ P a.s. as $n \to \infty$, where $H = (\partial_{ij}^2 \alpha(\theta^*) : 1 \le i,j \le d)$. Because θ^* is the unique global minimum of $\alpha(\cdot)$, H is positive definite, and consequently H_n^{-1} exists for n sufficiently large, and $H_n^{-1} \to H^{-1}$ P a.s. as $n \to \infty$. Hence,

$$\theta_n^* - \theta^* = H_n^{-1}(\nabla\alpha(\theta^*) - \nabla\alpha_n(\theta^*)).$$

But

$$\nabla\alpha(\theta^*) - \nabla\alpha_n(\theta^*) = -\frac{1}{n}\sum_{i=1}^{n} D_i,$$

where $D_i = 2(f(X_i) - g(\theta^*, X_{i-1}))\nabla g(\theta^*, X_{i-1})$. Now, $(D_i : i \ge 1)$ is a stationary sequence of square-integrable martingale differences, and consequently the martingale central limit theorem (CLT) (see Ethier and Kurtz (1986)) yields

$$\sqrt{n}(\nabla\alpha(\theta^*) - \nabla\alpha_n(\theta^*)) \xrightarrow{d} N(0, C)$$

as $n \to \infty$ where $N(0,C)$ is a d-dimensional multivariate normal random variable having covariance matrix $C = ED_1D_1^T$. We have therefore established the following CLT for θ_n^*.

Theorem 1 *Under A1-A2,*

$$\sqrt{n}(\theta_n^* - \theta^*) \xrightarrow{d} H^{-1}N(0, C)$$

as $n \to \infty$.

The above analysis presupposes that $g(\cdot)$ can be easily evaluated, so that θ_n^* can be computed without difficulty. As indicated earlier, we are especially concerned with problems in which $g(\cdot)$ is computed via simulation, thereby introducing additional error into our estimator of θ^*; this is the subject of Section 4.

4 GRID-BASED SIMULATION

Clearly, the conditional expectations associated with $g(\theta)$ can easily be computed via simulation of X. Specifically, suppose that $(W_j(i,\theta,x) : i,j \ge 1,\ \theta \in \Lambda,\ x \in S)$ is a collection of independent random variables in which $(W_j(i,\theta,x) : i,j \ge 1)$ is identically distributed with common distribution $P_\theta(f(X_1) \in \cdot \mid X_0 = x)$. We let $\tilde{P}(\cdot)$ and $\tilde{E}(\cdot)$ denote the probability and expectation operator associated with the probability space that supports the $W_j(i,\theta,x)$'s and the process X. Then,

$$g(\theta, x, i, m) = \frac{1}{m}\sum_{j=1}^{m} W_j(i, \theta, x)$$

is an estimator of $g(\theta, x)$. Furthermore, $\alpha_n(\theta)$ can be calculated numerically via the Monte Carlo estimator

$$\alpha_n(\theta, i, m) = \sum_{x,y}(f(y) - g(\theta, x, i, m))^2 \pi_n(x, y). \quad (5)$$

We note that, computationally speaking, $\alpha_n(\theta, i, m)$ requires only that $g(\theta, x, i, m)$ be calculated for states $x \in \{ X_0, \ldots, X_{n-1} \}$; this observation can result in significant computational savings when $|S|$ is (very) large.

However, it is clearly impossible to compute the function $\alpha_n(\cdot, i, m)$ over the entire parameter space Λ. Instead, one needs to restrict attention to a finite subset of Λ. Our approach will be to generate $\alpha_n(\cdot, i, n)$ on a uniform grid (hence, the term "grid-based simulation"). The grid will then be successively refined as more information becomes available on the likely location of the minimizer of $\alpha_n(\cdot)$.

More specifically, the iteration proceeds as follows. Suppose that at iteration i, we have a "guess" $\theta_n^*(i-1)$ available as to the likely location of some point in the set $\operatorname{argmin}\{ \alpha_n(\theta) : \theta \in \Lambda \}$. For $z > 0$, let

$$I(z) = \{ (i_1, \ldots, i_d) : i_j \in \mathcal{Z},\ |i_j| \le z,\ 1 \le j \le d \}.$$

For δ positive, we then proceed to generate $\alpha_n(\theta, i, m_n(i))$ over the grid points $\theta \in \Lambda_n(i)$, where

$$\Lambda_n(i) = (\theta_n^*(i-1) + n^{-(i+1)\delta} I(n^{2\delta})) \cap \Lambda.$$

We next select $\theta_n^*(i)$ to be any point in the set of global minimizers of $\{ \alpha_n(\theta, i, m_n(i)) : \theta \in \Lambda_n(i) \}$, and move on to the next iteration. The algorithm is initiated by setting $\theta_n^*(0) = 0$, and is terminated at iteration k with the final computed parameter estimator $\widehat{\theta}_n = \theta_n^*(k)$. Our choice for the sequence $\{ m_n(i) : i \ge 1 \}$ is $m_n(i) = [n^{4(i+2)\delta}]$ (where $[\cdot]$ denotes the greatest integer).

Let $\|\cdot\|$ be the norm on $\mathbb{R}^d$ defined by $\|x\| = \max_{1 \le i \le d}|x_i|$. Our main mathematical result of this section is the following.

Proposition 3 *Assume A1-A2. For each fixed* $k \ge 1$ *and* $\delta > 0$,

$$\tilde{P}(\|\theta_n^* - \theta_n^*(i)\| \le n^{-i\delta},\ 1 \le i \le k) \to 1$$

as $n \to \infty$.

Proof. Let $\mathcal{G}_{ni} = \sigma(X_0, \ldots, X_1,\ W_j(l,\theta,x) : \theta \in \Lambda,\ x \in S, j \geq 1,\ l \leq i)$. Then,

$$\tilde{P}\left(\| \theta_n^* - \theta_n^*(i) \| \leq n^{-i\delta},\ 1 \leq i \leq k\right) \tag{6}$$
$$= \tilde{E}\Bigg[\prod_{i=1}^{k-1} I(\| \theta_n^* - \theta_n^*(i) \| \leq n^{-i\delta})$$
$$\cdot \tilde{P}(\| \theta_n^* - \theta_n^*(k) \| \leq n^{k\delta} \mid \mathcal{G}_{n,k-1})\Bigg].$$

On the event $\{ \| \theta_n^* - \theta^*(k-1) \| \leq n^{(k-1)\delta} \}$, the convex hull of $\Lambda_n(k)$ contains θ_n^*. Hence, there exists a point $\theta_{nc}^*(k) \in \Lambda_n(k)$ for which $\| \theta_n^* - \theta_{nc}^*(k) \| \leq n^{-\delta(k+1)}$. We next observe that a sufficient condition for $\theta_n^*(k)$ to be within $n^{-\delta k}$ of θ_n^* (in the norm $\| \cdot \|$) is that $\theta_{nc}^*(k)$ have a strictly smaller objective value (with regard to the objective function $\alpha_n(\cdot, k, m_n(k))$ than all those points $\theta \in \Lambda_n(k)$ such that $\| \theta_n^* - \theta \| > n^{-k\delta}$. We now proceed to establish that this event occurs with high probability under our hypotheses.

Observe that for any positive deterministic $\tilde{a}$, there exists deterministic n_0 such that for $n > n_0$,

$$\{ \alpha_n(\theta, k, m_n(k)) > \alpha_n(\theta_{nc}^*(k), k, m_n(k)),$$
$$\theta \in \Lambda_n(k),\ \| \theta - \theta_n^* \| > n^{-k\delta} \}$$
$$\supseteq \{ \alpha_n(\theta) > \alpha_n(\theta_{nc}^*(k)) + \tilde{a}n^{-2k\delta},\ \theta \in \Lambda_n(k),$$
$$\| \theta - \theta_n^* \| > n^{-k\delta} \}$$
$$\cap \{ |\alpha_n(\theta) - \alpha_n(\theta, k, m_n(k))| \leq n^{-2(k+1)\delta},$$
$$\theta \in \Lambda_n(k) \}.$$

Hence, for n sufficiently large,

$$\tilde{P}\Bigg(\alpha_n(\theta, k, m_n(k)) > \alpha_n(\theta_{nc}^*(k), k, m_n(k)), \tag{7}$$
$$\theta \in \Lambda_n(k), \| \theta - \theta_n^* \| > n^{-k\delta} \mid \mathcal{G}_{n,k-1}\Bigg)$$
$$\geq I\Bigg(\alpha_n(\theta) > \alpha_n(\theta_{nc}^*(k)) + \tilde{a}n^{-2k\delta}, \theta \in \Lambda_n(k),$$
$$\| \theta - \theta_n^* \| > n^{-k\delta}\Bigg)$$
$$\cdot \prod_{\theta \in \Lambda_n(k)} \tilde{P}\Bigg(| \alpha_n(\theta) - \alpha_n(\theta, k, m_n(k)) |$$
$$\leq n^{-2(k+1)\delta} \mid \mathcal{G}_{n,k-1}\Bigg).$$

(We used above the fact that $\theta_{nc}^*(k)$ is $\mathcal{G}_{n,k-1}$ measurable and the independence of the $W_j(i,\theta,x)$'s.) Since θ_n^* is a minimizer of $\alpha_n(\cdot)$, $\nabla\alpha_n(\theta_n^*) = 0$ so Taylor's theorem (see, for example Sen and Singer (1993)) yields

$$\alpha_n(\theta) = \alpha_n(\theta_n^*) + \frac{1}{2}(\theta - \theta_n^*)^T H_n(\xi_n(\theta))(\theta - \theta_n^*),$$

where $\xi_n(\theta)$ lies on the line segment connecting θ and θ_n^*, and $H_n(x)$ is the Hessian of $\alpha_n(\cdot)$ evaluated at x. Now, as asserted earlier, $H_n(\cdot)$ is a continuous matrix function which converges uniformly on Λ, as $n \to \infty$, to the limiting continuous matrix function $H(\cdot)$, where $H(\cdot)$ is the Hessian of $\alpha(\cdot)$. Furthermore, θ^* is the unique global minimizer of $\alpha(\cdot)$, so $H(\cdot)$ is positive definite in a neighborhood of θ^*. Now, the minimal and maximal eigenvalues of a positive definite matrix are continuous functions of their matrix argument. As a consequence, the minimal and maximal eigenvalues of $H_n(\cdot)$ converge uniformly to the corresponding eigenvalues of $H(\cdot)$. Thus, it is evident that there exists positive constants ϵ, $\tilde{c}$, $\tilde{d}$ such that for any $x \in \mathbb{R}^d$, $\| \theta - \theta^* \| < \epsilon$,

$$\tilde{c}\| x \|^2 \leq x^T H_n(\theta)x \leq \tilde{d}\| x \|^2$$

for n large enough $\tilde{P}$ a.s. (We also use here the fact that our norm can be bounded above and below by constant multiples of the Euclidian norm.) Now, for $k \geq 2$, the diameter of $\Lambda_n(k)$ shrinks as $n \to \infty$. Consequently, on $\{ \| \theta_n^* - \theta_n^*(k-1) \| \leq n^{-\delta(k-1)} \}$, and for n large enough,

$$\tilde{c}\| \theta - \theta_n^* \|^2 \leq \alpha_n(\theta) - \alpha_n(\theta_n^*) \leq \tilde{d}\| \theta - \theta_n^* \|^2 \tag{8}$$

for $\theta \in \Lambda_n(k)$, $k \geq 2$. In view of the fact that $\| \theta_{nc}^*(k) - \theta_n^* \| \leq n^{-\delta(k+1)}$ it follows from (8) that for $k \geq 2$, there exists a $\mathcal{G}_{n,1}$ measurable random variable N_k that is finite $\tilde{P}$ a.s. such that

$$I\Bigg(\alpha_n(\theta) > \alpha_n(\theta_{nc}^*(k)) + \tilde{a}n^{-2k\delta}, \tag{9}$$
$$\theta \in \Lambda_n(k),\ \| \theta - \theta_n^* \| > n^{-k\delta}\Bigg)$$
$$= I(n \geq N_k)$$

on $\{ \| \theta_n^* - \theta_n^*(k-1) \| \leq n^{-\delta(k-1)} \}$ (if we choose a small enough δ. For $k = 1$, we use the fact that $\alpha_n(\cdot)$ converges uniformly to $\alpha(\cdot)$ outside any ϵ–neighborhood of θ_n^* and use the estimates (8) inside the ϵ–neighborhood, to arrive at (9).

Turning now to the second factor on the right–hand side of (7), observe that the $W_j(i,\theta,x)$'s are a family of uniformly bounded random variables (bounded by $\max(| f(x) |) : x \in S)$. It is easy to see that

$$\tilde{P}\Big(\mid \alpha_n(\theta) - \alpha_n(\theta,k,m_n(k)) \mid \le n^{-2(k+1)\delta} \mid \mathcal{G}_{n,k-1}\Big) \tag{10}$$
$$\ge \prod_{x\in S} \tilde{P}\Big(\mid g(\theta,x) - g(\theta,x,k,m_n(k)) \mid \le \tilde{b}n^{-2(k+1)\delta}\Big)$$

for $\tilde{b}$ chosen small enough and deterministic (see (5)). But for $p > 0$,

$$\tilde{P}\Big(\mid g(\theta,x) - g(\theta,x,k,m_n(k)) \mid \le \tilde{b}n^{-2(k+1)\delta}\Big) \tag{11}$$
$$= 1 - \tilde{P}\Big(\mid g(\theta,x) - g(\theta,x,k,m_n(k)) \mid > \tilde{b}n^{-2(k+1)\delta}\Big)$$
$$\ge 1 - \tilde{b}^{-p}n^{2(k+1)\delta p} \cdot \tilde{E} \mid g(\theta,x) - g(\theta,x,k,m_n(k)) \mid^p$$
$$\ge 1 - \tilde{b}^{-p}n^{2(k+1)\delta p}c(p) \cdot \max\big(\mid f(x) \mid^p : x \in S\big) m_n(k)^{-p/2}$$
$$= 1 - \tilde{b}^{-p}c(p)n^{-2\delta p} \max\big(\mid f(x) \mid^p : x \in S\big)$$
$$\triangleq 1 - rn^{-2\delta p};$$

the Burkholder inequality was applied in the second inequality; see Hall & Heyde (1980), p. 23. Application of (7), (9), (10), and (11), together with repeated conditioning in (6), yields the inequality

$$\tilde{P}\left(\| \theta_n^* - \theta_n^*(i) \| \le n^{-i\delta},\ 1 \le i \le k\right)$$
$$\ge \tilde{P}(N_1 \le n, \ldots, N_k \le n) \cdot (1 - rn^{-2\delta p})^{|S|\cdot|\Lambda_n(k)|\cdot k}$$
$$\ge \tilde{P}(N_1 \le n, \ldots, N_k \le n) \cdot (1 - rn^{-2\delta p})^{|S|\cdot(2n^{2\delta}+1)^d k}.$$

By choosing p sufficiently large and letting $n \to \infty$, we obtain the desired result. □

We note that by choosing $k\delta > \frac{1}{2}$, Proposition 3 and Theorem 1 combine to yield a CLT for $\widehat{\theta}_n$.

Theorem 2 *Assume A1-A2 and suppose* $k\delta > \frac{1}{2}$. *Then*

$$\sqrt{n}(\widehat{\theta}_n - \theta^*) \xrightarrow{d} H^{-1}N(0,C)$$

as $n \to \infty$, *where* H *and* C *are as in Theorem 1.*

In terms of the computational effort required to calculate $\widehat{\theta}_n$, note that the i'th iteration requires simulation at $(2n^{2\delta} + 1)^d$ points. Each simulation at the i'th iteration, for a given point, requires $n^{4(i+2)\delta}$ replications. Thus, the total work required at iteration i is of the order of $n^{2\delta d+4(i+2)\delta}$. Summing over the k iterations, we conclude that the total work is of order $n^{2\delta d+4(k+2)\delta}$. But k and δ can be chosen arbitrarily, subject to the constraint $k\delta > \frac{1}{2}$. Hence, by (for example), choosing the number of iterations k large, and $\delta = (\frac{1}{2} + \eta)/k$ for η positive, we note that we can make the exponent $2\delta d + 4(k+2)\delta$ as close as we wish to 2. Thus, roughly speaking, the computational effort required to compute $\widehat{\theta}_n$ is of order n^2.

This should come as no surprise. In the limit, an accuracy of order $n^{-1/2}$ in the location of the minimizer requires that we perform "function evaluations" that have accuracy n^{-1} (because of the locally quadratic structure of the objective function). To obtain simulations of accuracy n^{-1} requires a run-length of order n^2.

While the analysis of this paper is asymptotic, it does suggest that in implementing grid-based simulation, it is important to slowly refine the grid (i.e. k large), and that using a course grid (δ small) reduces the impact of dimensionality considerations (i.e. the impact of d being large).

ACKNOWLEDGMENTS

This collaborative work began during the first author's National Science Foundation sponsored visits to Stanford University. The work of the second author was supported by the Army Research Office under Contract No. DAAL03-91-G-0319.

REFERENCES

Billingsley, P. 1961. *Statistical Inference for Markov Processes.* Chicago: University of Chicago Press.

Bridges, E., K. B. Ensor and J. R. Thompson. 1992. Marketplace competition in the personal computer industry. *Decision Sciences* 23:467-477.

Cooke, J. R. and L. A. Stefanski. 1994. Simulation-extrapolation estimation in parametric measurement error models. *Journal of the American Statistical Association* 89:1314–1328.

Diggle, P. J. and R. J. Gratton. 1984. Monte Carlo methods of inference for implicit statistical models. *Journal of the Royal Statistical Society Series B* 46:193–227.

Duffie, D. and K. J. Singleton. 1993. Simulated moments estimation of Markov models of asset prices.

Econometrica 61:929–952.

Ensor, K. B.. 1994. Properties of simulation based estimators of stochastic processes. *Proceedings of the Thirty-Ninth Conference on the Design of Experiments. ARO Report No. 94-2*, 15-22.

Ethier, S. N. and T. G. Kurtz. 1986. *Markov Processes: Characterization and Convergence.* New York: John Wiley & Sons.

Hall, P. and C. C. Heyde. 1980. *Martingale Limit Theory and its Application.* New York: Academic Press.

Keane, M. P. 1994. A computationally practical simulation estimator for panel data. *Econometrica* 62:95–116.

Klimko, L. A. and P. I. Nelson. 1978. On conditional least squares estimation for stochastic processes. *The Annals of Statistics* 6:629–642.

Maa, J.-F., Pearl, D. K., Batoszyński, R., and Horn, D. 1993. Simulation–based optimization and estimation. *ASA Proceedings of the Statistical Computing Section*, 102–106.

Lee, L.-F. 1992. On efficiency of methods of simulated moments and maximum simulated likelihood estimation of discrete response models. *Econometric Theory* 8:518–552.

McFadden, D. 1989. A method of simulated moments for estimation of discrete response models without numerical integration. *Econometrica* 57:995–1026.

Pakes, A. and D. Pollard. 1989. Simulation and the asymptotics of optimization estimators. *Econometrica* 57:1027–1057.

Sen, P. K. and J. M. Singer. 1993. *Large Sample Methods in Statistics. An Introduction with Applications.* Chapman & Hall.

Thompson, J. R., B.W. Brown, and E. N. Atkinson. 1987. SIMEST An algorithm for simulation–based estimation of parameters characterizing a stochastic process. In *Cancer Modeling*, eds. J. R. Thompson and B. W. Brown,387–415. Marcel Dekker.

AUTHOR BIOGRAPHIES

KATHERINE B. ENSOR is Associate Professor in the Department of Statistics at Rice University. She joined the faculty at Rice University in 1987 after receiving her Ph.D. in Statistics from Texas A&M University. Her research interests include time series analysis, spatial processes, environmental statistics, and estimation for stochastic processes based on simulation methods. She serves on the editorial board of *The Journal of Statistical Computation and Simulation* and *Communications in Statistics.*

PETER W. GLYNN received his Ph.D. from Stanford. University, after which he joined the faculty of the Department of Industrial Engineering at the University of Wisconsin-Madison. In 1987, he returned to Stanford, where he currently holds the Thomas Ford Faculty Scholar Chair in the EES/OR Department. He was a co–winner of the 1993 Outstanding Simulation Publication Award sponsored by the TIMS College on Simulation. His research interests include discrete-event simulation, computational probability, queueing, and general theory for stochastic systems.

Proceedings of the 1996 Winter Simulation Conference
ed. J. M. Charnes, D. J. Morrice, D. T. Brunner, and J. J. Swain

SIMULATION BASED OPTIMIZATION

Alexander Shapiro

School of Industrial & Systems Engineering
Georgia Institute of Technology
Atlanta, GA 30332, U.S.A.

ABSTRACT

In this talk we consider a problem of optimizing an expected value function by Monte Carlo simulation methods. We discuss, somewhat in details, the stochastic counterpart (sample path) method where a relatively large sample is generated and the expected value function is approximated by the corresponding average function. Consequently the obtained approximation problem is solved by deterministic methods of nonlinear programming. One of advantages of this approach, compared with the classical stochastic approximation method, is that a statistical inference can be incorporated into optimization algorithms. This allows to develop a validation analysis, stopping rules and variance reduction techniques which in some cases considerably enhance numerical performance of the stochastic counterpart method.

1 INTRODUCTION

Consider the optimization problem

$$\min_{x \in S} f(x) \tag{1}$$

of minimization of a function $f(x)$ over a set $S \subset \mathbb{R}^m$. Suppose that the objective function $f(x)$ is given as an expected value $f(x) := \mathbb{E}_P\{g(x,\omega)\}$ with respect to a probability measure (distribution) P defined on a sample space $(\Omega, \mathcal{F})$. One can try then to solve the optimization problem (1) by a Monte Carlo simulation. A classical approach to such an optimization is the stochastic approximation (SA) method (e.g., Kushner and Clark, 1978). In case for every $\omega \in \Omega$ the function $g(\cdot,\omega)$ is convex, not necessarily differentiable, one can use a closely related stochastic subgradient (stochastic quasigradient) method (Ermoliev, 1988). In order to apply these methods only calculation of the gradient (in the convex case calculation of a subgradient) of the functions $g(\cdot,\hat{\omega})$ at iteration points for generated realizations (sample paths) $\hat{\omega}$ is required.

Drawbacks of the SA method are well known - slow convergence, absence of a good stopping rule, difficulty in handling constraints. In this talk we discuss an alternative approach to the above optimization problem which became known as the stochastic counterpart (or sample path) method. In the stochastic counterpart method a (large) sample $\omega_1, ..., \omega_n$ is generated and the function $f(\cdot)$ is approximated by the corresponding average function

$$\hat{f}_n(\cdot) = n^{-1} \sum_{i=1}^{n} g(\cdot, \omega_i).$$

This leads to the (stochastic counterpart) approximating problem

$$\min_{x \in S} \hat{f}_n(x), \tag{2}$$

which is solved by (deterministic) techniques of nonlinear programming. For such an approach to be applicable one needs a mechanism to evaluate the approximating functions $\hat{f}_n(x)$, and their derivatives, for all x in a given region in $\mathbb{R}^m$. There are basically two types of techniques which are available for that purpose, namely the likelihood ratios (LR) (also called the score function) method and the sample path generation.

Let us observe at this point that the SA and stochastic counterpart (SC) methods are asymptotically equivalent provided the expected value function $f(x)$ is smooth (differentiable), the "true" optimal solution x_0 is an interior point of the feasible region S and the SA method is implemented with the *optimal* choice of the corresponding step sizes (Shapiro, 1996). This should be not surprising in view of asymptotic optimality (efficiency) properties of SA estimators (Nevl'son and Has'minskii, 1978). It was discovered recently that by choosing relatively large step sizes and by averaging the obtained SA iterates it is possible (in the smooth case) to achieve asymptotic efficiency in an automatic way (Polyak, 1990). Therefore, for smooth problems, from the asymptotic point of view both methods are comparable.

There is, however, a number of advantages in the SC method which (when applicable) make this method computationally attractive. The involved constraints can be treated quite efficiently by (deterministic) optimization algorithms. A statistical inference for the SC method is available (e.g., Shapiro, 1996). This statistical inference allows to develop (statistically based) stopping rules, validation techniques and to construct confidence intervals and, in some cases, confidence regions for the optimal value and optimal solutions, respectively (cf. Shapiro and Homem-de-Mello, 1996). Furthermore, the SC method can be used together with variance reduction techniques which in some cases lead to significant improvements in the computational efficiency.

2 AN EXAMPLE OF $GI/G/1$ QUEUE

In this section we discuss an example of $GI/G/1$ queue whose service times depend on a parameter vector x. Although relatively simple this example demonstrates various aspects of the method. Let Y_i be the time between arrivals of the $(i-1)^{\text{th}}$ and i^{th} customers, and for a fixed value of x, let $Z_i(x)$ be the service time of the ith customer, $i = 1, 2, \ldots$, and let $G_i(x)$ denote the ith sojourn time (i.e. the total time spent by the ith customer in the queue). We assume that the first customer arrives at an empty queue and that for every $x \in S$ the queue is regenerative with the expected number of customers served in one busy period (regenerative cycle) being finite. Note that a recursion relation between the sojourn times is given by the Lindley's equation

$$G_i(x) = Z_i(x) + [G_{i-1}(x) - Y_i]_+. \tag{3}$$

Under standard regularity conditions, the long-run average functions

$$\hat{f}_n(x) = n^{-1} \sum_{i=1}^{n} G_i(x)$$

converge (pointwise) w.p.1, as $n \to \infty$, to the expected value (mean) steady state sojourn time $f(x)$. Consider the optimization problem

$$\min_{x \in S} f(x) + \psi(x), \tag{4}$$

where $\psi(x)$ is a (deterministic) cost function. By using Lindley's equation (3) a recursive relation between the gradients of the sojourn time functions can be derived and consequently the optimization problem (4) can be numerically solved by SA type algorithms (see, e.g., Chong and Ramadge (1993), L'Ecuyer and Glynn (1994), and references therein).

An alternative approach is to generate a large sample of interarrival and service times and then to solve a constructed approximation of (4) by deterministic methods of nonlinear programming. Suppose that the service times can be generated in the form $Z_i = \eta(U_i, x)$, where $U_1, \ldots$, is a sequence of iid random numbers and $\eta(u, x)$ is a known function. For example, if the service times have an exponential distribution with mean x, then one can take $\eta(u, x) = -x \ln u$ with U_i having the uniform (0,1) distribution. After a sample of $Y_1, \ldots, Y_n$, and $U_1, \ldots, U_n$, is generated the long-run average functions $\hat{f}_n(x)$ and their gradients can be calculated, at any given point x, by using Lindley's equation (3), separately for every busy period, and consequently the corresponding sample path approximation of the program (4) can be constructed. It is worth noting that constructed in that way, the approximating functions $\hat{f}_n(x)$ are typically *nondifferentiable*, piecewise smooth functions. Moreover, in some cases when the interarrival and service times distributions contain atoms, these nondifferentiabilities are carried over to the steady state, and hence the mean steady state function $f(x)$ is not everywhere differentiable (Shapiro and Wardi,1994).

A somewhat different approach to construction of a stochastic approximation of $f(x)$ is based on a LR transformation. Denote by $\tau(x)$ the number of customers served in one busy period. It is well known in the renewal theory of regenerative processes that, for a given x and $\tau = \tau(x)$, $G_i = G_i(x)$, the mean steady state sojourn time $f(x)$ can be written in the form

$$f(x) = \frac{\mathbb{E}\{\sum_{i=1}^{\tau} G_i\}}{\mathbb{E}\{\tau\}}, \tag{5}$$

provided both expectations in the right hand side of (5) exist.

Suppose now that the service times $Z_1, \ldots$, are iid and have a (common) pdf $p(z, x)$ depending on the parameter vector x. Then it is possible to show (Asmussen and Rubinstein, 1992, Rubinstein and Shapiro, 1993) that $f(x)$ can be represented in the form

$$f(x) = \frac{\mathbb{E}_{p_0}\{\sum_{t=1}^{\tau^*} G_t^* L_t(x)\}}{\mathbb{E}_{p_0}\{\sum_{t=1}^{\tau^*} L_t(x)\}}, \tag{6}$$

where $Z_1^*, \ldots$, is an iid random sample of service times generated from a (fixed) pdf $p_0(z)$, G_t^* and τ^* are the sojourn times and the number of customers served in one busy period, respectively, corresponding to the service times Z_t^*, and $L_t(x)$ is the likelihood ratio process

$$L_t(x) = \prod_{i=1}^{t} \frac{p(Z_i^*, x)}{p_0(Z_i^*)}.$$

Typically one takes the dominating pdf in the form $p_0(z) = p(z, x_0)$ for a chosen value x_0 of the parameter vector.

By generating n busy periods (regenerative cycles) and replacing the expectations in (6) by their sample averages, one obtains a (regenerative) LR estimator $\tilde{f}_n(x)$ of $f(x)$. Note that once a sample of the service and interarrival times is generated and the corresponding sojourn times $G_1^*, \ldots,$ are calculated, the function $\tilde{f}_n(\cdot)$ is given explicitly through the corresponding LR process $L_t(\cdot)$.

It should be noted that the above LR construction depends on two crucial assumptions, namely on identifiability of the regenerative cycles and a possibility of moving the parameter vector x into the corresponding pdf $p(z, x)$. In more complex situations it may happen that the regenerative cycles are too long or cannot be easily identified (the system is never empty). Note also that in order to keep the variance of the obtained LR estimators under control the reference value x_0 of the parameter vector, used in the dominating pdf $p_0(z) = p(z, x_0)$, should be carefully selected (see Rubinstein and Shapiro, 1993, for a discussion of this problem).

3 OPTIMIZATION ALGORITHMS

An important issue in an implementation of the SC method is the choice of a deterministic optimization algorithm. In case the considered problem is convex, it is suggested in Plambeck, Fu, Robinson and Suri (1996) to use a bundle type optimization procedure. An alternative approach is to apply a steepest descent method with an approximate line search, say by using Armijo step sizes (cf. Wardi, 1990). In this respect let us make the following observations.

Usually calculations of the average-function $\hat{f}_n(x)$ value, and its gradient $\nabla \hat{f}_n(x)$, at an iteration point $x = x_k$, are expensive with the time effort is proportional to the used sample size n. By the Central Limit Theorem, the involved stochastic error is inversely proportional to $\sqrt{n}$. Therefore, typically, in simulation based optimization one cannot hope to achieve an accuracy often attainable in deterministic optimization. Increasing the sample size does not help much since in order to improve the accuracy, say by one digit (10 times), one would need to increase the sample size 100 times.

Often the average-functions $\hat{f}_n(x)$ are piecewise smooth nondifferentiable functions. Consequently reliable estimates of the second order derivatives $\nabla^2 f(x)$, of the expected value function $f(x)$, are either unavailable or are too expensive to calculate. Therefore applicable algorithms are usually first-order optimization routines. In any case it does not make sense to try to solve the constructed stochastic counterpart problem (2) with high accuracy. Especially at first iterations, when the iterates are far away from the optimal solution, the sample size can be relatively small and can be increased gradually as the algorithm proceeds. In this respect steepest descent type algorithms offer more flexibility in updating the sample sizes than bundle type methods. It is shown in Shapiro and Wardi (1996) that, under mild regularity conditions, such steepest descent type algorithms produce iterates which converge with probability one to the set of "true" optimal solutions as the sample sizes are increasing to infinity.

4 VALIDATION ANALYSIS AND STOPPING RULES

Suppose that we are given a point x^* which is suggested as an approximation of an optimal solution x_0 of the program (1). Can we evaluate a quality of this approximation? Closely related to this question is a choice of stopping criteria for a considered algorithm. In this section we discuss some statistical tests for validation of optimality of the solution x^*.

Suppose that the expected value function $f(x)$ is *differentiable* at the point x^*. Also, assume that the feasible set S is defined by constraints as follows

$$S = \left\{ x \in \mathbb{R}^m : \begin{array}{l} c_i(x) = 0, \ i = 1, \ldots, k, \\ c_i(x) \geq 0, \ i = k+1, \ldots, l \end{array} \right\},$$

where $c_i(x)$ are (deterministic) continuously differentiable functions. By the first-order optimality conditions we have that if x_0 is an optimal solution of the problem (1), then (under a constraint qualification) there exist Lagrange multipliers λ_i such that $\lambda_i \geq 0$, $i \in J(x_0)$, and

$$\nabla f(x_0) - \sum_{i \in I(x_0)} \lambda_i \nabla c_i(x_0) = 0, \tag{7}$$

where $J(x) = \{i : c_i(x) = 0, \ i = k+1, \ldots, l\}$ denotes the index set of inequality constraints active at x and $I(x) = \{1, \ldots, k\} \cup J(x)$. Consider the polyhedral cone

$$C(x) = \left\{ z \in \mathbb{R}^m : \begin{array}{l} z = \sum_{i \in I(x)} \alpha_i \nabla c_i(x), \\ \alpha_i \geq 0, \ i \in J(x) \end{array} \right\}.$$

Then the optimality conditions (7) can be written in the form $\nabla f(x_0) \in C(x_0)$.

Suppose now that the gradient $\nabla f(x^*)$ can be estimated by a (random) vector $\gamma_n(x^*)$ such that $\gamma_n(x^*) \to \nabla f(x^*)$ w.p.1 as $n \to \infty$ and $\gamma_n(x^*)$ has (approximately) a multivariate normal distribution with mean vector $\mu = \nabla f(x^*)$ and a covariance matrix Ω_n. For example, if the gradient of $f(x)$ is estimated by the gradient $\gamma_n(x^*) = \nabla \hat{f}_n(x^*)$ of the corresponding average function, then asymptotic normality of $\gamma_n(x^*)$ follows by the Central Limit Theorem. Note that in this case the covariance matrix Ω_n can be consistently estimated from the same sample.

By using the estimator $\gamma_n(x^*)$, we can test the hypothesis:

$$H_0 : \nabla f(x^*) \in C(x^*)$$

against the alternative

$$H_1 : \nabla f(x^*) \notin C(x^*).$$

In order to test the (optimality-conditions) hypothesis H_0 one can use the following procedure (cf. Shapiro and Homem-de-Mello, 1996). Suppose that the covariance matrix Ω_n is nonsingular, and hence is positive definite, and that a consistent estimator $\hat{\Omega}_n$ of Ω_n is available. Then

$$T = \min_{z \in C(x^*)} (\gamma_n(x^*) - z)^T \hat{\Omega}_n^{-1} (\gamma_n(x^*) - z)$$

is an asymptotic analogue of the Hotelling's test statistic which is used in multivariate analysis. It is possible to show that if all Lagrange multipliers corresponding to the inequality constraints active at x^* are positive (*strict complementarity condition*), then the test statistic T has approximately (asymptotically) a noncentral chi-square distribution with $m-s$ degrees of freedom, where

$$s = \text{card}(I(x^*)) = k + \text{card}(J(x^*)),$$

("card(I)" denotes cardinality of the set I) and the noncentrality parameter

$$\kappa = \min_{z \in C(x^*)} (\nabla f(x^*) - z)^T \Omega_n^{-1} (\nabla f(x^*) - z).$$

In particular, under H_0 we have that $\kappa = 0$ and hence the null distribution of T is (asymptotically) central chi-square with $m - s$ degrees of freedom. Therefore for a calculated value T of the test statistic we can calculate the corresponding p-value, that is $p = \text{Prob}\{\chi^2_{m-s} \geq T\}$. This p-value gives an indication of the quality of the suggested solution x^* with respect to the stochastic precision. A large (close to one) p-value means that such precision was reached, so the algorithm cannot proceed further, whereas a small (close to zero) p-value indicates that either the current solution is far from the optimal or the deterministic error starts to dominate.

The test statistic T can be calculated at each (at some) iteration points and can be used in developing stopping criteria for considered algorithms. Numerical experience reported in Shapiro and Homem-de-Mello (1996) suggests that this test statistic is useful although alone does not give a good stopping rule and should be combined with other criteria. It may happen that the covariance matrix $\hat{\Omega}_n$ of $\nabla \hat{f}_n(x)$, at a current iteration point $x = x_k$, is large and consequently the corresponding T-value is small although, in fact, x_k is far away from x_0. In that case the algorithm reached its stochastic precision and either the sample size should be increased or some variance reduction procedure should be implemented.

On the other hand a reason for large T-value can be a small value of the covariance matrix $\hat{\Omega}_n$. In that case the deterministic error of the implemented optimization algorithm dominates the corresponding stochastic error and the algorithm can be stopped by a deterministic criterion.

It is also possible to construct a confidence region for $\nabla f(x_k) - P(\nabla f(x_k))$, where $P(\nabla f(x_k))$ denotes a point of the cone $C(x_k)$ closest to $\nabla f(x_k)$, with respect to the metric induced by the inverse of the covariance matrix. This can give an indication of the order of the stochastic error and hence to help in the choice of the generated sample size.

REFERENCES

Asmussen, S. and R.Y. Rubinstein. 1992. The efficiency and heavy traffic properties of the score function method in sensitivity analysis of queueing models. *Advances of Applied Probability* 24: 172-201.

Chong, E.K.P. and P.J. Ramadge. 1993. Optimization of queues using an infinitesimal perturbation analysis-based stochastic algorithm with general update times. *SIAM J. Control and Optimization* 31: 698-732.

Ermoliev, Y. and R.J.B. Wets, (eds). 1988. *Numerical techniques for stochastic optimization.* Springer-Verlag, Berlin.

Kushner, H.J. and D.S. Clark. 1978. *Stochastic approximation methods for constrained and unconstrained systems.* Springer-Verlag, New York, NY.

L'Ecuyer, P. and P. Glynn. 1994. Stochastic optimization by simulation: convergence proofs for the GI/G/1 queue in steady-state, *Management Science* 40: 1562-1578.

Nevel'son, M.B. and R.Z. Has'minskii. 1976. *Stochastic approximation and recursive estimation.*

American Mathematical Society Translations of Mathematical Monographs, Vol. 47, Rhode Island.

Plambeck, E.L., Fu, B.R., Robinson, S.M., and R. Suri. 1996. Sample-path optimization of convex stochastic performance functions, *Mathematical Programming, Series B*, to appear.

Polyak, B.T. 1990. New stochastic approximation type procedures, *Automat. i Telemekh.* 7: 98-107.

Rubinstein, R.Y., and A. Shapiro. 1993. *Discrete Event Systems: Sensitivity analysis and stochastic optimization by the score function Method.* John Wiley & Sons, New York, NY.

Shapiro, A. 1996. Simulation based optimization - convergence analysis and statistical inference. *Stochastic Models*, to appear.

Shapiro, A., and T. Homem-de-Mello. 1996. Simulation-based likelihood ratios approach to two stage stochastic programming with recourse, preprint.

Shapiro, A. and Y. Wardi. 1994. Nondifferentiability of the steady-state function in discrete event dynamic systems. *IEEE transactions on Automatic Control* 39: 1707-1711.

Shapiro, A. and Y. Wardi. 1996. Convergence analysis of stochastic algorithms. *Mathematics of Operations Research*, to appear.

Wardi, Y. 1990. Stochastic algorithms with Armijo step sizes for minimization of functions, *Journal of Optimization Theory and Applications* 64: 399-417.

AUTHOR BIOGRAPHY

ALEXANDER SHAPIRO is a Professor in the Department of Industrial and Systems Engineering at Georgia Institute of Technology. He received a Ph.D. degree in applied mathematics from the Ben-Gurion University (Israel) in 1981. His research interests are focused on stochastic programming, sensitivity analysis and multivariate statistical analysis.

Proceedings of the 1996 Winter Simulation Conference
ed. J. M. Charnes, D. J. Morrice, D. T. Brunner, and J. J. Swain

SAMPLE-PATH SOLUTION OF STOCHASTIC VARIATIONAL INEQUALITIES, WITH APPLICATIONS TO OPTION PRICING

Gül Gürkan
A. Yonca Özge
Stephen M. Robinson

Department of Industrial Engineering
University of Wisconsin–Madison
1513 University Avenue
Madison, WI 53706-1539, U.S.A.

ABSTRACT

This paper shows how to apply a variant of sample-path optimization to solve stochastic variational inequalities, including as a special case finding a zero of a gradient. We give a new set of sufficient conditions for almost-sure convergence of the method, and exhibit bounds on the error of the resulting approximate solution. We also illustrate the application of this method by using it to price an American call option on a dividend-paying stock.

1 INTRODUCTION

This paper shows how to use the technique of sample-path optimization (SPO) to solve stochastic variational inequalities, which include as special cases systems of nonlinear equations. This capability extends the range of application of SPO, since in some important cases the functions to be optimized are difficult to deal with, but their gradients can be approximated. In this section we review the SPO technique and give an example of a case in which its usual form is difficult to apply. The remainder of the paper then shows how to adapt the method to variational inequalities, and gives a numerical example illustrating an application of the new method.

To describe the usual form of SPO, we introduce an extended-real-valued stochastic process $\{L_n(x) | n = 1, 2, \dots\}$ where $x \in \mathbb{R}^k$. For each n and x, $L_n(x)$ is a random variable defined on the common probability space $(\Omega, \mathcal{F}, P)$; it takes values in $\mathbb{R} \cup \{+\infty\} \cup \{-\infty\}$, though we generally exclude $+\infty$ by requiring the function to be *proper* (for maximization): that is, never $+\infty$ and not everywhere $-\infty$. We can use the extended value $-\infty$ to model constraints on x, by setting $L_n(x) = -\infty$ for infeasible values of x. In the following we write $L_n(\omega, x)$ when we want to emphasize dependence of $L_n(x)$ on the sample point ω.

We assume the existence of a deterministic function L_∞ such that the L_n converge pointwise to L_∞ with probability one. A convenient interpretation is to regard the $L_n(x)$ as estimates of $L_\infty(x)$ computed by a simulation run of length n. Such convergence is often justified by regeneration theorems in the case of dynamic systems, or by the strong law of large numbers in the case of static systems. We wish to find (approximately) a constrained maximizer of $L_\infty(x)$, under the condition that one cannot observe L_∞ but only the L_n.

The method in its pure form is very simple: we fix a large simulation length n and the sample point ω (representing the random number streams used in the simulation), compute a maximizer x_n^* of the resulting deterministic function $L_n(\omega, \cdot)$, and take this point as an estimate of a maximizer of L_∞.

This form of the method was proposed in Plambeck *et al.* (1993, 1996) for use with infinitesimal perturbation analysis (IPA) gradient estimates; the key point is that IPA – when it applies – gives *exact* gradients of the L_n, so that one can apply the powerful technology of constrained deterministic optimization. Convergence of the general method is analyzed in Robinson (1996). Similar ideas were proposed by Rubinstein and Shapiro (1993) for use with the likelihood-ratio (LR) method, and these methods are also closely related to the retrospective optimization proposals of Healy and Schruben (1991) and to M-estimation and other techniques. Robinson (1996) gives a brief survey of these and other ideas similar to SPO that have appeared in the literature; a summary of the method's properties can be found in Gürkan *et al.* (1994). Gürkan (1996) and Gürkan and Özge (1996) show how SPO can be applied to the buffer allocation problem in certain tandem production lines. Plambeck *et al.* (1996) and Gürkan (1996) report extensive numerical results on the performance of the method on fairly large systems with various constraints on the variables.

In these experiments the method performed at least

as well as the stochastic approximation method or its single-run optimization variant (Meketon (1983), Suri and Leung (1989), Leung (1990)) in cases to which the latter two methods were applicable. It was also effective for problems with inequality constraints, in which stochastic approximation can have difficulty in maintaining feasibility; see Appendix F of Plambeck *et al.* (1996) for an example. In addition, the method does not require a predetermined choice of step size, thereby avoiding a potential difficulty of the stochastic approximation approach; see Fu and Healy (1992) and L'Ecuyer *et al.* (1994).

Existing sufficient conditions for the convergence of SPO can be found in Robinson (1996). A brief summary of these conditions is as follows: we require L_∞ to be a proper deterministic function with a nonempty, compact set of maximizers, and the L_n to be (with probability one) proper, upper semicontinuous functions such that the sequence $\{-L_n\}$ *epiconverges* to $-L_\infty$. Epiconvergence is a type of convergence often used in optimization, which is independent of the usual pointwise convergence. Kall (1986) gives a good review of epiconvergence and its connections to other ideas of convergence. Under the conditions just mentioned, Robinson (1996) shows that with probability one, for large n the set M_n of maximizers of L_n will be nonempty and compact, and the distance from *any* point of M_n to *some* point of the set M_∞ of maximizers of L_∞ will be small. Thus, for large enough n, if we maximize L_n we are guaranteed to be close to a maximizer of L_∞.

Unfortunately, for some problems found in applications it is difficult or impossible to maximize L_n. For example, Figure 1 shows the graph of L_{50} for an option-pricing problem described further in Section 3. This function is an average of step functions, and even though it would look better if we used more than 50 replications, it would still be discontinuous and extremely unpleasant to maximize. On the other hand, Figure 2 shows (also for 50 replications) the graph of a function f_{50} that is quite smooth. The f_n, which we shall show in Section 3 how to compute, have the property that f_n almost surely converges pointwise, as $n \to \infty$, to the *derivative* dL_∞ of L_∞.

Therefore, in this case a reasonable strategy seems to be to try to find a zero of f_n for large n, in the hope that it might be close to a zero of dL_∞, which under suitable conditions would be a maximizer of L_∞. The theory that we shall present in Section 2 shows when this procedure can be justified. Thus, by using this new theory we can apply SPO to this problem, whereas the version requiring maximization of the L_n would have been very difficult or impossible to apply.

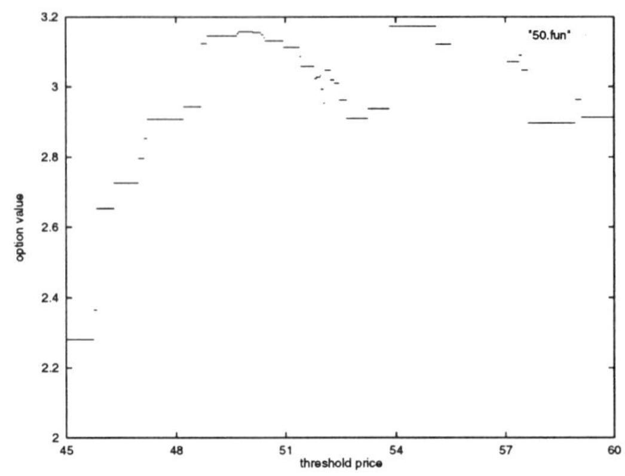

Figure 1: Sample option value

The rest of this paper has three numbered sections. Section 2 briefly introduces stochastic variational inequalities, and builds a theoretical framework to justify applying SPO to solve them. Particular cases covered by this theory include the first-order optimality conditions for nonlinear-programming problems, and this in turn includes unconstrained optimization by computing a critical point of the gradient. In Section 3 we illustrate the latter case by applying the method to price an American call option on a dividend-paying stock. We summarize what we have done in Section 4, and then conclude with acknowledgments and references.

2 VARIATIONAL INEQUALITIES AND SIMULATION

In this section we introduce the variational inequality problem, briefly review its significance, and indicate

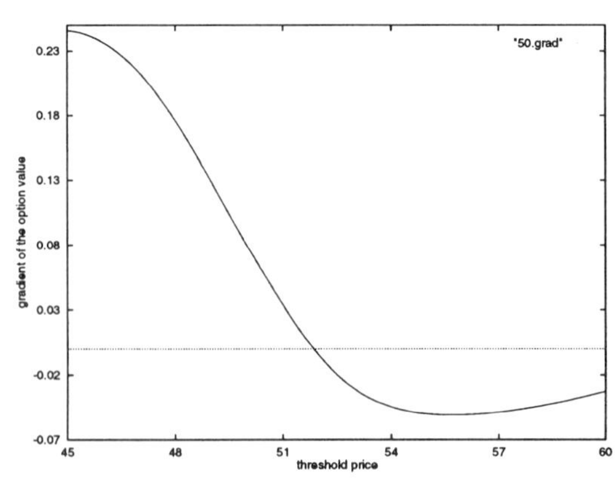

Figure 2: Sample gradient

how in a very special case this problem reduces to that of solving systems of nonlinear equations. Then we show how this problem can arise in a stochastic context, and we extend the SPO approach to provide a simple algorithm to solve such problems. Finally, we state a theorem giving conditions under which this approach is justified.

The ingredients of the (deterministic) variational inequality problem are a closed convex set C (in general, a subset of a Hilbert space, but here a subset of $\mathbb{R}^k$) and a function f from an open set Φ meeting C to $\mathbb{R}^k$. The problem is to find a point $x_0 \in C \cap \Phi$, if any exists, satisfying

$$\text{For each } x \in C, \quad \langle x - x_0, f(x_0)\rangle \geq 0, \tag{1}$$

where $\langle y, z\rangle$ denotes the inner product of y and z. Geometrically, (1) means that $f(x_0)$ is an inward normal to C at x_0.

The problem (1) models a very large number of equilibrium phenomena in economics, physics, and operations research. A survey of some of these can be found in Harker and Pang (1990). In problems arising from applications the convex set C is frequently polyhedral. As a simple example, consider the problem of solving systems of k nonlinear equations in k unknowns: to put this in the form (1) we need only take $C = \mathbb{R}^k$, so that the requirement is to find $x_0 \in \Phi$ such that $f(x_0) = 0$. Another very commonly occurring special case arises from the fact that the first-order necessary optimality conditions for a nonlinear-programming problem with continuously differentiable objective and constraint functions can be written in the form (1).

However, not all variational inequality problems arise from optimization, just as not all systems of nonlinear equations do. For example, in models of economic equilibrium, the lack of certain symmetry properties results in a model that is said to be *non-integrable;* in such a case, it is not possible to find the equilibrium prices and quantities by substituting an associated optimization problem for the variational inequality. For discussion of an actual model of this type that was heavily used in policy analysis, see Hogan (1975). The theory that we shall develop does not require any symmetry properties, so it applies to non-integrable models. In fact, one of the possible future applications that we have in mind is the solution of stochastic economic equilibrium models involving expectations or steady-state functions; in such cases one might hope that simulation together with gradient-estimation techniques could provide an effective alternative to discrete scenario representations of uncertainty, with their associated data-management problems.

We are concerned here with stochastic variational inequalities: that is, problems of the form (1) in which the function f is an (unobservable) expectation or steady-state function, and in which we can only observe f approximately by computing a sequence $\{f_n\}$ of functions (*e.g.,* by simulation) with the property that the f_n almost surely converge to f in a sense that will be made precise below. We shall show how one can use the sample-path technique on this problem when the set C is polyhedral.

Our technique will be first to convert the variational inequality (1) into an equivalent equation which, however, will involve a function that is continuous but is generally nonsmooth even if f is smooth. Then we introduce some technical terminology that will permit us to state the main theorem. That theorem gives conditions under which solutions of variational inequalities involving the approximate functions f_n will converge to solutions of the problem that we really want to solve. We conclude this section by commenting on some special cases, one of which involves the numerical example that we present in Section 3.

To convert (1) into an equivalent equation, we introduce the *normal map* induced by f and C, defined by $f_C(z) = f(\Pi_C(z)) + z - \Pi_C(z)$, where Π_C is the Euclidean projector on C. The function f_C is then well defined on $\Pi_C^{-1}(\Phi)$, and it is clearly continuous if f is. Further, it is easy to show that if x_0 satisfies (1) then $z_0 = x_0 - f(x_0)$ satisfies

$$f_C(z_0) = 0, \tag{2}$$

and that if (2) holds then $x_0 = \Pi_C(z_0)$ satisfies (1). Therefore for purposes of computation (1) and (2) are equivalent; we shall work with (2). For additional information and further references on normal maps, see *e.g.* Robinson (1995).

The new form of SPO works in the setting of a vector-valued stochastic process $\{f_n(x) \mid n = 1, 2, \ldots\}$, and a vector-valued function $f_\infty(x)$, where the parameter x takes values in $\mathbb{R}^k$. Again, for all $n \geq 1$ and all $x \in \mathbb{R}^k$, the random variables $f_n(x)$ are defined on a common probability space $(\Omega, \mathcal{F}, P)$. The aim is to find a point x_0 satisfying (1) for $f = f_\infty$, provided that one exists. We proceed by fixing a large simulation length n and the sample point ω, finding a point $x_n^*(\omega)$ solving (1) for $f = f_n$ (again, provided that one exists) and taking this point as an estimate of x_0. We shall give conditions below that ensure the existence of these quantities.

In the particular special case of an unconstrained simulation optimization problem, C would be $\mathbb{R}^k$ and f_∞ would be the gradient of the limit function L_∞.

By finding a zero of f_∞ (a critical point of L_∞), we are solving the first-order necessary optimality conditions, which is the approach taken by most deterministic optimization algorithms. Of course, such a critical point may or may not be an maximizer, and additional work is generally necessary to determine whether the appropriate second-order conditions hold. In the unconstrained case this amounts to checking for negative definiteness of the Hessian at the critical point. In some cases, notably when the problem is unconstrained and the objective function is known to be locally concave near the critical point, no second-order analysis is needed.

It turns out that *continuous convergence* plays a crucial role in the convergence analysis of this method. This notion of convergence is equivalent to uniform convergence to a continuous limit on compact sets. For an elementary treatment of the relationship between different types of functional convergence, see Kall (1986), and for a comprehensive treatment of continuous convergence and related issues, see Rockafellar and Wets (1996).

Definition 1 *A sequence f_n of extended-real-valued functions defined on $\mathbb{R}^k$* converges continuously *to an extended-real-valued function f_∞ defined on $\mathbb{R}^k$ (written $f_n \xrightarrow{c} f_\infty$) if for any $x \in \mathbb{R}^k$ and any sequence $\{x_n\}$ converging to x, one has $f_n(x_n) \to f_\infty(x)$. A sequence of functions from $\mathbb{R}^k$ into $\mathbb{R}^m$ converges continuously if each of the m component functions does so.*

The fact motivating our use of continuous convergence is that if $f_n \xrightarrow{c} f_\infty$ and $x_n \to x$, and if for each n, x_n solves (1) with $f = f_n$, then x solves (1) with $f = f_\infty$. This is very easy to check directly from (1). Thus, *if* the problems with $f = f_n$ have solutions, and *if* those solutions converge, then their limit will solve the problem with $f = f_\infty$.

We present next a theorem, for the case in which C is polyhedral, giving sufficient conditions for these approximate solutions to exist and to converge. In order to state this theorem we need a technical property that generalizes nonsingularity. We explain it very briefly here, and then mention some important special cases in which the property is easier to understand. For additional examples in more general cases we refer to Robinson (1995).

For this explanation we need some definitions. First, the normal cone $N_C(x)$ of C at x is the set

$$\{y^* \mid \text{for each } c \in C, \quad \langle y^*, c - x\rangle \le 0\}$$

provided that $x \in C$, and it is empty otherwise. If $x \in C$ then the *tangent cone* of C at x, written $T_C(x)$, is the polar of $N_C(x)$: that is, the set of all y such that $\langle y^*, y\rangle \le 0$ for each $y^* \in N_C(x)$. Second, the *critical cone* defined by C and a given point $z \in \mathbb{R}^k$ is

$$K(z) = T_C(\Pi_C(z)) \cap \{y^* \in \mathbb{R}^k \mid \langle y^*, z - \Pi_C(z)\rangle = 0\}.$$

Now fix any z and write $K = K(z)$. As K is polyhedral it has only finitely many faces; for each nonempty face F the normal cone of K takes a constant value, say N_F, on the relative interior of F. Then the set $\sigma_F = F + N_F$ is a nonempty polyhedral convex set of dimension k in $\mathbb{R}^k$. The collection $\mathcal{N}_K$ of all these σ_F for nonempty faces F of K is called the *normal manifold* of K. In each of these σ_F the projector Π_K reduces to an affine map (generally different for different σ_F). If M is a linear transformation from $\mathbb{R}^k$ to $\mathbb{R}^k$, then we say that the normal map M_K is *coherently oriented* if in each σ_F the determinant of the affine map obtained by restricting M_K to σ_F has the same nonzero sign.

As a simple illustration of this property, we can consider the case in which K happens to be a subspace (the "strict complementary slackness" situation in the optimization literature). Then the coherent orientation requirement reduces to nonsingularity of the section of M in K. In particular, if $C = \mathbb{R}^k$ (the case of nonlinear equations), then $K(z) = \mathbb{R}^k$ for each $z \in \mathbb{R}^k$, and then $\mathcal{N}_{K(z)}$ has only one cell, namely $\mathbb{R}^k$ itself. Then M_K is coherently oriented exactly when M is nonsingular. In general, the coherent orientation condition is a way of generalizing nonsingularity to the case of a nontrivial set C.

Theorem 1 *Let Φ be an open subset of $\mathbb{R}^k$ and let C be a polyhedral convex set in $\mathbb{R}^k$. Let x_0 be a point of Φ, and suppose f_∞ is a function from Φ to $\mathbb{R}^k$. Let $\{f_n \mid n = 1, 2, \dots\}$ be random functions from Φ to $\mathbb{R}^k$ such that for all $x \in \Phi$ and all finite n the random variables $f_n(x)$ are defined on a common probability space $(\Omega, \mathcal{F}, P)$. Let $z_0 = x_0 - f_\infty(x_0)$ and assume the following:*

a. With probability one, each f_n for $n = 1, 2, \dots$ is continuous and $f_n \xrightarrow{c} f_\infty$.

b. x_0 solves the variational inequality (1) defined by f_∞ and C.

c. f_∞ has a strong Fréchet derivative $df_\infty(x_0)$ at x_0, and $df_\infty(x_0)_{K(z_0)}$ is coherently oriented.

Then there exist a positive number λ, a compact subset $C_0 \subset C \cap \Phi$ containing x_0, a neighborhood $V \subset \Phi$ of x_0, and a set $\Delta \subset \Omega$ of measure zero, with the following properties: for $n = 1, 2, \dots$ and $\omega \in \Omega$ let

$$\xi_n(\omega) := \sup_{x \in C_0} \|f_n(\omega, x) - f_\infty(x)\|,$$

and

$$X_n(\omega) := \{x \in C \cap V \mid \text{for each } c \in C,\ \langle f_n(\omega, x), c - x\rangle \geq 0\}.$$

For each $\omega \notin \Delta$ there is then a finite integer N_ω such that for each $n \geq N_\omega$ the set $X_n(\omega)$ is a nonempty, compact subset of the ball

$$B(x_0, \lambda\xi_n(\omega)) = \{x \mid \|x - x_0\| \leq \lambda\xi_n(\omega)\}.$$

In the rest of this section we discuss some work in the literature related to Theorem 1 and point out similarities and differences. First, King and Rockafellar (1993) gave various convergence and asymptotic properties for solutions to stochastic generalized equations, of which the variational inequalities that we treat here are a particular case. Among other things they proved under certain regularity conditions that the limit superior of a localization of the perturbed solution set was nonempty and contained in the solution set of the limit problem; this corresponds in our case to a proof that the solutions of the problems involving the f_n have cluster points, and that each such point solves the limit problem. They also showed under additional assumptions that the solutions of the perturbed problems converged in probability to the solution of the limit problem, with the speed of convergence dependent on the goodness of the approximation. Our conditions are in some ways more restrictive than theirs, but we are able to establish almost-sure convergence and to bound the error of the approximate solutions by a constant multiple of the error in the approximation. Under considerably stronger assumptions than ours, they also obtained asymptotic results about the distribution of errors, a question which we do not treat at all here.

Shapiro and Wardi (1994, 1995) used conditions somewhat similar to those of Theorem 1 (in the special case $C = \mathbb{R}^k$) to prove the convergence of stochastic gradient descent algorithms. However, there are significant differences: for example, they made no assumption that the limiting function had a nonsingular derivative, but they derived no bounds on the resulting solutions. Also, they assumed that the functions involved were gradients or elements of generalized gradients. As we have indicated above, we have in mind applications in which it is important that the functions *not* be restricted to be gradients.

Shapiro (1995) investigates the convergence of approximate solutions of quite general problems by introducing a "stopping criterion" to measure the quality of approximation. However, he assumes the existence of solutions to the approximating problems, whereas under the hypotheses of Theorem 1 we can prove the existence of such solutions as well as bounding their distance from the true solution. On the other hand, Shapiro (1995) also gives results on the asymptotic distribution of solutions, as did King and Rockafellar (1993). We do not give such results.

As we have shown in Theorem 1, we can find approximate solutions of stochastic variational inequalities by a variant of sample-path optimization, provided that a generalized nonsingularity condition holds. Moreover, as $n \to \infty$ the distance of the approximate solutions from an exact solution is bounded above by a constant multiple of the uniform norm of $f_n - f_\infty$ on a compact set containing the true solution. In the next section we illustrate an application of this general result.

3 APPLICATION: OPTION PRICING

In this section we present a small numerical example illustrating how to apply SPO to the pricing of an American call option on a dividend-paying stock. Such an option is a derivative security giving the holder the right to *exercise* the option (by buying an underlying asset) at the *exercise price* K at any time up to the *expiration date* T. Under certain assumptions about markets, including the absence of transaction costs, it is never optimal to exercise an American call option prematurely (*i.e.*, before T) unless the stock pays a dividend during the life of the option contract (see Hull (1993) and Stoll and Whaley (1993)). In such a case, in order to obtain the right to receive the dividend it may be optimal to exercise the option just prior to an *ex-dividend date* t_1, provided that the stock price then exceeds a *threshold price.* Note that the choice of a threshold price is under the control of the option holder. We can value the option by finding its greatest expected return over all possible choices of threshold price.

Early examples of using Monte Carlo simulation in option pricing can be found in Boyle (1977), Hull and White (1987), and Johnson and Shanno (1987). In these studies, simulation was employed as a "black box" without providing any sensitivity information. More recently, Fu and Hu (1995) and Broadie and Glasserman (1996) have improved this approach by using gradient estimation methods. Welch and Chen (1991) formulated the pricing of an American call option on a dividend-paying stock as an optimization problem. Fu and Hu (1995) adapted this viewpoint and applied stochastic approximation to value this option.

Following the notation and discussion in Fu and Hu (1995), we define the variables that will be used

in the rest of this section as follows :
S_0 = the initial stock price,
S_t = the stock price at time t,
r = the annualized riskless interest rate (compounded continuously),
σ = volatility of the stock price process,
D = the dividend amount,
J_T = the net present value of the return on the option.

Among these variables S_t and J_T are random and the rest are deterministic. We assume that after the ex-dividend date the stock price drops by the dividend amount: *i.e.*, $S_{t_1^+} = S_{t_1^-} - D$. The sample option value for the American call is then given by

$$J_T = e^{-rT}[I_{\{S_{t_1^-}>s\}}(S_{t_1^-} - K)e^{r(T-t_1)} + I_{\{S_{t_1^-}\le s\}}(S_T - K)_+] \quad (3)$$

where I_A is the indicator function of the set A in the sense of probability theory, and $(\cdot)_+ = \max\{\cdot, 0\}$ (the "plus function"). We assume that except for the downward jump at the ex-dividend date the stock price changes continuously according to a function $h(\omega; S, t, r, \sigma)$ which gives the stock price at time t net of the present value of the dividend, given the stock price S at time 0 and the random vector ω. In this work we use the particular form of h that arises when the stock price fluctuations are described by the lognormal distribution, namely

$$h(\omega; S, t, r, \sigma) = Se^{(r-\sigma^2/2)t+\sigma\omega\sqrt{t}}, \quad (4)$$

where ω is a standard $N(0,1)$ random variable. For $N(0,1)$ random variables ω_1 and ω_2 we have

$$S_{t_1^-} = h(\omega_1; S_0 - De^{-rt_1}, t_1, r, \sigma) + D$$

and

$$S_T = h(\omega_2; S_{t_1^+}, T - t_1, r, \sigma).$$

We construct the functions L_n by drawing n samples from the standard normal distribution and averaging the resulting values of J_T. For a fixed, finite set of samples, L_n is therefore a deterministic step function of the threshold price s. As can be seen in Figure 1, the L_n are unsuitable for optimization. We therefore applied the general approach of Section 2 (with $C = \mathbb{R}$) by using an unbiased gradient estimator for $dE[J_T]/ds =: f_\infty$ provided in Fu and Hu (1995) to construct approximating functions f_n. Figure 2 shows a representative of f_{50}, which is quite well behaved. We next discuss briefly the construction of the f_n and some of their properties.

In Shi (1995) a discontinuous perturbation analysis (DPA) gradient estimator is derived for general step functions of one variable, and in Fu and Hu (1995) a smoothed perturbation analysis (SPA) gradient estimator is derived specifically for the gradient of $E[J_T]$ with respect to various parameters of the option. Each of these approaches results in the same unbiased gradient estimator for $dE[J_T]/ds$, which is shown in (5). Denoting by p the probability density function of ω and setting $\tau = T - t_1$, we have

$$(\frac{dJ_T}{ds})_{PA} = e^{-rT}\frac{\partial h^{-1}(y^*)}{\partial s}p(h^{-1}(y^*)) \quad (5)$$
$$[(h(\omega; s - D, \tau, r, \sigma) - K)_+ - (s - K)e^{r\tau}]$$

where $y^* = (s - D; S_0 - De^{-rt_1}, t_1, r, \sigma)$ and the shorthand notation $h^{-1}(y^*)$ stands for

$$h(\cdot\,; S_0 - De^{-rt_1}, t_1, r, \sigma)^{-1}(s - D).$$

For $y > 0$ and the choice of h given by (4), we have

$$h(\cdot\,; S, t, r, \sigma)^{-1}(y) = \frac{1}{\sigma\sqrt{t}}[\ln\frac{y}{S} - (r - \sigma^2/2)t],$$

while for $s > D$ the partial derivative in (5) is

$$\frac{\partial h^{-1}(y^*)}{\partial s} = \frac{1}{(s - D)\sigma\sqrt{t_1}}.$$

For $s \in (D, +\infty)$, the expression in (5) is therefore a continuous function of the threshold price s.

We constructed the approximating functions f_n by averaging the values of $(dJ_T/ds)_{PA}$. One can show that $f_n \xrightarrow{C} f_\infty$ and that f_∞ is continuously F-differentiable. Hence if there is an s^* with $dE[J_T(s^*)]/ds = 0$ and $d^2E[J_T(s^*)]/ds^2 < 0$, then Theorem 1 shows that for sufficiently large sample sizes n, the f_n will have zeros and these zeros will be close to a locally unique maximizer s^*.

In this example we used for the function h the particular form (4) arising from the geometric Brownian motion, since this is a popular stock price process in the finance community. Other forms could also be used, provided that they satisfy conditions to ensure the unbiasedness of the gradient estimate; see Fu and Hu (1995) and Shi (1995).

We now present the results of a small numerical experiment. The data used are taken from Stoll and Whaley (1993): $S_0 = 50.0$, $K = 50.0$, $T = 90.0$, $t_1 = 60.0$, $D = 2.0$, $r = 0.10$, $\sigma = 0.30$. This is a problem with only one dividend; hence one can use the analytical formula to find the optimal threshold price and the option value. Stoll and Whaley (1993) found these to be $s^* = 51.82$ and $E[J_T] = 2.93$.

The same problem is considered in Fu and Hu (1995); they applied stochastic approximation and reported that "convergence within a penny of the actual option value is achieved within 1000 simulations." Fu and Hu (1995) used 9 replications, which means that their total simulation budget was 9,000.

We applied SPO by finding a zero of f_n using the nonsmooth Newton method of Qi and Sun (1993). Results are presented in Table 1. The true solution is found within one penny with a simulation budget of 7,000 (two iterations, each with a sample size of 3,500). The table also gives results for larger sample sizes n. In addition, we used a very large sample size of 500,000 and found the exact solution (reported above). In each case the method converged in 2 iterations. Thus, on this problem SPO performed at least comparably to stochastic approximation.

Table 1: Results of SPO for Option Pricing

n	Option value	Threshold price	# Iter
35E2	2.937	51.850	2
50E2	2.931	51.803	2
10E4	2.939	51.828	2

4 CONCLUSIONS

In this paper we have shown how to use a simulation-based method called sample-path optimization (SPO) for solving stochastic variational inequalities. We presented a new set of sufficient conditions for the convergence of the method, and by applying these to the special case of nonlinear equations we showed how the method can be applied to the pricing of certain American call options. As an illustration we presented the results of a small numerical experiment. On that small problem SPO performed at least as well as stochastic approximation. However, the SPO technique appears to be applicable to a considerably larger class of problems than is stochastic approximation. Therefore, the sufficient conditions that we have given here may help in the solution of problems that are difficult to handle with current techniques.

ACKNOWLEDGMENT

The research reported here was sponsored by the U. S. Army Research Office under grant number DAAH04-95-1-0149, by the Air Force Office of Scientific Research, Air Force Materiel Command, USAF, under grant number F49620-95-1-0222, and by the North Atlantic Treaty Organization under grant number CRG 950360.

REFERENCES

Boyle, P. P. 1977. Options: a Monte Carlo approach. *Journal of Financial Economics* 4:323–338.

Broadie, M., and P. Glasserman. 1996. Estimating security price derivatives using simulation. *Management Science* 42:269–285.

Fu, M. C., and K. Healy. 1992. Simulation optimization of (s, S) inventory systems. In *Proceedings of the 1992 Winter Simulation Conference,* ed. J. J. Swain, D. Goldsman, R. C. Crain and J. R. Wilson, 506–514. Piscataway, NJ: Institute of Electrical and Electronics Engineers.

Fu, M. C., and J.-Q. Hu. 1995. Sensitivity Analysis for Monte Carlo Simulation of Option Pricing. *Probability in the Engineering and Informational Sciences* 9:417–449.

Gürkan, G. 1996. Performance optimization in simulation: Sample-path optimization of buffer allocations in tandem lines. Ph.D. Dissertation, Department of Industrial Engineering, University of Wisconsin-Madison, Madison, WI.

Gürkan, G., and A. Y. Özge. 1996. Sample-path optimization of buffer allocations in a tandem queue – Part I: Theoretical issues. Preprint, Department of Industrial Engineering, University of Wisconsin-Madison, Madison, WI.

Gürkan, G., A. Y. Özge, and S. M. Robinson. 1994. Sample-path optimization in simulation. In *Proceedings of the 1994 Winter Simulation Conference,* ed. J. D. Tew, S. Manivannan, D. A. Sadowski and A. F. Seila, 247–254. Piscataway, NJ: Institute of Electrical and Electronics Engineers.

Harker, P. T., and J.-S. Pang. 1990. Finite-dimensional variational inequality and nonlinear complementarity problems: A survey of theory, algorithms, and applications. *Mathematical Programming* 48:161–220.

Healy, K., and L. W. Schruben. 1991. Retrospective simulation response optimization. In *Proceedings of the 1991 Winter Simulation Conference,* ed. B. L. Nelson, W. D. Kelton, and G. M. Clark, 901–906. Piscataway, NJ: Institute of Electrical and Electronics Engineers.

Hogan, W. W. 1975. Energy policy models for Project Independence. *Computers & Operations Research* 2:251–271.

Hull, J. C. 1993. *Options, Futures, and Other Derivative Securities, 2nd ed.* Englewood Cliffs, NJ: Prentice-Hall.

Hull, J. C., and A. White. 1987. The pricing of options on assets with stochastic volatilities. *Journal of Finance* 42:281–300.

Johnson, H., and D. Shanno. 1987. Option pricing when the variance is changing. *Journal of Financial and Quantitative Analysis* 22:143–151.

Kall, P. 1986. Approximation to optimization problems: An elementary review. *Mathematics of Operations Research* 11:9–18.

King, A. J., and R. T. Rockafellar. 1993. Asymptotic theory for solutions in statistical estimation and stochastic programming. *Mathematics of Operations Research* 18:148–162.

L'Ecuyer, P., N. Giroux, and P. W. Glynn. 1994. Stochastic optimization by simulation: numerical experiments with the $M/M/1$ queue in steady-state. *Management Science* 40:1245–1261.

Leung, Y. T. 1990. Single-run optimization of discrete-event simulations. Ph.D. Dissertation, Department of Industrial Engineering, University of Wisconsin–Madison, Madison, Wisconsin.

Qi, L., and Sun, J. 1993. A nonsmooth version of Newton's method. *Mathematical Programming* 58:353–367.

Meketon, M. 1983. A tutorial on optimization in simulations. Unpublished tutorial presented at the 1983 Winter Simulation Conference.

Plambeck, E. L., B.-R. Fu, S. M. Robinson, and R. Suri. 1993. Throughput optimization in tandem production lines via nonsmooth programming. In *Proceedings of 1993 Summer Computer Simulation Conference,* ed. J. Schoen, 70–75. San Diego, CA: Society for Computer Simulation.

Plambeck, E. L., B.-R. Fu, S. M. Robinson, and R. Suri. 1996. Sample-path optimization of convex stochastic performance functions. Accepted by *Mathematical Programming.*

Robinson, S. M. 1995. Sensitivity analysis of variational inequalities by normal-map techniques. In *Variational Inequalities and Network Equilibrium Problems,* ed. F. Giannessi and A. Maugeri, 257–269. New York: Plenum Press.

Robinson, S. M. 1996. Analysis of sample-path optimization. Accepted by *Mathematics of Operations Research.*

Rockafellar, R. T., and R. J-B Wets. 1996. *Variational Analysis.* Berlin: Springer-Verlag (forthcoming).

Rubinstein, R. Y., and A. Shapiro. 1993. *Discrete Event Systems: Sensitivity Analysis and Stochastic Optimization by the Score Function Method.* Chichester and New York: John Wiley & Sons.

Shapiro, A. 1995. Simulation based optimization – Convergence analysis and statistical inference. To appear in *Stochastic Models.*

Shapiro, A., and Y. Wardi. 1994. Convergence analysis of stochastic algorithms. To appear in *Mathematics of Operations Research.*

Shapiro, A., and Y. Wardi. 1995. Convergence analysis of gradient descent stochastic algorithms. Preprint, School of Industrial and Systems Engineering, Georgia Institute of Technology, Atlanta, GA.

Shi, L. 1995. Discontinuous Perturbation Analysis of Discrete Event Dynamic Systems. Preprint, Department of Industrial Engineering, University of Wisconsin - Madison, Madison, WI.

Stoll, H. R., and R. E. Whaley. 1993. *Futures and Options.* Cincinnati: South-Western Publishing Co.

Suri, R., and Y. T. Leung. 1989. Single-run optimization of discrete event simulations – an empirical study using the $M/M/1$ queue. *IIE Transactions* 34:35–49.

Welch, R. L., and D. M. Chen. 1991. Static optimization of American contingent claims. *Advances in Futures and Options Research* 5:175–184.

AUTHOR BIOGRAPHIES

GÜL GÜRKAN is Research Assistant in the Department of Industrial Engineering, University of Wisconsin–Madison. Her research interests include stochastic and deterministic optimization, simulation, and stochastic processes. She is a member of INFORMS.

A. YONCA ÖZGE is Research Assistant in the Department of Industrial Engineering, University of Wisconsin–Madison. Her research interests include deterministic optimization, stochastic optimization, and simulation. She is a member of INFORMS.

STEPHEN M. ROBINSON is Professor of Industrial Engineering and Computer Sciences at the University of Wisconsin–Madison. His research interests are in quantitative methods in managerial economics, especially deterministic and stochastic optimization and methods to support decision under uncertainty. He is a member of INFORMS and of the Mathematical Programming Society, as well as of SIAM and IIE.

Proceedings of the 1996 Winter Simulation Conference
ed. J. M. Charnes, D. J. Morrice, D. T. Brunner, and J. J. Swain

SOME SUBJECTIVE VALIDATION METHODS USING GRAPHICAL DISPLAYS OF DATA

Robert G. Sargent

Simulation Research Group
College of Engineering and Computer Science
439 Link Hall
Syracuse University
Syracuse, New York 13244, U.S.A.

ABSTRACT

Subjective methods for operational validity are presented that use graphical displays of histograms, box plots, and behavior graphs. These methods allow the data to be correlated, have any statistical distribution, and be limited in the number of observations. Model data are used for the reference distribution (instead of a theoretical distribution such as the t or F) and for reference to compare the system data against. These methods are very general and can be used in validating different types of models.

1 INTRODUCTION

Determining that a model is valid for its intended purpose is critically important in developing a model and doing this is called the validation process. (See, e.g., Sargent (1996) for a discussion of the validation process). One step of the validation process is performing operational validity. Operational validity (Sargent 1994) is defined as "determining that a model's output behavior has sufficient accuracy for the model's intended purpose over the domain of the model's intended applicability." Various methods and techniques are used in conducting operational validity and they are applied either subjectively or objectively.

The major attribute determining which methods and techniques to use in performing operational validity is the observability of the problem entity being modeled. (A problem entity is some system (real or proposed), idea, situation, policy or phenomena.) A problem entity is (i) observable if it is possible to collect a reasonable amount of data on its operational behavior, (ii) partially observable if the amount of data that can be collected on its operational behavior is limited, and (iii) unobservable if no data can be collected on its operational behavior. The reason that observability is so important is because it is usually necessary to compare the model and problem entity behaviors (outputs) for different experimental conditions from the model's domain of intended application in order to establish a high degree of confidence that a model is valid.

It is preferable to make the comparisons of the behavior data between the problem entity and the model in objective ways, which generally is through the use of statistical tests. However, it is often not possible to use statistical tests because (i) the assumptions of the tests cannot (or only with extreme difficulty) be satisfied, (ii) the number of observations from the problem entity is insufficient, or (iii) the behavior of the problem entity is changing, e.g, is highly nonstationary. In these situations, other approaches for comparing the data are desirable.

We present some subjective approaches for performing operational validity that uses graphical displays for comparison of system and problem entity behavior data. We believe that the use of graphical displays is the most effective way of presenting the data for subjective comparison. (Furthermore, we also believe that the use of tables for comparing data is highly ineffective.) While the approaches we present are applicable to different types of models and problem entities, we orient the presentation of our approaches and the discussions about them towards the validation of stochastic discrete event simulation models. (This assumes that the problem entities are also stochastic.)

The remainder of this paper is organized as follows: Section 2 contains some different ways of displaying data graphically, Section 3 discusses some approaches for performing operational validity that use graphical displays of data, and Section 4 gives the summary.

2 GRAPHICAL DISPLAYS

In this section we present three different ways of displaying data graphically: histograms, box plots, and behavior graphs. These three types of displays (or graphs) allow data to be statistically dependent (i.e.,

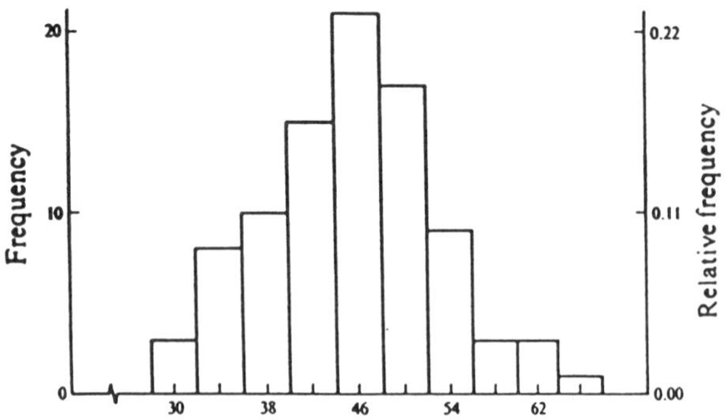

Figure 1: Histogram

correlated), which often occurs in behavior data of systems and simulation models.

2.1 Histograms

Histograms are used to display data for frequency and relative frequency distributions. To construct a histogram, the data are grouped into nonoverlapping intervals (or classes) of usually equal width. The number of intervals is usually between seven and fifteen, depending on the total number of observations. The area of each interval (class) is proportional to the number of observations in each interval. If the interval widths are equal, the height of each interval of a histogram of a frequency distribution is equal to the number of observations in the interval, and the height of each interval of a histogram of a relative frequency distribution is equal to the number of observations in the interval divided by the total number of observations. See Figure 1 for an example of a histogram.

The data plotted in a histogram should be identically distributed. (Note: Prior to putting data in a histogram, the data should be plotted in a time-sequence to ensure that there is no time trend in the data (Johnson 1994), which would cause the data to be not identically distributed.) The data may be statistically independent or dependent (i.e., correlated) (Box, Hunter, and Hunter 1978). The data plotted may be the observations collected on the system (or from the simulation model) or may be some function of partitioned subsets of the observations such as sample means.

2.2 Box Plots

A box plot (Johnson 1994) or box and whisker plot (Walpole and Myers 1993) depict the three quartiles and the two extreme values of a set of data. A rectangle box encloses the 25th percentile (lower quartile)

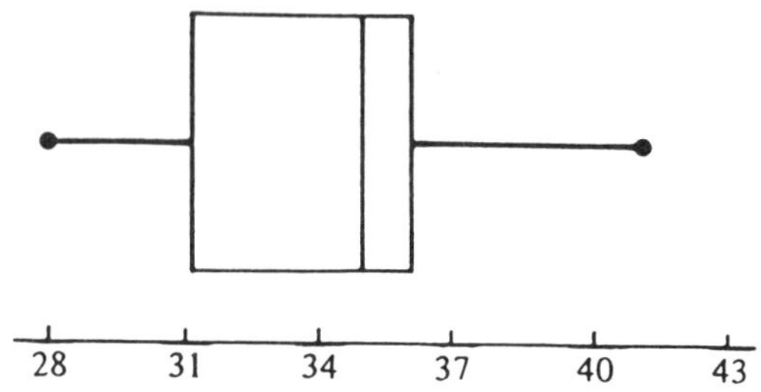

Figure 2: Box Plot

and the 75th percentile (upper quartile) and a line sectioning the box displays the 50th percentile (median). A whisker at each end of the box extends to an extreme value. See Figure 2 for an example.

The data used for box plots should be identically distributed. The data may be statistically independent or dependent (i.e., correlated.). The data used may be the observations themselves or some function of partitioned subsets of the data such as sample means.

2.3 Behavior Graphs

A behavior graph shows the relationship between two entities such as parameters, variables, and functions of random variables by plotting paired data on the two entities. The data points plotted may be the observations themselves, which may be few in number, or may be functions of partitioned subsets of the observations, which may be based on a large number of observations. Some examples of different types of relationships that may be shown in behavior graphs are relationships between two deterministic variables, between two functions of a single random variable such as its mean and standard deviation, between a parameter and some function of a random variable, and between a function of one random variable and a function of another random variable.

Figures 3 and 4 contain two examples of behavior graphs. They are from Anderson (1974) and are based on observations collected on an interactive computer system. Figure 3 contains data showing the relationship between the mean and the standard deviation of system reaction time. One can readily see that there is a linear relationship between them. Each data point is based on thirty independent observations. Figure 4 shows the relationship between the average length of a time slice and the total number of disk accesses based on twenty (correlated) observations per data point. A linear relationship between these two entities can be seen in the behavior graph.

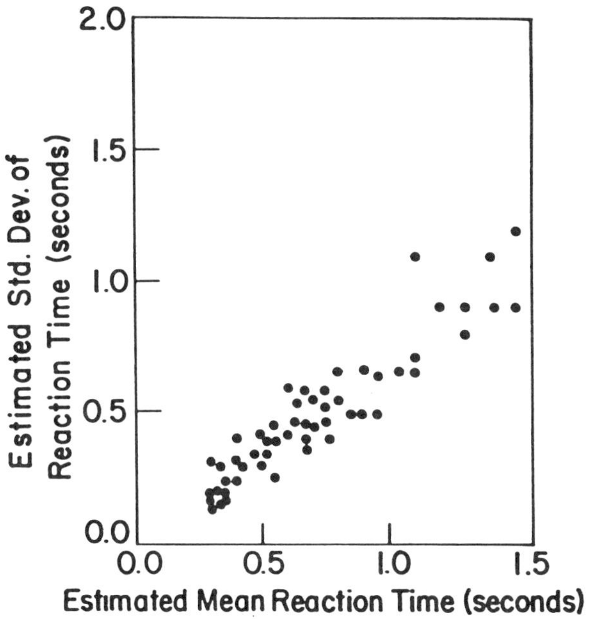

Figure 3: Mean vs. Standard Deviation

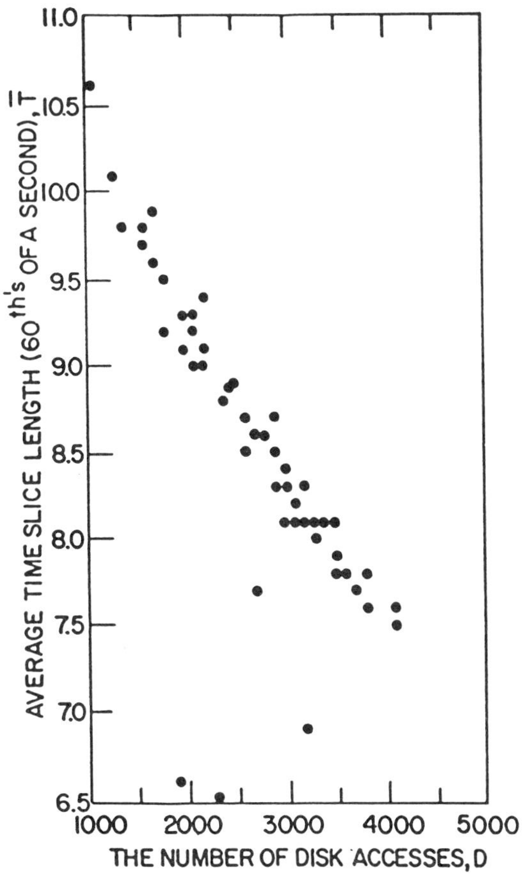

Figure 4: Average Time vs. Disk Accesses

3 VALIDATION APPROACHES

In operational validity, comparisons are made whenever possible between data from the problem entity and the model to determine if the model has sufficient accuracy for the model's intended purpose over the model's domain of intended applicability. It is preferable to do this in some objective way, which means using some type of statistical test if the system (or simulation model) is stochastic. However, this is often not possible because (i) the assumptions required of the statistical tests cannot be satisfied (or only with great difficulty) and/or (ii) there are insufficient data available from the system. Most (simple) statistical tests require the data to be independent, which is often not true of data collected from a system or simulation model. In addition, many statistical tests require the data to have a normal distribution, which is also usually not the case for data collected from a system or simulation model. Furthermore, even when the data does satisfy the statistical requirements, statistical tests frequently cannot be used for operational validity because there is insufficient number of observations from the system to obtain "meaningful" results from the statistical test. For example, the length of a confidence interval developed to compare the means of some model and system behavior variable can be to long for any practical usefulness. This can be caused by an insufficient number of system observations. Thus, it is often the case that statistical tests cannot be used for operational validity. We present in this section some approaches that can be used for operational validity, which eliminates the requirements of independence in the data, have no distributional requirements on the data, and can be used with limited number of system observations.

Statistical tests use a theoretical reference distribution such as the t or F distribution (Box, Hunter, and Hunter 1978; Johnson 1994; Walpole and Myers 1993). In the approaches presented for operational validity in this paper, data generated from the model are used for the reference distribution (or reference) instead of a theoretical distribution. (See Box, Hunter, and Hunter (1978) for a discussion on using data as a (external) reference distribution instead of a theoretical reference distribution.) The data generated from the model for use as a reference distribution (or reference) is displayed in one of the graphical displays described in Section 2 along with the data from the system. These two sets of data are compared subjectively to determine whether the model has sufficient accuracy for its intended purpose. This comparison can be make by the model development team and/or by experts using face validity or Turing tests.

In making comparisons between the model and system data to decide model validity, two types of incorrect decisions can be made. One is to reject a valid model as being invalid and the other is to accept an invalid model as being valid. The first type of incorrect decision is a type I error and is the model builder's risk, and the second type of incorrect decision is a type II error and is the model user's risk (Balci and Sargent 1981). The type II error is extremely important and should be minimized.

3.1 Histograms as Reference Distributions

Histograms of data generated from a model can be used as reference distributions for making comparisons between model and system data for operational validity. For each entity of interest (e.g., an output random variable or sample mean of some output random variable), a sufficient number of observations is generated from the model to have at least fifty data points to be placed in a histogram. The model data to be placed in each histogram need only be identically distributed. Similarly, for each entity of interest, system observations are used to obtain system data points. The system data points must also be identically distributed for each entity of interest. The number of system data points for each entity of interest is usually just a few and may be only one. For each entity of interest, a histogram of the model data, for use as the reference distribution, and a histogram of the system data are placed in the same figure to be compared subjectively to aid in deciding whether the model's entity of interest has sufficient accuracy for the model's intended purpose.

As an example, consider a simulation project by Lowery (1996). A simulation model was developed to predict the mean (average) number of beds used daily (census) in specific hospital units. Operational validity was performed to determine whether the simulation model's mean census (usage) of beds were within the required accuracy of four beds for large hospital units (i.e., units having a large number of beds). Since there was a day a week effect, only Mondays will be considered here. There were 24 system observations (weeks) available on Monday census for the unit that we consider and these observations are correlated. The data entities of interest to be compared must be determined. We select two for illustration purposes: a 24-week average daily (Monday) census and a 4-week average daily census. Observations were generated from the simulation model to obtain fifty 24-week average daily census (for Mondays) data points. A histogram of the 24-week average daily census is in Figure 5. This histogram is the reference distribution for the one system data point of a 24-week average daily census. One can readily see that this system data point lies within the reference distribution. Figure 6 contains a histogram of fifty 4-week average daily census (for Mondays) data points developed from observations from the simulation model for use as a reference distribution. The system observations are used to create six 4-week average daily census and a histogram of them is also placed in Figure 6. One can readily observe that the system histogram lies within the reference distribution. Thus, based on these two figures the simulation model can be judged to have sufficient accuracy with respect to mean daily census for this hospital unit on Mondays. Note that the only assumptions required for this approach are (i) the data compared must be on the same entity of interest, (ii) the data from the model have to be identically distributed, and (iii) the data from the system have to be identically distributed.

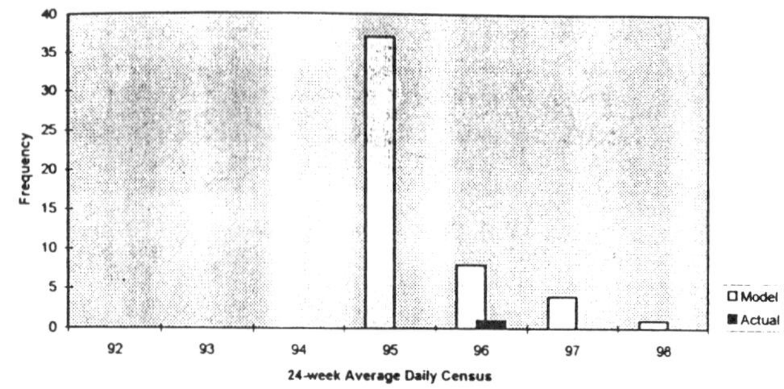

Figure 5: 24-weekly Average Daily Census

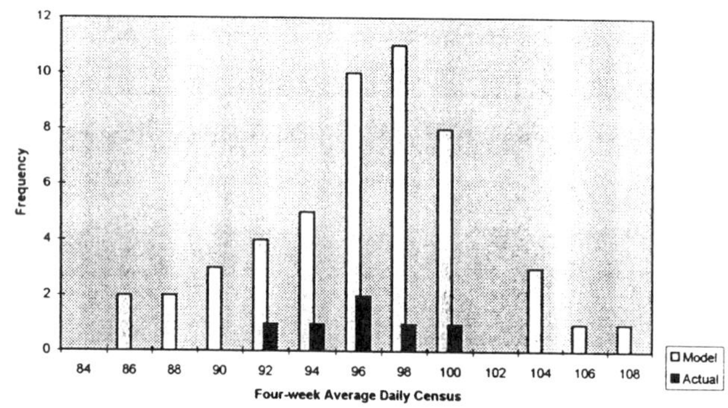

Figure 6: 4-week Average Daily Census

3.2 Box Plots as References

Box plots of data generated from a model can be used as references for making comparisons between model and system data for operational validity. For each entity of interest (e.g., an output random variable or sample mean of some output random variable), a sufficient number of observations is generated from the model to have at least fifty data points to develop a box plot. The model data points used to develop the box plot need only to be identically distributed. Similarly, for each entity of interest, system observations are used to develop system data points, which must be identically distributed. The number of system data points should be at least ten (hopefully) and preferably thirty or more. For each entity of interest, a box plot of the model data to be used as a reference, and a box plot of the system data are placed in the same figure to be compared subjectively. (To assist in making comparisons of means using box plots, a rule of thumb given in Walpole and Myers (1993) may be helpful: "a rough guideline is that if the 25th percentile line for one sample exceeds the median line for the other sample, there is strong evidence of a difference between the means." Note that a comparison can show that a model is invalid but cannot *prove* that a model is valid.)

An example of a box plot used in performing operational validity is given in Figure 7. The box plots are Sunday census observations for the same hospital unit discussed above. The model box plot, which is the reference, is developed from fifty observations (Sunday census) generated by the simulation model. The system box plot is developed from the 24 observations (Sunday census) collected on the hospital unit. In comparing the two box plots, it appears that the model has more variability in its Sunday census than the hospital unit. Regarding the mean census, it is this author's opinion that this pair of box plots shows insufficient evidence to judge that the model's mean census is not within four beds of the hospital's mean census; i.e., there is insufficient evidence to reject the model as being invalid. (In performing operational validity on this model, several different comparisons were made. This is an example of one of them.)

We note that box plots only require identically distributed data. Box plots are extremely effective in communication and are thus effective for conveying information on model validation and for helping with model acceptability (Sargent 1996). For example, a pair of box plots for each day of the week for the hospital unit discussed above could be put into a single graph to be used for communication to users regarding the validity of this simulation model. The use of box plots for operational validity may require more system data than the use of histograms, which were discussed above.

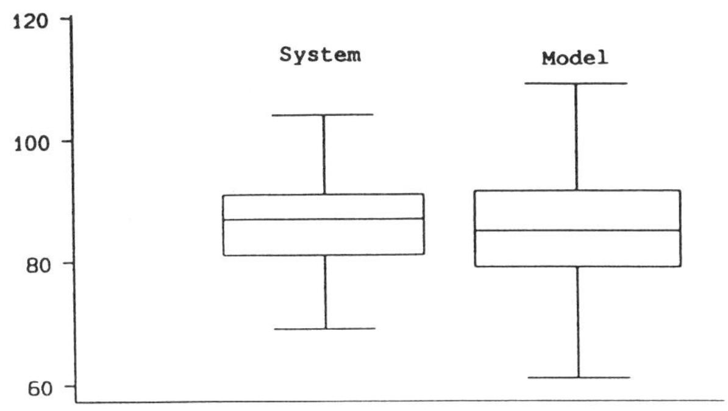

Figure 7: Box Plots of Sunday Census

3.3 Behavior Graphs as References

In performing operational validity, comparisons of different behavior relationships occuring in the system should be made to those occuring in a model. We suggest the use of behavior graphs as one approach to doing this. It is often difficult to use objective validation methods such as statistical tests because the system/model behavior may be nonstationary, may operate over a large portion of the application domain, and data may be correlated. Behavior graphs avoid the use of statistical tests by using subjective analysis and model data as references.

Behavior graphs can be used to show different types of relationships as discussed in Subsection 2.3. Different types of relationships require different amounts of system data. If operational validity is being performed on a deterministic model of a deterministic system, deterministic relationships are used and these generally require only a few observations. In performing operational validity on a stochastic model of a stochastic system, a large number of system observations are often needed. For example, stochastic simulation models of computer and communication systems generally have model and system relationships developed from a large number of observations from the model and from the system.

To illustrate behavior graphs, we consider a simulation model of an interactive computer system in Anderson and Sargent (1974) where behavior graphs were used to validate the model. Three behavior graphs they used are presented in Figures 8, 9 and 10. The relationship between the mean and standard deviation of reaction time is shown in Figure 8. One can readily observed that the same linear relationship occurs in both the model and the system. Figure 9 contain relationships for both the maximum observed value and the average value of reaction time versus the total number of disk accesses. Each data point

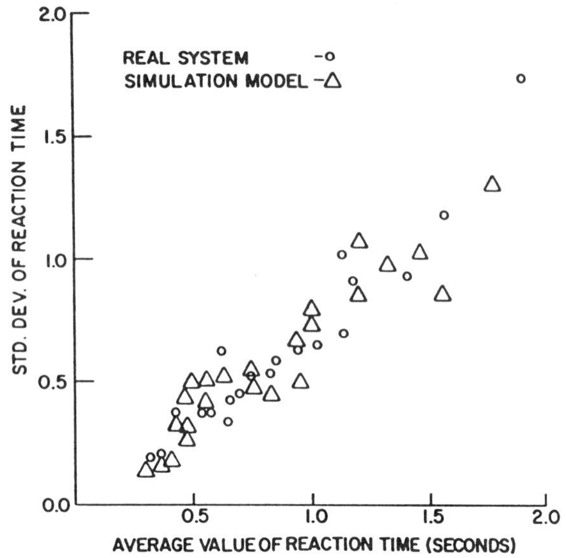

Figure 8: Mean vs. Standard Deviation

is from (or represents) five minutes of computer system time. We observe that these model and system relationships are similar with the exception that the system has more variability than the model. Figure 10 contains the relationship of average response time versus average background queue length. One can readily observe that these model and system relationships are similar except for two system data points, which is important to ask why". (For details on these graphs and the validation of this simulation model, see Anderson (1974) and Anderson and Sargent (1974).)

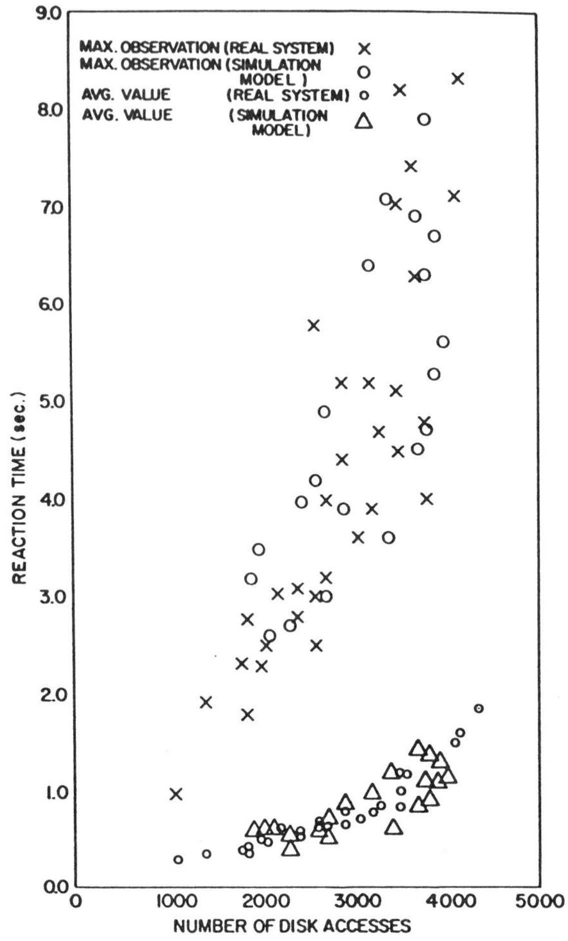

Figure 9: Reaction Time vs. Disk Accesses

4 SUMMARY

Three different types of graphical displays were presented that have minimal assumptions required of the data. Methods for operational validity that use these graphical displays were described. An important feature of these methods is that model data is used for the reference distribution (or reference) instead of a theoretical (statistical) distribution for the system data to be compared against. The graphical methods presented should provide significant help in performing operational validity since the use of graphs is the most used approach in performing operational validity (Sargent 1996).

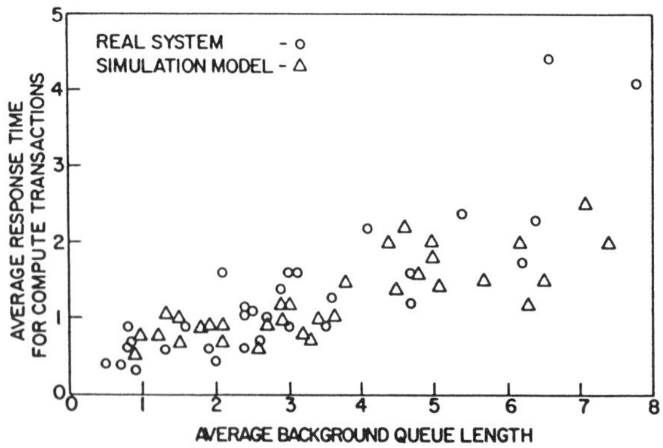

Figure 10: Response Time vs. Queue Length

ACKNOWLEDGMENT

This author acknowledges Julie Lowery for providing the figures from the hospital simulation study used in this paper.

REFERENCES

Anderson, H. (1974). ***An empirical investigation into foreground-background scheduling for an interactive computer system.*** Ph. D. thesis, Syracuse University.

Anderson, H. and R. Sargent (1974). An investigation into scheduling for an interactive computer system. *IBM journal of research and development 18*(2), 125–137.

Balci, O. and R. Sargent (1981). A methodology for cost-risk analysis in the statistical validation of simulation models. *Comm. of the ACM 24*(4), 62–71.

Box, G., W. Hunter, and J. Hunter (1978). *Statistics for experimenters.* John Wiley and Sons.

Johnson, R. (1994). *Miller and Freund's probability and statistics for engineers* (5th ed.). Prentice Hall.

Lowery, J. (1996). Design of hospital admissions scheduling system using simulation. In *Proceedings of the 1996 Winter Simulation Conference*, J.M. Charnes, D.J. Morrice, D.T. Brunner, and J.J. Swain (eds.).

Sargent, R. (1994). Verification and validation of simulation models. In *Proceedings of the 1994 Winter Simulation Conference*, J.D. Tew, S. Manivannan, D.A. Sadowski, and A.F. Seila (eds.), Lake Buena Vista, FL, pp. 77–87.

Sargent, R. (1996). Verifying and validating simulation models. In *Proceedings of the 1996 Winter Simulation Conference*, J.M. Charnes, D.J. Morrice, D.T. Brunner, and J.J. Swain (eds.).

Walpole, R. and R. Myers (1993). *Probability and statistics for engineers and scientists* (5th ed.). Macmillan Publishing Company.

AUTHOR BIOGRAPHY

ROBERT G. SARGENT is a Professor at Syracuse University. He received his education at the University of Michigan and has published widely. Dr. Sargent has served his profession in numerous ways and has been awarded the TIMS College on Simulation Distinguished Service Award for long-standing exceptional service to the simulation community. His research interests include the methodology areas of modeling and discrete event simulation, model validation, and performance evaluation. Professor Sargent is listed in *Who's Who in America*.

Proceedings of the 1996 Winter Simulation Conference
ed. J. M. Charnes, D. J. Morrice, D. T. Brunner, and J. J. Swain

VALIDATION OF TRACE-DRIVEN SIMULATION MODELS: REGRESSION ANALYSIS REVISITED

Jack P. C. Kleijnen
Bert Bettonvil
Willem Van Groenendahl

Department of Information Systems and Auditing (BIKA)/Center for Economic Research (CentER)
School of Management and Economics (FEW)
Tilburg University (KUB)
5000 LE Tilburg, NETHERLANDS

ABSTRACT

For the validation of trace-driven simulation models this paper recommends a simple statistical test that uses elementary regression analysis in a novel way. This test concerns a (joint) null-hypothesis: the outputs of the simulated and the real systems have the same means and the same variances. Technically, the differences between simulated and real outputs are regressed on their sums, and the resulting slope and intercept are tested to be zero. This paper further proves that it is wrong to use a naive test that regresses the simulation outputs on the real outcomes, and hypothesizes that the resulting regression line gives a 45° line through the origin. The new and the old tests are investigated in Monte Carlo experiments with inventory systems. The conclusion is that the new test has the correct type I error probability, whereas the old test (falsely) rejects a valid simulation model substantially more often than the nominal alpha level. The power of the new test increases, as the simulation model deviates more from the real system.

1 INTRODUCTION

This paper is the companion paper of Kleijnen, Bettonvil, and Van Groenendaal (1996), which has been accepted (conditionally) by *Management Science.* Both papers concern a novel test for the validation of trace-driven simulations. The *Management Science* paper estimates the statistical performance of that test, using a Monte Carlo study of single-server queueing simulations (namely, M/G/1), whereas this paper illustrates that performance through single-item inventory simulations (see §3 for details). Moreover, because of page restrictions Kleijnen, et al. (1996) covers only parts of the original working paper; this paper includes other parts of that working paper, and adds recent references.

The remainder of this section answers the following questions:
(i) What is meant by validation?
(ii) What has the literature to say about validation?
(iii) What is the contribution of this paper?
(iv) How is this paper organized?
Hasty readers may skip the next two subsections (§1.1 and 1.2).

1.1 Definition of Validation

This paper uses the following definition in the classic textbook by Law and Kelton (1991, p. 299): '*Validation* is concerned with determining whether the conceptual simulation model (as opposed to the computer program) is an accurate representation of the system under study'. To illustrate some validation issues, consider the following practical problem.

The management of a inventory system wants to control the total costs of their system, which consists of stock-carrying costs, ordering costs, and lost-sales costs (no backordering). To solve this problem, the Management Science/Operations Research (MS/OR) specialists build a simulation model that represents this inventory system. Before using that model to advise management, the MS/OR experts wish to validate their model; that is, determine whether the model is an accurate representation of the real inventory system.

Obviously, validation should not aim at a *perfect* model: the perfect model would be the real system itself. So, validation is interpreted here as comparing *data* on the real and the simulated systems.

Those data pertain to *inputs* and *outputs*; for example, customer demand per day and order lead times (which are stochastic inputs) and total inventory costs per day (which measures the output).

Comparing the output data of the real and simulated systems makes more sense if both systems are observed under *similar circumstances*: the analysts should not compare total costs during a period that includes a *long* lead time in the real system with the costs during a simulated period with a *short* lead time: the former period has more lost sales than the latter period has.

Hence, for validation purposes the analysts should feed real-world input data into the model, in *historical* order (assuming such data are available indeed). This is called *trace driven* simulation in computer performance modeling; we shall use this term throughout this paper. Law and Kelton (1991, p. 316) call this the 'correlated inspection approach' (after this validation phase, 'production runs' will follow). After running the simulation program, the analysts obtain simulation output; they compare that output with the historical output of the existing system.

Note: After this trace-driven validation, the analysts should use the historical input data to develop a (sub) model for the input. For example, they may specify a particular type of distribution (say, the Gaussian distribution) for the demand variable, possibly incorporating autocorrelations and time trends. After estimation of the parameters of that distribution, they may apply goodness-of-fit tests to verify whether this distribution gives an adequate approximation of this input. See Kleijnen (1974, pp. 68-69).

Note: Regression analysis of trace-driven simulations must be distinguished from the following situations. Van Groenendaal and Kleijnen (1996, figure 1), for example, make a scatter plot of predictions versus realizations. The two coordinates of a point use the same deterministic input; different points, however, correspond with different inputs. Hence points have different expectations and variances! So it is nonsense to test the hypothesis of equal means and variances respectively (see equation 1). It seems reasonable to compute the coefficient of determination R^2 (not ρ^2 or β_0 and β_1; see equation 2), to quantify the percentage of variation 'explained' by the model. There is no statistical test statistic for R^2; it is a mathematical (not a statistical) measure; see Kleijnen (1987, p. 193). Also see Mitchell (1996).

1.2 Literature on Validation

General discussions on validation of simulation models in MS/OR can be found in all textbooks on simulation, for example, Banks and Carson (1984), Law and Kelton (1991, pp. 298-324), and Pegden, Shannon, and Sadowski (1990, pp. 133-162). A well-known article on validation is Sargent (1991). A new monograph is Knepell and Arangno (1993). Recent survey articles are Balci (1994), including 102 references, and Kleijnen (1995), including 61 references. There are also many publications outside MS/OR, for example, in agriculture (see Mitchell 1996, and Muchow and Bellamy 1991) and in the earth sciences including hydrology, geochemistry, meteorology, and oceanography (Oreskes, Shrader-Frechette, and Belitz 1994). These contemporary publications all agree that it is essential to further develop the theory on validation, because of its great importance in the practice of MS/OR.

Unfortunately, the literature gives neither a standard theory on validation, nor a standard 'box of tools'. The literature does give a plethora of philosophical theories, statistical techniques, and software practices. The emphasis of the present article is on statistical techniques.

It might be argued that statistical techniques are not appropriate in validation. Statistical techniques, however, have the advantage of yielding reproducible, objective, quantitative data about the quality of a given simulation model.

Some authors (for example, Law and Kelton 1991, p. 319) claim that, when using statistical techniques, *hypothesis tests* are inappropriate; instead they advocate confidence intervals. Hypothesis tests, however, are closely related to confidence interval procedures; see Conover (1980), Kleijnen (1974), and also Law and Kelton (1991, p. 320). Moreover, the null-hypothesis on the means of (say) X and Y may be formulated as H_0: $E(Y) = E(X) + \delta$ where δ is not necessarily zero (δ depends on the purpose of the model; also see Mitchell 1996). This paper, however, concentrates on tests with $\delta = 0$. Such tests may easily overlook 'small' differences between the real and the simulated systems (a most powerful test is the t test for either independent or dependent X and Y, provided the assumption of normality holds; distribution-free tests may be surprisingly powerful; see again Conover 1980 and Kleijnen 1974.) However, some lack of power is acceptable, if in practice only 'large' differences are important. Anyhow, testing is only part of the whole validation process (again see the references above).

Unfortunately, experience shows that the correct use of mathematical statistics in MS/OR is less simple than might be expected. It is easy to apply the wrong statistical techniques: there is much statistical software, but that software does not warn against abuse, such as violations of statistical assumptions. On hindsight the correct use of statistics may seem easy. Indeed, Balci (1995) states: 'False beliefs exist about testing ... testing is easy ... no training or prior experience is required' (also see Mitchell 1996). This paper will provide a case in point: the wrong regression test has been used in many simulation publications (see the references in the next subsection, §1.3).

1.3 Contribution by This Paper

This paper is meant to contribute to the practice and the theory of validation (but it gives no panacea). It discusses how to validate trace-driven simulation models, emphasizing the familiar statistical technique of regression analysis, but advocating a novel test (regressing differences of simulated and real responses on sums).

Validation was interpreted above as *comparing* real and simulated outputs. More specifically, the analysts may compare the total costs (stock-carrying plus ordering plus lost-sales costs), averaged over all real and simulated days respectively.

Many years ago, Aigner (1972) already pointed out that it is wrong to expect unit slope and zero intercept, when regressing the simulated on the real outputs. He, however, focussed on econometric simulation models; he did not give the statistical test we shall propose in this paper. Years after Aigner, Harrison (1990) rediscovered that many authors still propose this bad intuitive idea. Harrison, however, discussed farming systems and synthetic models (including autocorrelations), not trace-driven discrete-event simulations; he does not propose the test we shall develop in this paper. Mayer, Stuart, and Swain (1994) challenge Harrison (1990), concluding that the old test is valid for their type of models (with autocorrelations).

Note: Aigner (1972) states that the intuitive idea dates back to Cohen and Cyert (1961). Harrison (1990, p. 184) refers to some more publications that apply this idea. Lysyk (1989) also uses this concept. Recently, the same idea was proposed in Kleijnen (1995, p. 155). So it seems high time to get rid of this concept, and to propose a better analysis. This is exactly the topic of this paper!

The essential assumption of the new test is *normally and independently distributed (n.i.d.)* outputs of the real system and the simulated system respectively. In practical simulations, however, output data may be non-stationary and autocorrelated. Unfortunately, most practitioners are familiar only with elementary statistical procedures that assume identically and independently distributed (i.i.d.) variables. Fortunately, it might be possible to derive i.i.d variables in simulation, so that it is correct to apply elementary statistical theory; for example, Law and Kelton (1991) give many examples of i.i.d. inputs and outputs, in their discussion of validation. Anyhow, in practice, simulationists often use the n.i.d. assumption, as is illustrated by the many applications (of the old test) referenced above. In general, *terminating* simulations (see Kleijnen and Van Groenendaal 1992, pp. 187-190) may give i.i.d. outputs, as is illustrated by the following queueing example.

The real and simulated systems should be observed under similar circumstances (see above); hence waiting times on a busy day in the real system should not be compared with waiting times of a simulated slow day. Those busy days may occur on (say) Saturdays. Suppose the simulation study concentrates on these days, because complaints are then most outspoken. Then there is still variation: some busy Saturdays are busier than others are. Obviously the busiest Saturday (of all Saturdays in the sample) should be compared with the busiest simulated day. These Saturdays may be assumed i.i.d.

Normal distributions may be explained by limit theorems. For example, trace data are summarized by one or a few statistics such as the average and selected quantiles. The queueing examples in Kleijnen et al. (1996) demonstrate that low traffic loads lead to normal output distributions for the average throughput time per day in terminating simulations. High traffic loads, however, give non-normality. This non-normality can be removed through transformations such as the Box-Cox transformation (which includes the logarithmic one); see Hoyle (1973). In practice the analysts can indeed test for non-normality: they can generate a large sample of simulated days.

This paper gives an academic *example* of the new and old tests, namely the validation of inventory simulation models. These models are derived from data provided by a *Monte Carlo laboratory* that uses academic simulation models (details will follow in §3). Surprisingly, the creation and use of such a laboratory seems novel in the research on *validation*.

Based on these experiments, this paper will give the following *conclusions*. (i) The old test (falsely) rejects a valid simulation model substantially more often than the nominal alpha level, whereas the new test has the correct type I error probability. (ii) The power of the new test increases, as the simulation model deviates more from the real system. (iii) Whereas in the queueing simulations in Kleijnen et al. 1996) the output should be transformed logarithmically, in the inventory system the original and the transformed outputs give statistical performances that are very close. This holds for both the novel and the old tests.

1.4 Organization of This Paper

§2 discusses the regression analysis of simulated and real outputs in trace-driven simulations. It proves that the old test is wrong. As an alternative this section proposes to test the hypothesis that means and variances of real and simulated outputs are equal. That hypothesis is tested through a novel regression test. §3 discusses a laboratory for studying various validation tests; this laboratory uses inventory simulation. §4 discusses future research. §5 gives conclusions.

2 REGRESSION ANALYSIS

This section summarizes Kleijnen et al. (1996). Let Y_i and X_i denote simulated and real outputs respectively in observation i, with $i = 1, ..., n$; capital letters denote random variables. *Trace-driven* simulation means that X_i and Y_i are dependent; it is realistic to suppose that the linear correlation coefficient is *positive*: $0 < \rho_{xy} \leq 1$. Assume the n pairs (X_i, Y_i) are i.i.d. Finally, assume these pairs have a bivariate *normal* distribution. Denote the means by μ_x and μ_y, and the variances by σ_x^2 and σ_y^2.

We propose the following *stringent validation* requirement (assuming positive correlation between real and simulated responses): a simulation model is valid if the real and the simulated systems have *identical means* (say) μ, and *identical variances* (say) σ^2:

$$H_0\text{: } \mu_x = \mu_y = \mu \text{ and } \sigma_x^2 = \sigma_y^2 = \sigma^2 . \quad (1)$$

Because of the well-known relationships

$$\beta_1 = \rho_{xy}\sigma_y/\sigma_x;\ \beta_0 = \mu_y - \beta_1\mu_x \quad (2)$$

Equation (1) is equivalent to $\beta_1 = \rho_{xy}$ (> 0) and $\beta_0 = \mu(1 - \rho_{xy}) < \mu$.

An *ideal, utopian* simulation model has $X_i = Y_i$ or *perfect fit*: $\rho_{xy} = 1$. Hence, the *old* test hypothesizes that fitting the regression model $y = \beta_0 + \beta_1 x$ gives $\beta_0 = 0$ and $\beta_1 = 1$. However, if and only if $\rho_{xy} = 1$, Equations (1) and (2) together give $\beta_0 = 0$ and $\beta_1 = 1$. So in practice *the old test is erroneous*; the empirical data in §3 show that this error is serious indeed.

Note: In the previous century Galton discovered that $\rho_{xy} < 1$ causes what he called 'regression to mediocrity' in his study on parents' and children's heights; see Larsen and Marx (1986, p. 447).

Suppose the real and the simulated means are positive; in practice this condition holds for inventory costs and waiting times. This gives $0 < \beta_1 < 1$ and $0 < \beta_0 < \mu$.

Note: An application of the old test is provided by Lysyk (1989). He indeed finds an estimated slope significantly smaller than unity, and a significantly positive intercept. Since he expects a unit slope and a zero intercept, he tries to explain this phenomenon away. Another recent example is Kozempel, Tomasula, and Craig (1995, p. 231).

The *novel* test of the joint hypothesis in Equation (1) accounts for *dependence* between X and Y, as follows. Compute the n i.i.d. differences (say) $D_i = X_i - Y_i$ and sums $Q_i = X_i + Y_i$. Regress D on Q:

$$E(D|\ Q = q) = \gamma_0 + \gamma_1 q. \quad (3)$$

It is easy to prove that a common variance of the correlated normal variables X and Y implies zero correlation between their differences and sums, D and Q (this result is due to Pitman and Morgan back in 1939; the standard F test for equality of two variances does not apply; see Kleijnen 1987, p. 99). This zero correlation implies that in Equation (3) $\gamma_1 = 0$; common means of X and Y imply $E(D) = 0$ or in Equation (3) $\gamma_0 = 0$ (see equation 2). Hence the hypothesis in Equation (1) gives

$$H_0\text{: } \gamma_0 = 0 \text{ and } \gamma_1 = 0. \quad (4)$$

The analysts should *simultaneously* test the joint hypothesis in Equation (4), with an experimentwise error rate α_E not exceeding the prespecified value α. This joint test can use an F statistic; for the general formula see any textbook on regression analysis or standard regression software; for the specific formula see Kleijnen et al. (1996). (A conservative alternative to this F test is provided by the t test for $\gamma_0 = 0$ and $\gamma_1 = 0$ respectively, combined with Bonferroni's inequality.) There is an analogous F statistic for the hypothesis $\beta_0 = 0$ and $\beta_1 = 1$ in the old test.

Note: The joint hypothesis in Equation (1) may be rejected because the first sub-hypothesis ($\mu_x = \mu_y$) or the second sub-hypothesis ($\sigma_x^2 = \sigma_y^2$) is rejected. Hence, a less stringent validation requirement is that the real and simulated means are equal, but their variances may differ (the variances are then treated as nuisance parameters; the Taguchi approach, however, does consider the variance to be an important performance measure). The hypothesis of equal means can be tested by the well-known paired t test or a distribution-free test; see Conover (1980), Kleijnen (1987) and Mayer et al. (1994). This variance heterogeneity may give a slope b_1 that is lower or higher than one, even if $0 < \rho < 1$; see Equation (2). Yet $0 < \beta_I < 1$ still holds if (but not if and only if) $\sigma_x > \sigma_y$ (this condition means that the simulation reduces the variability, possibly because it does not account for idiosyncrasies in the real system). Common means μ implies for the intercept $\beta_0 = \mu - \beta_1\mu = \mu(1 - \beta_1)$ (see equation 2). So a simulation model with $\mu_x = \mu_y$ and $\sigma_x \geq \sigma_y$ gives simulated responses that -when regressed on real responses- result in a slope less than unity and in a positive intercept (smaller than the average simulation response).

3 AN INVENTORY LABORATORY

To illustrate the validation issues discussed in the preceding sections, it might seem illuminating to apply the old and the new tests to (say) an inventory system in practice. Suppose historical data on inputs (demands or lead times) were collected, and used to drive the simulation model, followed by the statistical tests of the preceding section. Suppose further that the simulation model were not rejected. What lesson would have been learned from such a case study? Maybe this result would only mean that the tests have not enough power. In other words, more might be learned initially from applying the statistical validation tests to a number of examples with *known* properties, so that it is possible to conclude whether rejecting a simulation model is correct or not! So instead of studying a real system, we construct a Monte Carlo laboratory that represents the following single-item inventory systems.

Demand per day is n.i.d. with mean 500 and standard deviation 50. Initially, lead time is a shifted Poisson with mean 5, that is, to a constant lead time of one we add a Poisson variable with mean four (this mean, however, will be changed below). There are lost sales (no backorders). If physical stock plus receivable orders drop below the reorder point (say) ROP, then an order is placed. ROP is selected such that a prespecified service level is realized. The order size is selected such that the total inventory costs are minimized; the formulas for ROP and EOQ (economic order quantity) are rather complicated; the exact specification, however, is unimportant for this study so we refer to Kleijnen and Van Groenendaal (1992, pp. 95-96). ROP and EOQ are fixed by the cost parameters (stock- carrying costs \$1 per day per unit, lost-sales costs \$100 per unit, and ordering costs \$10,000 per order).

Trace-driven simulation means that the simulation and the real system share some input; we decided to use the same *demand* history. If we used a trace with data on both demand and lead times, then the simulation model would be *perfect* (see §2). So we suppose that the analysts do not know the historical lead times; instead they use a distribution function for these times.

We assume *terminating simulation*: a simulation run ends after 365 days have been simulated. Each run starts with the same initial stock; no orders are on their way. (On hindsight, it would have been more realistic to take the stock at the end of one year as the initial stock for next year; but this example is only a laboratory anyhow.)

The real and simulated outputs are the real and simulated total inventory costs per year (denoted by X and Y). We suppose that the analysts have $n = 10$ years of data available (we also experimented with 25 years of data, but this situation gave the same qualitative conclusions). A higher sample size n increases the power of the validation tests. We use three classic values for the *type I error rate* of the validation tests: α is 0.01, 0.05, and 0.10. Obviously, a higher α increases the power of the validation tests. To reduce information overload and to save space, we shall present results for a single α value, namely 0.10.

Note: To decrease both the type I and the type II error probabilities, the analysts might increase the sample size n (also see Balci 1994). In practice, however, the number of observations on the real system is usually fixed (and small).

We take 500 macro-replications to estimate the performance of the tests; by definition, each macro-replication either rejects or accepts a specific simulation model (resulting in a binomial variable).

Some of our experiments give estimates of the *type I* error of the validation tests: in these experiments we have the analysts use the correct lead time distribution with the correct parameter (so the mean is 5). Yet their simulation is not perfect, since they use *pseudorandom number streams* that differ from the steams used by the 'real' system when sampling lead times (not demands). Both streams use Turbo Pascal's standard generator (multiplier 134775813, additive constant 1, multiplier 2^{32}). Hence the real and simulated output variables X and Y have the same distribution so the hypothesis in Equation (1) holds. Yet the realized outputs x and y differ, so the realized correlation r_{xy} is smaller than one. The probability of a validation test rejecting this (valid) simulation model should be α.

Our computer programming ensures that lead times and demands use *non-overlapping* pseudorandom number streams (the demand history, once sampled, is saved). We select the seed through the computer clock.

To study the *type II* error of the validation tests,

we create a *gap* between on one hand the 'reality' of the laboratory and on the other hand the simulation model of the analysts. This gap implies that the means and/or variances of the real and simulated outputs X and Y are different. There are infinitely many ways to create such gaps; we select a few ways, as follows.

We have the analysts use the *wrong mean lead time*. Rather arbitrarily, we have them use a mean lead time ranging between three and seven. We increase means with *steps* of size 0.2.

Both the old and the new validation tests operate on the same data (X_i, Y_i); this improves the comparison of the two tests.

This design turns out to make the laboratory give clear results: see Figure 1. This figure shows that the old test (falsely) rejects a valid simulation model substantially more often than the nominal α value, whereas the new test has the correct type I error probability.

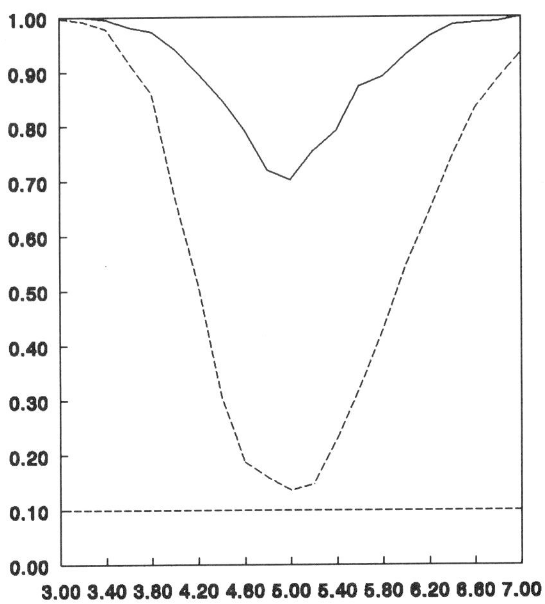

Figure 1: Estimated Probability of Rejecting the Simulated System, as a Function of the Deviation between the Means of Simulated and Real Lead Times, for the New and Old Validation Tests with $\alpha = 0.10$

Note: The power of any statistical test can be maximized over the whole domain of the parameters being tested, by simply always rejecting the null-hypothesis (that is, the simulation model is always rejected). Obviously, such a procedure is inferior. Therefore the first condition for any test is that its type I error probability is acceptable.

The figure further shows that as the analysts' error increases, the power of the test increases too. This (good) behavior is found for all type I error rates α and sample sizes n studied (but not displayed in the figure).

The figure is *asymmetric*: a mean simulated lead time (say) one unit too high has a very different effect on the total inventory costs than has a mean lead time one unit too low (lost sales against stock carrying). We also studied mean simulated lead times that are wrong on a logarithmic scale; they gave a more symmetric figure (not shown).

Notice that the analysts make *no specification errors*, when specifying the distribution type of the demands. Kleijnen et al. (1996) do study specification errors in their queueing simulations.

The analysts might apply a *normalizing transformation* to the outputs, such as the Box-Cox transformation. Such a transformation may make the new test better realize the prespecified type I error probability. In this inventory laboratory, however, the logarithmic transformation hardly affects the estimated performance of the two tests: the output is a sum, so apparently some limit theorem applies.

Note: The old validation test showed a certain 'perverse' behavior in the queueing simulations reported in Kleijnen et al. (1996); also see Harrison (1990, p. 187). In the inventory simulations, however, this behavior is not found.

4 FUTURE RESEARCH

Topics that require more research are:

(i) The novel test assumes *n.i.d.* observations on the real and simulated outputs (and so does the old test). How can this assumption be satisfied in simulations with *autocorrelations* and *time trends*? Autocorrelations might be removed through batching and similar approaches, which are popular in simulation (see Kleijnen and Van Groenendaal 1992, and also Mayer et al. 1994, pp. 99-100). Time trends might be removed through techniques used in econometrics; also see Barlas (1989, p. 68), who gives a system dynamics example that seems to allow subjective graphical analysis only, since the time series (simulated and real) show 'highly transient, non-stationary behavior'. Also see the use of 'differencing' in Box and Jenkins (1976, pp. 378-379). In other words, the academic examples in this paper and its companion paper (Kleijnen et al. 1996) need to be supplemented with practical applications.

(ii) A specific type of non-normality, namely *binary output variables* may be important in practice. An example is the probability of buffer overflow.

(iii) The proposed statistical test of trace-driven simulations is only part of the total validation and verification (V & V) process. This test needs to be incorporated in this total process.

5. CONCLUSIONS

This article focussed on *statistical hypothesis tests* for the validation of *trace-driven* simulations. It proved that it is *wrong* to expect unit slope and zero intercept when regressing simulated on real outputs. Therefore this paper applied a *novel* test: regress the differences of simulated and real responses on their sums.

Both tests were evaluated and illustrated by applying them in a *Monte Carlo laboratory* with academic inventory systems.

These experiments gave the following *conclusions*. The *old* test rejects a *valid* simulation model substantially more often than the α value indicates. The novel test does not reject a valid simulation model too often (its type I error probability equals the nominal value α). The power of the new test increases, as the simulation model deviates more from the real system.

REFERENCES

Aigner, D.J. (1972), 'A note on verification of computer simulation models', *Management Science*, 18, 11, 615-619.

Balci, O. (1995), 'Principles of simulation model validation, verification, and testing', *International Journal in Computer Simulation.*

Balci, O. (1994), 'Validation, verification, and testing techniques throughout the life cycle of a simulation study', in: *Annals of Operations Research*, 1994.

Banks, J., and Carson J.S. (1984), *Discrete-event System Simulation,* Prentice-Hall, Englewood Cliffs, New Jersey.

Barlas, Y. (1989), Multiple tests for validation of system dynamics type of simulation models, *European Journal of Operational Research*, 42, 1, 59-87.

Box, G.E.P. and Jenkins, G.M. (1976), *Time series analysis; revised edition*, Holden-Day, San Francisco.

Cohen, K.J. and Cyert, R.M. (1961), 'Computer models in dynamic economics', *The Quarterly Journal of Economics*, 75, 112-127.

Conover (1980)

Harrison, S.R. (1990), 'Regression of a model on real-system output: an invalid test of model validity', *Agricultural Systems*, 34, 183-190.

Hoyle, M.H. (1973), 'Transformations; an introduction and a bibliography. *International Statistical Review*, 14, 2, 203-223.

Kleijnen, J.P.C. (1995), 'Verification and validation of simulation models', *European Journal of Operational Research*, 82, 1, 145-162.

Kleijnen, J.P.C. (1987), *Statistical Tools for Simulation Practitioners,* Dekker, New York.

Kleijnen (1974), *Statistical techniques in simulation, volume I,* Marcel Dekker Inc., New York.

Kleijnen, J.P.C., Bettonvil, B., and Van Groenendaal W. (1996), Validation of simulation models: a novel regression test. *Management Science* (accepted).

Kleijnen, J.P.C., and Van Groenendaal W. (1992), *Simulation: a Statistical Perspective*, Wiley, Chichester, United Kingdom.

Knepell, P.L. and Arangno D.C. (1993), *Simulation validation: a confidence assessment methodology,* IEEE Computer Society Press, Los Alamitos (California).

Kozempel, M.F., Tomasula, P. and Craig, J.C. (1995), 'The development of the ERRC food process simulator', *Simulation; Practice and Theory*, 2, 4-5, 221-236.

Larsen, R.J. and Marx M.L. (1986), *An Introduction to Mathematical Statistics and its Applications.* Prentice-Hall, Englewood Cliffs, New Jersey.

Law A.M., and Kelton W.D. (1991), *Simulation Modeling and Analysis; Second Edition*, McGraw-Hill, New York.

Lysyk, T.J. (1989), 'Stochastic model of Eastern spruce budworm (lepidoptera: tortricidae) phenology on white spruce and balsam fir', *Journal of Economic Entomology*, 82, 4, 1161-1168.

Mayer, D.G., Stuart, M.A., and Swain, A.J. (1994), Regression of real-world data on model output: an appropriate overall test of validity. *Agricultural Systems*, 45, 93-104.

Mitchell, P.L. (1996), Misuse of regression for empirical validation of models. *Agricultural Systems* (accepted).

Muchow, R.C. and Bellamy, J.A., editors (1991), *Climatic Risk in Crop Production*, CAB International, Wallingford.

Oreskes, N., Shrader-Frechette, K. and Belitz, K. (1994), Verification, validation, and confirmation of numerical models in the earth sciences, *Science*, 263, 641-646.

Parks, S.K. and Miller K.W. (1988), Random number generators: good ones are hard to find. *Communications of the ACM*, 31, 10, 1192-1201.

Pegden C.P., Shannon R.E. and Sadowski R.P. (1990), *Introduction to Simulation using SIMAN,*

McGraw-Hill, New York.

Sargent, R.G. (1991), Simulation model verification and validation, *Proceedings of the 1991 Winter Simulation Conference*, 37-47.

Van Groenendaal, W. and Kleijnen J.P.C. (1996) Regression metamodels and design of experiments, *Proceedings of the 1996 Winter Simulation Conference* (edited by J.M. Charnes, D.J. Morrice, D.T. Brunner, and J.J. Swain).

AUTHOR BIOGRAPHY

JACK P.C. KLEIJNEN is Professor of Simulation and Information Systems in the Department of Information Systems and Auditing; he is also associated with the Center for Economic Research (CentER). Both the Department and the Center are within the School of Management and Economics of Tilburg University (KUB) in Tilburg, Netherlands. He received his Ph.D. in Management Science at Tilburg University. His research interests are in simulation, mathematical statistics, information systems, and logistics. He published six books and more than 130 articles; lectured at numerous conferences throughout Europe, the USA, Turkey, and Israel; he was a consultant for various organizations; and is a member of several editorial boards. He spent some years in the USA, at different universities and companies. He was awarded a number of fellowships, both nationally and internationally.

BERT BETTONVIL is Assistant Professor of Simulation in the Department of Information Systems and Auditing of the School of Management and Economics of Tilburg University (KUB). He received his Ph.D. at this university; the topic was sequential bifurcation. His main research interest is simulation, especially its programming and statistical aspects.

WILLEM J.H. VAN GROENENDAAL is Assistant Professor in the Department of Information Systems and Auditing, and Fellow at the Center for Economic Research (CentER). Both the Department and the Center are within the School of Management and Economics of Tilburg University in Tilburg, Netherlands. He is also a consultant for several international development agencies. He has a Ph.D. from Tilburg University. His research interests are in simulation, decision support, and investment analysis.

Proceedings of the 1996 Winter Simulation Conference
ed. J. M. Charnes, D. J. Morrice, D. T. Brunner, and J. J. Swain

ON THE ANOMALY OF RAN1() IN MONTE CARLO PRICING OF FINANCIAL DERIVATIVES

Akira Tajima
Syoiti Ninomiya
Shu Tezuka

Tokyo Research Laboratory
IBM Research
1623-14, Shimotsuruma, Yamato-shi, Kanagawa-ken, 242, JAPAN

ABSTRACT

Recently, Paskov reported that, the use of a certain pseudo-random number generator, ran1(), which is given in Numerical Recipes in C First Edition, makes Monte Carlo simulations for pricing financial derivatives converge to wrong values.

In this paper, we trace Paskov's experiment, investigate the characteristics and the generation algorithm of the pseudo-random number generator in question, and explain why the wrong convergences occur. We also present a method for avoiding such wrong convergences. A variance reduction procedure is applied together with the method to obtain more precise value, and the effectiveness is examined.

1 INTRODUCTION

Monte Carlo methods are effective for pricing some financial derivatives, especially when the price cannot be calculated analytically because it depends on the historical movement of the underlying variable or on multiple underlying variables (Hull 1993). For example, the payoffs of look-back options depend on the maximum or minimum prices of underlying assets during the lives of the options. The prices of such options are therefore difficult to calculate analytically, but easy to estimate by Monte Carlo simulation. On the other hand, there are two major drawbacks to using Monte Carlo methods in finance. One is that they cannot be smoothly applied to pricing American options, which can be exercised early, and the simulation therefore requires some optimal strategy for the exercise. This problem has been studied by Tilley (Tilley 1993) and Broadie and Glasserman (Broadie and Glasserman 1995). The other is the amount of computation time. Sometimes hundreds of thousands of simulation paths are necessary to obtain the required accuracy. Further, for hedging purposes, the sensitivities of the result to small changes in the initial values of the underlying variables are important in the financial domain.

In recent years, low-discrepancy sequences have been used in this domain, to improve the convergence speed. Paskov (Paskov 1994) investigated the effectiveness of low-discrepancy sequences in pricing a Collateralized Mortgage Obligation (CMO). He also reported that the use of a certain pseudo-random number generator, – ran1(), which is given in Numerical Recipes in C, First Edition (Press et al. 1988) –, makes Monte Carlo simulations converge to wrong values. Morokoff pointed out that in high dimensions, low-discrepancy sequences are no more uniform than random sequences (Morokoff and Caflisch 1994). Consequently, the Monte Carlo method with "random" sequences is still effective. Further, it is important not only to regard ran1() as a random number generator with a poor performance, but also to investigate why the simulation converged to wrong values in Paskov's experiment.

The objectives of this paper are to identify why the wrong values are obtained with ran1(), and to present a method for avoiding such wrong convergences. In section 2, we trace Paskov's experiment by using a pass-through Mortgage Backed Security (MBS), which is simpler than CMO in the sense that it has no tranches. In section 3, we investigate the cycle and the generation algorithm of ran1() and show why the Monte Carlo simulations converge to wrong values. We also present a method for avoiding the problem, and examine the effectiveness of a variance reduction procedure for obtaining a more precise result. Finally, we summarize this paper in section 4.

2 PRICING MORTGAGE-BACKED SECURITIES (MBS)

In this section, we trace Paskov's experiment (Paskov 1994), in which Monte Carlo simulations for estimating the present value of Collateralized Mortgage Obligation (CMO) converged to wrong values when the random number generator ran1() (Press et al. 1988) was used.

2.1 MBS Specification

Paskov does not give the specification of the tranches of CMO, so we simplify the problem into a pass-through MBS which with no tranches. The underlying pool of mortgages has a maturity of thirty years, and the a cash flow occurs every month; in other words, there are 360 cash flows, and the dimension of the problem is also 360. Let C be the monthly payment on the underlying pool of mortgages.

For $1 \leq k \leq 360$,

r_k – the appropriate interest rate;
w_k – the percentage pre-paying;
$a_{360-k+1}$ – the remaining annuity;
u_k – the discount factor.

Then, a_k is given by

$$a_k = \sum_{i=0}^{n}(1 + \frac{1}{1+r_0}),$$

where r_0 is the current monthly interest rate.

C and a_k are assumed to be constants. r_k, w_k, and u_k are stochastic variables, defined below. The interest rate follows the lognormal model, and the variable r_k is computed as

$$r_k = K_0 e^{\sigma\xi_k} r_{k-1} = K_0^k e^{\sigma\sum_{i=1}^{k}\xi_i} r_0,$$

where $\xi_k (1 \leq k \leq 360)$ is an independent normal random variable $N(0,1)$. Then, the discount factor u_k is given by

$$u_k = \prod_{i=0}^{k-1}(\frac{1}{1+r_i}).$$

The prepayment model w_k is a function of r_k, and is computed as

$$w_k = K_1 + K_2 arctan(K_3 r_k + K_4).$$

The cash flow in month k is

$$M_k = C(1-w_1)\cdots(1-w_{k-1})(1-w_k+w_k a_{360-k+1}).$$

Thus, the present value of the security is the sum of the present values of the cash flows in all the months,

$$PV = \sum_{k=1}^{360} M_k u_k(\xi_1, \ldots, \xi_{360}).$$

In this experiment, we set the constants as follows:

$$\begin{array}{ll} K_0 = 1/1.020201 & r_0 = 0.075/12 \\ K_1 = 0.24 & \sigma = 0.02 \\ K_2 = 0.134 & C = 2000 \\ K_3 = -261.17 \times 12 & a_0 = 0 \\ K_4 = 12.72 & w_0 = 0 \end{array}$$

To generate normal random variables $N(0,1)$ from uniform random numbers [0,1), we used the inverse method given in RISK (Moro 1995). Obviously, 360 numbers are necessary for one path of Monte Carlo simulation. When we use pseudo-random generators such as ran1(), we assign the numbers from one generator to 360 months sequentially. On the other hand, in the case with low-discrepancy sequences, we assign the i-th dimension of the sequence to the i-th month ($1 \leq i \leq 360$). Consequently, if we focus on the correlation problem among 360 variables, the serial correlations matters most with pseudo-random generators, whereas the independence among dimensions matters most with low-discrepancy sequences.

2.2 Pricing using Ran1()

Figure 1 shows the result of Monte Carlo simulations. The simulation using Sobol sequence converges to the correct value. But simulations using ran1() converge to three different values, all incorrect. This result is consistent with Paskov's study (Paskov 1994). For comparison, a result using drand48(), a system-supplied random number generator of IBM xlC, is also presented. For details of Sobol sequences, see (Sobol 1967).

3 RAN1() IN NUMERICAL RECIPES IN C, FIRST EDITION

In this section, we investigate the characteristics of ran1() in (Press et al. 1988), and identify why the simulation converged to wrong values.

3.1 The Ran1() Algorithm

The ran1() algorithm is based on three linear congruential generators. Linear congruential generators generate a sequence of integers between 0 and $m-1$ by means of the recurrence relation

$$I_{j+1} = aI_j + c \ (mod\ m).$$

If m, a and c are properly chosen, the period of the sequence will be equal to m.

In ran1(), one generator ix_1 with $m_1 = 259,200$, $a_1 = 7,141$, and $c_1 = 54,773$ is used for the most significant part of the output, another generator ix_2

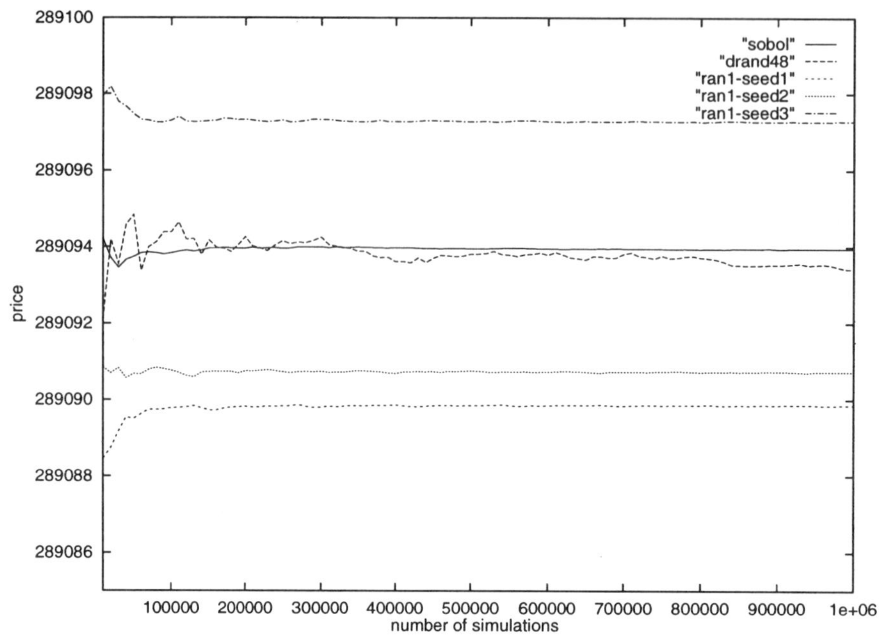

Figure 1: Monte Carlo Simulations using Sobol and ran1() with Three Seeds

with $m_2 = 134,456$, $a_2 = 8,121$, and $c_2 = 28,411$ for the least significant part, and the other generator ix_3 with $m_3 = 243,000$, $a_3 = 4,561$, and $c_3 = 51,349$ for the shuffling, using 97 buckets. The shuffling will prevent serial correlations of the output, which have a negative effect on the result of Monte Carlo method.

The float value $x[0,1)$ is generated by

$$x = (ix_1 + ix_2 \times \frac{1}{m_2}) \times \frac{1}{m_1}. \quad (1)$$

The ran1() algorithm is as follows:

Initialization: Fill the 97 buckets with float values.

Per Call:

1. Select a bucket by using the value generated by ix_3
2. Return the value in the bucket as output
3. Fill the bucket with the value generated by ix_1 and ix_2

3.2 Discrepancy of Points Generated by Ran1()

We measure the discrepancy of the points generated by ran1(). The discrepancy can be viewed as a quantitative measure for the deviation from uniform distribution, and is related to the error of the numerical integration (Niederreiter 1992). For N points $\mathbf{v}_0, \mathbf{v}_1, \ldots, \mathbf{v}_{N-1}$ in $[0,1]^k$, $k \geq 1$, and a subinterval $\mathbf{u} = [0, u_1) \times \cdots \times [0, u_k)$, where $0 < u_h \leq 1$ for $1 \leq h \leq k$, we define the L2-discrepancy as

$$T_N^{(k)} = (\int_{[0,1]^k} (\frac{A(\mathbf{u}; N)}{N} - V(\mathbf{u}))^2 d\mathbf{u})^{1/2},$$

where $A(\mathbf{u}; N)$ is the number of values of $n, 0 \leq n < N$, with $\mathbf{v}_n \in \mathbf{u}$, and $V(\mathbf{u})$ is the volume of $\mathbf{u}$.

Let $\mathbf{x}_n$ be a vector of 360 dimensions whose i-th component is used for the i-th dimension of the n-th path of the simulation. We assign values as

$$\mathbf{x}_n = (a_{360(n-1)+1}, a_{360(n-1)+2}, \ldots, a_{360(n-1)+360}),$$

where a_i is the i-th output of ran1(). We calculate the L2-discrepancy of $\mathbf{x}_n$ (Figure 2). The value never falls below 2^{-230}, which is reached after 4,000 points. The move of the value has a pattern with a cycle of about 10,000 or 11,000 points. Consequently, $\mathbf{x}_n$ has a cycle, and the accuracy of the integration does not improve after one cycle.

3.3 Floyd's Algorithm

We have observed that the vector generated from the output of ran1() seems to have a cycle. In this section, we examine whether ran1() has a short cycle.

To obtain the cycle we apply Floyd's algorithm (Knuth 1981). Suppose that we have a sequence of integers $X_0, X_1, X_2, \ldots$, in the range $0 \leq X_n < m$ formed by a rule $X_{n+1} = f(X_n)$. Then X_n is periodic, in the sense that there exist numbers λ and μ

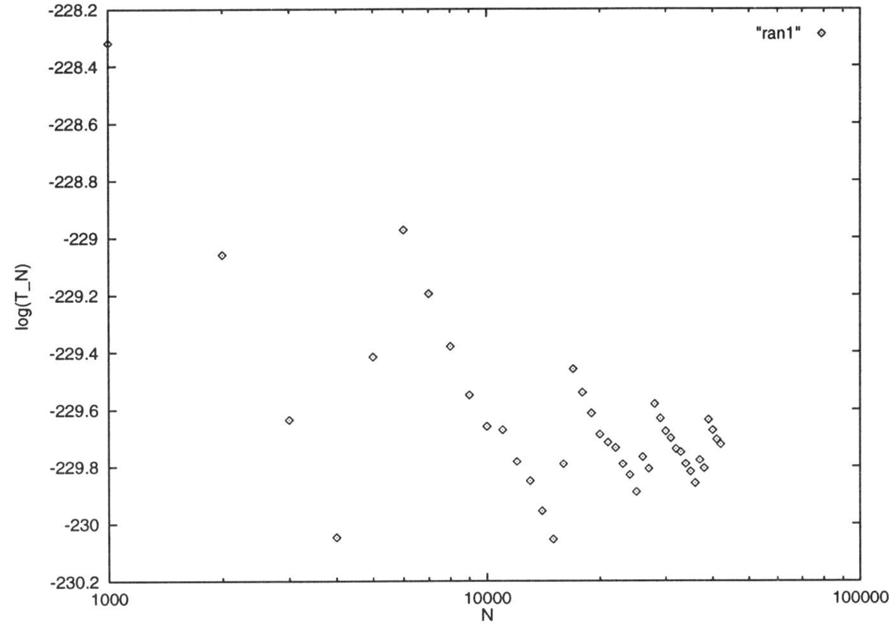

Figure 2: The L2-discrepancy of the Points Generated by ran1()

for which the values $X_0, X_1, \ldots, X_\mu, \ldots, X_{\mu+\lambda-1}$ are distinct, but $X_{n+\lambda} = X_n$ when $n \geq \mu$. That is to say, μ is the length of the transition state and λ is the period of the stationary state. Then there exists a value $n > 0$ such that $X_n = X_{2n}$ and the smallest such value of n lies in the range $\mu \leq n \leq \mu+\lambda$. Once n is found, μ is the smallest i that satisfies $X_i = X_{n+i}$. If none of the values of X_{n+i} for $0 \leq i \leq \mu$ is equal to X_n, $\lambda = n$; otherwise, λ is the smallest such i.

Using the algorithm, we examine the cycle of the sequence generated by ran1(). The algorithm can only be applied to sequences which is formed by a rule $X_{n+1} = f(X_n)$, and therefore satisfies $X_{n+\lambda} = X_n$ for large n. But ran1() does the shuffling using multiple buckets and does not correspond to the condition in the sense that it uses a series of historical values to generate another output, as well as the last output. Here we consider a vector $\mathbf{S}_n$ that consists of all the state variables of ran1(), including ix_1, ix_2, ix_3, and the contents of the buckets, instead of X_n. Then we can say that $\mathbf{S}_{n+1} = f(\mathbf{S}_n)$, and can apply the algorithm.

Unfortunately, the cycle is too large, more than 10^{10}, and we could not obtain it by computer simulations. Anyway, it is very large in comparison with the number necessary for the Monte Carlo simulation, 360×10^6, and does not seem to affect the result of Monte Carlo simulations. We therefore focus on the most significant part of the output of ran1(), that is, we modify ran1() to keep ix_2 equal to zero and apply the same algorithm. This time we obtain the cycle given in Table 1.

Table 1: μ and λ of the Most Significant Part of ran1()

	seed	μ	λ
#1	-386834056	490	3888000
#2	-1947093824	477	3888000
#3	-1505612680	735	3888000
#4	-965150217	444	3888000
#5	-1907586699	527	3888000
#6	-1548523014	650	3888000
#7	-829591941	472	3888000
#8	-1967060894	413	3888000
#9	-1997660300	367	3888000
#10	-158225988	549	3888000
average		512.4	3888000

3.4 The Cycle when Applied to MBS

We continue to examine the cycle of ran1() with the precision $1/m_1$, that is, we ignore the least significant part of the value. From the generation algorithm in Section 3.1, we can see that if all the 97 buckets are visited and the values are all replaced at least once, the effect of the initialization disappears and the generator goes into the stationary state. In the stationary state, the cycle of the sequence C is equal to the least common multiple of m_1 and m_3.

As in Section 3.1, $m_1 = 259,200 = 2^7 \times 3^4 \times 5^2$ and $m_3 = 243,000 = 2^3 \times 3^5 \times 5^3$, and therefore $C = 2^7 \times 3^5 \times 5^3 = 3,888,000$. This coincides with the result in Section 3.3. Here, $C = 3,888,000 = 360 \times 10800$; in other words, the cycle is a multiple of the dimension of the problem. Thus, when applied to the pricing of an MBS with 360 dimensions, there exists a cycle of 10,800 for each dimension. If we consider the vector $\mathbf{x}_n$ in Section 3.2,

$$\mathbf{x}_n = \mathbf{x}_{n+10800}$$

holds for $n > \mu/360$. In other words, a Monte Carlo simulation for an MBS has only 10,800 paths. This coincides with the result obtained in Section 3.2.

We have observed that the number of paths is limited in the stationary state; we therefore estimate the value of μ, the length of the transition state. If we assume that the buckets in ran1() are visited at random instead of through the value of ix_3, the expected number necessary to visit all the buckets, which is equal to the expected μ, can be calculated as the coupon collector's problem (Motwani and Raghavan 1995),

$$E[X] = n\sum_{i=1}^{n}\frac{1}{i} = 97 \times \sum_{i=1}^{97}\frac{1}{i} = 500.23.$$

This theoretical value is consistent with the experimental results given in Table 1. Anyway, one path requires 360 random numbers, and it takes only a few paths to reach the stationary state. Hence the effect of the initial instability is subtle and can be ignored when examining the wrong convergence.

To confirm that the cycle of 10,800 is the cause of the wrong convergence, we compared two Monte Carlo simulations, one a simulation of 1,000,000 paths using ran1(), and the other a simulation of 10,800 paths using the simplified version

$$x = (ix_1 + 0.5) \times \frac{1}{m_1}$$

instead of equation (1). That is to say, we approximate the least significant part by $0.5 \times 1/m_1$.

Figure 3 shows the results with three different seeds. The full simulation converges to the value obtained with the first 10,800 paths which is represented as a horizontal line. The amplitude is caused only by the least significant part.

3.5 Avoiding the Cycle

Here we examine whether we can avoid the incorrect convergence. We have observed that the inaccuracy is caused because the cycle of ran1() is a multiple of the dimension of the problem, 360.

As the cycle of the most significant part of ran1() $C = 3,888,000 = 2^7 \times 3^5 \times 5^3$, Monte Carlo simulation converges to the correct value, if the dimension of the problem is a prime. A simple way to realize this is to skip several random numbers after every path. Namely, for the vector $\mathbf{x}_n$, we assign values as follows:

$$\mathbf{x}_n = (a_{(360+\alpha)(n-1)+1}, a_{(360+\alpha)(n-1)+2}, \ldots, a_{(360+\alpha)(n-1)+360})$$

where α is the number of skipped random numbers. Figure 4 shows the result of Monte Carlo simulations using ran1() with the skip numbers 13, 41, and 143, which give the primes 373, 401, and 503 when added to 360. Compared with drand48(), we can say that the anomaly of ran1() is removed.

Further, to check whether they converge to the correct value precisely, we also apply a variance reduction procedure, the antithetic variable technique (Hull 1993). That is to say, after calculating one path by using a normal random number vector $\mathbf{f}_n$, which is generated from the uniform random number vector $\mathbf{x}_n$ as in Section 2.1, we also calculate another path by using $-\mathbf{f}_n$. Figure 5 shows the result obtained by using this antithetic variable technique. In this figure, the X-axis represents the number of paths, and not the number of random number vectors. That is to say, the number of random number vectors generated is the half of the number of paths, and is equal to 500,000.

From these experiments, we conclude that if we apply some appropriate techniques, the correct value can be calculated with a good convergence speed by using ran1().

4 CONCLUSION

We have investigated why the random number generator ran1() in (Press et al. 1988) caused wrong convergences when used for pricing financial derivatives.

The main reason is that the cycle of ran1() is a multiple of the number of dimensions of the problem, 360. Like other multiples of 12, this is a common number of dimensions in the financial domain, because a year has 12 months, and the same phenomenon will occur if the generator is used for pricing other derivative securities.

We have also presented a method for avoiding wrong convergences. Combined with the antithetic variable technique, ran1() can give the correct value with a good convergence speed.

We would like to examine what kinds of tests used to evaluate random number generators detect this problem with ran1().

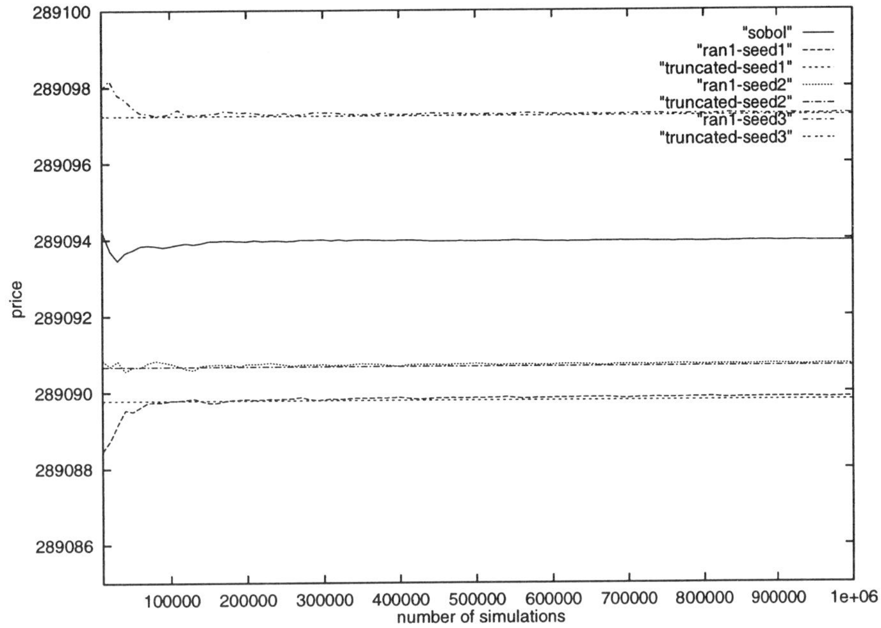

Figure 3: Simulation Almost Converges to the Value of the First 10,800 Trials

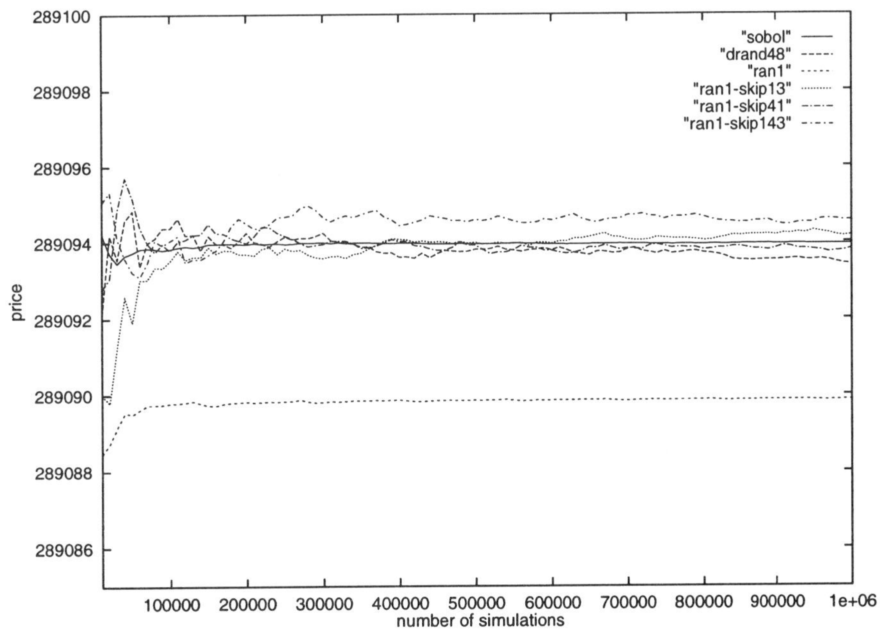

Figure 4: ran1() with Skips after Every 360 Random Numbers

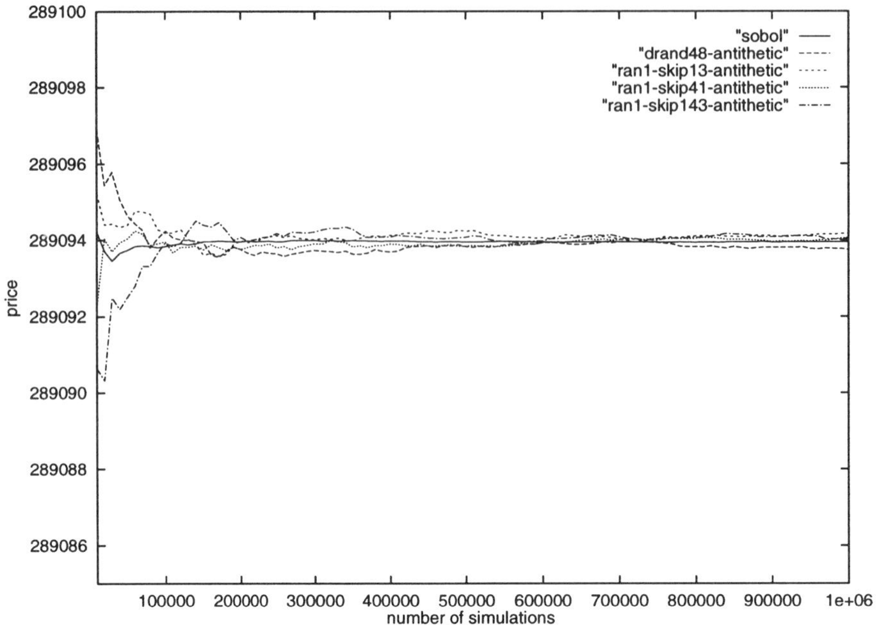

Figure 5: ran1() with Skips and Antithetic Variable Technique

REFERENCES

Broadie, M., and P. Glasserman. 1995. Pricing American-Style Securities Using Simulation. working paper, Columbia University.

Hull, J. C. 1993. *Options, Futures, and Other Derivative Securities, Second Edition.* Prentice Hall International.

Knuth, D. E. 1981. *The Art of Computer Programming, Second Edition, Vol. 2, Seminumerical Algorithms,* Addison-Wesley.

Moro, B. 1995. The Full Monte. *RISK,* Vol. 8, No. 2.

Morokoff, W. J. and R. E. Caflisch. 1994. Quasi-Random Sequences and Their Discrepancies. *Journal on Scientific Computing,* Vol. 15, No. 6.

Motwani, R. and P. Raghavan. 1995. *Randomized Algorithms,* Cambridge University Press.

Niederreiter, H. 1992. *Random Number Generation and Quasi-Monte Carlo Methods,* CBMS-NSF Regional Conference Series in Applied Math., No. 63, SIAM.

Paskov,S. H. 1994. Computing High Dimensional Integrals with Applications to Finance. Technical Report CUCS-023-94, Columbia University.

Press, W. H., S. Teukolsky, W. Vetterling, and B. Flannery. 1988. *Numerical Recipes in C, First Edition,* Cambridge University Press.

Sobol, I. M. 1967. The Distribution of Points in a Cube and the Approximate Evaluation of Integrals. *USSR Computational Mathematics and Mathematical Physics,* 7.

Tilley, J. A. 1993. Valuing American Options in a Path Simulation Model. *Transactions of the Society of Actuaries,* Vol. 45.

AUTHOR BIOGRAPHIES

AKIRA TAJIMA is a researcher in IBM Research Tokyo Research Laboratory. His research interests include low-discrepancy sequences, numerical integration, and their applications in financial engineering.

SYOITI NINOMIYA is a researcher in IBM Research Tokyo Research Laboratory. His research interests include pseudo random number generation, low-discrepancy sequences, and numerical integration.

SHU TEZUKA is manager of Exploratory Technologies and Applications group at IBM Tokyo Research Laboratory. He received a doctoral degree in Engineering from University of Tokyo in 1986. His research interests include random number generation, quasi-Monte Carlo methods, combinatorial optimization, discrete event simulation, and financial engineering. He is the author of the monograph, *Uniform Random Numbers: Theory and Practice*, Kluwer Academic Publishers, 1995. In 1995, he was awarded the *12th Inoue Prize for Science* on his significant contributions to uniform random number generation.

Proceedings of the 1996 Winter Simulation Conference
ed. J. M. Charnes, D. J. Morrice, D. T. Brunner, and J. J. Swain

USING NULL EFFECTS ESTIMATORS IN THE ANALYSIS OF SIMULATION OUTPUT

Lee W. Schruben
School of OR&IE
Cornell University
Ithaca, NY 14853

David Goldsman
School of ISyE
Georgia Institute of Technology
Atlanta, GA 30332

Douglas J. Morrice
MSIS Department
The University of Texas at Austin
Austin, Texas 78712-1175
End to End Simulation Department
Schlumberger Austin Research
Austin, Texas 78720

ABSTRACT

In a simulation experiment there are certain *null effects* whose true values are known to be zero. This paper investigates various estimators of the null effects. We apply these estimators of zero in two ways. First, null effects estimators are used to construct confidence intervals. We derive a null effects version of the batch means confidence interval that has certain desirable properties. Second, null effects estimators are used as control variates in variance reduction schemes. The achieved variance reductions are modest, but we present some interesting applications; for example, the output from one simulation model can be used to compute control variates for the output of another.

1 INTRODUCTION

This paper deals with various ways that simulation output can be used to estimate *null effects*, i.e., quantities whose true values are known to be zero. We subsequently apply these estimators of zero in two ways.

First, we will use the null effects to develop a confidence interval estimator (c.i.e.) for the mean of a stationary simulation output process. This c.i.e. is similar to the familiar batch means (BM) c.i.e., but it is obtained via a different route. (Indeed, this paper is not intended to change the way in which simulation experiments are run and analyzed but rather to present an alternative way of thinking about simulation output analysis!) The null effects interval estimation procedure does have desirable properties, one of which should facilitate in the selection of a batch size.

Next we show how null effects estimators can be used as control variates (c.v.'s) in variance reduction schemes. In one scheme, null effects estimators serve as internal c.v.'s and yield estimators of the mean that have smaller variance than the sample mean. A second application uses the output from a "control" simulation to compute a c.v. for an estimator from another simulation. The true mean of the control system need not be known. A third variance reduction scheme indirectly uses null effects c.v.'s to improve the estimator of the process mean.

2 BACKGROUND AND NOTATION

This section gives some preliminary definitions and examples. We shall regard the stationary stochastic sequence $Y_1, \ldots, Y_n$ generated by running a discrete-event simulation program as a vector $\boldsymbol{Y} \equiv (Y_1, \ldots, Y_n)'$ in R^n. This paper demonstrates the potential advantages of viewing the simulation output as consisting of different components of orthogonal basic vectors. In order to facilitate the study of various estimators of the variability of the simulation output, $\boldsymbol{Y}$ will be projected onto different coordinate systems that are determined by sets of basic vectors.

Sometimes a basis may be intentionally chosen to be *incomplete*; i.e., the basis only spans a proper subspace R^b of R^n ($b < n$). We denote a basis for R^b as b basic vectors, $\boldsymbol{\omega}_i \equiv (\omega_{i,1}, \ldots, \omega_{i,n})'$, $i = 1, \ldots, b$, each in R^n. The i-th component of $\boldsymbol{Y}$ in this basis is the orthogonal projection of $\boldsymbol{Y}$ and is given by

$$Y_i^\omega \equiv \frac{\boldsymbol{\omega}_i'\boldsymbol{Y}}{\boldsymbol{\omega}_i'\boldsymbol{\omega}_i} = \frac{\sum_{j=1}^n \omega_{i,j} Y_j}{\sum_{j=1}^n \omega_{i,j}^2}.$$

If $E[Y_i^\omega] = 0$, we call Y_i^ω a *null effect.* We will see that choosing a basis with components that are null effects (referred to here as a *null effects basis*) has certain advantages.

2.1 Example 1: Batch Means

BM is a popular method of estimating c.i.e.'s for $\mu \equiv E[Y_j]$. The BM basis $\boldsymbol{\omega}_i$, $i = 1, \ldots, b$, is not a null effects basis; its components are

$$\omega_{i,j} = \begin{cases} 1 & \text{if } (i-1)m+1 \le j \le im \\ 0 & \text{otherwise,} \end{cases}$$

with $m = n/b$ (assume b divides n). This basis is incomplete if $b < n$. Further, the subspace spanned by the basis changes radically when b changes (with fixed n); see Section 3.1.

2.2 Example 2: Harmonic Analysis

In order to identify specific periodic oscillations in the output vector, $\boldsymbol{Y}$ can be projected onto the Fourier basis (Bloomfield, 1976). The basic vectors are (assuming n is even):

$$\omega_{i,j} = \begin{cases} (2/n)\cos(2\pi ij/n), & i = 1,\ldots,\frac{n}{2} \\ (2/n)\sin(2\pi(i-\frac{n}{2})j/n), & i = \frac{n}{2}+1,\ldots,n. \end{cases}$$

Since $\boldsymbol{\omega}_i'\boldsymbol{\omega}_k = 0$ for $i \ne k$ and $\sum_j \omega_{i,j} = 0$ for all i (Chatfield 1980, p. 130), the Fourier basis is orthogonal and $E[Y_i^\omega] = 0$ for all i for a stationary sequence $\boldsymbol{Y}$. So the Fourier basis is a complete null basis.

2.3 Example 3: The Walsh Basis

The vectors in the Walsh basis are the columns in the design matrix for a full factorial experimental design with $n = 2^k$, including all additive effects and interactions (Beauchamp 1975, Sanchez 1986). For example, if a simulation run produces 8 observations, the (complete) Walsh basis is

$$\begin{aligned}
\boldsymbol{\omega}_1 &= (1,1,1,1,1,1,1,1)' \\
\boldsymbol{\omega}_2 &= (1,1,1,1,-1,-1,-1,-1)' \\
\boldsymbol{\omega}_3 &= (1,1,-1,-1,-1,-1,1,1)' \\
\boldsymbol{\omega}_4 &= (1,1,-1,-1,1,1,-1,-1)' \\
\boldsymbol{\omega}_5 &= (1,-1,-1,1,1,-1,-1,1)' \\
\boldsymbol{\omega}_6 &= (1,-1,-1,1,-1,1,1,-1)' \\
\boldsymbol{\omega}_7 &= (1,-1,1,-1,-1,1,-1,1)' \\
\boldsymbol{\omega}_8 &= (1,-1,1,-1,1,-1,1,-1)'
\end{aligned}$$

The variable $\boldsymbol{\omega}_1$ corresponds to the mean effect; $\boldsymbol{\omega}_2$, $\boldsymbol{\omega}_4$, and $\boldsymbol{\omega}_8$ correspond to the main effects in a saturated $n = 2^3$ factorial design with an additive model. The ordering of the Walsh basis is by *sequency* (i.e., $\boldsymbol{\omega}_i$ has $i-1$ sign changes). For stationary $\boldsymbol{Y}$, the Walsh basis is orthogonal and $E[W_i] = 0, i = 2,\ldots,b$, where $W_i \equiv Y_i^\omega$ for all i. So the Walsh basis is a null effects basis.

3 APPLICATIONS

3.1 Confidence Interval Estimation

Here we discuss two c.i.e.'s for μ based on batch means and Walsh components.

3.1.1 Batch Means c.i.e.'s

The BM $100(1-\alpha)\%$ c.i.e. for μ is

$$\mu \in \bar{Y}_n \pm t_{\alpha/2,b-1}\sqrt{\frac{\sum_{i=1}^b(\bar{Y}_{i,m}-\bar{Y}_n)^2}{(b-1)b}}, \qquad (1)$$

where the i-th batch mean $\bar{Y}_{i,m} \equiv Y_i^\omega = \sum_{j=(i-1)m+1}^{im} Y_j/m$, $i = 1,\ldots,b$, the sample mean $\bar{Y}_n \equiv \sum_{i=1}^n Y_i/n = \sum_{i=1}^b Y_i^\omega/b$, and $t_{\gamma,\nu}$ is the upper-γ quantile of the t-distribution with ν degrees of freedom.

3.1.2 Walsh c.i.e.'s

A c.i.e. for μ based on the Walsh components is

$$\mu \in W_1 \pm t_{\alpha/2,b-1}\sqrt{\frac{\sum_{i=2}^b W_i^2}{b-1}}. \qquad (2)$$

The asymptotic validity of this c.i.e. is discussed next.

3.1.3 Asymptotic Justification

The Walsh components, W_i, $i = 1,\ldots,b$, can be formed from non-overlapping batches of $\boldsymbol{Y}$. For instance, if $b = 2$, then $W_1 = (\bar{Y}_{1,n/2}+\bar{Y}_{2,n/2})/2$ and $W_2 = (\bar{Y}_{1,n/2}-\bar{Y}_{2,n/2})/2$. Further, when *$b$ is a power of 2 and evenly divides n*, the BM and Walsh c.i.e.'s, (1) and (2), are *equal*. We prove this "n-dimensional Pythagorean theorem" for the special case of $b = n$ ($b \le n$ is similar, but tedious): Observe that for the Walsh basis,

$$\sum_{i=1}^n \omega_{i,j}\omega_{i,k} = \sum_{i=1}^n \omega_{j,i}\omega_{k,i} = \boldsymbol{\omega}_j'\boldsymbol{\omega}_k = \begin{cases} n, & j = k \\ 0, & j \ne k. \end{cases}$$

Then

$$\sum_{i=1}^n W_i^2 = \frac{1}{n^2}\sum_{i=1}^n \boldsymbol{Y}'\boldsymbol{\omega}_i\boldsymbol{\omega}_i'\boldsymbol{Y} = \frac{1}{n}\sum_{i=1}^n Y_i^2.$$

So

$$n\sum_{i=2}^n W_i^2 = \sum_{i=1}^n Y_i^2 - nW_1^2 = \sum_{i=1}^n (Y_i-\bar{Y}_n)^2,$$

and thus, (1) equals (2). □

Consider $\boldsymbol{Y}$ as a scaled continuous-time function by defining $Y(t) \equiv Y_{\lfloor nt \rfloor}$, $0 \le t \le 1$, where $\lfloor \cdot \rfloor$ is the greatest integer function. The projection basis now consists of b continuous-time functions, $\{\omega_i(t), 0 \le t \le 1\}$, $i = 1, \ldots, b$, with $\omega_i(t) \equiv \omega_{i,\lfloor nt \rfloor}$. Then

$$Y_i^{\omega} \equiv \frac{\boldsymbol{\omega}_i'\boldsymbol{Y}}{\boldsymbol{\omega}_i'\boldsymbol{\omega}_i} = \frac{\int_0^1 \omega_i(t)Y(t)\,dt}{\int_0^1 \omega_i^2(t)\,dt}.$$

We apply the weak convergence arguments of Schruben (1983) and Glynn and Iglehart (1990) to obtain the asymptotic i.i.d. normality of the W_i's: With b fixed, the Walsh basis functions, $\{\omega_i(t), 0 \le t \le 1\}$, are constant over non-overlapping intervals. If the smallest such interval is selected as a batch size, then under general conditions, the resulting batch means converge weakly to i.i.d. normality; i.e., $\sqrt{m}(\bar{Y}_{i,m} - \mu) \xrightarrow{\mathcal{D}} \text{Nor}(0, \sigma^2)$, $i = 1, \ldots, b$, where $\sigma^2 \equiv \lim_{n\to\infty} n\text{Var}(\bar{Y}_n)$ and $\xrightarrow{\mathcal{D}}$ denotes convergence in distribution as $m \to \infty$. Since the Walsh basis components are orthogonal linear combinations of asymptotically i.i.d. normal batch means, they are also asymptotically i.i.d. normal; i.e., $\sqrt{m}W_i \xrightarrow{\mathcal{D}} \text{Nor}(0, \sigma^2/b)$, $i = 2, \ldots, b$.

3.1.4 Practical Considerations

Recall that when b is a power of two and evenly divides n, (1) and (2) are equal. In fact, based on empirical experimentation (not reported here), the performance characteristics of the BM and Walsh c.i.e.'s are essentially the same in general.

As is well known by simulation users, the main difficulty in using BM lies in determining the proper b to use (with n fixed) so that the assumption of i.i.d. normal batch means is approximately met. Schmeiser (1982) effectively argues that we rarely need $b > 30$. Sequential algorithms for finding an appropriate b (e.g., Fishman 1978, Fishman and Yarberry 1995, and Law and Kelton 1982) usually start with a particular value of b and reduce it while performing statistical tests of correlation and/or normality on the batch means.

A possible advantage of the Walsh c.i.e. procedure is that the Walsh basis is *constructive* as b increases; i.e., previous Walsh basic vectors remain in the basis as more are added. This contrasts with the BM basis which can change radically as b changes (with n fixed). Perhaps this observation could be used as the foundation for a batch size algorithm based on Walsh components.

We also see that the Walsh basis uses all of the output to estimate each W_i; the W_i's may converge to normality more quickly than do batch means. Indeed, Q-Q plots against normality have indicated consistently better performance for the W_i's (in terms of normality) than for the $\bar{Y}_{i,m}$'s.

3.2 Variance Reduction

Since the null effects have known means of zero, they make natural c.v.'s to use in variance reduction schemes for simulation. We give some applications.

3.2.1 Null Effects c.v.'s in a Single Run

When there is positive serial correlation in the stationary output series $\boldsymbol{Y}$, it is possible to construct a linear estimator of $\mu = E[Y_j]$ that is superior to the sample mean W_1. Several such estimators have been proposed (Halfin 1982) but usually in rather restrictive situations that require knowledge of the covariance structure of $\boldsymbol{Y}$. Our estimators require little knowledge of this structure. We can use c.v.'s of the form $C \equiv W_1 + kW_i$, $i = 2, \ldots, b$. Of course, $E[C] = E[W_1] = \mu$. The constant k is chosen in the hope of minimizing $\text{Var}(C)$. For $k > 0$, $\text{Var}(C) < \text{Var}(W_1)$ iff $\text{Cov}(W_1, W_i) < -k\text{Var}(W_i)/2$.

To illustrate, let $n = 4$ and $C = W_1 + kW_3$. Then $W_1 = (Y_1 + Y_2 + Y_3 + Y_4)/4$ and $W_3 = (Y_1 - Y_2 - Y_3 + Y_4)/4$. If we define $\gamma_j \equiv \text{Cov}(Y_i, Y_{i+j})$ for all j, some easy algebra gives $\text{Cov}(W_1, W_3) = (\gamma_3 - \gamma_1)/8$. For many stochastic processes, γ_j is decreasing in j; then $\text{Cov}(W_1, W_3) < 0$, and so small enough k will yield $\text{Var}(C) < \text{Var}(W_1)$. Now consider a first-order autoregressive [AR(1)] process, $Y_i = \mu + \alpha(Y_{i-1} - \mu) + \epsilon_i$, $i = 1, 2, \ldots$, with i.i.d. $\text{Nor}(0, 1-\alpha^2)$ ϵ_i's, and initialized in steady state. For the AR(1), $\gamma_j = \alpha^{|j|}$ for all j. We find $\text{Var}(W_1) = (1+\alpha)(2+\alpha+\alpha^2)/8$ and $\text{Var}(C) = (1+\alpha)[2+\alpha+\alpha^2 - 2k\alpha(1-\alpha) + k^2(2-\alpha)(1-\alpha)]/8$. $\text{Var}(C)$ is minimized with respect to k when $k = \alpha/(2-\alpha)$, in which case $\text{Var}(C) = (1/2)(1+\alpha)/(2-\alpha)$ and $\text{Var}(C)/\text{Var}(W_1)$ is between 0.964 and 1 for $\alpha \in [0, 1]$; so the variance reduction is modest.

3.2.2 Null Effects External c.v.'s

In external c.v. applications where one (stationary) simulation is used to produce c.v.'s for a second simulation, the expected value of the control simulation's variate must be known. Null effects estimators can therefore be used as c.v.'s when little is known about either simulation. The effectiveness of the approach depends on there being positive serial correlation within each of the individual output series (which is typical of many simulated systems) and positive correlation between outputs from the two simulations (as can often be induced by using common pseudo-random number streams as input).

3.2.3 Indirect Use of Null Effects c.v.'s

Here we indirectly use null effects c.v.'s to improve the estimator of the mean. The rationale for selecting the form of the c.v. used is as follows. Suppose a "reasonable" hypothesis involving μ is H_O versus H_A, and we know that H_O is true for the time series $\boldsymbol{Y}$ that is used to estimate μ. After estimating μ, we *alter* the time series by computing a null effect (which makes H_A true). The test statistic for the hypothesis test should be highly correlated with an indicator variable of whether H_O or H_A is true (else the test would not be very powerful). We should therefore be able to use the test statistic computed from the altered time series as a c.v. for the estimator from the original series.

For instance, consider the AR(1) process with $n = 4$. Suppose the altered output is $-\boldsymbol{\omega}_2'\boldsymbol{Y} = (-Y_1, -Y_2, Y_3, Y_4)'$. This altered series of four observations could be considered to contain severe initialization bias. A reasonably powerful test for initialization bias is based on the area under the so-called *standardized time series* (Schruben, Singh, and Tierney 1983). This test statistic is $\widetilde{Y} \equiv (3Y_1 + Y_2 + Y_3 + 3Y_4)/2$, for which $E[\widetilde{Y}] = 4\mu$. Thus, an unbiased c.v. estimator for μ is $C \equiv (\widetilde{Y} - \bar{Y}_n)/3$. Some algebra shows that $\mathrm{Var}(C) < \mathrm{Var}(\bar{Y}_n)$ iff $\gamma_0 - \gamma_2 < 2(\gamma_1 - \gamma_3)$ iff $\alpha > 1/2$. However, the variance reduction from this example is again modest. For example, when $\alpha = 0.9$, we obtain a variance reduction of about 4%. This is consistent with the variance reductions found by Halfin when the covariance structure is known and an optimal linear estimator is used.

4 CONCLUDING REMARKS

We have demonstrated that viewing a simulated time series as a vector in n-dimensional space can present useful insights when different bases for the space are considered. Relationships between the BM and Walsh bases yield alternative procedures for estimation and variance reduction.

Some subjects for future research: Techniques for consistent estimation of the Walsh spectrum (perhaps similar to spectral windowing) could be applied to BM estimation, and new procedures for selecting a batch size might be engineered using the Walsh basis. Also, although we illustrated some applications of null effects c.v.'s, no attempt were made to generalize the results, optimize the control, or use multiple c.v.'s — all ideas that might improve the results.

Certainly, the use of a different basis for the simulation output vector is one way of viewing the data differently, and it can lead to some entertaining new ideas.

ACKNOWLEDGEMENTS

This research was supported by National Science Foundation Grant No. DMI-9622269. The third author gratefully acknowledges the financial support of the CBA/GSB Faculty Research Committee of the College of Business of the University of Texas at Austin and the End to End Simulation Department at Schlumberger Austin Research.

REFERENCES

Beauchamp, K. G. 1975. *Walsh Functions and Their Applications.* New York: Academic Press.

Bloomfield, P. 1976. *Fourier Analysis of Time Series: An Introduction.* New York: John Wiley and Sons.

Chatfield, C. 1980. *The Analysis of Time Series: An Introduction.* New York: Chapman and Hall.

Fishman, G. S. 1978. *Principles of Discrete Event Simulation.* New York: John Wiley and Sons.

Fishman, G. S. and L. S. Yarberry. 1995. An Implementation of the Batch Means Method. Technical Report, Department of Operations Research, University of North Carolina, Chapel Hill, NC.

Glynn, P. W. and D. L. Iglehart. 1990. Simulation Output Analysis Using Standardized Time Series. *Mathematics of Operations Research* 15:1–16.

Halfin, S. 1982. Linear Estimators for a Class of Stationary Queueing Processes. *Operations Research* 30:515–529.

Law, A. M. and W. D. Kelton. 1982. Confidence Intervals for Steady-State Simulation, II: A Survey of Sequential Procedures. *Management Science* 28:550–562.

Sanchez, P. J. 1986. On the Correspondence Between Walsh Spectral Analysis and 2^k Factorial Experimental Designs. In *Proceedings of the 1986 Winter Simulation Conference* (ed. J. R. Wilson, J. O. Henriksen, and S. D. Roberts), 306–312. Institute of Electrical and Electronics Engineers, Piscataway, New Jersey.

Schmeiser, B. W. 1982. Batch Size Effects in the Analysis of Simulation Output. *Operations Research* 30:556–568.

Schruben, L. W. 1983. Confidence Interval Estimation Using Standardized Time Series. *Operations Research* 31:1090–1108.

Schruben L. W., H. Singh, and L. Tierney. 1983. Optimal Tests for Initialization Bias in Simulation Output. *Operations Research* 31:1167–1178.

AUTHOR BIOGRAPHIES

LEE W. SCHRUBEN is a Professor in the School of Operations Research and Industrial Engineering at Cornell University. He received his undergraduate degree in engineering from Cornell University and his Ph.D. from Yale University. His research interests are in the statistical design and analysis of simulation experiments and in graphical simulation modeling methods.

DAVID GOLDSMAN is an Associate Professor in the School of Industrial and Systems Engineering at the Georgia Institute of Technology. His research interests include simulation output analysis and ranking and selection. He is the Simulation Department Co-Editor for *IIE Transactions*, and the Past-President of the INFORMS College on Simulation.

DOUGLAS J. MORRICE is an Associate Professor in the Department of Management Science and Information Systems at The University of Texas at Austin. He is currently on sabbatical leave in the End to End Simulation Department at Schlumberger Austin Research. Dr. Morrice received his undergraduate degree in Operations Research at Carleton University in Ottawa, Canada. He holds an M.S. and a Ph.D. in Operations Research and Industrial Engineering from Cornell University. His research interests include discrete-event and qualitative simulation modeling and the statistical design and analysis of large-scale simulation experiments. Dr. Morrice is a member of INFORMS and the Council of Logistics Management. He served as Secretary for the INFORMS College on Simulation (1994-1996) and is Co-Editor of the *Proceedings of the 1996 Winter Simulation Conference.*

Proceedings of the 1996 Winter Simulation Conference
ed. J. M. Charnes, D. J. Morrice, D. T. Brunner, and J. J. Swain

TWO-STAGE PROCEDURES FOR MULTIPLE COMPARISONS WITH A CONTROL IN STEADY-STATE SIMULATIONS

Halim Damerdji

Department of Industrial Engineering
North Carolina State University
Raleigh, NC 27695-7906, U.S.A.

Marvin K. Nakayama

Department of Computer and Information Science
New Jersey Institute of Technology
Newark, NJ 07102-1982, U.S.A.

ABSTRACT

Suppose that we have k different stochastic systems, where μ_i denotes the steady-state mean of system i. We assume that the system labeled k is a control and want to compare the performance of the other systems, labeled $1,2,\ldots,k-1$, relative to this control. This problem is known in the statistical literature as multiple comparisons with a control (MCC). Independent steady-state simulations will be performed to compare the systems to the control. Two-stage procedures, based on the method of batch means, are presented to construct simultaneous lower one-sided confidence intervals for $\mu_i - \mu_k$ ($i = 1, 2, \ldots, k$), each having prespecified (absolute or relative) half-width δ. Under the assumption that the stochastic processes representing the evolution of the systems satisfy a functional central limit theorem, it can be shown that asymptotically (as $\delta \to 0$ with the size of the batches proportional to $1/\delta^2$), the joint probability that the confidence intervals simultaneously contain the $\mu_i - \mu_k$ ($i = 1, 2, \ldots, k-1$) is at least $1-\alpha$, where α is prespecified by the user.

1 INTRODUCTION

Suppose there are k different systems (i.e., stochastic processes), where system i has (unknown) steady-state mean μ_i and (unknown) asymptotic variance σ_i^2. We allow for the different systems to have different probability distributions governing them, and so the variances may be unequal. We assume that system k is a control and want to compare the mean performance of this system to that of each of the systems $i = 1, 2, \ldots, k-1$. For example, in a redesign setting, system k may be the existing system, and there are $(k-1)$ design alternatives under consideration. We assume here that larger performances are better than smaller. (The case when smaller performances are preferable can similarly be considered.)

Independent simulations of the various systems are run to construct simultaneous confidence intervals for $\mu_i - \mu_k$, $i = 1, 2, \ldots, k-1$. This is known as *multiple comparisons with a control* (MCC). In this paper, we propose some two-stage procedures based on the method of batch means for constructing lower one-sided MCC confidence intervals having prespecified (absolute or relative) half-width δ. It can be shown (see Damerdji and Nakayama 1996) that asymptotically (as $\delta \to 0$ and the batch sizes are proportional to $1/\delta^2$) the probability that the $\mu_i - \mu_k$, $i = 1, 2, \ldots, k-1$, simultaneously lie in the respective MCC confidence intervals is at least $1-\alpha$ (which the user prespecifies).

The problem of MCC has been previously studied only for the case of i.i.d. normals. For example, see Tamhane (1977), Dudewicz and Ramberg (1972), Dudewicz, Ramberg, and Chen (1975), and Dudewicz and Dalal (1983). For a review of these and other procedures, the reader is referred to Chapter 7 of Hochberg and Tamhane (1987) and Goldsman and Nelson (1994). None of these papers covers relative-width confidence intervals, as we do here.

The rest of the paper has the following organization. In Section 2 we define the notation used and state an assumption on the processes being simulated. The procedures are presented in Section 3, while Section 4 contains a brief discussion on how to specify values for the parameters needed to run our procedures.

2 NOTATION AND ASSUMPTIONS

For system i ($i = 1, 2, \ldots, k$), let $\mathbf{Y}_i = \{Y_i(t) : t \geq 0\} \in D[0,\infty)$ be a real-valued stochastic process representing the sample path of system i, where $D[0,\infty)$ is the space of right continuous functions on $[0,\infty)$ having left limits (for more details on $D[0,\infty)$, see Ethier and Kurtz 1986 or Glynn 1990). A large number of stochastic systems encountered in engineering can be modeled as such. It is assumed here that $\mathbf{Y}_1, \mathbf{Y}_2, \ldots, \mathbf{Y}_k$ are mutually independent. (In a simulation, this means that for all i and j with $j \neq i$, the sample paths of systems i and j are generated using non-overlapping streams of uniform random numbers.) Also, let $\mathbf{Y} = (\mathbf{Y}_1, \mathbf{Y}_2, \ldots, \mathbf{Y}_k)$ and $Y(t) = (Y_1(t), Y_2(t), \ldots, Y_k(t))$.

To establish the results, we need to assume that the process $\mathbf{Y}$ satisfies a functional central limit theorem (FCLT). More specifically, let "$\Rightarrow$" denote weak convergence (see Billingsley 1968), and then assume the following:

A1 *There exist a nonsingular $k \times k$ matrix $\Sigma = (\sigma_{i,j} : i, j = 1, \ldots, k)$ with $0 < \sigma_{i,i} < \infty$, $i = 1, 2, \ldots, k$, and a constant $\mu = (\mu_1, \mu_2, \ldots, \mu_k) \in \Re^k$ such that*

$$X_\delta \Rightarrow \Sigma B$$

as $\delta \to 0$, where B is a k-dimensional standard Brownian motion, $X_\delta = (X_{1,\delta}, X_{2,\delta}, \ldots, X_{k,\delta})$, and

$$X_{i,\delta}(t) = \frac{1}{\delta}\left(\frac{\int_0^{t/\delta^2} Y_i(s)ds}{1/\delta^2} - \mu_i t\right), \quad t \geq 0,$$

for $i = 1, 2, \ldots, k$.

Here, Σ is a diagonal matrix with ith diagonal entry $\sigma_i^2 \equiv \sigma_{i,i}$. For each $i = 1, 2, \ldots, k$, the constant μ_i appearing in A1 is precisely the steady-state mean of the process $\mathbf{Y}_i$. Also, σ_i^2 is the asymptotic variance of $\mathbf{Y}_i$.

A large number of stochastic systems that admit a steady state will also satisfy Assumption A1. For example, this assumption holds if the process $\mathbf{Y}$ satisfies any of the following: $\mathbf{Y}$ is regenerative and satisfies suitable moment conditions (see Glynn and Whitt 1987); $\mathbf{Y}$ is a martingale process (see Chapter 7 of Ethier and Kurtz 1986); $\mathbf{Y}$ satisfies appropriate mixing conditions (see Chapter 7 of Ethier and Kurtz 1986); or the $\mathbf{Y}(t)$ are associated (see Newman and Wright 1981). All these processes are weakly dependent, in the sense that events far apart in time are almost independent. Processes with long-range dependency may not satisfy the assumption.

3 THE MCC PROCEDURES

The basic idea of our two-stage MCC procedures is as follows. In the first stage we run independent simulations of the different systems. The method of batch means with $m \geq 2$ batches is applied to the output of each system, thus yielding an estimate of the variance of the first-stage sample mean. This is used to compute the total number of batches needed for each system, and so, additional simulation may need to be run in the second stage. (For each system, the total number of batches times the length of a batch will of course be the total simulation time that must be allocated to that system.) Statistics are then updated and the simultaneous MCC confidence intervals constructed. The two-stage batch means algorithm for constructing lower one-sided, absolute-half-width MCC confidence intervals is as follows:

Procedure 1

1. Specify the desired absolute half-width δ of each MCC confidence interval (where δ is small); the desired confidence level $1-\alpha$; and the number of initial batches $m \geq 2$. Let $\gamma \equiv \gamma(m, k, \alpha)$ solve Rinott's (1978) integral. (Wilcox 1984 presents tables for γ. Note that our notation differs from that used in Wilcox's tables. In particular, our m and γ correspond to his n_0 and h, respectively.)

2. First stage: Independently simulate systems $i = 1, 2, \ldots, k$, with run lengths $T_i = T_i(\delta)$, each proportional to $1/\delta^2$. For each system i, group the output into m (non-overlapping) batches, each of size T_i/m, and compute
$$Z_{i,j} = \frac{1}{T_i/m}\int_{(j-1)T_i/m}^{jT_i/m} Y_i(s)\, ds, \quad j \geq 1,$$
which is the sample mean of the jth batch.

3. For each system $i = 1, 2, \ldots, k$, compute the sample variance of the m batch means from the first stage as
$$S_i^2 = \frac{1}{m-1}\sum_{j=1}^{m}\left(Z_{i,j} - \frac{1}{m}\sum_{k=1}^{m} Z_{i,k}\right)^2.$$

4. For each system $i = 1, 2, \ldots, k$, compute the total number of batches to collect as
$$N_{a,i}(\delta) = \max\left\{m, \left\lceil\left(\frac{\gamma S_i}{\delta}\right)^2\right\rceil\right\},$$
where $\lceil x \rceil$ represents the smallest integer greater or equal to x.

5. Second stage: For each system i, if $N_{a,i}(\delta) \geq m+1$, simulate independently that system to collect the additional (non-overlapping) batches $m+1, \ldots, N_{a,i}(\delta)$, each of size T_i/m. Compute then the batch means $Z_{i,m+1}, \ldots, Z_{i,N_{a,i}(\delta)}$.

6. For each system $i = 1, 2, \ldots, k$, compute the overall sample mean as

$$\hat{\mu}_{a,i} = \frac{1}{N_{a,i}(\delta)} \sum_{j=1}^{N_{a,i}(\delta)} Z_{i,j}.$$

7. Simultaneously construct the absolute-precision, lower one-sided MCC confidence intervals

$$I_{a,i}(\delta) = [\hat{\mu}_{a,i} - \hat{\mu}_{a,k} - \delta, +\infty)$$

for $\mu_i - \mu_k$, $i = 1, 2, \ldots, k-1$.

Damerdji and Nakayama (1996) establish the following result.

Theorem 1 *Assume that Assumption A1 holds. If Procedure 1 is used, then*

$$\lim_{\delta \to 0} P\{\mu_i - \mu_k \in I_{a,i},\ i = 1, 2, \ldots, k-1\} > 1 - \alpha.$$

Theorem 1 establishes that our MCC confidence intervals are asymptotically valid (in the sense that the simultaneous coverage probability is at least $1 - \alpha$).

Procedure 1 above is for the case when we want MCC confidence intervals having a pre-specified absolute half-width. However, in certain settings, we may desire the interval for $\mu_i - \mu_k$, $i = 1, 2, \ldots, k-1$, to have half-width that is, say 5%, of $\mu_i - \mu_k$. Thus, we now develop a two-stage procedure for relative-width, lower one-sided, MCC confidence intervals.

Procedure 2

1. Specify the desired relative half-width of each MCC confidence interval δ (where δ is small); the desired confidence level $1 - \alpha$; and the number of initial batches $m \geq 2$. Let $\gamma \equiv \gamma(m, k, \alpha)$ solve Rinott's (1978) integral.

2. In the first stage independently simulate systems $i = 1, 2, \ldots, k$, with run lengths $T_i = T_i(\delta)$, each proportional to $1/\delta^2$. For each system i, group the output into m (non-overlapping) batches, each of size T_i/m, and compute the first m batch means $Z_{i,1}, Z_{i,2}, \ldots, Z_{i,m}$.

3. For each system $i = 1, 2, \ldots, k$, compute the sample mean and the sample variance of the m batch means from the first stage as

$$\tilde{\mu}_i = \frac{1}{m} \sum_{j=1}^{m} Z_{i,k}$$

and

$$S_i^2 = \frac{1}{m-1} \sum_{j=1}^{m} (Z_{i,j} - \tilde{\mu}_i)^2,$$

respectively.

4. For each system $i = 1, 2, \ldots, k$, compute the total number of batches to collect as

$$N_{r,i}(\delta) = \max\left\{m, \left\lceil \left(\frac{\gamma S_i}{\delta \tilde{\mu}_k}\right)^2 \right\rceil \right\},$$

assuming $\tilde{\mu}_k \neq 0$.

5. For the second stage, independently simulate systems $i = 1, 2, \ldots, k$, if needed, to collect additional (non-overlapping) batches $m+1, m+2, \ldots, N_{r,i}(\delta)$, each of size T_i/m. Compute the batch means $Z_{i,m+1}, \ldots, Z_{i,N_{r,i}(\delta)}$.

6. For each system $i = 1, 2, \ldots, k$, compute the overall sample mean as

$$\hat{\mu}_{r,i} = \frac{1}{N_{r,i}(\delta)} \sum_{j=1}^{N_{r,i}(\delta)} Z_{i,j}.$$

7. Simultaneously construct the relative-precision, lower one-sided MCC confidence intervals

$$I_{r,i}(\delta) = [\hat{\mu}_{r,i} - \hat{\mu}_{r,k} - \delta|\hat{\mu}_{r,k}|, +\infty)$$

for $\mu_i - \mu_k$, $i = 1, 2, \ldots, k-1$.

The relative half-widths of the above intervals are relative not with respect to $|\hat{\mu}_{r,i} - \hat{\mu}_{r,k}|$ but rather $|\hat{\mu}_{r,k}|$. Damerdji and Nakayama (1996) establish the validity of the following theorem:

Theorem 2 *Assume that Assumption A1 holds and that $\mu_k \neq 0$. If Procedure 2 is used, then*

$$\lim_{\delta \to 0} P\{\mu_i - \mu_k \in I_{r,i}(\delta),\ i = 1, 2, \ldots, k-1\} > 1 - \alpha.$$

4 SPECIFYING VALUES FOR PARAMETERS

To use Procedures 1 and 2 in practice, the practitioner must specify values for several parameters. These include the desired (absolute or relative) half-width δ, the run length of the first stage T_i (which must be proportional to $1/\delta^2$) for each system, and the number of initial batches m.

Because of the similarity of Procedures 1 and 2 and the two-stage stopping procedures developed by Nakayama (1994) for the single-system setting, it is reasonable to assume that appropriate values for the parameters of Nakayama's (1994) algorithm are also valid for our new procedures. Nakayama (1994) suggests that one should choose $5 \leq m \leq 15$ and $\delta < 0.025$. However, as Nakayama (1994) notes, selecting a reasonable value for T_i given δ is a delicate matter. In the case when simulating queueing systems, though, Nakayama (1994) proposes using some of the results of Whitt (1989a,1989b); for more details, see Sections 5 and 6 of Nakayama (1994).

ACKNOWLEDGMENTS

The work of the second author was supported in part by NJIT Grant 421180.

REFERENCES

Billingsley, P. 1968. *Convergence of Probability Measures.* New York: John Wiley.

Damerdji, H. and M. K. Nakayama. 1996. Two-Stage Procedures for Multiple Comparisons with the Best in Steady-State Simulations. In preparation.

Dudewicz, E. J. and S. R. Dalal. 1983. Multiple comparisons with a control when variances are unknown and unequal. *American Journal of Mathematics and Management Sciences* 4:275–295.

Dudewicz, E. J. and J. S. Ramberg. 1972. Multiple comparison with a control: Unknown variances. *Ann. Tech. Conf. Amer. Soc. Quality Control* 26:483–488.

Dudewicz, E. J., J. S. Ramberg, and H. J. Chen. 1972. New tables for multiple comparisons with a control (unknown variances). *Biometrische Zeitschrift* 17:437–445.

Ethier, S. N. and T. G. Kurtz. 1986. *Markov Processes: Characterization and Convergence.* John Wiley, New York.

Glynn, P. W. 1990. Diffusion approximations. Chapter 4 of *Handbooks in Operations Research and Management Science, Vol. 2, Stochastic Models*, ed. D. Heyman and M. Sobel. Elsevier Science Publishers B. V. (North-Holland).

Glynn, P. W. and W. Whitt. 1987. Sufficient conditions for functional-limit-theorem versions of $L = \lambda W$. *Queueing Systems: Theory and Applications* 1:279–287.

Goldsman, D. and B. L. Nelson. 1994. Ranking, selection and multiple comparisons in computer simulation. In *Proceedings of the 1994 Winter Simulation Conference*, ed. J. D. Tew, S. Manivannan, D. A. Sadowski, and A. F. Seila, 192–199. Piscataway, New Jersey: Institute of Electrical and Electronics Engineers.

Hochberg, Y. and A. C. Tamhane. 1987. *Multiple Comparison Procedures.* John Wiley, New York.

Nakayama, M. K. 1994. Two-stage stopping procedures based on standardized time series. *Management Science* 40:1189–1206.

Newman, C. M. and A. L. Wright. 1981. An invariance principle for certain dependent sequences. *Annals of Probability* 9:671–675.

Rinott, Y. 1978. On two-stage selection procedures and related probability-inequalities. *Communications in Statistics—Theory and Methods* A7:799–811.

Tamhane, Y. 1977. Multiple comparisons in model I: One-way ANOVA with unequal variances. *Communications in Statistics—Theory and Methods* A6:15–32.

Whitt, W. 1989a. Planning queueing simulations. *Management Science* 35:1341–1366.

Whitt, W. 1989b. Simulation run length planning. In *Proceedings of the 1989 Winter Simulation Conference*, ed. E. A. MacNair, K. J. Musselman, and P. Heidelberger, 106–112. Piscataway, New Jersey: Institute of Electrical and Electronics Engineers.

Wilcox, R. R. 1984. A table for Rinott's selection procedure. *Journal of Quality Technology* 16:97–100.

AUTHOR BIOGRAPHIES

HALIM DAMERDJI is an assistant professor in the Department of Industrial Engineering at North Carolina State University. His research interests are in simulation and applied stochastic processes. He is a member of INFORMS.

MARVIN K. NAKAYAMA is an assistant professor in the Department of Computer and Information Science at the New Jersey Institute of Technology. Previously, he has held positions in the Management Science/Computer Information Systems Department of the Graduate School of Management at Rutgers University, and at the IBM Thomas J. Watson Research Center in Yorktown Heights, New York. He received Ph.D. in Operations Research from Stanford University. His research interests include applied probability, simulation output analysis, gradient estimation, rare-event simulation, and random-variate generation.

Proceedings of the 1996 Winter Simulation Conference
ed. J. M. Charnes, D. J. Morrice, D. T. Brunner, and J. J. Swain

PARALLEL SIMULATION BY TIME SEGMENTATION: METHODOLOGY AND APPLICATIONS

Mehdi Hoseyni-Nasab

Department of Industrial Engineering
University of Wisconsin–Madison
1513 University Avenue
Madison, WI 53706, U.S.A.

Sigrún Andradóttir

School of Industrial and Systems Engineering
Georgia Institute of Technology
Atlanta, GA 30332, U.S.A.

ABSTRACT

In this paper we present a time parallel simulation approach and discuss conditions under which this approach is applicable. Our approach involves distributing the available processors among segments of the time horizon of the simulation. We show that under certain conditions, sample paths of the system generated by a common sequence of potential events will couple (i.e., become identical) with probability one. This property will be exploited to efficiently combine the information collected on different segments of the sample path and generate a complete valid sample path. We show that the expected coupling time of the system (i.e., the amount of time required for all the sample paths of the system to couple) is essential to the efficiency of the approach. We apply our parallel simulation approach to a class of Markovian queueing networks and investigate the efficiency of the method by providing bounds and estimates for the expected coupling times.

1 INTRODUCTION

The need for efficient simulation of discrete event systems has created a rapidly growing demand for parallel simulation methods that can effectively distribute the computational load of a simulation experiment among multiple processing units. Several parallel simulation methods have been studied and implemented in recent years. Depending on the strategy these methods choose to make concurrent use of the available processors, most parallel simulation methods fall into one of the following categories: *distributed simulation* methods, *multiple replications* methods, and *time parallel simulation* methods.

In distributed simulation methods the system under study is partitioned into subsystems and the available processors are appropriately distributed among the subsystems. Processors simulate the subsystems and communicate with each other to guarantee that the generated sample paths are valid (see Fujimoto (1993) for more details on distributed simulation).

In multiple replications methods, every processor generates sample paths of the system independently of the other processors. Then, the results obtained by different processors are averaged to provide estimators for the performance measures of interest (see Heidelberger (1988) for more details on the multiple replications approach).

A variety of time parallel simulation approaches have been suggested in recent years for parallel simulation of different classes of discrete event systems. Greenberg, Lubachevsky, and Mitrani (1990) and Chen and Serfozo (1995) take advantage of representations of the dynamics of the systems under study in terms of recurrence relations to achieve parallelism. Heidelberger and Stone (1990) propose a trace partitioning approach to perform parallel trace driven simulation of a cache, and Lin and Lazowska (1991) suggest a parallel simulation method applicable to a class of partially regenerative systems.

Andradóttir and Ott (1995) have recently proposed a time parallel simulation approach, namely the time segmentation approach, that can be applied to a variety of queueing networks. They have applied this method to Markovian queueing systems that consist of either loss stations or communication blocking stations. They have investigated the efficiency of the approach by providing bounds and estimates for the expected coupling times of the systems under study. The results of Andradóttir and Ott (1995) have been extended and strengthened by Hoseyni-Nasab and Andradóttir (1996a,b). They have shown that the time segmentation method is effectively applicable to different classes of queueing networks.

In this paper we describe how the time segmentation method can be applied to simulate discrete event systems whose sample paths satisfy certain constructibility conditions on parallel processors. We also present analytical and numerical results indicating the validity, applicability, and efficiency of the approach in the context of parallel simulation of a family of Markovian queueing networks containing loss and communication blocking stations.

The outline of the paper is as follows: In Section 2 we present the time segmentation parallel simulation method. In Section 3 we show that the time segmentation method is applicable to a class of Markovian queueing networks and investigate the efficiency of the approach by providing bounds for the expected coupling times. In Section 4 we further study the efficiency of the approach by estimating the expected coupling times for a number of queueing systems and providing numerical results directed towards understanding the dependence of the expected coupling times on the parameters of the system. We conclude the paper by a few remarks in Section 5.

2 THE PARALLEL SIMULATION APPROACH

In this section we describe the time segmentation method for parallel simulation of discrete event systems and discuss conditions under which this approach is applicable.

Suppose S is a discrete event system and suppose that we would like to generate long sample paths of a simulation model of S. Let $N(t)$ be a function of time t that uniquely determines the state of the system at time t. Now suppose we would like to use P processors concurrently to generate a sample path of S. Let T be the length of the required sample path, partition the interval $[0, T]$ into P subintervals $[t_0, t_1), [t_1, t_2), ..., [t_{P-1}, t_P]$, and assign one processor to each subinterval. Furthermore, suppose it is possible to generate valid sample paths of the system in parallel using a common sequence of potential events (i.e., given a sequence of events, a scheduled event in the sequence will be executed in a sample path if the event is feasible for the sample path). We start the simulation of the sample path on the first subinterval from the true initial state of the sample path. In other subintervals, the corresponding processors start simulating two sample paths, sample paths l and u (call them the bounding sample paths), using a common sequence of potential events. Let $N_l(t)$ and $N_u(t)$ denote the state of the system at time t in sample paths l and u, respectively. Suppose sample paths l and u can be chosen and the simulation can be conducted in such ways that by the time sample paths l and u couple (i.e., the first time T_c when $N_l(T_c) = N_u(T_c)$), any other sample path that has been constructed with the same sequence of potential events as l and u will have coupled with sample paths l and u. Finally, suppose that if two sample paths that have been constructed with the same sequence of potential events couple, then they will stay coupled for the remainder of the simulation period. For $i = 2, ..., P$, if there exists T_c^i such that $T_c^i = \inf\{t \in [t_{i-1}, t_i) : N_l(t) = N_u(t)\}$ (T_c^i is the coupling time of sample paths l and u on the interval $[t_{i-1}, t_i)$), then the above assumptions clearly imply that $N_l(t) = N(t) = N_u(t)$, for all $t \in [T_c^i, t_i)$. Therefore the information collected from either one of the bounding sample paths on the interval $[T_c^i, t_i)$ is valid data for the true sample path as well. Also, note that if sample paths l and u in the interval $[t_{i-2}, t_{i-1})$ are coupled prior to the end of the subinterval, then the real initial state of the sample path on the interval $[t_{i-1}, t_i)$ is the same as the terminal state of the coupled sample paths on the interval $[t_{i-2}, t_{i-1})$. Therefore, we can redo the simulation of the sample path on the interval $[t_{i-1}, T_c^i)$ with the true initial state (using the same sequence of potential events as the bounding sample paths of the interval) and collect the required data. If sample paths l and u in interval $[t_{i-2}, t_{i-1})$ are not coupled prior to time t_{i-1}, then we repeat what we did to sample paths l and u by initiating two sample paths in the interval $[t_{i-1}, t_i)$ starting at the initial states $N_l(t_{i-1})$ and $N_u(t_{i-1})$, which are the terminal states of sample paths l and u on the interval $[t_{i-2}, t_{i-1})$, respectively. By going through the above process (possibly repeatedly) and putting together the data collected on all the subintervals, we will finally generate a complete observation of the true sample path on the interval $[0, T]$. Note that the simulation is conducted in such a way that the sequence of potential events used for generating a complete observation of a true sample path is the same as the sequence obtained by putting together the sequences of potential events on the consecutive subintervals of the simulation.

As we mentioned before, the magnitudes of the coupling times are essential to the efficiency of the method. Note that if the bounding sample paths of a subinterval couple, then the part of the sample path generated after the coupling time on that subinterval is absolutely valid. Therefore, in general, the shorter the coupling times of the bounding paths the more efficiently we can construct the true sample path. In the best case scenario (i.e., when the bounding sample paths of all subintervals couple on the first run of the simulation), the total simulation time will be of order $\frac{2T}{P}$ or less. In the worst case scenario (i.e., when no bounding sample paths ever couple), the total simulation time will be of order T; that is the same order of magnitude as the simulation time of a single processor simulation of the system.

In the next section we will show that the time segmentation approach is applicable to a large class of queueing networks and investigate the efficiency of the method when applied to these systems.

3 APPLICABILITY AND EFFICIENCY OF THE APPROACH

The parallel simulation approach described in Section 2 is applicable to a variety of discrete event systems. In this section we show that the time segmentation approach can be applied to simulate a family of queueing networks with loss and communication blocking stations. We will also investigate the efficiency of the approach by studying the magnitude of the expected coupling times of the systems under study. Before stating our results we need to define the system under study and introduce some notation.

Suppose S is a network of n queueing stations. For $i = 1, ..., n$, let s_i and B_i denote the number of servers and the buffer capacity at station i, respectively, and let $C_i = B_i + s_i$. Suppose the service times at all servers of station i are independent and exponentially distributed with rate μ_i, $i = 1, ..., n$, and let jobs arrive from the outside to stations $1, ..., n$ according to independent Poisson processes with rates $\lambda_1, ..., \lambda_n$, respectively. For $i = 1, ..., n$, suppose p_{ij} is the probability that a job will attempt to join the queue at station j, immediately after a service completion at station i, $j = 1, ..., n$, and let $p_{i,n+1}$ denote the probability that a job will depart from the system upon a service completion at station i. Furthermore, suppose every station $i \in \{1, ..., n\}$ operates under one of the following two mechanisms:

1. Station i is a *loss* station if arriving jobs (from other stations in the system or from the outside) that find a full buffer at station i leave the system immediately.

2. Station i is a *communication blocking* station if arriving jobs from the outside that find a full buffer at station i leave the system immediately, and arriving jobs from other stations in the system that find a full buffer at station i get rerouted (independently of the past), after undergoing another service completion in the station of their most recent service completion.

The definition of communication blocking given above is different from the standard definition used in the literature. In the standard communication blocking mechanism, jobs will not start receiving service at a station unless there is room for them in the station that they are visiting next. However, it is not clear how the standard communication blocking mechanism works when we have queueing stations with multiple servers, multiple sources of arrivals, or multiple destinations for the departing customers. When the service times are exponentially distributed, the two definitions of communication blocking are equivalent if every station has only one server, one source of arrivals, and one destination for the departing customers.

For all $t \geq 0$, let $N(t) = (N_1(t), ..., N_n(t))$, where $N_i(t)$ is the number of customers at station i at time t, $i = 1, ..., n$. It is clear that $\{N(t)\}$ is a continuous-time Markov chain. Our goal is to show that the time segmentation approach can be used to simulate long sample paths of the Markov chain $\{N(t)\}$ efficiently.

For $i = 1, ..., n$, let T_i^a, T_i^s and T denote the time of the next arrival at station i, the time of the next potential service completion at station i, and the elapsed simulation time. Suppose we can generate four independent sequences $\{U_i^a(k)\}$, $\{U_i^s(k)\}$, $\{U_i^d(k)\}$, and $\{U_i^r(k)\}$ of independent $U[0,1]$ distributed random variables, for $i = 1, ..., n$. These sequences will be used to generate the sequence of interarrival times at station i, the sequence of potential service completion times at station i, the sequence of numerical marks that will be used to determine whether a potential service completion is an actual service completion at station i, and the sequence of destinations of the jobs that are departing station i, respectively. Finally, for $i = 1, ..., n$, let n_i^a, n_i^s, n_i^d, and n_i^r be the number of uniform random numbers from the sequences $\{U_i^a(k)\}$, $\{U_i^s(k)\}$, $\{U_i^d(k)\}$, and $\{U_i^r(k)\}$ that have been used to generate observations of the corresponding random variables, respectively.

We are now ready to state our algorithm. The empty sum in the following algorithm is defined to be zero. Algorithm 3.1 is a combination of Algorithms 2.1 and 3.1 in Andradóttir and Ott (1995).

Algorithm 3.1

Step 0: *Initialization.*
For $m = 1, ..., M$, select initial states N_i^m such that $0 \leq N_i^m \leq C_i$, $i = 1, ..., n$. For $i = 1, ..., n$, generate $U_i^a(1)$ and $U_i^s(1)$, and let $T_i^s = -\frac{\log(U_i^s(1))}{s_i\mu_i}$ and $T_i^a = -\frac{\log(U_i^a(1))}{\lambda_i}$. Let $T = 0$, $n_i^a = 1$, $n_i^s = 1$, $n_i^d = 0$ and $n_i^r = 0$, $i = 1, ..., n$. Go to Step 1.

Step 1: *Identifying the next event.*
Let $T^* = \min_{i=1,...,n}\{T_i^a, T_i^s\}$. If $T^* = T_i^a$, where $i \in \{1, ..., n\}$, then let $e^* = i$ and go to Step 2. Otherwise, if $T^* = T_i^s$, where $i \in \{1, ..., n\}$, then let $e^* = i$ and if station i is a loss or communication blocking station, go to Step 3 or 4, respectively.

Step 2: *Arrival at station e^*.*
Let $T = T^*$. For $m = 1, ..., M$, if $N_{e^*}^m < C_{e^*}$, then let $N_{e^*}^m = N_{e^*}^m + 1$. Let $n_{e^*}^a = n_{e^*}^a + 1$ and $T_{e^*}^a = T^* - \frac{\log(U_{e^*}^a(n_{e^*}^a))}{\lambda_{e^*}}$. Go to Step 1.

Step 3: *Departure from a loss station e^*.*
Let $T = T^*$ and $n^d_{e^*} = n^d_{e^*} + 1$. If $U^d_{e^*}(n^d_{e^*}) \leq \max_{m=1,...,M}\{N^m_{e^*}\}/s_{e^*}$, then let $n^r_{e^*} = n^r_{e^*} + 1$ and let $j \in \{1, ..., n+1\}$ be such that $\sum_{k=1}^{j-1} p_{e^*,k} < U^r_{e^*}(n^r_{e^*}) \leq \sum_{k=1}^{j} p_{e^*,k}$ (if $U^r_{e^*}(n^r_{e^*}) = 0$, then let $j = 1$). Otherwise, let $j = 0$.

- *Departure from the system.*
  If $j = n+1$, then for $m = 1, ..., M$, if $U^d_{e^*}(n^d_{e^*}) \leq N^d_{e^*}/s_{e^*}$, then let $N^m_{e^*} = N^m_{e^*} - 1$.
- *Departure at e^* and arrival at j.*
  If $1 \leq j \leq n$, then for $m = 1, ..., M$, if $U^d_{e^*}(n^d_{e^*}) \leq N^m_{e^*}/s_{e^*}$, then let $N^m_{e^*} = N^m_{e^*} - 1$. If also $N^m_j < C_j$, then let $N^m_j = N^m_j + 1$.

Let $n^s_{e^*} = n^s_{e^*} + 1$ and $T^s_{e^*} = T^* - \frac{\log(U^s_{e^*}(n^s_{e^*}))}{s_{e^*}\mu_{e^*}}$. Go to Step 1.

Step 4: *Departure from a communication blocking station e^*.*
Same as Step 3, except for the following:

- *Departure at e^* and arrival at j.*
  If $1 \leq j \leq n$, then for $m = 1, ..., M$, if $U^d_{e^*}(n^d_{e^*}) \leq N^m_{e^*}/s_{e^*}$ and $N^m_j < C_j$, then let $N^m_{e^*} = N^m_{e^*} - 1$ and $N^m_j = N^m_j + 1$.

Algorithm 3.1 generates multiple sample paths of the Markov chain $\{N(t)\}$ simultaneously. To generate the potential service times at station i for all the sample paths, we generate exponentially distributed service times with rate $s_i\mu_i$. Since the number of busy servers at station i may be less than s_i, we determine whether a potential service completion is an actual service completion depending on the number of busy servers in station i of each sample path through *thinning by rejection.* Our approach is related to the methods proposed by Cassandras and Strickland (1989) and Vakili (1991) for using a common sequence of potential events in parallel simulation of multiple discrete event systems.

Suppose we are generating two sample paths of the system, sample paths A and B, using Algorithm 3.1. Let $N^A_i(t)$ and $N^B_i(t)$ denote the number of jobs at station i at time t in sample paths A and B, respectively, for $i = 1, ..., n$. Let $T^{A,B}_c$ be the coupling time of sample paths A and B, that is $T^{A,B}_c = \inf\{t \geq 0 : N^A_i(t) = N^B_i(t), i = 1, ..., n\}$. Let sample path 1 be the sample path that initiates at the state where all servers are idle and sample path 2 be the sample path that initiates at the state where all buffers are full. Finally, let $T^{1,2}_c$ denote the coupling time of sample paths 1 and 2.

The following two propositions show that sample paths of the queueing system S generated by Algorithm 3.1 satisfy the conditions that are required for the time segmentation approach to be applicable. The proofs of Propositions 3.1 and 3.2 are given in Hoseyni-Nasab and Andradóttir (1996a).

Proposition 3.1 (Monotonicity Property) *Suppose S is a queueing system of loss and communication blocking stations. Let A and B be two sample paths of S generated concurrently by Algorithm 3.1 and suppose $N^A_i(0) \leq N^B_i(0)$, for $i = 1, ..., n$. Then $N^A_i(t) \leq N^B_i(t)$ for $i = 1, ..., n$ and all $t \geq 0$.*

Proposition 3.2 *In S, if $\{N(t)\}$ is an irreducible Markov chain, then $E\{T^{1,2}_c\} < \infty$. Also, for any two sample paths A and B we have $T^{A,B}_c \leq T^{1,2}_c$ (and hence $E\{T^{A,B}_c\} < \infty$).*

Proposition 3.1 indicates that sample path B that is initially above sample path A will never go below sample path A. This clearly implies that once two sample paths couple, they will stay coupled afterwards. Proposition 3.2 indicates that all sample paths generated by Algorithm 3.1 will couple with probability one and no later than the coupling time of sample paths 1 and 2. Moreover, the expected coupling time of all sample paths is finite. Considering the discussion in Section 2, Propositions 3.1 and 3.2 guarantee the validity of the approach. It is clear that sample paths 1 and 2 serve as the bounding paths on every subinterval of the simulation and the simulation of a long sample path of the queueing system S can be carried out by assigning the available processors to subintervals of the time horizon of the simulation and following the procedure described in Section 2.

As we discussed in Section 2, the magnitude of the expected coupling time of the system is essential to the efficiency of the time segmentation approach. In particular, it is important to understand how the expected coupling time of the system changes with respect to changes in the parameters of the system. The following proposition is one of our results that indicates that, under certain conditions, the expected coupling time of the queueing network S grows no faster than linearly with respect to the number of stations in the system. The proof of Proposition 3.3 can be found in Hoseyni-Nasab and Andradóttir (1996a).

Proposition 3.3 *For the queueing network S, suppose $C_i \leq C < \infty$, for $i = 1, ..., n$. Moreover, suppose that the Markov chain $\{N(t)\}$ is irreducible and that there exists a real number p such that $0 < p < 1$, and that for every station i at least one of the following two conditions is satisfied (for all $n \geq 1$):*

1. *$\lambda_i \geq p$ and $\frac{\lambda_i}{\lambda_i + s_i\mu_i + \sum_{j\neq i} p_{ji}s_j\mu_j} \geq p$; or*

2. $\mu_i \geq p$ *and* $\frac{\mu_i p_{i,n+1}}{\lambda_i + s_i \mu_i + \sum_{j \neq i} p_{ji} s_j \mu_j} \geq p.$

Let T_c *denote the coupling time of sample paths 1 and 2. Then*

$$E\{T_c\} \leq nM,$$

where M *is a quantity depending on* C *and* p.

In addition to the dependence of the expected coupling times on the number of stations in the system, we have studied the dependence of the expected coupling times on the buffer sizes, traffic rates, routing probabilities, and the types of the stations (loss vs. communication blocking) in the system. Our results indicate that the time segmentation approach can be efficiently used to simulate a large variety of queueing systems. More details on this issue can be found in Andradóttir and Ott (1995) and Hoseyni-Nasab and Andradóttir (1996a,b).

4 NUMERICAL RESULTS

In this section we present our numerical results aiming at better understanding the dependence of the expected coupling times of queueing networks on the parameters of the systems under study.

Our first set of numerical results has been obtained by simulating a network of n tandem communication blocking queueing stations with $s_i = 1$, $B_i = 10$, $\mu_i = 1$, for $i = 1, ..., n$, $\lambda_1 = 1$, $\lambda_i = 0$, for $i = 2, ..., n$, $p_{n,n+1} = 1$, $p_{i,n+1} = 0$, for $i = 1, ..., n-1$, and $n \in \{1, 2, 5, 10, 20, 100\}$. Note that for $n > 2$ this queueing system does not satisfy the conditions of Proposition 3.3. All clock times are independent and exponentially distributed. In each simulation two bounding sample paths are generated using Algorithm 3.1, one starting with all servers idle and one starting with all buffers full. The confidence intervals for the expected coupling times are obtained by generating 100 independent replications of the two sample paths. The results of our experiments are presented in Table 4.1.

The results in Table 4.1 suggest that for this class of queueing networks, the expected coupling times do not grow linearly with respect to the number of stations in the system. The growth rate of the expected coupling times appears to be beyond linear but no more than quadratic. This indicates that the conditions of Proposition 3.3 are at least to some extent necessary. To further investigate this issue we conducted simulations of two different variations of the queueing system of our first set of simulations. We slightly modified the parameters of the system in such a way that the modified systems satisfy the conditions of Proposition 3.3. Tables 4.2 and 4.3 show the results of simulations of two sets of queueing networks

Table 1: Dependence of the Expected Coupling Times of Tandem Networks of Communication Blocking Stations on the Number of Queueing Stations in the System when $\lambda_1 = 1, \lambda_i = 0$, for $i > 1$, $p_{n,n+1} = 1$, and $p_{i,n+1} = 0$, for $i < n$

Number of Stations, n	Coupling Times (95% Confidence Interval)
1	34.91 ($\pm$ 2.92)
2	66.09 ($\pm$ 5.60)
5	174.23 ($\pm$ 12.97)
10	435.92 ($\pm$ 26.64)
20	1,286.84 ($\pm$ 92.71)
100	12,061.88 ($\pm$ 772.24)

Table 2: Dependence of the Expected Coupling Times of Tandem Networks of Communication Blocking Stations on the Number of Queueing Stations in the System when $\lambda_1 = 1$, $\lambda_i = 0.1$, for $i > 1$, $p_{n,n+1} = 1$, and $p_{i,n+1} = 0$, for $i < n$

Number of Stations, n	Coupling Times (95% Confidence Interval)
1	34.91 ($\pm$ 2.92)
2	69.44 ($\pm$ 5.17)
5	106.13 ($\pm$ 6.40)
10	132.29 ($\pm$ 4.89)
20	159.18 ($\pm$ 6.20)
100	181.39 ($\pm$ 3.65)

that are different from the previous example in that $\lambda_i = 0.1$, for $i > 1$ (in Table 4.2), and $p_{i,n+1} = 0.1$, for $i < n$ (in Table 4.3), respectively. It is clear that even for large n these networks satisfy the conditions of Proposition 3.3. Tables 4.2 and 4.3 indicate that for the modified systems the expected coupling times grow sublinearly with respect to n. A thorough study of the conditions that affect the behavior of the expected coupling times is a subject of our current research.

5 CONCLUDING REMARKS

In this paper we presented a method for parallel simulation of discrete event systems, namely the time segmentation method, that is applicable to several classes of queueing networks. We presented our approach in a general setting for discrete event systems whose sample paths satisfy certain constructibility conditions. We showed that the efficiency of the approach is closely related to the expected coupling time

Table 3: Dependence of the Expected Coupling Times of Tandem Networks of Communication Blocking Stations on the Number of Queueing Stations in the System when $\lambda_1 = 1, \lambda_i = 0$, for $i > 1$, $p_{n,n+1} = 1$ and $p_{i,n+1} = 0.1$, for $i < n$

Number of Stations, n	Coupling Times (95% Confidence Interval)
1	34.91 (± 2.92)
2	59.18 (± 5.04)
5	114.86 (± 6.24)
10	150.75 (± 6.08)
20	184.42 (± 5.53)
100	284.79 (± 4.22)

of the sample paths of the system under study. We applied the time segmentation method to a family of Markovian queueing networks containing loss and communication blocking stations and investigated the validity and efficiency of the method through a variety of numerical and analytical results.

Our current research is directed towards understanding the behavior of the expected coupling times with respect to changes in the parameters of the systems under study, investigating further applications of the time segmentation approach, and studying the performance of the time segmentation approach in comparison with other parallel simulation methods.

ACKNOWLEDGMENT

This research was supported in part by the National Science Foundation under grants DDM-9210679 and DMI-9523111.

REFERENCES

Andradóttir, S., and T. Ott. 1995. Time Segmentation Parallel Simulation of Networks of Queues with Loss or Communication Blocking. *ACM Transactions on Modeling and Computer Simulation* 5(4):269–305.

Cassandras, C. G., and S. G. Strickland. 1989. Sample Path Properties of Timed Discrete Event Systems. *Proceedings of the IEEE* 77(1):59–71.

Chen, L., and R. Serfozo. 1995. Parallel Simulation by Multi-Instruction, Longest-Path Algorithms. Preprint.

Fujimoto, R. M. 1993. Parallel and Distributed Discrete Event Simulation: Algorithms and Applications. *Proceedings of the 1993 Winter Simulation Conference*, eds. G. W. Evans, M. Mollaghasemi, E. C. Russell, and W. E. Biles, 106–114.

Greenberg, A., B. Lubachevsky and I. Mitrani. 1990. Unboundedly Parallel Simulation Via Recurrence Relations. *Proceedings of the Conference on Measurement and Modeling of Computer Systems*, 1–12.

Heidelberger, P. 1988. Discrete Event Simulation and Parallel Processing: Statistical Properties. *SIAM Journal on Statistical Computing* 9(6):1114–1132.

Heidelberger, P., and H. Stone. 1990. Parallel Trace Driven Cache Simulation by Time Partitioning. *Proceedings of the 1990 Winter Simulation Conference*, eds. O. Balci, R. P. Sadowski, and R. E. Nance, 734–737.

Hoseyni-Nasab, M., and S. Andradóttir. 1996a. Efficiency of Time Segmentation Parallel Simulation of Finite Markovian Queueing Networks. Working Paper.

Hoseyni-Nasab, M., and S. Andradóttir. 1996b. Parallel Simulation of Transfer Lines by Time Segmentation. Working Paper.

Lin, Y., and D. Lazowska. 1991. A Time Division Algorithm for Parallel Simulation. *ACM Transactions on Modeling and Computer Simulation* 1 (1):73–83.

Vakili, P. 1991. Using a Standard Clock Technique for Efficient Simulation. *Operations Research Letters* 10(8):445–452.

AUTHOR BIOGRAPHIES

MEHDI HOSEYNI-NASAB is a Ph.D. student in the Department of Industrial Engineering of the University of Wisconsin–Madison and a guest student in the School of Industrial and Systems Engineering of the Georgia Institute of Technology. His research interests include parallel simulation and applied probability. He is a member of INFORMS and IIE.

SIGRÚN ANDRADÓTTIR is an Associate Professor in the School of Industrial and Systems Engineering at the Georgia Institute of Technology, on leave from the University of Wisconsin – Madison where she has held a faculty position since 1990. She received a B.S. in Mathematics from the University of Iceland in 1986, an M.S. in Statistics from Stanford University in 1989, and a Ph.D. in Operations Research from Stanford University in 1990. Her research interests include stochastic optimization, simulation, and stochastic processes. She presently serves as Associate Editor for *IIE Transactions, Stochastic Models,* and the *ACM Transactions on Modeling and Computer Simulation.* She is a member of INFORMS.

Proceedings of the 1996 Winter Simulation Conference
ed. J. M. Charnes, D. J. Morrice, D. T. Brunner, and J. J. Swain

EXPERIMENTAL EVALUATION OF CONFIDENCE INTERVAL PROCEDURES IN SEQUENTIAL STEADY-STATE SIMULATION

Donald C. McNickle
Krzysztof Pawlikowski
Gregory Ewing

University of Canterbury
Christchurch
NEW ZEALAND

ABSTRACT

Sequential analysis of simulation output is generally accepted as the most efficient way for securing representativeness of samples of collected observations. In this scenario a simulation experiment is stopped when the relative precision of estimates, defined as the relative width of confidence intervals at an assumed confidence level, reaches the required level. This paper deals with the statistical correctness of the methods proposed for estimating confidence intervals for mean values in sequential steady-state stochastic simulation. We formulate basic rules that should be followed in proper experimental analysis of coverage of different steady-state interval estimators. Our main argument is that such analysis should be done sequentially. The numerical results of our preliminary coverage analysis of the method of Spectral Analysis (SA/HW) and Non-overlapping Batch Means are presented, and compared with those obtained by traditional, non-sequential approaches.

1 INTRODUCTION

Sequential analysis of simulation output is generally accepted as the most efficient way for securing representativeness of samples of collected observations [see, for example, Law and Kelton (1992)]. In this scenario a simulation experiment is stopped when the relative precision of estimates, defined as the relative width of confidence intervals at an assumed confidence level, reaches the required level.

This paper deals with the statistical correctness of the methods proposed for estimating confidence intervals of mean values in sequential steady-state stochastic simulation. The main analytical problems of such simulation studies were discussed in Law (1983) and Pawlikowski (1990). They are caused by correlations between events observed during typical simulated processes. At least a dozen methods have been proposed for estimating confidence intervals of mean values from series of correlated observations collected during such simulation. A survey of the methods until 1990 can be found in Pawlikowski (1990). Newer proposals include those by Fox et al. (1991), Goldsman and Kang (1991), Howard et al. (1992). So far only a few implementations of these methods in an automated sequential simulation framework have been reported [see for example Fox et al. (1991), Heidelberger and Welch (1983), Pawlikowski et al. (1994), Rego and Sunderam (1992), Yau and Pawlikowski (1993)] and incorporated in some simulation packages. The methods are based on different approximations and their quality should be assessed by analysing the properties of the final confidence intervals they generate. A good method should produce narrow and stable confidence intervals, which should of course be valid, ie. they should contain the true value of the estimated performance measure (with the correct probability). Theoretical studies of various estimators of confidence intervals, reported before 1990, are surveyed in Pawlikowski (1990). Newer results can be found for example in Kang and Goldsman (1990).

Unfortunately, no satisfactorily exhaustive comparative studies of these methods have been reported yet, and it is difficult to find a good method for a specific range of applications. Additionally, most studies relate to non-sequential simulation experiments run on single processors. Very little is known about quality of these methods in sequential simulation, and in fast concurrent sequential simulations based on Multiple Replications in Parallel (MRIP), when multiple processors cooperate in production of data – see Pawlikowski et al. (1994).

The theoretical studies of confidence intervals reveal general conditions which have to be satisfied to ensure the validity of the final confidence intervals, but the correctness of any practical implementation of a specific method also has to be tested experimentally. In this paper we formulate a new methodology of such experimental studies of the methods used in sequential stochastic simulation for determining the final precision of results, and present the results of our comparative studies of two selected methods: SA/HW (the method of Spectral

Analysis in its version proposed in Heidelberger and Welch (1981), and the classical method of (non-overlapping) Batch Means, both in sequential simulations on single processors and in sequential simulations on multiple processors in MRIP scenario. Further directions of research in this area are indicated in the Conclusions.

2 EXPERIMENTAL ANALYSIS OF COVERAGE

In any performance evaluation studies of dynamic systems by means of stochastic discrete-event simulation the final estimates should be determined together with their statistical errors, which are usually measured by the half-width of the final confidence intervals. Restricting our attention to estimators of means, let us assume that we estimate theoretical mean $\mu = EX$ by

$$\overline{X}(n) = \frac{1}{n}\sum_{i=1}^{n} x_i$$

where $x_1, x_2, \ldots, x_n$ are observations collected during simulation. Then, one should also determine

$$P(\overline{X}(n) - \Delta \leq \mu \leq \overline{X}(n) + \Delta) = 1 - \alpha$$

i.e. the confidence interval (c.i.) of μ, at a given confidence level $1-\alpha$, $0<\alpha<1$. Δ is the half-width of the c.i. Typically, this will be $\Delta = t_{\kappa,1-\alpha/2}\hat{\sigma}[\overline{X}(n)]$, where $\hat{\sigma}^2[\overline{X}(n)]$ is an estimator of the variance of $\overline{X}(n)$, with κ degrees of freedom and $t_{\kappa,1-\alpha/2}$ is the $(1-\alpha/2)$ quantile of the Student t-distribution.

Various estimators of $\hat{\sigma}[\overline{X}(n)]$ have been proposed. This in sequel has created the need for a quality assessment of these estimators.

Let us note that in an ideal case the final c.i. would contain μ with the probability $1-\alpha$, or equivalently, if an experiment is repeated many times, one would expect to have μ contained in about $(1-\alpha)100\%$ of final confidence intervals. *Coverage of confidence intervals* is defined as the frequency with which the final confidence intervals $(\overline{X}(n) - \Delta \leq \mu \leq \overline{X}(n) + \Delta)$ contain the true value μ. While some interesting results have been achieved in theoretical studies of coverage [see eg. Glynn (1982), Kang and Goldsman (1990), Schruben (1980)], experimental analysis of coverage is still required for assessing the quality of practical implementations of methods used for determining confidence intervals in steady-state simulation. Of course, such analysis is limited to analytically tractable systems, since the value of μ has to be known.

As for any other point estimate, the coverage can be determined together with its c.i. :

$$\left(c - z_{1-\alpha/2}\sqrt{\frac{c(1-c)}{n_c}}\ ,\ c + z_{1-\alpha/2}\sqrt{\frac{c(1-c)}{n_c}}\right) \quad (1)$$

where c is the coverage, $z_{1-\alpha/2}$ is the $(1-\alpha/2)$ quantile of the standard normal distribution and n_c is the (suitably large) number of replicated experiments in the coverage analysis. In this study, n_c is at least 600, so the use of a normal approximation seems to be justified.

Then, an estimator of $\hat{\sigma}^2[\overline{X}(n)]$ used for determining the c.i. of μ can be considered as valid, ie. producing valid $100(1-\alpha)\%$ confidence intervals of μ, if the upper bound of the confidence interval of the coverage c in Equation (1) equals at least $(1-\alpha)$; see Sauer (1977).

Results of experimental coverage analysis have been reported in many publications, although majority of these results are related to simulations on single processors. Very little is known about coverage of estimators that could be used in parallel simulation executed in the MRIP scenario [Pawlikowski et al. (1994)]. It is strange, but while sequential simulation is generally recognised as the only way of producing results with the required precision since "*..no procedure in which the run length is fixed before the simulation begins can be relied upon to produce a c.i. that covers the true steady-state mean with the desired probability level*" [Law and Kelton (1982), Law and Kelton (1992)], even the original advocates of sequential simulation have applied non-sequential (fixed-sample size) approaches in their simulation studies of coverage. Certainly, if one accepts the arguments for the sequential approach as the only practical one then also such meta-simulation experiments as those for coverage analysis should be run sequentially!

In addition, most reported results on coverage were based on 50-200 replications [see for example Adam (1983), Fishman (1978), Heidelberger and Welch (1983), Kelton and Law (1984), Law and Carson (1979), Law and Kelton (1982), Lavenberg and Sauer (1977), Sauer (1979), Schriber and Andrews (1981), Schruben (1983)], which obviously puts in question the statistical representativeness of such experimental data. In all these cases, the estimates of coverage were based on only a few bad confidence intervals, ie. ones which did not contain μ.

This issue could be solved by requiring that the coverage can be estimated only after a minimum number of bad confidence intervals has been recorded.

It is also generally known that sequential steady-state simulation can produce very inaccurate estimates if the stopping criterion is only accidentally temporarily satisfied. Sensible practise is to ensure that estimates do not come from simulation runs that are too short. Thus, this effect should be similarly treated, and eliminated, in coverage analysis.

Recognising the significance of all these three factors, we have applied the following rules in experimental analysis of coverage of interval estimators from stochastic simulation:

R1. Coverage should be analysed sequentially, ie. analysis of coverage should be stopped when the relative precision (the relative half-width of c.i.) of the estimated coverage falls below an assumed level.

R2. An estimate of coverage has to be calculated from a representative sample of data, ie. the coverage analysis can start only after a minimum number of bad confidence intervals has been recorded.

R3. Results from simulation runs that are abnormally short should be not taken into account.

Details of our implementation of these rules for studying the quality of the final steady-state interval estimators of means in traditional simulation on single processors as well as fast concurrent simulation on multiple processors is discussed in the next section.

3 NUMERICAL RESULTS

Implementing the rules of coverage analysis formulated in the previous section, we must select (i) the minimum number of bad confidence intervals, Nmin, which have to be recorded before the sequential analysis of coverage can start, and (ii) the minimum sufficient length of simulation to produce valid steady-state estimates. The way in which we approached these problems is illustrated in Fig.1 and 2.

The results presented there are for SA/HW, the method of Spectral Analysis in its version proposed in Heidelberger and Welch (1981). Our implementation of this method for simulations on single processors followed exactly procedures specified in Pawlikowski (1990), including the procedure described there for detecting the length of the initial transient period. Its generalisation for simulations in the MRIP scenario was described in Pawlikowski et al. (1994).

All reported results were obtained by stopping simulations when the final steady-state results reached a (relative) precision of 5% or less, at the 0.95 confidence level. All series of replicated simulations were executed using strictly non-overlapping sequences of pseudo-random numbers generated by a linked sequence of congruential generators listed in Law and Kelton (1992).

The results presented in Fig.1 and 2 were obtained by running multiple independent replications of sequential steady-state simulation of $M/D/1/\infty$ queuing system on P=1 and 4 processors, respectively. The estimated parameter is the mean time that a customer spends in the queue. In all four cases the analysis of coverage was initiated after observing N_{min} bad confidence intervals. This happened after about 300 independent replications in the case of $N_{min}=30$ [Fig.1(a) and 2(a)] and after about 2000 independent replications in the case of $N_{min}=200$ [Fig.1(a) and 2(a)]. At this stage the lengths of executed simulations were analysed, and the results obtained from simulation runs shorter than L_{min} (one standard deviation below the average number of observations needed to stop sequential simulation with the required precision at the assumed confidence level) were discarded (hence the sudden improvement in coverage). L_{min} was also later used as the criterion for rejecting/accepting results from any additional replication in cases where sequential analysis of coverage had to be continued.

One can see that in all four cases filtering out too short simulation runs removes significant noise. On the other hand, continued instability of the coverage observed after $N_{min}=30$ bad confidence intervals have been collected shows that, possibly due to strong asymmetry of the sample distribution, many more than 30 bad confidence intervals had to be recorded to secure representativeness in the analysed data. This conclusion, on the basis of similar results we obtained for other queuing systems, suggested that many more replications were needed than used in previous studies. For this reason, in our further analysis of coverage we assumed $N_{min}=200$.

The results of sequential coverage analysis for SA/HW in simulations executed on a single processor, as well as for simulations in MRIP scenario for P=2 and 4 processors, are presented in Fig.3 (a)-(c), respectively. The performance of the SA/HW method improves with the number of processors or, equivalently, the number of independent simulation engines, used in the MRIP scenario. Another attractive feature of SA/HW is its good performance when simulating heavily loaded systems, ie. in the region where other methods usually fail. The "safe" degree of parallelisation for SA/HW has yet to be determined.

For comparison, Fig. 3(d) shows results one could obtain applying traditional fixed-sample-size approach, and estimating the coverage on the basis of first 200 replications. It is evident that the results obtained from traditional analyses of coverage cannot be considered as reliable.

Coverage analysis of different methods proposed for estimating confidence intervals of mean values in sequential steady-state stochastic simulation is illustrated here by the results obtained for the method of non-overlapping batch means (BM); see Fig.4 (a) and (b). The method was implemented following procedures specified in Pawlikowski (1990), including the procedure described there for detecting the length of the initial transient period. In the case of simulations executed on a single processor, SA/HW and BM offer similar (bad) coverage. When more processors are used under MRIP,

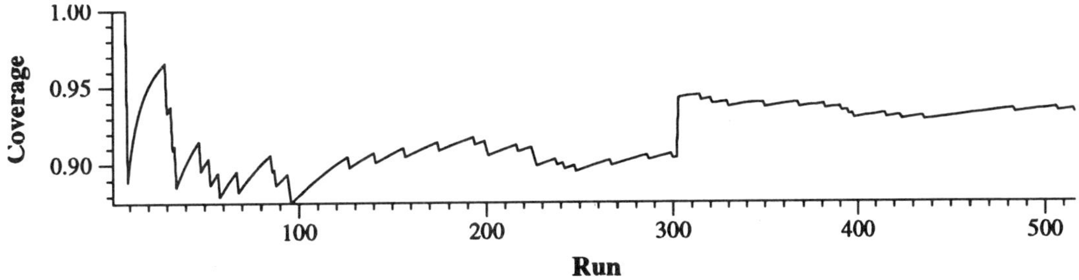

(a) N_{min} =30

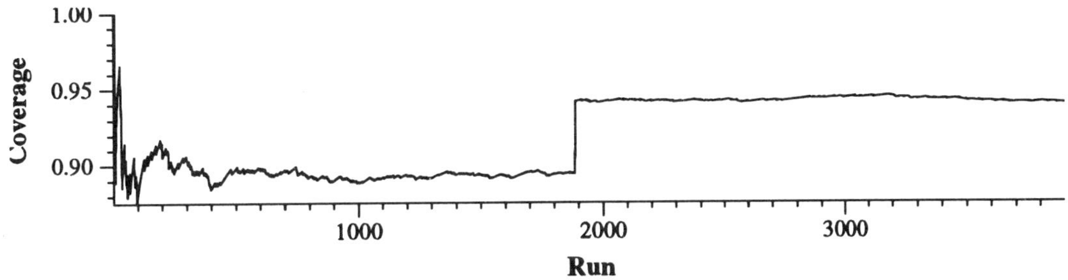

(b) N_{min} =200

Figure 1: Coverage as a Function of Sample Size for SA/HW in Steady-State Simulation of an M/D/1/∞ Queuing System for ρ=0.5, P=1 processor

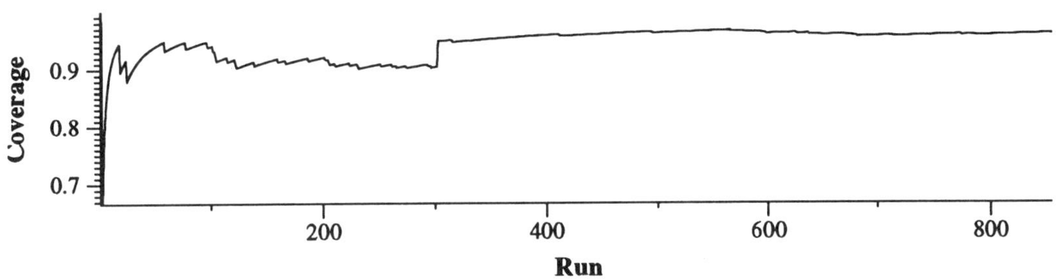

(a) N_{min}=30

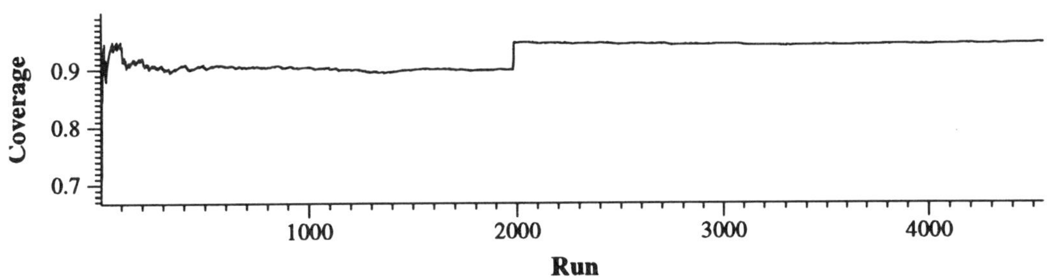

(b) N_{min}=200

Figure 2: Coverage as a Function of Sample Size for SA/HW in Steady-State Simulation of an M/D/1/∞ Queueing System for ρ=0.5, P=4 processors

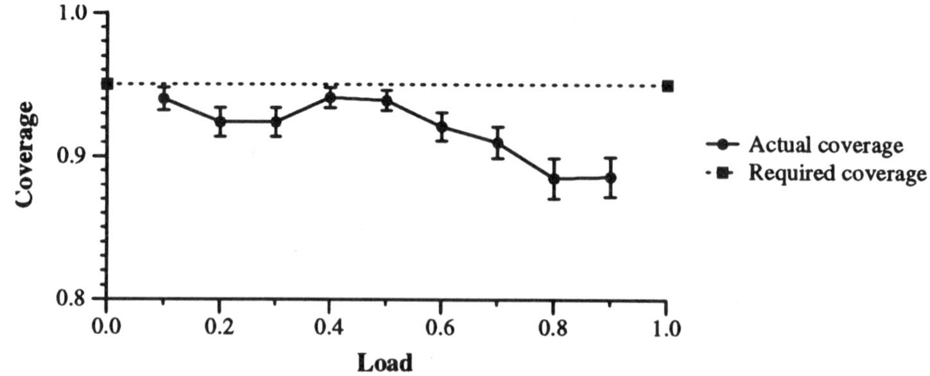

(a) Sequential Analysis, P=1, N_{min} =200

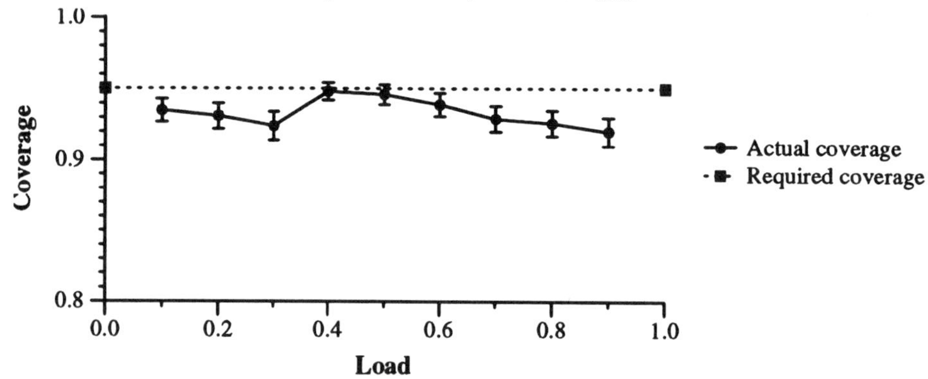

(b) Sequential Analysis, P=2 N_{min}=200

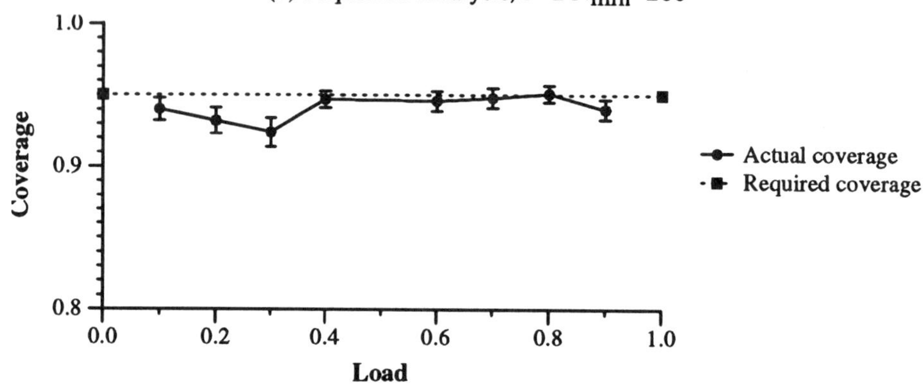

(c) Sequential Analysis, P=4, N_{min}=200

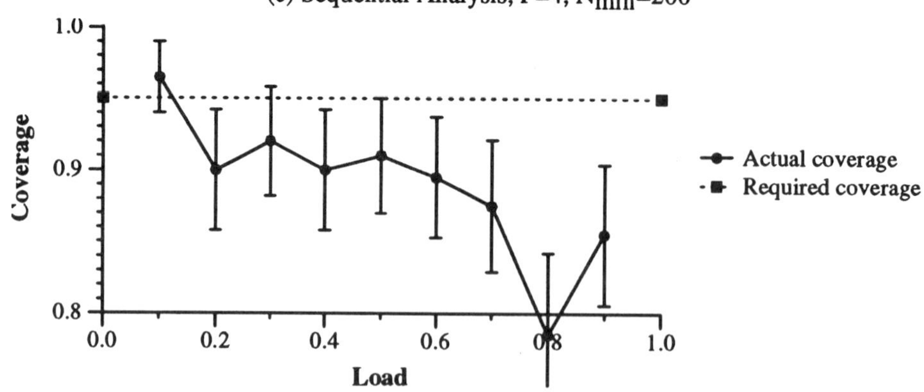

(d) Fixed-Sample-Size Analysis based on 200 Replications

Figure 3: Coverage of SA/HW in Steady-State Simulation of an M/D/1/∞ Queueing System

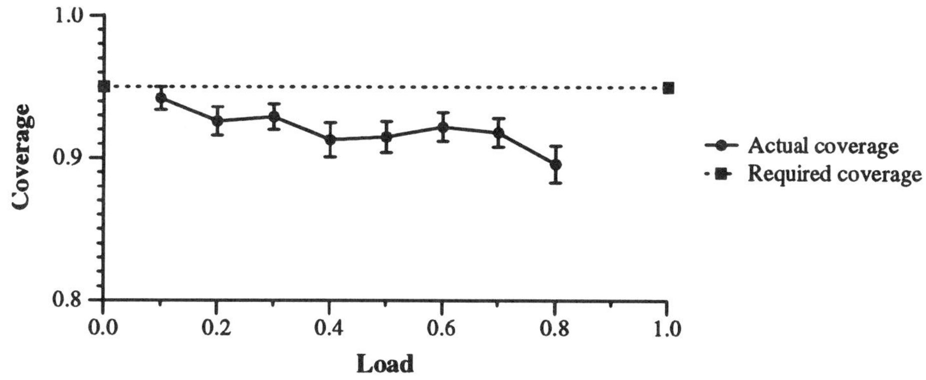

(a) Sequential Analysis, P=1, N_{min}=200

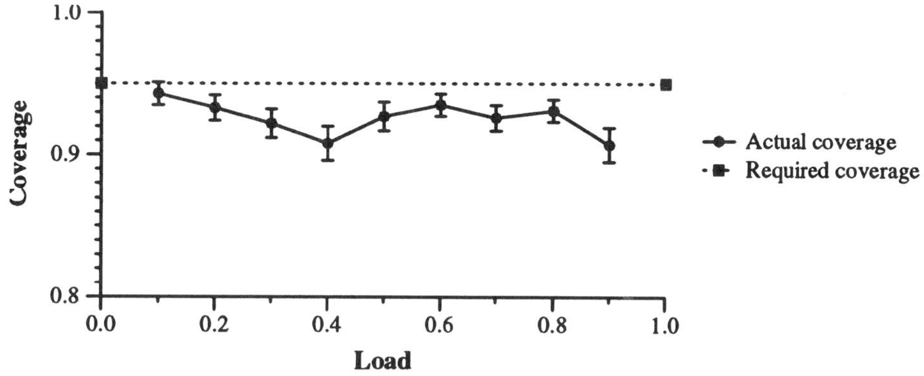

(b) Sequential Analysis, P=2, N_{min}=200

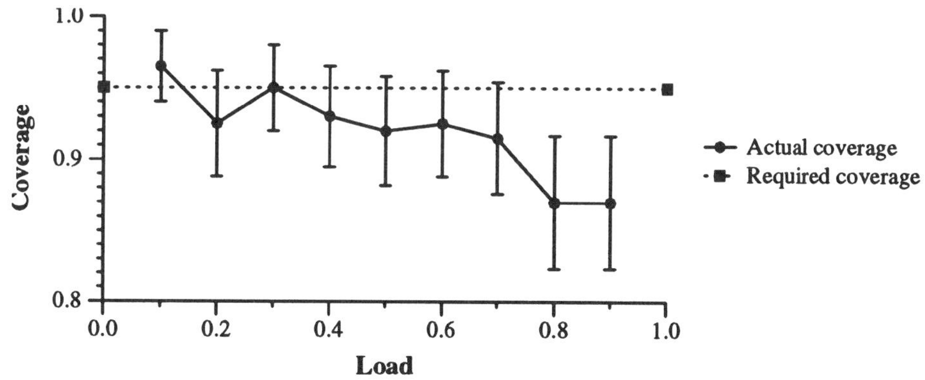

(c) Fixed-Sample-Size Analysis based on 200 Replications

Figure 4: Coverage of BM in Steady-State Simulation of an M/D/1/∞ Queuing System

using SA/HW one can obtain much better results. This has been consistently observed both when simulating M/M/1/∞ and M/D/1/∞ buffers. Fig. 4(c) again shows results one could obtain applying the traditional fixed-sample-size approach, estimating the coverage on the basis of first 200 replications. These results are much too optimistic if compared with the ones obtained on the basis of representative samples from sequential analysis of coverage. Our experimental comparative studies of different methods proposed for estimating confidence intervals of mean value in sequential steady-state stochastic simulation are continued.

4 CONCLUSIONS

We have formulated basic rules that should be followed in proper experimental analysis of coverage of different steady-state interval estimators. Our main argument is that such analysis should be done sequentially. The numerical results of our preliminary coverage analysis of the method of Batch Means and Spectral Analysis (SA/HW) have been also presented and compared with those obtained by traditional, non-sequential approach. As advocated in Law (1983), to draw more general conclusions about performance of interval estimators used in various methods of sequential steady-state simulation one needs to consider a number of different simulation models. Unfortunately until now no such standard set of reference models for coverage analysis has been wider adopted, in spite of that the issue being raised already in 1981 in Schriber and Andrews (1981).

REFERENCES

Adam, N. R. 1983. Achieving a Confidence Interval for Parameters Estimated by Simulation. *Management Science* 29: 856-866.

Fishman, G. S. 1978. Grouping Observations in Digital Simulation. *Management Science* 24: 510-521.

Fox, B. L., D. Goldsman and J. Swain. 1991. Spaced Batch Means. *Operations Research Letters* 10: 255-263.

Glynn, P. W. 1982. Coverage Error for Confidence Intervals Arising in Simulation Output Analysis. In *Proceedings of the 1982 Winter Simulation Conference*, IEEE Press, 369-375.

Goldsman, D., and K. Kang. 1991. Cramer-von Mises Variance Estimators for Simulations. In *Proceedings of the 1991 Winter Simulation Conference*, IEEE Press, 916-920.

Goldsman, D., et al. 1986. Large and Small Sample Comparisons of Various Variance Estimators. In *Proceedings of the 1982 Winter Simulation Conference* 278-284. IEEE Press.

Heidelberger, P., and P. D. Welch. 1981. A Spectral Method for Confidence Interval Generation and Run Length Control in Simulation. *Communications of the ACM* 25: 233-245.

Heidelberger, P., and P. D. Welch. 1983. Simulation Run Length Control in the Presence of an Initial Transient. *Operations Research*. 31: 1109-1144.

Howard, R. B. et al. 1992. Confidence Intervals for Univariate Discrete-Event Simulation Output Using the Kalman Filter. In *Proceedings of the 1992 Winter Simulation Conference* 586-593.

Kang, K., and D. Goldsman. 1990. The Correlation Between Mean and Variance Estimators in Computer Simulation. *Transactions of the IIE* 22 (1): 15-23.

Kelton, W. D., and A. M. Law. 1984. An Analytical Evaluation of Alternative Strategies in Steady-State Simulation. *Operations Research* 32 (1): 169-184.

Lavenberg, S. S., and C. H. Sauer. 1977. Sequential Stopping Rules for the Regenerative Method of Simulation. *IBM Journal of Research and Development* 21: 667-678.

Law, A. M., and J. S. Carson. 1979. A Sequential Procedure for Determining the Length of a Steady-State Simulation. *Operations Research* 27: 1011-1025.

Law, A. M., and W. D. Kelton. 1982. Confidence Intervals for Steady-State Simulations, II: A Survey of Sequential Procedures. *Management Science* 28 (5): 550-562.

Law, A. M. 1983. Statistical Analysis of Simulation Output Data. *Operations Research* 31 (6): 983-1029.

Law, A. M., and W. D. Kelton. 1992. *Simulation Modeling and Analysis*. New York: McGraw-Hill.

Pawlikowski, K. 1990. Steady-State Simulation of Queueing Processes: A Survey of Problems and Solutions. *ACM Computing Surveys* 22 (2): 123-170 (plus Corrigendum, *ACM Computing Surveys* 224, 409).

Pawlikowski, K., V. Yau and D. McNickle. 1994. Distributed Stochastic Discrete-Event Simulation in Parallel Time Streams. In *Proceedings of the. 1994 Winter Simulation Conference* 723-730. IEEE Press.

Rego, V. J., and V. S. Sunderam. 1992. Experiments with Concurrent Stochastic Simulation: the EcliPSe Paradigm. *Journal of Parallel and Distributed Computing* 14: 66-84.

Sauer, C. H. 1979. Confidence Intervals for Queueing Simulations of Computer Systems. *ACM Performance Evaluation Review* 8 (1-2): 46-55.

Schriber, T. J. and R. W. Andrews. 1981. A Conceptual Framework for Research in the Analysis of Simulation Output. *Communications of the ACM* 24 (4): 218-232.

Schruben, L. W. 1980. A Coverage Function for Interval Estimators of Simulation Response. *Management Science* 26: 18-27.

Schruben, L. W. 1983. Confidence Interval Estimation Using Standardized Time Series. *Operations Research* 30: 1090-1108.

Yau, V., and K. Pawlikowski. 1993. AKAROA: a Package for Automatic Generation and Process Control of Parallel Stochastic Simulation. In *Proceedings of the. 16th Australian Computer Science Conference* 71-82. Australian Computer Science Communications.

AUTHOR BIOGRAPHIES

DON MCNICKLE is a Senior Lecturer in Management Science in the Deparmtent of Management at the University of Canterbury. His research interests include queueing theory, networks of queues and statistical aspects of stochastic simulation. He is a member of INFORMS.

KRYS PAWLIKOWSKI is an Associate Professor of Computer Science at the University of Canterbury. His research interests include quantitative stochastic simulation, and performance modelling and evaluation of telecommunication networks. He received a PhD in Computer Engineering from the Technical University of Gdansk, Poland in 1975 He is a Senior Member of IEEE.

GREG EWING is a research associate in the Department of Computer Science at Canterbury. He recently completed his Ph.D. on Georgraphic Information Systems. His research interests include languages, 3D graphics and graphical user interfaces.

Proceedings of the 1996 Winter Simulation Conference
ed. J. M. Charnes, D. J. Morrice, D. T. Brunner, and J. J. Swain

THE RESTART/LRE METHOD FOR RARE EVENT SIMULATION

Carmelita Görg
Friedrich Schreiber

Communication Networks
Aachen University of Technology
D-52056 Aachen, GERMANY
[cg|fschr]@comnets.rwth-aachen.de

ABSTRACT

Using the LRE-algorithm (LRE: Limited Relative Error) for the evaluation of simulated data yields the stationary distribution function of an investigated random sequence and additionally the so-called local correlation coefficient, which represents relevant correlation evidence to be included in the error measure for controlling the simulation run time. In this paper a simplified LRE-algorithm is used to evaluate discrete sequences like the occupancy of finite buffer queueing systems $G/G/1/N$. It is shown how this algorithm is combined with the RESTART-method for an efficient rare event simulation.

A multi-stage RESTART/LRE-algorithm has been implemented as part of a stochastic simulation system and its performance has been verified by extensive simulations of the reference system $M/M/1/N$, whose properties including the local correlation coefficient can be described by analytical formulas. Approximate formulas for the optimal number of stages and number of trials are given. The new algorithm has been successfully applied to several finite buffer queueing systems, including the $SSMP(2)/D/1/N$ system with a correlated input stream, in order to gain by simulation the distribution function of the occupancy including very low loss probabilities in the order of 10^{-9} and even 10^{-21}, which are relevant for the performance of switching nodes in the ATM broadband network.

1 INTRODUCTION

The quantitative investigation of random systems by stochastic simulation is an important tool for telecommunication engineers. How to assure the *reliability* of simulated results is a problem of interest and depends on the statistical evaluation method used to control the simulation run time. Since run time is always limited, there are limitations for locating in a simulated sequence a *rare event* with sufficient accuracy.

This problem has become an issue of current interest in the performance investigation of switching nodes of the ATM broadband network, which can be modeled by a single server, finite buffer queueing system $G/D/1/N$ with a correlated general input process "G". System design has to guarantee that the event "cell loss" is a rare event with a probability in the range of $10^{-9} \cdots 10^{-10}$ (Le Boudec, 1991). How can this requirement be investigated and verified by simulation without excessive run time?

The relevant subfield within the various methods for variance reduction, see (Kleijnen, 1974), is *importance sampling* which originated as a method for more efficient Monte Carlo integration (Kahn and Marshall, 1953) and has been applied in a quite different way, called *importance splitting* in (Shahabuddin, 1995), to the statistical investigation of Markovian sequences by (Bayes, 1970) and has been further developed and made known under the name RESTART by (Villén-Altamirano and Villén-Altamirano, 1991) ("REpetitive Simulation Trials After Reaching Threshold").

As shown in (Schreiber and Görg, 1994) this method can be combined with the so-called LRE-method, see (Schreiber and Görg, 1996), which has been designed to resolve the often debated problem, see e.g. (Bratley et al., 1987), how to evaluate correlated sequences. This combination of two methods – named the RESTART/LRE-method – provides rare event simulation with a reliable error and run length control.

In this paper the RESTART/LRE-method with multiple stages is investigated and it is shown how the optimal number of stages provide a dramatic increase of speed-up in comparison to straightforward and single stage simulations. Examples are given of simulations that would have taken several years and even millions of years that could now be reduced to give good results within one day.

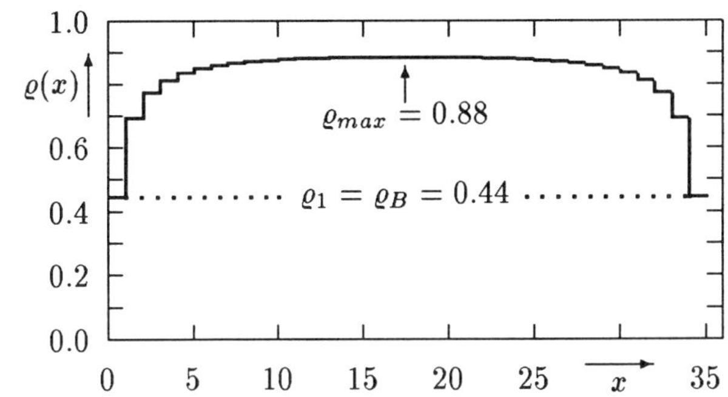

Figure 1: Queueing System M/M/1/34: Local C.C. $\varrho(x)$ of Occupancy x ($\eta = 0.8$)

2 REFERENCE MODEL M/M/1/N

It is absolutely necessary to verify a new simulation method using reference queueing models whose relevant properties can be described by analytically derived exact formulas. A basic reference model for the rare event simulation is the single server queueing system M/M/1/N-FIFO with a finite buffer size N, whose local correlation properties have been recently investigated (Schreiber, 1994). With η denoting the traffic load and the maximum occupancy $B = N+1$, the discrete random variable "occupancy β" is described by the following equations for the stationary complementary distribution function (compl. d.f.) $G(x) = 1 - F(x)$, for the loss probability P_L, and the local correlation coefficient (local c.c.) $\varrho(x)$ for the interval $i-1 \leq x < i$ and $i = 1, \cdots, B$, see (Schreiber, 1994):

$$\left.\begin{aligned} G(x) = G_i = \sum_{\beta=i}^{B} P(\beta) = \frac{\eta^i - \eta^{B+1}}{1-\eta^{B+1}}; \\ G_0 = 1; \quad G_B = P_L = \frac{\eta^B(1-\eta)}{1-\eta^{B+1}}; \end{aligned}\right\} \quad (1)$$

$$\varrho(x) = \varrho_i = 1 - \frac{(1-\eta)(1-\eta^{B+1})\,\eta^i}{(1+\eta)(1-\eta^i)(\eta^i - \eta^{B+1})}\,. \quad (2)$$

Analytical investigations and simulations of other, less elementary single server queueing systems with finite buffer size have shown that the character of the flat maximum curve $\varrho(x)$ of system M/M/1/N is representative for several finite systems, see Figure 1.

3 THE LRE-ALGORITHM

The LRE-algorithm (LRE: Limited Relative Error) has been designed to establish the distribution function (d.f.) $F(x)$ and the local correlation coefficient (local c.c.) $\varrho(x)$ of a stationary x-sequence with *a priori* unknown properties. The LRE-algorithm II given in (Schreiber, 1988) is an extension of the first LRE-algorithm I, which was defined for the evaluation of *independent* sequences. The algorithm requires, among other things, the finding of the location and size of the x-intervals and the detection of discrete points. The procedure for performing automatically this general, rather complicated evaluation task can be substantially simplified by applying the LRE-algorithm III (Schreiber and Görg, 1996), whose evaluation task has been restricted to the measurement of discrete random sequences. When, for instance, the occupancy and loss probabilities of a queueing system G/G/1/N are to be investigated, the discrete character of the random variable "occupancy β" with buffer size N, maximum occupancy $B = N+1$, and the exact range $\beta = 0, \cdots, B$ is known.

For deriving the complementary distribution function $G(x) = 1 - F(x)$ of the arrival occupancy β of a system G/G/1/N, the values of β are registered exclusively *prior* to the end of each interarrival time τ_a, that is at all instants "immediately before a new item arrives". By applying this rule the random state behavior of the system is associated to an embedded Semi-Markov chain with $B+1$ nodes.

3.1 The local correlation coefficient $\varrho(x)$

We now assume that β is generated by the random state sequence of a recurrent $(k+1)$-node Markov chain with $k = B = N+1$, whose complementary d.f. $G(x) = 1 - F(x)$ exists as shown by Figure 2. Then we can split at any point x on the real axis this chain into two parts and define an "$F(x)$-equivalent" 2-node Markov chain (Schreiber, 1987a), whose transition probabilities $p_0(x)$ and $p_1(x)$ determine its correlation coefficient $\varrho(x) = 1 - [p_0(x) + p_1(x)]$ for $i-1 \leq x < i$.

According to Equation (6a,b) in (Schreiber, 1987a) the function $\varrho(x)$ represents a well defined (first order) correlation measure for the $(k+1)$-node chain and is called the local c.c. $\varrho(x)$ due to its dependence on the location x (Ding and Schreiber, 1990). After the chain has performed n state changes (trials) the posterior mean $\tilde{\varrho}(x)$ can be obtained by relating the number of transitions a_i and $c_i \approx a_i$, respectively, across the separation line at x to the total number of observed events r_i in the left range $\beta = 0, 1, \cdots, i-1$ resp. $v_i = n - r_i$ in the right range $\beta = i, i+1, \cdots, B$, see (3). If the large sample conditions (Schreiber, 1987b):

$$n \geq 10^3; \; (r_i, v_i) \geq 10^2; \; (a_i, c_i, r_i - a_i, v_i - c_i) \geq 10$$

are fulfilled, the following posterior formulas express the *posterior compl. d.f.* $\tilde{G}(x)$, the *posterior mean occupancy* $\tilde{\beta}$, the *posterior local c.c.* $\tilde{\varrho}(x)$ with *correla-*

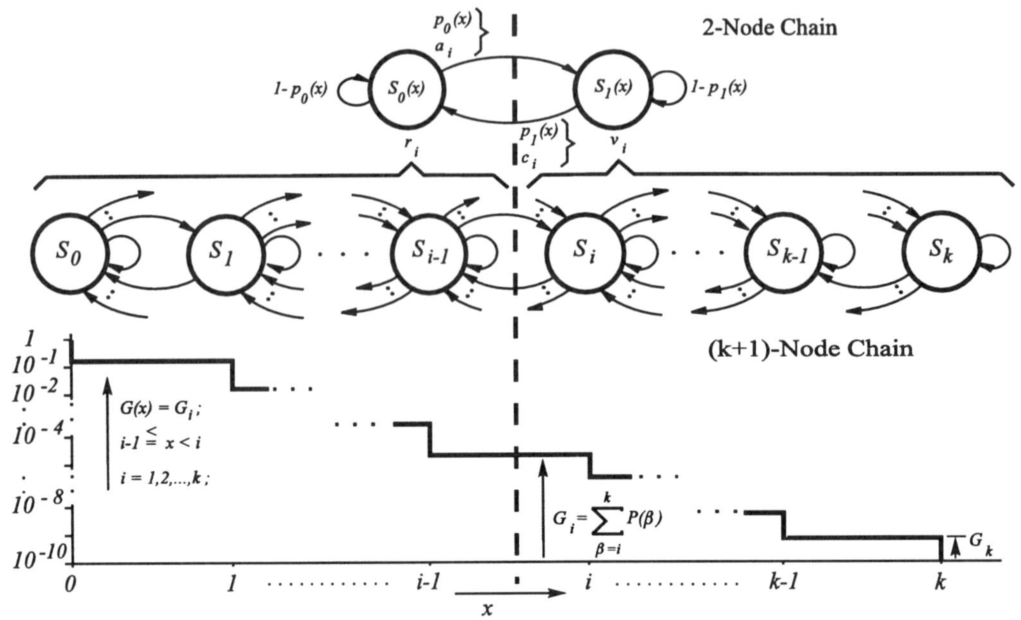

Figure 2: Discrete $(k+1)$-Node Markov Chain and the "$G(x)$-equivalent" 2-Node Markov Chain

tion factor $cf(x)$, and the *relative error* $d_G(x) \equiv d(x)$ for $i-1 \leq x < i, i = 1, \cdots, B$, see (Schreiber, 1988):

$$\left.\begin{array}{lllll} \tilde{G}(x) & = & \tilde{G}_i & = & v_i/n\,;\ \tilde{\beta} = \frac{1}{n}\sum_{i=1}^{B} v_i\,; \\ \tilde{\varrho}(x) & = & \tilde{\varrho}_i & = & 1 - \dfrac{c_i/v_i}{1 - v_i/n}\,; \\ cf(x) & = & cf_i & = & (1+\tilde{\varrho}_i)/(1-\tilde{\varrho}_i)\,; \\ d(x) & = & d_i & = & \left[\dfrac{1 - v_i/n}{v_i} \cdot cf_i\right]^{1/2}\,. \end{array}\right\} \quad (3)$$

Compared to conventional batch means and confidence interval methods, the relative error $d(x)$ has the great advantage that it contains correlation evidence, namely the measured values $\tilde{\varrho}_i$, and that it can be continuously reduced by evaluating the evidence of as many more trials as are needed to meet the error condition $d_i \leq d_{max}$, with d_{max} being the prescribed error limit. The total needed number of trials n is determined by d_{max} and by the smallest G-value to be established, namely $G_{min} = \tilde{G}_B$ in the present case; with $d_B = d_{max}$ we obtain from (3):

$$n = \frac{(1-\tilde{G}_B)\, cf_B}{\tilde{G}_B\, d_{max}^2} \approx \frac{cf_B}{\tilde{G}_B\, d_{max}^2}; \quad cf_B = \frac{1+\tilde{\varrho}_B}{1-\tilde{\varrho}_B}.$$

4 THE RESTART/LRE-ALGORITHM

Verifying the loss probability $P_L = G_B = 8.1 \times 10^{-5}$ of the reference system M/M/1/34 for $\eta = 0.8$ by simulation would afford $n = 3.2 \times 10^6$ trials for an error limit of $d_{max} = 0.1$ and, assuming a production rate of 10^4 trials per second, a simulation run time of $T = 5.3$ minutes. By raising the buffer size from $N = 34$ to $N = 85$ we achieve for $\eta = 0.8$ a loss probability $P_L = 9.26 \times 10^{-10}$ and a simulation with again $d_{max} = 0.1$ would now afford $n = 2.8 \times 10^{11}$ trials and the (prohibitive) run time $T = 324$ days. The demand of a reduced 5 % error limit $d_{max} = 0.05$ would increase the simulation run time for $N = 34$ to $T = 21.4$ minutes and for $N = 85$ to $T = 3.56$ years.

This is the typical situation where straightforward simulation is no longer possible and must be replaced by a simulation based on the so-called RESTART-principle (Villén-Altamirano and Villén-Altamirano, 1991). In the following this principle will be combined with the evaluation and error control methods of LRE-algorithm III.

The main object of simulation will be the distribution function of the occupancy β of system G/G/1/N with the "rare event" being given by the loss case $\beta = N+1 = B$ and the discrete $(B+1)$-node Markov chain Figure 2 representing a model for generating the random β-sequence to be investigated. The main idea is to identify intermediate states or stages of the simulation and to *restart* the simulation in these states thus obtaining conditional probabilities that can be converted to the original distribution function using results of the previous stage.

The following three subsections describe the RESTART/LRE-algorithm with one stage, the generalization to m stages or intermediate points is given in section 5.

4.1 LRE-Evaluation for one stage

In its simplest version the RESTART/LRE-algorithm for one stage is performed by two sequential LRE-runs.

First LRE-Run: The first LRE-run establishes the left part of $G(x)$ up to a certain "Intermediate Point" $i = I$. The level $G_I = P(\beta \geq I)$ of this interval belongs to a "less rare event" $\beta \geq I$ compared to the level $G_B = P(\beta = B) = P(\beta > N) = P_L$ of the "rare event" $\beta = B$: $1 \gg G_I \gg G_B$. To perform the restart-feature of the second LRE-run correctly information about the intermediate state has to be collected. For instance, when a transition $I-1 \rightarrow I$ is registered, the value of the *residual service time* τ_R of the job or message in service is stored and used as a τ_R-generator for producing a random variable τ_R according to the distribution expressed by $\tilde{F}(\tau_R)$ in the second LRE-run.

To execute the first LRE-run an error limit d_{1max} has to be prescribed and a level G'_I of the intermediate point I resp. interval $I-1 \leq x < I$ has to be determined in accordance with section 4.3. G'_I can be given as a parameter or it can be approximated by the program. Alternatively the intermediate point I can also be given as a parameter.

The first LRE-run needs n_1 number of trials:

$$n_1 = \frac{(1-\tilde{G}_I)\, cf_I}{\tilde{G}_I\, d_{1max}^2} \approx \frac{cf_I}{\tilde{G}_I\, d_{1max}^2}; \quad cf_I = \frac{1+\tilde{\varrho}_I}{1-\tilde{\varrho}_I}. \quad (4)$$

The intermediate point I is related to the following posterior quantities for $I-1 \leq x < I$ and the stopping criterion $d_I \leq d_{1max}$, compare (3):

$$\left.\begin{aligned} \tilde{G}_I &= v_I/n_1; \quad \tilde{G}_I \approx G'_I; \\ \tilde{\varrho}_I &= 1 - \frac{c_I/v_I}{1 - v_I/n_1}. \end{aligned}\right\} \quad (5)$$

Second LRE-Run: The second LRE-run establishes the right part of $G(x)$ for $i \geq I$ "under condition $\beta \geq I$" to be enforced by the RESTART-mechanism. This means that we now have to deal with the *conditional* compl. d.f. $G(x|\beta \geq I)$, which is related to the unconditional compl. d.f. $G(x)$ for $i-1 \leq x < i; i = I+1, \cdots, B$ as follows:

$$G(x) = G_I \cdot G(x|\beta \geq I) = G_i = G_I \cdot G_{i|\beta \geq I}. \quad (6)$$

The advantage of this expression is that the conditional levels $G_{i|\beta \geq I}$ are raised by the factor G_I^{-1} (e.g. $G_I^{-1} = 10^4$) compared to the unconditional levels G_i of the original function $G(x)$: $G_{i|\beta \geq I} = G_i \cdot G_I^{-1}$ and it is this effect, which enables us to measure successfully the rare event tail of the compl. d.f. $G(x)$.

The application of the RESTART-mechanism is quite simple: whenever the "next β-value" β_n represents a state $\beta_n < I$, a reset $\beta_n := I$ and after that a restart in state I is enforced. Unavoidably, these restarts cause deviations from the original random process involved, but any detrimental effects on the simulation result can be assumed as negligible due to the applied "service time correction": for each restart the current service time τ_b is replaced by a randomly chosen value of the residual service τ_R associated to transitions $I-1 \rightarrow I$, whose empirical d.f. $\tilde{F}(\tau_R)$ has been derived during the first LRE-run.

With the simplified notation $\tilde{\varrho}_B := \tilde{\varrho}_{B|\beta \geq I}$ we obtain in correspondence to (5) statements for the posterior conditional *loss probability* for $B-1 \leq x < B$ and the stopping criterion $d_{B|\beta \geq I} \leq d_{2max}$:

$$\left.\begin{aligned} \tilde{P}_{L|\beta \geq I} &= \tilde{G}_{B|\beta \geq I} = v_B/n_2; \\ \tilde{\varrho}_B &= 1 - \frac{c_B/v_B}{1 - v_B/n_2}. \end{aligned}\right\} \quad (7)$$

This simplification is justified by the fact that the posterior value $\tilde{\varrho}_B$ of the local c.c. $\varrho(x)$ of interval $B-1 \leq x < B$ at the right end is practically the same for simulation with or without RESTART. The number of restarts need no special consideration, because it represents a random quantity being controlled by the second LRE-run. Assuming $\tilde{G}_{B|\beta \geq I} \ll 1$ the final number of trials n_2 for the second LRE-run is, compare (4):

$$n_2 \approx \frac{cf_B}{\tilde{G}_{B|\beta \geq I}\, d_{2max}^2}; \quad cf_B = \frac{1+\tilde{\varrho}_B}{1-\tilde{\varrho}_B}. \quad (8)$$

4.2 Error Considerations

After both LRE-runs have been performed the levels $\tilde{G}_i$ of the *unconditional* compl.d.f. $\tilde{G}(x)$, see (6), for $i = I+1, \cdots, B$ are:

$$\tilde{G}_i = \tilde{G}_I\, \tilde{G}_{i|\beta \geq I}; \quad \tilde{P}_L = \tilde{G}_B = \tilde{G}_I\, \tilde{G}_{B|\beta \geq I}. \quad (9)$$

A careful investigation of the correct error expression for the posterior statements $\tilde{G}_i$ must take into account that all levels $\tilde{G}_i$ of the first LRE-run and $\tilde{G}_{i|\beta \geq I}$ of the second LRE-run represent mean values of the posterior random variables G_i and $G_{i|\beta \geq I}$, respectively, which – under the present large sample conditions (Schreiber, 1987b) – are normally distributed. A straightforward analysis proves that the relative error d_i of the unconditional level $\tilde{G}_i$, see (9), is given by:

$$d_i^2 = d_I^2 + d_{i|\beta\geq I}^2 = d_{1max}^2 + d_{i|\beta\geq I}^2 .$$

Due to the stopping criterion (5) we have set here with good approximation $d_I = d_{1max}$. In the same way we can set $d_{B|\beta\geq I} = d_{2max}$ due to the stopping criterion (7) and find for $i = B$ the relative error d_B of the unconditional *loss probability* $\tilde{P}_L = \tilde{G}_B$, which must be equal to a prescribed error limit d_{max} for the result $\tilde{G}_B$:

$$d_B^2 = d_{max}^2 = d_I^2 + d_{B|\beta\geq I}^2 = d_{1max}^2 + d_{2max}^2 .$$

The error limits d_{1max} and d_{2max} of the first and second LRE-run must be chosen in accordance with this equation. The error control in both LRE-runs safeguards that the condition $d_i \leq d_{max}$ is fulfilled for all levels $\tilde{G}_i$, $i = 1, \cdots, B$ including the unconditional levels $\tilde{G}_i$, see (9). This means that the prescribed value d_{max} is the effective error limit for the whole RESTART/LRE-algorithm.

4.3 Minimal Simulation Run Time

We replace in (4) and (8) the measured posterior quantities $\tilde{G}_I$, $\tilde{\varrho}_I$ and $\tilde{G}_{B|\beta\geq I}$, $\tilde{\varrho}_B$ by the corresponding variables G_I, ϱ_I and $G_{B|\beta\geq I}$, ϱ_B. Applying the relation $G_{B|\beta\geq I} = G_B/G_I$, $G_B \ll G_I$ which follows from (6) for $i = B$, we may therefore express the *total number of trials* $n = n_1 + n_2$ of both LRE-runs as a function of level G_I associated to the intermediate point I.

Using identical error limits for the first and second LRE-run $d_{1max} = d_{2max} = d_{Smax}$ we obtain the number of trials from $d_B = d_{max} = \sqrt{2}\, d_{Smax}$:

$$n(G_I) = n_1 + n_2 = \frac{2}{d_{max}^2}\left[\frac{cf_I}{G_I} + \frac{cf_B}{(G_B/G_I)}\right];$$

$$cf_I = \frac{1+\varrho_I}{1-\varrho_I}; \quad cf_B = \frac{1+\varrho_B}{1-\varrho_B} .$$

This function assumes its minimum n_{min} at the "optimal value" $G_I^\star$ of the variable G_I

$$\left.\begin{aligned} G_I^\star &= \sqrt{\frac{cf_I}{cf_B}} \cdot \sqrt{G_B}; \\ n_{min} &= n(G_I^\star) = \frac{4\sqrt{cf_I\, cf_B}}{\sqrt{G_B}\, d_{max}^2} . \end{aligned}\right\} \quad (10)$$

For system M/M/1/N with $N = 85$ and $\eta = 0.8$ we have $G_B = 9.26 \times 10^{-10}$, $\varrho_B = 0.44$, and $cf_B = 2.6$. Due to the flat maximum of the function $\varrho(x)$ in Figure 1 the local c.c. is nearly constant $\varrho_I \approx \varrho_{max} = 0.89$ and therefore also the correlation factor is nearly constant $cf_I = 17.2$. For $d_{max} = 0.1$ we compute from (10) and (1): $G_I^\star = 2.56\sqrt{G_B} = 7.78 \times 10^{-5}$, $n_{min} = 8.7 \times 10^7$ trials, and the optimal intermediate point $I_{opt} = \ln[G_I^\star + \eta^{B+1}(1 - G_I^\star)]/\ln(\eta) \approx 42$.

Assuming a production rate of 10^4 trials per second the minimal simulation run time would be $T_{min} \approx$ 2.4 hours. By comparing these values of n_{min} and T_{min} to $n = 2.8 \times 10^{11}$ trials and $T = 324$ days, which have been computed for a straightforward simulation of the same problem at the beginning of this section, we recognize the enormous potential of simulation speed-up by the RESTART/LRE-method.

The level G'_I of the intermediate point I to be prescribed in advance of the first LRE-run should be chosen as close as possible to the value of $G_I^\star$. But, under the conditions of normal simulation tasks, the optimal value $G_I^\star$ in (10) is *a priori* usually unknown because it depends on the values of the correlation factors cf_I, cf_B and of the loss probability $P_L = G_B$ itself, which become available only *a posteriori* as results of the statistical evaluation. For a series of RESTART/LRE-runs concerning the same simulation object the value of G'_I for each subsequent run can often be chosen closer to the optimum, because the values of cf_I and cf_B become known with increasing accuracy by the evaluation results of the previous runs. Therefore level G'_I must be chosen as a more or less good estimation of $G_I^\star$ and the needed number of trials will usually be $n > n_{min}$. In case that nothing is *a priori* known, it is recommended to use in (10) the values $cf_I = (1+\varrho_{max})/(1-\varrho_{max})$ and $cf_B = (1+\varrho_B)/(1-\varrho_B)$ of queueing system M/M/1/N, which can be computed for a given traffic load η from (2). Nevertheless, even in case of a relatively unfavorable choice of G'_I the essential simulation speed-up advantage of the RESTART-method is maintained; see also section 5.2 in (Villén-Altamirano and Villén-Altamirano, 1991).

5 THE MULTI-STAGE APPROACH

The extension of the RESTART/LRE-algorithm to m intermediate points $I_0, I_1, \cdots, I_{m-1}$ with G-levels: $1 \gg G_{I_0} \gg G_{I_1} \gg \cdots G_{I_{m-1}} \gg G_B$ is a natural extension of the above described algorithm with a single stage. The m intermediate points are also called stages leading to $m+1$ LRE-runs. This type of extension has also been envisaged in (Villén-Altamirano et al., 1994) and (Villén-Altamirano and Villén-Altamirano, 1994). The decisive advantage of the combination with the LRE-algorithm is that analytical formulas can be derived.

This leads under very general assumptions to the following optimal G-levels $G_{I_i}^\star$ ($i = 0, \cdots, m-1$) and minimal number of trials n_{min} :

$$G^{\star}_{I_i}(G_B) = \left(\frac{\prod_{j=0}^{i} {cf_{I_j}}^{m-i}}{\prod_{j=i+1}^{m} {cf_{I_j}}^{i+1}} G_B^{i+1} \right)^{1/(m+1)} ;$$

$$n_{min}(G_B) = \frac{m+1}{d^2_{Smax}} \left(\frac{1}{G_B} \prod_{i=0}^{m} cf_{I_i} \right)^{1/(m+1)} . \qquad (11)$$

To derive an approximation for the optimal number of stages m_{opt}, the correlation factor cf_{I_j} in (11) for $j = 0, \cdots, m-1$ is approximated by a fixed correlation factor named cf_M. The function $cf(x) = (1+\varrho(x))/(1-\varrho(x))$ is also a flat maximum curve with the maximum cf_{max}, if this is true for $\varrho(x)$ as shown in Figure 1, so that choosing the maximum for cf_M ($cf_M := cf_{max}$) is a good approximation and upper limit at the same time. Similar considerations as given in section 4.2 lead to an approximation of the error limit d_{Smax} in each run as a function of the overall error limit d_{max}:

$$d^2_{Smax} = (d^2_{max} + 1)^{1/(m+1)} - 1 \approx \frac{d^2_{max}}{m+1} .$$

Using these approximations the optimal number of stages m_{opt} follows from the necessary condition $(\partial n/\partial m = 0)$ for an optimum of n_{min} of which the nearest integer is taken for simulation runs:

$$m_{opt} \approx \frac{1}{2} \ln \left(\frac{cf_B}{cf_M G_B} \right) - 1 . \qquad (12)$$

Figure 3 shows the approximation of the optimal number of trials $n(m)$ on a logarithmic scale as a function of the number of stages m for the reference system $M/M/1/N$. The approximation is also an upper limit. Additionally a lower limit is shown.

It should be noted from (12) that the optimal number of stages does not depend on the maximum error level d_{max}. For smaller values of d_{max} the number of trials increases, but the optimum remains fixed.

A significant gain in terms of the total number of trials needed is already achieved for one intermediate point ($m=1$), which can be seen from the comparison of values with a straightforward simulation ($m=0$). Further improvements can be achieved when using the optimal number of stages. In the example in Figure 3 the relative speed-up $n(0)/n(1)$ is about 11 when comparing the straightforward simulation to the simulation with one intermediate point. In comparison with the optimal number of intermediate points ($m_{opt}=3$ in Figure 3) a relative speed-up $n(0)/n(3)$

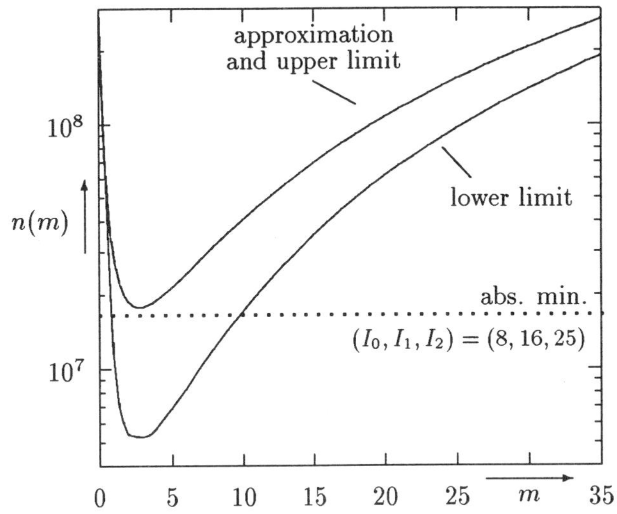

Figure 3: Queueing System M/M/1/34:
Total Number of Trials $n(m)$
versus Number of Intermediate Points m
(traffic load $\eta = 0.8$; max. error $d_{max} = 0.01$)

of about 18 is achieved. For this example it was possible to calculate the absolute minimum by exhaustive search, which is also given in Figure 3. For systems with very low loss probabilities the speed-up is much higher, so that now systems with, e.g. a loss probability of 10^{-20} can be simulated in affordable time, see section 6.

6 SIMULATION RESULTS

The LRE III and the RESTART/LRE-algorithm were implemented as part of a simulation system (Görg et al., 1991). Comprehensive RESTART/LRE simulation runs were performed and checked against theoretical and straightforward simulation results. Single server queueing systems of the type $M/G/1/N$-FIFO and $G/D/1/N$-FIFO were investigated.

In ATM simulations the service time is deterministic, so that systems of type $G/D/1/N$ are of interest. Figure 4 shows the systems $M/D/1/N$ and $M/M/1/N$ compared to a system SSMP(2)/D/1/N with a correlated Special Semi-Markov input Process SSMP(2) described in (Ding and Schreiber, 1990), (Ding, 1991), (Herrmann, 1993), and (Herrmann, 1994). The input process SSMP(2) distinguishes two states, in each of these states geometrically distributed interarrival times are generated with parameter q_1 and q_2. The transition probabilities between the states are indirectly given by the correlation coefficient κ and the stationary state probability P_1. In the example sys-

Table 1: Comparison of Simulated and Theoretical Values for $\eta = 0.4$ and $d_{max} = 0.03$

	M/M/1/21			M/M/1/50		
	sim.: $m = 1$	sim.: $m_{opt} = 9$	theoretical	sim.: $m = 5$	sim.: $m_{opt} = 22$	theoretical
$\tilde{G}_B$ resp. G_B	$1.091\ 10^{-9}$	$1.031\ 10^{-9}$	$1.056\ 10^{-9}$	$2.98\ 10^{-21}$	$3.02\ 10^{-21}$	$3.04\ 10^{-21}$
$\tilde{\varrho}_B$ resp. ϱ_B	0.291	0.280	0.286	0.292	0.256	0.286
$\tilde{n}$ (n straightforward)	$3.42\ 10^8$	$2.70\ 10^6$	($18.94\ 10^{11}$)	$5.14\ 10^8$	$1.39\ 10^7$	($6.57\ 10^{23}$)
Speed-Up $n/\tilde{n}$	$5.5\ 10^3$	$701\ 10^3$	–	$1.3\ 10^{15}$	$473\ 10^{15}$	–

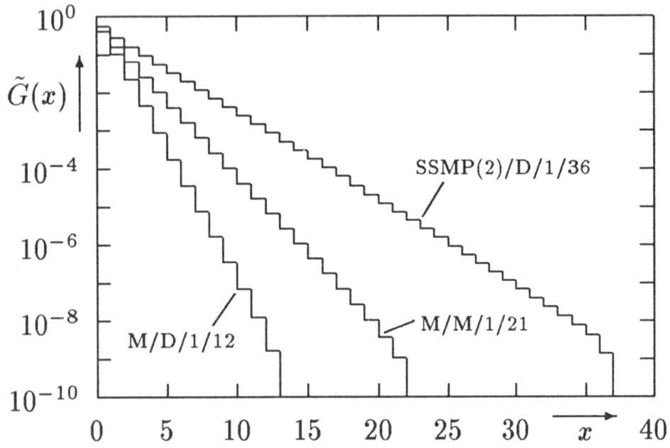

Figure 4: RESTART/LRE Simulation: Compl. Distribution Function $\tilde{G}(x)$ of Occupancy x ($m = 1$, $\eta = 0.4$, $G_B = P_L \approx 10^{-9}$, $d_{max} = 0.03$)

tem SSMP(2) the following parameters were used: $q_1 = 0.9$, $q_2 = 0.3$, $\kappa = 0.6$, $D = 3$, and P_1 was adapted according to the traffic load η.

The influence of the arrival and service time distributions on the compl. d.f. $G(x)$ of the occupancy x and the loss probability $P_L = G_B$ for a fixed traffic load of $\eta = 0.4$ is shown in Figure 4. The buffer size N for each system was chosen in such a way that the loss probability is approximately 10^{-9}. The M/D/1 system only needs a buffer size of $N = 12$ to fulfil the requirement of $P_L \approx 10^{-9}$, the M/M/1 system needs $N = 21$, whereas the SSMP(2)/D/1 system needs $N = 36$ due to the strong effect of correlated arrivals. The simulated compl. d.f. $\tilde{G}(x)$ of system M/M/1/N is in full agreement with the theoretical function $G(x)$, see (1). The SSMP(2)/D/1/N results were also checked successfully against the theoretical results given by (Herrmann, 1993). The differences are too small to be shown in the diagram.

Table 1 compares two reference systems with a loss probability of about 10^{-9} and 10^{-21}. The number of trials needed for the multi-stage simulations in comparison with the straightforward and single stage simulation are given. The simulation results are compared to the theoretical values. The main result concerning the loss probability $\tilde{P}_L = \tilde{G}_B$ resp. $P_L = G_B$ shows the conformity of simulation and theory.

If the quotient of B/m is small (e.g. $B/m < 3$) the resulting values for $\tilde{\varrho}_{I_i}$ are lower than in the original process which reduces the number of trials $\tilde{n}$ in the actual simulation. n gives the theoretically needed number of trials for a straightforward simulation. The speed-up factor is $n/\tilde{n}$. As mentioned in section 4 a straightforward simulation of these cases is practically impossible. Assuming 10^4 trials/sec results in 6 and 2×10^{12} years for the examples in Table 1. With the RESTART/LRE speed-up the simulations were all performed *within one day*.

7 FINAL REMARKS

In this paper the multi-stage RESTART/LRE-algorithm was described and the speed-up in comparison to straightforward and single stage simulations was discussed for representative examples.

The RESTART/LRE-algorithm as described here can be extended and adapted to other application fields, e.g. *multi-server* systems G/G/s/N. Another extension is the investigation of rare event details of more general stationary x-sequences having a pure continuous or a mixed continuous and discrete character. For this purpose the (more complicated) LRE-algorithm II (Schreiber, 1988) must be combined with the RESTART-method. Considering more complex systems, such as queueing networks, several open issues need to be resolved. The RESTART/LRE-method needs well defined intermediate states, an algorithm to generate these states, and the statistics for evaluating the original distribution function.

The results using the RESTART/LRE-algorithm are well established and will be further investigated to make them available in practical simulation tools.

ACKNOWLEDGEMENTS

The authors thank O. Fuss and W. Olzem, Aachen, for supporting these investigations by simulation work and Dr.-Ing. C. Herrman for helping with the *mutual verification* of simulation and theoretical results.

REFERENCES

Bayes, A. (1970). Statistical techniques for simulation models. *Australian Comp. Journ.*, 2:180–184.

Bratley, P., Fox, B., and Schrage, L. (1987). *A guide to simulation.* Springer, New York.

Ding, W. (1991). A unified correlated input process model for telecommunication networks. In *13th Int. Teletraffic Congress*, Copenhagen, Denmark. North-Holland.

Ding, W. and Schreiber, F. (1990). Local correlation properties of random sequences generated by queueing system M/M/1. *AEÜ*, 44:384–389.

Görg, C., Kluth, B., and Salewski, U. (1991). SIC – a system for stochastic simulation in C++. In *Modelling and Simulation 1991, Proceedings of the 1991 European Simulation Multiconference*, pages 413–414.

Herrmann, C. (1993). Analysis of the discrete-time SMP/D/1/s finite buffer queue with applications in ATM. In *IEEE INFOCOM'93, San Francisco, 1993, Session 2a.3, pp. 160-167.*

Herrmann, C. (1994). The local correlation properties of some discrete-time finite queues with semi-Markovian input. *AEÜ*, 48:244–255.

Kahn, H. and Marshall, A. (1953). Methods of reducing sample size in Monte Carlo computations. *J. Oper. Res. Society of America*, 1:263–278.

Kleijnen, J. (1974). *Statistical techniques in simulation*, volume I. Marcel Dekker, New York.

Le Boudec, J. (1991). An efficient solution method for Markov models of ATM links with loss priorities. *IEEE J. Select. Areas Commun.*, 9:408–417.

Schreiber, F. (1987a). The empirical stationary distribution function of Markovian correlated random sequences. *AEÜ*, 41:257–263.

Schreiber, F. (1987b). Measurement of the stationary state probabilities of the 2-node Markov chain. *AEÜ*, 41:117–122.

Schreiber, F. (1988). Effective control of simulation runs by a new evaluation algorithm for correlated random sequences. *AEÜ*, 42:347–354.

Schreiber, F. (1994). Local correlation properties of the finite buffer queueing system M/M/1/N. *AEÜ*, 48:63–67.

Schreiber, F. and Görg, C. (1994). Rare event simulation: a modified RESTART-method using the LRE-algorithm. In *Proceedings 14th Int. Teletraffic Congress*, pages 787–796. North-Holland.

Schreiber, F. and Görg, C. (1996). Stochastic simulation: a simplified LRE-algorithm for discrete random sequences. *AEÜ*, 50.

Shahabuddin, P. (1995). Rare event simulation in stochastic models. In *Proceedings of the 1995 Winter Simulation Conference*, pages 178–185, Arlington, Virginia, USA.

Villén-Altamirano, M., Martínez-Marrón, A., Gamo, J., and Fernández-Cuesta, F. (1994). Enhancements of the accelerated simulation method RESTART by considering multiple thresholds. In *Proceedings 14th Int. Teletraffic Congress*, pages 797–810. North-Holland.

Villén-Altamirano, M. and Villén-Altamirano, J. (1991). RESTART: a method for accelerating rare event simulations. In Cohen, J. and Pack, C., editors, *Queueing, Performance and Control in ATM*, pages 71–76. Proc. 13th Int. Teletraffic Congress, Copenhagen, North-Holland.

Villén-Altamirano, M. and Villén-Altamirano, J. (1994). RESTART: A straightforward method for fast simulation of rare events. In *Proc. of the 1994 Winter Simulation Conference*, pages 282–289.

AUTHOR BIOGRAPHIES

CARMELITA GÖRG received the Dipl.-Inform. degree in computer science from the University of Karlsruhe, Germany, in 1976, and the Dr. rer. nat. degree from Aachen University of Technology, Germany, in 1984 for her dissertation in the field of queueing strategies for communication networks. From 1985 until 1989 she worked as a consultant in the field of communication networks. Since 1989 she has been working as a group leader at the Communication Networks Institute, Aachen University of Technology. Her research interests are: performance analysis of communication networks, stochastic simulation, rare event simulation, high speed networks, and new services in telecommunication networks.

FRIEDRICH SCHREIBER studied communication engineering from 1948–1956 at the Universities of Karlsruhe, Rhode Island/USA, and Stuttgart, where he received the Dr.-Ing. degree for his dissertation in the field of magnetic materials. From 1956–1971 he worked at the Central Laboratories of Siemens AG in Munich. From 1971–1990 he was full professor and is now professor emeritus at the Aachen University of Technology, Faculty of Electrical Engineering. His research interests are: measurement of user behavior in data networks, optimal queueing strategies for communication networks, stochastic simulation techniques, quasi-ideal random generators, and new statistical evaluation methods for correlated random sequences.

Proceedings of the 1996 Winter Simulation Conference
ed. J. M. Charnes, D. J. Morrice, D. T. Brunner, and J. J. Swain

A GRADIENT APPROACH FOR SMARTLY ALLOCATING COMPUTING BUDGET FOR DISCRETE EVENT SIMULATION

Chun-Hung Chen
Hsiao-Chang Chen

Department of Systems Engineering
University of Pennsylvania
Philadelphia, PA 19104-6315, U.S.A.

Liyi Dai

Department of Systems Science and Mathematics
Washington University
St. Louis, MO 63130, U.S.A.

ABSTRACT

Simulation plays a vital role in analyzing many discrete-event systems. Usually, using simulation to solve such problems can be both expensive and time consuming. We present an effective approach to smartly allocate computing budget for discrete-event simulation. This approach can smartly determine the best simulation lengths for all simulation experiments and significantly reduce the total computation cost for obtaining the same confidence level. Numerical testing shows that our approach can obtain the same simulation quality with one-tenth the simulation effort.

1 INTRODUCTION

In order to efficiently manage and operate large man-made systems such as communication networks, traffic systems, and automated manufacturing plants, it is often necessary to apply extensive simulation to study their performance since no closed-form analytical solutions exist for such problems. Collectively, these types of systems are known as Discrete Event Systems (DES) (Ho 1991). Unfortunately, using simulation to solve such problems can be both expensive and time consuming due to their massive search space and their evolution in time according to complex man-made rules and the influence of random occurrences. In industry, with pressure to meet certain system specifications and only a limited budget to carry out necessary simulations, the limitations of traditional simulation technology can either delay a project or force it to go over budget.

Suppose we want to compare n different discrete-event systems (designs or alternatives), we do T simulation replications for all n designs (or alternatives). Totally, we need nT simulation replications. The simulation results become more accurate when T increases. If the accuracy requirement is not low (T is not small), and if the total number of designs in a decision problem is not small (n is large), then nT can be very large, which may easily make total simulation cost extremely high and preclude the feasibility of a simulation approach. To reduce total simulation time, one can either develop more efficient simulation technology or use faster computers to reduce the simulation time of each simulation experiment. In this paper, we present another approach to improve the overall simulation efficiency.

Our ideas are as follows. Intuitively, some bad designs can be discarded before completing all of the T replications. We don't have to waste efforts on simulating bad designs and so reduce overall simulation time. Then the question is how to systematically do this? When? And which designs? Ideally, we want to optimally choose the number of simulation replications for all designs to minimize the total simulation cost, while obtaining the desired confidence level. In fact, this question is equivalent to optimally decide which designs will receive computing budget for continuing simulation. Figure 1 illustrates the ideas by comparing a typical solution to this problem with the conventional approach using equal simulation lengths. Chen (1995) formulates this question and obtained promising preliminary results using very simple heuristics. In this paper, we will further discuss it and compare two approaches, one of which utilizes the gradient information.

To optimally allocate computing budget, first of all, one must have an efficient way to estimate the confidence level based on the results of the completed simulation. Further, one must have easy ways to anticipate how the confidence level will change if some computing budget is allocated and additional simulation replications are completed.

Goldsman and Nelson (1994) provide an excellent survey on current approaches (e.g., Goldsman, Nelson, and Schmeiser (1991), Gupta and Panchapakesan (1979), and Law and Kelton (1991)) to estimating simulation confidence level. In addition, Bechhofer, Santner, and Goldsman (1995) give a systematic and more detail discussion on this issue. Those approaches are mainly suitable for problems not having large number of designs

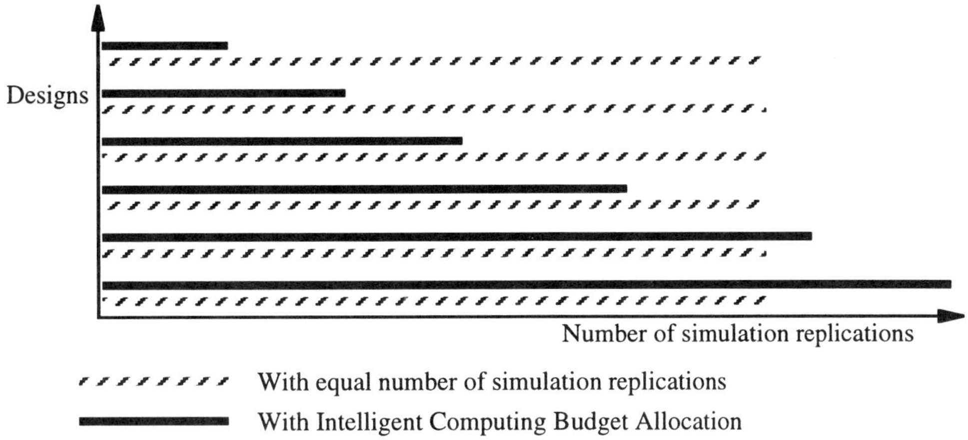

Figure 1: Comparison of Simulation Budget Allocations for Obtaining the Same Confidence Level

(e.g., Goldsman and Nelson (1994) suggest 2 to 20 designs). For real-life DES problems, the number of designs can grow up to numerous orders of magnitude easily. Chen (1996) presents a feasible way to quantify confidence level for *large* DES simulation (the "large" refers to large search space). Further, when the approach in Chen (1996) is applied the sensitivity information of the confidence level with respect to simulation replication numbers can be easily obtained, which will provide the basis to determine how to allocate computing budget among designs in this paper.

Section 2 describes the notation used in this paper and a brief overview of Chen (1996)'s approach to quantify confidence level for problems with large search space. In Section 3, we will define the "optimal computing budget allocation" problem. Two major difficulties in solving this problem will be pointed out. Since it is difficult to find an optimal solution for the computing budget allocation problem, and it is impractical to spend lot of time in finding the optimal solution, we propose a sequential approach to overcome these two difficulties in Sections 4 and 5. We call this approximation *smart computing budget allocation* scheme. Numerical testing in Section 6 shows that using this approach to smartly allocate computing budget can reduce the total computation time by about ten times for a 1000-design example. Section 7 concludes this paper.

2 A FEASIBLE APPROACH TO QUANTIFY CONFIDENCE LEVEL FOR PROBLEMS WITH LARGE SEARCH SPACE

Chen (1996) provides a simple approximation approach to quantify confidence level for problems with large search space and also provide some useful sensitivity information of the confidence level with respect to simulation replication numbers, which will provide the basis to determine how to allocate computing budget among designs in this paper. Denote

n: the total number of designs,

T: the length of simulation, the number of replication, or the total number of samples,

$\hat{J}_j(t)$: the t-th sample of the performance measure of design j,

$\bar{J}_j(T)$: the sample mean of design j, namely, $\bar{J}_j(T) = \frac{1}{T}\sum_{t=1}^{T}\hat{J}_j(t)$, and

J_j the performance measure, or more specifically, the mean performance measure of design j, i.e., the mean of $\hat{J}_j(t)$.

Assume that

i) $\hat{J}_j(t)$ is i.i.d. for all t,

ii) the simulations for designs i and j, $i \neq j$, are independent. Thus, $\hat{J}_i(t)$ and $\hat{J}_j(t)$ are independent.

For steady-state simulation, the sample $\hat{J}_j(t)$ may not be independent of $\hat{J}_j(s)$ for $s \neq t$. One possible way is to place the "raw" data in a few large batches, and work with these few batches as if they were independent (Banks, Carson, and Nelson 1995). As the strong law of large numbers, with probability 1,

$$\bar{J}_j(T) \to J_j, \text{ as } T \to \infty.$$

Without infinitely long simulations or infinite number of simulations replication, the sample mean $\bar{J}_j(T)$ is an approximation to J_j. We refer to the sample mean $\bar{J}_j(T)$ from one finite simulation experiment as an *observed* performance measure for a particular design's simulation. Let o_j be the index of the design having the j-th largest *observed* performance measure. With these notations, we have

$$\bar{J}_{o_1} > \bar{J}_{o_2} > \cdots > \bar{J}_{o_n}.$$

The traditional optimization approaches are to find the design with maximum performance measure. (Without loosing generality, we only consider maximization problems in this paper.) However, even the simulation cost for a good estimation of J_i could be very high, especially for complicated systems. Instead of insisting on picking the best design, Ordinal Optimization (Ho, Sreenivas, and Vakili 1992) concentrates on finding good enough designs and reduces the required simulation time dramatically. Comparing the observed performance measures at short simulation length T, we can select the observed best design (o_1) or the observed top-r designs (o_1, o_2, , .., o_r), and then ask what is the probability that at least one of the observed top-r designs actually belongs in top-k. This is crucial to Ordinal Optimization, although estimation of such a probability is very difficult for problems having large n.

Chen (1996) adopts the Bayesian model to analyze such confidence probability. Under the Bayesian model, J_i is treated as a random variable and has a prior distribution which describes the knowledge or the subjective belief about J_i before any sampling. The posterior distribution is updated after we observe the samples $\{\hat{J}_j(t), t=1,..,T\}$. The posterior distribution $p(J_j \mid \{\hat{J}_j(t), t=1,..,T\})$ summarizes the statistical properties of J_j given the prior knowledge and sampling information. When simulation stops, the statistical properties is described by the posterior distributions. We can estimate the probability that J_i is in some specific region, e.g., $\Pr\{J_j>0 \mid \{\hat{J}_j(t), t=1,..,T\}\}$, or compare two designs, e.g., $\Pr\{J_i-J_j>0 \mid \{\hat{J}_i(t), \hat{J}_j(t), t=1,..,T\}\}$. For notational simplicity, we denote $\tilde{J}_j$ as the posteriori

$$J_j \mid \{\hat{J}_j(t), t=1,..,T\}.$$

Namely, $\Pr\{\tilde{J}_j>0\}$ represents $\Pr\{J_j>0 \mid \{\tilde{J}_j(t), t=1, .., T\}\}$. The posterior distribution $p(\tilde{J}_j)$ illustrates what value J_j may be, based on samples $\{\hat{J}_j(t), t=1,..,T\}$ and the prior knowledge. With some normal assumptions, the posterior

$$\tilde{J}_j \equiv J_j \mid \{\hat{J}_j(t), t=1,2,\cdots,T\} \sim N\left(\frac{1}{T}\sum_{t=1}^{T}\hat{J}_j(t), \frac{\sigma_j^2}{T}\right)$$

Chen (1996) also shows that the Confidence Probability

CP1 ≡ Pr{At least one of designs o_1, o_2, .., o_r actually belongs in top-k}
≥ Approximated Confidence Probability

$$\text{ACP1} \equiv \prod_{j=r+k}^{n} \Pr\{\tilde{J}_{o_1} > \tilde{J}_{o_j}\},$$

and that

CP2 ≡ Pr{The true performance of the observed best design is not worse than β fraction of the performance of the true best design}
≥ Approximated Confidence Probability

$$\text{ACP2} \equiv \prod_{j=2}^{n} \Pr\{\tilde{J}_{o_1} > \beta\tilde{J}_{o_j}\}.$$

While CP1 and CP2 are very difficult to obtain, ACP1 and ACP2 can be computed very easily, and therefore will be used to approximate CP1 and CP2, respectively. Numerical testing shows that they can provide reasonably good approximation. Furthermore, since ACP1 and ACP2 are lower bounds of CP1 and CP2, we are sure that confidence level is sufficiently high when ACP1 or ACP2 is high enough. Although the definitions of CP1 and CP2 are different, the formulas for ACP1 and ACP2 are quite similar. For easy explanation, without loss of generality, we will only consider the simple case that

$$\text{ACP} \equiv \prod_{j=2}^{n} \Pr\{\tilde{J}_{o_1} > \tilde{J}_{o_j}\}$$

in the latter discussion, i.e., how to smartly allocate computing budget for obtaining satisfactory ACP.

3 PROBLEM DEFINITION

We follow the problem formulation given by Chen (1995). Let T_j be the simulation length, or the number of samples, of design j. If simulations are performed on a sequential computer and with simulation length T for all designs, the computation cost can be approximated by $T_1 + T_2 + \cdots + T_n = nT$. However, to ensure that ACP is larger than some value, we don't need to restrict ourselves to $T_1 = T_2 = \cdots = T_n$, and may choose different simulation lengths for different designs. This means T_i may not be equal to T_j for $i \neq j$. Furthermore, we can choose T_j for all j such that the total computation cost is minimized, while guaranteeing that ACP is greater than some satisfactory level. More specifically, we are considering the following problem.

$$\textbf{(P)} \quad \min_{T_1,\cdots,T_n} (T_1 + T_2 + \cdots + T_n),$$
$$\text{s.t. ACP} \geq \text{(a satisfactory level)}.$$

There are two major difficulties in solving (P):

Difficulty 1. ACP($T_1, T_2, \cdots, T_n$) can be computed only after doing simulations until $T_1, T_2, \cdots, T_n$. Before performing simulations until $T_1, T_2, \cdots, T_n$, how can we predict or estimate the ACP at $T_1, T_2, \cdots, T_n$?

Difficulty 2. $T_1, T_2, \cdots, T_n$ are integers. Even if we have techniques to estimate ACP at $T_1, T_2, \cdots, T_n$, an extremely large combinatorial space must be searched to find a solution to (P), especially when n is large.

Note that the purpose of solving (P) is to further reduce computation cost for obtaining a desired confidence level. We should not exert too much effort solving (P) during simulation. Otherwise, the additional cost of solving (P) will cancel the benefits of computing budget allocation. Hence, we need to solve (P) very efficiently, even if this means obtaining a sub-optimal solution. Efficiency is more crucial than optimality in this application.

4 A SEQUENTIAL APPROACH

This section presents a sequential procedure to overcome the difficulties in solving (P). To Optimally allocate computing resource, it is equivalent to determine which designs we should do more simulation. We will sequentially decide it, although this is usually not an optimal solution any more.

Before doing simulation there is neither knowledge about ACP nor a basis for choosing T_j. First, all designs are simulated until length t_0 to obtain statistical information about sample means and sample variances. Then we try to determine how to further allocate computing budget using available statistical information. When simulation is stopped at t_0, the posterior distribution of design j is

$$\tilde{J}_j \equiv \tilde{J}_j(t_0) \equiv J_j \mid \{\hat{J}_j(t), t = 1, 2, \cdot\cdot, t_0\} \sim N(\frac{1}{t_0}\sum_{t=1}^{t_0} \hat{J}_j(t), \frac{\sigma_j^2}{t_0})$$

At this moment, we have some ideas about each design and then can decide which designs are worthy of being allocated more computing budget. To determine how to further allocate computing budget, we have to be able to know how the ACP will change if some computing budget is allocated to some designs (Difficulty 1). More specifically, based on statistical information at t_0, we want to anticipate the posterior distribution at $t_0 + \Delta$, where Δ is a positive integer. To do this, we assume the sample mean and variance at t_0 are near those at $t_0 + \Delta$, and approximate the posterior distribution at $t_0 + \Delta$ using the *estimated* posterior distribution

$$N(\frac{1}{t_0}\sum_{t=1}^{t_0} \hat{J}_j(t), \frac{\sigma_j^2}{t_0 + \Delta})$$

Note that the denominator of variance portion is $t_0 + \Delta$ rather than t_0. This approximation will be satisfactory assuming t_0 is large enough and if Δ is not too large. On the other hand, we don't want to choose t_0 too large, or we will defeat the purpose of this approach. Using the estimated posterior distributions, we can estimate the ACP at $t_0 + \Delta$ using the statistical information at t_0, and call it the "Estimated ACP" or EACP.

Similarly, when simulation proceeds until $T_1, T_2, \cdots, T_{i-1}, T_i, T_{i+1}, \cdots, T_n$, we can also use the available information to estimate how ACP will change if design i is given additional budget Δ, i.e., EACP($T_1, T_2, \cdots, T_{i-1}, T_i + \Delta, T_{i+1}, \cdots, T_n$). This is accomplished by using the estimated posterior distribution

$$N(\frac{1}{T_i}\sum_{t=1}^{T_i} \hat{J}_i(t), \frac{\sigma_i^2}{T_i + \Delta})$$

for design i.

Now it is feasible to predict the ACP when the change of T_i's are not large. A possible sequential approximation approach to solving (P) is as follows. Since ACP will become larger as simulation proceeds, we sequentially add computing budget by b each time until ACP reached some satisfactory level (say P_{sat}). In order to minimize to the total computation cost, at each step, this budget b is allocated among some designs such that the EACP is maximized. Thus, at step k, k=1,2,3,..,

$$\textbf{(P-}k\textbf{)} \quad \max_{\tau_1^k, \cdots, \tau_n^k} \text{EACP}(T_1^k + \tau_1^k, T_2^k + \tau_2^k, \cdots, T_n^k + \tau_n^k),$$
$$\text{s.t. } \tau_1^k + \tau_2^k + .. + \tau_n^k = b \text{ and } \tau_i^k \geq 0 \text{ for all } i.$$

More specifically, the sequential algorithm is

<u>**A Sequential Algorithm for Smart Computing Budget Allocation**</u> **(SCBA)**

Step 0. PERFORM SIMULATION until length t_0 for all designs,
$k \leftarrow 0$,
$T_1^k = T_2^k = \cdots = T_n^k = t_0$.

Step 1. If ACP($T_1^k, T_2^k, \cdots, T_n^k$) $\geq P_{sat}$, stop, otherwise, go to Step 2.

Step 2. Solve (P-k),
$T_i^{k+1} = T_i^k + \tau_i^k$, for i = 1, .., n,
$k \leftarrow k+1$,

Step 3. PERFORM SIMULATION until ($T_1^k, T_2^k, \cdots, T_n^k$); go to Step 1.

Remarks:

1. Obviously, the computing budget allocated by this sequential approach is not the optimal way. As we discussed before, efficiency is more important than optimality. Otherwise, the additional cost of determining computing budget allocation may cancel its benefits.

2. b is the one-time increment of simulation budget. Small b means small step size and therefore will increase the total number of solving (P-k). On the other hand, large b may waste computing budget and result in a larger ACP.

5 SOLVING PROBLEM (P-k)

The next question is how to *efficiently* solve (P-k). While solving (P-k) is easier than solving (P), again, efficiency is much more important than optimality here. In this paper, we test two *quick and dirty* approaches to allocate the given computing budget b for obtaining large increment of ACP.

In the first approach, m designs are chosen and then the computing budget is equally distributed to them (each design has b/m). For each design, we calculate the anticipated increment of ACP if computing budget b/m is allocated to it. Then those designs are chosen if their anticipated ACP increments are among top-m.

Approach 1. Choose a positive integer m, and let $\Delta = b / m$ (assume Δ is an integer)

Step 1. For $i = 1, .., n$, calculate $D_i \equiv$ $\text{EACP}(T_1^k, T_2^k, \cdots, T_{i-1}^k, T_i^k + \Delta, T_{i+1}^k, \cdots, T_n^k)$ $- \text{ACP}(T_1^k, T_2^k, \cdots, T_{i-1}^k, T_i^k, T_{i+1}^k, \cdots, T_n^k)$.

Step 2. Find the set $S(m) \equiv \{ i : D_i$ is within the top-highest-$m\}$

Step 3. $\tau_i^k = \Delta$, for all $i \in S(m)$.

Approach 2. In the second approach, instead of equally allocating computing budget among some m designs, we apply steepest-descent method (Luenberger 1984) to solve (P-k). We do the following approximation to estimate the gradient of ACP with respect to T_i.

Lemma 1. Suppose the random variable $X \sim N(\Delta, \sigma^2)$, where $\Delta > 0$. Then $\Pr\{X<0\} \le \exp(-\frac{\Delta^2}{2\sigma^2})$.

<pf> Using Chernoff bounds (Ross 1994), we have

$$\Pr\{X<0\} \le \inf_{t<0} M(t) = \inf_{t<0} \exp(\frac{\sigma^2 t^2}{2} + \mu t).$$

Choose $t = -\frac{\mu}{\sigma^2}$, we have the minimum, i.e.,

$$\Pr\{X<0\} \le \exp(-\frac{\Delta^2}{2\sigma^2}). \qquad \text{Q.E.D.}$$

With this lemma,

$$\text{ACP} = \prod_{j=2}^{n} \Pr\{\tilde{J}_{o_1} > \tilde{J}_{o_j}\} = \prod_{j=2}^{n} (1 - \Pr\{\tilde{J}_{o_1} - \tilde{J}_{o_j} < 0\})$$

$$\ge \prod_{j=2}^{n} \left(1 - \exp(-\frac{\Delta_{1j}^2}{2\sigma_{1j}^2}) \right) = \text{ACP*},$$

where $\Delta_{1j} = \bar{J}_{o_1} - \bar{J}_{o_j}$ and $\sigma_{1j}^2 = \frac{\sigma_{o_1}^2}{T_{o_1}} + \frac{\sigma_{o_j}^2}{T_{o_j}}$.

For notation simplicity, let $s = o_1$. The gradient of ACP* with respect to T_i are as follows:

if $i \ne s$,

$$\frac{\partial}{\partial T_i} \text{ACP*} = \frac{\Delta_{1i}^2 T_s^2 \sigma_i^2}{2(T_s \sigma_i^2 + T_i \sigma_s^2)^2} \exp\left(-\frac{\Delta_{1i}^2}{2(\frac{\sigma_s^2}{T_s} + \frac{\sigma_i^2}{T_i})} \right)$$

$$\bullet \prod_{\substack{j=1 \\ j \ne s, j \ne i}}^{n} \left(1 - \exp(-\frac{\Delta_{1i}^2}{2(\frac{\sigma_s^2}{T_s} + \frac{\sigma_j^2}{T_j})}) \right)$$

if $i = s$, $\frac{\partial}{\partial T_s} \text{ACP*} =$

$$\sum_{i=2}^{n} \left\{ \frac{\Delta_{1i}^2 T_{o_i}^2 \sigma_s^2}{2(T_s \sigma_{o_i}^2 + T_{o_i} \sigma_s^2)^2} \exp\left(-\frac{\Delta_{1i}^2}{2(\frac{\sigma_s^2}{T_s} + \frac{\sigma_{o_i}^2}{T_{o_i}})} \right) \right.$$

$$\left. \bullet \prod_{j=2, j \ne o_i}^{n} \left(1 - \exp(-\frac{\Delta_{1i}^2}{2(\frac{\sigma_s^2}{T_s} + \frac{\sigma_{o_j}^2}{T_{o_j}})}) \right) \right\}$$

To avoid spending much time in iteratively finding the solution of (P-k), we only do a very limited numbers of iterations when applying steepest-descent method. Different numbers of iterations are tested in Section 6.

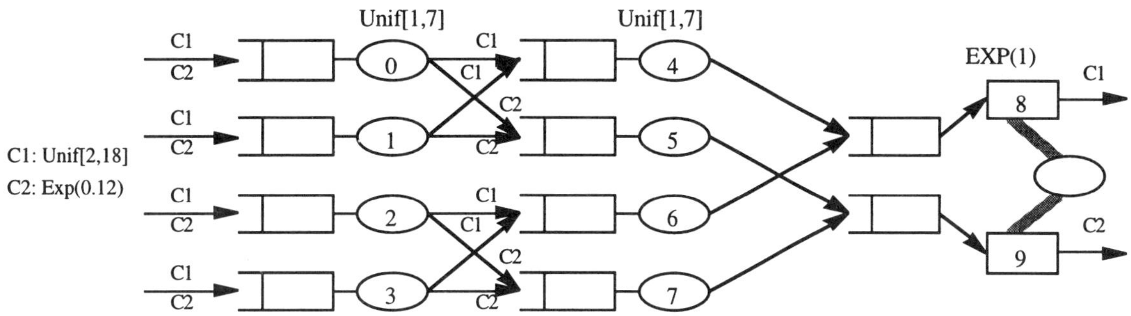

Figure 2: 10-Node Network with Priority, Interruption, and Shared Server

6 NUMERICAL TESTING

Two examples are tested. Example 1 is a steady-state simulation, and Example 2 is a terminating simulation.

Example 1. We consider a 10-node network (Please see Figure 2). Such a network could be the model for a large number of real-world systems, such as a manufacturing system, and a communication or a traffic network. For details about this example, please refer to Chen and Ho (1995). We consider the problem of optimally allocating 22 buffer units among the 10 different nodes for maximizing the throughput. Priority, interruption, blocking and multi-classes are included in this network. We denote the buffer size of node i by B_i. We set some constraints for symmetry reasons: $B_0 = B_1 = B_2 = B_3$, $B_4 = B_6$, and $B_5 = B_7$. In addition, B_8 & $B_9 \geq 1$. These constraints limit our search space to n=1000 different configurations.

The Standard Clock method (Chen and Ho 1995 and Vakili 1991), which is an efficient technique for DES simulation, is used to simulate this system. The computation cost for one design is roughly proportional to the number of clock ticks (for Standard Clock method, one event is generated at each clock tick). We define the computation cost as

$$\frac{1}{1000}\sum_{j=1}^{1000}\text{[the number of clock ticks when the simulation of design j is stopped]},$$

The $\frac{1}{1000}$ is used to rescale the cost. The computation cost for determining the smart computing budget allocation is so small as compared with the simulation cost that we ignore this portion. In this testing, we consider CP1 = Pr{At least one of the observed top-3 designs actually belongs in top-3} and CP2 = Pr{The true performance of the observed best design is not worse than 99.6% of the performance of the true best design}. We test Approach 1 and set $t_0 = 5$, $m = 25$, and $b = 125$. We repeat this testing 50 times. Each run has a different random seed. We consider the computation costs for different satisfactory confidence levels. Tables 1 and 2 contain the testing results for CP1 and CP2 respectively. The computation costs in these two tables are the average costs in the 50 testing runs.

Table 1: Speedup with the SCBA Method for ACP1

P_{sat}	without SCBA	with SCBA	Speedup
50%	24200	3952.5	6.12
60%	29100	4378.7	6.64
70%	45700	4670.0	9.78
80%	65600	5775.0	11.35

Table 2: Speedup with the SCBA Method for ACP2

P_{sat}	without SCBA	with SCBA	Speedup
50%	23000	3737.5	6.15
60%	29400	4105.0	7.16
70%	37300	4480.0	8.32
80%	54400	5446.2	9.98

Example 2. To further compare these two approaches, we test a simple single-node queue. The interarrival time is ~ Uniform[0.1, 1.9]. We consider 10 designs with different service times, which are Uniform[0.1, 1.85-i*0.05] for design i, i=1,..,10. Suppose we are interested in the average system time of the customers served between time 0 and time 10. Although the derivations in Sections 2 ~5 focus on maximization, we only need to reverse their inequality signs in order to apply to this minimization problem. We set b = 12. and t_0 = 10. 10,000 independent experiments are done to estimate the average cost for using different approaches. We consider CP = Pr{The observed best design is actually the true best design}. Table 3 shows the average total numbers

of simulation replications for obtaining the confidence level P_{sat} when setting $t_0 = 10$, and Table 4 is for $t_0 = 5$.

Table 3: Average Total Number of Simulation Replications for Different SCBA Approaches (T_0=10)

P_{sat}	60%	70%	80%	90%
App. 1 (m = 4)	172.4	230.3	350.0	608.4
App. 1 (m = 3)	170.5	225.9	322.5	542.7
App. 1 (m = 2)	170.0	226.5	322.5	511.9
App. 1 (m = 1)	176.6	228.9	324.6	509.6
App. 2 (1 itrn)	167.2	221.1	324.1	525.9
App. 2 (2 itrns)	162.2	212.0	307.7	498.6
App. 2 (4 itrns)	163.3	213.4	313.4	511.6
Without SCBA	327.1	488.2	796.4	1458.

Table 4: Average Total Number of Simulation Replications for Different SCBA Approaches (T_0=5)

P_{sat}	60%	70%	80%	90%
App. 1 (m = 4)	127.6	186.3	305.6	576.9
App. 1 (m = 3)	127.6	183.6	287.0	507.1
App. 1 (m = 2)	129.9	182.4	275.4	470.0
App. 1 (m = 1)	139.9	192.3	283.6	470.1
App. 2 (1 itrn)	130.1	185.9	290.4	496.2
App. 2 (2 itrns)	122.3	175.3	273.1	467.3
App. 2 (4 itrns)	125.9	185.3	285.5	492.1
Without SCBA	305.2	475.5	790.0	1462.

From Tables 3 and 4, we have the following observations:

- t_0 may affect the performance quite significantly, particularly when P_{sat} is small. How to choose an appropriate t_0 is problem-specific. This remains to be investigated.
- Different choices of m's obtain different performances. It is interesting to note that large m works better for low P_{sat}, while small m performs well for high P_{sat}. We conjecture that there exist some better ways which dynamically change m through simulation.
- For the steepest descent method, two iterations in each sequential optimization step works better than others. We need more testing to justify this. Ideally, we may gradually change the number of iterations through simulation to optimize the performance.
- The time savings factor of using SCBA increases as P_{sat} increases. This makes sense since we have more space to manipulate the allocation of computing budget when P_{sat}, the confidence level requirement, is higher.
- When $t_0 = 5$, Approach 2 with two iterations can reduce computation effort by 68% for P_{sat} = 90%. We believe that this time savings factor will be even larger if higher confidence level is required.

Since ACP is a lower bound of the confidence level CP and we use ACP to determine computing budget allocation, people may be concerned with the ending CP (actual confidence). In this testing, CP = Pr{The observed best design is actually the true best design} can be obtained numerically by calculating (total number of simulation experiments in which the observed best design is actually the true best design) / 10,000. Table 5 shows the numerical results of CP's for $t_0 = 5$.

Table 5: CP for Different SCBA Approaches (t_0=5)

P_{sat}	60%	70%	80%	90%
App. 1 (m = 4)	.629	.714	.815	.920
App. 1 (m = 3)	.630	.724	.817	.920
App. 1 (m = 2)	.634	.720	.813	.912
App. 1 (m = 1)	.632	.717	.804	.902
App. 2 (1 itrn)	.627	.724	.818	.918
App. 2 (2 itrns)	.627	.714	.816	.917
App. 2 (4 itrns)	.632	.713	.811	.919
Without SCBA	.696	.780	.879	.951

Our approaches stop simulation when ACP is no less than P_{sat}. We anticipate that the ending CP will be higher than P_{sat} since ACP is a lower bound of CP. Table 5 shows that CP's are not much higher than ACP's except the case in which SCBA is not used. The reason is that without using SCBA, all designs receive computing budget so that the simulation quality improvement is higher at each step. Consequently the ending CP's can be much higher than required level.

7 CONCLUDING REMARKS

In this paper we present two approaches to smartly allocate computing budget for DES simulation. Preliminary numerical testing shows that we can significantly reduce total computation cost. For real-time application problems, we have only a limited computing budget to carry out simulation. The SCBA can be applied to these problems to maximize the utilization of limited budget and obtain higher confidence level. In particular, from Table 1, the computation cost is 65,600 units for ensuring ACP1 > 80% without SCBA. On the other hand, with SCBA the computation cost is only 5,775 units. This implies that 5,775 SCBA computation units can obtain the same simulation quality as 65,600 computation units without SCBA. Application of the SCBA algorithm can obtain the same simulation quality with one-tenth the simulation effort.

In this paper, we compare two simple approaches using an example. The gradient approach performs slightly better than the other. While which approach is surprier needs more testing to justify, we fiirmly believe that there exists some more sopfisticaed way to accomplish better performance. We will test more examples in the future. The approaches using second order information is also one of the ongoing research topics.

ACKNOWLEDGMENTS

The work in this paper is partially supported by the University of Pennsylvania Research Foundation. The author would like to thank Dr. Russell R. Barton, Dr. Douglas J. Morrice, and the anonymous referees for their helpful suggestions and valuable comments.

REFERENCES

Banks, J., and J. S. Carson, B. L. Nelson. 1995. *Discrete-Event System Simulation*, Prentice-Hall.

Bechhofer, R. E., T. J. Santner, and D. M. Goldsman. 1995. *Design and Analysis of Experiments for Statistical Selection, Screening, and Multiple Comparisons*, John Wiley & Sons, Inc.

Casella, G., and R. L. Berger. 1990. *Statistical Inference*, Wadsworth.

Chen, C. H. 1995. "An Effective Approach to Smartly Allocate Computing Budget for Discrete Event Simulation," *Proceedings of the 34th IEEE Conference on Decision and Control*, 2598-2605.

Chen, C. H. 1996. "A Lower Bound for the Correct Subset-Selection Probability and Its Application to Discrete Event System Simulations," To appear on *IEEE Transactions on Automatic Control.*

Chen, C. H., and Y. C. Ho. 1995. "An Approximation Approach of the Standard Clock Method for General Discrete Event Simulation," *IEEE Transactions on Control Systems Technology*, Vol. 3, #3, 309-317.

Goldsman, G., and B. L. Nelson. 1994. "Ranking, Selection, and Multiple Comparison in Computer Simulation," *Proceedings of the 1994 Winter Simulation Conference*, 192-199.

Goldsman, G., B. L. Nelson, and B. Schmeiser. 1991. "Methods for Selecting the Best System," *Proceedings of the 1991 Winter Simulation Conference*, 177-186.

Gupta, S. S. and S. Panchapakesan. 1979. *Multiple Decision Procedures: Theory and Methodology of Selecting and Ranking Populations*, John Wiley.

Ho, Y. C. (Editor). 1991. *Discrete Event Dynamic Systems*, IEEE Press.

Ho, Y. C., R. S. Sreenivas, and P. Vakili. 1992. "Ordinal Optimization of DEDS", *Journal of Discrete Event Dynamic Systems*, 2, #2, 61-88.

Law, A. M. and W. D. Kelton. 1991. *Simulation Modeling & Analysis*, McGraw-Hill, Inc.

Luenberger, D. G. 1984. *Linear and Nonlinear Programming*, Addison-Wesley.

Ross, S. 1994. *A First Course in Probability*, Prentice Hall.

Vakili, P. 1991. "A Standard Clock Technique for Efficient Simulation, "*Operations Research Letters*, Vol. 10, 445-452.

AUTHOR BIOGRAPHIES

CHUN-HUNG CHEN is an Assistant Professor of Systems Engineering at the University of Pennsylvania, Philadelphia, PA. He received his Ph.D. degree in Simulation and Decision from Harvard University in 1994. His research interests cover a wide range of areas in Monte Carlo simulation, optimal control, stochastic decision processes, ordinal optimization, perturbation analysis, and their applications to manufacturing systems. Dr. Chen won the 1994 Harvard University Eliahu I. Jury Award for the best thesis in the field of control. He is also one of the recipients of the 1992 MasPar Parallel Computer Challenge Award.

HSIAO-CHANG CHEN is a Ph.D. candidate at the Systems Engineering Department, University of Pennsylvania. He received a double B.S. degree in Mathematics and Computer Science from the Eastern Michigan University in 1992, and he received an M.S. degree in Systems Science and Mathematics from Washington University, St. Louis in 1994. He is working on developing efficient approaches for discrete-event simulation.

LIYI DAI is an assistant professor in the Department of Systems Science and Mathematics at Washington University, MO. He received the M.S. degree from the Institute of Systems Science, Academia Sinica, Beijing, China, in 1986, and the Ph.D. degree from Harvard University in 1993. His research interests include discrete event dynamic systems, simulation, stochastic optimization, communication systems, and singular systems. He has coauthored over 30 papers in various journals and is the author of Singular Control Systems (Berlin: Springer-Verlag, 1989). Dr. Dai is listed in Who's Who among Asian Americans and is a recipient of the NSF CAREER award.

Proceedings of the 1996 Winter Simulation Conference
ed. J. M. Charnes, D. J. Morrice, D. T. Brunner, and J. J. Swain

DISCRETE STOCHASTIC OPTIMIZATION VIA A MODIFICATION OF THE STOCHASTIC RULER METHOD

Mahmoud H. Alrefaei

Department of Industrial Engineering
University of Wisconsin - Madison
1513 University Avenue
Madison, Wisconsin 53706, U.S.A.

Sigrún Andradóttir

School of Industrial and Systems Engineering
Georgia Institute of Technology
Atlanta, Georgia 30332, U.S.A.

ABSTRACT

In this paper, we present a modification of the stochastic ruler method for solving discrete stochastic optimization problems. Our method generates a stationary Markov chain sequence taking values in the feasible set of the underlying discrete optimization problem. The number of visits to every state by this Markov chain is used to estimate the optimal solution. Unlike the original stochastic ruler method, our method is guaranteed to converge *almost surely* to a global optimal solution. We present empirical results that illustrate the performance of our method, and we show that these results compare favorably with empirical results obtained using the original stochastic ruler method.

1 INTRODUCTION

Developing efficient methods for solving discrete stochastic optimization problems is an important area of research because such methods could be used to solve important problems in many fields, including manufacturing systems, management science, and logistics. The discrete stochastic optimization problem can be represented as follows:

$$\min_{x \in \mathcal{S}} f(x) = E[h(x, Y_x)], \qquad (1)$$

where $\mathcal{S}$ is a discrete set, and it is assumed to be finite throughout this paper, h is a deterministic real-valued function, and Y_x is a random variable that depends on the parameter $x \in \mathcal{S}$. For simplicity define the random variable $H(x) = h(x, Y_x)$ for all $x \in \mathcal{S}$, and let $\mathcal{S}^* = \{x^* \in \mathcal{S} : f(x^*) \leq f(x) \text{ for all } x \in \mathcal{S}\}$ denote the set of global solutions to the optimization problem (1).

Approaches that have been designed to solve the discrete stochastic optimization problem include ranking and selection procedures and multiple comparisons procedures. These procedures can be efficient when the number of alternatives is small. Goldsman, Nelson, and Schmeiser (1991) present demonstrations of three methods for discrete stochastic optimization (interactive analysis, ranking and selection, and multiple comparisons) on an airline-reservation-system simulation problem. For more details on ranking and selection and multiple comparisons methods, see Bechhofer, Santner, and Goldsman (1995).

Recently Andradóttir (1995, 1996) has proposed two methods for solving discrete stochastic optimization problems. These two methods involve generating Markov chains. The state that is visited most often by these Markov chains is used as an estimate of the solution. She has shown that these two methods converge to an optimal solution almost surely. Alrefaei and Andradóttir (1995) have proposed another method for discrete stochastic optimization that resembles the simulated annealing algorithm. They use the same criterion that Andradóttir (1995, 1996) used to estimate the solution. They have shown that their method converges to a global optimal solution almost surely. Gelfand and Mitter (1989) proposed to solve the discrete stochastic optimization problem using the simulated annealing algorithm. They show that this algorithm converges to a global optimal solution when the evaluations of the objective function values include noise that is normally distributed with mean zero and a small variance. Lee (1995) generalizes some of the results of Gelfand and Mitter (1989) to the case when the evaluations of the objective function values include noise which is not necessarily normally distributed. Fox and Heine (1995) have also investigated the application of simulated annealing to solve discrete stochastic optimization problems. When the set of feasible solutions is very big, the ordinal optimization approach proposed by Ho, Sreenivas, and Vakili (1992) can be used to narrow the focus on a smaller manageable set that contains a "good enough" (near-optimal) solution with high probability. Then the other methods we have mentioned could be used to find a solution over this smaller set.

Yan and Mukai (1992) proposed a method for discrete stochastic optimization that is called the *stochastic ruler method.* This method involves comparing the observed objective function values $H(x)$, where $x \in \mathcal{S}$, with a uniform random variable (the stochastic ruler) defined on the range of the observed objective function values. This uniform random variable is used as a scale against which the observations $H(x)$ are measured. This method uses an increasing sequence of observations per iteration and it requires some restrictive assumptions on the neighborhood structure. Yan and Mukai show that their method converges in probability to a global optimal solution. In this paper, we use the idea of the stochastic ruler method and propose a variant of this method that appears to behave better than the original method in practice and also has been shown to converge under more general conditions. Instead of using an increasing sequence of observations per iteration, we draw only a fixed number of observations per iteration, and we use a criterion that is similar to the criterion of Andradóttir (1995, 1996) for estimating the solution. Furthermore, unlike the original stochastic ruler method, our proposed method converges almost surely to a global optimal solution (the original method is only guaranteed to converge in probability). We provide a numerical example that verifies the validity of our method and we compare the performance of our method with that of the original stochastic ruler method.

This paper is organized as follows: In Section 2 we give a brief review of the original stochastic ruler method. In Section 3 we introduce our modified stochastic ruler method. In Section 4 we implement our method to solve a discrete stochastic optimization problem and we compare the results with the performance of the original stochastic ruler method. And finally in Section 5 we give some concluding remarks.

2 THE ORIGINAL STOCHASTIC RULER ALGORITHM

The stochastic ruler method is a random search method that was proposed by Yan and Mukai (1992) to solve the discrete stochastic optimization problem (1). To motivate our new method, in this section we give a brief review of the stochastic ruler method. To proceed we need the following definitions and assumptions that were presented by Yan and Mukai (1992).

Assumption 1 *Assume that* $E\{H(x)^2\} < \infty$, $\forall x \in \mathcal{S}$.

Definition 1 *For each* $x \in \mathcal{S}$, *there exists a subset* $N(x)$ *of* $\mathcal{S} \setminus \{x\}$, *which is called the set of neighbors of* x.

Assumption 2 *For any* $x, x' \in \mathcal{S}, x'$ *is reachable from* x*; i.e., there exists a finite sequence* $\{n_i\}_{i=0}^{l}$ *for some* l, *such that* $x_{n_0} = x, x_{n_l} = x'$, *and* $x_{n_{i+1}} \in N(x_{n_i})$, $i = 0, 1, 2, \ldots, l-1$.

Definition 2 *A function* $R : \mathcal{S} \times \mathcal{S} \to [0,1]$ *is said to be a transition probability for* $\mathcal{S}$ *and* N *if*

1. $R(x, x') > 0 \Leftrightarrow x' \in N(x)$, *and*
2. $\sum_{x' \in \mathcal{S}} R(x, x') = 1$, $\forall x \in \mathcal{S}$.

Assumption 3 *The neighbor system* N *and the transition probability* R *are symmetric, i.e.,*

1. $x' \in N(x) \Leftrightarrow x \in N(x')$, *and*
2. $R(x, x') = R(x', x)$, $\forall x, x' \in \mathcal{S}$.

Assumption 4 *The parameters* $a, b \in \Re$ *are selected to cover the range of the observed objective function values* $H(x)$, *where* $x \in \mathcal{S}$.

Assumption 5 *A sequence* $\{M_k\}$ *of positive integers satisfies* $M_k \to \infty$ *as* $k \to \infty$.

Instead of solving the original minimization problem (1) directly, Yan and Mukai (1992) solve the following alternative maximization problem

$$\max_{x \in \mathcal{S}} P(x, a, b), \tag{2}$$

where $P(x, a, b) = \text{Prob}\{H(x) \leq \Theta(a, b)\}$, for all $x \in \mathcal{S}$, and $\Theta(a, b)$ is a uniform random variable on the interval (a, b). The random variable $\Theta(a, b)$ is called the *stochastic ruler* and it is used as a scale against which the observations $H(x)$ are measured. Yan and Mukai (1992) have shown that the solution of the alternative maximization problem (2) is also a solution of the original minimization problem (1) provided that the interval (a, b) is chosen large enough (see Assumption 4). Therefore, they propose the following algorithm to solve the alternative maximization problem (2):

Algorithm 1 (Yan and Mukai (1992))

Parameters: $N, R, \{M_k\}, a, b$.

Step 0: *Select a starting point* $X_0 \in \mathcal{S}$ *and let* $k = 0$.

Step 1: *Given* $X_k = x$, *choose a candidate* Z_k *from* $N(x)$ *with probability distribution*

$$P\{Z_k = z | X_k = x\} = R(x, z), z \in N(x).$$

Step 2: *Given $Z_k = z$, draw a sample $h(z)$ from $H(z)$. Then draw a sample θ from $\Theta(a,b)$. If $h(z) > \theta$, then let $X_{k+1} = X_k$ and go to Step 3. Otherwise draw another sample $h(z)$ from $H(z)$ and draw another sample θ from $\Theta(a,b)$. If $h(z) > \theta$, then let $X_{k+1} = X_k$ and go to Step 3. Otherwise continue to draw and compare. If all M_k tests, $h(z) > \theta$, fail, then accept the candidate Z_k and set $X_{k+1} = Z_k = z$.*

Step 3: *Set $k = k+1$ and go to Step 1.*

Remark 1 Note that Step 2 can be rewritten as: Given $Z_k = z$, set

$$X_{k+1} = \begin{cases} Z_k & \text{with probability } p_{z,k}, \\ X_k & \text{with probability } 1 - p_{z,k}, \end{cases}$$

where

$$\begin{aligned} p_{z,k} &= [P\{H(z) \le \Theta(a,b)\}]^{M_k} \\ &= [P(z,a,b)]^{M_k}. \end{aligned}$$

Under Assumptions 1 through 5, Yan and Mukai (1992) have shown that the Markov chain $\{X_k\}$ generated by Algorithm 1 converges in probability to a global optimal solution of the discrete stochastic optimization problem (1).

In practice, the sequence $\{M_k\}$ plays a very crucial role in the convergence of Algorithm 1. In particular, if the sequence $\{M_k\}$ is selected to increase very rapidly then the algorithm could converge to a local optimal solution and stay there. On the other hand, if the sequence $\{M_k\}$ is selected to increase very slowly then the algorithm tends to take a longer time to settle down. To ease these difficulties, in the next section we present a new variant of the stochastic ruler algorithm that does not require an increasing sequence of observations to be drawn at every iteration. Furthermore, the new method can handle more general neighborhood structures than the original stochastic ruler method.

3 THE PROPOSED MODIFICATION OF THE STOCHASTIC RULER ALGORITHM

In this section we present a new variant of the stochastic ruler method (Algorithm 1) to solve the discrete stochastic optimization problem (1). This new method- requires only a finite number of observations per iteration instead of an increasing number of observations per iteration. Moreover, in our new method we relax Assumption 3 to include more general neighborhood structures and transition probabilities. To proceed, we use the following transition probability that also satisfies Definition 2. Let

$$R(x,x') = \frac{R'(x,x')}{D(x)}, \tag{3}$$

where $R' : \mathcal{S} \times \mathcal{S} \to \Re$ is a function such that $R'(x,x') > 0 \Leftrightarrow x' \in N(x)$, and $D : \mathcal{S} \to \Re$ is another function such that $D(x) = \sum_{x' \in N(x)} R'(x,x')$, $\forall x \in \mathcal{S}$. Now we use the following assumption instead of Assumption 3:

Assumption 6 *Let $R(x,x')$ be defined as in equation (3) and let N satisfy Definition 1. Then we assume that:*

1. *$x' \in N(x) \Leftrightarrow x \in N(x')$, and*
2. *$R'(x,x') = R'(x',x)$, $\forall x, x' \in \mathcal{S}$.*

As an example that satisfies Assumption 6 consider $R(x,x') = \frac{1}{|N(x)|}$ for all $x' \in N(x)$. Note that this choice of transition probability R only satisfies Assumption 3 if $|N(x)| = |N(x')|$ whenever $x' \in N(x)$.

Our modified stochastic ruler method is given below. Note that $\{X_k\}$ is the Markov chain that is generated by the algorithm, $V_k(x)$ is the number of times the Markov chain $\{X_k\}$ has visited state x up to time k, and $D(x)$ is defined in equation (3). Then the state X_k^* that maximizes $\frac{V_k(X_k)}{D(X_k)}$ is used as an estimate of the solution of the discrete stochastic optimization problem (1).

Algorithm 2

Parameters: N, R, M, a, b.

Step 0: *Select a starting point $X_0 \in \mathcal{S}$. Let $V_0(X_0) = 1$, and $V_0(x) = 0$, for all $x \in \mathcal{S}$, $x \ne X_0$. Let $k = 0$ and $X_k^* = X_0$.*

Step 1: *Given $X_k = x$, choose a candidate Z_k from $N(x)$ with probability distribution*

$$P\{Z_k = z | X_k = x\} = R(x,z) = \frac{R'(x,z)}{D(x)},$$

where $z \in N(x)$ and $R'(x,z)$ and $D(x)$ are defined in equation (3).

Step 2: *Given $Z_k = z$, draw a sample $h(z)$ from $H(z)$. Then draw a sample θ from $\Theta(a,b)$. If $h(z) > \theta$, then let $X_{k+1} = X_k$ and go to Step 3. Otherwise, draw another sample $h(z)$ from $H(z)$ and draw another sample θ from $\Theta(a,b)$. If $h(z) > \theta$, then let $X_{k+1} = X_k$ and go to Step 3. Otherwise, continue to draw and compare. If all M tests, $h(z) > \theta$, fail, then accept the candidate Z_k and set $X_{k+1} = Z_k = z$.*

Step 3: *Let $k = k+1$, $V_k(X_k) = V_{k-1}(X_k)+1$, and $V_k(x) = V_{k-1}(x)$, for all $x \in \mathcal{S}$, $x \neq X_k$. If $\frac{V_k(X_k)}{D(X_k)} > \frac{V_k(X^*_{k-1})}{D(X^*_{k-1})}$, then let $X^*_k = X_k$; otherwise let $X^*_k = X^*_{k-1}$. Go to Step 1.*

Remark 2 Note that Step 2 can be rewritten as: Given $Z_k = z$, set

$$X_{k+1} = \begin{cases} Z_k & \text{with probability } p_z, \\ X_k & \text{with probability } 1-p_z, \end{cases}$$

where $p_z = [P\{H(z) \le \Theta(a,b)\}]^M = [P(z,a,b)]^M$.

The next theorem has been proved by Alrefaei and Andradóttir (1996). It guarantees that the sequence $\{X^*_k\}$ generated by Algorithm 2 converges almost surely to a global optimal solution of the discrete stochastic optimization problem (1).

Theorem 1 *Under Assumptions 1, 2, 4 and 6 the sequence $\{X^*_k\}$ generated by Algorithm 2 converges almost surely to an element of $\mathcal{S}^*$ (in the sense that there exists a set A such that $P(A) = 1$ and for all $\omega \in A$, there exists $K_\omega > 0$ such that $X^*_k(\omega) \in \mathcal{S}^*$ for all $k \ge K_\omega$).*

4 NUMERICAL APPLICATION

In this section we implement Algorithm 2 to solve a simple discrete stochastic optimization problem, and compare its performance with that of Algorithm 1 on the same problem. In equation (1), let $\mathcal{S} = \{1, \ldots, 10\}$, and let $f(x) = E[Y_x]$ for all $x \in \mathcal{S}$, where Y_x is a uniform random variable on the interval $f(x) \pm 0.5$, for all $x \in \mathcal{S}$, and $f(1), \ldots, f(10)$ are $0.3, 0.7, 0.9, 0.5, 1.0, 1.4, 0.7, 0.8, 0.0$, and 0.6, respectively. See Figure 1 for the graph of the function. Let $a = -0.5, b = 1.9$, and let X_0 be uniformly distributed on the set $\mathcal{S}$.

We will apply Algorithm 1 to solve this optimization problem with a number of different choices of the parameters M, $\{N(x) : x \in \mathcal{S}\}$, and $\{R(x,x') : x \in \mathcal{S}$ and $x' \in N(x)\}$ that satisfy the conditions of Theorem 1 (so that Algorithm 2 is guaranteed to converge). In particular, we use two different values of the constant M: $M = 1$ and $M = 2$. Also, we use three different neighborhood structures $\{N(x) : x \in \mathcal{S}\}$. The first neighborhood structure is given by

$$N(x) = \begin{cases} \{2\} & \text{if } x = 1, \\ \{9\} & \text{if } x = 10, \\ \{x-1, x+1\} & \text{otherwise.} \end{cases} \tag{4}$$

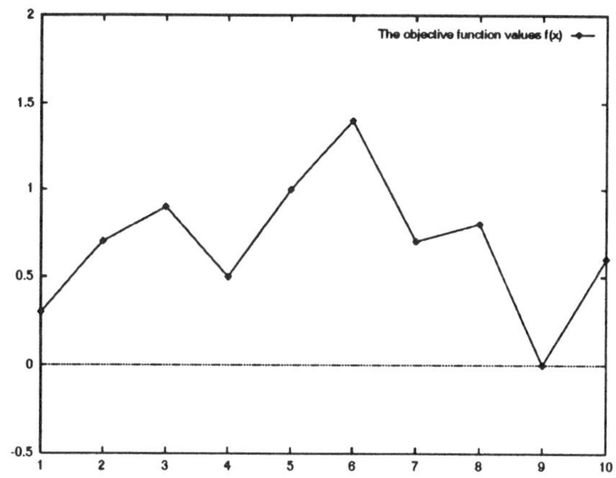

Figure 1: The Objective Function Values $f(x)$, where $f(x) = E[Y_x]$ and Y_x is a Uniform Random Variable on the Interval $f(x) \pm 0.5$.

The second neighborhood structure is given by

$$N(x) = \begin{cases} \{2,3\} & \text{if } x = 1, \\ \{1,3,4\} & \text{if } x = 2, \\ \{7,8,10\} & \text{if } x = 9, \\ \{8,9\} & \text{if } x = 10, \\ \{x \pm 1, x \pm 2\} & \text{otherwise.} \end{cases} \tag{5}$$

The third neighborhood structure is given by

$$N(x) = \mathcal{S} \setminus \{x\}, \tag{6}$$

for all $x \in \mathcal{S}$. In all cases, we let $R(x,x') = \frac{1}{|N(x)|}$, for all $x \in \mathcal{S}$ and $x' \in N(x)$. Note that in the third neighborhood structure (6), we have only one global minimum at $x = 9$. On the other hand, in the first neighborhood structure (4) we have three local minima at $x = 1, 4$, and 7 and one global minimum at $x = 9$, and in the second neighborhood structure (5), we have two local minima at $x = 1$ and $x = 4$ and one global minimum at $x = 9$. Since the first and second neighborhood structures result in more local minima, we expect that Algorithm 2 will converge more slowly in these settings than when the third neighborhood structure is used.

Note that for the first neighborhood structure (see equation (4)) it is impossible to select a transition probability R that satisfies Assumption 3. Therefore, Algorithm 1 is not guaranteed to converge when the first neighborhood structure is used. Also, for the second neighborhood structure (see equation (5)) it is difficult to select a transition probability R that satisfies Assumption 3. Therefore, it is difficult to apply Algorithm 1 using the second neighborhood structure. However, we implement Algorithm 1 using the third neighborhood structure defined in equation (6).

In this case, we let $M_k = \lfloor \log_5(k+10) \rfloor$ for all k; this choice of the sequence $\{M_k\}$ satisfies the guidelines specified by Yan and Mukai (1992).

Figures 2 and 3 show the results obtained by applying Algorithm 2 to solve this optimization problem with the choices of parameters described above. In particular, the figures show the number of observations of the objective function vs. the number of replications that have converged to the global solution out of a total of 100 replications that were used.

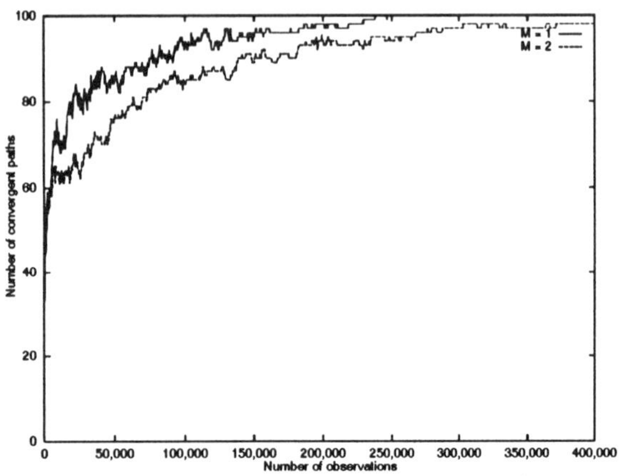

Figure 2: The Performance of Algorithm 2 as a Function of M when the Neighborhood Structure $\{N(x)\}$ given by Equation (4) is used.

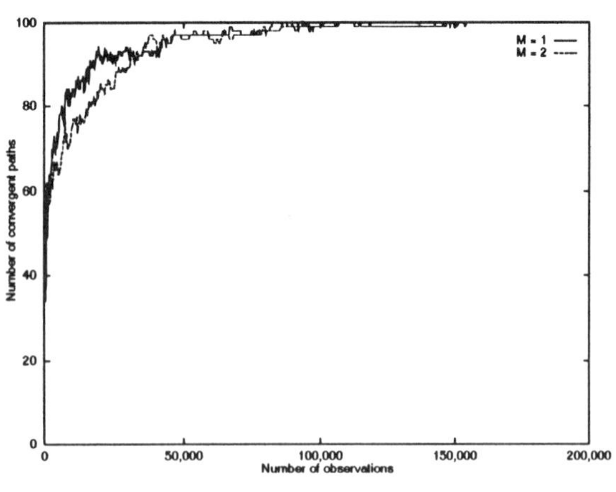

Figure 3: The Performance of Algorithm 2 as a Function of M when the Neighborhood Structure $\{N(x)\}$ given by Equation (5) is used.

Figure 4 includes a comparison between Algorithms 1 and 2 when the third neighborhood structure is used. Note that for Algorithm 2 all 100 replications converge to the true optimizer after 10,000 observations have been drawn. Furthermore, 90 of these replications converge after only 2,000 observations have been made. On the other hand, the number of convergent replications for Algorithm 1 does not exceed 60 replications after 50,000 observations have been obtained. In fact, in another experiment we completed 100 replications of 10 million iterations of Algorithm 1 (which required about 40 million observations per replication on the average) and we did not get convergence for all 100 replications.

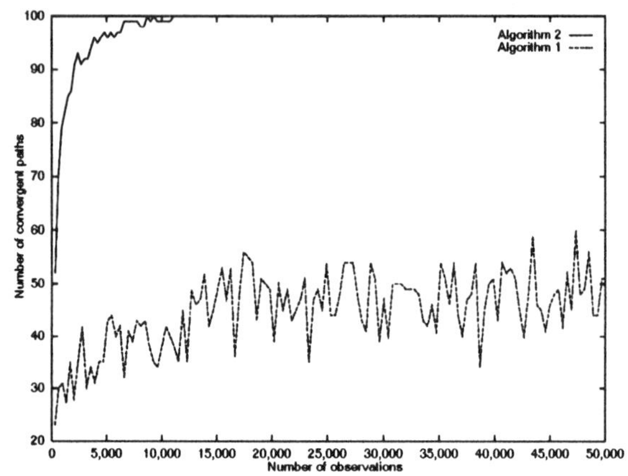

Figure 4: A Comparison Between Algorithms 1 and 2 when the Neighborhood Structure $\{N(x)\}$ given by Equation (6) is used, $M = 1$ in Algorithm 2, and $M_k = \lfloor \log_5(k+10) \rfloor$ for all k in Algorithm 1.

From the results given in Figure 4, we conclude that in this example Algorithm 2 converges much faster than Algorithm 1. Also, as expected, it appears to be better to have larger neighborhoods $N(x)$ where $x \in \mathcal{S}$, and smaller number of observations M in Algorithm 2 (the results for the neighborhood structure (6) are better than those for the neighborhood structures (4) and (5), and the results for $M = 1$ are better than those for $M = 2$).

5 CONCLUSION

In this paper, we have presented a new search method designed to solve the discrete stochastic optimization problem (1). This method resembles the stochastic ruler method of Yan and Mukai (1992). Unlike the original stochastic ruler method, the new method requires only a finite number of observations to be made at every iteration instead of using an increasing sequence of observations per iteration. Furthermore, this new method can handle more general neighborhood structures. It has the feature that it spends most of the time near the optimal solution. Also, unlike the original stochastic ruler method, our method converges *almost surely* to a global optimal solution.

In practice, in our example, it seems to be better to use a small number of observations per iteration to give the method more freedom to move around the state space aggressively and locate the solution very quickly. Also, it is better to use larger neighborhood sets when there are many local minimal solutions. However, we expect that may be better to use smaller neighborhood sets if the objective function is unimodal or has a fair amount of structure. In our example, the new method shows better performance than the original stochastic ruler method. In order to further investigate the efficiency of the proposed method, we are interested in implementing our method to solve more realistic discrete stochastic optimization problems (such as the buffer allocation problem). We are also interested in investigating the application of our method to solve discrete optimization problems in both transient and steady-state simulation.

ACKNOWLEDGMENTS

This research was supported in part by the National Science Foundation under grants DDM–9210679 and DMI–9523111. In addition, the research of the first author was supported in part by the Jordan University of Science and Technology.

REFERENCES

Alrefaei, M. H., and S. Andradóttir. 1995. A new search algorithm for discrete stochastic optimization. In *Proceedings of the 1995 Winter Simulation Conference*, ed. C. Alexopoulos, K. Kang, W. R. Lilegdon, and D. Goldsman, 236–241. Institute of Electrical and Electronics Engineers, Piscataway, New Jersey.

Alrefaei, M. H., and S. Andradóttir. 1996. A modification of the stochastic ruler method. Working paper.

Andradóttir, S. 1995. A method for discrete stochastic optimization. *Management Science* 41:1946–1961.

Andradóttir, S. 1996. A global search method for discrete stochastic optimization. To appear in the *SIAM Journal on Optimization*.

Bechhofer, R. E., T. J. Santner, and D. M. Goldsman. 1995. *Design and Analysis of Experiments for Statistical Selection, Screening, and Multiple Comparisons*. New York: Wiley.

Fox, B. L., and G. W. Heine. 1995. Probabilistic search with overrides. *The Annals of Applied Probability* 5:1087–1094.

Gelfand, S. B., and S. K. Mitter. 1989. Simulated annealing with noisy or imprecise energy measurements. *Journal of Optimization Theory and Applications* 62:49–62.

Goldsman, D., B. L. Nelson, and B. Schmeiser. 1991. Methods for selecting the best system. In *Proceedings of the 1991 Winter Simulation Conference*, ed. B. L. Nelson, W. D. Kelton, and C. M. Clark, 177–186. Institute of Electrical and Electronics Engineers, Piscataway, New Jersey.

Ho, Y. C., R. S. Sreenivas, and P. Vakili. 1992. Ordinal optimization of DEDS. *Journal of Discrete Event Dynamical Systems* 2:61–88.

Lee, J. 1995. Faster simulated annealing techniques for stochastic optimization problems, with application to queueing network simulation. Ph.D. Thesis, North Carolina State University, Raleigh.

Yan, D., and H. Mukai. 1992. Stochastic discrete optimization. *SIAM Journal on Control and Optimization* 30:594–612.

AUTHOR BIOGRAPHIES

MAHMOUD HASAN ALREFAEI is pursuing a joint Ph.D. degree in mathematics and industrial engineering at the University of Wisconsin – Madison. He received his B.S. and M.S. degrees in mathematics from Yarmouk University in Jordan. His research interests include discrete stochastic optimization and simulated annealing. He is presently a member of INFORMS and AMS.

SIGRÚN ANDRADÓTTIR is an Associate Professor in the School of Industrial and Systems Engineering at the Georgia Institute of Technology, on leave from the University of Wisconsin – Madison where she has held a faculty position since 1990. She received a B.S. in Mathematics from the University of Iceland in 1986, an M.S. in Statistics from Stanford University in 1989, and a Ph.D. in Operations Research from Stanford University in 1990. Her research interests include stochastic optimization, simulation, and stochastic processes. She presently serves as Associate Editor for *IIE Transactions*, *Stochastic Models*, and the *ACM Transactions on Modeling and Computer Simulation*. She is a member of INFORMS.

Proceedings of the 1996 Winter Simulation Conference
ed. J. M. Charnes, D. J. Morrice, D. T. Brunner, and J. J. Swain

VARIANCE PROPERTY OF DISCONTINUOUS PERTURBATION ANALYSIS

Leyuan Shi

Department of Industrial Engineering
University of Wisconsin-Madison
Madison, WI 53706

ABSTRACT

In recent article we presented a Discontinuous Perturbation Analysis (DPA) to deal with discontinuous sample performance functions. DPA is applicable to derivative estimation with respect to most threshold type of problems in Discrete Event Dynamic Systems (DEDS). In this paper, we examine the variance property of DPA method through a few examples. We will demonstrate that a better variance reduction can be obtained via DPA approach even if IPA Infinitesimal Perturbation Analysis (IPA) is applicable. A numerical example is also provided to compare DPA with other derivative estimation techniques.

1 INTRODUCTION

In the traditional simulation setting, one simulation corresponds to an estimate of the performance measure at one value of the parameter. None of the existing variance reduction methods could change this simple fact. But the driving force behind the development of Perturbation Analysis and Likelihood Ratio methods is the question of whether or not there is more information obtainable from that one simulation in the sense of estimating performance at other values of the parameter. From efforts made to tackle this issue came the development of Perturbation Analysis and later Likelihood Ratio method, which, simply put, is a means by which we compute derivative estimates from a single simulation (or actual observation) of the system.

Perturbation Analysis (PA) was initiated by Ho et al. (Ho and Cao, 1991), (Cassandras,1993) in the context of production lines in manufacturing systems. Its infinitesimal version, Infinitesimal Perturbation Analysis (IPA), was for the sensitivity analysis of throughput in closed queueing network. Since then, it has been successfully applied to many different systems. There also has been a considerable effort to extend the theory of IPA to handle situations where IPA fails (Brémaud and Vázquez-Abad, 1992), (Fu and Hu, 1992), (Gong and Ho, 1987), (Cassandras, 1993), (Shi, 1996), (Dai and Ho 1994). Likelihood Ratio (LR), also known as Score Function (SF) estimator has been proposed recently in the area of importance sampling (Glynn, 1987), (Reiman and Weiss, 1989), (Rubinstein, 1989). The relations between PA and LR have been discussed in (L'ecuyer, 1990).

To give the basic idea of IPA and LR, let us consider a stochastic system with a parameter θ. Suppose the performance measure of the system is

$$J(\theta) = E_\omega[L(x(t,\theta,\omega))].$$

where x is a random variable whose c.d.f. (cumulative distribution function) is $F(\theta,x)$ depending on θ. To simplify the discussion, we assume that x is a scalar. Thus, we have

$$J(\theta) = \int_{-\infty}^{\infty} L(x)dF(\theta,x). \tag{1}$$

Let $\xi = F(\theta,x)$. Then ξ is a random variable uniformly distributed on (0,1], and x is a function of ξ:

$$x = x(\theta,\xi) = F^{-1}(\theta,\xi) = inf\{x : f(\theta,\xi) \geq \xi\}. \tag{2}$$

Equation (1) is equivalent to

$$J(\theta) = \int_0^1 L(x(\theta,\xi))d\xi. \tag{3}$$

On the other hand, let $f(\theta,x)$ be the p.d.f. (probability density function), then (1) can be written as

$$J(\theta) = \int_{-\infty}^{\infty} L(x)f(\theta,x)dx. \tag{4}$$

Taking the derivatives of both sides of (3) and (4), we obtain

$$\frac{\partial}{\partial\theta}J(\theta) \quad = \quad \frac{\partial}{\partial\theta}\int_0^1 L(x(\theta,\xi))d\xi$$

$$= \int_{-\infty}^{\infty} \frac{\partial}{\partial \theta} L(x(\theta,\xi))d\xi$$
$$= \int_{-\infty}^{\infty} \frac{\partial}{\partial x} L(x) \frac{\partial x}{\partial \theta} d\xi$$
$$= E[\frac{\partial L(x)}{\partial x}\frac{\partial x}{\partial \theta}]. \quad (5)$$

and

$$\frac{\partial}{\partial \theta} J(\theta) = \frac{\partial}{\partial \theta}\int_{-\infty}^{\infty} L(x) f(\theta,x)dx$$
$$= \int_{-\infty}^{\infty} L(x)\frac{\partial}{\partial \theta} f(\theta,x)dx$$
$$= \int_{-\infty}^{\infty} L(x)\frac{\partial}{\partial \theta}[\ln f(\theta,x)] f(\theta,x)dx$$
$$= E[L(x)\frac{\partial}{\partial \theta}\ln f(\theta,x)], \quad (6)$$

respectively, provided that the two operations, "E" and "$\frac{\partial}{\partial\theta}$" are interchangeable in each case. Equation (6) also requires $f(\theta,x) \neq 0$. If the interchangeability in (5) or (6) holds, then

$$\frac{\partial L(x)}{\partial x}\frac{\partial x}{\partial \theta} \quad (7)$$

or

$$L(x)\frac{\partial \ln f(\theta,x)}{\partial \theta} \quad (8)$$

is an unbiased estimate of $\partial J(\theta)/\partial\theta$. Equation (7) corresponds to the Infinitesimal Perturbation Analysis (IPA) estimate, and Equation (8) is called the Likelihood Ratio (LR) estimate of $\partial J(\theta)/\partial\theta$.

In general, the IPA estimate has a smaller variance, enjoys the nice property of convergence with probability one and a certain amount of robustness with respect to uncertainties in the underlying distribution (Ho & Cao, 1991), but requires more restrictive conditions than the LR estimate. On the other hand, the LR estimate usually has a bigger variance, but applies to a wider range of systems since the requirement for interchangeability of two operators, "E" and "$\frac{\partial}{\partial\theta}$", is much milder.

As we can see in (2) that we use inversion to represent the random variable. It is well-known that there exists more than one way to represent a parametric family of random variables or stochastic processes with given marginal probability distributions. Different representations often give rise to different sample path derivatives. The problem of determining which sample path derivative has a smaller variance when multiple strongly consistent and unbiased sample path derivatives exist is important in practice. In this short paper we use an example to demonstrate that variance reduction can be obtained when we resort DPA approach. We also show that there is a trade-off between accuracy and computation effort.

2 SAMPLE PATH DERIVATIVE AND ITS VARIANCE

Sample path derivatives can be used as derivative estimates of performance measures in stochastic discrete event systems such as computer networks and communication systems. These derivative estimates can be calculated based on a single sample path using the technique of IPA or other derivative estimate techniques. Usually, there exists more than one way to represent a parametric family of random variables or stochastic processes with given marginal probability distributions. Different representations often give rise to different sample path derivatives which have the same mean but different variances. To see this, let us define the sample path derivative of a parametric random variable.

Consider a parametric family of cumulative distribution functions (c.d.f.'s) $\{F(x;\theta) : \theta \in \Theta\}$. A family of random variables $\{X(\theta)\}$ defined on a common probability space (Ω, F, P) is a representation of $\{F(x;\theta) : \theta \in \Theta\}$ if $X(\theta)$ is a representation of $F(x;\theta)$ for every $\theta \in \Theta$, i.e.,

$$Prob(X(\theta) < x) = F(x;\theta).$$

Given a family of $\{X(\theta) : \theta \in \Theta\}$ defined on (Ω, F, P). For simplicity of notation, we assume that $\theta \in \Theta = [a,b]$. Denote

$$\delta X(\theta,h) = \frac{X(\theta+h) - X(\theta)}{h},$$

where $\theta, \theta+h \in \Theta$.

Definition 1 $\{X(\theta)\}$ is a.s. differentiable at $\theta \in \Theta$ if there exists a random variable $Y(\theta)$ such that

$$\lim_{h\to 0} \delta X(\theta,h) = Y(\theta) \quad a.s.$$

We also call $Y(\theta)$ the sample path derivative of $X(\theta)$ and denote $dX(\theta)/d\theta = Y(\theta)$.

Definition 2 $\{X(\theta)\}$ is a.s. continuous (piecewise differentiable, differentiable) on Θ if $P(X(q)$ is continuous (piecewise differentiable, differentiable) on $\Theta) =$ 1.

Definition 3 $\{X(\theta)\}$ is *admissible* (with respect to θ) if $\{X(\theta)\}$ is a.s. continuous and piecewise differentiable on Θ, a.s. differentiable for every $\theta \in \Theta$ and $E[\sup_{\theta\in\Theta} dX(\theta)/d\theta] < \infty$.

Definition 4 if $\{X(\theta)\}$ is admissible and if $\{X(\theta)\}$ is also a representation of $\{F(x;\theta)\}$, then we say it is an *admissible* representation.

Given a parametric family of c.d.f.'s $\{F(x;\theta)\}$, there may exist more than one admissible representation. To illustrate this, we first define the inverse transform for a c.d.f. $G(x)$ to be

$$G^{-1}(u) = \sup\{y : G(y) \leq u\}.$$

It is well-known that $G^{-1}(U)$ is a representation of $G(x)$, where U is a uniform random variable over [0,1]. It then follows that $\{F^{-1}(U;\theta)\}$ is a representation of $\{F(x;\theta)\}$.

Example 1. Suppose $F_i(x) = 1 - e^{-\lambda_i x}$ $(x \geq 0, i = 1,2)$. Let $F(x;\theta) = \theta F_1(x) + (1-\theta)F_2(x)$, where $0 < \theta < 1$. Then it is easy to verify that $\{F^{-1}(U;\theta)\}$ is an admissible representation of $\{F(x;\theta)\}$. On the other hand, note that under the condition that $0 \leq U \leq \theta$ (or $\theta < U \leq 1$), $\frac{\theta - U}{\theta}$ (or $\frac{U-\theta}{1-\theta}$) is also uniformly distributed over [0,1]. Therefore,

$$\left\{ I_{[0\leq U\leq\theta]}F_1^{-1}(\frac{\theta-U}{\theta}) + I_{[\theta\leq U\leq 1]}F_2^{-1}(\frac{U-\theta}{1-\theta}) \right\}$$

is also an admissible representation.

Now let ξ be another uniform random variable over [0,1] which is independent of U. We can verify that

$$\{I_{[0\leq\xi\leq\theta]}F_1^{-1}(U) + I_{[\theta\leq\xi\leq 1]}F_2^{-1}(U)\}$$

is also a representation of $F(x;\theta)$. But **??** is not an *admissible* representation because it is not a.s. continuous on Θ. In the next section we will see that variance reduction can be obtained by using such a representation.

Lemma 1(Hu, 1993) If $\{F^{-1}(U,\theta)\}$ is an admissible representation, then

$$\frac{dF^{-1}(U;\theta)}{d\theta} = \frac{\partial F(x;\theta)/\partial\theta}{\partial F(x;\theta)/\partial x} |_{x=F^{-1}(U;\theta)},$$

and

$$Var\left[\frac{dF^{-1}(U;\theta)}{d\theta}\right] \leq Var\left[\frac{dX(\theta)}{d\theta}\right],$$

where $\{X(\theta)\}$ is any admissible representation of $\{F(x;\theta)\}$.

The lemma tells us that if inversion gives admissible representation, its corresponding sample path derivative has the smallest variance. In other words, inversion is the best representation in all **admissible** representations.

3 VARIANCE REDUCTION VIA DPA APPROACH

Many experiments have shown that IPA, when applicable, produces smaller variance of estimation error (based on fixed simulation length) than others (Glasserman & Yao, 1991). However IPA fails for most threshold (or structure) type of problems such as routing problem and queue length. The reason is mainly due to the sample path discontinuity with respect to the parameter in question, i.e., infinitesimal change in parameter value produces finite change in the sample path of the system. There has been a considerable effort to extend the theory of IPA to handle the situation where IPA fails (Shi, 1996),(Fu and Hu, 1992),(Gong and Ho, 1987), (Brémaud and Vázquez-Abad, 1992), (Dai and Ho 1994).

In (Shi, 1996), we propose another alternative approach called Discontinuous Perturbation Analysis (DPA) which tries to deal with discontinuity of sample performance functions, especially for threshold type of parameters or structure parameters. In this approach we use the step function to represent a discontinues sample performance function. We then introduce the d-function to sample path derivative. With this format, we are able to provide a framework for constructing the derivative estimation of certain classes of discrete event systems. In the same paper we also showed that the DPA formulation derived from this approach is very closely related with other exiting methods such as SPA (Smooth Perturbation Analysis) (Gong and Ho, 1987), RPA (Rare Perturbation Analysis) (Brémaud and Vázquez-Abad, 1992) and LR. In some cases, the SPA formula can be derived from DPA formula. A major result of the DPA approach can be summarized as the following:

Let

$$L(\theta,\omega,n) = L_i^k(\theta,\xi,\eta),$$

if $g^{k-1}(\theta) < \xi_i \leq g^k(\theta), k = 1,...M,$
and $0 = g^0(\theta) < g^1(\theta) < .. < g^k(\theta) < .. < g^M(\theta) = 1,$

Theorem (Shi, 1996) Assume that for all i and k, $L_i^k(\theta,\xi,\eta)$ satisfy
i) $L_i^k(\theta,\xi,\eta)$ is uniformly differentiable with respect to θ,
ii) $L_i^k(\theta,\xi,\eta)$ is a continuous function of ξ,
then we have

$$\begin{aligned}\frac{\partial}{\partial\theta}E[L(\theta,\omega,n)] &= E[\frac{\partial}{\partial\theta}L(\theta,\omega,n)\mid_{IPA}] \\ &+ E[\sum_{i=1}^{n}\sum_{k=1}^{M}(L_{i\theta}^k - L_{i\theta}^{k+1})\frac{dg_i^k(\theta)}{d\theta}]\end{aligned}$$

where $L_{i\theta}^k = L_i^k(\xi_i = g_i^k(\theta))$.

In the following we will provide an example to illustrate that when both IPA and DPA are applicable, DPA formula will provide a smaller variance than IPA formula, i.e., inversion is not the best representation in all **possible** representations.

Example 2. Consider a simple cyclic queueing network consisting of two servers and only one customer.

The service time distribution for server one is deterministic S1=m and for server two is $F(\theta, S_2) = \theta(1-e^{-\lambda_1 S_2})+(1-\theta)(1-e^{-\lambda_2 S_2})$. The state process of this system can be decomposed into regenerative periods, each consisting of one busy period and one idle period for each server. The mean length of a regenerative period is

$$E[L(\theta)] = E[S_1 + S_2(\theta)] = \mu + \frac{\theta}{\lambda_1} + \frac{1-\theta}{\lambda_2}.$$

Its derivative with respect to θ is

$$\frac{\partial}{\partial\theta}E[L(\theta)] = \frac{1}{\lambda_1} - \frac{1}{\lambda_2}.$$

The IPA estimate of $\frac{\partial}{\partial\theta}E[L(\theta)]$ is

$$\begin{aligned} e\,|_{IPA} &\equiv \frac{\partial}{\partial\theta}E[L(\theta)]\,|_{IPA} \\ &= \frac{\partial}{\partial\theta}[S_1 + S_2(\theta)] \\ &= -\frac{\partial F(\theta, S_2)/\partial\theta}{\partial F(\theta, S_2)/\partial S_2} \\ &= \frac{e^{-\lambda_1 S_2} - e^{-\lambda_2 S_2}}{\theta\lambda_1 e^{-\lambda_1 S_2} + (1-\theta)\lambda_2 e^{-\lambda_2 S_2}}. \end{aligned}$$

The LR estimate of $\frac{\partial}{\partial\theta}E[L(\theta)]$ is

$$\begin{aligned} e\,|_{LR} &\equiv \frac{\partial}{\partial\theta}E[L(\theta)]\,|_{LR} \\ &= (\mu + S_2(\theta))\frac{\lambda_1 e^{-\lambda_1 S_2} - \lambda_2 e^{-\lambda_2 S_2}}{\theta\lambda_1 e^{-\lambda_1 S_2} + (1-\theta)\lambda_2 e^{-\lambda_2 S_2}}. \end{aligned}$$

On the other hand, as we mentioned in the previous section that $S_2(\theta)$ can also be represented as

$$\{I_{[0\le\xi\le\theta]}F_1^{-1}(\eta) + I_{[\theta\le\xi\le 1]}F_2^{-1}(\eta)\}$$

That is

$$S_2(\theta) = \begin{cases} -\frac{1}{\lambda_1}\ln(\eta) & \text{if } 0 \le \xi \le \theta \\ -\frac{1}{\lambda_2}\ln(\eta) & \text{if } 1 \ge \xi > \theta \end{cases}$$

where ξ and h are two independent random variables with uniform distribution over [0,1]. Clearly $S_2(\theta)$ is not an admissible representation.

Using Discontinuous Perturbation Analysis (DPA) (Shi, 1996), we can use the step function to represent $S_2(\theta)$. We have

$$S_2(\theta) = -\frac{1}{\lambda_1}\ln(\eta) + (\frac{1}{\lambda_1}\ln(\eta) - \frac{1}{\lambda_2}\ln(\eta))H(\xi-\theta).$$

where

$$H(\xi-\theta) = \begin{cases} 0 & \text{if } \xi < \theta, \\ 1 & \text{if } \xi \ge \theta. \end{cases}$$

Then

$$\frac{d}{d\theta}S_2(\theta) = -(\frac{1}{\lambda_1}\ln(\eta) - \frac{1}{\lambda_2}\ln(\eta))\delta(\xi-\theta),$$

where δ-function is defined by

$$\delta(\xi-\theta) = \begin{cases} \infty & \text{if } \xi = \theta, \\ 0 & \text{if } \xi \ne \theta. \end{cases}$$

and for any continuous function $f(\xi)$

$$\int_0^1 f(\xi)\delta(\xi-\theta)d\xi = f(\theta).$$

To be able to calculate the sample path derivative we need to remove δ-function, hence the DPA estimate of $\frac{\partial}{\partial\theta}E[L(\theta)]$ is

$$\begin{aligned} e\,|_{DPA} &\equiv \frac{\partial}{\partial\theta}E[L(\theta)]\,|_{DPA} \\ &= E_\xi[\frac{dS_2(\theta)}{d\theta}] \\ &= \frac{1}{\lambda_2}\ln(\eta) - \frac{1}{\lambda_1}\ln(\eta). \end{aligned}$$

It is easy to verify that

$$E[e\,|_{DPA}] = E[e\,|_{IPA}] = E[e\,|_{LR}] = \frac{1}{\lambda_1} - \frac{1}{\lambda_2}.$$

However, their variances differ significantly. In fact,

$$Var[e\,|_{DPA}] = (\frac{1}{\lambda_1} - \frac{1}{\lambda_2})^2,$$

and

$$\begin{aligned} &Var[e\,|_{IPA}] \\ &= \int_0^\infty \frac{(e^{-\lambda_1 S_2} - e^{-\lambda_2 S_2})^2}{\theta\lambda_1 e^{-\lambda_1 S_2} + (1-\theta)\lambda_2 e^{-\lambda_2 S_2}}dS_2 \\ &\quad -(\frac{1}{\lambda_1} - \frac{1}{\lambda_2})^2, \end{aligned}$$

and

$$\begin{aligned} &Var[e\,|_{LR}] \\ &= \int_0^\infty (\mu + S_2)\frac{(\lambda_1 e^{-\lambda_1 S_2} - \lambda_2 e^{-\lambda_2 S_2})^2}{\theta\lambda_1 e^{-\lambda_1 S_2} + (1-\theta)\lambda_2 e^{-\lambda_2 S_2}}dS_2 \\ &\quad -(\frac{1}{\lambda_1} - \frac{1}{\lambda_2})^2. \end{aligned}$$

Proposition 1. When $2\lambda_1 \le \lambda_2$, we have

$$\begin{aligned} \lim_{\theta\to 0} Var[e\,|_{IPA}] &= \infty, \\ \lim_{\theta\to 0} Var[e\,|_{LR}] &= \infty. \end{aligned}$$

Proof:

Table 1: Comparison of the Variances of DPA and SPA

θ	$\rho(\lambda=1)$	N_r/N_c*	$dE[T]/d\theta$	DPA Mean(Var)	SPA Mean(Var)
0.2	0.62	100/500	1.4157	1.39619 (0.100597)	1.37187 (0.109193)
0.2	0.62	200/500	1.4157	1.40181 (0.099802)	1.39485 (0.129505)
0.5	0.65	30/1000	-0.798639	0.79770 (0.022511)	-0.81832 (0.026849)
0.5	0.65	200/500	-0.798639	-0.7951 (0.053081)	-0.7952 (0.072534)
0.7	0.57	30/1000	-1.78983	-1.71851 (0.034805)	-1.73729 (0.033423)
0.7	0.57	40/1000	-1.78983	-1.7359 (0.0332479)	-1.74807 (0.031902)

*N_r: Number of replications. *N_c: Number of customers simulated per replication.

We only prove when $\lambda_1 = 1$ and $\lambda_2 = 2$.
When $\lambda_1 = 1$ and $\lambda_2 = 2$ we have

$$\int_0^\infty \frac{(e^{-\lambda_1 S_2} - e^{-\lambda_2 S_2})^2}{\theta\lambda_1 e^{-\lambda_1 S_2} + (1-\theta)\lambda_2 e^{-\lambda_2 S_2}} dS_2$$
$$= \int_0^\infty \frac{e^{-2S_2} - 2e^{-3S_2} + e^{-4S_2}}{\theta\lambda_1 e^{-S_2} + (1-\theta)\lambda_2 e^{2S_2}} dS_2$$
$$= \int_0^\infty \frac{e^{-2S_2}}{\theta\lambda_1 e^{-S_2} + (1-\theta)\lambda_2 e^{2S_2}} dS_2$$
$$+ \int_0^\infty \frac{-2e^{-3S_2} + e^{-4S_2}}{\theta\lambda_1 e^{-S_2} + (1-\theta)\lambda_2 e^{-2S_2}} dS_2.$$

It is easy to see that the second term is bounded by a finite number. Now let us take a look the first term.

Since

$$\int_0^\infty \frac{e^{-2S_2}}{\theta\lambda_1 e^{-S_2} + (1-\theta)\lambda_2 e^{-2S_2}} dS_2$$
$$= \int_0^\infty \frac{1}{\theta\lambda_1 e^{S_2} + (1-\theta)\lambda_2} dS_2 \quad (\text{let } u = e^{S_2})$$
$$= \int_0^\infty \frac{1}{\theta\lambda_1 u + (1-\theta)\lambda_2} . \frac{1}{u} du$$
$$= \frac{1}{(1-\theta)\lambda_2} \left[\ln((1-\theta)\lambda_2 + \theta\lambda_1) - \ln(\theta\lambda_1)\right] .$$

That is

$$\lim_{\theta\to 0} Var[e \mid_{IPA}] = \infty.$$

Use the same, way we can also prove that

$$\lim_{\theta\to 0} Var[e \mid_{LR}] = \infty.$$

Figure 1 gives the comparison of the variance of the DPA, IPA, and LR estimates when we choose $\lambda_1 = 1$ and $\lambda_2 = 2$. We can see that the variance of the DPA estimate is a constant and the variance of IPA and LR increases when θ approaches to zero.

This example provides some insight in sample path derivative techniques. Usually, when IPA is applicable, we will no longer look for new derivative estimate formula, given the believe that IPA would be the most efficient derivative estimation method. Little has been done in previous research to compare the IPA estimator with other methods except for LR. This topic certainly deserves further research. In the following we give another example to show that there is a trade-off between accuracy (smaller variance) and computation effort.

Example 3. An M/G/1 queue is simulated. The service time is given as the follows

$$S_i = \begin{cases} 0.8 - \xi_i & \text{if } 0 \le \xi_i \le \theta, \\ \xi_i & \text{if } 1 \ge \xi_i\, \theta, \end{cases}$$

where $0 < \theta \le 0.8$.

The performance measure we are interested in is the mean system time. Since S_i is discontinuous with respect to θ, IPA will not work. There has been a considerable effort to extend the theory of IPA to handle situations where IPA fails. DPA is one of such efforts. Another well-known method is Smooth Perturbation Analysis (SPA) (Ho and Gong, 1986). Here we compare these two methods.

For this problem, the DPA estimate formula is

$$\frac{\partial}{\partial\theta} E[L] \mid_{DPA} = \sum_{i=1}^{M} (L_{+i\theta} - L_{-i\theta}),$$

where M is the total number of customer simulated and the SPA estimate formula has the following form

$$\frac{\partial}{\partial\theta} E[L] \mid_{SPA} = \frac{1}{\theta} \sum_{i\in A} (L - L_{-i\theta}),$$

where $A = \{i : \xi_i > \theta \text{ and } 1 \le i \le M\}$.

Table 1 shows that overall the DPA estimate formula provides a slightly smaller variance than SPA. On the other hand, even though the DPA estimate has smaller variance than the SPA, the DPA estimate needs a slightly more computation effort than SPA (Shi, 1996). This clearly shows that there is a trade-off between accuracy and computation effort.

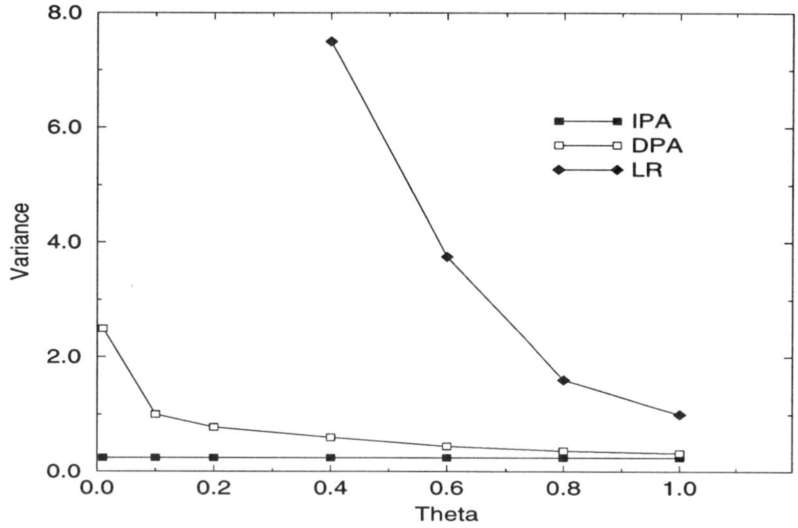

Figure 1: 90% Comparison of Variance of IPA, LR, and DPA

REFERENCES

Brémaud, P. and Vázquez-Abad F., 1992. On the pathwise computation of derivatives with respect to the rate of point process: The phantom RPA method. *Queueing Systems* 10: 249-270

Cassandras, C.G., 1993. *Discrete Event Systems: Modeling and Performance Analysis.* Irwin, March.

Dai, L. and Ho, Y.C., 1994. Structural infinitesimal perturbation analysis (SIPA) for derivative estimation in discrete event systems. Submitted to *Journal of Discrete Event Dynamic Systems.*

Fu, M.C. and Hu, J.Q., 1992. Extensions and generalizations of smoothed perturbation analysis in a generalized semi-markov framework. *IEEE Trans. on Automatic Control* 37: 1483-1500.

Hu. J.Q., 1993. Sample path derivatives: strong consistency and variance properties. Manuscript.

Glasserman, P., 1991. *Gradient Estimation Techniques for Discrete Event Systems.* Kluwer, MA.

Glasserman, P and Yao, D., 1991. Algebraic structure of some discrete event systems and their applications. *Discrete Event Dynamic Systems: Theory and Applications*, 1: 7-36.

Glynn, P.W., 1987. Likelihood ratio gradient estimation: an overview. *Proc. of 1987 Winter Simulation Conference* 366-375.

Gong, W.B. and Ho, Y.C., 1987. Smoothed perturbation analysis of discrete-event dynamic systems. *IEEE Transactions on Automatic Control* 33: 827-838.

Ho, Y.C. and Cao, X.R., 1991. *Discrete Event Dynamic Systems and Perturbation Analysis.* Kluwer Academic.

L'Ecuyer, P., 1990. A unified view of the IPA, SF, and LR gradient estimation techniques. *Management Science* 36: 1364-1383.

Reiman, M.I. and Weiss, A., 1989. Sensitivity analysis for simulation via likelihood ratios, *Operations Research* 37: 830-844.

Rubinstein, R. Y., 1989. Sensitivity analysis and performance extrapolation for computer simulation models. *Operations Research* 37: 72-81.

Shi, L, 1996. Discontinuous perturbation analysis for discontinuous sample performance functions. To appear in *IEEE Trans. on Automatic Control.*

AUTHOR BIOGRAPHY

LEYUAN SHI is an assistant professor in the Department of Industrial Engineering at University of Wisconsin-Madison. She holds a B.S. degree in mathematics from Nanjing Normal University, Chian (1982) and an M.S. degree in applied mathematics from Tsinghua University, China (1985), and M.S. and Ph.D degrees in applied mathematics from Harvard University (1990,1992). Her research interests include modeling, analysis, and optimization of discrete event systems, discrete-event simulation, and sensitivity analysis.

Proceedings of the 1996 Winter Simulation Conference
ed. J. M. Charnes, D. J. Morrice, D. T. Brunner, and J. J. Swain

FUZZY GRAPH BASED METAMODELING

Klaus–Peter Huber
Michael R. Berthold

IRF (Prof. D. Schmid), University of Karlsruhe
P.O. Box 6980, 76128 Karlsruhe, Germany

Helena Szczerbicka

Dept. of Computer Science, University of Bremen
P.O. Box 330440, 28334 Bremen, Germany

ABSTRACT

Analysis of simulation models has gained considerable interest in the past. However, their complexity still remains a considerable drawback in practical applications. One promising concept is the building of auxiliary models (*metamodels*) for different analysis goals. We present an efficient algorithm that constructs a metamodel only from simulation data, so no a priori knowledge has to be included. It will be shown that the resulting system approximates real valued functions with an adjustable precision. In addition the data can contain fuzzy patterns or values with a corresponding confidence–interval. This is especially well suited for simulation data due to its stochastic character. The metamodel is represented in form of a Fuzzy Graph which allows the analyst to directly extract easy to interpret if–then–rules. Application of this method to a real world token bus model is shown in detail.

1 INTRODUCTION

The use of modeling and simulation techniques has gained considerable influence for the development or the optimization of different systems. Unfortunately the complexity of the resulting simulation systems increases with the complexity of the real systems. This leads to several drawbacks: simulation becomes a highly time consuming task which makes it impossible to perform interactive simulations and in addition the analysis of the resulting simulation model is extremely complicated. Therefore there is a need for methods that help to analyze the behavior of complex models, e.g., to improve the validation process. In general two directions exist to solve this problem.

- *modelbased*: This means that the model can be analyzed based on knowledge about the concrete model structure. For example in petri net models the construction of a reachability graph helps to find places in the petri net that will never be reached.
- *databased*: If the model is executable, it is possible to perform experiments with the model e.g., by stochastic simulation of a queuing network model. The result is a huge database describing values of model parameters (factors) and corresponding values of the output parameter of interest. Analysis of dependencies between factors and the output parameters helps to build a simpler model or to better understand the behavior.

Due to the high complexity of models in real applications often there does not exist an analytical solution, so databased methods have to be used. One way is to analyze the data with statistical means like correlation analysis to find some unknown dependencies. Another way is to construct an auxiliary model, which is simpler and easier to handle than the original model. This auxiliary model is built using the example data generated by experimenting with the model and it should have the same behavior as the model.

In Blanning (1975) computation of partial derivatives was used to obtain the so–called *metamodel* from the simulation data. Depending on the task of analysis there may exist different kind of metamodels (see Figure 1), for example to analyze the parameter sensitivity (Blanning 1975; Huber and Szczerbicka 1994), to find bottlenecks (Kleijnen and Standridge 1988), or to optimize models (Yerramareddy et al. 1992). In all cases the extraction of a metamodel helps to reduce the complexity of the model that is being analyzed. Building a metamodel can be done including a priori knowledge or without it. If some underlying dependencies of parameters are known a priori, the metamodeling process can take these into account and generate a model where this information is incorporated. If this is not the case, the metamodel may be constructed only from the data set.

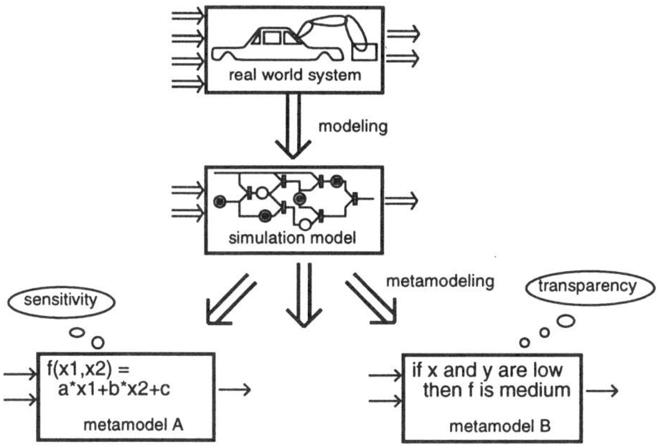

Figure 1: The Metamodeling Process

In this paper we present a metamodeling approach that only uses the data and constructs a so-called Fuzzy Graph. This allows an easy interpretation of the input-output behavior since the metamodel can be represented with fuzzy rules. Analysis of the rules helps to uncover the underlying dependencies between factors and one output parameter and the rule base can also be used as a fast simulator.

2 METAMODELING AND FUNCTION APPROXIMATION

Here, we concentrate on models where the behavior can be described by a function

$$f_{model}(\vec{x}) = f_{model}(x_1, \cdots, x_n) = y \qquad (1)$$

if n factors $x_1, \cdots, x_n$ are considered and y is the output parameter of interest. In practice this function can not be extracted explicitly from the model description and is therefore unknown. To build a suitable metamodel means to perform m experiments with the model to get m data points $(x_1^i, \cdots, x_n^i, y^i)$ $(1 \leq i \leq m)$ and to build a metamodel representing a function $f_{meta}(x_1, \cdots, x_n) = \tilde{y}$ with $\tilde{y}$ being the approximated value of y. If the data is generated by a stochastic simulation process y is a stochastic variable. There may exist a few different values of y for the same input parameters, e.g., by simulation with different random number streams. This should be taken into account when metamodeling. To validate the quality of the metamodel the difference between the observed y values and the approximated values $\tilde{y}$ can be evaluated.

Several approaches to build function approximators for metamodeling are known. They can be categorized into *statistical approaches* that use the data to adapt a special kind of function with statistical means and *Machine Learning Algorithms* that use the data to train a network or to generate a rule set.

2.1 Statistical Approaches for Metamodeling

Ad hoc methods use handfitted curves or graphical approaches where the response of the output to a changing input is represented graphically (Blanning 1975). Other methods use piecewise linear approximations (Meisel and Collins 1973) or linear regression models with a least square approach to define the parameters (Kleijnen 1979). Regression functions are very popular since the resulting models are easy to handle and interpret, and statistical methods like t-tests and F-tests can be used to validate the quality of the model (Friedman and Pressman 1988). For dealing with nonlinear behavior metamodeling regression functions are often defined as:

$$f_{meta}(\vec{x}) = \beta_0 + \sum_{j=1}^{n} \beta_j \cdot x_j + \sum_{i,l} \beta_{i,l} \cdot x_i \cdot x_l + \epsilon \quad (2)$$

with ϵ representing an error term and $\beta_{i,l}$ representing some user-defined in pair dependencies. To find a "best fitting" metamodel the regression parameters β_0, β_j and $\beta_{i,l}$ can be obtained by minimizing the mean square error due to the example data. The resulting function can be used for approximation or for analysis e.g., to obtain some information about the sensitivity of each factor.

In practice often the behavior of a simulation model can not be described with such first-order models. For example in Friedman and Pressman (1988) it is shown that the dependence between the factor *mean average numbers of customers* and the distribution parameters in a simple M/M/1-model can not be adequately described with that kind of functions. To solve this problem deep background knowledge about the model behaviour must be used. So for analysis of complex simulation models new approaches are required that build a metamodel without such a priori knowledge.

2.2 Machine Learning Approaches for Metamodeling

To avoid the need for background knowledge nowadays several methodologies have been proposed that make use of algorithms from the machine learning area. Prestructured Neural Networks are trained (Hurrion 1992; Pierreval 1996) or rule learning algorithms (Pierreval 1992) are used to build a metamodel using data from the simulation model. Although approximation with neural networks is done in a satisfying way this approach makes it hard to extract knowledge from the metamodel. Other approaches make use of rule learning algorithms to avoid this problem. In Pierreval (1992) preprocessed (i.e., cleaned and digitized) data from the simulation model was used to directly extract decision rules and in Huber and Szczerbicka (1994) a similar approach was used for sensitivity analysis. Both concepts are based on methods that learn a classifier but not a function approximator. Therefore they are restricted to applications where the output parameter is not continuous.

Another kind of learning techniques that are especially well suited to deal with "noisy" or "stochastic" data originate from the Soft Computing or Fuzzy Systems area (Zadeh 1994). They offer an easy way to model soft data points, for example values with a corresponding confidence interval. Unfortunately, most known methods that learn fuzzy systems from data have severe limitations in this context. Some require an a priori defined set of rules that is just fine tuned during training (Uebele, Abe, and Lan 1995), others construct the ruleset during training but the result depends heavily on the order of training examples (Simpson 1993). In Wang and Mendel (1991) a simple algorithm has been proposed that divides the feature space into partitions using a grid and assigns one rule to each tile. More sophisticated algorithms (Higgins and Goodman 1993) try to divide individual attributes step by step using an increasing number of membership functions. These methods tend to be very restrictive because the feature space is split into too many tiles. Each tile represents one rule but for most of these rules no evidence was encountered during training. For the purpose of metamodeling it is of much more interest, however, to find only a few rules that cover a large portion of the feature space.

In this paper a new approach is proposed that allows the usage of *Fuzzy Graphs* (Zadeh 1994) to represent the discovered knowledge. The main advantages are the automatic and fast construction of the Fuzzy Graph based on data examples and a straightforward knowledge extraction. Since the constructed Fuzzy Graph is represented with if–then–rules the metamodel is easy to understand. Additionally, the model behavior can be approximated. This makes it possible to perform fast simulation experiments with the metamodel. The following section describes how these Fuzzy Graphs are generated and how they can be used for function approximation. An example in section 4 demonstrates how the proposed algorithm models an artificial function.

3 FUZZY GRAPHS

The central idea of a Fuzzy Graph is to represent a function with means of Fuzzy Logic instead of mathematical equations. The domains of the input parameters and the domain of the output parameter are described with so–called linguistic variables represented by individual membership functions. The mapping from input to output is defined by a set of Fuzzy Points. Well–known techniques like the center–of–gravity calculation allow one to approximate real valued functions based on this Fuzzy Graph. In this paper, a Fuzzy Graph consists of a collection of Fuzzy Points that can be represented as if–then–rules. Other ways to represent Fuzzy Graphs are described for example in Zadeh (1994). The main advantage of the Fuzzy Graph concept is the very compact and easy to understand representation of a function.

The algorithm presented in this section automatically constructs a Fuzzy Graph based on a set of training examples. The training examples can feature "soft" values y, which enables the usage of targets with confidence intervals. Fuzzy Graphs built by this approach use a fixed granularization on the dependent variable which means that the user defines the membership functions of the output variable y before training. This is helpful to focus automatic generation of the Fuzzy Graph on specific regions of the output value and to weaken constraints (and therefore the evolving number of rules) on regions with a low focus of attention. The partitioning of the input variables is determined from the training examples. The algorithm we use is derived from a constructive Neural Network training algorithm that builds a specific type of networks with locally active hidden units (Berthold and Diamond 1995). In this paper, a modified version of that algorithm is used to automatically find a set of Fuzzy Points that describe the training data (Huber and Berthold 1995).

Output granularization has to be defined before training starts. Figure 2 shows an example of a one–dimensional Fuzzy Graph constructed by the proposed algorithm. The output granularization is known

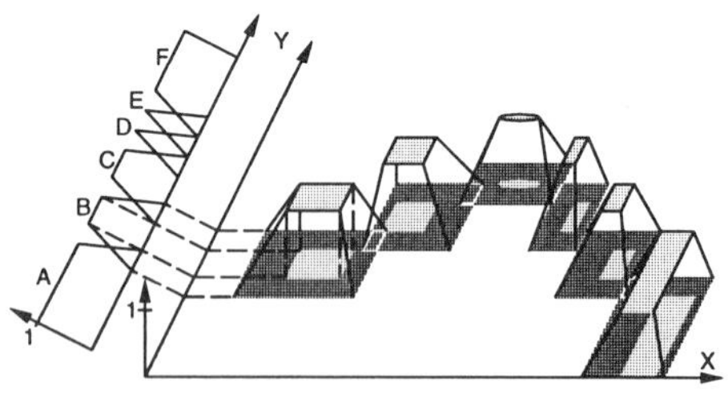

Figure 2: A One–dimensional Fuzzy Graph

a–priori: in this case 6 regions with individual membership functions are used. The used algorithm then places fuzzy points in the input space (X) to approximate the training examples. Before training starts, the output variable Y was already partitioned into 6 regions (or classes) with user defined membership functions. The Fuzzy Graph consists of fuzzy points with individual membership functions for the input variable X which are determined during training.

The constructed Fuzzy Graph consists of a collection of Fuzzy Points where the *core*–region represents the smallest area where patterns of the class were found, and the larger *support*–area contains no patterns of conflict. Therefore patterns in the core–region are given a membership value of 1 while in the support–region the membership value declines linearly to 0 along its boundaries. This leads to the membership functions illustrated in Figure 3. Each

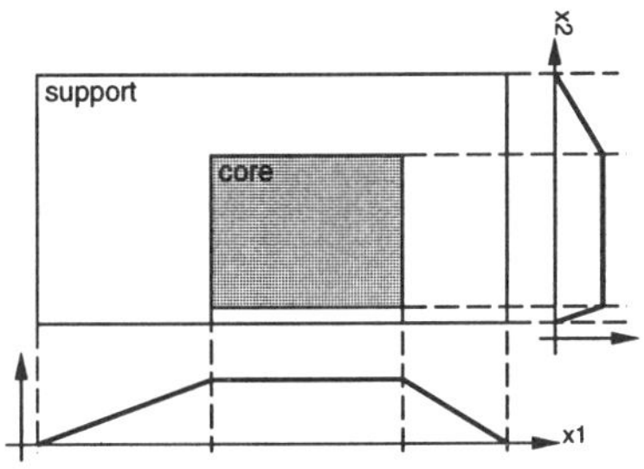

Figure 3: A Two–dimensional Fuzzy Point

such Fuzzy Point represents exactly one if–then–rule that can be described as:

IF	x_1	$\in [b_1, c_1] \subset (a_1, d_1)$
AND	$\cdots$	
AND	x_n	$\in [b_n, c_n] \subset (a_n, d_n)$
THEN		y is of class k
(weight: w)		

The input parameters are restricted through core $[b_i, c_i]$ and support–regions (a_i, d_i) with $a_i < b_i \leq c_i < d_i$ and each rule corresponds to one class k of the output parameter. Additionally, the rule weight shows the number of training examples that are covered by the core of the rule and is therefore an indication for the reliability of the rule.

3.1 Automatic Fuzzy Graph Construction

Construction of the Fuzzy Graph (or training) requires input patterns with a corresponding output. The output value (or *target*) can be defined as a soft value, using an individual membership function μ_T. For practical applications these soft targets can originate from stochastic simulation experiments. If there exist sharp targets a singleton can be used as the corresponding output membership function. From this soft target membership function μ_T the values μ_k for each class k ($1 \leq k \leq c$) are computed, using pre–defined membership functions for each class of y. The membership values for all classes are computed using a fuzzy and–operator (min) between the target and the class membership set. Training is then performed using these target membership values. This leads to training patterns consisting of an input vector $\vec{x} = (x_1, \cdots, x_n)$ with the corresponding target $\vec{\mu} = (\mu_1, \cdots, \mu_c)$ ($0 \leq \mu_i \leq 1$), where c denotes the number of classes.

The method makes sure that each training pattern is covered by a rule (Fuzzy Point) of the class with the highest membership value and that rules of classes with membership values = 0 do not cover the pattern. This is useful to tolerate moderately noisy patterns or small oscillations along class boundaries, as will be demonstrated later.

The training algorithm we use is based on three steps that introduce new rules when necessary and adjust the core– and support–regions of existing rules. The whole training process usually takes only about 4–5 training cycles until the structure of the Fuzzy Graph automatically stops to change. Two conditions will hold for all training patterns $(\vec{x}, \vec{\mu})$ after training: There is at least one rule of class k with the highest membership value μ_k which has $\vec{x}$ inside its core. And for all classes k with $\mu_k = 0$, $\vec{x}$ lies outside of their support area. This leads to a rule base where each pattern in the training set is covered by an appropriate rule and not covered by those of conflicting classes. The complete training algorithm is described in Huber and Berthold (1995).

3.2 Fuzzy Graph Function Approximation

To use the constructed Fuzzy Graph for function approximation the output values have to be fuzzified due to the predefined membership functions for the output parameter. These output membership functions have to be defined a priori by the user and can incorporate knowledge about the systems behavior and/or areas of interest. An example is illustrated in Figure 4.

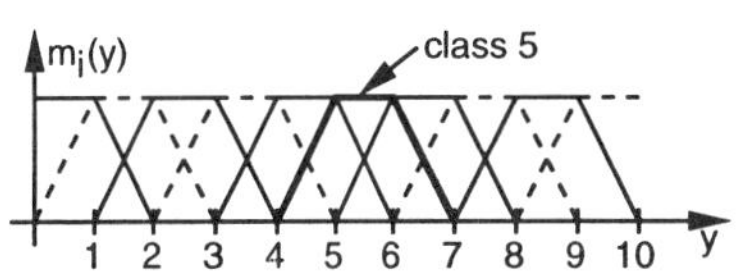

Figure 4: Fuzzification with 10 Output Classes

Based on this equidistant 10 class output–granularization 2000 randomly generated training points of an one–dimensional function were used for construction of a Fuzzy Graph. On the top of Figure 5 the Fuzzy Graph is shown and on the bottom the resulting function approximated by the Fuzzy Graph is depicted.

Especially within the context of analysing simulation data with stochastic output variables it is of interest how well a certain amount of uncertainty can

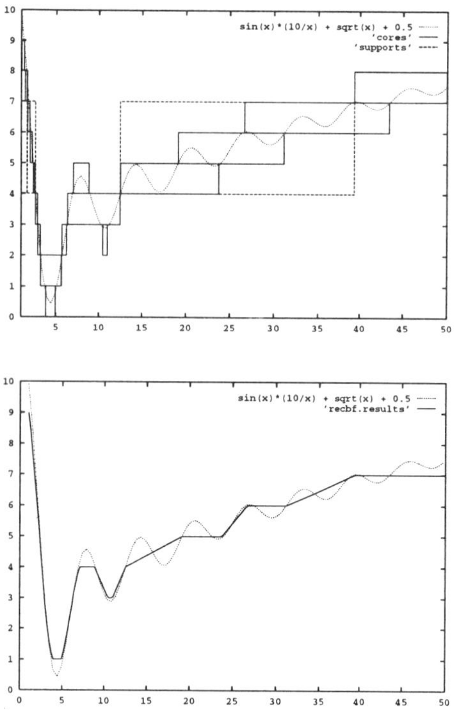

Figure 5: The Fuzzy Graph (top) and the Resulting Approximation (bottom)

be handled by this approach. A series of experiments was conducted with a specific amount of noise added to the training data to investigate the behavior of the Fuzzy Graph, for more details see Berthold and Huber (1996).

These experiments show that the resulting Fuzzy Graph approximates the original function well, with a specific degree of accuracy. In regions containing "noise", the Fuzzy Graph ignores the oscillations and tends to produce plateaus. The degree of noise tolerance depends mainly on the width of the membership functions for the output parameter. So the amount of smoothing can be controlled by the output fuzzification. Using more and finer membership functions results in higher precision but then the system tends to follow the data points very closely, and generates a lot of rules to model also the noise.

Good approximation performance helps to improve the reliability of a metamodel that is built with the presented Fuzzy Graph approach. In addition the metamodel can serve as an efficient simulation tool. Another advantage of the Fuzzy Graph metamodel is its easy to understand representation. This will be shown in the next section.

4 METAMODELING A TOKEN BUS

To demonstrate how the presented approach can be used to find rules in a real world simulation model, a token bus system was chosen. This system belongs to the class of field bus systems, i.e., a special type of communications systems, designed to connect machines and computers in a manufacturing environment. Important requirements in this area are "real time" facility, high flexibility, and low costs. In this section the analysis will mainly focus on the real time facility of the model, that is its capability to answer each request within a limited time. To guarantee this property for the given simulation model a metamodel will be built using the presented method and its behavior depending on different parameter settings will be explored.

The modeled token bus system corresponds to the seven level architecture of the ISO/OSI communication standard. Figure 6 shows the structure of the system. Many details like different message priori-

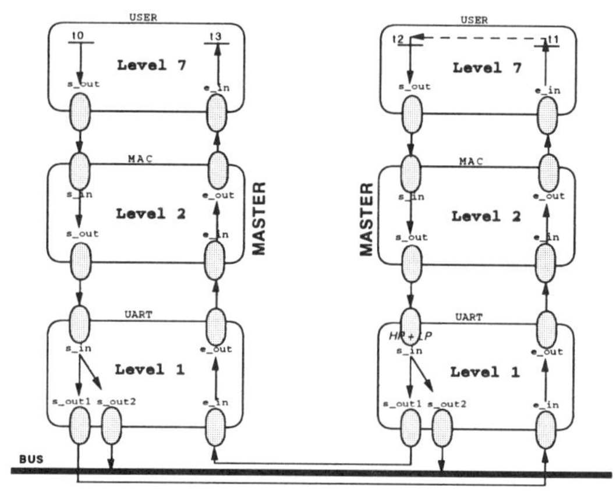

Figure 6: The Model Structure

ties and the token handling had to be taken into account when modeling the system with a queuing network model. The model was then implemented with a commercially available simulation environment. To illustrate the complexity of the underlying queuing network the internal structure of the UART module (Universal Asynchronous Receiver Transmitter) with its interfaces is illustrated in Figure 7. Since each station is modeled by four different modules the whole model consists of more than two hundred different queues and several hundreds connections. Due to this complexity of the internal structure a conventional analysis of this model is extremely time consuming and complicated.

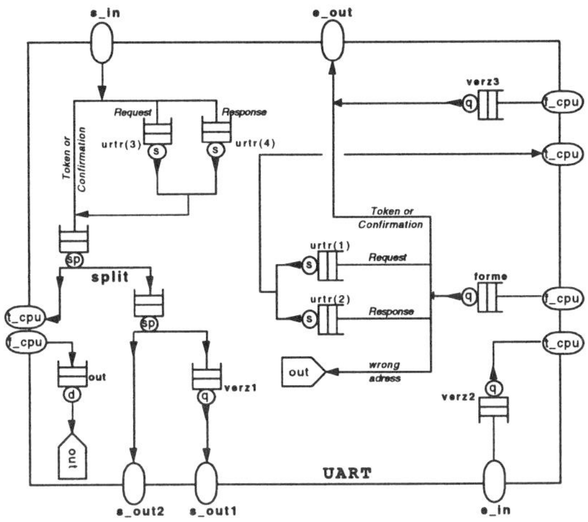

Figure 7: The Internal Structure of an UART

4.1 Metamodel Construction

Due to the large number of parameters (20 input and 10 output parameters) of the complete model the example analysis presented in this section will focus on the response time between two master stations in dependence of a selection of parameters of interest. It is desired that this response time always stays below an upper bound to guarantee that reaction of the system is always in time. Four input parameters were chosen while the other parameters remained fixed:

- *average time for execution (cpu1)*: describes the performance of the CPU module of station 1, i.e., the average time required to execute one command. This value is varied within 0.1 (fast) — 3.4 (slow).
- *workload rate (workload)*: describes the average idle time between two requests, this value is varied within 0.02 (low idle time, high workload) — 1.0 (low workload).
- *maximum target-rotation-time (trt)*: maximum allowed time to process the token. This parameter controls the time each station has to send messages, values were set within $(0.01, 0.4)$
- *number of additional stations (stations)* represents the background workload on the network. Many additional stations communicating over the network will increase the traffic on the network: $(1, 15)$

Since the construction of the metamodel only depends on the training data the set of examples has to be representative. For this the planning of simulation experiments must be done carefully. In regression analysis there exist some efficient techniques for experimental design because the assumption that the model behaves linear (or in a linear way) allows to make only few experiments to determine the regression parameters (Kleijnen and Standridge 1988). With Fuzzy Graph based metamodeling no assumption about the underlying model function is made, so the selection of representative data points is not straight forward. One way is to use a randomized setting of input parameters, a more sophisticated way is to use the information of the generated fuzzy graph for iterative selection of new simulation experiments. First ideas of this new approach are described in Frank and Huber (1996). In our application a full factorial design is not possible, therefore we used randomized settings for the input parameters. In the token bus experiment we performed some simulation experiments to validate the model and to see if the model behaves stable. 352 simulation experiments were performed where the input parameter values were varied randomly within the given intervals. The averaged response time (rt) was measured within $(0.088, 9.75)$. Each simulation experiment was repeated five times with a different random number stream of the simulation tool. From these five values the minimum, the maximum, and the average were taken and a triangular target membership function was generated (see Figure 8). With the

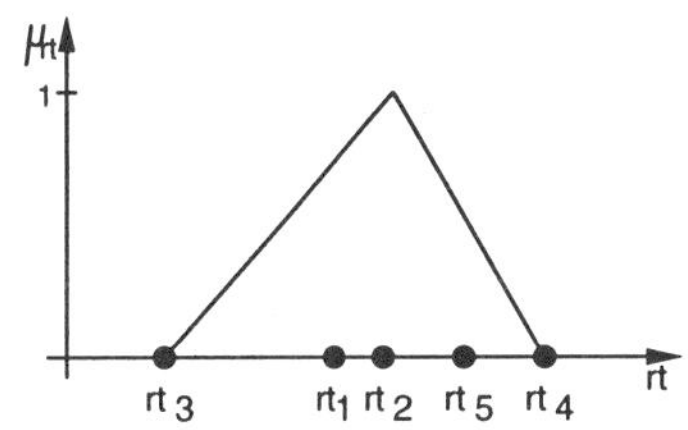

Figure 8: Generation of a Soft Target

Fuzzy Graph approach these four-dimensional data-vectors with their corresponding target membership function were used for training. Since the main focus of attention were fast responses (i.e., low values of rt) the membership functions for low values are defined finer than those for bigger values (see Figure 9). Three series of experiments were performed with two, five and ten membership functions.

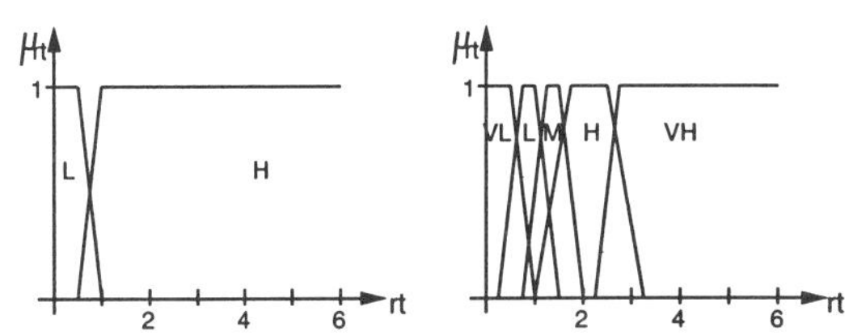

Figure 9: Two Used Types of Membership Functions

Fuzzy Graph construction required about 10 seconds on a SUN Sparc10 workstation. No training parameters besides the a priori definition of the output membership functions had to be considered or tuned. While a simulation run takes about 200 seconds the propagation of a new parameter set through the Fuzzy Graph is completed within fractions of a second, resulting in an increase in speed of two orders of magnitude. As expected the metamodel can be used for much faster simulation.

4.2 Metamodeling Results

To judge the reliability of the complete rulebase the quality of the whole metamodel can be analyzed by computing the mean relative error of the approximation of the metamodel. For this analysis an independent dataset that was not used for training, the so called cross–validation set, has to be used. The dataset of 352 training vectors was split into one tenth for testing and nine tenth for training and using each tenth once for testing ten cross–validation runs were performed. The average error on the corresponding testdata was 4.4% ± 1.0% (32 rules in average) with two, 4.1% ± 1.2% (59 rules in average) with five, and 3.3% ± 0.9% (68 rules in average) with ten memberships for the output. This approximation quality is sufficient because the primary goal of the presented approach is the extraction of few understandable rules instead of achieving minimal approximation errors.

One of the resulting rule bases from an experiment with two output classes was used for further analysis about the model behavior. In this case the classes are labeled *low* and *high*. Since the main focus of analysis are parameter settings which result in a low response time, rules of class `L = low` were investigated. From 29 rules 16 belong to this class and according to the rule weight two of the most important rules were:

$$\begin{array}{lll}
\text{IF} & \textit{cpu1} & \in [0.11, 1.69] \quad \subset (-\infty, 1.70) \\
\text{and} & \textit{workload} & \in [0.03, 0.99] \quad \subset (-\infty, +\infty) \\
\text{and} & \textit{trt} & \in [0.01, 0.39] \quad \subset (-\infty, +\infty) \\
\text{and} & \textit{stations} & \in [5, 15] \quad \subset (4, +\infty) \\
\text{THEN} & \multicolumn{2}{l}{\text{class low } (rt \in [0.0, 0.5] \subset (-\infty, 1.0))} \\
\text{(weight:} & \text{116)} &
\end{array}$$

$$\begin{array}{lll}
\text{IF} & \textit{cpu1} & \in [0.18, 2.16] \quad \subset (-\infty, 2.17) \\
\text{and} & \textit{workload} & \in [0.03, 0.98] \quad \subset (-\infty, +\infty) \\
\text{and} & \textit{trt} & \in [0.26, 0.39] \quad \subset (0.25, +\infty) \\
\text{and} & \textit{stations} & \in [1, 15] \quad \subset (-\infty, +\infty) \\
\text{THEN} & \multicolumn{2}{l}{\text{class low } (rt \in [0.0, 0.5] \subset (-\infty, 1.0))} \\
\text{(weight:} & \text{24)} &
\end{array}$$

Both rules demonstrate how the core always covers a confident subset of the support–region. The core of the first rule covers the whole range of two parameters, namely *workload rate* and *target–rotation–time*. It is only limited into one direction on the other two parameters, indicated by a support region having finite boundaries. The *performing time of CPU1* has to be below 1.70 and the *number of additional stations* above 4. This indicates that a certain amount of computation power together with some background stations guarantees fast responses no matter what settings are chosen for *workload rate* and *performing time.* In addition the weight of this rule can be used to judge its reliability. The weight indicates the number of patterns that were covered by this rule during training. In the case of the first rule shown above 116 training patterns fall inside its core region. This means that about 36% of all training patterns are covered by this rule, indicating a high reliability. Rules with low weight on the other hand might be indicators for outliers, irregularities in the dataset or regions of high sensitivity, i.e., regions where small changes of the attributes result in large variations of the output.

Another question of interest is the influence of some parameters considering the output class. In this model the parameter *target–rotation–time* is of importance. The first rule indicates that the *target–rotation–time* has no influence on the response time if the CPU is fast (below 1.7) and at least 5 background stations exist. The second rule above is an indication that if the CPU is not very slow (below 2.2) a *target–rotation–time* above 0.25 also leads to low response times. In this case this parameter influences the response time.

It can also be of interest to find "bad" examples, i.e., regions where the response time is very high. These indicate for parameter settings that are to be avoided. For example, the rule with the highest weight for the class *response time = high* was:

$$\begin{array}{lll}
\text{IF} & \textit{cpu1} & \in [2.71, 3.39] \quad \subset (2.70, +\infty) \\
\text{and} & \textit{workload} & \in [0.09, 0.98] \quad \subset (-\infty, +\infty) \\
\text{and} & \textit{trt} & \in [0.14, 0.39] \quad \subset (0.13, +\infty) \\
\text{and} & \textit{stations} & \in [1, 15] \quad \subset (-\infty, +\infty) \\
\text{THEN} & \multicolumn{2}{l}{\text{class high } (rt \in [0.5, 1.0] \subset (1.0, +\infty))} \\
\text{(weight:} & \text{38)} &
\end{array}$$

This rule indicates that if the CPU is very slow and the *target–rotation–time* is above a certain value the response time is high no matter what workload is considered (represented by background stations and the time between requests). Therefore if the system includes a slow CPU module the *target–rotation–time* should be set carefully if high response times are to be avoided. Since only 85 training patterns are of class *high* the weight of 38 is an indication for a high reliability also of this rule.

These results illustrate the applicability of the Fuzzy Graph approach for metamodeling tasks. Since the approximation error is acceptable the metamodel seems to be reliable and it can be used for new simulation experiments. The example rules delivered helpful information about dependencies between factors and the output of interest.

5 CONCLUSIONS

In this paper a new metamodeling approach based on Fuzzy Graphs has been proposed. With this approach it is possible to use not only real valued but also soft targets for training. This is especially well suited for the analysis of simulation models due to the stochastic character of the simulation data. It was demonstrated that the Fuzzy Graph approach is able to build suitable metamodels that can be used as a fast simulator. In addition the extracted rules deliver meaningful information about the relation between input parameters and the output, a very helpful information when analyzing simulation data of complex simulation models like the used token bus. Since the presented method is easy to handle and no parameters are needed the application of Fuzzy Graph metamodels allows to analyze complex simulation models straightforward.

REFERENCES

Berthold, M. R. and J. Diamond. 1995. Boosting the performance of RBF networks with Dynamic Decay Adjustment. In G. Tesauro, D. S. Touretzky, and T. K. Leen (Eds.), *Advances in Neural Information Processing Systems*, 7, Cambridge, pp. 521–528. MIT Press.

Berthold, M. R. and K.-P. Huber. 1996. Automatic construction of fuzzy graphs for function approximation. In *Proceedings of the NAFIPS*, 319–323.

Blanning, R. W. 1975. The construction and implementation of metamodels. *Simulation 24*, 177–184.

Frank, E. and K.-P. Huber. 1996, September. Active learning of soft rules for system modelling. In *Proceedings of the Fourth European Congress on Intelligent Techniques and Soft Computing.*

Friedman, L. W. and I. Pressman. 1988. The metamodel in simulation analysis: Can it be trusted? *Journal of the Operational Research Society 39*(10), 939–948.

Higgins, C. M. and R. M. Goodman. 1993. Learning fuzzy rule-based neural networks for control. In *Advances in Neural Information Processing Systems*, 5, California, pp. 350–357. Morgan Kaufmann.

Huber, K.-P. and M. R. Berthold. 1995. Building precise classifiers with automatic rule extraction. In *IEEE International Conference on Neural Networks*, 3, pp. 1263–1268.

Huber, K.-P. and H. Szczerbicka. 1994. Sensitivity analysis of simulation models with decision tree algorithms. In *Proceedings of the European Simulation Symposium ESS'94*, Volume 1, pp. 43–47.

Hurrion, R. D. 1992. Using a neural network to enhance the decision making quality of a visual interactive simulation model. *Journal of the Operational Research Society 43*(4), 333–341.

Kleijnen, J. 1979. Regression metamodels for generalizing simulation results. *IEEE Transactions on Systems, Man and Cybernetics 9*(2), 93–96.

Kleijnen, J. P. and C. R. Standridge. 1988. Experimental design and regression analysis in simulation: An FMS case study. *European Journal of Operational Research 33*, 257–261.

Meisel, W. S. and D. C. Collins. 1973, July. Repromodeling: An approach to efficient model utilization and interpretation. *IEEE Transactions on Systems, Man, and Cybernetics SMC-3*(4), 349–358.

Pierreval, H. 1992. Rule-based simulation metamodels. *European Journal of Operational Research 61*, 6–17.

Pierreval, H. 1996. A metamodeling approach based on neural networks. *International Journal of Computer Simulation 6*(2).

Simpson, P. K. 1993, January. Fuzzy min-max neural networks – part 2: Clustering. *IEEE Transactions on Fuzzy Systems 1*(1), 32–45.

Uebele, V., S. Abe, and M.-S. Lan. 1995, February. A neural-network-based fuzzy classifier. *IEEE Transactions on Systems, Man, and Cybernetics 25*(2).

Wang, L.-X. and J. M. Mendel. 1991. Generating rules by learning from examples. In *International Symposium on Intelligent Control*, pp. 263–268. IEEE.

Yerramareddy, S., D. K. Tcheng, S. C.-Y. Lu, and D. N. Assanis. 1992, June. Creating and using models for engineering design a machine-learning approach. *IEEE Expert*, 52–59.

Zadeh, L. A. 1994, November. Soft computing and fuzzy logic. *IEEE Software*, 48–56.

AUTHOR BIOGRAPHIES

KLAUS–PETER HUBER received his M.S. degree in Computer Science in 1988. Until 1992 he has been at Daimler–Benz AG, Stuttgart, working on expert systems, and is now a research assistant at the University of Karlsruhe.

MICHAEL R. BERTHOLD received his M.S. degree in Computer Science in 1992. In 1992 he has been a visiting researcher at Carnegie Mellon University, Pittsburgh, and joined Intel's Neural Network group in 1993. He is now a research assistant at the University of Karlsruhe.

HELENA SZCZERBICKA, Ph.D. 1982 in Computer Science from University of Warsaw, Poland. Since 1985 with the University of Karlsruhe and since 1994 professor of Computer Science at the University of Bremen, Germany. Associate member at the McLeods Institute of Simulation Sciences, co–editor of "Frontiers in Simulation" of the SCS.

Software/Modelware Tutorials

Proceedings of the 1996 Winter Simulation Conference
ed. J. M. Charnes, D. J. Morrice, D. T. Brunner, and J. J. Swain

INTRODUCTION TO WORK FLOW MODELING WITH BPSIMULATOR®

Don Lazzari
Systems Modeling Corporation
504 Beaver Street
Sewickley, Pennsylvania 15143, U.S.A.

Dr. Robert Crosslin
Technology Economics International
11212 Stephalee Lane
Rockville, Maryland 20852, U.S.A.

ABSTRACT

This paper highlights the features and functions of BPSimulator®, a simulation analysis tool focused on work flow and business processes. Based on Systems Modeling Corporation's general-purpose simulation system, Arena®, BPSimulator leverages all of the power and flexibility of Arena while providing the business analyst with an intuitive interface and a short learning curve. Like Arena, it is a comprehensive system that addresses all phases of a simulation project from input data analysis to the analysis of simulation output data.

1 INTRODUCTION

Recent advances in simulation technology have created a greater awareness and use of simulation as a decision support tool. Many top-level decision makers are now more aware of the potential benefits of simulation. However, even with the many important advances that have been made over the past several years, there are still many cases where business processes are being designed and implemented without the benefit of simulation. A very small percentage of systems that could benefit from simulation are actually simulated; the primary reason for this is the high level of effort required to employ simulation technology successfully. The key to making simulation technology more widely used is to make the tools significantly easier to learn and use without sacrificing modeling power and flexibility.

BPSimulator is focused on bringing the use of simulation to broad a new classes of users; users who may find much decision support benefit in simulation tools but who may not have the background or, perhaps more importantly, the time to master a comprehensive, general-purpose simulation tool. Its application focus addresses the needs of the business process analyst: activity-based costing, data input/output from Microsoft® Excel®, and the ability to import from process mapping tools like BPWin® from Logic Works and Visio® from Visio Corporation modeling constructs for worker efficiency and rework.

The key idea behind BPSimulator is to present the user with an easy-to-master tool set that still provides all of the robust simulation capability necessary to model work flow and business processes successfully and accurately. For advanced users who require an extra level of modeling sophistication, BPSimulator can run with Systems Modeling's general-purpose simulation system, Arena, thus allowing the advanced user access to Arena and SIMAN modeling constructs as well as C and FORTRAN subroutines.

2 OVERVIEW

BPSimulator provides an integrated framework for building business process simulation models. Tools for input data analysis, model building, interactive execution, animation, model verification, and output analysis are all contained within the bounds of BPSimulator.

2.1 Model Window

The model window is where new models are created, existing models are modified, animations are developed and models are executed. Since it is a native 32-bit Windows® application, BPSimulator takes full advantage of the easy-to-use Microsoft® Windows® operating system interface by including such components as 3-D controls, toolbars, status bars, tool tips, a context-sensitive help system, True-Type fonts, document windows, and common dialogs. Like any Windows application, model windows can be moved, resized, enlarged to full screen, or minimized to an icon. You can also have multiple model windows open in the work area at the same time. When a model window is first open, all of the toolbars are automatically displayed. The modeler can essentially design his or her own workspace by moving the toolbars around or turning them off based on his or her own preference.

The model window is where models and their associated animations are developed. This is done by: selecting modules from the BPSimulator tool set, placing them in the model window, responding to prompts in a dialog that appears when the user double-clicks on the module's handle and connecting modules to form a logical flow. When most modules are placed, basic animation constructs (e.g., queues, stations and resource icons) are placed simultaneously.

3 THE BPSIMULATOR TOOL SET

The BPSimulator contains 10 modules used for model development. Each module represents one or more modeling constructs used to depict a component of a system. The crux of the BPSimulator tool set is the Activity module. Within the Activity module (see Figure 1), a user can define the activity itself, which resources are needed to perform the activity, the processing time, inbound and outbound decision logic and various cost and rework parameters all by pointing and clicking with the mouse. The Activity module also makes it easy to model a system in a hierarchical fashion by allowing the user to specify whether an activity is a parent or child of a parent activity. When placed, the Activity module also brings with it entry and exit animation stations and icons to represent the processing state of the Activity and its associated queue. Real-time animation for any of the default statistics collected at each Activity can also be displayed simply by flipping switches.

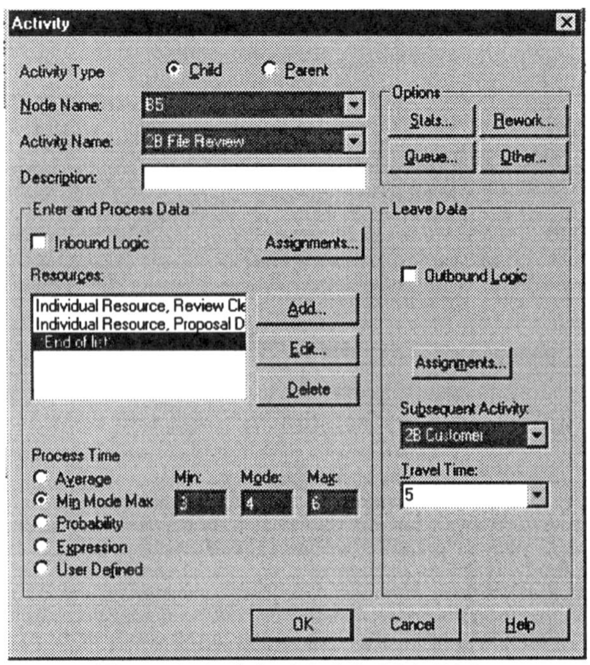

Figure 1: Activity Module Main Dialog

4 INTEGRATION WITH EXCEL®, BPWIN® AND VISIO®

Since a majority of business analysis data resides within spreadsheets, BPSimulator provides several ways to leverage work performed in other products.

For input data, BPSimulator provides OLE functionality allowing the user to embed an Excel spreadsheet directly into the model window. This not only provides easy access to Excel for editing input data; it also gives the user a single repository to view all input data at a glance. If the modeler has set up critical modeling values as variable names instead of "hard coding" specific values in the menu prompts, data for a simulation run can be read directly into the model through BPSimulator's Variables module, thus giving the modeler an added layer of flexibility in the model he or she constructs.

After collecting statistics from a simulation run, the user has the option of displaying the summary report directly in Excel and with the help of a utility included with BPSimulator, the entire report parses correctly into Excel's familiar cell format. Once in Excel, the user can take advantage of the graphical representation and data manipulation that Excel affords.

Since a majority of business process analysis projects begin with process mapping, BPSimulator can leverage the work already performed in some process mapping tools. With BPWin, IDEF modeling tool by Logic Works, information stored in the IDEF diagram can be imported directly into BPSimulator, essentially creating a simulation model from the IDEF model. Flowcharts created with the BPSimulator stencil in Visio can also be imported into BPSimulator, again resulting in a simulation model being built from a static process mapping tool.

5 INPUT ANALYZER

The graphical, menu-driven Input Analyzer provides users with the capability of fitting process or performance data to statistical distributions. Users can evaluate data downloaded from a file and may display the input data in the form of a histogram. The data may then be fitted to one of 14 commonly referenced distributions. Users unfamiliar with the data characteristics may choose to do a best fit analysis by selecting the Fit All option. This will perform statistical tests and choose the distribution that has the minimum mean square error. Additionally, the expression required as input to the Arena model will be provided and can be copied directly from the Input Analyzer and pasted into your Arena model.

Other options provided allow the user to modify parameters related to the input distribution to identify alternative input scenarios.

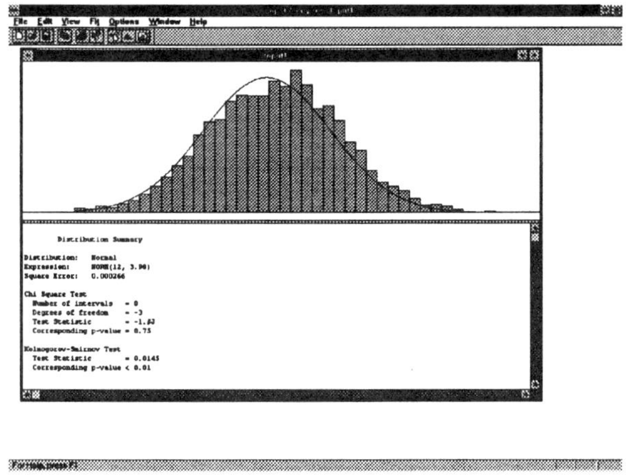

Figure 2: Curve Fitting with the Input Analyzer

6 OUTPUT ANALYZER

The graphical, menu-driven Output Analyzer provides users the capability to evaluate the statistical reliability of the results generated from their simulation model. Commonly used output data testing procedures such as correlograms, t-tests, data filters, moving averages, analysis of variance, and confidence intervals are included. These tests allow users to define and evaluate system performance or to compare the results from two or more simulation runs that evaluate different scenarios. The Output Analyzer is integrated within Arena such that a user can execute a simulation run, then select output analysis options that will display system performance characteristics.

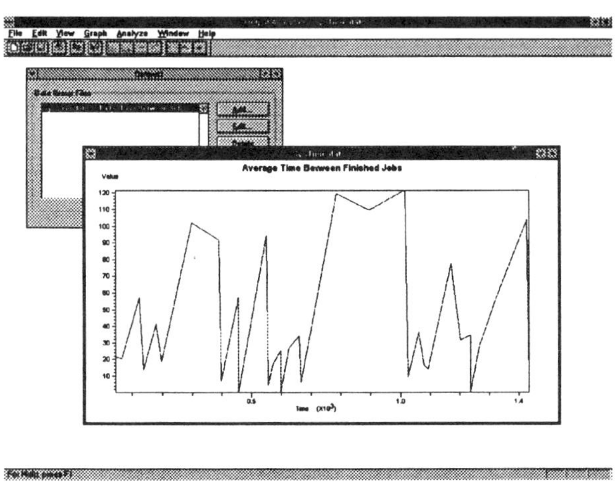

Figure 3: Performance Plot with Output Analyzer

7 SUMMARY

BPSimulator is an object-based, hierarchical modeling tool that offers the business analyst the power of simulation analysis without the learning curve and start up costs typically associated with simulation. It gives the business process analyst a decision support tool that combines the capabilities and power of a simulation language with an easy user interface for model building. BPSimulator is a comprehensive analysis tool that spans the scope of a simulation project from input data analysis to the analysis and presentation of model results.

REFERENCES

Hammann, J. and Markovitch, N. (1995) "Introduction to Arena" in Proceedings of the 1995 Winter Simulation Conference, IEEE, Piscataway, NJ.

Pegden, C.D., D. Davis (1992), "Arena: A SIMAN/Cinema-Based Hierarchical Modeling System" in Proceedings of the 1992 Winter Simulation Conference, IEEE, Piscataway, NJ.

Technology Economics International and Systems Modeling Corporation (1996), *BPSimulator User's Guide,* Sewickley, *PA*

Systems Modeling Corporation (1995), *Arena Template Reference Guide*, Sewickley, PA.

AUTHOR BIOGRAPHIES

DON LAZZARI is responsible for Systems Modeling's global work flow business. Don has a Bachelor of Arts and a degree in Writing from the University of Pittsburgh. Before joining Systems Modeling, Don held various sales and marketing positions with Matthews International. He also spent several years as a freelance writer/producer for KDKA-TV and a number of Pittsburgh area hospitals.

ROBERT CROSSLIN, Ph.D., is President of Technology Economics International, a Business Process Reengineering training and consulting firm based in Rockville, MD. He has consulted with companies and governments in both the United States and Europe. Prior to his founding TEI, Dr. Crosslin was a professor of Information Systems at The American University in Washington, DC.

Proceedings of the 1996 Winter Simulation Conference
ed. J. M. Charnes, D. J. Morrice, D. T. Brunner, and J. J. Swain

SIMULATION FOR HIGH-SPEED PROCESSING

David T. Sturrock
Glenn R. Drake

Systems Modeling Corporation
504 Beaver Street
Sewickley, Pennsylvania 15143, U.S.A.

ABSTRACT

High-speed processing lines are fundamental to the manufacturing processes of several industries, including food and beverage, consumer products, and pharmaceuticals. A *high-speed process* is defined here as a process where the volume and rate of jobs is such that discrete modeling and analysis cannot be done effectively. Traditionally, simulation analysis of high-speed processing lines has required a series of aggregations and simplifying assumptions to address the large volumes encountered. These approximations often cause the models to be poor predictors of real-world systems.

HiSpeed$im is a powerful, new approach for analyzing high-speed systems. It incorporates an interface and algorithmic engine specifically designed for simulating the complex control logic and equipment of high-speed processing. HiSpeed$im also leverages the powerful simulation platform of SIMAN/Arena. This combines the ease of use of a simulator and the speed of a custom-written product with the complete power and flexibility of the Arena general-purpose tool.

1 INTRODUCTION

High-speed processes typically take place at hundreds, sometimes thousands, of entities per minute. Typical examples include cans or bottles in food processing, pills in pharmaceuticals, or cigarettes in the tobacco industry. Simulation studies of these systems usually have many of the same modeling objectives as studies of other industries (e.g., system design, configuration, capacity analysis and planning). However, modeling high-speed processes also involves additional modeling obstacles due to the high speeds and large volumes of entities. Discrete entity approaches often overwhelm traditional simulation languages, resulting in significantly slower execution times.

There are two popular approaches to overcome the above problem: continuous equations and aggregation. Modeling a system with *continuous equations* typically has the advantage of very fast execution speeds. However, the complexity of mathematical equations increases rapidly for all but the simplest of models. Thus, the detail necessary to solve real problems can often not be represented by a set of equations. In addition, the lack of an interface between the system of equations and a discrete component often makes it impossible to model all aspects of the system (e.g., incoming supply or outgoing product flow).

The second approach, *aggregation*, works by allowing each logical entity to represent a quantity or lot of real entities. For example, each entity might represent 24 cans or 500 candies. Aggregation improves model execution speed and generally retains good modeling flexibility. However, sacrifices must be made in model accuracy and detail. For example, if each entity represents 500 candies, then all processing must be rounded to the nearest 500 candies. Furthermore, it may be very difficult to implement complex control logic. For example, continuing the problem above, it would be difficult to model the placement of sensors on a conveyor line to determine when there is a backup of 400 candies or when the conveyor is backed up to a point 10 feet from its end.

Clearly, another approach is needed. The simulation of high-speed processes requires a tool designed to handle the high volumes of entities, yet still able to retain modeling detail and flexibility.

2 MODELING APPROACH

HiSpeed$im is a collection of modules specifically designed for modeling high-speed operations. Each module incorporates all aspects of a particular system element (e.g., a machine). The modeling approach is illustrated in Figure 1.

Basically, you place and connect modules to represent the elements of the system. Then you fill in the descriptive information in the module dialogs. A good approach is to start simple and then, as your study develops, select options to specify more advanced features (e.g., sensors, failures, etc.). Model animation is automatically generated with each module, but these may be customized to your exact system.

The dialogs are self-contained and designed with terminology for high-speed applications. Thus, through a point-and-click graphical interface, you can develop models of sophisticated systems without writing any code. Finally, if the "built-in" operation of HiSpeed$im does not entirely address your operation, you can interface to modules of other products, such as the Arena template, to model the discrete portions of your logic exactly.

3 SYSTEM ELEMENTS

The main system elements in HiSpeed$im include Machines, Conveyors, Operators, and other devices such as Split and Merge.

3.1 Machines

Machines are the starting point for most models. This is where most of the actual processing or entity conversion takes place. After placing a machine, its description, run parameters and options can be accessed via the dialog illustrated in Figure 2.

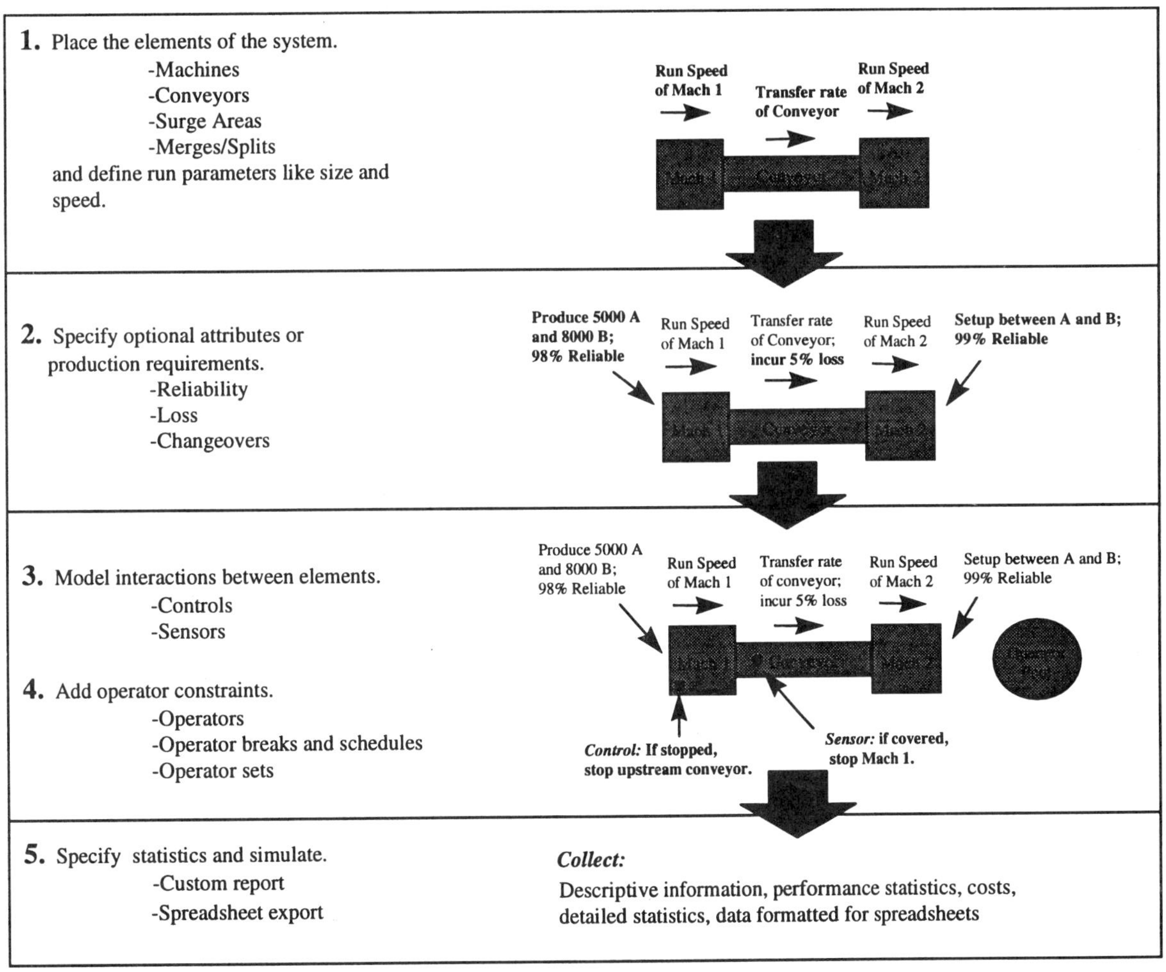

Figure 1: HiSpeed$im Modeling Approach

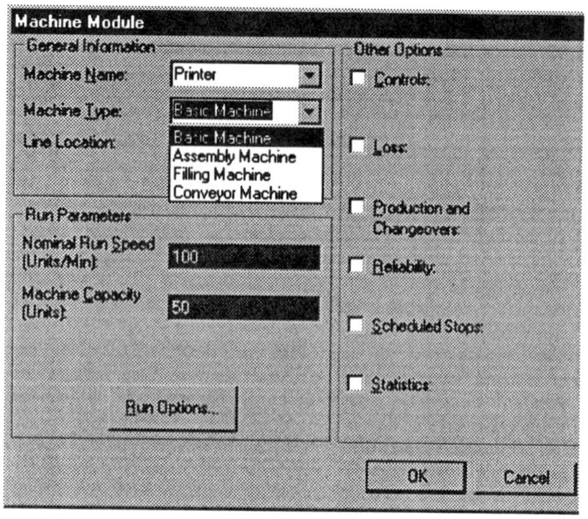

Figure 2: Machine Module Dialog

HiSpeed$im supports four types of machines:

•Basic Machines for standard processing;

•Assembly Machines for merging input from two or more lines (e.g., packaging processes);

•Conveying Machines for processes that involve significant product movement (e.g., washing or drying);

•Filling Machines for combining fluids with containers.

Depending on the machine type, the run parameters will change to prompt only for the appropriate information. For example, *Nominal Run Speed* applies to all machines, but the meaning of *Machine Capacity* changes with the machine type. Conveying machines require a capacity in terms of length, width, and density, while basic machines only require the number of units. The *Run Options* for machines include the ability to specify acceleration and deceleration parameters.

A machine's operation can be further customized by selecting from six optional areas: Controls, Loss, Production and Changeovers, Reliability, Scheduled Stops, and Statistics. *Controls* allows you to detect sets of conditions, such as a production levels or flow rates, and to take selected actions when such conditions occur (see Figure 3). For example, when a machine's production rate is decreased, it may in turn dictate that an upstream machine or conveyor slow down. Or, every time a target production level has been met (e.g., 1000 units), an entity might be generated to execute discrete logic such as operator intervention or custom statistics collection.

Loss allows you to specify product that will be lost (spoilage) due to routine production or special events such as changeovers and stoppages. *Production and Changeovers* allows you to specify the products produced at a machine, the effect of changeovers, and whether operator intervention is required. *Reliability* allows you to specify expected downtimes, repair times, and repair operators. *Scheduled Stops* specifies the times when a machine is scheduled to be out of service such as for breaks or preventive maintenance. *Statistics* allows you to define the statistical reports for machines.

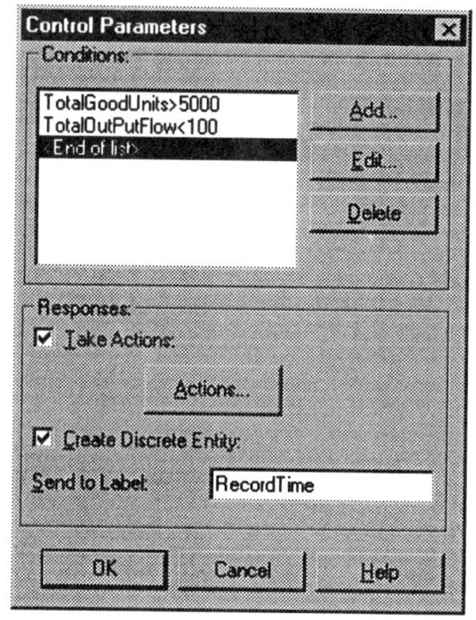

Figure 3: Control Parameters Dialog

3.2 Conveyors

Conveyors are the devices that move product between machines. Conveyor modules are placed the same way as machine modules. Their dialogs also have similar layouts and options (e.g., Controls, Reliability, etc.). HiSpeed$im supports two types of conveyors: accumulating and non-accumulating. In particular, accumulating conveyors have an additional option called *Sensors*. An unlimited number of sensors may be placed at specified positions along the conveyor. When a sensor is triggered, it can take actions identical to those described with controls above. Sensor triggering can be finely controlled. You can specify if a sensor should be covered or uncovered, the minimum time of the covering (uncovering), and whether additional conditions are necessary before taking actions.

3.3 Operators

Operators are the personnel who operate and maintain the equipment. Within equipment modules, operator involvement can be specified optionally for activities such as repairs, stoppages, and changeovers. They may also be controlled by discrete logic. Operators are defined in the Operators module (see Figure 4).

Operators may have a fixed capacity or follow a schedule. For example, you might require four operators for the day shift and only one for the other shifts. Operators may also have random breaks defined that affect their availability.

Operators can be grouped into sets (e.g., a set of electricians). When operators are required, you can select members from sets using operator selection rules. Each operator also has a skill factor defined. This value establishes an operator's proficiency in relation to the "average" proficiency. Skill factors may be used to adjust the delay times of operator activities.

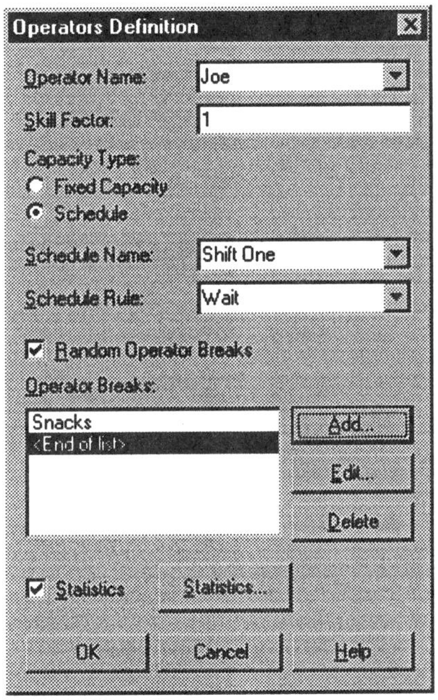

Figure 4: Operators Dialog

3.4 Other Devices

HiSpeed$im contains other system elements focused toward fulfilling specific needs. *Split* and *Merge* modules allow product flow to be separated and combined. The *Switch* module is similar to Split, but allows more sophisticated product or count-based flow control.

The *Surge Area* module facilitates the modeling of various types of surge or temporary accumulation areas. This module includes similar control options to those of machines and conveyors.

The *Palletizer* module represents palletizers, depalletizers, and combined units as well as the associated storage areas. In addition to the standard machine control options, it has options for controlling the storage area using a discrete interface. This allows discrete system logic such as pallet removal or creation.

4 MODEL OUTPUT

Model output takes several forms. The most obvious is animation. Animation of the machines and conveyors is automatically created as the logic is defined. This animation can also be customized by drawing your own pictures or selecting from supplied libraries. You may also import existing drawings from CAD packages. Business graphics like plots and histograms may be added to supplement the animation.

A second form of model output is summary report information. This report contains information about the system and its performance. The actual contents are customized by options in the appropriate modules. For each system element, you can choose to report descriptive information, production data summary, equipment performance summary, cost data, and detailed state statistics. Figure 5 illustrates part of a typical report for a machine.

Another output option is to export the summary report information to another program. This facilitates the use of a spreadsheet program to further sort or study the data and/or generate more extensive graphical analysis.

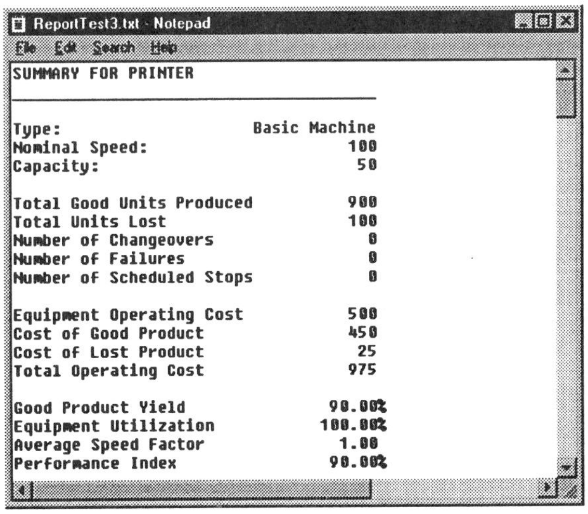
ReportTest3.txt - Notepad

File Edit Search Help

SUMMARY FOR PRINTER

Type:	Basic Machine
Nominal Speed:	100
Capacity:	50
Total Good Units Produced	900
Total Units Lost	100
Number of Changeovers	0
Number of Failures	0
Number of Scheduled Stops	0
Equipment Operating Cost	500
Cost of Good Product	450
Cost of Lost Product	25
Total Operating Cost	975
Good Product Yield	90.00%
Equipment Utilization	100.00%
Average Speed Factor	1.00
Performance Index	90.00%

Figure 5: Summary Report

5 SUMMARY

The benefits of simulation are well known in the high-speed processing industry, including reduced startup time and the security of making better, well informed decisions. Even a small improvement to system performance can realize tremendous gains. Unfortunately, the barriers to effective use of simulation are also well known—particularly the difficulty of modeling and the slow execution speeds.

HiSpeed$im is a new approach to modeling high-speed systems that overcomes the common difficulties. Because it is specifically designed for this industry and its challenges, difficult and complex systems can be modeled easily without writing code. Its intuitive interface allows non-simulationists to use simulation tools effectively.

HiSpeed$im is only one part of Systems Modeling's solution for high-speed process simulation. A related product is currently being developed for modeling liquid processing, storage, and flow. For example, modules will be available for modeling elements such as tanks, pipes, and valves with appropriate sensors, controls and actions. This product will seamlessly integrate with HiSpeed$im.

HiSpeed$im is the latest in a series of products from Systems Modeling to advance the state-of-the-art in simulation.

REFERENCES

Hammann, J., Markovitch, N. 1995. Introduction to Arena, in *Proceedings of the 1995 Winter Simulation Conference*, C. Alexopoulos, K. Kang, W. Lilegdon, D. Goldsman, Eds. IEEE, Piscataway, NJ.

Pegden, C.D., R.E. Shannon, and R.P. Sadowski 1995. *Introduction to Simulation Using SIMAN*, McGraw-Hill, New York, NY.

Systems Modeling Corporation 1995. *Arena User's Guide*, Sewickley, PA.

Systems Modeling Corporation 1995. *SIMAN Reference Guide*, Sewickley, PA.

Zepf, P.J. 1993. *How to Analyze Packaging Line Performance*, Institute of Packaging Professionals, Herndon, VA.

AUTHOR BIOGRAPHIES

GLENN R. DRAKE is an engineer with the Vertical Markets Team for Systems Modeling Corporation. This team is responsible for all aspects of vertical or specialty markets in simulation. Glenn has recently worked in the areas of manufacturing, business processes and real-time control. Glenn received his B.S in I.E. in 1994 and his M.S. I.E. in 1996 from Texas A&M University, College Station. He has a special interest in simulation-based real-time control.

DAVID T. STURROCK is the Vertical Markets team leader for Systems Modeling Corporation. Dave received his B.S.I.E. from The Pennsylvania State University with concentrations in manufacturing and automation. He joined Systems Modeling in 1988 after more than 11 years of manufacturing experience. Dave has applied simulation techniques in the areas of transportation systems, scheduling, plant layout, capacity analysis, process design, and real-time control. Dave is a member of IIE, SME, AMA, APICS, and SHS.

Proceedings of the 1996 Winter Simulation Conference
ed. J. M. Charnes, D. J. Morrice, D. T. Brunner, and J. J. Swain

ARENA® SOFTWARE TUTORIAL

Nancy A. Markovitch
David M. Profozich

Systems Modeling Corporation
504 Beaver Street
Sewickley, Pennsylvania 15143, U.S.A.

ABSTRACT

The Arena modeling system from Systems Modeling Corporation is a flexible and powerful tool that allows analysts to creates animated simulation models that accurately represent virtually any system. First released in 1993, Arena employs an object-oriented design for entirely graphical model development. Simulation analysts place graphical objects—called modules—on a layout in order to define system components such as machines, operators, and material handling devices. Arena is built on the SIMAN simulation language. After creating a simulation model graphically, Arena automatically generates the underlying SIMAN model used to perform simulation runs.

The graphical modules used by simulation analysts to create models are provided "off-the-shelf" with Arena. These modules can also be custom designed by the end user to produce a modeling environment that is tailored to a specific application area (e.g., welding, cross-dock, etc.) or a specific vertical market (e.g., semiconductor, textiles, etc.). The resulting collection of user-created modules is contained inside an Application Solution Template (AST) that can be shared by any licensed Arena user.

1 THE PRIMARY TEMPLATE — ARENA

The Arena template is the core collection of more than modules provided as part of the general Arena system. It was designed to provide a general-purpose collection of modeling features for all types of applications. In addition to providing core features for resources, queueing, inspection, system logic, and external file interface, the Arena template provides modules specifically focused on specific aspects of manufacturing and material handling. For manufacturing, it contains modules that incorporate such features as machine downtime and maintenance schedules. For material handling applications, modules exist for representing conveyors (synchronous and asynchronous) and various types of transportation devices.

Three panels compose the Arena template: the Common panel, containing modules representing fundamental simulation processes such as arrivals, service, and departures; the Support panel, containing supplemental modules for specific actions and decision logic; and the Transfer panel, whose modules are used to model the transfer (or flow) of entities through the system. Modules from these panels may be combined in the same model or may be used with other ASTs.

In order to develop a simulation model using the Arena template, the user simply picks a module, places it in the model, and then is prompted for the necessary information. For example, when placing the Server module from the Arena template, the user is asked for such information as how long entities spend at the server, the server's operating schedule, and where entities should go. After responding with the appropriate information, the user closes the dialog to accept the completed module.

Animation is automatically included with many of the modules in the Arena template to allow for rapid development of both a simulation model and accompanying animation. Graphics symbols that are automatically provided when placing a module from the Arena template can be changed with Arena's built-in graphics tools (similar to CAD systems) or can be replaced with icons from Arena's symbol library or from external applications (e.g., clip art).

2 ANIMATION

Arena animations can be run either concurrent with the executing simulation model or in post-process mode. Animations can be created in several ways: they can be created entirely using Arena's graphics drawing tools, they can be created from AutoCAD or other .DXF file formats, they can be created by using other Windows®-compliant drawing systems that can be pasted into Arena layouts, or any combination of the above.

Arena's drawing tools include all standard CAD objects (e.g., rectangle, ellipse, arc, text, etc.) and provides virtually unlimited color selection. Arena's interface with .DXF file formats was developed to allow for a direct import of CAD drawings to provide the animation background and dynamic icons. Once the drawings have been imported to Arena, the user can take advantage of Arena's drawing tool to embellish the layout. Dynamic paths contained in the CAD drawing can be used directly in Arena by "promoting" the paths. .DXF files containing dynamic icons can also be used in Arena.

Arena includes various animation options for real-time display of model statistics. For example, the user can place dynamic plots, histograms, levels, and time clocks directly within a simulation in order to illustrate system status as the model performs. This information is displayed on a real-time basis as well as on a post-process basis in the Arena statistical summary report.

3 FLOWCHART MODEL DEVELOPMENT

Arena was designed to make creating simulation models an entirely graphical process. All system behaviors are represented by using the graphical modules described above. For system logic such as IF/THEN/ELSE-type branching and queue selection rules, the user creates a flowchart of his system by placing the appropriate graphical modules on the Arena layout and directly connecting these modules. Non-animated modules from the Arena template that make up the majority of this flowchart can be freely interspersed with modules that contain animation.

4 MODEL JUMP-START WIZARD

When first starting a simulation model in Arena, the user may opt to utilize the Model Jump-Start Wizard. This technology allows for the rapid description of key system parameters such as the number of stations, entity or product information, and other data relevant to a model. After answering a series of questions regarding the new model, the Model Jump-Start Wizard rapidly creates a base-level simulation model using this information.

The Model Jump-Start Wizard was designed for users who are new to simulation and Arena, as well as for those who are more experienced. More experienced users typically use the Wizard to build a base level application quickly that they will substantially embellish using modules from the Arena template or other AST's outside of the Wizard itself.

5 CUSTOM DEVELOPMENT OF AST'S WITH ARENA PROFESSIONAL EDITION

A user can create a customized collection of graphical modules by using the features contained in the Professional Edition version of Arena. For example, a user working in the automotive industry may wish to develop a custom AST that contains modules for simulating welding lines quickly. User-created modules have focused dialogs, animation, and modeling functionality. Arena's hierarchy allows for the creation of higher-level modules from those contained in the Arena template, the SIMAN template, and/or other AST's.

Various AST's exist on the market today—some of which have been created and commercialized by Systems Modeling Corporation, others that were developed by other organizations who are licensed users for Arena Professional Edition. Currently available AST's include the BPSimulator, Call$im, and MP$im from Systems Modeling. Other organizations have developed AST's for such applications as automotive, production, process/chemical systems, forest resource management, and railroad operations.

6 SIMAN

The core technology of Arena is the SIMAN simulation language. The modules contained in the Arena template were created using SIMAN's modeling blocks as their components. This hierarchical concept prevails throughout the Arena template and other AST's that can be used within Arena. SIMAN blocks are made available to all Arena users in the SIMAN template.

SIMAN modules provide the user with increased flexibility and increased control of detailed system logic. Those users who have become accustomed to writing SIMAN code directly in a text editor are able to do so within Arena. In this case, Arena provides an option for directly recognizing this code, which is contained in a file external to the Arena graphical modeling environment.

7 INTEGRATION VIA ODBC, OLE

Arena is a Microsoft® Windows® 95 and Windows® NT compliant product. The entire product was developed using Microsoft's Foundation Classes (MFC) and is written in object-oriented Visual C++™. This allows users to leverage Microsoft's open architecture to integrate external data and applications with Arena models. Arena is also Microsoft® Office compliant that means that it utilizes all of the standard user interface options (e.g., toolbar buttons, function keys, etc.) that are in use in all Microsoft® Office products.

Arena's support of OLE (Object Linking and Embedding) allows the user to embed other technologies such as Excel® spreadsheets, Microsoft® Word files, clipart, and Microsoft® PowerPoint® presentations within simulation models.

Arena's support of ODBC (Open Database Connectivity) allows the user to integrate all database systems that are compliant with Microsoft's ODBC standard. Model data contained in products like FoxPro® database, Excel®, Oracle®, Informix®, and many others can easily be read into an Arena model without taking the time to enter it manually. For example, if a user has an Excel® file containing information on a manufacturing process plan, it can be directly imported to an Arena model via an ODBC wizard.

8 INPUT/OUTPUT ANALYZERS

Arena contains additional tools that are valuable for successfully conducting entire simulation projects. The Input Analyzer is useful for determining an appropriate distribution for input to an Arena model. The Input Analyzer allows the user to take raw data (e.g., time studies on process breakdowns or historically based order level information) and fit it to a statistical distribution. This distribution then can be incorporated directly into your model.

The Output Analyzer is used to display and analyze model data after the simulation run (or runs) has been performed. Graphical display options include plots, correlograms, histograms, and more. Multiple replications can be displayed on a single chart or can be lumped together for display of the aggregate performance over multiple runs. The Output Analyzer also provides analysis features such as confidence intervals, one-way analysis of variance, and comparisons (of multiple systems). Both the Input and Output Analyzers are directly available on the Arena Tools menu.

9 SCENARIO MANAGER

In order to manage the execution of multiple simulation runs, Arena's Scenario Manager provides an integrated environment for easy setup. The Scenario Manager allows the user to specify multiple executions (scenarios) of a model to be run on a batch basis. For example, once a valid model has been developed accurately representing the user's system, it is often useful to find out what the effects are on that system based on whether input1, input2, or input3 is used. The output from these runs can then be analyzed via output reports or within the Output Analyzer.

10 REAL-TIME CONTROL/MONITORING

An extension to Arena (called Arena RT) is available for the purpose of using a simulation model to interact with external client applications. This interaction is performed via an online messaging system. For example, the simulation model might contain aggregate-level system logic that sends tasks in real-time to a facility's shop floor control system. In this case, Arena's client might be a messaging queue that interfaces directly with PLCs. After completion of this operation (automated or manual), a message is sent back to the model so that the simulation can be updated and further instructions can be issued. During the execution of the model, the simulation and actual shop floor could operate concurrently. The animation could serve as a real-time monitoring device.

In addition to providing real-time control and monitoring of the shop floor, Arena RT provides the ability to spawn a concurrent model session so that the shop floor manager can look ahead at system performance while the factory is operating. This allows for the change of system configuration parameters in order to improve performance.

11 CONCLUSION

This paper has provided a basic overview of the Arena simulation system. A more complete review of the product can be arranged by contacting Systems Modeling Corporation. Also, additional information can be obtained on vertical market AST's offered for use in conjunction with Arena.

Arena is currently installed in more than 1,000 sites worldwide. It has been successfully used to simulate many types of manufacturing systems (e.g., automotive, aerospace, electronics, consumer goods, etc.) as well as applications in BPR, health care, call centers, warehousing, distribution, and more.

Arena has been certified by Microsoft® as a fully compliant Windows® 95 and Windows® NT application.

AUTHOR BIOGRAPHIES

NANCY A. MARKOVITCH is an Engineer for the Arena Products Team at Systems Modeling Corporation. Her responsibilities include technical phone support, in-house product support and training, technical sales support, marketing support, technical writing and review, software testing, software training, and simulation modeling. Nancy also serves as Systems Modeling's Training Coordinator. In this role, she oversees all details of the Standard Arena Training Course, including course content, materials, instructor training and scheduling, and facility setup. Nancy holds a bachelor's degree in Industrial and Systems Engineering from Youngstown State University. She is a member of Institute of Industrial Engineers and the Society of Women Engineers.

DAVID M. PROFOZICH is Vice President of sales and marketing for Systems Modeling Corporation. With SM since 1987, David directs the sales and marketing staff in two U.S. locations. He received his B.A. from Pennsylvania State University and an MBA from the Joseph Katz Graduate School of Business at the University of Pittsburgh. He has been published in *Manufacturing Systems Magazine* and *Managing Automation Magazine* on the use of simulation in manufacturing.

Proceedings of the 1996 Winter Simulation Conference
ed. J. M. Charnes, D. J. Morrice, D. T. Brunner, and J. J. Swain

MEDMODEL® - HEALTHCARE SIMULATION SOFTWARE

Diana F. Carroll

PROMODEL Corporation
Healthcare Systems Division
13333 Blanco Road, Suite 302
San Antonio, Texas 78216 U.S.A.

ABSTRACT

In response to the expressed needs of MedModel users, PROMODEL Corporation has made substantial improvements in its MedModel healthcare simulation software. Beginning with its inception in 1993, MedModel was specifically designed to be simple to use and tailorable to the needs of healthcare managers, engineers and clinicians. As such, it provides a basis for the comprehensive evaluation of large, complex problems which are representative of healthcare systems in general. This paper serves as a preliminary examination of some of MedModel's comprehensive features and capabilities. Special emphasis is placed on the new and modified features of MedModel 3.0. Preemption logic, pre and post shift logic, and new statements and functions have been carefully designed to solve unique hospital and healthcare specific simulation problems.

1 BACKGROUND

Hospital systems analysts are constantly faced with the difficult challenge of making sense, along with recommending courses of action, out of the various hospital systems they are chartered to study. This means they must recognize the types of systems found within a hospital, what each system is actually doing, what causes delays and bottlenecks, which actions are efficient and which are not and how the adoption of new policies, technologies or changes to existing structures overall will affect that system. The difficulty in arriving at this understanding derives from the very complexity of the systems under study. First, they are human systems subject to the vagaries of human actions and second, the healthcare process has too many interrelated and highly varied steps involved in even the simplest of healthcare processes to watch them all at once. Accordingly, the analyst has usually made decisions relying on massive databases that record and correlate all manner of clinical and administrative information. Although accurate in recording information on workload, staffing, patient mix, capacity and policy as they relate to cost and quality, the database is often unable to provide the true correlation among these factors and the dynamic interaction of each upon the other is usually lost. Moreover, the database approach relies heavily on past performance and is less helpful for predicting the result of anticipated changes to the system.

Finding an analytical tool that can handle the exhibited complexities of healthcare systems has proven to be a daunting task. What has been needed is a tool which allows the investigator to quickly and efficiently model the gamut of healthcare activities from patient admission to disposition and one that would be as effective for modeling a single activity (clinic, emergency department, nursing unit, radiology) as it would for a linked network of activities (ambulatory care, surgical services, food services, etc.).

Moreover, the problem of finding such a tool actually extends beyond simulation itself since simulation long ago proved its value to the manufacturing sector and has been used to evaluate process problems in healthcare as well. However, pure manufacturing is anything but an accurate reflection of what occurs in a healthcare setting. And, as such, simulators and simulation languages designed with the manufacturing environment in mind rarely contain the constructs and algorithms necessary to handle unique healthcare issues.

2 THE DEVELOPMENTAL FOCUS

Analysis of the hospital environment leads to the recognition that there are five broad categories of systems operating within its walls. Each has its own characteristics and each challenges the analyst attempting to develop a system model with its unique pattern of behaviors. These broad systems are: emergency departments, operating suites, nursing units, ambulatory patient care, and ancillary services. Using this knowledge, developmental activity was actually focused on two separate planes. First, there was the requirement to specifically understand the simulation needs and attributes of each of these broad hospital systems. Second was the focus on current and projected micro-computer and software technology that could be used to actually enhance the operational characteristics of the simulation itself.

2.1 Healthcare/Hospital Systems Needs

MedModel's genesis was prompted by the need for a simulation package that could adequately model each of these broad systems and yet be flexible enough to cross between them on models that link areas of more than one system. It needed to have constructs and algorithms that would assist the analyst in modeling healthcare and hospital systems. Based upon knowledge of the hospital and requests from scores of healthcare professionals, constructs and mechanisms that allowed for complex pathing, transparent relationships, multiple and repetitive activities and great operational variety were deemed crucial to a simulation software designed to solve even the most routine healthcare problems.

More importantly however, developmental concern went beyond simply capturing the essence of a healthcare system. Rather, it focused on ensuring that specific constructs represented precisely what was required to model the system of interest and did so with as much statistical validity as possible. For example, an initial concern for all five hospital systems surrounded the fact that patients and their records (doctor's notes, laboratory test results, etc.) often took different operational paths but they were, in fact, always linked to each other by the very fact that they were all a part of the patient's entire medical "persona". As such, one could not complete a process without the other being a part of the result. When the lab result is delayed, so is the patient's treatment and so on. Accordingly, specifically matching one with the other at various times in a simulation was often critical to the performance of a model.

Because these issues are often unique, how MedModel handles them is of specific interest.

2.1.1 Matching Entities

Since the treatment of a patient may well depend on the return of a lab or x-ray result and, conversely, the movement of the test result may depend on the availability of the patient, the ability to match patients and their supporting medical documents is a necessary capability. In the Acute Minor Illness Clinic shown as Figure 1, each patient has had a lab test done and waits in the patient waiting room for the results to be made available. Although one could assume that the longest waiting patient should be given the next available lab result, this is not necessarily the case. It has been shown that assuming that the longest waiting patient must be the next in the line for the next available lab result will yield waiting times that are significantly different statistically from those that correctly match an individual patient with his or her exact lab result.

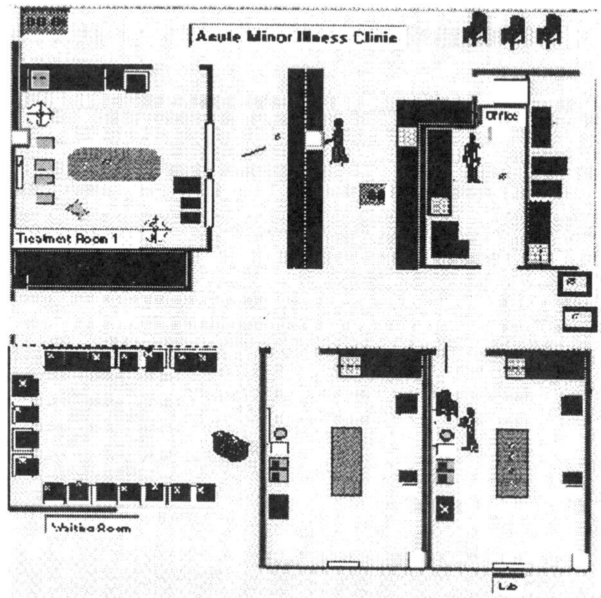

Figure 1: Acute Minor Illness Clinic with Lab Test

MedModel easily handles the requirement to keep identically "owned" entities together by using the built-in "matching" construct provided in the software. Shown below, the lab result is directly matched with its patient based upon a unique ID established early in the model and automatically transferred as an attribute to any entity "cloned" from the patient (Figure 2).

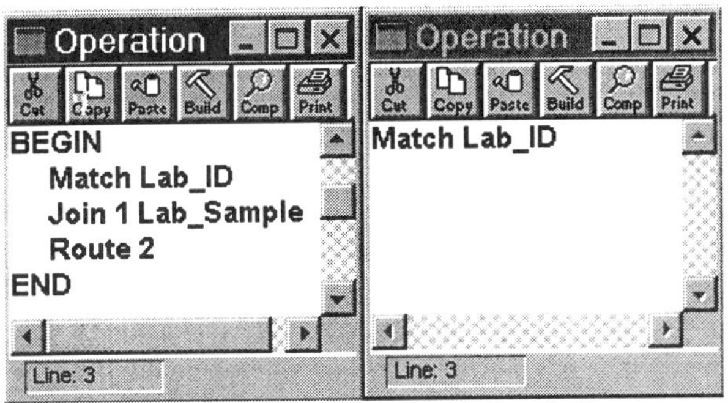

Patient Process Lab Result Process

Figure 2: Program Syntax for Lab Result Matching

2.1.2 Preempting Locations and Resources

Emergency Department modeling requires, among other items, a sophisticated and versatile capability to preempt rooms that are currently occupied by less acute patients as well as take staff resources from the work they are currently doing and moving them to an acute patient requiring immediate treatment. Also essential is the ability for the staff member to "remember" where he or she was in the treatment of the first patient and pick up from that point upon return.

In the case of the ED below (Figure 3), patients are placed in one of many exam rooms or may be placed in a trauma or cardiac room if their condition or lack of other space requires. But if a more serious trauma/cardiac patient should arrive, the room must be made immediately available to the arriving patient.

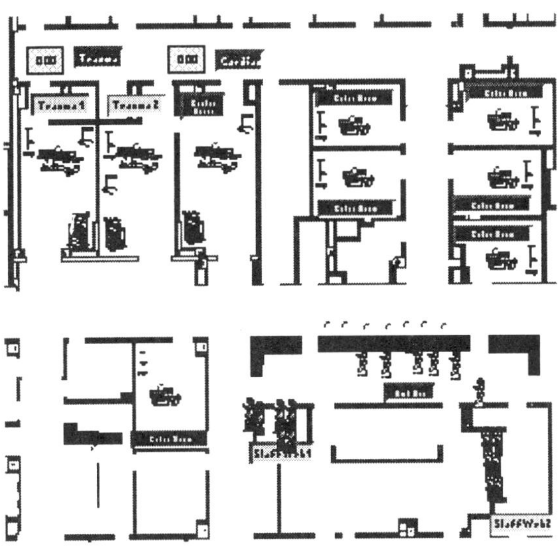

Figure 3: Emergency Department

To handle this situation, MedModel has an entire hierarchy of preemption constructs that have the ability to preempt locations from the currently occupying patient (Figure 4) and take resources from another patient when the new patient's condition requires immediate assistance (Figure 5). MedModel can even set up an entire system of preempting levels so that a patient with an acuity level of 1 will be preempted by a patient of acuity 2, an acuity 3 will preempt the acuity level 2 patient and so on. By merely filling in the priority level for the capturing of a location or resource, and applying the preemption hierarchy levels, MedModel will automatically preempt from one patient to another and remember where in the course of treatment the clinician left the patient. MedModel 3.0 adds preemption process logic to allow you to completely control what happens to the preempted patient rather than limiting you to program defaults. A preemption process record can be defined to postpone the actual preemption of a location or resource until the current entity explicitly releases it or to route the preempted entity to an alternate location (Figure 6).

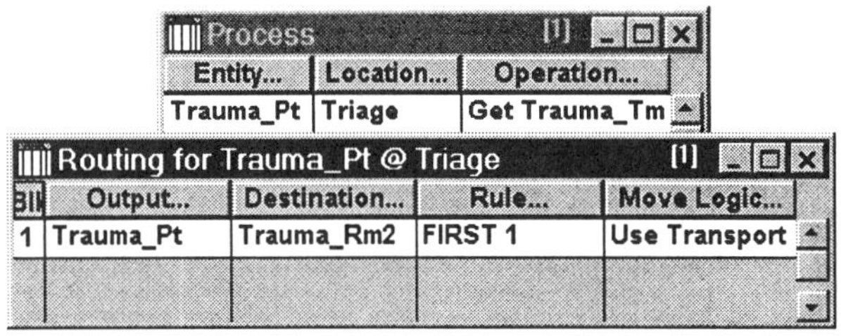

Figure 4: Preemption Priority to Capture a Location

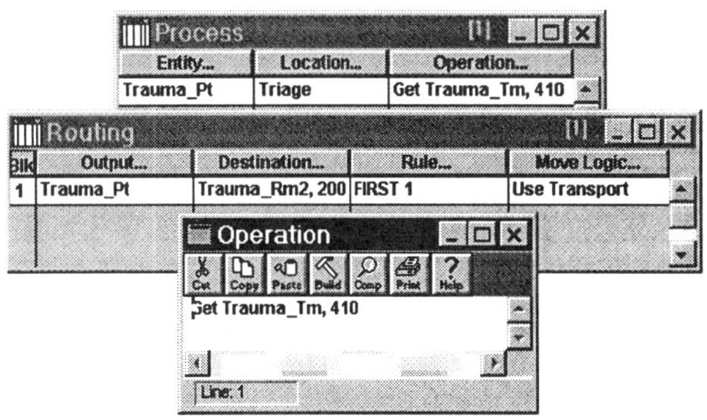

Figure 5: Preemption Priority to Capture a Resource

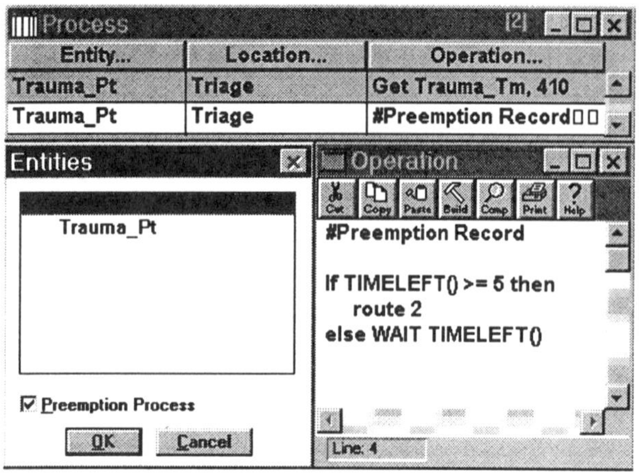

Figure 6: Preemption Process Logic

2.1.3 Multi-Use Subroutines and Macros

Healthcare in general and hospital systems in particular are characterized by many locations capable of having the same activity performed in each area. Examples such as non-specialized operating room suites, clinic exam rooms and inpatient rooms are but a few examples. A quick and effective way to model like activities at different locations is essential to any simulation software professing to accurately emulate healthcare activities. Not only must they be quick to use but they too, must be able to recall repetitive activities. In the nursing unit model depicted in Figure 7, all the patient rooms are identical and can house any of the patients assigned to the unit. Therefore, similar activities such as medication administration, meal serving, etc. can be handled as the same set of process logic.

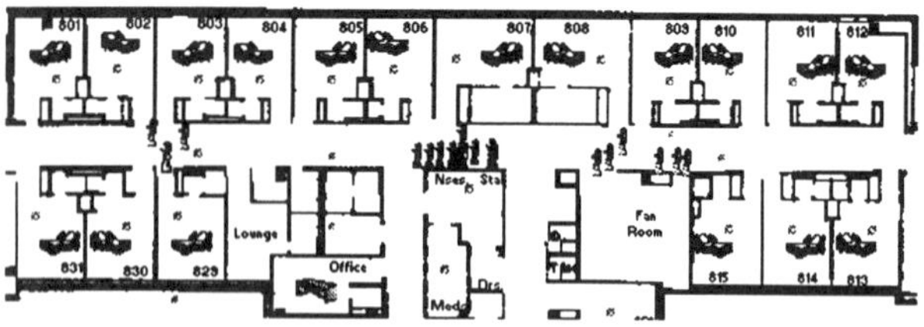

Figure 7: Nursing Unit

MedModel allows you to copy one set of logic statements from one room to another, but better yet is the ability to write the logic once as a subroutine or macro and then just use the name of the subroutine/macro whenever you wish its logic to be used.

For example, a patient who is feeble or comatose requires the nursing staff to roll the patient every four hours. This takes two staff members to accomplish the task but is only done for patients who can't turn themselves. Since it isn't known what type of patient may be in a particular room, MedModel simply allows you to write the logic to a subroutine. And since many activities in a nursing unit are time-based rather than activity-based, the subroutine can also be run independently of other activities that are occurring in the model. The subroutine logic for the scenario described above could look like that shown in Figure 8. Notice that the subroutine can also call a macro from within the subroutine logic.

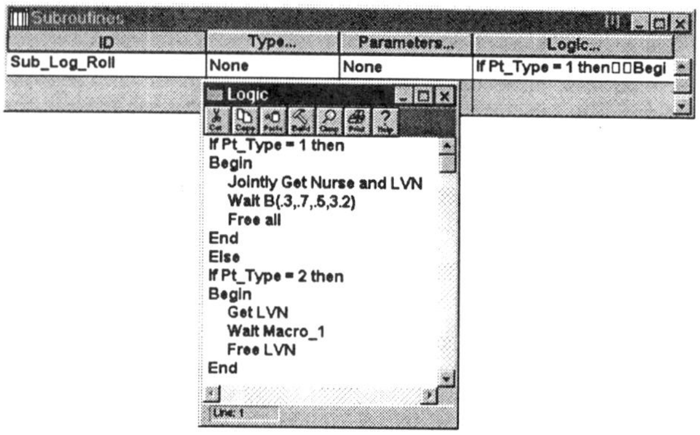

Figure 8: Subroutine and Macro Usage

2.1.4 Shifts

Certainly not the least of the problems handled deftly by MedModel is that concerning the management of resources and their accurate arrival and departure times. Nearly all healthcare workers are assigned to recurring shifts. Activity planning is often accomplished based on the knowledge of who will be at work during certain hours of the day. In fact, task responsibility is often changed from one hospital worker to another when certain shifts do not have a particular type of resource assigned. MedModel allows you to create any number of shifts of any length or days of the week along with break periods to represent breaks or meals (Figure 9). Single or multiple resources may be assigned to any shift. Locations may also be assigned to shifts which facilitates closing certain areas of the hospital during certain hours of the day. In MedModel 3.0, we have now added the ability to change shift assignments as the model runs. For example, a resource may work the day shift one week and the night shift the following week (Figure 10). In addition, logic can now be used to control whether a resource goes on break or off-shift (Figure 11). This solves one of the problems previously encountered of resources going off-shift when the clinic still had patients remaining in the clinic.

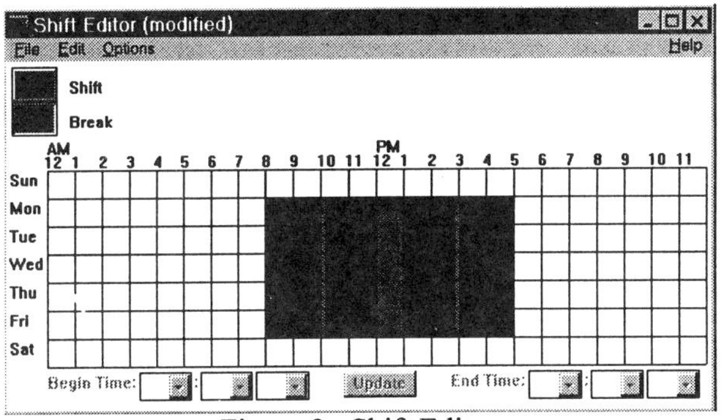

Figure 9: Shift Editor

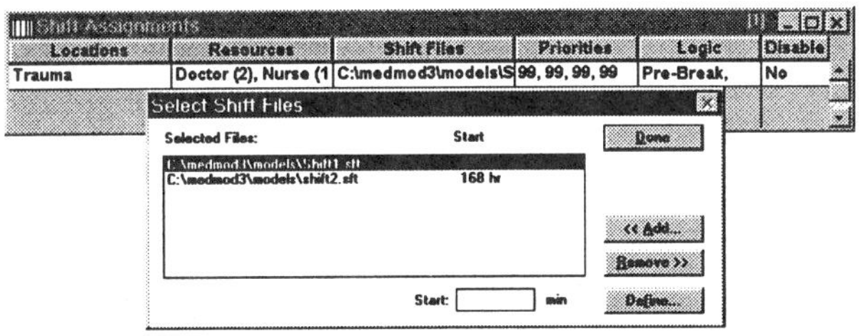

Figure 10: Shift Assignment

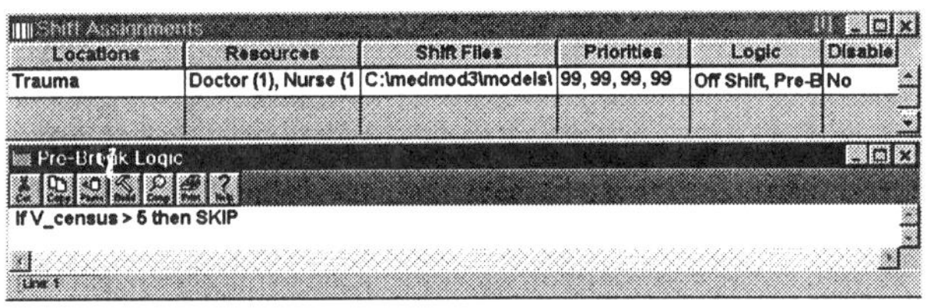

Figure 11: Shift Logic

2.1.5 In General

Of course, what has been addressed above is only a small part of what MedModel is capable of doing. Because the modeling of healthcare processes is almost always characterized by the need for complex process logic due to the seemingly endless variety of patients and activities, MedModel is equipped with an impressive collection of pre-programmed constructs that handle the vast majority of related problems. In other words, constructs that handle issues like the simultaneous but conditional use of different members of a healthcare team or the requirement to preempt certain medical activities when higher priorities come along, are built in. That makes it easier for the modeler to concentrate on modeling rather than on developing complex expressions to represent common activities.

2.2 Technology

To facilitate ease of use while simultaneously capturing extremely large-scale models, MedModel is based on object oriented programming using C++ and runs entirely under Microsoft Windows (including Windows 3.1, Windows for Workgroups 3.11, Windows 95, and Windows NT). In addition, MedModel is a true 32-bit program thereby eliminating nearly all size and memory constraints. This means that a model's size is limited only by the amount of memory available in your hardware and that MedModel takes full advantage of synchronized windowing capability, as well as dynamic data exchange offered by Windows. Naturally, larger and faster machines produce ever more dramatic results when building and running models. In fact, although any standard 486DX machine with VGA graphics and 8 Megabytes of RAM should suffice, a Pentium machine with Super VGA and 16-32MB RAM will show a dramatic speed-up of work.

But there is more to "technology" than the operating system itself. Use of software features available through existing programs that emphasize the capability to make full use of system capacity is also essential. Accordingly, MedModel avails itself of this technology by being nested in Microsoft Windows.

What makes MedModel unique then are the features that it calls on to make healthcare system modeling in general and hospital modeling specifically far easier. Coupled with responsive and versatile animation, MedModel provides a basis not only for rapid model building but ease of validation as well. The use of animation, or being able to "see" what's going on as the model progresses through its run-time, significantly enhances the modeler's ability to recognize whether the modeled system is operating correctly.

2.2.1 "Point and Click" Approach

Because MedModel was written under Windows, model definition depends, to a large extent, on nothing more than using a mouse to identify, select or place appropriate components of the model. The MedModel Statement Builder has been replaced in 3.0 by a new Logic Builder. The Logic Builder is much more comprehensive and powerful to assist you in creating valid logic and expressions. It is now a mode-less dialog which can remain open as you build your logic. Since it is fully menu-driven, the modeler need only "click" on an appropriate field to select an element for entry of information. In essence, a model can be built in its entirety simply by placing representative icons on the screen and then selecting different descriptive fields to define movement, relationships and activities between and among model entities and locations.

2.2.2 Custom Icons

MedModel also comes with an impressive library of colorful, pre-designed healthcare system icons representing everything from hospital-specific patients, staff members, material and treatment fixtures to instruments, used surgical trays, baby bassinets and the like. MedModel is also accompanied by an icon editor that enables the modeler to design any manner of icon desired using an almost limitless array of colors and shapes. This enables you to use the icons provided in the library or draw new icons in the graphic editor, MedModel allows for the import of Windows metafiles or .PCX clipart to be used as graphics too. The value of this capability transcends simple esthetics and allows the modeler to represent, as accurately as possible, specifically what is occurring in the hospital environment.

2.2.3 Automatic Processing and Path Entries

Unlike the majority of simulation software packages, MedModel presents the modeler with the capability to design and construct a model using nothing much more than the mouse. This is especially useful when identifying movement and processing steps. In this case, the modeler need only click on the succession of locations to which a patient may move and the required movement entries are made automatically. This holds true for both resource (i.e. doctor, nurse, technician, etc.) movement as well as processing. To create the required pathing for a clinical patient for example, the modeler need only draw a line from the clinic entry to the different locations the patient might go. In this manner, people can be made to walk down hospital corridors, through doors, up stairs and the like. Paths may be bi- or uni-directional. MedModel automatically enters all distances required to complete pathing and calculates correct movement times based on the distance and speed of the moving entity or resource.

3 CONCLUSIONS

MedModel is one of the most powerful tools available to use in analyzing the healthcare environment. It represents an opportunity for all elements of the healthcare sector to get involved with analytical tools that are far simpler, easier to use and yet more powerful than those previously available. Of greater significance however, is the fact that because of the comprehensive nature of the tool, assumptions and short-cuts that have routinely characterized healthcare and hospital simulations, are no longer necessary. Rather, it is now possible to model complex healthcare systems accurately and with confidence in the results.

AUTHOR BIOGRAPHY

DIANA F. CARROLL is the Senior Healthcare Simulation Consultant for PROMODEL Corporation. She has a B.A. degree in Psychology from Angelo State University and a M.B.A. degree in Information Resources Management from Webster University. She spent twenty-two years in the United States Air Force, fourteen in healthcare administration as a Medical Service Corps officer. Ms. Carroll's affiliations include the American College of Healthcare Executives, the Society for Health Systems of the IIE, the Healthcare Information Management Systems Society and the National Association for Female Executives.

Proceedings of the 1996 Winter Simulation Conference
ed. J. M. Charnes, D. J. Morrice, D. T. Brunner, and J. J. Swain

SIMULATION MODELING AND OPTIMIZATION USING PROMODEL

Deborah Benson

PROMODEL Corporation
1875 South State Street
Orem, Utah 84097, USA

ABSTRACT

The ProModel Optimization Suite is a powerful yet easy-to-use simulation tool for modeling all types of manufacturing systems ranging from small job shops and machining cells to large mass production and flexible manufacturing systems. ProModel is a Windows based system with an intuitive graphical interface and object-oriented modeling constructs that eliminate the need for programming. It combines the flexibility of a general purpose simulation language with the convenience of a data-driven simulator. In addition, ProModel utilizes an optimization tool called SimRunner which performs sophisticated "what-if" analysis by running automatic factorial design of experiments on the model, providing the best answer possible. This tutorial provides an overview of the ProModel Optimization Suite and presents its modeling, analysis, and optimization capabilities.

1 OVERVIEW OF PROMODEL

ProModel is a simulation and animation tool designed to quickly yet accurately model manufacturing systems of all types. Engineers and managers find the manufacturing oriented modeling elements and rule-based decision logic extremely easy to learn and use. Users are particularly delighted when they discover that ProModel is capable of modeling their most complex systems. Because it provides such an intuitive and straightforward approach to modeling, it is also attractive to professors in engineering and business programs who are interested in teaching modeling and analysis concepts without having to teach computer programming.

While most systems can be modeled by selecting from ProModel's complete set of modeling elements (e.g. resources, downtimes, etc.) and modifying the appropriate parameters, complete programming capability is also provided if needed for modeling special situations. Built-in language features include if-then-else logic, boolean expressions, variables, attributes, arrays and even access to external spreadsheet and text files. For those who prefer coding complex logic using a programming language such as C, Pascal or Basic, external subroutines may be dynamically linked to the model and called from anywhere inside the model at runtime. In this way, ProModel allows systems analysts and simulation experts to use the tools they are most comfortable with to provide total flexibility.

ProModel also provides several built-in distribution functions which, in conjunction with streams, return random values according to a statistical distribution. To aid the user in selecting an appropriate distribution for a data set, Stat::Fit is included with ProModel. Stat::Fit is a curve fitting software which fits analytical distributions to user data.

Model development is completely graphical and object-oriented. To the extent possible, all input is provided graphically with information being grouped by object type and presented in a "spreadsheet-like" format for quick and intuitive access. For example, when you define a machine you can define its icon, capacity, downtime characteristics, input and output rules, desired output statistics, etc. ProModel complies with GUI standards which means that individuals familiar with other standard Windows programs such as word processing or spreadsheets will have no trouble learning how to use ProModel. This data input approach minimizes the learning curve for beginners and maximizes the efficiency for modifying large and complex models.

A unique feature in ProModel is the ability to bring up a pop-up menu depending on the current context that prompts the user in defining any statement or expression. This enables any expression or statement to be entered using only the mouse. It also eliminates the need to remember variable or other element names that you wish to reference by allowing you to select them from a list box.

Quick and convenient online documentation is available through ProModel's integrated Help system and online tutorials. The Help system uses the Windows Help system which allows maximum flexibility for looking up anything from command syntax to descriptions of model building modules. ProModel also provides tutorials which contain quick lessons on how to build models, how to run models and access output reports and how to model various applications with the software.

To further reduce model development time, ProModel provides model merging capabilities to allow several individuals to be working separately on different sections of a large model. Additionally, frequently defined cells or even commonly used decision logic may be stored as submodel templates which eliminate the need to "re-invent the wheel" with every model. These templates can even have specially designated parameters which may be changed by the user.

Animation development is integrated with the model definition. A major drawback of many simulation software products is that animation development is independent from simulation model development. This makes it time consuming and inconvenient for engineers to use animation as a validation/verification tool. ProModel integrates system definition and animation development into one function. While defining routing locations, conveyors, AGV paths, etc., you essentially develop the animation layout. The layout screen is a virtual screen which can be scaled to an actual factory layout.

Simulation results are informative and may be displayed in tabular or graphical form. Many simulation software products require special commands to generate statistics that are difficult to interpret for non-simulationists. ProModel allows quick and convenient selection of reports and provides automatic tabular and graphical reports on all system performance measures. Output reports from several simulation runs can even be compared on the same graph.

ProModel runs on any standard 486 computer with Windows 3.x, Windows NT or Windows 95 operating systems. Most engineers, managers, and professors have easy access to IBM or compatible computers with VGA graphics capabilities. ProModel does not require any special graphics cards or special monitors. This makes it convenient and cost effective for companies and academic institutions that have standard microcomputers. ProModel also runs on LANs (Local Area Networks) .

2 MODELING ELEMENTS

The modeling elements of ProModel provide the building blocks for representing the physical and logical components of the system being modeled. Physical elements of the system such as parts, machines, or resources may be referenced either graphically or by name. Names of modeling elements may be any word containing up to 80 alphanumeric characters. Following is a brief description of each of these elements (see Figure 1).

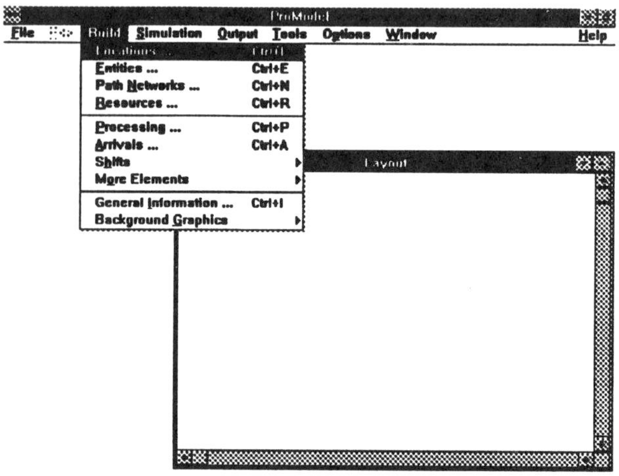

Figure 1: ProModel's Modeling Elements

2.1 Locations

Routing locations are fixed places in the system (e.g. machines, queues, storage areas, work stations, etc.) to where parts or entities are routed for processing, storage or simply to make some decision about further routing. Routing locations may be either single unit locations (e.g. a single machine) or multi-unit locations (e.g. a group of similar machines performing the same operation in parallel).

Routing locations may have a capacity greater than one and may have periodic downtimes as a function of clock time (e.g. shift changes), usage time (e.g. tool wear), usage frequency (e.g. change a dispenser after every n cycles), change of material (e.g. machine setup) or based on some user defined condition. Routing locations may be assigned input and output rules. Input rules are used for selecting what entity to process next while output rules are used for ranking entities (i.e. FIFO, LIFO, user-defined) in a multi-capacity location.

Two special types of locations that provide movement as well as performing holding and operation functions are queues and conveyors. A queue mimics the behavior of waiting lines, including the movement of entities through the line. Conveyors are accumulating or non-accumulating and have a particular speed and load spacing. Conveyors may be configured together to provide conveyor networks.

2.2 Entities (or parts)

Parts or entities refer to the items being processed in the system. These include raw materials, piece parts, assemblies, loads, WIP, finished products, etc. Entities of the same type or of different types may be consolidated into a single entity, separated into two or more additional entities or converted to one or more new entity types.

Entities may be assigned attributes that can be tested in making decisions or for gathering specialized statistics. The graphic of an entity can be changed as a result of an operation to show the physical change during the animation.

2.3 Path Networks

Path networks are optional and define the possible paths that entities and resources may travel when moving through the system. Path networks consist of nodes connected by path segments and are defined graphically with simple mouse clicks. Multiple path networks may be defined and one or more resources and/or entities may share the same network. Movement along a path network may be defined in terms of distance and speed or by time. Path distances are automatically computed based on the layout scale defined by the user.

There are three types of path networks: passing, non-passing, and crane. A passing network is used for open path movement where entities and resources are free to overtake one another. Non-passing networks consist of single-file tracks or guide paths such as those used for AGVs where vehicles are not able to pass. Crane networks define the operating envelope and interface points for bridge cranes.

2.4 Resources

A resource may be a person, tool, vehicle or other object that may be used to:

- transport material between routing locations.
- perform an operation on material at a location
- perform maintenance on a location or other resource that is down.

Resources may be either static or assigned to a path network for dynamic movement. A special type of dynamic resource is a crane. Built-in decision rules can be used for allocation of resources and for prioritization of part pick-up and delivery. Motion characteristics of resources such as empty and full speeds, acceleration, deceleration, pickup and delivery time can also be specified.

2.5 Processing (or routing)

This element defines the processing sequence and flow logic of entities between routing locations. The operation or service times at locations, resource requirements, processing logic, input/output relationship, routing conditions, and move times or requirements can be described using the Processing element.

Operation times can be defined by constants, distributions, functions, attributes, subroutines, etc. or an expression containing any combination of these. Operation logic can include IF-THEN-ELSE statements, loops, nested statement blocks and subroutine calls. Resource related statements such as GET, USE, JOINTLY GET with boolean expressions and built-in operation statements such as ACCUM, JOIN, GROUP greatly simplify otherwise complex logic in describing the processing requirements. Built-in and user-defined routing rules provide flexibility for modeling all types of routing conditions.

2.6 Arrivals (or production schedule)

Deterministic, conditional or stochastic arrivals can be modeled using this element. External files including production schedules or arrival data can be read into ProModel in the Arrivals element. Built-in or user defined distributions or spreadsheet created data can be used to define inter arrival times and quantities.

2.7 Shifts (or work schedules)

A powerful feature is the ability to define custom work and break schedules through ProModel's Shifts module. Work and break schedules are defined graphically by time of day and day of the week. Resources or locations are then assigned to a specific shift schedule. In addition, you can define shift and break logic that controls location and resource behavior when they go off-line and what happens once they are off-line.

3 ADDITIONAL MODELING ELEMENTS

ProModel provides additional modeling elements used in statements and expressions to define special decision and operating logic in a model. These elements include variables, attributes, functions and user defined-distributions.

There are several types of logic elements that may be defined by the user. Like model elements, names given to these elements may be up to 80 characters in length. Figure 2 shows the menu for accessing these additional elements. Some of these elements will be discussed in this section.

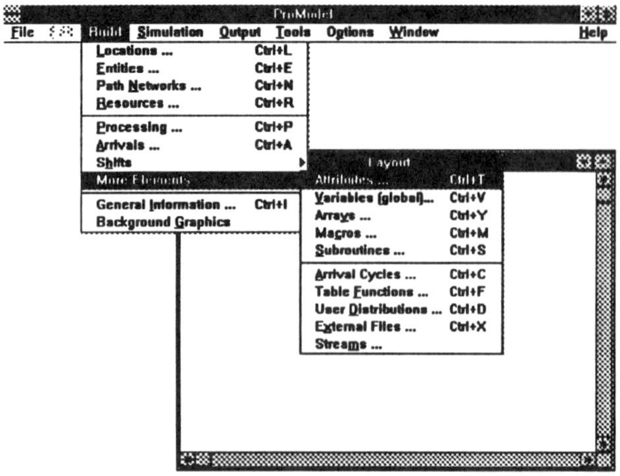

Figure 2: More Elements Menu

3.1 Attributes

Attributes for entities and locations can be defined. They may contain real or integer values. Location, resource and entity names may also be assigned to attributes.

3.2 Variables

Variables are used for decision making and statistical reporting. The value of a variable may be monitored over time and displayed at the end of the simulation as a time series plot or histogram. Variables may hold integer or real values. Local variables may also be used for quick convenience when defining logic.

3.3 Arrays

An array is a matrix of variables representing multiple values. An array may be one dimensional or multi-dimensional.

3.4 Macros

A macro is a complex expression or set of expressions that can be defined once and used multiple times as part of a logic statement (i.e. processing, scheduling, downtime logic). Macros can be helpful in situations where the same bit of logic is repeated in many places in the model.

3.5 Subroutines

A subroutine is a user defined block of statements that can be passed values when called and optionally return values when completed. Subroutines may be referenced in the place of any value or logic. A complex operation performed several places in a model may be appropriately defined by a single subroutine.

3.6 Arrival Cycles, Table Functions and User Defined Distributions

Arrival patterns, table functions or user distributions may be defined. For example, a user defined distribution which returns an operation time of 5 minutes 30 percent of the time and an operation time of 8 minutes 70 percent of the time can be expressed in the User Distributions element.

3.7 External Files

One of the powerful features of ProModel is the ability to read data from external text or spreadsheet files or write data to external files. For example, operation times (even in the form of expressions) from an EXCEL spreadsheet file can be read into ProModel using this element.

4 GRAPHICS

Graphics in ProModel are realistic and easy to create. Visually realistic animation helps simulation to become an effective communication vehicle between engineers and managers. ProModel comes with an extensive library of graphics with provision to create and add other graphics to the library. ProModel's graphics editor comes with a complete set of drawing tools and a full spectrum of color selections. Scaling, rotating, copying and many other editing features are available. You can even import drawings from other graphics packages. With little effort you can develop quick and simple 2D layouts, or, with little extra effort, 3D perspective layouts. CAD layout drawings (e.g. AutoCAD) can also be brought in to use as the model background. The Graphics Editor is shown in figure 3.

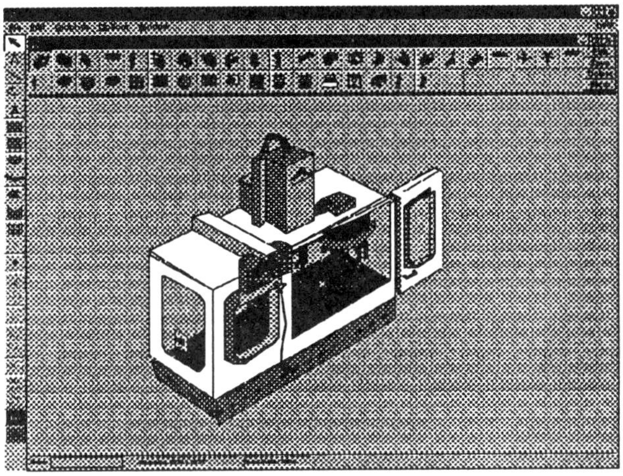

Figure 3: ProModel's Graphic Editor

5 SCENARIOS AND RUNTIME INTERFACE

The runtime interface (RTI) is a convenient and controlled environment for modifying selected model parameters (capacities, operation times, etc.) without having to change the model data directly. It also provides an experimental environment which permits multiple scenarios to be defined and simulated. The RTI can be accessed at the beginning of a simulation run for making modifications for a single run or for saving alternative modifications as scenarios for doing multiple scenario analysis.

6 RUNNING THE SIMULATION & ANIMATION

Models can be run for a specified length of time or until all entities have been processed. Multiple replications may also be specified. Models may be run with or without animation. The animation is very smooth and it maintains great resolution at any zoom factor.

The Views feature allows you to define and then quickly and easily access specific areas of the model layout. Selecting a view scrolls the layout window and adjusts the zoom so you see a specific region of the layout regardless of the layout window's size. Once a view is defined, you can select it while running the simulation or change the view from within the logic.

During the simulation, you can query the status of resources or the current value of any logic element. Source level debugging allows you to trace each action with the option to display the actual source statement that triggered the action. Trace statements may be confined to a single area of a model. Status lights for locations change colors to help you see various states such as busy, idle, down, etc.

The animation screen is a virtual screen which means the animation layout is limited only by memory in your computer. By turning off the animation, you can speed up the simulation, run for a while and turn the animation back on. The simulation clock resolution can be expressed in terms of hours, minutes, or seconds with a clock resolution of .00001 a second.

7 OUTPUT REPORTS

By choosing the statistics for resources, locations, entities, variables, etc., users can customize the output reports. The statistics are written to a results database which can be exported to spreadsheets. Additionally, graphical reports of the outputs can be displayed, printed or plotted. These graphs can be individual or comparative pie charts, histograms, time-series plots, etc. See Figure 4 for a sample location state graph.

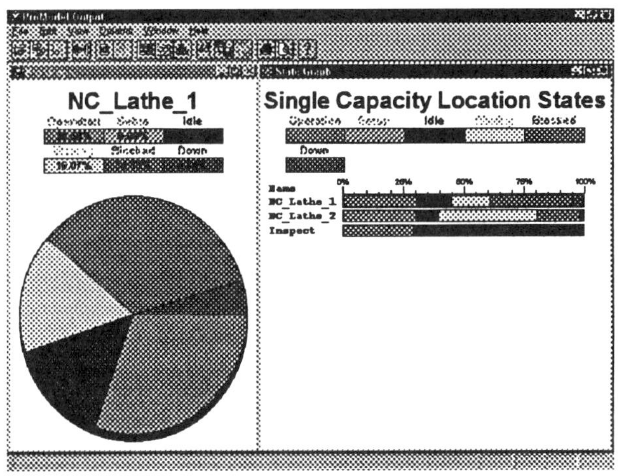

Figure 4: A Location State Graph

8 OPTIONS

One of the most accommodating features of ProModel is the provision for customizing the modeling environment to suit the preferences of the user. Under the Options menu, you can zoom in or out, display a grid, refresh the screen, set default directories for model files, icon library files, or model output files. You can set and save default settings (i.e. window sizes, appearance).

ProModel also allows you to choose editing preferences for model development. Default fonts and sizes for text and a choice of long or abbreviated menu lengths are available. Prompts and added explanations can also be activated which may assist the novice user. The expert user can benefit from choosing options that streamline the editing process.

9 OPTIMIZATION

SimRunner Optimization is an add-on capability for ProModel simulation software. SimRunner turns your simulation model into an answer machine by giving you the best answer possible while saving you the most time. The optimization process takes an existing ProModel simulation model and performs sophisticated "what-if" analysis and optimization automatically.

For each optimization project, you tell SimRunner which model to analyze/optimize, which input factors to change, and how you measure system simulation performance. You can run two types of projects in SimRunner: Stage One Optimization Analysis, and Stage Two Simulation Optimization.

A Stage One Optimization Analysis is a factorial design of experiments that tells you whether or not changing a given input factor is affecting the objective function, and the significance of the effects. A Stage Two Simulation Optimization is a multi-variable optimization that tries various combinations of input factors to arrive at the combination that provides the best objective function value.

The SimRunner Output report generates three types of data and reporting for you:

1. a data report - for import into spreadsheets
2. an analysis report - for text and word processing reports
3. charts - for graphical display of results.

Each of these can be printed or exported to other applications using common clipboard functions.

10 CONCLUSIONS

Until recently, manufacturing companies have not fully benefited from simulation in making continuous improvements because of the time, programming expertise, and cost involved in getting useful results. ProModel is designed for manufacturing companies to fully achieve the benefits of simulation technology at an affordable price. ProModel is directed toward making simulation a standard tool in the hands of engineers, managers and systems analysts just as spreadsheet software is in the hands of accountants and financial analysts.

REFERENCES

Harrell C. R. 1990. Trends in Manufacturing Simulation. In *Proceeding of Autofact Conference*, eds. A. Adlard, 21-31. Detroit, Michigan.

Harrell C. R. and K. Tumay. 1992. ProModel Tutorial. In *Proceedings of the Winter Simulation Conference*, eds. J. Swain, D. Goldsman, R. Crain, J. Wilson, 405-409. Arlington, VA

Harrell C. R. and J.J. Leavy. 1993. ProModel Tutorial. In *Proceedings of the Winter Simulation Conference*, eds. G.W. Evans, M. Mollaghasemi, E.C. Russell, W.E. Biles, 184-189. Los Angeles, CA

Baird S.P. and J.J. Leavy. 1994. Simulation Modeling Using ProModel for Windows. In *Proceedings of the Winter Simulation Conference*, eds. J. D. Tew, S. Manivannan, D. A. Sadowski, A. F. Seila, 527-532. Arlington, VA

AUTHOR BIOGRAPHY

DEBORAH BENSON is the Director of ProModel Support Services at PROMODEL Corporation. She has worked for PROMODEL Corporation since September 1992 and has held responsibilities in training, sales, and consulting. She authored the documentation for ProModel software, versions 2.0 and 3.0. She received a B.S. in Industrial Engineering from Arizona State University and a Master of Engineering Management from Brigham Young University. She currently serves as Treasurer for her local chapter of IIE.

Proceedings of the 1996 Winter Simulation Conference
ed. J. M. Charnes, D. J. Morrice, D. T. Brunner, and J. J. Swain

SIMULATION USING GPSS/H

Robert C. Crain

Wolverine Software Corporation
7617 Little River Turnpike, Suite 900
Annandale, Virginia 22003-2603, U.S.A.

ABSTRACT

GPSS/H is a tried-and-true simulation tool whose user base, both commercial and academic, continues to grow despite the presence of many "new" simulation technology trends. The process-interaction world view combines with the advanced features available in GPSS/H to make one of the most powerful and flexible tools available, *capable of handling the largest simulation projects* with ease, yet still providing exceptionally high performance.

The following sections provide an overview of GPSS/H and its process-interaction world view, a discussion of model-building interfaces including the advantages and pitfalls of graphical modeling environments, and a summary of advanced and recently-added GPSS/H features. Finally, the special-purpose simulator is discussed, along with features of GPSS/H which make it ideal for use as the engine in such a simulator.

1 INTRODUCTION

The widespread success of GPSS/H stems both from the superiority of its original design and from years of improvements and enhancements. GPSS/H requires some programming-style effort, but does so within a natural modeling framework that can be readily used without extensive programming experience. It is equally well suited for modeling simple systems and for modeling large, complex systems.

Although many new simulation tools have been introduced over the past decade, they are often designed for a limited set of applications. In strong contrast, GPSS/H continues to be one of the most *general*, *flexible*, and *powerful* simulation environments currently available. GPSS/H is presently applied worldwide in modeling manufacturing, transportation, distribution, telecommunications, hospitals, computers, logistics, mining, and many other types of queuing systems.

2 PRODUCT OVERVIEW

GPSS/H is a discrete-event simulation language. As is true with most languages, models are developed in text files. With GPSS/H, the text files are subsequently compiled *directly into memory* and executed. Rapid prototyping and iterative model development are encouraged by exceptionally fast compilation and execution.

GPSS/H uses the intuitive and natural *process-interaction* approach to modeling. The modeler specifies the sequence of events, separated by lapses in time, which describes the manner in which "objects" flow through a system. A GPSS/H model thus resembles the structure of a flowchart of the system being modeled. This intuitive modeling approach contributes greatly to the ease and speed with which simulation models can be built.

After the model has been built, the process representation is executed by GPSS/H, and the activities of "objects" are automatically controlled and monitored.

2.1 GPSS/H Process Representation

An "object" in a GPSS/H model might be a patient, a telephone call, or any other type of discrete entity. The representations of these entities in GPSS/H are called *transactions*. As the model executes, many transactions may be flowing through the model simultaneously—just as many "objects" would be moving through the real-world system. In addition, multiple transactions can execute GPSS/H model statements at the same instant in time without any special action required of the modeler. The execution of a process-interaction simulation model is thus akin to a multi-threaded computer program. This differs greatly from the single-threaded, sequential execution of most general-purpose programming languages.

The focus of many simulation projects is the use of system resources such as people, machines, conveyors, computers, physical space, and so on. In a GPSS/H simulation model, transactions ("objects") compete for the use of these system resources; as transactions flow through the process representation, they automatically queue up when unable to gain control of a necessary resource. The modeler does not need to specify the transaction's waiting time or its queuing behavior. Hence, the passage of time in a GPSS/H model can be represented *implicitly*, as in the case of a part waiting for a machine to be free, or *explicitly*, as in the case of a part being processed by a machine.

As is the case in most real-world systems, a GPSS/H model may consist of multiple processes operating simultaneously. Furthermore, each process may in some way affect the other processes in the system. For example, two parallel manufacturing processes may converge to a single inspection point where they are competing for a single resource—the inspector. GPSS/H provides the capability for multiple parallel processes to interact with each other automatically. Transactions ("objects") may be sent between processes; they may control or share common resources; or they may influence the (global) operation of all processes.

3 GRAPHICAL MODELING — GOOD AND BAD

Often, the power and ease-of-use of a simulation environment are confused with the model-building interface provided by the tool. This may be comprised of menus and data forms, or—as in the case of GPSS/H—it may consist of text entry, or it may be a combination of the two.

The latest "fashion" in simulation technology is directed at trying to model systems visually. Icons are placed on the computer screen to represent system components, and then the operating characteristics of each component are specified by moving through a series of menus and data forms. One inherent advantage of this approach is that even novices can build simple models quickly—*although not necessarily accurately*.

Building models of complicated systems, however, requires more than simply placing icons on the screen. To model many processes, a limited programming environment must be provided. For example, a part routing based on a time-dependent math equation cannot be represented visually. As a result, models of complex (real world) systems created using the visual approach often require the modeler to create substantial amounts of programming code in addition to the visual representation.

3.1 Developing and Editing Models

Graphical modeling tools can force their users to make the model fit within a rigid framework bounded by the available menus and forms. The advantage of such a framework is that it tends to steer even a beginning modeler through the model-building process. The disadvantage is that currently available frameworks are rarely versatile enough to *accurately* model complicated systems.

Additionally, large visual models can become very cumbersome to view, edit, and document. Large models can be comprised of many "screens" of icons, many of them with associated program code reachable only by going through multiple levels of menus. Editing—or even just browsing—these models forces the user to navigate through a labyrinth of icons, menus, click-buttons, data fields, and code segments.

3.2 What Defines an Easy-to-Use Simulation Tool?

A tool's ease-of-use cannot be evaluated meaningfully on the presence or absence of a single characteristic. Whether a tool is "easy-to-use" is determined by the *combination* of general characteristics and specific features that are frequently used in model development. Simulation software should be selected based on how well it is suited to the detail and complexity of the specific type of model to be developed.

The ease-of-use associated with a simulation tool can mean—among other things—easy to learn, easy to use repetitively, easy to use when modifying models, easy to use when building simple models, or easy to use when building large, complex models. Tools that are claimed to be "easy-to-use" often fall short when modeling complex, real-world systems.

Graphical model building, often touted as a breakthrough in ease-of-use, springs from attempts to apply to simulation recent trends in the design of computer interfaces used primarily for word-processing, spreadsheets, database access, and the like. Although a graphical user interface is well suited for many kinds of tasks, it is not always practical for developing simulation models—especially in circumstances where programming is necessary to define the operations of complex processes.

There is a point at which a graphical environment can present its user with more barriers than advantages. For example, proponents of graphical modeling techniques frequently claim shorter model development time, but this is primarily because users of graphical tools tend to build simpler models. As was discussed in section 3.1, creating and editing complex models with graphical tools can require more time than creating and editing such models in a text-based environment.

Wolverine's future product offerings can be expected to include the use of graphical interface techniques for model building, but not simply for the sake of appearance. Careful consideration is being given to how this technology can be used to bring real advantages to modelers of diverse and complicated real-world systems.

4 IMPORTANT FEATURES OF GPSS/H

Several unique characteristics make Wolverine's GPSS/H an excellent choice for a general simulation environment. A key feature of GPSS/H is the *conceptual flexibility* to model a wide range of different types of systems: any system that can be described as a process flow, with objects and resources acting upon each other, can be modeled. This may include people on a mass transit system, tasks in an office environment, or data flow within a computer network.

Definition flexibility is also provided within the language: complex math formulas, expressions, and constants can be used virtually anywhere in the model. To promote model readability, elements and entities may be specified by names instead of numbers. Basic simulation output data, such as queuing and service statistics, are *automatically* provided without any programming.

GPSS/H is available for PCs and SUN SPARC workstations. On the PC, GPSS/H Professional runs as a true 32-bit application under DOS, Windows 3.x, Windows 95, OS/2, or Windows NT, providing tremendous speed as well as model size that is limited only by the computer's available memory. Running under Windows, OS/2, Unix, and Windows NT, GPSS/H uses virtual memory, which allows model size to exceed the amount of physical memory installed in the machine.

4.1 GPSS/H File and Screen I/O

The file and screen I/O built into GPSS/H provide a variety of ways to get data into a model and to produce custom output. GPSS/H can read directly from the keyboard or from text files, and it can write directly to the screen or to text files. The GETLIST statement and the BGETLIST block read integer, character, and double-precision floating-point data. Data files are free-format (values on each line are simply separated by blanks), and special actions may be specified for error and end-of-file conditions.

Customized output is generated using the PUTPIC statement and the BPUTPIC block. These use a very intuitive "picture" type of format specification, which follows the "what you see is what you get" convention. Special provisions are included to allow easily formatted tabular output. Character strings can also be manipulated using built-in capabilities.

4.2 Scripting Language for Experiment Control

The results produced by a single run of a simulation model can only provide single estimates of random variables that may be subject to wide variations. Careful experimental design, using multiple runs, is *essential* to accurately predict the behavior of the model outputs. GPSS/H provides the tools to build a complete experimental framework.

A complete *scripting language* is available to construct experiments and control model execution. Experiments can be automated with DO loops and IF-THEN-ELSE structures. Statistics collection may be totally or selectively reset, and/or data values assigned, both during a model run and before or after each run in a series of runs. The experimental specifications and parameters, like any other model data, can be read in from a data file or from the keyboard if desired.

4.3 Statistically Robust Random-Number Streams

The need to provide multiple *independent* streams of random numbers for use in different parts of the model (or in the same parts for different runs) is very important, particularly after a model is largely complete and the modeler is concentrating on validation and the running of experiments. The *indexed Lehmer* random number generator provided in GPSS/H was designed and implemented specifically to provide exceptionally simple, straightforward control of the random number streams used in a model. Modelers can easily specify any number of streams and guarantee that they will be independent (that they will not be autocorrelated due to overlap). GPSS/H also *automatically* detects any accidental overlap, providing an extra measure of protection to users.

4.4 Validation and Debugging

The GPSS/H Interactive Debugger conveniently provides for rapid model development and verification. Simple debugger commands are used to control a model's execution and examine its status. Functions are provided to "step" through the model, to set breakpoints and traps that interrupt model execution based on multiple criteria, and to return to a previously saved state of the model. Almost all data values can be examined, including local data, global data, transaction attributes, entity statistics, and array data values.

The debugger provides a "windowing" mode that displays source code, model status, and interactive user input as the model runs.

A modeler can interrupt a long-running model at any time and use the debugging features to make sure that everything is running correctly before resuming execution.

The GPSS/H debugger has almost no effect on execution speed. Because of this, many modelers use the debugger as their everyday run-time environment for GPSS/H.

5 FEATURES OF GPSS/H RELEASE 3

GPSS/H is continually improving and evolving. Numerous enhancements, under development as of this writing, will be discussed in the tutorial session. Persons unable to attend may obtain the latest information by contacting Wolverine Software Corporation.

Some of the more significant additions to Release 3 of the widely-used GPSS/H Professional version are:

- The BLET Block and the LET Statement can now be used to assign a value to ***any*** GPSS/H data item. Unless you need the rarely used range-type assignments, *there is no longer any reason to use* the ASSIGN, SAVEVALUE, and MSAVEVALUE Blocks. The newly-extended BLET Block provides a single, straightforward syntax for assigning values to all GPSS/H data items. For example, using BLET to assign a value of 1 to the Transaction parameter named ALEX is quite intuitive:

  ```
  BLET        PF(ALEX)=1
  ```

  Using indirect addressing, such as assigning a value to the Parameter specified by the number given in PF(ALEX), is similarly intuitive, yet is not likely to be written by accident:

  ```
  BLET        PF(PF(ALEX))=1
  ```

- GPSS/H now supports built-in random-variate generators for 23 additional statistical distributions (26 in all).

- GPSS/H Professional now comes bundled with ExpertFit™, the highly-regarded distribution-fitting software from Averill M. Law and Associates.

- Release 3 of GPSS/H Professional supports user-written external routines in both C and FORTRAN. Although it is rarely necessary to go "outside" GPSS/H when developing a model, it can be helpful in special situations. For example, it might be desirable to use scheduling software from the real system as a component of the simulation model. Similarly, a modeler might want to use pre-existing computational code, or need to write extremely complex computational routines that can become somewhat cumbersome as GPSS/H Blocks. Other special situations might involve the need to interface with non-ASCII data files, or to develop a specialized user-interface.

- CHECKPOINT and RESTORE statements allow a model to save its state at a predetermined point during execution, then make repeated runs using that state as the starting point. In many cases, CHECKPOINT and RESTORE can be much easier to use than the traditional READ and SAVE statements.

- The SYSCALL statement and the BSYSCALL Block, which take an operating system command line as an operand, allow a running GPSS/H model to *shell out* to the operating system to perform the specified command. SYSCALL and BSYSCALL are especially useful when using existing programs to perform data analysis during model execution or between simulation runs. The models can communicate with the external programs through data files. The ability to shell out to the host operating system has also been implemented in the GPSS/H Interactive Debugger. In order to use this feature, one merely types a "$" followed by the operating system command at the debugger command line prompt.

- The operations that can be performed on Transactions in a User Chain have been extended. New SCANUCH and ALTERUCH Blocks allow examining and changing the Parameters of such Transactions without having to UNLINK and reLINK them. They operate on User Chains in exactly the same way as SCAN and ALTER operate on Groups.

- Floating-point Parameters can be examined and/or modified during operations on both User Chains and Groups.

6 BUILDING A SIMULATOR USING GPSS/H

Earlier in this paper, we contrasted the capabilities of visual-based modeling tools with those of languages. Regardless of which approach is used, the modeler must still build from scratch a model that represents the physical system of interest. Modeling complex systems *correctly* requires intimate knowledge of both the simulation software and the system under study. However, not everyone who can benefit from using simulation has the time or the training necessary to build simulation models.

As a result, a third type of modeling-tool, the *special-purpose simulator*, has emerged as a way of providing simulation capabilities to users with little or no modeling experience. Special-purpose simulators are most commonly developed under circumstances where:

- a single model development effort can benefit multiple users
- modeling expertise can only be obtained from indirect sources such as external consultants

In these cases, an experienced modeler develops the model, freeing the end-user from learning modeling and simulation-software skills.

The special-purpose simulator is a custom-built analysis tool designed by an experienced simulation-model builder. At its heart is a data-driven model of a specific system or set of similar systems. The simulator provides its user with a method to easily modify model parameters, define experiments, run tests, and get results. A simulator is usually comprised of a data-entry front end, a simulation engine, and an output browser. The simulation engine runs a parameterized model which accepts user-specified data at execution time. Combining these tools brings the power of simulation analysis into the hands of the non-simulationist.

6.1 Data-Entry Front End

The front end is the means by which the user of a special-purpose simulator modifies the run parameters without changing the underlying model. This may take several forms, the most basic and rarely used of which involves manually editing a text file. In another approach, the model itself prompts the user for input from the keyboard as the model executes. Still other designs require modifying data by using an external spreadsheet or database program. No matter which approach is used, the purpose of the front end is to conveniently produce a data file which can be used by the simulation model as it executes.

A more advanced approach integrates a customized front-end data-entry program, a simulation engine, and an output browser under a single outer *shell* (Figure 1). Typically created using a general-purpose programming language or a tool such as Visual Basic, the shell may be menu-driven. Data-entry "windows" and dialog boxes guide the user through the process of specifying parameters, running the model, and viewing the output. The shell may also provide built-in help facilities and data "range-checking" (*e.g.* verifying that all operation times are non-negative before executing).

6.2 Simulation Model

The most important component of the special-purpose simulator is the underlying model. Since the end user is generally prevented from modifying the model, this component determines the maximum flexibility offered by the simulator. It must be generic enough to accept a broad range of inputs, and it must be updated periodically to insure that the model *remains* valid.

A static simulation model can be produced and its design frozen when the simulator is initially created, or model code can be generated "on-the-fly" every time that the model parameters are modified by a user. In either case, user input is not limited to operating-parameter values — it can also alter logic embedded deeply within the model. For example, based on a user-specified value, the model could select one of three different order-picking algorithms that have been pre-coded into the model.

6.3 Simulation Engine

The simulation engine runs the model and generates output. There are several features to look for when selecting the engine.

Most importantly, the language used for the engine must be flexible enough to handle the demands that a generalized model places on the software. Flexibility is crucial in the areas of file input, file output, and control logic within the model. Execution speed is also a primary concern. The faster a model executes, the better—time executing a model is often down-time for the user. GPSS/H's speed and built-in flexibility make it the ideal simulation engine for a special-purpose simulator.

An excellent example of the use of GPSS/H as a simulation engine is given in Coughlan and Nolan(95).

6.4 Output Browser

The output browser displays the data generated by the model in an easy, understandable form. If the simulator's user has limited experience in simulation-modeling, the standard-style statistical reports provided by the engine may be difficult to decipher. Custom-formatted output, including summary statistics, should *always* be used to present simulator results.

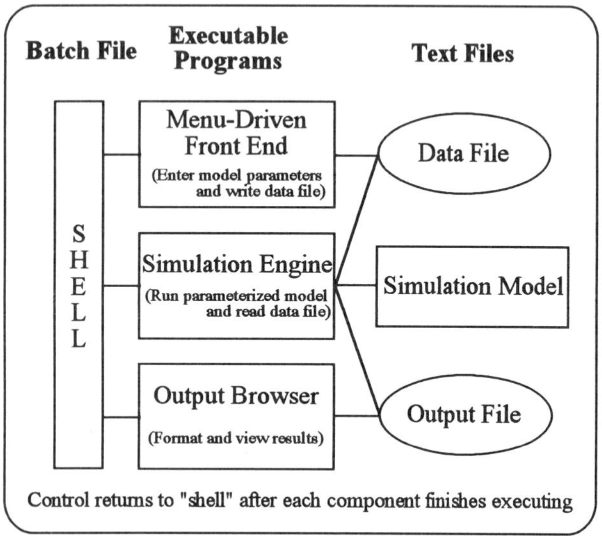

Figure 1: Components of a Special Purpose Simulator

Statistical analysis of the output can be performed directly by the shell program, or by a specialized statistical software product. For *special-purpose simulators* running under Microsoft Windows, SIMSTAT, from *MC^2 Analysis Systems,* reads and analyzes standard output data generated by GPSS/H.

Animation is yet another form of simulation output. Animating a generalized model can sometimes present obstacles. Accounting for variations in resource numbers and capacities, flow and routing-patterns, and physical layout dimensions makes animating a generic model more difficult than animating a specific model. However, a basic animation helps confirm model validity to the non-simulationist. High quality animations can be generated by coupling a GPSS/H model with Proof Animation™, a general-purpose animation tool.

6.5 Run-Time Versions Provide an Economical Simulation Engine

A simulator is generally developed for a single application, where it is intended to be used by many people. However, each user must have a copy of the simulation software in order to execute the model. For a simulator used by dozens or even hundreds of users, the cost of the simulation software may render a project too expensive. Wolverine's Run-time GPSS/H offers a solution to this problem.

Run-time GPSS/H is identical to Wolverine's 32-bit GPSS/H Professional, except that it can only run models which previously have been specially compiled with the regular Professional version. The run-time version allows economical distribution of high-performance GPSS/H-based simulators.

Security is another important feature provided by the run-time version. Since only *pre-compiled* models can be run, the end user cannot view or edit the model "source" code. The user has access only to the data files used by the front-end and the output browser; hence, confidential models can be safely distributed. *Even further security* can be obtained by producing special "project-specific" pre-compiled models that can only be run by a specially designated group of users.

SUMMARY

GPSS/H has a strong history of success in both commercial and academic environments. The product continues to evolve in functionality and to grow in use. Although GPSS/H uses a more traditional text-based model definition, it continues to break new ground in ease-of-use and modeling flexibility.

REFERENCES

Banks, J. 1991. Selecting simulation software. In *Proceedings of the 1991 Winter Simulation Conference,* ed. B.L. Nelson, W.D. Kelton, and G.M. Clark, 15-20. Piscataway, New Jersey: Institute of Electrical and Electronics Engineers.

Banks, J, J.S. Carson II, and J.N. Sy. 1996. *Getting started with GPSS/H.* 2d ed. Annandale, Virginia: Wolverine Software Corporation.

Coughlan, K.L., and Paul J. Nolan. 1995. Developing special purpose simulators under Microsoft Windows. In *Proceedings of the 1995 Winter Simulation Conference,* ed. C. Alexopoulos, K. Kang, W.R. Lilegdon, and D. Goldsman, 969-976. Piscataway, New Jersey: Institute of Electrical and Electronics Engineers.

Henriksen, J.O., and R.C. Crain. 1989. *GPSS/H reference manual.* 3d ed. Annandale, Virginia: Wolverine Software Corporation.

Law, A.M., and W.D. Kelton. 1982. *Simulation modeling and analysis.* New York: McGraw-Hill Book Company.

Schriber, T.J. 1991. *An introduction to simulation using GPSS/H.* New York: John Wiley & Sons.

Smith, D.S., D.T. Brunner, and R.C. Crain. 1992. Building a simulator with GPSS/H. In *Proceedings of the 1992 Winter Simulation Conference*, ed. J.J. Swain, D. Goldsman, R.C. Crain, and J.R. Wilson. 357-360. Piscataway, New Jersey: Institute of Electrical and Electronics Engineers.

Wolverine Software Corporation. 1996. *Using Proof Animation.* 2d ed.. Annandale, Virginia: Wolverine Software Corporation.

AUTHOR BIOGRAPHY

ROBERT C. CRAIN joined Wolverine Software Corporation in 1981. He received a B.S. in Political Science from Arizona State University in 1971, and an M.A. in Political Science from The Ohio State University in 1975. Among his Wolverine responsibilities is that of chief developer for PC and workstation implementations of GPSS/H. Mr. Crain is a Member of IEEE/CS and ACM. He served as Business Chair of the 1986 Winter Simulation Conference and General Chair of the 1992 Winter Simulation Conference.

Proceedings of the 1996 Winter Simulation Conference
ed. J. M. Charnes, D. J. Morrice, D. T. Brunner, and J. J. Swain

THE POWER AND PERFORMANCE OF PROOF ANIMATION

James O. Henriksen

Wolverine Software Corporation
7617 Little River Turnpike
Annandale, Virginia 22003, U.S.A.

ABSTRACT

The Proof Animation™ 3.x product family provides tools for animating a vast array of applications. There are versions of Proof Animation to meet any need, from small and mid-sized projects to large scale applications. The Proof product family runs on readily available, inexpensive PC hardware. Proof is ASCII-file-driven and features a general-purpose design. Its vector-based geometry provides a large animation canvas with the ability to zoom in or out while maintaining crisp, clear images. Proof's features include post-processing for maximum performance, built-in drawing tools and CAD import/export for ease of creating animation layouts, dynamic bar graphs and plots for displaying statistics, multiple-window display, and a unique presentation-making capability. Proof's open architecture makes it ideally suited for serving as an animation engine for models written in a wide variety of simulation and programming languages. Proof's superior power and performance assures smooth, realistic motion for animations regardless of their size, complexity, or application.

1 INTRODUCTION

Animation software has been around for many years; the concept is not new. Animation is often considered a requirement to a simulation study because it is a meaningful, proven way to show results to an audience with varied backgrounds. As more animation products enter the market, a user must make sure that the product purchased can meet the task at hand, as well as any future projects. Some animation software is limited to a single type of application. Moreover, some of these animation packages may only be useful for a specific type of problem *within* that application. In almost any application, there are many areas that benefit from simulation and animation. These can range from capacity analysis to scheduling, staffing, and more. Some application-specific animation software may not be able to handle the full range of problems even within a *specific* application. Buying single-application animation software can be costly, especially if a user does projects within a wide range of applications.

Proof Animation is a powerful general purpose animation tool. It is not tied to a specific area or application. Proof Animation can be used to animate the full range of applications from areas such as Business Process Reengineering to the classic applications such as healthcare, manufacturing, and traffic. The size of the system to be animated is not an issue when using Proof Animation. There are versions of Proof Animation that can be used to animate small to mid-size systems as well as a version to handle the largest systems.

A user may want to add animation to an already existing model. Since time is rarely unconstrained, he or she does not want to have to change any existing and valid model logic to make the animation look realistic. Some animation tools are so limited that it is difficult or even impossible to produce an animation that can mimic the motion of the real-world system. Wolverine Software Corporation's goal has always been to design animation software that is easy to use and able to produce realistic animations that help sell the results of simulation projects.

Another issue in animating existing models is interoperability. Can the animation software work with the existing simulation software? Proof Animation can be paired with most simulation and programming software. The user can continue to use the modeling tool with which the existing models were written.

Proof Animation 3.1 has an advanced feature set unparalleled in its ability to meet the challenges of animating today's complex systems while keeping ease of use a top priority. In the following sections we describe the products, discuss their design, give an overview of how they are used, and describe the advanced features.

2 THE PROOF ANIMATION FAMILY

All of the Proof Animation products run on readily available and inexpensive PC hardware. All run as 32-bit applications that require only a 386 or better CPU, a math coprocessor, and at least a VGA-compatible video card. Because this basic configuration is very widely available, the portability of animations developed with Proof is maximized. It is very likely that this hardware configuration will be available at any site.

The Proof Animation family runs under DOS and can also be launched as a full-screen application under Windows 3.x, Windows 95, Windows NT, or OS/2. A set of Windows icons are supplied with each of the Proof Animation products to provide single-click launching. The following products comprise the Proof Animation family:

- PROOF ANIMATION
  Proof Animation is the basic animator. Memory size is fixed and limited. Includes built-in CAD import/export feature. Good for small to mid-sized animations.

- PROOF PROFESSIONAL
  Proof Professional exploits all available extended memory for animating large systems. Includes built-in CAD import/export feature.

- RUN-TIME PROOF PROFESSIONAL
  Run-time Proof Professional runs developed animations or presentations, but has no animation *development* capabilities. It provides a low cost way to run different scenarios with a fixed layout file prepared using Proof Professional or Proof Animation.

- STUDENT PROOF ANIMATION
  The student version of Proof Animation is included with the *Using Proof Animation* text. Size and playing time limitations are imposed; otherwise it is identical to Proof Animation.

- PROOF ANIMATION DEMO MAKER
  Demo versions of animations can be prepared under a licensed copy of Proof Animation or Proof Professional containing the Demo-Maker add-on. Copies of the executable demo files can be reproduced and distributed free of charge and viewed by anyone. No licensed copy of Proof is needed to *view* a demo prepared with the Demo Maker.

3 MAXIMUM PERFORMANCE DESIGN

3.1 The General-Purpose Approach

One of the ways in which the performance of Proof Animation is maximized is its general-purpose design. A general-purpose animation product can be used without being tied to a specific simulation or programming language. While built to work easily with Wolverine's GPSS/H simulation software, Proof Animation also provides affordable and powerful animation software to users who develop models in other simulation and programming languages.

Most animation software from other vendors is directly coupled to their simulation software. In other words, one cannot use *their* animation software without also using *their* simulation software. In some cases, the simulation and animation software are sold only as a pair, so both must be purchased regardless of the needs of the user. The suggested advantage of the coupled approach is that because the animator has direct access to the simulation events, development of the animation is supposedly simplified. However, the real advantage is felt by the vendors. They sell more software since their users do not have an option to pick and choose based on their needs. Moreover, a user of the coupled software has little or no control over what information is passed to the animation; therefore, he or she may actually have to alter the modeling approach used in the simulation to achieve the desired appearance and level of detail for the animation.

Another disadvantage of the tightly coupled simulation/animation package is cost. Sole sources tend to be expensive. Vendors of these tightly coupled packages often claim that their approach is the *only* way to add animation to a simulation. Proof Animation has proved that wrong. The number of success stories using Proof Animation with other software continues to grow. Furthermore, a benefit of the mix-and-match strategy for software purchases is that the selection can be based on optimal functionality and price.

3.1.1 ASCII Input Files

Proof Animation can be used with other software because of two major design features. One is that Proof Animation is driven by ASCII files. Therefore, any software capable of writing ASCII text files can be used with Proof Animation.

Proof Animation requires two ASCII input files to run an animation: the layout file and the trace file. The layout file describes the geometric details of the background over which objects move, provides geometric definitions and properties for such objects, and defines logical paths along which the objects move.

Ordinarily, layout files are produced at least in part by using Proof Animation's drawing tools; however, the layout file command set specifications are published so programs can easily be written to generate layout files. For example, some users have written front ends for their simulation models that allow different system design parameters to be specified for each run. Based on these parameters, different geometric configurations are written and incorporated into a layout file. The new layout appears on screen when Proof Animation is invoked.

The trace file contains a time-ordered sequence of commands such as CREATE, DESTROY, PLACE, PLOT, MOVE, SET SPEED, SET COLOR and many more. This file provides Proof Animation with information on when, where, and what to create, destroy, place, plot, etc. Trace files are free-format, and the commands are easily learned and used. They provide exactly the kind of flexibility necessary to easily be integrated with the simulation model logic. Any language that can produce formatted ASCII output can write a trace file.

3.1.2 Post-Processor Animation

The second of the two design features that make Proof Animation compatible with other software is that it is a post-processing animation package. Post-processing means that it runs *after* the simulation has executed. Both the layout and trace files must exist before invoking Proof Animation. They cannot be written and read concurrently.

Two great advantages result from the post-processing approach. First, PC hardware is not shared between the simulation and the animation. This leaves the entire CPU for running the animation. Second, it provides the abilities to jump back and forth in time during the animation playback, to speed up or slow down the viewing speed, or show all or a specific portion of an animation. These features make it easy to investigate unusual system behavior or highlight points of interest.

3.2 Vector-Based Geometry

In the Proof Animation product family all layout file information is based on vector geometry. Vector-based descriptions are automatically mapped into the screen's pixels to build the image. One of the advantages of this approach is that layouts can be much larger than a single screen. With the ability to zoom in or out and pan, larger layouts are easily navigated to show the overall layout or zoomed in to whatever level of detail is necessary. Vector-based geometry also provides the ability to have moving objects realistically *rotate* around corners instead of the sliding effect to which other animation packages are limited.

Another advantage of vector-based geometry is that many CAD packages are capable of producing standard vector-based .DXF files. In many cases, a CAD drawing already exists for the system to be animated. If that is the case, the effort of redrawing an entire layout can be avoided. Proof Animation's built-in CAD Import/Export feature provides the capability to convert industry-standard .DXF files into Proof Animation layout files, *and vice versa*. Credibility of the study is enhanced when viewers see a familiar CAD drawing of the system integrated into the animation. These advantages maximize the power of the animation by giving a user total flexibility on the detail and complexity of the drawing.

Instead of vector geometry, animation packages may use a pixel-oriented approach for drawing. With the ability to manipulate individual pixels, one can produce detailed images. However, this level of detail is time consuming to draw and can often be lost because of the scale at which the animations are viewed. Some other disadvantages of pixel-orientation are: (1) pixel-oriented images cannot be rotated; (2) layouts are often confined to single-screen images. Some animators offer multi-screen operation; however, the individual screens are disjoint and independent, unlike Proof Animation's single, continuous canvas; (3) zooming in on pixel images magnifies the jagged edges inherent in all such images. When a zoom in is performed in Proof Animation, the vector-based image maintains its crisp and clear appearance. Lines continue to look like lines. If a zoom in is performed on a pixel-based animation layout, the effect of the jagged edged image makes a line look more like a stairway.

3.3 Smooth Motion

The maximum performance design of Proof Animation achieves very smooth motion. Proof Animation maintains this smooth motion by updating or refreshing the screen 60-70 times per second. Other software can often sustain refresh rates of only 5-10 updates per second. The ultimate purpose of an animation is to achieve a realistic depiction of the system being studied, allowing the audience to gain confidence in the results of the simulation study. Objects that move smoothly across the screen are more realistic than those that jump across the screen.

3.4 Color and Resolution Options

Beginning with Release 3.0, all versions of Proof, including the student version, now provide for operation in 256-color mode in a variety of screen resolutions. Prior to Release 3.0, "standard" versions of Proof operated in 16-color mode at 640x350 resolution. 256-color mode and higher screen resolutions were provided as no-cost options only to users of Proof Professional. Support was limited to about a half dozen video chip families.

The "normal" operating mode for Proof is 256-color mode, and the "normal" resolution is 640x480. Video hardware which supports this mode and resolution is very commonplace. Proof can also run at 800x600 and 1024x768 resolutions, provided enough video memory is available. These resolutions are now even available on some *laptop* machines. For older hardware with less video memory, Proof can run using 640x400 resolution. For very old hardware, Proof can still be run in its original 640x350, 16-color mode.

256-color animations are *much* more attractive than 16-color animations. The expanded color palette contains 24 foreground colors and 8 background colors. The background colors consist of 7 layout colors and 1 backdrop color. The background colors do not interfere with the foreground colors and therefore give a user much more flexibility when drawing the static background portion of the layout. With more colors from which to choose, the background can be drawn with more detail without sacrificing the color integrity. In addition, in 256-color mode, objects can be multi-colored. (In 16-color mode, objects must be monochrome.)

4 NAVIGATING PROOF'S MODES AND MENUS

Proof Animation is organized into seven menu-driven *modes*. Each mode is a collection of closely related functions. Switching among these functions is very easy. Usually, a single mouse click is all that is required. Switching among *modes* is also easily done, but it implies major changes of context. For example, running an animation and drawing a layout are vastly different activities. Each mode has one or two main menu bars at the top of the screen. Clicking on main menu items invokes the options for the lower level tools. The seven modes are summarized as follows:

- RUN MODE
  This is the mode in which animations are viewed. It provides menu tools for starting and stopping an animation, changing views, controlling viewing speed, jumping ahead and back in time, and more.

- DEBUG MODE
  This mode provides tools for stepping through an animation by individual events or time commands and examining the resulting movement. Information pertaining to an individual object can be obtained by clicking on the object with the mouse.

- DRAW MODE
  This mode contains the drawing tools used for creating the layout background. Tools are provided for drawing static elements such as lines, arcs, text, fills, etc. Dynamic elements such as messages, bars, plots, and layout objects are defined in Draw Mode.

- CLASS MODE
  This mode is used for defining object classes. A class serves as a template for creating both the dynamic objects that move around a layout and layout objects that generally remain stationary. The template determines an object's size and shape and other initial properties such as default speed and color. An animation will usually contain multiple object classes. For example, an animation of a hospital might contain classes that represent doctors, nurses, equipment, and patients.

- PATH MODE
  This mode is used for defining fixed paths. A path is a perfect mechanism for describing for guided, directional movement such as conveyors. The geometry or route of a path is easily defined by clicking on existing lines and arcs of a layout. Tools are also provided for defining path speeds, circularity, and accumulation status. Accumulating paths provide automatic queuing for objects that pile up at the end of the path.

- PRESENTATION MODE
  This mode is used for running scripted presentations. Scripts can include static bitmap slides and snippets of animation, separated by special effect segues such as screen fades and dissolves.

- SETUP MODE
  This mode is used for examining and altering infrequently changed configuration data. For

example, the color palette can be customized or the mouse speed changed in this mode.

5 CREATING ANIMATIONS AND PRESENTATIONS

5.1 Drawing the Layout

The first step in developing an animation is to draw a layout. If a .DXF formatted CAD drawing of the system is available, a user can begin by importing the drawing into a Proof Animation layout file. This is done using the built-in CAD import/export utility. Once imported, the drawing can be examined by layer or by line style. Specific layers and line styles can be deleted from the drawing as desired. When you save the resulting drawing, it is saved as a Proof Animation layout file. The original .DXF file remains intact.

If a user does not have a CAD drawing or prefers to draw using a computer, the drawing tools provided in Draw Mode are easy to use. Although it is mouse-oriented, Draw Mode also allows keyboard input, so if a user needs to draw a line of a specific length at exactly a certain angle, he or she can enter these specifications *numerically,* and the geometry will appear on the screen. To help in drawing scaled, accurate layouts, a visible grid is turned on automatically when Draw Mode is entered. For additional aid in drawing, Proof Animation has the Snap-to-Grid option. This option is also *on* as the default setting. Snap-to-Grid limits the drawing of layout elements from grid point to grid point, thus eliminating the chance of small gaps between the endpoints of *seemingly* connected lines. Other snap options which help draw accurate layouts are Snap-to-Endpoint which *magnetically* attracts the mouse cursor to the ends of lines and arcs, and Snap-to-Tangent which quickly finds points of tangency between lines and arcs. All of these options can be turned on or off by the user during the drawing session.

5.2 Defining Object Classes

Once the background of the animation is drawn, the second step in developing an animation is usually to define one or more object classes. This is done in Class Mode. Objects and object classes are among the most important constructs in Proof Animation. A class provides the geometric description of the individual objects that move throughout the animation. The class definition also includes the initial properties such as physical clearances, color, and speed of the individual objects. Each animation will usually have a collection of object classes.

It is helpful to think of an object class as the template from which the individual objects are made. An individual object is based on the single geometric description of a particular object class. There can be an arbitrary number of *objects*, such as widgets, in the system at once, but there need be only *one* widget object class.

Motion and color-changing commands in the trace file operate on *objects*. The drawn background components, produced in Draw Mode, cannot be moved or changed. If dynamic changes in background elements are required, the appropriate components must first be defined as object classes and can then be created and positioned directly in Draw Mode. Objects that are created and placed in the layout while the user is drawing the background are called *layout objects*. Layout objects enable a user to scale and position the objects into the layout while having the background components visible as reference points. While the animation is running, layout objects can be manipulated using trace file commands. For example, if an idle machine is shown as green and a busy machine as red, the machine must first be defined as an object class. Objects from that class can be created and placed as part of the layout file, and their color can be changed while the animation is executing.

5.3 Defining Paths

Proof Animation provides two kinds of motion: absolute and guided. Absolute motion, specified by the MOVE trace file command, causes an object to be moved between two points. Guided motion always occurs along a fixed route, called a path. If objects will follow guided motion, such as travel on conveyors or along guide wires, the next step in the animation development is to use Path Mode to define one or more paths.

Paths are comprised of lines and arcs that represent the route that the objects will follow. This underlying geometry must first be drawn using Draw Mode or imported from CAD. The logical path segments are then defined *on top* of the existing lines and arcs. A single line or arc can be part of one or more paths. Once defined, paths are saved as part of the layout file.

Using paths is very simple because Proof Animation does all the work. The most commonly used trace file path command is PLACE *object*ID ON *path.* Once an object is placed on a path, it will follow that path until it visually comes to rest at the end of the path or until it is PLACEd elsewhere or DESTROYed. All objects traveling on the same path can be stopped simultaneously and resume movement at a later time.

Paths provide outstanding animation power in response to a single trace file command.

Accumulating paths provide even greater power for animating paths on which queuing can take place. On accumulating paths, Proof Animation reflects physical reality by *visually* queuing objects when bottlenecks occur. This often makes a simulation model of the system much simpler to construct, because such queuing need not always be explicitly represented in the model code. Most systems contain some accumulation. This property can be used to represent certain types of conveyors, cars at a traffic signal, bank lines, and more. Paths play an especially important part in transportation, product flow, and material handling animations.

5.4 Writing the Trace File

The next step in the animation development is producing the animation trace file. Trace files consist of very readable ASCII commands. Trace files are time ordered. That means the specific animation events take place between TIME commands. Consider the following portion of a trace file:

```
TIME 34.6
CREATE PLANE 1
PLACE 1 ON RUNWAY3
SET 1 SPEED 75
TIME 52.8
```

It is very easy to visualize the results of these commands. At time 34.6, an object with an id number of 1 is created with geometry and properties inherited from class PLANE. This object will appear on screen at the beginning of a path named RUNWAY3 and begin moving along the path. The speed at which object 1 will move is set to 75 units of distance per unit of simulated time. These units are user-determined, e.g. feet and seconds. Proof Animation will continue reading trace file commands until it reads the TIME 52.8 command, signaling the end of the events that are to begin at time 34.6. It is very easy to produce simple trace files with any ASCII editor.

For most applications, it is impractical to create trace files by hand. Using a simulation model or program to generate the trace file is usually the *only* viable approach. In order to produce a trace file, output statements are inserted into the simulation model to write the appropriately formatted commands. The Proof Animation trace file command set has been designed to be easily generated. Any language with the ability to write a formatted ASCII file is capable of producing a trace file.

5.5 Building a Presentation

As an optional final step, a professional looking presentation can be built using Proof Animation. Presentation Mode lets users display scripted sequences consisting of bit-mapped screen images or slides, full animations, and/or segments selected from full animations. These presentation elements can be linked together using fades, dissolves, and other special effects, to produce a polished presentation. This is done by writing a simple ASCII presentation script file. Complete presentations can be viewed without ever exiting Proof Animation. This eliminates the awkwardness of switching back and forth between the computer and other display media during a presentation.

Slides can be created directly in Proof Animation or any software package capable of exporting industry-standard .PCX image files. There are many such packages available, and virtually all of them can produce very high-quality charts, graphs, and slides. Proof Animation can both read and write these .PCX screen images. It is very straightforward to save Proof Animation screen images as .PCX files and incorporate them into presentations as slides.

Presentations can be developed so that slides and animations appear on the screen for a defined amount of time. The viewer does not have to interact with the computer for the presentation to continue. Presentations can also be developed to continue once a key or mouse button is pressed, giving the viewer or presenter ample time to comment on what is currently on the screen.

When developing the presentation, a user can choose to highlight areas of interest within the animation by using different views or sound to draw the viewer's attention to particular aspects of the animation. Presentations can incorporate selectable menus defined by the presentation developer. These menus can be set up by topic, giving the viewer or presenter complete control and flexibility of what to show.

6 THE PROOF PROFESSIONAL ADVANTAGE

Proof Professional offers obvious advantages because it is limited only by the total memory available on a computer. No artificial memory limits are imposed; therefore, large-scale animations can be run on the PC.

Although Proof Professional runs on 386-based machines with a math coprocessor and VGA display, Pentium-bases PCs are the best platforms for running large-scale animations.

7 THE ADVANCED FEATURE SET

Advanced features make Proof Animation unparalleled in its capabilities. Keeping the user in mind, many of these features were added because of direct requests from our customers. This feature set is described below:

7.1 Multiple-Window Display

The animation screen can be divided into separate windows. Within each window, the view can be independently manipulated using zooms, pans and rotations to include a portion or all of the animation canvas.

With this feature users can maintain a window that keeps updating statistics in constant view while panning and zooming to different areas of the layout. For example, a user may choose to create a layout object that represents a clock to display the current system time and place it in a separate window that is always in view.

7.2 Displaying Dynamic Data

Along with bars, dynamic plots can be incorporated into a layout as a way of displaying statistics. A plot can be used to show any type of data displayed on X-Y axes. Plots, like bars, are defined and placed in the layout using Draw Mode. The data are displayed in the form of line segments that can be added, erased or changed via the new PLOT trace file command. A single plot can have many different types of data plotted simultaneously with each curve plotted in a different color. Plots offer a unique way of displaying changes over time. A viewer can look at a plot and see the differences in the value of the statistic as the animation progresses.

Textual messages can be incorporated into object classes. Each object created from that class will carry its own messages. The text displayed in the messages can be changed using the WRITE trace file command. Messages can be used to visually differentiate objects that are otherwise identical in appearance.

7.3 Features Affecting Object Movement

When an object is placed on a path, that object is positioned on the path at the object's hotpoint. The hotpoint of every object class is the point (0,0) as viewed in Class Mode. Placement of the hotpoint within the class is determined by how the geometry is drawn around that point. For example, the class can be drawn so that (0,0) is in the center, on an edge, or even outside of the geometry. Unlike compact objects, long, thin objects rounding corners of paths can exhibit a *fishtailing* effect, especially when the hotpoint is near the leading edge of the object

Proof Animation provides the ability to supply a second point of attachment to object classes. This point is called the rear guide point or RGP. The optional RGP is defined as a property of a class in Class Mode. It is input as a negative number which serves as the RGP's displacement from the hotpoint. Both the RGP and hotpoint will remain connected to a given path as long as such connections are physically feasible

Another feature that can affect the movement of objects is the ability to specify a negative speed or travel time for objects traveling on non-accumulating paths. Specifying a negative speed causes the object to travel in the reverse direction on a given path. This gives users the ability to have an object back into an area and pull out or vice versa. This is especially useful in animations that contain traffic of any type.

The PLACE...ON commands for objects entering paths has a new option called SQUEEZE. When SQUEEZE is used with a PLACE...ON, all objects on the path *behind* the point at which the new object will enter are pushed back, allowing sufficient room for the new object. Objects can be SQUEEZEd at the end or at any offset of a path as well as before or after a particular object already residing on the path. This feature is ideal for animating conveyors that are loaded manually.

The addition of the ATTACH trace file command lets the user connect objects to one another. All the objects then follow the movement of the leader object. There is a single level of attachment. For example, Object 2 can be attached to Object 1, but Object 3 cannot also be attached to Object 1 while Object 2 is still attached. To create a string of objects, a user can simply attach 2 to 1 and then attach 3 to 2, and so on. The movement of all of the objects is controlled by the first object in the string, in this case, Object 1. The trace file commands need only manipulate the leader object, and all the other attached objects will follow. Speed is also determined by the leader of the chain of objects. This new feature makes it simple to animate multi-car trains or vehicles in tow. The DETACH command is used to break the chain of objects at any point of attachment.

7.4 Features That Enhance The Trace File

A JUMP option has been added to the TIME command. If JUMP is used, the trace file is automatically fast-forwarded until a TIME command is encountered with a value that is equal to or greater than the value specified after the JUMP keyword. For example, a TIME JUMP 100 command issued in the trace file will fast forward the animation until it executed a TIME 100 or higher TIME command. The animation then begins to run at the specified viewing speed, processing trace file commands normally. This is a way to skip large warm-up periods in an animation without needing to stop and enter a fast-forward time via the Run Mode menu.

There is an option in the file menu that allows the writing of an *abridged* form of the trace file. When creating an abridged trace file, a full version of the trace file must first exist. A user determines the start and end times of the trace file that will be incorporated into the abridged version. This is especially useful in preparing animations for presentations. The abridged trace file contains the information needed to immediately update the state of the animation to the start time specified in the abridged trace file and continue running normally until the user-specified end time is reached. The abridged trace file eliminates any delays that would be incurred because of the additional trace file commands that are executed during a standard fast-forward.

8 SUMMARY

Wolverine Software's Proof Animation has set the standard for maximum power and performance. Some of Proof Animation's many unique features include the ability to show statistics using bar graphs and plots, create presentations, built-in CAD import/export, drawing tools, smooth motion, and a unique multi-windowing display.

Proof Animation is not tied to a specific application. There are features that make it an ideal choice for the animation of systems like computer networks, health care, transportation, reengineering, manufacturing, and more while maintaining ease of use.

An animation benefits a user in every phase of the study: verification, validation, presentation of results, and the overall system design process. Proof Animation's unmatched features make it the perfect tool for each of these phases regardless of the application.

REFERENCES

Earle, N.J. and Henriksen, J.O. 1994. Proof animation: reaching new heights in animation. *Proceedings of the 1994 Winter Simulation Conference*, eds. J. Tew, S. Manivannan, D. Sadowski, and A. Seila, 509-516 Institute of Electrical and Electronics Engineers, Piscataway, New Jersey.

Wolverine Software. 1995. *Using Proof Animation (Second Edition)*. Annandale, Virginia: Wolverine Software Corporation.

AUTHOR BIOGRAPHY

JAMES O. HENRIKSEN is the President of Wolverine Software Corporation. He is a frequent contributor to the literature on simulation. Mr. Henriksen served as the Business Chairman of the 1981 Winter Simulation Conference and as the General Chairman of the 1986 Winter Simulation Conference. He has also served on the Board of Directors of the conference as the ACM/SIGSIM representative.

Proceedings of the 1996 Winter Simulation Conference
ed. J. M. Charnes, D. J. Morrice, D. T. Brunner, and J. J. Swain

AN INTRODUCTION TO SLX™

James O. Henriksen

Wolverine Software Corporation
7617 Little River Turnpike, Suite 900
Annandale, VA 22003-2603, U.S.A.

ABSTRACT

SLX is Wolverine Software's "next generation" simulation language. SLX builds on the strengths of Wolverine's GPSS/H (Henriksen & Crain, 1996). It provides powerful simulation capabilities in a modern, C-like language framework. SLX provides a multiplicity of layers, ranging from the SLX kernel, at the bottom, through traditional simulation languages, e.g., GPSS/H, in the middle, to application-specific language dialects and extensions at the top. This paper provides an overview of SLX for readers who are already familiar with simulation. An earlier paper (Henriksen 1995) presented key concepts of the architecture of SLX. As of this writing, SLX has been heavily used by a number of alpha testers. Insights gained from our interactions with the software testers will be presented.

1 INTRODUCTION

The most important characteristic of SLX is its layered architecture. The success of SLX's layered approach depends on several factors:

A. The layers are well-conceived. In developing SLX, we have had the luxury of drawing on years of experience with GPSS/H. A great deal of SLX is based on GPSS/H. In some cases, source code from GPSS/H has been directly "lifted" for use in SLX. In other cases, we have modified, simplified, or adapted GPSS/H algorithms. In a few cases, we have eliminated pitfalls and shortcomings of GPSS/H. The end result is an extremely well-designed system.

B. The layers are not too far apart. Many other languages provide multiple layers, but typically there are wide gulfs between the layers. For example, a language might provide flowchart-oriented building blocks as its primary modeling paradigm, but also provide for "dropping down" into procedural languages such as C or FORTRAN. The problem with this approach is that there are only two layers, and they are too far apart. One must become familiar with many details of the C or FORTRAN *implementation* of the simulation language to be able to add C or FORTRAN extensions. Even worse, virtually none of the error checking and other safeguards provided in the simulation language are available in C or FORTRAN. The SLX kernel language is a powerful, C-like language, so users of SLX almost never find it necessary to drop down into a lower-level, more powerful language. Furthermore, the SLX kernel language includes complete checking to prevent errors such as referencing beyond the end of an array and using invalid pointer variables. The layers above the SLX kernel exploit kernel capabilities in straightforward ways. Transitions from layer to layer are very smooth.

C. The mechanisms for moving from layer to layer are very powerful. These mechanisms are *abstraction* mechanisms. A "higher level" entity provides a more abstract description than a "lower level" entity. Lower level implementation details are hidden at the upper levels. SLX provides both data and procedural abstraction mechanisms. Like C, SLX provides the ability to define new data types, and to build objects which are aggregations of data types. The procedural abstraction mechanisms of SLX are extremely powerful. SLX provides a macro language and a statement definition capability which allows introduction of new *statements* into SLX. (The SLX-hosted implementation of GPSS/H makes heavy use of the statement definition feature.) The definitions of macros and statements can contain extensive logic, including conditional expansion, looping, optional arguments, lists of arguments, etc. In fact, such definitions are actually *compiled* by SLX, allowing use of virtually all kernel-level statements. Macros and statement definitions offer far more than simple text substitution.

In the sections which follow, selected features of the SLX kernel are presented. Following the presentation of the SLX kernel, SLX's extensibility mechanisms are illustrated. Finally, brief presentations of how SLX has been used and how SLX is taught are presented.

2 SLX KERNEL FEATURES

The number of primitives required to support simulation is surprisingly small. Implementing some of these primitives in a general form, however, can be very difficult. Features such as SLX's generalized *wait until* are extremely difficult to implement. Not surprisingly, this feature has rarely appeared in other simulation software. Paradoxically, some of the features which are the most difficult to implement are the most easily understood. In the remainder of Section 2, we will present some representative features, to illustrate the functionality, ease-of-of-use, and *ease-of-learning* of SLX.

2.1 Objects and Pointers to Objects

The very mention of the word "object" inspires a wide range of expectations and emotions, due to the widespread influence of object-oriented programming (OOP). SLX has been influenced by OOP and incorporates some OOP ideas, but *SLX is not a truly object-oriented language*. Such a statement is a rarity in this day. Many products claiming to be object-oriented are far from it. In fact, although we do not claim that SLX is object-oriented, it is probably more so than some products for which OOP architecture is claimed.

In SLX, objects are used in two ways. *Passive* objects are used for modeling entities which have no "executable" behavior. For example, a parking lot could be modeled as a passive object. GPSS/H Facilities, Queues, and Storages are implemented as passive SLX objects. *Active* objects have executable behavior patterns. Customers in a supermarket are a good example of entities that would probably be modeled as active objects. SLX active objects are roughly equivalent to GPSS/H transactions. Some entities can be modeled either as active objects or passive objects. For example, a simple server with a FIFO queue can be modeled as a passive object. Its behavior depends solely on the requests made for it by active objects. (This is the way Facilities work in GPSS/H.) For more complicated servers, an active object may be more appropriate. Consider a butcher in a model of a supermarket. In a simple queueing model, the butcher can be represented as a passive object, responding to requests for service one customer at a time. In a more realistic model, the butcher would have a more complex behavior pattern, cycling through activities of cutting meat, arranging products in refrigerators, interacting with the deli department, taking breaks, etc. Such behavior would require modeling the butcher as an active object.

SLX objects can have a number of *standard properties*. All standard properties are comprised of explicitly identified sections of executable code. The *initial* property is invoked when an object is created. The *final* property is invoked when an object is about to be destroyed. The *report* property is invoked by the *report* statement; it is used for the obvious purpose. The *actions* property specifies the behavior pattern for an active object. It is invoked by the *activate* statement (discussed in the next section). A sample object follows:

```
object customer
  {
  integer  number_of_items;

  initial
    {
    number_of_items =
    rv_uniform(stream1, 10, 20);
    }

  actions
    {
    ...
    }

  report
    {
    print (ME, number_of_items)
    "* exiting, items purchased: *\n";
    }
  };
```

Objects are created by using the *new* operator, which returns a pointer to the newly created object:

```
pointer(customer)      cust;
cust = new customer;
```

The initial property of the customer is executed before the pointer to the new customer is assigned to `cust`. The initial property can be thought of as an easier-to-use counterpart to a C++ constructor. All uses of pointers are validated to ensure that pointers always point to objects of the proper type (or contain a NULL value). Use counts are maintained for all objects. If pointer1 and pointer2 are pointer variables, an assignment statement of the form "pointer1 = pointer2" causes the use count for the object pointed to by pointer1 to be decremented and the use count for the object pointed to by pointer2 to be incremented. Storage for an object is released if and only if its use count goes to zero.

2.2 Active Objects

An active object has an *actions* property. When the *activate* verb is used, a *puck* is created for the object and placed on the active puck list; i.e., the puck is placed in a ready-to-execute state. Pucks are the schedulable entities in SLX. Scheduled time delays and state-based delays, e.g., waiting for a server to become available, are puck-based operations. Thus manipulation of pucks is the basic mechanism by which a collection of objects experiences events over time. Pucks embody the means of achieving simulated parallelism.

Two kinds of parallelism can be described in SLX. Large-scale parallelism consists of interactions among active objects. For example, in a model of a freeway toll booth system, active car, truck, and bus objects would all interact. Small-scale or *local* parallelism consists of parallel activities carried out by the same object. For example, a customer entering a bank may decide that (s)he will wait in line for no longer than two minutes. At the end of that time, if the customer has not been served, (s)he will renege (exit the system without having been served.)

At first glance, the distinction between large-scale and local parallelism might appear to be somewhat arbitrary. What's large-scale, and what's local? In SLX, a precise distinction is made based on the SLX verbs which are used. Objects are activated by using the *activate* verb. In most cases activate is used in conjunction with new:

```
activate new customer;
```

The above statement creates a new customer object, creates a puck for it and places the puck in the active puck list, poised to execute the first statement in the customer object's actions property. The puck which executes the "activate new" statement continues executing. The new customer will compete with other pucks (according to their respective priorities) after the current puck has gone as far as it can in the actions property at the current instant of simulated time.

Local parallelism is described by using the *fork* statement:

```
fork
  {
  offspring actions
  }
parent
  {
  parent actions
  }
```

Execution of the fork statement creates an additional puck for the currently active object. The newly created puck is placed in the active puck list, poised to execute the offspring actions clause of the fork statement. The parent puck executes the optional parent actions clause. Unless otherwise specified, both pucks continue execution with the next statement following the fork statement. An example of the fork statement is given in Section 2.4.

2.3 Time Advance

Time advance is provided by the advance statement, modeled on the ADVANCE block of GPSS/H, e.g.,

```
advance rv_expo(stream2, 10.0);
```

The advance statement removes the puck from the active puck list and places it on the future event list, scheduled to resume execution after the specified time delay has taken place.

2.4 Control Variables & Wait Until

In SLX, state-conditioned delays are modeled using *control* variables and the *wait until* statement. The keyword "control" is used as a prefix on SLX variable declarations:

```
control integer count;
control boolean repair_completed;
```

The "control" keyword tells the SLX compiler that at each point at which the value of the control variable is changed, a check must be made to see whether any pucks in the model are currently waiting for the variable to attain a particular value or range of values. Such waits are described using the *wait until* statement:

```
wait until (count > 10);
wait until (repair_completed);
```

Compound conditions are allowed as well:

```
wait until (count >= 10
  or repair_completed
  and not repairman_busy);
```

SLX also supports *indefinite* (user-managed) waits. Three steps are required to implement an indefinite wait. First, the puck which is going to wait must be made accessible to other pucks. This is usually done by placing the puck into a set. Second, the puck executes a wait statement with no "until" clause. Finally, at a subsequent point in simulated time, another puck executes a *reactivate* statement to reactivate the waiting puck.

Let us consider an example which illustrates the use of the fork statement in conjunction with *wait until*. Assume that customer objects flowing through a model reach a point where they are willing to wait a maximum of two minutes for service. If they are not served within two minutes, they exit the system; i.e., they renege.

```
object customer
  {
  control boolean reneged;
  actions
    {
    ...
    fork
      {
      advance 2.0;    // max wait time
      reneged = TRUE;
      terminate;
      }
    parent
      {
      wait until (reneged
         or server available);
      if (reneged)
      terminate;
      }
    ...
    }
  };
```

In the above example, a Boolean control variable is used within the customer object for communicating "renege" status between two pucks which share the same object. At the point at which the customer begins waiting for service, it forks, creating a second puck. The offspring puck executes the logic enclosed in braces immediately following the fork statement. The original puck executes only the logic contained within braces following the "parent" clause. The offspring puck undergoes a two-minute delay, sets reneged to TRUE, and terminates itself. The parent puck waits for either the server to become available or for the two minutes to elapse. When it comes out of the `wait until`, it must distinguish which of these two possibilities has taken place. If the two minutes have elapsed, reneged will be TRUE, and the parent puck will terminate itself. If not, the parent will continue executing.

The sharing of a single data area makes communication between the two pucks trivial. The "reneged" variable shared by the two pucks is all that is needed to accomplish the communication required in this example. Note that if there are multiple customers active at a given time, each customer will have its own data area, so the "reneged" status for one customer cannot be confused with that of another.

In many simulation languages, operations such as reneging are difficult to implement. Because of this, languages sometimes include operations such as reneging as built-in features. Unfortunately if the language designer's concept of reneging does not exactly match your requirements, you're stuck. In SLX, carefully designed rock-bottom primitives allow you to build your own capabilities if none of the ones provided by others meet your needs.

2.5 Sets

SLX includes the capability for defining and manipulating sets of objects. Sets can be homogeneous (comprised of objects of a single type) or universal (comprised of objects of arbitrary types.) Homogeneous sets can be ranked in ascending or descending order on one or more attributes of objects of the type comprising the set. Some examples of set definitions follow:

```
set(widget) ranked FIFO  fifo_set;
set(widget) ranked LIFO  lifo_set;;
set(*)  universal_set;
set(job) ranked(ascending due_date,
descending priority)   job_set;
```

An iteration construct is provided for looping through the members of a set:

```
pointer(widget) w;
for (w = each widget in fifo_set)
  {
  ...
  }
```

If the subject set is a universal set, `for each` selects only objects of the specified type, skipping over any objects which are of other types. Arbitrarily complex forms of iteration can be built from lower-level first, last, successor, and predecessor primitives:

```
w = first widget in widget_set;
w = last widget in widget_set;
w = successor(w) in widget_set;
w = predecessor(w) in widget_set;
```

Using these primitives, *any* form of set iteration can be achieved.

3 EXTENSIBILITY FEATURES

SLX was designed to be an extensible platform on which a wide variety of higher level simulation applications could be built. In this section we will briefly present an example of how the extensibility mechanisms can be used to build new features out of old ones.

Suppose we wish to build a simple telephone book. Furthermore, assume that each entry in the book contains only a first name, a last name, and a telephone number. An entry in the phone book can be described as an SLX object:

```
object  book_entry
  {
  fstring(20)       first_name,
                    last_name;

  int               phone_number;
  }
```

Suppose that we wish to retrieve entries from the phone book both by name and by number. We could construct two SLX sets, one ranked by name and one ranked by number:

```
set(book_entry)ranked(ascending
  last_name, ascending first_name)
    phone_book;

set(book_entry) ranked (ascending
  phone_number)
    reverse_phone_book;
```

The name-sorted phone book could be printed as follows:

```
pointer(book_entry) name;
for (name = each book_entry in
  phone_book)
    print( name -> last_name,
           name -> first_name,
           name -> phone_number)

"*****, *****: *******\n":
```

Of course, the printing of phone books is best left to the telephone company. For the rest of us, the most common use of the phone book is to look up numbers. For the police, a relatively common use is to look up the name associated with a number. The SLX kernel contains no "look up" primitive; however, one can be constructed easily from existing language features. The following approach can be used to look up a phone number, given first and last names. First, copy the first and last names into a dummy book_entry object. We call this object query_book_entry; Next, place the query_book_entry object into phone_book. Since phone_book is ranked by ascending last and first name, and entries with identical keys are inserted into a set in FIFO order, we know that query_book_entry will be placed into phone_book immediately following the entry we wish to look up. Thus, the desired entry will be the predecessor of query_book_entry in phone_book. If the name we are looking up is not in the phone book, either the predecessor of query_book_entry will be NULL, or it will contain the wrong names. These conditions are easily detected. Finally, the query_book_entry can be removed from phone_book.

While each of the steps outlined above is straightforward, we'd like to have a less cumbersome way of issuing queries. SLX provides a statement definition facility which allows us to construct a "retrieve" statement. The definition of our retrieve statement is shown in Figure 1. The first line of the definition is a prototype which specifies the components of the retrieve statement. Names preceded by a pound sign ("#") represent components that are supplied by the user for each use of the statement. Braces ("{}") are used to enclose a group of specifications. The notation ",..." indicates that the immediately preceding component can be repeated as many times as desired, with repetitions separated by commas. The "@" in front of the "from" keyword tells SLX to ignore the usual meaning of "from" and treat it as a keyword of the retrieve statement. ("from" is a reserved word in SLX.)

The definition section specifies the mapping of the retrieve statement into lower-level SLX statements. Within the definition section, the *expand* statement is used to specify the lower-level SLX code that is to carry out the retrieval operation. The expand statement specifies one or more lines of output which is injected into the SLX compiler's input stream. A list of expressions can be supplied to be edited into the generated lines of output. Within an output line, groups of adjacent "#" symbols are replaced by edited values.

With one very important exception, this approach is similar to the use of *macros* in many languages. In most languages, the statements which are available to specify the internal logic of a macro are either very limited and use a syntax different from the host language, or they comprise a comparatively weak subset of the host language. In SLX, the "macro language" is SLX itself. Only a handful of statements are excluded from use within an SLX statement definition. For example, simulation constructs such as *wait until* or *advance* have no meaning during compilation of a program. Apart from these obvious restrictions, most of the rest of the SLX language can be used. For example, it is even possible to read a file as part of the process of expanding a statement!

The example in Figure 1 makes use of a local integer variable, i, which is used to iterate through the list of comma-delimited "#key = #value" items. The iteration is terminated when an empty value of #key[i] is encountered. This is an SLX convention. If a list of N items is provided, #key[1...N] will have non-empty values, and #key[i] will be empty for i > N.

A sample invocation of the retrieve statement and its expansion is shown in Figure 2. This example illustrates the power of SLX's statement definition facility. The retrieve statements are easy to read. If you didn't know that the retrieve statement was built using the statement definition facility, you would probably think that it was a built-in SLX statement. This example illustrates the extensibility of SLX. It's very easy to reshape the language by adding new statements which are specific to the kinds of problems you are working with. Whether you need a capability which is not present in basic SLX, or whether you just want a more expressive way of specifying complex logic in a more compact form, the statement definition facility can be of great use.

```
statement retrieve #ptr = #otype ( { #key = #value },... ) @from #set ;
    definition
        {
        int     i;

        expand                          "{\n";

        for (i=1; #key[i] != ""; i++)
                expand(#otype, #key[i], #value[i])

                                        "query_#.# = #;\n";

        expand(#otype, #set)            "place &query_# into #;\n";

        expand (#ptr, #otype, #set)     "# = predecessor(&query_#) in #;\n";

        expand(#otype, #set)            "remove &query_# from #;\n";

        expand(#ptr, #ptr, #key[1], #value[1])
                                        "if (# != NULL)\n  if (# -> # != #";

        for (i=2; #key[i] != ""; i++)
                expand(#ptr, #key[i], #value[i])

                                        " or # -> # != (#)";

        expand(#ptr)                    ")\n     # = NULL;\n";
        expand                          "}\n";
        }
```

Figure 1: The Definition of a "Retrieve" Statement

```
retrieve e = book_entry(last_name=lname, first_name=fname) from phone_book

    {
    query_book_entry.last_name = lname;
    query_book_entry.first_name = fname;
    place &query_book_entry into phone_book;
    e = predecessor(&query_book_entry) in phone_book;
    remove &query_book_entry from phone_book;
    if (e != NULL)
      if (e -> last_name != lname or e -> first_name != (fname))
         e = NULL;
    }
```

Figure 2: A Sample Expansion of the "Retrieve" Statement

4 EXPERIENCE WITH SLX TESTERS

As of this writing SLX has been intensively used by a select group of testers around the world. Interacting with the testers has resulted in a number of improvements to SLX and has yielded some interesting insights into how SLX is (and will be) used. The projects tackled by SLX testers have all been fairly intense efforts. Included among these have been a half-dozen masters degree thesis projects and one very large scale, real-world transportation model.

Two modest surprises have come out of the testing. First, the extent to which users have made direct use of SLX kernel-level features has exceeded our expectations. Perhaps biased by our years of experience with GPSS/H, we had expected that most users would prefer to use higher-level features, only occasionally resorting to kernel-level features. Virtually all the testers have made heavy use of low-level scheduling primitives such as wait until and fork. A second modest surprise has been the extent to which the testers have exploited SLX's statement-definition capabilities. We had envisioned the statement definition capability as being of interest primarily to builders of higher-level simulation packages, but nearly everyone who has touched SLX has made use of its statement definition facility.

Taken together, these two surprises demonstrate that we have achieved our goal of extensibility. SLX is easily adapted to different purposes by different users. With the ability to incorporate new statements into the language, the extended language used by the developer of a package for manufacturing applications is quite different from the extended language used by someone tackling transportation problems. Yet the two are built on a common foundation, and the underlying software is hardened and tempered by exposure to widely differing usage patterns.

5 TEACHING SLX

The architecture of SLX has potentially profound implications for teaching simulation. The usual approach to teaching simulation is to "dive in" at an intermediate level by providing an easily understood collection of building blocks and exploring some well-motivated examples. Students of simulation who tackle real-world applications sooner or later reach a point at which they have to go back and build a foundation under their knowledge; i.e., they have to learn how things really work. Depending on exactly when the foundation-building process takes place, students may have already developed usage patterns which ignore some of a language's capabilities and misuse others. For example, self-taught users of GPSS/H will almost always favor an "active-object, passive-server" world-view, even though the language is quite capable of expressing an "active-server, passive object" world-view. For users of very high-level simulation packages, especially graphically based model-builders, the foundation-building may *never* take place. Whether this is good or bad is a matter of religion. Advocates of the very high-level approach think this is good, while their more conservative counterparts are appalled by the danger of doing too much with too little knowledge.

In SLX, the number of kernel constructs which directly support simulation is very small. Depending on what one counts as a simulation feature, the number ranges from roughly 8 to 12. Our experience with GPSS/H has proven that this is a small enough number of building blocks for beginners to readily absorb. For example, we have seen many times that so-called "9-block GPSS/H" is easily mastered and quite powerful.

However, even with 9-block GPSS/H, students quickly reach a point at which foundation-building is necessary. With SLX, a bottom-up approach is feasible. For example, consider modeling a barbershop, a traditional introductory one-line, single-server queuing model. In a beginner's model, the barbershop runs from 9:00-5:00, at which time it summarily shuts down, ignoring the customer (if any) who is in the barber chair at that time and ignoring customers (if any) in the queue. In a second model, more realistic shutdown conditions can be implemented. At 5:00 the door to the shop is closed, and the barber does not leave until the current customer and all customers in the queue at 5:00 have been served. In SLX, this condition is easily expressed as a compound "wait until" condition, e.g., "wait until (queue empty and server idle)." Thus, SLX's wait until feature is well-motivated and easily understood at a very early stage of model building. In SLX, wait until is the foundation of *all* forms of state-based events. Thus mastery of wait until yields enormous benefits.

SLX kernel-level simulation primitives are *exposed,* i.e., they can be used *directly.* In most simulation software, primitives are bound into impenetrable higher-level features. For example, in GPSS/H there are at least five building blocks which internally utilize the equivalent of wait until. Some of these blocks have many external variations. Thus, students of GPSS/H must master the external variations *and* learn how the underlying `wait until` mechanism works. In SLX, it's easier to learn the general mechanism first. Wait until is both an SLX primitive and a fundamental modeling concept. Thus, by teaching/learning `wait until`, we can kill two birds with one stone.

As of this writing, an introductory SLX textbook is under development. A preliminary (at least) edition of the book will be available by the time of the 1996 Winter Simulation Conference. The book will use the approach outlined above for simultaneously exposing fundamental simulation concepts and their implementation in SLX. Given the enormous popularity of graphically-based, fill-in-the-blanks modeling tools, our approach to teaching SLX entails a degree of risk. We're going back to teaching the basics at a time when much of the rest of world is moving in the opposite direction. Our approach is motivated by a year's experience with our SLX testers and over twenty years' experience in helping users learn to use our simulation software. Our experience is biased by years of exposure to people who still believe it's important to know what they're doing.

6 CONCLUSIONS

SLX is a well-conceived, layered simulation system. Users of the upper layers can ignore lower layers. However, if their requirements are not met at a given level, they can move down one or more levels, without exerting extraordinary effort and without losing protection against potentially disastrous errors. Developers, who are used to working down among the lower layers, have at their disposal powerful extensibility mechanisms for building higher layers for use by themselves or others. The efficacy of the approaches offered by SLX has been demonstrated through intensive use. The benefits for teaching simulation derived from being able to simultaneously expose language primitives and fundamental modeling concepts remain to be demonstrated as of this writing. However, we are excited by SLX's great potential.

REFERENCES

Henriksen, J.O. 1995. An Introduction to SLX. In *Proceedings of the 1995 Winter Simulation Conference*, ed. C. Alexopoulos. 502-509. Institute of Electrical and Electronics Engineers, Piscataway, New Jersey.

Henriksen, J.O., and R.C. Crain. 1996. *GPSS/H reference manual*, fourth edition. Annandale, VA: Wolverine Software Corporation.

AUTHOR BIOGRAPHY

JAMES O. HENRIKSEN is the president of Wolverine Software Corporation. He is a frequent contributor to the literature on simulation and has presented many papers at the Winter Simulation Conference. Mr. Henriksen served as the Business Chairman of the 1981 Winter Simulation Conference and as the General Chairman of the 1986 Winter Simulation Conference. He has also served on the Board of Directors of the conference as the ACM/SIGSIM representative.

Proceedings of the 1996 Winter Simulation Conference
ed. J. M. Charnes, D. J. Morrice, D. T. Brunner, and J. J. Swain

A TUTORIAL ON PROSIM™: A KNOWLEDGE-BASED SIMULATION MODEL DESIGN TOOL

Madhavi Lingineni
Bruce Caraway
Perakath C. Benjamin

Knowledge Based Systems, Inc.
1500 University Drive East
College Station, Texas 77840, U.S.A.

Richard J. Mayer

Department of Industrial Engineering
Texas A&M University
College Station, Texas 77843, U.S.A.

ABSTRACT

This paper introduces PROSIM, a knowledge-based simulation design tool that automatically generates simulation models from process models. Domain experts who are familiar with the various aspects of the system may be interviewed and process models can be developed based on the evidence collected using PROSIM. The built-in simulation design support environment of PROSIM enables building simulation models automatically from process models. A purchase order processing system example is described to demonstrate the utility of PROSIM.

1 INTRODUCTION

The increasing complexity of organized enterprises has enhanced the utility of simulation as a managerial problem-solving and analysis tool. While specific areas in simulation modeling technology have shown rapid advances in the recent past, there are a few areas which would benefit from further research. Specifically, current simulation practice 1) affords little support for the initial analysis and model design tasks which are largely qualitative in nature, 2) involves the unproductive use of both the domain expert's and the simulation analyst's time in the modeling process, and 3) suffers from lack of widespread acceptance by decision makers due to the relatively long lead times and sophisticated skills needed for the effective use of simulation modeling techniques. It is the purpose of this tutorial to introduce PROSIM, a knowledge-based simulation model design tool that automatically generates WITNESS™ simulation models from IDEF3-based process models, as a possible solution for the aforementioned shortcomings.

2 PROCESS MODEL GENERATION USING PROSIM

PROSIM enables a modeler to capture processes or situations as an ordered sequence of events or activities. It is designed to capture process descriptions of the precedence and causality relations between events/situations in a form that is natural to domain experts in an environment. One of the primary goals of the tool is to provide mechanisms for expressing a domain expert's knowledge about how a particular process works.

PROSIM uses "scenarios" as the basic organizing structure for establishing the focus and boundary conditions for the process description. This feature is motivated by the tendency of humans to describe what they know in terms of an ordered sequence of observed activities within the context of a given situation.

The basic syntactic unit of PROSIM graphical process descriptions within the context of a scenario, is the Unit Of Behavior (UOB) represented by a rectangular box. Each UOB represents a specific view of the world in terms of a perceived state of affairs or state of change relative to the given scenario. UOBs are connected to one another via junctions and links. Junctions provide the semantic facilities for expressing synchronous and asynchronous behavior among a network of UOBs. Links represent temporal precedence, object flow, and relations between UOBs. Each UOB can have associated with it both descriptions in terms of other UOBs (a decomposition of a parent UOB) and a set of participating objects and their relations (an elaboration).

Let us consider a situation of an analyst interviewing a purchase department manager to capture main activities involved in generating purchase orders at the company. The purchase department manager is mainly responsible for the processing of purchase request forms, finding existing suppliers for the part, inviting bids from potential suppliers, receiving bids, evaluating bids, placing purchase orders, and acquiring the material. First, the initiating department (typically design or process planning) formally requests the needed material. If the requested material has existing suppliers, a purchase order can be placed immediately. On the other hand, if the requested material is new, or the existing supplier no longer produces the part, then the purchase department has to invite bids from potential suppliers, receive the bids, evaluate the bids, and place a purchase order to the selected supplier. The process flow diagram that depicts the sequence of activities involved in the purchase order generation process can be captured using PROSIM and is shown in Figure 1.

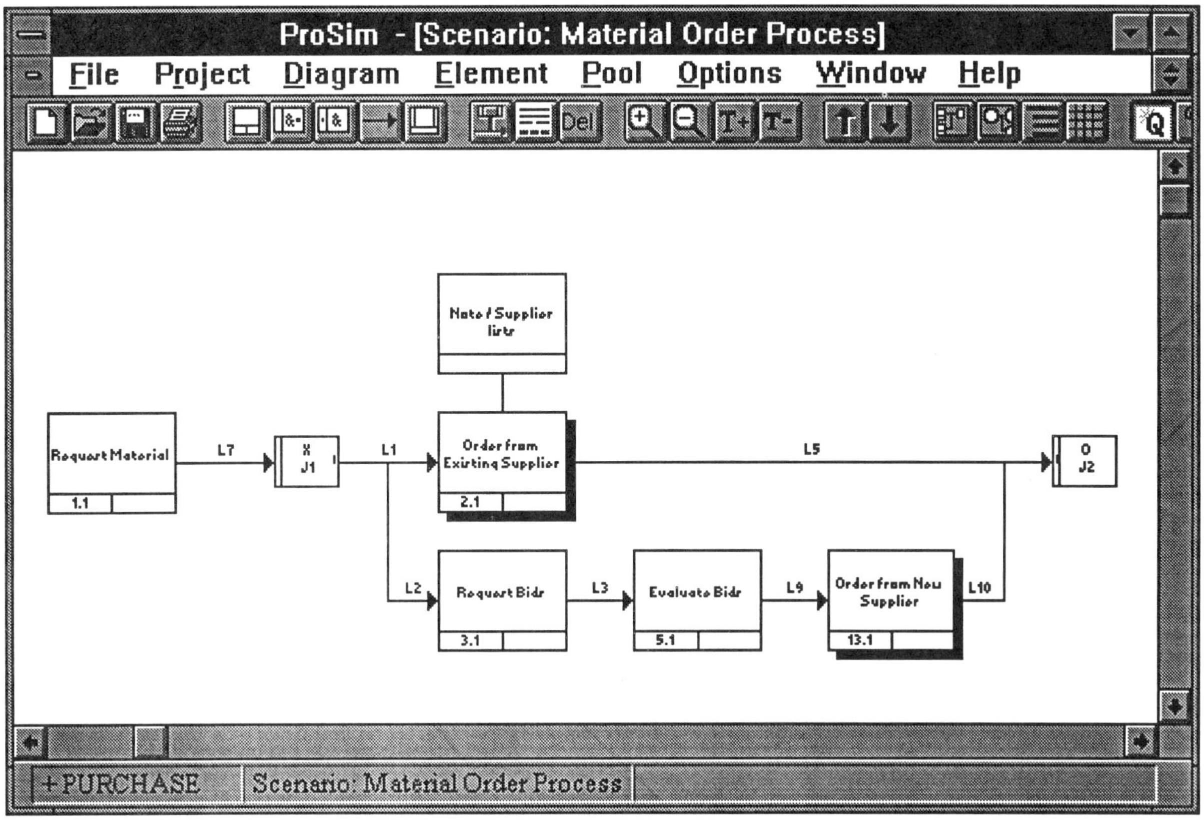

Figure 1: A Process Flow Diagram of the Purchase Order Process

3 GENERATING SIMULATION MODELS USING PROSIM

The intelligent support provided by PROSIM reduces the expertise required to design effective simulation models and perform simulation experiments. It provides an environment in which users can develop designs for discrete event simulation models from process descriptions and supports the conceptual modeling activities at various stages of the model design process. PROSIM can be used to support the analysis and design of manufacturing systems, business systems, logistics systems, command, control, communication, and intelligence systems, concurrent engineering systems, service systems, and information systems. It is a critical tool for realizing the monetary benefits of business process re-engineering and activity based costing.

PROSIM has a built-in simulation model design support environment. In other words, it has an expert system programmed to provide expertise at a level comparable to that of a simulation modeling expert. Process flow diagram constructs, UOBs, objects, and junctions, are required to be detailed within PROSIM in order to generate simulation model. Consider the *request for bids* UOB in the example process flow diagram in Figure 1. The real world objects that participate in the process of requesting bids are "Technical Package," "Purchase Department Manager," "Potential Supplier List," "Bid Package," and "Building 1." A "technical package" contains the technical data of the part to be ordered. The technical data of the part may include engineering specifications, tolerances, material to be used, special tools and jigs required to manufacture the part. A "Purchase Department Manager" requests bids by sending a technical package along with the bid package to potential suppliers. The "Potential Supplier List" consists of details of all the potential suppliers such as "supplier code," "supplier name," "supplier address," "supplier manufacturing capabilities." The "Bid Package" consists of information about the requirements of the bid responses, financial details, and other contractual details. This activity is performed in the location, "Building 1."

In PROSIM, a dedicated resource is categorized as either a *location* or a *queue*. The location label is provided whenever a non-waiting activity is represented by the UOB. If the UOB is used to represent a waiting activity, the dedicated resource is associated as a queue. Flow objects that participate in UOBs are called *entities*.

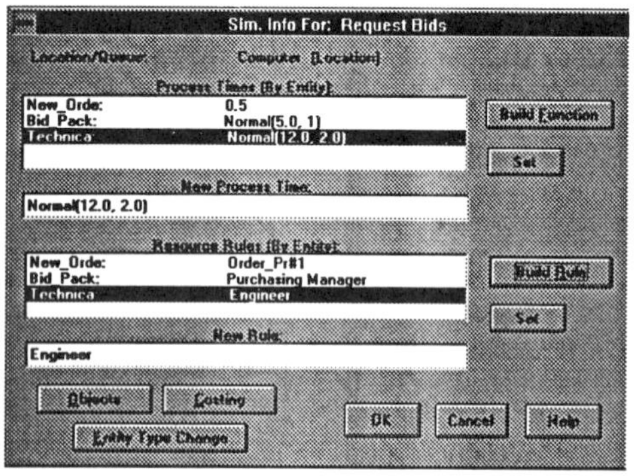

Figure 2: Dialog Box for Detailing a Unit of Behavior

3.1 UOB Detailing

The dialog box shown in Figure 2 is used to record the process times for the *Request Bids* UOB. Process times are recorded for the entities that participate in the process. The names of the entity objects that participate in this UOB are New Material Request Form (labeled "New Order"), "Bid Package," and "Technical Package." The process time for "New Order" is a constant of 0.5 days. The process time for "Bid Package" is a normal random variable with a mean of 5 days and a standard deviation of 1 day. The process time for "Technical Package" is a normal random variable with a mean of 12 days and a standard deviation of 2 days. The Objects such as "computers," "personnel manager" are categorized as *resources* in PROSIM. Resource objects, as well as other objects, can be shared between different UOBs in a PROSIM model.

3.2 Entity Detailing

Entities flow between UOBs in a scenario. The inter-arrival time distribution is an item of information required to simulate the behavior of entities in a system. Figure 3 displays how the details of "New Material Request Form" are recorded in a PROSIM model. Observe that the inter-arrival time for this entity is a negative exponential random variable with a mean of 10.0 time units. The arrival point is the first process visited by the entity as it flows through the scenario.

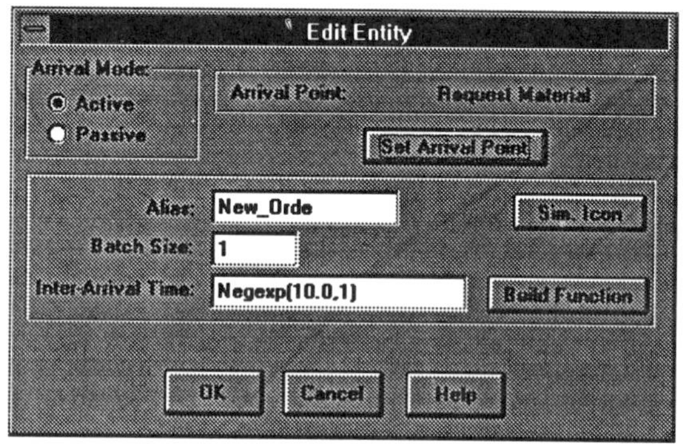

Figure 3: Entity Detailing

3.3 Decision Logic Specification

Junctions in PROSIM provide a mechanism to specify the logic of process branching. Different junction types are supported in PROSIM to aid in capturing the semantics of branching in real-world processes. Junctions support the description of 1) a process that splits into two or more process paths, or 2) two or more process paths converge into a single process. Junctions are classified in three different ways. First, they are classified according to the logical semantics involved: *AND* (&), *OR* (O), and *exclusive OR* (X). Junctions are further classified as either *fan-in* or *fan-out*, based on whether they represent a *convergence* or a *divergence* in the logic of the process description. They are also classified based on the coordination of the timing of the associated UOBs as either *synchronous* or *asynchronous* (KBSI 1992a).

An exclusive OR (X) junction is used to model decision logic. In our purchase order processing example, an exclusive OR junction is used to represent the logic that activity "request material" is followed exactly by one of the following two activities: 1) "order from an existing supplier", or 2) "request bids from new suppliers" with probabilities of 0.8 and 0.2, respectively (Figure 1).

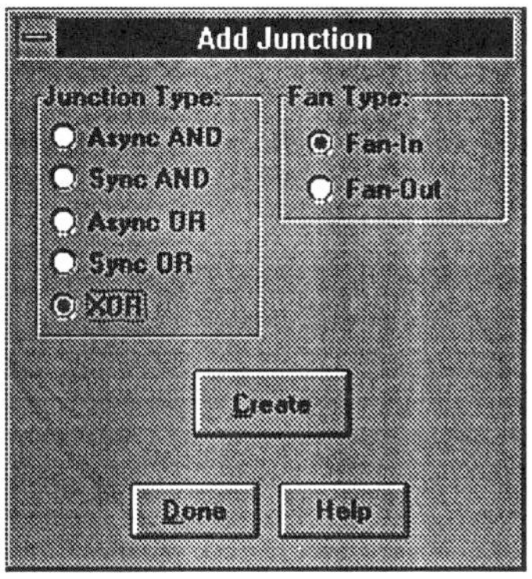

Figure 4: Adding a Junction

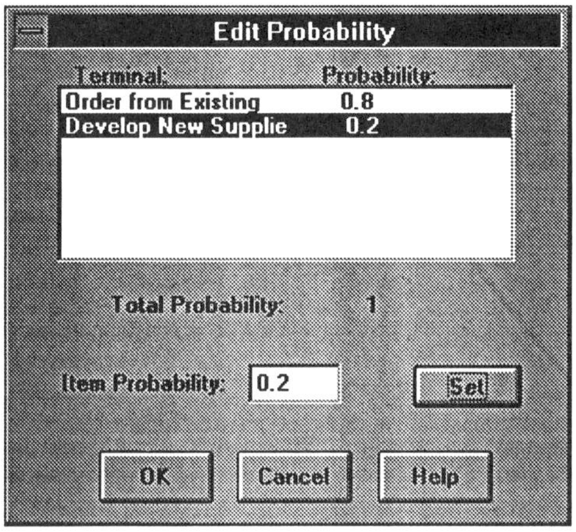

Figure 5: Junction Detailing

3.4 Performance Metric Specification

The goal in designing a simulation model is to generate data to answer a set of questions. Examples of questions include, "What is the total process time?", "What is the resource utilization?". PROSIM provides support for designing instrumentation to help answer these questions. In other words, PROSIM helps you select appropriate performance metrics that will automate the generation of output data to answer the questions. For example, the performance metric used to answer the first question is *Process Time* for the UOBs "preparation of bid package," "preparation of technical package," "filling in new material request form," and "filling in the repeat material request form." The PROSIM dialog boxes used to specify these performance metrics are shown in Figures 6 and 7.

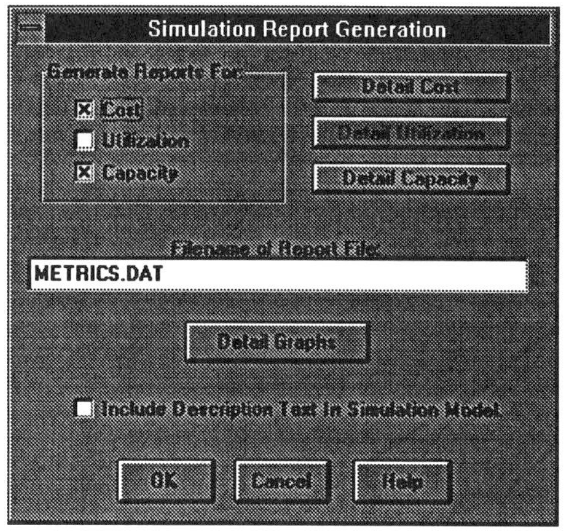

Figure 6: Simulation Performance Metrics Specification

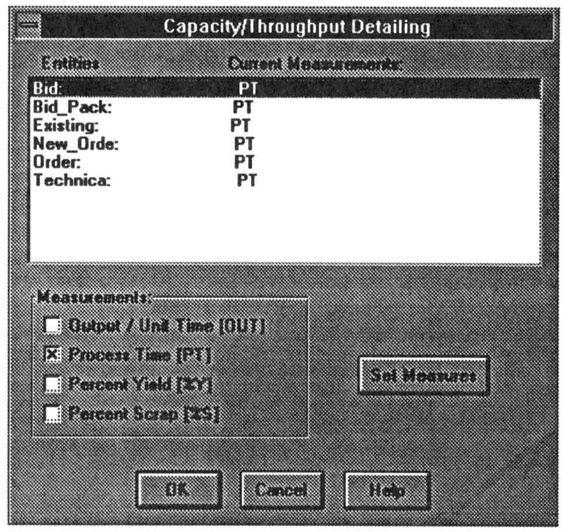

Figure 7: Metrics to Compute Process Time

3.5 Model Verification

In the context of PROSIM, model verification is defined as the process of ensuring that the minimum amount of information required to generate a WITNESS simulation model exists in the process model. Model verification is performed using the *Model Verification* option in the Diagram menu in PROSIM. A more detailed description is given in the *PROSIM User's Manual and Reference Guide* (1995).

3.6 Simulation Model Generation

PROSIM simulation model design specifications are now transformed to an equivalent representation in the WITNESS Input Command Language. This is done by using the *Build Simulation Model* option in the PROSIM

File menu. After the model has been generated, the next step is to open the application and load the model using the *Read Commands* option in the Model menu (see AT&T ISTEL 1992). After the WITNESS model has been loaded, the user can run the model and generate the desired performance metrics. WITNESS provides state-of-the-art capabilities for simulation-based analysis and experimentation.

4 POTENTIAL BENEFITS OF USING PROSIM

PROSIM has the following potential benefits: 1) it enables novice simulation modelers to develop and use complex simulation models, 2) it encourages and promotes the use of simulation as a decision support tool among a wide group of users from a variety of application domains, 3) it substantially reduces time and effort needed to develop simulation models, and 4) it facilitates the capture and retention of simulation model design rationale.

5 SUMMARY

This paper described salient features of PROSIM, a simulation model design tool. An example of purchase order processing system is provided to demonstrate the procedure of building process models from domain experts' descriptions and then generating a simulation model from this process model.

REFERENCES

AT&T ISTEL Limited. 1992. *WITNESS user manual*, Release 4.0, Cleveland, OH.

Knowledge Based Systems, Inc. (KBSI). 1992a. IDEF3 method report. Prepared for the U.S. Air Force Human Resources Laboratory, Contract No. F33615-C-90-0012.

Knowledge Based Systems, Inc. (KBSI). 1992b. Issues in knowledge-based assistance for simulation model design from IDEF3 descriptions. Technical Report for NSF SBIR Phase II Grant No. III-9123380.

Knowledge Based Systems, Inc. (KBSI). 1995. *PROSIM user's manual and reference guide*, Version 2.1, College Station, TX.

AUTHOR BIOGRAPHIES

MADHAVI L. LINGINENI received her Master of Science degree in Industrial Engineering from Texas A&M University in August 1993. She is currently working as a Systems Analyst at Knowledge Based Systems, Inc. Her research interests include business process reengineering, database design, simulation, system modeling, and reverse engineering.

BRUCE E. CARAWAY received his Bachelor of Science and Master of Science degrees in Industrial Engineering from Texas A&M University in 1990 and 1994, respectively. He joined Knowledge Based Systems, Inc. in January of 1993 as a Systems Analyst. He is currently working on a project at Tinker Air Force Base which seeks to baseline and identify process improvements on the E-3 AWACS Programmed Depot Maintenance (PDM) line using the IDEF methods. He is also working on the development of a simulation and shop floor control system for the E-3 PDM line.

PERAKATH C. BENJAMIN received his Master of Science degree in Industrial Engineering from the National Institute for Training in Industrial Engineering in 1983. He received his Ph.D. in Industrial Engineering from Texas A&M University in May 1991. As Vice President (Innovation and Engineering) at Knowledge Based Systems, Inc., Dr. Benjamin manages and directs research KBSI's research and development projects.

RICHARD J. MAYER received a Master of Science in Industrial Engineering from Purdue University in 1977. In 1988, he received a Ph.D. in Industrial Engineering from Texas A&M University. From 1984 to 1989, he was Project Manager and Principal Investigator on thirty-nine funded research efforts in the Knowledge Based Systems Laboratory. He founded Knowledge Based Systems, Inc. in 1988 and has received funding for applications in engineering design assistance, systems analysis and concurrent engineering methods and tools. Currently, he is an Associate Professor of Industrial Engineering at Texas A&M University.

Proceedings of the 1996 Winter Simulation Conference
ed. J. M. Charnes, D. J. Morrice, D. T. Brunner, and J. J. Swain

AWESIM: THE INTEGRATED SIMULATION SYSTEM

A. Alan B. Pritsker
Jean J. O'Reilly

Pritsker Corporation
8910 Purdue Road, Suite 600
Indianapolis, Indiana 46268, U.S.A.

ABSTRACT

AweSim is a totally new general-purpose simulation system, released in June of 1996. AweSim takes advantage of the latest in Windows® technology to integrate programs and provide componentware. AweSim includes the Visual SLAM simulation language to build network, discrete event, and continuous models. Network models require no programming yet allow user-coded inserts in Visual Basic or C. Discrete event and continuous models can be created using the object-oriented technology of Visual Basic, C or Visual C++ and can be combined with network models. This tutorial will demonstrate the process of using AweSim's componentware and provide examples of user interfaces that allow integration with other Windows applications both directly and through the AweSim database.

1 INTRODUCTION

AweSim is a general purpose simulation tool used to model systems as diverse as manufacturing, transportation, communication/information systems, military operations, health care, and banking. Network (process-oriented) models are built from a set of powerful, flexible building blocks which accomplish functions typically required to simulate any system. Should user-written extensions to a network model be necessary, they are developed in C or Visual Basic and easily linked to the basic system. Continuous variables may also be used whenever this is the most convenient way to represent system elements. This ability to construct models of unlimited size using the approach most suited to the problem gives AweSim utmost flexibility in model input.

Flexibility in model output is achieved through AweSim's relational database, which makes it easy to create custom reports using common tools such as Dbase, Access, and Excel.

2 MODEL INPUT

An AweSim model can consist of several components: networks, run controls, user-written inserts, data files, animations, and notes for documentation. These components are organized and accessed through the AweSim executive window as shown in Figure 1.

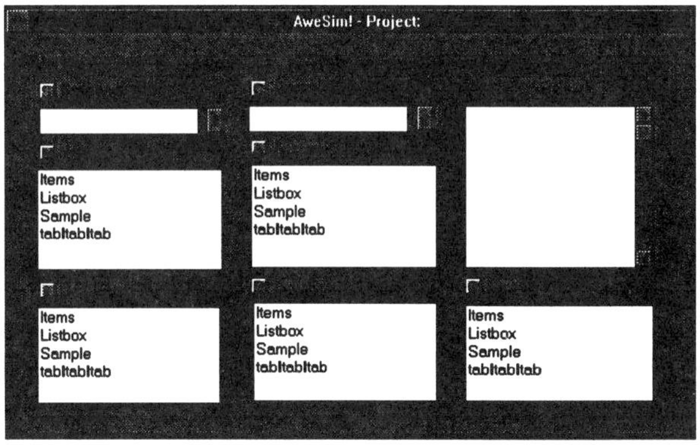

Figure 1: AweSim Executive Window

AweSim incorporates the Visual SLAM modeling methodology, which is a new simulator based on the powerful and proven modeling concepts used in the SLAM II modeling language. The basic component of a Visual SLAM model is a network, or flow diagram, which graphically portrays the flow of entities (people, parts or information, for example) through the system. A Visual SLAM network is made up of "nodes" at which processing is performed, connected by "activities" which define the routing of entities and the time required to perform operations. Visual SLAM nodes provide for such functions as entering or exiting the system, seizing or freeing a resource, changing variable values, collecting statistics, and starting or stopping entity flow based on status conditions.

The network segment shown in Figure 2, for example, shows a CREATE node, which generates entities according to an exponentially distributed interarrival time, connected by an ACTIVITY (represented by a direct branch) to an ASSIGN node.

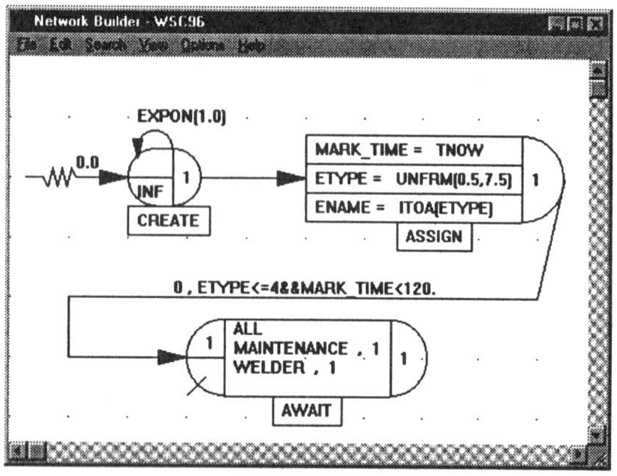

Figure 2: Graphical Network Builder

At the ASSIGN node, the newly created entity's attributes (characteristics) are assigned. In this example, the attribute MARK_TIME holds the entity's time of arrival, the ETYPE (entity type) attribute is randomly assigned to be an integer between 0 and 7, and an attribute called ENAME is the character representation of entity type. The ASSIGN node is followed by an AWAIT node at which two resources, MAINTENANCE and WELDER, are allocated to the newly arrived entity. The activity taking the entity from the ASSIGN node to the AWAIT node is taken only if entity type is less than or equal to 4 and arrival time is less than 120. This example is meant to suggest the flexibility provided by a rich set of random variables and intrinsic functions and by the ability to use complex expressions throughout a network model.

A network is built interactively in AweSim by selecting symbols from a graphical palette and locating them with the mouse. The symbol's parameters are specified by filling out a form, as illustrated by the ACTIVITY form shown in Figure 3. On-line error checking is performed upon completion of the form so that input errors can be corrected immediately. AweSim also facilitates model building by providing context-sensitive help and search capabilities (Figure 3).

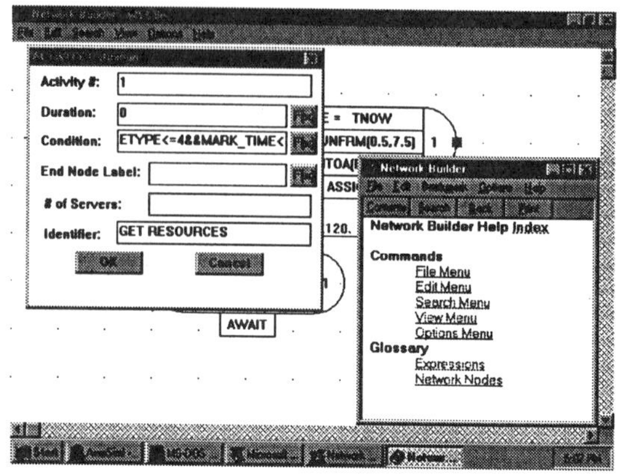

Figure 3: Network Builder dialog box and Help window

In addition to the Network Builder, AweSim includes a forms-based builder for developing run controls (simulation run length, output options, when to clear statistics, queue ranking rules, etc.) and text builders for User Data files and Notes for documentation. The User Insert builder is either a text builder, if AweSim is installed for use with user inserts written in C, or the Microsoft Visual Basic development environment.

3 MODEL OUTPUT

AweSim provides the ability to compare output from various scenarios both graphically and textually. A report "browser" allows alternative textual outputs to be compare side by side. Graphically, output may be viewed within AweSim in the form of bar charts, histograms, pie charts, and plots. Bar charts can be used to display the value of a statistic across scenarios. It is possible to view multiple windows of graphical output at the same time, as shown in Figure 4. Graphical and textual information from the AweSim database can be exported to other Windows packages such as EXCEL or Word for additional analysis and for documentation.

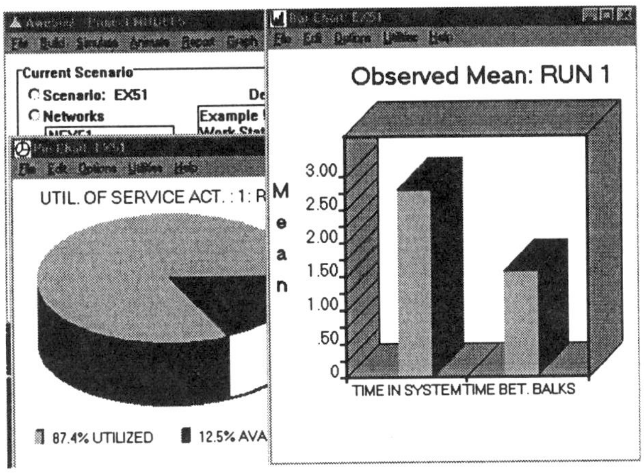

Figure 4: Multiple report windows

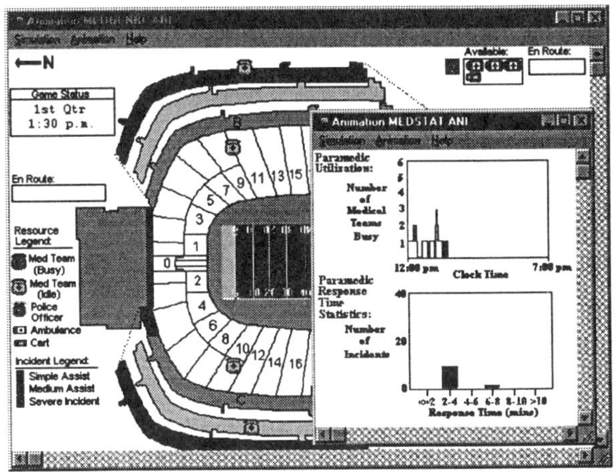

Figure 5: Animation with multiple windows

4 ANIMATION

With the AweSim animator, one may develop and display multiple animations for a single scenario. For example, the modeler can create one animation of a system at an aggregate level and another at a department level, side by side in separate windows. The two views may then be displayed by associating the animations with the current scenario and running the simulation. Animation constructs, called actions, correspond to elements in an AweSim network model. For example, the ACTIVITY action shows movement of a symbol. It requires that the modeler define a symbol, a graphical path location where movement of the symbol will be shown, and the number of an activity in the AweSim model to which to tie the movement. Other actions are provided to display resource status, items in a queue, variable values, and other system status conditions. The symbols manipulated by the animator are of two types: graphical items one wants to display or move, and the background on which they will appear. These symbols are stored in standard Windows bitmap format, allowing them to be exchanged between programs using the Windows clipboard.

5 INTERACTIVE EXECUTION

The AweSim Interactive Execution Environment (IEE) provides an interface with an executing simulation. The modeler may examine, modify, save, or load the current system status using the IEE. The IEE supports the debugging of a model under construction and verification of a completed model. The analyst may use the IEE to develop and analyze alternative control strategies for a system or to demonstrate the operation of the model to non-modelers. The IEE control panel (Figure 6) lets you advance through the model step-by-step or by setting breakpoints. At any point you may examine variable values, statistics, or queue entries or save the system status in order to reload and experiment with alternative scenarios.

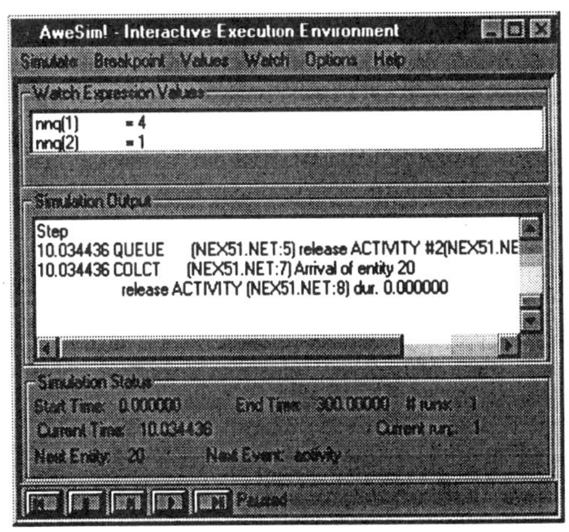

Figure 6: Interactive Execution Environment

6 INTEGRATION OF AWESIM! WITH OTHER SOFTWARE

AweSim was designed to integrate easily with other Windows applications. AweSim is built on a relational database which is accessible with standard tools such as Dbase, Access, FoxPro and Excel. Input data is easily moved from an Excel worksheet to the AweSim input tables. Output data is stored in AweSim output tables, available for creating custom reports using a favorite tool. In addition to standard output data, raw data from the simulation can be stored in standard database or Excel format for analysis, manipulation, or presentation. Data used to create AweSim output graphics can also be exported "on the fly" to an output file for use in any tool accepting comma-delimited input.

An AweSim animation can use graphics created from other programs. As mentioned in the discussion of animations, the graphical elements manipulated by the animator can be created using CAD, drawing or paint programs and loaded into AweSim by using the Windows bitmap format. The output charts and plots created by AweSim can be exported via the clipboard to other applications. For example, a pie chart created by AweSim may be copied to the clipboard and pasted into a word processing document describing the results of the model.

7 SUMMARY

AweSim is a brand new simulation support system which replaced the popular SLAMSYSTEM software in 1996. Written over the last two years, it takes advantage of the latest Windows technology in order to provide a simulation support system able to interface with a variety of familiar tools. It incorporates the Visual SLAM modeling methodology, a modern replacement for the proven SLAM II modeling language. AweSim is distributed by Pritsker Corporation, which offers regularly scheduled training classes as well as applications support.

AUTHOR BIOGRAPHIES

A. ALAN B. PRITSKER is Chairman of the Board of Pritsker corporation. He obtained a Ph.D. from the Ohio State University in 1961. He is a member of the National Academy of Engineering. Dr. Pritsker served twice as a member of the Board of Directors of the WSC and as Board Chairman in 1984 and 1985.

JEAN O'REILLY is a member of the technical staff at Pritsker Corporation. She holds an M.S. in Applied Mathematics from Purdue University. Since joining Pritsker Corporation in 1978, Ms. O'Reilly has been involved in software development and in a variety of consulting projects. She is currently responsible for technical support, training and quality assurance for the SLAM product family.

Proceedings of the 1996 Winter Simulation Conference
ed. J. M. Charnes, D. J. Morrice, D. T. Brunner, and J. J. Swain

MAKING BETTER MANUFACTURING DECISIONS WITH AIM

Julie N. Ehrlich
William R. Lilegdon

Pritsker Corporation
8910 Purdue Road, Suite 500
Indianapolis, Indiana 46268, U.S.A.

ABSTRACT

FACTOR/AIM (AIM) is a simulation system designed specifically for use in manufacturing decision support. AIM has been successfully applied to engineering design, scheduling, and planning problems within numerous manufacturing organizations.

Unlike language-based simulation products which require you to learn specific syntax and then abstract your system to fit this syntax, AIM uses the language of manufacturing. Example AIM components include machines, operators, materials, parts, jobsteps, process plans (routings), and conveyors. In addition, a comprehensive set of pre-defined manufacturing rules is available to you. Using AIM you can quickly and accurately build a model of any manufacturing process on your PC. Spending less time on modeling means more time to use the model to help you make decisions to improve your manufacturing operations.

This paper provides an introduction to AIM including AIM modeling constructs, the use of AIM for capacity engineering, planning and scheduling, and costing with AIM.

1 USING AIM TO MAKE MANUFACTURING DECISIONS

AIM is designed to help you make decisions regarding your manufacturing organization's productive capacity. When you build an AIM model, you develop a more thorough understanding of how your system operates and its capacity. You can use the model to investigate a variety of issues, for example to determine the impact of a proposed change, without affecting production. This enhances your ability to manage the system, control its capacity, and make better decisions regarding its operation. Which in turn improves profitably and your ability to predictably deliver quality product to your customers. These issues of predictability, profitability, and quality face every manufacturing organization today. Figure 1 shows the functional breakdown of the capacity management decision areas.

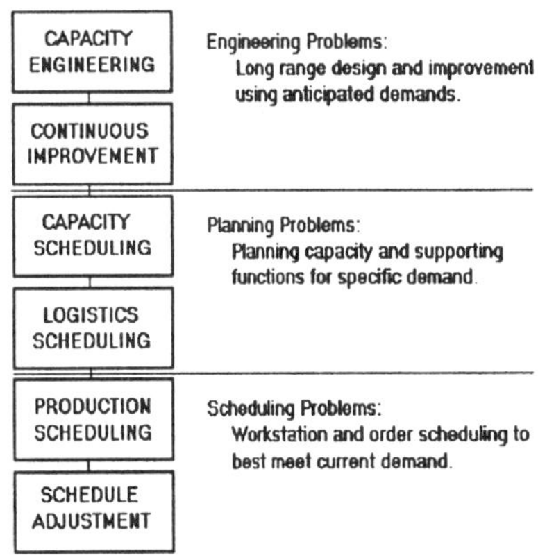

Figure 1. Capacity Management Decision Areas

Engineering problems focus on long-term questions regarding system design and continuous improvement. Planning problems address capacity issues including evaluating the impact of changing product mix or demand. Scheduling problems seek solutions to daily issues including on-time order completion, priority changes, and unplanned changes in resource availability.

Most products focus on one capacity management decision area. With AIM you build a single model and use it to support engineering, planning and scheduling decisions. For a product to support manufacturing design, planning, and scheduling, it must:

- make it easy to develop an accurate model of the manufacturing process,
- support loading complete data on the manufacturing system including the current system status and new orders to be processed,
- produce accurate, detailed equipment, material, and personnel schedules, as well as statistical performance information, and

- provide complete results in a timely manner -- when they are needed to make a decision.

AIM addresses all of these requirements. In addition, AIM results are presented in a manner which is easily understood by shop floor personnel – who most likely will not be familiar with traditional simulation analysis terminology.

1.1 Capacity Engineering Decisions

One key use of AIM for capacity engineering is in designing a new manufacturing facility. Most manufacturers would never consider building a new product without first designing a prototype, and then testing and refining that prototype before production begins. The same is true for a new manufacturing facility. Building the facility on the computer using AIM, before you buy equipment and begin construction, can save millions of dollars. By buying only the equipment you need, and ensuring that your facility design can produce at the anticipated demand rate, you minimize risk and capital expenditures.

Once a facility is in operation, AIM can be used to evaluate system improvements. You can analyze the impact of facility changes like: adding a new piece of equipment, changing the material handling system layout, reducing work-in-process buffer, etc. You can identify potential problems and correct them using your AIM model, instead of perturbing the real system with sub-optimal approaches. This translates into real monetary savings, improved employee relations, and increased customer satisfaction.

1.2 Planning Decisions

You can use an AIM model to address planning issues. Suppose you get a call from a new customer who wants you to make a new – though similar product. Wouldn't it be great to know the effect of this request before agreeing to do the job and setting a delivery date? AIM can help you determine the effect of adding a new product or increasing volume of existing product.

Your AIM model can also help you in logistics planning. In order to maintain a manageable schedule for the manufacturing floor, materials need to be ordered and procured at the proper time. You can use your AIM model to test material procurement policies such as reorder points, delivery procedures, etc., in order to manage your inventory better.

1.3 Scheduling Decisions

AIM is used to schedule operations for smaller scale production facilities, for individual departments within a larger facility, or for facilities requiring a PC based scheduling solution. Pritsker's FACTOR/PM and OrderLinX products were designed to fill the niche for large scale scheduling applications.

1.3.1 Data Integration for Scheduling

A key factor in the success of any scheduling system is the ability to integrate the scheduling software with existing production data systems. An accurate schedule depends upon precise production system status at the beginning of the schedule window, and complete data on new orders. Although it is possible to enter this required data manually, it is usually not practical to do so. Using AIM you can import information from other sources with speed and ease.

The AIM model is stored in an open, relational database. In-process status and new order data are loaded into the AIM database using custom data transfer procedures. The data connections usually associated with an AIM scheduling application are shown below in Figure 2.

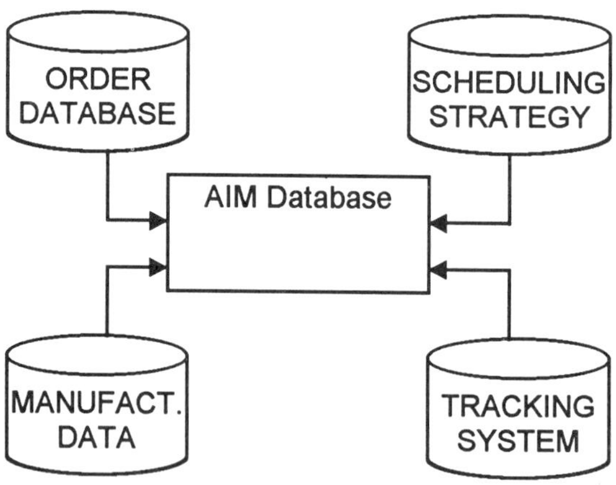

Figure 2. AIM Data Connections

1.3.2 The AIM Scheduling Process

The first step in the AIM scheduling process is to transfer information on system status into the AIM database. Then an order book containing all new orders to be processed during the scheduling interval, (e.g., shift, day, week), is imported to drive the production schedule. The scheduler executes the simulation and reviews a summary of the order, resource, and schedule performance. These reports allow the scheduler to detect potential scheduling problems and adjust the parameters of the AIM model. The new alternative is then executed and performance of the two alternatives is compared. This process is repeated until a feasible, desired schedule is achieved. Detailed schedules are generated for the components of interest (e.g., equipment, personnel, materials), and distributed to the proper personnel such as operators or automated cell controllers.

Schedules can also be viewed in an interactive graphical GANTT chart form (see Figure 3), using the AIM GANTTs Module.

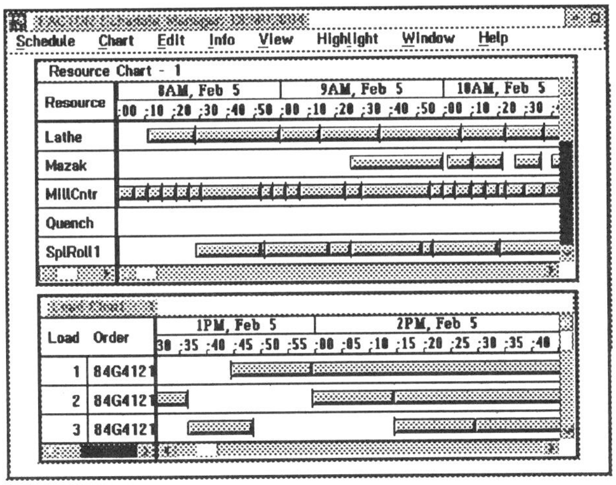

Figure 3. AIM GANTTs User Interface

The GANTTs Module is a graphical schedule display tool which provides a convenient mechanism to review an AIM schedule. A separate GANTT chart display for orders, jobs. and resources allows the scheduler to quickly review critical information. In addition, an extensive network of detailed information on the schedule and the AIM model is provided.

2 AIM ORGANIZATION

AIM is organized into four modules as shown in Figure 4 below. The AIM Executive controls data collection for a simulation and provides database and alternative management capabilities. It also invokes all the other AIM modules.

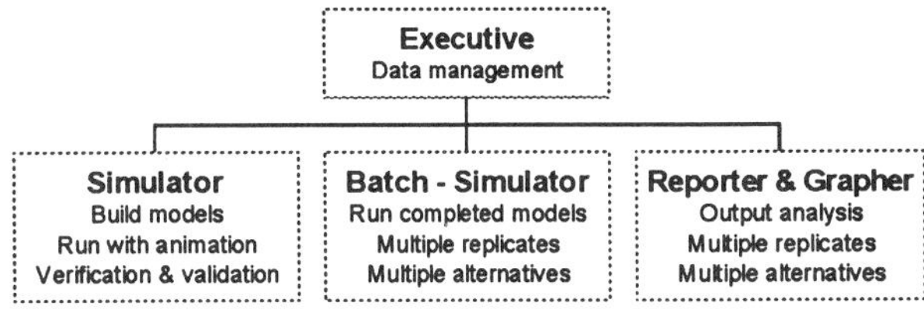

Figure 4. AIM Organization

The Simulator module is where you build your AIM model and run it with animation. It also contains summary report and graph capabilities to support model verification and validation. The Batch Simulator supports alternative evaluation through multiple replicates and multiple alternative runs. The Reporter displays textual reports on model performance. The Grapher displays plots, pie charts, bar charts, and histograms, summarizing model performance. Both the Reporter and Grapher support alternative and replicate comparison.

3 BUILDING AND RUNNING AN AIM MODEL

You create a model by pointing and clicking to locate graphical components (e.g. machines, conveyors) on the AIM facility window. Figure 5 shows an example model in the AIM facility window. This model contains machines, operators, and work-in-process buffers.

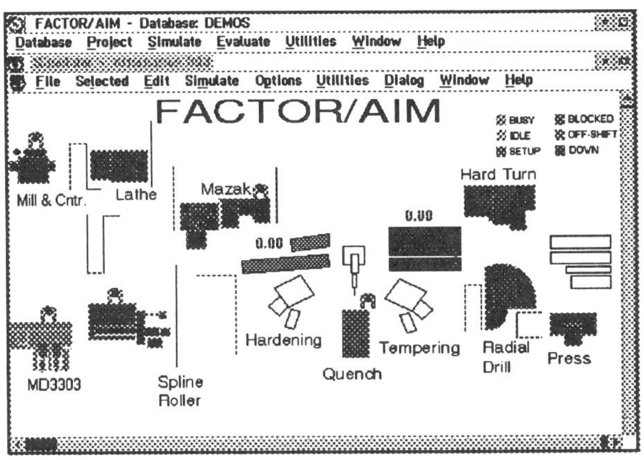

Figure 5. Sample FACTOR/AIM Facility Window

Once the graphical layout is complete, you add your parts and define their process routings. You can further refine the model as needed, using the information dialog boxes provided for each AIM component. For example, you can enable overtime for your operators by simply clicking a check box in a resource dialog box.

Another way to create a model in AIM is to load part or all of your model directly into the AIM database tables. Loading can be done using the database manager or using customized data import utilities. If you have existing information in electronic form, this can be a time-saving option.

Animation of your AIM model is automatic. As you create the layout of your system on the facility window, you are providing the location information needed to animate the model. If you load your resource and material handling information directly into AIM, you will need to provide location information for your layout.

Building and running a model in the AIM Simulator module is a very interactive process. You can build part of the model, run it, make changes, and run again. In addition, the Simulator has extensive support for model verification and validation including:

- automatic animation which allows you to watch your model as it simulates
- dynamic graphs of system conditions (e.g. resource queue length) which you can view while the simulation is running,
- reports and graphs summarizing system performance which you can view any time you pause the simulation,
- a network of status information boxes providing detained information on the state of each model component, and
- an events viewer which allows you to view all scheduled events and step through them while your model is running.

You can also run your model in batch mode, without animation, using the Batch Simulator module. You can request runs of multiple replicates and multiple alternatives for analysis purposes.

You have complete control over the type of data collected during an AIM simulation run. Summary and raw data outputs exist for every AIM component. All simulation output is stored in the AIM database. An extensive set of standard output reports and graphs are provided to analyze the system's performance. Figure 6 shows an example graph.

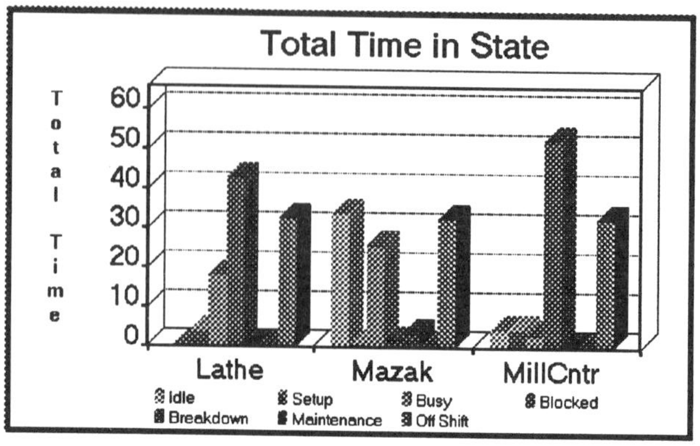

Figure 6. FACTOR/AIM Output Graphics

AIM reports and graphs make it easy to compare different potential configurations of your system – that is different AIM alternatives. Stochastic models can be built in AIM representing the random nature of the manufacturing process. AIM output reports and graphs are designed to reflect the stochastic nature of the analysis. AIM can also create schedule reports and communicate schedules graphically using a GANTT chart, as shown previously in Figure 3.

4 AIM MODELING COMPONENTS

AIM components correspond closely with the actual components of your manufacturing system. This reduces the amount of abstraction that is required to build models, and the amount of time it takes to build a model. Sample AIM modeling components include:

Machines	Parts
WIPs	Orders
Operators	Process Plans
Fixtures	Material Handling Equipment

The following sections provide more information on AIM components, and give a brief overview of the AIM modeling approach.

4.1 Types of AIM Components

You build an AIM model by defining the capacity constraining resources and the parts that flow through these resources. The resources are most commonly machines, operators, work-in-process areas, and the material handling equipment. Each resource component is represented directly in AIM, and has its own graphical location and appearance. Animation of resources is automatic.

You can control the availability of resources by assigning them to shift, breakdown, and maintenance schedules. The sample editor shown in the Figure 7 is used to define and access shifts, breakdowns, and maintenances for the machine,.

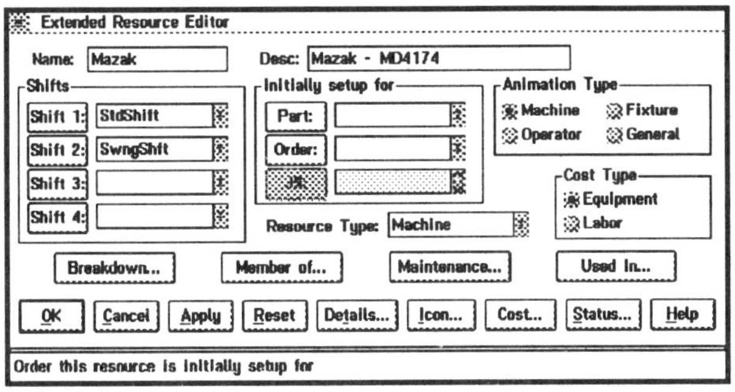

Figure 7. Sample Machine Resource Editor

The AIM process plan component captures the logic of your manufacturing system. You can think of a process plan as a flowchart describing the movement of parts or part families through their manufacturing process. You can easily translate your existing part routing sheets into AIM process plans.

An AIM process plan is actually a collection of jobsteps describing the sequence of operations performed on a part. These parts flow through the manufacturing facility as loads. At each jobstep, the load seizes the resources which are required to perform the operation. For example, a lathe machine and an operator are required for the job of setting up the lathe. Other resources which are no longer needed by the load are released. A sample of the types of jobsteps available in AIM include:

- Setup
- Operations
- Produce
- Assemble
- Move Between
- Add to Material
- Remove from Material
- Assign a Variable
- Accumulate/Split
- Inspection
- Kanban
- User Defined

Loads in AIM are created by several different means, including:

- Loading an existing order log or book directly into the AIM database. With this approach, you can compare the performance of your manufacturing operations under different configurations using actual production loading. You can also import system status information along with your order book in order to make planning and scheduling decisions.
- Modeling pull oriented manufacturing systems. You describe when new parts or sub-assemblies are needed by the system, for example when an inventory depletes to a threshold level. Pull orders generate new loads that represent the required parts or sub-assemblies.

Several other methods of load creation exist in AIM. An AIM model can contain any combination of these load creation methods.

4.2 Adding Manufacturing Details

You control the behavior of each AIM component using a robust library of pre-defined actions and rules. For example, you can control how a specific machine chooses the next job to start when it completes a job using AIM's extensive library of resource sequencing and selection rules. Examples of these rules follow.

- Adjusted Dynamic Slack
- Earliest Due Date
- Earliest Release Date
- FIFO
- High Attribute Value
- High Priority
- Large Load
- LIFO
- Least Av. Dynamic Slack Per Remaining Jobsteps
- Least Dynamic Slack
- User Defined
- Least No. of Jobsteps
- Least Processing Time
- Least Static Slack
- Longest Any Jobstep
- Longest Current Jobstep
- Low Attribute Value
- Low Priority
- Shortest Current Jobstep
- Smallest Load
- Min Setup Time over First N Loads
- etc...

AIM contains over 160 locations where you can choose from alternative actions or rules in order to refine component behavior and capture the details of your manufacturing operation. In most cases this information appears on the component's detail page. The resource detail page is shown in Figure 8 below.

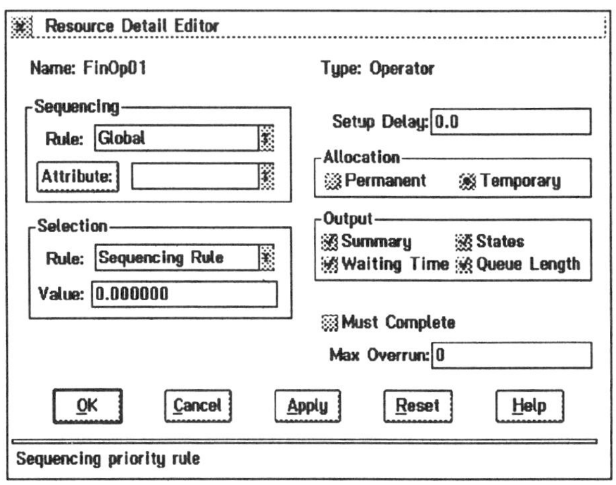

Figure 8. Sample Resource Details

If the built-in rules do not represent the logic of your system, you can create your own custom rule. You write custom rules in the C language and compile and link them without leaving AIM. AIM makes it easy for rules to be created by an expert user, and then used by less experienced modelers.

4.3 AIM Modeling Summary

You model in AIM by describing your production process using manufacturing terms. Figure 9 shows a summary of the major AIM modeling components and their relationships.

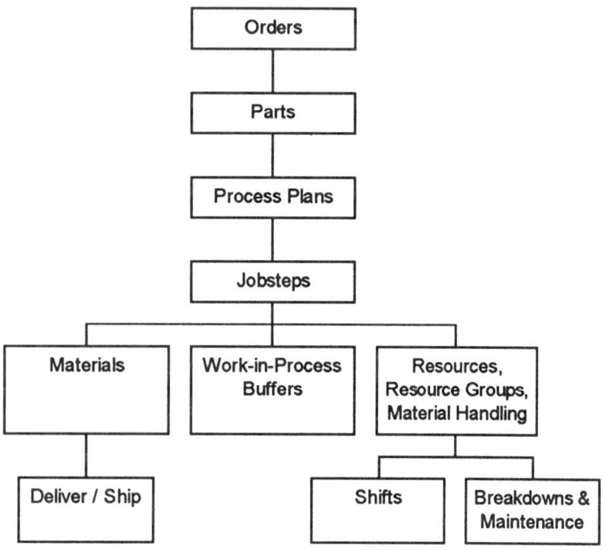

Figure 9. FACTOR Modeling Components

5 AIM COST MODELING

To this point you have seen how you can build a model in AIM which captures the details of your manufacturing process. You have also seen that this model can be used in the broad areas of capacity engineering, planning and scheduling. Now AIM also includes financial modeling capabilities.

The latest release of AIM, Version 6.0, extends its manufacturing simulation capabilities to include the modeling of manufacturing costs. This makes great sense. An AIM model already captures the details of how parts are processed. It is a logical extension to also capture the costs of these operations.

AIM Version 6.0 modelers can define product costs with simple fill-in-the-blank forms. These costs fall into four general classifications; operating costs, overhead costs, inventory costs, and lateness costs. Operating and overhead costs are defined for the resources in your facility – machines, labor, material handling devices. Inventory costs are associated with raw material, work-in-process buffers, and finished goods storage. Lateness costs are penalties associated with not completing orders on time.

Separate output reports and graphs are available in Version 6.0 to communicate the outcome of the cost simulation. The average operating expense per part, the value of inventory, and total expense incurred are just a sample of the reports and graphs available in Version 6.0. Utilizing the custom reporting of FACTOR/AIM it is possible to generate cash flow reports and pro forma income statements from the model.

5.1 Operating Costs

Operating costs are costs directly tied to the production of parts. Examples of operating costs include cost of labor to setup and operate a lathe; cost of fuel and lubrication to run a gantry crane; and cost of supplies used at a washing station. AIM tracks the operating costs consumed by a part as it flows through the shop. It also tracks the operating costs incurred by each machine, operator, fixture, and other resources. These costs can be reported for the system as a whole, by part type, or for each resource individually.

5.2 Overhead Costs

Overhead costs are costs that cannot be directly related to the processing of individual parts, Examples of overhead cost include: management costs, engineering costs, rent, capital equipment expenditures, and depreciation. Overhead costs can be defined for the same resources as operating costs, and they are evaluated and assigned to parts and resources at the end of the simulation.

5.3 Inventory Costs

Inventory costs reflect the value of parts in work-in-process storage, as well as the value of raw and finish goods inventory. Inventory carrying costs and turnover rates are also tracked and reported by AIM.

5.4 Lateness Costs

Lateness costs are incurred when product cannot be delivered on time. They can be used to quantify the negative effects of lateness including expedited shipping costs, contract penalties, and the cost of customer dissatisfaction.

6 THE AIM TUTORIAL

The tutorial at the Winter Simulation Conference will provide details about building an AIM model and scheduling with AIM. The presentation will include details about modeling components, example output reports and a demonstration of AIM. The modeling process will be discussed in detail, as will the use of the AIM model for making better manufacturing decisions in the areas of capacity engineering, planning and scheduling. In addition, upcoming product enhancements will be demonstrated.

REFERENCES

Grant, Floyd H. (1987), Scheduling and Loading Techniques. *Production and Inventory Control Handbook*, Second Edition, Green, J. H., Ed., American Production & Inventory Control Society.

Grant, Floyd H. (1986), Production Scheduling Using Simulation Technology, *Advanced Manufacturing Systems Conference*, IFC (Conferenced) Ltd., 129-138.

FACTOR AIM Modeling Reference (1995), Pritsker Corporation, West Lafayette, IN.

FACTOR AIM Output Analysis Reference (1996), Pritsker Corporation, West Lafayette, IN.

FACTOR AIM Database Tailoring Guide (1996), Pritsker Corporation, West Lafayette, IN.

FACTOR AIM Cost Modeling Guide (1995), Pritsker Corporation West Lafayette, IN.

FACTOR AIM GANTT Charts User's Guide (1995), Pritsker Corporation West Lafayette, IN.

MacFarland, D. G. and F. H. Grant (1987), "Shop Floor scheduling and Control using Simulation Technology," *Shop Control '87*, Cincinnati, OH.

AUTHOR BIOGRAPHY

JULIE N. EHRLICH is a Systems Consultant for Pritsker Corporation. She is responsible for FACTOR/AIM support and training. Julie has a Bachelors Degree in Forest Management from Purdue University and a Masters in Operations Research from the University of Georgia. Since joining Pritsker she has worked in the consulting, quality assurance, and documentation areas.

WILLIAM R. LILEGDON is a Vice President for Pritsker Corporation. He is responsible for the promotion, development direction, and delivery of Pritsker's engineering software and services. He received a Bachelors Degree in Industrial Engineering from Purdue University. Since joining Pritsker Corporation he has worked in the consulting, development, support and training areas. William led the development of SLAM II/PC, SLAMSYSTEM, and FACTOR/AIM software products. He has published numerous papers and articles on simulation products and their application.

Proceedings of the 1996 Winter Simulation Conference
ed. J. M. Charnes, D. J. Morrice, D. T. Brunner, and J. J. Swain

AUTOSTAT™: OUTPUT STATISTICAL ANALYSIS FOR AUTOMOD™ USERS

John S. Carson II

AutoSimulations
1827 Powers Ferry Road
Bldg 17, Suite 100
Atlanta, Georgia 30339, U.S.A.

ABSTRACT

AutoStat™ is an extension package for AutoMod™ and AutoSched™ models that provides complete support for simulation model experimentation and statistical analysis of outputs. Within a menu-driven, point and click environment, AutoStat provides assistance for setting up runs, automated execution of runs and consolidation of outputs across replications and scenarios. AutoStat automatically sets random number seeds to achieve statistically independent replications; it also sets factor levels (input parameters) to realize desired scenarios, without having to modify the underlying model.

AutoStat has a number of advanced features such as datafile factors, responses (or performance measures) read from custom user output files, and user-defined combination responses (linear combinations of other responses). AutoStat offers several statistical methods, including confidence intervals, a ranking and selection procedure, design of experiments, and warm-up determination. Outputs can be displayed and printed in tabular and graphical formats and exported to spreadsheet and other analysis software.

AutoStat saves simulation modelers considerable time and automates the effort of setting up and making runs, managing the input and output files, consolidating results and conducting analyses.

1 INTRODUCTION

The experimentation phase of simulation modeling and analysis is critical to the success of any simulation modeling project. AutoStat provides assistance for all parts of the experimentation phase, including:

- Setting up model runs
- File management for the model's input and output files
- Defining the desired scenarios to be run
- Automatic execution of desired scenarios, including control of random number streams, setting of factor levels (input parameters), and error tracking
- Consolidation of outputs across scenarios and across statistical replications
- Statistical calculations across replications, including sample means and standard deviations
- Statistical methods, including:
    - Confidence intervals
    - A ranking and selection procedure
    - Design of experiments
    - Warmup determination
- Tabular and graphical display and printing of results
- Export of tabular outputs to spreadsheets and other analysis software

AutoStat works with both AutoMod and AutoSched models. AutoMod is a general purpose simulation package with 3-D animation that provides pre-programmed, material handling simulators for conveyors and vehicle systems and special support for manufacturing systems. AutoSched is a simulation-based, finite-capacity production scheduling tool.

AutoStat provides all the tools you need for setting up runs and conducting a proper statistical analysis, in an easy-to-use point and click, menu-driven environment. AutoStat provides warm-up determination for steady-state analyses, single and multiple-comparison confidence intervals for the estimation and comparison of mean system performance measures, the design of experiments to determine the significant factors, and a ranking and selection procedure ("select the best") for finding the "best" system. Statistical results can be displayed in both tabular and graphical formats. Responses, or system performance measures, that can be analyzed include all those in the standard AutoMod output report plus responses printed to custom output reports. In addition, you can define a response as a linear combination of other, previously defined responses, without modifying the underlying model.

In support of experimentation and analysis, AutoStat provides automated set-up and execution of model runs, file management and consolidation of outputs across runs. You define the factors or input parameters that you want to vary. Each set of factor values defines a different scenario or system alternative of interest. Possible factors include any standard AutoMod input parameter (such as number of vehicles or operators, conveyor speed, or vehicle speed), any user-defined variable's initial value, or any user input data file (such as a production schedule or daily orders).

After you identify the factors you want to vary, define the scenarios of interest, and specify the number of replications desired, AutoStat will automatically make all the runs for all desired scenarios and archive the outputs for later analysis and display of comparative results. For each run, AutoStat automatically sets the seeds for all random number streams, and sets user-specified values for factors, including data file factors, without modification to the underlying model.

In support of debugging and verification, AutoStat allows previous runs to be repeated interactively, so that suspicious outputs can be investigated and the model verified and errors corrected. When repeating a run for debugging purposes, AutoStat automatically duplicates all previous settings, including random number seeds and factor values. Models with errors can be corrected and previous runs re-made for verification purposes.

If your model has probabilistic or statistical components, such as random arrivals or random processing times, then you need AutoStat for conducting a statistical analysis to determine whether observed differences in a response between two scenarios or two system alternatives are real differences or are merely due to random fluctuations. If your model is deterministic, then AutoStat can still save you considerable time and effort in setting up and making runs across numerous scenarios.

After AutoStat makes all the runs you request, you select the runs to compare, choose the type of analysis or graphical comparison, and AutoStat will extract the relevant outputs, do the statistical analysis and display the comparative results automatically. AutoStat can compare alternatives on the basis of any standard AutoMod output or any user-computed output written to a custom output report. Responses can be added after runs have been completed and analyses conducted on these responses, without having to re-run the model.

AutoStat makes it relatively simple to setup and run a large number of comparative simulations. With a few selections with the mouse, you can get results comparing the model's performance measures across any number of alternative scenarios. Results can be displayed in graphical and tabular form. In fact, without using the statistical capabilities at all, AutoStat can save you a tremendous amount of time during the debugging and verification phases of modeling, and thus help to make your model more robust. In addition, when setting up and making a set of runs for comparing alternatives, AutoStat can save you a tremendous amount of time compared to using batch files and manually managing all of the input and output files.

The remainder of this paper is organized as follows: Section 2 covers the basic concepts for setting up runs in AutoStat, namely, responses, factors, configurations and sample types. Section 3 explains the menu system. Section 4 covers the statistical procedures available in AutoStat. Section 5 discusses the use of AutoStat during model debugging and verification, and Section 6 discusses it use during experimentation. Section 7 provides a summary.

2 BASIC CONCEPTS

To use AutoStat to make runs for the purpose of experimentation or model verification, you begin by defining responses, factors, configurations and sample types. First, you define the responses or measures of system performance; second, you define the factors, or input parameters, that will be varied to generate alternative scenarios or system designs to be compared. A set of factors is grouped into a configuration, or experimental framework, to define the desired scenarios to be simulated. For each factor in a configuration, you then define the desired value or range of values, and AutoStat automatically generates the requested runs. You also must specify the sample type, namely the warmup period and runlength, plus the number of statistical replications. Then, AutoStat takes over, makes all requested runs, and archives the results for later analysis. This section discusses these basic concepts.

2.1 Responses

A response is a model output, usually a measure of system performance. For example, typical responses include machine or worker utilization, average and maximum queue size, and throughput. A response can be any output statistic captured by AutoMod or AutoSched and printed to any of the standard output reports.

AutoStat also allows user responses and combination responses. Any output statistic printed to a custom output report is called a user response. AutoStat can be "taught" to read a custom output report and "find" the desired output to be used as a response. To achieve this, the custom output report must contain a "snap separator", that is, a unique string that identifies the beginning of the custom report for the warm-up period (if any) and the beginning of the report for the steady-state portion of the run. For models without a warmup period, the snap separator can be any string at the top of the output file. The custom report also must contain one or two unique strings of characters, each usually a header or title, before the desired statistic, either on the same or a preceding line as the statistic. AutoStat is then told to look on the same line as, or a specified number of lines after, the unique character string, and to read a specified column from that line (counting columns starting at 0). You must also tell AutoStat whether the statistic is an average, a total (such as counting the occurrences of some event), a minimum or a maximum, so that AutoStat can handle any subsequent analyses properly.

Identification of AutoSched responses has been automated even further. Standard AutoSched output files are displayed in an edit table format, similar to a spreadsheet, with rows and columns. Clicking with the mouse on a desired element identifies the output in that row and column as a response.

Combination responses are new responses defined as linear combinations of other responses. This allows responses such as cost functions or averages to be computed by AutoStat without changing the model in any way.

For each run made, AutoStat saves the complete AutoMod report plus all custom reports requested to an archive in a compressed format (to minimize usage of hard disk space). Because of the archiving of all outputs, additional responses can be defined after a set of runs have been completed, without having to re-run the model, and additional analyses conducted.

2.2 Factors

A factor is a model input parameter that can be changed for the purpose of defining a scenario to be simulated. Usually, a set of factors set to particular values defines a scenario or alternative system design of interest. Examples of factors include the number of workers, product flow, machine speeds, production schedules and operating rules, order files, conveyor speeds, and vehicle performance specifications (speeds, accelerations).

In an AutoMod model, the following items can be factors:

- Capacity of a resource, counter, block or any other constrained entity.
- Speed or other performance parameter of a vehicle in an AGVS, ASRS, Bridge Crane, Power & Free or other vehicle system.
- Capacity of a vehicle (number of loads that can be transported).
- Speed of a conveyor section.
- Size of a load (important for conveyor models).
- Initial value of a variable.
- External input data file.

When any of these parameters is designated as a factor and assigned values, AutoStat will automatically override their value in the model (called the base value) without any more effort on your part than choosing the factor and typing the desired factor values.

Operating logic can be made a factor by using an input variable as a flag, for example to indicate which set of operating rules to use, and having the model select which rules to use based on the variable's value. In this way, alternative operating rules or logic can be compared against each other.

As mentioned, an external input data file can be defined as a factor. Examples of potential data file factors include production schedule files, product routing files, order files, machine processing times and downtime data. To use a datafile factor, first you create two or more alternative input files. Then by point and click you tell AutoStat that the input file is a variable factor and specify the alternative file names. For example, suppose a model of a distribution center reads a file of daily orders called "order.d". You obtain samples of order files for 5 different days over the past year, calling them "order.1", "order.2", ..., and "order.5". You tell AutoStat that "order.d" is a datafile factor and specify the names of the alternate order files. When such a file is used as a factor, AutoStat automatically copies the alternate files to the filename hard-coded in the model (saving the original and restoring it after all runs are completed). No changes to the model itself are required.

2.3 Configurations

Although you may have selected a large number of potential factors within AutoStat, in a particular phase of the project you may desire to experiment with only a subset of the selected factors. AutoStat uses the concept of a "configuration" to designate a set of factors to be varied for a particular analysis. The other factors are given a constant value, either the value hard-coded in the model or any other desired value. A configuration thus provides an experimental framework for a group of experiments.

For example, you may have defined 10 potential factors, but for an initial set of experiments you may desire to vary only 3 of these. You would define a configuration with the desired 3 factors designated as "variable" factors and the other seven set to a constant value. As another example, after some initial experimentation, you may find that some factors have no appreciable effect on system responses or an acceptable value becomes apparent. At that point, these factors become constants and the remaining smaller set become the variable factors. The configuration concept is thus an organizing principle for managing an on-going set of experiments.

2.4 Sample Types and Setups

After defining the factors and responses, you then define the sample type or run conditions. For example, for a particular model, it might be appropriate to have an 8 hour warm-up period and a 40 hour run. (For steady-state simulations with an initialization phase, AutoStat can assist in determining the appropriate warm-up period, as described in Section 4.1.) For a terminating (non-steady state) simulation, you simply define the desired runlength and a zero-length warmup period.

To setup the sample runs for a configuration, you define the value or range of values for each variable factor. You can setup runs one at a time by specifying values for each variable factor in a configuration, or you can setup multiple runs at once by defining the set of values for each factor and have AutoStat generate all combinations of factor values. Finally, you define the desired number of statistical replications, usually a minimum of 3. Each statistical replication of a scenario uses the same factor values but uses a unique stream of random numbers to guarantee statistical independence and random sampling.

2.5 Making Runs

After you define at least one factor, one response, one configuration, one sample type and set up the desired sample runs, you are ready to make the model runs. By clicking on the Run menu item "Do all runs", AutoStat will automatically make the requested number of independent replications of all the scenarios that you have set up.

For example, suppose that in a model of a production line, the following variables are the experimental factors, with factor values as given:

Number of Packing Machines: 3, 4, or 5

Number of Operators per Coloring Machine:

1 operator per 1 machine

1 operator per 2 machines

Production Schedule (a datafile factor):

"SCHED1.DAT"

"SCHED2.DAT"

AutoStat will automatically make the runs for each requested combination of factor values, and save the standard AutoMod output report and all custom reports that contain responses. In this example, for one replication per scenario and using all combinations of factor values, each set of runs amounts to $3 \times 2 \times 2 = 12$ runs. With 3 replications, AutoStat would make 36 runs. Multiple replications are needed by most statistical procedures.

3 MENUS AND MECHANICS

The MS-DOS and Unix versions of AutoMod before Version 8.0 work slightly differently than the Windows™ version (Version 8.0 and higher). The earlier versions work as follows. After building a model in AutoMod, from the project directory you enter AutoStat by typing

astat *model_name*

As in AutoMod, the right mouse button brings up the main menu. The first panel is the Model menu for opening models, saving AutoStat data, and quitting. The next four panels are the Responses menu, the Factors menu, the Configurations menu, and Samples menu, all used for setting up runs. The last panel is the Runs menu. With these five menus, you setup all AutoStat experiments and can get many of the benefits of using AutoStat in terms of data and scenario management.

The other menus cover the statistical and display capabilities:

1. Warmup Determination:
   Setup warmup runs, display Welch moving average plots of responses over time.
2. Confidence Intervals:
   Single and comparison intervals, the "select the best" ranking and selection procedure.
3. Design of Experiments:
   Setup DOE runs, graph and print DOE results.
4. Graphs:
   Plot of responses versus a factor
   Confidence interval plot of response for each configuration

Version 8.0 and higher of AutoMod are true Windows™ products for the PC, and are based on the Motif™ graphical user interface on Unix machines. The Windows version of AutoMod and AutoStat offers the same functionality as described here; the main difference is that menus are on the standard Windows menu bar instead of being the traditional AutoMod pop-up "card stack" menus. AutoStat is opened using the AutoStat icon in the AutoMod program group, and a model is opened from the Model Menu.

4 STATISTICAL PROCEDURES IN AUTOSTAT

In this section, we describe the statistical capabilities of AutoStat. These are divided between Warmup Determination in Section 4.1 and the analyses in Sections 4.2 - 4.5. The types of analyses offered include confidence intervals, a 2^k design of experiments, and a ranking and selection procedure called "Select the Best". (For more background on statistical methods in simulation, see Banks, Carson, and Nelson (1996) and Law and Kelton (1991).)

4.1 Warmup Determination

If you are interested in the steady-state behavior of a system and not the warmup or initial transient caused by starting the simulation in an empty and idle condition or some other state not representative of steady state, then you need to determine the length of the warmup or initial transient phase. After determining a reasonable length for the warmup period, you will run the simulation for a reasonably long period after the warmup phase, so that system performance statistics (responses) can be collected over a period that is close to steady state conditions. It is recommended that the steady-state phase be at least twice as long as the warm-up period, since the warm-up period is usually related to the (maximum) length of time it takes for parts to get through the system. Having a warm-up period that is too short, failing to load the system to a state representative of steady state, or failing to run the steady-state phase long enough, could result in performance statistics that are biased low due to the empty and idle initial state at time zero and that are not good estimates of the system's steady state behavior.

For warm-up determination, you should make a long run of one or more scenarios, breaking the run into a large number of short snaps. As a rule of thumb, we recommend 100 snaps. In AutoMod, a snap is a reporting period; standard output reports are written to a file at the end of each snap, statistical accumulators are reset, and the simulation continues. If you tell AutoStat to make 100 snaps for warmup determination, you will have a plot based on 100 response values, one response value for each snap period.

You should verify that all parts of the system become loaded and process a reasonable number of loads during the specified runlength. To determine whether it is reasonable to assume that the model has reached a steady state, you should examine all key responses and how fast, or whether, they are converging to a steady state. In addition, you may want to define a number of special responses for detecting that all subsystems, or all entity types, have been successfully loaded. For example, you may want to define responses for waiting times at the end of any product routing, or throughput by product type, to be sure that the whole system is loaded or that low volume or slow moving product has reached a steady state.

There is no known reliable method for automatically determining, by a statistical or mathematical calculation, when (or whether) a model has reached steady state. The best methods are graphical and depend upon your judgement and your knowledge of the actual system.

To assist in determining the length of the warmup period, AutoStat provides a moving average plot of any selected response over the snaps. To reduce random fluctuations or noise, the plot is averaged in two ways: across replications and within replications. The across-replication averaging is automatic. The within-replication average is the moving average plot, defined by a user-adjustable window size.

A moving average is the average over several adjacent snaps of the response value for those snaps. The number of adjacent snaps used to compute the moving average is called the window size. The purpose of the moving average is to smooth the plot in order to detect the underlying trend or low frequency fluctuations over time by dampening the random noise or high frequency fluctuations. Having a large number of short snaps makes the plots more meaningful, easier to interpret, and also allows for more flexibility in choosing a window size. For example, if you request 100 snaps and a window size of 1, the plot will consist of 100 response values. If you request a window size of 3, then snaps 1-3 are averaged, snaps 2-4 are averaged, ..., and snaps 98-100 are averaged, resulting in 98 values to be plotted.

To determine a reasonable warmup period, you must use not only the moving average plots with a suitably chosen window size, but also your knowledge that all areas of the system have reached a reasonable loading. It should be kept in mind that some configurations may not have steady state, usually indicated by a response value that continues to rise in value the longer the model is simulated. The AutoStat moving average plot can assist in determining whether a steady state is reached at all.

After you have determined the warmup period, you setup a sample type with the warmup and steady-state runlengths desired, and AutoStat will automatically ignore the responses over the warmup and use responses over the steady state for all analyses.

4.2 Confidence Intervals

A confidence interval for a single scenario is used to estimate long-run mean value of a response. It measures the effect on the estimate of mean response due to inherent statistical variation in the model. As more replications are made, the width of the confidence interval decreases, indicating greater certainty in the estimate in a statistical sense. (A short confidence interval does not, of course, say anything about a model's validity.)

A confidence interval for the difference in response between two scenarios is used to make a judgement regarding whether, and how much of, the observed difference is due to systematic differences in the two scenarios versus being due to mere random variation or noise.

To compute a confidence interval for a single scenario, AutoStat computes the overall average response over all replications, computes the sample standard deviation of the responses, and the standard error of the overall average response. Then it computes the confidence interval at the desired level of confidence (usually, 95%). The confidence interval is an interval around the overall average response value that estimates uncertainty in the estimate of mean response; it provides an estimate of the amount of variation in the overall sample average response due to statistical variation in the inputs for the particular runlength used.

A confidence interval should be used only when the response of interest is an average; it should not be used for maximums or minimums. This is because of the statistical theory (normality theory and the Central Limit Theorem) on which the computations are based. In fact, AutoStat will not compute confidence intervals for statistics of type minimum or maximum.

Confidence intervals are perhaps more useful when comparing the response of two (or more) scenarios. Suppose two alternative system designs (A and B) of a widget production facility were simulated for 4 replications and average widget time in system was compared. Suppose, further, that AutoStat reports a 95% confidence interval on the difference in time in system between systems A and B as:

3.2 ± 3.8 hours, or

-0.8 to 7.0 hours.

Since the interval contains zero, and thus overlaps both negative and positive ranges, no conclusion can be drawn as to whether system A or B is better, at least on a statistical basis. However, if a 4 hour improvement for widget time in system was of no practical importance, it can be concluded that neither system is better than the other. On the other hand, if a 2 hour difference was of importance, then more replications would be needed to reduce the width of the confidence interval so that a more precise estimate could be obtained. With a sufficient number of replications, the confidence interval should lie entirely to the left, or entirely to the right, of the value 2, in which case strong conclusions can be drawn.

If a tighter estimate of average difference were desired, how many more replications would be needed? In general, to decrease the confidence interval width by half, it is necessary to make 4 times the original number of runs. For example, for the widget example with 4 runs, it would take 12 more runs for a total of 16 to get a confidence interval of roughly half the original width (that is, of roughly $\pm 3.8/2 = 1.9$ hours in width).

4.3 Design of Experiments

One method for determining which factors have a statistically significant effect upon any response is to apply AutoStat's design of experiments. Statistical significance for a factor means that with high confidence the observed variation in mean response as the factor is varied from a low to a high setting is more than can be accounted for by mere random variation; that is, the observed variation of the response is larger than what would be expected due to the random variation alone. Thus, the factor is judged to have a significant effect on the response value. With this procedure, you can quickly eliminate from further experimentation many factors that have little or no effect on the important system performance measures, and use future computer and analysis time more profitably by concentrating on the factors most likely to have a large effect on system response.

AutoStat offers the so-called 2^k design; each factor is allowed two values, a "low" and a "high" setting. AutoStat offers the full factorial design, requiring all combinations of factor levels to be simulated. For example, for $k=5$ factors, a total of $2^k = 32$ runs are needed for each replicate of a full factorial design. AutoStat also offers several lower resolution designs, allowing a trade-off between number of runs and the possible inferences that can be drawn reliably.

4.4 Select the Best

AutoStat also offers a "Select the Best" ranking and selection procedure, one of the subset selection procedures presented in Law and Kelton (1991).

The goal is to select the best system, defined as the one with the minimum or maximum value of some specified response, from among k systems. For example, the best system might be the one with maximum throughput or minimum average time in system per part produced. The procedure can select either the single best system, or a subset containing the best system. The subset size is designated by m. For $m = 1$, the single best system is desired. For $m > 1$, the procedure will select a subset of size m containing the best system.

Besides specifying k and m, a user specifies the minimum probability of correct selection, P^*, and an indifference amount, d^*. Typically, $P^* = 0.95$, and the goal is $P(CS)$ = Prob(Correct Selection) $\geq P^*$.

The choice of indifference amount, d^*, depends on engineering or practical considerations. If the response of a system is very close to the response of the best system, then we do not care which of the two systems is chosen as best. More precisely, if the two system responses are within d^*, then we are "indifferent" to which one is chosen as best. With this setup, we are guaranteed that $P(CS) > \geq P^*$, where correct selection is defined as the best or any system within the indifference amount of the best.

The ranking and selection procedure in AutoStat is a two-stage procedure. First, an initial sample is taken of at least 5 runs of each configuration being compared. This initial sample is used to provide preliminary sample means and sample variances, and to determine the number of additional samples needed to select the best subset with the specified probability of correct selection. Note that P^* and d^* can be selected after the first-stage of sampling; the advantage of this is that AutoStat computes and displays the number of additional samples needed for a specified m, P^* and d^*. In general, the final sample size, N_i, required for each configuration i gets larger as subset size m or indifference amount d^* decreases, or as probability of correct selection P^* increases. It is recommended that P^* be set at 95% and m and/or d^* be made larger if the required sample size is too large for the available computer time.

Theoretically, the "Select the Best" procedure could be used in a factorial experiment to automatically find the combination of factor values that results in an optimum response. But in practice, with even a moderate number of factors and factor values, the procedure may require such a large number of replications as to exceed available computer time. Therefore, it is recommended that as many factors as possible be eliminated (and set at some appropriate constant value) by using either the Design of Experiments or multiple comparisons, or perhaps Select the Best with a moderately large subset size. Then Select the Best can be used with a small number of the most significant factors to determine the optimum combination of factor values.

4.5 Graphical Displays

AutoStat provides graphical displays of one or more responses versus a factor, and plots of confidence intervals for a selected set of configurations. In addition, the results of the Design of Experiments can be displayed graphically, namely, the confidence intervals for the main and interaction effects.

5 USING AUTOSTAT DURING MODEL TESTING AND VERIFICATION

Even before using its statistical capabilities, AutoStat can be of great assistance during model testing and verification. In brief, AutoStat provides a quick way to setup and run a large number of tests overnight or over a weekend, and a quick and convenient way to view the resulting outputs for face validity.

Key parameters can be varied easily by selecting them as factors. In fact, in a brief 10 to 15 minutes for setting up runs, almost any number of test runs can be setup and started. For verification purposes, you should make at least 3 statistical replications and run as wide a range of scenarios as possible, given the available computer time.

When you return to view the results, most likely AutoStat will have finished all the runs. If not, simply click on the Cancel button to abort the remaining runs. Completed runs will be saved and canceled runs can be completed at a later time if desired. For each run made, the usual AutoMod message window is on the screen; it will tell you whether the run completed the requested runlength or stopped with a modeling error. Of course, your model may also print messages to the message window, perhaps reporting some logical error detected during the run.

Using the Display Samples menu item on the Sample menu, you can display a table of all responses over all runs. A quick visual scan of this table will usually indicate whether the responses are consistent within a scenario. For example, looking at throughput for the whole system and key subsystems will tell you whether the model has a "lock-up" condition indicating a modeling error. The purpose here is not a detailed analysis, but rather to incorporate model verification into model development in an ongoing fashion and to provide more robust model testing by simulating over a wide range of factor settings.

If a run ends with an error or some unexpected output, you can use AutoStat to easily repeat that particular run in normal interactive mode to find the cause of the problem. You may use the debugger, watch the animation, or use any other debugging technique. Choose Repeat Run on the Runs menu, select the run you want to repeat, and AutoStat will automatically set the factor values and the random number seeds, so that you can repeat the previously made run. After finding suspected errors and modifying the model to fix them, you can again use AutoStat to repeat the previous run in order to verify the model's correctness.

Usually, after modeling errors have been fixed and the model has been satisfactorily verified, you should delete all runs made for verification and debugging purposes. After all, the model has probably changed during this period (to fix errors) and experimentation should be done with the final verified model.

The main purpose of using AutoStat in these early phases of a project before experimentation begins is as a quick test that a model runs to completion, and that the outputs (responses) are reasonable, over a wide range of possible inputs. This provides much more robust testing over a wider range of conditions than can be achieved by making one run at a time manually.

6 AUTOSTAT AND EXPERIMENTATION

AutoStat was designed to provide a set of basic statistical procedures to assist during the experimentation phase of a project. The data management and scenario management capabilities allow you to have more time to concentrate on the analysis itself instead of the tedious details of setting up runs and keeping track of all the output files.

The main purpose of the statistical procedures in AutoStat is to allow you to conclude with reasonable confidence that observed differences in response between two (or more) system designs or scenarios are due to a true systematic difference and not merely due to random noise. When a model has one or more input variables defined as a statistical distribution, say, for example, a time to failure that is exponentially distributed with mean of 6 hours, then random noise in the responses is unavoidable. Each independent replication uses a different stream of random numbers, the result being observed variation in the output responses. If there is a difference in the response of two system designs (or scenarios), is it a true systematic difference or merely noise? The statistical procedures built into AutoStat can assist in making a sound judgement.

Since AutoStat saves the standard AutoMod output report and all designated custom reports (in compressed form to minimize disk space usage), you can even add a new response after making the simulation runs. This comes in especially handy when your boss or customer requests a new or different measure of system performance; it avoids the necessity of repeating runs previously made.

7 SUMMARY

As with any analytic tool, it's up to you to interpret the simulation results properly and determine the practical significance of the factors on the responses. To leave more time for analysis, AutoStat can save you a tremendous amount of time by making it easier to setup and make any desired number of test runs and experimental runs. AutoStat will automatically save all the desired outputs and present them to you with a mouse click, in both tabular and graphical formats. For statistical analysis, AutoStat finds the desired outputs across any number of scenarios, does all the tedious computations, and presents the statistical results in tabular and graphical formats with a few clicks of the mouse.

REFERENCES

AutoStat User's Guide, *AutoMod User's Manual, Volume II*. 1993. Bountiful, UT: AutoSimulations.

Banks, J., J.S. Carson, and B.L. Nelson. 1996. *Discrete-Event System Simulation*. Second Edition. Upper Saddle River, NJ: Prentice-Hall.

Law, A.M. and W.D. Kelton. 1991. *Simulation Modeling and Analysis*. Second Edition. New York: McGraw-Hill.

AUTHOR BIOGRAPHY

JOHN S. CARSON, Senior Simulation Analyst, has been with AutoSimulations since 1994. He has over 20 years experience in simulation in a wide range of application areas, including manufacturing, distribution, warehousing and material handling, transportation and rapid transit systems, port operations and shipping, and medical/health care systems. His current interests center on the simulation of transportation systems of various kinds (traditional material handling to rapid transit), bulk and liquid processing, and software for simulation output analysis. He has been an independent simulation consultant, and has taught at Georgia Tech, the University of Florida, and the University of Wisconsin-Madison. He is the co-author of two textbooks, including the widely-used Discrete-Event System Simulation, Second Edition (1996). He holds a Ph.D. in Industrial Engineering and Operations Research from the University of Wisconsin-Madison, and is a member of IIE and INFORMS.

Proceedings of the 1996 Winter Simulation Conference
ed. J. M. Charnes, D. J. Morrice, D. T. Brunner, and J. J. Swain

AUTOMOD TUTORIAL

Matthew W. Rohrer
AutoSimulations
655 E. Medical Drive
Bountiful, Utah 84010, U.S.A.

ABSTRACT

The AutoMod™ simulation system differs significantly from other systems because of its ability to deal with the physical elements of a system in physical (graphical) terms and the logical elements of a system in logical terms. AutoMod also offers advanced features to allow users to simulate complex movement (kinematics and velocity) of equipment such as robots, machine tools, transfer lines, and special machinery. All graphics are represented in 3-D space with unlimited viewing control including: translation, rotation, scale, light-sourced solids, perspective, and continuous motion viewing.

The AutoMod CAD features are used to define the physical layout of manufacturing, material handling, and distribution systems. Unlike most other simulation software, the powerful AutoMod graphical interface accurately captures the physical constraints of distance, size, and space in 3-D.

AutoMod is really two programs. The build portion is for the physical and logical model definition. After the user has defined the physical and logical components of the model, it is then compiled into an executable model, where the simulation and animation run concurrently. The executable model is fully interactive, it can be stopped at any instant in simulated time to view statistics and model status.

INTRODUCTION

AutoMod combines the ease of use of a simulator with the power and flexibility of a simulation language. Simulators usually address specific real-world problems with predefined constructs, making model definition quick and easy. The AutoMod movement systems aid users in defining the movement of material either manually or by automated equipment. Material movement systems include:

- Transporter (path/vehicle systems such as fork trucks, AGVS, and human movers)
- Conveyors (including belt and roller types)
- Automated Storage and Retrieval Systems (AS/RS)
- Robots
- Bridge Cranes
- Power and Free Chain Conveyors

To define material movement systems, the user simply creates geometric entities such as paths and stations, then inputs the operating parameters such as velocity and acceleration. AutoMod automatically creates the corresponding model logic. 3-D animation is created automatically as well, providing a realistic picture of how a facility will operate. Model animation can be viewed from any angle or perspective, providing visualization capabilities unmatched in other simulation systems (see figure 1).

Figure 1. AutoMod Animation

A big difference between AutoMod and other simulation software is that users deal with geometry in a graphical way and create logic using a fourth-generation simulation language. This approach is intuitive to engineers who understand how their facilities operate but who do not have experience with simulation tools. Model output includes 3-D animation, automatically collected statistics, and business graphs. AutoMod provides users with a rich environment that facilitates thorough understanding and analysis of manufacturing systems.

The animation provided in AutoMod models is concurrent, meaning that the graphic pictures are running in step with the simulation model. The model execution environment is interactive, allowing the user to stop and start the simulation, run without animation in an accelerated time scale, and select objects from the animation screen to gather more information. Statistics can be viewed at any time during a simulation run. These model execution features make it easier to verify and validate models of complex systems and provide a forum for communication about system performance among project team members.

AUTOMOD INTERFACE

An AutoMod model consists of one or more systems. A system can either be a process system, in which control logic is defined, or a movement system. Each model must contain one process system and may contain any number of movement systems. Processes can contain complex logic to control the flow of either manufacturing materials or control messages (represented by loads), to contend for resources or to wait for user-specified times. Loads can move between processes with or without using movement systems.

Loads that flow through the process logic have the ability to claim and release resources, enter and leave queues, be added to and removed from order lists, change the value of variables, counters, and load attributes, create a new load or kill an existing load, read from and write to external files, and determine the next process. All interarrival and event times can be represented by deterministic values or be derived randomly from one of several statistical distributions. The AutoMod interface is window-oriented, utilizing menus, dialog boxes, selection lists, and mouse-based editor for developing process logic.

AUTOMOD WORLD VIEW

An AutoMod model is made up of a collection of systems. Any number of movement systems can be defined. A process system connects the movement systems to the logical flow of products. In the process system, loads move between processes (locations), and compete for resources (equipment, operators, and queues). The load is the active entity, executing action statements that are connected to the processes. Typical action statements give users the ability to:

- Use machines/operators
- Move into queues
- Clone new loads
- Change load types
- Wait on user-defined delay lists (order lists)
- Increment/decrement counters
- Set variable values
- Read from data files
- Send to other processes
- Make conditional tests

The process system is the backbone of an AutoMod model, providing the general-purpose simulation features required to model real-world problems. While material movement is important, it is not critical in manufacturing. No value is added to a product while it is being moved around a manufacturing facility. Value-added operations are performed by machines, processes, and workers in the facility. The AutoMod process system is where the value-added operations and control logic are defined.

The AutoMod process system includes a simulation language based on action statements. Action statements combine the power of a structured language with the ease of use of English-like, manufacturing-oriented syntax. AutoMod models are not limited in any way, so model logic can be of any size and complexity. The power and flexibility of the AutoMod language makes it easy to model almost any real-world situation. Figure 2 is an example of a process procedure:

```
begin Pfixit arriving procedure
   move into FixQ
   if FixedYet = 1 then send to die
     else begin
     choose a resource from among
       Picker(1), Picker(2)
     whose current loads is minimum
     save choice as RecIdle
     use RecIdle for uniform 10, 5 min
     set FixedYet to 1
     send to SizeWeigh
   end
end
```

Figure 2. Sample Procedure Logic

LOADS

Loads are the active entities in AutoMod and can be created in many ways, including deterministic or probabilistic generation. Their interarrival rate can be read from an ASCII data file or attached to a statistical distribution. Predefined random distributions can be used to fit most real-world random events. Loads are given types, such as “RedCar” or “PartA,” and they can have attributes such as color, stock keeping unit (SKU), priority, and makespan. Attributes can be accessed and modified in AutoMod action statements. Loads have 3-D shapes and dimensions like most other entities in AutoMod.

RESOURCES

Resources in AutoMod are used to represent machines, operators, fixtures, containers, and any other finite-capacity objects. There are two categories of resource state. The first is the working category (busy or idle), and the second is the availability category (up or down). During the animation, state colors indicate the status of each resource. Statistics are automatically collected for every resource in a model.

Loads use resources for specified processing times. These times can be deterministic or probabilistic, using built-in statistical distributions to simulate randomness. Processing times are either part specific or are applied to all parts using a particular resource. Resources can also be preempted if a higher priority load needs immediate attention.

Resources can have downtimes, which are defined by MTBF (mean time between failures) and MTTR (mean time to repair). These times can use statistical distributions to represent random failures. Though MTBF is calculated by model-simulated time by default, it can be based on parts processed as well as machine running time.

QUEUES AND ORDER LISTS

Because an AutoMod model is graphical as well as logical, loads must be located somewhere in the physical space of the model. AutoMod has two types of physical space: movement systems and queues. **Queues** in AutoMod are both physical and logical. They can have capacities ranging from one to infinity. When a queue has reached its capacity, the next load trying to enter that queue must wait until there is space available. Loads are sorted in queues by using order lists. Queue contents can be shown dynamically in the animation, and loads can be "stacked" in any direction.

Loads at a queue or process may be sorted or delayed until they are explicitly ordered to leave. To determine which action should take place next, the loads place themselves on order lists. An order list is not a physical entity like a queue, but a logical element that provides a way to sort loads that have been delayed for any reason. To remove a load from an order list, another load must execute an order action. Loads can be ordered to move to another process or order list, or to simply continue where they left off in their process logic. Order lists can be sorted by load priority or other load attributes, either in ascending or descending order.

VARIABLES AND COUNTERS

Data may be stored in an AutoMod model using **variables**. Variable values are changed using the "set" action as follows:

```
set Vtime to 123.456
```

Variables can be used in calculations or can be logically compared to other variables. In addition to integer, real, and string types, variables can also be used to store

- processes
- queues
- resources
- order lists
- counters
- loads
- locations

By setting a variable to another AutoMod type, such as a process, elegant approaches can be used to construct flexible and concise model logic. AutoMod also supports the notion of arrayed entities, making it easier to model real-world systems that have some repetition.

Counters are similar to variables, except that they can only be positive integer values. Counters have a maximum capacity, making them useful for global traffic control. When a load tries to increment a counter that is filled to capacity, the load will be delayed until another load decrements the counter. Statistics for counters are collected automatically.

TRAFFIC LIMITS AND BLOCKS

AutoMod has a powerful means of controlling the number of entities that can be either in processes or occupying physical space. Process **traffic limits** control the number of loads that are either on their way to or are in a process. **Blocks** control the number of entities occupying a physical space, making them very useful for preventing transporter vehicle collisions.

Blocks may have any capacity, but the capacity is generally set to one. Transporter vehicles (AGVs, fork trucks, etc.) and loads increment blocks automatically when moving through the physical space defined by the block. Loads can also claim blocks in process logic, as directed by the user. Blocks can have any shape, including combinations of cylinders, cubes, cones, or frustrums.

TEXT LABELS

AutoMod provides the ability to use **text labels** to enhance model communication. Labels may be added to any location in the model's physical space, and they can either be static or dynamic during the simulation. Labels can rotate when the animation view changes or can be attached to a fixed position on the screen.

TABLES

Tables in AutoMod supply the user with the means to collect statistics on any model entity and to classify those statistics for better understanding of their distribution. Tables automatically provide average, standard deviation, maximum, and minimum for all values entered in the table. Users can define the number of "bins" or categories, and AutoMod adds values to the appropriate bin when directed by the user in the tabulate action statement.

GRAPHS

Graphs in AutoMod are easy to define and update with the animation. Graph types include bar charts, pie charts, and timelines. Figure 3 shows a typical timeline business graph. Any model entity can be attached to a graph, including transporter vehicle velocity, number of loads on a conveyor section, or average utilization of a machine. Graphs can be printed or plotted to a variety of supported output devices, and graph displays can be controlled using the AutoMod language.

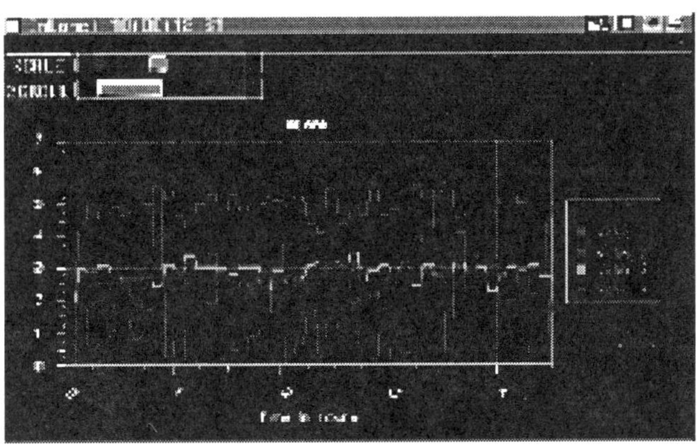

Figure 3. AutoMod Timeline Business Graph

RUN CONTROL

The AutoMod **Run control** allows users to define the warm-up and steady-state periods for the model by resetting time-persistent statistics. Reports can be printed for any run control period, or "snap." Business graph output can be automatically created, as well as post-processed animation records. These animation records can be played back after the simulation run with AutoView™.

Run control also provides an entity tracing capability that gives the model builder an event-by-event account of the model run. Tracing information is useful in verifying and validating a model.

PICTURE CONSTRUCTION IN 3-D

Both dynamic and static objects can be displayed during a model run. Dynamic objects represent loads, resources, queues, and statistics.

The static layout is the background graphics of the plant. It may contain column lines, aisle markings, and walls. Labels can identify specific areas in the facility.

There are several ways to create a layout of the system to be modeled. AutoMod comes with a three-dimensional graphics editor that allows the user to construct objects from standard graphic primitives. Cone, box, hemisphere, trapezoid, frustrum, cylinder, arc, vector (list), set, text, and triad are primitives that can be selected, placed, and scaled to create any static entity in the facility.

AutoMod also has the ability to import CAD information through the IGES standard. The acronym "IGES" stands for the Initial Graphics Exchange Standard. IGES is an industry standard exchange format for translating graphic data from one CAD system to another. Any IGES file of a plant layout that was created in a CAD system can be easily imported into AutoMod.

RUN-TIME ENVIRONMENT

In keeping with the AutoMod interactive features, the run-time environment provides the user with complete control of the model. The model can be viewed with the animation on or run with the animation off. AutoMod uses concurrent animation; the simulation progresses as the animation picture is being updated. With animation off, the simulation doesn't update graphics, but still performs all statistical calculations. The user can suspend the simulation at any instant to review statistics, take resources down, set break points or alarms, or control the view of the animation without constraint.

VIEW CONTROL

AutoMod provides a comprehensive, flexible, and easy-to-use method of interacting with a model during model execution. If the simulation project is in the experimentation phase where only parameter changes are made and the model needs to be re-run several times, AutoMod provides the ability to batch mode runs without animation. Whether the need is for a highly interactive mode or a batch mode, AutoMod lets you do it.

USER INTERACTION

AutoMod provides advanced debugging and trace facilities. A model can be single-stepped at any time during the animation. Also, the ability to set breakpoints and alarms allows the user to suspend the simulation when a certain event occurs or when a specific clock time is reached.

AutoMod also provides comprehensive reports. The reports can be displayed on request at any time during the animation. Printed versions of the reports can also be specified during model development.

AutoMod automatically keeps track of many statistics. These automatic reports are linked to specific entity types, such as

- Movement systems
- Processes
- Queues
- Resources
- Order lists

Figure 4 is an example of a transporter vehicle report. Vehicle states are tracked during the entire model run and reports are generated automatically.

Reports can be sorted alphabetically or numerically for easier analysis. The user can also develop and generate custom reports from within process procedures.

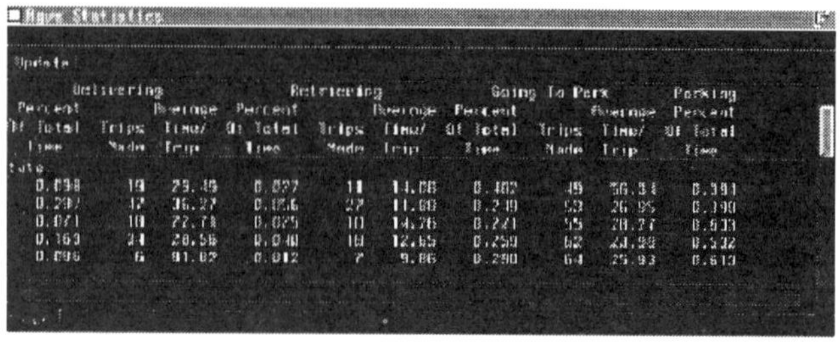

Figure 4. AGVS Statistics Report

SUMMARY

The AutoMod simulation system provides the ability to define the physical elements of a system using CAD-like graphics and the logical portion of the system using a powerful procedural language. A typical user can be three to ten times more productive using AutoMod in comparison to using any other simulation language. The accuracy and degree of detail with respect to movement systems is unequaled.

AutoMod allows the construction of very large, complex models. In fact, the AutoMod language has proven that the larger the project, the more benefits AutoMod has over alternative approaches.

AutoMod provides real-time, 3-D graphic animation. There are no limits to the views or the size of graphics. The degree of animation realism is also unmatched, as AutoMod provides light-sourced solid graphics with Z-depth sorting, showing all entities in correct relation to one another on the screen.

Enhancing the AutoMod capabilities are the following extensions and utilities:

- AutoSched™ - Provides a powerful manufacturing template for capacity planning.
- AutoView - A post-process animation package that allows you to create a directed "walk-through" of the model by panning, zooming, and moving back and forth in time and space. The animation view can be attached to a moving model entity like a load or vehicle, providing a realistic view from the "inside" of a model.
- AutoStat™ - Provides enhanced analysis of the statistics generated by AutoMod by calculating minimums, maximums, confidence intervals, Design of Experiments (DOE), and warm-up period to determine model steady state.

REFERENCES

AutoSimulations, Inc. 1989, *AutoMod User's Manual.*
AutoSimulations, Inc. 1989, *AutoMod Lessons Guide.*

BIOGRAPHY

MATT ROHRER, Vice President of Products, joined AutoSimulations, Inc. in 1988. Serving as a Simulation Analyst for five years, Mr. Rohrer completed simulation projects in distribution, manufacturing, and material handling. As a user and developer of AutoMod, he has contributed to the enhancement of the product to make AutoMod the most powerful simulation product available. His main interest is in extending the use of simulation technology beyond its traditional application in planning and design. Mr. Rohrer received a BS in Engineering from the University of Utah in 1983.

Proceedings of the 1996 Winter Simulation Conference
ed. J. M. Charnes, D. J. Morrice, D. T. Brunner, and J. J. Swain

AUTOSCHED TUTORIAL

Matthew W. Rohrer
AutoSimulations
655 E. Medical Drive
Bountiful, Utah 84010, U.S.A.

ABSTRACT

The AutoSched™ finite capacity planning and scheduling tool helps you increase throughput, reduce in-process inventory, and increase equipment and personnel utilization. AutoSched efficiently schedules all of the constraints in your factory, such as shift schedules, work setup rules, batching, preventative maintenance, machine efficiency, and operator skill classes.

With AutoSched, you can identify your bottleneck resources and apply intelligent operating rules to improve performance. One manufacturer reduced its average cycle time from 25 to 5 days, and reduced in-process inventory to one-fifth previous levels using AutoSched. Scheduling with AutoSched allows you to set accurate commit dates, based on the current mix of work, available resources, and existing commit dates.

AutoSched uses the AutoMod™ simulation system, AutoSimulations' proprietary simulation technology. AutoMod provides the discrete event simulation engine, 3-D graphics, and material handling constructs that are the foundation for AutoSched.

Built-in task selection rules in AutoSched allow you to model how machines and personnel select tasks to increase factory performance. You can also easily customize AutoSched with an engineering-oriented, English-like language that provides the flexibility to create custom task selection rules.

In addition to scheduling, the AutoSched planning tool can be used for capacity analysis. Simulation technology makes it possible for you to quickly and easily consider what if questions such as, What new equipment do we need to buy for producing the new product? Simply add the new equipment to the model, rerun it, and review the results. Effects of capital acquisitions, process changes, and changing product mixes can be evaluated prior to implementation.

INTRODUCTION

The AutoSched manufacturing capacity planning and scheduling tool is a template built on the AutoMod simulation engine. For most manufacturing problems, the AutoSched spreadsheet interface can be used to build models without programming.

Typical data includes part routings, machine mean time between failures, and setup requirements. In an AutoSched model, the equipment (station) is the intelligent decision maker, employing task selection rules to determine the processing sequence. AutoSched provides extensive reports, including Gantt charts for all model entities.

Because AutoSched is built on the foundation provided by AutoMod, 3-D animation is generated automatically (see figure 1). Users can define material movement systems and connect them to AutoSched models, adding the details of material movement between work operations. If the system being modeled has detailed rules of operation that are not part of standard AutoSched, the AutoMod language can be employed to represent complex algorithms.

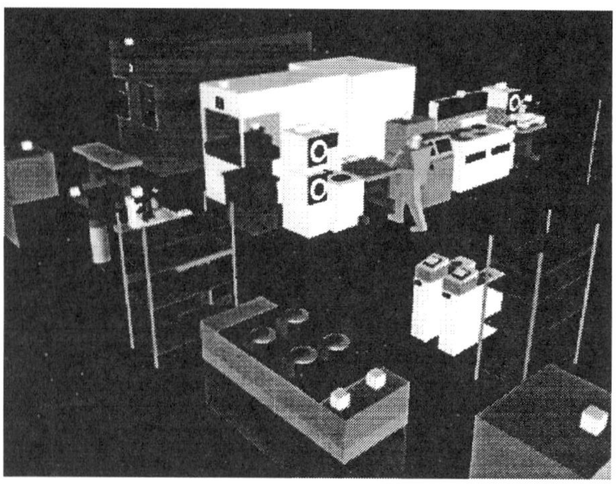

Figure 1. AutoSched Animation

AUTOSCHED INTERFACE

An AutoSched model is created mostly by populating AutoSched spreadsheets, called **edit tables**. The file format in AutoSched is flexible, with column order defined by the column headings. Definitions for all input fields are provided in the table-specific help, and data discrepancies tell the user when a data input error has been made. Manufacturing models can be built faster and with greater accuracy in AutoSched than with other simulation and scheduling products. The AutoMod foundation provides support for material movement systems as well as 3-D animation for better communication and visualization between engineers, production personnel, and managers. Figure 2 shows some of the AutoSched user interface.

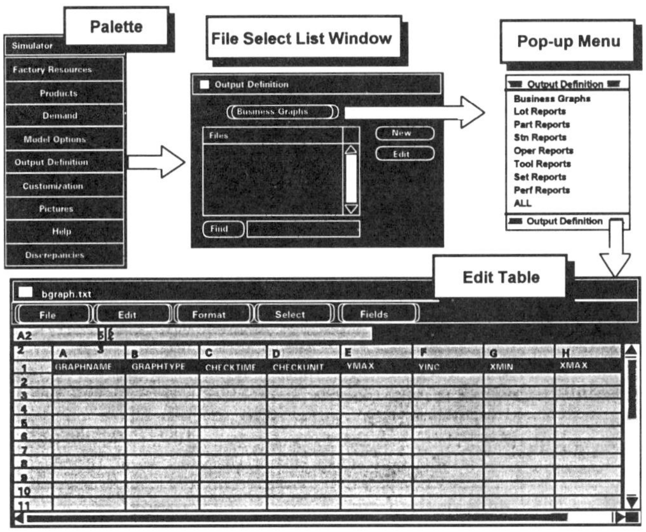

Figure 2. AutoSched User Interface

AUTOSCHED WORLD VIEW

AutoSched represents a manufacturing system as a number of **stations**. Stations can be machines, work benches, assembly positions, or any location where work is performed on a product. A station is made up of a machine with input and output queues, task rules, and downtime definitions.

A group of stations that performs essentially interchangeable work is called a **family**. Every station belongs to a family, even if the family contains only a single station. Families share a common input queue and work list for parts waiting for service from one of the stations in the family.

When a station completes its current task, it scans the **family work list** (FWL) to find another part to work on. If there is more than one part on the FWL, the station applies a task rule to make the "best" choice. Task selection rules include first in, first out (FIFO), same setup, shortest processing time, etc. There are 120 standard rules in AutoSched. Operators can also employ task selection rules to make decisions. Other factory resources, such as operators, are modeled with the same ease as stations.

Each station can have one or more calendars associated with it. **Calendars** specify when stations are unavailable for work.Calendars may be attached to factory resources to define preventive maintenance times, random downtimes, and shifts.

Lots flow between station families in a AutoSched model. Lots consist of a quantity of pieces of a given type of part and flow according to a routing that you define using manufacturing terminology.

The parts moving through a system have **routings** associated with them that define the processing sequence. Routings include the station, processing time, and setup requirements for each part at each processing step for the part. Yield and rework fields may also be included at any step. The demand on a manufacturing system is described in an **order** file. The part type, number of parts per lot, start time, and due date are defined as an order.

Task selection rules are the criteria the station uses to determine which lot to work on next. The task selection rule allows the station to either pick a lot from the potential parts or to wait for a better choice. The status of operators, tools, components, and other constraints may be considered in a task selection rule.

DATA REQUIREMENTS

To use AutoSched, you must provide three basic types of input:

- Factory resources (stations, operators, tools, storages, rules, calendars)
- Products (parts, BOM, routings, setup matrix, purchased parts, movement itineraries)
- Demand (factory orders, preventative maintenance, work orders)

AutoSched is extremely flexible in that it is only necessary to provide the data that is currently available. For example, if you do not want to include the detail of modeling human operators, you do not have to include it.

In addition, AutoSched has defaults for almost all of the possible features.

The data required for AutoSched can be entered from existing data bases or spreadsheets, or it can be entered through the powerful AutoSched edit tables. An edit table allows the user to input or modify data in a stand-alone or integrated database fashion.

PICTURE CONSTRUCTION IN 3-D

Both dynamic and static objects can be displayed during model execution. Dynamic objects represent stations, operators, and lots.

The static layout is the background graphics of the plant. It may contain column lines, aisle markings, and walls. Labels can identify specific areas in the facility.

There are several ways to create a layout of the system to be modeled. AutoSched comes with a three-dimensional graphics editor that allows the user to construct objects from standard graphics primitives. Cone, box, hemisphere, trapezoid, frustum, cylinder, arc, vector (list), set, text, and triad are primitives that can be selected, placed, and scaled to create any static entity in the facility.

AutoSched also has the ability to import CAD information through the IGES standard. The acronym "IGES" stands for the Initial Graphics Exchange Standard. IGES is an industry standard exchange format for translating graphic data from one CAD system to another. Any IGES file of a plant layout that was created in a CAD system can be easily imported into AutoMod.

TOOLS FOR EVALUATING RESULTS

GRAPHS

Graphs in AutoSched are easy to define and update with the animation. Graph types include bar charts, pie charts, and timelines. Figure 3 shows a typical timeline business graph. Any model entity can be attached to a graph, including transporter vehicle velocity, number of loads on a conveyor section, or average utilization of a machine. Graphs can be printed or plotted to a variety of supported output devices, and graph displays can be controlled using the AutoMod language.

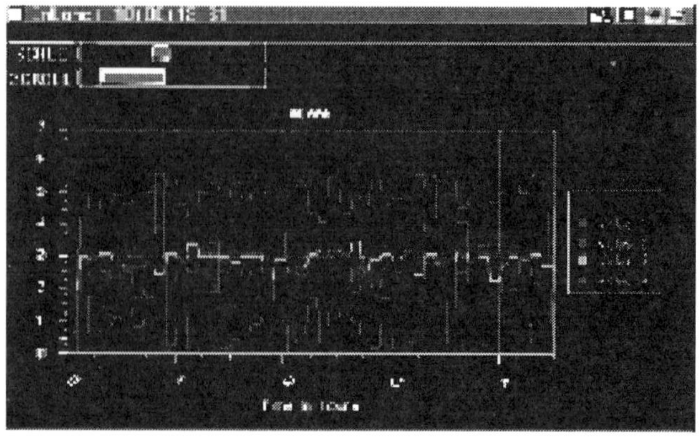

Figure 3. AutoSched Timeline Business Graph

RUN CONTROL

The AutoSched **Run control** allows users to define the warm-up and steady-state periods for the model by resetting time-persistent statistics. Reports can be printed for any run control period, or "snap." Business graph output can be automatically created, as well as post-processed animation records. These animation records can be played back after the simulation run with AutoView™.

Run control also provides an entity tracing capability that gives the model builder an event-by-event account of the model run. Tracing information is useful in verifying and validating a model.

REPORTS

AutoSched provides extensive reports on all defined entities. Station states are tracked, including percent of time:

- Processing
- Setting up
- Off shift
- Conducting preventive maintenance
- Down
- Idle
- Waiting for an operator
- Waiting for tool

Reports are presented in edit table format, and can be sorted on any column. They can also easily be exported to other spreadsheet and databases for evaluation.

AUTOSCHED BENEFITS

- **Data Organization** - You simply organize the data to define the model; you do not program.

- **Real-world Decision Orientation** - Stations, operators, and tools, rather than orders, make the decisions about what to work on next. This reflects how decisions are made in the real world, i.e., by operators who can look at the entire system.

- **Calendar Capability** - The simulation clock is converted to a calendar clock (month, day, year, hour, minute, and second). You can define an unlimited number of calendars and attach them to equipment and personnel. Calendars include information such as scheduled maintenance and holidays.

- **Schedule Diagnostics** - AutoSched provides Gantt charts and business graphs to help you visually interpret the schedules (see figure 4).

- Because AutoSched is based on AutoMod, **animation** aids in the understanding of scheduling dynamics.

- **Flexibility** - With AutoMod and AutoSched, the same tool can be used for:
  - Factory simulation
  - Finite capacity planning and analysis

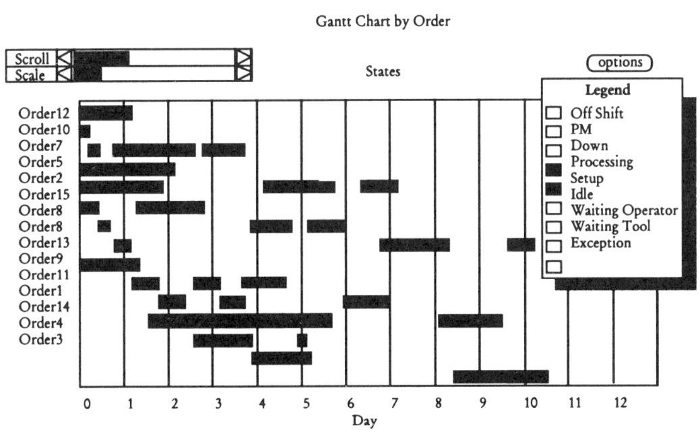

Figure 4: Sample Gantt Chart

SUMMARY

AutoSched is a powerful capacity planning and scheduling tool that allows manufacturing and process engineers to operate their facilities with greater efficiency. Because it is simulation based, AutoSched supports very detailed models where factory performance and model performance are within a few percentage points. This level of accuracy is required by many scheduling applications.

AutoSched provides an easy to use spreadsheet interface and flexible input file format that facilitate rapid model development. The AutoMod language can be used to customize models to achieve a higher level of modeling accuracy. The AutoMod foundation also provides 3-D graphics and material handling system constructs.

AutoSched has been used in many manufacturing operations to gain operating efficiency and to provide daily decision support.

REFERENCES

AutoSimulations, Inc. 1989, *AutoMod User's Manual.*
AutoSimulations, Inc. 1989, *AutoMod Lessons Guide.*

BIOGRAPHY

MATT ROHRER, Vice President of Products, joined AutoSimulations, Inc. in 1988. Serving as a Simulation Analyst for five years, Mr. Rohrer completed simulation projects in distribution, manufacturing, and material handling. As a user and developer of AutoMod and AutoSched, he has contributed to the enhancement of AutoSimulations' products to make them the most powerful simulation and scheduling products available. His main interest is in extending the use of simulation technology beyond its traditional application in planning and design. Mr. Rohrer received a BS in Engineering from the University of Utah in 1983.

Proceedings of the 1996 Winter Simulation Conference
ed. J. M. Charnes, D. J. Morrice, D. T. Brunner, and J. J. Swain

Advanced Uses for Micro Saint Simulation Software

Catherine E. Drury
K. R. Laughery

Micro Analysis and Design, Inc.
4900 Pearl East Circle,
Suite 201 E
Boulder, Colorado 80301, USA

ABSTRACT

Simulation is fast becoming *the* software tool for the 1990's. More companies each year are taking advantage of the ease of use of simulation software and see the benefits to using this tool. Unfortunately, there are still many people in Manufacturing, Health Care, Process Reengineering and Human Factors industries who are not using any sort of simulation software. Even in the information age of the 1990's it is still hard for potential simulation users to keep up with the changing market of computers and simulation software. In this paper we will focus on the changes in simulation software over the past ten years, the ease of use of the 1990's simulation software, and the broadening market for simulation software. Then, we will illustrate some of these changes by providing a step by step approach to using Micro Saint simulation software Version 2.0.

1 INTRODUCTION

Discrete event simulation has been a standard technique in the analysis of manufacturing systems for the past 30 years. More and more industries each year are seeing the advantage of using simulation software to analyze their systems. Today's rising costs and competitive pressure have led to an increased awareness in the value of simulation. This interest extends to other areas of industry as well. The adoption of total quality management (TQM) and process management techniques have created new challenges for managers and engineers. Process definition, quality measurement and control, process re-design, employee workload, safety and productivity are all areas where simulation is now being used to evaluate and improve efficiency.

Decision makers within organizations are applying simulation technology to a wider variety of problems. Consequently, the need for flexible, general purpose simulation tools capable of addressing all of these needs has increased. For example, Micro Saint 2.0, a discrete-event simulation tool, was developed as an efficient and cost-effective tool for simulating the complexities of systems within manufacturing, health care, retail, government, human factors, process reengineering and the service industry. The problems being analyzed range from process control and resource utilization to military maintenance procedures and human performance. What makes it unique is Micro Saint's focus on usability. To expand the base of users, we must make the products easier to learn and use. Therefore, usability is the driving force behind Micro Saint's design.

In this paper, we discuss how simulation is being made more usable by describing Micro Saint, how models are built with Micro Saint, and some of the applications of Micro Saint. This paper represents a "snapshot" in time of a simulation product designed for usability. We continue to look for and implement ways to improve the product's usability.

2 MICRO SAINT'S UNDERLYING METHODOLOGY

Micro Saint, developed in 1985, specifically for modeling human-based complex systems. The target users for the product were not simulation experts or computer programmers. Rather, we targeted individuals with little exposure to either simulation or computer science. This required a modeling approach that was much different from other simulation products that were available at the time. The engineers at Micro Analysis and Design, Inc. chose to develop Micro Saint implementing a methodology known as task network modeling. In task network model, activities are represented in a diagram as nodes, and the arrows between the nodes represent the sequence in which the activities are performed. A simple task network is shown in Figure 1.

This approach allowed users to develop models using the same techniques that they would use to define a flow diagram of the activity. Each activity, whether it be a human activity or machine activity, is defined using the same method. This minimizes the complexity of the user interface and eliminates the need for programming blocks specific to an application.

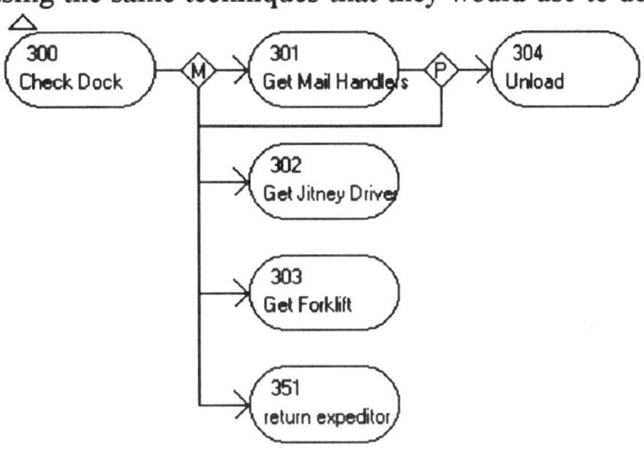

Figure 1: Sample Network Diagram

A Micro Saint model may be composed of many networks, each a sequence of tasks to be performed by a human, a series of processes that define an organization, or a series of machines in a manufacturing plant. Networks are composed of either additional lower-level networks or "tasks." Tasks represent the lowest level in the model and have specific parameters (timing information, conditions for execution, beginning and ending effects relating the task's performance to other system activities). An example of the window in Micro Saint for defining task information is presented in Figure 2. The following sections explains the task parameters in more detail. Micro Saint is now available on Windows 95, Windows NT, Windows 3.X, Macintosh, and UNIX platforms.

2.1 Task Timing Information

The task "mean time" is the average time that a task takes to complete. For example, if the task represents a human activity such as "transfer patient to recovery room," then the mean time to execute is the average time that it takes to transfer the patient. If the task represents a machine in a manufacturing process, then the mean time to execute is the average processing time for the machine. In many cases, the execution time is not constant. Rather, the elapsed time falls within a range of values that can be represented by a time distribution. Micro Saint supports more than 21 distribution types including normal, rectangular, exponential, gamma, Wiebull, Poisson, and others. In addition, users may enter parameters that control the spread of the distribution.

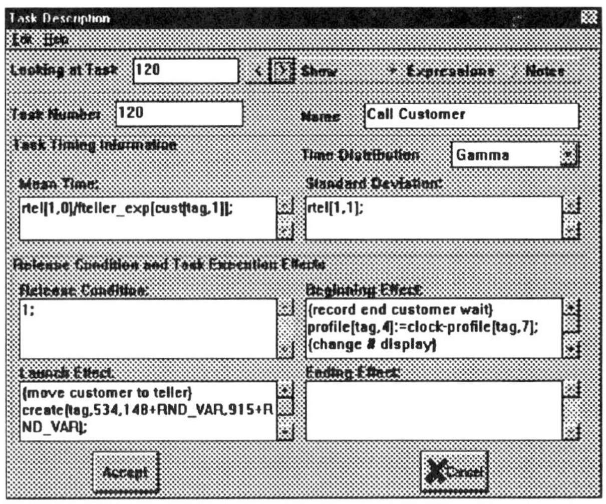

Figure 2: Task Description Window in Micro Saint

Alternatively, the mean time may be determined by the current state of the system or by an attribute of the process itself. In human performance modeling, the mean time to perform a task may be influenced by such conditions as how long the human has been working, the skill level of the human, or the current workload. In an insurance claim processing model, the time it takes to process a claim may be determined by the type of claim or the location of the client.

2.2 Conditions for Execution

Often, there are situations where a task cannot begin executing until certain conditions are met. A customer cannot make a transaction, even though the queue is empty, until a bank teller is available. A task may have resource requirements or other constraints (i.e., time of day, part type) that dictate when the task may begin executing. In Micro Saint, users enter a Boolean (logical) expression in the "release condition" field to control the execution of tasks. The release condition expression may be as simple as "teller <> 0", or it may be a complicated expression where several conditions are evaluated such as "(clock > 8 & clock < 16) & (clerk <> busy)." Entities moving through the network, whether they be patients, parts, or claim forms cannot be released into a task for processing until the release condition for the task evaluates to true.

2.3 Beginning and Ending Effects

These define how the task's performance affects other aspects of the system being modeled. For example, the current state of the system may change when a task begins or ends. For example, when a machine begins processing a part it becomes "busy" and is not available to another part until it has finished. The user would define the following expressions in Micro Saint to define this condition:

Release Condition: busy == 0; (A test for logical equality is represented as "==" in Micro Saint.) This keeps an entity (part, patient, etc.) from moving into the task when the task is "busy" processing another entity.

Beginning Effect: busy := 1; (Assignment statements are represented as ":=" in Micro Saint.) This sets the busy flag to TRUE so that the next entity cannot enter the task. As long as the task is executing, the busy flag will remain equal to "1".

Ending Effect: busy := 0; When the task finishes executing, the ending effect is evaluated and the busy flag is set to 0. Now, when the release condition is evaluated, the condition will be true and the next entity can enter the task.

The relationship between the release condition and the beginning and ending effects provides a general, yet powerful mechanism for users to define complex behaviors within the system they are modeling. Users may define variables that are specific to their system and manipulate the value of the variables as needed so that they can accurately represent their system. They do not have to compromise the accuracy of their model by relying on pre-defined "blocks" within the modeling tool nor do they have to learn a complex programming language in order to obtain the level of control required.

A feature that greatly increases Micro Saint's power is the "parser" that evaluates algebraic expressions. It provides the mathematical power of computer programming languages such as FORTRAN or C, but eliminates the need for compiling the model before running it. One of the biggest advantages of the parser is that it allows users to interactively change the value of model parameters while the model is executing. For example, the user can increase the number of resources available or change the execution time for a task while the model is executing to see what the overall effects on the system would be.

3 MODEL DEVELOPMENT WITH MICRO SAINT

The process of building a Micro Saint model involves two separate but interrelated steps. First, the user must define the structure of the task network. This is done by selecting the appropriate tool from the tool palette and placing the object in the drawing space of the network diagram. The Micro Saint tool palette and drawing space is presented in Figure 3.

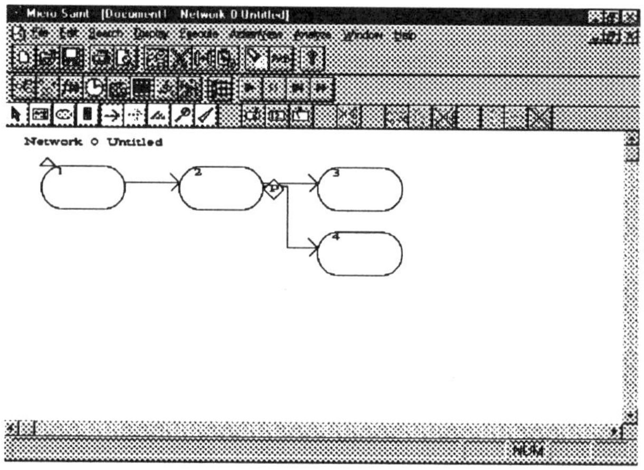

Figure 3: Micro Saint Tool Palette

Micro Saint uses a standard windows "point and click" approach to define the network objects. Using the mouse to "double-click" on an object will open it so that information specific to the object may be entered. Figure 2 above is an example of the Task Description window that is opened when the user double-clicks on a task (i.e., node in the network).

Task sequencing is defined by clicking and dragging with the mouse from the first task to the following task(s). A diamond shaped "decision icon" appears on the network diagram when more than one following task is defined. Users must enter the conditions that control the branching when there is more than one following task. Micro Saint provides the following decision types to ensure that all real-world situations may be represented in the model:

Probabilistic - The following task conditions are evaluated and the next task to execute is determined by the relative probabilities of all tasks listed. Probabilistic decisions allow only one of the following tasks to execute.

Multiple - The following task conditions are evaluated and all of the tasks whose conditions evaluate to non-zero will execute.

Tactical - The following task conditions are evaluated and the next task to execute is the task whose condition evaluates to the highest value.

Variables and algebraic expressions can be used in the branching logic and the value of the variables can be changed by conditions in the model. This gives the user complete control and manipulation of the network flow. Figure 4 is an example of a tactical decision that controls the flow of patients in an emergency room model through the network based on the seriousness of their injury.

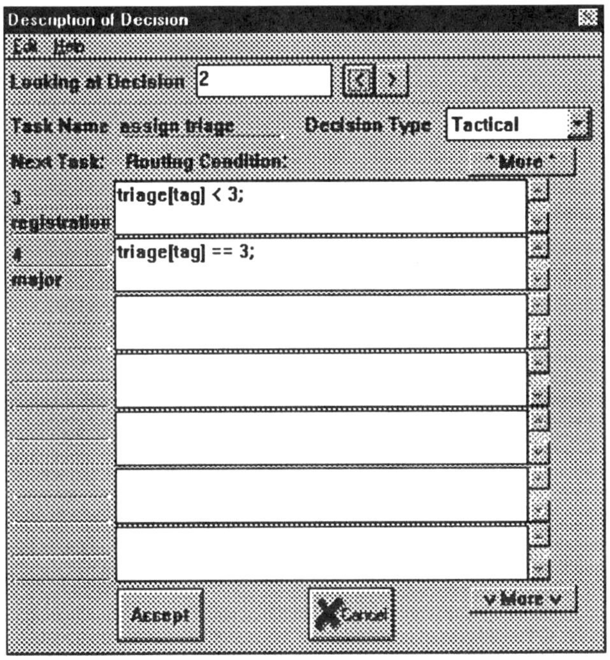

Figure 4: Micro Saint Tactical Decision

All of these features work to provide an environment for the model developer that is easy to learn and easy to use. Once the basic concepts are understood, any system or process can be modeled using Micro Saint. In addition, users can build models at any level of complexity. Some applications may require a very low-level, detailed definition while for others, a high-level definition is sufficient.

4 AUTOMATIC OPTIMIZATION WITH MICRO SAINT

Version 2.0 includes OptQuest, a model optimization function. OptQuest allows users to automatically search for optimal solutions to complex systems. The concept of optimizing solutions is very simple and very useful. Most systems being studied with simulation have constrained resources. These resources represent design variables that are critical to decision makers. Typical questions related to resources are:

- How many machines of each kind should we have in our shop in order to maximize the throughput without exceeding our budget?
- How many tellers and supervisors should we employ in our bank in order to maximize their utilization while also minimizing expected waiting time for the customer?
- What delivery frequency and batch size should we use in our just-in-time system in order to minimize inventory and production costs?

OptQuest allows for the creation of intelligent search procedures capable of finding optimal or near-optimal solutions to complex problems while only evaluating a small fraction of the possible alternatives. OptQuest is the result of implementing these cutting-edge technologies by customizing their operation to optimization to your simulation model. The Optimizer searches for the best possible answer to your optimization problem. OptQuest allows you to efficiently find high performance parameter settings for your most complex simulation model.

5 ANALYSIS AND RESULTS WITH MICRO SAINT

People build models to provide insight to, or answer specific questions about, a system or process. Some information can be gained by watching the Micro Saint model run. Micro Saint's symbolic animation capability provides an animated view of the network diagram as the model is running.

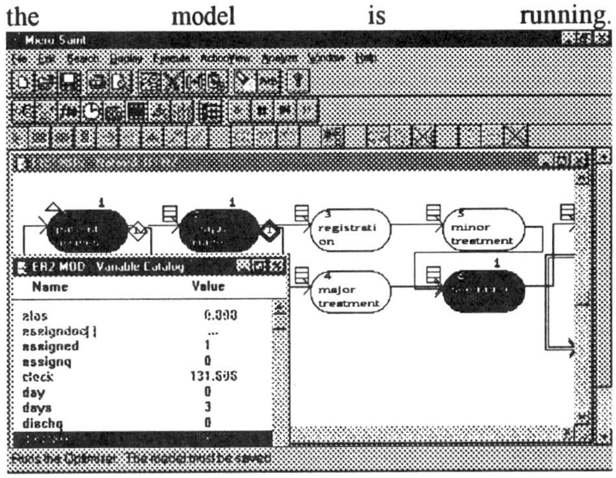

Figure 5: Micro Saint Symbolic Animation

Users can watch as entities flow through the network or wait in queues before being processed. This type of animation is particularly useful in debugging the model. An example of an animated network is shown in Figure 5. Micro Saint also provides an iconic animation capability called ActionView (see Figure 6) that allows users to build a realistic "picture" of their model.

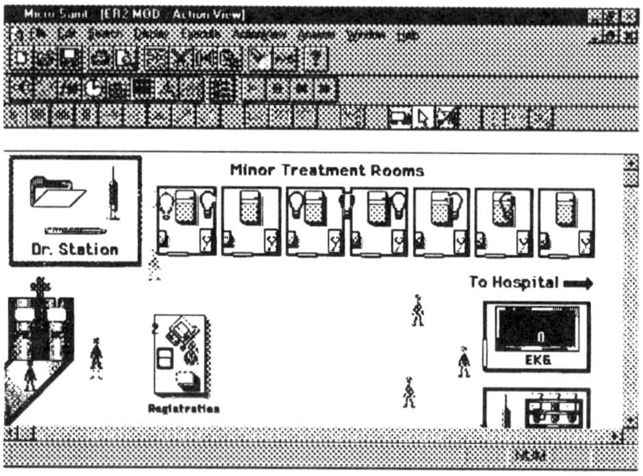

Figure 6: Micro Saint ActionView Animation

In addition, data can be collected at any time during the model run. A new feature of Version 2.0 includes automatic queue data collection. By turning on the data collection, and running a Micro Saint model, you can then at the end of the run automatically analyze queue data on the length of the queue, the average wait time, the maximum and minimum wait times (see Figure 7).

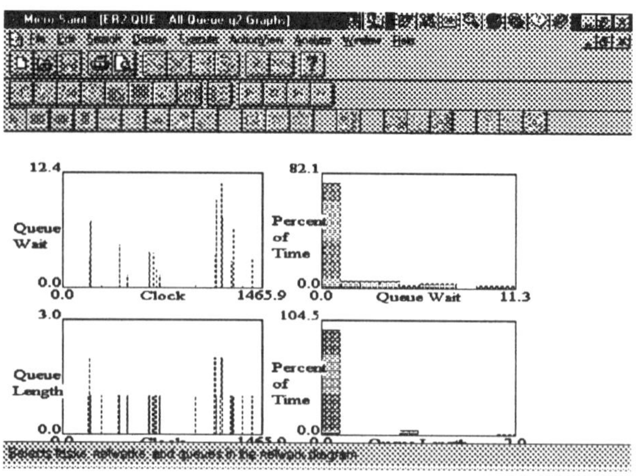

Figure 7: Micro Saint Automatic Data Collection Graphs

In addition you can create snapshots for specific information on the changing state of model variables. Sometimes it is sufficient to save the state of the system at the end of the run. However, in order to gain insight into the dynamic aspects of the system users can "take snapshots" of the model variables any time during run. These "snapshots" of data can be analyzed using the graphing capabilities within Micro Saint or imported into another statistical analysis package. Through the results of the simulation analysis and the insights gained, users can assess the relative merits of alternative solutions as well as predict their impact that leads to a better understanding of the costs and benefits.

6 MANUFACTURING APPLICATIONS

Micro Saint is used in the manufacturing industry to look at throughput, cycle time, the optimal schedules and in new factory layout. Major companies have found Micro Saint is easy to use and very flexible making it a choice for companies that need a simulation tool to use in different departments such as the work cell analysis and the reengineering of the accounting department.

The new Optimization feature in Micro Saint appeals to manufacturing facilities. For example, if a company is doing a simulation of their inventory system, the reorder point, the order quantity and the time between inventory reviews are the variables that that define the system. The optimizer can help define the optimal inventory policy with respect to minimizing a function of total expected cost.

For example, a Fortune 100 company had a cellular manufacturing layout that they felt could be made more efficient. They were about to make substantial capital improvements in manufacturing system equipment and wanted to make sure the new equipment was integrated with the old equipment in the best possible way. The company used Micro Saint to model their manufacturing system and reach their goals of finding a cellular layout that would meet their production goals. This study required a sophisticated model because the production goals were spread across 40 different product types , and capacity for inventory was limited, indicating a need for a JIT design. The study was limited to three weeks in length.

In the process of building the model, the customer gained an understanding of the system that showed them many areas for potential improvement. The project did not attempt to discover the "optimal design" for the cell rather it only identified one of an entire family of solutions that would work. More models were built in order to simulation the other production line cells. These models allowed the customer to consider tradeoffs based on operator utilization balanced against the cost of the equipment, operator training, installation requirements and process time. The bottom line was the manufacturer was able to increase production by 71% and increase profit per unit by 142%. This resulted in an increase in the annual profits of the company by over 400% on that single production line. The company reduced costs and improved quality with a small investment in simulation.

7 HEALTH CARE APPLICATIONS

In the health care industry, Micro Saint has been used to look at emergency room flow, ambulatory services, OBGYN units, pharmaceutical processes and patient file flow. Customers often choose health care facilities on the basis of the quality of service they receive. Every hospital or health care organization must answer questions about cost and quality. One way hospitals are able to look at cost versus quality is by using simulation. In one instance, an Ambulatory Surgery Department, the department that cares for patients both pre-operatively and post-operatively, had run out of space. The question was to build a new facility or redesign the existing space without changing the surgery schedule. Micro Saint was used to simulate the flow of patients from ambulatory surgery to surgery to recovery and back to ambulatory surgery. A management engineer simulated the effects on many different alternatives for routing patients through the system and maximized the utilization of the facilities.

The hospital staff was so pleased with the recommendations for the ambulatory surgery facilities that the management engineer was asked to help evaluate the surgery schedule on a daily basis. The changes recommended as a result of the simulation models allowed the hospital to increase the number of patients that can be scheduled for ambulatory surgery, therefore increasing profits while improving the quality of service being offered to patients.

This is just one example of how Micro Saint can help improve the patient care while reducing costs at a hospital.

8 HUMAN FACTORS AND ERGONOMICS APPLICATIONS

The use of simulation to analyze human factors and ergonomics is a relatively new application of simulation. However, we at Micro Analysis and Design have found simulation to be extremely useful in evaluating decisions such as "how many crewmembers do we need to fly an airplane?" and "what will the availability of trucks be for a given number of maintainers of given skill types?" Additionally, we have developed simulation techniques to answer these questions as a function of equipment design factors.

For example, during the early design phases of the Army Comanche helicopter, one of the key objectives was that the aircraft be piloted and operated entirely by one human. Central to that objective was the question, "Can one person do it all?" We were called upon to examine this issue with respect to four alternative helicopter cockpit designs. Specifically, the purpose of this project was to determine whether a one-person cockpit design would cause the operator's workload to reach unacceptable levels during the mission.

Micro Saint computer models were developed for each of the cockpit alternatives considered. Variables and modeling constructs were included in these models to track operator workload demands in the visual, auditory, cognitive, and psychomotor aspects of the operator's tasks. Using these models, simulation experiments were conducted under several experimental conditions corresponding to different cockpit designs. For each design, operator workload was predicted and the tasks driving workload were identified. The advantages of each of the alternatives were identified. The report we prepared discussed the relative advantages and disadvantages of each design as well as the projected technological requirements that would be required to support a one-person cockpit. Ultimately, we recommended that a one-person cockpit was not feasible within the technical and other design constraints.

This is one of many examples of human engineering questions we have addressed with Micro Saint.

9 BUSINESS PROCESS REENGINEERING APPLICATIONS

Business Process Reengineering is a term that, in effect, applies to all the areas previously discussed. In many cases the process the way it is today is looked at critically and in depth, the reengineering project is defined, objectives are set and a team is assembled. Then questions are asked by a team as to why a certain process is done in a certain way. The phrase, "because it has always been done that way" is a clue that closer scrutiny is warranted. To be competitive in today's environment one has to look beyond the limited scope of the same four walls. In evaluating the process many hard questions unfold. How many people needed to get the job done? Is there any cross training going on so one person can be more productive?

At this point in most BPR analyses, a process map is built. The process map is in turn used to identify the data requirements for what is the heart of the understanding phase, the process model. The process model is, essentially, a task network. When the data requirements are defined and the sources of the required data have been identified, a task analysis is conducted to verify the process map and begin the development of a process simulation model of the current process. In this case, the simulation takes an additional role - the means of process model verification and validation.

For example, in a hospital model, we gathered all the information for an obstetric unit, built a model and validated the model -- only then did we find out the process that was given to us by doctors was not the actual way the nurses went about their tasks! If we had not validated the model with simulation, the model would have been incomplete.

The final phase of the process reengineering methodology is the solution phase. Given the insights that were obtained from the process modeling effort, and the reengineering objectives, this phase searches for opportunities for radical improvement in cost savings, cycle time, service and quality. During the final phase we continue to look for ways to make the new process easier. The final phase also allows the Micro Saint user to play "what if" with the potential process modifications and test our initial hypotheses about how to reengineer the process. The decision to implement an identified reengineering solution is done by looking at the results of the process models and process objectives.

Business Process Reengineering takes on many different approaches. A Fortune 500 company has recently used Micro Saint to look at their underwriting process. The process was studied and modeled. In the modeling process the modelers found the claims were stacking up in the mailroom and sometimes up to a week before getting to the underwriter. After modeling the process flow and building a working model, the insurance company found the time it takes for a claim to be filed until the customer receives the check could be cut from ten to five days. They could accomplish this goal by using an express mail service. The net result is happy customers.

10 SUMMARY

In this paper, we have focused on the Micro Saint methodology and the underlying principles of modeling with Micro Saint in a variety of different industries. We have not attempted to cover all of the software features for building models, controlling a simulation, and generating and analyzing data. If you would like to know more, please contact Micro Analysis and Design.

What we have shown is that Micro Saint is a powerful tool for evaluating the dynamic aspects of systems within a wide variety of application areas. Micro Saint's primary strength is that it has an intuitive, graphical interface that allows users to quickly develop models that accurately represent their system. Users are then able to evaluate "what if" scenarios with a variety of inputs find the optimal solution. Simulation technology has been used by industrial engineers in a manufacturing context since the 1960's but only recently has been applied to more "white collar" applications. Regardless of the application area, the results are the same: better decisions can be made, money can be saved, productivity can be increased, and customers can receive a higher level of service.

AUTHOR BIOGRAPHIES

CATHERINE E. DRURY is the Marketing Director of Micro Analysis and Design, Inc. She has a Bachelor's of Science in Marketing from Miami University of Ohio. Currently, she is working towards her MBA at the University of Colorado at Boulder.

K. RONALD LAUGHERY JR., received his Ph.D. in Industrial Engineering from the State University of New York at Buffalo. He established Micro Analysis and Design in 1981, managing contracts for the development of computer modeling and simulation languages, the design and evaluation of training simulators, the development of supporting technologies for constructive and distributed simulations, and the development of tools for the Army MANPRINT program. Additionally, he participates in developing a number of simulation models for the Army, Air Force, and private industry.

Proceedings of the 1996 Winter Simulation Conference
ed. J. M. Charnes, D. J. Morrice, D. T. Brunner, and J. J. Swain

CSIM18 - THE SIMULATION ENGINE

Herb Schwetman

Mesquite Software, Inc.
3925 West Braker Lane
Austin, TX 78759-5321, USA

ABSTRACT

A simulation engine is the collection of components, features and support functions which are crucial to the implementation of an efficient discrete event simulation model. Furthermore, this model can be embedded in a larger application. A good simulation engine combines efficiency, functionality and completeness to enable a model builder to construct a comprehensive, customized model of either a specific system or a class of systems. This paper presents CSIM18, a simulation engine which supports development of applications with efficient, embedded simulation models on a variety of system platforms.

1 INTRODUCTION

Many application developers are finding a need to embed a simulation model into their application. Applications which help users design or configure complex systems for which performance is a major design goal are examples of such applications. This paper presents CSIM18, a simulation engine which has the features that developers find useful as they add simulation models to their applications.

An example of such an application would be a database design and layout program. In many cases, a database management system (DBMS) is used with a high performance on-line transaction processing (OLTP) system. A major goal of the system is to give "quick" responses to each kind of transaction in the stream of incoming transactions. The design program then needs to help the designer layout the database so that the response time requirement can be met. In order to evaluate a design, the designer needs to obtain estimates of transaction response times for the design just completed. A simulation model of the OLTP system with the DBMS as specified in the design program could provide the necessary estimates of the response times for each kind of transaction as the operation of the system is simulated in a realistic fashion. In this example, the design program is the larger application, and the model of the OLTP system and the DBMS is the embedded simulation model. In many cases, it would be appropriate and expedient to implement the embedded model with a simulation engine.

As this application is being designed and implemented, and the need for a simulation engine determined, a natural question is the one of "make or buy" - should the project implement their own simulation engine or should they examine the software marketplace, to see if an appropriate simulation engine can be purchased? Given the economics of software development, it is usually advantageous to buy a software component, if one can be found that (1) fills the need, (2) is compatible with the other software tools and components being used, (3) is supported and (4) is "reasonably" priced.

This paper first describes a simulation engine and presents the features which distinguish a good simulation engine. It next presents CSIM18 and shows why it fulfills the need for a simulation engine in many applications. The paper concludes with some examples of applications which have incorporated CSIM18 (or its predecessors) as the simulation engine for their models.

2 SIMULATION ENGINES

A simulation engine is a set of objects and methods which are used to construct simulation models which are embedded in applications. This is in contrast to a simulation environment which is itself the complete application. In many cases, an application is tailored to specific domain. In fact, an application can be quite complex, and the embedded simulation model is really a just minor component, usually used to help evaluate some aspect of the application.

The key point is that the simulation engine is used to build a model which is embedded in a larger application. The model is not the goal; the model is used to support the goal (the complete application). Some examples of applications with embedded models will be given in a later section.

There are several features which characterize a "good" simulation engine; these include:

a. the engine is a *complete* simulation package; it can support all of the important discrete-event simulation paradigms, including the process-oriented paradigm;
b. a model implemented with the engine can be an *accurate representations* of the "real" system; in particular, important features of the real system, including complex component interrelationships, unusual scheduling rules, and varying workload profiles can all be represented in the model;
c. models of almost any *size* can be created with the engine;
d. the engine is *efficient*; models constructed with the engine execute in an "reasonable" amount of time; in addition, both large and small models execute efficiently;
e. models based on the engine are *compact*;
f. models based on the engine are written in *standard programming languages*; in particular, developers do not have to learn a new language which is peculiar to the simulation model;
g. the model is *embeddable*; the model can be easily integrated with the rest of the application;
h. the engine has an *easy-to-use* interface; and
i. the model can be retargeted to *multiple platforms*; the model should not be an impediment to moving the application to different systems.

There are some additional features which could simplify the task of the application developer. These include:

a. *Automatic run-length control*: methods for determining when accurate results are achieved
b. *Random number streams* and *random number functions*: routines for generating streams of varying values from many different probability distributions, and
c. *Data collection* and *reporting routines*: routines for collecting data and generating useful ports.

When an application developer is evaluating simulation engines, these and other factors will be used to help select the best engine for use in the project.

It should be noted that a graphical user interface (GUI) was not listed as an important feature for a simulation engine. It is assumed that the larger application has its own GUI and that a GUI specific to the simulation model is inappropriate and not needed.

3 CSIM18

CSIM18 is a complete simulation engine. Earlier versions of CSIM18 are described in a number of articles (Schwetman 1986, 1988, 1990, 1994, 1995). These predecessors to CSIM18 have been used to model a large variety of systems, including communications systems [Edwa92], computer systems, transportation systems, microprocessors, and fault tolerant systems.

CSIM18 is a library of classes, functions, procedures and header files which enable developers to implement efficient models of complex systems. The models are written in either the C or C++ programming languages. The library is available on many different system platforms, including PC's with either Windows 3.1, Windows 95, Windows NT, OS/2 Warp and Linux operating systems. It is also available on almost all UNIX workstations, including Sun SPARC (SunOS and Solaris), DEC Alpha (with OSF/1), HP PA (with HP/UX), IBM RS/6000 (with AIX) and SGI workstations. There is also a version for the Power Mac (with the Metroworks C++ compiler).

CSIM18 supports process-oriented, discrete-event simulation models. In this kind of model, the complete system is represented by a collection of simulated resources and a collection of processes which compete for use of these resources. Typically, such a model is used to provide estimates of the performance of the real (modeled) system. The goal is to use the model to "try out" different system configurations or different workloads or different resource scheduling rules, to find the "best" configuration (or workload or schedule) with respect to the performance goal. In most cases, it is much "cheaper" to try out configurations in a system model than it is to use the real system.

The specific classes (simulation components) provided with CSIM18 (Mesquite 1996) include:

a. Processes: a CSIM process is an independent thread of execution; the active entities of a model are implemented as processes; since many processes can appear to be executing simultaneously, processes can mimic the activities of entities operating in parallel;
b. Facilities (these are called "active" resources in some simulation systems) - resources are typically "used" by processes in the model; facilities have one or more servers; processes queue up for access to a server;
c. Storages (these are called "passive" resources in some simulation systems) - a storage is initialized with a number of storage units; processes can "allocate" some subset of the storage units;
d. Events - events are used to synchronize actions of different processes, and
e. Mailboxes - mailboxes are used to interchange information between processes.

In addition, CSIM18 provides classes (simulation components) to assist the with collection of data in a model; these include:

a. Tables - used to collect real valued (double) data values and to report on their statistical properties;
b. Qtables - used to collect integer valued, time depend-

ent data values (e.g. queue lengths) and to report on their statistical properties;

c. Meters - used to measure the flow of entities past a point in the system and, also, to measure the times between successive passages of entities past this point; and

d. Boxes - used to collect data on time spent in a specified range of process activities.

A standard report can be generated for each facility, storage, table, qtable, meter and box. Histograms and confidence intervals can be requested for each table, qtable, meter and box. In addition, "inspector" functions are provided to give programs access to every data item collected for all of the structures with report capabilities. The developer can use these inspector functions to produce reports tailored for the specific application.

A "run-length-control" feature can be associated with the mean value of a table, qtable, meter or box (Mesquite 1995). This feature allows the modeler to specify a "stopping rule" for execution of the model. The stopping rule is based on achieving a specified level of confidence for some output statistic. A CPU time limit is also specified, so that the model will cease execution (with possibly inaccurate results) when this time limit is exceeded. This run-length-control feature could be very important for applications which depend on achieving accurate results in an efficient manner from the embedded simulation model.

CSIM18 has some other features which will assist the implementation of embedded models. These include:

a. the "rerun" statement: used to "tear down" a model, so that a new model can be constructed by the program;

b. the "reset" statement: used to discard results collected so far from all of the data collection facilities in a model; this used to allow for a "warm up" interval before actual data collection is started; and

c. model debugging aids: a selective event trace can be used to give details on selected (or all) aspects of the execution of the model; for some platforms, an interactive execution monitor lets the developer examine, interactively, many parts of the model and also lets the developer "step through" execution of the model.

Models produced with CSIM18 can execute very efficiently. A test program has been written and executed on many different platforms. The test consists of a model of an M/M/1 queue with 50% server utilization operating for 5000 arrivals and 5000 departures. The model requires 17,485 simulation events. The following tables shows the timings for executing this model on several platforms:

System:	Dell PC	SPARC	RS/6000
Processor:	Pentium 133	SStation5	PowPC 601
OS:	Win95	Solaris	AIX
Compiler:	MS VC++ 4.0	g++ 2.7	xlc xx
User CPU sec:	0.430	1.580	0.450
Events per sec:	40,662	11,066	38,856

CSIM18 is delivered with a complete set of documentation consisting of a Users' Guide (for either the C or C++ version) and a introductory ("Getting Started") manual, along with installation instructions.

CSIM18 is the newest version of a series of CSIM simulation software toolkits. CSIM was originally developed in 1985 at MCC. The preceding version, CSIM17, was developed and sold by Mesquite Software, Inc. (under the terms of a license agreement with MCC) in 1994. CSIM18 incorporates suggestions, enhancements and improvements from many sources including customers and developers using CSIM17.

4 EXAMPLES

Several projects have used previous versions of CSIM18 as a simulation engine for embedded models. This section describes some of these projects.

4.1 VisSim/Discrete Event

Visual Solutions, Inc., sells VisSim (Visual Solutions 1995), a simulation package for developing continuous, discrete, multi-rate and hybrid system models and for performing dynamic simulations. VisSim incorporates an outstanding graphics user interface which is used to create diagrams of the systems being simulated. Each model consists of a collection of components connected by flexWires. Components can perform many kinds of activities such as evaluating a mathematical function, generating a random value, performing an arithmetic or logical test, and producing different forms of animation. VisSim add-ons allow users to find optimal solutions to mathematical programming problems, model fuzzy-logic systems and do real-time control functions.

Some of the VisSim users expressed a need for discrete-event simulation capabilities in the VisSim modeling building environment. Visual Solutions responded to this need by engaging Mesquite Software to design and implement VisSim/Discrete-Event (Visual Solutions 1996), an add-on to VisSim which lets modelers use the VisSim GUI to construct, test and execute discrete-event simulation models. CSIM17 (the predecessor to CSIM18) is the simulation engine which "powers" VisSim/Discrete-Event.

Briefly, new VisSim blocks which represent most of the CSIM classes and methods are added to VisSim. These blocks are interconnected by VisSim flexWires. The model consists of one or more of "process diagrams". Process diagrams represent creating processes (the source block), terminating processes (the sink block) and many different kinds of process activity blocks (such as useServiceStation, allocateStorage, waitEvent and computeResult). Other blocks represent serviceStations (active resources), storages (passive resources), events and tables. Some of the standard VisSim blocks are used to present output results (display blocks, meters and strip charts) and perform some kinds of arithmetic operations. Compound blocks, which are aggregations of blocks, can be used to represent submodels.

VisSim/Discrete-Event is a powerful tool for creating discrete-event simulation models with a block diagramming interface. The models can represent many styles of process behavior and can be used to model many different kinds of systems. The internal simulation engine, CSIM17, when combined with the VisSim GUI enabled this new modeling environment to be created with a reasonably modest effort (about 2/3 of a man-year).

4.2 ArchGen

CAE + Corporation has developed and is selling a CAD system called ArchGen (Jain 1995, 1996). ArchGen captures, validates and optimizes system IC specifications. It includes a CSIM model for functional and performance simulation with clock edge accuracy. CSIM17 is the simulation engine used in ArchGen.

CAE+ decided to use CSIM because they needed a model which could be "connected" to the rest of the ArchGen components, and they needed a model which has a high rate of executing basic simulation activities (the IC designer is usually sitting at the workstation, waiting for the model to produce its estimates of chip performance). Mesquite Software has worked closely with CAE+, to insure that the functionality and performance needed by ArchGen would be available in CSIM.

4.3 *SPE•ED*

L&S Computer Technology, Inc. sells *SPE•ED* (Smith and Wong 1994), a tool for modeling, at a high level, the functionality and behavior of large software systems and for providing estimates of the performance of these systems in operation. *SPE•ED* supports the methods and techniques of performance engineering, to help software designers translate their designs to efficient implementations. L&S Computer Technology selected CSIM as the simulation engine because it was a "excellent fit" with the needs and capabilities of *SPE•ED*. In addition, CSIM was available on the platforms that L&S needed

4.4 Others

A large computer manufacturer introduced a large scale multiprocessor system and a parallel relational database management system (RDBMS). The system supports both high-volume on-line transaction processing and large, complex decision support queries. The manufacturer decide to offer a configuration planning and management application, to support both layout of the database and performance tuning of the transaction processing software. This application incorporated a simulation model of the multiprocessor system along with the parallel RDBMS and the transaction processing software. CSIM was chosen to be the basis for the simulation model. The key reasons for selecting CSIM were that CSIM could be easily integrated into the rest of the application, CSIM provided the necessary functionality for implementing the model, and CSIM provide efficient execution of the model. The efficient execution of large models turned out to be important, because some the configurations suggested for customers were very large and the transactions being simulated could be quite complex.

A research project at a large university built a software system for designing and evaluating fault-tolerant computer systems. It used CSIM as the simulation engine for the models of failures and repairs in the systems being designed.

A consulting firm built a model of inventory management system for a warehousing operation. They used CSIM as the basis of the model used to "try out" different management rules. The application was written in Visual Basic for execution on a PC with Windows. The Visual Basic application connected to the model which was implemented as a DLL written in Visual C++ and the CSIM17 library.

5 SUMMARY

CSIM18 is a simulation engine. It can be used to implement simulation models which will be embedded in domain specific applications. Rapid development of customized applications has been greatly enhanced by the emergence of application development environments such as Microsoft Visual C++, Borland C++, Microsoft Visual Basic and other similar products. One project has recently developed an application based on an Excel spreadsheet, with a link to a CSIM model.

The predecessor of CSIM18, CSIM17, has been successfully used to develop realistic models of many different kinds of complex systems. Some examples of applications incorporating CSIM models have be given in this paper. CSIM18 offers several new features which will be essential to developers. The functionality, compactness and efficiency of CSIM18 mean that models based on CSIM18 will enhance the surrounding applications.

ACKNOWLEDGMENTS

CSIM is copyrighted by Microelectronics and Computer Technology Corporation (MCC). CSIM17 and CSIM18 are supported and marketed by Mesquite Software, Inc. under a license from MCC. Dr. Jeff Brumfield developed the new data collection and presentation functions, including the run length control algorithm, in CSIM18.

REFERENCES

Edwards, G. and R. Sankar. 1992. Modeling and simulation of networks using CSIM. *Simulation* 58(2): 131 - 136.

Jain, P. 1995. A comprehensive pre-RTL design methodology. In *Proceedings of the 1995 International Verilog HDL Conference*. IEEE Computer Society Press.

Jain. P. 1996. Specification validation of embedded system ICs. In *Proceedings of Embedded Systems Conference 1996*. Miller Freeman, Inc.

Mesquite Software, Inc. 1995. Confidence intervals and run length control in CSIM18. Internal Report. Austin, TX.

Mesquite Software, Inc. 1996. *CSIM18 Users Guide*. Austin, TX.

Schwetman, H. 1986. CSIM: A C-based, process-oriented simulation language. In *Proceedings of the 1986 Winter Simulation Conference*, ed. J. Wilson, J. Henriksen, and S. Roberts, 387 - 396. Washington, DC.

Schwetman, H. 1988. Using CSIM to model complex systems. In *Proceedings of the 1988 Winter Simulation Conference*, ed. M. Abrams, P. Haigh, and J. Comfort, 246 - 253. San Diego, CA.

Schwetman, H. 1990. Introduction to process-oriented simulation and CSIM. In *Proceedings of the 1990 Winter Simulation Conference*, ed. O. Balci, R. Sadowski, and R. Nance, 154 - 157. New Orleans, LA.

Schwetman, H. 1994. CSIM17: A simulation model-building toolkit. In *Proceedings of the 1994 Winter Simulation Conference*, ed. J. Tew, S. Manivannan, D. Sadowski, A. Seila, 464 - 470. Orlando, FL

Schwetman, H. 1995. Object-oriented simulation modeling with C++/CSIM17. In *Proceedings of the 1995 Winter Simulation Conference*. ed. C. Alexopoulos, K. Kang, W. Lilegdon, D. Goldsman, 529 - 533. Washington, D.C.

Smith, C and B. Wong. 1994. SPE evaluation of a client/server application. In *Proceedings of Computer Measurement Group*. Orlando, FL.

Visual Solutions, Inc. 1995. *VisSim users' guide*. Westford, MA.

Visual Solutions, Inc. 1996. *VisSim/Discrete-Event user's guide*. Westford, MA.

AUTHOR BIOGRAPHY

HERB SCHWETMAN is founder and president of Mesquite Software, Inc. Prior to founding Mesquite Software in 1994, he was a Senior Member of the Technical Staff at MCC from 1984 until 1994. From 1972 until 1984, he was a Professor of Computer Sciences at Purdue University. He received his Ph.D. in Computer Science from The University of Texas at Austin in 1970. He has been involved in research into system modeling and simulation as

Proceedings of the 1996 Winter Simulation Conference
ed. J. M. Charnes, D. J. Morrice, D. T. Brunner, and J. J. Swain

ALPHA/SIM SIMULATION SOFTWARE TUTORIAL

Kendra E. Moore
John E. Brennan

ALPHATECH, Inc.
50 Mall Road
Burlington, MA 01803-4562, U.S.A.

ABSTRACT

ALPHA/Sim is a general-purpose, discrete-event simulation tool. ALPHA/Sim allows a user to graphically build a simulation model, enter input data via integrated forms, execute the simulation model, and view the simulation results, within a single graphical environment. In this paper, we introduce ALPHA/Sim and describe how to use ALPHA/Sim to build, simulate, and analyze a simple manufacturing system. In addition, we briefly describe some advanced features and list some sample applications.

1 INTRODUCTION

ALPHA/Sim is a general-purpose, discrete-event simulation tool. With ALPHA/Sim you can graphically build a simulation model, enter input data (timing delays, routing rules, initial conditions, and other data) via integrated forms, execute the simulation model, and view the simulation results, within a single graphical environment.

ALPHA/Sim provides a hierarchical modeling capability that allows models to be built from the bottom-up, top-down, or both. Models can be built without seeing or writing a single line of code; it is also possible to link to external software. ALPHA/Sim automatically collects statistics on populations (queues), delays, activity rates, and attributes.

ALPHA/Sim has been used in a wide number of applications including computer hardware systems, manufacturing systems, queuing systems, and military command and control. ALPHA/Sim currently runs on the PC (Windows NT) and Sun Workstation (SunOS and Solaris under the X Window System or Motif).

The modeling paradigm used in ALPHA/Sim is based on Petri nets (PNs). PNs were developed in the early 1960s to model concurrent operations in computer systems. Over the years PNs have been extended and applied to a wide range of systems characterized as being concurrent, asynchronous, distributed, parallel, and stochastic. PNs are a mathematical and graphical modeling tool. As a mathematical tool, PNs can be used to set up state equations, algebraic equations, and simulation models. As a graphical tool, PNs provide a visual modeling technique.

In this paper, we present a brief overview of PNs and describe how to use ALPHA/Sim to implement a simple manufacturing model. Specifically, we describe how to build the graphical model and define attributes, token types, timing delays, decision rules, output attribute definitions, and statistics collection. In addition, we briefly describe some advanced features and list some sample applications.

2 PETRI NETS

Petri nets (PNs) are a graphical and mathematical modeling technique originally developed by C.A. Petri in the early 1960s to characterize concurrent operations in computer systems (Petri 1962). PNs have been extended to capture many important aspects of large-scale systems, including attributes, timing relationships, and stochastic events (Moore and Lynch 1990, Moore et al. 1986, Murata 1989, Peterson 1981). The greatest appeal of PNs is their conceptual simplicity.

PNs consist of four primitive elements (tokens, places, transitions, and arcs) and the rules that govern their operation (Figure 1). PNs are based on a vision of *tokens* moving around a network. Tokens appear as dots and represent the objects or entities in a system. *Places* are shown as circles and represent the locations where objects await processing. Location can be either a physical location (e.g., the queue where a message waits to be processed) or a state (e.g., an idle resource). *Transitions* appear as rectangles and represent processes or events (e.g., processing a message or machining a part). Finally, *arcs* represent the paths of objects through the system. Arcs connect places to transitions and transitions to places; the direction of the path is indicated by an arrowhead at the end of the arc.

(a) Token (b) Place (c) Transition (d) Arc

Figure 1: Depiction of PN Primitives

PN firing rules specify the behavior of transitions; i.e., the conditions under which processes or events can occur. Three rules govern transition firing:

1. When all upstream places are occupied by at least one token, the transition is *enabled.*
2. Once enabled, the transition *fires.*
3. When a transition fires, exactly one token is removed from each upstream place and exactly one token is placed in each downstream place.

Figure 2 depicts these rules for a transition (Assemble) with two upstream places (Part A, Part B) and one downstream place (Assembly).

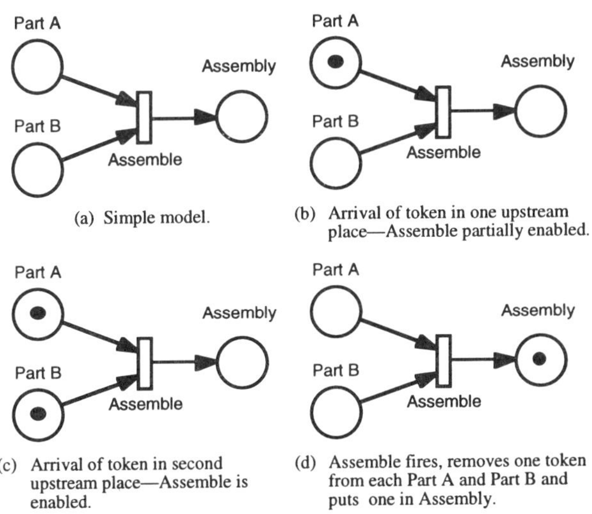

(a) Simple model.

(b) Arrival of token in one upstream place—Assemble partially enabled.

(c) Arrival of token in second upstream place—Assemble is enabled.

(d) Assemble fires, removes one token from each Part A and Part B and puts one in Assembly.

Figure 2: Transition Firing

Timing rules are associated with transitions and represent the time required to complete some activity. A timing rule may be stochastic, based on an assigned probability function, a computed value, or a constant. *Decision rules* are associated with places and resolve cases where more than one transition is enabled by the same token or set of tokens. There are three types of decision rules: priority, probability, and constructed. The *priority* decision rule (shown in Figure 3) states that, if all other firing rules are met, the token will leave by the path with the highest priority. The *probability* decision rule states that if all other firing rules are met, the token will select a path based on assigned probabilities. The *constructed* decision rule allows the user to specify the conditions under which the token will select a particular path, given that the firing rules are met.

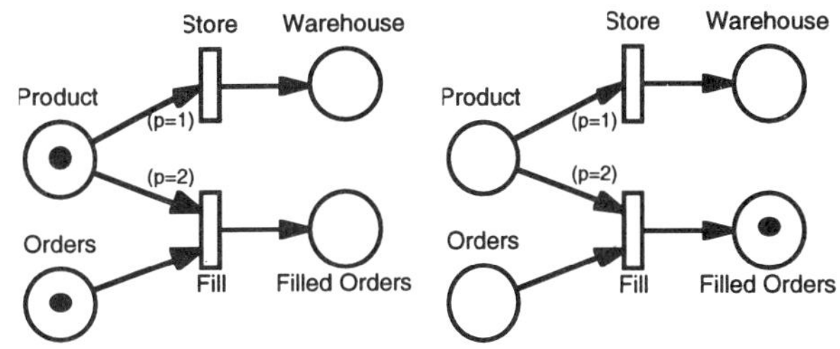

(a) Both transitions enabled; Fill will fire since its priority is higher than the priority of Store.

(b) Fill fires, consuming tokens from Product and Order, and putting one token in Filled Orders.

Figure 3: Effect of a Priority Decision Rule

Attributes on tokens are used to specify a set of characteristics associated with a token (e.g., size, type, priority, identity, etc.). The values of the attributes may be changed at transitions. They can also be used to determine timing and decision rules. Finally, the values of the attributes can be passed to external algorithms and the results incorporated into the PN model.

There are two other types of arcs, in addition to the standard arc, which provide complex transition logic. The *enable* arc is depicted as a line with a solid filled circle at the end where the arrowhead normally appears. The enable arc enables a transition when the upstream place has a token, but does not consume the token (the token remains in the upstream place). The *inhibit* arc is depicted as a line with a hollow circle at the end where the arrowhead normally appears. The inhibit arc disables a transition when the upstream place has a token in it (the token is not consumed along the inhibit arc).

Box nodes are used to encapsulate portions of a PN model and to provide a hierarchical modeling capability. Box nodes are used to group or cluster PN fragments that relate to various subsystems, functions, or organizational units.

3 BUILDING MODELS WITH ALPHA/Sim

With ALPHA/Sim you can: build and debug your models graphically; build models from the top-down, bottom-up, or both; easily modify model parameters and structure; navigate through the model; monitor results at any point in the simulation run; and save any model component for reuse in other models.

In the remainder of this section we illustrate how to use ALPHA/Sim to implement a simple manufacturing system which produces two types of parts. Type 1 parts are turned, milled, and plated, in that order; Type 2 parts are turned and milled. Input parameters include part mix and part processing times. Output parameters include buffer sizes (queue lengths), machine utilization, part latency, and throughput.

3.1 Drawing the Graphical Model

We begin by drawing the graphical model. Figure 4 shows an ALPHA/Sim screen. The screen has a menu bar at the top, an icon palette on the left, and a drawing window with scroll bars. We create the graphical model by using the mouse and icon palette to drop icons in the drawing window and connect them with arcs. Icons are automatically assigned default names (see Figure 4); these can be changed to be more meaningful.

Figure 5 shows the complete manufacturing model. The place-transition combination in the upper left corner periodically generates new parts into the place labeled Lathe_Q. Arriving parts wait in the Lathe_Q, until the Lathe becomes available. They are then turned and enter the Mill_Q, where they wait for the Mill to become available. Once the parts are milled, they are passed to the Plate_Q. Since Type 2 parts are done, they are immediately routed to the stock of Finished_Type_2 parts; meanwhile, Type 1 parts remain in the Plate_Q until the Plating_Machine becomes available. Once plating is complete, the Type 1 parts enter the Finished_Type_1 stock.

3.2 Defining Token Types

Once we have built the graphical model, we can define the token types using the Token Type Edit Form. For this model, we will use two token types: Part and Machine. Figure 6 shows the Token Type Edit Form for the Part token type. This form contains a field for identifying the token type and allows us to define the attributes associated with the token type. Each attribute definition consists of a name, class, type, and an initial range.

The attribute's class is a scalar (single) value, an array of values, or a matrix of values. If an attribute's class is array or matrix, we must also specify its size (rows and columns). The attribute's type refers to its format. Valid types include: Boolean, integer, real, string, or another (previously defined) token type. If the type is not another token type, we have the option of specifying an initial range for the attribute. Table 1 summarizes the token type definitions for the manufacturing model.

3.3 Place and Transition Forms

Once we have defined the token types, we can use the place and transition forms to assign token types to places and specify timing, routing, and other logical rules.

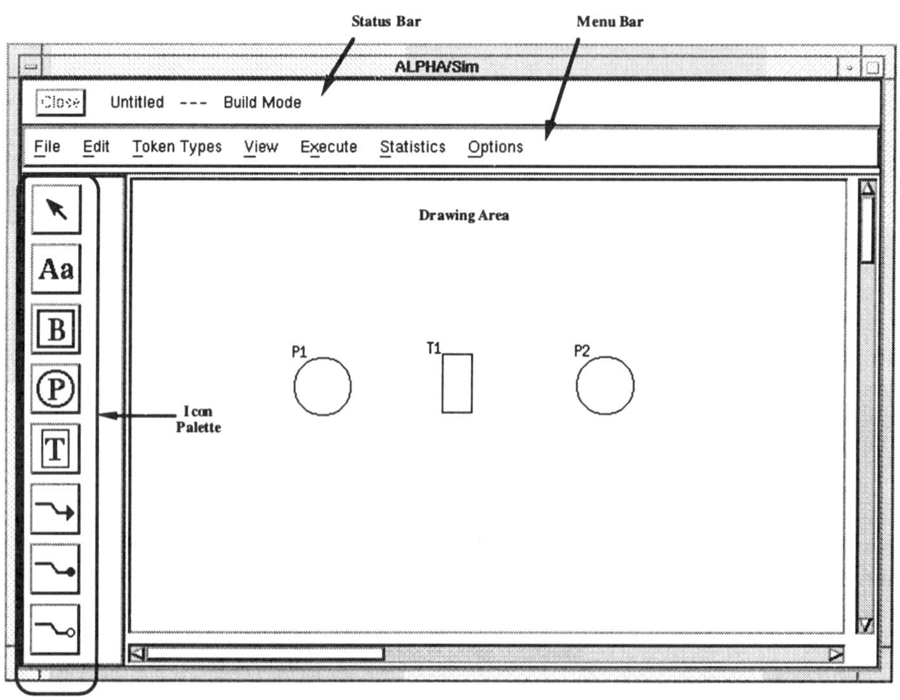

Figure 4: Sample ALPHA/Sim Screen

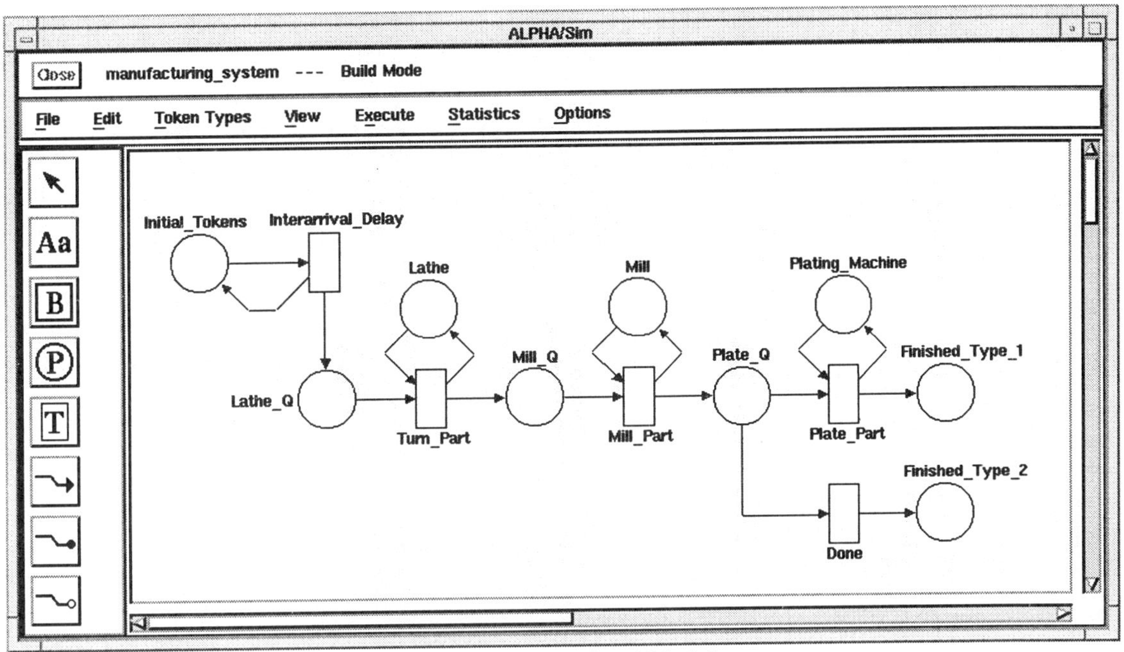

Figure 5: Completed Model

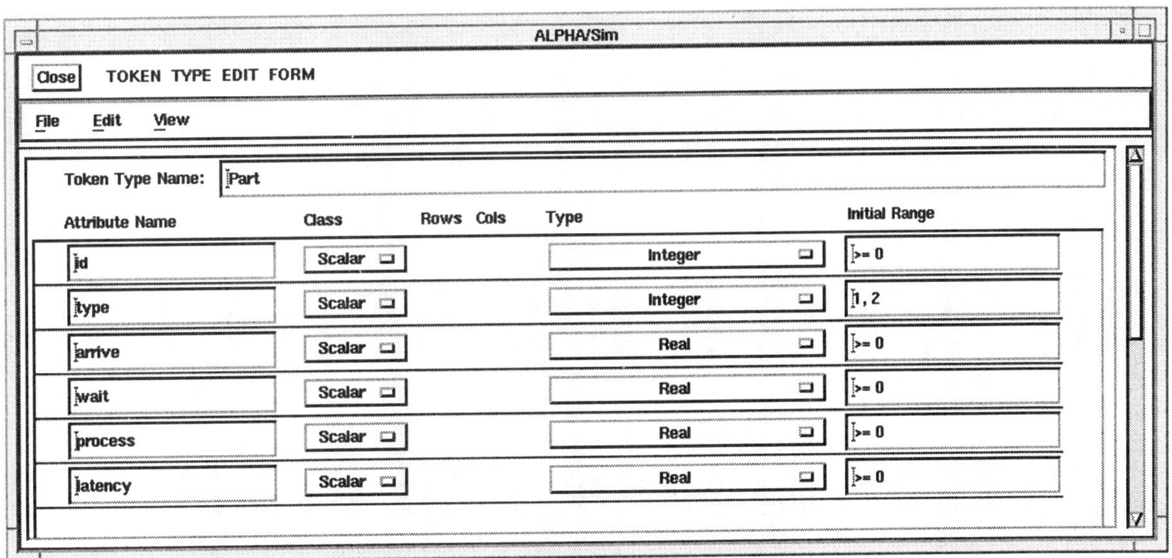

Figure 6: Token Type Edit Form for the Parts Token Type

Table 1: Token Type Definitions for Model

Token Type	Attributes	Class	Type
Part	id	Scalar	Integer
	type	Scalar	String
	arrive	Scalar	Real
	wait	Scalar	Real
	process	Scalar	Real
	latency	Scalar	Real
Machine	id	Scalar	Integer
	type	Scalar	String
	p_time	Scalar	Real

Figure 7 shows a sample place form. The top of the place form lists the input and output transitions, and allows us to specify the token type and the number of initial tokens. The middle of the form allows us to specify statistics collection and set the queuing order (FIFO, LIFO, or ascending/descending on an attribute value). The bottom of the form allows us to set decision rules (priority, probability, or constructed) for routing tokens out of the place.

Figure 8 shows a sample transition form. The left side of the transition form lists the input places; clicking on one of these places opens the input token profile displaying the input token type definition. Similarly, the right side of the form lists the output places and clicking on one opens the output token profile. Ordinarily, the input attribute values are mapped to the output attribute values; however, we can assign new values to these attributes using the area below the output token profile.

The center of the transition form is used to set a timing rule and to specify statistics collection. Regarding the timing rules, we can choose None, Selected Distribution, or Constructed. If we choose Selected Distribution, we are prompted to select one of the available distributions and provide the appropriate parameters. Table 2 lists these distributions and their parameters. If we choose Constructed, we can enter an expression utilizing other distributions or attribute values. The language used for the expression is English-like; e.g., the timing rule in the Interarrival_Delay transition is:

```
IF (Initial_Token.type = 1)
   exponential(10)
ELSE exponential(5)
```

"Initial_Token.type" is the value of the attribute "type" on the token coming from the place "Initial_Token". The lower left corner of the transition form is used to set specific enabling or inhibit logic (using attributes) or conditions for stopping the simulation.

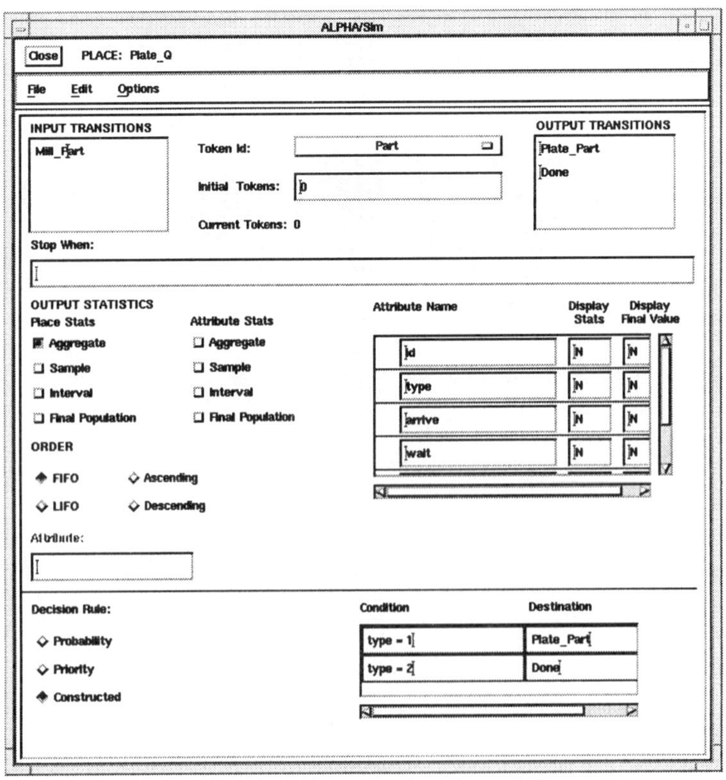

Figure 7: Place Form for the Plate_Q Place

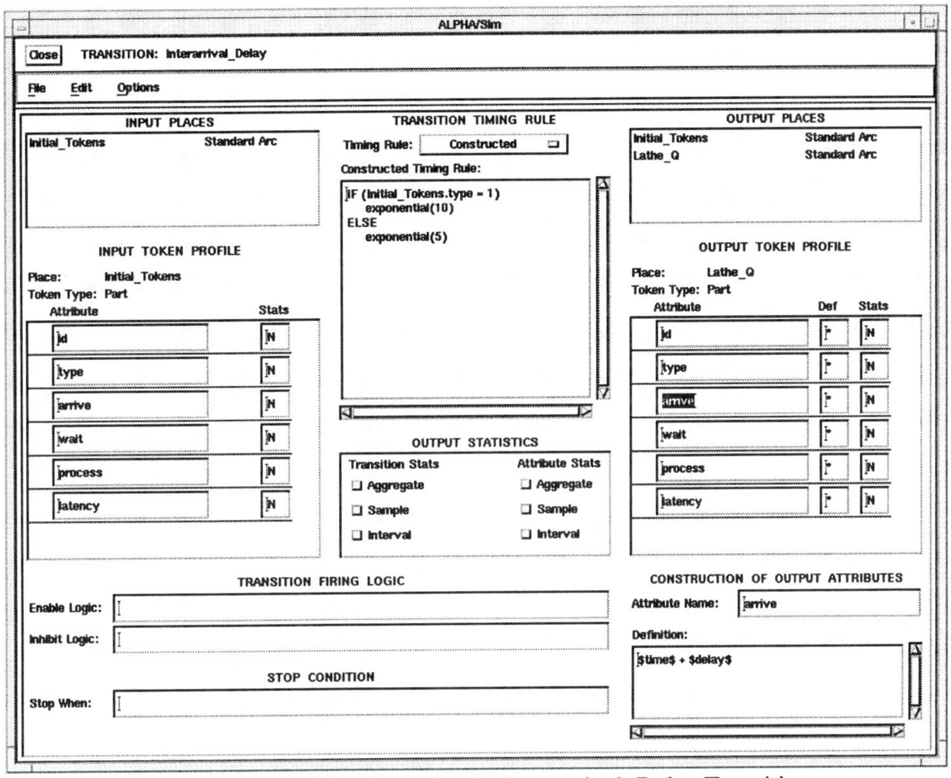

Figure 8: Transition Form for the Interarrival_Delay Transition

Table 2: Built-In Timing Distributions

Distribution	Parameters
Constant	Value
Exponential	Mean
Gamma	Alpha, Beta
Normal	Mean, Std Dev, Min, Max
Triangular	Min, Mode, Max
Uniform	Min, Max

3.4 Specifying the Model Logic via the Forms

We use these forms to associate the token types with places, and specify the initial tokens, the decision (routing) rules, timing rules, and output attribute definitions. First, we associate token types with places using the Token Id option menu in the Place Forms. At this time, we set two initial tokens in Initial_Token (one for each type of part), and one initial token in each of the machine places (Lathe, Mill, and Plating_Machine). Table 3 lists the token type assignments and initial populations for each place in the model; Table 4 lists the initial values of the attributes for those places with initial token populations.

Next, we specify a constructed decision rule for the place labeled Plating_Q to route Type 1 and Type 2 parts. Figure 7 shows the decision rule used for routing parts as it appears in the Plate_Q place form. Alternatively, we can use enabling logic in the Plate and Done transitions. If desired, we can set a queuing order in the Lathe_Q and Mill_Q places.

Table 3: Token Type Assignments and Initial Populations

Token Type	Places	# Init. Tokens
Part	Initial_Token	2
	Lathe_Q, Mill_Q, Plate_Q, Finished_Type_1, Finished_Type_2	0
Machine	Lathe, Mill, Plating_Machine	1

Table 4: Initial Token Values

Place	Attributes	Initial Value
Initial_Token	type	1 or 2
	all others	0
Machine	id	Unique integer
	type	"lathe", "mill", "plating"

Finally, we set various output attribute definitions to collect information on the machines and on individual parts as they pass through the system. ALPHA/Sim provides four system variables that can be used in expressions; these are $time$ (the current simulation time), $delay$ (the timing delay of an individual transition firing), $count$ (the number of times a specified transition has fired), and pop (the current number of tokens in a place). Since we are interested in the queuing, service, and system times for the customers, we will make use of the $time$ and $delay$ variables. Table 5 lists the expressions that are used in output attribute definitions; Fig. 8 shows a sample output attribute definition in a transition form (lower right). In addition, we can turn on statistics collection for places (average population), transitions (firing rates), and attributes, using the place and transition form statistics panels (see Figs. 7 and 8).

Table 5: Sample Output Attribute Definitions

Transition	Attribute	Definition
Interarrival_ Delay	id	\$count\$
	arrive	\$time\$ + \$delay\$
Turn	wait	\$time\$ - arrive
	process	\$delay\$
Mill	wait	\$time\$ - arrive - process
	process	process + \$delay\$
Plate	wait	\$time\$ - arrive - process
	process	process + \$delay\$
	latency	\$time\$ - \$delay\$ - arrive
Done	latency	\$time\$ - arrive

3.5 Controlling the Simulation Run

Additional forms are available to set the simulation run time, the number of replications and random number seeds, and statistics collection preferences. ALPHA/Sim has facilities for collecting aggregate, interval, and sample statistics. At runtime, ALPHA/Sim checks all expressions to make sure that there are no errors and executes the simulation. The results can be observed on-screen or sent to a file for further analysis. The simulation can also be run in batch mode.

4 ADVANCED FEATURES

ALPHA/Sim incorporates a number of additional features. These include: functions, enable and inhibit logic, stop when conditions, boxes, show tree, and various printing and file handling features. ALPHA/Sim includes over thirty built-in mathematical functions as well as arithmetic and logical operators that can be used in timing rules, decision rules, output attribute definitions, and other expressions. In addition, it is possible to incorporate user-defined functions and interact with external code. Enable and inhibit logic can be used in transition forms to specify which combinations of tokens will cause a transition to fire. Stop when conditions are logical expressions that can be used to halt the simulation if a specified condition is reached. Boxes provide a hierarchical modeling capability. Show tree allows you to view a model's hierarchy in a tree structure and provides an easy way to navigate through a model. The graphical model and the information contained in the forms can be printed out to a laser printer or sent to a file.

5 SAMPLE APPLICATIONS

ALPHA/Sim and its predecessor, Modeler, have been used to develop a wide array of discrete-event simulation models. These include computer components and systems (e.g., Ethernet system (Brennan, Walenty, and Moore 1995), client-server system, and high-speed disk systems), manufacturing systems (Moore and Gupta 1996), large-scale military command and control systems (Moore and Lynch 1990), and business process reengineering and workflow models for a charter air cargo and passenger service system.

The client-server system consists of several data processing nodes connected via a local area network (LAN). The model evaluates the impact of changing the number of hardware components and their capabilities on throughput and latency for individual processes. It also identifies bottlenecks in the system, thereby indicating good candidates for increasing capacity.

The charter air cargo and passenger service model depicts the workflow for a thirty person office responsible for handling and scheduling domestic and international transportation. This workflow is unique in that the staff's activities are frequently interrupted by higher priority tasks and phone calls or delayed due to communications delays. The model was used to determine the impact of automation and task redefinition on staffing requirements and throughput.

6 SUMMARY AND CONCLUSIONS

In this paper, we described a general-purpose, discrete-event simulation software tool called ALPHA/Sim. With ALPHA/Sim you can: build and debug your models graphically; build models from the top-down, bottom-up, or both; easily modify model parameters and structure; navigate through the model; monitor results at any point in the simulation run; and save any model component for reuse in other models.

With ALPHA/Sim's graphical modeling and simulation environment it is possible to develop and exercise simulation models without having to see or write a line of code. The graphical interface allows you to design the model using the mouse and icons. Integrated forms provide the means for specifying logic and input parameters for the model. ALPHA/Sim also provides the ability to interface with external software.

We described how to use this tool via a simple example of a manufacturing system. This example illustrates the key features of ALPHA/Sim. In addition, we briefly listed some of the advanced features of the tool. We also listed a number of sample applications and briefly described two of these, namely a client-server performance model and a business process workflow model.

ACKNOWLEDGMENTS

ALPHA/Sim is a trademark of ALPHATECH, Inc. Motif is a registered trademark of Open Software Foundation, Inc. Sun OS is a trademark of Sun Microsystems, Inc. Sun Workstation is a registered trademark of Sun Microsystems, Inc. UNIX is a registered trademark of UNIX Systems Labs, Inc. Windows NT is a trademark of Microsoft Corporation. X Window System is a trademark of the Massachusetts Institute of Technology.

REFERENCES

Brennan, J.E., M.E. Walenty, and K.E. Moore. 1995. Simulating a UNIX processor using Petri nets and ALPHA/Sim. In *Proceedings of the 1995 Summer Computer Simulation Conference*, ed. T. I. Ören and L. G. Birta, 63-69. Society for Computer Simulation, San Diego, CA.

Moore, K.E. and S.M. Gupta. 1996. Petri net-based analysis and simulation of traditional and flexible kanban control policies, ed. K. Stelson In *Proceedings of the Japan-USA Symposium on Flexible Automation*. American Society of Mechanical Engineers, New York, NY.

Moore, K.E. and J.P. Lynch. 1990. Stochastic, timed, attributed Petri net (STAPN) modeling of antisubmarine warfare C^3 architectures. In *Proceedings of the 1990 Symposium on Command and Control Research*, 311-325. SAIC, McLean, VA.

Moore, K.E., R.R. Tenney, P.A. Vail. 1986. Systematic evaluation of command and control systems. Technical Report TR-284, ALPHATECH, Inc., Burlington, MA.

Murata, T. 1989. Petri nets: properties, analysis and applications. *Proceedings of the IEEE* 77: 541-580.

Peterson, J. L. 1981. *Petri net theory and the modeling of systems*. Englewood Cliffs, NJ: Prentice-Hall.

Petri, C.A. 1962. *Kommunikation mit automaten*. Institut für Instrumentelle Mathematik, Schriften des IIM Nr. 3, Bonn, Germany. Also, English translation, "Communication with automata." Technical Report RADC-TR-65-377, vol. 1, Suppl. 1, January 1966, Griffiss Air Force Base, NY.

AUTHOR BIOGRAPHIES

KENDRA E. MOORE is the Simulation Products and Services program manager at ALPHATECH, Inc., in Burlington, Massachusetts. She received her MS degree (1989) in operations research from the Department of Industrial Engineering and Information Systems at Northeastern University in Boston, Massachusetts. She is currently pursuing a PhD in operations research at Northeastern; her research is focused on using Petri nets to model flexible manufacturing systems. She has a MA degree (1985) in philosophy of religion from Columbia University in New York City, and a BA degree (1981) in philosophy and religion from Stephens College in Columbia, Missouri. Since joining ALPHATECH in 1985, Ms. Moore has been actively involved in developing and applying simulation and modeling techniques and tools. Her areas of interest are Petri nets, discrete-event simulation, optimization, manufacturing systems, business process reengineering, and performance analysis.

JOHN E. BRENNAN is a senior simulation engineer at ALPHATECH, Inc., in Burlington, Massachusetts. He holds a BS degree (1981) from Catholic University in Washington, DC and an MS degree (1989) in operations research from the College of Business and Management at the University of Maryland at College Park. Since joining ALPHATECH in 1994, he has been involved in modeling and analysis of hardware and software architectures to support business process reengineering. Prior to joining ALPHATECH, Mr. Brennan was an engineer with the American Red Cross Biomedical Services division involved in systems design and analysis to support process improvement efforts in the collection, testing, manufacture, and distribution of blood and blood products. His areas of interest include simulation, decision support systems, process improvement, and health care applications.

Proceedings of the 1996 Winter Simulation Conference
ed. J. M. Charnes, D. J. Morrice, D. T. Brunner, and J. J. Swain

BUSINESS PROCESS MODELING WITH SIMPROCESS

Maya Binun

CACI Products Company
3333 North Torrey Pines Court
La Jolla, CA 92037, U.S.A.

ABSTRACT

This paper gives an overview of business process modeling with SIMPROCESS, its applications, unique features, basic and advanced modeling constructs, and benefits.

1 WHAT IS SIMPROCESS?

SIMPROCESS is a hierarchical and integrated process simulation tool that radically improves your productivity for process modeling and analysis. SIMPROCESS is designed for BPR and IT professionals of industrial and service enterprises who need to reduce the time and risk it takes to service customers, fulfill demand, and develop new products.

Unlike other tools, SIMPROCESS integrates process mapping, hierarchical event-driven simulation, and activity-based costing into a single tool. The architecture of SIMPROCESS provides an integrating framework for ABC. The building blocks of SIMPROCESS, namely processes, resources, and entities (flow objects), bridges ABC and dynamic process analysis. ABC embodies the concept that a business is a series of inter-related processes, and that these processes consist of activities that convert inputs to outputs. The modeling approach in SIMPROCESS manifests this concept, builds on it by organizing and analyzing cost information on an activity basis.

2 HOW DOES SIMPROCESS WORK?

SIMPROCESS uses four easy steps to model your processes.

2.1 Step #1. Create Process Model

Creating a process model using SIMPROCESS is as easy-as 1-2-3. First, you graphically select activities and processes from the tool palette and link them using connectors to define the process flow. Second, you define the entities (or flow objects) and resources used in the model. Third, you customize the behavior of your model and create a realistic model of your business process by filling in the dialog boxes.

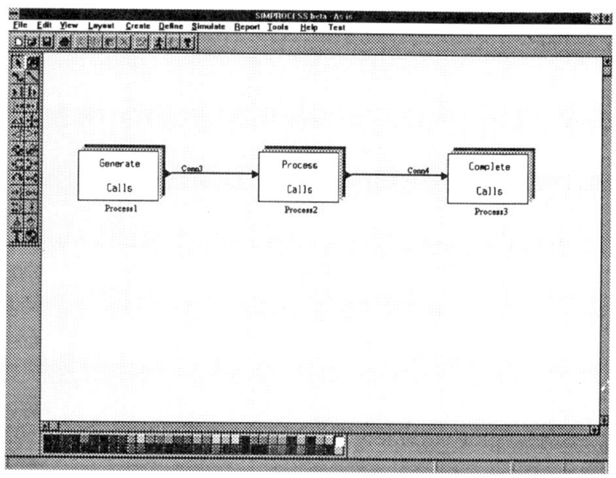

Figure 1: Create Process Model

2.2 Step #2. Simulate The Process

Before simulating your model, you select the performance measures of interest. For example, you may be interested in throughput and cycle time reports for entities, activity costs for processes, and utilization reports for resources. When you run the simulation, SIMPROCESS automatically verifies your model and begins advancing the simulation clock. During the simulation, SIMPROCESS provides you with an animated picture of the flow that helps you visualize your process in motion. You can also have SIMPROCESS generate real-time graphs, letting you view key performance measures, during the simulation.

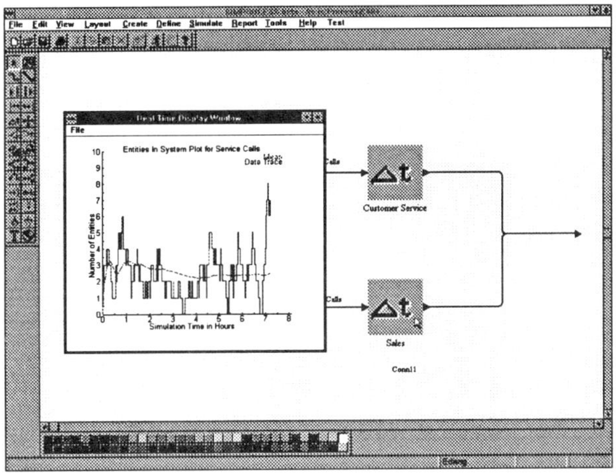

Figure 2: Simulate the Process

2.3 Step #3 - Analyze the Results

When the simulation is over, you can bring up your model results and analyze the performance measures of interest. In addition to the automatically generated cycle time, throughput, waiting time, resource utilization and cost reports, SIMPROCESS allows you to generate custom reports for tracking service levels or in-process inventory. All SIMPROCESS reports can be viewed in graphical or tabular form, and printed or exported to other software packages.

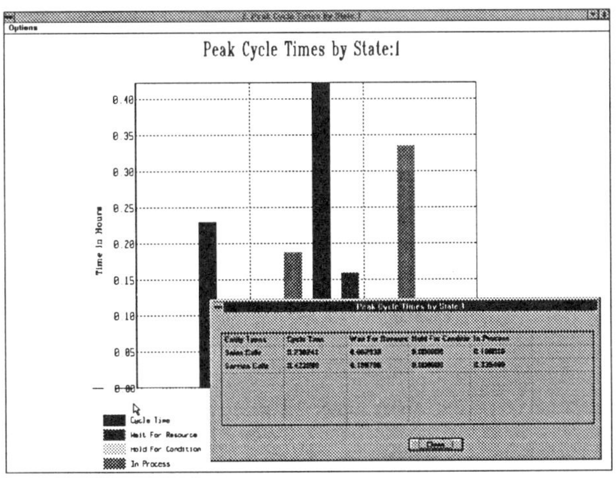

Figure 3: Analyze the Results

2.4 Step #4. Evaluate "Alternatives"

The primary purpose of using SIMPROCESS is to evaluate alternative business decisions. To facilitate this important decision support activity, SIMPROCESS provides two unique functions called "Alternative Sub-processes" and a "Design of Experiment". The "Alternative Sub-processes" function allows you to create alternative representations of a business process utilizing its object-oriented interface. The "Design of Experiment" allows you to define which parameters you wish to experiment with or compare. Using these powerful functions, you setup your business alternatives and let SIMPROCESS do the rest. When your simulations are run, you can compare the alternatives and choose the one that maximizes your service levels and profits.

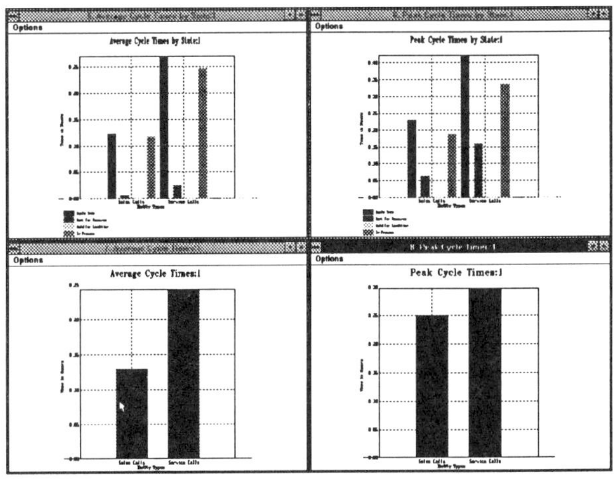

Figure 4: Evaluate Alternatives

3 WHAT MAKES SIMPROCESS UNIQUE?

SIMPROCESS is the first integrated tool that is specifically designed for business process modeling and analysis. SIMPROCESS combines process capture and event-driven simulation with activity-based costing. Here are some of the other features that distinguish SIMPROCESS from other modeling and analysis tools.

3.1 Hierarchical, Event-driven Simulation

A key distinguishing feature of SIMPROCESS is its hierarchical process modeling capability. SIMPROCESS is based on CACI's breakthrough simulation technologies MODSIM and SIMOBJECT. These underlying technologies provide hierarchical and event-driven simulation capabilities that had been previously unavailable for modeling large scale applications.

Unlike the hierarchical representations of processes using attached diagrams or files, SIMPROCESS offers true hierarchy based on object-orientation. This enables you to decompose a process into as many levels of detail as required. The Process construct allows you to create the hierarchy. A process may have several sub-processes and activities. For example, an inspection process that consists of a BRANCH activity and two DELAY activities can be defined as a hierarchical process template.

3.2 Activity-based Costing

The breakthrough activity-based modeling paradigm and resource constructs of SIMPROCESS offer a natural fit for its powerful activity-based costing engine. Activity-based costing is designed into SIMPROCESS, unlike other simulation tools, giving you automatic cost reports. One of the major challenges in successful implementation of ABC is finding the appropriate level of detail for the business process analysis. The organization of business processes is critical to reorganizing the cost data into activity pools. The hierarchical modeling approach of SIMPROCESS facilitates this organization and accommodates varying levels of detail for ABC analysis.

3.3 Methodology-independent Process Modeling

Some of the process modeling tools force you to learn a new methodology such as IDEF or systems dynamics. SIMPROCESS provides you with a flexible process modeling environment that is adaptable to any methodology that you choose. The process documentation features of SIMPROCESS are far superior to typical flowcharting or simulation tools because it is specifically designed for business process modeling and analysis.

3.4 Activity-based Modeling

SIMPROCESS utilizes a breakthrough activity-based modeling paradigm. Activity modeling constructs are the objects at the lowest level of the SIMPROCESS hierarchy and they are used for modeling the behavior of a process. Activities are non-decomposable. The default SIMPROCESS palette bar contains 18 built-in activity blocks: Generate, Dispose, Delay, Assemble, Branch, Merge, Batch, Unbatch, Split, Join, Transform, Copy, Gate, Assign, Synchronize, Get Resource, Free Resource, Replenish Resource. These activities can be connected or embedded into processes using simple flowcharting techniques making process documentation quick and meaningful.

In addition, you can create your own re-usable activity templates and add them to your palette bar. You can even customize these built-in activity blocks to represent the operational characteristics of your own business processes.

3.5 Powerful Resource Modeling Constructs

Accurate resource modeling requires model building blocks that allow you to define the hierarchical, shared and consumable behavior of real-world resources. In the real world, the performance of business processes is usually constrained by the limited availability of resources or by resource interdependencies. SIMPROCESS allows you to define the costs, capacity, usage, and interdependencies associated with resources. It automatically keeps track of the resource's utilization and costs.

Another unique aspect of SIMPROCESS is its ability to represent the hierarchical nature of resources using Departments and Workgroups. Schedules and Downtimes can be modeled to mimic the dynamic behavior of resources.

3.6 Advanced Modeling Constructs

Most BPR tools that provide simulation functionality use simplistic modeling constructs. When modeling the dynamics of real-world business processes, powerful functions such as attributes, expressions, system variables, and IF-THEN-ELSE logic are needed. SIMPROCESS provides these advanced modeling functions and eliminates the limitations of simplistic tools. Advanced modeling functions are very important for realistically modeling the complex behavior of business processes. These functions of SIMPROCESS differentiate it from other BPR tools that are based only on simplistic modeling functions. The advanced modeling functions combined with programming capabilities provide the power and flexibility to accurately analyze the dynamic behavior of real-world business problems.

3.6.1 User-defined Attributes

User defined attributes allow you to attach tags to entities that travel through the model. Attributes can be used for conditional branching, sequencing, or decision making in expressions or logic. For example, order quantity may depend on which customer the order came from. Furthermore, order processing time may depend on the size of each order. By defining an attribute named "order quantity" and writing an expression that multiplies "order quantity" by "process time per order," you can accurately model the order processing delay activity in your model.

3.6.2 System Attributes

These variables keep track of the states of model elements so that expressions or logic can be written to deal with complex business situations. For example, the number of customers waiting for a service representative is a system-state variable that changes over time. In a typical service process, the number of servers would be increased if the customers in line reach a certain number.

3.6.3 User-defined Expressions

When modeling real-world business processes, you will inevitably come across situations which will require modeling functionality that is not built-into a BPR tool. SIMPROCESS provides you with the capability to write user-defined expressions and use them for modeling or customizing the performance measures. For example, service level may be an important performance measure for a business that is trying to fulfill orders within a 48 hour window from the time orders are placed. Using global variables, user-defined attributes and user-defined expressions, you can compare the cycle time for each order with the 48 hour target and calculate the service level of the process.

3.6.4 Built-in Programming Environment

Although the built-in functionality and the ability to customize models with attributes, expressions, and distributions offer plenty of power for modeling most business processes, there may be situations where you will need added flexibility that can only be achieved by a programming environment. For these complex business process applications, SIMPROCESS gives you a built-in, C-like programming environment.

3.6.5 Built-in & User-defined Distributions

SIMPROCESS provides you with 14 standard probability distributions and allows you to create your own empirical distributions based on raw data. One of the unique features of SIMPROCESS is its built-in data analysis (curve fitting) function. This capability allows you to feed a number of data points into the software and get the distribution that best represents the data set.

3.6.6 Reusable Modeling Templates

SIMPROCESS lets you create reusable templates from its basic building blocks and activities. For example, you can define an inspection process template using three SIMPROCESS activity blocks such as DELAY (inspection activity), BRANCH (probabilistic outcome of inspection), and DELAY (rework activity). Then, you can attach an inspection icon to represent this process on the palette and save the inspection template in a library file. Every time you need to model an inspection process, you simply load the library, click on the inspection icon, drop it in your model diagram, and double-click on it to customize its parameters. One extremely powerful use of this feature is in creating templates that represent the operational behavior of automated process equipment.

Customized libraries can be made available to a cross functional team for reuse. This capability is a tremendous advantage for organizations that want to capture and maintain the most valuable asset of a business -- the process knowledge.

3.6.8 Event Logs

Event logs are built-in features for tracking custom performance measures. Timestamp event logs can be used for monitoring statistics such as makespan and in-process inventory. Recorder event logs can be used for monitoring statistics such as arrival and departure rates.

3.7 Comprehensive Statistical Analysis

SIMPROCESS contains a comprehensive set of statistical tools. This suite of tools can be used to fit input data to a distribution, to design and run experiments, and to analyze the results from multiple simulation runs. They are fully integrated into the SIMPROCESS modeling environment so that you can conduct extensive analysis within a single tool rather than multiple software products from multiple vendors. You do not have to be an expert in statistics to take advantage of these sophisticated capabilities.

3.8 Design of Experiments

The Experiment Manager is an integrated tool for designing, executing, and analyzing statistical simulation experiments with SIMPROCESS. This tool facilitates the "what if" analysis by systematically modifying the factors that impact the performance metrics and evaluating the results on those performance metrics. For example, you can set up 6 experiments where the number of available resources are varied from 5 to 7 and the interval between calls are varied from Light to Heavy for a Help Desk. This approach ensures that you perform a complete analysis of the business processes and minimizes the time to find the optimum solutions to the business problems.

3.9 Data Analysis Tool

The integrated Data Analysis tool is powerful function to analyze data and select the best distribution to fit that data. For example, the random nature of calls arriving at a Help Desk or the random nature of processing times can be captured by the Data Analysis tool and represented in your SIMPROCESS model. By modeling the data and representing the randomness associated with the data, you are more likely to increase the accuracy of your process analysis.

3.10 Multi-platform Support

SIMPROCESS is one of the few BPR tools that runs under Windows 95, Windows NT, and UNIX operating systems. Your models are fully compatible across platforms—giving you the flexibility to run large simulations on faster computers and take advantage of your existing hardware. As hardware and operating system technology moves forward, SIMPROCESS will take advantage of them and protect your investments.

4 SIMPROCESS APPLICATIONS

SIMPROCESS provides a rich array of integrated functions for modeling and analysis of business processes. From customer service to product development, from administrative to production processes, for every business process, SIMPROCESS allows you to visualize and evaluate the results of process changes before you commit expensive resources, time and money. Below are three applications where SIMPROCESS was used for effective business decisions.

4.1 An Order Fulfillment Application

A major Fortune 100 industrial enterprise was faced with a problem in one of its computer products business due to high inventory and low service levels. The proposed solutions included reduction of channel inventory and building product to order. SIMPROCESS was used in the reengineering project that resulted in a 50 percent reduction in inventory and a 63 percent increase in service level.

4.2 A Licensing Process Application

A State Government hired CACI to help reengineer its business processes that suffered from long service time and high cost per transaction. The proposed solutions included implementation of an automated workflow and imaging system as well as a full service counter. SIMPROCESS helped analyze the alternative solutions that increased throughput time from 80 days to 56 days, and reduced cost from $70 to $46 per transaction.

4.3 An Engineering Change Order Application

A major European automotive manufacturer was trying to shorten the time required for design changes in its product development process. The design changes originated in Europe and were implemented in South America. The business processes involved resources from designers to process engineers to purchasing agents. First, SIMPROCESS was used to create process maps and simulations of the AS-IS process. Then, TO-BE alternatives including policy changes and workflow automation were simulated to determine impact on cycle times.

ACKNOWLEDGMENTS

This paper was adapted from the SIMPROCESS Getting Started manual.

AUTHOR BIOGRAPHY

MAYA BINUN is a software developer for CACI Products Company in La Jolla, CA. She holds a BS degree in Chemical Engineering from Caltech and an MS degree in Engineering Sciences from UCSD. Prior to coming to CACI she has spent 5 years as a Process Engineer in pharmaceutical industry. Her efforts at CACI have been directed towards making SIMPROCESS an industry standard for business process simulation.

Proceedings of the 1996 Winter Simulation Conference
ed. J. M. Charnes, D. J. Morrice, D. T. Brunner, and J. J. Swain

CONDUCTING EXPERIMENTS WITH EXPERIMENT MANAGER

Michael Angel

CACI Products, Inc.
3333 N. Torrey Pines Ct.
La Jolla, CA 92037

ABSTRACT

This tutorial demonstrates how to conduct experiments and perform analysis with Experiment Manager, CACI's tool for design and analysis of simulation experiments. Experiment Manager works in conjunction with SIMPROCESS and provides a quick and object-oriented approach to designing simulation experiments. The tutorial will describe how to conduct a full factorial experiment with 2 factors and 3 levels each. These experiments will be performed using a model constructed in SIMPROCESS and the results will be analyzed using such features of Experiment Manager as box plots, confidence intervals, and time series plots.

1 INTRODUCTION

Experiment Manager is a set of statistical software tools integrated with the SIMPROCESS modeling tool. SIMPROCESS is a hierarchical icon based simulation modeling tool designed specifically for business process re-engineering (BPR). Experiment Manager extends the analyst's ability to conduct statistical data analysis and experiments using SIMPROCESS. Experiment Manager:

- Provides statistical software tools tailored to the statistical simulation environment
- Provides statistical software tools for all phases of a statistical simulation experiment
- Increases the accuracy of conducting an experiment
- Decreases the time to find the solution to the problem under study
- Interacts with the user through a unique easy to use object-oriented graphical user interface
- Organizes experiments
- Archives experiments
- Automates tedious data collection and data entry tasks such as collecting and merging SIMPROCESS statistical output

Section 2 of this paper is an overview of the Experiment Manager. It describes the functions and features of the major components of the Experiment Manager as well as some high level architectural considerations. The rest of the paper primarily focuses on the design and analysis of an experiment using the Experiment Manager Tool (EMT) component of Experiment Manager. Section 3 describes the specification of an experiment. Section 4 describes the execution and analysis of the experiment.

2 OVERVIEW OF THE EXPERIMENT MANAGER

There are two integrated experiment components in the Experiment Manager tool set: the Data Analysis Tool (DAT) and the Experiment Manager Tool (EMT). Integrated with SIMPROCESS, these tools provide a tightly integrated end-to-end environment for conducting statistical simulation experiments.

DAT is used to analyze data and construct a data model of the random process that generated the data. DAT helps you find a probability distribution that approximates the true distribution by analyzing a data set of measurements obtained from the random process. The parameters of the probability distributions supported in SIMPROCESS are computed using the techniques of maximum likelihood estimation (MLE). The analyst selects a distribution based on

Table 1. SIMPROCESS and Experiment Manager Functionality

	SIMPROCESS	DAT	EMT
Model Development	• Graphical specification of logical processes • Interpretive specification of logical processes • Specification of data model for modeling elements • Specification of performance measures	• Data Analysis • Modeling of Random Processes	• Factor View of Model
Verification & Validation	• Animation • Trace Files	• Goodness of fit	• Statistical validation
Experimentation & Analysis	• Single scenario execution • Time Series analysis • Single scenario analysis	• Data analysis of statistical output • Residual analysis of statistical tests	• Design of experiments • Execution of experiments • Collection and merging of statistical output • Regression analysis • 2k analysis Statistical comparison of alternatives • Transformations • Confidence intervals
Results & Presentation	• Extensive reporting mechanism • Single run statistical output • Model archive	• Box Plot • Standard Error • Confidence Interval plot • Empirical and probability distributions • Numerical results	• Box plot comparisons of scenarios • Standard error comparisons • Confidence interval comparisons • Time series plot • ANOVA tables • Hypothesis tests • Numerical results • Experiment archive

goodness-of-fit ranking or prior knowledge of the process that produced the data set. The selected probability distribution is then used to model the random process in SIMPROCESS.

EMT is used to design, execute, and analyze statistical simulation experiments. EMT helps the analyst select the factors (variables) of interest for the experiment and specify the scenarios (values for the variables) for execution. EMT executes all of the specified scenarios automatically collecting the statistical output for each scenario. EMT provides statistical tests for screening factors, developing quantitative relationships between important input variables and the performance of the system, and determining the best performing system of many.

The tightly integrated environment removes the drudgery of moving data from one tool to another and from the formats of one software tool to the formats of other software tools. This increases the accuracy of

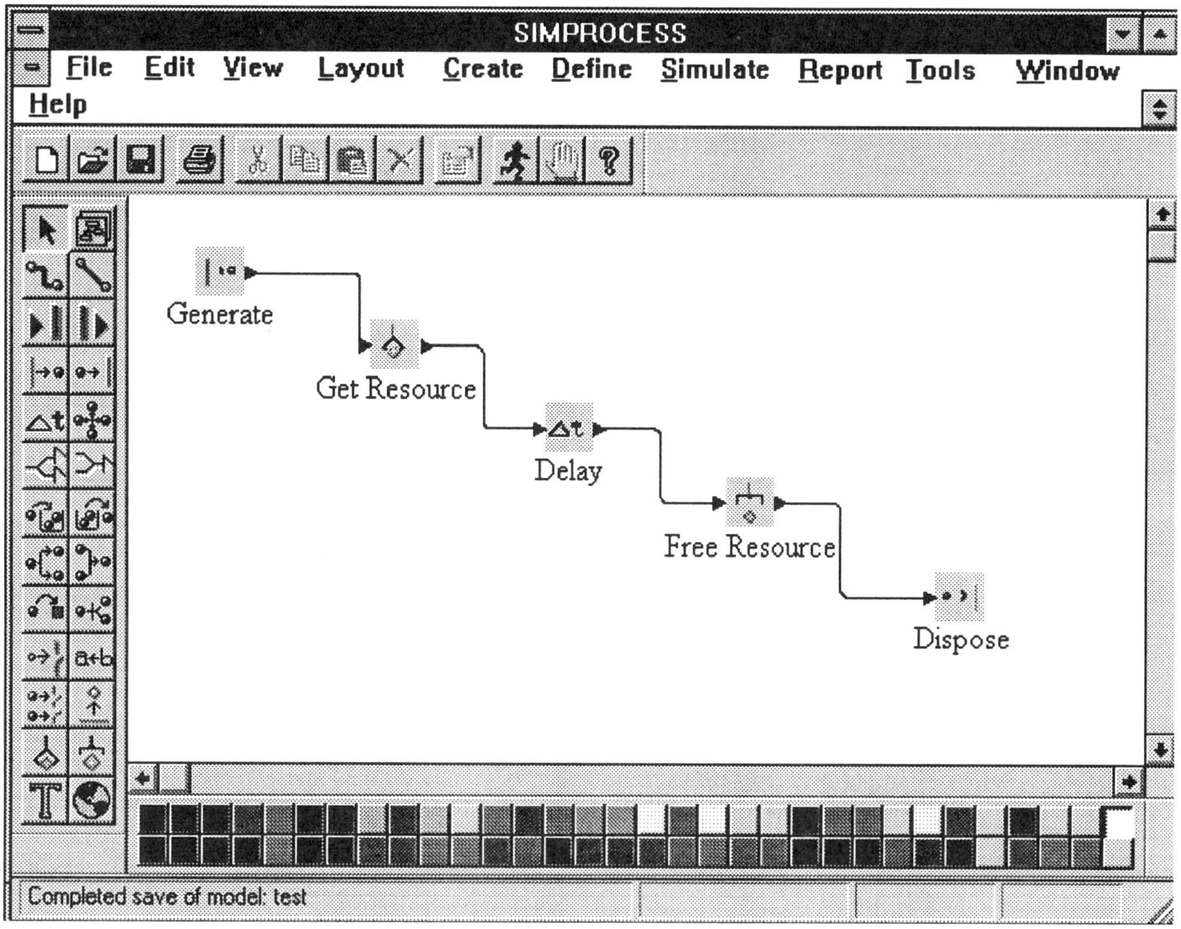

Figure 1. Simple Queuing Model

experimental information and decreases the time to find the solution to the problem under investigation. For example, data models constructed using DAT are automatically available to any SIMPROCESS modeling element that models some behavior as a random process.

Statistics collected during the execution of an experiment are automatically collected and even merged with existing data if necessary. Output graphs and reports automatically use the correct labels of SIMPROCESS modeling elements.

The end-to-end environment provides support for all of the major steps of a statistical simulation experiment including model development, verification & validation (v&v), experimentation, analysis and presentation of results. An end-to-end environment significantly shortens the turnaround time for obtaining statistically significant and presentable results. For the model development phase, SIMPROCESS provides graphical and interpretive specification of logical processes and DAT provides for data analysis and modeling of random processes. For the v&v phase, SIMPROCESS provides animation and trace files to verify the logical processes while EMT provides statistical procedures for validating a simulation model. For experimentation and analysis, SIMPROCESS provides single scenario execution and analysis capabilities and EMT provides multiple scenario execution and analysis capability. Finally, SIMPROCESS, DAT, and EMT all provide extensive graphing and reporting capabilities.

Table 1 shows the integrated capabilities of SIMPROCESS for all phases of the statistical experiment process. Each column of the table corresponds to a SIMPROCESS tool; each row to a phase of the statistical simulation experiment; and each cell to a set of capabilities provided by a tool for that phase of the statistical simulation experiment.

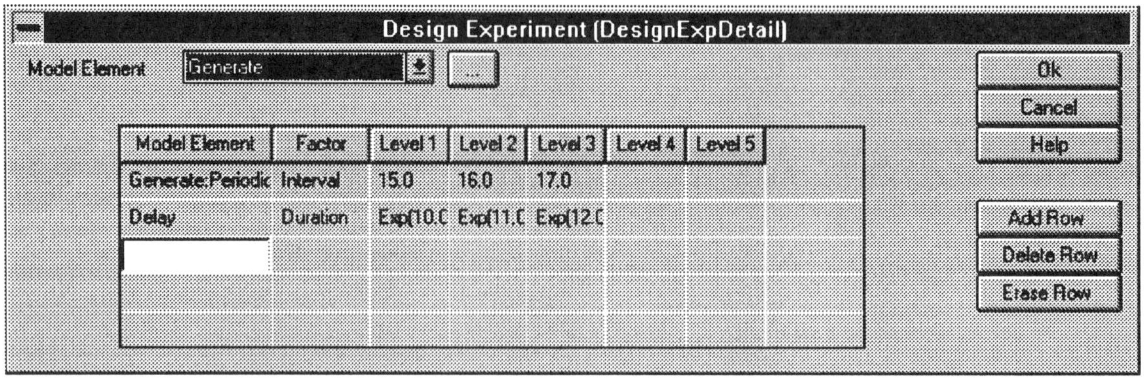

Figure 2: Specification of Factors and Levels

3 THE SPECIFICATION OF AN EXPERIMENT

EMT presents a different view of the model that makes experimentation easy. EMT views the model as a collection of factors for experimentation; SIMPROCESS views the model as a logical representation of the processes of a system. A factor is an input variable to the simulation model that could influence the output response. For example, the duration of a delay influences the total time required by an entity to move through a system whereas the particular bitmap used to represent the delay in the graphical view of the model does not. Note that although model elements are elementary from the SIMPROCESS perspective, they may have many behaviors from a Design of Experiment (DOE) point of view that affect the statistical output.

In SIMPROCESS, a simulation model is built by specifying the interactions between predefined model elements. Model elements include activities, resources, and entity types. Figure 1 shows the SIMPROCESS graphical representation of a simple queuing model used to demonstrate the design, execution, and analysis of an experiment using EMT. The interarrival times of entities is specified by the generate activity "Generate." Entities queue up for a single resource. The resource "Server" is made available to the first entity in the "Get Resource" queue. The use of the resource is a service that requires some length of time to occur. The service itself is modeled as a simple delay in the "Delay" activity. The server is freed for use by the next entity in the "Free Resource" activity. The entity exits the system in the "Dispose" activity. The statistical output of interest is the total time in the system. The total time in the system is the time an entity spends waiting in the queue plus the time it takes for service.

A statistical simulation experiment is initiated by creating an experiment object. An experiment object is a named object used as the organizational unit for managing experiments. The new experiment object manifests itself in the user interface as an experiment editor. The editor provides the interface for designing the experiment and performing analysis.

The design of the experiment is the specification of different input configurations of interest and the statistics used to measure their effect on performance. In EMT, the specification of different input configurations used to run a simulation is a two step process. The first step of the specification is the selection of factors and specification of levels - specific values for the factors. The second step of the specification is constructing the input configuration for individual simulation runs. The two-step specification of the input configuration is a deliberate design decision. Many types of statistical analysis of experimental results require a complete design i.e. a design that is executed at every combination of levels. A single combination of levels is a scenario or in DOE terminology a design point. Let's imagine an experiment that requires a complete design for four factors with each factor at four levels. There are 4*4*4*4 possible combinations of levels for the four factors. The complete design has 256 design points. In EMT, the user must specify four levels for each of the four factors. This is 4+4+4+4 or 16 level specifications in the first step of the process. The complete design is created simply by using the All Combinations option of the Scenarios command on the experiment editor. If the specification were a one step process, the user would be required to individually specify 256 scenarios rather than 16 levels to specify a complete design.

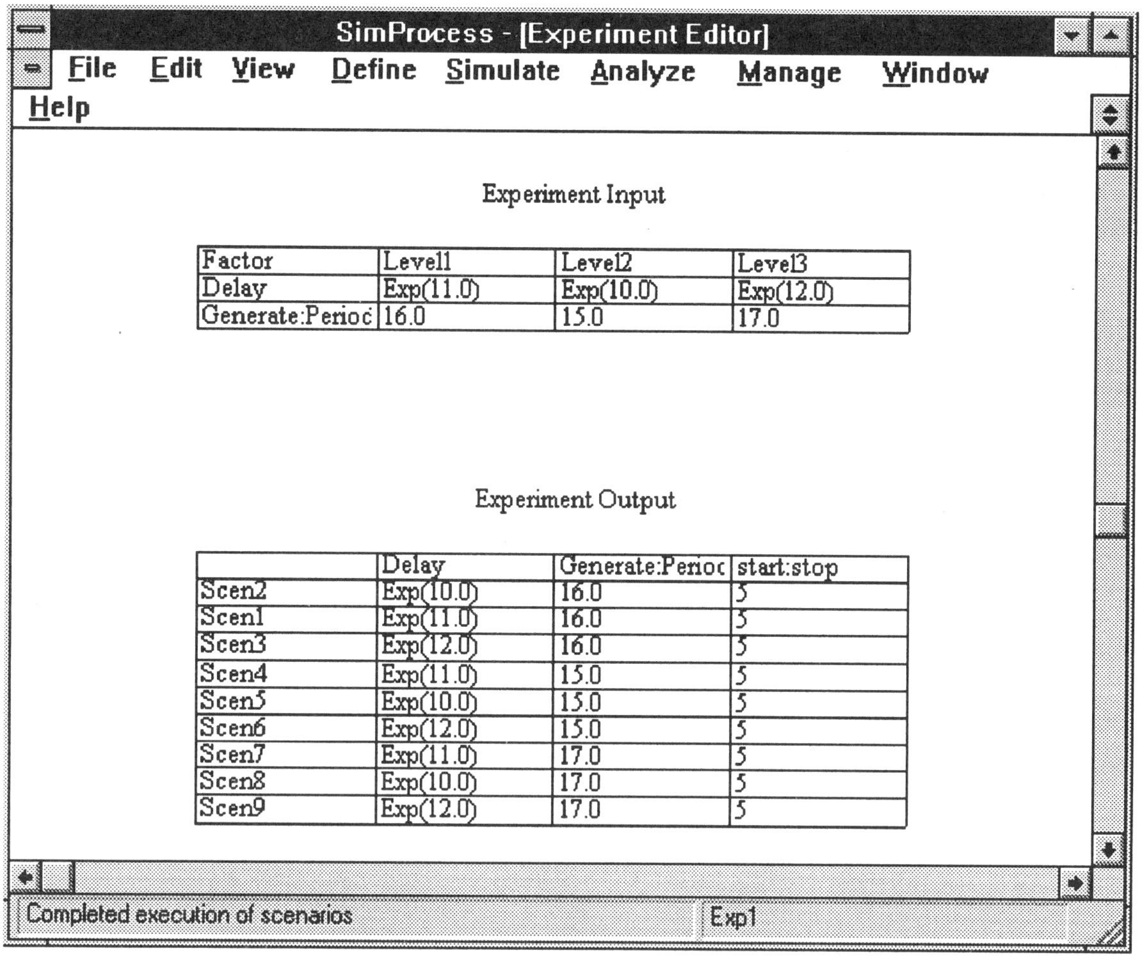

Figure 3: Experiment Input and Output Tables.

Figure 2 shows the spreadsheet-like user interface used to specify which factors to vary for a particular simulation experiment and what levels to use for the selected factors. Above the table is a context sensitive input selection list and text entry box. The first column of the table is used to specify the model element. Model elements are named objects in the simulation model that have at least one behavior designated as a factor. Selecting a cell in the model element column causes a list of selectable model elements to appear in the input selection list. The experimenter selects a model element for experimentation from the input selection list. Selecting the cell to the right of a selected model element in the factor column causes a list of factors to appear for the previously selected model element in the input selection list. Context sensitive spreadsheet data entry is used to correctly specify model elements and factors.

Each factor has a domain that is appropriate for the modeled behavior. Factors that model random processes have a sample domain. The sample domain is the set of supported distributions for SIMPROCESS. EMT domains also include real, integer, and enumeration domains. Levels are specified by selecting the next available cell after the selection of a model element and factor. The domain of the factor defines appropriate values for levels of the factor. Valid level specification is ensured by context sensitive spreadsheet data entry.

Because SIMPROCESS is a hierarchical simulation tool, any hierarchical process in a model is a potential factor. The domain of the hierarchical process is the enumeration of named alternative subprocesses. A level of the hierarchical process is an individual subprocess. The enumeration can be used to create an experiment that compares alternate subprocesses or configurations of the model.

The end result of the first step of the specification is a completed experiment input table as shown in Figure 3. In this experiment, two factors have been selected: "Generate:Period1 Interval" and "Delay Duration". Factor "Generate:Period1 Interval" has levels 15.0, 16.0, and 17.0. Factor "Delay Duration" has levels Exp(15.0), Exp(16.0), and Exp(17.0).

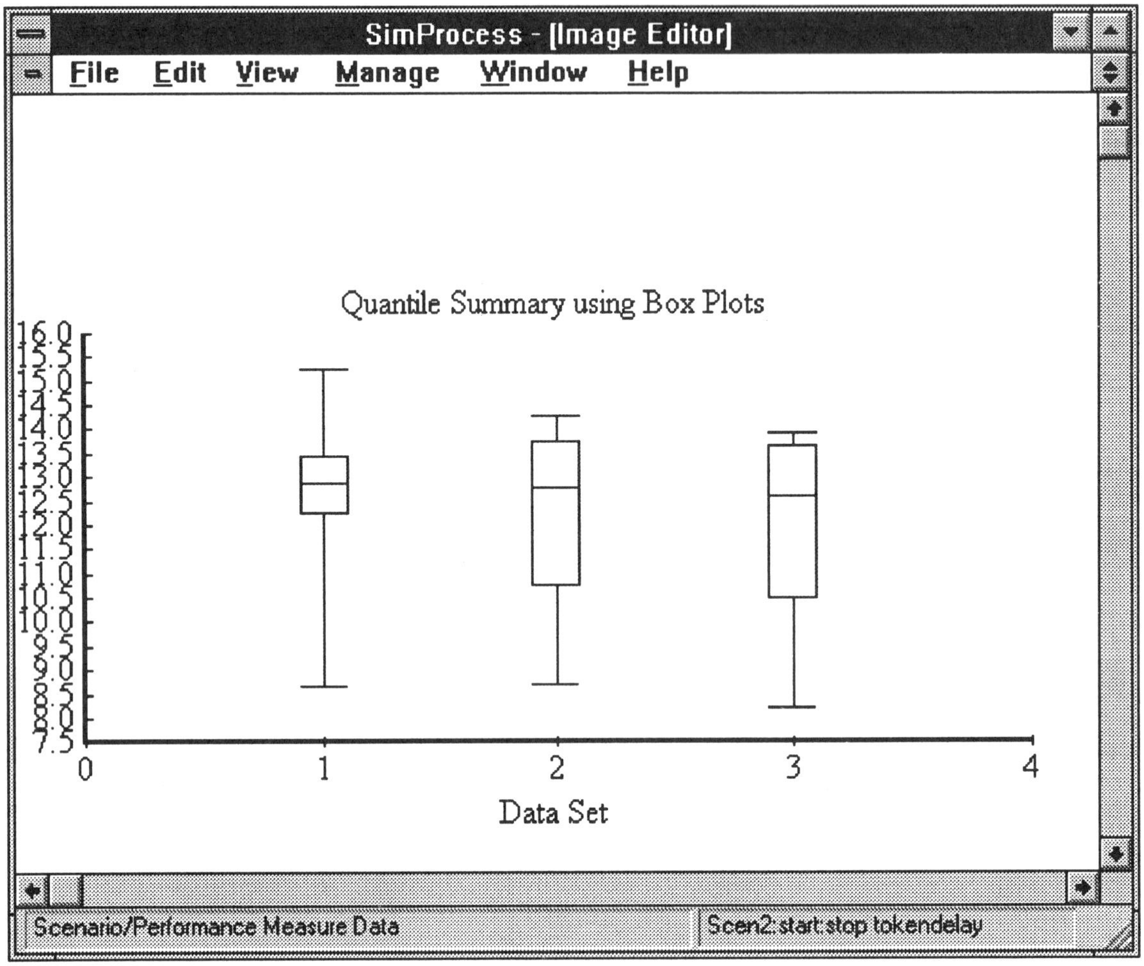

Figure 4: Box Plots of Selected Results

Exp is the standard abbreviation for an exponentional distribution. The parameter is the mean of the exponential distribution.

The second step in the design of the experiment is the specification of scenarios. There are nine possible scenarios to be executed as shown in Figure 3. If the design is complete all combinations are created by simply selecting the "scenario all" option from the experiment editor menu. If specific scenarios are necessary, they can be created by selecting an appropriate level for each factor in the experiment input table shown in Figure 3 and selecting the "scenario single" option from the experiment editor menu. Note the ability to select table entries on the user interface. The ability to select entries or "objects" is an inherent capability of the Experiment Manager object oriented user interface.

Figure 3 also shows the experiment output table. A named scenario has been created for each created scenario and a list of the levels that define the scenario. The performance measure start:stop is also shown in the table. Experiment Manager determines which statistics and reports are enabled in the SIMPROCESS model. A performance measure entry is created for each enabled SIMPROCESS statistic. Note that the cell entries beneath the start:stop performance measure indicate if statistical information has been collected. Prior to the execution of an experiment, the number is set to 0 indicating that no data points have been collected.

4 EXECUTION AND ANALYSIS OF AN EXPERIMENT

The run parameters used to execute the scenarios are specified in the SIMPROCESS editor. The run parameters include the number of replications, the run length of each replication, and the warm-up time. Either a terminating or batch-means experiment is specified. In the case of terminating simulations, EMT assumes that the specified warm-up period is long enough so that any bias is removed from the resulting observations. In the case of non-terminating simulations, EMT assumes that the replication lengths are of sufficient size to remove significant correlation from the observations. The number of replications determines the number of observations that will be recorded during execution of each scenario for each performance measure that is collected.

Scenarios are executed by selecting the run option from the menu of the experiment editor. The default action is to execute all of the scenarios. Specific scenarios can be executed by selecting the appropriate scenario entries in the experiment output table displayed by the experiment editor and then selecting the run option.

The initial seed values used for the random number streams for each scenario are identical by default. The current value of each random number stream is recorded with each scenario at the end of the execution. If more replications are required for the statistical analysis of an experiment, the experimenter can request that EMT merge results from subsequent executions of the scenario with already collected results. EMT will restart the experiment at the last value of each random number stream and automatically merge the new output results with existing data.

Figure 3 shows the experiment output table after execution of all nine scenarios. The cells beneath the performance measure column labeled “start:stop” indicate that 5 data points have been collected for each scenario. EMT uses the average value of a performance measure obtained at the end of each replication as a single estimate of the expected value of the performance measure. Statistical analysis of these estimates assume that the observations obtained from each replication are iid (independent and identically distributed).

Statistical output from multiple scenarios can be simultaneously viewed by selecting a performance measure entry in the experiment output table or by selecting one or more data sets and then selecting the desired operation from the experiment editor menu. EMT’s object oriented interface provides the experimenter with maximum flexibility in selecting data sets for viewing and statistical analysis. Figure 4 shows box plots of three selected data sets for graphical comparison of the performance of three scenarios for the start:stop statistic.

The EMT toolkit also includes statistical analysis tools for performing ANOVA, regression, transformations, time series analysis, and model validation.

5 FUTURE DIRECTIONS

The most interesting future direction for EMT is in the area of optimization. New advances in optimization algorithms, continuing improvement in hardware performance and greater expectations from commercial products are making the marriage of simulation and optimization not just a possibility but a requirement. Not only will SIMPROCESS and Experiment Manager provide “canned” optimization algorithms, but the environment can be extended to include custom algorithms. Other directions for growth include expanded statistical analysis particularly in the area of time series analysis. Time series analysis of simulation output can be used to determine appropriate warm-up times or replication lengths.

AUTHOR BIOGRAPHY

Michael Angel is a software developer for CACI’s Advanced Simulation Laboratory in La Jolla, CA. His efforts at CACI have focused on developing simulation modeling tools including the Experiment Manager.

Proceedings of the 1996 Winter Simulation Conference
ed. J. M. Charnes, D. J. Morrice, D. T. Brunner, and J. J. Swain

MODSIM III - A TUTORIAL

Alasdar Mullarney

CACI Products Company
3333 North Torrey Pines Court
La Jolla, CA 92037, U.S.A.

ABSTRACT

This tutorial introduces the MODSIM III language, showing how its simulation "world view" together with its object-oriented architecture and built in graphics contribute to successful simulation model building.

1 INTRODUCTION

Discrete systems simulation is one of the few truly new capabilities offered by the computing revolution. Computer simulation models provide us with a fascinating means to develop insight into the behaviors of the complex non-deterministic systems which surround us - in communications networks, transportation logistics, business and manufacturing processes, to name a few. These systems are costly to develop and modify. Feasibility analysis and performance prediction through simulation can greatly reduce the risk of failure or wasteful expense.

Simulation is an exploratory technique. We develop a model which represents our best understanding of a proposed system or modification. We run the model in the hope that it will confirm our design intuition. If it does not we endeavor to understand why, and use this knowledge to refine the model. Ultimately, we will use the model to explain and justify decisions affecting the real-world system.

So, simulation serves three principal roles: to help us articulate a coherent description of a system; to validate its dynamic behavior to our satisfaction; and, then to communicate this to colleagues and decision makers. These steps are commonly iterative. It is usually wise to begin with a high-level abstraction of the system, allowing the model to evolve as we gain greater understanding of system behavior. Furthermore, the system itself may change in this rapidly changing environment, requiring changes to the model.

For the simulation study to be effective, the model behavior should be presented in an understandable way, typically using animation. Ideally, domain experts should be involved in construction and validation of the model. Finally, the whole exercise must be completed within a useful time-frame, or it will serve little purpose.

Before embarking on model development, it may be prudent to look for a suitable, off-the-shelf, domain-specific simulator. Frequently, however, unique characteristics or fidelity requirements dictate the need for a custom model. The characteristics of simulation modeling, then, make some special demands of the model development environment. The modeler needs help in the conceptualization of systems with dynamic, interacting, probabilistic behaviors. The model should be understandable, maintainable, and should lend itself to incremental addition of detail. The environment should support graphical interaction for scenario development and animated model execution.

This tutorial will introduce the benefits of using an object oriented language specially developed for simulation, and enhanced with comprehensive graphics support. MODSIM III combines CACI's experience with simulation programming over three decades with advances in software engineering to offer the most productive environment for the development of large, complex, evolutionary, custom models.

Examples of MODSIM III's simulation features, and the benefits of object-oriented architecture, are demonstrated below using code fragments from a hypothesized airport/airspace planning model. Such a model might be concerned with the representation of aircraft, flight duration, air traffic controllers, runway allocation procedures, and so on. To be of any interest, such a model must represent multiple aircraft in flight concurrently, and delays due to contending requests for resources, such as runways.

2 DEFINITION BLOCK—AN INTERFACE DESCRIPTION

In support of modular program construction, objects in MODSIM III are described in two separate blocks of code. The Definition block describes the object type by declaring its variables and methods. This is the object description as it will be referred to by other objects in the simulation, and it provides the formal interface specification. An example of a Definition block for an aircraft object is shown below.

```
Aircraft = OBJECT;
  BestCruise : INTEGER;
  InFlight : BOOLEAN;
  ASK  METHOD SetCruise (IN speed:INTEGER);
  TELL METHOD Fly (IN distance:INTEGER);
END OBJECT;
```

The Definition block for an aircraft object declares the variables and methods that aircraft objects use in the simulation model. The information the aircraft knows is contained in its variables. In this simple case, the aircraft is responsible for the management of two variables, which represent its state:

* BestCruise - the optimal speed to cruise at for given conditions

* InFlight - whether or not the aircraft is actually in flight.

3 IMPLEMENTATION BLOCK—WHAT THE METHODS DO

The aircraft behaviors are described in its methods. These methods are named in the object description provided by the Definition block. The logic of what they do and how they affect the state variables of the object are described in the Implementation block, shown below:

```
OBJECT AircraftObj;

  ASK METHOD SetCruise (IN speed:INTEGER);
  BEGIN
    BestCruise := speed;
  END METHOD;

  TELL METHOD Fly (IN distance:INTEGER);
  BEGIN
    InFlight := TRUE
    WAIT DURATION distance/BestCruise;
    END WAIT;
    InFlight := FALSE;
    OUTPUT ("Arrived Safely at", SimTime);
  END METHOD;

END OBJECT;
```

The behaviors that objects can perform are the methods described in the Implementation block. In this case the aircraft is capable of the behaviors described in the following two methods:

* ASK METHOD SetCruise - When the aircraft is requested to perform this behavior, it registers the new value for its optimal cruising speed, instantaneously, that is, simulation time does not elapse.

* TELL METHOD Fly - When requested to perform this behavior, the aircraft calculates the required flight time to cover this distance at its cruising speed. This particular activity then pauses in execution until this period of time has elapsed within the simulation model, before completing the remainder of the behavior - in this case printing a notification that it has arrived safely. Unlike ASK methods, TELL methods are used to describe behaviors that elapse simulation time. While this method is paused, waiting for time to pass, other methods of other objects may be executing.

A key benefit of using MODSIM III in building complex simulations is the easy modeling of these behaviors. In a large model, many objects will have behaviors that must take account of the passage of time. Often, these behaviors will be concurrent, or overlapping in time. For example, our model will want to represent multiple "instances" of the aircraft object type. These instances can be created as needed; each can be given its own identifier, has its own state variables and can execute its methods as requested.

For a simple example of concurrent behaviors, let's look at how an aircraft dispatcher in our model might order two aircraft to fly to different destinations:

```
ASK JumboJet TO SetCruise(600);
TELL JumboJet TO Fly(3000);

ASK Biplane TO SetCruise(100);
TELL Biplane TO Fly(200) IN 1.0;
```

Using TELL methods, the flight times of both the JumboJet and Biplane aircraft can be modeled concurrently.

In this example, the aircraft object named JumboJet will elapse 5 hours flying a distance of 3000 miles at 600 mph. One hour after the JumboJet takes off (... IN 1.0), the Biplane aircraft will take off and fly 200 miles at 100 mph. It will complete its flight two hours before the JumboJet arrives at its destination. MODSIM III is responsible for sequencing the execution of the methods of both object instances, including the pauses to represent the flight times, so that the events of taking off and landing are played out in the correct order in the model. ASKing the object does not elapse any simulation time.

4 TIMING AND INTERACTION

Besides executing concurrently, time elapsing behaviors may interact. To make the model more realistic, we want to consider the effect of changing the cruising speed of an aircraft while it is in flight - perhaps in response to a change in weather conditions. Such a change invalidates the original computation of flight time, and a new arrival time must be determined based on the new cruising speed and the distance remaining. Let's look at how the logic, or implementation, of the methods of our aircraft objects can be refined to incorporate this modified behavior. The method which is responsible for registering a change in cruising speed can INTERRUPT the time-elapsing method, Fly, if appropriate. On recognition of this INTERRUPT, the remaining time to WAIT is reevaluated. To see the changes that we've made, compare this code with the original Implementation block for the aircraft object, presented earlier.

```
OBJECT AircraftObj;

 ASK METHOD SetCruise (IN speed:INTEGER);
 BEGIN
  BestCruise := speed;
  IF InFlight
   INTERRUPT SELF Fly;
  END IF;
 END METHOD;

 TELL METHOD Fly (IN distance:INTEGER);
 BEGIN
  InFlight := TRUE;
  WHILE distance > 0.0
   speed := BestCruise;
   start := SimTime;
   WAIT DURATION distance/BestCruise;
   ON INTERRUPT
    elapsed := SimTime-start;
    distance := distance-(elapsed*speed);
   END WAIT;
  END WHILE;
  InFlight := FALSE;
  OUTPUT ("Arrived Safely at", SimTime);
 END METHOD;
```

The aircraft's cruise speed can now be changed while in flight - the arrival time will be recomputed each time this occurs.

Look at how the Fly method describes the entire flight from take off to landing, allowing multiple speed change events, in a logical activity description. Contrast this with multiple, disconnected, event sub-routines in a conventional programming language which does not support the concept of time-elapsing behaviors.

Because MODSIM III provides you with a rich set of features to manage the complex scheduling, interaction and synchronizing of time-elapsing behaviors, you get increased readability and consistency in your models, factors that translate directly to increased productivity and maintainability.

Unlike making calls on a subroutine library, MODSIM III understands the meaning of these simulation features. Thus it can diagnose inadvertent misuse early - for example, WAIT statements are not allowed in ASK methods that are always instantaneous. Not only does such checking save time in building and running a model, but it can help avoid debugging subtle logic errors in simulations with complex interactions.

These specialized features for modeling concurrent and interacting behaviors distinguish MODSIM III as a simulation model development tool. In addition, MODSIM III includes a rich collection of simulation building block objects. These library objects are designed to fulfill many common simulation modeling requirements. MODSIM III uses the power of object oriented software architecture to allow these pre-built library objects to be readily adapted to special needs.

Consider contention for resources, an issue which is at the heart of many discrete system simulations. Specific allocation policies are a basis for common behavior. Objects incur delays in competing for resources; they queue for resources on some priority basis; they may choose to abandon requests after a time-out interval. Every simulation model will want to report to some degree on measurements of resource utilization, waiting time statistics, and so on.

MODSIM III provides a prebuilt Resource object as one of many objects in its simulation support libraries. In our airport model, for example, runways are clearly a resource. We could use an instance of ResourceObj taken directly from MODSIM III's library to model runway allocation, enqueueing and dequeueing the aircraft on a first-come-first-served basis, and recording statistics.

We need to make one important change, however. To avoid the danger of wake turbulence effects, it is important that a light aircraft not use a runway immediately following a large aircraft; it should delay a short time to allow wake vortices in the air to dissipate. This is where inheritance comes in. It allows us to describe a Runway object in terms of the existing ResourceObj provided by

MODSIM III. We only need to specify the differences between the new RunwayObj and ResourceObj.

Inheritance is one of the chief benefits of object oriented software construction, and the basis for providing libraries of useful objects which can be readily adapted to specialized needs.

In the example below, we have imported a resource management object from the MODSIM III library, defined an enumerated variable called AircraftCategory and show the Definition block for Runway. By declaring our Runway object to be derived from the library-supplied resource management object, it inherits all the built-in capabilities for enqueueing requests and maintaining utilization statistics. The Give method is declared as overridden, meaning that a different implementation, for just this method, will be substituted in the Implementation block (not shown). The Runway object also has an extra variable to 'remember' the last aircraft type. Our specialized implementation logic can now be designed to impose appropriate delays before giving the runway to aircraft of different categories.

```
FROM ResMod IMPORT ResourceObj;
TYPE
  AircraftCategory = (Light, Heavy);
...
Runway = OBJECT(ResourceObj);
  lastuse : AircraftCategory;
  OVERRIDE
    TELL METHOD Give(IN number : INTEGER);
END OBJECT;
...
```

The Runway object, derived from MODSIM III's resource management object has been customized to meet special modeling requirements.

Inheritance provides a disciplined way to selectively modify and extend object characteristics. As a specification mechanism, it maintains a clear distinction between those properties which continue to be available unchanged, and those enhancements designed to meet special needs - this is very valuable as software evolves through versions and upgrades.

New object types, derived through inheritance from existing objects, continue to conform to common interfaces, but incorporate additional capability. This is an excellent match to the evolutionary nature of successful simulation models; with increasing understanding of the system comes a desire to add details in areas of special focus.

The reuse of libraries of pre-built objects holds out the promise of real productivity gains in software development. Without a means to adapt such objects to special needs, this promise is rarely fulfilled. The extensibility offered by inheritance, coupled with the modular separation of interface definitions from actual implementation code are the mechanisms needed to support practical reuse of object libraries.

Object orientation offers other benefits to model development. The controlled access to object data structures through the object methods is just what is needed to build robust objects which can be the basis of reuse. Look back at the modified aircraft object implementation: any request to change the aircraft speed can now ensure a reevaluation of the flight time-which is faithful to the way things happen in the real world.

Taken together, support for object modeling concepts, along with concurrent time based behaviors, are what make MODSIM III an effective simulation productivity tool.

5 GRAPHICS AND SIMULATION

Through inheritance, the objects in your simulation can acquire a rich set of graphical properties and behaviors. You can use this to provide an interactive, graphically managed model that speeds up analyses and produces easy-to-understand results. Adding graphics is easy. You use a graphical editor to configure the appearance of icons, menus, dialog boxes and presentation charts. Minimal code then connects these to the entities and variables in the model. Adding graphics can enhance the appeal of a model in three principal areas:

5.1 Graphical Scenario Layout

Interactive graphical editing lets you define a scenario to simulate by selecting icons from the palette, positioning them on the screen, and configuring parameters through dialog boxes.

5.2 Dynamic Analysis

With a scenario on the screen, you can begin the simulation and see an animated picture of the system under study. In addition, you can study plots that are drawn while the simulation is running. You can pan and zoom on areas of special interest. These results, shown dynamically, will suggest alternatives that can be tried immediately. Interacting with the model in this way increases understanding of the system under study and speeds your analysis. Often errors that may have otherwise been difficult to find, will be obvious. Dynamic analysis contrasts sharply with the old iterative approach to simulation, where the following steps were repeated: prepare data, simulate, examine results, modify data, simulate, ...

5.3 Communication of Results

Through animation, you can dramatize the effect of alternative system configurations, spot unexpected behavior, and back up your recommendations. It's the best way to sell your ideas.

6 DEVELOPMENT ENVIRONMENT

MODSIM is a complete development environment. The MODSIM III simulation support, Compilation Manager, Object Manager and the Debugging Manager provide a complete environment for the successful development of advanced models.

6.1 Compilation Manager

The MODSIM compilation manager automatically determines which modules have been edited since the last compilation and recompiles only those modules and any other modules that depend on them. No make files are required.

6.2 Object Manager

MODSIM III provides a browsing tool for MODSIM III objects, variables and procedures. The Object Manager provides a concise representation of complex objects including aggregation of fields and methods and inheritance diagrams. For object types, all the fields and methods are displayed. Methods are followed by a condensed parameter list; fields have their types indicated. You can see all the attributes and capabilities this object has either defined or acquired through inheritance.

Further detail is provided by an inheritance diagram of this object, the module in which it is defined, any replaceable types it has declared and any type substitutions that have been made in ancestor objects. Clicking on a displayed ancestor selects this object type for browsing. A record is kept of all objects visited, making it easy to return to a previously browsed object. Browsing methods show you which ancestor originally defined the method, which ancestor implementation the object will invoke and a full parameter list.

6.3 Debugging Manager

Selective runtime checking of object referencing, invalid parameters, array bounds, and memory use are invaluable aids to software development. Models are large, complex programs, and debugging support for both simulation and programming errors are crucial to success. With debugging support enabled, a runtime error automatically drops you into debugging mode, allowing you to see where the error occurred and letting you examine variwhere the error occurred and letting you examine vari-ables. A traceback shows you the calling chain that led to the current method or procedure, so you can browse up and down the execution stack examining the sequence of procedure and method calls that preceded the error. The debugger supports a wide range of capabilities that are essential to interactive symbolic debugging. In addition it has special knowledge of MODSIM III's simulation constructs and can display the pending list, simulation time, and memory usage information.

7 MODSIM III AVAILABILITY

MODSIM III is developed and supported by CACI Products Company. MODSIM programs, including graphics, are completely portable across PC's running Windows95 or Windows NT, and all major UNIX platforms, with straightforward recompilation.

MODSIM III is available to your organization for a free trial in your environment, on your computer. CACI provides everything you need for a complete evaluation at your site: training, software, documentation, sample models and immediate support when you need it. In addition, CACI regularly offers time-tested training courses.

For more information contact us at:

CACI Products Company
3333 North Torrey Pines Court
La Jolla, CA 92037

Tel: 619-457-9681
Fax: 619-457-1184
Email: modsim@caciasl.com

AUTHOR BIOGRAPHY

Alasdar Mullarney is the Chief Technology Officer at CACI Products Company. During his 16 years with CACI he has made significant contributions to CACI's SIMSCRIPT II.5 simulation language and graphics. He has drawn on this experience to lead the design and development of MODSIM III. He received his Ph.D. from Trinity College, Dublin in 1976.

Proceedings of the 1996 Winter Simulation Conference
ed. J. M. Charnes, D. J. Morrice, D. T. Brunner, and J. J. Swain

INTRODUCTION TO THE WITNESS VISUAL INTERACTIVE SIMULATOR AND OLEII AUTOMATION

W. Bruce Thompson

AT&T Istel, Inc.
790 Turnpike Street
N. Andover, MA 01845, U.S.A.

ABSTRACT

The paper describes AT&T Istel's WITNESS Visual Interactive simulator. Benefits of using WITNESS are described and illustrated by explaining the steps required to build any WITNESS model. OLEII Automation as it pertains to WITNESS is also introduced.

1 INTRODUCTION

Witness is a comprehensive discrete event and continuous process simulator. It is designed to allow a person knowledgeable about the process under study to rapidly, incrementally, and accurately build, debug, validate, verify and exercise complex models. Model building time has been reduced by incorporating color coded status icons for each element, allowing portions or all of previously built models to be used (Sub-Models), user definable element library,(Designer Elements), the intelligent use of a graphical interface, form based modeling elements, pre-defined Input/Output rules, and built-in error checking. Modeling accuracy has been attained through the creation of an Action Language, a host of system and user-defined functions, user interactivity and Object Linking and Embedding (OLEII).

2 BUILDING A WITNESS MODEL

Using WITNESS' more advanced methods to construct a simulation model greatly reduces the simulation practioner's model building time. However, they do require a basic knowledge of how WITNESS works. WITNESS models are built in 3 steps:

1) Define the elements to be included in the model
2) Display the necessary elements
3) Detail the elements.

2.1 Defining Model Elements

In order to construct a simulation model, the user must first tell WITNESS what types of elements the model will contain. WITNESS contains 22 pre-defined elements that fall into 3 categories: 1) Physical, 2) Logical, and 3) Reporting. The user would enter the Define mode through a menu bar, select the element type to be defined, and provide a name and quantity required for the element.

2.1.1 Physical Elements

Table 1 lists the Physical elements that are available in both manufacturing and generic terms.

Table 1: WITNESS Physical Elements

Manufacturing	***Generic***
Parts	Entities
Machines	Activities
Module	Module
Labor	Resources
Tracks	Tracks
Vehicles	Vehicles
Buffers	Queues
Conveyors	Conveyors
Shifts	Shifts
Fluids	Items
Tanks	Store
Processors	Process
Pipes	Stream
Module	Module

2.1.2 Logical Elements

Table 2 lists the Logical elements. These elements provide the modeler greater control over execution of the model. Logical elements include:

Table 2: WITNESS Logical Elements

Variables	Attributes
Functions	Distributions
Files	Part Files

2.1.3 Reporting Elements

Table 3 lists the Reporting elements that allow user selected information to be graphically displayed while the model is running. Reporting elements include:

Table 3: WITNESS Reporting Elements

Histograms	Timeseries
Pie Charts	

In addition to these graphical reporting elements, the variables previously mentioned in the logical elements section can also be displayed.

2.2 Displaying Model Elements

Elements are displayed by clicking on the Display option from the menu bar and selecting the element to be displayed or, if the element is already displayed, the modeler may double-click on the required element, with the right mouse button, to bring up the Display form for that element. Once an element is selected, a sub-menu specific to the element type is presented. As an example, the sub-menu for a queue would contain selections for the Queue Name, Entity-in-the-Queue representation, and Icon graphic. The user would choose an option and place the chosen item accordingly. Should the initial placement of elements be unsatisfactory, a net is available to move an individual or group of elements to a new position.

2.3 Detailing Model Elements

Detailing a model involves the entering of cycle times, setup conditions, capacities, etc. Again, from the menu bar, a Detail option may be selected or, if the element is already displayed, the modeler may double-click on the required element, with the left mouse button, to bring up the Detail form for that element.

Each element type has a specific Detail form. The form for a Buffer/Queue would contain items such as the Name, Capacity, Delay Time (for black box analysis), Input Position, Output Position, as well as further conditions for Input and Output positions. The user would simply enter the relevant information into the appropriate field.

2.3.1 Input/Output Rules

WITNESS provides 9 pre-defined rules for Pushing or Pulling Parts/Entities through the model. At a basic level, elements such as Machines/Activities would either Pull from within the Input Rule editor or Push from within the Output Rule editor. At a more advanced level, IF/ELSEIF logic or User Defined Functions may be employed to move the Parts/Entities.

2.3.2 Action Language

Few, if any, models are straight forward enough to use only Push/Pull rules. More likely, movement of Parts/Entities would be contingent on conditions in some other part of the model. In addition, Parts/Entities are not always vanilla flavored. These items might also have lot sizes, specific routings or failure percentages, to name a few. WITNESS' Action Language allows these types of data to be added to the model at the beginning or end of each type of event. For example, a pass/fail determination needs to be made at the completion of a test. FOR/NEXT and WHILE/ENDWHILE constructs allow the user to change or monitor entity specific attributes or global variables thereby allowing greater control over model creation.

2.4 Basics Summary

So, in a nutshell, the steps necessary to create a simple model of an activity with an entity going through it would be to:

Define the entity: A
Define the activity: M1
Display the activity
Detail the activity
<u>Input Rule</u>: Pull A from World (Pre-defined WITNESS location)
<u>Output Rule</u>: Push to Ship (Pre-defined WITNESS location)
<u>Cycle Time</u>: 1

Click on Run on the menu bar and the model is running. Had we forgotten the Cycle Time, WITNESS would have displayed a screen prompting the user for the

Cycle Time. Had we forgotten an Output Rule, WITNESS would have displayed M1 in a magenta color indicating the activity is blocked. This makes it very easy to get the model debugged and running. If we had different entities going through M1, each of which had a different Cycle Time, we could have defined an attribute for the entity called CYC_T. When the entity was created, we could have assigned a value to CYC_T and referenced the attribute directly in the Cycle Time field of M1's Detail screen, etc., etc. Oh, by the way, all this could have been performed without resetting the model back to zero. WITNESS' Interactiveness is built into the software.

3 ADVANCED MODEL BUILDING

The following sections describe the techniques used to rapidly build a complex, user friendly model.

3.1 Designer Element Libraries

The menu bar has a DESIGNER selection. Clicking on this selection brings up a Designer Window. Once in the window, the user Defines an element. WITNESS will use the first three characters to create a sequentially numbered name each time the element is used in the model. For instance, ACT might be the name I give an activity. When the activity is used in the model, WITNESS would append 001 to it making the activity name ACT001. Once defined, the element needs an icon representation and positions for Entities and Labor, if necessary.

As above, clicking on Display on the menu bar brings up the display sub-menu for the element. From there, the items would be displayed in the Designer window.

Finally, if there are any Details of the element that would be common when placed in the model, these Details would be filled in on the Detail form of the element. Once the Design of the element is complete, adding a new element of this type to the model requires only that the element be clicked on once in the Designer Window, at which point the cursor will turn into a crosshair, and the user clicks at the point in the model at which he wants the element to appear.

3.2 Sub-Models

Sub-Models are pieces of other models that the modeler might want to use over. Creating a Sub-Model requires clicking on File, Save, Sub-Model, from the menu bar and then selecting which elements will be included in the Sub-Model. Selection may be from a list of elements, by mouse or by placing a net around a group of elements. Once saved, any reference to elements not contained in the Sub-Model will be listed at the completion of the save. Using the Sub-Model requires clicking on File, Open, Sub-Model and selecting the appropriate Sub-Model file. The user would then be prompted for what to do in case an element name in the Sub-Model matches an element name in the model into which it is placed. Options are to Prompt, Change or Ignore. If Change is selected, further selections are Prompt, Replace the First Letter, or Add a First Letter.

3.3 Module Elements and Hierarchical Modeling

When bringing a Sub-model into the model proper, the user is asked if he/she wants it assigned to a Module element. Assuming the answer is yes, the elements in the Sub-model would be assigned to an icon. Once assigned, the Module can then be entered into the Designer Elements window and brought into the model by simply clicking. The beauty of this assignment is that the model being presented needs only show the Module icon. The detail associated with the specific module is in a different part of WITNESS' virtual screen. Double clicking on the Module icon will take the user to that part of the screen.

3.4 Object Linking and Embedding - OLEII

WITNESS is capable of acting as an OLEII server. This means it can be run from an OLEII controller application. Microsoft Excel 5.0, Microsoft Access and Microsoft Visual Basic are examples of OLEII controllers. What is so good about this?

First, many firms maintain process data in spreadsheets. It saves model building time to use this data directly. Secondly, many employees know how to use a spreadsheet. Not many employees in a line decision-making authority have the time or desire to go in to a simulation model and make changes to see how to handle changes to the process. By using Excel as a front end to the simulation, the user can change the entrees in the spreadsheet, run the model, and have the necessary results of the model reported back to Excel. To the user, the simulation proper is akin to an invisible calculator. In this way, the proven benefits of simulation can be brought to a larger user base, not to mention where the decisions need to be made. The details of employing OLEII in a simulation are beyond the scope of the tutorial

3.5 Experimentation

WITNESS has a built-in feature that allows the user to automatically change the random number streams and re-run the same model, or a number of different models, to statistically determine the model's performance. Creating an experiment requires defining a Situation/Scenario, determining the Warm-up period, Run Length, Continuous vs. Restart data collection, how the random numbers will be exercised, what the output file name will be, and running the experiment.

Defining the Situation/Scenario requires that the scenario be named and, if applicable, assign a model file to be run for that scenario. If the experiment will be running inside of the model, no model name would be required. For the Warm-up period, assume a Timeseries has been placed in the model to look at the number of shipments every hour. Assume also that number appears to level off after 2 hours of simulation time. Two hours would be used as the Warm-up. The Run Length will be 1 week with data collection occurring at the end of each replication. Exercising the random numbers requires a little background.

Because simulation deals with stochastic systems, more than one run must be made to ensure that the value of a metric being reported upon is not caused by some strange combination of random events. As an example, if someone asked you how long it takes to get to work, you would probably answer with a range of values, perhaps 30 - 45 minutes. The exact number depends on the random events of weather, time of morning, time of year, etc. With simulation, these random events are represented by random number streams.

WITNESS has 999 unique random number streams numbered 1 through 999. Each model run would use a different set of random numbers. The easiest way of changing the numbers is with the Stream Offset. This option will use a random number stream that is a certain number higher than the one referenced in the model. For instance, assume I use the following distribution as a cycle time in a model:

LOGNORML (2, .15*2, 35)

The mean is 2, the standard deviation is 15% of the mean and the random number stream is 35. If I use a Stream Offset of 2 for this iteration of the experiment, the actual number stream being used would be 37. Exercising the model 6 times by simply changing the Offset is very convenient.

One thing remains. Where are the system statistics (i.e. Time in system, Total Throughput, etc.) going to go? Although there are 6 options, output from these iterations would normally go to a Comma Separated Values (CSV) file. This file format can be directly imported by a number of software programs for statistical analysis. For our purposes, a CSV file will be used as it will be input to WITNESS' statistical analysis program XA. Now the user has to simply run the experiment. The model will run 6 times and the output from each run will be placed in the output file WINSIM.CSV

4 XA

Packaged with WITNESS, although a separate program, is XA. XA will take a CSV output file from WITNESS, and calculate a Student's 't' statistic for each measurement over all runs of the experiment. It will also calculate 99%, 95%, and 90% confidence intervals using the 't' statistic. Measurements for which statistics are generated are user selectable.

5 SUMMARY

WITNESS is a discrete event, Visual, Interactive simulator. It has been designed to quickly and accurately create a simulation. Real-time color coded element status indications, combined with an extremely flexible graphic environment greatly reduce the time required to build a model

Basic models are built using a Define, Display, Detail cycle with pre-defined modeling elements and Input/Output rules. More complex models would incorporate IF/THEN/ELSEIF Input/Output rules, Object Linking and Embedding and the WITNESS Action Language. Modeling time can be significantly reduced by saving portions of previous simulations as Sub-Model for use in future simulations. In addition, elements common to the modeler's environment may be incorporated into a Designer library for quick Definition, Display and Detail.

Also described were the OLEII capabilities, how to set up experiments from within WITNESS and lastly, how to analyze the data from the experiments using the XA statistical analysis package.

AUTHOR BIOGRAPHY

W. BRUCE THOMPSON is a Sr. Technical Consultant with AT&T Istel, Inc. He received his BS degree from Northeastern University in 1981 and has been involved with simulation, primarily in the context of manufacturing, for the last 10 years. In addition to publishing articles on simulation in various trade magazines, he has presented at a variety of conferences, including Winter Simulation Conferences '90 - '95, APICS Conference '95, and AUTOFACT '92. His responsibilities for AT&T Istel include training, consulting, customer support, and Editor of the company's newsletter, *The Interactive Modeler*.

Proceedings of the 1996 Winter Simulation Conference
ed. J. M. Charnes, D. J. Morrice, D. T. Brunner, and J. J. Swain

BESTFIT, DISTRIBUTION FITTING SOFTWARE BY PALISADE CORPORATION

Linda Jankauskas
Sam McLafferty

Palisade Corporation
31 Decker Road
Newfield, NY 14867, U.S.A.

ABSTRACT

BestFit is distribution-fitting software for Microsoft Windows that finds the statistical distribution function that best fits a data set. BestFit provides a flexible, easy-to-use environment for analysis that allows users to focus on their data, not their software.

Users of simulation software can use distributions produced by BestFit to define uncertainty in their simulation models. BestFit helps you find the best representation of randomness in your model, making your simulation results more accurate. Users of @RISK, ProModel or any other simulation software will find BestFit an invaluable tool for defining uncertainty.

1 OVERVIEW OF THE SOFTWARE

BestFit's goal is to find the distribution that best fits your input data. BestFit does not produce an absolute answer, it identifies a distribution that most likely produced your data. For a given distribution, BestFit looks for the parameters of the function that optimize the goodness-of-fit, a measurement of the probability that the input data was produced by the given distribution.

Since its release in 1993, BestFit has undergone several upgrades. Each upgrade incorporated new features and capabilities directly requested by BestFit users. Palisade Corporation is dedicated to responding to its customers needs by offering new features and compatibility in a timely manner.

1.1 Entering a Data Set

BestFit allows three kinds of data: sample, density and cumulative. Methods for bringing your data into BestFit include: typing it directly into the input sheet, copying it from another Windows application, opening a text file containing the data or creating a link between BestFit and an Excel or Lotus 1-2-3 spreadsheet.

1.2 The BestFit Calculation

For each distribution type, BestFit makes a first guess of the best distribution parameters using maximum-likelihood estimators (MLEs). The MLEs of a distribution are the parameters of that function that maximize the likelihood of the distribution given a set of observational data.

BestFit optimizes the parameters calculated from the MLEs using the Levenberg-Marquardt method, an algorithm that maximizes the goodness-of-fit between a data set and a distribution function. This method takes a first guess of the parameters (the MLEs) and varies each parameter slightly until a good fit is found.

1.3 Graphs in BestFit

A good way of interpreting the results of a BestFit calculation is to visually assess how well a distribution agrees with the input data. Four graphs are available for this purpose: comparison, difference, P-P and Q-Q.

Comparison Graphs: The comparison graph superimposes the input and result distributions on the same graph, allowing you to visually compare them either as density or cumulative curves. This graph allows you to determine if the best fit distribution is well matched to the input data in specific areas. For example, it may be important to have a good match around the mean or in the tails.

Difference Graphs: The difference graph displays the absolute error between the input and result distributions. This error is defined as the difference between the input and result probability for each input value of X. Comparing the magnitude of the error to the magnitude of the result gives you an idea of the extent to which the result deviates from the input.

Probability-Probability Graphs: Probability-Probability (or P-P) graphs plot the distribution of the input data (P_i) against the distribution of the result ($F(x_i)$). If the fit is good the plot will be nearly linear.

Quantile-Quantile Graphs: Quantile-Quantile (or Q-Q) graphs plot the plot percentile values of the input distribution (x_i) against percentile values of the result ($F^{-1}(P_i)$). If the fit is good the plot will be nearly linear.

1.4 The BestFit Statistical Report

Basic statistics (mean, variance, mode, etc.) for each distribution are reported and can be compared to the statistics of the input. For each result, goodness-of-fit values and the corresponding confidence intervals are reported. These statistics measure how good the distribution fits the input data and how confident you can be that the data was produced by the distribution function.

The targeting option in BestFit compares percentile values and probabilities between distributions and the input data. For example, if the 5th and 95th percentile values are important to you, you can easily compare them in the BestFit statistical report.

Statistical results, including distribution functions, can be transferred to other Windows applications using the clipboard. In addition, results can be saved to a text file for export to other operating systems.

1.5 System Requirements and Compatibility

BestFit is available in 16-bit and 32-bit versions. The 32-bit version requires Microsoft Windows 95 or Windows NT. The 16-bit version requires Windows 3.1, Windows NT or Windows 95. Both versions will run under the minimum system requirements for the operating system. A multi-user, network version of BestFit can be used on any server-based PC network and is available in 16-bit and 32-bit versions.

2 HOW BESTFIT WORKS

For a given distribution, BestFit looks for the parameters of the function that optimize the goodness-of-fit. BestFit follows these steps when finding the best fit for your input data:

1. For each distribution type selected, parameters values are estimated using MLEs (maximum-likelihood estimators).

2. The fit is optimized using the Levenberg-Marquardt method (optional).

3. The goodness-of-fit is measured for the optimized function using chi-square, Kolmogorov-Smirnov and/or Anderson-Darling tests.

4. All functions are compared and the one with the lowest goodness-of-fit values is considered the best fit.

5. Detailed statistics (mean, variance, skewness, etc.) are reported for the input data and all results.

6. Goodness-of-fit statistics and their confidence intervals are reported for each distribution function.

2.1 Probability Distributions

BestFit offers 26 distribution functions. The density and distribution (when available) used for each function is defined in Evans, et al. (1993) and Law and Kelton (1982).

Table 1: Distributions Available in BestFit

Beta	Lognormal,
Binomial	Lognormal2
Chi-Square	Negative Binomial
Error Function	Normal
Erlang	Pareto
Exponential	Pearson Type V
Extreme Value	Pearson Type VI
Gamma	Poisson
Geometric	Rayleigh
Hypergeometric	Student's t
Inverse Gaussian	Triangular
Logistic	Uniform
Log-Logistic	Weibull

2.1 Maximum-Likelihood Estimators

As defined by Law and Kelton (1982), the MLEs of a distribution are the parameters of that function that maximize the likelihood of the distribution given a set of observational data. For any density distribution $f(x)$ with one parameter α and a corresponding set of observational data X_i, an expression called the likelihood may be defined:

$$L = \prod_{i=1}^{n} f(X_i, \alpha) \qquad (1)$$

To find the MLE, use (1) to maximize L with respect to α:

$$\frac{dL}{d\alpha} = 0 \qquad (2)$$

and solve (2) for α.

This paper was presented at the 1995 Winter Simulation Conference in Crystal City, VA.

BestFit calculates the maximum-likelihood estimators for each distribution using formulas defined in Evans et al. (1993) and Law and Kelton (1982).

2.2 Levenberg-Marquardt Method

BestFit can determine the best fit of a data set simply from its MLEs or it can optimize the parameters using the Levenberg-Marquardt method. The Levenberg-Marquardt Method is an algorithm used to maximize the goodness-of-fit between a data set and a distribution function. As explained by Press, et al. (1988), this method takes a first guess of the parameters of the distribution function, and then varies each parameter slightly until it finds a good fit.

The Levenberg-Marquardt method, which is a non-linear least-squares routine, takes an iterative approach to try to minimize the goodness-of-fit statistic (chi-square). The Levenberg-Marquardt method does not find the absolute minimum for chi-square; rather, it finds a local minima. The success of this method depends on the initial parameters used. Therefore, a good first guess (i.e., one produced from a well-defined input) will produce a good result, while a poor first guess might not provide a useful result.

The goodness-of-fit test used by BestFit for optimizing a distribution is the chi-square test. This test was chosen because it is the most frequently used and the fastest to calculate.

2.3 Goodness-Of-Fit Statistics

The process of calculating MLEs and optimizing the chi-square value gives a best guess for each distribution. BestFit then measures the quality of each fit using goodness-of-fit measurements.

2.3.1 What is Goodness-of-Fit?

Formally, goodness-of-fit is defined as the probability of the data given the parameters. In other words, the goodness-of-fit statistic tells you how probable it is that a given distribution function produced your data set. The goodness-of-fit statistic is usually used in a relative sense by comparing the values to the goodness-of-fit of other distribution functions.

BestFit offers three goodness-of-fit tests: chi-square, Kolmogorov-Smirnov and Anderson-Darling. The chi-square is the most common, but the others may supply more detailed information about your distribution.

2.3.2 Chi-Square Test

The chi-square test is the most common goodness-of-fit test. It can be used with any type of input data (sample, density or cumulative) and any type of distribution function (discrete or continuous).

A weakness of the chi-square test is that there are no clear guidelines for selecting intervals. In some situations, you can reach different conclusions from the same data depending on how you specified the intervals (number of classes).

The chi-square statistic used by BestFit, as defined in Law and Kelton (1982), is:

$$\chi^2 = \sum_{i=1}^{n} \frac{(P_i - p_i)^2}{p_i} \tag{3}$$

In (3) we define the probability values as

P_i = *the observed probability value for a given histogram bar*

p_i = *the theoretical probability that a value will fall with the X range of the histogram bar*

2.3.3 Kolmogorov-Smirnov Test

The Kolmogorov-Smirnov test does not depend on the number of intervals, which makes it more powerful than the chi-square test. A weakness of the Kolmogorov-Smirnov test is that it does not detect tail discrepancies very well.

The Kolmogorov-Smirnov statistic used by BestFit, as defined in Law and Kelton (1982), is:

$$D_n = \sup\left[\left|F_n(x) - \hat{F}(x)\right|\right] \tag{4}$$

In (4) we define the parameter values as

n = *total number of data points*

$\hat{F}(x)$ = *the hypothesized distribution*

$F_n(x) = \frac{N_x}{n}$

N_x = *the number of* X_i*'s less than x.*

2.3.4 Anderson-Darling Test

The Anderson-Darling test is very similar to the Kolmogorov-Smirnov test, but it places more emphasis on tail values. It does not depend on the number of intervals.

The Anderson-Darling Statistic used by BestFit, as defined in Anderson and Darling (1954), is:

$$A_n^2 = n\int_{-\infty}^{+\infty}\left[F_n(x) - \hat{F}(x)\right]^2 \Psi(x)\hat{f}(x)dx \qquad (5)$$

In (5) we define the parameter values as

$$\Psi = \frac{1}{\hat{F}(x)\left[1 - \hat{F}(x)\right]}$$

$\hat{f}(x)$ = *the hypothesized density function*

$\hat{F}(x)$ = *the hypothesized distribution function*

$$F_n(x) = \frac{N_x}{n}$$

N_x = *the number of* X_i*'s less than x.*

2.4 Confidence Levels and Critical Values

For the goodness of fit tests, the critical value determines whether you should accept or reject a fitted distribution at a given confidence level (α). The method of statistical hypothesis testing allows you to use a structured analytical method to make a decision regarding your BestFit results. This method allows you to control or measure the uncertainty involved in the decision.

Typically, the critical value depends on the type of distribution fit, the number of data points, and the confidence interval. Most critical values for the Anderson-Darling and Kolmogorov-Smirnov goodness-of-fit statistics have been found by Monte Carlo studies (see Chandra, et al. 1981, Stephens 1974 and Stephens 1977). BestFit reports both the smallest confidence value that passes the null hypothesis test (has a critical value greater than the goodness-of-fit statistic) and the critical values for selected confidence intervals.

2.4.1 Critical Values for the Chi-Square Test

For the chi-square test, BestFit calculates the critical value by finding the target point for the 1-α percentile of a chi-square distribution with N-1 degrees of freedom (N is the number of classes).

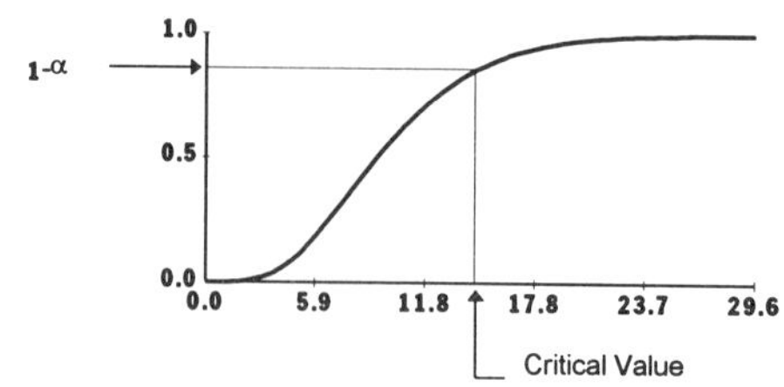

Figure 1: A Chi-Square Distribution With 10 Degrees of Freedom

When the calculated chi-square statistic is larger than the critical value, the null hypothesis should be rejected (the distribution is not a good fit). This particular measurement of the critical value is a conservative test, it does not take into account that the parameters were estimated from the data. The actual confidence is greater than or equal to α.

2.4.2 Critical Values for the Kolmogorov-Smirnov and Anderson-Darling Tests

The critical values for the Kolmogorov-Smirnov and Anderson-Darling tests were calculated using Monte Carlo simulation techniques as explained by Stephens (1974), Stephens (1977) and Chandra et al. (1981).

Unlike the chi-square test, where the critical value is the same for all distributions, the Kolmogorov-Smirnov and Anderson-Darling tests include special cases for the Normal, Exponential, Weibull and Extreme Value distributions. For all other distributions, the critical values are estimated using an "all parameters known" test that is more conservative than the tests for specific distributions. BestFit marks estimated confidence levels with an asterisk (*) in the statistical reports.

For the Kolmogorov-Smirnov and Anderson-Darling tests, BestFit does not compare the critical value to the actual test statistics. Instead, BestFit modifies the critical value using formulas defined by Stephens (1974), Stephens (1977) and Chandra et al. (1981). These modifications take the number of data points into consideration when evaluating confidence levels.

2.4.3 Interpreting the Critical Value

The tests described in this article are very sensitive to the magnitude n (the number of data points). If n is small, the goodness of fit will only measure large difference between the input data and the distribution function. As n increases, the modified test statistics increase and the null hypothesis will be rejected more often. The results produced by these tests should be considered "guidelines" in selecting a fit. Always

This paper was presented at the 1995 Winter Simulation Conference in Crystal City, VA.

evaluate the results by comparing statistics and graphs before accepting or rejecting a fit.

3 ADVANCED FEATURES OF BESTFIT

BestFit includes many advanced features that make distribution fitting easier and more powerful.

3.1 Linking BestFit to Spreadsheet Data

BestFit can directly link to a range of data in a spreadsheet. Instead of creating a copy of your data in BestFit, the project is linked directly to the source of the data and automatically updates every time the data changes in the spreadsheet.

BestFit acts like an add-in by displaying a new toolbar icon in Microsoft Excel or Lotus 1-2-3 and creating a new BestFit command in the Tools menu. To use BestFit as an add-in, simply highlight a range of data in your spreadsheet and click on the BestFit command. BestFit opens with a link to the selected data, allowing you to immediately run an analysis.

3.2 Filtering

BestFit's filtering option defines outliers in your data which will be ignored during a BestFit analysis. Analyze a subset of your data without creating a new project file.

For example, you may wish to only analyze X values greater than zero. Or, you may wish to filter out tail values by only looking at data within a few standard deviations of the mean.

3.3 Transferring Results to Your Spreadsheet

All BestFit results, including graphs and statistical reports, can be transferred to a spreadsheet or other software at the click of a button. Statistical reports can be copied to the Windows clipboard and pasted into other software packages for further customization.

BestFit can transfer the data points of a graph to Lotus 1-2-3 or Excel and create a graph in the spreadsheet's native format. Or, an image of the graph can be copied to the Windows clipboard and pasted into other software packages.

3.4 Subjective Probability Assessment

To generate a probability distribution in the absence of historical data, BestFit interacts with RISKview (distribution viewing software from Palisade Corporation). RISKview can be purchased by itself or bundled with BestFit.

RISKview generates a generalized BETA distribution from subjective input using two special functions: BETA-SUBJECTIVE and BETA-PERT. The BETA-SUBJECTIVE function accepts estimated minimum, most likely, mean and maximum values as inputs. The function gives you the flexibility of a generalized function like the TRIANGULAR while producing a more realistic distribution. The BETA-PERT function uses the classic Project Management PERT function to create a BETA distribution using estimated minimum, most likely and maximum values as inputs.

RISKview offers other distribution assessment tools including distribution drawing and smoothing, function previews, statistics, target values and other subjective probability techniques.

REFERENCES

Anderson, T. W., and D. A. Darling. 1954. A test of goodness of fit. *American Statistical Association Journal.* 1954:765-769.

Chandra, M., N.D. Singpurwalla and M.A. Stephens. 1981. Kolmogorov statistics for tests of fit for the extreme value and weibull distribution. *Journal of the American Statistical Association.* 76:729-31.

Evans, Merran, Nicholas Hastings and Brian Peacock. 1993. *Statistical Distributions.* 2d ed. New York: John Wiley & Sons, Inc.

Law, A. M., and W. D. Kelton. 1982. *Simulation modeling and analysis.* 2d ed. New York: McGraw-Hill Book Company.

Press, W. H., B. P. Flannery, S. A. Teukolsky, and W. T. Vetterling. 1988. *Numerical recipes in C: The art of scientific computing.* New York: Cambridge University Press.

Stephens, M.A. 1974. EDF statistics for goodness of fit and some comparisons. *Journal of the American Statistical Association.* 69:730-737.

Stephens, M.A. 1977. Goodness of fit for the extreme value distribution. *Biometrika.* 64:583-588.

AUTHOR BIOGRAPHIES

LINDA JANKAUSKAS is a Software Engineer at Palisade Corporation. She is involved in developing software for decision analysis, including BestFit, RISKview and DecisionTree.

SAM MCLAFFERTY is the President of Palisade Corporation. He has been developing risk analysis software for over ten years and is currently involved in developing software for decision analysis including @RISK and TopRank.

This paper was presented at the 1995 Winter Simulation Conference in Crystal City, VA.

Proceedings of the 1996 Winter Simulation Conference
ed. J. M. Charnes, D. J. Morrice, D. T. Brunner, and J. J. Swain

AN OBJECT-ORIENTED PHONE CENTER MODEL USING SIMPLE++

Gerald A. Levasseur

AESOP Corporation
233 South Wacker Drive
Sears Tower, Suite 9604
Chicago, Illinois, 60606, U.S.A.

ABSTRACT

A demonstration model and Application Object Template (object library) was created to show how SIMPLE++ simulation software can be applied to service industry telephone call handling centers. In addition, this example was designed to show modeling techniques that can be used to take advantage of some key object-orientation concepts to quickly create highly flexible simulation models. Finally, some features of the SIMPLE++ simulation package are illustrated.

1 INTRODUCTION

The object-oriented simulation package SIMPLE++ was first commercially introduced in Germany in February of 1992, and AESOP GmbH was tasked to further develop and market the software and related products and services. In North America, AESOP Corporation is responsible for SIMPLE++ licensing and related services. Since its creation, SIMPLE++ has been applied in a variety of industries including automotive manufacturing, rail transportation, electronics assembly and fabrication, pharmaceuticals, hospitals, chemical plants, semi-conductor facilities, material handling, and warehousing. Application functions include scheduling, production planning and control, design support, and capital equipment justification. Users include corporations, governments, universities and research institutions worldwide.

Geuder presented some of the most important object-oriented features of SIMPLE++ in two recent papers (Geuder 1995a, Geuder 1995b). The two most notable concepts that will be used in this example are inheritance and hierarchy. Inheritance is the ability of a child (a specific instance of a class) or a sub-class to maintain the properties or functionality of a parent (class). Hierarchy is the structure where multiple layers of inheritance are possible.

1.1 Problem Description

The Phone Center Demonstration Model presented here illustrates how SIMPLE++ can be used to distribute calls to telephone customer service sites, such as those shown in Figure 1. Such sites are widely used in financial service, travel, customer support, catalog and mail order businesses to answer customer calls. In this model, different types of calls arrive at a central distribution site, with each type of call following its own arrival distribution.

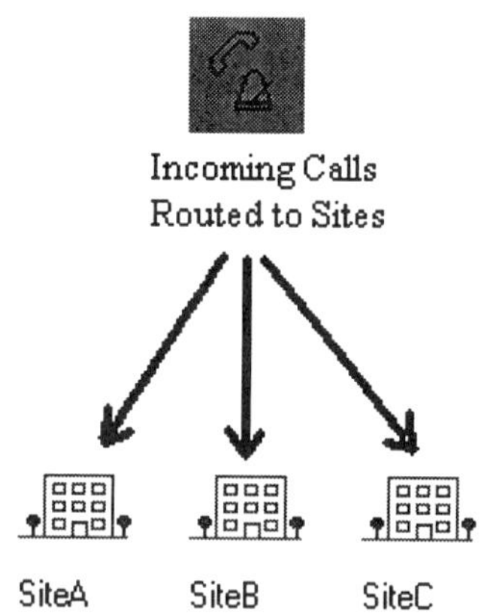

Figure 1: Incoming Calls are Distributed to Sites

For each type of call, there is one type of agent able to answer it. Upon a call's arrival, a decision is made to distribute the call to a specific phone site where the agents are located. There can be any number of sites, as long as there is at least one. One of three primary criteria may be selected to determine which site a call should be routed to:

1. Proportion of Agents at Site. The number of agents at a given site that can handle an incoming type of call is divided by the total number of agents at all sites that can handle the call. This proportion corresponds to the probability the call will be routed to a given site.

2. *Shortest Queue Length.* The site with the shortest queue length for a given type of call is selected. If there are no queues, then a secondary criterion may be selected.

3. *Number of Free Agents.* The site with the greatest number of free agents that can handle a given type of call is selected. If there are no free agents, a secondary criterion may be selected.

Secondary criterion may be selected in the event the first criterion conditions do not exist. This criterion can be any of the three primary criteria; however, if the secondary and primary criteria selected are the same, the secondary criterion will be ignored. If the primary criterion was Proportion of Agents at Sites, no secondary criterion is required. If the primary criterion was Shortest Queue Length and there are no queues, the secondary criterion will be used. If the primary criterion was Number of Free Agents and there are no free agents at any site, the secondary criterion will be used.

Once a call is distributed to a site, the call is given to the appropriate agent. If all agents are busy, the call will be placed in a queue until an agent becomes free. Calls exit the queue based on the First-In-First-Out (FIFO) rule. When a call is completed, appropriate statistics are collected and the caller exits the system.

1.2 Concepts Demonstrated

In addition to showing the potential application of SIMPLE++ in the service industry, this model demonstrates the following SIMPLE++ features and modeling techniques:

Hierarchy and Inheritance. This model shows how hierarchy and inheritance can be used to simplify modeling and contribute to modeling flexibility.

SimTALK. SimTALK is the built-in SIMPLE++ simulation language. This language is accessed by means of methods (objects that contain SimTALK coding).

Anonymous Identifiers. In SIMPLE++, anonymous identifiers are used to indirectly refer to an object without having to know in advance what object it is. This is especially useful because identical methods can be used in different models or objects (AESOP GmbH 1995). These anonymous identifiers can be used to greatly enhance the flexibility of a SIMPLE++ model, especially when used in conjunction with its object-orientation features.

Modeling Flexibility. This demonstration model will show how constructors and destructors, when used in conjunction with anonymous identifiers and other SIMPLE++ features, give the modeler a powerful tool in creating flexible models. SIMPLE++ constructors activate specified methods when a given object is inserted into a model, while destructors activate specified methods when a given object is removed from a model.

Running Multiple Models Simultaneously. Because all SIMPLE++ models are encapsulated objects, several models can be run at the same time by placing them inside another object. This demonstration model will show an example of how to run two models at once. A typical use may be to directly compare two dispatching criteria to see the effects side-by-side.

Other SIMPLE++ Features. This paper will also show some other SIMPLE++ features: Dialogue Boxes, Tables, and Plotters. Dialogue boxes are used so modelers can create a way for the end users to easily make choices while hiding model complexity. SIMPLE++ tables can hold different types of data, including tables within tables and even objects. Finally, plotters are used to graphically view performance statistics over time.

2 THE PHONE CENTER MODEL

Objects shown in the top five rows of the Application Object Template in Figure 2 are basic objects. These objects come with every SIMPLE++ package. The user may use these objects directly in a model, or may create an Application Object Template from them. In this example, the Application Object Template consists of the objects created for this demonstration model, and is found in the bottom four rows of the template.

To start the model, the user double-clicks on the Demo object in the Application Object Template. This will cause the CriteriaDialog box to open. The dialogue box feature in SIMPLE++ is used to hide model complexity when desired, so the ultimate end user does not need to understand the inner workings of a model to operate it. In this example, the user may select the previously described primary and secondary criteria for dispatching calls to sites.

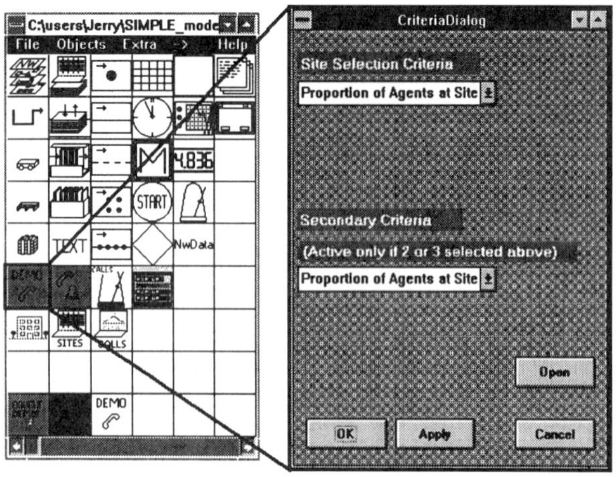

Figure 2: The Application Object Template for the phone center demonstration model (left), and a dialogue box used to select decision criteria (right).

After the criteria are selected, the user should select the Open button to open the Demo frame. A frame is simply an empty object that is used to hold other objects. In this example, the Demo frame holds an entire model. Once Open is selected, the frame shown in Figure 3 will appear, showing the contents of the frame Demo and revealing the structure of the model. The default behavior of double clicking on a frame is to open it directly; however, in this demonstration example the modeler added the command to open the dialogue box before opening the frame.

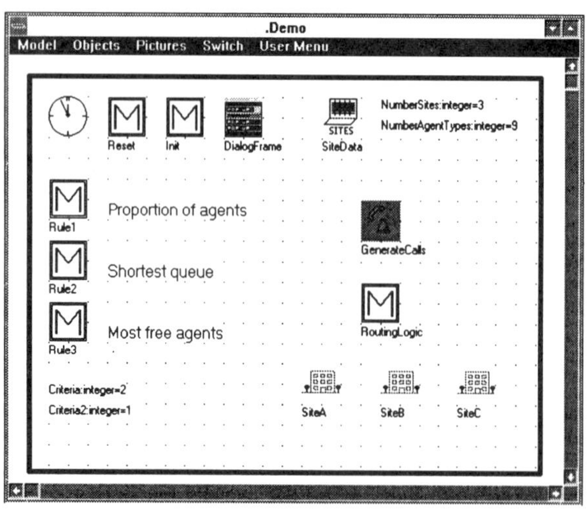

Figure 3: The Demonstration Model.

2.1 Inside the Demo Object

The EventController, represented by a clock icon in Figure 3, is an instance of the basic object that is used to control the model's simulated time. Before running the model, the user should open the Event Controller and run the Reset and Init options. This will call all methods in the model named Reset and Init, respectively, which are written by the modeler. In this example, they clear all data inserted into tables by previous runs while leaving user entered data intact. The Init methods in this example will calculate the total number of agents and the proportion of agents at each site, and will write this information to the SiteData table and other appropriate tables. Reset and Init are also excellent ways to set the initial conditions of a simulation run.

The SiteData table shown in Figure 3 is opened in Figure 4. This table contains the phone sites in the model, as well as tables-in-a-table that show the overall numbers of agents at all sites. The objects in the SiteName column are automatically entered into the table whenever a new site is inserted into the model. The user names the embedded table (called Number here) and enters the total number of agents at this site in the Total column. The Reset and Init methods fill in the TotalAllSites and Proportion columns, which contains the total numbers of agents at all sites and the proportion of the total at this site, respectively.

NumberSites and NumberAgentTypes shown in Figure 3 are NwData objects. NwData objects contain network data that can be used anywhere in the frame. In this example, the user does not need to do anything with these objects -- they are set automatically by Init. When Init methods are activated via the Event Controller, they check the dimension of the SiteData table. All variables in the model that refer to the number of sites are based on this dimension. Thus, any loops or counters referring to the number of sites are automatically re-adjusted, even if a site is added or deleted during a simulation run.

Criteria and Criteria2 are also NwData objects. These show the primary and secondary criteria selected in the dialogue box. The user does not need to do anything with these objects, but may change the criteria directly from here.

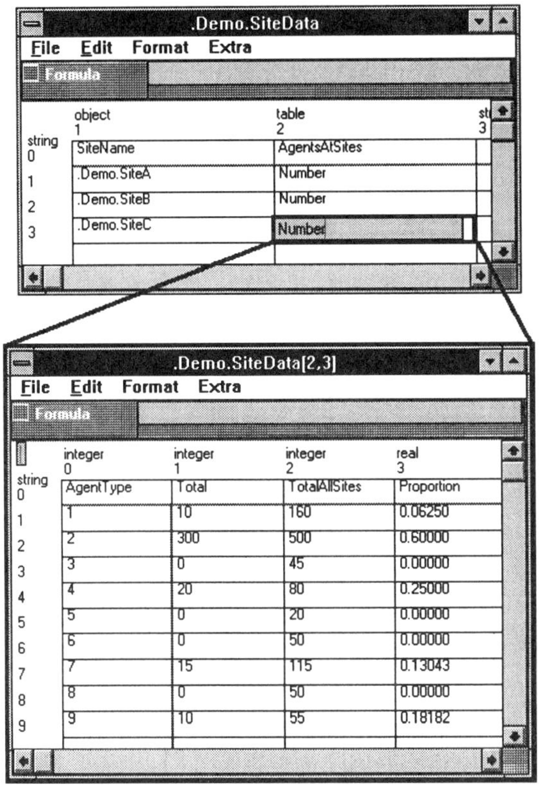

Figure 4: SiteData tables. Note they can contain a variety of types of data, including objects and nested tables.

2.2 Call Generation

The GenerateCalls object shown in Figure 3 is opened in Figure 5. This object contains nine generators, each which send calls to the RoutingLogic method. A generator is a basic object that calls a given method according to a specified distribution. Each of the generators in this example has its own distribution function that simulates each of the nine incoming call types. The user may change these distributions in the GenerateCalls dialogue box. There are nine built in distributions, or the user may define a formula, read a value from a table, or simply set a constant. In addition, any parameter for a distribution can be a variable. In the generator shown in Figure 5, the method RoutingLogic is activated according to a uniform distribution between 5 and 23 seconds. The anonymous identifier *location* refers to the frame the generator's current frame is in. If all generators follow identical distributions, the user may activate inheritance to use the parent's (located in the Application Object Template) distribution. Alternatively, the inheritance can be switched off to set individual values.

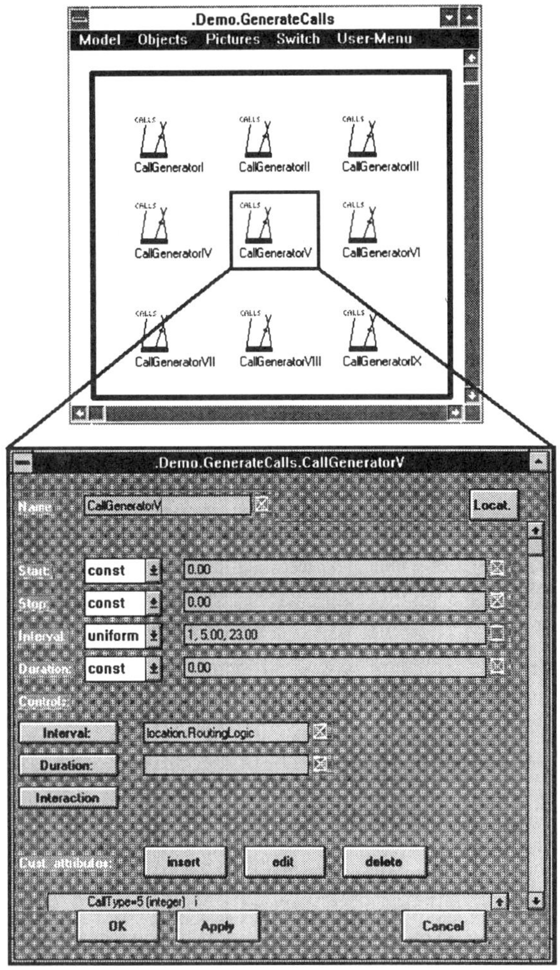

Figure 5: GenerateCalls and a Generator's Dialogue Box.

Because of SIMPLE++ object-orientation, one could easily modify this object and not worry about how it would effect the rest of the model. For example, the modeler could easily replace the GenerateCalls object and its nine generators with another object that contained only a single generator, a method and some tables that accomplished the same end result. That would be useful if the modeler wished to generalize the number of agents at some future date.

2.3 Routing Logic Methods

The RoutingLogic object shown in Figure 3 is a method. This method sends each in-coming call to other methods that perform Rule1, Rule2, or Rule3 depending on the primary criteria selected. Figure 6 shows this method's coding written in SimTALK. While the scope of this paper does not allow a detailed discussion of SimTALK, the reader can see how it may be activated. A full description of SimTALK is contained in the SIMPLE++ reference manual (AESOP GmbH, 1995).

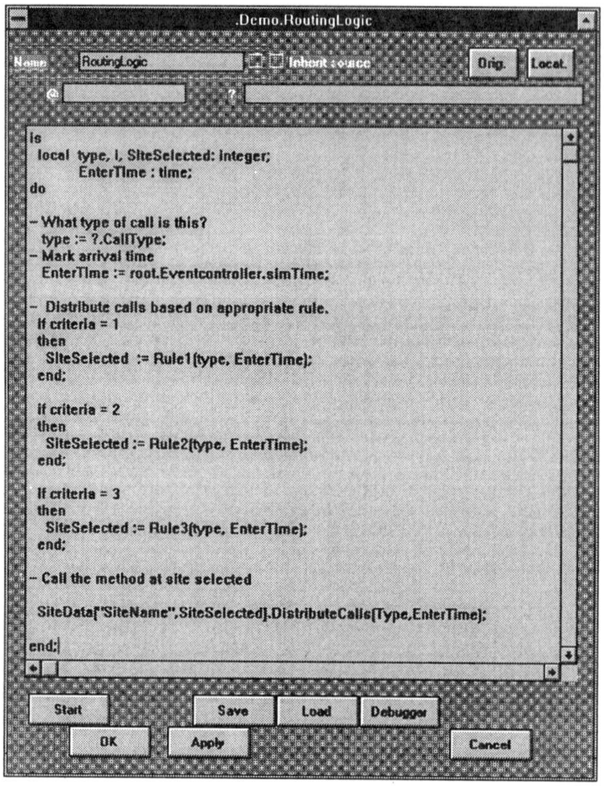

Figure 6: A Sample Method and SimTALK

The question mark used in the coding is an anonymous identifier that refers to the object that called this method, in this case a generator. Thus, this method can be called from anywhere in the model without even knowing what object called it. Also note the last line of coding before the end statement directly calls the site method DistributeCalls based on the object found in the SiteData table. By using tables to hold the object, any object with a method called DistributeCalls can be referenced, even if the object was inserted into the model after the simulation was started.

The rule methods select the sites according to the criteria previously described. If a secondary criterion is required, they will call the appropriate rule method. After the site has been selected, they will return an integer to RoutingLogic indicating which site has been selected. RoutingLogic then calls the DistributeCalls method that is located within the selected site's frame, giving it the type of call and the time it entered the system.

2.4 Inside the Site Object

An example of the Site frame is shown in Figure 7. When a call comes in from the RoutingLogic method in the Demo frame, it will go to the selected site's DistributeCalls method. Note that the site is directly referenced by the object in the table, and not by a string or integer variable. The method DistributeCalls decides if the call will go to a free agent immediately, or to a waiting list if no agents are free. If an agent is free, it decrements the number of free agents for this type of call in the Agents table, then it calculates the time the call will take. It will next calculate statistics and record them in the appropriate tables. Finally, after the call is complete, it will call the method Tally.

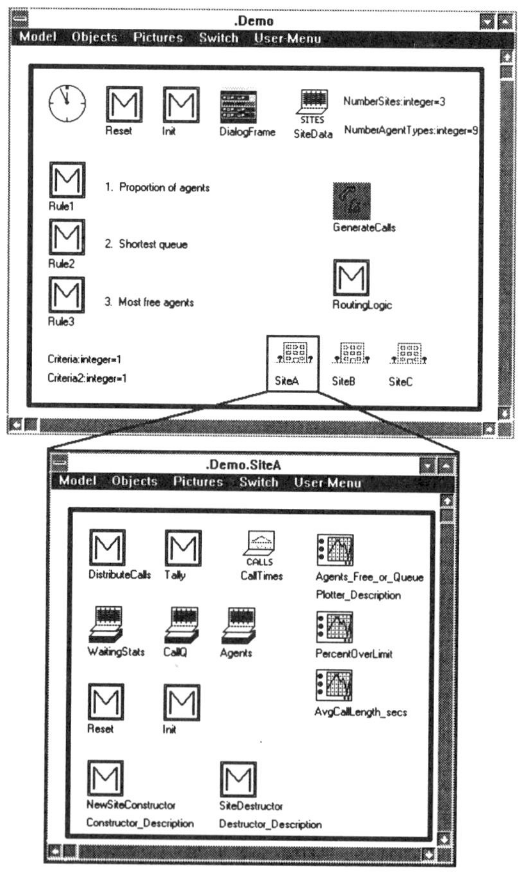

Figure 7: The Site Frame Contains the Objects Located in Each Site.

If no agents are free, DistributeCalls places this call in the CallQ table, which contains queues for each type of waiting call. It also updates the current count of waiting calls.

The method Tally is called when an agent becomes free. It adds one agent to the counter in the table Agents for this type of call. If there is a queue for this type of agent, it will take the next waiting call out of the queue and then call the method DistributeCalls. This time DistributeCalls will be able to send the call to an agent because an agent is free.

Figure 7 also shows an excellent example of hierarchy. Each site in the Demo object is a child of the site object in the Application Object Template. When the modeler wishes to make a change to the sites, only the object in the template needs to be modified -- all of the children in the model will automatically inherit those changes. In addition, issues such as repeated names are not a problem. For example, all sites have tables in them called SiteData, but each is unique because each is in a different child of the site object. The CallTimes table in SiteA is distinguished from the CallTimes table in SiteB by virtue of its location. Likewise, a modeler adding a new site does not need to remember to add the table because each site will inherit the table from the parent. Thus, the behavior of each child can be easily modified and tested.

The time a call will take once it is answered by an agent is assumed to follow a type of truncated normal distribution. The parameters this distribution will use is set by the user in the CallTimes table shown in Figure 8. The mean and standard deviation is set by the user, and the method DistributeCalls will set any generated values below zero equal to zero. For this demonstration model, the normal distribution was selected because most people are familiar with it, even though it is unlikely to be the distribution selected in an actual application. To make all distributions the same for each site, the user may set them in the Application Object Template CallTimes table and turn on the Inherit Contents option in each site's table in the Demo model. To set the distributions individually for each site, the user may change them in the Demo model with the Inherit Contents option off. This allows the greatest flexibility and ease of data input.

2.4.1 Plotters in the Site Object

This model contains three plotters in each site. They are:

Agents_Free_or_Queue: The number of free agents, or if there are no free agents, the queue length (represented as negative numbers).

AvgCallLength_secs: The average length of a call. This is defined as the time the call arrives until the call has been completed, including any time in spent queues.

PercentOverLimit: The percent of calls that take over a user-defined time limit. A sample plot is shown in Figure 9.

Here, the hierarchical structure assures that each plotter will plot the variables associated with the site it is in, even if this site did not exist when the run was started. No changes to the model are required for the plotters when adding a new site.

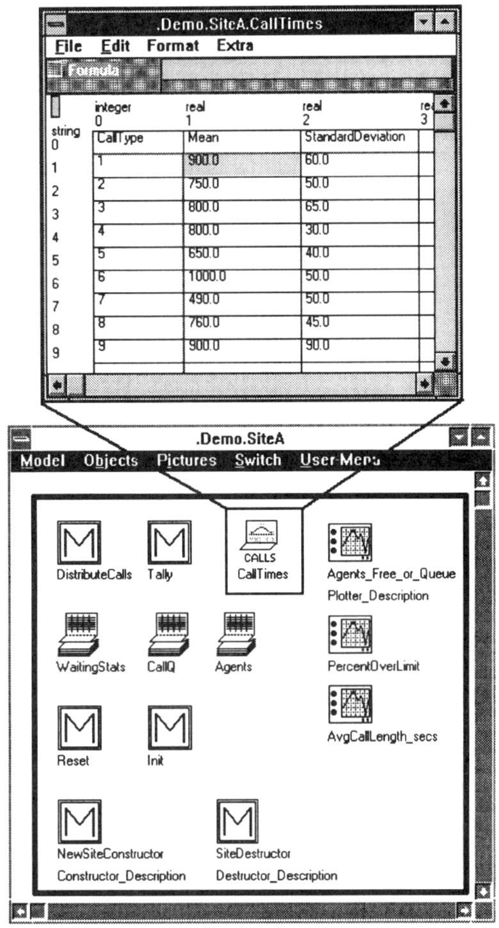

Figure 8: The CallQ Table Keeps Track of Waiting Calls

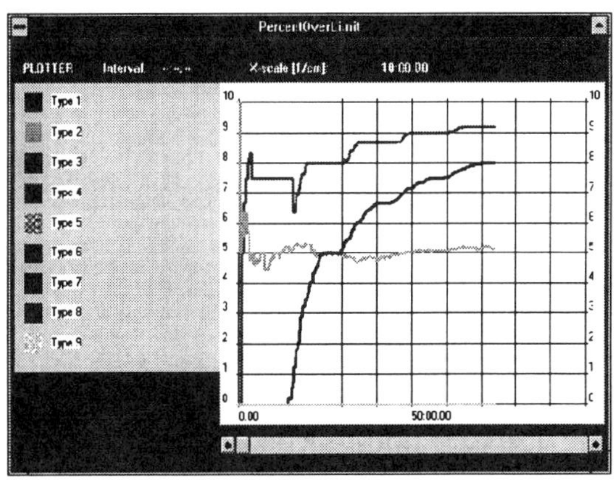

Figure 9: A Sample Plotter Output

2.4.2 Tables in the Site Object

In addition to the CallTimes table previously described, there are three other tables used in each site object. The user does not need to interact with the tables in the sites, but may open them to view data.

The WaitingStats table shown in Figure 10 is used by the model to collect data for the plotters. The columns NumberOfCalls, TotTime, Avg, OverLimit, and PercentOverLimit are filled in as the model runs. The Limit column is filled in before the model is run. If the user does not wish to use the values set in the Application Object Template's parent model, they can switch off the inheritance and set their own values at individual sites. Alternatively, they can change the parent's values to suit their needs.

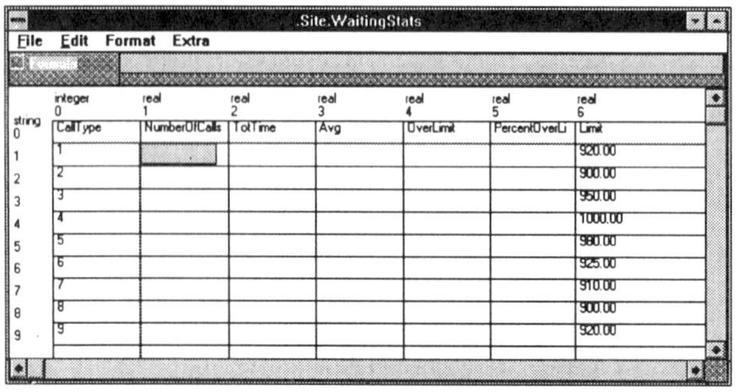

Figure 10: The WaitingStats Table

The CallQ table in Figure 11 keeps track of the queues for each type of call. In this example, the third type of agent has a queue of five calls waiting, each with its arrival time recorded in a queue file. The first type of agent has no queue, so the queue file is empty.

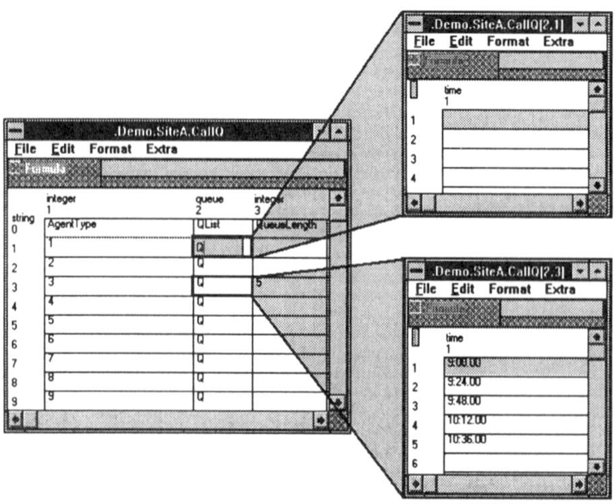

Table 11: The CallQ Table

Finally, the Agents table shown in Figure 12 gives the total number of agents and the number available. The total is set by the Init function when the model is started, and the number available is updated continuously as the model runs.

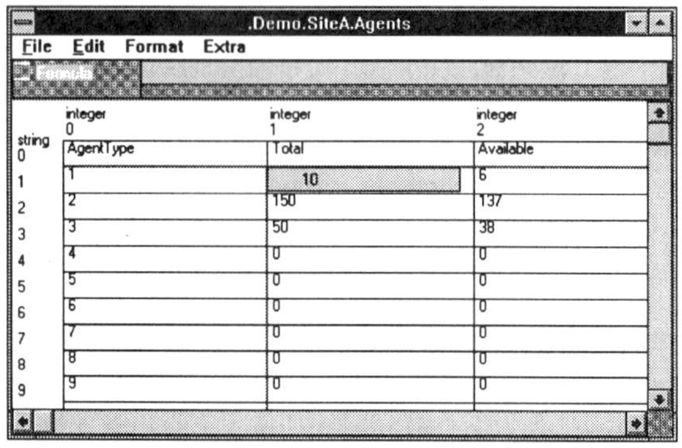

Figure 12: The Agents Table

3 MODIFYING THE PHONE CENTER MODEL

The Demo model in the Application Object Template can be easily modified or incorporated into other models. For example, sites can be easily added or deleted, or two instances of this model can be placed into a larger model to run both at once.

3.1 Adding and Deleting Sites.

The number of sites is controlled by the user, and the model is changed using the methods NewSiteConstructor or SiteDestructor shown in Figure 7. To add a site, the user just needs to add a child of the parent in the Application Object Template. This is done simply by clicking on the parent, then clicking in the Demo frame to add the child. In this demonstration model, whenever new site is added to a model as in Figure 13, a constructor modifies the model to automatically adjust for this new site. It does this by adding the new site to the SiteData table. The only action the user needs to take is to add the number of agents at the new site into the nested AgentsAtSites table. The operating system's output window will prompt the user of the required changes. Other than that, the entire process is fully automatic and no model modifications are required.

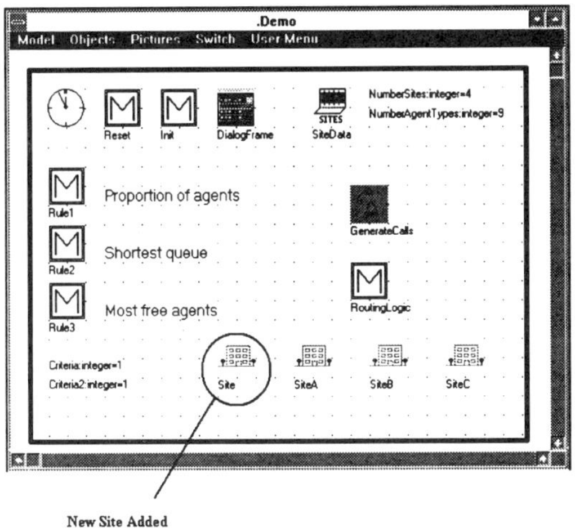

Figure 13: Constructors Used to Add a Site

If a site is removed from the model, a destructor automatically adjusts the model so that the model is fully operational without any user changes. Because each site is an object, no change to the model coding is required. The user simply needs to delete the site object from the model, and the computer will automatically re-calculate all of the data in the tables. If, however, deleting a site results in no agents of a given type being left in the model, the computer will prompt the user to add more agents of that type.

3.2 Running Multiple Models Simultaneously

The Application object Double Demo contains two Demo models, as shown in Figure 14. These can be run at the same time, each with its own number of sites, site selection criteria, number of agents, or other changes. This allows direct comparison of two strategies or options. In this example, hierarchy was used to treat two complete models as objects within a larger model.

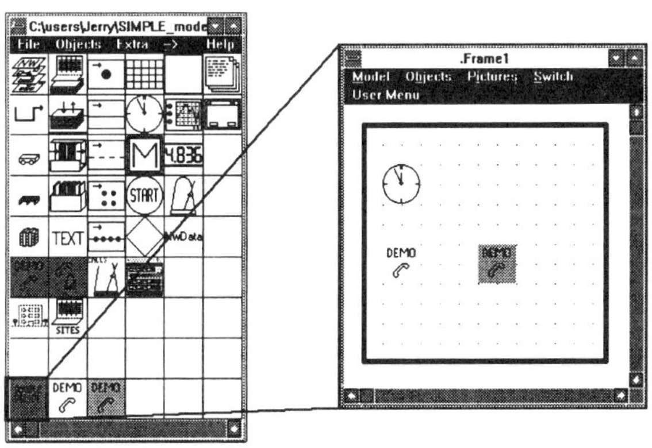

Figure 14: Two Models Can be Run Simultaneously

4 CONCLUSIONS

Object orientation, especially when combined with other SIMPLE++ features, can be used to create highly flexible models. In this paper, a service industry example was presented, but the concepts may be applied to any discrete-event simulation model.

REFERENCES

AESOP GmbH. 1995. *SIMPLE++ reference manual version 3.1.* Stuttgart, Germany.

Geuder, D. 1995a. Modular application objects: closing the gap between flexibility and ease of use. In *Proceedings of the 1995 European Simulation Conference*, ed. F. Breitenecker, I. Husinsky, 37-40. Eurosim Congress '95, Vienna, Austria.

Geuder, D. 1995b. Object Oriented Modeling with SIMPLE++. In *Proceedings of the 1995 Winter Simulation Conference*, ed. C. Alexopoulos, K. Kang, W. Lilegdon, D. Goldsman, 534-540. Arlington, Virginia

AUTHOR BIOGRAPHY

GERALD A. LEVASSEUR recently joined AESOP Corporation as their Chief Consultant. He received a BIE degree from Cleveland State University in 1985, an MSE from the University of Washington in 1990, and a Ph.D. degree in Industrial Engineering from the University of Washington in 1994. Jerry taught at the University of Tennessee at Chattanooga for two years, and has industrial engineering experience as a civilian employee of the Navy. He is a member of **SCS**, **IIE**, **ASQC**, and **APICS**.

Proceedings of the 1996 Winter Simulation Conference
ed. J. M. Charnes, D. J. Morrice, D. T. Brunner, and J. J. Swain

STRATEGIC LAYOUT PLANNING AND SIMULATION FOR LEAN MANUFACTURING A LAYOPT™ TUTORIAL

Eric S. Grajo

Production Modeling Corporation
3 Parklane Boulevard, Suite 910 West
Dearborn, Michigan 48126, U.S.A.

ABSTRACT

LayOPT™ is an innovative facilities layout analysis and optimization software package which can be used by layout planners and engineers in the optimal solution of single and multiple floor facility layout problems. LayOPT is ideal for the re-design of the overall layout for an existing facility or the development of a block layout for a new building. It can be applied to manufacturing, warehouse, office, and various service facility layout problems. This tutorial describes the LayOPT methodology and presents an overview of the software's features and capabilities.

1 INTRODUCTION TO LAYOPT

LayOPT is a facilities layout optimization software package which can be used by layout planners and engineers to solve single and multiple floor facility layout problems. It can be applied to manufacturing, warehouse, office, and various service facility layout problems. LayOPT allows the layout planner to generate alternative layout plans quickly and easily and to find the optimal layout among these alternatives.

LayOPT is an improvement algorithm that starts with an existing block layout, and given the flow and cost data, attempts to improve it by exchanging the locations of defined departments. While several available improvement algorithms perform basically the same function, many are severely limited by the kinds of exchanges they could perform. Not uncommon among them are constraints either on the layout itself or on the type of exchanges that could be considered (for instance, only equal area or adjacent departments). Consequently, results obtained from these algorithms may lead to inferior solutions. LayOPT overcomes these limitations and generates solutions that are generally 50% to 80% more efficient than their starting points.

LayOPT is a Windows-based software system with all the amenities of a user-friendly interface, including pull-down menus, toolbars, status bars, user-defined window sizes, and an on-line help system. It comes with a User's Guide/Reference and a Training Manual.

1.1 The LayOPT Algorithm

LayOPT is an improvement algorithm for developing alternative and efficient block layouts from an initial block layout provided by the user. The initial layout is typically either the existing or a proposed layout. In the absence of an initial layout, one may also be randomly generated by the program. The seed algorithm used in LayOPT is based on the algorithm developed by Bozer, Meller, and Erlebacher (1994).

LayOPT's algorithm is a steepest-descent, two-way exchange optimization routine. In each iteration, the algorithm picks the department pair whose exchange leads to the largest reduction in the objective function. It then automatically exchanges the pair to proceed to the next iteration. The objective function minimized by the LayOPT algorithm is the sum of the parts flows multiplied by the appropriate costs and expected distances between all department pairs with non-zero parts flow between them.

1.2 LayOPT Terminology

Before we describe the steps in creating and analyzing a layout problem in LayOPT, a brief discussion of commonly used terms is presented below.

1.2.1 Building and Grids

The building is the actual physical facility whose layout is being determined. It may consist of a single floor or more than one floor, including partial floors (mezzanines or loft areas with limited vertical access). The building footprint defines the actual shape (rectangular or irregular) of the building when viewed from the top. To represent a non-rectangularly shaped building in LayOPT, the smallest enclosing rectangle is drawn to surround the building footprint. As in most computer-based layout algorithms, layouts in LayOPT are represented as a matrix. Each element of the matrix corresponds to a grid square (or simply, grid) of specified area.

1.2.2 Floors and Lifts

A floor in LayOPT corresponds to a floor level in a facility. It may either be an actual floor or a mezzanine. Lifts are vertical material handling devices that allow for the transport of materials and people from one floor to another in a multi-floor facility. These may include elevators, vertical reciprocating conveyors (VRCs), gravity chutes, etc.

1.2.3 Departments

Departments are the smallest planning units that comprise a facility. A *department* may be an actual process department, a product cell, a single machine, a group of machines, or any part of the facility (offices, cafeteria, rest rooms, shipping and receiving docks, etc.) whose location relative to other departments is being determined.

A fixed department is one whose location is designated in the initial layout and is not allowed to change during optimization. Fixed departments are not considered when evaluating pairwise departmental exchanges. A free department is one whose location is allowed to change during optimization. All free departments are considered when evaluating pairwise departmental exchanges.

1.2.4 Obstacles

An obstacle is a general term used to describe any unusable part of a facility. Such obstacles include building columns, safety areas, freight elevators, etc. Obstacles may also be used to depict non-existing building areas as when representing a non-rectangularly shaped building with an enclosing rectangle, or when representing a mezzanine with an entire floor.

1.2.5 The Space Filling Curve

The space filling curve (SFC) is a device entered by the user and used by LayOPT to construct the layout. The SFC is a continuous line that visits all the assignable grids in a particular floor in the facility. To construct the layout, a set of departments whose areas fit in the floor defined by the SFC is sequenced along that curve. An SFC guarantees that no department is split because a separate curve is used for each floor and, within each floor, the curve visits the *neighbors* of a grid before visiting other grids.

Space filling curves can be drawn through virtually any building with obstacles and fixed departments as long as an obstacle or fixed department does not split the building into two or more disjointed pieces.

1.2.6 Flows and Weight Factors

Flow represents parts, information, or people that are transported from one department to another. These may be in the forms of actual part pieces, pallets, boxes, racks, bins, etc. The From-To chart is a table containing flow values from one department to another in a form similar to a mileage chart.

Weight factors represent either the actual unit costs associated with the individual flow values in the From-To chart or simply relative weights. Relative weights (on a user-defined scale) may represent such intangible factors as difficulty of the material moves, environmental and safety issues on various parts flows, or preferences for specific material handling containers. LayOPT minimizes either the layout cost expressed in dollars per unit time or the overall weighted sum of parts travel in the facility.

1.2.7 Layout Assessment

Layout assessment is the value of the objective function calculated by LayOPT : the sum of the flows multiplied by the appropriate weights and expected distances between all department pairs with non-zero parts flow between them. A lower layout assessment generally means a better layout.

2 METHODOLOGY

Creating a project in LayOPT involves defining the basic input data required to describe an initial layout and execute an optimization run, namely: (a) the building or facility, (b) departments and departmental properties, (c) flow and cost values, and (d) an initial departmental block arrangement.

2.1 Defining the Building

LayOPT can capture virtually any building shape and size. The grid size used in the matrix representation of the building shape is entirely dependent on the needs of the user. In general, a finer grid size can more accurately represent departmental areas and shapes, though, at the expense of faster computational run times. The opposite is true for a larger grid size -- less precision but faster run times. Depending on the grid size selected by the user, LayOPT can represent a department or an obstacle 5 ft. by 5 ft. in size, or even smaller.

Multi-level facilities are easily represented by assigning multiple floors to the building. In this regard, floor separation heights may or may not be the same across all floors. Vertical material handling equipment to transport parts, people, or products to and from departments on different floors can be placed anywhere on a floor and are assumed to service all defined floors. LayOPT concurrently optimizes the layouts of all the floors, i.e., each floor's layout is not developed independently of the others.

2.2 Defining Departments

A department is uniquely identified by its number and description. Defining a department also includes specifying its area requirements in square distance units (sq. ft. or sq. meters) or grids and designating it as free or fixed. Additionally, floor restrictions may be defined -- floors where the department may not be located at any time during and after the optimization run.

LayOPT does not limit department shapes to rectangles or squares. Departments can assume virtually any shape as long as they remain a single piece (i.e., not disjointed) and do not contain an enclosed void. Individual department shapes can be controlled through a user-defined shape parameter, one which compares the department's perimeter-to-area ratio to that of the ideal shape, e.g. a square.

2.3 Drawing an Initial Layout

Drawing the initial layout involves doing four basic steps: (a) drawing obstacles, (b) placing fixed departments, (c) drawing the space filling curve, and (d) placing free departments.

The user has unlimited freedom in designating unusable space in the facility. Defining obstacles is the first step in drawing the layout because grids assigned to obstacles may not be visited by the space filling curve.

Another set of grids not visited by the space filling curve is that of fixed departments. Fixed departments should also be drawn before the space filling curve. While fixed departments never move during the optimization run, they are nevertheless different from obstacles because flows exist between them and the other departments.

A space filling curve may then be drawn through the rest of the grids, i.e., those not assigned to either obstacles or fixed departments. The user may experiment with several types (in terms of orientation, density, or path) of curves either to obtain alternative layouts or to correct some department shapes.

To complete the layout, the free departments are then drawn either by sequencing them automatically along the defined space filling curve or by laying them out individually. See Figure 1. Boundaries may also be drawn to outline individual department block footprints.

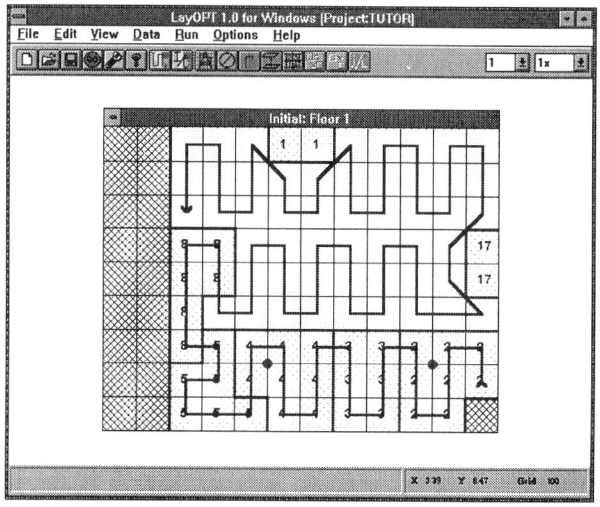

Figure 1: Partially Completed Initial Layout

2.4 Defining Weight Factors and Flows

The objective function minimized by LayOPT is the sum of parts flows multiplied by unit costs and the distances over which these parts travel. Having defined the departments and an initial layout from which distances can be calculated, the From-To chart may be entered directly by the user. Parts flow values entered in this chart may be expressed in different units, e.g., pieces, pallets, trips, racks, etc. However, they should represent parts flow between departments over the same time period.

The user may also enter from-to information using process flow charts. Here, the user graphically specifies departments visited in sequence by each subassembly or component that make up the end products. The quantity of each component that goes into the end product as well as transfer batch sizes are defined for each department pair in the process sequence. LayOPT then automatically generates the From-To chart from these process charts.

In addition to the From-To chart, the user also supplies a cost or weight factor matrix which represents either: (a) the actual unit cost of moving the parts between the departments, or (b) user-defined relative weights. The resulting layout assessment can therefore be either the total material movement cost or the weighted sum of parts travel in the facility.

2.5 Executing an Optimization Run

A LayOPT optimization run can be executed in one of two modes: automatic and interactive. In its true steepest-descent two-way exchange form, the automatic run searches all possible departmental exchange pairs based on space, shape, floor, and other restrictions and selects the best exchange at each iteration. The department location exchange is made and it proceeds iteratively until no further improvement in the objective function is realizable, at which point the optimal layout has been reached.

The interactive run mode allows the user to execute incremental 2-way departmental exchanges on the current layout. LayOPT presents the top 10 exchanges at each iteration in descending order of layout assessment savings and lets the user select the most appropriate (not necessarily the best) exchange. See Figure 2. Once the user selects one of the top ten exchanges, LayOPT switches the locations of those two departments and proceeds to the next iteration. A new list of top ten exchanges is then constructed from the 'new' layout. The user interactively participates in this process until no further exchanges can reduce the current layout assessment.

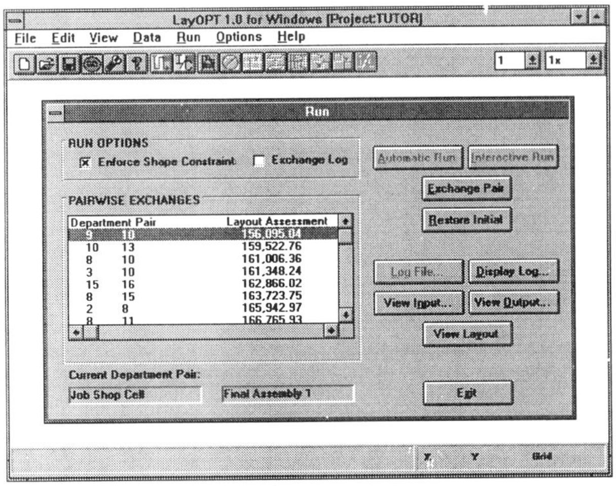

Figure 2: Interactive Run

2.6 Massaging the Final Layout

It is expected that the user will perform multiple runs starting with different initial layouts or space filling curves. This enables the generation of several alternative layouts that in general, satisfy the overall layout development objective. During these runs and in the resulting layouts, some departments may assume irregular shapes. LayOPT allows the user to 'massage' the layout by smoothing out or refining department shapes and make them more practical and usable.

Department shapes can be adjusted through either of the following layout massaging features: (a) even (no area loss/gain) grid exchange between departments, or (b) grid reassignment from one department to another.

Departmental input and output points can also be assigned anywhere within departmental boundaries to designate actual parts pick-up and drop-off points. This allows for a more realistic calculation of the layout assessment summary when presenting actual material handling costs for the final layout.

3 LAYOPT AND SIMULATION

For many facilities layout design or improvement tasks, conventional practice suggests arranging departments in a new plant in the same manner as has been done in similar plants in the past. While this easy approach saves time, it nevertheless runs the risk of inheriting the inefficiencies and shortcomings of the previous layout. On the other hand, most simulation studies attempt to determine satisfactory plant operating parameters based on an existing layout. Increasingly, analysts all too quickly jump to the micro-level simulation of a proposed system laid out inefficiently and perhaps show 10%-15% improvements in WIP levels or material handling costs. In the process, they may have missed a larger window of opportunity to save up to 80% in material movement and handling costs had they first optimized the layout.

The ills of *simulating a bad layout* are numerous and common. In many simulation studies, the objectives are operational and local: (a) how many carriers should a power and free system have to adequately service the needs of the assembly area? (b) how many forklifts does the shipping and receiving area need? (c) what is the best speed to run a particular transfer conveyor? (d) which of two possible temporary staging areas should WIP be sent to and what should the capacity of those staging areas be?

All these objectives focus on the one thing that lean manufacturing is all about: reducing non-value added time spent on material handling or storage systems. To this end, the layout on which the simulation is based should have already addressed the issue of minimizing parts travel distance and cost between departments. No amount of simulating can directly point out that a particular department should not be located where it is at. For instance, a simulation model might recommend six instead of seven forklifts to service the plant but it cannot directly show that by simply moving one department to another location, only two forklifts will be needed. By doing a layout optimization study before any discrete event simulation, it may even show that forklifts are not the most cost-effective way to transport the materials.

What the simulation model can provide, though, is an effective scheme for allocating work among available forklifts. Operational issues such as work scheduling, traffic flow control, break scheduling, and manpower allocation are best addressed by a discrete event model that considers both system capacity constraints and time or resource requirements. In this regard, it is important to emphasize the significance of layout planning in the macro-level, strategic phase (department location and space allocation) and that of simulation in the micro-level, operational phase (resource scheduling and utilization). With LayOPT, developing an optimal layout ensures that the overall cost of moving parts and finished goods between departments is the lowest feasible. It then allows simulation to address the operational issues more effectively, assured that the layout is already the most cost-efficient it can be.

The goal of lean manufacturing is the shortest, most cost-effective product-to-customer time possible. By one measure, that means the smallest, if not zero, non-value added portion in the product manufacturing cycle time. With that common goal, layout optimization and simulation are two complementary tools indispensable to any plant layout and productivity improvement task.

4 SUMMARY

This tutorial presented LayOPT as a Windows-based facilities layout analysis and optimization software package. It was developed to assist layout planners and engineers in solving single and multiple floor facility layout problems. LayOPT overcomes the limitations of current algorithms and generates solutions that are generally 50% to 80% more efficient. With its unique and state-of-the-art, as well as state-of-the-practice optimization engine, LayOPT empowers you to explore many quality alternatives to your facilities layout design challenges.

REFERENCES

Bozer, Y. A., R. D. Meller, and S. J. Erlebacher. 1994. An Improvement-type Layout Algorithm for Single and Multiple-floor Facilities. *Management Science* 40:918-932.

AUTHOR BIOGRAPHY

ERIC S. GRAJO is a senior consultant at Production Modeling Corporation, an industrial engineering and simulation consulting firm based in Dearborn, Michigan. He specializes in computer simulation and facilities layout planning for various manufacturing and material handling concerns in the automotive, electronics, food, pharmaceutical, chemical, and service industries. He led the team responsible for the development and application of LayOPT, a facilities layout optimization software package. Mr. Grajo has a Master's degree and has completed the course work for a Ph.D. in Industrial and Operations Engineering at the University of Michigan.

Proceedings of the 1996 Winter Simulation Conference
ed. J. M. Charnes, D. J. Morrice, D. T. Brunner, and J. J. Swain

TAYLOR II MANUFACTURING SIMULATION SOFTWARE

Cliff B. King

F&H Simulations, Inc.
P.O. Box 658
Orem, Utah 84059-0658

ABSTRACT

Taylor II is a menu-driven simulation package mainly used in manufacturing, warehousing, and material handling. It is developed for the analysis and quantitative evaluation of complex processes, especially those with a dynamic character. A lot of applications in different industries show that there is an increasing need for simulation tools. This paper will demonstrate the process of building, analyzing and presenting models of real world systems with Taylor II.

1 INTRODUCTION

Companies spend a lot of time and money to obtain highest performance from machines and plants. Besides expensive and complicated techniques, there are catchwords like flexibility, just-in-time, lean production, etc. All these terms are based on the optimization of delivery, reliability, throughput times, stocks, utilization: i.e. costs. The problem is often it is nearly impossible to make a quantitative approach to these characteristics. Two factors are responsible: complexity and uncertainty. With an increasing number of components being in relationship to each other, it becomes more and more difficult to describe the system mathematically. When the second factor, uncertainty, is involved, even analytical methods fail. Uncertainties include: machine breakdowns, varying cycle times and batch sizes, waste and reject, etc. Throughput times, buffer sizes, the effects of resources influencing each other cannot be determined with conventional methods. Here, at least, you have to resort to assumptions, experience and intuition, although a better insight would be desirable. At least both possibilities of a bad system dimension, too high and too low, cause much costs. For these problems simulation can be a powerful tool to justify decisions or even prevent you from making wrong decisions.

Taylor II is a package for the simulation of discrete event systems. In general, you can simulate all kinds of systems in which discrete entities are to be processed, transported and stored (queuing systems).

Usually these entities are parts, tools, pallets, etc. This shows that Taylor II is not limited to specific industries. Besides manufacturing there are applications in distribution centers, harbors, airports, etc. Even in administration, simulation can be useful. Think of people or documents (information) which can be represented by entities too.

Taylor is a Dutch product, developed by F&H Simulations B.V. since 1986. The package is installed in more than 2000 companies and educational institutes. In the middle of 1996, the package, now called Taylor II for Windows received a complete new structure and integrates all necessary functions for a simulation study. There is one platform for modeling simulation, animation, analysis and presentation. In the following, all important functions will be described.

2 BUILDING A MODEL

Working with Taylor II usually starts with building a model. All model building is menu driven. A model represents a real world system which is to be studied. In practice, this representation is often simplified. The most important aspect when doing this is to build a model as exactly as necessary and not as detailed as possible. What is the reason for this? The reason is not the least performance of simulation packages, Taylor II or others, but an efficient, economical and objective-oriented way of working.

A model in Taylor II consists of four fundamental entities being related to each other: elements, jobs, routings and products. On each element, one or more operations can take place. The three basic operations are processing, transporting and storing; in Taylor II they are called jobs. Jobs are characterized by a cycle time (cycle times for storage are naturally 0), which can be random. The entities using the jobs are called products. Products can be defined freely and represent parts, tools, pallets, people, etc. They take place on an element and are sent to another element when the operation is finished. Therefore, you need a description of how the products flow through a model. This description is stored in routings. Routings consist of a number of stages (routing records). At each stage it is defined which job-element is used and what the next stage is.

2.1 Layout and Routings

Defining a layout is the first step when building a model. A layout contains a number of different element types that you can position in different sizes and in any place on your screen. The following element types are available in Taylor II: Inout, Machine, Buffer, Conveyor, Transport, Path, Aid(operator), Warehouse and Reservoir. Choosing a type depends on the function in the real world system. A machine is the general purpose representation for any kind of operation so that a machine could be a robot, a mill, a counter, etc. Inout elements represent entries and exits (sink/sources) in a model. They can generate and 'eat' products freely. A buffer is for storage. Transports and conveyors represent continuous and discontinuous transportation. Aid elements are generally assigned to other elements. An aid could be an operator. Warehouses and reservoirs are buffers with some special functions. In a warehouse you can place products at a specific location, which is described by a number of cells horizontally and vertically. In a reservoir there is the possibility to define open/close control levels. Each element placed in the layout gets one job automatically.

When necessary, additional jobs can be assigned later. Besides the cycle time, a job also describes how an operation is done concerning input batch, additional resources, etc.

Now one or more routings are to be defined. By selecting the elements one after another you describe the path the products follow in the model. In most models the routing roughly reflect the flow of products because there are a number of product types taking (nearly) the same path and you can specify differences later. In some applications you have to handle a large number of product types all using different routings, i.e. in a job shop. In this case, Taylor II offers the modeler to import process routing descriptions from external files.

2.2 Detailing the Model

The second step is to detail the model. This means entering parameters describing the behavior of the model that differ from the default settings. Usually only some of the parameters are needed, so it is not necessary to fill a lot of masks before starting the model for the first time. Most of the parameters are related to the fundamental entities discussed above. Element parameters describe the behavior of the elements: what kind of element it is, stochastic (breakdowns) and deterministic (shifts, pauses), availability, how many products it can store, fixed and variable costs, how it selects its next task, etc. The job parameters are similar.

Many systems seem to be quite simple. When looking at details, one often finds different kinds of rules and strategies to make decisions. That is the reason why implementing these decision points takes most of the modeling time. An important aspect in this phase is the way a simulation tool supports the user. Two items have to be discussed: high flexibility for implementing control logic and easy to test right working of the model (verification). Besides special functions for verification the on-line animation in Taylor II plays a large part. When the layout and the routings are finished, animation is available immediately. All changes in the model can be seen on the screen at once. The user is able to detail and to test the model step-by-step until the model runs according to his wish.

Every company producing simulation software is in a dilemma when developing a concept for control logic. The software should be easy to use but, the system should also allow all possible constructions. These two aspects are hard to combine. Distributors go different ways from fixed implemented strategies or decision tables to interfaces for programming languages or combinations of them. Besides a number of default strategies, Taylor II uses a macro language called TLI (Taylor Language Interface). TLI is an easy to use programming language that allows for modifying the model's behavior powerfully in combination with simulation-specific predefined and user-definable variables. TLI is used at element, job and routing level. There is also the possibility to use TLI interactively during a simulation run to make queries and updates. An interface to own routines written in C, Basic, Pascal, etc. is also available. Typical situations in which the routines have to be modified are loops (i.e. rework), selecting elements as a result of a certain status, assembly, disassembly, etc. Taylor II can send products from each stage in the model to each other in each quantity. For this purpose the user has direct access to all addresses in a routing and replaces the (in most cases) fixed values by expressions of the following form:

select number *with* condition *from* list *order* expression *quantity* list *location* list

These so called 'select' statement are used for sending and receiving products. The italic parts in this structure can be values or other expressions. Selections can also be nested. This offers many possibilities for order picking, assembly and complex guiding and receiving strategies. Some examples:

select 1 from 3,4,5

One product will be sent to machine 3, 4 or 5.

select 1 from 3,4,5 order - utilization [L]

The same as above, but the machine with the lowest utilization has the highest priority.

select maximum[1,F] with elqueue[L]<5 from 3,4,5 quantity 4-elqueue[L]

An (express) order uses the maximum number of machines. Only machines with a stock less than 5 are taken into consideration.

Elements can be described by fixed acceptance rules for picking products or other TLI-user-definable rules and strategies for the entry and exit of products. Examples are:

Entry condition: plan from file

The element chooses the next order from a list stored in a disk file.

TLI-entry condition: time|600<120

A conveyor accepts products only for the first 120 seconds in a 600 second interval, (| stands for modulo). The expression is true for 120 seconds.

Product types are described by a code (in combination with an icon) and by size (important for conveyors). Sometimes it is necessary that products get individual information or signs. Information could be a delivery date, priority, state of work, etc. For this purpose, products are coupled with attributes. Attributes can also work as a pointer to a matrix in which i.e., cycletimes are stored. This enlarges the amount of possible information enormously. Querying and updating of the attributes are done at the jobs by evaluating triggers. A trigger is one (or more) simple lines of text with a TLI-expression. Triggers can be evaluated nearly every time an event is scheduled. They can influence products (with their attributes) or the model status itself. An example is:

Trigger on Entry: curcycle[J]:=att2[C]

The cycle time of the order is stored in the second attribute and is assigned when the product enters the job.

Assume that a model is built with Taylor II. Now starts the actual simulation, apart from the (little) simulation when testing the model.

3 SIMULATION

In stochastic processes random numbers play an important part. Taylor II comes with different random number generators which normally work independently from each other. In some experiments you may want to study alternatives of a system. Then the models should run with identical conditions and for this purpose it could be necessary that all generators generate the same sequence of random numbers.

Taylor II is fully event-oriented. this means that the time between two events (i.e., beginning and ending of an operation) needs no CPU time. An internal event list makes sure that all changes in status are evaluated.

For every simulation run, the start situation is free to define. Products can be stored in any quantity at any place in the model. To reduce the warm-up period (moment of time that a model runs with stable status) you can make a warmstart. For a number of various runs you can define a batch run.

During simulation you can zoom, pan and rotate freely, stop, make modifications and then continue. A model can be simulated during a certain period of time or until a special (user-defined) condition is met.

The time representation is fully user-definable (i.e. week 8; tue; 8:30,2) and is displayed by an analog and digital clock. With the single-step-mode you can trace your model in detail or you use the conditional single-step for watching a part of the model only.

While simulating, the screen output can be interactively changed, ranging from no information (as fast as possible), via statistics and simple animation to full animation. With full animation you see each product moving from stage to stage or queuing in the system. Taylor II automatically keeps track of a large number of common statistics like utilization, waiting times, etc. Default and user-defined statistics can be displayed as numbers or as dynamically changing graphs. Especially determining the end of the warm-up period. Taylor II can also keep track of any kind of variables you want and display them after the simulation run.

4 ANALYSIS

For a correct description of a model, a number of input data is required like arrival times, cycletimes, breakdown failures, etc. Often these data exist in protocols, databases, etc. But the question is: is there enough data to characterize a process well or are they only a part of reality? That means a detailed analysis with statistical methods should be done first. Taylor can read in datasets and analyze them. Two important parameters are the average value and the standard deviation. another point is the statistical distribution which fits best. A built in routine automatically makes a distribution suggestion for the dataset and displays this distribution depending on the type, in continuous or discrete form on the screen.

Typical information you get from a simulation study is utilizations, throughput times, production per period and costs. The results are available in different ways:

Tabular reports: contents and form are predefined, but may be changed by the user.

TLI reports: containing a mixture of explanation, results and illustration or giving specific results.

Queries: all results can be queried with TLI interactively (including minimum, maximum, average and standard deviation).

Predefined graphs: wait time histogram, queue graph, status diagram, utilization pie, etc.

User-defined graphs: representation of any kind of data like throughput times, costs, etc.

Graphs can be displayed in over 10 types of general purpose representation: histograms, bar graphs, x-y plots, pie charts, etc. including averages (whole and cumulative), standard deviation and so on. All reports and graphs can be sent to screen, disk file and printer or you can export these data for use in other programs.

5 ANIMATION

In the past animation has been treated as a nice but more or less unnecessary extra. Taylor II has an extensive built-in animation module.

There are several reasons. First, similar to the discussion in Section 2.2, the animation can be very useful (but generally not replace) for verifying a model. For this purpose the 2-D animation is available as soon as routing is created. With little additional work, 3-D animation is available too. They are useful when giving presentations to management or customers (often more convincing than rough numbers or in the educational sector. A number of ready-to-use 2-D and 3-D libraries come with the package.

With the built-in paintbox you can define your own icon-sets or draw background illustrations. The 3-D paint module allows you to change exiting or to add 3-D element icons. Background drawings are automatically translated into vector files for sealing and rotating. Existing drawings can be imported into the HPGL format.

6 ADDITIONAL FEATURES

In addition to the points above there are some special functions useful (not essentially necessary) for the daily work. When simulating a lot of alternative models with different parameters and settings you could lose your overview. For this purpose, Taylor II generates a model documentation (text file) describing the whole model. another important aid is the context-sensitive online help with index and page search facility.

Besides stochastic arrival patterns, Taylor II can process external files (ASCII-file) with an unlimited number of deterministic product arrivals including arrival time, quantities and product attributes. This offers the possibility to simulate with real data. Special arrival patterns, think of season influences of daily peaks, can be generated with genlink. Genlink generates an external file by taking samples from different distributions.

External files can also be used to define deterministic machine planning or routing specification (see Section 2.1). The handling with ASCII-files becomes easier with the integrated text editor. DDE (Dynamic Data Exchange) links exist for windows base products that allow such links.

The ability to created self-running presentations allows you to build models which explain themselves. With a runtime module you can give all models to other people that are participating in a simulation study.

Taylor II version 4.0 includes several new that have been specially designed for windows 95. The Taylor Language Interface has been extended to include the us of functions. There are also three new modules included. These are Advanced Statistics (with an autofit module which tests distributions to fit data and an experiment module), runtime Development Kit (allowing you to create customized applications), and the Animator (special 3-D animation module with shaded animation).

7 SUMMARY

Taylor II integrates all functions necessary for a simulation study and combines in a simple manner high flexibility with ease-of-use without making concessions to functionality. many applications in different industries; manufacturing, transportation, distribution, material handling, show this. Fast modeling and online animation are the concept for the use not only by simulation experts. Taylor II is used more and more in education institutes for their practical courses in engineering.

Continuous customer feed back and our own consulting department enable F&H to keep improving Taylor II. This experience is also an advantage for the support and training courses that we offer.

AUTHOR BIOGRAPHY

CLIFF B. KING is the current director of customer services at F&H Simulations, Inc. He is a member of the Institute of Industrial Engineers. Formerly he was the program manager of Morton International's Japanese automotive airbag programs. He received his B.S. in Mechanical Engineering from Brigham Young University.

Proceedings of the 1996 Winter Simulation Conference
ed. J. M. Charnes, D. J. Morrice, D. T. Brunner, and J. J. Swain

TIME MACHINE--THE DYNAMIC WORKSHEET

William B. Nordgren

F&H Simulations, Inc.
P.O. Box 658
Orem, UT 84059-0658, U.S.A.

ABSTRACT

Time Machine is a menu-driven simulation package mainly used for process re-engineering. It is developed for the analysis and quantitative evaluation of processes especially those with a dynamic character. A lot of applications in different industries show that there is an increasing need for simulation tools for processes such as paper flows, staffing, and customer service. This paper will demonstrate the process of building, analyzing and presenting models of real world systems with Time Machine.

1 INTRODUCTION

Competitive companies spend a lot of time and money to obtain highest performance from their employees and resources. Besides expensive and complicated techniques, there are buzzwords like flexibility, BPR, lean production, etc. All these terms are based on the optimization of delivery, reliability, customer satisfaction, and costs. The problem faced by management is that it is often impossible to make a quantitative approach to these characteristics. Two factors are responsible. The first is complexity. With an increasing number of components being in relationship to each other, it becomes more and more difficult to describe the system mathematically. The second factor is uncertainty. When uncertainty is involved, even analytical methods fail. Uncertainties can be classified as breaks, varying process times and batch sizes, waste and reject, etc. Throughput times, queue sizes, the effects of resource allocation cannot be determined with conventional methods. Many managers have to resort to assumptions, experience and intuition, to solve problems. For these types of problems simulation can be a powerful tool to justify decisions or even prevent you from making wrong decisions.

Time Machine is a package for the simulation of discrete event systems. In general, you can simulate all kinds of systems in which discrete entities are to be processed, and stored (queuing systems).

Usually these entities are parts, paper, action items, people, etc. This shows that Time Machine is not limited to specific industries. Types of applications include processing of people, paper work, phone calls, staffing, harbors, airports, and administration.

Time Machine is a Dutch product, developed by F&H Simulations B.V. since 1993. The package is installed in more than 500 companies and educational institutes. In the middle of 1996, the package, now called Time Machine for Windows received a complete new structure and integrates all necessary functions for a simulation study. There is one platform for modeling simulation, animation, analysis and presentation. In the following all important functions will be described.

2 BUILDING A MODEL

Working with Time Machine usually starts with building a model. All model building is menu driven. A model represents a real world system which is to be studied. In practice, this representation is often simplified. The most important aspect when doing this is to build a model that emulates the system correctly while minimizing the details. The reason for this is to build an efficient, economical and objective-oriented way of working.

A model in Time Machine consists of four fundamental entities being related to each other: elements, jobs, routings and products. On each element, one operation can take place. The two basic operations are processing, and storage; in Time Machine they are called jobs. Jobs are characterized by a cycle time (cycle times for storage are naturally 0), which can be random. The entities using the jobs are called products. Products can be defined freely and represent parts, tools, paper, people, etc. They take place on an element and are sent to another element when the operation is finished.

Therefore, you need a description of how the products flow through a model. This description is stored in routings. Routings consist of a number of stages and can be placed as connections between elements graphically in the same manner as creating a flow chart.

2.1 Layout and Routings

Defining a layout is the first step when building a model. A layout contains a number of elements that you can position in different sizes, and in any place, on your screen. The following two element types are available in Time Machine: World and Element. Choosing a type depends on the function in the real world system. An element is the general purpose representation for any kind of operation or processing point, or queue location. World elements represent entries and exits (sink/sources) in a model. They generate and 'eat' products. Each element placed in the layout gets one job automatically.

Besides the cycle time, a job also describes how an operation is done concerning input batch, additional resources, etc.

By selecting the elements one after another you describe the path the products follow in the model. In most models the routing roughly reflects the flow of products just as in defining a flow chart. In some applications you have to handle a large number of product types all using different routings i.e. in a paper work model. In this case, Time Machine offers the modeler the ability to specify how each product type is routed through the model by using the select statement.

2.2 Detailing the Model

The second step is to detail the model. This means that the modeler has to enter parameters describing the behavior of the model that differ from the default settings. Usually one only needs some of the parameters so that it is not necessary to detail before starting the model for the first time. Most of the parameters are related to the fundamental entities discussed above. Element parameters describe the behavior of the elements: what kind of element it is, stochastic (breakdowns) and deterministic (shifts, pauses), availability, how many products it can store, fixed and variable costs, how it selects its next task, etc. The job parameters are similar.

Many systems seem to be quite simple. When looking at details, one often finds different kinds of rules and strategies to make decisions. That is the reason why implementing these decision points takes most of the modeling time. An important aspect in this phase is the way a simulation tool supports the user. Two items have to be discussed: high flexibility for implementing control logic and easy model verification. Besides special functions for verification the on-line animation in Time Machine plays a large part. When the layout and the routings are finished, animation is available immediately. All changes in the model can be seen on the screen at once. The user is able to detail and to test the model step-by-step until the model runs according to his desire.

Besides a number of default settings, Time Machine uses a macro language called TLI (Taylor Language Interface). Since Time Machine is a simulation program developed from Taylor II, models built in Time Machine will run in Taylor II. Therefore Time Machine utilizes the same macro language as Taylor II. TLI is an easy to use programming language that allows one to modify the model's behavior powerfully in combination with simulation-specific predefined and user-definable variables. TLI is used at element, job and routing level. TLI can be used interactively during a simulation run to make queries and updates. Typical situations in which the routines have to be modified are loops (i.e. rework), selecting elements as a result of a certain status, assembly, disassembly, etc. Time Machine can send products from each stage in the model to each other in any quantity. For this purpose the user has direct access to all addresses in a routing and replaces the (in most cases) fixed values by expressions of the following form:

select number *with* condition *from* list *order* expression *quantity* list *location* list

These so called 'select' statement are used for sending and receiving products. The italic parts in this structure can be values or other expressions. Selections can also be nested. This offers many possibilities for order picking, assembly and complex guiding and receiving strategies. Some examples:

select 1 from 3,4,5

One product will be sent to element 3, 4 or 5.

select 1 from 3,4,5 order - utilization [L]

The same as above, but the element with the lowest utilization has the highest priority.

select maximum[1,F] with queue[L]<5 from 3,4,5 quantity 4-queue[L]

An (express) order uses the maximum number of elements. Only elements with less than 5 products are taken into consideration.

Elements can be described by fixed acceptance rules for picking products or other TLI-user-definable rules and strategies for the entry and exit of products. Examples are:

TLI-entry condition: time|600<120

An element accepts products only for the first 120 seconds in a 600 second interval, (| stands for modulo). The expression is true for 120 seconds.

Product types are described by a code in combination with an icon. Sometimes it is necessary that products get individual information or signs. Information could be a delivery date, priority, state of work, etc. For this purpose, products are coupled with attributes. Attributes can also work as a pointer to a matrix in which i.e., cycletimes are stored. This enlarges the amount of possible information enormously. Querying and updating of the attributes are done at the jobs by evaluating triggers. A trigger is one (or more) simple lines of text with a TLI-expression. Triggers can be evaluated nearly every time an event is scheduled. They can influence products (with their attributes) or the model status itself. An example is:

Trigger on Entry: curcycle[J]:=itemcode2[C]

The cycle time of the order is stored in the second attribute and is assigned when the product enters the job.

Assume that a model is built with Time Machine. Now starts the actual simulation.

3 SIMULATION

In stochastic processes random numbers play an important role. Time Machine comes with different random number generators which normally work independently from each other. In some experiments you may want to study alternatives of a system. The models should run with identical conditions and for this purpose it could be necessary that all generators generate the same sequence of random numbers.

Time Machine is fully event-oriented. This means that the time between two events (i.e. beginning and ending of an operation) needs no CPU time. An internal event list makes sure that all changes in status are evaluated.

For every simulation run, the start situation is free to define. Products can be stored in any quantity at any place in the model. To reduce the warm-up period (moment of time that a model runs with stable status) you can make a warmstart. For a number of various runs you can define a batch run.

During simulation you can zoom and pan, stop, make modifications and then continue. A model can be simulated during a certain period of time or until a special (user-defined) condition is met.

The time representation is fully user-definable (i.e. week 8; tue; 8:30,2) and is displayed by an analog and digital clock. With the single-step-mode you can trace your model in detail or use the conditional single-step for watching a part of the model only.

While simulating, the screen output can be interactively changed, ranging from no information (as fast as possible), via statistics and simple animation to full animation. With full animation you see each product moving from stage to stage or queuing in the system. Time Machine automatically keeps track of a large number of common statistics like utilization, waiting times, etc. Default and user-defined statistics can be displayed as numbers or as dynamically changing graphs. Time Machine can also keep track of any kind of variables you want and display them after the simulation run.

4 ANALYSIS

For a correct description of a model, a number of input data is required like arrival times, cycletimes, breaks, etc. Often these data exist in protocols, databases, etc. But the question is: is there enough data to characterize a process well or are they only a part of reality? That means a detailed analysis with statistical methods should be done first. Taylor can read in datasets and analyze them. Two important parameters are the average value and the standard deviation. Another point is the statistical distribution which fits best. A built in routine automatically makes a distribution suggestion for the dataset and displays this distribution depending on the type, in continuous or discrete form on the screen.

Typical information you get from a simulation study is utilizations, throughput times, production per period and costs. The results are available in different ways:

Tabular reports: contents and form are predefined, but may be changed by the user.

TLI reports: containing a mixture of explanation, results and illustration or giving specific results.

Queries: all results can be queried with TLI interactively (including minimum, maximum, average and standard deviation).

Predefined graphs: wait time histogram, queue graph, status diagram, utilization pie, etc.

User-defined graphs: representation of any kind of data like throughput times, costs, etc.

Graphs can be displayed in over 10 types of general purpose representation: histograms, bar graphs, x-y plots, pie charts, etc. including averages (whole and cumulative), standard deviation and so on. All reports and graphs can be sent to screen, disk file and printer or you can export these data for use in other programs.

5 ANIMATION

In the past animation has been treated as a nice but more or less unnecessary extra. Time Machine has a built-in animation module.

There are several reasons. First, similar to the discussion in Section 2.2, the animation can be very useful (but generally not replace) for verifying a model. For this purpose the 2-D animation is available as soon as routing is created.

With the built-in paintbox you can define your own icon-sets or draw background illustrations. Background drawings are automatically translated into bitmaped easy access. Existing bitmaped drawings can be imported into backgrounds as well.

6 ADDITIONAL FEATURES

In addition to the points above there are some special functions useful (not essentially necessary) for the daily work. When simulating a lot of alternative models with different parameters and settings you could lose your overview. For this purpose, Time Machine generates a model documentation (text file) describing the whole model. Another important aid is the context-sensitive online help with index and page search facility.

Time Machine version 2.0 includes several new features that have been specially designed for windows 95. The Taylor Language Interface has been extended and with autofit you can curve fit your existing data to prepare data for use in your model

7 SUMMARY

Time Machine integrates all functions necessary for a simulation study and combines in a simple manner high flexibility with ease-of-use without making concessions to functionality. Many applications in different industries, manufacturing, fast foods, banking, and management, have benefited from simulation. Fast modeling and online animation are the concept for the use not only by simulation experts. Time Machine is designed to be used by managers and is not more complicated to use that a good spreadsheet or flowcharting software.

Continuos customer feed back and our own consulting department enable F&H to keep improving Time Machine.

AUTHOR BIOGRAPHY

WILLIAM B. NORDGREN is President of F&H Simulations, Inc. U.S.A. He received his Master of Science in Computer Integrated Manufacturing (CIM), and Bachelor of Science in Manufacturing Engineering from Brigham Young University. Bill was co-founder of ProModel Corporation and served as Vice President of Customer Services and Training from 1988 until 1992. In 1993, Bill jointed F&H Simulations B.V. and founded F&H Simulations, Inc. U.S.A. Through this union Taylor II Simulation Software was introduced into the United States, Canada, Mexico and South America. He is listed in the 22nd, 23rd, 24th and 25th editions of Who's Who in the West. He has done simulation projects for several large corporations including Andersen Consulting, Black & Decker, Boeing, Eastman Kodak, IBM, Pillsbury, Thiokol, TRW and Whirlpool.

Proceedings of the 1996 Winter Simulation Conference
ed. J. M. Charnes, D. J. Morrice, D. T. Brunner, and J. J. Swain

MODELING WITH EXTEND™

David Krahl

Imagine That, Inc.
6830 Via Del Oro, Suite 230
San Jose, CA 95119, USA.

ABSTRACT

This document presents an overview of the Extend modeling environment, emphasizing Extend's ability to model a wide range of systems. Extend is a general purpose graphically oriented discrete event and continuous simulation application with an integrated authoring environment and development system.

1 INTRODUCTION

For many years there has been a perceived dichotomy in simulation software between simulation languages and simulators. The languages were viewed as more powerful and general purpose, while simulators focused on ease of use and were generally limited to a specific industry. Extend bridges these two types of programs in one easy-to-use, yet flexible software program. It exists as:

- A stand-alone simulation tool which can be used to create complex discrete event and continuous models without programming.
- A simulation authoring package where model interfaces can be easily created, without programming, to enhance productivity and ease of use.
- A development environment for building customized models of unique types of systems. The programming environment allows the modeler to create a simulator for a specific industry.

2 EXTEND'S MODELING ENVIRONMENT

Before looking into how Extend can be used to build models, it is helpful to understand the Extend modeling environment.

Extend models are constructed with library-based iconic blocks. Each block describes a step in a process or a calculation. Blocks reside in libraries. Each library represents a grouping of blocks with similar characteristics such as Discrete Event, Plotters, Electronics, or Business Process Reengineering. Blocks are placed on the model worksheet by dragging them from the library window onto the worksheet. The flow is then established between the blocks.

There are two types of logical flows between the Extend blocks. The first type of flow is "items," which represent the objects that move through the system. Items can have attributes and priorities. Examples of items include parts, patients, or a packet of information. The second type of logical flow is "values," which will change over time during the simulation run. Values represent a single number. Examples of values include the number of items in queue, the result of a random sample, and the level of fluid in a tank.

Each block has connectors that are the interface points of the block. Figure 1 shows the connector symbols for the value and item connectors.

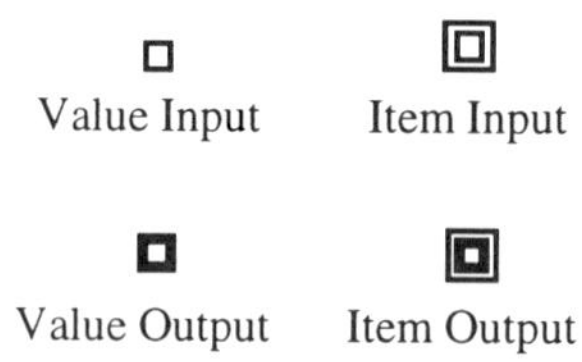

Figure 1: Value and Item Connectors

Connections are lines used to specify the logical flow from one connector to another. Double lines represent item connections and single lines represent value connections.

A model of a single server, single queue system would have the following form:

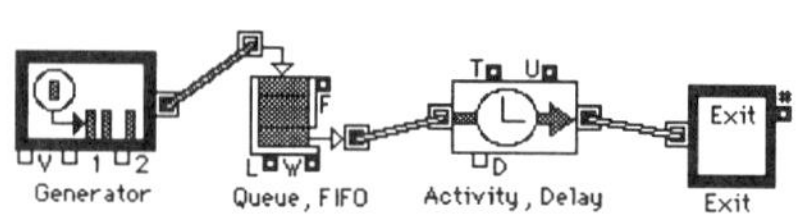

Figure 2: A Single Server Single Queue Model

The block on the far left represents a Generator which periodically creates items. Following this is a Queue block that holds items until requested by the next block. The Activity Delay represents a limited capacity of one processing unit and delays an item for a fixed amount of time. The last block in the model is an Exit block that removes items from the system.

An enhancement to this model would be to specify that the delay in the Activity Delay is determined by a specific random distribution. This can be done by connecting the output of an Input Random Number block to the delay connector (labeled "D") on the Activity Delay block as in Figure 3:

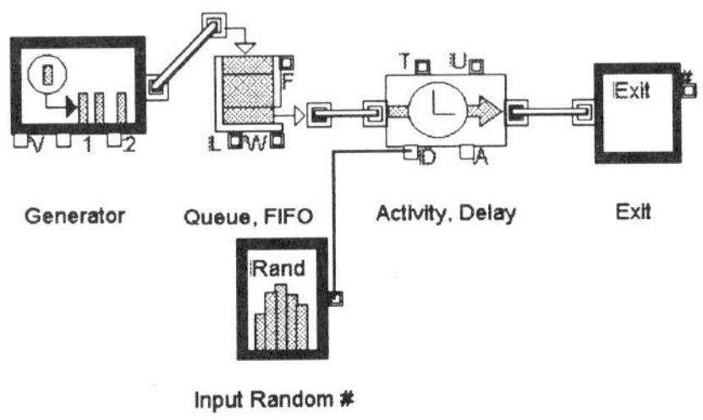

Figure 3: A Model With Random Process Times

Another feature that can be added to the model is a Discrete Event Plotter which graphically displays, in this example (Figure 4), the contents of the Queue. The Discrete Event Plotter value input connector will be connected to the Queue's length (labeled "L") value output connector as follows:

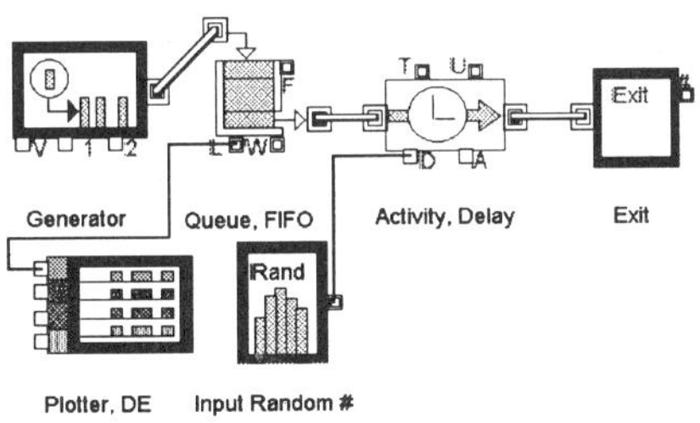

Figure 4: Discrete Event Plotter Added to Model

Simulation parameters such as the number of runs and simulation end time can be specified in the Simulation Setup dialog item under the Run menu. The simulation can then be run by selecting the Run Simulation menu item from the Run menu.

During the run, the current simulation status is displayed in a bar near the bottom of the monitor screen. This displays the estimated time before the run will be completed, the current simulation run time, the number of simulation steps completed so far, and the current simulation run number.

Once the simulation run has completed, the results of the simulation are reported within the blocks. Double clicking on each block reveals the information collected from the simulation run. For example, double clicking on the Queue FIFO block opens a dialog which shows the following information about the state of the Queue FIFO block:

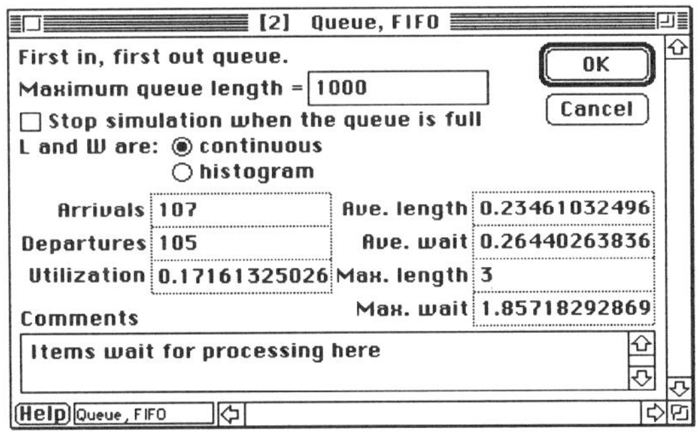

Figure 5: Dialog of Queue FIFO

The Plotter block shows the number of items stored in the Queue FIFO over time in both graphical and tabular format:

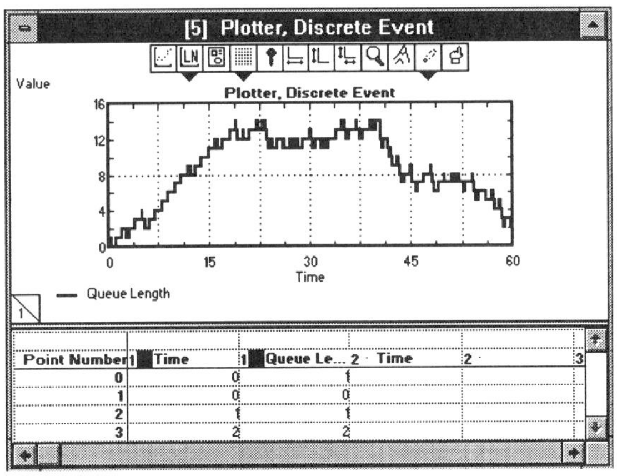

Figure 6: Plot of Queue Length

Simulation results may be stored in a table, plotted, cloned to a different area of the worksheet, exported to another program such as a spreadsheet or database, displayed in an animation, or even used to control some aspect of the outside world through external device drivers.

3 STANDARD EXTEND LIBRARIES

The standard Extend libraries include constructs for

discrete event modeling, results plotting, generic calculations, electronics design, interprocess communication, and utilities. For discrete event modeling, the most commonly used standard libraries are the Discrete Event, Generic, and Plotter. Additional, optional, discrete event libraries include the Business Process Reengineering and Manufacturing libraries.

Extend supports the following general modeling functionality for discrete event modeling:

- Attributes - Unique variables which are local to the items moving through the simulated system.
- Priorities - A unique value, local to a given item, which can be used to rank items in a queue or interrupt items in process.
- Values - The number of items represented by a single item. Setting a value will create clones of an item when that item arrives to a queue, resource, or exit block.

3.1 Discrete Event Library

The Discrete Event library contains blocks which are specific to modeling discrete event systems. In a discrete event model, the clock will update at intervals dictated by the individual events in the system. The discrete event blocks pass items to one another through their item connectors. If a discrete event block is unable to receive an item it rejects it and the item waits until it receives a downstream request.

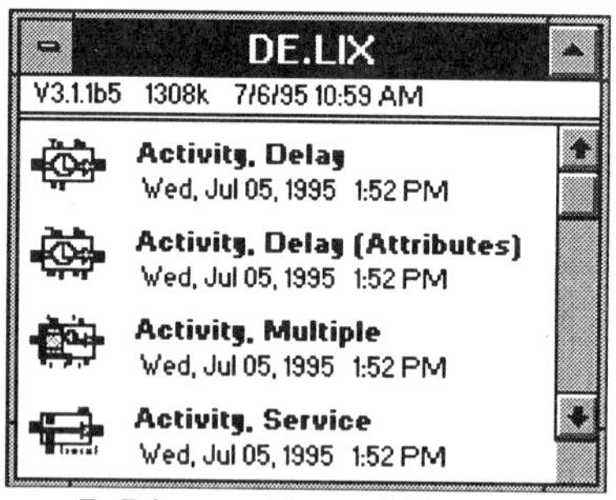

Figure 7: Discrete Event Library Window

The most commonly used block types in the Discrete Event library are as follows:

- Activities - Time delays
- Batching - Combining of items
- Resources - Limits capacities
- Decisions - Chooses alternate paths

3.2 The Generic Library

The Generic library is used for both continuous and discrete event modeling. In the continuous mode, calculations are performed at each evenly spaced clock step. In the discrete event mode, calculations are made in response to a request (message) from a discrete event block.

When used with the Discrete Event library, the Generic library is typically used to provide values for inputs or operate on the value outputs of the Discrete Event blocks. Typical examples of using the Generic library in this mode include using a Decision block to compare the length of two queues or using an Input Random Number block to generate a random time delay for an Activity.

There are a number of classes of Generic blocks. These include: mathematical calculation, integration, file operations, logical calculations, integration, statistical calculations, error reporting, simulation events, accumulation, and threshold detection.

3.3 Other Libraries

In addition to the above libraries, Extend also includes libraries for statistics, animation, plotters, utilities, electronics, filters, digital circuits, controls, dynamic data exchange (DDE), and DLLs or XCMDs. Libraries are available from third party developers for control systems, paper manufacturing, neural networks, biology, and signal processing.

4 CUSTOMIZING EXTEND

The above discussion illustrates the highly graphical and interactive nature of Extend. However, Extend can also take the shape of the model application. Interfaces, components, and graphics can be used which tailor the model to a specific application area.

The most visible aspect of a custom model is the user interface. By modifying an existing interface or creating a new one, the simulation modeler is able to create a model which can be exercised by someone more familiar with the system than with the simulation tool. Models can be built that fit naturally into the conceptual framework of the person using the model. In Extend, this can be done with a range of tools, including:

- Hierarchical modeling - Models can be subdivided into logical components.
- Dialog cloning and the Notebook - Consolidate critical parameters and results to a central location. This can be used to create a custom user interface.
- Interactive controls - Add sliders and switches which provide interactive control of the model

- Pictures and animation - The model is the interface.
- User messages - Users can be prompted for information at any point in the simulation.
- Block development environment - Create modeling components for a specific environment.

5 SAMPLE EXTEND MODELS

The following sections will present a number of applications which utilize Extend's features to create models in different areas. These models come from users with a variety of backgrounds and simulation experience levels. What is presented here is only a small subset of the areas where Extend is currently being applied.

5.1 Management Flight Simulator

The Deloitte & Touche Consulting Group Management Flight Simulator (a model for experimenting with different management strategies) is used internally for training sessions. This model, developed by Rick Blaiklock of their Ottawa office, focuses on the relationship between the economic factors of: staffing levels, sales effort, competition, employee training, and market research.

In the training session, participants run a consulting operation for a total of ten years of simulated time. At the end of each six month period, operational decisions are entered into the model and the economic results are plotted over time. Figure 8 illustrates the input panel for the decisions regarding the staffing levels of the model. Similar panels are used for the other business factors.

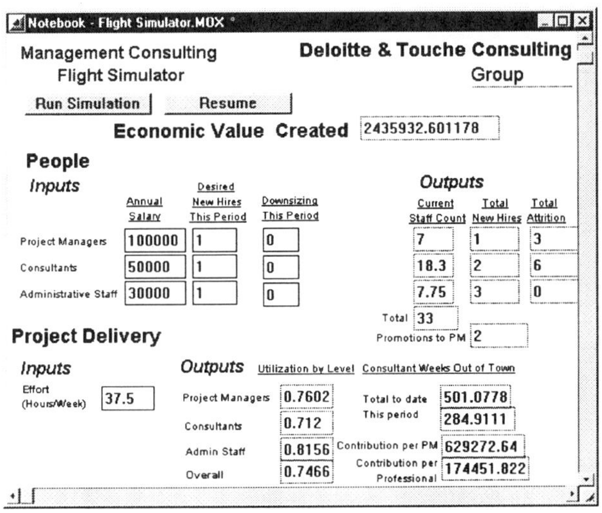

Figure 8: Management Flight Simulator

Here, dialog cloning is used extensively to create a user interface for model users who are not necessarily proficient in simulation. Because of this, the class participants are able to use a sophisticated model with only a brief introduction to the inner-workings of the model.

5.2 Economic Value Added Model

The Computer Aided Process Improvement (CAPI) Balanced Scorecard Economic Value Added Model (EVA) is an analysis tool designed to provide the backbone and baseline for reengineering and utility operations improvements. This allows executives to see the impact of decisions they have made on the balance sheet. The model is designed around the organizational framework of a large utility company operating via Strategic Business Units (SBUs). Of the four SBUs, three are revenue-generating and one is dedicated to cross-functional overhead processes. The revenue-generating SBUs are: Generation, Transmission, and Distribution, while the overhead processes are grouped into Services. The CAPI EVA model is capable of calculating the following:

- Economic Value Added (EVA) for each SBU and for the Corporation
- Return on Invested Capital (ROIC) for each SBU and for the Corporation
- Measured Operating Income (MOI) for each SBU and for the Corporation
- Average Capital for each SBU and for the Corporation
- Total Revenue for each SBU -- monthly and accumulated YTD
- KWH (kiloWatt hours) System Sales -- monthly and accumulated YTD
- Off System Sales -- monthly and accumulated YTD
- By Product Sales -- monthly and accumulated YTD
- Physical Plant and Equipment (PP&E) Maintenance -- monthly and YTD
- Total Cost to Generate Power -- monthly and accumulated YTD
- Total Labor Costs -- monthly and accumulated YTD
- Cost of Sales -- monthly and accumulated YTD

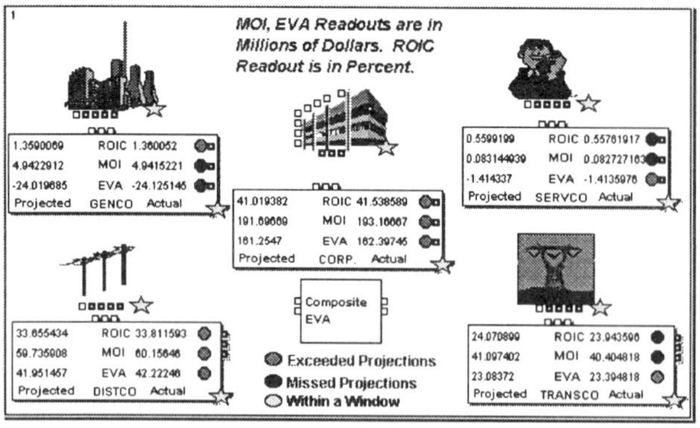

Figure 9: CAPI Economic Value Added Model

To build this model, standard Extend blocks were grouped together into hierarchical blocks (H-blocks). These were then placed in libraries. Models are built by selecting the hierarchical block from the library, placing it on the model worksheet, connecting other blocks, and filling in model parameters. By using this method a custom library was created without programming that incorporates the expertise of the library developer.

5.3 Bulk Processing

Another application of Extend is a model of a bulk processing system using blocks from Simulation Dynamics Inc. Bulk Flow Library. The blocks used to build the model in Figure 10 can also be used to build detailed models of any type of system that moves bulk quantities of materials.

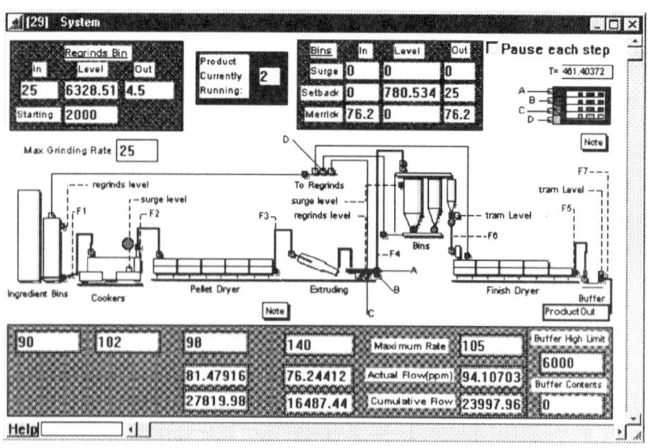

Figure 10: Bulk Processing Model

In this example, a library of blocks has been programmed to provide a simulator for a specific industry. Engineers are able to use the bulk processing architecture to build, run, and modify models without the burden of learning the intricacies of a general purpose simulation language. At the same time, Extend's flexibility remains should the need arise to model a system component which is beyond the scope of the existing Bulk Processing library.

The architecture is extremely applicable to production systems in industries such as pulp and paper, food processing, bottling lines, oil refining, high speed plastics molding, and any other application in which production is termed in rates instead of items.

5.4 Neural Networks

Another interesting application of Extend is the Neurolab library of blocks from Mikuni Berkeley. This library allows modelers to build neural networks by combining various network components.

Figure 11 shows a control network for an inverted pendulum. The control objective is to balance the pendulum with desired cart position. The two-layer feed-forward network is the compensator that improves the performance of the state feedback controller by adapting to the nonlinear dynamics and parameter changes of the inverted pendulum.

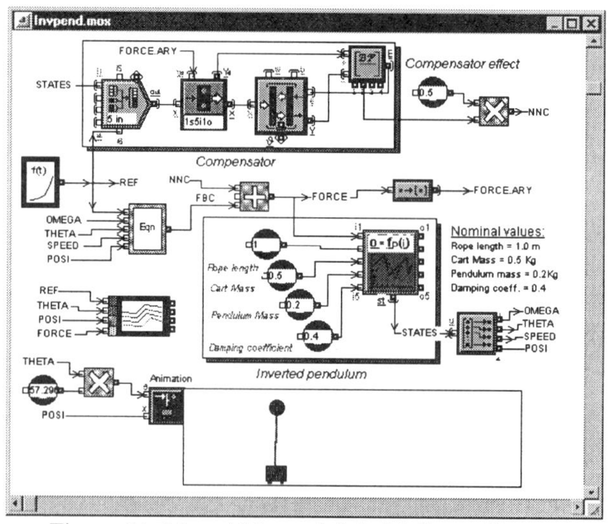

Figure 11: Neural Network Model Controling an Inverted Pendulum

With the Neurolab product, simulation has been opened up into neural network research and evaluation. Neurolab is applicable in any area where nonlinear mapping and adaptive capability are integral parts of problem solving. This is another example of providing a simulation software product for a specific area of interest through Extend's development environment.

6 SUMMARY

It is easy to focus on any one of Extend's features such as ease of use or graphical interface as its main attraction for simulation modelers. However, this view tends to eclipse the true power of Extend to provide modelers the ability to create unique models for their area of application. Extend's applicability and extensibility are unmatched in the simulation industry. As a result, Extend has become the simulation tool of choice for thousands of engineers and analysts ranging from simulation newcomers to experts.

REFERENCES

Computer Aided Process Improvement. 1996. *The Balanced Scorecard Economic Value Added Model.* Cincinnatti, OH

Imagine That, Inc. 1995. *Extend Software Manual.* San Jose, CA.

Deloitte & Touche. 1996. *Management Consulting Flight Simulator.* Ottawa, ON.

Krahl, David. 1995. Building End User Applications With Extend, In *Proceedings of the 1995 Winter Simulation Conference,* ed. C. Alexopoulos, K. Kang, W. R. Lilegdon, D. Goldsman, 413-419. IEEE Piscataway, NJ.

Mikuni Berkeley R&D Corp. 1995. *Neurolab User's Manual.* Richmond, CA

Simulation Dynamics. 1996. *Making Model.* Maryville, TN.

AUTHOR BIOGRAPHY

DAVID KRAHL is responsible for the technical support and block development for Extend at Imagine That, Inc. Mr. Krahl received a BS in 1986 in Industrial Engineering from the Rochester Institute of Technology and a MS in Project and Systems Management from Golden Gate University in 1996. He has performed consulting, technical support, and development for a wide range of simulation Products.

A SYSTEMATIC APPROACH TO FACTORY LAYOUT AND DESIGN WITH FACTORYPLAN, FACTORYOPT, AND FACTORYFLOW

David P. Sly

Cimtechnologies Corporation
2501 North Loop Drive, Suite 700
Iowa State University Research Park
Ames, Iowa 50010, U.S.A.

ABSTRACT

Since the finished quality of every layout project depends on a successful combination of products, processes and tools, it is essential to select each of these carefully. A good project will always begin with an analysis of the production volumes of the products with common process sequences and tool requirements in order to create manufacturing families. For each of these families you can select the "best manufacturing practices" that need to be used and re-engineer the processes and tooling to fit the desired throughput and inventory requirements. Only after you know the desired manufacturing practices to be used for each product family is it appropriate to begin developing layouts around a systematic flow and non-flow evaluation process. This systematic approach to layout design combines with Factory PLAN/OPT and FLOW software tools working in AutoCAD to create the most efficient factory designs in record time.

1 PRODUCT ANALYSIS

A good layout project begins by evaluating the major product family divisions and determining the appropriate levels of detail for classification. For example, a garden equipment manufacturer may select major component subassemblies used in multiple final product lines for classification, instead of the actual product lines themselves. You will then create a pareto chart similar to that shown in Figure 1 with products listed in decreasing order from left to right according to production volume. You will next want to reclassify some of your products into families that have very similar process routings and update the pareto chart. Strive for a pareto chart in which 80% of your products make up less than 20% of your production volume. If this ratio is not possible, you may choose to create multiple pareto charts based on highly different product types and manufacturing processes. Part quantity reduction is often another benefit of this type of analysis. Obviously the fewer unique products you make, the less your manufacturing complexity.

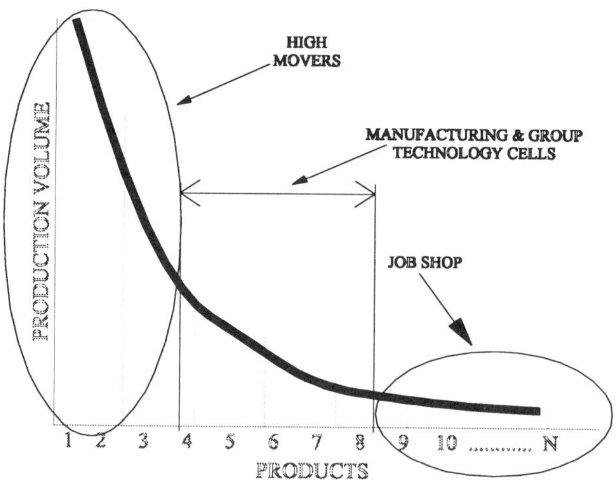

Figure 1: Products/Volume Pareto Chart

2 MANUFACTURING PRACTICES SELECTION

The next stage of your analysis will involve the selection of the appropriate manufacturing practices for groups of your product families. For example, your high volume products with very common processes will be good candidates for product focused manufacturing systems. Product focused systems include assembly and machining lines in which all of the necessary equipment is located in the same sequence as the manufacturing process and is often dedicated and balanced to the pace of the line.

Your very low volume products will be easy to identify and will be candidates for "Job-shop" facility arrangements in which equipment is often located according to shared tooling or operator needs. These low volume products are likely candidates for outsourcing, and therefore you should reevaluate their manufacturing profitability before proceeding with the layout project.

Finally, your "in-between" volume products will be likely candidates for manufacturing cells, group technology cells, and focused factories. These products often share some of their manufacturing processes with other products and thus will need to be reclassified accordingly. This "in-between" group of products will always be the most difficult to design manufacturing practices and layouts for; however, they also often represent the greatest opportunity for reductions in cost, throughput time and inventory. FactoryFLOW was designed to aid in this layout-oriented classification process by allowing you to color code common process flows with the desired equipment used in order to systematically create manufacturing cells from lists of products and tools.

Once you have properly classified your product families into desired manufacturing practices, you should undergo a thorough review of the manufacturing processes in each family. A layout is totally dependent on process sequences and equipment availability. Now is the time to create an efficient process that can translate into a productive layout. Skipping this opportunity will often result in layouts that fail to reduce throughput times and inventory, since you are doing little more than rearranging the furniture within the plant. While you cannot receive the benefits from JIT, Focused factories or manufacturing cells without creating a layout that supports these manufacturing practices, you likewise cannot benefit from cellular and focused layouts operating with traditional manufacturing practices.

3 EQUIPMENT SELECTION

Proper equipment selection is just as critical as proper manufacturing process design, and is often the step most ignored by layout planners. Sharing tooling among dissimilar product families and processes will result in significant penalties to efficient layouts, throughput times, and inventory. Eliminating shared tooling is therefore the most important goal in equipment selection. Too often inefficient layouts are created to provide flows from many different product areas in the plant to low cost equipment. If this equipment were duplicated and placed within different zones, the resulting inventory and throughput savings would likely far exceed the cost of the additional equipment.

Another important tool selection decision involves the use of expensive flexible equipment or inexpensive dedicated equipment. Expensive flexible equipment should only be used in cells with a high degree of process variability, and not used to join dissimilar cells. Therefore, the primary goal in tool selection is once again to isolate the processes and tooling in the cells from one another via dedicated low tech tooling whenever possible. Inventory and throughput times are reduced more by product-oriented manufacturing cells than from process-oriented Group Technology cells.

4 MATERIAL FLOW ANALYSIS

Material Flow is the primary activity that drives a factory layout. Good layouts have smooth and short flows with a minimum of backtracking and crossover. There are several different types of layout configurations available, and a typical factory will consist of several layout types among the different product families. Figure 2 illustrates common linear, S, U and L configurations often used in Product flow and cellular applications.

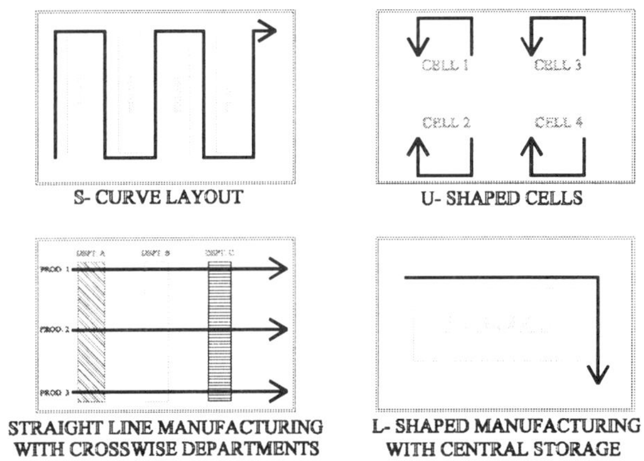

Figure 2. Layout Configurations

Material flow diagrams illustrate material moves. FactoryFLOW can generate a variety of diagrams from the same set of product data, allowing the user to focus on different aspects of flow. The key diagram from which all others are computed is the product flow diagram (Figure 3). This diagram shows different products/subassemblies/materials/processes in different colors and with line thickness according to either number of trips or cost. This diagram can be presented with actual paths for better numerical evaluation and aisle congestion analysis, or Euclidean paths for better visual evaluation of workcenter interrelationships. One key of the product flow diagram is that it is intelligent: a user can click on a line to find out what it represents. In addition, if users rearrange the equipment in the AutoCAD drawing, they need only select CALC again and FactoryFLOW can find all of the new locations, regenerate the flow diagram, and recompute the costs, distances, intensities, and time. It is this iterative approach with instant graphical and quantitative feedback that makes quantitative layout evaluation feasible.

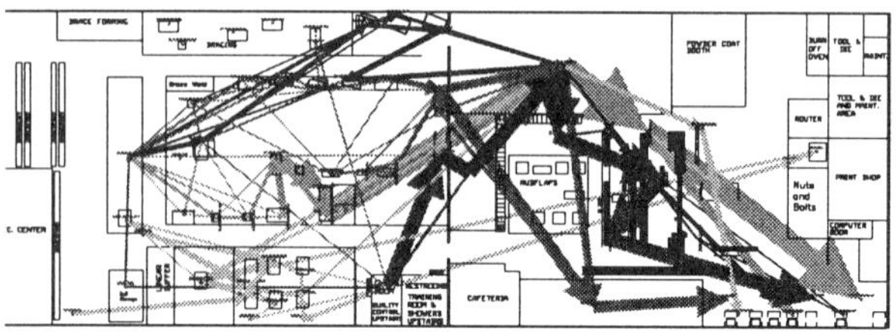

Figure 3. Product Flow Diagram

5 RELATIONSHIP ANALYSIS

Another important consideration in any layout analysis are the non-flow factors such as noise, dirt, contamination, supervision, safety, shared tooling, and so on. These activity relationships are best defined by listing all of the unique activities in a relationship chart (Figure 4) and categorizing the closeness affinity for each activity pair in a team meeting. You can enter these relationships into a spreadsheet or directly into FactoryPLAN's relationship editor in order to quickly get high quality relationship charts and diagrams complete with layout scores of your facility. It is important to evaluate the qualitative non-flow factors independent of the flow factors, even if it is desired to perform a relationship analysis that involves both flow and non-flow constraints.

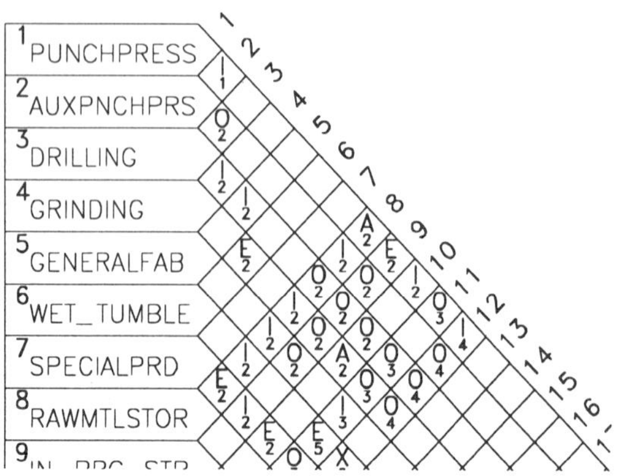

Figure 4. Activity Relationship Chart

FactoryPLAN also allows users to evaluate the layouts according to material flow intensities between activities. These flow intensities can come from FactoryFLOW studies, production analysis spreadsheets, or even be entered directly into FactoryPLAN via the supplied editor. Entering the flows directly is often the quickest and easiest for small facilities of less than 250k square feet, or manufacturing facilities with few dominant material flows, where using tools like FactoryFLOW may be overkill. FactoryPLAN can diagram non-flow relationships or flow-oriented relationships independently, or FactoryPLAN can aggregate these two kinds of relationships together using user-supplied weighting factors in order to generate layout diagrams and scores that best represent all relevant equipment and department adjacencies. Figure 5 shows an example of a relationship diagram.

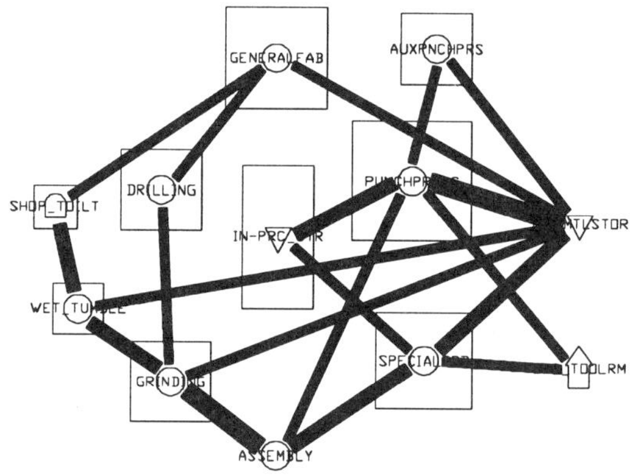

Figure 5. Activity Relationship Diagram

6 LAYOUT OPTIMIZATION

Once all of the processes and tooling requirements have been determined, and the flow and non-flow relations between them identified, optimization tools like FactoryOPT can be employed. FactoryOPT uses a spanning tree algorithm to generate a near-optimal arrangement of activities in a block layout based on flow and non-flow relationship data. You can interact with this arrangement at the spanning tree node diagram level or after FactoryOPT generates a block layout in AutoCAD. Once a good arrangement has been generated, FactoryPLAN and optionally FactoryFLOW can be used to diagram and score additional layout alternatives.

FactoryOPT can generate layouts with up to 256 unique activities using up to 128 different algorithm combinations. FactoryOPT can receive flow, non-flow, or aggregated flow/non-flow relationships as input. FactoryOPT works entirely inside AutoCAD along with

FactoryPLAN. FactoryOPT is based on the spanning tree algorithm developed for SPIRAL by Marc Goetschalckx from Georgia Tech. Figure 6 shows an AutoCAD-based spanning tree diagram created by FactoryOPT.

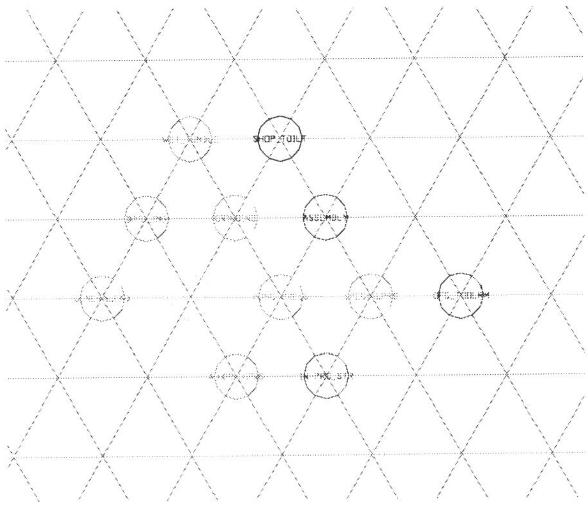

Figure 6. Spanning Tree Diagram

7 CONCLUSION

Successful layout projects are based on solid process and equipment definitions combined with a systematic flow and non-flow diagramming, evaluation, and benchmarking process. Layout projects that do not begin with thorough product, process and equipment evaluations or do not contain detailed relationship and material flow studies often result in design teams constantly bouncing from issue to issue with no end in sight. Software tools like FactoryFLOW, FactoryPLAN, and FactoryOPT can make short work of diagramming, scoring, and presenting layout alternatives within a systematic design framework. Such tools are becoming as necessary to factory layout designers as word processors are to typists.

ACKNOWLEDGMENTS

FactoryCAD, FactoryPLAN and FactoryFLOW are registered trademarks of Cimtechnologies Corporation. AutoCAD is a registered trademark of Autodesk, Inc.

AUTHOR BIOGRAPHY

David P. Sly is President and Founder of Cimtechnologies Corporation and original author of the FactoryCAD, FactoryPLAN, and FactoryFLOW products. Mr. Sly has been involved with industrial facilities layout and design for over a decade, performing consulting and development projects for John Deere, Ford, GM, SEMATECH, AT&T, and many other fortune 100 manufacturers worldwide. Dave received his bachelors and masters degrees in Industrial Engineering as well as his MBA from Iowa State University. He is a registered professional engineer in the state of Iowa, and is a member of the Society of Manufacturing Engineers, Institute of Industrial Engineers, and the Society for Computer Simulation.

Proceedings of the 1996 Winter Simulation Conference
ed. J. M. Charnes, D. J. Morrice, D. T. Brunner, and J. J. Swain

EXPERTFIT: TOTAL SUPPORT FOR SIMULATION INPUT MODELING

Averill M. Law
Michael G. McComas

Averill M. Law and Associates
P.O. Box 40996
Tucson, Arizona 85717

ABSTRACT

In this paper, we explain the important role of simulation input modeling in a successful simulation study. Two pitfalls in simulation input modeling are then presented and we explain how any analyst, regardless of their knowledge of statistics, can easily avoid these pitfalls through the use of ExpertFit, the Windows-based successor to the UniFit II input modeling package. We use a set of real-world system data to demonstrate how the package automatically specifies, evaluates, and ranks candidate probability distributions, and then assists an analyst in deciding whether the "best" candidate probability distribution provides an adequate representation of the data. If no candidate probability distribution provides an adequate fit, then ExpertFit can define an empirical distribution function. In either case, the probability distribution can be automatically expressed in the analyst's simulation software. We then consider the general case of selecting a probability distribution in the absence of data. As an example, we show how ExpertFit can be used to create busy-time and downtime models for machines that are subject to random breakdowns.

1 THE ROLE OF SIMULATION INPUT MODELING IN A SUCCESSFUL SIMULATION STUDY

In this section we will describe simulation input modeling and show consequences that might result if this important, but sometimes neglected, activity is performed improperly. We then suggest that with the use of ExpertFit any simulation analyst can perform simulation input modeling more quickly and with greater accuracy than would otherwise be possible.

1.1 The Nature of Simulation Input Modeling

One of the most important activities in a successful simulation study is that of representing each source of system randomness by a probability distribution. For example in a manufacturing system, processing times, operating times before failure, and repair times for a machine should usually be modeled by probability distributions.

In this paper, we use the phrase "simulation input modeling" to mean the process of choosing a probability distribution for each random component of the system under study and expressing this representation in a form that can be used with the analyst's choice of simulation software. In Sections 2 and 3 we will demonstrate how an analyst can easily and accurately choose an appropriate probabilistic representation using the ExpertFit software.

1.2 Two Pitfalls in Simulation Input Modeling

The authors have identified a number of pitfalls that can undermine the success of simulation studies (Law, McComas, and Vincent 1994, Law and Kelton 1991, Law and McComas 1989). Two of these pitfalls relate directly to simulation input modeling and are summarized in this section.

1.2.1 Pitfall Number 1: Replacing a Distribution by its Mean

Simulation analysts have sometimes replaced an input probability distribution by its mean in their simulation models. This practice may be caused by a lack of understanding on the part of the analyst or by lack of information on the actual form of the distribution (e.g., only an estimate of the mean of the distribution is available). Such a practice may produce completely erroneous results, as is shown by the following example.

Consider a manufacturing system consisting of a single machine tool at which jobs arrive to be processed. Suppose that the mean interarrival time of jobs is one minute and the mean processing time is 0.99 minute. Suppose further that the interarrival times and processing times actually have an exponential distribution. Then it can be shown that the long-run mean number of jobs waiting in the queue is *approximately 98*. On the other hand, suppose we were to follow the dangerous practice of replacing a source of randomness with a constant value. If we assume that each interarrival time is *exactly* one minute and each processing time is *exactly* 0.99 minute, *then each job is finished before the next arrives and no job ever waits in the queue*! The variability of the probability distributions, rather than just their means, has a significant impact on the congestion level in most queueing-type (e.g., manufacturing) systems. In Section 2 we shall show how the use of ExpertFit makes choosing an appropriate probability distribution a simple and easy process.

1.2.2 Pitfall Number 2: Incorrect Modeling of Random Machine Downtimes

The largest source of randomness for many manufacturing systems is that associated wth random machine downtimes. An analyst is often faced with representing in a simulation model the random machine downtimes of a machine that has not yet been purchased. Data concerning the actual downtime behavior of machine tools is, thus, unavailable and the analyst must rely on estimates of reliability provided by vendors and engineers. Suppose, for example, that a vendor claims that a machine tool will be down 10 percent of the time, but is unwilling or unable to provide more information on its operating time before breakdown and its repair time. Given the limited available information, some simulation analysts account for downtimes by simply reducing the machine processing rate by 10 percent. Law and McComas (1989) compare this practice to a more accurate model that we describe in Section 3. Although the two modeling approaches led to similar results for an average throughput measure of performance, the use of the reduced-production-rate model led to large errors with regard to measures such as average time in system and maximum number of jobs in queue. Accurate estimation of the latter performance measures is essential in many simulation studies. Thus, serious errors can result if an incorrrect, simplified approach is taken. We will show in Section 3 how easy it is to obtain a more accurate model of random machine downtimes using ExpertFit.

1.3 Advantages of Using ExpertFit

With the assistance of ExpertFit *any* analyst, regardless of their prior knowledge of statistics, can avoid the two pitfalls introduced above. When system data are available, a complete analysis with the package takes just minutes. The package identifies the "best" of the candidate probability distributions, and assists the analyst in deciding whether the fit is good. If none of the candidate distributions provides an adequate fit, then an empirical distribution function can be created by ExpertFit. In either case, the representation of system randomness can be automatically expressed in the analyst's choice of simulation software. Appropriate probability distributions can also be selected when no system data are available. For the important case of machine breakdowns, ExpertFit will determine appropriate busy-time and downtime probability distributions that match the system's behavior, *even if the machine is subject to blocking or starving.*

2 USING ExpertFit WHEN SYSTEM DATA ARE AVAILABLE

We now consider the case where an analyst has system data corresponding to the source of randomness to be represented in the simulation model. Our intention is to highlight the capabilities of ExpertFit. A demo disk with detailed commentary on program operation is available at no charge from the authors.

Three types of analyses are available for selecting probability models. In addition to the analysis of system data, there are two analysis types available when no system data are available (see Section 3). We have designed ExpertFit to embody our years of experience in selecting appropriate simulation input models. The user interface features multi-tabbed folders that correspond to the recommended steps in an analysis. Each tab organizes the appropriate options in a way that reflects our recommended analysis approach. Each option has default configuration settings that make it easy for an analyst to do any statistical procedure. All graphs are designed to assist in meaningful comparisons and to minimize possible analyst misinterpretation. For example: a) multiple models can be plotted on the same graph, b) error graphs are automatically scaled so that the visual size of an error reflects the severity of the error, and c) whenever possible, error bounds (safety limits) are displayed. These software features make it easy for any analyst to perform accurate and thorough analyses of data sets, regardless of their prior knowledge of statistics. On the other hand, the user interface is completely flexible so that an experienced analyst can easily access the full set of available tools for performing comprehensive and complete data set analyses in any order desired.

A data analysis is done using a folder with four tabs. The first tab has options for obtaining and displaying the features of a data sample; an analyst can read a data file, manually enter or edit a data set, paste in a data set from the system clipboard, as well as perform a variety of transformations. Once a sample is available, an analyst can create a number of graphical and tabular sample summaries, including histograms and plots designed to assess the randomness of the observations.

The data set we have chosen for this example consists of part processing times provided to us by a major automobile manufacturer.

At the second tab models can be fit to the sample. For the recommended guided fitting option, the basic information required by ExpertFit to begin the fitting and evaluation process is a specification of the range of the underlying random variable. For many data sets like the example processing times, the underlying random variable can be characterized as being greater than zero with no definite upper bound. ExpertFit responded to our choices by fitting distributions with ranges starting at zero and distributions whose lower endpoint was estimated from the data itself. These candidate models were then automatically evaluated. After a few seconds the result screen shown in Figure 1 was displayed.

ExpertFit fit and ranked 26 candidate models, with the three best-fitting models listed on the screen along with their scores. The displayed scores are calculated by a proprietary evaluation scheme that is based on our 18 years of research in this area, including the analysis of 35,000 computer-generated data sets. Results from the heuristics that we have found to be the best indicators of a good model fit are combined and the resulting numerical evaluation is normalized so that 100 indicates the best possible model and 0 indicates the worst possible model. These scores are *comparative* in nature and do not give an overall assessment of the quality of fit. ExpertFit provides a separate *absolute* evaluation of the adequacy of fit provided by the best-ranked model. This

Relative Evaluation of Candidate Models

Model	Relative Score	Model Range
1 - Inverted Weibull	100.00	Larger than 0
2 - Gamma(E)	92.00	Larger than 24.79809
3 - Log-Logistic(E)	90.00	Larger than 24.79809

26 models are defined with scores between 0.00 and 100.00

Absolute Evaluation of Model 1 - Inverted Weibull

Based on a heuristic evaluation, there is no current evidence for not using the primary model. If you are doing simulation, then the primary model will probably provide a good representation for your data. However, we recommend further confirmation of the primary model.

Additional Information Concerning Model 1 - Inverted Weibull

Result of an Anderson-Darling goodness -of-fit test at level .1	Do not reject
"Error" in the model mean relative to the sample average	-.09670 = .26%

Figure 1: Evaluation of Candidate Models for the Processing-Time Data

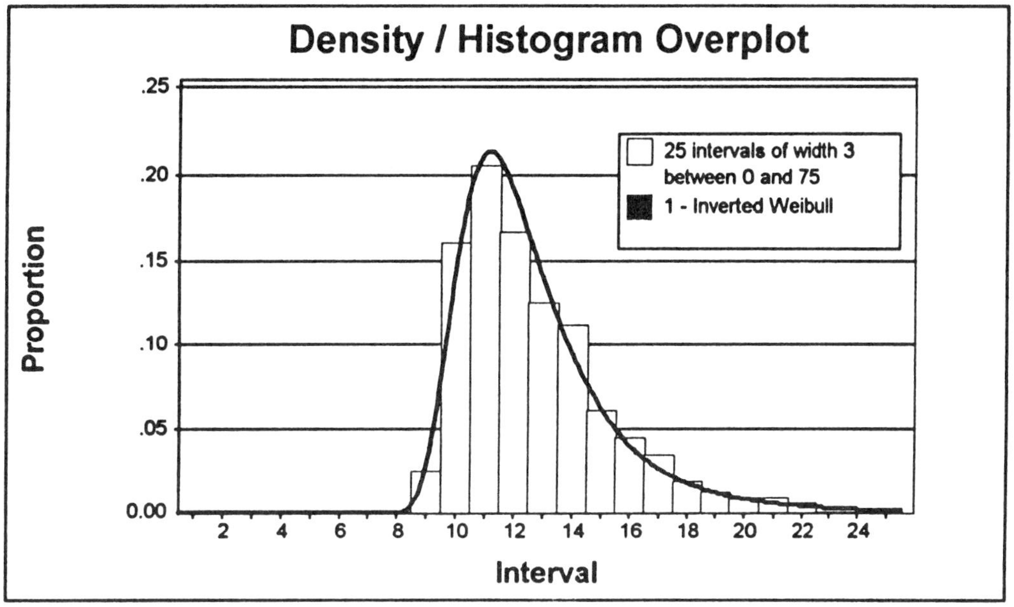

Figure 2: Density/Histogram OverPlot for the Processing-Time Data

absolute evaluation is critical because, perhaps, one third of all data sets are not well represented by a standard distribution. ***Furthermore, ExpertFit is the only software package that provides such an absolute evaluation.***

In Figure 1 we see that the inverted Weibull distribution (range starts at zero) is the best model for the processing-time data. Although the inverted Weibull distribution may be unfamiliar to you, it is can be used in most simulation packages since it can be generated as the inverse of a Weibull random variable. It should also be noted that ExpertFit completed the entire analysis without further input from the analyst; only the range had to be specified.

After guided fitting, an analyst is automatically transferred to the third tab at which specified models can be compared to the sample to assess the quality of fit. Among our favorite comparisons are the density/histogram overplot and the distribution function differences plot, which are shown in Figures 2 and 3, respectively. In the former case, the density function of

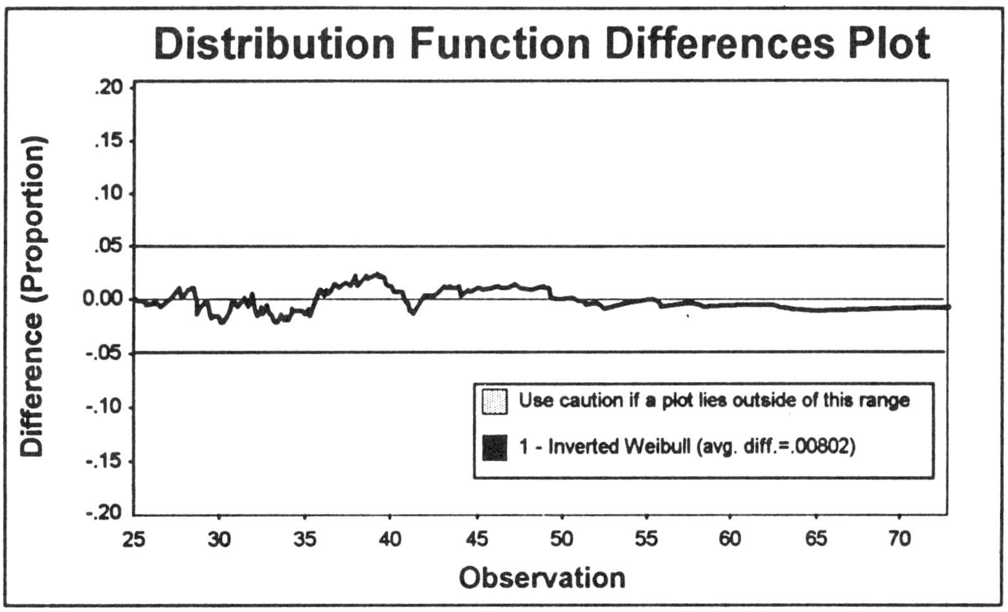

Figure 3: Distribution Function Differences Plot for the Processing-Time Data

Simulation Software	Representation
GPSS/H 3	RVIWEIB(<stream>,6.272056, 32.834140)
ProModel	InvWeibull(6.272056, 32.834140, <stream>, 0.000000)
Taylor II	1./weibull(0.028324, 6.272056)
WITNESS	1./WEIBULL(6.272056, 0.030456, <stream>)

Figure 4: Simulation Software Representation of the Inverted Weibull Distribution

the inverted Weibull distribution has been plotted over a histogram of the data (a graphical estimate of the true density function). This plot indicates that the inverted Weibull distribution is a good model for the observed data. The distribution function differences plot graphs the differences between a sample distribution function (a graphical estimate of the true distribution function) and the distribution function for the inverted Weibull distribution. Since these vertical differences are small (i.e., within the horizontal error bounds), this also suggests that the inverted Weibull distribution is a good representation for the data. Note that tab 3 also allows the analyst to perform several goodness-of-fit tests such as the chi-square test.

ExpertFit includes an option in tab 4 that allows one to display the representation of the inverted Weibull using different software packages. We show in Figure 4 the representations for four of the software packages supported by ExpertFit.

With some data samples, no candidate model provides an adequate representation. In this case we recommend the use of an empirical distribution function. One useful feature of ExpertFit is that in addition to using all of the sample values in the simulation software representation, it is possible to reduce the amount of required information through the use of a histogram-based empirical distribution function. We show a histogram-based representation (with 20 intervals) for two simulation software packages in Figure 5.

3 USING ExpertFit WHEN NO DATA ARE AVAILABLE

Quite often a simulation analyst must model a source of randomness for which no data are available. ExpertFit provides two analysis modes for this situation -- modeling of general activity times using triangular or beta distributions and modeling of random machine downtimes, for which we provide an example in this section. ExpertFit supports accurate modeling of systems with or without significant blocking or starving. For the example in this section, we will assume that the machine of interest is never blocked or starved.

Consider a machine that has an efficiency of 0.9; that is, it is actually producing parts 90 percent of the time. When the machine goes down, the average downtime is 60 minutes. However, the minimum downtime is 10 minutes. This information is specified to ExpertFit through a sequence of easy-to-use menus. After all of the required information has been specified, the average number of downs (actually the average number of busy-time/downtime cycles) per 8-hour shift is calculated by

Simulation Software	Representation
Arena	CONT(0.0000,24.800000, 0.0322,27.185000, 0.1576,29.570000, 0.3183,31.955000, 0.4791,34.340000, 0.5981,36.725000, 0.6945,39.110000, 0.7942,41.495000, 0.8457,43.880000, 0.8778,46.265000, 0.9068,48.650000, 0.9421,51.035000, 0.9550,53.420000, 0.9711,55.805000, 0.9807,58.190000, 0.9839,60.575000, 0.9904,62.960000, 0.9968,65.345000, 0.9968,67.730000, 0.9968,70.115000, 1.0000,72.500000)
AutoMod	continuous(0.0000:24.800000,0.0322:27.185000,0.1576:29.570000, 0.3183:31.955000,0.4791:34.340000,0.5981:36.725000,0.6945:39.110000, 0.7942:41.495000,0.8457:43.880000,0.8778:46.265000,0.9068:48.650000, 0.9421:51.035000,0.9550:53.420000,0.9711:55.805000,0.9807:58.190000, 0.9839:60.575000,0.9904:62.960000,0.9968:65.345000,0.9968:67.730000, 0.9968:70.115000,1.0000:72.500000)

Figure 5. Simulation Software Representation of the Empirical Distribution Function

Simulation Software	Busy-Time and Down-Time Representations
SIMSCRIPT II.5	GAMMA.F(540.000000, 0.700000, <stream>) 10.000000+GAMMA.F(50.000000, 1.400000, <stream>)
AweSim	GAMA(771.428571, 0.700000, <stream>) 10.000000 + GAMA(35.714286, 1.400000, <stream>)

Figure 6: Simulation Software Representations of Busy-Time and Downtime Models

the package to be 0.8. This makes sense since the average length of a busy-time/downtime cycle is 10 hours. A menu then allows various characteristics of the busy-time and downtime distributions to be displayed. We show the simulation software representations for two packages in Figure 6.

4 CONCLUSION

ExpertFit can help you develop more valid simulation models than if you use a standard statistical program, an input processor built into a simulation package (language or simulator), or hand calculations to determine input probability distributions. ExpertFit uses a sophisticated algorithm to determine the best-fitting distribution and, furthermore, has 43 built-in distributions. On the other hand, a typical simulation package contains roughly 10 distributions.

ExpertFit can represent most of its 43 distributions in 33 different simulation packages such as Arena, AutoMod, AweSim, COMNET III, FACTOR/AIM, GPSS/H, ManSim/X, MedModel, Micro Saint, MODSIM III, OPNET Modeler, ProModel, SES/*workbench*, SIMAN V, Simple++, SIMSCRIPT II.5, SLAMSYSTEM, Taylor II, and WITNESS, even though the distribution may not be available in the simulation package itself.

REFERENCES

Law, A.M. and W.D. Kelton. 1991. *Simulation Modeling and Analysis*, Second Edition. New York: McGraw-Hill.

Law, A.M. and M.G. McComas. 1989. Secrets of Successful Simulation Studies. *Industrial Engineering* **21**: 28-31 (May 1989).

Law, A.M., M.G. McComas, and S.G. Vincent. 1994. The Crucial Role of Input Modeling in Successful Simulation Studies. *Industrial Engineering* **26**: 55-59 (July 1994).

AUTHOR BIOGRAPHIES

AVERILL M. LAW is President of Averill M. Law & Associates, Inc. (Tucson, Arizona), a company specializing in simulation model building, training, and software. He has been a simulation consultant to more than 100 organizations, including General Motors, IBM, AT&T, General Electric, Nabisco, Xerox, NASA, the Air Force, the Army, and the Navy. He has presented more than 250 simulation short courses in 16 countries, and delivered more than 100 talks on simulation modeling at technical conferences.

He is the author (or coauthor) of three books and more than 35 papers on simulation, manufacturing, communications, operations research, and statistics, including the textbook *Simulation Modeling and Analysis* that is used by approximately 50,000 people worldwide. His series of papers on the simulation of manufacturing systems won the 1988 Institute of Industrial Engineers' best publication award. He is the codeveloper of the ExpertFit software package for selecting simulation input probability distributions, and he has developed several simulation videotapes. Dr. Law wrote a regular column on simulation for *Industrial Engineering* magazine from 1990 through 1991.

He has been a tenured faculty member and has taught simulation at the University of Wisconsin and the University of Arizona. Dr. Law has a Ph.D. in industrial engineering and operations research from the University of California at Berkeley.

MICHAEL G. MCCOMAS is Vice President for Consulting Services of Averill M. Law & Associates, Inc. He has considerable simulation modeling experience in a wide variety of application areas, and is the coauthor of seven published papers on simulation. His educational background includes an M.S. in systems and industrial engineering from the University of Arizona.

Modeling Methodology

Proceedings of the 1996 Winter Simulation Conference
ed. J. M. Charnes, D. J. Morrice, D. T. Brunner, and J. J. Swain

A TRACE-BASED APPROACH FOR MODELING WIRELESS CHANNEL BEHAVIOR

Giao T. Nguyen
Randy H. Katz

Computer Science Division
University of California at Berkeley
Berkeley, CA 94720-1776, U.S.A.

Brian Noble
Mahadev Satyanarayanan

School of Computer Science
Carnegie Mellon University
Pittsburgh, PA 15213-3891, U.S.A.

ABSTRACT

The loss behavior of wireless networks has become the focus of many recent research efforts. Although it is generally agreed that wireless communications experience higher error rates than wireline, the nature of these lossy links is not fully understood. This paper describes an effort to characterize the loss behavior of the AT&T WaveLAN, a popular in-building wireless interface. Using a trace-based approach, packet loss information is recorded, analyzed, and validated. Our results indicate that WaveLAN experiences an average packet error rate of 2 to 3 percent. Further analysis reveals that these errors are not independent, making it hard to model them with a simple two-state Markov chain. We derive another model based on the distributions of the error and error-free length of the packet streams. For validation, we modulate both the error models and the traces in a simulator. Trace-driven simulations yield an average TCP throughput of about 5 percent less than simulations using our best error model.

1. INTRODUCTION

With the proliferation of portable computers and wireless networks in recent years, many researchers have focused on designing better mobile systems. This has led to a growing interest in characterizing the loss behavior of many wireless technologies, including wireless LANs and packet radio networks. While it is well known that wireless links have higher error rates than their wired counterparts, the detailed characteristics of wireless errors are not well understood due to their inherent dependence on complex radio wave propagation. Characterizing the loss behavior is an important problem since it is one of the few key parameters that affect all levels of the network stack.

Due to the large number of protocols at different network layers, it is often infeasible to build and measure all of them. A better alternative is to trace the behaviors of existing implementations and apply the lessons learned to evaluate new designs. From wireless network traces (Noble, Nguyen, Satyanarayanan, and Katz 1996), we can improve our understanding of the wireless channel behavior, develop realistic models, and validate them. Traces also provide us a way to reproduce realistic network and mobile environments for comparing protocols and algorithms.

The focus of this paper is on the tracing and modeling of wireless channel errors. Although the uniform bit error rate model has often been used to simulate lossy links, it is inadequate to capture our measurements of in-building wireless errors. Our goal is to produce a realistic model that is not too difficult to implement. The resulting model contains several mathematical expressions for the error and error-free length extracted from many packet error traces. Validation of this model involves comparing the simulated TCP throughput with other models and traces.

The remainder of this paper is organized as follows. In section 2, we provide some background on wireless errors and the testbed for this study. Our trace-based approach consists of three phases, described in three separate sections along with the associated data. Section 3 describes the trace collection and presents the packet error rate data. In section 4, the errors are further analyzed and modeled. Finally, we discuss the validity of our models in section 5.

2. BACKGROUND

Wireless errors are mainly caused by the inability of the receiver to distinguish the transmitted signal from the background noise. The radio propagation patterns and their effect on receiver signal-to-noise ratio (SNR) have been studied for both the in-building and outdoor environments (Andersen, Rappaport, and Yeshiva 1995), (Cox and Leck 1975), (Eckhardt and Steenkiste 1996). The detrimental effect of wireless errors on reliable transport protocols has resulted in many efforts to improve the performance of TCP (Bakre and Badrinath 1995), (Balakrishnan, Seshan, and Katz 1995).

Since wireless errors highly depend on the environment and the network device, characterizing them requires some knowledge of the testbed. The following subsections describe our testbed and the wireless interface that we use in this study.

2.1 Testbed

The measurements presented in this paper were collected from the BARWAN wireless overlay network, which is built from many commercially available networking technologies. The available network interfaces include 915 MHz and 2.4 GHz AT&T WaveLAN, Metricom, CDPD, infrared, and DirecPC. The mobile hosts (MH) are IBM Thinkpad laptops and the base stations (BS) are Dell PCs. Both the mobile hosts and base stations are x86 computers running a common UNIX operating system, BSD/OS version 2.1. Although the base stations and mobile hosts can have multiple wireless interfaces, we only evaluate the 915 MHz WaveLAN in this study. Having complete control over this network allow us to obtain many device-specific data that are valuable for channel modeling.

2.2 WaveLAN

WaveLAN is a commercial wireless network interface operating in the 902-928 MHz ISM (Industrial, Scientific, and Medical) band (AT&T 1993), (Kohno, Meidan, and Milstein 1995). This direct sequence spread spectrum modem has a spreading factor of 11 chips per bit and a maximum bandwidth of about 2 Mbps. WaveLAN does not employ any forward error correction or retransmission at the device level, which makes it appropriate for studying wireless errors.

Medium access control (MAC) in WaveLAN resembles that of an Ethernet. Based on CSMA, it constantly monitors the state of the wireless channel. Before each transmission, the MAC must wait until the end of the current transmission plus a small delay known as the WaveLAN Inter-Frame Space (WIFS). Since it is difficult to detect a collision while transmitting, WaveLAN implements a collision avoidance scheme, which is different from the CSMA/CA being proposed IEEE 802.11 committee (Chen 1994). After the WIFS, the MAC waits for a random number of *antenna slots* less than an initial back-off value. If the channel becomes busy before its slot, the MAC doubles its back-off window and continues to monitor the channel for the next contention period. This back-off process is repeated for a maximum of 15 times before the packet is dropped.

The WaveLAN link is slightly asymmetric due to differences in implementation of the base station's ISA interface and the mobile host's PCMCIA counterpart. The ISA adapter has an Intel 82586 CSMA/CD LAN controller, while the PCMCIA version has the newer 82593 chip. Although both interfaces have 0.5 nanosecond bit-times and 32-bit inter-frame space, the length of the antenna slot for back-off purpose is significantly different: 12 bit-times for ISA and 32 bit-times for PCMCIA. To partially make up the longer antenna slot, the PCMCIA controller sets its initial back-off window to 16 slots (512 bit-times) instead of the ISA's 32 (384 bit-times). The larger average delay for PCMCIA leads to lower throughput.

Like many other wireless interfaces, WaveLAN provides access to the receiver signal strength information. The three signal measurements for each packet are signal level, noise level, and signal quality. The signal level is a 6-bit measurement derived from the receiver's automatic gain control (AGC) setting at the beginning of each transmission. Similarly, the noise level is measured immediately after the end of the transmission, when no WaveLAN interface within detectable range can transmit. The signal quality is the 4-bit measurement derived from the information used for antenna selection at the beginning of packet reception.

3. TRACE COLLECTION

Reproducing accurate loss behavior for wireless network is challenging because the quality of the wireless channel can vary dramatically over time and space. To handle this problem, we use a trace-based approach consisting of three phases: trace collection, analysis, and validation. In the trace collection phase, a large number of traces are collected for many different scenarios. The analysis phase involves extracting the data of interest, such as packet errors, and modeling them. In the validation phase, the models are simulated and compared with the traces and measurements.

This section describes the collection phase. To collect traces for many different scenarios, we implement a general purpose trace engine and workload generator. The following two subsections describe the trace recording process and workload generation. We then investigate the effects of traffic parameters including packet size and transfer rates. Finally, we examine the effect of distance on packet error rate.

3.1 Trace Recording

The network tracing facility for both the base station and the mobile host is transparent to the applications, as shown in Figure 1. It contains two key components: the *trace agent* and *trace collector*. The trace agent resides in the kernel where it can record data that is either inaccessible or expensive to obtain at the user level. Kernel hooks are added to the network device drivers to pass each received packet and device information to the trace agent. These data are periodically extracted from the kernel buffer and saved to disk by the user-level trace collector.

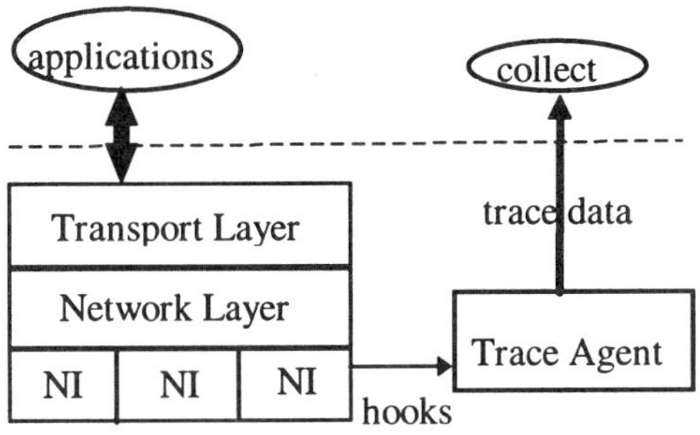

Figure 1: Network Trace Collection

When the trace collection program is executed, it opens a network tracing pseudo-device to start the kernel trace agent. The kernel trace agent then puts the wireless network interfaces (NI) into promiscuous mode to capture all packets, similar to the Berkeley Packet Filter (McCanne and Jacobson 1993). Device-specific data such as signal level, noise level, signal quality, and error status are also recorded for post-processing and analysis.

3.2 Workload Generation

To generate traffic for error measurements, UDP is preferred over TCP because it has no error-recovery and connection establishment mechanisms. Each UDP packet is specially formatted to include information for error detection, such as a sequence number. For our modeling of packet errors, we decided to hold transmission rate and packet size constant such that our models can be converted into time-based ones. The effects of these two parameters are discussed in the following subsections.

To simplify the analysis process, each trace captures a single traffic stream. We also pay special attention to reduce the interference of the trace workload and other traffic. Although the UDP packets need not be received for the sender to make progress, a standard kernel would send out a "port unreachable" error message in reply to each UDP packet whose destination port has no user-level listener. A kernel hook is added to suppress these interfering packets while collecting traces.

3.3 Effect of Transmission Rate

Due to the asymmetric nature of WaveLAN, and most wireless networks in general, each direction of the link is traced independently. Our measurements indicate that the base station can send large UDP packets of more than 1000 bytes at the maximum throughput of about 1.6 Mbps. Using a similar packet size, the PCMCIA interfaces only achieve the maximum throughput of about 1.2 Mbps. This lower transmission rate is mainly due to the differences in implementation of the radio interfaces discussed in section 2.2. Due to limited space, we only present the analysis for the higher bandwidth down-link (base station to mobile host).

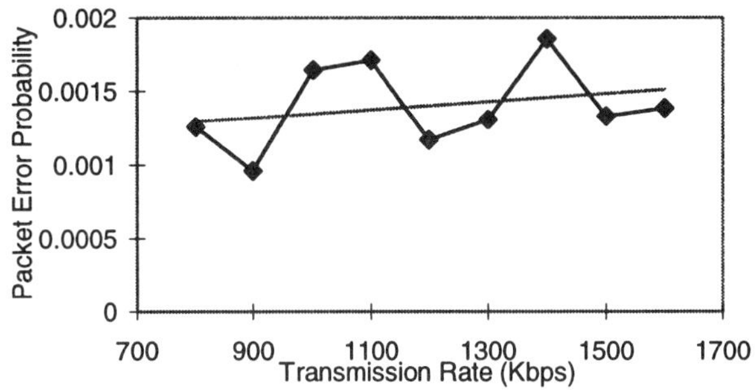

Figure 2: Packet Error Rate vs. Transmission Rate

To investigate the effect of the transmission rate, we use UDP streams with rates from 0.8 to 1.6 Mbps. The packet size and the distance are fixed at 1400 bytes and about 70 feet, respectively. Each data point in Figure 2 represents a single 1000-second trace. This plot reveals little correlation between the packet error probability and the transmission rate. Since the higher transmission rate does not lead to an increase in error probability, we will use nearly maximum transmission rates to capture more channel errors.

3.4 Effect of Packet Size

A similar experiment is conducted to examine the effect of packet size on error rate. In this experiment, the transmission rate is held constant at 1.5 Mbps while the packet size is varied from 100 bytes to 1400 bytes. To avoid fragmentation, packet size is chosen to be less than WaveLAN's maximum transfer unit of 1500 bytes.

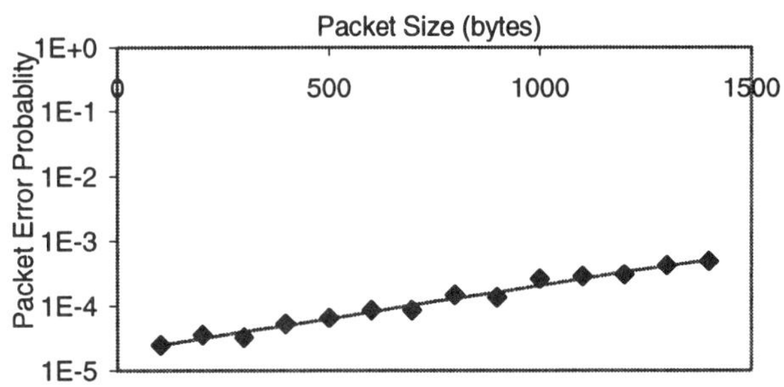

Figure 3: Packet Error Rate vs. Packet Size

The resulting packet error rates are plotted on the semi-log graph in Figure 3. Since the points in the graph lie on a straight line, we infer that the packet error rate increases exponentially with the packet size. Our regression analysis shows that packet error rate doubles for every 300-byte increment of the packet size.

The results of this experiment does not directly influence the choice of packet size for error modeling in section 4. For the data to be modeled, we choose to fix the packet size at 1400 bytes because we believe it is more representative of our WaveLAN network.

3.5 Effect of Distance

Another parameter that is generally believed to have high correlation with signal level and error rates is the distance between the sender and receiver. In free space, the power of electromagnetic radiation varies inversely with the square of distance, making distance an ideal indicator of signal level as well as loss rate. In practice, the imperfect propagation environment can render the inverse-squared relationship useless.

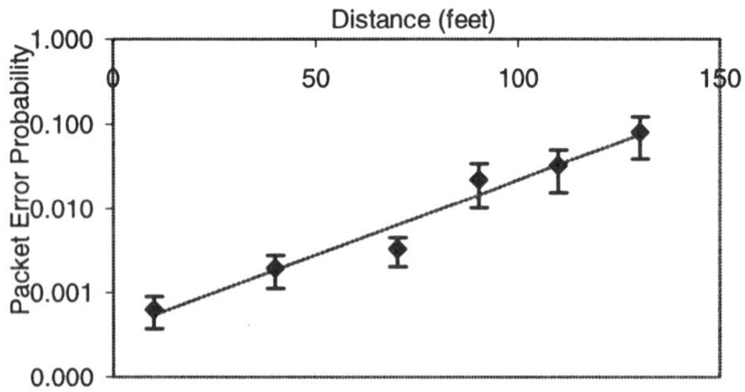

Figure 4: Packet Error Rate vs. Distance

Figure 4 plots the packet error rate measured at various distances in our building. Each data point represents UDP transfer totaling about 10^8 bytes over three 1000-second traces for a single room. We infer from the graph that packet error rate is an exponential function of distance. The regression analysis reveals that packet error rate doubles for every increase of 17 feet.

Table 1: Parameters of Distance Experiment

	packet error prob		error length		error-free length	
Distance	mean	stdev	mean	stdev	mean	stdev
10	0.000634	0.000229	1.000	0.000	1749.59	732.72
40	0.001955	0.000725	1.848	0.227	1023.43	334.40
70	0.003271	0.001086	2.005	0.210	574.56	148.45
90	0.021822	0.010412	2.428	0.018	122.89	58.99
110	0.032171	0.014786	2.540	0.482	87.26	39.75
130	0.080431	0.037123	2.298	0.290	28.90	8.82
average	0.024735	0.031554	2.054	0.564	549.37	699.39
mobile	0.032711	0.019324	2.365	0.089	84.60	37.36

Table 1 summarizes the data for the distance experiment. It contains the mean and standard deviation of the three key packet error characteristics: error rate, error length, and error-free length. The distance of less than 20 feet represents the single room scenario, where the sender and receiver have line-of-sight (LOS). The "average" row contains the averages of the data collected for all the distances. Data in the "mobile" row are collected while the experimenter is moving at the speed of about 5 feet per second inside our building. The packet error probability for this case is about 30% higher than the average, presumably because the presence of mobility produces more errors.

The error length is defined to be the number of packets that are lost consecutively. Similarly, the error-free length is number of packets that are successfully received between two adjacent bursts of error. This can also be translated into the inter-arrival time of errors. The average error length is 2 to 3 packets for most distances. The only noticeable exception is the same-room LOS scenario, which produces only single-packet error bursts. The average error-free lengths for different distances vary by two orders of magnitude and have large standard deviation.

4. MODELING WIRELESS ERRORS

The loss characteristics of wireless channels have been empirically observed to be bursty due to various fading effects. As the result, evaluating wireless network protocols with a uniform error model will likely produce inaccurate results. We investigate several models to capture the burstiness of wireless errors. Starting with the Fritchman binary error model (Fritchman 1967), we develop two models for our trace data. In the following subsection, we describe the two-state Markov model. Following that is the analysis of the error and error-free distributions to improve the two-state model.

4.1 Two-state Markov Model

The basic error model contains two states: error and error-free, each having its own distribution. When the channel is in the error state, any packets sent would be either lost or corrupted. The opposite is true for the error-free state. Being a Markov model, the duration of staying in each state can be expressed in term of transitional probabilities, as shown in Figure 5. This is the most simple form of the multiple-state Fritchman model.

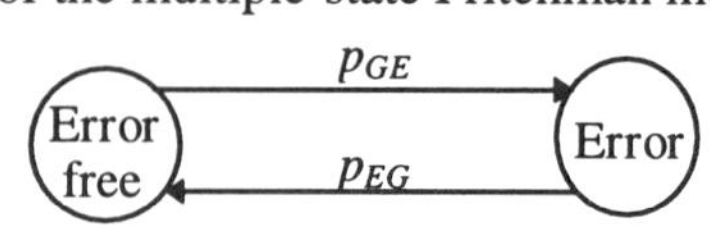

Figure 5: Two-state Error Model

In our derivation of parameters for the two-state Markov model, we label the error state E and error-free state G. The associated transitional probabilities, p_{GE} and p_{EG}, can be computed from the average error-free length

L_G and error length L_E. Since both L_E and L_G are geometrically distributed, the transitional probabilities can be expressed by the following formulas (Jain 1991):

$$p_{EG} = \frac{1}{L_E} \quad \text{and} \quad p_{GE} = \frac{1}{L_G}$$

The state transitions in the Markov model are memory-less. This nice property allows the transitions to be decided on an individual packet basis. If the length or duration of staying in each state is needed, the inverse of the cumulative distribution function (CDF), *F(x)*, can be used. By substituting the *1-F(x)* of the geometric distribution with a random number u uniformly distributed from 0 to 1, the length x of staying in a state with the leaving probability p is:

$$x = \frac{\log(u)}{\log(1-p)}$$

Table 2: Parameters for Two-state Markov Model

	Error	Error-free
probability	0.38201434	0.00600982
length (pkts)	2.61770277	166.394284
stdev of length	2.95928568	496.899532

Table 2 provides the error and error-free length averaged from about 30 1000-second traces that are used for modeling purpose. The transmission rate and packet size are fixed at 1.5 Mbps and 1400 bytes respectively, while the distance changes. Since the standard deviations for these parameters are relatively large compared to the means, this model will likely be inaccurate. This motivates us to find more accurate models without sacrificing complexity. The following section describes one such model for the same set of trace data.

4.2 Improved Two-state Model

Although the two-state Markov model can describe burstiness more accurately than the uniform error model, it cannot capture all loss behavior. The main limitation is the underlying assumption that the length distribution for each state is geometric. Our results indicates that neither the error and error-free length distributions is geometric.

The error length distribution is plotted on a semi-log graph in Figure 6. The dash line is the geometric fit for this data set. Our key observation is that almost 90 percent of all error bursts have length of less than 4 packets. This suggests that we should split error length distribution into two segments. Each segment is fitted with an exponential curve (straight lines on the semi-log graph). Since the best-fit curves do not intersect the coordinate (0,1), exponential terms must be multiplied by a constant factor e^k. The resulting CDF and its inverse that expresses the discrete length of error are:

$$1 - F(x) = e^{-x/a} e^k$$

$$x = \lceil a(k - \ln(1 - F(x)) \rceil$$

Since these expressions are derived from those of the exponential distribution, we will label them *Exponential**. The parameters a and k for each segment can be separately obtained using linear regression on x and natural log of *1-F(x)*. For this type of segmented curves, the splitting points must also be specified. These are the 1-CDF values since we are interested it the inverse function. The parameters for the error length distribution are presented in Table 3.

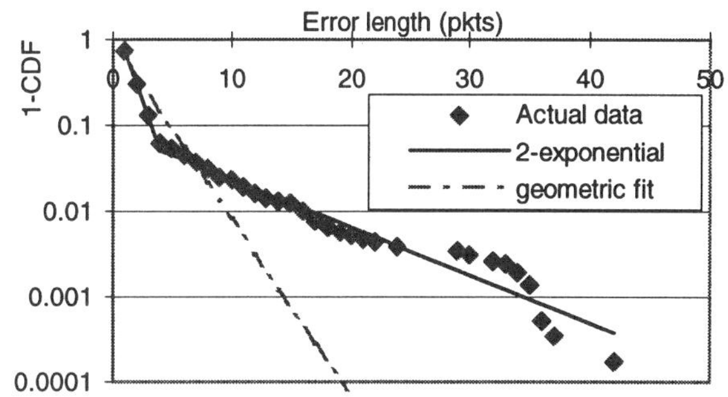

Figure 6: Error Length Distribution

The characteristic error-free length distribution is significantly different from that of the error length. In Figure 7, the 1-CDF is plotted on a log graph to provide more detail of the at the lower range. To avoid having an excessively complicated model, we decided to break this curve into 3 segments: 1 to 37, 37 to about 330, and the rest. The data points for the first two segments seem to fall on two different straight lines. Since the points of these segments fall on straight lines in this graph, they can be expressed by the following formula:

$$1 - F(x) = x^{-a} k$$

This expression describes the Pareto distribution. The inverse function that describes the error-free length is:

$$x = \left\lceil \frac{1}{k}(1 - F(x))^{-\frac{1}{a}} \right\rceil$$

The parameters a and k for the first two segments can be separately obtained by applying linear regression on the natural log of both x and *(1-F(x))*.

The third segment of the error-free length distribution is a concave curve. On a log graph, this type of curve implies exponential decaying function. Therefore, we use the *Exponential** expressions presented earlier to describe this segment. In Figure 7, our best fit curve with 3 segments is close enough to the actual distribution that it is almost invisible.

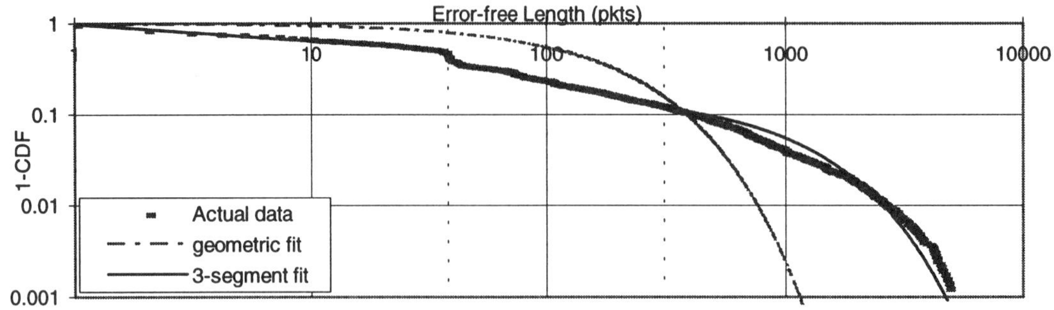

Figure 7: Error-free Length Distribution

Table 3: Parameters of the Improved Model for the Down-link

BS to MH	segment	Distribution	1-CDF	a	k	R2
Error-free	1 to 37	Pareto	1	0.183398	0.977663	0.974367
	38 to 330	Pareto	0.504178	0.566806	3.053316	0.996406
	331+	Exponential*	0.113339	925.719378	0.109955	0.933672
Error	1 to 3	Exponential*	1	1.152952	0.562514	0.999606
	4+	Exponential*	0.050955	7.753400	-2.460913	0.938160

Table 4: Parameter of the Improved Model for the Up-link

MH to BS	segment	Distribution	1-CDF	a	k	R2
Error-free	1 to 36	Pareto	1	0.217217	1.032932	0.911180
	37 to 300	Pareto	0.474259	0.780791	7.011119	0.993424
	301+	Exponential*	0.079019	441.524510	0.666545	0.935806
Error	1 to 3	Exponential*	1	1.152952	0.562514	0.999606
	4+	Exponential*	0.042026	5.144594	-2.391950	0.928936

Table 3 summarizes all the computed parameters for our improved model. This model requires 13 parameters: five pairs of *a* and *k*, and three 1-CDF values. For each pair *a* and *k* obtained from the regression of the 1-CDF values, there is an associated coefficient of determination R^2. This value is the ratio of the regression sum of squares (SSR) to the total sum of squares (SST), which is an indication for the quality of fit. Our model fits the 1-CDF of the experimental data with coefficients of determination greater than 90%.

Wireless loss behavior for the up-link closely resembles that of down-link. The improved two-state model for the up-link channel also contains 2 *Exponential** curves for the error length and a combination of 2 *Pareto*, 1 *Exponential** for the error-free length. Table 4 provides the model parameters computed for the up-link. Since the error and error-free distributions are very similar for both directions, we have higher confidence that the errors really occur in the wireless channel.

5. VALIDATION

Validation is the last of the three phases for our trace-based approach to modeling. Since our primary use of the channel models is evaluating the impact of the errors on higher-layer network protocols, we choose TCP throughput as the metric for validating our models. The approach that we take is validating the models in the network simulator *ns* (McCanne and Floyd 1996). In the following subsections, we describe the simulation of errors in *ns* and provide a quantitative comparison of all the results.

5.1 Simulating Packet Errors

To evaluate the error models for WaveLAN, we made several key enhancements to *ns*, in particular, the wireless channel and WaveLAN MAC protocol. We also implement the error models to generate errors for the channel.

Our simulation setup consists of 2 WaveLAN nodes. A 1000-second file transfer using TCP is done from the sender to the receiver. In this setup, we simulate three error models: uniform packet error rate, two-state Markov, and our improved two-state. The uniform packet error rate of 0.01549 is computed from the average error and error-free length. Other parameters for the two-state Markov and the improved model are taken from Table 2 and Table 3.

Trace modulation contains two steps. The first step is extracting the time and duration of the errors from the traces. In the second step, the errors are replayed in the channel at the simulated time corresponding to the traces. For the validation purpose, we collect 10 new traces from various locations and replay the errors in the simulation setup described above. Next, we average the TCP throughputs for all the trace-driven simulations and compare it with the throughputs of the error models.

5.2 Quantitative Comparison

Table 5: TCP Throughput for Error Models

	Throughput (Mbps)	relative difference
Uniform	1.5134	0.213429
Markov	1.4054	0.126841
Improved	1.3071	0.047995
Trace	1.2472	0.000000

Table 5 lists the TCP throughputs for the three models and the trace modulation. Having the trace-driven simulation throughput as the reference, the relative difference is computed for each model. The improved model produces throughput of about 5% higher than that of trace-driven simulation. The Markov and uniform error models differ by 13% and 21%, respectively. The higher TCP throughput for the Markov model confirms that the error lengths generated by the this model (shown in Figure 6) are shorter than the real ones.

6. CONCLUSION

We have showed that trace-based approach is valuable for modeling wireless errors. The large number of traces that we have collected provide a good basis for developing and validating several wireless error models. From the trace data, we derive the parameters for two variations of the two-state error model. Although the Markov model is much simpler, it produces TCP throughput that differs by 13% from the traces. The improved model with 13 parameters significantly improves the accuracy, only 5% from the traces. While the simple Markov model may be adequate for some applications, the higher accuracy of the improved model will justify its complexity for others.

Although it is our main goal to provide the best channel error model to study all network protocols, the results obtained thus far are more suitable for evaluating higher-level network protocols. This is partly due to our immediate interest in TCP performance evaluations. We believe it is possible to improve this model to study link-level and MAC protocols by using smaller packet sizes in tracing.

7. FUTURE WORK

Using a similar trace-based approach, we intend to characterize the behavior of the 915 MHz Metricom packet radio network and its interactions with the co-located WaveLAN network in the near future. Another goal is to improve our understanding of the mobility effects on various network characteristics such as errors and latency.

ACKNOWLEDGMENTS

The authors wish to thank Hari Balakrishnan, Bruce Mah, and Venkata Padmanabhan for many helpful suggestions in tracing and modeling of wireless networks. We also wish to thank Todd Hodes and Ann Chervenak, who offered comments on early drafts of this paper.

This research was supported by the Defense Advanced Research Project Agency (DARPA) and the Air Force Material Command (AFMC) under contract numbers DAAB07-95-C-D154 and F196828-93-C-0193, and the State of California MICRO Program. Additional support was provided by AT&T, Hughes Aircraft, IBM Corp., Intel Corp., and Metricom. The views and conclusions contained here are those of the authors and should not be interpreted as necessarily representing the official policies or endorsements, either express or implied, of AFMC, AT&T, DARPA, Hughes, IBM, Intel, Metricom, Carnegie Mellon University, the University of California, the State of California, or the U.S. Government.

REFERENCES

AT&T Global Information Solutions Company. 1993. *Architecture Specification for WaveLAN Air Interface.*

Andersen, J. B., T. S. Rappaport and S. Yeshiva. 1995. Propagation measurements and models for wireless communications channels. *IEEE Communications Magazine*, 42-49.

Bakre, A., and B. R. Badrinath. 1995. I-TCP: Indirect TCP for mobile hosts. In *Proceedings of the 15th*

International Conference on Distributed Computing Systems, 136-143.

Balakrishnan, H., S. Seshan, and R. H. Katz. 1995. Improving reliable transport and handoff performance in cellular wireless networks. *Wireless Networks*, 469-481.

Chen, K-C. 1994. Medium Access Control of Wireless LANs for mobile computing. *IEEE Network*, 50-63.

Cox, D. C., and R. P. Leck. 1975. Distributions of multipath delay spread and average excess delay for 910 MHz urban mobile radio paths. *IEEE Trans. Ant. Prop.*, 23: 206-213.

Eckhardt, D., and P. Steenkiste. 1996. Measurement and analysis of the error characteristics of an in-building wireless network. In *Proceeding of the ACM SIGCOMM.*

Fritchman, B. D. 1967. A binary channel characterization using partitioned Markov Chains. *IEEE Transactions on Information Theory*, 221-227.

Kohno, R., R. Meidan, and L. B. Milstein. 1995. Spread spectrum access methods for wireless communications. *IEEE Communication Magazine*, 58-67.

Jain, R. 1991. The art of computer systems performance analysis: techniques for experimental design, measurement, simulation, and modeling. *Wiley*, 491.

McCanne, S., and S. Floyd. Network Simulator. http://www-nrg.ee.lbl.gov/ns/

McCanne, S., and V. Jacobson 1993. The BSD packet filter: a new architecture for user-level packet capture. In *Proceedings of the 1993 Winter USENIX Technical Conference.*

Noble, B., G. Nguyen, M. Satyanarayanan, R. H. Katz. 1996. Mobile network tracing. *RFC Draft.*

AUTHOR BIOGRAPHIES

GIAO T. NGUYEN is a graduate student in Electrical Engineering and Computer Science at the University of California at Berkeley. His research interests are in the areas of computer networks, distributed and mobile computing. He received a Bachelor of Science in Computer Science and Engineering from the University of California at Los Angeles in 1994.

RANDY H. KATZ is a leading researcher in computer system design and implementation. His research experience has spanned numerous disciplines. He has written over 120 technical publications on CAD, database management, multiprocessor architectures, high performance storage systems, and video server architectures. Professor Katz developed the concept of Redundant Arrays of Inexpensive Disks (RAID), now a $10 billion industry segment. Katz's recent research has focused on wireless communications and mobile computing applications. He is a Fellow of the ACM and a Fellow of the IEEE.

BRIAN NOBLE is a Ph.D. candidate in Computer Science at Carnegie Melon University. His research interests include distributed systems, file systems, data repositories, mobile computing, and the measurement and evaluation of deployed systems. He received a Bachelor of Science in Electrical Engineering and Computer Science from the University of California at Berkeley in 1991, and a Master of Science in Computer Science from Carnegie Mellon in 1993.

MAHADEV SATYANARAYANAN is a professor of Computer Science at Carnegie Mellon University. His current research addresses the problem of information access in mobile computing environments. The Odyssey mobile computing platform and the Coda File System are outcomes of this work. Earlier, he was a principal architect and implementor of the Andrew File System, which focused on issues of scale, performance and security. Later versions of this system have been commercialized and incorporated into the Open Software Foundation's DCE offering. Satyanarayanan received the Ph.D. in Computer Science from Carnegie Mellon, after Bachelor's and Master's degrees from the Indian Institute of Technology, Madras. He is a member of ACM, IEEE, Usenix, and Sigma Xi, and has been a consultant and advisor to industry and government.

Proceedings of the 1996 Winter Simulation Conference
ed. J. M. Charnes, D. J. Morrice, D. T. Brunner, and J. J. Swain

PARALLEL SIMULATION ENVIRONMENT FOR MOBILE WIRELESS NETWORKS

Winston Liu
Ching-Chuan Chiang
Hsiao-Kuang Wu
Vikas Jha
Mario Gerla
Rajive Bagrodia

Computer Science Department
University of California, Los Angeles
Los Angeles, California 90095, U.S.A.

ABSTRACT

A parallel simulator has been designed for the evaluation of wireless, multihop, mobile networks. This paper describes the process of porting the simulator from a sequential to a parallel environment. Parallelization is critical in large radio networks, where the complexity of radio propagation models, channel access schemes and interference patterns makes sequential simulation very time consuming – in the order of several hours for 100 nodes experiments. With parallel execution, speedups of up to tenfold have been observed on a 16 processor SP/2, making large network simulations viable.

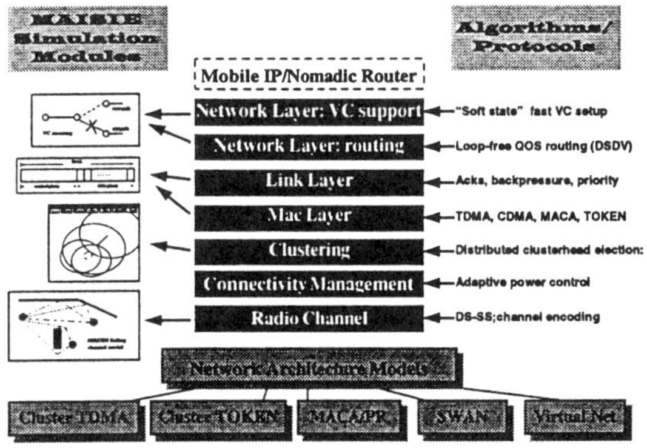

Figure 1: Wireless Network: Protocols and Models

1 INTRODUCTION

The rapid advancement in portable computing and wireless communications technology has led to significant research and investment to create an infrastructure to support mobile users. Such an infrastructure is beneficial to civilian applications such as disaster relief operations as well as military applications such as tactical mobile units. The need for such an infrastructure underlies the focus of the UCLA GloMo project. A major contribution of this project is the development of an instant wireless infrastructure which supports multimedia multi-hop wireless communications in presence of node mobility and failures. Protocols designed for this kind of environment are complex to evaluate analytically due to various factors such as complex channel access protocols, interference effects, traffic input patterns, node mobility , channel propagation properties, and radio characteristics. To alleviate the prohibitive cost of protocol implementation via trial and error, simulation offers a way to evaluate the protocol design before the actual implementation takes place. In simulating such an environment, the simulation execution time is usually directly proportional to the level of detail in the model and the size of the network that one wants to study. Given the complexity of the radio environment, sequential simulation of networks with thousands of nodes requires several days. To make the design more interactive, it is imperative to reduce run time without sacrificing the level of detail in the model. A promising solution is parallel simulation. This paper describes a parallel environment to support rapid prototyping of mobile, multimedia protocols. Sections 2 and 3 describe the wireless network architecture and the radio channel model which have been implemented in the parallel simulator. Section 4 reviews the Maisie parallel simulation environment and discusses the process of porting the model to a parallel architecture. Section 5 reports the results.

2 NETWORK ARCHITECTURE

The network architecture under study is a wireless, mobile, multihop architecture. Unlike cellular systems, there are no fixed base stations connected by a wireline network. The main motivation for mobile wireless multihopping is rapid deployment and dynamic reconfiguration. When the wireline network is not available, as in battlefield communications and search and rescue operations, multihop wireless networks provide the only feasible means for ground communications and information accesses. Figure 1 shows the typical protocol layers of the wireless network. In the GloMo project, various alternatives were developed for each layer, leading to several different network architectures as shown at the bottom of Figure 1. The protocol choices at each layer have been mapped into Maisie simulation modules, yielding a modular simulator structure ideally suited for the comparison of multiple architectures. The architecture evaluated in this case study is the Cluster Token. Its key layers are described in the following subsections.

2.1 Clustering

In multihop, mobile wireless networks, the aggregation of nodes into clusters controlled by a cluster head provides a convenient framework for the development of important features such as code separation (among clusters), channel access, routing, and bandwidth allocation (Gerla et al. 1995). Using a distributed algorithm with a cluster, a node is elected to be the cluster head. All nodes within transmission range of the cluster head belong to this cluster. That is, all nodes in a cluster can communicate with a cluster head and (possibly) with each other. The complexity and overhead of clustering rests in the selection of the cluster head. The most important criterion is stability. Frequent cluster head changes adversely affect the performance of other protocols such as scheduling and allocation which rely on it.

In our clustering algorithm, only two conditions cause the cluster head to change. One is when two cluster heads come within range of each other. The other is when a node becomes disconnected from any other cluster. This is an improvement (in stability) over existing algorithms which select the cluster head every time the cluster membership changes. Figure 2 shows a clustering example.

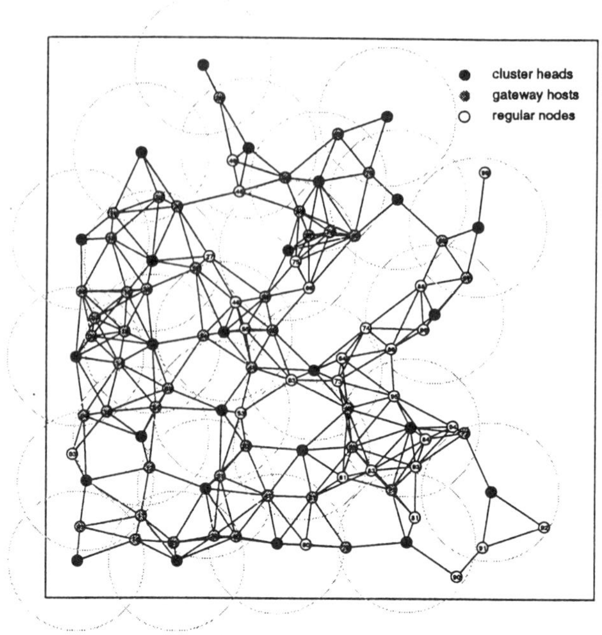

Figure 2: Clustering among 100 Hosts, Transmission Power=100

2.2 MAC layer

Clustering provides an effective way to allocate wireless channels among different clusters. Across clusters, we enhance spatial reuse using different spreading codes (i.e. CDMA). Within a cluster, we use a clusterhead controlled token protocol (i.e. polling) to allocate the channel among competing nodes. The token approach allows us to give priority to clusterheads in order to maximize channel utilization and minimize delay. A cluster head should get more chances to transmit because it is in charge of broadcasting within the cluster and of forwarding messages between mobile hosts which are not "connected".

The channel is assumed to be time slotted. In each time slot, only one node has the permission to transmit. In some cases the permission token may be lost. One such case occurs when the node with permission moves outside the cluster. Another case is when the host is a gateway (which belongs to more than one cluster). The gateway might be tuned to a different code (i.e. different cluster), thus missing the permission token which is then lost. To overcome these problems, the cluster head reissues the permission token after timeout. Figure 3 shows an example of slot scheduling (and gateway conflicts) in a 5 cluster network. Slot synchronization across clusters is assumed

2.3 Routing

Our model assumes shortest path routing. Routing tables are updated using a distance vector routing algorithm (Bertsekas et al. 1987). Each node maintains a routing table and broadcasts it to its neighbors whenever it gets the channel access permission. It updates its routing table when it receives the routing tables from its neighbors. In a time slot, the node broadcasts its routing table first and then transmit the data packet (if any).

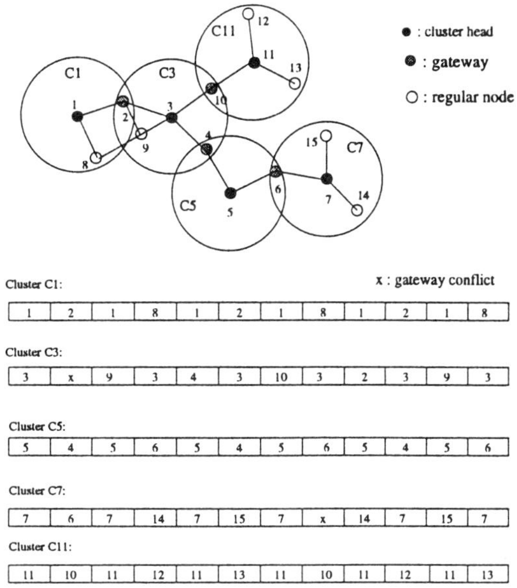

Figure 3: Example of Channel Scheduling

3 RADIO CHANNEL MODEL

An accurate radio channel simulation model is important not only for designing modulation and coding schemes that improve channel efficiency, but also for investigating the impact of channel fading on existing networking algorithms, such as clustering, routing and power adjustment. At present, most radio network protocol simulations are using the free space channel model which basically assumes that attenuation is only a function of transmitter-receiver distance. However, the radio channel characteristics are much too complex to be modeled by simple distance functions. Thus, the results are inaccurate.

To overcome this limitation, our simulator includes a rather sophisticated radio channel model which is an extension of the SIRCIM statistical impulse response model (Rappaport et al. 1990).

The radio channel is characterized by three propagation parameters: free space loss, multipath fading and shadowing. All these parameters are supplied by SIRCIM. SIRCIM provides impulse response characteristics which account for multipath fading. SIRCIM outputs include for example : the distribution of the number of multipath components in a particular multipath delay profile; the distribution of the number of multipath components etc.

SIRCIM provides impulse response at the *signal* level which is suitable for radio designs. For network performance evaluation purposes, we are more interested in received power at the *packet* level. Assuming that the channel is a Direct Sequence-Spread Spectrum channel, we can derive the mean signal power by performing convolution of the spread spectrum random chip sequence with the impulse response of the simulated channel.

Furthermore, in a mobile radio environment, we must model the fluctuation of received power caused by change of positions. To this end, the correlation between received power at different positions must be known. SIRCIM provides only *small-scale* spatial and temporal correlations. We have augmented SIRCIM to account for *large-scale* correlation as well.

In addition to multipath fading, the SIRCIM accounts for the shadowing effect caused by diffraction of radio waves around sharp edges. Shadowing, the slow fading, has been characterized in the literatures by roughly a log-normal distribution, with a standard deviation that depends on the roughness. A common assumption is that shadowing is independent from one location to another. Unfortunately, this assumption is not valid in a dynamic model with mobile users,in which location dependent correlation must be accounted for. In our simulation, we include the correlation model for shadow attenuation developed in (Gudmundson 1991).

The SIRCIM channel module is invoked (with a function call) every time a packet is transmitted in the network. Repeated computation of attenuation is required because the nodes are mobile and therefore change their relative position continuously.

4 SIMULATION ENVIRONMENT

A general purpose parallel environment for wireless network simulation has been designed at UCLA. The simulator is being built on an existing process-oriented, parallel simulation language called Maisie (Bagrodia and Liao 1994). A Maisie program is a collection of entities, each of which represents a specific object in the physical system and may be created and destroyed dynamically. Entities communicate with each other using timestamped messages. Every entity is associated with a unique message-buffer. The language provides an asynchronous send primitive to deposit a message with a current or future timestamp in the message buffer of a destination entity. Blocking and non-blocking receive primitives are also provided by the language to allow an entity to remove messages from its message buffer. The receive construct can be used to remove selective messages from the buffer such that a message is removed only when it is ready to be processed by the destination entity. Appropriate use of selective receives help in the generation of concise model descriptions.

Event scheduling constructs in Maisie are integrated with the communication (send and receive) primitives. Thus transmission delays in a physical

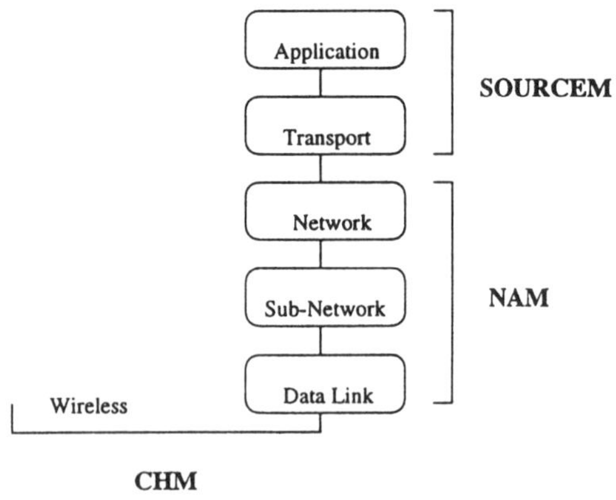

Figure 4: Networking Layers and Their Models

network can be modeled simply by incrementing the timestamp on the message when it is sent from the source to the destination node. A receive statement can optionally specify a timeout interval, where time is measured using the simulation clock; such a statement may be used to simulate the passage of time corresponding to activities like servicing of a job in a model, or transmission of a message in a network.

A Maisie program can be interfaced with a number of simulation algorithms: a splay-tree based implementation of the global event-list algorithm for the sequential implementation, a null-message or conditional event implementation of the parallel conservative synchronization algorithm (Jha et al. 1993), or a space-time implementation of the optimistic synchronization algorithm (Bagrodia, Chandy, and Liao 1991).

4.1 Model

The wireless simulator is being used to evaluate the stability and performance of network algorithms as a function of the radio characteristics, network control, and mobility patterns. A detailed description of the simulator and its use in evaluating mobile networks is described in (Bagrodia et al. 1995). This section recapitulates the simulator and describes the modifications that were necessary for efficient parallel execution. The simulator modules that have been used to construct the model described in this paper include: Application Traffic Models (SOURCEM), Network Algorithm Models (NAM), and Channel Models (CHM). Figure 4 depicts the primary layers in the simulator and describes the interface between them.

The SOURCEM components are used to generate the traffic load for the network. The NAM components are broken down into network layer protocol models such as IP, and wireless subnetwork control such as clustering, and mobility control such as power control, logical link control, and media access control. The clustering, cluster-head election, routing and channel access schemes described in the previous section were simulated in the NAM modules. The CHM components are used to provide a set of free space and fading channel models to simulate the wireless medium. The channel models include the effect of multi-path fading, shadowing, and interference for both indoor and outdoor transmissions. The model described in this paper was developed to study the relative stability of the clustering algorithm, and average number of hops and packet delay between a given source and destination pair under both free space and fading channel assumptions.

4.2 Reducing Simulation Execution Time

Simulations of complex systems such as the wireless network described in this paper can quickly become computationally intractable as the number of wireless nodes and the detail of the channel and other component models is increased. Some techniques that have been used to reduce the execution time of the model include: model aggregation, cloning components, hierarchical modeling and parallel model execution

Aggregation (Sanadidi 1993) or coarse grain modeling is an effective method if a component model's accuracy is fairly independent of the amount of detail used in modeling the component itself. Cloning (Sanadidi 1993) is useful in cases where there are many simulated components which tend to have the same behavior throughout the simulation. Hierarchical modeling (Bagrodia et al. 95) combines coarse and fine grain models to reduce simulation time in an iterative fashion. Each of the preceding methods reduces the execution time for the model by reducing the level of detail and hence potentially increasing the inaccuracy in the model. In contrast, parallel execution, if effective, can be used to reduce the execution time without reducing its accuracy. The next section describes the primary modifications that were made to the model to support its parallel execution.

4.3 From Sequential To Parallel Simulation

Parallel execution of the model basically requires that the model be subdivided into N partitions, such that $N \geq P$, where P is the number of processors available for execution of the model. In this paper we assume $N = P$. Each partition executes on a unique processor. The execution across partitions is synchronized through a parallel simulation algorithm. As the synchronization is handled transparently by the Maisie run-time system, in theory it should be easy to modify a sequential model for parallel execution. However, as described in a companion paper in this proceedings, there are a number of pitfalls that must be avoided in order to get an efficient parallel implementation. In the rest of this section, we first describe the modifications that were made to the initial model (Bagrodia et al. 1995) to yield a **correct** parallel implementation. Subsequently, we describe the modifications that were made to improve its parallel performance.

The first step in the parallelization effort was to eliminate global variables as the distributed memory architecture of SP/2 does not provide coherency support. This appears to be trivial, but can be complex if the model assumes some form of global view of the system for a distributed application and allows global variables to be modified and read at will. The best solution to preserve good performance in the parallel execution, is to redefine the global data structures in a distributed manner. For instance, there may be a global data structure which stores information regarding the VC connection status among all wireless nodes. Such a data structure should be distributed such that a wireless node is only aware of the VC connection status of itself and its immediate neighbors.

The next step is to eliminate unnecessary synchronizations in the model. In the sequential model, an entity was used to advance the time for all other entities in the system by broadcasting a message at the end of each slot interval. Since the time advance (equal to slot duration) is fixed, each entity can advance its local clock asynchronously eliminating the need for a centralized entity. Note that if the slot duration was determined based on some global property of the system, the decentralization would be considerably harder to implement.

For parallel execution, the simulation must be free of zero delay cycles in the communication topology. A model is said to contain a zero delay cycle if it is possible for an entity to receive a causally-ordered message with the same timestamp as another message sent by the entity (Misra 1986). Parallel execution of such models can either lead to deadlocks or instability. A sufficient condition to avoid zero delay cycles is to ensure that the receive time of a message is always greater than its send time; alternatively, it is possible to avoid such cycles by ensuring that an entity does not transmit a message with the same timestamp as an incoming message.

Upon the completion of these changes, the simulation was executed in parallel using the conservative run-time version of Maisie which is based on Chandy-Misra's null message parallel simulation algorithm (Misra 1986). A null message is a synchronization message which advances the clocks of neighbor entities. The initial result yielded slow-downs as opposed to speed-ups! Measurements showed that this was due to an extremely high "null message" ratio (defined as the number of null messages sent for every real message). This poor lookahead was certainly unexpected because the system utilizes fixed slots to synchronize within each cluster. Fixed slot length should itself serve as a good indication of lookahead. The poor lookahead in the initial model can be explained by the following two reasons.

First, the previous intuition of a good lookahead was based on the assumption that messages are only exchanged at the time that a slot starts or ends, but never in the middle. In reality, a number of messages such as routing updates, token arrivals, and data packets may arrive any time within the slot. The order in which these messages are processed sometimes will impact network performance; therefore, lookahead is not as simple as specifying a fixed slot time. Second, the original lookahead was poor because a number of the output messages are conditional, meaning that the entity has not enough information locally to deduce when it will send a message.

The following changes were made to the model to improve its lookahead and the null message ratio. (a) We exploited the fact that given the current state, although an entity does not know which message it will send next, it does know the time at which the message will be sent (if any). (b) Messages, whose sending and receiving time can be set to the start or the end of a slot without affecting the correctness of the simulation, are modified to take advantage of this behavior. By doing this, we were able to derive better lookahead upon receipt of a message. These changes in the model yielded a null message ratio in the range of 4.5 to 5.5 down from 25 in the first attempt. It is usually desirable to have a null message ratio of less than 1, but for the radio broadcast channel , this is the best that we are able to achieve given the fact that wireless network utilization is optimized when each node has on average 6 neighbors (Kleinrock et al. 1978).

Another issue of importance that may affect parallel performance is partitioning. In selecting partitions, it would be desirable to exploit fully both locality, and load balancing. Locality is achieved if neighboring nodes are placed on the same processor, thus reducing the communication overhead compared to neighboring nodes being placed on different processors. Load balancing (i.e. uniform distribution of entities among processors) is important as well because we want to fully utilize all the available computational resources. In our application, there are several strategies: random balanced, local balanced, and clustered balanced. Random balancing assigns enti-

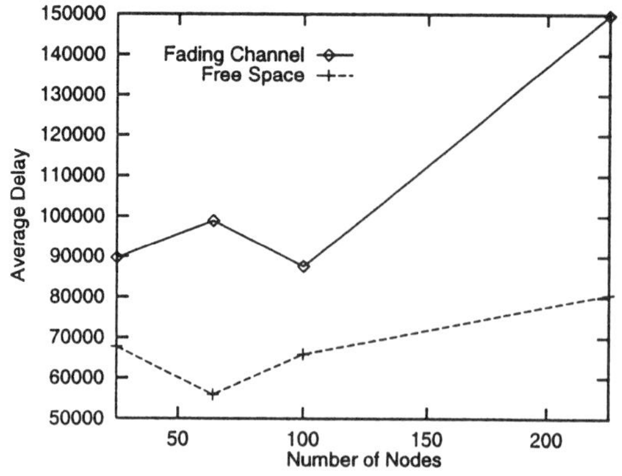

Figure 5: Network Performance

ties randomly to processors with the constraint that each partition has an approximately equal number of nodes. Local balancing tries to assign neighboring entities to the same processor while assigning an approximately equal number of entities to each processor. Clustered balancing first clusters the nodes into super nodes and then assigns the super nodes among different processors. The experiments reported in this paper use the first two partitioning techniques.

5 RESULTS

This section presents the results of parallel execution using Maisie conservative run-time simulator. We are studying the convergence of the distributed clustering algorithm, and throughput, average number of hops and end to end packet delay under both free space and fading channel assumptions. The experiments reported in this paper assume light load in the network with minimal mobility and were executed in exclusive mode on an IBM SP/2 multicomputer. Each node of this architecture is a RS/6000 processor with 128MB of RAM and a peak rating of 225 Mflops. The interconnection bandwidth is 40 Mbps and the hardware latency in the switch is 500 ns. Including software overheads, measurements show a minimum message latency of 30 μs. In order to ensure that the simulation has reached steady state, the baseline simulation time was established to be 250000 for all sets of runs.

5.1 Experiment Configurations

We consider four network configurations with 25, 64, 100, and 225 nodes randomly deployed in a square area (as in Figure 2) using a uniform probability distribution. The experiments serve two purposes, namely, to evaluate network performance and at the same time assess the efficiency of the parallel simulation implementation. Regarding network performance, Figure 2 illustrates a typical solution of the of the distributed clustering algorithm in the 100 nodes example. This solution is reached very rapidly after a few message exchanges among neighbors. Figure 5 shows the average end to end delay for both free and fading channel, for various network sizes. As expected, fading channel delay is higher due to packet loss and retransmission caused by fading.

To evaluate parallel simulation efficiency, the network examples are executed both sequentially, and in parallel on 1, 2, 4, 8, 12, and 16 processors respectively. We carry out three experiments. The first experiment uses random balanced partitioning while the second uses the more intelligent local balanced partitioning. Both experiments use a free space channel model. The third experiment uses the same partitioning strategy as the second, but employs a fading channel model which substantially increases the computational load.

5.2 Speed Ups

For free space channel model experiments, Figure 6, relative speed ups of parallel execution versus sequential execution. The speedup measure is derived by dividing the sequential execution time by the parallel execution time. In both experiments, the speed up generally improves with the number of nodes since a large number of nodes yields better load balancing across processors. The only glitch is for the random balanced partitioning experiment with two processors. This particular random balanced partitioning strategy uses a counter, which is incremented as an entity is created. In this strategy, entities are placed alternatively on one or the other processor in the order in which they were created. Since entities are sequentially deployed in the square grid as they are created, it follows that two successive entities are likely to be neighbors, and yet will fall in two different partitions. This is an example of conflict between locality and load balancing, which leads to high interprocess message cost and degraded performance. Next, we consider local balanced partitioning.

Table 1 compares the locality of messages in the two partitioning schemes used. Locality is defined as the number of messages (packets) that are sent within a processor, expressed as a percentage of the total number of messages. As expected, the local balanced partitioning provides significant improvement in locality. As a result, better speedups are obtained, as shown in Figure 7. Despite the improvement, however, the speed ups are not very high because of a high null message overhead (as explained at the end

Number of nodes	25	64	100	225
Random	11.9	14.4	11.8	19.3
Local	21.3	36.4	40.7	53.4

Table 1: Locality(%) of messages for the two partitioning schemes used, as the number of nodes are changed. The nodes are partitioned and mapped onto 16 processors.

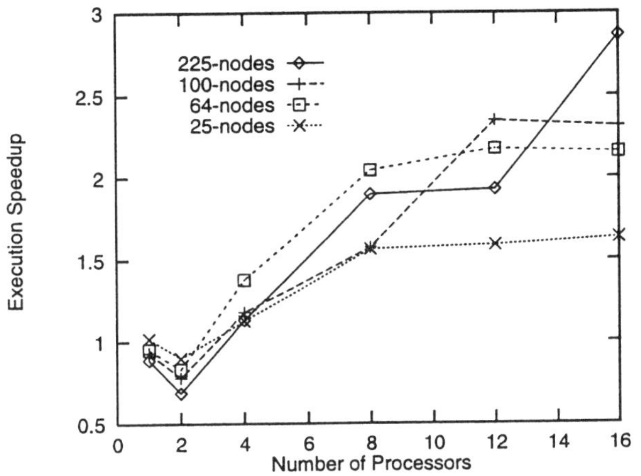

Figure 6: Free Space Channel with Random Balanced Partitioning

of this section).

For the fading channel model experiment, we only used local balanced partitioning since it performed better in the previous set of experiments. In Figure 8, the speed-up again increases with the number of processors and wireless nodes. In some cases, single processor parallel execution outperforms the sequential execution which uses a global event list. As discussed in (Jha et al. 1993), the primary reason for this is the lower context switching overhead. The higher speed up yielded by the fading channel model with respect to the free space model (10.4 vs 3.4) is due to the higher computation versus internode communication overhead. It is expected that future, more advanced radio network models will exhibit an even greater computation versus communication load ratio, due to sophisticated adaptive antennae and spread spectrum encoding schemes. In this experiment, we noticed that the 64 nodes case generally have higher speed up than the 100 nodes case except for the case with 12 processors. This is because for the 12 processors case, the 64 nodes can not be divided evenly among the processors.

Despite improving the lookahead, we found that NMR, the number of null messages sent for every real

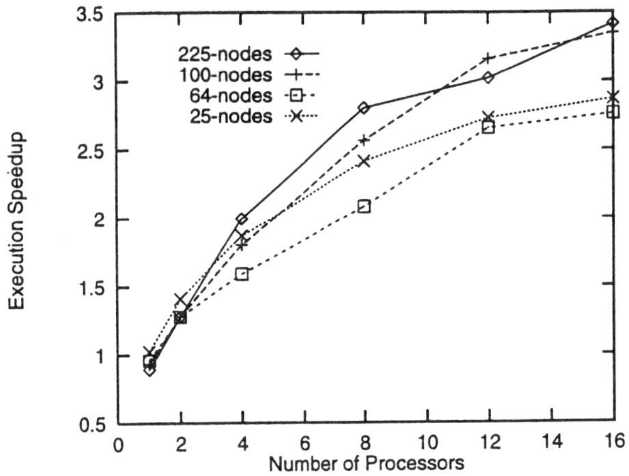

Figure 7: Free Space Channel with Local Balanced Partitioning

message, was relatively high - around 5.5-6.0. Since the computation associated with each real message (packet) is small in case of the experiments without the fading channel model, a high NMR means that the time spent in processing null messages is more than that spent in processing the real messages, resulting in relatively modest speedups. When the fading channel model is introduced, the computation associated with each real message increases substantially. Thus, even though the NMR is still high (around 6.0), the percentage of time spent in processing null messages is reduced, resulting in much better speedups (as good as 10 on 16 processors). Overall, the speed up is very encouraging considering the fact that we are able to reduce execution time from 3+ hours (sequential execution) down to 20 minutes using 16 processors for a network with 225 nodes.

6 CONCLUSION

A parallel simulator has been developed to study a number of protocols at different layers related in the design of mobile wireless networks using Maisie, a general purpose parallel simulation language. The message-passing paradigm used by Maisie provided a natural and simple way to model such networks. To study scalability issues, Maisie offers a parallel simulation approach which reduces execution time while preserving model accuracy. Results for static topology have shown that speed-up tend to increase as the number of nodes in the simulation increases. For the 225 nodes configuration with fading channel model, a speed-up factor of 10 was achieved. This dramatically reduces the turn around time of a simulation run.

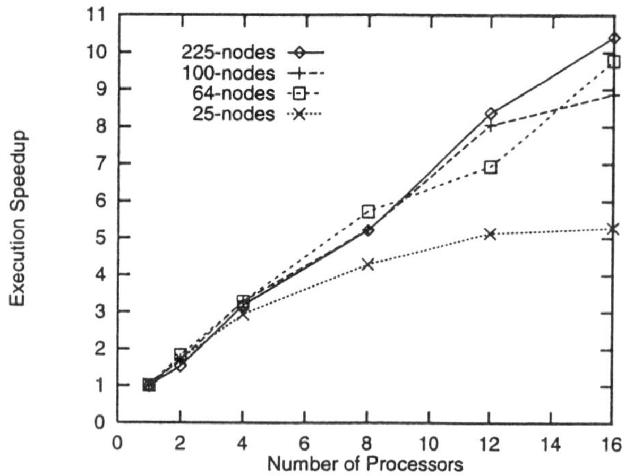

Figure 8: Fading Channel with Local Balanced Partitioning

In porting the simulation, we learned the following lessons: (a) A model designer should not assume the availability of a global snapshot of the system state. This is not an unreasonable constraint since no global snapshot exists in distributed applications. It is just an artifact to simplify the model. (b) Conditional messages are often unavoidable due to the behavior of the protocol that one tries to study. Thus, programmers must take great care to minimize the null messages that may result. One solution is to break up a single receive statement which may receive many different types of conditional messages into a sequence of receives if the arrival order of the various conditional messages is known. (c) Partitioning strategy used to decompose the network was found to affect the performance of the parallel simulator. (d) The general advice to programmers is that the simulator should be written from the beginning as a parallel program. The sequential execution should be considered as a special case. To our knowledge, this is the first parallel simulation environment that supports rapid prototyping of mobile wireless protocols. Work is in progress to parallelize the mobility model, to compare tradeoffs between conservative and optimistic run-time, and to study the behavior of the speed-up factor with respect to network load, degree of mobility and partitioning strategies.

ACKNOWLEDGEMENTS

This research was supported by the U.S. Dept. of Justice/FBI, ARPA/CSTO under contract J-FBI-93-112, by ARPA/CSTO under Contract DABT-63-94-C-0080 “Transparent Virtual Mobile Environment”, and by Intel under project “QoS Wireless Networks”.

REFERENCES

Bagrodia, R., K. M. Chandy, and W-T. Liao. 1991. A unifying framework for distributed simulations. *ACM Trans. on Modeling and Computer Simulation*, 1(4):348-385.

Bagrodia, R. M. Gerla, L. Kleinrock, J. Short, and J. T. Tsai. 1995. A Hierarchical simulation environment for Mobile Wireless Networks. *In proceedings of the 1995 Winter Simulation Conference*, December.

Bagrodia, R. and W-T. Liao. 1994. Maisie: A Language for the Design of Efficient Discrete-Event Simulations. *IEEE Transactions on Software Engineering*, 20(4):225-238.

Bertsekas,D. and R. Gallager. 1994. *Data Networks*, Prentice-Hall.

Gerla, M. and J. T. Tsai. 1995. Multicluster, mobile, multimedia radio network. *Wireless Networks*.

Gudmundson, M. 1991. Correlation Model For Shadow Fading In Mobil Radio Systems. *Electron. Lett.*, 2145-2146.

Jha, V. and R. Bagrodia. 1993. Transparent implementations of conservative algorithms in parallel simulation languages. *In proceedings of the 1993 Winter Simulation Conference*, December, 677-686.

Kleinrock, L. and J. Silvester. 1978. Optimum transmission radii for packet radio networks of why six is a magic number. *Nat. Telecom. Conf.*

Misra, J. 1986. Distributed Discrete-Event Simulation. *Computing Surveys*, 18(1), March.

Rappaport, T. S. and S. Y. Seidel. 1990. SIRCIM: Simulation of Indoor Radio Channel Impulse Response Models, *VTIP, Inc.*.

Sanadidi, M. 1993. A Comprehensive Performance Model of a Massively Parallel Processor. *Proceedings of AT&T Database Day*, October.

AUTHOR BIOGRAPHIES

RAJIVE L. BAGRODIA is an Associate Professor in Computer Science at UCLA.

CHING-CHUAN CHIANG is currently a Ph.D candidate in Computer Science at UCLA.

MARIO GERLA is currently a Professor in Computer Science at UCLA.

VIKAS JHA is a research engineer in Computer Science at UCLA.

WINSTON W. LIU is currently a Ph.D. candidate in Computer Science at UCLA.

HSIAO-KUANG WU is currently a Ph.D. candidate in Computer Science at UCLA.

Proceedings of the 1996 Winter Simulation Conference
ed. J. M. Charnes, D. J. Morrice, D. T. Brunner, and J. J. Swain

A CORBA FACILITY FOR NETWORK SIMULATION

Chien-Chung Shen

Bellcore
331 Newman Springs Road
Red Bank, New Jersey 07701, U.S.A.

ABSTRACT

Contemporary software development environments for discrete-event simulation have adopted either a language-based approach or a library-based approach. Although these approaches have advantages such as type checking and optimized code generation provided by the former and familiar programming environments facilitated by the latter, they suffer from the inherent limitations of model portability and interoperability, which may result in inflexible models and higher model development cost. This paper proposes a CORBA discrete-event simulation facility to facilitate portable and interoperable simulation models. The proposed facility is defined by a CORBA IDL interface which defines operations for object definition, inter-object communication and event scheduling. Based on the given IDL interface definition, different vendors could supply different products (at different costs) by using different simulation algorithms, different programming language, or on different operating system or hardware platforms. With respect to the simulation models, they see a consistent interface across all products. The paper presents the simulation facility IDL interface, describes its prototype implementation in C++, and illustrates its usage by a bounded-buffer producer-consumer example.

1 INTRODUCTION

Contemporary software development environments for discrete-event simulation have adopted either a language-based approach or a library-based approach (Martin and Bagrodia 1995). In either case, programmers are provided with a set of model definition primitives together with a set of parallel programming primitives for object definition and inter-object communication and synchronization. These primitives are provided either as language extensions or as functions implemented as library routines. Although these approaches have advantages such as type checking and optimized code generation provided by the former and familiar programming environments facilitated by the latter, they suffer from the following two inherent limitations.

- **Portability.** Simulation models developed using one simulation language might not be easily ported to another simulation language. The programmers will be required to learn new language constructs and perhaps an entirely new set of program development tools.
- **Interoperability.** Components of a simulation model are required to be programmed in the same host language dictated by the simulation language or library routines used. It is, so far, infeasible that a simulation model consists of interoperable components written in different programming languages, or running on different operating systems or hardware platforms.

These limitations may potentially lock simulation application development into a particular (vendor) environment and result in inflexible models and higher model development cost.

However, these restrictions are not inherent in simulation application development alone. In fact, one of the most difficult tasks challenging information industry today is to enable application interworking and construction in a distributed, heterogeneous, and multi-vendor environment.

Confronted with the challenge, the Object Management Group (OMG) was founded in 1989 to develop a set of standards, using *object technology*, to facilitate distributed computing, which guarantee application interoperability and portability. Among them, the *Common Object Request Broker Architecture (CORBA)* (OMG 1993) has emerged as a *de facto* standard for distributed object computing.

At its core, CORBA defines the facilities required to allows a client object to *transparently* invoke the services offered by CORBA-compliant objects (or CORBA objects for short) running on any machine in a heterogeneous distributed environment. The remote CORBA object are available across different operating systems (UNIX, Windows, OS/2, MVS), and different programming languages, such as C++, C, Ada, and Java, with many others to follow.

Each CORBA object has an *interface* that defines the services it offers to its clients, and this interface is defined in an Interface Definition Language (IDL) specified by the CORBA standard. The CORBA IDL is not a programming language and it does not replace the use of programming languages. Instead, the IDL's only role is to define interfaces which consist of operations available to the clients of the interfaces.

The advantage of using CORBA IDL is that it allows a CORBA object to define its interface (services) in a *declarative* fashion, which is independent of the programming language used to implement the object itself, or the programming language used to implement the clients of the object. In particular, the language used to implement CORBA objects need not be the same as that used by clients, and of course, the clients that invoke a given CORBA object need not all be implemented in the same programming language. For example, a client object written in Ada need not be aware that a CORBA object it is invoking is implemented in C++. In this case, the IDL definition of the object interface is translated, through an appropriate IDL compiler, automatically into Ada for the benefit of the client, and into C++ for the benefit of the implementer of the interface.

In this paper, we propose a CORBA discrete-event simulation facility using the message-based approach (Bagrodia, Chandy, and Misra 1987). The facility exports an interface, defined in CORBA IDL, consisting of operations for object definition, inter-object communication and event scheduling, and implements the interface using a discrete-event simulation algorithm. Simulation models utilizing the facility shall first ***bind*** to the facility (as an object) and then invoke the exported operations to obtain runtime support for executing simulation models.

Based on the proposed interface definition, different vendors are encouraged to supply different products (at different costs) by using different simulation algorithms, different programming language, or on different operating system or hardware platforms. However, with respect to the simulation models, they see a consistent interface across all products.

The remainder of the paper is organized as follows. We first present an overview of CORBA in the next section. The concept of message-based discrete-event simulation is introduced in Section 3. Based on that, we define a discrete-event simulation facility and present its CORBA IDL interface definition in Section 4. Section 5 describes its prototype implementation. Application of the facility using a bounded-buffer producer-consumer example is described in Section 6, and Section 7 is the conclusion.

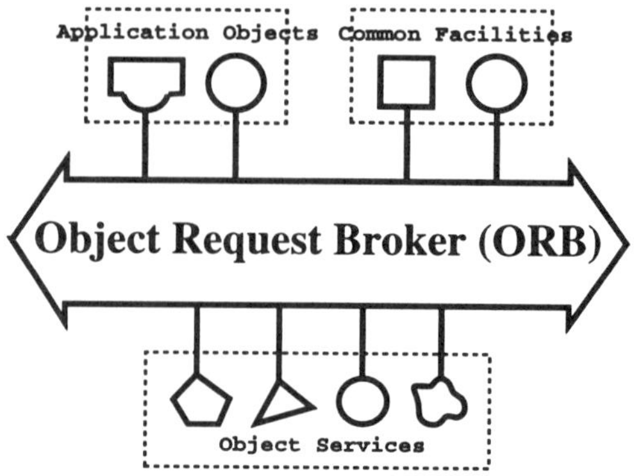

Figure 1: Object Management Architecture (OMA)

2 DISTRIBUTED OBJECT COMPUTING WITH CORBA

To enable distributed application interworking and construction, OMG defines the *Object Management Architecture (OMA)* Reference Model (OMG 1992) which identifies and characterizes the components, interfaces and protocols necessary to guarantee interoperability and portability of distributed applications in heterogeneous environments.

2.1 The OMA Reference Model

The OMA defines a reference model (Figure 1) for distributed object computing. Within the reference model, the most important component is the Object Request Broker (ORB). An ORB provides the basic mechanism for transparently making requests to and receiving responses from objects located locally or remotely without the client needing to be aware of the mechanisms used to communicate with the objects. As such, the ORB forms the foundation for building applications constructed from distributed objects and for interoperability between applications in distributed heterogeneous environments.

Every entity in the reference model is modeled as an object. These objects communicate with each other via the ORB. According to their functionality, objects are categorized into three groups:

- **Object Services** comprise a collection of services (interfaces and objects) that provide basic functions for using and implementing objects.

Examples are naming and event services.

- **Common Facilities** provide a collection of commonly-found services useful in many applications. In contrast to Object Services which are general purpose and application-domain independent, Common Facilities are services which are often application-domain specific and typically provide functionality directly to end-users. The simulation facility is an example.
- **Application Objects** are objects specific to particular end-user applications. Examples are simulation and banking applications.

2.2 The CORBA Architecture

The CORBA specifies a concrete description of the interfaces and services that must be provided by compliant Object Request Brokers. CORBA is composed of the following five major components.

- **ORB Core.** The ORB Core provides the basic communication capability between objects. It supports two different ways in which clients can make requests to objects: *static invocation* via interface-specific stubs and skeletons compiled from IDL interface definitions, and *dynamic invocations* via the Dynamic Invocation Interface. No matter which of these methods is used by a client to make a request, the ORB Core locates the desired object, activates it if it is not already executing, and delivers the request to it. The object performs the requested service and returns any result back to the ORB Core which then returns it to the client.
- **Interface Definition Language (IDL).** In CORBA, object services are defined as *interfaces* in IDL, a language-independent declarative language. IDL supports (multiple) inheritance in which derived interfaces inherit the operations and types defined in their base interfaces.
- **Interface Repository (IR).** The IR provides persistent storage for IDL interface definitions. Its primary function is to provide the interface information necessary to issue requests using Dynamic Invocation Interface.
- **Dynamic Invocation Interface (DII).** Some applications, such as browsers, require the capability of sending requests to objects without having compile-time knowledge of their interface definitions. The DII allows run-time discovery of interfaces from an IR, dynamic creation and invocation of requests to objects.

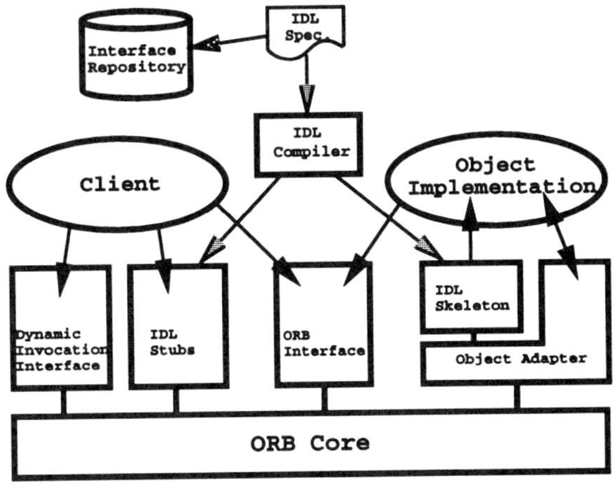

Figure 2: The CORBA Components

- **Object Adaptor (OA).** An object adaptor provides the means by which various types of object implementations can utilize ORB services, such as object method invocation, activation and deactivation of objects and implementations.

Figure 2 shows how the various CORBA components working together to facilitate distributed object computing. We assume that a client wants to invoke a service supported by an object. An IDL definition file is created to describe the interface (services) the object provides, which is stored in the IR as well as compiled into an IDL Stub and an IDL Skeleton. The client can initiate a request by calling the IDL Stub. Alternatively, the client looks up the IR and dynamically creates and invokes a request using DII. In either case, the request is directed to the ORB Core. The ORB Core locates the object implementation and then delivers the request to the OA managing that object. The OA feeds the request into the IDL Skeleton where it is then passed to the object implementation. Any return values are passed back through the IDL Skeleton and OA to the ORB Core. Then, depending on the original call, the ORB Core returns the value either through the IDL Stub or the DII to the client.

3 MESSAGE-BASED DISCRETE-EVENT SIMULATION

In this section, we review the basic concept of message-based discrete-event simulation which provides a more natural paradigm for simulating distributed systems, and therefore better serves as the foundation for the definition of a simulation facility.

In message-based simulation, each physical entity is abstracted by an *logical object* (*lo*), and interactions among the entities, called *events*, are represented by message communications among the corresponding

```
clock = 0;
Initialize event-list;
while (execution not terminated) do
{  fetch next tuple (m_i, s_i, d_i, t_i) from event-list;
   if (m_i is a timeout message) then clock = t_i;
   deliver m_i to d_i for simulation;
}
```

Figure 3: Message-Based Simulation Algorithm

lo. (Symbol *lo* represents both singular and plural forms.) A message-based simulation algorithm uses two data structures (Misra 1986): a *simulation clock* and an *event-list* (Figure 3). The simulation clock gives the time up to which the physical system has been simulated. The event-list is a partial order of tuples; a tuple is represented by (m, s, d, t), where m represents a message, s and d are the source and destination *lo* for m, and t is a timestamp. The partial order is typically based on the timestamp and ensures that events are simulated in the order of their dependencies. At every step of the simulation, the algorithm selects the tuple with the *smallest* timestamp, say (m_i, s_i, d_i, t_i), removes it from the event-list, and delivers m_i to d_i. Multiple tuples with the same timestamp may be handled in an arbitrary order, or be ordered deterministically using transparent sequence numbers to reflect their dependencies. The simulation of m_i by *lo* d_i may generate additional messages which are added to the event-list.

During the execution of a simulation program, the simulation clock advances in a *monotonic non-decreasing* manner through the timestamps associated with each tuple. Note that the simulation clock is completely decoupled from the physical processor clock. The physical time required to simulate a message does not have any effect on the simulation clock. How is the timestamp assigned to a message? When a message is generated, it is timestamped with the current value of the simulation clock — with one exception. We define a special message called a *timeout* message. The timeout message is scheduled by an *lo* for delivery to *itself* at a *future* time and is typically used to simulate the time of a simulation step that would be required by the physical entity to execute the corresponding operational step. An *operational step* refers to the statements executed by a physical entity to process a message received by it, and a *simulation step* models the activities that would be executed by the corresponding operational step. For example, consider a file-handler entity. On receiving a *read* request for the file, a physical (operational) file-handler will read the appropriate record from the file and return it to the requesting entity. If the file-handler is abstracted by a logical object, on receiving a *read* request, the *lo* estimates t, the time required for the corresponding physical entity to read the file and schedules a timeout message to itself t time units later. As the timestamp on all messages other than the timeout message refers to the current value of the simulation clock, the simulation time advances only when a timeout message is delivered to an *lo*.

4 SIMULATION FACILITY

In light of the above description, we define the `DESFacility` interface (Figure 4), in CORBA IDL, for message-based discrete-event simulation, which conists of operations for object definition, inter-object communication and event scheduling.

```
typedef string  ObjName;
typedef string  MsgType;
typedef any     MsgContents;
typedef long    Time;
typedef boolean BGuard;

struct WCItem {
  MsgType     msg_type;
  BGuard      bguard;
  Time        duration;
};
typedef sequence<WCItem> WaitCondition;

struct Message {
  ObjName     source;
  ObjName     sink;
  MsgType     msg_type;
  MsgContents contents;
};

interface DESFacility {
  void role(in ObjName myself);
  void enroll(in ObjName myself);
  void resign(in ObjName myself);
  void send(in Message msg);
  void receive(out Message msg,
               in  ObjName myself);
  void waituntil(out Message       msg,
                 in  ObjName       myself,
                 in  WaitCondition wc);
  void hold(in Time tm, in ObjName myself);
  Time now();
};
```

Figure 4: Interface Definition (DESFacility.idl)

Object Definition. Operations `role`, `enroll`, and `resign` are used to define participating simulation objects, and to demarcate their beginning and ending of execution.

Inter-Object Communication. Objects communicate with each other using buffered message-passing (asynchronous communication), where every object has a unique message-buffer. An object sends a message to another by invoking a `send` operation. Each message, with structure defined by `Message`, contains sender, receiver, message type, and message contents information. A message is deposited in the message-buffer of its destination object on the invocation of a `send` operation, and carries a timestamp which corresponds to the simulation time at which the corresponding `send` operation is invoked. An object accepts messages from its message buffer by invoking a `waituntil` operation. The `waituntil` operation takes a sequence of wait-condition items as an input argument to determine which message will be accepted by the object. Each wait-condition item specifies a message-type, say m_i and a boolean value, say b_i, which is said to be *enabled* if the message buffer contains a message of type m_i and b_i is true, and m_i is called an *enabling* message. The enabling message with the earliest timestamp is removed and delivered to the object. If all wait-condition items are disabled, the object is suspended for a *maximum* duration of *simulation* time equal to `duration` of a wait-condition item with `msg_type` equal to `timeout`. If omitted, a default wait-condition item with `msg_type` equal to `timeout` is set with an arbitrarily large `duration` value. If no enabling message is received in the `duration` interval, the object is sent a `timeout` message. An object must accept a `timeout` message that is sent to it. A separate `receive` operation is also provided to accept the message with the earliest timestamp from the message-buffer, regardless of its message type. It will block the invoking object until a message is available.

Event Scheduling. The `hold` operation enables the invoking object to schedule a timeout message for delivery to itself `tm` time units from now. It is used to specify the simulation time used by a simulation step. The `now` operation lets the invoking object to read the current value of the simulation clock.

5 PROTOTYPE IMPLEMENTATION

In this section, we describe a C++ implementation of the simulation facility in a multi-threaded version of the commercial ORB Orbix (IONA 1995). (Orbix is a Registered Trademark of IONA Technologies Ltd.)

We adopt the *wait-until* simulation algorithm

```
clock = 0;
Initialize event-list;
while (execution not terminated) do
{  fetch next tuple (m_i, s_i, d_i, t_i) from event-list;
   if (m_i is not accepted by d_i) then
       store m_i in temp_queue;
   else
   {  if (m_i is a timeout message) then clock = t_i;
      deliver m_i to d_i for simulation;
      merge temp_queue with event-list;
   }
}
```

Figure 5: Wait-Until Simulation Algorithm

(Franta 1977), as shown in Figure 5, to implement the simulation engine, in which an object may delay acceptance of a message based on its state such that messages are not necessarily delivered in the partial order specified by the event-list. Each *lo* may specify a wait-condition which restricts the types of message that it is willing to receive; a message is delivered to the destination *lo* only if it satisfies its wait-condition. The simulation clock is advanced through the timestamps associated with the timeout messages in the event-list.

In the C++ implementation, the `DESFacility` interface is implemented as the `DESFacility_i` C++ object (Figure 6) exporting the defined operations. Upon receiving an invocation for operation `send`, the `DESFacility_i` object inserts the message into the event list according to increasing timestamp order. On receiving an invocation for operations `waituntil` and `receive`, a thread is created implicitly to check if there is any enabling message. It will block the invoking object in case there is no enabling message. Finally, upon receiving an invocation for operation `hold`, a thread is created implicitly to first deposit a timeout message in the event list. It then block until the scheduled timeout message becomes deliverable.

As simulation facility is defined in CORBA IDL, it does not dictate its implementation. The simulation facility could very well be implemented using a parallel simulation algorithm to take advantage of parallel processing environment. With respect to its clients, the implementation is totally transparent, except for potential performance improvement.

6 AN EXAMPLE

As an example, we consider the simulation of a *bounded-buffer producer-consumer* system (Bagrodia and Shen 1991). The consumer object repeatedly re-

```
#include "DESFacility.hh"

struct Event {
  Message       msg;
  Time          tstamp;
  struct Event  *lst;
  struct Event  *nxt;
};

class DESFacility_i :
      public virtual DESFacilityBOAImpl
{
private:
  void enqueue(const struct Message& msg,
               Time tstamp);
public:
  Time          sim_clock;  // sim clock
  struct Event  eventList;  // event list
  DESFacility_i(char *marker); // constructor
  virtual ~DESFacility_i();    // destructor
  virtual void role(const ObjName myself);
  virtual void enroll(const ObjName myself);
  virtual void resign(const ObjName myself);
  virtual void send(const Message& msg);
  virtual void receive(Message& msg,
                       const ObjName myself);
  virtual void waituntil(Message& msg,
                     const ObjName myself,
                     const WaitCondition& wc)
  virtual void hold(Time tm,
                    const ObjName myself);
  virtual Time now();
};
```

Figure 6: DESFacility C++ Implementation Class

quests a data item from the buffer using a `request` message, waits to receive a `data` message, and invokes a `hold` operation to simulate data consumption. The producer object waits to receive a `free` message from the buffer which indicates that the buffer has a free slot, invokes a `hold` operation to simulate the generation of a data item, and sends the item to the buffer via a `data` message. The buffer object repeatedly invokes a `waituntil` operation that accepts a `request` message only when it is not empty and a `data` if it is not full.

Figures 7, 8, and 9 list the C++ source code for the producer, consumer, and buffer object, respectively, and Figure 10 depicts their interaction with the `DESFacility` object to obtain the discrete-event simulation service.

```
main (int argc, char **argv)
{
  DESFacility  *p;
  Message      msg_s, msg_r;
  int i, n, now, t_wait;
  p = DESFacility::_bind("sim:simSrv", "");
  p->enroll("producer");
  now = t_wait = 0;  n = Q_LENGTH;
  msg_s.source   = new char[20];
  strcpy(msg_s.source, "producer");
  msg_s.sink     = new char[20];
  strcpy(msg_s.sink,   "buffer");
  msg_s.msg_type = new char[20];
  strcpy(msg_s.msg_type, "data");
  for (i = 0; i < MAX_ITEM; i++) {
    if (n == 0) {
      now = p->now();
      p->receive(msg_r, "producer");
      n++;  t_wait += p->now() - now;
    }
    p->hold(rand()%MAXPRODUCE, "producer");
    n--;  p->send(msg_s);
  }
  now = p->now();
  printf("Producer utilization == %f\n",
       ((float)(now - t_wait)/(float) now));
  p->resign("producer");  p->_release();
}
```

Figure 7: The Producer Code

7 CONCLUSION

At the time when the computing industry is demanding portable and interoperable distributed applications, of which simulation is a notable example, CORBA has emerged as a *de facto* standard for distributed object computing. A CORBA-compliant object request broker serves as a *software bus* to facilitate communication among distributed objects, independently of their languages, operating systems, or hardware platforms. In addition, the OMA has depicted a reference model where application-specific services are provided as Common Facilities.

The paper describes a CORBA discrete-event simulation facility. The facility exports an interface, defined in CORBA IDL, consisting of model definition operations, and implements the interface using a discrete-event simulation algorithm.

The main thrust of the paper is not to propose a new simulation language or a more efficient parallel simulation algorithm, but to introduce a new paradigm for facilitating discrete-event simulation us-

```
main (int argc, char **argv)
{
  DESFacility *c;
  Message     msg_s, msg_r;
  int i, t, t_used, now;
  c = DESFacility::_bind("sim:simSrv", "");
  c->enroll("consumer");
  now = t_used = 0;
  msg_s.source   = new char[20];
  strcpy(msg_s.source, "consumer");
  msg_s.sink     = new char[20];
  strcpy(msg_s.source, "buffer");
  msg_s.msg_type = new char[20];
  strcpy(msg_s.source, "request");
  for (i = 0; i < MAX_ITEM; i++) {
    c->send(msg_s);
    c->receive(msg_r, "consumer");
    t_used += (t = rand() % MAXCONSUME);
    c->hold(t, "consumer");
  }
  now = c->now();
  printf("Consumer utilization == %f\n",
         ((float)(t_used) / (float)now));
  c->resign("consumer");  c->_release();
}
```

Figure 8: The Consumer Code

ing the newly emerged CORBA technology and to motivate more research activities and product development in this direction.

ACKNOWLEDGMENTS

This research was funded by the Army Research Laboratory under Cooperative Agreement No. DAAL01-96-2-0002."

REFERENCES

Bagrodia, R. L., K. M. Chandy, and J. Misra. 1987. A message-based approach to discrete-event simulation. *IEEE Transactions on Software Engineering* 13(6):654–665.

Bagrodia, R. L. and C.-C. Shen. 1991. MIDAS: Integrated design and simulation of distributed systems. *IEEE Transactions on Software Engineering* 17(10):1042–1058.

Franta, W. R. 1977. *The process view of simulation.* New York: Elsevier North-Holland Inc.

IONA. 1995. *Orbix: Programmer's Guide.* IONA Technologies Ltd.

Martin, J. M. and R. L. Bagrodia. 1995. COMPOSE: An object-oriented environment for parallel discrete-event simulation. In *Proceedings of the 1995 Winter Simulation Conference*. 163–166, Washington, DC.

Misra, J. 1986. Distributed discrete-event simulation. *Computing Survey* 18(1):39–65.

OMG. 1992. *Object Management Architecture Guide.* Object Management Group and X/Open.

OMG. 1993. *The Common Object Request Broker: Architecture and Specification.* Object Management Group and X/Open.

AUTHOR BIOGRAPHY

CHIEN-CHUNG SHEN is a Member of Technical Staff in Bellcore Applied Research. He received B.S. and M.S. degrees in computer science from National Chiao Tung University, Taiwan, in 1982 and 1984, respectively, and a Ph.D. degree in computer science from UCLA in 1992. His research interests include distributed network control and management for SONET, ATM, WDM and wireless networks, distributed object computing, and network simulation.

```
main (int argc, char **argv)
{
  DESFacility *b;
  Message     msg_s_p, msg_s_c, msg_r;
  WaitCondition wc;
  int q, i, v, tstart, tend;
  b = DESFacility::_bind("sim:simSrv", "");
  b->enroll("buffer");  tstart = b->now();
  msg_s_p.source   = new char[20];
  strcpy(msg_s_p.source, "buffer");
  msg_s_p.sink     = new char[20];
  strcpy(msg_s_p.sink, "producer");
  msg_s_p.msg_type = new char[20];
  strcpy(msg_s_p.msg_type, "free");
  msg_s_c.source   = new char[20];
  strcpy(msg_s_c.source, "buffer");
  msg_s_c.sink     = new char[20];
  strcpy(msg_s_c.sink, "consumer");
  msg_s_c.msg_type = new char[20];
  strcpy(msg_s_c.msg_type, "data");
  wc._length = 2;  wc._maximum = 2;
  wc._buffer = new WCItem[2];
  wc._buffer[0].msg_type = new char[20];
  strcpy(wc._buffer[0].msg_type, "data");
  wc._buffer[1].msg_type = new char[20];
  strcpy(wc._buffer[1].msg_type, "request");
  for (i = q = v = 0;;) {
    wc._buffer[0].bguard = (q < Q_LENGTH);
    wc._buffer[1].bguard = (q > 0);
    b->waituntil(msg_r, "buffer", wc);
    if (strcmp(msg_r.msg_type, "data") == 0) {
      tend = b->now();  v += q*(tend-tstart);
      tstart = tend;  q++;  break;
    } else {          // "request"
      b->send(msg_s_c);  tend = b->now();
      v += q*(tend-tstart); tstart=tend; q--;
      if ((++i) < MAX_ITEM)
        b->send(msg_s_p);
      else
        break;
    }
  }
  printf("The avg # of items in buf: %f\n",
         (float) v / (float) b->now());
  b->resign("buffer");  b->_release();
}
```

Figure 9: The Buffer Code

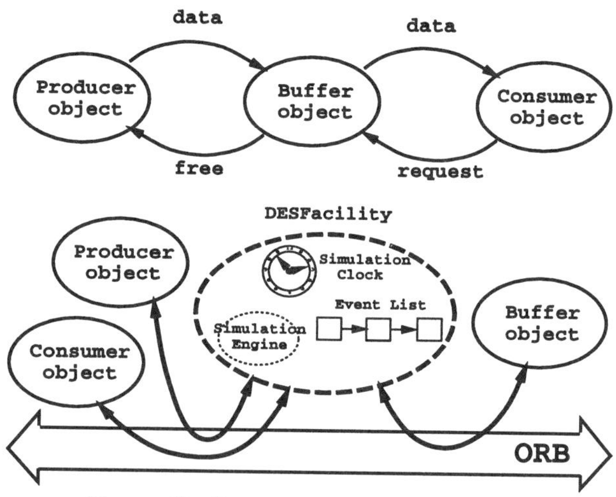

Figure 10: Prototype Implementation

Proceedings of the 1996 Winter Simulation Conference
ed. J. M. Charnes, D. J. Morrice, D. T. Brunner, and J. J. Swain

PERFORMANCE COMPARISON OF HIGH LEVEL ALGEBRAIC NETS DISTRIBUTED SIMULATION PROTOCOLS

Karim Djemame
Mohamed Bettaz

Computing Science Institute
University of Constantine
Constantine 25000, ALGERIA

Dennis C.Gilles
Lewis M.Mackenzie

Computing Science Department
University of Glasgow
Glasgow G12 8QQ, Scotland, UK

ABSTRACT

This paper addresses the problem of developing distributed simulation techniques to analyze ECATNets. ECATNets (Extended Concurrent Algebraic Term Nets) are a kind of High-Level Algebraic Nets used for specifying various aspects of distributed and parallel systems. Their most distinctive feature is that their semantics is defined in terms of rewriting logic. The conservative and optimistic approaches of Distributed Discrete Event Simulation (DDES) are used as the starting point to discuss a simulation framework for studying the behaviour of ECATNet models. The ECATNet model to be simulated is partitioned into several connected subnets. The various subnets are simulated in parallel by several Logical Processes. Next we develop two distributed simulation protocols to execute discrete-event simulations of ECATNets.

1 INTRODUCTION

Petri nets (Murata 1989) are an important graphical and mathematical tool applicable to many systems. They are a promising tool for describing and studying information processing systems that are characterized as being concurrent, asynchronous, distributed, parallel, nondeterministic and/or stochastic. ECATNets are a kind of High-Level Algebraic Nets. They are proposed as a way for specification, modeling and validation of applications from the area of communication networks, computer designs and other complex systems (Bettaz et al. 1993a, Bettaz et al. 1993b). They are built around a combination of three formalisms. The first two formalisms constitute a net/data model, and are used for defining the syntax of the system, in other terms to capture its structure. The net model, which is a kind of advanced Petri net (Jensen and Rozenberg 1991), is used to describe the process architecture of the system; the data model, which is an algebraic formalism (Ehrig and Mahr 1985), is used for specifying the data structures of the system. The third formalism, which is a rewriting logic (Meseguer 1992), is used for defining the semantics of the system, or in other words to describe its behaviour. According to this logic, the system behaviour may be explained by formal reasoning. Transforming our logic into a rewriting system (Bettaz and Mehemmel 1993) may be used for rapid prototyping and automatic proving of a system under design. However, the achieved models had two drawbacks: the occultation of the problem of time and a bad exploitation of the parallelism inherent to the studied models. In a previous paper (Djemame and Bettaz 1995), we presented an approach for the treatment of time in ECATNets, by introducing in each transition a firing delay to perform the operations "remove/deposit tokens". Simulation of these models can not only perform their validation, but can evaluate their performances as well. Simulation can also be amenable to parallel execution in order to exploit the inherent parallelism of these models. It is worth mentioning that we are dealing with the parallelism at two levels: the inter-module level, where the parallelism is achieved by partitioning the "initial" models w.r.t. a "separation of concern" strategy; the intra-module level, where the detection of the parallelism is permitted by the use of our rewriting logic.

In Distributed Discrete Event Simulation (DDES), the system being modeled is viewed as being composed of some number of Physical Processes (PPs) that interact at various points in simulated time. The simulator is constructed as a set of Logical Processes (or simply LPs), one per Physical Process. Logical Processes exchange timestamped event messages to interact. However, relationship between events may exist, so concurrent execution of these events must be synchronized, otherwise *causality errors* can occur. Distributed Discrete Event Simulation falls into two categories (Fujimoto 1990): *conservative* and *optimistic*. In the following, we consider synchronization protocols based on the Chandy-Misra scheme (CM) described by (Misra 1986) (conservative), and on Time Warp described by (Jefferson 1985) (optimistic, based on the *virtual time* paradigm). The conservative mechanisms strictly avoid the possibility of any causality error ever occurring by forcing LPs to block as long as there is the possibility for receiving messages with lower timestamp. The optimistic mechanisms use a detection and recovery approach, this means that causality errors are detected and a rollback mechanism is invoked to recover. Recently, (Chiola and Ferscha 1995) proposed a new DDES protocol called *probabilistic*, a performance efficient compromise between the two classical approaches.

The contributions of distributed simulation in the area of Petri nets and reported in the literature include (Ammar and Deng 1991, Nicol and Roy 1991, Thomas and Zahorjan 1991, Chiola and Ferscha 1993). They all deal with Timed and/or Stochastic nets. In (Chiola and Ferscha 1993), Petri net structural analysis is exploited for the efficient implementation of DDES techniques using both approaches: *conservative* and *optimistic*. To the best of our knowledge, no attention has been given to high-level Petri nets distributed simulation. We are only aware of the work in progress by Schöf on parallel simulation of THOR nets (Timed Hierarchical Object-Related Nets) (Schöf et al. 1995). It is worth mentioning that concerning simulation techniques for high-level nets with arc inscriptions, the enabling test and the firing operations are substantially more complex.

This paper is organized in the following way: in section 2, we review some basic notions about ECATNets, and illustrate their use by a complete example. Section 3 describes synchronization protocols based on the conservative and optimistic approaches for ECATNets distributed simulation. Section 4 presents some empirical results derived from the implementation of DDES strategies for ECATNet models. Some concluding remarks and perspectives of the research for future developments are given in section 5.

2 ALGEBRAIC TERM NETS

2.1 Basic Concepts

The graphical representation of a generic ECATNet is given by Figure 1. **IC**, **DT** and **CT** are multisets of (equivalence classes of) terms. The terms are defined by an algebraic specification of an abstract data type given by the user. **TC** (Transition Condition) is a boolean expression which may contain variables occurring in IC (Input Condition), DT (Destroyed Tokens) and CT (Created Tokens). Each place is associated with a capacity **C(p)** defined as a multiset of closed (equivalence classes of) terms. The marking **M(p)** of a place p of the net, which is itself a multiset of closed

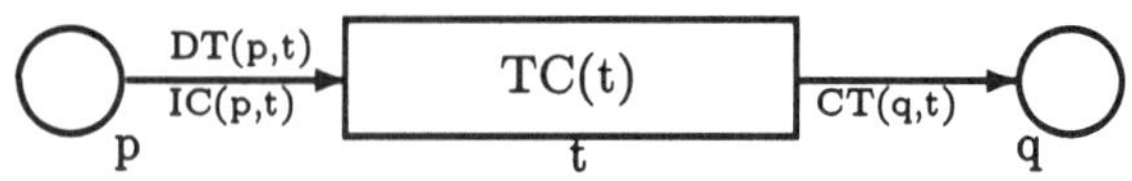

Figure 1: A Generic ECATNet

terms, is defined w.r.t. the capacity (which may be infinite).

A transition t is fireable when various conditions are simultaneously true. The first condition is that every IC(p,t) for each input place p is enabled. The second condition is that TC(t) is true. Finally the addition of CT(q,t) to each output place q must not result in q exceeding its capacity when this capacity is finite. When t is fired, DT(p,t) is removed from the input place p and simultaneously CT(q,t) is added to the output place q. Transition firing and its conditions are expressed by rewrite rules which are strongly depending on the form of the syntactic notation used for representing IC. Those rewrite rules (metarules) together with a set of deduction rules define a rewriting logic which gives the semantics of the net. They act as a parallelizing compiler which tries to find sequences of "code" which may be executed in parallel. Examples on concrete instantiations and practical use of our metarules may be found in (Bettaz et al. 1992, Bettaz 1993, Bettaz and Maouche 1993, Bettaz et al. 1994).

2.2 Aspect of Time

In (Djemame and Bettaz 1995) we have chosen to introduce in each transition a (marking) related rate. This will lead to take into account a firing delay "d" to perform the operations "remove/deposit tokens". This choice allows to preserve the incremental approach used for defining ECATNets and preserves the semantic framework defined in terms of rewriting logic. Thus, *firing times* as defined for Timed Petri nets are offered leading to modeling activities duration. Note that there will be two kinds of transitions. *Timed* transitions are specified by annotating each timed transition t_i with a firing rate λ_i. Let $\lambda : \mathrm{T} \rightarrow \mathcal{R}$ assigns firing delays λ_i to T-elements $t_i \in \mathrm{T}$. Zero delays are associated with transitions that are called *immediate*. ECATNets enriched with temporal specification are then suitable to discrete simulation. This is an important step in their quantitative performance evaluation.

2.3 An Example of Modeling with ECATNets

This example, borrowed from (Bettaz 1993), deals with the behaviour of the Ethernet transmitting station. It comprises four modules, each module is specified by an ECATNet model. The first module deals with the functions of formatting and transmitting starting. The second module is relative to the functions of transmission with success and acknowledgement. The third module is relative to the retransmission function. The fourth module treats essentially the functions of collision handling and acknowledgement (Figures 2-5).

The transmitter station transmits one frame at a time. The user is not allowed to request the transmission of a new frame before receiving the ackowledgement of the previous one. The formatting function starts when a token of type "d,s,data" is deposited in place FROM_USER. This token is considered as a primitive transferred from the user layer to the MAC layer for requesting the transmission of data "data", from a source "s" to a destination "d". The frame "d.s.data.fcs" is then deposited as soon as it is composed in a transmission register (TRANS_REG). The formatting function is consisting of the concatenation of sequences of bits corresponding to the addresses "d" and "s", to the data "data", and to the error control sequence "fcs" previously computed. On the other hand, the MAC layer is listening to the medium (CARRIER_SENSE) in order to avoid any collision occurring with a current transmission. The place CARRIER_SENSE is an interface between the MAC layer and the physical layer. When the medium becomes free (a token "false" is present in place CARRIER_SENSE), it waits a certain amount of time corresponding to the inter-frame spacing delay. Then, considering that the transmission may terminate with success (deposit of a token "false" in place BUSY_CHANNEL and a token "true" in place SUC_TRANS), it takes possession of the medium (CHANNEL_ACCESS) and the transmission starts (deposit of a token "true" in INIT_TRANS).

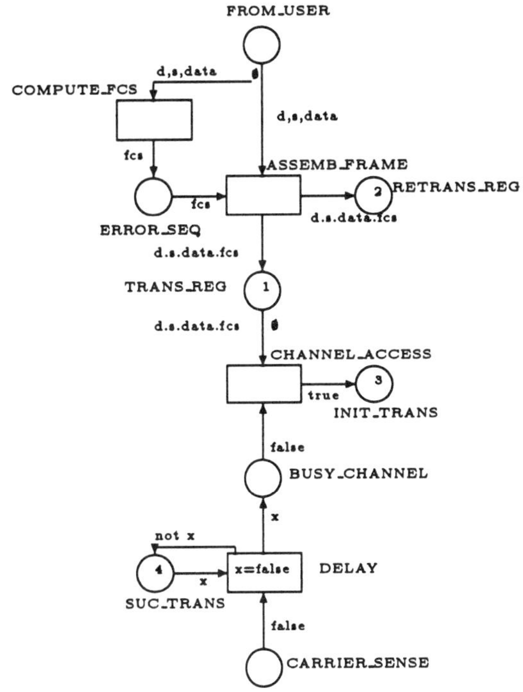

Figure 2: Starting of Transmission Module

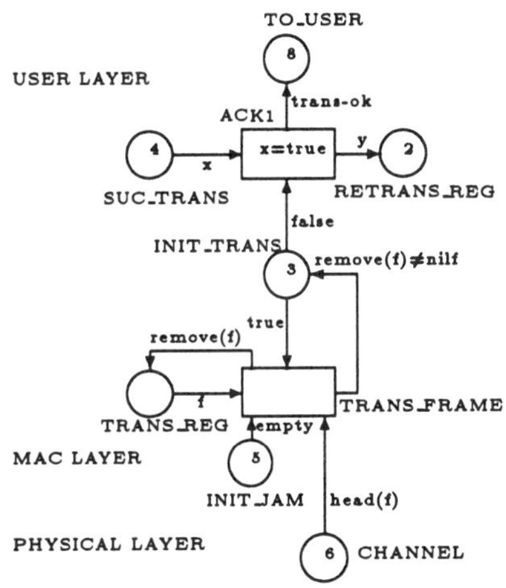

Figure 3: Transmission With Success Module

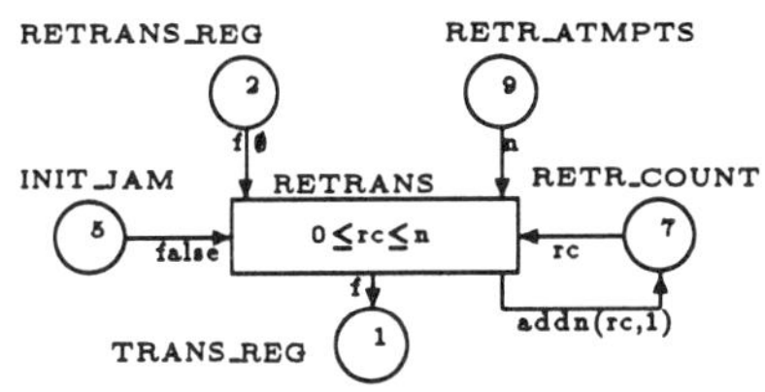

Figure 4: Retransmission Module

3 A DISTRIBUTED SIMULATION FRAMEWORK FOR ECATNETS

For each LP, we identify three components: the work partition assigned to it, its communication behaviour and its simulation engine are the constituent parts of the distributed simulation framework.

3.1 Net Partitioning Performance Impact

Starting from the "separation of concern" strategy and according to the conservative and the optimistic mechanisms, details about the protocols developed using such partitioning are found in (Djemame and Bettaz

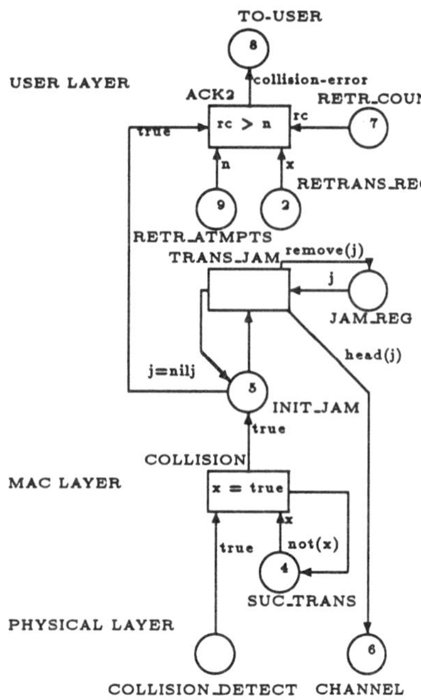

Figure 5: Collision Handling Module

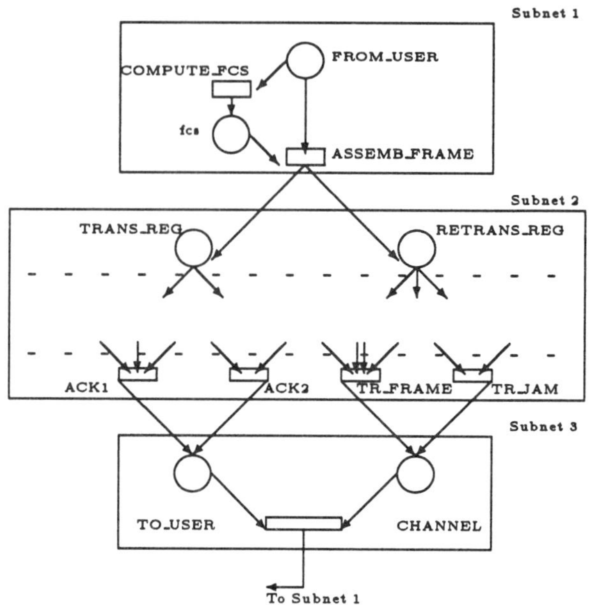

Figure 6: ECATNet Partition

1995). However, it has been seen that for performance reasons the generality of a partition should be limited in such a way that conflicting transitions together with all their input places should always reside in the same LP (Nicol and Roy 1991, Chiola and Ferscha 1993). Such partitioning avoids the implementation of a distributed conflict resolution algorithm and minimises overhead.

Separation of Subnets The partitioning has to be related to the firing rule : an LP should be a set of transitions along with their input places such that local information is sufficient to decide upon the enabling and firing of any transition. If P, T and A are respectively the set of places, the set of transitions and the set of arcs of the "intial" ECATNet model, the partition is a set of n subnets such that:
$ECATNet_i = (P_i, T_i, A_i, \lambda_i)$ where $\cup P_i$ = P, $\cup T_i$ = T, $A_i \subset (P_i \times T_i) \cup (T_i \times P_i)$, i=1..n.

λ_i is the set of rates of local transitions in T_i. We did not consider the multisets of terms IC, DT, CT and the boolean expression TC. They are defined as in the "initial" model and appear in the graphical representation of the subnets.

Definitions A place $p_i \in P_i$ in LP_i is said to be a member of the set of *input places* (IP_i) of LP_i if there exists a transition $t_j \notin LP_i$ which p_i is an output place. A transition $t_i \in T_i$ in LP_i is said to be a member of the set of *output transitions* (OT_i) of LP_i if there exists a place $p_j \notin LP_i$ which t_i is an input transition. A *communication arc* is an arc connecting a place (transition) in LP_i to a transition (place) in LP_j.
For each *output transition* t in subnet i, define :
$P_{t_{out}}$ = list of places to indicate which (input) places are related to t and the kind of relation that exists (CT).

Example Let's take the example of section 2.3, the modular specification of the Ethernet transmitting station. The partition is a set of three subnets, each subnet is simulated by an LP. Places and transitions are partitioned among the subnets as follows (Figure 6):
Subnet 1: 2 places, 2 transitions;
Subnet 2: 11 places, 8 transitions;
Subnet 3: 2 places, 1 transition;

TRANS_REG and TO_USER are input places in $subnet_2$ and $subnet_3$ respectively. In $subnet_1$, ASSEMB_FRAME is an output transition. (ASSEMB_FRAME, TRANS_REG), (ASSEMB_FRAME, RETRANS_REG), (TRANS_FRAME, CHANNEL) are communication arcs.

3.2 The Communication Interface

The decomposition of ECATNets requires an interface among ECATNet partitions preserving the behavioural semantics of the whole ECATNet. Such an interface has to be implemented by an appropriate protocol among the partitions according to the simulation strategy. If there is a communication arc from an *output transition* t to an *input place* p, there exists a unidirectional channel between them. This is motivated by (output) transitions which have to interact with their (input) places for sending tokens when these transitions fire. We can map the set of arcs ($T_k \times P_l$) interconnecting different subnets to the channels of the communication interface. We define I_k = ($CHANNELS$,m) of $subnet_k$ to be the communication interface with $CHANNELS = \bigcup_{i,j} ch_{i,j}$ where $ch_{i,j}$ = (LP_i,LP_j) is a set of directed channel from LP_i to LP_j corresponding to the arcs $(t_k,p_l) \in (T_k \times P_l)$ carrying messages of type m.

3.3 Chandy-Misra's Strategy

The simulation engine implements the simulation strategy. The simulation of events is performed in *virtual time* according to their causality. The data structures according to the conservative approach, a local virtual time (LVT) (representing an accumulated value of firing time in LP_i), a list of events (EVL) ordered by time of occurrence, input queues (IQ) (one queue per each input channel, which collects recently arrived messages ordered by time), and output queues (OQ) (one queue per output channel, which keeps messages to send, ordered by time) have to be maintained.

3.3.1 Lookahead Computation

Loohahead is required in conservative mechanisms to avoid deadlock situation (Misra 1986). It is the process' ability to predict what will happen, or more importantly, what will not happen in the simulated time as regards to its behaviour and when next it may affect other processes. It prevents incorrect computations from propagating too far ahead into the simulated time. Lookahead provides a "window" such that all events with timestamps in this window can be executed safely and without further communication between LPs. To highlight lookahead that exists in Petri nets simulation, if a transition starts firing at $Tsim$, an LP can predict exactly when the tokens generated by this firing are deposited ($Tsim+d$, where d is the firing delay associated with t). Lookahead comes directly from the ECATNet structure, and is the increment firing time of succeeding timed transitions in the LP plus the firing time of the transition of the output border of the LP. The value of lookahead can be established for a pair of transitions in each subnet by a static analysis of the subnet's structure.

3.3.2 Types of Messages

The form of the messages exchanged between LPs is (Type_message, source, destination, timestamp, type_token). *Source* and *destination* are either a place (in LP_i) or a transition (in LP_j). Type_token is the abstract data type of the token moved among subnets. The causality of events is preserved over all LPs by sending timestamped token messages of type **Tokens_Deposited**(t,p,TT,CT) in non-decreasing order. This message is carrying Created Tokens when t in LP_i fires leading to a deposit of tokens in place p in LP_j. **Tokens_Deposited** (t,p,TT, Null) is a Null message which is sent for synchronization purpose. It is a promise not to send a new message timestamped earlier than TT.

3.3.3 LP's Behaviour

The conservative approach allows only the processing of safe events, firing of transitions up to LVT for which the LP has been guaranteed not to receive messages with smaller timestamps. The behaviour of the conservative simulator is to process the first event of EVL if there is no token message in one of the IQ_is with smaller timestamp, or to process the token message with the minimum token time in IQs. Each input queue IQ_i has a clock CC_i associated with it that is equal to either the timestamp of the message at the head of the queue if the queue contains a message, or the timestamp of the last received message if the queue is empty. The LP blocks as soon as the minimum timestamp of messages in IQs is not larger than the occurrence time of the first event in EVL (if IQ_i becomes empty, the value of CC_i is changed to 0). The firing of a transition t is as follows: if t $\in$ (OT_i) in LP_i, then a bf Tokens_Deposited message is generated and inserted in the corresponding output queue (OQ). If t has an output place in LP_i, it schedules an event **End_Firing** of t. A Null message is also deposited for every output border transition in the corresponding OQ.

3.4 Time Warp Strategy

In order to simulate an ECATNet partition, the data structures of an LP we maintain with an optimitic simulator according to Time Warp are an input queue (IQ) which collects recently arrived messages (positive and negative), an output queue (OQP) which contains the *positive* messages to send, an output queue (OQN) which contains the *negative* copies of the messages recently sent (*antimessages* for unsending the originals), and an event stack (ES) which records all state variables such that a past state can be reconstructed in case of a rollback. The form of the entries is (t_i,LVT,M) where t_i is the transition that has fired at time LVT yielding a new marking M.

3.4.1 Types of Messages

Tokens_Deposited(t,p,TT,CT(t,p)) is a message carrying Created Tokens when t $\in OT_i$ fires leading to a deposit of tokens in place p in LP_j. The timestamp of this message is the accumulated firing time of transition t. **Tokens_Cancelled** is used in the rollback mechanism needed for synchonization. In this message, TT indicates which message should be cancelled.

The simulation engine's main task is not only to synchronize the LPs simulating the various subnets by controlling the timestamp (TT) of each message and the local virtual time (LVT), but also to implement the functions of the communication arcs. An LP processes messages in the input queue (IQ) by checking the sign (positive or negative) and the timestamp of each one (these messages are ordered by TT, the head of the queue corresponds to the smallest TT). Messages with timestamp > LVT are inserted in IQ. In case of a positive message, it is inserted in timestamp order, otherwise (the sign is negative) it annihilates the positive message in IQ previously sent. If the message is a *straggler* (timestamp of the message < LVT), the LP must roll back and restore a valid state. As for the conservative simulator, the processing of the first event in EVL or the first message in IQ generates either new events in EVL or output messages.

3.4.2 Message Cancellation

When an LP rolls back, it first inserts the straggler message into IQ and updates LVT. The state at (new) time LVT is restored. If rollback is applied with *aggressive cancellation*, all messages in OQN with token time > LVT are annihilated by removing them from OQN and sending them. All incorrect computation is undone by poping out all the records prematurely pushed in ES. The simulator can also apply *lazy cancellation*. In the case reevaluation yields exactly the same positive messages as already sent before, the new positive message is not resent. This will prevent unnecessary message transfers as well as possibly new rollbacks in other LPs.

3.4.3 Global Virtual Time

The *Global Virtual Time* (GVT) is considered as the virtual clock for the system as a whole. The knowledge of (GVT) reduces past state savings in ES. Any message in an input or output queue whose virtual time is < GVT can be discarded.

4 SIMULATION RESULTS

In the absence of a parallel computer, our parallel programs have been implemented in a network of (Sun Sparc) workstations. All our code is written in C plus MPI (Message Passing Interface) (MPI 1995). We have decided to run our initial tests for distributed simulation on discrete-event ECATNet model of the Ethernet transmitting station. The simulation has been running for 10,000 simulated time units. We have considered the simulator's implementations of the ECATNet partitioning in a parametrization: $\lambda_{DELAY} = 1.0$, probability of occurrence of a collision = 0.5, and N = 1, 2, 3, 4 processors respectively. Process 0 is a special process whose task is to start the simulation by assigning LP_i, i=1..3 to a dedicated processor and to compute GVT.

We faced two primary problems with the conservative approach. The first problem is related to cyclic models. A Null message sent out by one Logical Process could possibly circulate through a series of other Logical Processes and arrive back at the original sender at the time it was sent (eg. a Null message generated after transition DELAY in LP_2 fires). In some cases, exponential firing times are used, and because these have a minimum delay of zero, models must be modified for use on distributed simulation. To cope with this situation, a firing transition identifier was introduced in each Null message generated by a timed transition. If the Null message arrives back at the original sender at the time it was sent, it is simply discarded. Deadlock is avoided because there are no cycles in which the collective timestamp increment of messages traversing these cycles is 0. The second problem is related to the number of Null messages exchanged between LPs. Since a large number of transitions in LP_2 are immediate, there is no need to generate new Null messages when these transitions fire if there is no timed transition among the succeeding transitions up to the output border, i.e the accumulated firing time is 0, and this does not change lookahead.

With Time Warp, we faced the situation where a straggler positive message changes the marking in LP_2 but does not cause the enabling of any new event in the past. In such case, LP_2 does not have to rollback. The simplified mechanism which has been used to recover was an appropriate insertion of firings made on ES, and the top of ES was copied considering a potential change in the marking.

The results show that communication time between LPs is quite important. LP_1 and LP_3 contain one single input transition in the output border, whereas LP_2 contains four. LP_2 is the process with the largest event processing time because of its large number of transitions (8) and places (11) leading to an important number of events to schedule. The performances of the conservative simulator (Chandy-Misra's approach with deadlock avoidance) and the optimistic simulator

Table 1: LPs Execution Profiles (CM Approach, Number of Processors = 4)

Logical Process	Simulation Steps	Positve Messages	Null Messages	Time Evt. Processing	Time Communication	Time Blocking	Time Term. Protocol
1	20001	20002	10001	11.442	347.310	290.862	0.010
2	75000	20000	10001	66.773	282.757	213.071	0.027
3	20000	10000	10001	03.496	356.383	326.377	0.010

Table 2: LP_2 Execution Profile (CM Approach)

Number Proc.	Simulation Steps	Positve Messages	Null Messages	Time Evt. Processing	Time Communication	Time Blocking	Time Termin. Protocol
1	75000	20000	10001	79.703	819.843	162.407	1.067
2	75000	20000	10001	49.680	378.169	91.005	0.089
3	75000	20000	10001	83.855	281.944	200.557	0.410
4	75000	20000	10001	66.773	282.757	213.071	0.027

Table 3: LP_2 Execution Profile (TW, Lazy Cancellation)

Number Proc.	Simul. Steps	Number Rollbacks	Nb.Messages Pos./Neg.	Time Evt. Process.	Time Communic.	Time Rollback	Time Blocking	Time Term. Protocol
1	100001	5001	29997	123.722	861.796	2.385	0.461	1.008
2	85074	9950	20004	68.803	318.494	2.389	110.952	0.049
3	85125	9910	20002	73.184	281.583	2.357	219.943	0.054
4	85048	9970	20004	47.406	255.075	1.577	210.614	0.014

Table 4: Processing Time, CM vs TW

Nb.Proc.	Proc.Time (CM)	Proc.Time (TW)
1	15 mn 28 s 680	17 mn 49 s 944
2	07 mn 35 s 695	07 mn 15 s 698
3	06 mn 42 s 482	06 mn 46 s 962
4	06 mn 19 s 871	05 mn 40 s 175

(Time Warp, lazy cancellation) are shown in Tables 1-4. We have observed speedup of 2.4 and 3.1 using 3 and 4 processors respectively for the Ethernet transmitting station ECATNet model using our simulator. Speedup is reported by comparison with the distributed simulation code running on a single processor (Figure 7).

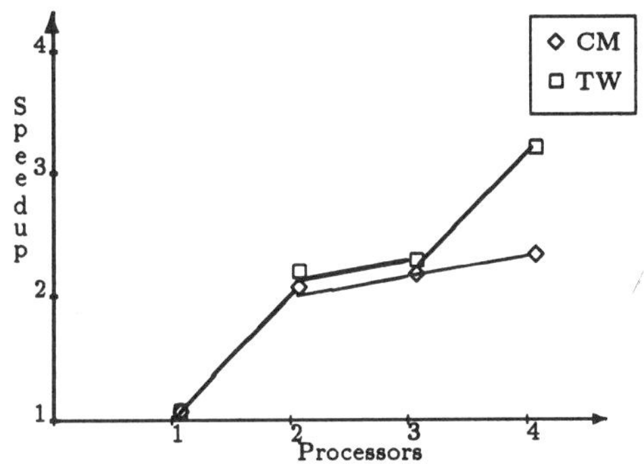

Figure 7: Speedup (CM vs TW)

5 CONCLUSION

Compared with other attempts on parallel or distributed simulation of Petri nets, the methods presented in this paper differ in at least one of the following points: ECATNets are high-level algebraic nets, the state of the net is distributed, and the simulation techniques have to respect timed transitions. The protocols we proposed are expected to improve the ability of efficiently simulate the behaviour of systems modelized by ECATNets over a period of time. An analysis of the structure of the ECATNet model had to be taken into account so that transitions sharing input places are always assigned to the same LP.

The design of a high-level algebraic net concurrent simulator is under investigation. Distributed Simulation is based on spatial decomposition whereas Concurrent Simulation (Jones et al. 1989) is based on temporal decomposition, which actually could be an alternative approach to the application of multiple processors to discrete event simulation. Our objective is a comparable study in order to clearly assess the performance of both approaches (Concurrent versus Distributed).

ACKNOWLEDGMENTS

K.Djemame is supported by a grant from the Algerian Ministry of Higher Education and Scientific Research, and the British Council under the Anglo-Algerian SATELLITE Programme.

REFERENCES

Ammar H. and Deng S. 1991. Parallel Simulation of Petri Nets using Spatial Decomposition. In *Proceedings of the IEEE International Symposium on Circuits and Systems*, pages 826-829, Singapore, jun.1991.

Bettaz M., Maouche M., Soualmi M. and Boukebeche M. 1992. Using ECATNets for Specifying Communication Software in the OSI Framework. In *Proceedings of ICCI'92*, pages 410-413. IEEE.

Bettaz M. 1993. Specification Hautement Compacte et Modulaire de l'ETHERNET: la Station Emettrice. In *Proceedings of CFIP'93*, Montreal, HERMES, Paris. (in french).

Bettaz M. and Maouche M. 1993. How to Specify Non Determinism and True Concurrency with Algebraic Term Nets. *Lecture Notes in Computer Science*, 655:164-180. Springer-Verlag.

Bettaz M. and Mehemmel A. 1993. Modeling and Proving of Truly Concurrent Systems with CATNets. In *Proceedings of Euromicro Workshop on Parallel and Distributed Processing*, pages 265-272.

Bettaz M., Maouche M., Soualmi M. and Boukebeche M. 1993a. Compact Modeling and Rapid Prototyping of Communication Software with ECATNets: a Case Study. *Simulation Series*, 25(1), 149-154. SCS and IEEE.

Bettaz M., Maouche M., Soualmi M. and Boukebeche M. 1993b. Protocol Specification using ECATNets. *Networking and Distributed Computing*, 3(1):7-35. Hermes, Paris.

Bettaz M., Maouche M., Soualmi M. and Boukebeche M. 1994. On Reusing ATNet Modules in Protocol Specification. *JSS*, 27(2):119-128, nov 1994.

Chiola G. and Ferscha A. 1993. Distributed Simulation of Timed Petri nets: Exploiting the Net Structure to Obtain Efficiency. *Lecture Notes in Computer Science*, 691:146-165. Springer-Verlag.

Chiola G. and Ferscha A. 1995. Performance Comparable Design of Efficient Synchronization Protocols for Distributed Simulation. In *Proceedings of MASCOTS'95*, pages 59-65, Durham, North Carolina, jan 1995. IEEE.

Djemame K. and Bettaz M. 1995. On the Parallel Simulation of ECATNets. Technical Report, Computing Science Institute, University of Constantine, Algeria, jun 1995.

Ehrig H. and Mahr B. 1985. Fundamentals of Algebraic Specifications. *EATCS Monographs on Theoretical Computer Science*, Springer-Verlag.

Fujimoto R. 1990. Parallel Discrete Event Simulation. *Communications of the ACM*, 33(10):31-53, oct 1990.

Jefferson D.R. 1985. Virtual Time. *ACM Transactions on Programming Languages and Systems*, 7(3):404-425, jul 1985.

Jensen K. and Rozenberg G.(Ed.). 1991. *High-Level Petri Nets*. Springer-Verlag, Berlin.

Jones D.W., Chou C.C., Renk D. and Bruell S.C. 1989. Experience with Concurrent Simulation. In *Proceedings of the 1989 Winter Simulation Conference*, pages 756-764, MacNair E.A., Musselman K.J., P.Heidelberger (Ed.).

Meseguer J. 1992. Conditional Rewriting Logic as a Unified Model of Concurrency. *Theoretical Computer Science*, 96, pages 73-155.

Misra J. 1986. Distributed Discrete Event Simulation. *ACM Computing Surveys*, 18(1):39-65, mar 1986.

Message Passing Interface Forum. 1995. MPI: A Message-Passing Interface Standard. Technical Report CS-93-214, University of Tennessee, jun 1995.

Murata T. 1989. Petri Nets: Properties, Analysis and Applications. *Proceedings of the IEEE*, 77(4):541-580, apr 1989.

Nicol D. and Roy S. 1991. Parallel Simulation of Timed Petri Nets. In *Proceedings of the 1991 Winter Simulation Conference*, pages 574-583, B.Nelson, D.Kelton, G.Clark (Ed.).

Schöf S., Sonnenschein M. and Wieting R. 1995. Efficient Simulation of THOR Nets. *Lecture Notes in Computer Science*, 935:412-431. Springer-Verlag.

Thomas G.S. and Zahorjan J. 1991. Parallel Simulation of Performance Petri Nets: Extending the Domain of Parallel Simulation. In *Proceedings of the 1991 Winter Simulation Conference*, pages 564–573, B.Nelson, D.Kelton, G.Clark (Ed.).

AUTHOR BIOGRAPHIES

KARIM DJEMAME completed an M.S. in computer science at the University of Constantine, Algeria, in 1991. He is currently a Ph.D. student on leave at the University of Glasgow.

MOHAMED BETTAZ is a professor in the institute of computing science at the university of Constantine, Algeria. His research interests include introducing formal specifications in the life cycle of software engineering. He is a member of IFIP.

DENNIS C.GILLES is an emeritus professor of computing science at the University of Glasgow.

LEWIS M.MACKENZIE is a lecturer in the department of computing science at the University of Glasgow. His research interests emphasize high performance networks and scalable parallel architectures.

Proceedings of the 1996 Winter Simulation Conference
ed. J. M. Charnes, D. J. Morrice, D. T. Brunner, and J. J. Swain

A PERFORMANCE ANALYSIS MODEL FOR DISTRIBUTED SIMULATIONS

David B. Cavitt
C. Michael Overstreet
Kurt J. Maly

Department of Computer Science
Old Dominion University
Norfolk, Virginia 23529-0162. U.S.A.

ABSTRACT

Distributed simulation has proved to be a cost effective technique for studying and understanding complex real-world systems. Many distributed simulations need to incorporate hundreds or even thousands of processors, across both local and wide area networks. As the size and complexity of distributed simulations increase so do the demands on the hardware and software resources that provide simulation services, causing performance bottlenecks that limit the effectiveness of distributed simulation. Different abstractions of performance are needed depending on who is analyzing the distributed simulation and for what reason. This paper proposes a framework for identifying the factors affecting performance and provides a technique for associating the performance factors with high-level system metrics that describe the behavior of the physical and logical resources and services used in the design and implementation of distributed simulations. Dynamic and static analysis of the performance information provides feedback on the execution of the simulation and can provide meaningful information as a guide in making decisions about the configuration and control of the available hardware and software resources for distributed simulation exercises.

1 INTRODUCTION

Distributed simulation is an effective technique for systems studies in research, acquisition, and training environments. For many applications, distributed simulations consist of using large numbers of workstations integrated into local-area and wide-area networks. To improve the benefit of distributing a simulation's computations and run-time environment it is necessary to understand the simulation's performance characteristics in the context of the distributed environment. The concept of distributed simulation performance is multifaceted and not easily characterized. Concurrent execution of simulation models and replicated resources and services are based on abstract models of computation, and performance characterizations of this system can be difficult to understand. The need for greater realism results in demand for resources that often exceeds the nominal capacity (both spatially and temporally) of the workstations currently used as simulation engines. When assessing distributed simulation performance, the goal of simulation speedup dominates most of the literature (Fujimoto 1990, Carothers and Fujimoto 1994, Falsafi and Wood 1994). Number of entities per workstation (where an entity is defined as some simulated real-world object) is also used as the singular metric of performance in Vrablik and Richardson (1994) and the ModSAF 1.4 Reverse Engineering Report (1995). Speedup and entities per workstation however are not sufficient for a full characterization of distributed simulation performance. This is especially true when distributed simulation is used in training and other man-in-loop environments. Performance information relates not just to the speed of computation, but to the efficiency and effectiveness of the simulation in utilizing the many shared resources and services within the distributed environment. Performance information characterizing concurrency, scaling, replication, and coordination can help characterize the simulation's ability to meet the many goals of the simulation study.

Further complicating the notion of performance, the utility of information depends on the needs of the intended audience. Different phases of the simulation life-cycle require people in different roles, each requiring different kinds of information relating performance with simulation goals. The roles include that of model developers and simulation programmers, configuration and experimentation planners, and system analysts or program managers. Performance information gathered and used by these people during the various phases of the simulation life cycle differentiates between the distributed simulation's different physical and logical components (i.e., hardware, O/S, application, models, experimentation). A systems analyst or project manager will generally be interested in high-level performance information on the simulation's capabilities, utilizations, and bottlenecks as they relate to the simulation study's goals. A model developer or programmer can use some of the same high-level information, but will most likely need additional measured data to derive lower-level information. Configuration and experimentation planners also need high-level information that allows them to assess the performance of the simulation. With the recognition that the distributed environment imposes certain constraints on the simulation's processing requirements, configuration and experimentation planners must have both low-level and high-level performance information that allows them to understand the affects of computer, network, and other architectural components.

Typical hardware performance information in distributed simulations, like CPU and network utilization, are easy to identify and measure using traditional tools. Obtaining performance information specific and meaningful to each of the life-cycle roles requires alternative methods, deriving the information from the composition of data characterizing the local run-time performance of individual services and resources, and global data on the structure and performance of the entire distributed simulation environment. The perception of both local detail and global structure provides both coarse and fine-grained views of distributed simulation performance. The performance information is represented at a level of detail that is sufficient for both the design and development, and the configuration and execution of distributed simulations.

Traditionally, most performance measurement and analysis methods of distributed simulations have been application specific. The facilities for measuring and analyzing the performance are generally tightly-coupled to the structure of the simulation and are intrusive during run-time execution, adversely impacting simulation performance measurements. System metrics are typically based on a set of criteria that indicate achievement or not of a particular performance goal, but do not articulate the reasons for the observed performance and behavior. Speedup or the number of simulated entities per workstation are examples of such metrics. The number of components that make up a distributed simulation and the complexity of interaction among the components make identifying all the performance factors a difficult task. A well-defined performance model establishes a relationship among these factors and the constructs and abstractions used in the simulation's design and implementation. The fundamental resource and service components of simulations, model characteristics, and distributed systems, provide a basis for identifying the performance model abstraction. The level the performance information is presented at is determined by the role and intended use of the information. The goal is to provide performance information to different decision makers - in terms of choices available to that individual - which could for example improve performance or indicate potentials for additional model actions. The intended use of the performance information defines what metrics are necessary to develop an understanding of the simulation's performance.

This paper discusses the preliminary concept and design of a distributed simulation performance model. The model describes the framework we are using to guide the process of identifying metrics that provide a specification of distributed simulation performance. The key components of the performance analysis model are still under development. An initial implementation of the model has provided data for evaluating the processing overhead of the performance measurements and verified the utility of aggregating and presenting meaningful performance information to decision makers. Section 2 defines the performance model abstraction and the semantics of the performance information as it relates to the physical and logical components of distributed simulation. Section 3 discusses the use of the performance model, the presentation of performance information, and gives examples of its use with ModSAF, a distributed wargaming simulation built by Loral Advanced Distributed Simulation and used for military training and doctrine development. Section 4 summarizes the proposed model.

2 PERFORMANCE ANALYSIS MODEL

Performance information typically has both space and time dimensions and can be considered the specification of the simulation's execution. To specify the execution we must describe where and when in the distributed simulation certain behaviors exist, which correlates with the space and time dimensions. The performance model represents the space dimension by using three sets of logical components that are a part of distributed simulations; distributed resources, simulation services, and simulation model characteristics. Distributed resources are considered objects or entities that utilize some key characteristics of distributed systems (such as concurrency). Simulation services are those logical components used in the architecture of present day discrete event simulations (such as event schedulers). The model characteristics are those components used to define and represent a simulation model (such as the entity attributes). Each set of components makes up a single dimension and when combined make up a three dimensional discrete space of factors effecting the performance of distributed simulations. Each factor is a function of one component from each dimension. Performance information associated with each factor is derived during the simulation exercise from the combined effect of each of the components. Simulation metrics establish a mapping between the performance factors defined by the model and the simulation's physical and logical resources and services. The performance information associated with each factor can consist of many metrics. A fourth dimension, time, represents the changing phases of simulation performance as time passes during the simulation exercise. Figure 1 is a graphical representation of the space and time dimensions of the performance model. Each shaded cube represents all possible factors that affect the simulation's performance at any instant in time. This representation attempts to define a framework from which a full spectrum of simulation metrics can be defined and used for performance analysis.

2.1 Space and Time Components

As described above, the three dimensions representing the space of performance factors consist of a set of distributed simulation components. The components represent the distributed simulation's resources, services, and model characteristics and define where and why in the simulation's execution space a certain behavior or performance factor is observed. One of the primary purposes of developing this performance model is to present a unified framework for describing the performance of distributed simulation. The challenging aspect of defining the model is to make sure the enumerated sets of components comprising the

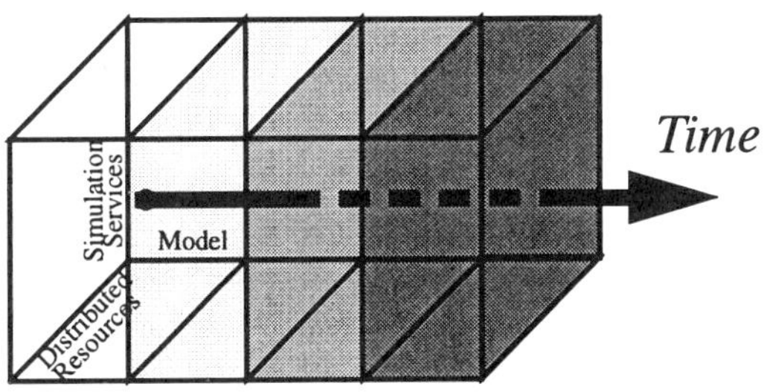

Figure 1: Representation of Performance Factors

performance model encompass the full spectrum of physical and logical components used in the distributed simulation's design and implementation. Figure 2 shows the three dimensions and the corresponding components used to define the model's performance factors.

The discrete space representation of the performance model provides a well-defined abstraction of logical components affecting the performance of distributed simulations. An important point about the representation is that since each dimension in space represents an enumerated set of components, no notion of magnitude or value is associated with the dimension (as is done in vector analysis). The location of each component in its dimension is merely a recognition of the existence of that component as a contributing factor to simulation performance. The discrete space representation results in a set of sixty potential performance factors, each factor being a function of the combined effect of distributed resources, simulation services, and model characteristics. Assessing simulation performance at any instant in time can be obtained by analyzing the three-dimensional discrete space model. A broader understanding of simulation performance can be gained by examining the changes to the discrete space model throughout the simulation exercise. Thus the time dimension is included as part of the model, relating the run-time characteristics of the distributed simulation to changes in performance. The resulting discrete space-time representation provides a high-level abstraction for identifying and communicating the spatial and temporal aspects of distributed simulation that effect performance.

The model's set of distributed resource components are derived from properties of distributed systems: concurrency, scaling, replication, and coordination. To benefit from the use of workstations in current network topologies, distributed simulation environments must be designed to incorporate, exploit, and compensate for these properties. Although these components interact, each can be considered separately to achieve an overall design objective. Concurrency, the parallel execution of simulation and model components, is an artifact of physically separate computers providing simulation resources and services (e.g., the coexistence of simulation entities). Scaling entails increasing the number of shared or replicated resources, services, and simulation entities with a goal of having no single simulation resource or service being in restricted supply. Replication, the existence of multiple copies of simulation data, is a mechanism to increase reliability and fault tolerance (a qualitative goal), and achieve better performance as it relates to higher availability of shared resources and services. Coordination synchronizes the timing and ordering of simulation events including the proper ordering of human intervention with the

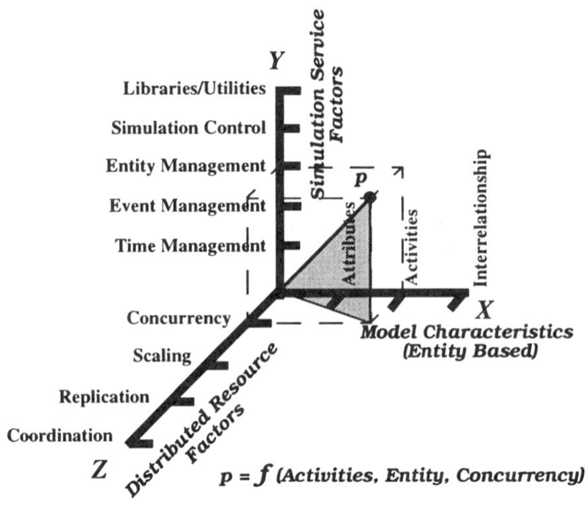

Figure 2: Performance Factors Representation

simulation (where applicable). Each of these components is used to define performance factors of distributed simulation.

Another dimension of the performance factor space deals with simulation service components. The model's service components are the functional components that are part of discrete event simulations. The services include the management of events and event schedulers, entity management, simulation administration and control, and support routines and libraries (such as random number and random variate routines). Distributed simulations generally have multiple schedulers executing concurrently on each of the computers participating in the simulation. The schedulers must invoke events in some coordinated or synchronized manner, assuring either correct program behavior or the recovery from errors induced by uncoordinated activity. The schedulers should ideally also execute efficiently, independent of the number of events that must be serviced as the size of the simulation experiment is scaled up or down. The aggregate cost of maintaining and accessing multiple and replicated events lists and other data structures can be significant and is considered along with each of the distributed resource components. Entity management in distributed simulation is affected by concurrency, scaling, replication, and coordination. The simultaneous creation, execution, and deletion of entities on separate computers relates the concurrency resources to the entity management services and the ability to increase the number of entities in the simulation relates the scaling component. Similarly, the replication and coordination requirements of the simulation affect the management of the simulation entities.

The control of distributed simulations (e.g., start-up, termination, cleanup operations) also affects run-time behavior and must be considered when assessing simulation performance. Additionally, for some simulations graphical displays and high levels of user interaction and intervention can be intrusive to the simulation and its impact on performance must be understood. Lastly, the library and utilities associated with distributed simulations are diverse and application dependent. However, the affect on performance can be dramatic depending on the amount of concurrency and replication, and whether or not the data associated with the libraries and utilities are shared.

The service components, when considered with the distributed resource components, define where in the distributed simulation's execution space a certain performance factor can be observed. The simulation's model characteristics make up a third dimension in the performance factor space and are combined with the resource/service pairs of components to define sources of performance information. The model characteristics are defined by entity attributes that describe some real-world object, entity activities that are the modeled behaviors of the entity, and entity interrelationships that are defined as any dependencies, interactions, or shared behaviors that exist among entities. The concurrent simulation of entities can cause resource and service contention, especially as the number of entities is scaled up. When an interrelationship exists between entities the replication and coordination components can significantly affect performance.

2.2 Simulation Metrics

The performance information associated with each factor defined in the space-time representation can be derived from run-time measurements of the distributed simulation and from static code analysis of the simulation code. Simulation metrics map the components of the space-time representation to physical and logical services and resources used in the distributed simulation environment (i.e., CPUs, network interfaces, event schedulers). The metrics become a specification of the simulation's execution expressed in terms of these services and resources; time-varying functions that characterize the distributed simulation's performance. The domain of metrics includes time, rate, utilization, reliability, and availability for each of the services and resources that contribute to a performance factor. The time metric domain is related to the throughput or responsiveness of the simulation in providing the service. The rate metric domain is related to the productivity or the frequency with which the simulation provides the service. The utilization metric domain is related to the utilization or the amount of the service the simulation uses or needs. The reliability metric domain relates to the correctness with which the simulation provides the service and its ability to meet the deadlines when providing the service. The availability metric domain relates to the simulation's ability to provide the service as required. These metric domains are applicable to each computer participating in the simulation and provide information on individual machines' performance. The same domains are used to define a global view, relating the concurrency, scaling, replication, and coordination components of the performance model with the overall progress and performance of the distributed simulation

Table 1 shows one possible set of simulation metrics that map the entity management services of a distributed simulation with the distributed resource components. The model characteristics are implicit within the definition of the metrics. The column labeled "Autonomous Simulation" shows metrics for the service as that metric pertains to each individual computer participating in the simulation exercise. The metrics relating the distributed resource components to the entity management service are listed in their respective columns and provide a global view of performance. The metrics shown primarily address the processing times associated with simulation entities. The processing time is the simulation work cycles spent executing entity behaviors and activities as well as the modification of entity attributes. The term "node" refers to any specified computer participating in the distributed simulation exercise. The time and rate domains are generally inversely related, the former giving the average amount of time the service takes for each entity type and the latter specifying the average frequency that the service is provided. The global metrics specify the performance across all computers participating in the simulation exercise. The individual metrics provide a more fine-grained view of the performance of specific computers. The reliability metrics assume certain deadlines or thresholds exist for processing entities and the metrics associated with the scaling component of the distributed resources assumes that certain nominal capacities exist. These deadlines and capacities could be defined by simulation requirements, previous empirical studies, or set arbitrarily.

The sample metrics present performance information at a consistent level of detail and are at a sufficiently high-level of abstraction, should allow an intuitive understanding of performance as it relates to the resources and services that are part of distributed simulation. Low-level simulation metrics such as transmission rate for a network interface or the average search time for getting the next event off of the event list are not presented. Low-level performance data, however, might be needed for deriving higher-level performance information.

The physical resources that make up a distributed simulation environment include CPUs, memory and buses, disk drives, network interfaces and transmission media. The operating system functions and interfaces provide access to these resources. The simulation services that request the resources include event schedulers, event-list management routines, clock and timer routines, random number generators, user interface and input/output routines, database routines, and others. The measurement of these physical and logical resources contributes data used to derive simulation metrics. The data can include count data, frequency data, and trace data. Count and frequency data of the physical and logical resources is considered low-level in terms of information content and it is used to derive higher-level simulation metrics that are defined by the performance model. Trace data may be low-level or high-level (containing actual measures of performance). For many distributed simulations, especially long running ones, the volume of measured data can be prohibitive to store, so processing low-level data into high-level trace information can have a distinct advantage (smaller storage requirements and reduced network utilization). The disadvantage of saving only high-level trace data is that it is difficult or impossible to recreate the analysis since content is lost once the data are processed into higher-level performance information.

3 USE OF MODEL

A primary concern of distributed simulation exercise and configuration planners is how to best utilize the available hardware and software to achieve the goals of the simulation exercise or experiment. The performance model provides a framework for specifying the simulation's performance at a level of abstraction that is suitable for making utilization decisions. Once the simulation is capable of providing the raw performance data, the metrics can be derived and the associated performance information analyzed to identify major performance problems. The global simulation metrics are used to observe overall performance. If a bottleneck exists somewhere, a finer-grained analysis can be done by isolating the components of the performance model that are contributing to the bottleneck, and analyzing the individual simulation metrics associated with each factor. After observing and analyzing the performance information, any changes in system configuration can be formulated, and further analysis done to assess the impact on performance. This kind of analysis can be an iterative process, repeatedly looking at performance information to identify exact reasons for performance problems. Why and where there are problems in the simulation's execution is just one aspect of using the performance

Table 1: Mapping of Service and Resource Components to Simulation Metrics

Simulation Service	Metric Domain	Autonomous Simulation (Individual Metric)	Distributed Resources (Global Metrics)			
			Concurrency	Scaling	Replication	Coordination
Entity Management	Time	Avg. Time To Process An Entity	Avg. Time To Process An Entity Per Node	Est. Δ in Time to Process An Entity Per Node With Entity Addition/Deletion Variance in Time To Process An Entity Per Node	Avg. Time To Process Replication Per Node	Avg. Time To Process Entity Coordination Tasks
	Rate	Avg. Number Of Entities Processed Per Unit Time	Avg. Number of Entities Processed Per Unit Time Per Node	Est. Δ in Number of Entities Processed Per Unit Time w/Entity Addition/Deletion Variance in Number of Entities Processed Per Unit Time	Avg. Number of Replications Processed Per Unit Time	Avg. Number of Entity Coordination Tasks Per Unit Time
	Utilization	Total Number of Entities Processed	Avg. Number of Entities Processed Per Node	Variance in Number of Entities Processed Per Node	Total Number of Replicated Entities	Total Number of Entity Interactions
	Reliability	Percentage of Total Run-Time Entity Processing Misses Deadlines	Avg. Number of Entities Missing Deadlines Per Node	Est. Δ in Number of Entity Additions/ Deletions To Deadline Variance in Percentage of Total Run-Time Entity Processing Misses Deadlines	Percentage of Total Run-Time Replication Processing Misses Deadlines	Avg. Number of Entity Interactions Using Stale Data
	Availability	Avg. Elapsed Time Entity Spends Waiting To Be Processed. Percentage Of Total Run-Time Spent Processing Entities.	Concurrency Index (Variance In The Number of Entities Processed Per Unit Time For All Nodes)	Est. Allowable Number of Entity Additions/Deletions	Replication Index (Variance In The Number of Replications Processed Per Unit Time For All Nodes)	Percentage of Total Run-Time Spent Processing Entity Coordination Tasks

model. In what phase of execution does the simulation's performance become limiting? This question is answered by analyzing the simulation metrics over time. The performance model provides the framework for analyzing independently, the different phases of the simulation's execution.

As an example of how the model can be used, consider an implementation of Loral's ModSAF, a Distributed Interactive Simulation (DIS) used for combat training and military doctrine development. A primary goal of ModSAF is to replicate the behavior of simulated military units, vehicles, and weapons systems to a level of realism sufficient for training military personnel (ModSAF Software Architecture Design and Overview Document 1995). Meeting this goal requires the simulation to heavily populate the virtual battlefield, simulating entity on entity combat engagements in an autonomous manner while allowing a user to maintain supervisory control of a large number of military units or individual vehicles.

Examine the availability metric at the bottom of Table 1. The proposed metric estimates the allowable number of entities that can be added to the existing simulation workload. The performance analysis model identifies this metric by the interdependencies of the entity management component of the simulation services (schedular), a scalability component of the distributed systems resources (time available for additional processing), and the entity attributes, activities, and interactions characterizing simulation models. An initial version of the performance model has been implemented by instrumenting ModSAF. During an experiment, data points were measured and derived from three aspects of the execution space of the DIS simulation; 1) the number of vehicles simulated on a single workstation, 2) the percentage of vehicle processing time spent handling the DIS interprocessor communication data packets called Protocol Data Unit (PDUs), and 3) the amount of time available for additional vehicle processing, referred to as slack, within the time constraints defined for the update rates of vehicle state data (Foster et al. 1994). Figure 3 shows a plot of the data. Note that the curve has not been smoothed, revealing the variations in required processing times dependent upon vehicle behavior.

The availability metric provides an estimate on the number of vehicles that can be added to a simulation exercise while still achieving acceptable simulation performance. This metric may be a linear or non-linear function of the slack time and the per vehicle processing time. Referring to Figure 3, we can see with 25 simulated vehicles approximately 140 milliseconds of slack exists (given a 2 Hz update rate for vehicle state). Assuming a simple linear function for calculating the availability metric, then a vehicle processing time of 14.4

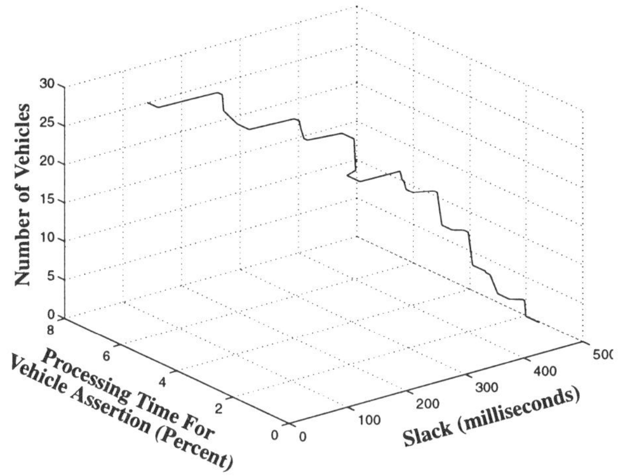

Figure 3: Relationship Among Number of Vehicles, Vehicle Assertion Processing, and Slack Available for Vehicle Processing

milliseconds per update means it is possible to add approximately 9 more vehicles and still meet the required 2 Hz update rate. This information could be derived at run-time and used for example by an analyst or decision maker who requires a greater number of vehicles to enhance the realism of a particular ModSAF simulation exercise.

This example serves to illustrate how a performance model that presents performance information at a high-level of abstraction can be used to guide analysts and configuration personnel in assessing distributed simulation performance. The implementation shows the feasibility of data collection and no sophisticated understanding was required of the simulation's hardware or software processing requirements (i.e., CPU or function timing data) nor any specialized tools required.

4 SUMMARY

The need to assess the limiting factors in the performance of distributed simulation lends itself to the development of a performance analysis model. Development of the model requires the identification of performance problems by understanding why the simulation performs the way it does, where the bottlenecks in performance occur, and when the effective performance of the distributed simulation is no longer adequate. This paper has discussed a performance model for distributed simulation. The need to specify performance using a high-level of abstraction, makes the proposed model suitable for performance assessment during all phases of the simulation life cycle. The model components are defined using a service/resource paradigm that maps the actual physical and logical components used in distributed simulations to the resource and service abstractions of the performance model. The model defines metrics that characterize distributed simulation, specifies what aspects of the simulation and distributed environment are to be observed, the analysis to be performed, and clarifies the semantics and use of the analysis results. Examples of the use of the model, while somewhat artificial in terms of scenario, were representative of the utility of a high-level abstraction for distributed simulation analysis.

An initial implementation of the performance analysis model has demonstrated the ability to collect performance data at a low level of intrusiveness, and can be used to derive metrics which help decision makers during the design of ModSAF training exercises. Future work will involve the refinement and validation of the performance model components, a complete definition of simulation metrics to fully map the set of service and resource components with the model characteristics, and the design and implementation of statistical techniques for presenting meaningful performance information usable throughout the simulation life-cycle. Attempts will be made to validate the use of the performance model during actual simulation exercises using the ModSAF simulation. Areas for future research include exploring the use of static code analysis for identifying resource and service requirements of the simulation as well as potential interactions and dependencies between simulation entities. Open research issues also include the design, use, and assessment of predictive techniques to guide decisions on how to better utilize hardware and software resources used in distributed simulation.

ACKNOWLEDGMENTS

Support for this research was funded by BMH Associates, Inc., Norfolk, VA, under grant number BMH-01N61339-95-C-0030.

REFERENCES

Carothers, C. D., and R. M. Fujimoto. 1994. Effect of Communication Overheads on Time Warp Performance: An Experiment Study. In *Proceedings of the 8th Workshop on Parallel and Distributed Simulation*, ed. D. K. Arvind, R. Bagrodia, Jason Yi-Bing Lin, 118-125, IEEE Computer Society Press, Los Alamitos, California.

Falsafi, B., and D. A. Wood. 1994. Cost/Performance of a Parallel Computer Simulator. In *Proceedings of the 8th Workshop on Parallel and Distributed Simulation*, ed. D. K. Arvind, R. Bagrodia, Jason Yi-Bing Lin, 173-182, IEEE Computer Society Press, Los Alamitos, California.

Foster, L., P. Maassel, and D. McBride. 1994. The Characterization of Entity State Error and Update Rate for Distributed Interactive Simulation. In *Proceedings of the 11th DIS Workshop on Standards for the Interoperability of Distributed Simulations*, September 1994, 61-73, Institute for Simulation & Training, Orlando, Florida.

Fujimoto, R.M. 1990. Performance of Time Warp Under Synthetic Workloads. In *Proceedings of the SCS Multiconference on Distributed Simulation*, ed. David Nicol, 23-28. Society for Computer Simulation, San Diego, California.

ModSAF 1.4 Reverse Engineering Report. 1995. Technical Report. Applied Research Laboratories, University of Texas at Austin.

ModSAF Software Architecture Design and Overview Document (SADOD) 1995. LADS Document Number 94070v1.0. Loral Advanced Distributed Simulation, Cambridge, Massachussetts.

Vrablik, R., and W. Richardson. 1994. Benchmarking and Optimization of ModSAF. In *Modular Semi-Automated Forces: Recent and Historical Publications*, LADS Document No. 94007 v.1.0, 141-147. Loral Advanced Distributed Simulation, Cambridge, Massachussetts.

AUTHOR BIOGRAPHIES

DAVID B. CAVITT is a doctoral student at Old Dominion University, Norfolk, Virginia. He received a BS degree in Computer Science at Old Dominion University. He has 9 years of experience in the use and development of simulations for military and engineering applications. His research interests include modeling and simulation, performance analysis, and distributed systems. David Cavitt is a member of ACM, and IEEE CS.

C. MICHAEL OVERSTREET is an Associate Professor of Computer Science at Old Dominion University. He is immediate past chair of the Special Interest Group in Simulation (SIGSIM) of the ACM. He received his B.S. from the University of Tennessee in 1966, an M.S. from Idaho State University in 1968, and an M.S. and Ph.D. from Virginia Polytechnic Institute and State University in 1975 and 1982 respectively. He has been a visiting research faculty member at the Kyushu Institute of Technology in Japan. His current research interests include model specification and analysis, distributed simulation, high performance networking, support for interactive instruction, and static code analysis in support of software maintenance tasks. He is currently a principal investigator in tasks funded by ARPA, ICASE at NASA Langley, and the National Science Foundation. Dr. Overstreet is a member of ACM, and IEEE CS.

KURT J. MALY received the Dipl. Ing. degree from the Technical University of Vienna, Austria, and the M.S. and Ph.D. Degrees from the Courant Institute of Mathematical Sciences, New York University, New York, NY. He is Kaufman Professor and Chair of Computer Science at Old Dominion University, Norfolk, VA. Before that, he was at the University of Minnesota, both as faculty member and Chair. He also is Visiting Professor at Chengdu University of Science and Technology, People's Republic of China and is Honorary Professor at Hefei University of Technology, PRC. He was a member of Board of the Microelectronic and Information Sciences Center, Minneapolis. His research interests include modeling and simulation, very high-performance networks protocols, reliability, interactive multimedia remote instruction, Internet resource access, and software maintenance. His research has been supported by DARPA, NSF, NASA, CIT, ARPA and the U.S. Navy among others. Dr. Maly is a member of the IEEE Computer Society and the Association for Computing Machinery.

Proceedings of the 1996 Winter Simulation Conference
ed. J. M. Charnes, D. J. Morrice, D. T. Brunner, and J. J. Swain

A TESTBED FOR PARALLEL SIMULATION PERFORMANCE PREDICTION

Alois Ferscha
James Johnson

Institut für Angewandte Informatik
Universität Wien
Lenaugasse 2/8, A-1080 Vienna, AUSTRIA

ABSTRACT

The overwhelming complexity of influencing factors determining the performance of parallel simulation executions demands a performance oriented development of logical process simulators. This paper presents an incremental code development process that supports early performance predictions of Time Warp protocols and several of its optimizations. A set of tools, N-MAP, for performance prediction and visualization has been developed, representing a testbed for a detailed *sensitivity analysis* of the various Time Warp execution parameters. As an example, the effects of various performance factors like the event structure underlying the simulation task, the average LVT progression per simulation step, the commitment rate, state saving overhead, etc. are demonstrated. We show how the scenario management features provided by the N-MAP tool can be efficiently utilized to predict performance sensitivities. For the particular example, the Time Warp protocol, though highly involved, N-MAP was able to predict the performance sensitivity that was measured from the full implementation executing on the Meiko CS-2.

1 INTRODUCTION

A huge variety of Time Warp (TW) parallel simulation (PS) protocols have been proposed in the literature, with a primary focus on correctness issues in their presentation. Performance aspects of the protocols, due to an overwhelming interweaving of influencing factors, have mostly been studied on the basis of abstractions (*models*) of the target execution platform, the TW implementation and the simulation problem. In most of the literature, performance analysis is used to motivate the optimization of the TW protocol with respect to one or more of these concerns, or to assert the performance gain obtained with optimizations. Specifically the question on the relative qualities of TW protocols (in protocol optimization studies) has often been raised, but general rules of superiority cannot be formulated since performance cannot be sufficiently characterized by models.

In the first category of TW performance analysis, all of the three influence factors (target platform, TW implementation, simulation problem) are implicitly or explicitly abstracted into *models*, mostly stochastic models. As an example, in the analysis by Gupta et. al. (1991) event processing time is assumed to be exponentially distributed, time stamps of messages are Poisson distributed in virtual time, the destination logical process (LP) for each message is randomly chosen and equally likely for every LP, etc. Performance investigations based on analytical models often fail to achieve a satisfactory accuracy due to unrealistic assumptions in the modeling process itself, as well as simplifying assumptions that make the evaluation of those models tractable (e.g. symmetry, homogeneity, $M \rightarrow M$ property). Since performance evaluation based on analytical models is prone to modeling errors, only relative trust can be placed in the results obtained (See Gupta et al. 1991 and Akyildiz et al. 1993 for exceptions).

As an improvement, performance investigations have been conducted upon full TW implementations, but operating under synthetic workloads, thus defining a second category of TW performance analysis. In this category, the analysis is based on a *real system* as far as the hardware and TW software is concerned, but the simulation task to be executed in this environment is still an abstraction (*model*) of the real simulation task – leaving just a single source of potential performance analysis bias due to modeling errors. The most prominent synthetic workload model for TW is PHOLD (Fujimoto 1990) which has been widely used in the community to demonstrate the *performance sensitivity* of TW implementations to the (event) structure of the simulation task. Another approach where a mixture of a real simulation task and a model of the execution environment is studied is trace driven simulation. Here, the behavior of a hypothetical environment is studied under the real load.

Finally, in the third analysis category, even full TW implementations with real workload sets often prohibit performance comparisons if different implementation strategies were followed or different target platforms were selected for the execution. Here, performance analysis potentially suffers from an "incomparability" dilemma addressed in (Ferscha and Chiola 1995). There we have developed a performance comparable implementation design, which isolates hardware from implementation and TW protocol related performance influences. A maximum of source code reuse gained from a conditional compilation implementation technique allows for a further reduction of incomparability to an acceptable level.

1.1 The need for performance predicition

In an existential discussion within the parallel and distributed simulation community (Fujimoto 1993), the availability of *performance prediction methods* for PS protocols has been pointed out as being critical for the future success and general acceptance of PS methods in practice (Lin 1993). For example, it is important for a simulationist to be able to evaluate the suitability of certain PS protocols for a specific simulation task *before* substantial efforts are invested in developing sophisticated PS codes. Another aspect is the early evaluation of the anticipated performance of a new PS protocol being developed.

None of the three performance analysis categories above appears adequate for these demands. Models, although able to give fast predictions, are too vague to be meaningfully applied. Analysis that considers full implementations, although accurate, just brings to light the design flaws committed earlier, which are irrevocable or very hard to repair at the time when major part of the development work is already done. In this paper we present a *performance prediction testbed for Time Warp*, N-MAP, designed and implemented to support performance engineering endeavors from the early design phase of TW protocols in order to avoid late and costly re-engineering. The performance engineering activities in a supervisory role escort the development of TW, ranging from *performance prediction* in the early development stages, to *measurements* of performance metrics of the preliminary or final program in the testing, debugging and tuning phase. Implementing TW *incrementally* in N-MAP, i.e. starting from a code skeleton and providing more and more detailed program code towards the full implementation, allows for very early performance based design decisions, systematic investigations of performance sensitivities using an automated scenario manager, and a maximum of code reuse when trying different TW optimizations using an automated version manager. In the next chapter we briefly recall TW performance factors. In Section 3 we develop a TW skeleton in N-MAP, upon which - as a demonstration - we investigate the performance effects of global virtual time (GVT) computation, throttling the optimism via the available memory, and the choice of the size of the checkpointing interval. We have executed the performance scenarios on the Meiko CS-2, and present the results in Section 4.

2 TW PERFORMANCE FACTORS

Collected arguments on the TW performance characteristics and influences have been reported in (Ferscha 1996). (See also Rönngren et al. (1993) for performance issues related to the implementation of Time Warp and some of its protocol optimizations, or Das et al. (1994) for a TW implementation with a minimal amount of event processing overhead.) Here we enlist some of them, outlining the overwhelming complexity of the performance issue:

- **Simulation Task** The structure of events underlying the simulation task exhibit properties such as persistency, concurrency, mutual exclusion, synchronization, causal connectedness, etc., which determine the potential TW performance. TW optimizations often utilize these properties.
- **Partitioning** A paramount TW performance factor is how the global simulation task is decomposed into LPs, and how these are assigned to processors.
- **Target Hardware Raw Performance**: processor speed, communication latency, memory hierarchies/size, cache levels, etc.
- **Communication/Synchronization Model** The target hardware together with several layers of system software influence performance via the routing strategy, policies for multicasting, scattering, buffering, etc.
- **Implementation related Optimizations** "Tricky" implementations of e.g. memory allocation at run time, active messages, interrupts, data referencing etc. can considerably accelerate TW simulations.
- **Protocol related Optimizations** e.g. aggressive/lazy cancellation, lazy re-evaluation, rollback filtering, infrequent/incremental state saving, cancelback, artificial rollback, time windows/buckets, GVT calculation, fossil collection and many others.

- **Partitioning related Optimizations** like the balancement of inter- and intra-LP load, the event-per-message ratio, rollback prevention by blocking, etc. have severe performance impact. Usually, information necessary for partitioning decisions is not available statically, claiming for methods optimizing the execution performance at runtime.

- **Simulation Engine** Yet another source of potential accelerations is the organization and implementation of the event list (binary heaps, splay trees, calendar queues, skip lists, etc.), other data structure manipulations (input queue, output queue, state stack, etc.), time progression, random number generation, etc.

3 IMPLEMENTING TW IN N-MAP

The N-MAP toolset aims at providing the software developer with an integrated environment for the development of performance efficient parallel programs. Starting from a rough description of the program's algorithmic structure in the form of skeletal code, the program is iteratively and incrementally refined by providing a more and more detailed description of the program's component behavior and execution time requirements in each successive development step. Under the constant supervision of performance prediction tools, the thus emerging program source code is tuned and modified in the direction of the most promising implementation strategy ultimately yielding a fully functional, performance efficient parallel program.

In the early development phases of TW implementations, the software developer is confronted with the difficult problem of choosing the TW implementation strategy which is most efficient and suited to the specific simulation task at hand from an huge wealth of possible implementation strategies which, each in its turn, offer a wide spectrum of possible performance optimizations. As a testbed for the development of parallel simulation protocols, N-MAP provides tools for performance prediction which allow the developer to determine the influence of performance critical factors and detect sources of performance loss. Furthermore, a meaningful TW performance analysis demands an in-depth investigation of all performance influencing factors and their interrelationship. For this, the N-MAP *scenario management* comes into play (see Figure 1).

At any point in program development, any variable used in the program source may be declared as a *mutable* and is subsequently handled by the N-MAP scenario manager. Each such mutable may then be assigned a unique value, or a selection of values to be systematically altered in different scenarios by the scenario editor. N-MAP then automatically creates a suite of simulation/execution runs by taking the Cartesian product over all mutable values in which each point in the resulting space represents a specific setting of mutable values.

For gathering performance data, N-MAP defines a standard set of *responses* which represent common performance metrics (e.g. execution time; busy, idle and communication time; packets/bytes sent/received etc.) which may be chosen for inclusion in the scenario. In addition to the standard responses, N-MAP also allows the definition of new, application specific responses.

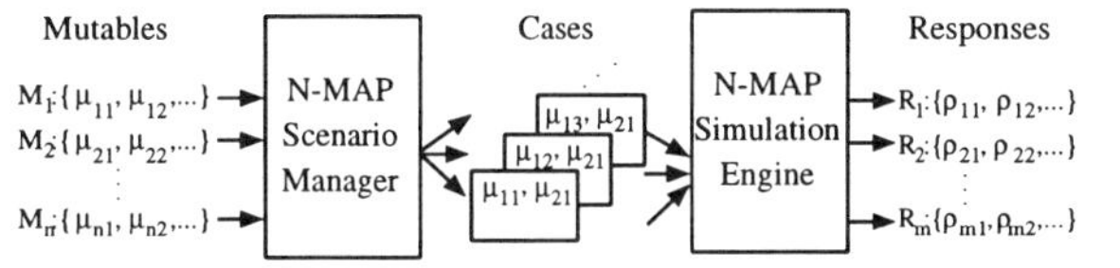

Figure 1: N-MAP Scenario Management

Contrary to other modeling techniques which, as a final result of modeling efforts, yield a *model* of program performance, the N-MAP methodology produces *operable source code* which may be translated into native C code for a variety of parallel platforms as well as for simulated execution on the local uniprocessor by means of N-MAP's built-in translation capabilities.

In the following, a general purpose TW simulator for timed Petri Nets is developed in the N-MAP environment. The principle features of the N-MAP toolset and methodology are demonstrated.

3.1 The Time Warp Task Structure

In the N-MAP environment, program behavior is described in the form of a *task structure specification* (TSS) which defines the sequential stream of task and communication calls to be performed on each processor. The syntax of the TSS is basically that of C with extensions for representing parallelism. The TSS for the TW implementation under investigation is shown in Figure 2.

The code segment labeled `LP` defines `MAXP` logical processors `lp(0)` through `lp(MAXP-1)` operating in an SPMD mode of execution. The `for` loop which encapsulates the whole of the TW source code serves to gather performance data over the number of simulation runs specified in the mutable `RUNS`. Each run begins with the initialization of the simulator and model (i.e. Petri net) in the code segment `INIT`.

During initialization the net description file (specified in the mutable `NETFILE`) is read and partitioned

```
process lp(i) where (i=0:MAXP-1;)                                    LP
{
/* Run several simulations one after the other                       */
for (run=0; run<RUNS; run++)
{
/* Initialize simulator and model */                                 INIT
initialstate(NETFILE,EVENT_POOL);
/* Insert initial internal events into EVL                           */
while(ie=next_ie()) chronological_insert(ie,EVL);
/* Log this state on state stack                                     */
log_new_state();
/* Main simulation loop  */                                          LOOP
while (GVT < ENDTIME)
  {
  /* Read all messages from the input buffer */                      INPUT
  while (m=next_IB())
    {
    /* Straggler message?                                            */
    if (ts_less_than_LVT(m))
      {
      /* Invoke rollback?                                            */
      dual=dual_exists(m,IQ);
      if ((positive(m) && !dual) || (negative(m) && dual))
        {
        /* Rollback to earliest state before timestamp */
        LVT=restore_earliest_state_before(ts(m));
        /* Generate antimsgs resulting from rollback   */
        generate_antimessages(LVT);
        /* Insert antimessages into OQ                 */
        while(ee=next_ee()) chronological_insert(ee,OQ);
        }
      }
    /* Insert message into IQ or annihilate                          */
    if (!remove_dual(m,IQ)) chronological_insert(m,IQ);
    }
  /* Was a GVT packet received? */                                   GVT/FOSS
  if (gvt_packet_received)
    {
    advance_GVT(); /* Calculate GVT estimate                         */

    /* Perform emergency fossil collection if memory                 */
    /* exhausted and checkpointing is used                           */
    if (percent_events_used()>MEMORY_LIMIT&&CHECKP_INTERVAL)
      incr_fossil_collection();
    else
      fossil_collection();
    }
  /* Loop back if insufficient memory */                             GUARD
  if (percent_events_used()>MEMORY_LIMIT) continue;
  /* Get the next event to process from EVL or IQ                    */
  /* If no events to process loop back  to INPUT                     */
  if (!e = get_first_EVL_or_IQ()) continue;
  /* Set LVT to timestamp and event and simulate */                  SIM
  LVT = ts(e);
  simulate_occurrence_of(e);
  /* Insert the internal events generated into the EVL               */
  while(ie=next_ie()) chronological_insert(ie,EVL);
  /* Remove pre-empted internal event from the EVL                   */
  while(ie=next_preempted_ie()) remove_event(ie, EVL);
  /* Insert external events into output queue only if                */
  /* dual does not exist (lazy cancellation)                         */
  while(ee=next_ee())
    if (!dual_update(ee,OQ)) chronological_insert(ee,OQ);
  /* Log this state incrementally onto the stack  */                 LOG
  /* if checkpointing used and checkpoint not yet reached */
  if (!checkpoint && CHECKP_INTERVAL) log_incr_state();
  /* Log the complete state information onto the stack               */
  /* if checkpoint reached or checkpointing not used                 */
  if (checkpoint || !CHECKP_INTERVAL) log_new_state();
  /* Fill the output buffer */                                       OUTPUT
  fill_OB(LVT);
  /* Send out messages in the output buffer                          */
  send_out_contents(OB);
  }
/* Free up all memory used during simulation */                      INIT
clean_up();
}
/* Print out performance data                                        */
print_results();
}
```

Figure 2: The Task Structure Specification for TW

among the processors and a pool of free events allocated on each processor. The number of available events (specified in the mutable `EVENT_POOL`) remains constant during simulation thus providing a means for limiting optimism through the limitation of available memory. Model initialization produces a list of initial internal events which are inserted into the event list (EVL) and the resulting initial state is subsequently logged on the state stack. The main simulation loop is then executed until GVT reaches the value contained in the mutable `ENDTIME`.

In the `INPUT` segment, incoming messages are read from the input buffer (IB) and their time stamps are checked against the current LVT. In the case of a causality violation, the rollback mechanism is invoked to restore the first consistent state prior to the time stamp of the straggler message and antimessages generated in the course of the rollback are inserted into the output queue. Finally, each incoming message is inserted into the input queue (IQ) or annihilated.

If a GVT calculation packet has been received during the `INPUT` phase, the `GVT/FOSS` code segment is executed. Using the information contained in the packet, the LP calculates a new GVT estimate, updates its own information in the GVT packet and forwards it to its successor (the GVT algorithm implemented is described in detail below). Based on the new GVT estimate, fossil collection is then performed in the state stack.

The following segment `GUARD` checks if sufficient memory (specified in the mutable `MEMORY_LIMIT`) is available to perform the local simulation of the next event. If not, the simulator loops back to `INPUT` to await the arrival of further messages or GVT calculation packets until sufficient memory is freed through the occurrence of a rollback (arrival of a straggler message) or fossil collection (arrival of a GVT calculation packet). In the case that no events are scheduled for local simulation in the IQ or EVL (i.e. the partition has become depleted of tokens), the simulator also loops back to the `INPUT` segment.

The occurrence of the next scheduled local event is simulated in the `SIM` segment and LVT is set to the occurrence time of event in the model. The model returns three lists of events to the simulator: 1) a list of the new internal events (transition firings) resulting from the occurrence of the event in the model which are to be scheduled for future simulation, 2) a list of previously scheduled events (transition firings) which have now been pre-empted by the occurrence of the event and 3) a list of new external events (token arrivals in other partitions) which must be sent to the respective LPs. The state of the simulator is updated to reflect the occurrence of the event by inserting the internal events into the EVL and removing the pre-empted events. External events are inserted into the OQ or alternately annihilated if a dual message is present in the OQ (lazy cancellation).

The current state of the simulation is saved in the `LOG` segment by copying the EVL to the state stack as well as the state variables used by the model. If checkpointing is enabled (mutable `CHECKP_INTERVAL` set to a value > 0), state information is saved incrementally on the stack between checkpoints and the complete state information only every `CHECKP_INTERVAL` simulation steps.

Finally, messages stored in the OQ having time stamps less or equal to the current LVT are moved to the output buffer (OB) and sent to the respective

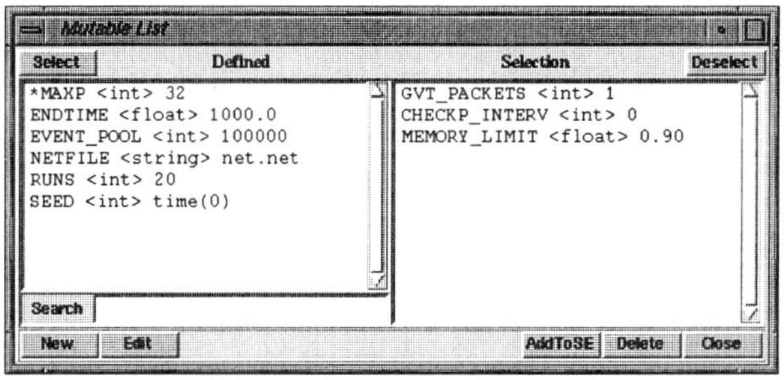

Figure 3: N-MAP Mutables List

LPs in the **OUTPUT** code segment and the simulator loops back to the **INPUT** phase. The current simulation run terminates when GVT reaches **ENDTIME** and all memory used by the simulation is freed before the next run begins. Upon program termination, the performance data gathered over all runs is averaged and the results printed.

3.2 Mutables

The mutables used in the TSS are shown in the *Mutables List* (Figure 3). Here, mutables may be assigned a unique value (left list) or alternately, a subset of mutables selected for variation in a scenario (right list). The following list summarizes the mutables defined in the TSS:

- **MAXP**: number of processors.
- **ENDTIME**: GVT at which simulation terminates.
- **EVENT_POOL**: number of events allocated on each processor during initialization.
- **NETFILE**: name of the Petri net description file.
- **RUNS**: number of simulation runs to execute.
- **SEED**: random number generator initialization.
- **GVT_PACKETS**: number of GVT packets.
- **CHECKP_INTERVAL**: steps between checkpoints.
- **MEMORY_LIMIT**: amount of memory available.

3.3 Execution Time Requirements

A schematic representation of the algorithmic structure and execution time requirements of the TW implementation is shown in Figure 4. The tasks defined in the TSS are grouped according according to their requirement types and the primary run-time factors influencing their execution times.

The simulation of the occurrence of an event in the model as well as GVT calculation are assumed to have fairly *constant* execution times. The group of tasks

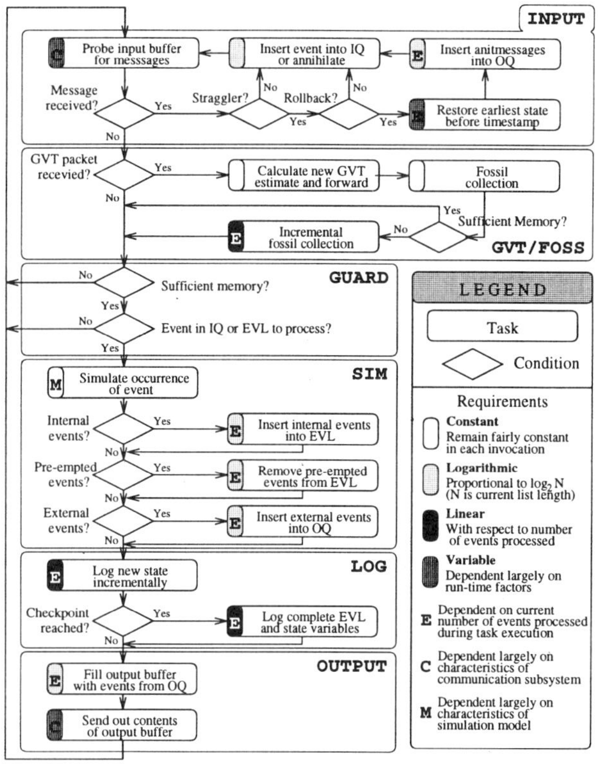

Figure 4: Algorithmic Structure and Execution Requirements of the Time Warp Simulation Protocol

involving list operations can be assumed to be *logarithmic* in their requirements since sorted lists have been implemented as "skip lists" (Pugh 1990) which have optimal element insertion/deletion costs of the order $O(\log_2 N)$, N being the current list length. Furthermore, skip lists have low overall execution times.

A further group of tasks are characterized by *variable* execution requirements since their execution times are dependent largely on run-time factors, e.g. tasks which involve interaction with the communication system or state restoration which is dependent on the "severity" of the rollback. The last group of tasks which deal with state saving and fossil collection can be considered to have *linear* requirements with respect to the number of events processed in the invocation of the task. These linear requirements stem primarily from the fact that during state saving each individual event to be saved must be duplicated and placed on the state stack. Similarly during fossil collection, each event must be freed individually and returned to the pool of free events.

Judging from the structure of the TW implementation under investigation, it appears as if optimization efforts should be concentrated primarily on reducing state saving/fossil collection costs as well as lowering the probability and severity of rollbacks.

3.4 Limiting Available Memory

As a mechanism to limit the optimism in TW, the potential performance gain of memory based throttling (Das and Fujimoto 1994) is investigated, here in the context of a distributed memory implementation of TW. In this approach, whenever an event occurrence is to be simulated, an artificial throttle (mutable **`MEMORY_LIMIT`**) will delay the execution for a constant amount of CPU time. For example, a value of 0.7 for **`MEMORY_LIMIT`** will cause the effect of throttle to set in when 70% of available memory has been used. Thus, the willingness of the simulator to execute the next event occurrence is related to the degree of memory exhaustion.

3.5 Checkpointing Interval

The checkpointing method implemented here saves a complete copy of the event list and state variables every **`CHECKP_INTERV`** simulation steps. Between checkpoints, only those events which have been added to or removed from the event list and state variables which have changed as a result of the simulation of an event are stored on the state stack in a form suitable to allow for subsequent restoration of any intermediate state in the case of a rollback.

The cost and memory requirements for incremental state saving in the case of Petri net simulation (Chiola and Ferscha 1993) is much lower than for complete state saving since each simulation step (transition firing) alters only a small portion of the event list and state variables. Especially when a large number of transitions in the partition have been scheduled for firing in the event list (resulting most likely from a large token population), incremental state saving costs can be considerably lower. For very large checkpointing intervals, however, the coast forward costs incurred during rollbacks may serve to mitigate the positive effect of incremental state saving.

3.6 GVT Calculation

	LP 0	LP 1	LP 2	LP 3		LP N-1	GVT
min LVT	16.754	15.732	14.562	15.921	...	16.054	15.732
Message Receipt Confirmations: from LP 0	—	3	—	1		7	
from LP 1	7	—	5	9	...	—	
from LP 2	—	4	—	12		—	
from LP 3	—	11	2	—		8	
			⋮			⋮	Increment
from LP N-1	15	—	12	—	...	—	5

Figure 5: Structure of a GVT Packet

The GVT calculation method implemented uses one or more *GVT packets* which circulate on complete, closed pre-defined paths among the processors. Figure 5 shows the structure of a GVT packet. To account for unprocessed "in transit" messages, outgoing messages are stamped with a sequence number for each output channel (Lin and Lazowska 1990). Upon receipt of a GVT packet, each processor posts message receipt confirmations in the GVT packet on a per channel basis by writing the sequence number of the last packet received on that channel which does not break the complete series of sequence numbers. Each processor can then calculate the set of unconfirmed messages by inspecting the message receipt confirmations posted in the GVT packet by the other processors and then calculate the minimum of its present LVT and the time stamps of all unconfirmed messages (`minLVT`) and post this in its respective column in the GVT packet. A GVT estimate is calculated by taking the minimum of all `minLVT`s in the GVT packet before forwarding the GVT packet to its successor on the path.

The *increment* field of the GVT packet contains an integer which is relatively prime to the total number of processors, **`MAXP`**, and which each processor adds to its own processor number (modulo **`MAXP`**) to determine its successor processor in the path. Each GVT packet in the network has a different increment so that no two GVT packets circulate along the same paths. Thus, in a network of 8 processors, the GVT packet with increment `1` travels on the path 1-2-3-4-5-6-7-0-1 whereas the packet with increment `5` is forwarded along the path 5-2-7-4-1-6-3-0-5.

In a network of **`MAXP`** processors and using one GVT packet, each processor must wait **`MAXP-1`** GVT calculation steps until it again receives the GVT calculation packet. By allowing 2 GVT packets to circulate in the network, the latency between the receipt of GVT packets is reduced and the frequency of GVT progression is increased on each processor thus allowing for more frequent fossil collections. For large processor counts or for simulations with large memory requirements, a more frequent GVT calculation (i.e. more GVT packets) may be advantageous despite increased communication costs. In other cases, however, a single GVT packet may provide for best performance.

4 SCENARIO EXECUTION ON THE CS-2

Figure 6 shows N-MAP's *Scenario Editor*. The mutables which have been chosen for variation in the scenario (`GVT_PACKETS`, `CHECKP_INTERV`, `MEMORY_LIMIT`) are displayed in the left listbox of the window. Each mutable is assigned a list of values which it is to assume in the scenario in a separate dialog window. The values may be given as a simple white space sep-

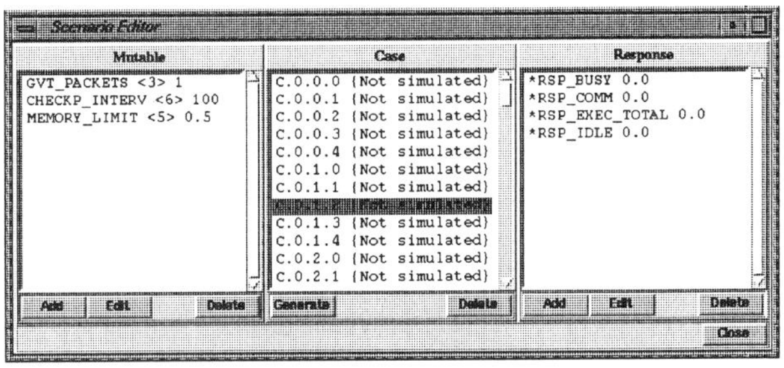

Figure 6: N-MAP Scenario Editor

arated list of values or alternately as a range of values by specifying an upper and lower bound and an increment which may be additive or multiplicative. The following values have been chosen for investigation:

Mutable	Values
GVT_PACKETS	1 2 4
CHECKP_INTERV	0 100 200 300 400 500
MEMORY_LIMIT	0.1 0.3 0.5 0.7 0.9

Figure 7: Scenario Variation of Mutable Values

The Petri net chosen as the simulation task (Figure 8) consists of 64 "regions" linked together in a ring fashion. Each region R_i can be considered to emulate the behavior of a more complex Petri net partition containing one input place P_i, one output transition T_i and a subnet Ψ_i. Both T_i and Ψ_i are infinite servers with exponentially distributed service times $\lambda_{ext} = 0.1$ and $\lambda_{int} = 1.0$ respectively. Initially, each place P_i contains $n = 8$ tokens.

The results of scenario execution are shown in Figure 9. The best results are obtained for the mutables setting `GVT_PACKETS=1`, `CHECKP_INTERVAL=100` and `MEMORY_LIMIT=0.5`.

The use of checkpointing (`CHECKP_INTERVAL>0`) is the most important factor contributing to performance improvement. The resimulation cost incurred during rollback or are obviously insignificant compared to the state saving and fossil collection costs

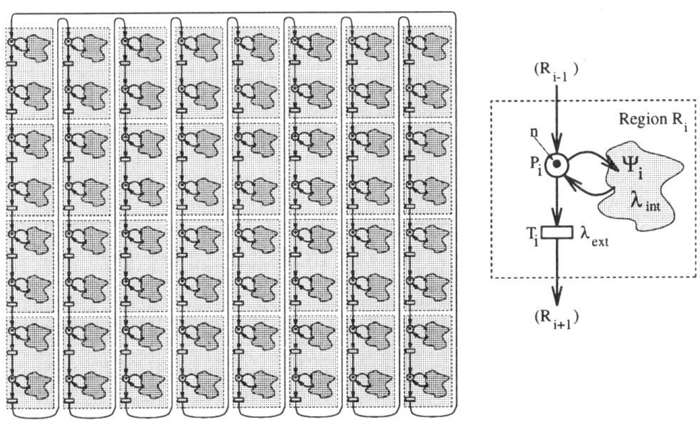

Figure 8: A Petri Net Simulation Task Consisting of 64 "Regions" Mapped onto 32 Processors

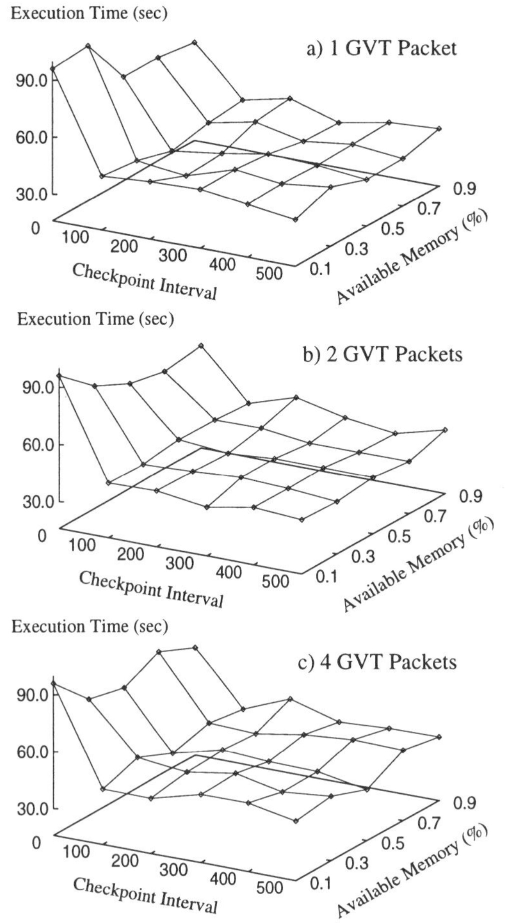

Figure 9: Results of Scenario Execution on CS-2

avoided through the use of checkpointing. The actual checkpointing interval used (100-500) appears to have little effect on performance. Point investigations for the mutable settings `GVT_PACKETS` = 1, `MEMORY_LIMIT` = 0.5 with `CHECKP_INTERVAL` = 1000, 2000, 5000 showed significant performance loss only after an interval of 2000.

In most cases, performance drops only slightly by using more `GVT_PACKETS` although an increased number of packets, providing for a more frequent GVT calculation, does serve to improve performance in cases of high memory requirements (`CHECKP_INTERVAL=0`) and where memory is very limited (`MEMORY_LIMIT` < 0.5). For this particular simulation task partitioned and mapped onto 32 processors, however, the GVT calculation frequency achieved with one GVT packet appears quite sufficient.

5 CONCLUSIONS

The availability of *performance prediction methods* and *tools* for parallel simulation protocols is without any doubt critical for the future success and general acceptance of parallel simulation in practice. For a simulationist it is of utmost importance, to to be able to evaluate the suitability of a certain parallel simulation protocol for a specific simulation task, for a certain multiprocessor system and a certain operational environment *before* substantial programming efforts are invested.

A performance prediction methodology and set of tools, the N-MAP testbed, has been developed, easing performance engineering endeavors of PS protocols from the early design phase in order to avoid late and costly re-engineering. As a testbed, N-MAP in a fully graphical user interface offers very early performance based implementation design decisions, systematic investigations of performance sensitivities using an automated scenario manager, and a maximum of code reuse when trying different TW optimization using an automated version manager. N-MAP is publicly available.

To demonstrate some of the features of the testbed, we have investigated the performance effects of GVT computation, of throttling the optimism via the available memory, and of the choice of the size of the checkpointing interval in a distributed memory implementation of Time Warp. Performance scenarios were defined and executed on the Meiko CS-2.

ACKNOWLEDGEMENTS

This work was partially supported by the Oesterreichische Nationalbank under grant No. 5069, and the Human Capital and Mobility program of the EU under grant CHRX-CT94-0452 (MATCH).

REFERENCES

Akyildiz, I. F., L. Chen, R. Das, R. M. Fujimoto, and R. F. Serfozo. 1993. The effect of memory capacity on Time Warp performance. *Journal of Parallel and Distributed Computing*, 18(4):411–422.

Chiola, G., and A. Ferscha. 1993. Distributed simulation of timed Petri nets: Exploiting the net structure to obtain efficiency. In *Proceedings of the 14th Int. Conf. on Application and Theory of Petri Nets 1993*, ed. M. Ajmone Marsan, 146–165. Lecture Notes in Computer Science, Springer Verlag, Berlin.

Das, S., R. Fujimoto, K. Panesar, D. Allison, and M. Hybinette. 1994. GTW: A Time Warp system for shared memory multiprocessors. In *Proceedings of the 1994 Winter Simulation Conference*, ed. J. D. Tew and S. Manivannan, 1332–1339.

Das, S., and R. M. Fujimoto. 1994. An adaptive memory management protocol for Time Warp parallel simulation. In *Proc. of the 1994 ACM Sigmetrics Conf. on Measurement and Modeling of Computer Systems*, 201–210. ACM.

Ferscha, A. 1996. Parallel and distributed simulation of discrete event systems. In *Parallel and Distributed Computing Handbook*, ed. A. Y. Zomaya, 1003–1041. McGraw-Hill.

Ferscha, A. and G. Chiola. 1995. Adaptive distributed simulation of Petri net models. In *Proceedings of 1995 Summer Computer Simulation Conference (SCSC '95)*.

Fujimoto, R. M. 1990. Performance of Time Warp under sythetic workloads. In *Proceedings of the SCS Multiconference on Distributed Simulation*, ed. D. Nicol, 23–28.

Fujimoto, R. M. 1993. Parallel discrete event simulation: Will the field survive? *ORSA Journal of Computing*, 5(3):218–230.

Gupta, A., I. Akyildiz, and R. Fujimoto. 1991. Performance analysis of Time Warp with multiple homogeneous processors. *IEEE Transactions on Software Engineering*, 17(10):1013–1027.

Lin, Y-B. 1993. Will parallel simulation research survive? *ORSA Journal of Computing*, 5(3):236–238.

Lin, Y-B. and Lazowska, E. 1990. Determining the global virtual time in a distributed simulation. In *1990 International Conference on Parallel Processing*, (III)201–209.

Pugh, W. 1990. Skip lists: A probabilistic alternative to balanced trees. *Communications of the ACM*, 33(6):668–677.

Rönngren, R., R. Ayani, R. M. Fujimoto, and S. R. Das. 1993. Efficient implementation of event sets in Time Warp. In *Proceedings of the 7th Workshop on Parallel and Distributed Simulation*, ed. R. Bagrodia and D. Jefferson, 101–108. IEEE Computer Society Press, Alamitos, Calfornia.

AUTHOR BIOGRAPHIES

ALOIS FERSCHA is an Associate Professor at the Department of Applied Computer Science and Information Systems at the University of Vienna, Austria. His current research interests include performance modeling and prediction, computer aided performance engineering of parallel software, distributed simulation and neural networks.

JAMES JOHNSON has been a Research Assistant the Department of Applied Computer Science and Information Systems at the University of Vienna since 1994. His research interests are in tools for computer system performance analysis.

Proceedings of the 1996 Winter Simulation Conference
ed. J. M. Charnes, D. J. Morrice, D. T. Brunner, and J. J. Swain

LOGICAL PROCESS SIZE IN PARALLEL SIMULATIONS

Fang Hao
Karen Wilson
Richard Fujimoto
Ellen Zegura

College of Computing
Georgia Institute of Technology
Atlanta, GA 30332-0280

ABSTRACT

Most existing synchronization protocols require that the simulation application be partitioned and mapped to logical processes to make it suitable for parallel execution. Assuming the simulation models some number of physical components, an important design question is how many components should be mapped to each logical process? This is a non-trivial question because logical process "size" affects the efficiency of the synchronization protocol, load balance, and approach for implementing shared state variables, as well as the efficiency of the event processing loop within the parallel simulator. This question is studied in the context of a Time Warp-based parallel simulator. Results of two experimental studies are described that compare the performance of parallel simulators using different logical process sizes. One study uses synthetic workloads; the other uses an Asynchronous Transfer Mode (ATM) network. These results show that the choice of logical process size can have a significant effect on performance, and the optimal size depends on model size, the number of available processors, and detailed semantics of the model itself.

1 INTRODUCTION

It is well known that long execution times severely limit the effectiveness of discrete event simulation techniques in modeling moderate and large sized systems. One approach to reducing execution time is to use parallel discrete event simulation (PDES) techniques to execute the simulation model (Fujimoto 1990b, and Nicol and Fujimoto, 1994). This is a particularly attractive approach when simulating large networks because such networks inherently contain large amounts of spatial concurrency. Here, we are primarily concerned with *optimistic* synchronization algorithms such as Time Warp (Jefferson 1985) that utilize a rollback and recovery procedure to synchronize the computation.

Traditional PDES approaches such as Time Warp define the simulation as a collection of logical processes (LPs) that execute concurrently, and communicate by exchanging timestamped *events*, or *messages*; these two terms are used synonymously here. For example, in an ATM network simulation, logical processes must be defined to model components in the network, e.g., traffic sources, switches, and multiplexers. Thus an important design question is: How should the simulation model be partitioned into logical processes? Is it better to define many small logical processes, e.g., a single switch might be modeled as several logical processes, or should one define "large" logical processes that include several components (e.g., several switches)? As elaborated upon later, this is a non-trivial question because of considerations such as load balancing, shared state, and interactions between the synchronization algorithm and the way the network has been partitioned. We refer to this question as the *logical process size* problem.

Here, we are especially concerned with *small-granularity* simulations containing only a modest amount of computation within each event, e.g., a few hundred machine instructions. Simulations of telecommunication networks are often small-granularity simulations because typical events might do little more than update queue length state variables and schedule new events. It is particularly important that parallel simulator overheads be minimized for these simulations because only a modest amount of overhead computation for each event could significantly degrade performance.

After considering a synthetic workload, we focus on simulation of ATM networks. Modern high-bandwidth networks must provide acceptable quality of service to an increasingly diverse set of applications and traffic sources ranging from low speed data transfers to high quality high-definition television (HDTV) distribution. ATM is one method for accommodating diverse applications on an integrated network. This diversity among sources has a direct impact on the design of high-speed networks and their expected performance. Traditional traffic assumptions are not applicable at the ATM *cell* (i.e., 53 byte fixed size packet) level (Das et al. 1994). Thus, the analytical approach to performance modeling is generally numerically intensive and often approximate. Further, the approximate analytical models require validation that is typically performed by simulations. In any case, simulation is an indispensable tool when it comes to testing the performance of a system over a wide variety of traffic loads and scenarios. However, simulations of high-speed networks require excessively long execution times, since network designers and researchers require simulations of large networks, and often have to collect statistics over minutes and hours of real time operation.

In the next section we review related work in accelerating the execution of simulations of telecommunication networks. Issues and tradeoffs concerning the partitioning of network simulations are then discussed and evaluated experimentally. A case study is described comparing the performance of *two* parallel simulations of the same ATM network model, one using "large" LPs, and the second using "small" ones. The performance of these two simulations of the same network model on a common parallel simulation platform are then compared experimentally.

2 RELATED WORK

A considerable amount of work in the circuit design community has been concerned with defining *partitioning algorithms* that divide circuits into subcircuits for simulation on parallel computers, e.g., see Davoren (1989), and Nandy and Loucks (1992). There, the partitioning problem is typically posed as decomposing a large circuit into N sub-circuits, where each sub-circuit is simulated on a different processor. These algorithms usually assume it is best to operate with only one logical process per processor, and do not directly address the question of how logical process size affects parallel simulation performance.

A goal of the work described here is development of a flexible, widely-applicable, parallel simulation tool for modeling telecommunication networks. This goal has been pursued by other researchers as well. For example, Earnshaw and Hind (1992) report up to an order of magnitude speedup in simulating B-ISDN networks. Gaujal et al. (1993) report an order of magnitude speedup in simulating call routing using application specific parallel simulation techniques, and report simulator performance exceeding 3,000,000 simulated calls per minute. Turner and Xu (1992) report success in speeding up telephone switching network simulations using a variation of Time Warp. More recently, work at the University of Calgary is developing a parallel simulation toolkit for modeling telecommunication networks (Unger and Xiao 1994). Not all report success, however. Phillips and Cuthbert (1991) report that their Transputer-based conservative simulations ran more slowly than a sequential simulator. Tallieu and Verboven (1991) also report disappointing performance in simulating an Ethernet. The central contribution of our results is in addressing the LP size question in the context of parallel simulators such as those described above. Although the experimental results presented in this paper apply exclusively to Time Warp simulations, some of the qualitative tradeoffs discussed in this work also apply to conservative simulation algorithms.

3 LOGICAL PROCESS SIZE

The physical system is assumed to be composed of a collection of *components*, where each component is informally defined as the most primitive element whose behavior is explicitly represented in the simulation model via state variables and events. Perhaps a more pragmatic definition of a component is the smallest portion of the simulation model that could be represented by a logical process, given the level of detail required to meet the goals of the simulation study. The components of the cell-level ATM simulation model described later include traffic sources, traffic sinks, and elements of each switch (e.g., output ports).

The central question to be addressed here concerns the number of components that should be included in a single logical process to maximize performance. The advantages of defining "large" logical processes that contain many components include:

- *efficient shared state.* State variables accessed by simulators for different components within the *same* logical process may be directly accessed by those components using local memory references.
- *event combining.* If it is known at model development time that an event will occur at several different components at the same instant in time, and these components are all modeled within a single LP, then the simulator need only schedule a *single* event at the LP.
- *process-oriented simulations.* Process-oriented simulators usually define each LP to contain a

single thread of execution that can "pause", e.g., to wait for another event. This is in contrast to event-oriented simulators where each event results in a procedure to be executed, and simulated time only advances between procedure calls. Implementation of process-oriented simulations typically requires a runtime stack to be maintained for each LP, and light-weight threads to switch execution among the LPs. Simulations using a small number of large LPs will typically require much less memory than simulations using a large number of small LPs. Due to reduced memory management overhead (e.g., caches operate more efficiently for small programs), large LP simulations may yield better performance as well as lower memory consumption.

The efficiency of the event scheduling mechanism may be affected by LP size, depending on the implementation of the parallel simulation executive. Event scheduling involves (1) locating the pending event list for the LP receiving the message, and (2) enqueueing the message into this data structure. Interprocessor communication will also be required if the sending and receiving LPs are mapped to different processors, however, this affects large and small LP simulators equally, assuming the same mapping of components to processors is used. If (1) above is faster when an LP schedules an event for itself compared to scheduling an event for a different LP, using large LPs offer some benefit because there will typically be more of these "self-scheduled" events. However, it is not a priori clear that self-scheduled events are necessarily faster than scheduling events between different LPs on the same processor. We conjecture that for most parallel simulation executives, large LPs offer little or no advantage over small LP simulators with respect to event scheduling because locating a destination LP should not be a time consuming operation in a well designed simulation executive. For example, in Das et al. (1994) locating the destination LP is a constant time operation (essentially, indexing into an array), independent of the destination or the number of LPs. Further, because a large LP including several components will typically have more pending events than a small LP, one would expect longer enqueue times, (2) above, giving the edge to small LP simulators.

In addition to scheduling events, each processor must repeatedly select the next event to be processed, typically by selecting the event with the smallest timestamp among those LPs mapped to the processor. We refer to the computation to perform this task as the *event selection overhead.* Event selection typically involves (1) selecting an LP to execute next, and (2) selecting the smallest timestamped event within that LP. Use of large LPs will typically reduce (1) because there are fewer LPs to manage, but at the cost of increasing (2), because there will usually be more pending events within each LP. If data structures with similar access times (e.g., $O(logN)$ insertion and deletion time priority queues) are used for maintaining the list of runnable LPs and the list of pending events within each LP, LP size may not significantly affect event selection overhead. Similarly, if both lists are combined into a single priority queue data structure as described in Das et al. (1994), LP size will not affect the event selection overhead at all.

"Small" logical processes do offer some significant advantages over large LPs:

- *greater potential concurrency.* Concurrency is limited by the number of LPs, so smaller LP simulators may offer more potential concurrency.
- *more flexible load distribution.* Better workload balance may be obtained because the granularity of computation that may be moved from one processor to another is finer.
- *potentially lower state saving overhead.* Optimistic protocols require snapshots of the LP's state to be made to enable rollback. Large LPs will typically contain more state than smaller ones, necessitating more state copying.
- *less rolled back computation or artificial blocking.* If two different, independent components are combined into a single larger LP in a Time Warp simulation, a rollback in one component may unnecessarily cause rollback events in the second component. Similarly, in a conservative simulation, blocking one component may unnecessarily delay processing events in another component within the same LP.

From the above discussion, we see that it is not a priori clear whether large or small LPs will yield the best performance, or how much difference this question makes with respect to overall performance. It is clear that the answer to this question depends on the simulation model and implementation details of the underlying simulation executive.

4 SYNTHETIC WORKLOAD

To quantitatively evaluate how different factors affect performance, a synthetic workload program called PHOLD was used (Fujimoto 1990a). The simulation model consists of a collection of entities, that send messages to each other. Each entity sends message to

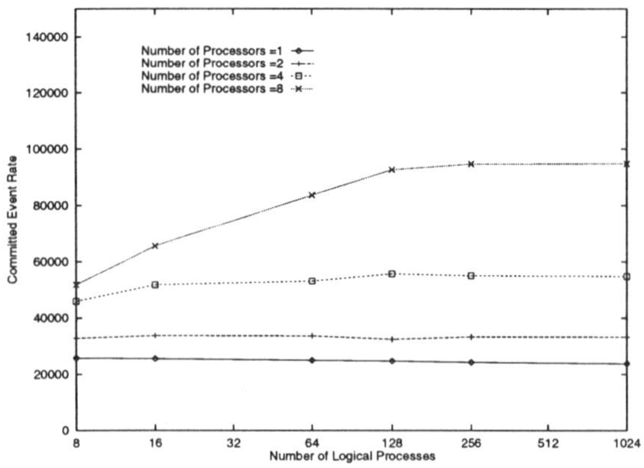

Figure 1: Events committed per second–PHOLD

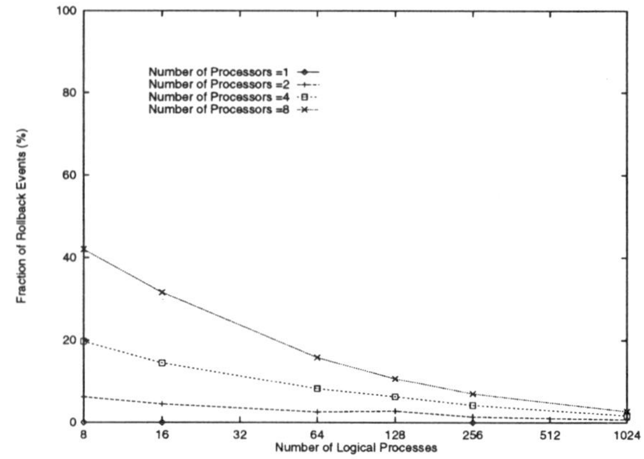

Figure 2: Rollbacks–PHOLD

another entity after processing each event. The destination is selected from a uniform distribution. The time stamp increment is selected from an exponential distribution. The entities are partitioned into logical processes so that each LP contains the same number of entities. The workload of the PHOLD simulator is specified as the message population, i.e., the number of unprocessed events that exist in a sequential execution of the simulator at any instant.

To emulate the effect of shared state, state update messages are added to the PHOLD model. The entities are clustered into a number of groups. All entities in the same group share state variables. The shared state is updated at simulated times following a Poisson distribution. In each group, one "master" entity is chosen that is responsible for sending update messages to the other entities in the group. If all entities in a group are in the same logical process, only one update message needs to be sent to implement each update, i.e., the master entity only sends one update message to itself. However, if the entities in a group span several logical processes, multiple update messages must be sent, i.e., the master entity must not only send an update message to itself, but also send one update message to each LP containing entities within the group.

4.1 Experiments without State Updates

The first set of experiments were conducted to evaluate the effect of rollbacks on the performance of PHOLD without state updates, under different LP sizes. The performance of PHOLD, as measured by the number of events committed per second with different numbers of logical processes (from 8 to 1024) and processors (1, 2, 4, or 8), are shown in Figure 1, where the message population is 64.

We observe that the message event rate increases with the number of logical processes, i.e., the fine-grained simulator produces better performance than the coarse-grained simulator. The principal reason is the coarse-grained simulator is prone to having more unnecessary rollbacks than the fine-grained simulator. We also observe that this effect is more pronounced when the number of processors is increased. This is because increasing the number of processors results in more optimistic execution, accentuating the rollback behavior.

Figure 2 shows the fraction of processed events that are rolled back. This graph confirms that the number of rollbacks increases significantly as the LP size increases (fewer LPs).

However, when the workload is heavy (e.g., message population is 1024), we observe that the number of logical processes does not have a dominant effect on the message event rate. This is because the heavy work load reduces the rate of advance of each LP and thus reduces the number of rollbacks.

4.2 Experiments with State Updates

A second set of experiments were conducted to evaluate the effect of state updates on performance. The performance was measured by determining the number of "effective" committed message events per second, where the number of "effective" events is defined as the number of total committed events minus the number of update events. The latter term represents an additional overhead incurred by simulations using small LPs. Figure 3 shows performance with different numbers of logical processes(from 8 to 1024), processors(4 or 8), and mean state update intervals (equal to 1 or 10 times the timestamp increment, or no update messages). The message population is 64.

We observe that: (1) In general, the effective event rate increases, and then declines, as the number of

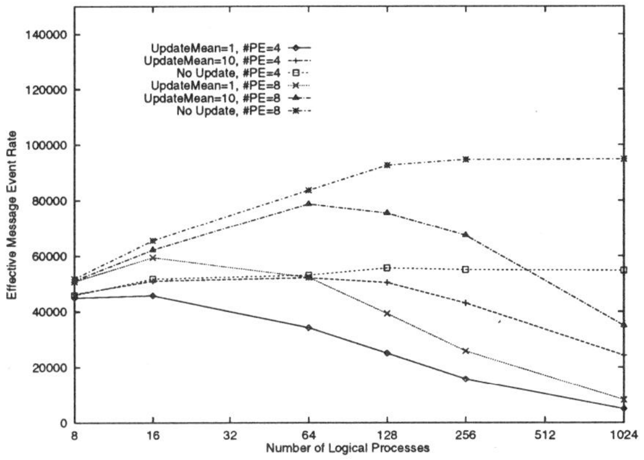

Figure 3: Effective events per second–PHOLD

logical processes is increased. This is because when the number of LPs is small, the number of update messages is small, so rollback is a dominant factor affecting performance; however, when the number of LPs is large, the number of update messages is large, so the overhead for update messages becomes dominant. (2) Performance is more sensitive to changes in the number of LPs when the mean value of the update interval is smaller, because more update messages are generated.

Again, when the workload is heavy (e.g., message population is 1024), performance is less sensitive to changes in the number of LPs, similar to what was observed in the previous experiments.

5 CASE STUDY: ATM SIMULATION

The simulation of large networks of ATM switches presents an interesting case study for optimal selection of logical process size. A single ATM switch is commonly abstracted as a set of output queues, one per output link. A large network of switches can then be partitioned into logical processes spanning a variety of process sizes, including (1) one process per output queue, (2) one process per switch, and (3) multiple switches per process. For this case study, we focus on the first and third approaches; our *fine-grained* (small LP) implementation uses one process per output queue while our *coarse-grained* (large LP) implementation uses multiple switches (and hosts) per process, where each host represents a single source of network traffic, and a single sink.

In order to compare the results from these two partitioning approaches, we maintain the following characteristics across the two implementations: (1) A switch is represented by a set of output queues, one per switch output link. (2) We use a common topology for the large network of switches, which is taken from an existing IP network called *Sesquinet*. The network consists of 94 switches and 80 hosts. Each host is represented by a source and a sink. (3) Traffic is generated using a uniform geometric bursty traffic model. The destination of each burst of cells is selected from a uniform distribution. All sources behave in a similar fashion. (4) The routing is shortest path, computed using unit weights on all links. (5) The simulators only model forwarding of ATM cells from one component to another. A static set of circuits is assumed, i.e., no new circuits are constructed during the simulation, and circuits are never torn down. (6) To emulate the effect of shared state (e.g., due to creating or destroying circuits during the simulation that cause routing table entries to change), both simulations use a common state update model. Specifically, state updates are generated using a Poisson process, and affect the state of a single switch. (7) Both simulators use identical mappings of logical processes to processors in the sense that the set of components modeled by LPs mapped to a single processor is the same for both simulators. The fine-grained simulation used here does not take advantage of the fact that the LPs used to model a single switch could be distributed over multiple processors.

We now describe the details particular to each of the two implementations used to study process granularity.

5.1 Fine-Grained Implementation

The fine-grained implementation uses a large number of small LPs, associating each host and each output queue in a switch with a separate LP. The fine-grained implementation of the Sesquinet has 436 LPs.

The fine-grained simulation uses copy state saving exclusively. Each state update event (pertaining to a single switch) results in the generation of additional state update events that are sent to other LPs modeling different output ports of the switch; the events are required to ensure the different copies remain consistent. This strategy for updating state is sometimes called "push" processing, in contrast to "pull" processing where LPs do not maintain local copies, but rather, state is explicitly requested by each LP via "query events" when needed (Wieland and Jefferson 1989). Push processing is more efficient when state updates occur less frequently than state queries.

5.2 Coarse-Grained Implementation

The coarse-grained implementation creates a small number of large LPs, grouping multiple sources together into source LPs, multiple sinks together into sink LPs and multiple switches together into switch

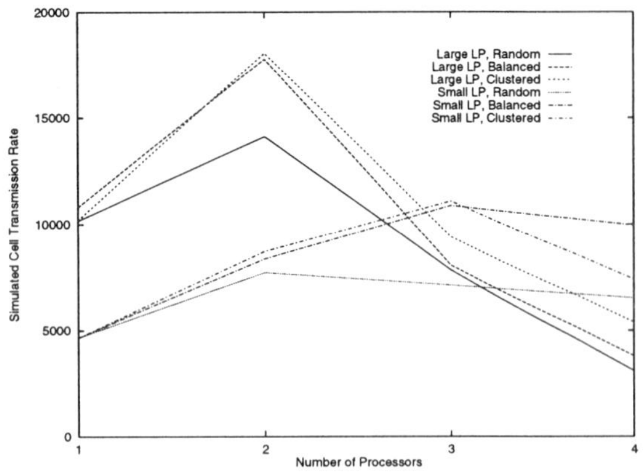

Figure 4: Simulated cell transmissions–ATM

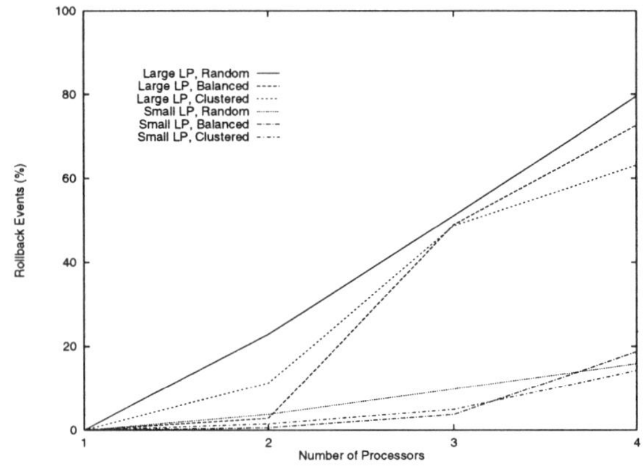

Figure 5: Rollbacks–ATM

LPs. In the coarse-grained implementation of the Sesquinet there is one source, one sink, and one switch LP per processor.

The implementation of the source model results in an event at each source during each time unit that determines whether the source turns on or off. The synchronous nature of this model results in many events across the different sources occuring at the same simulated time. The event handling routine in the coarse-grained simulator uses event combining (discussed earlier) and only schedules one event for all of the sources modeled by the LP. By contrast, the fine-grained simulator described earlier must schedule separate events for each source LP.

The coarse-grained simulator uses copy state saving for variables that are modified with each event and incremental state saving for other state variables. Because incremental state saving is somewhat more expensive than copy state saving, this situation favors fine-grained LPs. State updates are handled more efficiently in large-grained LPs, since all of the modified state is confined to a single LP.

6 PERFORMANCE RESULTS

The two parallel simulators were executed on a four processor shared-memory Sparc multiprocessor. Each processor is a 50 MHz HyperSparc. The performance metric used in this study is the number of simulated cell transmissions (sending a cell over a single communication link) per second. The Time Warp executive used in this study is described in Das et al. (1994). The first set of experiments does not consider state updates. The effect of the frequency of state updates on performance is then studied in a second set of experiments.

6.1 Experiments without State Updates

A sequence of experiments were performed using the following partitioning algorithms: (1) *Random mapping:* each switch or host node in the network is randomly mapped onto a processor. (2) *Balanced random mapping:* each node is randomly mapped onto a processor, with the constraint that each processor must be assigned an equal, total number of switch and host nodes. (3) *Clustering mapping:* the network is partitioned to balance the workload (as measured in the number of hosts and switches mapped to each processor) and to reduce interprocessor communications. The partitioning was performed manually.

The performance of the two simulators under different partitioning strategies with light traffic load is shown in Figure 4. The probability that a source is in the "on" (transmitting) state is 0.06; similar results were observed when the traffic load is high (e.g., the "on" probability is 0.67). Across the experiments, we observe that (1) The coarse-grained simulator far outperforms the fine-grained simulator for one or two processors. (2) The fine-grained simulator's performance approaches, and in some cases surpasses the coarse-grained simulator for three or four processors. (3) Among the three mapping strategies, clustering mapping yields the best performance, and random mapping the worst performance. Moreover, the partitioning algorithm does have a significant affect on performance in both simulators. (4) All of the simulators exhibit some degree of performance degradation as the number of processors increases, as elaborated upon below.

The principal reason the coarse-grained simulator outperforms the fine-grained simulator when using a small number of processors is because it can exploit event combining. This significantly reduces the amount of computation that must be performed to generate cells at the source LPs. Because simulator events contain relatively little computation, the overhead for scheduling and processing events significantly degrades the performance of the fine-grained simulator, which must schedule a new event for each source each clock tick.

A significant performance degradation was observed in the coarse-grained simulator as the number of processors is increased. This is because use of large LPs makes the simulator more sensitive to rollbacks, a phenomenon observed in the experiments using synthetic workloads. Figure 5 shows the fraction of processed events that are rolled back. As noted earlier, rolling back a large LP introduces unnecessary rollbacks in the computation, which in turn can result in unnecessary anti-messages and additional rollbacks in other LPs. We observed that both the small- and large-grained LPs yielded more rollbacks as the number of processors increased, as would be expected, however, the increased number of rollbacks was far more pronounced in the coarse-grained simulation. This ultimately led to the fine-grained simulation outperforming the coarse grained simulation, although the fine grained simulator using up to four processors never outperformed the coarse grained simulator using only two processors. It is possible that the fine grained simulation could yield the best performance overall if more processors were available for these experiments.

A second factor affecting performance with four processors is the presence of operating system daemons running on one processor of the four. The processor mappings assume identical processors, and thus allocate processing equally. We observed more severe performance degradations for the small-LP simulation than the large-LP simulations when all four processors were used (The experiments with fewer processors avoid the one executing the daemons). While in principal one could isolate the machine to avoid such interference, such a stand-alone configuration is not what one would typically encounter in practice. A more pragmatic approach is to adjust the processor mapping to compensate for this effect. Further experiments are are required to study this question, and to optimize performance of both simulators for increased numbers of processors.

Clustering mapping yields the best performance because it reduces the amount of inter-processor communications. This partitioning algorithm also yielded fewer rollbacks. The balanced random mapping yielded better performance than the simple random mapping. Balancing the workload reduces the number of rollbacks, as one would expect. Overall, simple mapping strategies such as random distribution of LPs to processors yielded poor performance for both the large and small grained simulators, especially as the number of processors was increased.

6.2 Experiments using State Updates

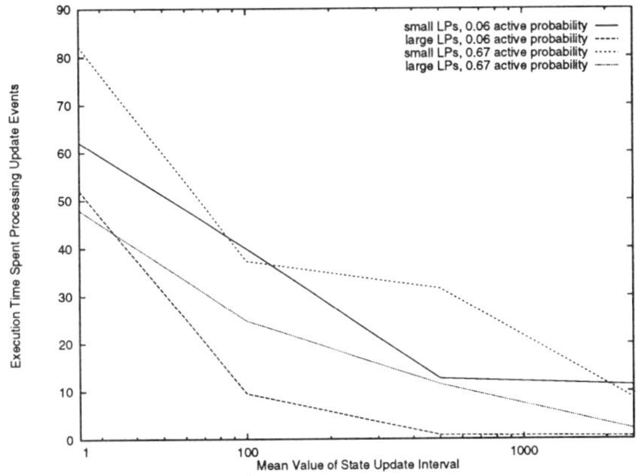

Figure 6: Time spent processing update events

A second set of experiments were performed to evaluate the effect of state updates on the relative performance of the two simulators. The same traffic generation model was used as in the experiments without state updates. As mentioned earlier, state updates were generated within each switch with an exponentially distributed time between updates. Experiments used 25, 100, 500, and 2500 as the mean values of the state update interval, where five units corresponds to the amount of time required to transmit one cell over a link. Each set of experiments were executed on one to four processors using the clustering approach to network partitioning.

The time spent performing state updates of the two simulator on four processors is shown in Figure 6; similar data was observed for different numbers of processors. The execution time of the small-grained simulation without state updates was 92 and 171 seconds for the lightly and heavily loaded networks, respectively. Execution times for the large-grained simulator were 59 and 116 seconds, again for the lightly and heavily loaded networks, respectively. As expected, as the state of each switch is updated more frequently (moving to the left in Figure 6), the fine-grained simulator incurs more state update overhead than the coarse grained-simulator. This overhead represents a significant portion of the simulation computation at high update frequencies. It is perhaps noteworthy that the performance of the fine-grained simulator would be degraded even further if LPs modeling a single switch were distributed across multiple processors because interprocessor communication would then be required with each state update.

7 CONCLUSIONS AND FUTURE WORK

This study addressed the question of partitioning simulations into logical processes, and whether small-grained or large-grained LPs yielded better performance. For the test cases that were examined, the central conclusions from this study are: (1) The most dominant effect favoring large-grained logical processes in these experiments was their ability to combine events containing the same timestamp into a single event. (2) Large grained logical processes suffered from being relatively sensitive to rollback as the number of processors increased, especially when the workload is low. Unnecessary rollbacks due to aggregating too many components into a single LP can have significant, detrimental effects in Time Warp simulations. (3) Partitioning the simulation to avoid state variables that are shared between LPs results in somewhat higher overheads for state updates, particularly if the state variables are updated very frequently.

Work is continuing to collect additional data across a variety of other test cases, e.g., other network topologies, source traffic models, and traffic loads. with the goal of developing an efficient, flexible parallel simulation tool for modeling ATM networks. Simulations using small sized LPs are able, in principal, to achieve better load balance because portions of a switch may be distributed over more than one processor, an aspect not explored in the results described here.

ACKNOWLEDGEMENTS

This work was supported by NSF Grant Number CDA-9501637 and NCR-9527163.

REFERENCES

Bae, J., and T. Suda. 1991. Survey of traffic control schemes and protocols in ATM networks. *Proceedings of the IEEE*, 79(2).

Das, S., R. Fujimoto, K. Panesar, D. Allison, and M. Hybinette. 1994. GTW: A Time Warp system for shared memory multiprocessors. In *1994 Winter Simulation Conference Proceedings*, pages 1332–1339.

Davoren, M., 1989. A structural mapping for parallel digital logic simulation. In *Proceedings of the SCS Multiconference on Distributed Simulation*, volume 21, pages 179–182. SCS Simulation Series.

Earnshaw, R. W., and A. Hind., 1992. A parallel simulator for performance modelling of broadband telecommunication networks. In *1992 Winter Simulation Conference Proceedings*, pages 1365–1373.

Fujimoto, R. 1990a. Performance of Time Warp under synthethic workloads. In *Distributed Simulation*, volume 22, pages 23–28. SCS Simulation Series.

Fujimoto, R. M. 1990b. Parallel discrete event simulation. *Communications of the ACM*, 33(10):30–53.

Gaujal, B., A. G. Greenberg, and D. M. Nicol. 1993. A sweep algorithm for massively parallel simulation of circuit-switched networks. *Journal of Parallel and Distributed Computing*, 18(4):484–500.

Jefferson, D. R. 1985. Virtual time. *ACM Transactions on Programming Languages and Systems*, 7(3):404–425.

Nandy, B., and W. Loucks. 1992. An algorithm for partitioning and mapping conservative parallel simulation onto multicomputers. In 6th *Workshop on Parallel and Distributed Simulation*, volume 24, pages 139–146. SCS Simulation Series.

Nicol, D. M., and R. M. Fujimoto. 1994. Parallel simulation today. *Annals of Operations Research*, 53:249–286.

Phillips, C. I., and L. G. Cuthbert. 1991. Concurrent discrete-event simulation tools. *IEEE Journal on Selected Areas in Communications*, 9(3):477–485.

Tallieu, F., and F. Verboven. 1991. Using Time Warp for computer network simulations on transputers. In *Proceedings of the 24th Annual Simulation Symposium*, volume 21, pages 112–117. IEEE Computer Society Press.

Turner, S., and M. Xu. 1992. Performance evaluation of the bounded Time Warp algorithm. In 6th *Workshop on Parallel and Distributed Simulation*, volume 24, pages 117–128. SCS Simulation Series.

Unger, B., and Z. Xiao. 1994. The fast parallel simulation of telecommunication networks. In *Proceedings of New Directions in Simulation for Manufacturing and Communications*. The Operations Research Society of Japan.

Wieland, F., and D. R. Jefferson. 1989. Case studies in serial and parallel simulation. In *Proceedings of the 1989 International Conference on Parallel Processing*, volume 3, pages 255–258.

AUTHOR BIOGRAPHIES

FANG HAO is a Ph.D. student in the College of Computing at Georgia Tech.

KAREN WILSON received the M.S. degree in Computer Science at Georgia Tech in 1995. She is currently employed by Digital Equipment Corporation.

RICHARD FUJIMOTO is a professor in the College of Computing at Georgia Tech.

ELLEN ZEGURA is an assistant professor in the College of Computing at Georgia Tech.

Proceedings of the 1996 Winter Simulation Conference
ed. J. M. Charnes, D. J. Morrice, D. T. Brunner, and J. J. Swain

EVENT SENSITIVE STATE SAVING IN TIME WARP PARALLEL DISCRETE EVENT SIMULATIONS

Sven Sköld
Robert Rönngren

Simulation Laboratory
Department of Teleinformatics
Royal Institute of Technology
Stockholm, SWEDEN

ABSTRACT

In this paper we present *event sensitive state saving* as a novel idea for sparse state saving in Time Warp synchronized parallel discrete event simulations. This state saving model is aimed at efficient simulation of models where the execution time or granularity for different types or classes of events typically has a large variance. The event sensitive state saving mechanism is sensitive to which class of event the previously executed event belongs, and decide when to save simulation state based on this information. We present an analytical analysis and compare this new state saving model with the performance of an existing adaptive state saving model as well as the traditional Time Warp state saving model with fixed state saving interval.

1 INTRODUCTION

The most well known optimistic parallel discrete event simulation mechanism is Time Warp, proposed by (Jefferson 1985). Any parallel simulation mechanism should guarantee that the simulation is synchronized such that the output is free from effects generated by causality errors. Time Warp uses a detection and recovery mechanism to address these issues. In Time Warp, a causality error is detected when a timestamped message arrives at a logical process (LP) and the timestamp of the message, i.e. the message receive time, is less than the time of the last processed event at that LP. The event that causes the causality error is called a straggler event. To recover from an erroneous state, the LP is rolled back in time to a state before the straggler event. Before resuming the execution, any possible effects from all messages that have been sent with timestamps greater than the straggler timestamp need to be undone. This is achieved by resending negative versions of previously sent messages, so called anti-messages, which will annihilate the original messages and thus undo the effect from the original messages. However, these anti-messages might possibly act as straggler messages themselves and hence cause further rollbacks at other LPs. Despite what may appear as a complicated mechanism, Time Warp has proved to be a promising mechanism for parallel simulation (Fujimoto 1990). However, there exist some performance issues related to Time Warp which still merits further attention by the research community.

One potential performance bottleneck of Time Warp is the need to periodically save the state for each LP to later be able to roll back to earlier states. Without an effective memory manager combined with an efficient state saving method, the amount of time as well as memory needed to save states could severely degrade performance. Furthermore, it might also put a limit on the maximum size of the models possible to simulate due to excessive memory consumption. During the last decade, two methods to reduce the state saving overhead has been proposed: incremental state saving (Bauer, and Sporrer 1993) (Palaniswamy, and Wilsey 1993) and sparse or infrequent state saving (Lin et al. 1993). In this paper we focus on sparse state saving methods.

Traditionally in Time Warp simulations, the state or state variables are saved after or before the processing of each event at an LP. This is usually referred to as *copy state saving*. However, one can observe that a state can be restored by reloading an earlier state and re-executing or *coasting forward* all events in between. Thus one could potentially save execution time and memory by not saving the state after each event, but instead save the state of an LP at some specific time interval which is greater than one - hence the name infrequent or *sparse state saving*.

Several methods for sparse state saving has been proposed. To our best knowledge these models are, without exceptions, based on the assumption that the execution time of events in coast forward can be approximated with a mean execution time, i.e. with an underlying assumption that the execution time, or granularity, show little variance. However, there exist many important classes of simulations where this obviously is not a good or even valid assumption. Consider for example battlefield simulations or simulations of mobile communication systems where many events are simple position updates for mobile entities. These events in general have very low granularity. However, these models typically also feature very large grain events which occur when entities in battlefield simulations enter into combat or when mobile units in a personal communication system perform handovers or interference measurements.

In this paper we propose an *event sensitive state saving* method for sparse state saving. This method takes into consideration the granularities of different event types or classes in the simulation. We present a simple analysis and compare this new state saving method to the performance of an adaptive state saving method (Rönngren, and Ayani 1994) as well as a traditional Time Warp simulator with a fixed state saving interval. We use two simple mobile communication network models as benchmarks.

This paper is organized as follows: Section 2 gives a short introduction to sparse state saving in Time Warp; other related work in this area is summarized in Section 3; Section 4 describes the event sensitive state saving approach; Section 5 shows our experimental results with this new state saving model; and finally Section 6 concludes the paper and summarizes this work.

2 SPARSE STATE SAVING

Sparse or infrequent state saving approaches are based on the observation that one simulation state can be recreated from an earlier state by re-executing the events between the two states. Thus, the simulation state does not need to be saved or checkpointed after the processing of each event. Hence, we can trade the overhead of saving state after each event for the overhead of restoring a state by coasting forward, or re-executing, the events in between the latest previously saved state and the current state (Preiss, Loucks, and MacIntyre 1994). For state restoration purposes and to assert the simulation can progress it is sufficient that at least one earlier state before the straggler event has been saved as well as all subsequent events that have been processed but not checkpointed, see the simulation snap shot example in Figure 1.

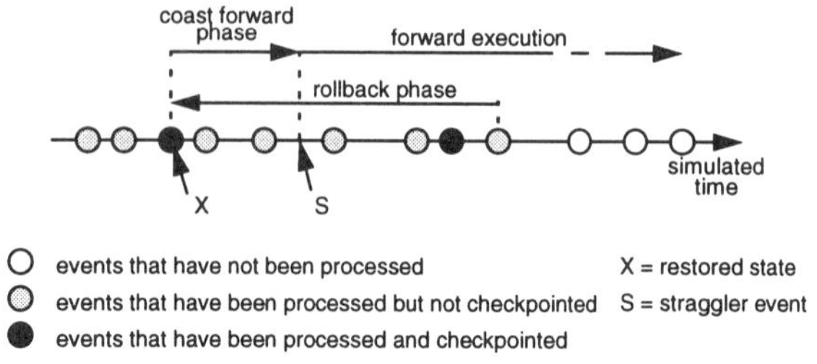

Figure 1: Simulation Snap Shot with $\chi = 5$

When a straggler event occurs, the saved state closest before the straggler event in simulation time (X in Figure 1) is recovered and the simulation coasts forward (without re-sending any messages) to the time of the straggler (S in Figure 1) where normal forward execution is resumed. We can therefore divide the simulation progress cycle of an LP into three different phases (Lin et al. 1993): forward execution phase; rollback phase; and coast forward phase.

If the frequency of state saving is reduced, the time to save states is also reduced and hence the cost for the forward execution phase is reduced. However, this reduction is at the expense of possibly longer rollbacks and of longer coast forward phases which increase the state restoration time. An appropriate state saving interval in a sparse state saving method should obtain a low cost from the forward execution phase while maintaining a low cost for the rollback and coast forward phases. Hence, this choice is dependent on the relations between the time to save the state of an LP; the time to coast forward or re-execute an event; the number of rollbacks; as well as the number of events involved in a rollback.

3 RELATED WORK

As the use of the Time Warp mechanism for optimistic synchronization of parallel discrete event simulations has increased, many researchers have presented different sparse state saving methods which differ in how to select the state saving interval. The approaches can be divided into two groups: *fixed sparse state saving*, where the chosen state saving interval is kept constant during the entire simulation, and *adaptive sparse state saving*, where the state saving interval is dynamically adapted to changes in monitored simulator parameters.

3.1 Fixed Sparse State Saving Method

In the simplest method, all LPs use the same static state saving interval, i.e. the simulation state for each LP is always saved after every χth event execution, where χ is the state saving interval. These methods work best when the simulation model is homogeneous, i.e. all LPs are identical and do not change behavior over time. The question is how to select an optimal state saving interval χ. (Lin et al. 1993) developed a manual regression model which works well for homogeneous models.

3.2 Adaptive Sparse State Saving Methods

(Fleischmann, and Wilsey 1995) performed an empirical study on sparse state saving approaches and present a heuristic algorithm for adapting to changes in the simulation model execution behavior. This method measures the time to execute N events and compares this measurement with the previous measurement. If the execution time has increased "significantly", the state saving interval is decreased by one otherwise it is increased by one.

(Rönngren, and Ayani 1994) developed an adaptive state saving method where the simulator dynamically changes the state saving interval for each LP depending on the individual LP's rollback behavior. Using this approach, a near optimum choice of the state saving interval can be selected at run time. For each LP, the monitored parameters are the number of processed events R_{obs} (including events that have been rolled back) and the number of rollbacks k_{obs} during an observation period. It is then possible to calculate the state saving interval χ_{min}, see Equation (1), which minimizes the sparse state saving overhead for the LP. In this equation, δ_s is the mean time needed to save the simulation state and δ_c is the mean execution time for an event in the coast forward phase. Equation (1) is then used to iteratively refine χ while adapting to changes in the rollback frequency R_{obs} / k_{obs}.

$$\chi_{min} = \left\lceil \sqrt{2\frac{R_{obs}\delta_s}{k_{obs}\delta_c}} \right\rceil \tag{1}$$

Recent work (Rönngren et al. 1996) shows that this adaptive method gives the best results when comparing the state saving performance for various sparse state saving techniques when simulating large, realistic cellular communication systems.

4 EVENT SENSITIVE STATE SAVING

We observe that previous work in estimating the optimal state saving interval typically depend on an estimation of the *mean* time for execution or coast forward of an event (δ_e or δ_c) or the estimation of the time to save state and time to coast forward (δ_s / δ_c) ratio. However, for these estimations to be optimal the variance should not be large, i.e. all types of events should have similar performance when comparing execution and coast forward times. Intuitively, even if there exist simulation models where the event execution and coast forward time variations are negligible, this can not generally be true. Clearly the coast forward time could be reduced if it was possible to reduce the probability of having to coast forward over large granular events. We therefore claim that this variance in event characteristics is an important factor when selecting state saving interval in a Time Warp synchronized discrete event simulation.

If we divide all possible types of events in a simulation model into a set of event classes, where members of each class exhibit a similar behavior when it comes to execution and coast forward times, we can treat each class as "individual events" without loosing generality. For a general simulation model the execution time for different event classes in an LP might vary considerably. We therefore propose a new sparse state saving method, called *event sensitive state saving* (ESSS), where we take this variance into consideration when selecting the optimal state saving interval.

4.1 The ESSS Approach

To further reduce the state saving overhead in sparse state saving methods we target the coast forward phase in the simulation progress cycle (see Figure 1). The time to execute this phase depends on the length of the state saving interval χ as well as the individual coast forward times for the events in the coast forward phase. The mean number of events in the coast forward phase is (χ-1)/2 under the assumption that state saving points and stragglers are independent. If we denote the *mean* coast forward time for all event classes as δ_c the average state saving overhead due to one coast forward phase is estimated as:

$$T_{coh} = \delta_c(\chi - 1)/2 \tag{2}$$

In traditional sparse state saving models, the state of an LP is saved after or before the execution of *any* event in a simulation run. From the coast forward point of view, this might not be the best approach. If the model under simulation is described by events where the coast forward time for different events has high variance, it should be more efficient to save the state after events that is *expensive*, i.e. with long coast forward times, compared to after events with short coast forward times. By employing a state saving technique that favors state saving after expensive events we could reduce the mean δ_c, i.e. it would be possible to reduce the *effective* δ_c that will add to the coast forward overhead T_{coh} in Equation (2).

The goal of the ESSS approach is therefore to implement a sparse state saving model where state saving is more probable after expensive high granularity events, than after events with lower granularity.

4.2 Execution Time Model

To understand the effects on the state saving overhead in sparse state saving we use Rönngren's execution time model (Rönngren, and Ayani 1994) where the simulation execution time T for an observation period when adopting a state saving interval χ is estimated as:

$$T(\chi) = R_{obs}\frac{\delta_s}{\chi} + k_{obs}\delta_r + k_{obs}\delta_c\frac{(\chi - 1)}{2} + R_{obs}\delta_e \tag{3}$$

In this expression, R_{obs} is the number of observed event executions (committed and rollbacked events), δ_s is the average time to save one state of the LP, k_{obs} is the number of observed rollbacks, δ_r is the average time to restore the state of the LP, δ_c is the average execution time for an event in a coast forward phase, and δ_e is the average execution time of an event in normal forward execution. The first term in the sum refers to the estimated time to save states. The last term is the time to execute events during forward execution, and together with the first term it reflects the time to execute the forward execution phase. The second term corresponds to the time to perform rollbacks and restore old states, i.e the rollback phase, and the third term estimates the time spent in coast forward phases, c.p. Equation (2).

The state saving overhead which depend directly on the state saving interval χ is term 2 and 3 in Equation (3):

$$T_{oh}(\chi) = R_{obs}\frac{\delta_s}{\chi} + k_{obs}\delta_c\frac{(\chi-1)}{2} \tag{4}$$

4.3 The ESSS Model

In the ESSS approach we classify each event class i in an LP by the following data and parameter estimations: f_i, the relative frequency of event class i, i.e. the observed number of executed events of class i divided by the total number of observed events; and δ_{ci}, the mean coast forward time for class i events.

Events that belong to an event class with high coast forward times or high granularity are called *heavy* events and correspondingly, events that belong to an event class with low coast forward times or low granularity are called *light* events. In Equation (3) and Equation (4), the execution time is estimated for an χ which does not explicitly account for whether the events in the coast forward phase are light or heavy. To be able to take this difference in event characteristics into consideration when deciding when to save state (provided that events of different classes are evenly distributed in time and events of the same class are not lumped together), we introduce a simulation tuning parameter: p_i, the probability of state saving after events of class i where $0.0 < p_i < 1.0$.

Thus, the average state saving interval χ_{ESSS} for the ESSS model with N event classes can be described as:

$$\chi_{ESSS} = 1/\left(\sum_{i=1}^{N} p_i f_i\right) \tag{5}$$

and the average coast forward event execution time δ_{cESSS} (provided that at least one p_j is less than one, i.e. the simulation state is not saved after each event which would imply copy state saving) as:

$$\delta_{cESSS} = \left(\sum_{i=1}^{N} (1-p_i)\ f_i\ \delta_{ci}\right)/\left(\sum_{j=1}^{N} (1-p_j) f_j\right) \tag{6}$$

Our goal is to reduce δ_{cESSS} in Equation (6) by saving heavy events with higher probability than light events, i.e. for heavy events we would like a high p_i and for light events a low p_i. By this approach it is possible to reduce the re-execution overhead in the coast forward phase.

4.4 Analytical Evaluation of ESSS

To evaluate the ESSS approach we compare the ESSS model to a sparse state saving (SSS) model with a fixed state saving interval given by minimizing $T_{oh}(\chi)$ in Equation (4), see Equation (1). We choose first to analytically study a simple model with the following parameters:

We have two event classes with events from each class equally frequent, i.e. N=2 and f_1=f_2=0.5. Events of class i=2 are considered heavy with a heavy factor HF, i.e. $\delta_{c2}=HF\delta_{c1}$. To find an estimate on how well the ESSS model performs we compare the state saving overhead when varying p_1 and p_2. The mean coast forward time for the ESSS model is described by Equation (6) and for the SSS model the mean coast forward time is:

$$\delta_{cSSS} = \sum_{i=1}^{N} f_i\ \delta_{ci} = \frac{\delta_{c1}+\delta_{c2}}{2} = \frac{\delta_{c1}(1+HF)}{2} \tag{7}$$

A comparison can be performed by estimating the simulation state saving overhead during an observation interval for ESSS and SSS respectively using Equation (4):

$$\frac{T_{ohESSS}(\chi)}{T_{ohSSS}(\chi)} = \frac{\left(\frac{\delta_s}{\chi} + \frac{k_{obs}}{R_{obs}}\delta_{cESSS}\frac{(\chi-1)}{2}\right)}{\left(\frac{\delta_s}{\chi} + \frac{k_{obs}}{R_{obs}}\delta_{cSSS}\frac{(\chi-1)}{2}\right)} \tag{8}$$

An important Time Warp simulation measurement is the efficiency parameter E which shows the ratio between the total number of committed events and the total number of processed events (both committed and rolled back events). This parameter together with the average rollback length RL, i.e. the average number of events a straggler causes the simulation to roll back across, can be used to describe the rollback frequency k_{obs} / R_{obs}:

$$\frac{k_{obs}}{R_{obs}} = \frac{1-E}{RL} \tag{9}$$

Furthermore we assume that the time to save the entire LP state δ_s in this simple model is the same as the time to re-execute or coast forward over a light event, i.e. $\delta_{c1} = \delta_s$. The state saving interval χ that minimizes the state saving overhead for SSS can be estimated by Equation (1) with δ_c equal to δ_{cSSS}. This gives the χ_{SSS} estimation:

$$\chi_{SSS} = \left\lceil \sqrt{2\frac{RL}{(1-E)}\frac{\delta_s}{\delta_{cSSS}}} \right\rceil \approx \left\lceil \sqrt{2\frac{RL}{(1-E)}\frac{2}{(1+HF)}} \right\rceil \qquad (10)$$

and the simulation state overhead comparison in Equation (8) can be estimated as:

$$\frac{T_{ohESSS}(\chi)}{T_{ohSSS}(\chi)} = \frac{\frac{\delta_s}{\chi_{ESSS}} + \frac{(1-E)}{RL}\delta_{c_{ESSS}}\frac{(\chi_{ESSS}-1)}{2}}{\frac{\delta_s}{\chi_{SSS}} + \frac{(1-E)}{RL}\delta_{c_{SSS}}\frac{(\chi_{SSS}-1)}{2}} \qquad (11)$$

If we for this analytical model vary the heavy factor *HF* and efficiency *E*, for a fixed rollback length *RL* of 2, we get the results in Figure 2 and 3 respectively. In Figure 2 we can see that the possible relative gain in reduced state saving overhead when using the ESSS state saving model is increased as the heavy factor is increased, i.e. a higher variance in the event coast forward time has a larger impact on state saving performance. For the situation when $\delta_{c2}=8\delta_{c1}$, the optimum choice of the state saving probabilities are p_1=0.0 and p_2=1.0, i.e. states are only saved after the heavy event (i=2). Performance improvement in the overhead directly associated with the state saving for the ESSS model compared to SSS is in this case close to a factor two.

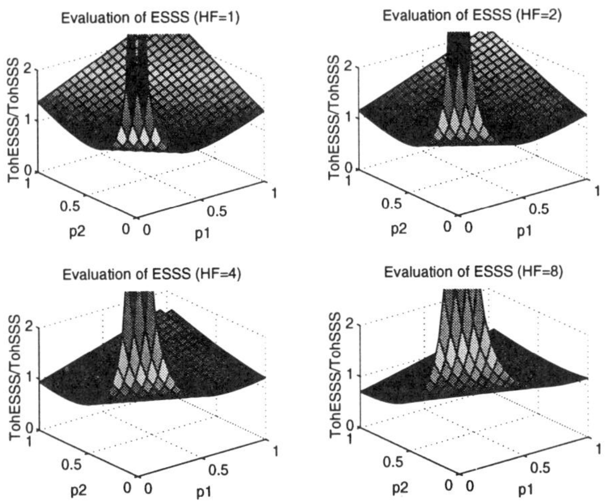

Figure 2: Analytical Estimation of State Saving Overhead for ESSS compared to SSS when varying the Heavy Factor *HF* (E=0.8, f_1=f_2=0.5)

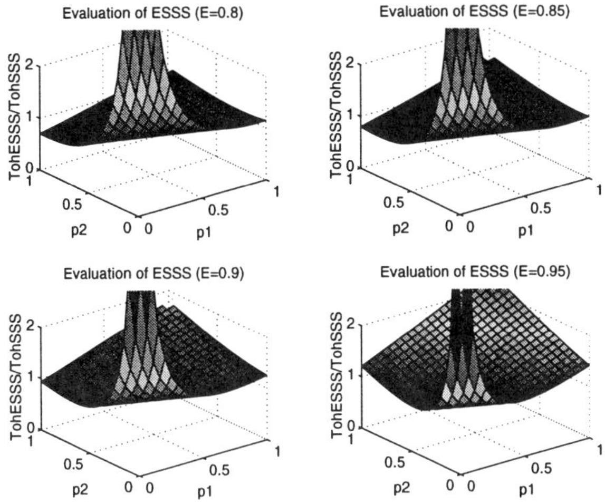

Figure 3: Analytical Estimation of State Saving Overhead for ESSS compared to SSS when varying the Efficiency *E* (HF=8, f_1=f_2=0.5)

When we vary the efficiency in the simple analytical example (see Figure 3) we can see that for lower efficiency (around 80%), i.e. for cases when the LPs are performing frequent rollbacks, the performance improvement is larger than for higher efficiencies. This also indicates that our attempt of targeting the coast forward phase could be a promising approach.

Summarizing the results from our analytical evaluation of ESSS, we claim that the ESSS model has a good potential to reduce the state saving overhead in sparse state saving approaches for simulation models which show a large variance in coast forward execution times.

5 PERFORMANCE EVALUATION

The impact of the ESSS model has been evaluated by implementation of the ESSS model on a Time Warp simulator targeted to shared memory multiprocessor workstations. The simulator is written in C and uses aggressive direct cancellation (i.e. anti-messages are sent immediately upon rollback) and static assignment of LPs to processors. GVT calculations as well as fossil (garbage) collection are performed once every second and on demand. Synchronization primitives are supplied by the p4 macro library (Butler, and Lusk 1994). The experiments were performed on a SUN multiprocessor workstation with 4 processors and 128 MB internal memory.

As simulation models we have selected a synthetic model, in which we could modify the parameters and hence the model behavior, as well as a real world simulation model which describes an existing physical system.

5.1 Experiments with a Synthetic Model

The synthetic model is a modification of the PHOLD model (Fujimoto 1989) which is a closed stochastic queueing model with a fully connected net of 64 nodes. Two types of equally probable messages are forwarded within the net. One type of message is labelled *light* and triggers an event that takes 30 microseconds to process, the other is labelled *heavy* and the time to execute the corresponding event is *HF* times longer. The message population is constant and messages are routed with 20% probability to the sending process and with 25% probability to one of eight so called *hot spots* that randomly move around in the network. Otherwise messages are equally likely to be routed to any of the other nodes. This model could be seen as a simple model of a mobile communication system with 64 nodes and 8 mobile terminals moving around in the system. The lighter messages trigger mobile position updates, whereas the heavier messages correspond to resource allocation events which typically are more costly.

Each node's state size is 1Kbyte and the time to save that state is approximately the same as the coast forward time for the lighter event, i.e. $\delta_s/\delta_{clight} = 1$. In the experiments we measured the event rate, the efficiency and the rollback length for various combinations of the two state saving probabilities p_1 (light event) and p_2 (heavy event). In each experiment approximately 300,000 events were committed, and each experiment was run three times.

For the first set of experiments, the heavy factor *HF* is eight and in average there is one light and one heavy message per node. The efficiency for this model is around 80%, i.e. this synthetic model is very similar to one of the analytical models studied in Section 4.4.

From the analytical analysis we would expect to get the best performance for the situation when states rarely are saved after light events (p_1 is close to zero) and almost always are saved after heavy events (p_2 is close to one). From the simulation results for the first set of experiments with the above described synthetic model, see Figure 4, we can see how varying the state saving probabilities affect the event rate, i.e. committed events per second. The results show that for this synthetic model the performance is improved when state is rarely saved after light events and more probably after heavy events, as would be expected from the analysis in Section 4.4. However, the best performance is not for $(p_1, p_2) = (0.0, 1.0)$ but for (0.0, 0.6), i.e state should not be saved after every heavy event, but approximately after every second heavy event even though the performance difference is very small compared to $(p_1, p_2) = (0.0, 1.0)$.

If we compare the ESSS performance for the synthetic simulation model and the first set of experiments with the

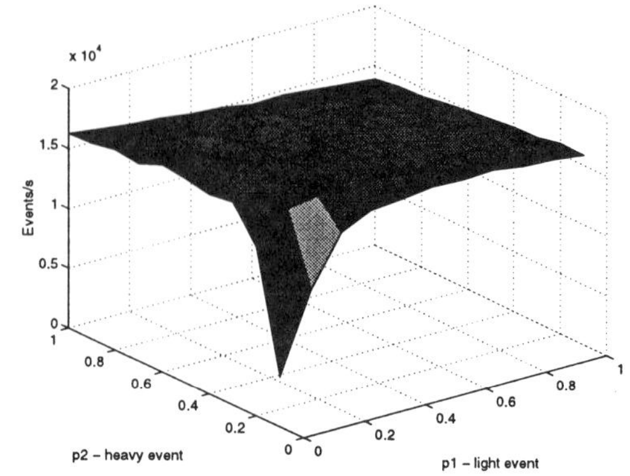

Figure 4: Simulation Results from the PHOLD Model with Hot Spots (f_1=f_2=0.5, *HF*=8)

best simulation performance when employing other sparse state saving techniques, we get the results in Table 1. However, these results show that the ESSS model does not outperform the sparse state saving method with fixed state saving interval of 4. This contradicts what we might have expected after studying the analytical analysis in Section 4.4. One factor contributing to this is the high variance in the actual state saving interval for the ESSS model compared to sparse state saving with fixed χ.

Table 1: Comparing Simulation Performance for the PHOLD Model with Hot Spots (f_1=f_2=0.5, *HF*=8)

State saving model	Mean χ	Efficiency [%]	Rollback length	Event rate [events/s]
Copy SS	1	79.82	1.85	14,073
Sparse SS	4	85.39	1.75	17,403
Adaptive SS	4.39	84.85	1.76	16,683
ESSS (p_1, p_2)=(0.0, 0.6)	3.32	83.97	1.78	17,223

For this PHOLD simulation model, rollbacks are typically short, only one or two events are rolled back, and even the rather high rollback frequencies of approximately 10% do not introduce high state saving overhead compared to the time needed for normal forward execution. The ESSS model shows better performance than Rönngren's adaptive state saving method, which partly can be explained by the overhead introduced by the adaptability in Rönngren's model. The traditional copy state saving that saves simulation state after each processed event is about 20% slower than the ESSS model as well as the sparse state saving method with χ=4.

For a second set of experiments, the heavy factor *HF* is sixty and in average there are two light and one heavy messages per node, i.e. f_1=2/3 and f_2=1/3. The efficiency for this model is lower than for the first set of experiments, around 75%. The reason to study this synthetical model is its resemblance with the real model to be studied in Section 5.2.

The results from the second set of experiments show that for a model with high variance in event granularity, the best performance is when state is rarely saved after light events and more frequently saved after heavy events. For this model the best performance is for (p_1, p_2) = (0.0, 1.0), see Figure 5.

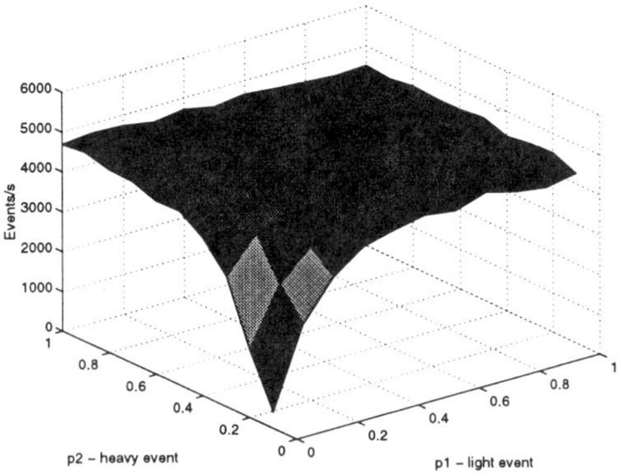

Figure 5: Simulation Results from the PHOLD Model with Hot Spots (f_1=2/3, f_2=1/3, HF=60)

A performance comparison for different state saving methods for the second model is found in Table 2. This PHOLD model also has short rollbacks, about two events in each rollback phase. However, the variance in event granularity is considerably higher for this model which results in increased rollback frequencies as well as decreased optimal state saving intervals. The ESSS model shows better performance than the other state saving methods for this set of experiments. The maximal event rate measured for ESSS is approximately 4% better than the event rate for Rönngren's adaptive method.

Table 2: Comparing Simulation Performance for the PHOLD Model with Hot Spots (f_1=2/3, f_2=1/3, HF=60)

State saving model	Mean χ	Efficiency [%]	Rollback length	Event rate [events/s]
Copy SS	1	74.57	2.09	4,558
Sparse SS	3	78.16	2.03	4,542
Adaptive SS	2.87	78.00	2.09	4,590
ESSS (p_1, p_2)=(0.0, 1.0)	2.49	75.49	2.12	4,772

5.2 Experiments with a Real World Model

The third simulation model is a mobile telecommunication network simulator which models a simple "ring" highway with uniformly distributed base stations. All cells are of equal size and have the same number of channels available. Calls from mobile users arrive according to a Poisson process and have exponentially distributed call durations. An initiated call is placed at a uniformly distributed random position on the highway. Mobility is modeled as a constant truncated Gaussian distributed random speed in one of the two possible directions. Each cell is simulated as an LP. The arrival of calls are generated by a central generator LP, and handover between cells is attempted by sending messages between cell LPs.

The model used for the performance evaluation include 20 base stations, each with 10 available channels, and each in a cell with size 2000 meters. Calls arrive at the rate of one call per second and with an average holding time of 120 seconds. The mobile terminals move with an average speed of 90 kilometers/hour. After an initialization phase, the mobile terminals will be approximately uniformly distributed over the cells.

The two most frequent types of events are *position updates* and *resource allocations* which account for about two thirds and one third of all events respectively. A resource allocation (heavy event) takes approximately 60 times longer than a position update (light event) due to the computation needed to calculate signal to interference ratios when allocating channel resources. For this model, the time to save the state of an LP is approximately the same as the coast forward time for the lighter event, i.e. $\delta_s/\delta_{clight}=1$. In the experiments we measured the event rate, i.e. the number of committed events per second for various combinations of the two state saving probabilities p_1 (light event) and p_2 (heavy event). In each experiment 300,000 events were committed, and each experiment was run three times.

Comparing the ESSS performance for this model with the best performance for the other three sparse state saving approaches, we get the results in Table 3:

Table 3: Comparing Simulation Performance for Mobile Communication Model

State saving model	Mean χ	Efficiency [%]	Rollback length	Event rate [events/s]
Copy SS	1	88.38	116.38	9,464
Sparse SS	12	87.46	129.39	10,150
Adaptive SS	12.70	87.48	126.72	10,089
ESSS (p_1, p_2)=(0.05, 0.15)	10.06	87.58	137.11	10,124

These results show that the ESSS model detects the difference in event granularity and hence the best performance for the ESSS model is for a lower mean state saving interval compared to the sparse and adaptive state saving methods. The choice of parameters (p_1, p_2) = (0.05, 0.15) indicates that the ESSS optimal performance is reached when the probability of saving state after light events are lower than after heavy events. This is in accordance with the analytical model. For this simulation model the ESSS, sparse and adaptive state saving methods reach about the same peak event rate. This is somewhat contradictory to the results from the second set of experiments with the synthetic model with high variance in event granularity. The reason for this is the difference in model geography. The synthetic, fully connected, model has a larger fan-out than the one dimensional real model which results in a higher number of rollbacks and shorter rollback lengths for the synthetic model. This difference indicates that the ESSS model has a larger potential for models with many and short rollbacks. This can be understood by considering the fact that most rollback recoveries start with a coast forward phase. In the case of ESSS there is a lower probability that the first event in a coast forward phase is a heavy event when ESSS is used compared to the other methods for sparse state saving. Thus for a constant efficiency ESSS is favored by a higher number of short rollbacks as compared to cases with fewer and longer rollbacks.

6 CONCLUSIONS

In this paper we have introduced a novel approach for sparse state saving in Time Warp synchronized discrete event simulation, event sensitive state saving (ESSS). This method is based on the observation that the time to execute the coast forward phase in the simulation progress cycle is dependent on the re-execution time of the coast forwarded events. For many important types of simulation models such as battle field simulations and simulations of mobile communication systems there is a significant difference in the coast forward cost for different types of events. This has not been captured in previously proposed methods for sparse state saving. In contrast, the ESSS state saving method uses information on the coast forward cost for individual events to decide when to save or checkpoint a state of logical processes.

In this paper we have presented an analytical analysis of the potential benefits from using an event sensitive approach. This analysis indicates that overhead due to sparse state saving in some cases can be reduced by nearly 50% when using the ESSS approach compared to other methods presented in the literature. Results for synthetic as well as realistic simulation models show that the ESSS model is a promising approach for simulation models where event granularity or execution time has large variance. ESSS is in particular favored by short rollbacks which often occurs in real well behaved simulation models. Our experimental results show that the ESSS approach could result in a more than 4% over all performance improvement to other state saving methods.

The result from this work shows that an efficient state saving method which takes more of the simulation modelling details (such as event granularity) into consideration is likely to improve state saving performance. We plan to continue this work and implement adaptive versions of the proposed method for sparse state saving.

REFERENCES

H. Bauer, and C. Sporrer. 1993. Reducing Rollback Overhead in Time Warp Based Distributed Simulation with Optimized Incremental State Saving. In *Proc. of the 26th Annual Simulation Symposium*, 12-20.

R. Butler, and E. Lusk. 1994. Monitors, messages, and clusters: the p4 parallel programming system. *Parallel Computing*: 20.

J. Fleischmann, and P. A. Wilsey. 1995. Comparative Analysis of Periodic State Saving Techniques in Time Warp Simulators. In *Proc. of the 9th Workshop on Parallel and Distributed Simulation*, 50-58.

R. M. Fujimoto. 1989. Time Warp on a Shared Memory Multiprocessor. *Transactions of the Society for Computer Simulation*, Vol. 6, No. 3:211-239.

R. M. Fujimoto. 1990. Parallel Discrete Event Simulation. *Comm. of the ACM*, Vol. 33, No. 10:30-53.

D. R. Jefferson. 1985. Virtual Time. *ACM Transactions on Programming Languages and Systems*, Vol. 7, No. 3:404-425.

Y.-B. Lin, B. R. Preiss, W. M. Loucks, and E. D. Lasowska. 1993. Selecting the Checkpoint Interval in Time Warp Simulation. In *Proc. of the 7th Workshop on Parallel and Distributed Simulation*, 3-10.

A. C. Palaniswamy, and P. A. Wilsey. 1993. An Analytical Comparison of Periodic Checkpointing and Incremental State Saving. In *Proc. of the 7th Workshop on Parallel and Distributed Simulation*, 127-134.

B. R. Preiss, W. M. Loucks, and I. D. MacIntyre. 1994. Effects on the Checkpoint Interval on Time and Space in Time Warp. *ACM Transactions on Modeling and Computer Simulation*, Vol. 4, No. 3:223-253.

R. Rönngren, and R. Ayani. 1994. Adaptive Checkpointing in Time Warp. In *Proc. of the 8th Workshop on Parallel and Distributed Simulation*, 110-117.

R. Rönngren, M. Liljenstam, J. Montagnat, and R. Ayani. 1996. A comparative study of state saving mechanisms for Time Warp synchronized parallel discrete event simulation. In *Proc. of the 29th Annual Simulation Symposium*, 5-14.

AUTHOR BIOGRAPHIES

SVEN SKÖLD is a Ph.D. student in the Department of Teleinformatics at the Royal Institute of Technology, Stockholm, Sweden. His research interests include efficient simulation techniques and in particular parallel discrete event simulation.

ROBERT RÖNNGREN is an associate professor in the Department of Teleinformatics at the Royal Institute of Technology, Stockholm, Sweden. His research interests include sequential and parallel discrete event simulation.

Proceedings of the 1996 Winter Simulation Conference
ed. J. M. Charnes, D. J. Morrice, D. T. Brunner, and J. J. Swain

EVENT SCHEDULING SCHEMES FOR TIME WARP ON DISTRIBUTED SYSTEMS

Eunmi Choi

Department of Computer Science
Michigan State University
East Lansing, MI 48824-1027, U.S.A.

Dugki Min

Department of Computer Engineering
KonKuk University
MoJinDong KwangJinGu Seoul, KOREA

ABSTRACT

To design an efficient event scheduling strategy for Time Warp on distributed systems, we study two major factors. The first factor is the amount of computation to be processed in a processor between two consecutive communication points. The computation granularity must be small enough to prevent Time Warp from progressing events over-optimistically, while large enough to minimize the degrading effects of performance due to the communication overhead. The second factor is the balancing progress of logical processes in a processor. Within the range of the given computation granularity, logical processes assigned to a processor progress at different speeds and this may cause more rollback situations. Considering these two factors, we propose two event scheduling schemes: the *Balancing Progress by Execution Chance (BPEC)* scheme and the *Balancing Progress by Virtual Time Window (BPVTW)* scheme. The BPEC scheme controls the progress of a process by limiting the number of chances for the process to schedule events, while the BPVTW scheme controls the progress by using a virtual time window, which allows events to be executed only if their timestamps are within the interval of the virtual time window. We have performed experiments on a cluster of DEC Alpha workstations to demonstrate the performances of both schemes.

1 INTRODUCTION

In simulating large VLSI circuits on distributed memory systems, one of the most important considerations for the Time Warp (Jefferson 1985) is to schedule events. Although the number of processors in distributed-memory systems is much smaller than that of the logical processes in VLSI circuits, the local memory size of each processor is so large that many logical processes can be allocated to a processor. The ratio of the number of logical processes to the number of physical processors is called the *LP ratio.* It represents the number of assigned logical processes per processor. For the simulation of VLSI circuits in which hundreds of thousands of gates are integrated in a chip, the LP ratio on distributed-memory systems of hundreds of processors is in the range of a couple of thousands. In addition to the large LP ratio, because of the optimistic characteristic of Time Warp, many processes allocated in a processor are likely to be active with many unprocessed events. Among the unprocessed events of active processes, the processor decides which event is executed first. The order of event execution determines the progress of the processes, and therefore it can significantly affect the performance of Time Warp on distributed-memory systems.

In this paper, we investigate two major factors in order to design an efficient event scheduling strategy for Time Warp on distributed-memory systems. The first factor is the amount of computation to be processed in a processor between two consecutive interprocessor communication points. This quantity is called the *computation granularity.* By adjusting the computation granularity to be larger than a certain degree, we can prevent the communication overhead from dominating the overall simulation performance. A large granularity, however, could increase the number of rollbacks, wasting the simulation time for recomputations. That is, while processors execute a large number of events without communication, processes could propagate many immature events and therefore the protocol should recover the earlier system state from the wrongly-advanced state when an event arrives whose timestamp is smaller than the current LVT (Local Virtual Time). A considerable amount of overhead can occur in the rollback situations because the past state of each process should be saved and recovered by re-computations. As a consequence, the appropriate computation granularity might be small enough to prevent Time Warp from progressing events over-optimistically, while keeping large enough to minimize the communication overhead. The second factor is the balancing progress of logical processes in a processor. Some logical processes may advance too far ahead of others, possibly leading to inefficient usages of memory and excessive rollbacks, especially for large application problems. By adjusting the granularity, we can control the degree of optimism appropriately as described in the first factor. However, within the range of the given computation granularity, the logical processes assigned to the same processor would progress at different speeds and this may cause more rollback situations. If we control the speed of each process by giving more chances of executing events to slower processes, we may reduce the frequency of rollback situations. We study two schemes of balancing the progress of processes; one of which is to restrict a process to execute events no more than the given number of times, and the other is not to allow an event to be executed if its timestamp is far ahead of other events in different processes. The details of those schemes are studied in Section 3.

The following section describes the simulation model and the terminologies used in this paper. In Sections 3, the event scheduling schemes are presented with the issues of computation granularity and balancing progress of processes. Section 4 shows the experimental results on a distributed system. The concluding remarks are presented in Section 5.

2 THE SIMULATION MODEL

Time Warp has been studied by several researchers recently on distributed-memory systems. Carothers et al. (1994) studied the effect of communication delay on a distributed system when the computation granularity is fixed to the event granularity. Preiss and Loucks (1995) studied memory management techniques on a distributed system. This paper focuses on event scheduling issues of the optimistic protocol on distributed-memory systems. Our implementation of Time Warp on distributed-memory systems is called *Distributed Optimistic Protocol (DOP).* The DOP executes events in an event-driven and optimistic fashion.

2.1 Rollback Handling

To handle the rollback situation, we use a cancellation mechanism called the *immediate cancellation* (Chung and Chung 1991). In this cancellation mechanism, the process that receives a straggler re-starts to process from the point of the straggler's timestamp, by ignoring all messages sent so far and all the already-processed events. The cancellation scheme applies to the input event queue where a straggler arrives. In the immediate cancellation, there is no antimessage to correct the already-sent wrong events unlike the aggressive cancellation mechanism (Jefferson 1985) and the lazy cancellation mechanism (Gafni 1988). At this point, it is the same as the direct cancellation mechanism (Fujimoto 1989) on shared memory systems.

2.2 Data Structure for Event Manipulation

In parallel discrete event simulation, the events that are generated but not yet executed are pending in the input queues. Each processor schedules the pending events according to the event scheduling strategy. The implementation of data structures of event queues and the event scheduling strategy have crucial influences on the simulation performance.

In our implementation, each logical process has its own event queue where all unprocessed events are pending and the already processed events are stored together. This type of implementation with an individual event queue per process is better than one central event queue shared by all processes in a processor, especially when the LP ratio is large. If all processes share one event queue together in a processor, it takes times to search and manipulate all pending events and rollbacked events. The data structure of the individual event queue per process is also easily adaptive to the situation of migrating a process to another processor by a dynamic load management mechanism. In each process's event queue of our implementation, all the events are sorted in the increasing order of timestamps. Instead of using separate output event queues, using a single event queue which can be used as both input and output event queues can reduce the queue manipulation time when a straggler arrives, by eliminating the corresponding event moving and queue handling from output event queues.

Our event scheduling schemes described in the next section basically follow the 'smallest-timestamp-first' selection scheme. Since searching all assigned processes linearly for the smallest timestamped event takes a lot of time in a large-scale application, we use a central priority queue structure for scheduling events in a processor. Several data structures of the priority queue have been proposed such as the calendar queue (Brown 1988) and the skew heap (Jones 1989). In this paper, we use the heap structure for the priority queue. Each item in the priority queue has a timestamp as the key to be sorted and a process index number. After each processor evaluates processes and determines whether they are active or not, unprocessed events are selected from the active processes according to the event scheduling scheme, and enqueued with the timestamps into the priority queue by the processor scheduler. From the scheduling heap, each processor dequeues the smallest one from the top and executes the event on the corresponding process.

2.3 GVT Computation

The GVT (Global Virtual Time) computation is necessary in optimistic protocol for fossil collection and for checking the termination condition. The GVT on distributed-memory systems is obtained by computing the minimum value of LVTs from all processes and the timestamps of events in the in-transit messages, which are sent by a sender processor and not yet received by the receiver processor, where the LVT of a process is determined by the smallest unprocessed event in a process. Since each processor maintains many processes, PVT (Processor Virtual Time) is computed as the minimum value of LVTs among the processes in a processor with considering the in-transit messages, and it is used to compute the GVT. Thus, the GVT can be computed as the minimum value of PVTs from all processors. We use a token passing mechanism to collect PVT values asynchronously from processors and compute the GVT value.

3 EVENT SCHEDULING

As one of the most popular simulation applications, VLSI circuits involve a large number of objects performing a small amount of computation for each communicating event. A VLSI circuit usually has several tens to more than hundreds of thousands of gates (or logical processes). The manipulation of an event requires around a hundred machine instructions; its event granularity is very small. In contrast, distributed memory systems consist of a much smaller number of big processors and the interprocessor communication latency is large. Thus, each processor should schedule the unprocessed events of a large number of assigned processes efficiently in order to achieve good performance. As we mentioned in Section 1, there are two important considerations we should make as we design an event scheduling strategy: the computation granularity and the progress balancing.

3.1 Computation Granularity

Computation granularity is the amount of computation to be processed by a processor between two consecutive interprocessor communication points (Choi and Chung 1995). This quantity is also called the *grain size*. If there is little overhead of interprocessor communication, the computation granularity is not a matter of concern. In this case, the ideal strategy is to perform an interprocessor communication for each event execution in order to exchange information as soon as possible. Frequent communications, however, may cause excessive rollbacks due to the tendency of overly optimistic execution in Time Warp. Therefore, a mechanism for restraining the overly optimistic execution is required in this case. The implementations of Time Warp on shared memory systems use this 'communication-per-event' style for the communicating pattern. As an exception, Das et al. (1994) discuss about the possibility of improving performance by processing a batch of k events, where k is the size of the batch. They claim that the batch processing approach reduces queue management overheads somewhat.

In distributed-memory systems, the communication latency is long, and therefore the computation granularity must be an important issue to be studied. An experimental study (Carothers et al. 1994) has investigated the effects of communication overhead on the performance of Time Warp, considering two types of simulation applications; one with a large event granularity and the other with a small event granularity. In the study, the computation granularity is fixed to the event granularity without allowing several events to be executed all at once. According to their performance results, the communication latency in distributed computing environments can significantly degrade the efficiency of Time Warp in the application problems which contain a large number of logical processes with a small event granularity. In contrast, for applications having large grained events, the communication latency appears to have little effect on the performance of Time Warp.

It is clear that a substantial amount of computation should be performed between interprocessor communications in order to yield appropriate performance results on distributed systems. In VLSI circuit simulations which have a small event granularity, a large computation granularity could be achieved by handling a batch of events together between communication points. However, the batch processing of a number of events with infrequent communications may unbalance the progress of logical processes in a processor from those of other processors. In this case, there is a high possibility that the incoming events from other processors become stragglers. The computation granularity, thus, should be tuned properly, depending on the characteristics of simulation application problems and the used computer systems. The optimal granularity should be large enough that the simulation performance cannot be degraded by the communication overhead, and small enought that rollback situations cannot be excessive.

3.2 Balancing Progress of Processes

The major characteristic of Time Warp is that each process can asynchronously execute its unprocessed events without considering causality effects. Due to this characteristic, some logical processes may advance too far ahead of others, possibly leading to the inefficient use of memory and excessive rollbacks. This situation might happen particularly for a large-scale simulation application with a small event granularity, such as VLSI circuits. By using the scheduling scheme of balancing the progress of logical processes, we can prevent the simulation from propagating incorrect computations too far ahead and reduce seriously excessive rollback situations. Two balancing schemes are presented in the following subsections.

3.2.1 Balancing Progress by Execution Chance (BPEC)

The BPEC is a scheme that balances the progress of logical processes. It limits the number of chances (opportunities) of executing events per logical process. In this scheme, we have two controllable parameters: one parameter for the computation granularity and the other parameter for balancing progress.

As the first parameter to control the computation granularity, this scheme sets the maximum number of events that can be executed by a processor between communication points. This quantity is called the *maximum batch size (MBS)*. Within the limit of the MBS, the processor scheduler selects events from logical processes through the central scheduling priority queue. That is, each processor selects and executes events up to the limit of the MBS. The scheduler counts the number of events executed in the processor since the last communication point while selecting events from the priority queue. Once the number of executed events reaches MBS, the processor stops scheduling events and performs the interprocessor communication. The second parameter is the maximum number of chances needed to execute events per logical process. When the scheduler selects events from the logical processes through the priority queue, it also counts the number of executed events per logical process. The purpose of counting events per process is to balance the progress of processes. By giving the appropriate execution chances to processes, we can achieve the benefit of using parallel optimistic protocol. As well as boosting only processes far behind, the overall progress of processes can advance by optimistically executing all processes. The maximum number of chances that a logical process can execute events between interprocessor communications is called the *Balancing Progress Chances (BPC)*. If a logical process no longer has unprocessed events or it has already executed as many events as the BPC, then the logical process is not allowed to enqueue its events into the priority queue.

By controlling the maximum allowable chances of executing unprocessed events per logical process, the BPEC scheme not only balances the relative speed of the processes' progress, but also controls the computation granularity. In the case that the MBS is set to be larger than the optimal value of grain size, the computation granularity can be adjusted by using the proper BPC value. When all logical processes are individually controlled by the BPC, the total number of executed events in a processor cannot increase too much, and the granularity between two communication points will be adjusted properly. The BPC is also able to control the event schedule by giving an equal number of execution chances to the logical processes. Instead of only executing events on a specific logical process, every process has the same chance to execute its events. Thus, the BPEC scheme contributes to progress balancing of logical processes and reducing the consequent rollback situations. Due to the characteristic of confining the strict upper bound of the number of events that can be executed by a processor between interprocessor communications, this scheme is a static method of controlling computation granularity.

3.2.2 Balancing Progress by Virtual Time Window (BPVTW)

Unlike the BPEC scheme, the BPVTW does not have the strict upper bound of execution chances for each logical process before interprocessor communication. Rather, the relative progress of logical processes is controlled with regard to the timestamp of events. That is, the scheduler prevents a process from executing events if the process is far ahead of the other processes in virtual time. For this purpose, we employ a virtual time window whose base is the GVT such that logical processes can execute only events having timestamps within the interval of the virtual time window. This virtual time window is called the *Balancing Progress Window* and the size of the window is denoted by BPW. Thus, the Balancing Progress Window has the interval $[GVT,\ GVT + BPW]$.

In this scheme, the overall progress is controlled by each logical process not by each processor. As in the BPEC scheme, the processor scheduler selects the smallest timestamped event from the central scheduling queue. The scheduler, however, does not count the number of executed events. When a logical process has an unprocessed event and its timestamp is in the interval $[GVT,\ GVT + BPW]$, the event is enqueued into the central scheduling queue. If there is no unprocessed event whose timestamp is in the interval $[GVT,\ GVT + BPW]$, the process does not have any chance to execute events until the next cycle after communication. In other words, the scheduler handles events until there are no available unprocessed events in the scheduling queue. Once the scheduling queue is empty, the processor communicates with other processors.

The computation granularity in the BPVTW scheme is indirectly controlled by the BPW value. Since there is no static limitation of the maximum number of events to be executed by a processor per communication, we do not employ the MBS. Instead, the BPVTW scheme concerns the real progress of processes with their timestamps in order to balance the progress of logical processes. The computation granularity is consequently adjusted when the scheduler restricts scheduling events according to the BPW value. Because there is not a limited number of executed events, the actual computation granularity may vary depending on the progress of the simulation. In this sense, the BTVTW scheme is considered a dynamic method of controlling the computation granularity.

A similar concept has been proposed as *Moving Time Window* (MTW) (Sokol et al. 1989). Although the MTW mechanism didn't provide considerable improvement on some cases (Fujimoto 1990), as it will be shown from the experimental results in Section 4, the BPVTW scheme achieves a good performance improvement for large-scale and small-event-granularity applications on parallel and distributed systems due to its ability to control the computation granularity. In addition, for those applications with a large LP ratio, the BPVTW scheme provides the balancing effects as well as control over the degree of optimism.

4 EXPERIMENTAL RESULTS

This section presents the experimental results on a cluster of six DEC Alpha workstations interconnected by a DEC GIGASwitch through FDDI. The DEC Alpha 3000 workstation has a 133 MHz clock rate and a 32 MB memory. The GIGAswitch supports a peak data transfer rate of 200Mbits per second. As benchmark circuits, we use several circuits from the ISCAS89 benchmark suite. In the circuits, we have the D f/f clocking interval set to be the same as the input clocking interval. The logical processes are randomly partitioned into processors. For parallel processing and the interprocessor communication on the distributed system, we use the PVM (Parallel Virtual Machine) 3.3 (Oak Ridge National Laboratory 1994). Time Warp is implemented on the distributed system as a master-slave model.

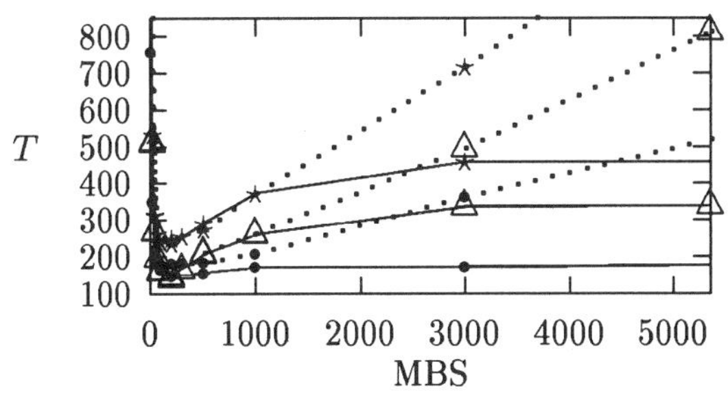

Figure 1: Total Execution Time (T) with BPEC Scheme for S38584 (Curves with Stars), S35932 (Curves with Triangles), S38417 Circuits (Curves with Dots)

4.1 Performance Measurement with the BPEC Scheme

This subsection explains the experimental results when the BPEC scheme is applied as the event scheduling technique to Time Warp on the distributed system.

4.1.1 Effects of the MBS

To study the effect of varying the MBS, we have performed experiments with three different circuits. Figure 1 shows the total execution time for the S38584 (curves with stars), S35932 (curves with triangles), and S38417 (curves with dots) circuits as a function of the MBS. The solid curves represent the case of $BPC = 1$ and the dotted curves show the case of $BPC = \infty$. The number of input vectors is 50, and the BPC is set to 1 and the unlimited value. The three circuits have the following LP ratios; 6733, 5357, and 6074, respectively.

As we mentioned in the previous section of the BPEC scheme, the computation granularity is controlled by values of the MBS and the BPC. To focus on only the effects of the MBS, in this subsection we consider only the dotted lines where the BPC has the unlimited value in Figure 1. As shown by the dotted lines, the total execution time can be minimized when the MBS is tuned properly. When the MBS is near to 1, the total execution time increases very sharply as the MBS decreases. In this situation, the computation granularity determined by the MBS is so small that the simulation spends most of the time in the frequent communications. It implies that

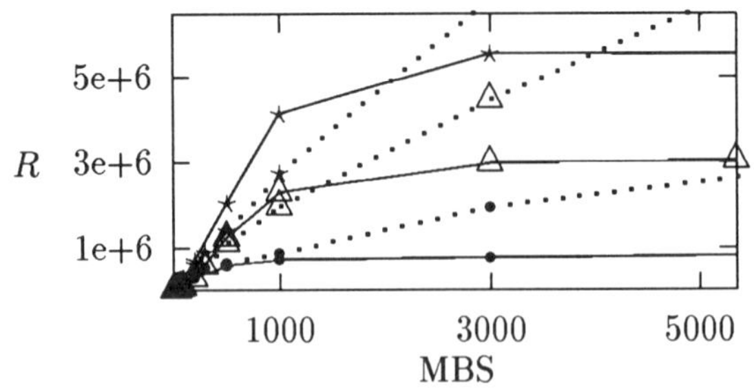

Figure 2: Number of Rollbacks (R) with BPEC Scheme for S38584 (Curves with Stars), S35932 (Curves with Triangles), S38417 Circuits (Curves with Dots)

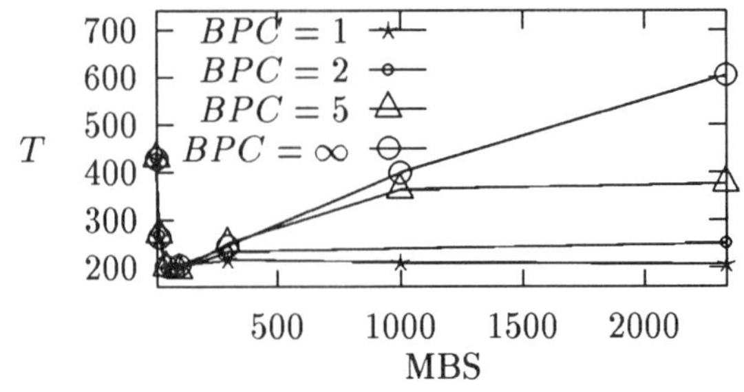

Figure 3: Total Execution Time (T) with BPEC Scheme for S13207 Circuit

the MBS should be larger than such a small value and that a substantial amount of computation could be performed between interprocessor communication points. The appropriate MBS will minimize the degrading effects of the communication overhead. In the extreme case of MBS = 1, the computation granularity is the same as the event granularity.

On the contrary, the total execution time increases linearly as the MBS increases after 200, approximately. It is because a large computation granularity may increase the number of rollbacks. To observe this relationship, Figure 2 shows the number of rollbacks varying the value of the MBS. Again, the dotted lines are for the case when the BPC is set to the unlimited value and the solid lines are when the BPC is set to 1. As mentioned before, we focus on the dotted lines in this subsection to focus on the effect of the MBS. In the dotted lines, the number of rollbacks is very small when the MBS is near to 1 because the overall simulation advances in a conservative fashion. As the MBS increases, the number of rollbacks increases almost linearly. This result implies that the larger the computation granularity is, the higher the probability is that immature events are propagated to other processors.

As a consequence, the appropriate value of the MBS should be as small as possible, but larger than a certain value so that the communication overhead can be reduced as much as possible. As for the cases in Figure 1, the optimal performance occurs when the MBS is around 200.

4.1.2 Effects of the BPC

In this subsection, we consider all the curves in Figure 1 to study the effects of the BPC. When the MBS is much larger than the optimal value, the simulation performance depends on the BPC value. Figure 3 shows the simulation results with the S13207 circuit. In the simulation, 100 input event vectors are used. Like the circuits in Figure 1, the curves in Figure 3 also reach steady performance when the BPC has small values, such as 1, 2, and 5. At this steady state, the BPC value controls the computation granularity because of the limited number of executed events per process. With a small BPC, processes have limited chances to execute events although they have more unprocessed events. Thus, a fast process is prevented from executing too many events so that the process cannot go too far ahead of the others, and we are then able to obtain the balancing effect with a small BPC value. In the interval between the optimal and the steady states, the computation granularity is controlled not only by the BPC, but also by the MBS. The depth of the valley of each curve at the optimal performance is deeper as the value of the BPC becomes larger. As shown in Figure 3 for the S13207 circuit, the simulation results with the BPC = 1 produce better performances than when the BPC is larger.

4.1.3 Effects of the LP Ratio

In order to study the effects of the LP ratio, we stage a set of simulations with three variations of the S35932 circuit. The experimental results are given in Figure 4. The D35932 circuit and the T35932 circuit are the circuits that are constructed by connecting the original S35932 circuit double and triple times, respectively. Therefore, the T35932 circuit has around a hundred thousand logic gates. Since the two circuits have logical gates multiple times of the S35932 circuit, they have multiple times higher LP ratios than the S35932 circuit. For all three curves, the BPC is fixed at 1. As before, each curve shows the optimal and the steady performances as the MBS varies. However, unlike the S13207 circuit which has the steady performance with the optimal performance on all the ranges

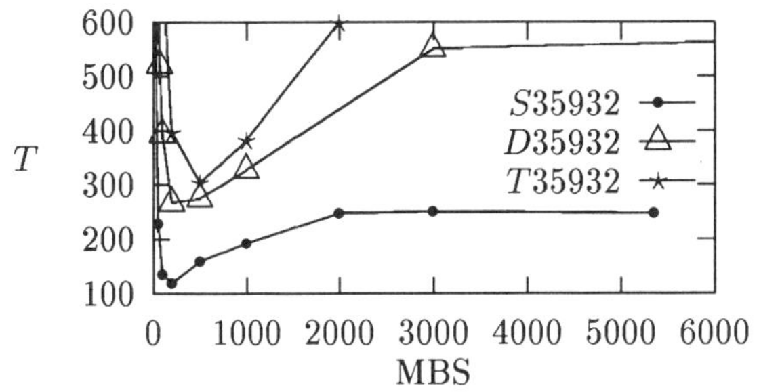

Figure 4: Effect of Granularity on Total Execution Time (T) by Varying LP ratio

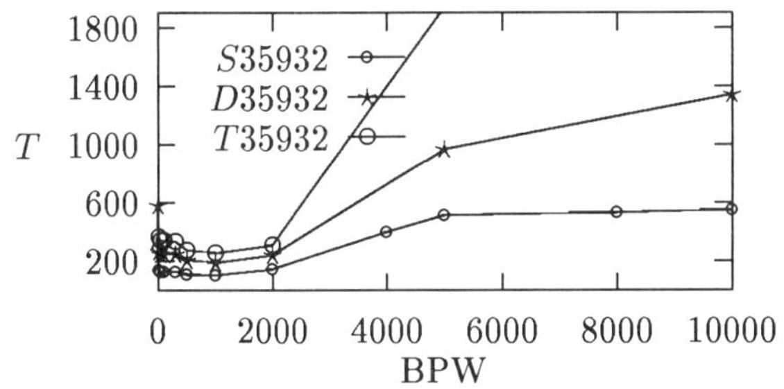

Figure 5: Total Execution Time (T) with BPVTW Scheme for S35932

of MBS when the BPC is 1, the steady performances of the three circuits in Figure 4 are much higher than their optimal performances. The larger the LP ratio is, the larger the difference is. Interestingly, even if tens of thousands of logic gates are assigned in a processor, the simulation shows reasonably good performance on distributed systems if the MBS is tuned appropriately. These results imply that the effects of computation granularity on the performance become more significant as the LP ratio becomes large enough.

4.2 Performance Measurement with the BPVTW Scheme

Additional sets of experiments have been performed with the BPVTW scheme on the distributed system. Figure 5 presents the experimental results in terms of the total execution time, varying the BPW. The three circuits, S35932, D35932 and T35932, are used, which are described in Section 4.1.3.

In the figure, all three circuits show good performances up to the BPW value of 2000. Unlike the BPEC scheme which shows sharp increases in total execution time for small values of BPC, the BPVTW scheme shows good performance even with very small values of BPW. It is because each process adds the unit delay of virtual time in executing events, and many events exist in the narrow virtual time interval. The interval of BPW which contains the optimal performances is quite wide in comparison to the sharp optimal range in the BPEC scheme. Thus, the progress balancing between logical processes with the virtual time window is effective and works well in the wide interval of BPW. This is because the BPVTW scheme concerns the actual progress of processes with their timestamps.

On the contrary, when the BPW is larger than 2000, the curves of total execution time increase with different rates as the BPW increases. In this larger window interval, the virtual time window does not take an important role of balancing progress of logical processes because there are many executed events in this interval. Also, the computation granularity is not restrained by the BPW. Thus, too many unprocessed events are executed, propagating lots of immature events. The larger the LP ratio is, the more unprocessed events are generated. In the figure, the T35932 circuit shows the worst performance when the BPW is larger than 2000. The D35932 circuit is worse than the S35932 circuit. Therefore, when the BPVTW scheme is used as the event scheduling method, the BPW must be tuned properly. Otherwise, the simulation performance will be as bad as if there were no limit of the MBS and the BPC in the BPEC scheme.

As discussed in Section 3, the BPVTW scheme does not control the computation granularity in a direct way as in the BPEC scheme. Rather, by using the virtual time window per logical process, the computation granularity is indirectly controlled while the progress of logical processes are balanced. In order to observe the effects of the BPW on the computation granularity, the average number of executed events between interprocessor communications are also shown in Figure 6. In the figure, where the simulation shows the almost optimal performance, the average number of executed events per communication is around 400 or 700, depending on the circuit. This amount is similar to the optimal grain sizes for the same circuits as shown in Figure 4.

5 CONCLUSION

In this paper, we propose two efficient event scheduling schemes on the optimistic protocol: the BPEC scheme and the BPVTW scheme. The BPEC scheme limits the number of executed events per logical process between interprocessor communications by using the BPC parameter. The BPVTW scheme controls

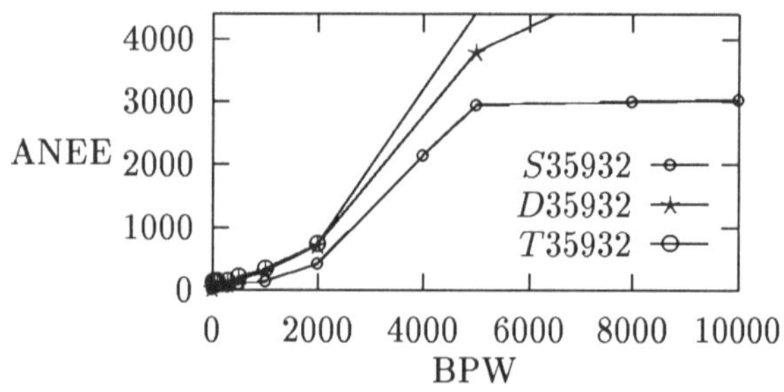

Figure 6: Average Number of Executed Events ($ANEE$) with BPVTW Scheme for S35932

the progress of processes in virtual time by using the BPW parameter, and prevents processes from executing events far ahead of other processes.

We experimented with the DOP considering the design issues of the control of the computation granularity and the event selection schemes on a cluster of DEC Alpha workstations. According to our experimental results, both schemes show good performances when the computation granularity is adjusted properly. Otherwise, communication overheads or excessive rollback situations may significantly degrade the simulation performance. The results also showed that the progress balancing among logical processes have significant effects on obtaining the optimal performance. In addition, it is shown that the LP ratio and the characteristic of the simulated circuit affect the simulation performance.

REFERENCES

Brown, R. 1988. Calendar Queues: A Fast O(1) Priority Queue Implementation for the Simulation Event Set Problem. *Communications of ACM*, 1220–1227.

Carothers, C. D., R. M. Fujimoto, and P. England. 1994. Effect of Communication Overheads on Time Warp Performance: An Experimental Study. In *Proceedings of the 8th Workshop on Parallel and Distributed Simulation*, 118–125.

Choi, E., and M. J. Chung. 1995. An Important Factor for Optimistic Protocol on Distributed Systems: Granularity. In *Proceedings of the 1995 Winter Simulation Conference*, 642–649.

Chung, Y., and M. J. Chung. 1991. Time Warp for Efficient Parallel Logic Simulation on a Massively Parallel SIMD Machine. In *Proceedings of the Tenth Annual IEEE International Phoenix Conference on Computers and Communications*, 183–189.

Das, S, R. M. Fujimoto, K. Panesar, D. Allison, and M. Hybinette. 1994. GTW: A Time Warp System For Shared Memory Multiprocessors. In *Proceedings of the 1994 Winter Simulation Conference*, 1332–1339.

Fujimoto, R. M. 1989. Time Warp on a Shared Memory Multiprocessor. In *Proceedings of 1989 International Conference on Parallel Processing* 3:242–249.

Fujimoto, R. M. 1990. Parallel Discrete Event Simulation. *Communications of ACM* 33:30–53.

Gafni, A. 1988. Rollback Mechanisms for Optimistic Distributed Simulation Systems. In *Proceedings of the SCS Multiconference on Distributed Simulation* 19:61–67.

Jefferson, D. 1985. Virtual Time. *ACM Transactions on Programming Languages and Systems* 7:404–425.

Jones, D. W. 1989. Concurrent Operations on Priority Queues. *Communications of ACM*, 132–137.

Preiss, B. R., and W. M. Loucks. 1995. Memory Management Techniques for Time Warp on a Distributed Memory Machine. In *Proceedings of the 9th Workshop on Parallel and Distributed Simulation*, 30–39.

Sokol, L. M., B. K. Stucky, and V. S. Hwang. 1989. MTW: A Control Mechanism for Parallel Discrete Simulation. In *Proceedings of the 1989 International Conference on Parallel Processing* 3:250–254.

The Oak Ridge National Laboratory. 1994. *PVM 3 User's Guide and Reference Manual.*

AUTHOR BIOGRAPHIES

EUNMI CHOI is a Ph.D. candidate in computer science at Michigan State University. Her current research interests include parallel asynchronous protocols, parallel logic simulation on parallel and distributed systems, and parallel and distributed algorithms. She received an M.S. in computer science from MSU in 1991, and a B.S. from Korea University in 1988. She is a member of ACM and the IEEE Computer Society.

DUGKI MIN received a B.S. degree in industrial engineering from Korea University in 1986, an M.S. degree in 1991 and a Ph.D. degree in 1995, both in computer science from Michigan State University. He is currently an Assistant Professor in the Department of Computer Engineering at Konkuk University. His research interests include parallel and distributed computing, distributed multimedia systems, and computer simulation.

Proceedings of the 1996 Winter Simulation Conference
ed. J. M. Charnes, D. J. Morrice, D. T. Brunner, and J. J. Swain

VIRTUAL REALITY AND SIMULATION: AN OVERVIEW

Robert Macredie, Simon J.E. Taylor, Xiaoning Yu and Richard Keeble

Department of Computer Science and Information Systems
Brunel University
Uxbridge, UB8 3PH, ENGLAND

ABSTRACT

Virtual environments can be used as an effective tool for training and education. This paper cites many recent examples of such systems and divides current work into virtual environment development and the development of supporting technology for distributed access. A novel supporting technology for distribution is introduced.

1 INTRODUCTION

A survey of the recent literature indicates that current research focusing on training is concerned with the development of virtual environments and the development of supporting technology to distribute access. Examples concerning the development of virtual environments show the wide range of applications used for training purposes.

Suzuki et al (1995) describe a driving simulator which allows a user to drive through a virtual town. ROADNET (Katz 1995) is a proposed system which utilises distributed interactive simulation technology to support a network of vehicle simulators. The paper describes the architecture and functionality of ROADNET. Similarly Stevens and Neal (1995) propose the use of distributed interactive simulation to support air traffic control simulation with the National Simulation Capability programme of the Federal Aviation Administration of the USA.

Mizumaki et al (1995) discuss an indoor distribution simulation system based on virtual reality technology. The system is designed to give the user the feeling of touring a substation for routine inspection. It is proposed that the validation of equipment operating procedures, training of unskilled workers and work schedule preparation can be improved by the use of this system.

Merril, Raju and Merril (1995) present a prototype virtual reality surgical simulator for laser iridotomy training. The aim of the work is to help physicians acquire the necessary skills without patient risk. Delingette et al (1995) present a craniofacial surgery simulation testbed which makes extensive use of virtual reality techniques. Virtual reality has been used to train surgeons in new procedures and can determine their competence level before going 'live' (Ota et al 1995). Zeigler et al (1995a and 1995b) discuss a virtual reality medical training system for arthroscopy based on computer graphics and virtual reality techniques.

Coleman (1995) presents the view that leaders in the medical profession suggest that accreditation as a trained physician in the field of endosurgery must begin with a training period in a simulated field. The aim is to develop manual dexterity skills before participating as a clinical assistant to an experienced operator. The Virtual Clinic is introduced as a highly realistic and interactive training system which offers surgeons the chance of gaining clinical expertise before encountering an actual surgical event.

As summarised by Zeltzer, Pioch and Aviles (1995) the US Navy sponsors the Virtual Environment Technology for Training programme which addresses key aspects of this technology. The initial prototype used for evaluation is based on training the officer of the deck on a submarine.

Vegas is a virtual reality application dedicated to the experimentation with evacuation scenarios (Sims 1995). The main goal is to communicate understanding as to the process of a disaster by the use of visualisation to present data generated by the simulator. It is hoped that it will allow users to understand fire and smoke spread. The system has been used to predict egress times from an underground subway station and evacuation times from a cross-channel ferry, as well as from several buildings. Bethel, Jacobsen and Holland (1995) discuss the combination of simulation with virtual reality and visualisation tools to aid training in environmental remediation. Stansfield et al (1995) focus on virtual reality to train people for small team, close quarters operations.

Stansfield et al (1995) present research aimed at developing a distributed, shared virtual reality simulation. Situational training of inspectors and escorts under programs to verify compliance with nuclear non-proliferation treaties is being used to investigate the efficacy of this technique.

Hollands and Mort (1995) discuss the use of virtual reality to visualise the simulation of manufacturing systems. It is observed that people unfamiliar with the simulated processes quickly understand the dynamics and the logic involved, and those familiar with the real system can immediately relate to the simulation output. The authors indicate that it seems probable that many manufacturing systems will not be suitable for, or benefit from, modelling using virtual reality techniques. However, with the provision of affordable virtual reality tools it is hoped that experience will show if the general manufacturing simulation community will find such features worth the extra modelling effort.

Bienvenue, Curtis and Thakkar (1995) investigate if educational researchers can use virtual environments in the classrooms of the USA for training and learning, and if teachers can look at virtual reality as being a discovery tool rather than a means to demonstrate principles.

Wakefield and O'Brien (1995) discuss the creation of a set of virtual reality and simulation tools for the detailed planning of construction activities.

Bowen-Loftin (1995) reports on the use of virtual environments for the training of NASA astronauts and ground-based personnel for a variety of activities such as repairing the Hubble Space Telescope. It is argued that this technology offers significant cost savings and increased training throughput. The move towards a shared virtual environment is discussed.

Macedomia (1995) describe a network software architecture for solving the problem of scaling very large distributed simulations. Logical partitioning of the virtual environment is accomplished via associated spatial, temporal and functional characteristics of entity classes. A vehicle simulation is given as an example of a practical application. Pullen (1995a and 1995b) discusses Distributed Virtual Simulation, a new technology that extends virtual reality to the networked environment. Future requirements of networking resources are discussed to supporting this proposed system. Faigle et al (1995) discuss how high performance computing can further enhance virtual environments.

The complexity and risks associated with the operation of the power transmission and distribution systems have increased the importance of personnel training for such systems. Garant et al (1995) present a virtual reality training simulator prototype to support this activity. The simulator is implemented on the basis of a distributed client-server architecture.

Mastaglio and Callahan (1995) present a virtual environment called the Close Combat Tactical Trainer (CCTT) which is claimed as the first fully distributed interactive simulation compliant training system.

On-going work at Brunel University is currently developing underlying technology to support distributed virtual environments. In this paper we present a novel approach to supporting distribution.

2 TOWARDS A DESKTOP VIRTUAL ENVIRONMENT

Typically, current systems are designed based on the following basic design principles, including:

· there is no central computer for event scheduling or conflict resolution,

· autonomous simulation nodes (computers) are responsible for maintaining the state of one or more simulation entities,

· there is a standard protocol for communicating "ground truth" data; receiving nodes are responsible for determining what is perceived, and

· simulation nodes communicate primarily changes of their state. Dead reckoning is used to reduce communications processing.

A system designed on the above principles suffers from high communication volume.

Object oriented programming is known to be a programming paradigm which deals with complex systems in such a way that maintainability, extensibility and reusability are accommodated (Maffeis 1993). Object orientation provides a natural basis for concurrency since the independency required for concurrent execution is made possible as each object contains its own data and procedures. Inter-object co-ordination is achieved by message passing.

One approach to addressing the problems mentioned above is to propose a new architecture which takes advantages of the features of the object oriented programming in the hope of reducing the volume of messages. In this paper, we present a prototype architecture with an object oriented structure.

This prototype object oriented architecture is composed of non-autonomous simulation nodes. With an appropriate user interface, the nodes appear as an integrated virtual environment to the user. Simulation entities are modelled as simulation entity objects which are distributed at different nodes and are migrated dynamically between the nodes during the simulation run. This mechanism of object management presents the possibility of reducing message volume. Communication efficiency can therefore be improved by two measures. One is to dynamically group closely related simulation entity objects to a specific node. Alternatively, a point to point communication approach can be used instead of broadcast.

The system is currently implemented on a network of IBM compatible PCs, with facilities providing real time communication, an interactive display and interface media.

3 THE SYSTEM ARCHITECTURE

As an object oriented simulation system, the basic system elements are objects located at different machines. These machines are called nodes and are connected by the network. There are two types of objects in our prototype. One class of object is used to manage system facilities such as keyboard and mouse operation, network operation and display operation. These objects are usually supplied by system support packages. The other class of object is used specifically for simulation. In this paper we focus on the class of simulation objects.

At any single node, there are three major objects: the message handling object, the situation display object and the simulation object.

3.1 The Message Handling Object

The message handling object controls the messages passed between local objects and the network. Unlike current approaches which broadcast all messages on the network (Pullen, 1995), it uses both broadcast and point to point approaches to send the messages to keep the number of redundant messages as low as possible. It broadcasts all messages about the updated states of local entity objects to other nodes via the network. Control is achieved by a point to point communication approach. The message handler also receives the messages from outside the node and distributes them to the proper local objects.

3.2 The Situation Display Object

With update messages, the situation display object is used to display the states of all entity objects distributed at the nodes. A function may be used in this object to restrict some object state display according to the different training scenarios. The display updates with simulation time advance. The update messages are from the local simulation object and the message handling object. The messages from the local simulation object include all changes of its derivative entity objects' states. The messages from the message handling object are the update messages from all other entity objects located at the other nodes. On some occasions "Dead Reckoning" algorithms can be used in this object to reduce the transmission of state updates.

3.3 The Simulation Object

The simulation object at a node contains a series of entity objects which represent physical entities in the real world (a simulation object controls its derivative entity objects simulation in each time loop). It is responsible for creating and deleting its derivative entity objects. It transfers the messages from its derivative entity objects to its other derivative entity objects or to the other objects located at the other nodes via the computer network. Another important task of the simulation object is to control the local simulation time advance. More detail about entity object management and simulation time management will be introduced in the rest of the paper.

4 SYSTEM MANAGEMENT

4.1 Object management

1) Entity object

In our prototype of system, entities in the real world are modelled by entity objects which are located at different nodes in the system. Each entity object has a unique identity that distinguishes it from other entity objects. During each simulation time loop, each entity object is activated by different messages and the entity object states can be updated. At each node, there is a table recording distribution of all entity objects in the system. When an entity object is created, deleted or transferred to an other node, a message is sent to all nodes in the system to inform them of this change and update the entity objects' distribution tables.

There is an owner relation between the entity objects. When we say the entity object A is the owner of the entity object B, it means the entity object A controls the entity object B in them system. The ownership can be changed. Each entity object has at most one owner. The ownership is used to judge if the entity object should be migrated or not. To reduce control messages passed over the network, any entity object should be located at the same node as its owner object. An attribute in the object is used to describe the ownership between entity objects.

2) Object list

At each node, a list is used to link all entity objects located at local nodes. During each simulation time loop, all entity objects are activated in the list order by control messages which are transferred from the simulation object. With the entity objects location changing, the list is dynamically changed by the system. When a new entity object is created or moved to a node from another node, it is automatically added at the end of the list as a result of its constructor. When an entity object in the list is deleted or moved out to another node, it also automatically removes itself from the list using itsdestructor.

3) Object Migration

Reducing the messages passed via the network and keeping the system load balanced can improve the system efficiency. This can be realised by object migration. Here we only discuss entity object migration.

One of the basic functions of entity object migration is to group entity objects together which have common ownership. Control messages are passed between entity objects. These messages are used by one entity object to control another entity object. They are only passed between the entity objects which have common ownership. Since messages passed locally are more efficient than the messages passed via the network, keeping control local can improve system efficiency.

Ownership can be changed during the simulation. When the owner of one entity object is changed, the entity object is immediately moved to the node where the new owner object is located.

Object migration also facilitates load balancing. Entity objects can be created, deleted, or moved dynamically during the simulation session. At each node, there is an entity object list recording the entity objects located at that node and an entity object location table recording the location of other entity objects at other nodes. A detection object is used at each node based on the above list and table to detect the load states of nodes in the system. When the load of one node is heavy while there is a node with no entity object located, some entity objects can be moved to the idle node to maintain load balance.

4.2 Time Management

Time management is used to control the advancement of simulation time during the simulation session. The simulation object at each node controls time management. To control the simulation time advance, three different time notions should be distinguished in time management.

Tli refers to the earliest message timestamp in the *i*th entity object. Tl refers to the earliest message timestamp in all entity objects. So

$$Tl = \min \{ Tli \mid i=1,...n\}$$

Here *n* is the number of entity objects in the simulation.

Wallclock time Tw is a scaled real time. A scale factor *f* is used to compress or expand real time. It is variable. Different scenarios can set different values of scale factor *f*. So

$$Tw = f * Tr$$

where Tr is the current real time

Simulation current time Tc is used for each simulation loop time advance.

$$Tc = \min \{Tl, Tw\}$$

When Tc = Tl, the entity objects process their own messages with time stamp Tl and change the related states. The other objects only change states which are related to time. When Tc = Tw, all objects only update the states which are related to time. Usually, the logical time is earlier than the wallclock time, but sometimes, there could be delay caused by nodes processing or network communication processing. So an adjusting value Ta should be added to the wallclock, i.e.

$$Ta = Tdelay+Tprocess,$$

where Tdelay is message delayed time comparing with its timestamp, Tprocess is the time the message object needs to process the message. So, the adjusted wallclock time is

$$Tw' = Tw+Ta$$

At each node, there is a message list in timestamp order. Each message is processed when the timestamp is the same as simulation time.

4.3 Message Management

Similar to the classes mentioned above, there are two major kinds of messages in the system. One kind of message serves the operating system, i.e. opening or closing a window, expending or reducing the size of window, moving the mouse, etc. The other kind of message is used for simulation execution, i.e. entity objects state update, an event happening, moving an entity object, etc.

For efficient communication, a broadcast approach is used for entity object state update and a point to point communication approach is used to transmit control messages between entity objects which are distributed at different nodes.

The simulation object at each node distinguishes which messages are only used for local entity objects and which messages should be transferred to the message handling object.

Message order affects the whole simulation event order, especially concurrent events. All messages have a time stamp marking effective time. All entity objects will process the messages in time stamp order. This guarantees the processing of concurrent events in the correct order. All message handling objects in the system run in parallel and are synchronised. In each time loop, message handling objects send all messages created by local entity objects to other nodes. When a node has no messages to broadcast, it broadcasts a time advance request message to inform other nodes that it has nothing to send and requests time advance. At the same time, each message handling object holds all messages until it has received messages from all nodes and then transfers them to the display object and simulation object in timestamp order.

5 Conclusion

This paper has introduced virtual reality and simulation within a training context as being a field which concerns interface design and distributed access. We have described a novel architecture which will be used to underpin future in virtual environments.

REFERENCES

Bethel, W., Jacobsen, J. and Holland, P. 1994. Site remediation in a virtual environment. Proceedings of the SPIE, The International Society for Optical Engineering, 2178: 78-87.

Bienvenue, L. A., Curtis, D. H. and Thakkar, U. 1995. Virtual environments in K-12 learning and discovery: a grand challenge in education? Computer Graphics, 29(4): 43-4.

Bowen-Loftin, R. 1994. Virtual Environments for Aerospace Training. In Proceedings of WESCON/94, Idea/Microelectronics, 384-7. IEEE, New York, USA.

Coleman, J. E. 1994. Virtual reality in medicine. In Proceedings of VR '94, the Fourth Annual Conference on Virtual Reality, 169-73. Mecklermedia, London, UK.

Delingette, H., Subsol, G., Cotin, S. and Pignon, J., 1994. Virtual reality and the simulation of craniofacial surgery. In Proceedings of Montpellier '94. 3rd International Conference. Interface to Real and Virtual Worlds, 399-408. EC2, Nanterre, France.

Faigle, C., Fox, G. C., Furmanski, W., Niemiec, J. and Simoni, D. A. 1993. Integrating virtual environments with high performance computing. In Proceedings of the IEEE Virtual Reality Annual International Symposium, 62-8. IEEE, New York, USA.

Garant, E., Daigle, A., Desbiens, P., Okapuu-von Veh, A., Rizzi, J.-C., Shaikh, A., Gauthier, R., Malowany, A. S. and Marceau, R. J. 1995. A virtual reality training system for power-utility personnel. In Proceedings of the IEEE Pacific Rim Conference on Communications, Computers and Signal Processing, 296-9. IEEE, New York, USA.

Hollands, R. J. and Mort, N. 1995. Manufacturing systems simulation: mixed mode and virtual reality simulation. In Proceedings of Fourth International Conference on Factory 2000 - Advanced Factory Automation, 651-7. IEE, London, UK.

Katz, W. 1994. ROADNET - distributed interactive simulation applied to driver training, city planning and transportation research. In Proceedings of the Twenty-Sixth Annual Summer Computer Simulation Conference, 936-41. SCS, San Diego, California, USA.

Macedomia, M. R., Zyda, M. J., Pratt, D. R., Brutzman, D. P. and Barham, P. T. 1995. Exploiting reality with multicast groups: a network architecture for large-scale virtual environments. In Proceedings of the Virtual Reality Annual International Symposium '95, 2-10. IEEE Computer Society Press, Los Alamitos, California, USA.

Maffeis, S. 1993. "ELECTRA - Making Distributed Programs Object-Oriented".: Proceedings of the Symposium on Experiences with Distributed and Multiprocessor System IV, USENIX. San Diego, California, USA.

Mastaglio, T. W. and Callahan, R. 1995. A large-scale complex virtual environment for team training. Computer, 28(7): 49-56.

Merril, G., Raju, R. and Merril, J. 1995. Changing the focus of surgical training. VR World, 3(2): 56-8, 60-1.

Mizumaki, Y., Morita, S., Asano, K. and Kamiji, N. 1995. Power substation simulation system using virtual reality technique. Transactions of the Institute of Electrical Engineers of Japan, 115-C(2): 267-72.

Ota, D., Loftin, B., Saito, T., Lea, R. and Keller, J. 1995. Virtual reality in surgical education. Computers in Biology and Medicine, 25(2): 123-37.

Pullen, J.M. and D. C. Wood. 1995. "Networking Technology and DIS", Proceedings of the IEEE, 83(8): 1156-1167.

Pullen, J. M. 1994a. Networking for distributed virtual simulation. In Proceedings of INET'94, the Annual Conference of the Internet Society, held in conjunction with JENC5, the 5th Joint European Networking Conference, 241-7. Plattner, B. R. and Kiers, J. P. A. The Internet Society, Reston, Virginia, USA.

Pullen, J. M. 1994b. Networking for distributed virtual simulation. Computer Networks and ISDN Systems, 27(3): 387-94.

Sims, D. 1995. See how they run: modeling evacuations in VR. IEEE Computer Graphics and Applications, 15(2): 11-3.

Smith, A. B. 1995. Using virtual reality for the simulation of infrared environments for human training. In Proceedings of the 33rd Annual Southeast Conference, 101-9. ACM, New York, USA.

Stansfield, S., Shawver, D., Miner, N. and Rogers, D. 1995. An application of shared virtual reality to situational training. In Proceedings of the Virtual Reality Annual International Symposium '95, 156-61. IEEE Computer Society Press, Los Alamitos, California, USA.

Stansfield, S., Shawver, D., Rogers, D. and Hightower, R. 1995. Mission visualisation for planning and training. IEEE Computer Graphics and Applications, 15(5): 12-14.

Stevens, R. J. and Neal, W. J. 1994. Distributed air traffic management simulation: the next application-domain for DIS. In Proceedings of the Twenty-Sixth Annual Summer Computer Simulation Conference, 828-33. SCS, San Diego, California, USA.

Suzuki, T., Yasuda, T., Yokoi, S. and Toriwaki, J. 1994. Construction of virtual town and driving simulation in it using graphics workstation. Journal of the Institute of Television Engineers of Japan, 48(10): 1318-25.

Wakefield, R. R. and O'Brien, J. B. 1994. A 'virtual reality' type simulation system for construction automation system development. In Proceedings of Automation and Robotics in Construction XI, the 11th International Symposium on Automation and Robotics in Construction, 239-47. Chamberlain, D. A. Elsevier, Amsterdam, Netherlands.

Zeltzer, D., Pioch, N. J. and Aviles, W. A. 1995. Training the officer of the deck. IEEE Computer Graphics and Applications, 15(6): 6-9.

Ziegler, R., Fischer, G., Muller, W. and Gobel, M. 1995. Virtual reality arthroscopy training simulator. Computers in Biology and Medicine, 25(2): 193-203.

Proceedings of the 1996 Winter Simulation Conference
ed. J. M. Charnes, D. J. Morrice, D. T. Brunner, and J. J. Swain

ENHANCING SIMULATION EDUCATION WITH INTELLIGENT TUTORING SYSTEMS

Simon J.E. Taylor

Department of Computer Science and Information Systems
Brunel University
Uxbridge UB8 3PH, ENGLAND

Julika Siemer

Department of Information Systems
London School of Economics and Political Science
Houghton Street
London WC2A 2AE, ENGLAND

ABSTRACT

The demand for education in the area of simulation is in the increase. This paper describes how education in the field of simulation can take advantage of the virtues of intelligent tutoring with respect to enhancing the educational process.

For this purpose, this paper gives an overview of what constitutes the objectives and the content of a comprehensive course in discrete event simulation. The architecture of an intelligent tutoring system is presented and it is discussed how these sophisticated learning aids offer individualised student guidance and support within a learning environment. The paper then introduces a prototype intelligent tutoring system, the simulation tutor, and suggests how the system might be developed to enhance education in simulation.

1 INTRODUCTION

Simulation provides the pedagogical challenge of educating students in a wide range of problem understanding skills (Paul and Hlupic, 1994). A recent approach to the use of computers as tutors is that of intelligent tutoring systems (Kaplan and Rock, 1995). Intelligent tutoring systems provide helpful guidance and adaptation to the student by exploring and understanding the needs of the individual student. This paper proposes the use of intelligent tutoring to enhance education in simulation. For this purpose, we present an overview of what constitutes the learning objectives and content of a comprehensive course in simulation. The paper continues to describe the architecture of intelligent tutoring systems and how these sophisticated learning aids can provide individualised tutoring. The paper presents a prototype intelligent tutoring system, the simulation tutor, and suggests how the system might enhance the delivery of skills and knowledge in simulation.

2 EDUCATION IN SIMULATION

An example of a typical course in discrete event simulation is taught at the Department of Computer Science and Information Systems, Brunel University, England. The course is dedicated to the objective of educating the student in problem understanding, and is a final year option open to B.Sc. Computer Science and B.Sc. Computing in Business students. This is achieved through an overview study of simulation approaches, and aims to provide the student with practical experience in different simulation languages and environments used in industry. The main modes of delivery of the course are lectures and practical case studies. The underlying themes ask questions such as, 'How may a simulation model be built, using an appropriate simulation tool, which represents the best understanding of the system under study?' and 'Using such a model, how may we experiment with it to help us understand more about that system?' Accordingly, the course objectives include the management of the simulation process, the development of simulation skills and promotion of awareness of the place of simulation relative to other disciplines (orientation).

2.1 Management of the Simulation Process

This concerns the key educational themes of model development and experimentation. Model development addresses the means by which a simulationist can confidently develop and subsequently demonstrate to the owner of the system the validity of a model developed for that system. Experimentation concerns the formulation of the simulation experiments to be performed on the simulation model.

2.2 Development of Simulation Skills

The student is taught a variety of techniques for this purpose. For example, during the course, the student is introduced to activity cycle diagrams, automatic program generators, visual interactive simulation modelling, simulation program structures, simulation software tools, the handling of stochastic input and output of the model, and issues relating to model confidence. This course of study is backed up with a workshop program to reinforce these practical skills, so giving the opportunity for the student to apply what he or she has learnt in order to concretise the knowledge delivered via the lecturing programme. For example, the student discovers activity cycle diagrams and then is given the opportunity to pursue a simple case study to model a system in terms of this technique. An automatic program generator is then introduced to give the student some experience in transferring the model into a simulator. Case studies of increasing complexity are introduced to address other aspects.

2.3 Simulation Orientation

Simulation Orientation is intended to present simulation as being part of a larger field concerning techniques relevant to decision making. The reason for this issue being addressed is exactly why simulation has found itself as part of other courses ranging from Operational Research to Management Science. It is important that the student is aware that simulation is not an isolated discipline and has strong links to other fields.

This course has proven to be extremely popular amongst undergraduates. While gratifying that this is the case, the size of the student cohort undertaking this scheme of study has increased the pressure placed on the educators responsible for course delivery. This is especially true for the practical case study elements of the course which require significant tutor-student interaction. These were some of the factors motivating a study launched to examine methods to alleviate this pressure. The most interesting possibility which arose from this study was that of a recent innovation in computer-based training which takes advantage of Artificial Intelligence. This is termed *intelligent tutoring systems.*

3 INTELLIGENT TUTORING SYSTEMS

The objective of intelligent tutoring systems (ITSs) research is to provide a new plateau of instructional capability by integrating artificial intelligence into computerised learning systems. Intelligent tutoring systems provide helpful guidance to the student and are intended to complement socalled conventional computerbased training systems with the features of an equivalent "online" human tutor. It is therefore the goal of an ITS to make the computer-based teaching process more adaptable to the student by exploring and understanding the student's special needs and interests, and by responding to these as a human tutor does. In order to provide this adaptability to the student an ITS makes use of three knowledge models. These are the *tutoring model*, which regulates the instructional interactions with the student, the *domain model*, which contains the knowledge about the domain to be taught, and the *student model*, which represents any information about the student. Adding the fourth component, the *user interface*, completes an ITS (Kaplan and Rock, 1995). To better understand the means by which ITS can aid education in simulation, we now review the four components.

3.1 The Domain Model

The domain model of an ITS contains the knowledge about the subject area to be taught. The ITS uses its domain knowledge to reason about and solve a problem which has been set for, or by, a student. This knowledge has to be represented in such a way that it supports reasoning that resembles the human problem-solving process within the teaching domain. Furthermore, different knowledge representations may be required to support the application of different teaching strategies. One rather pragmatic way of looking at what the domain model represents is to consider it as being a computer-based training tool and postponing decisions which have to be made concerning knowledge representation, etc. As will be seen, it is this approach which has given rise to our current implementation cycle.

3.2 The Tutoring Model

An ITS should exhibit the various tutoring characteristics encountered in the classroom. These are encapsulated in the tutoring model. It must therefore have control over the selection and sequencing of material to be presented to the student. The system must also be able to respond to questions concerning the subject material and must be able to apply learning strategies to determine when students need help and what kind of help is required. For this purpose the tutoring model incorporates different teaching strategies (O'Neil, Slawson and Baker 1991).

Teaching strategies are used to present material and depend on the subject matter and the instructional objectives of the ITS. A teaching strategy determines the style of material delivery that is employed in order to lead the student and to indicate the times at which intervention is required. Many ITSs apply different teaching strategies in different teaching situations. Reviews have been made by Siemer, Taylor, and Elliman (1995) and Elsom-Cook (1991).

An ITS should ideally include a selection of different

teaching strategies for intervention. The strategy may be chosen according to the peculiarities of a tutorial situation, such as the student's needs and preferences, experience and the domain of discourse (Angelides and Tong 1995).

3.3 The Student Model

To carry out intelligent tutoring a tutor has to have a good understanding of the student being taught. For this reason an ITS uses a student model to represent the student's emerging knowledge and skills of the subject matter. An ITS may use its student model to analyse the input of the student during tutoring interaction. The student's input may come in a variety of forms ranging from answers to questions posed by the ITS or moves taken in a game, to commands delivered within an editor. This information can sometimes be complemented by the student's educational history.

The most common approach is the representation of the student's knowledge in the form of an overlay model. Here the student's knowledge is represented as a subset of an expert's knowledge. As the student moves from his initial state of knowledge towards mastery, parts of knowledge are added to this expert knowledge subset. To determine the student's knowledge state the system generally refers to the same knowledge base using different interpreters. However, the overlay model allows only missing parts of knowledge to be represented. To represent misconceptions, i.e. a different conception of some part of knowledge, the overlay model is usually complemented by a educational bugs catalogue. The bugs catalogue is a library of common errors made by students. It represents typical deviations a student can make from the ideal expert behaviour. Student diagnosis is carried out by matching the student's faulty behaviour against the bugs in the bugs catalogue. From this remedial action can be taken by the system.

3.4 The User Interface

The user interface is generally recognised as a fourth component of an ITS architecture (Woolf and Hall 1995). As the student interacting with the ITS is working in an area he or she does not understand well the system has to ensure that any complications the student is exposed to when operating the system are kept to a minimum. A well designed interface can add considerably to the way in which the student will conceptualise the domain being taught (O'Malley 1990).

The user interface determines how students interact with the ITS (Bonar 1991). The teaching material may have to be presented in different formats, such as text, graphics, animation or video (Alpert, Singly and Carroll 1995) depending on the needs and requirements of the current situation, such as the teaching strategy employed or the nature of the material to be presented. A well designed human interface allows the ITS to present instruction and feedback to students in a clear way. At the same time it can provide students with tools and techniques to state problems and hypotheses to the ITS.

4 STUDENT-CENTRED TUTORING

An ITS employs the knowledge of its three models to provide studentcentred tutoring, i.e. the adaptation of the tutoring process to the student's needs and preferences. In order to provide this adaptability for a student, an ITS should apply suitable teaching strategies and presentations for each subject matter unit as needed, choosing the form that is most beneficial to the student for a particular instructional situation (Miller and Lucado 1992). The selection of an appropriate teaching strategy and its presentation requires knowledge about the needs of a student. The student model may provide information about the student's experience with particular remedial strategies and different ways of presentation. Accordingly, it may provide information on approaches that have proven to be successful or unsuccessful within earlier interventions, and the tutoring process may adapt accordingly.

A further common discriminating factor to consider in order to distinguish between students with different needs, at a stage where the creation of a full cognitive model has not yet been achieved, is the differentiation between student advancement stages. A system may, therefore, adjust its tutoring to the advancement stage of a student. A fundamental idea in education is that as students learn more about a subject, one can teach them progressively more advanced ideas. Accordingly, a novice may require different tutoring from an advanced student. Whilst the competent student may be able to appreciate and integrate shallow and subtle tuition, the novice student might require a detailed explanation of intermediate ideas (Silverman 1992).

A further issue of student-centred tutoring is the issue of proactive instruction and reactive instruction (Alpert, Singly and Carroll 1995, Silverman, 1992). An ITS may provide adaptability to the student by providing both help that may be invoked by the student or by the system. An advanced student, for example, may recognise his need for help and may decide to activate system help. However, when the student is a novice, or when the teaching domain is broader, system-invoked help seems more appropriate. The student may require intervention when the student makes a mistake without realising it, or when the student does not know what to do next. A student who has problems with the English language, for example, may appreciate a system-activated spellchecker or grammar-checker that assists him incrementally during the task as each difficulty occurs. Similarly, the PAT (Pump Algebra Tutor) system (Barker 1995) offers student-invoked help. The student has to ask for intervention by the system when he feels that help is required for the mathematical problem he has to solve. Attempts have been made to bring the two modes together by establishing a balance according to the student's individual preference (Moyse and Elsom-Cook 1992, Winkels 1992, Milheim and Martin 1991). The MoleHill system which teaches Smalltalk programming (Singley, Carroll and Alpert 1993), for example, offers both active and reactive instruction to its users.

ITSs have been introduced thus far as a novel development in computer-based training. These offer individualised student guidance and support within the learning environment. In order to demonstrate how these educational benefits can be used to enhance simulation education the next section introduces a prototype ITS for simulation which has been developed as part of this research.

5 THE SIMULATION TUTOR: AN INTELLIGENT TUTORING SYSTEM FOR SIMULATION

The simulation tutor is an ITS currently under development at Brunel. Given the quite heavy investment in development that an ITS requires, a prototyping approach has been taken. This has used a "paper-ITS" to form the basis for evaluation and definition of ITS interaction and the simulation tutor. A paper-ITS is essentially the domain model and user interface implemented using conventional computerbased training technology (in this case Asymmetrix Toolbook (Asymmetrix Corporation, 1994)). The tutor and student models are emulated by using a research assistant. To provide a forum for discussion and ongoing development, the domain model has been decomposed into sub-components representing the major elements of the simulation course (Paul and Balmer 1993). These are:

1. **The need for simulation.** This is first considered to make it clear to the student as to the reasons for when simulation should and should not be used. This follows the course text and uses examples generated by the educator and by the class in open discussion.
2. **Discrete event simulation.** The goal of this lecture is to introduce the students to the key terminology of discrete event simulation and its constituents. Again the course text is used and is backup by worked examples indicating how each elements works and its consequences.
3. **Different modelling approaches.** Following on from 2., different modelling approaches are considered. Worked examples are given.
4. **Activity cycle diagrams and the three phase approach.** This modelling technique is selected as one technique to study in depth. Further terminology and examples are introduced.
5. **Case study I.** The first of two case studies is introduced. Students are taken through different stages of model development via the first simple case study.
6. **Visual interactive simulation.** A simple simulation environment and features are introduced with respect to the earlier elements of the course. The importance of animation in simulation is demonstrated.
7. **Model testing and validation.** Following on from 6., issues of model testing and validation are covered.
8. **Case study II.** A more complex case study is attempted. This is used to summarise the previous elements of the course.
9. **Discrete event simulation software and languages.** This reviews the current state of the art.
10. **Sampling methods.** Sampling methods are reviewed.
11. **Planning and interpreting discrete event simulation experiments.** Again a review of the subject area.
12. **Summary.** The course is pulled together with respect to the learning objectives.

Initial work performed on the simulation tutor concerned the first four elements. Essentially, the domain model was an animated version of the course text. This added an extra dimension to the usual forms of course support and focused on relating the components of the simulation program to the flow of entities through a system. Interaction with the tutor and student models was performed via a research assistant who discussed the learning needs with the representative student groups.

6 EVALUATION OF THE PROTOTYPE

The prototype was evaluated with a group of representative students who used the system. The students had recently completed part of the simulation course at Brunel University. They were asked to interact with the system under supervision. The experiment was supported by a questionnaire which was developed based on suggestions made by Schneiderman (1987) to gain feedback from the students about the way they perceived the needs of tutoring and learning.

The results of the evaluation show that the simulation tutor has the potential of offering adaptive tutoring to the student. The tutor coordinates the information from its knowledge models to provide adaptability to the needs and preferences of the players. In general comments alluded to good structuring and presentation of the teaching material. Informational content was adequate, given that it was supported by course texts and lecture notes. However, more effort should have been made in the presentation of the graphics. Video should be included in the next release. The use of animation to demonstrate the working of the simulation model and the simulation itself was good. Students were keen to have model elements linked to the physical system via video sequences.

In regard to the user interface users expressed a general liking of the use of a windows environment with easy to use, self-explanatory buttons and features. The context-sensitive help facility providing hints and analogues was useful as was the use of a mapping system; the students responded well to being able to see where they were in relation to the other parts of the course. There was, however, general dissatisfaction with the lack of response that the prototype provided. This was to be expected as the tutoring parts of the system had not been fully implemented. Once this was explained to the students, the response was that if the role of the tutor was documented correctly, then the intelligent tutor would be more than satisfactory.

The overall response to the system was encouraging, and the feedback provided by the students suggests ideas to be incorporated into the next version of the prototype.

6 CONCLUSIONS

This paper has reported on an attempt to introduce ITS techniques into simulation education. An initial investigation has shown that ITSs could improve the educational process by providing individual guidence and student centred learning. Further work is currently addressing other elements of the simulation course and the implementation aspects of the simulation tutor.

ACKNOWLEDGEMENTS

The authors would like to thank Devika Kapoor for her help in implementing the prototype.

REFERENCES

Alpert, S.R., M.K. Singley, and J.M. Carroll. 1995. 'Multiple multimodal mentors: delivering computer-based instruction via specialized anthropomorphic advisors.' *Behaviour and Information Technology*, 14(2): 69-79 .

Angelides, M.C. and A.K.Y. Tong. 1995. 'Implementing multiple tutoring strategies in an ITS for music learning.' *Journal of Information Technology*, 10(1): 52-62.

Asymmetrix Corporation. 1994. *Asymmetrix Multimedia Toolbook Kit.* Asymetrix Corporation.

Barker, D. 1995. 'Seven New Ways to Learn Carnegie Mellon University is changing the way in which teachers teach and students learn.' *Byte*, 20(3): 54-55.

Bonar, J.G. 1991. Interface Architectures for Intelligent Tutoring Systems. In *Intelligent Tutoring Systems: Evolutions in Design*. ed. H.L. Burns , J.W. Parlett and C.L. Redfield. 35-68. Lawrence Erlbaum Associates, Hillsdale, New Jersey,

Elsom-Cook, M.T. 1991. Dialogue and Teaching Styles. In *Teaching Knowledge and Intelligent Tutoring*. ed. P. Goodyear. 61-84. Ablex Publishing Corporation, New Jersey.

Kaplan, R. and D. Rock. 1995. New Directions for Intelligent Tutoring. *AI Expert.* 10(2): 3040.

Milheim, W. D. and B. L. Martin. 1991, Theoretical Bases for the Use of Learner Control: Three Different Perspectives. *Journal of ComputerBased Instruction.* 18(3): 99-105.

Miller, M.L. and Lucado, S.R. 1992. Integrating Intelligent Tutoring, ComputerBased Training, and Interactive Video in a Prototype Maintenance Trainer. In *Intelligent Instruction by Computer*. ed. M.J. Farr and J. Psotka. 127-150. Taylor & Francis, New York.

Moyse, R. and M. ElsomCook. 1992. Knowledge Negotiation: An Introduction. In *Knowledge Negotiation*, ed. R. Moyse and M.T. ElsomCook. 1-20. Academic Press, London.

O'Malley C. 1990. Interface Issues for Guided Discovery Learning Environments. In *Guided Discovery Tutoring: A Framework for ICAI Research*. ed. M. Elsom-Cook. 24-41. Paul Chapman Publishing, UK.

O'Neil, H.F., D.A. Slawson, and E.L. Baker. 1991. Design of a DomainIndependent ProblemSolving Strategy for Intelligent Computer Assisted Instruction. In *Intelligent Tutoring Systems: Evolutions in Design.* ed. H.L. Burns, J.W. Parlett, and C.L. Redfield. 69-104. Lawrence Erlbaum Associates, Hillsdale, New Jersey.

Paul, R.J. and D. Balmer. 1993. *Simulation Modelling.* Chartwell Bratt.

Paul, R.J. and V. Hlupic. 1994. Designing and Managing a Masters Degree Course in Simulation Modelling. In *Proceedings of the 1994 Winter Simulation Conference*. ed. S. Manivannan J.D. Tew. D.A. Sadowski and A.F. Seila. 1394-1398. ACM Press.

Schneiderman, B. 1987. *Designing the User Interface Strategies for Effective Human Computer Interation.* Addison Wesley, California.

Silverman, B.G. 1992. *Critiquing Human Error, A Knowledge Based HumanComputer Collaboration Approach.* Academic Press, London.

Siemer, J., S. J. E. Taylor and A. D. Elliman. 1995. Intelligent Tutoring Systems for Simulation Modelling in Manufacturing Industry. *International Journal of Manufacturing System Design.* 2(3): 165-175.

Singley, M.K., J.M. Carroll, and S.R. Alpert. 1993. Incidental Reification of goals in an intelligent tutor for Smalltalk. In *Cognitive Models and Intelligent Environment for Learning Programming.* ed. E. Lemut, B. du Boulay, and G. Dettori. 145-155. SpringerVerlag, Berlin.

Winkels R. 1992. Explorations in Intelligent Tutoring and Help, *Frontiers in Artificial Intelligence and Applications Series*, IOS Press, Amsterdam.

Woolf B.P. and Hall W. 1995. Multimedia Pedagogues Interactive Systems for Teaching and Learning. *IEEE Computer*. May: 74-80.

AUTHOR BIOGRAPHIES

SIMON J.E. TAYLOR is a Lecturer and the Senior Tutor in the Department of Computer Science and Information Systems at Brunel University. His interest in simulation began with an in-depth study into the links between simulation methods and parallel simulation. Since then he has become an active researcher in simulation and contributes regularly to the Centre for Applied Simulation Modelling at Brunel University. His interest in Intelligent Tutoring Systems arose from experience in lecturing large classes and a chance meeting with his co-author.

JULIKA SIEMER is a Lecturer in the Department of Information Systems at the London School of Economics, a post to which she was appointed in July 1995. She studied computer science at Hildeshiem University, Germany, and holds an M.Sc. and a Ph.D. in Information Systems, both from the London School of Economics. She has four years of experience in researching in the area of intelligent tutoring systems and has developed a full-scale intelligent tutoring system for a business gaming-simulation.

Proceedings of the 1996 Winter Simulation Conference
ed. J. M. Charnes, D. J. Morrice, D. T. Brunner, and J. J. Swain

MODEL DEVELOPMENT AND HCI

Mike Pidd
Department of Management Science
The Management School
Lancaster University
Lancaster LA1 4YX
UK

ABSTRACT

Much of the progress within discrete simulation has gone hand-in-hand with general developments in computing. Recent years have seen software developers putting great efforts into improving the user interfaces of discrete simulation systems. This too parallels developments elsewhere. This paper considers what further benefits there might be for users and developers of simulation software from more careful attention to interface design.

1 INTRODUCTION

As computers have grown more powerful, so have users' expectations of how they might be used. Today's school students now assume that computers will be fast, friendly and easy to use. Many of them, certainly in the USA and Western Europe, are video game experts, connoisseurs who are easily bored with experiences that do not live up to their expectations. (See Microserfs, Coupland (1996), for amusing examples of this). These people will be the next generation of simulation modellers and users and they may not be satisfied with the interfaces available in today's discrete simulation software. Thus, it seems important that the designers and vendors of simulation software start to take the question of user interface design very seriously. This paper presents the basic requirements for simulation software that stem from a particular view of modelling that is argued elsewhere (Pidd, 1996a, 1996b). These are linked to current theories and ideas about HCI (Human Computer Interaction or the Human Computer Interface) to present desirable features of future discrete simulation software.

1.1 Historical Background

It is probably true to say that developments in discrete simulation software have proceeded hand-in-hand with general developments in computer software. There have been a few cases in which discrete simulation software has led the field, examples being object orientation in Simula (Dahl and Nygaard, 1966) and the introduction of interactive program generators by Clementson (1991) in his ECSL system. However, in most cases, discrete simulation software developers have been willing to pick up and use whatever tools and ideas appear within the computing community. Examples of this being the use of animated graphics for visualisation by Hurrion (1976) and the use of source level debuggers to support program development.

The purpose of this paper is to examine some of the links between simulation modelling and what is now known as HCI. Given that the idea of simulation is to use a computer as a dynamic model of some situation or system, then it would seem that the question of appropriate HCI is an important one. This paper begins by considering the different ways in which discrete simulation software is provided, looks at the different roles which people play in developing and using simulation models and then speculates on how things might be improved. The question at issue is, what can simulation users and developers learn from the field of HCI? In addition, is there anything that the HCI community can learn from discrete simulation?

2 THE USE OF DISCRETE SIMULATION SOFTWARE

2.1 Types of Discrete Simulation Software

Different authors have their own ways of discussing the various types of discrete simulation software that are now available. Law and Kelton (1991) divide this software into two main groups, those which require programming and the rest. The latter they describe as simulators. The former category includes general purpose programming languages and libraries such as C++ or FORTRAN, as well as dedicated simulation programming languages such as the SIMSCRIPT family. A rather more finely drawn categorisation is found in Pidd (1992), who has no less than seven categories, ranging from 'do-it-yourself' (in a general purpose programming language), through to Visual Interactive Modelling Systems (VIMS) such as Witness (Lanner Systems, 1996) and Microsaint (Micro Analysis and Design, 1992), via flow diagram or block diagram systems such as HOCUS (Szymankiewicz, McDonald and Turner, 1988) and GPSS.

This paper uses a simple tripartite classification as follows.

1. Software that requires all users to develop some true program code. This includes general purpose programming languages as well as simulation programming languages. It also includes those languages that provide support for visualisation and those that do not.
2. Visual Interactive Modelling Systems which are primarily based around some kind of visual motif for virtually all of their functions. In these systems, examples of which include Witness, Microsaint, ProModel and AutoMod, a visual interface is used for model building, controlling simulation runs and for interaction as a simulation run proceeds.
3. Layered systems in which the user can operate at a number of different levels, such as visual interactive modelling, automatic program generation, direct coding and low level bit twiddling. A few systems currently available incorporate some of these facilities and an example would be ARENA.

2.2 Groups of People Involved in Simulation Projects

Though this conference primarily focuses on those people who conduct the technical aspects of simulation projects, there is quite a range of people who may be involved across the set of activities that make up such projects. These may be different people or might be different roles that are conducted by one or more people. Examples include the following.

1. *Modellers*: whose main role is to understand the system being simulated and to capture that understanding in a model that is implemented in some software system or other. Thus the software interface must offer support to modellers as they develop, usually in a stepwise manner, their abstract and symbolic models.
2. *Programmers*: who have the task, in some projects, of taking model conceptualisations and realising them in program code. In an ideal world this would enable the programmers, though coding, to operate at a level which relates to the problem being modelled, rather than just the computing aspects of the work.
3. *Project managers*: who may be responsible for ensuring that the project meets the needs of the clients, is on-time and is properly conducted in a technical sense. This group of people may wish to establish standards that enable projects to be properly managed and consistently presented to clients.
4. *Customers*: who are footing the bill for the project and whose questions are being addressed in the project itself. This concern of this group may be to ensure that they get value for money and this may be aided by suitable HCI considerations.
5. *Users*: who may not be the same as the customers, but who may need to use a simulation model on one or more occasions to address issues of interest. This group may not be expert in simulation and may have limited expertise in computing. Thus the interface needs to be very supportive and should build on their previous experience and expertise.

In addition, of course, there may be software vendors keeping an eye on what is happening.

2.3 Discrete Simulation Modelling

A computer simulation involves experimentation on a computer-based representation of some system or other. This particular conference is interested, mainly, in discrete event simulation which is suited to systems made up of objects that change state and display dynamic behaviour. That is, those which change state through time. Thus, to build a discrete event model, the analyst must attempt to tease out and represent a number of features of the system being modelled. These are as follows.

1. *The main entities of the system*: that is, those classes of object whose behaviour seems important to the operation of the system. Note that word *seems*, if the modeller were absolutely sure of the important entities and if this were not a subject for thought, debate and argument, then it might not be necessary to model and simulate the system at all.
2. *The logical operations of these entities*: that is, the significant behaviour in which the classes of object engage as they persist, or appear, or disappear within the system being modelled. Once again, the previous sentence includes a weasel word, *significant*. This suggests that the modeller must make judgements about what features of the entities' behaviour deserves inclusion within the model.
3. *The logical interactions of the entities*: systems would be very simple to model and to simulate if the objects within them did not engage with one another. It is these complicated interactions through time that determine how the system itself will appear to behave and some of these interactions may be hard to tease out or may occur only rarely.
4. *The statistical distributions*: the entities will engage in activities, alone or whole other object classes

and these activities may occur at irregular intervals and for time periods that cannot be precisely determined in advance. It is normal to model these durations and intervals using statistical distributions that are believed to be good representations of the observed behaviour.

Thus an important HCI task is to ensure that the interface offers support on this model building, especially that it eases the gradual development of a model.

3 BASIC IDEAS OF HUMAN COMPUTER INTERACTION/INTERFACE

3.1 Interface Design

The main concern of HCI is to ensure that computer software and hardware is well suited to its users and to the tasks that they wish to perform. Early work on HCI tended to focus on the specialised needs of people such as pilots, engineers running large, complex systems and others such as air traffic controllers who worked with computers in situations of great pressure, risk or danger. More recently, the emergence of the Apple MacIntosh and other windowing-type operating systems has brought some of the benefits of the this early work to users of general purpose computer systems. HCI includes the design of hardware as well as software, but for the purposes of this paper, HCI is restricted to the software that enables users to interact with computer systems. This software interface should reflect the characteristics of the people who wish to use the system and the tasks and functions that they have in mind as they approach the computer system.

Lansdale and Ormerod (1994) suggest that the development of good software interfaces requires the co-operation of two groups of people, designers and psychologists. The designer must ensure that the software is able to achieve the tasks which the user may require of it. The psychologist must focus on the user to try to understand how the user might wish to conduct these tasks. This co-operation occurs during task analysis which aims to understand the logical structure of complex tasks, to identify the information needed by the users to carry out these complex tasks, to understand the activities conducted by users with the interface and to appreciate the conditions that may affect their performance.

On the same theme, Norman (1990) suggests that the interface designer must bear three things in mind if the interface is to be well-designed.

1. The types of people who may wish to use the system.
2. The tasks that these people wish to accomplish.
3. The tools that are most appropriate in enabling them to do those tasks.

This is an extension of what Norman (1988) terms 'user-centered design'.

3.2 User-centered Design

Norman (1988) suggests that designers should take account of seven principles if they wish to achieve a user-centered design.

1. *Use both knowledge in the world and knowledge in the head.* Knowledge in the world is that which is available externally, especially visibly and which will suggest what the function of some system component may be. Knowledge in the head is that which the user is able to internalise as they become practised and proficient and for which few cues are needed. Hence, at its simplest level, there should always be enough clues on-screen for the naive user to know what to do and yet the expert user should also feel comfortable and should not be frustrated by a having to step through a mind-numbing routine of protocols to achieve some simple end.
2. *Simplify the structure of tasks.*: This requires the designer to take account of the limitations of human memory. This is often conceptualised as *short term memory*, which is often believed to be capable of holding 7±2 items and *long term memory*, which seems to develop conceptual frameworks within which new experiences are considered. Thus, software must not overload short term memory, and it should be a good fit with the users' conceptual frameworks. Software that clashes markedly with a user's likely conceptual framework is likely to be difficult to learn and will be error prone in use.
3. *Make things visible.* Designers should struggle to ensure that sufficient information is visible to the user to enable them to do what they wish, within the limitations of the software. This clearly relates to the first principle of using knowledge in the world - it requires the designer to ensure that suitable cues are available to the user.
4. *Get the mappings right.* This principle requires the designer to ensure that the user can understand the links between the options, the information presented and the tasks that they wish to perform. The links between the users' intentions and the actions open to them in the software need to be made very clear. It also relates to the need to ensure that users have appropriate conceptual models, otherwise they may misunderstand the function of the system and its components.
5. *Exploit the power of constraints*: Constraints are those are restrictions that are deliberately built into the software to ensure that dangerous or bizarre actions

cannot be attempted. This might mean, on occasions, frustrating the user who wishes to cut corners.

6. *Design for error.* This principle requires the designer to assume that the user may make mistakes. It is vital, therefore, to ensure that, if mistakes and slips occur, the behaviour of the software is safe and helpful to the user. It is unreasonable and downright stupid to assume that even experts never make mistakes.

7. *When all else fails, standardise.* The idea of this final principle being to limit the amount of learning that the user must achieve in order to accomplish their required tasks.

Clearly, these principles are inter-dependent and overlap with one another. But their overall message is very clear. Focus on the user and make things easy for them. This requires the designer to know who the likely users will be and to understand what they may wish to do.

3.3 Some Principles of Learning

Kay (1990) builds on Bruner (1962) in suggesting that human cognition is made up of three interlinked aspects as follows, from the simplest through to the more abstract. As learning occurs, an individual moves through these levels.

1. *The enactive mentality.* In this, there is a focus on direct experimentation, on trying things out. It relates to very concrete thinking. Expressed in computer interface terms, this requires the user to be aware of exactly where they are in the system and exactly what is open to them at that point. Given that Bruner (op cit) argues that this is the way in which many of us start to learn about new things, this suggests that the interface be designed to ensure that its exploratory use can do no harm.

2. *The iconic mentality.* In this, the individual tries to recognise, compare and configure different aspects of their surroundings. This is rather more abstract than an enactive approach in that it suggests that a person may wish to change the world in some way or other and to make simple inferences. In computer interface terms, this means that the user needs recognisable images and objects whose function is clear. This should then reduce the need for learning by enactment and experimentation.

3. *The symbolic mentality.* In this, the individual tries to tie together chains of reasoning in a way that may be very abstract. In essence, the person has conceptual models which are applied to the task at hand. There is thus a need to see the connect between different symbols and to understand how they may be linked to achieve particular tasks. In discrete simulation terms, the interface should support modelling.

In these terms, simulation modelling operates with a symbolic mentality, but Kay(op cit) points out that this will not be achieved unless the other two levels are in place. That is, the ability to conduct symbolic reasoning sits on iconic and enactive foundations. Thus, useful discrete simulation software must make provision for all three.

3.4 The Effect of the Interface

Finally, it is important to bear in mind the comment made by Lansdale and Ormerod (op cit). Though they are insistent that good, user-centered design must be based on a good fit between task, technology and the user, they argue that one other aspect is of crucial importance. This is that the users' goals may not be static through time. There are two reasons for this, one of which may not be obvious.

The obvious point is that the task itself may change for reasons quite unconnected with the software. Its scale might increase (e.g. air traffic controllers might need to deal with a much more crowded airspace) or other changes may take away much of the task (e.g. decisions on the sharing of air space amongst national traffic controllers). Thus a sensible task analysis needs to look at possible changes in the task.

The less obvious point, and the one that matters for discrete simulation, is that the nature of the interface may change the task itself. To use a non-simulation example, consider word processors. They have two main virtues. The first is that they enable any trained user to produce work that is well-presented, hence this conference can provide presentation guidelines that enable authors to meet their requirements for camera-ready copy. The second is that it changes the nature of the task by allowing people to adopt a different approach to their writing. Writing with pen and ink requires a sentence, a paragraph even, to be composed in the head before it is committed to paper. This forces the pen-and-paper writing style to be thoughtful and considered (in the opinion of Stoll (1995)) or ponderous, in the opinion of others. Using a word processor, especially one with a good outliner, allows the user to hammer away at the keyboard with a stream of ideas that can later be expressed as lucid prose. Hence the software, and its interface, can change the nature of the task.

4 IMPLICATIONS FOR DISCRETE SIMULATION SOFTWARE

4.1 User-centered Design

Norman's seven principles can, perhaps, be reduced to four when viewed within the use of today's window-centric world. These are as follows.

a) Simplify the structure of tasks.
b) Get the mappings right.
c) Exploit the power of constraints.
d) Design for error.

This is not to say that today's windowing operating systems are perfect, far from it. However, adherence to their norms at least helps to enforce Norman's other three principles. A user who is familiar with other windowing software will at least have useful knowledge in the head and will know where to look for knowledge in the world. Such systems are also essentially visual (which presents severe problems for some users of computers with disabilities) and they provide some form of standardisation.

Principles a) and b) above relate to Kay's point about symbolic, iconic and enactive mentalities. That is, they relate to the need to ensure that the conceptual model of the software is abundantly clear in its interface and for it to fit with that assumed by the user. Those of the user are provided, in part at least, by the conceptual frameworks of their long term memory. The interface and its metaphors need to be consistent, clear and must make sense to the user, but this is not easy.

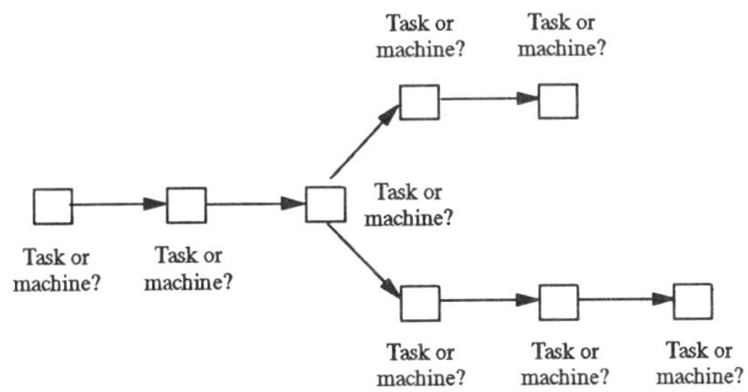

Figure 1: Task network or machine network

Consider for example, the metaphor that is usually employed in Visual Interactive Modelling Systems (VIMS). In these, the user develops a model by placing icons on-screen and by linking them with directed arcs in a network of some kind, see Figure 1. Though it may not be obvious, there are at least two ways in which this can be done - and this will not be obvious to the naive user.

1. Make this a task network. This is the approach employed in software such as Microsaint (op cit). Each node represents a task and each of these tasks may employ several resources which are committed during the duration of the task. Thus, for example, a medical consultation may require the co-operation of a patient, a doctor, a nurse and some specialised equipment. These resources are freed at the end of the task. One of these resources is modelled as the system entity which flows through the task network.
2. Make this a machine network. This is the approach commonly taken on manufacturing-oriented VIMS such as Witness (op cit). Each node represents a system resource that will be occupied as a system entity (usually an item being manufactured) passes through it. The node represents a machine that may well be capable of performing several tasks.

These distinctions will be not be obvious to a naive user from the interface nor may they be obvious from simple examples. The user may suddenly be brought to the jarring realisation that their conceptual model is a task network (or vice versa) and the simulation software assumes a machine network (or vice versa).

Principles c) and d) relate to the need for safety. They aim to minimise the risk that the user may make slips (accidental errors and abuses) or errors (deliberate actions that turn out to be misconceived). The interface and its metaphors must be clearly designed with this in mind. In the case of discrete simulation software, if designers wish to take these ideas seriously, then serious issues about various statistical aspects of the model and the simulation need to be taken into account. As a simple example, most VIMS permit the use of Normal distributions, but even sophisticated users may sometimes forget that Normally distributed variates range from plus to minus infinity. In most cases, users and modellers are probably assuming bounded Normal distributions. Whether this is what they will get may not be clear. If it is what they get, the bounds may not be stated anywhere that is easily accessible to the user.

4.2 Needs of Different Groups

It is probably true to say that early simulation software systems concentrated on the needs of modellers and programmers. The idea being to give maximum support to the technical aspects of the task in hand. That this effort has been successful is evident in the widespread use of VIMS and the growing acceptance of layered systems and approaches. Though the early, first-generation user interfaces of these systems now seem somewhat crude, the latest generation does seem to have taken some of the lessons of user-centered design to heart - assuming, that is, that the technical people are the users.

But what of the other groups of people involved in a simulation project? When a simulation model is used or viewed by someone who is not the developer, the user is modelling - comparing their conceptual model of the system with the perception of what the simulation model is doing. How can this be supported?

The most common approach has been to improve the quality of graphics and visualisation. Welcome though this is, it does carry some problems. Last year the author visited a state-of-the-art engineering company who were planning the development of a highly expensive flexible manufacturing line. As part of this, they developed a discrete simulation which they were keen to show off. The graphics and visualisation were stunning with 3 dimensions, correct proportions, panning and zooming ..etc.. But, the simulation itself was, in analytical terms, probably a waste of time. Simple calculations showed where the bottlenecks would occur and, given the discrete nature of the system components, it was pretty obvious what design would be needed. The author has a strong suspicion that this simulation may have hindered a sensible decision making process rather than supporting it. Though graphics and visualisation are important, it must be remembered that their role should be a support to people in their thinking. Sometimes a simple representation may be a better support to thinking than a complex one.

As with user-centered design for the analyst, so too for the other groups, the basic questions concern the tasks to which the different users wish to put the system. It may be that this implies that well designed systems should offer support to these users as they experiment with simulation models. If Bruner and Kay are correct, then these users will wish to enact scenarios in an attempt to learn about the model and to make inferences about the system being simulated. If appearances are deceptive, perhaps superb visualisation is not the be-and-end-ali for non-technical users?

REFERENCES

Bruner E. (1962) *On knowing: essays for the left hand.* Harvard Univ Press, Cambridge, Mass.

Clementson A.T. (1991) *The ESCL Plus system manual.* AT Clementson, The Chestnuts, Princes Road, Windermere, Cumbria, UK.

Coupland D. (1995) *Microserfs.* Flamingo, London.

Dahl O. and Nygaard K. (1996) SIMULA - an Algol-based simulation language. *Comm ACM,* 9, 9, 671-678.

Hurrion R. (1976) The design, use and requirements of an interactive visual computer simulation language to explore production planning problems. PhD Thesis, University of London.

Kay A. (1990) Interview. In Laurel B. and Mountford S.J. (1990) *The art of HCI design.* Addison-Wesley, Reading, Mass.

Lanner Systems (1996) *WITNESS User manual.* Lanner Systems, Redditch, Worcs, UK.

Lansdale M.W. and Ormerod T.C. (1994) *Understanding interfaces: a handbook of human computer dialogue.* Academic Press, New York.

Law A.M. & Kelton W.D (1991) *Simulation modelling & analysis,* second edition. McGraw-Hill, New York.

Micro Analysis & Design (1992) *Getting started with Micro Saint for Windows.* Micro Analysis & Design Simulation Software Inc., Boulder CA.

Norman D.A. (1988) *The psychology of everyday things.* Basic Books, New York.

Norman D.A. (1990) Interview. In Laurel B. and Mountford S.J. (1990) *The art of HCI design.* Addison-Wesley, Reading, Mass.

Pidd M. (1992) *Computer simulation in management science,* third edition. John Wiley & Sons Ltd, Chichester.

Pidd M. (1996a) *Tools for thinking: modelling in management science.* John Wiley & Sons Ltd, Chichester.

Pidd M. (1996b) Five simple principles of modelling. In *Proceedings of the 1996 Winter Simulation Conference,* San Diego CA.

Stoll C. (1995) *Silicon snake oil: second thoughts on the information superhighway.* Doubleday, New York.

Szymankiewicz J, McDonald J. and Turner K. (1988) *Solving business problems by simulation.* McGraw-Hill, Maidenhead, Berks, UK

AUTHOR BIOGRAPHY

MIKE PIDD is Professor of Management Science in the Management School of Lancaster University in the UK. He has written two books on simulation, of which the best known is Computer Simulation in Management Science (John Wiley). He has just completed a new book for the same publisher (Tools for thinking: modelling in management science) which addresses the issues often raised by critics of rational modelling. His current interests in computer simulation include object oriented methods.

Proceedings of the 1996 Winter Simulation Conference
ed. J. M. Charnes, D. J. Morrice, D. T. Brunner, and J. J. Swain

HCI AND SIMULATION PACKAGES

Jasna Kuljis

Department of Mathematical and Computing Sciences
Goldsmiths, University of London
New Cross, London SE14 6NN
UNITED KINGDOM

ABSTRACT

Computer-based simulation modelling is one of the domains that is particularly demanding in terms of user interfaces. Issues that influence the 'usability' of such systems are examined. Several representative systems were investigated in order to generate some general assumptions with respect to those characteristics of user interfaces employed in simulation systems. There is a need for simulation systems that can support the developments of simulation models in many domains, which are not supported by contemporary simulation software. Many user interface deficiencies are discovered and reported. Proposals are made on how user interfaces for simulation systems can be enhanced to match better the needs specific to the domain of simulation modelling, and on how better to support users in simulation model developments.

1 WHAT IS HCI?

To users, the interface is the system (Hix and Hartson, 1993). Computer systems often lack good user interfaces for a variety of reasons, including the lack of a good user interface design methodology and the lack of good tools to implement a user interface. An interface is often the single most important factor in determining the success of a system (Larson, 1992). It is also one of the most expensive. Smith and Mosier (1984) conducted a survey of people concerned with information systems design who on average estimated that 30 to 35 percent of operational software is required to support the user interface. Bobrow et al. (1986) claim that the user interface often constitutes one third to one half of the code of typical knowledge-based systems. This claim is reinforced by Myers and Rosson (1992) who argue that anywhere from an average of 48% to a maximum of nearly 100% of the code for an interactive system is now used to support the user interface.

There is no agreed upon definition of the range of topics which form the area of human-computer interaction. The following definition is one from ACM SIGCHI (1992): "Human-computer interaction is a discipline concerned with the design, evaluation and implementation of interactive computing systems for human use and with the study of major phenomena surrounding them". Benyon and Murray (1988) make an important distinction between the terms Human-Computer *Interaction* and Human-Computer *Interface.* Interaction includes all aspects of the environment such as the working practices, office layout, provision of help and guidance, and so on. The interface is the part of a system with which the user comes into contact physically, perceptually or cognitively. We are mostly concerned with the interface part of simulation systems.

In this paper we examine the usability and appropriateness of such approaches when dealing with simulation software development. We aim to examine user interfaces for discrete event simulation packages. In particular, to investigate issues that influence 'usability' of simulation systems. Usability has many meanings to many different people. Software vendors often claim that their products have attributes such as high level of software usability, or 'user friendliness'. However, this terminology mostly indicates that software has Windows like GUI. There is no generally agreed definition of usability. A definition proposed by the International Standards Organization (ISO) and listed in Booth (1989) states: "The usability of a product is the degree to which specific users can achieve specific goals within a particular environment; effectively, efficiently, comfortably, and in an acceptable manner." A more operational definition is given by Shackel (1991) who suggests that any system should have to pass the usability criteria of effectiveness, learnability, flexibility, and user attitude.

Shackel's four distinguishable and quantifiable dimensions effectiveness, learnability, flexibility, and attitude are not mutually exclusive in the sense that measures of, for example, effectiveness can, at the same time, also give some indication of system learnability. Effectiveness refers to levels of user performance, measured in terms of speed and/ or accuracy, in terms of proportion of task(s), proportion of users, or probability of completion of a given task. Flexibility refers to variations in task completion strategies supported by a system. Learnability refers to the ease with which new or occasional users may accomplish certain tasks. Attitude refers to user acceptability of the system in question. There are some problems, however, in setting appropriate numerical values against usability goals. This approach would be more appropriate when the usability goals are set during the design stage of requirements specification than as an evaluation criteria once a system is finished. Although quantitative data is required to accurately assess the usability of a system, it is qualitative information that informs designers how to change an unusable system. Our objectives are both to assess the usability of current simulation systems and to identify usability defects.

The usability evaluation of the representative simulation system was carried out using structured walkthrough (Booth, 1989), i.e. we worked through a series of tasks the user might be expected to perform looking for sources of potential difficulties. We have based the examination on simulation software for personal computers partly because the issues of interaction are not dependent on the computer platform and partly because the PC platform predominates among commercial simulation systems. Therefore, the results and findings can be generalised across the whole spectrum of simulation software regardless of the host system.

Table 1: Main System Characteristics

	XCELL+	Taylor II	ProModel	Micro Saint	Witness	Simscript II.5
Application area	Manufacturing	Mainly manufacturing	Mainly manufacturing	General purpose	Mainly manufacturing	General purpose
Software type	Data-driven simulator	Data-driven simulator	Data-driven simulator	Data-driven simulator	Data-driven simulator	Language
Interaction style	Menus and fill-in forms	Menus and fill-in forms	Windows GUI	Windows GUI	Windows GUI	Windows GUI
Interaction devices	Keyboard	Keyboard and mouse	Keyboard and mouse	Keyboard and mouse	Keyboard and mouse	Keyboard and mouse
Navigation facilitated using	Function keys	Mouse, arrow keys, and function keys	Mouse, arrow keys, and F1	Mouse, arrow keys, and function keys	Mouse, arrow keys, and F1	Mouse, arrow keys, and F1
Terminology	Manufacturing	Manufacturing	Manufacturing	No specific domain	Manufacturing	No specific domain
Screen layout	Overcrowded	Good	Good	Good	Poor	Good
No. of colours	12	16	64	16	16	64
Use of colours	Hideous	Good	Good	Good	Too extensive	Good

2 HCI FEATURES IN EXAMINED PACKAGES

We have reviewed six discrete simulation systems: XCELL+, Taylor II, ProModel for Windows, Micro Saint for Windows, WITNESS for Windows, and Simscript II.5 for Windows. We examine the following interaction characteristics of these systems: input-output devices employed, interaction styles, and use of graphics. We are also interested in: type of simulation system, application areas, hardware platform, operating system, and hardware requirements. Table 1 provides a summary of the main characteristics of each system.

We test each of the six listed systems on the task of developing a small queuing model (a bank). The test is performed by a user with a high computer literacy, low domain knowledge, and no knowledge of any of the six simulation systems. The ability to accomplish the task was based solely on consulting the user manuals and the available on-line help (i.e. without formal training). We try to identify which of the general usability principles are applied and also establish where the usability defects are in each examined system. When examining the user interface we are particularly interested in three aspects: firstly, how the user interface for a particular system aids the user in a model development process; secondly, can the user modify the existing interface to either accommodate the user's own preferences or to adjust the modelling environment to the needs of a particular model; and thirdly, does the system facilitate user interface development.

2.1 Simulation Environments

The provision of completely self-sufficient simulation environment is important for the following reasons:

- it can reduce the development time,
- it can support application consistency,
- it can aid the developers throughout the development cycle,
- it can support model completeness,
- it can provide checks of model validation.

The experience we have gained during this research convinced us that the model development process is generally not well supported. Five of the six examined simulation packages are data-driven simulators (XCELL+, Taylor II, ProModel for Windows, Micro Saint, and Witness for Windows) that use some sort of diagramming tools for representation of model logic, and one is a simulation language (Simscript II.5 for Windows). All six systems provide modelling environments. Two systems are general purpose (Micro Saint, Simscript II.5) whereas the other four are manufacturing or mainly manufacturing. Graphic elements for the representation of the model logic are pre-defined for all simulators, and cannot be changed for two of them (XCELL+ and Micro Saint). Names of model elements (i.e. machines, parts) are pre-defined and cannot be changed for any of the simulators, although the user can provide labels for individual instances of elements to describe better the domain related elements (i.e. an element machine can be labelled 'clerk' or 'bank teller' in a bank model using Taylor II). Similarly, all examined simulators use fixed, pre-defined, and unmodifiable attribute names (fields). Data entry is usually facilitated through pre-defined, unmodifiable fill-in forms which use the system's own element names, attribute names (fields), which usually have default values provided. Data validation is not a common facility (available in Taylor II and ProModel for Windows).

Background drawing tools are rarely facilitated (Taylor II, ProModel for Windows, Simscript II.5 for Windows), as is importing graphics from other applications (ProModel for Windows, Micro Saint, Simscript II.5 for Windows). Icon editors are more common (not provided in XCELL+ and Micro Saint), even though the majority of them only provide elementary drawing capabilities. The user is rarely allowed to control statistics collection (only in Micro Saint and Simscript II.5) and the way the statistics is displayed (only Simscript II.5 gives complete freedom). Report customisation is rarely allowed. If this facility is provided, only a limited set of options can be exercised. On-line help, if available, usually does not extend to anything more than an overview of basic system concepts. Context sensitive help is scarce and good context-sensitive help is almost non-existent (the exception is ProModel for Windows). On-line help for error messages is not available on the examined systems. The customisation of the modelling environment is virtually an unknown commodity. A limited customisation is offered only in ProModel for Windows. The development of separable user interfaces for particular simulation problems is possible only in Simscript II.5 for Windows, which facilitates user interface development by providing templates for menus, fill-in forms, and several types of graphs that can be then tailored to suit the problem.

2.2 Data Input

It is apparent that data input part of the system is considered as less important than, for example, the visual simulation part. Most of the papers on simulation systems only briefly mention the data input capabilities of systems, if at all. However, there is room for a great deal of improvement in the domain of data input and/or model specification that would improve existing simulation systems. We have already mentioned that data validation is supported in only two of the six examined simulation systems. None of the systems offers database capabilities for keeping multiple variations of a model. Data input forms, if available, are generally poorly designed. There is no help provision for individual data fields. Importing data files is supported in four of the examined systems. The format of imported data is usually an ordinary ASCII text file. Therefore, there is much to be improved in the way the simulation data is communicated to the systems.

Table 2 summarise user interfaces for data input/model specification for the systems examined. Most of the systems keep data in text files that can be accessed and modified from other environments. None of the examined systems provides database facilities. Simulation languages, like SIMSCRIPT II.5 for example, can provide most of the data input features (menu-driven system, data-input forms, on-line help, data validation, etc.) but at the expense of an extensive time-consuming programming effort. Some of the systems provide limited input error checking and model verification facilities.

Table 2: Data Input/ Model Specification

	XCELL+	Taylor II	ProModel	Micro Saint	Witness	Simscript II.5
Model logic representation	Diagramming tools	Diagramming tools	Diagramming tools	Diagramming tools	Diagramming tools	Program
Graphic elements	Pre-defined cannot be changed	Pre-defined can be changed	Default or user selected	Pre-defined cannot be changed	Pre-defined can be changed	Not provided
Model elements	Pre-defined	Pre-defined	Pre-defined	Pre-defined	Pre-defined	User defined
Element names	Up to 10 characters	Up to 8 characters	Up to 80 characters	Up to 20 characters	Up to 8 characters	User defined
Attribute names	Pre-defined cannot be changed	Pre-defined cannot be changed	Pre-defined cannot be changed	Pre-defined cannot be changed	Pre-defined cannot be changed	User defined
Default values provided	Yes	Yes	Yes	No	Yes	Can be programmed
Fill-in forms design	Not applicable	Good	Good	Well balanced	Poor and often confusing	Not applicable
Importing files supported	No	Yes	Yes	No	Yes	Yes
Data validation supported	No	Yes	Yes	No	No	Can be programmed
Model validation supported	Partially	No	No	No	No	Can be programmed

2.3 Visual Simulation

Visual programming tools are standard features in all visual interactive simulation (VIS) systems, and drawing tools are very common. Dynamic icons and animation are supported by most visual simulation systems. The interactive change of the simulation parameters and of the speed of animation whilst the simulation is being executed, are also often provided. Panning and zooming is another quite common facility. Most of the current simulation systems have some form of visual animation of a simulation run. Animation speed in simulation systems is commonly made adjustable by their users.

There are some problems with the animation speed that are not envisaged by the software developers. Simulation software is built for a particular hardware configuration and therefore for a particular processing speed (in MHz). The speed of animation (moving icons) is dependent on the computer processor speed. Hardware developments are much faster than software developments and by the time simulation software, based on a particular configuration, has reached the market it may well happen that the market has already adopted much faster computers. The user will probably install software on a much faster computer than it was intended for. Even though the user may have a facility to change animation speed, the slowest available speed may still be too fast for an animation observer. Table 3 provides a summary of some of the user interface features relevant to the design of simulation experiments.

Table 3: Simulation Experiment

	XCELL+	Taylor II	ProModel	Micro Saint	Witness	Simscript II.5
Background drawing tool	No	Yes	Yes	No	No	Yes
Icon editor	No	Yes	Yes	No	Limited capabilities	Yes
Importing graphics supported	No	No	Yes	Yes	No	Yes
Zooming supported	No	No	Yes	No	Using virtual windows	No
Panning supported	No	Limited (with arrow keys)	Yes	No	Limited	No
Interactive speed change	Yes	Yes	Yes	Yes	Limited to 3 speeds	Can be programmed
Interactive time change	Yes	Yes	Yes	No	No	Can be programmed
Interactive change of other simulation parameters	Yes	Yes	Yes	No	No	Can be programmed

Table 4: Simulation Results

	XCELL+	Taylor II	ProModel	Micro Saint	Witness	Simscript II.5
Control of statistics collection	System	System User partially	System User can only reduce the set	User	System	Modeller
Default statistics provided	Yes	Yes	Yes	Yes	Yes	NA
Graphics supported	No	Yes	Yes	Yes	Yes	Yes
Graph types	NA	Histogram, Pie chart Bar, Line, Scatter, Gantt, Area graph	Pie chart, Plot, Histogram, Bar, Line, Vertical line, and Step graph	Bar, Line, Scatter, and Step graph	Histogram Time series	Histogram, Pie chart, Line graphs, Dynamic bar graphs, Trace and X-Y plots, Dials, Meters
User defined presentation	No	Partially	No	Partially	Partially	Yes
Flexibility of presentation	None	Small	Small	Small	Small	Great
Tabular form statistics	Yes	Yes	Yes	Yes	Yes	Yes
Exporting files supported	Yes	Yes	Yes	Yes	Yes	Yes
Printing statistics tables	Yes	No	Yes	Yes	Yes	Yes
Printing graphs	NA	No	Yes	Yes	No	No

2.4 Simulation Results

Besides the standard reports for commonly occurring performance statistics (e.g. utilisation, queue sizes and delays, and throughput) some of the newer software allows the development of tailored reports. The user can choose the form of representation (e.g. textual output, table, representational graph, bar diagram, pie chart, histogram, time series), colours, labels, etc. In most of the VIS software, partial statistical results can be viewed during the simulation run. Graphs are updated dynamically during the simulation run. Some of the software has facilities to save the statistics into files that can then be printed. Rarely is there a facility to print the whole screen at any point of a simulation run or when the statistics are displayed/presented on the screen. Table 4 gives a summary of presentation of simulation results capabilities of the reviewed software.

Table 5: Printed Manuals

	XCELL+	Taylor II	ProModel	Micro Saint	Witness	Simscript II.5
Tutorial	Not provided	Not thorough enough	Not provided	Yes	Partially covered in the only manual.	Not provided
Getting started	Part of the User Guide	Not provided	Yes	Yes	Not provided	Not provided
User guide	Yes	Yes	Yes	Yes	Yes	Yes
Reference manual	Not provided	Not provided TLI reference manual provided	Yes	Not provided	Not provided	Yes
Index	Only XCELL+ glossary	Global for all manuals	System concepts	System concepts	System concepts	System concepts
Terminology	Manufacturing	Manufacturing Technical	Manufacturing	General simulation	Manufacturing	General simulation. Technical

2.5 User Support

As a rule documentation is highly erratic, written in a technical jargon and rarely organised in any structured manner. If it contains an index it usually requires the user to know the exact terminology used to be able to find the topic of interest. Similarly, the installation procedures are generally badly documented and off-putting All of the systems examined, except Simscript II.5, require a security. Considering the high cost of simulation software it makes some economic sense from the vendors point of view. However, from the customers point of view it shows a basic mistrust in customers' honesty and adds to the general unfriendliness of the simulation systems. User manuals are usually poorly written. Of the six simulation systems we have examined only two provide Tutorial (Taylor II and Micro Saint), three provide Getting Started (XCELL+, ProModel for Windows, and Micro Saint), two provide Reference manuals (Pro Model for Windows and Simscript II.5). An index is provided in all of them except XCELL+, but it usually lists only system concepts using a particular simulation system's terminology. Examples, if provided, are not followed throughout the development process and are therefore of not much use. The summary of characteristics of printed manuals is given in Table 5.

On-line help very rarely provides help for all facilities and the available tools in the simulation environment. Context-sensitive help is a rare commodity (it is only provided in Taylor II and ProModel for Windows) and is almost unheard of for system messages (i.e. error messages). An on-line tutorial is provided only in ProModel for Windows. Demonstration disks are provided for three of the examined systems Taylor II, ProModel for Windows, and Micro Saint, where ProModel provides a professional and carefully thought out product. Table 6 provides a summary of on-line user assistance offered in the six examined simulation packages.

Table 6: On-line User Assistance

	XCELL+	Taylor II	ProModel	Micro Saint	Witness	Simscript II.5
Model examples provided	Yes	Yes	Yes	Yes	Yes	Yes
Help type	One text screen	Isolated text screens	Hypertext	Limited hypertext	Isolated text screens	Hypertext
Navigation through help facilitated using	Not applicable	Mouse and arrow keys	Mouse point and click on link nodes.	Mouse point and click on link nodes.	Mouse point and click on cross reference buttons	Mouse point and click on link nodes.
Help text	Short information on system	Complete version of printed material	Differs from printed manuals	Differs from printed manuals	Differs slightly from printed manuals	Differs from printed manuals
Index of topics	Not applicable	Yes	Yes	Yes	Yes	Yes
Search facility within help	Not applicable	Searches for a first occurrence of a given string	Yes	Searches for a first occurrence of a given string	Not provided	Yes
Tutorial	Not provided	Not provided	Interactive lessons	Not provided	Not provided	Not provided
Context-sensitive help	Not supported	Only relevant page	Yes	Not provided	Not provided	Not provided
Help on help	Not provided	Not provided	Extensive	Limited	Not provided	Yes
Printing help text supported	No	No	Yes	Yes	No	Yes
Demonstration disk	Not provided	Elementary	Professional	Yes	Not provided	Not provided

3 USABILITY OF SIMULATION PACKAGES

Effectiveness of the system can be hindered if there are: defects in navigation through the system, problems in screen design and layout, inappropriate terminology, inappropriate feedback or complete lack of feedback, problems with modality, inconsequential redundancies, and problems in matching with user tasks. Learnability can be impeded if there are: defects in navigation, problems in screen design and layout, inappropriate terminology, inappropriate feedback or complete lack of feedback, and problems in matching with user tasks. Flexibility is impeded if there is no user control over the system and if the system imposes the order of the steps in a task. User attitudes towards the system can be seriously affected by any of the above usability defects.

The usability defects were identified in examined simulation systems. ProModel for Windows has only a few problems, which are its terminology and visual objects that are appropriate solely for the manufacturing domain. XCELL+, Witness for Windows, and Simscript II.5 have serious defects in navigation through the system. Feedback is inadequate in all five packages except in ProModel for Windows. Consistency of a system is assessed based on the degree to which the system performs in a predictable, well organised and standard fashion. XCELL+ has an inconsistent use of function keys, Taylor II an inconsistent use of interaction devices, and Simscript II.5 suffers from the inconsistency across its system modules. Modality is the state of the system operation that the user selects to perform a particular function. Only ProModel for Windows and Micro Saint provide clear feedback on the mode that the system is currently in. The four other systems either use modal dialogue which require the user to respond before any action can be taken or do not provide a clear indication in which mode the system is in. Only ProModel for Windows gives the user the feeling of being in control.

There is no provision, or if there is it is marginal, for user interface customisation. Only ProModel for Windows provides a minimal customisation of modelling environment. There is a facility for guiding an inexperienced user through the necessary steps of model development. Experienced users can choose the order in which to perform the steps of tasks. There is the possibility to create new simulation systems for limited domains with a custom made interface appropriate for the model domain. These capabilities are currently limited to bespoke programming that requires a substantial development effort. Sometimes user interface development can be facilitated using an object-oriented approach that reduces development time.

Many authors argue that the advantages of VIS include better validation, increased credibility (and hence model acceptance), better communication between modeller and client, incorporation of the decision maker into the model via interaction, and learning via playing with the VIS. However, there is little published empirical evidence to substantiate these claims. In addition, animation can be used to enhance a model's credibility and, according to Law and Kelton (1991), it is the main reason for animation's expanding use. Swider et al. (1994) feel that animation can provide convincing evidence that model behaviour is representative of the system under study. Cyr (1992) see advantage of using animation in its ability to demonstrate problems with the model itself which would otherwise be difficult to detect. Kalski and Davis (1991) point out that summary statistics sometimes do not show the active interactions of processes in a system, and they advocate the use of animation as an aid to the analyst in identifying the system status.

There are many animation proponents in the simulation community, especially the software vendors. However, there is very little published empirical evidence which would suggest how to design effective animation. Carpenter et al. (1993) conducted an experiment with 47 subjects to examine how well the animation communicated the operation of the simulation model. They considered combinations of three aspects of animation - movement, detail of icons, and colour. The results suggested that movement of icons is more important than their detail or colour in communicating the behaviour of a simulation model with moving entities. Swider et al. (1994) used 54 subjects to obtain objective and subjective measures in determining which combinations of animation presentation and speed were best for displaying violations of model assumptions. Based on the results of this study, Swider at el. (1994) recommend: the use of pictorial display with moving icons for simulation models with moving entities; the facility to set the presentation speed to make discrete differences visible; and to avoid overloading the user with too much visual information.

The results of the above two studies are not surprising and they match our intuition and common-sense. However, their importance is in substantiating our intuitive judgement with some more concrete evidence. Animations with moving icons are often used in current simulation systems even though presentation of animation is not often well thought about. Ideally, it may seem desirable to present information on the screen that has characteristics similar to the objects we perceive in the environment. The visual system could then use the same processes that it uses when perceiving objects in the environment. Factors that contribute towards the meaningfulness of a stimulus are the familiarity of an item and its associated imagery. The graphical representation of constructs for different applications should give definite information about the type of model component it represents, such as waiting queues, customers or servers in queuing systems or stores, or suppliers in store keeping systems (Kämper, 1993). Stasko's (1993) animation design recommendations state that animation should provide a sense of context, locality, and the relationship between and after states. Furthermore that the objects involved in an animation should depict application entities and that the animation actions should represent the user's mental model. If these recommendations were followed, the effectiveness, learnability, and the enthusiasm of a wider user population to use simulation systems might increase.

The eye-catching, appealing nature of animation can tempt designers to apply too many facets to an interface. Animation is, however, another attribute in which the often quoted design principle "less is more" does apply. Nevertheless, if the screen design is kept clean, simple, and well organised some redundant information can be quite useful to the user. The moderation principle is something that many simulation system developers should learn about. User interfaces that have screens crowded with too many objects, large numbers of offensive colours and incompatible colour schemes is more of a rule than an exception.

An essential aid in model development can be facilitated by selecting model components which are relevant to the model builder's modelling requirements. Our proposal (Kuljis 1994, 1995) is that simulation environments should provide model developers with the following:

i) Several pre-defined problem domains.
ii) A facility to create new problem domains.
iii) A facility to design and/or choose graphical representations for elements in a problem domain.
iv) A facility to set default values for a problem domain.
v) A facility to set defaults for statistical data collection.
vi) A facility to set defaults for the graphical presentation of simulation results.

REFERENCES

ACM Special Interest Group on Computer-Human Interaction Curriculum Development Group. 1992. *ACM SIGCHI Curricula for human-computer interaction.* Technical report, New York: Association for Computing Machinery, Inc.

Benyon, D. and D. Murray. 1988. Experience with adaptive interfaces. *The Computer Journal* 31(5): 465.

Bobrow, D.G., S. Mittal and M.J. Stefik. 1986. Expert systems: Perils and promise. *Communications of the ACM* 29(9): 880-894.

Booth, P. 1989. *An introduction to human-computer interaction.* Hove: Lawrence Erlbaum Associates.

Carpenter, M.L., Bauer Jr., K.W., Schuppe, T. F. and Vidulich, M. V. 1993. Animation: What's essential for effective communication of military simulation model operation? In *Proceedings of the 1993 Winter Simulation Conference,* 1081-1088. Institute of Electrical and Electronics Engineers, Los Angeles.

Cyr, R.W. 1992. Using animation to enhance a marine-terminal Monte Carlo simulaton. In *Proceedings of the 1992 Winter Simulation Conference,* 1000-1003. Institute of Electrical and Electronic Engineers, Arlington, Virginia.

Hix, D. and R.H. Hartson. 1993. *Developing user interfaces. Ensuring usability through product & process.* New York: John Wiley & Sons, Inc.

Kalski, D. R. and Davis, A. A. 1991. Computer animation with CINEMA. In *Proceedings of the 1991 Winter Simulation Conference,* 122-127. Institute of Electrical and Electronic Engineers, Phoenix, Arizona.

Kämper, S. 1993. A systematization of requirements for the user interface of simulation tools. *SAMS* 11 (2): 107-119.

Kuljis, J. 1994. User interfaces and discrete event simulation models. *Journal of Simulation Practice and Theory* 1(5): 207-221.

Kuljis, J. 1995. Usability of manufacturing simulation software. *International Journal of Manufacturing System Design* 2(2): 105-120.

Larson, J.A. 1992. *Interactive software. Tools for building interactive user interfaces.* Englewood Cliffs: Prentice Hall.

Law, A.M. and D.W. Kelton. 1991. *Simulation modeling and analysis.* 2nd ed., New York: McGraw-Hill

Myers, B. A. and M. B. Rosson. 1992. Survey on user interface programming. In *Proceedings of HCI Conference on Human Factors in Computing Systems,* 195-202. New York: ACM.

Shackel, B. 1991. Usability - context, framework, definition, design and evolution. In *Human Factors for Informatics Usability,* eds. Shackel, B. and S. Richardson. Cambridge: Cambridge University Press.

Smith, S.L. and J.N. Mosier. 1984. The use interface to computer-based information systems: A survey of current software design practice. *Behaviour and Information Technology* 3(3): 195-203.

Stasko, J.T. 1993. Animation in user interfaces: Principles and techniques. In *User Interface Software,* Bass, eds. L. and P. Dewan. Chichester: John Wiley & Sons, pp. 81-101.

Swider, C. L., Bauer Jr., K. W. and Schuppe, T. F. 1994. The effective use of animation in simulation model validation, *Proceedings of the 1994 Winter Simulation Conference,* 633-640. Institute of Electrical and Electronic Engineers, Orlando, Florida.

AUTHOR BIOGRAPHY

JASNA KULJIS is a Lecturer in the Department of Mathematical and Computing Sciences at Goldsmiths, University of London. She has a B.Sc. in Mathematics from Zagreb University, a M.S in Information Science from Pittsburgh University and a PhD in Information Systems from the London School of Economics. She is currently researching into human-computer interfaces to modelling systems, having previously spent over a decade as a researcher, consultant and teacher at the University of Zagreb Computer Centre and seven years as a Lecturer at the Roehampton Institute London. She has published widely in many aspects of computing, and has undertaken consultancy for a variety of government agencies and companies in Croatia, as well as for the Department of Health in the U.K.

Proceedings of the 1996 Winter Simulation Conference
ed. J. M. Charnes, D. J. Morrice, D. T. Brunner, and J. J. Swain

SOFT SYSTEMS METHODOLOGY AND SIMULATION MODELING

Brian Lehaney

Faculty of Business
University of Luton
Luton, Beds, LU1 3JU, UK

Ray J. Paul

Department of Computing and IS
Brunel University
Uxbridge, UB8 3PH, UK

ABSTRACT

Simulation is both popular and powerful, but reportage of simulation case studies indicates that in many cases process is treated cursorily, and end-user-acceptance of final models is not forthcoming. Whilst texts often proclaim the importance of process, this is usually left to the discretion of the modeler. A range of problem structuring (or soft) methodologies have been developed to address process issues. However, these can be both slow and unwieldy. This paper outlines a case which utilises the principles of soft methodologies in a relatively quick and dirty approach to process.

1 INTRODUCTION

Simulation provides managers with a powerful means to assess the demands on resources created by variable patterns of arrivals and service rates, such as those experienced in hospital outpatients departments. Analytical techniques such as queuing theory may not be of help in such situations, if their assumptions are too rigid or unrealistic or the situation is too complex. However, simulation does not provide a panacea, and much depends upon the way in which it is used. In order to ascertain the appropriate system specification, and model parameters, and in order to assess the value of the simulation, process of discussion an debate must be undertaken. Indeed, that very discussion and debate may provide resolution in itself.

Despite the possibility of good intentions, simulation textbooks tend to ignore, or mention only fleetingly, the important processes of problem formulation and logical model development. As Paul and Balmer (1993, p5) note 'Experience of this process of model formulation is not easy to provide in the context of the essentially artificial 'practical' exercises in either a textbook or academic course'. In fact simulation texts give no real guidance as to how this process should be undertaken.

In recent years problem structuring (soft methods) has flourished as an area of academic and practitioner interest. A range of methodologies has been developed, including Soft Systems Methodology (Checkland, 1981, Wilson, 1984). These draw from the following principles:

- emphasis on problem identification, problem structuring, and problem resolution, rather than on problem solution;
- acceptance of multiple problem-perspectives;
- belief that the researcher will affect the situation;
- consensus and participation rather than imposition;
- continual re-evaluation;
- no automatic acceptance of existing structures;
- involvement of those being researched in the research process;
- concern to elicit repressed or minority views;
- challenging approaches to 'norms'.

Whilst the approaches vary in many ways, the arguments in their favour are that they provide:

- structured approaches to problem identification;
- increased sense of model ownership, and hence increased likelihood of model confidence;
- increased probability of implementing recommendations;
- a reference framework which may be particularly useful in tricky situations, especially for the less experienced analyst;
- improved communication through the use of a known modelling methodology;
- an important reminder of the process to authors who are writing up case material - process is often omitted in case articles.

Soft Systems Methodology is advocated as the most suitable for combining with simulation because:

- each approach provides important features missing from the other;
- the two can be linked in a truly complementary manner;
- SSM is probably the best known soft method;
- SSM can be used to undertake the important problem structuring phase of simulation, assisting in identifying system boundaries and system activities;
- SSM creates an activity-based model which can be converted to an activity cycle diagram;
- simulation depicts a range of options which assists in assessing their feasibility and desirability;

- simulation produces dynamic models which can be used to investigate interactions;
- SSM enables 'hidden' activities to be identified, by means of issue-based root definitions;
- the use of SSM reduces the chance of implementation failure which results from misunderstandings.

2 WHAT IS SOFT SYSTEMS METHODOLOGY?

Soft systems methodology (SSM) is an approach to modelling developed by Checkland (1981) and Wilson (1984), as an alternative to traditional 'hard' approaches, which are based on the prescriptive use of techniques, for clearly defined problems. Such techniques are of little use in the initial analysis of a human activity system, which is likely to have problem areas that are not clearly identified, and which are unstructured. Soft system methodology enables the people involved in running a system (Actors), those responsible for controlling it (Owners), and those who receive its benefits (Customers), to participate in the process of developing a system model, which is likely to encourage acceptability of the model. SSM may be used to aid the identification of system boundaries and system activities, particularly in complex systems. This is particularly useful in a case where a 'hard' technique may eventually be applied, such as in the simulation of an out- patients system. Activity areas within the National Health Service (NHS) are typically 'messy', and particularly suited to an SSM approach, and to the application of simulation.

The general seven stages of soft systems methodology are shown in Figure 1. The methodology is iterative in approach, and is not prescriptive as to a starting point. The unstructured problem situation (1) is described by a rich picture (2). This is used to search for relevant systems which are described by a root definition (3) and conceptual model (4). Although other forms of root definition are possible, those which are both consensus-based, and issue-based are the most interesting and most useful. A root definition describes a system in terms of the following components.

Owners, the roles of monitoring and taking appropriate control actions;

Actors, the roles of running the system;

Customers, the beneficiaries (or victims) of the system;

Transformation, the system inputs converted to outputs;

Weltanschauung, the world view, the perspective taken in defining the system;

Environment, the constraints on the system.

The above can be formed as the mnemonic CATWOE, and may be used to deter the modeler from omitting important system elements. Any omissions should be deliberate and rational.

3 A CONCEPTUAL COMBINATION OF SIMULATION WITH SSM

Figure 2 shows how SSM and simulation may be combined. Phase 1 comprises the early stages of SSM, in which finding out about the problem situation is undertaken, the problem situation is expressed (rich picture), and the system is described (root definition). From the root definition, a conceptual model is formed which contains the minimum set of activities needed to support the root definition. The conceptual model is compared with the system (Phase 1a), and, where appropriate, the root definition is changed, and a new conceptual model developed. If the activity-level of the conceptual model is too broad, selected activities are expanded until the appropriate level of resolution is reached. As primary task root definitions are used, it is likely that the systems identified within the conceptual model will match those of the organisation. Following temporary entities through the system may result in systems which do not match those of the organisation and which may not have owners. At this stage, if such systems are identified, it may be that control actions are taken. If the organisation appoints an owner of a newly-identified system, for example, this may be sufficient to address any problem situations which have arisen.

If Phase 7 indicates that model output and system output do not match as well as is desired, the route of change must be through Phase 4, as adjusting the model ad hoc is likely to lead to self-fulfilling validation prophecies. (The modeler must make judgements regarding minor changes.) The route through Phase 4 preserves the integrity of the model, and mimics the 'Rich-Picture-to-Root-Defintion-to-Conceptual-Model-to-Rich-Picture-to-Root Definition' circle of SSM, which should not be broken by direct input from the Rich Picture to the Conceptual Model. (The latter should be built solely from the root definition.)

If the conceptual model development has been undertaken rigorously and with 'good' participation, it should be unnecessary to revisit the early SSM stages in the first iteration. These should, however, be revisited as part of the overall process, as once policy changes have been explored and implemented, the system, being dynamic, has led to another investigation. The cycle then continues.

The concept proposed is appealing in many ways, but it is not without its difficulties. As Mingers (1992) notes, despite Checkland's assurances that SSM is time-independent, it is in fact time consuming. A useful approach is one which enables an acceptable simulation model to be built quickly, but within a 'quick and dirty' framework which embodies the principles of soft methodologies.

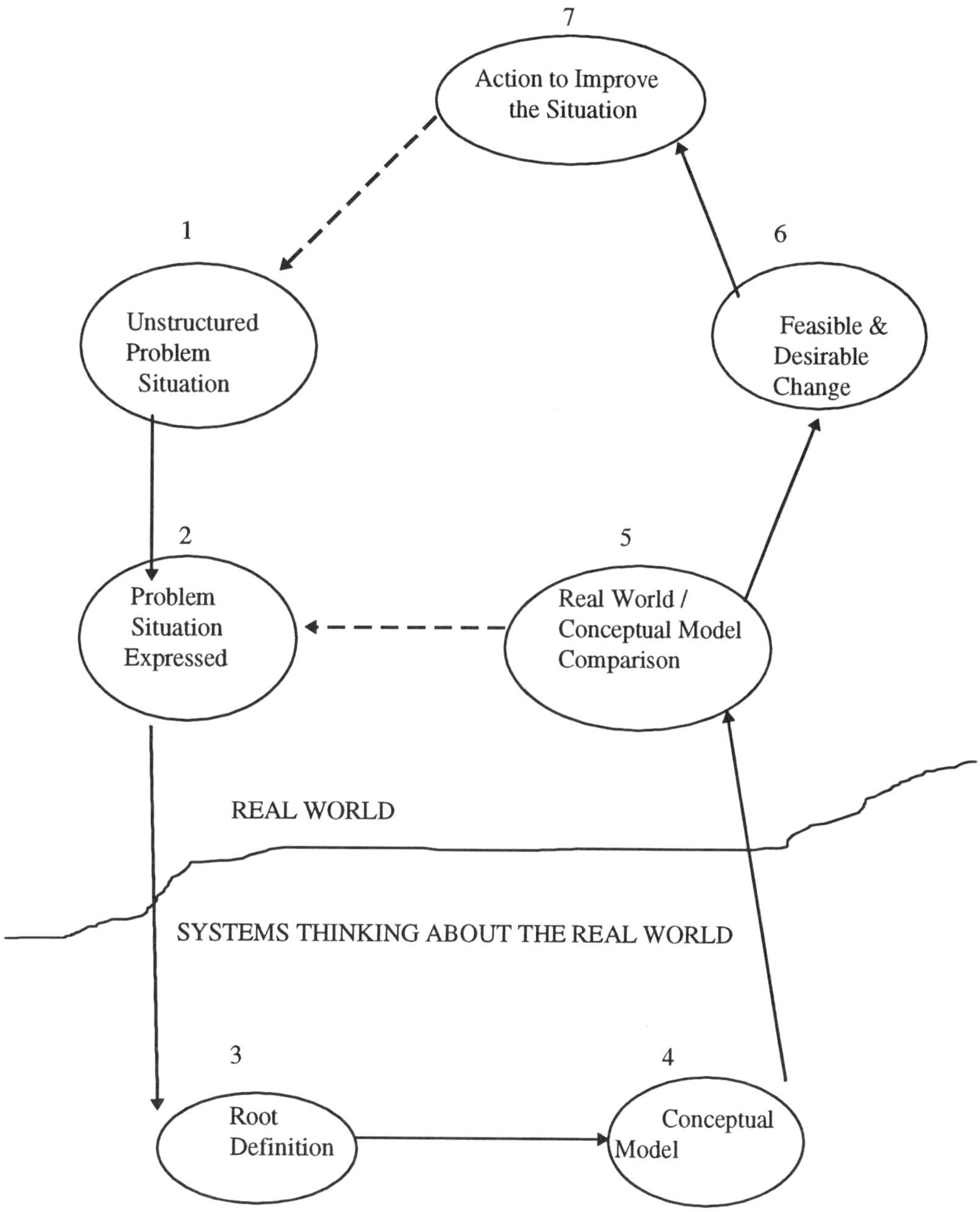

Figure 1: The SSM Seven Stage Process

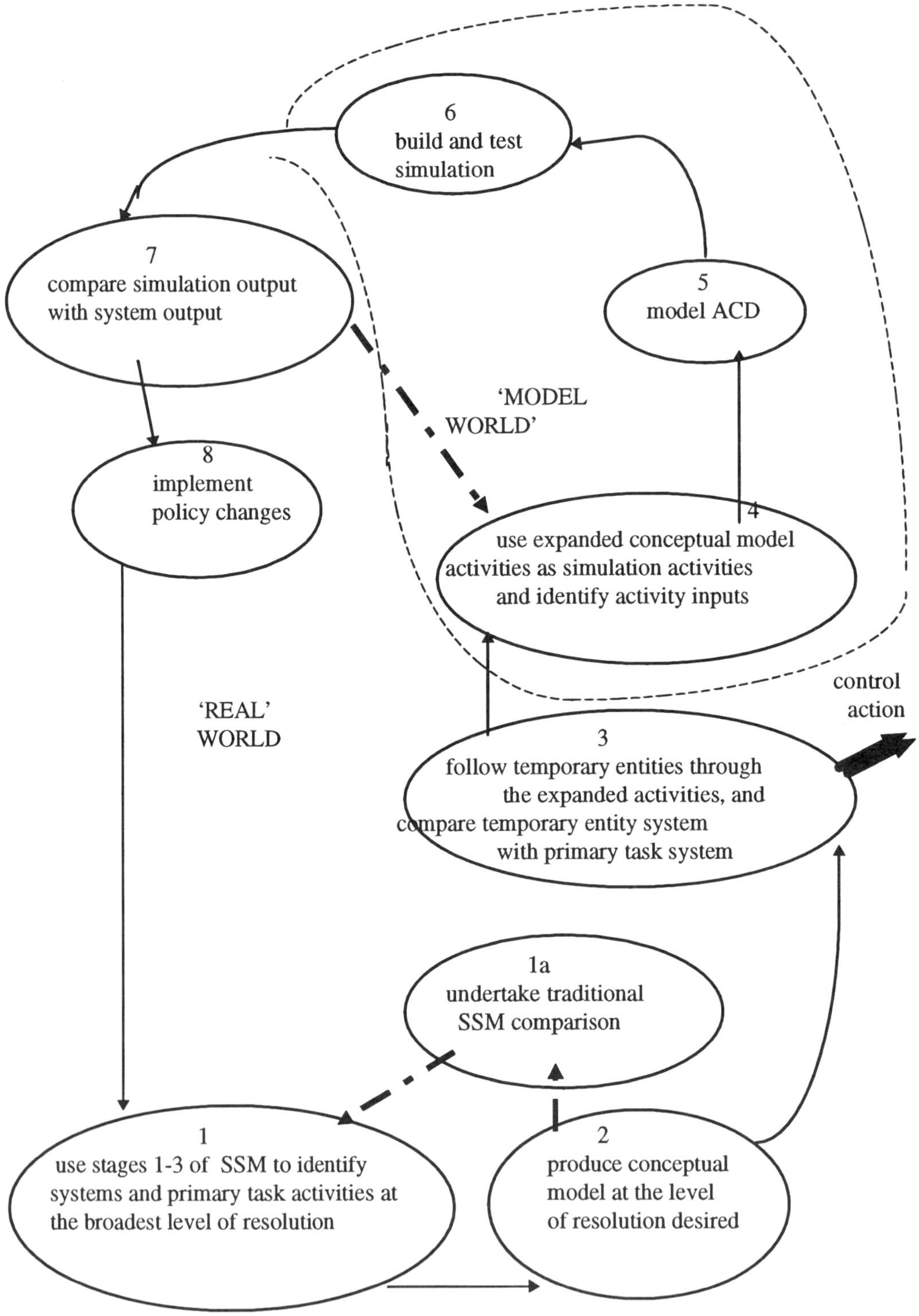

Figure 2: A Conceptual Combination of Simulation with SSM

Table 1: Simulation Modelling Process

The detailed methodology outlined below should be accompanied by critical reviews of the overall project, its process, its timing, and its outcomes to-date, by means of scheduled meetings of analysts and stakeholders, and by any other means considered to be useful by the analysts and stakeholders. The frequency and duration of meetings may be adjusted as the project develops and in order to ensure that analysts and stakeholders have sufficient opportunity to cover any major issues relating to the project which they wish to raise. Some stages may be undertaken in a different order to that shown below, and some stages may be undertaken in parallel. The terms 'model' and 'system model' refer to diagrams, numbers, and words that describe the analysts' and stakeholders' views of the operations of the system (e.g., flowcharts). This is distinct from a 'computer simulation', although the latter will be based on a 'system model'.

A analysts FC first contact KS key stakeholders ORS other relevant stakeholders
The First Contact below has been reassigned as a key stakeholder after the first two stages.

Actors	Actions
A, FC	agree the broad nature and scope of the project
A, FC	establish the initial key stakeholders
A, KS	establish the broad area of investigation
A, KS	re-establish the key stakeholders
A, KS	establish the nature and scope of the project
A, KS	agree working objectives
A, KS	agree an initial project process and timetable
A, KS	establish other relevant stakeholders
A, KS	agree initial model of system boundaries and system activities
A	build initial computer simulations
A	run initial computer simulations using 'rough' data estimates
A, KS	evaluate initial computer simulations
A, KS	establish other relevant stakeholders
A, KS	decide roughly how the system could be best examined 'on the ground'
A, KS, ORS	inform ORS of the nature and scope of the project
A, KS, ORS	refine the process of 'on the ground' system examination
A, KS,	agree a timetable to examine the system 'on the ground'
A, KS, ORS	examine the system 'on the ground'
A, KS, ORS	collate information
A, KS	refine the system model
A, KS	refine the nature and scope of the project
A, KS	refine the working objectives
A	refine the computer models
A	run the refined computer models using 'rough' data estimates
A, KS	establish confidence in results
A, KS	refine data requirements
A, KS, ORS	collect and collate additional data
A	run refined computer simulation using refined data
A, KS	experiment with scenarios and analyse results
A, KS	determine the external and internal criteria which policy should address
A, KS	assess which policy options are both feasible and desirable
KS	establish how policy is to be implemented (practical, technical)
KS	implement policy
A, KS	assess the success of policy in meeting internal and external criteria
A, KS	reassess the nature and scope of the project

4 A QUICK AND DIRTY CASE

A hospital outpatients department was the subject of a recent investigation. Table 1 shows the process. The discussions were essentially focus groups geared to patient throughflow. The key stakeholders were identified easily and very quickly. From this identification the rest follows.

Table 1 may appear to propose a sequential approach to a complex problem. In fact many of the activities in Table 1 may be undertaken in parallel, and the process is iterative. These activities can be grouped as a conceptual model, the root definition for which is as follows:

A system owned by a group of key stakeholders, run by analysts, key stakeholders, and other stakeholders, who use simulation modelling as an aid to develop and implement operational policy which meets internal and external criteria.

Using the CATWOE mnemonic:

Customer:	unspecified
Actors:	analysts, key and other stakeholders
Transformation:	to develop and implement operational policy
Weltanschaung:	by using simulation modelling as an aid, operational policy may be developed and implemented
Owners:	key stakeholders
Environment:	internal and external criteria

The system which is being simulated is the outpatients department. The root definition and conceptual model here relate to a system of resolution which utilises that simulation model, but is not itself being simulated. If it were to be simulated (not impossible) then another overarching root definition and conceptual model for a system of resolution would be needed. The important point is that instead of the simulation being used simply to help solve a single specific problem at an operational level, it is now part of a system to resolve on-going complex situations at a strategic level.

The above process has been used successfully to develop a simulation of an outpatients department. The term 'successfully' is used here to denote that the end-users have welcomed the recommendations resulting from the investigation, and are taking control actions (patient rescheduling) accordingly. The above process is now seen as an important part of their monitoring and control procedures.

5 CONCLUSIONS

Simulation literature is sparse in the extreme with regard to process, and the principles of soft methods are useful for anyone planning or examining an investigation involving simulation.

SSM is conceptually the most appropriate soft approach to combine with simulation, but it is time consuming and unwieldy.

A quick and dirty approach to developing simulation models which utiilises the principles of soft methods may encourages end-user confidence in the early stages, and if an over-arching 'soft' framework is utilised, this confidence may be further fostered and maintained.

REFERENCES

Checkland, P.B. 1981. *Systems Thinking, Systems Practice*. Chichester: Wiley.

Mingers, J., and Taylor, S. 1992. The use of soft systems methodology in practice. *Journal of the Operational Research Society*. 43(4): 321-332.

Paul R.J., Balmer D. 1993. *Simulation Modelling*. Bromley: Chartwell Bratt.

Wilson B. 1984. *Systems: Concepts, Methodologies, and Applications*. Cichester: Wiley.

BRIAN LEHANEY is Principal Research Fellow in the Department of Business Systems at the University of Luton. His research focuses on simulation model development and building end-user confidence.

RAY J. PAUL is the first UK Professor of Simulation, and is also Head of the Department of Computing and Information Systems at Brunel University. Amongst other activities, his research centre (CASM) explores numerous aspects of discrete event simulation.

Proceedings of the 1996 Winter Simulation Conference
ed. J. M. Charnes, D. J. Morrice, D. T. Brunner, and J. J. Swain

REDUNDANCY IN MODEL REPRESENTATION: A BLESSING OR A CURSE?

Richard E. Nance

Systems Research Center
Virginia Polytechnic Institute
and State University
Blacksburg, VA 24061-0251 USA

C. Michael Overstreet

Computer Science Department
Old Dominion University
Norfolk, VA 23529-0162 USA

Ernest H. Page

C3 Modeling & Simulation Center
Mitre Corp.
7525 Colshire Drive
McLean, VA 22102-3481 USA

ABSTRACT

With the intent of dispelling the prevailing negative connotations associated with redundancy, we argue that redundancy can effect benefits in model specification as opposed to model execution. Sources of redundancy are classified as accidental or intentional, and several examples are given for each. The comparative benefits and detriments are discussed briefly, and for the most interesting source of redundancy – that induced by a modeling methodology, we demonstrate that automated elimination of redundancy can actually improve the execution time. Although the set of models investigated is small, these results are encouraging for researchers in modeling methodologies using automated model diagnosis.

1 BACKGROUND

As computing technology has advanced, the problems perceived as solvable have grown more challenging, and the consequences have been models of ever-increasing size and complexity. Often, model development efforts focus on the edge of what is currently technically feasible. The need for control and discipline in the specification of such models, accompanied by a recognition of the needs of humans for assistance in performing such complex tasks, have led to the emergence of modeling methodologies and supporting environments. Recognition of the importance of modeling methodologies and computer-assisted support is evident in the prescriptive modeling community through efforts such as ANALYZE (Greenberg 1983; 1987; 1993), GAMS (Bisschop and Meeraus 1982; Brooke, Kendrick and Meeraus 1992) and Structured Modeling (Geoffrion 1987, 1992a, 1992b). Within the discrete event simulation community, the maturation of research in methodology-based support environments is seen in KBSim (Rothenberg 1989), Knowledge Based Simulation (KBS) (Baskaran and Reddy 1984), MODSYN (Rozenblit and Huang 1991), and the Visual Simulation Environment (VSE^{TM}) (Balci et al. 1995). VSE^{TM} is distinct in that it has evolved from research projects originating in 1983 to become a commercial product. Concurrently, simulation language vendors have expanded their software assistance for simulation activities, e.g., SLAMSYSTEM supporting SLAM II and Arena supporting SIMAN.

The consequences of this more expansive view of what modeling support is needed are summarized in (Nance 1994) with particular attention given to the Conical Methodology, the first attempt to provide life-cycle support to simulation activities (Nance 1981, 1987). A major consequence is a shift from program-centric modeling paradigms, where a simulation programming language (SPL) is considered as the only representational form, to model-centric paradigms in which the modeling process encompasses multiple abstraction levels and the focus is directed toward the transformation of model representations through iterative specification.

Model specification is that set of activities performed by a modeler (or modeling team) to create model representations that can can be communicated to humans. A model specification that is executable is considered a model implementation. Note that model specification can employ multiple representations, each at a different level of abstraction. One or more model specification languages can be used in these activities. While a model implementation (in a SPL or any other executable form) is a model specification, the reverse is not true. Since a model specification describes what behavior is to be exhibited whereas an implementation describes how that behavior is to be produced, the former is inherently more abstract than the latter.

The term "redundancy" typically induces negative connotations – something that is unnecessary and exacts a penalty by its inclusion, e.g., superflous constraints in a mathematical program, unexecuted and obsolete code in a simulation program. However, both examples pertain to the model execution, not the model specification. Maintaining unused storage and compiling unexecuted code is wasteful, but redundancy can prove beneficial in model specification. The objectives of this paper are to identify how redundancy is created in simulation model specifications, describe the beneficial influences of redundancy, and demonstrate that the objectionable influences might not be as perceived.

2 REPRESENTATIONAL REDUNDANCY

Redundancy can take several forms; the precise form of interest here is representational redundancy, which we define as the inclusion of any symbols in a model specification that are not required to fulfill the objectives for which the model is developed. "Symbols" can refer to icons in a graphical specification, textual strings used as attribute identifiers, or objects following the object-oriented paradigm. Methods that are never invoked or procedures that remain unexecuted provide additional examples. From this point, all uses of the term "redundancy" are intended to be limited to the "representational" form.

Careful distinctions must be drawn at this point: symbols external to the model specification but included to attain model development objectives are not redundant. An example is the use of embedded assertions to assist in testing and validation with the intent of assuring model correctness as an objective. In contrast, symbols internal to the model specification to support the attainment of objectives such as model reuse do exemplify redundancy.

Given the usual negative associations with redundancy, a typical reaction is that simulation models developed in practice provide few examples. We maintain that such a view can be overly optimistic. In a related paper (Nance, Overstreet and Page. 1996), we identify four major sources of redundancy in model specifications, and offer a number of examples. Because of space constraints, only a few examples are given herein, with the potential sources classified as accidental or intentional.

2.1 Accidental Redundancy

The most prevalent examples of accidental redundancy are those that are tool-induced. In particular, SPLs commonly use defaults on certain functions that force attributes to be carried along even if not used. Consider the TALLY and ACCUMULATE statements in Simscript II.5. Counters are required for computing the ten statistics available in each and are maintained even if the statistic is not accessed. A similar default decision of 12 parameters for a GPSS transaction imposes accidental redundancy if less than 12 are actually used.

Redundancy can be induced accidentally by the modeling methodology. Use of an object-oriented methodology can cause attributes imparted to the inheriting (created) objects to be redundant. For example, the class *vehicle* with an attribute *number_of_axles*, originally intended to distinguish trucks from cars, might be expanded through concatenation to define the class motorized vehicles, which is assigned the attribute type with enumerated values *motorcycle, car, light truck, bus, heavy truck,* and *tandem.* The inherited attribute *number_of_axles* might be superfluous but remain in the model description.

2.2 Intentional Redundancy

A common occurrence of redundancy can be attributed to the objective of reusability. For example, the reuse of an aircraft object to represent a satellite would cause several attributes related to maneuverability and speed changes to become redundant. However, the value of modifying the reused object simply does not argue for doing so.

Potentially, the most prevalent intentional redundancy can be observed in simulations using animation of output behavior. A host of attributes are attached to objects to "jazz up" the output, in the attempt to facilitate validation or promote model acceptance and credibility. Judged strictly from the objectives of the study they are unnecessary. One could assign the cause of redundancy here as stemming from the desire to utilize application domain knowledge to make the model reflect reality beyond that which is required.

Methodology-induced redundancy is clearly the most interesting. Intentionally including redundant description in the form of attributes, methods or even objects as part of the proper way to construct models seems counter-intuitive. However, the use of multiple descriptions to combat complexity is argued by Padmanaban et al. (1995). Incorporating descriptions from multiple domain experts is required "to provide more information than a single description" (Padmanaban, Benjamin, Mayer 1995, p.719). Clearly, the inability to partition these multiple descriptions must lead to redundancy.

The Condition Specification (CS) (Overstreet and Nance 1985), especially in its use within the Conical Methodology, forces even more obvious redundancy through its insistence on the declaration of relational attributes that relate one object to another. An indicative attribute, which conveys some useful information about the object (given the study objectives), can also serve a relational role, and often each of two interacting objects is assigned an attribute with the two attributes being so strongly dependent that one is unnecessary. Fortunately, automated model simplification techniques can be applied to recognize the dependency and eliminate the redundancy.

3 BALANCING THE NEGATIVES AND THE POSITIVES

The argument is put forth in this paper that, contrary to the usual definition of the term, redundancy, at least with respect to model representation, is not entirely bad. Of course, neither is redundancy entirely meritorious. So, can the negatives and positives be properly balanced? An obvious answer to this question seems to be "yes" and that this balancing is the purview of the modeler – such balancing, after all, *is* the essence of modeling. But upon closer examination this answer seems insufficient. As noted in Section 2, the modeler is not the sole mechanism for inducing representational redundancy. The tools that the modeler employs, as well as the modeling methodology itself, may be contributors to redundancy. Does a modeler have dominion over these factors apart from being free (generally) to select from a variety of methodologies and tools? The answer is not clear.

An example of tool-induced redundancy, in the form of default statistical collection routines, in Simscript II.5 and GPSS is given in Section 2.1. Tool-induced redundancy through the use of defaults is commendable as long as: (1) the defaults are reasonable, and (2) overriding the defaults is not prohibitive in time and effort. The tool developer's motive is laudable: provide convenience to the modeler. That convenience should not be attained by making departures from the defaults costly however. In both of the languages noted above the defaults are clearly reasonable. On the other hand, overriding default behaviors in these languages can be difficult.

Animation introduces some controversy as well. Some see it as a costly selling (marketing) gimmick; while others believe that it is essential to validation and model credibility and acceptance. The answer probably lies in the degree to which the animation developer attempts to mimic reality. Beyond a given point, no benefit is likely to accrue.

Redundancy created in the attempt to accomplish study objectives such as reusability is rarely to be criticized. The savings in reuse for instance clearly offset any costs incurred from storage loss or added computation. The inclusion of redundant documentation within the simulation model – internal documentation to promote maintainability as a study objective – seems eminently justifiable. On the other hand, redundant documentation could potentially *increase* the cost of maintenance since a single modification in design might require multiple changes in the documentation.

Redundancy in the representation emanating from the model development methodology probably requires the most careful defense. For what purpose are the instances of redundancy created? Is the redundancy limited to specifications that are not executable? How great is the penalty on execution performance if redundancy is produced in the implementation? In the section that follows we provide a demonstration of how methodology-induced representational redundancy can be circumvented to permit efficient model execution.

4 EXECUTION IMPROVEMENT: AN UNEXPECTED BENEFIT

The role of the CS, as with any specification language, is to operate at a level of abstraction above the implementation details; While the CS does not specify a particular time-flow mechanism (e.g., activity scan or event scheduling), such mechanisms are easily provided. Since the actions to be taken whenever a condition is satisfied can be prescribed in a level of detail similar to a programming language such as C, we have experimented with direct execution of action clusters to measure the performance benefits of diagnostic analysis through elimination of methodology-induced redundancy. The transformation required for direct execution of action clusters (DEAC) is minor, and a variety of algorithms for DEAC simulations can be defined; see (Page 1994, pp. 189-194).

Examples of CS model specifications may be found in many of the references cited herein. The most recent extension of the CS, to provide a more complete specification for parallel execution, is given in (Page 1994). In most of these sources, model specification is effected in the context of model development under the Conical Methodology. For medium- to large-scale models, the processes of model definition and model specification are intimately connected as the model evolves through successive elaboration and refinement. Selection of the four models used here is based on three criteria: (1) published in the literature, well known or easily understood; (2) not so lengthy as to require excessive space, and (3) provide among them a range in size and complexity for examining that factor. The Single Server System is too widely used to give a single reference. The Harbor Model appears in (Buxton and Laski 1963 and Schriber 1974). The MVS Computer Systems is from (Balci 1988). The Machine Repair has appeared in several published works, going back to (Nance 1971).

The DEAC algorithm uses the Action Cluster Incidence Graph (ACIG) described in (Overstreet 1982, Overstreet and Nance 1985, Overstreet and Nance 1986, Nance and Overstreet 1987a, Nance and Overstreet 1987b, Puthoff 1991). Space limitations restrict complete discussion of the graph, but the ACIG distinguishes between models actions whose occurrence is based strictly on the value of the simulation clock (called time-based actions), actions whose occurrence is based only on the values of model attributes other than the simulation clock (called state-based), and those actions which depend on both time and state (called mixed). In essence, model behavior is driven by testing to see if the condition controlling an action is true, and executing the action if so. Inefficient implementations may make many unnecessary tests of model conditions.

Figure 1 presents the DEAC algorithm used in these studies. This algorithm assumes a CS with no mixed ACs. A list, $\mathcal{A}$, of scheduled alarms is maintained as well as a list of state-based action clusters, $\mathcal{C}$, whose conditions should be tested within the current context of model execution. Note that the execution of the termination AC causes an exit from the simulation proper.

While the algorithmic construction of a minimal ACIG (which would produce the minimal number of tests) for any model specification is unsolvable (Overstreet 1982), we have identified techniques for simplifying ACIGS which, though not guaranteeing minimality, are effective and their benefits are illustrated in this section.

The eliminated edges in the ACIG represent modeling relationships that are specified following the CM so that concepts can be captured in a communicative form without regard to issues of efficiency, either in representation or execution. Redundancy abounds, but by recognizing redundant or unnecessary relationships prior to the implementation, no execution penalty is incurred. Note that eliminations ranging from 50% to 78% portend considerable savings in both programming effort and execution time.

4.1 Execution Results

For each of four models, we execute versions based on two ACIGs. The naive version (called "before" below) uses all state-based ACs as successors for each AC and provides a base case for comparison. The second version (called "after" below) incorporates the benefits of analysis by reducing the successor sets for individual Action Clusters based on analysis.

Table 1 presents data about each of the models, the effectiveness of the analysis techniques, and the effect of this analysis on performance.

The first set of data show the simplification of

Let $\mathcal{A}$ be the list of scheduled alarms.
Let $\mathcal{C}$ be the set of state-based clusters whose conditions should be tested immediately.
Let σ_{i_S} be the set of state-based successors for action cluster σ_i (where $1 \leq i \leq |$ ACs $|$).

Initially

```
∀σ_i, set σ_iS (from ACIG)
C = ∅
A = (initialization AC, initialization clock time)
```

Simulate

```
while (true) do

    clock ← time given by FIRST(A)
    while (clock = time given by FIRST(A)) do
        let σ_a be the AC corresponding to
        FIRST(A);
        remove FIRST(A)
        perform actions of σ_a
        add σ_aS to C

        while (C ≠ ∅)
            remove σ_c ← FIRST(C)
            if condition of σ_c is true
                perform actions of σ_c
                add σ_cS to C
            endif
        endwhile
    endwhile

endwhile
```

Figure 1: DEAC Algorithm for a CS without Mixed ACs

the ACIGs for each of the four sample models. The number of edges which can be removed (that is, are redundant) obviously depends on model structure; in some models, it is possible that no edges could be deleted. For these models, from 50% (6 of 12 for the single server queue) to 78% (91 of 116 for the MVS computing system) are eliminated.

Table 1 also includes counts of the total number of condition tests made during a typical run of each version of each of the four models. These data also indicate one aspect of the differences among the models: the frequency counts on number of tests vary by two orders of magnitude.

A significant indication of the benefit of the analysis is the "Percent True Tests" data. The analysis (data flow and expert system) is an attempt to determine a priori that certain tests need not be performed since they must come out false. So the percent of the tests which evaluate to true is a measure of the efficiency (or lack of wasted effort) of the implementation. Thus improving the percent of true tests for 8.5% to 19.3% (as is the case for the Harbor Model) indicates a more efficient implementation.

The last set of data in Table 1 present execution time data for each version of the four models. All runs are on a SPARCstation 5, running Solaris OS with each run consisting of 20 replications of each model (with different random number seeds so behaviors are not identical. The count and percentage data are from a single run (not the 20 replications); behaviors of each replication are similar and show little variation in counts or percentages.

Not included in this table is an actual count of tests which evaluated to *true* for the different versions of each model since for one model all versions must generate exactly the same count: correct implementation would require that any version perform actions whenever the model is in a state where the condition for the action is *true*. Each version for a single model must have identical counts for *true* tests; only the count of false tests can vary. This is the case with these runs.

Much of the data of Table 1 are represented in Table 2 as a percentage improvement. It compares the two versions of each model by computing $|before - after|/before \times 100$ for each metric. This shows, for example, that the speed-up varies from 27.3% to 53.2% among the four models, and that more than twice as many tests (220.5%) evaluate to *true* for the optimal compared to the base case for the Machine Repair model.

From this small number of models, it is inappropriate to draw general conclusions on the percent improvements likely with these techniques; thus we include no averages since it is unlikely these data represent "average" behavior. But since these models are not preselected in anticipation of these techniques working well with them, we do surmise that this approach is generally effective with a wide class of models.

Table 1: Performance Measures

Metric	Vers.	Sing. S. Q.	MVS Sys.	Har.	Mach. Rep.
ACIG	Before	12	116	67	29
Edges	After	6	25	26	10
Total	Before	12K	1,121K	30K	30K
Tests	After	6K	263K	14K	9K
% True	Before	17.7	8.7	8.5	11.2
Tests	After	33.3	33.9	19.3	35.9
Time	Before	2.2	215.7	4.6	6.2
(secs)	After	1.6	68.8	2.9	2.9

Table 2: Improvement of Optimal over Base: $|before - after|/before \times 100$

Metric	Single Serv. Q	MVS Sys	Harbor	Mach. Repair
Exec. Time	27.3	68.1	37.0	53.2
Num. Tests	50.0	76.5	53.3	69.1
% True	88.1	289.7	127.1	220.5

A comment on the effectiveness of our analysis for these four models: the techniques did find every edge which was valid to delete in the ACIG of each model. That is, for each of the remaining edges, at some point in an execution, the test it suggested might need to be performed did evaluate to *true* at least once. Put another way, if the test had not been performed (to check to see of another action should occur), the implementation would have been invalid. We do not expect the analysis to this effective for all models, but the completeness in these four cases is encouraging.

5 CONCLUDING SUMMARY

We continue to explore techniques and benefits of static analysis of model specifications. In this paper, we have demonstrated how one assumed negative aspect of redundancy, namely poor execution, can sometimes be ameliorated by analysis, so that the redundancy sometimes useful in model specifications to support analysis need not result in poor run-time performance.

REFERENCES

Balci, O. 1988. The implementation of four conceptual frameworks for simulation modeling in high-level languages, In: *Proceedings of the 1988 Winter Simulation Conference*, 287-295.

Balci, O., A. I. Bertelrud, C. M. Esterbrook, and R. E. Nance. 1995. A picture-based object-oriented visual simulation environment, In: *Proceedings of the 1995 Winter Simulation Conference*, 1333-1340, Arlington, VA.

Baskaran, V. and Y. V. Reddy. 1984. An interactive environment for knowledge-based simulation, In: *Proceedings of the 1984 Winter Simulation Conference*, 645-651.

Bisschop, J. and A. Meeraus. 1982. On the development of a general algebraic modeling system in a strategic planning environment, *Mathematical Programming Study*, 20:1-29.

Brooke, A., D. Kendrick and A. Meeraus. 1992. *GAMS: A user's guide, release 2.25*, San Francisco, CA: The Scientific Press.

Buxton, J. N. and J. G. Laski. 1963. Control and simulation language, *Computer Journal*, 5, 194-199.

Geoffrion, A.M. 1987. An introduction to structured modeling, *Management Science*, 33:547-588.

Geoffrion, A. M. 1992a. The SML language for structured modeling: levels 1 and 2, *Operations Research*, 40:38-57.

Geoffrion, A. M. 1992b. The SML language for structured modeling: levels 3 and 4. *Operations Research*, 40:58-75.

Greenberg, H. J. 1983. A functional description of ANALYZE, *ACM Transactions on Mathematical Software*, 9(1):18-56.

Greenberg, H. J. 1987. ANALYZE rulebase, In *Proceedings of NATO ASI: Mathematical Models for Decision Support*, 229-238, Berlin: Springer-Verlag.

Greenberg, H. J. 1993. *A computer-assisted analysis system for mathematical programming models and solutions: a user's guide for ANALYZE*, Boston, MA: Kluwer.

Nance, R. E. 1971. On time flow mechanisms for discrete event simulation, *Management Science*, 18(1):59-73, September.

Nance, R. E. 1981. Model representation in discrete event simulation: the conical methodology, Technical Report CS81003-R, Department of Computer Science, Virginia Tech, Blacksburg, VA.

Nance, R. E. 1987. The conical methodology: a framework for simulation model development, methodology and validation, *Simulation*, 19(1):38-43. Orlando, FL: The Society for Computer Simulation.

Nance, R.E. and C. M. Overstreet. 1987a. Diagnostic Assistance Using Digraph Representations of Discrete Event Simulation Model Specifications, *Transactions of the Society for Computer Simulation,* 4(1):33-57, January.

Nance, R.E. and C. M. Overstreet. 1987b. Exploring the forms of model diagnosis in a simulation support environment, In *Proceedings of the 1987 Winter Simulation Conference,* Atlanta, GA, December 14-16, 590-596.

Nance, R. E. 1994. The conical methodology and the evolution of simulation model development, *Annals of Operations Research* 53:1-46.

Nance, R. E., C. M. Overstreet, and E. H. Page. 1996. Redundancy in model specifications for discrete event simulation, work in progress.

Overstreet, C. M. 1982. Model specification and analysis for discrete event simulation, Ph.D. Dissertation, Department of Computer Science, Virginia Tech, Blacksburg, VA, December.

Overstreet, C. M. and R. E. Nance. 1985. A specification language to assist in analysis of discrete event simulation Models, *Communications of the ACM,* 28(2):190-201, February.

Overstreet, C. M. and R. E. Nance. 1986. World view based discrete event model simplification, In *Modelling and Simulation Methodology in the Artificial Intelligence Era*, Elsevier Science Publishers (North-Holland), 165-179.

Padmanaban, N., P. C. Benjamin and R. J. Mayer. 1995. Integrating multiple descriptions in simulation model design: A knowledge based approach, In *Proceedings of the 1995 Winter Simulation Conference*, 714-719, Arlington, VA.

Page, E. H. 1994. Simulation modeling methodology: principles and etiology of decision support, Ph.D. Dissertation, Department of Computer Science, Virginia Tech, Blacksburg, VA, September.

Puthoff, F. A. 1991. The model analyzer: prototyping the diagnosis of discrete-event simulation model specification, M.S. Thesis, Department of Computer Science, Virginia Tech, Blacksburg, VA, September.

Rothenberg, J. 1989. The nature of modeling, RAND Note N-3027-DARPA.

Rozenblit, J. W. and Y. M. Huang. 1991. Rule-based generation of model structures, in Multifaceted Modeling and System Design, *ORSA Journal on Computing* 3(4):330-344.

Schriber, T. J. 1974. *Simulation using GPSS*, John Wiley and Sons.

AUTHOR BIOGRAPHIES

RICHARD E. NANCE is the RADM John Adolphus Dahlgren Professor of Computer Science and the Director of the Systems Research Center at Virginia Tech (VPI&SU). Dr. Nance is also Chairman of the Board of Orca Computer, Inc. He received B.S. and M.S. degrees from N.C. State University in 1962 and 1966, and the Ph.D. degree from Purdue University in 1968. He has served on the faculties of Southern Methodist University and Virginia Tech, where he was Department Head of Computer Science, 1973-1979. Dr. Nance has held research appointments at the Naval Surface Weapons Center and at the Imperial College of Science and Technology (UK). Within ACM, he has chaired two special interest groups: Information Retrieval (SIGIR), 1970-71 and Simulation (SIGSIM), 1983-85. He has served as Chair of the External Activities Board and several ACM committees. He is the US representative to IFIP TC7. The author of over 100 papers on discrete event simulation, performance modeling and evaluation, computer networks, and software engineering, Dr. Nance has served on the Editorial Panel of Communications of the ACM for research contributions in simulation and statistical computing, 1985-89, as Area Editor for Computational Structures and Techniques of Operations Research, 1978-82, and as Department Editor for Simulation, Automation, and Information Systems of IIE Transactions, 1976-81. He served as Area Editor for Simulation, 1987-89 and as a member of the Advisory Board, 1989-92, ORSA Journal on Computing. Dr. Nance was the founding Editor-in-Chief of the ACM Transactions on Modeling and Computer Simulation, 1990-1995. He served as Program Chair for the 1990 Winter Simulation Conference. Dr. Nance received a Distinguished Service Award from the TIMS College on Simulation in 1987. In 1995 he was honored by an award for "Distinguished Service to SIGSIM and the Simulation Community" by the ACM Special Interest Group on Simulation. He is a member of Sigma Xi, Alpha Pi Mu, Upsilon Pi Epsilon, ACM, IIE, and INFORMS.

C. MICHAEL OVERSTREET is an Associate Professor of Computer Science and Graduate Program Director for Computer Science at Old Dominion University. He is immediate past chair of the Special Interest Group in Simulation (SIGSIM) of ACM. He received his B.S. from the University of Tennessee in 1966, an M.S. from Idaho State University in 1968, and an M.S. and Ph.D. from Virginia Polytechnic Institute and State University in 1975 and 1982. He has been a visiting research faculty member at the Kyushu Institute of Technology in Japan. His current research interests are in model specification and analysis, distributed simulation, high performance networking, remote instruction technologies, and static code analysis in support of software maintenance tasks. He is currently a principal investigator in tasks funded by the National Science Foundation, DARPA, and ICASE at NASA Langley. Dr. Overstreet is a member of ACM, and IEEE CS.

ERNEST H. PAGE is a Member of Technical Staff at the C3 Modeling and Simulation Center and Mitre Corp. in McLean, VA. He received his Ph.D. in Computer Science in 1994 from Virginia Polytechnic Institute and State University (VPI&SU). He received B.S. and M.S. degrees in Computer Science from VPI&SU in 1988 and 1990. His research interests include discrete event simulation, parallel and distributed systems, and software engineering. He is a member of ACM, ACM SIGSIM, IEEE CS, SCS, and Upsilon Pi Epsilon.

Proceedings of the 1996 Winter Simulation Conference
ed. J. M. Charnes, D. J. Morrice, D. T. Brunner, and J. J. Swain

EFFICIENT SIMULATION MODEL GENERATION USING AUTOMATIC PROGRAMMING TECHNIQUES

Jae-Wook Lee

Department of Electrical Engineering
Yonsei University
Seoul, Korea

Sungho Kang

Department of Electrical Engineering
Yonsei University
Seoul, Korea

ABSTRACT

Generating simulation models is a knowledge intensive, time consuming, and error-prone task in implementing a simulator. The main purpose of this research was to find an easy, fast, and reliable way to generate simulation models and model library. To solve this problem, the Automatic Element Routine Generation System (AERO) is developed as an efficient way for automatic model development using domain specific automatic programming techniques. Behavioral and structural models are generated from Boolean equations, truth tables, HDL descriptions, schematic diagrams, or incomplete specifications. Results show that the system could greatly reduce the cost of simulation model generation for CAD systems and, consequently, reduce the design cycle considerably.

1 INTRODUCTION

It is clear that simulation, at various levels, is an essential and commonly used method for verifying the design of digital systems. A simulation model is a program code which represents the functional behavior of an element and is written in a simulator's target language. Simulation models are created and added into a simulator library. The performance of a simulator highly depends on the number of models in the model library and the efficiency of the models. Since a simulator can only simulate systems constructed of the primitives in the model library, if there are not sufficient models, a large system may not be handled by the simulator. Also, if the model is written in an inefficient way, the speed of the simulator is slow.

Therefore, it is very important to find an effective way for developing a model library as well as for generating efficient models. Until recently, simulation models were developed through a knowledge intensive design process which is time consuming and error prone. Additionally, it is not easy to predict the performance of functional elements. To overcome these difficulties, an automatic way of model generation is required. Over the life time of a simulation system, the development of the simulation model library is the largest programming effort. Therefore, automatic model generation is clearly necessary and very significant, since it has the potential of saving hundreds of man-years of code development.

The concept of automatic code generation of functional models was initially introduced in (Szygenda 1979). Several attempts (Han 1991, Yang 1991) have been made to apply domain specific programming techniques (Barstow 1985) in the automatic model generation area, with great ease, accuracy, and efficiency. Domain specific automatic programming techniques allow a user to describe programs using informal and imprecise terms of domain, and produce efficient and reliable programs.

In this research, to eliminate inefficiency of the previous works, a new simulation model generation system, called AERO (Automatic Element Routine Generation System), is developed. Also, the AERO can generate simulation models from various descriptions for various simulation algorithms such as logic, fault, error simulators. This system can be used either when developing new simulators or for upgrading existing simulators, as a model generation tool.

2 CONFIGURATION OF AERO

To implement efficient domain specific automatic programming techniques, the following must be carefully considered. Firstly, knowledge of the application domain must be kept in some form of rule base to guide the program synthesis. It can be classified into domain independent knowledge base about implementing code syntax and domain specific knowledge base about application domain. Secondly, the system must provide the efficient user interface to get specifications and knowledge. Thirdly, in order to handle the users' incomplete or imprecise specifications, the system must be capable of figuring out the users' intention.

To generate simulation models effectively, the following must be considered. Firstly, the system must generate high performance models since models affect the overall performance of a simulator. Secondly, the system must verify generated models, since the correctness of the models means the accuracy of a simulator. Thirdly, the system must generate models which can easily interface with various simulators, with minimal modification. Also, the system should generate concise models, in order to minimize the usage of memory. Finally, the generated models for sequential devices must have some mechanisms to allocate data for each instance of the element, including internal memory states.

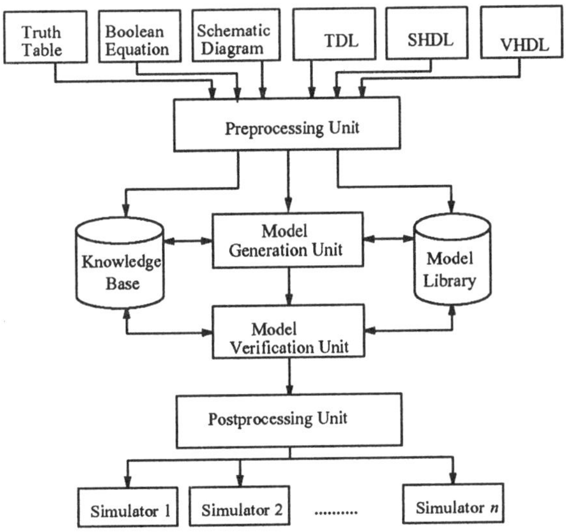

Figure 1: Configuration of the AERO

The AERO (Automatic Element Routine Generation System) can automatically generate models from various descriptions, using domain specific automatic programming techniques. As shown in Figure 1, the AERO consists of several subsystems, i.e., Preprocessing Unit, Automatic Code Generation Unit, Model Verification Unit, Postprocessing Unit, Knowledge Base and Model Library. This system can generate models written in the C programming language from truth tables, boolean equations, schematics, and hardware description languages.

3 PREPROCESSING UNIT

Figure 2 shows a block diagram of the Preprocessing Unit. The Preprocessing Unit includes parsers and translators which have interactive user interface facilities.

From the schematic description, a net list is generated with the help of the OCT environment (Spickelmier 1989). A netlist can also be generated from the hardware description language, Simulation Automation System (SAS) Hardware Description Language (SHDL) (Kang 1991), Tegas Description Language (TDL), and VHDL. These descriptions are parsed, and passed onto the Netlist Generator, which generates the levelized netlists. SHDL and VHDL are also used to describe behavioral domain models. Parsed behavioral descriptions include control flows and equations.

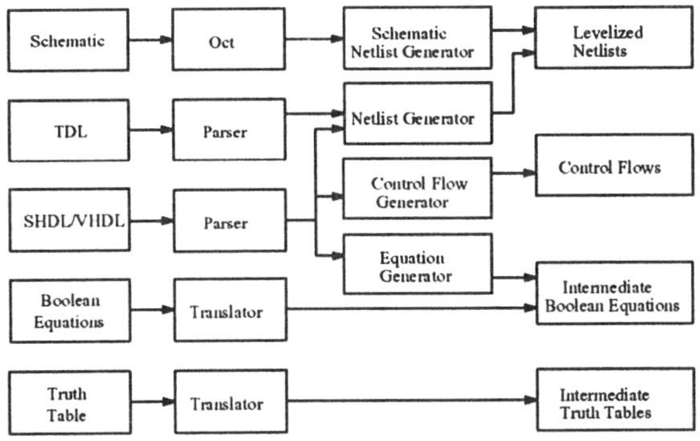

Figure 2: Preprocessing Unit

When truth table descriptions, which are one of the easiest and most natural means to describe a reasonable size circuit, are entered, they are translated into internal data structures. If the Boolean equations, or the gate implementation, of a circuit is known, it is convenient to use equations to describe the circuit. These equations are translated into internal data structures.

Interactive user interface displays the existing models which can be used to construct new models, or to store the generated models and the corresponding model specification into the model library. By grouping similar models into the same class, users can derive the characteristics and transformation rules and attach them onto this class. This knowledge will be used by the incomplete information handler and the input verifier to complete the missing part of the input specification and to perform the consistency check. The series of questions are shown in Figure 3.

4 MODEL GENERATION UNIT

Using the information from the Preprocessing Unit, the Model Generation Unit (MGU) synthesizes models. From the synthesized models, it can generate 'C' models according to user's command. The structure of the MGU is shown in Figure 4.

The general simulation models can be classified into structural and behavioral models. A structural

1. What's the model group name?
2. What's the model name?
3. How many states are desired for this model? [2-10]
4. Do you want to specify bit representations?
5. What's the type of your input specification? [a-h]
   (a) truth table (b) boolean equations
   (c) schematics (d) TDL
   (e) behavioral SHDL (f) behavioral VHDL
   (g) structural SHDL (h) structural VHDL

Figure 3: Questionnaire of Interactive User Interface

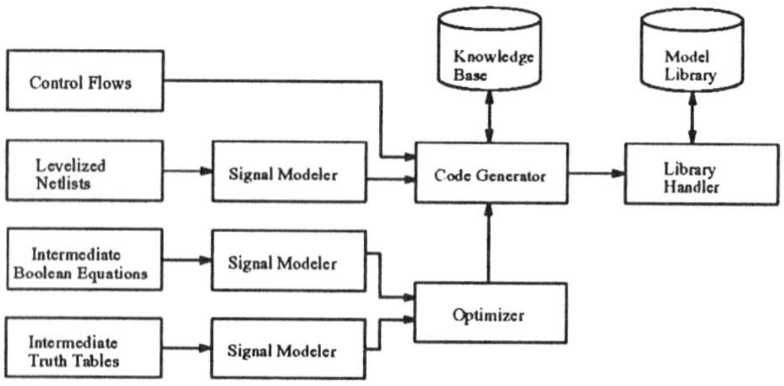

Figure 4: Model Generation Unit

model describes a circuit by connecting low level primitives. A behavioral model is used to describe the functional behavior of a circuit. This system can generate both structural and behavioral models. The simulation model consists of entity declarations, architectural bodies, and configuration declarations. An entity represents a part of a well-defined hardware design. An architecture specifies the relationship between the inputs and the outputs of a design entity, and a configuration specifies the binding of components to entities. With the information about input and output signal names, entity parts can be completed.

If the description is in the form of truth tables or Boolean equations, there are no primitives in the descriptions. However, 2 input AND, 2 input OR, and NOT gates are used to complete an 'architecture' part of structural C models. If the description is in the form of schematics or HDLs, then the information about primitives used in the descriptions is used for the structural models. The remainder of 'architecture' part is filled out using net list information, control flow information, and equation information which is synthesized using domain rules and given specifications.

According to the bit representations of a specific simulator, a signal modeler generates an enhanced table or equations. The signal modeler is used to dynamically customize AERO according to the user specified logic values and bit representations, so that the AERO process becomes very flexible in terms of generating models, which comply with various signal models. For any k-valued signal model, k distinct logic values can be represented in the computer by sets of bits with at least $\log_2 k$ bits in the set. For example, simulators like TEGAS use two word representations for three value simulation. Hence, two variables are used for each variable defined in the original truth table. Let us assume that we want to generate a simulation model for a two input AND gate, for three value logic; where the three state values are L, H, and X, for low signal value, high signal value and unknown value, respectively. The signal modeler may assign; bit0 = 0 and bit1 = 0 for L, bit0 = 0 and bit1 = 1 for H, and bit0 = 1 and bit1 = 1 for X. A C model based on the above bit assignments is shown in Figure 5.

```
AND2(i, o)
int i[2][2], o[1][2];
{
    int a0, a1;
    int b0, b1;
    int c0, c1;

    a0 = i[0][0];
    a1 = i[0][1];
    b0 = i[1][0];
    b1 = i[1][1];

    c0 = (a0&b0);
    c1 = ((a1&b1)|(a1&b0)|(a0&b1));

    o[0][0] = c0;
    o[0][1] = c1;
}
```

Figure 5: 2 Input AND Model for 3 Values

For more accurate simulation, 5 logic values (0, 1, U, D, E) are usually used. To represent 5 value, 3 bits are used, where logic L is represented by bit pattern 000, logic H is represented by bit pattern 110, the state U (for signal going-up) is represented by bit pattern 010, the state D (for signal going-down) is represented by bit pattern 100, and the state E (for error or unknown value) is represented by bit patterns 001, 011, 101, or 111. However, the encoding for the given value is not unique and the users should be able to decide on their own selections. A C model based on the above bit assignments is shown in Figure 6.

As an example, consider path delay fault sim-

```
OR2(i, o)
int i[2][3], o[1][3];
{
    int a0, a1, a2;
    int b0, b1, b2;
    int c0, c1, c2;

    a0 = i[0][0];
    a1 = i[0][1];
    a2 = i[0][2];
    b0 = i[1][0];
    b1 = i[1][1];
    b2 = i[1][2];

    c0 = (a0|b0);
    c1 = (a1|b1);
    c2 = ((¬a0&b2)|(¬a1&b2)|(a2&¬b0)|(a2&¬b1)
      (a2&b2)|(a0&¬a1&¬b0&b1)|(¬a0&a1&b0&¬b1));
    o[0][0] = c0;
    o[0][1] = c1;
    o[0][2] = c2;
}
```

Figure 6: 2 input OR Model for 5 Values

ulation (Kang 1994) using 4 values. Logic L is represented by bit pattern 00, logic H is represented by bit pattern 10, logic X is represented by bit pattern 11, and logic Z is represented by bit pattern 01. The evaluation routine for 2-to-1 MUX is shown in Figure 7.

When equations or truth tables are entered, it can generate the enhanced equations directly from the given equations, or can generate minimized equations using optimization routines. The optimizer generates optimized equations using the Quine-McCluskey's algorithm. If the description is given in equation form, the equations are transformed into a table form to permit the use of the same algorithm.

The code generator generates the simulation model from the model name, input pin names, output pin names, signal names, the optimized equations, and other information using domain rules. The code generation algorithm is based on a template matching mechanism (Biermann 1984).

5 MODEL VERIFICATION UNIT

When simulation models are automatically generated, it is necessary to verify the correctness of these models. Verification of models is accomplished in a similar fashion to the verification of design in that a simulator is used to verify new models. Some difficulty exists with respect to the determination of consistency among models at various levels.

Simulation at various levels is used in order to verify the correctness of models. The simulator can handle multi-level primitives according to the level

```
MUX21(i, o)
int i[3][2], o[1][2] ;
{
   int a0, a1, b0, b1, c0, c1, d0, d1 ;

   a0 = i[0][0] ;
   a1 = i[0][1] ;
   b0 = i[1][0] ;
   b1 = i[1][1] ;
   c0 = i[2][0] ;
   c1 = i[2][1] ;

   d0 = (¬c0&a0)|(c0&b0)|
      (c1&(a0|b0|(¬a1&b1)|(a1&¬b1)));
   d1 = (¬c0&a1)|(c0&b1)|
      (c1&(a1|b1|(¬a0&b0)|(a1&¬b0)));

   o[0][0] = d0 ;
   o[0][1] = d1 ;
}
```

Figure 7: 2-to-1 MUX for Delay Fault Simulation

of the primitives. The basic configuration is shown in Figure 8. From the given descriptions of models,

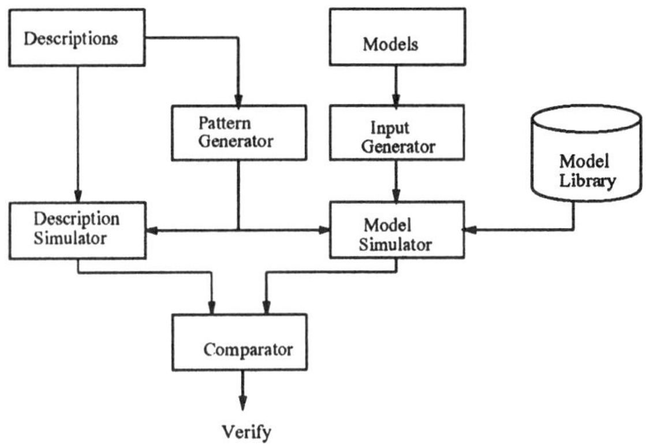

Figure 8: Model Verification Unit

expected behaviors are extracted and stored. This process is straightforward, to directly provide an environment for simulating a given model. If the user wants to see the results for each simulation pattern, they can be provided. If the user specifies the number of simulation patterns, then, according to the given number, the pattern set is generated. Also, if the user provides a simulation pattern set, it can be used for simulation. Then, according to the size of a described circuit, a random pattern set, or an exhaustive pattern set, may be generated by the pattern generator. Then, the simulation results are passed on to the comparator. After comparison, if all results are the same, the model is said to be correct. However, if any of results is not the same, an error message is provided and the simulation patterns in disagreement are provided to the user for further consideration.

6 MODEL LIBRARY AND KNOWLEDGE BASE

Model Library has several sub-libraries to store the same device using different levels of descriptions. The library handler is used to display all existing models and specifications. Also, it is used to place new specifications into the library. It stores and retrieves the circuit descriptions of various types and puts the generated models into the library. It is also used to link the structural models to their components and to add the generated elements to the specific simulators.

The Knowledge Base contains domain specific knowledge, used in model generation and verification. The knowledge stored in the Knowledge Base is classified into several categories including;

1) information about specific simulators, such as, logic values, memory state handling, how to call element routines, etc;

2) information about input descriptions, such as how to describe truth table specification, boolean equation specification, and HDLs; and

3) information about model verification.

7 POSTPROCESSING UNIT

The Postprocessing Unit generates an i/o interface for specific simulators. Since every simulator has its own way of using simulation models, the postprocessor should be capable of handling the variation. In order to do this, it requires various information about simulators, including; how to call an element routine, how to keep the state variables, etc. This information is stored in the Rule Base.

8 RESULTS

Using the AERO, many models were generated using ISCAS benchmark circuits (Brglez 1985). The results of model generation are shown in Table 1, including; circuit sizes, model generation times, number of simulation patterns, and simulation times. The circuit size is the number of gates, being assumed that the circuit was flattened into a gate level, if it is not at the gate level. Although model generation time increases according to the circuit size, the automatic model generation time is acceptable because of its speed and accuracy. For simulation purpose, 1024 random patterns were generated and used.

Simulation was performed using the generated models to prove thier efficiency and accuracy. The 3 value parallel fault simulation results are shown in Table 2. These results are derived using the PARSIM simulator (Kang 1990) For these results, 1024 random patterns are used.

Also, the mixed level path delay fault simulation results are shown in Table 3. For these results, 1000 random path delay faults and 1024 random patterns are used for benchmark circuits (Brglez 1989). In the table, the number of detected paths for robust

Circuit	# of Gates	Generation Time [sec]	# of Patterns	Simulation Time [sec]
c432	160	0.233	1024	7.013
c499	202	0.317	1024	14.317
c880	383	0.600	1024	19.600
c1355	546	0.833	1024	26.833
c1908	880	1.067	1024	35.067
c2670	1269	1.817	1024	47.817
c3540	1669	2.250	1024	64.250
c5315	2307	3.433	1024	106.433
c6288	2416	4.083	1024	163.083
c7552	3513	4.833	1024	261.833

Table 1: Results of Model Generation

Circuit	Faults	Fault Coverage[%]	CPU Time
c432	524	95.99	0.009 sec/pat
c499	758	97.49	0.032 sec/pat
c880	942	98.20	0.026 sec/pat
c1355	1574	89.07	0.183 sec/pat
c1908	1879	94.40	0.185 sec/pat
c2670	2747	81.61	0.580 sec/pat
c3540	3428	94.04	0.311 sec/pat
c5315	5350	95.73	0.534 sec/pat
c6288	7744	97.32	1.021 sec/pat
c7552	7550	91.08	1.345 sec/pat

Table 2: Results of Fault Simulation

Circuit	ROB	SNR	WNR	CPU[sec]
s713	556	144	242	51.14
s820	72	193	106	145.08
s832	131	185	131	152.03
s953	510	63	25	159.85
s1196	226	451	187	196.68
s1238	3	494	244	233.01
s1423	180	37	110	231.39
s1488	152	174	147	204.87
s1494	147	138	125	205.70
s5378	383	51	51	465.36

Table 3: Results of Delay Fault Simulation

test, strong non-robust test, and weak non-robust test, and simulation time for each circuit are represented.

As a final example, the design error simulation (Kang 1992) results are shown in Table 4. In the table, the number of design errors, simulation coverage metric, and simulation time for each circuit, are represented. For these results, 1024 random patterns are used.

Circuits	Errors	SCM [%]	CPU [sec]
c432	2489	97.99	2.15
c499	3012	98.24	3.60
c880	5171	98.36	6.73
c1355	7380	89.30	18.02
c1908	8775	94.21	10.43
c2670	13728	83.78	22.35
c3540	19130	94.88	38.75
c5315	33874	94.94	25.42
c6288	33472	92.46	33.67
c7552	38777	89.34	47.25

Table 4: Results of Design Error Simulation

These results show that automatic model generation is efficient, and the model generation time of the AERO is superior to that of experienced human programmers. Also, the results show that the generated models are efficient and accurate. If users are careful when writing the specifications, the generated code can approximate to the efficiency of the code written by a human programmer. Additionally, the generated models from the AERO can be used for various simulators.

9 CONCLUSION

The main purpose of this research was to find an easy, fast, and reliable way to generate simulation models and model libraries. To solve this difficult problem, a simulation model generation system using domain specific automatic programming techniques, is developed. The AERO can be used to construct simulation model libraries when developing new simulators or when upgrading existing simulators. The results prove that the time required to generate functional models is reduced considerably. In addition, since the designer needs not worry about the details of low level coding, the chance of errors in the design cycle can be significantly reduced. Therefore, this system could greatly reduce the cost of model generation for CAD systems and, consequently, could reduce the design cycle considerably.

REFERENCES

D. Barstow. 1985. Domain-Specific Automatic Programming. *IEEE Trans. on Software Eng.*

A. Biermann, G. Guiho and Y. Kodratoff. 1984. An Overview of Automatic Programming Construction Techniques. *Automatic Programming Construction Techniques.*

F. Brglez and H. Fujiwara. 1985. A Neutral Netlist of 10 Combinational Benchmark Circuits and a Target Translator in FORTRAN. *Proc. of ISCAS*: 695-698.

F. Brglez, D. Bryan and K. Kozminski. 1989. Combinational Profiles of Sequential Benchmark Circuits. *Proc. of ISCAS*: 1929-1934.

C. Han, S. Kang and S. Szygenda. 1991. AFMG: Automatic Functional Model Generation System for Digital Logic Simulation. *Proc. of ASIC Conf.*

S. Kang, and C. Han. 1990. *PARSIM Manual.* Univ. of Texas, Austin.

S. Kang and S. Szygenda. 1992. Modeling and Simulation of Design Errors. *Proc, of ICCD*: 443-446.

S. Kang, B. Underwood and O. Law. 1994. Path-Delay Fault Simulation for a Standard Scan Design Methodology. *Proc. of ICCD.*

R. Spickelmier. 1989. *Release Notes for Oct Tools Distribution 3.0.* Univ. of California, Berkeley.

S. Szygenda. 1979. Simulation of Digital Systems: Where We Are and Where We May Be Headed. *Computer Aided Design*: 41-54.

H. Yang and S. Szygenda. 1991. A Domain Specific Automatic Programming System for Element Routine Generation. *Proc. of SSC.*

AUTHOR BIOGRAPHIES

JAE-WOOK LEE is a M.S. student in the Department of Electrical Engineering at Yonsei University, Korea. He received a B.S. degree in Electrical Engineering from Yonsei Univesity in 1996.

SUNGHO KANG is an Assistant Professor in the Department of Electrical Engineering at Yonsei University, Korea. He received a B.S. degree from Seoul National University, Korea and the M.S. and Ph.D. degrees in Electrical and Computer Engineering from the University of Texas at Austin, respectively. He was a senior staff engineer at the Semiconductor Systems Design Technology, Motorola Inc., a research scientist at Schlumberger Laboratory for Computer Science and a post doctoral fellow at the University of Texas at Austin. Dr. Kang's research interests include Simulation, Design Verification, VLSI CAD, Testing and Design for Testability.

Proceedings of the 1996 Winter Simulation Conference
ed. J. M. Charnes, D. J. Morrice, D. T. Brunner, and J. J. Swain

PROGRESS IN MODULAR SIMULATION ENVIRONMENTS

Charles R. Standridge
James F. Kelly
Thomas Kelley
Jack Walther

Department of Industrial Engineering
Florida A & M University - Florida State University College of Engineering
2525 Pottsdamer Street
Tallahassee, FL 32310, U.S.A.

ABSTRACT

How each simulationist can design and implement software tailored for each particular simulation project is addressed by modular simulation environments. The requirements of such environments are derived from the needs of four distinct types of users. Inter-tool modularity deals with how data flows between a non-homogeneous set of software tools that can be changed on an ad hoc basis. Both simulation specific and widely applicable software tools may be used. The organization and management of simulation inputs and results to achieve this goal is important. Intra-tool modularity has to do with supporting simulation project tasks in a modular fashion. Modular modeling is well established. Possibilities for modular animation and the modular use of widely applicable tools, specifically spreadsheets, are discussed. An example modular simulation environment is given.

1 INTRODUCTION

Ideally, each simulationist would be able to select the set of software tools to use on each simulation project (Standridge and Centeno, 1994). The selection would be based on the particular requirements of that project. The tool set would contain both simulation specific tools such as model builders and simulation engines as well as tools with wide applicability such as word processors, statistical analysis packages, and spreadsheets.

Modular simulation environment concepts seek to provide the standard by which ad hoc collections of software tools can be used together to perform a simulation project. These concepts specify how simulation related data flows between tools in a general way so that heterogeneous software can work together. The concepts address how tools can be tailored for use on particular problems and how general purpose tools can be tailored for application in simulation.

Modular simulation environments are based on an approach similar to that of computer operating environments such as Microsoft Windows for personal computers and X-Windows for Unix-based work station computers. The windowing systems provide the design, structure and mechanisms for tool integration. Windows and X-Windows provide a user interface standard for software tools as well as standard mechanisms for sharing information between tools.

Some benefits of modular simulation environments are as follows:

1. High flexibility for end user selection of simulation software. This supports simultaneous use of simulation software from multiple software providers as well as locally developed tools.
2. Use of widely applicable software with which the end user is already familiar.
3. Inclusion of tools such as spreadsheets, word processors and presentation graphics generators that have not traditionally been a part of simulation environments.
4. Use of standard windows techniques for sharing information between tools.
5. Management and selective use of simulation inputs, results, and models.
6. Definition of a standard for simulation environment structure and user interface. Because of its flexibility, end users may generally adopt such a standard and tool builders may find developing software compatible with its requirements helpful.

This paper describes the design and initial implementation of modular simulation environments as well as an example application. Requirements for modular simulation environments are based on their four different types of users. These user types are derived from those proposed by Standridge and Centeno

(1991). The flow of information between tools is described. The use of modular modeling concepts to tailor simulation specific and general purpose software tools for particular applications is discussed. Modular modeling for network simulation languages is reviewed (Standridge, 1995).

2 SIMULATION ENVIRONMENT OVERVIEW

Traditional simulation environments have three major components: (1) a fixed set of software tools prescribed by the environment designers and implementers, (2) a database management system that transparently to the environment users controls the flow of data between the software tools, and (3) a user interface that gives access to all environment capabilities. Each software tool uses existing information in the database and adds the results of its own operations to the database.

TESS (Standridge, 1985; Standridge and Pritsker, 1987) was an early, pre-graphical user-interface, simulation environment built on this strategy. Software tools provided for building SLAM II network models, editing sets of SLAM II control statements, constructing animations, making statistical computations, graphing data, and reporting data. Simulation results could be collected automatically from SLAM II, GPSS/H, and MAP/1 simulations. A command language served as the user interface. Selective querying of the database was supported. A database subprogram library allowed computer programs to store and retrieve data from the TESS database. The database manager organized simulation inputs and outputs.

Balci and Nance (1987, 1992) have used a similar strategy in designing and implementing a simulation model development environment. This work has lead to the visual simulation environment (Balci et al., 1995). A visual user interface supports the development and simulation of visual models that may be hierarchical and object oriented. The collection of tools includes a visual editor, a library of existing component models, a static model analyzer, a model dynamics tester, a simulator, an analyzer for simulation results, a multi-media learning support system, and an evaluator for accessing the credibility of a simulation study.

Centeno and Standridge (1991) describe an information based simulation environment constructed using the same approach. Its distinguishing features include the description of a manufacturing system of interest as information stored in the database. Alternative models can be generated by querying the database. In addition, a knowledge-based user interface is proposed to guide the user through the steps of a simulation project.

Modular simulation environments are distinct from these existing environments in that the set of software tools are not pre-determined, the use of widely applicable software is encouraged, and the user-interface relies on the standards set by existing graphical computer operating system interfaces. Database management must accommodate the open-ended nature of the environment.

3 USER TYPE REQUIREMENTS

Modular simulation environment capabilities are based on the needs of four user types whose relationship is given in Figure 1. The user types represent typical roles. In any particular situation, the same individual may take on multiple roles or many individuals may participate in the same role.

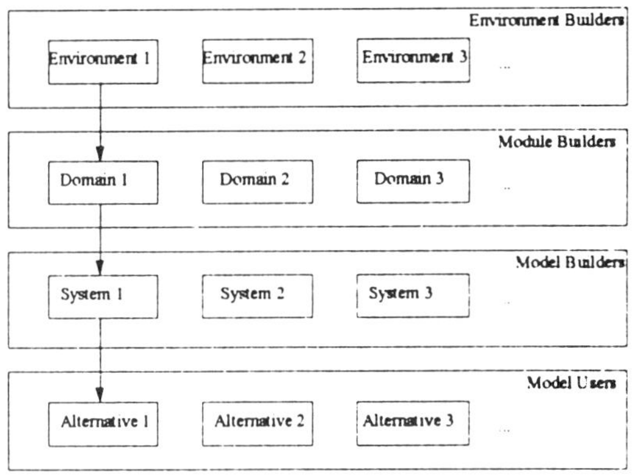

Figure 1: Modular Simulation Environment User Types

An environment builder gathers the software tools that form a simulation environment. The builder can construct multiple environments in this way. For example, an environment builder could select the SLAMSYSTEM simulation environment, the PROOF animation tool, and the EXCEL spreadsheet to comprise the environment.

In addition, an environment builder tailors the engine of any tool selected. SLX (Henriksen, 1995) and YANSL (Joines and Roberts, 1994 and 1995) are simulation engines. For example, the environment builder could construct a new transaction creation mechanism in SLX that reads the time of the next arrival and its attributes from a file as opposed to the traditional specification of a random or constant time between transaction creation. The extensibility capabilities of SLX would be used to define the new creation statement and encode the logic. The new statement could then be used as would any other statement.

A module builder deals with one environment. The environment is tailored multiple times, once for each application domain of interest.

Consider a module builder concerned with discrete parts manufacturing. Using the capabilities of the simulation language selected by the environment builder, the module builder could construct modules for modeling individual work stations, material handling between work stations, serial lines and the like. Spreadsheet macros for processing observations of performance measure values could be written. Such macros could compute confidence intervals using the method of replicates, help find the truncation point in a steady state simulation, and graph simulation performance measure observations. Interfaces to these capabilities would "speak the language" of the domain. Thus, the simulation environment would more closely "use the language" of the application domain.

A model builder is concerned about multiple systems in one domain. Using the domain specific modeling modules provided by the module builder, a model builder constructs a simulation model of any particular system in the domain. Other tools tailored by the module builder for the domain can be further tailored by model builder for the system. For example, a tool for graphing time series of observations could be labeled as graphing the number of parts in a buffer. The range of the number of batches considered in computing a confidence interval from one series of observations could be specified.

A model user evaluates multiple alternatives about one system. The alternatives are described using input data to the simulation model constructed by the model builder. Other tools tailored by the model builder are used to examine, compare, and analyze performance measures resulting from simulations. For examples, spreadsheet macros can be used to graph performance measure values and perform statistical analysis.

4 INTER-TOOL MODULARITY

One type of modularity in a simulation environment has to do with how data flows between tools. This is referred to as inter-tool modularity. It must accomplished in such a way that new tools can be added to the environment on an ad hoc basis. No modification of existing tools can be required though the ability to export data and capabilities for tailoring existing functionality can be exploited.

Figure 2 shows the tool set of the first demonstration modular simulation environment. Existing commercial tools are shown in capital letters. Both simulation specific and widely applicable tools are included. Tools developed to achieve inter-tool modularity are shown in lower case.

Modular Simulation Environment Demonstration		
SLAMSYSTEM		PROOF
EXCEL		ACCESS
Inputs	Result Collection	Result Processing
Tool Attachment	Tool De-attachment	

Figure 2: Demonstration Modular Simulation Environment Tool Set

SLAMSYSTEM is used for modeling and simulation activities. It is interesting to note that another simulation environment can be used within a modular simulation environment. PROOF provides animation capabilities. EXCEL spreadsheets are used for model input value entry and for processing simulation results. ACCESS databases can be used to organize, manage, examine, and present simulation results. A text editor helps with input value entry.

Template-based tools have been developed specifically for organizing, specifying, and optionally entering model input values within a modular simulation environment. Multiple data sets can be input to a model. A template or format is defined for each input data set. A list of templates needed for a particular simulation is maintained. A template tool provides for defining the format of one input data set, listing the input data sets needed for simulation input, and entering values.

Simulation result collection can be specified. Generic tools for organizing and collecting results are provided. The capability to extract values from each simulation language of interest must be coded. This has been done for SLAM II.

Results processing has to do with making simulation results ready for input to other software tools. It includes joining together results collected from the simulation of multiple alternatives. This facilitates the comparison of alternatives.

Finally, a new tool can be attached to the environment. This involves describing the input data formats it accepts and the format of the output data it produces if these are non-standard. Standard formats include text (ASCII) files, spreadsheet files, and database files. Existing tools may be removed.

Multiple data sets can be input to a model. A template or format is defined for each input data set. A list of templates needed for a particular simulation is maintained. A template tool provides for defining the format of one input data set, listing the input data sets needed for simulation input, and entering values.

Figure 3 shows the flow for input data. The TEMPLATE tool is used to describe the data. Values can be entered using any one of a variety of tools, including a spreadsheet such as EXCEL, a text editor such as NOTEPAD, or the TEMPLATE tool. This illustrates how a variety of tools can be employed to accomplish the same purpose in a modular simulation environment. Which tool is employed depends on each particular user.

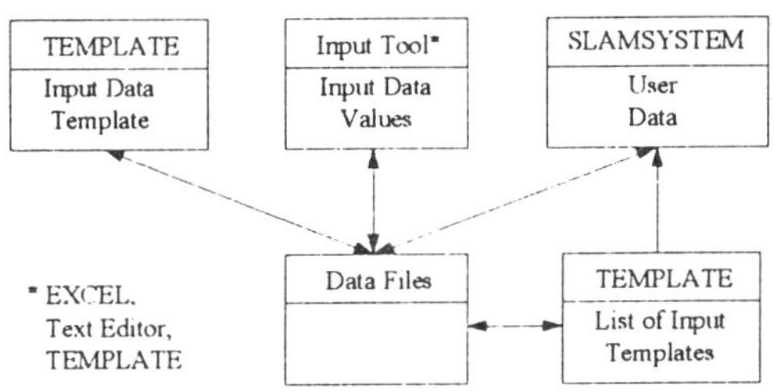

Figure 3: Flow of Simulation Input Data

The TEMPLATE tool also is used to define the data input requirements of a simulation model. This definition is a list of the templates for which values must be supplied in order to run the simulation. Running a simulation requires creating a list of the specific input data sets, one corresponding to each required template. This list defines the simulation input for one particular simulation run. Again, this input data list is created using the TEMPLATE tool. The input data list and the input data files are supplied to SLAMSYSTEM in its standard form for user data.

Within an input data set, values can be organized into rows and columns. Each row gives new values for the variables represented by the columns. Alternatively, each row can correspond to a different variable. A prompt at the beginning of the row specifies the input requirements.

The TEMPLATE tool is used to define the simulation results collected. Multiple result sets can be gathered for each simulation run. A template is given for each set. The template shows the name of the performance measure as known in the simulation result set and the corresponding name in the simulation language. For example, the performance measure, #INBUFFER, could correspond to the SLAM II function NNQ(1). A list of the simulation result sets to be gathered is created using the TEMPLATE tool.

Figure 4 shows the flow of information for the collection of simulation results. Each result template is defined using the TEMPLATE tool. A list of the all of the result templates desired for each simulation run is prepared. Based on the result template list, the result collector interacts with a simulation run to gather the performance measure values of interest. The result collector consists of a generic part and an interface to a particular simulation engine. Results can be stored in an ACCESS database for further processing.

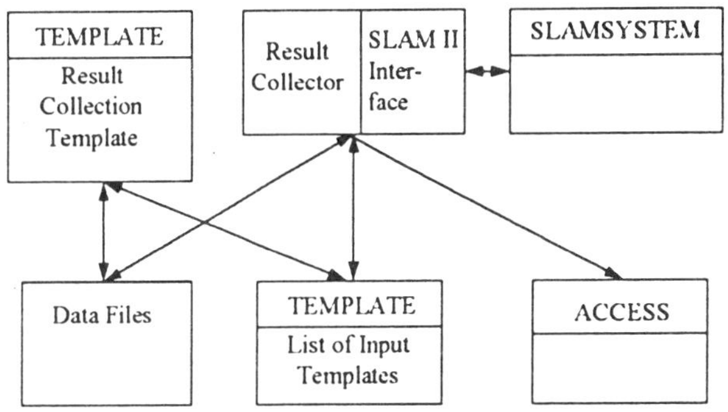

Figure 4: Simulation Result Collection

Result processing has to do with preparing simulation performance measure values gathered by the result collector for processing by other tools. Values of different performance measures from one alternative (different result sets) or values of the same performance measure from different alternatives can be organized together. EXCEL and PROOF can be used to analyze and display performance measure values.

Result processing is summarized in Figure 5. The TEMPLATE tool is used to specify which simulation results from which alternatives are of interest. The result processor produces a file of these result which be input to another tool. Data files collected from the simulation can be used directly in other tools if no re-organization of the simulation results is needed. This show the flexibility of a modular simulation environment is meeting user requirements.

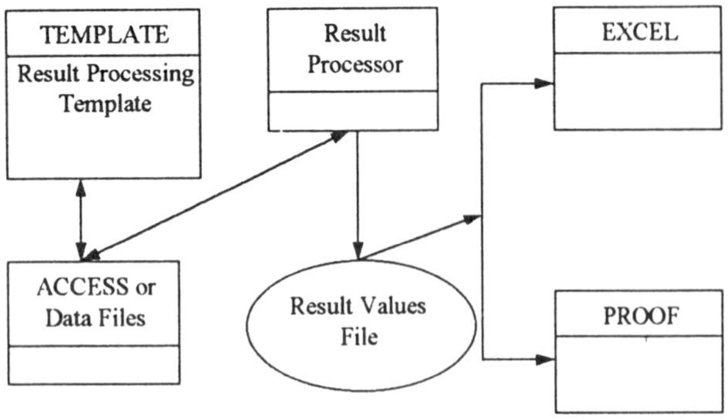

Figure 5: Simulation Result Processing

The organization of input values and simulation results is a fundamental aspect of achieving inter-tool modularity. Input data are labeled as follows:

Scenario -- What alternative is being described.
Variable -- Identifying name of an input quantity.
Value -- As specified by the model user

Note that there are multiple scenarios, versions, possible for each input data set. For example, there could be an input data set for the routes of parts through a job shop. Each scenario would represent a different possibility for routing jobs through the shop.

In addition, that there are multiple input data sets. Thus, a simulation alternative is defined by a model user by specifying the scenario of each input data set to be employed.

Performance measure values resulting from a simulation are labeled as in the following way.

Scenario -- What system alternative was simulated.
Type of values -- Time-persistent, observed, trace, end of simulation only.
Replicate -- ID number
Variable -- Identifying name of an performance measure.
Time or time interval -- When observed
Value -- As observed

For example, the following data sets could be collected: queue lengths, resource utilization, part time in the system, and throughput. The first two are time persistent, the third observed, and the last end of simulation only.

For result processing, the utilization of a particular resource and the corresponding queue length could be joined in a single data set. Alternatively, the queue lengths corresponding to a particular resource from each of two scenarios could be joined for comparison of the two alternatives.

5 AN EXAMPLE MODULAR SIMULATION ENVIRONMENT

Abrams, Standridge, et al. (1995) describe a simulation model of local caching policies. Files are retrieved via the World Wide Web to support distance learning. Caching allows the files to remain local, avoiding multiple retrievals of the same file. Caching policies tell what files to replace when a new file tries to enter a full cache.

A modular simulation environment supports this model. There are two input data sets: simulation options and experiment specification. Typical simulation options are the number of replications of each experiment, internet transmission time for files, and the name of the file listing the names of the files giving the files transmitted on the internet. This input data set in organized with a different variable and its value on each line. The experiment specification includes parameters such as the cache size, the replacement policy, and the parameters of the replacement policy. It is organized in a tabular format with one experiment per row. Columns correspond to experiment parameters.

The TEMPLATE tool is used to describe the two files. The Windows notepad editor is used to enter values. This shows the flexibility of a modular simulation environment. Either the TEMPLATE tool, the notepad editor, or a spreadsheet could have used to enter the values. The model users are computer scientists who are comfortable using the notepad editor.

The primary simulation results are gathered at the end of the simulation run (type end of simulation only) into one set. Results include the experimental parameter values and performance measures such as the hit rate, the percent of files found in the cache when requested.

Statistical analyses, such as regression and ANOVA, are used to determine which of the experimental parameters significantly affect the hit rate. These statistical analysis capabilities exist within a spreadsheet. Since only one result set is collected, it is stored by the results collector in a file format acceptable for direct spreadsheet input.

6 INTRA-TOOL MODULARITY

The second type of modularity has to do with how each tool does its work. Modular modeling concepts have long been discussed and are well implemented. See Zeigler (1990), Cota and Sargent (1992), and Sanderson et al. (1992) for basic definitions, concepts, and example implementations.

Standridge (1995) discusses modularity of network simulation languages and describes a modular network language, ModNet. This paper defines the characteristics of a module and the communication between modules. Module parameters are passed to a receiving module via a construct that resembles calling a subprogram in a general purpose programming language. Performance measure values are made public, that is accessible to all modules. Signals are public quantities that indicate that an event of note has taken place in a module. Other modules may respond to that event.

ModNet places bounds on the common entities of a network language. transactions and resources. Transactions may not cross module boundaries. Transactions may not be cloned. The need for cloning is satisfied by using a different module. A resource is defined and controlled by one module. This module may pass a resource name as input to another module to use. For example, a module may define and control a worker resource. The worker may perform tasks at two distinct work stations. Each work station is represented by a module. The name of the worker resource is passed to each work station module by the defining and controlling module.

ModNet is being implemented in the SLX simulation engine. An environment builder uses the extensibility properties of SLX to define the fundamental modeling capabilities available in ModNet. A module builder develops ModNet modules from which a model builder constructs models.

Consider the possibilities for the use of modularity concepts in other tools. A model user would like a spreadsheet interface and functionality tailored to the system under study. This could be constructed as follows. The environment builder is unlikely to have the capability of changing the spreadsheet engine, its core functions. The module builder would write spreadsheet macros that generate graphs and perform statistical computations generic to the domain of interest. For a example, a spreadsheet macro could graph the number of parts in a buffer versus simulation time. Another macro could use the method of batches to estimate a confidence interval concerning the mean number of parts in the buffer. The model builder could further tailor these macros to refer to the specific buffers in the model and specify the number of batches used in the confidence interval computation. Thus, the model user could simply ask for the graph of the number in the drill press buffer and the confidence interval concerning the mean.

Figure 6 shows one possible organization for animating a simulation. An animation is generated by a script that specifies a time ordered list of changes to a scene starting with an initial frame. Changes include modification to object attributes such as color or movement of objects. This specification allows the animation engine to produce a set of frames that show the time dynamics specified by the script.

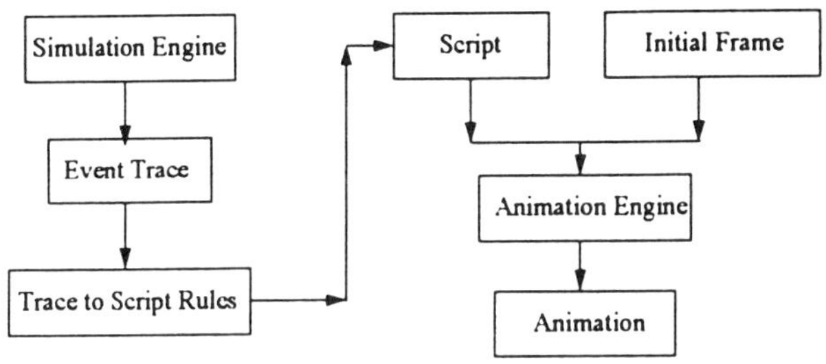

Figure 6: An Organization for Simulation Animation

The script can be based on the event trace produced by a simulation engine. A set of rules can be used to map the events of the simulation into a script that produces the animation of a simulation. In the case of a simulator such as ProModel, these rules are implicit in the specification of the model. In other words, the rules are embedded in the simulation engine. In the case of a simulation environment such as SLAMSYSTEM, the rules must be explicitly provided by a user.

Modular animation is achieved as follows. An initial frame is associated with each model module. The trace to script rules are specified. Thus, the animation for each instantiation of that module is specified. The combination of the animation of each module results in the animation of the model.

Animation of each module can be made optional. For example, an animation of the main module only gives a high level overview of the model. The animation can be expanded to include only those modules referenced by the main module to increase the level of detail. An animation of all modules shows all of the detail of the model.

7 SUMMARY

Current progress in the area of modular simulation environments has been discussed. Requirements for these environments arise from the four distinct types of users that take part in a simulation activity. Inter-tool modularity has to do with how data flow between heterogeneous tools. Intra-tool modularity is concerned with performing tasks in a modular fashion. Modular modeling is well established. Concepts of modularity can be applied to other tools such as spreadsheets and animators. The organization of simulation model inputs and outputs in a modular simulation environment is discussed. An overview of a modular simulation environment for a simulation of local caching of World Wide Web files is given.

REFERENCES

Balci, O. and R. E. Nance. 1987. Simulation model development environments: a research prototype. *Journal of the Operational Research Society* 38:753-763.

Balci, O. and R. E. Nance. 1992. Simulation model development environment. In *Proceedings of the 1992 Winter Simulation Conference*, ed. J. J. Swain, D. Goldsman, R. C. Crain, and J. R. Wilson, 726-735. IEEE, Piscataway, New Jersey.

Balci, O., A. I. Bertelrun, C. M. Esterbrook, and R. E. Nance. 1995. A Picture-Based Object Oriented Visual Simulation Environment. In *Proceedings of the 1995 Winter Simulation Conference*, ed. C. Alexopoulos, K. Kang, W. R. Lilegdon, and D. Goldsman, 1333-1340. IEEE, Piscataway, N.J.

Cota, B. A. and R. G. Sargent. 1992. A modification of the process interaction world view. *ACM Transactions on Modeling and Computer Simulation* 2: 109-129.

Centeno, M. A. and C. R. Standridge. 1991. Modeling manufacturing systems: an information-based approach. In *Proceedings of the 24th Annual Simulation Symposium*, ed. A. H. Rutan, 230-239. IEEE Computer Society Press, Los Alamitos, California.

Henriksen, J. O. 1995. An introduction to SLX. In *Proceedings of the 1995 Winter Simulation Conference*, ed. C. Alexopoulos, K. Kang, W. R. Lilegdon, and D. Goldsman, 502-509. IEEE, Piscataway, N.J.

Joines, J. A. and S. D. Roberts. 1994. Design of object-oriented simulations with C++. In *Proceedings of the 1994 Winter Simulation Conference*, ed. J. D. Tew, S. Manivannan, D. A. Sadowski, and A. F. Seila, 157-165. IEEE, Piscataway, N.J.

Joines, J. A. and S. D. Roberts. 1995. Design of object-oriented simulations in C++. In *Proceedings of the 1995 Winter Simulation Conference*, ed. C. Alexopoulos, K. Kang, W. R. Lilegdon, and D. Goldsman, 82-89. IEEE, Piscataway, N.J.

Sanderson, D. P., R. Sharma, R. Rozin, and S. Treu. 1991. The hierarchical simulation language HSL: a versatile tool for process-oriented simulation. *ACM Transactions on Modeling and Computer Simulation* 1: 113-153.

Standridge, C. R. 1985. Performing simulation projects with the extended simulation system (TESS). *Simulation* 45: 283-291.

Standridge, C. R. and A. A. B. Pritsker. 1987. *TESS: The Extended Simulation Support System*. New York: Halsted Press.

Standridge, C. R. and M. A. Centeno. 1994. Concepts for modular simulation environments. In *Proceedings of the 1994 Winter Simulation Conference*, ed. J. D. Tew, S. Manivannan, D. A. Sadowski, and A. F. Seila, 657-663. IEEE, Piscataway, N.J.

Standridge, C. R.. 1995. Modular modeling for network simulation languages: concepts and examples. In *Proceedings of the 1995 Winter Simulation Conference*, ed. C. Alexopoulos, K. Kang, W. R. Lilegdon, and D. Goldsman, 736-743. IEEE, Piscataway, N.J.

Ziegler, B. P. 1990. *Object oriented simulation with hierarchical modular models.* New York: Academic Press.

AUTHOR BIOGRAPHY

CHARLES R. STANDRIDGE is an associate professor in the Department of Industrial Engineering at the Florida A&M University - Florida State University College of Engineering. He led the development of the Simulation Data Language (SDL) and of The Extended Simulation Support System (TESS) for Pritsker Corporation. His current research interests are modular simulation environments and health system engineering. He is co-developing a simulation model of local file caching of internet files as well as models of new family therapy processes.

JAMES F. KELLY is a graduate student in the Department of Industrial Engineering at Florida A&M University - Florida State University College of Engineering. He is writing his masters thesis on inter-tool modularity in modular simulation environments.

THOMAS KELLEY is a graduate student in the Department of Industrial Engineering at Florida A&M University - Florida State University College of Engineering. He is writing his masters thesis on intra-tool modularity in modular simulation environments focusing on spreadsheets and animation.

JACK WALTHER is a graduate student in the Department of Industrial Engineering at Florida A&M University - Florida State University College of Engineering. He is writing his masters thesis on intra-tool modularity in modular simulation environments focusing on the implementation of ModNet in SLX.

Proceedings of the 1996 Winter Simulation Conference
ed. J. M. Charnes, D. J. Morrice, D. T. Brunner, and J. J. Swain

FIVE SIMPLE PRINCIPLES OF MODELLING

M. Pidd
Department of Management Science
The Management School
Lancaster University
Lancaster LA1 4YX
UK

ABSTRACT

As discrete simulation software gets easier to use it is tempting to assume that modelling also gets easier, unfortunately this is not the case. This paper suggests some simple principles of modelling that can be easily applied in discrete simulation, whether the model is implemented in a visual interactive modelling system or in a programming language. The modelling principles are used to suggest desirable features that might be included in simulation software.

INTRODUCTION

Since discrete simulation first became a practical proposition in the 1950s, it is probably fair to say that its development has gone hand in hand with general developments in computing. As computers have offered more bang per buck, so simulation models have been able to grow bigger and more complicated. In addition, as software has become easier to use, it has been possible for non-specialists to set about the task of building discrete simulation models.

The last decade has seen the development of powerful Visual Interactive Modelling Systems (VIMS) such as ProModel, Witness and many others. Using a graphical user interface, a modeller is able to both develop and run a simulation model in a point and click mode. Though VIMS are apparently targeted at non-specialists, they also make life much easier for specialist simulation analysts, who are able to develop straightforward models very quickly. In common parlance, this is 'simulation in the small'.

But some simulation models still need to be implemented as bespoke software, whether using programming systems such as MODSIM or in general purpose languages such as C++. This is the case for applications in which the logic of the model is tortuous and/or those in which there are severe computational demands due to model size or the need for fast experimentation. Such applications of 'simulation in the large', which may consist of programs with tens of thousands of lines of code or more, seem likely to persist whatever the development of VIMS.

This paper discusses a few practical principles which are relevant to simulation modelling, whether in the small or in the large. They are not intended to contradict the rigorous approaches suggested by various authors, most notably DEVS as developed by Zeigler (1976, 1984) and his colleagues. Instead they are intended to sit alongside them, nagging away at the modelling conscience of the analyst. Five such principles are discussed here, some of them could be subdivided, but that would break the 'seven, plus or minus two' rule of memorability. They assume that simulation modelling is a practical and yet intellectually hard task, one which would benefit from careful thought and planning (Pidd, 1996). They also assume that a simulation modeller will have basic technical competence in simulation modelling.

PRINCIPLE 1: MODEL SIMPLE, THINK COMPLICATED

In cybernetics, the science of control, is a tenet which is often known as the principle of requisite variety, due to Ashby (1956), which might seem to support the view that complex systems require complicated models. The principle of requisite variety can be stated in many ways, one of which is that 'variety must match variety'. Its origin lies in work in designing control systems that are to operate in complex environments. In its simplest form it suggests that, to be effective, a control system must be able to match the system which it is controlling. Thus, if a furnace can get too cool as well as too hot then the control system should include some way of detecting and responding to low temperatures as well as to high ones. Stated in this form, the principle of requisite variety is almost a truism. Ashby took it rather further than this commonsensical notion and developed a mathematical theory of its use.

What of its applicability to simulation modelling? Must a model be as complicated as the reality which is being modelled? Thankfully, the answer is no - for a reason which may not be initially obvious, but that is illustrated in figure 1. This shows that models are not just built, they are also used - which might be rather obvious, but is vitally important. It is crucial that the variety of the model and the user(s) combined can match that of the system being modelled but this does not mean that either one of the two components (model and user) must separately be able to do so. Systems theorists (Checkland, 1981) speak of emergent behaviour, which is the properties of a system that are not apparent in its components. In the context of modelling it is important to realise that the model and the user form a system. Requisite variety is an emergent property of that human:model system and not of its component.

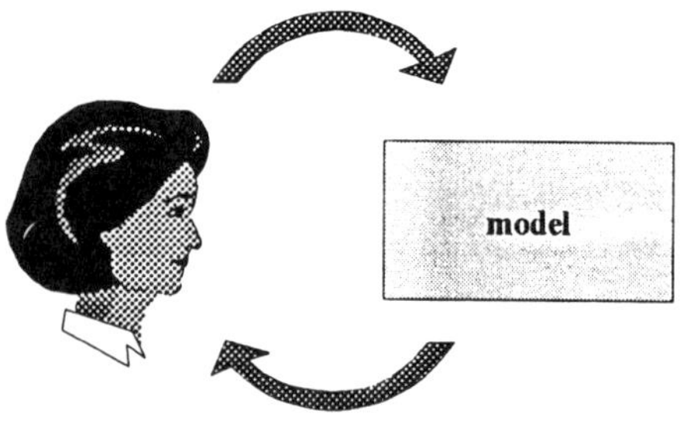

Figure 1: Requisite Variety in Modelling

The implication of this idea is captured in the aphorism 'model simple, think complicated'. That is, a simple model can be supplemented by highly critical thinking and rigorous argument and analysis. This can, of course, be taken to a ludicrous extreme. For example, a verbal model might be. "The world is hideously complicated and dangerous when things go wrong." No amount of critical thinking or analysis could take such a verbal model much further forward. At the opposite extreme, if a model itself does have requisite variety then there might be no need for much critical thinking. In such cases, the model could be built into a computer system which is used for day-to-day operation and control with no human intervention. Indeed, as Ashby pointed out in the principle of requisite variety, such complexity is essential if the model is to provide full and automatic control. Between these two extremes lies the realm within which most discrete simulation is conducted. The models are neither trivial nor fully requisite. Instead, they embody simplifications and aggregations.

The idea of simplicity needs to be linked with a suggestion from Little (1970) that a decision model should be easy to manipulate. As an analogy, there is a world of difference between the car driver and the skilled motor engineer. Most of us can drive cars whilst having only the vaguest of ideas about how the car works. We can do this because we have been trained to drive and also because the car itself gives us rapid feedback about our driving. We go off the road or hit other vehicles if we steer wrongly. A motor car is, after some training, easy to manipulate and it meets our preference for personal mobility. In a similar way, a model which is easy to manipulate (for example, its user interface might be made to resemble a spreadsheet) and which produces results which seem relevant, will be used. Thus simplicity has a second aspect, ease of use.

There are, of course, occasions when this metaphor collapses. In the world of travel, if we need to traverse the Pacific then driving across is not a viable option. Instead, we climb aboard a jet aircraft and surrender control to the aircraft crew with their own expertise. Similarly, there are some models which require the user to place their trust in the skills of the analysis team, because only they fully understand its workings. But users should only do so if the model produces results and insights which are relevant and appropriate to their situations. It is the joint responsibility of the analyst and the users to ensure that this is the case. Complicated models have no divine right of acceptance.

This idea that we might 'model simple, think complicated' brings us to the idea that models are 'tools for thinking'. It would be wrong to interpret that phrase as being 'tools to replace thinking'. Instead, they are tools to support and extend the power of thinking. Thus, a complicated model which is poorly employed may be worse than a simple model used as a tool for careful thought.

PRINCIPLE 2: BE PARSIMONIOUS, START SMALL AND ADD

The problem with the first principle of simplicity is knowing how simple or how complicated to be, and there is no general answer to this. Instead, like an army approaching a well-defended city at night, we use a little stealth and cunning. This is to employ the *Principle of Parsimony* which I have long found useful in computer simulation modelling (Pidd, 1984). In its more memorable form, this principle is sometimes known as *K.I.S.S.*, an acronym which stands for (*K*eep *I*t *S*imple, *S*tupid). The idea is that models should, ideally, be developed gradually; starting with simple assumptions and only adding complications as they become necessary. Instead of attempting, from the outset, a wonderful model which embodies every aspect of the situation in a realistic form we begin with something manageable, which may have unrealistic assumptions. The intention being to learn what we can from this simple model and then to refine it gradually, wherever this is necessary. Powell (1995) calls the same approach 'prototyping', this carrying the idea that it is best to quickly develop a working model, even if it is imperfect. It can be refined, or even abandoned, later. Starfield et al (1990) provide some interesting insights into this approach as applied in general mathematical modelling.

An example of this approach is described by Miser and Quade (1990a) who discuss a model that might be used to estimate how much of a runway would be visible to pilots through partial cloud cover. This might be an important consideration for an aircraft coming in to land in cloudy conditions. There are various types of cloud, but at a height relevant to a plane seeking visual contact with an airport runway, only some need be considered. The first question is, how could these be represented in a model? As before, the principle of parsimony suggests that starting simple might be the best approach, and thus some simple geometric shape has much to commend it. One possibility is to treat discrete clouds as if they were circular disks. As Miser and Quade point out, using circular disks in this way is not ridiculous as the relevant types of cloud do tend to be compact. It is also a helpful simplification, since the geometry of circular disks is well understood.

Cloud cover is often measured as a percentage of the sky that is visible from points on the ground and its pattern may be specific to the location of the airport. For example, nearby hills or ocean may cause certain patterns to predominate. These patterns could then be modelled at different heights by the use of the simplified circles to represent the clouds, and in this way it should be possible to estimate the visibility of the runway from different points in the sky. Whether this model is close enough to the likely 'real' distribution of cloud cover is the crucial question. This could be at least partially assessed by having an aircraft fly past different points under known cloud conditions and then attempting to compare the actual visibility with that predicted by the model. If the model is found to be too simple, then it can be further refined by, for example, using shapes which are closer to the actual shape of clouds. For instance, each cloud could be modelled as a set of circles which overlap (producing shapes which resemble Mickey Mouse's face in one form). Again, this refinement might be chosen because the geometry of circles is simple and thus the model is tractable.

Hence the principle of parsimony requires us to develop models that are, initially, too simple for the task in hand. Ideally, the basic structure of the model will be correct, but even this need not be so. The modeller then gradually refines or replaces the model, all the time bearing in mind the first principle of simplicity. The idea being to add nothing unnecessary to the model.

PRINCIPLE 3: DIVIDE AND CONQUER, AVOID MEGA-MODELS

This is common advice given to anyone trying to understand how a complex system operates. Powell (1995) calls this decomposition, as well as divide and conquer. Raiffa (1982, quoted in Miser and Quade, 1990b) has the following to say.

"Beware of general purpose, grandiose models that try to incorporate practically everything. Such models are difficult to validate, to interpret, to calibrate statistically and, most importantly to explain. You may be better off not with one big model but with a set of simpler models.."

Raiffa's point is partially related to the first two principles above, but also relates to the need to build a model from components, each of which should be themselves developed parsimoniously.

This is, it should be noted, not the same as advising software developers to use modular designs in the their computer programs, laudable though this advice may be. The suggestion is, rather, that the best way to face up to complexity is to break it down into manageable chunks, all the time making sure that the inter-relationships between the sub-models are well-understood. In really large scale simulation modelling, which might involve a sizeable team of modellers, then this is the only way to proceed. It requires careful project management, to ensure that each participant is operating as part of the team, but it is a very fruitful way to proceed.

PRINCIPLE 4: DO NOT FALL IN LOVE WITH DATA

A common failing of students when learning about modelling, is to insist that progress cannot be made unless there is some (or more) data available. Their assumption being that examination of the data will provide some clues that will extend their understanding. This may well be a mistake, even though exploratory data analysis is a very valuable technique. Modern statistical software or spreadsheets enable the rapid plotting and summary of large amounts of data and from this analysis, patterns may be quickly gleaned. Sometimes these patterns exist, but at other times they are, like beauty, in the eye of the beholder. Nevertheless, exploratory data analysis has much to commend it as an approach. It is, however, no substitute for careful thought and analysis.

Some of the dangers and pitfalls that await the unwary in their treatment of data are discussed below, but there is a fundamental point that should not be missed. This is that the model should drive the data collection and not vice versa. This means that the analyst should try to develop some ideas of the model and its parameters and from this should think about the type of data that might be needed. One of the problems with some case-style teaching is that the cases are often intended to be self-contained. That is, the students know that all the data they may need is available in the papers issued with the case. This is quite unlike real life in which data must be requested, justified and collected before it can be analysed. Data is not free, its collection has a cost as does its interpretation and analysis.

How then should the use of data be linked in to the parsimonious, gradual approach advocated earlier in this paper? If circumstances permit, then the best approach would be to develop a simple model and then to collect data to parameterise and test it. It may then be clear that the simple model is fine for the intended purpose or it may be that the model needs to be refined - which may need more data or different data. This new data will have a cost which should enable some rough cost:benefit calculation to check whether its collection and analysis will be worthwhile. And so on, until some point is reached at which the costs outweigh the benefits.

Of course, these ideal circumstances may not pertain and it may be necessary, especially when acting as a fee-charging external consultant to set up a complete data collection exercise at the start of the work. If this can be resisted then it would be a good idea to do so.

Within this fourth principle, there are a number of specific issues than can be addressed. These are almost principles in their own right, but are usefully gathered under this sceptical fourth principle.

1. Data mining and data grubbing: Data-mining is a term which has recently entered the language of statisticians and refers to attempts to develop statistical models from available data. Powerful computer packages are used to search for patterns in the data. A debasement of this is "data-grubbing", which some use as a term of abuse for approaches in which many different data series are unthinkingly collected and then read as data files by one of today's powerful statistical packages. These packages allow complicated analyses to be conducted on the data in a very short space of time. Regressions can be tried, linear and non-linear, other types of multivariate plastic surgery can be applied to the data and the original data series can be transformed by taking logarithms and the like. Within a simulation context, there are friendly packages to enable modellers to fit probability distributions to data that has been collected. This can all be done very quickly, by someone who knows very little about the statistical tools being used in the computer software. This can be like an attempt to bake a cake by collecting ingredients that look interesting, then mixing them together until boredom sets in, popping the resulting melange into the oven and waiting for the smoke to appear. Cakes cooked in this way should only be given to people with whom scores need to be settled.

This criticism of data-grubbing should not be interpreted as a call for a ban on friendly, powerful statistical packages. They are much too useful for that and they take the drudgery out of statistical modelling. However, they should not be a substitute for thought. Also, just because data is available it should not be assumed that it is useful.

2. Data is useful in model building: This criticism of data-grubbing and of a reliance on available data might be interpreted, wrongly, to imply that modelling is best carried out in an abstract way. It is certainly not the intention that data should be ignored in this way. It might, therefore, be helpful to divide data and information into three groups. First, there is preliminary or contextual data and information. One practical approach to problem structuring is to use the easy to remember questions of What, Why, When, Where, How and Who (Kipling's honest working men) as a useful guide in preliminary investigation. Clearly, the results of these questions may be qualitative or quantitative. In the latter case it may be necessary to conduct significant analyses on the data that is so produced. But this data is collected with a view to understanding more about the context of the problem, rather than the development of a detailed model. It may not be unusual for this preliminary analysis to reveal enough insights for there to be no real need to take a project any further.

The second type of data is that which might need to be collected and analysed in order to develop the model in some detail. This is model parameterisation, or model realisation (Willemain, 1995). But, as is repeatedly stated here, the model structure should drive the data collection and analysis, not the other way round. The third type of data is discussed under the heading 'Avoid using the same data to build and to test a model'.

3. *Beware of data provided on a plate*: An old adage amongst Management Information Systems professionals is that information is data plus interpretation. One feature of modern organisations is that computer systems collect almost every conceivable type of data about what is happening - except the data which you really need for a particular model. For modelling purposes, data is best ordered a la carte rather than table d'hôte. For example, in attempting to develop models for production control, it may be necessary to produce sub-models of customer demands for the products being made. Most companies monitor their sales by using data produced by the sales order processing systems that are used to take orders and issue invoices. Thus, the obvious way to get hold of demand data might be to take it from the customer order files, but there are at least three reasons why this might be a mistake.

The first is that such systems often only record the actual despatches to customers and this may be as much a reflection of the available finished stock as it is of the actual demand from customers. They may request one thing, but the company may be unable to supply and they may thus go elsewhere or may accept a substitute. The second reason is that, if the customer suspects that a required item is out of stock (due to past experience) they may not even request the item. Finally, the whole idea of implementing a new production system might be to make the company more attractive to customers who normally order from other suppliers.

Hence, for all of these reasons, the data from a sales order processing system might be treated with some caution. In this and other cases there may be no substitute for proper and well-organised data collection if a useful model is to be constructed. It may also be possible to take existing data and massage it in such a way as to account for some of its shortcomings. Thus, a small scale data collection exercise might be used to reveal the discrepancies in the full, system-produced data which may then be modified to take account of this. However, it must be born in mind that such data massage implies that a model of the data itself is being used as part of the modification process.

4. *Data is just a sample*: It is also important to remember that, in the vast majority of cases, data is just a sample of what could be used or might be available. This is true in a number of dimensions. First, the time dimension, this being the simplest to understand. When data is being used to build or to test a model then that data will have been collected at a particular time and over a certain period. If we say that the data is believed to be representative then we are implying that it is representative of a larger population which displays some regularity through time. We are not expecting someone to come along later and surprise us with data which differs drastically from that which we have already obtained. Nevertheless this may happen and is always a risk which is clear when we realise that the results of a model may be used to extrapolate into the future. The future may simply differ from the past, that is the population from which the data sample comes may behave differently in the future. Data is also a set of observations and this is the second aspect to note in the realisation that data is a sample of what might be obtained given enough time and other resources.

5. *Avoid using the same data to build and to test the model*: Most models used in management science make use of data in one form of another. The model will have parameters that must be given values. This process of parameterisation is usually based on data, whether specially collected or generally available. The trap to avoid, if possible, is the use of the same data to parameterise the model and then to test it. As an example, suppose that a company wishes to understand the sales pattern for one of its products. Careful data collection proceeds and they eventually have a time series of their weekly sales over the last 2 years. A management scientist then develops a forecasting model which is intended to suggest how sales might behave, given certain assumptions, over the next few months. The idea being that the model might be used, say, monthly each time with an updated data series, to suggest likely future sales.

There are a number of types of forecasting model that could be employed, and they have in common the fact that they are based on the analysis of historical data. Under skilful hands, computer programs are used to estimate the parameters of equations which lead to a model which is a good fit to the recent historical data. The goodness of fit can be expressed in standard statistical terms. But this goodness of fit is an evaluation of the degree to which the model fits the historical data. It is also important, where possible, to test how well the model predicts what might happen in the future. A tempting short-cut is to quote the goodness of fit statistic from the parameterisation exercise but this might be a mistake. A better approach, is shown in figure 2, in which the available data has been divided into two sets. The earlier data is used to parameterise the model and then the second set is used to test the model. This test set has not been used in the parameterisation but is being used as a surrogate future. Goodness of fit measures can be used to assess how well the model predicts this surrogate future.

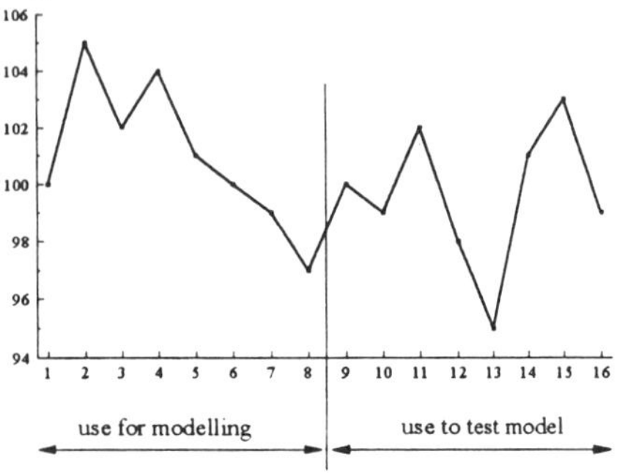

Figure 2: Use of Data in modelling

PRINCIPLE 5: MODEL BUILDING MAY FEEL LIKE MODELLING THROUGH

Because a simulation model is the result of an attempt to represent some part of reality so that action may be taken or understanding may be increased, it might be thought that model building is a linear and highly rational process. There have been few attempts to investigate this issue, as it relates to management science, but the evidence suggests that modelling is not a linear nor a smooth process. Instead, people seem to 'muddle through', making use of insights, perhaps taking time away from the modelling, trying to look at things from different perspectives and so on. This does not, of course, imply that modelling must be done this way, but it may well indicate how successful analysts actually operate.

A fascinating attempt to investigate this issue is reported by Willemain (1994). He gained the co-operation of a group of twelve experienced modellers, a mixture of practitioners and academics. All had completed graduate work in OR/MS and their average years of experience since finishing graduate school was over 15 years. They were not, in any sense novices. This group were asked to do two things. First they were given the chance to describe themselves, to express their views on their own approaches to modelling, to say what experience they had in modelling and to capture something important in a short, personal modelling story. Their self-descriptions and experience showed them to work mainly in areas in which they "pursue specific objectives towards fundamental changes in complex, existing systems". Also that they "develop a unique model for each problem, though all their models involve extensive computation". From this we can conclude that their work could in no way be described as 'airy-fairy', for their concerns seem down-to-earth.

Of their approach to actually building and developing models, Willemain (1994) summarises their responses thus. They "develop their models, not in one burst, but over an extended period of time marked by heavy client contact". Also, they are "guided by analogies, drawing and doodling, they develop more than one alternative model, in each case starting small and adding". Thus many of their claimed approaches are a good fit with the first four principles of modelling discussed in this paper.

In a second paper, Willemain (1995) reports an experiment with the same twelve people who were given modelling tasks and were asked to think aloud as they spent 60 minutes figuring out how best to develop suitable models. This is clearly an artificial task for two reasons. First it compresses their activity into just 60 minutes, when their expressed preference was to work "over an extended period of time". Secondly, the request to think aloud as they worked might distort their normal patterns of work. Despite these reservations, this thinking-aloud protocol reveals some interesting issues.

In analysing the tapes of their thinking aloud, Willemain (op cit) classifies their concerns under six headings.

- *The problem context*: which he relates to problem structuring as defined by Pidd and Woolley (1980). That is, the exercise which aims to gain a sufficient understanding of the problem to proceed to some form of formal modelling.
- *The model structure*: which he takes to be the process of deciding what category of model to use and of analysing data prior to actually building it.
- *Model realisation*: this is the process of parameter estimation for a model and/or calculation of results.
- *Model assessment*: which is deciding whether the model will be valid, useable and acceptable to a client.
- *Model implementation*: which is working with the client so as to gain some value from the model.

The tapes show that about 60% of the modellers' time was devoted to model structure, that is, what would be regarded as the core of model building.

About 30% of the time was divided equally between concerns about problem context and model assessment, with similar time spent on each issue. Just 10% was devoted to model realisation and almost none to questions of implementation.

As the study gave the modellers just 60 minutes to work on a problem it should be no surprise that so little time was devoted to model realisation or implementation. But what is significant is that so much time was spent on thinking about problem context and model assessment. What is also very important is the fact that the time spent on these three major concerns was scattered throughout the modelling session. The modeller kept picking up a concern for a while, dropping it, and then returning to it. Presumably this would be even more marked were it possible to follow how they operate over a much longer time period in their 'real' work.

Perhaps it is an exaggeration to say that modellers muddle through. However, it is equally an exaggeration to assert that modelling proceeds as a linear step-by-step process. It did not in Willemain's study and it probably does not in most of our experience. The other principles presented here should be used to provide some order to the muddling through. It is quite normal for a modeller to think in parallel whilst working on a model. Discussing ill-defined problem solving in general, Holyoak (1990) discusses how people tend to operate in parallel lines of thought and how they are continuously restructuring their ideas. A concern to understand the problem context goes hand in hand with a desire to produce a model which will be useful, as well as technically correct.

IMPLICATIONS FOR SIMULATION SOFTWARE

What then are the implications of these five principles for developers of simulation software? In answering that question, it must be re-emphasised that they assume that the modeller has some basic technical competence. Likewise, this brief final section assumes that the software is basically well designed and implemented, at least in technical terms. Over and above this basic competence, what help can software developers and vendors give to simulation modellers? Perhaps the best way address this is to look as at each of the five principles in turn.

1. Model simple, think complicated: This suggest something that is already well known to software developers, that modelling systems need to support the process of critical thinking and analysis as well as the process of model building. This might, perhaps, focus on the need to support experimentation, both formal and graphically based. At a basic level this implies support for common file formats to permit straightforward use of analysis packages. At a more sophisticated level it might mean the availability of database type support for analysis across a whole range of runs.

2. Be parsimonious, start small and add: This suggests the need for software that is properly layered, with top levels devoted to a VIMS sitting above a simulation code, which itself draws on libraries and may translate into C++ or similar. The reason for this being that the modeller needs some way of developing rapid prototypes that are either enhanced or discarded. Thus the VIMS may be used for this initial development, but the ability to translate the VIMS into a simulation code or into C++ means that a simple model can be enhanced in ways that are outside the scope of the VIMS. In addition, if the simulation code and its base language are object oriented, the early models could be code-based rather than in the VIMS and might draw on sensible class libraries.

3. Divide and conquer, avoid mega-models: If models are to be built up from smaller, relatively independent models then the modeller needs to operate within some paradigm that supports this. Two possibilities spring to mind, both of which have been found to be practically useful. DEVS and its extensions (Zeigler, 1976, 1984) was proposed with this need partly in mind. Object oriented systems provide an alternative way to attack this problem, as they require any model to be atomised into relatively independent components. Finally, the wide-spread use of windowing operating systems and the current vogue for 'network computers' means that it may, in future, be easier to develop communicating applets than has been the case until recently.

4. Do not fall in love with data: If it is easier for a modeller to develop rough-cut or prototype models then, possibly, an infatuation with data is less likely. Thus, the comments made above under the first three principles offer some support. But it seems unlikely that software developers can offer much more than this. It may even be that packages which ease (for example) the selection of input distributions may make things worse - though this most certainly need not be the case. It seems that thorough training is the only way to face up to this principle with all its variants.

5. Model building may feel like modelling through: The suggestions made above should help in this principle too, but another suggestion might also do so. When someone is muddling through with their modelling, then ease of use and simple interface design seems very important as does software that is fun to use. If the modeller has also, as is commonly the case, several tasks and modelling projects underway at the same time, then the question of interface support seems very important. The interface needs to be standard, predictable, fast to use and satisfying in its response. Not too much to ask, surely?

REFERENCES

Ashby R. (1956) An introduction to cybernetics. Chapman and Hall, London.

Checkland P.B. (1981) Systems thinking, systems practice. John Wiley & Sons, Chichester.

Holyoak K.J. (1990) Problem solving. In Osherson D.N. and Smith E.S. (1990) An invitation to cognitive science, volume 3: Thinking. MIT Press, Cambridge, Mass.

Little J.D.C. (1970) Managers and models: the concept of a decision calculus. Management Science 16, B466-B485.

Miser H.J. and Quade E.S. (1990a) Validation. In Miser H.J. and Quade E.S. (1990) Handbook of systems analysis: craft issues and procedural choices. John Wiley & Sons Ltd, Chichester.

Miser H.J. and Quade E.S. (1990b) Analytic strategies and their components. In Miser H.J. and Quade E.S. (1990) Handbook of systems analysis: craft issues and procedural choices. John Wiley & Sons Ltd, Chichester.

Pidd M. (1996) Tools for thinking: modelling in management science. John Wiley & Sons Ltd, Chichester, forthcoming.

Pidd M. (1984) Computer simulation in management science (1st edition). John Wiley & Sons Ltd, Chichester.

Pidd M. and Woolley R.N. (1980) A pilot study of problem structuring. Jnl Opl Res Soc, Vol 31, pp1063-1069.

Powell S.G. (1995) The teacher's forum: six key modeling heuristics. Interfaces 25, 4, 114-125.

Raiffa H. (1982) Policy analysis: a checklist of concerns. PP-82-2. International Institute for Applied Systems Analysis, Laxenburg, Austria

Starfield A.M., Smith K.A. and Bleloch A.L. (1990) How to model it: problem solving for the computer age. McGraw Hill Publishing Company, New York.

Willemain T.R. (1994) Insights on modelling from a dozen experts. Ops Res 42, 2, 213-222.

Willemain T.R. (1995) Model formulation: what experts think about and when. Ops Res 43, 6, 916-932.

Zeigler B.P. (1976) Theory of modelling and simulation. John Wiley & Sons Inc.

Zeigler B.P. (1984) Multi-facetted modelling and discrete event simulation. Academic Press.

AUTHOR BIOGRAPHY

MIKE PIDD is Professor of Management Science in the Management School of Lancaster University in the UK. He has written two books on simulation, of which the best known is Computer Simulation in Management Science (John Wiley). He has just completed a new book for the same publisher (Tools for thinking: modelling in management science) which addresses the issues often raised by critics of rational modelling. His current interests in computer simulation include object oriented methods.

Proceedings of the 1996 Winter Simulation Conference
ed. J. M. Charnes, D. J. Morrice, D. T. Brunner, and J. J. Swain

SIMULATION: "WHAT TO DO WITH THE MODEL AFTERWARD"

Daniel J. Muller
AutoSimulations
655 E. Medical Drive
Bountiful, Utah 84010, U.S.A.

ABSTRACT

Once the simulation analysis is completed, what becomes of the simulation model? Often all the efforts and costs incurred during the simulation project are forgotten once the results of the simulation are determined.

Simulation models can have a second life, in which they provide training, educating, and the means to assist the decision making process. Although simulation projects often involve the key individuals necessary for defining, analyzing, and evaluating the simulation models, they often overlook the individuals whom the simulation results will greatly impact. This paper discusses the need for education in the simulation process and some approaches for structuring models to be used for training and educational use.

1 INTRODUCTION

This paper presents various concepts for structuring and using simulation models as an educational and training tool. Using simulation for this purpose may play a greater role in a project's success than the simulation model development and resulting analyses. The effectiveness of any simulation project should always include the education of individuals who will be directly or indirectly affected by the simulation's outcome.

Education and training are the key to model acceptance. Using simulation for training purposes can often provide key insights and play a major role in the acceptance of the physical system changes prior to implementation. The first sections of this paper present the reasons for education in the simulation process. The final sections present different approaches on how to structure and apply a simulation model for future use after experimentation is completed.

2 THE NEED FOR EDUCATION

Every simulation model requires "selling" in the simulation project life cycle, such as selling the concept of simulation or selling the simulation results. Since simulation models often cross the lines between functional departments, many individuals are often affected by the project's outcome. The key to success lies in the information exchange between the interested "stakeholders" in the project. Figure 1 displays a sample of various project stakeholders.

There are many factors that can alter the success of a simulation project, such as:

1 - internal politics
2 - previous simulation successes and failures
3 - fear of change
4 - lack of communication

All of these influences can be reduced by educating the proper individuals. Communication during a simulation project is required throughout the simulation effort between the model builder and the stakeholders. This is notably important during the early stages of a project. Not only does the analyst require accurate logic and data from various sources, but the interested stakeholders require feedback on the potential project's impact on their operations.

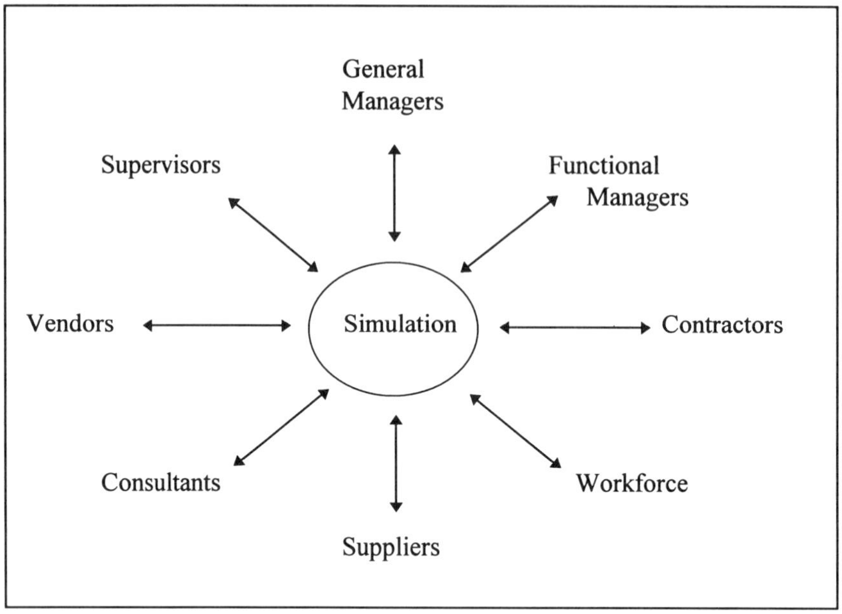

Figure 1: Various Simulation Project Stakeholders

The following section describes how simulation can be used to educate different parties based on individual responsibilities.

3 CONVEYING SIMULATION BENEFITS

When the boundaries of a simulation project are clearly defined, it should be the responsibility of the project manager to coordinate the efforts of the necessary individuals. The objectives and requirements of the simulation should be presented and documented. Key points addressing each person's role and the impact the project could have on their respective operations should be highlighted.

In educating these key individuals, certain concepts should be stressed relative to the simulation effort. Some of the points listed below can serve as a guideline in the education process:

1. Preview System Changes.
   Through the use of graphical simulation, everyone can observe the impact of the expected system's operations, thus reducing the uncertainty associated with a new concept.

2. Acceptance of Change.
   Simulation bridges the gap between the unknown and the real world. Individuals can foresee the effects of change on their respective systems, thus allowing them to accept and understand change.

3. Insurance Policy.
   Simulation provides an insurance policy against making a costly mistake when large capital expenditures are involved.

4. Project Feasibility.
   Simulation can be used to evaluate the feasibility of a concept to ensure the equipment performance and design criteria are accurate.

5. Project Confidence.
   Simulation provides a level of confidence necessary to take the project results towards implementation.

The results of the simulation project may result in the implementation of new and automated equipment, operating philosophies, operator task assignments, or material handling procedures. Subsequent to the implementation of these system changes, the education of employees is required to assure a successful operation. Using the simulation model to assist in the training and education of key individuals on the effects of system changes can greatly assist at this phase of the project.

4 DESIGNING INTERACTIVE MODELS

Prior to developing any simulation model, it is important to plan for model-user interactiveness. Many individuals associated with the simulation effort are not simulation modelers, nor do they necessarily understand the simulation process. Their vested interests lie in how the project results will impact their respective areas. By planning ahead, a model can be developed to support the education, training, and transfer of information to these of key individuals. By planning user involvement, an analyst can increase the likelihood of model acceptance.

In developing a model for individuals who may not fully understand the simulation process, the need to create easy interactive capabilities should be a high priority. Creating an interactive model for the sole use of demonstration and execution by non-simulation experts is well worth the effort. Providing this "hands-on" involvement will lead to increased model confidence and acceptance. The following are some examples of building user interactiveness:

1. Menuing System.
   The simulation model could include a menuing system that permits the user to specify certain performance data. This provides the user with the ability to run what-if scenarios for specific events to understand the cause-and-effect relationships between model components.

2. Data Files.
   Information used in the simulation model can be developed in data files which are read into the simulation model. These ASCII data files should be well documented so that the contents of the files are easily understood. This provides an easy way for a non-simulation expert to run experiments and what-if scenarios.

3. Model Interactive Dialogues
   A non-simulation expert can be easily trained to use a modeling interface or dialogue box that prompts the user to make a decision in the operation of the simulation. The user can then understand the cause and effect relationships of the system operations and unforeseen variability.

4. Equipment/Resource Failures.
   The model could include an interactive equipment or resource window for causing a resource breakdown. A user could see the effects of breakdowns on the system's performance and the resulting system's responses.

The above items can greatly improve the ease of use of a simulation model. Basic interactive planning does not require a large effort during the model development phase.

5 THE ROLE OF ANIMATION

The following paragraphs summarize the role of animated simulation models relative to various stockholders, for when education plays a key role in the project's success.

1. Management Acceptance.
   Since most simulation models are developed to prove a new approach, concept, or facility design change, the major concern is whether the system will meet its design requirements. Since major capital expenditures are often involved, simulations provide management the needed level of assurance for validating or invalidating the designs. Graphical models provide the visual means of presenting the system's findings to various levels of management in an efficient and expedient manner.

2. Equipment Venders.
   Graphical simulation models can be used to educate the equipment vendors on how their systems will perform and interact under various conditions. Besides validating or correcting the equipment quantities, simulation models can help a vendor analyze the performance characteristics such as reliability, throughput and maintenance requirements necessary for a successful implementation. Through the use of graphics, vendors can visually see the effects of equipment performance and quantity on the surrounding systems.

3. Operator Feedback and Acceptance. Simulation models can be used to educate employees on their roles in support of system changes. Animated models can help employees accept design and system changes that will alter their operations and/or responsibilities.
   Involving operators in the simulation process may supply key input and feedback on improving the system or identifying potential problems with the new concepts. This approach can only increase the acceptance of any new designs upon implementation at the shop floor level. A system that is not well received by the people responsible for its operation is destined for failure.

6 EDUCATIONAL AND TRAINING USES

The statistical output and project results can often hide the true potential of using simulation for educating and training. An animated model can be used in a variety of ways to convey different objectives. The following paragraphs briefly describe how an animated simulation model may be used for training and education.

1. Cause and Effect Relationships. An animated simulation model is an excellent way to educated people on the cause and effect relationship between system components. Since functional areas often divide individuals' responsibilities, a simulation model can be used to educate individuals across numerous physical boundaries to assist in facility improvements. This provides a quick way to communicate the effects of change within each area relative to the other.

2. System Response Requirements. Simulation models can be utilized to educate and train individuals on response measures associated with various scheduled and unscheduled system events. A model can be used to test various preventive maintenance plans to help minimize system and equipment downtime. Also, simulation can be used to evaluate the best response during equipment failures events. An animated model can be used to educate operators and management on the effects of schedule changes and failures through "hands-on" model interactiveness.

3. Videotape. Videotapes of animated simulation models can be used to educate and train individuals. Videotapes permit the analyst to highlight key performance features of a system, and they accommodate a large viewing audience. Since many large capital expenditures involve management at various corporate locations, a videotape can assist in bridge the distance and communicate the results. Also, a videotape can assist in defining and describing the material flows and functions associated with each area of a new design. This videotape can then be used to educate new employees prior to the physical system startup.

7 PITFALLS

Often the above approaches seem good in practice but are limited by time and schedules. Building a model with interactive capabilities does not usually add a large amount of model development time to a simulation effort. What usually takes the most effort is coordinating the schedules and allowing enough time in the project cycle to include all the key people. Various schedules, locations, and everyday responsibilities may reduce the amount of educating and training that can be applied to the project.

Another potential problem is the involvement of too many individuals in the simulation effort. As more individuals become aware of the simulation effort, often they will want to customize the effort to meet their internal objectives. This will always tend to increase the scope, schedule, and cost of simulation effort.

8 SUMMARY

Education and training play a major role in every simulation project. Whether enough time exists to involve all interested and affected individuals and departments is project-dependent. The greater the communication and involvement in the simulation model across all levels of management and operations, the greater the success in the project's acceptance can be expected. Developing a user-friendly interactive model for experimentation, what-if analysis and demonstration purposes will allow an analyst to perform quick what-if demonstrations. User-friendly models also allow non-simulation experts to participate in the project. Scheduling and coordinating the right individuals relative to maintaining the focus and objectives of the simulation effort is a difficult task. Simulation efforts tend to increase in size and scope as more individuals are involved. A good simulation project plan should always allow for the necessary time to educate key project members. The amount of time required is dependent upon which obstacles exist that could limit the project's success.

AUTHOR BIOGRAPHIES

DANIEL J. MULLER is a Senior Simulation Analyst with AutoSimulations. He received a Bachelor degree in Industrial Engineering from Georgia Tech in 1985, and he received his M.S. degree in Industrial Engineering from the University of Pittsburgh in 1991. He performs full time simulation consulting services for AutoSimulations' Mid-Atlantic region. He was the WSC' 94 exhibits chair. He is a member of IIE.

Proceedings of the 1996 Winter Simulation Conference
ed. J. M. Charnes, D. J. Morrice, D. T. Brunner, and J. J. Swain

GOAL DRIVEN SIMULATION INTELLIGENT BACK ENDS: A STATE OF THE ART REVIEW

Louis A. Molina
Carlos Gandarillas
Martha A. Centeno

Department of Industrial and Systems Engineering
College of Engineering and Design
Florida International University
Miami, Florida 33199, U.S.A.

ABSTRACT

Goal driven simulation (GDS) seeks to automate many of the output analysis and experimental design tasks of a simulation study. Theoretically, its use allows the reallocation of the simulation expert to other tasks. GDS capabilities include determining parameters to change, suggesting a rate of change, and testing these changes against a pre-established set of goals. Realizing GDS, however, requires the integration of techniques such as object oriented design, knowledge based systems, and neural nets. Before achieving this integration, there are still several issues to resolve including the type of interaction these techniques would have among themselves. This paper explores several of the issues concerning the realization of goal driven simulation systems, their impact on the simulation modeling methodology, how GDS works, and the need for its development.

1 INTRODUCTION

Simulation is the process of designing a computer model of a real system, and conducting experiments with this model, to understand the behavior of the system and/or evaluating various strategies for the operation of the system (Centeno and Shannon, 1995). Activities involved in a simulation study require the modeler to posses knowledge in various disciplines, such as modeling, probability and statistics, computer programming, and process analysis (Alsugair and Chang, 1994). In many instances, the modeler finds herself or himself not an expert in one or two of these areas; thus, the scope of the study is limited, or its delivery date gets extended. Of the various tasks in a simulation study, the experimentation and output analysis are the ones that lead to a real solution, but they are also among the most repetitive and time consuming tasks after model verification (Ford and Schroer, 1987). Unless the modeler has extensive experience with the type of systems being modeled, s/he find her/himself testing a large set of alternatives before a desired performance goal is achieved.

Goal driven simulation (GDS) incorporates the use of a knowledge based expert system in conjunction with a simulation language to achieve an optimal solution, within a set of desired objectives. Thus, GDS would reduce the time needed for experimentation and analysis, and it would allow the shifting of some modeler resources to other tasks in the simulation study. The latter has led us to believe that GDS research shall experience a significant growth during the next decade.

The need for GDS has been reflected in works in different fields such as the analysis of flexible manufacturing systems (FMS) (Kopacsi and Kovacs, 1993), tutoring of simulation users (Touran, 1990), manufacturing of electronics (Ford and Schroer, 1987), and the general work by Shannon, Mayer and Adelsberger (1985). These works clearly describe that two of the areas of the simulation modeling process that need great improvement are the design of experiments and output analysis areas. Traditional simulation languages fall short of providing adequate support; however, these authors point out that the integration of knowledge-based techniques with traditional tools will lead to the development of an "ideal" simulation modeling environment.

Advances in hardware and simulation software have allowed simulation modeling to be utilized in very diverse sectors. In fact, many simulation software companies offer packages targeted to specialized areas within industry. However, such expansion has not been paralleled by similar advancements in automated output analysis (Centeno and Shannon, 1995). It is not until recently that some simulation packages have begun to include capabilities for the analysis of data the simulation model uses and/or produces.

In Section two, we discuss some of the reasons why GDS is needed. In Section 3, we review various research efforts in this area and discuss a framework that we believe contains the minimum elements required for a GDS simulation system. In Section 4, we discuss the impact that GDS development will have on the simulation modeling methodology itself.

2 THE NEED FOR GDS

To better understand the need for GDS, it is necessary to first understand what the current simulation methodology can do. In this context, simulation modeling is a process that involves the designing, building, verification, and validation of a model, with which the modeler will experiment to explore a variety of "*what*-if" scenarios. Simulation can provide estimates for a variety of measures of performance, and it can evaluate the effects of changes to the operating conditions. However, traditional simulation modeling methodology can not optimize, it can not describe or deal with system characteristics that have not been previously described to it, and it can not solve problems just provide information from which solution may be inferred. Thus, it is clear that the power of this technique could greatly benefit from technique that will enable it to optimize and seek automatically for solutions.

The GDS philosophy offers a framework in which a variety of data modeling and knowledge managing techniques could be incorporated. GDS will search for the appropriate set of inputs that would yield measures of performance within a desired interval; thus, enabling simulation optimization. The GDS approach is only restricted by the knowledge contained in the knowledge base. Further, under GDS, the simulation environment would automatically run multiple alternatives, over a wider search space, using smarter search techniques, while taking into consideration the objectives of the study. Thus, the model user will have appropriate support when doing the analysis of several alternatives to obtain trends or patterns of the system's performance.

GDS also requires that domain knowledge be part of the knowledge base. This knowledge enables the GDS to support the modeler even when the complexity of the system under study grows because it will guide him or her on how to modify parameters to achieve desired goals.

In summary, the need for the GDS approach is paramount because

- ✓ traditional simulation methodology can not optimize.
- ✓ the complexity of simulated systems has grown.
- ✓ solutions tend to be dependent on the expertise of the model user.

Current commercial simulation languages do not support optimization methodologies based on domain knowledge that relieves the analyst from multiple scenarios. SimRunner©, a package that works with ProModel, performs optimizations on models for this language (Akbay, 1996). However, this is not domain knowledge based search, but statistical searches over a multidimensional parameter field. In this sense, the package falls short from providing the user with a pseudo-intelligent search.

Similarly, other packages such as SIMAN/ARENA support only tools that aid the analyst in performing the study of each scenario. However the capability of creating an integrated model and knowledge base is not found in the current versions. GPSS and AUTOMOD excel in other areas but fall short in the optimization process.

3 GOAL DRIVEN SIMULATION

In traditional simulation analysis, the user conducts experiments by varying inputs to the model or by changing the model itself. On the other hand, goal driven simulation (GDS) drives itself to achieve a set of desired goals (Shannon and Prakash, 1990). Under GDS, the user inputs the simulation model, its capabilities, his priorities, and goals of the simulation study. Goals are the desired value(s) for one or more of the systems measures of performance. Target values to be met may include average time in the system, resource utilization, and average number in queue. GDS would determine what parameters should be changed, and test these parameters, to ensure that they lead to the desired goals.

Research in the area of goal driven simulation has focused on two major areas. These are 1) building new simulation systems altogether, and 2) building hybrid systems. The first area calls for a redefinition and expansion of the simulation modeling methodology to incorporate the strengths of other disciplines such as artificial intelligence (AI); however, in many instances, the resulting simulation system has re-invented the wheel in the sense that they have had to duplicate what traditional simulation packages do well in order to attain minimal value added. This situation has been mostly caused by the incompatibility of AI-based tools with more traditional programming languages. Fortunately, software developments are taking us to a point in which robust AI-based tools can truly be combined with powerful traditional simulation languages.

The second area has seen a little bit more success because it calls for the addition of modules to traditional simulation packages, so as to have a simulation modeling environment that operates under the GDS philosophy.

The addition of intelligent back-ends (IBE) responsible for assessing the outputs of the simulation model against the user-provided goals seems to be a very promising alternative for GDS (Shannon and Prakash, 1990). Such back ends could operate in one of two ways:

- ✓ *Series*: The IBE is triggered as soon as the simulation model has finished generating the data of the first experiment. Then, the IBE assesses such output against the goals. If the outputs fall within the prescribed range, then the experiment is complete; otherwise, the IBE will utilize its knowledge base to decide how to modify the input parameters, so as to steer the simulation model towards the desired goal. Once the model provides new outputs, the IBE repeats the process for as long as needed.
- ✓ *Parallel*: Both the back end and the simulation model are running simultaneously. The simulation model produces a series of traces from which estimates for the measures of performance are continuously updated. The IBE is attentive to such estimates and, as soon as it can determine that the values of these estimates are heading outside the target, it will halt the simulation run, modify the input parameters, and re-start the simulation run.

In both cases, the IBE should have the capability of recognizing a situation where the goal(s) can not be met, no matter how the IBE changes the input parameter set, so as to avoid for the process falling in an "infinite loop."

In the parallel mode, the GDS environment must be capable of on-line triggering of the inference engine of the IBE at strategic points in time. The simulation model would still process events using the *next-event* time advance approach, but it will have a *special* event (that occurs at "some" points in time) which call the IBE engine right before the simulation clock is advanced. Thus, allowing for an *on-line* modification of input parameters (Farimani-Toroghi and Peck, 1990)

The evolution of expert systems tools has made it possible to embed knowledge-based systems (KBES) into other software. Similarly, data and knowledge representation techniques have also advanced (e.g. object oriented concepts are readily available in many languages). These advancements make the second of the two areas of GDS more promising. In fact, these advancements should help us move forward from the efforts preceding GDS (such as that by Bengu and Haddock (1986), Haddock and O'Keefe (1990)) to fully functional GDS environments.

It is worth noting, however, that at the heart of GDS is the incorporation of analysis and optimization techniques that are glued in such a way that the IBE evaluates various alternatives and relates them to the study objectives. Tompkins and Azadivar (1995) propose the use of genetic algorithms to solve the optimization of some simulation models.

Based on our previous discussion, a GDS system is formed by three main components: 1) A knowledge-based expert system, 2) an *open* simulation language, and 3) an object oriented programming language. Each component is composed of a various modules depending on its role of each component. Figure 1 gives a schematic representation of a GDS framework.

The simulation language for such a system would be preferably an object oriented system. This allows the description of the different parts of the model by means of properties that can be easily modified. This allows tangible changes in the representation of the system by removing and adding complete objects (Farimani-Toroghi and Peck, 1990). Among existing simulation languages, SIMAN and SLAM have proven to be capable of interacting with external tools (Ford and Schroer, 1990). However, with the advent of shared platforms and enhanced communications protocols, most software developed using object oriented design and an object oriented programming language (OOPL) would be able to be linked to other packages.

The contents of the KBES include historical runs, priority rules, statistical knowledge, capability rules, and model representations. These modules can be defined as follows

- ☑ <u>Historical runs</u> enable the collection of historic information of the system. The results from the modification of parameters and the output from the simulations are transformed into information expressed as cause-and-effect rules that could be use in later consultations.

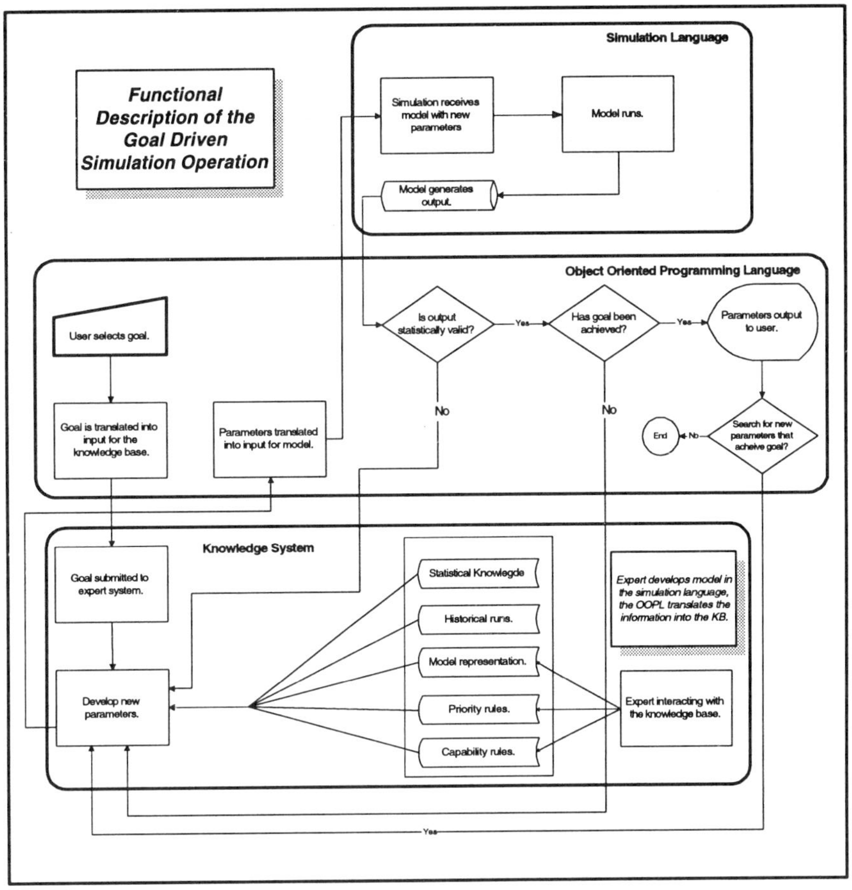

Figure 1: Proposed Framework for a GDS System

- ☑ *Priority rules* bound the search space according to a user given ranking of the input parameters. For example, a priority rule could guide the IBE to maximize the utilization of cheaper resources before increasing the capacity of the more expensive ones.
- ☑ *Statistical Knowledge* is needed to assess the closeness of outputs to the target, as well as to identify the changes required.
- ☑ *Capability rules*, or system capacity rules, define the allocation schemes of available system resources. These type of rules limits the search space to values that are realistic to the system under study.
- ☑ *Model representations* describe the model in terms of operational rules, including precedence relationships, entity flows, and any physical constraints fixed to the system. Works in this area include that by Benjamin et al. (1994) and Farimani-Toroghi and Peck (1990). The first work presents the application of an Air Force module call IDEF3 where descriptions of the model were successfully captured. The second work uses the production rule schema to represent some of the operational characteristics of the system.

Under GDS, the process is initiated by the user who seeks to satisfy a goal, bound by the restrictions previously encoded in the knowledge base. A model of the system is coded in the simulation language. It is then verified and validated. The goal requirements are passed to the knowledge base, which in turn would be accessed by the KBES to generates the model's parameters. These parameters are fed to the model via the OO user interface, and the simulation runs are triggered. Once this is obtained, runs can be executed to determine if the goals have been met. The historical component stores the results of the different runs in order to base the decision making process on actual performance of the model under different parameter settings. This removes the analyst from a great deal of repetitive calculation.

The use of a reasoning mechanism in GDS provides several advantages. The use of model specific information can describe the model using cause and effect relationships that relate the inputs to the outputs of the model. These inputs can then be manipulated to obtain a goal. The use of this approach, is what differentiates GDS from statistical methodologies or exhaustive studies of different scenarios.

4 IMPACT OF GDS ON THE SIMULATION METHODOLOGY

The simulation modeling methodology as described by Banks et al. (1996) would undergo a fundamental change if GDS were to be implemented. The study of several modified models would no longer be necessary to obtain the desired parameters. By avoiding the output analysis, and driving the simulation intelligently to the optimum scenario, there is no longer need for excessive hours of computational effort. The steps in a simulation study that would be impacted are experimental design and analysis of outputs.

4.1 The Experimental Design Stage

Under the experimental design stage, the user will not need to test the alternatives that possibly satisfy the desired goal. Instead, the user would input the goal and the system will provide a solution, or solutions, that satisfy the goal. Before any runs can be made, the interface would translate these goals into input for the KBES, which will establish what measures of performance need to be tracked to initiate the search towards the goal. This will initiate the search for a valid solution based on the information in the knowledge base. No longer will the model user be confronted with the situation in which a complete set of production runs are done just to find out during the analysis stage that a variable was left out of the data collection process.

4.2 Production Runs and Analysis

After the model is run, and given that the results of the model are statistically significant,, the outputs are analyzed and compared to the goal. If they meet or exceed the requirements of the goal, they are reported to the user. The user may then decide to have the system search for other parameters which yield the desired goal. In the case where the goal has not been achieved, the information is translated into historical knowledge for the expert system, triggering a new search for parameters. The new parameters are integrated into the simulation model for new runs to be performed.

4.3 The new role of the simulation model

The use of GDS not only will improve the cycle for studies, but will also change the notion of the model as a disposable entity. By creating a framework around the model, it becomes a dynamic entity. It will be allow modifications that are reflected in the knowledge base. This provides a tool that can continuously be monitored to reflect the performance of the real system. As changes in the parameters are varied, the system will provide feedback on expected performance. The investment in such a system will also allow the users to modify any physical representation of the model, that may be planned in the future.

5 SUMMARY

GDS are geared towards minimizing the need of human interaction during the analysis of output information; thus, reducing the risk of error. The traditional simulation methodology requires the analyst to perform what-if analysis. This trial and error approach requires the user to perform repetitive tasks such as statistical analysis of output. The focus of the modeler is shifted from the statistical output analysis to capability analysis. Capability analysis would encompass the study of all possible values for the assignment of resources within the system. In addition, users other than the modeler would have access to a system that would yield answers to "what-if" questions, without requiring from them an in-depth technical knowledge to develop a complete simulation study.

However, to realize GDS, further research is needed in the areas of classification of the types of systems to simulate, domain knowledge acquisition, and the integration of heterogeneous tools that may be rooted in heterogeneous platforms.

REFERENCES

Alsugair, A. and D. Y. Chang. 1994. A Goal Driven Approach to Discrete Event Simulation. In *Proceedings of the First Congress Held in Conjunction with A/E/C Systems*. 1: 515-522.

Akbay, K. 1996. Using Simulation Optimization to Find the Best Solution. IIE Solutions. 28(5):24-29.

Banks, J., J. S. Carson, II, and B. L. Nelson. 1996. *Discrete-Event System Simulation*. New Jersey: Prentice Hall.

Bengu, G. and J. Haddock. 1986. A Generative Simulation Optimization System. *Computers in Industrial Engineering*. 10(4): 301-313.

Centeno, M. A. and R. E. Shannon. 1995. Integrating AI and Databases in Application to Simulation. *Advances in Databases and Artificial Intelligence* 1: 311-332.

Erraguntla, M., P. C. Benjamin and R. Mayer. 1994. An Architecture of a Knowledge-Based Simulation Engine. In *Proceedings of the 1994 Winter Simulation Conference,* eds. J.D. Tew, S. Manivannan, D.A. Sadowski, and A.F. Seila, 673-680..

Farimani-Toroghi, S. and S. Peck. 1990. A Framework for Combined Simulation and Knowledge Based Systems. In *Expert Systems, Artificial Intelligence, Neural Networks, Vision, Manufacturing Modeling and Simulation, Proceedings of the Annual Pittsburgh Conference,* 903-907. SCS, San Diego, California.

Ford, D. R. and B. J. Schroer. 1987. An Expert Manufacturing Simulation System. *Simulation* 48: 193-200.

Haddock, J. and R. M. O'Keefe. 1990. Using Artificial Intelligence to Facilitate Manufacturing Systems Simulation. *Computers in Industrial* Engineering. 18: 275-283.

Kopacsi, S. and G. L. Kovacs. 1993. A Knowledge Based Simulation System for Flexible Manufacturing. In *Proceedings of the IEEE International Conference on Robotics and Automation,* 858-863. IEEE, Los Alamitos, Ca.

Lenat, D. B. 1995. Artificial Intelligence. *Scientific America.* 273 : 80-82.

Murray, K. J. and S. Sheppard. 1988. Knowledge-Based Simulation Model Specification. *Simulation* 50 : 112-119.

Romero, N. R. and L. Interrante. 1944. An Expert System Using Simulation Measures of Effectiveness. In *IEEE International Conference on Systems, Man, and Cybernetics* 3 : 2201-2206.

Schena, G. D. 1994. Objective-Driven Simulation as Tool for Plant Design- Application to Cyclone Selection with Cost-Minimization Function. *Transactions. Section C, Mineral Processing and Extractive Metallurgy* 103 : C155-C162.

Shannon, R. E., R. Mayer, and H. Adelsberger. 1985. Expert Systems and Simulation. *Simulation* 44 : 275-284.

Shannon, R. E. and S. Prakash. 1990. Goal driven Simulation Systems. In *1990 International Industrial Engineering Conference Proceedings.* 345-350.

Tibbitts, B. R. 1993. Flexible Simulation of a Complex Semiconductor Manufacturing Line Using a Rule-Based System. *IBM Journal of Research and Development* 37 : 507-521.

Tompkins, G. and F. Azadivar. 1995. Genetic Algorithms in Optimizing Simulated Systems. In *Proceedings of the 1995 Winter Simulation Conference,* eds. C. Alexopoulos, K. Kang, W.R. Lilegdon, and D. Goldsman, 757-762.

Touran, A. 1990. Integration of Simulation with Expert Systems. *Journal of Construction Management* 116 : 480-493.

AUTHOR BIOGRAPHIES

MARTHA A. CENTENO is an assistant professor in the Department of Industrial and Systems Engineering at Florida International University. She received a B.S. degree in Chemical Engineering from ITESO University (Mexico), an M.S. degree in Industrial Engineering from Louisiana State University in 1985, and a Ph.D. degree in Industrial Engineering from Texas A&M University in 1990. Her research interests are in the areas of intelligent simulation, smart data analysis, and engineering education. Dr. Centeno is a member of Alpha Pi Mi, IIE, INFORMS, SCS, ASA, and ASEE.

CARLOS GANDARILLAS is a graduate student at Florida International University, where he is pursuing a Master of Science degree in Industrial Engineering. His research interests are in the area of smart simulation. He is expected to graduate in August of 1997. He received a Bachelor of Science degree in Industrial and Systems Engineering from Florida International University in 1995. He is a member of Alpha Pi Mu.

LOUIS A. MOLINA is a graduate student at Florida International University, where he is pursuing a Master of Science degree in Industrial Engineering. His research interests are in the area of goal driven simulation and knowledge acquisition, from databases, for statistical modeling. He is expected to graduate in August of 1996. He received a Bachelor of Science degree in Industrial and Systems Engineering from Florida International University in 1994 and is a member of Tau Beta Pi and IIE.

Proceedings of the 1996 Winter Simulation Conference
ed. J. M. Charnes, D. J. Morrice, D. T. Brunner, and J. J. Swain

COMPARISON OF SIMULATION ENVIRONMENTS THROUGH ANALYTIC HIERARCHY PROCESS

S. Cem Karacal

Department of Mechanical and Industrial Engineering
Southern Illinois University
Edwardsville, IL 62026-1805

Terrence G. Beaumariage

Department of Industrial and Manufacturing Engineering
Oregon State University
Corvallis, OR 97331

Zeynep A. Karacal

Electronic Data Systems
Communications Industry
13736 Riverport Drive
St. Louis, MO 63146

ABSTRACT

This paper describes the application of analytic hierarchy process to evaluate a new methodology for simulation against conventional approaches. The study utilizes structured pair-wise comparison of analytic hierarchy process to reach a scale of preference between the two simulation modeling approaches. Both tangible and intangible aspects of modeling and simulation are combined to create a comprehensive platform for comparing simulation environments.

1 INTRODUCTION

This paper describes the application of the Analytic Hierarchy Process (AHP) to evaluate various aspects of the framework and the object-oriented methodology developed by Karacal and Mize (1996a, 1996b) against conventional simulation approaches. In simple terms, AHP is a multi-objective, multi-criteria decision methodology that utilizes structured pair wise comparisons among similar aspects of alternatives to reach a scale of preference. It is especially powerful when the problem has many aspects that are hard to quantify. In literature, most of the studies focused on comparing simulation environments or languages are based on tangible and measureable criteria such as execution speed, graphics capability, model size and complexity (Wallace, 1987). Although there were some approaches that tried to evaluate simulation systems through qualitative considerations, they were based on a set of disjoint, usually conflicting criteria. Several aspects of a simulation study such as ease of modeling and model effectiveness are difficult to measure. A comparision of environments based on a single or few number of aspects may lead to a narrow perspectived conclusions. The objective of this work is to unify tangible and intangible aspects of a simulation study through AHP to form a common platform for comparing traditional simulation approaches and the developed methodology.

2 ANALYTIC HIERARCHY PROCESS

The AHP process consists of a systematic approach based on breaking the decision problem into a hierarchy of interrelated elements. Applications of the methodology includes several diverse areas from economics to health care planning and energy policy (Wallace, 1987). A more comprehensive application of AHP for simulation environment evaluation purposes can also be seen from Beaumariage's (1990) study which compares object-oriented simulation environments against traditional environments such as SLAM and SIMAN.

Zahedi (1986) summarizes the AHP procedure in terms of four steps:

1) *Break the decision problem into a hierarchy of interrelated problems*: top level being the macro decision objective such as selecting the best alternative. The lower levels contain attributes which contribute to the quality of this decision. The next lower levels represents the increased details of these attributes. The bottom of the hierarchy contain decision alternatives or selection of choices. Figure 1 illustrates the standard format for AHP decision model.

2) *Provide the matrix data for pairwise comparision of the decision elements* : to express judgements in pairwise comparisions, the following scale of absolute values must be provided: 1, equal (weight); 3, moderate; 5, strong; 7, very strong; 9, extreme; 2, 4, 6,

8, for compromise; reciprocals are used for the inverse comparisions. The elements in the next hierarchical level are arranged in the form of a matrix and pairwise judgemental values are assigned in satisfying the decision element of the present level for which the comparision matrix is built. Similarly, elements in the next level down are subjected to pairwise comparisions for a particular decision element in previous level and values are assigned.

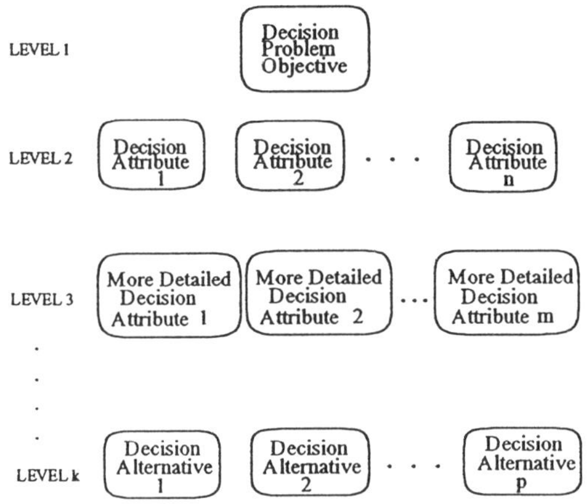

Figure 1: Standard Format of an AHP Decision Model

3) *Solve the pairwise comparision matrices for the eigen values and eigen vectors in order to estimate the relative weights of the decision elements*: The pairwise comparision values produce a ratio scale (a class of numbers whose ratios remain the same when each of them multiplied by a constant) of weights of the relative importance. AHP assumes that the evaluator does not know the actual weights, represented with vector W. Therefore the observed pairwise relative weights matrix, A, contains inconsistencies. The matrix A has rank 1.

$$A \cdot W = n \cdot W$$

where n is the eigenvalue and W is the eigen vector of A.

$$\hat{A} \cdot \hat{W} = \lambda_{max} \cdot \hat{W}$$

where $\hat{A}$ is the observed pairwise comparisions matrix, λ_{max} is the largest eigenvalue of $\hat{A}$, and $\hat{W}$ is the estimation of W. λ_{max} may be considered the estimation of n. The closer the value of computed λ_{max} is to n, the more consistent are the observed values of $\hat{A}$. As a result, two measures called Consistency Index (CI) and Consistency Ratio (CR) are defined as follow:

$$CI = \frac{\lambda_{max} - n}{n - 1} \quad \text{and} \quad CR = \frac{CI}{ACI} * 100$$

where ACI is the average index of randomly generated weights for a matrix of similar size. A CR value of 0.10 or less is considered acceptable, otherwise to resolve the inconsistencies in pairwise comparisions, the values of matrix $\hat{A}$ must be reassesed. Saaty (1988) recommends four different methods to estimate the values of W one of which is normalizing each column by dividing the elements of each column by the sum of the column and adding the elements in each resulting row and dividing this sum by the number of elements in the row. This is the process of averaging over the normalized columns.

4) *Aggregate the relative weights of the decision elements to obtain a rating for decision alternatives* : The last step of the process is the aggregation of the relative weights through the hierarchy by weighting relative values and summing the totals for each decision alternative and normalizing the results to sum to 1.

3 COMPARISION OF SIMULATION ENVIRONMENTS THROUGH AHP

The preliminary AHP model developed by the authors was discussed, critiqued, and iterated by a five member AHP study group formed at Industrial Engineering department of Oklahoma State University. Once the levels, the major aspects, and the criteria were finalized in terms of a set of nodes, the definition of linkages between the nodes is acomplished through an iterative process. Next, the resulting preference matrices were formed and weighted by the group, again in an iterative manner. The following section gives a summary of the resulting levels, major aspects, criteria and assessed weight matrices. The numbers before each aspect/criteria designate the node number.

Level 1 : Definition of the problem

1.1 - Best simulation approach : The problem on hand is the selection of the best simulation modeling methodology and the resulting model. The methodology in this context is interpreted as the whole process of conceptualizing and representing the system in terms of a simulation model using the underlying structures. The objective of the study is to reach a measure of preference between traditional simulation methodology based on existing simulation and programming languages and the new approach based on developed formalism and object oriented environment. In the new approach, the modeling is done by visualizing the system under study in terms of independent modular software constructs that represent the physical, logical, communication aspects of each system entity. These basic building blocks are designed using the sets and state-space representation scheme of the developed formalism. The formalism and the software also provide for nonprogrammed decision making capabilities through the use of knowledge-bases and artificial intelligence techniques. More detailed information on the new approach can be obtained from references Karacal and Mize (1996a and 1996b).

Level 2 : Main Aspects

2.1 - Model effectiveness : This is the model's capability of being used as a realistic decision support tool. That is to say, how closely the model expresses the real system in terms of the aspects that can be represented and the performance measures that can be obtained . In addition, the model's ability to manage change, extension, reusability, and detail level are also considered as part of this aspect. This node links to node 1.1.

2.2 - Model developer's potency and modeling effort : This aspect of the decision problem addresses the capabilities that are associated with the model developer and the effort required to build a model. The model developer's activities heavily depends on the conceptualization of the model, and the tools and facilities provided by the modeling environment. The lower level criteria are evaluated either in terms of increasing the modeler's capability or decreasing the modeling effort required. This node also links to 1.1.

2.3 - Model execution performance : This is basically the time required to experiment with the model. This aspect is considered as one of the main factors due to rapid deterioration of model execution time performance as layers of knowledge-based systems are added to the model in the developed methodology. This node links to 1.1.

2.4 - Model's degree of correspondence to the real system : This aspect is very important for the model's acceptance as a valid tool for gaining insight about the real system. Depending on the desired level of detail in the system to be represented, this aspect evaluates how accurately the real system can be expressed in the model. Similar to the other nodes of this level, this node also links to 1.1 to allow the relations defined at lower levels to factor into the final result.

Level 3 : Criteria Considered

3.1 - Formal modeling structures / modeling methodology : This criterion covers the underlying structures of the simulation paradigm and the modeling methodology dictated by those structures. In a sense, it is the science base of the simulation that gives the ability to answer questions like, how does one develop a model and why? This node links to model developer's potency and modeling effort, which is node 2.2.

3.2 - Model flexibility : This is the model's ability to express different aspects of the system as well as ease of model alteration and extension. The capability of developing models with different levels of detail without major model overhauls is also part of the flexibility criteria. This links to nodes 2.1, 2.2, and 2.4.

3.3 - Output provisions : This criterion represents the versatility of the data and the information from a simulation run. This includes the data collection facilities on physical and logical aspects of the system being modeled. This criterion has a strong influence on model effectiveness and therefore is linked to node 2.1.

3.4 - Execution speed : The computer time required to run the simulation model represents the execution speed. This criterion interacts only with node 2.3, which is the model execution performance aspect of a simulation.

3.5 - Physical, information, and control components : This is the simulation environment's ability to represent physical, information, and control components of the system under study in a modular fashion. This criterion increases the validity of the model, thereby promoting the credibility of the whole simulation study. This criterion links with all the aspects defined at level 2, except model execution performance. Therefore, it links to nodes 2.1, 2.2, and 2.4.

3.6 - Primitive modeling constructs : These are the basic building blocks of the model. The modularity and the variety of the constructs along with their expressiveness bring significant advantages to the whole simulation process. The primitive modeling constructs affect model developer's potency and model's degree of correspondence to the real system, and hence is linked to nodes 2.2 and 2.4.

3.7 - Non-programmed decision facilities : This criterion represents the simulation's ability to employ a hierarchical set of non-programmed decision support modules within the model. This is where mimicking the behavior of intelligent system entities that drive the entire operation of the real system comes into play. This criterion is considered to have an impact on all aspects defined at level 2, including model execution performance, and is linked to nodes 2.1, 2.2, 2.3, and 2.4.

Level 4 : Alternative Simulation Approaches

4.1 - Conventional simulation approach : This is the traditional approach to simulation. The system is mainly conceptualized in terms of physical components with no explicit information or control modules attached. The logic that governs the model behavior is implicitly expressed through a set of generic (not system relevant) and abstract modeling constructs. Modeling in this approach is analogous to developing a computer program in a simulation language.

4.2 - Formalism and OOP based simulation framework: This approach includes the new modeling methodology and its underlying formalism. In this approach, the system (manufacturing system) is perceived in terms of a set of interacting physical, information, and control components. These highly uniform, modular and alterable components are the basic model building blocks. Simulation modeling in this approach is the process of tailoring these default intelligent/nonintelligent constructs and defining the linkages among them to accurately represent a particular system's behavior.

Figure 2 shows the AHP hierarchical diagram. Tables 1 through 12 show the original pairwise weights of the AHP matrices agreed on by the study group.

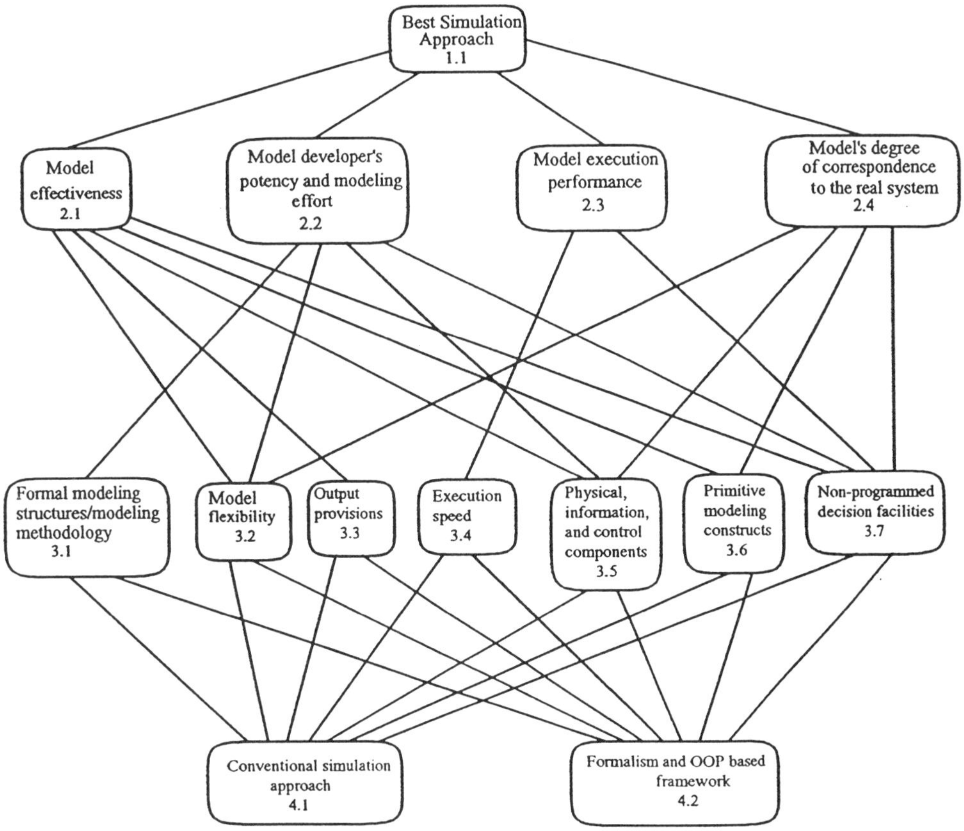

Figure 2 : AHP Hierarchical Diagram

Table 1: Node 1.1 Best Simulation Approach

Links from Lower Level:

1) Node 2.1 - Model effectiveness
2) Node 2.2 - Model developer's potency and modeling effort
3) Node 2.3 - Model execution performance
4) Node 2.4 - Model's degree of correspondence to the real system

Original weights

Row \ Col	1	2	3	4
1	1.000	5.000	8.000	3.000
2	0.200	1.000	6.000	0.250
3	0.125	0.167	1.000	0.167
4	0.333	4.000	6.000	1.000

Table 2 : Node 2.1 Model Effectiveness

Links from Lower Level:

1) Node 3.2 - Model flexibility
2) Node 3.3 - Output provisions
3) Node 3.5 - Physical, information, and control components
4) Node 3.7 - Non-programmed decision facilities

Original weights

Row \ Col	1	2	3	4
1	1.000	0.250	0.500	0.200
2	4.000	1.000	3.000	3.000
3	2.000	0.333	1.000	3.000
4	5.000	0.333	0.333	1.000

Table 3 : Node 2.2 Model Developer's Potency and Modeling Effort

Links from Lower Level:

1) Node 3.1 - Formal model structures and modeling methodology
2) Node 3.2 - Model flexibility
3) Node 3.5 - Physical, information, and control components
4) Node 3.6 - Primitive modeling constructs
5) Node 3.7 - Non-programmed decision facilities

Original weights

Row \ Col	1	2	3	4	5
1	1.000	0.333	0.333	0.333	3.000
2	3.000	1.000	4.000	2.000	3.000
3	3.000	0.250	1.000	0.333	4.000
4	3.000	0.500	3.000	1.000	4.000
5	0.333	0.333	0.250	0.250	1.000

Table 4 : Node 2.3 Model Execution Performance

Links from Lower Level:

1) Node 3.4 - Execution speed
2) Node 3.7 - Non-programmed decision facilities

Original weights

Row \ Col	1	2
1	1.000	8.000
2	0.125	1.000

Table 5 : Node 2.4 Model's Degree of Correspondence to Real System

Links from Lower Level:

1) Node 3.2 - Model flexibility
2) Node 3.5 - Physical, information, and control components
3) Node 3.6 - Primitive modeling constructs
4) Node 3.7 - Non-programmed decision facilities

Original weights

Col / Row	1	2	3	4
1	1.000	0.143	0.143	0.125
2	7.000	1.000	4.000	0.500
3	7.000	0.250	1.000	0.333
4	8.000	2.000	3.000	1.000

Table 6 : Node 3.1 Formal Modeling Structures/ Modeling Methodology

Links to Lower Level:

1) Node 4.1 - Conventional modeling approach

2) Node 4.2 - New modeling paradigm

Original weights

Col / Row	1	2
1	1.000	0.143
2	7.000	1.000

Table 7 : Node 3.2 Model Flexibility

Links from Lower Level:

1) Node 4.1 - Conventional modeling approach

2) Node 4.2 - New modeling paradigm

Original weights

Col / Row	1	2
1	1.000	0.333
2	3.000	1.000

Table 8 : Node 3 Output Provisions

Links from Lower Level:

1) Node 4.1 - Conventional modeling approach

2) Node 4.2 - New modeling paradigm

Original weights

Col / Row	1	2
1	1.000	0.333
2	3.000	1.000

Table 9 : Node 3.4 Execution Speed

Links from Lower Level:

1) Node 4.1 - Conventional modeling approach

2) Node 4.2 - New modeling paradigm

Original weights

Col / Row	1	2
1	1.000	5.000
2	0.200	1.000

Table 10 : Node 3 Physical, Information, and Control Components

Links from Lower Level:

1) Node 4.1 - Conventional modeling approach

2) Node 4.2 - New modeling paradigm

Original weights

Col Row	1	2
1	1.000	0.111
2	9.000	1.000

Table 11 : Node 3 Primitive Modeling Constructs

Links from Lower Level:

1) Node 4.1 - Conventional modeling approach

2) Node 4.2 - New modeling paradigm

Original weights

Col Row	1	2
1	1.000	0.333
2	3.000	1.000

Table 12 : Node 3 Non-programmed Decision Facilities

Links from Lower Level:

1) Node 4.1 - Conventional modeling approach

2) Node 4.2 - New modeling paradigm

Original weights

Col Row	1	2
1	1.000	0.143
2	7.000	1.000

The next step in the AHP procedure was the calculation of the relative weights of the decision elements. A set of spreadsheets are developed and are used to calculate the weights for each of the above matrices along with matrix consistencies. Then, after checking on the consistencies, and reassesing the assigned matrix values in an iterative manner, these relative weights are aggregated through a series of matrix calculations to yield a solution to the problem. Table 13 shows the resulting final weights.

Table 13 : Final Weights

Conventional simulation approaches	0.203
New simulation approach	0.797

The results of final weights obtained from AHP clearly indicate that the new simulation approach is preferrable to the conventional approach in terms of the aspects and criteria considered in the AHP study. The conclusion reached in this AHP study coincides with Beaumariage's (1990) results for object-oriented modeling approach, which were obtained using a different set of factors/criteria.

4 SUMMARY AND CONCLUSIONS

The example system modeled for validation of the formalism and the software developed shows the potential of the framework, even though the manufacturing system modeled is a simple one and the software is a prototype. This new approach allows the simulation analyst to collect and analyze certain types of data such as performance of different non-programmed control schemes and/or objects, and information processing capabilities of different entities and/or hierarchical levels. Traditional simulation modeling approaches do not allow studying various aspects of typical decision problems using a multi-criteria analysis. The formalism and the modeling methodology also dictate a uniform model. This indicates that different model developers who conceptualize a system in the same structured manner can come up with almost identical models. In contrast, the highly acclaimed and so called flexibility of the traditional approaches result in different models for the same real system if different model developers are involved in the model building process.

The final weights obtained from the AHP study is due to significant advantages provided by the new simulation approach. The formal structures provide the basic constructs for modeling information, control, and physical aspects of the system independently and simultaneously, which is a considerable improvement over the traditional approach. The interactive use of several knowledge-bases during the simulation allows more realistic and versatile representation of real systems.

REFERENCES

Beaumariage, T. G. 1990. Investigation of an Object Oriented Modeling Environment for the Generation of Simulation Models. Ph.D. Dissertation, School of IE and Management, Oklahoma State University, Stillwater OK.

Karacal, C., J. Mize. 1996a. A Formal Structure for Discrete Event Simulation, Part I : Modeling Multiple Level Systems, will appear in *IIE Transactions*.

Karacal, C., J. Mize. 1996b. A Formal Structure for Discrete Event Simulation, Part II : Object Oriented Software Implementation for Manufacturing Systems, will appear in *IIE Transactions*.

Saaty, T. L. 1988. *Decision Making : The Analytic Hierarchy Process*. Pittsburgh: RWS Publications.

Wallace, J. C. 1987. The Control and Transformation Metric : Toward the Measurement of Simulation Model Complexity. *Proceedings of the 1987 Winter Simulation Conference*, Atlanta, GA 597-603.

Zahedi, F. 1986. The Analytic Hierarchy Process - A Survey of the Method and its Applications. *Interfaces* 16:4: 96-108.

AUTHOR BIOGRAPHIES

S. CEM KARACAL is an Assistant Professor of Industrial Engineering at Southern Illinois University at Edwardsville. He received his B.S. degree in IE from Middle East Technical University, Ankara, Turkey, in 1982, and M.S. and Ph.D. degrees in Industrial Engineering from Oklahoma State University, in 1986 and 1991, respectively. He has four years of experience in academia, industry, and consulting. His primary areas of interest are object oriented modeling, AI applications in simulation, simulation optimization, and semiconductor manufacturing scheduling. He is a member of IIE, SME, Alpha Pi Mu, and Tau Beta Pi.

TERRENCE BEAUMARIAGE is a faculty at department of IME at Oregon State University. He holds BS (Rochester Inst. of Tech.), MS (Oklahoma Stat Univ.), and Ph.D. (Oklahoma State Univ.) degrees, all in Industrial Engineering. His interest are in simulation, manufacturing, quality control, and artificial intelligence. He was an NSF Graduate Fellow and is a member of IIE, NSPE, ASQC, ASEE, Alpha Pi Mu, and SCS.

ZEYNEP A. KARACAL is an Advanced Systems Engineer at Communications Industry of Electronic Data Systems (EDS). She received her B.S. degree in IE from Middle East Technical University, Ankara, Turkey, in 1982, and M.S. in Computer Science in 1988, and Ph.D. degree in Industrial Engineering in 1995, both from Oklahoma State University. She has two years of experience in consulting. Her interest areas are simulation optimization, software engineering, work force management, and database design. She is a member of Alpha Pi Mu, and Tau Beta Pi.

Proceedings of the 1996 Winter Simulation Conference
ed. J. M. Charnes, D. J. Morrice, D. T. Brunner, and J. J. Swain

ABSTRACTION MECHANISMS IN DISCRETE-EVENT INDUCTIVE MODELING

Hessam S. Sarjoughian

AI & Simulation Research Group
Electrical and Computer Engineering
University of Arizona
Tucson, AZ 85721-0104, U.S.A.
hessam@ece.arizona.edu
http://www-ais.ece.arizona.edu/~hessam

Bernard P. Zeigler

AI & Simulation Research Group
Electrical and Computer Engineering
University of Arizona
Tucson, AZ 85721-0104, U.S.A.
zeigler@ece.arizona.edu
http://www-ais.ece.arizona.edu/~zeigler

ABSTRACT

The power of abstraction lies in its ability to deal with "lack" of knowledge. In this regard, success in modeling and simulation rests on discovering useful abstractions that can support objectives of modeling. In our treatment, we refer to "data abstraction" as opposed to "structure simplification" since we consider a system's behavior rather than its structure. A system's behavior can be represented as time-varying input/output segments. Given the behavior of a causal, time-invariant system, we define some basic *abstraction mechanisms* to support inductive modeling. The basis for these abstraction mechanisms are a set of *general assumptions* which allow *consistent* abstraction of IO segments. Then, given these assumptions and non-monotonic reasoning paradigm, capable of handling them, we try to tackle the fundamental problem of insufficient knowledge in the realm of inductive modeling. In this way, by making useful abstractions, we can predict a system's unobserved behavior according to a well-defined framework of discrete-event inductive modeling.

1 INTRODUCTION

Vast amounts of observed data from various systems are continually collected with the expectation that they will assist us in understanding their underlying structure and behavior. An extensive body of research has been devoted to finding ways to predict a system's future behavior based on its previously observed behavior — Biermann and Feldman (1972), Zeigler (1976), Klir (1985), Michalski, Carbonell, and Mitchell (1986), Grossberg (1988), Cellier (1991), Omlin, Thornber, and Giles (1996) are some efforts. Here we do not discuss abstraction in deductive or abductive modeling paradigms. Hobbs and Moore (1985), Weld (1992), Kuipers (1994), and Fishwick (1995), among others, discuss abstraction in these settings.

In Sarjoughian (1995a) a Discrete-event Inductive Reasoner (DIR) has been developed based on the concepts from systems theory, Wymore (1993), and non-monotonic reasoning, McCarthy (1990), Davis (1990), Ginsberg (1993). While the former provides us simple means to formulate a well-defined structure representing a system's observed behavior, the latter supports the more powerful means necessary to reason with aforementioned structures in the absence of complete knowledge. Supposing that a causal, dynamic, time-invariant system's finite observed behavior can be partitioned into IO segments, we identify a well-defined set of assumptions which support prediction of the unobserved behavior of a system. An example would be when we have an unobserved input segment for which we would like to find its output segment. The problem is trivial if the input segment and its output segment are in the repository of observed IO segments. However, if no such input segment is available, it becomes essential to abstract certain features of the candidate input segment so that it is "equivalent" to one of the observed input segments for which we know its corresponding output segment.

To employ abstractions, Giunchiglia and Walsh (1992), and homomorphism, Wymore (1993), it becomes necessary to make assumptions (and thus non-monotonic reasoning) based on which unobserved IO segments can be predicted.

In our discussion, we do not argue the DIR's foundations nor its methodology. Moreover, we exclude the presentation of the overall architecture and the implementation of the DIR. Instead, we show the underlying mechanisms of a part which corresponds to defining abstraction mechanisms and how they would facilitate well-defined prediction of IO segments. Here, it suffices to say that the discrete-event inductive reasoner is comprised of a repository of observed IO segments, an inference engine and a logic-based truth maintenance system. Forbus and de Kleer (1993) describe and discuss thoroughly logic-based truth maintenance systems. For more discussion of the above and details of what follows refer to Sarjoughian (1995a), Sarjoughian and Zeigler (1995b).

2 ITERATIVE IO FUNCTION OBSERVATION STRUCTURE

Given input generator segments (Ω_G) and output generator segments (Ψ_G), the associated input/output generator segment set $IOspace_G$ is defined as:

$$\begin{aligned} IOspace_G = \ & \{(\omega_G, \psi_G) \mid (\omega_G, \psi_G) \in IOspace, \\ & \omega_G \in \Omega_G, \psi_G \in \Psi_G, \\ & IOspace_G \subseteq IOspace, \\ & dom(\omega_G) = dom(\psi_G)\} \end{aligned}$$

where $IOspace$ is a collection of IO trajectories (ω is an input trajectory and ψ is an output trajectory):

$$\begin{aligned} IOspace = \{(\omega, \psi) \mid (\omega, \psi) \in (X,T) \times (Y,T), \\ dom(\omega) = dom(\psi).\} \end{aligned}$$

Then, given an IOFO specification, Zeigler (1976), an *iterative IOFO specification* for a causal, time-invariant IO function observation structure can be defined as:

$$G_F = \langle T, X, S_i, Y, IOspace_G, F_G, \gamma_G \rangle$$

where

T	time base
X	input value set
Y	output value set
S_i	set of initial states
$IOspace_G$	causal, time-invariant input/output segment generator set
F_G	IO function generator set
γ_G	final state hypothesizer

with the following constraints:

$$F_G : S_i \longrightarrow \text{partial } IOspace_G,$$
$$\gamma_G : S_i \times IOspace_G \longrightarrow S_i.$$

The interpretation of the function γ_G is that it maps a given initial state and an input/output segment pair into another initial state.

To obtain an output segment ψ_G for an input segment $\omega_G \in \Omega_G$, an initial state $s_i \in S_i$, and a given data set F_G, the *output segment matching function* is defined as:

$$\eta_G : S_i \times \Omega_G \times F_G \longrightarrow \Psi_G.$$

where the elements of F_G are pairs of IO generator segments.

The interpretation of the above function is that if there exists $s_i \in S_i$ and $f : s_i \mapsto (\omega_G, \psi_G) \in F_G$ then

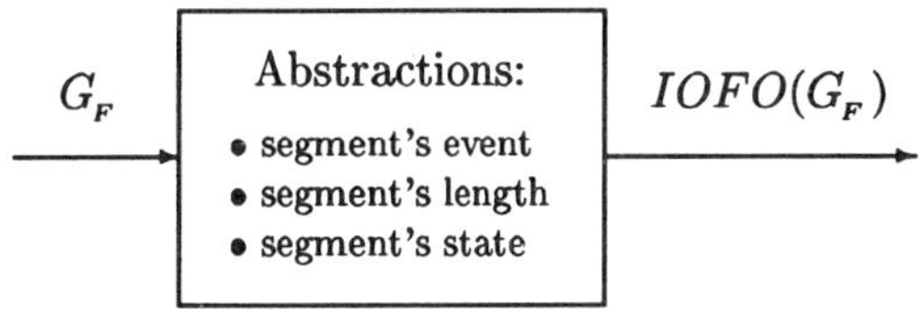

Figure 1: Role of Assumptions in Deriving $IOFO(G_F)$ From G_F.

$\eta_G : (s_i, \omega_G, f) = \psi_G$. For example, using the output segment matching function η_G and the special form of the *final state hypothesizer function* $\gamma_G : \Psi_G \rightarrow S_f$, the final state can be determined by consecutive application of γ_G followed by η_G. That is, $\gamma_G(\eta_G(s_i, \omega, f)) = \gamma_G(\psi) = s_f$ or using the general form of the final state hypothesizer, $\gamma_G : S_i \times IOspace_G^+ \rightarrow S_i$, we have the general form for combining η_G and γ_G:

$$\gamma_G(s_i, (\omega, \eta_G(s_i, \omega, f))) = \gamma_G(s_i, (\omega, \psi)) = s_i'.$$

This iterative IOFO specification can be specialized into Discrete-event IOFO specification (DEVF).

3 ASSUMPTION-BASED ITERATIVE IOFO SPECIFICATION

Once we have iterative IOFO, its *free* iterative specification is $IOFO(G_F) = \langle T, X, S_i, Y, IOspace_G^+, F_G^+ \rangle$, where F_G^+ is the set of all nonempty finite concatenations of elements of F_G. Likewise, $IOspace_G^+$ is constructed from $IOspace_G$.

However, if F_G does not represent a system's complete IO behavior, then it may be necessary to *compose* trajectories from segments, some of which have to be predicted based on some belief set which we denote as an *assumption set* (see Figure 1). To compose two segments ω_1 and ω_2, they may be concatenated as $\omega_2 \circ \omega_1$. However, unless the final state of ω_2 is the same as the initial state of the ω_1, we are forced to make $\omega_2 \circ \omega_1$ a hypothesis. That is, ignoring a mismatch between the final state of ω_2 and the initial state of the ω_1 results in a hypothesized trajectory. Given an input segment, we define the assumption set to allow abstractions on *length, event* and *state.* In this setting, the specification based on the assumption set is called *assumption-based iterative IOFO.*

For example, suppose we have two input segments $\omega_{i,j}, \omega_{k,\ell} \in \Omega_G$ (where they are contiguous w.r.t. to time) and the initial state s_i is associated with $\omega_{i,j}$. Given the extended output segment matching function,

$$\eta_G^* : S_i \times \Omega_G^+ \times F_G^+ \longrightarrow \Psi_G^+,$$

defined by:

$$\eta^*_G(s_i, \omega_{i,j} \circ \omega_{k,\ell} \circ \ldots \circ \omega_{n,q}, f_i, f_k, \ldots, f_n) = \eta_G((\ldots \gamma_G(\eta_G(\gamma_G(\eta_G(s_i, \omega_{i,j}, f_i)), \omega_{k,\ell}, f_k)), \ldots), \omega_{n,q}, f_n)$$

we have:

$$(\omega_{i,j}, \psi_{i,j}) \in F_G(s_i),$$
$$(\omega_{k,\ell}, \psi_{k,\ell}) \in F_G(s_k),$$
$$f_i : s_i \mapsto (\omega_{i,j}, \psi_{i,j}) \in F_G,$$
$$f_k : s_k \mapsto (\omega_{k,\ell}, \psi_{k,\ell}) \in F_G, \text{ and}$$
$$\eta^*_G(s_i, \omega_{i,j} \circ \omega_{k,\ell}, f_i, f_k) = \psi_{i,j} \circ \psi_{k,\ell}.$$

In the above composition (i.e., $\psi_{i,j} \circ \psi_{k,\ell}$), the final state of $(\omega_{i,j}, \psi_{i,j})$ is the same as the initial state of $(\omega_{k,\ell}, \psi_{k,\ell})$. If a composite output segment $\psi_{i,j} \circ \psi_{k,\ell}$ can be generated by applying η_G and γ_G on individual segments, then we have:

$$\eta_G(\gamma_G(\eta_G(s_i, \omega_{i,j}, f_i)), \omega_{k,\ell}, f_k) = \psi_{i,j} \circ \psi_{k,\ell}.$$

What this says is that if $\gamma_G(\eta_G(s_i, \omega_{i,j}, f_i)) = \gamma_G(\psi_{i,j}) = s'_k$ and the initial state associated with $\omega_{k,\ell}$ is s_k, then $s'_k = s_k$ must be satisfied in order for η^*_G to hold. In the next section, we discuss how to overcome the restrictiveness of *equality* that is required in composing two IO segments can be used to so that $s''_k = s_k$.

3.1 IO Segment Pair Types

The IO function generator set F_G is simply a database containing pairs, each comprised of an initial state associated with an IO segment pair. That is,

$$F_G : S_i \longrightarrow \text{partial } IOspace_G,$$
$$f = (s, g) \in F_G \quad \text{where} \quad g = (\omega, \psi).$$

In this form, no final state is assigned to any IO segment pair. Instead, the quasi-state identification function γ_G is specified in G_F to hypothesize about them. For any particular set of final states, we can suppose that every IO segment has both an initial state s_i and a final state s_f (i.e., $(s_i, s_f, (\omega, \psi))$). We now begin with a classification of all possible ways in which an input segment might be represented.

Let us denote an input segment as (s_i, ω), an output segment as (s_f, ψ), and an IO segment pair as $((s_i, \omega), (s_f, \psi))$. An input segment's representation, without considering its initial and final states for now, can be categorized into several types depending on whether events occur at one or both of its initial and final time-points; likewise, for output segments. For example, suppose we have an input segment ω and an output segment ψ with duration *dt*. We associate t_i with the initial time-point of the input/output segment pair and t_f with its final time-point. All segments are assumed to be of the discrete-event type since we confine our discussion to discrete-event systems. Thus,

- Input segment type 1:

$$\omega(t) = null\_event \quad \text{for} \quad t_i \le t \le t_f$$

- Input segment type 2:

$$\omega(t) = \begin{cases} input\_event & \text{for} \quad t = t_i \\ null\_event & \text{for} \quad t_i < t \le t_f \end{cases}$$

- Input segment type 3:

$$\omega(t) = \begin{cases} null\_event & \text{for} \quad t_i \le t < t_f \\ input\_event & \text{for} \quad t = t_f \end{cases}$$

- Input segment type 4:

$$\omega(t) = \begin{cases} input\_event & \text{for} \quad t = t_i \\ null\_event & \text{for} \quad t_i < t < t_f \\ input\_event & \text{for} \quad t = t_f \end{cases}$$

Likewise, output segments are of the same 4 types. Then, considering input and output segments together, we require that each conform to one of the following 4 types out of all 16 possible combinations of IO segment pairs.

- IO segment type 1:

$$\omega(t) = null\_event \quad \text{for} \quad t_i \le t \le t_f$$

$$\psi(t) = null\_event \quad \text{for} \quad t_i \le t \le t_f$$

- IO segment type 2:

$$\omega(t) = \begin{cases} input\_event & \text{for} \quad t = t_i \\ null\_event & \text{for} \quad t_i < t \le t_f \end{cases}$$

$$\psi(t) = null\_event \quad \text{for} \quad t_i \le t \le t_f$$

- IO segment type 3:

$$\omega(t) = null\_event \quad \text{for} \quad t_i \le t \le t_f$$

$$\psi(t) = \begin{cases} null\_event & \text{for} \quad t_i \le t < t_f \\ output\_event & \text{for} \quad t = t_f \end{cases}$$

- IO segment type 4:

$$\omega(t) = \begin{cases} input\_event & \text{for} \quad t = t_i \\ null\_event & \text{for} \quad t_i < t \leq t_f \end{cases}$$

$$\psi(t) = \begin{cases} null\_event & \text{for} \quad t_i \leq t < t_f \\ output\_event & \text{for} \quad t = t_f \end{cases}$$

The rationale for this restriction is the following: The trajectories for single-input, single-output systems as well as a class of multi-input multi-output systems, can be partitioned in several ways. Clearly, IO trajectories for a system ought to be partitioned with respect to one another. If there are no time-points at which both an input event and output event occur, then the partitioning of IO trajectories results in IO segments having the above proper types. This is not a restriction *per se* since inductive reasoning should operate on models having coarse granularity. A trajectory may be partitioned at time-points where either an input event or an output event occurs. Thus, partition points (time instants along a trajectory where an event occurs) should occur only at time-points where either an input event or an output event occurs, exclusively.

Having decided on the proper representations for input and output segments, we can determine the representation of *Complete IO segments.* That is, any IO trajectory pair (partitioned according to assumption set-I) would result in IO segment pairs represented as:

$$((s_i, (x_{val}, dt)), \quad (s_f, (y_{val}, dt)))$$

where dt is the duration of IO segments and $x_{val} \in \{nil, input\_event\}$ and $y_{val} \in \{nil, output\_event\}$.

Hence we can reformulate the earlier specification of G_F in terms of $((s_i, (x_{val}, dt)), (s_f, (y_{val}, dt)))$ as:

$$\widehat{G_F} = \langle T, X, S, Y, IOspace_G, \widehat{F_G} \rangle \qquad \text{where}$$

T	time base
X	input value set
Y	output value set
S	set of states
$IOspace_G$	time-invariant IO segment generator set
$\widehat{F_G}$	partial IO function generator set

and

$$\widehat{F_G}: \quad S_i \times PJN(IOspace_G, 1) \rightarrow S_f \times PJN(IOspace_G, 2)$$
such that $S_i \subseteq S$, and $S_f \subseteq S$ or

$$\widehat{F_G} = \{((s_i, (x_{val}, dt)), \quad (s_f, (y_{val}, dt))) \; s_i, s_f \in S, \; (x_{val}, dt) \in PJN(IOspace_G, 1), \; (y_{val}, dt) \in PJN(IOspace_G, 2)\}.$$

3.2 Input Segments Equivalence

The purpose behind the iterative IOFO specification is to support predictability. Given *candidate input segment*, the above assumption set is to be used to reason about IO segments contained in $\widehat{F_G}$ such that its output segment can be predicted. An unobserved input segment is called a candidate input segment. A *concrete* input segment (or IO segment), however, refers to one that is observed. We need to be specific about what the assumption set is, and what it entails given $\widehat{G_F}$.

Suppose we are given an input trajectory partitioned into a finite number of sequential segments and for which we would like to find its corresponding output trajectory. For example, suppose one of its input segments is $(s_i', (x_{val}', dt'))$, and there exists an input/output segment $((s_i, (x_{val}, dt)), (s_f, (y_{val}, dt))) \in \widehat{F_G}$. If the input segment $(s_i', (x_{val}', dt')$ is *equal* to the input segment $(s_i, (x_{val}, dt))$ (i.e., $x_{val} = x_{val}'$, $dt = dt'$, and $s_i = s_i'$), then it is trivial to determine its corresponding output segment.

However, the hope of composing the output trajectory is dashed if no equal input segment can be found in the database. For instance, given the three sequential input segments $(s_1', (x_1', dt_1'))$, $(s_2', (x_2', dt_2'))$, and $(s_3', (x_3', dt_3'))$, no output segments will be found for the second or third input segments if no output segment can be found for the first. That is, given the candidate input segment $(s_1', (x_1', dt_1')) \notin PJN(\widehat{F_G}, 1)$, (i.e., there exists no IO segment pair $((s_i, (x_{val}, dt)), (s_f, (y_{val}, dt))) \in \widehat{F_G}$ in the database, such that $x_{val} = x_1'$, $dt = dt_1'$, and $s_i = s_1'$), then no output trajectory can be obtained.

The notion of equality, of course, is *too strong* for inductive modeling. It becomes imperative to speak of *equivalence* instead. (Our usage of the term equivalence is different from the one used in Gill (1962) where various notions of (deductive) equivalence are defined for finite-state memory machines.) Otherwise, we have to limit our claims of prediction to trajectories that can be composed from the IO segments found in $\widehat{F_G}$ only. It is impossible to find in the database equal input segments for all imaginable new input segments. This underlies the need for defining equivalence between two input segments. We use the

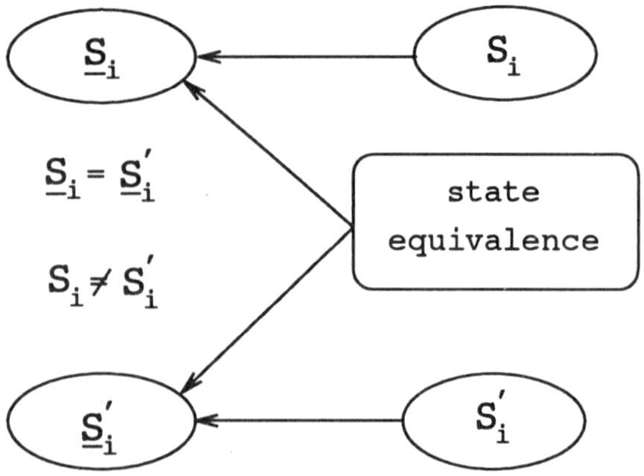

Figure 2: Abstraction of Two Concrete States Into Their Respective Abstracted States

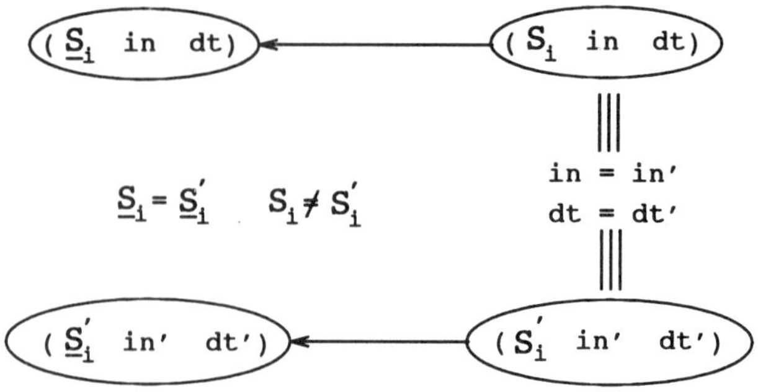

Figure 3: Abstraction of Two Input Segments Into Their Respective Abstracted Input Segments

term equivalence to indicate that, even though two input segments are not equal, we can *think* of them as being equal.

Given input segments $(s_i', \omega') \notin PJN(\widehat{F_G}, 1)$ and $(s_i, \omega) \in PJN(\widehat{F_G}, 1)$, when they can be *considered* to be equal? That is., when they can be called *equivalent*. Given the two input segments $(s_i, (x_{val}, dt))$ and $(s_i', (x_{val}', dt'))$, three primitive types of equivalences are possible. They are based on *length-equivalence*, *input-equivalence*, and *state-equivalence*. Each equivalence type ignores the inequality in one of three aspects: length (or duration), initial state, or input.

In our earlier example, we had $(s_i, (x_{val}, dt))$ and $(s_i', (x_{val}', dt'))$, where $s_i' \neq s_i$, $x_{val}' = x_{val}$, and $dt' = dt$. Consequently, we could assume state-equivalence in order to consider these two segments as being equivalent. How can this formally be accomplished? One approach is to turn the inequality $s_i' \neq s_i$ to an equality between their respective abstracted states $\underline{s}_i$ and $\underline{s}_i'$. Now, we treat the state-equivalence between s_i and s_i' as an *assumption*. Then, we can use $s_i' \neq s_i$ (fact) along with the state-equivalence assumption (belief) to construct two abstract states $\underline{s}_i$ and $\underline{s}_i'$ from the concrete states s_i and s_i' where $\underline{s}_i' = \underline{s}_i$ (refer to Figure 2). The term *fact* is restricted in the sense that its truth value is fixed and cannot be subjected to revision. The term *belief*, however, may change its truth value. Another difference between these is that, whereas a belief can be converted to a fact, the converse is not true. Every piece of data is either a fact or a belief, exclusively.

Hence the use of the state-equivalence is substantiated by treating the inequality of two states as a fact while using the state-equivalence assumption between them as a belief. Having two abstracted states $\underline{s}_i' = \underline{s}_i$, as well as two concrete input segments where $(s_i, (x_{val}, dt)) \neq (s_i', (x_{val}', dt'))$, then the abstract candidate input segment $(\underline{s}_i', (x_{val}', dt'))$ and the abstract observed input segment $(\underline{s}_i, (x_{val}, dt))$ are equal (see Figure 3). That is

$$(\underline{s}_i, (x_{val}, dt)) = (\underline{s}_i', (x_{val}', dt')).$$

Now, with well-defined semantics, we can simply use the state-equivalence knowing that in fact we are using inequality of s_i and s_i' together with the state-equivalence assumption. Thus, we may say that the two concrete input segments are equivalent.

Definition 1 *Two input segments $(s_i, (x_{val}, dt))$ and $(s_i', (x_{val}', dt'))$ are called:*

1. *length-equivalent iff*

$$x_{val} = x_{val}', \quad \underline{dt} = \underline{dt}', \quad s_i = s_i';$$

2. *input-equivalent iff*

$$\underline{x}_{val} = \underline{x}_{val}', \quad dt = dt', \quad s_i = s_i',$$

3. *state-equivalent iff*

$$x_{val} = x_{val}', \quad dt = dt', \quad \underline{s}_i = \underline{s}_i'$$

where $\underline{dt}, \underline{dt}', \underline{x}_{val}, \underline{x}_{val}', \underline{s}_i, \underline{s}_i'$ are abstractions of $dt, dt', x_{val}, x_{val}', s_i, s_i'$, respectively.

That is, despite the presence of an inequality in each of the above equivalences, the two input segments are *believed* to be equal. Hence, given two input segments, either the *length*, the *input*, or the *state* can be ignored in terms of their corresponding equivalences. Various combinations of the above equivalences comprise one form of the assumption set.

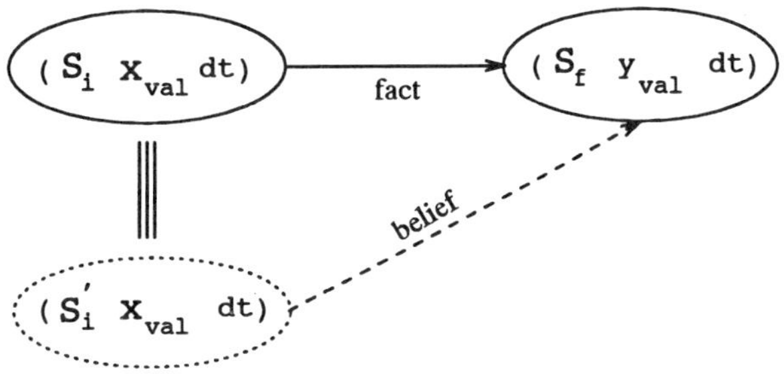

Figure 4: Constructing an Unobserved IO Segment

3.3 Predicting IO segments

In the above example, state-equivalence is supported by abstraction of states. In general, we have:

$$\begin{aligned}(s_i, (x_{val}, dt)) &\equiv (s_i', (x_{val}', dt')) \\ s_i \neq s_i', &\quad \underline{s}_i = \underline{s}_i'; \\ dt \neq dt', &\quad \underline{dt} = \underline{dt}'; \\ x_{val} \neq x_{val}', &\quad \underline{x}_{val} = \underline{x}_{val}'\end{aligned}$$

where one or more of the inequalities may be used in generating equivalence between two unequal input segments.

Although the concrete candidate input segment $(s_i', (x_{val}', dt')) \notin \widehat{F_G}$, we have seen its abstraction can be equal to the abstraction of an observed input segment. The equality between these two abstractions can be used to construct a new unobserved IO segment (cf. Figure 4) with the concrete candidate input segment and the output segment of the observed IO segment. That is, we can construct (predict)

$$((s_i', (x_{val}', dt')), \quad (s_f, (y_{val}, dt))).$$

To predict an output trajectory for an input trajectory, it is essential to partition the input trajectory into candidate input segments. The assumption on a segment's length is devised to partition an input trajectory using one of three possibilities. We have specialized length-equivalence to be one of *longest,exact,* and *all.* The first choice, *exact,* is used when any candidate input segment's length must be equal to one of the input segments from the repository. The second choice, *longest,* prefers the candidate input segments with the longest length which match input segments from the repository. The third choice, *all,* is defined to allow candidate input segments of any length.

To deal with incorrect hypothesized IO segments, the machinery of the non-monotonic reasoning provides well-defined means to retract such hypotheses once there exists support to do so.

4 AN EXAMPLE

Suppose we have observed two IO trajectories from a FIFO (first-in-first-out) discipline queue (refer to Figure 5). Then the discrete-event inductive reasoner is able to predict an output trajectory, given the candidate input trajectory. To generate the predicted output trajectory, an assumption set which is comprised of two assumptions is chosen. The first prescribes how the input trajectory is to be partitioned with respect to the IO segments in the repository. The second allows the inequality of the input events between a candidate input segment and an input segment from the repository of available IO segments. The predicted output trajectory is not completely correct. (Other possibilities exist based on alternative assumptions.) Instead of predicting output event c to occur at time 3, it is predicted to occur at time 2. This is due to length abstraction. Also, the output event is predicted as b instead of a, due to input event abstraction. When there was no matching input segment in the repository, (i.e., input event c and duration 1, and input event a and duration 7), it was inevitable that assumptions would be made when predicting output segments. Without making assumptions, it would have been impossible to predict anything given the observed IO segments one and two.

5 RELATED WORK/CONCLUSIONS

The importance of abstraction is undisputed in modeling and simulation. The inductive modeling research programs that we are aware of are based on techniques from artificial intelligence, probability theory in many different forms, neural networks, and genetic algorithms. Of these, our approach falls into the artificial intelligence arena in that it allows the introduction and manipulation of abstractions. We showed that once a well-defined structure is available, with the aid of powerful non-monotonic reasoning paradigm, abstractions can be formalized and manipulated to deal effectively with one of the fundamental issues in inductive modeling — qualification problem.

None of the other inductive modeling methodologies can support making explicit assumptions about a system's critical features such as state, inputs, and outputs. In an earlier effort, Dietterich (1984), given some observed behavior of a system, discussed how partial theories can be constructed and revised. However, this work lacks a framework. Hence, it provided no canonical representation for capturing a system's observed behavior. Also it did not establish the important role of non-monotonic reasoning in inductive

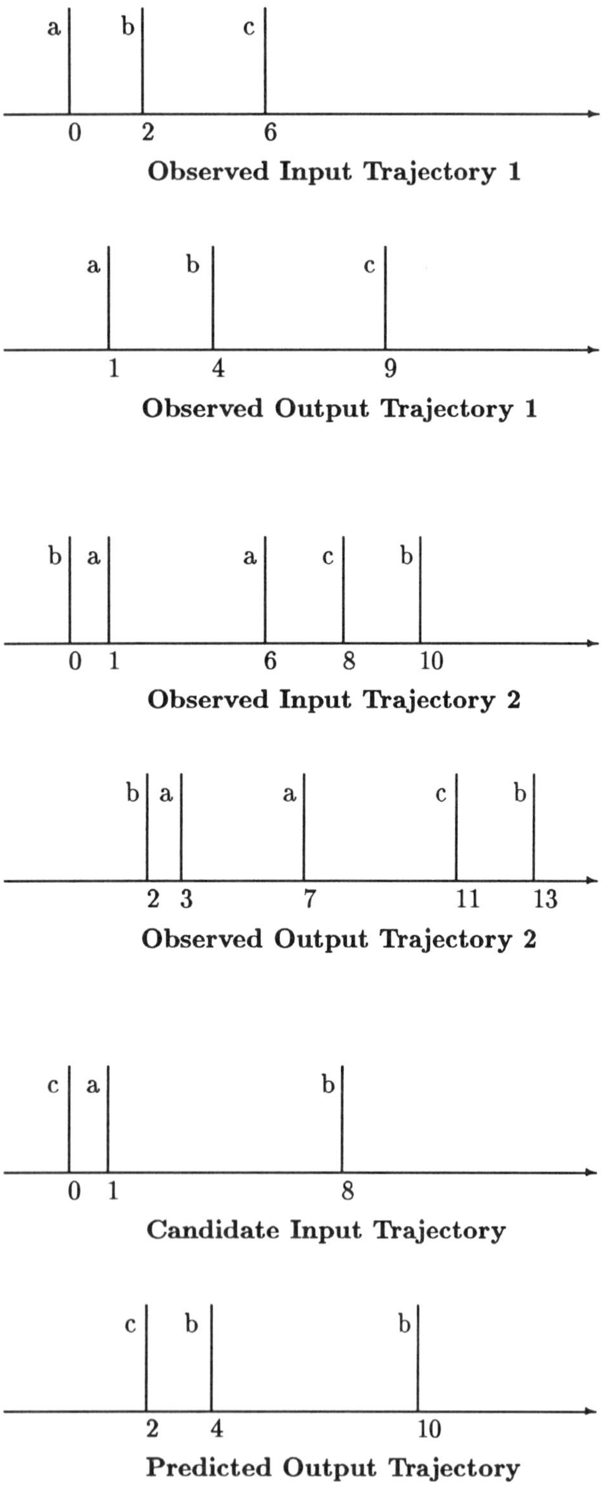

Figure 5: FIFO Discipline Example

modeling. Consequently, no abstraction mechanisms were introduced and integrated to support explicit reasoning with the observed data based on abstraction (assumptions).

Our concern is with causal/time-invariant dynamic systems. Other research efforts in inductive modeling that have focused on such systems are the works of Biermann and Feldman (1972), Dietterich (1984), Klir (1985), and Cellier (1991) although none of these takes into account the role of abstraction as we have.

In this light, our approach to inductive modeling differs fundamentally from those that are based on probability theory, neural networks, as well as other artificial approaches. We cannot engage in a discussion of AI-based inductive modeling approaches. However, it suffices to indicate that artificial intelligence approaches either represent states implicitly or contain no state variables; or just deal with static systems.

In conclusions, DEVS inductive modeling framework has been developed as part of the DEVS framework, Zeigler (1976, 1984, 1990), with the objective of laying a foundation based on concepts from systems theory and artificial intelligence. It supports abstractions representation and reasoning with them such that lack of data can be explicitly dealt with. The predictably of a system's behavior can benefit greatly by employing useful and effective abstractions. Our work discussed here suggests research in several directions: applying DIR to other types of systems and domains, studying issues related to handling large amounts of data, supporting the use of domain-specific knowledge, and formally characterizing abstraction (i.e., introduction of assumptions and their manipulation) in more depth.

REFERENCES

Biermann A. W. and J. A. Feldman, 1972. On the Synthesis of Finite-State Machines from Samples of Their Behavior, *IEEE Transactions on Computers*, Vol. 21, pp. 592–597.

Cellier F. E., 1991. *Continuous System Modeling*, Springer-Verlag, New York.

Davis E., 1990. *Representation of Commonsense Knowledge*, Morgan Kaufmann.

Dietterich T. G., 1984. Learning about Systems that Contain State Variables, *AAAI*, pp. 96–100.

Fishwick P. A., 1995. *Simulation Model Design and Execution: Building Digital Worlds*, Prentice Hall.

Forbus K. D. and J. de Kleer, 1993. *Building Problem Solvers*, MIT Press.

Genesereth M. and N. J. Nilsson, 1987. *Logical Foundations of Artificial Intelligence*, Morgan Kaufmann.

Gill A., 1962. *Introduction to the Theory of Finite-State Machines*, McGraw-Hill, New York.

Ginsberg M. L., 1993. *Essentials of Artificial Intelligence*, Morgan Kaufmann.

Giunchiglia F. and T. Walsh, 1992. A Theory of Abstraction, *Artificial Intelligence*, Vol. 57, pp. 323–389.

Grossberg S., 1988. *Neural Networks and Natural Intelligence*, MIT Press, Cambridge.

Hobbs J. R. and R. C. Moore (editors), 1985. *Formal Theories of the Commonsense World*, Ablex Publishing Corporation.

Klir G. J., 1985. *Architecture of Systems Problem Solver*, Plenum Press.

Kuipers B. J., 1994. *Qualitative Reasoning: Modeling and Simulation with Incomplete Knowledge*, MIT Press.

McCarthy J., 1990. *Formalizing Common Sense: Collected Papers of John McCarthy on Commonsense Reasoning*, edited by V. Lifschitz, Ablex Publishing Corporation.

Michalski R. S., J. G. Carbonell and T. M. Mitchell (editors), 1986. *Machine Learning: An Artificial Approach*, Vol. 2, Morgan Kaufmann.

Omlin C. W., K. K. Thornber, and C. L. Giles, 1996. Fuzzy Finite-state Automata can be Deterministically Encoded into Recurrent Neural Networks, Tech. Report UMIACS-96-12, University of Maryland.

Sarjoughian H. S., 1995a. Inductive Modeling of Discrete-event Systems: A TMS-based Non-monotonic Reasoning Approach, Ph.D. Thesis, University of Arizona, Department of Electrical and Computer Engineering.

Sarjoughian H. S. and B. P. Zeigler, 1995b. Inductive Modeling: A Framework Marrying Systems Theory and Non-monotonic Reasoning, P. Antsaklis, et. al. (Editors), *Hybrid Systems II*, LNCS 999, pp. 417-435, Springer Verlag.

Weld D., 1992. Reasoning About Model Accuracy, *Artificial Intelligence*, Vol. 56, pp. 255-300.

Wymore W. A., 1993. *Model-based Systems Engineering: An Introduction to the Mathematical Theory of Discrete Systems and to the Tricotyledon Theory of System Design*, CRC, Boca Raton.

Zeigler B. P., 1976. *Theory of Modeling and Simulation*, John Wiley and Sons.

Zeigler B. P., 1984. *Multi-Facetted Modelling and Simulation*, Academic Press.

Zeigler B. P., 1990. *Object-Oriented Simulation with Hierarchical, Modular Models: Intelligent Agents and Endomorphic Systems*, Academic Press.

AUTHOR BIOGRAPHIES

HESSAM S. SARJOUGHIAN earned his M.S. and Ph.D. degrees in Electrical and Computer Engineering from the University of Arizona in 1989 and 1995. His interests are Modeling and Simulation Methodologies, Artificial Intelligence, Software Engineering, and Intelligent Control with applications in engineered/natural systems. Presently, he is working on a Computer-Aided Business Engineering project funded by the Armstrong Research Laboratory, US Air Force. He has worked at Allied-Signal Aerospace company and International Business Machine.

BERNARD P. ZEIGLER is a professor of Electrical and Computer Engineering at the University of Arizona, Tucson. He has published over two hundred journal and conference articles on modeling and simulation, knowledge-based systems, and high-autonomy systems. His books include *Theory of Modelling and Simulation* (Wiley, 1976), *Multifaceted Modelling of Discrete Event Simulation* (Academic Press, 1984), and *Object-Oriented Simulation with Hierarchical, Modular Models: Intelligent Agents and Endomorphic Systems* (Academic Press, 1990). He has a new book which will appear in 1996 to be published by Springer-Verlag, NY: "Objects and Systems". Zeigler's research has been supported by federal agencies including NSF, NASA, USAF, and the US Army, as well as industrial sponsors including Siemens, McDonnell Douglas, and Motorola. He was elected as Fellow of the IEEE for his innovative work in discrete event modelling theory. In 1995, Zeigler served on a National Research Council committee to suggest directions for information technology in the 21st Century US Army and is currently a member of an NRC committee given a similar task by the US Navy, focusing on modelling and simulation. Zeigler is currently editor-in-chief of the Transactions of The Society for Computer Simulation. He received his B. Eng. Physics from McGill University, 1962, M.S.E.E. from MIT, 1964, and Ph.D. from the University of Michigan in 1968.

Proceedings of the 1996 Winter Simulation Conference
ed. J. M. Charnes, D. J. Morrice, D. T. Brunner, and J. J. Swain

PARAMETER IDENTIFICATION METHODS FOR METAMODELING SIMULATIONS

Don Caughlin

Mission Research Corporation
Colorado Springs, Colorado, 80903
donc@mrccos.com

ABSTRACT

A metamodel is a mathematical approximation of the system relationships defined by a high fidelity model or simulation. This paper presents methods that support new procedures that expanded the set of available metamodel representations beyond the traditional least squares formulation and added the capability to use dynamical metamodels. These methods compliment a new taxonomy of metamodel structures and procedures that separated the metamodeling process into a set of sequential decisions based on *a priori* information. This work was supported in part by The USAF Rome Laboratory Contract F30602-94-C-0110.

1 INTRODUCTION

In Caughlin (1994a) we introduced a framework for the application of System Identification techniques to develop suitable metamodels for tactical combat simulations used by the Department of Defense. We filled in the framework with concrete definitions and identified specific issues associated with the representation of dynamical systems.

Caughlin (1996) presented procedures based on this framework that allowed the separation of the metamodeling process into a set of sequential decisions based on *a priori* information. This paper presents specific methods that are consistent with the new framework and procedures to support the generation of metamodels.

The paper is organized as follows: Section 2 introduces metamodels; Section 3 covers the parameter identification methods; and Section 4 summarizes the paper.

2 METAMODELS

A model is a structure that can be used for understanding the behavior of a system Vemuri (1978). The model can be a physical structure such as a wind tunnel model used to determine the aerodynamics of an aircraft, or it could be a conceptual model represented by interactions, a system of equations, or a simulation.

A metamodel is a mathematical approximation of the system relationships defined by another, more detailed model (in our case – a tactical simulation).

There are two general metamodeling techniques: these are the "Direct" and "Inverse" methods. Direct metamodels are developed by applying basic principles to generate a more abstract (approximate) version of the original model. Inverse modeling begins with the input-output data generated by the high fidelity model or simulation and develops the metamodel from the data.

We presented new procedures in Caughlin (1996) that tailor the "Inverse Problem" to generate metamodels of simulations. We separated a complex procedure into two general steps. The first part of the process defined the problem. The second step was the metamodeling process. This step determined the metamodel set and generated the metamodel. We now discuss techniques for identifying the parameters of the selected representation.

3 PARAMETER ID METHODS

There are many taxonomies used to categorize identification methods. Methods can be referred to as off-line or on-line. Also, they can be classified as either open-loop or closed-loop methods. Further classification can be made as nonparametric, frequency domain, and as parameter identification methods.

Parameter identification methods are used when the candidate model is defined by a set of parameters.

The combination of model, error criterion, and

numerical method has lead to an overwhelming myriad of "identification methods." Most of the above techniques, however, can be classified by two elements of the identification method: the form of the identifier and the criterion of fit.

The form of the identifier defines the "experimental setup" or the manner in which the estimates are generated and compared. The criterion of fit establishes both the cost function and the method of its minimization. A summary discussion of these elements is included in Caughlin (1996). Details are found in Caughlin (1995).

If we categorize the identification method by the form of the identifier and the criterion of fit, we can reduce the many identification methods to four approaches: Prediction Error and Correlation, Maximum Likelihood, Optimization, and Approximation Techniques. We will now present some of the techniques that result.

3.1 Prediction Error and Correlation Approaches

Let the prediction error be given by $\varepsilon(t,\theta) = y(t) - \hat{y}(t|\theta)$ with $y(t)$ the output of the simulation and $\hat{y}(t|\theta)$ the output of the metamodel (θ is the parameter vector). A "good" model will have small prediction errors. There are two general approaches to define a measure of ε. The first is to define a norm that measures the size of ε and minimize that norm. This leads to the prediction error method (PEM). Another measure of ε is to require that ε be uncorrelated with past data. This is the correlation approach which contains the instrumental-variable (IV) method which we discuss in Section 3.1.4.

3.1.1 General Description of The Prediction Error Method

Filter the prediction sequence $\varepsilon(t,\theta)$ using a stable linear filter $L(q)$:

$$\varepsilon_F(t,\theta) = L(q)\varepsilon(t,\theta)$$

where q^{-1} is the backward shift operator defined as $q^{-1}u(t) = u(t-1)$.

This filtering acts like frequency weighting and can remove or enhance selected properties of the model. Then, using either a fixed or weighted (possibly time varying) norm:

$$V_N(\theta, D) = \frac{1}{N}\sum_{t=1}^{N} l\left(\varepsilon_F(t,\theta),\theta,t\right)$$

define the estimate $\hat{\theta}_N$ by the minimization:

$$\hat{\theta}_N = \hat{\theta}_N(D) = arg\min_{\theta\in D}\{V(\theta, D)\} \qquad (1)$$

where D is the set allowed by the model.

In general, PEM is a technique that approximates (smoothes) the empirical transfer function estimate to the model transfer function with a weighted norm corresponding to the model signal-to-noise at the frequency in question.

3.1.2 Specific PEM Methods

While "equation 1" can be solved numerically in the general case, specific methods are obtained as special cases with special selections of the filter $L(q)$ and the scalar valued norm function $l(\cdot)$.

Least Squares. If the predictor is linear, the prediction error becomes $\varepsilon(t,\theta) = y(t) - \phi^T(t)\theta$ where $\phi^T(t)$ is the vector of regressors that depends on the selected model structure. Also if $L(q) = 1$ and $l(\varepsilon) = \frac{1}{2}\varepsilon$, then the norm becomes:

$$V_N(\theta, D) = \frac{1}{N}\sum_{t=1}^{N}\frac{1}{2}\left[y(t) - \phi^T(t)\theta\right]^2$$

This is the least squares criterion for linear regression. The performance measure $J = \varepsilon^T\varepsilon$, was based on the view that all errors are equally important. Weighted least squares weights the errors and is based on the criterion $J = \varepsilon^T W\varepsilon$. Other versions of the least squares criterion are the Best Linear Unbiased Estimator where the weight is equal to the inverse of the measurement noise.

If the variance of the parameters is known (or assumed), then we can further improve on the Best Linear Unbiased (Gauss-Markov) Estimator. This improvement is called the minimum variance estimator and includes the variance of the parameters in the normal equations.

Ridge Regression. The aim of another modification of ordinary least squares - ridge regression - is the reduction of the mean square error (Press, 1986). This is accomplished by the addition of a symmetric matrix K to the regressor to improve the numerical conditioning of the estimator.

Chi-Square. In Chi-Square fitting, we assume that each data point y_i has a measurement error that is independently random and distributed as a normal distribution around the true model. Suppose that the standard deviation is the same for all points; then the probability of the data set is the product of probabilities of each point:

$$P = \Pi\{\exp\left[-\frac{1}{2}\left(\frac{y_i - y(x_i)}{\sigma}\right)^2\right]\Delta y\}$$

Maximizing this is equivalent to maximizing its logarithm, or minimizing the negative of its logarithm:

$$\left[\sum \frac{[y_i - y(x_i)]^2}{2\sigma^2}\right] - N \log \Delta y$$

Since N,σ, and Δy are all constants, minimizing this equation is equivalent to minimizing:

$$\sum_{i=1}^{N} [y_i - y(x_i; \theta_1 \dots \theta_M)]^2$$

If each data point has its own standard deviation, however, the probability of the data set is modified by considering σ_i in place of σ (see Press, 1986 for details).

Eigenstructure Realization Algorithm. The Eigenstructure Realization Algorithm (ERA) is included under the PEM methods because this algorithm uses the least squares approach to directly identify the Markov parameters of a steady state Kalman filter.

Consider a discrete, time-invariant multivariable linear system:

$$\begin{aligned} x_{t_{i+1}} &= A(\theta)x(t_i) + B(\theta)u(t_i) + M(\theta)w_d(t_i) \\ y(t_i) &= C(\theta)x(t_i) + D(\theta)u(t_i) + \nu(t_i) \end{aligned} \quad (2)$$

An observer for the above system can be developed that will be as stable as desired and the resulting Markov parameters will be the Markov parameters of the observer (Juang, 1993). The system Markov parameters can be extracted from the observer parameters. The major assumption is that of ergoticity.

Choose p such that $mp > n$ (where n is the number of states and m is the number of outputs) and, beginning at the p+1 measurement, let:

$$y = [y(p+1)\ y(p+2)\ y(p+3) \dots y(k-1)]$$

From the definition of the Kalman Filter we have:

$$\overline{Y} = \left[\, D\ C\overline{B}\ C\overline{A}\overline{B} \dots C\overline{A}^{k-1}\overline{B} \,\right]$$

with

$$\begin{aligned} \overline{A} &= A + MC \\ \overline{B} &= [B + MD, -M] \end{aligned}$$

and

$$U = \left|\begin{array}{cccc} u(p+1) & u(p+2) & \dots & u(k-1) \\ \begin{bmatrix} u(p) \\ y(p) \end{bmatrix} & \vdots & \dots & \begin{bmatrix} u(k-2) \\ y(k-2) \end{bmatrix} \\ \vdots & \begin{bmatrix} u(p) \\ y(p) \end{bmatrix} & \dots & \begin{bmatrix} u(k-3) \\ y(k-3) \end{bmatrix} \\ \vdots & \ddots & \vdots & \vdots \\ \begin{bmatrix} u(0) \\ y(0) \end{bmatrix} & \begin{bmatrix} u(1) \\ y(1) \end{bmatrix} & \dots & \begin{bmatrix} u(k-p-2) \\ y(k-p-2) \end{bmatrix} \end{array}\right|$$

When $C\overline{A}^k\overline{B} \approx 0$ for $k > p$, the system $y = \overline{Y}\,U$ can be solved for $\overline{Y}$ using a weighted least squares. Once the observer Markov parameters are determined, the system parameters must be extracted. After extracting the system Markov parameters from the observer, we can recover the state space model by the ERA. Define the following $r_1 \times s$ block data matrix:

$$H(\tau) = \begin{bmatrix} Y_\tau & Y_{\tau+1} & \dots & Y_{\tau+s-1} \\ Y_{\tau+1} & Y_{\tau+2} & \dots & Y_{\tau+s} \\ Y_{\tau+2} & Y_{\tau+3} & \dots & Y_{\tau+s+1} \\ \vdots & \vdots & \ddots & \vdots \\ Y_{\tau+r_1-1} & Y_{\tau+r_1} & \dots & Y_{\tau+r_1+s-2} \end{bmatrix}$$

The order of the system is determined by the singular value decomposition of $H(0)$:

$$H(0) = U\Sigma V^T = U_1 S_1 V_1^T$$

where Σ are all of the singular values. S_1 is an $n \times n$ diagonal matrix of positive singular values that are retained and n will become the order of the system:

$$\begin{aligned} A &= S_1^{-1/2} U_1^T H(1) V_1 S_1^{-1/2} \\ B &= S_1^{-1/2} V_1 E_m \\ C &= E_1^T U_1 S_1^{-1/2} \end{aligned}$$

where $E_r^T = \left[I_{r\times r}\ 0_{r\times(r_1-m)m}\right]$ and $E_m^T = \left[I_{m\times m}\ 0_{m\times(r_1-m)m}\right]$. The observer gain can be extracted in a similar fashion.

3.1.3 Correlation Approaches

Ideally the prediction error $\varepsilon(N, \hat{\theta})$ for a "good" model should be independent of past data Z^{N-1}. If $\varepsilon(N, \hat{\theta})$ is correlated with past data, then there is more information available in the data. A true test of the correlation of $\varepsilon(N, \hat{\theta})$ and Z^{N-1} requires testing every nonlinear transformation of $\varepsilon(N, \hat{\theta})$ with all possible functions of Z^{N-1}. This is not feasible.

We can, however, select a finite dimensional vector sequence $\{\zeta(t)\}$ derived from Z^{N-1} and force a certain transformation of $\varepsilon(N, \hat{\theta})$ to be uncorrelated with this sequence (Ljung, 1987). In general we can accomplish this by filtering the prediction errors:

$$\varepsilon_F(N, \hat{\theta}) = L(q)\varepsilon(N, \hat{\theta})$$

choosing a sequence of correlation vectors:

$$\{\zeta(t, \hat{\theta})\} = \{\zeta(t)(t, Z^{N-1}, \hat{\theta})\}$$

and a function: $\alpha\left(\varepsilon_F(N, \hat{\theta})\right)$ for computing:

$$f_N(\hat{\theta}, Z^{N-1}) = \frac{1}{N}\sum_{t=1}^{N} \zeta(t, \hat{\theta})\alpha\left(\varepsilon_F(N, \hat{\theta})\right)$$

and then finding $\hat{\theta}_N$ such that $f_N(\hat{\theta}, Z^{N-1}) = 0$.

3.1.4 Instrumental-Variable (IV) Method

If we define $\varepsilon(N,\hat{\theta})$ above to be $\varepsilon(N,\hat{\theta}) = \left[y(t) - \phi^T(t)\hat{\theta}\right]$ we can expand the sequence of the correlation vectors to include model dependent parameters by:

$$\{\zeta(t,\hat{\theta})\} = K_u(q,\hat{\theta})u(t)$$

where $K_u(q,\hat{\theta})$ is a $d \times m$ matrix filter and $L(q)$ is of dimension $p \times p$. With $dim\zeta(t) = dim\hat{\theta} = d \times p$, we have the instrumental-variable (IV) method:

$$\hat{\theta}_{IV} = \left[\zeta(t,\hat{\theta})^T X\right]^{-1} \zeta(t,\hat{\theta})^T y$$

If we allow $dim\zeta(t) > d$ and a minimum norm solution for $f_N(\hat{\theta}, Z^N)$, we have the extended IV method.

3.2 Maximum Likelihood Approaches

If we consider independent, identically distributed measurements, and if an efficient estimate (unbiased estimate with finite covariance such that no other unbiased estimate has a lower covariance) exists, it can always be found through maximum likelihood approaches. Although the maximum likelihood estimate will be biased for small samples, it will provide the unique minimum variance estimate attaining the Cramr-Rao lower bound if this is possible (Maybeck, 1982).

The objective is to provide a parameter estimator that does not require complete *a priori* parameter statistics yet still allows the inclusion of *a priori* knowledge. Unlike the best linear unbiased estimate provided by appropriately weighted least squares, this method propagates the probabilistic information in time and directly allows the inclusion of known statistical information.

The key to the identification algorithm will be the residuals of the state estimator, and the most significant drawback of the maximum likelihood approaches is the lack of theoretical knowledge on the behavior of the estimates for small sample sizes.

The following discussions are limited to linear-time invariant (discrete time) systems. Nonlinear effects can be included by appropriately modifying the prediction equations in either of two ways. First, nonlinear system effects can be directly included in the propagation of the state. Second, nonlinear measurements (with linear propagation) can be handled with an extended Kalman filter model.

Beginning with a linear time-invariant discrete state space model (see equation 2), there are a number of conditional probability destiny functions that could be used for the likelihood function. Variations include fixed length versus growing length functions, specification of *a priori* statistics, use of the initial conditions, and the sensitivity of the estimate on the identified parameters. The most appropriate density function is:

$$\begin{aligned} f_{x(t_i),Z(t_i)|\theta} &= f_{x(t_i)|Z(t_i),\theta} f_{Z(t_i)|\theta} \\ &= f_{x(t_i)|Z(t_i),\theta} \Pi_{j=1}^{i} f_{x(t_i)|Z(t_i),\theta} \end{aligned}$$

Minimization of the likelihood function with this density results in the state predicted by the Kalman-Bucy filter, but there is no closed form solution to compute the partial derivatives.

3.2.1 Full Scale Estimator

A full scale estimator can be derived that minimizes the likelihood function in an iterative process. This estimator uses the last N observations to identify v uncertain parameters in the system and input matrices A and B. (Note: Uncertainty in these parameters could not be separated from uncertainties in C and D. Consequently, the assumption is that C and D are known and the uncertainty is A and B.)

The iterative estimator for minimization of the likelihood equation:

$$\left.\frac{\partial L\left[\hat{\theta}, Z^N\right]}{\partial \hat{\theta}}\right|_{\hat{\theta}(t_i)=\theta_*(t_i)}$$

using the method of "steepest descent" is:

$$\hat{\theta}(t_i) = \hat{\theta}(t_i) - \left[\frac{\partial^2 L\left[\hat{\theta}, Z^N\right]}{\partial \hat{\theta}^2}\right]^{-1} \left[\frac{\partial L\left[\hat{\theta}, Z^N\right]}{\partial \hat{\theta}}\right]$$

To use this algorithm, the Hessian (second derivative matrix) must be of full rank. Using a technique called "scoring," we can approximate the Hessian with the conditional information matrix. However, considering the propagation of the values in time, incorporation of measurements, and the summation over the last N residuals, the implementation of the above equations is quite complex. Even with the approximations, the full scale estimator requires a large number of calculations (see Maybeck, 1982).

3.2.2 Modified Maximum Likelihood (MMLE)

In the modified maximum likelihood formulation, A, B, C, D, and M are estimated and used with the error covariance, P, to determine the Kalman gain, K, from an approximation based on the Ricatti equation. To provide a parameter estimator, we consider the measurement equation. Since we have assumed a Gaussian error model, the Conditional Probability Density Function (CPDF) for the measurement becomes:

$$P(z_i|z_{i-1},\theta) = \frac{1}{[(2\pi)^m det P]^{1/2}} \exp\left\{-\frac{1}{2}\tilde{z}_i^T(P)^{-1}\tilde{z}_i\right\}$$

where $P = E\{\tilde{z}\tilde{z}^T\}$ with dimension $m \times m$ and $\tilde{z} = z_i - \hat{z}$ is the innovations process (residuals) computed by the Kalman filter (where all of the matrices could be functions of θ).

Assuming a constant innovations covariance, use of a steady state filter results in a constant filter gain. This allows the CPDF to be written as:

$$P(z|\theta) = \Pi_{i=1}^{N} \frac{1}{[(2\pi)^m det P]^{1/2}} \exp\left\{-\frac{1}{2}\tilde{z}_i^T(P)^{-1}\tilde{z}_i\right\}$$

There are two approaches to the solution depending on whether *a priori* information is used.

Maximum Likelihood (ML) Estimation. Given the above CPDF, the ML LLF becomes:

$$\begin{aligned} LLF(\hat{\theta}) &= \frac{1}{2}\sum_{i=1}^{N}\{\tilde{z}_i^T(P)^{-1}\tilde{z}_i\} \\ &+\frac{N}{2}\log\det(P) + \frac{Nm}{2}\log 2\pi \end{aligned}$$

A necessary condition at the minimum is that $P = E\{\tilde{z}\tilde{z}^T\}$ must equal the sample innovations covariance (Goodwin, 1977). Therefore, since P has dimension $m \times m$, the first term in the LLF becomes $Nm/2$, and the minimization is reduced to a minimization of the determinant of the sample innovations covariance matrix.

When P is known, the LLF can be minimized by minimizing the following cost function:

$$J(\hat{\theta}) = \frac{1}{2}\sum_{i=1}^{N}\{\tilde{z}_i^T(P)^{-1}\tilde{z}_i\}$$

This minimization is usually carried out using a Gauss-Newton method using the first and second gradients of the cost function.

Maximum A Posteriori (MAP) Estimation. In the MAP estimator, we continue to require that $\hat{P} = \frac{1}{N}\sum_{i=1}^{N}\tilde{z}\tilde{z}^T$ but add the term $-\log P(\hat{\theta})$.

Assuming that θ is normally distributed with a covariance Σ:

$$\begin{aligned} -\log P(\hat{\theta}) &= \frac{1}{2}\left(\hat{\theta}-\theta\right)^T \Sigma^{-1}\left(\hat{\theta}-\theta\right) + \\ &\quad \frac{1}{2}\log\left((2\pi)^m \det\Sigma\right) \end{aligned}$$

the LLF_{MAP} becomes:

$$\begin{aligned} LLF_{MAP}(\hat{\theta}) &= \frac{1}{2}\sum_{i=1}^{N}\left\{\tilde{z}_i^T(\hat{P})^{-1}\tilde{z}_i\right\} \\ &+\frac{1}{2}(\hat{\theta}-\theta)\Sigma^{-1}(\hat{\theta}-\theta) \end{aligned}$$

which adds a quadratic term that biases the estimates toward *a priori* values.

3.3 Optimization

Often we are unable to formulate the problem such that a suitable prediction equation is available. Therefore we must resort to either a "nonlinear state space model" or a "simulation model." In these situations, where we are unable or unwilling to consider a linearized or perturbation approach, the best we can do is take the output of the model, incorporate it into a "cost" function, and adjust the model parameters to optimize (minimize) that function.

There are several "standard" numerical procedures that are used to search for the minimum of a function. These are the iterative optimization methods: successive approximation, Newton's method, the Gauss-Newton algorithm, to name a few.

In addition, there are several programs that are specifically designed to perform parameter estimation.

pEst. A minimum mean square error parameter estimator, pEst is an interactive program for the parameter estimation of nonlinear dynamic systems (Maine and Iliff, 1981). This program solves a vector set of time-varying, finite-dimensional, ordinary differential equations that are separated into a continuous-time state equation and a discrete-time measurement equation:

$$\begin{aligned} \dot{x} &= f\left[x(t), u(t), \theta\right] \\ z(t_i) &= g\left[x(t_i), u(t_i), \theta\right] \end{aligned}$$

pEst uses three separate minimization algorithms (steepest descent, modified Newton-Raphson, and

Davidod-Fletcher-Power) to minimize the following cost function:

$$J(\hat{\theta}) = \frac{1}{2n_N n_z}\sum_{i=1}^{n_t}\left[z(t_i) - \hat{z}(t_i)\right]^T W \left[z(t_i) - \hat{z}(t_i)\right]$$

where n_N equals the number of data points, and n_z is the number of response variables.

Simulated Annealing. Using statistical mechanical theories, an optimization technique called "simulated annealing" provides a new option to directly process nonlinear, discontinuous, stochastic functions (Ingber, 1993). Given data and a cost function, it will globally optimize that function by emulating the physical annealing process to arrive at a global minimum. See Ingber (1990) and Caughlin (1994b) for a description on how to use Adaptive Simulated Annealing.

3.4 Approximation Techniques for Identification

3.4.1 Stochastic Approximation

Stochastic approximation may be regarded as the application of gradient methods to stochastic problems. It is a scheme for successive approximation of a sought quantity when the observations involve random errors due to the stochastic nature of the problem. The main advantage is the simplicity of the implementation and the fact that prior knowledge of the noise statistics are not necessary.

Stochastic approximation can be applied to any problem which can be formulated as a regression in which repeated observations are made. This approach is an exact analog of the deterministic gradient procedure.

3.4.2 Spline Approximation

Polynomials are excellent approximating functions when a smooth function is to be approximated locally. Any such smooth piecewise polynomial function is called a spline, and they are commonly used for fitting data.

The typical use for the spline approximation is to construct a piecewise polynomial to fit data. An exact fit involves interpolation; an approximate fit uses least squares (minimum mean square error) approximation. To explain the structure and advantages of the spline, consider a truncated Taylor series (expanded about x_0 where D^i is the ith derivative):

$$\sum_{i=0}^{n}\frac{(x-x_0)^i}{i!}D^i f(x_0)$$

This polynomial should provide a satisfactory approximation for $f(x)$ if the function is sufficiently smooth and x is sufficiently close to x_0. But, if the function must be approximated over a larger interval, the degree of the polynomial may have to be unacceptably large.

The alternative to a higher order polynomial is to subdivide the interval into sufficiently small intervals such that, on each interval, a polynomial with a relatively low degree can provide an adequate approximation.

The construction of a series of splines over an interval is a stable and straightforward mathematical procedure. At the breakpoints, derivatives are continuous. At the end points, two conditions are possible. In the "natural" cubic spline, the second derivative is zero. In the "not-a-knot" end condition, the jump in the third derivative is zero.

Once developed, the spline can be evaluated, integrated, differentiated, augmented, or cut.

3.4.3 Canonical Variate Analysis

Another approximation technique is canonical variate analysis. The canonical variate method is a prediction error approximation technique that optimally predicts future responses based on a reduced order state space system (Larimore, 1989).

In the statistical literature, the canonical variate problem is one of maximizing the correlation between two sets of variables. Here we will use the technique to chose nonlinear combinations of past data to predict the future data by considering the fact that the conditional expectation is an optimal projection in Hilbert space. We optimally select k linear combinations of the past data for prediction of the future.

Observations coming from the behavior we desire to model are separated into the past p(t) of a vector process and the future f(t) of another vector process. They are assumed to be jointly stationary:

$$\begin{aligned} p^T &= (y^T(t), y^T(t-1), \ldots, u^T(t), \ldots)^T \\ f^T &= (y^T(t+1), y^T(t+2), \ldots, y^T(t-l))^T \end{aligned}$$

where the vector process $p(t)$ can include both inputs and outputs.

The optimal k^{th} order linear predictor $\hat{f}(t)$ of the past is measured by the prediction error:

$$E\left\{\| f - \hat{f} \|^2_{\Lambda^{-1}}\right\} \equiv \left\{(f - \hat{f})^T \Lambda^{-1} (f - \hat{f})\right\}$$

where Λ is arbitrary positive semidefinite, so that Λ^{-1} is a quadratic weighting matrix that is possibly singular. The CVA problem is to determine

$c(t) = J_k p(t)$ and $d(t) = L_k f(t)$ (a function of reduced order memory) such that the prediction error is minimized.

The connection between CVA and metamodeling is not direct and much of the literature is very confusing or misleading. First, recall that the metamodel is a reduced order model that is the result of an optimal projection of the higher order model onto a subspace of reduced dimensions. It can be shown that projection operators on a Hilbert Space of nonlinear functions can be expressed as a conditional expectation (Larimore, 1989). It can also be shown that eigenvectors of this conditional expectation have a common eigenvalue which is equal to the squared maximal correlation. If a process has a rational power spectrum (i.e., is a finite order Markov process) then there are a finite number of nonzero canonical correlations between the past and future outputs (Larimore and Baillieul, 1990).

The solution to the canonical variate problem is expressed by putting the covariance structure of the past and future data in a canonical form such that in this new basis the norm of the weighted prediction error is the sum of squares. This is equivalent to finding J and L such that:

$$\begin{aligned} J\Sigma_{pp}J^T &= I_m \\ L\Lambda L^T &= I_n \end{aligned}$$

$$J\Sigma_{pf}L^T = Diag\{\gamma_1 \geq \gamma_2 \geq, \ldots, \geq \gamma_q \geq 0, \ldots, 0\}$$

where $\Sigma_{pp}, \Sigma_{ff}, and \Sigma_{pf}$ are the covariance matrices of past, future, and cross covariance of the past and future data defined by:

$$\Sigma = \begin{pmatrix} \Sigma_{pp}\Sigma_{pf} \\ \\ \Sigma_{fp}\Sigma_{ff} \end{pmatrix}$$

with $Diag\{\gamma_1 \geq \gamma_2 \geq, \ldots, \geq \gamma_q \geq 0, \ldots, 0\}$ a diagonal matrix with the singular values on the diagonal. Since the past and future basis in the new basis are orthonornal and uncorrelated, the singular values are also the correlations between the canonical variates p and f.

In a linear system, independent variables are orthogonal. For nonlinear systems, stochastic independence is required. The maximal correlation is defined by:

$$\rho(p, f) = \sup_{p,f} \rho(p(y), f(y)) = \sup_{p,f} E\{p(y), f(y)\}$$

with $\| p \| = 1$ and $\| f \| = 1$.

If $\rho(p, f) = 0$, then $p(y), f(y)$ are statistically independent. Therefore, to find the optimal combination of past data to predict the future, we want the maximal correlation.

Determining this structure requires multiple steps. First, given the past and future vectors, the mean is removed to meet the constraints of the alternating conditional expectation (ACE) algorithm that will be used to determine the maximum correlation between transformed input and output variables c and d (Breiman and Friedman, 1985). Then a (Σ_{pp}, Λ) singular value decomposition of Σ_{pf} will determine a J and L such that after the transformations $c(t) = J_k p(t)$ and $d(t) = L_k f(t)$ and the covariances $\Sigma_{cc} = \Sigma_{dd} = I$.

3.4.4 State Space Reconstruction

Our final approximation technique, state space reconstruction generates a state space model from an optimal prediction of the future states from linear combinations of the past. Given the data from CVA, or any other identification method, we can use these predictions to parameterize a state space system for any order $k < q$ via a least squares regression.

Assume the following state space system:

$$\begin{aligned} x_{t_{i+1}} &= A(\theta)x(t_i) + B(\theta)u(t_i) + M(\theta)w_d(t_i) \\ y_{t_i} &= C(\theta)x(t_i) + D(\theta)u(t_i) + \\ &\quad O(\theta)w_d(t_i) + \nu(t_i) \end{aligned}$$

Define $m_{t_{i+1}} = J_k p(t_{i+1})$ and $M_t = J_k p(t)$. The state space system above expresses $(x_{ti+1}\ y_t)$ as a linear combination of $(x_t\ u_t)$. We can replace the predicted value of $x_{t_{i+1}}$ and $m_{t_{i+1}}$ with x_t and m_t and express $(m_{t_{i+1}}\ y_t)$ as a linear combination of $(m_t\ u_t)$. With this substitution, all of the data is available for a least squares fit of the two data sets leading to:

$$\begin{pmatrix} AB \\ CD \end{pmatrix} = cov\left\{\begin{pmatrix} m_{t_{i+1}} \\ y_t \end{pmatrix}, \begin{pmatrix} m_{t_i} \\ u_t \end{pmatrix}\right\} cov^{-1}\left\{\begin{pmatrix} m_t \\ u_t \end{pmatrix}, \begin{pmatrix} m_t \\ u_t \end{pmatrix}\right\}$$

with the prediction error given by:

$$S = \begin{pmatrix} S_{11}S_{12} \\ S_{21}S_{22} \end{pmatrix} = cov\left\{\begin{pmatrix} m_{t_{i+1}} \\ y_t \end{pmatrix}, \begin{pmatrix} m_{t_{i+1}} \\ y_t \end{pmatrix}\right\}$$

$$- \, cov\left\{\begin{pmatrix} m_{t_{i+1}} \\ y_t \end{pmatrix}, \begin{pmatrix} m_t \\ u_t \end{pmatrix}\right\} \cdot cov^{-1}\left\{\begin{pmatrix} m_t \\ u_t \end{pmatrix}, \begin{pmatrix} m_t \\ u_t \end{pmatrix}\right\} \cdot cov\left\{\begin{pmatrix} m_{t_{i+1}} \\ y_t \end{pmatrix}, \begin{pmatrix} m_t \\ u_t \end{pmatrix}\right\}$$

so that

$$\begin{aligned} Q &= S_{11} \\ O &= S_{21}S_{11}^{-1} \\ R &= S_{22} - S_{21}S_{11}^{-1}S_{12} \end{aligned}$$

4 SUMMARY

This paper presented methods that support new procedures to expand the set of available metamodel representations beyond the traditional least squares formulation and added the capability to use dynamical metamodels. These methods compliment a new taxonomy of metamodel structures and procedures that allow separation of the metamodeling process into a set of sequential decisions based on *a priori* information.

REFERENCES

Breiman, L and J.H. Friedman. 1985. Estimating Optimal Transformations for Multiple Regression and Correlation, pp 580-598, *Journal of the American Statistical Association*, Vol. 80, No. 391, September.

Caughlin, D. 1994a. A Metamodeling Approach to Model Abstraction, *Proceedings of the 1994 IEEE Dual-Use Technologies & Application Conference*. SUNY Institute of Technology, Utica/Rome, New York.

Caughlin, D. 1994b. An Evaluation of Simulated Annealing for Modeling Air Combat Simulations, *Proceedings of the 1994 IEEE Dual-Use Technologies & Application Conference*. SUNY Institute of Technology, Utica/Rome, New York.

Caughlin, D. 1995. *Final Report, Modeling Techniques and Applications, Volume I*. USAF Contract F30602-94-0110, Rome Laboratory/IRAE, 32 Hangar Rd, Griffis AFB, NY 13441-4114.

Caughlin, D. 1996. New Procedures to Metamodel Simulations, *Proceedings of the 6th Annual Conference on AI, Simulation, and Planning in High Autonomy Systems*, March 1996.

Goodwin, Payne. 1977. *Dynamic System Identification*. New York: Academic Press.

Ingber, A, L. 1990 Draft of Statistical Mechanical Aids to Calculating Term Structure Models. *Journal Phys. Rev.* Vol 42.

Ingber, A, L. 1993. Simulated Annealing: Practice versus Theory. Reprint from *Journal Mathl. Comput. Modeling*. December.

Juang, J, et.al. 1993. Identification of Observer/Kalman Filter Markov Parameters: Theory and Experiments. pp 320-329, *Journal of Guidance*, Vol 16, No. 2.

Larimore, W. E. 1989. System Identification and Filtering of Nonlinear Controller Markov Processes by Canonical Variate Analysis. Final Report for AF Office of Scientific Research, October 27, 1989.

Larimore,W. E and J. Baillieul. 1990. Identification and Filtering of Nonlinear Systems Using Canonical Variate Analysis. pp 635-640. *Proceedings of the* 29th *Conference on Decision and Control.*

Ljung, L. 1987. *System Identification: Theory for the User*. New Jersey: Prentice-Hall.

Maine, R. E. and K.W. Iliff. 1981. Formulation and Implementation of a Practical Algorithm for Parameter Estimation with Process and Measurement Noise. *SIAM Journal of Applied Mathematics*, Vol. 41, No. 3,

Maybeck, P. S. 1982. *Stochastic Models, Estimation, and Control, Vol 2*, New York: Academic Press.

Press, W. H., S. A. Teukolsky, W. T. Vetterling, and B. P. Flannery. 1986. *Numerical Recipes in FORTRAN*. New York: Cambridge University Press.

Vemuri, V. 1978 *Modeling of Complex Systems*, New York: Academic Press.

AUTHOR BIOGRAPHY

DON CAUGHLIN is Chief Scientist at Mission Research Corporation, Colorado Springs. He received a B.S. in Physics from the Air Force Academy, an MBA from the University of Utah, and M.S. and Ph.D. degrees in Electrical Engineering from the University of Florida. His research interests include system identification, pattern recognition, and intelligent control. Dr. Caughlin has over 28 years experience as an experimental test pilot, research scientist, program manager, and was also Associate Dean of the School of Engineering at the Air Force Institute of Technology. He currently holds an appointment as Research Associate Professor at the University of Colorado at Colorado Springs. He is a senior member of IEEE and AIAA and a member of the Society of Experimental Test Pilots.

Proceedings of the 1996 Winter Simulation Conference
ed. J. M. Charnes, D. J. Morrice, D. T. Brunner, and J. J. Swain

DYNAMIC MODEL ABSTRACTION

Kangsun Lee
Paul A. Fishwick

Department of Computer and Information Science and Engineering
University of Florida
Bldg. CSE, Room 301
Gainesville, FL 32611

ABSTRACT

While complex behavior can be generated through simple systems, as in chaotic and nonlinear systems, complex systems are found where a systems study contains multiple physical objects and interactions. Through the use of *hierarchy*, we are able to simplify and organize the complex system. Every level within the hierarchy may be refined into another level. System abstraction involves simplification through structural system representation as well as through behavioral approximations of executed model structure. There has been little work on creating a unified taxonomy for model abstraction. We present such a taxonomy and define two major sub-fields of model abstraction, while illustrating both sub-fields through detailed examples. The introduction of this taxonomy provides system and simulation researchers with a way in which to view and manage complex systems.

1 INTRODUCTION

Real world dynamic systems involve a large number of variables and interconnections. Abstraction is a technique of suppressing details and dealing instead with the generalized, idealized model of a system. The need of abstract models and traversing levels of abstractions are essential as complex models are used in practice. Computational efficiency and representational economy are main reasons of using abstract models in simulation (Fishwick 1987; Fishwick 1989; Zeigler 1972) and well as in programming languages (Berzins et al. 1986; Booch 1991).

Although many diverse areas employ abstraction methods, no agreed-upon taxonomy has been developed to categorize and structure them with underlying characterization of a general approach. Our goal is to clarify how abstraction methods relate to each other under a uniform taxonomy. We define system abstraction to be one of two types: *behavioral* or *structural*. In most cases, one should explore both of these methods when constructing systems. For instance, when a system is first being designed, one should construct it hierarchically, with simple system types at first, graduating to more complex model types later. Structural abstraction corresponds to this iterative procedure (Fishwick and Lee 1996; Fishwick 1996a; Fishwick 1996c). After creating the hierarchy, we may want to isolate abstraction levels, so a level can be executed apart from the rest of hierarchy with no detailed internal structure. This is where the behavioral approaches are employed. In depth discussions of each abstraction technique follow in the subsequent sections.

Our contribution is the formulation of a taxonomy capturing two types of abstraction, which have generally been overviewed in separate disciplines. Structural abstraction is found mostly in information on *design*, whereas behavioral abstraction is strewn across many fields of computer science and simulation. Through a unification in terminology, we demonstrate that structural and behavioral methods are complementary aspects of system abstraction. Structural abstraction is common in programming language development within computer science as well as in simulation. Behavioral abstraction is common in statistical analysis and automatic control where system abstractions are used in lieu of more complicated model-based transfer functions. Along with our discuss of the taxonomy, we present examples of each approach to complete the discussion.

The paper is organized as follows : we present the new system abstraction taxonomy with specific methods of each category in Section 2. Then we illustrate the abstraction types using two scenarios and show how abstraction methods perform in both linear and nonlinear system abstraction, in Sections 3 and 4. We close with a summary of the taxonomy and its advantages, with future goals to be achieved in Section 5.

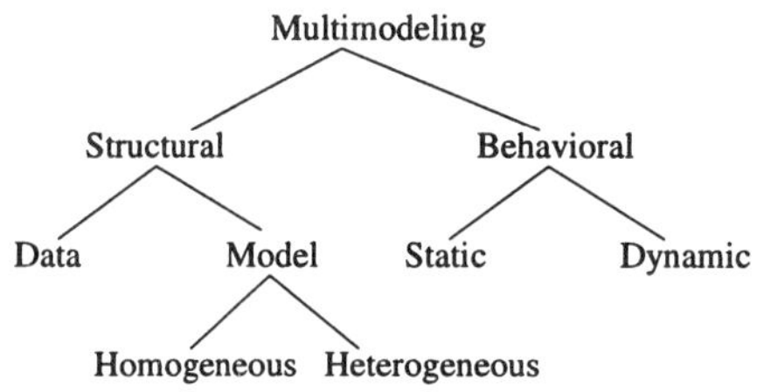

Figure 1: Proposed Taxonomy for Abstraction

2 ABSTRACTION TAXONOMY

Figure 1 illustrates our abstraction taxonomy. A system consists of data and model components. Data refers to values obtained either by observation or arbitrary assignment of values to model components. Model components, which serve as fundamental building blocks for models, take on the data values. Sample model components include state and event (Fishwick 1995). We sub-define structural abstraction of a system into ***data abstraction*** and ***model abstraction.***

- Data Abstraction : abstraction of input, output, time, parameter system values or time-dependent trajectories.
- Model Abstraction : abstraction of dynamical models.

Examples of data abstraction are symbolic value, statistic mean and variance, interval, ratio and fuzzy numbers. Data abstraction represents a way of compressing time-dependent information. In constructing a model, we further refine model abstraction to *homogeneous* and *heterogeneous* abstraction (Miller and Fishwick 1992; Fishwick 1991). For homogeneous-structural abstraction, dynamical systems can be abstracted with only one model type, depending on the level of information that one expects to receive from analysis. Specific model types are required at different abstraction levels. For example, one would not choose to model low-level physical behavior with a Petri net since a Petri net is an appropriate model type for a particular sort of condition within a system, where there is contention for resources by discretely-defined moving entities. Examples of homogeneous-structural abstraction are conceptual, declarative, functional, constraint and spatial modeling. Detailed discussion on each model type is shown in (Fishwick 1995). Models must be multi-layered so that different abstraction levels of the model respond to different needs of the analyst.

In heterogeneous-structural abstraction, different abstraction levels of a system are provided by allowing either homogeneous or heterogeneous model types together under one structure. To incorporate different levels together, we have constructed a multimodeling methodology (Fishwick 1991; Fishwick and Zeigler 1992; Fishwick 1993; Fishwick et al. 1994), which provides a way of structuring a heterogeneous and homogeneous set of model types together so that each type performs its part, and the behavior is preserved as levels are mapped (Fishwick 1988; Zeigler 1972; Zeigler 1990). Heterogeneous-structural abstraction is equivalent to multimodeling in the sense that we abstract a system structurally using homomorphic relationships of one level to another, providing multiple level abstractions. While the multimodel approach is sound for well-structured models defined in terms of state space functions and set-theoretic components, selecting system components in each level are dependent on the next-lowest level. This implies that we are unable to run each level ***independently.*** It is possible, to obtain *output* for any abstraction level but, nevertheless, the system model must be executed at the lowest levels of the hierarchy. A new definition and methodology are needed to better handle abstraction of systems and components.

Behavioral abstraction is where a system is abstracted by its behavior. We replace a system component with something more generic that approximates, to some degree of accuracy, the behavior of the system component at its refined levels. Therefore, discarding the refined levels that define a system component will still result in a complete behavioral description of a system (Fishwick and Lee 1996). By incorporating behavioral abstraction approaches into multimodeling allows each level to be understood independetly of the others. This is why we put multimodeling on the top of our taxonomy.

We have two approaches of specifying system behavior:

- Static approach : one takes a system and captures only the steady state output value instead of a complete output trajectory. The input value is defined to be the integral of time value over the simulation trajectory.
- Dynamic approach : one needs to associate time-dependent input and output trajectories.

System identification (Ljung and Soderstrom 1983; Johansson 1993) is to abstract a system by *mathematical models.* Modeling the system consists of selecting a general, parameterized mathematical representation and then tuning the parameters, so that behavior predicted by the model coincides with measurements from the real system. Parameter estimation procedure provides a search through parameter space, effectively, to achieve a close-to optimal

Table 1: Sample Abstraction Categories and Associated Techniques

Base Abstraction Type	Abstraction Technique
Data Abstraction	Symbolic Value Mean, Variance Interval, Ratio Fuzzy Number
Structural Abstraction	Conceptual Modeling Declaration Modeling Functional Modeling Constraint Modeling Spatial Modeling Multimodeling
Behavioral Abstraction	Regression System Identification Neural Network Wavelet Genetic Algorithm

mapping between the actual values of the system and the approximate abstract system. Commonly used parameter models are ARX, ARMAX, OE (Output Error) and BJ (Box-Jenkins) (Tan et al. 1995; Johansson 1993). Brief explanations of these models are shown in Section 3.2.2.

Neural networks have been established as a general approximation tool for fitting models from input/output data (Cynbenko 1989; Tang et al. 1991; Tang and Fishwick 1993). From the system identification perspective, a neural network is just another model structure (Ljung and Soderstrom 1983; Barron 1989). The inputs are linearly combined at the nodes of the hidden layer(s) and then subjected to a threshold-like non-linearity, and then the procedure is repeated until the output nodes are reached. Backpropagation, recurrent and temporal neural networks have been shown to be applicable to modeling an identification (Fishwick and Lee 1996; Mills et al. 1995). On the other hand, recently introduced *wavelet decomposition* achieves the same quality of approximation with a network of reduced size by replacing the neurons by "wavelons", i.e. computing units obtained by cascading an affine transform and multidimensional wavelets (Ahang and Benveniste 1992).

Table 1 summarizes the based categories along with some sample abstraction techniques discussed so far. Having defined the model abstraction taxonomy, we now proceed to illustrate the different abstraction techniques using the following example.

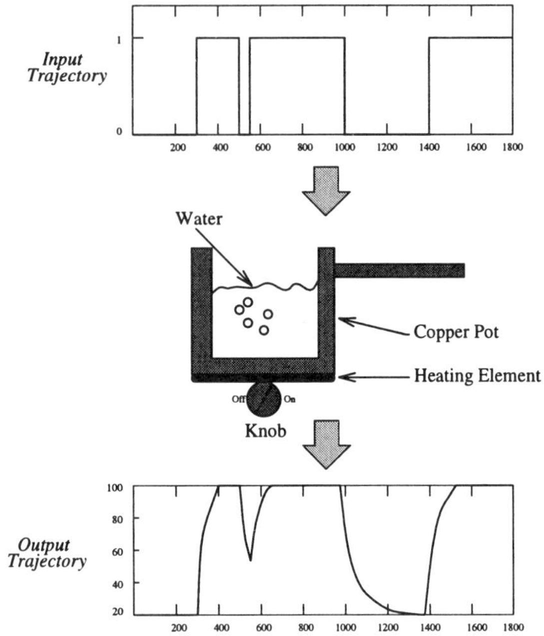

Figure 2: Boiling Water System

3 EXAMPLE I: BOILING WATER MODEL

Consider a pot of water in Figure 2. Here we show a picture of the boiling pot along with an input and output trajectory. The input reflects the state of the knob, which serves to specify external events for the system. The output defines the temperature of the water over time. Newton's law of cooling states that $Rq_h = \Delta T = T_1 - T_2$ where T_1 is the temperature of the source (heating element), and T_2 is the temperature of the water. q_h is heat flow. Since T_2 is our state variable we let $T = T_2$ for convenience. By combining Newton's law with the capacitance law, and using the law of capacitors in series, we arrive at:

$$k = \frac{C_1 + C_2}{RC_1C_2}$$
$$\dot{T} = k(T_1 - T)$$

3.1 Structural Abstraction

The structural approach to system abstraction for the boiling water is defined in a recent text (Fishwick 1995) where the boiling water is included as a sub-system within a system of two flasks and a human operator who mixes the flasks once the liquids are boiling. In the structural abstraction approach to systems, we first need to define our levels of abstraction and then choose which models types to use at each level.

We show part of the multimodel in Figures 3 and 4. The first model is a compressed version of all the

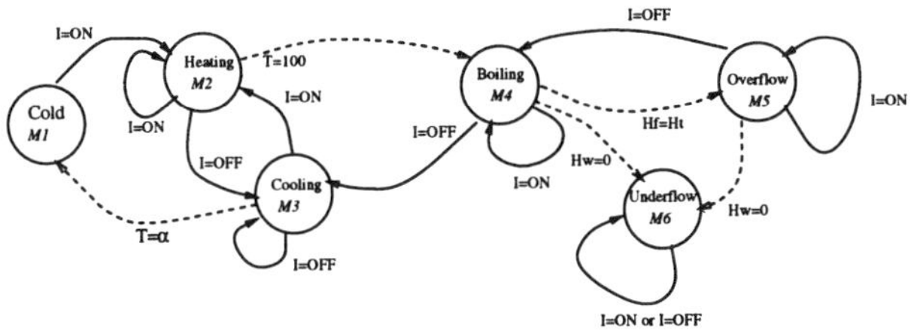

Figure 3: Six State Automaton Controller for the Boiling Water Multimodel

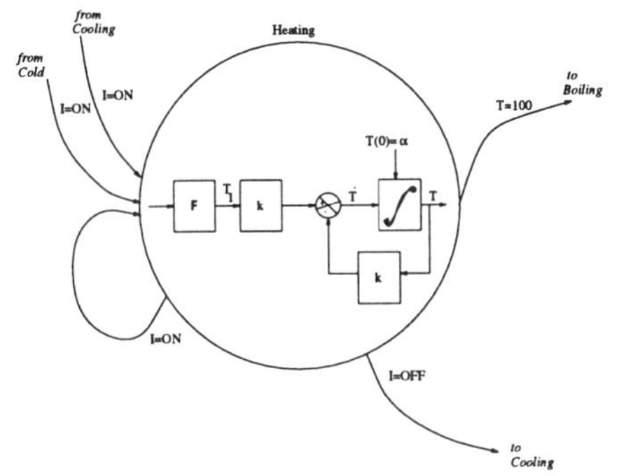

Figure 4: Decomposition of *Heating* State

hierarchy. Figure 4 shows Newton's law of cooling in a functional block form.

3.2 Behavioral Abstraction

3.2.1 Static Approach

In the static approach, we're interested only in the *final* (i.e., steady state) temperature of the water. Our two inputs are: (1) total amount of elapsed time for the input trajectory and (2) integral value of the input trajectory integrated over time. The output is the temperature of the water at the elapsed time. A graph of elapsed time versus temperature is shown in Figure 5. This information is obtained directly from the underlying simulation of the boiling water system. We chose a subset of all possible input time trajectories in such a way that some nonlinearity was introduced into the graph in Figure 5. This was done to challenge the behavioral parameter estimation methods in creating a good fit. This explains why Figure 5 contains a small area of discontinuity in the region between steady state temperature values of 20 and 40.

We will use linear regression to exemplify the static approach. In general, a polynomial fit to data in vectors x and y is a function p of the form:

$$p(x) = c_1 x^n + c_2 x^{n-1} + \ldots + c_d$$

The degree is n and the number of coefficients is $d = n + 1$. The regression coefficients $c_1, c_2, \ldots, c_n$ are determined by solving a system of simultaneous

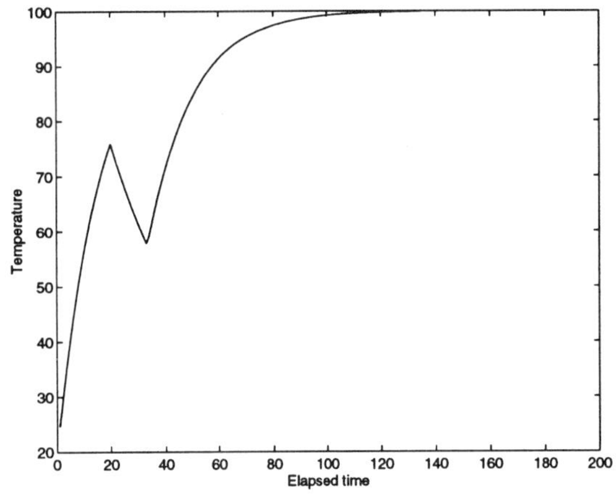

Figure 5: Elapsed Time vs. Temperature

linear equations: $Ac = y$ (Law and Kelton 1991). Figure 6 shows the result. The approximation is poor in the graph's central region because linear regression is done by polynomial fit, and so it generates a *monotonically increasing* function.

3.2.2 Dynamic Approach

In the dynamic approach, we're interested in time dependent behavior. In this case, we are concerned not only in the steady state temperature but also the way in which the temperature changes over time. For this approach, we chose a system with just one input and one output, both time-varying trajectories. The input is the input "knob off/knob on" trajectory and the output is the temperature trajectory.

The Box-Jenkins method is a frequently used linear system identification method in time series analysis (Tang, de Almeida, and Fishwick 1991; Tang and Fishwick 1993; The MathWorks 1991). Its structure is given by

$$y(t) = \frac{B(q)}{F(q)} u(t - nk) + \frac{C(q)}{D(q)} e(t)$$

with

$$\begin{aligned} y(t) &: \text{output signal} \\ B(q) &= b_1 + b_2 q^{-1} + \ldots + b_{nb} q^{-nb} \\ F(q) &= 1 + f_1 q^{-1} + \ldots + f_{nf} q^{-nf} \\ C(q) &= 1 + c_1 q^{-1} + \ldots + c_{nc} q^{-nc} \\ D(q) &= 1 + d_1 q^{-1} + \ldots + d_{nd} q^{-nd} \end{aligned}$$

The numbers nb, nc, nd and nf are the orders of the respective polynomials and q is the shift operator. The number nk is the number of delays from input to output. Figures 7 and 8 show the approximation result and the accompanying error. Successful identification of $y(t)$ depends on how well we guess the

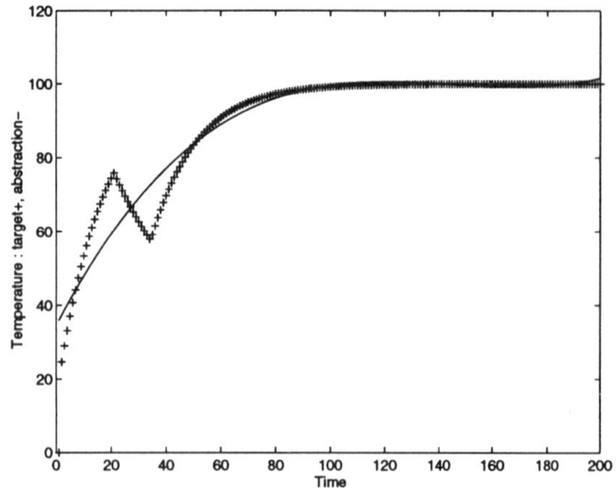

Figure 6: Linear Regression

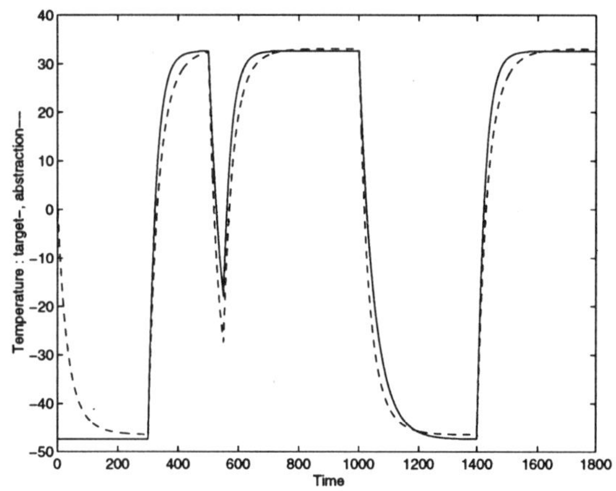

Figure 7: Box-Jenkins Method

values of nb, nc, nd, nf and nk. Heuristics and "expert rules," if available, aid us in choosing parameters. For example, too large a value for a parameter results in computational difficulties to generate $y(t)$, while too small a value results in a rough estimation. We often had to tune parameters by hand in order to get a good approximation.

4 EXAMPLE II: HEMATOPOIESIS MODEL

Though the abstraction methods discussed so far were good at linear system abstraction, non-linear system abstraction involves more complex behavior mapping. In this section, we show how these abstraction methods perform under non-linear conditions.

Our model deals with the regulation of hematopoiesis, the formation of blood cell elements in the body. Hematopoiesis is the process of blood creation in the body. White and red blood cells are produced in bone marrow. From the marrow they enter the blood circulatory system. As the oxygen level decreases in the body, there is a feedback back to the bone marrow–which produces more cells.

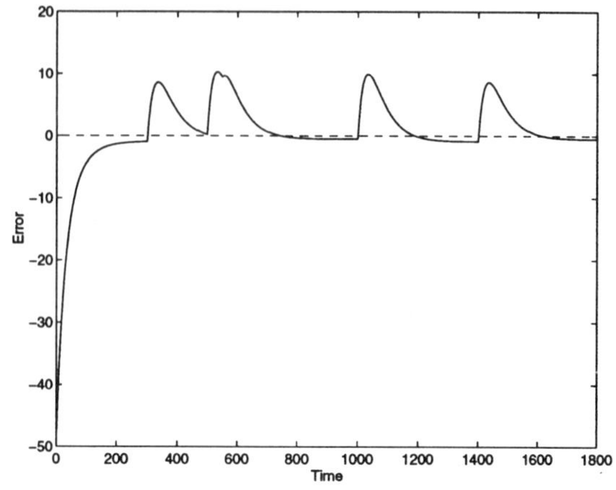

Figure 8: Abstraction Error in Box-Jenkins Method

Mackey and Glass (Mackey and Glass 1977) provide a delay model for hematopoieses of the following form:

$$\frac{dP(t)}{dt} = \frac{\lambda \theta^m P(t-T)}{\theta^m + P^m(t-T)} - gP(t)$$

where, λ : the flux of cells into the blood stream, P(t) : the concentration of cells(the population species) in the circulating blood ($cells/mm^3$), g : day^{-1}, cell loss rate per day and T : maturation delay.

We use λ as an input. Depending on the maturation delay T, we can generate different solutions. In a lower maturation delay, the system shows periodic behaviors, but, as the delay moves upward, nonperiod trajectories appear. Figure 9 shows a nonperiodic trajectory when the delay is 20.

Since we are interested in abstracting the time dependent behavior of cell concentration in the circulating blood, we will restrict out experiments to dynamic-behavioral abstractions. Also, to see how abstraction techniques perform under heavy nonperiodic and nonlinearity, we choose maturation delay 20. Now, the dynamic-behavioral abstraction of hematopoiesis model is to approximate equation (1) with a discrete model of the form where f is a nonlinear function to be estimated with order na:

$$P(t) = f(P(t-1), ..., P(t-na))$$

Small sampling period for the discretization makes the order of the discrete model very high due to the long dependence of $P(t)$ on $P(t-20)$, which results in numerical difficulties to compute the optimal function of f. Therefore, increasing sampling period is needed as long as the discretization is not too

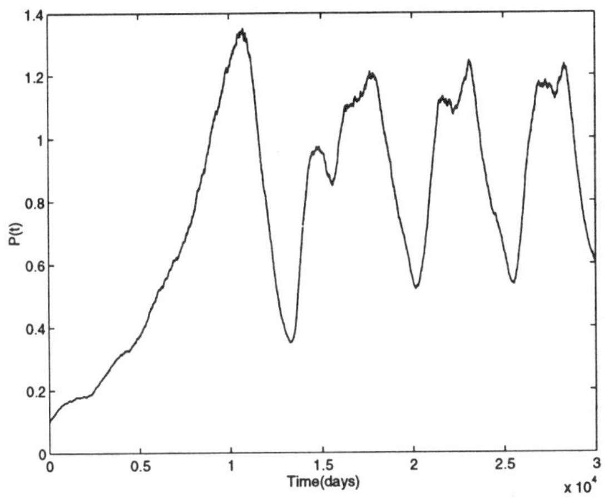

Figure 9: Cell Concentration vs. Time for Delay T = 20 Days

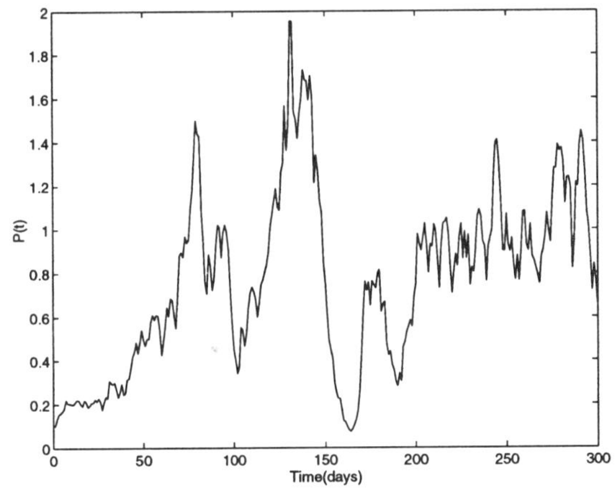

Figure 10: Hematopoiesis Model for Delay = 20 Days with Increased Sampling Period: Abstraction Target

rough. Figure 10 displays the time trajectory for the total concentration of blood cells when the sampling period is increased by 100, which introduces more nonlinearity and instability. We choose Figure 10 as the abstraction target and use λ for input.

ADALINE was developed by Widrow and Hoff (Widrow and Sterns 1985). Their neural network model differs from the perceptronq in that ADALINE neurons have a linear transfer function. The ADALINE network also enables the Widrow-Hoff learning rule, known as the Least Mean Square (LMS) rule, to adjust weights and biases according to the magnitude of errors.

The ADALINE neural network for the hematopoiesis model performs abstraction as shown in Figure 11. An ADALINE neural network takes initial weights and biases, an input signal and a target signal, and then filters the signal adaptively based on

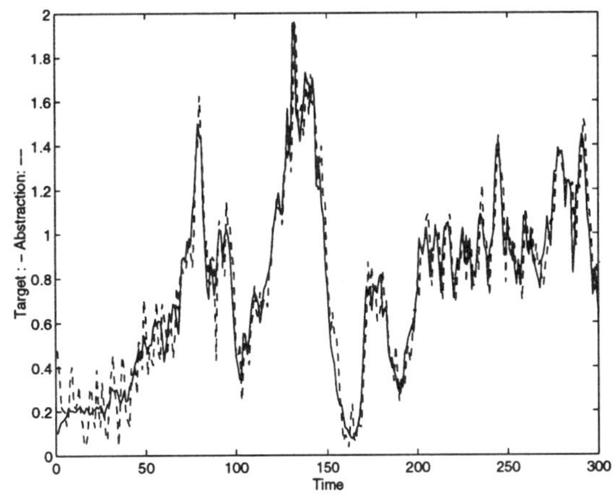

Figure 11: ADALINE Network for Hematopoiesis Model

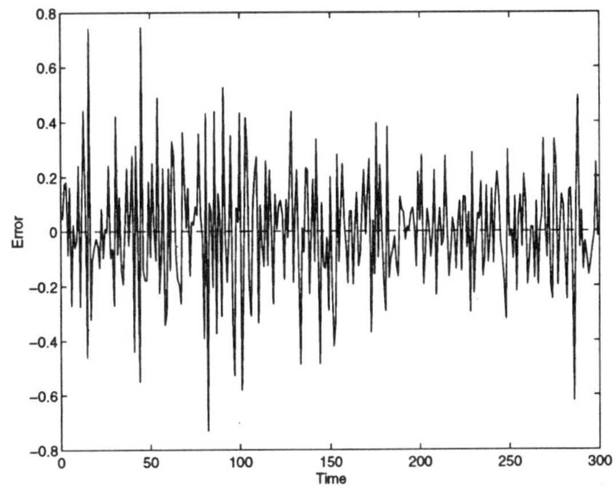

Figure 12: Abstraction Error in ADALINE Network

input delay and learning rate parameters. In most cases, input delay can be guessed by the modeled system itself. For hematopoiesis model, we know an output at time t is determined by 20 most recent inputs, which could be inferred by the the delay-differential equation (1). A proper learning rate is determined by repetitive trials until a good fit is achieved with fewer learning-stage perturbations.

5 CONCLUSIONS

We have presented a new taxonomy for model abstractions in dynamic systems. The taxonomy of abstraction types, with multimodeling at the top level, is constructed by *model engineering* perspective: when a system is first being developed, one should use structural abstraction to organize the whole system hierarchically with simple system types, and then graduate to more complex model types. Below the structural abstraction, each component is black-box with no detailed internal structure. Behavioral abstraction is used to represent those black-boxes by approximating the behavior of the system components. By combining structural and behavioral abstraction together, each level of abstraction is independent from the lower abstraction levels, so a level can be executed apart from the rest of the hierarchy. These two concepts: structural and behavioral abstraction are blended together to form a comprehensive taxonomy. In addition to the taxonomy, we discussed several abstraction methods according to the categories they belong to and showed how they perform in linear and nonlinear system abstractions. We felt it important to provide both linear and nonlinear models since one technique may fare well for one type of system and then poorly on the other.

Given that we have developed this taxonomy, a good question is "What to do with it?" We are developing a system called MOOSE (Fishwick 1996b), standing for multimodeling object oriented simulation environment, in which the taxonomy is to be applied. MOOSE models are constructed using a graphical user interface which begins with the user specifying an object oriented class hierarchy. This procedure takes advantage of structural abstraction. For exploiting behavioral model abstraction, our current plans are to provide two or three basic techniques and allow the user to choose which they would like. Moreover, we are developing a semi-automated approach to developing behavioral abstractions of multimodel components which can benefit most from the computational gain afforded by not having to simulate at the lowest level.

ACKNOWLEDGMENTS

We would like to acknowledge the following funding sources which have contributed towards our study of modeling and implementation of a multimodeling simulation environment: (1) Rome Laboratory, Griffiss Air Force Base, New York under contract F30602-95-C-0267 and grant F30602-95-1-0031; (2) Department of the Interior under grant 14-45-0009-1544-154 and the (3) National Science Foundation Engineering Research Center (ERC) in Particle Science and Technology at the University of Florida (with Industrial Partners of the ERC) under grant EEC-94-02989.

REFERENCES

Ahang, Q. and A. Benveniste. 1992. Wavelet Networks. *IEEE Transactions on Neural Networks 3*(6).

Barron, A. R. 1989. Statistical Properties of Artificial Neural Networks. In *Proceedings of the 28th IEEE Conference on Decision and Control*, pp. 280–285.

Berzins, V., M. Gray, and D. Naumann. 1986. Abstraction-Based Software Development. *Communications of the ACM 29*(5), 402–415.

Booch, G. 1991. *Object Oriented Design with Applications.* The Benjamin/Cummings Publishing Company, Inc.

Cynbenko, G. 1989. Approximation by Superposition of a Sigmoidal Function. *Mathematics of Control, Signals and Systems 2*, 303–314.

Fishwick, P. A. 1987. A Taxonomy for Process Abstraction in Simulation Modeling. In *Proceedings of the IEEE International Conference on Systems, Man and Cybernetics*, Volume 1, Alexandria, Virginia, pp. 144 – 151.

Fishwick, P. A. 1988. The Role of Process Abstraction in Simulation. *IEEE Transactions on Systems, Man and Cybernetics 18*(1), 18 – 39.

Fishwick, P. A. 1989. Abstraction Level Traversal in Hierarchical Modeling. In B. P. Zeigler, M. Elzas, and T. Oren (Eds.), *Modelling and Simulation Methodology: Knowledge Systems Paradigms*, pp. 393 – 429. Elsevier North Holland.

Fishwick, P. A. 1991. Heterogeneous Decomposition and Coupling for Combined Modeling. In *Proceedings of the 1991 Winter Simulation Conference*, Phoenix, AZ, pp. 1199 – 1208.

Fishwick, P. A. 1993. A Simulation Environment for Multimodeling. *Discrete Event Dynamic Systems: Theory and Applications 3*, 151–171.

Fishwick, P. A. 1995. *Simulation Model Design and Execution: Building Digital Worlds.* Prentice Hall.

Fishwick, P. A. 1996a. A Taxonomy for Simulation Modeling Based on a Computational Framework. *IIE Transactions on IE Research.* Revised and re-submitted May 1996.

Fishwick, P. A. 1996b. Extending Object Oriented Design for Physical Modeling. *ACM Transactions on Modeling and Computer Simulation.* (submitted for review).

Fishwick, P. A. 1996c. Toward a Convergence of Systems and Software Engineering. *IEEE Transactions on Systems, Man and Cybernetics.* Submitted May 1996.

Fishwick, P. A. and K. Lee. 1996. Two Methods for Exploiting Abstraction in Systems. In *Proceedings of the AI, Simulation and Planning Conference in High Autonomous Systems Conference*, pp. 257–264.

Fishwick, P. A., H. Narayanan, J. Sticklen, and A. Bonarini. 1994. Multimodel Approaches for System Reasoning and Simulation. *IEEE Transactions on Systems, Man and Cybernetics 24*(10), 1433–1449.

Fishwick, P. A. and B. P. Zeigler. 1992. A Multimodel Methodology for Qualitative Model Engineering. *ACM Transactions on Modeling and Computer Simulation 2*(1), 52–81.

Johansson, R. 1993. *System Modeling and Identification.* Prentice Hall, Englewood Cliffs, New Jersey.

Law, A. M. and D. W. Kelton. 1991. *Simulation Modeling and Analysis.* McGraw-Hill.

Ljung, L. and T. Soderstrom. 1983. *Theory and Practice of Recursive Identification.* MIT Press, Cambridge, Mass.

Mackey, M. and L. Glass. 1977. Oscillations and Chaos in Physiological Control Systems. *Science 197*, 287–289.

Miller, V. T. and P. A. Fishwick. 1992. Heterogeneous hierarchical models. In *Proceedings of the Conference on Artificial Intelligence X: Knowledge Based Systems*, Orlando, FL. SPIE.

Mills, P., M. Tade, and A. Zomaya. 1995. Identification and Control Using a Hybrid Reinforcement Learning System. *International Journal in Computer Simulation 5*, 109–126.

Tan, K., Y. Li, D. Murray-Smith, and K. Sharman. 1995. System Identification and Linearization Using Genetic Algorithms with Simulated Annealing. *Proceedings of the First IEE/IEEE International Conference on GA in Engineering Systems: Innovations and Applications*, 164–169.

Tang, Z., C. de Almeida, and P. A. Fishwick. 1991. Time Series Forecasting using Neural Networks vs. Box-Jenkins Methodology. *Simulation 57*(5), 303–310.

Tang, Z. and P. A. Fishwick. 1993. Feed-Forward Neural Nets as Models for Time Series Forecasting. *ORSA Journal of Computing 5*(4), 374–386.

The MathWorks, I. 1991. *System Identification Toolbox.* The MathWorks, Inc.

Widrow, B. and S. Sterns. 1985. *Adaptive Signal Processing.* Prentice-Hall.

Zeigler, B. P. 1972. Towards a Formal Theory of Modeling and Simulation: Structure Preserving Morphisms. *Journal of the Association for Computing Machinery 19*(4), 742 – 764.

Zeigler, B. P. 1990. *Object Oriented Simulation with Hierarchical, Modular Models: Intelligent Agents and Endomorphic Systems.* Academic Press.

AUTHOR BIOGRAPHIES

PAUL A. FISHWICK is an Associate Professor in the Department of Computer and Information Science and Engineering at the University of Florida. He received the PhD in Computer and Information Science from the University of Pennsylvania in 1986. He also has six years of industrial/government production and research experience working at Newport News Shipbuilding and Dry Dock Co. (doing CAD/CAM parts definition research) and at NASA Langley Research Center (studying engineering data base models for structural engineering). His research interests are in computer simulation modeling and analysis methods for complex systems. He is a senior member of the IEEE and the Society for Computer Simulation. He is also a member of the IEEE Society for Systems, Man and Cybernetics, ACM and AAAI. Dr. Fishwick founded the `comp.simulation` Internet news group (Simulation Digest) in 1987. He has chaired workshops and conferences in the area of computer simulation, and will serve as General Chair of the 2000 Winter Simulation Conference. He was chairman of the IEEE Computer Society technical committee on simulation (TCSIM) for two years (1988-1990) and he is on the editorial boards of several journals including the *ACM Transactions on Modeling and Computer Simulation*, *IEEE Transactions on Systems, Man and Cybernetics*, *The Transactions of the Society for Computer Simulation*, *International Journal of Computer Simulation*, and the *Journal of Systems Engineering*. Dr. Fishwick's WWW home page is `http: //www.cise.ufl.edu /~fishw ick` and his E-mail address is `fishw ick@ cise.ufl.edu`.

KANGSUN LEE received the B.S and M.S degree in Computer Science from Ewha Womans University, Korea in 1992 and 1994, respectively. She is currently a Ph.D student in the Computer and Information Sciences and Engineering department at the University of Florida, Gainesville. Her research interests are in Modeling methodology, Abstraction techniques and simulation. Kangsun Lee's WWW home page is `http://www.cise.ufl.edu/~kslee` and her E-mail address is `kslee@cise.ufl.edu`

Proceedings of the 1996 Winter Simulation Conference
ed. J. M. Charnes, D. J. Morrice, D. T. Brunner, and J. J. Swain

WEB-BASED SIMULATION: SOME PERSONAL OBSERVATIONS

Paul A. Fishwick

Department of Computer and Information Science and Engineering
University of Florida
Building CSE, Room 301
Gainesville, FL 32611

ABSTRACT

The web has a short history, but has grown at an exponential pace since its introduction six years ago. *Web-based simulation* represents the connection between the web and the field of simulation. Web-based simulation is not an existing field but rather an idea which represents an interest on the part of simulationists to exploit web technology. To further this cause, I present some issues and concepts in web-based simulation to serve as a back drop for a more formal discussion, and potentially the formation of a new simulation sub-area.

1 INTRODUCTION TO THE WEB

What is web-based simulation? First, we need to discuss a few terms before we can adequately address this question. But, before doing that, we need to say something about the nature of this topic. The web represents a rapid growth of Internet-based interaction over the past six years. Yet, the topic reflects more of a *technology-transfer* area than one with a solid theoretic underpinning. Rapid advances have taken place through practical application, engineering and entrepreneurial activity. A new web advancement will appear, almost overnight, as a working product to be tried by eager users. This situation—not unlike the gold rush in the mid-nineteenth century—suggests that, by default, an article on "web-based simulation" needs to focus on technology. Yet, as researchers, we collectively require a substantial knowledge of this technology if we are to apply the web to our disciplines. It is not healthy to ignore the web and remain fixated to either purely academic tomes or yesterday's approaches to simulation. My purpose in this article is to inform the reader about some of the possibilities with an applications and technology transfer focus. I have been involved as a heavy user of web technology and will impart some of my personal experiences with the web as they relate to simulation. I will unabashedly refer to specific packages and methods, not with the deliberate intention of presenting a biased view, but because this is not a survey paper and there are too many web-based methods to be able to develop a timely paper without making the content obsolete once it is in print. Indeed, novel web-based applications seem to crop up so frequently that information on web-based simulation will likely change as rapidly as these applications. In this sense, the article is somewhat anecdotal, and is meant to help simulation-interested readers who are beginning to explore the web, and possibly even simulation. Furthermore, in the spirit of the subject matter, all of our references will be universal resource locators (URLs) and a hypermedia copy of this document should be accessed to be able to link appropriately where required. The hypermedia document is available by pointing your browser at: `http://www.cise.ufl.edu/~fishwick/websim.html` as soon as you are near a PC or workstation. Beyond this unique URL, which serves as the gateway to all further citations and references, all other URLs will not be visible in this article for the sole reason that their employment is not possible without a browser. Nothing is gained from reading the printed URL, and with a browser, you will not require this document. So, I encourage you at the next opportunity to go to a browser at locate the previous URL. It is assumed that you will have access to a browser, as there are several free browsers available and some (such as Microsoft's Internet Navigator) come bundled with their operating systems (Windows 95 and NT), which in turn are bundled with all PCs. The PC is the most popular platform for web-based applications due to the huge number of PCs and the PC open architecture. Other supported platforms include Apple's MacIntosh and Unix workstations such as the ones made by Sun, Hewlett-Packard and Silicon Graphics. Most web software is currently packaged for Windows 95 first, then Windows NT and Windows 3.1, followed by Unix versions.

The Internet is a loosely connected world-wide network of heterogeneous computers. The World Wide Web (WWW) is a set of on-line hypermedia documents that reside on the Internet. Hypermedia documents are built using a language called hypertext markup language (HTML). HTML started as a subset of SGML (Standard Graphics Markup Language) but grew to absorb the needs of web authors. Access to the web is afforded using a web *browser* such as Netscape, Mosaic or Internet Navigator. When you use the browser, you are presented with a default *home page* which presents you with your first online document. This document will include *hyperlinks* which, when clicked using a mouse, link to another document. In this way, a document is connected to other documents in a sprawling network: there is no centralized control—just documents loosely connected to one another through links. Documents may contain more than the usual printed document:

- Text: various natural languages are supported.
- Images: typically GIF files seem to be the best supported but JPEG is also supported. Image converters exist for most platforms.
- Video: supported video types are MPEG and AVI, with the additional GIF89 that supports inline small video clips meant for small icon display.
- Audio: AU and WAV are supported formats, as well as digital standards such as MIDI.

Although web browsers support many exciting features, our purpose is to say something about how the web and the field of simulation relate to one another: web-bases simulation. Given a browser and the ability to write new web pages, what are the issues for simulationists? How can we exploit the new technologies and how will our field change as a result of the web?

2 HOW SIMULATION WILL BE AFFECTED

There are many aspects to simulation as there are for any scientific discipline. We will focus on the following aspects: 1) Education and Training, 2) Publications, 3) Simulation Programs. All three of these topics are interrelated. For the academic readers, the topics of education and publication are probably on the top of the list, and industrial simulationists will want to focus more on training and the practical methods for building simulations that use the web.

2.1 Education and Training

How do we currently educate our students using simulation? Most modern simulation packages today have graphical front ends (so called graphical user interfaces—GUIs). This allows a student to learn through virtual world exploration. The web will affect this process in several ways. First, a diskette or CD-ROM has a limited amount of local storage but the web provides us with effectively infinite storage. The "storage" is on the Internet and not limited to one's own machine. Therefore simulation packages which include help text and information about the pieces of the model, can have these pieces on the web so that they do not require local storage. This encourages some collaboration in that a company marketing their simulation tool, conveniently priced for educators, will have to think more *globally* about information and resources. It is not necessary for a company to re-build every piece of information about a manufacturing process, for instance, from scratch. Information on automated guided vehicles (AGVs), machine specifications, and automated conveyance mechanisms may already be located somewhere on the web. These sorts of devices are common in simulation programs built for manufacturing systems analysis. Why re-invent the wheel by building our own information databases on these manufacturing devices? The web encourages this sort of global view with a *re-use* of knowledge and information. While academics may revel in this kind of reusable knowledge, industry will have to rethink certain proprietary concerns: a company needs to make a profit to survive. They also need to produce quality simulation software, but this does not negate the use of the information of other web sites. Credit must be given where it is due, as usual, and it is not clear exactly how copyright, trademarks and patents will change as a result of the web phenomenon. It is too early to tell, but we need to address these issues now.

We have discussed the use of web sites in user documentation for a simulation program, but the program could just as easily be placed on the web as well. After all, is the program really that different from hypermedia "documentation?" We discuss this in more detail within this section. The most profound affect the web may have on our method of teaching simulation lies with the use of multimedia, since it is multimedia that we find in every web link. The web encourages distance learning more so than the typical simulation textbook. On any web page it is possible to include images and video of the instructor along with synchronized slides or overheads. This immerses the student in a synthetic learning environment that is more congenial than one they would get simply by reading a book or watching a videotape. Various technologies are around to aid us in creating multimedia web pages for simulation education. Before mentioning a few of them, it is worth defining the concepts of *client* and *server* since the web encapsulates these two roles. A server provides a client with information and services. The daily metaphors of servers and clients (whether going to a restaurant or visiting your doctor or lawyer) are the best way of understanding these roles. If you enter your browser and access a travel agency in Connecticut, you are able to do this because the travel agency has a computer which contains a web server. This web server services your needs by delivering multimedia documents to you via your browser. The relationship is directional: the agency is the server and you (or your machine) are the client. Every machine, potentially, has the ability to be both server and client so there is no reason why the agency might not use your machine as a server if you have it set up as such, delivering information required by the travel agency. Any machine can serve, then, as both client and server.

A *plug-in* is a software program to allow third party vendors (with respect to the browser designer) to provide multimedia service for web pages. For example, if I install the PDF reader from Adobe, Inc. into Netscape, I will then be able to include PDF files in one of my web pages, so when I access someone elses web page which contains a server capability delivering PDF files, my PDF plug-in will activate and I will be able to see PDF files. Plug-ins exist for numerous functions. A recent trend has been for commercial vendors to provide plug-ins for free to anyone, but to charge a fee for those wanting to serve clients. If you want, for example, to design your own PDF files so that others can access them over the web, this will involve a cost *to you*, and this is where the company generates its required revenue. If web software is free, it is free only in the sense that someone else is paying for you to use it in the same way that TV and Radio advertisers pay for the programming which you enjoy.

What plug-ins are useful to a simulation educator? When you have a need to deliver real-time video and audio to your students, then you will need to purchase the server software, while your students can freely obtain the client-side plug-in. For non-real time video and audio, many free players exist (e.g., MPEG and AVI for video and AU, MIDI for audio), and so most companies seem to be making a profit for providing a much-needed real-time audio-video stream, especially over low bandwidth connections such as via 28.8K modems. VDOLIVE and VDOPHONE are especially good at providing close-to real time audio and video. VDOLIVE is a method of "streaming" video and audio in the same way that a TV does. VDOPHONE is an Internet substitute for a videophone. Both of these packages employ wavelet compression to achieve excellent performance. Video frame rate is generally sacrificed where necessary to maintain uninterrupted audio. This is as it should be since "broken" video is acceptable, but this is not true for audio. Occasionally, another marketing technique is to provide a "personal server" for free. This provides you with the ability to use one stream, which is fine as long as two clients (i.e., students) do not attempt to access the video stream simultaneously. Many other plugins are useful as well, such as one that can display color slides, possibly in synchronization with the audio and video streaming plugins.

In the sense that our conferences represent "education and training," we need to make major overhauls of the way we do business. The use of teleconferencing will help us to interact prior to meeting at a conference. Personal meetings with each other cannot be replaced by electronic means, but new web tools such as CU-SEEME and VDOPHONE (section 2.2) have the means to allow us to communicate more effectively. E-mail, mailing lists and USENET news groups (including the new web-based news groups) are useful technologies. The USENET news group `comp.simulation` has been in operation for almost a decade and has helped to grow the discipline. The ELECSIM conference has been successful in forging ahead to permit us to communicate with one another without forcing cross-country trips to conferences.

2.2 Publications

We publish for one reason: to inform others of new or existing simulation research and applications. We write articles in everything from conference proceedings such as this one, and trade magazines to academic archival journals. The web is dramatically changing the way we publish, and it is changing so fast that many are rethinking how publications are delivered and purchased. Let's consider archival simulation journals first. If you publish an article in an archival journal, there is an accepted practice which is defined as follows:

1. *Refereeing*: you submit a manuscript to the journal's Editor in Chief. The editor then has an editorial board which is used to manage the review process for the paper. The idea is that, for your paper to be accepted, it should be sent to qualified reviewers who are experts in the sub-simulation area that represents your paper's focus. Moreover, these reviewers are

anonymous so as not to bias the review. Sometimes, conferences try to snip part of an author's submission so that the reviewer doesn't know who wrote the article. This procedure tends not to be practiced in the simulation community, and the procedure has flaws since self-reference (to establish background) in a paper generally links a specific person to the paper. Once the paper is reviewed, it can be rejected, accepted or subject to revision with provided comments and critique by the reviewers. A review process from time of submission to a decision can take anywhere from three months to six months.

2. *Publication*: The review process can be lengthy, and assuming that you have made the proper revisions to your paper, your paper may be accepted by the Editor in Chief on recommendation of a member of the board. The time from acceptance to the time your paper appears depends on two things: the periodicity of the journal and the current queue (backlog) of already accepted papers. Times from six months to two years are common.

3. *Archival*: Libraries typically archive journals so that others can search for your article (using a scientific index) for their own research.

This entire process is lengthy but there is good reason for much of the time taken to evaluate the paper. We, as scientists, also accept this method since it has proven to be the way to introduce original simulation knowledge to grow our information base. The web has changed the playing field - many new ideas are surfacing. Most of the simulation magazines and journals are making new inroads here. For example, both the ACM Transactions on Modeling and Computer Simulation (TOMACS) and the SCS Transactions on Simulation are rapidly transitioning to electronic form. Let's consider some issues:

1. *Literature Search*: a publication begins with a thorough literature search. This search is made easier on the web through search tools, but searching for the right information is still difficult. There is so much information that searches can become unwieldy. Often, searches on the web result in "missing link" problems: all information is distributed and so links change dynamically. Only evolved and maintained links remain in way not unlike the processes of genetic variation and fitness in living organisms. Libraries are changing their functions to help with information access and retrieval. There is still an art to locating just the right information and this is the task of an information specialist—the modern librarian.

2. *Electronic Access*: make everything electronic. The author submits the paper using an electronic medium such as Postscript, which is fast becoming a kind of pseudo-standard for page layout among journals. Referees are sent (either via mail attachments or using file transfer-ftp) a compressed Postscript file. Referees submit their reviews by e-mail and the author is informed of the decision by e-mail as well. All of this makes sense and is happening now. All citations use URL links instead of printing a bibliography. This allows the reader to jump directly to a cited paper, avoiding lengthy and sometimes unsuccessful literature searches.

3. *Refereeing*: more comments can be made available to an author if anyone is permitted to attach comments to an article being considered for publication. Why limit the comments to two or three people? We call this "reader refereeing": let those who are interested in reading an article make comments on it. One possibility is to have a combination of traditional and reader refereeing.

4. *Revenue*: who pays for the article? Before, everything was so simple - the entity which published the hardcopy of the journal received basically all the revenue. However, if the volunteers (editorial board) are doing all the work and do not need the publisher, then should simulation journals be free? Well, there are costs lurking everywhere: the cost of the server equipment and the cost of archiving journal articles. Someone has to pay for these costs. Costs have been translated from paper-printing into digital service and storage.

5. *Preprints*: remember that the goal of publishing is for you to impart information to others on your new ideas. If you can put your papers immediately online in your web page, and have these web pages indexed by web robots and *intelligent agents*, then you don't have to wait any longer to see your work "in print." It is in print immediately. The generally accepted practice is for authors to place preprints on the web, either in a home page or in a preprint area maintained by someone else, and to simultaneously submit the article for consideration by a journal. Whereas before, only a handful of people might ever thoroughly read your journal article, now anyone with a web browser can find it

using a search engine and read it. Yet, there is so much information on the web, that we want to narrow our search. As societies abandon paper in favor of web-publishing, there still will be *value* in doing your search for simulation papers via society servers first since that is where papers have been checked for quality.

2.3 Simulation Programs

The most exciting part of web-based simulation is in the simulations themselves. Operating on the web gives new meaning to the word *document*. A web document can contain videos, interactive simulations, images and audio in addition to the usual text that traditional documents contain. How do we embed simulations in documents? A simulation requires a model and a model requires computing hardware to execute the model. Some parts of a model can be executed on another machine and other parts are executed on your machine where your web browser is located. For simplicity, we'll assume that you are the one building the simulation and others want to execute it. To build a simulation which executes on your machine requires that someone else (the client) access your simulation (the server) and cause a program to execute. Typically scripting languages such as Perl are used to build a high level program that runs your simulation code once you have obtained all required simulation inputs and parameters, generally using a "form" capability. If many users are running your simulation simultaneously, you will need to make sure your Perl script is set up to handle this. Your computer resources may be taxed, depending on the number of readers wanting to run the simulation. The idea of Java and Javascript is to bundle a piece of code (called an applet) and send it to the client's computer to execute, not yours. This way, it makes no difference how many clients want to run your simulation at least from the perspective of model execution.

In our discussion so far, you have constructed a simulation which runs on your computer or sends applets to a client's computer. This is fine for single-user simulations, but what if you need to have a simulation where multiple users *interact* with each other? The best example of user interaction is found in multi-user gaming simulations such as DOOM and QUAKE. In such a simulation, a global server (one person's machine) is used to capture and maintain the state of the world. Each player makes moves and information is moved from client to server. Multi-User Dungeons (MUD) operate in a similar fashion, the difference being that the interaction among users may extend toward a cooperative relation to other players rather than a strictly competitive one. The MUD is also a basis for applying education through simulation within a virtual simulated "world." This represents a shared solution to multi-user simulation.

If the simulation model is spread over a network of computers, then we have a distributed solution. Each computer is associated with either a single physical entity or a set of entities. The distributed interactive simulation (DIS) initiative begun by the Department of Defense (DoD) is a good example of distributed model execution. The DoD focus is to have an interactive simulation capability mainly for training people, but increasingly other aspects of the military (such as "acquisition") are finding uses for DIS. Some of the computers in a DIS network may be running semi-automated forces (SAFOR) and others will have humans attached to often sophisticated human-computer interfaces so that the training exercise is made as realistic as possible. If there is no real-time human-computer interaction (HCI) necessary, we may still split our models and do parallel and distributed simulation (using conservative or optimistic time advance).

How does the web relate to all of this? Since the web is a collection of documents, an interactive simulation could ostensibly be entered from within a web document. A user locates a document, presses the mouse button on an image portraying a virtual world and then is immersed in a simulation. Most DIS simulations do not currently operate in this fashion, but there is no reason why this could not be done. After all, the web is all about storing and retrieving multimedia information, and simulation is a key *kind of information*. The virtual reality markup language (VRML) suggests a way to do this. VRML plugins allow one to construct 3D geometry and have it rendered in real time. A document contains a URL which points to a VRML file. The browser recognizes this particular information type and launches the VRML plugin. Then the user browses a 3D scene using whatever HCI devices are available.

With DIS and MUDs, we are only scratching the surface of what is possible. You have probably noticed that most companies currently have web sites. What if they were to take the products that they manufacture and put digital equivalents of these products on their web site? This would radically change the way in which we do modeling and simulation. One of the biggest reasons why simulation of large systems is expensive is that each one of us rebuilds model components instead of re-using what could be out there on the web. Customers will benefit from such web-based models since they can try out a product before purchasing it. This also goes for industrial customers as much as it does for individuals. For the original equipment manufacturers, this scheme will allow for them to sell more equipment since potential buyers can try out the equipment simply by linking the part of their simulation model which models the equipment to the manufacturer's web site. Manufacturers will have to spend more time building digital web-based models for their products, but this will be time well spent once readers and users locate the models.

We have done some preliminary studies of web-based simulation at the University of Florida. Figure 1 displays the home web page for a simulation of four disks and one CPU. A job enters the CPU and then is assigned a disk unit for file access. Then, this job loops a certain number of times. This looping simulates the time taken by an average program (i.e., job) which has need of CPU and file storage resources. The web page is organized in an overall linear fashion, starting with a paper title, author list and then proceeding with a picture of the system to be simulated. The square area of the screen uses an "ISMAP" feature which allows us (the authors of the web page) to perform a given action if the reader clicks on a certain region within this square. At the bottom of Figure 1, the user clicks on this to begin entering simulation input information. All input is entered using the FORMS feature of HTML. Once the the information has been entered, the reader can execute the simulation. Since our simulation uses the "server side approach," using a Perl script which invokes an executable on a SPARC Unix workstation, the simulation executes and produces time-dependent graphs for CPU and Disk resource queue sizes. The graphs are produced by dynamically creating an HTML web page from the simulation output, which is then available to the reader.

3 REDUCING THE COST OF SIMULATION

Simulation has always been a costly enterprise, and even though more inexpensive computer equipment is driving costs down, the ever-lurking cost makes itself known in all of our concerns: education, publication and simulation programs. We have just discussed reducing this cost by the re-use of objects within simulation models but there is a broader strategy: advertising. Everyone has questions about how money is to be effectively exchanged over the web. Revenue issues will affect the simulation community as much as any other, so we need to consider the transfer of money. Simulation products and services can be bought through traditional credit card means, and we have seen that many services (in the form of web products) are free for the client side of information delivery. The cost is on the server side. Many simulation services may be possible through inventive advertising methods such as including an advertiser's message in a web page accessed by simulation practitioners. Most web-search engines contain this feature. The web user benefits by using the search engine and the search engine service is paid for by the advertiser. Everyone wins. In this way, education, publication and simulation software may become cheaper as a result of embedded advertisements. We should be concerned about over-commercializing simulation knowledge; however, web advertisements are *less obtrusive* to the reader and provide direct links to the advertiser. Furthermore, by the web's nature, advertiser information tends to be more information-rich compared to other types of media such as television, radio and magazines. The advertising links are also interactive so there is the potential for providing quality information to readers. This means that simulation publications, educational products and interactive simulation programs of the near future might incorporate direct advertising to cover costs. The advertising is less blatant, more accommodating, and provides for a three-way benefit to the server, client and advertiser.

4 SUMMARY

We are in a kind of twenty-first century gold rush since everyone is rushing to put multimedia information on the web, but there is great uncertainty about how things such look and feel. How will simulation-related societies change and deliver their publications? Will companies expand their sales by offering simulation models of their products? One thing is for certain. We will never know any of these answers unless we experiment and take risks if necessary. The "right way to do simulation" will naturally emerge and evolve as we try various web-based simulation approaches. The worst possible approach is to sit back and adopt a "let's wait" attitude: simulation will be worse off as a discipline unless we move forward now to incorporate the web-based technologies.

ACKNOWLEDGMENTS

I would like to acknowledge the following funding sources which have contributed towards our study of modeling and simulation. (1) Rome Laboratory, Griffiss Air Force Base, New York under contract F30602-95-C-0267 and grant F30602-95-1-0031; (2) Department of the Interior under grant 14-45-0009-1544-154 and the (3) National Science Foundation Engineering Research Center (ERC) in Particle Science and Technology at the University of Florida

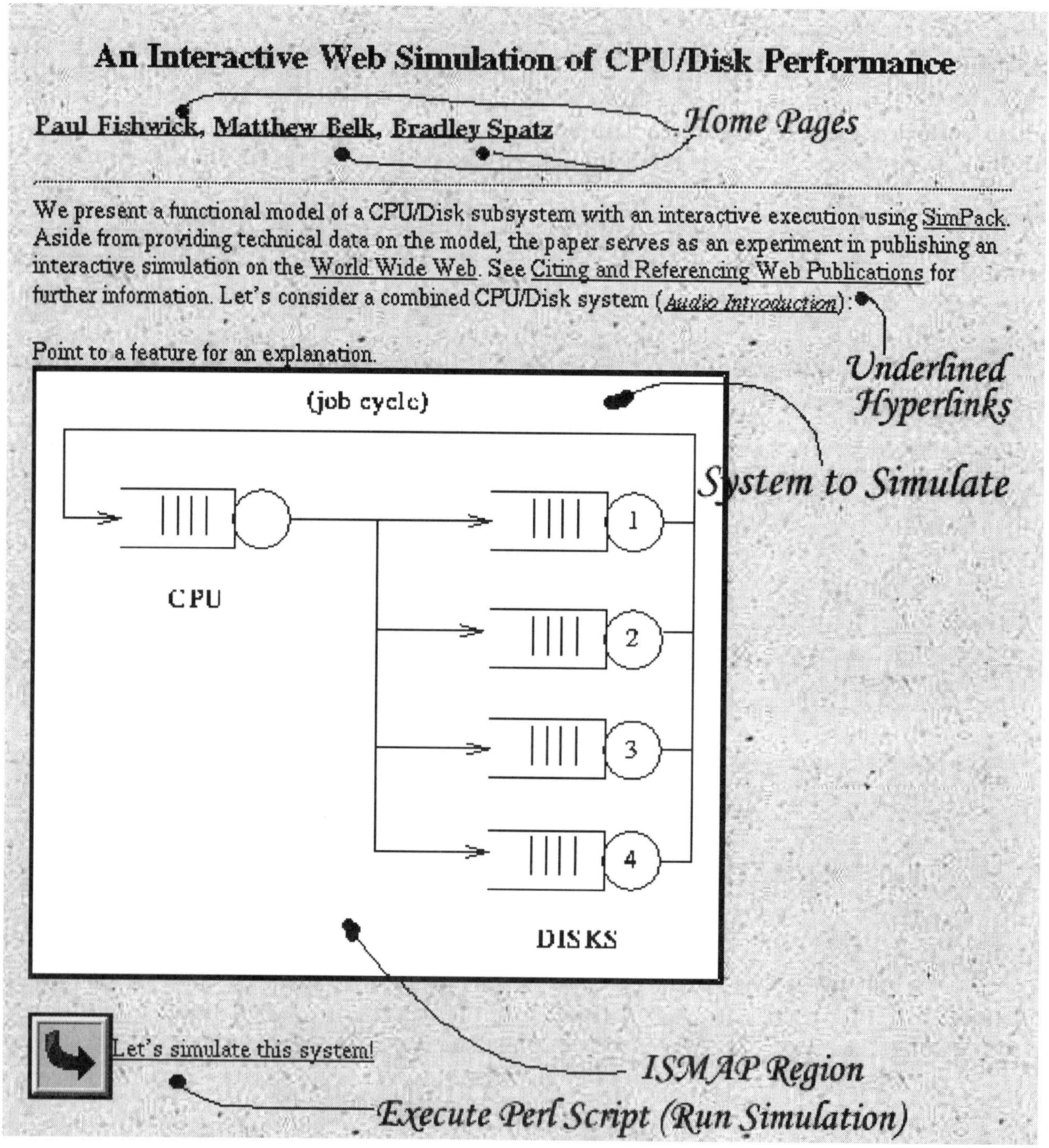

Figure 1: Web page of a CPU/Disk simulation.

(with Industrial Partners of the ERC) under grant EEC-94-02989.

AUTHOR BIOGRAPHY

PAUL A. FISHWICK is an Associate Professor in the Department of Computer and Information Science and Engineering at the University of Florida. He received the PhD in Computer and Information Science from the University of Pennsylvania in 1986. He also has six years of industrial/government production and research experience working at Newport News Shipbuilding and Dry Dock Co. (doing CAD/ CAM parts definition research) and at NASA Langley Research Center (studying engineering data base models for structural engineering). His research interests are in computer simulation modeling and analysis methods for complex systems. He is a senior member of the IEEE and the Society for Computer Simulation. He is also a member of the IEEE Society for Systems, Man and Cybernetics, ACM and AAAI. Dr. Fishwick founded the `comp.simulation` Internet news group (Simulation Digest) in 1987. He has chaired workshops and conferences in the area of computer simulation, and will serve as General Chair of the 2000 Winter Simulation Conference. He was chairman of the IEEE Computer Society technical committee on simulation (TCSIM) for two years (1988-1990) and he is on the editorial boards of several journals including the *ACM Transactions on Modeling and Computer Simulation*, *IEEE Transactions on Systems, Man and Cybernetics*, *The Transactions of the Society for Computer Simulation*, *International Journal of Computer Simulation*, and the *Journal of Systems Engineering*. Dr. Fishwick's WWW home page is `http://www.cise.ufl.edu/ ~fishwick` and his E-mail address is `fishwick@cise.ufl.edu`.

Proceedings of the 1996 Winter Simulation Conference
ed. J. M. Charnes, D. J. Morrice, D. T. Brunner, and J. J. Swain

DISCRETE EVENT SIMULATION ON THE WORLD WIDE WEB USING JAVA

Arnold H. Buss
Kirk A. Stork

Operations Research Department
Naval Postgraduate School
Monterey, CA 93943-5000, U.S.A.

ABSTRACT

This paper introduces Simkit, a small set of Java classes for creating discrete event simulation models. Simkit may be used to either implement stand-alone models or Web page applets. Exploiting network capabilities of Java, the *lingua franca* of the World Wide Web (WWW), Simkit models can easily be implemented as applets and executed in a Web browser. Java's graphical capabilities enable the rapid development of intuitive user interfaces. Java's use of interpreted bytecodes, while imposing a performance penalty, enable development of platform-independent models. The language's inherent internet-awareness make other possibilities, such as distributed simulation, much easier to implement.

1 INTRODUCTION

The Naval Postgraduate School Simkit was developed to provide simple simulation tools in an object oriented computer language supported on a wide variety of computing hardware. The goal was to provide simulation tools for the analyst and researcher that:

- are accessible to analysts without professional programming skills.
- are reliable enough for moderately sized projects.
- are capable enough for real problems.
- are conducive to rapid development of exploratory models.
- are low cost in money, programmer time and resources.
- promote code sharing and reuse.
- promote model sharing and reuse.
- allow for exploration of advanced simulation concepts, such as distributed simulation and remote entities

Examination of these objectives lead to the rejection of specialized high level languages that are typically expensive and require a workstation to run. General high level languages were explored in search of an accessible, available object oriented language, resulting in the selection of Java over other prime candidates such as Objective C, C++, Smalltalk and others. Perhaps most important, Java's internet capabilities make it the only candidate capable of creating simulation models that can be run on the Web itself.

The WWW presents many new possibilities for creation, distribution, and execution of simulation models (indeed, of *all* computer models). Some of the potential uses include the following.

Models can be created and posted to a Web site so any user with a Java-enabled browser can execute the model. Since the browser executes compiled (bytecode) classes, the source coded need not be revealed. Java capable browsers, such as Netscape Navigator 2.0 and Microsoft Internet Explorer 3.0, are becoming the standard, so such simulation models will be widely available for use.

With the wider distribution of running models, a substantially larger potential user base exists. The useful feedback from this user base could be orders of magnitude larger than possible with traditional methods of model distribution.

Modelers at different locations can develop parts of the same project and post their classes for the other members of the team. With Java's ability to import remote classes, new subproject classes could rapidly be created to interface with existing classes. Since the classes themselves, rather than a copy, would be used by the other team members for testing their code, the likelihood of errors creeping in by noisy copies is reduced. Posting compiled bytecodes also helps the reliability of the overall project. By accelerating the availability of prototype classes to the project, the entire project could become more robust as well as more timely.

Finally, Javasim supports the possibility of distributed simulation through Java's built-in internet awareness. Simulation models my be executed remotely, with different parts executing on machines that could be anywhere. Furthermore, even the client/server structure typically necessary for proper synchronization is also easily implemented in Java itself.

The remainder of this paper is organized as follows. In the next section we will give a brief overview of the Java language with emphasis on features that are key to Javasim. Next, we discuss Object-oriented design as applied to simulation programming. Section 4 describes the Simkit hierarchy and Section 5 illustrates its use in a simulation applet. Finally, conclusions and directions of future research are presented.

2 JAVA

The hype surrounding the Java language gives a casual observer the impression that it is only suitable for writing animated "applets," small programs that typically have cute but inconsequential animations. The explosion of such applets on the WWW, together with Java capability of the major Web browsers, support the notion that Java is not a "serious" language. In fact, Java is a powerful, well-designed language that is an excellent platform for developing complex, object-oriented applications.

Even a modestly comprehensive description of the Java language is obviously beyond the scope of this paper; for an excellent overview, see Cornell and Horstman (1996). However, several properties of Java made it the ideal platform for Simkit, as discussed above.

Java supports most of the language features for object-oriented programming, including classes, encapsulation, polymorphism, and inheritance. Unlike C++, Java does not support multiple inheritance, but rather borrows the protocol concept from Objective-C, called an "interface" in Java. An interface is simply the promise that a given class will implement a specified list of methods. Use of interfaces gives the modeler much more flexibility for extensions to base functionality than inheritance without giving up any power. Indeed, without interfaces the only way to combine different groups of method specifications is through multiple inheritance. It turns out that multiple inheritance is far too restrictive, particularly for the kind of flexibility required in simulation modeling. Interfaces are a critical feature in Javasim, and most of the basic abstract functionality are defined in interfaces (see Section 4).

3 DESIGN CONSIDERATIONS

There is considerable difference between programming in an Object-Oriented (OO)language and utilizing Object-Oriented design concepts (Bootch, 1996). Although an OO language may support OO design, it does not guarantee it. The design view we take is along the lines of Cox's concept of the "Software Integrated Circuit" (Cox, 1986). Cox's "software IC" concept. The fundamental components for a simulation software IC are those of the Event Graph (Schruben, 1995).

In designing Simkit, we attempted to break the act of creating a simulation model into its fundamental functional components. By utilizing available modules that provided functionality, such as Doug Lea's public-domain collections classes, we maintained a certain degree of modularity on the project.

A simulation language typically takes a certain "World View," the lens through which the underlying modeling paradigm views the model landscape. Major world views include Next-Event, Process/Resource Interaction, and Activity Scanning (Law and Kelton, 1991). The Process world view appears to be the most common among commercial languages (GPSS, SIMAN, SIMSCRIPT, etc). Implementations of the Process view involve an Event List operating in the background. Thus, while the user creates the model using high-level block diagrams, the language translates the resulting structure into a Discrete-Event model.

The Process world view allows the modeler to operate at a relatively high level of modeling. Relatively straightforward queueing-oriented models are easy to create and run. However, even modest departures from queueing models forces the modeler out of the pre-constructed blocks. Commercially available languages typically force the user to implement user-defined blocks in conventional programming languages, such as C or Fortran.

The design objective for Simkit was to achieve as much modeling ease as possible by enabling the construction of resident entities and blocking resources, while allowing the user to customize at a higher level than conventional programming. The latter was accomplished by incorporating a Discrete-Event approach based Lee Schruben's Event Graphs (1983, 1995). Thus, the user of Simkit has the choice of implementing a model using a Process orientation, a Discrete-Event Orientation, or some combination. The use of Event Graphs to create customized "blocks" is a powerful modeling capability.

4 OVERVIEW OF Simkit

Simkit consists of a programming class library for discrete event simulation written in the Java computer language. Simkit provides general code for simulation and contains specific code providing modeling capabilities of interest to the military analyst. We will addresses only the general simulation tools provided Simkit, with primary emphasis on Javasim, the core of Simkit.

Simkit consists of three Java packages, each containing a set of related classes that work together to provide functionality in a key area to support the model. Graphical User Interface elements are provided by the classes in the awt package. Utilities, such as data structures, random variate generation, and statistics collection, are implemented in the util classes. Event-driven simulation is facilitated by the classes in the Javasim package. The packages form a hierarchy, much like a class inheritance hierarchy, which is shown as a directed acyclic graph in Figure 1.

The Mil package is an extension to Javasim that provides tools for building simulation models with particular application to military problems. In addition to the intrinsic interest in such models, Mil can be seen as a prototype that demonstrates the extensibility of Javasim. The design of Javasim enables modelers to extend the language itself to incorporate their particular needs. This extensibility makes Javasim an attractive alternative to special-purpose simulation languages which tend to be more restrictive.

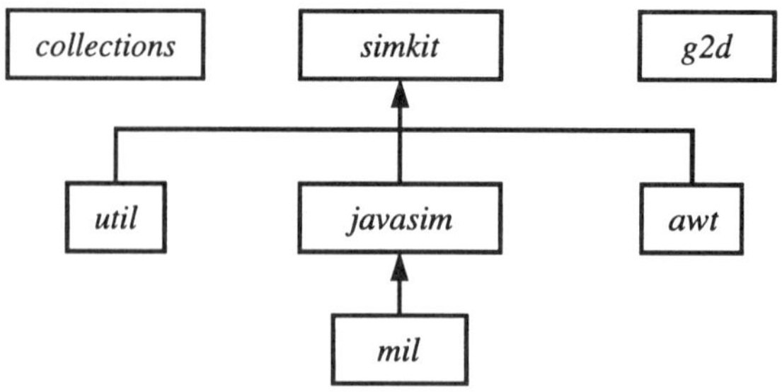

Figure 1: Simkit Package Hierarchy

Classes in one package can use classes in other packages. In fact, Javasim relies on classes in each of the other packages, each providing a distinct set of services to the programmer. We will now briefly describe each package. In the spirit of OO design, we have utilized existing classes to as great extent as possible, specifically the Collections and g2d packages.

4.1 Collections and G2d

The collections package, generously contributed to the public domain by Doug Lea, provides a very complete set of data structures. This package is completely independent of Simkit and is useful in many contexts. Simkit uses collections for most of its data structure needs. Complete documentation for collections is provided with the source code distribution available on the internet.

Charting capabilities are provided by the g2d (Graphics, 2-dimensional) package. Minor modifications were made to the original Graph2D classes provided by Leigh Brookshaw to give them their own identity and to simplify the programming interface for Simkit. The package itself is quite complete, and can be used independently of Simkit. Documentation for the original Graph2D is included with the source code distribution on the internet. The modified code making up the g2d package as shown here is also available on the internet, but the changes are slight. We recommend getting the most recent source code from the original author for uses other than Simkit. While not in the public domain, the g2d package is covered by the GNU General Public License (1991).

4.2 Awt

The classes in the awt package provide a limited set of pre-built graphical user interface elements used by Javasim. The main features of awt include simple windows to display tables and graphs. Awt relies on g2d for its lower-level functions. Several classes originally presented in the book Core Java (Cornell and Horstman, 1996) are used to provide formatted number output and self verifying text input boxes.

4.3 Util

The util package contains classes that give support functionality to the simulation model. These include data collection, random variate generation, and data structures that gather summary statistics on usage.

4.4 Javasim

Javasim is the core of Simkit and provides all the general discrete-event simulation facilities in Simkit. Although incomplete in comparison to commercial simulation environments, Javasim provides the most of the needed functionality for simulation modeling.

The structure of Javasim is shown in Figure 2. There are three main layers to the overall design: The core Javasim classes, an Extension layer, and a User layer. As Figure 2 indicates, Javasim consists of two

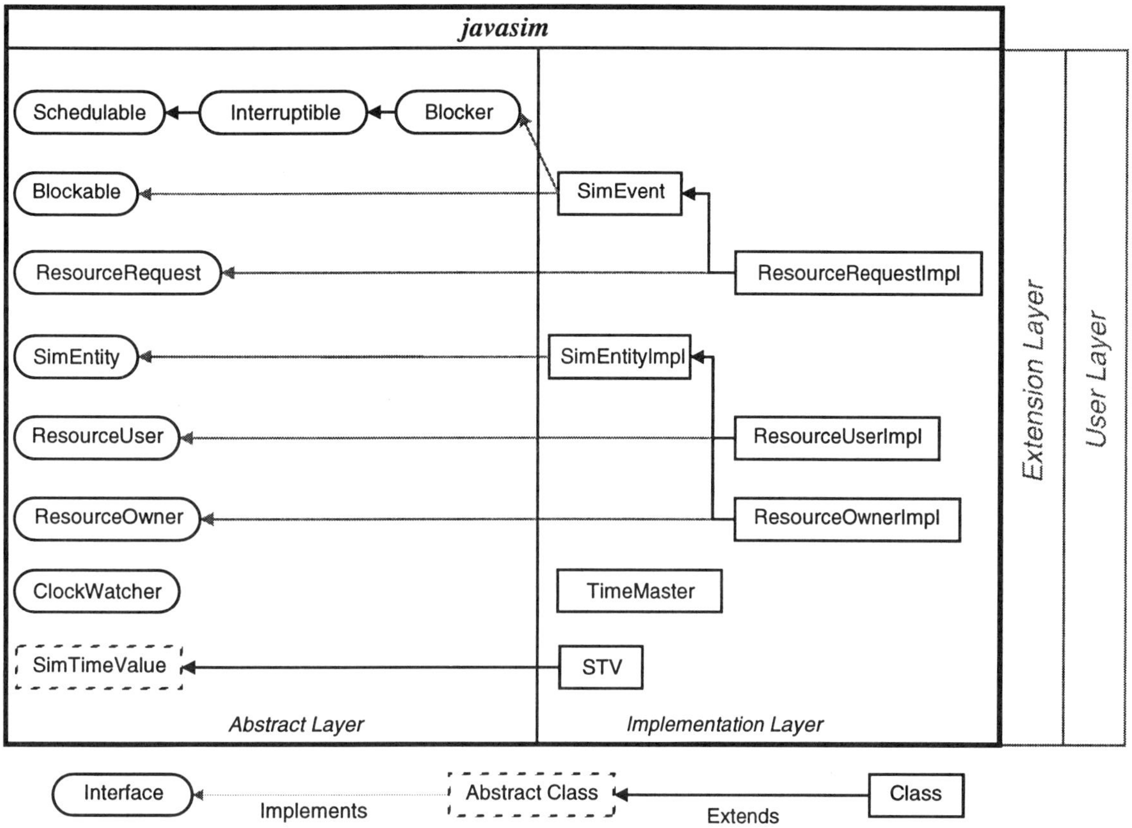

Figure 2: Javasim Design

components: an "Abstract" layer and an "Implementation" layer. The Abstract layer defines functionality mostly through Java Interfaces. Each interface only provides a list of methods that are to be implemented by classes in the Implementation layer. The relationship between interfaces and the classes that implement them is much weaker than the inheritance relationship. Specifically, a class which implements an interface does not have an "is-a" relationship to the interface. Rather, the class simply promises that it will have the interface's methods in some manner. Interfaces are thus identical to Objective-C protocols. The one "Abstract" class, SimTimeValue, provides the timekeeping in a robust manner.

The basic discrete event framework in Javasim is provided by the Timemaster class, which manages the event list. The execution and scheduling of events is done by the SimEvent and SimEntityImp classes, respectively. The Interruptable interface allows the possibility of events being cancelled. The STV class implements simulated time. These classes are sufficient to create any discrete-event simulation models.

The remaining three classes, ResourceRequestImpl, ResourceUserImpl, and ResourceOwnerImpl, together implement the concept of blocking resources, allowing some degree of creating Process/Resource models with Simkit.

Using interfaces rather than an inheritance hierarchy allows Javasim to easily incorporate the Extension Layer. While a modeler will not change anything in Javasim itself, there may still be a need for special-purpose, customized tools. For example, the Mil extension supports the ability for the modeler to easily create such entities as ships, airplanes, submarines, etc. Furthermore, such interactions between model entities, such as sensing and detection, can be created as as part of the language itself by use of the extension layer.

4.5 Mil

Mil is a simulation framework for simulating scenarios of interest to military analysts. In addition to being an example for the Extension layer of Simkit, Mil provides functionality for such military applications as ship-missile defense (Stork, 1996).

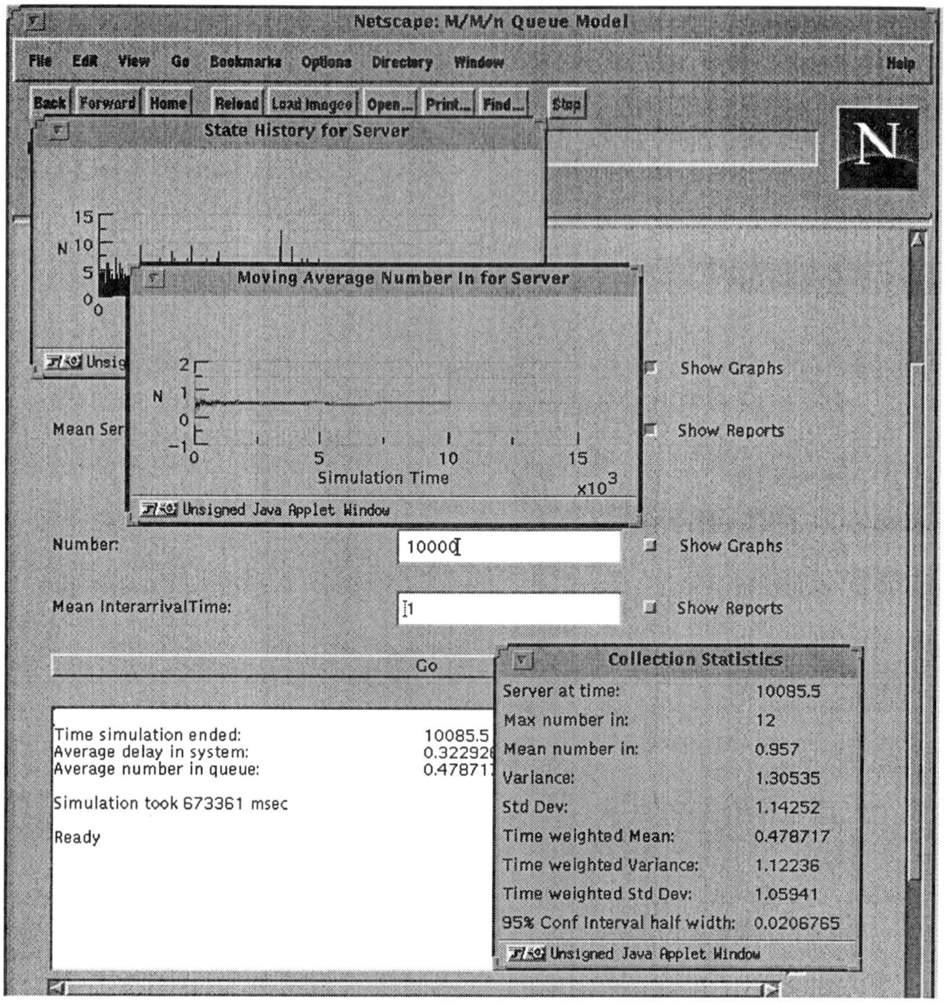

Figure 3: Output of MMn Applet

5 EXAMPLE

We illustrate a model in Simkit with the simple $M/M/n$ queue as an applet. A screen shot of the applet running in Netscape is shown in Figure 3. The user interface is easily implemented and allows parameter input from text boxes and execution with a simple button. Check boxes allow the user to decide at run time to display choices of summary statistics and graphs of output.

To illustrate how Simkit is be used for a Process/Resource view of the model, we show the source code for the ResCustomer class below.

```
class ResCustomer extends ResourceUserImpl {
  private double myServiceDelay;
  private Time myServiceRequestTime;
  private ResourceOwnerImpl myServer;
  private Histogram delayStat;
  private double meanST;

  public void setServiceTime(double d) {
    myServiceDelay = d;
  }

  public void getService(
                  ResourceOwnerImpl server,
                  Histogram reportTo) {
    myServer = server;
    delayStat = reportTo;
    myServiceRequestTime = TimeMaster.SimTime();
    getResources( myServer, 1L );
  }

  public void receiveResources(
                        ResourceRequest res){
    super.receiveResources(res);
    delayStat.getSample(
      ((TimeMaster.SimTime()).value()) -
```

```
        ( myServiceRequestTime.value()));
    ResEndService e = new ResEndService( this );
    e.waitDelay( new Time(myServiceDelay) );
  }

  public void doEndService() {
    returnResources(myServer);
  }
} // END CLASS Customer
```

The ResCustomer class provides the core functionality of the model. The methods are mostly self-explanatory, representing the customer requesting and subsequently receiving the server resource, waiting for the service time, and relinquishing the server. This class also illustrates the use of the Time class wrapper. All variables representing time in Simkit are declared to be an instance of Time. This abstraction of the concept of a time adds robustness to Simkit models.

6 CONCLUSIONS

Simkit is a class library written in Java that provides core functionality for the creation of simulation models. While relatively small, Simkit is nevertheless is sufficiently powerful for many modeling needs. Its user layer allows Simkit to be used as-is for creating simulations, and its extension layer allows language extensions at a level substantially higher than base-level code. Unique network-capable features of Java can be exploited to enable straightforward implementation of simulation applets and distributed models.

ACKNOWLEDGEMENTS

Support for this work from the Naval Postgraduate School is gratefully acknowledged. Portions of this work were part of the second author's Masters Thesis in Operations Research at the Naval Postgraduate School.

REFERENCES

Bootch, G. 1996. *Object Solutions: Managing the Object-Oriented Project*, Addison-Wesley, NY.

Cornell, G. and C. Horstman. 1996. *Core Java*, Sunsoft Press, Mountain View, CA.

Cox, B. 1986. *Object-Oriented Programming: An Evolutionary Approach*, Addison-Wesley, NY.

Flanagan, D. 1996. *Java in a Nutshell*, O'Reilly & Associates, Inc., Sebastopol, CA.

GNU General Public License. 1991. The Free Software Foundation, Cambridge, MA.

Law, A. and D. Kelton. 1991. *Simulation Modeling and Analysis, Second Edition*, McGraw-Hill, NY.

Schruben, L. 1983. Simulation Modeling with Event Graphs, *Communications of the ACM*, **26**, 957–963.

Schruben, Lee. 1995. *Graphical Simulation Modeling and Analysis Using Sigma for Windows*, Boyd and Fraser Publishing Company, Danvers, MA.

Stork, K. 1996. A Simulation Study of Countermeasure Effectiveness Against Anti-Ship Missiles. Masters Thesis, Department of Operations Research, Naval Postgraduate School, Monterey, CA.

AUTHORS BIOGRAPHIES

ARNOLD BUSS is a Visiting Assistant Professor of Operations Research at the Naval Graduate School. He received his PhD in Operations Research from Cornell University, and is a member of INFORMS, IIE, POMS, and MORS. His research interests include Object-Oriented simulation modeling, manufacturing applications, and project management.

KIRK STORK is a Lieutenant in the United States Navy. After enlisting in 1981 and completing the Navy's Nuclear Power Training, he attended the University of Washington and obtained his BSE and a commission in the Navy in 1989. He received his MSOR at the Naval Postgraduate School and is currently attending the Navy's Submarine Officer's Advanced Course in Groton, CT.

Proceedings of the 1996 Winter Simulation Conference
ed. J. M. Charnes, D. J. Morrice, D. T. Brunner, and J. J. Swain

JAVA-BASED QUERY DRIVEN SIMULATION ENVIRONMENT

Rajesh S. Nair
John A. Miller
Zhiwei Zhang

Department of Computer Science
University of Georgia
Athens, Ga 30602-7404, U.S.A.

ABSTRACT

The concept of Web-based simulation can be finally realized using Java. We have used Java to create a powerful simulation modeling library which is based on the process interaction paradigm. Java threads make it easy for us to implement each process (or active entity) as a thread. Our library, JSIM, supports both simulation and animation thus rendering the model developer's job easier. An important component of our approach is that we integrate our Java Simulator with a Database Management System (DBMS). The conceptual basis for this integration is an environment based on Query Driven Simulation, which allows simulation analysts to study the performance of systems simply by querying a database. Simulation inputs and outputs are stored in databases, and simulation models can be launched as a part of query processing. This integration also resolves a security restriction problem which Java imposes. Most browsers (e.g., Netscape Navigator) restrict Java applets from writing into a file which may be necessary to make simulation results available to the user. Our approach bypasses this problem by writing into a database. Databases are much less of a security concern since they already have substantial authorization mechanisms.

1 INTRODUCTION

Simulation has come a long way since its inception. An increasing number of real world concepts are being modeled and studied using simulation. Recently, simulation has been influenced by an increasingly popular phenomenon - the World Wide Web or WWW, as it is known. The popularity of the Web has propelled it into being a medium for all kinds of applications including simulation.

The concept of Web-based simulation can be finally realized using Java. Java is an object-oriented programming language for the Web. With Java, it is possible to have "executables" on the Web. This provides users with a dynamic environment as opposed to a largely static, text-driven environment that was existent prior to Java. Simulation users can finally view results over the Web in a dynamic environment complete with animation, sound and textual information.

Java is an important advance in Web technology, and one that is particularly important to simulation. Previously, interactivity or minimal dynamism on the Web were accomplished using Common Gateway Interface (CGI) scripts, typically coded in Perl or C/C++. Unfortunately, these scripts run on Web servers, thus limiting how dynamic the display on a remote Web browser could be. Java is fundamentally different in that it allows applets downloaded from a Web server to be run by the browser. Therefore, dynamism is only limited by the speed at which Java code can be executed. Since Java aims for universal portability, browsers provide a byte code interpreter to execute code produced by a Java compiler. Consequently, it is not as fast as purely compiled code (e.g., C/C++) but is faster than purely interpreted code (e.g., Perl). However, in the future it is expected that native code compilers will be available for Java to make its speed competitive with C/C++. It should be noted that since Java source code (.java) is compiled to byte code (.class), source code need not be made available to the Web as it must be for purely interpreted languages. Let us summarize by listing some of advantages of using Java to implement simulation models.

1. Simulation models implemented as Java applets can be made widely available. Anyone that you give permission to that has a Java capable Web browser (e.g., Netscape 2.0 or higher, HotJava 1.0 or higher) could potentially run your simulation model on their local machine.

2. The goal of universal portability means that one simply retrieves an applet and runs it. One does

not have to port to a different platform, or even recompile or relink.

3. Java applets run on a browser allowing a higher degree of dynamism.

4. Java has built-in threads making it easier to implement simulations following the process interaction paradigm.

5. Java has built-in support for producing sophisticated animations. Geometric objects can be readily created, moved, color changed and destroyed. In addition, icons/images (.gif, .jpeg) can be similarly used.

6. Some also see Java as a better C++. It is smaller, cleaner, safer and easier to learn than C++. Notably, it does not provide pointers and automates storage management through the use of garbage collection. Thus, many of the most common C++ errors would be avoided.

Java shows great promise for simulation and animation. We have used these strengths of Java to build a comprehensive library, called JSIM, to support both simulation and animation. One potential problem with Java involves making data persistent. For example, in Netscape Navigator 2.0, applets cannot read or write files at all (see http://www.javasoft.com/java.sun.com/sfaq). Our solution to this problem is to have applets read from and write to a database instead. In fact, a Database Management System (DBMS) is an integral part of our simulation environment. The integration of the simulation and database components is done following the notion of Query Driven Simulation (QDS).

In the rest of this paper, we overview the Query Driven Simulation environment, and then cover the three main aspects of the JSIM library: simulation, animation and database access. This is followed by an example simulation model and an example QDS query. Finally, conclusions and future work are presented.

2 QUERY DRIVEN SIMULATION

Query Driven Simulation is based on the tenet that simulation analysts as well as naive users should see a system based upon QDS as a sophisticated information system which will be able to store or generate information about the behavior of systems the user wishes to study (Miller and Weyrich 1989,Miller et al. 1991a,1991b,Miller et al. 1996a, Miller et al. 1996b). This means that the user must be provided with an easy to use environment where he/she may trigger an action using a simple query language.

The QDS environment principally consists of three communicating processes and several data stores (see Figure 1). The processes include a Web browser (e.g. Netscape Navigator or HotJava), a Web server (e.g. Netscape Commerce Server) and a DBMS (e.g. m-SQL). The data stores consist of models, data and meta-data stored in databases or files accessible to the Web server. The Web server will also store the applets used in the QDS environment such as the QDS Applet, the Design Applet and potentially numerous model applets. The JSIM library simplifies the development of model applets. The QDS environment is aimed towards providing a friendly and easy to use interface to the user. A typical session involving a user and the QDS system can be explained as follows. When the user first contacts the Web server (e.g. httpd), they download the main control applet for the environment, namely the QDS Applet. The QDS Applet upon starting up gives the following options to the user.

1. Find Model Applet
2. Run Model Applet
3. Submit SQL Query
4. Save Results to DBMS
5. Update Model Index
6. Submit QDS Query
7. Design Model

All the options, excepting option 7, can be explained by outlining the functions performed by option 6. If the user elects to submit a QDS query, the QDS Applet first treats the query as a normal SQL query and submits it to the database engine. The query is run against the existing database and the results are returned to the QDS Applet. If the results are adequate, then they are presented to the user. Otherwise, the QDS Applet submits a query to the Model Index in the database to determine the relevant model. Once the entry for the model in the Model Index table is found, the QDS Applet extracts the parameters required by the model applet from the query itself. (For option 2, the QDS Applet requests the user for the parameters.) This is called "defining a scenario". The QDS Applet then requests the Web server for the model applet code. The Web server responds by providing the applet code which is then executed with the appropriate parameters. The model applets use the classes provided in the JSIM library (section 3.4). The results generated by the model execution may or may not be stored in the database

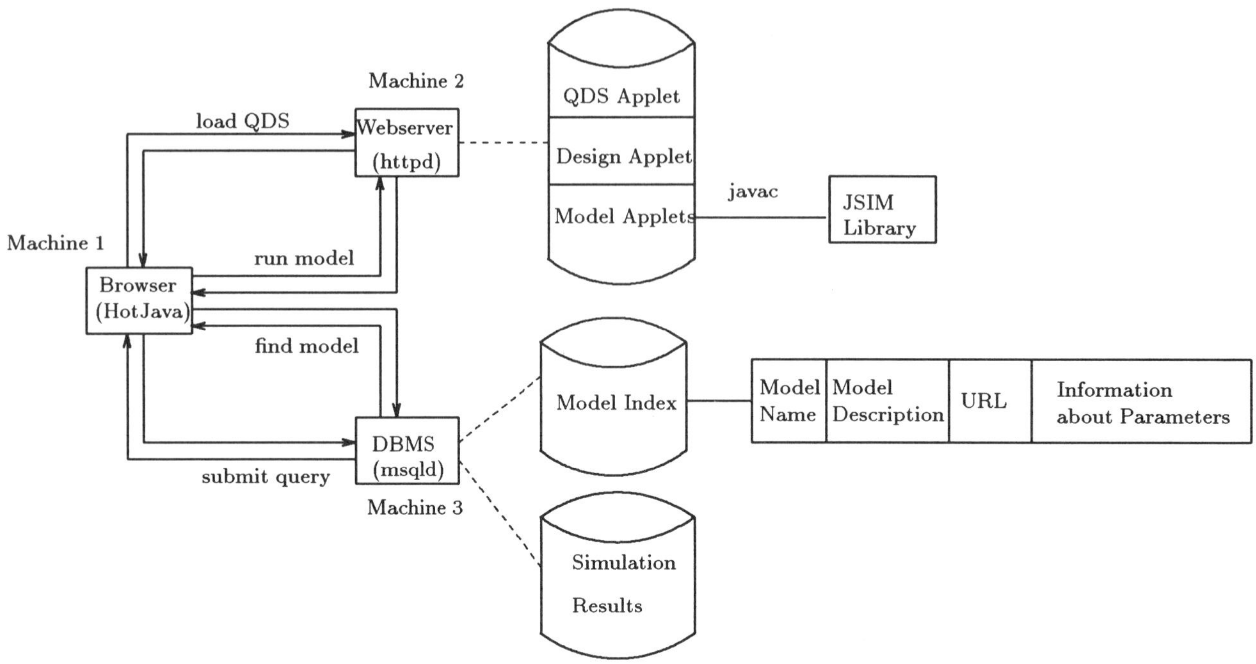

Figure 1 : The QDS Environment

depending on the user's choice. It is also the responsibility of the model applet to display the results of model execution to the user. Option 7, Design Model, is different from the other options in that it is meant for the simulation model builder rather than the user. A special applet called the Design Applet is invoked when the user selects this option. The Design Applet helps the model builder design a simulation model by providing him/her with a graphical interface in which a model can be built primarily by drag and drop operations. Currently, the Design Applet is not fully functional.

3 JSIM:THE JAVA SIMULATION LIBRARY

JSIM, written in Java, is a simulation toolkit with a library of Java classes. JSIM provides most of the features of SIMODULA developed by (Miller and Weyrich 1989). In our implementation, we have built simulation, animation, random variate generation, statistical analysis and database access facilities into the library. In building our library, we followed the process interaction paradigm. Java's built-in support for threads helped us in implementing a true process interaction paradigm. The simulation entities have been coded as threads and are truly independent and may perform actions on their own. Throughout the discussion, we consider a simple simulation example - that of a bank with one or more tellers and multiple customers. The customer arrival time is random as is the service time.

In the following subsection, we discuss our library in greater detail. We begin with the simulation component.

3.1 Simulation Facilities

We use the process interaction paradigm for simulation since we feel it is closer to real world concepts than the event scheduling approach. Java threads make it possible for us to realize the process interaction paradigm in a straightforward manner.

As can be seen from the bank simulation example of section 4, both customers as well as tellers are independent entities. Each simulation entity, namely customer, teller, etc., is coded as a Java thread. Java's support for multi-threaded programming makes it an ideal platform for an implementation of the process interaction paradigm.

Furthermore, we have provided for both virtual-time clock and real-time clock simulation capabilities. The virtual-time clock simulation is ideal for speedy simulation with little or no machine interference in simulation timings. It is particularly helpful if the real world events that are being simulated are long lived. Virtual-time clock simulation uses the so-called "next-event" strategy for advancing simulation time. Here, the clock value jumps to the time at which the next event is due to occur. It is usually implemented with a scheduler and a Future Event List (FEL). The scheduler first schedules an event by inserting it into the FEL depending on its activation time. It then transfers to the event at the front of the FEL by removing it from the FEL and updating the simulation clock to the time associated with that event. Real-time clock simulation, as the name suggests involves using a real world clock (system clock time obtained using the Java function System.currentTimeMillis) for simulation timing. It is well suited for real-time animation where simulation events are represented by corresponding animation events. In real-time clock simulation, the clock does not move in jumps but progresses smoothly. Entities are delayed by using sleep or suspend which are methods within the thread class.

We also provide for animation of virtual-time clock simulation as is explained in the next subsection.

3.2 Animation Facilities

As mentioned above, we provide for animation of both real-time clock and virtual-time clock simulations. To support real-time clock simulation, we built in animation capabilities into each of the simulation entities. Each simulation entity has its own *draw* method. We used a technique called double buffering to implement multi-thread animation capabilities. The double buffering technique is generally used to reduce flicker during animation. However, we have used this technique to our advantage in a totally different manner. A single *display thread* periodically calls every active simulation entity to draw itself into an offscreen buffer. After the offscreen buffer has been completely updated, the display thread repaints the screen by copying the offscreen buffer onto the active display buffer. This process continues until the end of simulation.

Animating the virtual-time clock simulation required a totally different approach. Here, simulation events could not trigger off corresponding animation events since the time frame is virtual. Instead, we adopted the following approach: the simulation writes every simulation event into a trace file stored in the database. The virtual-time clock animation thread then reads the trace file (after simulation ends) and performs animation. The virtual-time has to be converted to real-time for animation. There are methods to draw each simulation event. The advantage of using virtual-time clock simulation in animating is that one can fast forward, rewind, pause while watching the animation.

An added advantage of using the Web as the interface (Java applets) is that one has unlimited access to image sources and other resources needed for animation. For example, rather than having geometric shapes move around in a bank simulation, one could use images of peoples heads.

3.3 Database Access Facilities

The conceptual basis of integrating our simulation library with database access is QDS. Our simulation environment supports QDS in a natural and flexible manner. QDS is characterized by two basic ideas:

1) Storing simulation results and simulation models in a database: QDS emphasizes that simulation results, which are usually generated at a great computational cost must not be discarded, but instead be stored in a database. Simulation models are also required to be stored and retrieved systemically. Currently, we used the MiniSQL (mSQL) (Kimpton, T.R., 1995) database engine for our implementation. The reasons for using mSQL are as follows:

- mSQL, being a lightweight database engine that implements a subset of ANSI SQL, requires less memory and overhead than most relational DBMS engines. Nevertheless, it is powerful enough for our requirements.
- mSQL is the only DBMS which has a Java API, MsqlJava (Collins, D. 1996), at the moment this document is being written. We wanted database access to seamlessly integrate with the simulation library. This required that the DBMS provide an API rather than resorting to embedded SQL for database access.
- We expect that conversion from MsqlJava API to JDBC calls will be easy when JDBC is made available by JavaSoft and JDBC driver managers are made available by individual DBMS vendors.
- mSQL is free of cost and is available for download from http://Hughes.com.au/product/msql/.

We have designed our database to maintain a Model Index and store simulation results. The Model Index contains a list of all available models along with a brief description and information about the parameters the model requires (refer to Figure 1). It also contains a Uniform Resource Locator (URL) to the model applet. Each model is associated with a table in which it stores its results. These are the tables that are queried when a user issues either a normal query or a QDS query. The models, themselves, are stored as Java applets. These applets are invoked whenever a model needs to be executed.

2) Presenting a simple user interface: QDS envisages that even naive users must be able to use a simulation system based on QDS. We chose SQL as our user interface query language because SQL is both popular and simple. Currently, our user interface is functionally simple where the user can issue a query on one or more models (each model stored as a single table in the database) using one or more conditions. The available models and the stored data attributes can be obtained by querying the database. The models satisfying the query will be displayed to the user in a scrollable list. The user can select the model and the attributes he/she is interested in from the model list and can either run the model or issue a query. The user is also allowed to specify a conditional search which will serve as the where clause in the SQL query. The query is then sent to the database engine which will return the results if they exist in the database. If the results are not available in the database, the relevant simulation model is executed and the results are provided to the user.

The basic ideas of QDS also help overcome a problem posed by Java - that of providing persistence to simulation results. Java's in-built security restrictions as well as the security restrictions imposed by most Web browsers do not allow Java applets to write applet generated results into a file. We solved this problem by writing results into a database thereby shifting the responsibility of security from the browser to the database which often has elaborate security provisions. We have been successful in writing results from an applet into the mSQL database while working with the HotJava browser. The popular Web browser, Netscape, imposes a further restriction on database access by requiring that both the http server and the database server be running on the same machine.

3.4 The JSIM Class Library

The goals of creating the JSIM simulation library are:

- *to provide a set of utilities for building a working simulation*
  JSIM provides the basic utilities for creating a working simulation. Our implementation of a bank simulation illustrates the power of the simulation library.

- *to provide statistical or other data to be stored in the database*
  As a key component of query-driven simulation, the simulation component generates statistical data during model execution, which will then be stored in a database for future queries. As stated before, re-executing a simulation model with the same input parameters as before could be expensive if we do not store the previously generated data.

- *to provide facilities for animation*
  Java is a language for programming on the Web and is designed to deliver interactive content using applets embedded in HTML pages over a network of heterogeneous systems. Our simulation system can be used to animate the simulation process using applets.

- *to provide an instructional tool for simulation*
  The saying "it is better to see once than hear a hundred times" has been verified by psychological experiments (Humphreys and Riddoch, 1987). Visual facilities such as overhead projectors and models have been widely used for educational purposes and are proven to be effective, especially for explaining abstract concepts. For the people who are used to concrete thinking, this is also supported by research in interface design (Hutchins, Hollan and Norman, 1985). The rapid development of the Internet, especially the World Wide Web, places a huge impact on the way we live and learn. With animation, students are able to visualize and interact with simulation processes, which will help them learn better and reduce their learning curve.

- *to avoid learning a new simulation language*
  Java is an object-oriented programm-ing language and its syntax is similar to C++. Java is arguably simpler than C++ and is, therefore, easier to learn for novice programmers, easy to adopt for C++ programmers, yet powerful enough for complex tasks. With all its power and wide applications, Java becomes an alternative programming language to C/C++ and gains popularity with the rapid growth of Web technology. After having grasped Java, there is no need to learn other special simulation languages and simulationists can use JSIM to develop simulation models quickly.

The following are the classes in the JSIM library: *Display_Thread, Graph, Histogram, Plot, Model_Applet, Queue, FCFS_Queue, LCFS_Queue, Priority_Queue , Resource, Server, Simobject, Statistic, Transport* and *Variate*. Figure 2 shows the class inheritance hierarchy in JSIM. There are three types of queues derived from *Queue* in JSIM: FCFS_Queue, LCFS_Queue and Priority_Queue. The *Statistic* class implements methods for collecting, analyzing, testing, and displaying simulation results. The *Tally* method gathers sample statistics, while *Accumulate* gathers time-persistent statistics. In addition, the

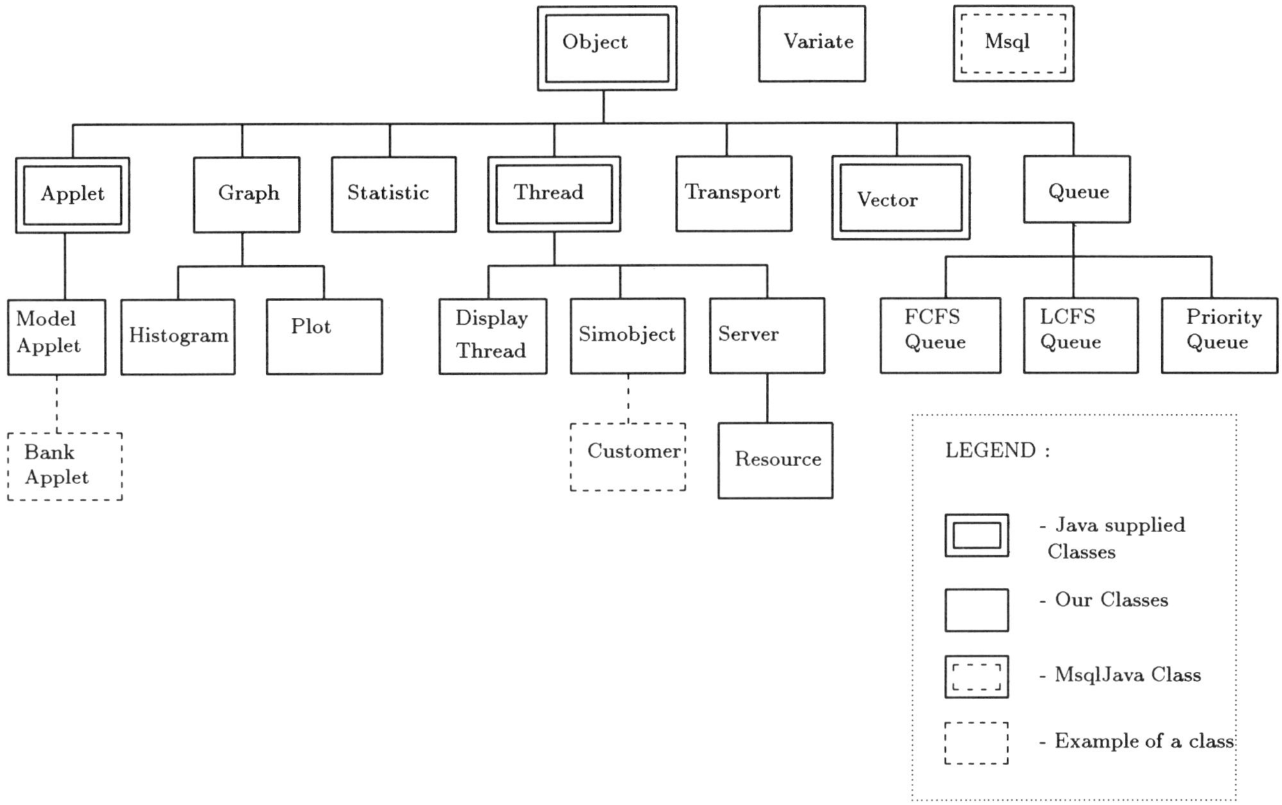

Figure 2: Class hierarchy in JSIM

Statistic class also provides methods for calculating mean, variance, standard deviation of sample statistics, e.g., mean of customer inter-arrival time. The class *Variate* provides procedures to generate random numbers and random variates. It has nine procedures for continuous random variates and four for discrete random variates.

The *Server* class, derived from *Thread*, is the service provider and has basic methods such as associating queues with the server and setting priorities to associated queues. The *Resource* class, derived from *Server,* allows user defined resources for simulation models. A resource consists of a number of service units (e.g., bank tellers), and a Queue feeding the service units. A process requests resources by calling *request*, and releases resources by calling *relinquish.* If a request for resources is not granted, the process making the request waits in the resource queue. The *Histogram* and *Plot* classes are responsible for generating histograms and plotting simulation results, respectively, and they are derived from *Graph.* The *Transport* class provides movement facilities to the simulation objects. Simulation objects can move only using transports or queues. The transport object is used to connect two key locations, for example, the head of a queue and a service station at a resource.

The *Simobject* class provides an abstraction for simulation entities, for example, the customer class is derived from the Simobject class in the bank simulation example of Figure 3. The Simobject class houses basic methods for joining a queue, claiming a resource, etc. The *Display* class implements the display thread that is responsible for animating the simulation.

4 AN EXAMPLE

We use a classical bank simulation to show the power and capabilities of our simulation environment. The Bank_Applet is an aggregation of five types of objects: one or more Servers (Tellers), a FCFS_Queue, a Transport, a Customer_Generator and many Customers. The relationships between these components are shown in Figure 3 which is an OMT diagram (Rumbaugh et al., 1991).

The Bank_Applet models an M/M/s queue where s represents the parameter Num_Tellers. Both the Customer and Customer_Generator are derived from the Simobject class. The Customer_Generator class is used to generate several Customer threads based on the specified Inter_Arrival_Time. The Customer thread and the Customer_Generator thread make use of the Variate and Statistic classes to maintain their

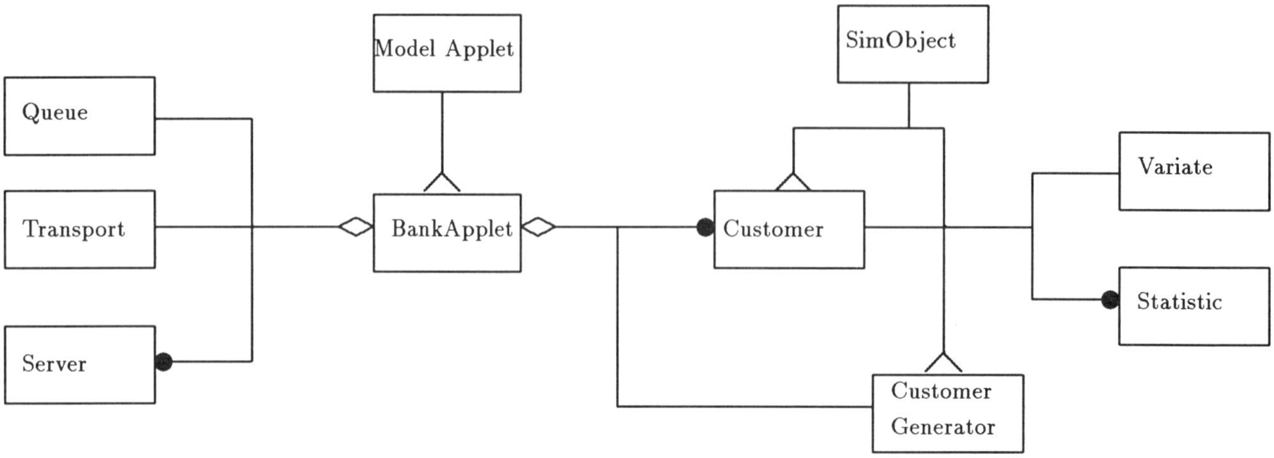

Figure 3 : Bank Simulation Example

statistics. The Transport class is used to connect the Queue to the service station at the Server.

When a Customer entity is first created, it tries to engage the Server. If the Server is busy with some other Customer, the Customer entity enters the Queue and waits its turn. If the Customer is able to engage the Server, it will be served for a period of time referred to as service time which is derived from an exponential distribution. After the service time is over, the Customer relinquishes the Server and leaves the bank (Customer thread terminates).

A typical QDS query for the above bank simulation model would be as follows:

```
SELECT Waiting_Time, Avg_Queue_Length
FROM Bank_Applet
WHERE Num_Tellers=1 AND Inter_Arrival_Time=3
```

In response to this query, the table corresponding to the bank model will be queried and the results returned if available. If the required results are not available in the table, then the QDS system retrieves the model information from the Model Index table. It then invokes the relevant model applet using the parameters given by the user in the query, namely, Num_Tellers = 1 and Inter_Arrival_Time = 3 (other parameters will take on their default values). The user is then given the results of this new model execution.

5 CONCLUSIONS AND FUTURE WORK

We believe that Web-based simulation will become an important technology for the future. The importance of the World Wide Web steadily increases; and now with Java, highly interactive and dynamic simulations/animations may be executed on the Web. Recognizing this, we have an ongoing project to develop sophisticated capabilities for Web-based simulation. Our QDS environment and JSIM library provide a widely available and easy-to-use facility for developing Web-based simulations and animations. The environment combines simulation, animation and database access exploiting the strengths of Java (e.g. it is a high-level, clean and safe object-oriented language and has multi-threading and sophisticated display/GUI facilities) as well as minimizing its weaknesses (e.g. lack of data persistence in applets).

As an ongoing project, we still have much work to do. The following are the major areas of interest for future work. We are currently incorporating advanced features such as batch/unbatch, split/join and routing features into our library. We will also be converting from MsqlJava API to JDBC API when a stable version of JDBC becomes available. This will allow us to use a variety of Relational or Object-Relational DBMSs. Another area of focus is the Graphical Model Designer. The Graphical Model Designer is intended to be a GUI- based model builder that a simulationist can use to quickly build models by drag and drop operations.

ACKNOWLEDGEMENTS

We would like to acknowledge Darry Collins for writing and making available the Java API to mSQL. We would also like to thank Vijayalakhsmi Natarajan, Department of Computer Science, University of

Georgia, who helped us in developing database access for our class library and Hongwei Zhao, Department of Computer Science, University of Georgia for contributing to the design of the graphical designer.

REFERENCES

Collins, D. (1996) *MsqlJava: A Java class library for mSQL.* URL: http://mama.minmet.uq.oz.au/msqljava/tutorial.html

Hutchins, E.L., Hollan, J.D. and Norman, D.A. (1985). Direct manipulation interfaces. *Human-Computer Interaction.* 1:311-38.

Humphreys, G.W. and Riddoch, M.J. (Editors) (1987). *Visual Object Processing: A Cognitive Neuropsychological Approach.* Hove, UK: Lawrence Erlbaum Associates.

Kimpton, T.R. (1995) *Mini SQL: A lightweight database server.* URL: http://Hughes.com.au/product/msql/manual.htm

Miller, J.A., Kochut, K.J., Potter, W.D., Ucar, E., and Keskin, A. (1991b). Query-driven simulation using active KDL: A functional object-oriented database system. *International Journal in Computer Simulation,* 1(1): pp. 1-30.

Miller, J.A., Potter, W.D., Kochut, K.J., Keskin, A., and Ucar, E. (1991a). The active KDL object-oriented database system and its application to simulation support. *Journal of Object-Oriented Programming,* 4(4): pp. 30-45.

Miller, J.A., Potter, W.D., Kochut, K.J., and Ramesh, D. (1996a). Object-oriented simulation languages and environment: a four-level architecture. *Object-oriented Simulation.* Zobrist-Leonard, Editors. IEEE Press (1996). (to appear)

Miller, J.A. and Weyrich, O.R. (1989). Query driven simulation using SIMODULA. *Proceedings of the 22nd Annual Simulation Symposium, Tampa, Florida.* pp. 167-181.

Miller, J.A., Weyrich, O.R., Potter, W.D., and Kessler, V.C. (1996b). The SIMODULA/OBJECTR query driven simulation support environment. *Progress in Simulation,* Vol. 3. Leonard-Zobrist, Editors. (to appear)

Rumbaugh, J., Blaha, M., Premerlani, W., Eddy, F. and Lorensen, W. (1991) *Object-Oriented Modeling and Design.* Prentice Hall, Englewood Cliffs, New Jersey.

AUTHOR BIOGRAPHIES

RAJESH S. NAIR is a graduate student in the MS program in the Department of Computer Science at the University of Georgia. He was awarded a University-wide Assistantship for the year 1995-'96. He received his BE degree in Computer Technology from the University of Bombay, India in 1994. His research interests include application of database systems and simulation environments. This paper is part of his ongoing research work toward the MS program.

JOHN A. MILLER is an Associate Professor and the Graduate Coordinator in the Department of Computer Science at the University of Georgia. His research interests include Simulation, Database Systems, and Parallel & Distributed Systems. Dr. Miller received the BS degree in Applied Mathematics from Northwestern University in 1980, and the MS and PhD in Information and Computer Science from the Georgia Institute of Technology in 1982 and 1986, respectively. During his undergraduate education, he worked as a programmer at the Princeton Plasma Physics Laboratory. Dr. Miller has been active in conferences in both simulation and database. He has been the Publication Co-Chair for RIDE (Research Issues in Data Engineering), President/General Chair of the Annual Simulation Symposium, and Coordinator of the Modeling Methodologies Track of the Winter Simulation Conference. He has also been a Guest Editor for the International Journal in Computer Simulation.

ZHIWEI ZHANG is a graduate student in the MS program at the Department of Computer Science, University of Georgia. He received his Masters in Library Science from the University of Maryland and a certificate in Library Science from Nanjing University, China. His research interests include Simulation and Database & Information systems.

Proceedings of the 1996 Winter Simulation Conference
ed. J. M. Charnes, D. J. Morrice, D. T. Brunner, and J. J. Swain

LANGUAGE BASED STATE SAVING EXTENSIONS FOR OPTIMISTIC PARALLEL SIMULATION

Fabian Gomes
Brian Unger

Department of Computer Science
The University of Calgary
2500 University Drive N.W.
Calgary, Alberta T2N 1N4
CANADA

John Cleary

Department of Computer Science
University of Waikato
Te Whare Wananga o Waikato
Private Bag 3105
Hamilton
NEW ZEALAND

ABSTRACT

One of the greatest challenges in making optimistic synchronization techniques such as Time Warp practical tools is making state saving efficient and easy to use. State saving is necessary so that when optimistic execution is found to be out of order, rollback can be used to recover an earlier execution state. Previous work has shown that the most robust and efficient technique for saving state is to incrementally save copies of small parts of the state at the point that they are modified. Unfortunately, this requires significant programmer intervention to insert additional code.

In this paper, C^{++} language extensions for transparent incremental state saving are presented. Operator overloading and type parameterization are used to incrementally save basic data types. Building on this, two new type-specifiers, "`recover`" and "`nonrecover`" are described. They allow a single declaration to specify, for example, that all the member variables of a class are to be state saved, and for all the resulting state saving calls to be automatically generated. Issues, including how these specifiers interact with class inheritance and function declarations are examined and solved.

1 INTRODUCTION

Computer simulation is a valuable tool for the design and analysis of complex systems. However, the simulation of many important systems requires massive amounts of processor time and memory. The goal of parallel discrete event simulation is the acceleration of sequential simulation execution through concurrent execution on multiple processors (Fujimoto 1990).

Optimistic synchronization has been proposed to maximally exploit the inherent parallelism within systems. Time Warp, a well known optimistic synchronization technique based on the virtual time paradigm (Jefferson 1985), has been shown to be capable of attaining impressive speedup. Optimistic methods are characterized by logical processes executing aggressively and independently of others. A causality error detection and recovery scheme based on rollback is used to assure causality in the asynchronous parallel execution.

Two serious problems with rollback implementation need to be adequately resolved. The first is the efficient saving of state information during forward execution. The second problem is to solve the first in a way which is transparent to the programmer, i.e. that doesn't substantially complicate model development.

State saving mechanisms can be categorized as copy state saving (CSS) or incremental state saving (ISS) (Bauer and Sporer 1993, Bruce 1995, Cleary et al. 1994, Steinman 1993). In CSS mechanisms, a checkpoint of the entire object's state is taken on each event. CSS mechanisms are processor intensive and consume a large amount of memory. An optimization of the CSS strategy is to reduce the frequency of checkpointing using periodic state saving (PSS) (Fleischmann and Wilsey 1995, Lin et al. 1993, Preiss, Loucks, and MacIntyre 1994, Rōnngren and Ayani 1994). Estimation of the optimal frequency of checkpointing, using static or adaptive schemes, is critical, as the object may need to rollback further if there is no state saved at the point of a synchronization error.

One major advantage of CSS is that it can easily be made transparent to the programmer. That is, no extra code is required to ensure that all the appropriate state is saved. Its major disadvantage is that it can easily become very expensive when the state size is large. This is particularly acute when each event has a small computational grain and only a small part of the state is modified.

On the other hand, in ISS mechanisms only information related to changes to the state are saved. Generally, ISS is robust in its performance. Its cost tends to be a constant fraction of the event execution time, so that it seldom causes unexpected costs. The cost of rollback is proportional to the distance rolled-back, and in principle this can be arbitrarily large. Practical experience however shows that in most Time Warp executions rollbacks are limited to a few events, so that this is not a problem (Fujimoto 1989, Gomes and Unger 1994, Xiao and Unger 1995).

The major difficulty with ISS is the need for programmer intervention, in explicitly saving a state prior to its modification. This code and the effort of inserting it is needed only for Time Warp, thus exposing the underlying synchronization protocol during the model design, development and validation phases. Also mistakes in the ISS code can lead to subtle and difficult to detect bugs. Clearly some easy transparent mechanism for implementing ISS is necessary if Time Warp is to become part of the normal simulation life-cycle.

In this paper, C^{++} language extensions to support transparent incremental state saving are addressed. Incremental state saving is automated without compromising efficiency. A simple ISS backtrail mechanism is presented in section 2. Section 3 details the use of parameterized types to define recoverable basic data types. They have the same semantics as basic data types, but have built-in incremental state saving for write operations. The specification for a recoverable type annotation semantics and its use is detailed in section 4. This uses two new keywords "**recover**" and "**nonrecover**" to enable a type-safe declaration of which state is to be saved. Finally, a summary is presented in section 5.

2 SIMKIT SYSTEM

SimKit is a C^{++} class library that is designed for very fast discrete event simulation (Gomes et al. 1995). The primary goal of SimKit is to provide an event-oriented logical process modeling interface that facilitates building application models for sequential and parallel simulation with high performance execution capabilities.

The parallel SimKit implementation is atop a Time Warp executive interface called *WarpKit*. SimKit specifies three classes called **sk_simulation**, **sk_lp** and **sk_event**. In a simulation execution, there is one instance of **sk_simulation** that controls the various phases of the simulation execution. **sk_lp** is an abstract base class that is used to derive application level logical processes. Logical processes communicate using objects derived from **sk_event**. A **send_and_delete()** primitive in the **sk_lp** class is provided for inter-LP communication.

In SimKit, an ISS backtrail mechanism is implemented. In this mechanism, a *backtrail* of address/value pairs is constructed during forward execution. An address/value pair, representing an image of the state, is referred to as a state log element (SLE). SLEs are recorded prior to a write operation, and are linked to form the backtrail, from which a past system state can be recovered. Recovery entails scanning the backtrail, from the most recent SLE recorded back to the one at the recovery point, and writing the old values back into the associated addresses.

On most architectures two backtrails, one allowing 4-byte data and the other allowing for 8-byte data are instantiated for each **sk_lp**. These two sizes efficiently cover most basic data types, including integer, pointer, double, long long, etc.

3 PARAMETERIZED STATE DATA TYPES

Incremental State saving can be automated in C^{++} using *parameterized types* and *operator overloading*. A parameterized type defines a new type in terms of (or parameterized by) another unspecified type. The following design uses the **template** construct, to define specialized state types from basic data types (Ellis and Stroustrup 1990, Stroustrup 1988). State variables declared using the state type templates are said to be *recoverable*. Each of these specialized recoverable types need to have the same semantics as the base data type but are extended to include state saving. Assignment related operators are redefined to record a backtrail entry prior to the actual write operation, using operator overloading. In C^{++}, write operations on an integer data type include assignment operations like =, *=, /=, \%=, +=, -=, and, prefix and postfix versions of ++ and --. Integers also masquerade as bit vectors so assignment operators of the form \&=, |=, \^{ }=, \~{ }=, <<= and >>= also need to be overloaded.

3.1 State Data Type Templates

Two templates State<T> and StatePtr<T> are used in the declaration of recoverable variables. The former is used in declaring recoverable variables of basic types while the latter in declaring pointer type recoverable variables. Sample source code for the above template class definitions are provided in appendix A.

Basic recoverable types are defined using **class State<T>** as follows:

```
class State<int> SInt1, SInt2( 100 );
```

Recoverable integers SInt1 and SInt2 are constructed having the semantics of their base type, i.e. int. SInt2 is initialized to 100 during construction. Such initialization does not cause SInt2 to be state saved.

An automatic conversion operator allows recoverable objects SInt1 and SInt2 to be used in place of an int base type, as in the expressions,

```
a_int = SInt2;
if ( SInt2 < 100 ) { ... }
SInt2 = 1 + a_int + SInt1 * 2;
```

The overloaded assignment operator in class State<int> records a backtrail entry for SInt2 prior to the assignment in the last statement.

Recoverable pointer data types are defined using class StatePtr<T> as follows:

```
class StatePtr< int > SPtrInt( & a_int );
class StatePtr< State<int> > SPtrSInt;
SPtrSInt = &SInt1;
```

In the first declaration, a recoverable pointer to an int data type is defined. SPtrInt is initialized to point to the variable a_int. No state saving of SPtrInt takes place during initialization within the constructor. In the second, a recoverable pointer to a recoverable integer is defined. SPtrSInt gets state saved prior to the assignment, in the third statement.

An overloaded de-reference operator in the class StatePtr returns the object pointed to by the StatePtr type object. Only in the second of the following assignments is a record entered into the backtrail. In the first case, the int type data pointed to is not recoverable and so is not state saved. In the second, as the object pointed to is recoverable, the overloaded assignment defined for the class State type object SInt1 is invoked.

```
* SPtrInt = 10;
* SPtrSInt = 10;
```

In the next example, a recoverable pointer SPtrSPtrSInt, to another recoverable pointer to a recoverable integer type variable is declared.

```
class StatePtr< StatePtr< State<int> > >
      SPtrSPtrSInt;
```

Now, consider the following assignment statements,

```
SPtrSPtrSInt = &SPtrSInt;
*SPtrSPtrSInt = &SInt2;
**SPtrSPtrSInt = 200;
```

In the first assignment SPtrSPtrSInt is saved, in the second SPtrSInt is saved, and in the third SInt2 is saved.

3.2 Template Design Issues

When designing these templates, a number of important design considerations were taken into account.

1. The result should always be safe - all modifications of a recoverable object should be trailed. A recoverable object may be modified directly by write operations or indirectly via a pointer de-reference. All write operations are identified and overloaded to record a backtrail entry prior to the write operation. Recoverable objects modified indirectly by de-referencing a class StatePtr are also saved, by overloading the de-reference operator to return a pointer to the recoverable object whose overloaded assignment operator is invoked.

2. As a consequence of the first point, the system should be type safe, so that state accessed via function parameters, or pointers is safe. All recoverable objects are defined using class State and class StatePtr, that are parameterized by a specified type. Incorrect type usage will generate compile-time error messages, for instance,

```
void fn( int * );
fn( & a_int );
fn( & SInt1 ); // Error
```

A compile time error message is generated in the second function invocation, as the actual parameter is a pointer to a class State<int>, where as the formal parameter expects a pointer to an int.

3. It should be possible for the user to force the point of state saving and this should be safe and efficient. A type conversion function Save() is defined in the templates. It does two things. It forces a state save operation on the object's variable and returns a reference to the variable of the original base type. This reference can be subsequently used to modify the variable without incurring state saving overheads (recall that it is only necessary to record a variable on the backtrail the first time it is modified in each event).

Consider the example:

```
int & temp = SInt1.Save();
for (int i=0; i<100;i++)
    temp++;
```

In the above example, a reference type variable temp is initialized to refer to the int data type at the location of SInt1. SInt1 is explicitly state saved only once prior to the entry into the for loop. The data of the recoverable object is modified within the loop without any records being entered into the backtrail.

4. Temporary variables on the stack must never be trailed as they may no longer exist upon rollback. In C^{++}, variables may be created on the stack due to pass by value parameter or declarations local to a function. Such variables have a temporary lifetime limited to the function execution. A rollback would cause an invalid value to be written into a stack location which once contained the temporary recoverable variable. In pass by value parameter passing, a compiler invokes an object's copy constructor to create a temporary object that is an exact duplicate of the actual parameter. Pass by value temporary variables are prevented by making the copy constructors private. Local declarations are prevented by using an execution time check in the default constructors. No local state type variables may be declared once simulation event processing begins.

An unfortunate side effect of making the copy constructor private, is that recoverable object construction and initialization using another recoverable object generates a compile-time error message, as in the case of `SInt3` declaration below:

```
class State<int> SInt3( SInt2 ); // Error
```

Alternatively, the programmer can use an assignment to initialize a recoverable object as a copy of another, as shown:

```
class State<int> SInt3;
SInt3 = SInt2;
```

5. It should be possible to dynamicly create recoverable objects. Dynamic state allocation is enabled by providing an overloaded **new** operator in each of the template classes. The **new** invokes a rollback sensitive memory allocator to dynamicly construct a recoverable object.

6. The mechanism must be totally transparent when executed on a non-optimistic system. In a non-optimistic system, the template just returns the base type making all the recovery declarations invisible. So no state saving is performed on write operations.

4 RECOVER TYPE SPECIFIER

One of the problems in using templates parameterized by basic types, is that the programmer must individually declare all state variables. This can be quite cumbersome and inconvenient. To alleviate this, two new keywords "**recover**" and "**nonrecover**" are used as type-specifiers.

In the simplest form, a "**recover int I**" declaration translates to "**State<int> I**". In the context of pointer data types, a declaration "**recover int * IP**" specifies that both the pointer and the object pointed to are recoverable. This declaration translates to "**StatePtr< State<int> > IP**". Declarations of the form "**State<int> * Ip**" and "**StatePtr< int > iP**" need to be explicitly represented as "**recover int * nonrecover Ip**" and "**nonrecover int * recover iP**", respectively.

A preprocessor front end to the C^{++} compiler has been developed for SimKit which uses the new keywords to generate C^{++} code wherein state variables are individually declared using the template classes.

4.1 Semantics of Recover Type Specifier

The **recover** and **nonrecover** keywords may be used in a number of contexts.

1. The first context is in association with a **class-key**. A **class-key** can be one of the **class**, **struct** or **union** keywords (Ellis and Stroustrup 1990). By default all variables do not have a state saving attribute. Consider the example declarations:

```
class z{...};
recover class A{...};
class B : class z{...};
recover class C : class z{...};
recover class D : nonrecover class A{...};
class E : class A{...};
```

Objects instantiated from `class A` will have all their variables recoverable. The **recover** keyword qualifying `class C`, forces all variables including those of the base `class z` to be recoverable in any objects instantiated from `class C`. Objects instantiated from the derived `class D` will have variables declared in `class D` recoverable and, because of the **nonrecover** qualifier, variables declared in its base class non recoverable. Objects of `class E` will only have variables declared within `class A` recoverable. In the case of pointer type member variables in a class, the use of **recover** as a `class` qualifier, specifies that both the pointer and the object pointed to are recoverable, unless explicitly over-ridden by the use of **nonrecover** in the pointer declaration.

2. The second use of the keyword is as an aggregate declaration within a `class`, `struct` or `union`. Members within a class may be grouped together as **recover** and **nonrecover** types, using a similar mechanism to the access specifiers `public`, `protected` and `private`. For instance, in the next example, all of `I`, `J`, `K` and `D` are recoverable, but none of `i`, `j` and `d`.

```
class A {
  recover:
    int I, J, K;
    double D;
  nonrecover:
    int i, j;
    double d;
};
```

The use of such aggregate declarations in classes needs some care. For example in the declaration:

```
class F {
    int x, y, z;
  recover:
    int I, J, K;
```

```
    double D;
  nonrecover:
    int i, j, k;
    double d;
};
class G : recover class F { ... }
```

the variables x, y, z, I, J, K, and D will all be recoverable in the context of **class G**. However, because i, j, k, d were explicitly declared to be non-recoverable, their type is not modified and they remain non-recoverable.

Note that the variables x, y and z may be recoverable or non-recoverable depending upon the context of their usage. To deal with this correctly, the implementation automatically creates two versions of the **class F**, one recoverable and the other not. The correct version is used depending upon the context.

3. In specifying recover type functions, the **recover** type-specifer is applied to the function return type as well as to all the arguments passed to the function. The example declaration

```
int * f1(int i, int *p, int **pp) recover;
```

translates to,

```
State<int> * f1(
 int i,
 State<int> *p,
 StatePtr< State<int> > *pp
 );
```

The translation of types here is rather different to that in the other contexts. Because C^{++} passes parameters by value and because these are allocated as temporaries on the stack, they must not be made recoverable. The rule used is that the "outer-most" type declaration is left non-recoverable and all types within this are made recoverable. So the declaration `int i` is left non-recoverable. The declaration `int *p` is changed to a non-recoverable pointer to a recoverable integer. The declaration `int **pp` is changed to a non-recoverable pointer to a recoverable pointer to a recoverable integer type.

4. A **recover** keyword may also be used to derive types using the **typedef** construct. The new type-name may then be used to declare recoverable variables of the new derived type. For instance,

```
typedef recover int SIntType;
SIntType SInt1, SInt2;
```

4.2 Sample State Declaration

An example using **recover** is shown in Figure 1. **class nm_output** defines an output buffered logical

```
class nm_output : public sk_lp {
 public:
    nm_output( ... );
    void process(const sk_event *);
    void initialize( );
    void terminate(void);
    afr_link *link(void);
    int id(void);
    int index(void);
    double average_occup(int,sk_time)
    double std_dev_occup(int,sk_time);
    int cells_dropped(int,int);
    int max_occupancy(int);
    int curr_occupancy(int);
    int cells_sent(int, int);
    int cells_received(void);
    double average_ctd(int);
    double std_dev_ctd(int);
private:
    nm_node *My_node;
    afr_link * My_link;
    int My_index, Id;
    sk_time Cell_transm_delay;
    RPT_STYLE Rpt;
    char *Report_1, *Report_2;
    void reset_stats(sk_time);
    void report(sk_time);
    void report_formatted(sk_time);
    void report_wsv(sk_time);
  recover:
    nm_buffer *Buffer;
    int Cells_received;
    int Link_state;
    int Cells_sent[AFR_N_CLASS][2];
    afr_tally State_ctd[AFR_N_CLASS];
};
```

Figure 1: Recoverable State Declaration Example

process in an ATM switch in the ATM-TN simulation model (Unger et al. 1995). It is a typical example of the modifications necessary to ensure correct state saving in this extensive body of software. Note that **recover** is used only once in the example. In 6000 lines of C^{++} header source files, 35 **recover** and 5 **nonrecover** keywords were needed to define the recoverable state in the simulation model.

5 CONCLUSIONS

Two significant barriers to the adoption of PDES techniques such as Time Warp are the execution cost of state saving and the difficulties of writing correctly state saved Time Warp models. Previous work has shown that incremental state saving can be a low cost robust state saving mechanism. This paper has addressed the second problem of ensuring that state is saved safely, correctly and with minimal programmer effort.

ACKNOWLEDGMENTS

This work has been supported by grants from the Canadian Natural Sciences and Engineering Research Council and by the New Zealand Public Good Science Fund. We would also like to thank Steve Franks for many useful discussions.

APPENDIX: TEMPLATE CLASS DEFINITION

```
template <class T> class State {
 private:
  T _datum;
  State(const State<T> &) { }
 public:
  State() : _datum(NULL) { ... }
  State(T t) : _datum(t) { ... }
  ~State() { ... }
  operator T() {  return _datum; }
  T value() { return _datum; }
  State<T>& operator=(T t)
  { ISSMgr.save( &_datum );
    _datum = t;
    return *this;
  }
  State<T>& operator=(const State<T> &t)
  { ISSMgr.save( &_datum );
    _datum = t.value();
    return *this;
  }
   // All other forms of assignments
   // e.g. +=, *=, ++, --, ...
  T& Save()
  { ISSMgr.save( &_datum );
    return _datum;
  }
  State<T>* operator&()
  {  return this; }
};

template <class T> class StatePtr{
 private:
    T* _datum_ptr;
    StatePtr(const StatePtr<T> &)
    { /* NO OP */ }
 public:
  StatePtr() : _datum_ptr(NULL) { ... }
  StatePtr(T* p) : _datum_ptr(p)
  { ... }
  ~StatePtr() { ... }
  operator T*() { return _datum_ptr; }
  T* value() { return _datum_ptr; }
  StatePtr<T>& operator=(T* p)
  { ISSMgr.save( &_datum_ptr );
    _datum_ptr = p;
    _datum_ptr = p; return *this;
  }
  StatePtr<T>&
      operator=(const StatePtr<T>& p)
  { ISSMgr.save( &_datum_ptr );
    _datum_ptr = p.value();
    return *this;
  }
   // All other forms of assignments
   // e.g. +=, *=, ++, --
  T* & Save()
  { ISSMgr.save( &_datum_ptr );
    return _datum_ptr;
  }
  StatePtr<T>* operator&()
  {  return this; }

  T& operator*()
  { return *_datum_ptr; }
};
```

REFERENCES

Bauer, H., and C. Sporer. 1993. Reducing rollback overhead in Time Warp based distributed simulation with optimized incremental state saving. In *Proceedings of the 26th Annual Simulation Symposium*, ed. J. Miller, 12–20. Arlington, Virginia.

Bruce, D. 1995. Treatment of state in optimistic systems. In *Proceedings of the 9th Workshop on Parallel and Distributed Simulation (PADS95)*, ed. Y-

B. Lin, and M. Bailey, 40–49. Lake Placid, New York.

Cleary, J., F. Gomes, B. Unger, X. Zhonge, and R. Thudt. 1994. Cost of state saving and rollback. In *Proceedings of the 8th Workshop on Parallel and Distributed Simulation (PADS94)*, ed. D. Arvind, R. Bagrodia, and Y-B. Lin, 24(1):94–101. Edinburgh, Scotland, U.K.

Ellis, M., and B. Stroustrup. 1990. *The annotated C^{++} reference manual.* Addison-Wesley.

Fleischmann, J., and P. Wilsey. 1995. Comparative analysis of periodic state saving techniques in Time Warp simulators. In *Proceedings of the 9th Workshop on Parallel and Distributed Simulation (PADS95)*, ed. Y-B. Lin, and M. Bailey, 50–58. Lake Placid, New York.

Fujimoto, R. 1989. Time Warp on a shared memory multiprocessor. *Transactions of the Society for Computer Simulation*, 6(3):211–239.

Fujimoto, R. 1990. Parallel discrete event simulation. *Communications of the ACM*, 33(10):30–53.

Gomes, F., S. Franks, B. Unger, Z. Xiao, J. Cleary, and A. Covington. SimKit: A high performance logical process simulation class library in C^{++}. In *Proceedings of the 1995 Winter Simulation Conference*, ed. C. Alexopoulos, K. Kang, W. Lilegdon, and D. Goldsman, 706–713. Arlington, Virginia.

Gomes, F., and B. Unger. 1994. Benchmarking Shared Memory Time Warp with Signalling System No. 7 Performance Model – 3. Project Report CPSC 601.24, Department of Computer Science, University of Calgary, Canada.

Jefferson, D. 1985. Virtual time. *ACM Transactions on Programming Languages and Systems*, 7(3):404–425.

Lin, Y-B., B. Preiss, W. Loucks, and E. Lazowska. 1993. Selecting the checkpoint interval in Time Warp simulation. In *Proceedings of the 7th Workshop on Parallel and Distributed Simulation (PADS93)*, ed. R. Bagrodia, and D. Jefferson, 23(1):3–10. San Diego, California.

Preiss, B., W. Loucks, and I. MacIntyre. 1992. Effects of the checkpoint interval on time and space in Time Warp. *ACM Transactions on Modeling and Computer Simulation*, 4(3):223–253.

Rōnngren, R. and R. Ayani. 1994. Adaptive checkpointing in Time Warp. In *Proceedings of the 8th Workshop on Parallel and Distributed Simulation (PADS94)*, ed. D. Arvind, R. Bagrodia, and Y-B. Lin, 24(1):110–117. Edinburgh, Scotland, U.K.

Steinman, J. 1993. Incremental state saving in SPEEDES using C^{++}. In *Proceedings of the 1993 Winter Simulation Conference*, ed. G. Evans, M. Mollaghasemi, E. Russell, and W. Biles, 687–696. Los Angeles, California.

Stroustrup, B. 1988. Parameterized types for C^{++}. In *Proceedings of the 1988 USENIX C^{++} Conference*, 1–18.

Unger, B., F. Gomes, Z. Xiao, P. Gburzynski, T. Ono-Tesfaye, S. Ramaswamy, C. Williamson, and A. Covington. 1995. A high fidelity ATM traffic and network simulator. In *Proceedings of the 1995 Winter Simulation Conference*, ed. C. Alexopoulos, K. Kang, W. Lilegdon, and D. Goldsman, 996–1003. Arlington, Virginia.

Xiao, Z. and B. Unger. 1995. Performance of ATM-TN Wnet model. TeleSim Project Report ATM1.1-PR, Department of Computer Science, University of Calgary, Canada: WurcNet Inc.

AUTHOR BIOGRAPHIES

FABIAN GOMES is a post doctorate fellow in the Department of Computer Science at the University of Calgary. He received his Ph.D. degree in Computer Science from the University of Calgary in 1996. His research interests are in parallel simulation, modeling ATM broadband networks, distributed systems and rollback based computing.

BRIAN UNGER is a professor in the Department of Computer Science at the University of Calgary and is the President of WurcNet Inc. His current interests include the modeling and simulation of ATM broadband networks, parallel and distributed simulation, and simulation in Java.

JOHN CLEARY is a professor in the Department of Computer Science at Waikato University. His research interests include parallel and distributed systems, logic programming, complexity theory applied to adaptive and learning systems, and parallel discrete event simulation.

Proceedings of the 1996 Winter Simulation Conference
ed. J. M. Charnes, D. J. Morrice, D. T. Brunner, and J. J. Swain

A PARALLEL GPSS BASED ON THE PARASOL SIMULATION SYSTEM

Felipe Knop
Edward Mascarenhas
Vernon Rego

Department of Computer Sciences
Purdue University
West Lafayette, IN 47907-1398, U.S.A.

ABSTRACT

Much of the research in parallel discrete-event simulation (PDES) has resulted in new experimental simulation languages or toolkits. Meanwhile, the simulation community continues to use existing (serial) commercial tools which are reportedly more powerful and flexible from a modeler's point of view. A possible way to increase the impact of PDES in the simulation community is to make existing simulation packages execute in parallel. Towards this end, we present a parallelization of the GPSS simulation language. We implement parallel GPSS as a GPSS-to-C++ translator and execute the transformed code with the help of the *ParaSol* parallel simulation system. The mapping from GPSS to *ParaSol* is simple because, unlike other parallel simulation systems, *ParaSol* is transaction oriented. On the other hand, because GPSS was not designed with parallelism in mind, there are GPSS constructs that can behave poorly in a parallel environment. We present details on the mapping, some of the challenges we faced in this task, and key solutions that we adopted to enhance parallelism.

1 INTRODUCTION

It has been said that research in parallel discrete-event simulation (PDES) has not made an impact on the simulation community (Fujimoto 1993). A possible reason for this is that extant PDES systems typically sacrifice model implementation effort for the promise of speedup. Application programmers are forced to use explicit message exchange constructs, lookahead information, and even state-saving/restoration procedures. Meanwhile, the simulation community continues to use existing (serial) commercial tools, which are reportedly more powerful and flexible from the modeler's point of view.

In this paper we describe a project to parallelize the GPSS simulation language (Bobillier et al. 1976). GPSS was chosen because of its widespread use and potentially large base of existing simulation programs. Since GPSS was not developed with parallel execution in mind, it offers a good test of the suitability of current PDES work for the parallelization of sequential simulation languages.

This work is based on the *ParaSol* project, aimed at developing a parallel simulation system based on mobile threads (Mascarenhas et al. 1995). *ParaSol*'s main design goals are ease of use – to minimize user-visible complexities of parallel simulation – and flexibility – to maximize the system's use across different application domains. We implement parallel GPSS as a GPSS-to-C++ translator and execute the transformed program using *ParaSol*. The mapping from GPSS to *ParaSol* is simplified because both are transaction oriented. But though, at first glance, both systems appear to have many similarities, there are dramatic differences as well. Because GPSS was not designed to run on a parallel machine, it has features that can make it behave poorly in a parallel environment. We present details on the mapping, some challenges that we faced in parallelizing GPSS, and some key solutions that we adopted in the implementation.

The remainder of the paper is organized as follows. Section 2 contains a review of related work on parallelizing sequential simulation languages. Sections 3 and 4 present overviews of *ParaSol* and GPSS, respectively. In section 5 we describe the most important aspects of the mapping from GPSS to *ParaSol*. In Section 6 we point out some problems caused by the sequential nature of GPSS, and in Section 7 we describe our solutions to these problems. Performance experiments are presented in Section 8. We conclude in Section 9.

2 RELATED WORK

Today's popular (sequential) simulation languages were not designed for parallel execution and, as a result, are difficult to parallelize. It should come as no surprise, therefore, that research in PDES systems invariably results in new simulation languages or toolkits. For examples of such systems see Bagrodia (1991), BoyanTech (1995), and Gomes et al. (1995). Typically, these systems offer an application interface consisting of a set of logical processes that exchange timestamped messages with one another, using either language constructs or function calls.

Tsai and Fujimoto (1993) describe a framework for parallelizing simulation languages. In this framework, a common layer implements a set of basic primitives that can be used to build compilers for existing languages. The common layer maintains two abstractions, *state variables* and *unprocessed events*, and standard operations that can be applied to these. For example, variables can be created, destroyed, read, and modified, and events can be scheduled, examined, and deleted. The parallel implementation of the shared variables and their operations is achieved through the use of *space time memory*, allowing variables to be accessed by entities that execute at different points in simulation time. The parallel implementation of the event list is obtained with mechanisms similar to those found in Time Warp. As a case study, the paper presents an implementation of the SIMSCRIPT II.5 language for a shared-memory environment. It is unclear whether the approach can be used to parallelize transaction-oriented languages such as GPSS for architectures that do not support shared-memory multiprocessing.

Nicol and Heidelberger (1994) describe a different approach. Here, the goal is to parallelize simulation tools with little or no modification, because these tools are generally large and complex. The idea is to use an existing tool to define sub-models that will run on different processors, with tool *extensions* defining interfaces between sub-models. The extensions incorporate all required communication and synchronization. Despite the apparent simplification, this implies that a sequential application will have to undergo some modification in order for it to run in parallel. In implementing tool extensions, no information about the sub-models is assumed, except that which is provided explicitly by the sub-model. This constrains parallel models in that they are forced to run under a conservative synchronization mechanism. The computation and dissemination of lookahead information – sometimes beneficial in conservative mechanisms – is encapsulated in the extensions. The proposed approach is applied to produce a library which extends the CSIM simulation system (Schwetman 1986). To enable extensions, without requiring significant changes, a tool will generally require features that may not be present in other simulation tools.

3 *ParaSol* OVERVIEW

ParaSol is a parallel simulation system based on the (active-transaction) process-interaction paradigm. Instead of using timestamped messages for communication and synchronization among logical processes (LPs), as done in existing systems, *ParaSol* relies on transactions that transparently migrate between LPs, to obtain a more powerful effect. Indeed, this transparency leads to greatly simplified model development for many simulation problems, and was an important design consideration.

ParaSol presents a programming environment that offers *transactions* – dynamic, computational units with some private data – and a set of *global objects*. Both transactions and objects are distributed among the physical processors hosting a simulation. Transactions, which are dynamically created and destroyed, usually spend their time either performing local computations or accessing objects. To access an object located at a remote process, a transaction *migrates* to the process where the object is located, thus enhancing locality. A transaction causes simulation time to pass when it executes a *hold* command (analogous to the `ADVANCE` block in GPSS).

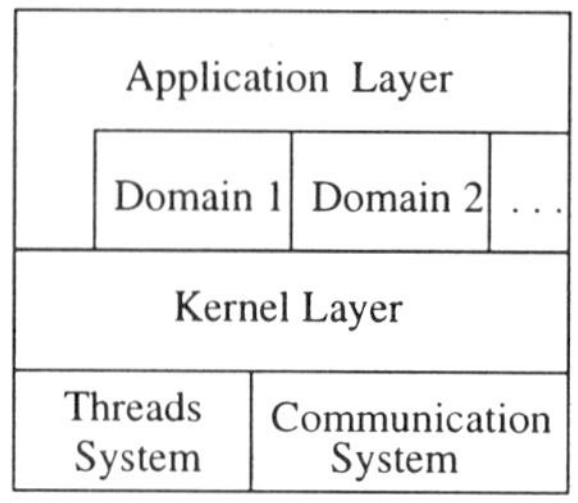

Figure 1: *ParaSol*'s Layered Architecture

ParaSol's architecture is shown in Figure 1. The kernel provides basic services, and the domain libraries support higher level services geared towards specific application domains. The kernel insulates the upper layers from most parallel simulation details, including transaction management, migration, communication, rollback, etc. The kernel is supported from below by the *Ariadne* threads system (Mascarenhas and Rego 1996) and a suitable communications subsystem, e.g., PVM (Sunderam 1990). Support for migratable threads is provided by the Ariadne system.

The kernel programming interface is represented by public methods of class `PSol`, the main simulation class in *ParaSol*'s C++ interface. An explanation of some of the kernel primitives will help clarify the description of the mapping from GPSS to *ParaSol*.

A transaction's execution environment consists of the local variables of the function currently being executed, and the local variables of all functions in its calling chain. Transactions are directly supported by the kernel layer through the use of threads. They are dynamically created with method `trCreate()`. As in other process-oriented simulation languages, simulation time progresses when transactions execute `trHold(double time)`, a primitive analogous to `hold()` in CSIM. If the simulation time is t before `trHold(x)` is executed, it will be $t + x$ after this primitive is executed. Other transactions may execute between time t and time $t + x$.

Transactions are suspended and resumed with primitives

```
int trSuspend(void)
int trResume(int tid, double delta_t)
```

If a transaction is to be suspended, layers above the kernel are responsible for storing the transaction's id, so that it can be used when the transaction is resumed.

In *ParaSol* the term logical process (LP) – slightly twisted in relation to the definition used in other systems – describes the *static* (albeit still active) part of the simulation model. While transactions migrate from LP to LP, LPs themselves remain largely static. Each logical process in *ParaSol* consists of a single thread – the "LP thread" – and objects hosted by the LP. An LP is created and bound to one specific (UNIX) process at the start of the simulation, using primitive `bindLP()`.

Transactions (threads) migrate from one LP to another, using primitive `trMigrate(int LPid)`. Migrations occur instantaneously in terms of simulated time; if a delay is required it can be added by invoking `trHold()`.

To make a C++ object usable as a *ParaSol* global object, two kernel primitives are invoked on its behalf. The first registers the object for state saving, causing methods `save()` and `restore()` to be called when the system is saving or restoring state (in this instance the kernel does not insulate the layers above from parallel simulation functions). The second primitive registers the object as a global object, allowing transactions throughout the system to locate the position (LP number) of the object. Global objects have one "original" copy, located at some process, and several proxies. In general there is one proxy at each of the other processes. When a transaction accesses a proxy's method, a check is made within the method to determine if the object is an original or a proxy. If the object is a proxy, the transaction migrates to the appropriate LP where the same method is invoked on the original object.

4 GPSS OVERVIEW

Like *ParaSol*, GPSS adopts an active-transaction process-interaction world view. A GPSS program consists of a sequence of "blocks" that represents the flow of transactions through the system. Transactions, which are created by the `GENERATE` block, progress from one block to the next unless routed to another block by a `TRANSFER` block.

A large number of permanent entities are provided to facilitate modeling. For example, *Facilities* are FIFO single-server queues. Block `SEIZE` is used by a transaction to "seize" a facility's server. Seizing an already seized facility forces a transaction to *block* until the facility is released. A transaction holding a facility releases it using `RELEASE`. *Storages* are similar to facilities but may hold more than one transaction at a time.

Transactions have access to local (per transaction) variables called "parameters" and global (system-wide) variables called "savevalues". There is also support for computational entities such as random number generators. Time advances through the execution of the `ADVANCE` block, which suspends the invoking transaction for a user-specified amount of simulated time (only integer numbers allowed). Transactions finish their execution by entering the `TERMINATE` block. At each such block, a global termination counter is decremented by some given amount. When the counter reaches 0, the simulation is terminated.

Some powerful features make GPSS useful in manufacturing applications. For example, the `SPLIT`, `MATCH`, `GATHER`, and `ASSEMBLE` blocks allow transactions to split into two or more transactions and later rejoin, effectively enabling the modeling of machine parts that split and take different paths.

```
GENERATE    FN$ARRIV
QUEUE       WAIT
SEIZE       BARB
DEPART      WAIT
ADVANCE     FN$SERV
RELEASE     BARB
TERMINATE   1
START       10
```

Figure 2: Example of GPSS Program

A small example of a GPSS program is given in Figure 2. This example (taken from Bobillier et al. (1976)) models a barber shop: a barber, represented by facility `BARB`, can serve only one customer at a time. Customers, represented by transactions, enter the shop and wait for service if the barber is busy. Customers are served in order of arrival. A GPSS queue—an entity that helps in generating queueing statistics—called `WAIT` is used to represent the waiting room. The `START` block sets the termination counter, indicating how many transactions must terminate for the simulation to finish. The transactions are created by the `GENERATE` block, with the time between creations given by function "ARRIV".

At the heart of the GPSS simulator we find *transaction chains* containing all transactions in the system. The *future events chain* (FEC), actually a simulation calendar, contains the list of transactions scheduled for execution at a future time, in general those that have entered an `ADVANCE` block. The *current events chain* (CEC) contains transactions due to execute at the current time, but also contains those that are blocked, waiting for conditions such as the release of a facility. The CEC is organized as a priority queue, with transactions having higher priority coming first. Among transactions with the same priority, a FIFO ordering is adopted. The CEC is always scanned completely in each simulation cycle, from beginning to end, before it receives new transactions from the FEC. The transactions in the CEC chain are executed until they enter `ADVANCE` or become blocked for some reason. Certain events, such as the release of a facility or storage, force the simulation engine to rescan the whole CEC, since some transactions on the CEC may have become ready to execute once again.

5 GPSS-*ParaSol* MAPPING

At a first glance, GPSS appears readily amenable to execution support from *ParaSol*. We implement the language as a GPSS-to-C++ translator, where the transformed program invokes primitives from the *ParaSol* kernel and a "GPSS domain" library.

Entities in GPSS are mapped as follows. Static entities such as facilities, queues, storages, and even savevalues are implemented as *ParaSol* global objects. As an example, the GPSS facility object, similar but not identical to the Facility object in *ParaSol*'s queueing domain, presents the outside world with methods such as `seize()` and `release()`, which implement the behavior of the `SEIZE` and `RELEASE` blocks, respectively. Other methods are provided to return facility statistics (e.g., facility utilization and queueing delay). These methods are required not only for printing final statistics but also for implementing GPSS's *standard attributes*. The GPSS facility object also contains state saving and restoration methods which are required by *ParaSol*'s optimistic synchronization protocol.

Not surprisingly, GPSS's transactions are actually *ParaSol*'s transaction threads. The execution path of GPSS transactions, which comprise the bulk of the GPSS input lines, is incorporated into the thread code. Threads then move from block to block as directed by the input program. Transaction attributes such as parameters and priorities are implemented via the *thread attribute area* supported by *Ariadne*. An alternative would be to place these transaction attributes on the thread's stack, but this would hinder the translation of GPSS blocks that allow one transaction to examine another's attributes. Time advances (`ADVANCE` block) are achieved through calls to the kernel `trHold()` primitive.

Transactions entering a `SEIZE` block, for example, are represented by threads executing

```
GPSS_Facility_ptr->seize()
```

If the transaction is "refused" by the `SEIZE` block because the facility is already in use, then the corresponding thread is placed in a queue inside the GPSS facility object. In terms of the domain-kernel interface, the queue is nothing more than a list of thread ids. The thread is suspended with the kernel's `trSuspend()` primitive. To resume the thread, it suffices to call the kernel's `trResume()` primitive with the thread id as parameter. When resumed (by another thread executing the `RELEASE` block), the thread eventually returns from `seize()`, now owning the facility.

Parallel execution of GPSS is achieved by first distributing the global objects among different LPs and then placing these LPs in distinct processes, as illustrated in Figure 3. This figure shows a transaction that executes a method at a proxy facility at LP #1. This forces it to migrate to LP #2 and access the same method at the real facility. The number of LPs used in the program is determined at compile time. This number gives an upper bound on the maximum parallelism possible. The assignment of LPs to processors is done at run time, based on the number of processors available; each processor may host more than one LP. When a thread accesses a remote object, the thread *migrates* to the LP where the object is located, as explained in Section 3. For example, inside `GPSS_Facility::seize()` (and also in many of the GPSS domain library methods), there is a test to determine whether the facility is remote. If it is, the thread migrates to the appropriate LP. Otherwise, the thread goes through the actual `SEIZE` procedure on the local object. A thread migrates along with its attribute area, which is required for the correct implementation of parameters, priorities and other GPSS transaction attributes.

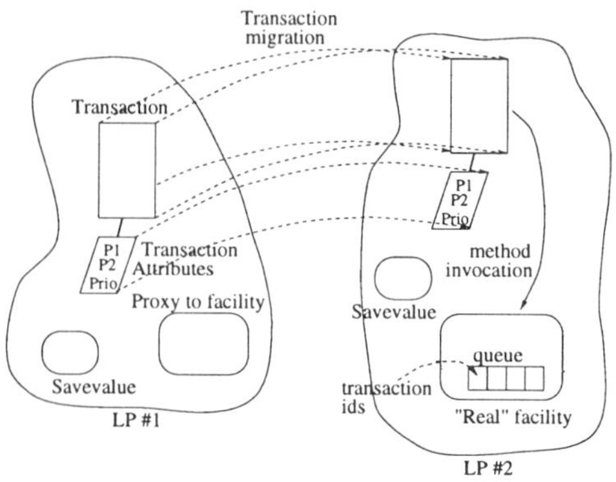

Figure 3: Transactions and GPSS Objects

```
ATHREAD generateBlk5(void)
{
  int intTmp;
  gpssP->advance(8);
  for(; ;) {
    pSolP->trCreate(transactionThread,
                              <stk sz> ...);
    intTmp = uniform_AB(40,60);
    if(intTmp > 0)
      gpssP->advance(intTmp);
  }
}
```

Figure 4: *ParaSol* Code Created for Block `GENERATE 50,10,8`

A GPSS program may have several `GENERATE` blocks, with each possibly specifying a different rate of creation for transactions. For each `GENERATE` block in the program, the translator creates a *generate thread*. Figure 4 shows an example of code executed by a generate thread. The example corresponds to the sequence `GENERATE 50,10,8`, which is supposed to create new transactions with inter-creation time uniformly distributed between 50-10 and 50+10 time units, with the first one created at time 8. In the figure, method `advance()` is nothing but a wrapper around `trHold()`. The reason for not having `trHold()` itself in the code is explained in Section 7.4. The generate threads, one for each `GENERATE` block, are created at the beginning of the simulation.

Termination requires the use of a centralized `GPSS_Termination` global object. Threads executing the `TERMINATE` block migrate to the LP holding the `GPSS_Termination` object, using the same mechanism as in `GPSS_Facility`. There, the termination counter, which is a member of the termination object, is decremented; when it gets to 0 all processes are informed about the termination. When `TERMINATE`'s operand is 0, the termination counter need not be decremented. As an optimization for this situation, the thread does *not* migrate to the termination object's location, since there is no need for it to access the termination object.

6 ISSUES INVOLVED IN PARALLELIZING GPSS

The main obstacle in parallelizing GPSS is the many instances where the language definition, and its natural sequential implementation, assumes the existence of centralized data structures. Typical examples are statements such as

```
      TEST E   Q8,15              (1)
      TRANSFER BOTH,HERE,THERE    (2)
VAR   VARIABLE Q1+Q2+Q3           (3)
```

In the first statement above, the transaction must block while the number of elements in queue number 8 is not 15. In a typical GPSS implementation, the transaction is left in the CEC (see Section 4) and is awakened every time the CEC is rescanned. This kind of blocking, called a "non-unique blocking condition", is one where the system does not know a priori what can be done to unblock the transaction. Consequently, there is no specialized queue for "transactions waiting for queue 8 to have 15 elements".

In the second statement, according to the language definition, the transaction must attempt to go to the block labeled `HERE` and, if the block is "refusing transactions" (for example, the block is a `SEIZE` and the facility is in use), the transaction must try the block labeled `THERE`. If the latter also refuses entry, the transaction must suspend itself inside the `TRANSFER` block and try again when the CEC is rescanned.

In the third statement, the transaction must obtain the number of elements in queues 1, 2, and 3 to compute the value of the variable.

To be able to parallelize GPSS, we must do away with centralized data structures such as the FEC and CEC. While the FEC is handled "automatically" by the PDES synchronization mechanism, the CEC still requires some suitable replacement. The existence of the CEC may severely impact performance, since it assumes that all transactions it holds will have to be rescanned every time some "significant" event, such as the release of a facility, occurs. While the rescan in a sequential setting is already recognized as inefficient (Bobillier et al. 1976), doing it in a distributed environment would certainly lead to bottlenecks: if the same scheme is adopted in a distributed implementation, the "significant" events would have to be reported to all processes, resulting in unacceptable communication cost.

Statement 3 above highlights the importance of object placement in the performance of parallel GPSS: if the three queue objects are located on different processors, then the thread implementing the transaction must migrate at least twice just to compute the value required in the example.

Statement 2, which like statement 1 presents a non-unique blocking condition, illustrates how much the language depends on centralized information: if the objects referred to by blocks `HERE` and `THERE` are located at different processes, then the transaction would have to migrate twice to determine the status of these blocks. Moreover, it would have to repeat the check every time some "significant" event occurs at any place in the system.

Another difficulty related to the CEC is its policy of considering transaction priorities when inserting transactions in the chain. Like many other systems, *ParaSol* does not have a CEC-equivalent. It inserts transactions in the calendar based only on their resumption times.

7 SOME SOLUTIONS

7.1 Object Placement

A judicious placement of objects may alleviate some performance problems by reducing the number of thread migrations. For instance, in statement 2, having the objects accessed by blocks `HERE` and `THERE` in the same process would help tremendously, as would having queues 1,2, and 3 in the same process at statement 3. Of course, applying this reasoning to all the blocks would result in a sequential simulation! It can be observed that having queues 1–3 in different processes is likely to be less damaging than having objects in `HERE` and `THERE` in different processes, since in the latter case the set of migrations may occur several times for each transaction that enters a `TRANSFER BOTH,...` block.

Since, in some applications, the user may be better equipped than the compiler to decide the location of some key entities, constructs are provided to allow manual placement of entities at either compile time or program load time.

7.2 Replacement of the CEC

The CEC and the mechanism used to handle non-unique blocking conditions must be altered for the parallel implementation. Suppose a transaction is blocked, waiting for the condition "queue 4 has 3 elements and storage 6 is full". In the sequential implementation, the transaction is placed on the CEC. Every time the CEC is rescanned, the transaction is resumed. The transaction then proceeds to check one more time if the two conditions are true. In the distributed setting, queue 4 and storage 6 may be located on different LPs, say LP1 and LP2, respectively.

Since a "global" CEC is not maintained, the following scheme is adopted when a transaction is about to block on a non-unique blocking condition (we use the example above to explain it):

- Thread T1 visits queue 4 at LP1 and places a notification record [T1,LP3] at queue 4
- Thread T1 then visits storage 6 at LP2 and places a record [T1,LP3] at storage 6
- Thread T1 moves to LP3 and suspends itself. LP3 may be LP1, LP2, or yet another LP.
- When the status of queue 4, for example, changes, the queue object (that is, the particular thread which, executing some method in the queue object, alters the queue's status) checks its notification records. Upon finding the entry [T1,LP3] there, it sends a messenger thread to LP3. Once there, the messenger confirms that T1 is still blocked and, if so, awakens it. T1 then visits queue 4 and storage 6 once again.

Although still expensive, this scheme is nevertheless cheaper than sending broadcast notification messages every time any entity changes state (the distributed equivalent of the CEC). For unique blocking conditions, the implementation is simpler and more efficient: transactions are suspended and placed in existing queues at the appropriate entities. As an example, consider the queue of transactions waiting to enter the facility entity, located within the GPSS facility object.

7.3 Space-Time Memory

Besides triggering expensive multiple migrations, constructs such as that shown in example 3 above (`Q1+Q2+Q3`) create havoc with optimistic synchronization protocols, since migrations may cause rollbacks at one or more LPs. On the other hand, in such constructs the queue objects themselves are not altered by the operations following the migration. Therefore, the following scheme may be adopted:

- Variables that are frequently accessed in a read-only fashion, such as the queue size in the given example, are stored in a "space-time" manner, not unlike the Tsai and Fujimoto approach (1993): not only the most current value, but also *all* past values are stored. An appropriate format for this is a time-ordered sequence of (t_i, v_i) pairs, where v_i is the variable's value and t_i is the simulated time at which the variable is loaded with the value v_i.
- If a thread migrates to an LP only to retrieve the value of some variable, then a special "read-only" migration is used that does not cause a rollback even if the destination LP is already in the future of the migrated thread. In this case, the migrated thread is scheduled as if it were not a straggler, and it then accesses a past value of the desired variable by examining the past history of values stored in the space-time memory.

To avoid causality problems, the migrating thread should refrain from altering the state of any object. If it decides to do so, the thread is required to update its migration status from "read-only migration" to the default "read-write" migration. Then, if the destination LP is in the future of the arriving thread, in terms of virtual time, a rollback procedure is immediately initiated.

The mechanism used for reclaiming the space-time memory is similar to that used in fossil collection: values saved before GVT (Global Virtual Time, Fujimoto 1990) may be discarded since they are no longer needed. Notice, however, that all manipulation of the space-time memory is done at the domain level, independently of the kernel. The kernel could have provided all support for space-time memory by incorporating it into the state saving mechanisms. Past values of variables would then be retrieved from system snapshots taken for state saving purposes. The drawback of this approach is that it would require the kernel to save state at *every* execution of an event, preventing the use of infrequent state saving mechanisms.

If keeping all values since the GVT in space-time memory is too expensive, the space-time memory may be configured to store only the last few values. This may be the case if each value requires a large amount of memory, or if the GVT is not computed frequently enough, If some older value that is not stored in space-time memory is requested, then the read-only migration is updated to the default migration.

Together with the read-only migration, the space-time memory can be used in the implementation of several GPSS constructs, including the savevalues and many of GPSS's "standard attributes".

7.4 Transaction Priorities and the CEC

It is possible to simulate the structure of the CEC (transactions ordered by insertion time and priority) without having to change *ParaSol*'s kernel. First, we observe that the *ParaSol* calendar, the equivalent of GPSS's FEC and CEC combined, already respects a FIFO ordering for threads having the same time: if a thread executes `trHold(x)`, and the current time is t, its resumption is scheduled as the *last* entry in the calendar having time $t + x$. Next, we can take advantage of GPSS's *integer* time advances and use *ParaSol*'s arbitrary time advances. If the thread priority changes from 0 to p ($0 \leq p < 128$) then the next `ADVANCE x` issued by the thread at time t will cause the thread to be resumed at *ParaSol* time $t + x - cp$, where $c < 1/128$ is a constant. In this way, the thread is resumed before all threads with priority smaller than p scheduled for $t + x$ (GPSS time).

The simulation of the CEC's structure is not perfect, though. The order of scheduling two transactions arriving at an LP from two different processors, with the same simulation timestamp, is essentially random, and depends on the real times at which the transactions arrive at the LP.

8 EXPERIMENTS: ENTITY MAPPING

We report results of initial performance experiments that measure the effect of the entity-to-LP mapping.

Our test program simulates a closed queueing network and is written in a GPSS subset. The network has 64 facilities, divided into four groups of 16 facilities in tandem. Jobs that leave one group can be routed to any other group. The GPSS code is shown in Figure 5. The `GENERATE` block at line 2 generates 32 transactions, with inter-generation time uniformly distributed between 80 and 120. These transactions begin execution at the line following the `GENERATE` block. Once created, transactions go with equal probability to blocks labeled `SWT1`, `SWT2`, `SWT3`, or `SWT4`. At each such block, transactions visit 16 facilities in series, after which they go back to the `TRANSFER` block at line 3. Blocks at lines 31–33 actually define the termination condition for this particular program: a transaction is created at time 160000. Upon being created at line 31, it executes the termination block

```
1        SIMULATE  1
2        GENERATE  100,20,,32  32 jobs generated
*                  once every U(80,120) units
3 ROUT   TRANSFER  0.250,,SWT1 With prob 1/4,
*                  transfer transactions to blocks
*                  SWT1, SWT2, SWT3, or SWT4
4        TRANSFER  0.333,,SWT2
5        TRANSFER  0.500,,SWT3
6        TRANSFER  ,SWT4
7 SWT1   ASSIGN    1,16    P1 <- 16
8        ASSIGN    2,1     Entities [1-16]
9 LOO1   QUEUE     P2
10       SEIZE     P2
11       DEPART    P2
12       ADVANCE   50,10   Service time: U(40,60)
13       RELEASE   P2
14       ASSIGN    2+,1    P2++
15       LOOP      1,LOO1  P1--;if(P1==0) goto LOO1
16       TRANSFER  ,ROUT

17 SWT2  ASSIGN    1,16    P1 <- 16
18       ASSIGN    2,17    Entities [17-32]
19 LOO2  QUEUE     P2
20       SEIZE     P2
21       DEPART    P2
22       ADVANCE   50,10   Service time: U(40,60)
23       RELEASE   P2
24       ASSIGN    2+,1    P2++
25       LOOP      1,LOO2  P1--;if(P1==0) goto LOO2
26       TRANSFER  ,ROUT
27 SWT3  ASSIGN    1,16    P1 <- 16
28       ASSIGN    2,33    Entities [33-48]
         [...]
29 SWT4  ASSIGN    1,16    P1 <- 16
30       ASSIGN    2,49    Entities [49-64]
         [...]
31       GENERATE  160000  create transctns every
*                          160000 time units     (*)
32       TERMINATE 1       termination transctn  (+)
33       START     1       set termination cnt to 1
34       END
```

Figure 5: A Closed Queueing Network Coded in the GPSS Subset

at line 32. Since the termination counter is now 0, the simulation is terminated.

How GPSS entities are mapped onto LPs is crucial to the performance of parallel GPSS. We have experimented with different mappings, in order to evaluate its effect on execution time. The ideal mapping for the given queueing network is to have each group of 16 tandem facilities and queues placed at a different LP: facilities 1–16 are placed at LP 0, facilities 17–32 at LP 1, and so on. Queues are treated identically. By doing this, we balance computation across processes. Also, we minimize communication, since transactions visit 16 facilities and queues at a time without the need for remote object access.

Like queues and facilities, GENERATE blocks and termination objects must also be mapped. All transactions created by GENERATE begin their execution at the LP hosting the generate thread (Section 5). Transactions executing TERMINATE with a non-zero operand must migrate to the LP that hosts the termination object. The GENERATE block on line 31 should be placed at the LP that contains the termination object. LP 0 is chosen for both. The generate block on line 2 can be placed at any LP, and LP 0 is chosen for this block.

Table 1 shows the results of executions using different entity-to-LP mappings. All four processors in a SUN SPARCstation 20, a shared memory multiprocessor, were

Table 1: Performance of a Closed Queueing Network Coded in GPSS with Several Different Entity-to-LP Mappings

	Mapping	Exec time	Num rollbacks	Msg read ops
1	Optimal mapping	52.7	1959	14907
2	GENERATE block (*) and TERMINATE block (+) in different LPs	53.5	2976	18515
3	Facilities and queues 13–16 at LP 1; facilities and queues 29–32 at LP 0	66.4	4755	33779
4	Facility and queue 10 at LP 1	66.6	3953	26819
5	Unbalanced: facilities and queues 17–32 at LP 0 (LP 1 hosts no entities)	82.3	4424	35653

used in the experiment. Execution times are in seconds. Approach 1 shows, as a reference, the performance of the optimal mapping. Approach 2 shows the performance of a near-optimal mapping; only the GENERATE block at line 31 is moved to LP 1. This seems to be a trivial variation of the optimal mapping, since only one transaction is created on line 31. In an optimistic environment, however, the sequence of events is more complex. Since all job transactions are generated at LP 0, LP 1 is initially idle and has no alternative but to schedule the only calendar entry it has, which is for the generate thread (line 31) at time 160000. Once scheduled, the generate thread creates a transaction that immediately migrates to LP 0 in search of the termination object. At LP 0, this transaction is placed in the calendar but not scheduled, since time at LP 0 is still far from the termination time. The GENERATE block on line 31 creates a transaction *every* 160000 time units, so that a new transaction is created at each of the virtual times 320000, 480000, etc.; each immediately migrates to LP 0. Eventually, job transactions migrate from LP 0 to LP 1, causing a rollback at LP 1. Several anti-thread messages (*ParaSol*'s equivalent of anti-messages) are sent from LP 1 to LP 0. The whole process is repeated whenever LP 1 finds itself idle. Though the effect on execution time, as shown in Table 1, is small, the effect on number of rollbacks is not. On a distributed-memory environment, execution times would suffer because of the larger communication delays.

Approach 3 shows a more significant departure from the optimal mapping: four facilities and queues are moved from LP 0 to LP 1, and vice versa. Not surprisingly, the execution time, number of rollbacks, and number of message operations are all higher than in the optimal mapping. In approach 4, the optimal mapping is changed by placing only one facility-queue pair (number 10)—originally at LP 0—at LP 1. The execution time is close to that given by approach 3. When compared to the optimal mapping, this indicates that even small changes may have a noticeable effect on execution time. After accessing facilities 1–9, transactions at LP 0 must migrate to LP 1 to access facility 10, and then return to LP 0 to access facilities 11–16. These extra migrations account for the increased communication/rollback overhead over the optimal mapping.

Finally, approach 5 shows a more radical departure: all entities at LP 1 are now assigned to LP 0, resulting in an unbalanced load. The execution time is greatly affected, with the processor hosting LP 1 becoming idle for the entire run.

9 CONCLUSION

This paper describes the implementation of a parallel GPSS simulation language, based on the *ParaSol* system. Because of GPSS's widespread use, this effort is one significant step towards making parallel simulation more practical and useful to the simulation community. *ParaSol*'s process-orientation enhances the mapping from GPSS to *ParaSol* , since GPSS is also process-oriented. GPSS, however, relies heavily on centralized data structures. These can hamper the performance of parallel applications. We describe some techniques to tackle bottlenecks arising in the parallel GPSS implementation, such as user-guided placement of simulation objects, the use of notification mechanisms to control communication costs in non-unique blocking conditions, and the use of space-time memory and read-only migrations to reduce the number of rollbacks.

At the present time, the implementation of a GPSS core is complete. Although we expect GPSS programs to run in parallel without changes, some programs may not possess a sufficient amount of inherent parallelism. These may require code rearrangement or may run efficiently only for select entity-processor mappings.

ACKNOWLEDGMENTS

This research was supported in part by ONR-9310233, BMDO-34798-MA, and NSF-CCR 9311862. The first author was supported by CNPq-Brazil, grant 260059/91.9.

REFERENCES

Bagrodia, R.L. 1991. Iterative design of efficient simulations using Maisie. In *Proceedings of the 1991 Winter Simulation Conference*, pages 243–247.

Bobillier, P., B. Kahan, and A. Probst. 1976. *Simulation with GPSS and GPSS V.* Prentice Hall.

BoyanTech, Inc. 1995. McLean, VA 22102. *CPSim 1.0 User's Guide and Reference Manual.*

Fujimoto, R.M. 1990. Parallel discrete event simulation. *Communications of the ACM*, 33(10):30–53.

Fujimoto, R.M. 1993. Parallel discrete event simulation: Will the field survive? *ORSA Journal of Computing*, 5(3):213–230.

Gomes, F., S. Franks, B. Unger, Z. Xiao, J. Cleary, and A. Covington. 1995. SimKit: A high performance logical process simulation class in C++. In *Proceedings of the 1995 Winter Simulation Conference*, pages 706–713.

Mascarenhas, E., F. Knop, and V. Rego. 1995. ParaSol: a multithreaded system for parallel simulation based on mobile threads. In *Proceedings of the 1995 Winter Simulation Conference*, pages 690–697.

Mascarenhas, E. and V. Rego. 1996. Ariadne: architecture of a portable threads system supporting thread migration. *Software – Practice and Experience*, 26(3):327–356.

Nicol, D. and P. Heidelberger. 1994. On extending parallelism to serial simulators. Technical Report ICASE Report No. 94–95, Institute for Computer Applications in Science and Engineering – NASA Langley Research Center. December.

Schwetman, H.D. 1986. CSIM: A C-based, process-oriented simulation language. In *Proceedings of the 1986 Winter Simulation Conference*, pages 387–396.

Sunderam, V.S. 1990. PVM: a framework for parallel distributed computing. *Concurrency: Practice and Experience*, 2(4).

Tsai, J.-J. and R.M. Fujimoto. 1993. Automatic parallelization of discrete event simulation programs. In *Proceedings of the 1993 Winter Simulation Conference*, pages 697–705.

AUTHOR BIOGRAPHIES

FELIPE KNOP, Ph.D., Computer Sciences Department, Purdue University, (August 1996), received a Masters degree in Computer Sciences from Purdue University in 1993 and a Masters degree in Electrical Engineering from University of São Paulo, Brazil, in 1990. His current research interests include parallel and distributed simulation, and multiprocessor operating systems.

EDWARD MASCARENHAS, Ph.D., Computer Sciences Department, Purdue University, (August 1996), received a Masters degree in Industrial Engineering from NITIE (Bombay, India), and a Masters degree in Computer Sciences from Purdue University (West Lafayette) in 1993. His research interests include parallel computation, distributed simulation, and multithreaded programming environments.

VERNON REGO is a Professor of Computer Sciences at Purdue University. He received his M.Sc.(Hons) in Mathematics from B.I.T.S (Pilani, India), and an M.S. and Ph.D. in Computer Science from Michigan State University (East Lansing) in 1985. He was awarded the 1992 IEEE/Gordon Bell Prize in parallel processing research, and is an Editor of *IEEE Transactions on Computers*. His research interests include parallel simulation, parallel processing, modeling and software engineering.

Proceedings of the 1996 Winter Simulation Conference
ed. J. M. Charnes, D. J. Morrice, D. T. Brunner, and J. J. Swain

USING ZPL TO DEVELOP A PARALLEL CHAOS ROUTER SIMULATOR

Wilkey Richardson

Department of Electrical and Computer Engineering
University of Arizona
Tucson, AZ 85721, U.S.A.

Mary L. Bailey

Computer Science Department
University of Arizona
Tucson, AZ 85721, U.S.A.

William H. Sanders

Center for Reliable and High-Performance Computing
Coordinated Science Laboratory
University of Illinois at Urbana-Champaign
Urbana, IL 61801, U.S.A.

ABSTRACT

This paper reports on our experience in writing a parallel version of a chaos router simulator using the new data-driven parallel language ZPL. The simulator is a large program that tests the capabilities of ZPL. The (parallel) ZPL program is compared with the existing serial implementation on two very different architectures: a 16-processor Intel Paragon and a cluster of eight Alpha workstations. On the Paragon, the simulator performs best when simulating medium- to large-sized routers, and on the Alpha cluster, it performs best when simulating large routers. Thus a user can choose the parallel platform best suited to the router size.

1 INTRODUCTION

There are some simulators that are good candidates for parallel simulation. These simulators are generally too slow on a sequential machine and are much faster on a parallel or distributed machine. The first criterion is often much easier to judge than the second, since the second is only truly verified after the implementation of a (good) parallel simulator. A sequential simulator has been implemented to simulate a chaos router, a randomizing, non-minimal adaptive router for multicomputers (Bolding and Snyder 1992, Konstantinidou and Snyder 1994). It exhibits the first characteristic, slowness, when simulating large routers. The router simulator has been used extensively on a sequential machine, and one version of the router node has been implemented in VLSI (Boulding et al. 1994). While the sequential simulator proved sufficient for many different router configurations, it proved to be impractical for very large router sizes. Thus it becomes a candidate for a parallel simulation, assuming the parallel version can run much faster than the serial one and, in this case, can simulate larger router configurations than are practical using the sequential simulator.

There are several issues to consider when implementing a parallel simulation. First is whether the simulation is event-driven or time-driven. Event-driven strategies are most appropriate when the activity of the system being simulated is uneven. This is not true in the chaos router simulator under heavy loads. Almost all router nodes are active during each time interval, so a time-driven simulation is preferred, eliminating the overhead of event management.

The second issue is the implementation language. Ideally, the simulator will be portable, implying the use of some type of standard parallel language or communication library. There are many parallel languages that could be selected. We chose to use ZPL, a new data-driven portable parallel language that is a sublanguage of a more general family of parallel programming languages A-ZPL designed by Lin and Snyder (1992, 1993). ZPL is attractive for this parallel simulation for three reasons. First, a data-driven model is appropriate given the time-driven simulation model. Second, ZPL is portable, so users can run the parallel simulation on various platforms. Third, ZPL is a sublangague of A-ZPL. Many simulators are event-driven and thus cannot use a data-driven language but require a more general parallel language such as A-ZPL.

This paper reports on our experience writing a parallel version of the chaos router simulator using this new portable data-driven parallel language ZPL. We compare the resulting parallel simulator with the serial one on two very different platforms: the Intel Paragon and a network of Alpha workstations. The Intel Paragon performs best for small- to medium-sized problems but runs into memory problems for the largest sizes. The Alpha network performs poorly for small- to medium-sized problems but performs well for larger-sized problems. In fact, it can simulate larger sized routers than can be simulated by the Paragon or by a single Alpha workstation. Thus we show that our original goal is met – the parallel simulator can run faster than the serial one on the Paragon for many router sizes, and we can simulate larger router nodes (mainly on the Alpha cluster due to memory limitations on the Paragon). Moreover, we believe that our experiences using ZPL are valuable to other parallel programmers who wish to use this new language. We believe that the parallel chaos router simulator is the largest ZPL program that has been written to date.

2 THE PARALLEL CHAOS ROUTER

The chaos router is a randomizing, nonminimal adaptive router for multicomputers. It dynamically selects communication paths depending on network traffic, thus bypassing network congestion. Moreover, messages do not necessarily take a minimal path between source and destination – they can be *derouted*, or steered off a minimal path, to avoid congestion. The chaos router uses randomization to eliminate the need for livelock protection, thus probabilistically ensuring that each message reaches its destination. We briefly describe the chaos router algorithm, focusing on those parts of the algorithm critical in the simulation; the router is fully described in Bolding and Snyder (1992).

The current router specification assumes that the communication structure is an n-dimensional torus, with a processor/router pair at each torus node. We further constrain the communication structure to be a two-dimensional torus, as in the hardware implementation. Each processor communicates with its corresponding router node; each router node communicates with both its corresponding processor and its four adjacent router nodes. A block diagram of a router node is shown in Figure 1.

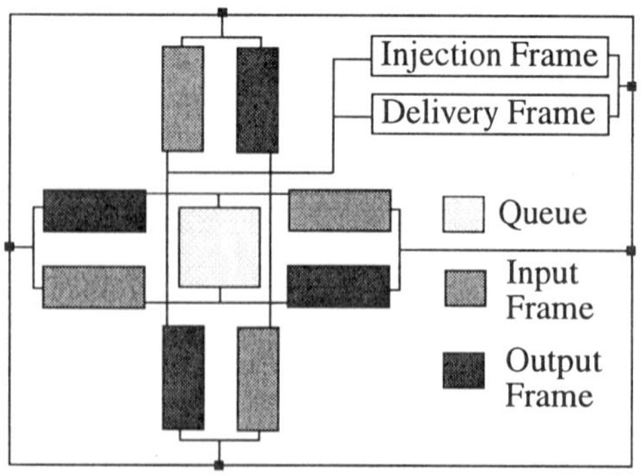

Figure 1: A Chaos Router Node

Messages enter the routing network through the Injection Frame and exit through the Delivery Frame. Both are actually FIFO buffers that connect the router to its corresponding processor. There are also two buffers associated with each of the four router communication channels: the Input Frame and Output Frame. When a message enters the router, either through the Injection Frame or an Input Frame, its header is read to determine which Output Frames and/or Delivery Frame can be used to profitably route the message. These desirable output buffers are then searched in a cyclical order, with the first free buffer receiving the message. If no desirable output buffer is free, then after a specified number of cycles the message is stored in a central Queue. Messages in the Queue have priority for accessing outgoing channel(s) when they become free. If a message is destined for the Queue, but the Queue is full, a message in the Queue is derouted. The message to be derouted is picked randomly.

3 THE ZPL IMPLEMENTATION

The parallel simulator uses the fact that there is a uniform torus of router nodes to create an array or ensemble of router nodes for the ZPL language. Ensembles in ZPL are automatically distributed across processors but are managed as a single entity in a ZPL program. Once the router ensemble was created, most of the ZPL simulation was a relatively straightforward implementation of the single-router node operation. Communication among the elements of an ensemble are facilitated by ZPL operators, including operators for "wrapping" data around the edges. It should be noted that even though the ZPL program is written in a purely SIMD style, the code is not executed in lock-step, as in an SIMD architecture. There are specific places in the code where synchronizations are performed; otherwise each processor runs its portion of the code asynchronously (Lin and Snyder 1993).

Our original implementation was quite slow; a 64-node router ran 16 times slower on a one-processor Paragon than the serial version! We thus focused initially on optimizing the ZPL code. There were four areas where we optimized: (1) communication, (2) memory usage, (3) procedure calls, and (4) statistics. The first two were needed mainly because of our naive translation of the serial router code into ZPL code. The third was needed because of the way that ZPL compiles procedure calls. The fourth was needed because of the way router statistics are computed, where global data are required. We will discuss each of these four areas, focusing on those aspects that might help other ZPL programmers.

In ZPL, there are two operators that are used to communicate data within an ensemble, the "wrap" and "@" (at) operators, both of which take an ensemble as their operand. In the chaos router, the data for each router node are encapsulated in a single ensemble, which we initially used as the operand for these operators. However, these operators invoke communication, so when we wrapped an ensemble node, the ZPL compiler packaged the entire router node (approximately 5KB) into a message and sent it to another processor. This resulted in much larger messages than were necessary. By modifying the code to use only the portion of the node that was needed, we reduced the message size to 24 bytes, resulting in a 47% decrease in execution time for a 64-node router.

We had to pay close attention to memory usage in the router simulation because ZPL does not support dynamically allocated data structures, which the serial router simulator uses extensively. In some places, it is easy to avoid the dynamic structures by using ensembles; in other places, it is more difficult. For example, each node has a queue of messages that are being injected into the router by the processor associated with this node. This queue is typically quite small, but it can grow large when the router is congested. For the ZPL implementation, we statically fixed the sizes of these data structures to be the largest size supported in the serial simulator. In the case of the injection queue, this resulted in a 4KB queue per node. As the number of nodes grows, this becomes a significant amount of memory, which is often in short supply. Thus we more carefully determined the appropriate size for each data structure, shrinking them whenever possible. In the case of the injection queue, the amount of memory was reduced to 200 bytes per node with no loss of functionality. This resulted in a 17% savings in execution time for a 64-node router.

The third problem area involved procedure calls. There are two types of procedures in ZPL, parallel and promoted procedures. A parallel procedure uses a parallel construct, references an ensemble, or performs I/O. All other procedures are promoted procedures. Parallel procedures cannot be called from promoted procedures. When a promoted procedure is called from a parallel procedure, the promoted procedure is called in parallel, once for each element in the ensemble found in the parallel procedure. We originally assumed that the optimal performance would be obtained using parallel procedure calls, since it would allow the ZPL compiler to exploit the natural data parallelism. In fact, this is not true. Because ZPL is not recompiled for different numbers of processors, the ZPL compiler must insert additional code each time an ensemble is accessed in order to dereference ensemble elements, making it expensive to unnecessarily access ensembles. This is never a problem in promoted procedures, since the dereferencing is done before the procedure is called. We thus reinspected the code, creating as many promoted procedures as possible. In addition, we subdivided many parallel procedures to make additional promoted procedures, thus reducing ensemble references. This reduces the special code required for ensemble accesses, resulting in less overhead. These procedure changes resulted in a surprising 16% decrease in execution time.

The final optimization involved our statistics collection, which are collected to better understand the performance of the chaos router. The serial simulator computes statistics on a global basis while it is running, combining information from all nodes. We had two options for the parallel simulation: to collect statistics on a per-node basis and combine the data at some intervals, or to collect statistics on a per-processor basis. We began with the first option, to collect statistics on a per-node basis, combining the data at a "reasonable" interval. We immediately faced a trade-off. The longer the interval was between combining data, the more memory the simulation used to store the data it was collecting. On the other hand, the combination basically was a global sum, which is an expensive, serial operation (even with language support from ZPL). We were not able to find a good point in this space. If we combined data too frequently, we had communication bottlenecks, and if we waited too long we used up a large amount of memory, causing thrashing. Thus we considered the second alternative: collecting data on a per-processor basis. Unfortunately, there is no provision for this type of operation in ZPL. After talking with the ZPL designers, we wrote some C subroutines for data collection that were linked in with the ZPL code. This provided a "hook" for our data collection and improved the performance of the parallel simulation. While this optimization is more evident for larger configurations of the chaos router, we still obtained a 5% improvement for a 64-node router.

These four optimizations reduced the run-time of the ZPL implementation dramatically. The ZPL implementation now runs only 2.38 times slower than the serial implementation on a one-processor Paragon. Figure 2 illustrates the improvement for each of the four optimizations for a 64-node router on the Intel Paragon. Note that some optimizations such as statistics will show greater

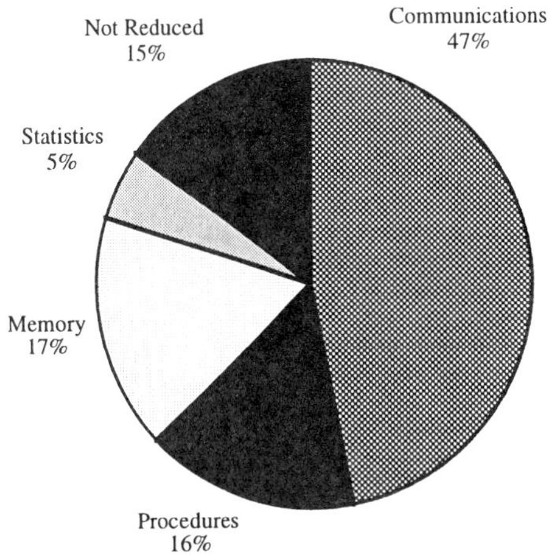

Figure 2: Improvement in Execution Time for ZPL Optimizations

improvement for larger router sizes. There is still room for improvement. For example, the ZPL compiler packages code that is passed between ensemble elements residing on the same processor as if it were being sent to another processor, increasing overhead. With additional compiler improvements and improvements in our code, we expect the execution time of the parallel code to continue to improve, although we don't expect the ZPL program to ever be as fast as the serial one.

In summary, our experience with ZPL has been quite positive. The language proved to be easy to use, with good documentation (Forman and Lin 1994, Lin 1994, Snyder 1994), and the compiler is reasonably efficient. The ZPL compiler was still under development during this project, and the compiler team was very helpful in fixing bugs that we found in the compiler and telling us when we were at fault.

4 EVALUATION OF THE SIMULATION

We now turn to the performance of the chaos router simulator on multiple processors. There are two main evaluation metrics that were used to measure the performance of the parallel simulator:

- The execution time of the parallel simulator and its relationship to the speed of the (existing) sequential simulator, and
- The portability of the parallel simulator and its ability to run on different architectures.

Performance tests were run on (1) a 16-node Intel Paragon and (2) a small network of Digital Alpha workstations (eight workstations). These are two very different architectures, with different computation/communication ratios. Thus they should provide useful insight into any architecture-related issues in performance.

Three parameters were varied in these experiments:

- Load, the number of router messages generated as a percentage of the rate that would saturate the network;
- Processors, the number of processors used in the parallel simulation; and
- Nodes, the number of router nodes in the simulation.

We first discuss the performance of the parallel simulator on each of these platforms, and then we discuss the impact of the platform on the simulation.

4.1 The Intel Paragon

The Intel Paragon used in these experiments has 16 processors, each with 32MB of memory. Below are the parameters used for these experiments.

Load:	20%, 40%, 60%, 80%
Processors:	1, 2, 4, 8, 16
Nodes:	16, 64, 256, 1024, 4096, 16384

Not all possible permutations of these parameters were run; the larger-sized experiments (nodes greater than 1024) were measured only on the larger numbers of processor due to performance problems. These will be discussed later in this section.

Below is a graph of the execution times for all of the experiments with a load of 80%. This load was chosen since it is the heaviest load. In all cases, the one-processor parallel version is somewhat slower than the serial version, so additional processors are needed to "break-even" or to show an improvement. Not surprisingly, the 16-node instance fails to run faster than the serial implementation, but all other versions do outperform the serial implementation, given sufficient numbers of processors. The 4096 instance was not run on the Paragon for one and two processors, due to thrashing (all processors shared the same disk, so performance seriously degraded when much swapping occurred). For similar reasons, the 16384-node instance was only run on 16 processors. This version was too large to be run on the serial simulator. As the instances grow, the relative performance increases, and we see greater improvements over the serial version. Moreover, we are able to run larger experiments on the Paragon than can be run on the serial version (the 16384-node instance). We now discuss specific issues in the experiments to illustrate performance issues in the parallel simulator.

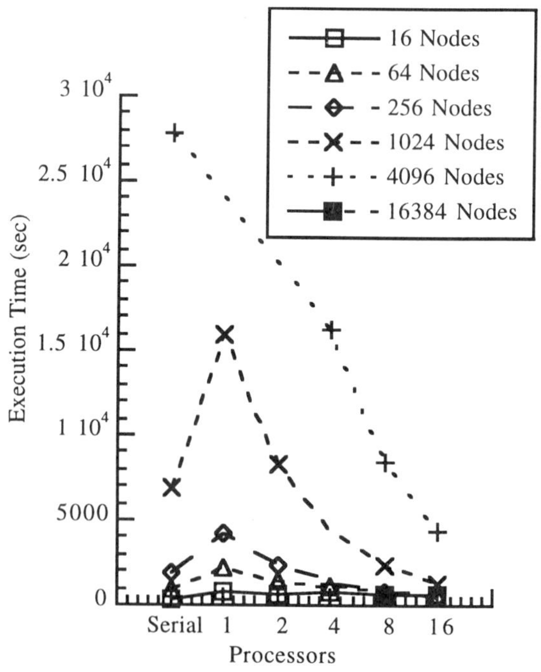

Figure 3: Execution Times on the Intel Paragon

The 16-node experiments were run to see how the communications affected the performance of the parallel simulator. Clearly a parallel machine is not necessary for this sized experiment. In particular, this instance stresses the communications of the Paragon, since there is a small computation/communication ratio. We had, in fact, expected to see performance degrade as the number of processors was increased. To our surprise this didn't happen. The execution time actually decreased slightly when going from 8 to 16 processors.

The 64-node instances were again used to calibrate the "break-even" point for the parallel simulation. Here we observed some improvement over the serial version above two processors. It is in the 256-node instances where the performance of the parallel simulator begins to show reasonable improvement over the serial simulator. Here, we gain an improvement of approximately 3.5 over the serial version (and an improvement of 8 over the one-processor parallel version). This trend continues in the larger instances.

We now address speedup. Figure 3 shows the speedup of the parallel simulator vs. the serial simulator (effective speedup), and Figure 4 shows the speedup of the parallel simulator vs. its one-processor instance. Not surprisingly, the speedup curves are better for the larger problem sizes, and the relative speedups are much better than the effective speedups. For the 4096-node instance, the largest instance recommended for the serial router, the Paragon with 16 processors can improve the execution time by a factor of 6.4, a significant improvement since the serial version takes almost 8 hours to complete.The 16-processor Paragon is a good choice for 256 or more nodes.

The relative speedup is more impressive than the effective speedup, due to the fact that the one-processor ZPL program is around 2.4 times slower than the serial version. As the problem size increases, the speedup curves come closer to the ideal speedup; in the best case, we achieve a speedup of 1.89 in the 1024-processor instance when we go from one to two processors. As we "optimized" the ZPL program, the relative speedups tended to remain constant, while the effective speedups improved, implying that if we make additional optimizations in the ZPL code we can expect to see corresponding improvements in the effective speedup.

Thus far, we have fixed the load to be 80%; what effect does changing the load have on the execution times? We have run all of the different instances discussed above with various loads, ranging from 20% to 80%. For all instances, the execution times increase as the load increases, but the speedups also increase with increasing load, probably due to better balance among the processors. As an example, Figure 5 shows a plot of the effective speedup of the 256-node router with various loads. This plot is typical. The effective speedups are similar for each load, but they do increase with increased load. We had actually expected more of an impact from processor load, with the 20% load showing much poorer performance.

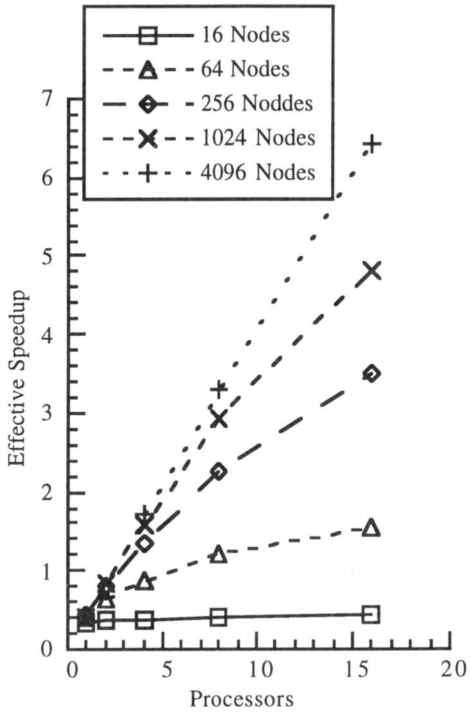

Figure 4: Effective Speedup of the Chaos Router Simulation on the Intel Paragon

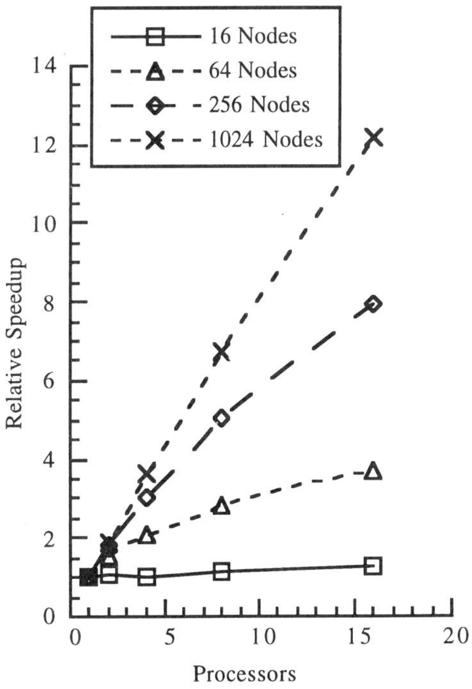

Figure 5: Relative Speedup of the Chaos Router Simulation on the Intel Paragon

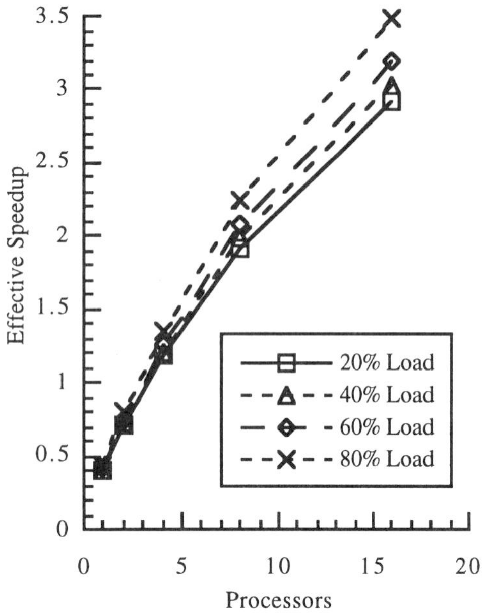

Figure 6: The Impact of Router Load on Effective Speedup Using 256 Nodes

4.2 The Alpha Cluster

The Digital Alpha cluster used in these experiments is a subset of the faculty workstation network of the Computer Science Department at the University of Arizona. The network is connected via an Ethernet and was not isolated from outside traffic during the runs, since we were not allowed to disturb the normal operation of the departmental system. However, we did make every effort to use idle workstations and run the experiments during "off hours." Due to the number and availability of workstations, we limited the experiments to eight workstations. All were identically configured with 64MB of memory. Below are the parameters used for these experiments.

Load:	20%, 40%, 60%, 80%
Processors:	1, 2, 4, 8
Nodes:	16, 64, 256, 1024, 4096, 16384

Not all possible permutations of these parameters were run; the largest-sized experiments (16384 nodes) were measured only on four and eight processors due to the length of time taken for the simulation.

The execution times for the larger experiments (256 nodes and greater) using 80% loads are shown in Figure 6. We don't show the 16- and 64-node instances here; they exhibited fairly large slowdowns due to the small number of nodes per processor (Richardson 1995). Note that we only gain performance improvements over the serial simulator in the largest two instances (4096 and 16384 nodes). This is likely due to several factors. First, the network is connected via an Ethernet bus, which only allows one message to be transmitted at a time. Thus all messages are serialized here, while in the Paragon, multiple messages can be transmitted simultaneously. Second, ZPL uses PVM to interface to the network. This means that it uses the TCP/IP protocol to send messages. While the TCP/IP protocol is robust, it incurs a nontrivial software overhead cost that must be borne by the parallel program. Finally, we cannot be completely sure that there was no other network traffic competing for the limited Ethernet bandwidth, although we believe that this effect was minimal.

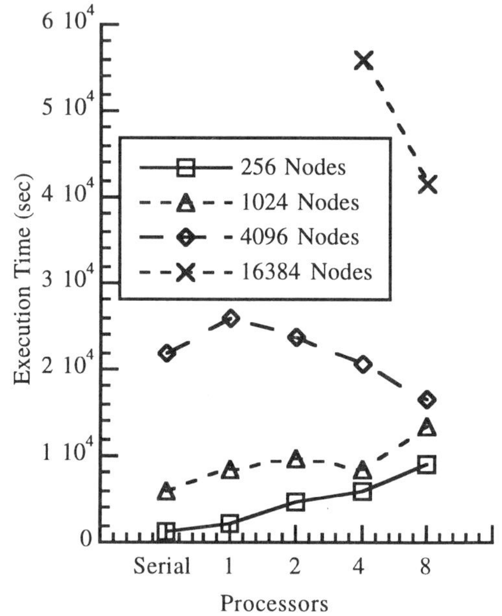

Figure 7: Execution Times on the Digital Alpha Cluster

The effective and relative speedups are shown in Figures 7 and 8, respectively. We do not include the 16384-node instances because we have no serial or one-processor baseline, since the problem size is too large for a uniprocessor. It is clear from the effective speedup curve that only the 4096-node instance has any speedup, and that the speedup here is minimal. Perhaps more interesting is that fact that this is true for the relative speedup too. It is not the slowdown of the parallel implementation that appears to be hurting performance, but rather the computation/communication ratio. For good performance on an Ethernet with this high-performance processor, we need lots of computation per processor.

The effect of nodes on speedup was not as clear for the Alpha as for the Paragon. The general trend still held – that an increased load tended to increase the effective parallelism. However, there were instances where this trend did not hold. For example, the effective speedup decreased from 0.71 to 0.43 when going from a load of

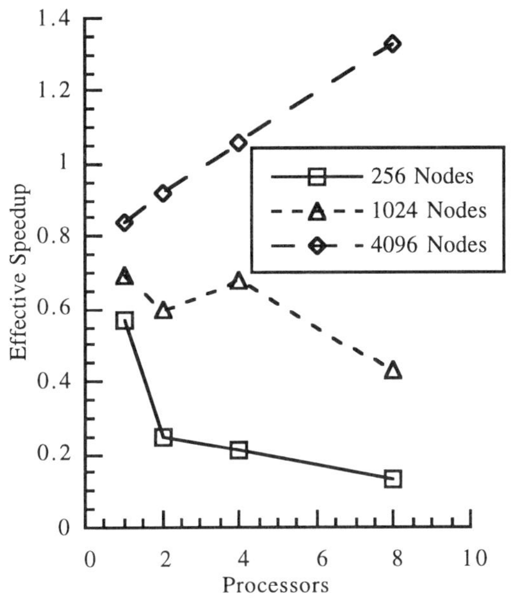

Figure 8: The Effective Speedup of the Chaos Router Simulation on the Digital Alpha

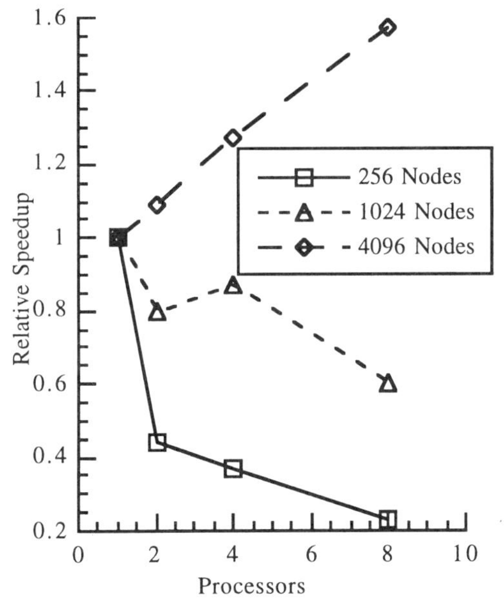

Figure 9: The Relative Speedup of the Chaos Router Simulation on the Digital Alpha

60% to 80% using 1024 nodes on eight processors. It is not clear whether these experiments were anomalous; we have not yet had enough access time to replicate the experiments sufficiently to insure that the measured differences are due to program effects rather than to interference from other Ethernet traffic. For the experiments where we achieved positive speedup, node sizes of 4096 and above, we saw none of the anomalies.

4.3 Comparing the Performance Results

The Alpha and Paragon provided very different architectural platforms on which to test the effectiveness of the parallel chaos router. The processor on the Paragon, the i860, is much slower than the Alpha processor, while the Paragon's interconnection network is much faster. Moreover, the Paragon's interconnection network essentially "matches" the communication patterns of the router, while the Ethernet serializes all communication.

These differences manifested themselves in the performance of the parallel simulation. On the Paragon, thrashing occurred more quickly than we had expected from the serial results. For example, we should have been able to run the 2048-node parallel simulator on a single processor without undue thrashing. One effect that we had failed to consider originally is that the parallel simulator uses substantially more memory than the serial version, because in the serial version memory could be dynamically allocated and pointers were often used to pass information. These were not possible in the parallel simulator because pointers cannot be used when the data are spread among multiple processors. Because each Alpha processor has twice as much memory as a Paragon processor, thrashing was not as evident on the Alpha cluster. Moreover, each Alpha has its own disk, while the Paragon processors all share a single disk array, which further exacerbates thrashing on the Paragon.

Overall, the two architectures complemented each other quite nicely. For small- to medium-sized problems, the Paragon had the best performance. This is likely due to its communication network. The mesh interconnection network maps well to this problem, and its speed is also faster than the Ethernet. This makes interprocessor communication much more efficient. However, for large-sized problems, the Paragon ran into memory limitations. In these cases, the Alpha cluster performed better. In fact, the Alpha cluster can run larger-sized router nodes than can run on a single Alpha. Each Alpha processor has more memory, and the computation/communication ratio is more favorable for the Ethernet for large router sizes. Thus the user can choose the best platform for the router simulation based on the router size, using the Paragon for medium-sized problems and the Alpha cluster for the larger problems.

5 CONCLUSIONS

In this paper, we discussed our experiences in using ZPL for a parallel chaos router simulator. ZPL was easy to use, although we did have to understand some aspects of the compiler in order to obtain more efficient performance. The lack of dynamic memory, especially for use with a single node, proved somewhat difficult, but we were able to use static memory with little difficulty. The ability to write C routines to be called from ZPL proved invaluable for gathering statistics about the performance of the chaos router. Moreover, ZPL's portability allowed us to test the parallel simulator on different architectures without rewriting the simulator.

While we achieved our original goal of a faster parallel simulator, the parallel simulation running on a single processor is still slower than the optimized serial simulator by a factor of approximately 2.4. We expect some improvement in this factor as the ZPL compiler matures, but the one-processor parallel implementation will remain slower than the serial one. This is because the parallel program has inherent overheads (as compared to C) due to the fact that pointers cannot be used for data that can be partitioned across multiple processors in non-shared memory architectures.

The two architectures tested here complemented each other nicely. The Paragon performed best for medium- to large-sized routers, likely due to the lower computation/communication ratio (it has a slower processor and a faster interconnection structure than the workstation cluster), and the fact that the interconnection structure mapped well to this problem. However, it was susceptible to thrashing for large routers. The workstation cluster appears to be limited by the Ethernet and PVM for medium-sized routers, but it could simulate really large routers that are not viable for either the Paragon or a single Alpha. With a faster interconnection such as an ATM network, we expect the performance of the workstation cluster to improve for medium-sized problems.

REFERENCES

Bolding, K., M. Fulgham, and L. Snyder. 1994. The Case for Chaotic Adaptive Routing. Technical Report UW-CSE-94-02-04, University of Washington.

Bolding, K., and L. Snyder. 1992. Mesh and Torus Chaotic Routing. In *Advanced Research in VLSI and Parallel Systems: Proceedings of the 1992 Brown/MIT Conferences*, 222–247.

Forman, G., and C. Lin. 1994. ZPL Tutorial. Technical Report, Computer Science and Engineering Department, University of Washington.

Konstantinidou, S., and L. Snyder. 1994. The Chaos Router. *IEEE Trans. on Computers* 43(12):1386–1397.

Lin. C. 1994. ZPL Language Reference Manual. Technical Report UW-CSE-94-10-06, Univ. of Washington.

Lin, C., and L. Snyder. Data Ensembles in Orca C. 1992. *In 5th Workshop on Languages and Compilers for Parallel Computing*, New Haven, CT.

Lin, C., and L. Snyder.1993. ZPL: An Array Sublanguage. In *Languages and Compilers for Parallel Computing*, U. Banerjee, D. Gelernter, A. Nicolau, and D. Padua, editors, 96–114. Springer-Verlag.

Snyder. L. 1994. A ZPL Programming Guide. Technical Report UW-CSE-94-12-02, University of Washington.

Richardson. G. W. 1995. *Evaluation of a Parallel Chaos Router Simulator*, Master's Thesis, Dept. Of Electrical and Computer Engineering, University of Arizona.

AUTHOR BIBLIOGRAPHIES

WILKEY RICHARDSON is a Member of the Technical Staff at Hughes Missile Systems Company working on real time simulation. He has previously been an engineer at Kaman Corp., Raytheon Co., and Sperry Corp. He received a BS in Computer Science and Mathematics from the University of Kentucky in 1982, and an MS in Electrical and Computer Engineering from the University of Arizona in 1995.

MARY L. BAILEY is a Senior Member of the Technical Staff at Rincon Research Corporation. She was previously an Assistant Professor in the Computer Science Department at the University of Arizona. She received a B.A. in Mathematics and Physics from Vanderbilt University, and an M.A. in Mathematics and M.S. and Ph.D. in Computer Science and Engineering from the University of Washington. She was Program Chair of PADS'95 and General Chair of PADS'96. She also served on the executive committee of ICCAD (1993-1996). Her research interests include parallel and distributed simulation, logic simulation, and special-purpose architectures.

WILLIAM H. SANDERS is an Associate Professor of Electrical and Computer Engineering and Research Associate Professor at the Coordinated Science Laboratory at the University of Illinois at Urbana-Champaign. His research interests include methods for performance, dependability, and performability evaluation, computer networks and protocols, and fault-tolerant computing. He is the developer of two tools for assessing the performability of systems represented as stochastic activity networks: METASAN and UltraSAN. UltraSAN has been distributed widely to industry and academia, and is currently being used at more than 100 universities, five companies, and NASA. His research awards include the Digital Equipment Corporation Incentives for Excellence Faculty Award 1989-1991. He is a member of the IFIP Working Group 10.4 on Dependable Computing and is also a member of Sigma Xi and Eta Kappa Nu academic honor societies.

Proceedings of the 1996 Winter Simulation Conference
ed. J. M. Charnes, D. J. Morrice, D. T. Brunner, and J. J. Swain

STEPS TOWARDS A BETTER INTERNAL GPSS MECHANISM

Ingolf Ståhl

Stockholm School of Economics
Box 6501
S-113 83 Stockholm, Sweden

ABSTRACT

This paper presents the internal procedures used by the micro-GPSS system. They are quite different from those used by standard GPSS, like GPSS/H. Instead of the Future and Current Events Chain, five different lists are used. The advantages of this approach is that queue statistics, both in front of servers and due to WAIT conditions, can be obtained in a far simpler manner, that programs with long waiting lines are executed more efficiently and that, in the case of equal priority, the transaction having waited the longest time in front of a server will always be the first to be served.

1. INTRODUCTION

Recent papers by T. Schriber and D. Brunner (1994 and 1995) have stressed the importance of understanding the internal procedures used by the software of discrete-event simulation for implementing the fundamental management of entities, involving such things as scheduling events and handling waiting lines. In these papers, the approaches taken in this regard by GPSS/H, SIMAN and ProModel are compared, with a focus on the two first languages.

A fundamental difference can be seen comparing GPSS/H and SIMAN. Both systems have their comparative advantages and disadvantages. As regards GPSS/H, a problem is that all events in a waiting stage are handled on **one** single list. This has, in the case of many waiting transactions, the disadvantage of a more cumbersome search procedure. This is only partially hidden by the efficient compilation done by the GPSS/H system. Another disadvantage is that the measuring of waiting line statistics requires separate pairs of blocks, QUEUE-DEPART.

Against this background, this paper wants to bring to attention that another GPSS version, micro-GPSS (**mG**), ever since its first beginning in the 1970s has used a completely different mechanism than the standard approach used by the IBM GPSS versions, by GPSS/PC and by GPSS/H. While these Standard GPSS (**SG**) versions use only two standard lists, or chains, for handling events and waiting lists, called the Current Events Chain (CEC) and the Future Events Chain (FEC), mG uses for these purposes five different types of lists.

2. SOME DIFFERENCES IN SYNTAX

In order to help the novice to micro-GPSS understand the presentation of the five different types of lists, we shall first present some of the fundamental differences in syntax between mG and SG. mG uses only 22 block types, five of which do not exist in SG, namely ARRIVE, GOTO, IF, LET and WAITIF. (See also Ståhl 1996b for more details.)

ARRIVE replaces QUEUE and has (almost) the same syntax; GOTO replaces TRANSFER, but with a simpler syntax; LET replaces ASSIGN and SAVEVALUE (and BLET in GPSS/H); IF replaces TEST and GATE blocks **with** an address, while WAITIF replaces TEST and GATE blocks **without** an address.

In contrast to TEST, IF works with a "straight logic", namely that we proceed to the address in the C operand, if the stated condition is **true** (not false as in SG).

WAITIF works in a similar way: we wait, prior to the WAITIF block, if, and as long as, the stated condition is **true**. The WAITIF condition can concern either a name in the A operand and a server status code in the B operand, or SNAs or constants in the A and B operands. The server status code in the B operand involves the letters EFNU; namely U (in **U**se) or N (**N**ot in **U**se) for facilities; E (**E**mpty), NE (**N**ot **E**mpty), F (**F**ull) and NF (**N**ot **F**ull) for storages. In the SNA case there is an SNA in at least one of the two operands.

3. THE FIVE MICRO-GPSS LISTS

The micro-GPSS processor works with the following five different types of lists:

1. **QLs**, Queue lists, one for **each** server, i.e. facility or storage, handling the waiting lines in front of these servers.

2. **WELs**, i.e. **W**AITIF **E**FNU Lists, where EFNU represents the server status codes E, F, NE, NF, NU, U, mentioned above. These lists contain the transactions waiting in front of a WAITIF block with server status code. The correspondence between such WE-lists and WAITIF blocks is discussed below in section 10.

3. One **WSL**, i.e. one **W**AITIF SNA **L**ist, where all transactions, waiting in front of a WAITIF block with an SNA in at least one operand, are placed.

4. One **FEL**, i.e. **F**uture **E**vents **L**ist, where all events that are to be executed at a time later than the present clock time are placed.

5. One **REL**, i.e. **R**eady **E**vent **L**ist, where all events that are ready to be executed at the present clock time are placed. All transactions on this list have earlier been on one of the other four lists.

The REL and the FEL are to some extent part of the same list, but we shall for pedagogical reasons first treat them as completely separate lists and in section 11 discuss the actual implementation.

On the QLs, the WELs and the WSL the transactions will as a main rule be sorted primarily on basis of priority, and in the case of equal priority on the basis of time of "true" entry into the list. In certain cases, transactions having just left a list will be put back at the **front** of the list, implying a "dummy" entry. The transactions on the FEL are on the other hand primarily sorted according to event time and only in the case of equal event times on the basis of priority.

On the REL, where all event times refer to the present simulation clock time, all transactions are primarily sorted according to priority. At equal priority, the following applies: The transactions just moved from the FEL are generally put last in each priority group. The remaining transactions are in each priority group sorted according to the entry time into the QL, WEL or WSL from which they have just been removed. Thus, in each priority group on the REL, the transactions brought in from these three types of lists will (almost) always be ahead of the transactions brought in from the FEL (with "entry time" = present clock time; see also section 11). The sorting according to priority is done in descending order, i.e. the transactions with a higher priority number are placed close to the front of the list, while the sorting according to time is done in ascending order, i.e. the lower times are closer to the front.

4. PHASES OF PROGRAM EXECUTION

The **Initialization Phase** is identical to that of GPSS/H (Schriber and Brunner 1994). Here the simulation clock is set to 0 and the initial GENERATE events are placed on the FEL. We here determine as an attribute of the transaction its NB (next block), in this case the block following the actual GENERATE block. This phase finally involves the placement of all events taking place at time 0 on the REL.

In the **Entity Movement Phase** all events on the REL are executed and brought through as many blocks as possible. In the process some events might be brought from a QL, a WEL or the WSL onto the REL. This EMP is described in detail in section 5. As will be discussed there, this phase continues until all events on the REL have been executed and the REL is empty, after which we move to the clock update phase.

The Clock Update Phase: If the FEL is empty, we go to the End Phase. If the FEL is not empty, we do the following: We remove the first item from the FEL and advance the clock to this time. We move this event to the REL, as well as all other events on the FEL that have this clock time as their move time, in such a way that they keep the same order as on the FEL. We then go back to the EMP.

In the **End Phase** all standard reports are printed and control is passed to a superloop in which the simulation program is possibly restarted.

5. THE ENTITY MOVEMENT PHASE

The EMP is subdivided into four steps:

Step 0: We remove the transaction at the front of the REL.

Step 1: We determine the event connected with the NB of this transaction and execute this event. We distinguish between the following types of events:

a **GENERATE**. An IAT is sampled and a new GENERATE event is put on the FEL to happen at the **move time** T= clock time + IAT. The event is put on the FEL, in its priority class, as the last event among the events to happen at time T. We next go to step 2, i.e. to proceed to the next block.

b. **ADVANCE**. The transaction is moved to the FEL with the move time set to the time of exit from the ADVANCE block. It is placed on the FEL according to the same rule as for GENERATE under point a. above. We next go to step 3, since this transaction cannot move further ahead.

c. An **LRR** event (i.e. LEAVE, RELEASE or RETURN) will carry out (i) the QR scheme (Queue Remove scheme) described in section 6, which will move one transaction, or, in the case of LEAVE, possibly several transactions, from the QL to the REL; (ii) the WER scheme, described in section 8, to see if we shall move transactions from one or several WELs and finally (iii) we go to step 2.

d. An **EPS** event (i.e. ENTER, PREEMPT or SEIZE). We carry out the SE scheme (see section 7) to see if the transaction may enter the block. If it **cannot enter**, we move the transaction into the QL of the server. If the transaction has just come from this QL, i.e. the transaction's JM (Just Moved)-switch is on, we move it to the front of the QL; else it is put as the last member of its priority group on the QL. We next go to step 3. If it **can enter**, we carry out the WER scheme, described in section 8, to see if we shall move transactions from one or several WELs. We finally go to step 2.

e. A **WAITIF EFNU** event, i.e. a WAITIF block in server mode. We here check whether or not the EFNU condition is true. If it is **true**, the transaction is moved to the WEL connected with the actual block (see section 10 below). If the transaction has just come from a WEL, i.e. the JM-switch is on, it goes to the front of the WEL. If the JM-switch is off, the transaction is placed as the last member of its priority group on the WEL. We finally go to step 3. If the EFNU condition is **not true**, i.e. the transaction does not have to wait, we remove the first transaction on the corresponding WEL, if this WEL is not empty, and move it to the REL. We finally proceed to step 2.

f. A **WAITIF SNA** event, i.e. a WAITIF block with an SNA. We here check whether the waiting condition is true. If this condition is **true**, the transaction is moved to the WSL. If the transaction has just come from the WSL, i.e. the JM-switch is on, it goes back to the WSL, to be placed on basis of original entry time into the WSL. If the JM-switch is off, the transaction is placed as the last member of its priority group on the WSL. We finally go to step 3. If the waiting condition is **not true**, we remove the first transaction on the WSL, unless WSL is empty, and move it to the REL. We finally proceed to step 2.

g. An **ASSEMBLE** event. A special process is carried out. We go to step 2 or step 3 depending on whether the transaction can proceed or is stopped.

h. A **SPLIT** event. A number of copies, determined by the A operand, of the transaction are put on the REL, with the NB determined by the B operand. The original transaction goes to step 2.

i. **GOTO** and **IF** without D operand. Since a new NB is determined on the basis of the address in the operands we go directly to step 2b (skipping 2a where a new NB is determined).

j. **TERMINATE**. We decrease the TC (Termination Counter) by the A operand value. If the TC becomes ≤ 0, we go to the End Phase. Otherwise we go to step 3.

k. **PRIORITY**. We change the transaction priority and go to step 3 to put the transaction on the REL, possibly in a new priority class. Another transaction on the REL might now have a higher priority.

l. All **other** events will be executed without having any effect on lists or the Next Block. We here proceed to step 2.

Step 2: The transaction can proceed forwards to another block at this clock time. We have three substeps: 2a. We increase NB by 1; 2b. We carry out the WS scheme (WAITIF SNA scheme) described in section 9, thereby possibly moving transactions from the WSL to the REL and 2c. We go back to step 1.

Step 3: We carry out the WS scheme. If REL is empty, we go to the CUP (Clock Update Phase). If REL is not empty, we remove the first event from the REL, establish its NB and go to step 1 with this event.

6. THE QR SCHEME

The QR (**Q**ueue **R**emove) scheme moves one or several transactions from the QL to the REL. The processor tests if there are any events on the QL associated with the server referred to in the LRR block. If the QL is not empty, we distinguish between two cases of action:

a. The case of either a RELEASE-RETURN block or of a LEAVE block when there is not any B>1 in the program, implying that the transaction freeing the server allows room for exactly **one** other transaction. In this case we remove the first transaction on the QL and move it to the REL, placing it, in its priority group, on the basis of the time of entry into the QL.

b. The case of LEAVE in a program where there is some B>1. We here have two subcases:

(i) The LEAVE block has B=1. We start the search in the QL from the front for the first transaction that has an ENTER with B=1. This transaction is then removed from the QL and moved to the REL, in the same way as under 6a.

(ii) The LEAVE block has a B>1. We start a search in the QL, from the front, determining for each transaction on the list, which all involve ENTER events, a value B_e of the B operand of this ENTER block. If $B=B_e$, the same action takes place as under (i) and we are finished with the QR scheme. If $B<B_e$, we continue to the next transaction. If $B>B_e$, we decrease B by B_e and move the transaction from the QL to the REL, but continue to the next transaction with this procedure, until we reach the last transaction on the QL or until B=0.

7. THE SE SCHEME

The SE (Server **Entry**) scheme deals with an EPS block. In the case of SEIZE, the transaction may enter if the facility is idle. In the case of ENTER, the transaction may enter if the free capacity of the server $\geq$ B. In the case of PREEMPT, the transaction may enter if the facility is idle. It may also enter if the facility is busy by a SEIZE block, or, in the case of PREEMPT A,PR, by a transaction with a lower priority. In the latter two cases we carry out a special pre-emption activity. It is outside the scope of this paper to discuss this activity.

In all of these three cases, if a transaction can enter the server, it proceeds to the next block. If it cannot enter, its behavior depends on if it has come to the REL from the FEL or from a QL. In the FEL case the transaction is put on the QL as the last member of its priority group. In the QL case it is put back at the front on the QL, i.e. at the position it just held.

8. THE WER SCHEME

The WER (**WE** list **R**emove) scheme will possibly move a transaction from one or several WELs to the REL. A certain server is connected with each LLR or EPS block. With each server there is, in turn, a connection to one or several WELs (this association is discussed further in section 10 below). Having established the server of the block, we next look through **all** WELs associated with this server. For each such WEL there is in turn associated a specific WAITIF condition. If this WAITIF condition is fulfilled, no action will be taken (i.e. waiting will continue). If the WAITIF condition is **no longer** true, we will remove the first transaction of the WEL and move it to the REL, where it is placed on the basis of the start wait time in this WEL.

9. THE WS SCHEME

The WS (**W**AITIF SNA) scheme is carried out, if there is at least one WAITIF SNA block in the program. We first check whether or not the WSL is empty. If the WSL is empty, we proceed to the next task. If the WSL is not empty, we start from the front and go backwards through the whole list. For each transaction we check whether or not the wait condition still holds. If it still holds we proceed to the next transaction. If the wait condition does **not** hold, we remove this transaction from the WSL and put it in its priority group on the REL, placing it here on the basis of time of entry into the WSL. In this way we continue until all transactions on the WSL have been examined.

10. WELS AND WAITIF EFNU BLOCKS

When discussing the relationship between WELs and WAITIF EFNU blocks, we shall distinguish between the following three cases as regards the A operand of the WAITIF block:

a. The A operand is a name. All blocks that have the same server name **and** the same one of the six status codes (E, F, NE, NF, NU or U) share the same WEL. Transactions having to wait at this WEL will, if they have the same priority, be sorted according to when they started to wait due to this specific condition.

b. The A operand is a parameter referring to a **facility**. We allow in this case both for the case of P*j*=U and P*j*=NU, for a certain number, *n*, of integer values of the parameter P*j*, thus allowing for *n* WELs of 1=U, .., *n*=U, and for *n* WELs of 1=NU,.., *n*=NU. Every block WAITIF P*j*=U will then refer to the first mentioned *n* lists and every block WAITIF P*j*=NU will refer to the second mentioned *n* lists.

c. The A operand is a parameter referring to a **storage**. Since this kind of usage is rare, we allow in this case only for **one** single WEL. The first WAITIF condition referring to E, F, NF or NE in the program will use a number of WELs and every WAITIF block with a parameter in the A operand and the same status code in the B operand will refer to these WELs. If there then, later in the program, is another WAITIF with a parameter in the A operand, but another status code in the B operand, then this WAITIF EFNU block will internally be treated as a WAITIF SNA block and the transactions waiting at this block will be waiting on the WSL.

This can be exemplified as follows: Assume we first have a block WAITIF P1=E and later a block WAITIF P3=F. Then the transactions waiting before the first block will do that on a number of WELs, for 1=E, 2=E, etc. The transactions waiting in front of the other block will be waiting on the WSL as long as R(P3) =0, i.e. as long as the remaining free capacity of the storage referred to by P3 is 0.

11. IMPLEMENTATION OF REL AND FEL

The REL and FEL together constitute an EL, Events List, with all transactions on the REL at the front of the EL and all transactions on the FEL at the end of the EL. Each transaction on this EL has a pointer to the preceding member and a pointer to the succeeding member, except that the first member has no predecessor and the last member no successor. We have one list pointer to the first member and one to the last member on this EL. We also have a special REL pointer, the

RELL (REL **Last**) pointer, pointing at the last member of the REL. The successor of this member is thus the first member of the FEL. The movement of all events taking place at the new simulation clock value in the CUP can now be done in a very simple fashion. We just have to establish, starting with the successor to the old RELL pointer, the last new member on the REL and then change the RELL pointer to point at this transaction.

When we bring in a new transaction from a QL, a WEL or the WSL to the REL, we will start to move it from the back of the REL, i.e. from the transaction at which the RELL pointer points. We will put it in front of any transaction with a lower priority and, in the case of equal priority, with a lower waiting time, i.e. with a higher entry time into the original QL, WEL or WSL. Since the transactions which in the CUP were brought in from the FEL all have 0 waiting times, the new transaction will, in the case of equal priority, be put in front of these FEL transactions, provided it has a waiting time >0.

An advantage with the joint approach is that all removals from this EL will take place from the front and can be done by the same procedure. When we remove the first transaction from the REL, the front of REL is of course the front of the EL. When we remove the first transaction from the FEL, the REL is empty and we hence remove the front transaction from the EL.

12. COMPARING WITH THE SG APPROACH

In Ståhl (1996a) we show that, except for four specific differences, to be mentioned below, the procedure presented above for mG will result in the same execution results as that of SG. We there compare the mG approach with the SG approach as presented by Schriber, in 1974 for GPSS/360 and in 1990 for GPSS/H.

We shall here only mention the following: The Initialization Phases are equivalent, with the only difference being that the events put on the FEC in SG are put on the FEL in mG. Likewise the CUPs are very similar. In both cases, we update the clock to the event time of the first transaction on the FEL/FEC and move **all** events happening at this updated clock time from the FEL/FEC to the REL/ CEC, before proceeding to the EMP. One, in reality unimportant difference, to be called **difference 1**, is that execution stops with a normal report in mG, when the FEL is empty, while it stops with an error report in SG. The End Phase is also similar, in that all standard reports are printed, but the actual reports vary in format.

All the obvious differences happen in the EMP. One difference between mG and most SG versions (GPSS/H and GPSS/V, but not GPSS/PC) deals with when the arrival of, for example, customer 2 at a GENERATE block is put on the FEC/FEL. In SG this is done when customer 1 is moved **out of** the GENERATE block, but in mG this is done when customer 1 moves **into** the GENERATE block. This difference, **difference 2**, will matter greatly if the GENERATE block is followed immediately by, for example, SEIZE. In GPSS/H, the introduction of an ADVANCE block with no operands between GENERATE and SEIZE would then make a big difference. In mG such an ADVANCE block would in this case not matter.

Such an ADVANCE block without any operand, will in other cases lead to another difference. Let us look at the following simple program in SG (GPSS/H). (For the corresponding mG code, see Ståhl 1996a).

```
        SIMULATE
NUM     GENERATE    20
        ASSIGN      1,N$NUM
BEGIN   ADVANCE
        SEIZE       RIDE
        ADVANCE     20
        RELEASE     RIDE
        TRANSFER    .5,,BEGIN
        TERMINATE
        GENERATE    101
        TERMINATE   1
        START       1
        END
```

Program example 1

The result of this program will be different in SG and mG. The first customer will in SG repeat his ride immediately, while in mG the second customer will immediately get a ride. The reason for this is that in mG, entry into the (BEGIN) ADVANCE block will always cause the transaction to be placed on the FEL and the next transaction to be taken from the REL or FEL. ADVANCE without an A operand hence does the same thing as BUFFER or YIELD in GPSS/H.

This is **difference 3**. It should be noted that if we add an A operand, e.g. 0.00001, to the ADVANCE block the results will be identical in mG and SG, with the second customer immediately getting a ride. Likewise, if we replace (BEGIN) ADVANCE with, for example, a simple assignment (ASSIGN VAL,0,XL in SG; LET X$VAL=0 in mG), SG and mG will also yield identical results, now with the just released first customer getting his second ride right away.

13. FIFO IN WAITING LINES IN MG, BUT FIFO ON THE CEC IN SG

In mG, the transactions blocked by an EPS-block referring to a specific server are, as noted, all put on a specific QL, where those transactions that have the same priority are sorted according to their time of entry into this QL. For the simple case of a facility, where all entries are made with SEIZE, and of storages, where all entries are made with ENTER with B=1, we will then always remove the first transaction from the QL. This implies, in the case of equal priority for all transactions, a straight FIFO-discipline as regards the waiting line in front of a server. A transaction will in this case never go ahead of another transaction that has waited a longer time in front of this **specific server**.

This FIFO discipline with regard to the server might, however, be violated in SG. This constitutes **difference** 4. This is due to the fact that the transactions on the CEC are never resorted while they remain on the CEC, except for the above mentioned case with PRIORITY. The only other way they can be resorted is by moving them to the FEC, e.g. to be have an exit from an ADVANCE A block scheduled, and then return at the exit from this block. Once they return from the FEC, they are put at the end of their respective priority group.

Let us as an example assume that a transaction, customer 1, at time 20, is first blocked trying to seize one facility, FACA, and upon being unblocked from this facility at time 25, **immediately** tries to seize another facility, FACB, which at this time is busy. Another transaction, customer 2, will at time 23, immediately when generated, try to seize facility FACB, which is busy already at this time.

In SG, customer 1 remains waiting for facility B on the position determined by having come to the CEC at time 20. Customer 2, who is transferred from FEC at time 23 to seize FACB B directly, without having to first use facility A, is then placed on the CEC **behind** customer 2. Since facility B was busy also at time 23, customer 2 starts waiting for facility B **earlier** than customer 1. However, when facility B becomes free, e.g. at time 30, the rescan starts from the beginning of the CEC and hence customer 1 gets the chance to move first to seize facility B that has now become idle, in spite of the fact that customer 2 has been waiting in front of facility B for a longer time.

In mG, customer 2 will be placed on the QL of FACB at time 23 and customer 1 on this QL at time 25. Customer 1 is hence placed behind customer 2 on this QL. When FACB becomes free at time 30, customer 2, who has waited the longest will be served.

The mG approach appears as more intuitively appealing, since it implies that the person having waited the longest time among those with equal priority in front of a specific server will always be served first. That some servers are needed jointly under some, but not all, circumstances does not immediately imply that there is a joint waiting line. It also appears easier to learn that each server has its own waiting line, with FIFO-discipline in case of equal priority. To understand the SG approach one must really understand the CEC approach.

It is furthermore easier to understand the idea of how to get a joint waiting line in mG than of how to get separate waiting lines in SG. To get a joint waiting line in mG in the example above, one just puts a block PRIORITY A, with A>0, implying a **higher** priority, between SEIZE FACA and SEIZE FACB. To get separate waiting lines in front of FACA and FACB in GPSS/H, one could also include a block PRIORITY A between SEIZE FACA and SEIZE FACB, but with A=0, i.e. implying **unchanged** priority. Why this is so also requires good understanding of the CEC mechanism.

The problem that waiting in SG is bound to the CEC and not the waiting lines is perhaps even more clear in program example 2, presented below in its GPSS/H form.

```
         SIMULATE
EXHIB    STORAGE      5
         GENERATE     10
         TEST GE      C1,100
         ASSIGN       1,C1
         ENTER        EXHIB
         ADVANCE      120
         LEAVE        EXHIB
         TERMINATE
         GENERATE     10
         ASSIGN       2,C1
         ENTER        EXHIB
         ADVANCE      120
         LEAVE        EXHIB
         TERMINATE
         GENERATE     400
         TERMINATE    1
         START        1
         END
```

Program example 2

We here have two kinds of visitors coming to an exhibition; invited guests and ordinary visitors. The ordinary visitors are allowed into exhibition building only after 100 minutes. We assume that all visitors stay for two hours. There can only be five visitors at a time in the small exhibition room, talking to the artist. Visitors of each type arrive 10 minutes apart. When a visitor enters the exhibition room, we save, in P1 for ordinary visitors and P2 for invited guests, the time that he started to wait at the entrance of the exhibition room.

In the SG case, an ordinary visitor who started to wait at the exhibition room at time 100 is allowed to go into the room before an invited guest who arrived and started waiting at time 60. In the corresponding program in mG this will not happen, but the people waiting at the exhibition room enter this in strict first come-first served order.

14. OTHER ADVANTAGES OF THE MG APPROACH

There are two other substantial advantages of the mG approach besides the strict adherence to the FIFO-principle dealt with in section 13. The first and most important one is the easy way in which gathering of queue statistics can be coded.

The gathering of queue statistics referring to a facility SERV would in SG require three blocks: QUEUE SERV, SEIZE SERV and DEPART SERV. In mG it is enough to write SEIZE SERV,Q. The same Q as B operand (or possibly C operand) can also be used in connection with ENTER.

We can now in micro-GPSS obtain queue statistics in a simplified fashion also in connection with a WAIT condition by giving the name of the queue as a C operand of WAITIF. Thus, the gathering of statistics in front of a GATE, which in SG would require e.g. the three blocks QUEUE QSTUD, GATE SNE STOCK and DEPART QSTUD, can in mG be handled by one single block WAITIF STOCK=E,QSTUD.

The saving in number of blocks is substantial. As an example, we have rewritten all programs in Schriber's famous "red book" from 1974 using mG. In many cases we have with the 22 blocks of mG been forced to use several blocks to replace a non-existing SG block, e.g. LOOP. Yet for the 29 programs presented in the book, the average number of blocks used is virtually the same (18.6 in SG and 18.8 in mG). This is mainly due to the fact that we have saved on average 1.24 blocks per program thanks to this simplified queue statistics gathering, based on separate QLs.

The other advantage is that the program execution of mG is in principle more efficient than that of SG. This is quite clear when comparing mG with SG systems such as GPSS/PC. It is less easily evident when comparing with GPSS/H, which is a compiled system and focused on rapid execution. Yet there are examples when mG, although interpretative and not so well coded for optimization, will execute faster than GPSS/H. This refers to cases with very long waiting lines, like in program example 3, presented here in GPSS/H.

```
          SIMULATE
          INITIAL     XL$MAX,2000
          GENERATE    1.8,0.6
          TEST NE     Q$SAL,XL$MAX,BYE
          QUEUE       SAL
          SEIZE       SAL
          DEPART      SAL
          ADVANCE     25,5
          RELEASE     SAL
BYE       TERMINATE
          GENERATE    1,0.6,,,1
          QUEUE       SUL
          TEST NE     Q$SUL,4,BYTT
          PREEMPT     SAL
          DEPART      SUL
          ADVANCE     1,0.5
          RETURN      SAL
          TERMINATE
BYTT      DEPART      SUL
          TERMINATE
          GENERATE    32000
          TERMINATE   1
          START       1
          END
```

Program example 3

Although GPSS/H is very efficient when skipping all transactions for which the Scan Indicator is on, the computer still has to investigate, in the EMP, **every** single transaction on the CEC, e.g. at RELEASE, to check the status of the Scan Indicator. In mG it is enough to remove the very first transaction from the QL of SAL. When used by a compiling system, focused on optimization, the separate QL approach should be faster also in the case of much shorter waiting lines. For better speed, one would in GPSS/H in this case have to use a more complicated and less-easy-to-learn LINK/UNLINK construction, allowing for a non-standard road to the more efficient approach with separate QLs.

Against this background we shall finally argue that the approach with separate QLs for each server, and separate WELs for most blocks of the WAIT type, should be made standard in future languages of the GPSS-type.

ACKNOWLEDGEMENTS

The ideas behind this paper, as well as the development of micro-GPSS, have been greatly influenced by Tom Schriber, not only through his publications, listed in the references, but also through his comments, by voice, mail and Email.

REFERENCES

Schriber, T. J. 1974. *Simulation Using GPSS*, Wiley, N.Y.

Schriber, T. J. 1990. *An Introduction to Simulation with GPSS Using GPSS/H*, Wiley, N.Y.

Schriber, T. J. 1995. How Discrete-Event Simulation Software Works. In *Eurosim'95 Simulation Congress,* ed. F. Breitenecker and I. Husinsky, 17-28. Elsevier, Amsterdam.

Schriber, T. J. and D. T. Brunner. 1994. Inside Simulation Software: How It works and Why It Matters.In *Proceedings of the 1994 Winter Simulation Conference,* ed. J.Tew, S. Manivanna, D. Sadowski, and A. Seila, 45-54. SCS, La Jolla.

Schriber, T. J. and D. T. Brunner. 1995. Inside Simulation Software: How It works and Why It Matters. In *Proceedings of the 1995 Winter Simulation Conference*, ed. C. Alexopoulos, K.Kang, W. Lilegdon and D. Goldsman, 451-456. Piscataway, New Jersey: Institute of Electrical and Electronics Engineers.

Ståhl, I. 1990. *Introduction to Simulation with GPSS: On the PC, Macintosh and VAX,* Prentice Hall International, Hemel Hempstead, U.K., 1990.

Ståhl, I. 1995. *Simulation Made Simple with micro-GPSS: A Short Tutorial with Seven Lessons,* Stockholm School of Economics, Stockholm.

Ståhl, I. 1996a. *Steps Towards a Better Internal GPSS Mechanism,* EFI Working Paper, Stockholm School of Economics, Stockholm.

Ståhl, I. 1996b. Teaching the Fundamentals of Simulation in a Very Short Time. In this volume.

AUTHOR BIOGRAPHY

INGOLF STÅHL is Professor at the Stockholm School of Economics, Stockholm, and has a chair in Computer Based Applications of Economic Theory. He was visiting Professor, Hofstra University, N.Y., 1983-1985 and leader of research project on inter-active simulation at the International Institute for Applied Systems Analysis, Vienna, 1979-1982. He has taught GPSS for twenty years at universities and colleges in Sweden and the USA. He has on the basis of this experience led the development of the micro-GPSS system. He is also consultant in simulation to Swedish banks and industry.

Proceedings of the 1996 Winter Simulation Conference
ed. J. M. Charnes, D. J. Morrice, D. T. Brunner, and J. J. Swain

ELIMINATING CANCELING EDGES FROM THE SIMULATION GRAPH MODEL METHODOLOGY

Ricki G. Ingalls

Manufacturing Strategy Group
Corporate Operations
Compaq Computer Corporation
20555 SH 249
Houston, TX 77070

Douglas J. Morrice

MSIS Department
The University of Texas at Austin
Austin, TX 78712
End to End Simulation Department
Schlumberger Austin Research
Austin, TX 78720

Andrew B. Whinston

MSIS Department
The University of Texas at Austin
Austin, TX 78712

ABSTRACT

Event Graphs and Simulation Graph Models provide a powerful and general modeling framework for discrete event simulation. Within this framework it has been shown that an event cancellation construct in the form of a canceling edge is a modeling convenience rather than a necessary modeling tool. As a result, very little work on the formal development of Event Graphs and Simulation Graph Models directly considers the canceling edge construct. However, to the simulation practitioner the modeling convenience and functionality provided by canceling edges is important. This is clearly demonstrated by the presence of event canceling edges in SIGMA, the commercial software implementation of Event Graphs. This paper proposes an extension of the Event Graph methodology that will completely eliminate canceling edges without loosing the functionality that canceling edges provide. Since this extension is formally developed within the Simulation Graph Model framework, it provides a general approach to handling event cancellation.

1 INTRODUCTION

Event Graphs (EGs) have included event cancellation constructs in the form of event canceling edges since their introduction (Schruben 1983). Subsequent papers also include the canceling edge construct (Sargent 1988, Som and Sargent 1989). After EGs were formalized and extended into Simulation Graphs (SGs) and Simulation Graph Models (SGMs), it was shown theoretically that event canceling edges were a modeling convenience but not a necessary modeling tool (Yücesan 1989, Yücesan and Schruben 1992). As a result, very little of the formal development work on EGs and SGMs directly addresses the issue of event cancellation. Recently, Savage and Schruben (1995) demonstrated that it is practically possible to model without event canceling edges.

One might conclude that event canceling edges, in particular, and event cancellation, in general, should be eliminated from the EG methodology. However, simulation practitioners clearly find the modeling convenience and functionality provided by event canceling edges, beneficial. This is demonstrated by the fact that the commercial software implementation of EGs, SIGMA (Schruben 1995, page 75) includes the event canceling edge construct.

Since event canceling has some benefit in terms of modeling convenience this paper does not advocate its elimination altogether. Instead, the event canceling edge construct is eliminated and replaced by a more general event canceling construct in the form of an edge execution condition. Essentially, the edge execution condition is tested at the moment the event is scheduled to be executed. If the condition is true, the event is executed; if the condition is false the event is discarded without execution. The edge execution condition retains the functionality and modeling convenience of canceling edge.

This paper extends EGs to include the edge execution condition. In addition, this construct is formally developed within the SG framework to provide an extension to SGMs. Its generality is demonstrated within this framework. The complete simulation execution sequence of the SGM extension is provided in algorithmic form. The results are demonstrated on an example from Savage and Schruben (1995).

The remainder of the paper is organized as follows. Section 2 provides a description of EGs and SGMs. Section 3 introduces the new EG construct that includes an edge execution condition, extends SGMs to include this new construct, and provides the algorithm that demonstrates the complete simulation execution sequence. Section 4 demonstrates the methodology on an example from Savage and Schruben (1995). The example is described using the canceling edge approach, the approach developed by Savage and Schruben (1995), and the approach developed in this paper. Section 5 provides a preliminary run-time comparison of computation effort required by the three approaches on the example provided in Section 4. Section 6 contains some concluding remarks.

2 ORIGINAL IMPLEMENTATION

Figure 1 depicts the basic EG construct. This construct is called a *scheduling edge*. The nodes labeled A and B represent events. The edge specifies that there is a relationship between the two events. More specifically, the construct can be interpreted as follows "if condition *(i)* is true at the instant event A occurs, then event *B* will be scheduled to occur *t* time units later" (Schruben 1983). The quantity *t* may assume the value zero, in which case *B* happens at the same instant as *A*. Note that it is possible (and often necessary) to specify an edge with no condition. The scheduling edge schedules the event vertex at the head of the edge in accordance with the time specification with the edge

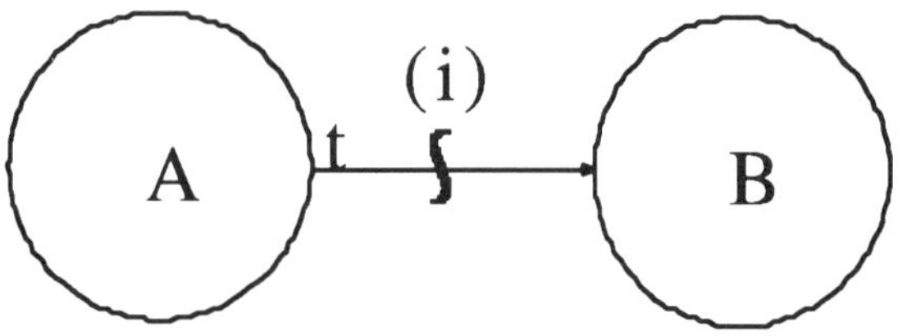

Figure 1: Scheduling Edge

Figure 2 depicts a second basic construct in EGs referred to as a *canceling edge*. The canceling edge has the following interpretation: "if condition *(i)* is true at the instant event A occurs, then the currently scheduled event *B* will be canceled *t* time units later" (Schruben 1983). In most instances, *t* equals zero for the canceling edge (Schruben 1995, page 76). The canceling edge is used to delete scheduled transactions that execute the vertex that is at the head of the edge. The canceling edge was added to the event graph framework to handle disruptions such as machine breakdowns.

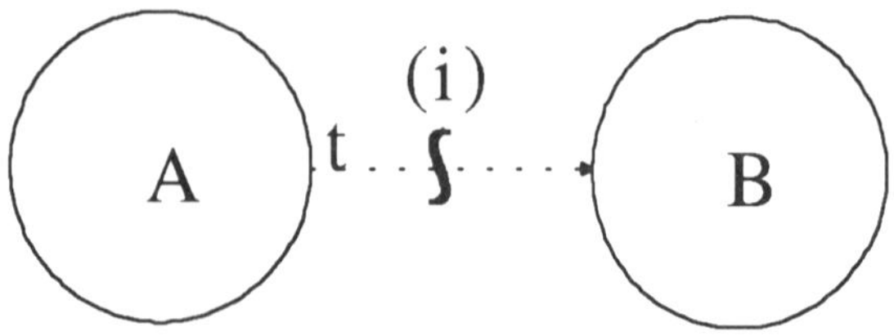

Figure 2: Canceling Edge

In the event graph, it is also possible to parameterize the event vertices and thus extend the basic constructs in Figures 1 and 2. Parameterization is accomplished through vertex parameters and edge attributes. A vertex parameter list is a string of state variables associated with a particular vertex. An edge attribute list is a string of expressions associated with a particular edge. These lists are used in scheduling or canceling specific instances of events. Using these parameters is analogous to passing values in subroutines in high-level programming languages. For example, Figure 3 provides an extension of Figure 1. The construct in Figure 3 is interpreted as follows: "if condition *(i)* it true at the instant event A occurs, then event *B(j)* will be scheduled to occur *t* time units later with parameter string *j*, equal to *k*" (Schruben 1995, page 79).

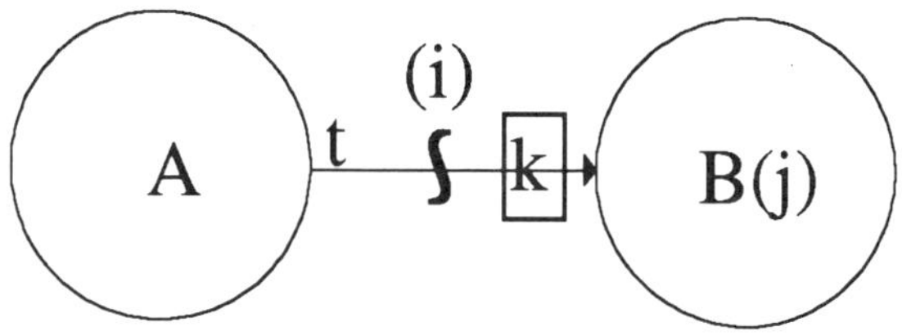

Figure 3: Vertex Parameters and Edge Attributes on a Scheduling Edge

Yücesan and Schruben (1992) and Schruben and Yücesan (1993) extend EGs to SGs and SGMs. This extension provides a formal graph theoretic framework for EGs. Within this framework, Yücesan (1989) shows that canceling edges and parameterization are modeling conveniences and do not augment the modeling capabilities of these graphs. As a result, very little of the formal development in SGs has focused directly on the canceling construct.

3 THE EDGE EXECUTION CONDITION

The basic construct in the modified event graph framework is shown in Figure 4. As before, the nodes labeled A and B represent events and the edge specifies that there is a relationship between the two events. However, the construct is now interpreted as follows: "if condition (i) is true at the instant event A occurs, then event B will be scheduled to occur *t* time units later. Event B will be executed *t* time units later with the state variables in array n set equal to the values in array k if condition (j) is true *t* time units later." This simple modification is the basis for eliminating canceling edges in the simulation graph methodology.

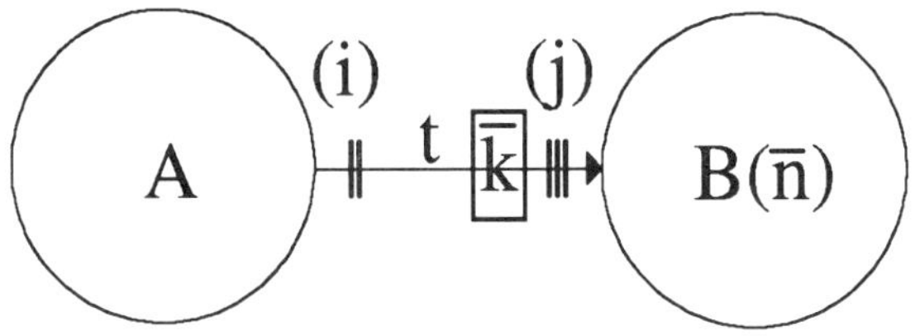

Figure 4: Scheduling Edge with an Execution Condition

3.1 Simulation Graphs Extension

Using the notation of Yücesan and Schruben (1992) and Schruben and Yücesan (1993), let $G=(V(G),E(G),\psi_G)$ be a directed graph where $V(G)$ is the set of vertices and $E(G)$ is the set of edges. G can be a multi-arc graph, meaning that more than one edge can connect the same two vertices flowing in the same direction. Vertices and edges can have functions and attributes assigned to them.

The function that is assigned to the vertices is:

(i) $\mathcal{F}=\{f_v : v \in V(G)\}$, the set of state transitions functions associated with vertex v.

The functions assigned to the arcs are:

(i) $C=\{C_e : e \in E(G)\}$, the set of *scheduling edge conditions*.

(ii) $X=\{X_e : e \in E(G)\}$, the set of *execution edge conditions*.

(iii) $\mathcal{T}=\{t_e : e \in E(G)\}$, the set of *edge delay times*, and

(iv) $\Gamma=\{\gamma_e : e \in E(G)\}$, the set of *event execution priorities*.

With the modifications described above, the Simulation Graph Model (SGM) is now defined as $\mathcal{L}=\{\mathcal{F},C,X,\mathcal{T},\Gamma,G\}$.

In contrast to Schruben and Yücesan (1993), there are no canceling edges in the graph. Instead, the set X, which is evaluated at the time that the event comes off the calendar to be executed, replaces the function of the canceling edge. If the condition X_v is true, then the vertex associated with the event notice is executed. If X_v is false, then the event notice is discarded. This implementation provides a precise way for canceling scheduled events in the methodology and in practice.

Although Schruben and Yücesan (1993) specifically detail the function of canceling edges in their terminology, their canceling edge step is not straightforward. The function of canceling edges is to search for and remove one or more *already scheduled* event notices from the event calendar, $\mathcal{L}$. The edge execution edge conditions permits the same functionality, i.e., the ability to not execute an event notice already scheduled on the calendar, without requiring a search of the events calendar. The edge execution approach does require some extra overhead associated with carrying more information on the event calendar and perhaps the need for additional state variables. However, in many instances, the overhead can be justified if it eliminates a costly search procedure.

3.2 Simulation Graph Execution Sequence

As with the framework in Schruben and Yücesan (1993), this framework also needs additional mechanisms to facilitate the execution of the model. The symbol τ will be used to represent the global simulation clock and $\mathcal{L}$, the event calendar. However, our definition of $\mathcal{L}$ will be different from the one specified in Schruben and Yücesan (1993). Our definition of $\mathcal{L}$ is an ordered set, $\mathcal{L}=\{(t_1,\gamma_1,v_1,e_1,a_1),(t_2,\gamma_2,v_2,e_2,a_2),\ldots\}$ where t_i represents the execution time, γ_i represents the execution priority, v_i is the vertex to be executed, e_i is the index of the edge that is being scheduled, and a_i are the values of the edge attributes. Each (t,γ,v,e,a) is an *event notice*. It is important to note that Schruben and Yücesan (1993) did not include the values of the edge attributes as part of the event notices. Edge attributes are included here since the parameter passing mechanism is intended to pass the values of the variables and not the references.

We also define the following sets:

(i) S_v : the set of state variables that can be changed at vertex v. (Since S is the set of all state variables, $S_v \subset S$).

(ii) P_v : the set of state variables in the parameter list of vertex v.

(iii) Φ_e : the set of state variables involved in the scheduling conditions on edge e.

(iv) ϑ_e : the set of state variables involved in the execution conditions on edge e.

(v) A_e : the set of state variables used to determine the values of a_e on edge e.

In addition to these sets, there is one conditional statement, ω, which is evaluated to determine if the simulation should stop. With these definitions, the execution of the Simulation Graph model is carried out as follows.

Initialize (Run Initialization)

Step 1. Initialize the global simulation clock. $\tau \leftarrow 0$.

Step 2. Insert one or more event notices into the event calendar: For simplicity, we will assume that each of these event notices could be executed at time 0. $\mathcal{L} = \mathcal{L} \cup \{(0,\gamma_1,v_1,e_1,a_1),(0,\gamma_2,v_2,e_2,a_2),...\}$.

Execute (execution of the model implementation)

Step 1. Remove the first event notice from $\mathcal{L}$. $\mathcal{L} = \mathcal{L} \setminus \{\ell | \ell = (t_1,\gamma_1,v_1,e_1,a_1)\}$. Event notice ℓ is removed from the calendar. This implies $t_\ell \equiv t_1, \gamma_\ell \equiv \gamma_1, v_\ell \equiv v_1, e_\ell \equiv e_1, \text{ and } a_\ell \equiv a_1$.

Step 2. If the execution edge condition, $X_{e_\ell}(\vartheta_{e_\ell}, a_\ell) = FALSE$, then go to Step 1 of execute.

Step 3. Update the simulation clock. $\tau \leftarrow t_\ell$.

Step 4. Assign the attributes to the parameters of the vertex. $P_{v_\ell} \leftarrow a_\ell$. If Y is the ith state variable in the vertex parameter list, i.e. $(P_v)_i = Y$, then $Y \leftarrow (a_\ell)_i$.

Step 5. Evaluate the state change. $S_{v_\ell} \leftarrow f_{v_\ell}(S)$.

Step 6. Schedule further events. For each edge, e_{ij}, emanating from v_ℓ, if $C_{e_{ij}}(\Phi_{e_{ij}}) = TRUE$ then evaluate $A_{e_{ij}}$ and assign the attribute value of the new event notice, k, $a_k \leftarrow A_{e_{ij}}$. Generate the inter-event time, $t_k = f(t_{e_{ij}})$, and schedule the event notice where $\mathcal{L} = \mathcal{L} \cup \{(\tau + t_k, \gamma_{e_{ij}}, v_j, e_{ij}, a_k)\}$.

Step 7. Terminate the execution of the simulation if any of the following conditions is satisfied:

(i) τ exceeds T_{stop}.

(ii) The simulation stopping condition, ω, evaluates TRUE.

(iii) $\mathcal{L}$ is empty.

Otherwise, go to Step 1 of Execute.

4 EXAMPLE

Savage and Schruben (1995) present an example in which they demonstrate a procedure to model canceling edges out of EGs. The same example will be used to demonstrate the edge execution condition approach.

The example is one of a single server system that has two types of customers. Of the two types of customers, customer type 1, has the highest priority. If a type 2 customer is being served when a type 1 customer arrives, the type 2 customer is preempted by the type 1 customer and will wait until all type 1 customers have been served. Figure 5 shows how this system can be modeled with a canceling edge (Savage and Schruben 1995).

4.1 With a Canceling Edge

For this system, the state variables are:

Q1 and Q2	The number of customers in each queue.
S	Server Status: 1 = available, 0 = busy, -1 = preempted

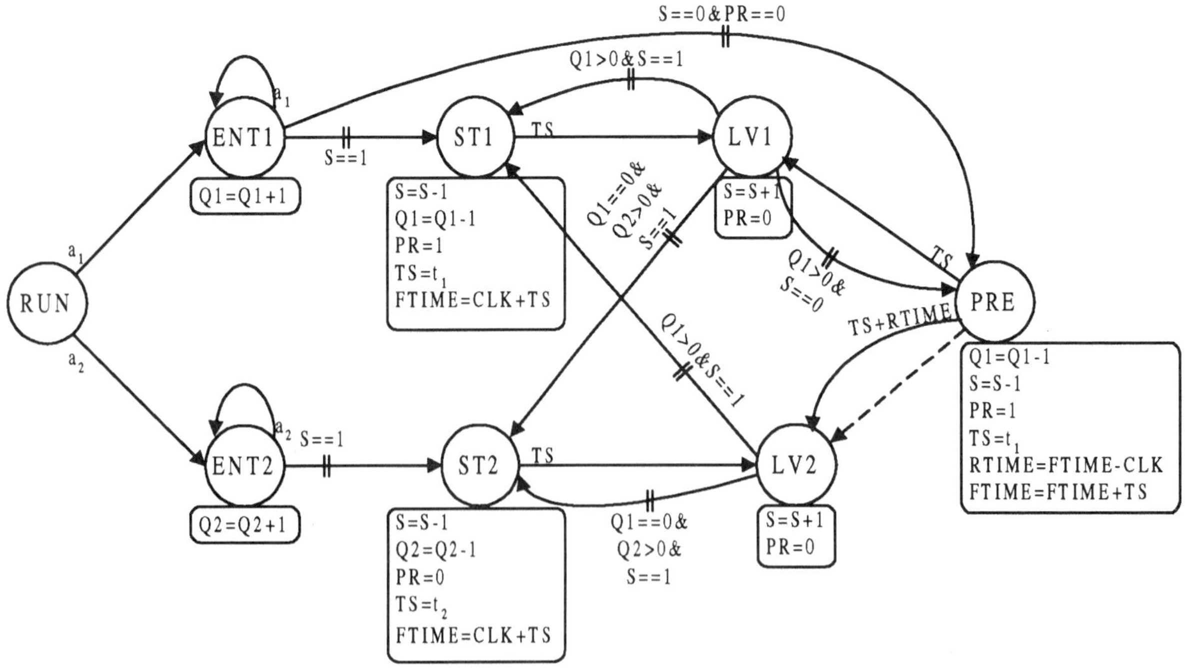

Figure 5: Single Server Preemptive System with a Canceling Edge (Savage and Schruben 1995)

PR If =1, then a priority customer is being served.
TS The length of time for service.
FTIME The finishing time of service (if not preempted).
RTIME The remaining time in the service.

In this system, each type of customer has its own arrival distribution, a_i. When a type 1 customer enters the system (ENT1), if there is a server available, the customer starts his service (ST1) and then finishes his service (LV1) after TS time has elapsed. When a type 2 customer enters the system (ENT2) and a server is available, then the type 2 customer starts his service (ST2) and is *scheduled* to finish his service TS time units later. However, if a type 1 customer enters the system while the server is busy with a type 2 customer, the type 1 customer preempts (PRE) the type 2 customer by taking the type 2 customer off of the future events list if the event the type 2 customer is scheduled to execute is LV2. The type 1 customer then is served for TS time units and then finishes his service (LV1). At the same time, the type 2 customer is scheduled to finish his service TS+RTIME units later. When a type 1 customer finishes service, one of three things can occur. First, if a type 2 customer was preempted (S==0) and there is another type 1 customer in the queue (Q1>0), then the type 2 customer that was previously preempted is preempted again and service is given to the new type 1 customer. Second, if there are new type 1 customers in the queue (Q1>0) and there was not a previous preemption (S==1), then the service is given to the next type 1 customer in the queue. Third, if there are no type 1 customers (Q1==0) and there are type 2 customers (Q2>0) and the server is available (S==1), then the service is given to the first type 2 customer in the queue.

4.2 Without a Canceling Edge

Figure 6 contains the Savage and Schruben (1995) model without canceling edges. They eliminate canceling edges by putting a check event (CHK) in the model that will not allow a type 2 customer to leave the system (LV2) until no type 1 customers are in the system (S==0). The check event is processed whenever a type 1 customer is scheduled to leave the system. The actual execution of this model as intended is problematic due to the number of zero-time events and the lack of priorities on the arcs. Regardless, this method of eliminating the canceling edge should work with minor modifications.

There are three basic problems with this solution. The first is that the modeler, and not the methodology, is responsible for eliminating the canceling edge. In many instance, this could be a challenging modeling exercise by itself. The second problem is that there is additional overhead in terms of scheduling the check event. The third problem is more philosophical, but one

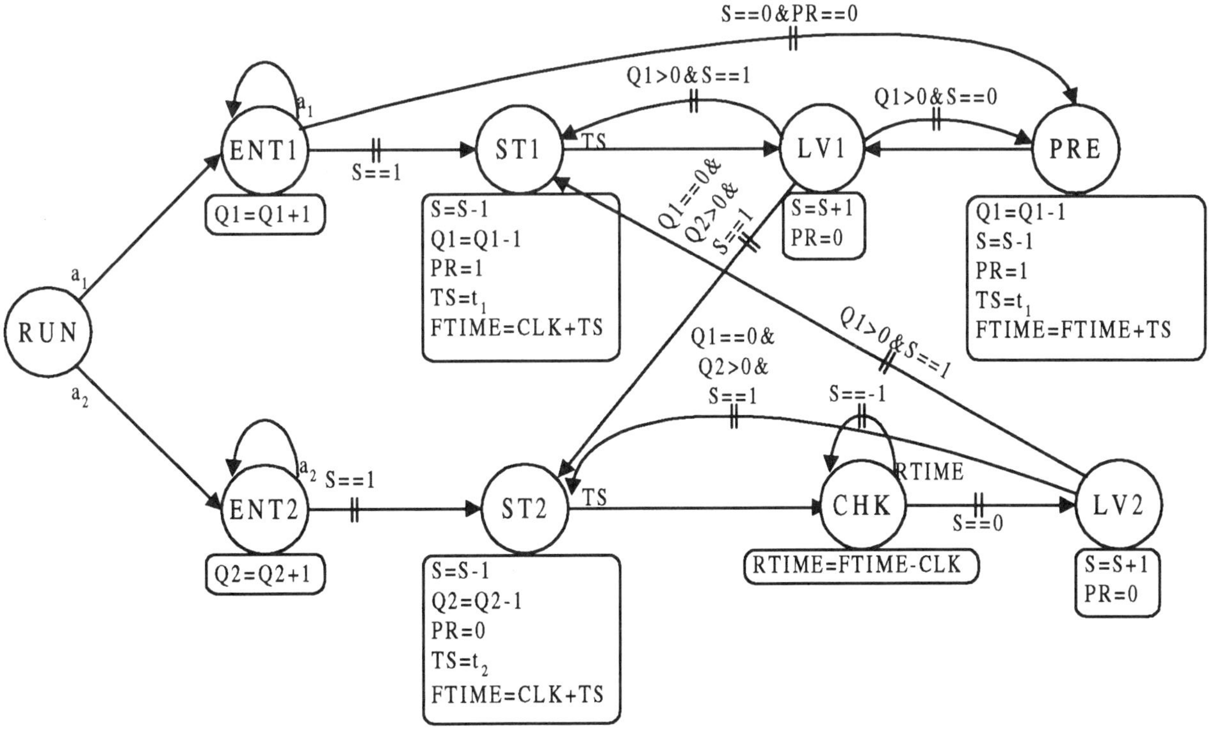

Figure 6: Single Server Preemptive System without a Canceling Edge (Savage and Schruben, 1995)

of the strengths of event graphs is that the events in the model usually correspond to events that occur in the actual system being modeled. A check event never actually occurs in the physical system.

4.3 With Execution Conditions

The model with execution conditions, as they are outlined in this paper, is very straightforward. Figure 7 shows the model as it would be with execution conditions.

In this system, the state variables are:

Q1 and Q2	The number of customers in each queue
S	Server status
P	The number of preemptions that have occurred
T	The type of customer being serviced
ST	The starting time in the service
RT	The remaining time in the service

In this system, each type of customer has its own arrival distribution, a_i. When a type 1 customer enters the system (ENT1), if there is a server available (S==1), the customer starts his service (ST1) and then finishes his service (LV1) after t_1 time has elapsed. When a type 2 customer enters the system (ENT2) and a server is available (S==1), then the type 2 customer starts his service (ST2) and is *scheduled* to finish his service TS time units later. When the event is scheduled, the number of preemptions that have occurred (P) is stored in attribute 1 (A(1)). When the event is scheduled to occur, if there have been no more preemptions (P==A(1)), then the type 2 customer leaves the system (LV2). However, if a type 1 customer enters the system while the server is busy (S==0) with a type 2 customer (T==2), the type 1 customer preempts (ST1P) the type 2 customer by incrementing P, the number of preemptions and storing the remaining service time for the type 2 customer (RT). This will not allow the type 2 customer to finish because P>A(1). In any case, when a type 1 customer finishes his service (LV1), then one of three things can occur. First, if there is a type 1 customer in the queue, then a new type 1 service is started. Second, if there is no one in the type 1 queue (Q1==0) and a type 2 customer has been preempted (RT>0), then the type 2 customer is restarted with completion scheduled for RT time units later. Again, the leave event (LV2) for the type 2 customer will only be executed if his is not preempted (P==A(1)). Third, if there is no one in the type 1 queue (Q1==0) and no one has been preempted (RT==0) and there is a type 2 customer waiting, then a new type 2 customer starts service (ST2).

5 RUN-TIME CONSIDERATIONS

In order to determine which of the three approaches is

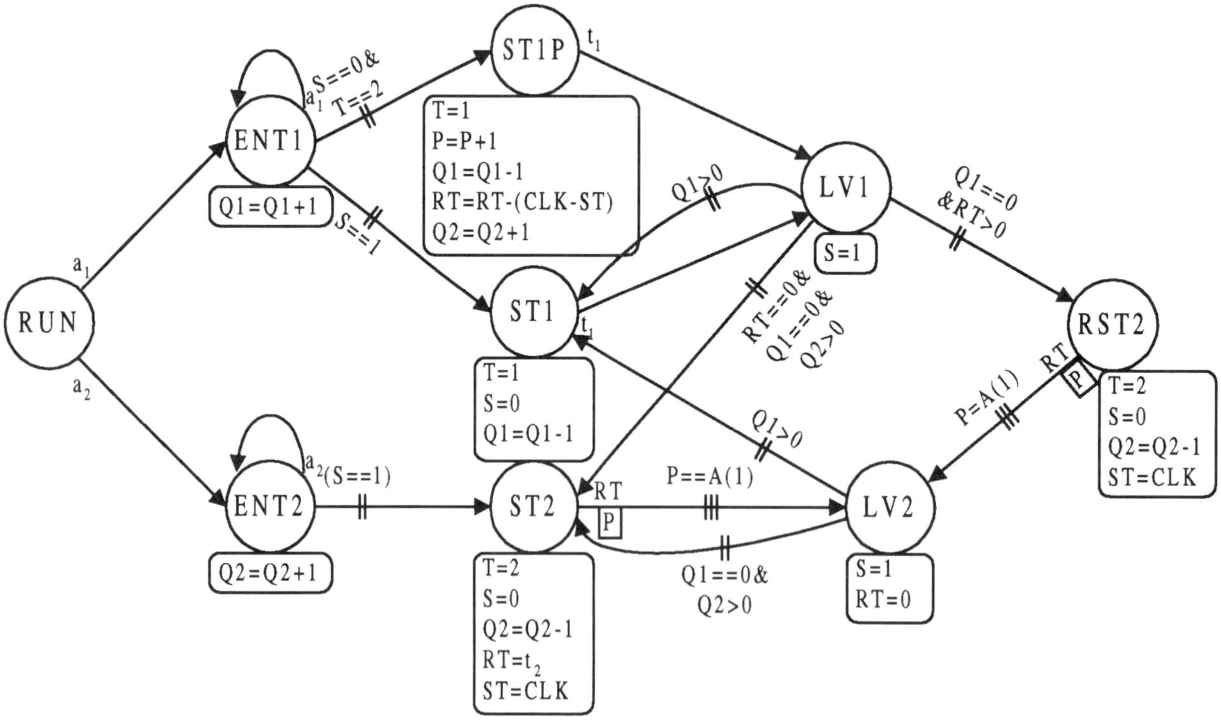

Figure 7: Single Server Preemptive System with Execution Conditions

more efficient, consider the run-time complexity of each of the three models. First, we define complexity as the number of times that an event is schedule or canceled since the management of the calendar is easily the most time consuming aspect of any discrete-event simulation.

To best gauge the run-time, let us pick a state that will be revisited, and thus could be called a cycle, in the simulation. Let us assume that the simulation is about to execute an ST2 event. If the simulation is about to execute an ST2 event, we know that there are no type 1 customers in the system. The complex part of this simulation is when type 1 customers preempt type 2 customers. Let us assume that n type 1 customers and m type 2 customers will enter the system before the next ST2 event is executed. Under these conditions, we can calculate the number of events that will executed between ST2 events. The different calculations on the number of events for each type of model are in Figure 8.

Events Scheduled or Canceled	Number of Occurrences
With Canceling Edge	
ST2	1
ENT2	m
ENT1,PRE,2*LV2,LV1	n
LV2	1
TOTAL	$5n+m+2$
Without Canceling Edge	
ST2,CHK	1
ENT2	m
ENT1,PRE,LV1,CHK	n
LV2	1
TOTAL	$4n+m+3$
With Execution Conditions	***(Worst Case)***
ST2	1
ENT2	m
ST1P,LV1,RST2	n
LV2	1
TOTAL	$3n+m+2$
With Execution Conditions	***(Best Case)***
ST2	1
ENT2	m
ST1P	1
LV1	n
ST1	n-1
RST2,LV2	1
TOTAL	$2n+m+3$

Figure 8: Complexity Calculations for Different Canceling Edge Elimination Methods

Assuming that scheduling events is the most time consuming task in a simulation, then the model with execution conditions should be able to outperform equivalent models that use either canceling edges or the approach suggest by Savage and Schruben (1995). At this point it is difficult to make a more definitive statement but these preliminary results look promising. A more thorough analysis of the computational requirements of these methods is the subject of ongoing research.

6 CONCLUSION

The implementation of execution conditions in the simulation graph methodology provides a standard, complete methodology for modeling interrupts. The implementation of execution conditions should be straightforward in any simulation software and be very useful for the simulation practitioner.

ACKNOWLEDGMENTS

We gratefully acknowledge the financial support of the CBA/GSB Faculty Research Committee of the College of Business Administration of the University of Texas at Austin. The second author gratefully acknowledges additional support from the End to End Simulation Department at Schlumberger Austin Research.

REFERENCES

Sargent, R. G. 1988. Event Graph Modeling for Simulation with an Application to Flexible Manufacturing Systems. *Management Science* 34 (10): 1231-1251.

Savage, E. L. and L. W. Schruben. 1995. Eliminating Event Cancellation in Discrete Event Simulation. In *Proceedings of the 1995 Winter Simulation Conference*, ed. C. Alexopoulis, K. Kang, W. R. Lilegdon, and D. Goldsman, 744-750. Institute of Electrical and Electronics Engineers, Piscataway, New Jersey.

Schruben, L.W. 1983. Simulation Modeling with Event Graphs. *Communications of the ACM*, 26(11): 957-963.

Schruben, L. W. 1995. *Graphical Simulation Modeling and Analysis Using Sigma for Windows.* Danvers, Massachusetts: Boyd & Fraser Publishing Company.

Schruben, L. W. and E. Yücesan. 1993. Modeling Paradigms for Discrete Event Simulation. *Operations Research Letters* 13: 265-275.

Som, T. K. and R. G. Sargent. 1989. Formal Development of Event Graphs as an Aid to Structured and Efficient Simulation Programs. *ORSA Journal on Computing* 1 (2):107-125.

Yücesan, E. 1989. Simulation Graphs: A Mathematical Framework for The Design and Analysis of

Discrete Event Simulations. Ph.D. Dissertation, School of Operations Research and Industrial Engineering, Cornell University, Ithaca, New York.

Yücesan E. and L. W. Schruben. 1992. Structural and Behavioral Equivalence of Simulation Models. *ACM Transactions on Modeling and Computer Simulation* 2 (1): 82-103.

AUTHOR BIOGRAPHIES

RICKI G. INGALLS is the Manager of Business Analysis in the Manufacturing Strategy Group at Compaq Computer Corporation in Houston, Texas. He has been involved in the application and development of operational modeling tools and techniques in the electronics industry for over 12 years. He has a B.S. in Mathematics from East Texas Baptist College, a M.S. in Industrial Engineering from Texas A&M University and is currently a Management Science Ph.D. candidate at the University of Texas at Austin. Prior to re-joining Compaq, he was on the technical staff of the Operational Modeling Group at SEMATECH, Manager of the Operations Analysis Group at Compaq, a consultant with the Electronics Automation Application Center of General Electric Co. and an Industrial Engineer with Motorola, Inc. His research interests include simulation methodologies, qualitative simulation, and simulation applications. He is a member of the Society for Computer Simulation.

DOUGLAS J. MORRICE is an Associate Professor in the Department of Management Science and Information Systems at The University of Texas at Austin. He is currently on sabbatical leave in the End to End Simulation Department at Schlumberger Austin Research in Austin, Texas. Dr. Morrice received his undergraduate degree in Operations Research at Carleton University in Ottawa, Canada. He holds an M.S. and a Ph.D. in Operations Research and Industrial Engineering from Cornell University. His research interests include discrete event and qualitative simulation modeling and the statistical design and analysis of large scale simulation experiments. Dr. Morrice is a member of the The Institute for Operations Research and Management Science (InfORMS) and the Council of Logistics Management. He served as the Secretary for the InfORMS College on Simulation (1994-1996) and is Co-Editor of the *Proceedings of the 1996 Winter Simulation Conference.*

ANDREW B. WHINSTON is a professor of both business and computer science at the University of Texas at Austin, where he is also a fellow of the IC2 Institute and director of the Center for Information Services Management, and holds the Hugh Roy Cullen Centennial Chair in Business Administration. His research deals with decision support systems theory, distributed AI, organization modeling and qualitative modeling. He received his Ph.D. in management from Carnegie Mellon University, Pittsburgh, PA, and is a member of the Institute of Management Science.

Proceedings of the 1996 Winter Simulation Conference
ed. J. M. Charnes, D. J. Morrice, D. T. Brunner, and J. J. Swain

PROGRAMMING COMBINED DISCRETE-CONTINUOUS SIMULATION MODELS FOR PERFORMANCE

J. Frederick Klingener

President, Brock Engineering, P.C.
Roxbury, Connecticut 06783, U.S.A

ABSTRACT

Continuous state variables in combined discrete-continuous simulation models (combined models) commonly represent physical quantities, such as fluid levels or temperatures, that are governed by physical laws, and these laws are expressed as differential equations of state. The combined simulation modeling program commonly integrates the differential equations numerically, in step with its computations that describe the evolution of the discrete events. In addition to the pitfalls familiar to numerical integration, special hazards due to the interacting discrete events may confront the analyst seeking high performance in a complex model.

This paper first discusses, in the context of discrete event modeling packages, some requirements for obtaining accuracy and speed in the numerical integration of the continuous variables in combined models, and second, it describes approaches that can be used to meet those requirements in selected commercial modeling packages.

1.0 BACKGROUND AND PURPOSE

The distinguishing feature of combined discrete-continuous simulation models (combined models) is the existence of continuous state variables that interact in complex or unpredictable ways with discrete events. The introduction of continuous variables into a discrete event simulation sets two tasks: 1) evaluation of the continuous variable itself often by numerical integration of a governing differential equation, and 2) assuring proper interaction among the continuous and discrete state variables. Figure 1 shows the three fundamental types of interaction, described by Pritsker [1968], that can occur between discrete and continuous variables: 1) a discrete event causes a change in the value of a continuous variable, 2) a discrete event causes a change in the relation governing the evolution of a continuous variable, and 3) a continuous variable causes a discrete event to occur or to be scheduled by achieving a threshold value.

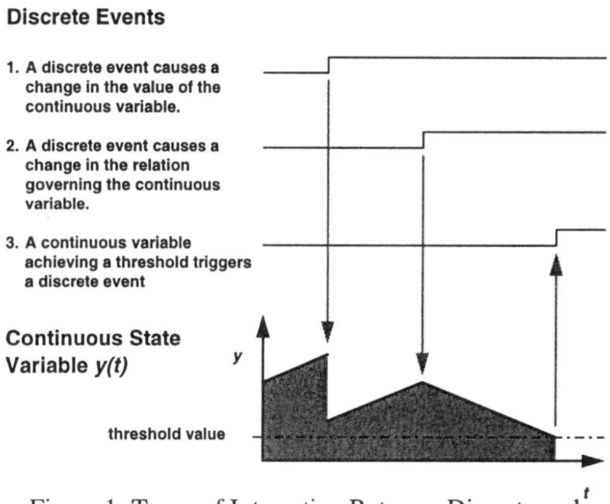

Figure 1: Types of Interaction Between Discrete and Continuous Variables

Combined modeling goes beyond the simple numerical integration of the continuous variables and specification of the particular interactions that may occur among continuous and discrete state variables; there is a fundamental mismatch between the way that time is advanced by a stack-based discrete event modeler and the way it is advanced by a marching continuous variable integrator. On the one hand, the common discrete event modeler is event-driven. It advances time asynchronously, that is it takes the execution time of an event from the top of a sorted stack, calculates its effect on the system state, schedules dependent events by placing them into appropriate places in the event stack, and advances the simulation clock to the next scheduled event. On the other hand, the common continuous variable solver is synchronous. It uses the system state at some time, t, to compute the value of a continuous state variable at some short time later at $t+\Delta t$. The solver then advances the simulation clock to $t+\Delta t$ and continues.

Press et al. [1992] describe classical techniques used to perform the synchronous calculation with speed and accuracy. In the methods they describe, there is a tradeoff between accuracy and speed, and the determinant of the tradeoff is the time step size. If the integration proceeds using a large number of small time steps, then the accuracy can be high at the expense of computation speed. Conversely, in a solution that uses a small number of larger time steps to obtain rapid execution, the accuracy may suffer. The adaptive step size methods described by Press et al. [1992] have been developed to manage this classical tradeoff, but these methods depend on the freedom to adjust the time step size, Δt, in response to the local conditions relating to the integration process itself. If the time step size is subject to outside constraints, such as might be required to detect and respond with sufficient resolution to discrete events, then performance may suffer.

The purpose of this paper is to describe techniques that an analyst can use to build high performance combined models by reconciling the mismatch between the views of time held by the discrete event modeler and the numerical integrator while preserving the theoretical bases and implementation methods of the classical methods of high performance numerical integration of continuous variables. The second section of this paper discusses first an overall strategy that enables the techniques, and second the details of managing the three types of interaction between discrete events and continuous state variables.

2.0 IMPLEMENTING NUMERICAL METHODS IN A COMBINED MODEL

The first step in adapting the numerical integrator to the discrete event environment is to recognize that the continuous state variable may be only piecewise smooth between the interacting events. A numerical procedure that seeks high accuracy should reflect this structure. The second step is to acknowledge that, given the system state when the simulation clock stands at time t, any value computed for a continuous variable y some time Δt in the future is speculative. It must be considered invalid until the simulation clock reaches $t+\Delta t$ without encountering any event or condition that affects the continuous variable or the relation that governs its evolution. The subsections that follow first describe an approach to the implementation of speculative computation. Then they outline strategies for programming the three types of interactions between discrete events and the piecewise smooth state variables.

2.1 Speculative Computation

As a minimum procedure, when the value of $y(t+\Delta t)$ is computed at time t, it must be considered speculative and saved locally or privately, then published for use by other processes only when it becomes valid. Apart from assuring consistency in the integration of the continuous variables, this strategy enables the techniques for managing the interactions between continuous and discrete variables described in the next three subsections.

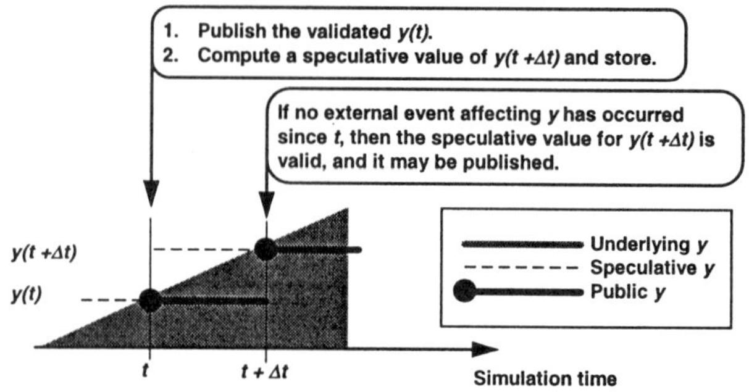

Figure 2: Strategy for Speculative Computation in a Discrete Model

Figure 2 shows a schematic of the method. In this figure and in the three figures that follow, a continuous state variable, y, is evaluated numerically as a function of time at increments of Δt, given an expression for its rate of change. The underlying value of $y(t)$ is shown as a gray line, the speculative (local) value of $y(t)$ is a dashed line, and the validated (public) value of $y(t)$ is a solid black line. When the simulation clock arrives at time t without encountering some invalidating event, the speculative value of $y(t)$ may be published or made available to other processes. Then, the system state at t may be used to project a speculative value of y at $t + \Delta t$. This speculative y, invalid until $t + \Delta t$, is stored privately or locally, and the clock may be advanced.

Figure 2 also illustrates a secondary point about combined models. Note that the solution for $y(t)$ can approach arbitrarily high accuracy (matching the underlying y) at t and again at $t + \Delta t$, but there will be an error at other times. For this reason, it is desirable to synchronize the numerical integration of related continuous variables. Under circumstances where this is not practical, it may be necessary to store enough of the state at time t to permit another process to reconstruct y at an intermediate time.

The speculative computation facilitates programming of the interactions between continuous variables and discrete events. In general, if an interacting discrete event happens between t and $t+\Delta t$, then the speculative value of $y(t + \Delta t)$ must be abandoned, a current value of y, must be calculated and published, and a new speculative value of y must be calculated and stored. The following subsections describe how interacting discrete events that happen between t and $t + \Delta t$ affect the subsequent calculations of $y(t)$ for the three types of interactions described in section 1.

2.2 Interaction Type 1: A Discrete Event Changes the Value of a Continuous Variable

Figure 3 shows a procedure for programming the numerical integration of a continuous variable in the vicinity of a Type 1 interaction, in which a discrete event causes a change in the value of a continuous variable. The objectives of the procedure are to manage the transition between piecewise smooth segments by the following steps: 1) terminate the segment that ends with the interrupt, 2) reset the continuous state variable to the new value, and 3) restart the integration with a new segment. Figure 3 shows that at t (the time step immediately before an interrupting event), the procedure is the same as that shown in figure 2 - the validated $y(t)$ is published and a speculative $y(t + \Delta t)$ is computed and stored. At t_I, in response to a discrete event that sets the value of y to y_I, the following steps are programmed: 1) the speculative value of $y(t + \Delta t)$ is abandoned, 2) y_I is published as the valid y, and 3) a revised speculative $y(t_I + \Delta t)$ is calculated and stored.

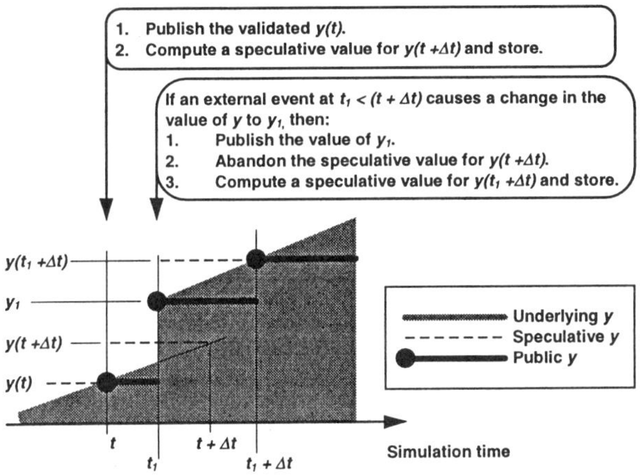

Figure 3: Strategy for Programming a Type 1 Interaction

2.3 Interaction Type 2: A Discrete Event Affects the Relation Governing a Continuous Variable

Interactions of the second type (a discrete event causing a change in the relation governing the evolution of a continuous variable) can be programmed in a similar way, as Figure 4 shows. The principal difference between this case and the Type 1 interaction is that in the Type 1 case, the value of $y(t_I)$ is specified by the interrupting process, while in this Type 2 case, the value of $y(t_I)$ must be calculated by the integrating process itself before publishing. The performance requirements of the particular model dictate the level of sophistication required in this calculation of $y(t_I)$ in the Type 2 case. In the simplest case, perhaps the value of $y(t_I)$ could be simply interpolated between $y(t)$ and $y(t + \Delta t)$, while a more demanding case might require prior storage of all of the required elements of the system state at t to enable a complete reconstruction of $y(t_I)$.

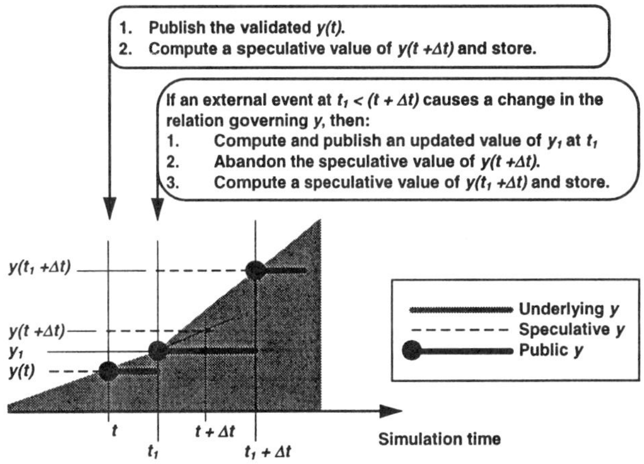

Figure 4: Strategy for Programming a Type 2 Interaction

2.4 Interaction Type 3: A Continuous Variable Achieving a Threshold Causes a Discrete Event

The processing of a Type 3 interaction (an interaction in which the continuous variable achieving a threshold causes a discrete event to occur) differs in two fundamental ways from the processing of Type 1 and Type 2 interactions: 1) the integration process itself identifies the time of the interaction, and 2) the integration process triggers discrete events. Figure 5 shows the strategy for programming a Type 3 interaction. At t, the speculative value $y(t + \Delta t)$ may be checked to see whether y has crossed the threshold. If no crossing is found, then the process may continue in the simple form shown in figure 2. If a crossing has occurred, then the integrating process must find the time of crossing by a method that gives suitable accuracy. Again, the performance requirements of the particular application guide the choice of a method. In the simplest case, the crossing time $t_{threshold}$ may be estimated by interpolating the value of $y(t)$ between t and $t + \Delta t$, while for more demanding applications, a root-finding procedure might be required.

Type 3 interactions additionally require that the modeling program provide special features not required for the other types of interactions. The numerical integrating process must be able 1) to trigger discrete events immediately (at $t_{threshold}$ in the example), 2) to

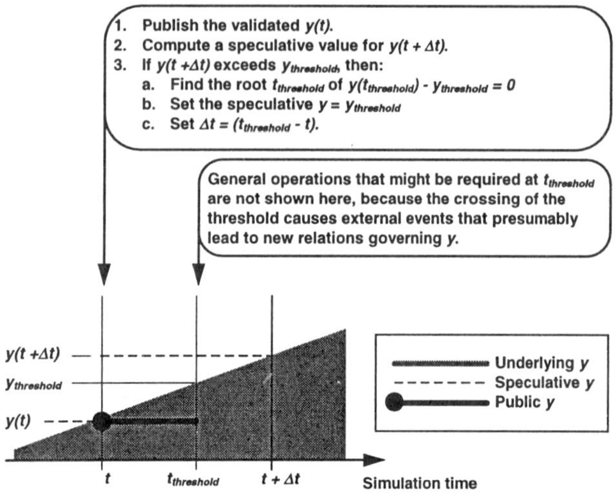

Figure 5: Strategy for Programming a Type 3 Interaction

cause discrete events to be scheduled, and 3) to reset or adjust other continuous variables and relations that govern them.

3.0 SUMMARY

Continuous state variables in combined models are composed of piecewise smooth functions of time, divided by discrete events into continuous segments. If synchronous integration methods are used, then the precision with which boundary events can be resolved depends on the selection of integrating time step - the smaller the step, the finer the resolution. To meet resolution requirements, the analyst may place an upper bound on the permissible time step size in addition to that dictated by the needs for accuracy in the integration process itself. This additional constraint may result in a performance penalty. Using asynchronous (event-driven) integration methods to divide the smooth pieces removes the need for the additional constraint

Figure 6 summarizes the differences between synchronous and asynchronous integration. In the case of synchronous integration, the fidelity with which the computed result reflects the interactions with discrete events clearly depends on the selection of step size. In the case of asynchronous integration, the continuous state variable is divided into its three piecewise smooth segments. In the case of the Types 1 and 2 interactions, the division reflects the "exact" time resolution of the discrete event scheduler, and in the case of the Type 3 interaction, the time resolution of the detection threshold can be made arbitrarily fine. Thus, fidelity at the segment boundaries is independent of the integration step size.

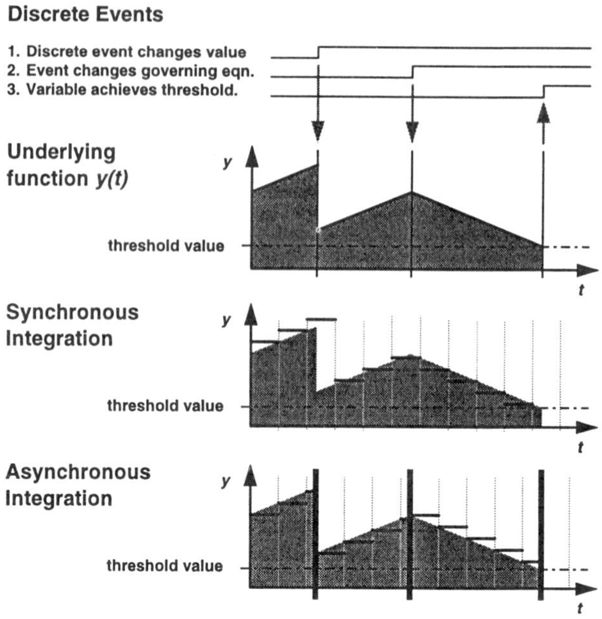

Figure 6: Comparison of the Results of Synchronous and Asynchronous Integration

4.0 IMPLEMENTATION

The earlier sections of this paper discussed the methods of speculative computation and asynchronous integration that are needed to adapt classical numerical integration methods, which are basically synchronous, to the asynchronous environment of the commercial discrete modeler. This section first discusses general programming features (data structures and function calling conventions) that are useful for implementing those methods. Then, for two commercial modelers, it discusses how the analyst can exploit programming features and how a simple example model can be programmed. An exploration of comparative performance or even preference is outside the scope of this paper.

4.1 General

The notions critical to adaptation of classical numerical integration methods to a discrete event environment are: 1) speculative computation described in section 2.1and 2) interruptible processes to enable the methods described in sections 2.2 through 2.4. To implement speculative computation, it is useful, though not necessary, for the modeler's data structure to embody some idea of locality or privacy to protect the speculative values of the continuous state variables until they can be validated. Further, the modeler should have some facility for interrupting the integration process so that the speculative values of continuous state variables can be modified to reflect changes due to discrete events that occur between computation and validation. This section discusses approaches to combined modeling that can be used with two modelers ProModel for Windows (ProModel Corporation, Orem UT) and Extend (Imagine That!, San Jose CA).

As a framework for discussion, the following simple system was constructed in both models: pallets containing massive steel objects arrive at a preparation area. When space is available, a pallet is loaded into a furnace, in which it is heated by radiation and convection, to a predetermined threshold temperature. Then the pallet is moved to a cooling area and then to disposal. The pallet temperature is a represented by a continuous state variable that follows the differential equations of heat flow. The variable is divided into smooth pieces by the charging and discharging events.

The model includes interactions of all three types, the arrival of a pallet in the furnace combines types 1 and 2 interactions by setting both the value of the continuous state variable and the relation governing it. The type three interaction occurs when the pallet temperature reaches the threshold temperature.

4.2 ProModel for Windows

ProModel for Windows (PMW) is a discrete event modeler that has no explicit support for modeling continuous processes. It does have features that an analyst can use to build integrators that implement the techniques that section 2 describes. PMW has a useful data structure that includes real and integer variables with either global or local scope. In addition, the user can define real or integer attributes for entities and/or locations. PWM provides at least two constructs that can be used to program numerical integrators: 1) subroutines launched by the `ACTIVATE` statement, and 2) operations code that is executed by arrival of an entity at a location. This section takes the latter approach.

Klingener [1995] describes a combined model built in PMW based on a synchronous Euler integration scheme using the approach. Although the methods were useful for illustrating the technique of embedding numerical integration schemes in entity operations code, they are not suitable for serious modeling. Adding such features as speculative computation, a robust integration scheme, adaptive time stepping, and Type 3 interactions no change to the basic structure of the model. However, managing interactions of Types 1 and 2 requires the creation, by interrupting events (using PMW's `ORDER` statement), of dummy entities that update the continuous variable as required, that arrange the orderly termination of the already running Operations code, and that flush invalid speculative results.

The programming is described here from the bottom up. Listing 1 shows the subroutine `derivl` that returns the rate of change of the charge temperature, given the simulation clock time and the state variables. Listing 2 shows the `dy_rk4` subroutine that implements the 4th order Runge-Kutta numerical integration method. It calls the `derivl` routine and returns the change in temperature during the integration time step.

Listing 1: Differential Change in Temperature

```
# real derivl(real time, real charge_temp)
#---------------------------------------------------------------
# uses global variables:
#    area -     heat transfer area (ft^2)
#    charge_weight              (lbf)
#    specific_heat              (BTU/lbf/degR)
#    Hconv - convect. h. t. coeff (BTU/hr/ft^2/degR)
#    Hrad  - rad.     h. t. coeff (BTU/hr/ft^2/degR^4)

RETURN (area/(charge_weight*specific_heat))*
       (Hconv*(Thigh-charge_temp) +
        Hrad*(Thigh**4-charge_temp**4))
```

Listing 2: Fourth Order Runge-Kutta Procedure

```
# real dy_rk4(real x, real y, real dx)
#---------------------------------------------------------------
# unadorned fourth order Runge_Kutta
# ref Press, et al. 2nd Ed. page 711 eqn 16.1.3

REAL k1,k2,k3,k4

k1=derivl(x     ,y      )*dx
k2=derivl(x+dx/2,y+k1/2)*dx
k3=derivl(x+dx/2,y+k2/2)*dx
k4=derivl(x+dx  ,y+k3  )*dx

RETURN k1/6 + k2/3 + k3/3 + k4/6
```

Listing 3 shows the core operations performed by the Entity `RK4` each time it enters the Location `integrators`. At the end of the operations block, `RK4` is `ROUTE`d either back to the beginning (using the `CONTINUE` rule) or, if an interrupt has started the integration of a new segment, to the `EXIT`. The listed procedure performs a 4th order Runge-Kutta integration with adaptive time stepping, speculative computation, and threshold detection. When the threshold condition is met, the type 3 interaction is effected by issuing a `SEND` statement to the Entity waiting at the `furnace` Location.

Numerical integration processes, turned off when not needed, can be spawned by `ORDER`ing a dummy entity to a real or dummy location. Listing 4 shows the operations typical for such an Entity, here `Intr2` at `integrators`. It performs the initialization appropriate to its type and then spawns the recurring process by routing another Entity `RK4` back to the same Location.

Listing 3: Asynchronous Numerical Integration Operations

```
#################################################################
#  OPERATIONS BLOCK FOR ENTITY RK4 AT LOCATION integrators     #
#    This block implements a simple fourth order Runge-Kutta   #
#    integration with adaptive time step control.              #
#    refer to Press, et al. Numerical Recipes in C, 2nd Ed.    #
#    pp. 710-716                                               #
#################################################################

#  DECLARE LOCAL VARIABLES
#----------------------------------------------------------------
INT retry                    # flag controlling adaptive stepper
INT too_slow, too_sloppy     # flags reflecting truncation error
REAL max_err=max_error_per_RK_step, min_err=max_err/100.0
REAL y_1step,y_2step,t_err   # interim results

#  SAVE THE CONTEXT IN ATTRIBUTES
#----------------------------------------------------------------
LocAttr1=CLOCK()             # use these later for intr handling
EntAttr1=CLOCK()

#  COMPUTE SPECULATIVE TEMPERATURE AND TRUNCATION ERROR
#----------------------------------------------------------------
DO BEGIN                     # until retry=False
  y_1step=Temp+dy_rk4(CLOCK(HR),Temp,time_step)
  y_2step=Temp+dy_rk4(CLOCK(HR),Temp,time_step/2)
  y_2step=y_2step+dy_rk4(CLOCK(HR)+time_step/2,y_2step,
                                                  time_step/2)
  t_err=y_2step-y_1step      # Press 2nd. p.715 16.2.2

  #  DECIDE WHETHER RESULT IS WITHIN ACCURACY SPECIFICATIONS
  #--------------------------------------------------------------
  IF abs(t_err)<min_err THEN too_slow=True
                        ELSE too_slow=False
  IF abs(t_err)>max_err THEN too_sloppy=True
                        ELSE too_sloppy=False
  retry=TRUE
  IF too_slow=True THEN time_step=2*time_step          # double
    ELSE IF too_sloppy=True THEN time_step=time_step/2   # half
                            ELSE retry=False       # just right
END UNTIL retry=False

# This completes the computation of the speculative result
# for the temperature at time_step in the future.
# Keep the best value in the (local) variable, y_2step:
y_2step=y_2step+(t_err/15.0)        # Press 2nd. p. 715 16.2.3

#  CHECK FOR THRESHOLD CROSSING (Type 3 Interaction)
#----------------------------------------------------------------
IF y_2step>=Tthreshold THEN BEGIN
  # interpolate time to crossing
  time_step=time_step*(Tthreshold-Temp)/(y_2step-Temp)
  WAIT time_step
  Temp=Tthreshold
  SEND 1 Workpiece TO cooling
                             # Output Destination  Rule
  ROUTE 2 BREAK END          # RK4    EXIT         FIRST 1

#  WAIT FOR SIMULATION CLOCK TO CATCH UP
#----------------------------------------------------------------
WAIT time_step

#  DECIDE WHETHER THERE HAS BEEN AN INTERRUPT
#----------------------------------------------------------------
IF LocAttr1=EntAttr1         # no interruption
 THEN BEGIN
  # Publish the speculative result as the global variable Temp
    Temp=y_2step
  # Then go back for another bite
                             # Output Destination  Rule
    ROUTE 1  END             # RK4    integrators  CONTINUE 1
 ELSE BEGIN                  # interruption
                             # Output Destination  Rule
    ROUTE 2 BREAK END        # RK4    EXIT         FIRST 1
```

Listing 4: Type 2 Interaction Operations

```
# Intr2 Type 2 Interaction Preamble
#----------------------------------------------------------------
# update the value of the continuous variable first
  REAL short_step
# calculate the size of the short step
  short_step=CLOCK()-LocAttr1
# call the integration routine for a new Temp.
  Temp=Temp+dy_rk4(CLOCK(),Temp,short_step)
# restart the integration
  time_step=0.05
# exit ROUTING:                Output Destination  Rule
#                              RK4    integrators  FIRST 1
```

4.2 EXTEND

Extend from Imagine That! is a visual simulation modeling language that has support libraries for either discrete or continuous systems. Combined models are constructed in the discrete mode. While Extend has library support for integration of continuous variables in discrete models, that support is based on a fixed time step size Euler method. In the context of issues discussed in this paper, that support lacks the speed, accuracy and stability that are required for high performance modeling.

However, Extend, as part of its C legacy, inherits a rich and expressive underlying structure along with its built-in ModL language. ModL has an expressive data structure that includes user-definable real, integer, string, and array types (though no structures). In addition, Extend's block structure supports message passing among processes and programming of the block responses to those messages. Figure 7 shows one approach to modeling a combined process in Extend. The required behavior of the furnace block (receiving an item, triggering the integration process, and passing the item on after receiving a release message from the integrator) may be obtained by modifying one of Extend's existing library blocks. The user-written integration block may be triggered either by a message to Intr2In or, if the model required that item attributes be passed to the integrator, the item itself could be passed. The numerical integration itself is coded into the integrator block script. When the threshold temperature was reached, then the integrator block sends a message to the furnace via the Intr3Out connection.

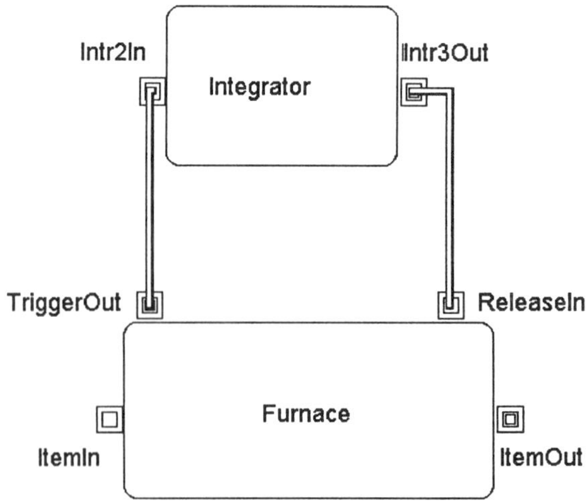

Figure 7: Block Interactions in Extend

4.3 CONCLUSIONS

Of the examined modelers none explicitly supports combined models with the features required for high performance. ProModel for Windows has no built-in support at all. Extend supports either discrete event or continuous models, but its support for continuous variables in combined models is suitable only for small or undemanding problems.

Although ProModel for Windows and Extend have different modeling paradigms, both have data structures and programming features that enable construction of robust combined models. The methods described implement speculative computation, a robust integration scheme, adaptive time step adjustment, threshold detection, and asynchronous restarts after interruption. The ability to use different time step sizes for processes that have different time characteristic and the ability to turn off numerical integration when it is not needed are crucial advantages of the approach.

REFERENCES

Klingener, J. Frederick, 1995, Combined Discrete-Continuous Simulation Models in ProModel for Windows. In *Proceedings of the 1995 Winter Simulation Conference*, ed. C. Alexopoulos, K. Kang, W. R. Lilegdon, D. Goldsman, 445-450. Institute of Electrical and Electronics Engineers, San Francisco, California.

Press, William H., Brian P. Flannery, Saul Teukolsky, William T. Vetterling. 1992. *Numerical Recipes in C - The Art of Scientific Computing*. 2d ed. New York: The Cambridge University Press.

Pritsker, A. A. B. 1986. *Introduction to Simulation and SLAM II*. 3d ed. West Lafayette, Ind: Systems Publishing Corporation.

AUTHOR BIOGRAPHY

FREDERICK KLINGENER is president of Brock Engineering, P. C. in Roxbury Connecticut. He received a B. S. in 1962 and an M.S. in 1966 in mechanical engineering from Carnegie Institute of Technology. He is a registered mechanical engineer in the state of Alaska. He has had broad experience in design, analysis, and fabrication of nuclear, robotic, and automotive mechanical systems. He has recently worked in cost, risk, and production analysis of the army's program to dispose of unserviceable chemical weapons. Mr. Klingener's current interests include data analysis of real and simulated manufacturing processes.

Proceedings of the 1996 Winter Simulation Conference
ed. J. M. Charnes, D. J. Morrice, D. T. Brunner, and J. J. Swain

THE DESIGN OF AN EFFICIENT SIMULATOR FOR THE PENTIUM PRO PROCESSOR

David A. Sykes

ObjectKnowlogy, Inc.
11 Oakview Drive
Greenville, SC 29605–1816, U.S.A.

Brain A. Malloy

Department of Computer Science
Clemson University
Clemson, SC 29634–1906, U.S.A.

ABSTRACT

The increasing size and complexity of computer systems has created the need to develop new techniques to facilitate the design and evaluation of developing architectures. The most common technique applied to the development of new architectures is simulation, which permits detailed accurate modeling of the architecture. However, there are several problems associated with using simulation in developing computer systems. First, traditional simulation approaches are not flexible enough to permit easy extension or modification of the model. While most simulators are parameterized to provide flexibility, parameterization alone is not adequate to permit the modification and extension required for developing architectures. Second, traditional simulation approaches are computation intensive, slowing development and prohibiting simulation of large programs. To address the first problem, we use an object-oriented design for our simulator to produce a flexible, extensible model. To address the second problem, we incorporate the *decode* step of processor simulation into the state of each instruction object, saving the time to simulate the decode during execution. We have implemented a prototype of our design and initial experiments with the object-oriented prototype, coded in C++, ran twice as fast as another prototype that we implemented using the traditional approach to simulation where the simulator was written in C.

1 INTRODUCTION

The increasing size and complexity of computer systems has created the need to develop new techniques to facilitate the design and evaluation of developing architectures. The most common technique applied to the development of new architectures is simulation, which permits detailed accurate modeling of the architecture (Malloy 1993; Philip 1992). The most common simulation approaches for computer systems are *trace-driven* and *execution-driven* simulation. Both of these approaches have been used successfully in developing and testing new architectures.

However, there are several problems associated with using simulation in developing computer systems. First, traditional simulation approaches are not flexible enough to permit easy extension or modification of the model. While most simulators are parameterized to provide flexibility (Malloy 1993; Philip 1992), parameterization alone is not adequate to permit the modification and extension required for developing architectures. Second, traditional simulation approaches are typically computation intensive, which slows development and prohibits simulation of large programs. Third, to provide accurate results and to avoid design errors, simulators must be validated; validation of simulators, especially simulators for developing architectures, is difficult and time consuming.

In this paper, we address the first two problems associated with traditional approaches to simulation of developing architectures. To address the first problem, we use an object-oriented design for our simulator to produce a flexible, extensible model. To address the second problem, we incorporate the *decode* step of processor simulation into the state of each instruction object, saving the time to simulate the decode during execution. We have implemented a prototype of our design and initial experiments with the object-oriented prototype, coded in C++, ran twice as fast as another prototype that we implemented using the traditional approach to simulation where the simulator was written in C. However, almost twice as much memory was required to execute the object-oriented simulator over the traditional simulator. Our current focus is only on execution-driven simulation. Further work will address the reusability of our design and the components it produced in the construction of trace-driven simulators.

The remainder of this paper is organized as follows. In the next section, we provide background about trace-driven and execution-driven simulation together with information about the target architecture of our simulator: the Pentium Pro Processor. In section 3, we present the object model for our simulator together with discussion about the advantages of our design. In section 4 we describe our prototype and in section 5 we discuss our experiments and give concluding remarks.

2 BACKGROUND

In this section, we provide background about simulators and processors. We begin by discussing two techniques applied to processor simulation: *trace-driven* and *execution-driven* simulation. We then overview recent trends in processor design with special emphasis on the Pentium Pro processor by Intel, selected because of its complexity and the challenge it introduces to simulator development.

2.1 Processor Simulation

The classic approach to simulating existing or developing architecture is *trace-driven simulation* where the simulation is based on a predetermined instruction sequence or *trace*. Trace-driven simulation requires information about branch targets and memory references made during the execution of the program. This technique is used extensively to evaluate uniprocessor cache performance and techniques have been developed for capturing trace information to include references from system calls and the operating system. However, the problem with applying the trace-driven approach to simulating developing architectures is that the simulated architecture must be similar to the architecture on which the trace was obtained; this is a significant problem when simulating an architecture that is in the development stage (Koldinger, Eggers, and Levy 1991). For a new architecture, the order of events, the latency of communications, and/or the number of processors might be completely different from the trace host, rendering the traces useless when evaluating timing-sensitive applications. Moreover, using traces obtained from a sequential system may be completely inappropriate for use in simulating a parallel system since the order of execution is usually dependent on the order in which different parts of a parallel program complete execution (Mukherjee and Bennett 1990). Previous research has shown that traces obtained on a multiprocessor can produce erroneous results when used to simulate a different multiprocessor (Dahlgren 1991). Even when the trace-driven approach is applied to uniprocessor simulation, traces are difficult to obtain and trace files are frequently prohibitively large (Ball and Larus 1992).

Because of the drawbacks of trace-driven simulation, recent trends reveal a proliferation of the use of *execution-driven simulation*, also referred to as instruction level simulation, register-transfer simulation or cycle-by-cycle simulation. In the execution-driven approach, the simulator actually executes an input program. Thus, the main actions of the simulator are expressed with a large **case** statement enclosed in a **while loop** that runs for the length of the simulation; on each iteration of the loop, an instruction is executed and each case option is an `op code` for the processor to decode and execute. Execution-driven simulation permits development of the software together with the hardware because the simulator isn't bound by the traces.

Although recent research has produced reliable execution driven simulators (Malloy 1993; Philip 1992), the classic, time-honored approach to processor simulation is trace-driven simulation.

2.2 Trends in Processor Design, The Pentium Pro

In the earliest stored-program computers, hardware was expensive so that early processor designers produced a machine with a single register for arithmetic instructions. Since all operations would accumulate in a single register, it was called an *accumulator*; the EDSAC computer was a single accumulator machine. This early design philosophy was replaced by more elaborate schemes, since hardware became progressively less expensive. Recent processor designs combine a lush supply of registers with multiple instructions issued on each execution cycle and elaborate memory hierarchies.

An extremely efficient microarchitecture, the Pentium Pro Processor by Intel, has recently been presented that combines innovations resulting in processor speeds that significantly exceed the 100 MHz Pentium Processor, though it is manufactured with the same semiconductor process that produced the Pentium Processor. The Pentium processor is superscalar, using five stages to extract high throughput. However, the Pentium Pro processor is capable of speeds in excess of 200 MHz. To achieve this, several design innovations were incorporated into the new processor including *dynamic execution*, *associative memory*, *branch prediction*, *branch recovery*, a *branch target buffer* and an *instruction pool* that permits three independent processor units to communicate. The instruction pool is implemented as *content addressable memory* called the *reorder buffer* (ROB).

Dynamic Execution technology can be summarized as optimally adjusting instruction execution by predicting program flow, using data flow analysis to choose the best instruction to execute and then executing the instructions speculatively, in a preferred order. To do this, the processor uses an in-order *fetch/decode* unit, an out-of-order *dispatch/execute* unit, an in-order *retire unit* and a *bus interface* unit that communicates with the off chip cache supporting as many as four concurrent cache accesses.

Figure 1 illustrates the important design features of the Pentium Pro Processor. At the top of Figure 1, is the system bus and the off chip cache or *L2 Cache*. The *Bus Interface* connects the processor with the system bus and the off chip L2 cache. The processor has an on-chip cache with an 8K byte instruction cache, shown in Figure 1 as *L1 I-Cache*, and an 8K data cache, shown in Figure 1 as *L1 D-Cache*. The *Fetch/Decode* unit fetches instructions from the L1 I-Cache while the *Dispatch/Execute* unit and the *Retire* unit fetches (stores) instructions from (to) the L1 D-Cache. All three units access the *Instruction Pool*. Figure 1 is an extremely simplified description of the Pentium Pro Processor, which requires detailed simulation to capture subtle but important nuances to facilitate processor development. A simulator for the Pentium Pro Processor must permit easy extension and modification together with efficient execution.

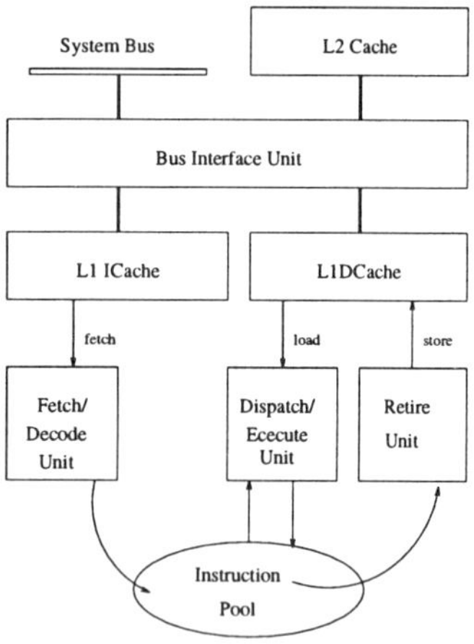

Figure 1: The Pentium Pro Processor Design

3 DESIGN OF THE SIMULATOR

In our brief description of the Pentium Pro Processor, provided in the previous section, we have tried to capture the flexibility and power of the processor. However, development and testing of such a processor requires the use of a simulator that is also powerful and flexible. The processor simulator that we present in this paper was motivated by the following goals:

- *efficiency,* particularly with respect to execution time since traditional simulators are CPU-intensive;
- *extensibility,* to permit addition of new kinds of instructions or simulation of new hardware components to study their impact without requiring large changes to existing hardware or software;
- *flexibility,* to permit easy adaptation to new algorithms; for example, adaptation of a new caching strategy or pipeline schedule to determine their effects; and
- *reusability,* such that we can build simulators for other computer architectures using existing software components.

We have chosen to use an incremental development approach. In starting our work, we have focused on simulators for simple architectures and are building on that work to address computers that have increasing amounts of sophistication. Our plans for future work include the development of an object-oriented framework for simulators. We are first focusing on building a set of components that can be reused to simulate a wide range of architectures—for example, an architecture containing multiple levels of cache memory, register arrays, and instruction pipelines including those that support out-of-order processing and in-order retirement of instructions. Our ultimate goal is a simulator for the Pentium Pro. The design that we present in this paper comprises the fundamental architecture. From that and its refinement over a series of increments, we will identify a more general design that will be captured in a framework from which we can design many simulators. We have already determined that framework will include both abstract classes and templates.

To address the goals listed above, we begin with the object-oriented programming paradigm to exploit its strengths for supporting extensible, flexible designs and reusable components. For reasons of efficiency, we adopted C++ as our implementation language, although our design is virtually language-independent.

3.1 Analysis of Computer Architectures

We began our development effort with an analysis of the domain of computer architecture. The object model for our initial design is illustrated using "unified method notation" in Figure 2. Although this model is in the development stage, we consider it sufficient for investigating a design for simulating a variety of computer architectures. In the model, we attempt to capture major object classes within the domain so that when we consider a specific computer architecture—for example a Motorola 68040 or the

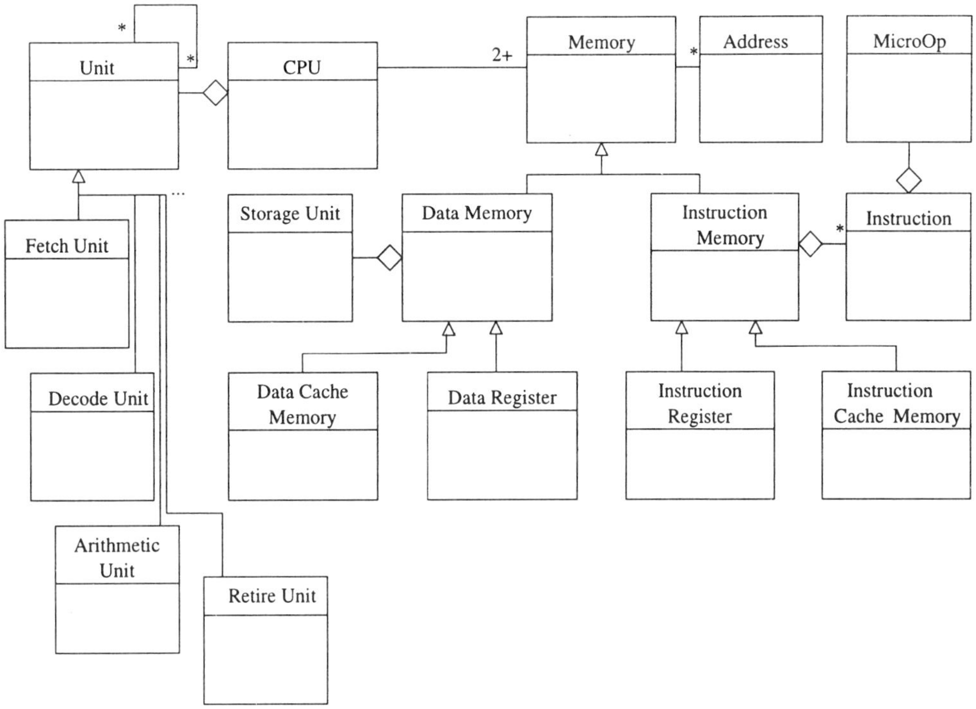

Figure 2: Analysis Object Model for the Computer Architecture Domain

Pentium Pro Processor—we will refine the classes in this model and likely introduce new relationships.

The object model is based on an assumption that data and instruction memories are separate, although there is no assumption as to whether this separation is logical or physical. Implicit in this assumption is that instructions can be executed only from instruction memory.

3.2 Design Overview

Our simulator architecture is based fundamentally on the separation of instruction memory and data memory. Instruction memory contains only instructions—never data—that we treat as objects. That is, instead of representing instruction memory as an array of bytes (or words) that are fetched and assembled into instructions that are then decoded, we represent instruction memory as an associative array of instructions (associated with addresses), each of which knows how to "execute" itself given a current processor environment.

There are at least three implications that arise from a design based on such assumptions:

1. The "decode" step of simulation is implicitly represented in the state of each instruction object, saving the steps in the simulator that interprets the bit pattern of a sequence of bytes retrieved from the memory array. A set of experiments we ran on a design prototype shows that the execution of a simulator using this technique is approximately twice as fast as another prototype we built in C using a traditional design, although about twice as much memory is required to run the simulator. We note that the prototype uses a very simple implementation of instruction memory: an array of pointers to instruction objects that yields fast access to instructions, but at a cost of more space. (the prototype is discussed further in Section 4.)

2. Writing into instruction store introduces inefficiency. In most circumstances, runtime modification of code is not performed by application programs so this is not a problem. Some processor architectures separate code and address spaces physically, making runtime modification of instructions impossible. However, some processors do allow writes to instruction memory and some algorithms—most notably, those that perform "bit block transfer" (BITBLT) in animated computer graphics applications—optimize their efficiency by, in essence, assembling a customized segment of code that executes its function faster than a data-driven approach would yield. Since, in our design, we essentially *pre*-decode instructions in instruction memory, then any write into a byte of that memory must re-decode the instruction stored there. This is pos-

sible in our design. We can treat the associative memory as a cache, such that any write to a byte at an address invalidates the instruction associated with that address. If it can be assumed that a sequence of writes to instruction memory will result in replacement of one or more whole instructions before they are executed, then construction (by decoding bytes) of the new instruction objects can be accomplished straightforwardly. Algorithms used by interpreters in compiling language constructs into a cache to achieve faster execution—for example,the caching of methods by a Smalltalk interpreter—can be adapted for use here in a way that is transparent to all objects outside those representing instruction memories. If that assumption is not valid, then a byte image of instruction memory must be maintained by the simulator and instruction objects in memory be reconstructed from that. [This would certainly be expensive in terms of both time and space.]

3. If a transfer of control is made to an address that corresponds to a byte that lies within—that is, after the first byte of—an instruction object, then a situation similar to the previous one arises and the equivalent of a byte image must be maintained and used to reconstruct instructions in memory.

We believe that these three situations are rare and have decided not to support them until a later version of our simulators, although we are keeping them in mind as we design.

4 DESIGN PROTOTYPE

We developed a prototype as a proof of concept for our design. Our prototype implementation supports simulation of a simple computer architecture that has byte addressing and single-word (two 8-bit bytes) instructions. The basic architecture is illustrated in Figure 3. All addresses are real and comprise twelve bits. Instruction and data memory are physically separate. Transfer instructions contain instruction memory addresses, while all other instructions contain data memory references. There is a single sixteen-bit accumulator (AC), a twelve-bit stack pointer register (SP) used for subroutine calls and returns, and a status register that describes characteristics of the contents of AC.

The prototype was built in C++ to prove the design concept and to collect rough performance characteristics. We purposely chose a simple architecture for our initial experiments so that we could evaluate our object-oriented design, devoid of nonessential considerations. We also built a corresponding simulator in C based on traditional simulator design so that we could directly compare the designs and their performance.

The design object model for the prototype is shown in Figure 4. It includes some details about the class specifications. In that model, Environment represents the computer configuration, including the registers in the CPU. We have not included processing units that perform arithmetic, for example, in the prototype, but future designs will include such components. We are currently evaluating the design vis-a-vis the benefits of this change as well as defining a CPU class. We expect the interface for the environment to contain operations that gain access to all system components and handle requests such as fetching instructions from code store and reading and writing registers.

In terms of an algorithm, the most significant change in the object-oriented version from the procedural version is the way in which instructions are fetched from code store, decoded, and executed. The high-level algorithms may be described as:

```
do {
  instPtr =
    this->fetchInstruction(PC.read());
  PC = PC + instPtr->size();
  instPtr->execute(*this);
} while (not done);
```

In a procedural version, all data used in the simulation is typically global to the main simulation loop and to all functions called by that loop. In the object-oriented version, `PC` is the program counter and `instPtr` is a pointer to the current instruction [object] to be executed, and `this` is the environment that is responsible for running a simulation (member function `go`). All data used in the simulation is defined in the environment object, but not to other objects it messages, thus the need to store return values—for example, into `instPtr`—and pass parameters to instructions—for example, `*this` to an instruction being `executed`.

Experiments with our prototype show that the object oriented simulator, coded in C++, ran twice as fast as another prototype that we implemented using the traditional approach to simulation where the simulator was written in C. Figure 5 illustrates these results. The table of Figure 5 lists three programs having complexity $O(n)$, $O(n^2)$, and $O(n^3)$. Conceptually, these correspond to programs that compute factorial, perform a sort, and perform a fast Fourier transform, respectively. On average, these programs executed 262,145, 16,787,457 and 67,110,145 instructions respectively. The $O(n)$ program required 97 clock units to execute using the traditional C simulator and only 56 clock units to execute using the

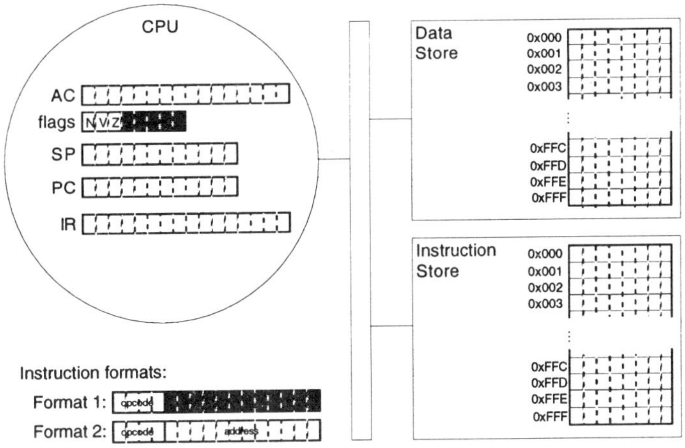

Instruction Set	
LDA addr	*load AC with data[addr]*
ADD addr	*add data[addr] to [AC], set flags*
STA addr	*store [AC] in data[addr]*
JSR addr	*jump to subprogram at address addr and save return address on stack*
RTS	*return from subprogram*
HLT	*stop execution*
BRZ addr	*branch on [AC] zero to addr*
BRN addr	*branch on [AC] negative to addr*
BRV addr	*branch on [AC] zero to addr*
JMP addr	*jump unconditionally to addr*
CMP	*boolean complement [AC]*
XOR addr	*bitwise xor data[addr] with [AC]*
AND addr	*bitwise and data[addr] with [AC]*
NEG	*negate [AC]*
LDR addr	*use the lower twelve bits of AC as a data address and load contents into AC*
NOP	*no operation*

Figure 3: Architecture for the Design Prototype

object-oriented C++ simulator. The other two programs experienced similar speedups. The graph in Figure 5 highlights this result.

From this data we have concluded that the storage of instructions as objects, which by their existence makes them "pre-decoded," speeds up execution. However, we can realize a benefit only for programs that involve looping because initialization of the object-oriented simulator requires more time.

5 CONCLUDING REMARKS

We have presented a design that exploits object technology to produce a flexible, extensible, efficient simulator for developing architectures. We have proven the concept of our design by the development of a prototype for a simple processor. We have compared the execution characteristics of that prototype implementation with the implementation of a simulator in C to determine the costs and benefits of our approach.

We are currently extending the prototype in two ways. First, we are modifying the analysis model to address the Pentium Pro Processor. Second, to further demonstrate the extensibility of our model, we are looking at our design to build an object-oriented framework. As we extend our design over a series of more and more complex processor architectures, ending at a simulator for a Pentium Pro, we will validate our design in terms of its reusability and extensibility.

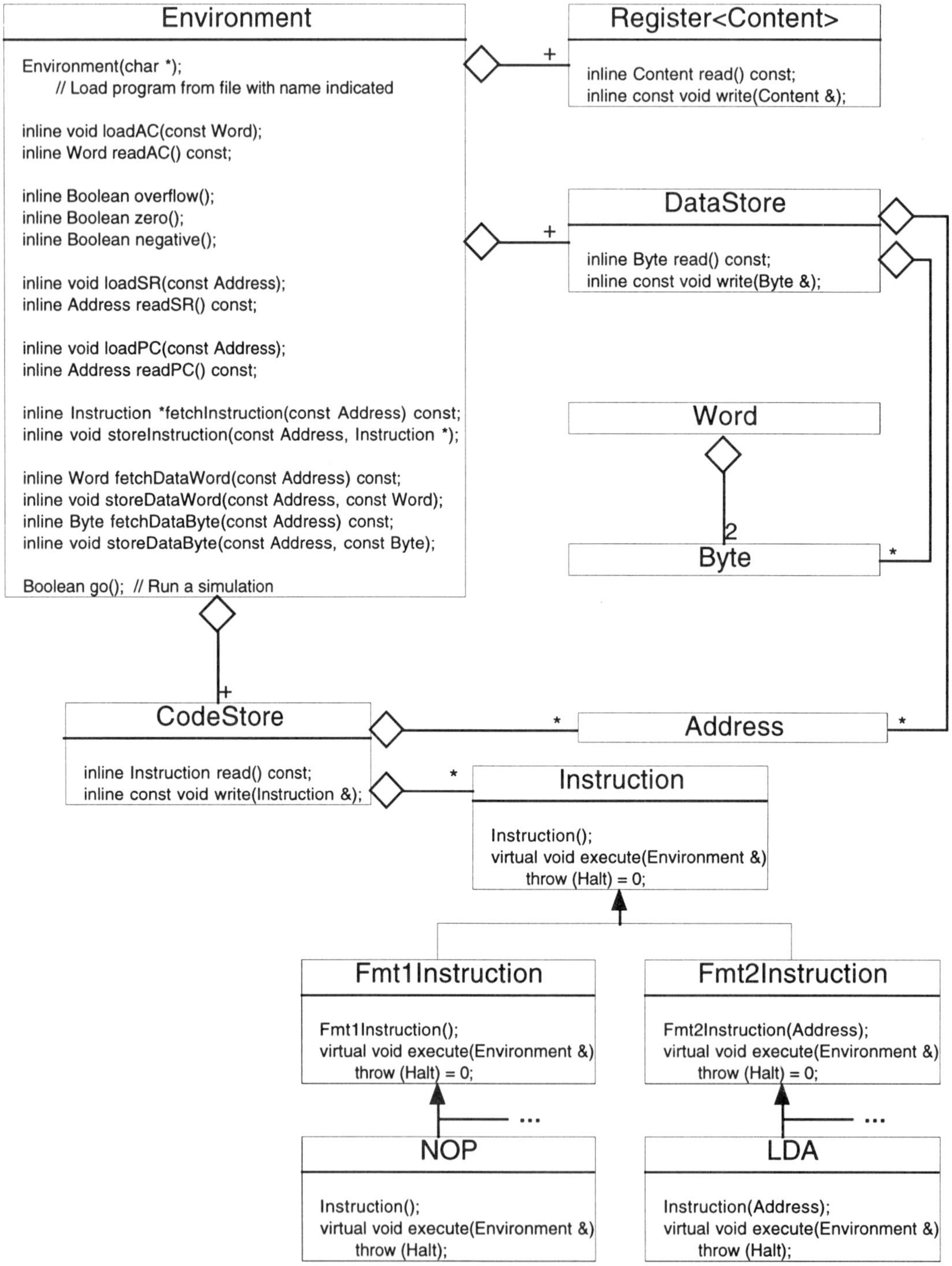

Figure 4: Design Object Model for Our Prototype

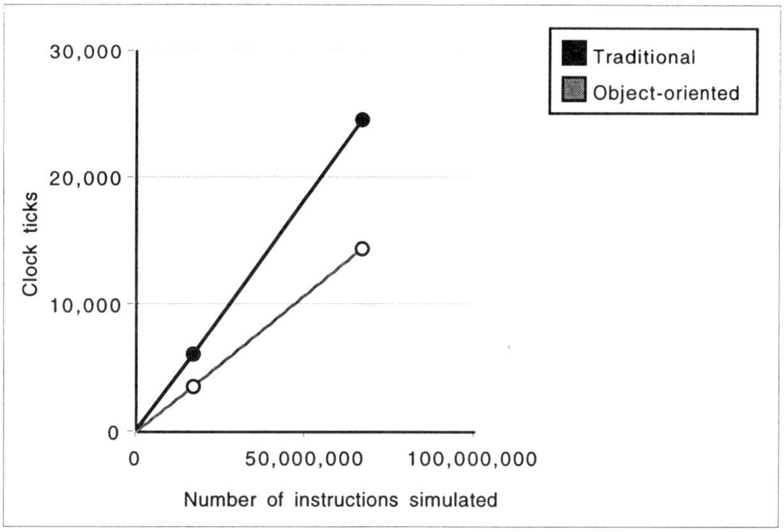

Program complexity	Instructions executed	Traditional time	Object-oriented time
$O(n)$	262,145	97	56
$O(n^2)$	16,787,457	6122	3602
$O(n^3)$	67,110,145	24556	14397

Figure 5: Experimental Results

REFERENCES

Ball, T. and Larus, J. R. January 1992. Optimally profiling and tracing programs. In *Proceedings of Symposium on Principles of Programming Languages,* 59–70.

Dahlgren, F. 1991. A program-driven simulation model of an MIMD multiprocessor. In *24th Simulation Symposium,* 40–49.

Koldinger, E., Eggers, S., and Levy, H. 1991. On the validity of trace-driven simulation for multiprocessors. In *Proceedings of 18th Annual Symposium on Computer Architecture,* 244–253.

Malloy, B. 1993. The validation of a multiprocessor simulator. In *Proceedings of the 1993 Winter Simulation Conference,* 625–631.

Mukherjee, R. and Bennett, J. 1990. Simulation of parallel computer systems on a shared memory multiprocessor. In *Proceedings of 23rd Hawaii International Conference on System Science,* 1:242–251.

Phillip, M. J. 1992. Performance issues for the 88110 RISC microprocessor. In *Proceedings of IEEE COMPCON,* 163–168.

AUTHOR BIOGRAPHIES

DAVID A. SYKES is a Principal Investigator with ObjectKnowlogy, Inc. He received a B.S. degree in computer science from Purdue University in 1972, an M.A. degree in computer sciences from the University of Texas at Austin in 1975, and a Ph.D. degree in computer science from Clemson University in 1995. His research interests focus on the design and testing of object-oriented software systems.

BRIAN A. MALLOY is an Associate Professor in the department of Computer Science at Clemson University. He received an M.S. and Ph.D. from the University of Pittsburgh in 1984 and 1991. His research interests focus on simulation techniques that exploit object technology and parallelization.

Proceedings of the 1996 Winter Simulation Conference
ed. J. M. Charnes, D. J. Morrice, D. T. Brunner, and J. J. Swain

HYBRID HIGH-LEVEL NETS

Ralf Wieting

Oldenburger Forschungs- und Entwicklungsinstitut für
Informatik-Werkzeuge- und Systeme (OFFIS)
Escherweg 2, D–26121 Oldenburg, GERMANY

ABSTRACT

This paper presents a new methodology for modeling and simulation of hybrid systems which is called Hybrid High-level Nets (HYNETS). The new methodology integrates three established modeling approaches into one language. High-level Petri Nets represent the basic discrete framework, differential algebraic equations (DAEs) are used to describe continuous systems behavior and object-oriented concepts improve the expressiveness and compactness of models. The paper explains basic ideas of the language and illustrates its usefulness with a small example.

1 INTRODUCTION

Modeling and simulation of hybrid systems, i. e. systems consisting of a mixture of discrete and continuous components, is a research area that becomes more and more interesting. This is due to the fact that most systems of real world applications are not purely discrete nor purely continuous and often both parts influence each other.

Consider for instance the small example in Figure 1: A level of liquid in a tank changes continuously dependent on the output of a pump and the opening degree of a valve. Two level indicators report to a control unit when a certain level is reached. The control unit reacts on these discrete events by sending instructions for regulating the pump and the valve in such a way that the tank level always stays in between level S_1 and level S_2.

Modeling methodologies which are able to describe these kinds of systems in a convenient way and which allow an efficient analyzation or simulation of models to get information about the systems' behavior are very useful in a wide range of application areas; examples are manufacturing systems, production lines, chemical plants, mechatronic systems, process control and ecological systems.

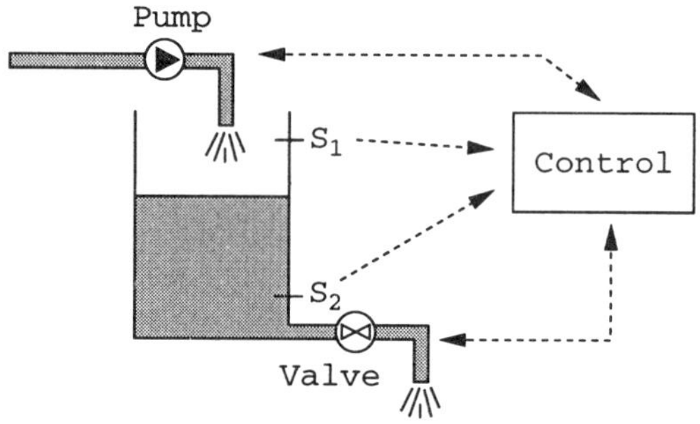

Figure 1: Example of a Hybrid System

This is the reason why many hybrid modeling methodologies have been and will be proposed. Most of the known approaches base upon continuous simulation languages which are extended by components allowing the specification of discrete actions as well. Cellier (1979) was one of the first who introduced a sound methodology for specifying and simulating hybrid systems. By his approach a systems description is divided into a continuous and a discrete part and the interaction is governed by an interface. Similar interface methods followed (cf. Göllü and Varaiya 1989). The main disadvantage of these approaches is the fact that all continuous systems behavior has to be specified completely by one single system of differential equations. A more recent methodology tries to overcome this shortcoming by using object-oriented concepts (Elmqvist et al. 1993).

In this paper I present a hybrid modeling methodology which, in contrast to the approaches mentioned above, is based upon a discrete language, namely high-level Petri nets. Before I explain the basic ideas of this approach in Section 3 and the evolution of hybrid models in Section 4, I give a brief overview of the different classes of Petri nets in the next section. In Section 5 I illustrate the capabilities of the new language. Finally, in Section 6 I summarize the main results and give some ideas on future objectives.

2 PETRI NETS

Petri nets were originally introduced by C. A. Petri (1962). Since then a lot of variants and extensions of Petri nets have been proposed and investigated. See Murata (1989) for a comprehensive introduction.

Due to their good properties in theoretical analysis, practical modeling, and graphical visualization of concurrent discrete systems, they are in use in a wide range of application areas. Especially high-level Petri nets (Jensen and Rozenberg 1991), which lead to more succinct and manageable models, have shown good suitability.

Nevertheless, Petri nets are basically discrete models. They are not inherently qualified for modeling continuous systems. In order to overcome this drawback but still preserve the good properties of Petri nets several continuous extensions have been proposed. David and Alla (1992) and Trivedi and Kulkarni (1993) independently introduced two formal hybrid modeling approaches on the basis of Place/Transition Petri nets. In both cases basic extensions relate to a new kind of real valued tokens which reside on continuous places. These tokens can be modified continuously by other net elements.

Both approaches have already shown good results for investigating small hybrid systems (cf. Petterson and Lennartson 1995). Thus, in order to improve the expressiveness and compactness of hybrid models it is obvious to integrate similar extensions into high-level Petri nets. First ideas from this conclusion have been presented in Wieting and Sonnenschein (1995). A more detailed description of the new approach is presented in Section 3.

A slightly different approach which is also based on a class of high-level Petri nets is presented in (Brielmann 1995). In this approach extended Predicate/Transition Nets serve as a basis and first order differential equations are used to describe continuous systems behavior. But in contrast to my proposal the differential equations are transformed into according difference equations before they are modeled with elements of the discrete Predicate/Transition Nets.

3 HYBRID HIGH-LEVEL NETS

In this section I will introduce the new modeling methodology of **Hy**brid High-level **Nets** (HyNets).

The basic framework of HyNets is represented by a special class of high-level Petri nets called **T**imed **H**ierarchical **O**bject-**R**elated **N**ets (Thorns) (cf. Schöf, Sonnenschein, and Wieting 1995). These nets already offer several reasonable features for modeling complex discrete systems. Two different timing concepts (delay times and firing durations) related to transitions allow the description of many different temporal processes and dependencies. Due to an appropriate hierarchy concept even large models can be structured into clear peaces. Finally, an object-oriented programming language, namely C++, is used to define complex objects and to inscribe transitions with activation conditions and firing actions, etc. In this way even complex discrete systems can be described in a compact and clear way.

In the following sections I explain how the concepts of Thorns can be enhanced to describe the behavior of continuous systems as well. Since most continuous systems can be described by differential algebraic equations (DAEs), it should be possible to specify ordinary first order differential equations in explicit form ($x' = f(x,y,t)$) and algebraic equations ($x = f(y,t)$) within the new language. Since C++ does not provide constructs for the specification of DAEs, this objective has to be achieved by the introduction of an enhanced inscription language.

3.1 Inscription Language

As pointed out above, the originally used inscription language of Thorns had to be enhanced to be able to specify DAEs. For this purpose a new **Hy**brid **o**bject-oriented **La**nguage called HoLa was developed (Majchszak 1996). The roots of HoLa are basically C++. Thus, the syntax is very similar and the basic object-oriented concepts like inheritance, polymorphism and dynamic binding are still present.

In order to distinguish continuously changing state variables from other variables in a HyNet model, HoLa offers the new elementary class `Real` next to standard classes like `Int`, `Double`, etc. Only objects or object attributes of type `Real` can be changed in a continuous way. Hence, only variables of this type may appear on the left hand side of a DAE.

Let us consider the example of Figure 1 again to illustrate, how a user can define a complex object type. Assuming the relevant attributes of the tank are its area A and the level l and density d of the liquid filled into it, a HoLa definition of the tank may look as follows:

```
class Tank {             Double Tank :: v(){
  Float    A;              return( A * l );
  Double   d;            }
  Real     l;            Double Tank :: p(){
  Double   v();            return( d * l );
  Double   p();          }
};
```

The definition shows a complex object type with the attributes mentioned above and two methods $v()$

and $p()$ where $v()$ calculates the liquid volume and $p()$ calculates the pressure on the bottom of the tank. Since the attribute l is of type `Real`, it can be modified by differential equations (see Section 5 for more details).

3.2 HyNet Structure

Following the customary notation for defining Petri Nets, I define the structure of a HyNet as a triple (P, T, A) satisfying the requirements below:

1. P is a final set of *places*.
2. T is a final set of *transitions* partitioned into two disjoint sets of discrete transitions T_D and continuous transitions T_C.
3. $T \cap P = \emptyset$ and $T \cup P \neq \emptyset$.
4. A is a final set of *arcs* partitioned into two disjoint sets of discrete directed arcs A_D and continuous undirected arcs $A_C \subseteq (P \times T_C)$. Furthermore, discrete arcs are partitioned into three disjoint subsets of standard arcs $A_S \subseteq (P \times T_D) \cup (T_D \times P)$, enabling arcs $A_E \subseteq (P \times T)$ and inhibitor arcs $A_I \subseteq (P \times T)$.

The semantics of these net elements is described in Section 4. The graphical representation is shown in Figure 2. As usual, places are represented by circles and transitions by rectangles. In order to distinguish continuous transitions from discrete ones, continuous transitions have two border lines.

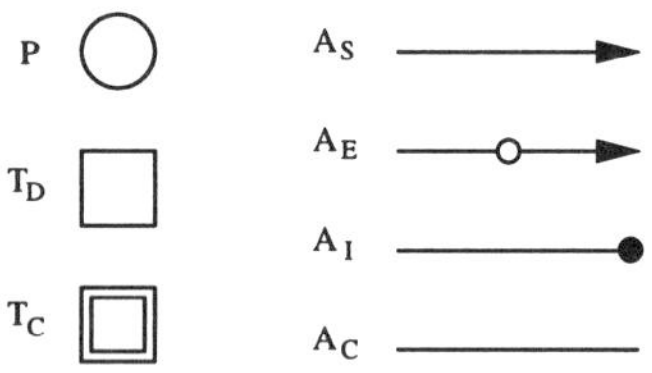

Figure 2: Graphical Representation

Figure 3 shows the allowed connections between places and transitions according to the definition above.

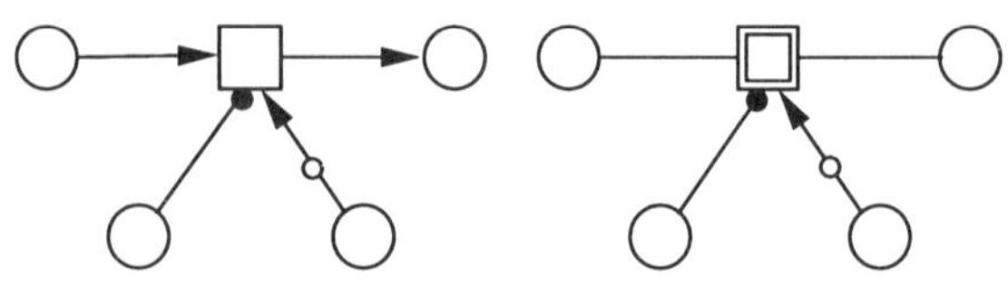

Figure 3: Possible Connections between Places and Discrete (a) and Continuous (b) Transitions

3.3 HyNet Inscriptions

The following inscriptions can be assigned to the net elements of a HyNet.

Places are labeled by a *place type* and a *place capacity*. The type of a place defines the kind of objects which may reside on a place. This type is either an elementary HoLA-class (`Bool`, `Char`, `String`, `Int`, `Long`, `Float`, `Double`, `Real` or `Token`) or a user defined complex HoLA-class e. g. `Tank`. In the case that a place type is defined by a complex class, objects of derived classes are allowed to reside on that place, too.

The place capacity specifies the maximum number of objects held by a place at each moment. It is defined by a positive integer or `Omega` where `Omega` represents infinite capacity.

Arcs have an *arc weight* and a *variable name*. The arc weight of a standard arc and an enabling arc is a positive integer which defines how many object have to reside on the incident place to enable the incident transition. The weight of inhibitor and continuous arcs is one per default.

The variable name of an arc is used to reference objects of a place from a transition. Together with the weight of an arc a set of variables is defined. For example, if an arc is labeled with the weight 2 and the variable name x, two variables $x.at(1)$ and $x.at(2)$ are defined. If the arc weight is one the variable is defined directly by the variable name. Variables may be used by the inscriptions of a transition (see below). Inhibitor arcs don't need a variable name.

Transitions have the most comprehensive inscriptions:

- The *firing capacity* of a transition is a natural positive number or `Omega`. It specifies how often a transition can fire in parallel with itself. Since a continuous transition modifies all objects on surrounding places at each time, it has an infinite firing capacity per default.
- Every transition can be labeled with an *activation condition*. This condition is used to make further requirements to objects enabling that transition. It has to be specified as a boolean HoLA expression without side effects over variables of incoming arcs. In order to avoid name clashes of variable names two arcs incident to a transition must not have the same name.
- Another important inscription of a transition is the *firing action*. In case of a discrete transition the action is specified in form of a sequence of HoLA statements and in case of a continuous transition it is defined by a system of algebraic and first order

differential equations respecting HoLA syntax. In both cases variables of incoming and outgoing arcs can be used, but it has to be assured that only variables of type `Real` appear on the left hand side of equations.

- Finally, discrete transitions have two more inscriptions defining their behavior in time: a *delay time* and a *firing time*. Both inscriptions are specified as HoLA expressions without side effects over variables of incoming arcs. The results obtained by an evaluation of these expressions have to be of type `Double`.

Now, the structure of a HyNet and all possible inscriptions are introduced. In the following I will describe the semantics of the different net elements and their interactions.

4 EVOLUTION OF HYNETS

Evolution of a HyNet means changing the state of a net. This is done by the firing of transitions or more precisely by the firing of occurrence elements. Before I explain which conditions have to be fulfilled to enable transitions and what happens if a transition fires I have to describe the terms of a state and an occurrence element.

An *occurrence element* is a transition combined with a set of objects that is bound to variables of the transition's ingoing arcs. It is denoted as follows: $[t : < v_1 = o_1, \ldots, v_n = o_n >]$ where t is a transition, v_i are variables of ingoing arcs and o_i are objects from preset places bound to the variables.

The *state* of a HyNet is given by the current time of the state, the markings of places and information about firing occurrence elements. The latter is discussed in more detail in Section 4.4.

A marking of a place consists of the set of objects residing on a place. All objects are distinguished by an unambiguous identification number — that's the reason why we can speak of sets of objects instead of multi-sets of objects. Furthermore, objects are labeled with a time-stamp indicating the time of their instantiation. This time-stamp may be greater than the current time of the state if the object will be produced by a discrete transition in the future (see Section 4.1). In this case an object is called invisible. With respect to these additional labels an object is completely described by its value, its identifier and its time-stamp. I omit identifier and time-stamp if they are not necessary in some context.

4.1 Discrete Transitions

Since the definition of HyNets bases upon the definition of Thorns, the activation and firing of discrete transitions is quite similar to that of Thorn transitions (cf. Schöf, Sonnenschein, and Wieting 1995). Thus, I only summarize some essentials for firing discrete transitions and illustrate the firing with a small example.

Figure 4 shows a small net with a discrete transition and four places. The inscription of the transition is given on the right side of the figure. Place $p1$ has the type Int, an infinite capacity and a marking consisting of three objects (identifiers and time-stamps are not shown).

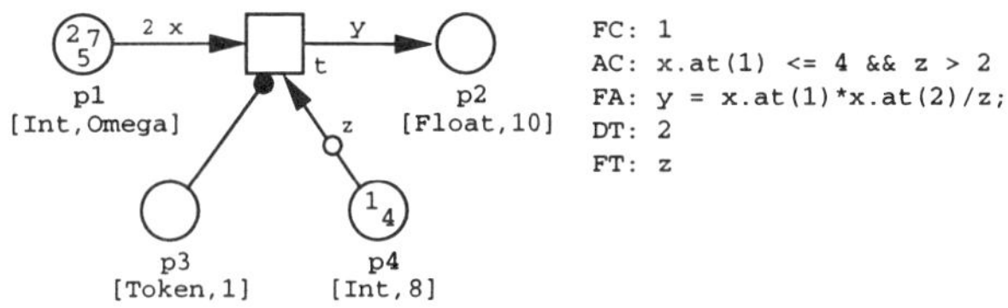

Figure 4: Discrete Transition and some Places

An occurrence element of a discrete transition will be delayed if the *precondition* of the transition is fulfilled. Therefore all places connected by inhibitor arcs have to be empty, every variable of an ingoing arc has to be bound by an object, and the activation condition has to be fulfilled by the bound objects.

Considering the example above, we only have two occurrence elements that fulfill the precondition: $[t : < x_1 = 2, x_2 = 5, z = 4 >]$ and $[t : < x_1 = 2, x_2 = 7, z = 4 >]$. These elements will be delayed.

The delay time of an occurrence element is determined by the according expression of the transition. If an occurrence element has been delayed for the whole delay time without interruption it is called completely delayed. Completely delayed occurrence elements begin to fire as soon as they are enabled. A discrete occurrence element is *enabled* if the precondition is fulfilled, the transition has free firing capacity, and places in the postset of the transition can receive the amount of objects produced by the firing of the transition.

Both occurrence elements of the example are completely delayed after two time units, but only one of them is enabled, because the transition t only has a firing capacity of one. This situation is called a conflict. Conflicts between discrete occurrence elements are solved by a nondeterministic choice — for more details about conflicts see Section 4.3. Let us assume that in this case $[t : < x_1 = 2, x_2 = 5, z = 4 >]$ is chosen.

At the *beginning of a firing* of an occurrence element all objects bound by variables of ingoing standard arcs are consumed from the corresponding preset places. Objects bound via enabling arcs are not consumed — these objects are only read by the transition. The firing capacity is decremented, the firing time is determined by the according expression, the firing action is executed, and produced objects are put on postset places. These objects get an unambiguous identifier and a time-stamp calculated by the current time plus the firing time.

In the example objects 2 and 5 are consumed from place $p1$ and a new object 2.5 is produced on place $p2$. The time-stamp of this object is calculated according to the current time. Notice: although this object is invisible in the current state, it already uses capacity of place $p2$.

At the *end of a firing* of an occurrence element only the firing capacity of the transition has to be incremented again. All produced objects get visible automatically, because the time of state equals the time-stamps of the objects now.

After four time units of firing, transition t of the example could fire again if there would be enough objects residing on place $p1$ then.

4.2 Continuous Transitions

The idea of continuous transitions is to change values of objects residing on places adjacent to the transition continuously. Only objects or object attributes of type `Real` can be modified. I explain the activation condition and the firing rule of a continuous transition on a small example, too.

Figure 5 shows a small net consisting of a continuous transition and four places together with the net inscriptions.

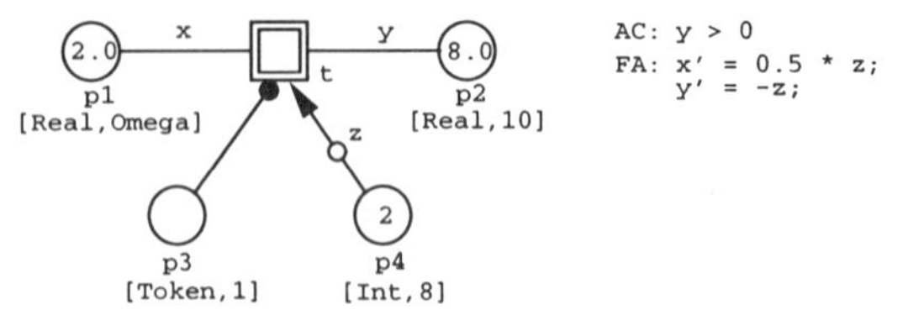

Figure 5: Continuous Transition and some Places

A continuous occurrence element is *enabled* under the same conditions as a discrete one. A precondition is not necessary because continuous transitions are not delayed. Thus, a continuous occurrence element is enabled if all places connected via inhibitor arcs are empty, every variable of an incident arc can be bound by an object, and the activation condition is fulfilled by the bound objects.

In the example of Figure 5 the occurrence element $[\,t : <x = 2.0, y = 8.0, z = 2>\,]$ is enabled.

An enabled continuous occurrence element *fires* as long as it is enabled. Firing of a continuous occurrence element means continuously changing the values of bound objects according to the equations of the transitions firing action — algebraic equations assign values to objects and differential equations change object values. The firing does not influence the object identification numbers and time-stamps, since only object values are changed and no new objects are created. There are only two reasons why a continuous occurrence element becomes disabled:

(1) the activation condition is no longer fulfilled or

(2) a discrete transition consumes an object bound by the occurrence element or produces an object on a place connected via an inhibitor arc.

Considering the example above the continuous occurrence element fires exactly for four time units. Then, the object 8.0 bound to the variable y becomes zero and the activation condition is no longer true. The marking of place $p1$ at this moment is 6.0.

4.3 Conflict Handling

Before I describe the firing rule of a HyNet, I will explain the handling and resolution of conflicts between occurrence elements. I omit reflections on enabling and inhibitor arcs because they are not relevant in this context. There are five basic cases of conflicts:

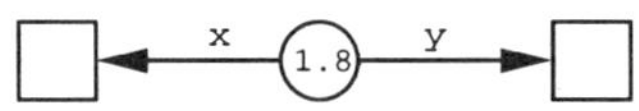

Case 1: Two or more enabled discrete occurrence elements want to consume the same object. This is the classical conflict. It is solved by a nondeterministic choice: only one transition can fire.

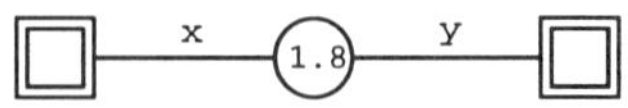

Case 2: Two or more enabled continuous occurrence elements want to change the same object or object attribute, respectively. In this case we have to distinguish three subcases depending on the firing actions of the transitions:

(a) Both transitions want to assign different values to the object by algebraic equations, e. g. $x = 3$ and $y = sin\, t$. This conflict is solved by a nondeterministic choice as in case 1.

(b) One transition wants to set the object value by an algebraic equation, e. g. $x = 3t$, and the other one wants to change the object value by a differential equation, e.g. $y' = 0.5\,y$. Here we give higher priority to algebraic equations.

(c) Both transitions want to change the object value by differential equations, e.g. $x' = 3$ and $y' = sin\ t$. In this case the object will be shared by the transitions. The new object value is defined by the sum of changes. The advantage of this definition is shown with the example of Section 5.

Since subcases (a) and (b) lead to states which are not well defined, a modeler should avoid such situations.

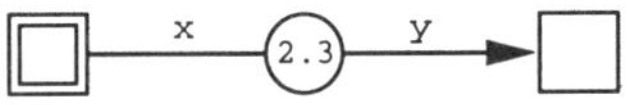

Case 3: An enabled discrete occurrence element wants to consume an object which is part of two or more enabled continuous occurrence elements. Here we give higher priority to the discrete occurrence element, i.e. the discrete transition can fire with the bound object and the according continuous occurrence element becomes disabled.

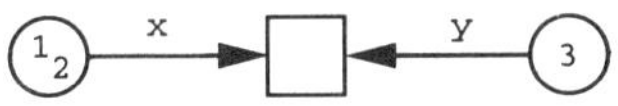

Case 4: A discrete transition is enabled by more than one occurrence element but it has not enough firing capacity to fire all of them at the same time. This internal conflict is solved in the same nondeterministic way as case 1.

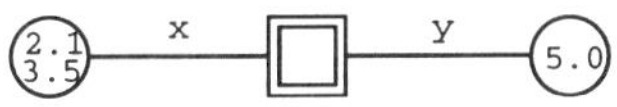

Case 5: A continuous transition is enabled by more than one occurrence element. Since continuous transitions have infinite firing capacity no conflicts result from lacking firing capacity. Here, a conflict can only arise if two or more occurrence elements want to assign different values to the same object by an algebraic equation, e.g. if the firing action is $y = x$ such a situation occurs. This kind of conflicts is solved as in case 2(a). Thus, in the example above, no assumptions can be made on the value of the object bound to y during the firing of the continuous transition (it may be 2.1 or 3.5).

These five cases represent the basic kinds of conflicts. More complex situations can be reduced to basic conflicts. Then, they are solved according to the specified rules.

The idea of sharing objects between continuous occurrence elements and giving discrete transitions higher priority over continuous ones was previously proposed by Le Bail, Alla, and David (1991) for hybrid Petri nets.

4.4 Firing Rule

As already mentioned at the beginning of this section, evolution of a HyNet means changing the state over time. This is done by the firing of discrete and continuous occurrence elements. The firing of a discrete occurrence element can be split into two discrete events, namely the beginning and the end of a firing. At these times, the state is changed by some discrete actions including the consumption and production of objects (cf. Section 4.1). Between such discrete events the state is modified continuously by enabled continuous occurrence elements (cf. Section 4.2). Within time periods where only continuous occurrence elements fire, no objects are consumed or produced, only object values are changed.

If we take the conflict handling of Section 4.3 into account the evolution of a HyNet can be formulated by the following algorithm:

1. Execute all actions belonging to discrete occurrence elements at the current time. This is done by interleaving actions which belong to enabled occurrence elements that start firing and occurrence elements that stop firing at this moment until no such element exists. Notice that the execution of these actions may enable or disable other (conflicting) occurrence elements and that time does not proceed in this step.
2. Determine all enabled continuous occurrence elements and change the values of bound objects according to equations specified by the firing actions until
   - a discrete occurrence element becomes enabled or finishes its firing or
   - a continuous occurrence element becomes enabled or disabled.
3. Continue with 1.

This algorithm implements a firing rule that always fires a maximum conflict free set of occurrence elements.

If in step two no continuous occurrence element is enabled no object values will be changed. Time elapses without any actions until one of the listed events occur. If none of these events will occur and no continuous occurrence element is enabled the algorithm does not proceed. The net is dead.

4.5 Evolution Graph

The evolution of a Petri net is usually represented by a reachability graph or coverability graph respectively (cf. Murata 1989). The nodes of these graphs represent markings and the edges relate to fired transitions responsible for the change of a marking.

A similar term, called evolution graph, was introduced by Le Bail, Alla, and David (1991) for hybrid Petri nets. Here, the nodes of a graph not only represent the marking of a net but also an invariant speed vector for continuous transitions and a vector of reserved marks. The edges between these nodes are labeled with transitions as for ordinary Petri nets.

The ideas of Le Bail et al. can be applied to an evolution graph for HyNets. With respect to the algorithm of Section 4.4 a HyNet evolution graph can be constructed as follows.

Every node of the graph represents a HyNet-state which is completely defined by

- the current time of state,
- the markings of all places including visible and invisible objects,
- a set of firing discrete occurrence elements together with the time specifying the end of their firing, and
- a set of firing continuous transitions.

As long as the firing rule algorithm stays in step two, a state (or a node of the evolution graph) is valid: time elapses and object values are changed continuously by the firing continuous transitions. A state changes to another state

- if the set of firing discrete occurrence elements changes, i. e. if a discrete occurrence element finishes firing or a new one starts firing, or
- if the set of firing continuous occurrence elements changes, i. e. if a continuous occurrence element becomes enabled or disabled.

The event responsible for a change of state can be used as a label of an edge in the evolution graph.

5 EXAMPLE

In this section I illustrate some features of HyNets with a small example. Figure 6 shows the HyNet model of the tank system introduced in Section 1. The elements on the left side describe the continuous behavior of the tank while the elements on the right model the discrete control unit.

The type of the place *Tank* on the left hand side of the figure is defined by the HoLA class presented in Section 3.1. Currently it holds a tank object with an area A of 8.0. The tank can be filled with liquid of density $d = 0.9$. The current liquid level of the tank is $l = 0.0$. The measures for these values have to be set properly according to physical laws.

In a similar way a complex object type `Valve` can be defined. A valve object has an attribute f for its maximum throughput, a real valued attribute o for its opening degree ($0 \leq o \leq 1$) and a method $t()$ which calculates the actual throughput ($f * o$). The place

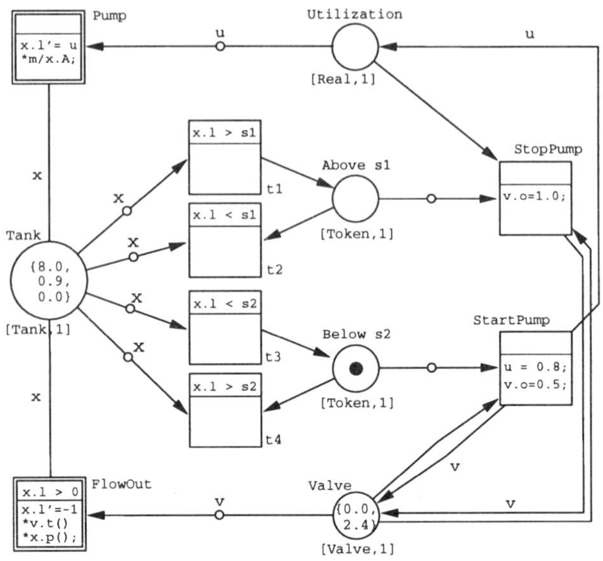

Figure 6: HyNet of the Example of Figure 1

Valve currently holds a closed valve with a maximum throughput of 2.4.

In the depicted situation only the discrete transition *StartPump* is enabled. Because it has a zero valued delay and firing time — like all other discrete transitions in this model — it fires immediately. Thereby the opening degree of the valve is set to 50% and an object 0.8 is produced on the place *Utilization*.

This place describes the power utilization of the pump. If it is marked, the continuous transition *Pump* can fire. Thus, in this state the tank level $x.l$ is changed continuously according to the differential equation $x.l' = u * m/x.A$ specified by the firing action of *Pump* (m describes the maximum output of the pump).

At the same time, the liquid level of the tank rises, the other continuous transition *FlowOut* is enabled. Then it fires continuously according to the equation $x.l' = -v.t() * x.p()$ where $x.p()$ specifies the pressure on the valve (cf. Section 3.1).

In this situation both continuous transitions change the liquid level of the tank concurrently. The level only rises if the output of the pump is larger than the throughput of the valve — let us assume this case.

The next event occurs when the level reaches the value $s2$. The discrete sensor transition $t4$ removes the token from the place *Below* $s2$. This has no further effects. Only when the level reaches $s1$ and the sensor transition $t1$ fires, the pump will be stopped and the valve will be opened to 100% by the firing of transition *StopPump*.

The following evolution of the net is straight forward. The sensor transitions $t1$ to $t4$ report values of the liquid level of the tank to the control transitions *StopPump* or *StartPump* and these transitions regulate the flow into and out of the tank. They do this in such a way, that, after an initialization phase, the liquid level always stays between $s1$ and $s2$.

6 CONCLUSION

In this paper, I have presented a new modeling methodology in which three established modeling approaches are combined.

- High-level Petri Nets represent the basic framework. They have a formal basis with a well-defined semantics and they provide an attractive graphical visualization of system models.
- Object-oriented concepts extend the expressive power of the methodology and lead to more succinct and manageable models.
- Differential algebraic equations allow the specification of continuous system behavior. Therefore many different continuous processes can be described conveniently.

Together, these concepts form a comprehensive and comfortable modeling approach capable of modeling complex hybrid systems in a wide range of application areas.

In the future, I want to show the suitability of HyNets by several larger case studies. Another important objective is the development of efficient simulation strategies which allow an extensive investigation of HyNets models by computer tools.

REFERENCES

Brielmann, M. 1995. Modelling Differential Equations by Basic Information Technology Means. Cadlab Rep. 7/95, Cadlab, Paderborn, Germany.

Cellier, F. E. 1979. Combined Continuous/Discrete System Simulation by Use of Digital Computers: Techniques and Tools. Dissertation, Swiss Federal Institute of Technology Zurich, Switzerland.

David, R., and H. Alla. 1992. *Petri Nets & Grafcet — Tools for modelling discrete event systems.* New York: Prentice Hall.

Elmqvist, H., F. E. Cellier, and M. Otter. 1993. Object-Oriented Modeling of Hybrid Systems. In *Proceedings of the European Simulation Symposium — ESS '93*, 31–41. Delft, The Netherlands.

Göllü, A., and P. Varaiya. 1989. Hybrid Dynamical Systems. In *Proceedings of the 28th Conference on Decision and Control*, 2708–2712. Tampa, Florida.

Jensen, K., and G. Rozenberg. 1991. *High-level Petri Nets — Theory and Application.* Berlin: Springer-Verlag.

Le Bail, B., H. Alla, and R. David. 1991. Hybrid Petri Nets. In *Proceedings of the European Control Conference*, 1472–1477. Grenoble, France.

Majchszak, R. 1996. Eine objektorientierte Beschriftungssprache höherer Petrinetze für hybride Modelle. Master's thesis, Fachbereich Informatik, Universität Oldenburg, Oldenburg, Germany.

Murata, T. 1989. Petri Nets: Properties, Analysis and Application. In *Proceedings of the IEEE* 77 (4): 541–580.

Petri, C. A. 1962. Kommunikation mit Automaten. Schriften des IIM Nr. 2, Institut für Instrumentelle Mathematik, Bonn, Germany. Also, English translation: 1966. Communication with Automata. Technical Report RADC-TR-65-377, Vol. 1, Suppl. 1, Griffiss Air Force Base, New York.

Pettersson, S., and B. Lennartson. 1995. Hybrid Modelling focused on Hybrid Petri Nets. In *Proceedings of the 2nd European Workshop on Real-time and Hybrid Systems.* Grenoble, France.

Schöf, S., M. Sonnenschein, and R. Wieting. 1995. Efficient Simulation of Thor Nets. In *Proceedings of the 16th International Conference on Application and Theory of Petri Nets*, ed. G. De Michelis and M. Diaz, volume 935 of *Lecture Notes in Computer Science*, 412–431. Turin, Italy.

Trivedi, K. S., and V. G. Kulkarni. 1993. FSPNs: Fluid Stochastic Petri Nets. In *Proceedings of the 14th International Conference on Application and Theory of Petri Nets*, ed. M. Ajmone Marsan, volume 691 of *Lecture Notes in Computer Science*, 24–31. Chicago, Illinois.

Wieting, R., and M. Sonnenschein. 1995. Extending High-level Petri Nets for Modeling Hybrid Systems. In *Proceedings of the IMACS Symposium on Systems Analysis and Simulation*, ed. A. Sydow, 259–262. Berlin, Germany.

AUTHOR BIOGRAPHY

RALF WIETING is a research assistant in the Department of Systems Modeling at the Oldenburger Forschungs- und Entwicklungsinstitut für Informatik-Werkzeuge- und Systeme (OFFIS). He received a diploma degree in computer science from the University of Oldenburg in 1992. His research interests include modeling and simulation of hybrid systems using high-level Petri Nets.

Proceedings of the 1996 Winter Simulation Conference
ed. J. M. Charnes, D. J. Morrice, D. T. Brunner, and J. J. Swain

A GENERIC ARCHITECTURE FOR INTELLIGENT INSTRUCTION FOR SIMULATION MODELLING SOFTWARE PACKAGES

Tajudeen A. Atolagbe
Vlatka Hlupic

Department of Computer Science and Information Systems
Brunel University, Uxbridge,
Middlesex UB8 3PH, UNITED KINGDOM

ABSTRACT

This paper describes an architecture for an intelligent interactive instructional simulation modelling environment. It revolves around the production of tutorial and courseware authoring for different simulation software packages with a generic user interface shell. The generic shell is a *"front end"* which provides a uniform graphical user interface to diverse simulation modelling software tutorials. An object oriented perspective is combined with tutorial activities based on the task classification structure to form the interactions between objects. The development environment brings together objects that are functionally coherent and allows them to share common resources within the shell.

1 INTRODUCTION

Organisations adopt different strategies for assessing training needs and ways of implementing them. Because of the increasing use of simulation modelling as an analysis tool and the complexity of developing simulation models, computer based training is a viable option for implementing simulation software training. The provision, maintenance, and continuous updating of courseware, both in academic and industrial establishments, will continue to require huge investments. If an organisation is to sustain its competitive position, full integration of training with business activities, and a framework within which the tutorials needs and strategies for their development are vital.

The Intelligent Simulation Training System (1991), provides a generic architecture for building an intelligent simulation software training system. It uses a generic modelling of simulation scenarios for drilling the learner. The system is domain dependent, and its portability is limited. Its domain knowledge is specifically for simulation modelling. The TOTS system (1987) provides a domain-independent intelligent tutoring shell for task oriented domains and is based on *"procedural network"*. The shell is domain specific and does not allow the user to interact directly with the target software. The EXCALIBUR (1987) produced a generic knowledge base system for learning. The system allows the learner to interact with the tutorial program(s).

Using simulation modelling software involves learning the model development skills, knowing what procedures exist for modelling exercises, and how to accomplish various conceptual model formation. It also involves recalling the procedures whenever a similar situation arises. It can be inferred that using a simulation software package requires analytical and cognitive skills.

The aim of this research is to develop an instructional simulation software environment and provide for the learner to interact directly with various simulation software packages. This method permits the learner to acquire procedural knowledge, which forms the foundation for cognitive skill acquisition (1992). The curricular consists of practical activities and illustrations, which help the learner to develop basic skills and knowledge for using the package.

An object oriented approach that provides a generic, direct manipulation graphical user interface for an interactive instructional simulation modelling development environment is presented. The software provides for cross platform integration and development of interactive tutorials for simulation software packages. The object oriented approach allows functionality of the simulation software package to be mapped directly into a tutorial structure.

The tutorial operations and learners mental model can be used to obtain the systems requirements which forms the basis for initial interface design. The architecture embraces systems for producing, authoring and delivery of tutorials for different simulation software packages with a shared graphical user interface. The shell is capable of seamless integration of audio, text and graphical images, which are used at various levels during instruction. The shell is a graphical user interface suitable for accessing heterogeneous instructional software and it can be utilised for verse delivery of tutorials. It embraces systems for the production and authoring for different simulation software packages with common front end across the packages. The "shell" also incorporates different learning aids, and learner interaction is based on direct manipulation dialogue.

Learning hierarchies proposed by Gagne (1965) can be employed to yield task classification structure. This method allows tutorial tasks to be represented in small modules thereby allowing portability and ease of maintenance. Alderman (1978) suggested that hierarchical structure of subject materials is suitable for a *"rule-by-practice"* learning method. The tutorial curriculum can be classified into hierarchical structure based on its functionality of the simulation software and its operational techniques.

The tool possesses three unique features: unique front end, text inputs which provides for a cross-platform courseware development, and an application software environment. These features provide facilities for courseware authoring and maintenance, and for application software to run independent of the tutorial. It can provide the learner with a content rich tutorial environment with an interactive guide.

2 TEACHING SIMULATION MODELLING SOFTWARE

There is a large gap between the functional behaviour of a simulation software package and the learner's perception of software operations. This misapprehension can be attributed to *"cognitive strain"* which users adopt as strategies designed to minimise cognitive load during problem solving and concept-attainment tasks (Bruner et al., 1956). The difficulty of usage can be exacerbated by the process of models representation and analysis. The functionality of the simulation software can be classified into a hierarchical task structure. The behaviour of simulation software can be depicted by constructing a conceptual model. Teaching operations of the software requires cognitive association of all instructional activities, structured into hierarchical array of sub-activities. We identified the following strategies for effective teaching of simulation modelling software: (1) tutorial tasks organised in a pedagogy structure, (2) minimising the cognitive load on the learner, (3) providing meaningful error messages during development, (4) provision of interactive tutorials, (5) use of animation and gaming-simulation as an alternative teaching strategy and (6) providing context sensitive help. The architecture provides for the development of procedural knowledge, direct mapping of structured curriculum tasks to an application model based on appropriate pedagogical structure.

The functionality of a simulation package and the tutorial operations can be classified into a pedagogy structure. The shell allow for direct mapping of simulation problem into predefined structure of the tutorial. The tutorial operations are represented by a root object, on which a hierarchical decomposed task object depends. The tutorial tasks of the software are arranged in pedagogy order, reflecting stages of building a simulation model and post simulation analysis. The sub-task level consist of diverse models of the software behaviour represented in varying levels of description.

An interactive tutorial that permits the learner to interact with simulated scenarios at certain levels during instruction can be provided within the shell. The architecture also uses animation to display the conceptual model of a scenario, with icons representing different activities. This method allows long-term transfer of skills and reinforces the proficiencies of using the software.

The user interface for a simulation software consists of data and functions objects, and their attributes. The learner activities involve manipulation of these objects. The object-oriented model consists of the following semantic elements: all objects are classified into types defined by their behaviour and inter-relation with other objects and their attributes. The behaviour of the objects can be defined by their operation, and their inherent characteristics.

The system provides the learner with an opportunity to exploit most features the simulation package such as ProModel PC, SIMFACTORY 11.5, WITNESS etc. A structured decomposition of the domain knowledge elements and functions are progressively presented as the learner advances through the curriculum. This technique enhances learner participation in building a simulation model of the problem. The shell also provides for unobtrusive "walk-through" stages of model development. Simulation model of a discrete event systems, passes through different activities and the system clock advances asynchronously. Constructing a simulation model involves depicting the activities and entities of the problems through time. This can be represented and presented during instruction by using a structured decomposition of the operations and taking the learner through stages of developing a simulation model.

3 SYSTEM ARCHITECTURE

An "*object oriented*" technique effectively captures and displays systems requirements during the development phases. The generic "*shell*" is represented by objects depicting various operations. The shell is based on the four layers architecture. These consist of the following: the presentation system, the application and tutorial module layer, knowledge representation and domain expert layer, and the main kernel.

1. The presentation layer consists of the graphical user interface (GUI), which consists of a dialogue system that communicates with all objects in the system. It also provides for direct manipulation of all objects on the GUI and provides feedback to the learner. It provides an effective and transparent interface across the shell.
2. The application module consists of a simulation modelling software package and online documentation provided by the vendor. It can be linked directly to the tutorial system and the courseware authoring environment, as shown in figure 1. The courseware authoring environment provides a platform for tutorial development and for courseware maintenance. It provides for components that can be used for integrating text, graphics, animation and for creating or importing dynamic multimedia objects. This is a hypermedia representation and other media objects can be integrated. A tutorial unit may consists of a combination of graphics, digital video images and animation.
3. The knowledge base contain structured domain knowledge, and simulation modelling problem, plus simulation modelling examples. The structured knowledge is organised into hierarchical task units. The control system is the domain dependent problem solving strategy that determines how the tutorial can be manipulated into the learner's requirements. The knowledge base is linked to a domain "*blackboard*" which contains all the instructional strategy and control of the tutorial. It executes rules which depend on the learners' behaviour during instruction.
4. The kernel consists of inheritance characteristics of all objects and their classes. It executes the codes and provides means for sharing objects and manages the memory. The kernel provide pre-emptive threads, and multiple threads operations. It allows for each tutorial unit to provides a pre-emptive "*threads*" to execute other units.

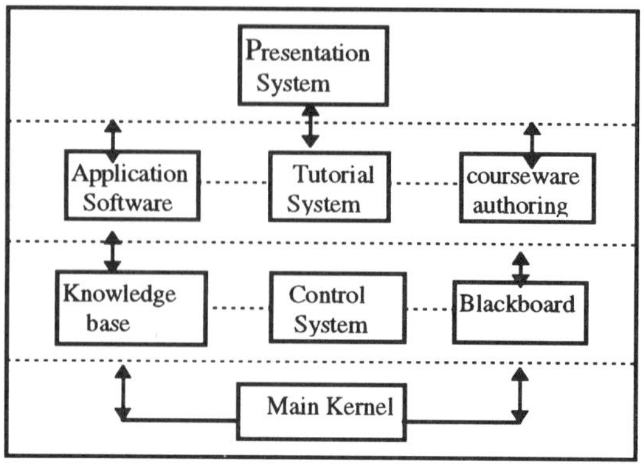

Figure 1: Schematic Representation of the Shell

Each object has defined characteristics which can be embedded into the software. The presentation system provides a uniform representation of the objects. It also provides a mechanism for assessing both the applications and the tutorials modules. The GUI allows the user to communicate directly with the system (1982), whilst the shell consists of a set of routines, which allows the learner to control the instruction. It reflects the general tasks classification structure and semantics for both the tutorials and the application software. The GUI consists of icons that represent objects which the user can manipulate while under instruction. By providing a unique front end, courseware designers can write diverse tutorials for different simulation software packages.

Research findings suggests that acquisition of procedural knowledge can be improved if learner can observe and interact with visual models of the abstract concepts (1982). Shneiderman, (1987) suggested that learners can learn 'quickly' by manipulating the interface object. Tutorial activities depend on the hierarchical structured tasks represented as objects.

The architecture exploits both hypermedia and artificial intelligent methods for courseware authoring. Angelides and Paul (1993), suggested that for an instructional tool to be effective, it should incorporate "*intelligent tutoring*" components. The artificial intelligence components can be embedded within the tutorial environment. The generic tool allows the user to suspend a tutorial unit and interact directly with the simulation software or to seek other method of learning i.e. to choose a demonstration option or observe the animation of the scenario. Some simulation software vendor provided tutorials, for example ProModel PC, permits the learner to interact with a subset of the real application software. These tutorials do not teach the compete functionality of the software. This method may result in a reduction of the learners' skills level development.

The instructional paradigm is based on different techniques of teaching a simulation software package. The tutorial units provides a "*step-by-step*" guide to constructing a simulation model and illustrating the use of all the components of the software. The tutorial objective is for the learner to develop a procedural knowledge for a specific simulation software package. This approach is similar to SOPHIE, I, II, III (1982) which is based on developing the users skills in troubleshooting electronic circuit. The systems are based on expert systems tutoring environment, but offer limited control to the user.

The control system provides for dynamic process communication between the "blackboard" and the event manager, as shown in figure 2. It allows the learner to use their own model and generates simulation models based on their tutorial activities. The event manager controls the operations of the software and the tutorial activities. The control system consists of inference mechanism which controls the structure of the program, and interpreter rules that are embedded in the system. It compares the learner input with the information in its knowledge base and decides which information satisfies the rules. This technique is used for accessing the tutorial network and it limits the search and retrieval time.

The *"kernel"* can execute only one tutorial at a time, but allows user interruption at any time. During interruption it "spawns" supplementary threads to perform other tasks. By separating the graphical user interface from the underlying instructional applications and application software, the shell can be easily transfer to other environments. Other advantages of this approach are versatility, code reusability and ease of updating, Booch (1990).

All the components of the shell send and receive events messages via a protocol. The controller performs this function and co-ordinates the scheduling of processing time between application system and the tutorial system. The protocol determines the order of the events based on priority values set in the program. User interruption triggers events which provides a multitasking environment and allocates the resources.

The blackboard comprises of the student model and a model generator. It can provide for communicating between the two models. The student model provides the learner with different problems based on the user *"overlay"* model. The "overlay" model describes the learner in relation to its interactions with the system and competence level. It maps problem domain into tasks classification structure of the tutorial. It also presents increasingly complex exercises based on the current tutorial unit level.

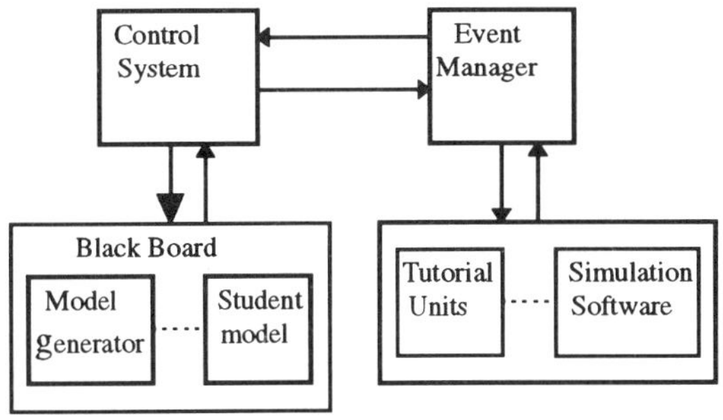

Figure 2: Inter-Process Dialogue

The model generator allows for a hierarchical structured tutorials domain knowledge representation. It provides a structured approach to building a simulation model and consists of combination of the following related phases: (1) identification of simulation concepts and software components, (2) formation of the relationship between concepts and software, (3) hierarchical approach to model representation and mapping problem domain into predefined structure of the tutorial, (4) the output analysis. It first identifies the tutorial needs of the learner and presents the tutorial as represented by rules. It employs the "overlay utilisation rule" to generate simulation model exercises and tutorials.

3.2 Simulation Model

The second layer of the architecture allows for interactive development of simulation models and the representation of the systems components. For a discrete-event simulation model, the state of the *"entities"* change at discrete times. The entity can be represented as objects, characterised by their attributes. An object transforms into different states of activities within an interval. All the objects relates to events, which have dynamic characteristics determined by the user. The *"entities"* advance through different states and behaviour asynchronously and depends on dynamic message routing.

An example of a simulation model of a scenario can be interrupted by the learner at any time. Learner interruption allows for manipulation of the attributes and allows the learner to suspend the current task or perform another operation. The scope and characteristic of the simulation model must be appropriate to the learner's tutorial level and should permit instructional control. Constructing a simulation model of a scenario involves the follows phases: hierarchical representation of the scenario, relating scenario to a classification structure depicting the functionality of the software, and components of the package. Post simulation analysis illustrates output collection, and statistical analysis.

4 CURRICULUM DEVELOPMENT

All simulation software packages possess different characteristics and functionality, they also differ in how entities are represented. Developing curriculum for a simulation modelling package should be a dynamic process set within a context, which unifies the simulation modelling theory and practices. The representation of the domain should consists of the tutorial units with specified internal structures that relate to the theory. It should allow the software to run independently of the tutorial. This representation is similar to that of a Lisp program, that captures the learners behaviour during instructions. It is essential to identify which paradigm will guide the way simulation theory is applied in the tutorial. The difficulties of writing tutorials for a simulation software package without relating to simulation theory is complicated. To minimise these obstructions, the following procedures were formulated:

1. The curriculum is decomposed so that all the contents and instructional strategy reflect the functionality of the software.
2. The curriculum relates to information on simulation modelling theory at appropriate level in the tutorial.
3. Developing models that can be supported by the simulation software
4. An hierarchical representation of building the model, demarcating conceptual representation and statistical analysis.
5. The functionality of the simulation software should be demonstrated in the tutorial.

This representation allows modularity and each composite part is independent of the other. The advantage of this architecture is its simplicity and object oriented nature, plus its usefulness as an integrated courseware authoring tool.

4. 1 Tutorial Units

Discrete event simulation software consist of three major components - library, control shell and report generator, Pidd (1992), and the tutorial units can structured to depict the functionality of the software and these components. The tutorial units can be represented as structured tasks and subtasks, elicited from the tutorial as a tree. The subtask unit portrays the characteristics of the software package and its functionality. Hierarchical task analysis can be applied to organise the operations involved in using the software into a task classification structure, Gagne (1965). The tutorial consists of problem solving operations, with various degrees of complexity within domain. Simulation modelling operations can be decomposed into a coherent sequence. The instructional paradigm is based on the combination of hierarchical task analysis, learner control technique and learner cognitive preference.

Decomposition of the tutorial tasks into structured units can minimise the cognitive complexity of the operations, Gagne (1968). A simulation modelling exercise can be decomposed into smaller subsystems and the curriculum task units can be represented as "tree" in an hierarchical structured network. This consists of several units, arranged in pedagogy array of tasks operations. A unit is represented as a node and it consists of several subtasks which can be depicted as leaves. The tree node can be linked a tutorial units, depicting a component of the software and the leaf nodes correspond to a command in the software. The contents of a node can easily be linked to other child nodes, as shown in figure 2, and addition instructions can be attached as leave at different levels.

The tasks are depicted as graphs with individual units, as nodes and prerequisite relations as links. Instructions are encapsulated on each node. The units are represented by a semantic network and each node represents a tutorial unit, which can be linked to other similar operation. The arcs represents the "*segment*" of tutorial unit.

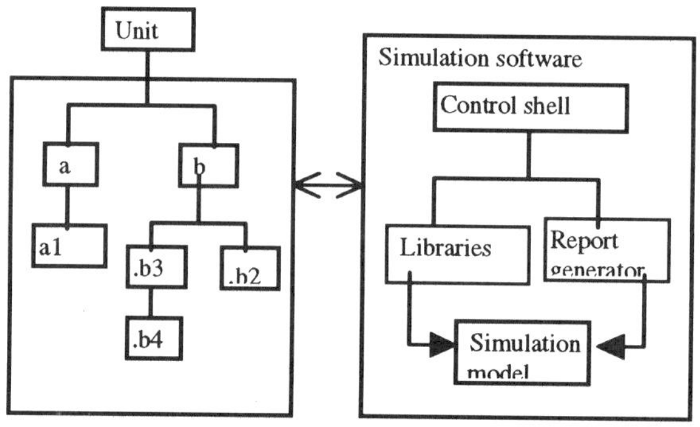

Figure 2: Structured Task Unit and Simulation Software Components

Each node in the structured tasks tree can represent a production rule, which controls the presentation frame. All the nodes possess attributes which indicate the names of up to two or more child nodes that are dependently linked. Navigation through the tutorial depends on the learners path, which can be directly mapped into the task structure.

Each unit is represented as object and every object has one or more activities. All the nodes consists of classes and subclasses. The tutorial can be represented into subtasks levels. These consists of a hierarchical model building process and other operations, which are represented as classes and subclasses in the tasks structure.

The sequence of instruction from a basic model representation to a complex simulation model and analysis employs the "*overlay utilisation rule*". This rule considers the learners' "*current knowledge*" before presenting the tutorial and can provide control. The learner control provides the choice of content, direct manipulation of instruction, priority of actions and tutorial scope. It also provides control for sequence selection, tutorial option and review. It incorporates commands which can allow the learner to navigate through the tutorial freely.

Tutorial knowledge for a simulation software package can be generated from training and user manuals, task classification structure and structured interviews with experienced users, plus case studies of diverse simulation projects. Knowledge is elicited from the tutorials after careful analysis of the task classification structure which is mapped

5 USER INTERFACE OBJECTS

The user interface objects provides an interactive, "*direct manipulation*" of the curriculum objects and all other functional components of the system. The user interface can support unlimited set of semantic messages and provides direct mapping of tutorial knowledge to the functionality of the software, to a representation on the GUI. All the object have a semantic constituent represented as sets of rules that can manipulate the objects on the interface. Some of the rules describe units of lessons referenced by unique name, each unit consists of sets of classes and objects stored as sets. The tutorial task decomposition is utilised as object hierarchies which depicts the tutorial functions and dialogue specification. This can be used to form a specification of the graphical user interface and depict the object oriented hierarchical structure. The GUI consists of data and object functions which can be inherited. Each application comprises of framework objects. The encapsulated object is specified by its behaviour and state. It permits control of data between objects. Some classes were incorporated for the specific needs of the tutorial. They include the following:

- Action-Class is a lower level class, which maintains a list of actions that are displayed to the learner when appropriate. Each window has an action associated with it. This class is responsible for freeing the memory associated with each action.
- The feedback-Class is a component of the intelligent tutoring system. It provides a content sensitive feedback to the learner. This depends on the type of error and the level of the learner in the task classification structure.
- The Navigation-Class helps the learner to keep up with the current position in the presentation program. It also monitors all the windows titles in the presentation. It allows the learner to navigate throughout the program. The navigation system address context, location, and progress indicator.
- The Windows-Class controls the graphical user interface portion of the sub-frame windows in the presentation. It refreshes the user interface when required, and options chosen by the user. The option can also be used to display textual multiple choice questions. Texts that cannot be displayed are scrolled.
- Model-Class controls all simulation model examples presented to the learner. The learner can use this option to interact with the simulation model to change any of the variables or use the instructional gaming system

These classes present a highly interactive choice of principles and methodologies which can be used by the courseware designer.

Polymorphism of a single task to refer to different sub-task and inter-related tasks, that can be satisfied by rules, represented by the knowledge elicitation method. The form of object orientation is as proposed by Booch (1990). All tasks and operations are modelled as objects, closely linked to the structures and behaviour of the package, incorporating single and multiple inheritance. All objects that have common characteristics and behaviours can by grouped together.

5.1 Portability to other platforms

The major factors for portability are modularization of instruction, with code reusability, file transfer, adaptation to different environment, and cross platform integration of simulation software packages. A "*transfer*" command from the courseware authoring option allows the text file and the interface to be transfer to diverse environment. Courseware authors can also import data from text files into the tutorials. Choice of development software is crucial for the development of the tool, as the primary objective is to produce a tool that will enable the transfer of both the interface and the files to a target environments. The only feasible programming languages are

Visual C++ or Visual Smalltalk. Visual C++ allows partial transfer of its code to other platform, but it is complicated and it has a long development cycles. Visual Smalltalk allows easier transfer of codes to major wider area and is easier to learn. Although both languages support object orientation, Visual Smalltalk has a better memory management and supports multiplatform development. Object Orientation methodology is being used in the design of the software but the development environment will be Visual Smalltalk.

6 CONCLUSIONS

The paper describes an architecture for an intelligent instructional system for simulation modelling software packages. It describes an integrated environment for teaching simulation modelling, courseware production and delivery. The curriculum development environment provides a dynamic process set within a context which unites the simulation modelling theory and the functionality of the software. The tutorial development framework allows for flexible courseware authoring and maintenance. The development environment allows the transfer of both the front end and the text file to diverse environments.

The direct manipulation of user interface object allows the tool to be adaptable and can simulate learning by allowing the user to manipulate objects of the system and adapt a preferred learning styles. It allows the learner to interact with the simulation software package whilst under instructional control. The tutorial of a simulation software can be tailored extensively to suit this methodology. The architecture provides for multiplatform utilisation. This can be achieved by structuring the tutorial into tasks units and by separating the tutorials from the rest of the application. The shell can manage cross platform transfer across the various hardware platforms of Smalltalk and can be implemented within a range of curricular.

The object oriented architecture ensures a uniform approach for analysis and design of the system, and consistencies of the system for courseware authoring and maintenance. The framework objects can be encapsulated, and allow control of data between all objects. Its inhabitant characteristics allows changes to be made to the tutorials when the software functionality is improved. The interactive development environment allows for the courseware author to visualise the functionality of the software and its components. This reduces the time required for analysis and development of the courseware and improves the courseware quality. It also helps to standardise the curriculum and improves usability of the system . The architecture may also provide a platform for development and implementation of such systems.

REFERENCES

Alderman, D.L. 1978. Evaluation of the TICCIT *Journal of Computer Assisted Instructional System in the Community College*. Final Report Vol. 1 Educational Testing Services, Princeton, NY

Anderson, J.R, Breuker. D, and Sandberg, J (eds.) 1989. Psychology and Intelligent Tutoring. Artificial Intelligence and Education, *Proceedings of the 4th International Conference of AI and Education, IOS*, Amsterdam,

Anglids M.C., and Ray Paul. 1993. Towards a Framework for Integrating Tutoring Systems and Gaming-Simulation. *In Proceeding of the Winter Simulation Conference, ed.* G.W. Evans, M. Mollaghasemi, E.C. Rusell, and W.E. Biles, 1281-1289. Institute of Electrical Engineers, Los Angeles, California.

Burton, R.R. and Brown, J.S. 1982. An Investigation of Computer Coaching for Informal Learning Activities. In Sleeman, D. and Brown, J.S. (eds.) *Intelligent tutoring systems*, Academic Press, London.

Shneiderman, B. 1987. *Designing the user interface: strategies for effective human computer interaction*, Addison-Wesley, USA

Bruner, J.S., Goodnow, J.J, Austin, G.A. 1956. *A study of thinking*. New York: Wiley.

Dede, C. 1986. A Review and Synthesis of recent Research in Intelligent Computer-Assisted Instruction. *International Journal of Man-Machine* Studies, 24 329-353.

Gagne, R.M. 1965. *Learning hierarchies.* New York: Holt, Rinehart, and Winston.

Booch, G. 1990. *Object oriented design with applications.* Benjamin/Cummings Publishing Company, Inc. USA.

Hartley, J.R and Tait, K. 1986. Learner Control and Educational Advice in Computer based Learning, The study-station concept. *Computers and Education*, 10, (2), 259-265.

Hartley, J.R. 1973. The Design and Evaluation of an Adaptive Teaching System. *International Journal of Man-Machine Studies*, 5, 421-436

Ohlsson, S. 1993. Impact of Cognitive theory on the practice of courseware authoring. *Journal of Computer Assisted Learning*. Vol. 9 No. 4, 194-221

Richard, T., Webb, G.I., and Craske, N. 1987. *Object oriented control for ICAL systems.* Technical Report, T., La Trobe University, Boundoora, Australia.

Richard, T., Webb, G.I., 1987. *Topic structuring in ECCLES: A simplified CAL authoring Methodology.* Technical Report 9, La University, Boundoora, Australia.

Patrick, J. 1992. *Training research and practice,* Academic Press, London.

Sleeman, D. and Brown, J.S. *Intelligent tutoring systems*, London Academic Press

Draman, M. 1991. A generic Architecture for Intelligent Simulation Training Systems, *In Proceeding of the Winter Simulation. Institute of Electrical Engineers.*

Mitchell, P.D. *EDSIM*. 1978. A Classroom in a Computer for Lesson planning Practise. In Meardy, J, (ed.). *Perspectives on Academic Gaming and Simulation*. pp. 191-204. Kogan Page.

Nuron Data. 1990. *Nextpert object user manual.*

Patrick, J. 1992. *Training research and practice,* Academic Press.

Pidd, M, 1992. *Computer simulation in management science,* Third Edition, John Wiley. London.

Poole, T and Szymankiewicz, J. 1977. *Using simulation to solve problems.* McGraw-Hill Book Company (UK) Ltd.

AUTHOR BIOGRAPHIES

TAJUDEEN ATOLAGBE works as Consultant with Microparadigm Research and Development Group. He is currently undergoing a Ph.D. in the interdisciplinary program in Simulation Modelling and Instructional Systems at the Brunel University. Holds an M.Sc. in Business Information Systems and initial background in Systems Training and Development. Current research interests include aspects of instructional systems, Simulation Modelling, HCI and Object Orientation methods.

VLATKA HLUPIC is a Lecturer in Simulation Modelling in the Department of Computer Science at the Brunel University. She holds a B.Sc.(Econ) and an M.Sc. in Information Systems from the University of Zagreb, and a Ph.D. in information Systems at the London Schools of Economics, England. She is researching into, and has published extensively, in simulation modelling software approaches to manufacturing problems. She has practical experience in the manufacturing and waste disposal industries, as well as having held a variety of teaching posts in England and Croatia. her current research interests are in manufacturing simulation, software evaluation and selection, and in simulators and simulation languages.

Military Applications

Proceedings of the 1996 Winter Simulation Conference
ed. J. M. Charnes, D. J. Morrice, D. T. Brunner, and J. J. Swain

THE JUDY THEATER SURVEILLANCE AND STRIKE SIMULATION MODEL

Douglas R. Brouse

Science Applications International Corporation (SAIC)
4001 North Fairfax Drive, Suite 300
Arlington, Virginia 22203, U.S.A.

William A. Cann

Science Applications International Corporation (SAIC)
4001 North Fairfax Drive, Suite 300
Arlington, Virginia 22203, U.S.A.

ABSTRACT

JUDY is a stand-alone constructive architecture-level simulation designed specifically to investigate the potential contribution of current and emerging systems, concepts, and technologies to the prosecution of theater targets, including ground, air, and ballistic. The simulation includes all phases of end-to-end target prosecution: surveillance, cueing, exploitation, command & control, sensor & weapon tasking, fire control, strike, and engagement. JUDY can address various theater mission areas, including:

- wide area ground surveillance/reconnaissance
- surveillance & strike of time critical targets
- theater missile defense (TMD)
- boost phase intercept (BPI)
- ground strike surveillance & planning (ATO support)
- imagery exploitation, target classification & ID
- real-time retargeting of strike packages
- air-breather surveillance & intercept (ADI, CMD)

A key feature of JUDY is the ability to model an integrated multi-mission architecture including the simultaneous operation of TMD, BPI, CMD, and Air Strikes. The JUDY model was designed to support architecture-level trade studies, examine parametric sensitivities, and provide first order insights into issues of system performance and concepts of operations (CONOPS). JUDY is a faster than real-time Monte Carlo discrete event simulation that explicitly models surveillance platforms, strike assets, C4I processes, point-to-point communications, and upwards of 100,000 individual threat vehicles. The individual threat entities operate in a geographic environment that includes: terrain elevation (DTED), road networks, cultural data such as foliage, and a simplified atmosphere. Multi-replication batch run times may range from 5-30 minutes on a Sun Sparc 20.

1 BACKGROUND

Soon after the conclusion of Desert Storm, the Defense Advanced Research Projects Agency (DARPA) established the WarBreaker facility to explore the potential contribution of current and emerging surveillance technologies, command and control concepts, and weapon systems to the prosecution of theater Time Critical Targets (TCTs) using Distributed Interactive Simulation (DIS) and Man-In-The-Loop (MITL) experiments. However, it quickly became clear there was an immediate need for a more traditional constructive simulation tool to support architecture analyses and performance-level trade-offs of WarBreaker technologies and concepts. Such a tool could be used for the conduct of quantitative studies and analyses while also serving the DIS component by helping to filter and define the scenarios and cases of interest for MITL experiments. This need spurred SAIC to design, develop, and employ the JUDY theater surveillance and strike architecture-level simulation model. Since its origin, JUDY has expanded and evolved to become a unique tool within the analysis community, capable of addressing a variety of mission areas.

2 MODEL OVERVIEW

The JUDY model is a Monte Carlo, discrete-event system simulation model of approximately medium fidelity. JUDY's primary focus is on the evaluation of architecture-level performance and CONOPS metrics among the Blue Forces with a mission against the Red forces. The JUDY model is designed as an entity based simulation to assess the performance of small numbers (dozens) of Blue Force surveillance and strike entities against large numbers (thousands) of Red Force threat entities. The Blue surveillance, C4I, and weapon elements are all explicitly modeled as well as the Red threat elements and targets. Functionality within JUDY is modeled at the necessary "performance level" required to evaluate the system and architecture impacts. For output, JUDY collects a wide variety of statistical metrics for evaluating defense performance and effectiveness including sensor utilization, platform utilization, sensor and weapon tasking, weapon expenditures, communications loading, and threat leakage. JUDY is written in FORTRAN 90 and consists of approximately 50,000 lines of code with 400 subroutines. User and technical documentation for JUDY is available (Brouse, Cann, and Kalbaugh 1996).

From the outset, JUDY was designed to be plug and play since alternative architecture and system design was the intended analysis regime. The highest level Blue entity is an element. An element is typed as an aircraft, ground site, aircraft, or satellite. All elements may have sensors, weapons, functions, and communications associated with them. At the extremes, a Blue architecture may consist of a single, centralized, multiple function element or many, distributed, single function elements.

The Red threats are designed to be a single entity but can represent some level of aggregation. The threats move across the terrain in deterministic or random motions. The movement can be on-road or off-road. Fixed ground targets such as bridges, C2 nodes, and IADS sites may also be specified. In addition, Red threats may be defined as aircraft, cruise missiles, or ballistic missiles.

The environment consists of a table top playing surface that can be overlaid with masks to represent cultural data, line-of-sight (LOS) data, road data and weather data. Two modes for LOS can be used: direct DTED computation or a LOS distribution derived from DTED. Road data is provided by Digital Chart of the World (DCW).

These three model segments (elements, threats, environment) comprise the basis of JUDY's modeling features and are described in more detail as follows.

3 BLUE ELEMENT DEFINITION

The element is the basic building block of a surveillance and strike architecture. The element is defined by its component definitions which are the platform type, the procedure (or function) definition, the sensor definition, the weapon definition and the communications definition.

As a platform, an element may be an aircraft, a ground site, or a satellite. Each platform is set to be either active or inactive at the onset of a simulation. In addition, each platform is given a tasking mode. The tasking modes control whether the platform will accept internal or external tasking and whether the platform will change its waypoints to accommodate the tasking and targeting.

The aircraft requires that a home base, takeoff time, flight waypoints and aircraft type be defined. All aircraft will start at the home base, takeoff at a given time, then traverse the waypoints as the simulation progresses. An aircraft's characteristics include speed, rate of turn, endurance, rate of climb, and turn around time. The takeoff time can be randomized to ensure different starting conditions for each replication. When the aircraft has reached the limit of its endurance it will return to base, unless it is currently tasked to perform a strike in which case it will return to base when the strike is complete. The aircraft will also return to base when it has used the last of its weapons.

A fixed ground element will be situated at its location throughout the entire simulation. The satellite element does not move and is geo-stationary over the scenario geography at all times.

The functionality of an element is defined by the procedures it has onboard. Each procedure allows the element to accept certain communications inputs, perform certain functions, and output communications messages to specified destinations. The functions which may be assigned to elements include:

- sense
- process
- detect
- recognition
- correlation
- data fusion
- sensor management
- weapon allocation
- ground target strike
- air target strike
- boost phase intercept strike
- data relay
- fire control radar

All necessary functions to complete an architecture and mission must be mapped to the elements. When using more than one element, the functional flow is defined by prescribing a destination element for each function's outgoing communications messages. An example functional flow for a theater missile defense mission is shown in Figure 1.

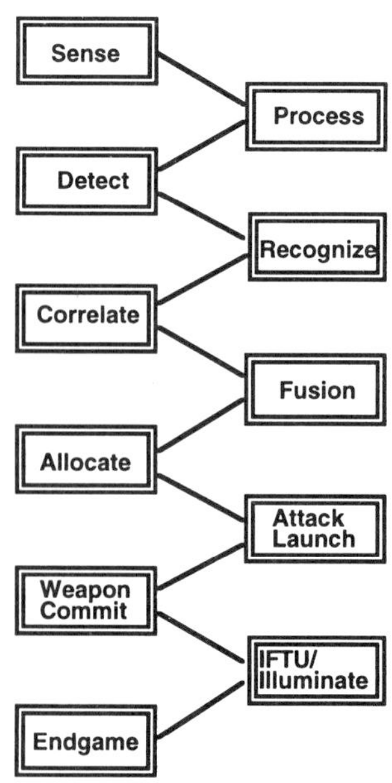

Figure 1: Example TMD Functional Flow

Each function has an inherent time, number of processors, and list of destinations. A function on an element can name its own element as its destination if the next function is onboard the same element.

Each sensor package used onboard an element is described by its sensor characteristics. Each sensor package can have multiple components and each component requires resources to operate. In addition, each sensor package is given a limited number of resources available at any one time. This allows control over the simultaneous operation of multiple sensor components. Sensor characteristics which define sensor operation include field of view constraints (azimuth, elevation, range), resolution, frequency, scan time, beamwidth, minimum detectable velocity, Pd, target location error, and other detailed radar parameters. Sensors which can be modeled in JUDY include MTI, Spot SAR, Strip SAR, ISAR, EO/IR imagery, space-based IR, airborne IR, and air defense surveillance and fire control radars.

Weapons in JUDY can be delivered from ground systems (i.e. SAMs and ATACMS) or air vehicles (i.e. AAMRAM and ground munitions). Weapons delivery is modeled using interceptor flyout fans or a constant velocity. Aircraft delivery systems may be enabled by on-board target acquisition sensor such a forward looking infrared (FLIR). If an aircraft does not immediately acquire the target, the aircraft may perform a search.

Communications within JUDY are point-to-point between elements. The communications capability of an element is defined by its baud rate and the function message sizes. Each element has transmit and receive message queues with maximum message buffer and queue disciplines. Message queues can be either FIFO (First In, First Out) or LIFO (Last In, First Out). Once a transmit or receive buffer limit has been exceeded, messages are lost. . Communications within JUDY may be LOS-constrained or non-LOS constrained

4 RED THREAT DEFINITION

JUDY is capable of modeling a variety of threat target types (ground, air, ballistic). Each threat target type is defined by its motion style, value, LOS type, and terrain following nature. JUDY allows these target motion styles:

- fixed (no motion)
- random motion on roads
- random motion off roads
- TCTs on roads
- TCTs off roads
- deterministic motion off roads
- air breathers (waypoints)
- ballistic missiles (flyout curve)

Target value is used to rank targets during strike allocation. The LOS type specifies whether LOS calculations are done with actual DTED or a statistical interpretation of DTED. Additionally, a target may travel at an absolute altitude or at an altitude above local terrain.

Movement types associated with road networks require additional road input data defined in detail in separate input files. JUDY has been used to model as many as 200,000 individual ground targets moving along a road network.

5 ENVIRONMENT DESCRIPTION

JUDY operates in a geographic environment that includes: terrain elevation, road networks, cultural data such as foliage, and a simplified atmosphere. The terrain elevation can be defined using DTED data or using a cumulative density function (CDF) which is a pre-processed statistical interpretation of minimum grazing angles based on DTED. It is also possible to run JUDY with a bald earth. The road networks are defined by a series of nodes (intersections) and links (connections).

Cultural data consists of a type or code for each grid point in the scenario geographic region. Cultural data codes exist for water, open land, forest or vegetation, and urban areas. In addition to cultural specifications for each grid point, a delimitation flag for each grid point may also be specified to exclude certain region from consideration. The simplified atmosphere includes a fraction or percent of grids affected by weather and an update time for each weather calculation..

6 MODEL OUTPUT

The JUDY model is capable of generating a wide variety of outputs which describe the event history of the simulation. If desired, each key event or activity may be traced throughout the simulation. In addition, JUDY collects and reports a vast array of descriptive statistics which provide measures of effectiveness for architecture performance. The JUDY simulation model generates three primary output files. First, JUDY can record each key event during a replication in a history file. Next, JUDY collects and reports statistics for each Monte Carlo trial or iteration in the replication statistics file. Lastly, JUDY collects and reports statistics over all replications in the grand summary statistics file.

Pertinent output statistics are gathered in all key functional areas such as surveillance, communications, battle management, weapons systems use, and defense effectiveness. The statistics collected include sensor utilization, targets detected and tracked, tasking response time, target revisit rates, target location certainty, area coverage. In addition, the model outputs strike mission effectiveness, targets killed, sorties flown, weapons fired, and threat keepout distance.

7 INTERFACE TO OTHER MODELS

JUDY may be linked through input/output with other associated tools to form an analysis chain suitable for addressing particular theater surveillance and strike issues. RASPUTIN may be employed on the front-end to provide credible threat vehicle deployments and movements. CTEM, which generates optimal strike platform allocations across the target set, may provide a master attack plan for JUDY in order to assess surveillance support to the ATO. JUDY may also be used with a mission planning tool which generates specific pre-planned sensor tasking instructions for associated surveillance assets. In the near future, animation of JUDY model event dynamics will be available.

8 SUPPORTED ANALYSES

The JUDY model was heavily employed as a major analytic component of DARPA's WarBreaker program and it has supported a number of other DARPA surveillance and technology studies. SAIC has used JUDY to explore and evaluate a variety of concepts including off-board targeting, wide area surveillance, and persistent/precision track, boost phase intercept, and integrated theater defense operations. JUDY has also been used to investigate HAE UAV performance issues and candidate CONOPS. JUDY has also been employed, in conjunction with other models, to assess the contribution of HAE in the planning and execution of the ATO process. JUDY has been used to support the Navy Real-time Targeting Study, Joint Precision Strike studies, and various other ADI and CMD analyses. Examples of analyses supported with the JUDY simulation are found in Brouse, Cann, and Kalbaugh (1995).

REFERENCES

Brouse, D. R., W. A. Cann, S. Kalbaugh. May 1996. JUDY Simulation Model User's Reference. SAIC Technical Report, Arlington, Virginia.

Brouse, D. R., W. A. Cann, S. Kalbaugh. March 1995. JUDY Theater Surveillance and Strike Simulation Model. SAIC Technical Briefing, Arlington, Virginia.

AUTHOR BIOGRAPHIES

DOUGLAS R. BROUSE is a Senior Operations Research Analyst with SAIC in Arlington. He holds a B.A. in Mathematics from Mansfield University and an M.S. in Industrial Engineering from Purdue University. Mr. Brouse has been involved in the development of the JUDY model and has employed the model heavily in various analyses. His current responsibilities include modeling and analysis of emerging concepts for theater surveillance and warfare.

WILLIAM A. CANN is a Senior Operations Research Analyst with SAIC in Arlington, Virginia. He holds a B.S. in Industrial Engineering and Operations Research from Virginia Polytechnic Institute and an M.S. in Operations Research from George Washington University. Mr. Cann is the originating author and developer of the JUDY model, its structure, and approach. His current responsibilities include modeling and analysis of emerging concepts for theater surveillance and warfare.

Proceedings of the 1996 Winter Simulation Conference
ed. J. M. Charnes, D. J. Morrice, D. T. Brunner, and J. J. Swain

THE JOINT TRAINING SYSTEM: A PILLAR OF JOINT READINESS AND AN AUTOMATION CHALLENGE

Joseph C. Barto, III
Lieutenant Colonel, United States Army
Joint Warfighting Center
Fort Monroe, VA 23651

Patrick D. Clark
Commander, United States Navy
Joint Warfighting Center
Fort Monroe, VA 23651

ABSTRACT

The Joint Training System (JTS) (Figure 1) defines a multi-step approach to identify training requirements, plan, execute, and assess joint training events. First, and foremost, the JTS is focused on warfighting. "The specific objective is to develop a joint training program that bolsters the combatant commanders' ability to execute the National Military Strategy while simultaneously maintaining high readiness of our forces as a pre-requisite to deterring aggression and responding to crisis. The desired end state is the improved readiness of joint forces, a training and exercise strategy better aligned with the National Military Strategy, improved interoperability, and a more stable process for optimizing the application of scarce resources (dollars, forces, time)." Successful implementation of the JTS depends upon automating the system's processes so that the development, sharing, and reporting of training information and products is conducted as efficiently and effectively as possible.

1 INTRODUCTION

Readiness is the right people, properly equipped, trained as a team capable of fighting and winning our nation's wars. That formula with its fundamental dependent relationships relates the commander's age old problem. "Given my assigned missions how do I best man, equip, and train my forces to successfully accomplish those missions; and if I have done everything I can and still determine that I can not accomplish the mission how do I tell my boss in terms so that he can help me." Consequently; training proficiency, equipment availability, and personnel resources are the three pillars of joint readiness. These

Figure 1: The Joint Training System

three pillars directly affect the creation and maintenance of a trained and ready force able to perform assigned missions.

Under Title X of the United States Code, the combatant commanders (e.g. the Commanders and Chiefs of US European Command, US Atlantic Command, US Central Command, US Southern Command, US Pacific Command, US Transportation Command, US Strategic Command, US Special Operations Command, and US Space Command) are responsible for assessing their own joint training proficiency. The Goldwater Nichols Act directed the Chairman, Joint Chiefs of Staff (CJCS) to develop a uniform system to assess the preparedness of the combatant commanders to carry out assigned missions. The emerging JTS as directed in the Joint Training Master Plan (JTMP) links joint training to joint readiness. The Universal Joint Task List (UJTL), developing Joint Mission Essential Task Lists (JMETLs), preparing task-based joint training plans, executing and evaluating task-based joint training, and assessing training proficiency based upon demonstrated performance establish a uniform system for joint training readiness assessment; and more importantly, a methodology to correct those deficiencies and validate the solutions within the same task-based training system.

Using this system, the combatant command, using JMET derived training objectives for individual training events conduct training evaluations tied directly to those training objectives. The UJTL provides a common framework to describe the full range of tasks which may be accomplished at the Strategic National, Strategic Theater, Operational levels of war and the primary purview of the Joint Force Commander and provides a link to the Service Tactical Task Lists. UJTL derived JMETLs are a comprehensive and authoritative standard with approved tasks, conditions, and standards. Since the JMETL reflects those essential tasks which must be accomplished in order to accomplish the mission, the training proficiency evaluation of those JMETs reflects and provides insights to the command's ability to accomplish the assigned missions. This linkage closes the joint training system cycle--the command's readiness to accomplish its missions.

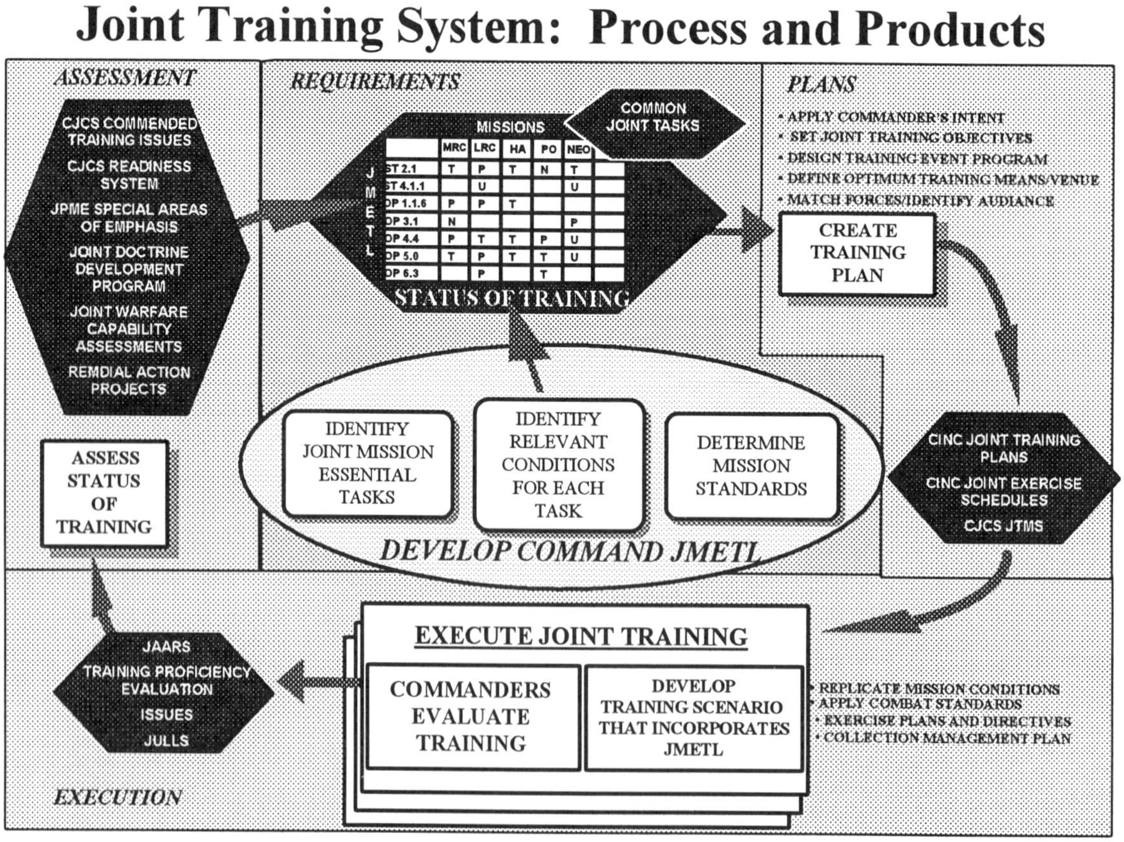

Figure 2. Joint Training System: Process and Products

2 The Joint Training System Processes and Products

There are specific processes and products associated with each phase of the JTS (see Figure 2). The four phases of the JTS are Requirements, Plans, Execution, and Assessment.

2.1 Phase I: Requirements.

The purpose of this phase is to install the methodology and tools to translate strategy to missions to tasks. The mission-to-task JTS revolves around clear statements of joint requirements. This phase uses major inputs including: analysis of operation plan (OPLAN, CONPLAN, functional plan) missions resulting from Joint Strategic Capabilities Plan (JSCP) planning tasks, joint doctrine, and the Universal Joint Task List (UJTL) as an interoperability tool.

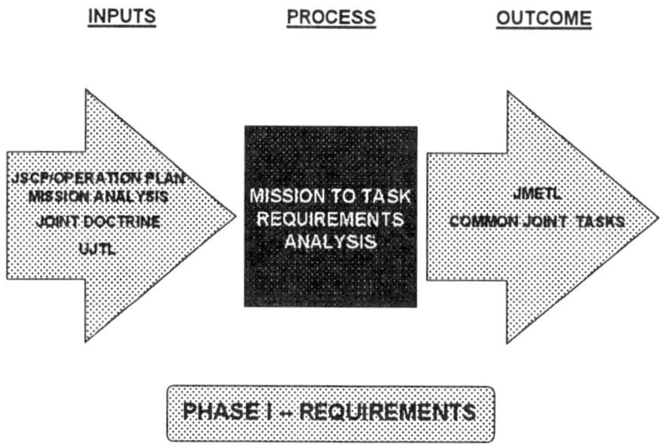

The end products are each combatant command's JMETL, and a multi-command list of common joint tasks.

2.3 Phase II: Planning.

Once command requirements are approved and training readiness assessments are considered, joint training plans and exercise schedules are developed to address the JMETL requirements. The objective is to incrementally install tasks, conditions and standards into joint training and exercise programs. The products of this phase are the CINC Joint Training Plans, CINC Joint Exercise and Training Schedules, and the CJCS Joint Training Master Schedule.

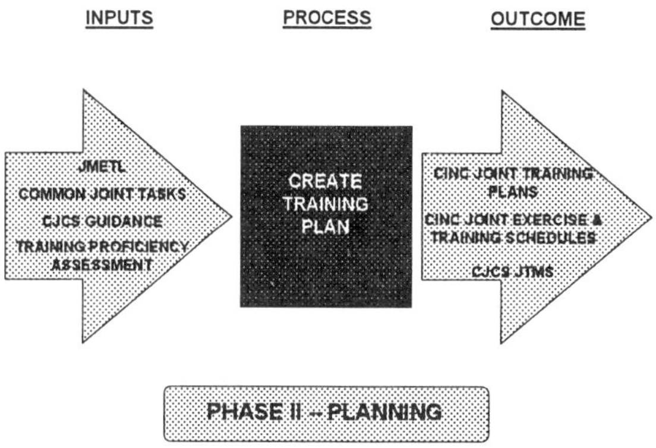

2.4 Phase III: Execution

The actual conduct and evaluation of joint and multinational training events and the support infrastructure to support a joint training event are the focus of the execution phase. Both collective joint training events and individual joint training events contribute to the personnel "train up" process in preparation for joint operations. Within this phase, discrete joint training exercises and events are planned, prepared, executed, and evaluated. Moreover, standardized training development tools and automated products assist trainers in executing JMETL-based training events. As part of the execution phase, commanders are responsible for systematically evaluating each training exercise or event to determine the level of training proficiency attained for each JMET-derived training objective. Evaluation is an internal command responsibility, intended to determine whether specific training objectives were met. In addition to the actual training received, the products of this phase are the Joint After Action Report (JAAR) task proficiency observations, command training proficiency evaluation, Joint Universal Lessons Learned (JULLs) and *Issues*.

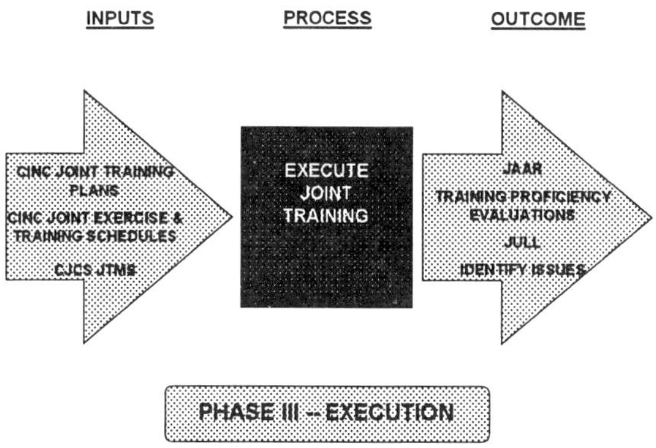

Ultimately, joint training depends on the efficient and effective execution of joint exercises and training events. Most of these activities fall under the CJCS Exercise Program consisting of those activities sponsored by the CJCS and those sponsored by the individual combatant commanders. These two categories can be further

2.4 Phase IV: Assessment.

The final phase of the JTS is the assessment process. While commanders evaluate training proficiency during every training exercise or event, the assessment phase allows the commander to use aggregated results from all training events to judge their commands overall mission capability. Assessments synthesize multiple training event evaluations with the commander's assessment of JMET proficiency. The assessment phase completes the joint training cycle. The products of this assessment serve to inform future training plans or, when high value issues are raised, near term training plans may require adjustment to focus on those critical shortcomings or deficiencies. This snapshot of a command's proficiency in accomplishing JMETs directly reflects the command's ability to perform assigned missions. Therefore, this assessment may be reported out by the command as input to CJCS and combatant command readiness reporting systems. Finally, systemic *Issues* requiring resolution outside the purview of the organization are defined, analyzed, corrected and returned to the joint community for validation.

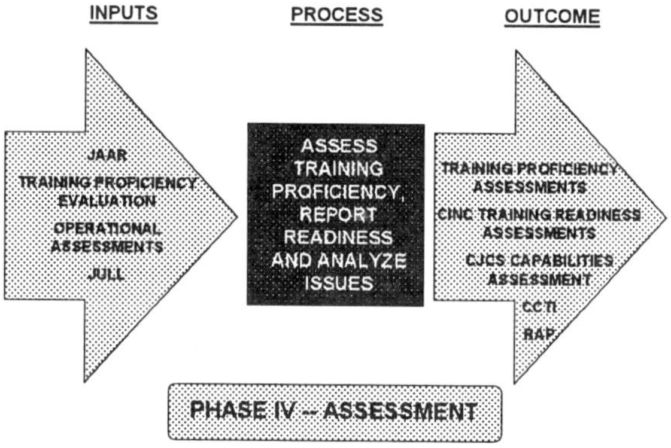

3. Uses of Joint Training Assessment Information.

The results of the commander's assessment dictates one or more of the following corrective actions:

3.1 Adjust the current Joint Training Plan:

If during the evaluation of individual training objectives during joint training events a shortcoming or deficiency is deemed immediately critical to mission accomplishment then the commander may elect to immediately revise training plans correcting that identified task shortcoming or deficiency. Commanders should understand that this decision may have drastic short term impacts on training events that are already well into the planning or preparation stages of the execution phase.

3.2 Inform future Joint Training Plans:

Training proficiency assessments are a key and essential portion of the joint training system. All things being equal commanders should focus their training resources and efforts on JMETL tasks assessed U (Untrained), P (Partially Trained), or N (Not Observed). Obviously, an assessment of T (Trained) in any given task or JMET indicates full capability to perform that JMETL based task, under established conditions, to the desired standard. T (Trained) is the goal. Theoretically, if the command is Trained in all the JMETs associated with a specific mission then the command can report that they are READY to perform that mission based upon demonstrated performance.

3.3 Report out Readiness to perform assigned missions.

Reports outside the command may come in many formats or purviews. All reporting is a request for assistance and not an admission of guilt or performance impotence. One format is the Joint Monthly Readiness Report (JMRR). For the MRC scenarios addressed in the JMRR, a joint training assessment is required to the CJCS as well as resource status in terms of personnel and equipment. This same information may be reported out to other agencies requesting status reports because it is based on demonstrated training proficiency against established standards. The Joint Warfighting Capabilities Assessment (JWCA) and the Remedial Action Projects (RAP) Program are two examples.

3.4 Report Out *Issues*.

Many identified deficiencies are addressed within the commands internal procedural or resource capabilities. However, those deficiencies outside the command's purview to correct are reported outside the command along with recommended corrections. The primary method to communicate deficiencies or shortcomings both within the command and to external agencies is through the Doctrine, Organization, Training, Materiel, Education (DOTME) construct. This analytical methodology allows for a comprehensive analysis to be conducted which considers all aspects of an issue. The result is a complete, coordinated and analyzed product leading the command and/or the joint community to comprehensive issue resolution. For example, if the organizational structure is changed; then the way that organization is employed doctrinally, trained, equipped, and its personnel educated is effected.

4. The Automation Challenge.

Over the next two years the CJCS has directed the full implementation of the JTS. Clearly, the JTS is an information intense system whose efficiency is largely dependent upon the rapid exchange and sharing information. Additionally, the installation of the JTS and its Automation Tools must not be so administratively burdensome that if fails simply from the weight of its complexity and manpower requirements. Finally, the automation effort must account for the needs of the user and focus on the products required and the allow for quick, reliable electronic communications around the world.

As the JTS processes are being validated, the automation effort is beginning under a program entitled the Joint Event Training Tool (JETT). (Figure 3) As the Program Management structure is organized and software integration and development occurs there are several key points to keep in mind.

4.1 Settle on a Program Structure and agree early on specific responsibilities. JETT has a Steering Group, a Configuration Control Board, a Program Manager, and an Executive Agent with specific technical expertise.

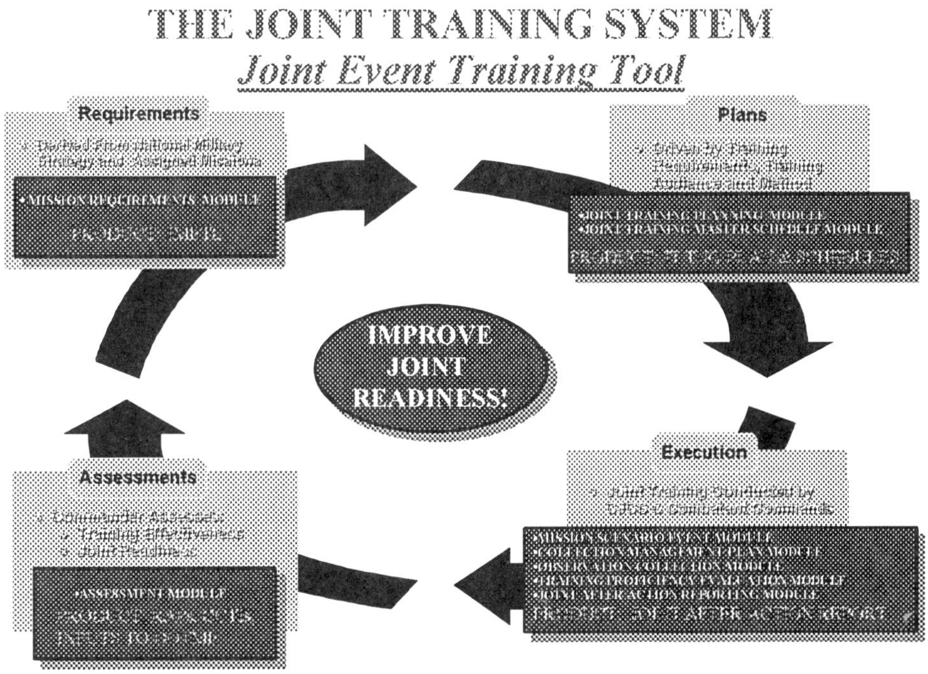

Figure 3: Joint Event Training Tool

4.2 Ensure a thorough understanding of the process before building automation software. The functionality must be fully developed before considering software alternatives. Designate a specific effort to focus, capture and update the functional requirements document. An unclear understanding of the required functionality makes every piece of software look great.

4.3 Agree to a technical approach early and hold the line. DoD has many and varied rules and regulations applying to software and software development. Get an expert and put them on the CCB.

4.4 Constantly stay aware of the user resources required. Education may be the most expensive resource and will dictate the success of the implementation plan. Ensure the entire life cycle is considered to include initial training, distribution, and life cycle costs. Putting some disks in the mail with installation instructions is not a fielding plan.

4.5 Develop an implementation and public relations campaign up front. Bring the customers in early and often and make them part of the developmental team. Create a sense of ownership among your users. Use of existing software that may have been developed locally helps. Do not discount anything a user brings to the table without a full evaluation.

REFERENCES:

Chairman Joint Chiefs of Staff Instruction 3500.01, Joint Training Policy, dtd 21 November 1994

Chairman Joint Chiefs of Staff Instruction 3500.02A Joint Training Master Plan 1998, dtd 8 December 1995

Chairman Joint Chiefs of Staff Instruction 3500.03 Joint Training Manual, dtd 1 June 1996

AUTHOR BIOGRAPHIES

JOSEPH C. BARTO, III is a Lieutenant Colonel in the United States Army currently assigned to the Joint Warfighting Center at Fort Monroe, Virginia. He is holds a B.S. from the United States Military Academy at West Point, NY and a MPA from James Madison University at Harrisonburg, VA.

PATRICK D. CLARK is a commander in the United States Navy currently assigned to the Joint Warfighting Center at Fort Monroe, VA. He holds a B.A. from the University of Wisconsin and a M.A. from the Naval War College at Newport News, RI.

Proceedings of the 1996 Winter Simulation Conference
ed. J. M. Charnes, D. J. Morrice, D. T. Brunner, and J. J. Swain

MODELING AND SIMULATION ENABLING TECHNOLOGIES FOR MILITARY APPLICATIONS

Alex F. Sisti
Steven D. Farr

US Air Force
Rome Laboratory
32 Hangar Road
Rome, NY 13441, U.S.A.

ABSTRACT

While there has been much attention paid lately to the field of Modeling and Simulation as a general tool to be used in specific and well-founded application areas, little has been done to address those broad technology areas that enable such application-oriented simulations to be more easily constructed, run and analyzed. This paper deals with the wide variety of research issues in simulation science being addressed by Government, academia and industry, and their application to the military domain; specifically, to the problems of the Intelligence analyst

1 INTRODUCTION

In defining the Air Force's New Vector for Modeling and Simulation, the Chief of Staff and Secretary of the Air Force stated that "We need to expand our involvement and investment in advanced simulation technologies to improve our readiness and lower our costs today, and prepare us to dominate the battles of tomorrow" (Air Force Laboratory Ad Hoc Working Group 1996). The rapid growth and expanding role of simulation in the Air Force and throughout DoD attest to the tremendous power and potential that this technology provides to nearly every aspect of the defense enterprise. The Air Force, in parallel with the rest of DoD, is making a substantial investment in the development and use of simulation to support improved decision making, training, systems development, and testing.

Historically, modeling and simulation emphasis within the Air Force has been on the end use product, (i.e., using existing simulation technologies to develop simulations which would be suitable for defined applications). However, as our investment base grows, and as we engage in major long term simulation projects, it becomes imperative that we are aware of emerging or developing simulation technologies that may contribute to or affect these or follow-on programs. Further, if there are significant enabling technologies in development which appear to have unique applicability and high payoff for the Air Force, we should track these developments and consider investing in them to ensure their eventual applicability to our needs.

As the Air Force's Center of Excellence for Research and Development in Command, Control, Communications, Computers and Intelligence (C4I), we have a strong commitment to technology transition and technology transfer, and our Modeling and Simulation programs reflect that commitment. While our charter calls for us to apply Modeling and Simulation tools and support to our operational customers, we also feel that the most successful applications should be, and are, rooted in basic and advanced research into simulation science itself. It is our very strong feeling that as an R&D lab, bent on maximizing efficiency and effectiveness, and on avoiding duplication, such a cradle-to-grave approach to systems development and acquisition is essential (Sisti 96). This paper will describe the foundational technologies in simulation science being addressed by Rome Laboratory basic research programs, and some successful applications of that technology toward solving real-world operational problems.

2 ENABLING TECHNOLOGIES IN SIMULATION SCIENCE

There is a tremendous amount of research being performed, both within and outside of the Department of Defense, in advancing and rethinking the theoretical foundations of simulation science, and in "borrowing" the theoretical foundations from other disciplines and reapplying them to traditional simulation. The technological scope of this body of research varies greatly, as does each area's level of maturity. The following sections describe some of the more important enabling technologies, based on the subjective metrics of degree of maturity, Air Force relevance and potential benefit to the simulation community in general. The interested reader may also refer to Appendix B in bibliographical reference (Air Force Laboratory Ad Hoc Working Group 1996) for more detailed writeups on these and other technology areas.

2.1 Hierarchical Simulation Construction

Hierarchical Simulation refers to the notion of a validated analytical hierarchy of models, consisting of four levels of modeling fidelity, ranging from most detailed (engineering level modeling) to most aggregated (theater/campaign level modeling). The basic idea of hierarchical simulation construction is to treat an engineering-level model -- say a radar - as being a "software zoom" of its more coarsely-modeled counterpart in a platform-level model; for example, an early warning radar on an aircraft. This platform-level model, in turn, can be viewed as a "zoom" of a rudimentarily modeled aircraft in a mission-level model; perhaps consisting of several aircraft in an air-to-air combat scenario. Finally, the highest level of the hierarchy is the theater- or campaign-level model, in which air-to-air combat is but one facet of the overall campaign (Sisti 1992).

If the objective of a particular simulation experiment is to assess the capability of an existing (or proposed) piece of equipment, or processing technique, obviously a detailed model of that equipment or technique is needed. Ideally (in order to assess its overall "battle worthiness"), it should be evaluated within the context of a theater-level simulation. Again, ideally (since accuracy is key, especially in a pre-Milestone 0 design alternative study), one would like for the entire system to be at the same level of detail as the engineering-level model of the component of interest. Since a simulation involving thousands of entities, each made up of many components, is not inconceivable or even rare, such a simulation is impossible today; or at least unfeasible. Fortunately, it is not even necessary. There are two alternatives being examined now, under the technical umbrella called "mixed fidelity simulation". The first – and this is probably the most prevalent approach taken by researchers -- is what we're calling "model integration". This involves actually replacing a coarsely-modeled function or entity with a more detailed version; the "software zoom" mentioned earlier. Note that this still involves mapping all the input/output associated with each version so that their interconnections to the rest of the system can be resolved. Also, some degree of Verification, Validation and Accreditation needs to be made to ensure that the codes being swapped are both modeling the same thing, albeit at different degrees of fidelity. "Model aggregation", on the other hand, deals with the idea that the level of detail -- complexity, numbers of inputs/outputs, etc. -- should be fairly consistent from component to component, throughout the theater-level simulation. But then how does that ensure that an individual component of interest -- say the radar mentioned earlier -- is modeled to the accuracy needed for a sensitivity study? The answer lies in the field of "model abstraction", described in more detail in the next section.

2.2 Model Abstraction

Model Abstraction is the intelligent capture of the essence of the behavior of a model without all the details (and therefore runtime complexities) of how that behavior is implemented in code (Frantz 1996). As there is no one "best" solution or approach, it is also one of the most diverse areas. It is as old as modeling itself, in that the intent of the modeler has always been to capture the essence of the behavior of the real-world entity or process, to whatever level of detail (complexity) he/she could afford, or was willing to wait. As a discipline, it aligns closely with, and complements, the concept of mixed fidelity simulation within a hierarchical framework. The basic premise is that the appropriate level of resolution and fidelity should be determined by the end-user requirements posed on the simulation, with the goal being to provide a representation of entities and their behaviors that is sufficiently detailed to support the intended end-use, yet lacks any unnecessary complexities. Avoiding such complexities reduces the developmental and computational requirements associated with the simulation, enables the representational focus to remain end-use motivated, and reduces the false security of "more is better".

Abstraction techniques range from lookup tables, where the entries are the outputs of many simulation runs of the detailed code; to performance curves and response surfaces; to mathematical "metamodeling" (Caughlin 1996), which tries to reduce the behavior of a model to some mathematical equation (itself a model), involving the crucial input set -- those factors to which an output of interest is most sensitive. Traditional study in this area has dealt with reduced order polynomials comprised of these input factors and, while more recent research applies another pervasive enabling technology area -- neural nets -- to represent reduced-order models; yet another involves the use of "rational functions" (Cassandras 1993).

One interesting advance in abstraction research has been the application and adaptation of the concept of "qualitative reasoning"; borrowed from the field of Artificial Intelligence. Qualitative Simulation, as this application is called, concerns itself with getting away from the idea of "exactness"; the mindset of traditional (quantitative) simulationists. Some ancillary topics of research in qualitative simulation are: fuzzy modeling, Random Set Theory, Possibility Theory, Rough Sets and Dempster-Shafer Theory, and ordinal optimization. The common factor is that all strive to represent "intermediate degrees of truth" (uncertainty) in such a way as to attain optimal answers, or ranges of answers, as opposed to an optimum answer to 10-decimal place precision.

2.3 Object-Oriented Simulation

Object Oriented programming, as a more natural and intuitive way of representing and coupling software components, is a well-founded software science paradigm. Extending the application of object-oriented techniques to the modeling (i.e., representation) and simulation of complex entities and systems, such as aircraft, sensors, weapons and C4I systems, would greatly facilitate their creation, insertion and interoperability. An object-oriented simulation enables the simulationist to more naturally represent the various types of relationships among these entities and systems, as well as their relationships with their human operators. The semantics of these relationships are particularly important, transcending well beyond what is currently supported in object-oriented programming languages today. Another area of research addresses the attempts of the simulation community to create object-oriented simulations without a clear engineering approach to building models. The emerging area of "model engineering" is as significant method in simulation science as was "software engineering" in the computer science field. Yet another potential area relating to this paradigm relates to the idea of a Virtual Model Repository, discussed later in this paper; specifically, how this repository can benefit from research pursuits in Object-Oriented database technology.

The Object-Oriented paradigm, as a computer science discipline, is quite mature, dating back over 25 years. However, the application of object-oriented principles to modeling and simulation of complex systems is still relatively new. Model Engineering holds much promise as a disciplined approach to simulation construction, and object-oriented database technology is as important and relevant a research area today as general database technology was in the 60s and 70s.

2.4 Simulation Paradigms

An important area of research in simulation science involves investigating new simulation paradigms. Whereas traditional approaches have been adequate for the body of problem-solving applications for which they've been used, the problems confronted by designers, analysts and decision-makers of today imply the need for new ways of representing the behavior of real-world entities/processes and their interactions, new ways of presenting simulation results and data, and new ways of executing a simulation experiment. The reasons are manifold; however, they can be reduced to two basic areas: in some cases, new paradigms are needed to take full advantage of advances in other technology areas (e.g., increased capabilities of hardware now allow problems of enormous complexity to be run), while in other situations, requirements have arisen due to an increased acceptance of simulation as a tool, and the user community's desire to extend its utility.

Traditional modeling paradigms (discrete-event system modeling, continuous system modeling, Monte Carlo simulations, etc.) evolved from the capabilities of the hardware, software and engineering/mathematical principles of their time. They were essentially children born of a bad marriage; that being the union between the domain expert (who didn't know how to program a computer) and the computer modeler (who didn't necessarily know the domain being modeled). However useful they had been, these paradigms are now giving way to new modeling "styles" that are more intuitive and natural (Object-Oriented Simulation, Qualitative/Fuzzy Simulation, Generative Analysis, Multimodeling/Multi-Faceted Modeling and paradigms based on Petri Nets and Neural Nets), quicker (Parallel/Distributed Simulation and Concurrent Simulation), more accurate (Hierarchical

Simulation/Mixed Fidelity Simulation) and more dynamic and interactive (Web-Based Simulation, System Dynamics modeling, Adaptive/Heuristic Simulation and Man/Hardware-in-the-Loop Simulation).

2.5 Distributed Processing/Distributed Simulation

Distributed Processing, as it relates to modeling and simulation, originally was employed as a way of speeding up simulation run time, by distributing the functions of a simulation among separate processors. This approach -- a conscious and deliberate attempt to break apart an existing piece of code and force a distribution -- has since given way to a new tact; that being the networking and interoperation of already distributed models, simulations and simulators. Starting with the SIMNET project in 1983, and continuing on with the ideas of Distributed Interactive Simulation (DIS) and High-Level Architecture (HLA), distributed simulation is an enabling technology that has only begun to show its tremendous potential in effecting the synthetic battlefield environments envisioned by the highest-level military planners and decision-makers.

2.6 Simulation Support Tools

In the classic sense, simulation has long been viewed as comprising three basic phases: simulation setup, simulation execution and post-run analysis/reporting. There has been considerable effort and money expended on the latter two phases; however, simulation design and setup as a technological discipline has gone virtually ignored by the research community (or more specifically, the research funding). As we note regarding most aspects of technology insertion, there have been some piecemeal and ad hoc improvement; e.g., graphical user interfaces and code-generating front-ends. For the most part, however, there is much to be gained by a concerted effort to apply some of the technology areas that are currently being pursued in research circles in the simulation community. The following sections discuss some of the more prevalent enabling technologies as they relate to designing, configuring and managing a simulation experiment.

2.6.1 Model Management Tools

Much has been written lately regarding the need for a "repository of models". There seems to be a general consensus of opinion that what is envisioned is not so much a physically manifested library, but rather a Virtual Model Repository (VMR); given the requirement to reuse as much legacy (read: validated) code as possible, and the fact that these models are independently owned and maintained, at sites distributed around the country. This Virtual Model Repository would consist of validated simulations, models, model components (at varying levels of resolution), data and inter-model coupling relationships. These pieces, either through conformance to model/data format standards or by judicious adherence to "wrapper" constructs, would be selected, instantiated and appropriately coupled to build the desired simulation; all by the Model Management System. In fact, a Model Management System in its most elemental form, is to a Model Base (our VMR) what a database management system is to a database. It eases, and to some extent automates, the user's ultimate model selection.

2.6.2 Resolution/Validation Management Tools

These tools assume the necessity for the proposition, approval and adoption of a definition of discrete levels of fidelity (the simulation community freely uses the term 'fidelity' interchangeably with 'resolution'); which currently are undefined. Basically, this would be a necessary precursory stage to a model's inclusion in the Virtual Model Repository (VMR). In order for the Model Management System to be of any utility, each component in the VMR must have a specific level of resolution associated with it, and it must have been validated as an accurate representation of the behavior of its real-world counterpart, at the prescribed level of fidelity.

There is a run-time aspect to this technology area as well. Earlier in the paper, we introduced the concept of mixed fidelity simulation as it relates to hierarchical construction. In such a framework, we may wish to construct large-scale simulations in which its entities may be modeled at differing levels of resolution. While that section primarily refers to the general paradigm of composing a simulation from building blocks of different (but appropriate) levels of fidelity, most prototype implementations of hierarchical simulation systems do not allow for dynamically changing resolution of entities, once configured. Dynamic Resolution Management represents an excellent research topic whose time has come. The essence of the concept is that at certain stages of the simulation -- either temporal or spatial -- it makes sense to transition from modeling an entity or process at one level of fidelity to representing it at another. The need for such a transition, called a "software zoom", could be either scripted or, more likely, could be dynamically deduced using a new concept called Significant Event Detection (Air Force Laboratory Ad Hoc Working Group 1996).

2.6.3 Experimental Frame

Closely aligned with Model Management Systems, but broader, is the idea of the Experiment Frame (Zeigler 1990). The vital component of the Simulation setup phase, an Experiment Frame enables a simulation designer or constructor to translate the objectives and issues to be addressed by a modeling effort into conditions under which experiments will be run with a model or real system. This technology would allow the user to completely set up an experiment involving multiple executions of a single simulation with parameters changed for each run, or execution of multiple simulations with varying parameters.

2.6.4 Simulation Design

Simulation Design (Fishwick 1995) is the overarching technology area which includes the concepts of experimental frames, Model Management Systems and other sub-phases of the pre-simulation run stage. As does its construction industry counterpart, it speaks to the notion that one can't jump into construction (in our case, a simulation) without a well thought-out blueprint.

2.6.5 Data Acquisition and Insertion

Arguably the "oldest" new idea, this technology area refers to users'/analysts' desire to imbue their simulations with live, or real-time, input data. Some examples of this are: scanning circuitry schematic drawings as a way of setting up appropriate simulations; scanning photographs, documents and other forms of imagery; or driving existing flight simulations with real (or recorded) flight path/waypoint data.

3 MODELING AND SIMULATION FOR THE INTELLIGENCE ANALYST: THE CONCEPTS ARE APPLIED

The scientific and technical intelligence (S&TI) community provides an excellent audience to introduce these developing technologies for several reasons. Intelligence analysts are typically experienced in the use of automated systems and, quite often, modeling and simulation applications. Given these user characteristics, the simulation development community is free to reduce emphasis on building initial prototypes that are significantly robust. For example, input parameter checking, extensive help utilities, and intuitive GUIs need not be implemented to demonstrate the operational suitability of the underlying simulation science. While these factors are necessary for final user acceptance, their development should not be the subject of scarce basic research and development dollars.

The S&TI organization most familiar to the authors is the Air Force's National Air Intelligence Center (NAIC) located at Wright-Patterson Air Force Base in Ohio. NAIC is responsible for analysis of current and projected foreign aerospace technology trends and capabilities. This broad area of responsibility dictates the need for group analysis and assessment. Consider, for example, the common scenario where the intelligence produced by the office responsible for analysis of a missile seeker is passed to the office with responsibility for the air-to-air missile itself. The flow of information continues to the office charged with analysis of an entire aircraft , and from there on to the office responsible for air-to-air engagement analysis. Each stage of analysis requires a different level of modeling fidelity, ranging from the detailed engineering level to the mission (few on few) level. All this must be performed while maintaining traceability back to the original intelligence source. The application of techniques such as hierarchical modeling and simulation support environments seem quite natural to the process described above.

Another interesting aspect of dealing with NAIC is the organization's utilization of support software. Examples include geographic information systems, relational databases, solid geometry modelers (e.g., CAD), automatic text retrieval systems, and desktop publishing packages. Modeling and simulation applications must eventually be seamlessly integrated with these support applications if they are to achieve any level of user acceptance. The opportunity to observe and understand the entire analysis process is critical in order to make decisions on which simulation technologies are mature enough for operational use.

3.1 Case Study: Modeling of an Integrated Air Defense System

In practice, the utilization of just one of the enabling technologies discussed earlier in this paper will rarely occur. Instead, a marriage of these techniques will be necessary in order to provide a simulation capability suitable for operational use. An example of this is found in an integrated air defense system (IADS) modeling effort currently under development for several US Air Force agencies. For the uninitiated, an air defense system consists of the components necessary to detect, track, identify, engage, and disable incoming airborne adversarial weapons systems. That statement alone should scope the complexity of modeling such a system; however, the problem is compounded when one considers active versus passive detection systems, radar jamming, alternate tracking and fusion algorithms, surface-to-air-missiles versus interceptor aircraft for engagement, and the command and control infrastructure. The approach to modeling these systems involves the application of model integration and abstraction techniques, petri nets, and a host of simulation support tools.

Already in existence are simulations capable of performing detection and engagement. Our challenge was to model the C2 components and to integrate all these codes to provide an end-to-end modeling capability. Existing models selected for our use are the FPG flight path generator, the Advanced Low Altitude Radar Model (ALARM), and the Ground Radar Clutter Estimator (GRACE). The Air-to-Air System Performance Evaluation Model (AASPEM) and the Enhanced Surface to Air Missile Simulation (ESAMS) will eventually be used for air to air engagement modeling and surface to air missile effectiveness modeling, respectively.

Model integration was achieved through the Digital Integrated Modeling Environment (DIME). The DIME is a simulation support environment that facilitates the integration of existing models. Inter-model communication in the DIME is achieved through the concept of a cradle. To integrate a model, the developer must develop a cradle that serves as the interface between the model and the DIME environment. The DIME also provides additional support functions such as report generation, plotting functions, and mapping.

The opportunity to introduce emerging technology was seized in the development of the command and control model. The underlying C2 model simulation is based on a Petri Net engine called Modeler. C2 nodes and data links are represented within Modeler. Detailed algorithms for tracking, fusion, and, eventually, weapon control are performed within a set of submodels. Representation of various C2 node types as objects allows for arbitrary configuration of node combinations and therefore, provides a means to represent any air defense system.

To date, DIME only supports non-reactive modeling (i.e., each model executes sequentially without interaction with other models). In our system, the flight path generation is performed first, ALARM executes once for each flightpath/radar pair, then the C2 model executes. Long term goals are to eventually allow dynamic modeling whereby models (including humans) interact during system execution. To this end, the C2 model was designed so as to not prohibit the man in the loop. An open systems man-in-the-loop version of AASPEM is also under development. Dynamic modeling also offers the chance to apply model abstraction and distributed processing techniques; especially where ALARM is concerned. The independent executions for the flightpath/radar pairings present a clear opportunity to distribute the processing load.

Model abstraction opportunities have been identified as applicable to the ALARM ground based radar model (Frantz 1996). For example, ALARM facilitates an *explicit assumption* concerning components of the returned signal strength. The computational complexity of the model is reduced by explicitly assuming that there is no significant contribution to the returned signal from individually selectable elements of the target such as rotor blades, jamming, clutter or atmospheric attenuation. Without consideration of these factors, the element of the code responsible for their calculation is not invoked. *Causal approximation* offers another abstraction opportunity. For example, a simplified model of standoff jamming could substituted for the more detailed model currently in place, particularly if the model is being used to compare alternative radar designs (rather than predict absolute performance).

4 SUMMARY

This paper endeavored to familiarize the reader with current topics of interest to researchers in the simulation community, and to demonstrate some selected Air Force programs where this research has been applied. We feel that of all the enabling technologies, Model Abstraction is probably the most important enabling technology area in simulation science today. The requirements for mixed fidelity composition, for representing models at varying degrees of resolution and for reducing the complexity of monolithic legacy code are also widely stated by all the services. There is much research under way, using a wide variety of approaches; however, it is a fairly immature discipline, with few actual prototype implementations in place. Nonetheless, the potential benefits of such research to the Air Force and the rest of the DoD, and to the simulation community in general, are enormous and far-reaching.

REFERENCES

Air Force Laboratory Ad Hoc Working Group; "Analysis Of Emerging Enabling Technologies With High Potential To Support The Use Of Modeling And Simulation In The Air Force (Interim Report). 1996.

Cassandras, C. G. 1993. "Discrete Event Systems - Modeling and Performance Analysis", Aksen Associates Publishers.

Caughlin, D. 1996. "Metamodeling Techniques and Applications", Rome Laboratory Technical Report.

Fishwick, P. A.. 1995. "Simulation Model Design and Execution - Building Digital Worlds", Prentice Hall Publishers.

Frantz, F. 1996. "Model Abstraction Techniques", Rome Laboratory Technical Report.

Sisti, A. F. 1992. "Large-Scale Battlefield Simulation Using a Multi-Level Model Integration Methodology", Rome Laboratory Technical Report.

Sisti, A. F., and Trott, K. C. 1996. "SYNBAD: A Distributed Interactive Simulation (DIS) Environment for C3I Capability Assessment", Rome Laboratory Technical Report.

Zeigler, B. P. 1990 "Object-Oriented Simulation with Hierarchical, Modular Models", Academic Press/Harcourt Brace Jovanovich.

AUTHOR BIOGRAPHIES

ALEX SISTI is an Electronics Engineer with the Intelligence Technology Branch, Signals Intelligence Division, the Intelligence and Reconnaissance Directorate at Rome Laboratory, Rome, NY. He has been actively applying his modeling and simulation research interests in a variety of Air Force programs since 1982. He received an A.A.S. degree in Accounting in 1975, a B.S. in Computer Science in 1982 and earned his M.S. in Computer Engineering from Syracuse University in 1989.

STEVE FARR is an Electronics Engineer with the Intelligence Technology Branch, Signals Intelligence Division, the Intelligence and Reconnaissance Directorate at Rome Laboratory, Rome, NY. Steve received his B.S. from the University of Connecticut in 1983, an MBA from Rensealear Polytechnic Institute in 1986 and an M.S. from Syracuse University in 1993. His professional responsibilities include the application of emerging technologies to specific modeling and simulation requirements. He has been actively applying his modeling and simulation research interests to numerous Air Force programs over the last 12 years.

Proceedings of the 1996 Winter Simulation Conference
ed. J. M. Charnes, D. J. Morrice, D. T. Brunner, and J. J. Swain

KNOWLEDGE ACQUISITION IN SUPPORT OF ARTIFICIAL INTELLIGENCE: AN EXAMPLE AND SOME LESSONS LEARNED

Martin S. Kleiner

Logicon RDA
113 Kingsway, Suite 102
Hampton, Virginia 22639, U.S.A.

Scott A. Carey

Logicon RDA
317 S. 2nd Street
Leavenworth, Kansas 66048, U.S.A.

ABSTRACT

The Defense Advanced Research Projects Agency (DARPA) Command Forces (CFOR) program has undertaken the development of intelligent agents to perform the function of command and control within advanced warfighting simulations. While service-related tactics and doctrine publications provide a general description of what a commander must do, they do not address the cognitive processes that are applied during command decision making. Furthermore, existing material is not in a format that lends itself to knowledge engineering. DARPA instituted a focused knowledge acquisition (KA) effort as a component of its Synthetic Theater of War (STOW) program to support the development of intelligent command entities (CEs).

This paper presents the methodology and products that have been created to support the development of Army CEs within the STOW CFOR program. Additionally, it provides some insights and lessons learned that may have applicability across a wide field of KA efforts.

1 BACKGROUND

Logicon RDA is currently under contract with DARPA to provide KA services and subject matter expert (SME) support for the STOW Advanced Concept Technology Demonstration (ACTD). This ACTD has been designated STOW-97 and will be conducted in support of U.S. Atlantic Command during exercise Unified Endeavor 98-1. Core simulation technologies that support STOW are under development by DARPA's Information Systems Office (ISO). These core technologies include Modular Semi-Automated Forces (ModSAF), Intelligent Forces (IFOR), and CFOR for the U.S. Navy, U.S. Air Force, U.S. Marine Corps, and U.S. Army components of a Joint Task Force (JTF). ModSAF provides platform-level representations of weapons systems and small units whose behavior is controlled through the use of rule-based tasks organized in task frames or by an intelligent agent, interacting with the vehicle-level ModSAF simulation through an entity control interface.

CE representations are being developed as part of the CFOR program. CEs exercise command and control over subordinates, which may be manned simulators, virtual simulations such as ModSAF and IFOR, or, in some cases, real manned combat vehicles. The first CEs will be the Army's heavy company team (Co/Tm) commander, and the attack helicopter company commander, as well as a limited battalion command capability.

A detailed description of the Army capabilities planned for STOW may be found in Feldmann (1995). The CFOR concept and technical reference model are described in Dahman, et al. (1994).

2 COMMAND FORCES COMMAND ENTITIES

The goal of the user model is to represent the discrete decision-making processes that a specific functional commander and, where appropriate, his staff use to plan and execute combat operations. In order to be effective over a wide range of applications, the model must be soundly based on doctrinal principles and take into account the following considerations:

- Command decision making involves both innovative and deductive reasoning (commonly referred to as the art and science of warfare).
- Command decision making is always highly contextual. This requires a clear definition of the variables that set the context (e.g., the construct of Mission, Enemy, Terrain/Weather, Troops and Time Available (METT-T)) and a method for weighting these variables.
- The model should incorporate human factors such as levels of aggressiveness and willingness to take risks.
- The model should accept the inputs and produce its outputs in the same form and format as the real-world entity that it is emulating.

The CE must be capable of both planning and executing an operation, in the sense that as the reality of an ongoing operation diverges from the plan the CE must recognize this and adapt its unit's behavior to deal with the situation. To date, several developers have demonstrated effective CEs, at a more than rudimentary level, within varied tactical situations.

3 DOMAIN DESCRIPTION

In order to fully understand the scope and intricacies required of the CFOR CEs it is first important to understand the real-world entities that they must emulate and the environment in which they must operate.

3.1 Organization

The Army is a hierarchical organization whose basic building blocks are individual soldiers assembled and organized into small groups. Each group has specific missions, functions and lines of authority. The most elemental group is the section or squad. In the infantry, the squad is the lowest-level element that can conduct the collective task of fire and maneuver. In units possessing major weapons systems, such as tanks, infantry fighting vehicles, or artillery pieces, the crews of one or more weapons system constitutes a section. Multiples of these elements (usually two to five) are then organized into platoons and then into company-level units (sometimes designated as batteries or troops depending on their branch or mission).

The full authority and responsibility of command and control decision making reside at the company level. This is the first level where the commander has complete responsibility for planning, organizing and executing the full range of tactical missions. The company commander is also responsible for the logistical, administrative and personnel support of the unit and all attached elements. Although the term *commander* is sometimes used at lower levels, such as tank commander, they have a much narrower (although still great) responsibility for fighting their weapons system and directing their crews.

Continuing up the organizational structure, a number of companies are organized into a battalion, with the battalion commander exercising authority over all of the company commanders. This is the first level where the commander is supported by an organized and assigned staff. A number of battalions constitute a brigade, with a brigade commander and staff, and thus up through division, corps, and Army levels.

3.2 Command Authority and Responsibility

While the Army is organized from the bottom up, its missions, command authority and command responsibility are delegated from the JTF commander and flow down to the company level. The company level is where the majority of actual fighting takes place during wartime.

A unique aspect of ground combat forces is that at each sequential level of command, down to the company, the commander is given a mission, a set of assets and a three-dimensional block of terrain. He is then given full responsibility and authority for this terrain, within which he is to employ his assets and accomplish the mission.

3.3 The Decision-Making Environment

Along with the commander's authority and responsibility comes a degree of autonomy that is unique. In some respects this autonomy is complete, as in the fact that no one, including his higher headquarters, can fire into, enter or otherwise effect change within the block of terrain, without explicit coordination and permission. In other respects, his autonomy is restricted, bounded and influenced by outside factors:

- The commander must plan and execute his actions in accordance with the mission guidance from higher headquarters (i.e., he may not retreat if his mission is to attack).
- The commander is expected to follow the commonly held principles of war, doctrine, tactics, techniques and procedures (he does not rush hastily ahead nor fall behind in an attack, thereby exposing his, or an adjacent unit's, flank).
- The commander is influenced by the enemy's actions (he can not execute a pursuit if the enemy executes an effective defense).
- The assets and terrain assigned to the commander enable or preclude certain actions (e.g., without bridging support he must find and use natural fording points on streams, thus limiting mobility and avenues of approach).

For planning purposes these influences have been collected and ordered into the factors of: Mission, Enemy, Troops, Terrain and Time, referred to as METT-T.

The company commander also possesses a higher-order knowledge that results from his training, experiences, and knowledge of his commanders and subordinates. This knowledge and the factors of METT-T form the basis for the commander's decision making.

As an analogy, the JTF HQ can be viewed as a force-generating element (and, in fact, force generation is a function that is attributed to high-level headquarters). Each of the intervening headquarters can be viewed, again simplistically, as a reduction and distribution gear, applying this force down to and across the battlefield. In the end, the force is applied by the weapons systems, both human and physical, that belong to the companies. To make use of this analogy though, one must realize that force is not applied evenly nor equally across the battlefield. The principles of war dictate that a commander both "Concentrate combat power at the decisive place and time" (mass) and "Allocate minimum essential combat power to secondary efforts" (economy). It is here that the cognitive capabilities of the intervening commanders, and their staffs, come into being. It is their knowledge of the higher commander's intent, the mission, the situation and the enemy (METT-T) that allows them to alter the "ratios" and paths of the individual gears, thus applying the maximum force at the opportune time and ensuring decisive victory.

Finally, it must be noted that the decision-making process is continuous and iterative. When a company commander receives a mission it is likely that he has already formulated a general concept or plan, based on a warning order or some other prior knowledge. As he receives more information, he considers various courses of action, choosing some and eliminating others. He goes through this process for each phase of his mission. He then formulates his decisions into a plan and issues it as an order to his platoons and attached support elements. As he executes the plan he reviews each decision and the factors that now exist in reality (METT-T). If the actual factors vary significantly from the planning factors that caused him to select a given course of action, he may decide to modify or choose a different course of action. If, during the planning phase, the commander has considered and analyzed the full realm of possibilities, this real-time tactical decision making becomes the substitution of a more appropriate (and previously devised) course of action to meet the circumstances of reality. While seemingly reactive, it is the result of proactive prior planning.

One critical note on command decision making, especially in combat, is that it must be conducted within the elements that Carl von Clausewitz refers to as the "fog and friction of war." Translated into modern problem solving terminology, this means that the commander is always dealing with the following issues:

- An incomplete and changing problem specification
- A constant requirement for situation awareness that is never fully satisfied
- The feedback of effects on both enemy and friendly forces, neither of which is entirely true nor complete
- The issue of a rapid tempo and fleeting opportunities, requiring that decisions must be made before complete decision criteria are available

In summary, the issue is succinctly stated in the following quotation by Helmuth von Moltke:

> The problem is to grasp, in innumerable special cases, the actual situation which is covered by the mist of uncertainty, to appraise the facts correctly and to guess the unknown elements, to reach a decision quickly and then to carry it out forcefully and relentlessly.

4 DEVELOPMENT OF A GENERALIZED UNIVERSAL KNOWLEDGE BASE

One of the basic precepts of the CFOR program is that there will be many agencies utilizing various approaches to developing the CEs. In some cases different agencies will develop representations of the same CE, and in other cases they will develop complementary CEs that will be required to interact with each other. Since all of the CEs will be required to work within an interactive environment that includes humans as well as constructive and virtual simulations, this demands that the CEs be based on a universal and doctrinally recognizable knowledge base.

To ensure doctrinal soundness, inherent validity, and a high degree of user acceptance, a decision was made to develop the knowledge base within the construct of the Army Training and Evaluation Program (ARTEP). The ARTEP, by type and echelon of unit, defines the missions, the collective tasks, the conditions and standards that Army tactical units are expected to perform and achieve. At the interface between each echelon the tasks of the lower element, from individual soldier to brigade level, are subsumed into the missions and tasks of the higher unit.

Utilizing the appropriate ARTEP Mission Training Plan (MTP) for the units to be modeled, the critical warfighting tasks are extracted and a discreet decision module is built around each of these collective tasks. This provides a set of clearly defined, doctrinally based, mission-related primary decision elements. Each module is produced in a standardized format and consists of:

- The decisions to be made
- The possible paths or interactions of these decisions
- The variables involved in the decision as defined by a doctrinally based set of factors; i.e., METT-T
- A list of inputs required to make the decision, as well as the sources for those inputs
- The specific outputs required from each decision, including the required form and format (i.e. orders, plans, messages, etc.)

Additionally, each module includes detailed graphics depicting the execution of the task in order to provide the knowledge engineers with a contextual visualization and understanding of the task.

While the individual modules are the fundamental building blocks of the command decision process, alone they fail to address the key elements of the higher-order process that occurs at the company level and above. Key elements of this higher, second-order decision-making process are:

- Decisions for one task are interrelated with and must be made in concert with the other tasks and decisions that constitute a mission, operation or campaign
- Collectively, as well as individually, the tasks must meet

the criteria of:
- Suitability (they accomplish the mission)
- Feasibility (the resources are realistically available)
- Acceptability (the pain is worth the gain)
- Individual task decisions must be part of a cohesive process that is conducted within and defined by the specific factors of METT-T

In order to accommodate the requirement to make these decisions on an interactive, mission-related basis a generic, but representative, tactical mission is developed and amplified from the planning phase through its execution. This higher, second-order decision process describes the interrelationships, interactions and dependencies of the subtasks and decisions within the context of an overall mission. This second-order process normally represents a mission that begins with a planning process and proceeds through preparation, execution (to include adaptation to changes), reconstitution, and transition into a subsequent mission. In keeping with the goal of adhering to a real-world form and function, this integrated event has been designated as a virtual field training exercise (vFTX) and mirrors the field Army's use of an FTX as a capstone training and evaluation tool for each echelon of command.

The first series of task modules and vFTX addressed the attack mission. These products were accompanied by a series of additional papers, designed to provide the computer scientist and software engineer with necessary background and contextual information. The papers provided a broad overview of the U.S. Army and then focused on the decision-making environment of the company team commander. They include information on:
- Task organization
- Threat tactics, doctrine and organization
- Terrain and weather effects
- Operational terms and graphics

These initial deliveries were followed by additional task modules and vFTXs covering defensive operations, company logistics operations, and the missions of delay, screen, and guard. The concept for developing the CEs is that they will initially be able to conduct a full, but very basic, mission from beginning to end. Their capabilities are then expanded in terms of both types of missions and the complexities of the missions.

4.1 Support for the Development Process

From the outset it was known that additional information and levels of detail would be required in order to translate the initial knowledge base into functioning CEs. At the initiation of each development agency's effort they are briefed on the concept and contents of the knowledge base and encouraged to initiate a dialogue with the SMEs. Questions received from the developers generally fall into two categories, either approach specific or general to the domain.

The first are questions that are directly related to or generated by the particular approach that a developer is taking. These often require the creation of products that are tailored to the developer and might be useful to only their approach. In some instances, however, the development of these products points out an area that represents a general information deficit. In cases where the issue of proprietary information arises, the issue is referred to the MITRE program integration office and, if necessary, a generic product is produced and distributed to all of the developers.

The second case involves questions that are common amongst all of the developers and arise out of vagueness or a lack of detail in the general knowledge base. More often than not, these points revolve around the most complex issues that represent the more advanced human behaviors such as course of action (COA) generation and bounding. In this particular case a tutorial was developed and presented to the developers and the program integration office collectively. While there was some concern that this tutorial did not reduce the COA generation process to a series of flowcharts and decision matrices, it did involve the developers in the very human process of developing and bounding COAs in a constrained environment. The end result was that the developers gained a much greater appreciation for the process and were able to implement it with significant success.

As the CEs continue to be developed in both breadth and complexity of mission, a continuous dialogue has developed between the developers and the SMEs. This aspect not only provides the developers with additional information but also expands the technical understanding of the supporting SMEs.

4.2 Testing and Evaluation of Command Entities

Once the development of the CEs was initiated, it was necessary to develop a method for evaluating their tactical behavior and performance in a realistic manner. Again, the real-world method was employed by adapting the evaluation portion of the ARTEP to measure the company teams' performance. Utilizing the real-world training and evaluation paradigm as a guide, it was understood that the virtual CEs would progress from a limited capability to a broader and more complex capability, mirroring their human counterparts. In order to accommodate this growth, a progressive training and evaluation regime was developed. At an early stage, a series of products was developed that laid out to the developers the performance criteria that was expected of their CEs. These products included a matrix, as shown in figure 1, and an ordered task listing:
- The matrix defines the possible universe within which the virtual company team commander would be expected to be tested. The variables consisted of the

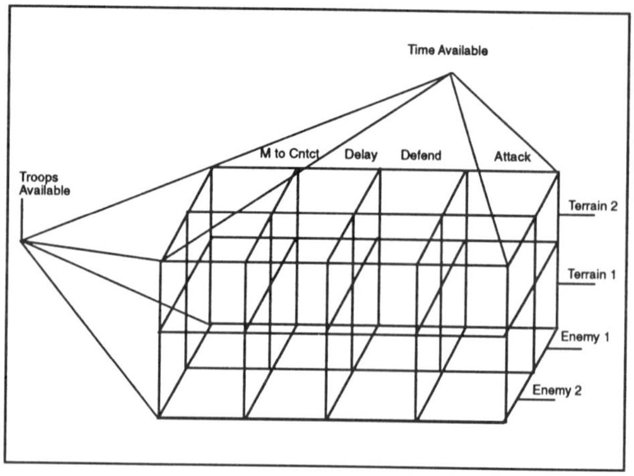

Figure 1: Command Entity Test

standard categories of METT-T, which were then broken down into clearly distinct sets with one to four variations of each factor.

- Mission
  - Attack
  - Defend
  - Delay
  - Movement to contact
- Enemy
  - Type 1 force ratios as expected for the type of operation
  - Type 2 force ratios either greater or lesser than expected
- Troops available
  - A general variable ranging from pure armor or mechanized infantry to a mixture including scouts and mortars. The number of platoons will range from two to six.
  - Terrain
- Time available
- A general variable that will fall within the practical range that would be encountered, but that might force a commander to choose between certain options within a mission, such as routes of march or the emplacement of various types or numbers of obstacles.
- A listing of tasks that is designed to accommodate growth from a very primary capability to a more advanced and adaptive level of competence. The tasks were placed in three categories by mission type:
  - Category 1. A discretionary selection of tasks within a mission type that constitutes an elementary vFTX within that mission. The ability to accomplish a vFTX consisting of these tasks within each mission type is necessary in order to constitute a rudimentary virtual Co/Tm commander.
  - Category 2. The compliment of tasks within a mission type that, combined with category 1 tasks and a wider range of the other variables, demonstrates an increased ability to properly react to unplanned for circumstances and constitute an "apprentice" virtual Co/Tm commander.
  - Category 3. The full range of tasks within a mission type constitutes a competent virtual Co/Tm commander.

The numbering of these tasks in no way constitutes a rank ordering of their importance to a commander, either virtual or human. This breakdown of tasks is simply a way of accommodating the development, growth and fair testing of candidate software.

4.3 Mission-to-Task Decomposition Methodology

Although the ARTEP prescribes a listing of all applicable tasks by mission, it does not present any distinct methodology for determining which tasks are required to accomplish any specific mission under a set of conditions. To date, the development of mission-related task strings for a given set of METT-T has been accomplished by highly experienced professionals and has fallen primarily within the purview of the "art of war". When the devolution process to lower units and the integration of supporting BOSs is undertaken, this process becomes very complex but is an absolutely critical component of commanding a unit.

To overcome this deficiency, an effort was launched to develop a clear and concise mission-to-task decomposition methodology. The goal of this effort was to provide the developers with a tool that could apply the variables at hand and consequently extract the specific set of tasks required to accomplish any given mission. While almost any task can occur in any given mission the ones that are actually required are generated by the factors of ETT-T as associated with the assigned mission. Although the tasks often, and sometimes must, occur sequentially, they are not necessarily mutually exclusive and many may occur concurrently as well as in varied sequences.

As work on the methodology progressed, it soon became apparent that there was some element of phasing that lay under and, along with the specific factors of ETT-T, influenced the selection and ordering of the tasks. Initial efforts were made at utilizing a geographically based phasing method. While this method proved to be effective in attack operations, which normally involve movement and an orientation on terrain or an enemy location, it proved unsuitable for application to defensive operations. Subsequently, a temporal-based phasing method was developed, but this too proved unsatisfactory when applied across all mission areas. At this point it became obvious that the phasing should not be coupled with one of the primary variables in the METT-T equation, but should rather be associated with general processes that occur within all missions.

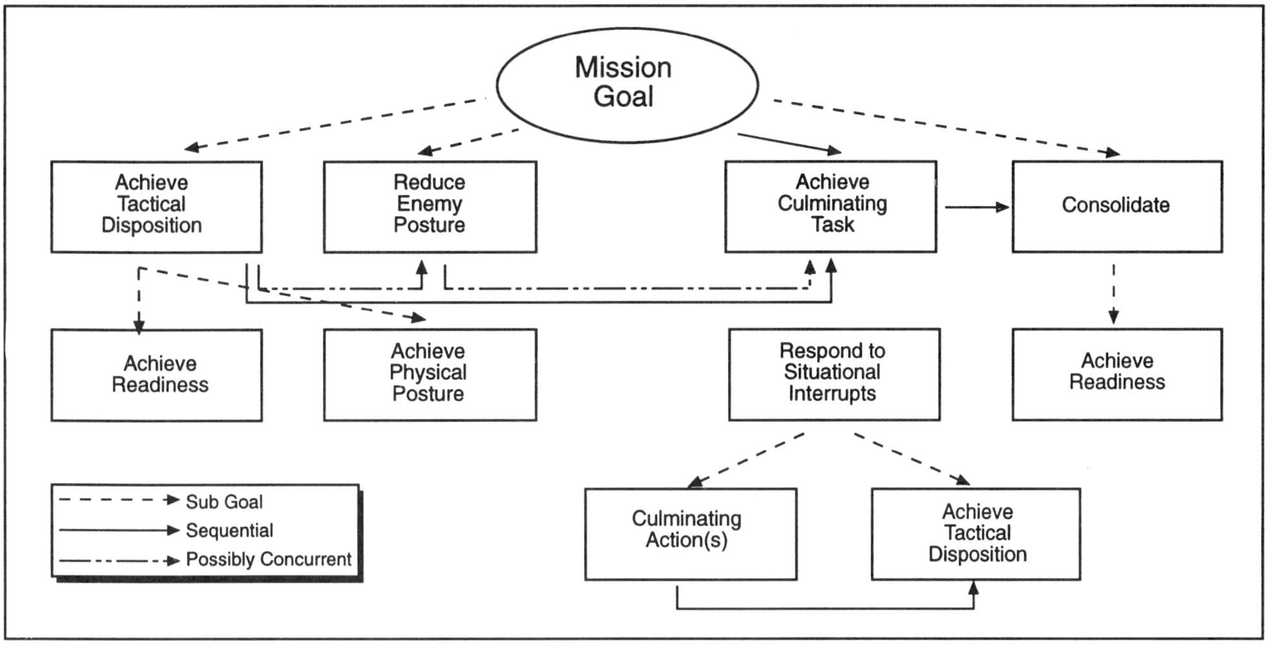

Figure 2: Universal Mission-to-Task Decomposition Model

After careful analysis it was determined that each mission within the ARTEP MTP can be decomposed into logical segments (see figure 2), each with a precondition (or set of preconditions) and a goal. These segments are:

- Achieve tactical disposition
  - Achieve readiness
  - Achieve physical posture
- Reduce enemy posture
- Achieve culminating task
- Consolidate

For any given mission, one culminating task is selected and represents the entry point into the process. Once the culminating task is selected, the reverse planning sequence is utilized in conjunction with the factors/variables of ETT-T to determine the specific string of tasks that will be required under the given conditions.

Situational interrupts may occur when the unit is influenced by unplanned factors, such as an enemy attack or impassable terrain. Situational interrupts may preclude the unit from accomplishing its mission.

An example of the company team attack mission-to-task decomposition is depicted in figure 3, including an assignment of missions to subordinate platoons. In the interest of brevity a listing of the referenced tasks has been omitted, but may be found in ARTEP 71-1 MTP.

This mission-to-task decomposition has been developed for all heavy battalion task force missions, for all heavy company team missions, and for a representative sampling of platoon-level missions. It has been used for the integration of combat support (CS) and combat service support (CSS) operations and is extensible to virtually all types and echelons of units conducting operations.

5 LESSONS LEARNED

5.1 The Role of Doctrine and Its Adversaries

In creating a knowledge base to support the development of intelligent CEs, doctrine is not merely useful or helpful, but is in fact absolutely critical. Military doctrine represents not only the group's previous experiences but its practitioner's hard won insights and, when formed properly, the innovative thinker's projections for future warfighting. As stated in the Army's capstone manual, "Doctrine touches all aspects of the Army. It facilitates communications between Army personnel no matter where they serve, establishes a shared professional culture and approach to operations, and serves as the basis for the curriculum in the Army school system" (FM-100-5, 1993).

If the knowledge base is not tightly tied to the service's doctrinal base, then it may only represent one individual's knowledge and experiences within a limited set of circumstances. An individual's experiences are more often than not anecdotal, explaining what happened, but not why it happened. They may often represent a very specific solution to a unique problem. Developers unfamiliar with the domain will then integrate these methods as the norm and thus limit the CE's ability to deal with the wide range of situations encountered. When the knowledge base is developed as an expansion of the service's doctrinal

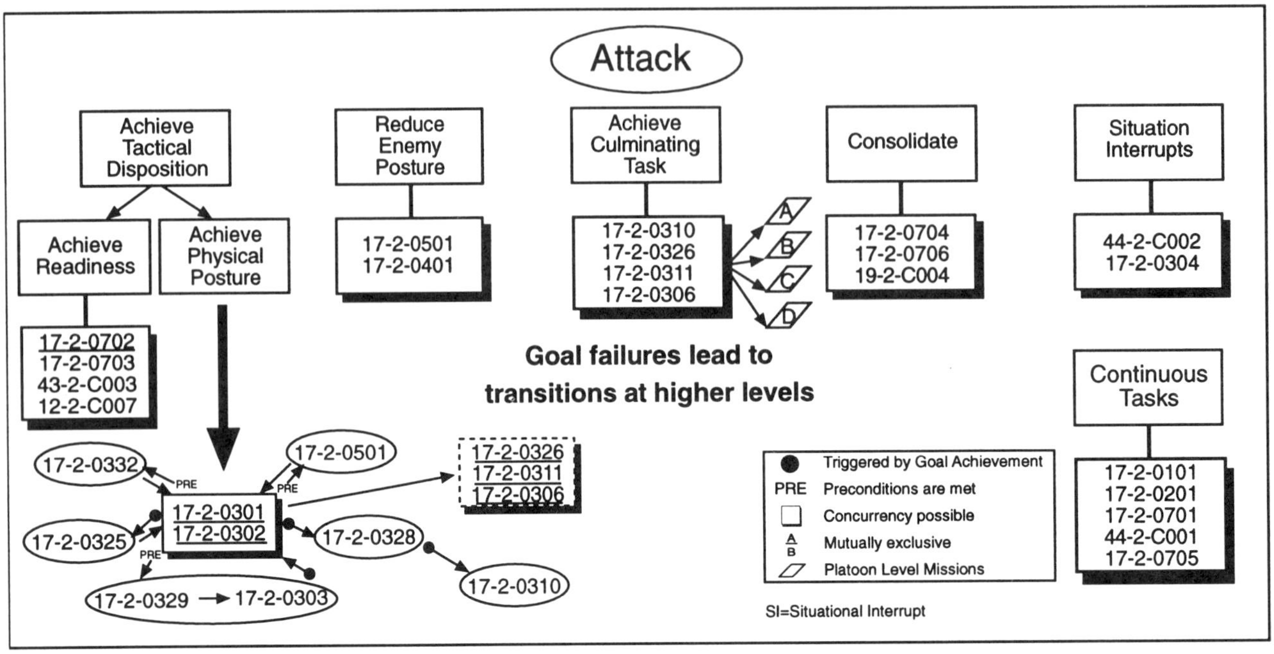

Figure 3: Company Team Attack Mission-to-Task Decomposition

base, the CEs become much more robust in their ability to deal with a host of circumstances.

A second benefit of utilizing the doctrinal base as the foundation for the CEs is that it significantly supports the validation of the end product. When the vast majority of the knowledge base is rooted in doctrine, there is little question as to its acceptability or why it is included. Additionally, when the developers of the CEs have become steeped in the terms, jargon, and processes that are used by the actual practitioners they are emulating, they are much better able to articulate their processes to the target audience and are much more accepted by them.

Unfortunately, doctrine has many adversaries, both within the services and in the development community. These adversaries hold the view that doctrine is simply a rigid set of rules and processes that represents the lowest common denominator for articulating a problem or finding a solution. In their view doctrine is a backward-looking, stifling influence that will inhibit initiative, preclude innovation, and eliminate aggressiveness. This precept is often held by those that have neither studied nor practiced doctrine in its true sense, and the battle to overcome their resistance is constant. As appropriately practiced or integrated into a CE, the use of doctrine meets the following criteria: "Never static, always dynamic, the Army's doctrine is firmly rooted in the realities of current capabilities. At the same time it reaches out with a measure of confidence to the future" (ibid.).

5.2 Combined Arms Integration

With the development of various CEs, it became apparent that a specific effort would have to be to made to ensure that these elements could perform as an integrated combined arms team on the virtual battlefield. This situation was anticipated because it accurately reflects the circumstances as they exist in reality. Although aware that they must operate in a combined arms environment, the Army's branches concentrate primarily on their specific areas of responsibility. It is the task of the combined arms integration centers and field commanders to ensure that the various processes, procedures and equipment operate in a collective and synergistic manner.

In the CFOR program the combined arms team's complexity came to light when the question was raised as to how the fire support CEs would integrate the maneuver CEs concept of operations. This requires the fire support and maneuver CEs to carry on a dialogue with the resulting fire support plans and products supporting the maneuver CE's plan. This requirement had not been actively anticipated by the majority of the CE developers and it was decided that they should be made aware of such issues sooner rather than later. To this end, supporting materials were developed and a seminar was conducted that detailed the integration and interaction that is required to conduct combined arms operations at the battalion task force level and below. This seminar used the same form and format as the previously developed products (i.e., decision modules, vFTX) and built on them where necessary. In particular, it expanded the mission-to-task methodology to the point where it is now capable of incorporating the required CS and CSS functionality along with the maneuver CEs requirements. The seminar was very interactive and proved a significant boost to stimulating the combined arms integration development.

5.3 Individual Knowledge, Experience and Personalities

While both domain SMEs and CE developers must possess significant knowledge and experience in their respective fields, it is often the individual personalities and attitudes that contribute to the project's success or difficulties.

While there are many successful military officers, only a small subset are well suited for supporting the development of intelligent CEs. First of all, their military service must have been a vocation, not simply an occupation or a job. This group has both the aptitude and inclination to spend the necessary hours of research required to articulate the decision-making processes that seemingly comes natural to them (e.g., one SME reported that while he could develop the detailed plan for occupying a battalion assembly area (ARTEP Task #7-1-3001) in less than half an hour, it took over a week to develop the decision paths and factors showing how and why this is done in a doctrinal manner.)

Secondly, most military officers are used to working in a hierarchical and authoritarian structure where decisions are made, orders are issued and there is little after-the-fact questioning. Working with development engineers (when it is done correctly) is a very collegial proposition in which almost everything is questioned, both to determine its necessity and validity, and also to understand the process being modeled.

Thirdly, an effective SME must be willing to absorb significant amounts of constructive criticism. Once an SME has completed a product, it receives a two and sometimes three-step review by peers and superiors. This review is detailed to the point that it addresses individual words in order to ensure doctrinal validity, eliminate anecdotal or personalized aspects, and remove ambiguity about the decision process being described. This process results in the highest quality product, but also requires a certain degree of kenosis on the part of the SME. Collectively these are major adjustments that many potential SMEs do not wish to make.

On the other hand, CE developers tend to take either a global or a specific approach to the development process. The first group actively seeks to learn all they can about the domain in which their agent will be operating. The second group, however, takes a tighter view of the domain and is characterized by the approach "don't confuse the issue with extraneous details, just define what I have to model." While this second group is very good at producing highly detailed representations of very specific behaviors, the first group, in general, seems to be much more successful at developing intelligent, adaptive agents capable of operating in a variety of situations.

ACKNOWLEDGMENTS

The authors wish to thank the Command Forces team at DARPA, MITRE, and NRaD, with special thanks to Dr. Lashon Booker, for their dedication, support and guidance.

REFERENCES

ARTEP 71-1-MTP, *Mission Training Plan For The Tank and Mechanized Infantry Company and Company Team* Headquarters, Department of the Army, 3 October 1988.

Dahman, Judith S., Salisbury, Marnie R., Booker, Lashon B., Seidel, David W., 1994. *Command Forces: An Extension of DIS Virtual Simulation.*

Feldmann, Peggy, CDR, USN, 1995, *STOW-97 Army Synthetic Forces.*

Field Manual 100-5, *Operations,* Headquarters, Department of the Army, 14 June 1993.

AUTHOR BIOGRAPHIES

MARTIN S. KLEINER is the director of Logicon RDA's Knowledge Acquisition program and has served in this capacity since its inception two years ago. He retired from the U.S. Army with the rank of colonel and has experience in the areas of maneuver warfare, intelligence, campaign planning, operational level targeting, and research and development. He received a baccalaureate degree in organizational behavior and management from the University of Houston and is a graduate of the U.S. Army War College.

SCOTT A. CAREY is the deputy task leader for Logicon RDA's Knowledge Acquisition program in support of the DARPA Command Forces project. He is a retired Army lieutenant colonel with more than twenty years experience as an armor officer. His last assignment before retirement in 1994 was as the senior command and control battlefield operating system subject matter expert and observer/controller for the Battle Command Training Program at Fort Leavenworth, Kansas. He has been in his current position for the past two years. Mr. Carey has a bachelor of science degree in education from the University of Maine and a master of science degree in business administration from Boston University.

Proceedings of the 1996 Winter Simulation Conference
ed. J. M. Charnes, D. J. Morrice, D. T. Brunner, and J. J. Swain

USING AFTER ACTION REVIEW SYSTEMS FOR EXERCISE PLANNING AND CONTROL

John A. Hixson

Logicon RDA
P.O. Box 607
Leavenworth, Kansas 66048, U.S.A.

ABSTRACT

After Action Review Systems—the training assessment component of simulation training exercises—provide vital feedback to the military training audience. This paper describes how AARSs serve another essential training function—exercise planning and control—by supplying archived data, in a variety of formats, arrays, and sequences, that can be used to develop plans and monitor exercise progress. This claim is based on Logicon's nine years of experience developing and using an AARS to support the United States Army Battle Command Training Program.

1 BACKGROUND

The U.S. Army fielded Logicon's After Action Review System (AARS) in January 1994 to assist the Battle Command Training Program (BCTP) in assessing simulation-supported training exercises for large units. The AARS software collects, stores and correlates selected military unit, event and ground truth data obtained from the Corps Battle Simulation (CBS), a constructive simulation that provides a realistic, stressful environment in which the training audience must behave (generate orders) as if they were in an actual battlefield situation. Commercial off-the-shelf software (COTS) provides the graphics, text and presentation capabilities to format and present the collected data. The AARS runs on a group of SUN model 5 workstations and 20 Sparcstations, organized into two local area networks and connected by a wide area network that links the exercise site with the National Simulation Center at Fort Leavenworth, Kansas.

Data collection and AAR preparation processes begin before the start of the exercise and continue through to the end of the exercise when the after action review is presented. Objective data, such as artillery missions fired, battle damage assessments, unit movements, and unit logistics status, is collected from the simulation and stored in the AARS database. Subjective data, from observation of the training audience conducting command and control tasks, is gathered from the simulation workstations and the tactical staffs by a group of workstation controllers and observers and forwarded to the analysts. Using these data, along with a comparison of standard Army doctrine for the particular type operation and the exercise unit's plan of operation, analysts determine key events and trends related to the unit's training objectives. These events and trends are then incorporated into presentation materials, including standard doctrinal templates, proposed exercise unit scheme of maneuver, battle summaries of significant events keyed to a sequence of timed situation snapshots and force ratio charts. These materials can either be presented as overhead slides or as part of a narrated automated sequence, using the AARS presentation subsystem. AAR facilitators and exercise controllers use these presentation materials during after action reviews and the exercise control meetings. These materials are also archived on the system and available for use during future exercise planning.

2 EXERCISE PLANNING AND DESIGN

The AARS has capabilities that can facilitate and improve the planning and design of a simulation-driven exercise. As figure 1 shows, a training exercise has three major components: planning, execution and control, and assessment (or feedback). Planning begins by determining on a scenario which events will accommodate the unit's training objectives. These objectives are translated into questions that will be answered by the outcome of the exercise. The questions also indicate which Battlefield Operating Systems (BOSs) will be of primary interest. The BOSs consist of a variety of functions that assist the commander in building and sustaining combat power. Commanders integrate and coordinate these functions to synchronize battle effects in time, space, and purpose. Each BOS, such as fire support, has been further defined by a series of subtasks that need to be accomplished during the curse of normal operations. They vary in importance, or criticality, depending on the type of operation and the unit training objectives. Once the exercise planner has identified the critical tasks, they are incorporated into an operational

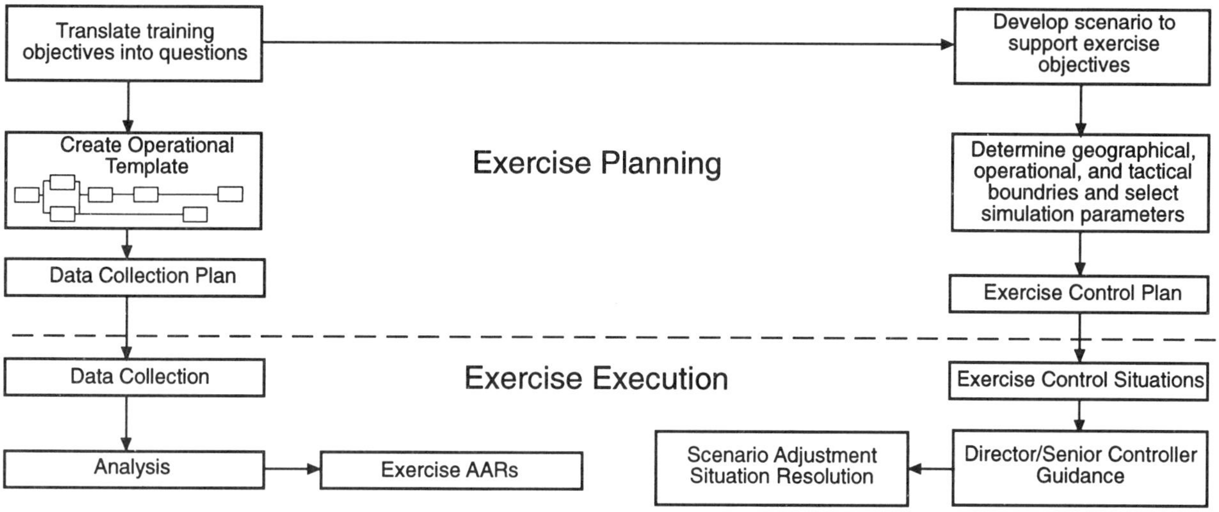

Figure 1: Exercise Planning and Execution

template of the tasks or activities involved in the operation of interest. From the template, we tentatively determine what products are to be created. This, in turn, provides a list of the data needed for preparation. The next step in planning is the data collection plan. Here we marry the critical tasks, the products selected for discussion, the necessary data items, and where and how the data will be collected.

Exercise execution and control proceed together. During execution, the unit attempts to put their plan into operation. Exercise control consists of monitoring the progress of their efforts and the results, through a process of identifying the key events associated with the unit training objectives, periodically reviewing the progress of exercise activities, determining when, where and how adjustments to the scenario need to be made, and the adjudication of any outcomes that are unrealistic or detrimental to the intent of the exercise.

AARSs are designed to store and access large amounts of data For example, the current CBS AARS has 18 gigabytes of available data storage, including archives of data needed to plan and design an exercise. The AARS provides:

- Operational and data collection templates
- Scenarios, key events, and control measures
- Parameter selection checks

Operational templates can be selected from a series of templates, one for each standard military operation, e.g., hasty attack, or deliberate attack, or movement to contact. These standard templates can be modified as desired. Data collection plan templates, archived from previous exercises, can also be accessed through the AARS libraries. These will assist the exercise planners in determining the MEL and the data collection plan.

A database of previous exercise scenarios, key events, and control measures provides information that must be inserted into the exercise through exercise control action, e.g., decisions and actions by higher command levels, certain types of intelligence data. This data is of material assistance in determination of the exercise layout, exercise control organization and functions.

The AARS also provides a means for checking the selection of simulation parameters. For example, exercise performance data, on selected combat and combat support systems, can be collected and analyzed during pre-exercise preparations to determine the reality of results. Additionally, the AARS Geographical Information System (GIS) permits the exercise planners and controllers to set and review the geographical operational and tactical limits of the exercise playbox.

3 EXERCISE CONTROL

During the exercise, the exercise director and the control cell monitor all aspects of the exercise to determine whether the scenario is on track and what corrections are required. The workstation controllers receive exercise data from the AARS, the control cell, the exercise director and senior observers, and C^4I system inputs. (C^4I system inputs will be fully integrated into simulation-driven exercises when the Army fields its digitized battle command and control system, ABCS (Army Battlefield Control System), which will transmit operational directives and reports. Exercise analysts and controllers will need access to this data to understand the flow of operations.)

The controllers watch for the key events they previously identified when they constructed the MEL. As the exercise unfolds, the control cell monitors these events and periodically briefs the exercise director on their progress so that any necessary changes or corrections can be made to attain the training objectives.

Figure 2 shows the interaction of exercise control and the AARS. During the execution of the exercise, data is collected by the AARS for use in the assessment of the exercise unit training. Many products prepared in the assessment process will be used both in the after action review and by exercise control. Exercise controllers use the products to monitor exercise unit actions and activities as well as those of the opposing forces (OPFOR) by using the following AARS capabilities:

- Operational situation snapshots
- Animation clips
- Correlation of forces and means (COFM)
- Battlefield Operating Systems (BOS) products

Operational situation snapshots show the position of the exercise and OPFOR units at particular times of interest. These snapshots can be tailored to show specific units and selected levels of command. Additionally, for field artillery cannon and missile units, target acquisition and counterfire radars, and electronic warfare emitters, the range capability of the system can be shown. Such snapshots are particularly useful to controllers in ascertaining whether or not the exercise unit is keeping pace with the flow of the exercise and accomplishing key events needed to meet their training objectives.

Animation clips, constructed in the GIS, are used to view and assess the progress of the battle, either in limited time blocks or over more extended periods. These clips can be viewed by the control cell on a Sparcstation monitor or on a much larger screen using the AARS presentation subsystem. These clips, along with an explanation of key events that have been completed, provide a comprehensive operations summary for the controller staff.

Correlation of forces and means (COFM) functionality is frequently used to monitor the strength ratio of the OPFOR in particular subareas of the exercise playbox where key events in the scenario are taking place. This ratio is based on type units and combat systems, and is computed automatically.

Automated BOS products—normally line, bar, area or pie charts—are used to assess and monitor the progress of key events in selected BOS areas. BOS products contain selected parameter statistical information, such as types of field artillery missions and numbers of rounds fired, that can be scheduled for periodic collection, display and rapid review, and that clearly indicates trends on a particular BOS. Additionally, the exercise control element monitors key events from the standpoint of reality and adverse impact on the exercise.

These AARS capabilities are in constant use during the exercise, not only to support the periodic control assessment briefings, but also to take more immediate corrective actions. For example, let's assume that the execution of a deep attack by the exercise unit, using attack helicopters, is a unit training objective. During the conduct of this operation, the exercise unit experienced what it believed to be an unusually heavy and unrealistic loss in aircraft to enemy antiaircraft action. The control element responds to this protest, using AARS to help assess this action, by collecting and presenting the necessary simulation data on weapon systems parameters, critical time periods, numbers of aircraft, routes, targets, opposing antiaircraft units locations and zones of acquisition and fire, friendly suppression fires and actions, and the resulting aircraft losses. This data is presented to the senior controller in formats that permit a quick review of the operation's specifics, both statistically and geographically. Based on his assessment of this information, the exercise controller will make a decision on whether to let the result stand or modify it in some fashion. This type of procedure is employed several times throughout the course of the exercise.

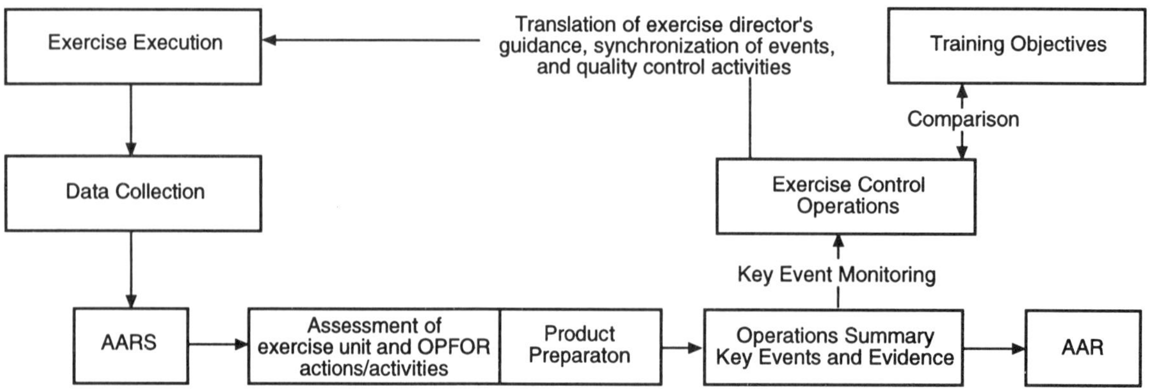

Figure 2: Interaction of Exercise Control and AARS

4 CONCLUSION

Although intended as a training feedback support system, the AARS has important planning and control uses in simulation training. To leverage the AARS's functionality, and avoid unsatisfactory results, training simulations should be designed with both training assessment and exercise planning and control requirements in mind.

AUTHOR BIOGRAPHY

JOHN HIXSON is the Logicon RDA task manager for the support and enhancement of the U.S. Army Battle Command Training Program, After Action Review System, at Fort Leavenworth, Kansas. He received a B.S. from the United States Military Academy, an M.A. in military history from Rice University, and is currently working toward a Ph.D. at the University of Kansas. His research interests are in the areas of analysis of military operations and the application of simulations to training. Mr. Hixson is a member of the Council on America's Military Past and the Association of American Military Historians.

Proceedings of the 1996 Winter Simulation Conference
ed. J. M. Charnes, D. J. Morrice, D. T. Brunner, and J. J. Swain

COMBAT SIMULATIONS IN U.S. ARMY TRAINING AND TESTING

Lawrence J. Dacunto
David J. Prybyla

Logicon RDA
3620 "A" 96th Street S.W.
Tacoma, Washington 98499-4430, U.S.A

ABSTRACT

The U.S. Army has invested billions of dollars in new command and control (C^2) automation systems and communications for transfer of associated information. Testing of these new systems is a very costly process that can be significantly enhanced by using simulations. Training tactical units in command post exercises can also be significantly improved if they are integrated with "go to war" systems. This paper explains how U.S. Army constructive simulations have been applied to testing of and training with C^2 systems, through automated interfaces conceived and developed by Logicon, Inc., contractor to the U.S. Army Electronic Proving Ground.

1 BACKGROUND

The U.S. Army exercises and evaluates tactical operations through a conceptual framework of seven battlefield operating systems (BOSs), shown in table 1. The Army is developing and fielding the Army Battle Command System (ABCS) to provide C^2 automation to support these BOSs. The first five BOSs in table 1 are closely related to ABCS major component systems, and the sixth (mobility and survivability) is supported by the Engineer C^2 System, architecturally a subsystem of the Maneuver Control System (MCS). The last BOS in table 1, C^2, is the process in which each of the other BOSs is executed; that is, C^2 is the synchronization of all BOSs at each battle command echelon (e.g. at a division level), referred to as the Force Level Control System.

Through a combination of manual and automated processes, these BOSs are executed at each battle command echelon. From battalion through higher echelons, each BOS becomes more complex and is supported by staffs of increasing complexity and size that acquire information, plan, and direct subordinate units to conduct combat actions. Information on the tactical situation is communicated to each command post (CP), or operational facility of larger CPs, via oral, written or graphic reports exchanged through communications systems, radars, and intelligence sensors, or direct observation by commanders and staff officers. Many of these information sources can only be present in a full operational environment, which includes properly equipped enemy player units; this is resource intensive, almost impossible to represent effectively in peacetime. Hence the Army must simulate such a combat environment to achieve realistic and efficient training, particularly for the higher echelon battle commanders and staffs.

2 SIMULATIONS IN SUPPORT OF TRAINING

To evaluate the state of training of a battle command echelon (e.g., a mechanized division battle command and staff organization), player CPs are observed either during a field

Table 1: Relationships Between BOSs and ABCS Component Systems

Battlefield Operating Systems	ABCS Component Systems
(1) Maneuver	Maneuver Control System (MCS)
(2) Fire Support	Advanced Field Artillery Tactical Data System (AFATDS)
(3) Air Defense (AD)	Forward Area AD C^2 System (FAADC^2)
(4) Intelligence	All Source Analysis System (ASAS)
(5) Combat Service Support (CSS)	CSS Control System (CSSCS)
(6) Mobility and Survivability	Engineer C^2 System
(7) Command and Control (C^2)	Force Level Control System

training exercise (FTX), in which subordinate organizations participate in associated field operations, or a command post exercise (CPX), in which only the command, staff and selected signal units are engaged in the exercise. The fewer units involved, both friendly and enemy, the fewer resources and more efficient the evaluation of the command echelon, but the greater the challenge of realistically providing information on the tactical situation to the players.

2.1 Standard Training Architecture

For evaluation of a division and its maneuver brigade command echelons, controllers are employed above and below the player echelons providing tactical information as would be directed or reported by the next higher (corps) and lower (battalion) units. These controllers must be instructed on the tactical information to be provided to insure evaluation objectives are achieved. The earliest method of providing these instructions was through manually prepared and executed Master Events Lists (MELs), a time-phased scripting of required controller actions. However MELs require high labor resources in preparation and execution. They also lack training realism because they cannot be modified during the exercise to accommodate unexpected actions by the player units and cannot easily integrate actions and reactions of a live opposing (i.e. enemy) force.

2.2 Use of Simulations

Originating as analytical tools and improving in validity and responsiveness with ever more powerful computer technology, combat simulations were employed beginning in the late 1970s to facilitate training of commanders and staffs. Known as constructive simulations, such combat resolution models receive inputs of opposing (red) and friendly (blue) player actions, wargame the exchange in cyclic updates (i.e., every so many minutes), and produce outputs of combat resolution in terms of battle results (e.g. casualties, battle damage, front line trace).

When a constructive simulation is used to drive an exercise, enormous efficiencies are gained through elimination of MEL preparation and reduction in numbers of controllers. Training realism can be enhanced through employment of controllers serving as a live opposing force (OPFOR). For added training efficiency, the combat simulation can be operated at a location distant from the player units, connected only by tactical, strategic or commercial communications.

Figure 1 depicts the training situation described above, complemented with a constructive simulation as well as a live OPFOR. Controllers for constructive simulations are commonly positioned within a Battle Simulation Center

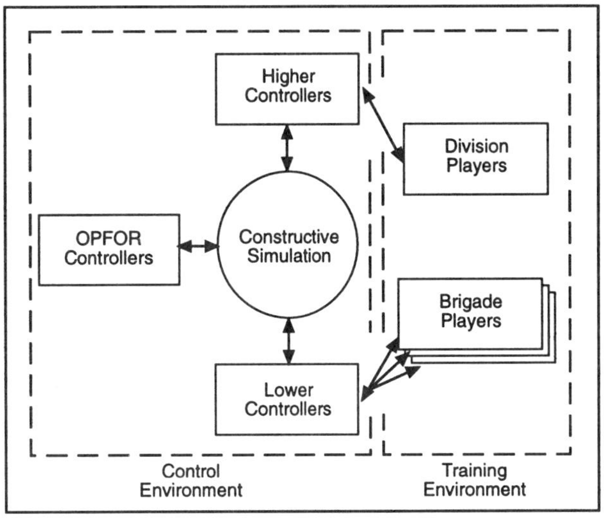

Figure 1: Simulation Supported Battle Exercise

(BSC), putting into the simulation their subordinate echelon actions and orders and reporting simulation outputs/ results to the player units in their field locations.

For a division-level exercise, battalion-level controllers would normally maneuver company level units in the simulation and report results at the battalion level to player brigade Tactical Operations Centers (TOCs). OPFOR controllers are usually positioned in a separate area of the BSC, with the ability to interface directly to the simulation. To implement this concept, computer devices are required for interfacing between the simulation and controllers. Complementary simulations are frequently linked, particularly for the exercise of logistic systems.

By the mid-1980s, constructive simulations were used regularly for training by the U.S. Army, and BSCs were fielded at U.S. Army corps and division installations throughout the world. An entire family of simulations (FAMSIM) was established, and models have been continually improved in both realism and responsiveness. FAMSIM includes the Corps Battle Simulation (CBS) for employment from brigade through corps level, Brigade/ Battalion Simulation (BBS) optimized for those two levels, Janus for battalion and below exercises, and supporting simulations: Combat Service Support Training Simulation System (CSSTSS) and, for intelligence support, the Tactical Simulation (TACSIM) and Battlefield Intelligence Collection Model (BICM).

3 SIMULATIONS IN SUPPORT OF TESTING

While constructive simulations were increasingly used to train battle commanders and staffs, tactical headquarters were targeted to receive increasing numbers and types of ABCS command and control automation systems that interoperate with communications, intelligence and other computer systems. Referred to collectively as C^4I systems, most were in stages of evolutionary development and fielding, requiring significant resources for operational testing, usually troop units in FTXs or CPXs.

3.1 Early C^4I System Testing

The earliest ABCS component systems (MCS and CSSCS) underwent tests based on MELs, but testers disliked the high labor intensity of preparing and executing MELs and their lack of training realism. Noting successful application of constructive simulations in the training environment, test officials directed that subsequent C^4I test exercises be driven by constructive simulation. In addition to being more efficient and realistic, this innovation provided a further technological opportunity. Since Logicon had been under contract to support testing of C^4I systems as well as development and operation of constructive simulations, our staff had experience to conceive the creation of automated exchanges between the combat simulation and the C^4I systems.

These interfaces, called Simulation Support Models (SSMs), were proposed to parse the simulation output into formatted C^4I system messages, and transmit these data into the appropriate point in the C^4I system. To a limited degree, the SSMs would take some C^4I system messages (outputs of the lowest and highest echelon blue players), parse into formatted simulation orders and automatically insert them into the combat simulation. Through these capabilities, Logicon proposed that the SSM interface would effortlessly create blue player activity of subordinate, lateral and higher units and even automated subsystems (e.g. radars) which are not actively being exercised.

3.2 SSM Development

Army officials at the Test and Experimentation Command (TEXCOM) and the former Army Tactical Command and Control System (ATCCS) Experimentation Site (AES) funded a proof of principle development because this approach offered resource savings in MEL preparation and execution as well as reduction of human errors. No longer would there be a need for numerous controllers, particularly at echelons below the player units for submission of bottom-up feeds into the players. An additional benefit was the creation of an audit trail for post hoc operational assessment. Data can be recorded for subsequent analysis of system "value added", such as comparing ground truth in the simulation with perceived truth in the player unit C^4I system.

In 1991, the first SSM was prototyped linking the U.S. Army's MCS and the Corps Battle Simulation (CBS), the official combat resolution model for tactical operations from brigade through corps levels. Although it does not portray all battlefield entities or events, CBS has provided sufficient volume and detail to successfully drive hundreds of brigade, division and corps training events since 1987. After successful use in MCS testing, further SSMs were developed for the CSSCS in 1992. By 1994, with additional funding from the Defense Modeling and Simulation Office (DMSO), SSMs were used in major interoperability tests for the five major components of ABCS supporting maneuver, fire support, air defense, intelligence and combat service support.

Because both MCS and the All Source Analysis System (ASAS) employ standard U.S. Message Text Formats (USMTF), one SSM supports both these systems, and one generic fire support SSM supports AFATDS as well as the systems it will replace. There is one additional SSM for CSS and another for air defense. In some exercises, intelligence play requires supporting intelligence simulations (TACSIM or BICM), which interface directly with CBS and hence need no SSM interface.

3.3 Database Synchronization

We have learned that the most complicated action required for simulations to support exercises is the creation and synchronization of databases for both the associated C^4I systems and the simulation. Most C^2 systems cannot operate without installation of an established database, and the manual creation of databases is labor intensive, time consuming and costly. As two and more C^4I systems were tested simultaneously, the need became increasingly apparent that databases of all C^4I systems had to be precisely synchronized not only among themselves, but also with those of the constructive simulation, as well as any of its feeder simulations (e.g. CSSTSS).

Not only do data elements need precise definition and description, essential in any computer system, but levels of data detail require the SSM to aggregate or de-aggregate detail to insure full congruence of databases. For example, some C^4I systems track general supply categories (e.g. cargo truck), others track by National Stock Number (NSN) or line item number (LIN), and still others require greater detail, such as lot numbers for ammunition. Hence effective use of constructive simulations as exercise drivers require all systems involved to have, at exercise start, precisely nested data that match actual requirements of the organizations and equipment of the player and notional units involved. During exercises, each exchange between simulation and C^4I systems requires data defined and parsed to the appropriate level of detail. The SSM performs this type of transaction thousands of times for each report cycle and each unit portrayed in CBS, thus decreasing the potential for data mapping errors and delays in processing times if this were performed by controller personnel.

To simplify this effort, Logicon has developed a database populator tool that crosswalks the Unit Task Organizations (UTO) and Order of Battle (OB) with data in the Vertical Force Accounting System (VFAS), the Commander's Tracked Items List (CTIL), approved unit planning factors, and CSSTSS data. Additional information is provided in both automated and semi-automated means from the unit or test organization. Using a combination of electronic feed and manual input of data between CBS and the C^{4}I systems, the data is synchronized prior to the start of the exercise. This database populator tool automates what would otherwise be a burdensome, manual process of determining, synchronizing and loading databases of both the C^{4}I and simulation systems.

Logistics data is the most voluminous, and the CSS SSM maps the CBS data to data in both the C^{4}I system (CSSCS) and a supporting logistics simulation (CSSTSS). Because CSSTSS does not portray some logistics information required by the CSSCS (e.g., unit-level resource data), CSSTSS information provided to CSSCS is supplemented with additional information from the CBS itself through the SSM, including data for food, fuel and ammunition resources on hand. The SSM uses CBS and CSSTSS as its simulation components. CBS unit data are de-aggregated and reformatted using a variety of tools to provide the appropriate message flow for each C^{4}I system. CSSTSS unit data mirrors CBS unit data. Through the SSM, CSSTSS provides to CSSCS Class I, II, III, V, and IX at the supply points, and data for Class VII, maintenance and personnel assets in units.

3.4 Architecture

Figure 2 shows the comprehensive picture of how simulations support testing or training with relationships among all the components described earlier. On the right side of the figure, player units in field CPs are employing C^{4}I systems driven by constructive simulations (CBS and CSSTSS). These are both influenced through SSM and controllers using CBS workstations. All data is synchronized not only at exercise start, but continually throughout the exercise.

It is important to note that no changes to the core CBS models or algorithms need to be made for SSM activities. The SSM interface software uses the standard CBS software package as delivered by the U.S. Army National Simulation Center (NSC). The SSM establishes CBS unit databases and task organization, model input orders if required, and utilities to extract information from CBS databases and messages. The information extracted is loaded into high resolution databases and used to generate messages transmitted as manual output to controllers or electronic data to the C^{4}I systems.

4 THE SSM IN SUPPORT OF TRAINING

Initially CBS and SSMs were used in technical testing of C^{4}I systems, and these tests normally involved, as C^{4}I system operators, either technicians or soldiers borrowed from troop units. However, when full scale operational tests began with troop unit commanders and staffs, officials realized that there were additional training benefits, because the CBS with SSMs:

- Permits player units to "train as they fight", allowing them to take precisely the same C^2 actions as they would in actual combat and avoiding "swivel chair" interfaces between simulation and C^{4}I systems.
- Creates more realistic training through automated displays on C^{4}I systems of such data as real-time air tracks and counterfire radar returns.

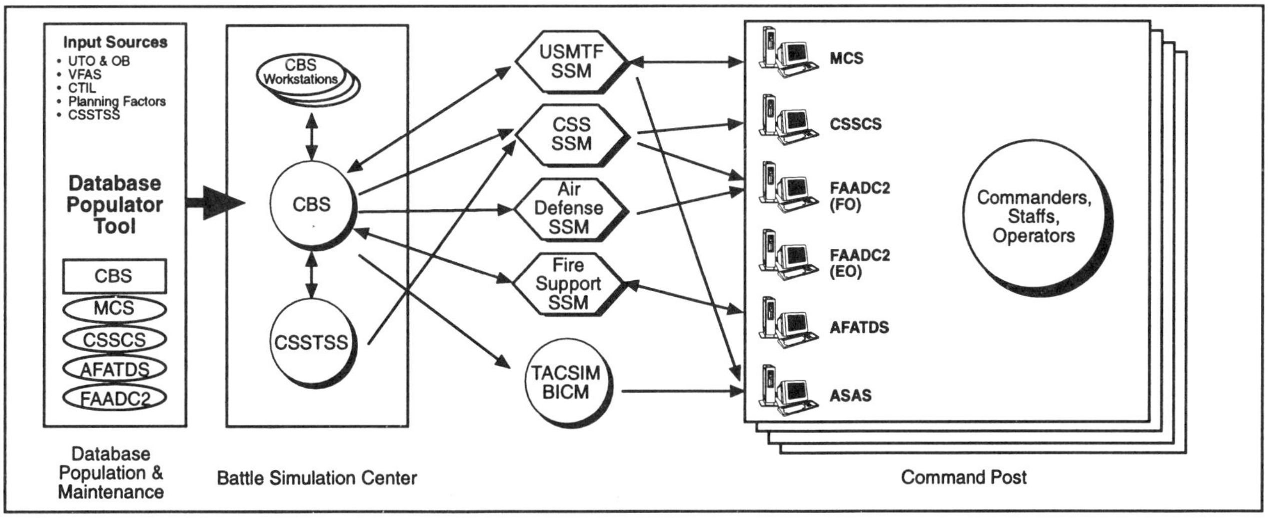

Figure 2: Simulation Supported Training and Testing

- Facilitates multi-echelon training. For example, in a division exercise, maneuver units could play down to battalion level (two echelons), but artillery and support commanders could play down to only one echelon, division artillery (DIVARTY) and division support command.
- Frees soldiers for training who would otherwise be filling increased controller requirements.
- Enhances interoperability of C^4I systems when all use synchronized databases.
- Allows troop units to participate in simulation-driven joint exercises through linkage of the SSM to the Joint Confederation of Models.
- Permits flexibility in exercise architecture by replicating simulated operational units.
- Induces greater user acceptance of C^4I systems by showing staff and commanders the benefits of the systems on live exercises.

5 FIRE SUPPORT TESTING AND TRAINING

To fully exercise the fire support BOS or to test the AFATDS automation system, all the type units in figure 3 would have to be fielded and replicated in full to assure sufficient loading and exercise of the fire support system. Consequently, test or training organizations employing AFATDS in conjunction with other ABCS component systems would have to field large formations of soldiers and equipment to generate the input necessary for evaluations of artillery units or the utility of AFATDS itself. (If field artillery units are exercised without other ABCS component systems, the Fire Support Automated Training System (FSATS) can be used for AFATDS stimulation.)

5.1 Fire Support SSM Utility

The fire support SSM stimulates AFATDS, in near real time, by transmitting CBS-generated resource and update messages, causing AFATDS to compose and transmit messages needed for data collection and analysis. The SSM accepts orders to fire artillery and inputs those orders into CBS for execution for all CBS-fired ammunition types.

In figure 4, all the elements from figure 3 which are depicted in clouds are represented by the fire support SSM, which receives CBS outputs of resources and other update messages and formats them into resources and messages, causing trainer/tester AFATDS systems to compose and transmit messages to player AFATDS systems. The fire support SSM can replicate those echelons and automated systems that would provide information to AFATDS

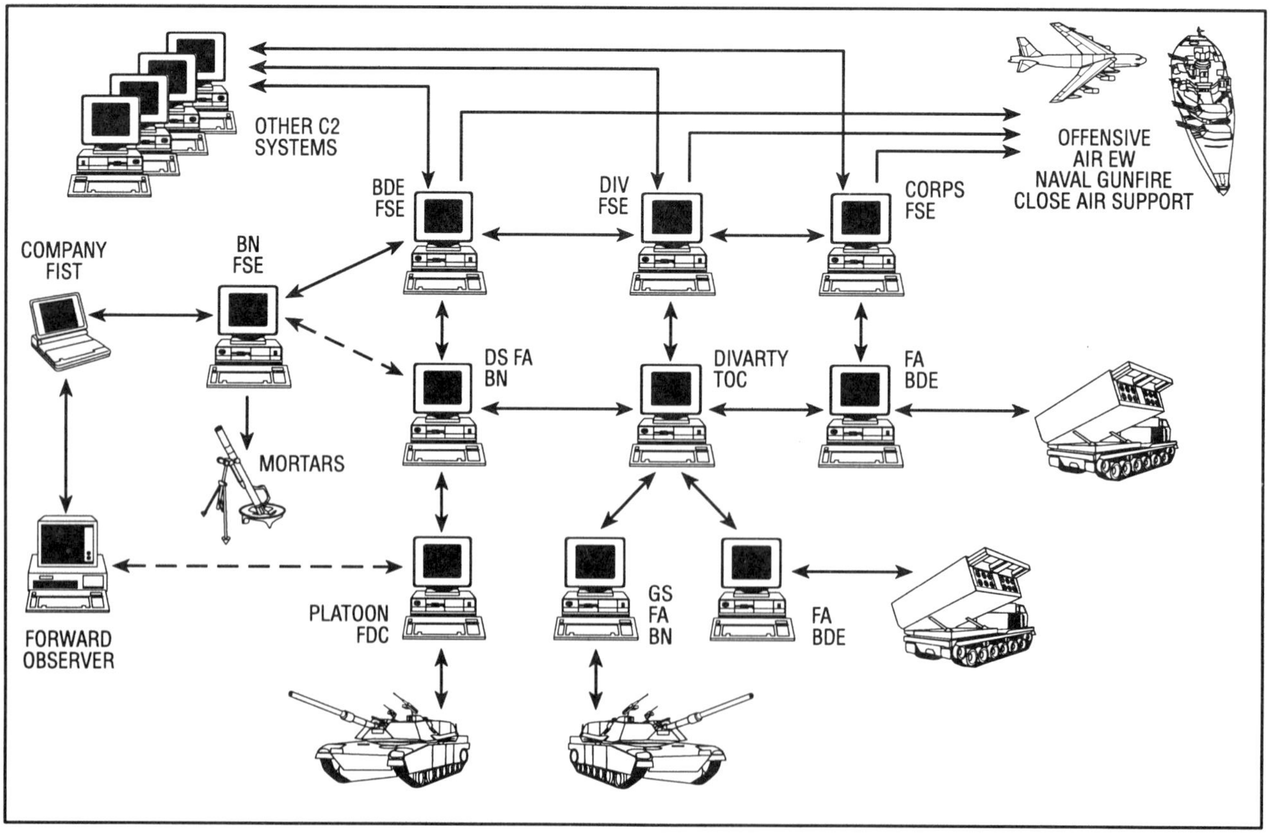

Figure 3: AFATDS in Tactical Employment

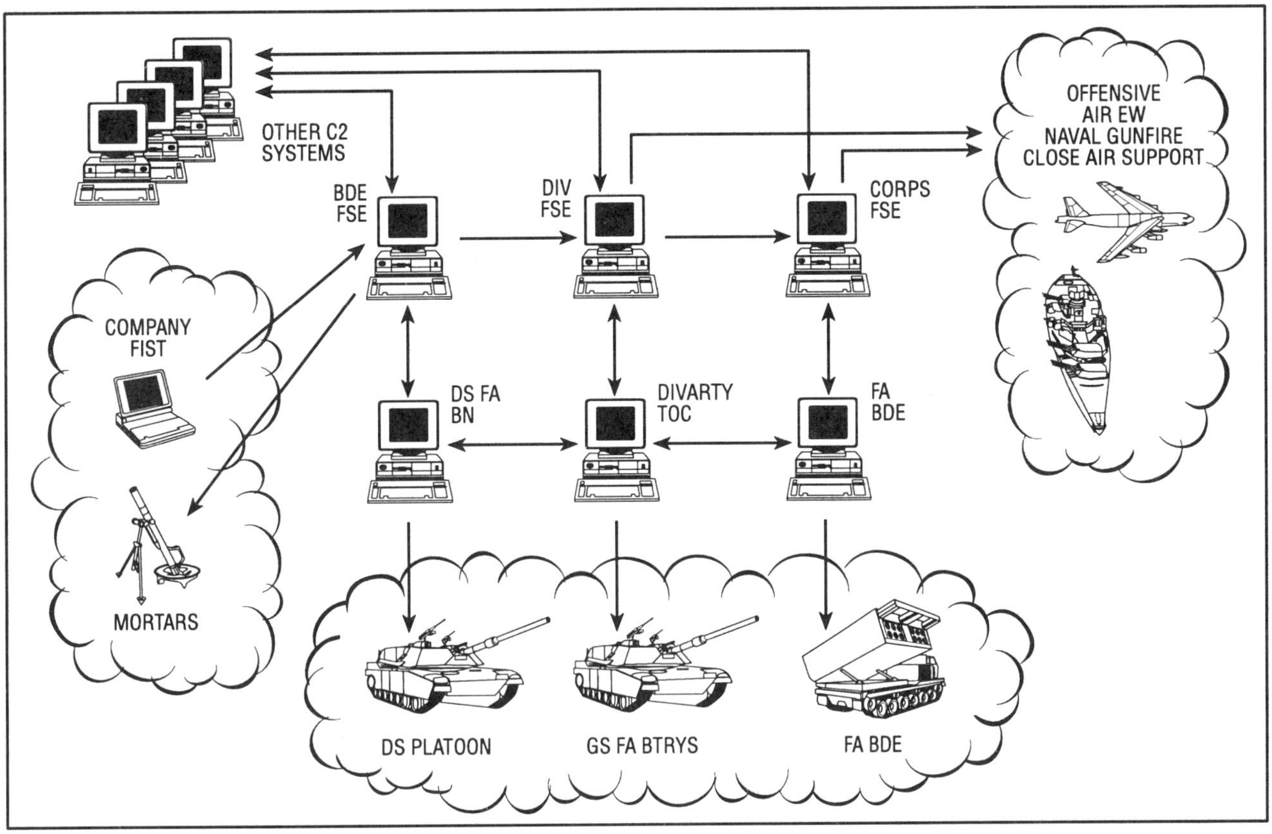

Figure 4: Fire Support SSM Employment

during tactical operations. These include computer systems within batteries and platoons, artillery and mortar forward observers using digital message devices, and the AN/TPQ 36 and TPQ 37 counterfire radar systems.

5.2 Fire Support SSM Functions

As one detailed example, when tracked CBS maneuver units (simulated, not player units) become engaged in combat, the SSM automatically generates a fire request as would be done by a forward observer accompanying the engaged unit, and transmits that fire request to the lowest responsible Fire Support Element (FSE) being played in the exercise. (Units become engaged through embedded CBS decision rules considering such factors as proximity to the enemy, intervisibility, enemy firing and weapons ranging.)

The FSE player, using AFATDS, receives the fire request precisely as though a live player had transmitted it, and makes disposition of the fire request. If the FSE approves it, he forwards an appropriate fire order via AFATDS to the direct support (DS) artillery battalion. Using their AFATDS, battalion players select a fire unit, compute a firing solution, and transmit the order to a subordinate battery; any of these steps after the FSE can be replicated by the SSM. When the SSM receives the order, it translates the fire order into a CBS input message, then injects the message into CBS for execution. Fire is promptly emplaced on the prescribed coordinates, and CBS attrits any enemy located at that position in the CBS. The SSM registers returns from the in-flight artillery rounds on enemy counterfire radars if they are active in the simulation and directed to the air corridor through which artillery rounds pass.

After the combat resolution cycle in which the mission is fired, the SSM decrements the firing unit's ammunition stock and captures the CBS message containing battle damage assessment and munitions expenditures. If this had been an "observed" fire mission, the SSM automatically extracts the battle damage assessment, formats it into a standard message and sends it to the FSE player, who will then act to end the fire mission by so notifying the artillery battalion. When they receive an end-of-mission order from the FSE, the artillery battalion staff terminates the mission with the firing battery, which in turn sends the doctrinal mission fired report to the battalion.

5.3 Fire Support SSM Stimulation

The fire support SSM provides the following stimuli to AFATDS:

- Tactical communications connections

- Doctrinal network architecture
- Requests for fire from observers and radars
- Movement updates for all artillery unit types
- Ammunition status for cannons, missiles
- Reports on incoming enemy fires
- Mission-fired reports
- General, unformatted unit status information
- Orders to fire for simulated fire units
- Battle damage assessment from observed fire

6 SSM EMPLOYMENT

The SSMs have been used on all ABCS system confidence demonstrations and interoperability tests for the past two years. In addition, they have been used by the Command and General Staff College (CGSC) during Prairie Warrior 94, 95 and 96, by the U.S. Army Forces Command during Roving Sands 95, and by 1st Cavalry Division during training exercises. This is a significant amount of usage because AFATDS and CSSCS have thus far only been fielded to the initial units that participated in system testing.

In addition, the AD SSM has been used by several corps air defense artillery (ADA) brigades and every divisional ADA battalion which has thus far been fielded the $FAADC^2$ system. Troop units appreciate the realism of live tracks displayed to system operators just as the tracks would be received from active radars and aircraft, but stimulated from relevant air activity in CBS.

The U.S. Army Battle Command Training Program (BCTP) has initiated efforts to integrate SSMs into its warfighter exercises, which are training evaluations of U.S. Army division and corps battle command organizations.

7 ONGOING INITIATIVES

At the Fort Lewis Field Office of the Electronic Proving Ground (EPG), work continues to refine these SSMs and insure compatibility with updated software versions for both the simulation model and the C^4I systems, since all these systems are in evolutionary development. Additionally, a feasibility study is underway, funded by EPG, for determining costs and benefits of similar interfaces to the Janus simulation, used at the lower Army command echelons.

Though interfaces between C^4I systems and constructive simulations have been enormously successful in the testing environment, exploitation lies ahead in the world of training. As units become equipped with increasing numbers and types of C^4I systems, U.S. Army field commanders will appreciate the training benefits cited above. They will likely encourage fulfillment of training materiel requirements for extensions of these interfaces among all available exercise drivers and C^4I components of the ABCS. Field commanders will no doubt also demand extension of SSM downward into virtual simulations such as tank crew trainers, and upward into Joint C^2 systems and constructive simulations. Further, a Logicon independent development project has served as proof of principle that these SSM interfaces can be extended into the U.S. Army Standard After Action Review System (STAARS), thereby enhancing training evaluations of C^4I -equipped units participating in the BCTP.

The SSMs provide troop units with an inexpensive stimulus for sustainment training for commanders, staffs and system operators. We foresee that the U.S. Army will soon direct that existing SSMs be documented for fielding to BSCs for easy access by supported troop units.

In Spring 1996, the U.S. Army awarded to Loral corporation a contract for development of WARSIM 2000, the eventual replacement of CBS and other FAMSIM models. Included in the WARSIM concept are interfaces to C^4I systems, like the SSM described herein, and automated interfaces to After Action Review Systems; some WARSIM work will be executed by Logicon as a subcontractor to Loral.

On a broader scale, the DMSO has an initiative underway to develop modular reconfigurable C^4I system interfaces among simulations and C^4I systems of the several U.S. armed services, and Logicon is also expected to be involved in this effort.

AUTHOR BIOGRAPHIES

LAWRENCE J. DACUNTO is manager of the Logicon C^3 Engineering and Fabrication Division supporting the Electronic Proving Ground Field Office at Fort Lewis, Washington. He received a master of science in management from Rensselaer Polytechnic Institute and, as a U.S. Army colonel, served as a brigade commander and later as director of C^3I at the Combined Arms Center, Fort Leavenworth, KS.

DAVID J. PRYBYLA is task manager for development of simulation technology in support of the Electronic Proving Ground Field Office at Fort Lewis, Washington. He received a master's degree in system management from University of Denver and another in education from Boston University. He has been working in U.S. Army simulations and training for over nine years, after concluding a career as a U.S. Army officer.

Proceedings of the 1996 Winter Simulation Conference
ed. J. M. Charnes, D. J. Morrice, D. T. Brunner, and J. J. Swain

A SYSTEM TO PROJECT INJURY AND ILLNESS INCIDENCE DURING MILITARY OPERATIONS

Christopher G. Blood
Ed O'Donnell
Dan Rotblatt

Medical Information Systems and Operations Research Department
Naval Health Research Center
P.O. Box 85122
San Diego, CA 92186-5122, U.S.A.

ABSTRACT

Modeling of medical resource requirements during military operations requires projections of disease and non-battle injury (DNBI) and wounded-in-action (WIA) rates. Historical data were extracted from unit diaries of infantry and support troop deployed during four previous combat engagements. A planning tool (FORECAS) was developed that uses the statistical distributions of DNBI and WIA incidence from previous operations to simulate injury and illness arrival rates for future scenarios. Output of the simulated data reflects the nuances of the empirical data.

1 INTRODUCTION

Forecasts of the expected injury and illness incidence in a combat theater is requisite to the medical resource planning of a military operation (Systems Research 1992 and Galarza 1987). Distribution characteristics of casualty incidence patterns may be derived by fitting theoretical distributions to illness and injury data of previous operations. These distributions can then provide a basis for modeling future disease and non-battle injury (DNBI) rates and WIA (wounded-in-action) incidence.

A medical casualty forecasting system (FORECAS) has recently been developed. The purpose of the FORECAS software is threefold: 1) to provide medical planners with estimates of the average daily rates of medical admissions that may be sustained in a given scenario, 2) to indicate the maximum daily patient loads that may be incurred and for which planning is necessary, and 3) to enhance understanding of the statistical properties of injury and illness rates for use in future modeling efforts. The utility of FORECAS as a planning tool goes beyond supplying the average daily admission rates for operations; FORECAS also reflects the 'pulse and pause' nature of combat operations in the inter-arrival times and the magnitudes of projected medical admission frequencies. By displaying the dispersion, the range, and other statistical properties, FORECAS graphically depicts the potential patient flow within a theater of operations--information that is critical to proper medical resource allocation.

Development of a casualty simulation model required the analysis of historical data to determine the appropriate probability distributions. Data were extracted from the unit diaries of Marine Corps battalions deployed (1) to Okinawa in 1945 (April 1 through June 30), (2) to Korea between February and June of 1951, and (3) to Vietnam between May and June, 1968. Additionally, medical incidence data from the 1983 Falklands conflict was extracted and analyzed. These time periods were selected because they were thought to be representative of potential future operations. All told, the data (Blood 1995 and Blood 1994) upon which this medical projection system bases its forecasts exceeds 3.5 million person-days.

Studies performed at the Naval Health Research Center (O'Donnell 1993 and O'Donnell 1994) found that DNBI rates among combat support troops follow a simple Poisson process that can be approximated by a normal distribution. Among infantry troops, however, greater fluctuations in DNBI rates were evidenced and this process is best represented by a lognormal distribution. Rates of WIA are characterized by a non-stationary Poisson process combined with the appropriate time series attributes (inter-arrival patterns). Further, there was found to be a significant amount of autocorrelation within both the DNBI time series data and the WIA data. Another significant aspect of the empirical rate analysis was the degree of correlation between the WIA rates and DNBI incidence among infantry troops (Blood 1993). Support troops DNBI data, in contrast, is a simple random variable that is independently and identically distributed.

2 USE OF THE FORECAS PROJECTION SYSTEM FORECAS

System minimum requirements: an IBM or IBM-compatible PC, an EGA to SVGA monitor, 640k of base memory, 500k disk space.

The first input screen employs a graphical user interface

to prompt the user for the information needed in the simulations. The screen displays the following options regarding illness or injury incidence to be projected: DNBI, WIA, DNBI and WIA.

The second input screen prompts the user for information pertaining to battle tempo. Again, the user makes inputs with the use of the left-side mouse button. This screen requests the degree of battle intensity of the proposed battle scenario. The combat intensity options are: NONE, LIGHT, MODERATE, HEAVY, and INTENSE. The "NONE" category represents a 'no combat' situation; though tensions may exist, no combat is taking place (an occupying force, for example). While DNBI arrivals will still occur in this situation, there will be no WIA. The other battle tempo options reflect the magnitude of injury/illness incidence evidenced in previous combat operations of varying intensities (Blood 1995).

The third input screen allows the user to specify the numbers and types of troops for the projected scenario. There are three categories of troops: INFANTRY TROOPS, SUPPORT TROOPS, SERVICE SUPPORT TROOPS. Combat troops refer to infantry personnel; Support troops represent intra-divisional support such as tank, artillery, light-armored infantry, and combat engineer units; Service Support personnel includes the Force Service Support Group (FSSG) and the Surveillance, Reconnaissance, Intelligence Group (SRIG). Below each of the three troop categories are buttons for specifying the number of that troop type deployed.

The fourth input screen prompts the user for two parameters regarding the operational theater and length of the proposed scenario. The user chooses between EUROPE, EAST ASIA, and SOUTHWEST ASIA as the region and then specifies expected duration of the engagement.

Upon clicking the right-side mouse button, the first of the graphical displays is presented. The graphical presentations consist of DNBI and/or WIA incidence line charts plotted across the temporal course of the simulated scenario. Initially, each graph depicts a single troop type. If the user has selected more than one troop type, there will be a graph for each troop category followed by a composite line chart across all troops categories.

Each arrival stream (DNBI, WIA) is depicted in different colors to enhance user viewing. Along the y (vertical) axis the scale shows the magnitude of the arrivals, while the x (horizontal) axis represents the temporal course of the operation (See Figure 1).

A statistical summary screen follows each graphical depiction of injury/illness incidence. The user-defined parameters are shown at the top of the screen. Below these parameters are descriptive statistics representing the casualty flow over the operation. For DNBI* and WIA incidence, the output shows:

-TOTAL NUMBER OF PRESENTATIONS
-TOTAL NUMBER OF ADMISSIONS
-Total number of diseases*
-Total number of nonbattle injuries*
-Total number of battle fatigue cases*
-DAILY ADMISSION AVERAGES
-MAXIMUM ADMISSIONS ON A SINGLE DAY
-AVERAGE RATE PER 1000 STRENGTH PER DAY
-MAXIMUM DAILY RATE PER 1000 STRENGTH
-MINIMUM DAILY RATE PER 1000 STRENGTH

Presentations represent the total number of patients expected to be admitted to a treatment facility. Admissions represent only those patients whose treatment is expected to last three days or longer.

Upon hitting the "enter" key, the system then proceeds to the next graphical presentation (line chart for the next troop category). This process repeats itself until all troop categories have been viewed or the user quits. After all the graphs and statistical summary screens have been viewed,

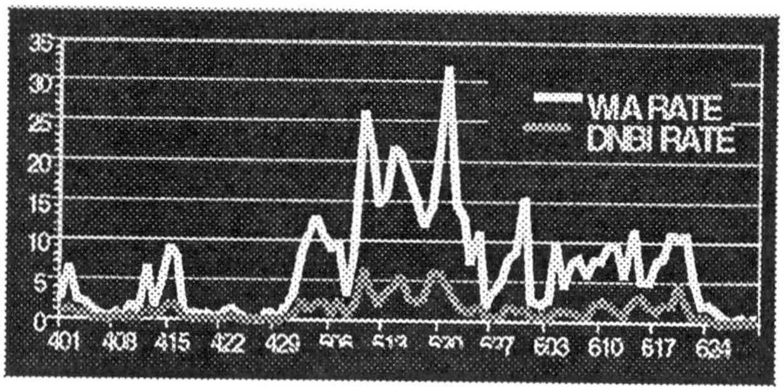

Figure 1: FORECAS depiction of daily patient load

the user is returned to the initial input screen to create another scenario or to quit the program.

3 STATISTICAL UNDERPINNINGS OF FORECAS

The historical data that was studied to form the basis for the simulation comes from Marine company and battalion unit diaries. These battalions were of varying strengths. Numbers of illnesses and injuries were computed per 1000 troop strength per day. This standardization of the data gives the output a uniform perspective which can be easily adjusted across various troop strength levels. The actual simulations are accomplished with the aid of uniform random number generators to start each of the rates.

3.1 DNBI Rates

The simulation of support troop DNBI rates makes use of variates drawn from a normal distribution. These variates historical data. The variates Y, are distributed as:

$$Y \sim N(\mu,\sigma^2)$$

Generation of the infantry troop DNBI can be derived from a lognormal approximation to a Poisson process. The variates from a lognormal distribution are distributed as:

$$X \sim LN(\mu,\sigma^2)$$

Generation of this variate is similar to the normal distribution. After deriving Y, the normal variate, the lognormal variate, X, is then generated:

$$X = e^Y$$

This results in an IID lognormal variate. This is than put into the equation:

$$X_t = \beta_0 + \beta_1 X_{t-1} + \beta_2 W_t + LN(\mu,\sigma^2)$$

where:

X_t = the infantry DNBI variate with the necessary auto and cross correlations.

β_i = constants used to obtain the necessary auto and cross correlations.

W_t = the WIA variate.

$LN(\mu,\sigma^2)$ = the IID lognormal variate

3.2 WIA Rates

Examination of individual battalions shows the historical data of the combat troop WIA rates are represented by single and batch arrivals in a non-homogeneous Poisson process. Representation of the 1000 troop strength per day can be simulated with the use of a continuous distribution. These variates are exponentially distributed.

$$W \sim \text{expo}(\beta)$$

where:

β is the estimated mean.

There is also a strong degree of autocorrelation within each of the combat troop series. Specifying that any of the combat troops series data as being independent is inappropriate. Generation of the series than involves producing a stream of variates that possess the needed serial correlation. Generating the series with autocorrelation then is:

$$W_t = \alpha_1 + \alpha_2 W_{t-1} + \alpha_3 W_{t-2} + \alpha_3 W_{t-2} + \text{expo}(\beta)$$

approximate the Poisson process which best represents the

where:

W_t = the WIA rate at time t.

α_t = constants used to develop the autocorrelation.

$\text{expo}(\beta)$ = the random number drawn from the computer.

The support troops and the extra-divisional troops both have less or no autocorrelation within each series so a simple exponential distribution is used to generate the interarrival times of each casualty event.

3.3 Validation of the Model

Analysis of the casualty stream data generated by FORECAS is warranted to ensure that the projection system accurately reflects the statistical trends evident in the empirical data. FORECAS also has a feature that creates an ASCII data set which can be imported into any common statistical software program to perform validation tests. Analyses were performed on the time series data in the form of serial and cross correlation tests. Past research found that support troop rates are identically and independently distributed -- consequently no autocorrelation should exist within each of these series and none was found. The combat troop DNBI empirical data however possessed a degree of autocorrelation which were accurately reflected in plots of the infantry troop DNBI simulated series.

Autocorrelation function graphs were also used to compare the simulated combat troop time series with the actual data and the plots indicated a significant degree of autocorrelation at the first day which decreases thereafter. This is consistent with the findings from analyses of the historical data. A test for cross-correlation between the DNBI and WIA series demonstrated a statistically significant finding similar to the empirical data. These analyses indicate that the simulations accurately depicted the statistical patterns evidenced in the empirical data.

4 CONCLUSIONS

Because the FORECAS projection system accurately simulates the medical admissions seen in previous military operations and because a great deal of consistency in admission patterns has been witnessed across operations in the past, this tool may prove useful in forecasting the medical requirements of future operations. Such a planning tool may likewise have similar utility in the civilian sector, to the extent that sufficient empirical data exists upon which to base projections.

REFERENCES

Blood CG, Gauker ED: The Relationship Between Battle Intensity and Disease Rates Among Marine Corps Infantry Units. *Military Medicine*, 1993, 158 (5): 340-344.

Blood CG, Gauker ED, et al.: *Projections of Battlefield Medical Casualties Among U.S. Marine Forces For Various Theaters of Operations*, NHRC Report No. 93-35. Naval Health Research Center, San Diego CA, 1995.

Blood CG, Gauker ED, et al.: Comparisons of Casualty Presentation and Admission Rates during Various Combat Operations, *Military Medicine*, 1994, 159 (6): 457-461.

Blood CG, Jolly RT: Comparisons of Disease and Nonbattle Injury (DNBI) Incidence Across Various Military Operations, *Military Medicine*, 1995, 160 (5): 258-263.

Burdick DS, Chu K, et al.: *Computer Simulation Techniques*. New York: John Wiley & Sons, Inc. 1966.

Flannery BP, Press WH et al.: *Numerical Recipes in C*. New York: Cambridge University Press. 1992.

Galarza JG: *Using the Deployable Medical Systems Clinical Data Base for materiel requirements*. Paper presented to Army Operations Research Symposium (AORS XXVI), Fort Lee VA, 1987.

Kelton DW, Law AM: *Simulation Modeling and Analysis*. New York: McGraw-Hill Book Company, 1982.

O'Donnell ER, Blood CG: *Distribution Characteristics of Marine Corps Casualty and Illness Rates*, NHRC Report No. 93-20. Naval Health Research Center, San Diego CA, 1993.

O'Donnell ER, Blood CG: *A Bivariate AutoRegressive Integrated Moving Average Analysis of Combat Troop Casualty Rates*, NHRC Report No. 94-2. Naval Health Research Center, San Diego CA, 1994.

Systems Research and Applications Corporation. *Preliminary software requirements for the medical planning and execution system of the Joint Operations Planning and Execution System* (SRAC Report No. 91-674-234.1), 1992.

AUTHOR BIOGRAPHIES

CHRISTOPHER G. BLOOD is Head of the Operations Research Division at the Naval Health Research Center (NHRC) in San Diego, California. He has been involved in analyzing ground and ship-based casualty incidence from past operations and is active in the development of medical planning tools to project casualty incidence in future scenarios. Mr. Blood is a member of the Military Operations Research Society.

EDWARD O'DONNELL is a former GEO Centers, Inc. Research Analyst who worked at NHRC on ground casualty projections for future scenarios. Mr. O'Donnell was the lead programmer in developing the first version of the FORECAS ground casualty planning tool.

DAN ROTBLATT is a Research Analyst working at the Naval Health Research Center for GEO Centers, Inc. Mr. Rotblatt's work focuses on projections of casualties and examining various factors underlying fluctuations in casualty incidence such as battle intensity, weapons inventories, and environmental parameters.

Proceedings of the 1996 Winter Simulation Conference
ed. J. M. Charnes, D. J. Morrice, D. T. Brunner, and J. J. Swain

THE OPTIMAL PLACEMENT OF CASUALTY EVACUATION ASSETS: A LINEAR PROGRAMMING MODEL

Scott C. Sundstrom
Christopher G. Blood
Serge A. Matheny

Medical Information Systems and Operations Research Department
Naval Health Research Center
P.O. Box 85122
San Diego, CA 92186-5122

ABSTRACT

Through the use of linear programming techniques, the optimal number and positioning of patient evacuation assets within a theater of operations may be determined to ensure the orderly transport of casualties from the front lines to third echelon medical treatment facilities. The Probabilistic Location Set Covering Problem has been chosen as the core module for a linear programming model to assist in these determinations. The Optimal Placement of Casualty Evacuation Assets (OPTEVAC) model prompts the user to enter the dimensions of the theater, troop deployment nodes, types of evacuation assets available, and preferred locations of medical treatment facilities. The OPTEVAC model then provides output as to the required numbers of ground and air ambulances as well as the optimal positioning of those evacuation assets and ambulance exchange points.

1 INTRODUCTION

Combat casualties in a theater of operations are treated at mobile medical facilities organized into a series of echelons with the facilites at the forward echelons having the greatest mobility but least surgical capability (Fleet Marine Force Manual 1990) The efficient evacuation of casualties from echelon to echelon is essential to ensure the wounded personnel reach a facility with the capability to render the required level of care. Distances between medical treatment facilities, as well as factors such as the type of terrain and mode of transportation, may all impact the evacuation process. Likewise, the number of treatment facilities deployed and where they are located greatly affects the casualty handling process and the ability to provide adequate casualty care.

In the Kuwaiti theater of operations, treatment facilities were spaced further apart than the distances considered under the existing medical evacuation doctrine; consequently, some casualties requiring evacuation were out of helicopter range or in locations not accessible to fixed-wing aircraft (Naval Medical Doctrine Center 1991) While casualties were relatively few in the Gulf Conflict, and the wounded personnel did receive needed treatment in a timely fashion, future operations may yield larger numbers of casualties with a greater potential for overwhelming the casualty evacuation system. Accurate assessments of the required numbers, types, and deployment locations of the evacuation assets ensure that wounded personnel are transported expeditiously from the point of injury through the various echelons of care to an Echelon III treatment facility, as dictated by the severity of the wound.

The focus of this report is the selection and development of an appropriate linear programming technique which will allow optimization of the number and positioning of evacuation assets at given treatment facilities in a theater of operations. Because inter-theater evacuation of patients is handled for all services by the U.S. Transportation Command (TRANSCOM), the present effort focuses on intra-theater casualty evacuation -- transporting wounded Marines from the point of injury, through Echelons I and II in the combat zone, and ending at an Echelon III level of care (e.g., Fleet Hospital). The methodology developed will be flexible enough to allow incorporation of treatment facilities and evacuation assets of other service branches within the combat zone, particularly those of the Army which may very well have forces interdeployed with the Marines.

The optimization planning tool will incorporate the projection methodology of the FORECAS system (O'Donnell and Blood 1994) which simulates daily casualty averages and maximum daily patient loads. The simulations generated by FORECAS are based on the statistical properties of empirical data observed from four previous ground combat operations (Blood and Gauker 1993A). In addition to patient flows, the optimization model will require data on troop strengths and deployment locations, terrain, size of theater, numbers and types of ground/air ambulances available, types and locations of

Echelon II and Echelon III treatment facilities, and length of evacuation delay. Using this input, the planning tool will compute the optimum numbers, types, and deployment coordinates of the evacuation assets, including ambulance exchange points.

A comprehensive examination of existing optimization models was conducted and the Probabilistic Location Set Covering Problem (PLSCP) (ReVelle and Hogan 1989 and ReVelle 1989) was determined to be the one that most accurately represented the demands and constraints needed to model the intra-theater evacuation of casualties. Other optimization models that were examined include the location set covering problem (LSCP), the maximal covering location problem (MCLP), the maximum availability location problem (MALP), and two derivatives of the chosen PLSCP: the α-reliable p-center problem and the maximum reliability location problem (MRLP). The PLSCP was chosen over the other models because it minimizes the total number of required evacuation assets within the theater, while at the same time incorporating randomness in vehicle availability. The vehicles of the other models investigated, are sited only on the basis of geographical coverage and not on the basis of availability. The assumption underlying these alternative models is that the vehicles are continuously available for evacuation of casualties. In low demand operations, this assumption is not unreasonable; however in high tempo military operations where frequent casualty calls keep evacuation assets on the roads, this assumption cannot be justified. Use of the PLSCP algorithms within the evacuation model (OPTEVAC) will allow medical planners to input the "alpha-level" at which the optimization model is to run -- a statistic which represents the percentage of the maximum daily casualty load for which the proposed assets will meet the demand. The maximum daily casualty load will be computed by adding the maximum projected wounded-in-action (WIA) incidence to the maximum projected disease and non-battle injury (DNBI) incidence during the operation. The alpha level is important because of the casualty "spikes" that occur during a combat operation. For instance, the planner may want to economize resources by deploying dedicated assets to ensure evacuation coverage for 80% of the maximum casualty load ($\alpha = .80$) and rely on "vehicles of opportunity" to transport the casualties that exceed the capabilities of these dedicated assets. Alternatively, if evacuation assets were programmed for the maximal projected daily casualty load, many assets would be used but a small percentage of the time, if at all.

Patient flows within the OPTEVAC model will be determined from the user-specified input of troop placement, with projections based on casualty distribution trends evidenced during previous combat operations. The patient stream is actually composed of two separate medical admission flows: WIA and DNBI. The daily WIA flow is generated by drawing a random deviate from an exponential distribution based on 'battle intensity'-specific averages and geographical theater considerations. Projected DNBI incidence is similarly derived using variates drawn from a lognormal distribution. The casualty generator also incorporates two other statistical patterns observed within the empirical data: cross-correlation between WIA and DNBI and autocorrelation within the WIA flows (Naval Medical Doctrine Center 1991 and Blood and Gauker 1993B).

The following section describes the linear programming model (PLSCP) chosen for optimizing casualty evacuation assets and provides a detailed analysis of how the problem is mathematically represented. Essential to accurate assessment of the numbers of evacuation assets required are variables relating to the specific types of assets (transportation mode, litter capacity, range, mobility, etc.) as well as factors related to terrain and weather. The scope of the present report, however, is the linear program to be used in optimizing evacuation assets; the parameters relating to environmental conditions and ambulance characteristics are the subject of a parallel effort.

2 THE MODEL

The OPTEVAC model will provide medical planners with the required number and optimal placements of evacuation assets to ensure sufficient casualty transport while minimizing oversupply of ground and air ambulances. The Probabilistic Location Set Covering Problem (PLSCP) linear program has been selected as the core module for OPTEVAC, and it has been modified to make it applicable to the optimization of casualty evacuation assets in a combat theater.

The PLSCP module, incorporated into the casualty evacuation application, seeks the positions of z evacuation assets (1) in a theater of operations which minimizes the number of vehicles (ground and air ambulances) required so that there is a 95% certainty (2) that the user-specified evacuation demands will be met. Mathematical formulation of the PLSCP for the casualty handling process is:

$$\text{minimize } z = \sum_{j \in J} x_j$$

subject to (1)

$$\sum_{j \in N_i} x_j \geq b_i \quad \forall i, \qquad x_j \text{ are integers,}$$

where b_i is the smallest integer satisfying

$$1 - \left(\frac{F_i}{b_i}\right)^{b_i} \geq .95 \tag{2}$$

Define:

$$F_i = \left(\frac{1}{24}\right) t \sum_{k \in M_i} f_k \tag{3}$$

where:

t = the average duration (hours) of a casualty call within the theater (average distance from a demand node to a treatment facility / weighted average speed of available assets, using the ratios of available assets);

f_k = frequency of casualty calls, or trips, at demand node k (casualty calls/day);

M_i = the set of demand nodes within S of demand node i.

Variables:

z = the total number of evacuation assets distributed over all of the facilities in the theater of operations ($\sum x_j$);

i,I = index and set of demand nodes;

j,J = index and set of treatment facility sites;

d_{ji} = the distance from facility site j to demand node i;

S = the distance standard within which a treatment facility is desired to be found;

N_i = $\{j|d_{ji} \leq S\}$; the set of facility sites within S of demand node i;

x_j = the number of evacuation assets at facility site j;

b_i = the minimum number of evacuation assets required within M_i .

The OPTEVAC planning tool consists of input screens which employ graphical user interfaces prompting the user for the information needed for the simulations. The first input that the planner must supply into the OPTEVAC model is the dimensions of the theater. An image of a grid with the user-specified dimensions will then be displayed on the computer screen immediately following this input. The next required input pertains to location of troop deployment. Following troop deployment input, the user will be asked to specify the numbers and types of medical treatment facilities and to indicate their locations within the theater. Other key inputs to the model will include factors such as battle intensity, geographical region, expected length of the operation, evacuation delay, "alpha-level" associated with the maximum projected daily casualty load, and types of evacuation assets available.

The standard PLSCP linear program, by itself, is not sufficiently robust to handle the demands of the OPTEVAC model, so modifications were required to fit it to the casualty evacuation model. The PLSCP was originally written to satisfy the demands of a city population in need of ground ambulances in a timely fashion. The urban transport problem assumed only one type of ambulance, which implied one speed and one litter capacity. In a military theater, however, there is a multitude of vehicles available, and each has it's own distinct speed, range, and litter capacity. Furthermore, within a combat theater, different types of vehicles need to be dispatched simultaneously.

To integrate more than one type of evacuation asset into the model, information regarding the capacities, speeds, and the numbers of each vehicle available for deployment must be included. Both ground and air ambulances are actively involved in the casualty evacuation process, so a revised algorithm is needed to determine the demand, or frequency of calls per day, of a given configuration of transportation assets. This algorithm, which represents the core of the OPTEVAC model, considers the placement and allocation of evacuation assets for varying litter capacities.

The model performs the necessary distance calculations between the troop deployment nodes and the medical treatment facilites and begins to place the various types of available vehicles at treatment facilites to solve for the optimal vehicle configuration. Different placements of troop nodes and treatment facilites will of course yield different numbers and combinations of each vehicle located at the facilities. Not always will all of the available vehicles be required for a given treatment facility configuration since the model finds the minimum number of required vehicles to handle the casualty flow. Similarly, the results may indicate that more vehicles must be made available than were originally entered to handle the casualty load.

The OPTEVAC program utilizes the PLSCP module to solve for the optimal number of evacuation assets given a specific configuration of treatment facilities. As part of the PLSCP model, the first task is to determine each of the b_i variables (minimum number of required assets within demand area), and there will be as many b_i variables as there are demand nodes. Each b_i can be solved after the planner has entered the aforementioned required inputs into the program, and solving for each b_i will be accomplished by using Newton's Method (Swokowski 1983). Newton's Method will be used to find the b_i's after the user has entered the inputs required to solve for F_i (total evacuation asset demand). The required b_i's will be inserted into the constraint matrix of the PLSCP and hence complete the system of equations. There will be "i" (# of demand nodes) equations and "j" (# of treatment facilities) variables in the constraint matrix.

The following casualty evacuation problem, solved by the Simplex Method (Hillier and Lieberman 1990), demonstrates that OPTEVAC can yield solutions for multiple casualty evacuation asset types. In this facile example, consider a 60-day operation in which there are five Echelon II facilities, three troop deployment nodes, and that the average WIA and DNBI rates are 5.77 per 1000 strength per day and 4.22 per 1000 strength per day, respectively. Also assume that 10 ground ambulances and 2 air ambulances are available for deployment. The first calculation performed is to determine the frequency of calls per day at each troop deployment node within the theater. The necessary algorithm must factor in the placement and allocation of evacuation assets for varying litter capacities. Therefore, the PLSCP algorithm needs to be modified such that separate frequencies ($f_{k,ground}$ and $f_{k,air}$) and separate demand areas' vehicles ($b_{i,ground}$ and $b_{i,air}$) are computed for each type of vehicle. Defined below is an application of the algorithm that calculates the frequency of calls per day, when multiple vehicle assets are being used for evacuation (vehicle specifications are also listed). For instance, with only two types of vehicles (e.g., a single ground ambulance type and a single air ambulance type) this algorithm computes $f_{k,ground}$ and $f_{k,air}$ as follows:

	Litter capacity	Speed*	# available
Ground ambulance	4	16 km/hr	10
Air ambulance	6	150 km/hr	2

* The speeds of the dedicated assets include the loading and unloading of patients (Planning for Health Service Support 1994).

Define variables:

a/g ratio =

$$\frac{\textit{air ambulance speed}}{\textit{ground ambulance speed}} = \frac{150\ km/hr}{16\ km/hr} = 9.375\ ,$$

g capac = ground ambulance capacity,
a capac = air ambulance capacity,
g avail = number of ground ambulances available,
a avail = number of air ambulances available,
tot avail = the total number of ground and air ambulances available,

$$\text{g factor} = \frac{g\ avail}{tot\ avail} \times (g\ capac)\ ,$$

$$\text{a factor} = \frac{a\ avail}{tot\ avail} \times (a\ capac) \times (a/g\ ratio)\ ,$$

$$\text{Z factor} = \frac{maximum\ daily\ casualties}{g\ avail + a\ avail}\ ,$$

$$f_{k,ground} = \frac{(Z\ factor) \times (g\ factor)}{g\ capac}\ ,$$

$$f_{k,air} = \frac{(Z\ factor) \times (a\ factor)}{a\ capac}\ ,$$

The sum of the frequency of calls from all demand nodes within the theater ($f_{k,ground} + f_{k,air}$) will be used to calculate F_i (3), which will then lead to the calculations of the b_i's. By way of the Simplex Method, the PLSCP module determines that 3 ground ambulances and 1 air ambulance are required within M_1 (demand area 1), 4 ground ambulances and 1 air ambulance are required within M_2 (demand area 2), and 4 ground ambulances and 1 air ambulance are required within M_3 (demand area 3) as well (symbolically, for ground ambulances: $b_{1,ground} = 3$, $b_{2,ground} = 4$, $b_{3,ground} = 4$, and for air ambulances: $b_{1,air} = b_{2,air} = b_{3,air} = 1$).

Referring to Figure 1, the corresponding mathematical setup of the PLSCP, when solving for the optimal number of ground ambulances in the example scenario, is the following:

$$\begin{aligned} &\text{minimize } z = x_{11} + x_{21} + x_{31} + x_{41} + x_{51} \\ &\text{subject to} \\ &x_{11} + x_{21} \geq 3 \\ &x_{21} + x_{31} + x_{41} + x_{51} \geq 4 \\ &x_{41} + x_{51} \geq 4 \end{aligned} \tag{4}$$

where x_j are positive integers,

and when solving for the optimal number of air ambulances in the example scenario, the corresponding mathematical setup is the following:

$$\begin{aligned} &\text{minimize } z = x_{12} + x_{22} + x_{32} + x_{42} + x_{52} \\ &\text{subject to} \\ &x_{12} + x_{22} + x_{32} + x_{42} + x_{52} \geq 1 \\ &x_{12} + x_{22} + x_{32} + x_{42} + x_{52} \geq 1 \\ &x_{12} + x_{22} + x_{32} + x_{42} + x_{52} \geq 1 \end{aligned} \tag{5}$$

where x_j are positive integers.

In the aforementioned equations, the first subscript represents the Echelon II treatment facilities, and the second represents the type of vehicle. The first equation of the ground ambulance constraint matrix may be interpreted as "troop deployment node 1 is within range of medical treatment facilities 1 and 2 by way of ground ambulance". Similarly, each equation in the air ambulance constraint matrix indicates that none of the troop deployment nodes are out of air evacuation range for each of the five Echelon II treatment facilities.

After solving the above linear programming problems, the optimal solution was found to be placement of two ground ambulances at treatment facility 1, one at treatment facility 2, one at treatment facility 3, three at treatment facility 4, and one at treatment facility 5, for a total of eight necessary ground ambulances ($x_{11} = 2$, $x_{21} = 1$, $x_{31} = 1$, $x_{41} = 3$, $x_{51} = 1$, and $z_{ground} = 8$). By inserting these results into the constraint matrix of the PLSCP for ground ambulances, it is clear that the results are indeed a minimum:

$$\begin{array}{lcl} 2+1 \geq 3 & & 3 \geq 3 \\ 1+1+3+1 \geq 4 & => & 6 \geq 4 \\ 3+1 \geq 4 & & 4 \geq 4 \end{array}$$

thus the constraint equations hold, and

$$z_{ground} = x_{11}+x_{21}+x_{31}+x_{41}+x_{51} = 2+1+1+3+1 = 8$$

Further, only a single air ambulance was required ($z_{air} = 1$) and could be placed at any Echelon II facility, though placing it in the center facility would be the logical choice.

In this sample scenario, the optimization algorithms were able to minimize the number of evacuation assets required in the operation. Ten ground ambulances and two air ambulances were available for the operation, but only eight land transports and one air asset were necessary to perform the evacuation demands of the defined scenario. OPTEVAC finds the optimum numbers of required dedicated assets for a given theater of operations, and in the example scenario, the planner saved on resources by 20% for ground ambulances and 50% for air ambulances. These results demonstrate that the OPTEVAC model allows the user to determine the minimum numbers and optimal positioning of patient transportation assets required to meet the casualty handling demands of the operation.

3 SUMMARY

The PLSCP model is the appropriate choice for modeling the efficient transport of casualties between intra-theater medical facilities. The model will be integrated into a software environment that will allow planners to project the optimal numbers and placement of assets needed for casualty handling in a theater of operations. The user will be able to input the dimensions of the theater, the quantity, types, and placement of medical treatment facilities, as well as the types of evacuation assets available. The OPTEVAC casualty evacuation model will then provide the optimal number and position of the needed ground and air ambulances.

REFERENCES

Blood, C.G., & Gauker, E.D. 1993A. *Comparisons of medical presentation and admission rates during various combat operations*. (NHRC Report No. 93-12). San Diego, CA: Naval Health Research Center.

Blood, C.G., & Gauker, E.D. 1993B. The relationship between battle intensity and disease rates among Marine Corps infantry units. *Military Medicine* 158: 340-344.

Fleet Marine Force Manual (FMFM) 4-50, Health Service Support. Department of the Navy, Headquarters United States Marine Corps, Washington D.C. September 1990.

Hillier, F.S., & Lieberman, G.J. 1990. *Introduction to Operations Research* (Fifth Edition). New York: McGraw-Hill Publishing Company.

Naval Medical Doctrine Center. Naval Health Service Support lessons learned in Operation Desert Shield-Storm. Bureau of Medicine and Surgery, June 1991.

O'Donnell, E.R., & Blood, C.G. 1994. *A simulation program to project casualty and illness rates*. (NHRC Report No. 94-2B). San Diego, CA: Naval Health Research Center.

Planning for Health Service Support, FM 8-55. Headquarters, Department of the Army, Washington D.C. September 1994.

ReVelle, C. 1989. Review, extension and prediction in emergency services siting models, *European Journal of Operational Research* 40: 58-69.

ReVelle, C., & Hogan, K. 1989. The maximum reliability location problem and α-reliability p-center problem: derivatives of the probabilistic location set covering problem, *Annals of Operations Research* 18: 155-174.

Swokowski, E.W. 1983. *Calculus with Analytic Geometry* (Alternate Edition). Boston: Prindle, Weber & Schmidt.

AUTHOR BIOGRAPHIES

SCOTT C. SUNDSTROM is an Operations Research Analyst for the Operations Research Division at the Naval Health Research Center (NHRC) in San Diego, California. He is employed as a contractor with GEO-CENTERS, INC., a private research and development company. His research interests include mathematical modeling, linear programming, and statistical distributions. Mr. Sundstrom received his B.A. degree in Applied Mathematics from the University of California at San Diego, and he is a member of the Military Operations Research Society (MORS).

CHRISTOPHER G. BLOOD is Head of the Operations Research Division at NHRC in San Diego, California. He has been involved in analyzing ground and ship-based casualty incidence from past operations and is active in the development of medical planning tools to project casualty incidence in future scenarios and the related medical resource requirements. Mr. Blood is a member of MORS.

SERGE A. MATHENY is a Mathematical Analyst at NHRC in San Diego, California. He works as a contractor with the research and development firm, GEO-CENTERS, INC. Currently, he is working on medical evacuation aspects of ground military operations. His interests include linear programming, numerical analysis, and probability distributions. Mr. Matheny has an M.S. degree in Applied Mathematics from the University of Arizona, and he is a member of MORS.

Proceedings of the 1996 Winter Simulation Conference
ed. J. M. Charnes, D. J. Morrice, D. T. Brunner, and J. J. Swain

SIMULATION OF CASUALTY SUSTAINMENT DURING NAVAL COMBAT OPERATIONS

Christopher G. Blood
Michael S. Odowick
Jeffrey S. Marks

Naval Health Research Center
P.O. Box 85122
San Diego, CA 92186-5122, U.S.A.

ABSTRACT

Accurate forecasts of the casualties likely to be sustained during naval warfare allow sufficient resources to be programmed to meet the medical demands of the operation while at the same time minimizing resource overallocation. A simulation tool called the shipboard casualty projection system (SHIPCAS) has recently been developed to assist planners in medical resource programming.

SHIPCAS is a forecasting tool that projects WIA (wounded-in-action), KIA (killed-in-action), and DNBI (disease and non-battle injuries) incidence among naval surface forces. By simulating ship strikes during naval operations, SHIPCAS provides medical planners with the injury and illness estimates required to assist in determining the needed medical resources. The SHIPCAS system models casualties afloat by allowing the user to define a specific scenario in terms of task force composition, expected battle intensity, and length of operation. The model then produces graphical and tabular information detailing the total number of casualties across the operation, the daily average number of casualties, the maximum daily number of casualties, and the casualty rates per 1000 strength per day. In addition to projecting the number of ships hit and the resulting casualties, SHIPCAS also provides estimates of the temporal points in the operation during which ship strikes are most likely.

1 INTRODUCTION

Casualty estimates are provided for two levels of care: presentations and admissions. Presentations represent all injuries and illnesses requiring admission to a medical treatment facility; admissions are the subset of medical presentations that are retained for treatment three days or longer.

Two separate components are essential to projecting shipboard casualty incidence: 1) calculation of the number of ships likely to be struck during an operation, and 2) estimation of the wounded and killed which would result from the ship strikes. In order to provide operationally-relevant projections, historical naval warfare data from eighty operations and 850 ship attacks in the western Pacific during World War II were extracted and analyzed in terms of the hit rates and casualty incidence. (Blood 1990 and 1992). The operations were segregated into five separate battle intensities (no combat, light, moderate, heavy, intense) and ship attack rates, WIA and KIA frequencies, and distributions of weapons and ship types were then examined for each battle tempo. Shipboard DNBI rate projections are based upon ship type; while DNBI rates were found to vary by size of ship (Blood and Griffith 1990), combat status had but a slight impact on illness incidence (Blood et al. 1992), (Blood and Nirona 1990).

2 USE OF THE SIMULATION TOOL

2.1 Designing a Scenario

Simulation of the casualties sustained afloat begins with defining a task force and setting the operational parameters. To choose the ship types involved in the scenario, TASKFORCE is selected under the EDIT option. From this sub-menu, selection of MAJOR COMBATANTS brings up a screen for selection of the warships to be included in the task force. The options provided include Destroyers, Carriers, Frigates, and Cruisers. In addition to the major combatants, the task force may also include auxiliary ships. The number of auxiliary ships are specified in the same manner as for the major combatants.

Once the composition and size of task force have been set, the user returns to the main menu. Then the operational scenario may be defined by selecting SCENARIO under the EDIT option. The resulting screen permits the user to assign a battle intensity level and to designate the length of the operation. The choices for battle intensity are None, Light, Moderate, Heavy, and Intense. Lastly, the user enters the length of the operation on this screen in terms of the number of days the operation will last.

2.2 Running the Model

After the required user input has been entered into the model, the simulation may be executed under the RUN option in one of two modes: RUN mode yields casualty data based on a single simulation of the user-defined operational scenario; ITERATE mode computes mean casualty statistics across a number of simulated operations. The "single" simulation mode provides 1) tabular information on the ships hit and the weapons involved, 2) a table of casualties sustained by ship type, 3) a graph projecting temporal points in the operation in which the ship hits occur, 4) a graph delineating casualty sustainment over time, and 5) summary statistics across the operation. "ITERATE" option mode requires the user to designate the number of iterations of the simulated scenario across which the casualty statistics will be calculated. The projection program then computes means across the iterations for the number of ship hits, frequencies and rates of WIA, frequencies and rates of KIA, maximum daily casualty load, and total operational casualty load. A tabular display of projected DNBI (disease and non-battle injury) frequencies and rates, for each ship type in the task force, is available in both RUN and ITERATE mode.

2.3 Viewing the Results

The VIEW option from the main menu allows the user to choose from HIT DISTRIBUTION, CASUALTY DISTRIBUTION, DNBI DISTRIBUTION, and SUMMARY STATISTICS sub-options to view the output of the model. The HIT DISTRIBUTION is available separately for major combatant ships and auxiliary vessels.

Additional information in the form of a graphical display of the number of hits that occurred each day of the operation can be obtained by selecting the GRAPH button.

The second choice of viewing options is the CASUALTY DISTRIBUTION screen. This window depicts WIA and KIA frequencies, and rates per 1000 strength per day for both combat and auxiliary ships. A graphical display of the projected casualties on a daily basis is available by selection of the GRAPH button

The DNBI DISTRIBUTION screen is similar to the CASUALTY DISTRIBUTION screen except that it provides tabular information on the disease and non-battle injuries projected to occur during the operation. The last option within the VIEW menu is the SUMMARY screen. This option provides a tabular display of summary statistics across the operation. These statistics include mean daily casualty frequencies and mean casualty rates per 1000 strength per day as well as the maximum single day casualty load and total operation casualty load.

3 STATISTICAL BASIS AND MODELING METHODOLOGY

3.1 Hit Rates and Battle Intensity

Ship hit rates were computed for each of the eighty WWII operations analyzed. Because large numbers of ships were involved in many operations and relatively few ships were actually struck, rates were computed as hits per 100 ship days ((Hits/Ships*Days)*100). The data across all operations yielded a range of 0.0 to 50.00 hits per 100 ship days. These rates were partitioned into five groups of near equal observations to represent the varying battle intensities required by medical resource models. The range and mean ship hit rate for each battle intensity level are:

	RANGE	MEAN
NONE	0.0000	0.0000
LIGHT	0.0295 to 0.2579	0.1739
MODERATE	0.2738 to 0.6095	0.4297
HIGH	0.7067 to 3.8462	1.0236
INTENSE	5.8824 to 50.000	11.7798

The first step that the SHIPCAS program performs is the simulation of the number of hits that occur for the user-designed scenario. Variates from a normal distribution are generated around the estimated parameters of the hit rate for the chosen battle intensity to determine the number of hits occurring during the operation. Statistically, this process is represented by equation (1):

$$H = (Y/100)S \quad (1)$$

where:

H = the number of hits (in the hypothetical operation),

Y ~ N(m,s), variates from a normal distribution,

m = mean number of hits per 100 ship days (from the empirical data),

s = standard deviation of the number of hits per 100 ship days (from the empirical data),

S = the number of days in the operation.

3.2 Timing of Hits

The next process is simulation of the temporal points in the operation when each of the hits occurred. Analysis of the empirical data allowed the computation of a mean inter-arrival time between hits (3.8 days). Random deviates based on this mean and drawn from an exponential distribution then provide projected days in the operation on which each hit occurs. Equation (2) symbolizes this process.

$$T \sim \text{expo}(b), \quad (2)$$

T = inter-arrival time,

b = estimated mean of the exponential distribution.

3.3 Ship Hit Determination

Examination of the distribution of the ship types that were attacked during the historical scenarios, as well as their overall presence, provides the foundation for the ship hit determinations within the SHIPCAS model. The relative risks of each ship type were computed to control for the amount of days of exposure. The distributions of hit percentage, total ship days percentage, and relative risk were:

Combatant	% of hit ships	Ship days %	Relative Risk
Destroyers	55.4	57.3	0.97
Carriers	15.7	12.3	1.28
Frigates	8.2	20.4	0.40
Cruisers	20.6	9.9	2.08

Auxiliaries	% of hit ships	Ship days %	Relative Risk
Cargo	7.1	6.2	1.14
Mine Sweepers	32.0	15.9	2.01
Torpedo Boats	30.6	24.5	1.25
Tank Landing	27.2	43.0	0.63
Transports	3.1	10.4	0.30

To account for the relative risk of ships in the user defined scenario, equation (4) shows how each empirically computed risk was weighted by the number of ships the user chose of that type:

$$r_k = n_k x_k \quad (4)$$

where:

k = each individual ship category (i.e., destroyers, carriers, etc.),

r_k= the weighted relative risk for ship type k in the user defined task force,

n_k= the number of ships of category k in the task force,

x_k = the relative risk of a ship of category k.

The probability of any individual ship in the task force being hit is then calculated in equation (5) by dividing the relative risk of a ship of that type by the aggregated weighted risks of all ship types:

$$Y_i = x_i / \sum_k r_k \qquad (5)$$

where:

i = the category of the target ship (i.e., destroyer, carrier, etc.),

Y_i = the probability of an individual ship of type i being hit,

x_i = the relative risk of a ship of category i,

k = each individual ship category (i.e., destroyers, carriers, etc.),

r_k = the weighted relative risk for ship type k in the user defined task force.

With the individual ship probabilities normalized to 100 percent, they are then aggregated to form a continuous distribution between 0 and 1. A uniform random variate is then chosen between 0 and 1 to determine which ship is struck for each hit during the operation.

3.4 Weapon Determination

The next step in the shipboard casualty simulation process is designation of the weapon type associated with each hit. Again using the empirical data, distributions of attacks by weapons, ships, and battle intensity levels were analyzed and percentage distributions were computed for 1) each weapon by ship combination, 2) weapons within each battle intensity level, and 3) overall weapon distribution. These percentage distributions were determined from a matrix of weapon probabilities for each weapon hitting each ship type during each battle intensity. For cells where no events have occurred, a value for each weapon type was determined using equation (6) as follows:

$$\mathbf{W_{mjk} = O_m + (I_{mj} - O_m) + (T_{mk} - O_m)} \qquad (6)$$

where:

m = weapon,

j = battle intensity,

k = ship type,

W_{mjk} = weapon m percent with battle intensity j and ship k,

O_m = overall proportion of weapon m striking any ship,

I_{mj} = proportion of weapon m striking during an operation of j battle intensity,

T_{mk} = proportion of weapon m striking ship type k.

Combining the probabilities of the different weapon and ship type combinations, as with ship type determination alone, yields a distribution between 0 and 1. A uniform random deviate is then chosen to determine the specific weapon that strikes the ship.

3.5 WIA and KIA

Empirical data (Blood 1992) indicating the numbers of WIA and KIA sustained in various attacks were used to calculate the average number of casualties for different "weapon by ship" combinations. Goodness of fit analysis indicated that shipboard WIA and KIA incidence are best represented by a Poisson process symbolized in equation (7). SHIPCAS, therefore, yields its wounded and killed projections by drawing a random number from a Poisson process based on the mean frequency of casualties for each weapon by ship combination.

$$X \sim \text{Poisson}(l) \qquad (7)$$

where:

X is the projected casualties (WIA or KIA) with estimated parameter l.

In addition to these casualties, analysis of the historical data provided information on casualties that occur during operations that are not dependent upon a ship being hit. These "background" casualties, which may result from events such as the firing of weapons, near misses, or defensive maneuvers, are derived by drawing a random deviate from a normal distribution surrounding the mean background casualties observed for that particular battle intensity. Once the total numbers of casualties are projected, the WIA and KIA rates per 1000 strength per day are computed based on the crew complements of each ship.

3.6 Disease and Non-battle Injuries

Studies have indicated that ship type is a significant factor in disease and non-battle injury incidence while combat status had little practical effect on DNBI incidence. Rates of DNBI, therefore, are averaged across the different ship types in the task force and provide the basis for the simulation of disease and non-battle injury rates. Mean rates of DNBI incidence are transformed into frequencies, based on the designated length of the operation and the crew complements, and DNBI projections are then generated by drawing a deviate from a normal random distribution. This quantity is then partitioned into a disease component and a non-battle injury component based on distributions observed in the empirical data (Blood et al. 1992).

4 CONCLUSION

The shipboard casualty projection system simulates casualties likely to be sustained during various naval combat scenarios. Run in "single" simulation mode, SHIPCAS accurately depicts the variability in ship hit rates and casualty incidence witnessed in previous naval engagements. Deriving projections from "multiple" iterations of a single scenario successfully approximates the mean ship hit rates and casualty sustainment of the empirical data. These forecasts may, in turn, be used as input to the types of models which determine the specific bed, supply, and health care personnel requirements (Systems Research 1992 and Galarza 1987). It is noted that these projections are based on ship structures and weapon systems which have since undergone significant technological advances. Currently, data from a subject matter expert panel of weapons system and ship structure engineers are being used to modify the SHIPCAS model projections to reflect the technological advances in these areas.

REFERENCES

Blood, C. G. 1990. Shipboard and ground troop casualty rates among navy and marine corps presonnel during world war II operations. Report No. 90-16. Naval Health Research Center, San Diego, CA.

Blood, C. G., and D. K. Griffith. 1990. Ship size as a factor in illness incidence among U.S. navy vessels. *Military Medicine* 155(7):310-314.

Blood, C. G., and C. B. Nirona. 1990. Outpatient illness incidence aboard U.S. navy ships during and after the Vietnam conflict. *Military Medicine* 155(10):472-476.

Blood, C. G. 1992. Analyses of battle casualties by weapon type aboard U.S. navy warships. *Military Medicine* 157(3):124-130.

Blood, C. G., W. M. Pugh, E. D. Gauker, and D. M. Pearsall. 1992. Comparisons of wartime and peacetime disease and non-battle injury rates aboard ships of the royal british navy. *Military Medicine* 157(12):641-644.

Galarza, J. G. Using the deployable medical systems clinical data base for material requirements. Presentation to *Army Operations Research Symposium* (AORS XXVI). Fort Lee, VA. October 1987.

Systems Research and Applications Corporation: Preliminary software requirements for the medical planning and execution system of joint operation planning execution system. TR-91-674-0234.1. January 1992.

AUTHOR BIOGRAPHIES

CHRISTOPHER G. BLOOD is head of the Operations Research Division at the Naval Health Research Center (NHRC) in San Diego, California. He has been involved in analyzing ground and ship-based casualty incidence from past operations and is active in the development of medical planning tools to project casualty incidence in future scenarios and the related medical resource requirements. Mr. Blood is a member of the Military Operations Research Society.

MICHAEL S. ODOWICK is a former GEO-Centers, Inc. research analyst who worked at NHRC on shipboard casualty projections for future scenarios. Mr. Odowick was the lead programmer in developing the first version of the SHIPCAS shipboard casualty planning tool.

JEFFREY S. MARKS is an Operations Research Analyst and Group Supervisor working at the Naval Health Research Center for GEO-Centers, Inc. Mr. Marks' work focuses on projections of casualties and examining the factors underlying fluctuations in casualty incidence. Mr. Marks is an adjunct professor in the College of Business at San Diego State University, a member of the American Statistical Association, and a member of the Military Operations Research Society.

Proceedings of the 1996 Winter Simulation Conference
ed. J. M. Charnes, D. J. Morrice, D. T. Brunner, and J. J. Swain

REENGINEERING THE UNITED STATES ARMY'S TACTICAL COMMAND AND CONTROL OPERATIONAL ARCHITECTURE FOR INFORMATION OPERATIONS

Robert G. Phelan Jr.
Michael L. McGinnis

Operations Research Center
United States Military Academy
West Point, New York 10996, U.S.A

ABSTRACT

During the past few years, the Army has undertaken a major effort to integrate computer, information, and communication technologies throughout the force. These technologies have made a profound impact on the Army's tactical command and control (C^2) processes, especially at division and below. This paper presents a computer simulation model to support the reengineering of the United States Army's staff organization at division and below for information operations. The simulation model under development will provide force planners with a tool for methodically conducting comparative analyses of specific alternatives for reengineering tactical level Army staffs to improve efficiency and effectiveness of staff work related to information operations.

1 INTRODUCTION

For the United States Armed Forces to win decisively on tomorrow's battlefields, military commanders must make well-informed, timely decisions. The high tempo of joint and coalition military operations during recent years has accelerated the military decision making processes at all levels of command as never before. The pace of military operations demands that military staffs provide commanders with timely information for making decisions. This is necessary to give our land combat forces sufficient time to carefully and thoroughly plan the military operation based on unambiguous commander's guidance.

The Army is presently attempting to meet the future information needs of military commanders by integrating communication, computer, and information technologies at all levels of the force. However, these initial efforts have had the unintended effect of overloading commanders and staffs with battlefield information, thereby degrading the military decision making process rather than making it more efficient. Overcoming these problems will require that the Army reengineer its command and control architecture currently based on a stove-piped military staff organized around traditional battlefield operating systems. Specifically, the military staffs from battalion through division must be reengineered for information operations that consists of the processes for collecting, processing, storing, retrieving, disseminating, displaying, protecting, and denying information.

We present work in sponsored by the Army Digitization Office (ADO), Headquarters, Department of the Army, to develop a computer simulation model for reengineering the current tactical level army staffs into staffs organized for information operations. The simulation model will permit Army force developers to evaluate alternatives for reorganizing military staffs based on measures of performance important to military commanders. We also discuss how the model can support development of the operational and system architectures of the Army Enterprise Strategy.

2 BACKGROUND

Battlefield information requirements of commanders focuses primarily on the status of friendly and enemy forces and the enemy's intentions and capabilities. As straightforward as this may seem, Operations Desert Shield and Desert Storm provide anecdotal evidence of how communication, computer, and information technologies have complicated information operations. For example, according to Lieutenant Colonel John Burke, Office of the Director for Information Systems for Command, Control, Communications and Computers, Department of the Army, approximately 700,000 telephone calls and 152,000 messages were exchanged daily between military units within the theater of operations using 35,000 radio frequencies. Managing the volume of information generated by fast-paced military operations is necessary to provide manage the type and amount of information flowing to military commanders.

Currently, tactical information generally flows along traditional lines of communication defined by traditional relationships between battlefield operating systems. Battlefield operating systems include maneuver, fire support, air defense, intelligence, mobility/counter-mobility, survivability, and combat service support. Standard operating procedures and doctrine generally determine the types of information exchanged between the battlefield operating system staff elements and command and control centers. Tactical Operation Centers (TOCs) serve as the command and control nodes for processing and analyzing information for commander from battalion through division. The staff elements acquire commander's critical information requirements (CCIR) and plan future operations. Key staff actions include developing, analyzing and briefing courses of action, performing intelligence preparation of the battlefield, and wargaming.

To be successful, military commanders and staffs at all levels must effectively manage battlefield information and the time available for planning and executing military operations. Through *FORCE XXI Joint Venture*, the Army is digitizing the battlefield and creating an integrated information architecture where battlefield information is shared horizontally and vertically across the force to enhance command and control. In the end, there is little doubt that information technology will accelerate the operational tempo by compressing the C^2 cycle for military operations. Figure 1 depicts the *I*nformation operations-*D*ecisionmaking-military *A*ction (IDA) loop of the command and control process (McGinnis, Technical Report FY95-6, 1995).

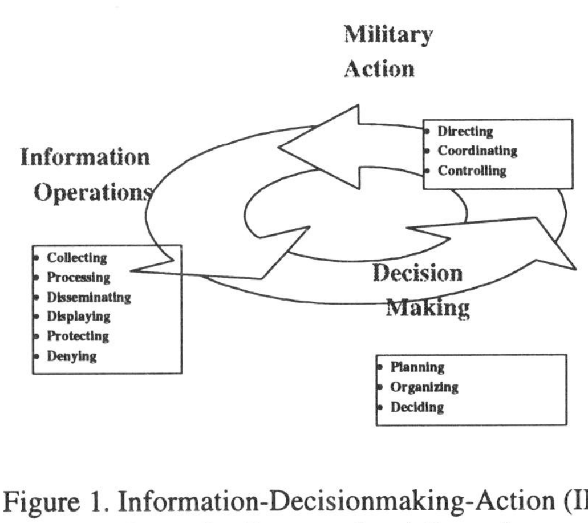

Figure 1. Information-Decisionmaking-Action (IDA) Loop for Command and Control

A major goal of FORCE XXI Joint Venture is to use the information technology, combined with reengineering the military staffs, to create a *relevant common picture* of the battlefield. This involves collecting, analyzing, and disseminating information about the terrain, weather and status and disposition of friendly and enemy forces that can be shared by all levels of command. A higher-level concept closely related to the relevant common picture is *dominant situational awareness* of the battle area (McGinnis et. al., Technical Report FY95-5, 1995). This involves interpreting and exploiting the relevant common picture of the battle area to predict likely enemy actions.

3 THE ARMY ENTERPRISE STRATEGY

In addition to Force XXI Joint Venture, the Army has simultaneously undertaken the development of a high-level data architecture to support future information operations called the *Army Enterprise Strategy*. This strategy involves designing, developing, and implementing the Army's operational, system, and technical architectures. When fully developed, the strategy will specify the Army's information system requirements and how the system will be integrated from the strategic to tactical levels of command. Part of architecture development is the requirement to provide interoperability between the Army's battlefield operating systems, other branches of military service, and allied forces as well.

The Army's operational architecture defines the context and framework for fielding Command, Control, Communications, Computers, and Intelligence (C^4I) systems that meet operational requirements. It defines the force elements and the requirement to exchange information between those elements.

The system architecture describes the systems required to support the operational architecture concept. It also defines C^2 nodes, connected through communications equipment to accomplish the information exchange.

The technical architecture identifies the rules, conditions, and constraints for the technical requirements of the common operating environment and for the data model. This is a preliminary step towards a seamless, interoperable architecture for exchanging information electronically across all levels of command: strategic, operational and tactical.

4 THE ARMY BATTLE COMMAND SYSTEM

An important project related to the Army Enterprise Strategy is the future Army Battle Command System (ABCS) that will link operations on the battlefield with the sustaining base. The design of this system will rely heavily on the operational, system, and technical architectures of the Army Enterprise Strategy discussed above. An important characteristic of the ABCS is seamless exchange information across all battlefield operating systems and levels of command.

Military operations in the 21st century will be conducted by projecting US forces centrally-based within the continental United States to crisis situations around the world. The nature of future operations will require that military units be ready to deploy with little or no notice. The forces will be expected to conduct joint military operations immediately upon arriving in the area of operations.

The ABCS will satisfy two near-term critical battle commander C^2 requirements: development of the Relevant Common (RCP) picture and interoperability. The ABCS will be the Army's integrated command and control system that will carry it into the 21st century It will provide future commanders with a versatile C^2 system supporting them through all force projection stages from mobilization and pre-deployment activities through operations and back to re-deployment and reconstitution. It is the physical realization of the Army Enterprise Strategy.

5 OPERATIONAL ARCHITECTURE

Recent advances in information technology have changed the way many activities conduct business. For example, the world wide web is providing users access to a wide variety of information, products, and services. The Army is currently researching the development of a web-like information architecture for conducting at division and below. The operational architecture will reflect force structure changes for conducting information operations. The operational architecture will also guide the development of the systems architecture by identifying operational requirements for equipment based on likely future operations.

The operational architecture will involve modeling current tactical operations "as is" to establish a baseline for modeling future operations. Operational modeling will include routine command and control (C^2) tasks and the information required to support those tasks. It will also model the flow of information through the C^4I network. The operational model will also simulate the information database that supports the C^2 and C^4I operations. Figure 2 illustrates the dynamics of the processes to be modeled.

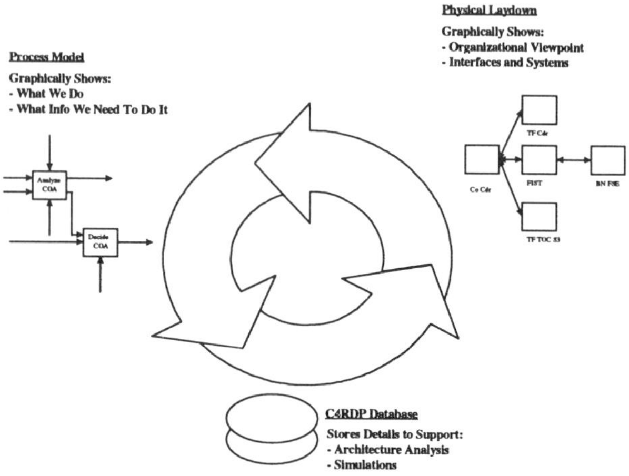

Figure 2. Operational Architecture Modeling

An activity model represents the activities performed by the tactical military units during military operations. For command and control modeling, the primary resource to be modeled is battlefield information used by the commander and staff for making decisions about the tactical situation. Battlefield information is broken down into two categories: critical and routine. The activity model is constructed by decomposing the C^2 processes into tasks and subtasks that must be accomplished by the tactical-level military unit.

The physical laydown of the tasks and subtasks provides a graphical representation of the organizational, C^2, and C^4I system relationships that exist within and between tactical organizations.

The primary source of model data and data exchanges for this study is the Command, Control, Communications and Computers Requirements Definition Program (C^4RDP) database maintained by the Army's Signal Center at Fort Gordon, Georgia. It contains over 130,000 approved doctrinal information exchanges that occur at all levels of command from the National Command Authority to the lowest tactical unit. It is the Army's primary source of information exchanged between combat, combat support, and combat service support units.

Two other models related to the development of the command and control activity model for the operational architecture are the Network Assessment Model (NAM) and the C^2NET model. The Network Assessment Model, also maintained by the Signal Center at Fort Gordon, is a communications system simulation model of various communication architectures. It models friendly and enemy force C^2 nodes, their communications equipment, and voice and data communication traffic.

The TRADOC Analysis Center (TRAC), Fort Leavenworth, Kansas, developed the C^2NET model in 1993 to support Force XXI. It is an interactive C^2 workflow model that uses petri-nets to simulate tactical-level command and control processes. C^2NET models military staff processes from Brigade to Corps level.

6 PROCESS MODELING

Functional decomposition of the tactical command and control process into tasks and subtasks is accomplished using the Integrated Definition Method (IDEF) modeling method, specifically, IDEF0 to model C^2 activities and processes. The size and complexity of C^2 systems made it necessary to model the process as a system of sub-systems. This will create a group of IDEF0 models that can be logically joined together based on information exchanges. IDEF3, used for process modeling, will incorporate IDEF0 modeling.

Preliminary work suggests that reengineering the Army's tactical-level operational architecture for information operations as shown in Figure 1 will require changes to the current C^2 system. This work will be accomplished in several steps. First, we diagram the current C^2 processes to depict relationships between activities where events occur sequentially and simultaneously, and to identify precedence conditions between the activities, if any. Next, we use subject matter experts to brainstorm ideas for changing the command and control processes based on reengineering for information operations.

Currently, tactical C^2 systems are organized around the seven battlefield operating systems. The viewpoint taken in our modeling is that of reengineering command and control for the benefit of the tactical commander.

Work-in-progress includes using IDEF0 to IDEF3 linkages to model changes to tactical-level command and control based on integrating new information technology into the process. Figures 3 and 4 depict a partial decomposition of the future command and control process system from the commander's viewpoint using IDEF. The modeling described here is a collaborative effort between the authors and the TRADOC Program Integration Office (TPIO), Fort Leavenworth, Kansas.

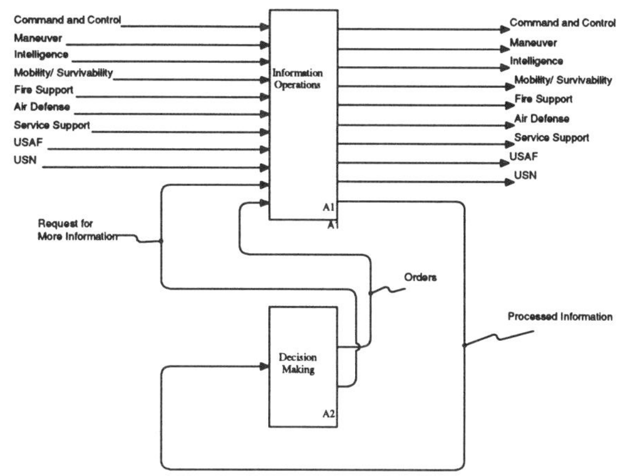

Figure 3. First Level Decomposition of C^2

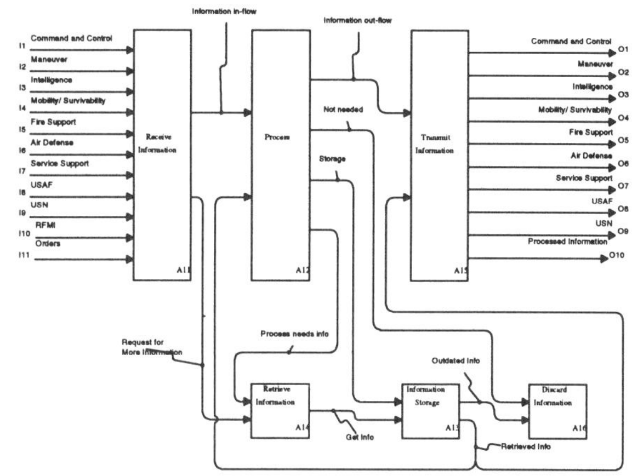

Figure 4. Decomposition of C^2 Information Operations

7 PROCESS SIMULATION

A process-oriented simulation under development will be used for conducting experimental tests to measure the relative effectiveness of the current C^2 system versus alternatives for future systems. Apparently, the reengineering the Army's tactical command and control processes using a discrete event, process-oriented simulation represents a new application of this modeling approach, at least for the Army. If successful, this methodology will (hopefully) be incorporated into existing IDEF0 process models to provide the Army with a flexible, dynamic tool for conducting analysis of competing force design alternatives. This section presents the analytical framework for modeling the C^2 system. The linkages between the current IDEF0 models and the discrete event process simulation model proposed here are also discussed.

Consider the simple C^2 information network depicted in Figure 5. This C^2 node consists of two M/M/1 queues where information arrives into the node with rate λ_1. The information is processed by Server 1 at a rate of μ and is then disseminated. It immediately joins queue 2 for more processing by Server 2 also at rate μ. When Server 2 is finished, the information departs the C^2 node. This model assumes no information is lost and each queue has infinite capacity.

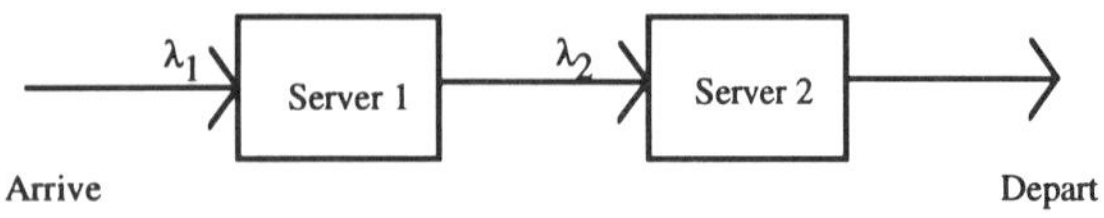

Figure 5. Simplified C^2 Node System

Although each server within this network has a service rate of μ_i associated with it, the information cannot be processed any faster than there is information in the system. Therefore, the rate out of Server *i* is 0 when that server is in state 0 and μ_i otherwise. The departure rate from Server 1 to queue 2 determines queue 2's arrival rate. In a steady state system this is also queue 1's arrival rate.

Next, consider a generalized network of *k* servers where information arrives to the C^2 node system and to each Server *i*, $i \in \{1,2,\ldots,k\}$, according to a Poisson process with rate r_i. The information then joins queue *i* until processed by the server. Then information is disseminated to the queue for Server *j*, $j \in \{1,2,\ldots,k\}$, with probability P_{ij} such that $\sum_{j=1}^{k} P_{ij} \le 1$. Alternatively, the information departs Server *i* with probability $P_{exit} = 1 - \sum_{j=1}^{k} P_{ij}$. Queuing networks of this type are commonly referred to as *Jackson Networks* (Walrand, 1988).

The arrival rates at any server within the network are characterized by the following flow conservation equations: $\lambda_j = r_j + \sum_{i=1}^{k} P_{ij}\lambda_i$ for $i \in \{1,2,\ldots,k\}$. $P_{ij}\lambda_i$ denotes the arrival rate to queue *j* for information departing from Server *i*.

Figure 6 depicts a two-server Jackson Network that enumerates the possible paths for disseminating information are enumerated. The values give the information flow rates along each path of the network.

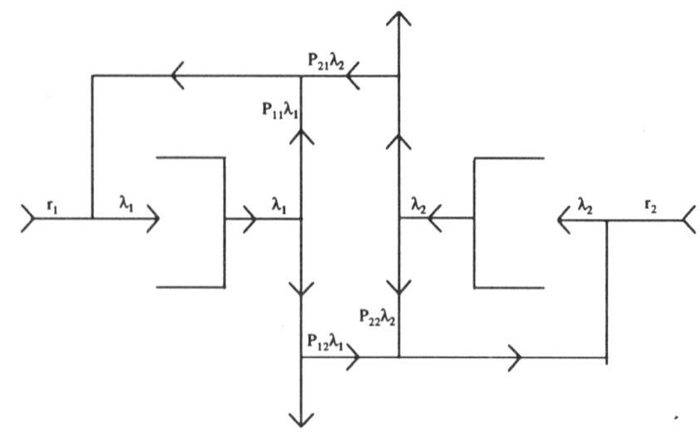

Figure 6. Expanded Jackson Network of a C^2 Node

In general, open Markovian networks can be described in terms of a Jackson Network. A probability routing matrix **P** describes the relative probability that information may take a particular route once it entered the C^2 system. If any $P_{ij} = 0$, then the information path defined from i to j does not exist. If $\mathbf{P} = \mathbf{0}$ then the network is a system of independent queues where information may arrive at any server of the C^2 node and all information departs the C^2 node upon completion of service.

Many aspects of the tactical command and control system can be modeled using Jackson networks. Each stovepiped element within the current C^2 system, and the associated staff functions may be modeled as a M/M/S queuing system. *S* represents the number of independent servers for processing information within a staff section. The C^2 node, also referred to as a tactical operations center (TOC), may be modeled as an interconnected series of Jackson networks comprised of staff sections representing each battlefield operating systems. Figure 7 diagrams the flow of information within a command and control node.

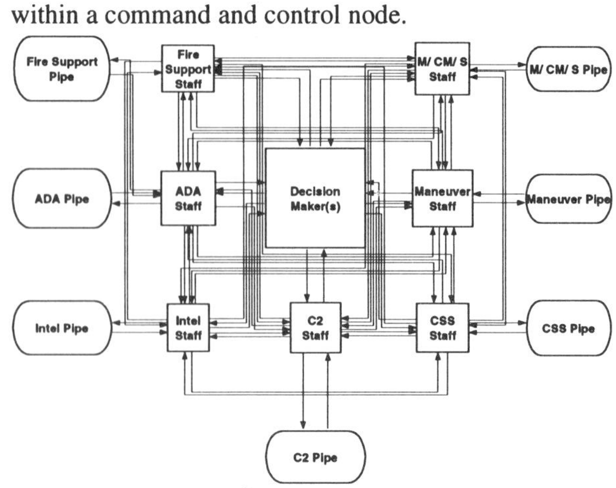

Figure 7. C^2 Node Information Flow

If we assume a Poisson arrival process for information that enters the C^2 node and exponentially distributed information processing times by each staff sections (i.e., server), then this C^2 node may be modeled as a Jackson Network. $\lambda_j = r_i + \sum_{i=1}^{k} P_{ij}\lambda_i$, $i \neq j$, describes the systems conservation of flow equations for information, where j represents either any battlefield operating system staff section or the military commander. For this system, we do not allow for the case where $P_{ij} = 0$. The only external arrival of information permitted is that designated for the commander. Stability of the C^2 system requires that $\mu_j > \lambda_j$ otherwise queue j would grow out of control although this is a fairly common occurrence in real-world situations.

Figure 8 depicts one representation of a future C^2 node reengineered for information operations. In this system, we require $\mu_j > \lambda_j$ for all functions except the military decision making cycle where human processors may be in the loop. Using a discrete-event process simulation, this C^2 system representation can be analyzed in terms of making tradeoffs between human and computer-based methods for processing information. The system can also be used to study the effects of information overload on humans or to determine acceptable arrival rates of information for human processors.

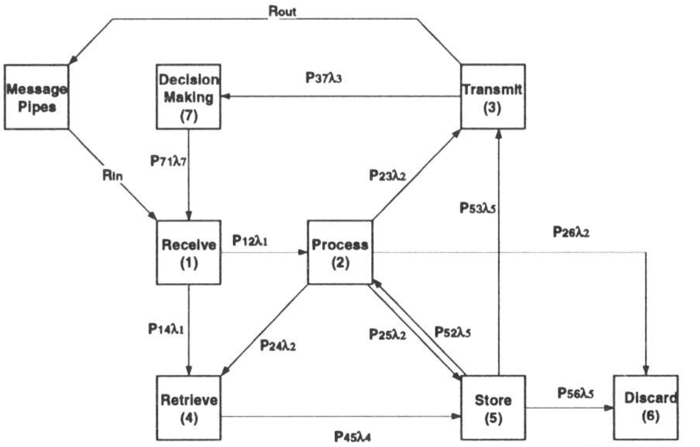

Figure 8. Framework for Analyzing Future C^2 System Information Flow Rates

Modeling either of the systems as a Poisson process assumes both exponential interarrivals and service times. However, there are no validated data sets that would to confirm this assumption. One experiment that will attempt to do this is called Prairie Warrior 1996 conducted at Fort Leavenworth. This experiment will study alternatives for organizing staff sections manned by students from the Army's Command and General Staff College and equipped with advanced computer and information technology. The Battle Command Battle Laboratory of Fort Gordon is designing experiments to record network traffic data during the exercise. It is expected that this data can then be used to determine interarrival and service time distributions for the staffs and staff functions under consideration. These distributions can then be used in the discrete event simulation model being developed for studying the C^2 system.

Figure 9 illustrated how the discrete event process model fits within the existing hierarchy of data and process models. Arrival and service rate estimates are taken from the C^4RDP and C^2NET respectively.

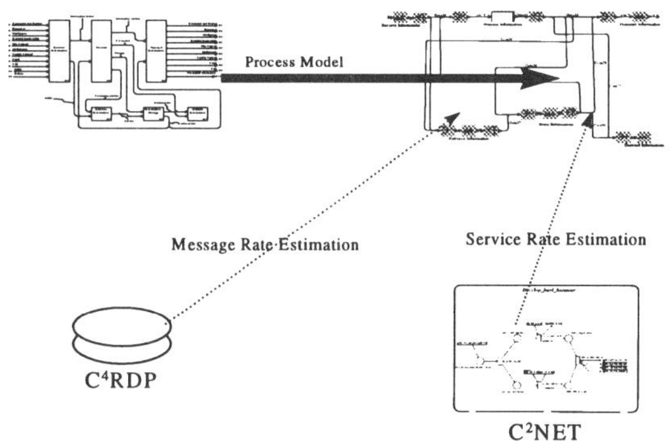

Figure 9. Linking the Discrete Even Process Simulation Model to Existing Data and Process Models

Currently, information message arrival and processing rates used for the discrete event process simulation model were obtained from the C^4RDP. The data reflects the flow of information to and from a "typical" division tactical operations center. The information exchanges were cross-referenced with the first level functional decomposition of the battlefield as defined by TRADOC's blueprint of the battlefield (TRADOC, 1993). Two important measures of information exchange provided by the C^4RDP that are also used as performance measure in the discrete event simulation are *speed of service* and *cost of failure*.

Speed of service reflects the amount of time that can be allowed to pass from when the information is sent to when it is received. This time varies from less than one second to over eight hours.

Cost of failure describes the impact on mission accomplishment from the originator's viewpoint if the transmission of the information is not completed. Categories of cost of failure are given in Table 1.

Table 1. *Cost of Failure* Categories

Indispensable	Causes complete mission failure if not satisfied 100% at any time
Critical	Causes complete mission failure if not satisfied over 75% over extended time periods
Essential	Has a negative impact on the mission if not satisfied over 75% over extended time periods

The next three tables give examples of the type of data available from the C^{4}RDP. For example, this includes mean interarrival times for messages arriving or leaving the fire support section. The Network Assessment Model uses similar data to estimate arrival rates and assumes exponential interarrivals.

Table 2. Division In-Flow Mean Time (hours/message)

System	Speed of Service	Critical	Essential	Indispensable
Fire Support	5-10 Sec	6.00		0.50
Fire Support	25-59 Sec	0.06		24.00
Fire Support	1-10 Min	0.22	0.41	0.17
Fire Support	10-60 Min	4.80	1.20	0.20
Fire Support	1-2 Hrs	24.00	0.32	2.40
Fire Support	2-3 Hrs	12.00		
Fire Support	4-8 Hrs	24.00		

Table 3. Division Internal Information Flow Mean Time (hours/message)

System	Speed of Service	Critical	Essential	Indispensable
Fire Support	5-10 Sec	8.00		
Fire Support	25-59 Sec	0.03		
Fire Support	1-10 Min	0.73		0.28
Fire Support	10-60 Min		0.47	2.67
Fire Support	1-2 Hrs		6.00	
Fire Support	2-3 Hrs		0.75	
Fire Support	4-8 Hrs			1.33

Table 4. Division Information Out-Flow Mean Time (hours/message)

System	Speed of Service	Critical	Essential	Indispensable
Fire Support	5-10 Sec	2.67		0.10
Fire Support	25-59 Sec	0.01		
Fire Support	1-10 Min	0.01	0.43	0.44
Fire Support	10-60 Min	0.50	0.63	6
Fire Support	1-2 Hrs		6	6

The data presented here, among other data, is used to estimate the internal flow rates and routing probabilities for the internal Jackson network. It has also been useful in the development of performance measures. However, more effort is required in this area. For example, the US Army Training and Doctrine Command (TRADOC) Headquarters is currently spending considerable effort to estimate the times that tactical units routinely spend to perform doctrinally approved command and control tasks with tactical operation centers from brigade to corps level for the C^{2}NET model.

The simulation package selected for the discrete event model is *SIMPROCESS*. This simulation software is currently being beta tested. The simulation software is a process oriented and permits dynamic modeling for simulating the stochastic flow of items through user defined processes. Figure 10 depicts a sample screen created for a future C^2 system using *SIMPROCESS*.

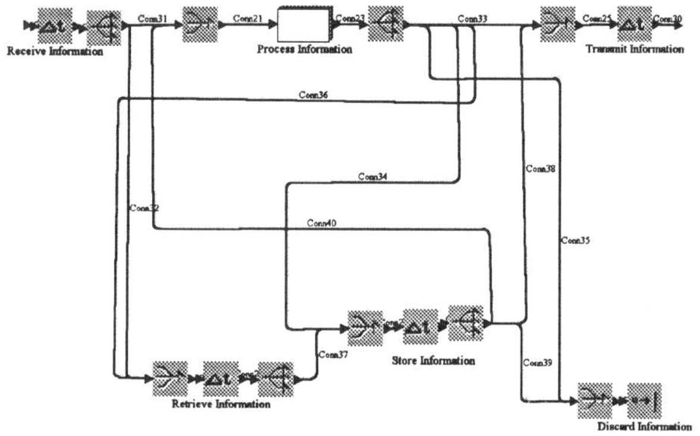

Figure 10. *SIMPROCESS* representation of a C^2 System

8 FUTURE WORK

Future work will focus on completing those tasks required to analyze different staff alternatives. The first is incorporating IDEF0 process models under development by TRADOC into the simulation model. The second is the incorporation of a data network simulation focusing on the internal movement of information. The third major area involves devising and incorporating an analytical memory model to refine the storage of information function.

Future operations may involve situations where commanders and their staffs are not physically collocated. A supporting data network will be constructed to pass information between systems and locations. The network topology and the physical medium will directly influence information flows. A network simulation model will be overlaid onto the process model.

Information storage and retrieval will also play a major factor in the development of both tactical situational awareness and the supporting data network's performance. For the Force XXI Army, it is imperative that each staff section use the most current and up to date data. Many of the problems facing our future systems are strikingly similar to today's parallel computing platforms. Raw and processed data may be stored in some central location or at each of the workstations. During our model development, we plan to apply current analytical parallel computing performance models to ascertain the impact of various data storage alternatives.

REFERENCES

Almasi, G.S. and Gottlieb, A. 1989. *Highly Parallel Computing*, The Benjamin/ Cummings Publishing Co. Redwood City, CA.

Banks, J. and Carson, J.S. 1984. *Discrete-Event System Simulation*, Prentice-Hall, Englewood Cliffs, NJ.

Blanchard B.S. and Fabrycky, W.J. 1990. *Systems Engineering and Analysis*, Prentice-Hall, Englewood Cliffs, NJ.

CACI Products Co. 1996. *SIMPROCESS User's Guide*, CACI Products Co., LaJolla, CA.

Campen, A.D., 1994. *The First Information War,* AFCEA International Press, Fairfax, VA.

Director of Information Systems for Command, Control, Communications and Computers, 1996. *Army Technical Architecture*, Headquarters, Department of the Army, Washington, D.C.

Josephson, E.H. and Macedonia, R.M. 1994. *Fighting Smarter: Leveraging Information Age Technology*, The Institute of Land Warfare, Association of the United States Army, Arlington, VA.

Kleinrock, L., 1975. *Queueing Systems, Volume 1: Theory*, John Wiley & Sons, New York, NY..

Law, A.M. and Kelton, W.D., 1991. *Simulation Modeling and Analysis*, 2nd Ed, New York: McGraw-Hill.

McGinnis, M.L, Phelan, R.G. 1995. "Using Computer Simulation to Reengineer Traditional Stove-Piped Army Staffs for Information Operations in the 21st Century." *Proceedings of the 1995 Winter Simulation Conference* (Washington D.C. Dec. 3-6). IEEE, Piscataway, N.J., 1158-1163.

McGinnis, M.L., Barbero, M.P., Hull, G. and Caraccilo, D. J. 1995. *Strategic Framework For Organizing Force XXI For Future Operations*, Technical Report FY95-4, Operations Research Center, West Point, NY.

McGinnis et al 1995. *Critical Battlefield Information for FORCE XXI Commanders*, Technical Report FY95-5, Operations Research Center, West Point, NY.

McGinnis et al 1995. *A FORCE XXI C2 System and Modular Force Structure*, Technical Report FY95-6, Operations Research Center, West Point, NY.

Potomac Systems Engineering, 1992. *Final Report - Command and Control Functional Area Model Data Study, Volumes 1-2*, Potomac Systems Engineering, Leavenworth, KS.

Sullivan, G.R. and Dubik, J.M. 1994. *War in the Information Age*, Strategic Studies Institute, U.S. Army War College, Carlisle Barracks, PA.

Sullivan, G.R., 1994. Unpublished message to the US Army Training and Doctrine Command (TRADOC) Headquarters, Fort Monroe, VA.

TRADOC 1993. *Blueprint of the Battlefield*, Pamphlet 11-9, Headquarters TRADOC, Department of the Army, Ft. Monroe, VA.

TRADOC 1994. *Force XXI Operations: A Concept for the Evolution of Full-Dimensional Operations for the Strategic Army of the Early Twenty-First Century*, Pamphlet 525-5, Headquarters TRADOC, Department of the Army, Ft. Monroe, VA.

Walrand, J. 1988. *An Introduction to Queueing Networks*, Prentice-Hall, Englewood Cliffs, NJ.

AUTHOR BIOGRAPHIES

MICHAEL L. MCGINNIS, the Director of the Operations Research Center at the United States Military Academy, West Point, New York, is currently attending the Naval War College at Newport, RI. He received a B.S. degree from USMA in 1977, an M.S. degree in Applied Mathematics and an M.S. degree in Operations Research from Rensselaer Polytechnic Institute in 1986, and a Ph.D. in Systems and Industrial Engineering from the University of Arizona in 1994. His current research interests are command and control, modeling and solving manpower requirement problems, and solving large-scale resource scheduling problems by exact and heuristic procedures.

ROBERT G. PHELAN JR. is assigned to the Operations Research Center at the United States Military Academy, West Point, New York. He received a B.S. from the University of Notre Dame in Aerospace Engineering in 1982, an M.E. in Manufacturing Engineering from Boston University in 1984, an M.B.A. from Boston University in 1993 and an M.S. in Systems Engineering from the University of Virginia in 1994. His current research interests are simulation of command and control systems and use of expert systems for resource scheduling.

Proceedings of the 1996 Winter Simulation Conference
ed. J. M. Charnes, D. J. Morrice, D. T. Brunner, and J. J. Swain

MANAGEMENT OF PERSONNEL POLICIES TO INCREASE THE STABILITY OF PATRIOT CREW MEMBERS AND THEIR FAMILIES: A SIMULATION APPROACH

Michael J. Kwinn Jr.
Robert G. Phelan Jr.

Operations Research Center
United States Military Academy
West Point, New York 10996, USA

ABSTRACT

The overwhelming success of the Patriot Air Defense weapon system during Operation Desert Storm has become a double edge sword for the system's enlisted crew members and their families. During the war, the Patriot system became the first proven means of defending against Tactical Ballistic Missile (TBM) attacks. The people and the governments of Israel, Saudi Arabia and the United States praised the Patriot crew members as true heroes. Unfortunately for the Patriot crew members, their success also meant that every country wanted the protection that the system offered. A battalion is now permanently stationed in Korea. Another battalion rotates every six months to Saudi Arabia. The increased number of deployments of the system and the crew members has had a dramatically negative impact on morale, retention and recruitment of Patriot crew members. In this paper, we present a simulation model which allows a personnel manager to vary personnel policies effecting Patriot crew members to determine the impact on the number and frequency of deployments and on unit readiness.

1 INTRODUCTION

In the early stages of Operation Desert Shield military planners determined one of the major threats to US and allied forces in Saudi Arabia prior to any offensive action was from Iraqi Scud missile attacks. The solution was to send Patriot battalions as soon as possible to attack and destroy the in-bound Scuds. This anti-ballistic missile defense capability had never before been proven in combat conditions. When the first Patriot system engaged, fired and destroyed the first Scud missile over Riyadh, the effectiveness of the system became legendary - and its crew members became nomads. Suddenly every country allied with the United States wanted Patriot coverage to protect against TBM threats.

The Patriot weapon system and the soldiers that man it were moving all over the globe. Soldiers who were stationed at Fort Bliss deployed to South Korea with their Patriot Battalion. They would return from that deployment only to be sent to Germany. Shortly after arrival to Germany, they would deploy for six months to Southwest Asia (SWA) with a different battalion. It was not unheard of for solders to be away from their families for 24-30 months out of the last 36. Retention rates were dropping dramatically and morale was at an all time low. The large number of deployments did not decrease with time. When the Army decided to make the battalion in Korea a permanent stationing, it did reduce the deployment time and the uncertainty of which battalion would deploy next [Costello, 1996]. It also increased the deployment time for the soldiers from 6 to 12 months.

Prior to the conflict in Southwest Asia (SWA), the Army personnel system was undergoing dramatic reductions in all areas. With the end of the Cold War, there was an almost universal cry for a "peace dividend". In the early 1990's, this "peace dividend" took the form of a massive reduction in the Army's personnel end-strength and a realignment of installations. At the end of fiscal year 1989, the Army's enlisted end-strength was 658,000 soldiers stationed throughout the world. By the end of fiscal year 1995, the enlisted end-strength was only 422,000 soldiers [Hersh, 1995]. According to the Army Chief of Staff, General Dennis Reimer, the Army may have to reduce even further to pay for modernization programs [Reimer Interview, 1995]. Army planners are currently analyzing options on how to reduce the Army end-strength by an additional 20,000-40,000 soldiers.

Senior leaders in the US Army Personnel Command (PERSCOM) demand from their branch managers immediate analysis of the almost daily changes to the personnel system. General Reimer, recently stated, "The models that we've had that have worked well during the Cold War don't adjust well to a changing situation...." [Adelsberger, 1995]. Currently, personnel managers base many of their decisions on the *output* of the current models. Personnel managers use most models as decision *support* tools as opposed to decision *analysis* tools.

In this paper, we present a simulation model developed for the Plans and Analysis Branch, Enlisted Personnel Management Division, US Army Personnel Services Command (EPMD, PERSCOM). We designed the model to assist personnel managers in analyzing the impact changes to the personnel system. This model allows managers and analysts to modify policies and determine the impact on personnel stability and unit readiness.

2 BACKGROUND

2.1 Types of Moves

When a soldier moves from one location to another, it is a Permanent Change of Station, (PCS) move which generally falls into one of five categories: Unit, Operational, Training, Rotational, and Accessions and Separations. The movement of a soldier as a result of a unit's activation or deactivation is considered a "Unit move". When a soldier moves between one unit between units both within the Continental United States (CONUS), this is an "Operational move". An exception to this is when the move is to or from a professional development school. This is considered a "Training move". Whenever a soldier is moved "over water" (ie. From Germany to the United States), it is counted as a "Rotational move". Moves which result from a soldier either entering or exiting the Army are considered "Accession and Separation moves".

The costs associated with all of these moves total over $1 Billion annually. Rotational moves and Accession and Separation moves account for over 90% of the PCS costs [Hix, 1995]. The number of moves, and therefore the cost, increased dramatically in recent years due to the draw down. The Army closely managed the voluntary separations by soldiers by MOS and grade but not by unit or geographical region. The result was a large increase in operational moves.

2.2 Needs Analysis

Another significant impact on personnel policy was the changing force structure; especially the reduction in the number of soldiers required in Europe. For years, soldiers knew that roughly every other assignment would be overseas, usually to Europe. This was due to the fact that in the 1980's, approximately 30% of the enlisted soldier authorizations in the Army were in Europe. By the end of FY95, the ratio was down to 13%!

This realignment of forces changed the Army's "comfortable" rotational scheme for soldiers moving from overseas to stateside assignments. Base closures and threats of even further force reductions created uncertainty for the Army's leaders, too. Each base-closure recommendation from the Base Realignment Commission (BRAC) further disrupted the personnel system.

We met with LTC James Thomas, Section Chief, Plans and Analysis section, Distribution Division, Enlisted Personnel Management Directorate, PERSCOM in September of 1995. One of the needs that he explained to us was that his superiors often asked his staff to conduct analysis on the effects of changing personnel policies. For example, they would ask about the effects on unit readiness if the Army extended the tour lengths in Germany. They were also very interested in the effects of reducing the number of Operational moves allowed in a given fiscal year. He stated that his office did not have an analytical tool to directly assist his staff in analyzing these types of problems. They best they could do is long-handed work with a calculator and a pencil. As the Army continues to operate under increasing tight budgets and to realign units, he anticipated that questions such as these would be asked more frequently.

We determined that he needed an analytical tool with which his staff could quickly and easily modify PCS policies and could determine the effects of these policies on the number of moves and the readiness of the units. The model would have to provide accurate results in a short amount of time.

3 MODEL DEVELOPMENT

3.1 Use of Simulation

The personnel system is very stochastic in nature. The number of soldiers promoted in a given month varies greatly. The number of soldiers who leave the service in a given month also varies greatly. For this reason, we wanted any analytical tool which we developed to account for this variability. We were trying to predict an outcome based on a given set of parameters (the current personnel picture) and rates which varied. We could not just provide our decision maker with one pat answer and not address that variability. This was one of the major reasons we chose to develop a simulation of the personnel system. By using simulation, we could replicate the stochastic nature of the personnel system while begin able to analyze each policy by running the simulation a number of times.

We decided to use the ProModel® software package for this simulation. This is a discrete-time simulation package. Each minute of simulation time represents one month of real time. ProModel® is an object-oriented simulation package which makes programming the simulation quite easily. Additionally, it provides an outstanding use of animation and a user can easily transform the statistical output into graphical format. Displayed in the figure below is the graphics of the simulation. The numbers within the boxes near the locations are the number of soldiers at that location.

Figure 1: ProModel display for PCS Simulation Model

The animation and the statistical analysis features of the software are extremely valuable when attempting to convince a decision maker on a proposal. The problem is that it can potentially limit the overall size of the model. We will address this later.

3.2 Proof of Principle

Simulating every MOS in the Army is a very large undertaking. The overall modeling effort would take a great deal of work. It would also require a large amount of memory unless we trade-off resolution for aggregation. Prior to launching a large modeling effort, we decided to ensure our efforts would meet the needs of our client. For this reason, the simulation model we develop in this report is a proof of principle using only one MOS. We chose to model 16T, Patriot Crewmember. We decided on this MOS for a number of reasons. First, it was a representative MOS in that a 16T could be stationed in a number of locations both stateside and overseas and could also be assigned to duties away from a Patriot unit, like Drill Sergeant. Second, all of the analysts were Air Defense officers so there was a level of familiarity with the 16T career patterns. Finally, we conducted an interim brief on the project to MG Costello, Chief, Air Defense Branch. He asked that we focus on this MOS due to the lack of assignment stability for these soldiers in light of the requirement for a rotating battalion to deploy to Southwest Asia (SWA).

3.3 Concept for Modularity

Since we were only working on one MOS, we wanted to make the model as modular as possible to allow for easy expansion to encompass all MOSs. This would also allow for adaptability to different policies or realignment of forces. This modular concept also made the initial model development a great deal easier and quicker.

3.3.1 Types of Locations

We categorized each location, or unit assignment, a 16T could be assigned to, into one of three types of locations: CONUS TOE, CONUS TDA and OCONUS. A CONUS TOE unit is a line Patriot unit in the Continental United States. A CONUS TDA unit could be a Drill Sergeant or Recruiting assignment, for example, in the Continental United States. Any assignment Outside the Continental United States is considered an OCONUS assignment for the purposes of our model.

The differences in the types of locations are the criteria personnel managers use to determine a soldiers eligibility to move and the type of move generated from that type of location. Movement out of a OCONUS or a CONUS TDA assignment is based on a soldier reaching a certain TOS. For an assignment to Germany, for example, generally when a soldier has been on station for 30 months, he or she will be looked at for reassignment. Assignments in Korea are normally 12 months and CONUS TDA assignments are 24-36 months in length. The difference here is that a move from a CONUS TDA assignment can be an "Operational move". Our model counts all moves to or from an OCONUS assignment as either Rotational or as an Accession or Separation move. Soldiers in a CONUS TOE unit are only moved when there is a requirement at another location, and usually only after 24 months TOS. If there is no requirement, a soldier remains at his or her current location.

3.3.2 Processing Blocks

We further break down the processing at the different locations into three blocks: Administrative, Orders, and Movement. The Administrative block is just what the title says. All administrative actions such as promotions, demotions and attritions are done in this block of code. In the Orders block, soldiers are "given" orders for reassignment to another location, or unit. Finally, in the Movement block, a soldier "PCSs" to his or her new duty location. We will discuss the actual coding of these blocks in a later section.

4 CHARACTERISTICS OF THE MODEL

4.1 Locations

For our specific model, we defined eight separate locations a soldier can "move" to or from. Soldiers can be assigned to the Advanced Individual Training location at Fort Bliss, a CONUS TOE unit at either Fort Bliss, Fort Hood, or Fort Polk. They can also be assigned to an OCONUS unit in Germany or in Korea. Finally they can be assigned to a CONUS TDA unit. We aggregated all the CONUS TDA assignments into one unit. We did this because our client was not interested in the cost of each move, simply the number of moves, by type. This allowed us to consider a move from the Wisconsin National Guard to Fort Bliss the same as a move from Washington, DC to Fort Polk, for example.

4.2 Entities

There is only one type of entity in our model: a soldier. We model each 16T individually. This requires more memory than aggregating the entities but allow for greater resolution as to the individual characteristics of a soldier. We account for the various characteristics of the soldiers by assigning different attributes to each entity.

4.3 Attributes

We defined four attributes for each individual soldier:

4.3.1 Skill Level (SL)

This is self-explanatory. As a soldier in this model is promoted to skill level 5 (E8), he or she leaves the system. This is because that soldier would then become a 16Z. This is an entirely different MOS and is managed separately.

4.3.2 Time on Station (TOS)

This is the number of months a soldier is at a particular location. This is incremented every month and is the key indicator to determine *when* a soldier is eligible for reassignment.

4.3.3 PCS Designator (PCS_CODE)

This a designator which is assigned to an entity to essentially, "put a soldier on orders" to another location. This is explained in the Movement Processing paragraph below.

4.3.4 Time Until Movement (TTM)

This is a value given to an entity to signify the number of months until that entity proceeds to its new assignment after being "put on" assignment instructions. Prior to putting a soldier on orders, this number is negative. After a soldier is put on orders, he or she is assigned a positive number based on the need of the gaining unit. We discuss the actual assignment algorithm later. An entity's TTM is decremented every time period (month). When it becomes zero, the entity moves to the new "unit".

4.4 Variables

Most of the variables we use in this model are defined to calculate which unit a solider ready for PCS should be assigned to. An assignment officer generally reassigns a PCS eligible soldier to the unit with the greatest "need". This is based on a priority fill plan. Under this plan, different units are designated to have a higher priority. Personnel managers attempt to keep these units at high fill percentages. To replicate the assignment process, we defined a great number of variables in this model. We defined each of the variables explained below for each skill level and location. For example, the actual number of skill level 1 soldiers at Fort Bliss is defined as Bliss_act1.

4.4.1 Target

Different units have different "fill priorities". Some units, such as rapid deployment units, have a high fill priority. Personnel managers try and maintain a fill level of over 100% for these units. This way, even with normal fluctuations in personnel levels, a high priority unit will never drop below 100% fill. There are "bill payers" for this over-manning. These are the low priority units. Personnel managers are willing to let these units fill levels drop below 100%, perhaps as low as 95%. To account for these different priority units, we used a "target" value instead of simply a unit's authorized strength. For example, if a unit is a high priority unit and has an authorized strength of 200 personnel, the target value would be 204 (200*102%). Similarly, if a low priority unit has an authorized strength of 100, its "target" could be 98 or even 95 depending on the priority. The usefulness of this definition will be more transparent when we discuss the "need" variable below.

4.4.2 Actual

The variable is simply the actual number of soldiers, by skill level, at a given installation. This variable is incremented when an entity arrives at a location.

4.4.3 Gain

This is the number of soldiers "on orders" for a given installation, by skill level. When an entity is assigned a PCS Code, thereby designating a location to where that entity will move, this variable is incremented. When the entity is considered and arrival (thus incrementing the actual counter at a location) this variable is decremented.

4.4.4 Loss

This is obviously just the opposite of a gain variable. When an entity is assigned a PCS Code, this variable is incremented as a loss at that location. Upon arrival a the new location, this variable is decremented.

4.4.5 Need

This is where all the variables come together to calculate a unit's "need" for a new soldier. The formula we use is simple: NEED = TARGET - ACTUAL - GAIN + LOSS. We calculate this for each skill level and each location. For example, the calculation of Fort Hood's need of skill level 1 soldiers is: HOOD_NEED1 = HOOD_TGT1 - HOOD_ACT1 - HOOD_GAIN1 + HOOD_LOSS1. Based on this calculation, a high priority unit would have a higher need even when its actual strength (measured against authorized) is higher than that of a low priority unit.

4.5 Processing

As we stated before, there are three "blocks" of processing in each location, Administrative, Orders, and Movement. When an entity arrives at a given location, the processing first increments the "actual" variable and decrements the "gain" variable for that location. A new "overall actual" value for that location is then calculated.

Once we have updated the variables properly, the entity enters a DO/UNTIL Loop. The entity remains in this loop until it is assigned a PCS_CODE of 1 (leave the system, or ETS) or its TTM = 0.

4.5.1 Administrative Processing

During any given month, a soldier either attrites, is promoted, demoted, or is continued at the same skill level another month. The rate at which soldiers in a given skill level and MOS complete one of these actions in a given month varies wildly depending on numerous factors. To simplify our calculations, we used parameters used in the Enlisted Loss Inventory Model, or ELIM. This is a widely recognized model with the personnel community. This model calculates a thirty-six month weighted average to compute rates for attritions, promotions and demotions by skill level and MOS. These values are given as deterministic parameters. We replicate the randomness of this process by generating a random number and comparing it to a "yardstick". This yardstick varies from 0 to 1. Within this yardstick are "bands" for attrition, promotion, demotion, and continuation as seen in the figure below:

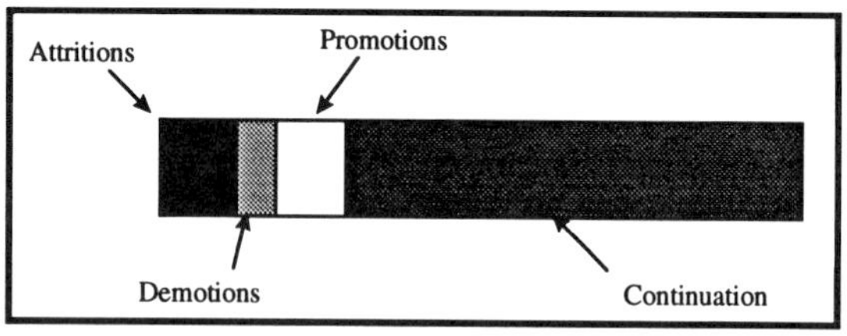

Figure 2: Personnel Action Bar

The model then compares the random value generated against this yardstick and depending on where it "lands", the appropriate personnel action is taken on the entity. For a promotion, the skill level is incremented and it is decremented for a demotion. In the case of an attrition, the entity is assigned a PCS_CODE = 1 and then leaves the system. This process of generating a new random number and then comparing a random number against a yardstick every time increment could allow an entity to be promoted one month, promoted the next month, demoted the next month, promoted again the net and then finally attrited (or any combination). This may not seem realistic for a given soldier. We are not concerned with this because though we are tracking individual entities, we are only interested in information from the aggregate. The individual promotions, demotions and attritions are unimportant to our analysis. We are merely interested in the fact that in a given installation, for a given month, for a given skill level and MOS, "roughly" a certain percentage will fall into various administrative actions. Refer to Annex B for the processing code in this block.

4.5.2 Orders Assignment Processing

Embedded with the Administrative Processing code is the Orders Assignment Processing. In the Army, when a soldier reaches a certain Time on Station (TOS) "threshold", an assignment manager begins to look at that soldier for reassignment. The trigger mechanism is different depending on the type of assignment or the location of the assignment. For example, if a soldier is assigned to Germany, an assignment manager will look for that soldier to redeploy based solely on that soldier's TOS, regardless of the readiness of his or her current unit or other units. The same can be said for most TDA assignments. These assignments last two or three years generally depending on the type of assignment. When those two or three years are over, the soldier moves to a new location. Again, the reassignment trigger is TOS.

For a soldier in a CONUS TOE unit, however, there is more involved. These soldiers will generally stay on station at least 24 months prior to being allowed to be reassigned elsewhere. Once this soldier attains that TOS threshold, he or she can be reassigned if there is a greater need elsewhere. In the case of moving the soldier to another CONUS TOE unit (thus generating an Operational move), this need must be much greater.

We attempted to account for these realities in our model. Once an entity reaches a given TOS threshold, it enters the orders processing block. Once in this block, the need for a given location is calculated for the given skill level. We do a pair-wise comparison and calculate the MAX NEED. This is the location with the greatest need for this skill level entity. For a OCONUS or a TDA location, the soldier is then assigned a PCS_CODE corresponding to the location with the greatest need. In the case of a CONUS TOE unit, the need at the MAX NEED location must be significantly greater than that at the current location. If it is just slightly greater, there is not need to move the soldier immediately. In this case, the entity is assigned a PCS_CODE of -1 (not on orders) and continues back to the administrative block for processing the next month with a higher TOS.

After an entity is assigned a PCS_CODE >1, the process then calculates the Time to Move (TTM) for the entity. AS the MAX NEED increases, the TTM decreases in a step-wise manner. If the MAX NEED is high, the entity will move in two months. If the need is low, the entity will move in six or more months.

In our simulation, we also must account for a soldier electing to remain at an overseas location for another tour. This is termed a Consecutive Overseas Tour (COT). For example, a soldier in Germany can request to have another 36 month tour immediately after his initial 36 month tour is completed. A soldier can request waivers to reduce the length of either the initial or subsequent tour. With these waivers, the overall tour length could vary from 48 months to 72 months. Under another program, soldiers can also extend their tours beyond the regular 36 months for months ranging from 2 to 24. We account for this in the model by again comparing the random number generated in the administrative block with a COT rate. [Black, Jul 96] Once an entity "accepts" a COT, his TOS is adjusted based on a distribution to reflect the variability in the length of the extension or new tour length.

After the entity is assigned a PCS_CODE and a TTM, it returns to the administrative block and continues to be compared to be attrited, promoted, demoted or continued. The TTM is decremented every month in this block. Upon reaching zero, the soldier moves to the movement processing.

4.5.3 Movement Processing

We liken the movement of entities around the simulation to a monorail system. After leaving the location processing, the simulation analyzes the PCS_CODE attribute of the entity. The entity is then moved to the location alluded to by this attribute. Prior to movement, the PCS_CODE is reset to -1, the TTM is reset to -1, The TOS is reset to 0, and the "actual" and "loss" variables for that skill level at the location are decremented. If the movement is between two CONUS TOE locations or a CONUS TOE and a CONUS TDA location, the move is an operational move. In this instance, the OPER_MOVE variable is incremented. Finally, the entity is told to wait either one (for a move between CONUS units) or two (for a move overseas) time increments to make the move.

4.6 Arrivals

Entities arrive to the system in only one location: Fort Bliss AIT. We do not manage the arrivals or the processing at this location. This model is not interested in recruiting rates or retention rates at Basic or AIT. Each entity which arrives at the AIT location, moves to the unit with the greatest need for skill level one soldiers immediately. We establish the arrivals at this location to correspond with the AIT graduation rates we received from the ELIM output [Hersh, 1995]. These rates can be modified once the interface of the model has been established to fully validate the model.

5 VERIFICATION OF THE MODEL

We have run the model a number of times under various initial conditions. The model seems to run well. Entities are being promoted as shown in the figure below. In this figure, the lines represent the actual number of soldiers, by skill level at Fort Bliss at a given time. Note that the time scale along the x-axis is in hours. This simulation package does not allow "months" as a time increment. We used minutes to represent months. Therefore, one hour of simulation time would equal 60 months of "real" time. As you can see, soldiers are being promoted and are attriting and/or moving out of the location. This is a graph taken directly from the ProModel® statistics package.

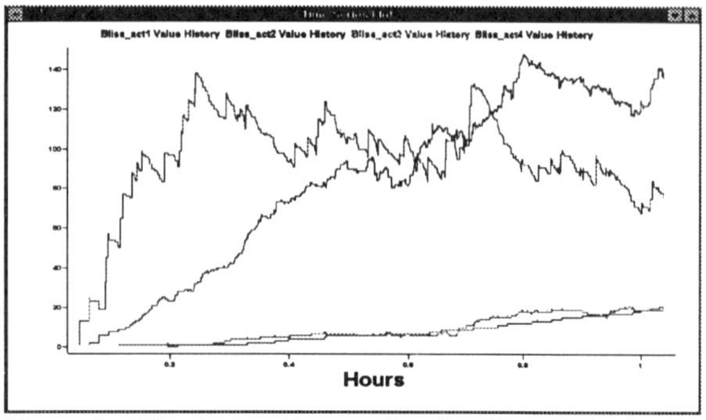

Figure 3: Simulation Output Graph of Fort Bliss by SL

Figure 4, below, shows the number of soldiers at the different locations. Note that these numbers begin to level out after about 30 months. This shows that the system is stabilizing. This verifies that the model is operating as we intended and should continue to do so when we modify the initial conditions.

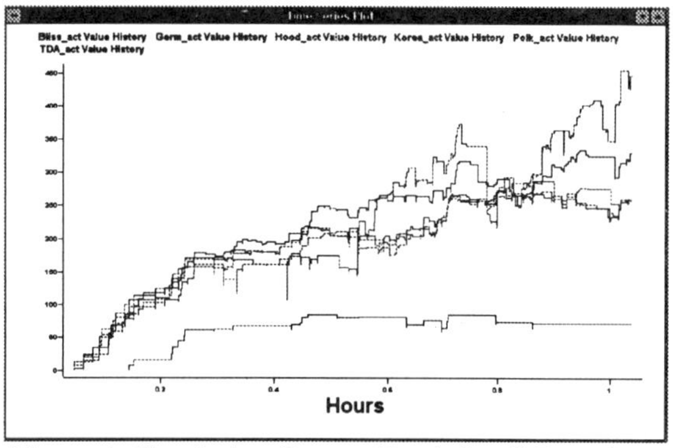

Figure 4: Simulation Output Graph for all Locations

6 VALIDATION AND FUTURE DEVELOPMENT

In order to fully validate the model we must begin with realistic numbers and other data. A front-end interface with the Enlisted Master File must be developed to do this. This should be an easy step for an individual with access to the database and programming skills. The database must be queried and sorted into 16T soldiers. The programmer must then modify this data so that it is in the format required by ProModel. For example, for each soldier with a Fort Bliss Unit Identifier Code (UIC), assign the entity with a 2 in the Location column.

Once this interface is established, we must modify the rates, the TOS thresholds and the TTM values to obtain realistic annual operational move numbers. We should also see the correct number of promotions, demotions and attritions. Since the model seems to be operating correctly, the only modifications should be in these rates. Once fully validated, the model can be used in analysis, but only for 16T MOS.

In order to use this simulation for evaluation of PCS policies for the entire Army, the model should be expanded to include all MOSs. On the surface this seems an easy task. To expand to different MOSs, simply assign another attribute to each entity delineating that entities MOS. The problem becomes computational. The sheer number of soldiers in the Army requires that some aggregation be done in order to expand this model. The question becomes: what do you aggregate? As you aggregate, you lose some resolution in the model. This will be a problem. The use of ProModel exacerbates the problem. ProModel is an easy-to-use, visual simulation package. The problem is that it requires a great deal of memory to use the animation and the other features of the package. A user might be able to do less aggregation with a different simulation package.

7 USE OF THE MODEL IN ANALYSIS

This simulation was designed to assist a personnel manager, or analyst, in determining the effect of changing TOS policies on the number of moves and level of readiness of a unit. After modifying the TOS thresholds to reflect the policy recommendation, a user simply runs the simulation. The simulation will track the number of operational moves required under this new policy. The user can then compare the number of moves under the current policy with the number under the proposed policy.

The user can also view the impact of the policy on readiness of a unit. After running the simulation, the user views the statistics from the run. By analyzing the graphs, (which is extremely simple to create in ProModel®), a user can see where the drops are in units readiness, how deep these drops are and how long the system will take to recover completely.

For example, suppose a decision maker recommends a policy which increases the tour lengths in Germany from 36 to 42 months. An analyst would first run the model under the current 36 month TOS policy and obtain the baseline results. Then the analyst would re-run the model after changing the TOS threshold for Germany. The analyst would simply have to compare the results of the two runs to determine the impacts of such a policy. One would expect that this would initially increase the number of operational moves because soldiers would not return from overseas on their "projected" rotation cycle. There may also be a noticeable drop in readiness of CONUS units. The analyst could easily show a decision maker (using the graphs in the ProModel® statistics package) the effects of this policy after one, two or five years.

8 CONCLUSIONS

This proof of principle is a worthwhile analysis tool for analyzing PCS policies for 16Ts. There is a long way to go, however, to make it a worthwhile analysis tool for the entire Army's personnel picture. Due to the modular design approach, this job will be considerably easier. The size of the model will be a problem but should not stall efforts to expand this model. In this time of ever-reducing budgets and personnel drawdowns, personnel managers need better analysis tools to provide meaningful answers to decision makers. This is a first step in that direction. It should not be the last. This project, or a similar effort, should be expanded to include all Army MOSs. We can no longer afford to analyze policies *after* we implement them.

ACKNOWLEDGMENTS

We wish to acknowledge the Assistant Secretary of the Army, Financial Management and Comptroller for funding this project. We also acknowledge the Plans and Analysis Branch, Enlisted Personnel Management Division, Personnel Services Command for their willingness and foresight to improve their models and their analysis.

REFERENCES

Adelsberger, Bernard, *Troop budget needs fixing*, Army Times, Dec 4, 95

Army Training Requirements and Resources System (ATRRS) Selected Actual Summaries run, 16T FY94-96 Actual Grads, 27 Mar 96.

Bonder, Seth, *Impact of the New Global Environment on US National Security Planning - Challenges to the OR Community*, International Transactions in Operational Research Vol. 1, No. 1, Mar 94.

Costello, John, Major General, State of the Air Defense Artillery, slides from briefing to faculty, USMA, 5 Mar 96.

General Research Corporation (GRC), *Military Occupational Specialty Level System (MOSLS) Advanced Training Course*, course packet, 13 Mar 96.

Hersh, Doug, Office of the Deputy Chief of Staff, Personnel, unpublished message on Army Enlisted End-Strength data, 28 Sep 95.

Hix, Mike, *Assignment Stability: An In-Progress Review*, RAND briefing, 28 September 1995.

Interview with General Dennis Reimer, Army Chief of Staff on Nov 21, 1995, Army Times, Dec 4, 95.

Kwinn, B.T., Muhammad, A., *Continuous Time Simulation Approach to Developing US Army Enlisted Personnel Policy*, work in progress, submitted for 1996 MORS Conference.

McGinnis, M. L., Kays, J. L., and Slaten, Pamela, *Computer Simulation of US Army Officer Professional Development*, Winter Simulation Conference Proceedings, Dec 1994.

McGinnis, M. L., and Fernandez, E., *Initial Entry Training for the United States Army: Optimal and Heuristic Scheduling Procedures*, IEEE Decision and Control Proceedings, 1995.

Operations Management Division, EPMD, PERSCOM *Assignment Nomination Subsystem*, Briefing Packet, 10 Oct 95.

Office of the Deputy Chief of Staff, Personnel, MOSEL run, 20 Oct 95.

Park, Chan S., *Contemporary Engineering Economics*, Third Edition, Addison-Weslsley Publishing Company, New York, May 1994.

Stover, E., Tucker, B. and Zopelis, J., briefing to MG James Costello, *Enlisted PCS Simulation Model*, 5 Mar 96.

Thorpe, Robert, *Enlisted Distribution Target Model*, Information Paper, Planning and Analysis Branch, EPMD, PERSCOM, Sep 1, 95.

AUTHOR BIOGRAPHY

MICHAEL J. KWINN, JR. is an Analyst in the Operations Research Center at the United States Military Academy (USMA) at West Point. He received a BS from USMA in 1984, and an MS in Systems Engineering from the University of Arizona. The University of Arizona has accepted him into their Ph.D. program. His research interests are personnel management systems and force structure analysis through the use of optimization and simulation models.

ROBERT G. PHELAN JR. is an Analyst in the Operations Research Center at the United States Military Academy (USMA), West Point, New York. He received a BS from the University of Notre Dame in Aerospace Engineering in 1982, an M.E. in Manufacturing Engineering from Boston University in 1984, an MBA from Boston University in 1993 and an MS in Systems Engineering from the University of Virginia in 1994. His current research interests are simulation of command and control systems and use of expert systems for resource scheduling.

Proceedings of the 1996 Winter Simulation Conference
ed. J. M. Charnes, D. J. Morrice, D. T. Brunner, and J. J. Swain

MEASURING THE EFFECTIVENESS OF SIMULATION-BASED TRAINING

Michael L. McGinnis

Department of Systems Engineering, USMA
West Point, New York 10996

George F. Stone III

National Simulation Center, US Army
Orlando, Florida

ABSTRACT

The relationship between measures of training performance and the achievement of training goals is often difficult to quantify. This continues to be an obstacle to the development of curricula for simulation-based training systems. These include both constructive and virtual computer simulations for training. This paper discusses the development of training effectiveness measures as they relate to the design of instructional, simulation-based training systems featuring automated tools for planning training.

1 INTRODUCTION

Recent advances in computer technology have substantially increased the use of simulation-based training by industry, academia, and government agencies. The users of simulation-based training have found it to be an efficient, effective means of training personnel in certain tasks and sustaining the proficiency of those tasks as well. This is especially true for virtual and computer simulation training systems being developed to train personnel as a substitute for dangerous (risky) and resource intensive real-world training. As the future cost of simulation-based training systems decreases, it is expected that they will become a prominent feature of the workplace. During the past several years, the Army has been researching new training performance measures for simulation-based training systems. Agencies involved in this effort include the National Simulation Center, Fort Leavenworth, Kansas, the US Army Simulation, Training, and Instrumentation Command, Orlando, Florida, and the Operations Research Center at the US Military Academy, West Point, New York.

This paper suggests a methodology for measuring the effectiveness of military simulation-based training. Although simulation-based training involves both individuals and teams, our paper is oriented towards team training. Section 2 briefly summarizes previous work in this area. Section 3 outlines a methodological approach for conducting simulation-based training and measuring its effectiveness. Section 4 presents an elementary, yet illustrative, example of a training effectiveness measure developed for simulation-based training.

2 BACKGROUND

A survey of the literature reveals previous studies attempting to define measures of training effectiveness (MOTE) (see Turnage, Houser, & Hofmann, 1989 and Tannenbaum, Cannon-Bowers, Salas, & Mathieu, 1993). Much research focused on specific quantitative or qualitative performance measures within narrowly defined training domains (Spurgin, Moieni, & Orvis, 1993). Others developed methods for conducting and evaluating simulation-based training. The latter generally addressed the training processes, actions, and procedures followed by trainers in conducting simulation-based training.

One of the first systematic training models developed was the Instructional Systems Design (ISD) model. ISD was used during World War II by the U.S. military to train soldiers in aircraft recognition. Recently, Sloman (1994) synthesized the major components of the ISD model into a high-level framework for evaluating training effectiveness. Figure 1 summarizes the stages of Sloman's model.

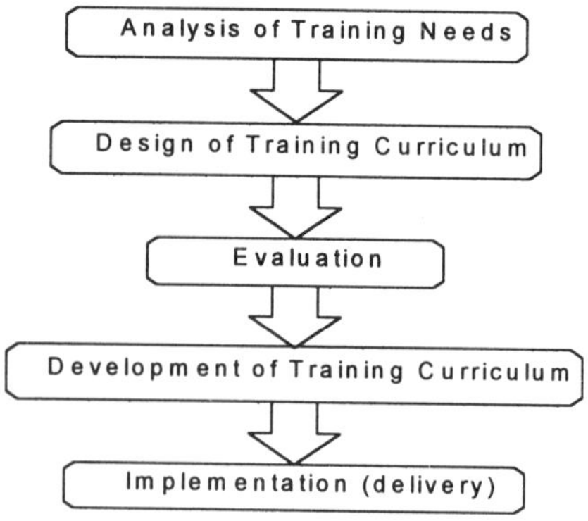

Figure 1. Sloman's ISD systematic training model.

In Sloman's model, training needs derive from a thorough analysis of the training requirements. From this, appropriate and relevant training curricula may be designed and developed. Next, the training curriculum is implemented within a simulation training program. Finally, the training personnel evaluate the proficiency of the team undergoing simulation training. Sloman suggests that the effectiveness of future simulation training will depend on how well the five components of his ISD model are integrated into training curricula.

According to Williams, Reynolds, Carolan, Anglin & Shrestha (1989), there are several important factors to consider when integrating simulation training into an organization's training program. First, the simulation training must be linked to strategic objectives of the organization. Second, it must also be directly related to the functions and processes of the organization. Lastly, training system development must be perceived by system users as relevant to real-world tasks. In this way training curricula will provide information and training for accomplishing real-world tasks.

3 METHODOLOGY FOR MEASURING THE EFFECTIVENESS OF MILITARY SIMULATION-BASED TRAINING

The approach presented here for conducting and measuring the effectiveness of simulation-based training generally follows the ISD process described above in Section 2. The six major components of our process are: (1) training task identification; (2) training proficiency evaluation; (3) training task prioritization; (4) identification of simulation training support; (5) simulation training execution; and (6) feedback.

Identification of military training tasks for simulation-based training curriculum has benefited from at least twenty years of extensive Army analysis in this area. Previous work determined many tasks for real-world missions that Army units may be assigned. These plans, called *mission training plans* (MTPs), specify the conditions, tasks, subtasks, and standards that Army organizations must meet in order to accomplish combat and peacetime missions. For example, mission training plans may identify the tasks required of an infantry unit to defend key terrain from enemy capture, or a quartermaster unit's tasks for setting up and operating a laundry and shower site for units in the field. Each mission training plan identifies the conditions for performing mission related tasks. They also subdivide major tasks into subtasks with standards for accomplishing each one. Work by the agencies noted in Section 1 included modifying the tasks and subtasks from mission training plans, and other sources, for use in real-world simulation-based training.

The second component is the assessment of a training unit's proficiency at performing mission related training tasks against a measurable standard. It is important that task proficiency be evaluated in a fair, unbiased manner that accurately reflects the training unit's ability to perform key and essential tasks for accomplishing specific missions. Task evaluation may be performed by the team undergoing training or by an outside, independent evaluator. Recent efforts to automate the task proficiency evaluation process within a Training Exercise Development System (TREDS) are discussed by Crissey, Stone, Briggs, & Mollaghasemi, (1994) and McGinnis & Stone (1995).

In most cases, task proficiency standards are measurable, fixed criteria. Performance outcomes are usually assessed as pass or fail (i.e., go or no go). This helps remove subjectivity and ambiguity that might otherwise bias training performance measurements. See Gonzalez & Ingraham (1994) for further discussion of the application of measurable standards to simulation-based training. Next, we aggregate task and subtask ratings into an overall determination (measure) of training proficiency (see Section 4 below). Based on this assessment, the military unit is rated as either fully trained (T), partially trained (P) but needs practice, or untrained (U).

The third component, prioritization of tasks and subtasks, is a preliminary step to identifying and scheduling training support and resources, i.e., the fourth component, for executing simulation-based training as accomplished by the fifth component (see above). Task prioritization also helps training managers select training scenarios that meet the training goals of the military unit(s) undergoing training. The training task rankings, reflecting the value of each training task to mission accomplishment, and training performance scores obtained during simulation training execution are used to determine training proficiency scores.

The unit receives feedback during an after action review of training performance and lessons learned. Training evaluations provide the military unit(s) with quantitative results in the form of an overall score that benchmarks their performance during training. Training results are also used for adjusting task prioritization to reflect updated training priorities. This marks the start point for the next iteration of the simulation-based training process.

4 FORMULATION OF AN ELEMENTARY SIMULATION-BASED TRAINING EFFECTIVENESS MEASURE

The effectiveness measure presented here for simulation-based training is partially based upon multiple attribute utility theory and decision making methods. See Clemen (1990), Hwang & Yoon (1981), and Saaty (1980) for discussions of the theory and methods.

As explained above, our methodology requires that tasks be identified in advance of training. Following this, each task is ranked (weighted), relative to the others, by the military unit undergoing training. The weights reflect the importance of each task to mission accomplishment. Precise measurable standards are best for evaluating the training proficiency of the military unit. Figure 2 depicts a weighted sequence of training tasks.

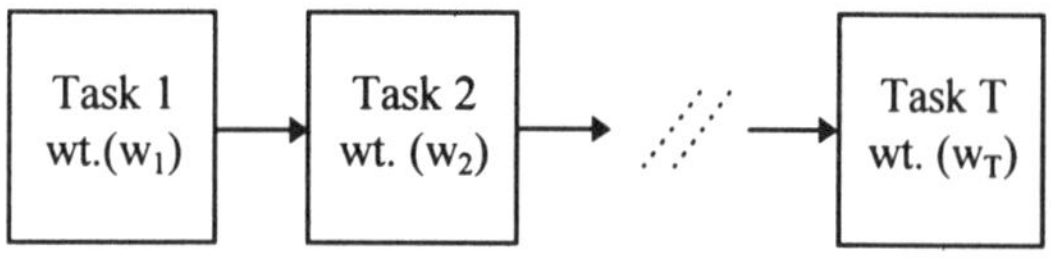

Figure 2. Training tasks with weights

Next, a training task tree is constructed by breaking down each top level task *t* into lower level subtasks. The highest level tasks, level 1, are usually described in broad, general terms. These high level tasks are the ones that must be successfully completed for the military unit to achieve mission success. At each level, task decomposition leads to more basic, specific subtasks. Working down the tree, lower level subtasks generally reflect "how" the parent task is to be accomplished. Figure 3 illustrates the hierarchical structure of a partial training task tree for a single level 1 task *t*, decomposed through *N* subtask levels with *L* subtasks at each level.

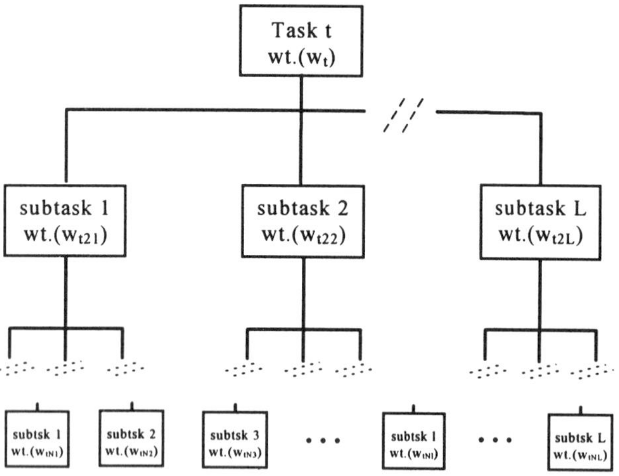

Figure 3. A hierarchical task tree.

Weights must be determined for each subtask. As mentioned above, they represent the importance of subtasks to mission or parent task accomplishment. The combined weights of all subtasks branching from a parent task must sum to one. They may be derived mathematically using various methods or be elicited as subjective assessments from unit personnel.

The tree is solved, branch by branch, starting at the lowest level and working up. A linear combination of weights *W* and scores *S* is computed for each branch. These are rolled up to the next branch and added to the linear combination of weights and scores for the parent node. This process continues for all tasks *L*, branches *M* (within a level), and levels *N* until a total score is computed for the highest level task (level 1). Mathematically, this is given by

$$\sum_{i=1}^{N} \sum_{j=1}^{M} \sum_{j=1}^{L} W_{ijk} S_{ijk} .$$

Training proficiency score computations for a hypothetical training proficiency task tree may be illustrated using a simple additive procedure. We note, however, real-world situations may demand a more sophisticated, robust method. Here, we compute the overall task proficiency score for a level 1 task expanded two levels deep. Two branches emanate from the first level to the second level. Three branches emanate from each of the second level nodes.

In practice, task scores may have different units of measurement such as time or distance. Before these task scores can be used to compute a task proficiency score they must be converted into unit-less values called *utiles* (see Clemen, 1990). For this example, we assume all real-world task scores have been converted into utile values as given in Table 1. The utile values, in this case, range between 0 to 10 yielding a utopian value for the level 1 task of 32. Table 1 gives the weights and scores for each subtask, as well as, the combined and aggregated scores computed for each level.

Level	Task	Weight	Task Score	Combined Score	Aggregate Scores
1					**18.2**
2	1	0.4	7	2.8 (+4.8)=	**7.6**
	2	0.6	8	4.8 (+5.8)=	**10.6**
3	11	0.3	1		
	12	0.2	10		
	13	0.5	5	4.8	
	21	0.3	8		
	22	0.3	6		
	23	0.4	4	5.8	

Table 1. Training proficiency computation results.

The hypothetical results reveal that the military unit achieved a raw task proficiency score of 18.2 for the level 1 task. This establishes a training proficiency benchmark for this task of approximately 50% of the utopian value.

Training measurement parameters such as task conditions, standards, weights, and utile ranges may vary by training unit, simulation site, or training scenario. Obviously, task proficiency scores are sensitive to changes in these training effectiveness parameters. Therefore, care must be taken to properly account for such changes when making comparative analyses of a military unit's training proficiency over time using the methodology presented above.

5 CONCLUSIONS

At this time, none of the Army's simulation-based training facilities of Combined Armed Tactical Trainer (CATT) family have yet been built. The first to be developed will be the Close Combat Tactical Trainer (CCTT) for training armor and mechanized forces at battalion and below. The first CCTT facility is currently under construction and is expected to be operational in 1997 or 1998. Therefore, the potential value of the methodology presented in this paper is unknown at this time. Future work will include testing and evaluating the methodology and various training effectiveness measures in simulation-based training experiments to be conducted at a prototype facility located at the US Army Simulation, Training, and Instrumentation Command. Experimentation involving potential future system users will hopefully provide training system developers with feedback for improving both the simulation training systems under development and the methods for measuring training effectiveness.

There is also a high potential for transferring the military simulation-based training technology and training methodology to non-military applications. For example, a System for Training Emergency Personnel (STEP), proposed previously by McGinnis (1994) for training civilian emergency personnel, is very similar to TREDS task identification and the Instructional Systems Design (ISD) model discussed previously. Finally, it is hoped that this work will motivate future research efforts in this area that will lead to improved measures of simulation-based training effectiveness and fielding the best training system possible.

REFERENCES

Clemen, R.T. 1990. *Making Hard Decisions: An Introduction to Decision Analysis.* Belmont, CA: Duxbury Press.

Crissey, M., Stone, G.F., Briggs, D., & Mollaghasemi, M. 1994. Training exercise planning: Leveraging data and technologies. *Proceedings of the 16th Interservice/Industry Training Systems and Education Conference.* Washington, DC: National Security Industrial Association.

Gonzalez, A. & Ingraham, L. 1994. Computer-based exercise progression in simulation-based training. *IEEE Transactions on Systems, Man, and Cybernetics* 24(6): 863-874.

Hwang, C. & Yoon, K. 1981. *Multiple Attribute Decision Making Methods and Applications: A State-of-the-Art Survey.* New York: Springer-Verlag.

McGinnis, M.L. 1994. Unpublished Technical Report: *The system for training emergency personnel (STEP).* The United States Military Academy, West Point, New York.

McGinnis, M.L. and Stone, G.F. 1995. *Recent initiatives to develop decision support systems for command and control (C2) and training planning.* Industry, Engineering, and Management Systems (IEMS) 1995 Proceedings: 584-589.

Saaty, T.L. 1980. *The Analytical Hierarchy Process.* New York: McGraw Hill.

Sloman, M. 1994. *A handbook for training strategy.* Brookfield, Vermont: Gower.

Spurgin, A., Moieni, P., & Orvis, D. 1993. Some thoughts on the use of data for assessing training effectiveness. *Proceedings of the 1993 Simulation Multiconference on the International Simulators Conference.* San Diego, CA: Society for Computer Simulation: 579-584.

Tannenbaum, S., Cannon-Bowers, J., Salas, E., & Mathieu, J. 1993. Technical Report 93-011: *Factors that influence training effectiveness: a conceptual model and longitudinal analysis.* Orlando, FL: Naval Air Warfare Center Training Systems Division.

Turnage, J., Houser, T., & Hofmann, D. 1989. *An assessment of performance methodologies for collective military training.* Orlando, FL: University of Central Florida.

Williams, K., Reynolds, R, Carolan, T., Anglin, P., & Shrestha, L. 1989. Technical report TR89-035: *An evaluation of a methodology for cognitively structuring and adaptively sequencing exercise content for embedded training.* Orlando, FL: University of Central Florida, Institute for Simulation and Training.

AUTHOR BIOGRAPHIES

MICHAEL L. MCGINNIS directs the Operations Research Center at the United States Military Academy (USMA), West Point, New York. He received a B.S. degree from USMA in 1977, M.S. degrees in Applied Mathematics and Operations Research from Rensselaer Polytechnic Institute in 1986, a Ph.D. in Systems and Industrial Engineering from the University of Arizona in 1994, and a M.A. in National Security and Strategic Studies from the Naval War College in 1996. His applied research interests are military command and control, personnel forecasting, and resource scheduling. E-mail: fm0768@exmail.usma.edu.

GEORGE F. STONE III is United States Military Academy (1980) graduate with a B.S. in General Science. He also received a M.S. in Industrial Engineering from Texas A&M University in 1989, and more recently his doctorate in Industrial Engineering at the University of Central Florida. Major Stone has had numerous field artillery assignments including two battery commands in West Germany. Major Stone is a former Assistant Professor in the Systems Engineering Department at West Point and is currently the Deputy TRADOC Systems Manager for WARSIM 2000. His research areas include simulations, operations research/systems analysis, systems engineering and training.

Proceedings of the 1996 Winter Simulation Conference
ed. J. M. Charnes, D. J. Morrice, D. T. Brunner, and J. J. Swain

CLOSING THE GAP BETWEEN SIMULATION AND COMBAT COMPUTER SYSTEMS

Roger D. Smith
Mystech Associates
Orlando, Florida 32833

ABSTRACT

As simulations have evolved, interoperability between them has emerged as a fundamental technique for increasing their applicability and minimizing the cost of development and maintenance. From totally independent systems, through manual interfaces, automated interfaces, messaging standards, control standards, and architectural standards - the field has been transformed from a set of independent programs into a loose confederation working together to maximize each others contributions. The next step in this evolution is the inclusion of real-world combat computers in the simulation federations, particularly command and control computers.

This paper will define the steps that must be taken to support this integration with a large number of simulation systems. It will explore interoperability projects that are taking place within the simulation and C4I system communities, describing architectures such as the simulation High Level Architecture, the Modular Reconfigurable C4I Interface, the Defense Information Infrastructure - Common Operating Environment, and the Joint Military Command and Information System. Interoperability methods for the generic integration of systems built to a common architecture will be proposed and some argument given to its viability.

1 SIMULATION INTEROPERABILITY

Though simulations have been used by the military for analysis and training for centuries, it was not until the advent of computer networks that the issue of integrating multiple heterogeneous systems emerged. Prior to that, it was naturally assumed that each simulation was a virtual world unto itself, and that information exchange was a laborious, human-driven process. Analytical simulations were the first to crack the "interoperability barrier" by creating simulations that were specifically designed to accept the output of another model as the input for beginning a different stage of analysis.

The virtual community created the Simulator Networking (SIMNET) systems with their associated protocols for joining multiple copies of the same simulator into a common synthetic world. Those these were intended for homogeneous simulations at a single location, they pointed the way for more ambitious projects. Distributed Interactive Simulation (DIS) protocols followed, adding the ability to join a large variety of simulations and to do so across wide area networks.

As this was developing the constructive community was creating the Generic Data System (GDS) to bridge the gaps between constructive simulations. These allowed widely different simulations to exchange world data and deposit it in a common data repository and analysis system. This was followed by the Aggregate Level Simulation Protocol (ALSP) which created a more dynamic environment and provided distributed management functions.

The High Level Architecture (HLA) promises to join both of these communities and allow them to work together at a level not attainable before. Simulations from many fields: staff training, crew training, live training, analysis, engineering, and testing will be able to take advantage of a single architecture for distributed simulation.

The next step in this evolution seems to be the integration of real world C4I systems into the synthetic world. Projects like the Modular Reconfigurable C4I Interface (MRCI), Tactical Simulation System (TACSIM), and other experiments are illustrating the fact that these command systems can be extensions of the distributed simulation paradigm. TACSIM supports dedicated interfaces between simulations and intelligence analysis systems and MRCI promises to extend this to the generic level available to simulations through the HLA Run Time Infrastructure (RTI). With this continuous expansion of the dimensions of interoperability it is probable that the integration of other forms of combat computer systems will follow: radar trackers, radios, communication stations, operational sensor systems, and even combat aircraft. As real systems, particularly C4I systems, settle into the use of standard message formats and data exchange methods, the level of interoperability between these different fields will continue to increase.

The era of dedicated, stand alone simulation models is coming to a close. The class of problems that can be addressed with today's technology far exceeds the ability of a single model to capitalize upon. Distributed, cooperative simulation solutions that can still operate independently are going to be the systems of the foreseeable future.

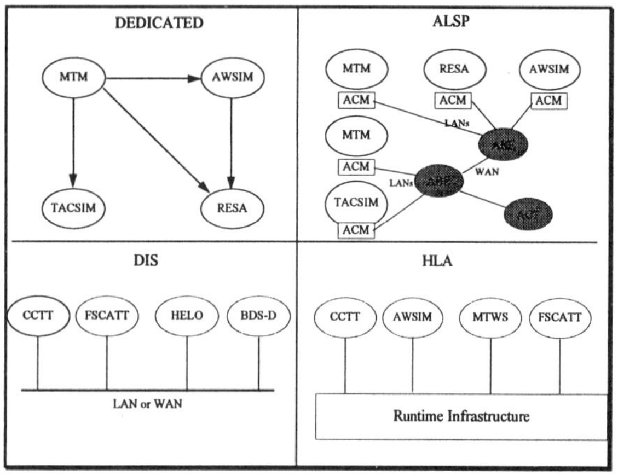

Figure 1. Simulation Interoperability Techniques

A topological illustration of each of these levels of interoperability is shown in Figure 1.

2 C4I SYSTEM INTEROPERABILITY

Within the military command and control communities, standardization programs have been evolving in response to operational and budgetary issues very similar to those encountered in the simulation community. As computerized systems were developed for command, control, communications, and intelligence a variety of different hardware architectures, data representations, and network standards created an environment in which each system was electronically isolated from the others around it. Each was unable to exchange data except through direct human intervention. Standardization first occurred within each service and is now extending across the services and into allied systems.

2.1 Messages

The most basic characteristic of information is the ability to communicate it in an understandable form. Messages generated at one source must be correctly understood by a number of recipients. The Joint Interoperability of Tactical Command and Control Systems (JINTACCS) program is one of the most successful programs at creating standard messages which can be understood by humans and read by computers. These standards dictate the "punctuation" of the message, eliminating the military message equivalents of sentence fragments, run-on sentences, and mis-abbreviation. JINTACCS messages are recognizable by their abundant use of the "/" character as a punctuation symbol.

The United States Message Text Format (USMTF) organizes JINTACCS by defining a set of standard messages. The information contained in these creates the equivalent of "sentences" for military topics. These are grouped together to make "paragraphs" which describe all of the information that may be known about an enemy unit or that may be needed to serve a friendly unit. Multiple paragraphs can be placed in single message to allow the exchange of specific information on a large number of battlefield entities.

2.2 Network

Standard messages require a standard messaging system to transmit them between a variety of equipment types. In 1994, the proof-of-concept for the Global Command and Control System was fielded. This allowed computers from many locations to exchange data by both "pushing" it to other systems and "pulling" it from systems, creating the seeds for a globally distributed set of data that is completely accessible from anywhere in the world. The system allows other computers to join the network through the use of Common Operating Environment (COE) standards and components. By adhering to the COE standards or reusing the COE components, any C4I system can join the command network and exchange data with systems already resident on the network. The COE standards describe how systems must interact with each other, similar to the Internet protocols which allow a network to span the world without stumbling on the barriers of language, telephone line, computer architecture, and electric power differences.

2.3 Operations

The Command and Control Reference Model (C2RM) is an attempt to further standardize military communications by creating a standard taxonomy that describes all events and states in military operations, particularly command and control (Figure 2). The C2RM standardizes the phases of combat and the types of activities that are occurring in each of these phases. It formulates the steps through which all decisions are made and when information should be exchanged.

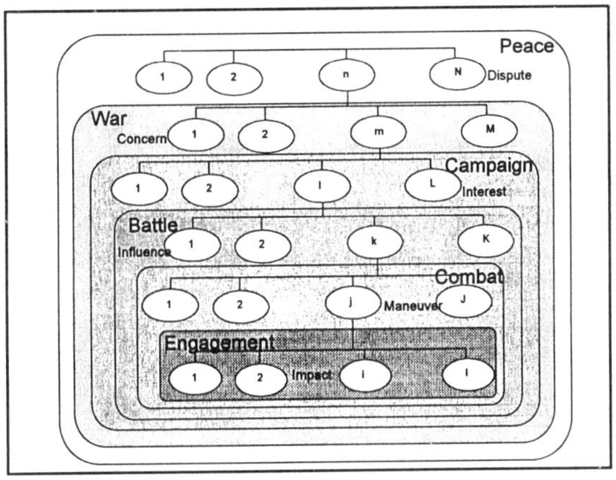

Figure 2. C2 Operations Standard

2.4 International

The International Command and Control Systems Interoperability Project (IC2 SIP) is attempting to extend interoperability beyond US systems into those of our allies in Germany and France. Between different countries there are the barriers of language, message format, and computer architecture which make combined operations in Iraq, Somalia, and Bosnia very difficult. Assuming that the content of military operations and communications is similar, this project converts messages from one country into the language and format of another. The messages being translated are very data intensive and contain little "free text", as a result the system is less a language translator than a data reformatter.

IC2 SIP uses a concept like Inter-lingua for data description, a language to which any other language can be translated and from which any language can be synthesized. This allows an "Appliqué" message in US Variable Message Format to pass through a decomposition template, be converted into the common data format, fed into a composition template, and reformatted as a French "SIR" or German "GeFuSys" message (Figure 3).

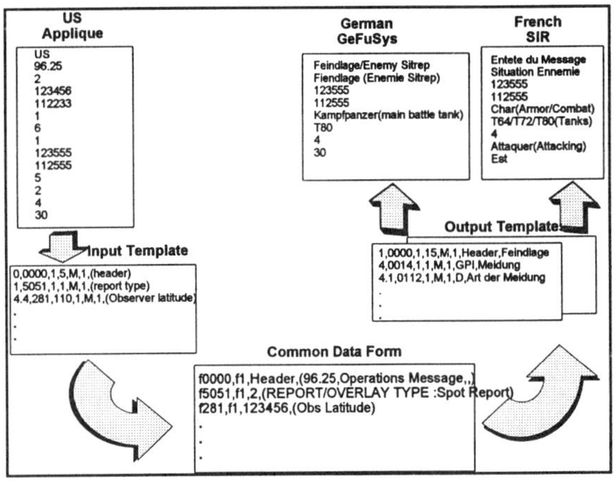

Figure 3. International C4I Interoperability

3 BRIDGING THE GAP

Constant standardization has created an optimistic atmosphere in which military program managers expect to join all combat computer systems - be they command and control, maintenance, or simulation. As these interfaces develop, it is clear that the method for computer translation and communication is quite similar to that used in interpersonal interfaces. The liberal insertion of translators, templates, and bounding of expression make the computers seem less magical than one might have hoped, but they are performing tasks for which humans have yet to develop concise, efficient methods in any domain - computer, spoken, or written.

3.1 Special Cases

The excitement for interoperability between simulations and command and control computers is based upon the operational success of dedicated interfaces in use today (Figure 4). These include the ability of the Tactical Simulation System (TACSIM) to generate intelligence reports about a synthetic battlefield and transmit them directly into intelligence processing systems using the USMTF messages described above. The Warrior, All Source Analysis System (ASAS), Enhanced Tactical Users Terminal (ETUT), and Enhanced Processing and Dissemination System (EPDS) are all used by intelligence analysts to manage and process reports

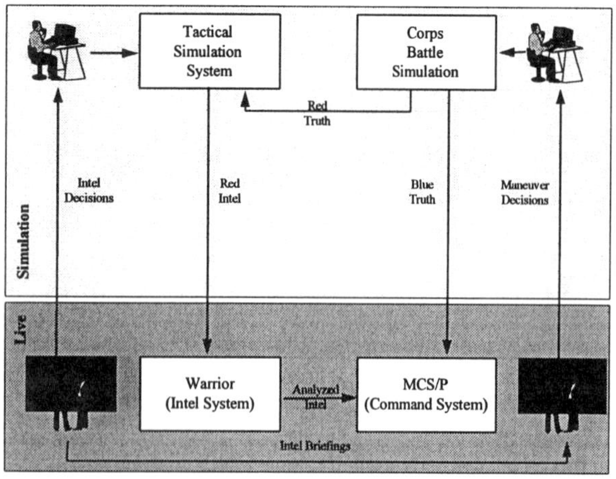

Figure 4. Current Simulation-C4I Interfaces

on enemy activities and TACSIM can feed the same reports to these systems during training exercises. This allows the analysts to work with simulated data in the same form, quantity, and with the same interface that he/she uses in a real conflict.

The Corps Battle Simulation (CBS) has also been able to create interfaces with the Maneuver Control System/Phoenix (MCS/P) and Advanced Field Artillery Tactical Data System (AFATDS) computers. A commander can enter firing orders and plans into the AFATDS system and transmit these into the simulation rather than to real-world artillery systems. He may also receive force status information from CBS through the simulation/C4I interfaces. In both cases the orders can be read, acted upon, simulated, and the results disseminated without requiring the intervention or translation of a human operator. These efficiencies reduce the man-power and the time required to prepare for training exercises.

Similarly, the Air Warfare Simulation (AWSIM) has been extended to accept Air Tasking Orders (ATOs) from the Contingency Theater Automated Planning System (CTAPS). This allows Air Force commanders to prepare daily mission orders in their combat system and send it directly to the simulation. Personnel do not have to be trained to understand the simulation interface and order syntax, a skill that is not useful to combat soldiers.

Each of these interfaces is a dedicated instance which can not easily be extended to other systems. But the concepts are being converted to standard architectures and formats in both the simulation and command and control arena. Future simulation systems are being designed to interface with the training audience purely through their native computer systems.

3.2 Design Standards

The Warfighter Simulation 2000 (WARSIM 2000), currently under development, must reduce the man-power required to operate an exercise by 33% in the near-term and 67% long-term. Much of this savings is being accomplished by interfacing the simulation with the training audience through their native equipment, eliminating a layer of "role players" that currently translate real-world orders into simulation orders and simulation reports into real-world reports. These "role players" make up a significant number of training personnel during an exercise and are usually extracted from the training audience, missing the opportunity to practice their skills. WARSIM will interface with the following communications systems: Enhanced Position Locating and Reporting System (EPLRS), Single Channel Ground and Airborne Radio System (SINCGARS), and the Mobile Subscriber Equipment (MSE), providing connectivity to a host of information processors for maneuver, artillery, logistics, air defense, and intelligence operations.

The Air Force's National Air and Space Warfare Model (NASM) is being designed to operate in a similar manner, achieving the same savings, and providing the same training improvements. It will exchange orders, reports, and plans through a host of real combat systems, such as the Contingency Theater Automated Planning System (CTAPS), Joint Deployable Intelligence Support System (JDISS), Special Operations Forces Planning and Rehearsal System (SOFPARS), and Tactical Data Information Link (TADIL).

The Navy's Battle Force Tactical Trainer (BFTT) has opted for a slightly different interface to accomplish the same objectives. The Joint Military Command and Information System (JMCIS) is a system and messaging standard for many command and control systems and BFTT will participate in that network as if it were a real command node. All data will be exchanged through the JMCIS standard and the differences between simulation and command systems will be obscured.

4 DOMAIN SPECIFIC INTERFACES

To accomplish the type of interoperability that has been discussed above it is essential that standards emerge and be used in both the simulation and combat computer arenas.

4.1 Simulation Architectures

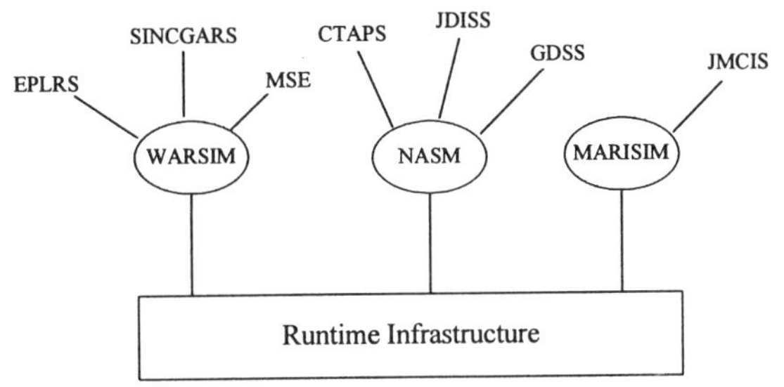

Figure 5. New Simulation-C4I Interfaces

The first experiments in simulation interoperability relied upon dedicated interfaces between two simulations and were followed by standards designed to support the functionality necessary to share a synthetic battlefield. This same transformation is beginning to occur with regard to interfaces between simulations and C4I systems. The next generation of training simulations are pushing for seamless interfaces to C4I systems and they plan to accomplish this by creating dedicated interfaces directly from the simulation to the C4I systems (Figure 5). This provides connectivity to C4I systems only through a specific host simulation. To support generic interoperability the interface to the C4I system must be through a standard protocol which can be accessed directly by any simulation adhering to the standard.
To create simulation interoperability the High Level Architecture uses the Runtime Infrastructure (RTI), allowing applications on a variety of host computers to access a standard set of services for exchanging data with other applications connected to it. Extending this bridging technique will allow any simulation to access any C4I system through standard RTI services. The Defense Modeling and Simulation Office (DMSO) is developing the Modular Reconfigurable C4I Interface (MRCI) to serve as a common gateway application to C4I systems architectures (Figure 6). It will operate as an RTI member and support the exchange of information between compliant simulations and C4I systems. However, the design calls for the integration of simulation specific software into the real C4I systems which is a significant limitation to the ability to implement and scale the solution.

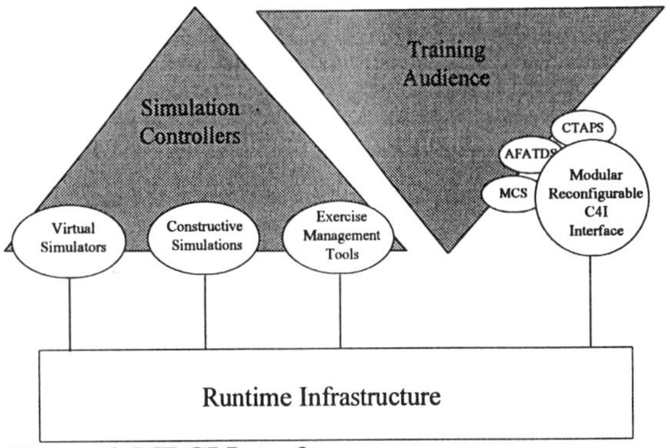

Figure 6. MRCI Interfaces

4.2 C4I Architectures

Within the C4I community, the standard emerging most prominently is the Global Command and Control System (GCCS), of which the Joint Military Command and Information System (JMCIS) is an instance. Once interfaces are built for these standards the simulation community will be able to access any of the processing, communication, analysis, and decision support systems operating on these networks. As computerized combat systems continue to grow they will adhere to the GCCS standard and interoperabilty with simulation systems will be achieved without special consideration.

4.3 Bridging Architecture

The final step would be to replace the MRCI with a more general, non-intrusive gateway that joins the simulation and the C4I standard protocols. New simulations will not be required to understand C4I specific services, but rather, a single interface will provide the bridging mechanism for connecting to all RTI compliant simulations.

Similarly, the Joint Simulation Systems (JSIMS) project proposes a single architecture for all command staff trainers. Within this architecture there exists a component which specifically models the physical performance of pieces of equipment. This component also supports the surrogate representation (ghosting) of real-world equipment. This capability specifically targeted at the ability for a simulation to communicate with internal models with out having to differentiate between real equipment and simulated equipment. The methods required to export or import a message across the simulation-C4I boundary are encapsulated in the equipment component, allowing the simulation and the C4I system to operate naturally without special consideration for the receiving equipment. This begins to support the identification of a non-intrusive bridge between the two domains which understands both communication standards, but does not interfere with the operations of systems in either world (Figure 7). Such a Simulation-C4I Gateway could serve any simulation adhering to the RTI and any C4I system adhering to the COE. Remote systems on both sides could join the exercise without having to be configured for participation.

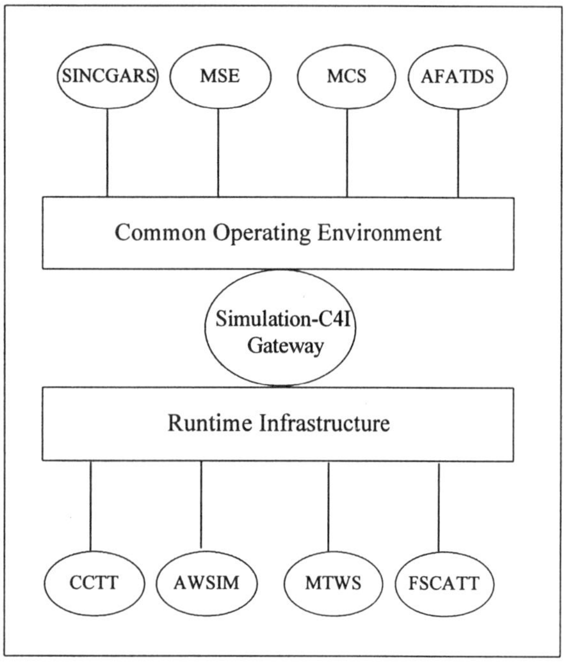

Figure 7. Bridging Two Architectures

5 CROSS DOMAIN ARCHITECTURES

Though the military is going to great lengths to join simulations and C4I systems, in the future, simulation capability may be built directly into real-world C4I systems. The current trend to provide decision support capabilities to commanders through their C4I systems is already placing simulation-like capabilities within their combat systems. A joint simulation-C4I architecture could allow the creation of a system from the pieces needed for their mission (real or training), whether they come from a simulation or tactical source.

Creating a cross domain architecture will require a joint requirements analysis process to identify the commonality between simulations and C4I systems. This must be followed by a process of requirements alignment in which similar, but not identical requirements, are matched and a system designed to meet both of them, or to modify them into a common requirement. From this set of requirements a common simulation-C4I architecture can be developed. It must be able to accommodate both the simulation and C4I specific requirements in order to serve both communities. Finally, the internal and external interfaces, be they simulation or C4I, must be defined.

6 CONCLUSION

There are more similarities between simulation systems and combat systems than there are differences. If we focus on these similarities we will find opportunities for integration and cooperation that are not immediately evident otherwise. In both cases, the user of the system is a soldier who is trying to do his/her job or improve his/her ability to do that job. Realizing that the user is one-in-the-same soldier leads to a design, development, and procurement approach between systems that is common and tends to support interoperability rather than thwart it.

Simulations have had a measurable effect on the military's ability to perform effectively in combat. A joint architecture will begin closing the temporal gap between training events and performance events. As simulations become an integral part of the battlefield it will be possible to conduct just-in-time training for contingencies or fine nuances which were not predicted and planned for in earlier training. For this to occur simulations must become part of the military table of organization and equipment (TOE), they must become a go-to-war system.

REFERENCES

Allen, Gary and Smith, Roger. 1994. TACSIM: Intelligence Training for Tomorrow's Battlefield. *Military Intelligence Professional Bulletin*. Fort Huachuca, Arizona.

Ben-Natan, Ron. 1995. *CORBA: A Guide to Common Object Request Broker Architecture*. McGraw Hill. New York, NY.

Dahmann, Judith and Jefferson, Mark. 1996. The Modular Reconfigurable C4I Interface. *Summary Report - 14th Workshop on Standards for the Interoperability of Distributed Simulations*. Institute for Simulation and Training. Orlando, FL.

Defense Modeling and Simulation Office. 1996. High Level Architecture - Object Model Templates. DMSO. Alexandria, VA.

Defense Modeling and Simulation Office. 1995. Department of Defense High Level Architecture for Modeling and Simulation, Version 0.2. DMSO. Alexandria, VA.

Defense Modeling and Simulation Office. 1995. HLA Management Plan: High Level Architecture for Modeling and Simulation. DMSO. Alexandria, VA.

Joint Simulation System Project Office. 1996. JSIMS Domain Specific Software Architecture. http://www.jsims.mil/lfs_dssa.html.

Joint Simulation System Project Office. 1996. JSIMS Architecture Implementation Plan. http://www.jsims.mil/lfs_aip.html.

McGarry, Stephen. 1995. The DoD High Level Architecture (HLA) Run-Time Infrastructure (RTI) and Its Relationship to Distributed Simulation. *Summary Report - 13th Workshop on Standards for the Interoperability of Distributed Simulations*. Institute for Simulation and Training. Orlando, FL.

Mowbray, Thomas J. and Zahavi, Ron. 1995. *The Essential CORBA: System Integration Using Distributed Objects*. John Wiley & Sons Inc. New York, New York.

Smith, Roger D. 1995. The Conflict Between Heterogeneous Simulations and Interoperability. *Proceedings of the 1995 Interservice / Industry Training Systems and Education Conference*. Albuquerque, NM.

Smith, Roger D. 1996. *Proceedings of the Electronic Conference on Interoperability in Training Simulation*. Mystech Associates. Falls Church, VA.

Tiernan, Tom. 1996. Modular Reconfigurable C4I Interface Program Plan. DMSO Briefing. http://www.dmso.mil/projects/hla/mrci/.

US Army Communications-Electronics Command. 1996. International Interoperability Today: The International Command and Control Systems Interoperability Project. Unpublished Document. Fort Monmouth, New Jersey.

US Army Communications-Electronics Command. 1996. The Command and Control Reference Model for Modeling, Simulations, and Technology Applications. Fort Monmouth, New Jersey.

US Army Command and Control Systems. 1993. Army Tactical Command and Control System Message Catalog. ACCS-A3-500-004. Fort Monmouth, New Jersey.

US Army Command and Control Systems. 1994. Army Tactical Command and Control System Specification. ACCS-A1-100-006A. Fort Monmouth, New Jersey.

Wilson, Annette L. and Weatherly, Richard M. 1994. The Aggregate Level Simulation Protocol: An Evolving System. *Proceedings of the 1994 Winter Simulation Conference*. Orlando, Florida.

AUTHOR BIOGRAPHY

ROGER D. SMITH is the Chief Simulation Scientist and a Corporate Technical Director for Mystech Associates. He is actively involved in designing, developing, and fielding simulations, having contributed to JSIMS, WARSIM, BLRSI, ADST, TACSIM, AWSIM, ALBAM, and others. He is very active in the simulation community serving as the Vice Chairman for the ACM Special Interest Group on Simulation, the General Chairman for the Electronic Conferences on Training Simulation, and as a member of the editorial boards *of ACM Transactions on Modeling and Computer Simulation* and the *International Journal of Computer Simulation, Modeling, and Analysis*. His research interests include embedded simulations in operational equipment, integrated simulation-C4I systems, and problems in aggregation/disaggregation.

Proceedings of the 1996 Winter Simulation Conference
ed. J. M. Charnes, D. J. Morrice, D. T. Brunner, and J. J. Swain

EFFICIENT DATA CONSISTENCY IN HLA/DIS++

Sudhir Srinivasan

Mystech Associates, Inc.
5205 Leesburg Pike, Suite 1200
Falls Church, VA 22041, U.S.A.

ABSTRACT

Simulation exercises based on the DIS paradigm are faced with a severe scalability limitation due to the use of time-out updates or "heartbeat" messages for maintaining consistency despite message loss. Since they carry information that is redundant and useless in most cases, these messages represent a waste of network bandwidth and computational resources. This paper presents an elegant solution that eliminates heartbeats without compromising consistency. Rather than using additional schemes and software components as others have done, the solution proposed here consists of careful use of reliable and unreliable communications based on a key insight into the communication properties of DIS-based simulations. Not only does our solution eliminate heartbeats, it also minimizes additional resource consumption by using reliable communications minimally. Finally, due to its simplicity, it can be easily incorporated into HLA-compliant simulations.

1 INTRODUCTION

A distributed simulation consists of a set of simulators distributed geographically and interconnected via a system of networks (local and wide-area), simulating a common application, typically a battle. The Distributed Interactive Simulation (DIS) paradigm and protocols define a common way for simulators to exchange information so as to maintain a consistent view of the battlefield at each simulator. In this paper, we focus on one aspect of these protocols, namely, the *time-out update*, also known in the literature as the *heartbeat*. This data consistency feature of the DIS paradigm requires each simulator to periodically send an update on the state of each *entity* it simulates, even if the state of the entity has not changed since the last update. The DoD High Level Architecture (HLA) (McGarry, Weatherly and Wilson, 1995), which is the chosen architecture for the next generation of DIS, referred to as DIS++, provides a *minimum rate* communication service to be used by DIS-based simulations to implement heartbeats. Clearly, heartbeats impose a severe scalability limitation. With the advent of large exercises such as STOW (Calvin et.al., 1995) which are expected to simulate on the order of hundreds of thousands of entities, the scalability problem due to the heartbeat has become critical to the point of making such exercises infeasible. Consequently, there is a severe need for a scalable solution.

While reliable communication can eliminate the need for heartbeats, it introduces problems of its own (excessive bandwidth consumption and potentially higher communication latency). We describe a new solution to the problem that combines unreliable and reliable communications in a simple and elegant way to yield an effective, scalable solution. The main features of our solution are:

- it eliminates the need for heartbeats;
- it uses reliable communication only when required, thus minimizing the number of acknowledgment messages and retransmissions;
- it is simple to implement — specifically, the HLA provides the required primitives, so that our solution can be incorporated trivially by HLA-compliant simulations.

Due to the power and simplicity of our solution, we strongly believe it can solve the scalability problem for large-scale distributed simulation exercises.

2 THE PROBLEM

DIS (Dahmann and Wood, 1995) refers to two things: a *paradigm* for interoperability among distributed real-time simulators, and an IEEE standard set of *protocols and data structures* that implement this paradigm. With the advent of the HLA as directed by the DoD Modeling and Simulation Master Plan (DMSO, 1995), it is expected that the DIS community will adopt the HLA as the architecture for DIS++, the next generation of DIS.

Irrespective of the implementation (protocols vs. HLA), the basic DIS paradigm is expected to be an important technology in future distributed simulation exercises. Essentially, a distributed simulation based on the DIS paradigm consists of a set of geographically distributed simulators tightly synchronized to a global wall-clock time. The simulators maintain a consistent view of the battlefield by exchanging ground truth such as state updates (e.g. position and velocity) and interactions (firing of a weapon) and modifying their local databases accordingly. To reduce the number of state updates, DIS uses a mechanism called *remote entity approximation* (or *dead reckoning*, DR) wherein simulated entities use a common algorithm to estimate the trajectory of other entities based on the most recently received updates. With this scheme, state updates are generated only when the actual trajectory of an entity differs from the trajectory estimated by the DR algorithm by some threshold amount. In the rest of this paper, "DIS" refers to the DIS paradigm rather than the protocols, unless specified otherwise.

DIS does not assume a reliable communication medium underlying the simulators. In other words, it is possible for messages (in particular, state updates) to be lost. To compensate for lost updates, DIS includes a *time-out update* mechanism, commonly known as the *heartbeat*. Under this mechanism, an entity must generate an update if a given amount of time has elapsed since the last one, even though its state or appearance have not changed. The periodicity of these heartbeats is adjusted (typically, 5 seconds) so that the error due to lost updates is within acceptable bounds. Thus, the primary use of the heartbeat mechanism is to maintain data consistency in the face of lost messages. Heartbeats have also been used to detect the removal of an entity from an exercise — an entity that has not sent a heartbeat for a certain number of time-outs is considered deleted.

Heartbeats impose a severe scalability limitation on DIS exercises. This scheme worked well for the original SIMNET protocols because SIMNET exercises were on a small scale (few tens of entities). DARPA's STOW program expects to conduct exercises involving on the order of 100,000 entities by the year 2000 (Calvin et.al., 1995). A large number of these entities are expected to be in a steady state most of the time (i.e. their state and appearance change very infrequently). Unfortunately, DIS requires these entities to generate updates at every time-out. Clearly, this will place excessive demands on the network (for transporting messages) and processors (for generating and consuming the messages). This (mostly redundant) traffic is projected to constitute a significant portion of the total traffic in large-scale exercises. The problem will only worsen as DIS++ incorporates new kinds of entities such as dynamic terrain and environmental effects (smoke and weather). Consequently, there is a significant need for a scalable solution that eliminates or minimizes the use of heartbeats.

3 RELATED WORK

To date, only two efforts have addressed the problem: the STOW real-time information transfer and network (RITN) project (Van Hook, Calvin and Smith, 1995) and the DIS lite & query protocol (Taylor, 1995).

The STOW approach is based on a key characterization of the state of DIS entities into *active* and *quiescent* states. A quiescent entity is one that generates only heartbeats (we defer a formal definition to Section 4.2 since our solution is based on these definitions as well). The main idea behind STOW's *consistency protocol* is to reduce the number of heartbeat messages generated by combining the updates for *all* quiescent entities at a simulator into a single message at each timestep. The protocol requires simulations to effectively maintain a shared *consistency list* that is used to track the set of quiescent entities and their versions. Using the information in the consistency lists, entities receiving a heartbeat update can determine the quiescent entities for which they do not have current information and request this information using a *negative acknowledgment* (NACK) scheme. All entities that do not have current information send a NACK message indicating the missing pieces of information to the emitter of the updates which responds with the missing information. Since a large number of NACK's could be generated in a short period of time (NACK's follow a bursty pattern), the protocol incorporates a *NACK suppression* scheme using back-off in which a sender of a NACK waits a random amount of time before sending its NACK. If the NACK of another sender is received during this back-off period, the sending of the NACK is aborted.

Clearly, the STOW approach is fairly complex (the consistency list is a shared data structure with a single writer and multiple readers, the implementation of which is non-trivial; so also the NACK suppression scheme). Recognizing scalability limitations of the consistency protocol, STOW has proposed *consistency agents* to off-load some of the consistency tasks from the simulations. A key factor to good performance is quiescence detection. Simulations must be able to accurately determine when entities have become quiescent or active. Quiescence is usually detected when an entity has been generating heartbeats for some period of time. This threshold period must be chosen very carefully since it introduces a performance trade-off: if quiescence is detected too often, the overhead of the consistency protocol will degrade performance; on the other hand, if quiescence is not detected often enough, the benefits of the protocol are not observed. To simplify this problem, the STOW program adopted a weaker definition of quiescence where entities are considered quiescent only when they are stationary (trading off some performance in the process). It is important to note that the consistency protocol requires heartbeats to be *broadcast* to all simulations (whereas state updates would be multicast to only those simulations that require them). As such, the scalability of the STOW scheme improves upon that of DIS only in that it is limited by the number of simulations rather than the number of entities. While this represents a significant improvement, we believe it will prove insufficient as exercises that involve large numbers of simulations are planned.

The DIS-Lite & Query Protocol (DLQP) is based on an analysis of the contents of the *entity state protocol data unit* (ESPDU), the most often used unit of data exchange in DIS. DLQP proposes alternate "lightweight" PDU's that would be used instead of the ESPDU's, to reduce the bandwidth consumed. Further, it defines "minimal" versions of these new PDU's that would be used for the heartbeats, again, to reduce bandwidth consumption. Finally, it proposes increasing time-out intervals to reduce the number of heartbeats. A query protocol is used to fill in gaps of information, much like the NACK scheme of the consistency protocol. DLQP uses a suppression scheme for queries similar to that used for NACK's in the consistency protocol. Since DLQP is based on the structure and contents of the ESPDU, it is not clear how it will be extended to DIS++ which will use the HLA interface specification for data exchange, rather than the ESPDU. A more detailed comparison of DLQP with the STOW consistency protocol is given in (Smith and Van Hook, 1996).

4 TOWARDS A NEW SOLUTION

At first thought, it seems the solution to the heartbeat problem lies in reliable communications: if all messages were sent reliably, there would be no need for heartbeats. Unfortunately, reliable communication has its own costs. The main drawback of reliable communication is that reliability is achieved using acknowledgment messages and retransmission of lost messages which increases the bandwidth consumed, compared to unreliable communication. This introduces an interesting trade-off that is an open problem worthy of investigation: on the one hand, the reliable communications for *all* updates increases the bandwidth consumption; on the other hand, the use of heartbeats with unreliable communications also increases bandwidth consumption — which is better? Ironically, while reliable communication protocols are designed to alleviate the loss of messages, it is precisely that situation that causes the most problems for these protocols. It is easy to reach a state where the loss of a message causes a sender's retransmission window to fill up and thus halt further sending of new messages (thereby increasing the latency of communication for those messages). This second problem is severe enough due to the real-time requirements of distributed simulations to make reliable communication *for all messages* infeasible (at least at the time of writing — as we move towards more reliable communication media, reliable communications may prove to be the solution).

4.1 Requirements

From the above discussion, some combination of reliable and unreliable communication appears necessary, to solve the scalability problem of heartbeats. Ideally, the solution must utilize unreliable communications most of the time and use reliable communications only when required. The goal is to develop a solution that eliminates (or minimizes) the use of heartbeats while minimizing the use of acknowledgments and retransmissions.

Ideally, a scheme that eliminates heartbeats would be preferable. In practice, the DIS community has used heartbeats for a second purpose: detecting the loss of entities. An entity from which some specified number of contiguous heartbeats have not been received is deleted to prevent other entities from unrealistically engaging or dead-reckoning it. If heartbeats are eliminated, we must address this issue of "dropped entities" — i.e. how to detect entities that are either no longer operational or that cannot communicate with the rest of the simulation.

In keeping with the philosophy of the STOW RITN project (Van Hook, Calvin and Smith, 1995), we argue that the primary focus of efforts in this area should be on the data consistency aspects, for the following reasons:

- The issues of data consistency and detecting dropped entities were originally coupled because a single mechanism (heartbeats) could address both. *There is no other reason to link these two issues.* Since the heartbeat mechanism is proving inadequate to solve the data consistency issue, we argue that the two issues should be decoupled and independent solutions developed for each.
- The decoupling of the two issues allows the use of more efficient schemes (such as query-reply schemes using unicasts rather than multi- or broadcasts) to detect dropped entities
- Such outages are temporary in nature and during such outages, interactions are meaningless by definition (Smith and Van Hook, 1996). Schemes developed to detect such situations must focus on quick detection rather than recovery since no protocol can compensate for the semantic gap created by the sudden loss of entities.

4.2 Solution Strategy

Under the DIS paradigm, state updates must be emitted under three conditions:

- dead-reckoning threshold is exceeded
- appearance has changed
- time-out period has elapsed

An entity that emits updates under the first two conditions is said to be in an *active* state while an entity that emits updates only under the last condition is said to be in a *quiescent* state. Note, stationary entities (whose appearance is not changing) are quiescent by definition, but quiescent entities are not necessarily stationary — a moving entity that does not deviate significantly from its dead-reckoned path is also quiescent. Typically, entities transition between these states in the course of a simulation: an active entity could become quiescent; a quiescent entity could become active; a quiescent entity could move to another quiescent state (for example, by changing its appearance only).

To understand the solution we propose, we must first understand the nature of communications among DIS simulations. A very important requirement is low latency of communication. DIS simulations operate in real-time (typically with humans in the loop). To be realistic, the communication latency must be low enough (~ 100 ms) so that the occurrence of an event at one simulator and the notification of the same at another are effectively instantaneous. An important consequence of this real-time requirement is that it limits the benefits of reliable communications, as follows: the loss of a message implies the information in the message will not reach its destination at the time it would have reached if the message had not been lost. In a sense, we can say the damage is already done when a message is lost. Any remedial action (such as a retransmission) can only reduce the effect of this damage, but cannot eliminate it. Another factor that limits the benefit of reliable communication is the property that a new update makes any older update obsolete immediately. Thus, a retransmission by a reliability mechanism pays off only if it happens closer in time to the original lost message than to the next update. Recalling the definition of active and quiescent states of entities, this implies that reliable communications will not provide significant benefit when an entity is in an active state. Since the goal is to eliminate the heartbeats in the quiescent states, we have the following insight which is key to the solution proposed here:

> **Observation:** Reliable communications will have the most benefit if used while transitioning to a quiescent state.

Based on this observation, we describe next the solution we propose to solve the scalability problem due to heartbeats.

5 THE SOLUTION

The rationale behind our scheme (developed in Section 4) is summarized in Figure 1. In an active state, the entity generates new state updates often enough so that the time-out does not occur. Reliable communication does not provide significant benefit in this state because new updates (being generated frequently enough) subsume older updates. In other words, the time to retransmit the lost message (including timing out waiting for an acknowledgment) will typically be close to or greater than the time to the next new state update, significantly reducing or eliminating the benefit of the retransmission. In the quiescent state, we wish to eliminate the s completely. Since the heartbeats are generated primarily to ensure data consistency in the event some simulators do not receive some of these updates, our solution is to use reliable communication while transitioning to a quiescent state, thus eliminating the need for heartbeats in the quiescent state.

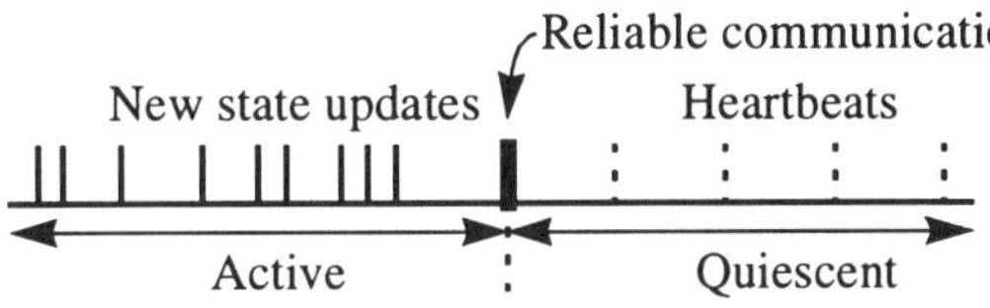

Figure 1: Rationale

To implement our solution, a simulation follows the state diagram of Figure 2. The upper entry in each state describes the state while the lower entry (in italics) describes the communication paradigm used in that state. In an Active state, the entity sends updates using unreliable communications (as is done in DIS now). The occurrence of a time-out is taken as an indication that the entity is transitioning to a quiescent state and the entity enters the Transition state where it sends an update using reliable communication. In the Transition state, one of two things can happen: if the entity generates a new update, then it is put back into the Active state and the update is sent using unreliable communication as before; if a second time-out is generated, the entity enters the Quiescent state where updates are suppressed. The entity comes out of the Quiescent state when it generates its next new state update, at which time it moves to the Active state. Note, it is possible to reduce the Transition and Quiescent states into a single state that sends only one message with reliable communication and suppresses all subsequent messages. With this optimization, a single time-out is needed to transition the entity out of the Active state — thereafter, the timer can be disabled until the entity re-enters the Active state. It is very important to note that the two classes of messages (reliable and unreliable) should follow different logical paths in the communications software of a simulation. This is required so that hold-ups in the reliable communication channel (such as a full retransmission buffer) do not interfere with the unreliable communications.

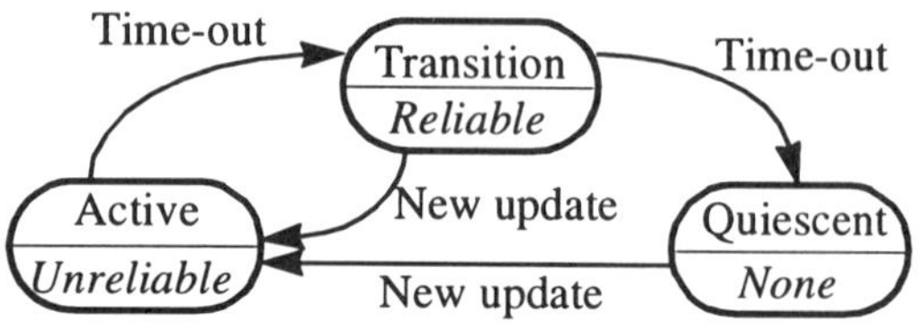

Figure 2: Implementation

The simple scheme we have proposed above has several powerful advantages:

- It eliminates heartbeat messages completely — the only update is in the transition state and this update is required to preserve the effect of the heartbeats (data consistency).
- It uses reliable communication only where it is most likely to be beneficial.
- It uses reliable communication at a minimum, resulting in a minimum number of projected acknowledgments and retransmissions (and thus, extra bandwidth consumed).
- Reliable communications do not interfere with unreliable communications, so that in an active state, the entity's updates can be delivered in a timely manner.
- It does *not* require quiescence detection; in a sense, the quiescence detection is built into the scheme (note also, we are using the stronger definition of quiescence here, where a quiescent entity need not be stationary).
- Any retransmissions required due to lost messages are handled by the communications software rather than the simulation, which will lead to better performance since the retransmissions can presumably be performed more efficiently at the communications layer.
- Since updates are usually sent to several receivers, the use of reliable communications has the benefit that retransmissions are sent only to those receivers that did not receive the original message. With the heartbeats of DIS, effectively, the retransmission is sent to *every* receiver of the original update.
- It can benefit single-entity simulators as well as those that simulate a number of entities, such as synthetic forces.
- It is relatively easy to implement.

The power of the solution may be enhanced by the following optimization. Recall, the solution utilizes the property that newer updates subsume older ones. Consider an entity that has just moved into the Transition state (Figure 2). Soon after, it generates a new state update, moving it back into the Active state. In such a case, it would be beneficial if the simulation had the capability to abort the reliable transmission of the message it sent in the Transition state (if such a transmission is in progress). While this capability may not exist with off-the-shelf communication software such as TCP, it is one that can be built in easily. This optimization could be useful in the case where an entity toggles back and forth between the Active and Transition states.

Theoretically, the scalability of this solution is limited only by the number of entities in the Transition state simultaneously (i.e. the number of entities using reliable communications at any time). In practice, this number is expected to be very low. Optimizations such as controlling time-out intervals and aborting unnecessary retransmissions (described above) can further improve scalability.

6 HIGH LEVEL ARCHITECTURE/DIS++

One of the advantages of the solution we have proposed here is its ease of implementation. This is especially true for simulations that are HLA-compliant. The HLA is a part of a common technical framework being developed by the Architecture Management Group (AMG) of the Defense Modeling and Simulation Office (DMSO) to promote interoperability and reuse among simulations. It consists of a set of compliance rules, an object model template (OMT) to describe the object models of simulations (and federations of simulations) and an interface specification describing the interfaces using which the simulations interact. The services described in the interface specification are provided by a *run-time infrastructure* (RTI). Among other things, the RTI provides communication services to the simulations. Currently (Architecture Management Group, 1996), the RTI is expected to provide both reliable and unreliable (called *best-effort*) communication. As such, the RTI provides the primitives needed to implement our solution. HLA-compliant simulations can simply choose the appropriate RTI service depending on the state they are currently in.

The RTI being developed currently by the Defense Modeling and Simulation Office is a prototype designed primarily for demonstrating the HLA concept in practice. As such, some aspects of its design are guided by schedule rather than by performance. Therefore, while our solution can be implemented using the RTI prototype, performance may not be satisfactory. We expect the performance to improve with future RTI implementations.

7 CONCLUSION

The use of time-out updates (or heartbeats) in the DIS paradigm poses a scalability limitation for distributed simulation exercises. This problem has become acute with the advent of large-scale simulation exercises such as STOW and alternate schemes are needed. We have proposed an elegant yet powerful solution to this problem. Our solution consists simply of judicious use of reliable and unreliable communications based on a key insight into the communication requirements of DIS simulations and does not require any additional software/hardware schemes. It completely eliminates the use of heartbeats. Further, it uses reliable communications only when required and in a way such that they need not interfere with the normal, unreliable communications used for state udpates. Consequently, it minimizes the use of acknowledgments and retransmissions. Overall, we believe it will significantly improve the scalability of distributed simulations. Finally, our solution is easy to implement, especially in HLA-compliant simulations.

REFERENCES

Architecture Management Group. 1996. Interface Specification for the HLA, Version 0.5. Working document available from URL: http://www.dmso.mil/docslib/hla/IF_Spec_v0.45.doc, Defence Modeling and Simulation Office, Alexandria, Virginia.

Calvin, J.O., J. Seeger, G.D. Troxel, and D.J. Van Hook. 1995. STOW Realtime Information Transfer and Networking system architecture. In *Proceedings of the 12th Workshop on Standards for the Interoperability of Distributed Simulations*, 343-353, Institute for Simulation & Training, Orlando, Florida.

Dahmann, J.S., and D.C. Wood. 1995. Editors, Special Issue on Distributed Interactive Simulation. *Proceedings of the IEEE*, Vol. 83, No. 8.

DMSO. 1995. DoD Modeling and Simulation Master Plan, Defense Modeling and Simulation Office, Alexandria, Virginia, October 1995.

McGarry, S., R. Weatherly, and A. Wilson. 1995. The DoD High Level Architecture (HLA) Run-time Infrastructure (RTI) and its relationship to distributed simulation. In *Proceedings of the 13th Workshop on Standards for the Interoperability of Distributed Simulations*, 595-606, Institute for Simulation & Training, Orlando, Florida.

Smith, J.E. and D.J. Van Hook. 1996. Comparison of consistency protocol vs. DIS-Lite. In *Proceedings of the 14th Workshop on Standards for the Interoperability of Distributed Simulations*, 875-884, Institute for Simulation & Training, Orlando, Florida.

Taylor, D. 1995. DIS-Lite and Query protocol. In *Proceedings of the 13th Workshop on Standards for the Interoperability of Distributed Simulations*, 697-703, Institute for Simulation & Training, Orlando, Florida.

Van Hook, D.J., J.O. Calvin, and J.E. Smith. 1995. Data consistency mechanisms to support distributed simulation. In *Proceedings of the 13th Workshop on Standards for the Interoperability of Distributed Simulations*, 797-806, Institute for Simulation & Training, Orlando, Florida.

AUTHOR BIOGRAPHY

SUDHIR SRINIVASAN is a research scientist at Mystech Associates, Inc., conducting research in parallel and distributed modeling and simulation. He received his Ph.D. in Computer Science from the University of Virginia in 1995 and Bachelor of Engineering also in Computer Science from the Bangalore University, India, in 1990. From August 1995 through February 1996, he was also a research associate at the University of Virginia, working on identifying fundamental issues in linking models at different levels of resolution. His main research interest is in parallel and distributed simulation, especially the High Level Architecture. He is supervising the ongoing development of a performance analysis tool for the HLA. His other interests are in parallel and distributed computing and networking. He has published several papers in conferences and journals and is a member of the ACM and the IEEE Computer Society.

Proceedings of the 1996 Winter Simulation Conference
ed. J. M. Charnes, D. J. Morrice, D. T. Brunner, and J. J. Swain

EVOLVING THE VV&A PROCESS FOR THE ALSP JOINT TRAINING CONFEDERATION

John A. Tufarolo
Ernest H. Page

The MITRE Corporation
1820 Dolley Madison Boulevard
McLean, Virginia 22102-3481

ABSTRACT

The Aggregate Level Simulation Protocol (ALSP) Joint Training Confederation (JTC) provides the environment for several large-scale computer-assisted command post exercises (CPXs) each year. The JTC consists of a collection of Service simulations and evolves according to an annual development cycle. The JTC environment poses several challenges to simulation model verification, validation and accreditation (VV&A). Some of these challenges are of a technical or fundamental nature, others are more programmatic. We briefly describe several of these challenges and the efforts currently underway to address them.

1. INTRODUCTION

The Aggregate Level Simulation Protocol (ALSP) began as a Defense Advanced Research Project Agency (DARPA) funded effort in the spring 1992 to investigate the feasibility of interconnecting existing constructive simulations to form a cohesive exercise environment (DARPA 1990). Subsequent to a successful demonstration of these capabilities in the fall of 1992, ALSP was fielded as the infrastructure for a premier joint exercise environment. Since 1992 this environment, the Joint Training Confederation (JTC), has grown from two primary simulations to eight and has supported several major exercises each year. For 1996 these include: RSOI and UFL, hosted by the Korean Battle Simulation Center, Seoul, Korea; Prairie Warrior, hosted by the National Simulation Center, Ft. Leavenworth, Kansas; and United Endeavor, hosted by the Joint Training and Analysis Simulation Center, Suffolk, Virginia. For a complete list of JTC-supported exercises refer to (Page et al. 1996).

Many aspects of ALSP and the JTC have been necessarily experimental in their nature. Although many of the technologies leveraged are mature ones, the JTC arena provides a novel context for their application. As a result, the processes and methodologies guiding the JTC have followed an evolutionary path. Many problems encountered have admitted solution, but many challenges remain.

The techniques and methodologies for verification, validation and accreditation (VV&A) are an example of this phenomenon. The techniques for VV&A of discrete event simulations are well studied and widely utilized (e.g., see (Balci 1994b)). Likewise, software testing methodologies are well known (e.g., see (Hetzel 1988)) and are the basis of a variety of standards (e.g., see (Department of Defense 1988)). However, the application of these techniques and methodologies within the JTC is not straightforward — a variety of technical and programmatic barriers to their application exist. The size and scope of the JTC is increasing to keep pace with the ever-expanding training requirements, yet the time and dollars allocated to testing promises to remain fixed. In order to maintain JTC acceptability (a function of sufficient validity), the process of engendering and quantifying JTC credibility must strive to be both flexible and optimal.

This paper describes the latest efforts in refining the VV&A processes within the JTC. Section 2 provides a review of the JTC development process. Evaluation of the impact on this process on VV&A is given in Section 3. Section 4 identifies some current initiatives to improve cost-effectiveness of VV&A in the JTC. Conclusions are given in Section 5.

2. THE JTC DEVELOPMENT PROCESS

Page et al. (1996) describe the VV&A practices for the ALSP JTC focusing on the testing process and the technical and operational challenges introduced through the Advanced Distributed Simulation (ADS) arena. A more general description of the JTC development process is given in (Miller and Zabek 1996).

The JTC development cycle comprises the transformation of user requirements and existing training simulations into a user-validated system (a "confederation" of simulations). A portrayal of the overall ALSP system

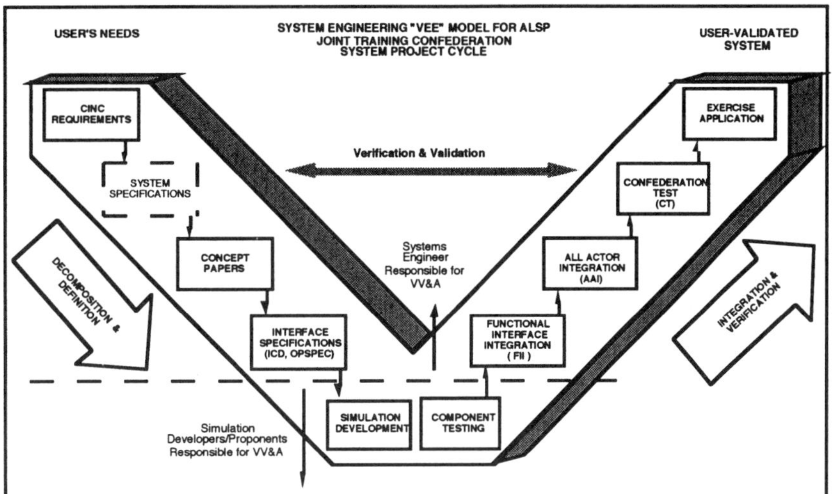

Figure 1: Vee Model

project cycle is presented in Figure 1, as viewed using a system engineering "vee" model adapted from (Forsberg and Mooz 1992).

A single iteration of the system life cycle begins with user needs (requirements) at the upper left and concludes with a user-validated system at the upper right. Descending the left side of the "vee" reflects decomposition and definition of the system. Ascending the right side of the "vee" represents system integration and verification. Each level of the model provides a correspondence between: (1) defined requirements and specifications, and (2) integration and verification events. In order to promote prudent system development, requirements and specifications should not be defined in a way that cannot be tested or verified. Similarly, integration and test events should have corresponding requirements and/or specifications they propound to verify and/or validate.

2.1 Overview

For the JTC, user needs are captured in the form of CINC/Service requirements. These high-level requirements are defined, agreed upon and prioritized for implementation. Based upon available funding (as well as the perceived availability of future funding) a collection of high-level designs are constructed in the form of concept papers. These concepts are evaluated against a variety of criteria and, if accepted, are refined into a set of model specifications and implemented within the various simulations comprising the JTC. Subsequently, the system of simulations is integrated, tested and evaluated. The "final" system consists of an accredited confederation (new versions of individual simulation software along with the ALSP Infrastructure Software) and associated documentation. It is this final system that is intended for deployment and subsequent use at exercise locations.

2.2 Design and Specification

From CINC/Service requirements, sets of *concept papers* are written. A concept paper describes the new or modified confederation functionality being proposed, identifies the relevant requirements, and suggests methods for achieving the desired capability. Concept papers outline the applicability of the concept to the JTC, impact on existing simulations, and estimated level of effort for affected simulations. These papers also provide a forum for bringing new simulations into the JTC.

Concepts are evaluated and approved through the Interface Working Group (IWG). The interested reader should refer to (Fischer 1994) for a description of the organization and management of the Joint Training Confederation.

Subsequent to concept approval, detailed designs are formulated. These designs span several documents: Interface Control Documents (ICDs) describe message syntax, content, and sequencing requirements. Actor Implementation Documents (AIDs) detail the ICD implementation within a specific simulation ("actor" in the parlance of ALSP). The Operational Specification (OPSPEC) provides the model definition in terms of objects and attributes as well as simulation-object ownership and behavioral "scenario descriptions" and serves as the primary communicative model representation.

Currently the dashed item in Figure 1 labeled "System Specification" does not exist explicitly for the JTC. Individual simulations maintain system specifications describing private objects and behaviors within their respective simulations. Ideally, these specifications should be consolidated to form a single system description describing objects and behaviors — both private and public — for the JTC *as one system*.

2.3 Development and Implementation

The simulation developers use the design documents to generate modified versions of their simulation software. By organizational agreement, each Service (or model proponent) has the responsibility of VV&A for its simulation (Fischer 1994). Similarly, each Service *reserves* responsibility for evaluating any code changes made with respect to either Service or JTC requirements. This important point is elaborated in Section 3.

2.4 Integration and Test

Following simulation development, integration and testing of the confederation takes place. Three test stages are identified: Functional Interface Integration (FII), All Actor Integration (AAI), and Confederation Test (CT). These tests are supported by participants in the JTC community.

Test plans created to support the AAI are intended to test the inter-simulation functionality. Formal reports from the AAI are not generated. However, results are informally promulgated to the simulation developers to permit refinement of their interfaces.

The CT is the pinnacle of the JTC development process, and serves as the user acceptance test. Testing is conducted by the user community, and formalized test reports are generated. Stress tests are added to provide a means to operate the JTC under larger, "exercise-like" loads. The CT culminates with a written test report presented to the Review Panel for approval. This report describes the tests performed at the CT and the results obtained. The Review Panel promulgates the official accreditation of the ALSP JTC.

2.5 Deployment

Confederation deployment includes delivery and employment of new simulation and ALSP Infrastructure Software at user sites. The JTC Systems Engineer provides support during "first-use" — the first major exercise to operate a newly accredited JTC. Associated with the JTC deployment, exercise centers typically conduct additional on-site testing of the JTC using their specific environment and data. This supplementary testing, to include Systems Engineer support if requested, is funded by the exercise sites.

3. EVALUATION

An evaluation of the VV&A process within the Joint Training Confederation must begin with a recognition that a singular "VV&A Agent" role is not defined for the JTC (Fischer 1994). Such an agent, if singularly defined, would require the omnipotence to establish and oversee the VV&A within the individual simulations as well as the VV&A of the confederation as a whole. In the current JTC management structure each Service (or model proponent) has the responsibility of VV&A for its simulation. The ALSP Systems Engineer assumes responsibility for VV&A of the system as a whole (see the demarcation of VV&A responsibilities in Figure 1). As Balci (1994a) indicates, a collection of valid subcomponents does not necessarily result in a valid system when these components are integrated. Therefore, validation of the whole requires validation beyond that of the parts. However, the limited authority of the Systems Engineer in this role implies that most of the testing undertaken to support the system-level validation activity is necessarily functional or "black-box".

3.1 The Vanilla Approach to Systems Testing

Adopting the term used by (Harel 1992) we describe the vanilla approach to systems testing (testing "in the large" as delineated in (Hetzel 1988)) as being comprised of four stages: (1) component test, (2) integration test, (3) system test, and (4) acceptance test (assuming the existence of a customer for the system). Figure 2 depicts the JTC test process superposed on the vanilla approach. Although the mapping is not as clean as the figure indicates, the absence of a component phase of system testing is evident. While lack of a component test may not directly affect validation, it potentially results in inefficiencies in the development cycle — since faults may go undetected until later in the cycle thus tending to increase the cost of their correction.

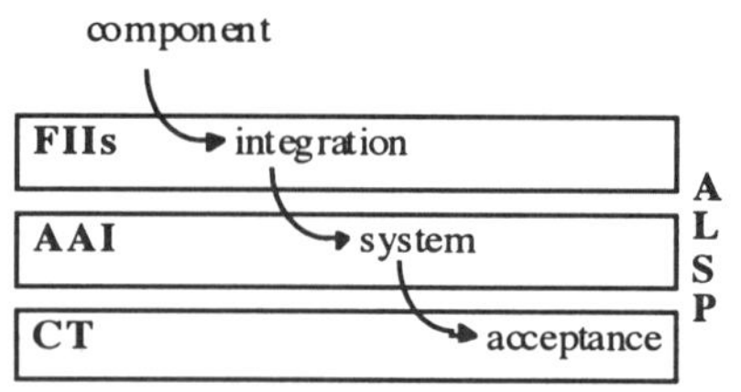

Figure 2: JTC Test Process Superposed on the Vanilla Approach

3.2 Problems Associated with a Traditional Approach

The addition of a component test phase to JTC development would provide the general vanilla framework, and as such should tend to increase cost-effectiveness. However, a number of complications (with respect to model validation and accreditation) are associated with the application of such a traditional testing approach within the context of the JTC. A few such complications are noted below.

- *The test system and fielded system differ in scale.*

The systems and personnel assembled to support and participate in a JTC exercise are extensive, typically involving thousands of people and hundreds of workstations over a month long period. In contrast, the 1996 ALSP JTC confederation test (user acceptance test) involved roughly 100 people for two weeks, and approximately 40 simulation workstations.

Both practicality and cost limit the size and scale of JTC testing. A test cannot cost a much as an exercise. In fact, testing costs arguably must be many orders of magnitude less than an exercise. As a result, the system that is used to support training is much larger in scale than the system tested.

- *The test system and fielded system differ by platforms.*

Exercises are usually held at military "simulation centers" (for examples, see section 1). Each center has a significantly different hardware, software, and network infrastructure. Therefore, test results are only applicable (for certain measures, e.g. performance) with respect to a particular infrastructure (the one used for the test) and, of course, subject to the caveat of the aforementioned scale problem.

- *The test system and fielded system differ by composition.*

The ALSP JTC may be viewed as a layered architecture. At the hub are a set of simulations (e.g. CBS, AWSIM, RESA, JECEWSI, TACSIM, CSSTSS, MTWS, and PSM for 1996 — see (Page et al. 1996) for details) that exchange information and coordinate the advance of simulation time over ALSP. Many of these simulations are interactive, i.e. they have user input terminals and battlefield situational displays. These devices are generally referred to as "controller workstations" and provide a layer around the simulation hub.

An additional layer of systems (referred to in the JTC community as peripheral software systems or PSS) are routinely attached to the JTC when employed at an exercise. These systems are influenced by actions and changes made within the JTC, and facilitate information exchange with the training audience. The training audience generally does not interact directly with workstations or PSS, but is (ideally) immersed in a wartime-equivalent environment. In some instances, the PSS provide information in a facsimile of real-world formats. A very small number of the PSS support linking real-world command and control (C^2) systems and command, control, communications, computers and intelligence (C^4I) systems to the JTC. More often the linkage requires a manual interface (so called “man-in-the-loop”) to convey information between the JTC and the training audience. The mélange of workstations, PSS, C^2/C^4I systems, and manual interface mechanisms provides the outer “external systems” layer of the JTC architecture.

During a CT, a system consisting of the JTC and a collection of external systems anticipated for the coming year's training exercises is tested. Unfortunately, the tested system is never identical to the one used at any exercise. Subsets of the simulations and external systems are selected, often based on political or cost rationale that supersede technical considerations. New external systems are occasionally used in exercises (to satisfy exercise requirements) that were not tested during a CT.

- *The test system and fielded system differ by code.*

The JTC development can be viewed as multi-level development life cycle. Each simulation component comprising the JTC has a separate life cycle (necessary to satisfy primary Service requirements) in addition to that of the JTC. These individual development life cycles must be coordinated to coincide with the JTC life cycle. Currently no formal mechanism exists to facilitate such coordination. Modifications to software can occur throughout the year (as dictated by the primary Service development cycle). By organizational agreement, configuration management authority falls to the Service.

- *The test system and fielded system differ by data.*

Many military simulations, including those in the JTC, are essentially "data driven," i.e. the outcome of any particular event in the model is sensitive (at varying levels of sensitivity) to values in the model database (see (Davis 1992) for design rationale). For the JTC simulations these databases include information regarding the types of battlefield equipment, order-of-battle (hierarchical) information, and details of the geographical region being played.

These databases differ greatly between exercises. One exemplary exercise includes next-generation (2010) weapons systems in a fictionalized European scenario. Another uses current weapon systems in a Korean defense scenario.

The results of any test are tightly coupled to the collection of databases underlying the simulations. As a result, database tests must be conducted prior to each exercise. However these tests also fail to be definitive. Database parameters may be changed during an exercise, typically to correct an inappropriate action observed by exercise controllers, or to ensure the exercise scenario evolves as required to meet training objectives.

- *The test system and fielded system differ by personnel.*

Several interfaces between the training audience and the JTC simulations have been automated (i.e. a direct link between C^2/C^4I systems and simulations). Many of them have not. This is the primary role of an exercise controller: to communicate with the training audience via real world mechanisms, and interact with the simulations using controller workstations.

There is also a level of critical personnel generally referred to as "technical control". These personnel operate the simulations, the ALSP infrastructure software, and the attendant systems and networks.

Both the exercise controllers and technical personnel vary from exercise to exercise. Each can introduce mistakes that unintentionally disrupt the exercise. Novice workstation operators often input unexpected values, producing equally unexpected events in the simulations. System and network personnel will adjust systems and networks differently, resulting in varying behavioral characteristics.

- *System behaviors are subject to evaluation/modification.*

A mechanism for fault tolerance is typically used in training exercises under the nomenclature "white-cells". White-cells are a group of personnel that behave as referees and are used to adjudicate questions regarding the state of the simulation. When an undesired outcome (in terms of training objectives) is detected in the state of the simulation, an exercise white-cell can intercede and realign the simulated conditions to better meet the objectives.

For example, during a recent exercise, an initial engagement involved a massive cruise missile attack by Opposing Forces (OPFOR) against friendly forces. As a result of operator errors during exercise set-up, many friendly forces did not have fire control for their air defenses. These units were virtually wiped out during the opening attack. Such an imbalance of forces threatened to seriously impact the training objectives for the exercise, so the referees interceded and the friendly force structure was replenished to the expected post-attack level.

3.3 Why Do Bad Things Happen at Exercises?

In "traditional" settings (i.e. simulation used for analysis) failure of the simulation is strictly a function of simulation invalidity. This isn't customarily the case in the interactive training simulation world. Numerous problems can be traced to causes outside the simulations themselves, including computer failure, network failure, and operator error. To a lesser extent, general software errors occur. Consequently, the items that a substantive portion of our testing efforts focus on, and the corresponding results that lead us to consider the simulations as "sufficiently valid" are frequently *not* contributing to exercise problems.

Although the testing efforts are valid and successful test results desirable, the downside is we end up with a confederation of valid simulations which can fail to support an exercise.

For example, in the first days of one exercise the network performance became significantly degraded such that the simulations were running so slow (at a ratio of 0.4:1 with real time) that the training audience was becoming affected. After several hours of investigation the problem was tracked to the fact that a workstation had been added to the network that morning and given the same IP address as one of the mainframes hosting the simulations. How VV&A can (or should) address problems like these is an open issue.

3.4 Is a Traditional Approach Infeasible?

The accumulation of factors noted above begs the question, "to what extent can model validity be determined during acceptance testing?" The answer seems to be a little, but only partially. The differences in scale, composition, data, and personnel demand that much of the validation activity may only be undertaken within the immediate context of an exercise — using the existent training system, data and personnel as the basis for testing. If this is true, then is there any value in conducting the traditional (component, integration, system, acceptance) tests? Should these tests be abandoned, and all test expenditures redistributed to pre-exercise efforts? Such course of action would seem unwise. The benefits gleaned from a disciplined, bottom-up testing approach extend beyond validation. They contribute to early error detection as well as verification, and provide feedback regarding the overall reliability of the system.

A cost-effective solution would seem to lie somewhere in the middle, as it often does; its precise location only identifiable by trial and error.

4. NEW DIRECTIONS

Although the challenges to the conduct of VV&A are numerous — some technical and others programmatic — affecting change within the process is necessarily slow. In addition to the natural resistance to change, the role of the JTC as a proven and reliable environment supporting training exercises demands that modifications to a process that works sufficiently well should be made with caution. There is wisdom to the philosophy "if it ain't broke don't fix it".

Several initiatives are underway for the 1997 development cycle.

- A test harness is being designed to support component testing at the system level. The harness will be "programmed" to stimulate an actor with respect to the interfaces it participates in (as described in the ICDs and AIDs).
- To address the sensitivity of model validation on the underlying model databases, a mechanism is being provided to allow each actor to *register* the values of key database elements with the ALSP infrastructure during the confederation initialization process. The database elements are enumerated attribute values defined at the confederation level — so-called *enumerations mapping*. A tool has been developed to collect this information and provide cross-referencing to facilitate the detection of incomplete entries or errant mappings. This facility could significantly improve what has been an entirely manual, labor-intensive and error-prone process. However, not all of the JTC actors can easily register their enumerations mapping. In some cases this mapping is resident in code rather than a separate database, thus making the creation of a routine that extracts this information a difficult (if not impossible) task.
- Another tool has been developed to monitor the message traffic and examine messages for invalid enumerations. Note that should each actor have the capability to register its enumerations mapping (as noted above) *and* if no database changes occur during an exercise, this tool would provide no additional benefit. However, in the absence of such capabilities and guarantees, this tool, while providing less information than an enumerations mapping cross-reference, can provide a valuable mechanism for quickly identifying certain common errors that occur during confederation operations.

5. CONCLUSIONS

Within the JTC development process, cost-effectiveness serves as the overriding objective. While organizational and bureaucratic factors tend to hinder early error detection and strict configuration management, the VV&A process within the JTC seems to overcome these difficulties. The track record of the JTC as a training vehicle serves as witness to this phenomena. While the size and scope of the JTC continues to expand to keep pace with ever-expanding training requirements, the time and money allocated to testing promises to remain fixed (or even diminish). Given that model validity is a function of cost versus benefit, the challenge to the Systems Engineer for the JTC is to maximize the benefits (i.e. the validity) in the face of fixed costs.

REFERENCES

Balci, O. (1994a). Principles of Simulation Model Validation, Verification, and Testing. Technical Report. TR 94-24. Department of Computer Science. Blacksburg, VA, June 10. (To appear in: *International Journal in Computer Simulation*).

Balci, O. (1994b). "Validation, Verification, and Testing Techniques Throughout the Life Cycle of a Simulation Study." *Annals of Operations Research*(53): 121-173.

Davis, P. K. (1992). Generalizing Concepts and Methods of Verification, Validation, and Accreditation (VV&A) for Military Simulations. R-4249-ACQ. The RAND Corporation. Santa Monica, CA.

Defense Advanced Research Project Agency (DARPA) (1990). Advanced Distributed Simulation Program. Project Report. Advanced Research Project Agency, March.

Department of Defense (1988). MIL-STD-498: Software Development and Documentation, United States Department of Defense.

Fischer, M. C. (1994). "Aggregate Level Simulation Protocol (ALSP) Managing Confederation Development." *Proceedings of the 1994 Winter Simulation Conference.*

Forsberg, K. and H. Mooz (1992). "The Relationship of Systems Engineering to the Project Cycle." *Engineering Management Journal* **4**(3): 36-38.

Harel, D. (1992). "Biting the Silver Bullet:Toward a Brighter Future for System Development." *IEEE Computer* **25**(1): 8-20.

Hetzel, B. (1988). *The Complete Guide to Software Testing.* Wellesley, MA, QED Information Sciences, Inc.

Miller, G. and A. Zabek (1996). The Joint Training Confederation and the Aggregate Level Simulation Protocol. *Phalanx.* **29:** 24-27.

Page, E., J. Tufarolo and B. Canova. (1996). "A Case Study of Verfication, Validation and Accreditation for Advanced Distributed Simulation." *ACM Transactions on Modeling and Computer Simulation* (Submitted for publication).

AUTHOR BIOGRAPHIES

JOHN A. TUFAROLO is a Senior Simulation Engineer for the MITRE Corporation in Reston, Virginia, where he is currently the MITRE project manager for the Synthetic Theater of War (STOW) Command Forces (CFOR) program. He has over 10 years experience in modeling and simulation, supporting development, testing, and application of military training and analysis simulations. Mr. Tufarolo has worked as a systems engineer and lead test planner for the Aggregate Level Simulation Protocol (ALSP) program, and currently serves as a member of the Defense Modeling and Simulation Office (DMSO) High Level Architecture (HLA) simulation management group. He received a B.S. degree in electrical engineering from Drexel University in 1985 and is currently an M.S. candidate in Systems Engineering at George Mason University. He is a member of the ACM, IEEE, and SIGSIM.

ERNEST H. PAGE is a member of the technical staff of the MITRE Corporation where he is currently working on the Aggregate Level Simulation Protocol (ALSP) program. He received the Ph.D., M.S. and B.S. degrees in Computer Science from the Virginia Polytechnic Institute and State University (VPI&SU) in 1994, 1990 and 1988 respectively. Dr. Page serves as a member of the Defense Modeling and Simulation Office (DMSO) Verification, Validation and Accreditation Technology Working Group (VV&A TWG) and VV&A Technology Support Team. He has been active with the High Level Architecture (HLA) initiative, working with both the Test and Interface Specification Working Groups. Dr. Page is the Secretary/Treasurer for the Association for Computing Machinery (ACM) Special Interest Group on Simulation (SIGSIM). His research interests include discrete event simulation, parallel and distributed systems, and software engineering. He is a member of ACM, SIGSIM, IEEE CS and Upsilon Pi Epsilon.

Proceedings of the 1996 Winter Simulation Conference
ed. J. M. Charnes, D. J. Morrice, D. T. Brunner, and J. J. Swain

AN EXPERIMENT IN SIMULATION INTEROPERABILITY

Gary N. Bundy
David W. Seidel
Ben C. King
Carl D. Burke

The MITRE Corporation
1820 Dolley Madison Blvd
McLean, Virginia 22102

ABSTRACT

The AWSIM Interoperability with ModSAF (AIM) program is investigating interfacing traditional constructive simulations with virtual simulations. The program addresses a number of issues involved in this interoperability. As a mechanism for exploring these issues, ModSAF is used to create a detailed window into an AWSIM exercise; modeling of AWSIM aircraft entering the window is transferred to ModSAF. A prototype was developed to investigate the practical aspects of the interface issues. Knowledge, algorithms, and software developed in this program will support future Air Force wargame development.

1 INTRODUCTION

The AWSIM Interoperability with ModSAF (AIM) program is investigating approaches and issues in interfacing traditional constructive simulations with virtual simulations.

The form of integration assumes a broad constructive simulation within which a virtual simulation portrays a detailed window (virtual playbox). In this case, the constructive simulation is AWSIM, an Air Force simulation widely used in theater-wide computer-aided exercises; and the virtual simulation is ModSAF, a detailed simulation used in conjunction with vehicle trainers. The mechanism for exchange is the Aggregate Level Simulation Protocol (ALSP), a mechanism used in theater training exercises.

When aircraft modeled in AWSIM cross into a ModSAF window, responsibility for modeling the aircraft is transferred to ModSAF. When the aircraft cross out of the window, modeling responsibility is passed back to AWSIM. AWSIM retains command and control responsibility for the aircraft regardless of where it is modeled.

1.1 Issues

Issues that form the framework for the AIM investigation include the following:

- **Deaggregation/Reaggregation**. When an aggregate unit in a constructive model is decomposed into its constituent entities, the entities should be properly placed on the battlefield with sufficient information to continue the mission of the unit when responsibility for modeling the entities is transferred to the virtual simulation. Similarly, when a unit reaggregates, the constructive simulation should receive sufficient information about the unit to reflect the results of battlefield activities that occurred in the virtual simulation.
- **Coherency**. Consistent views of the battlefield should be maintained in both constructive and virtual simulations, including the state of the entities on the battlefield, the behaviors and activities that they perform, the time being represented, and the terrain that forms the context of the battle. A seamless transition of modeling and interactions should be present across the boundary between the simulations.
- **Command and Control**. Units in both the constructive and the virtual simulations should be able to accept command and control directives in a consistent way.
- **Exercise Support**. Mechanisms should be provided to facilitate the management of the total exercise. This entails an understanding of the total state of the exercise and coordinated control of the component pieces.

1.2 Constraints

Several constraints drive the AIM project:

- **Practicality**. The results of the investigation should be valid in the world of military exercises and training.

  AWSIM and ModSAF are simulations that are in regular use.

  Entities used in the prototype are real-world aircraft types, missions, and environment.

- **Reuse**. Wherever possible, existing software and processes should be used. In addition, where future related software development is contemplated, software and processes developed under this program should be transferable to the new program.

  The Aggregate Level Simulation Protocol (ALSP) is an existing information-exchange mechanism. AWSIM was already adapted to use ALSP and ALSP already provides mechanisms for ground-truth data exchanges and time management.

- **Visibility**. The exchange of information between the simulations should be explicit and externally visible.

  Information is exchanged using a message-based medium.

- **Maintaining Model Credibility**. In order to maintain the credibility of each simulation, new modeling should be avoided. Similarly, new software elements that introduce modeling concepts should be avoided.

1.3 Implementation Approach

The approach chosen to implement AIM was to use an enhanced ALSP protocol (See Figure 1). Since AWSIM was already adapted to exchange information and regulate time using ALSP, much of the effort in adapting AWSIM was already accomplished. What remained was

- Adapt ModSAF so that it performs basic ALSP interface functionality (broadcast modeled aircraft characteristics, receive and portray aircraft modeled in AWSIM, and coordinate time via ALSP).
- Adapt ModSAF to perform AIM-specific functionality (coordinate hand-off from and to AWSIM via ALSP, process and generate weapons interactions, and coordinate command and control from AWSIM via ALSP).
- Adapt AWSIM to perform AIM-specific functionality (coordinate hand-off from and to ModSAF via ALSP and coordinate command and control to ModSAF aircraft via ALSP).
- Adapt the ALSP protocol to handle the new requirements presented by AIM (hand-off and command and control).

2 SIMULATION COMPARISON

AWSIM and ModSAF differ in how they perceive and model air warfare. This section describes elements that are relevant to bringing them together in AIM. Paragraphs below depict how objects are described in each of the simulations and how each simulation models relevant activities.

2.1 Object Description

In AWSIM and ModSAF, objects are represented using attributes. This representation comes into play when one simulation must internalize (ghost) an object modeled by the other and when modeling responsibility must pass from one simulation to the other. Table 1 compares significant data elements as they are represented in the two simulations

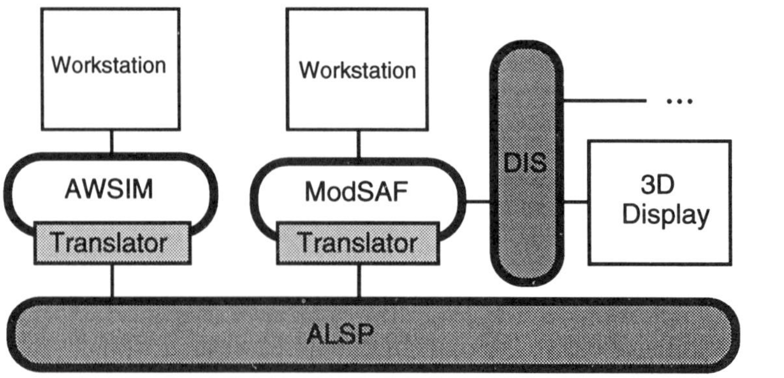

Figure 1. AIM Architecture

Table 1. Object Description Comparison

Data type	ModSAF	AWSIM
Entity represented	Individual aircraft	Flight of aircraft
Flight name	None	Used by AWSIM to track and control aircraft.
Location	3-dimensional distance from the center of the earth	Latitude, longitude, altitude above mean sea level
Attitude	Attitude of each aircraft.	Not modeled
Dead Reckoning parameters	Rate of change of positional and attitude data.	Course and speed.
Tail number	Marking text represents tail number	For reporting purposes only.
Formation	Formations can be specified	Not modeled
Sensor status	Most sensors are not modeled	Modeled
Weapon load	Quantity of each type of weapon on each aircraft	Aggregate quantity of each type of weapon for the flight.
Fuel	Maintained for each aircraft	An average fuel load for all aircraft in the flight is kept

2.2 Modeling

In AWSIM and ModSAF, activities of single entities and interactions between entities are modeled in different ways. These differences are significant when comparing similar activities that take place separately in each simulation (results of AWSIM air-to-ground attack versus results of ModSAF air-to-ground attack) and when attempting to integrate activities where participating objects are modeled in different simulations (air-to-air engagement between AWSIM and ModSAF aircraft). Table 2 compares activities as they are modeled in the two simulations.

3 RESOLUTION APPROACHES

The differences between the simulations that are described in Section 2 must be rectified if an interoperable federation of simulations is to exist. The most obvious and straightforward approach to accomplishing this is to add representations and modeling to each simulation to make it match the other. However, this is not the best approach for a number of reasons:

- Some abstractions are purposely built into a simulation (1) to permit its users to deal with the activity being modeled at a higher level (2) to ease scenario development or (3) to ease processing requirements, permitting large scenarios to be executed in relatively small computers.
- The cost to add detail to the more abstract simulation is excessive.
- Implementation of detailed modeling is often not required since what is evident is appearance at the point of interface.

Therefore, three approaches to the rectification present themselves:

- **Add modeling.** New modeling of activity can be added to a simulation. For example, in AWSIM, fuel could be modeled (consumed, replenished when refueled) for each aircraft in a flight individually. As discussed above, adding the modeling of a specific function is the most costly alternative.
- **Add representation.** New data elements can be added to a simulation. For example, in AWSIM, a flight formation could be kept. While this approach may be adequate, it often introduces simplistic, operationally inappropriate results.

Table 2. Battlefield Activity Comparison

Activity	ModSAF	AWSIM
Entity Update rate	Often enough that, using dead reckoning parameters, another simulation could calculate location and attitude within prescribed values (typically about once per second)	Selectable (once every 10 seconds to once per minute).
Ground Entities	ModSAF models ground vehicles • Permits better target selection for air-to-ground attacks. • Permits more detailed flight profiles • Permits more rapid reattack decisions	AWSIM models only SAMs, airbases, and search radars. In ALSP exercises, AWSIM passes target preferences in ALSP air-to-ground interactions; the ground simulation (CBS) selects specific targets.
Jamming	Not modeled	AWSIM models self protection and stand-off jamming
Terrain	Detailed terrain representation is part of ModSAF. Outside the terrain box, no earth representation exists. Current air implementations do not use terrain.	AWSIM does not model terrain • No map-of-the-earth flying • No terrain masking of radars and ground-based weapons • No terrain effect on detection of ground targets in complex terrain
Air picture	ModSAF has a complete picture only of its playbox.	AWSIM keeps a picture of the total battlefield through ALSP.
Command and Control	Direct from GUI	Future tasking can be input (ATO-like orders)
Time	ModSAF and DIS evolve time in very small "ticks" and lock time advance to the wall clock.	AWSIM advances time in fixed (10 second to 1 minute) increments. It attempts to synchronize simulation time to wall clock time but may fall behind if processing needs are heavy. In ALSP confederations, AWSIM regulates its time through ALSP.

- **Create information**. New information can be introduced at the time of exchange of data. For example, AWSIM could apply some algorithm to determine individual aircraft fuel state and broadcast that information. Alternatively, AWSIM could broadcast flight fuel state and ModSAF could apply an algorithm to determine individual aircraft fuel state. In general, it appears best to assign this role to the simulation with the most detailed modeling of the particular function—ModSAF in this case.

It must be noted that all of these approaches create information that is not present when the simulations operate separately. By doing so, any validation or accreditation of either of the simulations is abrogated; the process must be performed again against the simulations interoperating together.

4 AIM IMPLEMENTATION METHODOLOGY

Given the differences in representation and modeling described above, mechanisms were chosen to permit interoperability between AWSIM and ModSAF. Each of the mechanisms entail compromises between the desired seamless integration and the practical requirement to integrate without violent disruption to either simulation.

This section describes the compromises that have been made and resulting modeling discrepancies. The section is divided according to the interoperability issues described in Section 1.

Table 3. Deaggregation/Reaggregation Issues

Issue	Resolution	Resulting Anomalies
AWSIM doesn't model attitude	ModSAF creates attitude data for each aircraft	Aircraft attitude is jerky and unrealistic.
AWSIM doesn't model formation	ModSAF creates formation when it initially observes the flight.	Formations may be inappropriate. A rapid deaggregation, reaggregation, deaggregation sequence may lead to inconsistent formations.
ModSAF doesn't model Flights of aircraft	AWSIM passes flight data in ALSP message. ModSAF remembers and uses this data when reporting on the flight.	None
Units of measure differ between ModSAF and AWSIM	ALSP units of measure are used for information exchange.	None
AWSIM aggregates weapon and fuel loads	Each time a flight of aircraft newly appears in ModSAF, ModSAF splits loads onto aircraft.	A deaggregation, reaggregation, deaggregation sequence may lead to inconsistent loads.

4.1 Deaggregation/Reaggregation

When a flight of AWSIM aircraft enters ModSAF's window of interest, ModSAF begins to ghost the flight. When the flight enters ModSAF's window of control, modeling responsibility is handed-off to ModSAF—ModSAF must broadcast information about the flight over ALSP. When the flight of aircraft leaves the window of control, the hand-off mechanism is invoked to transfer modeling responsibility for the flight back to AWSIM. These transfers require private information that is not normally carried in ALSP update messages; the private data was incorporated into the update messages and the hand-off message exchange. These issues are described in Table 3.

4.2 Coherency

Consistent views of the battlefield should be maintained in both AWSIM and ModSAF, including the state of the entities on the battlefield, the activities that they are performing, the time being represented, and the terrain that forms the context of the battle. A seamless transition of modeling and interaction should be present across the window boundary. These issues are discussed in Table 4.

4.3 Command and Control

Units in both the constructive and the virtual simulations should be able to accept command and control directives in a consistent way. These issues are discussed in Table 5.

4.4 Exercise Support

Mechanisms should be provided to facilitate the management of the total exercise. This is particularly true in exercises, like AIM, with multiple simulations and complicated topologies, These mechanisms permit an understanding of the total state of the exercise and coordinated control of the component simulations. These issues are described in Table 6.

5 CONCLUSIONS

While the AIM investigation and prototype implementation are incomplete, several conclusions can be drawn from progress to date.

- Complete, seamless interoperability is not practical, and, given the strengths of each simulation and the intended use of the federated system, not desirable.
- Reasonable work-arounds can be found for most simulation disconnects. Most unreasonable anomalies exist because fundamental modeling is lacking from a simulation.
- Exercise support is inadequate for this combination of simulations. ModSAF and DIS do not provide facilities to incorporate necessary features.
- Verification, Validation, and Accreditation (VV&A) of separate simulations does not produce a verified, validated, and accredited federation of simulations.

Table 4. Coherency Issues

Issue	Resolution	Resulting Anomalies
ModSAF has a limited geographic playbox.	ModSAF models aircraft in its window of control and ghosts those in its window of interest.	Aircraft outside the window of interest are invisible to ModSAF objects.
AWSIM updates location less frequently than ModSAF.	ModSAF dead reckons aircraft between AWSIM updates.	Kinematics of AWSIM aircraft appear jerky and unrealistic to ModSAF observers.
ModSAF models ground vehicles; AWSIM does not.	To attack ground targets, AWSIM passes target preferences in ALSP air-to-ground interactions for the ground simulation to use to determine specific targets.	ModSAF permits better target selection for air-to-ground attacks.
AWSIM models self protection and stand-off jamming; ModSAF does not.	AWSIM jamming is not turned on in AIM.	Electronic warfare is not a part of AIM.
AWSIM uses logical time; ModSAF time is tied to wall clock time	ModSAF is assigned the role of time controller; it broadcasts its perception of time and AWSIM adheres to it.	If AWSIM workload is heavy, it will fall behind wall clock time and lose synchronization with ModSAF
AWSIM assesses interactions less frequently than ModSAF.	ModSAF aircraft firing at AWSIM aircraft cause ALSP interaction messages to be sent to AWSIM. AWSIM determines the results of the interaction when its next increment of modeling occurs (up to one minute away). It then sends a revised ALSP update message showing the new state of the aircraft or its destruction.	The delay in ModSAF's learning the result of the weapon shoot could be interpreted by ModSAF as a miss and cause further weapon firings.
SAMs and Air-to-air missiles are not modeled as objects in AWSIM	When a ModSAF-controlled aircraft fires at an AWSIM-controlled aircraft, an ALSP interaction message is sent from ModSAF to AWSIM. ModSAF calculates the number of weapons that could have hit the target aircraft and includes that number in the interaction message. AWSIM processes the message and determines whether the weapons hit.	Since AWSIM aircraft cannot maneuver in the same manner as ModSAF aircraft in reaction to attack, new pH/pK values are needed for AWSIM; the source of this data is currently unknown
	When an AWSIM-controlled aircraft fires at a ModSAF-controlled aircraft, an ALSP interaction message is sent from AWSIM to ModSAF. AWSIM calculates the number of weapons that could have hit the target aircraft and includes that number in the interaction message. ModSAF processes the interaction message and determines whether the weapons hit.	Incompatible aircraft parameters preclude a good understanding of hit probabilities and interaction results.

Table 5. Command and Control Issues

Issue	Resolution	Resulting Anomalies
An operator must be able to realistically order flights of aircraft	AIM adaptations of the ALSP protocol permit the AWSIM operator to send command and control orders to ModSAF-modeled aircraft. These messages are limited to AWSIM-implemented mission assignments and weapon delivery. However, since AWSIM is an accepted Air Force simulation, the orders should closely adhere to Air Force doctrine	None
An operator must be able to receive reports from flights of aircraft	ModSAF generates status reports describing a flight of aircraft to AWSIM. The frequency and content of these messages are the subject of on-going knowledge acquisition; current implementation is an estimate of the result	None

Table 6. Exercise Support Issues

Issue	Resolution	Resulting Anomalies
It is useful to monitor the health of the simulations and their communications	ALSP infrastructure software provides this facility for the constructive confederation, and consequently also for AIM	None
A mechanism is needed to assure that only one simulation owns (controls the update of) a particular attribute of an entity or the entity as a whole	ALSP protocol and supporting infrastructure software provide this mechanism	None
During long exercises, it is often necessary to pause exercise activity in all simulations simultaneously. This requires automated control over all simulation clocks. AWSIM and ALSP provide this control; ModSAF and DIS do not	No automated pause capability is provided in AIM.	AIM exercises must proceed to completion without pause in simulation time.
During long exercises, it is often necessary to periodically store the current state of the simulations outside of simulation time (time stops while state save occurs). AWSIM and ALSP provide a mechanism for synchronizing the state saves so that all states represent the same instant in simulation time; ModSAF and DIS do not.	No automated state save capability is provided in AIM.	No facility is available to recover from simulation crashes or to reverse operational decisions during the exercise.

ACKNOWLEDGMENTS

The AIM project is sponsored by the US Air Force's Electronic Systems Center (ESC). It builds on previous efforts in support of the Synthetic Theater of War (STOW) program that was sponsored by the Defense Advanced Research Projects Agency (DARPA).

REFERENCES

Ceranowicz, A. Z, J. E. Smith, A. J. Courtemanche, R. B. Calder, November 1992, *ModSAF Programmer's Guide*, Loral Advanced Distributed Simulation, Inc., Cambridge MA.

Defense Modeling and Simulation Office, July 1993, *Survey of Semi-Automated Forces*, Arlington VA.

DIS Steering Committee, October 1993, *The DIS Vision, A Map to the Future of Distributed Simulation*, Orlando FL.

Institute for Simulation and Training, February 1994, *Standard for Distributed Interactive Simulation—Application Protocols, Version 2.0 Fourth Draft*, Orlando FL

Institute for Simulation and Training, June 1993, *Communication Architecture for Distributed Interactive Simulation (CADIS) Guidance Document*, Orlando FL

Institute for Simulation and Training, June 1993, *Communication Architecture for Distributed Interactive Simulation (CADIS) Proposed IEEE Final Draft Standard*, Orlando FL

Institute for Simulation and Training, June 1993, *Communication Architecture for Distributed Interactive Simulation (CADIS) Rationale*, Orlando FL

Loral Advanced Distributed Simulation, Inc., February 1993, *ModSAF Software Architecture Design and Overview Document*, Cambridge MA.

Miller, G. R., A. R. Adams, D. W. Seidel, November 1993, *Aggregate Level Simulation Protocol (ALSP) 1993 Confederation Annual Report*, MITRE Corporation, McLean VA.

MITRE Corporation, April 1996, *Aggregate Level Simulation Protocol (ALSP) Technical Specification,* The MITRE Corporation, McLean VA (not in public domain).

MITRE Corporation, February 1996, *Aggregate Level Simulation Protocol (ALSP) 1996 Joint Training Confederation Operational Specification,* The MITRE Corporation, McLean VA (not in public domain).

MITRE Corporation, April 1995, *Aggregate Level Simulation Protocol (ALSP) Infrastructure Software User Manual,* The MITRE Corporation, McLean VA (not in public domain).

MITRE Corporation, March 1996, *AWSIM - ModSAF Interface Control Document*, McLean VA (not in public domain).

AUTHOR'S BIOGRAPHIES

GARY N. BUNDY is the technical leader for the AIM effort at The MITRE Corporation. He has been with MITRE for over five years. During that time he has supported DoD agencies in integrating military simulations as well as integrating military simulations with C4I systems.

DAVID W. SEIDEL is project leader for the AIM effort at the MITRE Corporation. For the past five years he has supported DoD agencies in integrating military simulations. He has over fifteen years of experience in development and application of military wargames.

BEN C. KING has been supporting modeling and simulation interfacing at the MITRE Corporation for five years. In addition to leading the technical effort on AIM, he implemented key portions of ALSP and Command Forces software.

CARL D. BURKE is has performed software development and Army systems analysis at the MITRE Corporation for ten years. His previous modeling and simulation experience includes support for the JANUS interface to BDS-D.

Proceedings of the 1996 Winter Simulation Conference
ed. J. M. Charnes, D. J. Morrice, D. T. Brunner, and J. J. Swain

A SIMULATION OF THE EVACUATION OF AMERICAN CITIZENS WITH AN OBJECT-ORIENTED, ANIMATED MODEL

Jeffrey E. Sumner
Eric A. Zahn

TASC, Inc.
2555 University Boulevard
Fairborn, OH 45342, U.S.A.

ABSTRACT

This paper highlights a model developed by TASC, Inc. under contract to the Defense Advanced Research Projects Agency (DARPA). This project models and simulates an evacuation of American citizens and other important foreign nationals from a destabilized foreign country by utilizing available U.S. military resources. This project demonstrates how different evacuation plans can quickly be compared analytically through computer simulation. The software chosen for the simulation and analysis is the Integrated Model Development Environment (IMDE), an object-oriented discrete-event simulation package. The implementation of the model features reconfigurable mission plans and detailed simulation-independent animations. This model was demonstrated at the 1995 Joint Warfare Interoperability Demonstration (JWID).

1 INTRODUCTION

The Department of Defense (DoD) has an ongoing goal to improve the quality and accuracy of crisis action planning in an increasingly constrained global environment. With decreasing manpower and resources, greater emphasis has been placed on tools to aid the high level military planners. When a crisis situation arises, and the National Command Authority (NCA) has determined that military intervention is required, a Joint Task Force (JTF) is assembled to cope with the situation. This paper highlights research in the application of simulation as a JTF planning tool.

Much of the current research and development in military support systems is presented each year at the Joint Warfare Interoperability Demonstration (JWID), which features the application of new technologies to a hypothetical scenario indicative of modern military engagements. For JWID 1995, the scenario was partly focused on a humanitarian assistance situation arising from a typhoon in the southwestern Pacific. During the relief efforts, civil unrest occurs in the fictional nation of Gompin, a nation with strong ties to the United States and home to a large number of U.S. citizens. The unrest forces the NCA to authorize a Non-Combatant Evacuation Operation (NEO), with the goal being to safely evacuate all U.S. citizens with minimal loss of life.

A JTF is assembled to plan and execute the NEO operation. To help the JTF with this planning process for the NEO, a simulation was created using IMDE, the Integrated Model Development Environment. The simulation takes into account all possible resources that could be used in the operation, and allows the JTF planners to enter a number of different evacuation plans. These plans include information about what resources to utilize, what times the evacuation order should be given, and other pertinent data. The simulation of the plan is animated so the JTF planners can visualize plan execution and identify any problems before a Course of Action (COA) is actually selected.

Section 2 of this paper describes the model in-depth, including the different requirements that need to be incorporated. Section 3 describes IMDE, the simulation package chosen for the project, how the model was implemented, and the animation capabilities. Section 4 narrates how the final project comes together, explaining what is involved in the final version of the project and what was seen at JWID '95. Section 5 lists recent improvements to the model; finally, Section 6 concludes the paper by summarizing the project and proposing enhancements.

2 MODEL REQUIREMENTS

JWID '95 focused on providing joint service commanders with the technological resources to deal with a disaster scenario. In this case, an island nation in the South Pacific has been hit by Typhoon Jennifer, an intense tropical storm. This nation, fictionally named

Gompin, was devastated; electrical and telephone services were eliminated for almost all of the island.

At the request of the Gompin government, the United States immediately begins mobilizing disaster relief efforts. The already strong rebel factions in Gompin seize the opportunity to challenge the strained Gompin government in the aftermath of the storm.

At this point in the disaster scenario, U.S. advisors anticipate that matters will worsen on the island and that it will become unsafe for any U.S. citizens to remain in Gompin. Intelligence reports indicate that riots and violence, directed towards anyone connected with the current government, may break out at any time. The decision is made by the NCA to assemble a JTF to evacuate all American citizens and designated foreign nationals from Gompin as quickly as possible with minimal civilian casualties.

2.1 Specifics of a NEO

The NEO involves many steps in its execution. First, a decision must be made as to how many planes and ships will be used in the operation. There are two major U.S. Air Bases in the region available for the operation, Kadena AB at Okinawa, and Yokota AB just west of Tokyo, Japan. Both have C-130s and C-141s ready to transport citizens from Gompin to Guam, another island which is a U.S. protectorate that has escaped major damage from Typhoon Jennifer.

There is also the issue of where the evacuation point(s) on Gompin will be. The Gompin government has guaranteed the use of an air strip to the U.S.; however, it is away from the majority of American citizens living in the capital city area. There is a commercial airport available for the capital city, but the JTF should not rely on that airport to be open due to the civil unrest. In addition, a smaller airport is available near the northwest part of the island; but intelligence suggests it has a short runway and cannot handle many of the larger transports. The one major Gompin seaport will also be considered.

Transportation of the citizens to the evacuation points is also another issue. The island contains a network of highways which the citizens will utilize to get to the evacuation points. The actual speed of the citizens will vary based on terrain and road conditions. Also, some highways may be impassable due to the rebel activities on the island; these roadblocks will also have to be modeled.

Once the American evacuees arrive at the evacuation points, they will either load onto an American plane or ship already present, or will wait for one to arrive. The times of departure for the planes and/or ships will be determined by an external plan file, which will be read in by the model. More specifics and requirements of the plan file will be detailed in section 3.5. The planes, once they arrive, will load the evacuees and leave for Guam. Each evacuee has a certain loading time that must be factored in; also, the issue of how long a non-full plane will wait at a port for other evacuees to arrive must also be addressed in the model.

The final result of the model will enable the JTF to enter different plans into the simulation and analyze the effectiveness of each plan. For instance, if the plan is to utilize only half of the planes available, keeping half of the force in reserve for other contingency operations, how much longer will the evacuation process take? The JTF can receive this information from the NEO simulation and then make a knowledgeable decision.

2.2 The Object Model Working Group

The DARPA JTF Advanced Technology Demonstration (ATD) is a demonstration of applied technology for the next generation command and control environment. This system includes an open architecture, infrastructure, various servers, and a wide array of applications. In order for the servers and applications to share information, they must also share the internal representations of the data. For instance, the situation server needs to pass the current situation to the planning application so the JTF planner can review the situation and act accordingly. The situation would be contained in a collection of objects, each with its own states and behaviors.

To facilitate widespread intercommunication, there needs to be a standard for the objects so each application knows the structure and capabilities of the objects. The Object Model Working Group (OMWG) is a DARPA subgroup in charge of developing a common schema that defines the common objects for use in the JTF-ATD. A skeleton schema is already available, with naming conventions and concept documentation. Figure 1 shows a simplified extraction of the OMWG hierarchy for two classes representing military aircraft, the C-130H and the F-18. Both descend from the Platform class, one of the key classes in the hierarchy (Carrico 1995).

One of the major objectives of the NEO model was to evaluate the OMWG for application in an object-oriented analysis model. The objects of the NEO simulation represent a small subset of the OMWG schema, but have provided valuable insight into the use and complexity of such a schema in modeling and simulation. The following section details the software that was chosen for the model and how the model was built to fulfill the requirements listed in this section.

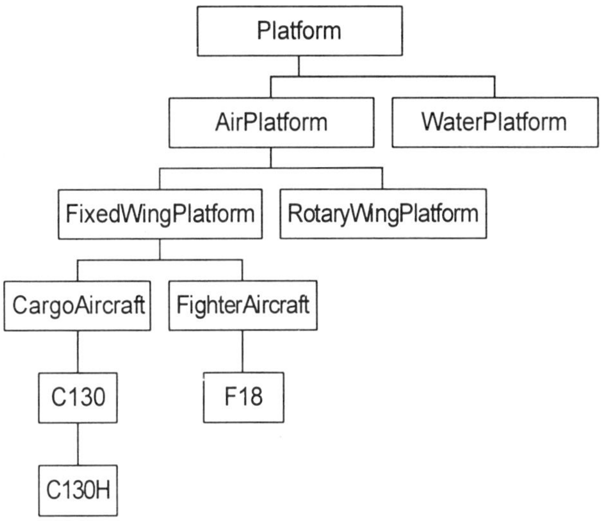

Figure 1: The OMWG Schema for the C-130H and the F-18

3 NEO Implementation

The development of the NEO model was performed by integrating several different software engineering tools and methodologies. This section briefly reviews several of these. First, a discussion of the Integrated Model Development Environment (IMDE) is presented. Following this is a description of the class selection and design process used in the architecture of the NEO model. Third is a discussion of how the Model-View-Controller (MVC) methodology was used, as well as the benefits received. Next is a discussion of the animation capabilities that were incorporated into the model. Last is a presentation of the plan file, which provides reconfigurable mission plans for rapidly developing and simulating various COAs.

3.1 An Overview of IMDE

IMDE is a domain-independent CASE tool used to develop and analyze object-oriented discrete-event simulation models. The USAF Logistics Research Division of Armstrong Laboratory has funded the development of IMDE at TASC, Inc. since 1990. It is a single application running under either UNIX or Windows NT, and directly interfaces with the Versant Object-oriented Database Management System (ODBMS). IMDE was chosen as the simulation package for this project due to its flexibility in modeling detailed situations, its object-oriented, modular development environment, and its graphical descriptions of the behavior of objects in the simulation.

IMDE is strongly based in object-oriented technology. All model parts in the simulation are defined as "objects" that have both variables that describe the state of an object, which are called attributes, and functions that specify its behavior, which are termed methods. By storing all objects in a distributed ODBMS, simulation objects can be shared and re-used in other simulations without recoding.

Objects from any field of study can be defined in IMDE. Models from the aerospace, business, education, and manufacturing fields have all been simulated and analyzed with equal success. The construction of simulation projects in IMDE can be done in a group environment, which allows multiple developers to create objects simultaneously and link them together. Finally, IMDE contains a comprehensive data analysis environment, where simple statistics, confidence intervals, hypothesis testing, and sensitivity analysis can be performed. Time trace plots of desired values can be shown quickly and easily.

The construction of models within IMDE is based on building objects and graphically assembling them to create complete simulation models. Templates for the design of these objects are called classes. For example, a C-130 class would be created that would contain all of the attributes and methods necessary to model an actual C-130 aircraft in the simulation. Several "instances" of this single class can be created in the simulation, each of which would model a specific C-130 aircraft. The definition of a class in IMDE includes detailing its attributes and methods. A graphical editor is employed to visually define all methods of any class. The editor allows a developer to define different behaviors of a single class with a separate flow chart that sequentially details the behavior logic. Flow charts are created by dropping nodes that represent simulation actions onto a canvas and connecting them in the correct order. IMDE uses these networks to generate source code in MODSIM II, an object-oriented simulation language, or C++ (Clark 1994).

Once all classes for a simulation have been defined, they can be linked together to form a model in IMDE's project editor. This editor allows the user to specify which entities will be present in the model and at what quantities. The project diagram illustrates the "has a" relationship of model parts, thereby representing the aggregation relationships of the model. With this framework in place, IMDE allows the user to further configure the model. Class substitution can also be performed, replacing objects in the model with other objects sharing the same inheritance hierarchy. For instance, if a modeler wants to determine the impact of using C-141 aircraft instead of C-130s, then a new C-141 class would be developed and substituted for the C-130 class in the simulation.

Also, lists of objects can be shortened or expanded in the project diagram. For example, one of the attributes of the Resource class in NEO is a list of aircraft. The initial number of aircraft that the Resource contains can be changed at this level, and a new simulation could be run without recompilation. Each component of the project can have input sets and statistics collection sets attached, detailing which attributes are parameters to be specified by the user, and which are noted for statistics collection.

Once the project diagram is complete, IMDE generates source code in MODSIM II or C++. This source code is then compiled and linked to create an executable simulation. All of this, including the running of the simulation and subsequent data collection, can be performed within IMDE.

At the conclusion of the simulation run(s), the data generated by the statistics collector is loaded into the database and can be analyzed within IMDE. As stated above, IMDE provides a complete data analysis package, calculating standard statistics as means, deviations, skews, and medians, as well as generating histograms, time traces, and scatterplots from the data.

3.2 Class Selection and Design

A main requirement is to have the class selection for the NEO model adhere to the OMWG schema. Therefore, a large number of classes were developed in order to match the OMWG schema and satisfy the needs of the NEO model. Since the number of classes involved is so large, this section briefly describes only the base level classes.

At the root level of the simulation is the JTF class. This class contains a list of Resources, a list of SafeHavens, and a list of HotSpots. The JTF class is responsible for reading in the plan and mission leg files, discussed later, and starting up the entire operation. A HotSpot is any area where an evacuation needs to occur. A HotSpot contains a road network that is used by the evacuees to travel to the evacuation points.

The OMWG schema defines a class called a Platform with serves as a parent to most vehicle classes. Therefore, all aircraft and ship classes defined in the NEO model descend from the Platform class. This allows certain operations to be performed by Platforms, such as traveling to specific locations, loading and unloading evacuees, etc., where it is irrelevant whether the actual vehicle is an aircraft or a ship.

The OMWG also defines a class called Port, which serves as a place of departure or arrival for Platforms. This Port class was used to model the actual air and sea ports that American citizens travel to in order to be evacuated. The SafeHaven class is descended from Port because it can not only serve as a port, it also serves as a location where evacuees can be transported. The Resource class models an air or sea base that houses Platforms, and is descended from SafeHaven. Each Resource contains a list of Platforms that may be used in the evacuation operation.

A class called PeopleGroup is used to model a generic group of people. An instance of this class represents a group of evacuees that travels together to the nearest Port and waits to be evacuated. Descended from the PeopleGroup class is a class named RebelGroup. This class models actions such as rebels taking control of a Port and destroying links in a road network.

3.3 MVC Architecture

NEO was developed under a powerful object-oriented methodology called Model-View-Controller (MVC). This architecture separates the responsibilities of software components into three different layers. The model layer handles all of an application's functionality. The view layer implements all feedback mechanisms used to present information to the user. Lastly, the controller layer handles all interaction that a user has with the application (Barkakati 1991).

NEO makes extensive use of the model and view layers. Less attention was allocated to the controller layer because all user interaction is handled inside of IMDE rather than during the actual execution of the simulation. The model layer was implemented through the development of the classes described in the last section. These classes provide the necessary functionality by modeling all of the real-world objects, such as planes, ports, evacuees, etc., that have a part in this simulation. These classes, and more importantly the interaction between them, provide for the accurate simulation of an actual evacuation operation.

The view layer of the NEO simulation provides the mechanism for displaying all information to the user during the execution of the evacuation operation. Several windows are displayed showing maps of the locale(s) where the evacuation is taking place as well as an expanded overview map which depicts the entire operation. Each object in the model layer contains a list of views that will need to display a graphical representation of the object. As the state of the object changes, the views are updated. For example, the current NEO simulation displays two windows, the evacuation location (Gompin) and an overview map of southwest Asia. A particular instance of a C-130 object would have a list of views that contains these windows.

As the C-130 travels, each view is updated. The C-130 will always appear on the map of southwest Asia, but it will also appear on the map of Gompin as it approaches that location. This provides a true synchronous "zoom-in, zoom-out" display. Since the different views operate independently of each other, the C-130 object does not have to coordinate how it is displayed in the different windows. Instead, the C-130 simply iterates through its list of views telling each one to update the C-130's representation given its new position.

The power of the object-oriented design and the MVC architecture is evidenced by the high flexibility of the system. Additional windows could be added as separate views. Appropriate backgrounds such as maps would have to be loaded, and coordinate systems would need to be set. Since the functionality and display responsibilities are separated, objects such as the C-130 would not need to be modified to display itself on the new views. Instead, the C-130 would simply have the new views added to its view list so that the new views could be told to update themselves as the state of the C-130 changes.

3.4 Animation

The animation capabilities of the NEO simulation provide the user with a top-down view of all activities occurring during the model's execution. The main window shows a scaled-out view of Southeast Asia. This map shows several areas of interest: United States air bases (Yokota and Kadena), a United States carrier fleet (Comphibron Eleven), an evacuation "safe haven" (Guam), and an evacuation site, or "hot spot" (Gompin). Air and sea networks show the routes that planes and ships will take during the simulation. This window shows the evacuation operation at a high level as planes and ships travel to the evacuation points, load evacuees, travel to the safe havens to drop off the evacuees, and return to base. Figure 2 shows this window.

For each hot spot, a "zoomed-in" window displays a map of the area, including locations of air/sea ports and road networks on which evacuees travel. These windows show several events occurring. At the beginning of the simulation, blue icons in the shape of a person appear at random places on the map, with each icon representing a group of approximately 100 evacuees. As the planes and ships approach the hot spot on the main window, they also appear on the hot spot window when they get close enough to display. This allows the user to see the behavior of the ships, planes and people groups in much greater detail as they arrive at the ports. An example of a hot spot window is shown in Figure 3.

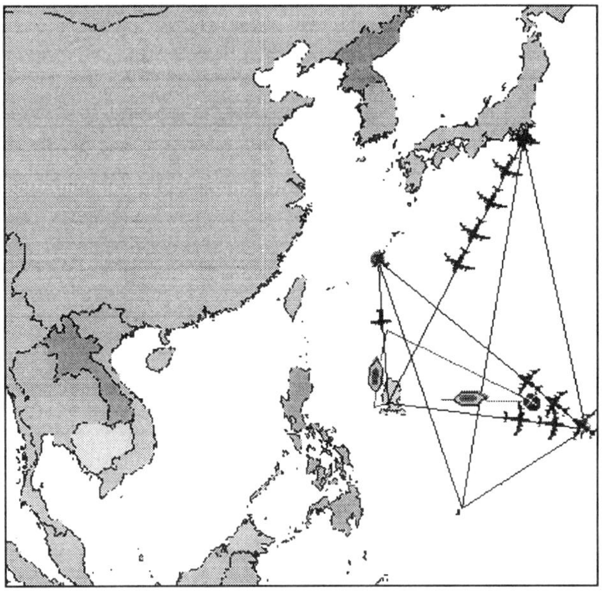

Figure 2: The Southeast Asia Window

The hot spot views can also show several other occurrences, depending on the actions specified in the plan file. Rebel groups, appearing as red icons in the shape of a person, can destroy links in a road network and also take over a port. In the latter case, the port is shown to be under hostile control by displaying a rebel flag over the port. The NEO simulation allows the planner to send in SH-60 helicopters carrying troops to regain control over the port. When this happens, a helicopter flies from one of the carriers, lands at the port, and returns to the carrier when control has been established. To signify control of the port, the rebel flag is replaced by an American flag.

Three other windows are displayed during the simulation. First, a summary window shows current counts of American citizens evacuated, number of missions generated, number of sorties generated, number of failed sorties, etc. Another window displays the current simulation time in hours and also provides a button to allow the user to pause and resume the simulation at will. Lastly, the third window allows the user to select several class attributes to display run-time statistical graphs during the simulation.

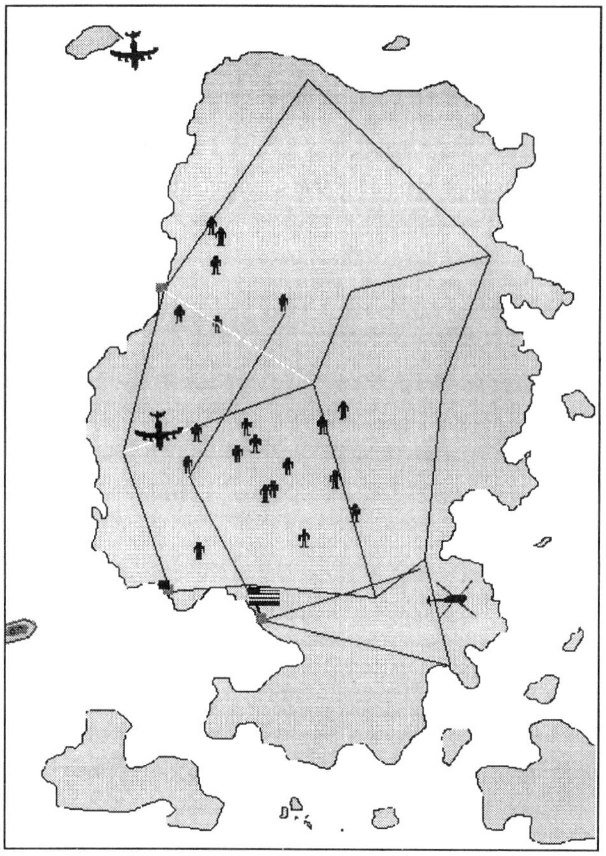

Figure 3: A Sample Hot Spot Window (Gompin)

3.5 Plan File

Since the purpose of the NEO simulation is to allow a JTF planner to determine how effective a plan will be in an actual evacuation operation, a textual plan file is used to convey to the model the types of events that will occur and the times that they will happen. There are several types of events that can be controlled through this file. For each event in the plan file, a time is specified that dictates when the event is to occur. First, the JTF object is told to perform a plan action which starts the simulation and performs any necessary initialization procedures. Next, the people in the area are told to evacuate at a specific time. At this time, each group of people determines the shortest possible path to an evacuation point and starts traveling toward that location. Also, helicopters such as the SH-60 can be told to travel to a hostile area and secure it from rebels.

The event definitions in the plan file pertaining to aircraft and ships are slightly different. These evacuation vehicles, or platforms, follow a specific pattern of departing from a set location (an airbase or somewhere at sea), traveling to an evacuation point to load passengers, transporting and unloading the passengers to a safe haven, and returning to the original location to end the mission. Therefore, these entries in the plan file specify the time that the mission will start, the number and type of platforms to use, the starting location, the evacuation point, and also the safe haven.

4 DEMONSTRATION OF THE MODEL

The NEO demonstration model was utilized at JWID '95 to model a real NEO situation. The scenario called for 4200 people to be evacuated. The two bases in the area, Yokota and Kadena, both had five C-130H aircraft available for use in the NEO. A fictional point near a popular city was used for the evacuation point in Gompin. The evacuation of the U.S. citizens on Gompin was to begin at the end of Day 4.

Two plans were used to compare and contrast to one another. The first plan attempted to evacuate the citizens using all five aircraft at each airbase, ten in all. The second plan was a more conservative one, keeping two C-130s back at each base in case there was a need somewhere else in the region, so six C-130s in all were used for the NEO mission. In both plans, the evacuation order by the JTF was given at the 96 hour mark; at that time, the people groups on Gompin moved to the port by using the shortest path. This path took into account any blocked highways and, in some cases, meant moving off the main highways for a majority of the time.

The two charts below highlight the differences between the two plans. In the first case, when utilizing all ten planes available, all evacuees were safely rescued within 50 hours of the evacuation order - about two days, as seen in Figure 4. In the second case, shown in Figure 5, the operation took a little over three days - 80 hours. This information can now be presented to whomever is coordinating the NEO in the JTF for their decision, or another plan could be created and then analyzed using this simulation.

5 POST-JWID EFFORTS

Since JWID '95, a great deal has been done to improve and enhance the NEO simulation. First, the capability has been added to allow the user to define multi-leg missions in more detail. Second, the model was enhanced to account for subsystem failure and maintenance activities on planes and ships. Third, the model was successfully converted from MODSIM to C++. The following sections describe each of these items in detail as well as the benefits realized from these modifications.

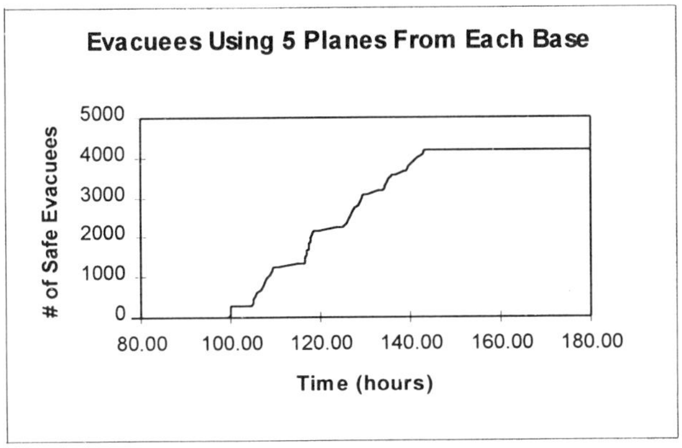

Figure 4: Time Trace of Evacuees with the First Plan

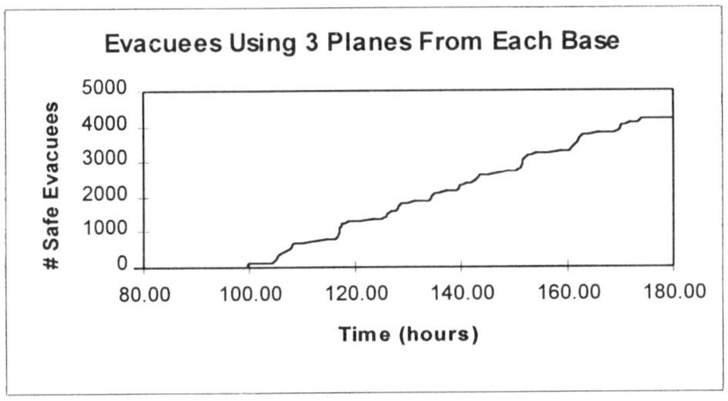

Figure 5: Time Trace of Evacuees with the Second Plan

5.1 Multi-leg Missions

Previously, any mission performed by an aircraft or ship comprised of exactly three legs: departing from a home base and arriving at an evacuation point to load evacuees, traveling to a safe haven to drop the evacuees off, and returning to the original base. With the development of a mission leg file, a JTF planner can create missions of varying lengths. For each mission, one line is entered into the file for each leg.

Several pieces of information are specified for each mission leg. One of the most important items is the name of the mission to which the leg belongs. Also, the name and quantity of the desired platform is specified. A mission leg also contains an originating location, an ending location, and an action. A relative departure time is included which specifies how long after the mission starts that the mission leg should begin. For example, if the plan file specified that a mission should begin at time 2500, and the mission contains two legs with relative departure times of 0000 and 0250, then the first leg would start at time 2500 and the second would start at 2750.

A mission leg also specifies a lead time and cancel time. The lead time specifies the amount of time needed to prepare for the leg, and the cancel time specifies the amount of time that a platform can delay departure before the leg is considered to have failed. Therefore, if the second mission leg in the previous example had a lead time of 0100 and a cancel time of 0200, the platform would start preparing for the leg at time 2650, and the leg would be considered to have failed if departure did not occur by 2950.

The main benefit of the mission leg file is the ability to define the structure of a mission as the JTF planner sees fit. The basic three-leg mission can still be used, but missions can now be made to have as many legs as desired. This provides a great deal more flexibility to the planner, which in turn greatly increases the possible options that can be simulated and evaluated. Also, since the mission leg file is read in at the beginning of the simulation, changes can be made to the file in order to see the effects on the operation without needing to recompile the model.

5.2 Platform Maintenance

An operation of any kind rarely follows a best-case scenario. Random elements and unforeseen events are always possible and even likely. Therefore, the possibility of failure and subsequent maintenance activities have been added to the NEO model. For each aircraft, a small number of highly fallible subsystems have been added. For example, the C-130 is currently comprised of five subsystems. In reality, the number of subsystems is much larger. Modeling all of these subsystems would drastically reduce the execution speed of the simulation. Therefore, the number was kept small enough to retain a high execution speed, but also large enough to show some noticeable effects of component failure.

For each subsystem, a mean time between failures (MTBF) and a mean time to repair (MTTR) are specified. At the beginning of the simulation, a failure time is determined by drawing a random number based on the MTBF. After each mission leg, the flying time is subtracted from the failure time. When the failure time reaches zero, the component has failed and the platform cannot continue on another leg until the proper maintenance has occurred. The amount of time needed for maintenance is determined by drawing a random number around the component's MTTR (Zahn et al. 1995).

The primary benefit of this modification is that the JTF planner can see how well a plan executes when random events such as component failure occur. Failure and maintenance could cause a platform to depart after the cancel time has passed, thereby generating a failed sortie. If too many sorties have failed due to required maintenance, then the planner may need to modify the mission legs and plan files in order to specify a more conservative flying schedule.

5.3 C++ Simulation and Animation

Since JWID '95, TASC has developed its own C++ simulation engine and graphics library. With this engine, TASC has modified the IMDE system to generate C++ source code as well as MODSIM II. The original MODSIM II version of the NEO model has been ported to C++ through the use of the new simulation engine and graphics library, which was beneficial because several advantages are realized by using C++. First of all, C++ is more widely known and recognized, which makes it easier for people familiar with C++ to understand how a model is constructed. Second, C++ is a more cost effective solution since it is possible to use public domain C++ compilers. Third, modelers now have the capability of incorporating third party libraries written in C++ into their models.

6 CONCLUSION

There are many high-technology software tools available to help military advisors in their decision making. One of these tools has been detailed in this paper, a model to simulate a possible NEO task. The flexibility of the model allows the user to analyze many different COAs before settling on an optimal choice. The animation involved in the simulation facilitates the decision process by graphically illustrating the outcome of a potential plan. With the use of IMDE, the object-oriented simulation package used to model the NEO, graphical comparisons can be made between two or more candidate COAs quickly and easily.

The conversion to C++, the addition of maintenance activities to the platforms and the addition of multiple leg missions have increased the power and fidelity of the base NEO model. Future enhancements may include expanding the maintenance capabilities by adding more subsystems to each platform, modeling the possible replacement of failed aircraft, and adding the functionality to graphically define road, sea and air networks.

ACKNOWLEDGMENTS

The authors would like to thank Mr. Patrick Clark and Capt Todd Carrico for their invaluable help in the writing of this paper. The research presented was sponsored by DARPA and the Logistics Research Division, Armstrong Laboratory, U.S. Air Force, Wright-Patterson AFB, OH 45433-7604 under Contract F33615-92-D-1052, DARPA Order Number 0008.

REFERENCES

Barkakati, N. 1991. *Object-Oriented Programming in C++*, Carmel: SAMS.

CACI Products Company. 1991. *MODSIM II Reference Manual.*

Carrico, T. 1995. The Object Model Working Group Document. Logistics Research Division, Armstrong Laboratory, U.S. Air Force, Wright-Patterson AFB, Ohio.

Clark, P. K., et al. 1994. Object-oriented Simulation with IMDE. In *Western Simulation Multiconference - Object-oriented Simulation Conference '94 Proceedings*, 131-136.

Zahn, E. A., N. J. Stute, and P. K. Clark. 1995. An Object-Oriented Simulation of Air Force Support Equipment Usage. *In 1995 Winter Simulation Conference Proceedings*, 1193-1199.

AUTHOR BIOGRAPHIES

JEFFREY E. SUMNER is a Staff MTS at TASC, Inc. He received his B.S. in Computer Engineering in 1990 and his M.B.A. in Project Management in 1995, both from Wright State University. He has been involved in numerous software development projects and has worked on the IMDE simulation tool for the past three years. His interests include object-oriented software development and discrete-event simulation.

ERIC A. ZAHN is a Staff MTS at TASC, Inc. He received his B.S. in Systems and Control Engineering in 1993 and his M.S. in Systems, Control and Industrial Engineering in 1994, both from Case Western Reserve University. He has utilized various simulation packages for the last five years, and has worked on the IMDE simulation tool for the last two. His interests include discrete-event and object-oriented modeling and simulation.

Proceedings of the 1996 Winter Simulation Conference
ed. J. M. Charnes, D. J. Morrice, D. T. Brunner, and J. J. Swain

DYNAMIC MEMORY USAGE IN PARALLEL SIMULATION: A CASE STUDY OF A LARGE-SCALE MILITARY LOGISTICS APPLICATION

C. J. M. Booth
D. I. Bruce
P. R. Hoare
M. J. Kirton
K.R. Milner
I. J. Relf

Defence Research Agency,
St. Andrew's Road, Malvern,
Worcestershire WR14 3PS, UNITED KINGDOM.

ABSTRACT

We report on a comparative study of the performance of shared and distributed memory parallel simulation algorithms on a large-scale military logistics simulation, and describe the nature of the application and its parallelisation in some detail. We demonstrate that the patterns of communication in the simulation were such that a standard implementation of Breathing Time Buckets (BTB) was unable to achieve any speed-up. New variants of BTB were designed and implemented, and we were able to achieve a speed-up of 7.4 compared to a critical path limit of 16.9. The logistics simulation contained a number of complex data structures and its parallel implementation was highly memory-intensive. The resulting patterns of memory access significantly degraded the shared memory simulation performance relative to the equivalent distributed memory version. These results cast doubt on the effectiveness of the current generation of shared memory parallel computers in dealing with optimistic simulations of large, dynamic scenarios.

1 INTRODUCTION

The emergence of relatively low-cost shared memory parallel computers has been a significant development in the last few years. Many manufacturers now offer multi-processor workstations with a global memory model. The advantage of a single address space is that data is immediately visible to each processor and message-passing can be implemented by simply passing data references. Although shared memory machines have distinct advantages over distributed memory ones, developing efficient shared memory simulators presents a number of problems, particularly for memory-intensive applications.

The logistics application consisted of about 16,000 lines of C++ code and simulated the activity of the British Army's Royal Logistics Corps during the first few days of a military campaign. There were three main features associated with the logistics application that made it worthy of detailed study in the context of parallel simulation. First, it was highly memory-intensive: the state associated with many of the simulation objects contained dynamic data structures of considerable complexity, thus generating a large overhead associated with saving and restoring state. Second, it exhibited poor lookahead by repeatedly generating events in the near future. Finally, there were a number of 'self-propelled' objects—objects that moved rapidly ahead in simulation time independently of all other objects in the system, needing to be repeatedly rolled back unless the optimism in the underlying simulator were constrained in some way.

A major aim of our research programme was to build general-purpose simulators capable of dealing with fully-interconnected objects with no pre-determinable lookahead. For these reasons, our work was based on optimistic simulation protocols: BTB (Steinman 1992) and Time Warp (Jefferson 1985). Our simulators were coded in C++ and all implemented the same simulation programmer's interface (SPI). Our shared and distributed memory computers were both SPARC-based and ran the same operating system (Solaris 2.3), enabling us to make a direct comparison between simulators.

The remainder of this paper is arranged as follows. In section 2, we describe the logistics simulation. Our shared and distributed memory simulators are described in sections 3 and 4, respectively; extensions to the standard BTB algorithm are detailed in section 5. We present our results in section 6 and summarise our work in section 7.

2 LOGISTICS APPLICATION

2.1 Major functional components

Our logistics model (Hoare et al. 1995) simulated the deployment of equipment during a military campaign. The re-supply model consisted of fixed first-line storage areas (HQ and Division), intermediate storage areas (Brigade and Battle Group) and front-line (Unit-level) dumps connected by a network of links as shown in Figure 1.

Storage areas consisted of a number of separate areas called circuits in which stacks of stores were arranged in rows. Trucks in a given storage area traversed the stacks collecting stores and were then assembled into convoys for dispatch to a target storage area or dump where they would unload. A dump consisted of a number of trucks in the process of being unloaded as stores were required. Trucks returned to their starting point when they were empty. The base personnel were also modelled in that they were required to rest for a minimum number of hours per day.

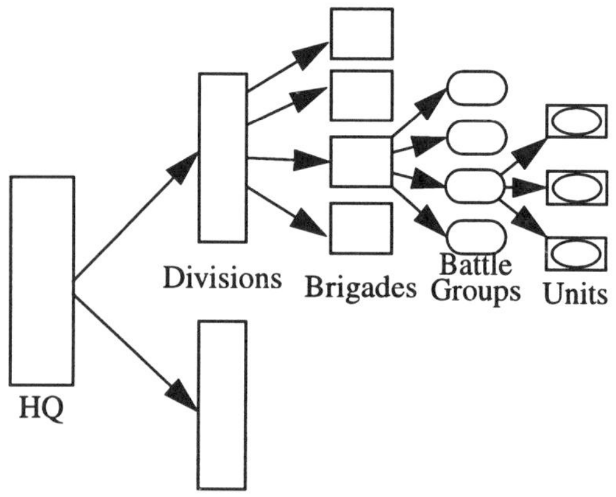

Figure 1: The Basic Layout of the Logistics Scenario

The model operated on a daily cycle; at the beginning of the simulation half of the trucks in the storage areas were already loaded awaiting dispatch. For each simulated day, the loaded trucks were dispatched to destinations to satisfy actual requirements and the remaining empty trucks were loaded with the predicted requirements for the next day. In the results we report on, the model was simulated for a period of two days.

We used two configurations based on Figure 1 which we will refer to as single-division and twin-division, in which the single HQ provided orders for one and two hierarchies respectively.

2.2 Simulation Objects

The storage areas and dumps were treated as simulation objects, and each had a significant amount of associated state and processing. For the results reported in section 6, the simulation objects were distributed across the processors in a round-robin fashion.

A major factor affecting the writing and debugging of large-scale optimistic simulations is state saving. The logistics application contained a number of complex data structures which needed to be recoverable in the event of a rollback. Neither full state saving nor incremental state saving offered attractive solutions—in particular, it was not possible to determine before an event was processed just how many elements in a particular data structure would change.

In order to minimise the programming effort associated with the saving of state, we developed a prototype data structure manager which maintained its own time-dependent state information. Its implementation as a C++ class hid the details from the model writer but was only a half-way solution to the problem as it only dealt with list types. We shall report elsewhere on our results using a version of logistics in which state saving was carried out transparently for all data types using persistent data structures (Bruce 1995).

2.3 Event Distributions

Within the hierarchical structure depicted in Figure 1, a simulation object only communicated externally with its immediate ancestor and descendants; communication between objects at the same level in the hierarchy did not take place. Excluding start-up messages, the ratio of intra- to inter-object messages was about 5:1.

Many of the events were concerned with the movement, loading and unloading of trucks around the Battle Group storage areas. In fact, for the single-division configuration, these types of event that the sixteen Battle Groups scheduled for themselves accounted for about 45% of the total number of generated messages.

Each of the 48 unit-level storage dumps independently generated a re-order event for itself every simulated half-hour. The unit-level storage dumps were therefore essentially self-propelled objects.

The major interaction between storage areas at different levels arose from the sending and returning of trucks: on the outward journey the trucks were filled with supplies, unloaded at pre-determined stacks and returned to the storage area that dispatched them. At the unit-level dumps, supplies were consumed at pre-determined rates. The units could also send urgent requests to the next level up in the hierarchy for supplies.

In Figure 2, we present the measured distribution of lookahead times for the single-division scenario, where we define the lookahead to be the difference between the receive and send times of an event. Over half of the events were generated with time stamps less than two simulated minutes into the future. At an even finer level of granularity, nearly ten thousand events were scheduled with times less than 0.01 simulated minutes. The events at thirty minutes corresponded to the re-order events fired off by the unit-level storage areas.

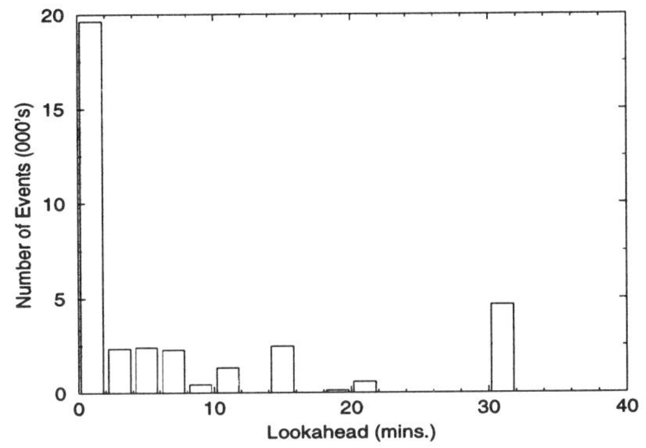

Figure 2: Distribution of Lookahead Times for the Single-Division Configuration

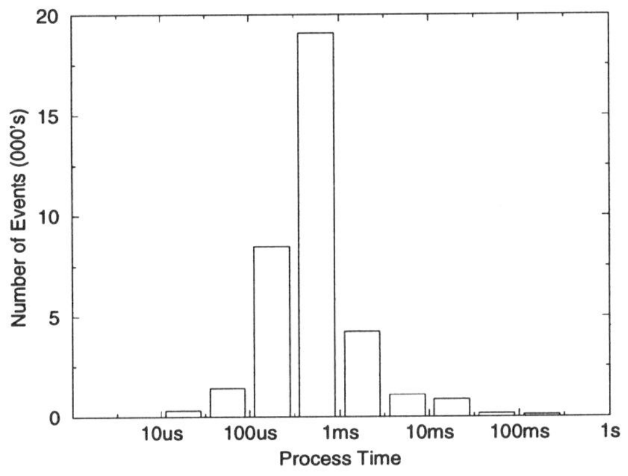

Figure 3: Distribution of Event Processing Times for the Single-Division Configuration

Figure 3 shows that the distribution of event processing times was very broad—ranging from about 10 μs to 1.0s—though the overwhelming majority resided in the range 0.1—10 ms. The larger granularity events were involved in searching a storage area for supplies—this search involved iterating over heavily-nested loops. The granularity of these types of event was very data-dependent and also led to the consequence that an event processed for the second time following a rollback sometimes required a significantly different amount of CPU time than was needed to process it initially.

Finally, we note that many of the events in the logistics simulation needed to pass lists of data values (for example, a truck contained a list of the pallets it was carrying). Due to the inherent difficulties in passing dynamic data structures between different memory spaces, these lists were implemented as fixed-size arrays. This made the copying and movement of the events straightforward, but added to the memory consumption of the simulation. In practice, we found the event sizes to be grouped around 1 Kbyte, 8 Kbytes and 16 Kbytes.

3 SHARED MEMORY SIMULATORS

3.1 Hardware

The shared memory computer we used was an 8-processor SPARCserver 1000 (SS1000) manufactured by Sun Microsystems. It contained 50 MHz SuperSPARC processors with 1 Mbyte of external cache per processor and a 64-bit wide packet switched XDBUS operating at 40 MHz, giving a peak bandwidth of 320 Mbytes/s to its 256 Mbyte memory. The Solaris threads library provided the necessary parallel programming support.

3.2 Memory Management

Our early performance experiments revealed that the dynamic memory routines of the compiler scaled very poorly. Since our parallel simulators made considerable use of dynamic memory we found it necessary to implement our own dynamic memory algorithm, which we did by means of a C++ `MemoryManager` class. Moreover, these experiments demonstrated the significant effect data locality could have on performance cf. (Fujimoto and Panesar 1995), and thus each processor had its own `MemoryManager`.

In the interests of access speed, memory within a `MemoryManager` was maintained using a set of free-lists. We used a BSD-style algorithm, in which a free-list was maintained for each power of 2. Memory returned by the `MemoryManager` was also aligned on cache-line boundaries and no block returned was smaller than the cache-line size (64 bytes in our case). These restrictions were designed to prevent the problems associated with false sharing (Torrellas et al. 1994) that can occur with multi-processor cache-based machines. Dynamic memory management on a given processor was also completely independent of all other processors: there were no shared variables and no concatenation of blocks or merging of free-lists. In addition, free-lists operated a 'last-in first-out' policy, which meant that memory returned from a free-list had a good chance of residing in the local cache of its processor.

In our current version, the `MemoryManager` class had no mechanism for preventing a net transfer of memory from one processor to another, which could be a problem if the simulation contained net source or sink objects.

3.3 Breathing Time Buckets

Despite requiring synchronisation at the end of each cycle, the Breathing Time Buckets (BTB) algorithm (Steinman 1992) has the advantages of simplicity coupled with a certain elegance and robustness. BTB consists of a number of distinct phases: event processing and local event horizon (LEH) exchange; global simulation time (GST) determination; event committal and exchange; sorting and merging event queues; local rollback and housekeeping. On the SS1000, we used the Solaris threads library to bind one 'main loop' thread to each processor.

Each processor maintained its own separate event queue and was responsible for a subset of simulation objects. Output messages from a given processor were inserted directly into the input queues of the appropriate nodes. GST was determined via a set of shared variables arranged in a tree structure. As each processor crossed its LEH it informed its parent of the crossing and wrote its LEH value into a single global variable if it was less than the value recorded thus far. Once the root node was informed that all LEH's were crossed, it set the global Boolean `allCrossed` to true, allowing all processors to move on to the next phase of the BTB cycle. We used a tree-based algorithm to ascertain when event exchange was completed, since this was substantially faster than using a shared counter.

3.4 Time Warp

Our implementation of Time Warp was similar in spirit to the BTB implementation described above. A 'main loop' thread was run by each processor. Again, there was no global input queue; output messages were inserted into an unsorted input buffer residing on the receiving processor. The input buffer was scanned by the receiver and messages extracted and placed into its pending queue in time stamp order. After an event was processed, it was inserted into the processed event queue associated with its simulation object. Rollback was implemented using direct cancellation (Fujimoto 1989).

We used a version of the global virtual time (GVT) algorithm described in (Fujimoto and Hybinette 1994). This algorithm makes use of the fact that on a shared memory computer there are no messages in transit and has the advantage of being able to operate without closing the simulation down. Also, there is no need to acknowledge messages or perform rounds of token passing.

4 DISTRIBUTED MEMORY SIMULATORS

4.1 Hardware

We used a 22-processor Meiko CS-2 containing 50 MHz SuperSPARCs with 1 Mbyte of external cache and 64 Mbytes of memory per processor. Two of the processors were I/O nodes with file systems mounted and connections to ethernet; the remaining 20 processors effectively operated as diskless workstations served by the I/O nodes. Each processor had its own dedicated interface to a high-performance communications network—a Meiko designed communications processor named Elan—which provided 50 Mbytes/s bisectional bandwidth over a physical link operating at 0.6 Gbit/s in each direction. For a 64 byte message, the latency (including transfer) and effective bandwidth between UNIX processes on different processors was 12 μs and 5.54 Mbytes/s; for 64 Kbyte messages the corresponding figures were 1.491 μs and 43.95 Mbytes/s.

The CS-2 also supported a global data space, and very efficient message transfers were achievable by the direct memory access (DMA) transfer procedures supported by the Elan processor. Also, there was direct support in hardware for processor synchronisation.

4.2 Breathing Time Buckets

The parallel BTB program comprised the same executable running on every processor. The BTB processes interacted via message passing and hardware synchronisation. In order to perform the LEH broadcast as efficiently as possible, we used the global memory provided by the CS-2. Thus when a processor crossed its LEH, it broadcast this value via low-level DMA's to all other processors. When all LEH's were crossed, each BTB process was able to determine independently the global simulation time.

During the event committal stage, each BTB process stored its generated events and placed them in an array of message buffers, according to their destination. Message sending was accomplished using the Elan Channel Protocol, which provided a fully robust yet very efficient technique for sending variable sized messages. After sending its messages, each BTB process performed a hardware synchronisation and blocked until all processors reached this point.

5 EXTENSIONS TO BTB

During the course of our work on logistics, we developed three variants on the basic BTB algorithm, which we named *limited*, *extended* and *limited-extended*. *Limited* BTB stops processing events as soon as its local event horizon (LEH) is crossed. This can significantly reduce the number of potentially costly rollbacks but reduces the optimism of the simulator. *Extended* BTB folds back into the event queue events generated for objects on the same processor. Such events are not in general considered for the LEH calculation and for applications like logistics, this can greatly reduce the number of BTB cycles required for a given end simulation time.

There is a slight subtlety in the implementation concerning the way in which *extended* BTB interacts with optimistically processed events. A processor may continue to process beyond its LEH in the hope that optimistically processed events will not need to be rolled back. Unfortunately, a local event may cause one of these optimistically processed events to roll back; if this is the case, the local event *must* contribute to LEH—if it did not, then any optimistically processed events that should have been rolled back may be incorrectly committed.

Another consequence of *extended* BTB is that in simulations which consist entirely of local events, no BTB cycles will be triggered and therefore no dynamic memory will be recovered. *Extended* BTB therefore requires an additional synchronisation mechanism, akin to GVT in Time Warp, in which memory recovery is forced at regular intervals.

As its name suggests, *limited-extended* folded events back into the event queue but stopped processing after the LEH was crossed.

6 RESULTS

The following results were obtained using simulators that had already demonstrated good speed-up on the queuing network application described in (Damitio et al. 1994). For example, the distributed memory BTB simulator achieved a maximum speed-up of 17.5 on 20 processors and achieved significant speed-ups over a range of parameters. There was also no significant difference in performance between the shared memory and distributed memory simulators using the synthetic application.

All of the speed-up figures quoted in the results below were calculated relative to an optimised sequential simulator.

6.1 Distributed Memory

The speed-up figures obtained for the two logistics scenarios (single-division and twin-division) on the CS-2 are shown in Figures 4 and 5 respectively. Standard BTB was unable to achieve any speed-up on the single-division configuration (and this result was repeated on the SS1000). The main reason for this was the large number of rollbacks that occurred as a result of objects sending messages a short distance into the future. These messages often triggered costly rollbacks, severely degrading the overall performance.

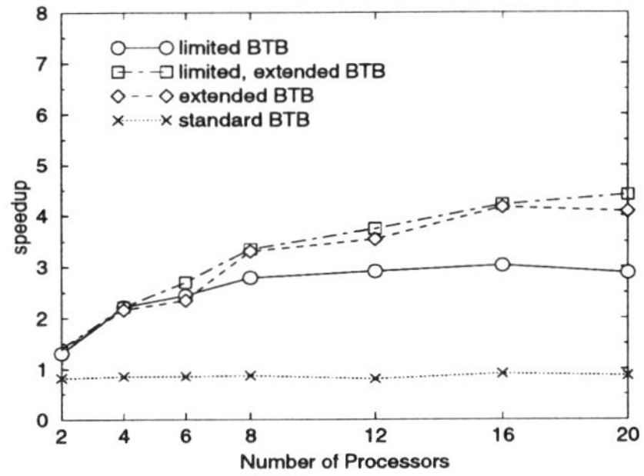

Figure 4: Speed-up for the Single-Division Logistics on the CS-2

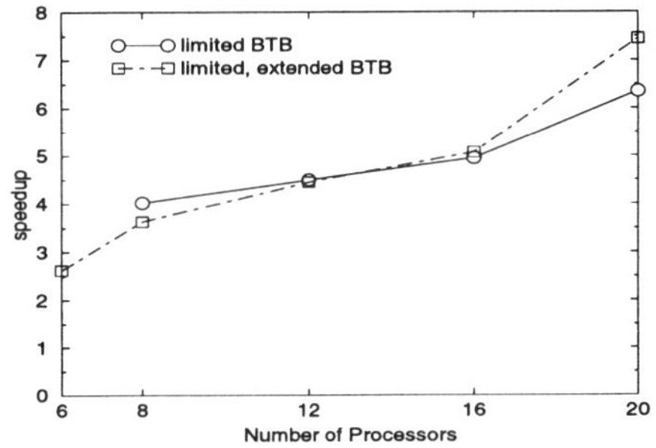

Figure 5: Speed-up for the Twin-Division Logistics on the CS-2

Overall, the best speed-up was obtained with *limited-extended* BTB; a speed-up of 4.1 was achieved on the single-division logistics (out of a maximum possible speed-up of 8.1) and 7.4 achieved on the twin-division scenario (out of a maximum of 16.9). Our maximum speed-up figures were obtained using a critical-path analysis (Wieland et al. 1992) in which the best possible completion time for a simulation was calculated (neglecting state-saving and message-passing overheads). The twin-division scenario was so memory-intensive that the *extended* BTB simulator would not run it to completion and the *limited* and *limited-extended* simulators could only do so using at least 8 and 6 processors respectively.

6.2 Shared Memory

The speed-up curves for single-division and twin-division logistics simulations on the SS1000 are shown in Figures 6 and 7 respectively. If we compare these results with those obtained on the CS-2, we see that the speed-ups achieved on 8 nodes were significantly worse than the equivalent CS-2 values. For single-division logistics, the speed-up was 2.3 on the SS1000 compared with 3.3 on the CS-2; for the twin-division we obtained a speed-up of 2.7

on the SS1000 and 4.0 on the CS-2. In terms of run-time, the distributed memory simulator consistently outperformed its shared memory counterpart by about 50%.

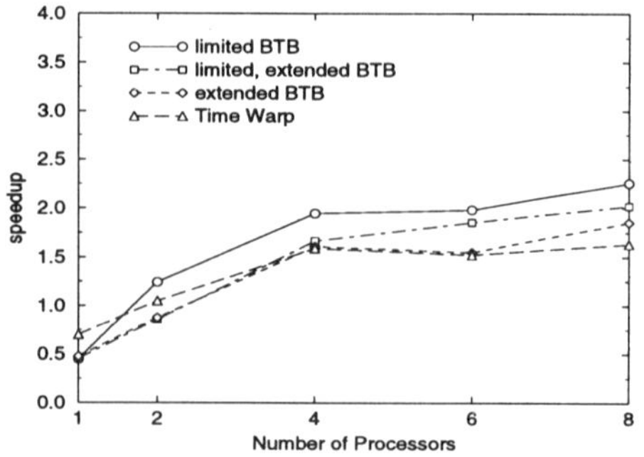

Figure 6: Speed-up for the Single-Division Logistics on the SS1000

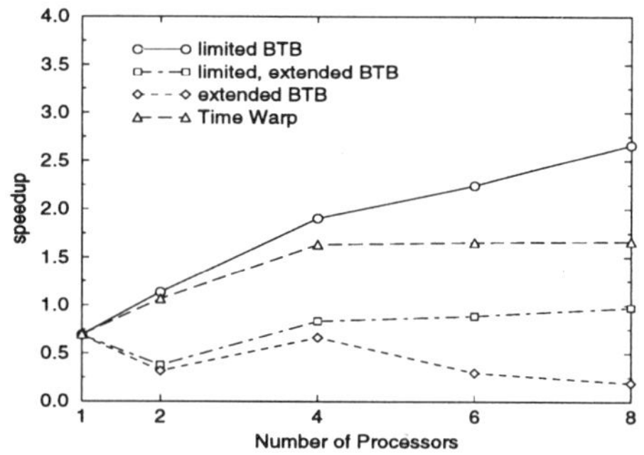

Figure 7: Speed-up for the Twin-Division Logistics on the SS1000

A breakdown of overheads for *limited* BTB is shown in Figure 8. 'wait LEH' represents the waiting time at the end of a cycle when a processor has reached its LEH, and is waiting for all other processors to synchronize; 'save state' is the time spent saving the state associated with each event; 'commit' includes memory recovery; the main constituents of 'housekeeping' are the merging and sorting of pending event queues; and 'wait exchange' is the time spent awaiting messages from other processors at the end of a cycle.

Much of the overhead associated with BTB was in the wait LEH time at the end of a cycle. The different BTB variants also exhibited quite different behaviour for the logistics application. *Limited* BTB (Figure 8) generated just over 1800 BTB cycles in the example shown and had no rollbacks; a substantial amount of time was spent waiting at the end of a cycle, but no time was wasted unnecessarily processing events. *Extended* BTB generated just 40 cycles for the same simulation, and it spent much less time waiting for inter-processor synchronisation; however, it also spent a considerable amount of time processing events that were subsequently rolled back (this was roughly 20% of the total number of events for *extended* BTB). The overall effect was that *limited* BTB consistently outperformed *extended* BTB.

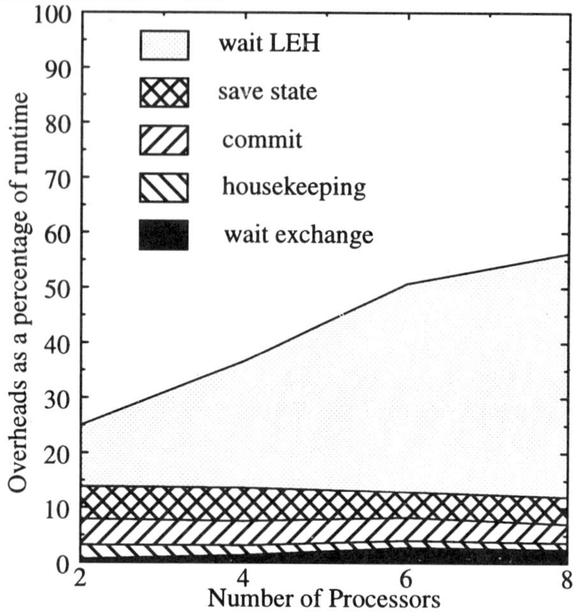

Figure 8: Breakdown of Overheads for the Single-Division Logistics Simulation on the SS1000 using *limited* BTB

6.3 Degradation of event processing time

When we analysed the breakdown of the overall run-time, we discovered that on the SS1000, the average time to process an event degraded significantly with increasing numbers of processors (Figure 9). Similar degradation was not observed on the CS-2 (Figure 10) and this is the main reason for the performance difference between our shared and distributed memory simulators.

There is also a difference between the event processing time for the sequential simulator and the parallel simulators running on one node. This is due to the overhead of the state-saving implementation, which is absent for the sequential simulator. Also, memory access times were slightly better on the CS-2, compared to the SS1000, which meant that the absolute run-times for the sequential simulator were consistently 10% less than on the SS1000.

Before we consider possible explanations for the degradation in event processing time on the SS1000, we should note that this behaviour is observed both for BTB and Time Warp; it is not specific just to BTB (or its variants). The event processing time is being averaged over committed events, so that a comparison is being made over exactly the same set of events. Also, the memory management allocator used was designed to avoid false sharing effects, in which two or more independent varia-

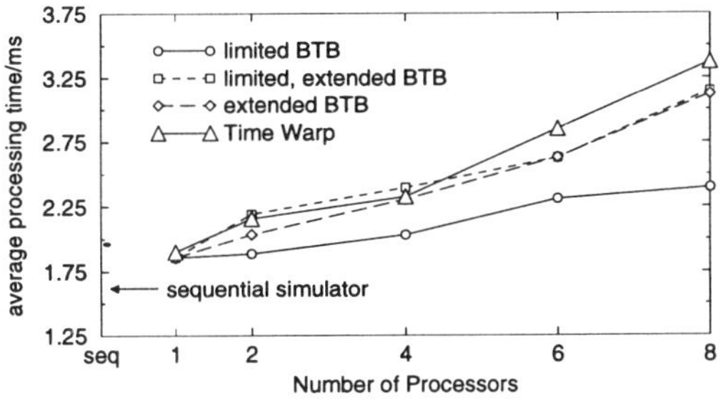

Figure 9: Average Event Processing Time for the Single-Division Logistics Simulation on the SS1000

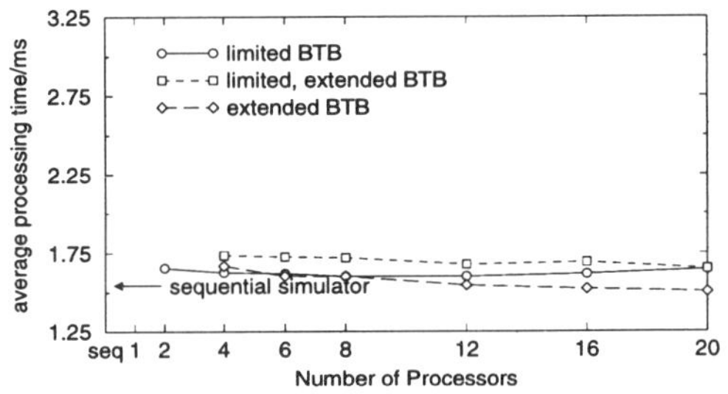

Figure 10: Average Event Processing Time for the Single-Division Logistics Simulation on the CS-2

bles share the same cache line. Moreover, it is worth bearing in mind that our synthetic queuing network application did not show any such degradation in event processing time.

Our hypothesis is that the shared memory performance is being affected by the bus traffic generated, since logistics is a very memory-intensive application, typically consuming 50 Mbytes or more in a typical run. One of the difficulties in programming shared memory machines is in analysing and diagnosing performance bottlenecks. Unfortunately, reliable and accurate tools for monitoring cache miss rates were not available, and therefore in order to test whether it was the bus traffic causing the problem we introduced random delays into our shared memory BTB simulator, immediately prior to an event being processed. The rationale here is that as the delay was progressively increased, the bus traffic density would be correspondingly reduced, as each processor spent more and more time waiting. The results of applying delays to the SS1000 BTB simulator are shown in Figure 11.

The average processing times for *extended* and *limited-extended* decreased significantly as the wait time was increased. This effect was less marked for *limited* BTB, but under normal conditions (i.e. without any artificially in-

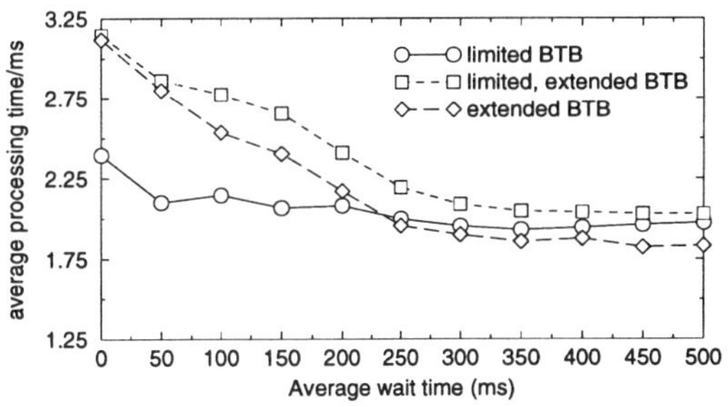

Figure 11: Average Processing Time as a Function of Wait Time on the SS1000 (all data points were generated using 8 processors with wait times normally distributed and variance = mean/2)

troduced waiting time), *limited* spends considerably longer waiting for synchronisation at the end of a cycle, compared to either *extended* or *limited-extended.* Under normal conditions it would therefore generate less bus traffic and hence be less prone to degradation in event processing time than either *extended* or *limited-extended.*

The explanation for the lack of degradation in event processing time for our synthetic queuing application (Damitio et al. 1994) is that the spin loop at the core of the event processing procedure did not generate any bus traffic. In fact, by artificially increasing the state size of each server object and modifying its event processing, we were able to demonstrate the same effect observed with the logistics application.

7 CONCLUSIONS

We have presented a study of the application of parallel simulation technology to a large-scale military logistics simulation and compared its performance using both shared memory and distributed memory computers. We have developed extensions to the basic BTB algorithm (*limited, extended* and *limited-extended*) and have managed to exploit roughly half of the available parallelism for a large-scale logistics simulation on the CS-2.

Comparing our shared and distributed memory simulators, we have shown that with a memory-intensive application like logistics, the speed-up obtainable can be affected significantly by specifically shared memory effects. Since optimistic simulations tend to make heavy use of dynamic memory, these results would seem to cast doubt on the capability of the current generation of multiprocessor workstations in dealing with optimistic simulations of large, dynamic applications.

ACKNOWLEDGEMENTS

It is a pleasure to thank Brian Roberts and Brian Merrifield for their long-standing support and encouragement.

REFERENCES

Bruce, D. 1995. "The treatment of state in optimistic systems", pp 40–49 in *Proc. Ninth Workshop on Parallel and Distributed Simulation*, Lake Placid, New York, 14–16 June 1995.

Damitio, M. et al. 1994. "Comparing the breathing time buckets algorithm and the Time Warp Operating System on a transputer architecture", pp 141–145 in *Proc. Modelling and Simulation (SCS European Simulation Multiconference)* 1994, Barcelona, Spain, 1–3 June 1994.

Fujimoto, R. M. 1989. "Time Warp on a shared memory multiprocessor", *Transactions of the SCS*, Vol. 6, No. 3, July 1989.

Fujimoto, R. M. and M. Hybinette. 1994. "Computing global virtual time in shared memory multiprocessors", Georgia Institute of Technology Technical Report, August 1994.

Fujimoto, R. M. and K. S. Panesar. 1995. "Buffer management in shared memory Time Warp systems", pp 149–156 in *Proc. Ninth Workshop on Parallel and Distributed Simulation*, Lake Placid, New York, 14–16 June 1995.

Hoare, P. R. et al. 1995. "The application of high performance parallel computing to military simulation", pp 115–119 in *Proc. Military, Government and Aerospace Simulation Conference*, Phoenix, Arizona, 9–13 April 1995.

Jefferson, D. R. 1985. "Virtual time", pp 404–425 in *ACM Transactions on Programming Languages and Systems*, Vol. 7, No. 3, July 1985.

Steinman, J. S. 1992. "SPEEDES: A multiple-synchronization environment for parallel discrete-event simulation", pp 251–286 in *International Journal of Computer Simulation*, Vol. 2, No. 3, 1992.

Torrellas, J., M. S. Lam and J. L. Hennessy. 1994. "False sharing and spatial locality in multiprocessor caches", pp 651–663 in *IEEE Transactions on Computers*, Vol. 43. No. 6, June 1994.

Wieland, F. et al. 1992. "A critical path tool for parallel simulation performance optimization", pp 196–206 in *Proc. 25th Hawaii International Conference on System Sciences*, Vol. II, Kauai, Hawaii, 7–10 January 1992.

AUTHOR BIOGRAPHIES

CHRIS BOOTH has a BA in Mathematics (1986) and an MSc in Computation (1987) from the University of Oxford. He is interested in adaptive synchronisation algorithms for PDES, and is currently working on the APOSTLE simulation language.

DAVID BRUCE has a BSc in Mathematics (1988) from the University of Kent at Canterbury. His research interests include parallel discrete event simulation, state saving for optimistic computation, programming languages and type systems and he is currently working on the APOSTLE simulation language and the DoD High Level Architecture (HLA) for simulation.

PETER HOARE has a Computer Science BSc and PhD from Royal Holloway, University of London. His research interests currently include distributed/parallel simulation, computer graphics and Internet protocols and systems. He is involved in the UK STOW programme as the technical lead on the DoD HLA.

MICHAEL KIRTON has BSc and PhD degrees in Physics, and before joining DRA held an IBM (UK) Fellowship. He has published over thirty scientific papers in fields ranging from noise in solid-state microstructures to parallel simulation.

ROY MILNER has a BSc in Mathematics (1983) from the University of Nottingham. His research interests include PDES on shared memory machines and he is currently working on the DoD HLA.

IAN RELF has a BEng in Engineering Electronics from the University of Warwick (1989). His research interests include the application of PDES to military simulations. He is currently working on a large-scale DIS air-ground simulation.

Manufacturing Applications

Proceedings of the 1996 Winter Simulation Conference
ed. J. M. Charnes, D. J. Morrice, D. T. Brunner, and J. J. Swain

A NEW PARADIGM FOR MANUFACTURING ENTERPRISE MODELING: REUSABLE, MULTI-TOOL MODELING

Dursun Delen
David B. Pratt
Manjunath Kamath

Center for Computer Integrated Manufacturing
School of Industrial Engineering and Management
Oklahoma State University
Stillwater, OK 74078, U.S.A.

ABSTRACT

The power of modeling of manufacturing systems can be enhanced through the application of reusable and pluggable modeling constructs. In this paper we discuss a framework for model reusability and how the seperation concept together with object oriented paradigm support this capability. We then discuss the importance of reusable and pluggable organizational controls in a modeling environment. Several important issues related to the implementation of reusability and pluggability are summarized. Examples are presented to illustrate our approach for implementing reusable and pluggable organizational controls.

1 INTRODUCTION

Advanced manufacturing technology, without a doubt, is a major corporate advantage in today's global market. Strategic application of advanced technology can markedly improve a manufacturer's product quality, responsiveness to customers, process control, process flexibility, and capital investment flexibility, all of which are determinants of global manufacturing competitiveness.

The stochastic nature of manufacturing systems, coupled with ever changing market behavior, makes it difficult for manufacturing engineers to analyze and design/redesign a complete discrete part manufacturing system. It is rarely feasible to do experiments with the actual system because such experimentation is often too costly or too disruptive to the system. In many cases, the "system" might not even exist, as in the case of the design of a new discrete part manufacturing system. Nevertheless, we may want to study it in its various proposed alternative configurations to recommend how it should be built in the first place. For these reasons, it is usually necessary to build a model as a representation of the system and conduct the analysis using this model.

It has been demonstrated many times that modeling, analytical or simulation, is vital for the design of complex manufacturing systems (Leung and Suri 1990; ElMaraghy and Ravi 1992; Mize et al. 1992; Suri and de Treville 1993). Simulation and analytical models are used to analyze the stochastic behavior of these systems under various scenarios. Nevertheless, modeling of such complex systems is not without its shortcomings. First, models have traditionally been viewed as single purpose, throw-away efforts. A model is built from scratch to address a particular problem or question, and then it is often discarded with little thought given to additional use. This single-use, throw-away mentality of modeling is very expensive, time consuming and wasteful. Second, lack of access to models by non-modeling specialists limits their usage and value. The availability of a modeling expert becomes a constraint on the usage and value-added contributions of the model.

The advent of the object oriented paradigm has provided an impetus for researchers to create modeling environments for building reusable simulation models (Bhuskute et al. 1992; Adiga 1989; Glassey and Adiga 1989). Designing for reusability involves identification of behaviors that are useful in more than one context. In general, this implies a system design which adheres rather strictly to the "one-component-one-function" doctrine (Glassey and Adiga 1989). If a component performs more than one function, its usage becomes limited to situations in which all of its functions are required. Alternatively, if a strict one-to-one functionality is maintained between component and function, the components truly become "building blocks" from which a total system model can be constructed (Pratt et al. 1994).

2 REUSABLE AND PLUGGABLE MODELING

Object oriented programming (OOP), a paradigm in which all program variables are represented as objects which communicate by means of message passing, is a significant advancement toward the development of multiple use, general purpose, plug-compatible models. OOP possesses four key concepts which facilitate this advancement: encapsulation, message passing, late binding, and inheritance (Budd 1991).

A key consequence of the reusability emphasis is the implementation of the separation concept (Pratt et al. 1994). The implementation of separation involves the creation of separate and distinct modeling primitives for physical elements, information flow, and control decisions. Traditional modeling approaches have not considered the separate and distinct modeling of physical, information, and control elements. For example, in many simulation languages, the constructs that are provided for information and control are frequently hard coded and dispersed into the model. This results in code that is hard to modify and difficult to use for multiple purposes.

Another advantage of the separation of physical, information, and control objects is that it allows the system modeler to think of these elements independently during model development. This provides a more natural modeling environment. In other words, when developing the physical model, the modeler need not be concerned with the information or control aspects. The process involves selecting the appropriate physical components without being constrained by concerns regarding how to model information flow. Similarly, information flow is considered without regard to physical objects. This independence facilitates the creation of models with a higher degree of integrity and greater flexibility relative to experimentation with the model.

Bhuskute et al. (1992) proposed a framework with reusability features for modeling and simulation of discrete part manufacturing systems. As suggested therein, a modeler can visualize each modeling object in terms of its physical, information, and control aspects. Visualizing physical and information components of a system as distinct elements is straightforward. In most cases, these elements are tangible and easily defined; for a manufacturing system, a lathe, a drill press, or an AGV is a physical component, whereas a bill of material or a routing sheet is an information component. Control components are potentially more difficult to grasp. When a control element interacts with the physical element it controls, it evaluates the state of the system on the basis of physical system status and other available information. Then an action is taken (i.e., a decision is made) based on an algorithm or a decision process. The decision is then communicated to physical, information or other control/decision components. The modeler can store not only the primitive objects but also the composite objects (i.e., a work station or an assembly department) in the object library, and couple them in a suitable manner to represent the real world manufacturing system.

This framework (Bhuskute et al. 1992) was initially designed for creating models of discrete part manufacturing systems which were to be exercised by a simulation tool only. A prototype modeling environment was developed in which the models created were highly reusable within a simulation context. From a broader perspective, the models created by this environment were not 'reusable' in the sense that they could not be exercised by any tool other than simulation. While this was a major limitation, the richness of the models created and the ease with which the models could be changed opened up new possibilities. The question was "Now that we have the ability to create and modify with ease detailed descriptions of complex manufacturing systems, how can we use these 'models' with other analysis tools?" Recognizing this opportunity, Duse et al. (1993) suggested that a modeling framework be designed where a model of the enterprise be created without keeping any tool or any specific problem as the context for model development. This led to the concept of a "base model."

3 BASE MODEL CONCEPT

A base model is an abstraction of a real world manufacturing system in the richest possible way (Duse et al. 1993). It acts as the substrate from which further abstractions can be made for providing tool-specific models. Thus the <u>modeling</u> activity is viewed as a separate and distinct activity from <u>problem solving</u> (which requires the use of specific tools and further abstractions). Thus, reusability with respect to tools is achieved through the concept of a "Base Model" and "Configurators/Translators" which configure/translate problem specific and tool specific models from the base model. More detailed discussion of *tool independent modeling* and the *base model* concept can be found in Kamath et al. (1995 and 1996). In the creation of a base model, a library of manufacturing modeling primitives, described earlier in this paper, is used. These primitives are classified into three types: physical primitives, information primitives, and control/decision primitives. A user can construct the manufacturing system specific base model by simply selecting appropriate modeling primitives and assembling them together using a software environment (Figure 1).

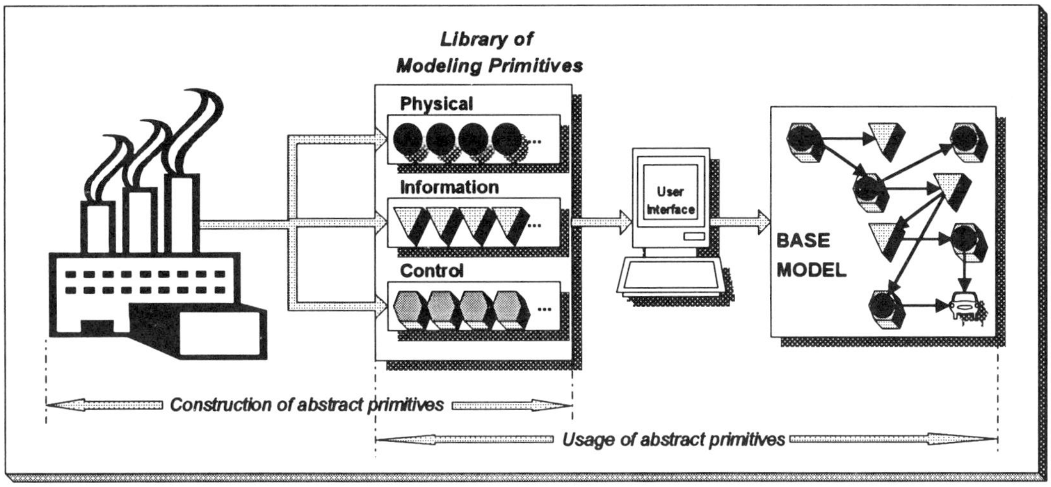

Figure 1: Construction and Usage of Abstract Modeling Primitives

A base model is never static. It evolves with the organization and is persistent in time. It is maintained as an ongoing activity just as company databases are created and updated on an ongoing basis. Maintaining the base model is thus a modeling activity for the sake of modeling and not pursued with an immediate specific purpose in mind, as is done in the traditional modeling approach which is purpose driven and tool specific. Given the existence of such a base model, one can derive tool-specific models, called execution models, to address a particular modeling problem or question at any time.

A comprehensive modeling environment should provide a variety of appropriate analysis, modeling, and optimization tools for solving problems which lie in different domains. For instance, rough-cut system design problems require aggregate estimates of performance measures and modelers usually employ analytical methodologies (e.g., queuing theory) for their solution (Suri and de Treville 1993). For determining detailed and accurate estimates of performance measures, modelers typically employ a simulation based approach. The base model, which is a tool independent representation of a specific manufacturing system, has all the information necessary to support the analysis tools available. Once an analysis tool is determined based on the specific experimental circumstances, the next step is to extract an execution model from the base model for the selected solver. Figure 2 illustrates this process.

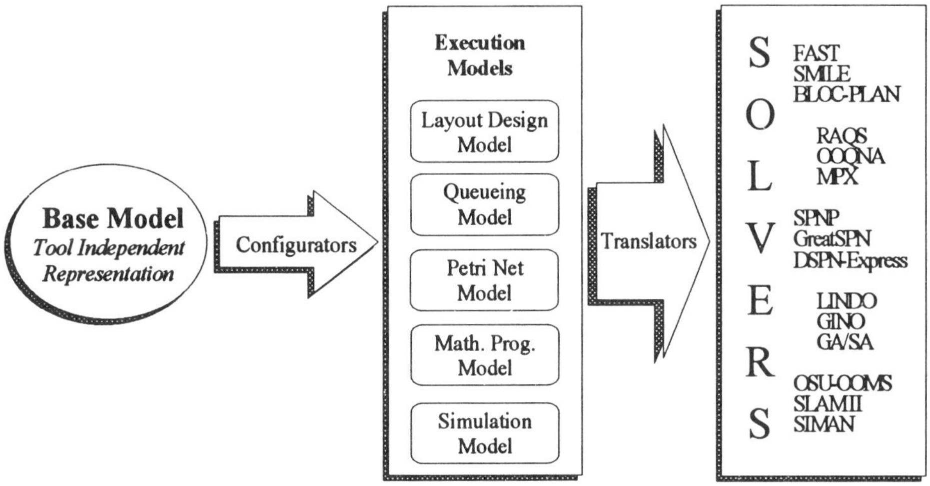

Figure 2: Tool Independent Model Representation

A new modeling framework has been developed based on the concept of the base model (Kamath et al. 1995). This framework, illustrated in Figure 3, includes both the construction phase and the utilization phase of the base model. The Manufacturing System (*1*) block in Figure 3 represents the real world manufacturing system. A Model Builder can construct the tool independent, generic, persistent Base Model (*3*) by using the Base Model Configurator (*2*). Once the base model is constructed a Decision Maker can seek answers to questions related to the manufacturing system through an Experimental Frame (*4*). Based on the nature of the problem to be solved and the complexities of the manufacturing system under study an appropriate Tool Specification (*5*) can be determined. With the tool specification in hand, a tool dependent configurator extracts the Execution Model (*6*) from the base model. Performing Analysis (*7*) will lead to either another experimental frame definition or Recommended Changes (*8*) for the system. As Changes are Implemented (*9*) within the manufacturing system, it is necessary to update the base model to reflect the changes.

The base model concept and modeling framework described in this section effectively address the first shortcoming of traditional modeling approaches, namely reusability. A proof-of-concept implementation of the modeling framework described above has been developed in the VisualWorks 2.0 development environment (ParcPlace 1994) which is based on the Smalltalk-80 language (Goldberg and Robson 1989). The user of the modeling environment can construct, save, and reuse a base model of a manufacturing system. The user can also analyze the manufacturing system performance using discrete-event simulation models, queueing network models, or Petri net models which are automatically configured by the modeling environment using the base model.

4 ORGANIZATIONAL CONTROLS

4.1 Reusable Controls

It is our experience that designing a framework to support the reuse of physical elements and information elements is relatively easy. Object oriented modeling environments have been developed by other researchers, examples of which are those developed by Glassey and Adiga (1989), Thomasma and Ulgen (1988), Narayanan et al. (1992), LeFrancois and Montreuil (1994). Most

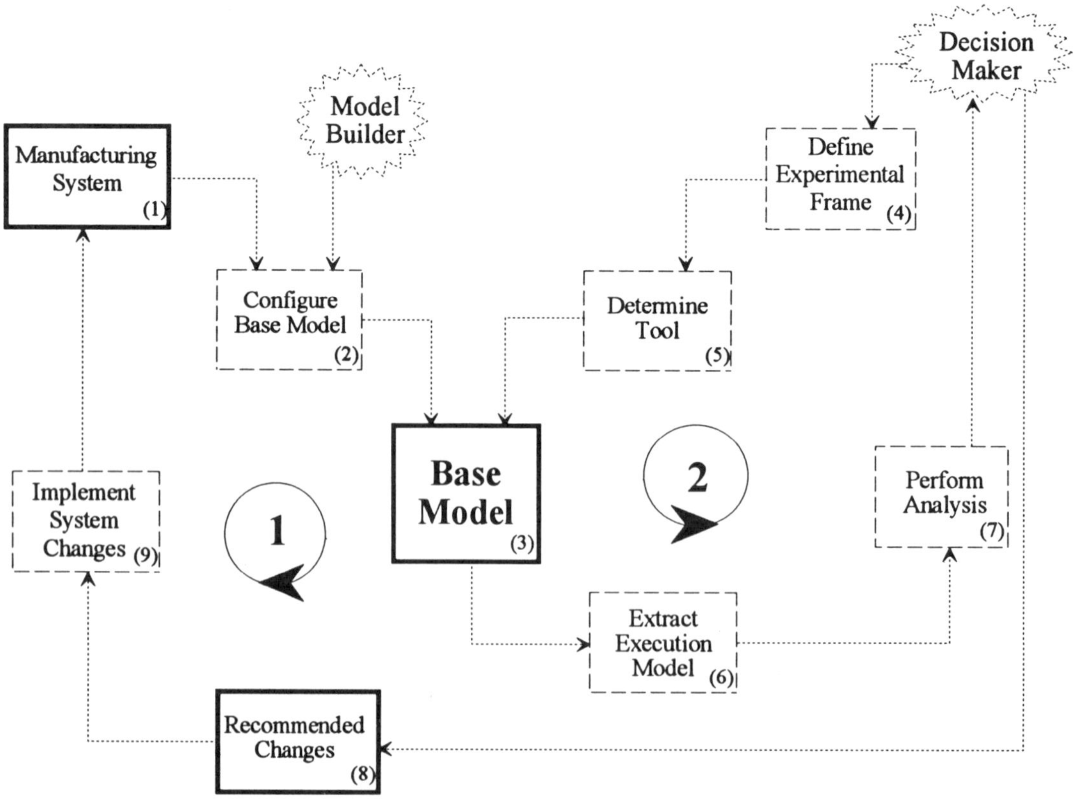

Figure 3: A New Framework for Manufacturing Systems Modeling

of the research efforts in the literature have been directed at the modeling of the physical elements and information elements. Furthermore, all of the efforts address simulation modeling only. The OOSIM group (Narayanan et al. 1992) and SmartSim/SmarterSim group (Thomasma and Ulgen 1988; Thomasma et al. 1990) have concentrated on lower level control for specifying material movements. Object oriented modeling and simulation of automated control in manufacturing has been proposed by Bodner et al. (1993). As per these authors, " ... the separation of control and plant resources in the model allows explicit representation of the factory control system, a critical feature of an automated system." However, in general, in the existing modeling environments, the control elements have been implicitly incorporated, and cannot be distinguished from other elements. That is, the control elements are interwoven with the other aspects of the model. This results in "forced hard coding" of control aspects. These modeling frameworks do not easily lend themselves to modeling organizational controls in a reusable way.

Achieving reusable organizational controls requires the modeler to consider the following two main issues:

1. *Identification of the decision points.* Decision points occur in three types of flows; the physical flow, the information flow, and the control flow. To identify the decision points, one must trace the threads of each of these flows. If one fails to recognize a decision point explicitly, then the logic for answering that decision-question can become hard coded, limiting the scope for achieving pluggable controls. The model developer must *be aware of default decisions.* These are the decisions that are not explicitly seen as decision making points, since a particular operating procedure is so ingrained that the concept of doing things differently cannot be recognized unless special attention is paid to such cases. For modeling, one needs to question every method to discover whether there exists a potential decision point.
2. *Identification of the set of questions that need to be answered at a decision point.* Granularity of the decision-question is a primary determinant of the degree of reusability that can be achieved. Composite questions (coarse granularity) will lead to logic that involves multi-dimensional decision making. Thus a change in logic along any one dimension may require re-coding of the complete logic. For example, consider the decision of customer order processing. The logic employed for answering this question may also involve determining the priority for the order if accepted, quantity of order accepted, and estimated delivery date based on which acceptance decision is made. While modeling, one needs to consider these aspects of the decision individually so that one does not group the logic for all these dimensions in one segment of code.

4.2 Pluggable Controls

Pluggability can be defined as a modeling capability that enables a modeler to replace a modeling object with a compatible one without changing the rest of the model. The modeling object can be a physical, information, or control/decision element. Since these elements can be grouped according to the levels in a manufacturing organization such as the machine level, the department level, and the plant level, we can potentially have pluggability at multiple levels. In our prototype implementation, we have achieved pluggability with respect to certain elements at different levels in the organization. Several characteristics of the OOP framework, such as polymorphism, inheritance, and encapsulation, simplify the implementation of pluggability.

Pluggability greatly facilitates the modeling of control logics in a manufacturing setting. The model user should be able to manipulate these control logics in order to find the 'optimum' set for a given physical and organizational configuration. Figure 4 presents a work station object for an illustration of pluggable control logics. The work station object has input and output buffers, a set of attributes (name, parent, location, footprint, status, reliability parameters) as well as a set of controller objects. The controller objects are of three functional classes; input queue controllers, output queue controllers, and resource controllers. Each functional controller class has a set of potential logics which can be used for answering a set of specific questions. For the input queue controller, these logics are grouped into four categories - part addition, part removal, primary part selection, and secondary part selection. In each of these categories there is a set of logics, as depicted in Figure 5, to answer the same question with different information usage. For each of the actions of part addition, part removal, and part selection, the model builder can substitute one logic element for another. In our prototype implementation, the user has the following choices.

- Add Checks (part addition) - *addAlways* (no restriction on adding a part), *capacity* (restriction on the queue size), *size* (restriction on the part size), *type* (restriction on the part type).
- Remove Checks (part removal) - *presence* (the only restriction is the presence of the part), and *productionOrderCompletion* (restriction is the

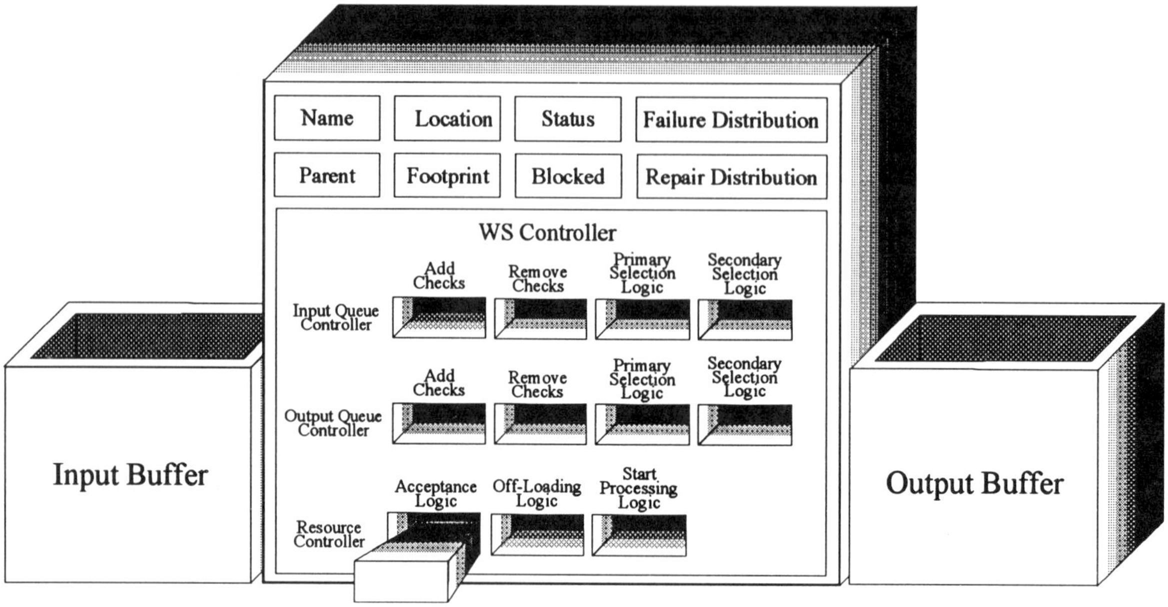

Figure 4: Pluggable Control Logics at the Workstation Level

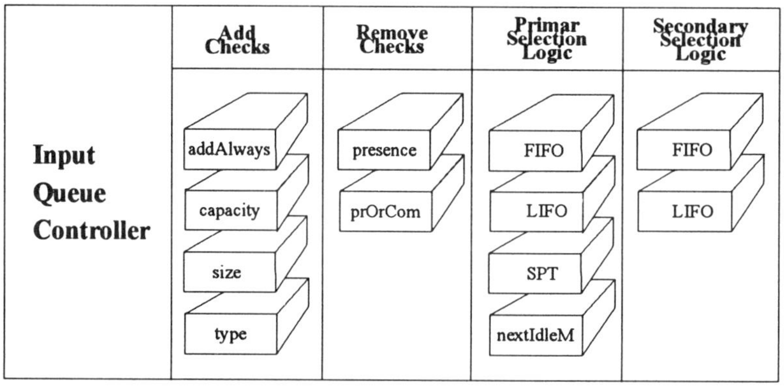

Figure 5: Alternative Logics for Input Queue Controller within a Workstation Controller

completion of the production order).

- Primary Selection Logic (part selection) - *FIFO* (first in first out), *LIFO* (last in first out), *SPT* (shortest processing time), and *nextIdleM* (select the part whose next processing machine is idle).
- Secondary Selection Logic (tie breaker) - *FIFO* (first in first out), and *LIFO* (last in first out).

In our control framework the internal implementation of a logic algorithm is transparent to the modeler so long as the controller can respond with appropriate feedback to a posed question. In order to answer the posed question, the logic algorithm might use complicated artificial intelligence techniques, simple calculations, or a set of if-then rules. Neither the user nor the model developer should be concerned about the internal structure of the control logics.

5 ONGOING RESEARCH

When this modeling environment is used for manufacturing system reconfiguration, the phases involved in responding to particular questions posed are often executed in an iterative fashion. Even the determination of the problem and the selection of the appropriate analysis/optimization tool requires extensive expertise. An intelligent advisor would be of great

An intelligent advisor would be of great benefit to non-modeling specialists (e.g., manufacturing managers) to determine a structured problem from a given set of symptoms and select an analysis/optimization tool to address such a problem (Delen et al. 1996). Furthermore, such an intelligent advisor can also help the user to analyze the output from several iterations and to determine if further investigation is needed. One of the main ongoing research efforts is directed at conceptualizing, designing, and implementing such an intelligent advisor.

ACKNOWLEDGMENTS

Support for this research is being provided by the National Science Foundation through Grant No. DMI-9400572. We also wish to acknowledge the support provided by the AT&T Foundation and the Oklahoma Center for the Advancement of Science and Technology (OCAST) during the formative period of this research. This work could not have progressed to its current level had it not been for the significant contributions of Prof. Joe Mize whose efforts are gratefully acknowledged. The efforts of former research assistants, especially Manoj Duse and Jagannath Gharpure, are also acknowledged.

REFERENCES

Adiga, S. 1989. Software Modelling of Manufacturing Systems: A Case for an Object-Oriented Programming Approach. *Annals of Operations Research* 17: 363-378.

Bhuskute, H. C., M. N. Duse, J. T. Gharpure, D. B. Pratt, M. Kamath, and J. H. Mize. 1992. Design and implementation of a highly reusable modeling and simulation framework for discrete part manufacturing systems. In *Proceedings of the 1992 Winter Simulation Conference*, ed. J. J. Swain, D. Goldsman, R. C. Crain, and J. R. Wilson, 680-688. Piscataway, New Jersey: Institute of Electrical and Electronics Engineering.

Bodner, D. A., S. D. Schneider, S. Narayanan, U. Sreekanth, T. Govindaraj, L. F. McGinnis, and C. M. Mitchell. 1993. Object-Oriented Modeling and Simulation of Automated Control in Manufacturing. In *Proceedings of the 1993 IEEE Conference on Robotics and Automation*, 3: 83-88. Atlanta, Georgia.

Budd, T. 1991. *An Introduction to Object-Oriented Programming.* Reading, Massachusetts: Addison-Wesley Publishing Company.

Delen, D., D. B. Pratt, and M. Kamath. 1996. A New Paradigm for Modeling of Complex Manufacturing Systems. In *Proceedings of The First Conference on Intelligent Manufacturing Systems*, ed. M. Ozbayrak, H. Taskin, E. Oztemel, 27-37. Sapanca-Sakarya, Turkey.

Duse, M. N., J. T. Gharpure, H. Bhuskute, M. Kamath, D. B. Pratt, and J. H. Mize. 1993. Tool-Independent Model Representation. In *Proceedings of the 2nd Industrial Engineering Research Conference*, 700-704. Norcross, GA.: Institute of Industrial Engineers.

Glassey, C. R., and S. Adiga. 1989. Conceptual design of a software object library for simulation of semiconductor manufacturing systems. *Journal of Object Oriented Programming* 2, 4: 39-43.

ElMaraghy, H. A. and T. Ravi. 1992. Modern Tools for the Design, Modeling, and Evaluation of Flexible Manufacturing Systems. *Robotics & Computer Integrated Manufacturing* 9, 4/5: 335-340.

Goldberg, A. and B. Robson. 1989. *Smalltalk-80: The Language*. Reading, MA.: Addison-Wesley.

Kamath, M., D. B. Pratt, and J. H. Mize. 1995. A Comprehensive Modeling and Analysis Environment for Manufacturing Systems. In *Proceedings of the 4th Industrial Engineering Research Conference*, 759-768 Norcross, GA.: Institute of Industrial Engineers.

Kamath, M., D. B. Pratt, D. Delen, S. Sivaramakrishnan, B. Krishnamoorthy, and R. Fernandes. 1996. An Advanced Modeling Environment to Support Rapid Analysis and Reconfiguration of Agile Manufacturing Systems. Submitted to *IIE Transactions.*

LeFrancois, P. and B. Montreuil. 1994. An Object-Oriented Knowledge Representation for Intelligent Control of Manufacturing Workstations. *IIE Transactions* 26, 1: 11-26.

Leung, Y. and R. Suri. 1990. Performance Evaluation of Discrete Manufacturing Systems. *IEEE Control Systems Magazine*, June, 77-86.

Mize, J. H., H. C. Bhuskute, D. B. Pratt, and M. Kamath. 1992. Modeling of Integrated Manufacturing Systems Using an Object-Oriented Approach. *IIE Transactions* 24, 3: 14-26.

Narayanan, S., D. A. Bodner, and C. M. Mitchell. 1992. Object-Oriented Simulation to Support Modeling and Control of Automated Manufacturing Systems. In *Proceedings of the 1992 Western Multi-Conference*, 1-15. San Diego, CA.: Society for Computer Simulation.

ParcPlace. 1994. *VisualWorks Release 2.0 User's Guide*. ParcPlace Systems, Inc., CA.

Pratt, D. B., P. Farrington, C. Basnet, H. Bhuskute, M. Kamath, and J. H. Mize. 1994. The Separation of Physical, Information, and Control Elements for

Facilitating Reusability in Simulation Modeling. *International Journal in Computer Simulation*, 4: 327-342.

Suri, R. and S. de Treville. 1993. Rapid Modeling: The Use of Queueing Models to Support Time-Based Competitive Manufacturing. *Operations Research in Production Planning and Control*, 21-30. Springer-Verlag.

Thomasma, T., and O. M. Ulgen. 1988. Hierarchical, Modular Simulation Modeling in Icon-Based Simulation Program Generators for Manufacturing. In *Proceedings of the 1988 Winter Simulation Conference*, ed. M. Abrams, P. Haigh, J. Comfort, 254-262. Piscataway, New Jersey: Institute of Electrical and Electronics Engineering.

Thomasma, T., Y. Mao, and O. M. Ulgen. 1990. Manufacturing Simulation in Smalltalk, *Object-Oriented Simulation*, ed. A. Guasch, 93-97. SCS.

AUTHOR BIOGRAPHIES

DURSUN DELEN is a Research Associate in the Center for Computer Integrated Manufacturing within the School of Industrial Engineering and Management at Oklahoma State University, Stillwater, OK. He received a B.S. degree in Industrial Engineering from the Istanbul Technical University, Istanbul, Turkey, in 1986, and an M.S. degree in Industrial Engineering from the Yildiz University, Istanbul, Turkey, in 1988. He is currently completing his Ph.D. degree in Industrial Engineering and Management at Oklahoma State University. He has three years of industrial experience in information systems analysis and design. His research interests include manufacturing systems modeling, discrete event simulation, object-oriented modeling, and expert systems. He is a member of Alpha Pi Mu, IIE, and INFORMS.

DAVID B. PRATT is an Assistant Professor in the School of Industrial Engineering and Management at Oklahoma State University, Stillwater, OK. He holds B.S., M.S., and Ph.D. degrees in Industrial Engineering from Oklahoma State University. He has twelve years experience in the petroleum, aerospace and pulp and paper industries. His research and teaching interests include manufacturing systems modeling, enterprise integration, and the strategic implications of CIM. He is a registered Professional Engineer in Oklahoma, a certified Fellow in Production and Inventory Management, and an ASQC Certified Quality Engineer. He is a member of IIE, NSPE, APICS, INFORMS, and ASQC.

MANJUNATH KAMATH is an Associate Professor in the School of Industrial Engineering and Management and Director of the Center for Computer Integrated Manufacturing at Oklahoma State University, Stillwater, OK. He received the B.Tech. Degree in Mechanical Engineering from the Indian Institute of Technology, Madras, India, in 1982, the M.E. degree in Automation from the Indian Institute of Science, Bangalore, India, in 1984, and the Ph.D. degree in Industrial Engineering from the University of Wisconsin-Madison, in 1989. His primary areas of interest are stochastic modeling and queueing theory, analytical performance modeling of manufacturing systems, object-oriented modeling and simulation of discrete-event systems and Petri nets. He is a member of IEEE, IIE, and INFORMS.

Proceedings of the 1996 Winter Simulation Conference
ed. J. M. Charnes, D. J. Morrice, D. T. Brunner, and J. J. Swain

ENTERPRISE MODELING WITHIN AN ENTERPRISE ENGINEERING FRAMEWORK

Donald H. Liles
Adrien R. Presley

Automation & Robotics Research Institute
The University of Texas at Arlington
7300 Jack Newell Boulevard South
Fort Worth, TX 76118, USA

ABSTRACT

The role of abstraction, or modeling, is a major element in Enterprise Engineering. Enterprise engineering deals with the analysis, design, implementation and operation of an enterprise. The Enterprise Engineer addresses a fundamental question: "how to design and improve all elements associated with the total enterprise through the use of engineering and analysis methods and tools to more effectively achieve its goals and objectives". This paper describes a describes a multi-view reference architecture for modeling an enterprise. It presents a modeling scheme under development which supports the architecture and acts as a tool for Enterprise Engineering.

1 INTRODUCTION

Enterprise Engineering is the body of knowledge, principles, and practices having to do with the analysis, design, implementation and operation of an enterprise. In a continually changing and unpredictable competitive environment, the Enterprise Engineer addresses a fundamental question: "how to design and improve all elements associated with the total enterprise through the use of engineering and analysis methods and tools to more effectively achieve its goals and objectives".

There are several world view assumptions present in enterprise engineering. The first assumption is that the enterprise can be viewed as a complex system. This is necessary because systems in organizations are systems of organized complexity. Complexity is the result of the multiplicity and intricacy of man's interaction with other components of the system. Secondly, the enterprise is to be viewed as a system of processes. These processes are engineered both individually and holistically. The final assumption is the use of engineering rigor in transforming the enterprise. The enterprise engineering paradigm views the enterprise as a complex system of processes that can be engineered to accomplish specific organizational objectives. In the Enterprise Engineering paradigm, the enterprise is viewed as a complex system of processes that can be engineered to accomplish specific organizational objectives. Enterprise Engineering recognizes the ever-changing organic nature of the enterprise, and therefore has a valid world view or paradigm (Liles et al. 1995).

These assumptions have implications in enterprise modeling. Models are abstractions of real life systems. Models are created to assist an analyst extract requisite details of the system in order to gain a better understanding of the complex system. An enterprise model is a symbolic representation of the enterprise and the things that it deals with. An enterprise model contains representations of individual facts, objects, and relationships that occur within the enterprise. According to Burkhart (1992), the description of these interactions is the basis for measuring how well the activities of the enterprise work together. The selection and design of enterprise processes for the effective cooperation is a prime objective of enterprise engineering. Enterprise models can assist the goal of Enterprise Engineering by helping to represent and analyze the structure of activities and their interactions.

This paper presents an enterprise modeling scheme which supports a process centered approach to the analysis and design of the enterprise. The scheme, currently under development, utilizes concepts of object-oriented and agent modeling techniques. An innovative feature of the scheme is the ability to present federated activity, process, business rule, organization, and resource views of the process. The paper will first discuss an enterprise modeling architecture which the scheme supports.

2 ENTERPRISE ARCHITECTURE

An enterprise architecture is a "blueprint" or "picture" which assists in the design of an enterprise. An enterprise architecture must define three things. First, what are the activities that an enterprise performs? Second, how should these activities be performed? And finally, how should the enterprise be constructed?

An enterprise is a collection of enterprise activities organized into a set of business processes which cooperate to produce desired enterprise results (Presley 1993). In the context of this paper, an enterprise activity is defined as any organized behavior which transforms inputs into outputs. It is proposed that enterprise activities are the basic building blocks of an enterprise and, furthermore, that enterprise activities become useful only when organized into business processes.

The architecture takes a systems view of an enterprise in which an enterprise is seen as a system which takes in inputs and produces outputs under some set of environmental constraints. Figure 1 shows several sets of enterprise activities (boxes) logically organized into business processes (shaded ellipses). The business processes are organized into an enterprise represented by the larger box. At this high level of abstraction, the enterprise itself is represented as a system which takes inputs and transforms them into output using available resources under the bounds of certain constraints.

The figure also shows the interaction of three process categories. It is proposed that business processes naturally fall into three categories: (1) those processes which transform external constraints into internal constraints, (2) those processes which acquire and prepare resources, and (3) those processes which use resources to produce enterprise results. In the figure, a category 1 process (BP8) takes constraints from the environment and through some process converts them into a set of constraints which will be used to constrain other processes. A category 2 process (BP7) likewise takes in resources from outside of the system and prepares them for use by a category 3 process (BP5).

An enterprise must be viewed from several perspectives if it is to be fully described and understood (Barnett 1994; ESPIRIT Consortium AMICE 1991). Previous work in the development of architectures by the Automation & Robotics Research Institute (Presley et al. 1993) describes a five view approach. The Business Rule (or Information) View defines the entities managed by the enterprise and the rules governing their relationships and interactions. The Activity View defines the functions performed by the enterprise (*what* is done) while the Business Process View defines a time sequenced set of processes (*how* it is done). The resources and capabilities managed by the enterprise are defined in a Resource View. Finally, the Organization View is used to define how the enterprise organizes itself and the set of constraints and rules governing how it manages itself and its processes.

This research proposes that unless all of these views are included, a comprehensive model of the enterprise is not possible. A comprehensive enterprise model addresses a major shortcomings of current modeling techniques: that of disjoint modeling of views. Traditional process modeling methodologies typically emphasize one aspect or view of an enterprise over others. Each aspect is modeled from vastly different

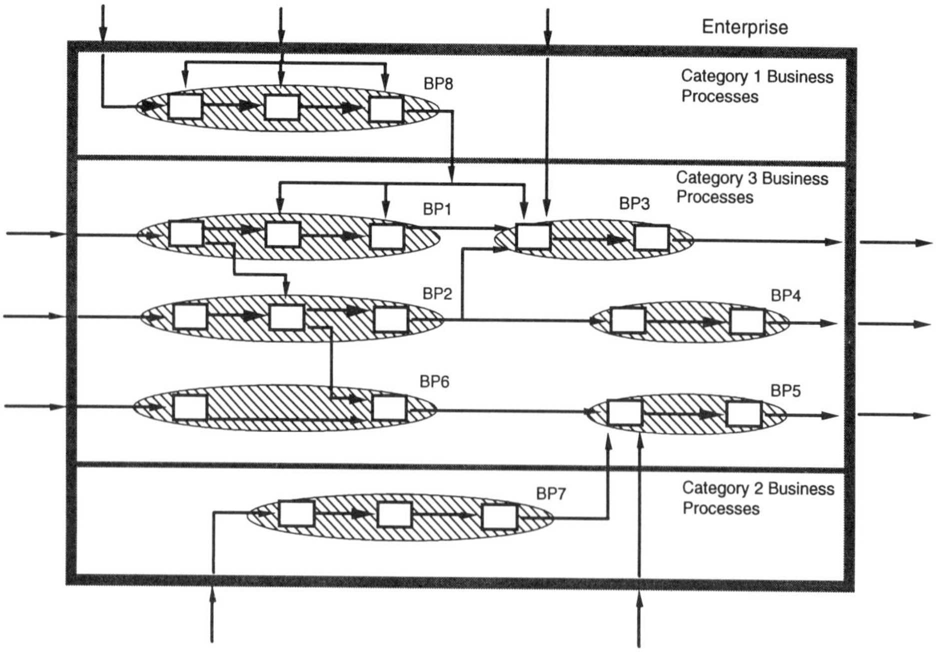

Figure 1: Enterprise as Collection of Business Processes

conceptual perspectives. Separation of the aspects of a business process into a number of separate models is an unnatural representation method. The distinct pieces of a business process exist and have meaning as a single unit.

This does not, however, mean that all these views must be present in all models. A model is an abstract representation of reality which should exclude details of the world which are not of interest to the modeler or the ultimate users of the model. Models are developed to answer specific questions about the enterprise. As long as the model answers the questions for which it was developed, only one or a few of these views may be sufficient.

3 ENTERPRISE MODELING SCHEME

In this section, the modeling scheme being developed to address the needs of the enterprise modeling architecture is discussed. The concentration of this research is on higher level enterprise processes, although the scheme is applicable at all levels of the enterprise. The model is developed by progressively specifying the five views, starting with the business rule view and ending with the organizational view. Feedback and refinement of previously defined views is accomplished to ensure consistency among the views.

The scheme is built upon the IDEF modeling methods (Mayer, Painter, and DeWitte 1992). IDEF is a suite of modeling methods which developed out of the Air Force's Integrated Computer Aided Manufacturing (ICAM) project in the 1980's. Each of the IDEF methods provides a set of modeling syntax and steps for describing a particular perspective of an enterprise. The IDEF suite provides for functional modeling (IDEF0), information modeling (IDEF1), data modeling (IDEF1x), systems dynamics modeling (IDEF2), process description capture (IDEF3), object oriented design (IDEF4), and ontology capture (IDEF5), among others. The modeling scheme being described makes use of IDEF0, IDEF3, and IDEF5.

A holon or agent based approach to identifying and representing the activities and resources is used. The term holon was first proposed by Arthur Koestler (Koestler, 1989) as the basic unit for modeling biological and social systems in his book *The Ghost in the Machine*. The term, according to Koestler, is intended to describe any entity which is at the same time "a whole unto itself, and a part of other whole(s)." Holons belong to structures which consist of self contained units capable of functioning independently but nevertheless are dependent on other units. This structure, called a holarchy, is a temporary assembly of holons which has a specific set of temporal goals and objectives. The strength of a holarchy lies in its ability to construct highly complex, resource efficient systems which are highly resilient to internal and external disturbances and are adaptable to changes in the environment. As can be seen, the structure of an enterprise can be considered to a holarchy.

3.1 Business Rule View

A business rule model identifies the objects of interest in a particular domain and their relationships. In this way, the business rule view is closely related to the concept of an ontology. According to Benjamin, et al. (1995), "an ontology is a description of the kinds of things, both physical and conceptual, that make up a given domain, their associated properties, and the relationships that hold among them as represented by the terminology in that domain". Within the IDEF suite, the IDEF5 ontology capture method (Benjamin et al. 1994) was developed for this purpose. It facilitates the collection of knowledge about physical and conceptual objects along with their associations. It provides facilities for diagrammatic representations of an ontology. In addition, a structured text language for detailed ontology characterization is present. Figure 2 shows a small part of a model for an enterprise. The circles represent kinds (roughly equivalent to class or type) with arrows representing relations. As can be seen both tangible things (Production Plan), concepts (Status) and activities (Support Production) can be defined as entities.

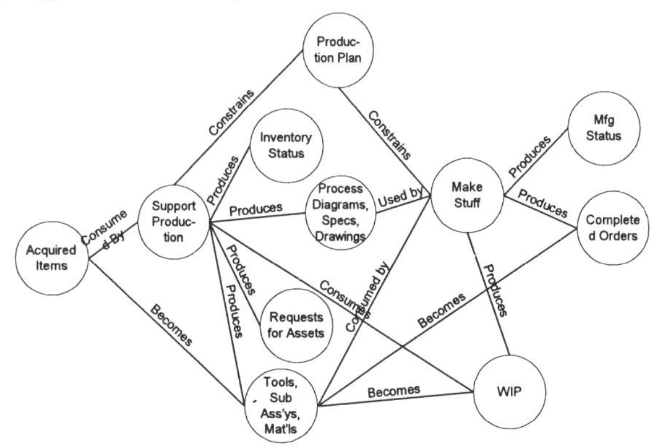

Figure 2: Partial Ontology Model

The ontology model can itself be thought of as the integrated model of the enterprise. The creation of this model is extremely important as the ontology model will form the basis for the other four views. Entities and relationships for all views are defined within the business rule view. Each of the views is built by extracting the entities and relationships particular to that view. This helps to ensure integration of the views and eliminates redundancy in building the views. For example, a resource entity is defined once. Properties and relationships for each view are progressively defined as the views are developed. Even as these new properties and relationships are defined, there will be only one instance and definition of the resource. This is contrasted to traditional techniques where the resource is defined in the activity view, redefined within an information view, etc.

The model is created through knowledge acquisition methods such as interviews and reviews of existing models and documents. The modeling scheme and method provides a limited set of modeling elements of predefined objects and relations for creating an enterprise model.

3.2 Activity View

An IDEF0 activity model is created based on the business rule model created in the previous step. IDEF0 is used to represent the functional (i.e., activity or process oriented) framework of a system. There are five elements to the IDEF0 functional model (see figure 3): the activity (or process) is represented by boxes; inputs are represented by the arrows flowing into the left hand side of an activity box; outputs are represented by arrows flowing out the right hand side of an activity box; the arrows flowing into the top portion of the box represent constraints or controls on the activities; and the final element represented by arrows flowing into the bottom of the activity box are the mechanisms that carry out the activity (Mayer et al. 1992).

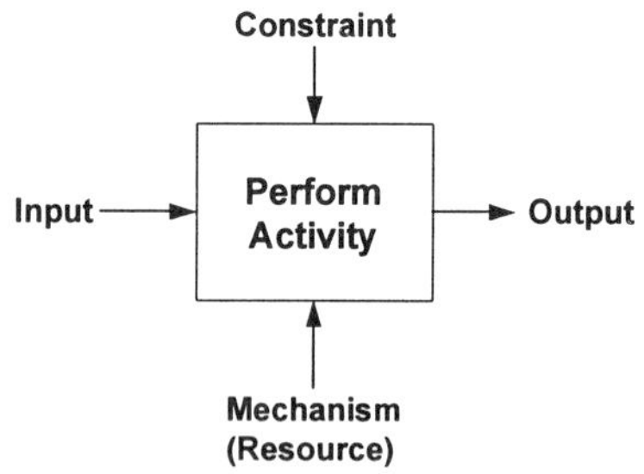

Figure 3: IDEF0 Representation

Another characteristic of the IDEF0 modeling technique is that each activity and the ICOMs can be decomposed (or exploded) into more detailed levels of analysis. This characteristic is especially useful in enterprise modeling where details about lower level activities can be captured, but at the same time, be hidden from models of the enterprise at higher, more abstract levels. This can be thought of as equivalent to the development of hierarchical simulation models.

The use of the IDEF5 model will enforce a level of structure to the standard IDEF0 process. The model is bounded by identifying what is to be included in the model. A process can be considered as a special relationship that exists among objects. In this scheme, however, processes are themselves defines as objects. This allows for the capture of additional knowledge and the ability make inferences about the processes. The identified processes are used to create a list of candidate activities. The objects and activities are defined in additional detail, with the scheme enforcing some properties which must be defined for the model. This will be done by defining a small core set of properties for activities, mechanisms, constraints, and objects manipulated by activities.

The identification and specification of the IDEF0 activities from the IDEF5 model is shown conceptually in figure 4. In this example, an activity called "Fill Orders" and the ICOMs associated with it are identified. The activity model created in this activity will be identical to a "standard" IDEF0 model which would show 3 to six such activity boxes and the interactions of the activities as defined by the flow of arrows among the activities.

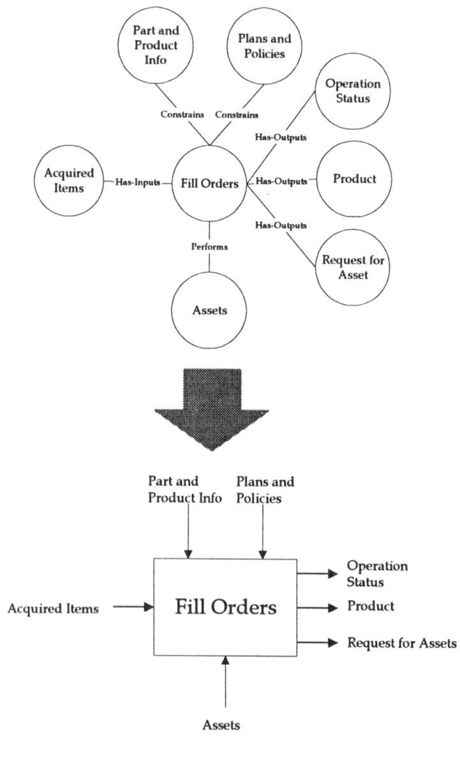

Figure 4: Deriving an IDEF0 Activity Box

At this point actual mechanisms are not identified, but rather, capabilities or kinds of mechanisms required are identified. As part of this activity the ontology specification for activity entities are updated. Properties defined include a description of the activity, and required capabilities.

3.3 Resource View

The Resource View defines two basic aspects: the resources necessary to accomplish activities (and therefore acts as a capability model), and how those resources are organized and "owned" by the organization. To form the resource view, actors and the process they perform are coupled, in essence forming the holons described earlier.

This coupling of entity and process allows for attributes and performance characteristics of interest to be represented in a more holistic manner. It is proposed that only when a particular process is assigned to a particular resource can the values of these performance characteristics be identified. This allows for increased reasoning to be applied. The assignment is done through user defined selection rules with the aid of available analysis tools. Once the agents to perform the activities are identified, the performance of the enterprise can be defined.

As with this the Activity View, the Resource View is built from the Business Rule View. Two classes are especially important here: Activity and Actor. Also important is the relationship of the two as expressed by attributes named Activity-Performance and Actor-Performance. Both of the performance attributes can be of two types: capability or actual performance. The performance capability required by an activity will be matched to the performance capabilities available in potential actors. These will be evaluated on a set of measures as determined by a decision maker to select the best alternative. Common kinds of performance include cost and cycle time.

There are many types of analysis techniques which could be applied. The goal of this research is to develop a modeling scheme to support both manual and automated analysis and design of enterprise processes. The ontology based approach is being pursued to specifically facilitate the use of automated reasoning tools. Automated analysis and design would be supported by advanced techniques such as artificial intelligence and simulation techniques, especially in the evaluation of resources and alternate ways of performing a process. The ontology based approach is closely related to object oriented methods of process modeling and analysis approaches described by several authors. In the area of simulation, for instance, Pratt, Mize et al. (Mize et al. 1992, Pratt, Mize and Kamanth 1993) describe an object oriented approach to building simulation models. The modeling method described here could be adapted or interfaced to work with such simulation approaches. The advantage to be gained is in the availability of a federated model from which various analysis techniques could be built.

3.4 Business Process View

This step creates the business process view for the actor described in the previous step. The business process view defines the time ordered sequence in which a process is executed. It essentially describes the behavior of the actor. The process view is not specified until the agent is defined because for the purpose of designing an enterprise, we are first interested in the activity view ("what is done"). Once that is defined, we are now interested in the behavior that allows for the activity to be accomplished. That is, knowing "what needs to be done?" for an enterprise, we are now interested in the "what can it do?" when selecting resources. The holon approach taken with this model in which each resource can perform an activity any way it seems fit, means that the "how does it do it" (internal behavior) is not of interest until the agent is defined. The behavior is defined by the set of internal processes used by the holon. For any agent, several process scenarios can be created. Examples include alternate ways of performing a process based on existing constraints, and the tracing of alternate inputs through the agent. The specification of the process model is accomplished by again examining the IDEF5 Business Rule Model. It may be necessary to develop additional IDEF5 submodels to define information which was not previously available. This process view is specified using the IDEF3 Process Description Capture Method (Mayer, Painter, and DeWitte 1992). The method consists of process flow diagrams and elaboration diagrams. One of the appealing aspects of IDEF3 is its relationship to the IDEF5 ontology capture method through the use of object state transition diagrams. Additionally, there are commercial tools available for the development of simulation models from IDEF3 process models. An example of an IDEF3 model for a process to ship a customer order is shown in figure 5.

3.5 Organization View

The organization view defines the reporting and constraint structures put in place to guide the performance of entities and activities. Both of these aspects is defined by examining the IDEF5 business rule model. A limited set of predefined kinds and relations are provided within the scheme to assist in developing this view. The modeler would, by identifying the instances of each kind, complete the model. The result of this activity is an Organization View specified in IDEF5. It is a simple matter to then display this information in a form familiar to the user, such as a standard organization chart.

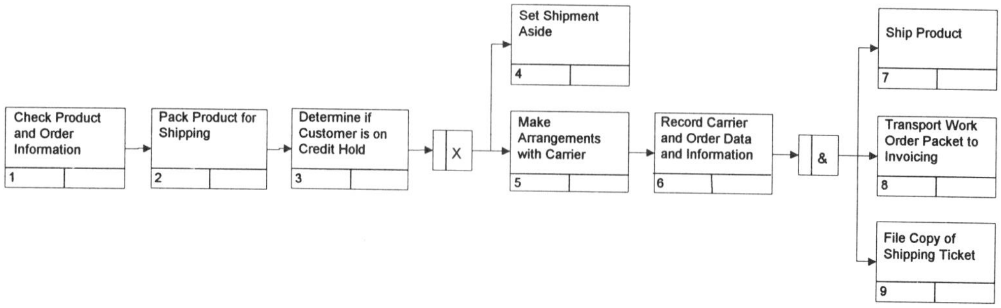

Figure 5: Example IDEF3 Flow Diagram

Many of the entities defined in the other views also participate in the organization view. However, several entities have been defined specifically to support the organization view. Included here are planning objects such as goals, plans, policies, and measures. The roles and positions agents in the enterprise play are identified. These are then linked to each other, activities, and planning objects through a set of management and organization links (Fox, Barbuceanu and Gruninger 1995).

4 CONCLUSION AND FUTURE DIRECTIONS

This paper has presented an architecture for modeling an enterprise and a representation scheme designed to support the architecture. The development of the representation scheme was guided by the need to facilitate both manual and automated analysis and design of enterprise processes. Research is being conducted into possible techniques with which the scheme could interface. The results of this research will provide a valuable tool for the modeling of enterprises.

Perhaps more importantly, this research is being conducted within the larger framework of the development of the discipline of Enterprise Engineering. A discipline has six basic characteristics: (1) a focus of study, (2) a world view or paradigm, (3) a set of reference disciplines used to establish the discipline, (4) principles and practices associated with the discipline, (5) an active research or theory development agenda, and (6) the deployment of education and promotion of professionalism (Liles et al. 1995). This research is furthering the goal of developing a tool for the application of the principles and practices of this emerging discipline. It is also a component of the active research program being conducted by the Automation & Robotics Research Institute in Enterprise Engineering.

ACKNOWLEDGMENTS

Research for this paper is funded in part by the National Center for Manufacturing Sciences, the State of Texas Advanced Technology Program Grant 003656-036 and by the National Science Foundation sponsored Agile Aerospace Manufacturing Research Center.

REFERENCES

Barnett, W., A.R. Presley, M.E. Johnson, and D.H. Liles. 1994. An Architecture for the Virtual Enterprise. In *Proceedings of the IEEE International Conference on Systems, Man, and Cybernetics*, 506-511. Institute of Electrical and Electronics Engineers. Piscataway, New Jersey.

Benjamin, P. C., R. J. Mayer, T. M. Blinn, C. Marshall, and R. Tye. 1995. Toward an Environment for Collaborative Enterprise Engineering. In *Proceedings of the Society for Enterprise Engineering (SEE) Conference*. Society for Enterprise Engineering, Kettering, Ohio.

Benjamin, P. C., C. Menzel, R. J. Mayer, F. Fillion, M. T. Futrell, P.S. DeWitte, and M. Lingineni. 1994. *Ontology Capture Method (IDEF5)*. College Station, TX: Knowledge Based Systems, Inc.

Burkhart, R. 199). Process-based Definition of Enterprise Models. In *Enterprise Integration Modeling: Proceedings of the First International Conference*, ed. C. Petrie, 229-238. Cambridge, Massachusetts: The MIT Press.

Fox, M. S., M. Barbuceanu and M. Gruninger. 1995. An Organisation Ontology for Enterprise Modeling: Preliminary Concepts for Linking Structure and Behaviour. In *Proceeding of the Fourth Workshop on Enabling Technologies: Infrastructures for Collaborative Enterprises*, West Virginia University.

ESPIRIT Consortium AMICE. 1991. *Computer Integrated Manufacturing: Open Systems Architecture*. Berlin: Springer-Verlag.

Koestler, A. 1989. *The Ghost in the Machine*. London: Arkana Books.

Liles, D.H., M.E. Johnson, L.M. Meade, and D.R. Underdown 1995. Enterprise Engineering: A Discipline? In *Proceeding of the Society for Enterprise Engineering Conference*. Society for Enterprise Engineering, Kettering, Ohio.

Mayer, R. J., M. Painter and P. DeWitte. 1992. IDEF Family of Methods for Concurrent Engineering and Business Re-engineering Applications. College Station, TX: Knowledge Based Systems, Inc.

Mize, J. H., H. C. Bhuskute, D. B. Pratt and M. Kamath. 1992. Modeling of Integrated Manufacturing Systems Using an Object-Oriented Approach. *IIE Transactions* 24:14-25.

Pratt, D. B., J. H. Mize and M. Kamath. 1993. A Case for Bottom-up Modeling. In *Proceedings of the 2nd Industrial Engineering Research Conference*, 430-434. Institute of Industrial Engineers, Atlanta, Georgia.

Presley, A.R., B.L Huff, and D.H. Liles. 1993. A Comprehensive Enterprise Model for Small Manufacturers. In *Proceedings of the 2nd Industrial Engineering Research Conference*, 430-434. Institute of Industrial Engineers, Atlanta, Georgia.

AUTHOR BIOGRAPHIES

DONALD H. LILES is Professor of Industrial and Manufacturing Systems Engineering at The University of Texas at Arlington, and Associate Director of the Automation & Robotics Research Institute. His primary research interest is in the emerging discipline of Enterprise Engineering. He is active in numerous technical and professional organizations, including the Institute of Industrial Engineers, Alpha Pi Mu, and the Society for Enterprise Engineering.

ADRIEN PRESLEY is a Research Industrial Engineer with the Automation & Robotics Research Institute (ARRI) of The University of Texas at Arlington (UTA). He is currently pursuing his Ph.D. degree from the Industrial and Manufacturing Systems Engineering department at UTA, where he also received his MS degree. His research interests are in enterprise modeling and analysis, management of technology, and artificial intelligence applications in manufacturing.

Proceedings of the 1996 Winter Simulation Conference
ed. J. M. Charnes, D. J. Morrice, D. T. Brunner, and J. J. Swain

AN OVERVIEW OF CIM ENTERPRISE MODELING METHODOLOGIES

Ganesh Kateel
Manjunath Kamath
David Pratt

Center for Computer Integrated Manufacturing
School of Industrial Engineering and Management
Oklahoma State University
Stillwater, Oklahoma 74078, U.S.A.

ABSTRACT

Computer integrated manufacturing (CIM) systems are increasingly being used as weapons by manufacturing enterprises in competitive business environments. The complicated nature of these systems and the high initial investment requirements have necessitated their accurate modeling. A number of models, modeling methodologies, and modeling tools have been developed and used for this purpose. We first present a brief overview of several CIM models as well as modeling tools and methods. Many of the models are said to emphasize only a part of the system. A concern in the research community is that these models must be integrated. We conclude the paper by examining the rationale and feasibility of integrating the different models and/or creating integrated models.

1 INTRODUCTION

Many manufacturing firms, zealous to maintain their competitiveness in the world class manufacturing environment, attempt to use state of the art technology in their operations. Advancements in computer and information technology have led to the development of computerized applications in many functional areas of manufacturing enterprises. As a result, we have a plethora of three letter acronyms starting with C (for Computerized) such as CAD, CAM, and CAE along with others such as MRP, MRP-II, and SFC which deal with different aspects of manufacturing. The design of such systems is generally developed in a fashion that emphasizes 'local' solutions mostly overlooking the overall organizational effectiveness. This together with the use of heterogeneous databases and incompatible computer operating systems, has led to 'islands of automation' which suffer from data inconsistencies and lack of control of the functional interactions between manufacturing application systems.

Traditionally, different functions of enterprises have been modeled with a focus on the performance evaluation of the individual departments (referred to in this paper as local optimization) rather than the effectiveness of the entire organization. But, a modern enterprise requires the departmental goals to be in tune with the organizational goals for its betterment and survival. Designers need tools that will allow them to evaluate the impact of local decisions on the performance of the total enterprise. However, many CIM models and modeling approaches tend to focus only on particular functional parts of the enterprise. Hence, one of the issues that we examine in this paper is the rationale and feasibility of integrating the different models and/or creating integrated models.

The rest of the paper is organized as follows. Several different definitions of CIM are provided in Section 2 as a basis for further discussion. Section 3 provides a general perspective on modeling with an emphasis on CIM modeling. CIM enterprise architectures, modeling tools and methods are briefly discussed in sections 4, 5, and 6 respectively. Section 7 discusses the issues pertaining to integration of models.

2 CIM ENTERPRISES-DIVERSE PERSPECTIVES

Jorysz and Vernadat (1990) state the objective of CIM as the appropriate integration of enterprise operations by means of efficient information exchanges within the enterprise with the help of information technology. This indicates that CIM includes all of the design and manufacturing functions of CAD/CAM, as well as all the business functions such as marketing and accounting. A comprehensive definition of CIM has been given by Doumeingts et al. (1995): "CIM refers to a global approach in an industrial environment which aims at improving industrial performances. This approach is applied in an integrated way to all activities, from designing to delivery and after sale, and uses various methods, means and techniques (computer and automatic) in order to simultaneously improve productivity, decrease costs, meet due dates, increase product quality, secure flexibility at local or global level in a manufacturing system, and involve every actor. In such an approach, economic, social and human aspects are at least as important as technical aspects."

Nagata et al. (1993) describe CIM as "an integrated system which combines the areas of production, marketing and R&D, to manage and operate them under a single management strategy with the support of computers so that the production operation can be efficient and flexible."

A few of the other definitions of CIM are: "The automation and integration of information, processes, and functions in a manufacturing environment, including customers and vendors, with the result being a closed-loop, functionally integrated manufacturing planning and control system" (Weston 1994). "CIM is the integration of computer based manufacturing process, drawing on a common database and communicating via some form of computer networks" (Bessant 1991).

The foregoing discussion indicates that most authors encompass a broad range of enterprise activities in the definition of CIM. So it is difficult and may be inappropriate to view CIM in isolation. It is not surprising to see that many enterprise modeling tools, methods and architectures are being used to model or are being structured to model CIM. Further evidence for this can be found in Ngwenyama and Grant (1994).

Generally the implementation of CIM involves high investment and complex system design issues. Although much research is being carried out regarding the development and implementation of CIM, many problems are yet to be addressed satisfactorily. Ngwenyama and Grant (1994) provide a list of these problems which includes lack of integration, islands of automation, sub-optimization of resources, ad-hoc development, lack of well-defined architecture, confusing definition of CIM and so on. *Many researchers have tried to address these problems by developing good models of CIM because models form a rational basis not only for designing new systems but also for learning about existing systems. Evaluation using models instead of the actual systems allows fast acquisition of knowledge and avoids the risk of costly disruptions to the real system. Models can be used for system optimization, performance analysis and prediction, gathering insight into the system or facilitating learning processes.*

3 CIM MODELING

Modeling is the process of creating an abstraction of a real world system which reflects the system properties to the desired degree of detail (Kochikar and Narendran 1994). Models filter out irrelevant details and represent only information essential to the task. The traditional approach to modeling begins with the identification of the purpose of the model. Then the modeler identifies all system features germane to the purpose, constructs and validates an abstract structure with sufficient scope and complexity so that it can aid in integration of system components and their interaction under suitable assumptions.

The purpose driven, tool dependent modeling described above has the inherent disadvantage that one has to construct different models for different purposes even though the system being represented is the same. Hence, Duse et al. (1992) suggest building robust, tool independent models to represent manufacturing enterprises from which one can extract the information relevant for a particular purpose. They refer to this model as a *Base Model.* This base model is a rich, robust, representation of the system consisting of physical, informational and control elements as well as their relationships.

The above base model concept is a new paradigm in the field of manufacturing systems modeling, and additional research is necessary before researchers fully implement this concept. Meanwhile, constraints such as cost, quality, flexibility, time, size, nature, and complexity make the designing and modeling of CIM a very complicated task and as such requires good modeling methodology and tools. The comprehensive definition of CIM provided by Doumeingts et al. (1995) indicates that CIM systems are remarkably complicated with a multitude of activities carried out by human beings and machines using information flows and control signals. Generalized models and well defined architectures are required to understand the system complexity at a manageable level. Since humans have "bounded rationality" they like to break (such) complicated systems into better manageable pieces for analysis, design and management (Little 1991). Humans prefer to abstract only that part of reality which may be relevant to them. Due to this tendency, most of the modeling approaches have separate models for different aspects of CIM. The above argument follows from systems theory. It is well known that any system, with a very few exceptions, consists of subsystems. These subsystems can be considered to be systems in their own right. Hence, they can be modeled independently. This may not allow us to have a 'big-picture' of the system and it is difficult to move from one model to the other (Savolainen et al. 1995).

The means that we use to model systems are referred to as *modeling tools*. An Entity Relationship (ER) diagram is an example of a tool used for representing the relationship between entities. The procedure to construct models may be termed as *methodology*. The result of using the modeling tools in a manner prescribed by the methodology is a representation or a *model*. In the case of CIM modeling this may be an architecture.

As a preparation for the further discussion about CIM enterprise modeling issues, a brief overview of existing architectures, modeling tools, and modeling methods is presented in the following sections.

4 CIM ENTERPRISE ARCHITECTURES

Architectures of CIM systems contain conceptual models as well as rules that help to translate the models into a working reality (O'Sullivan 1994). O'Sullivan defines architecture as "a body of rules that define those system features which directly affect the manufacturing environment into which the system is placed. These features include system configuration, component locations, interfaces between the system and its environment, and modes of operation." Most authors refer to two types of architecture. The first is a detailed collection of generic information management and automatic control tasks and their necessary functional requirements referred to as *reference* architecture. A *particular* architecture is the instantiation of a reference architecture.

A CIM reference model committee of Purdue University (1989) describes a reference model or architecture as a previously agreed upon or standard definitive document or conceptual representation of a system. The reference model defines requirements common to all implementations but is independent of the specific requirements of any particular implementation. Several reference architectures have been put forth by collaborative research projects, computer manufacturers and individuals. Some of the important architectures frequently referred in the literature are briefly discussed below.

CIM-OSA: CIM-OSA is an Open-System Architecture and has three levels of model derivation viz., requirements definition, design specification and implementation description, four views viz., functional, informational, resource, and organizational views, and three levels of instantiation viz., generic, partial and particular models (Jorysz and Vernadat 1990). This architecture guides the system designer in deciding what is to be implemented to achieve what is required.

ICAM Architecture: ICAM (Integrated Computer Aided Manufacturing) architecture uses tools such as IDEF0 (ICAM DEFinition - Zero) and IDEF1 (ICAM DEFinition - One). A Heirarchial decomposition method is used to define this architecture. The architectures generated in ICAM are not generally available publicly (O'Sullivan 1994).

CAM-I Architecture: The Computer Aided Manufacturing - International (CAM-I) architecture is a general representation of manufacturing enterprises. The functional decomposition method used to create this architecture is prescribed to indicate the details such as company policies and procedures, organizational structure and standards (Doumeingts et al. 1995).

NBS Architecture: The NBS (National Bureau of Standards - now, National Institute of Standards and Technology) architecture uses a hierarchical control approach with five levels of hierarchy viz., factory, shop, cell, workstation and machine. Each level can be broken into more activities. The decomposition is based on procedures, functions, or rules. NBS (NITS) developed this architecture so that manufacturing system vendors could develop products compatible for CIM (O'Sullivan 1994).

IMPACS Architecture: The IMPACS architecture uses IDEF0, Data Flow Diagrams (DFD), Graphe a Resultatis er Activities Intercies (GRAI) Grids and nets, IDEF1x and group technology. IMPACS outlines a cellular architecture. The production cells are controlled by software modules such as dispatcher, scheduler, mover, producer and monitor. These software modules are designed to be compatible with each other even when they are developed by different vendors. O'Sullivan (1994) states that the IMPACS architecture has been widely accepted among manufacturing software vendors as a useful and practical interpretation of the production management system.

5 MODELING TOOLS

Models are built using the modeling tools in the manner prescribed by modeling methodologies. *System modeling tools* refer to techniques used for diagrammatically representing functions or activities (O'Sullivan 1994).

IDEF Modeling Tools: IDEF0, IDEF1, IDEF1x, and IDEF2 are modeling tools developed by the ICAM (Integrated Computer Aided Manufacturing) project of the US Air Force at Soft Tech, Inc. These four main modeling tools, which are complementary to each other, provide functional, informational and dynamic models of the system. IDEF models are used primarily for requirements definition.

IDEF0 Models: IDEF0 is a comprehensive and expressive functional modeling language capable of graphically representing a wide variety of business, manufacturing and other types of enterprise operations to any level of detail. Three important features of IDEF0 technique are activity modeling graphics, gradual exposition of details and disciplined team work. The basic construction used in an IDEF0 model is a function block linked to other function blocks by inputs, outputs, mechanisms and controls. Links between the blocks may be either physical objects, such as material flow, or information flow. IDEF0 models have three important features viz., 'context' which indicates the position that the subject model takes up in the systems hierarchy, 'viewpoint' which refers to the perspective which the model adopts, and the 'purpose' which indicates the reason for existence for the model (Pandya 1995). IDEF0 has been widely used in industry because of its user-friendliness, computer support, rigor and conciseness, and well documented rules and procedures. The static nature of the models produced using IDEF0 has been cited as the major drawback of this tool. A comprehensive treatise on the state of the art of IDEF0 can be found in Colquhoun et al. (1993).

IDEF1 and IDEF1x Models: IDEF1 is a technique for modeling information requirements of a function, in terms of the structure of information. (An overly simplified example: What information should an invoice have?). It is based on the entity-relationship approach developed by Chen (1976). IDEF1x is an extension of IDEF1 and deals with the flow of information. IDEF1x is not widely used due to its drawbacks such as its inability to support composite entity types and the strict procedural requirements such as complete enumeration of attributes of entities before instantiation (Pandya 1995).

IDEF2 Models: IDEF2 models have the capability to describe as well as analyze a system (Banerjee and Al-Maliki 1988). However, IDEF2 is an unsupported simulation language (Young and Vesterager 1991). Hence, other simulation languages such as ARENA (Drevna and Kasales 1994) and PROMODEL (Baird and Leavy 1994) are more commonly used.

Structured System Analysis (SSA): (Gane and Sarson 1979). SSA is considered as a data flow approach to systems design. It is also quite effective in representing the flow of physical entities. This technique prompts the user to think in terms of what to accomplish than how. SSA uses hierarchical decomposition similar to IDEF0. However, it is more detailed and software oriented than IDEF0 and is the most preferred modeling tool for data flows (O'Sullivan 1994). This is also referred to in the literature as Data Flow Diagrams (DFD).

GRAI Grids and GRAI Nets: GRAI grids and GRAI nets are the tools developed by the GRAI Laboratories of France to model decision-flow processes in manufacturing environments. In the GRAI grid, various activities are modeled with respect to the decisions and information flows between them while the GRAI net delineates the decision making process itself.

6 MODELING METHODS

Modeling methods are guidelines to combine the modeling tools described above to model a particular system. Some of the methods used for modeling CIM enterprise systems include IDEF, SSADM, SADT, GIM, and CIM-OSA cube.

IDEF Method: This methodology prescribes the integration of IDEF0, IDEF1 and IDEF2 models which describe functional, informational and dynamic model of an enterprise, respectively.

Structured Systems Analysis and Design Methodology (SSADM): SSADM is a procedural framework which was developed specifically for use in system development projects. It uses three modeling tools: data flow diagrams (DFDs), logical data structures (LDSs), and entity life histories (ELHs) to provide function, data, and event views of the systems, respectively (Pandya 1995).

Structured Analysis and Design Technique (SADT): SADT (Ross 1977) is a system analysis and design methodology to represent the structure of the system diagrammatically. The representation begins with a diagram depicting the general description of a system. Modelling then progresses in a top down fashion. This method makes use of a number of graphical and textual tools such as activity diagrams, data diagrams, node lists, and data dictionaries to represent the structure of the system being addressed. The analysts are required to consider activity or function and data views of the system being modeled, thereby encouraging the creation of integrated enterprises. The methodology requires a functional model before considering the physical design. This methodology has computer support in the form of software packages viz., AUTOIDEF0 and SPECIFX.

GRAI Integrated Method (GIM): The ESPIRIT IMPACS project uses an approach which employs tools such as GRAI, IDEF0, IDEF1x, and Group Technology. This combination is called the GRAI Integrated Method (GIM).

CIM-OSA Method: CIM-OSA defines an integrated methodology to design, implement, operate, and maintain an enterprise. The CIM-OSA method deploys many well-accepted ideas and principles to design a CIM enterprise. These include functional decomposition in SADT; function/activity and information modeling used in IDEF, the entity-relationship model and the three schema approach from ANSI/SPARC development for data communications (Jorysz and Vernadat 1991).

Object-Oriented Approach: Object oriented concepts provide a new promising ontology for enterprise and CIM modeling. Ngwenyama and Grant (1994) and Kim et al. (1993) suggest ways to model CIM information systems using an object-oriented approach. The methodology proposed by Kim et al. (1993) consists of an analysis phase and a design phase. In the analysis phase, functional decomposition is employed to define the information flow among the manufacturing functions and their infrastructures. Decomposed functions are represented by functional diagrams. The functional diagrams are then transformed into an object-oriented information model consisting of a class dictionary and class relationship diagrams. Using these, a class dictionary consisting of function classes and entity classes is formed. These class dictionaries can then be translated to a specific data dictionary of an object oriented database management system.

7 MODELS, ARCHITECTURES, AND INTEGRATION

Consideration of the architectures and modeling methods described above reveals that almost all architectures do rely upon multiple views of manufacturing enterprises. The functional and informational views appear in most of the architectures. But most of these architectures have differing additional views. IDEF has a dynamic view. The CIM-OSA architecture has a resource view and an organization view. CAM-I depicts three more perspectives viz., management perspective, computer systems perspective, and physical structure perspective in addition to functional and informational perspectives.

Savolainen et al. (1995) note that a single model will not be able to capture all aspects of a CIM enterprise. Therefore, a complete model is composed of different views, aspects, or perspectives.

Weston (1994) suggests that CIM may be viewed from three different perspectives, viz., from an engineering perspective, from the viewpoint of information system and networking and from the operations view point. He encourages manufacturing organizations to examine CIM in the context of a total organizational perspective. He argues that a single dimensional CIM perspective most likely will produce results that are less than desired, particularly in a global competitive environment.

Brandimarte and Cantamessa (1995), however, note that while the aim of the CIM community is to develop an integrated manufacturing enterprise, the integration between the different domains and cultures which cooperate in this field is difficult to achieve. But it would be advantageous to the CIM community to integrate the different disciplines of CIM modeling, mainly the physical, communication (information) and operational dimensions.

Little (1991) observes that "we indulge in modeling myopia if we believe as system analysts that we can (or should) be building complete models of our system and setting all the control variables. Doing so misses the major opportunities for system improvement that are possible by finding new ways to improve the people on the front lines of the organization by giving them information, training and tools with which to improve our performance." He further comments "the building of more and more complicated models of systems using the same methodologies is likely to yield diminishing returns. Managers face dozens of different problems each day; not just late schedules and excess inventories but also issues such as key people being hired away, roofs that leak, customer dissatisfaction with products, and employee absenteeism. Thus a hundred different models are often needed, not one big model." This seems to be a very strong argument/opinion against building of a single integrated model.

At this point it may be worthwhile to discuss the differences, if any, between development of an integrated model and model integration. The first is the development of one 'big- picture' (global view) from which the detailed local models can be built by 'zooming-in'. Conventionally this may be akin to the top-down hierarchical approach. Most of the models discussed above fall in this category. The second is the pooling of the different models and tying them with some common thread. The argument in favor of the first approach (integrated model) is that the local models built with the big-picture in mind will not be striving towards local optimization, rather they would contribute towards the system effectiveness. But the biggest drawback here is that it would be a Herculean task to use this concept in existing systems which are already in operation. It may be easier to adopt this while establishing new systems. A serious challenge for the designers of such models is that those models should be flexible and must be robust to adapt to changing environments without losing focus.

The argument in favor of the second approach (model integration) is that this is a plausible integration methodology in the case of existing organizations. But the problem here is that each of the local models developed separately will be myopic and naturally strive towards local optimization. Another practical problem is to make these models communicate with each other, as these might have been developed on different platforms and may operate in different languages.

An attempt at *integrating distributed models* called the ENVISION architecture was presented by Heim (1994). He describes model integration as an alternative to aggregate refinement and decomposition. He provides a conceptual framework to integrate individually and independently functioning models to communicate with each other by acting as message passing objects in client-server relationships. This allows data sharing and coordination of the activities. This also reduces the programming effort and simplifies model validation. The author presents a prototype implementation of the model integration architecture.

Wang et al. (1993) challenge the approach of integrating distributed models built using different tools to represent the various viewpoints. They cite some of the latest developments in this area reported by Gay, et al. (1991). Gay et al. (1991) are said to have described the functional, informational, and dynamic perspectives of a flexible manufacturing system by applying IDEF0, IDEF1 and SLAM II, respectively. Wang et al. (1993) challenge these approaches using the following arguments: first, since the dynamic view point is being modeled using a language different from that used to represent functional and informational view, some actions have to be re-represented in building the dynamic version of the model. Second, the different modeling tools use different syntax. This may result in loss of information especially if different people have modeled the various views. Third, any change in the system will require a significant remodeling effort since the change in one is not automatically reflected in the other. This makes the modeling and remodeling process highly inefficient.

Wang et al. (1993) present a comprehensive methodology called IDEM *(Integrated system DEscription Model)* to overcome this difficulty. IDEM uses an extended IDEF0 methodology to build a functional view of the system. An information view of the model is described by adding an object oriented information model on to the previously defined functional model. A dynamic view of the model is created by associating time attributes to the functional model. The dynamic behavior of the model is governed by defining rule sets. IDEM is developed using a LISP based language called LOOPS.

The first International Conference on Enterprise Modeling Technology (ICIEMT) identified three types of approaches to the problem of syntactic and semantic model integration (Petrie 1992). They are described below.

Master Models are a single reference model from which all other models and instantiations are derived. This is similar to what has been referred to in this paper as an integrated model. *Unified models* are metamodels which translate between models. Generally these models may have similar structures. For example, all the models may be databases which can communicate with each other. *Federated models* are loosely coupled models. Here the individual models may not have same structure. In such a case the interface between the models has to be manually constructed. Hence these models are also referred to as interfaced models.

8 CONCLUSION

The literature reviewed suggests that there are various tools and methods to model CIM enterprises. There appears to be some consensus that there is a need to develop a generic approach to build CIM models which can represent the entire manufacturing organization while simultaneously allowing each function of the enterprise to maintain its uniqueness. Though the present day models do not appear to have completely achieved this character, some of the approaches appear to be quite promising.

ACKNOWLEDGMENT

We wish to acknowledge the support provided by the National Science Foundation for this research through Grant No. EEC-9527493.

REFERENCES

Baird, S. P., and J. J. Leavy. 1994. Simulation modeling using PROMODEL for Windows. In *Proceedings of the 1994 Winter Simulation Conference*, ed. J. D. Tew, S. Manivannan, D. A. Sadowski, A. F. Seila. 527-532. Institute of Electrical and Electronics Engineers, Piscataway, New Jersey.

Banerjee, S. K., and I. Al-Maliki 1988. A structured approach to FMS modelling. *International Journal of Computer Integrated Manufacturing*. 1(2): 77-88.

Bessant, J. 1991. *Managing Advanced Manufacturing Technology*. Manchester: NCC Blackwell,

Brandimarte, P., and M. Cantamessa. 1995. Methodologies for designing CIM systems: A critique. *Computers in Industry* 25: 281-293.

Chen, P.P. 1976. The entity-relationship model: Toward a unified view of data. *ACM Transactions on Data Base Systems*, 1(1), 1976.

CIM reference model committee. 1989. A reference model for computer integrated manufacturing from the viewpoint of industrial automation. *International Journal of Computer Integrated Manufacturing*. 2(2): 114-127.

Colquhoun, G. J., R. W. Baines, and R. Crossley. 1993. A state of the art of IDEF0. *International Journal of Computer Integrated Manufacturing*. 6(4): 252-264.

Doumeingts, G., B. Vallespir, and D. Chen. 1995. Methodologies for designing CIM systems: A survey. *Computers in Industry*. 25:263-280.

Drevna, M. J. and C. J. Kasales 1994. Introduction to ARENA. In *Proceedings of the 1994 Winter Simulation Conference*, ed. J. D. Tew, S. Manivannan, D. A. Sadowski, A. F. Seila. 77-80. Institute of Electrical and Electronics Engineers, Piscataway, New Jersey.

Duse, M., J. Gharpure, H. Bhuskute, M. Kamath, D. B. Pratt, and J. H. Mize. 1992. Tool independent model representation. In *Proceedings of the 2nd Industrial Engineering Research Conference*, 700-704. IIE, Norcross, GA.

Gane, C., and T. Sarson. 1979. *Structured Systems Analysis-Tools and Techniques*. Englewood Cliffs, NJ: Prentice Hall International.

Gay, R. K. L., R. Lim , and L. W. Shing. 1991. Using IDEF methodology for functional, information and dynamic modelling of a FMS. In *Proceedings of International conference on Computer integrated Manufacturing*. ed. B.S. Lim. 97-100. World Scientific, Singapore.

Heim, J. A. 1994. Integrating distributed models: the architecture of ENVISION. *International Journal of Computer Integrated Manufacturing*. 7(1):47-64.

Jorysz, R. H. and F. B. Vernadat. 1990. CIM-OSA part 1: total enterprise modeling and function view. *International Journal of Computer Integrated Manufacturing*, 3(3) & (4):144-156.

Kim C., K. Kim, and I. Choi. 1993. An object oriented information modeling methodology for manufacturing information systems. *Computers and Industrial Engineering*. 24(3):337-353.

Kochikar, V. P., and T. T. Narendran. 1994. On using abstract models for analysis of flexible manufacturing systems. *International Journal of Production Research* 32(10):2303-2322.

Little, J. D. C. 1991. Are there 'laws' of manufacturing? *Manufacturing Systems*. Washington D. C.: National Academy Press.

Nagata, T., Y Nagata, and H. Koshimitsu. 1993. Building a CIM system for compound plant: Utilization of distributed processing system. *Computers & Industrial Engineering*. 24(4): 561-569.

Ngwenyama, O. K., and D. A. Grant. 1994. Enterprise modeling for CIM information systems architectures: An object oriented approach. *Computers & Industrial Engineering*. 26(2): 279-293.

O'Sullivan, D. 1994. *Manufacturing System Redesign: Creating the Integrated Manufacturing Environment*, New Jersey: Prentice Hall.

Pandya, K. V. 1995. Review of modeling techniques and tools for decision making in manufacturing management. In *IEE Proceedings on Science, Measurement, and Technology* 142(5): 371-377.

Petrie, C. J. 1992. Introduction In *Proceedings of the First International Conference on Enterprise Modeling* Ed. Petrie, C. J., 1-17. Massachusetts: MIT Press.

Ross, D. 1985. Applications and extensions of SADT. *Computer* April: 25-34.

Savolainen, T., D. Beeckmann, P. Groumpos, and H. Jagdev. 1995. Positioning of modeling approaches, methods and tools. *Computers in Industry*. 25: 255-262.

Wang W., K. Popplewell, and R. Bell. 1993. An integrated multi-view system description approach to approximate factory modeling. *International Journal of Computer Integrated Manufacturing*. 6(3): 165-174.

Weston, Jr. "TED". 1994. Three dimensions of CIM, *Production and Inventory Management Journal*. First Quarter: 59-61.

Young, R.E. and J. Vesterager. 1991. An approach to CIM system development whereby manufacturing people can design and build their own CIM system. *International Journal of Computer Integrated Manufacturing*. 4(5): 288-299.

AUTHOR BIOGRAPHIES

GANESH PAI KATEEL is a Research associate in the Center for Computer Integrated Manufacturing in the School of Industrial Engineering and Management at Oklahoma State University, Stillwater, OK. He received the B.Tech degree in Mechanical Engineering from the Karnataka Regional Engineering College, India and Post Graduate Diploma in Industrial Engineering form the National Institute for Training in Industrial Engineering, Bombay, India. He has two years experience in the CNC machine building industry. He has taught both at undergraduate and graduate levels at Karnataka Regional Engineering College, India for four years. He was the Project Officer for the State of Karnataka, National Technical Manpower Information System, Govt. of India from 1990 to 1994. Currently he is completing his M.S. degree in Industrial Engineering and Management at Oklahoma State University. During 1995 - 96, he taught Production Management at the College of Business, Oklahoma State University.

MANJUNATH KAMATH is an Associate Professor in the School of Industrial Engineering and Management and Director of Center for Computer Integrated Manufacturing at Oklahoma State University, Stillwater, OK. He received the B.Tech. degree in Mechanical Engineering from the Indian Institute of Technology, Madras, India, in 1982, the M.E. degree in Automation from the Indian Institute of Science, Bangalore, India, in 1984, and the Ph.D. degree in Industrial Engineering from the University of Wisconsin-Madison, in 1989. His primary areas of interest are stochastic modeling and queueing theory, analytical performance modeling of manufacturing systems, object-oriented modeling and simulation of discrete-event systems and Petri nets. He is a member IEEE, IIE, and INFORMS.

DAVID B. PRATT is an Assistant Professor in the School of Industrial Engineering and Management at Oklahoma State University, Stillwater, OK. He holds B.S., M.S., and Ph.D. degrees in Industrial Engineering from Oklahoma State University. He has twelve years experience in the petroleum, aerospace and pulp and paper industries. His research and teaching interests include manufacturing systems modeling, enterprise integration, and the strategic implications of CIM. He is a registered Professional Engineer in Oklahoma, a certified Fellow in Production and Inventory Management, and an ASQC Certified Quality Engineer. He is a member of IIE, NSPE, APICS, INFORMS, and ASQC.

Proceedings of the 1996 Winter Simulation Conference
ed. J. M. Charnes, D. J. Morrice, D. T. Brunner, and J. J. Swain

SIMULATION IN THE DESIGN OF GROUND-BASED INTRABAY AUTOMATION SYSTEMS

Thomas Jefferson
Madhav Rangaswami
Gregory Stoner

Intel Corporation
Mail Stop CH3-84
5000 West Chandler Boulevard
Chandler, Arizona 85226 USA

ABSTRACT

Material transport within a semiconductor factory (fab) can be broken into two categories - interbay (bay to bay movement) and intrabay (within bay) movement.

Due to the fragile properties of semiconductors, there are risks associated with any movement of material. These risks range from misprocessed or damaged wafers, to throughput loss. These risks are increased during manual intrabay handling, where several operations may be taking place in a populated, confined area.

Intrabay automated material handling is a means by which the above-mentioned risks can be reduced or eliminated. Simulation can play a key role in the design of intrabay automation systems, in verifying that the system will not negatively affect throughput of an area.

A simulation model has been developed to analyze the effects of intrabay AMHS on a manufacturing bay. Model architecture is discussed, and three case studies demonstrate applications of the model.

1 INTRODUCTION AND OVERVIEW

Automated material handling systems (AMHS) in semiconductor factories can be separated into two classes - interbay and intrabay. Interbay systems are large, factory wide systems which move materials between bays or functional areas throughout the factory. Interbay systems are typically monorail-type movement systems, where vehicles move material using the monorail and interface with AS/RS machines (stockers) in areas where materials are processed (manufacturing bays). The advantages of simulation in the design of interbay systems have been documented in the past. Pillai (1989) showed how simulation can be used for design verification and sizing the different components of the total system. Nadoli and Rangaswami (1993) presented a modeling methodology for simulating interbay systems.

Intrabay material handling systems are specific to one processing area or bay. These systems, which move materials directly to and from tools within a bay and load and unload equipment, allow for better tracking of materials, more reliable material handling, and reduced overall system variability for even the most complex operations. Cost savings can also be realized by reducing the manpower required to run a bay, improvements in ergonomics and overall cleanliness.

An automation system typically consists of a shop floor controls system communicating with an AMHS and an automated equipment control system. Figure 1 graphically depicts this automation architecture.

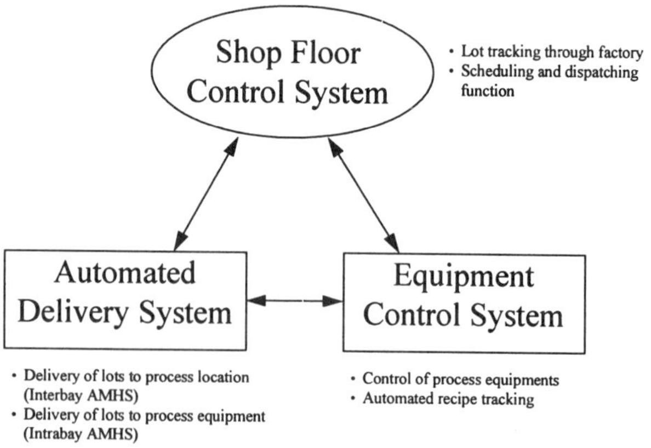

Figure 1: Intrabay automation architecture.

The shop floor control system communicates with the AMHS and provides information on lot delivery locations. In addition, an intrabay automation system requires a tight coupling between the delivery mechanism and the equipment control system.

In a full factory setting, these complex interactions are very difficult to predict; hence, requiring the use of a simulation or modeling tool. In addition, the manufacturing process may also include constraints whose effect on factory output become difficult to predict. At Intel, the Material Handling Systems Simulation Group simulation has built a suite of modeling tools that can be used to model and predict the behavior and performance of the AMHS in a full factory setting. An intrabay modeling methodology has been developed where a core model interacts with add-on components to create a flexible tool that can be configured, based on the type of processing equipment which is being automated, or the type of outputs which need to be collected. This approach has led to a tool which is both powerful and easy to implement. All the simulation models used within the group use AutoMod II (AutoSimulations, 1992) as the simulation tool environment.

The remainder of this paper will describe the modeling methodology and present applications, in the form or case studies. Section 3 details the model architecture, and discusses both the model core and the add-on approach to the model. Section 4 presents 3 case studies which demonstrate applications of the model. Section 5 lists conclusions, and areas of possible future study and application of intrabay automation.

2 AMHS SIMULATOR ARCHITECTURE

The Intrabay AMHS simulation model consists of a core model and four major sub-sections, or modules, graphically depicted in Figure 2, below.

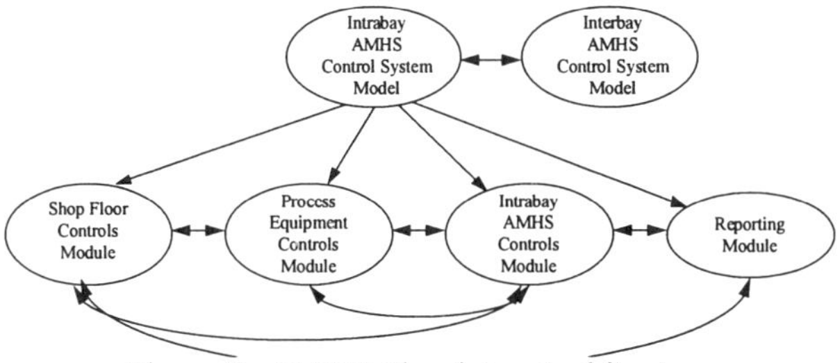

Figure 2: AMHS Simulator Architecture

2.1 Shop Floor Control Module

This module includes code which simulates the WIP scheduling rules that are built into the Shop Floor Controls Systems. Different WIP management rules, such as Theory Of Constraints (TOC), Just-In-Time processing (JIT), etc. can be simulated.

2.2 Process Equipment Module

This module includes details of the process equipment being modeled, such as its speed, setup time information, batching characteristics, equipment buffer configurations, and the amount of re-work being processed.

2.3 Intrabay AMHS Module

A typical Intrabay AMHS consists of a storage system and a transport system. This module simulates all the interactions that happen between the different AMHS components, including transport to the process equipment. It can also include AMHS reliability information, and can interface with an Interbay AMHS model to simulate lot movement between the different parts of the factory.

2.4 Reporting Module

This module includes software to provide output reports and charts. These reports include the standard reports provided by the simulation tool as well as custom reports which provide more information on the performance of the AMHS.

The simulation software in these modules has been written in such a way that it can be easily modified in the event of changes. For example, the Intrabay AMHS module can be changed to model the characteristics of different variations of AMHS, without modifying too much code. Another feature of the Intrabay model is that much of the information is present in different data files. These files can be easily changed to perform sensitivity analyses to understand the effects of different input parameters. This paper will now show how this modeling tool has been used to model three different variations of intrabay systems within Intel's wafer fabrication facilities.

3 CASE STUDIES

3.1 Simulation of Diffusion Intrabay

Figure 3 shows a typical layout of an intrabay AMHS in a diffusion processing bay.

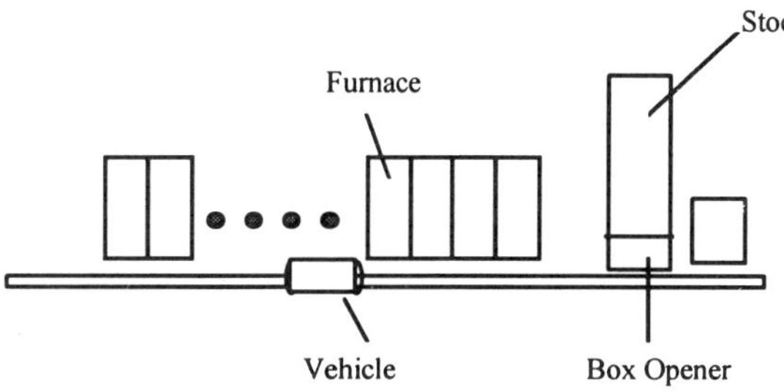

Figure 3: Diffusion Intrabay line

The system consists of an automated storage system (also known as a stocker) communicating with a ground-based robotic vehicle, capable of carrying multiple loads, which delivers lots to the diffusion process tool. The stocker also provides an interface with the Interbay delivery system. Diffusion operations are batch processes, with up to 6 lots being loaded into the tool. In addition, the tool has a buffer which can hold one extra batch.

The objectives of the simulation model were to determine the time needed to deliver a batches of lots to the furnaces at different factory wafer start loading conditions, and to determine the utilizations of the different AMHS components (stocker robot, stocker storage, and vehicle utilization). Figure 4 shows the variation of the batch delivery time with the loading of an Intrabay line.

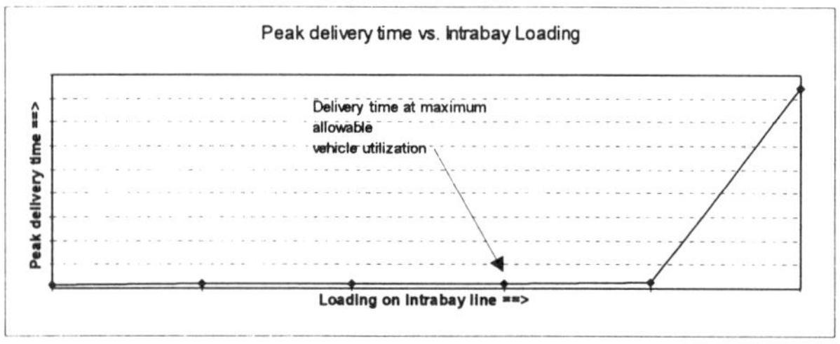

Figure 4: Batch Delivery Time to the Tool vs. Loading on Intrabay Line

Figure 4 shows that the batch delivery times remain constant up to a certain level, beyond which the delivery times increase almost exponentially. Figure 5 shows the utilization of the robotic vehicle.

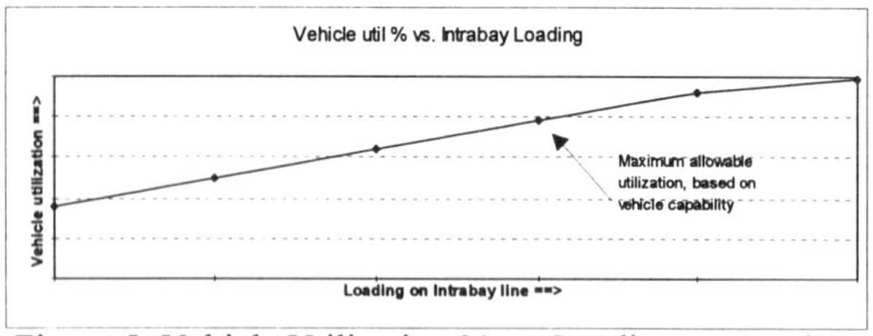

Figure 5: Vehicle Utilization % vs. Loading on Intrabay Line

Figure 5, when combined with Figure 4, shows that the delivery times begin to increase when the loading on the robotic vehicle increases beyond its capable limit. This defined the maximum allowable throughput capability of the Intrabay line.

After the system was released to production, the different AMHS components were monitored. The validity of the simulation model was confirmed when the delivery times seen with the actual equipment closely followed the estimates provided through the dynamic simulation model.

3.2 Intrabay in Lithography Processing

The process of lithography (litho) is the most critical processing in semiconductor manufacturing. Due to the intricate nature of litho processing, these tools are very expensive, costing millions of dollars. When investing such a sum in equipment, a high priority is placed on attaining its maximum utilization. Intrabay automation was seen as one way to help increase tool utilization and reduce equipment costs.

The model used to model diffusion intrabay automation was used as a baseline to simulate an automated environment. The diffusion model was viewed as the core model, as it had been proven to be an accurate simulator of actual diffusion operations. However, differences between diffusion operations and litho operations such as batch size, processing time, layout of equipment, and lot buffering led to additions to the model which built upon this core and added to the flexibility of the model.

The spin, expose, and develop processing sequence performed by a litho tool is shorter in duration than diffusion processing, and therefore the time to deliver materials (the time between processed lots) could negatively impact tool utilization. As a result, the number of lots at the tool waiting to be processed - the buffer size - becomes a factor in AMHS design. If a tool has too small a buffer, the tool could be starved for material to process, negatively affecting tool utilization.

Two new functions were added to the model to provide for the new requirements brought on by lithography processing. First, the way in which lots requested the vehicle was modified. The structure of the model was modified to work in the following manner. Once a lot arrives at the stocker, it first checks to see if there is a location available on a tool before attempting to move to the tool. If a location is available, the lot will claim it, and then initiate the sequence of steps needed to move from the stocker to the tool port. If no buffer location is available, the lot will continue to search for available locations, until one becomes available.

The second addition to the core model was to include tool processing time into the scope of the model. The addition of processing time enabled the utilization of the tool to be measured, and in turn the effect of buffer size on tool utilization could be observed. Each tool port (buffer) has its own queue, with a capacity of one lot, and lots in these queues attempt to obtain a resource (the tool), also with a capacity of one. A lot holds the resource for the processing time, then releases it to be claimed by other lots waiting to use that resource. It should be mentioned that, in keeping with the goal of flexibility in the model, all of the changes made to accommodate litho tool modeling are configurable. Also, the model can revert back to its diffusion format by changing input flags.

Once the above-mentioned 'add-ons' were complete, simulations were run to determine the effect of buffer size on tool utilization. Each simulation run had the same inputs for throughput level, tool processing time, and number of tools on the intrabay line, with buffer size the only variable. Four scenarios were simulated, with the buffer size ranging from zero to three. Results of this experiment can be viewed in Figures 6 and 7, below.

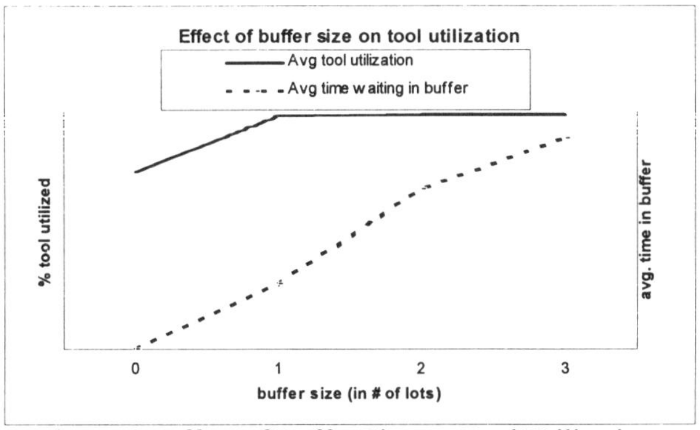

Figure 6. Effect of Buffer Size on Tool Utilization.

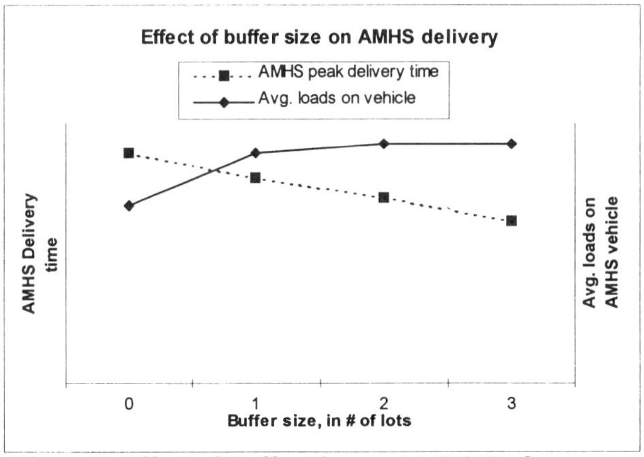

Figure 7: Effect of Buffer Size on AMHS Performance.

Figure 6 shows the results generated by the model to determine the effect of buffer size on tool utilization. Figure 6 also shows the average time spent in buffer as a function of buffer size. The analysis suggests that utilization of the tool will be significantly afftected if there is no buffering, and there is no gain in tool utilization by adding more than one buffer location.

Figure 7 shows the effect of buffer size on the ability of the automation system to transport lots to tools. Two metrics are shown, delivery time and the average lots on the AMHS vehicle. A decrease in delivery time was observed as a result of increasing the buffer size. This decrease can be explained as a result of the AMHS vehicle having a greater opportunity to carry more than one lot, which led to more than one lot delivery of materials for a typical cyclew, and thus shorter delivery times per lot.

From these results, it can be seen that although increasing the buffer size beyond one does not increase tool utilization, an increase will result in better AMHS performance and more 'flex' in the overall system. At this point in time, it is too early in the project to validate these results against an actual running system. However, the model used contains the same core software as the diffusion model, which has been validated against actual performance. The ability to re-use the core software of another model which has been validated is another example of the benefits of the core/add-on methodology of automation modeling. For the same reason, this approach has also been adopted for additional intrabay automation simulation studies, described in the next section.

3.3 The Value of Simulation in Analysis of Future Intrabay Systems

The analysis of future applications of intrabay automation has concentrated on the use of value engineering to perform analysis quickly, to keep up with changing process requirements. As in previous intrabay analysis, models were utilized to filter out the highest priority tools based on the benefits of automation, the return on investment, and the compatibility of automation. This filtering process is shown graphically in Figure 8, below.

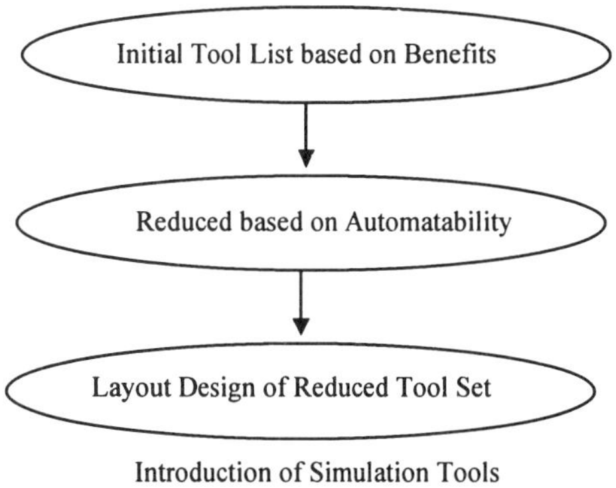

Figure 8: Filter Process Used in Determination of Functional Areas to Automate

Once the filtering process has been completed, the next steps are to once again look at the intrabay layout configuration and automation requirements. From a layout perspective, it is necessary to determine if the tools can be positioned for effective automation access, without violating safety, microcontamination, and ergonomic guidelines. These intrabay-inclusive layouts must also be re-verified to ensure that all process performance requirements are met. It is at this point that the value of the simulation model is exhibited.

In the first case study of diffusion, the performance requirement was based upon the batch delivery time from storage to tool port.. Then, in the case of lithography, analysis was delivery time based with an additional emphasis placed upon tool buffer requirements. In modeling new intrabay options, the initial focus was again on delivery time performance within the design parameters of the intrabay automation equipment. The key to success will be the rate at which the simulation core model, using existing add-ons, will be implemented for two different functional areas of the layout. As seen in Figure 9, the reduction in time to deliver the model results has greatly increased the flexibility to consider new process requirements for run rate and tool quantities.

Model	Simulation Software Development Time	Analysis at Implementation
Diffusion: Case Study #1	~ months	- delivery time
Lithography: Case Study #2	~ weeks	- delivery time + buffer, tool utilization
New Intrabay: Case Study #3	~days	- delivery time + buffer, tool utilization + ...

Figure 9: Time to Simulation Implementation

Using the methodology shown in Figure 8, it has been decided that two areas of processing, thin films and plasma etch, may be best suited for intrabay automation. The initial layout of these bays with intrabay automation are similar to the layouts seen in diffusion. Characteristically, thin films tools are larger in foot print and have faster run rates. Therefore, due to the nature of these functional areas, a variable number of tools per line must be considered in the layout verification process. (Figure 10).

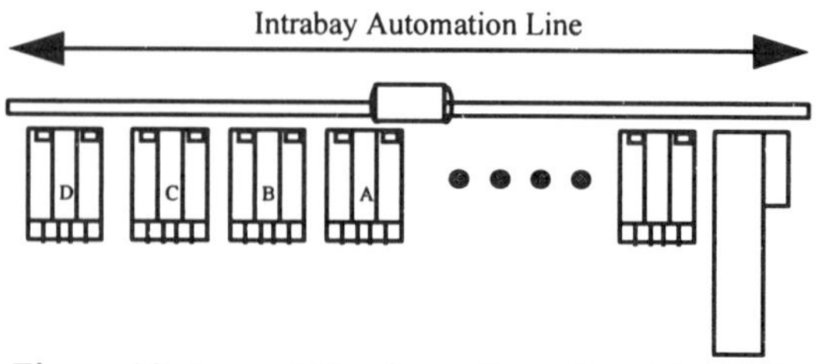

Figure 10: Layout Configurations Considered for Simulation

The purpose of the simulation study therefore is to verify the feasibility of each process tool and its layout when coupled with intrabay automation. The key indicators once again were batch or lot delivery time, and the throughput requirements of the intrabay vehicle.

Figure 12 exhibits that although the etch equipment modeled can support a larger number of tools per intrabay system, the faster run rate of the thin films tools requires more AMHS deliveries per system. As expected, the higher activity rate for the thin films case in turn resulted in slightly longer delivery times per system, as seen in Figure 11. Although the length of these bays is very similar, the tool interface requirements for automation have contrasting differences. These differences, coupled with manufacturing operating requirements, could lead to a modification of the layout to keep the batch delivery times within desired limits.

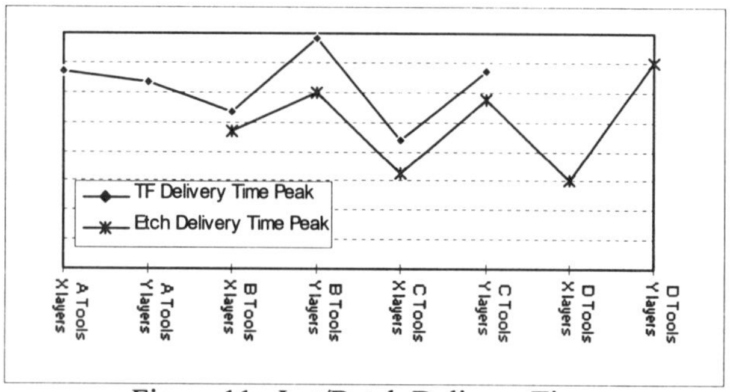

Figure 11: Lot/Batch Delivery Time

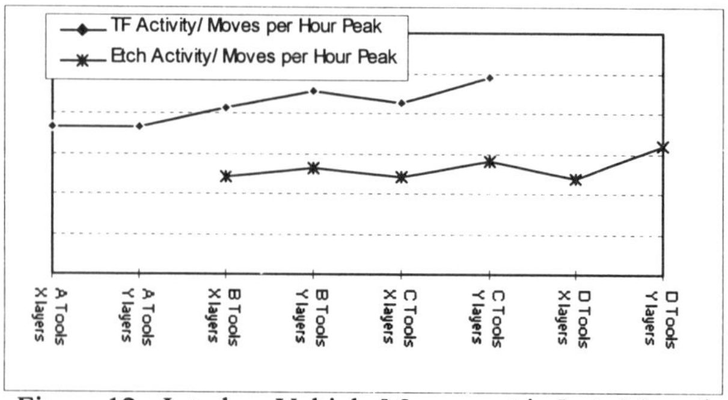

Figure 12: Intrabay Vehicle Movement in Lots Moved Per Hour

These types of sensitivity analysis will be performed through modification of existing simulation software or via the creation of new software modules. Additional data analysis will include the effect of operational assumptions on tool utilization. Ultimately as was done (or is going to be done) in the previous case studies, all model results will be compared to the actual field data after installation.

4 CONCLUSIONS AND FUTURE SCOPE OF WORK

4.1 Conclusions

The Intrabay simulation model has proved to be very useful in designing automated Intrabay systems within Intel's wafer fabrication plants. The same tool has been used to model different processing areas without any major simulation code modifications. We have successfully utilized this modeling approach to understand system behavior in three different processing areas. As an additional benefit of using this modeling approach, we have been able to reduce the modeling cycle time, from months to days. The reduction in development time can be attributed to the overall modeling methodology and the structured model interfaces that were built to support this methodology. We estimate that we can now develop simulators for a new vehicle control system in a matter of days.

4.2 Future Scope of Work

To date, we have utilized this simulation model for modeling 200mm toolset intrabay material movement. With the transitioning of 200mm to 300mm wafers comes new opportunities for automation and layout design. These options will likely be studied and evaluated in the existing 200mm arena, and will require data to support their consideration.

Therefore, the core and add-on approach to simulation will be called upon to incorporate modules for new controls software and new interface requirements for interbay material handling systems. Other possible modules will address additional changes that may be developed from the continual improvement process. These changes could include intelligent vehicle controls and work rule scheduling methodologies. Systems like these are already in development.

ACKNOWLEDGEMENTS

Many people have assisted us during the development of this work. The authors would collectively like to thank Devadas Pillai, Syed Kazmi, and Jeff Pettinato.

REFERENCES

AutoSimulations, Inc. 1992. AutoMod Reference Manual.

Pillai, D. 1989. Designing Automated Material Handling Systems for Large-scale Wafer Fabrication Automation. In *SME Semiconductor Manufacturing Conference.* Society of Manufacturing Engineers, Dearborn, Michigan.

Nadoli, G., and M. Rangaswami. An Integrated Modeling Methodology for Material Handling Systems Design. In *Proceedings of the 1993 Winter Simulation Conference,* ed, G.W. Evans, M. Mollaghasemi, E.C. Russell, and W.E. Biles, 785-789. Institute of Electrical and Electronics Engineers, Piscataway, New Jersey.

AUTHOR BIOGRAPHIES

THOMAS JEFFERSON is an Industrial Engineer with the Material Handling Systems Group at Intel Corporation, responsible for simulation and modeling of various automation systems. He received a B.S. in Industrial Engineering from the Rochester Institute of Technology in 1994. Current research interests include material handling concepts for next-generation manufacturing, whole factory modeling, and the use of simulation in the design of semiconductor assembly/test factory automation systems.

MADHAV RANGASWAMI is the manager of the Simulation Group within the Material Handling Systems Group at Intel Corporation. He received a B.Tech in Mechanical Engineering from Indian Institute of Technology, Delhi (India), an M.E. in Mechanical Engineering from the Indian Institute of Science, Bangalore (India) and an M.S. in Industrial Engineering from the University of Florida, Gainesville. He has worked in the field of computer simulation for the last 12 years, and has been involved with many automation projects within Intel. His current interests are in the fields of simulation, computer integrated manufacturing and factory scheduling systems.

GREGORY STONER is a Industrial Engineer in the Automated Material Handling Systems Simulation group at Intel Corporation. He received a B.S. in Engineering and a M.S. in Industrial Engineering from Purdue University, West Lafayette, Indiana. His research interests are in the fields of simulation, material handling and storage systems, and layout design.

Proceedings of the 1996 Winter Simulation Conference
ed. J. M. Charnes, D. J. Morrice, D. T. Brunner, and J. J. Swain

EVALUATION OF LOT RELEASE POLICIES FOR SEMICONDUCTOR MANUFACTURING SYSTEMS

Raka Sandell

MOS12 Die Manufacturing
Semiconductor Products Sector
Motorola Inc.
1300, North Alma School Road,
Chandler, Arizona- 85224, U.S.A

Krishna Srinivasan

Worldwide Enterprise Modeling Group
Motorola Manufacturing Systems
Motorola Inc.
1299, E. Algonquin Road
Schaumburg, Illinois 60196, U.S.A.

ABSTRACT

This paper describes a simulation based approach to evaluate and benchmark various lot release strategies in semiconductor manufacturing systems. A full-factorial experiment, with 5 factors, has been designed and executed for this purpose. A variation of the Bonferroni inequality is used to compute confidence intervals for differences in system performance for different operating strategies. The results show that no lot release policy dominates across all scenarios.

1 INTRODUCTION

Semiconductor manufacturing is generally characterized by investments in process technology which are over a billion dollars. The industry is characterized by extremely short product life-cycles, frequently decreasing profit margins and intense competition. In such a volatile scenario, maintaining a competitive advantage and remaining profitable, in operational terms requires minimization of cycle time and work-in-process inventory and the maximization of throughput. Cycle time is the time it takes for a product to go from start to finish. Cycle time reduction is important as it leads to a shorter order lead time and hence better customer satisfaction. Also, for the same level of throughput, a shorter cycle time results in a smaller work-in-process inventory that not only reduces the capital tied up, but also minimizes the inventory buffer that needs to be maintained at the downstream end of the plant. When the market demand changes and product designs become obsolete, such inventory may lose value. There is also a technological reason for reducing the cycle time. The shorter the period that the wafers are exposed to aerial contaminants while waiting for processing, the smaller is the yield loss. It is also important to reduce the variance of cycle time. Low variability in cycle time allows a more accurate prediction of production completion time and ability to meet due dates. This also facilitates improved downstream coordination of operations on completed wafers such as wafer probe, assembly and test.

The performance measures of interest for a product generally depend on the nature of business one is in. For make-to-order parts as in ASIC (Application Specific Integrated Circuit) factories, adhering to due dates is of paramount importance. It is also important to minimize the mean cycle time. Hence, a performance measure like mean + 3 × standard deviation, also known as 98% cycle time, is of great significance. For a make-to-market environment, like Dynamic Random Access Memory (DRAM) chip manufacturing, typical objectives are maximization of throughput while maintaining cycle time goals. For extremely expensive chips like microprocessors (pentium, power PC etc.), yield is an important criterion. Thus in all types of industries, the reduction of cycle time assumes great significance.

Different types of control policies are implemented to ensure that the factory's cycle time is minimized under different operating characteristics. Some examples of such policies include determining the instants of time when wafers or lots are to be brought into the factory, deciding how operators choose parts for processing given that there are many wafers to choose from, and determining the optimal schedules for machine maintenance. Certain rules are formally defined in order to address such operational problems in wafer fabrication facilities. These rules are referred to as scheduling policies.

In general, the scheduling policies used to exercise control over fab operations are of two types. In the first type, one can specify rules for determining when new lots are to be released into the plant. This is referred to as the release policy. Clearly, the release policy must meet some constraints such as maintaining an average release rate of lots. Examples of lot release policies are deterministic release and closed-loop release which are explained in subsequent sections. Second, for lots already in the plant, one has to decide which lot is to be processed next at each machine as it becomes available. These are referred to as dispatching policies. Such policies determine the sequence in which jobs in front of a machine are processed by the machine. Examples of such policies are Shortest Processing Time (SPT) and Earliest Due Date (EDD). The dispatching policy need not be a global rule i.e., the dispatching rule at different workstations may be different.

Many papers in literature have compared the cycle time performance of lot release and dispatching policies. Most of them have concluded that lot release policies have a much larger impact on fab performance than dispatching policies (Wein(1988), Glassey and Resende(1988a, 1988b)). In addition, these papers have concluded that one policy generally dominates over other release policies investigated. For example, Spearman and Zazanis(1992) conclude that the CONWIP release policy is more superior than open-loop release policies. Wein(1988) shows that the workload regulating rule is much superior to deterministic and Poisson release policies. In this study, we adopt the premise that lot release policies have a much larger impact than dispatching policies on fab performance. While earlier authors have tested the quality of policies on one manufacturing facility, this study attempts to provide a comparison of release policies across vastly differing manufacturing facilities. The study explores whether conclusions derived by other authors can be extended and held to be true under all types of manufacturing scenarios. Preliminary results showed that under different circumstances, the performance of policies tend to vary greatly. Hence, a full factorial experiment is constructed and executed to test the impact of common lot release policies on the cycle time performance of vastly differing manufacturing facilities.

As this study will strive to provide a quantitative comparison of the effectiveness of different lot release strategies, a basis for comparison has to be defined *a priori*. As described earlier, the objective of most scheduling policies for semiconductor manufacturing is the minimization of mean cycle time and average WIP and the maximization of throughput. Glassey and Resende(1988a) suggest guidelines for the comparison of scheduling policies. Scheduling policies can be compared based on a tradeoff curve plotted between the mean cycle time and mean throughput for a system. Figure 1 describes the cycle time D(t) as a function of the fab throughput t. It is a well known fact that as throughput approaches the capacity of the fab (i.e. the utilization of machines becomes very high) the average queuing time (cycle time) approaches infinity. As shown in Figure 1, scheduling policy SA is superior to policy SB for a given throughput t if DA(t) < DB (t), where DA (t) and DB (t) are the tradeoff curves for policies SA and SB, respectively.

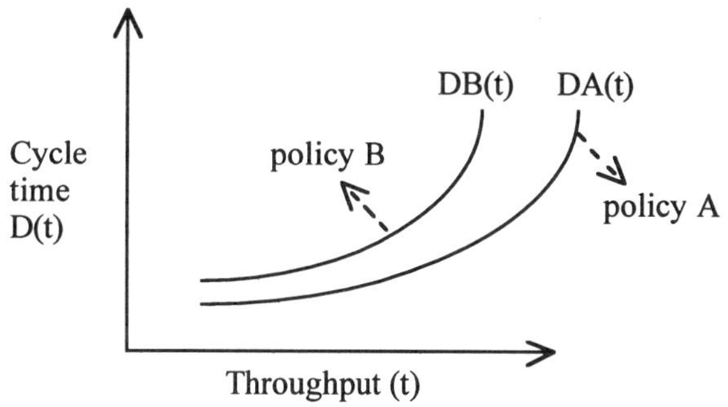

Figure 1: Scheduling policies: Basis for comparison

The experiment described in subsequent sections will use the above described scheme for testing release policies. The scheduling policies will be tested under different throughput rates for each of the system configurations in order to make the above comparison. A variation of the Bonferroni inequality will be used to create confidence intervals for the difference between the policies.

The paper is organized in the following manner. Section 2 provides an overview of past work in this area. Section 3 elaborates on the experimental design and the factors used in the experiment. Section 4 describes the use of a modified version of the Bonferroni inequality to compare release policies. The results are discussed in Section 5 and analysis of release policies for an actual system and conclusions are provided in Sections 6 ad 7 respectively.

2 LITERATURE SURVEY AND DEFINITIONS

This section describes recent forays in the area of dispatching and lot release policies for semiconductor manufacturing systems. Wein(1988) is concerned with assessing the impact that scheduling can have on the performance of semiconductor wafer fabrication facilities. The performance measure considered here is the mean cycle time for a lot of wafers. Several release policies and dispatching rules have been evaluated using a simulation model of a fictitious, single product semiconductor facility. Three versions of the model, which differ by the number of servers present at particular stations are used. Four different types of release policies were considered.

The first release policy evaluated is the deterministic input policy. Given the total number of wafers that need to be processed in a week, the average time between the entry of wafers of equivalent quantities is obtained from the ratio between the total time available and the number of wafers to be processed. In other words, if the total available time in a week is Z units and the total number of wafers that need to be processed is N, then the average inter-arrival time between wafers is $(Z \div N)$ units. For the deterministic release policy, the inter-arrival times are sampled from a constant distribution. The second type of release is the Poisson release policy. For the Poisson release, the inter-arrival times are sampled from an exponential distribution with mean $(Z \div N)$. The third release policy studied is the closed loop input policy where the number of lots in the system is held constant. The departure of a job/wafer from the system immediately triggers the arrival of a new job into the system. Here the throughput/ wafers started is controlled by the maximum WIP permitted in the system. The fourth release policy is the workload regulating input policy, which releases wafers in lots into the system whenever the total amount of work in the system for any bottleneck falls below a pre-specified level.

Wein(1988) showed that the workload regulating policy dominates the other three policies. As mentioned before, these results were applicable to a single product fictitious facility where the number of servers were perturbed and hence, by no means the experiment tested the robustness of the applicability of the result to other manufacturing configurations.

Glassey and Resende(1988a, 1988b) also carried out comparisons of the four release policies and another policy called the starvation avoidance policy. Based on experiments carried out with the fictitious one-product facility used by Wein(1988) and based on perturbations to the number of servers, they concluded that the starvation avoidance policy was superior to the other four policies. Here again, the authors had not tested the robustness of the result to alternate manufacturing configurations. Similar results were observed in Lu, Ramasamy and Kumar (1992).

Spearman and Zazanis(1992) demonstrated, through analytical means the dominance of the CONWIP release policy over Poisson release policy for a network of tandem exponential queues. The network comprises a set of machines where the process occurs in tandem (as in an assembly line) and each of the machines that comprise the network are single server systems where the service time is assumed to be sampled from an exponential distribution.

These results presented in the literature survey have not been clearly understood by the practicing community. Even though the workload regulating policy and the starvation avoidance policies have been shown to be quite effective in improving performance metrics, practitioners have found them impractical to implement them on the factory floor. This is primarily due to the difficulties involved in the detection of the bottleneck on a dynamic basis in a large sized factory. The bottleneck tends to shift constantly as a function of various random events and fluctuations in the product- mix. Therefore, policies such as workload regulation and starvation avoidance, which are based on bottleneck activity are not examined in great detail in this study. Instead, this study concentrates on simplistic policies like deterministic input, CONWIP and Poisson release which are easily implementable on the factory floor.

Among the three release policies described above - Poisson, deterministic and CONWIP, there lies a great deal of confusion with the understanding of the operation and overall relative effectiveness of CONWIP type closed-loop policies for different factory configurations. Many fab managers tend to assume on the basis of results reported by Spearman and Zazanis(1992) for tandem manufacturing lines that CONWIP is the panacea for all operational problems faced in their factories. In many cases, the authors have observed that the results presented by Spearman and Zazanis(1992) have been misinterpreted as a generalized proof of the superiority of CONWIP over all release policies for all types of performance measures. This contradicts results obtained by Wein(1988) and Glassey and Resende(1988a, 1988b) where, the CONWIP policy has been shown to be inferior to an open-loop policy like the deterministic release policy where the inferiority of a policy was measured based on the amount of increase in mean cycle time with an increase in average throughput. This study will attempt to resolve this confusion and make recommendations with respect to the relative performance of these release policies for different manufacturing configurations.

This study will show that one release policy does not dominate others for all types of manufacturing systems. In fact, the results that have been reported in all these studies have been based on different types of manufacturing systems. While those reported by Wein(1988) and Glassey and Resende(1988a, 1988b) are for a fictitious one-product semiconductor manufacturing facility, the results reported by Spearman and Zazanis(1992) are based on tandem (assembly line) manufacturing systems and hence might not apply to complex systems like semiconductor manufacturing systems. Such disparities in the base assumptions, the concomitant variations in the conclusions and the resulting confusion in the practicing community calls for a detailed experimental design that tests the results obtained using scheduling policies for multiple manufacturing systems.

3 EXPERIMENTAL DESIGN

The earlier section brought to light the confusion that exists over the relative effectiveness of the deterministic, Poisson and CONWIP policies for different types of manufacturing systems. In this section, a full-factorial experiment is designed which will permit the evaluation of performance of these policies for a wide spectrum of factory configurations.

3.1 Experimental Factors Perturbed

While designing a full-factorial experiment, it is important to intuitively determine a list of factors that are expected to affect the output response in a significant manner. In this case, the configurations used and types of analysis carried out by earlier researchers provide pointers to the factors that need to be perturbed in this experiment. In addition, it has been conjectured in past literature that the superiority of CONWIP type closed-loop policies may be attributed to the fact that there exists a negative correlation between the arrival process and the WIP in the system. A large WIP in the system limits the rate of arrival into the system while a low WIP level accelerates the arrival rate. In other words, variability in WIP translates to arrival variability. This phenomenon may prove favorable to performance measures of some systems. The factors that are considered for this study include those influence the system configuration and those that greatly impact the WIP in the system. These factors, their levels of perturbation and the respective reasons for inclusion are described below.

i) System configuration

This factor refers to the size of the system and is a function of the process flow and the number of work stations in the system. The inclusion of this factor is motivated by the fact that earlier researchers have published results based on experiments carried on different sets of systems. Therefore, the use of this factor will help determine if the size/type of system has any bearing on the relative ranking of the three lot release policies.

The system configuration factor is varied over three levels. Each level corresponds to a process flow and a corresponding tool list. The first system involves three machines and five process steps. The second system involves five machines and thirteen process steps, and the third system configuration comprises nine machines and twenty five process steps. A fourth system is simulated separately. It uses the process flow and tool list corresponding to an actual semiconductor manufacturing system. The process flows are shown in Figure 2, 3 and, 4 and the tool list characteristics are displayed in Tables 1, 2, and 3 respectively. The number of identical tools at each work station is computed by ensuring that the utilization of the station given a product start rate, is less than 1. The product start rate is derived from the average inter-arrival time of wafers. For the purpose of tool-set determination, the average inter-arrival time between lots of 25 wafers each is assumed to be 0.85 hours. This number has been found to be consistent with the scale of operation of typical medium to large-sized semiconductor manufacturing facilities.

Table 1. Tool List for 3 machine 5 step system

Tool no.	No. of passes	Tool TP (wafers/ hour)	MTBF (hours)	MTTR (hours)	No. of tools	Utilization
1	2	75	9	1	1	0.98
2	2	60	10	1	2	0.61
3	1	42	9	1	1	0.87

Sequence of stations: 1-2-1-2-3

Figure 2: Process Flow for 3 machine 5 step system

Table 2. Tool List for 5 machine 13 step system

Tool No.	No. of passes	Tool TP (wafers / hour)	No. of tools	Utilization
1	3	72	2	0.76
2	3	60	2	0.92
3	3	42	3	0.87
4	2	55	2	0.66
5	2	45	2	0.82

Sequence of stations: 1-2-1-2-3-1-2-3-4-5-3-4-5

Figure 3: Process Flow for 5 machine 13 step system:

Table 3. Tool List for 9 machine 25 step system

Tool No.	No. of passes	Tool TP (wafers / hour)	No. of tools	Utilization
1	3	72	2	0.76
2	3	60	2	0.91
3	3	42	3	0.87
4	3	55	3	0.66
5	3	45	3	0.82
6	3	34	4	0.81
7	3	36	4	0.76
8	2	40	2	0.92
9	2	82	1	0.90

1-2-1-2-3-1-2-3-4-5-3-4-5-6-7-4-5-6-7-8-9-6-7-8-9

Figure 4: Process Flow for 9 machine 25 step system:

ii) Release policy

This is the main focus of the experimental design. The

output response derived from the performance measures of interest are observed as a function of release policies under different system scenarios. As described earlier, this factor is varied over three levels.

The first level represents the deterministic release policy where jobs are released into the factory such that the inter-arrival time is a constant. The second level involves releasing jobs into the factory such that the inter-arrival times are sampled from an exponential distribution. The third level studies an implementation of CONWIP for releasing jobs into the factory. Here the system is modeled as a closed queuing network where the departure of a job from the system triggers a new arrival into it. Thus the coefficient of variation (the ratio between the standard deviation and mean of a random process) of inter-arrival time of jobs into the system is identical to the coefficient of variation of the departure process of the last machine.

iii) Start rate

The start rate of wafers is controlled by specifying the mean inter-arrival time of wafers into the system for deterministic and Poisson release policies. Variations in start rates alter the utilization of stations in the system. This factor enables the testing of robustness of the results as a function of factory congestion and equipment utilization. Typically this factor is varied over six levels.

The range of mean inter-arrival times used for this experiment is from 0.85 hours per lot of 25 wafers to 1.3 hours per lot in steps of 0.05 or 0.1. Specifically, the levels of the factors of inter-arrival time are 0.85, 0.9, 0.95, 1.0, 1.05, 1.1 and 1.2. These levels of inter-arrival times are meaningless with respect to CONWIP release policy. This is because, while the start rate for open-loop policies can be controlled by an external arrival rate, the start rate for closed-loop systems is controlled by altering the permitted WIP in the system. Sample preliminary simulation runs are required while ensuring that the start rate resulting from WIP fluctuations in CONWIP/ closed-loop systems is comparable with the start rates achieved by external release in open-loop systems using Poisson and deterministic policies. The different levels of WIP required to achieve average lot inter-arrival times equivalent to 0.85, 0.9, 0.95, 1.0, 1.05, 1.1 and 1.2 hours is determined in an iterative fashion.

iv) Variability in service time

The variability in service time is controlled by altering the probability distribution for service times. This factor is varied over two levels. The arrival stream for CONWIP systems are controlled by WIP fluctuations. The WIP fluctuations are in turn impacted by the variability in service time. Thus, this factor tests the validity of conclusions in the presence of large variability in WIP induced by variability in service time.

The service time is perturbed over two levels. These two levels are deterministic (or constant) service time and exponentially distributed service time with CV of the distribution 0 and 1 respectively.

v) Tool reliability

Semiconductor manufacturing systems are characterized by large variations in tool reliability due to the tight process specifications that need to be maintained on the factory floor. These large fluctuations in tool reliability greatly impact the resultant variability in processing time (Segal and Whitt (1988)). As described in the earlier section, the variability in service time impacts the variability in WIP in a significant manner and is considered an important factor in this experimental study.

Variations in tool reliability are captured through changes in the Mean Time Between Failures (MTBF) and the Mean Time To Repair (MTTR). The distributions of MTBF and MTTR considered are constant and exponential. This factor is varied over three levels. The table below shows the three levels of this factor considered for the experiment.

Table 4. Three levels of tool reliability

Tool No	LOW		MEDIUM		HIGH	
	MTBF (hours)	MTTR (hours)	MTBF (hours)	MTTR (hours)	MTBF (hours)	MTTR (hours)
	Const.	Const.	Expon.	Const.	Expon.	Expon.
1	9	1	9	1	4.5	1
2	10	1	10	1	5	1
3	9	1	9	1	4.5	1
4	10	1	10	1	5	1
5	8	1	8	1	4	1
6	10	1	10	1	5	1
7	8	1	8	1	4	1
8	11	1	11	1	5.5	1
9	7	1	7	1	3.5	1

3.2 Output Responses Studied

This study uses the basis for comparison suggested by Glassey and Resende(1988a) as described in the earlier section. The primary response of interest is cycle time, as a function of the factor levels. Cycle time is estimated using the mean and the half width of a 95% confidence interval. Half width is defined as the absolute value of the difference between one of the bounds of the confidence interval and the mean. Another response of interest is the coefficient of variation (CV) of the arrival process. The CV of arrivals for the deterministic and Poisson release policies are 0 and 1, respectively. But, for the CONWIP release policy, CV of arrivals is a function of the departure process and the WIP in the system. The CV of arrivals is estimated from the simulation experiment. The third response of interest is number of jobs processed during the simulation run which is a measure of the throughput obtained from the system.

3.3 Simulation Details

A full factorial experiment is carried out with the factors just described and the responses listed are estimated for each combination of the factors. The full-factorial experiment requires a total of 270 (3×3×5×2×3) simulation runs. The number of combinations is derived from the product of the number of levels of each of the factors considered.

Simulations were run using a C based discrete event simulator - Delphi (Chance(1995)), developed at SEMATECH. Simulations were typically run for 1000 days. Statistics were collected after 200 days in order to correct for initialization bias. The duration of the simulation run, number of replications and the truncation point were determined using recommendations made by Schruben *et al.*(1983).

4 ANALYSIS OF RESULTS

The results of the simulation runs for the design points for the experiment are not reported due to the volume of output data generated. To summarize this data, a modified version of the Bonferroni inequality is used. This is elaborated in the next section. Release policies are evaluated for a combination of system configuration, variability of service time and tool reliability. Among the responses obtained for different throughput rates, the mean cycle time pertaining to comparable throughput rates for two release policies, is used to reach a verdict on the relative performance of these release policies.

Comparison of cycle time for two policies can be made by computing the difference of the respective means. There is a caveat here. For the purpose of the comparison, it is necessary to devise a suitable methodology to test for the statistical significance of the differences and it is not sufficient to observe just a small difference in the mean cycle time irrespective of the width of its confidence interval. The primary reason for this fact is that the cycle time observed is typically a confidence interval and hence one does not have a 100 percent confidence interval for the value of cycle time. Thus, one has to compare two intervals, both of which are not 100% confidence intervals. A modified version of the Bonferroni inequality is used to establish confidence intervals with a certain probability, for the difference in cycle time for two systems.

Suppose that I_s is a $100(1-\alpha_s)$ percent confidence interval for the measure of performance μ_s (where s = 1,2,...,k). This can be stated as follows

$$P(\mu_s \in I_s) = 1 - \alpha_s \quad \text{for } s = 1, 2, \ldots, k$$

Then, Law and Kelton (1992) show that the probability that all k confidence intervals simultaneously contain their respective true measures satisfies

$$P(\mu_s \in I_s \text{ for all } s = 1,2,\ldots,k) \geq 1 - \sum_{s=1}^{k} \alpha_s$$

This result is known as the Bonferroni inequality. The above result is altered in the following manner in order to obtain a confidence interval for the difference between two means. In the case of this study, the mean and the half width of cycle time for two systems under the two release policies for a given system configuration and system variability has been obtained from the simulation experiment.

The primary idea used for comparison of two systems is as follows. If system A has a 95% confidence interval of (5.5, 6.3) for a particular performance measure (say μ_1) and system B has a 95% confidence interval of (8.1,9.5) for a measure (say μ_2) then the difference of the two measures ($\mu_1 - \mu_2$) has *at least* a 90% confidence interval of (8.1-6.3, 9.5-5.5) or (1.8, 4). This result is proved in the following manner.

Claim: If $P(\mu_1 \in (l_1, u_1)) = 1 - \alpha_1$ and
$P(\mu_2 \in (l_2, u_2)) = 1 - \alpha_2$ then
$P(\mu_1 - \mu_2 \in (l_1 - u_2, u_1 - l_2)) \geq 1 - \alpha_1 - \alpha_2$
where l_1 and u_1 are the lower and upper bounds respectively for the confidence interval for μ_1 and l_2 and u_2 are the lower and upper bounds, respectively, for the confidence interval for μ_2.

$P(\mu_1 - \mu_2 \in (l_1 - u_2, u_1 - l_2)) =$
$P(\mu_1 - \mu_2 \in (l_1 - u_2, u_1 - l_2) / (\mu_1 \in (l_1, u_1) \text{ and } (\mu_2 \in (l_2, u_2)))$
$+$
$P(\mu_1 - \mu_2 \in (l_1 - u_2, u_1 - l_2) / (\mu_1 \in (l_1, u_1) \text{ and } (\mu_2 \notin (l_2, u_2)))$
$+$
$P(\mu_1 - \mu_2 \in (l_1 - u_2, u_1 - l_2) / (\mu_1 \notin (l_1, u_1) \text{ and } (\mu_2 \in (l_2, u_2)))$
$+$
$P(\mu_1 - \mu_2 \in (l_1 - u_2, u_1 - l_2) / (\mu_1 \notin (l_1, u_1) \text{ and } (\mu_2 \notin (l_2, u_2)))$

It is apparent that the expression
$P(\mu_1 - \mu_2 \in (l_1 - u_2, u_1 - l_2) / (\mu_1 \in (l_1, u_1) \text{ and } (\mu_2 \in (l_2, u_2)))$

$= \quad P(\mu_1 \in (l_1, u_1) \text{ and } \mu_2 \in (l_2, u_2)) = 1 - \alpha_1 - \alpha_2$

The other three expressions are all greater than or equal to zero as they are all probabilities.

Thus,
$P(\mu_1 - \mu_2 \in (l_1 - u_2, u_1 - l_2)) = 1 - \alpha_1 - \alpha_2 +$ (a positive quantity)
So, $P(\mu_1 - \mu_2 \in (l_1 - u_2, u_1 - l_2)) \geq 1 - \alpha_1 - \alpha_2$

The following numerical example is used to explain this result. Consider two 95% confidence intervals for μ_1 and μ_2.
Let $P(\mu_1 \in (19.5, 20.5) = 0.95$ and $P(\mu_2 \in (17, 19)) = 0.95$.
Then $P(\mu_1 - \mu_2 \in (19.5 - 19, 20.5 - 17)) \geq 0.9$ or
$P(\mu_1 - \mu_2 \in (0.5, 3.5)) \quad \geq 0.9$

With respect to the results of the experiment, if the cycle time and the corresponding half-width are computed for two systems at a given throughput, then a 90% confidence interval for the difference in cycle time for the two systems can be obtained. If this interval does not contain 0 it implies that the difference between the two means is greater than 0 with at least 90% probability. Then, with at least 90% confidence it can be stated that one system clearly dominates the other in terms of the performance measure. In other words the probability that one system is superior to the other is at least 0.9.

The concepts stated in the earlier paragraph were used to statistically prove the superiority of one release policy over the others under different scenarios. If both the bounds in the confidence interval for the difference in means are negative then CONWIP is superior to deterministic policy for that configuration. If both the bounds are positive then the converse is true. In the following tables, the results of the application of this result to the experiment results are summarized. In Table 5, the number of experiments where the differences indicated the superiority of CONWIP over deterministic policy (C > D) and vice-versa (C < D) and number of instances where the difference was insignificant are reported. Table 6 shows the average 90% confidence lower and upper bounds for the difference in mean for the three instances. Table 7 displays the utilization of the bottleneck tool and Table 8 lists the average CV of arrival under CONWIP for each of the instances.

Table 5 CONWIP vs. Deterministic - comparison count

Name of system	C > D	D = C	C < D
3mac_5step	10	12	0
5mac_13step	4	16	10
9mac_25step	0	2	26

Table 6. Average lower bound, Average upper bound in each of the instances

Name of system	C > D	D = C	C < D
3mac_5step	(-1.6, -0.52)	(-0.49, 0.38)	-
5mac_13step	(-1.7, -1.1)	(-0.2, 0.04)	(0.54, 0.68)
9mac_25step	-	(-0.1, 0.5)	(1.19, 2.88)

Table 7. Average utilization of bottleneck in each of the three instances

Name of system	C > D	D = C	C < D
3mac_5step	0.86	0.67	-
5mac_13step	0.79	0.72	0.83
9mac_25step	-	0.68	0.75

Table 8. Average CV in each of the three instances for CONWIP release policy

Name of system	C > D	D = C	C < D
3mac_5step	0.62	0.51	-
5mac_13step	0.47	0.62	0.65
9mac_25step	-	0.58	0.74

5 DISCUSSION OF RESULTS

The conclusions observed from the summary results are classified on the basis of system configuration. In all the cases, the difference between CONWIP and deterministic release policies tends to be almost statistically insignificant for low values of utilization (for low throughput) of the work stations (from Table 7).

i) The 3 machine, 5 step system

CONWIP policy clearly dominates the deterministic release policy. This result can be inferred from the fact that in 10 of the 22 experiments run with this configuration, both the bounds are negative while none of the instances show the converse result. 12 cases show statistically insignificant difference, generally due to low utilization. These results hold true even when the CV of the arrival process is high (greater than 0.8). The arrival pattern, though highly variable, effectively uses information about the state of the system and results in lower cycle time than the deterministic release policy. Thus, for a small system, the total WIP in the system is a useful measure for controlling release of lots into the system if the objective is the minimization of mean cycle time. Both CONWIP and deterministic release policy dominate the Poisson release policy.

ii) The 5 Machine, 13 step system

This system has lesser number of instances where the CONWIP release policy dominates the deterministic release policy. In 4 experiments, the CONWIP policy behaves superior to the deterministic release, 10 of them show the converse result and 16 experiments resulted in statistically insignificant differences.

Table 8 shows that CONWIP dominates deterministic release for systems with low CV of arrival while the converse result is observed for high CV values. Middle values of CV go with statistically insignificant differences. This implies that for highly variable systems, the total WIP in the system is not a good measure to control release of wafers into the factory if the performance measure of interest is mean cycle time. But for systems which are less variable (resulting from medium level tool reliability or medium variation in service time), one observes smaller values of CV of release under CONWIP release and the differences between the mean cycle time under the two release policies tends to become insignificant. This indicates that the use of WIP in the system as a parameter to control release of wafers into the system neither aids or hinders the performance of the system. For systems with even further reduced variability (low level of tool reliability and deterministic service distribution) the CONWIP release behaves superior to the deterministic release policy. Here, the value of CV of arrival is very small. For such systems, the total WIP in the system is an useful measure to control the release of lots into the system.

Both CONWIP and deterministic release policies dominate the Poisson release policy.

iii) The 9 machine 25 step system

The deterministic release policy dominates CONWIP release policy for this system in most of the experiments. 26 instances result in positive lower and upper bounds and 2 result in statistically insignificant differences. None of the instances resulted in CONWIP dominating the deterministic release policy. The average coefficient of variation of the arrival process for the CONWIP release policy is 0.735.

Thus, the total WIP in the system may not be an useful measure for controlling release of lots into large systems if the performance measure of interest is mean cycle time.

As demonstrated by Spearman and Zazanis(1992) and Wein(1988) the CONWIP release policy dominates the Poisson release policy.

6 ANALYSIS OF ACTUAL SYSTEM

The purpose of this analysis is to quantitatively study tradeoffs between different types of release policies used in semiconductor manufacturing systems. Again, the release policies of interest are the deterministic release, CONWIP and Poisson release policies. The manufacturing system used for the initial simulation results consists of the process flow and tool set defined by process engineers at SEMATECH. The tool set comprises of 45 different work stations each of which consists of parallel identical machines. Machines are highly unreliable as typical reliability levels are around 80%. The process flow comprises of 250 process steps. The number of process steps is much greater than the number of work stations. This means that there is reentrant flow in the system. Assuming no scrap, the start rate is the desired output from the system. It is generally specified as the number of wafers processed in a week. For simulations performed, this value has been perturbed about 4500 wafer starts per week. The number of parallel identical tools at each work station is determined using a spreadsheet calculation by ensuring that the utilization of the work station is less than 1 assuming a start rate of 4500 wafers per week.

The performance measures of interest are the mean and standard deviation of cycle time. The simulations were run using ManSim for 1000 days, with an initial condition of zero work-in-process inventory in the factory. Statistics are collected after a period of 200 days (warm-up period) to account for initialization bias.

The results of the simulation runs are shown below. Tables 9 and 10 provide the mean and standard dev. of cycle time as a function of start rates for the three release policies.

Table 9. Throughput vs. Mean Cycle Time

Throughput	Cycle Time (in hours)		
(wafers per week)	Deterministic	CONWIP	Poisson
4606	562.3	586.3	601.2
4501	497.6	522.4	540.2
4269	418.8	441	458.7
4195	402.1	419.3	433.4
3897	358.3	369.9	379.3
3498	325	334.3	341.6

Table 10. TP vs. Standard Dev. of Cycle Time

Throughput	Standard deviation of Cycle Time (hours)		
wafers/week	Deterministic	CONWIP	Poisson
4606	23.9	17	24.2
4501	19.6	11.9	19.4
4269	17.1	11.4	18.2
4195	16.6	11.2	15.3
3897	15.3	12.6	14.9

The above results show that the mean cycle time using

deterministic release policy is lower than that for Poisson and CONWIP release policy. The standard deviation under CONWIP is lesser than that for the other two release policies.

7 CONCLUSIONS

The results of the previous sections showed that it is not possible to assume that one release performs better than another under all circumstances. A careful study is required for each system to study the relationship between the departure process of the last machine and the utilization of the bottleneck. Based on these initial studies, it appears that the deterministic release policy dominates CONWIP and Poisson policy for larger systems while the CONWIP is the superior policy for small systems. There is also some preliminary evidence on the correlation between superiority of deterministic policy, high utilization and high CV of arrival under CONWIP policy.

For ASIC (Application Specific Integrated Circuits) products, if there is a need to ensure on-time delivery (minimization of both earliness and tardiness), CONWIP seems to be an effective release policy. If minimization of mean cycle time is of greatest importance, then the deterministic release policy is recommended. For make-to-market products like DRAMs, the use of CONWIP will ensure maximum throughput without any WIP explosions in the factory. In other words, it will prevent overloading the factory without explicit consideration for its actual capacity. The yield implications of these release policies is not very obvious from these results. But Srinivasan, Sandell and Brown (1994) have shown that a low mean cycle time translates to smaller yield losses. Hence, the deterministic release policy which results in smaller mean cycle time might lead to smaller yield losses. In reality, the actual time spent at each of the process steps has to be known before such conclusions can be drawn.

REFERENCES

Chance Frank, DELPHI: A C-Based Manufacturing Simulator, Version 8, Level 27.

Glassey C. Roger, and Mauricio G.C. Resende, Closed-Loop Job Release Control for VLSI Circuit Manufacturing, *IEEE Transactions on Semicon-ductor Manufacturing*, Vol. 1, No. 1, pp. 36- 46, 1988a.

Glassey C. Roger, and Mauricio G.C. Resende, A Scheduling Rule for Job Release in Semiconductor Fabrication, *Operations Research Letters*, Vol. 7, No. 5, pp. 213-217, 1988b.

Law M. Averill, and W. David Kelton, *Simulation Modeling and Analysis*, McGraw-Hill Inc., Second Edition, 1991.

Lu C.H Steve, D. Ramaswamy, and P. R. Kumar, Efficient Scheduling Policies to Reduce Mean and Variance of Cycle-Time in Semiconductor Manufacturing Plants, *IEEE Transactions on Semiconductor Manufacturing*, Vol. 7, No. 3, pp. 374-388, 1994.

Schruben L., H. Singh, and L. Tierney, Optimal Tests for Initialization Bias in Simulation Output, *Operations Research*, Vol. 31, pp. 1167-1178, 1983.

Segal M., and W. Whitt, A Queuing Network Analyzer for Manufacturing, *Proceedings of 12th International Teletraffic Congress*, Torino, Italy, 1988.

Spearman L. Mark, and Michael A. Zazanis, Push and Pull Production Systems: Issues and Comparisons, *Operations Research*, Vol. 40, No. 3, pp. 521-532, 1992.

Srinivasan K., R. Sandell and S. Brown, Correlation Between Yield and Waiting Time, *SEMATECH Technology Transfer Report No. 94112635 A-XFR.*

Wein M. Lawrence, Scheduling Semiconductor Wafer Fabrication, *IEEE Transactions on Semiconductor Manufacturing*, Vol. 1, No. 3, pp. 115-129, 1988.

AUTHOR BIOGRAPHIES

RAKA SANDELL is a Manufacturing Systems Engineer at MOS 12 Die Manufacturing at Motorola in Chandler, Arizona, specializing in production planning, real-time dispatching, and modeling, analysis and control of semiconductor manufacturing systems. She has a B.S. degree in Electrical Engineering from the University of Delhi and an M.S. degree in Operations Research from the University of Texas, Austin.

KRISHNA SRINIVASAN is a Senior Modeling Engineer with the Worldwide Enterprise Modeling Group of Motorola Manufacturing Systems at Motorola's Corporate Offices in Schaumburg, Illinois. His areas of specialization include developing solutions for sales, operations and production planning problems, queuing networks and factory modeling. He has a B.S. degree in Mechanical Engineering from the Indian Institute of Technology, Madras and an M.S. degree in Operations Research from the University of Texas, Austin.

Proceedings of the 1996 Winter Simulation Conference
ed. J. M. Charnes, D. J. Morrice, D. T. Brunner, and J. J. Swain

AN INVESTIGATION OF OPERATING METHODS FOR 0.25 MICRON SEMICONDUCTOR MANUFACTURING

James F. Hallas
Texas Instruments
East Building,
Mail Station 385
P.O. Box 655012
Dallas, TX 75265 USA

Jane D. Kim
Carolyn Internicola
SEMATECH
2706 Montopolis Drive
Austin, TX 78741-6499 USA

Charles T. Mosier
School of Business,
Clarkson University
Box 5790
Potsdam, NY 13699-5790 USA

ABSTRACT

This study investigates a number of operational issues associated with the control of microprocessor fabrication facilities, specifically expanding the domain of previous research to investigate the effects of lot size, test wafer proportion, and tool productivity on wafer fabrication performance. Response variables included cost and production performance metrics.

1. INTRODUCTION

The elements required to support and plan for a semiconductor production facility (fab) process technology and tool set is naturally partitioned into two distinct and yet interrelated groups. The first, termed *factory architecture*, encompasses strategic and technical decisions made concerning the manner in which the fab accommodates and enables operation of the process tool set. Factory architecture elements include: shell structure, tool layout, air handling, material handling and movement, fluid distribution, and tool accommodation systems.

Temporally, factory architecture issues dominate in the initial facility planning stages through tool set installation and product qualification, after which the facility infrastructure is typically set. With a typical shell design life of 10 to 15 years (Burnett 1995), the factory architecture decision making must anticipate the needs of a variety of tool sets and process technologies. This requires a degree of flexibility and robustness in all architectural decisions associated with facility design. Further, although the facility has become a shrinking portion of the overall cost of a fab, an ineffectual design (i.e., one that does not anticipate tool and process requirements and is not well integrated with operational business practices) will severely limit the long-term profitability of the business.

The second group of elements, *factory operations*, refers to the business practices that govern the operation and control of daily life within the fab. Factory operations include items such as production planning and shop floor control, yield management, staffing and training polices, modeling and simulation, decision support systems, and equipment and process control. These operational decisions dominate from the time of tool installation and product qualification throughout the active life of the fab until refit or shut down. Operational business practices are critical in the context of the evaluation and selection of factory architectural elements since the operational practices determine to what extent the potential of the factory architecture, to enable the tool set and process technology, will be realized in production.

This study reports the results of a full factorial simulation experiment (27 experimental treatments) which investigating the effects of the following operational factors: **lot size**, **test wafer level**, and **tool productivity**. The lot size factor was investigated at three levels, i.e., 12, 24, and 48 wafers per lot. Test wafers were run as a non-revenue producing product consuming 0%, 10%, or 20% of aggregate lot starts. Tool productivity refers to the *speed* of the process equipment as measured in wafers per hour. To obtain the tool productivity factor levels, the nominal equipment speeds were varied ±10%.

A baseline simulation model with a 24 wafer lot size, 0% test wafers, and nominal tool productivity was developed using an equipment and product set previously defined by the SEMATECH capital productivity (CP) group (Rowe, et. al., 1995). This baseline model contained four process flows and 168 pieces of primary processing equipment. The primary performance metrics analyzed include cost/cm^2 (in the year 2001), cycle time, throughput, and overall equipment effectiveness (OEE). Statistical analyses of the results illustrated the significant impact data quality (with regard to tool productivity and test wafer requirements) can have on the ultimate cost and performance of a fabricator. Overall, there was minimal performance difference between 24 and 48 wafer lot sizes. Statistical significance was found for all anticipated main effects. The impact of factor interactions was minimal, although their graphical examination allows considerable insight into fab system dynamics.

The life cycle of a modern semiconductor fabrication facility is one of constant flux. It is not uncommon for the volume and mix of products intended for a new fab to drastically change over the 18 to 36 month planning and construction phase. Once in production, the market-driven requirements to implement product enhancements, die area shrinks, tool upgrades, and new process technologies are constant. These requirements demand that the architecture of the fab superstructure and support systems be closely synchronized with efficient business and operating practices. This synchronization is required to support process technologies and tool sets that are routinely strained beyond their original design parameters. The SIA reiterates this message in the statement, *The semiconductor factory requires the following two levels of integration: 1) the process flow with the set of equipment and procedures that comprise the process and device technology, and 2) the various disciplines required to build and operate a successful factory.* (SIA 1994, p. 143).

2. PREVIOUS RESEARCH

Research into operational issues associated with microprocessor fabrication has traditionally received less attention than the more technical issues association with the engineering and physics of the production transformation process required to fabricate microprocessors. In the current climate of intense competitiveness, these operational issues have increased in importance. For example, recently Hung and Leachman (1996) attempted to apply both simulation and mathematical programming approaches to the production planning problem in the microprocessor fabrication environment. Earlier, a closed loop job release control system for the microprocessor fabrication environment was presented (Glassey and Resende 1988).. A much more complete review of production planning approaches applied to the microprocessor fabrication arena can be found in focussed survey papers (e.g. Uszoy, et.al. 1992).

Investigations into the nature of microprocessor fabrication range from empirical studies (e.g., of the relationship between die yield and cycle time (Cunningham and Shanthikumar 1996) to analytic studies (e.g. again, die yield verses cycle time (Wein 1992)). More and more, researchers are focusing on semiconductor scheduling issues (e.g., Glassey and Petrakian 1989, Lozinski and Glassey 1988, Lu, et. al., 1994, Wein 1988) and industry practices (Hood, et. al., 1995, Leachman 1994).

3. THE EXPERIMENTAL DESIGN

3.1. The Tool Set

Similar to the product specification, SEMATECH CP 200 mm wafer diameter, 0.25 μm critical dimension tool database was adopted for this study. The capacity balanced line for the four logic process flows contains 168 primary processing tools (see Table 1). The capacity balancing was completed to achieve 20,000 wafers starts per month with a cycle time to raw processing time ratio of less than 2.5 for a 25 wafer lot size.

The parametric tool data contained in the database are representative of the SEMATECH estimate of mature production tools purchased in 1998 and operating in 2001 (Rowe, et.al. 1995). In an effort to more accurately reflect typical (rather than best-of-breed) manufacturing operations, selective tool parameters were degraded. Reliability parameters: mean time between failure (MTBF), mean time to repair (MTTR), E10 downtime percentages, and E10 engineering time percentages from the CP 0.5 μm tool database (Elahi, 1995) were superimposed over the 0.25 μm values. Preventive maintenance times also were lengthened by 50% to dynamically model the myriad of minor productivity losses and delays routinely encountered in fab operations (Rowe, et.al. 1995, p. 13). Other tool parameters, including setup and run-out times, were unchanged.

It should also be noted that critical dimension, overlay, and patterned defect inspection tools for the lithography sector are the only metrology tools represented in this simulation model. The omission of other metrology or in-line test equipment is a significant deviation from most current leading edge semiconductor manufacturing environments and will be addressed in future studies.

3.2. The Product

Four (200 mm wafer diameter, 0.25 μm critical dimension) logic products with associated process flows as specified in the SEMATECH CP program (Rowe, et. al., 1995) were the basis for this study.

The four products included in the simulation model are as follows:

1. **High performance logic** (suitable for advanced microprocessor fabrication).
2. **Vertically modulated well** (VMW) (featuring MEV implant and shallow trench isolation).
3. **Low power logic** (based on the VMW process).
4. An **application-specific integrated circuit** (ASIC) process (no deep ultraviolet (DUV) lithography).

Throughout the discrete event simulations, each of the four products was allocated one-fourth of the available product wafer starts.

Table 1: Tool Type Summary

Process Sector	# of Tools	Tool Type	# of Tools
Litho-graphy	36	DUV Litho (248nm)	5
		0.35μ I-Line	6
		Large Field I-Line	7
		Litho Apply	2
		CD Measure	7
		Overlay	9
CFM	3	Defect Inspection	3
Etch	24	Oxide/Nitride Etch	8
		Metal Etch	5
		Poly/Si Etch	3
		CDE	2
		W Plug Etch	3
		Oxide Etch	3
Surface Preparation	19	Vapor Pre Clean	1
		Critical Pre Clean	3
		CMP Post Clean	3
		Std Wet Clean	12
Thermal	29	LPCVD Deposition	9
		Ox and Anneal	14
		RTP	6
Implant	7	Hi-I Implant	2
		Med-I Implant	4
		MEV Implant	1
BEOL Films	44	Asher	16
		APCVD BPSG	1
		CAP Oxide	4
		PECVD	1
		Ti, TiN Deposition	5
		Sputter TACS	5
		W CVD	2
		CMP Oxide	6
		ILD Gap Fill	4
Misc	6	Parametric Tester	3
		Backside Grind	2
		Laser Scribe	1
Total	**168**		**168**

An additional product, i.e., test wafers, was also included in our model. The test wafer product followed the high performance logic process flow. The only deviation from the SEMATECH specification of process flows was the slight adjustment of processing times required for variations in the lot size, and the inclusion of material movement times. The lot size modification was instituted to facilitate the inclusion of lot size as an experimental factor. The addition of material movement times created new steps in the flows but had no effect on existing ones.

3.3. Simulation Model Details

The fab layout was developed based on the CP Rev. 2 data set, resulting in model requiring 51,905 ft^2 of primary cleanroom space.

3.4. Experimental Factors

The experimental factors of lot size, tool productivity, and test wafers are discussed below. A brief description of the supplemental factors, WIP, and material movement time also is included.

Lot Size: Lot size was examined at three levels: 12, 24, and 48 wafers per lot. It was assumed that process and transfer lot sizes are equal. The lot sizes of 24 and 48 wafers were selected to reflect the lot sizes of existing fabs. A lot size of 12 was investigated as one that may be of interest for future fabs that produce a wide variety of low volume products. Twelve wafer lots are also of particular interest for proposed 300 mm fabs.

A primary objective in the development of the fab layout model was to minimize product travel while maintaining proper relative location of tools. To minimize product travel between tools, the process flow was analyzed to determine which tools should be in close proximity to one another. Tools that are expected to remain in the fab for a long period of time were located near the center of the fab, and tools required more frequent repair, upgrading, or modification were placed on the periphery. After determining tool location within the bays, the optimal arrangement of the bays was determined.

The tool layout was generally in a format known as a farm arrangement, i.e., with homogeneous tools grouped together. In only a few instances, such as ashers, were homogeneous tool types dispersed throughout the fab. The latter method was chosen for this study to simplify the calculation of process times where both lot size and tool productivity required modification.

Tool Productivity: The three tool productivity factor levels were coded as **low**, **nominal**, and **high**. The central level, nominal, represents the tool productivity (or speed of the tool), measured in wafers per hour. The low level corresponds to slowing down the productivity of each tool in the line by 10 %. This was achieved by multiplying the tool process times by 1.1. Likewise, the high level of tool productivity corresponds to speeding up the tools by 10%, as measured in wafers per hour. This level was implemented by multiplying the process times by 0.9. This treatment was similar for both serial and batch tools.

The wafers per hour used in the nominal level of tool productivity represents a consensus between IC manufacturers and equipment suppliers, as documented by the SEMATECH CP program. The productivity consensus is meant to be indicative of best-of-breed tools purchased in the 1998 time frame. The 10% variation in tool *speed* to achieve the high and low factor levels were chosen to illustrate the impact of deviations from intended wafer per hour productivity rates. This situation is one that may arise when wafer per hour rates realized on tools in production do not correspond to the rates assumed when the original capacity and space planning was completed for the fab. Such *speed* differences may arise for several reasons: tool suppliers may not be selected when the facility is designed, tools may not perform as advertised, final tool selection may be based on criteria other than throughput, or tools may be required to run processes that differ substantially from those anticipated at the time of facility design.

Test Wafers: Test wafers were examined at three levels: **0%**, **10%**, and **20%** of total wafer starts. Uniform terminology regarding test, monitor, control, non-prime, and non-product wafers often is not used. In this paper, *test wafer* is assumed to include the preponderance of non-product wafers that must be started in the line, regardless of wafer ultimate conditions of usage. This definition is intended to be similar to the SIA roadmap usage of *monitor wafers*, which refers to desired ratios between product and total wafer starts/usage [SIA 1994, p.164].

There are several ways to model test wafers. In reality, test wafers require a wide variety of films and structures to fulfill their myriad functions. An accurate portrayal of test wafers was beyond the scope of this project. However, the most important factors to consider for test wafers are their impact on the cost of silicon purchased, and the capacity loss of tools required to process and run the test wafers. For this study, test wafer costs were incurred as a percentage of total product wafer starts. The concept of test wafers as a capacity loss on the tool set was also included.

This capacity loss was implemented by the creation of a fifth product type: test wafers. The test wafer product followed the same process flow as the high performance logic. The number of test wafer starts per month was calibrated to the total 20K per month of releases. The other four product types were adjusted to maintain an equal 25% mix for the remainder of the available monthly wafer starts.

The 0% level was chosen to represent the optimal potential of the system as well as to allow easy comparison to other studies that do not consider test wafers as a capacity loss. Ten percent was considered to represent best-of-breed operations for fabricators in 1995 and the Roadmap goal for 1998. Finally, 20% was selected as representative of the 1995 Roadmap goal [SIA 1994, p.164] and perhaps more typical current fab operations.

3.5. Supplemental Factors

Follow-up experimentation was conducted including various level of shop work-in-process (WIP) levels and various levels of transport time associated with the automated material handling system (AMHS). For a complete discussion of the results of the follow-up experimentation see Hallas, et. al., (1996).

3.6. Metrics

Four metrics were collected in this study:

- cost per square centimeter of silicon produced in the year 2001 (cost/cm^2),
- throughput,
- cycle time, and
- OEE.

The first three measures represent three of the five defined factory integration performance metrics itemized in the Roadmap [SIA 1995]. The other two SIA-listed metrics, yield learning curve and extendibility, both require an extended temporal analysis of a fab life cycle that was beyond the scope of this effort. The last primary metric, aggregate OEE, is reported for each experimental treatment. While OEE is typically a tool-based measure (Ames, et. al., 1995, Barber 1995), it is reported here as an aggregate line measure.

A full factorial experiment was conducted (3^3) on the primary experimental factors. Discrete even simulation was in conjunction with static cost calculations to model fab performance. Ten batch means (replications) of one year were analyzed, with the initial transient data (one year) discarded. Randomly selected initial seeds for the simulator random number generators were selected to support the observation independence assumptions associated with analysis of variance.

4. ANALYSIS OF RESULTS

Cost $/cm^2 SI: For cost, as well as other response variables, there was little impact on the system between 24 and 48 wafer lots (See Figure 1).

This is, to some degree, due to the fact that the line balancing was done using a 25 wafer lot size. However, it is not clear that rebalancing the line would substantially improve efficiencies for 48 wafer lots. Further, the 12 wafer lot size incurred a large setup penalty. This relationship between the different lot sizes appears to be due primarily to the shifting bottlenecks that result from the unbalanced lines. Since 24 and 48 wafer lots share nearly the same setup-independent bottlenecks, there is little performance difference between these two lot sizes. Because the 12 wafer lot models have a setup-dependent bottleneck, there is a significantly greater cost for its use.

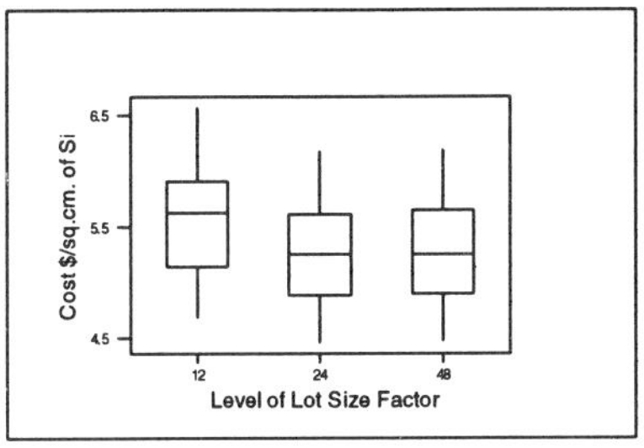

Figure 1: Lot Size Factor Effect on Cost $\$/cm^2$ SI

Alternatively, increasing tool productivity through all levels has a positive impact on cost from the increased throughput it allows (See Figure 2).

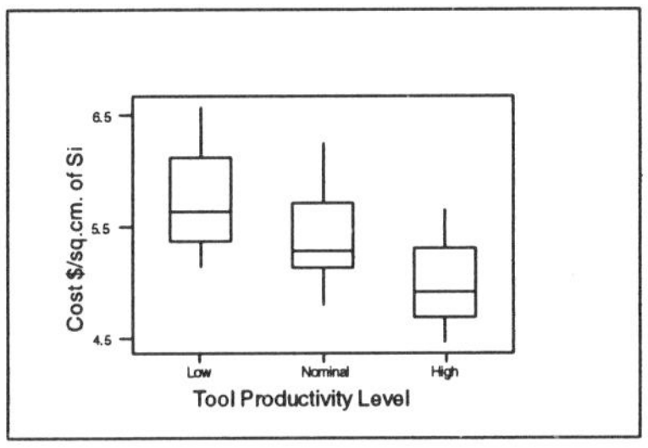

Figure 2: Tool Productivity Effect on Cost $\$/cm^2$ SI

Increasing test wafers negatively influences cost by using more production capacity for non-revenue producing wafers. (See Figure 3). As indicated in Table 2, all three factors are significant.

The statistical significance of the interactions may be, to some degree, due to the extremely long simulation runs which allow for very small within-group variation. Therefore, although most interactions are statistically significant, a focus on the statistically significant main effects may be appropriate.

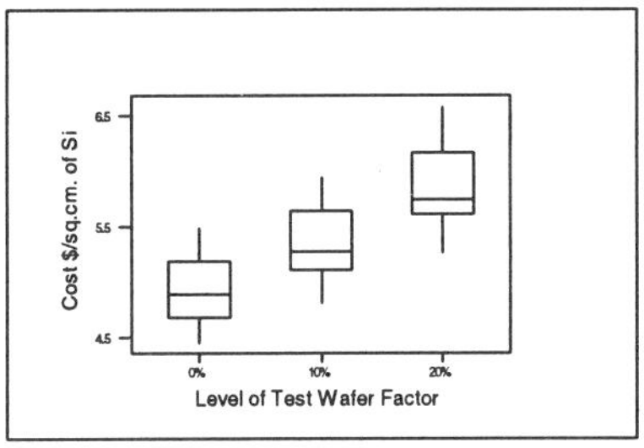

Figure 3: Test Wafer Effect on Cost $\$/cm^2$ SI

Table 2: ANOVA Results for Cost $\$/cm^2$ SI

Cost $\$/cm^2$ SI			
Source	**df**	**F**	**p-value**
<u>L</u>ot <u>S</u>ize (LS)	2	1066.77	4.7×10^{-121}
<u>T</u>ool <u>P</u>roductivity (TP)	2	4927.56	2.2×10^{-197}
<u>T</u>est <u>W</u>afer (TW)	2	7353.79	4.3×10^{-218}
LS×TP	4	16.00	1.2×10^{-11}
LS×TW	4	24.62	4.02×10^{-17}
TP×TW	4	23.78	1.3×10^{-16}
LS×TP×TW	8	1.18	0.3202

Throughput: Throughput of the simulated fab was measured in terms of wafers produced per week. All three factors show a substantial statistical impact on throughput (See Table 3).

Table 3: ANOVA Results for Throughput

Throughput			
Source	**df**	**F**	**p-value**
<u>L</u>ot <u>S</u>ize (LS)	2	1980.45	3.8×10^{-151}
<u>T</u>ool <u>P</u>roductivity (TP)	2	$1.0\times10^{+4}$	4.4×10^{-234}
<u>T</u>est <u>W</u>afer (TW)	2	$1.5\times10^{+4}$	2.9×10^{-255}
LS×TP	4	33.24	4.12×10^{-22}
LS×TW	4	2.66	0.033405
TP×TW	4	47.08	2.86×10^{-29}
LS×TP×TW	8	1.99	0.0483

Again, although the interaction effects are statistically significant, they explain much less of the variation than the main effects and are well behaved when plotted. As expected, throughput increases as tool productivity increases (See Figure 4).

Additionally, throughput decreased significantly as the number of test wafers increased (see Figure 5). When modeled as a direct incursion on capacity, the stark impact of the level of test wafers, a factor largely under engineering control, is clear.

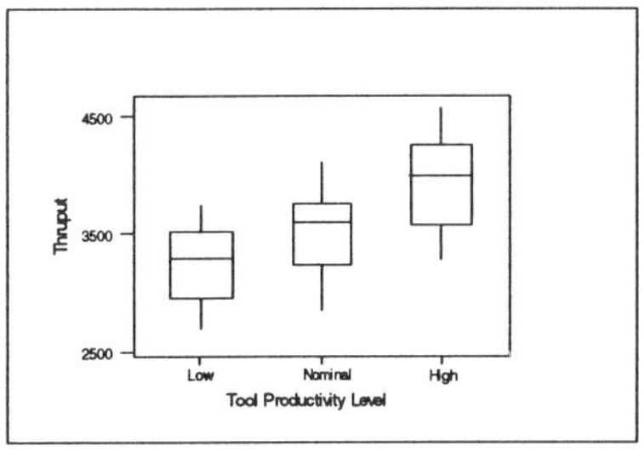

Figure 4: Tool Productivity Effect on Throughput

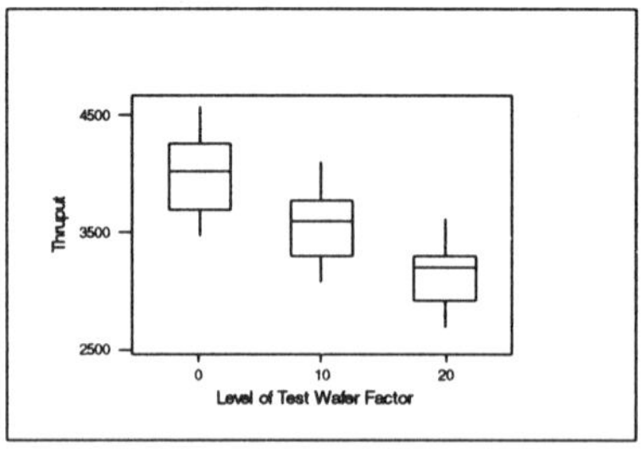

Figure 5: Test Wafer Effect on Throughput

Similar to the cost main effects, increasing from a lot size of 24 to 48 provides no significant benefit in throughput. In fact, in Figure 6 the impact is slightly negative.

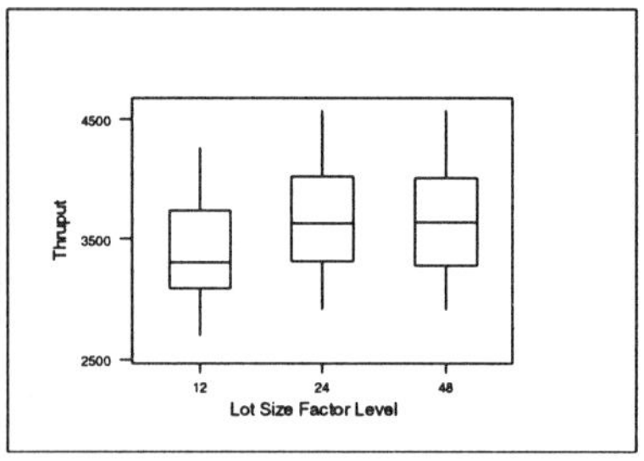

Figure 6: Lot Size Effect on Throughput

Much of the throughput loss seen in the 12 wafer lot size is explainable through the treatment of setups. This is discussed further in the cycle time section below. As expected, the throughput main effects are inverted relative to those of the previous metric, cost.

Cycle Time: Cycle time of the simulated fab was measurd in terms of hours. The percentage of test wafers had no impact on the cycle time of the four process flows (see Table 4).

This result can be seen by the small F value for test wafers and is logical due to test wafers following the same process flow as high performance logic. Otherwise, lot size (Figure 7) and tool productivity (Figure 8) trends follow that of cost and are inverted relative to throughput.

An examination of the 12 wafer lot size treatments will show that cost and cycle time increase while throughput and OEE (discussed later) decrease relative to treatments with 24 wafer lots.

The only counterintuitive result here is the cycle time seems to increase slightly when lot size changes from 24 to 48. This occurs due to the conservative setup assumptions used for this study. There are natural penalties that a small lot size would incur, such as more frequent sequence dependent setups due to increased switching between product and recipe queues, longer batching delays, and more congestion due to a higher number of lots in the line. However, in this study the 12 wafer lot size incurred further inefficiencies because only process times, and not setups, were modified for lot size changes. The setup frequency for some tools in lithography, CMP, and RTP have setups based on the number of lots processed. Therefore, treatments with 12 wafer lots incurred substantially more setups per wafer than 24 wafer lots (while 48 wafer lots received less). This is the cause of the observed cycle time increase for the 12 wafer lots.

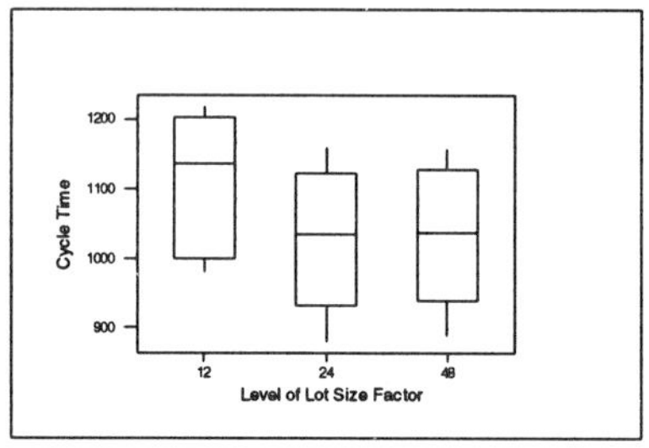

Figure 7: Lot Size Effect on Cycle Time

Table 4: ANOVA Results for Cycle Time

Cycle Time			
Source	**df**	**F**	**p-value**
Lot Size (LS)	2	1607.15	7.8×10^{-141}
Tool Productivity (TP)	2	8019.51	4.6×10^{-233}
Test Wafer (TW)	2	0.30	0.741092
LS×TP	4	21.01	6.8×10^{-15}
LS×TW	4	4.81	0.000945
TP×TW	4	0.82	0.513472
LS×TP×TW	8	0.96	0.467984

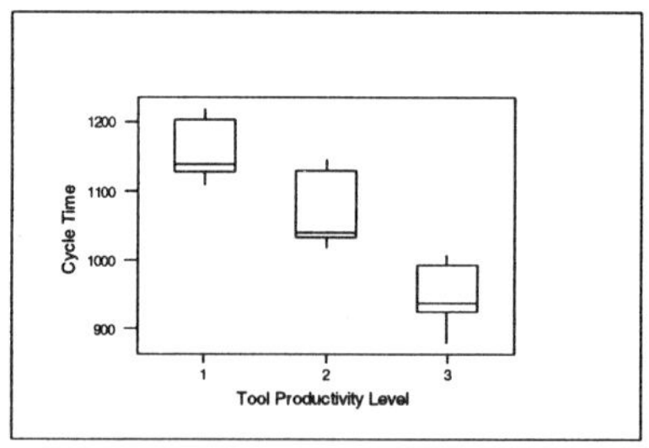

Figure 8: Tool Productivity Effect on Cycle Time

Although in many cases, significant interactions cause one to doubt the assessment of the signifiance of main effects, in this instance, the assessment of the significance of the test wafer factor seems appropriate.

OEE: OEE measures the amount of tool potential that is realized in production. Table 5 indicates significance of all main and interactive effects of this experiment. Alternatively, since tool productivity essentially alters the denominator for OEE (i.e., the anticipated potential of tools), it is practically unaffected by tool productivity, as can be seen on the main effects graph (see Figure 9).

Table 5: ANOVA Results for OEE

OEE			
Source	df	F	p-value
Lot Size (LS)	2	3134.29	3.1×10^{-174}
Tool Productivity (TP)	2	117.64	1.86×10^{-36}
Test Wafer (TW)	2	2.3×10^{4}	1.1×10^{-277}
LS×TP	4	36.79	4.85×10^{-24}
LS×TW	4	12.26	4.27×10^{-9}
TP×TW	4	3.81	0.005041
LS×TP×TW	8	2.99	0.003247

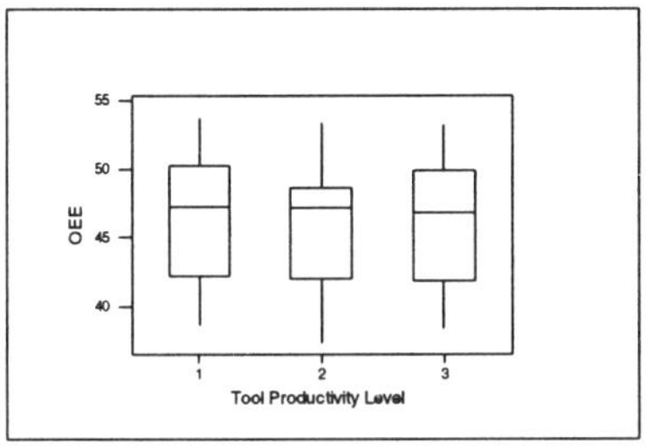

Figure 9: Tool Productivity Effect on OEE

In other words, changing the productivity (or speed) of a particular tool has little practical effect on its utilization and thus its OEE. This observation reiterates the need to ensure that a macroscopic view of the system is taken at all times to avoid myopic efforts at productivity improvement. Thus, although tool productivity is statistically significant in the model, the variation it explains is an order of magnitude less than the factors of lot size and test wafer percentage. Again, this statistical result may arises because of the cohesive simulation data provided by the extreme length of the replications.

Other than with the results associated with tool productivity, OEE results generally mimic those of throughput and follow the inverse of cost and cycle time (see Figures 10 and 11).

5. CONCLUSIONS AND EXTENSIONS

In general, results indicate minimal differences in performance between 24 and 48 wafer lot sizes, with significant differences in performance at 12 wafer lot sizes. Alternatively, clearly for specific markets or market conditions, 12 wafer lots still may be the preferable alternative. The impact that tool productivity has on system performance highlights the importance of data quality during the iterative strategic planning phase of a fab. Viewing test wafers as a capacity loss (in addition to the added material and logistical costs) demonstrated the direct impact that they can have on the bottom line. Lot size, tool productivity, and test wafer percentage were not found to interact materially with respect to measures. Overall, our results indicate an extreme sensitivity of a variety of fab performance measures to constraints determined by operating characteristics, such as raw processing time and setups, was shown. The interested reader is referred to the full paper [Hallas, et. al., 1996] for additional detail and discussion.

The results of follow-up experimentation indicated that inventory (WIP) level a large impact on fab performance, with results corresponding to those anticipated by queuing theory, while material movement time had a relatively negligible impact.

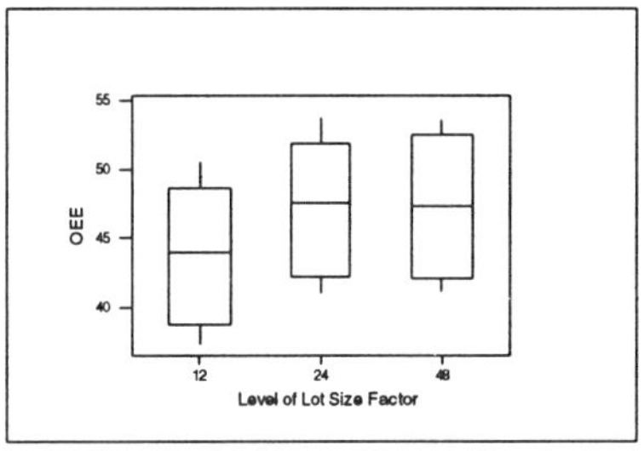

Figure 10: Lot Size Effect on OEE

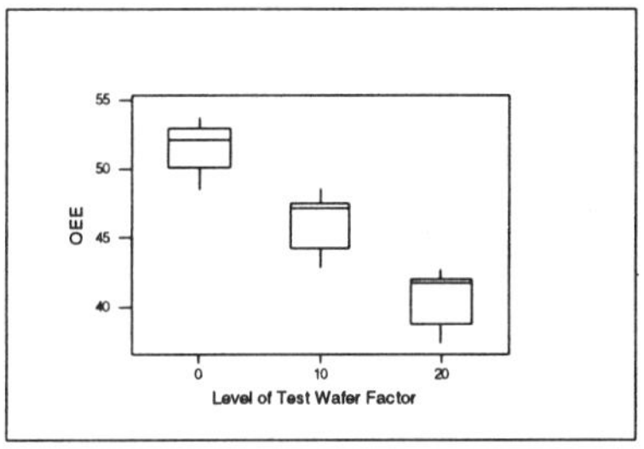

Figure 11: Test Wafer Effect on OEE

Several extensions are suggested by the results of this work. First, to provide a much needed validation, industry benchmarking analyses would be useful for comparing actual fabs with various treatments of an experimental design. Further, studies incorporating specific cost considerations in the layout design process (possibly seeking the minimum investment fabricator) would be useful. Possibly most pertinent to this study, the common practice of running various lot sizes concurrently within the fab should be investigated.

ACKNOWLEDGMENT

The authors wish to extend a special thanks to Steven Brown and Bill Jacobs for their guidance and support. In addition, we wish to thank Rick McKiddie and Scott Chandler for their technical contributions.

REFERENCES

Ames, V.A., J. Gililland, J. Konopka, R. Schnabl, and K. Barber. 1995. Semiconductor Manufacturing Productivity Overall Equipment Effectiveness (OEE) Guidebook Rev. 1.0. TECH. TRANS. #95032745A-GEN, SEMATECH, Austin, Tx.

Barber, K. 1995. Overall Equipment Effectiveness (OEE) Success Story. TECH. TRANS. #95032750A-XFR, SEMATECH, Austin, Tx.

Burnett, J. 1995. Microelectronic facility design for submicron ULSI chip manufacturing. Working paper. University of Arizona.

Cunningham, S.P. and J.G. Shanthikumar. 1996. Empirical Results on the Relationship Between Die Yield and Cycle Time in Semiconductor Wafer Fabrication. *IEEE Transactions on Semiconductor Manufacturing* 9:273--277.

Elahi, K. 1995. 0.50 Micro Baseline Fabrication Analysis. TECH. TRANS. #95042804A, SEMATECH, Austin, Tx.

Glassey, C.R. and M.G.C. Resende. 1988. Closed-loop job release control for VLSI, circuit manufacturing. *IEEE Transaction on Semiconductor Manufacturing* 1:36--46.

Glassey, C.R. and R.G. Petrakian. 1989. The use of bottleneck starvation avoidance with queue predictions in shop floor control. In *Proceedings of the 1989 Winter Simulation Conference*, 908--917.

Hallas, J.F., J.D. Kim, C. Internicola, 1996. An Experimental Investigation of Architectural and Operating Methods for 0.25 Micro Semiconductor Manufacturing. TECH. TRANS. #96013073A-TR, SEMATECH, Austin, Tx.

Hood, S., D. Miller, and D. Fronckowiak. 1995. User perspective on factory simulation. *Semiconductor Fabtech* 3:59--64.

Hung, Y-F and R.C. Leachman. 1996. A Production Planning Methodology for Semiconductor Manufacturing Based on Iterative Simulation and Linear Programming Calculations. *IEEE Transaction on Semiconductor Manufacturing* 9:257--269.

Leachman, R. 1994. The Competitive Semiconductor Manufacturing Survey: Second Report on Results of the Main Phase. Report CSM-08, UC Berkeley.

Lozinski, C. and C.R. Glassey. 1988. Bottleneck starvation indicators for shop floor control. *IEEE Transactions on Semiconductor Manufacturing* 1:147--153.

Lu, S.C.H., D. Ramaswamy, and P.R. Kumar. 1994. Efficient scheduling policies to reduce Mean and variance of cycle-time in semiconductor manufacturing plants. *IEEE Transactions on Semiconductor Manufacturing* 7:374--388.

Rowe, B., T. Tryba, K. Monnig, G. Escher, J. Caldwell, J. Greene, K. Elahi, and A. Alan. 1995. 0.25 Micron Process Equipment Requirements: Rev. 2.0-Closure. TECH. TRANS. #95042802A-ENG, SEMATECH, Austin, Tx.

Semiconductor Industry Association (SIA). 1994. *The National Technology Roadmap for Semiconductors.*

Uszoy, R., et al. 1992. A review of production planning and scheduling models in the semiconductor industry, Part I: System characteristics, performance evaluation, and production planning. *IEEE Transactions* 24:47--60.

Wein, L.M. 1988. Scheduling semiconductor wafer fabrication. *IEEE Transactions on Semiconductor Manufacturing* 1:115--130.

Wein, L.M. 1992. On the relationship between yield and cycle time in semiconductor wafer fabrication. *IEEE Transactions on Semiconductor Manufacturing* 5:156--158.

AUTHOR BIOGRAPHIES

JAMES F. HALLAS has earned graduate degrees in engineering and business from Clarkson University (Potsdam, NY). He has held manufacturing, research, and teaching positions with Raymond Corp. (Greene, NY), IBM (Burlington, VT), Clarkson University (Potsdam, NY), SEMATECH (Austin, TX), and International 300 mm Initiative (Austin, TX). He is currently a member of the start-up management team for Texas Instruments (Dallas, TX) DMOS 6 facility that will manufacture 0.25 and 0.18 micron logic devices.

JANE D. KIM graduated from the College of William and Mary in 1990 with a BS in Mathematics. She pursued graduate studies in Operations Research and Industrial Engineering at the University of Texas at Austin, during which time she worked at SEMATECH as a Factory Performance Analyst. She graduated from UT in 1996 with an MSE.

CAROLYN INTERNICOLA will be graduating from Rochester Institute of Technology in May of 1998 with a BS in Industrial Engineering. As part of here co-op program she has worked at SEMATECH as a Factory Performance Analyst and at DIGITAL Semiconductor as an Industrial Engineering Co-op.

CHARLES T. MOSIER is Associate Professor of Management at Clarkson University (Potsdam, NY). His research interests are in the area of manufacturing system management and design, with a focus on group technology and the application of simulation modeling to manufacturing problems.

Proceedings of the 1996 Winter Simulation Conference
ed. J. M. Charnes, D. J. Morrice, D. T. Brunner, and J. J. Swain

A COMPARISON STUDY OF THE LOGIC OF FOUR WAFER FABRICATION SIMULATORS

Scott J. Mason

Advanced Micro Devices
5204 East Ben White Boulevard, Mailstop 612
Austin, Texas 78741, U. S. A.

Paul A. Jensen

The University of Texas at Austin
Mechanical Engineering, Mailcode C2200
Austin, Texas 78712, U. S. A.

John W. Fowler

Industrial & Management Systems Engineering
Arizona State University
Tempe, Arizona 85287-5906, U. S. A.

ABSTRACT

Many semiconductor manufacturing companies use one of the following four simulators to aid in analyzing, planning, and operating their manufacturing facilities: Tyecin Systems' ManSim/X, AutoSimulations' AutoSched, Systems Modeling's Wafer Fabrication Template, or Chance Industrial Solution's Delphi, which Wright, Williams, and Kelly now licenses as Factory Explorer. A benchmark study of the four packages was conducted, using actual factory data. The packages each produce different results between replications of the data, due to the data's stochastic nature. The differences which exist between the simulation packages studied in this paper, along with the modeling assumptions required to model the data, caused differences in the experimental output metrics of interest: product cycle time and tool utilization parameters. Some possible reasons are each simulation package's batching logic, setup avoidance rules, and rework occurrence estimators.

1 INTRODUCTION

The semiconductor industry is a vital part of America's economy. With annual sales totaling billions of dollars, the manufacturers of these semiconductors work in a very competitive and lucrative arena. Though many different varieties of semiconductor chips exist, each manufacturer is faced with the enormous challenge of keeping up with this ever-changing industry's state-of-the-art technology. Products which are deemed advanced or new concepts can be outdated and replaced within two to three years. Companies which are best able to meet future customers' anticipated needs in a cost effective, timely manner will continue to thrive in the years to come.

There currently exists a variety of commercial, corporate, and university developed simulation and analytical tools for the analysis of wafer fabrication facilities (fabs). There have also been a number of strategies and rules developed to control the flow of semiconductor chips or product in a wafer fab. In order to investigate the possible merits of these strategies, many schedulers, planners, and manufacturing engineers have enlisted the aid of discrete event simulation and flow/queue analysis.

Without disrupting the current work-in-process in their own manufacturing facility, planners can perform "what-if" analyses using simulation and queuing models to find out whether a proposed change in the factory's schedule and/or toolset configuration will prove useful or profitable to the company in the long run. However, this type of analysis is a "point in time" look. Once the simulation analysis has been performed, it becomes obsolete as soon as the configuration or product mix of the factory changes. Indeed, this can occur on a daily basis in current fabs. Therefore, care must be taken to ensure the most current information from the fab is present in the simulation or queuing model.

Working with modeling and simulation practitioners from different semiconductor manufacturing companies around the world, it is evident that no standard semiconductor industry simulation package exists. However, the most widely used tools appear to be Tyecin Systems' ManSim/X, AutoSimulations' AutoSched, Systems Modeling's Wafer Fabrication Template, and Chance Industrial Solution's Delphi, which Wright, Williams, and Kelly now licenses as Factory Explorer.

This paper benchmarks the four factory performance analysis packages using actual semiconductor factory data. This data includes a list of the tools used in the semiconductor facility, the different types of products manufactured by the facility, as well as the process steps required to make each of the individual products. The same datasets are input into each package. Once the simulation models have been developed, simulation experiments are run under identical conditions in each package. After running the simulation experiments, a comparison of the output statistics of each of the packages is performed and explanations of both the similarities and differences in the results among the simulation software packages is given.

While there are many other attributes that could be used to compare the packages, only their predicted product cycle times and tool utilizations are examined. Other possible attributes include analysis run times, cost of the packages, ease of use, flexibility, etc. While these attributes were studied, they are not reported in this paper because results in these areas are generally obsolete by the time that they are published. This is due to the fact that most of the suppliers release at least two versions of their software each year. The authors have found, however, that the underlying way that basic elements (e.g. batching, setups, etc.) are handled does not change as often.

2 SIMULATION PACKAGE INFORMATION

The four simulation packages investigated were selected on the basis of their wide-spread use in the semiconductor industry, their ability to allow the user to build simulation models through flat ASCII text files, and their availability to the authors.

AutoSched 4.0 is based on another one of AutoSimulation's modeling packages, AutoMod. AutoSched provides added functionalities which are not present in the standard AutoMod software. However, a user must own the AutoMod software before attempting to run AutoSched. AutoSched allows the users to perform discrete event simulations by simply populating required data fields, such as process flow, tool set, and operator information. Standard "rules" exist for processing jobsteps, but the user does have the ability to write custom step instructions for each individual jobstep to further customize the simulation model. Used in combination with AutoStat, the user can run multiple replications of his or her simulation model to get a better idea of the long run average cycle time of a particular product or the utilization of a certain tool group (AutoSimulations, Inc. 1993).

Delphi 8.20 is a C-based simulator which allows the user to create discrete event simulation models by simply creating a process file and a routing file which contain process flow, tool set, and operator information. Extensive simulation run support is offered in Delphi, with the user having many different command line options at his or her disposal for configuring a particular simulation experiment run or set of experimental replications. Detailed queuing analyses are provided in the Delphi output report, along with product cycle times, queue lengths at each work station, rework information, setup occurrences, and other quantities of interest (Chance 1994). The Delphi simulation engine has been incorporated into Wright, Williams, and Kelly's Factory Explorer.

The ManSim/X 3.4 factory performance simulator was specifically developed to analyze the performance of wafer fabrication facilities by Tyecin Systems. A heavy semiconductor nomenclature exists in the software package which helps the model builder, who is familiar with the jargon of the industry, to build simulation models rapidly. Users can build discrete event simulation models by simply populating required data fields, such as product and process flow information, tool set data, and operator information. Many different selectable output reports are also available to the user which provide a variety of output statistics, including tool, operator, process, and cost information (Tyecin Systems, Inc. 1994).

WFT 1.23 is an application specific template which attaches to Systems Modeling Corporation's simulation tool suite, Arena. Arena allows the user to create simulation models by directly using the SIMAN and CINEMA languages. However, the WFT was developed to help users build simulation models which are specific to the semiconductor industry without having to know the SIMAN and CINEMA simulation languages. The user specifies both the pertinent data to be used in the model, such as resources, process flows, and operators, and the "rules" to be followed at each jobstep. By specifying the data and the model "rules," the WFT modeler can build a model which very accurately depicts almost any operation present in his or her fab. Also, in addition to performing discrete event simulations, WFT has the ability to perform flow and queue analyses on a simulation model, providing the user with a "rough" estimate of the performance of the fab (Systems Modeling Corporation, 1994).

3 SIMULATION MODEL DEVELOPMENT

While each of the simulation packages being investigated was created to simulate operations in a production environment, the way in which these operations are modeled varies among the tools. Many different types of processes exist in a semiconductor fab. A major challenge in trying to create a discrete event simulation model of a semiconductor fab is to accurately emulate the wide variety of operations which take place in the actual fab.

Many specialized aspects of a particular manufacturing process exist which may not be directly modeled in a particular simulation tool: specialized process steps, operator breaks, scheduled downtimes, product- and/or sequence-dependent setups, and company proprietary lot release policies. However, it is the modeler who must make the proper assumptions and decisions to most accurately represent his or her actual fab. Therefore, a modeler must thoroughly understand the following aspects of a factory's operation: load rules and batching, process step rules and delays, wafer rework and scrap, setup, lot releases, sequencing/dispatching rules, simulation clock type, and machine downtimes.

The amount of effort required by a modeler to create and implement his or her simulation model can often determine whether or not the simulation package will again be used in the future. While many simulation packages provide the user with a graphical user interface (GUI), these GUIs often can become tedious and inefficient for building large simulation models. Indeed, pointing and clicking a mouse thousands of times in order to populate dialog box fields is a very time-consuming task.

To this end, the developers of most simulation packages have provided their users with the capability to read in and create simulation models from an ASCII flat file(s) as an alternative to building the model using the program's GUI. The ASCII file model building capability may also prove useful to modelers who have access to their facility's shop floor control system. Data from this system may be downloaded and subsequently formatted by the modeler in order to obtain most of the inputs required by the simulation package.

3.1 Testbed Datasets

A testbed dataset format was developed to provide planners, researchers, and software suppliers with actual data and models that can be used to benchmark their control strategies and software (Feigin et al. 1994). The testbed includes actual manufacturing data from several wafer fabrication facilities, organized into a standard format. The data sets do not include real product names, company names, or other nomenclature that could serve to identify the source of the data.

The goal in development of the fab level data sets was to include the minimum information necessary to model a factory. In support of this goal, the data sets were kept as simple as possible, specifically by including data but not modeling assumptions. The resulting data is given in five ASCII flat files, containing product routings and processing times, rework routings, equipment information, operator data, and product starts information. A comments file also exists which gives information about the dataset, such as tool setup requirements and product-type descriptions.

To date, six testbed datasets have been compiled. In Mason (1995), four of the datasets are modeled in the four simulation packages mentioned above. This paper presents the results for two of these datasets: the simplest dataset and the most complex dataset.

3.2 Modeling Assumptions

Conversion programs were written to convert the testbed datasets into the proper ASCII file formats required by each of the four simulation packages. Since not all of the testbed datasets' fields map directly into all of the software packages, certain modeling assumptions were made to most accurately model the specified data. However, some assumptions were made globally, applying to each of the four testbed datasets as they were modeled in all four simulation packages. The two most important global assumptions are described below.

The time between successive machine failures was assumed to be exponentially distributed, as was the time required to repair a machine. Another global assumption was that product lot starts were made uniformly, with a constant amount of time elapsing between each successive lot release into the factory.

4 DESIGN OF EXPERIMENTS

After the testbed datasets had been properly converted into simulation models for each one of the four packages being investigated in this paper, simulation replications were performed with all four testbed datasets. Prior to performing a simulation replication, Delphi performs a capacity analysis and calculates the maximum component input rate for each of the different product types which are present in the simulation data. This value was used in all models as the maximum allowable input rate. It was decided to perform experiments with three different levels for wafer starts per week, in terms of a percentage of the maximum allowable rate: 55%, 75%, and 95% of the maximum component input rate.

Each simulation model was run for a period of 1,000 days, with 25% of the run (250 days) being used as a "warm-up" period to account for initialization effects of starting the model with an empty and idle fab. After viewing WIP vs. Time curves for each of the simulation models, it was decided that 250 days was a sufficient length for the warm-up period, as each of the simulation models had achieved steady-state by this time. Deterministic lot releases into the fab were used in each of the simulation models, with each product type having its own unique, constant interarrival time.

For each of the three start rate levels, twenty replications were performed so that long term averages could be calculated for product cycle time and resource utilization, the two main quantities of interest in this paper. Once these quantities were obtained, comparisons were made between the four simulation packages for the same wafers per week start rate.

5 EXPERIMENTAL RESULTS AND ANALYSES

For both of the datasets investigated, the experimental tool results are given for the three tools which had the most number of lots waiting to be processed at the work station, i.e. tools with the largest average waiting queues. The percentage of time each tool spent processing lots (busy), incurring a setup (setup), failing due to unscheduled machine breakdown (down), and waiting to process a lot (idle) is given in the experimental tool data table.

Then, the average cycle time for each product type over the 20 simulation replications is given for both of the start rate levels. Some possible reasons are given for the similarities and differences between the outputs of the simulation packages being investigated for both of the datasets.

5.1 Dataset Four

The fourth testbed dataset is comprised of seven products. Products 1, 2, and 5 all share an identical process routing, each following the same identical set of 92 process steps. The remaining four products, 3, 4, 6, and 7, share an identical process routing as well, with each product following a 19 process step flow. Therefore, there are a total of 111 process steps in this dataset, with a total of 35 different tool groups being present. There are no operators in this dataset, no probability of rework at any of the jobsteps, and setup is not required at any process step. This dataset is the smallest and simplest of all of the testbed datasets.

All of the simulation packages being benchmarked were able to successfully read-in and run this dataset. Table 1 shows the experimental tool utilization data

Table 1: Dataset Four Experimental Tool Data

	AutoSched						Delphi					
	Tool	Busy	Setup	Down	Idle	Avg Q	Tool	Busy	Setup	Down	Idle	Avg Q
55%	DFC4	50.6	0.0	7.7	41.7	0.5	DFB3	77.2	0.0	7.2	15.6	0.5
	DFB3	78.7	0.0	7.5	13.8	0.5	DFC4	50.6	0.0	7.8	41.6	0.5
	ASM2	35.5	0.0	14.3	50.2	0.3	ASM2	35.5	0.0	16.8	47.7	0.3
75%	DFC4	69.0	0.0	7.5	23.5	1.5	DFC4	69.0	0.0	8.0	23.0	1.4
	DFB3	87.6	0.0	7.6	4.7	0.8	DFB3	89.3	0.0	7.1	3.6	0.7
	ASM2	48.4	0.0	14.5	37.1	0.7	ASM2	48.4	0.0	16.4	35.1	0.7
95%	DFC4	87.4	0.0	7.3	5.3	5.8	DFC4	87.4	0.0	7.6	5.0	6.6
	ASM2	61.3	0.0	14.1	24.5	1.5	ASM2	61.3	0.0	16.3	22.4	1.7
	D1_9	60.3	0.0	12.2	27.5	1.1	QLES	76.0	0.0	11.8	12.2	1.4

	ManSim						WFT					
	Tool	Busy	Setup	Down	Idle	Avg Q	Tool	Busy	Setup	Down	Idle	Avg Q
55%	DFB3	93.1	0.0	6.7	0.2	1.9	DFC4	50.6	0.0	7.3	42.1	0.5
	DFC4	50.6	0.0	7.3	42.1	0.4	DFB1	48.8	0.0	7.3	43.9	0.3
	ASM2	35.5	0.0	14.2	50.3	0.3	ASM2	35.5	0.0	14.3	50.2	0.3
75%	DFB3	93.2	0.0	6.7	0.1	2.4	DFC4	69.0	0.0	7.3	23.7	1.3
	DFC4	69.0	0.0	7.3	23.7	1.2	QLES	64.7	0.0	10.8	24.5	0.7
	ASM2	48.4	0.0	14.1	37.5	0.5	ASM2	48.4	0.0	14.8	36.7	0.6
95%	DFC4	87.4	0.0	7.2	5.4	5.3	DFC4	87.4	0.0	7.6	5.0	6.7
	DFB3	93.3	0.0	6.6	0.1	3.0	QLES	82.0	0.0	10.7	7.4	3.6
	ASM2	61.4	0.0	14.0	24.6	1.3	ASM2	61.3	0.0	15.0	23.6	1.3

obtained from the replications of Dataset Four. The results show good agreement between the downtime percentages of the four simulation packages. As expected, the start rate of the simulation model does not have an effect on machine downtime, as this is a random variable which is independent of start rate. However, in some simulation packages, such as WFT, the user can model machine downtimes based on the total number of wafers processed by a machine. This type of downtime is dependent on start rate by definition and would increase with start rate, but was not used in this benchmarking study.

The majority of the tools in the top three bottleneck list have identical busy percentages as calculated by each simulation package. The factory's bottleneck, tool DFC4, is correctly identified by each of the packages at the 95% start rate level, with its busy, down, and idle percentages, along with average queue length being similar for all four of the simulation packages. Each package obtains similar results for the factory's bottleneck, DFC4, with the range of values obtained for the tool's average queue length being less than 1.5 lots between all of the packages, from 5.3 lots to 6.7 lots. In fact, the range of average queue lengths for each of the tools is very small, within one to one and one-half lots. The simplicity of this dataset, having no rework, no operators, and no setup, allows each of the simulation packages to compute similar output performance measures for the given input data.

Table 2 shows the experimental cycle time data obtained from the simulation replications of this dataset. Each of the simulation packages computes a similar cycle time for each of the different start rates being investigated. In both AutoSched and ManSim/X, it is evident that the tie-breaking rule used to select between identical lots at a given step in the process flow is FIFO, as the product types which have identical process routings experience increasing cycle times according to their position in the products list. Product 5 has a longer cycle time than does product 2 because product 2 lots are released into the factory before product 5 lots, therefore always being the "lead" product type. Delphi and WFT, however, do not seem to have this sort of tie-breaking rule, as identical product's cycle times are not arranged in ascending order. However, all four packages are comparable in terms of their estimations of cycle time.

5.2 Dataset Five

The fifth testbed dataset is made up of 21 different product types. The 21 different process flows range in size from 108 to 259 steps, with the total dataset containing 3,798 jobsteps and 85 different tool groups. This dataset is the largest of the testbed datasets studied in this paper. There are four operator groups present in this dataset, but no rework is required at any of the jobsteps. Setup is required on some of the tools in this dataset, with two of the tools, 19 and 66, requiring setup avoidance rules.

One type of setup avoidance rule specifies that if a lot arrives at the work station which requires the same setup that the machine is currently configured for, process that lot immediately. Another rule uses the above logic, but also places a maximum time delay on a lot, saying that if a lot has been waiting longer than this specified maximum time, re-setup the machine for the new lot and process that lot. This could play an important part in determining when particular lots begin processing and how fully the machine is utilized.

All of the simulation packages being evaluated successfully read this dataset into memory except the evaluated version of WFT, Version 1.23. WFT was unable to read Dataset Five into memory due to its size and/or the program's memory limitations. However, WFT's author, Systems Modeling Corporation, has

Table 2: Dataset Four Experimental Cycle Time Data

	AutoSched			**Delphi**			**ManSim**			**WFT**		
Product	**55%**	**75%**	**95%**	**55%**	**75%**	**95%**	**55%**	**75%**	**95%**	**55%**	**75%**	**95%**
1	152.6	167.0	198.2	148.5	162.5	204.1	150.9	162.8	195.1	148.1	161.9	217.3
2	157.0	170.6	201.3	148.5	162.4	204.1	155.8	164.8	198.0	148.3	161.9	217.3
3	23.0	24.1	24.8	24.8	26.2	28.4	21.4	22.2	23.7	25.4	27.3	32.9
4	24.5	25.5	26.4	24.6	26.3	28.3	22.5	24.0	26.3	25.4	27.1	32.4
5	163.6	175.5	205.5	148.5	162.6	204.1	157.4	167.1	200.6	148.5	161.9	217.4
6	25.7	26.3	27.9	24.7	26.2	28.3	22.5	24.8	27.6	25.6	27.3	32.6
7	26.5	27.6	28.8	24.7	26.2	28.5	23.6	26.5	29.7	25.2	27.0	32.3

All times are specified in hours.

stated that this size and/or memory limitation has been eliminated in later versions of the software package. Table 3 shows the experimental tool data obtained from the simulation replications of this dataset. Batching tools dominate this dataset, with all of the tools being identified as bottlenecks by AutoSched and ManSim/X belonging to this tool type. The only serial tool which is present on the bottleneck tool list is tool 19 on the Delphi list.

The setup required of tool 19 places it on Delphi's list, as this setup directly impacts the tool's average queue length. However, this tool does not appear in the lists of the other two packages. Tool 19 has a group-dependent setup associated with it which was modeled in Delphi, but not in AutoSched or ManSim/X, as group-dependent setups were not modeled in the latter two simulation packages.

A particular lot may be assigned a group ID which is then used for determining the need for a specific type of tool setup. This group-dependent setup is 45 minutes in duration, which definitely would increase the number of lots and the amount of time which they have to wait at the tool before it is properly setup. This type of setup obviously plays an important part in determining when particular lots are worked and how often the machine is busy processing actual lots.

The experimental cycle time data obtained from the simulation replications for Dataset Five are presented in Table 4. Surprisingly, AutoSched shows little variation in cycle time as the start rate is increased, especially when compared to the other simulation packages. The cycle times calculated by AutoSched for the 55% input rate case are very close to the values arrived at for the 95% case. While the AutoSched bottleneck tool lists show relatively small average queue lengths, the batching efficiencies, a measure of how "full" a batch is, are significantly increasing in AutoSched, allowing for more products to be processed at a time, thus keeping the overall product cycle time down. This, combined with the fact that group-dependent setups were not modeled in AutoSched and ManSim/X, keeps the two packages' cycle times below the results obtained in Delphi.

The ManSim/X and Delphi cycle time numbers increase significantly when going from the 75% to the 95% start rate. Setup on tool 19 is driving Delphi's cycle times up at higher input rates, especially due to the previously mentioned group-dependent setups, while tool 77's zero idle time is causing the increased cycle time in the ManSim/X results. The cycle time results also follow the expected trend based on the average queue sizes present in each of the simulation models according to Little's Law: increased queue lengths (WIP) leads to an increase in cycle time.

Another possible reason for Delphi having longer cycle times than the other simulation packages is that Delphi has both group- and sequence-dependent setup modeling capability. The testbed datasets used in this benchmarking study contain both types of setup. Consequently, both types of setup were modeled in Delphi. AutoSched's and ManSim/X's simulation models contain sequence-dependent setups, but not group-dependent setups. This suggests lower cycle times for these two packages due to the decrease in the amount of setups which must be performed.

Table 3: Dataset Five Experimental Tool Data

	AutoSched						Delphi						ManSim					
	Tool	**Busy**	**Setup**	**Down**	**Idle**	**Avg Q**	**Tool**	**Busy**	**Setup**	**Down**	**Idle**	**Avg Q**	**Tool**	**Busy**	**Setup**	**Down**	**Idle**	**Avg Q**
55%	77	91.8	0.0	5.0	3.1	1.4	32	96.0	0.0	4.0	0.0	2.1	32	95.8	0.0	4.1	0.1	3.9
	32	95.3	0.0	4.6	0.1	1.1	34	93.3	0.0	4.4	2.3	1.7	77	94.3	0.0	5.0	0.7	3.9
	75	82.8	0.0	5.0	12.2	1.0	89	96.6	0.0	0.0	3.4	1.5	75	90.6	0.0	5.0	4.4	2.2
75%	77	93.9	0.0	5.0	1.1	2.1	19	38.0	51.4	6.7	3.9	4.3	77	95.0	0.0	4.9	0.1	6.2
	75	87.4	0.0	5.0	7.6	1.6	32	95.9	0.0	4.0	0.2	2.7	32	95.6	0.0	4.1	0.3	5.6
	32	95.3	0.0	4.6	0.1	1.5	89	98.5	0.0	0.0	1.5	2.4	89	97.4	0.0	0.0	2.6	4.9
95%	77	94.6	0.0	5.0	0.4	3.2	19	48.1	45.1	6.7	0.1	12.2	77	95.1	0.0	4.9	0.0	9.2
	75	91.0	0.0	4.9	4.1	2.3	4	89.0	0.0	4.3	5.8	5.9	89	92.8	0.0	0.0	7.2	8.7
	32	95.6	0.0	4.4	0.1	1.9	77	94.4	0.0	5.2	0.4	4.3	32	95.1	0.0	4.1	0.8	7.7

Table 4: Dataset Five Experimental Cycle Time Data

Product	AutoSched			Delphi			ManSim		
	55%	75%	95%	55%	75%	95%	55%	75%	95%
1	163.3	164.9	166.0	272.6	303.8	377.7	173.2	192.5	272.2
2	177.6	180.1	181.6	329.2	358.2	428.2	190.9	207.1	246.0
3	166.2	170.0	172.6	285.5	319.4	398.7	181.2	202.3	287.6
4	198.4	198.4	202.2	342.7	375.2	452.2	221.5	237.9	286.6
5	183.2	185.8	188.5	345.9	390.3	488.7	200.9	223.0	327.2
6	197.0	198.1	199.6	418.1	460.4	557.0	223.9	238.3	285.9
7	200.5	202.0	205.8	331.6	371.4	466.4	222.5	246.3	313.5
8	210.7	212.9	214.5	399.7	436.9	530.5	233.7	251.5	298.1
9	196.6	197.7	197.6	266.2	301.1	384.1	220.3	242.6	354.7
10	154.0	156.8	159.6	304.6	343.4	434.3	171.2	191.9	277.8
11	283.8	291.4	303.2	369.1	405.9	493.9	397.8	452.6	532.9
12	176.2	179.3	183.6	441.4	488.5	615.6	198.3	222.2	284.2
13	292.5	302.3	315.9	291.0	325.9	409.7	390.6	442.2	540.8
14	187.9	190.5	193.3	350.6	384.8	468.3	210.4	225.6	267.8
15	283.3	293.4	307.0	423.0	471.8	598.4	359.3	395.4	454.8
16	351.1	361.6	379.7	498.6	551.9	689.6	502.7	583.1	758.1
17	306.9	316.4	330.7	445.9	493.0	622.7	401.3	452.6	542.2
18	99.6	101.4	101.9	247.6	272.5	335.6	107.7	120.0	147.4
19	104.0	109.2	110.3	213.8	243.1	302.4	116.9	135.4	204.1
20	103.6	103.8	106.0	254.9	280.9	343.4	113.2	125.4	156.7
21	110.8	111.4	113.2	208.0	233.4	295.8	119.9	133.6	197.3

All times are specified in hours.

6 CONCLUSIONS AND COMMENTS

Dr. George Box is widely credited with saying that, "All simulation models are wrong, some are useful." Simulation models are "wrong" in the sense that they are only computer representations of real systems. Through the use of dispatching rules, random number generators, and sampling from statistical distributions, simulation models attempt to imitate the actual dynamic processes found in a manufacturing facility. "Useful" can be interpreted as does the simulation model accomplish the tasks which the modeler intended it to perform? Does it answer the question(s) which the modeler was asking? The accuracy and functionality of a simulation model determines both how wrong is "wrong" and how useful is "useful."

A simulation package can produce different results between replications of the same model, due to its stochastic nature and the enormous number of events which occur in the model: lot moves, random number sampling, resource contingencies, lot rework and scrap, and forming of batches. The different ways in which each of the simulation packages studied in this paper handle these events, coupled with the modeling assumptions required to model the testbed datasets, caused differences in the experimental results. While the simulation packages' results are comparable for the smallest, simplest dataset, Dataset Four, a wide range of values for both product cycle time and machine utilization were observed among the software packages for Dataset Five.

Some possible reasons for these discrepancies which were discussed are each simulation package's batching logic, setup avoidance rules, and rework occurrence estimators. The results confirm that batching efficiency, a measure of how "full" each batch is, increases with start rate as expected. A noticeable increase in the batching efficiencies of the tools helped to keep the tool's busy percentage fairly constant and average queue lengths at the batching tools somewhat low. This also in turn can help to shorten the product's average cycle time as more lots experience concurrent processing with other lots at one or more places in the process flow. The setup avoidance rules used by the simulation packages dictate the frequency of required setups, thus affecting both machine utilization and overall product cycle time. Finally, the rework occurrence estimators used by the simulation packages, being either deterministic or probabilistic, directly affect the number of lots which must pass through a given job step and therefore, a given work station.

This benchmarking project was definitely a larger undertaking than at first expected, mostly due to the amount of time required to decipher the nuances of each simulation package and the huge amounts of data which must be accounted for in a simulation model. Due to the size of simulation model data, there is no doubt that simulation packages which allow the user to ASCII load-in simulation model data are the only practical way to build factory-sized simulation models.

Even after the model data has been read into the simulation package, great care must be taken during the large amounts of time necessary to check the data's integrity and the simulation model's assumptions into the software package. As described in section 3.1, various assumptions were required when building the simulation models for the software packages being studied in this paper. A careless oversight or misunderstanding of the simulation package's assumptions and/or methodologies can produce simulation results which are of no use to the modeler and might even have detrimental effects on a factory if used to make policy decisions.

This paper represents the culmination of one and one-half years of modeling, simulation, experimentation, and analysis. During this time, newer releases of each of the simulation packages evaluated in this paper have become available to modelers, in effect making the results in this paper obsolete. This paper is not intended to provide "the answer" for all of the questions associated with the modeling and simulation of manufacturing facilities. Rather, it is a point-in-time comparison study of the basic logic in four state-of-the-art simulation packages which have been used to model semiconductor fabs. The author encourages the reader to perform additional simulation experiments to gain a better understanding of both his or her own modeling needs and requirements, and to get a feel for the types of simulation packages which are available to modelers today.

REFERENCES

AutoSimulations, Inc. 1993. *AutoSched and the Simulator System.* Bountiful, Utah: AutoSimulations, Inc.

Chance, F. 1994. *Delphi: Integrated Capacity Analysis and Factory Simulation.* Version 8.20. Dublin, California: Chance Industrial Solutions.

Feigin, G., J. W. Fowler, J. Robinson, and R. Leachman. 1994. Semiconductor Wafer Manufacturing Data Format Specification. SEMATECH, Austin, Texas.

Mason, S. 1995. A Comparison Study of Four Factory Performance Simulators. Master's Report, Mechanical Engineering Department, The University of Texas at Austin, Austin, Texas.

Systems Modeling Corporation. 1994. *Wafer Fabrication Reference Guide.* Sewickley, Pennsylvania: Systems Modeling Corporation.

Tyecin Systems, Inc. 1994. *ManSim/X User's Manual.* Release 3.4. Los Altos, California: Tyecin Systems, Inc.

AUTHOR BIOGRAPHIES

SCOTT J. MASON is an Operational Modeling Engineer at Advanced Micro Devices (AMD) in Austin, Texas, specializing in cost and capacity modeling. He has a B.S. degree in Mechanical Engineering and an M.S. in Operations Research and Industrial Engineering, both from The University of Texas at Austin. Prior to working at AMD, he was an independent simulation consultant to SEMATECH.

PAUL A. JENSEN is the Cullen Trust Professor of Mechanical Engineering. He has B.S. and M.S. degrees in electrical engineering from the University of Illinois and the University of Pittsburgh, and the Ph.D. in operations research and industrial engineering from Johns Hopkins University. He has taught at the University of Texas since 1967. In addition to prior employment with the Westinghouse Electric Corp., he has practical experience from a number of consulting engagements. He is the co-author with J. Wesley Barnes of the book Network Flow Programming, the winner of the 1980 Book-of-the-Year Award from the Institute of Industrial Engineers.

JOHN W. FOWLER is an Assistant Professor in the Industrial and Management Systems Engineering Department at Arizona State University. Prior to his current position, he was a Senior Member of Technical Staff in the Modeling, CAD, and Statistical Methods Division of SEMATECH and an Adjunct Assistant Professor in the Graduate Program in Operations of the Mechanical Engineering Department at The University of Texas at Austin. He received his Ph.D. in Industrial Engineering from Texas A&M University and spent the last 1.5 years of his doctoral studies as an intern at Advanced Micro Devices. His research interests include modeling, analysis, and control of manufacturing (especially semiconductor) systems. He is a member of IIE, INFORMS, and SCS.

Proceedings of the 1996 Winter Simulation Conference
ed. J. M. Charnes, D. J. Morrice, D. T. Brunner, and J. J. Swain

APPLICATION OF A 2-STAGE GROUP-SCREENING DESIGN TO A WHOLE-LINE SEMICONDUCTOR MANUFACTURING SIMULATION MODEL

Theodora Ivanova

Lucent Technologies
9333 S. John Young Parkway
Orlando, FL 32819

Mansooreh Mollaghasemi

Department of Industrial
Engineering and Management
Systems
University of Central Florida
Orlando, FL 32816

Linda Malone

Department of Industrial
Engineering and Management
Systems
University of Central Florida
Orlando, FL 32816

ABSTRACT

The focus of the paper is on the application of an experimental design methodology to a semi-conductor manufacturing simulation model. A complex whole-line simulation model of a semiconductor fab is built. Seventeen input factors are set for investigation through a 2-stage group-screening experimental design. A multiple response regression metamodel is built to define the relationships between the significant input factors and the four response variables of interest. The combination of simulation modeling methods with experimental design and regression analysis techniques allows the development of a flexible tool for capacity analysis of a semiconductor manufacturing facility.

1 INTRODUCTION

As semiconductor companies look for ways to increase their competitiveness, many are turning to simulation modeling to help them control their facilities.

One of the major manufacturers of Application Specific Integrated Circuits (ASIC), has formed an operations research team whose main task is to create simulation models and to assist the company's management in making its future business decisions.

At present, simulation is the only tool that is capable of modeling the complex, often random nature of the semiconductor manufacturing environment. Simulation modeling, however, has certain drawbacks, such as the lack of optimization capability. Also, the simulation model is often referred to as a "black box", because the explicit relationships between its input and output parameters are typically unknown. That is why simulation modeling becomes most effective in combination with other analysis methods, such as experimental design and regression analysis.

Experimental design allows examination of the input factor effects on the system response variables. In cases where the effects of less than 11 input factors are studied, Biles (1984) recommends the application of fractional factorial designs for the simulation experiments. Research presented by Hood and Welch (1990, 1993) shows the application of fractional factorial Resolution III and IV designs in modeling the logistics of semiconductor manufacturing lines. In cases where more than 11 input factors are studied, the recommended type of design is a group-screening design. A 2-stage group-screening procedure was introduced by Watson (1961) and further developed for multiple-stage designs by Patel (1962) and Li (1962). Significant contribution to the group-screening design method has been made by Mauro and Smith with their numerous papers on the robustness and effectiveness of the method (Mauro and Smith 1982, 1984, and Mauro 1984).

Based on the experimental design results, regression analysis equations are built to define the relationships between the input factors and the measures of performance. The regression metamodel concepts were introduced to simulation by Kleijnen (1979). A long-term advocate for the implementation of multiple response regression metamodels to simulation output analysis is Friedman (1984, 1987, 1989).

Although group-screening design combined with regression metamodel analysis appears well suited for the analysis of large-scale semiconductor manufacturing simulation models, there is a limited number of papers dealing with this type of experimental design application. The objective of the present study is to build a whole-line simulation model and to estimate the future Work-In-Process (WIP) levels, cycle times and throughputs for two basic semiconductor products. Further, the most significant input factors for the production measures of performance are to be identified through the application of group-screening design to the simulation model.

The organization of the paper is as follows: in Section Two, an overview of the theoretical aspects of the group-screening experimental design and multiple response regression metamodels is presented. Section Three presents the whole-line simulation model definitions and the output analysis steps. Section Four includes the application of a 2-stage group-screening design to the simulation model and the multiple response regression metamodel analysis. Finally, Section Five summarizes the results from present research.

2 THEORETICAL BACKGROUND

2.1 Two-Stage Group Screening Design

Watson (1961) suggests that the k input factors in a model can be separated into g groups of f factors each, by any method. Each group is then considered as a single factor called group-factor. At the upper level of a group-factor, all factors in that group are at their high levels. The lower level of a group-factor is determined by setting all individual factors at their low levels. If a group-factor is found to be significant, a second stage of the design is set, where the original factors from the significant groups are tested individually. If after the first stage there is still a considerable number of important factors left in the experiment, further regrouping might be applied and the group-screening process will then have more than two stages (Li 1962 and Patel 1962). Kleijnen (1987) recommends keeping the "unimportant" factors at fixed levels during the next stages of the experiment.

2.2 Multiple Response Regression Metamodel

After every group-screening design stage, the most significant input factors are determined by the use of regression analysis. Multiple response regression metamodels relate each system response to the most significant input factors. The simplest multiresponse metamodel is the additive first-order (linear) model. Multiple linear regression equations are typically built after each experimental stage and a global F-test is used to analyze the hypothesis that all regression model coefficients, β_{km}, equal zero. Then, through individual t-tests, the significant input factors are determined. The insignificant factor deletion procedure is iterative, (i.e. one factor is deleted at each step, after which the t-tests are run again). After including only the significant group-factors, a linear regression metamodel is built. The significant input factors are used as individual or group-factors in the second stage of the group-screening design. At the second experimental stage, similar tests are performed to determine the significant input factors and so forth.

3 THE WHOLE-LINE SIMULATION MODEL

3.1 Model Assumptions and Definitions

The ManSim/X simulator, developed by Tyecin Systems Inc., was used to build the fab simulation model. ManSim/X has been specifically designed for capacity analysis and production planning of semiconductor manufacturing facilities.

The whole-line simulation is a model of a 6" semiconductor wafer fab with more than 250 machines and operators, grouped into multiple work areas. Two basic recipes for two products are included in the model. Different operational rules are used to control the interactions between the model elements.

3.2 Model Validation and Output Results

The model validation processes included variable reasonableness tests, conceptual and operational validity tests, comparisons with mathematical models, etc. Based on the WIP autocorrelation functions for five runs, a warm-up period of 90 days was determined. All statistical calculations used in further simulation runs were based on the truncated "steady-state" time series with a length of 270 days. An overall confidence level of 0.80 was set for four system measures of performance, namely cycle times for Product 1 and Product 2, and throughputs for Product 1 and Product 2. By using Bonferroni's inequality, the individual confidence level for each response was set at 0.95.

Queue size analysis showed that implanters, steppers and etchers are the three most critical production facility groups. The comparatively large queue sizes at these workstations, which form even in the case of stable WIP output time-series, remind of the danger that the workstations could easily become a fab "bottleneck" at certain conditions. Therefore, there was a need for further study to identify the factors which are significant for the performance of these three workstations and for the overall factory performance.

4 GROUP-SCREENING EXPERIMENTAL DESIGN

4.1 Group-Screening Design - Stage I

The objective of the group-screening experimental work was to determine the importance of certain input factors on the four simulation model responses, namely, the cycle times for Products 1 and 2, and throughputs for Products 1 and 2. Seventeen input factors were selected at the beginning of the experiment. Fifteen input factors are related to the three most critical wafer fab facility groups, namely implanters, steppers and etchers and two input factors are related to the overall fab performance. Following is a list of the input factors for Stage I screening process:

X1 = MTBF (Mean Time Between Failures) for steppers
X2 = MTBF for implanters
X3 = MTBF for etchers
X4 = MTTR (Mean Time to Repair) for steppers
X5 = MTTR for implanters
X6 = MTTR for etchers
X7 = Lot Dispatch Rule for steppers (the rule by which a lot is chosen from the queue in front of a machine)
X8 = Lot Dispatch Rule for implanters
X9 = Lot Dispatch Rule for etchers
X10 = Number of steppers
X11 = Number of implanters
X12 = Number of etchers
X13 = Operator/machine Ratio for steppers
X14 = Operator/machine Ratio for implanters
X15 = Operator/machine Ratio for etchers
X16 = Lot Release Rule (the rule which organizes the lot release into production)
X17 = Hot Lots percentage for both products.

The seventeen input factors were tested for significance through a group-screening design. By using factor grouping rules (Watson 1961), seven group-factors were formed at the first design stage, as shown in Figure 1.

A two-level fractional factorial 2_{IV}^{7-3} design with 16 design points and 5 replicates for each was planned. A full factorial design was set for the first 4 variables *(A, B, C, D)*. The rest of the design input variables *E*, *F* and G were defined as design generators, where

$$E = ABC \,;\; F = BCD \,;\; G = ACD \qquad (3)$$

The defining relation for this Resolution IV design is :

$$I = ABCE = BCDF = ADEF = ACDG = BDEG = ABFG = CEFG \qquad (4)$$

Table 1 defines the low and high levels for each group factor. The low level for each factor was chosen to be more constraining to the simulation model compared to the high input factor level. Trial runs were performed to make sure that the model is stable under the low factor level setting. Then, the high factor levels were set as an improvement over the base level for each factor.

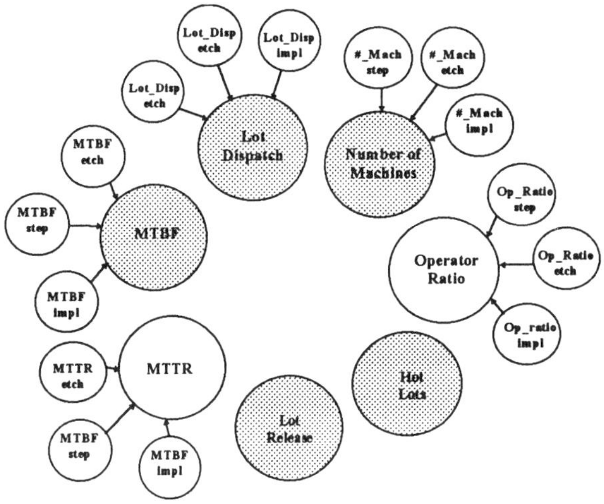

Figure 1: Group-Screening Design - Stage I

This method for setting the low and high factor levels ensures that there is sufficient resource capacity and that the model is stable for all experimental runs (Hood and Welch 1992).

Table 1: Group-Factor Levels

Group - Factor Description	Name	Low Levels (-1)	High Levels (+1)
MTBF	A	base	2 * base
MTTR	B	base	.5 * base
Lot Release Rule	C	FIFO	Fewest Lots at Next Queue
Number of Machines	D	base	base + 1
Operator / machine Ratio	E	base	1.5 * base
Lot Dispatch Rule	F	Random	Constant
Hot Lots	G	10%	5%

As a next step the JMP software, a product of SAS Institute Inc., was used to build a multiple response regression metamodel based on the Stage I results and to determine the significant group-factors. It was assumed that the simulation model output results could be generalized in a linear regression metamodel with no interactions between the group

screening factors and no quadratic terms. The qualitative nature of some of the input factors and the narrow range between the low and high factor levels for others, let us maintain the linearity assumption.
The *F*- test results appear in Table 2. All probability values *p* are less than 0.05; therefore, all four models are statistically significant. The regression coefficients for the four response variables were tested one at a time using *t* tests and are presented in Table 3. The shaded regression coefficients have *p* values of less than 0.05 and are considered statistically significant.

In conclusion, at the end of stage I of the group-screening design, input group factors MTBF, Lot Dispatch Rule, Number of Machines, Lot Release Rule and Hot Lots were declared statistically significant for at least one of the four output responses. Therefore, these group factors were further investigated in Stage II of the experimental design. On the other hand, the MTTR and Operator-to-Machine Ratio group-factors were found insignificant with respect to all responses and were dropped from the next experimental stage.

Table 2: Least-Squares Analysis Table by Response Variable - Stage I

Response Variable	Source	d.f.	Sum of Squares	Mean Square	F Ratio	Prob>F	RSquare
Cycle Time for Product 1 (hrs)	Model	7	21379.22	3054.17	7.86	0.0047	0.87
	Error	8	3108.655	388.58			
	Total	15	24487.88				
Cycle Time for Product 2 (hrs)	Model	7	17057.48	2436.78	7.944	0.0045	0.87
	Error	8	2453.95	306.74			
	Total	15	19511.43				
Throughput Product 1 (wafers)	Model	7	4407.69	629.671	15.939	0.0004	0.93
	Error	8	316.04	39.505			
	Total	15	4723.73				
Throughput Product 2 (wafers)	Model	6	5270.347	878.391	8.681	0.0025	0.85
	Error	9	910.67	101.186			
	Total		6181.018				

Table 3: Linear Regression Metamodel Coefficients - Stage I

			Input Factors							
			Intercept	MTBF	MTTR	Lot Dispatch Rule	Number of Machines	Operator/ Machine Ratio	Lot Release Rule	Hot Lots
			Regression Coefficients							
Response Variable	Average Value	Standard Deviation	$\beta 0$	$\beta 1$	$\beta 2$	$\beta 3$	$\beta 4$	$\beta 5$	$\beta 6$	$\beta 7$
Cycle Time for Product 1 (hours)	762.2	23.55	762.24	-18.94	-1.53	-14.23	-12.63	-3.47	-6.71	-4.8
Cycle Time for Product 2 (hours)	698.4	21.31	698.38	-19.06	-2.99	-11.01	-11.51	-0.74	-5.56	-5.08
Thruput for Product 1 (wafers)	1294.7	23.41	1294.7	2.23	-0.59	8.51	-0.59	-2.11	3.67	1.44
Thruput for Product 2 (wafers)	2183.8	26.45	2183.8	2.08	0.29	3.2	6.91	2.775	6.76	8.8

Note: The shaded cells mark the significant input factors regression coefficients.

4.2 Group-Screening Design - Stage II

In Stage II, the five significant group factors were separated into individual factors. The separation of the three significant group-factors (MTBF, Lot Dispatch Rule and number of machines) and the two single factors (Lot Release Rule and "hot lots") resulted in 11 individual input factors to be examined in the second stage of the experimental design, as follows:

A = MTBF for steppers
B = MTBF for implanters
C = MTBF for etchers
D = Lot Dispatch Rule for steppers
E = Lot Dispatch Rule for implanters
F = Lot Dispatch Rule for etchers
G = Number of steppers
H = Number of implanters
I = Number of etchers
J = Lot Release Rule
K = Hot Lots percentage for both products.

To obtain a Resolution IV experimental design and to minimize the number of simulation runs, a Plackett-Burman design with 24 runs was performed at this stage. Each run was replicated 5 times. A total of 120 simulation runs were performed, equivalent to almost 120 hours of computer run time. As in Stage I of the experiment, the global F-test for the model adequacy indicated that all four models are significant (see Table 4). Table 5 displays the regression coefficients for all input factors, where the significant input factor coefficients are shaded. As shown in Table 5, the MTBF at etchers, the Lot Dispatch Rule at implanters and etchers, the number of machines at steppers and implanters, and the Lot Release Rule have significant positive effects on the two cycle time variables, (i.e. cycle times decrease when these factors are set at their high levels - see Table 1). The "hot lots" percentage does not have a significant effect on the average cycle time for the products, which could be expected. Although the "hot lots" cycle time decreases, the cycle time for the "regular" lots increases, therefore, the average product cycle time does not change. Factors which have significant influence on the throughput levels are the MTBF at implanters, Lot Dispatch Rule on the implanters and etchers, number of machines in the steppers, etchers and implanters groups, the Lot Release Rules and the percentage of Hot Lots.

Table 4: Least-Squares Analysis Table by Response Variable - Stage II

Response Variable	Source	d.f.	Sum of Squares	Mean Square	F Ratio	Prob>F	RSquare
Cycle Time for Product 1 (hrs)	Model	10	22229.76	2222.98	11.49	< .0001	0.896
	Error	13	2514.34	193.41			
	Total	23	24744.11				
Cycle Time for Product 2 (hrs)	Model	10	25557.18	2555.72	11.23	< .0001	0.898
	Error	13	2958.51	227.58			
	Total	23	28515.69				
Throughput Product 1 (wafers)	Model	8	3362.83	420.35	8.26	0.0003	0.815
	Error	15	763.49	50.89			
	Total	23	4126.32				
Throughput Product 2 (wafers)	Model	12	17936.24	1494.69	20.67	< .0001	0.957
	Error	11	795.54	72.32			
	Total	23	18731.79				

Table 5: Linear Regression Metamodel Coefficients - Stage II

			Input Factors											
			Intercept	MTBF_ step	MTBF_ impl	MTBF_ etch	Lot Disp_step	Lot Disp_impl	Lot Disp_etch	Number_ Mach_ step	Number_ Mach_ impl	Number_ Mach_ etch	Lot Release	Hot Lots
			Regression Coefficients											
Response Variable	Average Value	Standard Deviation	$\beta 0$	$\beta 1$	$\beta 2$	$\beta 3$	$\beta 4$	$\beta 5$	$\beta 6$	$\beta 7$	$\beta 8$	$\beta 9$	$\beta 10$	$\beta 11$
Cycle Time for Product 1 (hours)	747	26.2	747	-1.32	-1.42	-7.26	-7.04	-4.15	-5.29	-8.62	-13.11	3.13	-11.8	-5.56
Cycle Time for Product 2 (hours)	684.1	22.1	684.1	1.12	-2.81	-8.72	-9.73	-0.7	-7.78	-11.52	-9.58	-0.33	-9.26	-3.46
Thruput for Product 1 (wafers)	1289.2	27.11	1289.2	0.41	1.27	-2.23	-0.28	4.65	3.27	0.56	1.72	5.03	3.34	0.81
Thruput for Product 2 (wafers)	2181.6	32.63	2181.6	-0.49	7.65	1.44	-2.08	-3.88	10.46	5.02	3.87	4.57	4.41	8.62

Note: The shaded cells mark the significant input factors regression coefficients.

Note that the decrease in the percentage of "hot lots" from 10% to 5% has a significant positive influence on the throughput of Product 2. The only factor which has no significance on any of the four response variables is the MTBF on steppers. The Lot Release Rule, on the other hand, has a significant influence on all four responses. A conclusion could be made that the higher the number of response variables, the harder it becomes to identify factors that are totally insignificant for all response variables.

5 CONCLUSIONS

A whole-line simulation model of an ASIC wafer fab was built and validated. This model is a flexible tool for capacity analysis of the semiconductor manufacturing facility. Additionally, a 2-stage group-screening experiment was designed to study the interactions between the input factors and the multiple measures of performance. The experience with performing group-screening design on a simulation model with multiple responses leads us to believe that although group- screening design is efficient in cases with a large number of input factors and one response variable, it is not as efficient when multiple response variables are involved. At the end of stage I, only two out of the seven group-factors were declared insignificant which brings us back to the still considerable number of eleven individual factors at the second stage. Therefore, it could be concluded that the greater the number of response variables, the less efficient the group-screening design method becomes.

REFERENCES

Biles, William E. 1984. Design of Simulation Experiment. *Proceed. of the 1984 Winter Simulation Conference*, 99 - 104.

Friedman, Linda Weiser. 1987. Design and Analysis of Multivariate Response Simulations: The State of the Art. *Behavioral Science*: 32, 138 - 148.

Friedman, Linda Weiser. 1984. Multivariate Simulation Output Analysis: Past, Present and Future. *Proceed. of the 1984 Winter Simulation Conference*, 277 - 279.

Friedman, Linda Weiser. 1989. The Multivariate Metamodel in Queuing System Simulation. *Computers and Industrial Engineering:* 16, No. 2, 329 - 337.

Hood, Sarah J. and Peter D. Welch. 1992. Experimental Design Issues in Simulation with Examples from Semiconductor Manufacturing. *Proceed. of the 1992 Winter Simulation Conference,* 255 - 313.

Hood, Sarah J. and Peter D. Welch. 1990. The Application of Experimental Design to the Analysis of Semiconductor Manufacturing Lines. *Proceed. of the 1990 Winter Simulation Conference*, 303 - 309.

Kleijnen, Jack P. C. 1979. Regression Metamodels for Generalizing Simulation Results. *IEEE Trans. on Systems, Man, and Cybernetics:* 9, No. 2, 93 - 96.

Kleijnen, Jack P. C. 1978 *Statistical Tools for Simulation Practitioners*, Marcel Dekker Inc.

Li, C. H. 1962. A Sequential Method for Screening Experimental Variables. *Journal of American Statistical Association:* 57, 455 - 477.

Mauro, Carl A. and Dennis E. Smith. 1982. Factor Screening in Computer Simulation. *Simulation*, 49 - 54.

Mauro, Carl A. 1984. On the Performance of Two-Stage Group Screening Experiments. *Technometrics*: 26, No. 3, 255 - 264.

Patel, M.S. 1962. Group-Screening with more than two stages. *Technometrics*: 4, 209 - 217.

Watson, G. S. 1961. A Study of the Group Screening Method. *Technometrics*: 3, No. 3, 371 - 388.

AUTHOR BIOGRAPHIES

THEODORA IVANOVA is a member of Technical Staff in the Manufacturing Systems Engineering Department at Lucent Technologies in Orlando. She received a B.A. in Electrical Engineering from St. Petersburg, Russia, and a M.S. degree in Operation Research from the University of Central Florida in Orlando, Florida. Her present research interests include simulation modeling and analysis, production planning and control of semiconductor manufacturing system, and experimental design. She is a member of IEEE.

MANSOOREH MOLLAGHASEMI received her Ph.D. in Industrial Engineering from the University of Louisville in 1991. She also holds a B.S. and an M.S. in Chemical Engineering from the University of Louisville. She is currently an Associate Professor in the Department of Industrial Engineering at the University of Central Florida. She is actively involved in funded research through Lucent Technologies, Rhone-Poulenc Rorer Pharmaceuticals, and the US Army. Her research interests involve simulation modeling and analysis, multiple criteria decision making, decision support systems, and multiple response simulation optimization. She is a member of IIE and INFORMS.

LINDA C. MALONE is an associate professor in the Industrial Engineering Department of the University of Central Florida. She earned a B.S. in mathematics from Emory and Henry College, a M.S. in mathematics from the University of Tennessee in Knoxville, and a Ph.D. in statistics from Virginia Polytechnic Institute and State University in 1979. She is an Associate Editor of the *Journal of Statistical Computation and Simulation.* Her major interests are in Design of Experiments, Response Surface Analysis, and Regression. She has a number of publications and has co-authored an introductory statistics text book.

Proceedings of the 1996 Winter Simulation Conference
ed. J. M. Charnes, D. J. Morrice, D. T. Brunner, and J. J. Swain

TRW/TED ACCELEROMETER WAFER PROCESS PRODUCTION FACILITY: MANUFACTURING SIMULATION

Lawrence S. Farey

193 S.W. Seminole Drive
Aloha, OR 97006

ABSTRACT

Simulation is a cost affective way to analyze and make improvements to dynamic complex manufacturing systems, such as semi conductor processing. The TRW/TED Accelerometer Wafer Process Production Facility is a semi conductor processing facility which makes automotive air bag sensors. A manufacturing simulation of the facility was developed to determine what changes could be made to the facility and/or personnel to maximize production of the sensor chips. A simulation model of the existing facility was developed and validated. Eight experiment models were built. The models included adding: lot starts, a second shift, personnel to the first shift, equipment, personnel and equipment, personnel and a second shift, changing process times, and eliminating process steps. Throughput was chosen as the performance criteria to determine the alternatives effectiveness.

The model for the existing conditions indicated a production throughput of 825,552 chips per year. Experiment model which added personnel and a second shift produced the greatest increase in throughput, 2,555,280 chips per year. This is an increase of 210% over the existing conditions. The conclusion is to add personnel to the first shift and add a second shift.

The conclusion was based on operational, technical and physical data, it did not take into consideration financial objectives. Using cost information in the analysis will give a more accurate solution.

1 MANUFACTURING SIMULATION

1.1 Why Simulation Is Needed In Manufacturing

Three key words indicate why simulation is an essential tool when attempting to analyze and improve a manufacturing system: complexity, uncertainty and cost.

Suppose we wish to determine the production capacity of a factory. The factory is defined as follows: produces a single product, has a straight line product flow, no machine breakdowns, no product rework, no operator interface, and no variation in process times. Using this definition a mathematical formula or spreadsheet program can be used to determine production capacity. Now suppose we want to add a second product to the factory leaving all other assumptions the same. We could use linear programming to optimize the production capacity of the factory. These models are static (not influenced by time) and deterministic (contains no random variables) mathematical models.

The problem with this scenario is most manufacturing systems are dynamic and stochastic in nature. Gogg and Mott (1992) state, "A dynamic system implies action. Factors which influence a system can change as time progresses (a manufacturing system is subject to part scheduling changes, equipment breakdowns, part defects, etc.). Stochastic suggests that these changes can vary indiscriminately."

Simulation models are excellent imitators of dynamic/stochastic systems because they are able to account for the affects of changes occurring within a system. They can analyze complex systems with nonlinear product flows and operator interface with machines. They can account for uncertainty by handling product rework, machine breakdowns, and variable process times.

Organizations must be able to react quickly to changes from the market place, changes in technology, and changes on the factory floor. Using a validated simulation model, developed with the proper software, a company can make decisions on the computer before implementing them on the factory floor. This is a very cost effective way of determining a course of action.

1.2 Simulation Software

There are numerous simulation software packages currently available. In choosing a package the user should look for three characteristics: speed, flexibility and animation. Taylor II software was chosen because it excels in all three characteristics.

Speed: Taylor II uses an optimized simulation algorithm that guarantees fast simulation runs. Simulations can be run for a fixed time or until a certain condition is reached.

Flexibility: Taylor II is a menu driven package that allows the user to graphically build models by positioning resources like machines, buffers, conveyors, reservoirs, warehouses, transports and labor on the screen. Details are added to the model by using pre defined parameters, distributions and Taylor interface language. Information on model runs is collected through several pre defined reports or through user created reports.

Animation: Animation servers two purposes: model verification (does the model behave as intended?) and presentation of the model and results. Because the animation is integrated into Taylor II and can be used instantly, it will verify (or refute) the model immediately.

Taylor II's combination of these three characteristics gives the experienced simulation user the flexibility and performance to model the most complex systems and present the results in a timely manner with as much detail as the project requires.

2 ACCELEROMETER FACILITY MODEL

2.1 Problem Definition

The production of the air bag sensors (semi conductor chips) developed by the accelerometer facility has been a success. Requests for the sensor by the automotive industry is increasing. To determine if these requests can be met, the maximum capacity of the facility must be established.

There area four methods of maximizing the capacity of the existing facility: add employees, add process equipment, add employees and process equipment, and change or eliminate process steps.

Changing the physical layout will not be considered since the original layout was developed taking into consideration process flow, cleanliness requirements, and air flow dynamics of the clean rooms.

Adding employees is restricted floor space in the clean rooms, the skill level required to operate the equipment, and machine cycle times. The machine cycle times prevent an employee from being assigned to only one machine or work station.

Equipment should be added at the bottleneck points. These points were determined through interviews with the accelerometer personnel and by observation. The bottlenecks occur at the Ultratech 1000 and the Dektak in lithography and the VWR ovens in metals.

Through experiments being conducted the accelerometer personnel feel the process can be changed as follows: the process time for the VWR oven can be reduced, the process time for the Dektak can be reduced, and the process steps for the sensor stops can be eliminated. With more experience and training the set up times for the Ultratech 1000 and the Dektak can be reduced.

2.2 Simulation Goal

Determine the maximum output of accelerometer parts (chips) from the TED Accelerometer Wafer Processing Facility at Space Park in Redondo Beach.

2.3 Criteria

This is the performance index of how the alternative's effectiveness will be evaluated. The performance criteria will be throughput (number of wafers processed per week).

2.4 Assumptions

1. The set up and process time values supplied by the accelerometer facility personnel accurately describe the operation of the facility.
2. One shift is 8 hours.
3. Each employee is considered proficient at operating all equipment in their area.
4. No personal fatigue or delays are considered for the employees.
5. There are 12 wafers in each lot, 2,100 chips on each wafer.
6. There is no scrap.
7. The equipment breakdown (MTBF) and repair time (MTTR) will use the Weibull distribution. Times are estimated by accelerometer facility personnel.
8. There is no starting WIP.

2.5 Input Parameters.

1. Quantity of equipment.
2. Quantity of labor.
3. Machine cycle times (with distributions).
4. Machine set up times.
5. Machine breakdown and repair times (with distributions).

2.6 Model Descriptions

Model 0 - Existing conditions: 12 lots in the system, one shift, two employees in lithography, two employees in metal, one employee in final assembly.

Model 1 - Increase lot starts from 12 to 16. All other conditions remain the same.

Model 2 - Add a second shift. Duplicate the first shift conditions.

Model 3 - Add a third employee to the lithography area and a third employee to the metal area. Assume the added employees have the same skill levels and are proficient at operating equipment as the existing employees. All other conditions remain the same.

Model 4 - Add one Ultratech 1000, one Dektak, and one VWR oven. All other conditions remain the same.

Model 5 - Combine models 3 and 4.

Model 6 - Combine models 2 and 3.

Model 7 - Change the process time for the VWR oven from 65 hours to 8 hours. Change the set up time for the Ultratech 1000 from 30 minutes to 15 minutes. Change the Dektak set up time from 10 minutes to 0 minutes and the run time from 68 minutes to 30 minutes. All other conditions remain the same.

Model 8 - Remove the process step for the sensor stops. All other conditions remain the same.

"Multiple model replications are always required when stochastic are involved. A general rule of thumb is to always perform at least three to five replications for each experiment." (Gogg & Mott 1992) Four simulation runs of 2000 hours each were performed for each experiment.

2.7 Results

The results of the simulation runs are ranked in descending order based on throughput.

Table 1: Wafer and Chips Throughput

Model	Ave. Wafer/Week	Ave. Chips/Year
6	23.40	2,555,280
2	15.96	1,742,832
8	12.18	1,330,056
5	9.90	1,081,080
3	9.87	1,077,804
7	9.60	1,048,320
4	8.34	910,728
1	8.02	875,784
0	7.56	825,552

2.8 Analysis

When stochastic are used statistical analysis should be performed to determine if there is a difference in the means of the simulation results. Comparisons will be made between models where the mean values in the results are "close" to each other. As an example, the average difference between model 2 and model 3 is approximately 6 wafers per week. These models will not be compared because from observation there is a significant difference between the two means. Statistical analysis is not required to conclude that model 2's throughput is superior to model 3's throughput.

A paired-t test will be used to compare close alternatives. The paired-t test uses the statistic for small sample test concerning mean, the one-sample t-test. The mean and standard deviation used in the calculations are found by using the difference between the two means being tested. The t value calculated is used to test the hypothesis that the two means are equal. The confidence level is 90%. The test requires the number of model replications performed for each alternative be equal. Common random number streams must be used with model replications.

If t>2.353, there is a statistical difference between the means and the model with the larger mean is superior. If t<2.353, there is no statistical difference between the means and one model is not superior to the other.

Comparisons will be made between: (1) model 0 and model 1, (2) model 0 and model 4, (3) model 3 and model 4, (4) model 3 and model 5, (5) model 3 and model 7, (6) model 4 and Model 5, (7) Model 4 and Model 7, (8) model 5 and model 7.

Table 2: t Value Comparison

Comparison	t value	Choice
1) model 0 and model 1	1.0395	none
2) model 0 and model 4	3.8060	model 4
3) model 3 and model 4	7.0952	model 3
4) model 3 and model 5	0.1901	none
5) model 3 and model 7	2.1828	none
6) model 4 and model 5	10.069	model 5
7) model 4 and model 7	8.3446	model 7
8) model 5 and model 7	5.0000	model 5

2.9 Conclusions

Simulation is intended to provide a reasonable estimate of the systems behavior. There are hundreds of scenarios (experiments) which can be simulated using the model. The scenarios chosen were determined by estimating which conditions could most likely be changed.

The goal of the simulation study was to determine the maximum output of accelerometer parts from the facility. The models are listed in order, from most desirable to least desirable, based on the selected performance criteria. (1) model 6, (2) model 2, (3) model 8, (4) model 5 or model 3, (5) model 3 or model 7, (6) model 4, (7) model 1.

The to maximize production based on throughput one employee should be added to both lithography and metals and a second shift should be added.

A test model with two employees added to both lithography and metal was run. The additional employees did not add to the throughput of the facility due to the dynamics of the product, equipment, and employees.

3 RECOMMENDATIONS

3.1 Financial Considerations

This simulation project used operational, technical and physical data. The conclusions were based on this data and did not take into consideration financial objectives. Therefore, it is important that potential solutions be evaluated using financial data. It is necessary to determine what financial data should be included in the simulation analysis. Generally, manufacturing simulations will impact operational costs (direct material, direct labor, and overhead). However, they may also impact general and administrative costs, capital costs, and depreciation.

An important objective for any organization is to reduce wait time and inventory carrying costs. Wait time is non-value added time and can dramatically add to the cost of production. Carrying costs have traditionally been hidden costs and are usually not included in the production decision making process. Simulation is the most effective tool for measuring and substantially reducing wait times and carrying costs.

Gogg & Mott (1992) state that costs can be integrated with simulation projects in three ways: full integration, partial integration, and external integration. Full integration requires all cost information be processed within the simulation. No additional data processing is required to obtain the cost results used for analysis and determining a course of action. Partial integration means some cost information is processed within the simulation. Other costs are processed externally and the results are added to the simulation cost results for analysis and action. External integration means all cost information is processed outside the simulation. The cost information is integrated with the simulation results for analysis and course of action.

Taylor II software is capable of allowing full integration of cost data by using fixed, variable and interest cost parameters for all elements in the model. Fixed costs represent the amount of money per time interval. The fixed costs are always counted, whether a machine is busy or not. A negative value can be entered to accumulate depreciation costs. Variable costs represent the amount of money per busy time interval. The variable costs are only counted over the period of time the element was busy. This can be utility costs and consumable material costs. Interest cost values represent an amount of money per time interval per product. The more products in the system and the longer they are stored, the higher the total interest cost. This parameter is used to accumulate wait time and carrying costs.

Since Taylor II software is able to perform full integration of costs, it is recommended that cost information be accumulated and added to the simulation models. The models should be run again and the results analyzed using throughput and cost data. This analysis will give a more accurate and complete solution of the problem of maximizing throughput by keeping the costs at a minimum. It will be counter productive to increase the throughput of the facility if the cost per part produced is increased in the process.

3.2 Increase Model Accuracy

In addition to output analysis, design of experiment and equipment justification analysis; the accelerometer facility model can be used for process control, shop scheduling, raw material scheduling and product cost analysis. If the model is to be used for additional analysis it is recommended that more accurate empirical data be gathered.

Two sets of data must be collected: employee data and machine data. Employee data consist of machine set up times and employee process times (attended machine cycle times). This data can be collected manually or semi-automatically. Machine data consists of machine process time (unattended machine cycle times), mean time between failure (MTBF) and mean time to repair (MTTR) or "down time". This data can be collected manually, semi-automatically or automatically. The cost and accuracy of data collection increases moving from manual to automatic.

Manual data collection is accomplished by using log sheets. The operator uses a watch and enters start and stop times on the log sheet for employee data and machine data. The problem with this form of data collection is reduced accuracy. The operator may be too busy to enter the time on the log sheet when the operation is finished or, in the case of an unattended machine, the operator may not be present when the machine cycle is finished. In these cases, a guess is made for the stop time. The MTBF is usually an estimate (if any information is recorded at all). The MTTR is usually recorded in a log. The majority of the information supplied by the accelerometer personnel was from log sheets.

Semi-automatic data collection can be accomplished using bar code readers to record employee data and machine data. The operator does not require a watch or writing instruments to record stop and start times. The start and stop times are recorded by using a bar code reader which uses the computers internal clock to record the time values. The MTBF must still be estimated.

Automatic data collection is used for machine data only. This is accomplished by using direct connection of the equipment to a computer. Many pieces of semiconductor process equipment have some form of computer connection (usually RS232). An interface program must be written to collect the information supplied by the equipment. If no computer connection is available, data collection devices can be retrofitted to the equipment. However, the expense for hardware and software development may be high.

3.3 Irwindale Facility

The simulation project evaluated the manufacturing process through the facility at Space Park. The simulation should include the process of component placement, bonding, final packaging and final inspection performed at TRW Irwindale. Incorporating the Irwindale operation into this model would allow an analysis of the total process for producing the air bag sensor device. It is recommended data be collected from the Irwindale facility and added to the current manufacturing simulation.

4 UPDATE

The model and original report were completed in the third quarter of 1994. The model is currently being used to make production related decisions as described in the first paragraph of section 3.2.

A bar code data gathering system and automatic data collection from process equipment have been implemented. The accelerometer management have indicated they will continue to modify and update the model as conditions change in the facility. The model will be used as a tool to help make production related decisions.

REFERENCES

Gogg, Thomas J. and Jack R.A. Mott. 1992. Improve Quality and Productivity with Simulation. Library of Congress Catalog Card Number:92-85591. ISBN 1-882229-0307.

AUTHOR BIOGRAPHY

LAWRENCE S. FAREY is a Senior Facility Engineer for Fujitsu Computer Products of America and principle of LSF Consulting. He received a B.S. degree in Industrial Engineering from Oregon State University in 1987 and he received a Masters degree in Business Administration from California State University, Fullerton in 1991. He is a licensed Engineer in the State of California and a Certified Plant Engineer. He is a member of the National Society of Professional Engineers, the Association for Facilities Engineering, and the Institute of Industrial Engineers.

Proceedings of the 1996 Winter Simulation Conference
ed. J. M. Charnes, D. J. Morrice, D. T. Brunner, and J. J. Swain

MANUFACTURING SIMULATION CONSULTANT'S FORUM

Chair
F. Bradley Armstrong

ABB Power T&D Company Inc.
Transmission Technology Institute
1021 Main Campus Drive
Raleigh, NC 27606-5202, U.S.A.

Panelists

Bennett Foster

Dupont
1007 Market Street
Wilmington, DE 19898, U.S.A.

Randall Gibson

Automation Associates, Inc.
512 Via de la Valle, Suite 209
Solana Beach, CA 92675, U.S.A.

Charles Kern

Simons Technologies
P. O. Box 1286
Atlanta, GA 30301, U.S.A.

Darrell Starks

Isys Corporation
253 Glenmary Avenue
Louisville, KY 40204, U.S.A.

ABSTRACT

Consultants are an important segment of the manufacturing simulation community because they are on the forefront of change and innovation. From their position they have much to offer the rest of the community. This session offers the panelists and audience the opportunity to share their thoughts and views on communication issues when conducting manufacturing simulation projects.

1 INTRODUCTION

This year's panel is organized around the theme: *Communication Issues in Manufacturing Simulation*. The intent is to share information on the different aspects of communication associated with simulation usage in manufacturing.

Webster's II New College Dictionary defines communication as: "The exchange of ideas, messages, or information, as by speech, signals, or writing". The dictionary also states that "in recent years the verb *communicate* has developed the meaning 'to express oneself effectively'."

In a strict interpretation, simulation can be considered as more of a communication tool than as an analysis tool because it does not provide answers to problems, but rather takes data and transforms it into information from which decisions can be made. This information is presented as an animation and statistical results. Over time, the swing has been toward animation as the format of choice, and this evolutionary process has had some interesting effects on the simulation project process.

2 THE PANEL

The panelists, as evidenced by their biographies, are a diverse group of experienced manufacturing simulation consultants. Each of the panelists has selected a question or statement related to the general theme and will moderate a discussion using their topic as a departure point. The remainder of this section lists the topics and a brief position or background statement.

2.1 Brad Armstrong: Simulation Project Team Communication

A basic value of simulation and animation as analysis tools is their ability to provide a common ground for people in different functional areas and organizational levels to exchange ideas and understand issues. Figure 1 illustrates the roles and defines the knowledge and perspective each type of team member contributes.

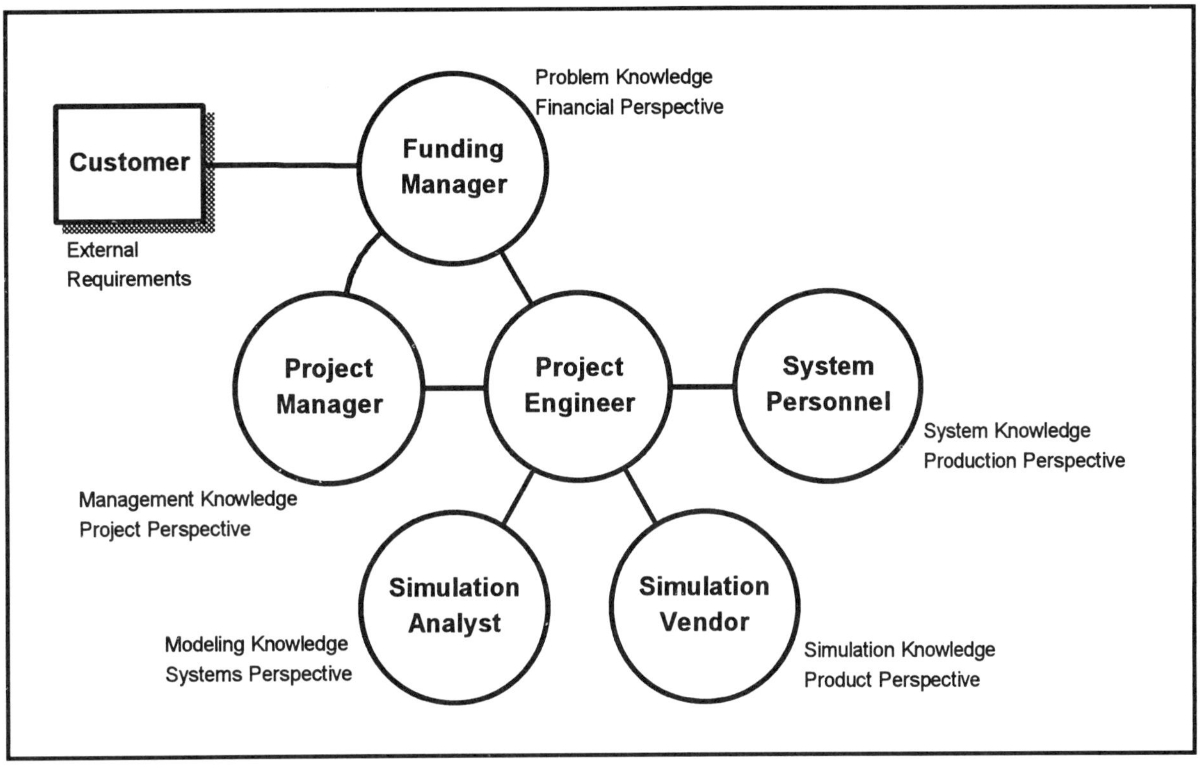

Figure 1: Representative Project Team

The structure of an organization lends itself to specialization with compartmentalized responsibilities and rewards. Simulation, and especially animation, allow individuals with different sets of knowledge and perspective to blur the organizational lines to achieve successful system solutions. Although each simulation project team differs in the number and type of people, I believe there are basic roles and responsibilities present in any team. In some cases, a single individual must play all of the roles; in other cases, each role will have one or more persons.

Even though animation is currently seen as the key to effective communication, assumptions have been the best method historically and still remain important. The problem with relying too heavily on an animation is that often the discussions they invoke are not documented. This approach results in group consensus during the meeting, but the lack of a permanent record promotes misinterpretation and faulty remembering that can cause significant problems later in the project. This effect is proportional to the number of people involved which, in turn, is proportional to the size and complexity of the system being modeled.

A detailed set of assumptions completely defines the project in a form understandable to all project team members. For the Modeler, simply stating an assumption is often not enough when conveying information about the model. Often the other team members will not understand the implications associated with a given assumption until the final results are presented with catastrophic results. To avoid this undesirable outcome, the assumption should be stated along with a justification for making the assumption, the possible consequences of using the assumption, and alternative assumptions that could be used. This greater level of detail should be used for key assumptions that either have serious consequences or are controversial. An example is shown in Figure 2.

Assumption: Equipment downtime is not important.

Justification: No downtime data is available and all of the line personnel (Joe, Betsy, and Sue) say down time rarely occurs.

Consequences: If down time is important, the model results will overestimate manufacturing capability and give optimistic results.

Alternatives: Collect down time data and evaluate its importance.

Figure 2: Example of a Fully Documented Assumption

2.2 Bennett Foster: The Model Purpose

Do you see both greater interest and less skepticism about modeling as a result of animation - and does this put even greater responsibility on the simulation developer to properly define model goals, scope, and level of detail at the onset of a modeling project?

Animation has taken simulation from a buyers market to a sellers market in some corners of the business world. The glitz of animation and menu driven data input tends to blunt the skepticism which many of us encountered, before the days of animation. However with this often ready acceptance of simulation modeling comes a possible trap. Years ago the client would always demand that the consultant be very specific on:

1. what the model would do
2. the type of questions the model could answer
3. the types of data that go into the model

Because the prospective user often wasn't sure that the kind of modeling we were describing could even be done. Now we may occasionally find a client who has decided that (s)he "wants one of those" - before the modeling work is fully laid out.

There are at least two different tasks facing the simulation consultant at the beginning of a simulation modeling project:

1. the responsibility (which has always existed) of properly defining modeling objectives, scope, and level of detail.
2. the need to manage expectations for a generation of users that has grown up with "Nintendo" and "Excel" (or "Lotus") - and automatically assumes that simulation models will come with a variety of advanced control logic and i/o features which are expensive (if not impossible) to develop inside current modeling tool and hardware environments.

2.3 Randall Gibson: Simulation Modeling: Caution - Some Programming Required"

The recent emphasis on ease-of-use by many simulation software vendors can be misleading, even dangerous. While it is true that many simulation software packages have become relative easier to use, i.e. to operate or employ, that must not be confused with the other activities involved in "using" the software. Activities such as the level of effort, training, and experience required to correctly specify, design, construct, debug, test and verify a model, and then produce correct results. For anything other than a trivial problem, these activities have in the past -- and still do -- require experience and the ability to program the solution.

A broad definition of programming would be the ability or discipline to think logically and organize complex activities into a structured, logical procedure or unified sequence of operations. Any real world system sufficiently complex to require or benefit from analysis using a simulation model will necessitate programming the system logic and interrelationships in order to produce a useful model.

Simulation practitioners, their managers and future clients will do well to keep this in mind, and not be lulled into false expectations about how quick and easy creating valid models will be.

2.4 Charles Kerns: The Role of Simulation Software as a Communication Aid

If we properly develop our product and services, if our client organization accepts this new technology, and if the way they do business, and the nature of the service they supply, radically change will any of us ever imagine that we could have done work any other way? My company is a simulation software supplier that has been spun-off from an Engineering firm with expertise primarily in the process industries. We supply a process simulator that our parent company and clients use throughout the design process. Our clients' design cycle includes the following phases:

1. Advanced Planning (an architectural technique)
2. Conceptual Design
3. Detailed Steady-state Mass Balance
4. Control Design
5. Transition to a dynamic model
6. Staging and Testing
7. Building an Operator Trainer
8. Installation and Start-up

In the past each of these phases results in a paper deliverable that is human tested and which is used to perform the next phase of the project. Problems accumulate and are fixed at the end of the project. By integrating these phases through a simulation based design tool we can build an operating model that can have layers added to it and be tested as the project proceeds, so that the system can be modeled and its performance can be tested and demonstrated.

Relying on humans to mentally model a system's performance, communicate problems, and develop and demonstrate solutions to others using dead paper models is difficult and costly. Using sophisticated simulation tools to demonstrate, communicate, and evaluate a system's performance has and will continue to be developed as technology develops. I have given you a glimpse into an area where developing simulation technology is helping create the future. We have come a long way. We have a long way to go. This and other uses of simulation are some of the areas where simulation community is helping to create the way we have been modeling for some time now.

2.5 Darrell Starks: Animation: The Double Edged Sword?

Some believe that animation has been the best advance and worst advance in the simulation technology area. Let us first address the idea that animation has caused problems in the area of simulation. Those of us who were brought up on simulation starting in the 1970's saw animation as challenging our statistically correct simulation analyses. By refusing to change our paradigm, we saw animation as corrupting the simulation user by allowing the user to make decisions based upon what the user saw on the screen instead of considering the massive amounts of output that a simulation generated. And to our credit we were correct in that many in our profession caused the users and decision makers to equate animation with simulation. We were correct in fearing that users and decision makers would use the one observation of the experiment from the animation to make critical decisions. Therefore animation was allowed to be misused initially because some of us refused to see the great benefits that animation would have provided.

That leads to the fact that animation has become one of the best features of simulation and the simulation project. First animation allows the model builder to more quickly verify and validate the models and thus allows the analysis phase of the project to begin more quickly and to be done in more depth. And the animation does in effect allow analysis. That is, animation provides "qualitative" analysis. It allows the user to visualize the system and perhaps eliminate some of the more outlandish scenarios. It allows the user "play" with the model in a visual environment and quickly view what might happen in a particular scenario. But after the user has established a set of feasible scenarios it is still the quantitative analysis using the output statistics over n replicates that must be used as the foundation for critical decisions.

And finally the most useful purpose of animation is that it allows for communication among all participants in a simulation project. It gives everyone, especially those not familiar with simulation and modeling, that "warm fuzzy" feeling that the Modeler has truly captured THE SYSTEM and therefore gives great credibility to the output statistics. Animation has allowed us in the simulation community to become the Merlin's of the 21st century using animation as our "crystal ball".

3 SUMMARY

For a consultant, using simulation and animation present a never-ending cycle of communication challenges. Starting with the need to sell the tool and ending with gaining acceptance of project results, a consultant often struggles with over- or under-selling.

The word "communicate" when talking about simulation immediately evokes thoughts of animation and two of the panelists focused on this aspect. However, there are other, broader communication issues. Two other panelists sent messages to the people who purchase simulation software and services. The first focused on software capabilities and the second on being clear about project goals. Lastly, there is the issue of the simulation team itself and its organization.

A common issue raised by the panel is the consultant's use of animation when communicating with untrained participants. There is no doubt that animation has drastically changed the focus and approach used when performing simulation projects. The question becomes whether the change is for the better or not. On the positive, it has made it possible to sell the need and results effortlessly. On the negative, it has led people to abandon other communication aids and techniques because they are seen as unnecessary.

As well-trained consultants who have spent considerable time and effort becoming specialists, we tend to have a protective view towards simulation and its usage -- maybe too much so. For the salesman in each of us, animation is dream come true. For the analyst in us, it is a potential nightmare and we suffer pangs of conscience over its potential misuse. Regardless, it is clear that although software alone may be able to give people a pretty picture, without a well-defined process to behind it, solutions are still hard to achieve and no amount of hype is going to change that fact.

REFERENCES

Webster's II New College Dictionary. Houghton Mifflin Company, 1995.

AUTHOR BIOGRAPHIES

F. BRADLEY ARMSTRONG is a Fellow Engineer at the ABB Power T&D Transmission Technology Institute. Prior to joining ABB, he founded the consulting firm Simulation Engineering Associates. He has also worked for Hughes Aircraft Company, Pritsker Corporation, and General Dynamics. He received a B.S. in Mechanical Engineering from the University of Texas at Austin and an M.S. in Industrial Engineering from Purdue University. He will receive an M.S. in Integrated Manufacturing Systems Engineering from North Carolina State University. He is a professionally registered engineer , a senior member of IIE and a member of SCS. His current interests include generic modeling, experimental design, and World Class Manufacturing.

BENNETT FOSTER has worked as an operations research consultant for over 20 years in DuPont. He holds B.S. and M.Eng. degrees from Virginia Tech in Industrial Engineering and Operations Research. He is a P.E. and holds CPIM certification from APICS. Bennett has taught simulation at Vanderbilt University, and continuing education simulation courses for AIChE and in DuPont.

RANDALL GIBSON is President of Automation Associates, Inc. (AAI), an independent engineering firm specializing in systems design, analysis and simulation modeling for commercial and industrial companies. AAI has special expertise in modeling manufacturing systems and operations, and has offices in San Diego and Seattle. The company has clients in a variety of industries, including electronics and computer manufacturing, semiconductor manufacturing, warehousing and distribution operations, marine terminal and port operations, traffic modeling, and restaurant operations re-engineering. Mr. Gibson was previously involved in the design and implementation of material handling automation systems as President of White Data Systems, Inc., in San Diego, now part of the Pinnacle Automation group. He holds B.A. degrees in Applied Physics/Information Sciences, and in Psychology from the University of California, San Diego, and an M.S. in Electrical Engineering and Computer Science from the University of California, Berkeley. He is a senior member of the Institute for Industrial Engineers and the Society for Computer Simulation.

CHARLES KERN is Senior Operations Research Analyst at Simons Technologies, Inc. (STI). At STI he is primarily responsible for discrete event product development. His major clients and customers are in the Food & Beverage, and Pulp & Paper industries. Prior to joining STI , he worked for Harnishfeger Engineers, Arthur D. Little, major clients including Caterpillar Inc, and the U.S. Postal Services. He has built models with numerous tools including SLAM II, ProModel, AutoMod II, Extend and IDEAS?. He received a B.S. In Mathematics from Northern Illinois, and a M.S. Operations Research from Georgia Tech. His current efforts are in mixed discrete event and continuous simulation. His outside interests include The History of Technological Innovation.

DARRELL STARKS is a senior consultant with Isys Corporation of Louisville and is currently a doctoral candidate at the University of Louisville. Mr. Starks has worked extensively in the application of simulation technology to real world applications while at McDonnell-Douglas Corporation, Pritsker Corporation, Biles & Associates, and Isys Coporation. Application areas include manufacturing, transportation, logistics, hospitals, and restaurants. Mr. Starks research interests are in the area of interfacing simulation to optimization techniques for real world application.

Proceedings of the 1996 Winter Simulation Conference
ed. J. M. Charnes, D. J. Morrice, D. T. Brunner, and J. J. Swain

ON-LINE KNOWLEDGE-BASED SIMULATION FOR FMS: A STATE OF THE ART SURVEY

Lissette R. Gonzalez
Marelys L. Garcia
Martha A. Centeno

Department of Industrial and Systems Engineering
College of Engineering and Design
Florida International University
Miami, Florida 33199, U.S.A.

ABSTRACT

Simulation modeling has been widely used to study many aspects of FMS design, planing and control. Yet, simulation modeling still offers other capabilities that are proven to be effective for the on-line control of FMS. An example of it is the training of neural nets off-line, so that it will later control the decision making process when the FMS is in operation. if the neural net runs out of knowledge, it turns back to a simulation model to learn new situations. In this paper, we review the various ways in which on-line knowledge-based simulation for FMS has been approached. This paper represents the early stages of on-going research efforts at Florida International University.

1 INTRODUCTION

A *flexible manufacturing system (FMS)* is a production system consisting of a set of identical and/or complementary numerically controlled machines, connected through an automated transportation system Tempelmeier (1993). FMSs have continuously gained importance not only in big factories but also in the small and medium serial production environments because in a properly organized FMS the production costs and delivery times are relatively low. However, the success of an FMS heavily depends on having effective real-time planning, scheduling, and control. Theoretically, "real-time" should mean providing a decision immediately. But, in practice, the speed at which the decision is needed is actually a function of various system parameters such as the magnitude of part processing times.

The design and operation of an FMS involve solving intricate operational problems in the areas of planning, scheduling, and control. These problems render real-time control of the system difficult to achieve. Some of the planning problems include the number part types to produce in the FMS, and the components and layout of the FMS itself. Scheduling problems include order releasing and sequencing, whereas control problems involve monitoring and controlling of machines. In addition, no two FMS are identical; thus, system decision making varies both in type and complexity Drake, et. al. (1995).

Computer-based simulation can be used in many parts of the life cycle of an FMS. Traditionally, simulation has been used for capacity planning, bottlenecks detection, and creating and testing manufacturing schedules. This means that the simulation modeling activity is done off-line. However, recent efforts have been geared towards using simulation for real-time decision making; thus, drifting from the conventional off-line approach. This new approach to using simulation for FMS analysis requires the handling of large amount of knowledge to prepare the simulation model and to evaluate its outputs. Thus, the combination of traditional simulation systems with knowledge-based systems seems to provide a feasible and effective approach to the decision making problem in an FMS environment. This combination has resulted in what has been called knowledge-based simulation Kopacsi et. al. (1993).

Section 2 provides a description of an architecture of a knowledge based, on-line simulation system. Section 3 survey the current status on the usage of knowledge-based on-line simulation systems as applied to an FMS scenario. This section also provides a review of recent developments in this field. Section 4 summarizes some of the areas and related issues that need further research, so as to fully exploit the use of knowledge based on-line simulation in the context of FMS.

2 STRUCTURE OF A KNOWLEDGE-BASED ON-LINE SIMULATION SYSTEM

On-Line simulation is a computerized system capable of performing both, deterministic and stochastic simulation in real-time (or quasi real-time) to monitor, control, and schedule parts and resources in a discrete part manufacturing system such as an FMS. The level of complexity of an on-line simulation system depends upon the following factors:

Freshness of Dynamic Data: Refers to the frequency by which data is collected from the FMS. This desired frequency level is determined by the elapsed time to conduct the simulation and the complexity of the simulation model.

Degree of Automation of Data Collection: Refers to the means of collecting the data from the FMS. This process could be fully automated or manual through the use of operators.

Degree of Reaction to Data from Shop-floor: Refers to those changes that would prompt a reaction in the on-line simulation system.

Frequency of Simulation: Refers to the frequency and number of times that a simulation would be conducted. This would depend upon the purpose for which the on-line simulation system has been implemented, for instance short-term planning.

Levels of Capability: Refers to the way the on-line simulation system is structured to handle the control decisions.

The Knowledge-based on-line simulation system provides the ability to control the FMS environment by integrating several knowledge-bases with a simulator and data communication software. This architecture can be used for modeling, simulating and analyzing several of the operational problems discussed in the previous section that characterize an FMS.

To integrate the on-line simulation principle and FMS, there are several means of interfacing the simulation model and the physical system. Figure 1 presents a typical viewpoint of this interfacing as given by Tayanithi, (1992) .

Figure 1 shows the major components of the knowledge-based on-line architecture: FMS emulation, nested databases. knowledge bases, on-line simulation (OLS), knowledge-based controller and supervisor interface.

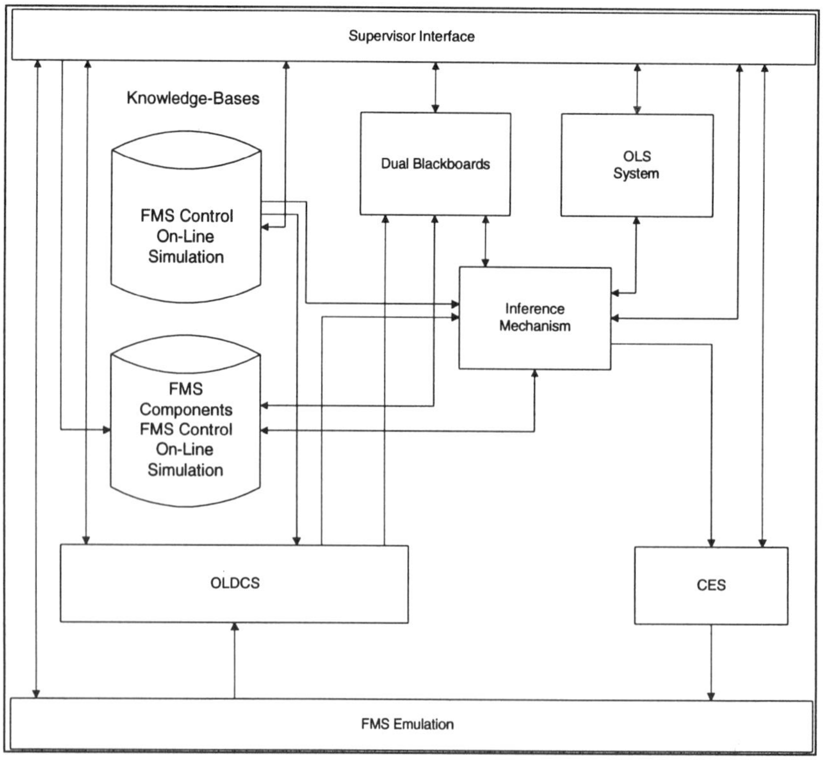

Figure 1: Knowledge-Based On-line Simulation Architecture

The emulated FMS is monitored by the knowledge-based controller; thus, detecting problem immediately when they occur. Upon detection, it would classify the control problems and check to see if a suitable solution resides in the KBs. In the event that no solution is found, alternative actions would be developed and evaluated using on-line simulation. The knowledge base would select the best alternative based on the results from the simulation.

The KB controller includes several inferential methods like forward chaining and blackboard architectures for storing data and improving the inferential process. It interacts with the emulated FMS, nested databases , knowledge bases, OLS, and supervisory interface. The emulated FMS and the KB controller are connected through an on-line data collection system (OLDCS) and a control execution system (CES), as depicted in Figure 1. The OLDCS collects information about the FMS status to allow the KB controller to monitor the system state variables, so as to make appropriate decisions. The CES executes the actions as described by the methods encapsulated in the control decision criteria Tayanithi et. al. (1992).

3 RELEVANT WORK

Many researchers have devoted considerable time to study and prototyping OLS systems in FMS scenarios. Tayanithi, et. al. (1992) presents a study regarding the complexity reduction during interruption analysis in a FMS using a knowledge-based on-line simulation system (KBOLS). Their effort focused on interruptions caused by machine breakdowns and rush orders. These interruptions prompt for changes in the loading scheduling decisions; thus, posing a high level of complexity in the decision making, even when the number of machines and part types in a FMS is small. They developed a new architecture that combines KB, simulation, and an on-line data acquisition techniques in order to perform the interruption analysis. Several experiments were conducted using this architecture to evaluate the improvements in terms of complexity reduction and response time. Results showed a significant reduction of complexity while making critical on-line decisions during the interruptions in a FMS. Another contribution of the study is that the developed architecture is capable of reducing the search space to only a few alternatives instead of evaluating every single one; thus, reducing the response time.

In a similar study done by Manivannan and Banks (1991), a framework for achieving real-time control was developed. The framework integrates data collection devices with a dynamic knowledge base and a simulation model. The framework is known as real-time knowledge-based simulation (RTKBS). The meaning of modeling control and analysis framework within the RTKBS is provided using a very abstract terminology. The problem of synchronizing various events within the FMS and their times of occurrence in both the simulation and the actual manufacturing cell is discussed. Response time reduction is accomplished by creating a dynamic knowledge base containing the results from previous simulations. The RTKBS system is applied to a manufacturing cell connected by a variety of material handling systems.

Szu-Yung David Wu et. al. (1988), proposes a control structure for flexible manufacturing cells that takes advantage of the Expert System technology and discrete event simulation. MPECS (Multi-pass Expert Control System) utilizes artificial intelligence and simulation in real-time control procedures to maximize the benefits provided by each method. The system breaks up a production cycle into various decision point periods. At each of these decision points, the current system status is assessed and simulations are conducted to test which control rule performs the best, among those alternatives generated by the expert system. One significant result from this study establishes that the performance of MPECS is considerably better than that of single-pass traditional methods that constantly use a single rule.

Smith et. al. (1994) examined the application of discrete event simulation for shop floor control of an FMS. In their architecture, namely RapidCIM, simulation is also used as a task generator to control the physical equipment of the shop floor. The way the system is configured provides for a clear separation of decision making and execution; thus, it introduces flexibility into the shop floor control system.

Harmonosky (1995) performed a very comprehensive review of recent developments in the field of knowledge based on-line simulation, focusing on those problems regarding real-time scheduling in an FMS. One of the studies reviewed is the one done by Krishnamurthi and Vasudevan (1991). In their work, they present a framework for a domain-based on-line simulation, which could be used as a general purpose decision support system. This indeed represents the biggest contribution of their study, since as the authors claim, the framework although domain-specific, is generic to all problems generated within such domain. The simulation continuously monitors the real systems so that it always reflects the current system state. This system constitutes an example of what Harmonosky identifies as continuos real-time scheduling, in which every decision regarding the task to be scheduled next is made as time moves forward in the physical system. This opposes the other type of real-time scheduling system, "exception", in which decisions are made whenever an event like a breakdown occurs.

Another study that Harmonosky discusses is the one performed by Duffe and Prabhu (1994). In their study, the authors exemplarize the use of simulation to evaluate local schedules generated by local controllers in an FMS environment. All entities in the system develop a local schedule, which is optimal among their alternatives. Furthermore, the best alternative among all local schedules is selected for execution posterior to the simulation evaluation of these optimal local alternatives. The authors sustain that this method of providing feedback to the entities, allowing them to select their own local schedule, provides faster scheduling decision making and evaluation, which in turns makes this method more attractive for real-time applications.

4 SUMMARY

It has been established that the complexity of the operational problems that characterize an FMS calls for an efficient tool that would facilitate decision making by providing a real-time control in a reasonable response time. A viable alternative to these problems has been the use of knowledge-bases in conjunction with on-line simulation implementations. Although a relevant amount of work has been undertaken in this regard, there are still areas for future research opportunities. One of the areas many of the reviewed authors have agreed on is the frequency by which data should be collected. This area is very important for the overall performance of the system since creating large databases has a negative effect in the response time.

In addition, another area for potential improvement involves simulation software developers. Simulation packages of the future should provide the programming capability to help resolve the issue of interfacing the simulation with the real system which is already being considered by commercial developers as per Harmonosky (1995).

REFERENCES

Drake, Glenn R., J. S. Smith, and B. A. Peters. 1995. Simulation as a Planning and Scheduling Tool for Flexible Manufacturing Systems. In *Proceedings of the 1995 Winter Simulation Conference,* ed. Alexopoulos, K. Kang, W. R. Lilegdon, and D. Goldsman, 805-812.

Harmonosky, C. 1995. Simulation-Based Real-Time Scheduling: Review of recent Developments. In *Proceedings of the 1995 Winter Simulation Conference,* ed. Alexopoulos, K. Kang, W. R. Lilegdon, and D. Goldsman, 220-225.

Kopacsi, S., N. T. Koussoulas, and D. Gavalcova. 1993. Knowledge Based Modelling, Simulation and Evaluation of Flexible Manufacturing Systems. *Knowledge Based Hybrid Systems* (B-11): 203-212.

Manivannan, S. and J. Banks. 1991. Real-Time Control of a Manufacturing Cell Using Knowledge-Based Simulation. In *Proceedings of the 1991 Winter Simulation Conference,* ed. Barry L. Nelson, W. David Kelton, and Gordon M. Clark, 251-260.

Manivannan, S. and J. Banks. 1992. Design of a Knowledge-Based On-Line Simulation System to Control a Manufacturing Shop Floor. *IIE Transactions* 24:72-83.

Stecke, K. E. 1985. Design, Planning, Scheduling, and Control Problems of Flexible Manufacturing Systems. *Annals of Operations Research* 3:3-12.

Tayanithi, P., S. Manivannan, and J. Banks. 1992. Complexity Reduction During Interruption Analysis in a Flexible Manufacturing System Using Knowledge-Based On-Line Simulation. *Journal of Manufacturing Systems* 12:153-169.

Tayanithi, P., S. Manivannan and J. Banks. 1992. A Knowledge-Based Simulation Architecture to Analyze Interruptions In a Flexible Manufacturing System. *Journal of Manufacturing Systems* 11:195-213.

Tempelmeier, H. and H. Kuhn. 1993. Flexible Manufacturing Systems. *Decision Support for Design and Operation.* Ed. John Wiley & Sons, Inc.

Whitney, C. 1985. Control Principles in Flexible Manufacturing. *Journal of Manufacturing Systems* 4:157-167.

Wu, S. and R. A. Wysk. 1988. Multi-pass Expert Control System - A Control/Scheduling Structure for Flexible Manufacturing Cells. *Journal of Manufacturing Systems* 7:107-119.

Wu, Zhiming. 1992. Combining of Expert System and Simulator for FMS Rescheduling. *Japan/USA Symposium on Flexible Automation* 2:1127-1130.

AUTHOR BIOGRAPHIES

LISSETTE GONZALEZ is a graduate student at Florida International University, where She is pursuing a Master of Science degree in Industrial Engineering. Her research interests are in the area of FMS modeling and optimization. She is expected to graduate in December of 1996. She received a Bachelor of Science degree in Industrial and Systems Engineering from Florida International University in 1995. She is a member of IIE.

MARELYS L. GARCIA is a graduate student at Florida International University, where She is pursuing a Master of Science degree in Industrial Engineering. Her research interests are in the area of smart simulation, smart data analysis, and human-computer interaction. She is expected to graduate in August of 1996. She received a Bachelor of Science degree in Industrial and Systems Engineering from Florida International University in 1994. She is a member of Tau Beta Pi, Alpha Pi Mu, and IIE

MARTHA A. CENTENO is an assistant professor in the Department of Industrial and Systems Engineering at Florida International University. She received a B.S. degree in Chemical Engineering from ITESO University (Mexico), an M.S. degree in Industrial Engineering from Louisiana State University in 1985, and a Ph.D. degree in Industrial Engineering from Texas A&M University in 1990. Her research interests are in the areas of intelligent simulation, smart data analysis, and engineering education. Dr. Centeno is a member of Alpha Pi Mu, IIE, INFORMS, SCS, ASA, and ASEE.

Proceedings of the 1996 Winter Simulation Conference
ed. J. M. Charnes, D. J. Morrice, D. T. Brunner, and J. J. Swain

CONTROLLING ACTIVITIES IN A VIRTUAL MANUFACTURING CELL

Michael Iuliano
Albert Jones

National Institute of Standards and Technology
Metrology Bldg., Room A127
Gaithersburg, MD 20899

ABSTRACT

Researchers at the National Institute of Standards and Technology are developing a virtual manufacturing cell. This cell will contain simulation models of a wide range of manufacturing equipment, processes, and systems. It will have commercial and prototype applications software which implement production functions from order release to final inspection. There will be an information base and a collection of interfaces which will provide the integrating infrastructure for these applications. These interfaces will be based on information models and exchange protocols, and will specify what information is shared across those applications and how it is exchanged. This paper describes the current virtual manufacturing cell, with special emphasis on the hierarchy we are developing to control activities within the cell.

1. INTRODUCTION

Since the mid 1980s, the National Institute of Standards and Technology (NIST) has been involved in research and development related to manufacturing systems integration. The Automated Manufacturing Research Facility (Simpson, Hocken, Albus, 1982) was designed and built as a development and demonstration testbed for a wide range of systems integration projects. Hierarchical control principles (Albus, Barbera, Nagel, 1981), together with contem- porary information management and communication strategies were used to design and build these integrated systems. The computer technology available at that time was used to implement these principles and strategies on top of a real manufacturing cell. As a result, the AMRF was subject to the same kinds of problems faced by factories across the country: difficult and time consuming development of new ideas; and, frequent breakdowns, software glitches and communications problems.

To overcome these equipment-related problems, NIST is collaborating with vendors, users, and university researchers to develop a virtual manufacturing cell. This cell will contain simulation models of manufacturing equipment, processes, and systems. These models will be used for off-line programming of individual pieces of equipment and on-line control of processes and systems. There will also be commercial and prototype applications software and a manufacturing information base. The applications software will implement production functions from order release to final production and inspection. A major effort will be the development of the interfaces between the applications and this information base. These interfaces will provide the methods for creating, updating, inserting, and extracting required information. Together, they will provide the integrating infrastructure for both the simulation models and the software applications.

This virtual manufacturing cell will provide a highly flexible and cost effective means for extending the earlier AMRF work in integration and control . We intend to investigate both traditional and emerging control strategies. We will examine a wide range of integration mechanisms including file transfer, messages passing, object request brokering, and databases . As it evolves, this virtual manufacturing cell will provide a place for testing the robustness of interface standards, and conformance of software products to those standards. It will be a vehicle for cooperative research and development on a wide range of manufacturing problems. Initially, the virtual manufacturing cell will be housed in a laboratory at NIST. Our long term goal, however, is to use the Internet and emerging web technologies to integrate models, applications, and information bases from locations around the country and the world.

This paper is organized as follows. Section 2 provides a short description of the cell including software applications and hardware platforms. Section 3 presents an overview of the proposed control hierarchy - a cell controller and several machine controllers. Section 4 gives details about the current capabilities of the controllers. Finally, Section 5 looks at the future evolution of the cell.

2. VIRTUAL MANUFACTURING CELL

The cell contains software packages which run on a wide range of hardware platforms and operating systems. These packages support the applications which implement the following functions: cell control, machine control, cell simulation, machine simulation, routing, operations planning, scheduling, NC verification, and shop floor data collection. Prototype applications under development at NIST include order entry, job routing, cell control, machine control, and process plan editing. Commercial software packages include:

- product data management, MATRIX™
- operations planning, ICEM™ PART
- shop floor simulation, QUEST™
- NC program simulation, VNC™
- scheduling, AUTOSCHED™

These packages run on a number of hardware platforms under a number of operating systems including: IRIX™ on SGI machines, SOLARIS™ on SUN workstations, and WINDOWS NT™ on desktop and notebook PCs.

We will show a high degree of interoperability by replacing these particular packages and applications with others that have the same functionality. This will give vendors, manufacturers, and university researchers the opportunity to test the capabilities of a single software application on vastly different simulated manufacturing systems, or different software applications on the same simulated manufacturing system. Beginning next year, we plan to add some new applications such as Manufacturing Resource Planning (MRP), Manufacturing Execution Systems (MES), inventory control, tool management, and quality control.

3. THE CONTROL HIERARCHY

As noted in section 1, NIST has been involved in the theory, implementation, and demonstration of hierarchical control systems for a number of years. The control system being implemented in the virtual manufacturing cell is based on the approach described in (Davis, Jones, Saleh, 1992). The system being controlled contains a number of simulated machines grouped into a single cell. These machines may be machine tools, robots, or coordinate measuring machines. The control hierarchy contains a cell controller and one machine controller for each machine. Each of the controllers in this hierarchy performs the same four functions: planning, scheduling, execution, and monitoring. The cell controller performs these functions to control the movement of parts around the cell. Each machine controller performs these functions to control the parts at a single machine.

Our goal is to have this virtual cell emulate a real factory as closely as possible. At the moment, there are some mnior differences (see Figure 1). In a real factory, the cell controller would issue a collection of commands to each machine controller. The machine controller, which could be a computer or a human operator, would issue commands to the commercial controller that controls the motions of the real equipment. Currently, the controllers in the hierarchy for the virtual cell behave in a slightly different way. First, the cell controller interfaces directly with the cell simulator. It does not issue commands directly to machine controllers. Second, machine controllers are embedded in the machine simulators. These controllers can be activitated in three ways: by the cell simulator, by an external operator, or by the workflow manager. Activation occurs whenever parts arrive at a machine to be processed. These behaviors will be modified in the coming year so that the controllers in the virtual cell behave more like controllers in a real cell.

Currently, this control hierarchy is supported by several prototype applications developed at NIST that implement the following functions: order entry, routing, and operations planning. The order entry application creates a file that contains the orders to be manufactured in the virtual cell. Each order contains the following information: Order ID, Part ID, Part Description, Order Size, Routing ID, Lot Size, Release Date, and Due Date. The routing application generates the cell level process plan. This plan contains all the possible routings through the cell for a given order. Each plan contains the following information: Routing ID, Part ID, Step ID, Step Description, Alternate Step ID, Next Step ID, Machine ID, and Processing Time. The operations planning application generates the machine level process plan. This plan contains all of the tasks that the machine is expected to perform and all of the data needed to perform those tasks. The structure and information content for this plan are described in detail in section 4.1. The output files generated by these applications are managed by MATRIX™.

3.1 The Cell Controller

As noted above, the cell controller is responsible for planning, scheduling, executing, and monitoring the movement of parts around the cell. The planning function generates a run-time routing that will enable each order to be completed by its due date. This plan contains: 1) the list of machines to be used in the production of each part, 2) precedence relations among those machines, and 3) expected duration at each machine. The generation of this plan involves the following steps.

1) retrieve the process plan (see below) for this order
2) parse the plan to identify all alternative routings
3) select feasible routings based on the current state of the system

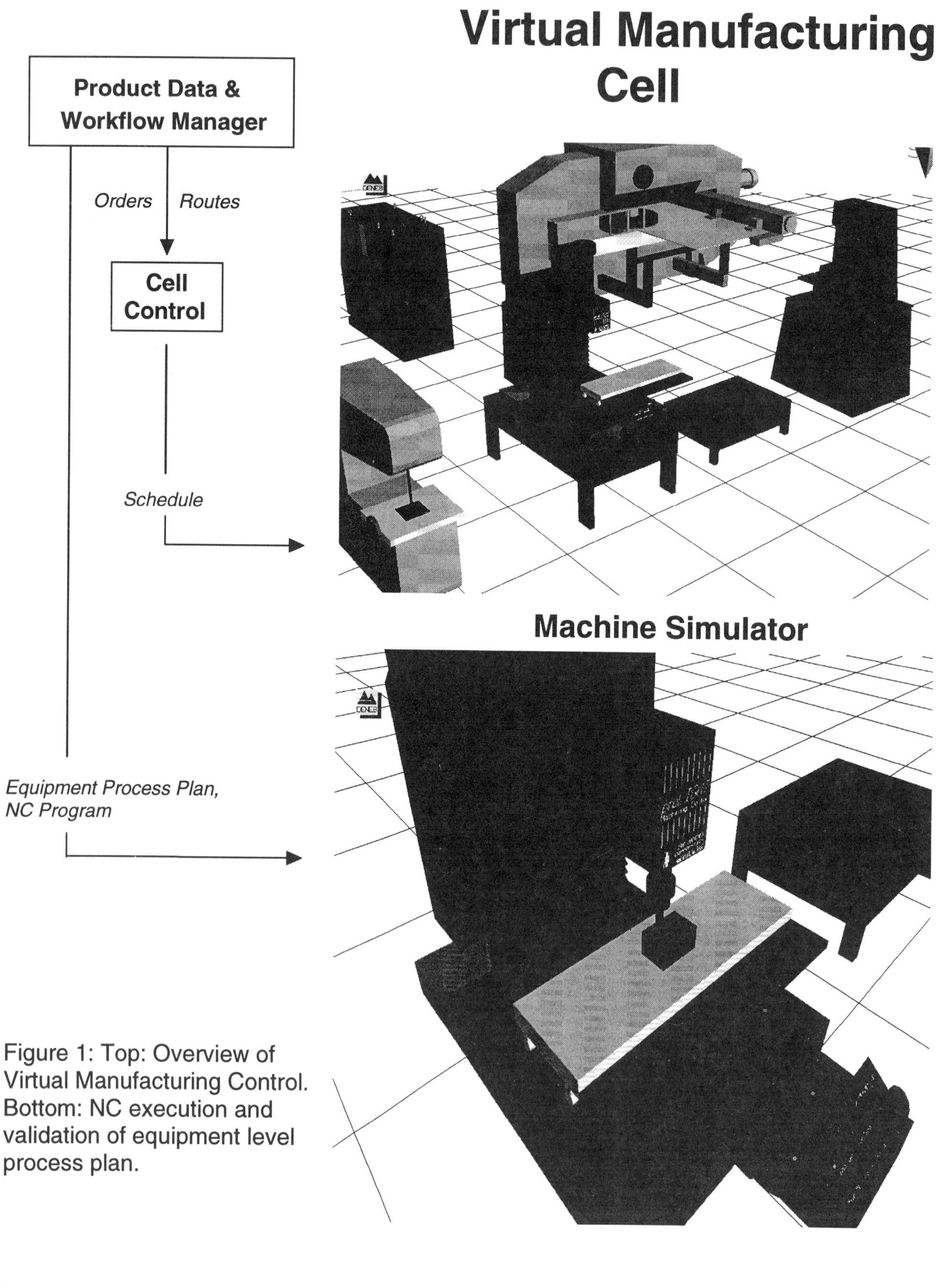

Figure 1: Top: Overview of Virtual Manufacturing Control. Bottom: NC execution and validation of equipment level process plan.

4) rank all feasible routings based on current performance criteria and process times
5) select routing which optimizes those criteria
6) pass this run-time routing to the scheduler.

Whenever conditions in the cell evolve to the point that the due date for this order cannot be met with the current routing plan, a new one must be created. The planner will repeat the preceding steps to select one of the remaining alternatives.

For each new order, the scheduling function uses its due date, routing, processing times, current cell status, and the current performance measures to determine predicted start and finish times at each machine in routing. The scheduling function also uses the updated state information from the monitoring function (see below) to determine whether or not to generate a revised schedule.

The execution function formulates the data package that drives the cell simulator. This package contains the orders, the routing for each order, and the dispatch list for each machine. The order file is used to create an initialization file for the simulation. Each time a job is finished at one machine, the simulator determines its next destination from the routing file. The dispatch list contains the exact sequence of jobs to be performed at every machine.

The monitoring function keeps an up-to-date model of cell status. This model can be compared to the predicted model that comes from the scheduler to determine if everything is "on-track". This comparison can be done automatically by the monitoring function or by a human operator. Whenever a problem occurs, the scheduler can be invoked using the model from the monitoring function.

3.2 Machine Controllers

Each machine controller is responsible for planning, scheduling, executing, and monitoring operations at a single machine. For each job that arrives at the machine, the planning function creates a run-time operations plan. This plan contains the tasks that must be performed in order to complete the machining, inspection, or material handling for that part. There are two types of scheduling that must be done by the machine controller: sequencing of parts at the machine and sequencing of tasks within the operations plan. As noted above, the sequencing of the parts at the machine is provided by the cell execution function. Since the operations plan contains only one feasible sequence of tasks at this time, no additional scheduling is required. The execution function formulates the data package that the machine simulator needs to execute the tasks in that plan. Since all of these tasks are executed automatically within the machine simulator, the only feedback occurs when all tasks have been completed. This feedback is sent to the cell simulator that forwards the appropriate the message to the cell level monitoring function.

4. MORE ON IMPLEMENTATION

4.1 The Cell Controller

In the current implementation, each cell level process plan contains only one feasible routing. The planning function simply retrieves this plan from MATRIX™, parses it and sends it to the scheduling function. The scheduling function, implemented in AUTOSCHED™ , computes the start and finish times at each of the machines in this routing.

The execution function formulates the data package that drives the cell simulator implemented in QUEST™. This data package contains a list of orders, the routing for each order, and the dispatch list for each machine. The order and the routing files are read into the cell simulator from MATRIX™. The order file is used to initiate the simulation. Each time a job is finished at one machine the cell simulator determines its next destination from the routing file. The dispatch list is derived from the schedule produced by AUTOSCHED™. It contains the exact sequence of jobs to be performed at every machine. This sequence overrides the normal queue selection rules used by the cell simulator to pick the next job.

The monitoring function maintains an up-to-date model of cell status in the relational database which is implemented in ACCESS™ . A preliminary information model describing the entities and their relationships can be found in (Jones, Riddick, Rabelo, 1996). Updates to this database are based on a collection of "change" messages issued by the cell simulator each time an event occurs in the simulation. These messages are written to a file which resides in a public directory on the network server. This file is retrieved by a file handler that parses the file and creates the necessary Standard Query Language (SQL) update queries. These queries are used to update the model in the database.

Machine controllers implement the planning, scheduling, and execution functions. The implementation of the scheduling function is discussed above. The planning and execution functions are implemented in the same prototype application within the simulation package, VNC™. This simplifies prototype development because of the native bindings with the machine simulator in the simulation environment. Moreover, since they are implemented sequentially within the same application, no feedback is needed on the completion of individual work elements within the plan. Hence, we have no need for a monitoring function at this time. In the following sections, we provide more details on the implementation of the planning and execution functions.

4.2 The Machine Controller

In the following sections we provide more details on the implementation of the machine controllers.

4.2.1 The Planning Function. As noted above, the planning function must retrieve and parse an operations plan. The current operations plan structure is based on work described in (Mclean, 1987). The plan consists of keywords, work elements, and associated attribute/value pairs. There are presently three keywords corresponding to the three sections in the plan: HEADER, RESOURCE, and PROCEDURE. The HEADER section contains administrative information about the plan such as the part_id, machine_id, date, revision_number, etc. The RESOURCE section lists all of the data files that appear in the PROCEDURE section. The PROCEDURE section can contain an arbitrary number of work elements, which are listed in a sequential manner indexed by a step number . These work elements are directives indicating a type of task to be completed.(see Figure 2).

Figure 2. Simple Operations Plan

```
HEADER
    part_id = test_part1
    plan_name=setup2
    machine_id = vmc100
    date=July, 6, 1996
    revision_number=7
    end_header

RESOURCES
    machine_file = WORKCELLS/VMC_100
    tool_file = TOOLS/TWIST_DRILLS_225
    fixture_file = FIXTURES/FIXTURE_75
    workpiece_file = WORK/WORKPIECE_1
    nc_program_file= NC/test_part1.cnc
    end_resources

PROCEDURE
Step 1  LOAD_TOOL
    tool_name = TWIST_DRILLS_225
    tool_id = T7
    slot = 7
    end_step
Step 2  LOAD_FIXTURE
    fixture_name = CLAMP
    fixture_id = 75
    location = x_axis
    x_offset = 152.4
    y_offset = 101.6
    z_offset = 44.45
    end_step
Step 3  LOAD_WORKPIECE
    workpiece_name = BAR_3x5x2
    workpiece_id = WORKPIECE_21
    location = Fixture
    x_offset = 0
    y_offset = 0
    z_offset = 0
    end_step
Step 4  LOAD_NC_PROGRAM
    nc_program_name= test_part1
    machine_id  = vmc100
    end_step
Step 5  RUN_NC_PROGRAM
    end_step
```

Each work element contains attribute/value pairs that represent the data files required to implement that particular work element. At the moment, we support LOAD_TOOL, LOAD_FIXTURE, LOAD_ WORKPIECE, LOAD_ NC_ PROGRAM, and RUN_NC_ PROGRAM. LOAD_TOOL requires a tool_name, tool_id, slot. LOAD_FIXTURE requires a fixture_ name, fixture_ id, location, x_offset, y_offset, and z_offset. LOAD_ WORKPIECE requires a workpiece_ name, workpiece_id, location, x_offset, y_offset and z_offset. LOAD_ NC_ PROGRAM requires an nc_program_ name and machine_id. RUN_NC_ PROGRAM has no attributes.

4.2.2 The Execution Function We are using the VNC™ simulation package to execute the operations plan. To accomplish this, simulation models of several entities are required including the machine, cutting tools, a fixture, a workpiece, and the machine controller. These models are generated in the simulation environment and saved into individual files that are managed by the product data manager.

For a machine, the model consists of the following: geometry of the physical parts of the machine, the degrees of freedom of the machine, the association of each machine physical part with each degree of freedom, the definition of the degree of freedom as translational/rotational and the direction of translation/rotation, the translation/ rotation travel limits, the axis translation/rotation travel speeds/ accelerations.

The model of the machine controller consists of a MIMIC™ configuration file that is the controller emulator within VNC™. This emulator is capable of reproducing the functionality of any CNC controller in the simulation environment. Unique configuration files have been created and configured for several machine tool controllers.

For a tool, the model consists of the following: geometry of the physical parts of the tool, the definition of which geometry is the cutting edge, and the definition of the name of the tool as it will be referred to in the simulation. The tool geometry is composed of two sub-geometries: a geometry of the shank and a geometry of the cutting edge. For fixtures and workpieces, the model consists of the geometry of the fixture/workpiece and the definition of a name.

Once all of these models have been retrieved into the environment, the NC program is executed. No feedback is provided until either an error occurs or the program runs to completion. If any errors are detected, they are reported back to the process planner, who must make modifications to the operations plan. Typical errors include missing files and a variety of cutting problems (for example, tool cuts into fixture). Once the simulation has run to completion, the final simulated part can be compared to the original designed part.

5. FUTURE PLANS

The virtual manufacturing cell will continue to evolve over the next several years. During that time we will focus on four areas: interoperability testing, capability testing, expanded functionality, and distribution over the Internet. Because of the focus on interoperability, we will spend considerable time interchanging software applications that implement the same functions. Furthermore, we can begin to measure the impact of various types of integration mechanisms including message passing, file transfer, databases, common object request brokers, and work flow management systems. The virtual manufacturing cell will provide a unique opportunity for testing the capabilities of the existing software applications and adding new ones. Manufacturers, vendors, and university researchers will be able to test the capabilities of a single software application on vastly different simulated manufacturing systems, or to test different software applications on the same simulated manufacturing system. We plan to add new software applications such as MRP, MES, inventory control, tool management, and quality control. Finally, we want to make the various testing capabilities of the virtual manufacturing cell available to other researchers around the world using emerging Internet tools such as JAVA (Gosling, Joy, Steele, 1996).

** No approval or endorsement of any commercial product by the National Institute of Standards and Technology is intended or implied by these selections.*

ACKNOWLEDGMENTS

Work described in this paper was sponsored by the US Navy Manufacturing Technology Program and the NIST Systems Integration for Manufacturing Applications (SIMA) Program and is not subject to copyright.

The authors wish to thank Chuck McLean for his advice on process plan work elements.

REFERENCES

Albus, J., Barbera, A., and Nagel, R., Theory and Practice of Hierarchical Control, *Proceedings of 23rd IEEE ComputerSociety International Conference*, 18-39, 1981.

Davis, W., Jones, A., and Saleh, A., A Generic Architecture for Intelligent Control Systems, Computer Integrated Manufacturing Systems, Vol. 5, No. 2, 105-113, 1992.

Gosling, J., Joy, W., and Steele, G., The JAVA Language Specification, *Addison-Wesley*, 1996.

Jones, A., Riddick, F., and Rabelo, L., Development of a Predictive/Reactive Scheduler using Genetic Algorithms and Simulation-based Scheduling Software, *Proceedings of AMPST'96*, 589-598, 1996.

McLean, C., Interface Concepts for Plug-Compatible Production Management Systems, *IFIP WG 5.7: Information Flow in Automated Manufacturing Systems,* North Holland, 307-318, 1987.

Simpson, J., Hocken, R., and Albus, J., The Automated Manufacturing Research Cell of the National Bureau of Standards, *Journal of Manufacturing Systems*, Vol. 1, No. 1, 17-32, 1982

Proceedings of the 1996 Winter Simulation Conference
ed. J. M. Charnes, D. J. Morrice, D. T. Brunner, and J. J. Swain

A SIMULATION-BASED CONTROLLER BUILDER FOR FLEXIBLE MANUFACTURING SYSTEMS

Fernando G. Gonzalez

Department of Electrical and Computer Engineering
University of Illinois at Urbana-Champaign
1406 W. Green Street
Urbana, Illinois 61801 USA

ABSTRACT

This paper discusses a new simulation and control prototype software package that can be employed as a controller for a simple FMS. In developing this software, we attempted to achieve a compromise between the programming flexibility of a general-purpose language like C++ and the convenience of a simulation language like SIMAN® (see Pegden et al. [1995]). In particular, we desire to provide standard programming features within the model such as recursive function calls while providing a single model which can be used to simulate and control a real-world system. To test the efficacy of the developed simulation tool, we developed a simulation model which can control a physical emulator of an FMS which has been constructed in our research laboratory.

1. INTRODUCTION

When writing a simulation model, one can either employ a dedicated simulation language like SIMAN or a general-purpose programming language like C++. The simulation language has the advantage of being easy to use which usually hastens model development. Consider the SIMAN® simulation language. It provides a set of basic modeling elements which can quickly be assembled to build your model. It also provides the essential software elements that are needed to execute the model and to perform the requisite statistical analyses. If one elects to program a simulation model using a general-purpose language, then one must also provide the software needed to execute these auxiliary functions.

However, simulation language like SIMAN have their limitations. SIMAN is not a general purpose programming language like C++, and therefore, cannot provide full modeling flexibility. One is limited to programming within the constraints of the set imposed by the provided modeling elements. For many modeling situations, the provided set of modeling elements is sufficient, and using a simulation language like SIMAN is an optimal choice over developing the simulation model in a language like C++. However, to model more complex systems like an FMS, the set of modeling elements provided by most simulation languages impose severe modeling constraints, particularly when one attempts to assess the impact of the control architecture upon the system. Using a general programming language becomes essential. (Davis et al. [1993] discussed in detail the numerous restrictions that entity flow-based simulation languages impose upon the modeling of an FMS. Mize et al. [1992] further discuss the inaccuracies that ensue from the use of current simulation languages to model an FMS.) Although SIMAN and other simulation languages do permit C subprograms to be included within the model, there is only so much that one can accomplish with these patches. Furthermore, these patches invariably make the model more difficult to verify.

This capability to control an FMS using simulation has already been demonstrated in the literature. Smith et. al. [1994] presented their RapidSIM effort where they enhanced SIMAN with additional constructs to make it possible to implement the physical control of equipment. They added a TASK block and modified the DELAY, ROUTE, MOVE and TRANSPORT blocks. The modeling approach discussed in this paper has some similarities to the approach discussed by Smith et al. It too employs a basic set of SIMAN-like modeling elements and introduces new blocks to manage the modeled system. The presented approach also expands the general modeling flexibility of the language to permit standard features such as recursive function calls to be employed in the modeling. This added flexibility is essential when modeling complex systems such as an FMS.

In this paper a set of objects are developed to provide the minimum necessary modeling elements that one must use in every simulation and controller regardless of the methodology being used. Care has been taken to ensure that the software does not impose on the methodology the designer may have. That is, whether one chooses to write the controller using an entity-flow based method or a message-flow based method, this software gives the flexibility to implement either one.

For this study, the modeled and controlled system will be a physical emulator for an FMS which we have constructed. The study will also demonstrate some new features which could be provided within simulation languages to increase their modeling flexibility.

For discussion purposes in this paper, an event represents the action taken between simulated or actual physical events. For example, a delay event is the action taken when a delay is finished, and a QUESEIZE event is the action taken when the resources needed for executing the event are allocated. Please note also that time does not advance while an event is being executed.

2. THE EXPLORED MODELING FRAMEWORK

The goal for this research effort was to develop a simulation approach using C++ which could provide the modeling convenience of a simulation language like SIMAN while providing the additional capability needed to control a real-world system. This paper concentrates first on defining a set of basic modeling elements using C++ and then adapts these blocks to model the controller of an FMS. Each included modeling element is represented as an object in C++. As with most simulation approaches, this approach is also event driven, and an execution object manages the simulation by the sequential processing of events. An event list is maintained where the scheduled events are stored in chronological order based upon the time that they are scheduled to occur.

In the development of this modeling environment, only a minimal set of modeling elements have been included. The decision of which elements were to be provided was based upon whether the object required an event to be scheduled on the event list or requires the allocation of resources. Since the proposed modeling environment will provide an increased capacity to include C++ programming elements, we decided to provide for the other modeling requirements, like the TALLY block, by defining a library of C++ functions. Some of the more trivial modeling elements like the COUNT, BRANCH and ASSIGN are not implemented since they can be trivially implemented in C++.

A simulation executive object manages the chronologically-ordered event list, controls the processing of the events and manages the allocation of resources. After each event is processed, the executive object pops the event with the smallest event time off of the event list and then invokes the proper function to manage its execution. Given this streamlined approach, there are two basic types of events that occur: QUESEIZE, where the event occurs as a result of resources becoming available and DELAY where the event occurs when a delay is finished. There are also other types of events that provide an expanded modeling flexibility for this simulation environment which are not considered in most simulation tools including the PENDING, GOSUB, and RETURNTO elements. These will be discussed later.

Each modeling element object is instantiated from an object subclass for the element type. A feature of all of these element subclasses is there ability to store (queue) the pointer of the entity which enters the element. Within a system's model, a given modeling element may be used several times. For example, a particular model may have several DELAY elements. Each use of an element within a model will assign a unique Label which will also be attached to every event that is instantiated using that element. For example, the seventh QUESEIZE employed in the model would be assigned the label QUESEIZE7. A DELAY32 label would be attached to the thirty-second DELAY element included in the model. Whenever an entity enters a given modeling element an event object of the particular type may be instantiated, depending on whether the element is of the type that needs to schedule an event. Among the event object's attributes are its event label (e.g. DELAY32) and the scheduled event time. For example, an event may be DELAY37, 3:15. This means that a DELAY37 event is to occur at 3:15. This simulates the completion of the process that is being modeled with the DELAY37 element.

To enhance the capability of the user to include C++ functions, either from the modeling tools library or a special user-provided function, the modeling tool makes a significant deviation from the approach employed in most simulation. Usually, the list of modeling elements included within the model specification are assumed to be sequentially attached to each other unless otherwise specified via modifiers such as DETACH or NEXT(Label). That is, the entity will usually flow from one model element to the next model element in the model's statement. Under our framework, this assumption is not made. Instead, the modeler can insert a set of C++ statements in-line between any two modeling elements. The C++ code can then perform any function upon the entity that is required, including tally statistics from its attribute list or return resources allocated to the entity. Note that neither of these actions has consequences upon the evolution of time. The modeling element that follows the last statement of the inserted C++ code will then receive the entity after the intervening C++ statement have been executed upon it. An event ends when the next modeling element is encountered. At this point, the entity is stored inside of the modeling element's object until the event associated with it occurs. When this event occurs, the entity is removed from the element's object and processing of the event starts. The following is an example of a DELAY37 event. Notice that the first step in the event is to remove the entity from the DELAY37 element and the last step is to store the entity inside of the next element's object.

```
case DELAY37:
    ent = Motor_d37.remove();
              .
              .
              .
    get_machine_q42.insert(ent);
    break;
```

Note: Motor_d37 is the name of the DELAY37 modeling element's object and likewise get_machine_q42 is the name of the QUESEIZE42 element's object.

The following provides a brief outline on how modeling element objects and the simulation executive object are implemented. The DELAY block is an object that consists of a priority queue used to store the position of the entity's pointer, a pointer to a random number generator that can generate random numbers from a prespecified probability distribution, and an event label. The priority among the entities in the queue is based upon the time that the entity is due to be removed from the DELAY block. That is, its order is determined by the entities' remaining delay time. This way the order of the entities in the queue is synchronized with the order of the DELAY events that are scheduled for that particular DELAY block. The random number generator is used to compute the delay time. Each entity entering a delay block is given a random number from this generator and is used as the delay time. When an entity is inserted into a DELAY block, the block generates the delay time, computes the time when the delay will finish, stores the pointer to the entity in its queue using the time it will finish as the priority and finally schedules a delay event for that particular DELAY block in the event list. The element label associated with the modeled element in the model statement is also stored as an attribute within the instantiated entity object. This permits the executive program to know at what location the entity object currently resides in model network.

The QUESEIZE element also contains a queue to store the pointer of the involved entity object and an element label, but no random number generator. When an entity is inserted into the QUESEIZE block, it simply puts the entity into the rear of the queue. No event is scheduled since the entity is waiting for resources to become available. After the execution of every event, the executive object checks each QUESEIZE element to see if any waiting entity needs the available resources to continue. If so, a QUESEIZE event is scheduled to occur immediately. That is, it schedules it into the event list with the current time as the time to occur.

The CREATE element is similar to the DELAY but functions automatically without an entity flowing into the element. When a create event occurs, it automatically creates a new create event object and schedules it on to the event list. Since a new create event is scheduled every time one occurs, there is always a create event scheduled to occur in the future. This models the constant flow of entities entering a system. Within the C++ library, there is also a function which can create a duplicate entity at any point while a given entity is flowing through the modeled network.

The resource object is used to manage the resources. Only the executive object can interact with this object. This private object is invoked by the QUESEIZE element when it either checks or seizes some resources or by the Release function when it releases some resources.

The executive object starts its cycle by processing the event object from the event list with the minimum scheduled event time. It begins the processing of each event object by removing it from the event list. The simulation time is then set to the event's scheduled event time. After the model executes the event, it is possible that resources may have been released, so the executive object checks each QUESEIZE element to see if any entity can now allocate the required resources. If so the corresponding QUESEIZE event is scheduled to occur at the current simulation time and is placed in the front of the event list. After the event object is processed, it is destructed, and the next event is removed form the event list thus starting a new cycle.

In Program 1, an example of a simulation using this developed simulation tool is provided. This model simulates a simple M/M/1 queue where the arrival rate is LAMDA and the service rate is MU. This simulation estimates the average queue size.

3. FURTHER MODEL ENHANCEMENTS

In addition to implementing the basic set of SIMAN like blocks, additional modeling elements not found in most simulation language have been added to enhance the software's capacity to provide expanded control over the flow of entities and to permit the model to function to as a controller of the real system.

3.1 Flow Control Enhancements

To provide enhanced control over the flow of entities, several new features have been incorporated. Each modeling element object contains three public functions. The first function is the Insert function. This function permits the modeler through the use of C++ code to insert an entity into any other modeled element object within the model. The inserted entity object would then begin to flow from its new location. The second function is the Remove function. It permits the modeler to remove an entity object from the modeled element object where it currently resides. The modeler can then Insert the element elsewhere or the entity will continue to

```
#define        SERVER   0

int            total_ent;
ResourceStruct  resources[50] = { "Server", 1};
RandomExp arrival(1/LAMBDA),
               service_time(1/MU);
QueSeize       server_q1("Server Que", SERVER,1);
Release        server_r1;
Delay          service_d1( &service_time);
Create         arrivals_a1(&arrival);
Tally          TIS_t1("time in sys.");

void
perform(int EventNumber)
   {
   Entity *ent;

   switch (EventNumber)
      {
      case CREATE1:
           if (get_TNOW() > 150)
               break;
           ent = arrivals_a1.create();
           ent->attr[0] = get_TNOW();
           server_q1.insert(ent);
           break;

      case QUESEIZE1:
           ent = server_q1.remove();
           service_d1.insert(ent);
           break;

      case DELAY1:
           ent = service_d1.remove();
           server_r1.release();
           TIS_t1.tally( get_TNOW() -
               ent->attr[0]);
           total_ent++;
           delete ent;
           break;

      default:
           cout << "Oops. Did a boo boo\n";
           break;
      }
   }

void
main()
   {
   InitializeResources(resources);
   total_ent = 0;
   sim( arrival.get() );
   PrintResults();
   }
```

Program 1: Sample Model for M/M/1 Queue

flow from the position where the Remove function was invoked. A third public function, Display, provides the pointers of the entity objects that currently reside at a given modeled element object. It returns a null pointer if no entity objects currently reside at that element.

There are times in a simulation that a section of the model needs to be repeated several times. For example, in our emulator of an FMS, the submodel that models the movement of the automated guided vehicle (AGV) is repeated several times through out the model, i.e. once to take the part to the machine and then once again to pick up the part and so on. Languages like SIMAN provide little support for recursive submodel calls. The submodels simply allow one to organize the model into groups of blocks. These submodels always return entities to a specific location. In order to support recursive submodel calls, two new blocks have been added. The GOSUB and the RETURNTO blocks allow a submodel to be called from anywhere in the model and always returns to the location where it was called, much like the recursive function calls in C or any general programming language.

The GOSUB block is used to call a submodel. An example is gosub(ent,MOVE,7), where ent is the pointer to the entity, MOVE is the name of the submodel and 7 is the return address or the return event. In this example, the entity is passed to the block via the variable ent that points to the entity, the first event in the submodel is called MOVE and the next event to execute when the submodel is done is called RETURN7. The RETURNTO element is the last block in the submodel. It sends the entity back to the return event. An example is returnto(ent), where ent points to the entity. Parameters can be passed to the submodel by using a space in the entity object called param. This is like the space for the attributes but is designated for parameter passing.

The ability to return the entity to the location where it was called from is provided by storing the return address inside of the entity itself. Each entity has a return address stack. Since there may be several entities within the submodel at the same time, each entity must hold there own return address unlike a general programming language that usually has only a single process operating and therefore, uses the system stack to hold the return address.

When GOSUB is called, it stores the entity in a holding area and pushes its return address onto the entities return address stack. It then sets a flag which signals the executive object that an event must be executed before finishing the cycle and provides it with a pointer to the next event, the first event in the submodel. The RETURNTO block works in much the same way, but it pops the return address off of the return address stack in the entity and uses this address to tell the executive object where to go to next. The following is an example of a section of a model where the recursive submodel called MOVE_AGV is called to move the AGV from the machine to the port. It has just finished turning the machine's motor needed to load the AGV with the part. Notice the passing of the parameters through the "param" spots 5 and 6. Param 5 holds the location of the AGV and param 6 hold the destination.

```
case PENDING35:
        ent = turn_motor_p35.remove();
        Motor_r9.release();
        ent->param[5] = PORT;
        ent->param[6] = MACHINE;
        gosub(ent,MOVE_AGV,7);
        break;

case RETURN7:
        ent = get_ent();
             .
             .
             .
case MOVE_AGV:
        ent = get_ent();
        move_from = ent->param[5];
        move_to =ent->param[6];
             .
             .
             .
        returnto(ent);
        break;
```

3.2 System Control Enhancements

The original desire for developing this simulation tool was to permit the simulation model to function as the controller for the system. To accomplish this goal, the controller must send commands to the physical machine to execute a task. A new model element called the PENDING block has been included to allow the model to synchronize itself with the hardware.

The execution of the PENDING event requires a major redefinition of the manner in which the executive object manages the simulation. First, the clock must run in real-time instead of simulated time. That is, instead of extracting the next event from the event list and setting the current time to the time that event is to occur as is done in simulation, the controller waits until that time arrives before it executes the event. So if the event is to occur at 5:00 the controller waits until 5:00 before it executes it as opposed to simply setting the current time to 5:00. The controller looks at the computer system's clock to get the current time. When a PENDING element is active, the simulation object must wait for a feedback signal from the hardware to indicate that the requested task has been completed. Hence the occurrence of the PENDING event cannot be scheduled. While the executive object is waiting for the PENDING event to occur, the real-time clock will continue to advance. The executive object also knows that there are additional scheduled events on the calendar. Since one or more of these events could occur before the PENDING event occurs, it is essential that the simulation executive object continuously compare the real-time clock against the earliest event time for an event object on the event list. Whenever the real-time equals the earliest event time, the associated event is processed.

When a machine is simulated, the DELAY block calculates the time it will be when the task is complete based on its duration and schedules a "task is finished" event for that time. However, when the machines actually exist and the model is to employ control of the system, the controller sends a signal to the machine to perform the task. It then waits until the machine signals that it has completed the task before the "task is finished" event is executed. The PENDING block in control mode does precisely that. However when a PENDING block is executed the event is put into a list of events that are pending for signals from the hardware before it can execute. This list is called the pending event list. So instead of scheduling events onto an event list where each event has an associated time of occurrence, the PENDING block puts events into a pending event list where there is no time associated with the events. In this list, the events simply wait until the hardware signals the controller that the requested task is done. The corresponding event is then pulled off of the list and executed.

Since the model is used for simulation as well as for control, the driver runs in two modes; simulation and control mode. In the simulation mode the clock runs in simulated time and the PENDING block operates in an identical fashion to the DELAY block. In this mode, there is no difference between the DELAY and PENDING block. In the control mode the system clock runs in real-time, and the PENDING block performs as indicated above. So one only sets the mode of operation, and the model either controls the system or simulates it. Note, in control mode one can still use the DELAY block. It always operates as a DELAY block but the delay is measured in real-time, not in simulated time. It is only the PENDING block that switches modes. This is useful when controlling an emulator where some of the machines exists and must be controlled while others are simulated. Also note that one must introduce the code to send commands to the hardware and provide checks so that the commands are not sent in simulation mode. This requirement is easily achieved in C++. The following example shows a section of the model communicating a command to turn the motor by 1 unit after seizing the motor.

```
case QUESEIZE9:
        ent = Motor_q9.remove();
        if (control_mode)
            send_msg("TURN");
        turn_motor_p35.insert(ent);
        break;
case PENDING35:
        ent = turn_motor_p35.remove();
```

4. THE FMS EMULATOR.

To develop an educational and research laboratory for the coordination of discrete-event systems, the Manufacturing Systems Laboratory at the University of Illinois has constructed an FMS emulator (see Davis et al. [1994]). To insure safety and economy, all physical processing has been omitted. We have also constructed physical analogs for the material handling systems (MHSs). The controller for this system is built using this package.

The schematic for the constructed FMS emulator is depicted in Figure 1. Figure 2 provides a photograph of the emulator. The emulator has four Processing Centers, numbered 1 through 4. Each Processing Center (PC) is a cell containing one primary subordinate processing resource (the Unit Process) and a dedicated MHS.

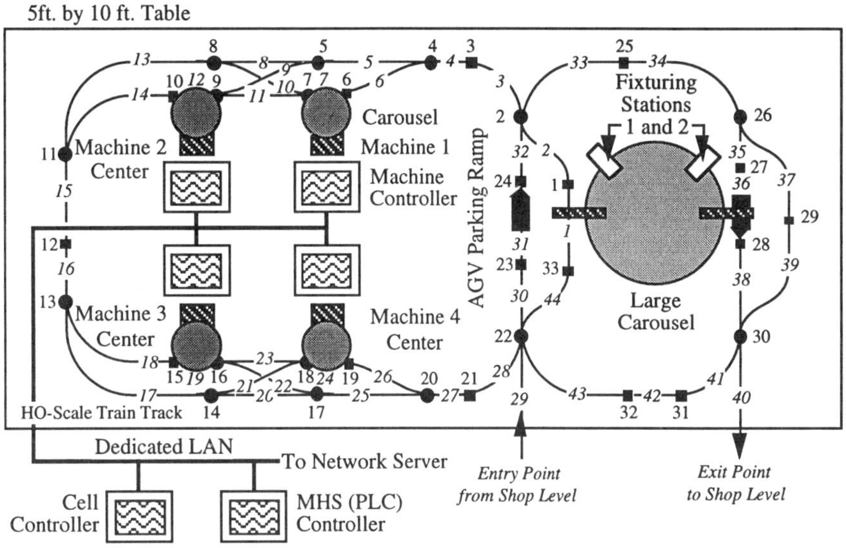

Figure 1: Schematic Layout for FMS Emulator

In Figure 3, we show the constructed PC. Its MHS is represented by a carousel with six electromagnets which

Figure 2.: FMS Emulator under Construction

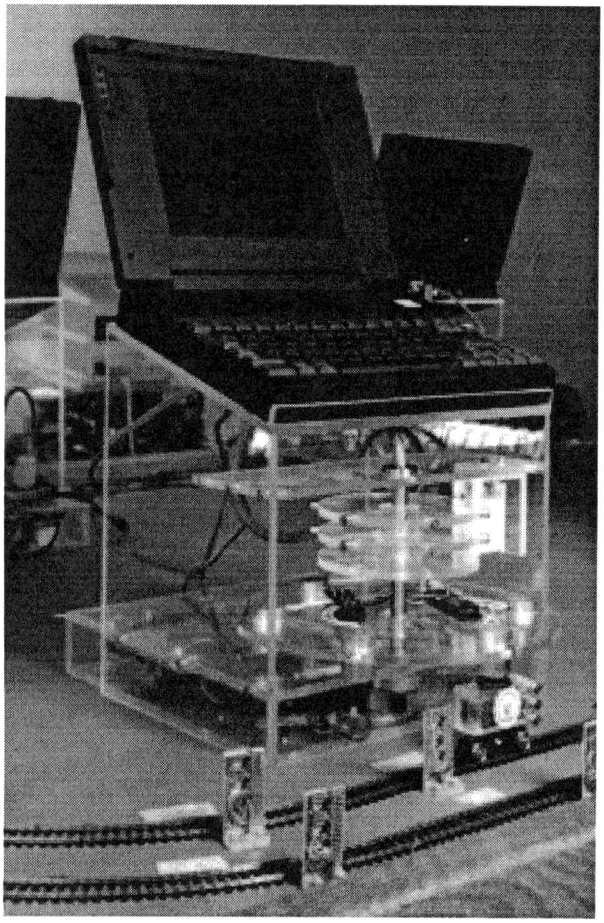

Figure 3: The Emulated Processing Center

hold the jobs (represented by color-coded steel washers) residing in either the input or output queues of the PC. The movement of the carousel is controlled by a dedicated Programmable Logic Controller (PLC). The PC controller is a lap-top UNIX workstation which is placed on a rostrum atop each PC. This controller is connected to the MHS PLC via a dedicated RS 232 link, and its display provides the summary status for the MHS.

The PC Controller summarizes each task as it is being implemented by the subordinate unit process. Future plans provide for the inclusion of another Unix computer at each PC to animate the operation of the unit processes. Software packages such as Deneb Robotics' Virtual NC can provide the desired animation for machining processes. (These packages are very expensive and are not critical to current implementation of the emulator.)

Within the emulated FMS (see Figure 2), another subordinate process is the Fixturing Center (FC) which also represents a cell. The structure and emulation of the FC is very similar to that of a PC. The FC has a dedicated MHS consisting of a primary carousel capable of holding sixteen jobs and two smaller carousels for loading and unloading jobs from the AGVs. The movement of these carousels is controlled by a dedicated PLC. The FC has two fixturing positions which represent the unit processes. A dedicated lap-top UNIX workstation provides the real-time status information for each fixturer. This same workstation also issue control messages to the dedicated PLC for the FC's MHS.

The final subordinate process is the Cell MHS. Automated Guided Vehicles (AGVs) are employed as the MHS and are emulated with a HO-scale electric trains. The train layout is diagrammed in Figure 1 and pictured in Figure 2. In this layout, there are over forty track segments which can be individually powered. Sensory switches are provided on each track segment to detect the presence of an AGV (see Figure 3). A Petri net has been developed to insure that no more than one AGV ever occupies a single track segment at a time.

A dedicated PLC receives directives from the Material Handling controller to move a given AGV from one location to another using the incorporated Petri net logic. The dedicated PLC returns the location of each AGV as it enters each track segment to the Material Handling controller. The Material Handling controller is also responsible for determining which of the pending material handling transfers will be processed and which AGV will be assigned to complete the requested transfer.

The cell controller is implemented by another UNIX workstation. It is connected to each of the subordinate PCs, FC and MHS controllers (and a network server) via an ether network. Various commands and feedback informations flow across this network. The role of the cell Controller is to orchestrate the flow of the entities of all types (jobs as well as supporting resources) among the subordinate processes within the cell.

5. CONCLUSION

Presented here is a simulation and control software that has the following properties:

1. The simulation is written in C++ providing the full modeling flexibility of a general purpose language.

2. The set of basic SIMAN like blocks are available to provide ease in modeling.

3. The package itself is a C++ library of routines and objects. The package is portable and may be run on any platform that has a C++ compiler.

4. This package supports subroutines not just submodels. Subroutines can be called from anywhere in the model

and returns to where they were called. You can even call them recursively. Submodels do not do this.

5. The package can run in "control" mode. In this mode, with the hardware drivers written, your simulation can actually control the system. The simulated-time becomes real-time and the event list includes a pending event list where the events are waiting for signals from the hardware drivers. You can write a controller for your system using SIMAN like blocks.

The system was also successfully tested on the physical FMS emulator. Future research is being done to build an intelligent controller for this emulator using real-time simulation and controller concurrently. That is, the controller will run real-time simulations concurrently in order to provide feed-forward information used for making intelligent control decisions.

REFERENCES

Davis, W. J., D. Setterdahl, J. Macro, V. Izokaitis, and B. Bauman. 1993. Recent Advances in the Modeling, Scheduling and Control of Flexible Automation. *Proc. of the 1993 Winter Simulation Conference*, eds. G.W. Evans, M. Mollaghasemi, E.C. Russell, W.E. Biles, 143-155, The Society for Computer Simulation, San Diego, CA.

Davis, W. J., B. Bauman, J. Macro and D. Setterdahl. 1994. Constructing an Emulator for Research and Education in the Control of Flexible Automation. *Proceedings of the ORSA Technical Section on Manufacturing Management Conf.*, eds. J. Buzacott and C.A. Yano, 151-157.

Mize J. H., H. C. Bhuskute, and M. Kamath. 1992. Modeling of Integrated Manufacturing Systems. *IIE Transactions* , 24(3):14-26.

Pegden, C.D., R.E. Shannon and R. P. Sadowski. 1995. Introduction to Simulation using SIMAN (second edition). McGraw-Hill, New York.

Smith, J. S., R. A. Wysk, D. T. Sturrock, S. E. Ramaswamy, G. D. Smith and S. B. Joshi. 1994. Discrete Event Simulation for Shop Floor Control. *Proc. of the 1994 Winter Simulation Conference*, Eds. J. D. Tew, S. Manivannan, D. A. Sadowski, and A.F. Seila, 962-969.

AUTHOR BIOGRAPHY

MR. FERNANDO G. GONZALEZ is doctoral student in Electrical and Computer Engineering at the University of Illinois. He received his B.S. in computer science and M.S. in electrical engineering at Florida International University. His current research addresses the real-time management of discrete-event systems. He is a recipient of an NSF Support for Under Represented Groups in Engineering (SURGE) Fellowship and a GTE Minority Fellowship..

Proceedings of the 1996 Winter Simulation Conference
ed. J. M. Charnes, D. J. Morrice, D. T. Brunner, and J. J. Swain

EFFICIENT STATE-INDEPENDENT SEQUENCING RULES FOR CERTAIN FLEXIBLE MANUFACTURING SYSTEMS

Arne Thesen

Department of Industrial Engineering
University of Wisconsin-Madison
Madison, WI 53706, U.S.A.

ABSTRACT

We introduce three state-independent methods for sequencing production of parts with random processing times on a simple production system such that productivity is maximized. The system of interest is one where parts of several different types are first processed on a common machine after which they proceed to separate, type-specific production lines. Referred to as *random*, *rotation*, and *most-behind* the rules are all based on easily computed estimates of the optimal product mix for the given part routings. Extensions of these rules are shown to give close to optimal performance in the case where blocking is present. The rules differs from conventional sequencing heuristics in that their performance is excellent over a wide range of system parameter values. The rules differ from optimal, state-dependent, scheduling rules in that no rule base or information about the current system states is needed.

1 INTRODUCTION

Consider the problem of determining the processing sequence of parts in a simple production system such that the number of parts produced per day is maximized. If all part routings and processing times are known in advance, then the consequences of all sequencing decisions can be predicted in advance, and the problem can be treated as a combinatorial optimization problem. In this case, a deterministic sequence of operations can be developed prior to the start of operations. On the other hand, if only the *distributions* of the processing times are known in advance, then optimal schedules cannot be developed in advance. Instead, state dependent scheduling rules leading to optimized expected performance can, under certain circumstances be developed. Based on assumptions of exponential service times and enumerable state spaces, these methods usually employ semi-Markov decision processes (SMDP, Howard, 1971) to develop specific scheduling decisions for each individual system state. Typical applications can be found in Seidmann (1988), Yih and Thesen (1991), and Chen (1992). Efficient algorithms for solving SMDP problems can be found in Stidham and Weber (1993). Given the need for real time information and the need to accommodate a large and complex rule base, the resulting scheduling rules are often of limited practical value. However they can be of significant value as benchmarks for evaluating more implementable heuristic rules (see, for example, Seidmann and Tenenbaum, 1994).

Many simple scheduling rules are *static* in the sense that priorities are established prior to production, and remain unchanged over the life of the process. Common sequencing rules such as *Shortest Processing Time First* (*SPT*) and *Longest Processing Time First* (*LPT*) are static rules. Harrison and Wein (1990) introduced a static rule that performs well for a class of small systems using a small number of circulating pallets. Chen and Thesen (1995) show that the resulting performance is very good, but never optimal.

Given the difficulties in implementing state-dependent scheduling rules, the scheduling rules that we have encountered most frequently in practice in systems with dynamically changing workloads are still of the simple FIFO variety. In this paper we show that significant improvements in performance are possible if more carefully prepared state-independent scheduling rules are used.

Our ultimate goal is to develop systems with performance equal or close to the one that is attainable by optimized state depended scheduling rules, but without the need for any real time decision making at all. Towards this end, we introduce the following three different management procedures:

- *Random* (the next job is selected at random according to a pre-computed distribution)
- *Most Behind* (the next job is the one most behind its current production target)

- *Rotation* (the next job is the next entry in a pre-computed sequence)

Each of these rules employs pre-computed parameters. These parameters are based on the product mix resulting from a relatively simple linear program (LP) optimizing the production rates when the effects of queuing delays and blocking are ignored. We will see that all rules perform exceptionally well when large buffer sizes are allowed. Modifications of these rules designed to avoid delays due to blocking extend this excellent performance to cases with small buffer sizes.

This paper is organized as follows. The production system of interest is described in section two, as is a simple LP algorithm for estimating an upper bound of productivity and the corresponding product mix. The three state-independent scheduling procedures are introduced in section three, as are extensions to deal with the problem of blocking. An evaluation is presented in section four. A summary and conclusions are given in section five.

2 THE PROBLEM

The goal of this study is to develop new simple sequencing rules for a stochastic production system first studied by Seidmann (1988) (Figure 1). Briefly, the production system is defined as follows. Parts of *n* different types are first machined on a shared CNC, then they proceed to dedicated assembly lines (one for each part type). Processing times are exponentially distributed and buffer spaces are limited. Parts can be processed in any sequence on the CNC and there is an infinite supply of raw material. Our objective is to determine, in real time, a sequence of products to be produced at the CNC such that the overall production rate is maximized. A table summarizing the key features of this problem is given in Table 1.

2.1 Product Mix Estimates

We will refer to the *product mix* as the set (*p*) of percentages (p_1, p_2 ,.., p_i) describing the ratio between the output of individual parts and total production. For the class of systems studied here, total production (measured in parts per hour) is known if *p* and the utilization of any one machine are known. We will refer to the *optimal product mix* (*P*) as the product mix that leads to the highest possible production rate. It follows that any scheduling rule that gives 100% utilization of at least one machine while at the same time ensuring that the optimal product mix is maintained, will give the highest possible (e.g. optimal) production rate. The problem is now reduced to one of finding the optimal product mix, and a method for ensuring 100 percent utilization of one machine.

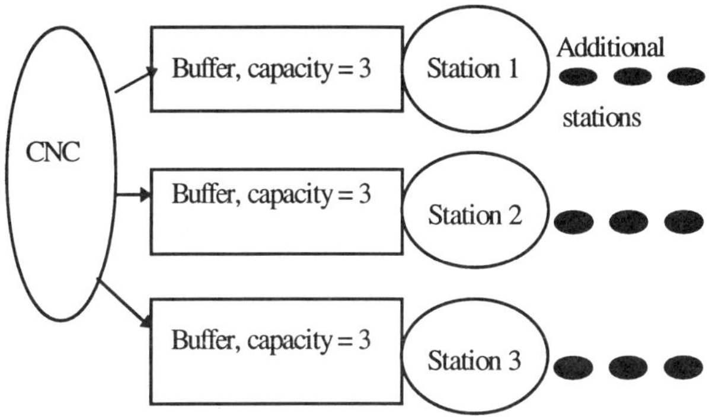

Figure 1: A FMS consisting of a CNC feeding parts of different types to three separate assembly lines

Table 1: Key elements of the problem of interest

Parts	Parts of *n* different types are produced. There is an unlimited supply of raw materials for all parts. All parts produced by the system can be sold.
Facilities	All parts are first processed at a single machining center. They then continue to separate assembly lines for each part type. The first station in each line is the bottleneck; hence subsequent stations need not be analyzed. Buffer capacities (b) in front of each line are identical.
Processing	Processing times are exponentially distributed random variables with the following parameters: m_i = Mean processing time at the machining center for parts of type *i*. A_i = Mean processing time at station 1 for parts of type i.
Control	*CNC*: Parts may be processed in any sequence. The identity of the part to be processed next is determined at the time processing starts. *Assembly lines*: Any part in the input buffer may be selected for processing.
Available Inform-ation	Three types of information are available: • Processing times (actual or expected) • System states (Buffer full/ not full, expected completion times) • Product mix (Target and/or actual)
Objective	Maximize parts produced per hour.

Table 2: The part routings used in this study

	Mean Processing Times						Upper bounds on throughput rates							
Rou ting	Machining Center (min.)			Assembly Lines (min.)			Parts per Hr.	Product mix (%)			Machine Utilization (%)			
	m_1	m_2	m_3	a_1	a_2	a_3		p_1	p_2	p_3	cnc	r_1	r_2	r_3
1	0.12	.24	.36	2.0	3.0	1.0	110	27	18	55	50	100	100	100
2	0.24	.48	.72	2.0	3.0	1.0	110	27	18	55	100	100	100	100
3	.375	.75	1.125	2.0	3.0	1.0	80	37	25	38	100	100	100	.50
4	0.6	1.2	1.8	2.0	3.0	1.0	60	50	37	17	100	100	100	17
5	1.00	2.0	3.00	2.0	3.0	1.0	45	67	33	0	100	100	75	0
6	.36	.24	0.12	2.0	3.0	1.0	110	27	18	55	50	100	100	100
7	.72	.48	0.24	2.0	3.0	1.0	110	27	18	55	76	100	100	.50
8	1.125	.75	.375	2.0	3.0	1.0	100	20	20	60	100	67	100	100
9	1.8	1.2	0.6	2.0	3.0	1.0	80	0	25	75	100	0	100	100
10	3.00	2.0	1.00	2.0	3.0	1.0	60	0	0	100	100	0	0	100

We first develop an estimator of P. Ignoring inefficiencies due to starvation and blocking caused by finite buffer sizes, random service times, and poor decisions, we formulate a linear program that maximizes the throughput rate for this system. The objective is to:

Maximize z = $\sum_1^n x_i$ (total output per hour)

Subj. to: $\sum_1^n m_i * x_i <= 60$ (Cell capacity)

$a_i * x_i <= 60 \quad i = 1,..,n$ (Line capacity)

Where:

- x_i = parts of type i produced per hour
- a_i = mean processing time at the first station of the ith assembly line (minutes)
- m_i = mean processing time at the cell center for the ith part type (minutes)
- n = Number of part types.

We define the *ideal product mix* is as the fraction of each type produced in the bound established above. Specifically, we define P_i as:

$$P_i = \frac{x_i}{\sum_{j=1}^{n} x_j}$$

Many simplifying assumptions were made above, and the ideal product mix may not be identical to the optimal product mix. However, since the solution of the LP given above is an upper bound, any schedule that is consistent with the solution to this LP must be optimal. Hence, a schedule resulting in the ideal product mix and in full utilization of at least one machine is optimal. We will see that this is easily achieved when the buffers are large (i.e. no blocking). We will also see, for the case when blocking is present, that the three different scheduling rules using the ideal P_i's as targets dominate the performance of other well known scheduling heuristics.

2.2 Scenarios

We will study the systems performance when producing parts using ten different part routings (Table 2). The upper bounds on throughput rates given in this table were established by solving the linear program given in the previous section using the appropriate parameters. All part routings use the same mean times at the assembly lines. However, to observe a number of different bottleneck conditions, five different ratios between processing times at the CNC and the assembly lines were used. The first five and the last five routings use identical processing times at the CNC, and they differ only in that processing times for part types one and three were reversed (i.e. 1,2,3 vs. 3,2,1). This was done to negate any arbitrary benefit from a given sequence established by the relative magnitudes of the processing times. Observe that the CNC is a bottleneck for scenarios 2 through 5 and 8 through 10, and that all three lines are bottlenecks only for scenarios 1, 2, and 6. Scenarios 5, 9 and 10 have less than three parts types in the ideal product mix. Buffer allocations are assumed in all cases to be identical for all machines.

2.3 "Optimal" Decision Rules

Seidmann (1988) showed that stochastic scheduling systems of the type discussed here can be modeled as a Semi-Markov Decision Processes (SMDP). Chen (1992) used a three-phase implementation of the value iteration algorithm to find recommended state transitions for each decision state for several variations of the example given above. Unfortunately Chen found that the state space increases exponentially with the number of buffer spaces, and he was unable to develop optimal schedules for problems significantly larger than the ones studied here. One useful result of Chen's research was the observation that the resulting optimal schedule can often be expressed in a simple decision table. He also showed that the rules developed this way can easily be extended to larger problems. The table used for scenarios one through four is shown in Table 3.

Table 3: Optimal state dependent decision rules for scenarios one through four.

CONDITIONS			ACTIONS		
Space Available In			Start Production of Part of		
Line A?	Line B?	Line C?	Type A	Type B	Type C
Y	I	I	Y		
N	Y	I		Y	
N	N	I			Y

3 DECISION RULES

Here we introduce three procedures for initiating production of parts of different types at the CNC such that a given fraction of parts of each type are produced. The *random* rule picks work pieces at random, using probabilities corresponding to the given product mix. The *rotation* rule uses the knowledge of the product mix to implement a repeating fixed sequence of parts to be produced. Finally, the *most behind* selects the next work piece as a part most under represented in the current mix. Note that these rules are not static since part priorities will change dynamically as production progresses.

3.1 State Independent Rules

Random. Under the random rule, we draw the next part at random according to the ideal product mix fractions (P_i). In other studies, a random rule has never been shown to be effective. In fact such rules are often included in studies as a neutral benchmark to determine the effectiveness of other rules. However, since the random rule used here includes information about the ideal product mix, it is quite effective. The expected throughput rate for a given set of P_i ´s is readily computed (although we will not be doing so here). One advantage of this rule is that it needs no information about the current state of the system. A disadvantage is that it takes no action to avoid full buffers. This leads to blocking when the system is operated with limited buffer spaces.

Rotation. Rotation sequences are constructed such that parts of the same type are evenly spaced in the sequence, and such that the ratio between the different parts in the sequence is identical to the given product mix. For example the sequence (1, 2, 1, 1, 2, 1, 3,...) implements a rotation schedule for the 50 %, 33 %, 17 % ideal part mix for scenario 4. The rotation schedules used for all scenarios are shown in Table 4. Again, the expected throughput rate for a given rotation schedule is readily computed. By placing a given part type at even intervals in the sequence, the assembly line buffer for that part is given time to clear before a new part of the same type is produced. Simulation experiments show that this feature causes this rule to perform better that the random rule in avoiding blocking when buffer capacities are limited. A disadvantage of this rule is the fact that (short) rotation sequences are difficult or impossible to develop for many product mixes.

Table 4: Rotation schedules

Cases	Product mix	Production sequence
1,2,6,7	27,18,55	3, 1, 3, 2, 3, 1, 3, 2, 3, 1, 3, ...
3	37,25,38	1, 3, 2, 1, 3, 1, 3, 2, ...
4	50, 33, 17	1, 2, 1, 2, 1, 3, ...
5	67, 33, 0	1, 1, 2, ...
8	20, 20, 60	3, 2, 3, 1, 3, ...
9	0, 25, 75	3, 3, 3, 2
10	0, 0, 100	3

Most behind. Under this rule, we multiply the present total output by the ideal product mix probabilities to obtain the current production target for each part type. This value is compared to the actual production. The most under-produced part is selected for production. The performance of this rule is quite similar to the one observed for the Rotation rule. However, sequences can be developed with equal ease for any product mix. A dis-advantage of this rule is that it may be difficult to collect the required data in real time. Presently, we are unable to estimate the expected throughput rate for a given rotation schedule without using simulation.

Table 5. Some state-independent rules

Rule	**Description**	**Real Time Information Needed**	**Avoids full buffers**	**Remarks**
Random	Pick next part at random, in	None	No	

	accordance with the ideal product mix			
Rotation	Pick part form a pre-computed fixed sequence.	None	Yes	Rotation sequences difficult to develop
Most Behind	Pick part most under represented in current product mix	Current product mix	Yes	Probably good for smaller buffers

3.2 State-Dependent Extensions

The simple sequencing rules introduced above perform well when the assembly lines have large buffer capacities. However, frequent delays due to blocking occur when buffer spaces are limited. Here we will discuss ways to avoid this problem while retaining the ideal product mix generated by these rules.

Unequal buffer allocations. Delays due to blocking can be substantially reduced by reallocating limited buffer spaces from poorly utilized buffers to more highly utilized ones. Thesen and Chen (1996) shows that optimal scheduling rules change when this is done, and they propose an algorithm for simultaneously establishing optimal buffer allocations and scheduling rules. Heuristic scheduling rules such as *Probabilistic Shortest Queue* (*PSQ*, Yao and Buzacott, 1984) and *Fastest Shortest Queue* (*FSQ*, Seidmann and Tenenbaum, 1994) that use information about currently available buffer capacities work very well when limited buffer spaces are optimally allocated.

Blocking avoidance. The simplest way to avoid blocking is not to start processing of a part at the CNC if it will be going to a full buffer. This may be achieved by excluding parts going to full buffers from consideration for processing at the CNC when the next work piece is selected. A problem with this strategy is that the resulting product mix may deviate from the ideal one for some rules. In fact, only the *most-behind* rule will self-correct for the distortions introduced by this scheme.

Anticipation. It is difficult to avoid blocking when all buffers are full. One efficient strategy is to select the part with the longest possible processing time. Also, process completion times can very often be determined with great accuracy when processing starts. Using this knowledge, the CNC scheduler may be able to select a part for processing, knowing that there will be room in the corresponding input buffer when processing at the CNC is completed. However, the resulting product mix may again be distorted.

4 EVALUATION

In this section we report on experiments designed to evaluate the decision rules discussed in the previous section. For reference purposes, the conventional *SPT* rule is also evaluated. Simulation using ProModel for Windows (Baird and Leary, 1994) was used to estimate the throughput rates resulting from the use of different decision rules. The batch means technique (10 batches, 10,000 parts per batch) was used to estimate a confidence interval for the throughput rate for each rule under each scenario. All confidence intervals have half widths less that 5% of the mean. Only the resulting means are shown in this report. The optimal throughput rates were obtained using a value iteration based SMDP algorithm due to Chen (1992).

4.1 State-Independent Rules

We first evaluated performance under the conditions of "weak blocking". In this case we assumed that 30 buffer spaces were available in front of each line. The results for scenarios one through five are shown in Figure 2. We see that performance within a few percent of optimal was observed in all cases. However, we also see that the conventional SPT rule performed rather poorly. The results for scenarios six through ten are quite similar.

The performance observed for these rules under scenarios one through five with three buffer spaces in front of each line is shown in Figure 3. We see that the *most behind* and *rotation* rules resulted in relatively good performance, while the random rule performed less well. In all cases did the optimal rule result in the best performance. An analysis of the simulation output indicates that the reason for the inferior performance of the random rule is that blocking is more likely to occur as full buffers are not avoided. The results are similar for scenarios 6 - 10.

4.2 State-Dependent Rules

Figure 4 gives the observed performance for all ten scenarios when the rules are modified to implement the state-dependent suggestions given for blocking avoidance and anticipation in the previous section. The difference in performance between the modified rules and the optimal rules is not statistically significant at the 5% level. A separate test was run to determine if performance improved when information about buffer availability at the *end* of processing at the CNC was used in place of knowledge about buffer availability at the *start* of processing. No significant improvement was found.

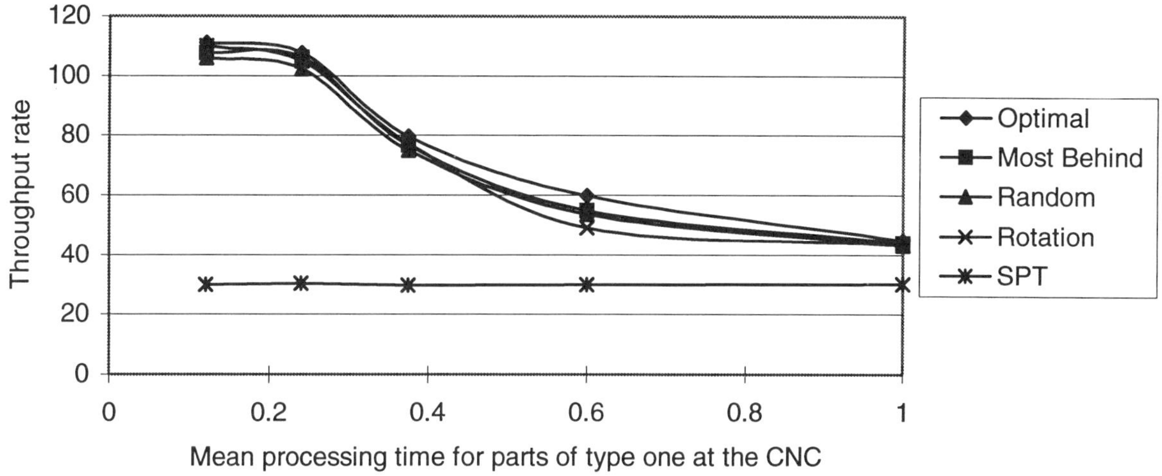

Figure 2: Performance of the simple scheduling rules for scenarios one through five under the condition of weak blocking. Thirty buffer spaces are available at each of the assembly stations

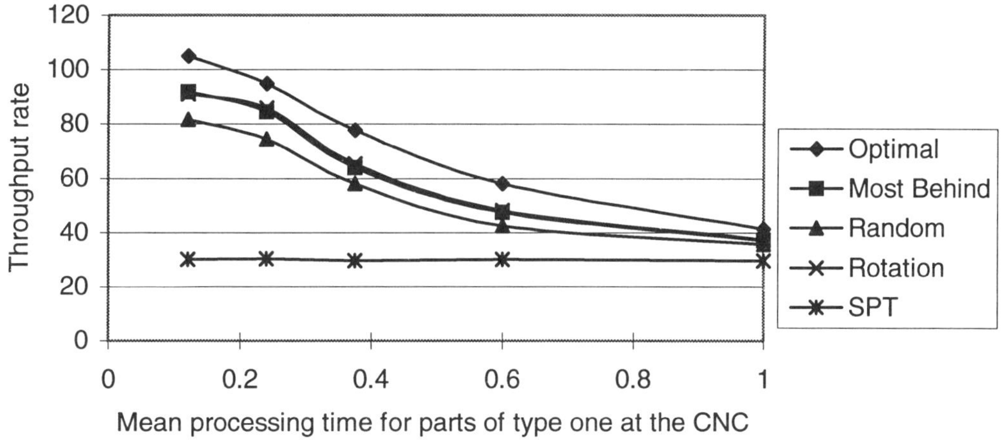

Figure 3: Performance of the simple scheduling rules for scenarios one through five under the condition of strong blocking. Three buffer spaces are available at each of the assembly stations

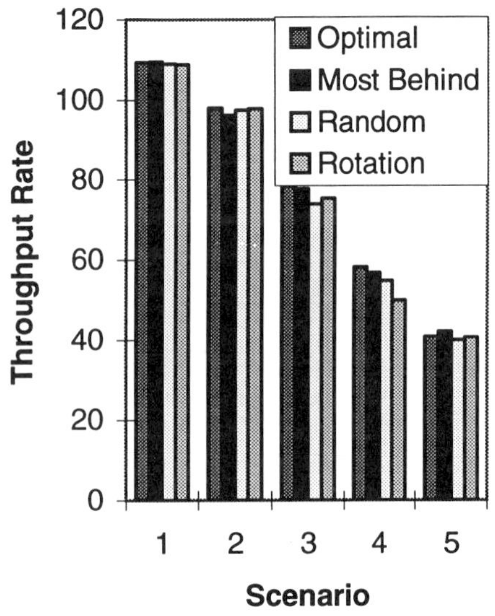

Figure 4: Throughput rate for state dependent rules. Three buffer spaces at each of the assembly stations

5 CONCLUSION

We have introduced three different sequencing rules all of which exploit information about an optimal product mix that is easily computed using information about part routings. The rules were shown to give good to excellent performance in a predictable manner for a wide range of scenarios. The choice of a rule for a given situation depends on the availability of buffer spaces, and on the information available to the scheduler. If the system has large buffer spaces and if no real-time information is available about buffer populations, then the random rule should work well. When buffer spaces are limited and real time information is available, then the *most-behind* rule should give close to optimal performance.

REFERENCES

Baird, S.P., and J.J. Leavy. 1994. Simulation Modeling Using Promodel for Windows. In *Proceedings of the 1994 Winter Simulation Conference*, eds. Jefferey D. Tew, Mani S. Manivannan, Deborah A. Sadowski, and Andrewe F. Seila, 527-532.

Chen, J.C., 1992, *State Dependent Scheduling for Manufacturing Systems*, Ph.D. Dissertation. Department of Industrial Engineering, University of Wisconsin-Madison.

Chen, L-Y., and A. Thesen. 1995. Performance Analysis of A Two-Station Closed Queueing Network Using Static and Optimal Sequencing Policies. *Technical Report 95-5*, Industrial Engineering, Univ. of Wisconsin at Madison.

Harrison, J. M., and L. M. Wein, 1990. Scheduling Network of Queues: Heavy Traffic Analysis of a Two-Station Closed Network. *Operations Research* **38**, 1052-1064.

Howard, R.A., 1971, *Dynamic Probabilistic Systems. Volume II: Semi-Markov and Decision Processes*, John Wiley & Sons, New York, New York.

Seidmann, A., 1988, "Optimal Dynamic Routing in Flexible Manufacturing Systems with Limited Buffers", *Annals of Operations Research*, 15, 291-311.

Seidmann, A., and A. Tenenbaum, 1994. Throughput Maximization on Flexible Manufacturing Systems. *IIE Transactions*, Vol. 26, No. 1, 90-100.

Stidham, S., and R. Weber,1993, "A Survey of Markov Decision Models for Control of Networks of Queues", *Queueing Systems*, 13, 291-314

Thesen, A., and J.T. Chen , 1996, "Jointly Optimized Buffer Allocation and Sequencing Rules for a FMS", *Proceedings of IASTED Conference on Modelling, Simulation and Optimization*", Gold Coast, Australia, May 6.

Yao, D.D., and J.A. Buzacott, 1984, "Modeling a Class of State-Dependent Routing in FMSs", *Annals of OR*, 3, 153-168

Yih, Y., and A. Thesen, 1991,. "Semi-Markov Decision Models for Real-Time Scheduling". *International Journal for Production Research*, Vol. 29, No. 11, 2331-2346.

AUTHOR BIOGRAPHY

ARNE THESEN is the Robert A. Ratner Professor of Industrial Engineering at the University of Wisconsin-Madison. He holds a MSIE and Ph.D. from the University of Illinois Urbana-Champaign, and he is a member of INFORMS and a Senior Member of IIE. He has authored or co-authored three textbooks in the areas of simulation and operations research, and he has published over 40 papers in the same areas. A former department chair, his current interests are in short term scheduling and heuristic optimization methods.

Proceedings of the 1996 Winter Simulation Conference
ed. J. M. Charnes, D. J. Morrice, D. T. Brunner, and J. J. Swain

SIMULATION SYSTEM FOR REAL-TIME PLANNING, SCHEDULING, AND CONTROL

Glenn R. Drake

Systems Modeling Corporation
504 Beaver Street
Sewickley, Pennsylvania 15143, U.S.A.

Jeffrey S. Smith

Department of Industrial Engineering
Texas A&M University
College Station, Texas 77843, U.S.A.

ABSTRACT

This paper introduces a framework for on-line simulation systems in the operational planning, scheduling, and control of manufacturing systems. Five basic concepts for software design of an on-line simulation system are identified and an example simulator is illustrated.

1 MOTIVATION

There has been a growing interest in real-time, "intelligent" shop floor control systems that use simulation technology to predict the future impact of short-term manufacturing decisions (Erickson *et al.* 1987, Wu and Wysk 1989, Harmonosky and Robohn 1991, Rogers and Flanagan 1991, Smith *et al.* 1994, Drake *et al.* 1995, Jones *et al.* 1995). Although recent researchers have examined the responsibilities and underlying issues of on-line simulation systems, there has been a tendency to view the simulation software as a "black box" and examine solely its interactions with other system components such as data collection devices, neural networks, expert systems, genetic algorithms, control mechanisms, and process planning functions. Thus, the design of the simulation system itself has been largely ignored and a framework for on-line simulation systems is still needed.

There is currently a large commercial market for simulation-based analysis in manufacturing. This market is illustrated by the demand for products such as Arena/SIMAN® (Pegden *et al.*, 1995), SlamSystem™ and FACTOR™ (O'Reilly, 1995), Witness™ and PROVISA™ , Extend™ (Krahl, 1995), and AutoMod/AutoSched™ . Some of these packages (e.g., Arena/SIMAN, Extend, Witness, SlamSystem) are primarily designed for and used in design applications (i.e., long-term, predictive analysis). Others, such as AutoSched, PROVISA, and FACTOR, are primarily designed for and used in short-term finite capacity planning and scheduling applications. Irrespective of their primary focus, it is often difficult to implement these systems in on-line planning, scheduling, and control applications due to one or more of the following reasons:

(1) The conceptual framework is geared exclusively towards use by humans and not the decision support software commonly found in shop floor control systems for computer integrated manufacturing (CIM).

(2) The system does not employ modeling constructs or a framework that facilitates the development of simulation models intended for real-time use. This includes inadequate support for modeling complex control logic, whereby either the necessary constructs are not available or the modeling of control logic is integrated with physical characteristics such that changes are difficult to implement. Additionally, many systems do not incorporate constructs for real-time task dispatching.

(3) The conceptual framework implicitly assumes that simulationists (i.e., modelers) and end-users are identical by providing a single primary interface or a set of "rigid" interfaces (e.g., a "scheduling" interface and a "modeling" interface) for constructing models and executing simulation experiments. However, in the context of on-line planning, scheduling, and control applications, end-users often include a *variety* of personnel (e.g., schedulers, capacity planners, managers) who were not directly involved in model development, have little or no simulation expertise, and who desire to use only a portion of the simulation tool.

The common inefficiencies of current commercial packages and the increasing interest in real-time planning, scheduling, and control indicate that an effective framework for on-line simulation systems is needed. When developing important concepts for such a framework, it is useful to examine the applications of on-line simulation technology.

2 THE USERS OF ON-LINE SIMULATION SYSTEMS

On-line simulation systems incorporate two powerful features: 1) the ability to reliably *predict* the future behavior of the shop floor given its current status, and 2) the ability to *emulate* and/or *dictate* the control logic of a manufacturing system. These two capabilities offer potential benefits to a variety of end-users in a manufacturing organization. Figure 1 shows an on-line simulation system integrated with several functional areas of a facility.

First, simulation can help *capacity planners* determine order lot sizes, release dates, and work calendars for resources by incorporating simulated scheduling constraints in their decision making. Second, once orders are released, simulation offers real-time *schedulers* the ability to evaluate strategies for sequencing jobs through the workcenters. In dynamic manufacturing environments, it may be advantageous to change the way a shop is controlled at certain points in time.

Third, once the desired operational strategy and part mix have been determined, simulation can interact in real-time with *shop floor management.* Work lists can be generated and distributed to shop floor execution or dispatched on-line in a task-by-task manner (Smith *et al.* 1994). Running parallel to the actual system in real-time allows the simulation system to keep track of current status and send feedback to the scheduling function on the current schedule's performance.

Fourth, by emulating current shop conditions, simulation can help the marketing and sales functions reliably predict order leadtimes and quote accurate due dates to customers. Using simulation in this manner can decrease the amount of work in process and add to the reputation of the organization by improving its ability to meet promised delivery dates (Rogers and Flanagan 1991). Finally, the simulation system must interact with engineering for the necessary data (e.g., process plans, layout information) to perform valid analysis. Simulation output can give feedback to the engineering function on the performance of a current design (e.g., the layout of a workcenter).

3 CONCEPTS FOR ON-LINE SIMULATION SYSTEMS

It is apparent that several functional areas of an organization might interact with on-line simulation technology. In response, five basic concepts for simulation systems in on-line planning, scheduling, and control are identified:

(1) The system should be multifaceted as suggested by Zeigler (1980). A multifaceted environment acknowledges that the system is approachable from different points of view by different users with different objectives at different times.

(2) The system should separate the logically distinct activities of on-line simulation-based planning, scheduling, and control. Thus, the functional stages of modeling physical structure, modeling control logic, simulation input, experimentation, output analysis, and task dispatching should be distinguished through a modular approach that recognizes the details and mechanisms required for each element.

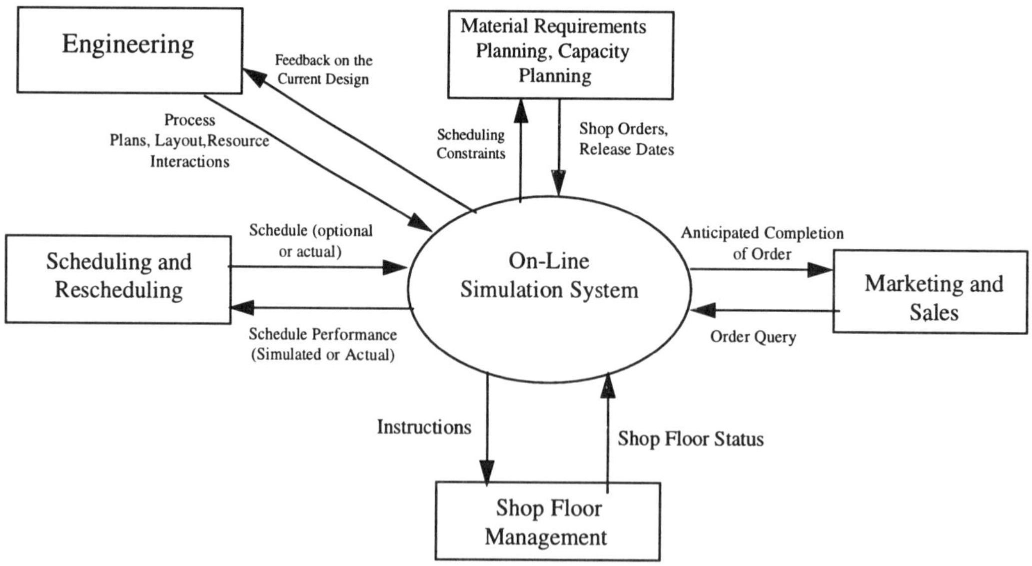

Figure 1: An On-Line Simulation System for Planning, Scheduling, and Control.

(3) The system should incorporate not only good human interfaces, but also explicit software interfaces to the activities of model development, simulation input, experimentation, output analysis, and task dispatching.

(4) The system should allow a model of a manufacturing system to be *re-used* for design, planning, scheduling and control with minimal modification within and between applications.

(5) The system should be flexible yet easy to use for modelers *and* end-users (Pinedo 1995). Flexibility within the simulation system is necessary for modeling complex control logic and reusing resident simulation models. The ease of use of the software is dependent on how well the skills and perspectives of end-users are recognized (i.e., a multifaceted environment).

4 GENERAL FRAMEWORK

When simulationists develop simulation models and turn them over to end-users for experimentation, they are delivering *simulators*. These end-users often include plant engineers, managers, and sales representatives with little or no simulation expertise. Though they may contribute input or functional specifications to the modeler(s) for project purposes, these personnel are usually not directly involved in model development. Instead, they require the simulator for daily decision making and task dispatching. If the same interface is used for these end-users as was used for the simulationists, then these end-users have access to simulation primitives that may be confusing no matter how "easy" the package is to use (Brunner and Crain 1991). Moreover, access to these primitives is generally unnecessary.

Figure 2 shows a general framework for on-line simulation systems in scheduling and control. In departure from traditional simulation software, the user interface is separated into two distinct environments: the *modeling environment* for simulationists and the *simulator environment* for end-users. The modeling environment focuses on the simulation constructs and logic necessary for developing simulation models intended for on-line use. Once a model is completed, it is "handed over" to the simulator environment which focuses upon the activities that comprise the real-time planning, scheduling, and control process (e.g., data entry, experimentation, output analysis).

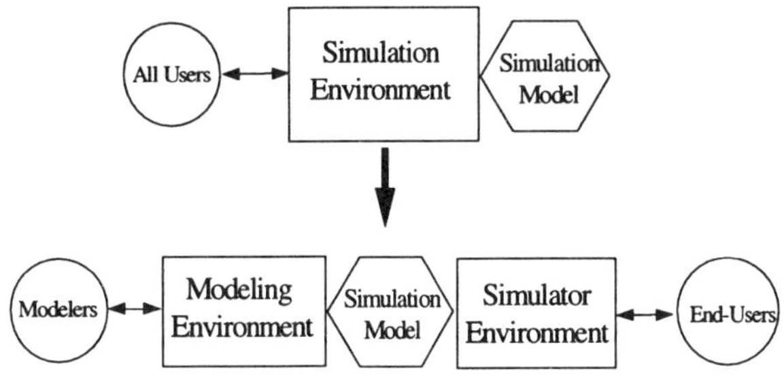

Figure 2: General Framework.

The simulation model developed for on-line planning, scheduling, and control applications can be decomposed into five primary functional elements. These functional elements are:

(1) *Physical Structure:* The definition of machines, people, tooling, transporters, and storage areas in the facility, as well as the layout and decision points of the manufacturing process.

(2) *Task Dispatching:* The simulator's I/O links with external components for task dispatching and the protocol of these messages.

(3) *Control Logic:* The operating rules (e.g., scheduling rules for equipment and people, batching rules), simulator run-time parameters (e.g., run length, output generation period), and exception handling (e.g., a dispatched task is not performed or is performed improperly).

(4) *Simulation Input:* The product definitions (e.g., routings, processing times, resource requirements, bills of materials), product demand (e.g., order quantities, due dates, start dates, priorities), resource relationships (e.g., resource groups), and work calendars (e.g., alternative shifts).

(5) *Simulation Output:* The start- and end-times for tasks in the system. These times may include additional "tag" information such as task codes, affected products, or utilized resources.

Figure 3 illustrates the functional elements of an on-line simulation model and their relationships with the modeling and simulator environments in Figure 2. Note that the modeling environment is separated into two distinct frames or interfaces: the *model frame* for defining the physical structure and task dispatching of the simulator, and the *rules frame* for developing the control logic or rule base of the simulator (e.g., alternative scheduling rules, batching rules, procedures for exception handling). Functions for loading simulation input from input databases and writing simulation output to output databases are automatically compiled during model development and do not require manual editing. End-users define simulation input and manipulate simulation output (e.g., generate Gantt charts) through the simulator environment.

The modeling environment's organization incorporates flexibility into the simulation system in two primary ways. First, it facilitates the development of simulators which separate the functional elements of physical structure, control logic, simulation input, and simulation output so that different scenarios (e.g., part mixes, scheduling rules, work calendars) can be readily "plugged in" and evaluated without model recompilation. Constructs for modeling task dispatching are incorporated with constructs for modeling physical structure such that the same model used for design, planning, and scheduling can be used for direct shop floor control. Thus, within the simulator environment, end-users exploit generic, reusable models designed for easy "what if" experimentation and automated execution. This concept is referred to as *flexible simulation* (Drake *et al.* 1995). The modularity of the rules frame facilitates the addition of new rules to the simulator's rule base.

Figure 3 shows the simulator environment separated into three distinct frames: the *entity frame* for simulation input (e.g., entering production data), the *experiment frame* for experimentation and execution (e.g., selecting operating rules from the rule base), and the *output analysis frame* for data manipulation and presentation (e.g., creating Gantt charts, order tracking). The *interactive montitor* provides a real-time interface to the on-line simulator tool. This monitor might include real-time animation, and could feasibly be detached from the simulator environment and implemented "stand-alone" at workcenters for real-time interaction (e.g., entering status and receiving tasks from the simulator).

Additionally, the proposed simulator environment incorporates software interfaces for the activities of simulation input, experimentation, output analysis, and real-time task dispatching. These software interfaces are illustrated in Figure 4. A *simulator language* to the manager of the simulation engine facilitates the automation of real-time planning, scheduling and control applications. The language incorporates commands for verifying simulator programs (i.e., debugging), loading simulation input from the entity frame, setting up and executing simulation experiments in the experiment frame, and calling analysis and presentation routines in the output analysis frame. Macros of commands might be created for end-users with different "views" or objectives (e.g., a scheduling macro, a marketing macro). To dispatch tasks on-line, the simulation system and shop floor management communicate through a *task initiation queue* (TIQ) and a *task completion queue* (TCQ) (Smith *et al.* 1994). Monitoring and tracking of tasks can also be implemented using the TIQ/TCQ interface. *Structured query language* (SQL) might be used to link with a variety of database management systems and extract simulation input (e.g., shop status, process plans) or forward simulation output (e.g., schedules, performance, expected completion times).

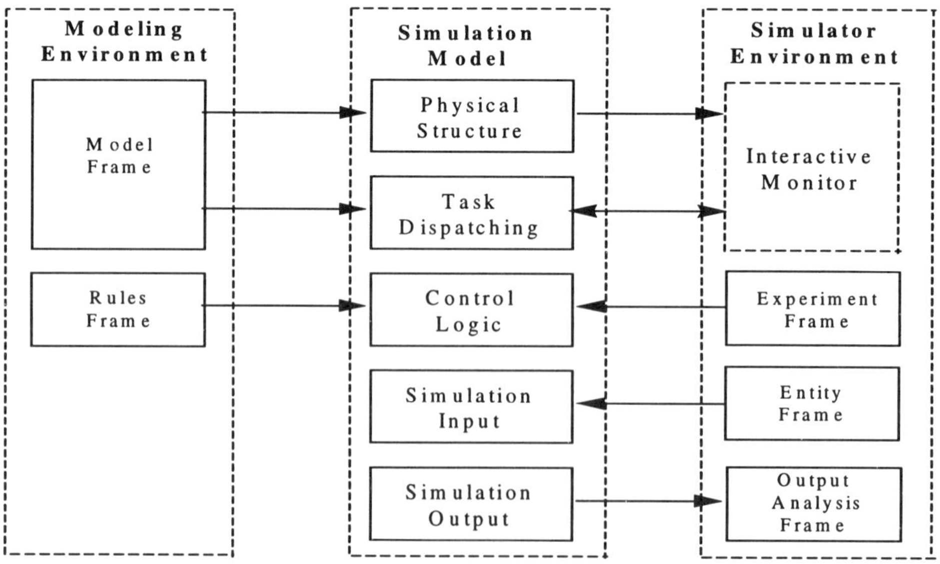

Figure 3: General Framework and Functional Elements of an On-Line Simulation Model.

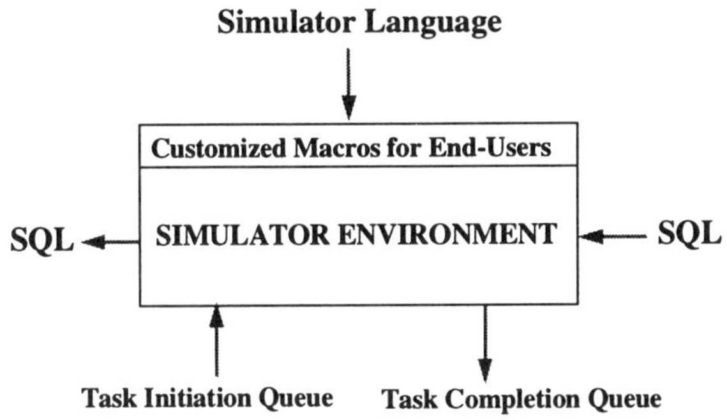

Figure 4: Software Interfaces to the Simulator Environment.

5 EXAMPLE OF A CONVENTIONAL ON-LINE SIMULATOR

To further motivate the general framework discussed in Section 4, a conventional SIMAN V (Pegden, *et al.*, 1995) simulator for a simple manufacturing system is now illustrated. The example system is shown in Figure 5. Six types of jobs arrive from Workcenter A to Workcenter B. Workcenter B is a small station that consists of two machines with an input buffer in front of each. Depending on the part type, jobs can be processed on Machine 1 only, Machine 2 only, or either of the resources. Jobs are forwarded to Workcenter C after processing. The operators at Workcenter B have three primary objectives. First, they want short lead times to aid on-time completions of orders at the factory level. Second, they want low levels of work-in-process (WIP) to minimize inventory costs. Third, they would like to forward anticipated completion times of jobs to the personnel of Workcenter C so that those operators can plan their operations. To meet their objectives, the operators at Workcenter B confront four major operating decisions. Three of these decision points are illustrated as question marks in Figure 5. The first decision entails incoming jobs with alternative routings, whereby one of the two machines must be selected. Machine 2 is the more expensive model and processing times are always shorter on this resource. The next two decisions involve the processing sequences on Machine 1 and Machine 2 (i.e., job dispatching from the input buffers). The last decision involves the work calendars of Machine 1 and Machine 2, whereby a fixed daily percentage of downtime is required for each resource for routine maintenance.

Due to the dynamic nature of the product mix released from Workcenter A, and the impact of the four operating decisions, it is considered advantageous to change the decision strategies of Workcenter B at certain points in time. Thus, to maintain short lead times and low WIP, strategies are alternated based on real-time shop conditions. Table 1 lists the processing times for the six part types manufactured in Workcenter B and the alternative strategies (i.e., decision rules) for machine selection and queue dispatching. These rules are sensitive to the workstation's dynamics and considered good candidates for improved performance. A model of Workcenter B was developed for on-line planning, scheduling, and control applications at Workstation B using the SIMAN V simulation language and environment. The simulator is generic in that different combinations of the decision rules in Table 1 can be assigned without model editing or recompilation. This simplifies "what if" experimentation by end-users unfamiliar with SIMAN or the model's structure. Once a satisfactory strategy is determined, task lists can be generated and distributed to the operators in Workcenter B and predicted completion times of jobs forwarded to the personnel in Workcenter C. SIMAN code for the model and experiment files is shown in Figures 6 and 7 respectively. X's mark the functional elements (i.e., physical structure (P), control logic (C), task dispatching (T), simulation input (I), and simulation output (O)) associated with each SIMAN construct. For example, Line 6 in Figure 6 defines some of the physical structure of the model as well as task dispatching.

Figures 6 and 7 illustrate the modeling effort required to develop a "user-friendly" simulator using the conventional Arena/SIMAN environment. Though an attempt was made by the modeler to textually partition the model and experiment files into their functional elements, the simulation system does not explicitly support a modular separation of these activities. Some general observations will now be made on the simulator's functional elements:

(1) *Physical Structure:* Lines 2 through 12 in the model file and Lines 3 through 6 in the experiment file define the layout and decision points of the process in Workcenter B. Within the proposed framework, these constructs would be defined in the model frame of the modeling environment.

(2) *Task Dispatching:* Lines 6, 11, and 52 of the model file define the I/O points for external communication and Lines 27 through 30 of the experiment file define the communication protocol. Within the proposed framework, these constructs would be defined in the model frame.

(3) *Control Logic:* Lines 13 through 48 in the model file and Line 25 in the experiment file define

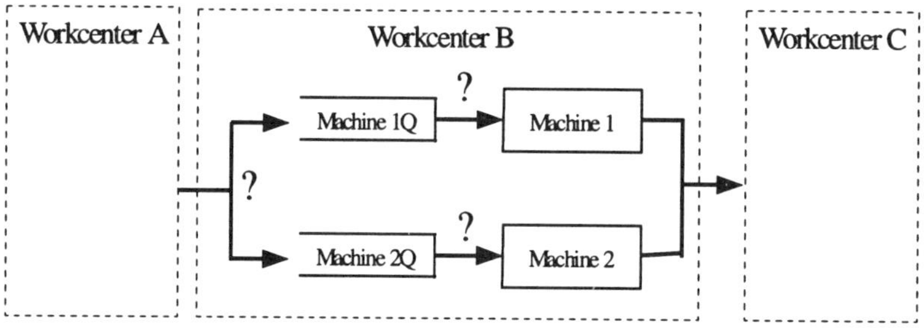

Figure 5: Example System.

the rule base. Within the proposed framework, these constructs would be defined in the rules frame of the modeling environment. End-users select rules from the rule base by assigning numerical values to the variables *Decision1*, *Decision2*, and *Decision3*.

(3) The assignment of work calendars to the machines requires the editing of Lines 3 and 4 in the experiment file. Within the proposed framework, all of these assignments would take place in the experiment frame of the simulator environment and not require model editing.

(4) *Simulation Input:* Lines 51 through 57 in the model file define code for reading production orders. Within the proposed framework, this logic would be automatically compiled during model development. In the experiment file, new orders are entered from the external file Orders.DB (Line 31 in Figure 7). The processing times for each part type are defined in Lines 13 and 14, and the work calendars for machines defined in Line 26. Entering new product types, processing times, or calendars requires modification of the model code. Within the proposed framework, this data is external to the model and is defined through the entity frame of the simulator environment.

(5) *Simulation Output*: Lines 59 through 66 in the model file define code for recording the start- and end-times of tasks in the system. This aggregate data is forwarded to the external file Output.DB (Line 32 in Figure 7). Within the proposed framework, these constructs would be automatically compiled during model development. Manipulation of the data (e.g., computing performance measures) would take place in the output analysis frame of the simulator environment. Different statistics/presentation macros might be applied by end-users with different objectives.

6 SUMMARY

The simulator presented in Figures 6 and 7 is "flexible" in that different scheduling rules can be defined in the model without recompilation. Also, the simulation output is aggregate such that multiple statistics can be calculated from the raw data by external functions. Thus, the output from the simulator is useful for a variety of end-users with different objectives. The generic characteristics of the model simplifies operational use of the tool by personnel unfamiliar with the SIMAN simulation language.

However, to change work calendars or enter additional production data *would* require editing of the model code. Developing flexibility for these instances necessitates additional effort by the modeler. For large models, it is likely the efforts to make user-friendly models could be considerable. The framework presented in this paper simplifies the development of generic simulation models for on-line use.

REFERENCES

Brunner, D. T., and R. C. Crain. 1991. GPSS/H in the 1990s. In *Proceedings of the 1991 Winter Simulation Conference*: 81-85.

Drake, G., Smith, J. S., and B. A. Peters. 1995. Simulation as a planning and scheduling tool for FMS. In *Proceedings of the 1995 Winter Simulation Conference*: 805-812.

Table 1: Operation Times and Decision Rules.

Processing Times (Min)

PartType	M/C 1	M/C 2
1	10	-
2	-	6
3	13	-
4	-	9
5	15	8
6	12	6

Rule Base
Alternative Machine Selection:
Always M1, Always M2, SNQ, STPT
Machine 1Q Dispatching:
FIFO,SPT
Machine 2Q Dispatching:
FIFO,SPT

Erickson, C., Vandenberge, A., and T. Miles. 1987. Simulation, animation, and shop-floor control. In *Proceedings of the 1987 Winter Simulation Conference*: 649-652.

Harmonosky, C. M., and S. F. Robohn. 1991. Real-time scheduling in computer integrated manufacturing: A review of recent literature. *International Journal of Computer Integrated Manufacturing*, **4**, 331-340.

Jones, A., Rabelo, L., and Y. Yuehwern. 1995. A hybrid approach for real-time sequencing and scheduling. *International Journal of Computer Integrated Manufacturing*, **8**, 145-154.

Krahl, D. 1995. An introduction to Extend. In *Proceedings of the 1995 Winter Simulation Conference*: 413-419.

O'Reilly, J. J. 1995. Introduction to SLAM II and SLAMSYSTEM. In *Proceedings of the 1995 Winter Simulation Conference*: 467-471.

Pegden, C. D., Shannon, R. E., and R. P. Sadowski. 1990. *Introduction to Simulation using SIMAN*, McGraw-Hill, Inc., New York.

Pindeo, M. 1995. *Scheduling: theory, algorithms, and systems*, Prentice Hall, Englewood Cliffs, NJ.

Rogers, P., and M. Flanagan. 1991. On-line simulation for real-time scheduling of manufacturing systems. *Industrial Engineering*, **23**, 37-40.

Smith, J. S., Wysk, R. A., Sturrock, D. T., Ramaswamy, S. E., Smith, G. D., and S. B. Joshi. 1994. Discrete event simulation for shop floor control. In *Proceedings of the 1994 Winter Simulation Conference*: 962-969.

Wu, D. S., and R. A. Wysk. 1989. An application of discrete-event simulation to on-line control and scheduling in flexible manufacturing. *International Journal of Production Research*, **27**, 1603-1624.

Zeigler, B. P. 1980. Concepts and software for advanced simulation methodologies. *IEEE: Simulation with Discrete Models: A State-of-the-Art View*, 25-43.

AUTHOR BIOGRAPHIES

GLENN R. DRAKE is an associate applications engineer for Systems Modeling Corporation. He received his B.S.I.E. in 1994 and his M.S.I.E. in 1996 from Texas A&M University.

JEFFREY S. SMITH is an assistant professor in the Industrial Engineering Department at Texas A&M University. His research interests are in shop floor control, manufacturing systems design, analysis, control, and simulation.

SIMAN Constructs	Comments	P	C	T	I	O
BEGIN,Yes,Yes;	Begin Experiment	X				
PROJECT,Example,GRD,,No;	Model Information	X				
RESOURCES: 1,Machine1,SCHEDULE(Daily);	Machine 1 Resource	X	X			
2,Machine2,SCHEDULE(Daily);	Machine 2 Resource	X	X			
QUEUES: 1,Machine1Q,LVF(Priority);	Machine 1 Queue	X	X			
2,Machine2Q,LVF(Priority);	Machine 2 Queue	X	X			
VARIABLES: PartExec:	Emulation Variable			X		
ReadExec:	Emulation Variable			X		
Counter:	Loop Variable		X			
SUM1:	Sums Variable		X			
SUM2:	Sums Variable		X			
NUM:	Part Identifier Variable					
OpTime(6,2),10,0,13,0,15,12,	Processing Times				X	
0,6,0,9,8,6:						
Decision1:	Alternative Machine Var.		X			
Decision2:	Queue 1 Dispatching Var.		X			
Decision3;	Queue 2 Dispatching Var.		X			
ATTRIBUTES: ProcessTime:	Processing Time				X	
OrderNum:	Order Identifier				X	
PartNum:	Part Type				X	
PartID:	Part Identifier				X	
Quantity:	Order Quantity				X	
Priority:	Queue Rank		X			
Again;	Order Variable				X	
EXPRESSIONS: QueueRules(2),TNOW,ProcessTime;	Queue Ranking Rules		X			
SCHEDULES: Daily,1*480,0*120,1*480,0*120,1*480	"Daily" Work Calendar				X	
TASKS: 1,ProcessPart,PartExec,	TIQ task for processing			X		
"process part %6.0f",PartID,IDENT:						
2,ReadOrder,ReadExec,	TIQ task forreadingOrders			X	X	
"Enter Order TGID=%6.0f",IDENT;						
FILES: 1,OrdersFile,"Orders.DB",SEQ,,Rewind,",":	Orders Database				X	
2,OutFile,"Output.DB",SEQ,,Rewind,;	Output Database					X
REPLICATE,1,0,1800,Yes,Yes;	Run Control Information		X			

Figure 6: Example SIMAN V Experiment File.

SIMAN Constructs	Comments	P	C	T	I	O
BEGIN;	Begin Model	X				
;************Physical Structure****************						
;Processing at Machine 1						
Mach1Q QUEUE,Machine1Q;	The Machine 1 Queue	X				
SEIZE,Machine1:NEXT(Out1);	Seize Machine 1	X				
In1 DELAY,ProcessTime,ProcessPart:NEXT(Out2);	Process the Part	X		X		
In2 RELEASE,Machine1:DISPOSE;	Release Machine 1&Leave	X				
;Processing at Machine 2						
Mach2Q QUEUE,Machine2Q;	The Machine 2 Queue	X				
SEIZE,Machine2:NEXT(Out3);	Seize Machine 2	X				
In3 DELAY,ProcessTime,ProcessPart:NEXT(Out4);	Process the Part	X		X		
In4 RELEASE,Machine2:DISPOSE;	Release Machine 2&Leave	X				
;************Control Logic of System********************						
;Alternative Machine Decision (Decision 1)						
Ent BRANCH,1:	Alternative Machine Rules		X			
IF,PartNum.EQ.1.OR.PartNum.EQ.3.OR.Decision1.EQ.1,Mach1:	Select Machine 1 In-Buffer		X			
IF,PartNum.EQ.2.OR.PartNum.EQ.4.OR.Decision1.EQ.2,Mach2:	Select Machine 2 In-Buffer		X			
IF,Decision1.EQ.3,SNQ:	SNQ Rule to Select		X			
IF,Decision1.EQ.4,STPT;	STPT Rule to Select		X			
;Smallest Number in Queue Rule (SNQ)						
SNQ BRANCH,1:	SNQ Rule		X			
IF,NQ(Machine1Q).EQ.NQ(Machine2Q),STPT:	STPT Tiebreaker		X			
IF,NQ(Machine1Q).GT.NQ(Machine2Q),Mach2:	Select Machine 2 In-Buffer		X			
ELSE,Mach1;	Select Machine 1 In-Buffer		X			
;Shortest Total Processing Time in Queue Rule (STPT)	STPT Rule					
STPT ASSIGN:Counter=1:			X			
ProcessTime=OpTime(PartNum,1):	Calculate the Total				X	
SUM1=ProcessTime;	Processing Time in the In-					
WHILE:Counter.LE.NQ(Machine1Q);	Buffer for Machine 1		X			
ASSIGN: SUM1=SUM1+AQUE(Machine1Q,Counter,1):			X			
Counter=Counter+1;						
ENDWHILE;			X			
ASSIGN:Counter=1:			X			
ProcessTime=OpTime(PartNum,2):	Calculate the Total				X	
SUM2=ProcessTime;	Processing Time in the In-					
WHILE:Counter.LE.NQ(Machine2Q);	Buffer for Machine 2		X			
ASSIGN: SUM2=SUM2+AQUE(Machine2Q,Counter,1):			X			
Counter=Counter+1;						
ENDWHILE;			X			
BRANCH,1:			X			
IF,SUM1.GT.SUM2,Mach2:	Select Machine 2 In-Buffer		X			
ELSE,Mach1;	Select Machine 1 In-Buffer		X			
;Machine 1 Queue Dispatching (Decision 2)						
Mach1 ASSIGN:Priority=QueueRules(Decision2):	Assign Queue 1 Rank		X			
ProcessTime=OpTime(PartNum,1):NEXT(Mach1Q);					X	
;Machine 2 Queue Dispatching (Decision 3)						
Mach2 ASSIGN:Priority=QueueRules(Decision3):	Assign Queue 2 Rank		X			
ProcessTime=OpTime(PartNum,2): NEXT(Mach2Q);					X	
;********Simulation Input and Output*********************						
;Reading Orders from Orders Database						
CREATE,1:,1;	Create One Entity				X	
loop1 DELAY,0,ReadOrder;	Send ReadOrder Task			X	X	
loop2 READ,OrdersFile:OrderNum,PartNum,Quantity,Again;	Read an Order				X	
DUPLICATE:Quantity,Id;	Create the Order				X	
BRANCH,1:	Read another Order?				X	
IF,Again.EQ.0,loop1:ELSE,loop2;	Assign Part Identifiers				X	
Id ASSIGN: NUM=NUM+1: PartID=NUM: NEXT(Ent);	Enter Part into System				X	
;Generating Output						
Out1 WRITE,OutFile:"S,1,%8.3f,%6.0f\n":	Write StartTime of job					X
TNOW,OrderNum,PartID:NEXT(In1);	at Machine 1					
Out2 WRITE,OutFile:"F,1,%8.3f,%6.0f\n":	Write EndTime of job					X
TNOW,OrderNum,PartID:NEXT(In2);	at Machine 1					
Out3 WRITE,OutFile:"S,2,%8.3f,%6.0f\n":	Write StartTime of job					X
TNOW,OrderNum,PartID:NEXT(In3);	at Machine 2					
Out4 WRITE,OutFile:"F,2,%8.3f,%6.0f\n":	Write EndTime of job					X
TNOW,OrderNum,PartID:NEXT(In4);	at Machine 2					

Figure 7: Example SIMAN V Model File.

Proceedings of the 1996 Winter Simulation Conference
ed. J. M. Charnes, D. J. Morrice, D. T. Brunner, and J. J. Swain

DYNAMIC RESCHEDULING OF A JOB SHOP: A SIMULATION STUDY

Anand S. Kunnathur
P.S. Sundararaghavan

ISOM Department
The University of Toledo
Toledo, OH 43606-3390, U.S.A.

Sriram Sampath

Cap Gemini America
25800 Science Park Drive, Suite 180
Beachwood, OH 44122

ABSTRACT

The development of a rule based Expert System (ES), driven by a discrete event simulation model, that performs dynamic shop scheduling is described. Based on a flowtime prediction heuristic that we have developed and base-line runs to establish the efficacy of scheduling strategies such as SPT, Critical Ratio, Total Work etc, a re-scheduling based dispatching strategy is investigated in a dynamic job shop environment. The results are discussed and analyzed.

1 INTRODUCTION

Global competition has enhanced the significance of manufacturing effectiveness. Better manufacturing schedules provide competitive advantage through reduced production cost by way of lesser inventories and increased productivity. Moreover, increasing global competition has driven many US companies into investing heavily in Advanced Manufacturing Technologies, like Flexible Manufacturing Systems (FMS) and Computer Integrated Manufacturing (CIM). These manufacturing systems have created a whole new set of operational problems which emphasize the importance of superior job scheduling strategies.

In this increasingly competitive manufacturing environment, simulation has gone from a tool of "last resort" to being viewed as an invaluable design and problem solving methodology (Shannon 1988).

Artificial Intelligence (AI), the technology that attempts to preserve domain intelligence (Knowledge Base) in order to use the same for decision making in the future, has matured enough to redirect the research in scheduling. There are several capabilities of AI that make this technology particularly suitable for scheduling (Shaw, Park, Raman 1990).

In this paper we report the development of a generic rule based Expert System (ES) which is driven by a front end simulation engine and which uses several heuristics developed by us. We also report on their performance.

The rescheduling heuristic developed in this work uses a flowtime prediction strategy. We have also investigated the effectiveness of this strategy to set due-dates for newly arriving jobs. The organization of the paper is as follows.

The next section reviews the relevant literature on expert system shells in general and specific ES shells developed for scheduling. Section 3 presents the mechanics of the ES shell, detailed problem description and notation. The shop and job characteristics are also described in detail. SIMAN-V and C's role in the model is described. The objective measures that have been used in the model are also explained.

Section 4 describes the simulation model and the scheduling rules that have been used to obtain the baseline data used in the ES shell. Section 5 describes the studies that have been conducted using the ES shell and includes a review of relevant job shop scheduling literature. A flowtime prediction heuristic, based on dynamic shop conditions, is developed in Section 6. Various scenarios are discussed where such prediction is useful and possible.

Section 6 also describes our re-scheduling experiments in detail. The re-scheduling heuristic that has been developed in this work is elaborated. Results from the heuristic are compared and analyzed with the baseline results.

2 LITERATURE REVIEW

Expert Systems Applications developed probably constitutes the major segment of Artificial Intelligence research that has been conducted in the past two decades. We discuss the job scheduling related ES in this review. Job Shop scheduling has conventionally been looked upon as a computationally complex mathematical problem. Recent research on the real-world domain of scheduling demonstrates that the computational complexity issue is a minor part of the job shop scheduling problem in the real world and that expert schedulers do exist. (Mckay, Buzacott, Charness and Safayeni 1989). Mathematically, a dynamic shop scheduling problem is "*np*-complete", (Morton, Pentico 1994) which justifies heuristic methods.

The first real prototype expert system for scheduling was ISIS, developed by Fox and Smith (1984). It uses backward interval scheduling and beam search, with an evaluation function defined by estimation of constraint violations. The OPIS system (Ow, Smith and Howie 1988) employs an opportunistic approach to improve upon ISIS. It selects the most appropriate strategy for scheduling opportunistically; the resulting flexibility achieved in problem solving results in better performance.

OPAL is a multi-knowledge-based commercial scheduling software developed by Bensana, Bel and Dubois in 1988. Shaw (1988) has developed a knowledge-based scheduling approach based on the problem-solving techniques developed in artificial intelligence. Wysk, Wu and Yang (1986) use multipass expert system to decide the appropriate scheduling rules based on information such as the current system status, scheduling objective and management goals.

Shaw, Park and Raman (1990) use a machine learning approach to perform intelligent scheduling. Based on simulation runs they determined the most effective dispatching rule for a set of system attributes. O'Keefe and Rao (1992) use a look-ahead simulation strategy to evaluate the performance of candidate dispatching rules in an FMS cell.

3 RULES BASED ES SHELL FOR SCHEDULING

3.1 General Shop Conditions

The ES shell developed in this work resides on a discrete event simulation model of a production shop. Most of the researchers in the past have used systems involving shops having anywhere between five and twelve machines. Conway (1965) performed an elaborate study on flow shop with nine machines and 9000 jobs simulated over 30 different rules. Muhlemann *et al.* (1982) conducted a study on a job shop consisting of 18 machines. Ahmed and Fisher (1992) used a five machine job shop to conduct their study on due-dates. Fry *et al.* (1988) used a ten machine job shop to conduct their study on processing time dispatching rules. Ragatz and Mabert (1994) describe a nine machine shop in the explorative study on due date assignment rules. The simulation model that we have developed is tested on a typical job-shop with no break-downs.

3.2 Dynamic Shop Design

The dynamic shop modeled in the discrete event simulation engine consists of ten machines and uses warm-up period to make the shop representative. The job arrivals are modeled with exponential inter-arrival time and the process times have been assumed to be normal with relatively low variance.

Table 1: Processing-time Distribution at Various Machines

Mac->	1	2	3	4	5	6	7	8	9	10
Mean **Std.D**	15 2	17 2.5	12 1	18 3	19 2.5	17 2	14 1.5	12 0.5	18 2.75	9 1

Table 1 gives the processing time distribution at each of the machines. The inter-arrival time of jobs is exponential with an average of 6/hour. In the Job Shop Model mode every job can have anywhere between 2 and 8 operations, i.e., the number of operations per job is drawn from a discrete probability distribution with a minimum of 2 and a maximum of 8. The visitation sequence is randomly selected (the next machine in the sequence is drawn from a discrete probability distribution with a minimum of 1 and a maximum of 10; every unvisited machine has an equal probability of being selected) with no machine being visited twice by the same job.

Due Date assignment is done internally using the Total-Work (TWK) rule for all the runs. The TWK (Total Work) rule is as follows:

$$d_j = A_j + k \sum_{i \in mj} p_i \qquad (1)$$

where d_j - Due Date of job j

A_j - Arrival time of j

p_i - Process time of the j in machine i.

k - Parameter than reflects the amount of expected delay a job will experience. The parameter k is 3,4,5 or 6 with equal probability.

m_j - Set of machines visited by job j

3.3 Objective Measures Studied

The following objective measures have been fused in the ES shell: **Average Flowtime, Average Tardiness and Percentage Tardy .** With the arrival rates and processing times assumed, the Dynamic Job Shop has an average machine utilization of 75.7%, with a maximum of 95% and a minimum of 45%.

3.4 EXPERT SYSTEM SHELL DESIGN

SIMAN-V™ has been used as the simulation modeling tool. All the experiments were done on a Gateway ® 2000 Pentium-66Mhz. The *jobs* in the dynamic shop are *entities* created in the SIMAN-V simulation model. SIMAN-V allows memory pointers to be accessed by C. The entire rule based scheduling module is written in C and compiled in WATCOM® C compiler. The C object module is linked with SIMAN-V runtime libraries to generate a final run-time module.

Entities (memory variables) are created by the discrete event simulation model based on the arrival distribution. The created entities are assigned job attributes, viz., process times, due date, predicted flow time and the machining sequence. The entities enter the shop where they are processed by the machines. The entity leaving the machine after completion of processing triggers the ES shell which schedules the next entity to the free machine. The entity gets processed by the appropriate machine as per the machining sequence. Upon completion of all required processing the entity leaves the system after updating the statistics.

The rule based scheduling shell dispatches jobs depending on the system characteristics and objective measures selected in the user interface by the user. The system currently uses simple IF..THEN..ELSE rules to schedule jobs. Experimental data from past simulation runs serve as a knowledge base to guide decision making when the ES is used to make accept/reject decisions and perform adaptive re-scheduling.

4 TESTING AND ENHANCEMENTS

4.1 Dispatch Heuristics Tested

The ES shell with the front end simulation engine is generic and hence can simulate a variety of job shops. This section describes the baseline or one-pass heuristics tested in the scheduling engine. The baseline runs have been conducted in order to test the ES shell. The results from the baseline runs have been analyzed here prior to the development of heuristics to address the objectives.

One-pass heuristics are simple heuristic procedures that build up a single complete solution a step at a time (Morton, Pentigo 1994). These are very useful primarily because these serve as simple solutions which meet specific shop objectives. These are iteratively used to build more sophisticated multi-pass heuristics or search heuristics that fit well in a rule based ES development paradigm.

- We have experimented with a number of one-pass heuristics after a thorough search of job shop scheduling literature. (Muhlemann 1982, Morton, Pentigo 1994) For convenience we have named these one-pass heuristics as *pure strategies*. Performance of objective measures from these pure strategies have served as fundamental experimental data to drive the ES shell. We have classified these pure strategies into ***simple dispatch heuristics*** and ***simple bottleneck heuristics***.

Simple dispatch heuristics profess a scheduling strategy that is based on the simple premise: *highest priority first*. Whenever a machine becomes available, we simply schedule the highest priority job currently available at that machine. Dispatch heuristics that have been used in this work:. *Static* priorities tried were First In First Out (FIFO), and *dynamic* priorities like Minimum Slack First, Slack Per Operation, Minimum Total Work First were also tried. ***Local*** priorities like Shortest Processing Time first and ***forecast*** priorities like Critical Ratio and Least Work Remaining have also been tried.(Standard definitions have been used.)

4.2 Bottleneck Heuristics Tested

Bottleneck heuristics are dispatch strategies where the bottleneck machine is identified from the set of machines available in the shop and the scheduling strategy at the bottleneck machine is different from the other machines.

A machine is defined as a *bottleneck* of the shop at time t if it has the maximum work-load among all machines at time t. In a dynamic environment the bottleneck tends to be dynamic depending on the jobs that are currently available in the system. In the simulation model the work-load is calculated for all machines whenever a new job enters the system or whenever a job leaves the system after completion of processing. The machine with the maximum work-load is deemed as the *bottleneck*. Scheduling at the bottleneck machine is different from the non-bottleneck machines. Because of the superior performance of SPT in comparison to the rest of the simple dispatch heuristics when it comes to flowtime criterion, all bottleneck machines have been scheduled on SPT. The non-bottleneck machines have been scheduled using four different options (LWR, EDD, TWK, Least Slack/opn)

4.3 Results from the Baseline Runs

The simulation model was validated based on the input parameters and the performance of objective measures. Using the run length analysis tools, viz., SIMAN output analyzer, available in ARENA, the Graphical User Interface that resides on SIMAN-V, the run length and warm-up period were determined after extensive experimentation. Using a moving average analysis of flowtime the initial condition bias (warm-up) was determined to be 20 days. The total run length was 80 days. Therefore data was collected and analyzed for 60 days.

The results from the baseline runs are tabulated in Table 2 and 3. The results of a paired difference t-test conducted between results obtained using SPT vs EDD, and SPT vs FIFO indicate clearly that there is a statistical difference between all the objective measure values between various dispatching rules.

Both simple dispatch and simple bottleneck heuristics are reported in Table 2. As can be seen from the Tables, SPT performs best on flowtime, followed by Bottleneck - SPT/Non-Bottleneck - LWR rule. However TWK rule performs the best when it comes to percentage tardy as objective measure. On the percent tardy criterion there is hardly any statistical difference between Bottleneck - SPT/Non-Bottleneck - LWR rule, Bottleneck - SPT/Non-

Bottleneck - TWK rule and SPT rule.

EDD, Slk/Opn and Critical ratio perform equally well when it comes to Average tardiness criterion. The most interesting result of the whole exercise is that all Bottleneck based rules perform well for all the objective measures taken together. As identified by other scheduling literature, SPT is probably the best when it comes flowtime criterion.

Table 2: Performance of dispatching rules

Characteristics	SPT	EDD	FIFO	TWK	MinSlk	LWR
Flowtime	314.84	333.88	353.69	325.42	333.88	350.35
Average Tardiness	110.11	59.57	84.55	123.18	59.57	137.00
Percentage Tardy	17.64	36.62	41.35	12.61	36.62	18.58

Table 3: Performance of dispatching rules

Characteristics	Slk/Op	C.Rat.	B-SPT N-EDD	B-SPT N-LWR	B-SPT N-TWK	B-SPT N-S/o
Flowtime	340.31	337.67	325.33	317.49	318.92	331.31
Average Tardiness	57.74	59.80	84.64	105.35	113.36	78.28
Percentage Tardy	35.28	37.55	23.15	19.48	16.34	22.18

Fry, Philipoom and Blackstone (1988) present a list of truncated SPT rules which attempt to capture the effectiveness of SPT in meeting the flowtime criterion but perform well with tardiness criterion. Our bottleneck based rules seem to do exactly that when one looks at all the criterion in total. Truncated rules are difficult to implement whereas bottleneck based rules are simple and easily implementable. We have seen what rules are appropriate for each of the objective measures. In the next section we describe studies conducted in dynamic job shop re-scheduling using the experimental data gathered. We first develop a flowtime prediction heuristic for the purpose of dynamic rescheduling.

5 DYNAMIC RESCHEDULING

5.1 Flowtime Prediction

Predicting flowtime in a dynamic environment is a difficult task primarily because of the fact that there is no definite method to estimate the waiting time of the job at various queues in a shop. However, knowing an estimate of flowtime before the acceptance of an order is important to give a delivery commitment to the customer. In this increasingly customer-driven market the probability of success of a firm in the market place is directly dependent on its ability to give and adhere to delivery commitments. Conventionally, researchers in dynamic scheduling have tried to address delivery commitments through improved due-date setting. We examine relevant literature on due-date setting in this section.

Eilon and Chowdhury (1976) used expected waiting times and current queue lengths to more accurately estimate job flow time. Weeks (1979), in a simulation study of a 24 machine job shop having labor and machine constraints concludes that rules based on the number of jobs in the system, provide good due date performance.

Ragatz and Mabert (1994) conducted a simulation analysis of due-date assignment rules. They report on a number of performance measures including: standard deviation of lateness, mean tardiness, and mean absolute missed due dates and made several interesting comments about the role of shop characteristics and status in affecting performance. We have developed a flowtime prediction heuristic in this paper, which uses both characteristic and shop status information.

Eilon and Chowdhury (1976) used expected waiting times and current queue lengths to estimate job flow time. Our flowtime prediction heuristic uses expected waiting times and current queue lengths. However our estimation uses the job's processing time, along with queue lengths. Ragatz and Mabert (1982) demonstrated a methodology to include both job characteristics and shop status information for setting due dates.

5.2 Flowtime Prediction Heuristic

Weeks (1979), in a simulation study of a 24 machine dual

constrained (machine and labor) shop, used a rule, based on the number of jobs in the system, to provide good due date performance. Even though Week's heuristic used information on total shop congestion as a basis for developing flow time estimates, it ignored the job characteristics. Ragatz and Mabert (1994) in their study found the Weeks rule to be performing rather poorly in comparison to other rules which use both job and shop characteristics.

The flowtime prediction heuristic developed in this work uses both job and shop characteristics. The heuristic assumes that the dispatching rule is SPT. (SPT has performed the best when it comes to mean flowtime criterion.) The heuristic is as follows:

5.3 Heuristic I

Suppose a new job arrives at time *t*:

Step-1. The percentile to which the simulated processing time of job belongs (based on the processing time distribution at each work-station) is determined at each work center.

Step-2. Using the actual queue size at each work-station at that point of time in the simulation run the potential location of the new job at various queues is ascertained, using the information in *step 1*, ignoring all other jobs that may be received after the receipt of this job.

Step-3. Using the information in *step 2* the projected waiting time of the job at each work-station is determined and summated over all the work-stations that the job would visit. Mathematically the waiting time at time *t* of the job at various queues can be expressed as,

$$\text{Waiting time } W^t = \sum_{j=1}^{n} (m \; x \; \overline{P}j) \qquad (2)$$

where,
m - number of *jobs* in front of the current *job*.

$\overline{P}j$ - Mean processing time at machine *j*.

n - Number of machines visited by the *job*.

Step-4. The waiting time determined in *step 3* is added to the TWK (total work) of the job and the whole value is multiplied by an experimental factor γ to give the predicted flowtime. Mathematically,

$$\text{Predicted flowtime} = \gamma(W^t + \text{TWK}) \qquad (3)$$

Initial simulation runs using $\gamma = 1$ were performed. A comparison of the average predicted flowtime and average actual flowtime resulted in a value of 1.75 for γ. This value is dependent upon the shop characteristics and reflects the degree of bottlenecking in the shop.

5.4 Flow Time Prediction Results

In order to investigate the effectiveness of the flowtime predicting heuristic the predicted flowtime was compared with actual flowtime. Ten replications of the simulation were conducted with SPT dispatching. A *paired difference t* test to examine whether there is a statistical difference between the distributions of the actual and predicted flowtimes was negative. The results indicate that there is no statistical difference between the two distributions. Therefore the flowtime prediction heuristic is effective in giving an estimate of the actual flowtime. Comparison of the averages indicated that this heuristic could certainly be used to serve as a tool for rescheduling.

5.5 Rescheduling

Historically there have been two approaches to scheduling: ***sequencing***, the approach that seeks to establish an order for all open jobs and ***dispatching***, the approach that provides a solution by the use of local rules for selection of one job from the list of available jobs at decision epochs. It has been reported that sequencing approach is more efficient than the dispatching approach in a pure static environment. (Matsura, *et al*; 1993) However in a dynamic environment it is practically impossible to adopt a total sequencing approach, simply because the problem cannot be solved satisfactorily. Therefore dispatching is probably the only solution. *Re-scheduling* as has been defined earlier is a goal driven strategy that attempts to involve shop characteristics, shop objectives and dynamic shop status information to perform effective dispatching.

Muhlemann, *et.al* (1982) investigated a rescheduling strategy for a 15 machine job shop. They concluded that frequent scheduling can produce significant improvements on shop performance. Matsura, *et.al* (1993) suggest that a rescheduling (referred as switching strategy by the authors) strategy is very efficient when there are uncertainties in the system. Garg and Wang (1989) demonstrate that a rescheduling based strategy is more effective in meeting shop objectives in a Flexible Manufacturing Cell (FMC).

Rescheduling can be triggered by a change in the shop conditions. The challenging exercise is therefore to identify shop conditions which could serve as a vehicle to trigger rescheduling. In the rescheduling heuristic developed in this work, variation of predicted mean flowtime from the actual mean flowtime serves as the trigger to perform rescheduling. The rescheduling strategy reported in this paper uses the flowtime prediction heuristic described in the previous segment. The rescheduling heuristic is as follows:

5.6 Heuristic II

Suppose a new job arrives at time *t*:

Let,

k - Jobs completed
n - Jobs in the system (Work in progress)
p - Proportion (w.r.t total number possible in the whole simulation run) of jobs completed in time t.
f_1 - Mean actual flowtime of completed jobs
f_2 - Mean predicted flowtime for all work in progress (W.I.P)
f_o^o - Mean adjusted flowtime based on experimental runs
f_3 - Mean flowtime estimate for $(k + n)$ *jobs*
f_o^t - Predicted flowtime mean at time t

then,

Step-1. The flowtime for the job is predicted based on the Heuristic-I described earlier and the mean predicted flowtime (f_2) for all work in progress (W.I.P) is recalculated.

Step-2. Whenever a job completes processing (time t) the following are computed:

f_3 - Mean flowtime estimate for $(k + n)$ *jobs*

$$= \frac{(f_2 n + f_1 k)}{(k + n)}$$

f_o^t - Mean predicted flowtime at time t

$$= f_o^o (1-p) + f_1 p$$

Based on straight simulation runs f_o^o is determined as follows:

$$f_o^o = \bar{p} \times \left(\bar{m} + \frac{\sum_{i=1}^{r} p_i q_i}{r} \right)$$

where $\bar{p}$ = Average processing time of all machines in the shop
$\bar{m}$ = Average number of machines visited by any job in the shop
q_i = Average queue length of the ith machine based on prior simulation runs
r = The total number of machines in the shop, 10 in this study.

For the shop and job characteristics used in the model f_o^o was determined to be 263.84 minutes.

Step-3. Decision to reschedule: The rescheduling parameter is determined based on the parameters computed in *step 2.* The rescheduling parameter (at time t) is given as:

$$R = \frac{(f_3 - f_o^t)}{f_o^t}$$

The decision to reschedule is then based on the following rule:

If R > Δ → RESCHEDULE
R ≤ Δ → HOLD OFF

where Δ is an experimental decision variable.

Note: Once a reschedule has been performed further reschedules are possible only after a simulation run for time T. This is done to ensure that one does not go overboard in rescheduling every other minute.

Δ and T are experimental variables used in this research to study their influence on the performance of the rescheduling heuristic.

5.7 Candidate Rules for Dispatching

The rescheduling heuristic described above describes the process of arriving at a decision to reschedule. An analysis of the baseline runs indicated that bottleneck based rules perform well in meeting overall objective measures. Therefore two out of our three candidate rules for dispatching are bottleneck based rules. The ES shell uses two sets of dispatching rules, one for the *flowtime minimization* criterion and the other for *tardiness minimization* criterion. The candidate rules for each of the criteria are as follows:

(i) Flowtime Minimization:

The following rules are the candidate rules for the flowtime minimization criterion; (a) SPT, (b) Bottleneck Machine - SPT and Non-bottleneck machines - LWR and © bottleneck Machine - SPT and Non-bottleneck machines - TWK. Whenever our heuristic advises a reschedule, the dispatching rule is changed to a different rule (by rotation) within the aforesaid list of candidate rules.

(ii) Tardiness Minimization:

Based on an analysis of the baseline runs, the following dispatch rules perform well, for minimizing average tardiness: (a) EDD; (b) Slk/Opn (ratio of total slack to number of operations remaining) and © Critical Ratio. Therefore these rules are the candidate rules for the average tardiness minimization criterion. Rescheduling is done within this set of candidate rules in rotation.

We experimented with various values for the variables Δ and T to investigate their influence in the performance of objective measures. Next, we analyze the performance of the rescheduling heuristic for both criteria.

Table 4: Objective Measures for various values of Δ
(Flowtime minimization criterion; T=4 Days)

Characteristics	Δ=0.05	Δ=0.10	Δ=0.125	Δ=0.15	Δ=0.20
Flowtime	318.96	318.96	315.30	315.30	315.86
Average Tardiness	114.30	114.30	106.73	106.73	108.76
Percentage Tardy	16.78	16.78	16.38	16.38	16.42
No. of Reschedules	10	8	7	5	1

Table5: Objective Measures for various values of Δ
(Average Tardiness minimization criterion; T=4 Days)

Characteristics	Δ=0.05	Δ=0.10	Δ=0.125	Δ=0.15	Δ=0.20
Flowtime	333.08	329.79	329.79	327.90	330.60
Average Tardiness	90.06	76.94	76.94	88.98	90.68
Percentage Tardy	25.35	29.41	29.41	23.24	24.50
No. of Reschedules	11	8	8	6	5

5.8 Rescheduling Results

The performance of the rescheduling heuristic for both the criteria are reported in Tables 4 and 5. The rescheduling decision factor, Δ, directly affects the performance of objective measures for both criteria. Shop performance is very sensitive to Δ. The decision factor has a direct influence on the number of reschedules until a certain value. This is because Δ measures the variation of average flowtime vis-à-vis the average flowtime under SPT scheduling. A lower value for Δ would therefore mean more frequent re-scheduling and lower average flowtime. For larger values for the decision factor, there is hardly any rescheduling.

Table 5 describes similar results as in the case of tardiness minimization. However, it was observed that rescheduling does not seem to be more effective than EDD when it comes to average tardiness minimization criterion. This is attributed the fact that the rescheduling trigger is flowtime based. Further studies can be conducted with a tardiness based rescheduling criterion.

6 CONCLUSION

We have seen in this chapter that a rule based system works well for implementing a rescheduling based strategy. The rescheduling strategy developed in this research performs effectively for the shop characteristics used. We believe that the rescheduling strategy could perform even better for a different set of shop characteristics. However this calls for extensive experimentation with this heuristic. We have seen that an optimal (which is dependent on the system parameters) number of reschedules does result in a superior overall scheduling strategy. The effectiveness of rescheduling in tardiness minimization was not as profound as in flowtime minimization. However this could be attributed to the rescheduling trigger, which in our case is flowtime based. Simulation driven, expert system implemented rescheduling as a strategy, appears to hold a lot of promise. For complete details the reader is referred to Sampath (1996).

ACKNOWLEDGMENTS

The work of Sriram Sampath (1996)during his master' s program was partially supported by the ISOM Academic Challenge Grant of the College of Business Administration at The University of Toledo.

REFERENCES

Ahmed, Imtiaz and W.W. Fisher. 1992. Due Date Assignment, Job Order Release, and Sequencing Interaction in Job Shop Scheduling. *Decision Sciences* 23: 633-647.

Bensana, E., Bel, and D. Dubois. 1988. OPAL: A milti-knowledge-based system for industrial job-shop scheduling. *International Journal of Production Research* 26 (5): 795-819.

Conway, R.W. 1965. Priority Dispatching and Work In

Progress Inventory in a Job Shop. *Journal of Industrial Engineering* 16: 123-130.

Conway, R.W., W.L. Maxwell, and L.W. Miller. 1967 *Theory of Scheduling*. Addison-Wesley.

Eilon, S., and I.G. Chowdhury. Due Dates in Job Shop Scheduling. *International Journal of Production Research* 14 (2): 223-237.

Fox, M.S., and S.F. Smith. ISIS - A Knowledge Based System for Factory Scheduling. *Expert Systems* 1: 25-49.

Fry, T.D., P.R. Philipoom, and J.H. Blackstone. 1988. A simulation study of processing time dispatching rules. *Journal of Operations Management* 7 (4): 77-92.

Garg, A., and H. Wang. On Simulation Run Length: A Step Towards Real Time Control in an FMS. *International Journal of Operations and Production Management* 10 (9): 74-88.

Matsuura, H., H. Tsobone, and M. Kanezashi. 1993. Sequencing, dispatching, and switching in a dynamic manufacturing environment. *International Journal of Production Research* 31 (7): 1671-1688.

McKay, K.E., J.A. Buzacott, N. Charness, and F. Safyeni. Can Expert Systems deal with the job shop scheduling domain. *WATMIMS working paper* No. 89-01.

Morton, T.E., and D.W. Pentigo. 1994. Heuristic Scheduling Systems, John Wiley & Sons.

Muhlemann, A.P., A.G. Lockett, and C. K. Farn. 1982. Job shop scheduling heuristics and frequency of scheduling. *International Journal of Production Research* 20 (2): 227-241.

O'Keefe, R.M., and R. Rao. 1992. Part input into a Flexible Flow System: An evaluation of look-ahead simulation and a fuzzy rule base. *The International Journal of Flexible Manufacturing Systems* 4: 113-127.

Ow, P.S., S.F. Smith, and R. Howie. A Co-operative Scheduling System. *Expert Systems and Intelligent Manufacturing*, edited by M.D. Oliff, North Holland: 70-89.

Ragatz, G.L., and V.A. Mabert. (1982) An Investigation of Flow Time predictability in a Single Machine Shop. *Proceedings*, Midwest AIDS: 154-156.

Ragatz, G.L., and V.A. Mabert. 1994. A simulation analysis of due-date assignment rules. *Journal of Operations Management* 5 (1): 27-39.

Sampath, Sriram. 1996. Unpublished M.S. Thesis, College of Business Administration, The University of Toledo, Toledo, OH 43606.

Shannon, R.E. 1988. Knowledge based simulation techniques for manufacturing. *International Journal of Production Research* 26 (5): 953-973.

Shaw, M.J. 1988. Knowledge-based scheduling in flexible manufacturing systems: An integration of pattern-directed inference and heuristic search. *International Journal of production Research* 26 (5): 821-844.

Shaw, M.J., S.C. Park, and N. Raman. 1990. Intelligent Scheduling with Machine Learning Capabilities: The Induction of Scheduling Knowledge. *Carnegie Mellon University — The Robotics Institute-Technical Report*.

Turban, E. *Decision Support Systems: Management Support Systems*, 2nd Edition, Macmillan.

Weeks, J.K. 1979. A Simulation Study of Predictable Due Dates. *Management Science* 25 (4): 363-373.

Wysk, R.A., S.D. Wu, and N. Yang. Multi-pass Expert Control System for Manufacturing Systems. *Proceedings of the Symposium on Real-time Optimization in Automated Manufacturing Facilities*, National Bureau of Standards, Gaithersburg, MD.

AUTHOR BIOGRAPHIES

ANAND S. KUNNATHUR received the master's degree in physics from the University of Delhi, the master's degree in mathematics from York University, Canada, and the Ph.D. degree in management science from the University of Tennessee. His research interests are in the areas of information systems, manufacturing management, and the interface between management science and functional areas of business. He has published in journals such as *Operations Research, TIMS* special issue on "Optimization in statistics," *SIAM Sci. and Stat. Computing, EJOR, Omega, Computers and Industrial Engineering, Information and Management*, and others.

P. S. SUNDARARAGHAVAN received the Bachelor's degree in chemical engineering from the University of Madras, the Post Graduate Diploma in management from the Indian Institute of Management in Calcutta, and the Ph.D. degree in management science from the University of Tennessee. He is a former Chairman of the ISOM Department at The University of Toledo and the founder of The University of Toledo Chapter of the APICS Society. He has published in journals such as *Naval Research Logistics, SIAM Algebraic and Discrete Methods, Decision Sciences, IIE Transactions, OPSEARCH, EJOR, Journal of Operations Research Society*, and others. His current research interests include sequencing and scheduling, simulation, production planning and control, and issues in advanced manufacturing technology adoption.

SRIRAM SAMPATH received the M.S. degree in Manufacturing Management from the College of Business Administration of The University of Toledo. He works for Cap Gemini in the area of manufacturing software development and customer support.

Proceedings of the 1996 Winter Simulation Conference
ed. J. M. Charnes, D. J. Morrice, D. T. Brunner, and J. J. Swain

EVALUATION OF A DISTRIBUTION CENTER TOW-LINE MATERIAL HANDLING SYSTEM THROUGH SIMULATION MODELING

Jay Bakst
Automation Associates, Inc.
13333 Bel-Red Road
Bellevue, WA 98005 U.S.A.

Joel Hoffner
E & E Corporation
225 West Second Street
Bethlehem, PA 18015 U.S.A.

Kris Jacoby
Random House, Inc.
400 Hahn Road
Westminster, MD 21154 U.S.A.

ABSTRACT

Random House, one of the world's largest and most distinguished publishing houses, embarked on a mission to improve its tow-line material handling system at its main distribution center located in Westminster, Maryland. A mechanical tow-line system, while a rugged low-tech but reliable solution, still comes with complicated management and control issues which require a simulation to evaluate and understand. A three person team designed, developed, and implemented a simulation model of a proposed modification to the existing transport system. The results were then analyzed and presented to company management.

This paper discusses the processes and solutions of this simulation including how our team approach improved the model, how the model identified significant variables in the implementation and how the model results were used to help justify the capital expenditure.

1. INTRODUCTION

THE RANDOM HOUSE DISTRIBUTION CENTER RECEIVES AND SHIPS HIGH VOLUMES OF PALLETIZED BOOKS EACH DAY ....

Random House is one of the world's largest and most distinguished publishing houses. The main distribution center for Random House is located in Westminster, Maryland. Each day the facility receives from 500 to 1,200 pallet loads of books from binderies and ships from 350,000 to 600,000 books to retail stores and wholesale outlets. These outgoing shipments are also palletized but may consist of a mix of cases and single books that are consolidated into case lots.

The Westminster distribution center has grown in stages and consists of three buildings totaling over 762,000 square feet. The first building was opened in 1966. The second, located adjacent to the first, was constructed in 1972. The third addition was dedicated in 1983 and is connected to the other two warehouses via a 565 foot covered bridge.

THE MATERIAL HANDLING SYSTEM AT THE RANDOM HOUSE DISTRIBUTION CENTER WAS BEING PUSHED TO ITS LIMITS ....

As the Westminster facility grew, so had the mechanical tow-line system which is the primary means for transporting pallets of books throughout the three buildings. The tow-line system uses an in-floor continually moving chain to transport both loaded and empty carts. Although the path of the tow-line system was appropriate when each of these buildings were built, the design of the system no longer supports the current method and volume of operation and Random House's plans for future growth.

The existing tow-line system consists of three independent loops which (1) rely heavily on transfer connections between buildings and (2) have a path which weaves extensively within pallet rack aisles. These two factors have contributed to reducing the efficiency of the operation, increasing overall pallet transport time, and reducing the predictability of pallet transport time.

RANDOM HOUSE EMBARKED ON A MISSION TO IMPROVE THE MATERIAL HANDLING SYSTEM THROUGHPUT TO SUPPORT CURRENT AND FUTURE NEEDS ....

The goal of Random House was to design and implement a material handling system for pallet transport between buildings that achieved the following goals:

1. support an anticipated 25% increase in volume within the next five to ten years
2. support the current and future pallet movement patterns

3. increase the reliability and predictability of goods being transported between buildings
4. reduce material handling costs and improve the efficiency of the operation

TOW-LINE SYSTEMS ARE RELIABLE WORKHORSES BUT LACK THE GLAMOUR OF HIGH TECH MATERIAL HANDLING SYSTEMS....

Tow-line systems, for movement of cart loads of material, have been in use for decades in distribution centers, freight docks, newspaper production and manufacturing operations. Although overshadowed by higher technology Automated Guided Vehicle Systems in recent years, tow-line transportation remains a rugged, reliable and low cost means of automatically moving and routing heavy, sometimes awkward, loads(Young, 1991). Specialized cart design in combination with the ability to program cart movement creates an efficient system(Bradt, 1971)

Tow-line movement occurs by means of a continuously moving chain that is installed below floor level. The chain has pusher dogs at regularly spaced intervals which capture and pull a pin that is, in turn, attached to a transportation cart. The cart is custom tailored in size and shape to the load that it handles.

After evaluating several different methods of pallet transport, including Automated Guided Vehicle Systems, Random House made the decision to modify and upgrade the existing tow-line system. Although a mechanical tow-line system is not as glamorous as other higher tech systems, it is a untiring workhorse. The tow-line system is capable of delivering a large number of heavy pallet loads to various points throughout the facility. In addition, the existing tow-line system is mechanically sound and has experienced very little downtime for repairs over the past 30 years.

A MAJOR CHANGE TO THE CONVEYANCE LOGISTICS WAS PROPOSED TO IMPROVE DISTRIBUTION CENTER OPERATIONS ....

To achieve the project goals and improve the system performance, a major redesign of the tow-line system was studied. The system layout was reconfigured to (1) provide a path from any point of origin to any destination that is less dependent upon transfers, (2) eliminate excessive looping of the path within the pallet racking which currently causes congestion in the aisle ways and increases cart transport time, and (3) provide overflow spurs to reduce the number of carts that recirculate in the system.

The result of the study was a proposal to simplify the three loop system into a single loop which skirts the perimeter of the buildings. This was done quite easily by abandoning the internal portions of the loops and adding some new track.

2. THE SIMULATION MODEL DESIGN

BEFORE PROCEEDING WITH A MAJOR CAPITAL EXPENSE, SIMULATION MODELING WAS USED TO ANALYZE AND VALIDATE THE PROPOSED CHANGE ....

Although management at Random House believed the performance of the redesigned system would be an improvement over the current system, they wanted assurance that the system would achieve the desired goals and that the additional system enhancements would be beneficial. Some key questions to be answered were:

1. Is the new single loop system capable of transporting up to 2,000 pallet loads per day using a load profile whereby the majority of moves are generated on the first shift?
2. Is the average transport time for a pallet load from origin to destination less than 45 minutes and is delivery time fairly predictable?
3. Are the added overflow spurs in buildings 1 and 2 beneficial in reducing the number of carts that recirculate? And if so, what is the proper overflow capacity?
4. Should the tow-line shut down if an overflow spur fills?
5. Is a capacity of 6 carts sufficient for the new transfer in building 2?
6. How many carts are required in the system and how should they be managed?

After reviewing the complexity and volume of peak day pallet move data and taking into consideration system variables such as system downtime, spur service times, and time to unload carts, a decision was made to develop a simulation model. A simulation model was the best means of determining the expected performance benefits and would assist in evaluating design parameters.

Model animation was planned to be an important means to evaluate the logical operation of the model during testing and also provide credible evidence to the management of Random House that the model emulated real life system events.

THE SIMULATION MODEL HAD THREE GOALS ....

The primary goal of the model was to determine the predictability of the transport time for loaded carts and to recommend the number of carts for each of two operating modes:

1. shutdown tow-line when an overflow fills
2. recirculate normally if an overflow fills

Transport time is defined as the elapsed time from the dispatch of a loaded cart at the origin to the arrival of the load at the destination. Wait times at the origin for an empty cart, as well as load and unload times and wait times at the destination for service, were not included in the transport time.

For purposes of the model, the load was considered dispatched when an empty cart was loaded and a destination assigned. Arrival at a destination was defined as arrival of a cart at either the destination spur or the group overflow spur associated with the intended destination.

The secondary goal of the model was to ascertain the impact of each of four parameters on the transport time of loaded carts and on the number of recirculating carts.

1. capacity of transfers
2. capacity of overflow spurs
3. capacity of the empty cart pools
4. time to service spurs and unload carts

A third goal was to derive a recommended quantity for each of the four parameters that was consistent with the desired performance at the level of operation for the busiest historical day at the distribution center.

A TEAM JOINTLY DEVELOPED THE MODEL AND EVALUATED THE RESULTS ....

The authors worked as a team to develop and utilize the model to assess the performance of the proposed system changes. Kris Jacoby represented Random House; Joel Hoffner provided the tow-line experience; and Jay Bakst wrote the simulation model using SIMAN for modeling and ARENA for animation.

However, rather than each of the participants having an isolated assignment, a team approach was adopted. Each had areas of prime responsibility, but the help of team members was repeatedly accepted and sought out.

Although Joel developed the design document and Jay the model itself, each had a role in reviewing and improving the design and validating the model. But the cooperation went deeper than that. Sharing our knowledge of the tow-line, modeling techniques and the distribution process enabled us to make appropriate simplifying assumptions that allowed use of conveyor constructs extensively and still get results in a reasonable time. Had we not done this, free path constructs would have been used with far less reliability and accuracy.

The teamwork extended to Kris identifying situations that caused the model to develop a gridlock condition which affected 25-30% of the iterations. This led to a speedy repair of the model.

THE SYSTEM LAYOUT WAS STRAIGHTFORWARD BUT CHALLENGING ....

The perimeter tow-line system design consists of a loop segment within Buildings 1 and 2 approximately 2100 feet long with 38 spurs; a loop segment within Building 3 approximately 2100 feet long with 19 spurs; and a bridge approximately 500 feet in length connecting the buildings together. The total distance around the perimeter of the buildings and through the bridge is 5200 feet. (see diagram 1)

Carts are 48 inches long and are transported on 20 foot centers on the main tow-line and on 6 foot centers on the transfers. There are currently 350 carts available for use in the system. The system may dispatch as many as 2000 loads within a day with the majority of the loads moved during the first shift (0730 to 1600).

Tow-line speed is 60 feet per minute. Each spur may hold a different number of carts. Multiple spurs with the same identity were aggregated in the model.

The transfer capacity at the entrance to the bridge in Buildings 1 and 2 was estimated to be 6 carts. The final capacity of the transfer in Buildings 1 and 2 was determined with the aid of the model. The capacity of the transfer at the entrance to the tunnel in Building 3 was constrained to 3 carts as this was an existing transfer that would be retained in the proposed system.

RULES FOR EMPTY CART MANAGEMENT WERE ESTABLISHED ....

In order to manage empty carts in the system, Empty Cart Pools were created. An Empty Cart Pool reflects carts stored in the immediate vicinity of the point where a previously loaded cart was unloaded. A maximum of 25 pools were created and maintained. The pools were associated with a group of spurs that are in close proximity to each other. This allowed easy placement of the cart into the pool and equally easy retrieval of an empty cart when needed.

When a load is initiated at a spur, an empty cart is consumed. The primary source for empty carts is a series of empty cart pools. Each pool is associated with

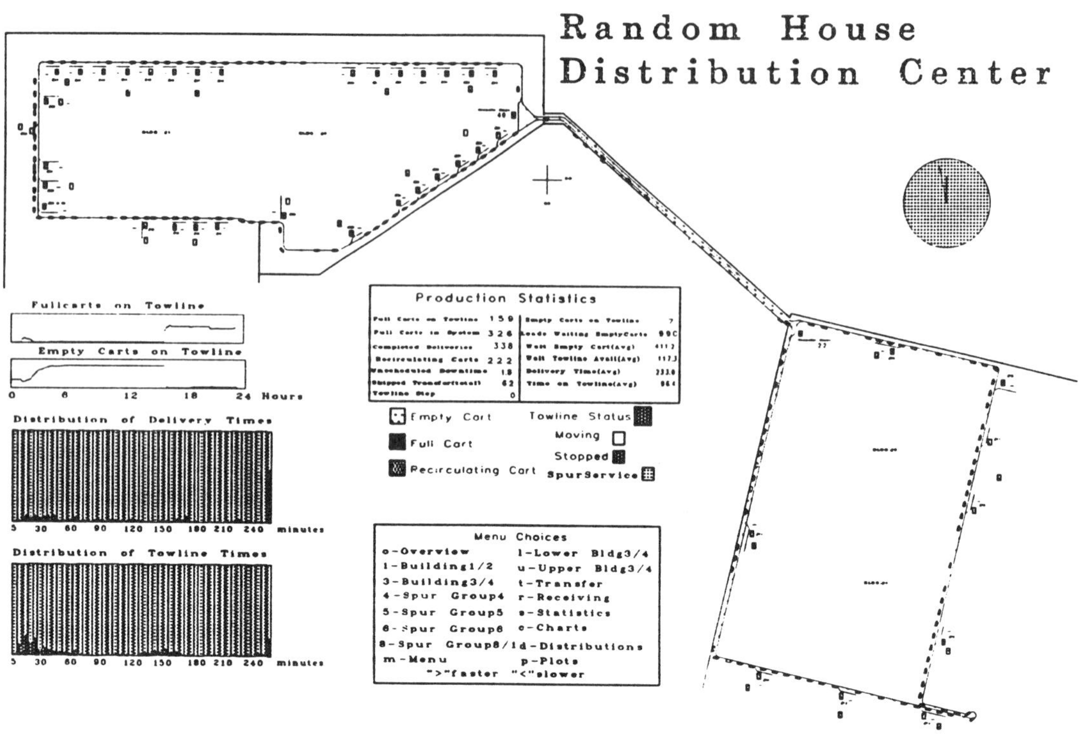

Diagram 1

Hit "m" for menu

a groups of spurs. If no carts are available from the empty cart pool, then the operator waits until an empty cart enters the spur area on the towline.

Empty cart pools were assigned a maximum number of carts and a reorder point. If the number of carts in the pool dropped below the reorder point and there was someone working in the area (initiating a load or servicing the spur), an empty cart was removed from the towline and stored in the empty cart pool. The limits for each pool were set via a menu entry prior to model execution.

SPUR AND EMPTY CART POOL DESIGN WAS CAREFULLY CONSIDERED ....

There are a total of 36 regular spurs in the model. In some cases, a number of these regular spurs, can have the same identity. In actual operation, a cart destined for this multiple spur configuration, will enter the first spur that has room. To simplify the model, these multiple spurs with the same identity were configured as a single spur with a capacity equal to the aggregate number carts in all the spurs. This reduced the complexity of the model with no loss of integrity or precision.

In addition to the 36 regular spurs, the new system design includes 5 group overflow spurs. The overflow spurs are associated with a series of spurs in the same general vicinity. The overflow spurs provide a second chance for a cart to arrive near its destination in a timely manner. This reduces the number of carts that recirculate in the system.

Empty cart pools are associated with sets of spurs in close proximity. The pools contain empty carts which are used to route a load from one of the spurs in the set to a remote destination. If no empty cart exists, the load must wait to be transported.

For ease of visual observation and ease of experimenting with various capacities, the spurs, transfers and empty cart pools were modeled as queues with menu selectable capacities. The spurs, transfers and empty cart pools appeared in the animation as small bar charts which showed the percentage full at a quick glance.

RULES FOR ENGAGING AND DISPATCHING LOADED CARTS WERE CONSTRUCTED ....

An activity file was provided by Random House with actual peak daily pallet move data. The file defined the point of origin, destination spur and the time of origin for each load. A load may originate at any spur. The destination for a load may be any spur except the origin. Loads must wait for an available empty cart before being initiated into the system.

Loaded carts are placed on the tow-line as the next empty 20 foot space passes the spur. Loaded carts move counter-clockwise around the tow-line system. Loaded carts take the shortest route to their destination unless a bypass transfer is full.

Loaded carts queue in a transfer if traffic on the main tow-line does not allow a merge from the transfer to occur. Loaded carts move from the main tow-line to the destination spur if room exists on the spur. If the spur is full, the cart bypasses the spur. If a loaded cart passes the destination spur, it will move into the overflow spur unless it is also full. If the overflow spur is full one of two actions can be chosen:

1. the loaded cart recirculates until the destination or overflow spur is able to accept the cart, or,
2. the tow-line stops until at least one cart is unloaded from the overflow spur

3. SIMULATION MODEL DEVELOPMENT

THE MODEL WAS DEVELOPED TO EMULATE THE SYSTEM CONSTRUCTION AND TO OBSERVE STATUS IN RELATION TO PERFORMANCE ....

The main path of the tow-line system weaves its way around the buildings. When the tow-line stops, it is necessary to stop all of the carts on the towline in their place and wait until the disturbance is over.

Second, the number of carts in a spur varies over a set range. When using the animation, the actual number of carts in a spur is less valuable than having an idea of the utilization of the spur.

Third, throughout the day, operators come to a spur to unload the carts and to "service" the zone. While they are in the area, several activities (taking off empty carts, loading the spur, emptying loaded carts) could take place.

The main line of the tow-line system was modeled as a non-accumulating conveyor. Push dog entities were placed on the conveyor and acted as carriers for empty and loaded carts. The spurs where the carts were diverted automatically were a combination of queues which stored the carts, and levels which displayed the current utilization of the spur. A similar construct was used for empty cart pools. Operator activity was displayed with global variables that changed with each change in the state of service activity.

The levels displaying the queue data and the global variables showing operator status provided a dynamic display which was extremely easy to understand. At all times while running the animation the relative number of carts in each of the spurs, the empty cart pools, and the overflow spurs was easily viewed.

MODEL INITIALIZATION, STARTUP AND STABILIZATION WAS EASED BY THE MODE OF OPERATION ....

During model development, concerns about initialization and stabilization were identified. A startup scenario was constructed to address these concerns. Empty carts were placed on the tow-line and in empty cart pools throughout the system. This was accomplished by running the conveyor at near infinite speed for the first few minutes and metering empty carts onto the tow-line from a single source spur. Likewise, empty cart pools were filled to their respective reorder points.

Fortunately, system activity in the initial hours of operation each day from midnight to approximately 7 AM is light. This allowed the system to reach equilibrium and emulate actual performance within the first hour of operation. Statistics were reset at one hour of model execution.

SYSTEM DOWNTIME WAS SCHEDULED TO REFLECT ACTUAL EVENTS ....

The tempo of operations in the Random House distribution center was often broken by scheduled operational break times and unscheduled tow-line shutdowns required for operating convenience.

For a model and project such as this, where dozens of different cases were evaluated, it was important to eliminate the effects of factors which are outside the bounds of the study. Modeling explicit breaks in operation by means of a fixed table of outage periods eliminated the possibility that random variations in these outage periods would inadvertently skew the results and therefore was more effective than randomly scheduled events. Actual historical downtime statistics were obtained and a file of outage periods was constructed.

MOVEMENT, LOADING AND UNLOADING TIMES FOR CARTS WERE BASED UPON ACTUAL HISTORICAL PERFORMANCE AND OPERATIONAL STATISTICS ....

Similarly, an activity file was developed and used as a basis for the workload in each of the runs. Scenarios for average and peak activity were developed. The transport time of the system and the number of cart recirculations were effected by the four major parameters listed earlier, each of which was modeled and specified individually.

Operator activities such as cart marshaling and servicing were developed by looking at dispatching logs and observation times. With these sources the servicing times represented realistic expectations.

MENU DRIVEN MODEL PARAMETER ENTRIES WERE DEFINED TO EASE MODEL EXECUTION....

The SIMAN simulation language was used to develop the model and the ARENA product was used for animation purposes. This is in keeping with the concerns identified by M. Seppanen(1995) for modeling industrial strength systems. Animation on selected cases provided the validation of cart movement and insight to the activities occurring in the model.

The nature of the analysis encouraged parameterized data sets which made it easier to modify the data in an orderly fashion and to assure complete coverage of the options to be tested.

A menu-based variable entry system developed by Automation Associates provided a vital means of storing multiple data sets, tracking the multiple model run scenarios and enabling several parameter runs to be made consecutively. Finally, the SIMAN model without any constraints and/or delay from animation enabled us to run many different scenarios quickly. Iterations were batched and run overnight.

MODEL OUTPUTS WERE SELECTED TO EMPHASIZE THE PERFORMANCE CRITERIA TO BE INVESTIGATED ....

The primary purpose of the model was clear: to evaluate predictability of transport within the facility. With this in mind, specific outputs were selected. The model provided average, minimum and maximum transport (flow) times for the carts dispatched. A frequency distribution showing number of carts for various transport times was used to interpret the variance of transport times.

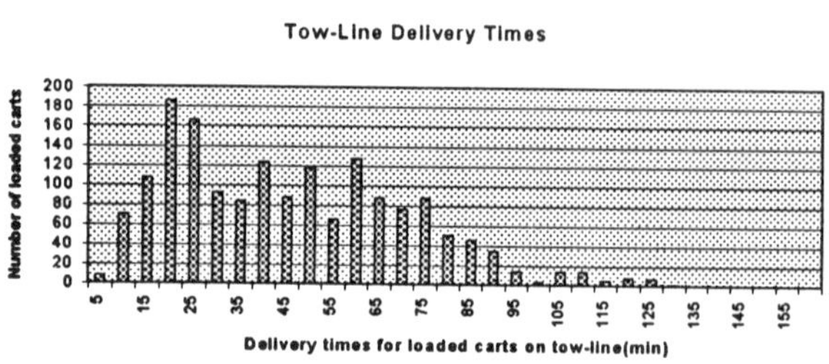

To aid in the validation and interpretation of the model results, statistics were displayed:

1. on the animation screen as the model was running
2. as end of run statistics
3. as graphics which tracked values over time and maintained a history throughout the run

The output processor was used to answer a few specific questions but comparing differences among many different runs was the main evaluation technique.

4. ANALYSIS

MODEL CASES WERE SELECTED TO PROVIDE A PARAMETRIC STUDY OF THE MOST SIGNIFICANT VARIABLES ....

We hypothesized that transport time was most significantly effected by a few variables. The test cases were designed to demonstrate the sensitivity of transport time to:

1. number of carts in the system
2. empty pool sizes
3. time to service a cart after arrival at the destination
4. overflow spur capacities
5. transfer capacities
6. shutdown versus recirculation upon spur overflow

MORE THAN 100 CASES WERE RUN AND THE RESULTS INTERPRETED ....

Numerous model runs were made by parametrically varying the variables that were selected as the most significant. Each series of runs varied one parameter over a representative range while holding the other parameters constant. The results were summarized on a spreadsheet for easy comparison and evaluation.

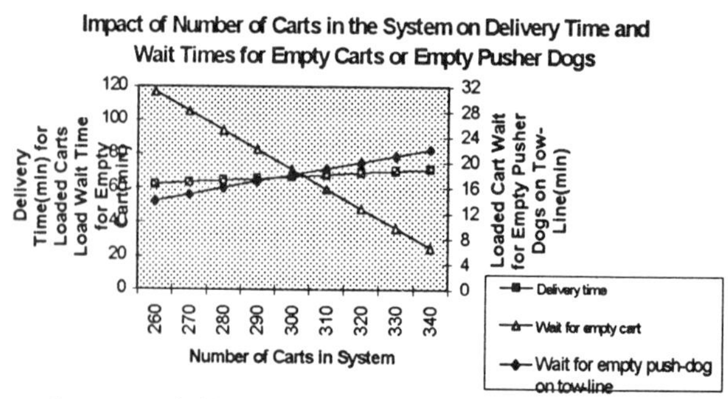

It was quickly determined that certain parameters, such as transfer capacity, even when varied over a wide range had very little impact on the dependent variable, transport time. For other parameters, such as number of carts in the system, transport time was shown to be highly sensitive to the parameter over a narrow range.

Interpretation of initial results guided the selection of additional runs for further evaluation. In other words, the design of the experiment was fluid and driven by results of each new case that was tested. The investigation converged on results in an iterative fashion with the analyst and the model working in a closed loop environment.

5. RESULTS AND CONCLUSIONS

MODEL RESULTS DEMONSTRATED THAT THROUGHPUT CAPACITY, TRANSPORT TIMES AND VARIANCE MET THE DESIGN CRITERIA ....

The capability of the tow-line system to deliver loads was found to be limited to approximately 2300 to 2500 loads per day using the load schedule profile provided by Random House wherein the highest concentration of loads occurred between 0700 and 1600 hours.

The system had no difficulty in handling the 1764 loads associated with the peak expected daily load assuming that spur service, cart loading, loaded cart placement on the tow-line and empty cart removal from the tow-line were as recommended.

The average transport time for a load was 45 minutes, once placed on the tow-line, and after eliminating abnormally long times for the few carts trapped on the line during a long scheduled shutdown at 0100 hours. This conclusion was based upon allowing carts to recirculate.

The average full delivery time which includes the wait for an empty pusher dog, was 53 minutes, also after eliminating abnormally long times for the few carts trapped on the line during the long scheduled shutdown at 0100 hours.

Substantial delays were incurred waiting for an empty dog after the cart is loaded. This condition was less significant as the number of loads scheduled per day decreases. This implied that personnel may be required to linger in an area after carts are loaded to wait for empty dogs to occur.

Approximately 90% of the carts were on the tow-line less than 80 minutes whereas the total delivery time was less than 95 minutes 90% of the time.

MODEL EXECUTION HIGHLIGHTED OPERATIONS CONSTRAINTS THAT WERE EXPECTED AND UNEXPECTED....

The model showed that the system ran more efficiently and delivered loads up to 10% faster if carts are allowed to recirculate when the overflow spurs fill as opposed to shutting down the tow-line. This was not in concert with the expectations of the team prior to modeling.

Increasing empty cart pools from the prescribed capacity was found to have no significant effect on system performance. Similarly, increasing transfer and overflow spur capacities from the design quantities had little effect on system performance. Recirculations due to a full transfer were minimal under all conditions of operation.

Decreasing the maximum service interval a moderate amount from 95 minutes to 60 minutes had no effect on system performance. Service times less than 60 minutes began to have a beneficial effect. This was expected.

THE MODEL BROUGHT A NEW LEVEL OF UNDERSTANDING TO THE DYNAMICS OF THE OPERATION IN RELATION TO THE NUMBER OF CARTS ....

The impact of the number of carts in the system and empty cart management was found to be profound. The total number of carts in the system must be carefully selected to optimize system performance. For the parameters in the model, 260 to 270 carts provided the best results. Too many carts caused gridlock in the system and deteriorated performance. Too few carts starved the system.

The system performance was highly impacted by empty cart management. The spurs downstream of the receiving area (spur 74) were frequently starved for empty carts. A strategy to deliver empties to those spurs must be developed. Increasing the empty cart pools in these areas provides initial relief, however, these larger pools have no impact after the system achieves steady state because empty carts are not available to replenish the pools.

USING SIMULATION PROVIDED THE MEANS TO QUANTIFY SIGNIFICANT RESULTS AND IDENTIFY MANAGEMENT ISSUES.

The results of the simulation showed that: (1) the size of the transfer lines played a distinct role in the model's usefulness and (2) the number of carts on the tow-line had a profound effect on performance.

The model also showed that the need to manage empty carts will be a substantial and difficult task. This has given management an important "heads up" in preparation of the new design implementation. With this forewarning, new empty cart strategies can be developed.

THE MODEL PAVED THE WAY FOR CAPITAL APPROPRIATION APPROVALS FOR THE SYSTEM MODIFICATIONS ....

Simulation answered significant questions as to the viability of the proposed design:

1. The new design would support the necessary number of pallet moves for current and future needs.
2. Standards for consistent delivery times could be developed.
3. The system should continue to run when an overflow spur is full.
4. A capacity of six(6) carts at the new transfer bypass system was sufficient to prevent loaded carts from traveling all the way around to other buildings.
5. Overflow spurs were beneficial in preventing carts from recirculating.
6. The system operated best within a range of 260-270 empty carts.

With these questions answered, the company approved the expenditure and is in the process of modifying the system. The system is expected to be completed in mid-1996.

REFERENCES

Bradt, L. Jack and Jack Dorrance, "Expanding Role for Tow-line Cart Systems", Automation Magazine, 1971.

Seppanen, Marvin S., "Developing Industrial Strength Simulation Models", Proceedings of Winter Simulation Conference 1995, pp936.

Young, Gene, "Gene Young's Tow-Line Labor Saver", Grower talks Magazine, 1991.

AUTHOR BIOGRAPHIES

JAY BAKST is the Division Manager for Automation Associates, Inc. in Bellevue, Washington. AAI is an independent engineering firm specializing in system design, analysis and simulation modeling. Jay holds his B.S. (Mathematics) and his M.S.(Operations Research) from Rensselaer Polytechnic Institute. Prior to joining AAI, he owned Technologies Services Group, a simulation modeling consulting firm. He is a member of the Institute of Industrial Engineers, the American Production and Inventory Control Society, a board member of the Kent Chamber of Commerce and a member of the Greater Seattle Chamber of Commerce.

JOEL HOFFNER is a principal in the consulting firm of E&E Corporation in Bethlehem, Pennsylvania. He has a BSEE from Lehigh University where he continues to serve on Doctoral Review Committees for Ph.D. candidates specializing in Operations Research and Manufacturing Engineering. Joel provides management, technical, and operational consulting for the steel and material handling industries world-wide.

KRIS JACOBY is an industrial engineer for the Distribution Operations of Random House Inc. Prior to joining Random House, she was a Senior Management Systems Analyst for the Radar Systems Group of Hughes Aircraft Company. Kris holds a B.S. in Industrial Engineering from the Milwaukee School of Engineering and a Certificate in Production and Inventory Control from California State University Domininquez Hills. She is a member of the Institute of Industrial Engineers and a member of the Institute of Packaging Professionals.

Proceedings of the 1996 Winter Simulation Conference
ed. J. M. Charnes, D. J. Morrice, D. T. Brunner, and J. J. Swain

MODELING OF CHAIN CONVEYORS AND THEIR EQUIPMENT INTERFACES

Ali K. Gunal
Shigeru Sadakane

Production Modeling Corporation
Three Parklane Boulevard, Suite 910 West
Dearborn, Michigan 48126, U.S.A.

Edward J. Williams

206-2 Engineering Computer Center, Mail Drop 3
Ford Motor Company
Dearborn, Michigan 48121-2053, U.S.A.

ABSTRACT

Chain conveyors are a specific type of conveyor often used in a variety of manufacturing and production applications, such as body and paint shops. These conveyors must typically interface with other types of conveyors such as cross-transfer conveyors, and also with other material-handling equipment such as lift tables and hold tables. Micromodeling of chain conveyors and their equipment interfaces requires close attention to numerous details. These details include not only static and operational properties of the chain conveyors themselves, but also the particulars of dimensional and operational interfaces of the conveyors and the equipment served by the conveyors, such as lift tables and the conveyor acceleration and deceleration ramps.

In this paper, we first delineate the situations in which micromodeling of material-handling equipment is appropriate. We then present an overview of conveyor types and terminology. Next, we describe the challenges of modeling chain conveyors accurately, and our recommendations for meeting these challenges within the framework of typical modeling tools and simulation-study contexts. As an example, we present details of these recommendations relative to the AutoMod modeling tool. In conclusion, we summarize these recommendations and indicate promising directions for further development of modeling techniques and enhancement of model-building tools.

1 MACRO- VS. MICRO-MODELING

Macro models are, by definition, overview models with a "coarse" level of detail. In contrast, micro models incorporate a "fine" (high) level of detail (Ülgen, Shore, and Grajo 1994). The appropriate level of detail for a particular model within a simulation study, and hence the decision of whether to build a macro or a micro model, properly depends on the objectives of the study, availability of data, credibility concerns, and constraints of model-development and computer-run time available (Law and McComas 1991). In view of this credibility concern, the modeler making the "micro model versus macro model" decision properly anticipates the task of validation by asking "Will our modeling team be able to use the model – in place of the system – to make the decisions required by the study objectives?" (Ruch and Kellert 1995).

Simulation may be applied to the study of material handling systems during any or all of four project phases: the conceptual phase, the detailed design phase, the launching phase, and the fully-operational phase (Ülgen and Upendram 1995). In the context of modeling material-handling systems, typical indications calling for development of a micromodel are requirements to minimize both global and local work-in-process levels and maximize utilization of material-handling equipment, plus availability of detailed dimensional, cycle-time, and downtime data relative to individual pieces of equipment, such as the aerial gravity conveyors and motorized roller conveyors compared in (Cerda 1995). Such modeling is frequently required due to the mathematical intractibility of operational questions involving material-handling equipment. For example, (Tsai 1995) verifies the "NP-hard" status of minimizing the likelihood of conveyor stoppage to complete assembly-station work, via mixed-model sequencing, when a conveyor serves even a single station with arbitrary processing times. In such a context, detailed modeling of conveyors becomes vital to address decision-making policy relative to both line balancing and product sequencing. The motivation to model conveyors, in particular, at a high level of detail increases in studies undertaken to assess competing conveyor management strategies, such as dynamic allocation of workpieces to conveyors which alternately merge with each other, diverge again, and must serve widely dispersed, dissimilar work cells (Laughery 1995). Macro models are frequently analyzed to optimize conveyor and overall system performance via selection of a workpiece to convey, selection of a conveyor, choice of waiting discipline among conveyors, or choice of waiting discipline among jobs awaiting a conveyor. Micro models of conveyor performance furnish required input to such macro models (Backers and Steffens 1983).

2 DEFINITIONS AND TERMINOLOGY OF CONVEYORS AND INTERFACE EQUIPMENT

Conveyors can transport a high volume of items over a fixed path at adjustable speed with little manual intervention. Many varieties of conveyors are in use, such as belt conveyors (endless belt), chute conveyors (metal slides), screw conveyors (large spiral contained in confining trough or tube), chain conveyors (endless chain), and roller conveyors (load carried on transverse rollers which are either gravity- or power-driven) (Sule 1988). For example, belt or roller conveyors can be readily configured to transport small, odd-shaped items (Gould 1994). Chain-driven conveyors with roller surfaces, the type of conveyor considered in this paper, are occasionally non-powered (gravity) or, more usually, powered (live). Such conveyors are relatively inexpensive, readily assembled and adjusted, and well suited for a wide range of loads, provided the materials being transported have, or are mounted on, a rigid riding surface (Allegri 1992). These chain-driven conveyors may accumulate material by use of slip or clutch mechanisms built into the rollers. The "roller flight" variety typically uses two parallel sections of chain supporting rollers on non-rotating shafts. Those rollers can turn under the material, permitting its accumulation (Gould 1993). Additionally, acceleration and deceleration sections may be appended to powered roller conveyors when precise material control is required, as at interface points, through inclines or declines, or around curves (K. W. Tunnell Company, Incorporated 1995). Recent advances in designs and materials have markedly increased speeds and decreased operating noise of powered roller conveyors, hence increasing their economic appeal (Witt 1995).

These conveyors characteristically interact with other material-handling equipment such as lift tables and hold tables. Lift tables provide a working or material-transfer surface at heights and positions chosen for ergonomic and operational advantage (Tompkins and White 1984). Hold tables are a passive variant of lift tables; unlike a lift table, a hold table cannot move a load perpendicularly to the direction of travel of a downstream conveyor to align the load for placement on that conveyor.

3 CHALLENGES OF MODELING CHAIN CONVEYORS

As predicted in (Muth and White 1979), work in conveyor modeling comprises both analytical models (limited to a low level of complexity but quantifying fundamental relationships among important conveyor parameters) and simulation models (allowing tradeoff studies during design and analysis of operational problems of existing conveyors). Numerous studies illustrate the wide applicability of simulation to the analysis of conveyor systems, such as designing and implementing a power-and-free conveyor system (Good and Bauner 1984), comparative macro evaluations of accumulating and non-accumulating conveyor systems (Henriksen and Schriber 1986), layout and flow path analysis of overhead conveyors (Foote et al. 1988), micromodeling of conveyors transporting extremely fragile workpieces forbidden to touch one another (Hopings 1988), and rapid assessments of proposed modifications to a power and free conveyor system in an automotive paint shop (Graehl 1992).

Chain conveyors are widely used in automotive assembly plants. High volume production requires cycle times in the order of a few seconds. For example, for a conveyor running at fifteen feet per minute and for pallets of length fifteen feet, a time loss of two seconds amounts to two jobs per hour and forty jobs per day over two ten-hour shifts. Depending on the sales price of a car, such a loss might be somewhere from $500,000 to $1,200,000 in lost revenue per day. On the other hand, the types of material handling equipment required to support such production might be very costly to install, operate, and maintain. Therefore, the system design must strive for balance between high throughput capacity and low redundancy in material handling equipment requirements. Consequently, it is necessary that simulation models of such material handling systems should represent sufficient detail of the parts movement to make accurate assessments of the adequacy of those systems and the overall production systems of which they are a component. Output results from micromodels of conveyor configuration and performance can then guide experimentation with macromodels of the manufacturing system.

A typical vehicle assembly plant would have an abundance of various kinds of conveyors. Chain conveyors constitute one of the most commonly used types of material handling equipment in many assembly plants. Modeling of those conveyors can be a challenging task. The following is a discussion of some of the issues in modeling of chain conveyors and the supporting material handling equipment that can be found in a typical assembly plant.

3.1 Production and Transfer Conveyors

These conveyor sections move pallets through production operations. They are also used for buffer storage purposes, holding pallets between production departments or during off-shift activities such as clean-ups. Furthermore, pallets are accumulated in a bank of conveyors before they are sequenced for the next operation.

There are essentially two types of chain conveyors: accumulating and non-accumulating. Accumulating conveyors, also referred to as roller flight conveyors, have the ability to hold pallets without stopping the power chain. A non-accumulating conveyor allows the pallets to stop only when the entire conveyor chain stops. In a typical setup, non-accumulating conveyors are used to move pallets through production processes (paint booths, wash rooms, etc.). Accumulating conveyors are used for other purposes mentioned previously (e.g., temporary storage, delivery, and resequencing). Regardless of the accumulation capabilities, most conveyors have special head and tail sections (speed-up sections) that are used to accelerate and decelerate pallets, respectively. Those sections are needed to adjust the speed of pallets as they move onto and off the conveyors, which typically have much lower chain speeds than the equipment with which they interface.

Pallets move on chain conveyors with a specified minimum distance between them. Depending on the speed of the conveyor, this distance may be different on various conveyors of a typical production setup. On an accumulating conveyor, the spacing between pallets might be different in accumulation from that between moving pallets. Also, once in accumulation, a pallet does not start moving again until the preceding pallet moves away to a distance defined as the moving space.

Most simulation software provides built-in constructs to model accumulating and non-accumulating conveyors. Simulation software that does not provide such constructs might require substantial amounts of overhead code to account for the operational characteristics of either type of conveyors for a typical production system. Even with built-in conveyor constructs, some of the operational characteristics might be challenging modeling tasks. A speed-up section, for example, consists of several smaller sections of the same conveyor that run at different speeds (from low to high speed or vice versa along the direction of travel). A pallet changes its speed when its center of gravity shifts from one portion of the speed-up section to the next. As there might be pallets of different lengths moving through the system, the time to get onto and off a conveyor differs among pallets. Some simulation software (such as AutoMod) allows precise calculation of these time delays by providing mechanisms to define such small sections of conveyor, the pallet size, and locations for speed changes. In other packages, the same effect can be achieved by calculating such delays prior to simulation and by explicitly representing them in the model. A further complication to modeling speed-up sections is the requirement that no accumulation be allowed on those sections, even on accumulating conveyors.

3.2 Cross Transfer Conveyors

Cross transfers are fast-moving chain conveyors with no accumulation capability. These conveyors are used to move pallets between different sections of production and transfer conveyors. A cross transfer conveyor typically interfaces several production and transfer conveyors. Incoming pallets are moved to one of several conveyors served by the cross transfer conveyor. There is a series of hold and lift tables along those conveyor sections to compensate for the lack of accumulation capability. Hold tables are used only for temporary storage purposes. Lift tables also act as the interface between the cross transfer and production conveyors, as they have the capability to push the pallets perpendicularly. Pallets move sideways on a cross transfer. A pallet moves along a cross transfer from one hold/lift table to another, stopping as necessary. A pallet stops on a table if the next table is occupied by another pallet. Once stopped at a table, the pallet is raised off (disengaged from) the conveyor chain to prevent its moving further (as the chain runs continuously on a cross transfer). Once the next table becomes available, the table lowers the pallet back onto (into reengagement with) the chain and the pallet starts moving along the conveyor again. Only after a pallet reaches the next table or clears a limit switch placed at a distance from the table, can another pallet move onto the table. Such control is required because cross transfer conveyors typically move at higher speeds and serve several chain conveyors. Typical simulation software constructs for conveyors provide no built-in support for such control logic. There is a clear need to model such logic to represent cross transfer conveyors accurately.

3.3 Lift and Hold Tables

These special tables are used along cross transfer conveyors. The main purpose of a lift table is to move pallets onto and off the cross transfer. A lift table has powered rolls to move pallets in a direction perpendicular to the direction of travel on the cross transfer. These rolls are used to move the pallets from/to production and transfer conveyors. A hold table is typically used to provide temporary storage on the conveyor to allow accumulation (cross transfers otherwise have no accumulation capability). Either type of table raises pallets from the conveyor chain to stop them from moving. Once the pallet is ready to move forward, the table lowers the pallet back onto the chain, and upon contact with the chain, the pallet starts moving again, as described in the previous subsection. Pallets do not stop at a table if they can continue to move forward (i.e., if the next table on the chain is available) and do not experience the cycle of going up and down at that table.

Availability of a table is usually signaled by the last pallet to visit the table. There is a clear limit switch placed at a distance from the table. This switch may be on the cross transfer conveyor, on an adjacent table, or on a production/transfer conveyor served by the cross transfer. Once the pallet reaches a clear switch, it sets the switch "off," indicating that it is now safe for its successor pallet to move onto the conveyor. Accordingly, any pallet waiting to get clearance to move onto the table sets the switch "on," flagging that the table is now assigned to its use.

In a simulation model, cross transfer conveyors can be treated as non-accumulating conveyors with discrete segments between hold/lift tables. Simulation packages with built-in conveyor constructs provide adequate support to model the movement between tables. However, additional steps should be taken to account for the up-down cycle of tables. Also, the model should replicate the logic to control the access to the tables. This effect can be achieved by treating the tables as resources and by capturing and releasing them at appropriate points.

3.4 Powered Roll Beds

Powered-roll beds are relatively short conveyor sections which use powered rolls for high-speed transfer of pallets. They are typically used between production/transfer conveyors and other higher-speed transfer equipment such as lift tables, turntables, and lifts (Sims 1992). They also provide buffer storage space for one pallet at a time. Those conveyors run at higher speeds than the others described above. Consequently, during transfers from/to power-roll beds, pallets experience speed changes. As indicated earlier, due to different sizes of pallets, the time to clear a power-roll bed differs among pallets. Consequently, a simulation model needs to represent the clearance time by accounting for the travel distances and the length of the pallets. With simulation languages such as AutoMod, it is possible to address such detail by utilizing built-in capabilities. With some other languages, up-front calculations are required for each power roll bed to determine the time it takes to move onto and off that power roll bed.

3.5 Turntables

Turntables are short conveyor tread sections mounted on a bearing surface; they can rotate around a vertical axis to reorient pallets in transit (Cahill 1985). Some production processes require the parts to be in a certain orientation. Movement on conveyors and cross transfers between them changes the pallet orientation. Also, transfer/production conveyor segments are sometimes at angles to each other. Consequently, in a typical production setup utilizing chain conveyors, there would be several turntables with different amounts of rotation (e.g., 45, 90, 180 degree turns). Once a pallet moves onto it, a turntable rotates and aligns itself with the new path that the pallet will follow. In most cases, a turntable does not start its rotation back to its loading position before the pallet clears a limit switch. Once the pallet clears the limit switch, the table rotates back to the loading position. The rotation time is typically proportional to the angle of rotation. Clearly, a simulation model should make sure that a pallet does not attempt to move onto a turntable before the turntable returns to the loading position. Turntables can be modeled in most simulation languages by using resources and by capturing and releasing those resources at appropriate points in time.

3.6 Lifts

In a typical production facility, there will be conveyors and production processes at different floors. Lifts (elevators) are used for transferring pallets vertically between different floors. Apart from the vertical travel time, there are delays in loading and unloading of lifts (Hudson 1954). Also, in most cases, there will be clearance limits for enabling/ disabling access to lifts. Due to various pallet sizes, those delays will differ among pallets. Consequently, the model should have either an explicit (in-travel distances, pallet sizes, and locations of clear limits) or an implicit (in-time delays and signals) representation of the delays that occur in the load-travel-unload-return cycle of lifts.

4 AN AUTOMOD EXAMPLE OF CHAIN CONVEYORS

AutoMod is an industrial simulation system combining the convenience of CAD-like drawing tools, an engineering-oriented modeling language, and accurate capture of distances, sizes, speeds, and accelerations as aids to building accurate simulation models supported by three-dimensional animation (Rohrer 1994a).

The example model is a small portion of a large body shop simulation built for an automobile manufacturer. Although small in scale, this example demonstrates some of the components mentioned above. Figure 1 (next page) depicts a CAD layout of the portion of the body shop, including three accumulating chain conveyors, two cross transfer conveyors, and several lift and hold tables.

AutoMod has detailed built-in constructs for modeling conveyors. The CAD drawing of the layout is imported into AutoMod as a static background and a conveyor system is superimposed upon it. Figure 2 (next page) displays the conveyor system used for this portion of the system. As Figure 2 shows, to capture the details of transfers between equipment running at different speeds, smaller conveyor pieces were slightly extended over the gaps between the actual conveyors. An "x" represents locations where loads stop and/or either set or reset clear switches. Also, the speed-up sections on chain conveyors are represented as separate conveyor pieces appended to the end/tail of conveyor segments. Furthermore, multiple stations were defined to represent the locations where the speed of a pallet changes. The stations in AutoMod are defined with a load alignment attribute that determines when a load is considered to be *at* that location. For example, if a station has a "trailing edge" alignment, then the load will be considered to have arrived at that station when the trailing edge of the load reaches it. Consequently, travel speeds and distances can be accurately captured in the model by appropriately laying out stations and selecting their alignment attributes accordingly.

The lift tables interfacing with conveyor segments are represented as smaller conveyor sections attached directly to the end of a chain conveyor segment. The cross transfer conveyors are also represented as separate conveyor segments laid atop the segments that represent lift/hold tables. To model transitions between lift/hold tables and the cross transfer conveyors, two stations are created at the same coordinates. One of the stations is then attached to the conveyor segment that represents the hold/lift table; the other, to the conveyor segment representing the cross transfer line. Then, by using the "move" command at appropriate points, the load representing the pallet is instantaneously transferred from one segment to the other, displaying a smooth movement in animation. The time delays in lowering and raising the load to/from the chain are explicitly represented in the model by using the "wait" command.

Clear limit switches are represented in the model as conveyor stations. Once a load reaches a station representing a clear switch, it releases a resource that corresponds to the area that the switch controls. Clearly, most of those stations have a "trailing edge" alignment. Loads trying to enter the area wait to capture the resource and start moving as soon as the previous pallet clears the area. For each lift table and each hold table, there is a corresponding resource in the model.

5 ANALYSIS AND RESULTS OF AUTOMOD EXAMPLE

The complete simulation model of the body shop contained many more conveyors, processing times at stations, and downtimes associated with processes and some of the material handling equipment. Analysis of this model was conveniently undertaken with AutoStat. AutoStat, working in conjunction with AutoMod, provides determination of initial transients, management of scenarios as a database, and a "Design of Experiments" feature (Rohrer 1994b).

Analyses performed with the base model indicated that the system as designed could not meet the target production. Detailed analyses showed that the portion of the system shown in Figure 1 was one of several problem areas. Part of the problem was due to the distance between the first and second lift tables on the cross transfer (right side of drawing). As indicated earlier, on a cross transfer conveyor, pallets do not move until the next lift/hold table is cleared by the previous pallet. In this case, because of the long distance, each pallet experienced a delay exceeding the maximum allowable cycle time. The problem was remedied by placing a clear limit switch at a distance sufficiently removed from the lift table to avoid collisions. Consequently, the cycle time at the table was considerably reduced, thereby increasing the throughput.

Another part of the problem was due to the flow of pallets through the area. Normal flow of pallets required that the pallets go up on the cross transfer to the first conveyor, and proceed to the cross transfer (right side of Figure 1). The top two conveyors were to be used as temporary storage only when the pallets could not be sent (e.g., due to equipment down time) to the next set of conveyors that are beyond the top part of the drawing. However, simulations showed that the system would back up faster because once the first conveyor is blocked, there was no other way to utilize the middle conveyor for temporary storage. An alternative to this routing scheme was to send the pallets to the last conveyor at the top of the surge as the normal flow and then to accumulate pallets on the bottom two

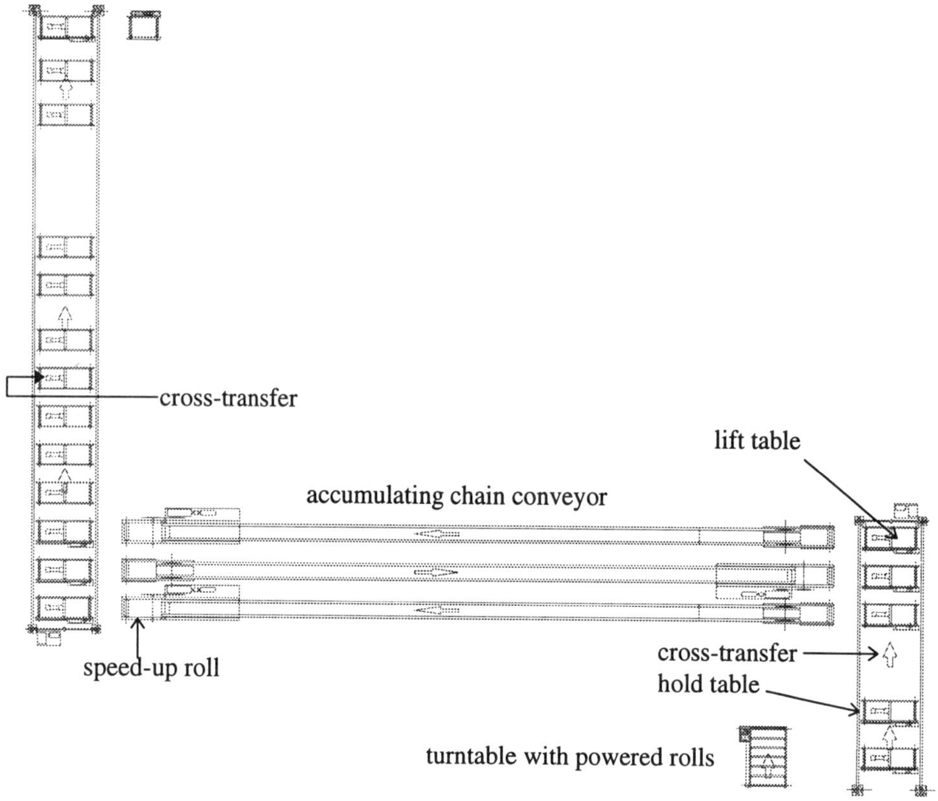

Figure 1: Actual Conveyor System, CAD Layout

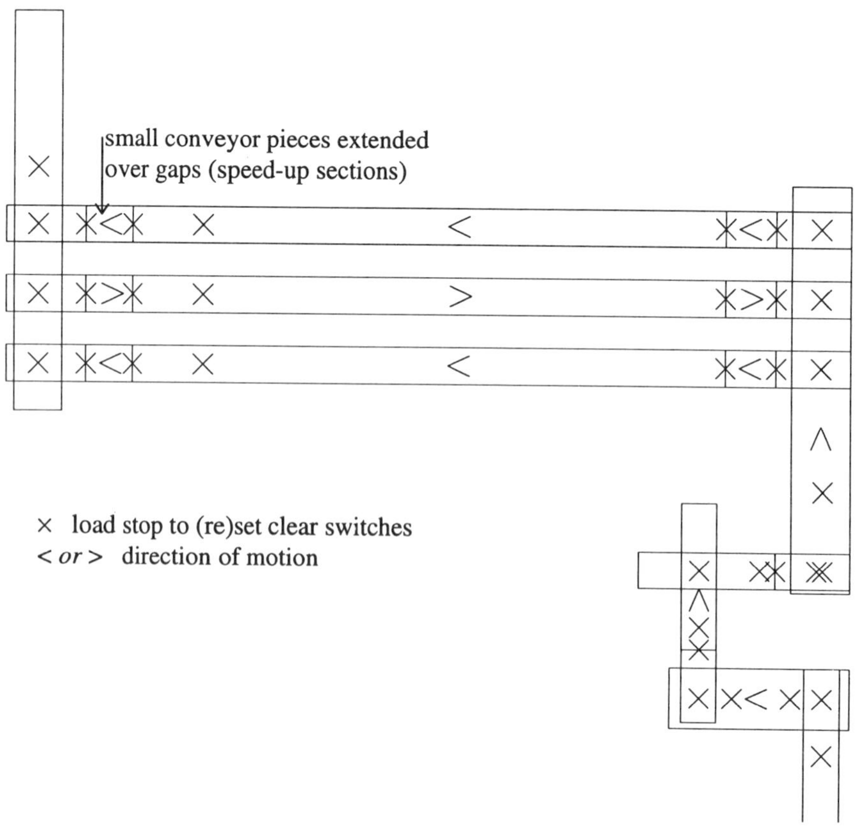

Figure 2: Model Conveyor System, AutoMod Representation

conveyors. Again, the model runs showed that this routing scheme worked better than the original and improved the overall throughput of the system.

Similarly, the complete model helped to identify several other problems with the conveyor system. By facilitating quick evaluation of the alternatives, the simulation model not only played a significant role in the detailed design of the system, but also provided quantitative support for managerial capital-investment decisions constrained by a tight project timetable.

In all these ways, results from micromodeling of conveyors, when included in macromodels of the production system, helped assess the degree to which material handling improvements spawned improvements in overall system performance.

6 SUMMARY

Particularly at the micromodeling level of detail, the accurate representation of conveyors and the equipment interfacing with them comprises numerous challenges. Meeting these challenges requires close attention to operational specifications and detail of the equipment, plus keen awareness of the capabilities of the simulation modeling tool chosen for use. This paper surveys these challenges and specific examples of them pertinent to a model including chain conveyors, cross transfer conveyors, lift and hold tables, and clear switches.

ACKNOWLEDGMENTS

Dr. Onur M. Ülgen, Professor, Department of Industrial & Systems Engineering, University of Michigan – Dearborn, and president, Production Modeling Corporation, has made valuable suggestions helpful to the presentation, organization, and clarity of this paper. Additionally, the cogent criticisms of two anonymous referees were especially helpful in these regards.

A preliminary version of this paper was presented at the June 1996 AutoSimulations annual users' group meeting (*AutoSimulations Symposium '96 Mountain Rendezvous Proceedings*, pages 313-319).

APPENDIX: TRADEMARKS

AutoMod and AutoStat are registered trademarks of AutoSimulations, Incorporated.

REFERENCES

Allegri, Theodore H., Sr. 1992. *Materials Handling Principles and Practice.* Malabar, Florida: Krieger Publishing Company.

Backers, R., and H. Steffens. 1983. Development of Strategies for Controlling Automated Material Flow Systems. In *Proceedings of the 1st International Conference on Automated Materials Handling*, ed. R. H. Hollier, 9-20.

Cahill, James M. 1985. Package-Handling Conveyors. In *Materials Handling Handbook*, 2d ed, ed. Raymond A. Kulwiec, 317-339. New York, New York: John Wiley & Sons, Incorporated.

Cerda, Carlos B. Ramírez. 1995. Performance Evaluation of an Automated Material Handling System for a Machining Line Using Simulation. In *Proceedings of the 1995 Winter Simulation Conference*, eds. Christos Alexopoulos, Keebom Kang, William R. Lilegdon, and David Goldsman, 881-888.

Foote, Bobbie L., A. Ravindran, Adedeji B. Badiru, Lawrence M. Leemis, and Larry M. Williams. 1988. Simulation and Network Analysis Pay Off in Conveyor System Design. *Industrial Engineering* 20(6):48-53.

Good, George L., and J. Thomas Bauner. 1984. On the Use of Simulation in the Design and Installation of a Power and Free Conveyor System. In *Proceedings of the 1984 Winter Simulation Conference*, eds. Sallie Sheppard, Udo W. Pooch, and C. Dennis Pegden, 425-428.

Gould, Les. 1993. Application Guidelines for Accumulation Conveyors. *Modern Materials Handling* 48(5):42-43.

Gould, Les. 1994. Conveyors Designed to Handle Small Products with Ease. *Modern Materials Handling* 49(3):44-45.

Graehl, David W. 1992. Insights into Carrier Control: a Simulation of a Power and Free Conveyor Through an Automotive Paint Shop. In *Proceedings of the 1992 Winter Simulation Conference*, eds. James J. Swain, David Goldsman, Robert C. Crain, and James R. Wilson, 925-932.

Henriksen, James O., and Thomas J. Schriber. 1986. Simplified Approaches to Modeling Accumulating and Nonaccumulating Conveyor Systems. In *Proceedings of the 1986 Winter Simulation Conference*, eds. James R. Wilson, James O. Henriksen, and Stephen D. Roberts, 575-593.

Hopings, Donald B. 1988. Simulation of Discrete Conveyor Systems. In *Proceedings of the 1988 Winter Simulation Conference*, eds. Michael A. Abrams, Peter L. Haigh, and John C. Comfort, 575-582.

Hudson, Wilbur G. 1954. *Conveyors and Related Equipment*, 3d ed. New York, New York: John Wiley & Sons, Incorporated.

K. W. Tunnell Company, Incorporated. 1995. Containerization. In *Standard Handbook of Plant Engineering*, 2d ed., ed. Robert C. Rosaler, 8-30–8-46. New York, New York: McGraw-Hill, Incorporated.

Laughery, K. Ronald. 1995. A Micro Saint Model of Conveyor Management Strategies. In *Proceedings of the 1995 Winter Simulation Conference*, eds. Christos Alexopoulos, Keebom Kang, William R. Lilegdon, and David Goldsman, 818-822.

Law, Averill M., and Michael G. McComas. 1991. Secrets of Successful Simulation Studies. *In Proceedings of the 1991 Winter Simulation Conference*, eds. Barry L. Nelson, W. David Kelton, and Gordon M. Clark, 21-27.

Muth, Eginhard J., and John A. White. 1979. Conveyor Theory: a Survey. *AIIE Transactions* 11(4):270-277.

Rohrer, Matthew. 1994a. AutoMod. In *Proceedings of the 1994 Winter Simulation Conference*, eds. Jeffrey D. Tew, Mani S. Manivannan, Deborah A. Sadowski, and Andrew F. Seila, 487-492.

Rohrer, Matthew. 1994b. AutoStat. In *Proceedings of the 1994 Winter Simulation Conference*, eds. Jeffrey D. Tew, Mani S. Manivannan, Deborah A. Sadowski, and Andrew F. Seila, 493-495.

Ruch, Stéphane, and Patrick Kellert. 1995. Validation of Manufacturing Systems Conceptual Models. In *Proceedings of the 1995 Summer Computer Simulation Conference*, eds. Tuncer I. Ören and Louis G. Birta, 400-406.

Sims, E. Ralph, Jr. 1992. Conveyors Are Often Taken for Granted by Material Handling Planners. *Industrial Engineering* 24(3):39-42.

Sule, D. R. 1988. *Manufacturing Facilities Location, Planning, and Design.* Boston, Massachusetts: PWS-KENT Publishing Company.

Tompkins, James A., and John A. White. 1984. *Facilities Planning.* New York, New York: John Wiley & Sons, Incorporated.

Tsai, Li-Hui. 1995. Mixed-model Sequencing to Minimize Utility Work and the Risk of Conveyor Stoppage. *Management Science* 41(3):485-495.

Ülgen, Onur M., John Shore, and Eric S. Grajo. 1994. The Role of Macro- and Micro-Level Simulation Models in Throughput Analysis. Paper presented at *1994 ORSA/TIMS Joint National Meeting*, 23-26 October, Detroit, Michigan.

Ülgen, Onur M., and Sanjay S. Upendram. 1995. The Role of Simulation in Design of Material Handling Systems. In *AUTOFACT Conference Proceedings Volume 1*, ed. Lisa Moody, 259-273.

Witt, Clyde E. 1995. Powered Roller Conveyors: Sound Engineering Creates Better Environment. *Material Handling Engineering* 50(13):41-46.

AUTHOR BIOGRAPHIES

ALI K. GUNAL is a Systems Consultant at Production Modeling Corporation. He received his Ph.D. degree in Industrial Engineering from Texas Tech University in 1991. Prior to joining PMC, he worked as an Operations Research Specialist for the State of Washington, where he developed a simulation system for modeling and analysis of civil lawsuit litigation. At PMC, he is involved in consulting services for the analysis of manufacturing systems using simulation and other Industrial Engineering tools. He is familiar with several simulation systems including AutoMod, ARENA, QUEST, ROBCAD, and IGRIP. He is a member of the Institute for Operations Research and the Management Sciences [INFORMS], the Institute of Industrial Engineers [IIE], and the Society of Manufacturing Engineers [SME].

SHIGERU SADAKANE holds a bachelor's degree in Industrial & Systems Engineering (University of Michigan - Dearborn, 1994). He is an Applications Engineer at Production Modeling Corporation, and highly familiar with simulation and facilities-layout optimization systems including AutoMod, WITNESS, LayOPT, QUEST, IGRIP, and SIMAN. He is a member of the Institute of Industrial Engineers [IIE] and the Society of Manufacturing Engineers [SME].

EDWARD J. WILLIAMS holds bachelor's and master's degrees in mathematics (Michigan State University, 1967; University of Wisconsin, 1968). From 1969 to 1971, he did statistical programming and analysis of biomedical data at Walter Reed Army Hospital, Washington, D.C. He joined Ford in 1972, where he works as a computer software analyst supporting statistical and simulation software. Since 1980, he has taught evening classes at the University of Michigan, including both undergraduate and graduate simulation classes using GPSS/H, SLAM II, or SIMAN. He is a member of the Association for Computing Machinery [ACM] and its Special Interest Group in Simulation [SIGSIM], the Institute of Electrical and Electronics Engineers [IEEE], the Society for Computer Simulation [SCS], and the Society of Manufacturing Engineers [SME]. He serves on the editorial board of the *International Journal of Industrial Engineering – Applications and Practice.*

Proceedings of the 1996 Winter Simulation Conference
ed. J. M. Charnes, D. J. Morrice, D. T. Brunner, and J. J. Swain

A SIMULATION TESTBED FOR INVESTIGATION OF TANDEM AGV SYSTEMS

Leon Bing-Shiun Chuang
Joseph A. Heim

Industrial Engineering, 352650
University of Washington
Seattle, Washington 98195, USA

ABSTRACT

A tandem, one vehicle per loop, approach to Automated Guided Vehicle System (AGVS) guide path design has been proposed as a more efficient and flexible alternative to traditional AGV system design. The tandem design can substantially simplify the design task while providing even further flexibility for changes in production system layout and operation. Tandem guide paths, however, have potential disadvantages because vehicle break-down, or even scheduled maintenance of vehicles, results in loop inaccessibility.

Previous work comparing the efficacy of traditional guide path configurations and tandem loop configurations provided only limited flexibility for experimentation. A tandem AGV system simulation testbed has been developed to facilitate investigation and analysis of a variety of loop reconfiguration strategies and algorithms. A particular focus of our current investigations is the concept of *Real-Time Loop Reconfiguration* (RTLR), a means of responding to single vehicle failures in tandem AGV systems. The RTLR can solve the potential loop inaccessibility problem and also enhance the flexibility of the overall AGV system design.

Results from initial studies used to verify the correct operation of the testbed are presented, and a brief outline of future applications for the tandem loop reconfiguration model are discussed.

1 INTRODUCTION

The automated guided vehicle (AGV) has the potential for improving the performance of the material handling tasks required in advanced manufacturing systems. However, good design of the AGV system plays an essential role in using the technology efficiently. The AGV system design effort is comprised of the route configuration, AGV dispatching rules selection, traffic control, and obviously, vehicle selection. In this paper we will focus on the design of the guide paths rather than the mechanical or electrical components of the systems.

The tandem guide path design approach has been recognized as a more efficient and flexible alternative to traditional guide path design, because it significantly reduces the complexity of the overall AGV system design effort (Bozer and Srinivasan 1991). The tandem AGV system configuration as illustrated in Figure 1, however, does present potential disadvantages: a loop (i.e., one closed segment of the AGV system guide path set) may become inaccessible because the AGV assigned to a single loop is unavailable. The loss of an AGV is usually a result of vehicle break-down or scheduled maintenance. But there are also opportunities to enhance the flexibility of the production system layout with tandem guide paths.

Previous ways of responding to the loss of vehicle-services has been to provide a stand-by AGV. Providing "spare" vehicles, however, can be an expensive way to provide continued material handling services to all loops. And, in some instances, this approach may not be practical, because vehicle configurations may not be consistent across all loops.

To compensate for this potential problem, we have developed what we call *Real-Time Loop Reconfiguration* (RTLR). RTLR is a means of responding to the loss of one or more vehicles (i.e., loop accessibility) and a way to dynamically change the control strategies for the AGV system based on product mix or production machinery availability.

The complex nature of AGV system design in this situation provides several opportunities to combine analytical and simulation-based methodologies. In this paper we will report on the simulation aspects of the project. The complex nature of AGV system design in this situation provides several opportunities to combine analytical and simulation-based methodologies. In this paper we will report on the simulation aspects of the project. Discrete event simulation techniques have been frequently used to design and evaluate alternative AGV systems, often because complex production-based systems become analytically intractable due to the interactions among the many elements of the overall manufacturing system. Simulation models can provide detailed analyses of the dynamic behavior of the system along time, as opposed, analytical approaches are generally limited to examinations of static performance characteristics for the system (Pegden, Shannon and Sadowski 1995). The dynamic aspects of simulation readily supports the types of investigation and analysis of the loop inaccessibility problem and the dynamic behavior of RTLR during operation of the AGV system.

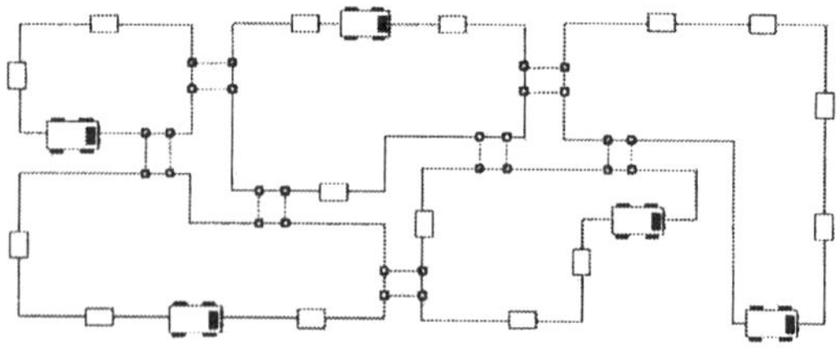

Figure 1: The Tandem AGV System Configuration

2 TANDEM AGV SYSTEMS

The tandem AGV system configuration was first proposed by Bozer and Srinivasan in 1991. The intention was to reduce complexity and the attendant difficulties in the traditional AGV system design stage caused by the need to prevent vehicle congestion and blockage when two or more vehicle routes must share the same guide paths; these problems do not exist in the tandem AGV system. Additionally, the AGV system dispatch rules are greatly simplified, since a transport request can only be made to the AGV traveling in the loop from which the request originates. The tandem approach to guide path management strongly supports group technology facility layout techniques: each tandem loop represents a manufacturing cell performing a certain associated series of processes.

2.1 Design Rules for Tandem AGV Systems

The tandem AGV configuration is defined as an AGV system in which all stations have been partitioned into non-overlapping, single vehicle closed loops with pick-up/drop-off points interfacing adjacent loops. Therefore, an AGV can travel only inside the non-overlapped loop to which it is assigned, eliminating vehicle congestion, blockage, deadlock or AGV assignment rules.

The term *loop inaccessibility* as used here means a tandem AGV loop that can not be reached because the AGV residing in this loop is unavailable. Unavailability of the AGV could be because of maintenance requirements (e.g., battery recharging), AGV breakdown, or the desire to add another loop (i.e., workcell to the guide path configuration set). Note that loop inaccessibility means that the machinery, equipment, tools, materials and work pieces in progress within the manufacturing cell cannot be reached; all the machines in that loop may still be functioning.

2.2 Real Time Loop Reconfiguration

The RTLR strategy is to pre-define alternative, overlapped reconfiguration guide paths which connect two adjacent loops. These reconfigurations allow the AGV in an adjacent loop to travel across to the inaccessible loop when necessary, thus making the inaccessible loop reachable once again. Naturally, we would expect that such reconfigurations or combining of loops will increase the utilization of the "borrowed," adjacent-loop AGV.

Rules for determining which adjacent loops should be reconfigured to provide transport services to the inaccessible loop is another aspect of this research which is being examined analytically; it will not be discussed in this paper. In the simulation testbed presented here, our rule for reconfiguration is to always use the shortest adjacent loop, therefore, as shown in Figure 2, inaccessible Loop-3 is always reconfigured with Loop-1 as Loop-3/1, likewise, inaccessible Loop-5 is always reconfigured wit Loop-4 as Loop-5/4.

3 THE CONCEPTUAL MODEL AND SIMULATION MODELING

Following the traditional simulation project steps (Law and Kelton 1991), this study was conducted first by identifying the problems of interest, constructing and understanding the conceptual model and then building the simulation model using an appropriate computer programming language. In the following sections, several of these steps are explained in detail.

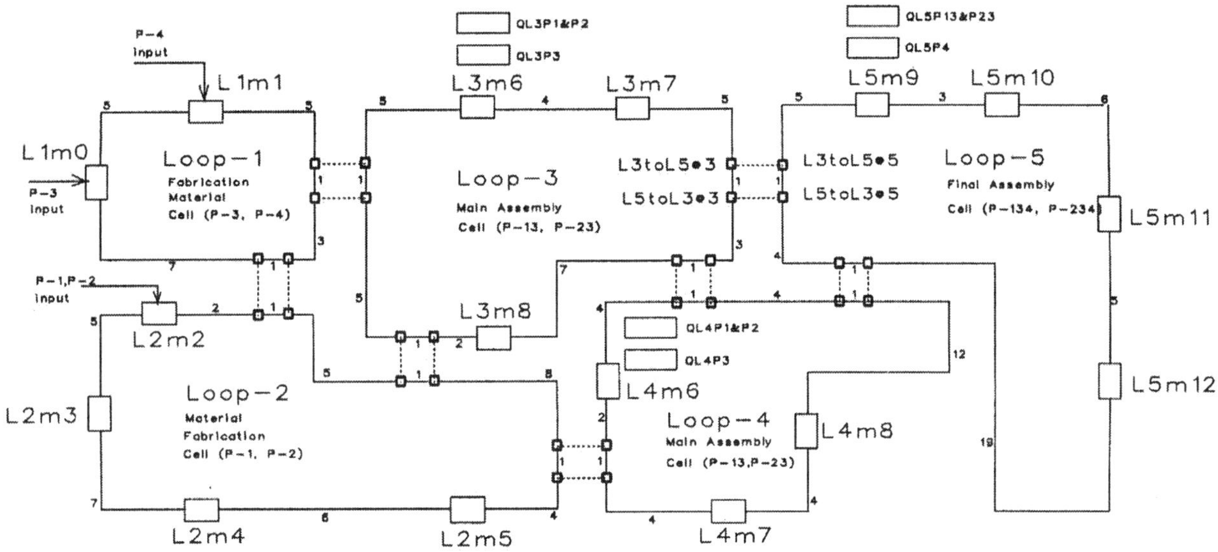

Figure 2: The Production System Layout of the Simulation Testbed

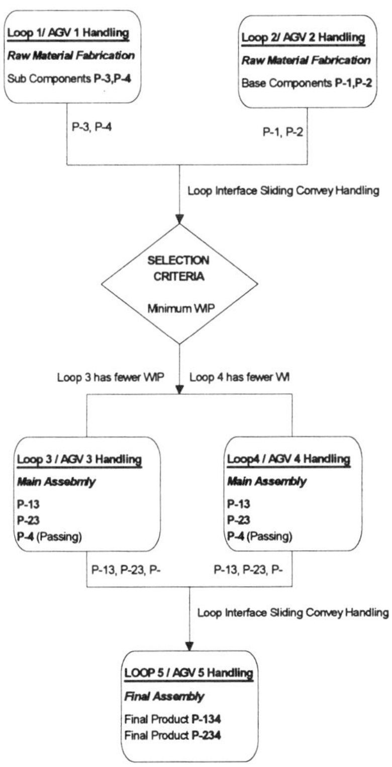

Figure 3: The Processes Followed to Make the Two Final Products, P-134 and P-234

3.1 Problem Statement and Planning

The objective of the initial study undertaken with the testbed was to characterize the impact on the performance of the system when a loop became inaccessible. We wanted to evaluate the effect of RTLR: could it minimize the loss caused by loop inaccessibility?Since this is the first introduction of RTLR, we also want to provide justification for using it in a tandem AGVS design and thus arises some interesting issues for possible future studies.

3.2 The Conceptual Model Understanding

The conceptual testbed model presented in this paper is an assembly manufacturing system laid out in a tandem AGV configuration following group technology plant layout guidelines. Figure 2 illustrates the production system layout of the simulation testbed.

Figure 3 explains the processes followed to make the two final products, P-134 and P-234. Two base components, P-1 and P-2, and two sub-components, P-3 and P-4, are sent to the *material fabrication* processes. P-1 and P-2 are then assembled with P-3 in the *main assembly* process; these two intermediate products, called P-13 and P-23, are then assembled with P-4 in the final assembly process, and the final products ,P-134 and P-234, leave the system. Each end product is composed of one base component P-1 or P-2, and two sub-components P-3 and P-4.

Material handling requests between machines within each tandem loop (i.e., manufacturing cell) are performed by the AGV assigned to the loop. Moreover, the material handling between loops is assumed to be performed by a sliding conveyor.

Fabricated components choose the main assembly loop -- Loop-3 or Loop-4 -- which has least work-in-process (WIP). Obviously, more comprehensive rules could be adopted for loop selection. See section 5.4 for further discussions.

3.3 Verification and Validation

Verification and validation are the two important parts for simulation modeling. Verification insures that the simulation model behaves as it was programmed to behave, and validation is necessary to convince we that the model properly represents reality.

We have verified our simulation program by exhaustively tracing entities through the system and examining, quantitatively, the flow time, waiting time and traveling time. Our results have given us confidence that the components follow the right assembly sequences from material fabrication through final assembly, and that the values obtained are what one would expect based on hand calculations. Animation has also been a useful auxiliary tool for visual verification of proper model operation.

Validation is more difficult because the relatively small systems we have constructed thus far would not justify AGVs in an actual production situation. The manufacturing assembly system we have developed, however, does reflect appropriate magnitudes: the dimensions of the production floor and loops, the vehicle traveling speed (100 ft/min.), production machinery operation times, and the handling time (30 seconds for AGVs) all conform with common industrial specifications.

4 EXPERIMENT DESIGNS

In order to observe the impact of the loop inaccessibility, as well as to justify to what extent the implementation of RTLR can compensate for the loop inaccessibility, a total nine experiments are conducted in three categories as listed below:

I. Normal Operation

- E1. System operating with no AGV or machine breakdown.

II. Loop Inaccessibility

- E2. Non-critical Loop-3 inaccessibility with periodic maintenance for AGV-3.
- E3. Critical Loop-5 inaccessibility with periodic maintenance for AGV-5.
- E4. Non-critical Loop-3 inaccessibility with unpredictable AGV-3 breakdown.
- E5. Critical Loop-5 inaccessibility with unpredictable AGV-5 breakdown.

III. Real Time Loop Reconfiguration (RTLR)

- E6. RTLR implementation of Loop-3/1 corresponding to E2.
- E7. RTLR implementation of Loop-5/4 corresponding to E3.
- E8. RTLR implementation of Loop-3/1 corresponding to E4.
- E9. RTLR implementation of Loop-5/4 corresponding to E5.

In the normal operation scenario, all AGVs are traveling in good condition at a constant speed of 100 feet/minute (fpm). The *critical loop* is defined as a loop that is unique to the system, and its loop inaccessibility will cause the system operation to fail immediately (i.e., the material fabrication loops and the final assembly loop in the testbed).

In contrast, a non-critical loop contains functions that are duplicated in another loop and, therefore, its loss will not stop the entire system. A main assembly loop in the testbed would be considered non-critical because there are two of them. The AGV maintenance schedule for all experiments is 400 minutes after 4000 minutes of operation. During AGV maintenance, the loop is inaccessible. A stochastic AGV break-down occurs once in each experiment and cause the loop to be inaccessible for 1440 minutes (24 hours).

5 ANALYSIS OF SIMULATION RESULTS

In order to evaluate the differences in system performance between experiments, three kinds of manufacturing measurement factors are used. The product associated metrics, the system associated metrics and the resource associated metrics.

The product metrics include product flowtime through the system, product waiting time for AGVs, product waiting time for machines, and the ratio of material handling time (including all time spent in waiting and material handling) to machine processing time. These are based on final products. For example, the product flowtime of P-134 is measured as the time interval between the input of base component P-1 and the output of the final product P-134.

System metrics include the throughput of the assembling system during a specific time period (in this study, 24,000 minutes with the first 2,000 minutes as warm-up period is simulated for all experiments), and the maximum quantity of work-in-process (WIP) of the actual products in system. We consider WIP to be the number of products in a manufacturing cell waiting for either AGV handling or machines processing. It's possible that the simulation model may generate more WIP than the physical capacity would permit; we ignore these constraints. The resource metrics used here include the AGV and the maximum machine utilization of each loop, which are estimated as the time a resource (either an AGV or a machine) is on duty over the total time stimulated.

The metrics were collected after the system performance had achieved steady state, so that the statistical bias of the warming-up period could be minimized.

5.1 Negative Impact of Loop Inaccessibility

Examining Table 1, you can see that the four product associated measurement factors, the product flowtime, the waiting time for AGVs, the time waiting for machines and the ratio of handling time to processes time, all increase when loop inaccessibility occurs. As expected, inaccessibility of the critical loop has more serious negative impact on performance than the non-critical loop in terms of these factors. The worst situation happens when the critical loop is not accessible with unpredictable AGV break-down which is 2.6 times of that in normal operation.

Table 1: Product Associated Measurement Factors

		Product Flowtime (minutes)	Waiting AGVs (minutes)	Waiting Machines (minutes)	Handling/ Processing (%)
I	E1	245.49	23.96	33.17	31.16
II	E2	301.45	29.33	88.83	43.94
	E3	334.23	45.16	100.79	49.44
	E4	398.68	29.02	181.38	57.61
	E5	646.78	98.25	360.25	73.87
III	E6	245.38	46.03	32.66	31.13
	E7	245.57	51.17	32.99	31.74
	E8	244.75	47.00	31.94	30.95
	E9	250.86	54.89	32.22	32.63

Figure 4 shows the product flowtime for Experiment 1 (E1), normal operation. Figures 5 through 8 illustrate the dynamic behavior of the system in terms of product flowtime measurement for the four loop inaccessibility experiments, E2, E3, E4 and E5. It is evident by examining Figures 5 and 7, that when there is inaccessibility to non-critical loop resources, the system can still yield continuous output. And also interesting to note are the two sets of product flowtime measurement output when the non-critical loop becomes inaccessible: one for those products blocked and delayed by the inaccessible non-critical loop (Loop 3), and one for those going through the other loop performing the same manufacturing activity as the inaccessible non-critical loop. However, in contrast, Figures 6 and 8 illustrate the consequences of critical loop failure. A discontinuous output is generated with an output gap equal to the length of loop inaccessibility; all the products are affected by loss of the critical loop.

Table 2 provides a list of system associated measurement factors, the throughput of the simulated assembly manufacturing testbed during a 24,000-minute period, and the maximum number of WIP for each tandem AGV loop/manufacturing cell recorded along simulation run. As mentioned previously, the output gap is caused by critical loop failure and results are a loss in terms of throughput for the tandem AGV system. In Experiment 5 (E5) the size of work in process becomes completely unacceptable: 221, as compared to 13 during normal operation.

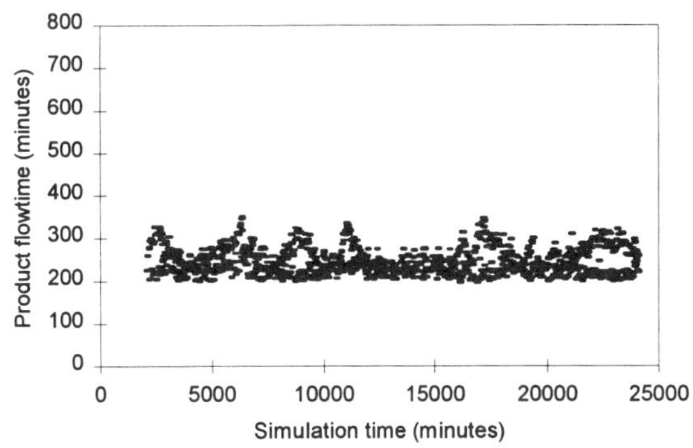

Figure 4: Product Flowtime of E1

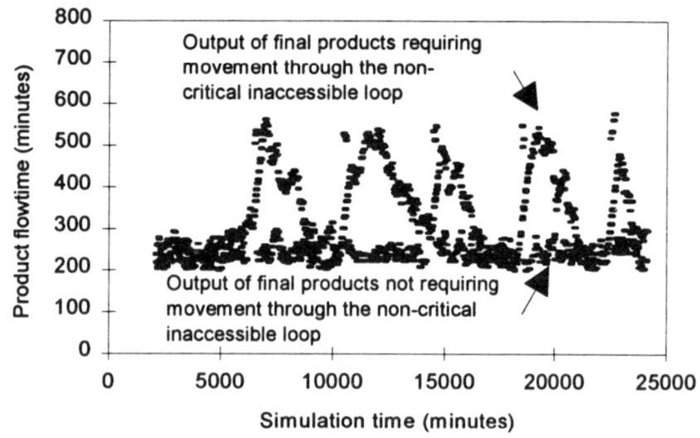

Figure 5: Product Flowtime of E2

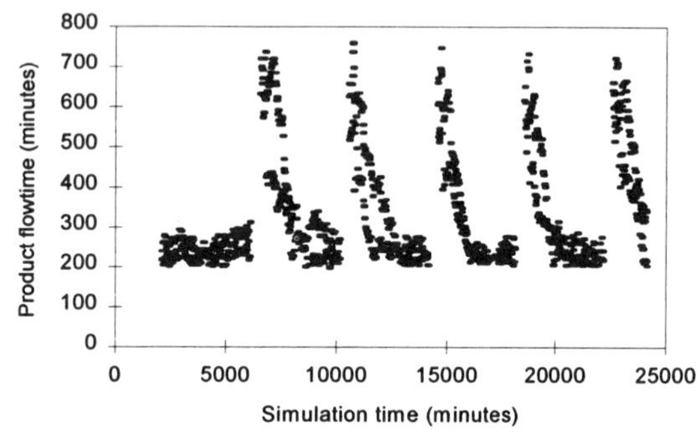

Figure 6: Product Flowtime of E3

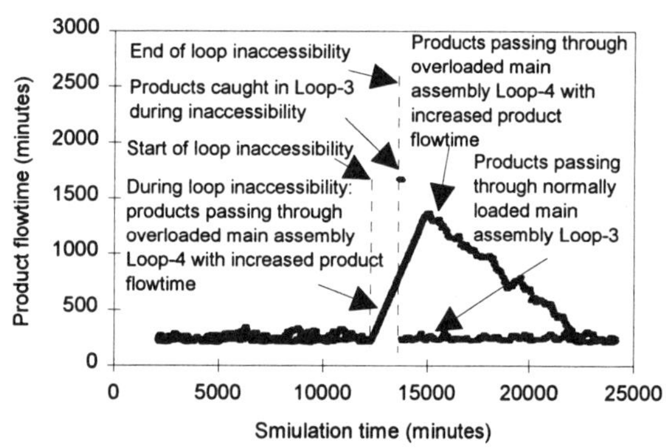

Figure 7: Product Flowtime of E4

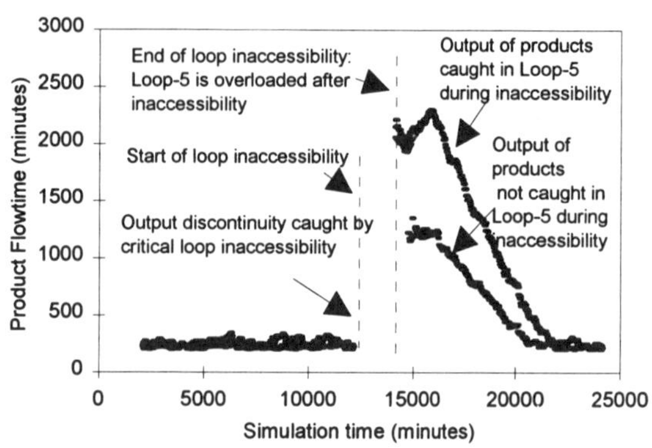

Figure 8: Product Flowtime of E5

Table 2: System Associated Measurement Factors

		Throughput 22,000 Minutes	Maximum WIP Recorded L1	L2	L3	L4	L5
I	E1	1223	1	7	12	13	9
II	E2	1219	1	6	12	23	9
	E3	1205	1	6	12	19	60
	E4	1204	1	7	12	61	9
	E5	1185	1	7	11	60	221
III	E6	1221	2	6	11	13	9
	E7	1224	1	7	13	12	14
	E8	1221	2	7	12	13	9
	E9	1223	1	7	12	17	15

The AGV utilization factor not only provides information about system balance during the design stage, but it can also help determine the appropriate number of vehicles necessary under varying conditions. In Table 3, AGV utilization among experiments seems to be only minimally influenced by loop inaccessibility. This implies a design with less loading can return to steady state without creating bottlenecks in the system due to overloaded AGVs.

Table 3: Resource Associated Measurement Factors

AGV Utilization

Maximum Machine Utilization

		L1	L2	L3	L4	L5
I	E1	28.97	54.62	71.43	32.26	60.89
		55.56	72.51	91.79	86.66	78.39
II	E2	29.11	56.68	66.18	36.99	61.18
		55.56	72.69	83.69	95.04	78.70
	E3	29.11	56.47	66.92	36.13	60.77
		55.56	72.36	85.89	91.96	78.04
	E4	29.10	56.40	66.80	36.15	61.19
		55.56	72.48	85.22	93.20	78.42
III	E5	29.09	56.38	67.13	36.18	60.80
		55.56	72.46	85.61	92.65	78.37
	E6	32.99	55.28	66.66	33.38	60.93
		55.56	72.66	90.36	88.29	78.65
	E7	28.95	54.30	71.84	36.95	57.01
		55.56	72.50	91.57	86.67	78.27
	E8	32.76	55.56	66.67	33.29	60.88
		55.56	72.64	90.85	87.84	78.65
	E9	28.95	54.58	71.37	37.35	57.12
		55.56	72.65	90.31	88.40	78.70

5.2 Compensating Effects of Implementing RTLR during Loop Inaccessibility

The results of the simulation experiments illustrate that implementing RTLR provides a compensating effect for the loss of a single loop guide path. From Tables 1 ,2 and 3, the metrics reflect no significant differences from those recorded in normal operation. Even for the critical loop inaccessibility situation, real time loop reconfiguration can make the system behave as if the system were under normal operation. Figures 9 through 12 show this in terms of product flowtime. Although the measurements of product waiting time for AGVs in E6, E7, E8 and E9 rise to an amount from 1.9 to 2.3 times of that in normal operation, the overall performance measurement in terms of product flowtime and the ratio of handling time to machine process time performs almost as well as it was in normal operation.

5.3 Factors Essentially Affecting Loop Inaccessibility

From observing the simulation generated results, one can find some important factors that essentially affect the impact of loop inaccessibility. In general, we can infer from the simulation results that the non-critical loop inaccessibly will cause congestion, and that the critical loop inaccessibility will cause an additional problem of discontinuity in processing flow. The time interval of the loop inaccessibility plays a more influential role in affecting the extent of the impact far more than the frequency of the loop inaccessibility.

5.4 Future Studies

Initial tests with relatively simple RTLR situations have been presented in this paper. There are, however, more complex and interesting issues for which further studies will help us understand when RTLR may be most appropriately applied as well as which strategies for loop reconfiguration are most likely to enhance the tandem AGV system flexibility. Some of these include:

1. RTLR can be used to help partition the manufacturing floor during the early stages of the production system design by examining the consequences of pre-defining "spare" guide paths for production system reconfiguration (i.e., redefining manufacturing workcells). In this way, a tandem AGV system with RTLR capability can reduce cost associated with an over partitioned floor (Harit 1995).

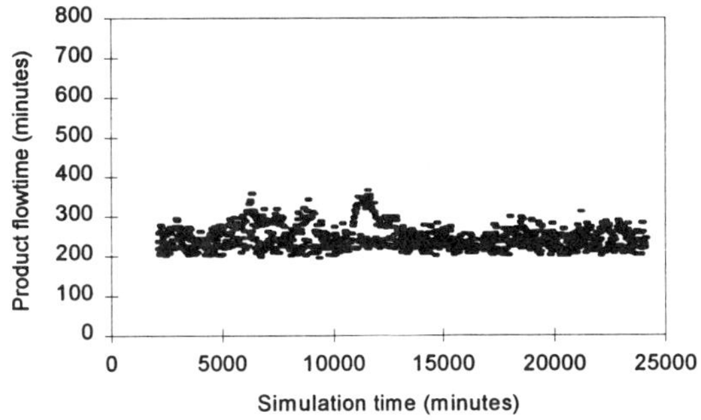

Figure 9: Product Flowtime of E6

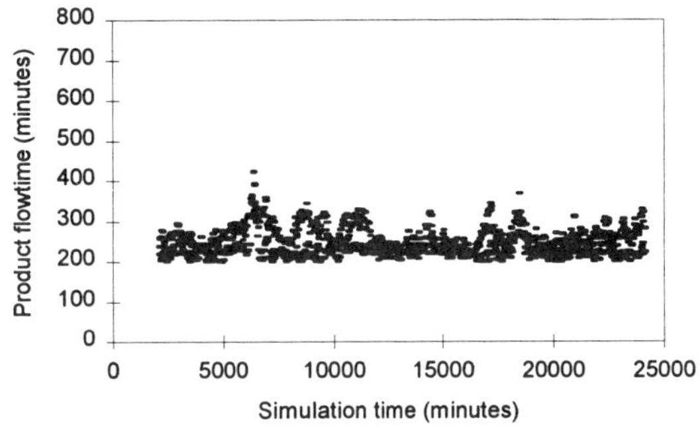

Figure 10: Product Flowtime of E7

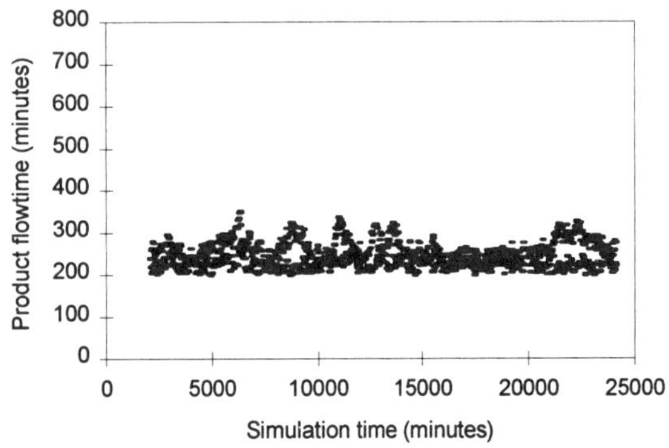

Figure 11: Product Flowtime of E8

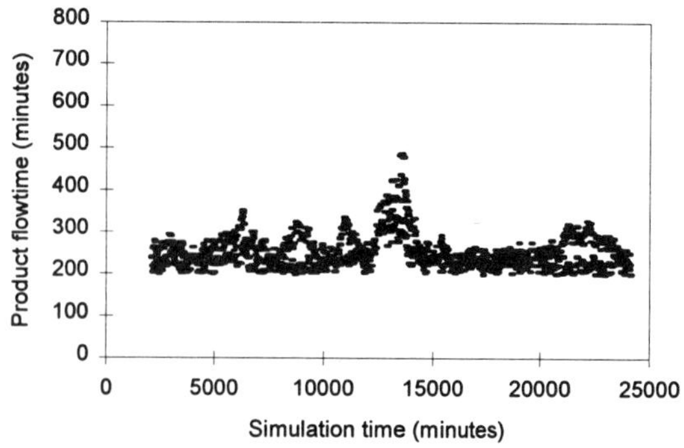

Figure 12: Product Flowtime of E9

2. In this paper, the "shortest AGV traveling guide path" rule is applied to select from among available adjacent loops (for example, while non-critical Loop 3 is inaccessible, Loop 1 which has the shortest guide path is selected among adjacent Loop 2, 4 and 5). But it is likely that many tandem systems would have more than two adjacent loops, and, because implementing RTLR will directly increase the material handling loads in the reconfigured loops, rules for selecting from among the adjacent loops should be developed to optimize the reconfiguration.

3. We have examined the results of scheduled maintenance in which vehicles are unavailable. It would be helpful to integrate RTLR and the creation of AGV maintenance schedules, thus reducing the impact on production output during required maintenance periods. The simulation testbed can provide an efficient mechanism for providing alternative solutions of efficient integration of vehicle maintenance schedule and RTLR .

4. RTLR can be used to dynamically (on-line) partition the manufacturing floor into independent AGV transport loops during system operation in response to actual resource utilization. By examining the consequences of various system reconfigurations with an embedded simulation model, a tandem AGV floor with RTLR capability could be reconfigured flexibly and dynamically to minimize resource idleness and maximize system performance.

6 CONCLUSIONS

AGV systems have been studied for many years to improve their capability and enhance their flexibility by providing better overall designs. We have demonstrated the use of simulation to support the application of Real-Time Loop Reconfiguration (RTLR) in tandem AGV system design; the results generated by the models indicate that RTLR can effectively compensate for loop inaccessibility in tandem loop guide paths. Furthermore, RTLR can be used to enhance the flexibility of tandem AGV systems by supporting the design of dynamically changing production process flows with pre-partitioned guide path alternatives.

REFERENCES

Bozer, Y. A. and M. M. Srinivasan. 1991. "Tandem Configurations for Automated Guided Vehicle Systems Analysis". *IIE Transaction* 23:1, p71-p82.

Bozer, Y. A. and M. M. Srinivasan. 1992. "Tandem AGV Systems: A Partitioning Algorithm and Performance Comparison with Conventional AGV Systems". *European Journal of Operations Research* , 63, p173-p191.

Choi, H., H. Jin and J. Lee. "Traditional and Tandem AGV System Layout: A Simulation Study". *Simulation* 63:2, p85-p93,1994.

Harit, S. 1995. "Partitioning Algorithm for Multiple Server Tandem Loop Configuration". Ph.D. dissertation, Department of Industrial Engineering, University of Arkansas.

Law, A. M. and W. D. Kelton. 1991. *Simulation Modeling and Analysis*, 2nd Ed, New York: McGraw-Hill.

Pegden, C. D., R. E. Shannon and R. P. Sadowski. 1995, *Introduction to Simulation Using SIMAN.* 2nd Ed. New York: McGraw-Hill.

AUTHOR BIOGRAPHIES

LEON BING-SHIUN CHUANG received his Masters degree from the Department of Industrial Engineering at the University of Washington in 1996. In 1992 Mr. Chuang received his Bachelor of Engineering in Aeronautical Engineering at the Tamkang University in Taipei, Taiwan. His research interest is manufacturing applications of simulation modeling.

JOSEPH A. HEIM is an assistant professor of Industrial Engineering at the University of Washington in Seattle, Washington. His teaching responsibilities include computer integrated systems, simulation, inventory management and scheduling. Heim's research focuses on coordination of complex collaborative tasks with distributed discrete event simulation models. Professor Heim is a member of ASEE, ACM, IIE and SME.

Proceedings of the 1996 Winter Simulation Conference
ed. J. M. Charnes, D. J. Morrice, D. T. Brunner, and J. J. Swain

SIMULATION OPTIMIZATION IN STRATEGIC LOCATION OF SEMI-FINISHED PRODUCTS IN A PULL-TYPE PRODUCTION SYSTEM

Farhad Azadivar, Junhong Shu, and Moyeen Ahmad

Department of Industrial and Manufacturing Systems Engineering
Kansas State University, Manhattan, Kansas 66506, U.S.A

ABSTRACT

In many manufacturing environments there are intermediate states of semi-finished products that are shared by several members of the product family. This is true, for instance, in machine tools manufacturing or manufacturing of several printed circuits from the same basic board. For these systems, processing parts ahead of time and locating them at strategic stages of the operation may improve the lead time with relatively low levels of work-in-process inventory. This is especially true when production stages do not have a dedicated resource and share a limited number of production equipment. In this paper a simulation model has been developed and interfaced with a simulation-optimization method to determine these levels. Furthermore, a method has been suggested to group optimization parameters such that fewer variables need to be optimized.

1 INTRODUCTION

In many manufacturing environments several parts or products are made from the same raw material in several stages of manufacturing operations. After each process, the product assumes a particular state that is closer to its final shape. The closer the product is to its final state the shorter the lead time becomes to transform it to a finished product. However, as the product gets closer to its final state, its flexibility for being transformed into a larger variety of final products is decreased. Raturi et. al. (1990) provide a number of good examples in machine tools industry where the proposed problem exists and a solution will help the manufacturer to better satisfy customer demands. For uncertain demand patterns for final products, a good compromise among the overall inventory carrying costs, service levels, and lead times might be achieved if inventories of semi-finished products were kept in various intermediate stages. This consideration will have even a better effect when the system operates in a pull-type production environment.

The optimum levels of work in-process in general pull-type manufacturing systems have been studied by several researchers including Ohta and Miyazaki (1988), and Bitran and Chang (1987). So and Pinault (1988) presented some analytical results for determining the optimum number of in-process inventories in a single product pull-type system. Azadivar and Lee (1985) showed the results for a multi-product flexible manufacturing system. A closely related work to this study is that of Garg and Lee (1994) where the operating characteristics of multi-product production lines are investigated. Using a modified linear production rule by Tang (1990), they suggest target levels for inventory at each stage to be set to cope with uncertainties in demand.

This paper provides a heuristic method for determining these target values for a given multi-stage production system. However, the problem under study here has an additional complexity in that the resources are shared for various stages of production. This puts an additional constraint on allocating the in-process inventories, as stages that need more scarce resources need to be supplied with larger levels of semi-finished products.

Since each process does not have its dedicated resource and has to share it with other activities and because of the stochastic nature of demands and processing times, Tang's linear production rule or other analytical methods would not be easy to apply here. Thus for this study, the system is modelled through computer simulation and is formulated as a constrained stochastic optimization problem. A modeling and optimization procedure has been developed to determine the near optimum levels of in-process inventories to keep at each stage of operation. The objective of the optimization is to minimize the average lead time for satisfying a set of uncertain demands while keeping the inventory levels at a reasonable values and maintaining an acceptable service level. In addition, a software package has been developed that facilitates application of this methodology. Through a graphical interface the user supplies inputs on the parameters and the operating conditions of the system. The program uses these inputs and provides near optimum values of the work in-process at each stage.

2 PROBLEM DEFINITION

To help in understanding the problem, consider the system shown in Figure 1. Four different products are produced from the same stock of bar through several operations utilizing four machines MC1, MC2, MC3, and MC4. In this figure, nodes represent the states of the product while arcs denote the processes of transforming a product from one state to another. Arcs are also marked by the manufacturing resource (e.g. machine) needed for the transformation. All processing times are random and are distributed according to given distributions. Orders for different products are also random and arrive at random intervals.

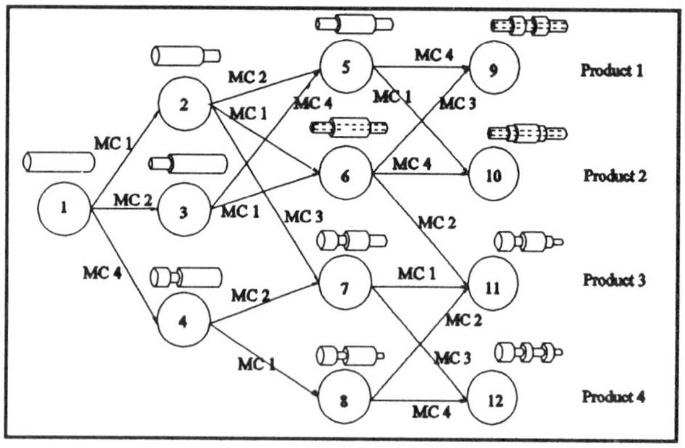

Figure 1: Production System for a Four Product Family

For the problem under study, it is assumed that the production period consists of just one cycle. The demand for products during this cycle is assumed to be uncertain with known probability distributions. It is also assumed that the lead time for obtaining the raw material is an order of magnitude larger than all due dates for orders. This means that if the plant runs out of raw material during a production cycle, there will not be enough time to order and receive the raw material in time and the corresponding orders will be lost. Thus all needed raw materials have to be ordered in advance. An example of this situation is that of a machine tool manufacturer ordering raw forgings for press brakes which would be customized for a variety of models. Further, it is assumed that the manufacturer orders as many sets of raw material as are sufficient to supply the expected value of the demand.

With these assumptions, the maximum service level, in terms of the fraction of the total demand that is satisfied can be achieved if all the stock is kept in the raw state at node 1. However, this may create long lead times because each order has to wait for the product to go through all stages of production. Alternatively, they can be processed ahead of time and kept at final states for each product. This will reduce the lead time to practically zero for orders that are satisfied, but some demands will go unsatisfied because raw materials needed are tied up in other products for which the demand did not turn out to be as expected. A solution between these extremes will result in a lead time in between while satisfying an acceptable portion of the demand.

3 PROBLEM FORMULATION

Before presenting the general formulation for the problem the following clarifications are in order:

1. Since the problem is considered in its general form with stochastic demand patterns and processing times, and the performance of the system is evaluated through computer simulation, the objective function and the constraints for this formulation are not available analytically. They can only be evaluated for a given set of decision variables and only through multiple runs of the simulation model.

2. The lead time has been defined as the average of the lead times for individual final products and is estimated for only those demands that are supplied. Thus, the demand for products that are not produced, due to lack of raw material, is assumed lost. This means neither the lead time nor the service level alone can be considered as the performance measure for the system. Otherwise, one could show a drastic reduction in the lead time by losing too many orders.

Based on these definitions and assumptions the problem in its general form can be stated as following:

$$\begin{array}{ll} \text{Minimize} & f(x_1, x_2, \ldots, x_n) \\ \text{S.T.} & g(x_1, x_2, \ldots, x_n) \geq b \end{array}$$

where, x_1, x_2, x_3 ....., x_n are the decision variables representing the levels of work-in-process inventory at different nodes, f is the average lead time for customers as a function of work-in-process inventories, g denotes the service level in terms of the fraction of the total demand that is satisfied, also as a function of the in-process inventories, and b is the minimum tolerable service level.

The lead time obtained from one simulation run is only one stochastic observation on the response of the system. This observation is noise corrupted and consists of the theoretical mean (expected value) and some random error. The expected value of this function can be optimized by optimizing the theoretical regression function of the response defined by:

$$y(x_1, x_2, \ldots, x_n) = E[f(x_1, x_2, \ldots, x_n) \mid \forall x_i].$$

Since this function is not known analytically, its values have to be estimated based on several observations made on its simulation model.

Service level is also a stochastic variable. This makes it mathematically incorrect to require the service level obtained from the simulation model, which is a random variable, to be compared to a constant value. A more correct way to deal with this constraint is for the decision-maker to specify a probability indicating the acceptable risk for violating a given constraint. If we denote the risk for violating the minimum tolerable service level as α, the modified problem can be rewritten as follows:

Minimize: $y(x_1, x_2, \ldots, x_n)$
S.T. $P[g(x_1, x_2, \ldots, x_n) \geq b] \geq (1 - \alpha)$

4 OPTIMIZATION METHODOLOGY

Azadivar (1992) provides an overview of the available simulations-optimization procedures. For this problem, SIMICOM algorithm developed by Azadivar and Lee (1986) has been employed. SIMICOM is developed for solving the optimization problems as defined above and is based on Box's (1965) complex search method. It can be interfaced with any simulation model automatically to seek the optimum.

4.1 Example 1

To apply and test the procedure, a simple system for production of three products, belonging to the same family, was considered. The network presentation of this system is given in Figure 2. The system is assumed to operate in a pull-type production environment. All products are manufactured in two stages and all the state changes are performed by two machines denoted by MC1 and MC2.

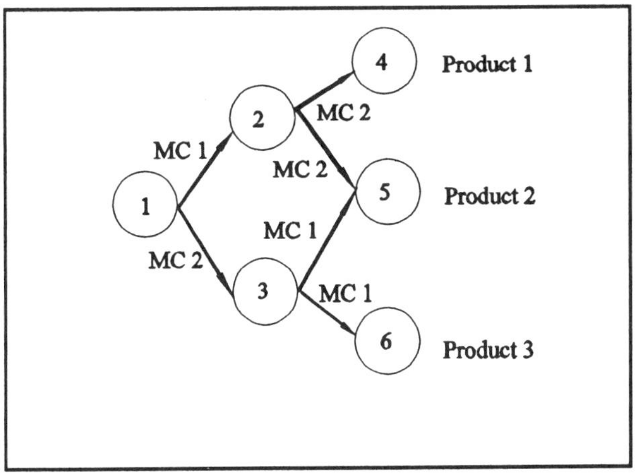

Figure 2: Three Product Production System for Example 1

Orders for the products can come in batches for any of the final nodes. If there are finished products waiting at these nodes, orders are met right away. However, if there is no product waiting, a set of orders is sent to one or more of the supplying nodes based on the workloads on the machines. A tally is kept on the work already assigned and not yet completed for each machine. For instance, if an order comes to node 5 (Figure 2), it can be either satisfied by semi-finished products at node 2 using machine 2 through process 2-5 (arc 2 5) or by those at node 3 using machine 1 through process 3-5 (arc 3 5). Depending on the workloads on these two machines (time needed to process the products that are waiting in the machine queue), the load will be assigned to the machine with the lighter load. Ties are broken arbitrarily.

This problem was solved for a case where orders arrive with interarrival times that are distributed normally with a mean of 30 and a standard deviation of 5 time periods. Each order consists of a batch of 10 that could be for products 1, 2, or 3 with probabilities of 0.5, 0.35 and 0.15 respectively. The production cycle consists of 600 time units with an expected total demand of 200 units for the planning period. This will be the number of sets of raw material that will be ordered. The processing times were assumed to be all normal with parameters (4,1), (3,1), (5,2), (4,1), (4,1) and (3,1) time periods for processes 1-2, 1-3, 2-4, 2-5, 3-5, and 3-6 respectively. The minimum desirable service level was specified as 90%. The resulting work-in-process inventories are given below:

Nodes	2	3	4	5	6
WIP	4	4	9	16	16

This solution results in an average lead time of 38.07 time periods with an average service level of 93%. It took 44 evaluations of the simulation model to reach this solution.

Here, since there were only 6 variables involved, the solution could be obtained rather easily. However, in many situations, decisions should be made on a large number of work-in-process values where many stages are involved. In those situations, the simulation-optimization requires a large number of simulation evaluations which in some cases will demand an infeasible amount of computer time. Besides, this will put a heavier burden on the management in controlling a large number of decision variables.

One way to alleviate this problem is to group the nodes in a few groups according to a suitable set of criteria. Then, rather than solving the problem for the work-in-process for each individual node, it can be solved for the same amount assigned to each node of a given group. This could reduce the number of variables into a manageable size. The quality of the solution obtained in this way may, of course, be inferior to the one for individual nodes, but the optimization efforts will be reduced significantly.

The following section covers a discussion and analysis of some grouping criteria with a conclusion on the most suitable one for a given situation.

5 CRITERIA FOR GROUPING OF NODES

In searching for common characteristics for nodes for grouping purposes, the following factors showed reasonable relevance.

5.1 Stages of Production (A)

The nodes can be grouped according to stages of production. Stages are defined according to the relative position of the nodes in the network. The initial raw material node is considered to be at the lowest stage (stage 1) and the final product nodes are considered to be at the highest stage. The nodes that follow stage 1 are at stage 2 and those following stage 2 nodes are denoted stage 3 and so on. To incorporate this factor in grouping the nodes each node is given a weight according to the stage it is at.

5.2 Flexibility of Nodes (B)

According to this criterion, the flexibility is defined as the fraction of final products that can be reached from a node. Consider the network shown in Figure 3 . Let us assume that when an order arrives the probabilities that it will be for products 1,2,3 and 4 are 0. 50, 0.25, 0.10, and 0.15 respectively. The fraction of the total expected demand for each of these products will be proportional to these probabilities. From node 9, only the demand for product 1 can be satisfied. The probability of an order being for product 1 is 50%. So, from node 9, 50% of the total demand can be satisfied. This gives node 9 a flexibility factor of 0.50. From node 6, we can satisfy any demand for products 1,2 or 3. The probability of an order being for products 1,2 or 3 is 0.85 (0.50+0.25+0.10). So, node 6 has a flexibility factor of .85. In this way all the nodes are assigned a flexibility factor. These factors for some of the nodes are shown in Figure 3.

5.3 Combined Grouping Criteria (AB)

If nodes are grouped according to their flexibility, the optimization program will automatically allocate higher levels of inventory to the most flexible nodes. By keeping higher levels of inventory in the most flexible nodes, acceptable service levels will be maintained but satisfactory lead times may not be achieved. If nodes are grouped according to stages, the optimization program will automatically maintain higher levels of inventory in the final stages. By keeping high levels of inventory in the final stages, the optimization program can reduce lead time but that might not allow for meeting service level requirements. Since none of the two rules satisfy both lead time and service level criteria, a combination of these rules was also examined.

6 SUITABLE GROUPING CRITERIA

As discussed before, grouping represents a tradeoff between the computational efficiency and the quality of the solution. By carefully selecting the suitable criterion for grouping, efficient designs can be achieved without sacrificing the quality of the solution significantly. To determine the suitable criteria for grouping, several experiments were performed.

The three criteria A, B, and AB for grouping were tested on a set of 3-stage network problems possessing various specific characteristics. The characteristics that were considered are as follows.

6.1 Overall Flexibility of the Network

Here, the overall flexibility of a network is defined as the average flexibility of each node in the network. Figures 4 and 5 demonstrate two systems with high and low overall flexibilities, respectively. Two networks with average flexibility of approximately 40% and 70% were examined.

6.2 Demand Patterns

Orders arrive at intervals distributed according to a distribution specified by the user. The probabilities for distributions of the orders among the products are also assigned by the user. If orders arrive with equal probabilities for all products we refer to the system as one with a uniform demand pattern. If probabilities for products are significantly different, the system is referred to as having nonuniform demand pattern. Two demand patterns; one with equal probability for each product and one with probabilities varying between 0.1 to 0.4 were considered.

6.3 Batch Sizes

Orders for parts can come individually or in batches. For the experiments in this study two batch sizes of 6 and 9 were tested.

6.4 Required Service Level

Two levels of 90% and 95% tolerable service levels were considered in conducting the experiments.

For each case, the inventory levels were determined using all the rules. The resulting lead time for each case was used as the measure of performance for comparison purposes. For each case, an analysis of variance was performed and using Fisher's Least Significant Difference (LSD) the significantly better (at 10% level) rule for each case was determined. The summary of conclusions from these experiments is as follows:

1) Rule B, grouping nodes according to their flexibility, performed well in all cases where the tolerable service level was at the lower level.

2) Rule B performed equally well for relatively high service level requirements for uniform demand patterns.

3) Rule B was also quite consistent when the network was highly flexible and demand patterns were uniform.

4) For high service level requirements and non-uniform demand patterns, Rule AB, grouping according to a combination of stage weight and flexibility, performed more consistently.

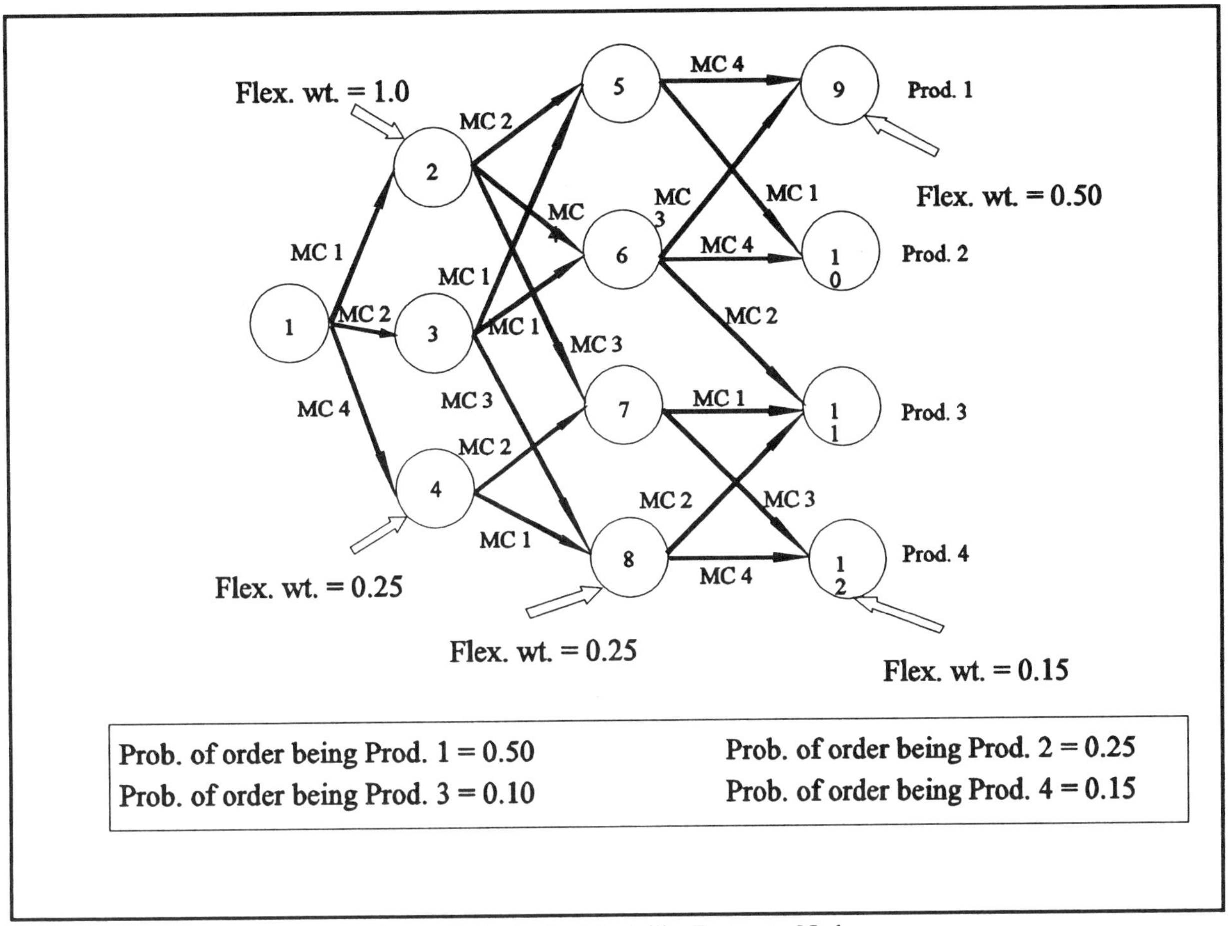

Figure 3: Assigning Flexibility Factors to Nodes

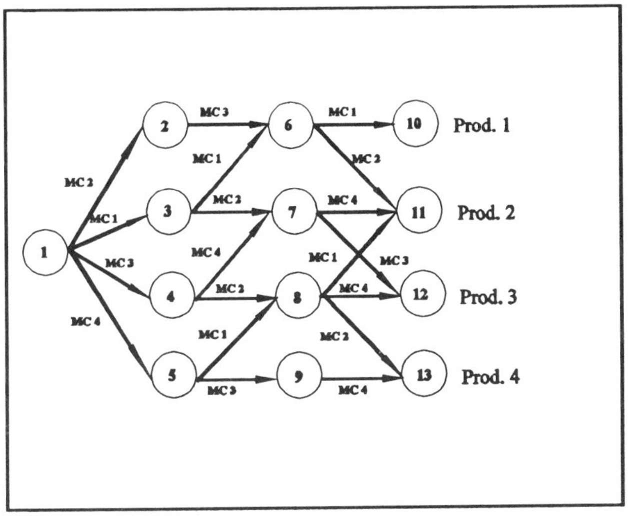

Figure 4: An Example of a High Flexibility Network

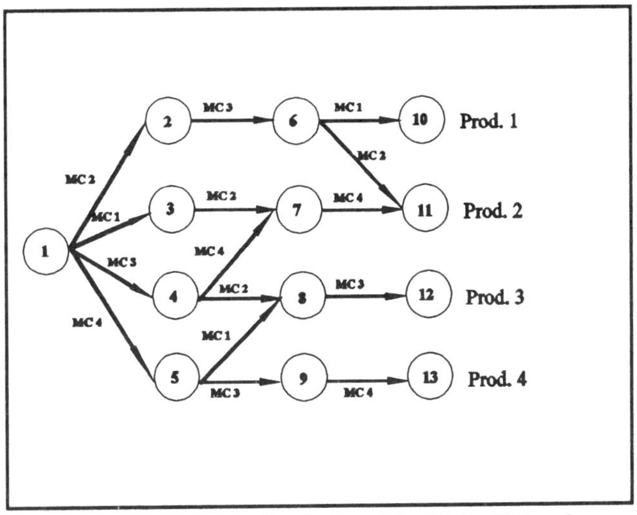

Figure 5: An Example of a Low Flexibility Network

5) Rule A, grouping the nodes according to their stages, never dominated the other two rules. Thus an argument could be made against using this rule by itself in any of the cases.

7 IMPLEMENTATION

A generic software has been developed that implements the procedure described in this paper automatically. Users provide values for the parameters of the system through an interactive interface. These include the number of final product types; the number and characteristics of each operation; possible routing for each product; resources required for processing each product; distribution of operation times; and the demand patterns for final products. A graphic interface has been developed through which these input parameters can be supplied by constructing the network and assigning resources and processing times to the arcs. The program automatically builds the simulation model, interfaces it with the optimization routine and determines the results based on the number of groupings specified by the user. Using this program the user can try various grouping criteria to choose the best for a given system.

8 CONCLUSIONS

This paper presented an application of simulation-optimization for strategically locating semi-finished products in a manufacturing system where several products are manufactured using the same raw material through several sets of manufacturing operations. The results showed that it is possible to service demand more efficiently and maintain reasonable levels of in-process inventory if the semi-finished products are manufactured in advance and are located in appropriate stages of operation and in appropriate quantities. The experience with this model showed that computation efforts for conducting such optimization process for relatively large problems is sometimes prohibitive. However, it was discovered that it may not be necessary to determine a separate optimum level for each variable individually. Depending on the structure of the system, many nodes represent characteristics by which they can be classified into various categories and the reasonable inventory levels could be specified only for each class. This drastically reduces the optimization efforts and the management requirements for the implementation of the results for a solution that still may be reasonably good to implement.

ACKNOWLEDGEMENT

This study was funded by a grant from the Advanced Manufacturing Institute of Kansas State University.

REFERENCES

Azadivar, F., 1992, A Tutorial on Simulation Optimization, *Proceedings of 1992 Winter Simulation Conference*,198-203.

Azadivar, F. and Lee, Y. H., 1988, An algorithm for Optimization of Discrete-Variable Systems by Computer Simulation, *Mathematics and Computers in Simulation*, **30**, 331-345.

Azadivar, F. and Lee, Y. H., 1986, Optimum Number of Buffer Spaces in a Flexible Manufacturing System, *Proceeding of the Second Conference on Flexible Manufacturing Systems*, Ann Arbor, Michigan.

Bitran, G. R. and Chang, L., 1987, A Mathematical Programming Approach to a Deterministic Kanban System, *Management Science*, **33**, 427-441.

Box, M. J., 1965, A New Method of Constrained Optimization and a Comparison with Other Methods, *Computer Journal*, **8**, 42-52.

Garg, A. and Lee, H., 1994, A Divergent Production Control Model with Stage-Dependent Lead Times, *Proceedings of the 3rd Industrial Engineering Research Conference*, 88-93.

Ohta, H. and Miyazaki, S., 1988, The Optimal Operation Planning of Kanban to Minimize the Total Operation Cost, *International Journal of Production Research*, **26**, 1605-1611.

Raturi, A.S., Meredith, J.R., McCutcheon, D.M. and Camm. J.D., 1990, Coping with Build to Forecast Environment, *Journal of Operations Management*, **9**, 2, 230-249.

So, K.C. and Pinault, S.C., 1988, Allocating Buffer Storage in a Pull System, *International Journal of Production Research*, **26**, 1959-1980.

Tang, C.S., 1990, The Impact of Uncertainty on a Production Line, *Management Science*, **36**, 12, 1518-1531.

AUTHOR BIOGRAPHIES

FARHAD AZADIVAR is a professor in the Department of Industrial and Manufacturing Systems Engineering at Kansas State University. He is also the director of the Advanced Manufacturing Institute, a Center of Excellence in Manufacturing research. His research interests are in modeling and optimization of manufacturing systems.

JUNHONG SHU is a Ph.D. candidate in the Department of Industrial and Manufacturing Systems Engineering at Kansas State University. His work is on optimization of maintenance policies in just-in-time manufacturing.

MOYEEN AHMAD is a graduate of Industrial and Manufacturing Systems Engineering from Kansas State University.

Proceedings of the 1996 Winter Simulation Conference
ed. J. M. Charnes, D. J. Morrice, D. T. Brunner, and J. J. Swain

THE USE OF SIMULATION IN THE OPTIMIZATION OF A CELLULAR MANUFACTURING SYSTEM

Stacy M. Sammons
Jeffery K. Cochran

Department of Industrial and Management Systems Engineering
Arizona State University
Tempe, AZ 85287-5906, U.S.A.

ABSTRACT

In the evaluation of a cellular manufacturing system, multiple factors should be considered, including the flow of the parts and the assignment of the workers. Simulation techniques can be effectively used in order to understand and optimize the behavior of the system. In this paper, a work cell is evaluated by means of discrete simulation using SLAM II. An optimization method is presented to determine the best alternative among proposed changes to the system. The structure of the approach allows the optimization to be performed with direction given by management. The results reveal that priority ranking of the parts, automated testing, and distributed job functions implemented together would produce the most significant improvement in the utilization of the team members.

1 INTRODUCTION

Simulation can be utilized in the manufacturing environment to predict system behavior or to conduct experiments and anticipate changes in the system before any adjustments are actually made. The model represents the system and allows for relatively easy modification and analysis. Objectives of the use of simulation to optimize a manufacturing system may include:

- Maximization of the utilization of the machinery and the workers.
- Minimization of the work-in-process in the system.
- Maximization of the production throughput of the system.

During the simulation process, many potential systems are often examined. It is beneficial to obtain the best solution in the fewest number of runs or calculations. When many design changes are being considered, it may not be feasible to run every possible combination, due to time or cost constraints. Therefore, an algorithm is needed to streamline the analysis.

Extensive research has been done on the role of simulation in cellular manufacturing. It can be used to ease the conversion from a traditional, functional layout to a work cell (Baran 1991). In addition, simulation can be used after implementation as a decision support tool to illustrate the effect of decisions on the system (Rahimifard and Newman 1995). The discussion has been primarily limited to analysis of the manufacturing system, rather than optimization. An optimum-seeking method is often used with simulation in order to determine the final preferred system from those under evaluation (Mollaghasemi and Evans 1994).

This paper describes how the flexibility offered by simulation can be utilized both to obtain savings during the decision process, and to involve management throughout the procedure of optimization. This approach is applied to a cellular manufacturing environment.

2 MODEL DEVELOPMENT

Hamilton Standard, a subsidiary of United Technologies Corporation, provides aircraft products for commercial and military applications. During the early 1990s, the company decided to pursue focused facilities with the intention of implementing continuous flow manufacturing techniques in a self-managed work team environment. The techniques implemented include the use of a cellular layout, point of use storage, and a visual pull system.

The assembly system features high performance "work teams" which are responsible for the product from the procurement of material through the shipping of completed components to the customer. The facility utilizes six work cell teams, each specializing in a different product line. A work cell is created by drawing people and equipment from their functional areas, and placing them in the same work area to reduce movement distances and allow the product to flow from one bench to another. Figure 1 shows a chart of the layout and the product flow for the work cell.

The employees at Hamilton Standard are cross-functional, capable of performing any of the functions within the team. Each team consists of six associates: one associate to purchase the materials, one associate to schedule production, one associate to receive parts, two associates to assemble and test the products, and one associate to ship the product to the customer. The team members currently work for six hours per day in their primary function, while the other two hours are used for administrative functions, such as accounting, filing paperwork, training, and attending meetings.

This paper focuses on the Outflow Safety Valve Cell Team which assembles three types of valves: a safety valve, an outflow valve, and a pressure regulating valve. The objective is to analyze the system and to optimize the utilization of the two team members who are involved in the assembly process, allowing them to accomplish other functions within the facility. The assembly of the valves can be classified as a small series production, due to the level of demand. Therefore, the focus is not on the throughput of the system. Rather, it is on the improvement of the use of work teams in the manufacturing environment.

A schematic of the system is shown in Figure 2. The entities of the system are the valves. They enter the system after an order has been placed, the necessary parts have arrived, and the production has been scheduled. The valves must wait on a rack for a team member to become available. Then, they move from bench to bench during the assembly process. There are five benches and a testing chamber used during the process. The same team member works on the valve from start to finish. If the valve does not pass testing, rework must be done before it is tested again. Finally, the valve leaves the cell to be shipped to the customer.

Data was collected on the system to specify input parameters and probability distributions. In addition, data was collected on the performance of the system to aid in validating the model. The team maintained detailed and up-to-date information about the operation of their work cell. By working with employees of the company familiar with the overall system, it was determined that the recorded values were representative of the way the cell was currently operating.

The daily starts and exits for the cell were obtained for six months, along with the work-in-process. Histograms were used to visually ascertain the apparent theoretical distribution of the data. The Kolmogorov-Smirnov (K-S) test was used to determine whether the actual distribution was significantly different from the apparent theoretical distribution that was assumed (Friedman and Friedman 1985).

The number of daily starts for the valves was compared to an exponential distribution. For a six hour work day, a new valve enters the system with the time between creations being exponential with a mean of 324 minutes. Of the valves that are ordered, approximately 40% are safety valves, 25% are outflow valves, and 35% are pressure relief valves. The safety valve had the shortest cycle time, and the pressure relief valve

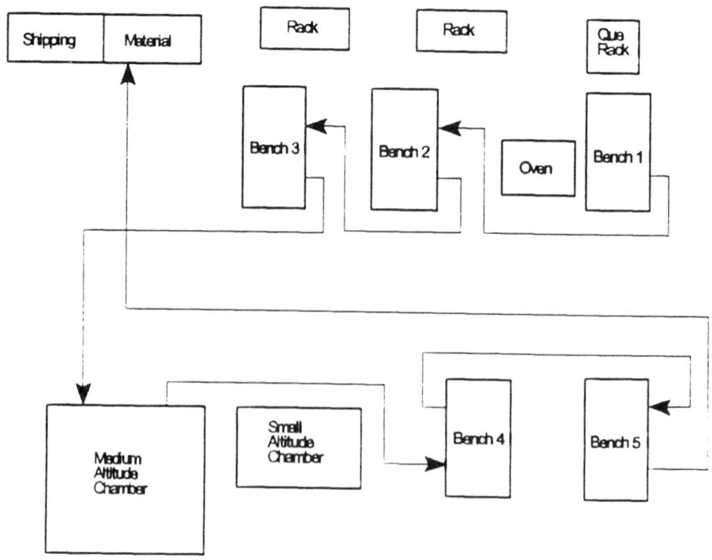

Figure 1: Product Flow for the Work Cell

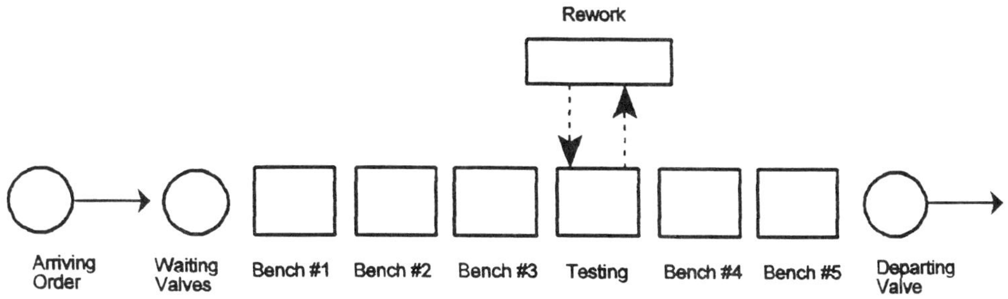

Figure 2: Schematic Diagram of Assembly and Testing Stations

measured the longest cycle time. It was necessary to determine the percentage of valves that required additional work because they did not pass the testing stage. The first time test yields demonstrated that approximately 92.5% of the valves passed testing the first time. When rework was required, the time was triangularly distributed with a minimum of 30, a mode of 50, and a maximum of 120 minutes.

3 ANALYSIS STRATEGY

A network model for the system was drawn as preparation for computer processing in SLAM II, Simulation Language for Alternative Modeling (Pritsker 1995). After entities are created, they are routed through probabilistic branching to specify the type of valve and the duration of the assembly operations. They wait for one of two team members to become available. Resources are used to model the team members, with the entities waiting in a file. The assembly benches are represented by activities in series. Statistics are collected on the cycle time for each of the valve types, and the entities are terminated. Statistics are also collected during the simulation on a time-persistent variable, which represents the number of entities in the system.

The main benefit of the network diagram is the ease with which it can be directly translated to SLAM code. Through the use of symbols, it allows the modeler to visualize the system and the flow of the entities. In addition, it provides a detailed picture of the organization of the model, which includes branching, the assignment of attributes to specific entities, and the collection of variables used for analysis.

It was necessary to establish that the computer program was executing as intended through model verification. A trace was performed to aid in the verification. This provides further detail about a specific time in the simulation. Additionally, a manual process of reviewing data inputs and outputs was done, ensuring that no significant discrepancies existed between expected and observed performance.

The process of establishing that a desired accuracy existed between the simulation model and the real system involved a comparison of the model output with the data that was obtained for the assembly system. Validation shows that the system can be represented by the model that was developed. The cycle times for the valves, the number of valves that were assembled in a given time period, and the average number of valves that were in the cell showed that the model represented the process in the work cell at Hamilton Standard. The organization of the model is directly comparable with the system's structure, and the model only has enough complexity to portray the important characteristics of the real system.

The process of establishing the experimental conditions for using the model consisted of developing an efficient experimental design to explain the relationship between the simulation response and the controllable variables and to make each simulation so that the most information could be obtained from the data. The utilization of the resources was affected by the work hours of the team members and the time between arrivals of the valves, among many other factors.

The initial conditions were established by using a pilot run to estimate a state which was representative of the long-term behavior of the system. Several runs were needed in order to accurately assess the system, due to normal variation. The use of a few long runs as opposed to many short runs generally produces a better estimate of the steady state mean, minimizing bias. Therefore, the simulation was run for six months with six hour work days.

4 APPLICATION OF THE MODEL

The simulation model was executed with the determined conditions to obtain output values for the system. The average cycle times for the three valves were the following: 257 minutes for the safety valve, 475 for the outflow valve, and 612 minutes for the pressure relief valve. Additional measures of performance included the total number of observations, the number of entities in the system, the time spent waiting for a resource, and the utilization of the resources. The average values from the results of the simulation runs are summarized in Table 1 for the original system.

An average of 142 valves were assembled in six months with a waiting time of 269 minutes. An average of 2.3 valves were in the work cell, either located at a bench or waiting for a resource. The utilization was approximately 72% for each team member. A value above 85% is usually considered high when variability is involved in the system.

Table 1: "As Is" Performance Measures

	Average Value
Throughput (parts / six months)	142
Work-In-Process (parts)	2.3
Waiting Time (minutes)	269
Utilization (% for team members)	143

The model was used to predict performance of the system. Design changes were made to the model, within realistic constraints, in order to attempt to improve the utilization of the team members. The first change was to establish a priority ranking in the file, rather than relying on the default, First-In First-Out (FIFO) ranking. A Low-Value-First (LVF) ranking was used on the attribute representing the type of valve. This ensured that the safety valve, the valve with the lowest cycle time, was placed first in the file, followed by the outflow valve and then the pressure relief valve.

The next change referred to the testing procedure used at Hamilton Standard. They were considering the use of fully automated test equipment, run entirely by computer. The duration of the activity for testing was adjusted to eliminate the variance involved.

The final design change involved the reallocation of resources. Rather than having each team member follow every step of the assembly process, they were divided. The objective was to maintain relatively equal utilization for the two employees, while separating their job functions. The first choice was to have one team member assemble only at the first three benches, and the second team member to test and finish the assembly. However, this was not effective because the job duties were unequal and the system backed up waiting for the first team member.

The model was adjusted so that the first team member was dedicated to the first two benches. The second team member was responsible for the remaining assembly and testing functions. This system was successful in providing an equal amount of work for the two employees. It has the advantage over the original distributed system of allowing the associates to concentrate on only a part of the assembly process, so that their experience level may improve at a faster rate.

Optimization can be accomplished with different methods. The most obvious method is enumeration, which consists of running every possible combination of the changes to the system. There are M^N combinations, where N is the number of modifications and M is the number of alternatives considered within each modification. In this case, there are eight combinations. The results for the utilization are shown in Figure 3 for each of the simulations. The utilization values are included with 95% confidence intervals representing the high and low values.

A more efficient method is shown in Figure 4, Figure 5, and Figure 6, which uses pattern search optimization. For three design changes, this method can be illustrated graphically by a cube. The front face represents the system with team members performing all of the assembly functions, while the back face represents the system with team members dedicated to specific functions.

The original system is represented by the box for FIFO priority, manual testing, and distributed functions. Starting with the original system, the adjacent boxes are simulated and compared to the original system. The best alternative is selected, and then the boxes adjacent to this new system are simulated. This is continued until the optimum is reached. The benefit is that there is only a need to run the model for alternatives that are next to each successive optimum.

Although this method only eliminated the need for one run in locating the optimum in this case; it would be more of an advantage for models with extensive design changes. Additionally, it allows management to be involved in each step. Management can direct the search and provide knowledge of other constraints that may affect the process. They can streamline the process and direct it to the areas that are most relevant. In this way, the pattern search optimization method takes advantage of past knowledge and experience.

5 CONCLUSIONS

A one-tailed hypothesis test was used to determine if the modifications made statistically significant improvements to the utilization. The three modifications of priority ranking in the file, priority ranking with automated testing, and priority ranking with automated testing and dedicated functions showed statistically significant improvements in lowering the utilization of the team members. The best value for the utilization

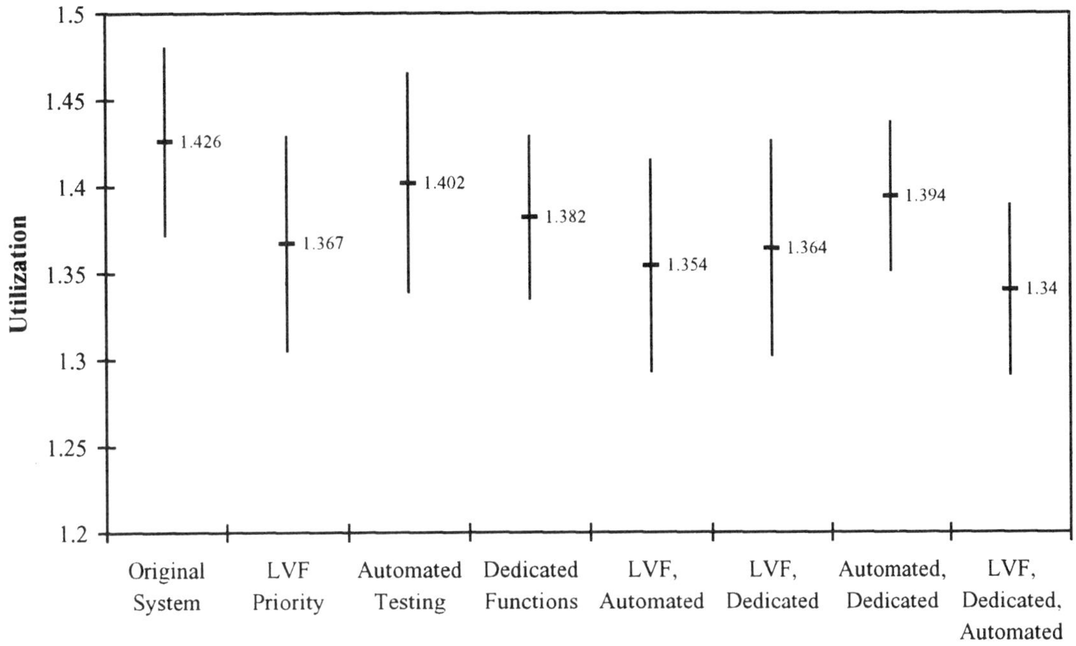

Figure 3: Effect of Design Changes on Utilization

	Distributed: Manual	Distributed: Automated	Dedicated: Manual	Dedicated: Automated
FIFO	**1.426**	1.402	1.382	
LVF	1.367			

Figure 4: Simulated Alternatives - Step 1

	Distributed: Manual	Distributed: Automated	Dedicated: Manual	Dedicated: Automated
FIFO	1.426			
LVF	**1.367**	1.354	1.364	

Figure 5: Simulated Alternatives - Step 2

	Distributed: Manual	Distributed: Automated	Dedicated: Manual	Dedicated: Automated
FIFO		1.402		
LVF	1.367	**1.354**		1.340

Figure 6: Simulated Alternatives - Step 3

occurred when all three modifications were made to the process.

Simulation was an effective tool for this analysis in anticipating changes in the work cell. It gave an evaluation of the system in terms of a performance measure, and allowed for easy modification and optimization of the system. The results for this process were as expected, and the pattern search optimization effectively demonstrated the significance of the design changes on the manufacturing cell.

Hamilton Standard is considering the establishment of additional facilities with work teams. The results of simulation are relevant to the process of learning how to effectively manage teams in a manufacturing environment.

REFERENCES

Baron, J. J. 1991. Tips on Tackling GT-Based Cells. *Manufacturing Engineering* 106:46-49.

Friedman, L. W. and Friedman, H. H. 1985. The Probability Distribution as a Performance Criterion When Comparing Alternative Systems. *Simulation*, 262-264.

Mollaghasemi, M. and Evans, G. W. 1994. Multicriteria Design of Manufacturing Systems Through Simulation Optimization. *IEEE Transactions on Systems, Man, and Cybernetics* 24:1407-1411.

Pritsker, A. 1995. *Introduction to Simulation and SLAMII*. John Wiley & Sons, New York.

Rahimifard, S. and Newman, S. T. 1995. The Role of Simulation in Operational Planning and Control of Flexible Machining Cells. In *Proceedings of the 1995 Winter Simulation Conference*, ed. C. Alexopoulos, K. Kang, W. R. Lilegdon, and D. Goldsman, 793-798.

AUTHOR BIOGRAPHIES

STACY M. SAMMONS is a graduate student in the Department of Industrial and Management Systems Engineering at Arizona State University. Her primary focus is in management of technology and information systems. In addition, she is attending the American Graduate School of International Management, and she received a B.S. degree in aerospace engineering from the University of Arizona in 1995. She is a licensed Engineer-In-Training in the state of Arizona.

JEFFERY K. COCHRAN is an associate professor of Industrial and Management Systems Engineering at Arizona State University, where he is also a director of the Systems Simulations Laboratory. Dr. Cochran was educated at Purdue University (BSE 73, MSNE 76, MSIE 82, PhD 84). His industrial experience includes positions at Battelle Northwest Laboratory, Los Alamos National Laboratory, and NASA's Laboratory for Applications of Remote Sensing. Professor Cochran has directed research projects for the National Science Foundation's I/UCRC for Manufacturing and Automation, the National Institute of Standards and Technology, and the Institute for Manufacturing and Automation Research. His primary research interests are simulation modeling and analysis, computer integrated manufacturing, operations research, and artificial intelligence. Dr. Cochran is the author of over fifty scholarly publications, and is a Professional Engineer.

Proceedings of the 1996 Winter Simulation Conference
ed. J. M. Charnes, D. J. Morrice, D. T. Brunner, and J. J. Swain

USE OF RESOURCE BLOCKS IN SLAM II FOR THE EVALUATION OF OPERATOR-MACHINE ASSIGNMENTS

Lynn S. Lachenmaier
Jeffery K. Cochran

Department of Industrial and Management Systems Engineering
Arizona State University
Tempe, AZ 85287-5906, U.S.A.

ABSTRACT

This paper examines the problem of assigning jobs to a set of manufacturing machine operators. This classic problem involves a set of semi-skilled personnel who can run subsets of machinery. The number of operators assigned to groups of machinery can be varied to optimize throughput. While the operator-machine problem has been studied in previous literature using a wide variety of analytical algorithms, we look at a solution that uses the RESOURCE block in SLAM II to simulate possible operator-machine combinations. A semiconductor fabrication facility was modeled using the operators as resources, and four alternate assignments were compared. Our concept for developing a simulation model of this problem in SLAM II proved to be an effective way to evaluate operator-machine assignments.

1 INTRODUCTION

A classic problem for manufacturing managers is the operator-machine assignment problem. This problem arises when there are more machines than operators, and each operator has been trained on only a subset of the available machines (Wittrock 1992). One operator can run more than one machine, and conversely, multiple operators can share the duties of running one machine or machine group. If an operator is assigned to too many machines, parts sit idle in one machine while the operator is busy with the others. This "machine interference" affects the overall system output (Stafford 1988).

In many facilities, there is no structured method of choosing the operator-machine assignment. The decision is typically left to the production supervisor and can vary from shift to shift within the same department. Decisions are made based on experience or informal rules of thumb (Jackman and Johnson 1993). This lack of structure can lead to inefficient combinations of operators and machines restricting the potential output of the facility.

This paper examines prior approaches to this problem and explores the possibility of comparing operator-machine assignments that have been determined acceptable to management as design questions in a SLAM II simulation model. The operators are modeled as resources in a SLAM II RESOURCE block. Four assignments from a semiconductor fabrication facility at Delco Electronics in Kokomo, IN are compared to illustrate the application of this concept.

2 BACKGROUND

Many approaches have been proposed to determine operator assignments that optimize system output. One method discussed in the literature is to create a mathematical representation of the system. Stafford (1988) shows how deterministic mathematical models of the operator-machine assignment problem can be solved both graphically and through the calculation of three relatively simple formulas. Using this method provides the manager with "K," the optimum number of machines to assign to any one operator. The formulas are easily computed, but assuming manufacturing processes to be deterministic limits the applications of this approach. Stafford's formulations also assume that all operators will be assigned the same number of machines.

More flexibility and realism are gained by modeling the system as probabilistic. Wittrock (1988) demonstrates the problem as a network flow problem. His more complex model deals with the concept that operators will have overlapping skill sets. The optimum number of jobs is not found per operator, but for teams of operators with identical skills. This approach finds an optimum number of operators to assign to a job, but this technique requires complex mathematical algorithms to maintain the robustness of its solutions. Queueing networks are popular in manufacturing applications, but do tend to have difficulty representing systems that involve merging parts together in sub-assemblies, such as the bonding of wafers in multi-stack semiconductor devices (Jackman and Johnson 1993).

Lutz et al. (1994) use a decision support system (DSS) to generate operator assignment schedules by solving a dynamic programming heuristic. However, this method relies on the assumption of a balanced production line (where the average processing times on all machines are the same).

3 MODELING THE OPERATOR-ASSIGNMENT PROBLEM WITH SLAM II

Many of the assumptions that are required for the mathematical models are not necessary for a simulation model of a manufacturing system. Operators, or servers, do not have to be assigned equal numbers of jobs. All machines can have probabilistic processing times with totally different underlying distributions. Merging sub-assemblies is easily done.

The operator-machine assignment problem can be tackled in the simulation language SLAM II by making use of the RESOURCE block and the AWAIT node (Pritsker 1995). Just as a machine can be a resource for which parts wait, the operators can also be coded as resources for waiting parts. Machine interference, then, is represented by parts sitting in the AWAIT node after a machine has become available while all the operators assigned to that machine are utilized elsewhere. Different design alternatives can then be tested by altering the operator RESOURCE block and changing the resource label (RLBL) specified in the AWAIT and FREE nodes.

This method does not aim to optimize the combination of operators and machines, but instead allows management to compare the relative effectiveness of assignments known to be acceptable under other constraints, such as cross-training, union rules, or operator limitations. The simulation model acts a source of feedback to the manager who wants to know "which of these several assignment combinations is best," without disrupting his personnel by trying them all in the real factory.

While this paper focuses on the methods used to represent operator assignment in SLAM II, other simulation packages also allow the user to specify which machines an operator can attend. MANSIM, for example, which does not require the level of programming of SLAM II, allows the user to input "operator skills" in a Windows menu format. The same goal of creating machine interference is accomplished.

3.1 System Description

This approach was tested on a real operator-machine assignment scenario at Delco Electronics. A fairly complex model was designed that captured the essential functions of an acid etch room in a semiconductor fab. The model demonstrated the processing of five part types through seventeen major operations. Two part types were sub-assemblies that are bonded to two of the other three parts. The seventeen operations consisted of twenty-eight individual machines. Eighteen operators were available to run jobs.

The operators were modeled in the RESOURCE block by listing the AWAIT nodes associated with a coded group of one or more operators (see Table 1). The groups were identified as "OPERA, OPERB," etc. For instance, to show that two operators (grouped as OPERA) would be assigned to jobs consisting of machines 1, 4 and 8, the RESOURCE code would look like "RESOURCE/OPERA(2),1,4,8."

Table 1: A SLAM II RESOURCE Block that Defines the Operator-Machine Assignments for One Possible Arrangement

OPERA	1	2	11			
OPERB	2	5	66	23	55	7
OPERC	1	9	17	59	63	
OPERD	1	25				
OPERE	1	53				
OPERF	1	51	57	61	15	
OPERG	1	19				
OPERH	1	33	43			
OPERI	1	35	45			
OPERJ	1	37	29			
OPERK	3	39	47			
OPERL	1	27	65			
OPERM	1	21				
OPERN	2	49				

In this way, any number of operators can be associated with any number of specific machines. Sharing of jobs is allowed; multiple jobs per operator are allowed. Operators can have different skill sets (the machines to which they could be assigned are limited), or they could all be identical and assignable to any machine.

The entities in the SLAM II model are lots of wafers that need processing through the seventeen workstations of the acid etch room. The product flows are modeled very simply by a series of AWAIT and FREE nodes (see Figure 1). When a part arrives at a workstation, it waits first for a machine to be available, then it waits for an operator from the group assigned to that job to become available. When the operator finishes her setup and measurements, she is freed, and in some cases, the machine runs for some time by itself. Machines that do not run without the full attention of the operator are modeled with a zero duration activity between the operator free and the machine free. Finally the machine is freed and the part proceeds to the next station. This pattern is repeated thirty times in the model to represent the flow of product through the entire system.

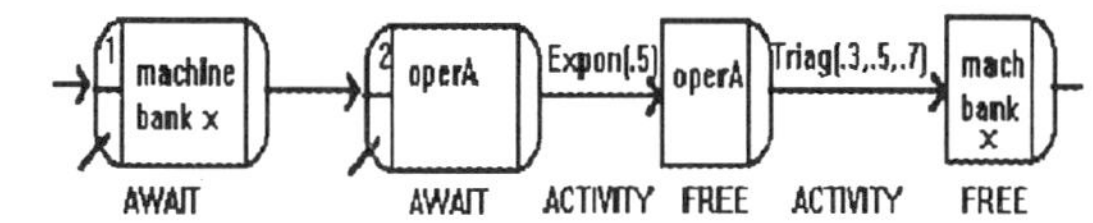

Figure 1: Illustration of a Portion of the Network Diagram Where a Part Arrives to a Given Workstation

Durations for operator time and machining time were derived from data collected by engineering. Raw data was used to fit probability distributions. In this system, different machines have different levels of variability and were described by totally different distributions. The values used are shown in Tables 2. From these tables, you can see that highly unique systems can be modeled with SLAM II.

Table 2: Probability Distributions Used as Machine Times (time a machine runs without the aid of an operator) and Operator Attention Times

MACH	DISTRIBUTION	OPER	DISTRIBUTION
AUTO	TRIAG(.3,.5,.7)	AUTO	EXPON(.471)
BAKE	TRIAG(.167,.5,1)	BAKE	EXPON(.127)
OXIDE	TRIAG(.075,.218,.433)	OXIDE	EXPON(.213)
STRIP	TRIAG(1.083, 1.083, 1.583)	STRIP	EXPON(.122)
METAL	EXPON(.042)	METAL	EXPON(.531)
CD	0	CD	EXPON(.5)
CONT	0	CONT	EXPON(.217)
PRINT	UNFRM(1.183,1.267)	PRINT	EXPON(.904)
PUNCH	0	PUNCH	EXPON(4.483)
PAD	EXPON(.1)	PAD	EXPON(.151)
BOND	UNFRM(.6, 1.033)	BOND	EXPON(2.392)
IR	0	IR	TRIAG(.217, 1.1, 13.383)
VIKING1	0	VIKING1	UNFRM(4.2, 7.6)
VIKING2	UNFRM(.833,1.234)	VIKING2	EXPON(1.628)
ECE	EXPON(6.501)	ECE	EXPON(2.939)
GLASS	UNFRM(.934,1.167)	GLASS	EXPON(.235)
CAV	UNFRM(2.723,3.133)	CAV	EXPON(.342)
KOH	TRIAG(1.05,1.991,3.717)	KOH	EXPON(.348)
OUT	RNORM(7.3, 2.8)		

The basic model was validated by conducting a Kolmogorov-Smirnov (K-S) test on the data used and by examining the initial summary reports generated by the system. Idle times of machinery and operators and system output were compared to historical data. While some pieces of the model were more true to life than others, the results maintained the general behavior of the acid etch room.

3.2 Design Alternatives

To examine the SLAM II model's use as a method of evaluating operator-machine assignments, four methods of operator assignment were tested. With eighteen operators and twenty-eight machines, more than 4.099 x 10^{31} (18!(18!-8!)) combinations are possible, even if we ignore job sharing, and restrict the maximum number of machines run by a given operator to two! However, there is no reason for us to examine more than a minuscule fraction of the possible assignments. In a real facility, a manager would probably want to compare his current practices to a handful of hypothetical alternatives suggested to him by his staff that he knows would satisfy his other constraints.

We reproduced this scenario by comparing four assignments. The two assignments used currently by first shift and second shift were compared, as well as two hypothetical arrangements. The first hypothetical assignment is based on raw labor point data; the other assumes full cross-training of the operators. The resource blocks representing each assignment is shown in Table 3.

Table 3: RESOURCE Blocks that Define the AWAIT Nodes (jobs) Assigned to Each Group of Operators

DesignAlt1

OPERA	1	2	11			
OPERB	2	5	66	23	55	7
OPERC	1	9	17	59	63	
OPERD	1	25				
OPERE	1	53				
OPERF	1	51	57	61	15	
OPERG	1	19				
OPERH	1	33	43			
OPERI	1	35	45			
OPERJ	1	37	29			
OPERK	3	39	47			
OPERL	1	27	65			
OPERM	1	21				
OPERN	2	49				

DesignAlt2

OPERA	2	2	11				
OPERB	1	5	66	23	55	7	
OPERC	1	9	17	59	63		
OPERD	1	25					
OPERE	3	53	51	57	61	27	65
OPERF	1	19					
OPERG	1	33	43				
OPERH	2	35	45				
OPERI	1	37	29				
OPERJ	3	39	47				
OPERK	1	21	15				
OPERL	1	49					

DesignAlt3

OPERA	1	2	11			
OPERB	2	5	66	23	55	7
OPERC	3	9	17	59	63	
OPERD	1	25				
OPERE	1	53				
OPERF	1	51	57	61	19	21
OPERG	1	15	27	65		
OPERH	1	33	43			
OPERI	1	35	45			
OPERJ	2	37	29			
OPERK	3	39	47			
OPERL	1	49				

DesignAlt4

OPERA	18	2	11	5	66	23
		55	7	9	17	59
		63	25	53	51	57
		61	19	21	15	27
		65	33	43	35	45
		37	29	39	47	49

Each alternative was measured by the number of “ships”, or total number of lots shipped to the customer from the system over a two week period. Our goal, then was to find the alternative that shipped the most lots. All four designs were compared at their steady state, as data from the warm-up period of the system was truncated and disregarded. Pilot runs of the original model were used to determine the proper number of runs to make of each alternative to compare them at a 90% confidence level within +/- 2 shipped units. Twenty runs were performed for each design alternative.

The programming time required to test a new arrangement after the base model was established was minimal. From our base model containing 342 lines of code, a new operator arrangement could be created in less than 30 minutes. The models ran in seconds.

4 ANALYSIS OF RESULTS

Data was collected from the runs of each design alternative. The maximum number shipped by the system in any two weeks was 39 units and the minimum was 0. The data for each operator assignment alternative is summarized below in Table 4.

Table 4: Number of Ships Generated by Each Design Alternative over Twenty Runs

	DesignAlt1	DesignAlt2	DesignAlt3	DesignAlt4
	18	10	33	16
	15	14	13	14
n	23	6	16	3
u	16	3	19	14
m	19	9	19	18
b	20	29	15	21
e	2	0	13	16
r	11	14	20	16
	21	9	17	17
o	20	15	18	19
f	16	1	25	18
	21	15	10	12
s	16	22	13	20
h	17	0	13	18
i	18	4	35	20
p	12	8	39	10
s	23	26	18	23
	13	15	21	22
	20	8	19	16
	21	27	28	18
Average	17.1	11.8	20.2	16.6
Std Dev	4.9	8.8	8.0	4.5

As you can see, the four design alternatives shipped different numbers of parts. Operator assignment method 3 generated the highest number of ships in a two week period on the average, and therefore could be the best alternative of our four tested. However, since our model is probabilistic and the values vary from run to run, we checked to see if the differences in output between the design alternatives were statistically significant. This was done by analyzing the experiment with Design-Ease software. Design-Ease created the graph shown in Figure 2 to demonstrate the variance in our results.

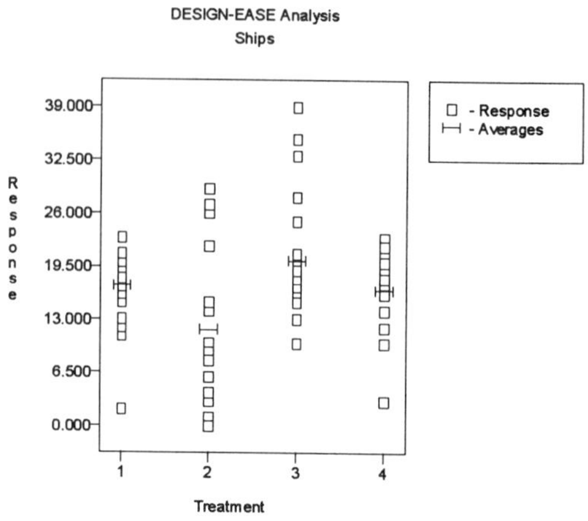

Figure 2: Interpretative Graph from Design-Ease Showing Means and Ranges of the Four Alternatives

An Analysis of Variance (ANOVA) was conducted to determine whether or not the four means were different. The null hypothesis tested was that the four mean ships were equal. The results of the ANOVA are shown in Table 5.

Table 5: ANOVA Results Adapted from Design-Ease

Analysis of Ships

SOURCE	SUM OF SQS	DF	MEAN SQ	F VALUE	PROB > F
MODEL	731.500	3	243.83	5.23	0.0025
RESIDUAL	3541.700	76	46.60		
COR TOT	4273.200	79			

Treatment	MEAN DIFF	DF	STD ERROR	t FOR H0 COEFF=0	PROB > |t|
1 vs 2	5.35	1	2.159	2.478	0.0154
1 vs 3	-3.10	1	2.159	-1.436	0.1551
1 vs 4	0.55	1	2.159	0.255	0.7996
2 vs 3	-8.45	1	2.159	-3.914	0.0002
2 vs 4	-4.80	1	2.159	-2.224	0.0292
3 vs 4	3.65	1	2.159	1.691	0.0950

The ANOVA table gives us an F value of 5.23, which is significant at a (1 - .0025) or 99.75% level of confidence. So this leads us to reject the null hypothesis that the four mean ships are the same. However, what we really want to know is which means are different, and which one is best. This can be figured by examining the last rows of the ANOVA sheet where a Least Significant

Difference test is performed. At a 90% confidence level ($\alpha = .1$), we can draw the conclusions outlined below:

- Mean ships for Design 1 are statistically higher than Design 2, but not significantly different than Designs 3 and 4.
- Mean ships for Design 2 are also statistically lower than Designs 3 and 4.
- Mean ships for Design 3 are slightly higher than Design 4 (90.5% confidence).

Therefore, we conclude with 90% confidence than Design Alternative 2 produces the least amount of ships, and is the worst of the four alternatives. Design Alternative 3 is probably the best strategy of job assignment, but more testing would have to be done to distinguish it from Design Alternative 1.

The results provided by the SLAM II simulation can be judged for reasonableness from a qualitative standpoint as well. It makes sense that design alternative 3 was the most efficient of the alternatives, since it was based on carefully collected labor point data. This data is calculated by the industrial engineer and tells the manager how many hours a shift each machine must be run in order to produce enough product to meet schedules.

Design alternative 2, on the other hand, was created in a less mathematical way. It is the current arrangement used in the area during the first shift. It makes sense that it is least efficient in terms of output, since it was constructed to balance out the effects of high absenteeism, feuding operators, and other unique managerial factors, rather than maximizing output alone.

One surprise to the researchers was the mediocre performance of design alternative 4. In this arrangement, all operators were considered to be fully cross-trained, and could rush to any job that needed an operator. Before performing the simulation runs, we expected this alternative to perform the best. However, since there are more machines than operators, machine interference still occurs. In the model, no particular priority was given to bottleneck operations. In the real fab, we would rather an operator wait around idle so he could immediately service a bottleneck machine when necessary. This is better than attending to a non-bottleneck machine and leaving the bottleneck machine idle for any period of time. Design alternative 3 recognizes bottlenecks and deals with them more efficiently.

5 FUTURE WORK

The work done to date with the SLAM II model has encouraged the management of the semiconductor facility to expand the model of the entire fab to represent operator-machine assignments. A model from the package MANSIM is being revised to represent operator allocation in the photolithography room, the acid etch room, and thin film deposition areas. Initial work shows that including current operator assignments results in a much more accurate simulation of the day-to-day operations of the facility. The next step is to test possible improved arrangements.

Initial attempts have also been made to test analytical queueing network models of similar systems. These were created and executed much more quickly than the simulation models, but generated less accurate results.

The usefulness of simulation in evaluating operator-machine assignments could also be greatly increased by combining it with mathematical optimization techniques. Integer programming models could be developed to generate test alternatives.

6 CONCLUSION

While many approaches exist for solving the operator-machine assignment problem, use of simulation allows the greatest flexibility for unique manufacturing environments. By making use of the RESOURCE block provided in SLAM II, many combinations of operators and machines can be tested to determine a suitable arrangement. This method is not limited to our test model of a rather complex acid etch facility, but could be used in a variety of manufacturing environments. Any facility that has more machines than operators could represent the operators as resources rather than servers. Test alternatives can be quickly generated and analyzed.

The general method of assigning operators to groups of machinery can also be applied in newer dynamic simulation languages. Windows based packages can be altered and tested even more quickly than SLAM II models.

Because this approach is not a mathematical algorithm or expert system, it is limited in the number of alternatives that could be realistically compared. A manager would not want to generate every possible operator-workstation combination. However, this restriction is not critical in a real manufacturing environment. While the number of *possible* assignment combinations may be large, the number of *feasible* combinations is much more manageable. A manager can use simulation to test a small number of arrangements he knows to be feasible with respect to the constraints of operator cross-training, aptitudes, and relative ability of the operators to work together. As we found alternatives that did not perform as expected, the manager may also be surprised to find out which alternative is most productive. By using a model to test expectations, the manager could save money by not over-training or over-hiring.

Implementation of this method would provide facilities with a way to evaluate the job assignments they have relied on in the past, and could encourage them to develop a standard arrangement across all shifts. A simulation model that included only the most important elements of the real system could be developed fairly quickly using machine processing data and work measurement standards for the operating personnel. Manufacturing plants with existing simulation models could adapt them to study operator-machine assignments.

Further work could be done to incorporate this evaluative tool with a mathematical method that would truly optimize the operator-machine assignment in relation to the constraints understood by management.

ACKNOWLEDGMENTS

The authors wish to thank Tyrone Edens, Dick Weddell, Bill Bradley, Fred Blessinger and Scott Schuster of Delco Electronics for their support and assistance in data collection. We also appreciate the efforts of the anonymous appraisers whose comments led to significant improvements in the final paper.

REFERENCES

Jackman, John and Eric Johnson. 1993. The Role of Queueing Network Models in Performance Evaluation of Manufacturing Systems. *Journal of the Operational Research Society*. 44(8): 797 - 807.

Lutz, C.M, K.R. Davis, and C.F. Turner III. 1994. Development of Operator Assignment Schedules: A DSS Approach. *Omega, International Journal of Management Science*. 22(2): 57 - 67.

Pritsker, A. Alan B. 1995. *Introduction to Simulation and SLAM II*.

Stafford, Jr. Edward F. 1988. An Optimal Solution Technique for the Operator-Machine Assignment Problem. *Production and Inventory Management Journal*. 29(3): 25 - 29.

Wittrock, Robert J. 1992. Notes: Operator Assignment and the Parametric Preflow Algorithm. *Management Science*. 38(9): 1354 - 1359.

AUTHOR BIOGRAPHIES

Lynn S. Lachenmaier is a graduate research assistant at Arizona State University, where she is working toward her MSE degree in industrial engineering. Her graduate work is being funded by Delco Electronics, Kokomo, IN, for whom she has worked one year as a manufacturing supervisor. She graduated from Purdue University in 1994 (BSIM).

Jeffery K. Cochran is Associate Professor of Industrial & Management Systems Engineering at Arizona State University, where he is also a Director of the Systems Simulations Laboratory. Dr. Cochran was educated at Purdue University (BSE 73, MSNE 76, MSIE 82, PhD 84). His industrial experience includes positions at Battelle Northwest Laboratory, Los Alamos National Laboratory, and NASA's Laboratory for Applications of Remote Sensing. Professor Cochran has directed research projects for the National Science Foundation's I/UCRC for Manufacturing and Automation, the National Institute of Standards and Technology, and the Institute for Manufacturing and Automation Research. His primary research interests are simulation modeling and analysis, computer integrated manufacturing, operations research, and artificial intelligence. Dr. Cochran is the author of over fifty scholarly publications, and is a Professional Engineer.

Proceedings of the 1996 Winter Simulation Conference
ed. J. M. Charnes, D. J. Morrice, D. T. Brunner, and J. J. Swain

EFFICIENT MODULE-BASED MODELING FOR A LARGE-SCALE AS/RS-AGV SYSTEM

Soemon Takakuwa

School of Economics
Nagoya University
Furo-cho, Chikusa-ku, Nagoya-shi, Aichi, 464-01 JAPAN

ABSTRACT

A method of modeling large-scale AS/RS-AGV systems is proposed in an attempt to describe the AS/RS precisely and flexibly in which a lot of items are stored and retrieved. The system considered in this study consists of the large-scale AS/RS with stacker cranes, the looped-track AGV system, aisle conveyors connecting these two systems, and incoming and outgoing conveyors. There are two major stages to model large-scale AS/RS-AGV systems efficiently and to execute simulation experiments. The first stage is to generate a simulation program by placing appropriate modules and inputting the selected parameters such as the number of banks, the number of bays, the number of levels of AS/RS, and the number of AGVs. The second stage is to perform simulation experiments. In the proposed model, every item number puts on each rack, the corresponding number of cases for each item, the time of arriving in the warehouse, and the time of departing from the system are stored and recorded. Generally, it takes a lot of time to build a complex program for a particular type of large-scale AS/RS-AGV system. It is found that simulation/ animation models can be made quickly, by using a module-based modeling method developed in this study.

1 INTRODUCTION

The modern warehouse must play the role not only of storage for raw materials, parts, and end products, but also of a dynamic inventory control for a smooth logistic system, such as procurement, production, inventory, sales, and distribution, by establishing the information system to update kinds and quantities of stored items. Recently, the automated storage and retrieval system (AS/RS) has been utilized together with AGVs in the above-mentioned fields. Performance analysis of an AS/RS is a complex problem. Some approaches exist for performing such an investigation (Pulat 1988, Takakuwa 1989).

With regard to modeling AS/RS, the degree of preciseness of modeling depends on the purpose of analysis. There are some reports on applications of simulation to model an AS/RS (Harmonosky and Sadowski 1984, Medeiros, Enscore, and Smith 1986, Muller 1989, Gunal, Grajo, and Blanck 1993). In the previous studies, the AS/RS is modeled precisely and realistically to behave as a real system (Takakuwa 1994, Takakuwa 1995).

Generally, when introducing a large-scale AS/RS-AGV system, there are a lot of alternatives. Management should decide particular specifications of the system, such as the specifications of AS/RS, the number of AGVs, and the number of incoming/outgoing conveyors, considering the frequency of handling items. For analyzing and comparing possible alternatives of the system, it is necessary to model and perform simulation experiments of a number of models that describe the alternatives. Because it takes a lot of time (possibly a couple of months even for skilled programmers) to build a complex program for a particular type of large-scale AS/RS-AGV system, analyzing and comparing alternatives is a time-consuming task. In this paper, an efficient modeling system is proposed to build simulation programs for the AS/RS-(looped-track) AGVs system, and the flexibility of the proposed system is described together with some analytical results explored by the previous studies.

2 LARGE-SCALE AS/RS-AGV SYSTEM

2.1 The Storage System and the Conveyor Interface

A general view of the AS/RS is shown in Figure 1. Storage and retrievals can be performed to/from the racks on either side of an aisle. The storage system is linked to the rest of the system through a peripheral aisle-conveyor system. Each aisle of the system is interfaced with the peripheral conveyor system through picking and dropping stations, which are called

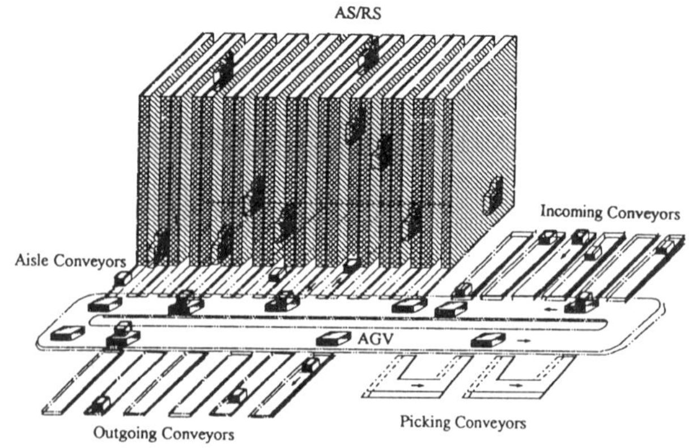

Figure 1: Layout of Large-Scale AS/RS-AGV System

the home position (HP), of a stacker crane at one end of the aisle.

Each aisle of the system is served by a stacker crane which can move in horizontal and vertical directions simultaneously. Once the stacker crane is positioned at a designated location for storage/retrieval to/from a certain rack, the consecutive operations of extending its shuttle into the rack, dropping/picking the item and pulling the shuttle back are performed. The similar movement applies to picking/dropping an item from/to the picking and dropping stations.

Every incoming item is first moved onto the home position at one end of an aisle, and then picked by the stacker crane, and finally moved to its destination rack. On the other hand, an outgoing item is picked from its rack in the system by the stacker crane, and is transferred to the home position. Once dropped on the home position, the item is moved onto the outgoing conveyor. Typically, these incoming/outgoing aisle conveyors are connected to the AGV system.

2.2 Control Logic of AS/RS

A detailed examination of the system requirements reveals several interesting issues that should be addressed for an accurate representation of the system. The control logic of the AS/RS is designated so that storage and retrievals are made to the rack in one of the four different ways shown in Figure 2.

The searching method imposes two related constraints on the simulation model. First of all, the model needs to know the required time to travel to a given rack from the home position or, in general, the required time to travel from a given position in the aisle to another. The travel time is given by the maximum

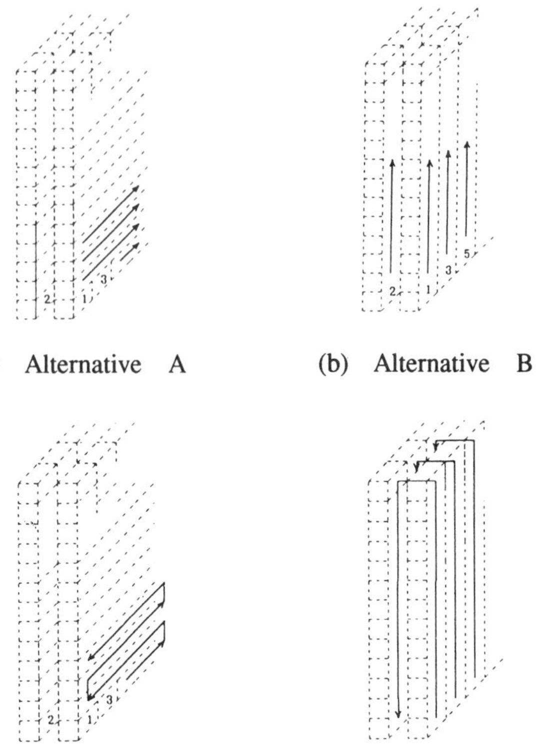

(a) Alternative A (b) Alternative B (c) Alternative C (d) Alternative D

Figure 2: Storing/Retrieving Policy

value of those needed by the movement of horizontal and vertical directions. Secondly, it should perform a search of the available racks to find the first available or open location. The AS/RS in question is required to examine more than thousands of racks in each operation for storage and/or retrievals. Thus, finding an efficient way of making such operations is a necessity for a simulation model that is of some practical value in making quick analysis of the system under varying conditions.

2.3 Movement of the AGV

Movement of AGVs in looped-track AGVs systems consists of several major procedures.

2.3.1 Requesting a Vehicle

When an item arrives at the position to be loaded and transferred by a vehicle, the item tries to request the nearest available vehicle. If there is no vehicle available at that time, it would try again when a vehicle become available.

2.3.2 Moving a Vehicle

There are three major types of AGV movements. The first type is on the assigned (or requested) vehicle going to its destination point. The second type is on the loading (i.e., carrying items) vehicle. The third type is on the unloading vehicle which is pushed ahead by the above-mentioned two types of vehicles.

2.3.3 Controlling Vehicles

The function of controlling the vehicles is performed in searching control points (stations) on the way to the destination control points. This function is also performed in checking whether loading/unloading vehicles are in the same path to the destination point. In case there are vehicles in the path, these vehicles are pushed ahead to the appropriate positions. Then the loading vehicle is moved to its destination.

2.4 Material Flow

Specifications of the system are summarized in Table 1. The system to be analyzed comprises three major subsystems, namely, the automated warehouse, the conveyance system and the handling system. There are two types of material flow, namely, incoming and outgoing, in the AS/RS system considered in this study. Each item enters at one of the arrival stations to the system. Then, the item will be conveyed to the incoming isle conveyor directly connected to the AS/RS for the designated high-rise rack lane, moving on one of AGVs. The AS/RS receives items solely from the arrivals stations.

After an incoming item reaches the warehouse gate, it must wait for a stacker crane to be transported to the assigned rack. Both incoming and outgoing items are handled by a stacker crane between the two high-rise rack lanes in the automated warehouse.

On the contrary, outgoing items leaving the warehouse are routed on a conveyance system toward an assigned departure station. After arriving at the station, the item leaves the system.

Table 1: Specifications of the System

Items	Parameters
Stacker Cranes:	
Horizontal direction	
Velocity	0.083 - 2.667(m/sec.)
Acceleration	0. 294(m/sec.^{2})
Vertical direction	
Velocity	0.083 - 1.050(m./sec.)
Acceleration	0.294(m./sec.^{2})
Loading/unloading time	13(sec.)
Incoming Conveyors:	
Velocity	1(m./sec.)
Length	5(m.)
Outgoing Conveyors:	
Velocity	1(m./sec.)
Length	5(m.)
Picking Conveyors:	
Velocity	1(m./sec.)
Length	10(m.)
AGVs:	
Velocity	0.500 - 1.667(m./sec.)
Acceleration	0. 490(m./sec.^{2})
Loading/unloading time	8(sec.)
Items to be handled:	
Number of kinds of items on each rack	1 - 5(kinds)
Number of cases put on each rack	1 - 20(cases)

3 MODULE-BASED MODELING SYSTEM

3.1 Efficient Modeling System for Large-Scale AS/RS-AGV System

The basic idea of the proposed procedure is based upon a flexible modeling system described in the previous paper (Takakuwa 1995). The system comprises ten major items, and each item has more detailed parameters to be set interactively. For example, the item of "AS/RS" has the parameters of the number of banks, levels and bays, travel time of stacker cranes, and the loading/unloading time of stacker cranes. By responding to the prompt for each parameter, appropriate figures may be inputted. After all required parameters are inputted, the corresponding simulation programs (the model and experiment programs) are generated automatically.

In the proposed procedure in this paper, on the other hand, there are two steps to execute simulation experiments. The first step is to generate simulation programs, by selecting and allocating the required modules of the Arena template for describing a particular AS/RS-AGV system. There are 23 modules for the particular type of the large-scale AS/RS-AGV system, as shown in Table 2. Although, in this study, the simulation is performed by SIMAN/Arena, the basic idea of this procedure can be applied to other simulation languages as well (Pegden et al. 1994).
The template system comprises the following major items;

(1) AS/RS.
(2) Inventory files.
(3) AGV system.

(4) Incoming conveyors.
(5) Outgoing conveyors.
(6) Picking conveyors.
(7) Aisle conveyors.

3.2 Building a Simulation Model

The movement system is modeled using the SIMAN/Arena simulation language constructs. In addition to the movement system, the external files are also used for generating requests for storage and retrievals, collecting various performance statistics, and monitoring the status of the system.

Requests for storage and retrievals are generated in the model based on the searching method described in Section 2.2. In addition to the typical model and experimental simulation program, five additional files are used for keeping information on storage and retrievals in the system. (A rack is identified by a particular combination of the bank number, the level number, and the bay number.)

Figure 3 shows an illustrative example of Arena template and a series of associated modules for the large-scale AS/RS-AGV systems. In Arena, simulation models are built by placing modules in a working area of the model window, providing data for these modules, and specifying the flow of entities through modules. A module defines the underlying logic that is applied when an entity is directed to the module, as well as the associated graphical animation, to depict the module's activities during a simulation run.

After a module has been placed in the model windows, its associated data may be edited. The module for defining the initial conditions inside AS/RS (named 'Zaiko Sakusei' in Table 2) is shown in Figure 4. Furthermore, each item has more detailed parameters to be set interactively. By responding to the prompt for each parameter, appropriate figures may be inputted. Another module for creating incoming/outgoing items (named 'Nyuko Shuko' in Table 2) is shown in Figure 5. This module is prepared for creating entities of incoming and outgoing items, and specifying case assignments.

To define the flow of entities among modules, either direct connections or station transfers may be used. A direct connection is formed by placing a connection from a module exit point to a module entry point. Entities that leave a module through an exit point transferred through the connection to the entry point with no time delay. A station transfer takes place when an entity leaves a module by means of a route, transport, or convey. In these cases, a station destination is specified and the entity is sent to the module that defines the station. Allocating stations for the looped track of AGVs is shown in Figure 6. Dummy stations might be

Table 2: List of Modules for Large-Scale AS/RS-AGV System

Module		Corresponding	Description
No.	Name	Subsystem	
1	Zaiko Sakusei	N/A	Assign the initial condition on each rack of AS/RS.
2	Nyuko Shuko	N/A	Create incoming and outgoing entities.
3	AisleConv	Aisle Conveyor	Define the specifications of aisle conveyors.
4	ASRS1	AS/RS	Define the specifications of each pair of high- rise rack lanes.
5	ASRS2	AS/RS	Define the specifications of the oveall AS/RS.
6	InCnvBegin	Incoming Conveyors	Specify the starting point of each incoming conveyor.
7	InCnvEnd	Incoming Conveyors	Specify the ending point of each incoming conveyor.
8	OutCnvBegin	Outgoing Conveyors	Specify the starting point of each outgoing conveyor.
9	OutCnvEnd	Outgoing Conveyors	Specify the ending point of each outgoing conveyor.
10	InAslBegin	Incoming Aisle Conveyor	Specify the starting point of each incoming aisle conveyor.
11	OutAslEnd	Outgoing Aisle Conveyor	Specify the ending point of each incoming aisle conveyor.
12	PickCnvBegin	Picking	Specify the starting point of each picking conveyor.
13	PickCnvEnd	Picking	Specify the ending point of each picking conveyor.
14	PickWorker	Picking	Specify the picking operation.
15	DummyStation	AGV	Set each dummy Station.
16	STV	AGV	Define the specifications of AGVs.
17	STVTrack	AGV	Set the looped track of AGVs.
18	Conveyors	Conveyor Related	Specify the specifications of conveyors.
19	STVFileInit	AGV	Initialize the file to control AGVs.
20	STVFileCheckSTV	AGV	Check and/or modify the file to control AGVs.
21	FreeSTV	AGV	Free the AGV.
22	Hikiate	AGV	Assign the AGV to be allocated.
23	Oidashi	AGV	Push ahead the unloading AGV.

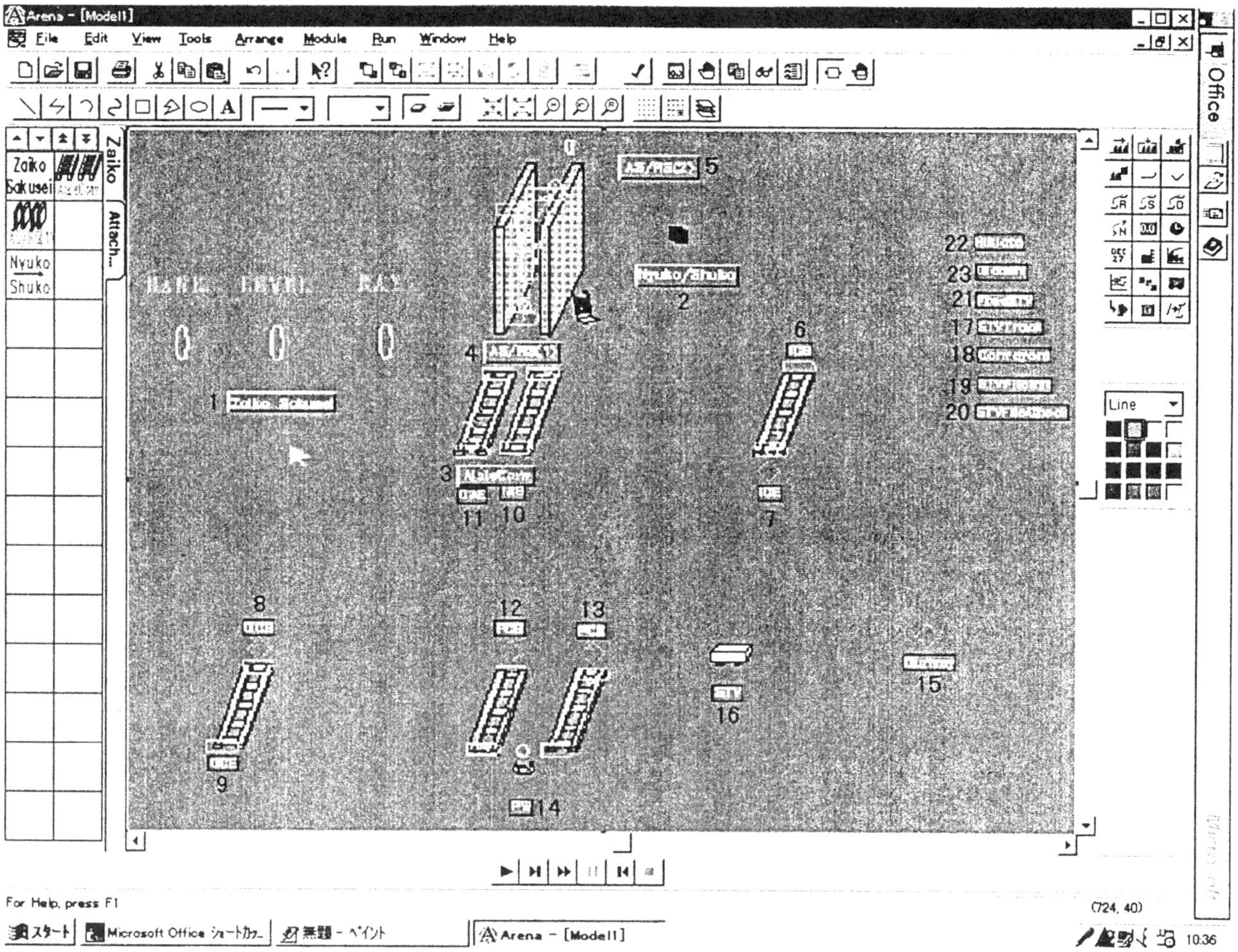

Figure 3: Arena Template and Modules

placed in the system, if necessary. Furthermore, AS/RS, aisle conveyors, incoming conveyors, outgoing conveyors, and picking conveyors are placed and allocated in the appropriate positions on the working area of the window, as shown in Figure 7. After all necessary modules are placed in an appropriate fashion for a particular model and the corresponding values are provided for their operands, a simulation run may be performed.

3.3 Advantages of the Proposed Approach

As mentioned in Section 1, it takes a lot of time (possibly a couple of months even for skilled programmers) to build a complex program for a large-scale AS/RS-AGV system. If the system developed in this study is utilized, it would take only a few hours even for a programming newcomer to develop a simulation program for the similar large-scale AS/RS-AGV system. Thus, the time required to build simulation programs could be reduced drastically, by utilizing the proposed module-based modeling method.

In addition, as the model is developed to be a completed one, the animation model is also built together with the model. When construction of the simulation model is completed, the corresponding animation can be run immediately. This is a great advantage of the module-based modeling method over the flexible modeling system developed in the previous paper.

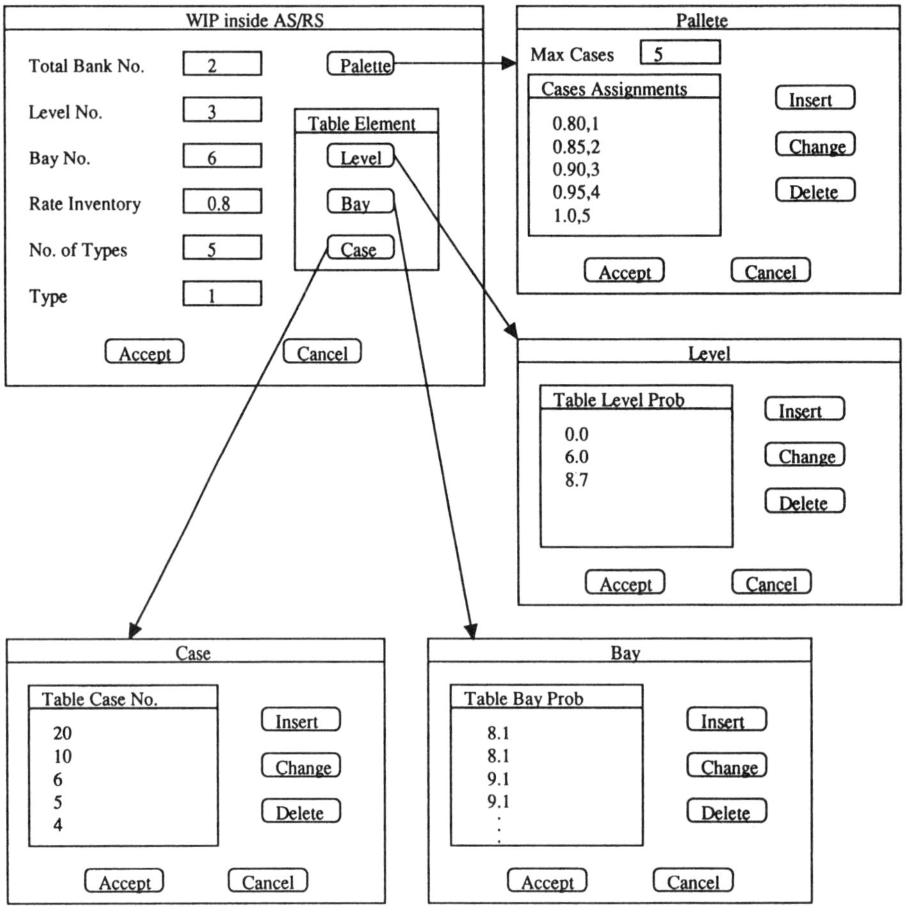

Figure 4: Module for Defining the Initial Conditions inside AS/RS

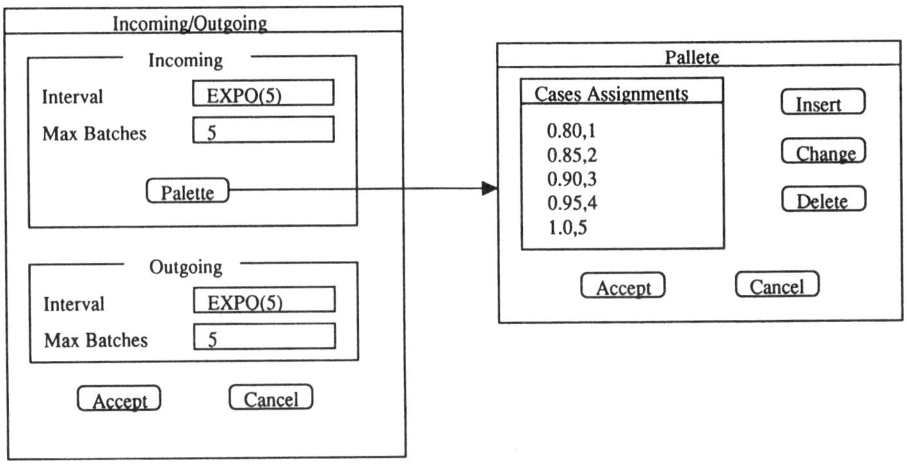

Figure 5: Module for Creating Incoming/Outgoing Items

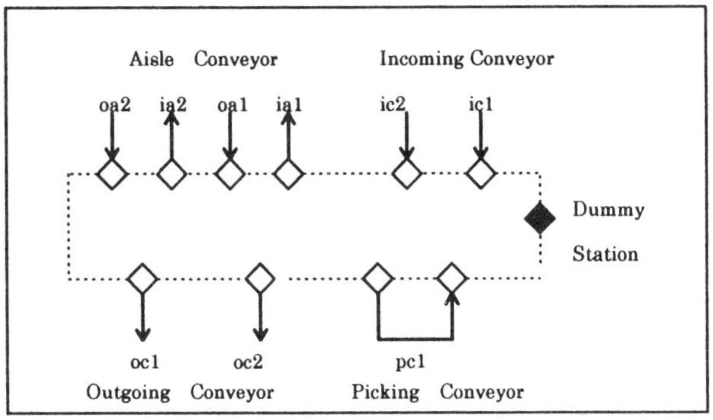

Figure 6: Allocating Stations for the Looped Track

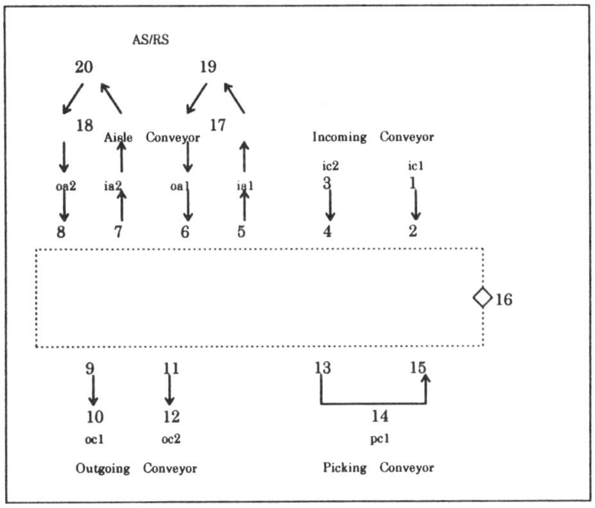

Figure 7: Allocating Appropriate Modules

4. MODELING FLEXIBILITY AND ANALYTICAL FINDINGS OF PREVIOUS STUDIES

4.1 Major Selected Alternatives

As described in Section 1, when introducing large-scale AS/RS-AGV systems, there are a lot of alternatives. Therefore, management will have to decide particular specifications of the system. In this section, some of the typical alternatives are reviewed and some important analytical results found in the series of the previous papers are summarized. The following three issues are considered for evaluation:

(1) Storing/retrieving policy.
(2) Overall layout on the conveyors.
(3) Arrangement of racks inside AS/RS.

4.2 Storing/Retrieving Policy

As mentioned in Section 2.2, there are four typical alternatives typically on the control logic of the AS/RS. The logic is designated so that storage and retrievals are made to the rack in the four different ways shown in Figure 2 (i.e., the alternatives A1 through A4).

According to the previous study, it is concluded that the alternatives A1 and A3 are more effective than other alternatives especially in installing more AGVs, regardless of the percentage of occupied racks (Takakuwa 1994). This may be because the travel time by the vertical movement is relatively longer than by the horizontal movement. In addition, the inventory level does not affect the total flow time. Therefore, the desirable storing/retrieving policy with the search by the horizontal direction basis is recommended.

4.3 Overall Layout on the Conveyors

In general, there are two stages, for management to decide, in installing the particular type of the system described in this study. First stage is determining the overall layout of the system. Relative locations of AS/RS, incoming conveyors, and outgoing conveyors might be different from each other for four alternatives on layout. Second stage is determining more detailed specifications of the system, such as the number of AGVs, the number of incoming and outgoing conveyors.

According to the previous study, the system can be operated more efficiently by locating the loading and unloading positions (i.e., the incoming conveyors and the AS/RS, and the AS/RS and the outgoing conveyors) are located as close as possible (Takakuwa 1993).

4.4 Arrangement of Racks inside AS/RS

In this section, alternatives which differ in the rack arrangement inside the AS/RS are examined for the purpose of evaluating efficiency of the handling. The total number of racks are same inside the AS/RS, whereas the numbers of the corresponding bank, bay, and level of three types of AS/RS are different from each other.

According to the previous study, it is concluded that the systems with more stacker cranes are more effective (Takakuwa 1995). This may be because of the difference of the number of stacker cranes; they might have enough stacker cranes to handle incoming and/or outgoing items under the current condition. Hence, even though the total number of racks are the same, if the number of stacker cranes are a few, the items cannot be handled efficiently in AS/RS-AGV systems.

5 SUMMARY

(1) In this paper, an efficient module-based modeling method is presented for generating simulation programs for complex and large-scale AS/RS-AGV systems.
(2) Flexibility of the proposed system is summarized. In addition, by utilizing the proposed module-based modeling method, the time to build simulation programs could be drastically reduced,
(3) Some important analytical results are reviewed on major alternatives of the large-scale AS/RS-AGV systems.

ACKNOWLEDGMENTS

The author wishes to thank Mr. Motoo Shinozuka and Mr. Masanori Tsujimoto of Daifuku Co., Ltd. for their valuable comments on AS/RS and materials handling systems. In addition, the author wishes to thank Mr. H. Tanaka and Mr. M. Maekawa for their assistance.

REFERENCES

Gunal, A,E. Grajo, and D. Blanck. 1993. Generalization of an AS/RS in SIMAN/CINEMA. In *Proceedings of the 1993 Winter Simulation Conference*, 857- 865. Institute of Electrical and Electronics Engineers, Piscataway, New Jersey.

Harmonosky, C.M. and R.P. Sadowski. 1984. A simulation model and analysis: integrating AGV's with non-automated material handling. In *Proceedings of the 1984 Winter Simulation Conference*, 178-183. Institute of Electrical and Electronics Engineers, Piscataway, New Jersey.

Medeiros, D. J., E. E. Enscore, and A. Smith. 1986. Performance analysis of miniload systems. In *Proceedings of the 1986 Winter Simulation Conference*, 606-612. Institute of Electrical and Electronics Engineers, Piscataway, New Jersey.

Muller, D. 1989. AS/RS and warehouse modeling. In *Proceedings of the 1989 Winter Simulation Conference*, 802-810. Institute of Electrical and Electronics Engineers, Piscataway, New Jersey.

Pegden, C. D., R. E. Shannon, and R.P. Sadowski. 1994. *Introduction to simulation using SIMAN.* 2nd ed., McGraw-Hill, Inc., New York.

Pulat, B.M. and P.S. Pulat. 1989. Performance analysis of automatic storage and retrieval system - a comparative approach. In *Proceedings of the 1988 Winter Simulation Conference*, 591-596. Institute of Electrical and Electronics Engineers, Piscataway, New Jersey.

Takakuwa, S. 1989. Module modeling and economic optimization for large-scale AS/RS. In *Proceedings of the 1989 Winter Simulation Conference*, 795-801. Institute of Electrical and Electronics Engineers, Piscataway, New Jersey.

Takakuwa, S. 1993. Design and cost-effectiveness analysis of large-scale AS/RS-AGV systems. In *Proceedings of the 1993 Winter Simulation Conference*, 851-856. Institute of Electrical and Electronics Engineers, Piscataway, New Jersey.

Takakuwa, S. 1994. Precise modeling and analysis of large-scale AS/RS. In *Proceedings of the 1994 Winter Simulation Conference*, 1001-1007. Institute of Electrical and Electronics Engineers, Piscataway, New Jersey.

Takakuwa, S. 1995. Flexible modeling and analysis of large-scale AS/RS-AGV systems. In *Proceedings of the 1995 Winter Simulation Conference*, 873-880. Institute of Electrical and Electronics Engineers, Piscataway, New Jersey.

AUTHOR BIOGRAPHY

SOEMON TAKAKUWA is a Professor of Economics at Nagoya University in Japan. He received his B. Sc. and M. Sc. degrees in industrial engineering from Nagoya Institute of Technology in 1975 and from Tokyo Institute of Technology in 1977 respectively. His Ph.D. is in industrial engineering from The Pennsylvania State University. He received Dr. of Economics from Nagoya University in 1996. He has prepared the Japanese edition of *Introduction to simulation using SIMAN.* He was awarded by Japan Foundation for the Promotion of Machine Tools their Prize in 1984, and by SCS, International for the Best Paper Award of "Modelling and Simulation 1992". His research interests are in the area of optimization of manufacturing systems and simulation.

Proceedings of the 1996 Winter Simulation Conference
ed. J. M. Charnes, D. J. Morrice, D. T. Brunner, and J. J. Swain

INCORPORATING FINANCIAL AND ERGONOMIC CONSIDERATIONS INTO MODELS OF MANUFACTURING SYSTEMS

Henry Herper

Otto-von-Guericke-University of Magdeburg
Department of Computer Simulation and Graphics
P.O. BOX 4120
D-39016 Magdeburg, Germany

Thomas Kagerl

Engineering Office for Project Management, Work Structuring and Plant Organisation
Schleinufer 14
D-39104 Magdeburg, Germany

Dietrich Ziems

Otto-von-Guericke-University of Magdeburg
Department of Logistics and Mechanical Handling Engineering
P.O. BOX 4120
D-39016 Magdeburg, Germany

ABSTRACT

This contribution describes a modular concept by which simulation results can be post-run interpreted under different aspects (multifunctional interpretation). The extension of classic simulation models for materials flow systems and manufacturing systems by economic, ecological and environmental aspects as well as the advantages and limits of this approach are discussed in this paper. A gear manufacturing was chosen as an example to demonstrate the application of a multifunctional interpretation.

1 INTRODUCTION

Classic simulators are able to model technical processes in manufacturing and material handling systems with a sufficient degree of details. They are used by factory planners to check and improve plans of technical systems. During the last years these tools were extended with components to support the modelling of flexible manufacturing systems, automated production lines and automated materials flow processes.

But today the consideration of worker, costs and environmental factors becomes more and more important in designing of technical systems. Criteria like qualification, work organisation, compensation of employees and work structuring must be more integrated into the planning process and for that reason also into the planning tools. At the same time economic and ecological questions gain more and more importance.

To support the assessment of such aspects it is necessary to provide and process additional data and methods for the application in computer-based planning tools. In such a way the reliability of planning results could be improved and qualified statements about ergonomics, economy and ecology of manufacturing systems and materials flow systems could be available already during the planning stage.

Conventional simulators only allow limited assessment of a materials flow system. Normally they are based on state variables of flows of material and resources in location and time. Such models do not fully suffice e.g. for the introduction of new work structures like team work. Often it is necessary to allow a view on simulated systems under different aspects. In this case the simulation results are to be prepared under various viewpoints and for different user groups. We refer to this process as multifunctional interpretation of simulation results. The problem analysis must clear up which questions of different user groups should be answered by the simulation system, how the additional result data could be generated out of the model data, which additional input data are to be provided and how these data are to process.

The classic form of a simulation study can deliver the planning team many necessary information about the materials flow and can show an insight into the process. The modelling of the behaviour of materials flow systems and manufacturing systems in location and time is the basis of the development of expanded models. Banks (1996) claims 80 to 85% of the time that an item is on the manufacturing floor is expended in material handling or waiting for material handling to occur. Therefore the planning tools are often concentrated on the modelling of these items, especially on the states of flowing objects and technical resources. Theses models can be generated very precisely for flexible manufacturing systems with a high level of automation. They deliver results which are well suited for the assessment of the planning. The strain of the employees, ecological or economic aspects are normally modelled only if they directly influence the production sequences or if they are requested as target figures or factors of evaluation of the modelling.

2 BACKGROUND

Today's simulation tools are equipped with extensive components for the assessment of states and behaviour of manufacturing systems and materials flow systems. The simulation results are delivered in a processed form as compressed result data. They include for instance the utilisation rate of transport devices and manufacturing facilities, the utilisation of storage space, or the throughput time of items or orders. These data allow the materials flow planner an effective assessment of technical systems, their performance and their performance limits. Besides most simulators offer primary result data. These are for instance trace protocols. The animation trace file is a special form of primary result data which allows a post run animation.

With the help of the animation components in today's simulation tools, the experienced user gets various possibilities, for instance to assess the modelled processes and to put the system states in a proper place due to their history.

Classic simulation models are state models of operands, operators and operations (process stages). The model structure and the kind of result data are determined by the aims of the simulation study. In multifunctional models not all result data must be generated by the simulation tool. Primary result data can be connected and interpreted with extended input data. The dataflow is shown in Figure 1.

The interpretation makes is possible to gain an extended evaluation under different aspects. Therefore it is necessary that the simulation model is designed, structured and detailed in a proper way. The methods described in this contribution concentrate on the form of post-run evaluation above explained to make simulation results accessible to a multifunctional interpretation.

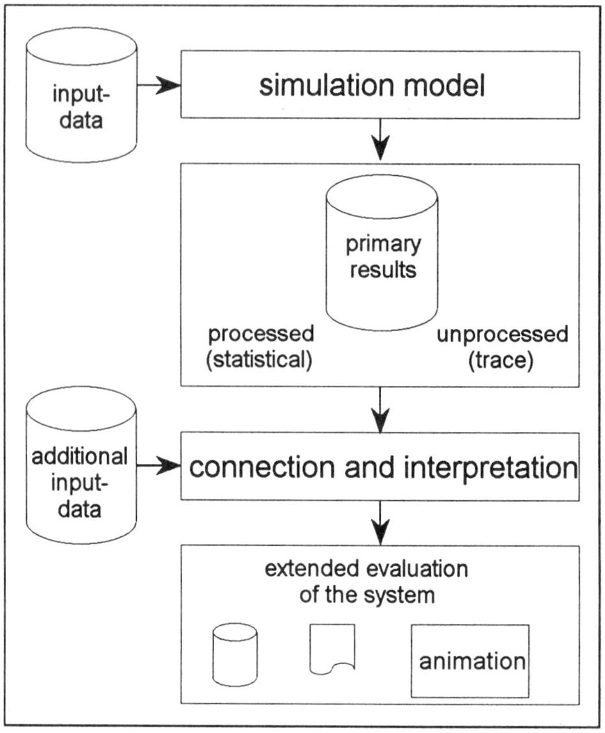

Figure 1: Data Flow

3 EXTENSION OF SIMULATION TOOLS

One aspect of the extension of simulation tools is for instance the detailed modelling of workers. In classic simulation models statements about single workers are only possible in an abstract way. Often only the utilisation of the worker is given indicated. Qualifications as the ability to carry out a production sequence at a workplace are normally defined in simulators, but only to distinguish between "The worker can make this job" and "The worker can not make this job".

But it is not enough to use only the state modelling of technical systems if it is necessary to assess different work structures concerning their demands on the worker. Components allowing a distinguished estimation of physical strain and qualification must be integrated.

Two different approaches are possible for the necessary extension of simulation tools. On the one hand there is the possibility to expand the simulators with further functions. The result would be specialised simulators. Some simulator developers follow this way for instance to model value added processes in much more detail. This approach is very expensive and demands an individual's own solution for each simulator. The second approach consists in a post-run combination and assessment of uncompressed result

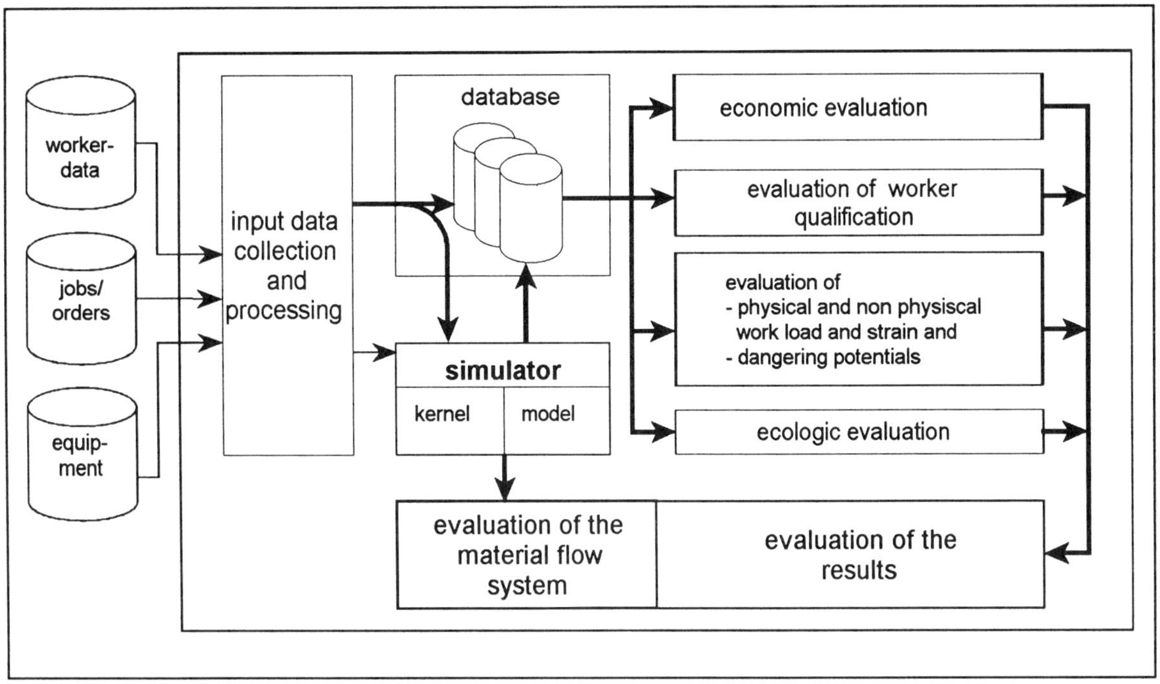

Figure 2: Components and Interfaces of the System

data produced by the simulator with additional information. Additional modules allow an assessment under various viewpoints. The structure and components of such an approach are shown in Figure 2.

3.1 STRUCTURE OF THE SYSTEM

The precondition for the winning of additional information is a simulation model that describes the analysed system very detailed. It must have a clear connection of all manufacturing sequences to machines, items and workers. The series of manufacturing sequences (processes) are to be recorded in trace protocols. Also the writing of trace protocols of changes in states of system components must be possible. Both forms of trace protocols constitute as primary result data the basis for the extended data evaluation.

The multifunctional interpretation system has the following basic functions:

- input data collection and processing,
- interfaces to the simulation model,
- assessments,
- help and documentation systems and
- data base interface.

The assessment modules can be chosen in relation to the aims of the simulation study. The following modules were developed in connection with research projects and simulation studies for the industry:

- assessment of load, strain and danger potential of workers,
- assessment of the value added by the manufacturing process and
- estimation of the environmental danger potential by the manufacturing process.

For the generation of additional information, additional input data are to be collected and to be processed. The generated data can be stored and managed in a data base.

The simulation model generates data which can also be stored in the data base. The internal assessment methods of the simulation tools are used for the assessment of the system behaviour in location and time. The user gets his other results about the system in the usual and accepted way from the simulation tool. The simulation tool creates also the animation corresponding to its implemented abilities.

The additional assessment tools deliver the extended interpretation and evaluation of the simulation results. The users get further aspects, assessment criteria and data for their decision making. Trace protocols generated by the simulator provide the basis for the estimation. All modules use the trace protocols by connecting the events and processes with additional data. Due to the modular concept chosen it is possible to integrate different assessment tools. The components run under a uniform user interface.

3.2 DATABASE

Data are managed in a relational data base. In this data base a redundancy free data management is aimed. The data formats are chosen in such a way that they allow a platform independent exchange. Therefore we use the ASCII format.

The data necessary for the extended result evaluation must be collected and committed to the evaluation components. The events or processes must be assigned to the trace protocol to connect the additional input data with the result data generated by the simulator.

Events and processes are the elementary components of the state model. Manufacturing sequences for the evaluation processes are built from these events and processes. Every working sequence consists of several working activities which are connected to a working task. These working sequences are not interruptive. They are processed by a worker or a team at a defined location, on a defined item with defined devices. The combination of working activities to a working sequence is realised in the stage of the additional data collection. There are two ways to determine the duration of a working sequence. On the one hand the duration is determined by input data, on the other hand the duration is computed by the simulator.

The application of this concept demands a trace file generated by the simulator with the following contents

- start,
- end,
- working location,
- working item and
- worker

for each working sequence. The combination of working sequences and primary simulation results is realised by a connection file which is represented by a matrix. Various evaluation components can generate multifunctional interpretations from these combined result data.

Another kind of data connection is the extension of state protocols generated by the simulator with additional data. These state protocols are necessary for the evaluation of data which can not be collected from working sequences.

3.3 INTERFACES AND RESULT DATA EVALUATION

The extended result data evaluation is also realised by a modular concept. This allows the integration of various evaluation modules related to the requirements of the users. The result data evaluation is post run. This method has the advantage that the computed simulation results can be interpreted under various viewpoints. The disadvantage is that the results of the

viewpoint	object	state	primary result	evaluation	result
	type: resource (operator)				
ecology	forklift truck	transport	duration duration	energy consumption/ period heat noise carbon dioxide emmision	energy consumption/ operation environmental load
economy		busy idle	duration duration	work costs waiting costs	cost / period idle cost/ period
ergonomy	worker	in work wait	duration duration	work load recreation time	stress level stress level
	type: issue (operand)				
economy	part	processing quality checking transportation	duration duration duration	increase in value yes increase in value yes (partially) increase in value no	increase in value/ period
ecology			duration	heat loss	energy demand

Figure 3: Examples for added costs and values

evaluation can not be used for the decision making during the simulation run.

The connection to different simulators was one of the main aims in the development of the interfaces. A high acceptance by a user can only be reached, if the modules work with his or her simulator and the additional costs for modeling are low. For different viewpoints and users we provide different modules. The functionality of the modules shall be described in the following paragraphs.

The module for the economic evaluation allows a detailed analysis of the cost behavior of the simulated system and the added value. For the usage of this module it is necessary that the user can combine every working sequence with its cost and added value (Figure 3). This kind of economic analysis is at this time directly supported by some simulators. The advantage of the external evaluation module is that the user can choose the form of the evaluation according to the available data. The module allows the calculation of direct costs (for example consumption of operating resources and energy) and indirect costs (for example floating capital).

Another module was developed for the estimation of load and strain of workers. The evaluation component for the physical work load and strain was developed in the project EMSIG and described in Ehrhardt (1994). Investigations in several firms have shown that the physical load and strain of workers becomes more and more important. The activities done by the workers include a lot of control tasks with a high level of responsibility. The estimation of mental load and strain of the worker is much more difficult than physical work load, because the individual differences are much higher. The evaluation component includes methods used in ergonomic investigations in firms.

The qualification of workers is very important for reengineering manufacturing systems. Materials flow simulators generate often only results about the utilization of workers. The developed evaluation module takes into consideration how well a worker is qualified for an activity. It compares the qualification of the worker with the requirements of the working process. The interpretation of the results allows the determination of necessary training for the workers already in the planning stage.

Many manufacturing sequences include danger potentials for the workers. The danger potential is caused by the working sequence but also by the substances used and by the environmental conditions. A special evaluation module supports the user in the estimation of kind and duration of danger potentials in the planning stage.

All Modules work separately on the same database. Often it is necessary to combine the results of the different evaluation models with each other and with the primary simulation results of the technical system. Therefore we developed a separate component which compresses and presents combined result data.

4 MODEL OF AN ASSEMBLY SYSTEM

The developed modules were tried in a simulation study supporting the reengineering of a gear manufacturing. The planning team consists of materials flow engineers, work structuring engineers and quality managers. Different variants of the system were analysed and evaluated under different viewpoints with the help of the simulation model. The variants differ in the chosen form of the working structures and layout of the material flow system.

The main goal in the restructuring of the manufacturing system is an increased performance with lower throughput times.

The simulation results allow statements about:

- performance of the system and the throughput time of items,
- cost and added value,
- demand of qualification,
- load and strain of workers and
- possible danger potentials.

For the development of the model we used the simulation language GPSS/H™ and for the animation the system PROOF™.

5 SUMMARY

Traditional simulation models for materials flow systems and manufacturing systems describe and insight the processes, technical performance and the system behavior in location and time using discrete events and states. The primary results are related to these goals.

The integration of additional aspects like ecological, economic and ergonomical evaluation demands extensions of the model. This can be done with two different approaches. The first one is the extension of the simulator and the other one is a post run interpretation. The post run interpretation is usable if the decision making is not influenced by the additional evaluation.

A multifunctional interpretation can be realised with various simulators which are able to generate trace files in the described form. The presented modular concept of evaluation modules and files allows the interpretation of primary simulation result data under various viewpoints and user interests. The evaluation methods were successfully used with result data of different simulation tools. This approach opens up several new application fields for traditional simulators and models.

REFERENCES

Banks, J., John S Carson, and Berry L. Nelson. 1996. Discrete-event system simulation. Upper Saddle River, New Jersey: Prentice Hall.

Ehrhardt, I., Henry Herper, and Hansjürgen Gebhardt. 1994. Modelling Strain of Manual Work in Manufacturing Systems. In: *Proceedings of the 1994 Winter Simulation Conference*, ed. J. D. Tew, S. Manivnnan, D. A. Sadowski and A. F. Seila,1044-1049. Institute of Electrical and Electronics Engineers, Piscataway, New Jersey.

AUTHOR BIOGRAPHIES

HENRY HERPER works as Assistent Professor in the department of Computer Simulation and Graphics at the University of Magdeburg. His areas of research are the modelling of logistical and manufacturing systems and their dispositions. He is a member of the ASIM and the GPSS-Users'-Group Europe.

Thomas Kagerl works as Project Manager in an Engineering Office for Project Management, Work Structuring and Plant Organisation. One of his recent projects has been the reengineering of a gear manufacturing.

Dietrich Ziems has been Professor at the University of Magdeburg (Germany), Department of Logistics and Materials Handling Engineering since 1992. He obtained his doctorate at the University of Magdeburg. He teaches planning of material handling systems, organisation and control of material flow processes and logistics. His research interests include applications of CAD-layout planning, simulation and animation of manufacturing, multimedial technologies in education. He is a member of the ASIM.

Proceedings of the 1996 Winter Simulation Conference
ed. J. M. Charnes, D. J. Morrice, D. T. Brunner, and J. J. Swain

A GENERIC SIMULATOR FOR MODELING MANUFACTURING MODULES

Bernard J. Schroer
Phillip A. Farrington
James J. Swain
Dawn R. Utley

University of Alabama in Huntsville
Huntsville, Alabama 35899 USA

ABSTRACT

This paper presents a simulator for designing and analyzing modular manufacturing lines. Included in this paper are a description of the simulator, a sample application, and the benefits of using this approach.

1 INTRODUCTION

Computer simulation has become a common and accepted tool in the design, layout, and analysis of manufacturing lines. Many companies will not install a new manufacturing line until the line has been evaluated using simulation. Management now sees simulation as inexpensive insurance against costly mistakes.

However there are still some pitfalls in using simulation. Probably the most serious pitfall is the time to build, verify and validate a simulation model. Model development takes time and generally more time than estimated and than often available. In addition, detailed data for the simulation model is often nonexistent, unreliable, and generally difficult to obtain.

Another pitfall in simulation is the need for an employee who knows how to build a simulation model, is experienced in using a simulation package, and has a general background in simulation and statistics. Quite often firms lack this expertise and cannot afford a full time simulation modeler.

One approach to reducing the pitfalls is the use of simulators. The advantages of simulators are relative ease of use, rapid model development, and changing model parameters.

2 BACKGROUND

In 1989 the University of Alabama in Huntsville (UAH) was awarded a contract by the State of Alabama to provide technical assistance to the apparel industry. As part of this contract, visits were made to over one-hundred apparel manufacturers in Alabama. A common response from these manufacturers was the need for assistance in converting their traditional manufacturing lines to modules (Gilbert, 1988; Kulers and Dewitt, 1990; Fralix, et.al., 1990; AAMA, 1987; Holoyda, 1992; Kron, 1987). As a result, a number of simulation models were written for designing and analyzing the proposed manufacturing modules (Schroer, 1990; Schroer, et.al., 1992, and Farrington, et.al., 1994).

Surprisingly, after developing these models, it was noticed that:

- Management wanted a quick design and analysis which resulted in the development of high-level models.
- The manufacturing modules were very similar for all the manufacturers.
- The manufacturing modules were generally U-shaped and did not have branches, such as diverging and converging lines.
- The simulation models were also similar and used identical features of the simulation package.

As a result of these observations, a generic modular manufacturing simulator (SSE5) was developed with partial funding from the NASA Marshall Space Flight

Center (MSFC), Schroer and Wang (1992). The inputs to the SSE5 simulator are:

- Number of stations and operators.
- Number of machines at each station.
- Cycle time for each station.
- Operator assignment (fixed at a station or moveable and if movable, the station routing sequence).

The SSE5 outputs are production, queue statistics, machine utilization, and operator utilization. The simulator is distributed through the NASA Marshall Space Flight Center technology transfer program. Since 1993, over 800 firms have requested copies of the simulator.

3 MODULAR MANUFACTURING SIMULATOR

Although the SSE5 was developed specifically for the apparel manufacturing industry, the feedback from a number of users indicated that it could be used in other domains such as electronics or small electromechanical assembly. As a result, the SSE5 simulator was modified to a generic Modular Manufacturing Simulator (MMS).

3.1 System Description

An overview of the Modular Manufacturing Simulator is given in Figure 1. The MMS can be used for designing and evaluating modular manufacturing lines with the following characteristics:

- The system consists of one line with an unlimited number of stations (all stations are in series).
- Each station may have unlimited number of machines with each machine performing the identical operation.
- There are an unlimited number of operators.
- There is unlimited space for work-in-process (WIP) in front of each station.
- There is always enough WIP in front of the first station so there is no delay waiting for parts.
- Some operators may be fixed at specific stations.
- Some operators may move between a given number of stations. The movement of operators is defined by a set of predefined rules.
- Work is done in lots of one part. However, it is possible to perform work in lots of more than one part by defining all parameters in terms of lots.
- There is no machine breakdown.

The MMS runs on a PC, with a minimu of 8Mb of memory, under the Windows operating system.

3.2 Operator Movement Rules

Each non-fixed operator moves based on one of the following rules: Maximum WIP Rule, Rabbit Chase Rule, and Push/Pull Rule. Figure 2 illustrates each of these rules for operator movement between three stations.

The WIP Rule is as follows:

- If the operator has worked for more than the time limit or has exceeded the part limit at the current station, or the WIP at the next station has exceeded the WIP limit, the operator will move to the station on the priority list with the largest WIP.

- If the maximum WIP is at the current station, the operator will stay at the current station and complete another part.

- If the station with the maximum WIP is busy, the operator will move to the station on the priority list with the second largest WIP.

- If two stations have the same WIP, the operator will move to the station with the higher priority.

- If all stations are busy, or there is no WIP at any station in the priority list, the operator will stay at the current station and make another part.

The Rabbit Chase Rule is as follows:

- If the operator has worked for more than the time limit or has exceeded the part limit at the current station, or the WIP at the next station has exceeded the WIP limit, the operator will move to the next station in the priority list.

- If the next station is busy or there is no WIP at the station, the operator will skip the station and go to the next station on the priority list.

- If the operator is at the last station on the priority list, the operator will try to move to the first station on the priority list.

- If all stations are busy or there is no WIP at any station, the operator will stay at the current station and make another part.

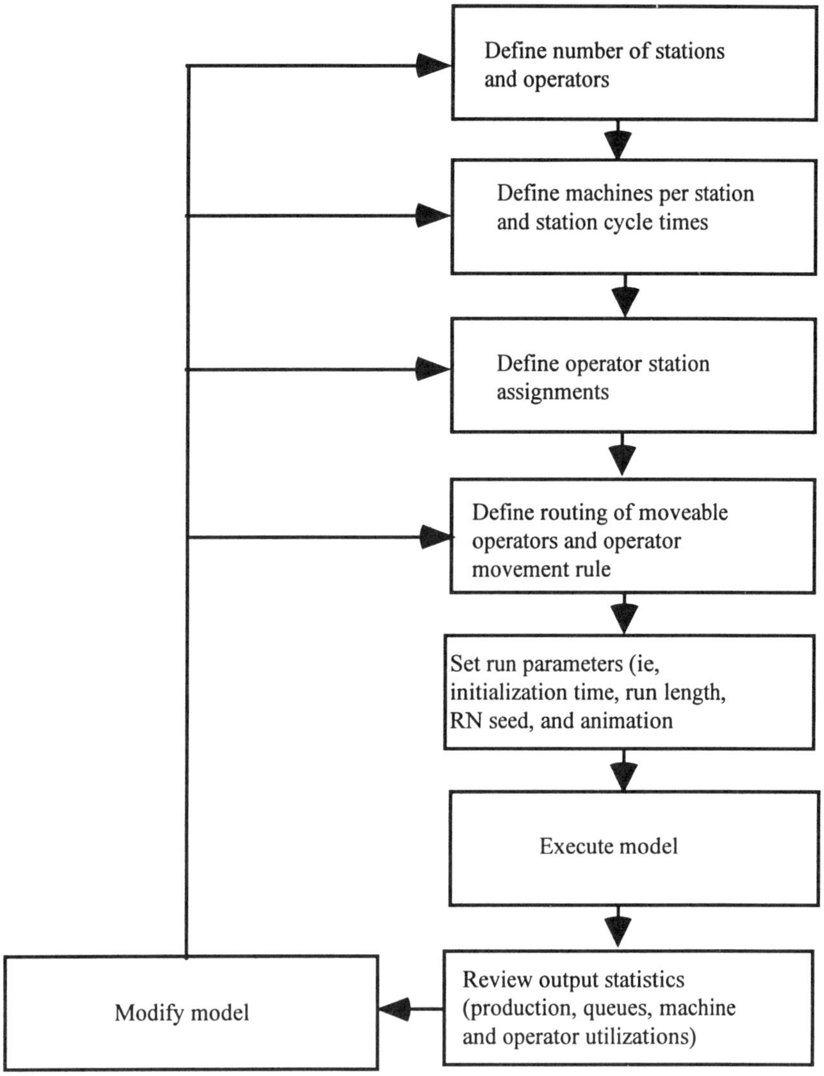

Figure 1: MMS system overview

The Push/Pull Rule is as follows:

- If the operator has worked for more than the time limit or has exceeded the part limit at the current station, or the WIP at the next station has exceeded the WIP limit, the operator will move to the next station on the priority list.

- If the next station on the priority list is busy or has no WIP at the station, the operator will skip the station and go to the next station on the priority list.

- If the operator is at the last station on the priority list, the operator will move back one station on the priority list. If this station is busy or has no WIP, the operator will move back two station on the priority list.

4 SAMPLE PROBLEM

Figure 3 is an example of a typical manufacturing line that can be evaluated using the Modular Manufacturing Simulator. Tables 1, 2, and 3 describe the inputs to the MMS. These tables are input through three MMS spreadsheets. The MMS outputs are typical simulation outputs and include:

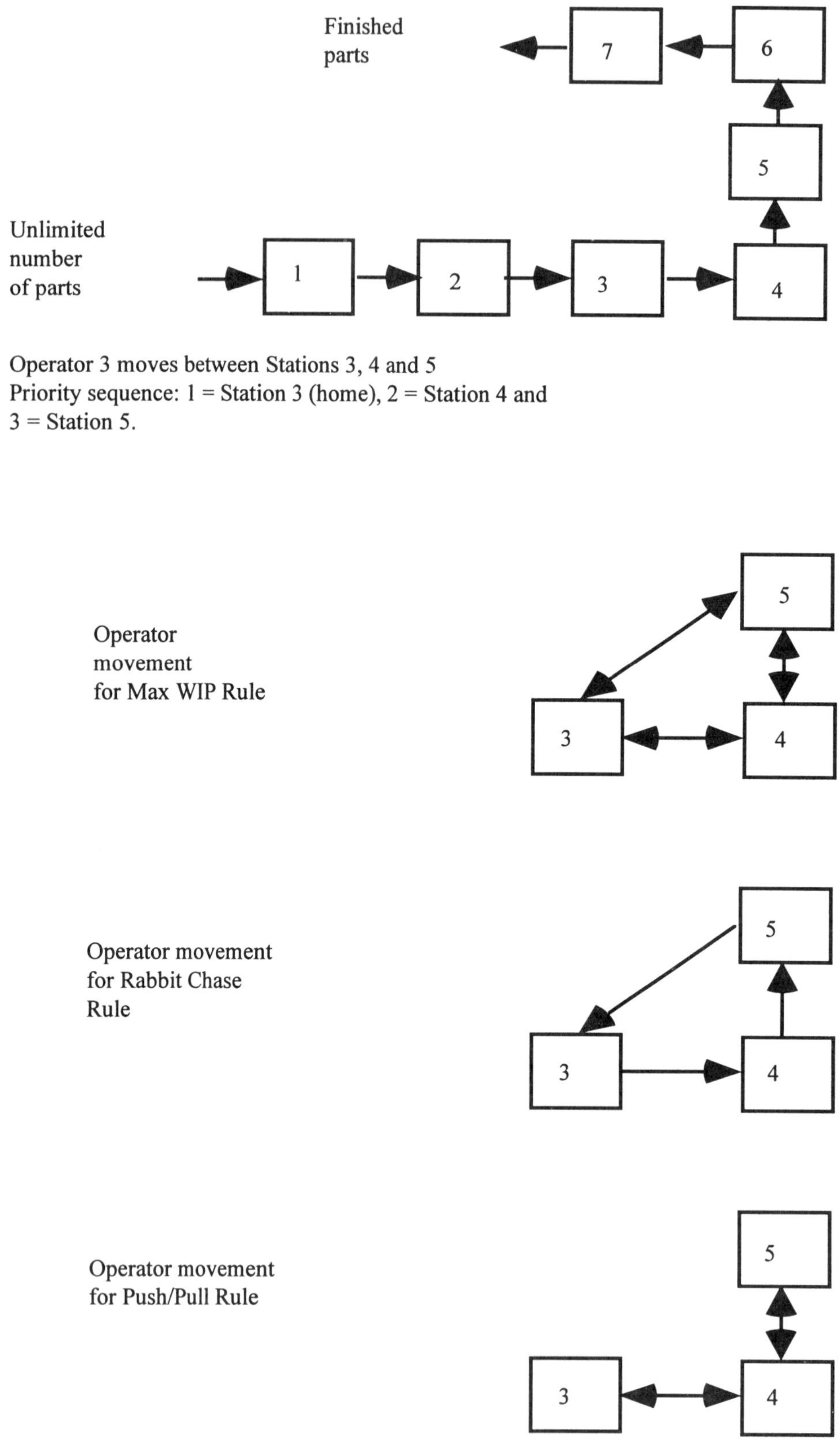

Figure 2: Example of operator movement rules

- Production
- Queues - number in and out, average content, average time in queue, minimum and maximum content, and current content
- Machine utilization - WIP, percent busy, and number of operations
- Operator utilization - percent at each station and percent idle

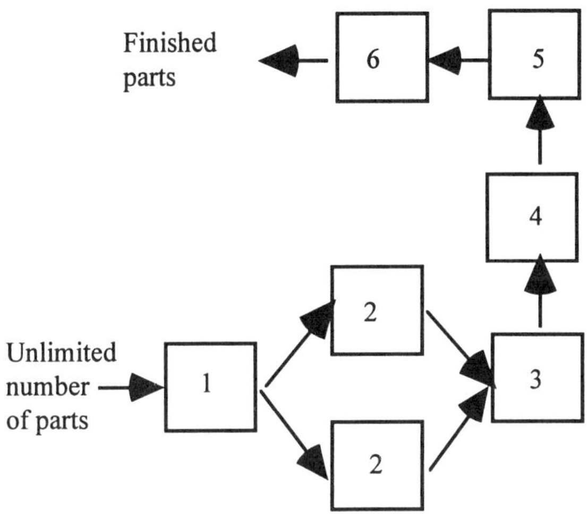

Figure 3: Typical manufacturing module

Table 1: Cycle time input

Station	Number of machines	Cycle time (minutes)
1	1	L(30,3)
2	2	L(60,12)
3	1	L(10,3)
4	1	L(12,2)
5	1	L(8,3)
6	1	L(10,3)
7	1	L(28,5)

L(x,y) = log normal distribution with mean x and standard deviation y

Table 2: Operator assignment input

Operator	Type	Station Assignment	Operator Efficiency (%)
1	Fixed	1	100
2	Fixed	2	100
3	Fixed	2	100
4	Moveable	3(home),4,5	3=100, 4=90, 5=90
5	Fixed	6	100

Table 3: Moveable Operator 4 input

Station	Priority	Time limit	Part limit	WIP limit	Movement Rule
3	1(home)	30	100	50	Rabbit Chase
4	2	30	100	50	Rabbit Chase
5	3	30	100	50	Rabbit Chase

5 CONCLUSIONS

In summary, the following observations can be made about the Modular Manufacturing Simulator:

o The MMS has been successfully used to develop models for apparel, electronics, and electromechanical manufacturing domains.

o A model of a manufacturing module could be developed and executed in less than fifteen minutes.

o Most firms had the necessary data available to construct the MMS model. For example, most firms had a data base of standard times. A good assumption is to use a log normal distribution for the standard times with the standard deviation expressed as a percentage of the mean. A simple sentivity analysi on the standard deviation can be done.

o The lack of animation was not really a drawback since the manufacturing modules are relative straightforward and easy to visualize.

The Modular Manufacturing Simulator has become an invaluable tool for supporting the State's Manufacturing Extension Program (MEP) which is funded in part by the National Institutes of Standards and Technology (NIST). The MMS offers the State MEP the ability to demonstrate the power of simulation to small and medium manufacturers and to actually model a client's manufacturing module quickly and at their manufacturer's location.

A followup evaluation by the NASA Marshall Space Flight Center has indicated that the SSE5 has been one of the major contributors to the economic impact of the MSFC technology transfer program (Tyson, 1996). One apparel manufacturer responded that the use of the simulator has resulted in an annual savings of $2M. These results indicated that the simulator is being used and having an impact on improving a manufacturer's productivity and competitiveness.

ACKNOWLEDGMENTS

This project has been funded in party by the Alabama Industrial Development Training, Alabama Department of Economic and Community Affairs, and the National Institutes of Standards and Technology's Manufacturing Extension Program.

REFERENCES

AAMA, 1987: *Getting Started in Quick Response*, American Apparel Manufacturers Association Arlington, VA.

Farrington, P.A., B.J. Schroer, J.J. Swain and Y.C. Feng, 1994, "Simulators as a Tool for Rapid Manufacturing Simulation," *Proceedings of the 1994 Winter Simulation* Conference, eds. J.D. Tew, S. Manivannan, D.A. Sadowski, and A.F. Seila, December 11-14, Orlando, Florida, pp. 994-1000.

Fralix, M., J. Langford, C. Carrere, and T. Little,1990: "Reflections on Modular," *Bobbin*, April, pp. 60-66.

Gilbert, C., 1988: "Tracking Modular Production," *Apparel Industry Magazine*, April.

Holoyda, O., 1992: "Quick Response Applications of Technology Enable U.S. Apparel Companies to Improve Competitiveness," *Industry, Trade, and Technology Review*, October, pp. 8-9.

Kron, P., 1987: "Pondering Modular," Apparel Industry Magazine, August, pp. 70-80.

Kulers, B. and J. Dewitt, 1990: "Modular Goes Mainstream," Apparel Industry Magazine, May, pp. 44-52.

Schroer, B. and J. Wang, 1992: *Simulation Support Environment for Modular Manufacturing Systems* SSE5, UAH Research Report 92-03, University of Alabama in Huntsville, October.

Schroer, B. and M. Ziemke, 1992: *Modern Apparel Manufacturing Systems and Simulation, Handbook.*, University of Alabama in Huntsville.

Schroer, B., J. Wang and M. Ziemke, 1992: "A Look at TSS Through Simulation," *Bobbin*, July, pp. 114-119.

Schroer, B., 1990: "Using Simulation Before Implementing a Unit Production System," *Bobbin*, November, pp. 110-114.

Tyson, T., 1996: "MSFC Technology Transfer Program," Presentation to the Technology Transfer Society, Huntsville, AL, January.

AUTHOR BIOGRAPHIES

BERNARD J. SCHROER is Associate Vice President for Research and Professor of Industrial and Systems Engineering at the University of Alabama in Huntsville. He holds a Ph.D. in Industrial Engineering from Oklahoma State Unversity. His professional affliations include SCS, NSPE, and the Technology Transfer Society. He is a Fellow of IIE.

PHILLIP A. FARRINGTON is an Assistant Professor of Industrial and Systems Engineering at the University of Alabama in Huntsville. He holds B.S. and M.S. degrees the University of Missouri, and a Ph.D. in Industrial Engineering from Oklahoma State University. His research interests include manufacturing systems design and system simulation. His professional affiliations include membership in ASEE, ASQC, IIE, and SME.

JAMES J. SWAIN is an Associate Professor in the Department of Industrial and Systems Engineering at the Univesity of Alabama in Huntsville. He received his B.A., B.S., and M.S. from the University of Notre Dame and a Ph.D. in Industrial Engineering from Purdue University. His research interests include Monte Carlo methods, statistical modeling, and output analysis. He has been active in the WSC and TIMS College on Simulation.

DAWN R. UTLEY is an Assistant Professor of Industrial and Systems Engineering at the University of Alabama in Huntsville. She holds a B.S.C.E. from Tennessee Technological University, an M.S.I.E. from the University of Tennessee and a Ph.D. in Industrial and Systems Engineering from the Unversity of Alabama in Huntsville. Her research interests include engineering management, organization structure, and quality improvement. She is a registered Professional Engineer and active in IIE, ASEE, and ASEM.

General Applications

Proceedings of the 1996 Winter Simulation Conference
ed. J. M. Charnes, D. J. Morrice, D. T. Brunner, and J. J. Swain

USING SIMULATION TO CRAFT A NATIONAL ORGAN TRANSPLANTATION POLICY

A. Alan B. Pritsker, Ph.D.

Pritsker Corporation
8910 Purdue Road, Suite 600
Indianapolis, IN 46268-1170 USA

O. Patrick Daily, Ph.D.

United Network for Organ Sharing
1100 Boulders Parkway, Suite 500
Richmond, VA 23225 USA

Kenneth D. Pritsker, Ph.D.

Strategic Reengineering Institute
2304 Eldridge Street, Suite 200
Pittsburgh, PA, 15217 USA

ABSTRACT

For many years, we have been able to use simulations for studying manufacturing and service systems, resulting in the ability to make better managerial decisions. As simulation methods have become more sophisticated, we have talked about the benefits of conducting large-scale simulations to support the selections of alternative public and corporate policies. This paper presents a description and analysis of what can be accomplished when simulations are used to support policy analysis and selection. The National Liver Allocation and Distribution Policy is used as the vehicle for presenting specific details of how simulation can be used to support policy selection.

1 THE CHALLENGE: CRAFTING A NATIONAL TRANSPLANTATION POLICY

Imagine the challenge of establishing a set of national policies that regulate the allocation of donated liver organs to patients in need of a liver transplant. At an operational level, the simulation task uses past data to make future projections of performance measures for different methods of allocating the liver organs to waiting patients. However, the diversity of constituents, the multiplicity of performance measures and the emotional public concerns create an environment in which decision making and policy setting is extremely complex.

The simulation effort was sponsored by the United Network for Organ Sharing (UNOS), a non-profit national organization which sets transplantation allocation and distribution policy. UNOS sets policy by involving a diverse set of constituents which include physicians, surgeons, transplant centers, organ procurement organizations, patients, families of patients, hospitals, insurance companies, the government's Division of Organ Transplantation (DOT), national law makers and the general public.

To help simulate the impact of policy changes, UNOS sought the assistance of the Pritsker Corporation to apply simulation methods in early 1995. Because of the complexity and seriousness of this project, the assignment quickly escalated into a testing ground for simulation's ability to help formulate and evaluate complex medical and societal concerns.

2 USING SIMULATION TO EVALUATE TRANSPLANTATION POLICY ALTERNATIVES

An early series of meetings between UNOS and Pritsker personnel resulted in an agreement on the individual responsibilities concerning potential policy alternatives, model development, data availability, performance measures and model validation.

UNOS, through its constituent-based committee structure, would be responsible for defining and evaluating alternative policies for transplantation allocation and distribution. The Pritsker Corporation would be responsible for building a model of the national transplantation network based on patient-status, patient requests, organs available, geographic location, distribution methods, and organ-patient matching. UNOS and Pritsker Corporation would jointly develop component models for survival, removal and relist functions, patient status transitions, and organ and patient arrival processes. The initial approach was to build, verify and validate a model with the transplantation community defining the performance measures, specifying the policy alternatives and then running the simulations to objectively evaluate the policy alternatives. The UNOS committee structure would then have the responsibility of selecting a policy, based in part on the results of the simulations, and to recommend a policy alternative to the UNOS Board of Directors. If the Board of Directors accepts a recommendation, it is broadcast for public comment prior to final adoption by the Board of Directors.

3 APPLYING SIMULATION IN A POLICY-CRAFTING CONTEXT.

Typically, policy-level decisions are the result of a complex organizational process that involve social, economic, psychological and political dimensions (Allison, 1971). This complexity increases exponentially in public policy settings due to diverse constituents, qualitative performance measures and difficulties of implementing policy changes (Bryson, 1988). Given these conditions, there is a question as to how simulation can play a productive role in the setting of policy.

Our experience suggests that if simulation can assist in operational decisions, the same approach can be utilized to foster a more productive policy-crafting environment. Policy-makers have been aggressively searching for tools to assist them in objectively studying alternatives (Mockler, 1991), yet simulation as a support tool has been sparingly used. Unlike the operational uses of simulation (where one manager wants to analyze the impact of adding one machine based on three quantifiable performance objectives), policy applications face several dilemmas that complicate the simulation effort, by:

1. Involving a broad set of decision-makers with diverse interests, conflicting objectives and different internal thought processes (Peters, Hammonds and Summers, 1974);

2. Balancing quantifiable and non-quantifiable performance measures, thereby creating a potential conflict over the relative benefits of proposed alternatives (Eccles, 1991); and

3. Understanding the issues from a broader perspective because of the more significant long-term implications, suggesting the need to evaluate the "down-stream" implications of policy decisions (Armstrong, 1982).

The crafting of a national transplantation allocation and distribution policy illustrates the differences between policy setting and operational decision-making. This paper describes the use of simulation to support the shaping of policy in this complex setting over a fourteen month period. The major challenge was to carefully integrate the scientific/quantitative methods associated with simulation into a organizational policy-setting process.

4 OVERVIEW OF THE MEDICAL SITUATION

Organs for transplantation are a scarce resource. The number of people who donate organs has not kept pace with increasing demand. Thousands of people wait for organs every year, with many dying while they are on an organ transplantation waiting list. The list of waiting patients continues to grow at a faster rate than the number of donors. How to allocate such a scarce and valuable resource is indeed complex. Much is at stake among the potential recipients, doctors, Organ Procurement Organizations (OPO), and transplant centers. The formulation of a comprehensive transplant allocation policy revolves around developing an organ distribution system that is equitable, efficient and effective.

5 THE TRANSPLANTATION SYSTEM ENVIRONMENT

Many issues arise with organ allocation. What may maximize medical benefit for one person may not necessarily result in equitable treatment for another. In pursuit of a policy to balance all aspects of this issue, UNOS organizes its activities through a Board of Directors and committees consisting of volunteers representing a broad cross-section of the transplantation community.

The national system for collecting data on patients and organs, called the National Organ Procurement Transplantation Network (OPTN), is operated by UNOS. Organ allocation policies are constantly reviewed and altered to reflect the advances being made in science, medicine and communications. These changes are regularly reviewed and modified to ensure equity among those in the transplantation community and to instill confidence in the community at large.

Because of the significance of the allocation process, the Federal Government has established a Division of Organ Transplantation (DOT) within the U.S. Department of Health and Human Services. DOT partially funds the National Organ Procurement and Transplantation Network and maintains oversight responsibility to ensure that medical utility and justice for potential recipients is maintained by organ allocation policies.

6 BUILDING A MODEL FOR COMPARING TRANSPLANTATION POLICIES

The objective of model building was to develop the UNOS Liver Allocation Model (ULAM) for comparing proposed alternative allocation policies. By focusing on comparing alternatives, system components which do not have an impact on the comparison of policies could be eliminated from the modeling effort. A secondary objective was to create a model that could be used and updated at UNOS headquarters to meet the future needs of the transplantation community (Garvin, 1993).

The model building activities followed the process described in *Simulation with Visual SLAM and AweSim* (Pritsker et al, 1996) and in the WSC Proceedings (Withers et al, 1993). A model specification document was written to describe the entities, events and component models for the arrival streams of donors and patients, the patient-status change process, the offering and acceptance of organs by doctors/patients and the relist and survivability functions relating to patient post-transplantation status. A specification for the reports and displays to be output from ULAM was also detailed. The policies to be evaluated initially using ULAM were limited to those prescribed in terms of patient health status and the geographical areas where patients are located. An Allocation Modeling Oversight Committee was created to review the specification and, after discussion and modification, the specification document was approved. This committee played a critical role in validating the basic structure of the model (Benveniste, 1972).

From the model specification document, a list of significant factors was developed that served as the foundation of the ULAM effort. Figure 1 provides an overview of ULAM which is modeled and run using Visual SLAM and AweSim. The modeling and simulation techniques used in building ULAM include: fitting donor and patient arrival processes using non-homogeneous Poisson processes (NHPP) having exponential rate functions which may include both a polynomial and some trigonometric components; fitting distributions to data on transition times between states of medical urgency; applying variance reduction techniques using common random number streams and prior information; organizing data structures for efficient file searching and ranking capabilities; using bootstrapping techniques for attribute sampling; building submodels employing biostatistical procedures such as Kaplan-Meier and logistic regression; and estimating the performance measures suggested by the Oversight Committee.

All waiting patients have been diagnosed as having end-stage liver disease. Fundamental to liver allocations is a patient's medical urgency status. The definitions of medical urgency status are:

Status	*Definition*
1	Patients must be in an Intensive Care Unit (ICU) due to acute or chronic liver failure and have a life expectancy of less than 7 days without a liver transplant.
2	Patients have been continuously hospitalized in an acute care bed for at least five days, or are ICU bound.
3	Patients require continuous medical care.
4	Patients are at home and functioning normally and liver transplantation would be an elective procedure.
7	A patient listed as Status 7 is temporarily in-active; however, the patient continues accruing waiting time up to a maximum of 30 days.

Status transitions are made between these states and to two other statuses: Death where the patient has died while waiting for transplant (Status 8); and Removal where the patient is no longer considered as part of the UNOS system (Status 9). When either of these status changes occur, the patient is taken off the waiting list and all relevant statistics are updated.

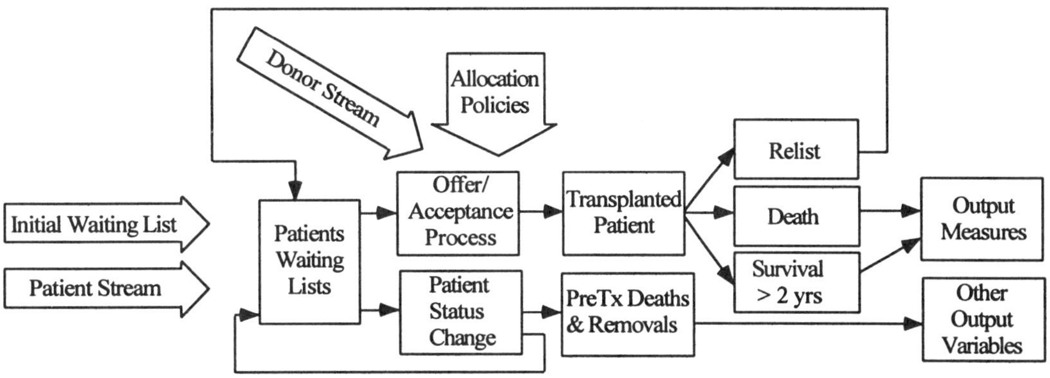

Figure 1. Overview of ULAM

Eligibility for allocation of a donated organ to a waiting patient is determined from blood type compatibility, ranges of organ weight and patient age. In allocating livers to patients in a geographical area, a ranking criterion is used. The ranking criterion is computed as the sum of points given to a patient based on the patient's status, blood type compatibility with the donor, and ranking with respect to other patients in the prescribed geographical area relative to amount of time waiting. This allocation method was included in ULAM and a model of the offering/acceptance of organs to patients was also included.

7 COMPARING NATIONAL LIVER ORGAN TRANSPLANTATION POLICIES

The initial use of ULAM in June 1995 discovered that the differences in alternative policies were clinically not large enough to recommend a change from the current policy. The first set of policies that were modeled allocated organs to the sickest patient first within different geographical areas. For example, in a local area such as an OPO which has 1 to 5 transplant centers associated with it, allocation would be to Status 1 patients first, then Status 2 patients and then Status 3 and 4 patients combined. Thus a local Status 3 patient could be allocated an organ before a Status 1 patient outside the OPO. Under these conditions, ULAM results showed that the trade-off between pre-transplant deaths (deaths occurring while people are waiting for an organ) and post-transplant deaths resulted in the total number of deaths being approximately the same for the policies considered (within the clinically significant difference). Prior to the modeling effort, the intuition of the transplantation community was that there would be clinically significant differences in the total deaths associated with policies concerning local, regional and national allocation methods.

The identification of this surprising result began the iterative process between the modeling effort and the transplantation allocation policy setters. As such, the policy-crafting process could be viewed as a discontinuous process (Hedberg and Johnson, 1977). To address the difference in expectation with regard to total deaths, a greater amount of discussion was held with the transplantation community and the model development effort intensified. This interaction produced a better understanding of the policy issues associated with transplantation (Morimoto 1973) and resulted in a greater level of confidence in ULAM's ability to compare policies.

Over the next several months, ULAM was enhanced by defining new status types; making geographical boundaries based on a percentage of total patients on the National Waiting List; devising new point systems to include population about the donor site; and using waiting time directly. In addition, restrictive capabilities were included in policies to allow a limit on the number of transplants for an individual patient and to require an identical blood type match between donor and patient.

In parallel with the modeling, a subcommittee of the Liver Committee was established to select performance measures which should be used for policy selection. The modeling team developed a categorization of performance measures according to utility and equity as viewed from medical, patient and system perspectives. The categorization of performance measures is presented in Table 1 and was a major step toward understanding and evaluating policies. The performance measure subcommittee selected a subset of these measures to be the basis for comparing policies.

8 NEW POLICY DEVELOPMENT AND POLICY SELECTION

The Liver Committee recommended a hybrid policy which maintains the local OPO structure while using the 20% of the National List as the definition of a regional allocation. The Liver Committee also recommended that the Board of Directors give consideration to retaining the current policy. The Allocation Committee developed its own recommendation by modifying the Liver Committee proposal. The change established two definitions of a local area based on percentages of the total National List; i.e., 5% of the National List for Status 1 and 2 patients and 2% of the National List for Status 3 patients. Both the Liver and the Allocation Committee policies involved allocating organs first to those acute patients who were, by distance, the closest 20% of all patients to the donor hospital.

After each new policy was suggested and upon approval of the appropriate committee, ULAM was enhanced in order to estimate the necessary performance measures. ULAM's flexible modeler and user interfaces provided direct access to outputs in the form of charts, plots, tables, and animations. Many of the proposed changes fit into the flexible organization that was developed for ULAM. However, some of the changes were more fundamental and created difficulties, for example, changing the definition of status types and the inclusion of both geographical boundaries and dynamic boundaries based on percentages of the national list. These changes were difficult because they required rebuilding component models since survival, relist and removal functions and transition matrices are all dependent upon the definition of medical status. In June 1996, the Board of Directors approved the change in definitions of medical status, approved the integration of

Table 1. Listing of Potential Liver Allocation and Distribution Performance Measures.

Medical Utility Measures	
MU1	Total (nonrepeated) transplants
MU2	Total deaths
	MU2.1 Pre-transplant deaths
	MU2.2 Post-transplant deaths
MU3	Total other patients removed
MU4	Percent of transplanted patients that survive
MU5	Total patients relisted
MU6	Number of life-years
MU7	Number of quality life-years
MU8	Size of the end waiting list
Patient Utility Measures	
PU1	Probability of receiving a transplant
PU2	Probability of dying on the waiting list[1]
PU3	Patient transitions from status at registration to other medical statuses
System Utility Measures	
SU1	Distance organs travel
SU2	Size of the end waiting list
SU3	Total costs of transplantation
Medical Equity Measures	
ME1	Total (nonrepeated) transplants[1]
ME2	Differences in percent of status types transplanted
Patient Equity Measures	
PE1	Differences in the probability of receiving a transplant
PE2	Differences in the probability of dying on the waiting list
PE3	Differences in waiting time across across categories
System Equity Measures	
SE1	Impacts on geographic areas or individual centers
SE2	Local use of organs and its impact on organ donation
SE3	Differentials in size of waiting lists

[1] Several measures may be classified as either utility measures or equity measures

medical statuses with their associated points for ranking, decided to maintain the geographical areas for distribution, and approved the priority allocation to acute patients based on a percentage of the total number of acute and chronic patients.

9 POLICY CRAFTING ISSUES

At the Winter Simulation Conference 1996, a description of ULAM will be presented as well as a chronological description of the interaction between the various UNOS committees and the support provided by the UNOS/Pritsker Team. A panel discussion will then be held dealing with issues that arise in using simulation to craft policy. Some of the issues are listed in Table 2 along with illustrative examples from the transplantation field. Many of these issues relate to the establishment of the demarcation lines between a system and its environment which is a key to successful modeling (Pritsker et al 1996).

10 CONCLUSION: SIMULATION'S EXCITING OPPORTUNITIES IN CRAFTING POLICY

Simulation has an outstanding track record in assisting operational decision-making by systematically estimating the performance implications of various alternatives. As companies today are examining broader, more large-scale changes, the application of simulation can help in the crafting of the best policy alternative.

The UNOS transplantation example illustrates how policy crafting involves broader issues with complex implications impacting diverse groups. Simulation's future application in supporting policy setting is highly dependent upon how modeling and simulation programs can foster meaningful executive communication and systematic understanding of the issues being considered.

For fourteen months, the UNOS/Pritsker Corporation team worked closely together to evolve a recommendation for a new national transplantation policy for livers based on delicately balancing of the intuitive solutions of the various participants and the analytic performance estimation of ULAM. During this time, many lessons were learned concerning how to best apply simulation methods into an organizational policy crafting context. These lessons can be grouped into three major areas:

Table 2 Issues Related to Crafting Policy

Issue	*Discussion*
1. Personal and organizational relations	Are geographical boundaries necessary in order to build personal relations and organizational ties?
2. Evaluation of benefits	How do you measure transplant center proficiency which is dependent on the experience of the surgical team, facilities available, rate of performing transplants and patient mix?
3. Ownership of assets	Does the local procuring organization have a claim to the organs they procure or are organs a national resource?
4. Distributions of a scarce life-saving resource.	Given that an organ is a scarce resource, should patients who have a higher probability of survival have preference in allocation?
5. Setting standards for policy statements.	If waiting time points are calculated based on registration time, how is a standard established for time of registration for terminally ill liver patients?
6. Evaluating downstream implications.	Are there ways to "game" the policy?
7. Can a national public policy be properly evaluated using a democratic process?	Can fair and effective policy be developed through a consensus process and by committees whose members have different points of view?

1 Reaching Consensus on the Goals, Methods and Approaches of Simulation

Policy issues, by definition, involve a broad spectrum of participants with diverse viewpoints, experiences and expectations. In order to have quantitative outputs accepted in such a decision-making environment, the various policy-makers must acquire a confidence in the simulation methods and models.

2 Specifying Definitions, Values and Measures

Policy selection requires the use of intuition, experience and judgment by those closest to the issue being evaluated. Conversely, simulation is based on precision, quantification and probabilistic reasoning. The key to blending these complementary approaches is to use qualitative information and simulation model outputs in such a way as to create a continuous dialogue that is aimed at improving policy setters intuition while improving the model's comparative ability.

3 Developing Policy Alternatives and Models Concurrently

Conceptually, a model's performance measures and policy selection criteria should be separate from the process of generating policy alternatives. However, to ensure a model can accurately capture the implications of the policy alternative, an iterative process between model specification, policy articulation and simulation testing is necessary. While the managing of this iterative process is difficult, the benefits in terms of model capabilities and output acceptance make the effort worthwhile.

The use of simulation to address policy-level issues offers new opportunities and challenges for our field. Our success at integrating the systematic science of simulation with the organizational process of policy development will make a vital contribution to effective policy selection.

ACKNOWLEDGMENTS

The authors wish to thank the other members of the UNOS/Pritsker Project Team for their contributions to the project. This includes Erick B. Edwards, Ph.D., and Ann M. Harper, UNOS; David L. Martin, Janet S. Reust, James R. Wilson, Ph.D., (consultant) and Michael E. Kuhl (consultant), Pritsker Corporation. The authors also wish to acknowledge the individuals who led the effort relating to policy analysis, evaluation and selection: Bruce Lucas, M.D., University of Kentucky; Margaret D. Allen, M.D., University of Washington Medical Center; Andrew Klein, M.D., The Johns Hopkins University; John P. Roberts, M.D, University of California, San Francisco; James F. Burdick, M.D., The Johns Hopkins University; Walter K. Graham, Executive Director, UNOS; and Ms. Judy Braslow, Director, Division of Organ Transplantation, H.R.S.A, D.H.H.S. We also appreciate the constructive comments and efforts of the more than 300 individuals who participated in committee meetings and who are an integral part of the policy selection process.

REFERENCES

Allison, G.T. 1971. *Essence of Decision: Explaining the Cuban Missile Crisis.* Boston, MA: Little, Brown.

Armstrong, J. S. 1982. The Value of Formal Planning for Strategic Decisions: Review of Empirical Research. *Strategic Management Journal.* 197-211.

Benveniste, A. 1972. *The Politics of Expertise.* Berkeley, CA: Glendessary Press.

Bryson, J. M. 1988. *Strategic Planning for Public and Non-Profit Organizations.* San Francisco: Jossey-Bass.

Eccles, R. G. 1991. The Performance Management Manifesto. *Harvard Business Review.* 61(1): 131-137.

Garuin, D. A. 1993. Building a Learning Organization. *Harvard Business Review.*

Hedberg, B. L. T. and S. A. Johnson. 1977. Strategy Formulation on a Discontinuous Process. *International Studies on Management and Organization* . VII 2: 88-109.

Kuhl, M. F., J. R. Wilson and M. A. Johnson. 1995. Estimation and Simulation of Nonhomogeneous Poisson Processes Having Multiple Periodicities. In *Proceedings, Winter Simulation Conference.* 374-383.

Mockler, R. J. 1991. A Catalog of Commercially Available Software for Strategic Planning. *Planning Review.* May/June: 28-35.

Morimoto, K. 1973. Notes on the Context of Learning. *Harvard Educational Review.*

Peters, J. T., K. R. Hammond, and D. A. Summers. 1974. A Note on Intuitive Versus Analytic Thinking. *Organizational Behavior and Human Performance.* 12: 125-131.

Phillips, M. G., Ed. 1996. *Organ Procurement, Preservation and Distribution in Transplantation.* 2nd ed. Richmond, VA: UNOS.

Pritsker, A. A. B. et al. 1995. Organ Transplantation Policy Evaluation. In *Proceedings, Winter Simulation Conference.* 1314-1323.

Pritsker, A. A. B. et al. 1996. *Simulation With Visual SLAM and AweSim.* New York and West Lafayette, IN: John Wiley and Systems Publishing.

Withers, B. D. et al,. 1993. A Structured Definition of the Modeling Process. In *Proceedings, Winter Simulation Conference.* 1109-1117.

AUTHOR BIOGRAPHIES

A. ALAN B. PRITSKER is the Chairman and CEO of Pritsker Corporation. He obtained a Ph.D. from The Ohio State University in 1961. He is a member of National Academy of Engineering. Dr. Pritsker served twice as a member of the Board of Directors of the WSC and as Board Chairman in 1984 and 1985.

O. PATRICK DAILY is Director for Research and Scientific Registry Operations at UNOS. He attained a Ph.D. from the University of Maryland in 1976. He has 28 years of experience in medical research, healthcare operations and senior level management. Dr. Daily was Commissioned Officer in the U.S. Navy Medical Department for 24 years. He has broad experience in government, medical research, multicenter data collection and healthcare management.

KENNETH D. PRITSKER is Managing Partner of the Strategic Reengineering Institute. He obtained a Ph.D. from the University of Washington in 1987. He taught strategic planning and policy analysis at the University of Pittsburgh and was a strategic planner for the Boeing Corporation from 1990 to 1993. The Strategic Reengineering Institute serves as an industry, academic and consulting consortium to expand traditional business process reengineering approaches to a macro-industry level.

Proceedings of the 1996 Winter Simulation Conference
ed. J. M. Charnes, D. J. Morrice, D. T. Brunner, and J. J. Swain

AN ONLINE, SIMULATION-BASED PATIENT SCHEDULING SYSTEM

Hans Manansang
Joseph A. Heim

Industrial Engineering Program
University of Washington
Seattle, WA 98195

ABSTRACT

In this research, the application of on-line simulation for patient scheduling decision support in a clinical environment is investigated. A prototype system was designed and constructed to demonstrate the concepts and to address various problems to be faced in implementation of the system.

1. INTRODUCTION

Delivering quality health care at a reasonable cost and in a timely manner has become a fundamental issue that must be addressed by all health care enterprises. Coordinating the administrative aspects of patient care delivery, however, is consuming more staff time and becoming increasingly complex. Detailed examination of Patient Care Coordinator (PCC) activities in a clinical service at the University of Washington Medical Center found that substantial time and effort are expended in coordinating patient and medical resources. The inefficient use of patient care professional time is mainly caused by complex issues concerning allocation over time of limited and highly interdependent resources, and uncertainty surrounding the prospect of treatment. Response to the dynamic allocation of resources above can be categorized as scheduling problem (Roberts 1981, Crowstone and Malone 1993). With an anticipated growth of out-patient service demand, the severity of these scheduling issues are expected to escalate.

A Mohs-micrographic controlled surgery for skin cancer treatment facility at University of Washington was selected as a case study for investigation of scheduling issues in health care organizations. The current approach to patient management is based on long-range calendar planning between patient and medical resources. However, because the real-time activities involved in coordinating health care delivery are subject to a great deal of variance and unexpected change, there is a need to study the dynamic behavior of the system and be able to prepare for the consequences of the changes to the patient treatment schedule.

In the research reported here, we examine a simulation modeling approach for scheduling patients in medical clinics. Traditional use of simulation to study the behavior of the system and address off-line decision making has been performed (Roberts, 1981 and Pardue, 1995). However, long-range scheduling developed off-line is static and will quickly lose its validity in a rapidly changing situation. Recently there has been an increasing interest in the development and use of simulation for real-time and continuous analysis in area of manufacturing, (Drake, Smith, Peters 1995 and Harmonosky 1990). This type of simulation is categorized as on-line simulation. Distinctively, an on-line simulation refers to continuous simulation run and update with data input and output from physical system in real-time. On-line simulation provides the advantages of a "look ahead" and "what-if" capabilities in real-time manner while continuously monitoring the system. The combination of both features provides continuous decision support to determine the next incident to occur and anticipate the actions to be executed (Harmonosky 1995).

The paper examines the coordination of multiple resource allocation and resource sharing dependencies among several closely related specialty clinics within the University of Washington Medical Center. We also report on development of a prototype for real-time scheduling using on-line simulation concepts.

2. PROBLEM DESCRIPTION

2.1. Workflow Overview

A Mohs-micrographic controlled surgery for skin cancer treatment includes the following basic treatments: 1) registration, 2) evaluation, 3) the Mohs surgery, and 4) reconstructive surgery. The initial visits by the patient are primarily concerned with administrative activities undertaken by a Patient Care Coordinator (PCC): gathering patient medical records, demographic data, and financial status (i.e., insurance policy/managed care). Once these activities are completed, the patient is scheduled for evaluation and/or examinations. Figure 1 illustrates the overall Mohs Treatment workflow for patients.

Evaluation, which includes detection, and test for malignancy, are performed between 1:00 PM and 3:30 PM, Monday, Tuesday and Wednesday. If the patient lesion is suspected of being cancerous, then a biopsy is performed. The tissue is sent to pathology laboratory for vigorous histological evaluation. Pathology results are available in about three days. If the biopsied tissue test result is a positive malignancy, the patient is scheduled for surgery.

The Mohs-micrographic controlled surgery hours of operation are 8:00 AM until 11:00 PM, Monday, Tuesday, Wednesday and Thursday. Once the patient arrives for surgery, the patient is assigned to one of the surgery suites and receives pre-biopsied treatment such as local anesthesia. Next, Mohs surgery is performed by removing a section (a section corresponds to a thin horizontal layer of the skin in a grid pattern) around the cancerous area. The layer then is examined microscopically by the histologist for evidence of cancer cells. During this investigation, the patient is taken into the post-treatment area, and the doctor will begin to treat another patient. If the inspection of the layer by the histologist is found to still contain malignant cells, the doctor continues to remove a layer

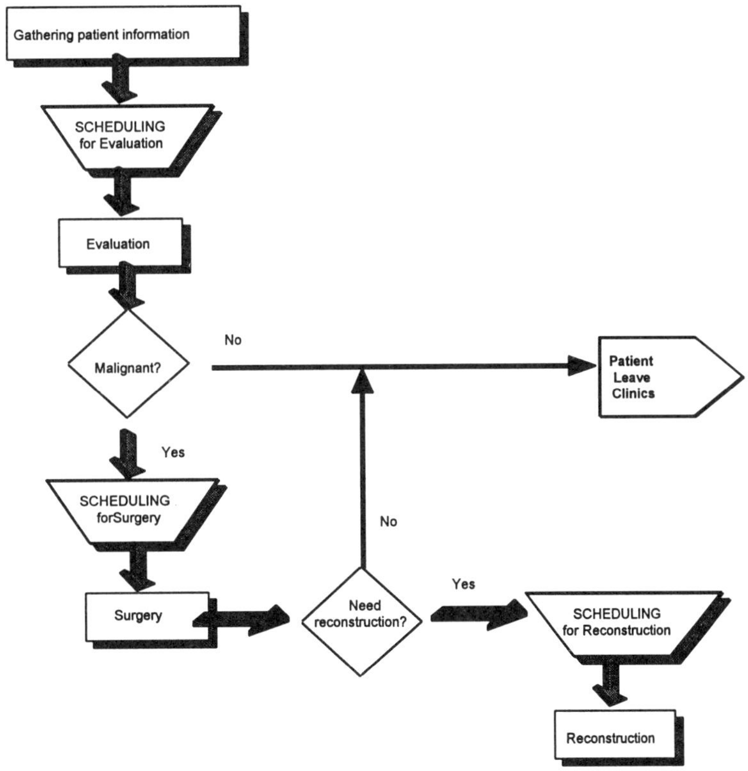

Figure 1. Mohs Treatment Workflow Overview

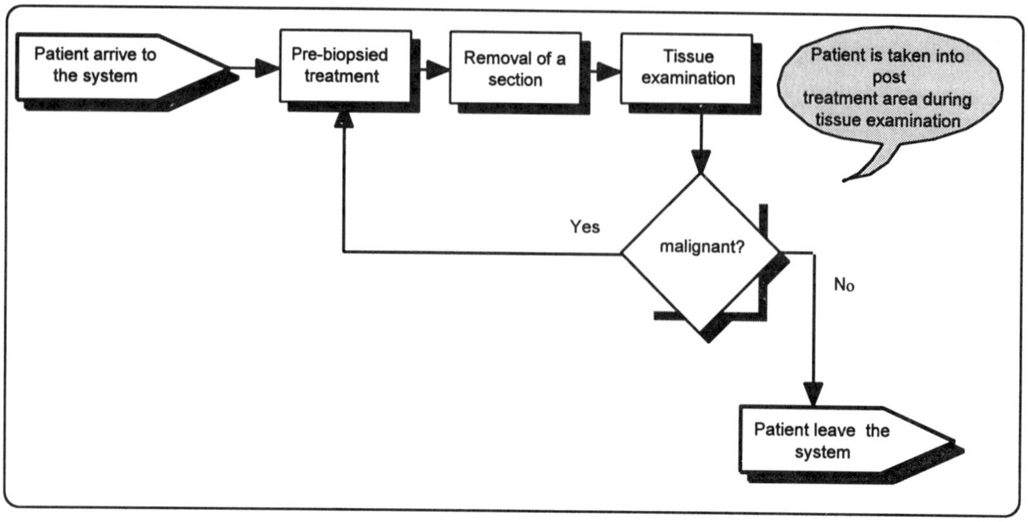

Figure 2. Mohs Surgery Process Flow

of tissue until no margin of malignancy exist (see figure 2).

A patient blocking situation may occur when the doctor has finished performing an excision, yet the histologist is still working on the previous patient's tissue. The number of sections to be removed are estimated during patient evaluation and subsequently used as the basis for scheduling time needed for the patient's surgery. Table 1 contains the treatment resources available.

Resource Description	Quantity
Mohs Doctor	1
Nurses	3
Histologists	1
Medical Assistance	1
Surgery Suites	2

Table 1. Mohs Treatment Resources

There are three types of carcinomas: basal cell carcinoma, squamous cell carcinoma, and malignant melanoma. Patients with malignant melanoma malignancy need to be treated in two sequential contiguous days. Patients with a larger lesion have to go through reconstruction in the Otolaryngology clinic. The hours of reconstruction are Tuesday and Thursday from 1:30PM to 3:30PM. Reconstructions usually are performed within two days after surgery and carried out in the operating room which is shared with other out-patient clinics. Currently, the activities in reconstruction including the interdependent scheduling for Mohs Surgery are not included in this study.

2.2. Scheduling Issues

For Mohs surgery, scheduling of patients is performed by a Patient Care Coordinator (PCC). Negotiation occurs in scheduling both the patient and the medical resources. To determine a schedule for surgery, the PCC has to estimate the processing time of the tasks, the time the patient has to arrive to the clinic and the availability of the medical resources.

The PCC also has to estimate the maximum time needed for the predicted number of sections to be removed for every patient. The maximum duration for all patients in a day contributes to the total completion time which should fit within the Mohs Treatment clinic hours of operation. Fitting this estimated duration of tasks to the hours of the operation is heavily dependent on the experience of the PCC. A delay in processing of one patient beyond the estimated time will cause delay in treatments for successor patients; hence, increasing the completion time and so the lateness of the Mohs surgery operation hours.

In more severe cases, a delay in Mohs surgery could postpone the schedule for subsequent reconstruction treatment. Reconstructive surgery is dictated by the availability of the resources which must be coordinated with Mohs surgery patients.

Complexities increase for patients with two contiguous days of surgery (in the case of malignant melanoma and reconstruction); any delay can cause the whole system to become congested. To avoid these circumstances, PCCs are forced to increase slack time in the Mohs surgery clinic schedule to accommodate the unexpected. Not only this relaxation of schedule is not optimal, but it also introduces significant clinic inefficiency. Externally, frequent changes of patient and medical resource schedule and duration of treatment require continuous schedule negotiation.

With the complexities of coordinating patient schedules to meet with the resource constraints in an efficient and adaptive response, the current physical system needs a way to *look-ahead* and perform *what-if* analysis in a real-time.

3. FRAMEWORK OF ON-LINE SIMULATION SCHEDULING SYSTEM

3.1. System Framework

In following section, we discuss the development of a real-time scheduling system using the concepts of on-line simulation. The generic scheduling system framework includes the following elements: the human user interface system, a database management system, and a discrete-events simulation engine. See Figure 3.

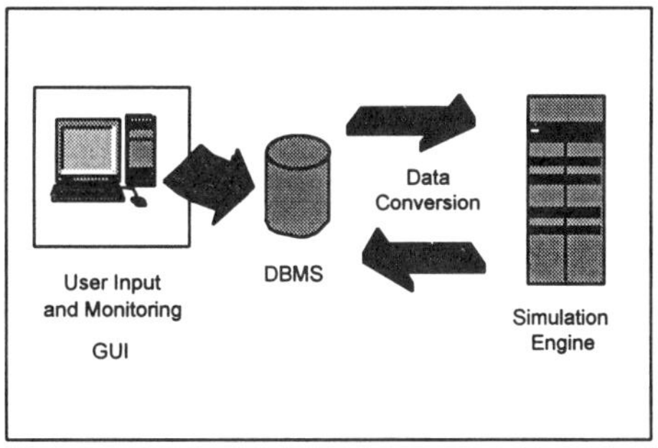

Figure 3. Patient Scheduling System Framework

The graphical user interface provides a means of entering an event schedule into the system. The database stores the value of variables such arrival times of the patients, the diagnosis of treatment, and, eventually, the time the patient leaves the system. The simulation engine provide the simulation capability to examine and predict the behavior of the system according to the variables stored in the database.

The value of major treatment variables are entered by the PCC : the arrival time to the Dermatology clinic, the number of sections to be removed, the type of malignancy (for each patient), and the availability of medical resources. When the user requests execution of the simulation model, the data is then feed to the simulation engine from the database. The simulation output then is stored back into the database and can be examined by the user in the monitoring window.

The simulation output provides statistical analysis on the departure time of the last patient, duration of each patient's treatments and characteristics of the system based on the selected patients. Using these *look-ahead* features, the PCC is able to evaluate decision alternatives on scheduling patients for the desired time period (i.e. the number of patients to be scheduled on a certain day, which patients are eligible based on number of sections to be removed, and when each patient should arrive to begin treatment).

The physical surgery system, in a general sense, can be characterized as a stochastic job shop machine problem with two processing systems (excision and inspection of the tissue); recirculating, blocking, release dates (corresponding to arrival time of the patient), and carcinoma type as constraints. The goal of the system may involve minimizing the makespan of individual queue delay and minimizing the total completion time for the collective patient per day. Analytical approaches to this problem are found to be extraordinarily complex (Pinedo, 1995).

To facilitate the stochastic nature and the complexities of the scheduling problems, the models are designed to conduct multiple simulation runs, each configured to provide the maximum amount of information to the PCC trying to construct a clinic schedule. The parameters for number of sections, patient arrival time, number of patients and type of malignancy can be set to deterministic values or be stochastically determined by the simulation engine during the run.

3.2. Prototype Implementation

In the construction of the prototype, Visual Basic, ACCESS and ARENA were used for the user interface system, the database management system and the simulation engine respectively. Visual Basic was chosen as the interface software due to its ease of programming and its capability to communicate directly with ACCESS databases. ARENA, a general purpose discrete-event simulation environment, provided a flexible modeling facility.

The model include a loading module to transfer data from the database to ARENA v1.28 and vice versa. This is performed through ASCII code conversion. The data from ACCESS is assigned to the specific variables at the beginning of a simulation experiment.

The user interface provide a single point of contact for system control and analysis. This includes: 1) appointment and event calendars graphical user interface for instant user input to be incorporated to simulation run and 2) statistics on the "what if" condition from the simulation result. Controls are provided at the interface that allow the user to browse the existing schedule, to edit the schedule, and to run a simulation. For each block appointment time, the interface will prompt a window for additional information on the number of sections to be taken and type of carcinoma. Additional programming is embedded in the GUI (graphical user interface) to insure that the surgery and reconstruction will be done within two days of one another (this limits late week Mohs surgery when reconstructive surgery is not possible). Data entered with the GUI are stored directly to the ACCESS database. When the simulation is run, ARENA will be started from within the Visual Basic program. The output from ARENA is stored in ACCESS and displayed at the request of the user.

Data on service duration needed in the simulation engine consists of duration of patient evaluation, time for removing lesion, time for tissue processing and length of time needed for anesthetizing the section of skin to be removed. These data are derived from interviews and observations of clinical workflow and patient scheduling activities.

4.0 DISCUSSION

4.1 Initial Results

To gain insight on the usefulness of the simulation-based scheduling concept, a comparison between two parameter sets was developed. The first used patient arrival time data obtained from clinic observation and data collection efforts. The second was conducted with arrival time set as the *earliest time of the day* policy. Each scheduled five patients. Both were subject to the same resource constraints and the estimated number of sections to be removed for each patient.

The output shows that the same results on total completion time are obtained from both parameter sets. However, the patients in the second setting (as expected) experienced longer queue delays. A logical explanation of this is that the fifth patient in the second setting arrives at 8:00 A.M. in the morning and waited for the other patients to be processed. Hence, a longer waiting time was experienced. This also shows that the data obtained from the PCC for the first setting are quality judgment based on PCC experience in the scheduling area.

4.2 Future Work

Further development of online simulation scheduling system has focused on creating a tighter integration with the current database and making the model more broadly available and comprehensive. The model will be accessible on a data communications network where patients, medical resources, and medical center staff can remotely perform *what-if* analyses. With adoption of an object-oriented approach to simulation scheduling, classes of models can be developed, and patient scheduling with the related clinics to Mohs surgery will be able to interact with one another. And finally, by using time compression in simulation, we hope that the models will provide closer to reality model fidelity (Miller, Sheth, Kochut, Wang and Murugan 1995).

4.3 Conclusions

One of the primary concerns in health care is efficient utilization of professionals and the technological resources provided in their support. In this paper we have discussed a surgery clinic in which the stochastic aspects of patient disease, diagnostic unknowns and the variance of daily clinic operations combine to create a complex and time-consuming scheduling task. We believe that many of the concepts of on-line simulation, borrowed from manufacturing, but tempered by the human aspects of the problem, can be effectively used to support scheduling decisions when treating patients. A prototype simulation-based system for testing some of those ideas has been developed and we have discussed our preliminary results. It appears that a combination of qualitative professional skills and comprehensive planning tools can substantially improve the service and efficiency of the health care system.

REFERENCES

Crowston, K. and Malone, T. W. November 1993. The Interdisciplinary Study of Coordination. *ACM Computing Surveys, 1994(March) 26(1), 87-119.*

Drake, Glenn R., Smith, Jeffrey S., and Peter, Brett A. 1995. *Simulation As a Planning and Scheduling Tool For Flexible Manufacturing Systems*. Department of Industrial Engineering Texax A&M University.

Harmonosky, Catherine M. 1990. *Implementation Issues Using Simulation for Real-Time Scheduling, Control, and Monitoring.* Proceedings of the 1990 Winter Simulation Conference.

Harmonosky, Catherine M. 1995. *Simulation-Based Real-Time Scheduling: Review of Recent Developments.*

Proceedings of the 1995 Winter Simulation Conference.

Miller, A. John, Sheth, P. Amit, Kochut J. Krys, Wang Xuzhong and Arun Murugan. 1995. *Simulation Modeling within Workflow Technology.* Proceedings of the 1995 Winter Simulation Conference

Pinedo, M. 1995. Scheduling : *Theory, Algorithms and Systems.* Prentice Hall International Series in Industrial and System Engineering.

Pardue, Harold. J., Clark, Thomas D. Jr. Cognetta, Armand B. Jr., 1995. *A System Analysis and Model of Real-Time Skin Cancer Treatment.* Proceedings of the 1995 Winter Simulation Conference.

Roberts, Stepen D. 1981. *Survey of the Application of Simulation to Health Care.* Society for Computer Simulation (Simulation Council).

AUTHOR BIOGRAPHIES

HANS MANANSANG is a Master of Science candidate in the Industrial Engineering Program at the University of Washington. He received his Bachelor of Science in Mechanical Engineering degree at Seattle University in 1990. His research interest is in health care and manufacturing application simulation modeling. He is a member of INFORMS.

JOSEPH A. HEIM is an assistant professor of Industrial Engineering at the University of Washington in Seattle, Washington. His teaching responsibilities include computer integrated systems, simulation, inventory management, and scheduling. Heim's research focuses on coordination of complex collaborative tasks with distributed discrete event simulation models. Professor Heim is a member of ASEE, ACM, IIE and SME.

Proceedings of the 1996 Winter Simulation Conference
ed. J. M. Charnes, D. J. Morrice, D. T. Brunner, and J. J. Swain

ANALYZING THE TEMPORAL ASSOCIATION BETWEEN HEALTH DISORDERS AND MEDICAL TREATMENTS USING PROBABILITY MODELS AND MONTE CARLO SIMULATION

Sheldon H. Jacobson

Department of Industrial and Systems Engineering
Virginia Polytechnic Institute and State University
Blacksburg, Virginia 24061-0118

Douglas J. Morrice

MSIS Department
The University of Texas at Austin
Austin, Texas 78712-1175
End to End Simulation Department
Schlumberger Austin Research Center
Austin, Texas 78720

ABSTRACT

Statistical analysis is often used in medical studies to provide evidence for an association between a treatment and a patient condition. The work in this paper is motivated by a recent medical study which analyzes the *temporal association* between a treatment (i.e., an implant procedure) and the rare autoimmune disorders, polymyositis and dermatomyositis (PM/DM). To address this association mathematically, this paper develops a probability model based on the multinomial distribution that can be used to make statistical inferences about the timing of incidences of a treatment/condition pair. This paper details an empirical study of this model using Monte Carlo simulation. It also describes some analytical results developed in subsequent research efforts. Data from the medical study illustrate the application of this model and its results.

1 INTRODUCTION

Medical studies often provide statistical data as evidence for an association between a treatment (e.g., medical procedure, drug, therapy) and a subsequent patient condition (e.g., disorder, syndrome, disease). Frequently reported statistics such as the number or proportion of a treatment condition pair are inherently *static* and fail to capture an additional dimension often found in patient data: the *timing* of the incidences of a patient condition after the treatment has been administered. The latter is referred to as *temporal association* (Cukier et al. 1993). Approaches designed to detect temporal association are potentially more powerful than static approaches in establishing connections between treatments and patient conditions. Whereas a static approach might consider the number of incidences in a given population of patients statistically insignificant, a temporal approach applied to the same data might find the timing of the incidences to be extremely rare. For example, while the total number of patients who develop a particular condition any time after receiving the treatment may be very close to the expected number, these patients may do so much sooner than expected.

This paper develops a mathematical model to study temporal association in a problem encountered in a medical study conducted by Cukier et al. (1993). Their study is designed to determine the existence of an association between the uncommon autoimmune disorders PM/DM and an implant procedure used to correct wrinkles due to aging, acne scars, and other superficial skin defects.

Other literature exists on the modeling of event timings in medical data. This research considers problems such as estimating failure time distributions, constructing regression models for censored data, and comparing survival time distributions between multiple groups of patients in clinical trials. (see, for example, Kalbfleisch and Prentice 1980 and Fleming and Harrington 1991). None of this work, however, directly addresses the temporal association problem considered in this paper.

To examine the temporal association, Cukier et al. (1993) conducted a retrospective cohort study that included approximately 345,000 patients who had received the implants over an eight year period. Of the 345,000, 9 patients developed one of the autoimmune disorders. Table 1 contains the timings of the disease diagnoses after the implants were given. A case is recorded in month 1 if it is diagnosed in the interval $(0, 1]$ months, month 2 if it is diagnosed in the interval $(1, 2]$ months, and so on. Months not included in the table had no new cases diagnosed.

Cukier et al. (1993) also construct a conditional probability distribution that reflects the temporal likelihood of a patient developing at least one of the autoimmune disorders in each one month post-implant period over a 96 month time horizon (96 months are considered because the maximum post-

Table 1: Timing of Diagnoses for the 9 Patients After the Implants Were Given

Month	1	2	4	6	7	8	25
Number of Cases	1	1	3	1	1	1	1

exposure observation period for any of the patients was eight years). This distribution is shown in Figure 1. These data reflect the likelihood of incidences of PM/DM in a patient population that grows from zero at the beginning of the study period to 345,000 patients after eight years, assuming the incidences were purely coincidental after the implants were given. Finally, Cukier et al. (1993) assess the likelihood of the incidence timings for the 9 patients who developed the disorders against the temporal probability distribution using Monte Carlo simulation.

The objective of this paper is to formulate a mathematical model to describe the problem studied in Cukier et al. (1993), and to present the results of a Monte Carlo study used to support the results in Cukier et al. (1993). The paper also describes some analytical results that extend the Monte Carlo simulation study. The remainder of the paper is organized as follows. In Section 2, the mathematical model is formulated. The model is not restricted to the particular problem studied in Cukier et al. (1993). In fact, it is general enough to study the temporal association between any patient condition and treatment. Section 3 describes the details of the Monte Carlo simulation model and the results used in Cukier et al. (1993). Section 4 describes some of the analytical results that extend this research. Section 5, summarizes the results and provides some concluding remarks.

2 MODEL FORMULATION

The problem described in Section 1 can be restated to facilitate the construction of the underlying probability model. Suppose the units in the sample are balls (rather than patients who develop the disorder) and the monthly cells on the time scale are urns. An equivalent problem formulation is that of randomly distributing N balls into Q urns where the balls act independently. The probability of any ball falling into the qth urn is u_q, $q = 1, \ldots, Q$, where $\sum_{q=1}^{Q} u_q = 1$. Let N_q denote the number of balls placed in urn q $(\sum_{q=1}^{Q} N_q = N)$. The underlying probability distribution for the $\{N_q, q = 1, \ldots, Q\}$ is the multinomial distribution, (see Johnson and Kotz 1977, page 108).

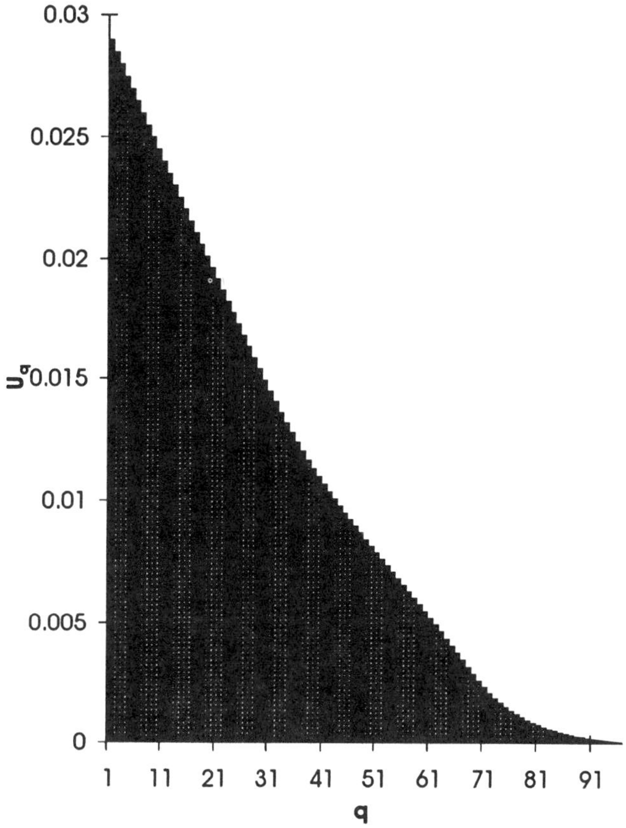

Figure 1: Temporal Probability Distribution of Patients Expected to Develop Coincidental Autoimmune Disorders

In the multinomial framework, there are various ways to characterize the temporal association between the disorders and the implant procedure. Cukier et al. (1993) identify the discrete random variable

$$R = \sum_{q=1}^{Q} q N_q \tag{1}$$

to be appropriate for their study and purpose, where R can assume the values $N, N+1, \ldots, QN$. Although other measures could be used to characterize temporal association, R is a natural quantity to consider for the medical study because it measures the mean latency period for a particular sample of patients. In other words, it can be interpreted as the cumulative number of patient-months until all patients in which the disorder occurred actually developed the disorder after receiving the implant treatment. For example, from the data in Table 1, the observed value of R is 61. In other words, it took a total 61 patient-months for all 9 patients to develop an autoimmune disorder after the implant treatment. Therefore, R is a way to compare different patient group configurations on the basis of temporal ordering.

In order to study the temporal association, one must characterize the probability distribution of R and then assess the likelihood of observed values of R being generated from this distribution. For example, Cukier et al. (1993) assess the likelihood of the disorders occurring in the 9 patients as soon after treatment as they did. Stated in terms of R, they assess the likelihood of a randomly selected 9 patient group (from the multinomial distribution with probabilities given in Figure 1) having a total patient-month value of less than or equal to 61, which is just the tail probability

$$P\{R \leq 61\}. \tag{2}$$

A small value for this tail probability provides evidence for a strong temporal association between the disorders and the implant treatment since this indicates that disorders occur much sooner than expected from the patient population.

Based on the properties of the multinomial distribution, certain aspects of the distribution of R can be easily quantified. The mean and the variance of R are

$$E[R] = \sum_{q=1}^{Q} q u_q N \tag{3}$$

and

$$Var[R] = \sum_{q=1}^{Q} q^2 u_q(1-u_q)N - 2\sum_{q}\sum_{<r} q r u_q u_r N. \tag{4}$$

Using the values for the $\{u_q\}$ shown in Figure 1, along with $Q = 96$, and $N = 9$, yields $E[R] \approx 229.056$ and $Var[R] \approx 3226.70$. An analytic upper bound for the tail probability in (2) can be obtained from the one-sided Chebyshev inequality (Ross 1988, page 352):

$$P\{R < 62\} \leq \frac{3226.70}{3226.70 + (229.056 - 62)^2} \approx 0.104. \tag{5}$$

Expression (5) illustrates that the tail probability , $P\{R \leq 61\}$ is bounded above by 0.104. This means close to 90% of all possible 9 patient groups would have a patient-month total of greater than or equal to 62. While this indicates that the observed R for the 9 patient group in Table 1 is somewhat rare, it is far from definitive since statistical significance in most studies is declared on a probability of 0.05 or less. In most instances, the Chebyshev inequality is quite conservative and does not provide a tight upper bound. In fact, this will be demonstrated in Section 4 to be the case.

Section 3 provides the details of the Monte Carlo simulation model that was used to generate the results reported in Cukier et al. (1993). We demonstrate that the model is able to produce a very precise estimate of the tail probability in (2).

3 A MONTE CARLO SIMULATION APPROACH

The relationship between the multinomial distribution and R results in a straightforward Monte Carlo simulation model for this problem. The simulation input was generated by the linear congruential pseudo random number generator RAND from Law and Kelton (1991), pages 449-450. According to the authors, this generator is portable and well-tested (see also Marse and Roberts 1983). In our Monte Carlo experiment, nine pseudo uniform random variates were transformed using inversion (Law and Kelton 1991, page 469) to produce a sample of nine random variates from the multinomial distribution. Then, R was computed from the multinomial sample. An estimate of the probability distribution for R was generated by performing $1,000,000$ replications of this experiment. The following algorithm summarizes the logic used in a simulation program written in FORTRAN:

Algorithm 1 *Initialize the Number of Samples, N, Q, and the $\{u_q\}$*

Compute the cumulative probability distribution, C_q, for u_q, $q = 1, 2, \ldots, Q$

For $I = 1$, Number of Samples

Set $F_R = 0$, for $R = N, N+1, \ldots, QN$ (initialize the distribution of observed frequencies for R to zero)

Set $R = 0$

Set $N_q = 0$ for $q = 1, 2, \ldots, Q$ to zero

For $J = 1, N$

Sample U (Uniform(0,1) random variate)

Set $K = 1$

While $U > C_K$ do

$K = K + 1$

EndWhile

$N_K = N_K + 1$

EndFor J

For $L = 1, Q$

$R = R + N_L L$

EndFor L

$F_R = F_R + 1$

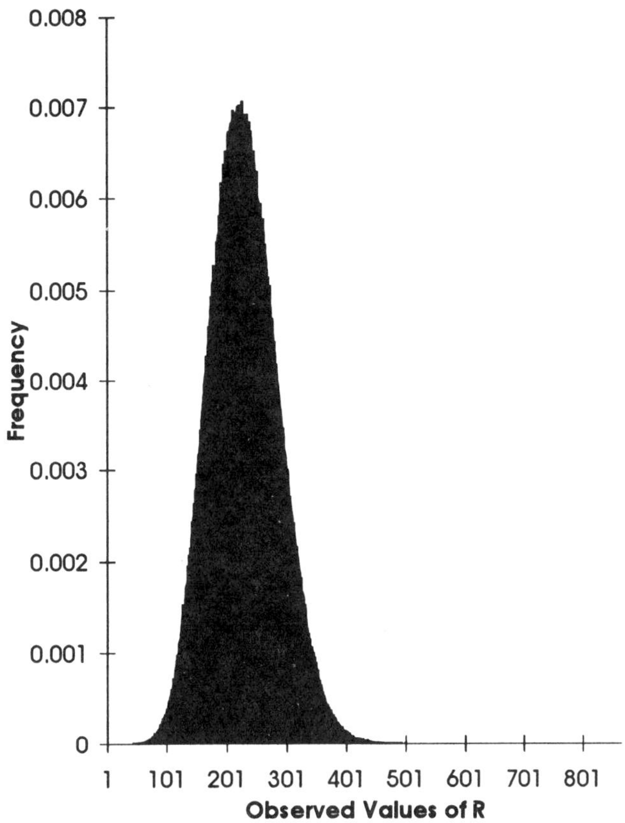

Figure 2: Observed Frequency for R from a Monte Carlo Simulation Run of 1,000,000 Samples

EndFor I

Calculate the cumulative sum of F_R *for* $R = N, N+1, \ldots, QN$

Output F_R *and the cumulative sums for* $R = N, N+1, \ldots, QN$.

Figure 2 is a distribution of the observed frequencies of values for R from the Monte Carlo simulation run. This plot (suitably standardized) serves as an unbiased estimate of the probability mass function (Law and Kelton 1991, page 362). It appears to be symmetric about the approximate mean value of 229.056 with very little probability in the tails beyond the value of $R \leq 61$.

The $P\{R \leq 61\}$ was estimated by,

$$\hat{P} = \frac{(\#\text{ of samples for which the observed R} \leq 61)}{1,000,000}. \tag{6}$$

The data illustrates that the probability of observing $R \leq 61$ is extremely low. Out of the 1,000,000 replications, 92 fall at or below the value of 61. Therefore, a point estimate for $P\{R \leq 61\}$ is $\hat{P} = 9.2 \times 10^{-5}$. Since the sample size is large and (6) is a proportion, large sample theory can be used to produce an approximate confidence interval for $P\{R \leq 61\}$. Using the procedure found in Bickel and Doksum (1977), page 160, an approximate $(1-\alpha) \times 100\%$, $(0 < \alpha < 1)$ confidence interval for $P\{R \leq 61\}$ is

$$\frac{S + \frac{k_\alpha^2}{2} \pm k_\alpha \sqrt{\frac{S(\eta - S)}{\eta} + \frac{k_\alpha^2}{4}}}{\eta + k_\alpha^2}$$

where S is the number of Monte Carlo simulation replicates observed in the tail, k_α is the $1 - \frac{\alpha}{2}$ percentile from the normal distribution, and η represents the total number of simulation replicates. In this example, $S = 92$, $k_\alpha = 1.96$ (for a 95% confidence level), and $\eta = 1,000,000$. Hence, the observed confidence interval is $(7.5024 \times 10^{-5}, 1.1282 \times 10^{-4})$. Bickel and Doksum (1977) state that the use of this confidence interval is satisfactory as long as the smaller of $(1,000,000)(P\{R \leq 61\})$ or $(1,000,000)(1 - P\{R \leq 61\})$ is at least 5. For the example in this paper, the value is estimated to be 92.

4 ANALYTICAL APPROACHES

The probability distribution for R can be determined analytically by addressing the following combinatorial problem: find all possible values of $\{N_q, q = 1, \ldots, Q\}$ such that

$$\sum_{q=1}^{Q} N_q = N \tag{7}$$

and

$$\sum_{q=1}^{Q} q N_q = R. \tag{8}$$

Once the possible combinations of $\{N_q\}$ are known, the probability that R equals some prespecified value can be calculated analytically using the probabilities from the multinomial distribution. Since we are interested in a cumulative tail probability, this procedure would have to be repeated for each fixed value of R. Total enumeration for all the $\{N_q, q = 1, \ldots, Q\}$ that satisfy the above two constraints for different values of R is quite difficult because the number of possible combinations of the $\{N_q\}$ grows exponentially in the size of Q, N, and R (see Jacobson and Morrice 1996). Hence a total enumeration approach is impractical for realistic sized problems such as the one considered in Cukier et al. (1993).

Fortunately, this problem can be solved without using total enumeration. In order to define this procedure, some additional notation is needed. Let n_q

be the total number of balls in the first q urns, $q = 1, 2, \ldots, Q$. By definition, $n_q = \sum_{i=1}^{q} N_i$. In addition, let

$$R_q = \sum_{i=1}^{q} iN_i$$

and

$$p_{q,n,r} = P\{R_q = r, n_q = n\}$$

for $q = 1, 2, \ldots, Q$, $n = 1, 2, \ldots, N$, $r = n, n+1, \ldots, qn$. Finally, let $V_{q,n,r}$ be the total number of patient samples (or vectors) of size n that are possible for the first q urns with $R_q = r$ and $n_q = n$ balls.

The problem is to determine the distribution for $R_Q (\equiv R)$, i.e., $p_{Q,N,r}$. The following theorem provides the basis for a recursive approach to calculating $p_{q,n,r}$.

Theorem 1 *The probability $p_{q,n,r}$ can be computed using the recursion*

$$p_{q,n,r} = \sum_{m=0}^{h(q,n,r)} p_{q-1,n-m,r-qm} \binom{n}{m} u_q^m \quad (9)$$

for $q = 2, 3, \ldots, Q$, $n = 1, 2, \ldots, N$, and $r = n, n+1, \ldots, qn$, where $h(q, n, r) = min\{n, \lceil r/q \rceil\}$. The initial boundary conditions for this recursion are

1. $p_{q,n,r} = 0$ *for* $q = 1, 2, \ldots, Q, n = 1, 2, \ldots, N$ *and* $r = 1, 2, \ldots, N-1$.
2. $p_{1,n,n} = u_1^n$ *for* $n = 1, 2, \ldots, N$.
3. $p_{q,0,0} = 1$ *for* $q = 1, 2, \ldots, Q$.
4. $p_{q,0,r} = 0$ *for* $r = 1, 2, \ldots, qN$ *and* $q = 1, 2 \ldots, Q$.

Proof: See Jacobson and Morrice (1996).

The recursion in (9) can be successively applied, starting with the values $q = 2$ and $n = 1$, until the desired values for $q(= Q)$, $n(= N)$, and r are reached. A recursive algorithm is described in Jacobson and Morrice (1996).

A corollary to Theorem 1 can be used to determine the number of vectors for each possible value of R.

Corollary 1 *The number of vectors, $V_{q,n,r}$, can be computed using the recursion*

$$V_{q,n,r} = \sum_{m=0}^{h(q,n,r)} V_{q-1,n-m,r-qm} \binom{n}{m} \quad (10)$$

for $q = 2, 3, \ldots, Q$, $n = 1, 2, \ldots, N$, and $r = n, n+1, \ldots, qn$, where $h(q, n, r) = min\{n, \lceil r/q \rceil\}$. The initial boundary conditions for this recursion are

1. $V_{q,n,r} = 0$ *for* $q = 1, 2, \ldots, Q, n = 1, 2, \ldots, N$ *and* $r = 1, 2, \ldots, N-1$.
2. $V_{1,n,n} = 1$ *for* $n = 1, 2, \ldots, N$.
3. $V_{q,0,0} = 1$ *for* $q = 1, 2, \ldots, Q$.
4. $V_{q,0,r} = 0$ *for* $r = 1, 2, \ldots, qN$ *and* $q = 1, 2, \ldots, Q$.

Proof: See Jacobson and Morrice (1996).

To solve the recursion in (9) for all values of r ($r = 1, 2, \ldots, QN$), in the worst case, QN recursions must be constructed for each value of r, where each recursion has at most $N+1$ parts in the summation. Therefore, the computational effort needed to determine the entire distribution for R is polynomial in Q and N, where the largest polynomial term is of order Q^2N^3. Solving the recursion in (10) for the number of possible vectors requires the the same amount of computational effort as solving the recursion in (9). In addition,

Lemma 1

$$\sum_{r=1}^{QN} V_{Q,N,r} = Q^N.$$

Proof: See Jacobson and Morrice (1996).

The recursive procedure derived from Theorem 1 was implemented in FORTRAN. This program can be used to characterize the entire distribution for R in the Cukier et al. (1993) study. The probability distribution is given in Figure 3. For $R \leq 61$, the $P\{R \leq 61\} = 1.0250 \times 10^{-4}$. This value validates the Monte Carlo simulation results reported by Cukier et al. (1993) and described in the last section, since the confidence interval produced by the simulation covers 1.0250×10^{-4}.

Convolving the distribution in Figure 1 with itself N times provides an alternative way to formulate the probability distribution of R. This follows since R is also the sum of sample of size N from the distribution in Figure 1. The following theorem provides a recursive procedure for calculating the probability distribution of R using this convolution formulation.

Theorem 2 *The probability $p_{q,n,r}$ can be computed using the convolution formula*

$$p_{q,n+1,r} = \sum_{i=\max(1,r-q)}^{r-1} p_{q,n,i} u_{r-i}$$

for $q = 1, 2, \ldots, Q$, $n = 1, 2, \ldots, N-1$, $r = n+1, n+2, \ldots, q(n+1)$. The initial boundary conditions for this formula are

$$p_{q,1,r} = \begin{cases} u_r & \text{if } r \leq q \\ 0 & \text{if } r > q \end{cases}$$

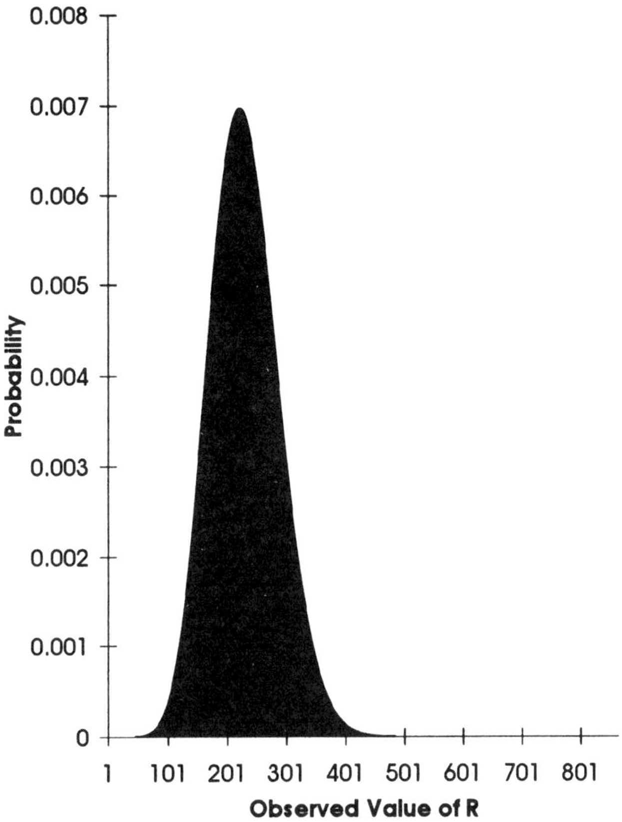

Figure 3: Probability Distribution for R

for $r = 1, 2, \ldots, QN$, $q = 1, 2, \ldots, Q$.

Proof: See Jacobson and Morrice (1996).

When the convolution formula in Theorem 2 is set up as a recursive algorithm, this algorithm requires polynomial computational effort in Q and N to generate the entire distribution for R. To obtain this distribution for Q alone, the largest polynomial term is of order Q^2N^2. However, if this distribution is required for all $q = 1, 2 \ldots, Q$, then the largest polynomial term is of order Q^3N^2.

The convolution formulation has the advantage over the formulation in Theorem 1 in computational speed if the distribution for R is required for the single value Q. However, if the distribution of R is required for all values of $q = 1, 2 \ldots, Q$, then the Theorem 1 formulation has the advantage over the convolution formulation in computational speed. This follows from the order of the largest polynomial terms for the two approaches.

Theorems 1 and 2 eliminate the need to perform a Monte Carlo simulation to calculate the probability distribution for R. In addition to being able to produce exact results, the algorithms associated with Theorems 1 and 2 are very fast relative to the simulation described in the last section. In particular, the simulation takes about 21.6 CPU seconds on the DEC AlphaServer 2100 4/275 to produce the data for Figure 2. On the same machine, the algorithm from Theorem 1 produces the data for Figure 3 in about 0.33 seconds of CPU time and the convolution approach produces the same results in about 0.14 seconds of CPU time. Perhaps the only advantage of the simulation over the results from Theorems 1 and 2 is that the simulation produces actual patient samples whereas the algorithms from the theorems do not produce such samples explicitly. However, sample counts are provided by the results in Corollary 1 and Lemma 1.

5 CONCLUSIONS

This paper develops an analytical model for describing the temporal association between disorders and treatments in medical studies. The model was inspired by a medical study conducted by Cukier et al. (1993). A complete description of the Monte Carlo simulation approach used by Cukier et al. (1993) has been provided. In addition, a recursive procedure and a convolution procedure that provide analytical solution to their problem have been described. The solution algorithms using these formulations are very efficient and can be used to solve realistically sized problems.

The model formulation presented in this paper is general and not restricted to the specific problem given in the cited medical study. Studies similar to Cukier et al. (1993) are common in the medical literature. Hochberg (1993) cites several examples and specifically discusses Cukier et al. (1993) and Bridges et al. (1993) as similar types of studies. The latter focuses on a possible connection between silicon breast implants and rheumatic disease.

ACKNOWLEDGEMENTS

The authors would like to thank Dr. Richard A. Beauchamp from the Texas Department of Health, Austin, Texas and Dr. Merwin W. Hemphill from the Texas Air Control Board, Austin, Texas for their help and advice on this research. The first author acknowledges support from the NSF (DMI-9409266, DMI-9423929) and the AFOSR (F49620-95-1-0124). The second author acknowledges the financial support of the CBA / GSB Faculty Research Committee of the College of Business and the End to End Simulation Department at Schlumberger Austin Research.

REFERENCES

Bickel, P. J. and K. A. Doksum. 1977. *Mathematical statistics: basic ideas and selected topics.* Oakland, California: Holden-Day.

Bridges, A. J., C. Conley, G. Wang, D. E. Burns, and F. B. Vasey. 1993. A clinical and immunological evaluation of women with silicon breast implants and symptoms of rheumatic disease. *Annals of Internal Medicine* 118:929–937.

Cukier, J., R. A. Beauchamp, J. S. Spindler, S. Spindler, C. Lorenzo, and D. E. Trentham. 1993. Association between bovine collagen dermal implants and a dermatomyositis or a polymyositis-like syndrome. *Annals of Internal Medicine* 118:920–928.

Fleming, T. R. and D. P. Harrington. 1991. *Counting processes and survival analysis.* New York: John Wiley and Sons, Inc..

Hochberg, M. C. 1993. Cosmetic surgical procedures and connective tissue disease: the cleopatra syndrome revisited. *Annals of Internal Medicine* 118:981–983.

Jacobson, S. H. and D. J. Morrice. 1996. A mathematical model for assessing the temporal association between health disorders and medical treatments. Technical Report, The University of Texas at Austin, Austin, Texas.

Johnson, N. L. and S. Kotz. 1977. *Urn models and their applications: an approach to modern discrete probability theory.* New York: John Wiley and Sons, Inc..

Kalbfleisch J. D. and R. L. Prentice. 1980. *The statistical analysis of failure time data.* New York: John Wiley and Sons, Inc..

Law A. M. and W. D. Kelton. 1991. *Simulation modeling and analysis.* 2nd ed. New York: McGraw-Hill, Inc..

Marse K. and S. D. Roberts. 1983. Implementing a portable FORTRAN uniform(0,1) generator. *Simulation* 21:135–139.

Ross, S. 1988. *A first course in probability.* 2nd ed. New York: Macmillan Publishing Company.

AUTHOR BIOGRAPHIES

SHELDON H. JACOBSON is an Associate Professor in the Department of Industrial and Systems Engineering at Virginia Polytechnic Institute and State University (Virginia Tech). Before joining Virginia Tech, he served for five years on the faculty in the Department of Operations Research at Case Western Reserve University. He has a B.Sc. and M.Sc. in Mathematics from McGill University, and a Ph.D. in Operations Research from Cornell University. He has served as the Advanced Tutorial Track Coordinator at both the 1994 and the 1995 Winter Simulation Conferences. He also served as the Doctoral Colloquium Coordinator at both the 1993 and the 1994 Winter Simulation Conferences. He served as the Treasurer for the InfORMS College on Simulation (1994-1996). His research interests include simulation optimization and sensitivity analysis, frequency domain approaches to analyzing simulation outputs, and issues related to the complexity of analyzing structural properties of discrete event simulation models.

DOUGLAS J. MORRICE is an Associate Professor in the Department of Management Science and Information Systems at The University of Texas at Austin. He is currently on sabbatical leave in the End to End Simulation Department at Schlumberger Austin Research Center in Austin, Texas. Dr. Morrice received his undergraduate degree in Operations Research at Carleton University in Ottawa, Canada. He holds an M.S. and a Ph.D. in Operations Research and Industrial Engineering from Cornell University. His research interests include discrete event and qualitative simulation modeling and the statistical design and analysis of large scale simulation experiments. Dr. Morrice is a member of the The Institute for Operations Research and Management Science (InfORMS) and the Council of Logistics Management. He served as the Secretary for the InfORMS College on Simulation (1994-1996) and is Co-Editor of the *Proceedings of the 1996 Winter Simulation Conference.*

Proceedings of the 1996 Winter Simulation Conference
ed. J. M. Charnes, D. J. Morrice, D. T. Brunner, and J. J. Swain

MODELING DISEASE PROCESSES FOR DRUG DEVELOPMENT: BRIDGING THE GAP BETWEEN QUANTITATIVE AND HEURISTIC MODELS

Pamela K. Fink
L. Tandy Herren

Medical Science Systems, Inc.
100 N.E. Loop 410, Suite 1350
San Antonio, Texas 78216, U.S.A.

ABSTRACT

Drug development is a complex and extremely risky process. It involves developing knowledge of the relevant biology at a variety of levels of detail in order to piece together a coherent model of the disease process and the potential point(s) of intervention. Ultimately, this model must be complete enough to support the prediction of effects and the explanation of clinical trial results. Unfortunately, such models of a disease process have been primarily developed and maintained in human minds. Consequently, such models are limited by a human's ability to store all of the needed information, structure it properly, and reason about its consequences involving both complex feedback and time dependencies. The following paper describes a knowledge-centered approach to the development of biological models that support the drug development process. This technique focuses on a multi-level, hierarchical approach that models various levels of the related biology using appropriate representation techniques based on both the type and availability of the knowledge.

1 THE DRUG DEVELOPMENT PROBLEM

New drug development is an extremely speculative endeavor. In order to bring a new drug to market, numerous hurdles must be overcome. Each hurdle involves gaining knowledge about how the drug works, under what situations it works, and whether or not it is safe. New drug development is typically motivated by a need or opportunity to intervene in a particular disease process to improve a patient's quality of life. This improvement could be based upon alleviation of symptoms, slowing the progression of symptoms, or actually curing the disease. However, though the success of a drug intervention can only be evaluated at the patient symptom level, current breakthroughs in potential drug interventions are at the sub-cellular level. Thus, for the drug development process to be successful (i.e., a drug has been identified, tested, marketed, used, and deemed effective and safe), several very complex linkages must be made:

1) the signs and symptoms of the disease must be linked to the basic biology that causes them,
2) an intervention must be identified that will alter the basic biology such that the disease signs and symptoms are altered in the desired manner, and
3) a chemical compound must be identified that alters the basic biology of a human in the desired way, but without any major undesirable effects.

To understand how a drug acting at the sub-cellular level can affect a disease's symptom complex, the individual cell behavior, the behavior of pools of cells, and the cellular interactions within a specific biological environment must be known. In addition, how cell behaviors and the various chemicals work to affect the symptom complex at the patient level must also be pieced together.

The problem, of course, is that there are a large number of unknowns and uncertainties in how all of the pieces fit together, which makes the process of drug development extremely unpredictable. In order to reduce the unknowns and increase the predictability, well-targeted laboratory experiments, animal testing, and clinical trials must be run to develop the knowledge needed to put together a coherent picture of the disease process and the possible intervention(s). Each stage of this development process contributes to understanding the biological processes involved at the various levels of detail, from basic biochemistry, up through cellular biology, on into organs and systems, and finally to the symptom complex as it relates to patient perceptions. Experimentation and analysis works to build confidence in the proposed intervention and, hopefully, reduces some of its unpredictability.

Humans faced with the challenge of developing new drugs evolve mental models of the relevant biological processes based on extensive experimentation and analysis. However, a disease process and the related biology are invariably complex, interrelated, and evolving over time. Even if all information about a process were available, it is still difficult for a human to effectively assimilate all of the data into a complete and coherent model or to identify all of its implications for disease progression. Thus, predictions based on such human mental models can be inaccurate and unreliable. Techniques that help to increase the knowledge and understanding of a particular disease, and that support the prediction of system behavior under specified conditions and over time, can reduce the uncertainty and risk involved with developing a new drug, and consequently reduce the time and cost.

One technique used to support the problem of predicting complex system behaviors has been the use of computer modeling and simulation. Numerous modeling and simulation methodologies exist. Most approaches used in the biomedical area have been either classic, numerically-based simulation techniques or knowledge-based, artificial intelligence techniques. These two approaches touch on the two ends of the knowledge spectrum needed to support drug development.

With numeric, or quantitative, approaches, sets of mathematical equations are developed that describe the behavior of a particular device or system. Such models are precise, but are only accurate if the device or system to be modeled is simple enough and understood well enough that it can be sufficiently and accurately described by a mathematical representation. Though such techniques have been used in numerous biological areas (Aliev and Saks, 1994) (Celada and Seiden, 1992) (Conolly and Kimbell, 1994) (Dalton et al., 1994) (Danis et al., 1995) (Hoang, 1995) (Pedley et al., 1994) (Robertson et al., 1994), the techniques do not scale up well to larger portions of human biology. They work well at the sub-cellular and chemical levels because the knowledge is usually available to model isolated portions at this level of biology with mathematical equations. However, such techniques are not well-suited for the higher, patient level models of the disease's symptom progression.

At the other end of the modeling spectrum are the heuristic-based approaches used to build knowledge-based and expert systems to model a particular disease at the patient level. Numerous systems have been developed over the years that can diagnose a particular class of disease given the signs and symptoms experienced by a particular patient along with any relevant test results (Shortliff, 1976) (Patil, Szolovits & Schwartz, 1981) (Weiss et al., 1978) (Chandrasekaran, 1983) (Miller, Pople, & Myers, 1982) (Kuipers, 1986). These systems have also been used to plan therapy regimens based on the diagnosis. Such approaches work well when the knowledge involved is at a higher, associative, heuristic, cause and effect level, but are not as effective as traditional techniques for modeling lower level, more detailed knowledge.

In the middle of the modeling spectrum are the qualitative modeling techniques developed primarily in the engineering, as opposed to the biological, domains (Bobrow, 1985) (Davis, 1983) (deKleer and Brown, 1984) (Forbus, 1984) (Genesereth, 1982) (Rieger & Grinberg, 1977) (Round, 1989). Many of these techniques have also been applied to modeling biology as a test of the approach (Uckun, 1992). These approaches provide a middle ground for representing a device or system explicitly as a set of behaviors when pure quantitative knowledge is not available or is not needed for the reasoning process being studied.

Thus, a large amount of modeling has been done in the biomedical field, but it has tended to touch primarily on only one point in the knowledge spectrum in any given modeling effort. Though some work has been done to link model-based and associative knowledge (Fink & Lusth, 1987), little work has been done to link the disparate levels of knowledge available at the sub-cellular level up through the cellular and system levels to the patient level in the biomedical arena (Karp & Frieland, 1989). In order to do this, a combination of modeling techniques must be used that can bridge the gap between quantitative models and heuristic models. The following sections discuss the various forms of knowledge involved in modeling the entire spectrum of biology needed to support drug discovery, describe how such a model is actually developed, and illustrate the technique with an example.

2 CHARACTERISTICS OF BIOLOGICAL KNOWLEDGE

Knowledge about disease processes is inherently complex, variable, and uncertain. As described in the initial section, it also spans numerous levels of detail, some of which are better understood than others. The various levels of knowledge needed to more fully understand how a particular chemical compound can influence a disease process are as follows:

1) Sub-cellular, chemical interactions that generate a local environment

2) Individual cell behaviors within an environment

3) Sets, or pools, of a specific cell type within an environment

4) Aggregations of different cell pools whose chemical products and behaviors create a particular environment

5) The environment generated by the aggregation of cell pools that manifests itself at the patient level as the disease's symptom complex

Notice that this list contains a feedback loop, since the environment listed in number one is aggregated to generate a global system environment in number four, which in turn influences numbers 2 and 3, and is manifested at the patient level in number five as the symptom complex. The complex interaction between the levels of biological knowledge makes it advantageous to represent and implement the knowledge independently within a hierarchy to accurately specify all relationships between levels.

To illustrate the types of knowledge included in this list, consider an example of otitis media. Otitis media is an infection of the middle ear. It is usually triggered by a viral infection of the epithelial cells in the nasal passages, as with a cold or the flu. As a result of the infection, the epithelial cell ciliary function is impaired and fluid builds up in the sinuses and associated passages including, potentially, the middle ear. This fluid environment is perfect for the proliferation of bacteria, which exacerbates the situation, eventually causing enough fluid to backup so that pain and swelling occur.

The knowledge about otitis media needed to build a disease model can be categorized according to the five level hierarchy described above. At the sub-cellular level (Level 1), the chemical environment consists of the classic pro-inflammatory cytokines such as IL-1, TNF, IL-2, IL-4, histamine, and numerous others generated by the pro-inflammatory set of cells. This chemical environment evolves over time based on the initial viral infection and inhibits the middle ear's drainage ability, which eventually causes the proliferation of bacteria. The cytokines and other chemicals found in the environment are the products of individual cells responding to specific chemical signals, a Level 2 phenomenon. Initially, the cells are responding to virally infected epithelial cells. Later, cells are responding to other cells that had initially responded in a cascade of events. Each of the individual cells can be aggregated into pools of the particular type of cell with a set of possible behavioral states. To model the cascade of events that represents the course of the disease, a pool of each cell type is necessary to represent the change of cells in the system from state to state (e.g. the probability, likelihood, or confidence that an epithelial cell is virally infected or virally immune). This represents knowledge at Level 3. The dynamics of the cell pools captures the progressive nature of the disease. Together, the pooled sets of cells produce certain chemicals based on cell states and the extracellular environment. This constitutes Level 4 knowledge. This, in turn, creates the extracellular environment, which feeds back to the various cell pools and generates the symptom complex, Level 5 knowledge.

The otitis media disease process just described is a superficial example of the kind of knowledge needed to properly and effectively model a disease in support of drug development. To truly model the process, sufficient detail must be provided so that the model can run as a simulation. For example, in order to model otitis media properly, sufficient knowledge of all of the cell types involved must be available, including what calls them into the area, what possible states they can be in, what environmental factors determine what state they are in, and what they do and produce when they are in any given state. Methods of acquiring and representing knowledge about diseases to build a viable model must surmount a number of obstacles and issues. These are as follows:

1) *The level of existing knowledge about a disease process and the basic supporting biology tends to be quite shallow*

The current state of knowledge about a disease process often is relatively shallow, existing only as a set of abstracted rules that are empirically derived. Thus, researchers often do not have information about exact causal relationships and only have empirical data on associations. For example, researchers may know that a drug affects the symptoms of a disease but may not know how it directly affects the biology that causes the symptoms. They may know that a particular cell is involved in the disease process, but do not know all of its effects on other cells in the area. Thus, though representation of such a large amount of knowledge needed for a model of a particular disease can be problematic, obtaining the needed knowledge to make a model work can be even more difficult.

2) *The size and complexity of human anatomy and physiology is large and variable*

The sheer complexity of the underlying biology of a disease process is staggering. Not only do aspects of anatomy vary from person to person but the basic biological processes are greatly modified by the background and history of a particular patient. The magnitude of knowledge that must be acquired and encoded can be prohibitive unless efforts are made to simplify and aggregate wherever possible and appropriate.

3) *Biological systems are based on self-generated feedback and control and are not static, but evolving, adapting systems*

Biological systems work because of the self-regulating feedback. They are, fundamentally, control systems. As illustrated above by the levels of knowledge needed to model a disease process, feedback is the foundation of the process. Though current simulation techniques can handle feedback, the level of complexity and the pervasiveness of the feedback in biological systems is beyond what is commonly modeled in other application areas.

4) *What is known about a particular disease process is often uncertain and contradictory*

Although the best scientific minds may have participated in model development and the best, most current research available was obtained, there are generally still large areas of uncertainty in the knowledge. Experts must synthesize knowledge from a wide variety of sources to bring to bear on the current modeling effort. Those sources may only be tangentially related or may be unrepresentative studies that provide insight without confirmation. For example, research on a particular cell type's behavior under certain conditions may be used to support the model even though the data from the research is based on an entirely different disease process than the one being modeled.

5) *The knowledge on which the model is built is constantly changing and evolving*

The biological sciences are constantly uncovering new information about disease processes. One of the goals of a biological model is to synthesize that knowledge to express a comprehensive theory about the disease. Once the model has been built representing the theory, the theory can be tested, refined, and validated. As new information emerges, the model can be altered to incorporate it.

As a result of these issues, not only is the actual modeling process difficult in a biological area, but the knowledge acquisition process is highly complex, as well. Thus, although at the lower levels of knowledge certain mathematical techniques may be appropriate and sufficient, at the higher levels less restrictive methods of representation are needed that allow for the complex feedback interactions and the large uncertainties in the knowledge. Modeling techniques that are forgiving of the lack and imprecision of knowledge are needed to effectively develop useful models in support of drug discovery.

3 THE KNOWLEDGE-CENTERED MODEL DEVELOPMENT PROCESS

Model-based reasoning is the use of the structural, behavioral, and/or functional knowledge of a system in order to diagnose, predict, and explain its behavior over time (Uckun, 1992). Techniques have evolved out of the area of computer science known as artificial intelligence, specifically knowledge-based systems development and qualitative reasoning (Bobrow, 1985) (Davis, 1983) (deKleer and Brown, 1984) (Forbus, 1984) (Genesereth, 1982) (Rieger & Grinberg, 1977) (Round, 1989), and have been used to model areas either not amenable to traditional numerical methods or for which the level of accuracy of a numerical model was not needed. Knowledge-based systems development techniques are used to acquire knowledge about the underlying model of a domain and to convert this knowledge into an interactive computer simulation. The form of the knowledge in the model is qualitative and/or heuristic, representing entities and relationships by relative values rather than by precise amounts.

Formal definitions of quantitative and qualitative modeling exist (Cellier, 1991) (Clancy, 1989) (Fishwick, 1989a), as do definitions of knowledge-based systems (Waterman, 1986). Much discussion has occurred around what constitutes quantitative vs. qualitative modeling and why one would use qualitative modeling techniques (Cellier, 1991) (Fishwick, 1989b). Knowledge-centered modeling attempts to merge these various techniques into a single, multi-level, hierarchical model. The knowledge available for the model can be precise mathematical relationships between entities or can represent abstracted rules provided by experts. The level of knowledge can vary from a high degree of precision to a high degree of abstraction within a single knowledge-centered model. The reason for the use of qualitative or heuristic modeling techniques in the drug development area is pragmatic --- the knowledge is simply not available in the detail and completeness needed at all levels of the biology required to build a working model. In addition, even if it were, the model would be so large and complex that aggregation at certain levels into qualitative models would still be required in order for the simulation to run in a reasonable period of time.

Knowledge-centered modeling has the following key attributes:

1) it is hierarchical,
2) it is modular,
3) it can model sufficient depth,
4) it can model sufficient breath, and
5) it is synergistic.

This technique allows for sufficient depth at the cellular and sub-cellular levels to model potential drug interventions at their source, while at the same time covering sufficient breadth of the relevant biology to model the symptom complex. Because of the hierarchical and modular nature of the resulting models, the simulations are sufficiently complex to allow for new and unique insights into biological processes.

The knowledge acquisition process used to build such a multi-level model results in the collection of many different pieces of data and information from many different sources. This process is time consuming and human intensive, often involving multiple domain experts over the course of many months. It follows both a top-down, structural approach and a bottom-up, data-driven approach. High level, general, textbook-oriented knowledge is collected as well as low level details on the latest in vitro laboratory experiments. The goal of knowledge acquisition is to determine which entities need to be represented in the model and the key relationships between the entities that drive the behavior of the system. Once the knowledge has been collected, the implementation process begins.

Implementation of a knowledge-centered model is hierarchical in nature. The process starts by identifying the symptoms that are of interest in the drug development effort. For example, in otitis media, the key symptoms are fluid behind the ear drum, pain, and fever. During the knowledge acquisition process, identified symptoms get mapped backwards into the biological environment that drives them and ultimately to the cellular and sub-cellular biology that causes the environment.

Once the territory is mapped with respect to the disease and its biological drivers, the top level of the model is segmented according to the major structural/functional components involved in the disease process. For example, in otitis media these major components might be the bacterial propagation and toxin productions, the middle ear area, the mucosal surface of the middle ear and eustachian tube, and the sub-mucosal connective tissue area.

Once the top level structural and functional relationships have been defined, more details are developed within each module to define its functionality. For example, within the mucosal surface of the middle ear module, a model of epithelial cells and their associated ciliary functions would be developed. This constitutes the Level 3, cell pool, knowledge in the model. It includes details on how virally infected epithelial cells behave, as well as what happens when they are exposed to pro-inflammatory chemicals from other cells responding to the viral infection. Finally, their behavior from exposure to bacterial outputs also has to be included. Such details must be worked through for every major component of the model. The knowledge is pulled not only from the set of human experts available to the project, but also from any literature that can be found on the subject, including any and all information on the specific cells and chemicals of interest. Again, this is done using both a top-down, structural and a bottom-up, data-driven approach.

As the knowledge is collected to the lowest level of detail needed, usually the cellular or sub-cellular level, implementation of the working model begins. Implementation is an iterative process in which levels of the model are built hierarchically, tested, and assembled into a working whole. Object-oriented techniques can be applied to speed the development effort and one hallmark of knowledge-centered modeling is how readily it adapts to this approach. The implementation process involves defining associations between entities and the qualitative relationships between them. These qualitative relations may be represented numerically and, at the lowest level of the model, usually are numerical values. However, these values are converted into fuzzy categories and it is these qualitative values that are actually passed around the model. Thus, knowledge-centered models include quantitative and qualitative values as well as heuristic associations.

Before a knowledge-centered model is complete, the basic biology resulting from the disease process that is now embodied in the model must be mapped back up to the symptom complex. Specific cell models are the bottom most layer in a hierarchical representation of the disease process at the cellular level. Anatomical features of the disease group the cells according to the point of their principle effects. Thus, some cells may have a large influence in one part of the anatomy but not in others. These groupings are necessary to localize effects and establish appropriate feedback between interacting cell types. This organization and grouping occurs as a natural result of building a knowledge-centered model and directly supports the symptom complex mapping. For example, the production of a specific cytokine by the cells in a given region can be mapped to the level of a particular symptom of the disease, such as the effect of histamine on the level of tissue plasma and consequently the level of edema in the area.

Once the model is functional and has a symptom complex output, it can be run as a simulation, tested, and validated against available biologic reference patterns. These reference patterns define, based on available research data, what the patterns of cell behavior should be under the disease conditions. They also provide details on the course of the symptom complex for the disease process itself. The model behavior is compared at all possible levels to available reference patterns. The model is modified and tuned so that its outputs fit within reasonable variation of the biologic reference patterns. This provides a baseline for model behavior from which model predictions about unknown scenarios (e.g. novel drug interventions) can be extrapolated with some confidence.

A knowledge-centered model constitutes a coherent theory of the disease process and underlying biology being studied. Once the theory has been validated to the best available data, the model's behavior can be explored and assessed for a variety of purposes. In the area of identifying new potential intervention opportunities, various pieces of the biology represented in the model can be altered to simulate a particular drug intervention. For example, in the otitis media scenario, an inhibitor of specific mediators of inflammation can be represented by decreased amounts of these mediators, such as complement and prostaglandins. Then the resulting symptom complex as output from the simulation can be evaluated for change from the untreated state (e.g. fluid levels decrease). To better understand confusing or conflicting clinical trial data for a potential new therapy, the model input data can be set to mimic the clinical trial situation with respect to type of patient and form of treatment. The resulting scenario can then be simulated, the model behavior examined at various levels, and the effects on disease outcome evaluated. This information can then be used to support the design of more effective clinical trials.

4 AN EXAMPLE - A MODEL OF OTITIS MEDIA

The top level components of an otitis media model include a cell model of the bacteria that can infect the middle ear, models of cells at the eustachian tube surface, models of cells in the submucosal connective tissue, and a model of how the cells and cell products drive patient symptoms. Anatomically, these represent layers in the middle ear. The bacteria invade the middle ear, irritate epithelial cells at the eustachian tube surface, and the immune system responds by sending inflammatory and immune cells into the submucosal connective tissue layer to counter the infection. Some inflammatory cells also appear at the eustachian tube surface to directly kill bacteria.

The underlying biology at the cell pool, individual, and chemical levels (levels 3, 2, and 1, respectively) is developed within each of these major components of the otitis media disease process. Detailed models of each of the cells involved in the disease process are necessary to represent the effect of disease processes at the cellular level. In otitis media, the cell types modeled include mucosal epithelial cells, PMN's, macrophages, mast cells, basophils, eosinophils, and a number of other immune cells. These cell models each include knowledge about the relative number of a cell type, the potential cell states, and the chemical production rates of the cells in each of the potential states. Information about the signals in the system that move cells between states and/or cause cells to generate their products complete the representation of each cell type. So, for example, a mucosal epithelial cell is modeled as being in one of several potential states, including resting, virally infected, virally immune, proliferating, and activated. Each of these states represents a distinct behavior, such as a specific chemical production profile and responsiveness to other chemicals. For example, resting epithelials do not produce anything, but are susceptible to viral infection and subsequent change of state, resulting in viral, IL-1, and interferon production. Resting epithelial cells can also become exposed to interferon and change state to a virally immune cell that still produces nothing, but is no longer susceptible to the state change to virally infected.

In addition to the knowledge about the basic biology, knowledge is represented about how cell products drive patient symptoms to model the observable effects of the disease. For example, the state of a given epithelial cell defines the ciliary function for which it is capable. Taken as a whole, along with glandular secretions in the area, the mucociliary function of the entire population of epithelial cells can be calculated to determine their ability to clear fluid from the middle ear. This ability to clear, along with glandular production rates, determines whether or not fluid backs up into the middle ear, and at what rate, to cause the key symptoms of a middle ear infection.

Finally, knowledge about treatment effects at the cellular level is necessary to model possible interventions in the disease process. These effects must also be mapped through the biology up to the patient level in order to experiment with the effectiveness of various therapies. Thus, for example, an antibiotic will increase the death rate of the bacteria, reducing the bacterial load in the area and causing an alteration in the patient's immune response. This alteration in immune response is manifested in a modification to the numbers of cells in various states and, consequently, the level of certain chemicals. This then influences the symptom complex. Other potential therapies can also be modeled in a similar fashion, allowing exploration and assessment of hypothetical treatment scenarios.

5 CONCLUSIONS

Drug development involves first finding a key locus of intervention in the biology of the disease which, if affected, alters the course of disease progression, and then testing this intervention in the laboratory and in clinical trials to verify its effect and assess its safety. The pathology of a disease is often so complex that it takes years of research to discover a leverage point that provides a cure or at least relieves the symptoms. Current approaches to drug development concentrate on standard laboratory experimentation to generate hypotheses and animal tests to further evaluate the hypotheses. This is a very labor intensive and time consuming process in which a positive outcome is not assured. It relies on discovering an insight, which happens in due course rather than on a fixed schedule. Testing a proposed intervention` once found also can take years of research to determine if its effect is significant enough and if its advantages outweigh any potential risks or side effects. The use of models to support this process can reduce time and risk by supporting exploration of the relevant biology in a consistent, coherent, and complete fashion. Such models can also serve as electronic laboratories for testing hypotheses quickly and with little risk. However, until recently, available modeling techniques were not powerful enough or flexible enough to capture sufficient portions of a disease process' biology to provide significant support. Knowledge-centered modeling provides this power.

REFERENCES

Aliev, M.K., and V.A. Saks. 1994. Mathematical modeling of intracellular transport processes and the creatine kinase systems: a probability approach. *Molecular and Cellular Biochemistry*. 133-134:333-346.

Bobrow, D.G., ed. 1985. *Qualitative Reasoning about Physical Systems*. MIT Press.

Celada, F. and P.E. Seiden. 1992. A computer model of cellular interactions in the immune system. *Immunology Today*. 13(2):56-62.

Cellier, F.E. 1991. Qualitative Modeling and Simulation: Promise or Illusion. *Proceedings of the 1991 Winter Simulation Conference*. WSC '91.

Chandrasekaran, B. 1983. On evaluating AI systems for medical diagnosis. *AI Magazine*. 4(2):34.

Clancy, W.J. 1989. Viewing Knowledge Bases as Qualitative Models. *IEEE Expert*. 4(2):9-23.

Conolly, R.B. and J.S. Kimbell. 1994. Computer simulation of cell growth governed by stochastic processes: application to clonal growth cancer models. *Toxicology and Applied Pharmacology*. 124(2):284-295.

Dalton, J.T., M.G. Weintjes, and J.L. Au. 1994. Effects of bladder resorption on pharmacokinetic data analysis. *Journal of Pharmacokinetics and Biopharmaceutics*. 22(3):183-205.

Danis, R.P., R.S. Moorthy, and J. Savage. 1995. Microhemodynamics of retinal collateral vessel formation. *Medical Hypotheses*. 44(2):103-109.

Davis, R. 1983. Reasoning from first principles in electronic troubleshooting. *International Journal of Man-Machine Studies*. 19(5):403-423.

deKleer, J. and J.S. Brown. 1984. A qualitative physics based on confluences. *Artificial Intelligence*. 24:7-83.

Fink, P.K. and J.C. Lusth. 1987. Expert systems and diagnostic expertise in the mechanical and electrical domains. *IEEE Transactions on Systems, Man, and Cybernetics*. 17(3):340-349.

Fishwick, P. 1989a. A Study of Terminology and Issues in Qualitative Simulation. *Simulation*. 52(3):5-9.

Fishwick, P. 1989b. Qualitative methodology in simulation model engineering. *Simulation*. 52(3):95-101.

Forbus, K. 1984. Qualitative Process Theory. Artificial Intelligence. 24(1-3):85-168.

Genesereth, M.R. 1982. Diagnosis Using Hierarchical Design Models. *Proceedings of the National Conference on Artificial Intelligence*. 278-283.

Hoang, K. 1995. Physiologically based pharmacokinetic models: mathematical fundamentals and simulation implementations. *Toxicology Letters*. 79(1-3):99-106.

Karp, P. and P. Friedland. 1989. Coordinating the Use of Qualitative and Quantitative Knowledge in Declarative Device Modeling. In *Artificial Intelligence, Simulation, and Modeling*. L.E. Widman et al. (eds.). John Wiley & Sons.

Kuipers, B. 1986. Qualitative Simulation. *Artificial Intelligence*. 29:289.

Miller, R.A., H.E. Pople, and J.D. Myers. 1982. INTERNIST-1: An Experimental Computer-based Diagnostic Consultant for General Internal Medicine. *The New England Journal of Medicine.* 307:468-476.

Patil, R.S., P. Szolovits, and W.B. Schwartz. 1981. Causal understanding of patient illness in medical diagnosis. *Proceedings of the Seventh International Joint Conference on Artificial Intelligence.* 893-899.

Pedley, T.J., et al. 1994. Gas flow and mixing in the airways. Critical Care Medicine. 22 (9 Suppl):S24-36.

Rieger, C. and M. Grinberg. 1977. The Declarative Representation and Procedural Simulation of Causality in Physical Mechanisms. *Proceedings of the Fifth International Joint Conference on Artificial Intelligence.* 250-256.

Robertson, M.J., et al. 1994. Agonist-antagonist interactions at angiotensin receptors: application of a two-state receptor model. *Trends in Pharmacological Sciences.* 15(10):364-369.

Round, A. 1989. Knowledge-Based Simulation. *The Handbook of Artificial Intelligence*, Vol IV. Barr, Cohan, and Feigenbaum (eds.). Addison-Wesley.

Shortliff, E.H. 1976. *Computer-Based Medical Consultation: MYCIN.* Elsevier, New York.

Uckun, S. 1992. Model-Based Reasoning in Biomedicine. *Critical Reviews in Biomedical Engineering.* 19(4):261-292.

Waterman, D.A. 1986. *A Guide to Expert Systems.* Addison-Wesley.

Weiss, S.M., C. Kulikowski, S. Amarel, and A. Safir. 1978. A Model-Based Method for Computer-Aided Medical Decision Making. *Artificial Intelligence.* 11(2):145-172.

AUTHOR BIOGRAPHIES

PAMELA K. FINK is Executive Director of Discovery and Technology Systems for Medical Science Systems, Inc. She received a B.A. degree in Mathematics and Philosophy from Eckerd College in 1989 and a Masters' and Ph.D. in Computer Science from Duke University in 1991 and 1993, respectively. Her research interests include issues in knowledge acquisition, representation, and utilization as they relate to problem solving and computer software implementation. Currently, she develops computer models of complex biological processes to support pharmaceutical and medical device research and development. She is a member of the American Association for Artificial Intelligence, the IEEE Computer Society, the Association for Computing Machinery, and the Society for Computer Simulation.

L. TANDY HERREN is a Staff Scientist for Medical Science Systems, Inc. She received a B.S. degree, an M.A. degree, and a Ph.D. in Psychology from Ohio State University. She also received an M.S. in computer science from Ohio State. Her current research interests involve biomedical simulation for therapy development, and knowledge acquisition and representation methodologies. She is a member of the American Association of Artificial Intelligence and the IEEE Computer Society.

Proceedings of the 1996 Winter Simulation Conference
ed. J. M. Charnes, D. J. Morrice, D. T. Brunner, and J. J. Swain

SIMULATING LYME DISEASE USING PARALLEL DISCRETE EVENT SIMULATION

Ewa Deelman
Boleslaw K. Szymanski

Department of Computer Science
Rensselaer Polytechnic Institute
Troy, NY 12180

Thomas Caraco

Department of Biological Sciences
State University of New York at Albany
Albany, NY 12222

ABSTRACT

Lyme Disease affects many people in the northeastern United States. One of the most important mechanisms that sustains the epidemic is the interaction between white-footed mice (*Peromyscus leucopus*) and deer ticks (*Ixodes scapularis*). When mice move around in their territory they carry diseased ticks to new locations. Our system simulates the different developmental stages of the tick through its active spring/summer period. The system uses the optimistic protocol for Parallel Discrete Event Simulation. In this paper, we present the model of the spread of the disease. We describe how we parallelize the problem, and we sketch a new global virtual time algorithm used in our system. We present performance benefits resulting from a parallel platform.

1 BIOLOGICAL OVERVIEW

Lyme Disease is caused by a spirochete (*Borrelia burgdorferi*) which is most commonly present in ticks (Barbour and Fish 1993, Miller *et al.* 1990). When an infected tick feeds on an animal or human, the spirochete may be transferred into the host's blood stream, causing an infection. Since the ticks are practically immobile, the spread of the disease is driven by their mobile hosts, such as mice and deer. An infected tick might infect a mouse, which in turn may infect an uninfected tick at a new location.

The life cycle of ticks encompasses several stages over a period of two years. The eggs hatch in the summer and become larvae. If a larva is successful in feeding on a host, it molts into a nymph. The nymphs that survive the winter, and that are able to find a blood meal during the spring, molt into adults. Adult female ticks deposit eggs. Ticks are assumed to hatch uninfected. The ticks in their larval and nymphal stages prefer to feed on mice whereas the adults prefer deer. Ticks that are unable to find a host die off. Since the mouse-tick interaction is at the heart of the Lyme disease cycle, our simulation is currently modeling only mice and ticks (Figure 1). The duration of the simulation is 180 days, starting in the late spring. This time is the most active for the ticks and mice. Mice, during that time, are looking for nesting sites and may carry ticks a considerable distance (Ostfeld *et al.* 1996).

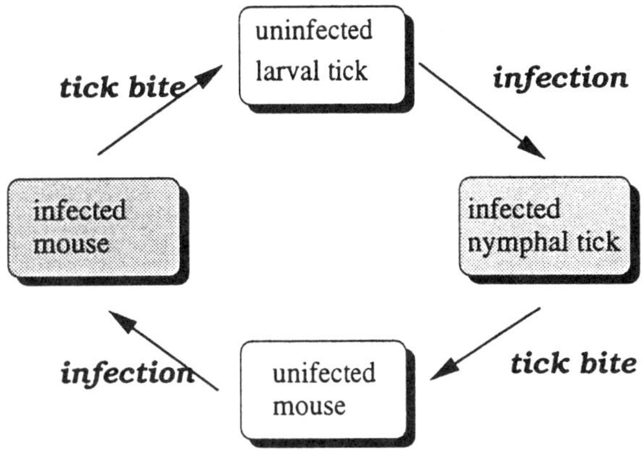

Figure 1: Infection Cycle

2 SIMULATION MODEL

Our goal is to understand the mouse-tick interaction at the lowest level, so we treat mice as individuals. We have chosen to use discrete event simulation (DES), because it lends itself well to individual-based modeling (Deelman, Caraco, and Szymanski 1996). It allows one to model the behavior of an individual through events that comprise the individual's life history. DES is also appropriate because of the temporal and spatial aspects of the physical system. We might have an area where there are no ticks or mice, and therefore no events. Time-step simulation methods might unnecessarily have to check the "empty" areas for activity. Also, DES can progress through simulated time quickly during episodes where the simulation objects are practically inactive, such as during the winter months.

Mice, as mentioned, are treated as individuals. We discretize the space, which results in a two-dimensional lattice, wrapped in both directions. The node size is assumed to be the size of a mouse home range ($400m^2$). Ticks and mice can be present at any node of the lattice. Ticks, because of their sheer number (as many as 1200 larvae/$400m^2$, Ostfeld *et al.* 1996), are viewed as background. Ticks are immobile. Mice move as a result of the occurrence of a *Move Event.* Other events associated with mice are:

1. *Disperse Event*, the mouse starts looking for a nesting site. This event triggers a new *Move Event.*
2. *Kill Event*, the mouse is removed from the simulation.
3. *Tick Bite*, when a mouse is in an area occupied by ticks, the mouse will be bitten by a tick with a predefined probability. The type of bite (larval or nymphal) depends on the type of ticks present in the area. When this event occurs, a number of ticks are removed from the location and placed on the mouse. This event creates the event *Tick Drop*, since after a blood meal, a tick drops off. If the mouse moves out of the area before the ticks drop off, the ticks will be deposited at the new location during the *Tick Drop* event.
4. *Tick Drop*, the ticks that are present on the mouse drop off. We assume that ticks molt just before the drop. Larvae that fed on the mouse drop off as nymphs. If the type of ticks on the mouse were nymphs, they drop off as adults. Also, a new *Tick Bite* of the same type (larval or nymphal) is triggered.

When a mouse starts moving in a certain direction, it will continue moving in the same direction until it settles at a nesting site. With each move, the mouse's survival probability diminishes. If a mouse cannot find an empty site within a certain number of steps, it dies. A mouse can also die of other causes, which are modeled by the random selection of the life span for each individual. Figure 2 shows a diagram of possible events associated with a mouse.

Although ticks are not modeled as individuals, their densities are updated at the time mouse events occur. It is possible that there are ticks at the lattice nodes where no mice are present. In this case the update of densities is done periodically. For simplicity, we update these densities during the *fossil collection*

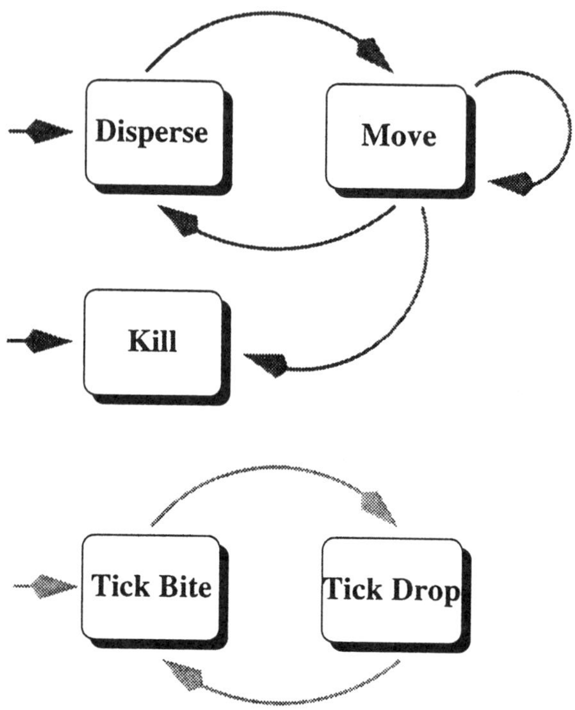

Figure 2: Possible Events for a Mouse

described below. We do not count all individual tick bites, since studies show that there can be as many as five individual larval bites per day. Accordingly, we combine multiple bites into one: ten larval bites or five nymphal bites at one time.

At the beginning of the simulation we have only nymphs that have over-wintered. They are then questing nymphs. At about the 90th day eggs hatch, larval ticks enter the simulation, and the number and type of ticks at each spatial node are updated. When a mouse is bitten by ticks, the number of ticks at the lattice node where the mouse is located is decreased by the number of ticks that bit the mouse. When the ticks drop off the animal, the tick densities at the lattice node are increased. We also make the assumption that when the mouse dies, the ticks (if any) present on the mouse die as well. We assume that mice are bitten by ticks as long as there are enough questing ticks on the node of the lattice. This assumption implies that there is a threshold for both larval and nymphal ticks below which we do not "notice" any new bites.

3 IMPLEMENTATION ISSUES

The model is implemented on a parallel platform, the IBM SP2, a MIMD machine (below, we present the results of runs on up to 16 processors). The implementation is written in C++ to take full advantage

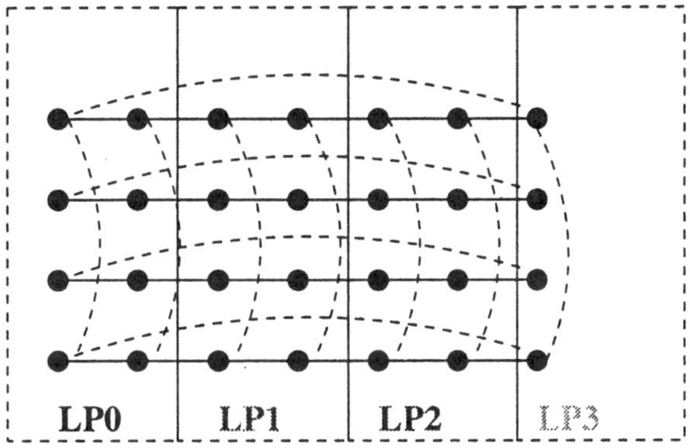

Figure 3: Strip Decomposition for 4 *LPs*

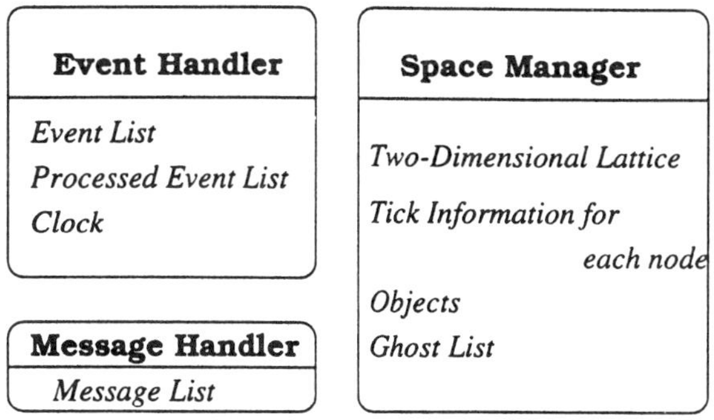

Figure 4: Components of an *LP*

of the object-oriented nature of DES. In Parallel Discrete Event Simulation (PDES), the physical system is divided into several physical processes. The behavior of each physical process is simulated by a Logical Process *LP* (Fujimoto 1990). In general, an *LP* consists of a state, event queue, and clock. The simulation progresses as events are removed from the queue and are processed. The event causes a state change, and the simulation clock advances. The MPI message passing library (Gropp, Lusk, and Skjellum 1994) is used for communication between *LPs*. We use a strip decomposition in the dominant direction to divide the space (Figure 3). The number of strips is equal to the number of processors. We assign one *LP* per strip. We chose this decomposition as the starting point for the system. As the research progresses, we will add other spatial decompositions.

Each *LP* is composed of three logical components: the *Event Scheduler*, which is responsible for processing events, the *Message Handler*, which carries out the communication between *LPs*, and the *Space Manager*, which oversees the movement of objects on the lattice (Figure 4). The *Space Manager* indicates to the objects if the space is occupied or not. Based on this information, the individual makes a decision whether to continue to move or to stay and, in the case of a mouse, to nest. The *Space Manager* also checks if the object is moving outside of the area assigned to the *LP*. If it does, control is given to the *Message Handler*, which sends the object, the *Move Event*, and all the future events associated with the object to the appropriate *LP*.

3.1 Optimistic Protocol Aspects

The major challenge in PDES is to make sure that causality between events is preserved. There are two major approaches to the problem: conservative (Chandy and Misra 1979) and optimistic (Jefferson 1985). In the conservative approach, causality errors are avoided by making sure that no *LP* processes an event for time t unless all events in the system with a timestamp less than t are already processed. In the optimistic approach, causality errors are allowed to occur. When an event with timestamp t_x is received by an *LP* with a clock of $t_c > t_x$, the *LP* has to restore its state to that just prior to time t_x and restart the computation from there. An event causing a rollback is known as a *straggler*. Obviously, the state of an *LP* has to be saved frequently. The more frequently the state is saved, the more memory is used; however, if the state is saved infrequently, the cost (in terms of time) of the rollback grows.

Our parallel simulation uses the optimistic protocol. Since the state of an individual *LP* consists of several lattice nodes (each with a number of individuals) and "background" information, it is prohibitively expensive in this application to save the state of the *LP* after each event, especially so because a single event affects at most only two lattice nodes (the *Move Event*). Therefore, we are using incremental state saving, described by Steinman (Steinman 1993). Each simulation event is augmented with information about the subset of state variables it has changed.

When an event is processed, we place it on the *processed list*. If the event causes a message to be sent to another *LP*, as in the case of a *Move Event* between two lattice strips, we place the message on the *message list*. We use aggressive cancellation; therefore, when a rollback occurs, *antimessages* are sent immediately to minimize the progress of the erroneous computation on other *LPs*. *Antimessages* are used to cancel the original messages that were sent between the logical time to which we have to roll back and the current logical time. The set of such messages is easily determined by scanning the *message list*. Next, we remove events from the *processed list* and "undo" them. When an event is undone, the data which were changed during its processing are restored. This is possible because each event is augmented with this information.

Processing a *Tick Bite* event involves calculating the number and type of ticks that will bite the mouse based on the number and type of ticks present at the lattice node where the mouse resides. We "place" that number and type of ticks on the mouse. If there are some infected ticks present in the biting sample, the mouse, if not already infected, will become infected with the spirochete. In order to be able to "undo" the bite, the *Tick Bite* event "remembers" the infection status of the mouse before the bite, as well as the number and type of ticks present at the lattice node. When the *Tick Bite* is undone, the ticks present on the mouse are returned to the lattice node in their original infection status. If the mouse was originally uninfected, it will be returned to a healthy state.

If a *Move Event* which involved a move from one *LP* to another is undone, we have to put back the object at the old location and place all the events that were sent out with it on the event list. This is done with the support of the *ghost list*. When an event causes the object to be removed from an *LP*'s space, for example, due to a move or death, the object and all its future events are placed on the *ghost list*. When the events are undone, it is very easy to restore the object and its events from that list. The original move to a new *LP* is cancelled during the *antimessage* phase previously mentioned.

3.2 Global Virtual Time

A very important part of optimistic protocols is the *global virtual time* (*gvt*) calculation. The *gvt* is the minimum time of all the *local virtual times* (*lvt*) of all the *LPs* and of all the timestamps of the messages in transit (Jefferson 1985). Since there are no events in the system with a time smaller than the *gvt*, all information that refers to events that happened before the *gvt* can be removed from memory. This property is very important because optimistic simulations use large amounts of memory to save state information necessary to support rollback. When reclaiming memory (also called *fossil collecting*), we can remove "old" events from the *processed list*, "old" messages from the *message list*, and "old" objects and their events from the *ghost list*.

The challenge of the *gvt* calculation is to capture information about the messages in transit. In our system, the simulation messages are augmented with information that the sender has about all the *LPs*. This knowledge includes not only the known *lvts* of all *LPs*, but also basic information about messages believed to be still in transit. We have designed a new *gvt* algorithm for our simulation system. It not only calculates the *gvt* but also gives us a good view of how the simulation progresses throughout the entire system. Our *Continuously Monitoring Global Virtual Time* (*CMGVT*) algorithm relies on two basic properties of our system: messages are received in the order in which they were sent, and each simulation message has a serial number. The first property holds in most parallel systems. The second only requires that an extra tag be attached to each message.

Each message and *antimessage* can be simply represented with four quantities: sender id, receiver id, serial number and timestamp. We call this representation of a message a *forcing*, since in optimistic simulations, a message might "force" a process to roll back. To indicate which messages are in transit, we combine two data structures: one quantitative—the *Message Matrix* (MM), which holds the knowledge an *LP* has about the messages sent by itself and its neighbors plus that *LP*'s knowledge of other processes' knowledge of messages sent by itself and its neighbors, and the second qualitative–the *Table of Forcing Vectors* (TFV), which indicates which messages are still unacknowledged. The messages in the table are described in terms of *forcings*. A sample MM and TFV are presented in Eq.(1) and Eq.(2) respectively. The data structures represent those of LP_1, and the system consists of four *LPs*.

$$MM_1 = \begin{bmatrix} 0 & 0 & 0 \\ 6 & 0 & 2 \\ 3 & 0 & 2 \\ 2 & 0 & 1 \end{bmatrix} \tag{1}$$

The columns of the MM refer to LP_1 (column 0, counting from 0) and its neighbors (here in the strip decomposition, each *LP* has two neighbors): LP_0 (column 1) and LP_2 (last column). The rows of the MM represent the knowledge each of the four *LPs* has about the number of messages LP_1 and LP_1's neighbors have sent. For example, the entry (1,0)=6 in Eq.(1) shows that LP_1 knows that LP_1 has sent out 6 messages. Entry (1,2)=2 shows that LP_1 knows that LP_2 has sent out 2 messages. The matrix also contains the knowledge other *LPs* have. Entry (2,0)=3 shows that LP_2 knows of 3 messages sent by LP_1. Since row 0 is 0, LP_1 does know anything about LP_0's knowledge. Also, because column 1 is 0, LP_1 does not know of any other *LP* that has any knowledge of LP_0's messages.

$$TFV_1 = \begin{bmatrix} 0 & [F(5,1,4), F(7,1,5), F(10,2,1)] \\ 15 & \\ 20 & [F(11,1,2)] \\ 36 & [F(10,1,3)] \end{bmatrix} \quad (2)$$

The TFV is used to keep track of all the messages still unacknowledged as well as the latest known *lvt* of all the *LPs* in the system. The table is indexed by the *LP* id. The first element in each row is the *lvt*. In Eq.(2) LP_1's view of the system is as follows: LP_1 is at time 15, LP_2 at 20, LP_3 at 36, and there is no knowledge about LP_0. The second entry in the table is the vector of forcings. When a message is sent to LP_x, a forcing $F(t, s, c)$, where t is the timestamp of the message, s is the sender's id, and c is the serial number of the message, is entered in row x. From Eq.(2) we see that LP_1 has sent two messages to LP_0: one at time 5 with serial number 4 and the other at time 7 with serial number 5. To LP_1's knowledge there is also a message sent by LP_2 to LP_0 in the system ($F(10, 2, 1)$). From Eq.(2) also notice that the first message sent by LP_1 ($F(t, 1, 1)$) was acknowledged, since it is not present in the TFV_1.

The *gvt* is easily calculated just by looking at the TFV. The *gvt* is simply the minimum of the *lvts* present in the table and of the timestamps present in the forcings. In the above example, the *gvt* is 0 since there is no information about the progress of LP_0.

When a message is sent, the MM and TFV are appended. When a message is received, the incoming information is compared with local data and the receiver updates its own information. The sender might, for example, indicate that an *LP* (LP_x) is aware of m messages sent by LP_y. If the receiver has a forcing for a message n sent by LP_y to LP_x, and $n \leq m$, then that forcing can be deleted, since that message is indirectly acknowledged by LP_x. Details of the algorithm and the proof of correctness are presented in Deelman and Szymanski 1996. Similar algorithms have been developed previously for systems in which logical time is monotonically increasing (Raynal 1996). We extended the idea of a matrix clock to be able to account for the time going backward as well as forward.

The *CMGVT* distinguishes itself from other *gvt* algorithms (Steinman *et al.* 1995) since it does not require special synchronization rounds in order to calculate the *gvt*. The *CMGVT* is computed by each process based on the information currently available to it. The *gvt* can be calculated when needed, for example, when an *LP* is about to run out of memory, or at some predetermined intervals. Here, we update the

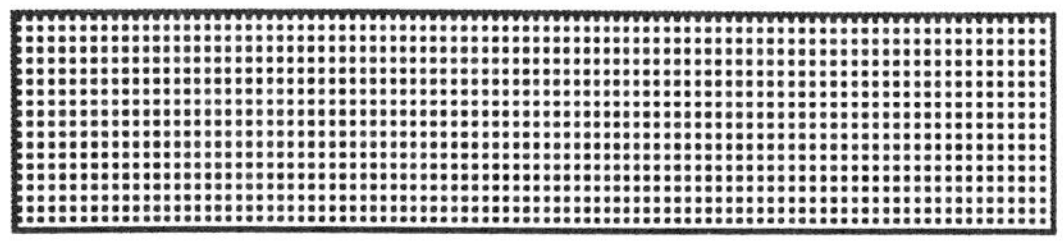

Figure 5: Mice on Day 1

gvt each time the difference between the *lvt* and the *gvt* goes beyond a threshold value *GVT_DIS* which is an adjustable parameter of the simulation. Obviously, as in most *gvt* algorithms, we calculate only an estimate of the *gvt*, since, to get an actual value, the simulation would have to be suspended. It is interesting to note that each *LP* can have a different estimate of the *gvt*, based on the information it has received from other *LPs*. A problem might arise when some *LPs* do not communicate often enough. In this case, if one *LP* has not received any new information from another, it just queries the noncommunicating *LP*. Even though at first glance it seems expensive to send additional information required by this protocol along with the simulation messages, we are able to achieve good results.

4 RESULTS

Initially we divide the space into as many sections as we have processors. We calculate a new *gvt* when the simulation gets ahead of the previous *gvt* by 10 days. For all the simulations we have the following initial conditions: mice are placed on every other node of the lattice (Figure 5) and ticks are positioned in a 20-node wide "band" in which every fourth row is populated with ticks, 15% of which are infected with the spirochete. Figure 6 shows the infected ticks at the beginning of the simulation. These initial conditions are interesting, because we can see the spread of the diseased ticks to areas previously devoid of them. The spread is caused by mice carrying diseased ticks to new locations and infected mice infecting uninfected ticks.

Figure 7 shows the progress of the disease on day 86 of the simulation. The initial "lines" of ticks are becoming "blurry" as the mice pick up infected ticks and move them to new locations. Figures 8 and 9 show the final configuration of mice and ticks, respectively. The data points at the top of Figure 9 represent infected ticks that the mice carried from the lower part of the figure. This is possible because the lattice wraps around its edges.

We are able to achieve good speedup for small data sets: 2,400 lattice nodes with 800 mice initially. The

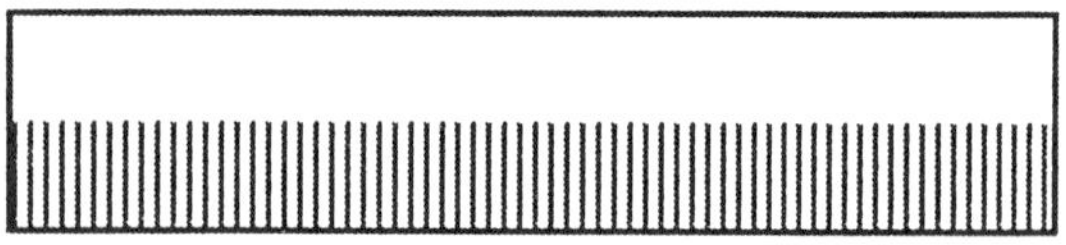

Figure 6: Infected Ticks on Day 1

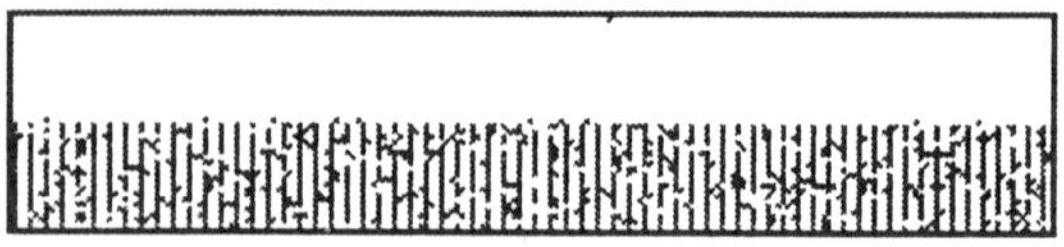

Figure 7: Infected Ticks on Day 86

results are shown in Figure 10. The speedup grows with the number of *LPs* for up to 10 processors. With 12 processors the communication overhead becomes large, decreasing the overall performance.

When the lattice size is increased to 32,000 nodes and 8,000 mice, with the same distribution of mice and ticks, the speedups are less impressive (Figure 11). This is caused by rollbacks whose cost is proportional to the size of the lattice for which each *LP* is responsible. With 4 processors, the lattice size per *LP* is large—8,000 nodes. When a rollback occurs, all the events that happened in the affected time in all of the 8,000 nodes have to be rolled back.

We investigate several methods of reducing the possibility and cost of rollbacks.

First, we increase the number of strips into which the lattice is divided, thus decreasing the area assigned to each *LP*. The processes are mapped by the job scheduler of the IBM SP2; therefore, a process is not aware if processes with which it exchanges data are run on the same processor as it is running. Hence, it always calls interprocess communication for such exchanges, slowing the execution. The results of dividing the problem into as many as 20 *LPs* on up to 16 processors are presented in Table 1. The best times for 4 processors are achieved when each processor has 5 *LPs*. Still, there is no speedup (the sequential time is 227 sec.). The speedup with 8 processors and 16 *LPs* is very small, around 1.6. The speedup with 12 processors is best when 20 *LPs* per processor are used,

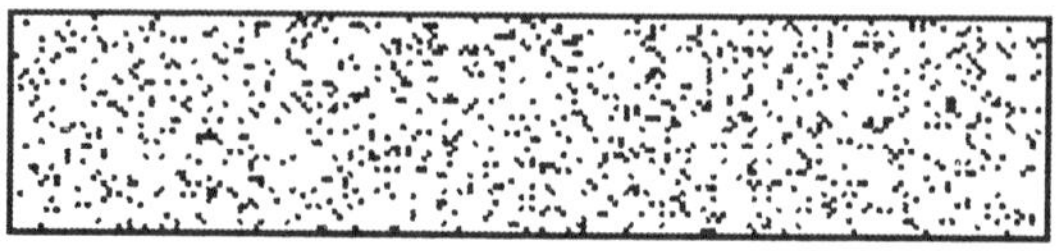

Figure 8: Mice on Day 180

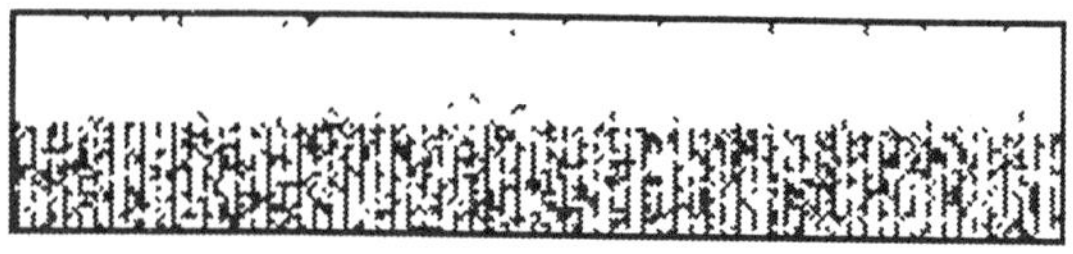

Figure 9: Infected Ticks on Day 180

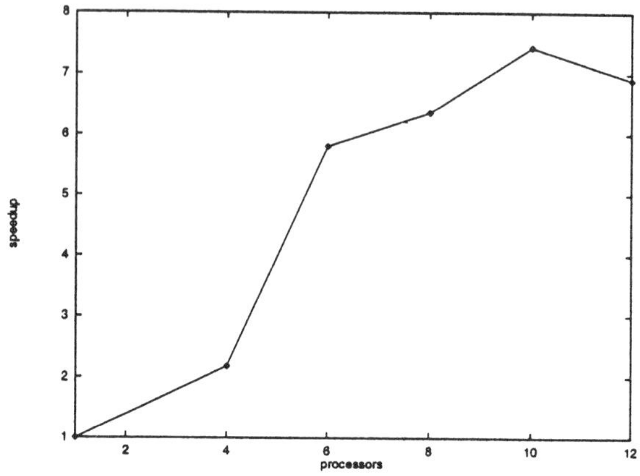

Figure 10: Speedup for Small Data Set

and is equal to 2.2. With 16 processors the speedup improves slightly to 2.8.

Another way to decrease the impact of rollbacks is to curb the optimism by allowing each *LP* to process events only within a limited time into the future. Toward this end, a process is allowed to advance only by 20 or 30 days ahead of the average *lvt* of others. Each process knows this value from the information sent by our *CMGVT* protocol. The results are shown in Table 2. The performance of this method was better than in the previous case only for 16 processors, with the best speedup of 3.7 for the 30-day time cap.

We also tried to combine both the use of multi-

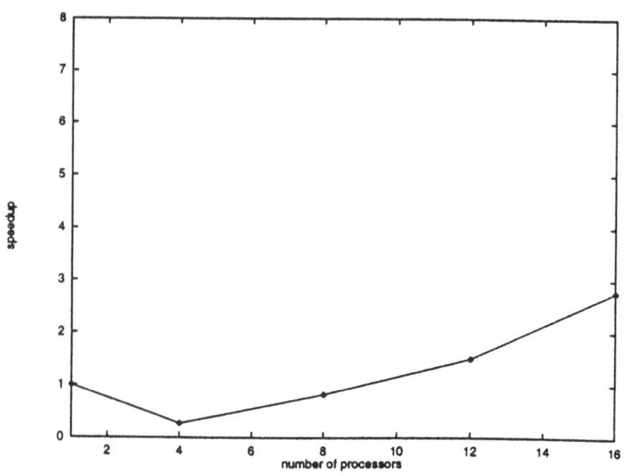

Figure 11: Speedup for Large Data Set

Table 1: Runtime in Seconds for Multiple LPs per Processor

Processors	Number of LPs 8	12	16	20
4	502.8	510.07	289.96	226.85
8	275.76	280.4	139.7	249.29
12	-	149.42	116	103.7
16	-	-	82	98.16

Table 2: Curbing the Optimism (time in sec.)

Processors	20 Days	30 Days
4	769.27	936
8	177.46	224
12	122	128.48
16	> 200	61.79

ple *LP*s per processor and curbing the optimism to processing only up to 30 days ahead of the average *lvt*. The results are shown in Table 3. The combined method resulted in an overall improvement over each of the component methods with the most significant improvement for 12 processors.

Table 3: Multiple LPs and Curbed Optimism (time in sec.)

Processors	Number of LPs 8	12	16	20
4	492.98	379.37	359.6	242.55
8	224.0	215.23	118.17	188.7
12	-	128.48	88.81	100.27
16	-	-	61.79	74.49

We checked if we were indeed reducing the number of rollbacks by using multiple *LP*s and reducing optimism. The dramatic results of a typical run on 8 processors are shown in Figure 12. With the increase of the number of *LP*s, the average number of rollbacks per *LP* decreases significantly. However, the number of rollbacks is not the only measure of performance. For example, the average number of rollbacks for 20 *LP*s is smaller than for 16 *LP*s , but the runtime is higher. This is because an increase in the number of *LP*s per processor intensifies the contention for the CPU, which slows the entire simulation. We have also noticed that the average number of rollbacks decreased as we curbed the optimism of the simulation. For 8 processors, 8 *LP*s, and an optimism cap of 30

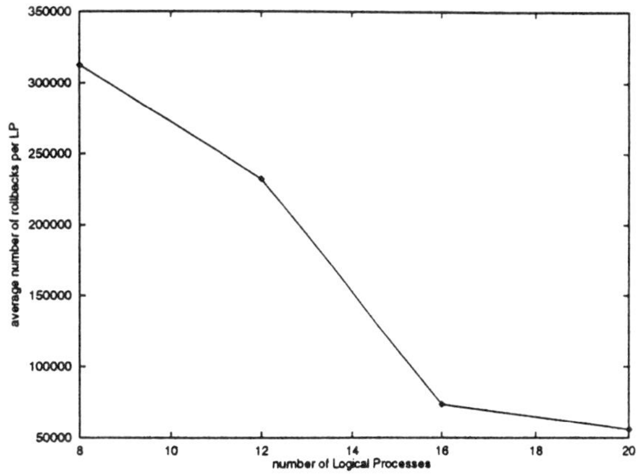

Figure 12: Average Number of Rollbacks for 8 Processors

days, we had on the average 136,355 rollbacks per *LP*; for a 20 day cap, the average was 92,770.

We investigated if the size of the messages had an adverse effect on the performance of the simulation. Our messages have the overhead of adding the *gvt* information. The size of the *Message Matrix* is $C_m p$, where C_m is the connectivity of an *LP* (up to 8 for spatially explicit two-dimensional problems), and p is the number of *LP*s. The size of the *Table of Forcing Vectors* is $c_t p$, where c_t is the maximum size of a *Forcing Vector* and is set by the system (the results presented here use $c_t = 10$). Thus, the additional message size is $C_m p + c_t p = O(p)$.

To see how the message size affects the performance of the simulation, we have increased the messages by an additional $.5(C_m + c_t)p$. The sender packs the message buffer up to its maximum size, but the receiver just reads out the original information. The results are shown in Figure 13. Increasing the message size had an unexpected effect: it actually improved the performance of the simulation! It reduced the number of rollbacks and the number of messages sent between *LP*s. This improvement might be caused by slowing down *LP*s that send many messages, or slowing down *LP*s that get ahead of others, thus allowing the slower *LP*s to catch up. These results encourage us to investigate methods for controlling the progress of the simulation in order to achieve better performance.

ACKNOWLEDGMENTS

This work was supported by the National Science Foundation under Grant BIR-9320264. The content of this paper does not necessarily reflect the position or policy of the U.S. Government—no official endorsement should be inferred or implied.

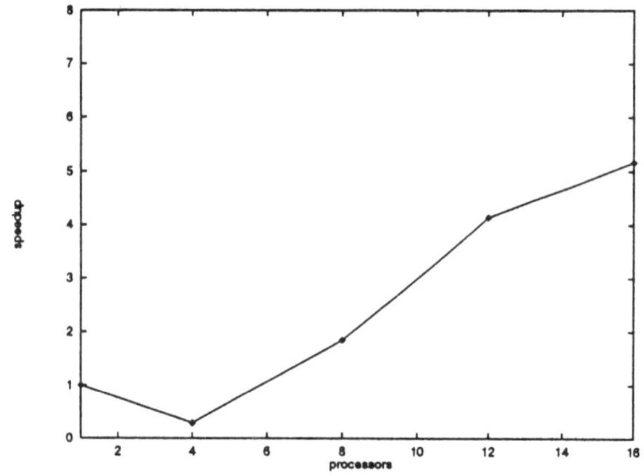

Figure 13: Speedup for Large Messages

REFERENCES

Barbour, A. and D. Fish. 1993. The biological and social phenomenon of Lyme disease. *Science* 260:1610-1616.

Chandy, K. M. and J. Misra. 1979 Distributed Simulation: A Case Study in Design and Verification of Distributed Programs. *IEEE Transactions on Software Engineering*, 5:440–452

Deelman, E., T. Caraco and B. K. Szymanski. 1996. Parallel Discrete Event Simulation of Lyme Disease. In *Pacific Symposium on Biocomputing* 191-202.

Deelman, E. and B. K. Szymanski. 1996 Continuously Monitored Global Virtual Time. *Department of Computer Science Technical Report 96-18*, Rensselaer Polytechnic Institute.

Fujimoto, R. M. 1990. Parallel Discrete Event Simulation. *Communications of the ACM* 33:31-53.

Gropp,W., E. Lusk and A. Skjellum. 1994. *Using MPI.* The MIT Press.

Jefferson, D. R. 1985. Virtual Time. *Trans. Prog. Lang. and Syst.* 7:404-425.

Miller, G. L., R. B. Craven, R. E. Bailey and T. F.Tsai. 1990 The epidemiology of Lyme disease in the United States 1987-1998. *Laboratory Medicine* 21:285-289.

Ostfeld, R. S., K. R. Kirsten and O. M.Cepeda. 1996. Temporal and Spatial Dynamics of *Ixodes scapularis* (Acari: Ixodidae) in a Rural Landscape. *Journal of Medical Entomology.* 33:90-95.

Raynal, M., and Mukesh Singhal. 1996. Logical Time: Capturing Causality in Distributed Systems. *IEEE Computer* 49.

Steinman, J. S. 1993. Incremental State Saving in SPEEDES using C++. In *Proceedings of the 1993 Winter Simulation Conference.* 687-696.

Steinman, J. S., C. A. Lee, L. F. Wilson, and D. M. Nicol. 1995. Global Virtual Time and Distributed Synchronization. *Workshop on Parallel and Distributed Simulation*, pages 139–148.

AUTHOR BIOGRAPHIES

EWA DEELMAN is pursuing her Ph.D. in the Department of Computer Science at Rensselaer Polytechnic Institute. She received a B.A. in Mathematics at Wells College and a M.Sc. in Computer Science at SUNY New Paltz. Her research interests include designing algorithms for parallel scientific computation and building tools and compilers for parallel platforms. Her Ph.D. work focuses on new paradigms for Parallel Discrete Event Simulation and their use in biological applications. email: deelmane@cs.rpi.edu, http://www.cs.rpi.edu/~deelmane

BOLESLAW K. SZYMANSKI is a Professor of Computer Science and a cofounder of the Scientific Computing Research Center at Rensselaer Polytechnic Institute, Troy, NY. He received a Ph.D. in Computer Science from the National Academy of Science in Warsaw, Poland, in 1976 and was a post-doctoral fellow at Aberdeen University in Aberdeen, UK. His research interests include language and compiler issues in large scale computing systems; analysis, design and verification of distributed and parallel algorithms; and simulation and modeling of computer and ecological systems. email: szymansk@cs.rpi.edu, http://www.cs.rpi.edu/~szymansk

THOMAS CARACO is an Associate Professor of Biological Sciences at the University at Albany, State University of New York. His research interests include the effects of behavior on population dynamics, and the landscape ecology of epidemics. Dr. Caraco is an editor of *Evolutionary Ecology* and an advisory editor for *Behavioral Ecology and Sociobiology.*

Proceedings of the 1996 Winter Simulation Conference
ed. J. M. Charnes, D. J. Morrice, D. T. Brunner, and J. J. Swain

DESIGN OF HOSPITAL ADMISSIONS SCHEDULING SYSTEM USING SIMULATION

Julie C. Lowery

VA Health Services Research & Development Field Program
P.O. Box 130170
Ann Arbor, MI 48113-0170

ABSTRACT

The ultimate objective of the research reported herein is to design an admissions scheduling system which can control hospital occupancy. This paper reports on the initial phase of the research--demonstrating the technical feasibility of developing a simulation model of patient arrivals to and discharges from a hospital, which can be used to design the scheduling system. The specific steps that were accomplished for demonstrating technical feasibility of model development were: (1) develop a hospital simulation model, which can be used for designing the scheduling system; (2) validate the simulation model in two test hospitals; and (3) use the model to design an improved scheduling system which reduces the variability in daily census.

1 INTRODUCTION

The control of inpatient bed occupancy has been cited as a source of potentially significant savings in hospitals, without requiring major changes in the structure of medical practice or the method of reimbursement for health care services (Martin, Dahlstrom, and Johnston, 1985; Griffith, Hancock, and Munson, 1973). Specifically, a reduction in the variation of daily occupancy can improve operational efficiency by reducing bed and personnel requirements and improving the utilization of specialized equipment (Redelmeier and Fuchs, 1993). The ultimate objective of the research reported herein is to design an admissions scheduling system which can control hospital occupancy. This paper reports on the initial phase of the research--demonstrating the technical feasibility of developing a simulation model of patient arrivals to and discharges from a hospital, which can be used to design the scheduling system. The specific steps that were accomplished for demonstrating technical feasibility of model development were: (1) develop a hospital simulation model, which can be used for designing the scheduling system; (2) validate the simulation model in two test hospitals; and (3) use the model to design an improved scheduling system which reduces the variability in daily census.

2 BACKGROUND

A number of simulation studies have been conducted which have investigated the effects of alternative scheduling systems on hospital performance measures, such as bed occupancy, number of turnaways, cancellations, and patient misplacements. The results from these studies suggest that such performance measures can be improved with alternative scheduling systems. These improved scheduling systems include such features as estimating the length of stay of patients prior to their admission (Robinson, Wing, and Davis; 1968); limiting the number of scheduled admissions (Jeang, 1990); scheduling elective surgery and medicine patients in a more uniform manner throughout the week (Butler, 1992); and instituting a series of scheduling parameters, including number of patients to schedule and number of beds to reserve for emergent admissions, which vary by day of week (Hancock and Walter, 1983; Hancock et al., 1976).

With the exception of Hancock and Walter's Admission Scheduling and Control System (ASCS), none of the recommended systems was ever actually implemented. Hancock and Walter claim that implementation of ASCS saved the eight hospitals in which it was implemented from $45,000 to $750,000 per year. The savings in underbedded hospitals were achieved primarily by treating additional patients without incurring a corresponding increase in the costs of staff, equipment, and beds. The savings in overbedded hospitals were achieved by reducing bed capacity without affecting the delivery of services. However, reports on implementation of the ASCS provide only sketchy data to support the claim that predicted and observed improvements in occupancy were actually due to a reduction in census variation, which was the purpose of the scheduling system (Hamilton, Hancock, and Hawley, 1975; Hancock and Walter, 1983; Johnston, Hancock, and Steiger, 1975; Yannitelli and Hancock, 1975). The results from their research indicate that ASCS holds considerable potential for improving the efficiency of the delivery of inpatient care; but additional research is needed to determine the relationship between the recommended scheduling system, a reduction in census variability, and either an increase in occupancy or reduction in costs. The simulation model discussed in this paper will be used to design a scheduling system like the one described by Hancock and Walter, and to evaluate its effect on census variability.

3 MODEL DESCRIPTION

A hospital simulation model was written in the simulation language GPSS/H (Henriksen and Crain, 1993). The model represents patient movement through multiple bed services based on actual patient "flow patterns" identified from each hospital's historical data on patient admissions, discharges, and transfers. Included in these patient flow patterns are admissions to and transfers among medicine (including dermatology and neurology), surgery, medical intensive care unit/coronary care unit (including medical/coronary stepdown unit and telemetry), and surgical intensive care unit (including surgical stepdown unit). The patient flow patterns identified from the two test hospitals were as follows:

Hospital A
Medicine
Medicine-->CCU/MICU-->Medicine
Medicine-->SICU-->Surgery
Medicine-->Surgery
CCU/MICU
CCU/MICU-->Medicine
Surgery
Surgery-->SICU-->Surgery
Surgery-->Medicine
SICU-->Surgery
Surgery-->SICU

Hospital B
Medicine
Medicine-->CCU/MICU-->Medicine
Medicine-->SICU-->Medicine
Medicine-->CCU/MICU
CCU/MICU
CCU/MICU-->Medicine
Surgery
Surgery-->SICU-->Surgery
Surgery-->CCU/MICU-->Medicine
SICU-->Surgery
SICU-->Medicine

A small proportion of patients at each site follows more lengthy flow patterns, involving multiple transfers among services. For purposes of model simplification, these flow patterns have been reduced to one of those listed above; this simplification did not adversely affect model validation.

The distribution of admissions across the possible flow patterns for each admitting bed service is one of the model inputs, and is based on historical workload. The above flow patterns are examples of patterns that patients *should* follow, given bed availability in the appropriate bed services. However, one of the objectives of model development is to investigate the consequences of high average occupancies. Therefore, the model also includes patient flow patterns in the event that a bed is not available in the desired service. That is, secondary patient flow patterns for each site were identified and modeled for patients who cannot follow the primary, or desired, flow patterns because of lack of available beds.

Theoretical input distributions were used for both emergent arrivals and patient length of stay, by major bed service. Use of theoretical rather than empirical distributions for these variables facilitates model implementation because a hospital can use summary data as model input, rather than have to perform a detailed analysis of patient-specific data. Specifically, theoretical distributions require the hospital to simply enter the distribution's parameters (e.g., mean and standard deviation) as model input; while empirical distributions require the hospital to develop a distribution from individual patient records. The exponential distribution was used to model emergency interarrival times, and the lognormal distribution was used for modeling length of stay. Both of these distributions have been used extensively in previous simulation studies of health care systems.

4 DESCRIPTION OF SCHEDULING SYSTEM TO BE MODELED

Once the simulation model is developed and validated for a given hospital, it can be used to design the scheduling system, which consists of various scheduling parameters and decision rules. The scheduling system is incorporated into the simulation model, then combinations of different values of the parameters are systematically tried as model input, along with the hospital's historic or projected values for the other input variables. Model output includes predicted values of hospital performance, which will be reviewed by hospital clinical and administrative staff. The values of the scheduling parameters which result in a desirable combination of values for daily census, overall occupancy, cancellations, and turnaways become the parameters of the scheduling system to be implemented in the hospital. If the hospital decides to implement the recommended scheduling system, it will consist of a simple software program which performs a series of calculations using the parameter values plus daily data on bed availability. The results from the calculations will be used by admitting personnel in their daily admitting and scheduling decisions.

The admitting decisions that must be made each day include the following:

1) How many elective admissions should be scheduled, by bed service, by day of the week?
2) Given current bed availability at a given point on a particular day of the week, how many patients (if any) should be called in?
3) Given current bed availability at a given point on a particular day of the week, how many scheduled admissions (if any) should be canceled (to ensure bed availability for emergency patients)?

The answers to the above questions ultimately determine a hospital's overall occupancy, as well as the number of cancellations and turnaways incurred.

A computer-based scheduling system should perform the required calculations to provide the answers to the above questions. To perform these calculations, the scheduling system requires data on current bed availability (for questions 2 and 3), as well as values of the scheduling parameters, which are day-of-the-week specific.

The scheduling parameters required for responding to question 1 are the day-of-the-week numbers of scheduled appointment slots. No further calculations based on current bed availability are required. For question 2, the research by Hancock and Walter has shown that two decision numbers should be determined: a "call-in allowance" (CIA), and a "call-in maximum" (CIM). Using these numbers, the answer to question 2 is calculated as follows: If the current number of empty beds > CIA, then the number of patients to be called in = empty beds - CIA, but should not exceed CIM. The call-in allowance is used to determine if the number of empty beds is sufficiently large that additional patients can be called in without adversely affecting bed availability for scheduled admissions later in the week. If enough beds are available that patients can be called in, then the call-in maximum sets a limit on the number of patients that can be called in, so that the distribution of discharges (and hence, daily census) is not adversely affected.

Finally, question 3 requires a value for the number of empty beds that must be reserved for emergency patients, or the "emergency reserve allowance" (ERA). The answer to question 3 is calculated as follows: If ERA > the number of empty beds expected after all scheduled patients have been admitted, then the number of scheduled patients to cancel is ERA - empty beds.

In summary then, the values of 28 different scheduling parameters must be determined for each bed service which falls under the guidelines of the proposed scheduling system: four decision numbers for each of the seven days of the week. The four decision numbers are (1) number of elective patients to schedule; (2) call-in allowance; (3) call-in maximum; and (4) emergency reserve allowance. The purpose of the simulation model is to help determine the values of the 28 scheduling parameters.

5 RESULTS

The simulation model that will be used for designing the scheduling system must be (1) easily applied to multiple hospitals; (2) valid--i.e., an accurate representation of the actual system being studied; and (3) able to demonstrate a reduction in census variability with a change in the scheduling system. The achievement of these criteria is discussed below.

5.1 Ease of Application to Multiple Hospitals

The model is easily customized for a given hospital by making changes to definitions of the following input variables:

- Number of major bed sections
- Number of beds in each section
- Average and standard deviation of patient length of stay in each section
- Arrival rates of emergency patients, by day of week
- Distribution of emergent admits, by treating specialty and patient flow pattern
- Distribution of elective admits, by treating specialty and patient flow pattern
- Identification of first and second alternative bed sections (used when all beds in desired section are full), by bed section

To apply the model to a new hospital, the hospital characteristics listed above are determined, and changes are made to the corresponding definitions, tables, or matrices in the model "definitions" section. These changes consist of simply editing numbers or unit identifiers. No changes need to be made to any of the program code describing patient movement. Thus, the model was easily customized to represent the two test sites included the research.

5.2 Validation

The validity of a simulation model is established by comparing model and system behavior. The methodology used in this research for performing the comparison was a graphical approach, which is the most commonly used method for the validation of simulation models (Sargent, 1994). Confidence intervals and hypothesis tests are also used, but less frequently. A graphical methodology, rather than hypothesis testing, was used for evaluating the validity of the admissions scheduling model because observations of daily census are autocorrelated, and because limited data are available from the actual system. When observations are autocorrelated, classical statistical tests based on identically and independently distributed observations are not directly applicable. Furthermore, because a simulation model is only an approximation of the actual system, the null hypothesis that model behavior and system behavior are the same will almost certainly be false (Law and Kelton, 1991, p. 312). While formal statistical tests may lead to the conclusion that a model is not valid, it may still be valid for the purpose for which it is intended. This is especially true for models that are designed primarily for comparing alternatives than for predicting absolute answers, as is the admissions scheduling model.

To compare system and model behavior, actual census data were obtained from the two test hospitals for a six-month time period. A six-month time period was selected because it is expected that this would be the amount of time for which the scheduling parameters would be set. (Every six months the performance of the system and the assumptions underlying the model would be evaluated and the parameters reset, if necessary.) Because the parameters are set by day of week, it is important that the model accurately predict census by day of week; hence, actual census by day of week was compared against model census by day of week. Furthermore, because of the potential problem of autocorrelation of census figures (even though the observations for a given day of the week are separated by six-day intervals), the average daily census for a four-week time period was used as the performance measure for comparison.

During a six month time period (26 weeks), six observations of the four-week average daily census (ADC) were available for each day of the week from both medical centers. These observations constitute the "actual" data. The distribution of these observations was compared with the distribution of 50 observations of four-week ADC data from the model. Any number of observations can be generated from the model; 50 was selected because it appears (from the graphs) to be a sufficient number for depicting the general shape of the distribution. A valid model should produce a distribution of observations that encompasses the observations from the actual system. Since the four-week ADC is a random variable, the six observations of actual data represent a sample from a distribution, which we want the model to be able to reproduce. Figure 1 is an example of the model and actual distribution of four-week average daily census data by day of week for Sunday at one of the test sites. Similar results were obtained for all days of the week, for both test sites, demonstrating that the samples of actual data fall well within the distributions from the model.

In addition to seeing that the distribution of actual observations falls within the model's distribution of observations, we want to see that the system's mean census is close to that of the model's mean census.

Figure 1: Hospital A
Sunday Model vs. Actual Census

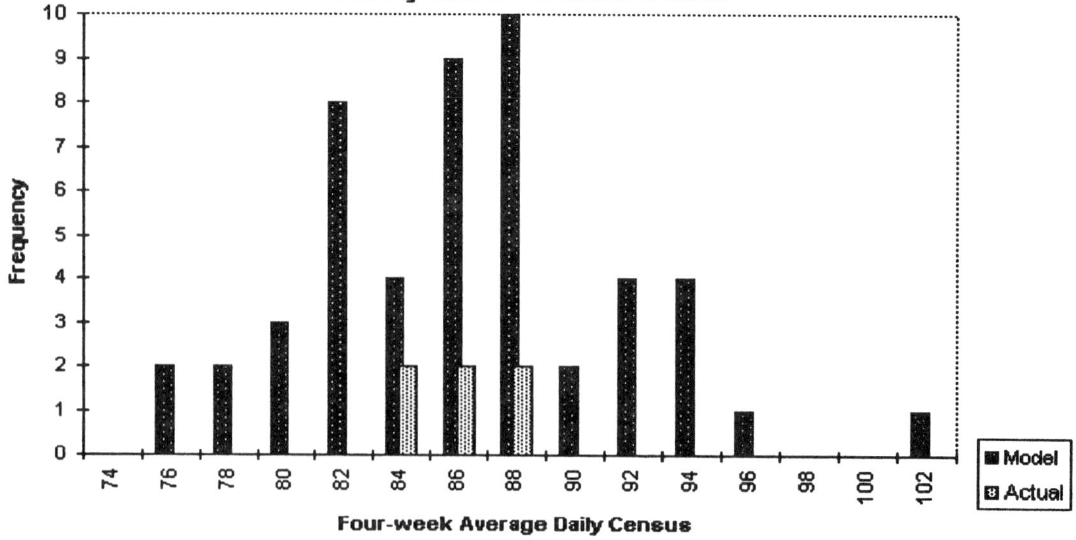

Therefore, we compared the mean from the sample of hospital observations with the model's "population" mean--i.e., the mean of the 50 four-week observations generated by the model. Table 1 shows the difference between model and actual means for all days of the week for hospital A. The difference ranges from 0.1 to 1.7. These differences are within the acceptable range established by the Chiefs of Medicine and Surgery at the test site, who stated that a credible model should predict within two beds of the actual census. Table 2 displays the differences for hospital B, all of which are also less than two beds. Based on these results and the graphical comparisons, the model is considered valid. The validation methodology and results are also discussed by Sargent elsewhere in these *Proceedings* (see "Some Subjective Validation Methods Using Graphical Displays").

Table 1: Difference between Model and Actual Mean Daily Census, Hospital A

	Model Mean[1]	Actual Mean[2]	Difference
Sun	85.4	85.3	+0.1
Mon	95.6	95.5	+0.1
Tues	98.9	98.1	+0.8
Weds	98.4	98.2	+0.2
Thurs	98.5	96.8	+1.7
Fri	84.7	83.8	+0.9
Sat	78.1	78.0	+0.1

[1]Based on 50 4-week observations.
[2]Based on six 4-week observations.

Table 2: Difference between Model and Actual Mean Daily Census, Hospital B

	Model Mean[1]	Actual Mean[2]	Difference
Sun	48.1	47.9	+0.2
Mon	55.8	56.3	-0.5
Tues	58.0	57.9	+0.1
Weds	60.5	60.3	+0.2
Thurs	61.6	61.9	-0.3
Fri	53.1	53.4	-0.3
Sat	46.9	47.2	-0.3

[1]Based on 50 4-week observations.
[2]Based on six 4-week observations.

5.3 Reduction in Census Variability with New Scheduling System

As indicated above, the purpose of the admissions scheduling model is to identify a set of scheduling parameters which results in a decrease in census variation from the current system. Therefore, if the simulation model is to be used for this purpose, it must be able to show a reduction in the variation in daily census between the current scheduling system and the proposed, improved scheduling system. For example, figure 2 presents a "box and whisper" plot of the variation in daily census at Hospital A for Wednesdays. The figure presents the results for five model replications of one year each (52 observations) under the current system, and five replications under the new scheduling system (i.e., one set of values of the scheduling parameters). The plots for all days of the week showed a modest, but observable, reduction in census variation under the new scheduling system for Mondays through Thursdays. The new scheduling system is most likely to affect these days of the week, because the majority of admissions on these days are scheduled and can be controlled. Additional experimentation with the simulation model to investigate the effects of other values of the scheduling parameters on census variability is currently being conducted.

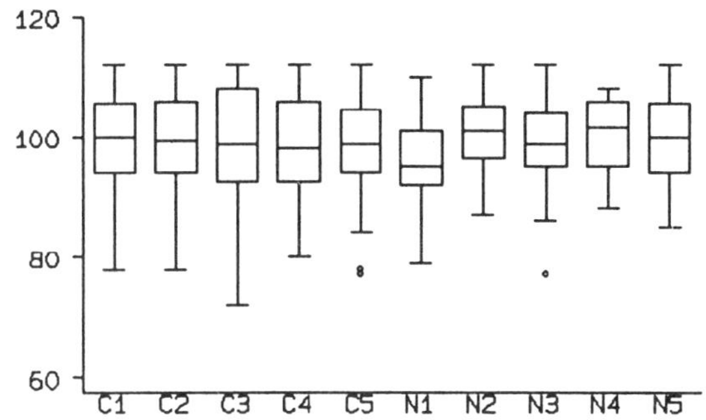

Figure 2: Current vs. New Scheduling System Wednesday Replications 1-5

6 CONCLUSION

The results from this research demonstrate the feasibility of developing a valid simulation model of patient arrivals to and discharges from a hospital, which can be used for evaluating the effects of alternative scheduling systems on census variability. The model is easily customized to multiple hospitals, is valid, and can demonstrate a reduction in census variability from modification of values of the scheduling parameters. Work is proceeding on using the model to design scheduling systems for the two test hospitals.

The hospital simulation model offers other benefits besides the design of an improved scheduling system. For example, it can be used for determining the appropriate allocation of beds between different specialties and subspecialties. One of the test sites, as well as the Michigan Voluntary Hospital Association, have expressed interest in using the model for this purpose. The simulation model also provides a mechanism for structuring the hospital's planning process. By requiring hospital clinical and administrative staff to project inpatient workload and case-mix, and to define these projections in the precise terms required for the model's input variables, the model facilitates the discussion and definition of assumptions necessary for the planning process. In addition, the knowledge gained about the admissions process while designing and validating the simulation model may prove to be invaluable to helping the clinical and administrative staff understand how the current system really works, as opposed to how everyone thinks it operates (Shannon, 1992, p. 66). In turn, the process of designing and validating the model may be of great value toward suggesting improvements in current procedures (Banks and Carson, 1984, p. 4), even if the proposed scheduling system is never implemented.

ACKNOWLEDGEMENTS

This research was funded by the Department of Veterans Affairs Health Services Research & Development Service.

REFERENCES

Banks J, Carson JS. *Discrete Event System Simulation.* Englewood Cliffs, NJ: Prentice-Hall, Inc., 1984.

Butler TW, Reeves GR, Karwan KR, Sweigart JR. Assessing the impact of patient care policies using simulation analysis. *Journal of the Society for Health Systems* 1992;3(3):38-53.

Griffith JR, Hancock WM, Munson FC. Practical ways to contain hospital costs. *Harvard Business Review* 1973;51(6):131-139.

Hamilton R, Hancock W, Hawley K. *The Admissions Scheduling and Control System: The Admissions Simulator.* Report 75-1, Bureau of Hospital Administration, The University of Michigan, February 1975.

Hancock WM, Walter PF. *The ASCS Admission Scheduling and Control System.* Ann Arbor, MI: Health Administration Press, 1983.

Hancock WM, Warner DM, Heda S, Fuhs P. Admission scheduling and control systems. In: Griffith JR et al., ed. *Cost Control in Hospitals.* Ann Arbor, MI: Health Administration Press, 1976.

Henriksen JO, Crain RC. *GPSS/H Reference Manual.* Annandale, VA: Wolverine Software Corporation, 1993.

Jeang A. Inpatient admission system model using SIMAN software. *Journal of Medical Systems* 1990;14(6):365-374.

Johnston C, Hancock W, Steiger D. *The First Three Months of Implementation of the Admission Scheduling and Control System in Delta Hospital.* Report 75-9, Bureau of Hospital Administration, The University of Michigan, November 1975.

Law AM, Kelton WD. *Simulation Modeling & Analysis.* New York: McGraw Hill, Inc., 1991, 759 pp.

Martin JB, Dahlstrom GA, Johnston CM. Impact of administrative technology on acute care bed need. *Health Services Research* 1985;20(1):63-81.

Redelmeier DA, Fuchs VR. Hospital expenditures in the United States and Canada. *New England Journal of Medicine* 1993;328(11):772-778.

Robinson GH, Wing P, Davis LE. Computer simulation of hospital patient scheduling systems. *Health Services Research* 1968;3(2):130-141.

Sargent RG. Verification and validation of simulation models. In *Proceedings of the 1994 Winter Simulation Conference.* Edited by JD Tew, S Manivannan, DA Sadowski, AF Seila. Baltimore: Association for Computing Machinery, 1994, pp. 77-87.

Shannon RE. Introduction to simulation. In *Proceedings of the 1992 Winter Simulation Conference.* Edited by JJ Swain, D Goldsman, RC Crain, JR Wilson. Baltimore: Association for Computing Machinery, 1992, pp. 65-73.

Yannitelli P, Hancock W. *Implementation of the ASCS into Charles Hospital.* Report 75-3, Bureau of Hospital Administration, The University of Michigan, June 1975.

AUTHOR BIOGRAPHY

JULIE C. LOWERY is Associate Director of the Center for Practice Management and Outcomes Research, VA Health Services Research and Development Field Program, Ann Arbor, MI. She has a BS in Microbiology and a Master's in Health Services Administration, both from the University of Michigan. She received her PhD in Health Services Organization and Policy from the School of Public Health, University of Michigan, where she specialized in operations research and information systems in health care.

Proceedings of the 1996 Winter Simulation Conference
ed. J. M. Charnes, D. J. Morrice, D. T. Brunner, and J. J. Swain

A SIMULATION MODEL FOR EVALUATING PERSONNEL SCHEDULES IN A HOSPITAL EMERGENCY DEPARTMENT

Gerald W. Evans
Tesham B. Gor

Department of Industrial Engineering
University of Louisville
Louisville, KY 40292

Edward Unger

Baptist Hospital East
Louisville, KY

ABSTRACT

This paper describes an Arena simulation model of a particular hospital's emergency department. The model allows the simulation of the process flows of 13 different types of patients, and is used to evaluate various feasible schedules for nurses, technicians, and doctors. The main performance measure used in the evaluation process is the average length-of-stay of patients in the emergency department.

1 INTRODUCTION

Recent years have seen an increase in the use of mathematical modeling techniques for the design and operation of health care facilities. For example, specialized simulation software packages such as Med Model (McGuire, 1994) and the hospital template of Arena (Drevna and Kasales, 1994) have recently been developed. For an overview of modeling techniques used in health care settings, see Pierskalla and Brailer (1991).

One of the main reasons for the increase in the popularity of modeling techniques in health care settings is pressure resulting from the desire to maintain a high level of quality for patient care while reducing (or at least not increasing) costs. Techniques of mathematical modeling, including those associated with simulation, statistical analysis, optimization, and multicriteria decision analysis, allow one to optimize the use of health care resources (e.g., through efficient scheduling of those resources) subject to various constraints.

This paper involves a discussion of a simulation model developed for an emergency department of a Louisville, Kentucky hospital. Emergency departments are especially important since the number of emergency room visits in the United States exceeds 86 million per year. In addition, approximately 30% of a hospital's admissions are through the emergency department and about 50% of a hospital's total revenue is generated from patients who use or come through the emergency department (see Hospital Statistics).

The main purpose of the simulation model, developed using the Arena software package, was to have a vehicle for investigating the desirability of various personnel schedules for the emergency department. See Draeger (1992), Iskander and Carter (1991), Kumar and Kapur (1989), Ladany and Turban (1978), and McGuire (1994) for additional examples of simulation models of emergency departments.

2 DESCRIPTION OF THE EMERGENCY DEPARTMENT

The emergency department under study consists of two areas: a full service emergency room (ER) and Emergency Care Express. Emergency Care Express is a "fast-track" emergency room service that treats minor medical emergencies from 11am to 11pm seven days a week. The focus of this study however was the full service ER which is operational seven days per week, 24 hours per day. This ER consists of 16 rooms, including two cardiac rooms, one ear-nose-throat (ENT) room, and one room for patients with behavioral health problems. The ER treats approximately 39,000 patients per year.

Concern about the ER was raised because the average patient time in the system was approximately 142 minutes, which was significantly greater than the standard industrial average of 120 minutes. Hence, the main objective of the study was to investigate various schedules for nurses, ER technicians, and doctors in order to reduce the average patient time in the system. Other objectives included providing management with better insight into the working of the ER and to determine how other factors (e.g., patient load) affected the average length of stay (LOS) of patients.

A patients enters the ER by one of two modes: Walk in or squad. Patients are assessed upon arrival by a nurse (RN) and urgency of care is prioritized. The RN initially assesses the patient and obtains baseline information such as vital signs and chief complaint. The physician then examines the patient and appropriate medication, laboratory tests or X-ray procedures are ordered. Patients receive a certain amount of direct care every hour during their length of stay based on their acuity. The activities performed by RN, technicians and physicians are of two types, one is direct and the other one is indirect. Direct treatments are hands-on type of treatment on patients. It includes assessing patient conditions, taking vital signs, applying dressings, hanging IV's, patient treatment etc. Whereas indirect time spent by the staff of the ER includes coordinating patient care with other departments, processing physician orders, handling communication, general administrative work, cleaning beds, assisting in hygiene etc. For physicians the indirect time includes communicating with other expert physicians for their comments and advice, and evaluating patient condition from X-rays and lab reports. Patients have a probability of leaving the ER for a period of time to go to ancillary departments such as X-rays etc. Patients leave the ER in one of the two ways: discharge to home or admission to the hospital. In either case they leave the model/system and they no longer require care from the ER staff.

For this study, patients were classified according to 13 significant patient groups. Data was collected to establish the fraction of patients in each group -- see Table 1. Associated with each group of patients was a particular process flow, as defined by a sequence of activities and associated resource types (doctors, nurses, technicians) required to perform those activities. For example, the process associated with patients who have cardiovascular disorders is shown in Figure 1. Note that this process flowchart includes some probabilistic branching, as is the case for most of the categories of patients.

Table 1: Categories of Patients

No.	DIAGNOSIS	PROBABILITY OF OCCURENCE (%)
1	Allergy	5
2	Cardio-Vascular Disorders	12
3	EENT	6
4	Gastro-Intestinal Disorders	13
5	Genito-Urinary Disorders	5
6	Gynec. & Obs. Disorders	5
7	Musculoskeletal Disorders	13
8	Neurological Trauma	7
9	Orthopedic	10
10	Psychiatric Disorders	4
11	Respiratory Disorders	10
12	Surgical Trauma	5
13	Others	5

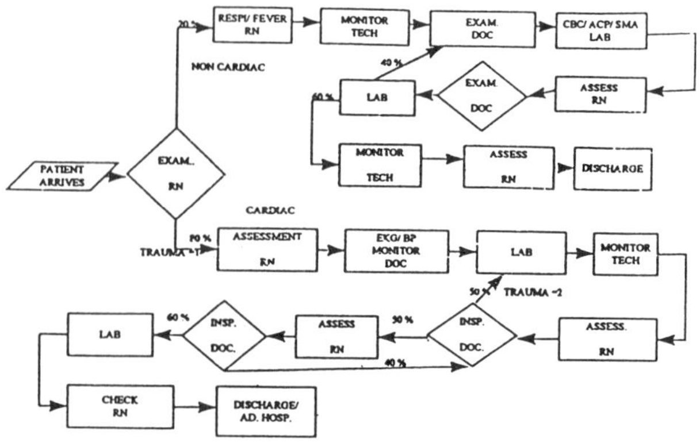

Figure 1: Process Flow for Patients with Cardiovascular Disorders

3 DATA COLLECTION

Data for building the model was collected on patient arrival rates; types of treatments for various categories of patients; resources (e.g., nurses, technicians, or doctors) and durations required for the various activities associated with the treatment of patients; possible schedules of nurses, technicians, and doctors; and indirect time spent by ER staff.

Some of this data was collected through the use of group discussions with some of the nursing staff. Other data (such as that associated with arrival rates for different types of patients) was collected from ER logs and patient records.

The current schedules used for the nurses, technicians, and doctors of the ER are given in Table 2. This data was used as input to the model for validation purposes.

Table 2: Current Personnel Schedules in the ER

RESOURCE	TIME					
	12mid-7a.m.	7a.m.-11a.m.	11a.m.-2p.m.	2p.m.-3p.m.	3p.m.-11p.m.	11p.m.-12 mid.
Nurses	4	4	6	6	7	4
Technicians	2	3	4	4	4	2
Doctors	1	1	1	2	2	2

4 SIMULATION MODEL DESCRIPTION

The development of the simulation model for the ER was accomplished by using Arena. In the model, the perspective used to describe the process is that of a patient flow in the

ER. The dynamic nature of the ER is captured by altering the time between arrivals of patients by time of the day. Patients arrive according to an interarrival distribution based upon the pattern identified for the time of day.

Parameters and attribute values (e.g., trauma level) are assigned upon arrival for each patient. Initially a trauma level value of 0 is assigned upon arrival and trauma values are changed in the model with improving or deteriorating condition of the patient with a value of 0 as lowest and 3 as the highest acuity level for trauma. Patient case type is assigned to each patient according to probability of occurrence of cases identified at the time of data collection. Specific patient care activities according to case type are grouped together and patients follow a process flow. Appropriate staff responsibilities are assigned at each node according to the flow chart made for patient type.

When a patient arrives, a trauma value is stored as an attribute value, and he is routed in the model according to a treatment type assigned with the help of a "branch" SIMAN node. The patient is routed in a sequence identified in the flow chart which was collected for each treatment type. The patient moves from one server to another, which represents appropriate treatment given to him. Hence, each server represents a treatment and the one resource: either RN, technician or doctor, selected to serve at that station. Inspection nodes are used when the decision is to be made about the condition of the patient and the next set of procedures. Finally when the patient passes through all treatment steps, he is routed to a depart node where all time-persistent statistics like length of stay etc. are collected.

Case type and acuity level determine the probability distribution applied in routing a particular patient through or around specific activities in the model. The priority of treatment is based on the acuity level of the patient. In the case of two or more patients attempting to capture one resource, the patient with the higher acuity is served first. In the case of a tie FIFO rule is used. Activity durations are dependent on patient treatment type and acuity level, and are represented by the appropriate distribution obtained from detailed observation data. In the case of lab and X-ray procedures, an average waiting time is built into the activity time since these locations (lab and X-Ray) are not modeled explicitly.

While modeling the system, exceptional cases and deaths in the ER are omitted, and the following additional assumptions are applied:

1. Each patient retains a single classification according to treatment, throughout his stay in the ER.
2. For the purposes of the model, the processing of patients according to treatment type was judged adequate to represent the process.
3. In the event that a staff person is busy with an activity at the end of his/her shift or during indirect time, that person finishes the activity before going off duty, as would actually occur in the ER.
4. For the purpose of simulation, the model was developed for the procedures done in the ER work area only. The triage work area where initial administrative work is done on a patient was not considered in the model due to lack of data; also, ER managers wanted to perform the analysis on only the internal working of the ER.
5. All staff resources are assumed to be available without any specific break time. In the real system, ER staff breaks are not scheduled, they occur whenever the staff gets free and senses there is a time to take a break.
6. It is not mandated that a single nurse treat a particular patient during the patients' stay.
7. Reneging patients, who choose not to wait after they were initially assessed, were not considered while modeling.
8. All persons of a particular resource type, were assumed to have the same level of expertise and skills.
9. The usage of a specialist doctor while treating a patient was not considered while modeling.

5 MODEL VERIFICATION AND VALIDATION

The model verification was performed using the trace and break option of Arena. The trace option records the movement of entities from node to node and the processing of entities at each node. The break option stops the simulation when a logical error occurs. The trace output report was examined to determine whether the entities were proceeding through the system as desired. After examining the trace it was determined that the computer program implementing the simulation model of the ER was executing properly.

Several steps were taken throughout the project to validate the model and build its credibility. These included an initial introduction of the ER managers and RN representative to the concept and objectives of simulation as an analysis tool. In identifying elements to be included in the model and determining appropriate levels of detail, the help of management engineering department was sought. Through the data collection process, first hand knowledge of the ER was gained across all shifts, including usual and exceptional patient cases. Those issues necessary to include in the model were prioritized and respectively addressed during development of the model.

The completed model was run on a pilot basis to perform first pass validation against actual data. The system performance measure monitored was total time spent by patients in the ER, which was of main interest to management in decision making. Relative frequency distribution of length of stay of patient in the ER (LOS) for the model's one week of activity was compared against actual results from patient observation data.

After initial validation, the ER managers were presented the model flow, the model assumptions and details, validation results and the output of the current model. Suggestions and corrections to the model details and assumptions were solicited from the ER managers.

Final model changes were made, the model was run with five replications and the results were validated against actual data. The average LOS of the simulation model was found to be 142 minutes. With 95% of confidence interval (CI), average LOS was found to be between 134 and 149 minutes. This compares to an actual average LOS for the ER of 142 minutes.

6 MODEL EXPERIMENTATION

Experimentation with the model is ongoing. Design variables and parameters with which one could experiment using the model include the mix of patient types; process activities, including durations of those activities; demand for the ER; and schedules for the nurses, technicians, and doctors. As was mentioned earlier, the emphasis in this initial effort was on the personnel schedules, as measured by the numbers of nurses, technicians, and doctors on duty in the ER for each hour of the simulation run.

In the production runs of the model 14 days of the operation of the ER were simulated, including a warm-up period of seven days. All statistics were cleared after the warm-up period but the entities were kept in the system in order to begin the simulation in a steady state mode.

Five different schedules, labeled A,B,C,D, and E, were input to the model for experimentation purposes. These schedules varied only in the numbers of nurses and technicians on duty from the actual system, and kept the numbers of doctors on duty the same as the actual system. See Table 3 for those five schedules.

Each of the five schedules was run for five replications, with the average LOS and a 95% confidence interval for average LOS computed in each case. These results are shown in Table 4. Note that Schedule D gave the lowest average LOS, but that additional replications would have to be made in order to establish significant differences between Schedule D and Schedules A and C.

Table 3: Experimental Personnel Schedules

RESOURCE	SCHEDULE	TIME			
		7a.m.-11a.m.	11a.m.-3p.m.	3 p.m.-11p.m.	11 p.m.-7a.m.
NURSES	A	4	6	7	5
	B	3	6	7	5
	C	5	6	7	5
	D	5	7	7	5
	E	5	7	7	4
TECH.	A	3	3	5	3
	B	4	5	5	3
	C	4	5	4	3
	D	3	4	5	3
	E	3	5	5	3

Table 4: Simulation Output Associated with the Experimental Schedules

SCHEDULE	AV. LOS	AV. LOS WITH 95% CI
A	137.2	131.9-142.5
B	143.3	140.5-146.3
C	138.8	135.3-142.3
D	136.9	134.0-139.9
E	143.3	136.3-149.7

7 CONCLUSIONS

A simulation model is an ideal tool to cope with the diversified and dynamic nature of a hospital's emergency department. The model developed for this work allowed the evaluation of various schedules for the nurses, technicians, and doctors with respect to such performance measures as average length of stay of the patients.

There are many extensions to this work which are currently being investigated. Examples of these extensions would include the development of an "optimization module" which would interface with the simulation model to optimize over a large number of potential schedules. Such an optimization module would consider several objectives simultaneously, such as length-of-stay, personnel costs, and desirability of various schedules as considered by the nurses, technicians, and doctors.

REFERENCES

Draeger, M.A., 1992. An emergency department simulation model used to evaluate alternative nurse staffing and patient population scenarios. In *Proceedings of the 1992 Winter Simulation Conference*, ed. J.J. Swain, D. Goldsman, R.C. Crain, and J.R. Wilson, 1057-1064. Piscataway, New Jersey: Institute of Electrical and Electronic Engineers.

Drevna, M.C. and C.J. Kasales, 1994. Introduction to Arena. In *Proceedings of the 1994 Winter Simulation Conference*, ed. J.D. Tew, S. Manivannan, D.A. Sadowski, and A.F. Seila, 431-436. Piscataway, New Jersey: Institute of Electrical and Electronic Engineers.

Hospital Statistics, 1989-91 Ed., American Hospitals Association, page 12.

Iskander, W.H. and D. Canter, 1991. A simulation model for a same day care facility at a university hospital. In *Proceedings of the 1991 Winter Simulation Conference*, ed. B.L. Nelson, W.D. Kelton, and G.M. Clark, 846-853.

Kumar, A.P. and R. Kapur, 1989. Discrete simulation application - scheduling staff for the emergency room. In *Proceedings of the 1989 Winter Simulation Conference*, ed. E.A. MacNair, K.J. Musselman, and P. Heidelberger, 1112-1120. Piscataway, New Jersey: Institute of Electrical and Electronic Engineers.

Ladany, S.P. and E. Turban, 1978. A simulation of emergency rooms. *Computers and Operations Research* 4: 89-100.

McGuire, F., 1994. Using Simulation to reduce length of stay in emergency departments. In *Proceedings of the 1994 Winter Simulation Conference*, ed. J. Tew, S. Manivannan, D.A. Sadowski, and A.F. Seila, 861-867. Piscataway, New Jersey: Institute of Electrical and Electronic Engineers.

Pierskalla, W.P. and D.J. Brailer, 1994. Applications of operations research in health care delivery. In *Handbooks on Operations Research and Management Sciences*, Vol. 6 6, S.M. Pollock et cl. edition, Elsevier Science, B.U.

AUTHOR BIOGRAPHIES

Gerald W. Evans is a Professor in the Department of Industrial Engineering at the University of Louisville. He received a B.S. degree in mathematics, and M.S. and Ph.D. degrees in industrial engineering, all from Purdue University. His research interests include multicriteria decision making and its interface with simulation modeling. He is a member of IIE, INFORMS and DSI, and served as Proceedings Editor for the 1993 Winter Simulation Conference.

Tesham B. Gor is a Ph.D. student in the Department of Industrial Engineering at the University of Louisville. He received his M.S. in Industrial Engineering from the University of Louisville, and B.E. in Mechanical Engineering from the University of Poona in India. His research interests are in the areas of simulation, process control, quality control, and wavelet theory. He was recently awarded the American International Education in Japan Fellowship to conduct his research work at the Toyohashi University of Technology in Japan. Mr. Gor is a member of IIE and SME.

Edward Unger is an industrial engineer working at Baptist Hospital East in Louisville, Kentucky. He received a B.S. degree in industrial engineering in 1993 and an M.S. degree in industrial engineering in 1996, both from the University of Louisville. His research interests include the applications of industrial engineering techniques in health care.

Proceedings of the 1996 Winter Simulation Conference
ed. J. M. Charnes, D. J. Morrice, D. T. Brunner, and J. J. Swain

USE OF SIMULATION TO TEST CLIENT-SERVER MODELS

Yogesh L Deshpande
Roger Jenkins

Faculty of Business and Technology
University of Western Sydney, Macarthur
P O Box 555, Campbelltown
NSW 2560, AUSTRALIA

Simon Taylor

Department of Computer Science and
Information Systems
Brunel University
Uxbridge, UB8 3PH ENGLAND

ABSTRACT

Simulation is used for many purposes: for example, to analyse a complex system, to visualise the functioning of a system, and to optimise or tune a system. While there is no limitation on the use of simulation, the general consensus is that an analytical solution, if one is possible, is always to be preferred to simulation as a methodology. In the field of information systems, client-server models exhibit a degree of complexity and richness not amenable to easy analytical solutions, except for some specific algorithms useful in limited contexts. Simulation could, therefore, be a good strategy to analyse the client-server systems and help in better implementation of feasible solutions. This paper examines the current state of client-server models and use of simulation in dealing with the problems encountered. The paper then compares the seven-layer OSI model for communications architecture and recommends that a similarly layered approach is likely to prove useful in simulating client-server systems. In the process, the paper also points out that the simulation models bring into a sharp focus the importance of software metrics, an area of vital importance in software development.

1. INTRODUCTION

Simulation is a powerful tool in analysing complex systems. It is used for a wide variety of purposes such as optimising or fine-tuning a system, visualising the functioning of a system, or simply for better understanding through the analysis of a system. While there are few limits on the use of simulation, the general consensus is that, for a given problem, an analytical solution, if one is possible, is to be preferred to simulation as a methodology. In practice, however, there are many situations and systems where interactions among the components of the system are not amenable to easy analysis and consequently analytical solutions are based on simplified models. Client-server (c-s) systems are a good example of these limitations and hence a field which could benefit from simulation studies.

The use of simulation to test the design of a client-server system is, however, not so straightforward. Simulation as a strategy for problem-solving requires a good understanding of simulation modelling and the computing and statistical aspects of simulation. Client-server systems themselves can be quite complex with possibly several options for the systems designer to choose from. Since client-server modelling is a relatively recent development, there is a shortage of experts in this area, widely acknowledged by the software industry. Additionally, the developers of such systems are generally under severe time constraints, unable to experiment sufficiently with the alternative designs and not necessarily familiar with simulation strategies or tools. There is thus a possibility that for a number of reasons, the systems designers may neglect simulation altogether or use it in a manner which may limit the usefulness of the results obtained, or even invalidate them.

The difficulties enumerated above notwithstanding, this paper joins a growing body of work which contends that there is much to be gained by the use of simulation in the client-server environment. Even the simplified models presented in this paper bring into a sharp focus the complex nature of layers of sub-systems, such as network topologies and communication protocols, on which a client-server system is based and about which a designer is forced to make assumptions. At the same time, a review of the literature in this area shows that simulation modelling is used either in a somewhat limited capacity, concentrating on sub-systems and algorithms or at a level of detail which makes it complex and expensive (Robinson 1994). There is thus an untapped potential for the use of simulation of the systems. The possible application of simulation strategy to client-server problem also points to the vital need of proper measurements of the client-server system parameters, an issue directly linked to software metrics and performance monitoring. The paper concludes that even a simple simulation model of a client-server system is beneficial to the system designers and developers.

The paper is organised as follows. Section 2 briefly describes the client-server models. Section 3 relates simulation work to client-server and finds that the modelling exercise in this area includes details of communication networks, hardware and processor characteristics, algorithmic analysis, query optimisation and file or database distribution. Section 4 then relates the client-server systems to the seven layers of the Open System International (OSI) model to bring out the hidden complexity of the models and to recommend simulation strategies in a similarly layered modelling approach. Section 5 describes a case study under way, using a commercially available simulation package. Section 6 deals with the question of statistical aspects of simulation and compares them with software metrics. Section 7 concludes the paper.

2. CLIENT-SERVER MODELS

Almost all writings on this subject start with a statement that there is no single definition of what is meant by client-server model of computing. Nevertheless there is a fairly good agreement about what does characterise them (Stallings and Slyke 1994, IESC Report 1995). The following working definition by Boar (1993), later adopted by the IES Committee (IESC) set up by the Australian Federal Department of Finance (1995) is relatively straightforward: "*A computing model whereby the application processing is partitioned across multiple processing platforms and where processors cooperate to complete the processing as a single integrated task*".

This definition emphasises the logical nature of the client-server model and not the physical one which would include the hardware and network components.

Client-server models belong to the class of distributed computing models. In theory, even one client and one server on the same hardware platform could qualify as a client-server system. In practice, though, there are likely to be many clients, based on multiple platforms, and being served by at least one server, connected through networks. Figure 1 illustrates a general client-server model with multiple clients connected to multiple servers via a local or wide area network.

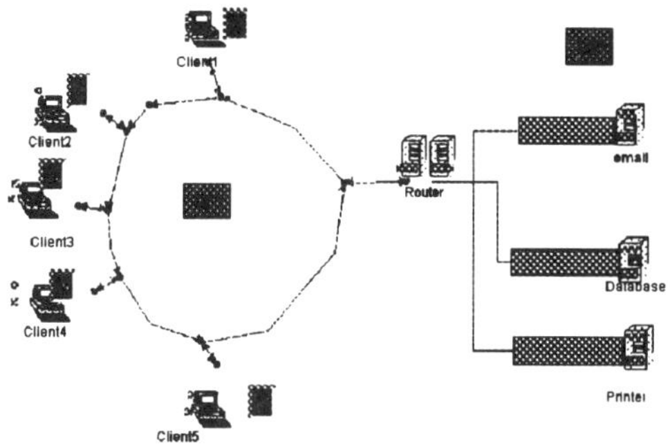

Figure 1: A simple client-server model

A more elaborate model put out by the Gartner Group (1994) has gained wide acceptance among computing practitioners. They define client-server by postulating five configurations through combinations of the three components into which any application may be subdivided:

- Presentation
- Processing
- Data management

These components may be distributed across the client(s) and the server(s) in five ways, as shown in Figure 2.

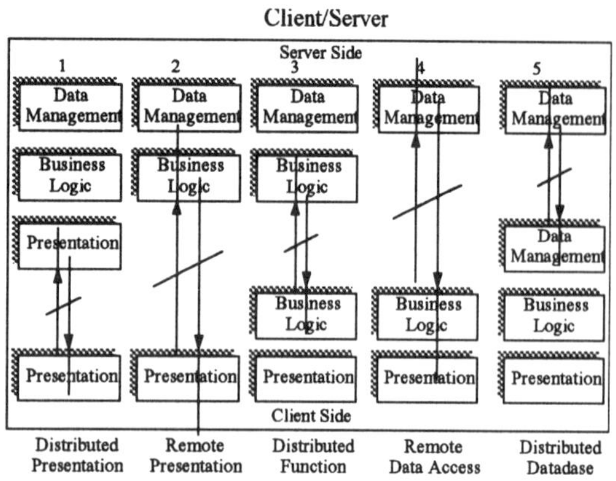

Figure 2: The Gartner Group models of client-server systems

The first two models, distributed and remote presentations, represent a situation when the server is the dominant partner and the client is either a dumb terminal or is there mainly to provide a user-friendly graphical user interface (GUI). The remaining three allow the client to carry out processing either partially or wholly on the client(s) and are of major interest here. In these situations, a designer is faced with many choices about how to split processing and data management functions between the clients and servers.

Consider a hypothetical application for development. Typically, it will contain tens, if not hundreds, of modules dealing with user interface, inputting data, validation of input data, queries, report generation, processing and output messages. The client-server architectural options will allow a split of these functional modules on one or more clients on the one hand and on one or more servers on the other. Furthermore, the server(s) will also be responsible for data management which may lead to distribution of databases and tables over many servers. The performance of the system will clearly depend, among other things, upon how these modules are distributed across all the clients and servers in the system.

A number of authors have written in detail on strategies and guidelines for such functional and data distribution (Booth 1981, Krantz 1995, among others). However, since the design of a client-server system is specific to a given application, such guidelines have been necessarily general in nature. Consequently, details of actual implementation of a system design and its effects on the system performance are also of general nature. Anecdotal evidence points to the designers making choices about splits in client-server functionality on the basis of experience, intuition and trial-and-error methods.

The foregoing description and discussion indicate the possibility, and desirability, of applying simulation strategies to test and improve client-server designs. The following sections elaborate on this theme.

3. SIMULATION AND CLIENT -SERVER

This paper is mainly concerned about using simulation modelling in a specific environment, viz: client-server systems. Familiarity with simulation methodology will, therefore, be assumed and the reader is referred to well-known titles such as Tocher (1963), Naylor et al (1966), Banks and Carson(1984), Pidd (1992), Szymankiewicz et al (1988) for further elaboration of simulation strategies. Any comments related to the simulation methodology in this paper should be read only in the context of client-server systems modelling and design and not as any critique contributing to the field of simulation.

Section 2 described the five generally accepted models of client-server computing. It also pointed out that the design of a client-server system is specific to a given application. For the discussion of use of simulation, for the rest of this paper, therefore, it is necessary to make certain assumptions which are as follows:

- There is only one server with more than one client;
- The client will have presentation and partial functionality while the server will carry out the remaining processing and data management, i.e. the third model in the Gartner Group's classification will be implemented;
- The application will be a database management;
- The systems designers want to reduce the response time to the minimum.

These assumptions are really meant to facilitate identification of characteristics of the client-server environment germane to simulation. As will be seen, any or all of these assumptions may be relaxed without adversely affecting the conclusions.

The assumptions made above are partial details of only a logical model. They do not at all consider the possible variations in hardware and software platforms, nor are the network and communication parameters identified. The view is of a systems designer who takes all the other parameters as given and is only concerned with decisions about how to split the functionality within the application across multiple clients and the server.

A typical working scenario of the system would involve the following:

- a client would present a user interface to validate input and generate requests (transactions) for the server to process;
- the transactions could take the form of simple queries, or updates to the database, or requests for detailed reports;
- the server would receive these transactions/ requests over the network and respond appropriately;
- the server would carry out database management tasks, as and when needed.

In order to simulate such a system, a simulation expert would want to know answers to questions such as:

- what are the possible types of transactions and what are their (statistical) distributions?
- do all clients follow the same distributions?
- how much processing time is needed on the client and how much on the server?
- what is the communication overhead and how is it measured?

Depending on the level of detail in a simulation model, there could be many more similar questions to which answers may be required.

Given the required details, a simulation model may be built and statistically valid experiments run to report on various combinations of functional division between the clients and the server.

The idea of using simulation in client-server environment has been explored by several researchers in much greater detail, as reported in literature (Smith 1990, Jain 1991, Gold 1993, McBeath and Keezer 1993, Shen and Butler 1994, Robinson 1994, Komatsu, Wakayama and Nose 1994). The level of detail, however, is a potential challenge to systems designers in general, as discussed below.

The main characteristic of the first three questions about a simplified client-server system raised above is that they are related to the application under consideration and not to the hardware/software platform on which the application will be built. They are also the questions of interest to any systems designer. The performance of a system based on this logical design could be analysed without much more data.

In the final analysis, however, performance of a given system will depend upon many factors, such as network topology, transmission protocols, client and server processor speeds, number of clients supported, choice of software packages and underlying operating systems. A systems designer is aware of these factors but has far fewer measurements available on which to judge the relative contribution of these factors and hence their likely effect on the system performance. Many of these factors belong to areas of specialisation about which a systems designer has only a vague model in his/her mind.

In the references cited above, for example, Shen and Butler (1994) include the network characteristics in their model-building, Smith (1990) and Robinson (1994) allow hardware characteristics, Komatsu, Wakayama and Nose (1994) incorporate internal processes of a computer system and models using the so-called Remote Procedure Calls (RPC) have also been reported.

The detailed simulation modelling would make the subsequent experiments and output analyses more valid than those based on a less detailed one. On the other hand, Robinson notes that the increased complexity is a drawback and, in many cases, simpler models may be adequate to address basic issues.

The foregoing analysis suggests that what is likely to be useful to a systems designer is the ability to include the levels of detail selectively and in such a way as to make modelling easier and more tractable. Section 4 suggests ways of doing that by utilising the familiar Open Systems Interconnection (OSI) model.

4. CLIENT-SERVER AND OSI

As very briefly discussed in Section 3, client-server modelling is complicated by the considerations of a large number of variables, ranging from software and hardware platforms through to networks and communications protocols. The OSI model, representing both client and server (Figure 3), is in fact a good guide to sort these factors in layers where they belong.

	Layer
1	*Physical*
2	*Data link*
3	*Network*
4	*Transport*
5	*Session*
6	*Presentation*
7	*Application*

Figure 3: OSI model of communications architecture

Reverting back to the simplified example of Section 3, it is seen that the first three questions about the system characteristics belong to the Application layer whereas the fourth properly formulated straddles two layers, Network and Communications. From the system designer's point of view, a model involving a (client) application layer connected to (server) application layer is easily understood. Such a model would then necessarily assume that the characteristics of the layers below remained unchanged and therefore allow exclusive consideration of allocation of functionality between a client and the server.

This approach logically leads to another step in the full exercise of simulation. Each layer, especially the Network and Communications Protocols can be independently modelled, in their own right, to a level of detail relevant to a given system. It may be then possible to draw upon work already done in these areas (e.g. Mitrani 1987, Law and McComas 1994, Komatsu, Wakayama and Nose, 1994, Rumekasten 1994, Shen and Butler, 1994). Since there will generally be more than one information system running on these platforms, these Network and Communications models could be directly utilised in other simulation exercises. The layered models also allow analysis of possible or proposed changes to specific characteristics of any given layer.

The layered approach need not follow the OSI model exclusively. IESC report (1995), for example (Figure 4), or TCP/IP protocol does not adhere to seven layers. As mentioned above, what is needed is a good model at the appropriate level and a systems designer may in fact be in the best position to decide upon the layer of interest.

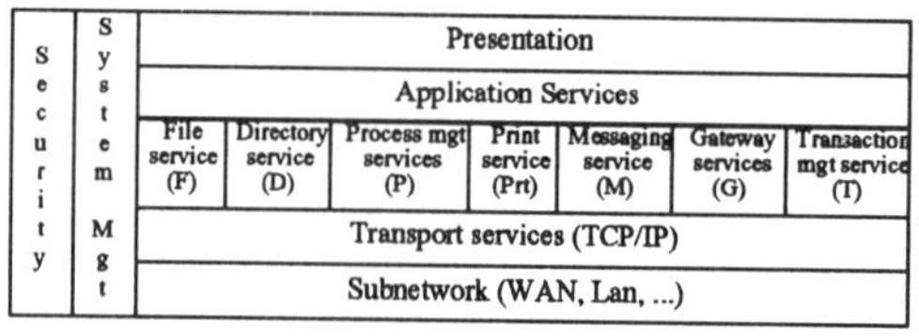

Figure 4: Layered model of client-server systems (Source: IESC Report 1995)

Work is in progress to implement this layered approach for the simple system described above. Even at an early stage, this modelling exercise has highlighted benefits, such as better understanding of issues, the role of Software Metrics and the possibility of using commercially available and inexpensive simulation packages. Section 5 describes the case study and discusses the relevant issues.

5. CASE STUDY

The case study is based on the assumptions listed in Section 3. The simplicity of the model meant that there was no need to write customised simulation package. The choice of the package was ProModel, for its ease of use, ability to animate the model, statistical support and the at least one author's familiarity with it. The first models were shown to a number of people who suggested a few, later obvious, improvements. The results of these attempts have been described in Sections 3 and 4 and led to the adoption of the layered approach to client-server systems modelling. There are now three separate models, one each dealing with the Application layer, the Network layer and the Communications Protocol layer.

To illustrate the nature of these models, consider how the system would deal with a transaction generated by a client. The broad steps involved in processing the transaction would be as follows:

- The client side would present the user interface, accept input and process it as per a given specification, passing it on to the Network layer for further processing.
- The Network would process the transaction and pass it on to the Communications system.
- The Communications layer on the client side would connect with that on the server side

and ensure that messages were passed on correctly.

- These messages would travel upwards to the Application layer of the server in the reverse fashion, processed and the output would travel back to the client in a similar manner.

Although this is a very simplified model, it has already proved to be useful in many ways.

Firstly, the exercise brought home to the people involved in the discussion that there are a number of possible views of the same system, all of them important, and that it is essential to develop a common language across the different expertise. The layers clarified to everyone exactly what was under discussion and needed further clarifications.

Secondly, it forced the participants to consider different ways of measuring the "traffic". A transaction viewed at the client Application layer goes through a few transformations before reaching the server Application layer, again as a complete transaction. The statistical distributions of the types of transactions and of the time it takes for these transactions to reach the destination are likely to be different at each layer. A larger question is thus raised about the appropriate measurements and collection of data in order to validate the simulation exercise.

Thirdly, the case study has highlighted the appropriateness of constructs available in the simulation packages. ProModel is primarily designed to aid in modelling manufacturing systems and uses icons and terminology of that environment. The success and speed of modelling depends substantially on the ability of experts participating in the study to interpret the terminology flexibly and not be diverted by the language used by the package. This is not always so obvious.

A further advantage which simulation modellers clearly recognise, was the effect of animating the model. As a generalisation, systems designers tend to think in more abstract terms, in algorithmic patterns and, at times, in static models of their systems. Animated simulation models of even the simplified models seems to generate much more interest than even extended discussions.

As mentioned above, this case study has raised a major concern about the measurements of any given system and the different points of view of simulation strategists and systems designers. Section 6 briefly discusses these issues.

6. STATISTICAL ASPECTS OF SIMULATION AND SOFTWARE METRICS

There is a lot of similarity between the approaches adopted in simulation modelling and systems design (Robinson 1994). However, after the logical design of a system model, the system developers proceed to its physical design straightaway, very often under severe time pressure. Validation of the models they may have built is static and through structured walkthroughs or meetings with the responsible clients or through proto-typing.

In contrast, simulation strategists attach a great deal of importance to relevant and accurate measurements, or data, before any experiments are conducted and models validated.

One of the early benefits of the case study, as mentioned above, was the focus on input and output data of the given system. From the point of view of validating the simulation model of a client-server system, data would be needed to cover as many of the variations in the systems parameters as possible. Depending on the level of modelling, such data would include measurements specific to the application, network and communications layers before simulation experiments could begin.

The conceptual modelling exercise and consequent attention to measurements also is extendable to post-implementation systems. Whatever assumptions are made in the pre-design stage and, data collected accordingly, the data capture and further validation tests can be continued throughout the systems life cycle, leading to more pro-active systems maintenance.

It may be pointed out that the model can only perform the task of highlighting the paucity and/or the (poor) quality of data about the system(s) under consideration. Changes in practice and improvements in the quality of data are matters for action only by the appropriate authorities.

7. CONCLUSIONS

Simulation modelling and experiments are used widely and for a variety of purposes. This paper joins a growing body of simulation work in the client-server environment. The paper analysed the client-server model to relate it to the seven-layers of OSI model and suggested that a similar, layer-based approach be used in the simulation exercise. The benefits of this approach are in terms of clearer understanding of the client-server models, possibility of using commercially available simulation packages, and a sharp focus on the issues of software metrics. The modelling exercise alone is worth the effort to clarify the issues to the software community. The client-server environment is still unfamiliar to many people and causes difficulties in understanding since it engages many rapidly-changing technologies. Simulation modelling makes the client-server models much more tractable and thus should prove to be of great benefit overall.

REFERENCES

Banks, J. and Carson, J.S. 1984. *Discrete-Event System Simulation*. Prentice-Hall:

Boar, B.H. 1993. *Implementing Client/Server Computing: A Strategic Perspective*, New York, NY: McGraw-Hill.

Booth, G.M. 1981. *The Distributed System Environment*. New York: McGraw-Hill.

Gartner Group. 1994. *Client/Server Computing in the 1990s*, Stamford,CT: Client/server Logical View Personal Advisory Service.

Gold, J. 1993. Simulation aims to predict software performance. *Software Magazine*. July: 15-19.

Information Exchange Steering Committee (IESC). 1995. *Client/Server Computing in the Australian Public Service: The Next Wave?* Canberra: Department of Finance, Australian Federal Government.

Jain, R. 1991. *The Art of Computer Systems Performance Analysis*. New York, NY: John Wiley and Sons.

Jiang, Y., Cao, J. and Gupta, G.K. 1992. *Dynamic Distributed Query Processing Through Query Migration*. In *Research and Practical Issues in Databases, Proceedings of the 3rd Australian Database Conference*, ed. B. Srinivasan and J. Zeleznikov, 197-209. Melbourne: World Scientific.

Komatsu, T., Wakayama, H. and Nose, J. 1994. Model Description in the INSYDE Simulator for Evaluating Large-scale Computer System Performance. In *Proceedings of the 1994 Winter Simulation Conference*, ed. J.D.Tew et al, 1272-1279, Boston: The Society for Computer Simulation.

Krantz, S. 1995. *Real World Client/Server*. Gulf Breeze, Fl: Maximum Press.

Law, A.M. and McComas, M.G. 1994. Simulation of Communications Networks. In *Proceedings of the 1994 Winter Simulation Conference*, ed. J.D.Tew et al, 166-170, Boston: The Society for Computer Simulation.

McBeath, D.F. and Keezer, W.S. 1993. Simulation in Support of Software Development. In *Proceedings of the 1993 Winter Simulation Conference*, ed. Evans, G.W., Mollaghesemi, M., Russell, E.C., Biles, W.E., 1143-1151. Baltimore, MD: Association for Computing Machinery.

Mitrani, I. 1987. *Modelling of computer and communication systems*. Cambridge: Cambridge University Press.

Naylor, T.H., Balintfy, J.L, Burdick, D.S. and Chu, K. 1966. *Computer Simulation Techniques*. John Wiley:

North, K. 1996. Performance Testing, ODBC, and Native SQL APIs. In *Dr Dobb's Sourcebook*, Vol 21, Issue 13, (January-February), 17-21.

Pidd, M. 1992. *Computer Simulation in Management Science*. John Wiley and Sons:

Robinson, J. 1994. Capacity and Performance Analysis of Computer Systems. In *Proceedings of the 1994 Winter Simulation Conference*, ed. J.D.Tew et al, 34-41, Boston: The Society for Computer Simulation

Rumekasten, M. 1994. Simulation of Heterogeneous Networks. In *Proceedings of the 1994 Winter Simulation Conference*, ed. J.D.Tew et al, 1264-1271, Boston: The Society for Computer Simulation

Shen, J. and S. Butler. 1994. Performance Modeling Study of a Client/Server System Architecture. In *Proceedings of the 1994 Winter Simulation Conference*, ed. J.D.Tew et al, 1280-1287, Boston: The Society for Computer Simulation.

Smith, C.U. 1990. *Performance Engineering of Software Systems*. Reading, MA: Addison-Wesley.

Stallings, W. and Van Slyke, R. 1994. *Business Data Communications*. 2nd edition. New York, NY: Macamillan.

Szymankiewicz, J., McDonald, J and Turner, K. 1988. *Solving Business Problems by*

Simulation. McGraw-Hill: Maidenhead, England.
Tocher, K.D. 1963. *The Art of Simulation.* English University Press: London.

AUTHOR BIOGRAPHIES

YOGESH L. DESHPANDE is a senior lecturer in the Department of Computing and Information Systems in the Faculty of Business and Technology at the University of Western Sydney, Macarthur. His research interests include discrete event simulation modelling, client-server systems, and information systems. He is a member of the Australian Computer Society, Computer Society of the IEEE, and the Operational Research Society.

ROGER JENKINS is a lecturer in the Department of Manufacturing and Quality Systems in the Faculty of Business and Technology at the University of Western Sydney, Macarthur. His research interests include discrete event and continuous simulation modelling and materials and resource planning.

SIMON TAYLOR is a lecturer in the Department of Computer Science and Information Systems at Brunel University. His research interests include simulation modelling, and distributed and parallel simulation.

Proceedings of the 1996 Winter Simulation Conference
ed. J. M. Charnes, D. J. Morrice, D. T. Brunner, and J. J. Swain

SIMULATING A PROPOSED INFORMATION DELIVERY SYSTEM

Eugenio M. Alvarez

Nine West Group, Inc.
9 West Broad Street
Stamford, Connecticut 06902, U.S.A

Walter Callahan

NewYork Power Authority
Indian Point Nuclear Power Plant Unit 3
Buchanon, New York 10511, U.S.A

Paul Mitchell
Raymond Telerole

New York Power Authority
123 Main Street
White Plains, New York 10601, U.S.A

Peter Hoefer

Lubin School of Business
Pace University
Pace Plaza
New York, New York 10038, U.S.A

ABSTRACT

This paper describes the results of a simulation analysis of hardware designs for a proposed data delivery system. The system is to serve clients of a large accounting firm. Once the required hardware components were identified, and their system parameters and response times estimated, the design of such a system required selection of components in a way to optimize specific performance characteristics. Simulation was used to understand the performance of the system, and to help with decisions on the components to be used.

1 INTRODUCTION

This story begins with one of us being involved with designing a system to give customers an interactive ability to retrieve information from a large accounting firm. In the next section we describe the design of this "InfoDelivery System," a data storage and acquisition system, intended to give users current information on time and expense records. The system design enables the user to access the information through on-line reports that are dynamically linked. The user starts with a report and "drills down" to the linked reports the user wishes to access. It is expected that this system will eventually replace numerous reporting systems currently in use.

Since speed and system reliability are vitally important to users of this system, basing the system on single components may not be acceptable. However, some duplicate components are expensive. This study will not describe the InfoDelivery System in detail, but will provide an analysis of the critical component mix required to access the system. In this context, the relevant variables are speed of access and cost.

2 SYSTEM DESIGN OVERVIEW

The InfoDelivery System contains three subsystems that link together to create the application as a whole: the client computer (PC), the network, and the mainframe database.

The speed of data retrieval is vitally important to the application. During the analysis of this system, customer "throughput time" was closely monitored. Early in the study, we learned that acceptable time was an average 20 seconds or less per customer. Thus, components had to be selected in a way that allows average throughput time of 20 seconds or less, while containing cost. The discussion in this paper mainly focuses on throughput time; the information provided to management enabled them to decide which design they could accept based upon cost and performance.

The critical hardware requirements and connections are described in Figure 1 (see next page). There are three components of interest associated with six variables (the variable names are capitalized in parentheses):

1. The PC client, which consists of
    A. A security module(s) (SECURITY)
    B. A data access layer(s) (DAL)
    C. A swap pool(s) (SWAP)
    D. An error handler(s) (ERRH)
2. The network (an ODBC communication layer(s))
3. The DB2 mainframe database(s).

We treated each of the six variables (SECURITY, DAL, SWAP, ERRH, ODBC and DB2) as discrete components that could be connected in parallel. Thus, if a queue formed for DB2, a second DB2 component could be added to reduce that queue. In effect, adding components was modeled as adding parallel identical servers. A technical session with

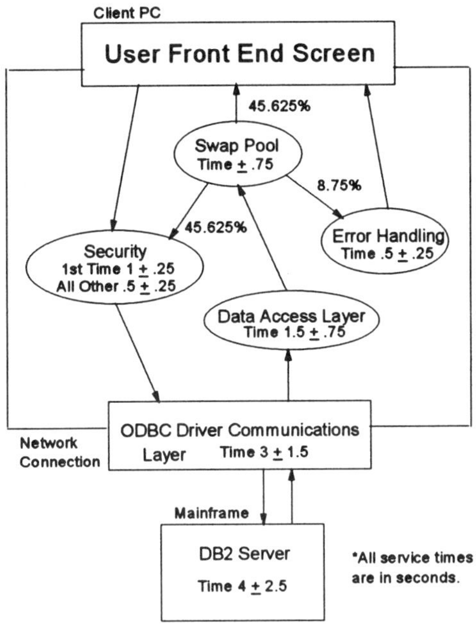

Figure 1: Hardware Requirements and Connections

management, not involving costs, led to the determination of the feasible values of the variables, as follows:

The value of SECURITY was determined to be feasible between 1 and 3;
the value of DAL was determined to be feasible between 1 and 2;
the value of SWAP was determined to be feasible between 1 and 2;
the value of ERRH was determined to be feasible between 1 and 2;
the value of ODBC was determined to be feasible between 2 and 3;
the value of DB2 was determined to be feasible between 1 and 3.

The cost per component varied greatly. The estimates are as follows:

Variable	Cost per Unit
SECURITY	$500.00
ODBC	$3,000.00
DB2	$50,000.00
DAL	$100.00
SWAP	$75.00
ERRH	$10.00

Clearly, the large variability of per unit cost affected the decision process.

Other parameters that are important for this paper were either estimated, or known to be, as follows:

Maximum number of users: 5,000.

Three periods of system load over a cycle (10 business days):

- Peak period - 5,000 users within a 16 hour (2 business day) period
- Normal period - 2,500 users within a 48 hour (6 business day) period
- Off period - 500 users within a 16 hour period (2 business day) period.

In an initial analysis, what is obvious was confirmed; the peak period use dominates the decision about component mix. This paper therefore focuses on peak period use, although the entire study has information on all periods.

The exponential distribution was used to model interarrival times for requests for service, as determined by the peak period requirements. Service times were modeled by the uniform distribution, with estimated mean and half-widths (in seconds) as follows:

SECURITY	first time	1, .25
	second time	.5, .25
DAL		1.5, .75
SWAP		1, .75
ERRH		.5, .25
ODBC		3, 1.5
DB2		4, 2.5.

Experiments using these parameters with certain different distributions did not show the decision options were very sensitive to those changes (although the throughput time averages increased with distributions containing more variation). Thus, the report contains performance based upon the uniform distribution, without compromising the decision making process.

3 THE SIMULATION MODEL AND EXPERIMENT

The systems described above can be easily modeled in any modern simulation language. We chose GPSS/H (Wolverine (1989)). We studied a base model for sensitivity to various model changes, and to determine the approximate number of replications necessary to have a reasonable error term in confidence intervals for throughput time (Schriber (1990)). Experts observed the simulated model performance, and suggested changes so the simulated system behaved like what they believed was an actual working system. With four of the six variables having two levels of treatment (SWAP, DAL, ERRH and ODBC), and the other two variables having three levels of treatment (SECURITY and DB2), a complete analysis consisted of evaluating the performance of 144 models (Box (1978), Madu (1993)). We used 25 replications for each model, looking at the peak performance period only. Five recommended designs, with mean throughput time and cost, are listed in Table 1. Note that there are statistical "ties" between some configurations (no statistical significance was detected). Using ANOVA, a significant difference was detected among the five configurations in Table 1. The last configuration is significantly slower (and cheaper) than the other configurations listed in Table 1; the first configuration is significantly faster (and more expensive).

4 CONCLUSION: DISCUSSION OF THE RESULTS

There were many model designs that met the less than 20 second average throughput requirement in a statistically significant sense. Some of those designs came close to the theoretical minimum of just more than 13.5 seconds. Table 1 lists five designs which distinguish themselves from other designs because of sample throughput ranking and cost.

The results show that for "optimum" performance, at least three thousand dollars must be added to the least cost design that meets the required performance criterion. The design with the "best" simulated average time, listed at the top of Table 1, is extremely costly because of the third DB2 component. Even though there is a significant difference among the performances of that design and other less costly designs (such as the middle three in Table 1), we do not recommend it because of the cost. There is also a significant difference among the throughput times for the last design listed on Table 1 and the four above it. We nevertheless suggest any of the last four, with a stipulation that other information on reliability and other practical measures be taken into consideration before selecting a particular model. For example, if there is suspicion that the error handlers will not be very reliable, given their cheap cost, we recommend purchasing more of them than may be needed.

ACKNOWLEDGEMENTS

We thank Dr.Christian Madu, Professor of Management Science and Research Professor at the Lubin School of Business, Pace University, for the suggestions he made to improving the analysis in this paper.

REFERENCES

Box, George E. P., William G Hunter and J. Stuart Hunter. 1978. *Statistics for Experimenters*. New York, NY. Wiley-Interscience.

Madu, Christian N., and Chu-Hua Kuei. 1993. *Experimental Statistical Design and Analysis in Simulation Modeling*. Westport, CT. Quantum.

Schriber, Thomas J. 1990. *Introduction to Simulation Using GPSS/H*. New York, NY. Wiley and Sons.

Wolverine Software Corporation. 1989. *GPSS/H Reference Manual*. 3rd Edition. Annandale, VA.

AUTHOR BIOGRAPHIES

EUGENIO M. ALVAREZ is the Data Administrator for Nine West Group Inc. He received a B.A. degree in Computer Information Science from Iona College, and an M.S. degree in Management Science from Pace University. He has previously worked for AT&T, the New York Power Authority, and most recently, KPMG Peat Marwick.

WALTER M. CALLAHAN is the Senior Accountant at the Indian Point #3 Nuclear Power Plant, a subsidiary of the New York Power Authority. He holds a B.A. degree in Social Sciences from SUNY Binhamton, and an M.S. degree in Management Science from Pace University.

Table 1: Selected Designs: Performance and Cost

Design*	Estimated Cost	Mean Throughput: Peak Period	Comments
(3,3,3,2,2,1)	$160,860	13.58 secs	Best simulated average time
(3,3,2,2,2,1)	$110,860	13.84 secs	
(2,3,2,2,2,1)	$110,360	13.86 secs	
(1,3,2,2,2,1)	$109,860	13.89 secs	
(1,2,2,1,1,1)	$106,685	15.77 secs	Least cost design with speed < 20 secs

*The six-tuple (A,B,C,D,E,F) indicates the number of components for A=SECURITY, B=ODBC, C=DB2, D=DAL, E=SWAP, F=ERRH.

PAUL R. MITCHELL is a Senior Engineer in the Performance Engineering group of the New York Power Authority. Previously, he was a senior engineer with Niagara Mohawk Power Corporation and a principal engineer with General Physics Corp. He received a B.S. in Mechanical Engineering from the University of Arizona in 1970, and an M.S. in Management Science from Pace University in 1996.

RAY TELAROLE is Manager of Technical Training for the New York Power Authority Power Generation Division. He received a B.A. degree in Humanities from St. Thomas Seminary College, and an M.S. degree in Management Science from Pace University. He has previously worked for the Utah Power and Light Company, and most recently General Physics Corporation.

PETER HOEFER is Professor of Management Science and Associate Dean and Director of Graduate Programs at Pace University's Lubin School of Business. He holds a Ph.D. degree in Mathematics from the City University of New York. In addition to administration, his interests include risk analysis and simulation.

Proceedings of the 1996 Winter Simulation Conference
ed. J. M. Charnes, D. J. Morrice, D. T. Brunner, and J. J. Swain

SIMULATOR FOR THE EVALUATION OF DISTRIBUTED NETWORK-SYSTEM PERFORMANCE

Toshio Komatsu
Yukihiro Nakamura

NTT Information and Communication Systems Laboratories
Nippon Telegraph And Telephone Corporation
1-2356 Take Yokosuka-shi Kanagawa 238-03 JAPAN

Junro Nose

NTT Software Corporation
4-40 Hon-cho Naka-ku
Yokohama-shi Kanagawa 231 JAPAN

ABSTRACT

The INSYDE simulator was developed as a tool for estimating the performance of large-scale, complex computer systems using graphical type figures to describe system operations. This paper provides an overview of an enhanced version of INSYDE that can be applied to distributed network systems. The model description method and a number of application examples are also presented. A major challenge in simulating distributed network systems that interconnect multiple computer systems over a network is how to describe the intricate, complex model conditions of computer systems including task processing and interrupts on the one hand, and network-related model conditions (e.g., specifying protocol, media access method, congestion control, broadcast communication, and so on) on the other in the most parsimonious fashion with the least amount of model description. In this work, the complex and intricate conditions of distributed network systems are simply described by creating a protocol media access method library, realizing packet splitting/merging and broadcast communication functions as nodes, adding system information that can be checked for congestion control, and multiplexing task processing.

1 INTRODUCTION

The simulator INSYDE was developed as a tool for evaluating the performance of large-scale, complex computer systems. The simulator has already been applied to estimate the performance of a number of actual system developed in-house, and the evaluation results were reflected in the system designs. Focusing on the flow of transaction processing, INSYDE describes system operations as processing flows in a flowchart-like fashion. Remaining open to modify and extend the model later, hardware and software resources conditions are separated out from the process flow and defined in table format. The descriptive power of an earlier version of INSYDE presented at the 1994 Winter Simulation Conference (See Komatsu, Nose, and Wakayama 1994) has been markedly enhanced by providing support for indirect indication of upper level equipment by files, introducing variables to node parameters, and by implementing definitions of variables in table format.

Now in our most recent work, we have extended INSYDE giving it the capability to very simply represent descriptions of complex and intricate conditions of distributed network systems.

One of the major difficulties of simulating distributed network systems that interconnect multiple computer systems across a network is how to represent intricate and complex computer model conditions (inter-task processing and interrupts) and network model conditions (specifying protocol, media access method, congestion control, broadcast communication, etc.) simply with the least amount of description. When computers are deployed in distributed topologies, this raises the issue that, even when the transaction flows are identical, if the processing computer systems are different, then a different number of process flows must be represented.

The extended version of INSYDE also represents the operations of distributed network systems as flowchart-like process flows, with the distinguishing characteristics of resources, number of distributed units, performance, and so on, represented in a graphical format and using tables. However, four extensions had to be implemented before network-specific conditions could be represented as process flows in the same way as computer systems:

(1) CSMA/CD(carrier sensitive multiple access with collision detection), FDDI(fiber distributed data interface), and other lower-layer protocol media access methods were organized as a library.

(2) Network-specific processes including packet splitting/merging and broadcast communication functions were defined as nodes.

(3) System information accessible within INSYDE was added to make it possible to describe congestion control, and the abilities to dynamically modify device

performance and to terminate transactions were implemented as nodes.
(4) Finally, the ability to represent buffer controls and other functions within devices was implemented.

Moreover, in order for the new version of INSYDE to represent distributed topologies with relatively few process flows, inter-task processing and variable definition capabilities were extended. Finally, INSYDE was extended so it could represent various communication processes between computer systems. Implementing these various methods, we ended up with 73 types of nodes for representing process flows, and 8 types of definition tables, giving us the ability to simply describe the complex and intricate conditions associated with distributed network systems using a combination of process flows and definition tables.

In the network field, there are a number of fairly good tools available that are capable of representing the minute conditions of networks in graphical and table format, but when computer systems are factored in together with networks, it is exceedingly difficult to represent a detailed internal process model of a computer system (See Bharath and Kermani 1984, Marsan et al. 1990, and Dupuy et al. 1990). There are quite a number of simulation tools available in the computer field as well, but none of these are very good at representing some of the most characteristic processes of computers such as interrupts, inter-task processes, and so on. Of course, computer-specific tools are also incapable of representing network model conditions (See Okada and Tanaka 1991).

These were some of the considerations motivating our efforts to extend the INSYDE system platform. The rest of the paper will be organized as follows. Section 2 will outline some of the unique attributes of distributed network systems. Section 3 will lay out the issues involved in model description, which will be followed in Section 4 with the approach to model description that we have adopted in the extended version of INSYDE. Finally, Section 5 will present a number of application examples and the results that were achieved using the enhanced INSYDE simulator.

2 DISTINCTIVE FEATURES OF DISTRIBUTED NETWORK SYSTEMS

A typical distributed network system configuration is illustrated in Fig. 1. Let us briefly consider four distinctive features of this topology in terms of how these features affect simulation.
(1) Topologies in which multiple computer system are interconnected by a LAN generally include a range of devices including processors and file storage devices and various software resources such as tasks, tables, and files. In addition, the variety and number of these hardware and software elements is usually quite large.

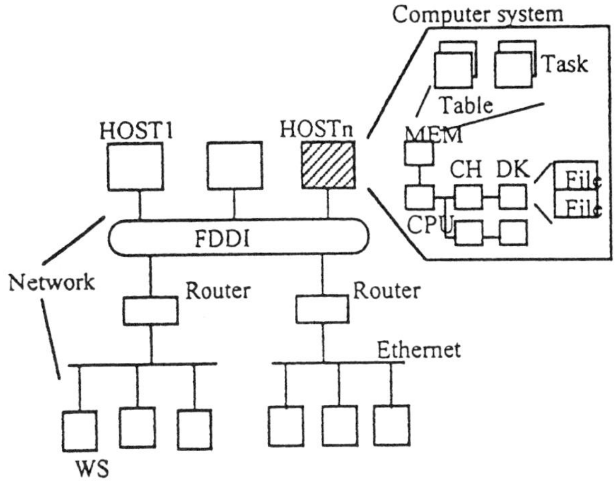

FDDI : Fiber Distributed Data Interface
WS : Work Station
MEM : Main Memory
CPU : Central Processing Unit
CH : Channel Device
DK : Disk Device

Figure 1: Overview of Distributed Network System

(2) The transactions flowing through networks and computer systems assume many different varieties and are multiplexed. And even when the same transactions are being processed, the number of resources accessed and the amount of processing will vary depending on the nature of the transaction.
(3) There are also various distinctive kinds of processing associated with networks such as processing associated with protocols, routing, broadcast communications, congestion control, and so on. Other kinds of distinctive network-related processing support inter-center communications.
(4) There is distinctive processing in computer systems to handle interrupts, exclusive control, inter-task communications, and other functions.

3 MODEL DESCRIPTION REQUIREMENTS

Having looked at the distinctive attributes of network systems, here we will list some of the requirements in order for INSYDE to simply represent the intricacy and logical complexity of networks and distributed network systems.

3.1 Requirements to Represent Networks

(1) Integrate the network model descriptive method with that of the computer model.
(2) In order to reduce the number of processing flows, if

the processing flows are the same, make it possible to represent them as a single processing flow, even though the number of LAN devices being accessed and amount of processing may vary.
(3) Enable LAN-specific functions such as retry control after collisions, congestion control, and broadcast communication to be represented by process flows.
(4) Enable buffer control and other detailed device internal operations to be described.

3.2 Requirements to Represent Distributed Systems

(1) Integrate the distributed network model descriptive method with that of the computer model.
(2) If processing flows are the same, make it possible to represent those flows as a single processing flow, even though the hardware and software configurations making up the computer system are different.

Table 1: Functions of Nodes for Processing Flow Description

Type of Node		Number	Functions
Time delay		9	These delay transactions with resources in the reserved or unreserved status.
Control	Start, End	8	These indicate the beginning or end of processing flow. For termination,continuous restart and queuing to other processes is possible,except when the transaction is deleted.
	Branch	14	Unconditional branching and branching according to various conditions is possible. Conditions can be set for the work area of each transaction or variable,etc.
	Processing Transfer	9	These transfer processing between system processes and task processes or between task processes.
	Resource management	13	These handle resource reservation and release. Multiple resources can be reserved or released simultaneously.
	Splitting , Merging	8	These split transactions into multiple processes that may proceed in the same direction or in different directions. It is possible to specify that the split transaction be complete reintegrated or partially re-interred.
LAN		4	These split/merge packets, support broadcast communication,etc.
User		1	This is defined by the user.
Macro		2	This calls a pre-defined macro process flow.
Measurement		3	These collect statistical data over a freely specified measurement interval.
Connection Node		2	These indicate the processing flow interconnection on input screen.

4 DISTRIBUTED NETWORK SYSTEM MODEL DESCRIPTION METHOD

4.1 Modeling Overview

In order to integrate the representation of networks with that of computer systems, we focus on the transaction flow of system operations, and describe them as processing flows. Processing flows are described in two tiers: as system processing flows covering an entire process from the time a transaction is created until it is terminated, and as groups of task processing flows that execute individual work processes.

Processing flows are described through the use of 73 different kinds of nodes as shown in Table 1. In addition, to facilitate unified modification and correction of models, the full range of model conditions including load conditions, device conditions, task and other software conditions, and LAN conditions are defined using eight tables that encompass types of items.

4.2 Network Description Method

In the extended version of INSYDE, lower-layer protocol network descriptions are provided in the form of a library, while higher-layer protocol network descriptions are supported with additional functions enabling processing flows to be simply described.

4.2.1 Creating a Lower-Layer Library

The various functional capabilities of the protocol media access method including CSMA/CD, token passing, and FDDI will be provided in the form of a library. Figure 2 shows that the access method is not directly described in the LAN node, but rather is defined using a LAN definition table. The detailed parameter items will differ depending on the access method.

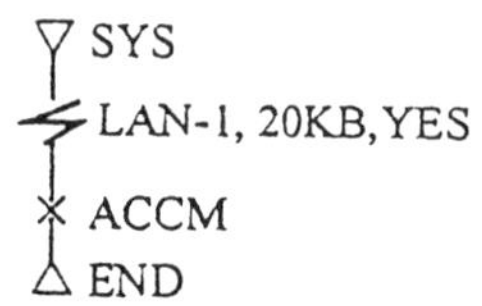

WW-XX,YY,ZZ: Media access and packet splitting
WW: LAN name
XX: LAN number
YY: Data length
ZZ: Specifies packet merging
× ACCM: Combines packet segments of the same generation

Figure 2: Example of Using a LAN Library

(1) CSMA/CD : The main defining parameters of the CSMA/CD method are slot size, frame interval, number of retransmissions, and cable length. When a frame is sent, frame collisions are detected. If a collision occurs, the frame is retransmitted a prescribed number of times according to the prescribed algorithm. If the number of retransmissions exceeds the prescribed value, statistical processing is invoked, the number of retransmissions is reset, and the frame is resent

(2) Token Passing : Both token bus and token ring topologies are supported by INSYDE's library. Here, the main defining parameters are THT (token-holding time) and the repeat extension bit. In order to allocate tokens in the proper order in the LAN interconnection node, the hierarchical position of the interconnection node must be specified. INSYDE approximately allocates tokens to interconnected nodes in the order in which LAN nodes are executed.
(3) FDDI : The token allocation rules and link delay time-related provisions are largely the same as in • for token passing. In addition, a TRT (token rotation time) value is employed to support both synchronous and asynchronous transmission. In order to achieve efficient transmission of files and other data, INSYDE provides that the token is not released until all the data have been transferred.

4.2.2 Higher-Layer Capabilities

Higher-layer capabilities that can be represented in the extended version of INSYDE include the following.
(1) Packet Splitting/Merging : The LAN node not only has media access capability, that is also endowed with a packet splitting function. The DIV node is also supported that only does packet splitting. The user can then specify whether packets split by either node are to be merged by an ACCM node. Using these nodes, packet splitting and merging can thus be represented in nested fashion as illustrated in Fig. 3. Since generation management is imposed when packets are split, there is no need to describe merged packets at the ACCM node in terms of merge packet number and generation parameters.

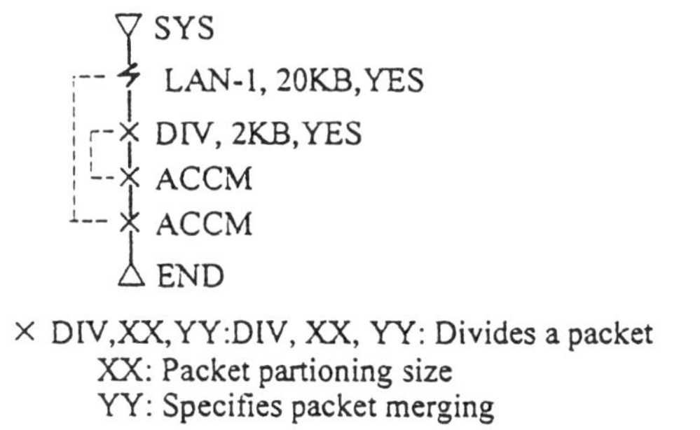

Figure 3: Example of Nested Packet Splitting/Merging

(2) Routing : In cases where the processing flow is the same (i.e., where the same kinds of devices are passed through the flow), Fig. 4 shows that the routing can be represented by a single process flow and variable definition tables even though the number of devices accessed and the amount of processing differs. One will observe in Fig. 4 that, in order to process split packets at a router, the current packet length for each transaction is set by the SET node.
(3)Broadcast Communication: Broadcast communications can be represented using the PARA, BROAD, and SPLT nodes, which correspond to different broadcast topologies. For example, where there are relatively few broadcast destinations, a topology can be represented with a branch splitting off at the PARA node. Or, in cases where there are many broadcast destinations and a simple flow is involved, the topology can be described with the BROAD node. In cases where processing will continue after the broadcast, a transaction that is first split and then continues in the same direction can be represented with the SPLT node.

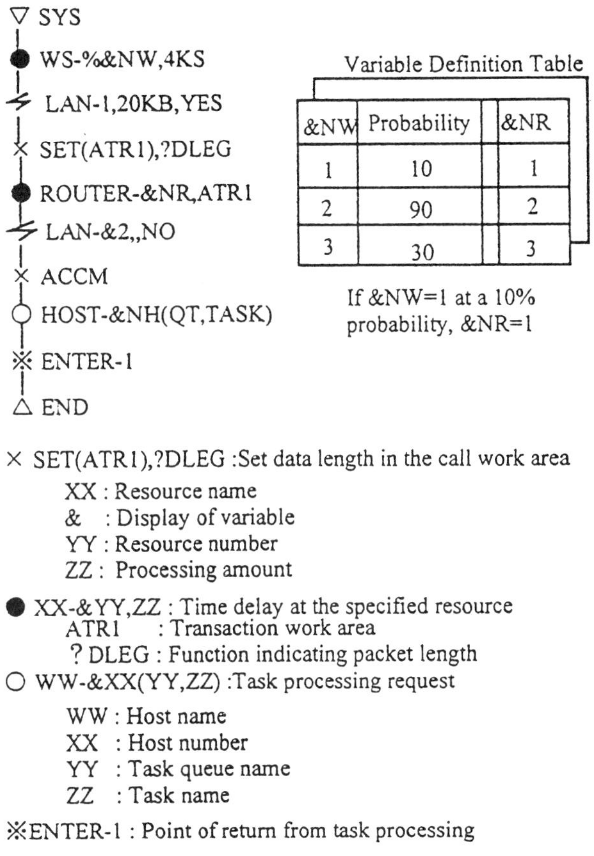

Figure 4: Example of Routing Description

(4) Congestion Control : In order to simply represent congestion control in a process flow, two additional things are required: system information that can be referenced, and nodes for terminating and dynamically modifying the performance of devices. These capabilities are illustrated in Fig. 5 where a stay situation is observed, device performance in incrementally modified, and the transaction is terminated.

4.2.3 Defining Detailed Device Configurations

In the process of investigating a new protocol or configuration of a router, bridge, switch, or other piece of equipment, there are many cases where a researcher wants to model the internal aspects of a device in detail. Figure 6 illustrates that a range of device characteristics can be easily defined with a set of queuing compartments and corresponding device definition tables. The management of input and output buffers can then be represented using the tables.

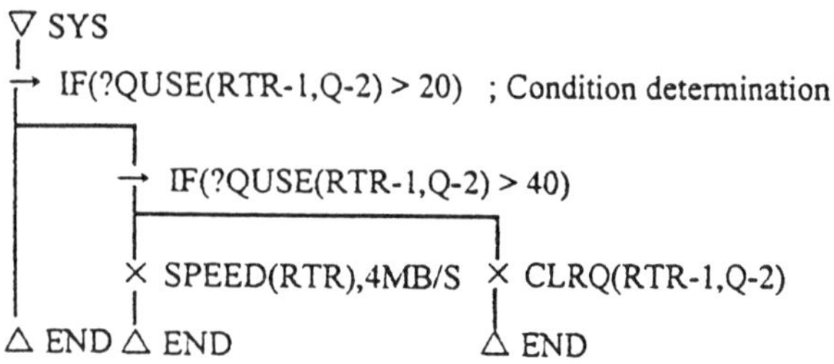

?QUSE(XX,YY) : Function indicating use state
XX : Device name
YY : Device queue name
× SPEED(XX),YY : Modify device performance
XX : Device name
YY : Performance value
× CLRQ(XX,YY) : Reduce queuing transactions
XX : Device name
YY : Devise queue name

Figure 5: Example of Congestion Control Description

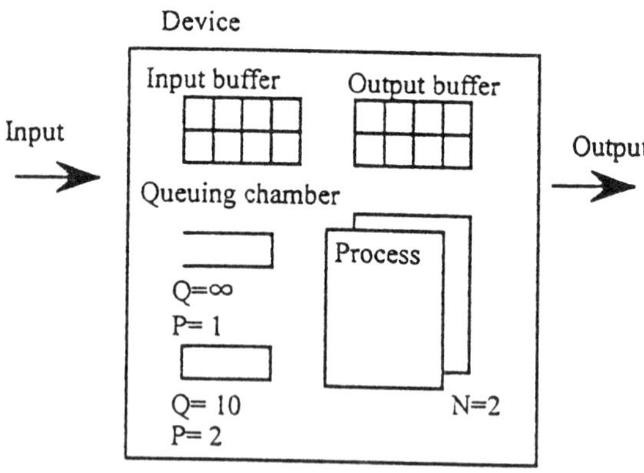

Q : Queuing capacity
P : Device acquisition priority
N : Degree of multiplexing

Figure 6: Example of Detailed Device Modeling

4.3 Distributed Network Description Method

In earlier work we significantly enhanced the descriptive power of INSYDE for modeling computer systems by indirect indication of upper level equipment by files, introducing variables as node parameters, defining variables in table format, and so on (See Komatsu, Nose, and Wakayama 1994). Building on this descriptive approach, we have extended the INSYDE environment to support multiple computer systems in distributed topologies as follows.

(1) Resource Identification Method
In the majority of cases where the computational load is distributed over a distributed computer system, each interconnected computer exhibits the same kind of processing flow. In order to reduce the number of processing flows in cases such as these, we have identified distributed equipment, tasks, tables, and other resources with a resource name and a resource number. Although the number of configured devices, the file configurations, and the number of tasks will vary for each computer system, and even though the processing computer systems will differ depending on the transaction, Fig. 7 shows that the processing flow can be represented by a single processing flow and set of variable definition tables.

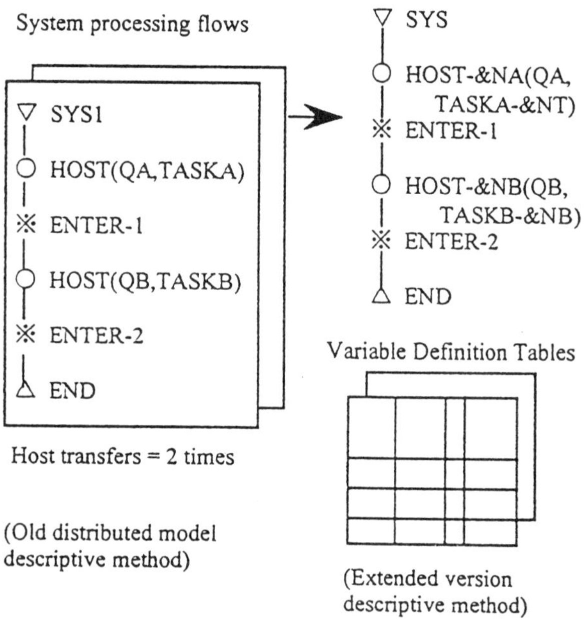

Figure 7: Reducing the Number of Processing Flows

(2) Communication Between Computer Systems
Since communications between computer systems involves communications moving in both directions, two types of communications are supported in the extended version of INSYDE. Cases where a response is required from the recipient of the communication and where no response is required are represented by LINK nodes and SEND nodes, respectively. Since each and every

communication is controlled by the initiator of the communication, there is no need for the response destination to be represented as a parameter in the LON node which indicates that a response has been returned to the communication initiator.

5 APPLICATION EXAMPLE

5.1 Model Conditions

We applied the extended version of INSYDE to a distributed network system having the same kind of topology as shown schematically in Fig. 1. The main conditions of the model were that the distributed system interconnected eight computers that were concurrently executing ten different kinds of jobs. Here we assume that the same kind of work would produce the same processing flow. Among the various kinds of work that were performed, the processing flow of one of the jobs is briefly described as follows.

A transaction was generated at a workstation. The transaction was divided into packets and sent via Ethernet, router, and FDDI to host computer system HOSTn where the packets were reassembled. The destination HOST varies depending on the nature of the transaction. Here n signifies the number of the HOST, and the transaction is processed by transferring the task from TASK1 to TASK2 at HOSTn. After the processing at HOSTn is completed, the packets are again split up and transferred to HOSTm over an FDDI link for further processing. Here, the value of m varies depending on HOSTn and the nature of the transaction. After the processing is completed, the transaction is terminated. The processing of each task involves CPU processing and file processing. Here, we have simplified the actual flow which was longer and more complex in order to facilitate the explanation.

5.2 Example of Model Description

As an example of the model description, portions of the system processing flow and task processing flow representing the job described in the previous section are shown in Fig. 8. In the workstation generating the transaction, since the device numbers will vary depending on the nature of the transaction, it is defined by variables. Descriptions are omitted, but since the HOST number, the task number, and so on are described using variables, the variables must be defined.

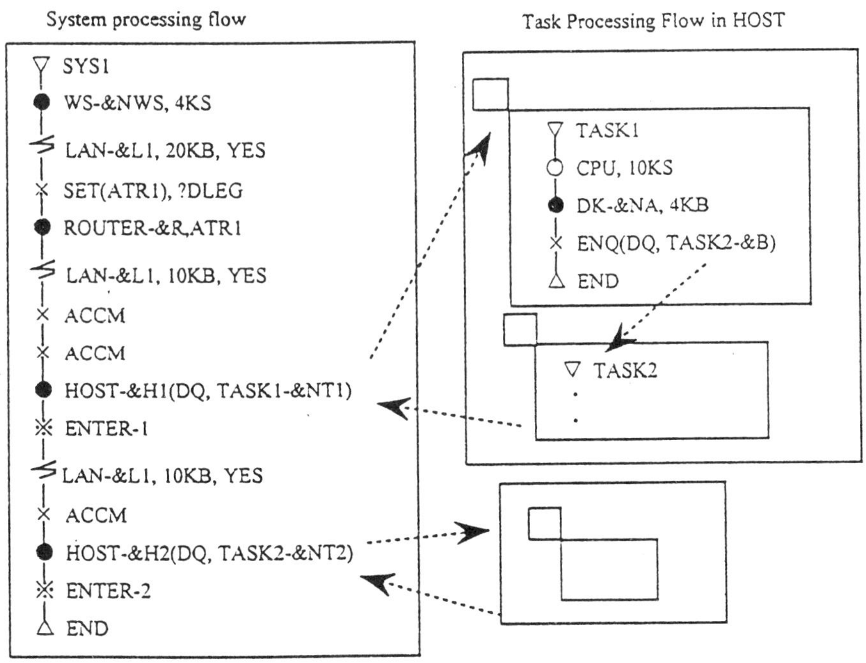

○ CPU,XX : CPU processing time, transaction extension
XX : Processing amount
× ENQ(XX,YY-ZZ) : transfer of task processing
XX : Task queue name
YY : Task name
ZZ : Task number

Figure 8: Example of Processing Flow Description

5.3 Application Results

Although we are not able to generalize since the results in any particular case will differ depending on the model conditions, as a broad indication we will compare results using the old version of INSYDE designed for computer systems and the new descriptive method outlined here in representing the work outlined above in Section 5.1.

5.3.1 Reducing the Processing Flow Number

Since there are two transfers to the host, the system process flow number described by the old version is given by the number of jobs × the number of hosts × the number of hosts, or 10 × 8 × 8 = 640. In the extended version of INSYDE on the other hand, since the host number is defined by variables, the transaction can be terminated with a system processing flow number equivalent to the job number of 10. Of course, the variable definitions have to be added to this. Since there are transfers of task processes, the task processing flow number using the old version is given by the number of jobs × the number of hosts × the number of hosts × the number of times the task processing is transferred = 1,280. In the expanded version, using variable definitions as in the previous example, the transaction can be terminated with a task processing flow number equal to the number of jobs × the number of times the task processing is transferred = 20. Although variables must be defined in order to reduce the processing flow number, and this increases the complexity somewhat, the variables are represented in a table format which reduces the complexity.

5.3.2 Number of Processing Flow Nodes

In the older version of INSYDE, packet splitting and merging must be represented using combinations of existing nodes. In cases where the processing amount is adversely affected by variables, programming is required to effect the splitting, and this complicates the description. Representing the protocol media access method also becomes more difficult. In the extended version of INSYDE, however, these various aspects can be represented using a small number of nodes; namely, LAN, DIV, and ACCM nodes.

5.3.3 Easy Visualization of Processing Flows

Although variable definitions are added in the extended version of INSYDE, there are fewer flows and fewer nodes per processing flow, which makes the extended version process flows much easier for users to visualize.

6 CONCLUSIONS

This paper has presented an overview of INSYDE, a powerful simulation tool for evaluating the performance of large-scale, complex distributed network systems using graphical figures and tables to represent system operations. We have also discussed the model description method and an application example. The extended version of INSYDE greatly simplifies the representation and the amount of description that is required to represent the complex and intricate conditions of distributed network systems by creating a protocol media access method library, realizing packet splitting/merging and broadcast communication functions as nodes, adding system information that can be referenced for congestion control, and multiplexing task processing.

The extended version of INSYDE not only provides a powerful tool for modeling and representing distributed network systems, it also features an extremely user-friendly man-machine interface. This user friendliness, for example, permits users to input other data outside of the actual numerical values and resource names with the click of a mouse, or to use the mouse to select a resource name that was previously defined as a table from a list.

We intend to follow up on this work by expanding the media access topology library and by developing other simple, intuitive simulation tools incorporating the powerful capabilities of INSYDE.

ACKNOWLEDGMENTS

The authors gratefully acknowledge the support and encouragement of their colleagues over the span of years that the INSYDE simulator was in development. The also express their appreciation to their coworkers for valuable discussions in formulating the model conditions of actual systems.

REFERENCES

BHARATH-KUMAR, K., and P.KERMANI. 1984. Performance Evaluation Tool(PET): An Analysis Tool for Computer Communication Networks, *IEEE JASC, Vol.SAC2, No.1*: 220-225.

DUPUY,A., J.SCHWARTZ, Y.YEMINI, and D.BACON. 1990. NEST:A Network Simulation and Prototyping Testbed, *COMMUNICATION OF THE ACM, Vol.33, No.10*: 64-74.

Komatsu,T., J.Nose, and H.Wakayama. 1994. Model Description in the INSYDE Simulator for Evaluating Large-Scale Computer System Performance, *Proceedings of the 1994 Winter Simulation Conference*: 1272-1289.

MARSAN, M.A., G.BALBO, G.BRUNO, and F.NERI. 1990. TOPNET:A Tool for the visual Simulation of Communication Networks, *IEEE JSAC, Vol.8, No.9*: 1735-1747.

Okada, Y., and Y.Tanaka. 1991. FES:A Toolkit System for the Development of Visual Interactive Simulators, *Trans.IPG Japan, Vol.32, No.6*: 766-776.

AUTHOR BIOGRAPHIES

TOSHIO KOMATSU is a Senior Research Engineer in the NTT Infonnation and Communication Systems Laboratories. He received B.S. degree in 1972 and M.S. in 1974 in electrical engineering from Kyushu Institute of Technology. Since joining NTT in 1974, he has been active in R & D on computer systems hardware architecture and computer systems performance evaluation techniques. He is currently interested especially in modelling and analyzing techniques and tools for distributed computing systems. He is a member of the Information Processing Society of Japan.

JUNRO NOSE is a Senior Manager in the Productivity Innovation Promotion Headquarters at NTT Software Corporation. He received B.S. degree in 1967 and M.S. in 1969 in electrical engineering from Kobe University. He joined NTT in 1969 and has been active in R & D on fault diagnosis technique, videotex communication system and systems evaluation techniques. He joined NTT Software Corporation in 1993 and is now responsible in the systems evaluation systems and services. He is a member of the Information Processing Society of Japan.

YUKIHIKO NAKAMURA is an Excutive Manager in the NTT Information and Communication Systems Laboratories. He received the B.S., M.S., and Ph.D. degrees in Applied Mathematics and Physics from Kyoto Univercity in 1967, 1969 and 1995 respectively. He joined NTT in 1969, from 1969 to 1980, he was engaged in R & D on computer systems hardware architecture, and since 1981 he has been engaged in R&D of high-level synthesis technology for parallel architecture design. He is a member of the technical committee of the ICCAD, ED&TC, ISSS and ASP-DAC.

Proceedings of the 1996 Winter Simulation Conference
ed. J. M. Charnes, D. J. Morrice, D. T. Brunner, and J. J. Swain

MODELING FILE-SYSTEM INPUT TRACES VIA A TWO-LEVEL ARRIVAL PROCESS

Peter P. Ware

Department of Computer
and Information Science
The Ohio State University
Columbus, OH 43210, U.S.A.

Thomas W. Page, Jr.

Department of Computer
and Information Science
The Ohio State University
Columbus, OH 43210, U.S.A.

Barry L. Nelson

Department of Industrial Engineering
and Management Sciences
Northwestern University
Evanston, IL 60208, U.S.A.

ABSTRACT

A method for analyzing, modeling and simulating a two–level arrival–counting process is presented. This method is particularly appropriate when the number of independent processes is large. The initial motivation for this method was the need to analyze and represent computer file system trace data that involves activity on some 8,000 files. The method is also applicable to network trace data characterizing communication patterns between pairs of computers.

Cluster analysis with a novel stopping rule is used to decompose the arrival process into groups. The resulting clusters can be characterized using the time between clusters, the time between arrivals within clusters, and the size of each cluster. Each of these three components is then analyzed as a univariate problem.

The effectiveness of this method is measured by comparing the output of a simulation driven by the original trace data to the output of the same simulation driven by the input model.

1 INTRODUCTION

The motivation for this work is the desire to drive simulations of distributed, replicated, file systems. Much like network traffic or memory reference patterns, workloads presented to file systems tend to be bursty and exhibit a high degree of locality. That is, a file which has been accessed recently tends to be accessed again soon. Further, accesses tend to come in *clusters*; there is a burst of activity, followed by a break, followed by another burst, etc. Figure 1 is the access pattern for one such file.

We can drive file system simulation with one of several available sets of trace data. However, there are a number of limitations to the use of traces. Traces are of finite length. If the events of interest in the simulation occur sufficiently infrequently (as they do in our application), then the length of the trace may be insufficient for quality results. Despite their limited size, traces are nevertheless voluminous and hence cumbersome. They are highly inflexible, representing only what happened over some specific interval of time, which may or may not be more widely representative. Thus, even when in possession of trace data, there is motivation to produce a synthetic workload which can capture important aspects of the recorded trace, but with greater flexibility. The work reported here is a step along the path to building such a synthetic trace generator for file accesses.

1.1 Data Characteristics

Figure 2 shows the inter–arrival times for our computer file system trace data (Hisgen 1990). This data totals over 2 Gigabytes, representing approximately 29 million events collected over a four day period from 114 workstations running a Unix-like operating system. This data appears very nearly exponentially distributed. However, this view from 10,000 ft. (an altitude from which the identity of the file being accessed is obscured) is too coarse for our simulation purposes and such a workload model would not capture the critical characteristics of the arrival process. If one zooms in closer to where one can observe the data decomposed into a separate process per file, a very different pattern emerges.

Figure 1 shows this arrival–counting process at a more detailed level (access to a single file), illustrating some important characteristics. The data appears to be generated by a two–level process: one process generating clusters of events, and within each cluster, a second process generating individual events. A sequence of independent, identically distributed interarrival times would do a poor job of representing such patterns.

In Figure 1, the vertical axis is a cumulative count of the number of accesses to the file and the horizontal

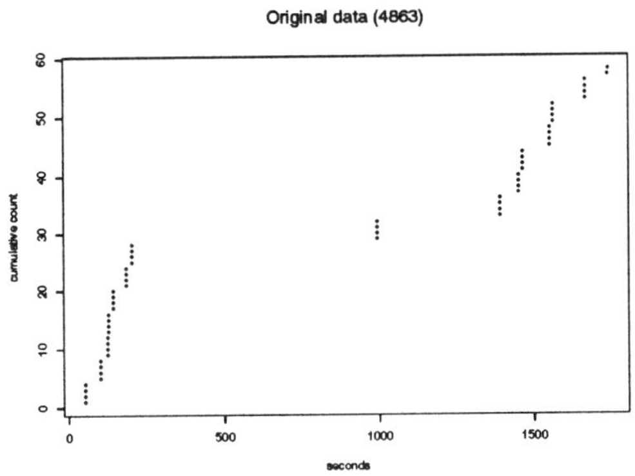

Figure 1: Several Clustered Events

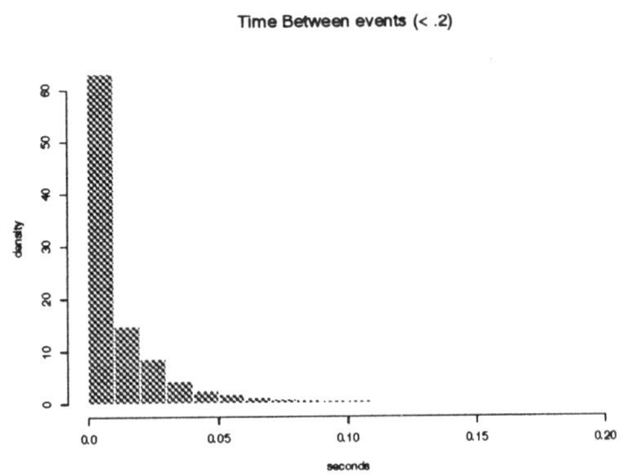

Figure 2: Probability Density Plot of Inter–arrival Times

axis is in seconds. Examining this figure, one could manually draw circles to cluster the data and model the time between the start of each cluster (the *inter–cluster* time) to form the first process. Then, within all the clusters, one could match a distribution to the time between each event (the *intra–cluster* time) to represent the second process within each cluster. A distribution for the *size* of each cluster can then be used to terminate the second process.

An automated method for doing such clustering is needed when a data set is composed of more than a handful of such fine–grained patterns. The data shown in Figure 2 is composed of nearly 8,000 such objects (0.73% of the samples (3038) are >= 0.2 seconds and are not displayed). Each file may have its own access pattern, some of which may be characterized by frequent long bursts, others by rare short flurries of activity, with still others showing little clustering. Manual clustering of such a large data set is impractical, yet no static parameters (say for maximum intra-cluster gap) yield good results over the entire non-homogeneous data set.

We adapt a well-known algorithm called *hierarchical clustering* with *single linkage*, to measure the distance between clusters. Although some of the literature on hierarchical clustering disparages this method, it has ideal characteristics for this type of data, as shown in Section 3.

The final (and critical) step in *hierarchical clustering* is to determine when clustering should be stopped (between the extremes of one cluster per event and one cluster containing all events). Frequently, such a decision is reached manually; however this is again not practical for such large data sets. Section 3.1 introduces a new stopping rule designed specifically for this application.

The methodology we propose uses the Johnson translation system to fit a distribution to the *inter–cluster times* and *intra–cluster times*, and the empirical CDF for the distribution of the *cluster size*. Each of these can be treated as an independent, univariate distribution if clustering successfully identified a two–level process.

The effectiveness of this methodology is evaluated in Section 5 by comparing the output of the same simulation driven by two representations of the same input process. The simulation is first driven by the actual trace data and then is driven by the arrivals generated from our two–level process. The output of the two simulations is compared to establish that the features of the trace data that most affect the simulation output are accurately represented by the input model.

2 BACKGROUND ON APPLICATION

The impetus for this work was the desire to answer the question, "How frequently do conflicting updates occur in a replicated data system using optimistic concurrency control?" A number of systems have been built with this principle (see for example Bayou from Xerox PARC (Terry, Theimer, Petersen, Demers, Spreitzer & Hauser 1995), Coda from CMU (Satyanarayanan, Kistler, Kumar, Okasaki, Siegel & Steere 1990) and Ficus from UCLA (Guy, Heidemann, Mak, Page, Popek & Rothmeier 1990)). However, due to differing architectures and environments, it has proved difficult to answer this question in a very general way using the actual systems. Hence, a flexible simulation was undertaken.

A replicated data object with n copies is modeled as a finite state machine with n "normal" states and one "conflict" state (see Figure 3). In state n, all replicas are mutually consistent; that is, all updates to the object have been applied to all replicas (in the same order) and hence the replicas contain the same value. If an update (write) operation occurs while the object is in state n, the update is initially applied to one of the replicas, and the model transitions to state 1. State 1 models the situation where only one replica contains the most recent data value. The update then propagates to the other replicas asynchronously, and the model transitions up through states 2, 3, etc., as each additional replica is informed, until mutual consistency is restored and the model returns to state n. In any state $i : 1 \leq i \leq n$, i of the n replicas contains the most up-to-date data, while $n-i$ contain stale data.

If an update arrives while the object is in any state i: $i < n$, two possibilities result: either the update is applied to one of the replicas which already contains the latest version, or to one that does not. This is the essential difference between weak consistency and the traditional strongly consistent algorithms. Conventional algorithms prevent conflicts by restricting updates to one of the up-to-date replicas. However, the cost of doing so may exceed the cost of dealing with the occasional conflict, if conflicts are sufficiently rare and/or easy to repair. If the replica selection algorithm chooses one of the i replicas that is already up to date, then the state transitions back to 1 and update propagation resumes spreading knowledge of the new update. However, if one of the $n - i$ replicas containing an old version is selected, then a state exists in which there is no longer a total ordering of file versions, an *update–update* conflict exists and the model transitions to the conflict state. When in the conflict state, C, update propagations cannot in general restore a correct, where "correct" is defined as one-copy-serializable, mutually consistent replicated data value. Any further updates to the object while in the conflict state will leave the file in the conflict state. Another type of operation called "repair" is required to restore a new dominant version which must then be propagated to all of the replicas before the file is once again mutually consistent. This operation is ignored.

If we were to assume that updates are generated by a Poisson process with rate λ, that the replica selection policy is equally likely to choose any replica for initial application of the update, and the time to propagate an update to any individual site is exponentially distributed with mean $1/\mu$, then the model becomes Markovian and is easily solved analytically.

As we have seen, however, a Poisson process is a poor model of the bursty arrival process typical of file system access, and it is precisely the pattern of rapid updates that makes conflicts more likely. Hence it is critical that the input process accurately capture this characteristic.

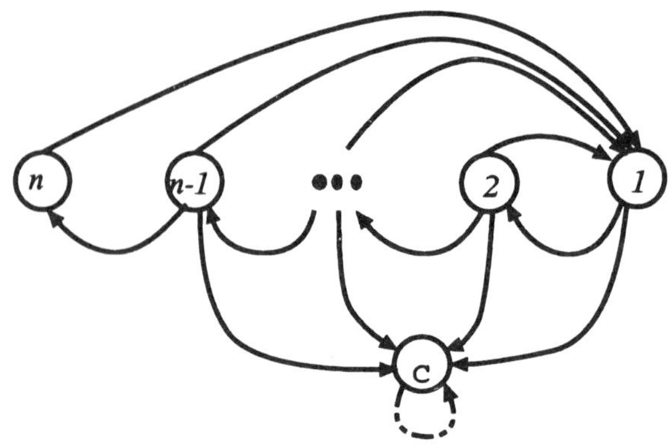

Figure 3: State Transition Diagram For An n Replica File

3 CLUSTERING

The hypothesis is that file access data (and many other types of workload) are effectively modeled by two–levels of processes; the first generating bursts, and the second generating events within bursts. In order to fit distributions for these processes for a given data set, it is necessary to "cluster" the data into bursts. We consider a form of agglomerative hierarchical clustering, as described, for example, in Jain (1991), but simplified by the fact that our application requires clustering in the time dimension only.

The data is in the form of an arrival process. Thus, each event has a time at which it occurs, and the events are totally ordered by their time. Let $T_1, T_2, \ldots, T_n$ be the sequence of access times for a particular file. The basic algorithm for agglomerative clustering is as follows:

1. Start with each event T_i in a cluster by itself and initialize the iteration counter i to 0.

2. Construct the inter–cluster distance array in which the jth element is the distance between cluster j and cluster $j+1$.

3. Find the smallest element of the distance array (if the smallest value is not unique, randomly choose one from the set of smallest distances). Say the kth distance is chosen. Combine clusters k and $k+1$.

4. Repeat steps 2 and 3 until all events are part of a single cluster (we examine earlier termination conditions below).

The above method produces a sequence of clusterings starting with each event in a cluster by itself, and terminating with all events in a single cluster. If this model is a good representation of the data set, then in between these degenerate cases lies one or more appropriate clusterings. The difficulty lies in automatically deciding which intermediate clustering is the best one. If we can determine that we have reached the point where any further iterations would combine clusters which should remain distinct, then we should terminate the algorithm and report the resulting clusters.

A second issue is how to compute distances between clusters. The statistical literature provides a large number of distance measures that are reasonable in various situations (Kaufman & Rousseeuw 1990). We employ one of the oldest measures, known as *single linkage* or *nearest neighbor*: The distance between two clusters is the minimum distance between any member of one cluster and any member of the other. Translated into our context, the "distance" between two clusters is the absolute value of the difference between the two closest event times. For example, if cluster j is $\{T_m, T_{m+1}, \ldots, T_{m+r}\}$, and cluster $j+1$ is $\{T_{m+r+1}, T_{m+r+2}, \ldots, T_{m+r+s}\}$, then the distance between them is $T_{m+r+1} - T_{m+r}$. Notice that employing single linkage implies that clusters will always be formed from sequential event times, a property that we obviously desire. In fact, the single linkage distance measure is criticized in the clustering literature for its tendency to form clusters that look like chains, but this is precisely the type of clusters for which we search.

3.1 Terminating the Algorithm

Define the random variable H to be the smallest value of the distance array chosen in each iteration. That is, H_i is the distance between the two clusters selected for merging in the ith iteration of the clustering algorithm; it is monotonically increasing as the distance is always positive. Thus, the slope of the H_i curve is always nonnegative.

Intuitively, if the data is truly bursty, the early elements of the H_i sequence should be relatively small since we are combining clusters which are part of the same burst. At some step we will merge the last pair of clusters that are indeed part of the same burst. The next value of H_i represents an attempt to combine two clusters which are separated by an inter–cluster gap which should be very much larger than the greatest intra–cluster gap. At that point the slope of the H_i curve turns dramatically upward. We need to be able to recognize this point in an automated fashion.

Many clustering stopping rules have been proposed in the statistical and application literature. For example, Milligan & Cooper (1985), describes and tests 30 such rules! Most of these rules are based on the relative variability within clusters to between clusters. Our method exploits the fact that our data is an arrival counting process. Specifically, we compute a finite–difference estimate of the second derivative of the H_i curve, and end clustering when this derivative is significantly different from 0. The following argument justifies this approach:

Suppose that the file–access process is completely random, meaning that no clusters physically exist. Such a process is often well modeled as a Poisson process with arrival rate λ events per unit time. In other words, the inter–event times $G_i = T_{i+1} - T_i$, $i = 1, 2, \ldots, n-1$, are independent and identically exponentially distributed random variables with mean $1/\lambda$. Because of the nature of the single linkage distance measure, the heights $H_1, H_2, \ldots, H_{n-1}$ will therefore be the order statistics (sorted values) of $G_1, G_2, \ldots, G_{n-1}$. Thus, the smallest inter–event gap will determine the first cluster, the second smallest gap the next cluster, and so on. Using known results for the order statistics of the exponential distribution (e.g., Devroye (1986))), the expected increase (finite–difference first derivative) of H_i over H_{i-1} is

$$\mathrm{E}[H_i - H_{i-1}] = \frac{1}{\lambda(n-i)}. \tag{1}$$

When n, the number of event times, is large, and i is not too close to n, then (1) will be nearly constant for values of i close together. Therefore, the local increase in H_i will be approximately linear, in expected value, and the second derivative approximately 0, for a process without clusters.

Our stopping rule looks for the behavior expected of a Poisson process. When the behavior of H_i departs significantly for what is expected, then we assume that we are no longer combining intra–cluster event times, and have started to combine clusters. Using the second derivative test avoids the need to specify a value of λ.

If we apply this rule to a Poisson event process, then it is likely to break the data up into a small number of very large clusters, since (1) does change rapidly for i near n. This implies that we should first subject our data sets to a Poisson–process test to determine if clustering is needed. If the hypothesis of a Poisson process is rejected, then we apply the clustering algorithm.

On the other hand, if we apply this rule to a deterministic, completely regular process, it will place all observations in a single cluster, since the second derivative will be zero at all stages. Therefore, a single–cluster outcome should alert us to a data set that represents a regular process. However, the methodology continues to work as the inter—arrival times within one large cluster are fitted with some univariate distribution.

4 MODELING AFTER CLUSTERING

There are four steps in modeling and generating data after the sample data is clustered:

1. DATA ANALYSIS: Extract the observed inter–cluster times, intra–cluster times, and cluster–size, as represented by the random variables $\{\hat{X}_1, \ldots, \hat{X}_n\}$, $\{\hat{Y}_1, \ldots, \hat{Y}_m\}$, and $\{\hat{D}_1, \ldots, \hat{D}_n\}$, respectively. Here n is the number of clusters and $m = \sum_{i=1}^{n} \hat{D}_i - n$;

2. MODELING: Fit distributions to the observed inter–cluster times, intra–cluster times, and cluster–size data. Our software uses the `Fittr1` package (Swain, Venkatraman & Wilson 1988) to determine the parameters of the Johnson translation system (Ord 1972, Johnson 1949) for each of the random variables.

3. GENERATION: Use the parameters selected by `Fittr1` for random–variate generation of X, Y, and D. (Though D is discrete, we fit a continuous distribution and then round to the nearest integer to obtain the cluster size.)

4. SYNTHETIC TRACE: Use the generated values to form a two–level process. After an inter–cluster gap interval X_i, a new process is started. This process generates D_i events with the time between events given by $Y_j, \ldots, Y_{j+D_i}$, and then terminates. Note that for the ith cluster, $j = 1 + \sum_{k=1}^{i-1}(D_i - 1)$.

At each interval X_i a new process is started. This process generates D_i events with the time between events given by Y and then terminates. Figure 4 shows this two–level process with the ith process generating $D_i = 4$ events with the time between events given by $Y_j, \ldots, Y_{j+2}$. The $i + 1^{st}$ cluster is shown starting X_{i+1} time units later.

Figure 5 shows data generated for file 4863.

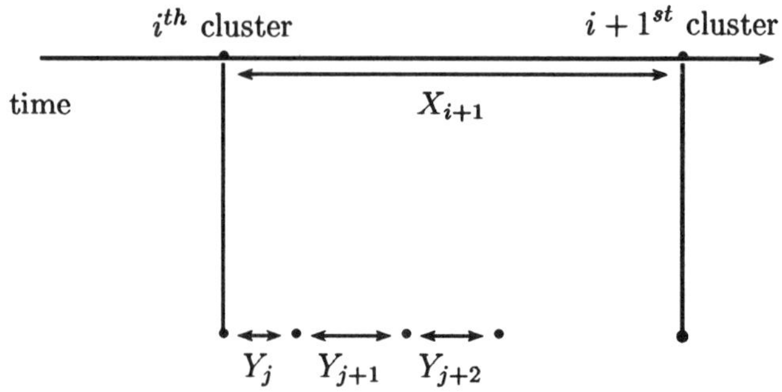

Figure 4: A two–level Process

5 VALIDATION EXPERIMENT

The purpose of this experiment is to test the hypothesis that the clustering method presented for generating file access traces which mimic the bursty behavior of the actual trace captures the characteristics of the trace data essential to the replicated file system simulation. If the results of driving a significant simulation experiment with the synthetic data are reasonably close to those generated with the original trace data, we will conclude that the hypothesis is validated.

5.1 Method

Twenty files out of 853 files that were accessed during the busiest two hour period of the data were selected at random. The clustering algorithm was run and then distributions fit to the inter-cluster gap, the intra-cluster gap, and the number of events per cluster. These distributions were then used to generate synthetic workloads for each file.

Next a simulation experiment was performed twice for each file: once driven by the actual trace data, and then driven by the corresponding synthetic workload model. A discrete event simulation of the replicated filing model presented in Section 2 was run repeatedly as μ (the rate at which updates are propagated from the most up-to-date replica) was varied from .001 to 1000. The number of replicas was fixed at 5. Each simulation run was terminated when the 95% confidence interval of the mean time to conflict (the mean time to absorption where conflicts are considered an absorbing state) was less than 5% of the mean. This results in a pair of curves for each file: one generated using the trace data and the other generated using the synthetic load model. Each curve plots the mean time to conflict as a function of the update propagation rate.

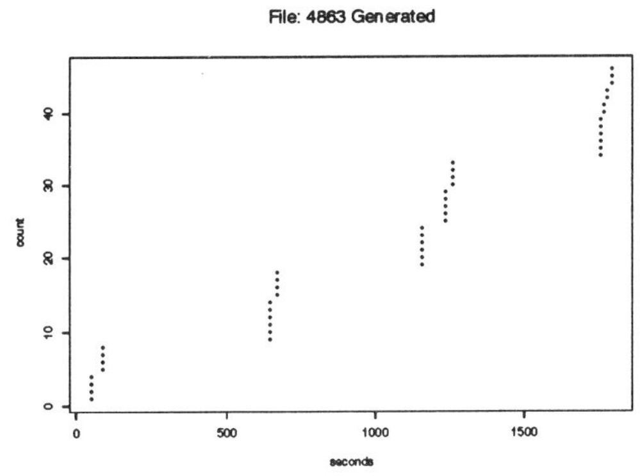

Figure 5: Generated Data Corresponding to File 4863 (see Figure 1)

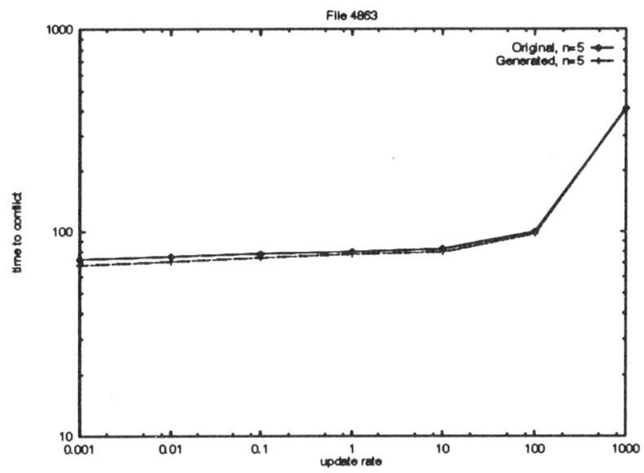

Figure 6: Mean Time to Conflict for the Original and Generated Data for File 4863

Table 1: Summary Statistics For The Inter-arrival Times Of The Original and Generated Data for File 4863

Statistic	Original	Generated
mean	29.66	29.95
std. dev.	116	114
median	0.0021	0.0032
coef variation	3.93	3.83
lexis ratio	459.642	439.98
skewness	5.35	5.33
count	57	3307

5.2 Single File Illustration

Figure 5 shows the results of generating data for a single file. Visually comparing this to Figure 1 suggests similar patterns of clustering. The summary statistics in Table 1 show the original data and the generated data are close according to several measures. The mean and standard deviation are within 1%. The other values show the distributions are similar. Note the wide variability of the inter–arrival times as shown by the large standard deviation (~ 116).

Figure 6 shows the results of running the simulation with both the empirical and synthetic trace data. The mean time to conflict as a function of update rate is close for both data sets, lending support to the hypothesis that the generated data captures important characteristics of the empirical trace.

5.3 Results

The results of running the experiments are presented in Table 2. This table gives the relative error expressed as a percentage when comparing the conflict rate of the simulation using empirical data versus generated data. Generally, the results are satisfactory. Most results were within 30% of the original data.

The results that differed more than 30% come from two sources of error. The first is a basic limitation in using empirical data: the inter-arrival times are fixed even though the data suggests that smaller or larger values may exist. In this particular experiment, the mean time to conflict is closely related to the minimum inter–arrival time. A smaller minimum inter–arrival time causes conflicts to occur more often. Since the generated data can have smaller inter–arrival times, the mean time to conflict tends to be smaller for the generated data, especially as the update rate increases.

The other source of error is introduced by the sensitivity of the method to any errors in the *cluster size* distribution. The size of the clusters affects how many of the intra–cluster arrival times are generated. Since these are the predominant cause of conflicts (the intra–cluster times are typically an order of magnitude smaller then the inter–cluster times), changing the number such events can drastically affect the mean time to conflict. For our eventual file system research, this is not problematic as the nature of file systems will tend to treat clusters as a whole. For other purposes, this aspect needs refining.

6 SUMMARY

This paper presents a methodology for analyzing, modeling and simulating certain arrival–counting processes. The method is appropriate when (1) the number of such processes is too large to analyze manually, and (2) the processes resemble two–level pro-

Table 2: Relative Errors at Each Update Rate for Each File

	Update rate (1/seconds)						
File	**.001**	**.01**	**.1**	**1**	**10**	**100**	**1000**
5503	-26.48	-12.77	9.67	1.97	0.16	0.00	0.00
5479	-22.44	-9.81	9.85	1.04	0.05	0.00	0.00
3781	-18.25	-12.68	-11.21	-11.85	-18.69	-4.99	0.24
5260	-5.72	-0.12	0.42	2.85	0.12	0.00	0.00
4863	-5.51	-5.96	-7.79	-7.72	-7.35	-3.60	-0.54
7187	-4.79	-2.74	1.66	-0.06	0.00	0.00	0.00
5570	-3.96	-5.46	-1.75	-0.07	0.00	0.00	0.00
0524	-1.84	-2.14	-1.33	-0.84	-4.03	-12.30	-3.81
5488	-1.26	2.03	3.87	-2.51	0.01	0.00	0.00
7324	-0.19	0.53	-2.18	0.48	0.11	0.00	0.00
5580	1.76	6.50	17.19	1.56	0.07	0.00	0.00
5155	4.66	15.03	24.04	2.22	0.16	0.00	0.00
7174	8.84	8.48	11.89	1.07	0.12	0.00	0.00
1087	32.94	33.39	33.10	29.38	33.67	33.48	1.45
5532	46.40	57.34	42.45	6.51	0.68	0.06	0.00
0098	52.78	53.17	42.52	-15.05	-5.57	0.27	0.05
6491	54.63	59.42	46.22	4.74	0.35	0.00	0.00
5043	161.63	160.16	42.54	-17.41	-9.91	-4.14	-0.07

cesses. Trace data from both computer file systems and from communications networks typically fit these characteristics. In the case of file system traces, the data is decomposed into separate traces for each file accessed; for network traffic, communications between distinct pairs of source and destination sites constitute an appropriate decomposition. The purpose of this input modeling effort is to be able to generate synthetic traces that exhibit similar "burstiness" to the original recorded data.

The method first uses hierarchical clustering to group the data into bursts. A novel stopping rule is presented which allows the method to use different stopping parameters for each file. This is in contrast to previous work (see for example Jain & Routhier (1986)) which required a single stopping rule be applied to the whole data set.

A simple simulation modeling effort is described whose goal is to predict the frequency of concurrent updates in a lazy-update propagation replicated distributed file system. The system is very sensitive to the burstiness of the update traffic. Hence the simulation is a good test of the degree to which the method captures this characteristic of the trace data.

Preliminary results lead us to believe that the method does a good job at modeling file system trace data.

6.1 Limitations and Follow-up Work

A limitation that should be noted is that no attempt has been made to model the time varying nature of the workload. It is well known that file system workloads are highly cyclical. For the purposes of our simulation application however, we are more interested in the worst case scenario which is represented by the busiest sections of the trace. Hence, no work is planned in this area.

As of yet, we have not attempted to model the mix of file operations. The experiment reported utilizes only update events. Further work is required to generate an appropriate distribution of event types (read, lookup, create, etc.). Similarly, identity of the site making the file system access has been ignored to this point.

A longer range goal of this work is to produce a flexible file system load generation tool. Given a set of trace data for an existing system, the tool should be able to generate events which mimic the behavior of the existing system. But further, it should be parameterized so that it can be tuned to generate predictable workloads for file systems for which no trace data are available, either because they do not yet exist, or because instrumenting them to gather the traces is infeasible. For example, one might wish to generate a trace for a file system which mimics an existing one, but has 10 times as many users accessing 5 times as many files. The work presented here is a step along that path.

REFERENCES

Devroye, L. (1986), *Non-Uniform Random Variate Generation*, Springer-Verlag, New York, NY.

Guy, R. G., Heidemann, J. S., Mak, W., Page, Jr., T. W., Popek, G. J. & Rothmeier, D. (1990), Implementation of the Ficus replicated file system, *in* 'USENIX Conference Proceedings', USENIX, pp. 63–71.

Hisgen, A. (1990), Dec firefly trace data, Data obtained from author on 8mm tape.

Jain, R. (1991), *The Art of Computer Systems Performance Analysis: Techniques for Experiment Design, Measurement, Simulation and Modeling*, John Wiley & Sons, Inc.

Jain, R. & Routhier, S. A. (1986), 'Packet trains - measurements and a new model for computer network traffic', *IEEE Journal on Selected Areas in Communications* **SAC-4**(6), 986–995.

Johnson, N. L. (1949), 'Systems of frequency curves generated by methods of translation', *Biometrika* **36**, 149–176.

Kaufman, L. & Rousseeuw, J. (1990), *Finding Groups in Data: An Introduction to Cluster Analysis*, John Wiley & Sons, Inc, New York, NY.

Milligan, G. W. & Cooper, M. C. (1985), 'An examination of procedures for determining the number of clusters in a data set', *Psychometrika* **50**(2), 159–179.

Ord, J. K. (1972), *Families of Frequency Distributions*, Griffin, London.

Satyanarayanan, M., Kistler, J. J., Kumar, P., Okasaki, M. E., Siegel, E. H. & Steere, D. C. (1990), 'Coda: A highly available file system for a distributed workstation environment', *IEEE Transactions on Computers* **39**(4), 447–459.

Swain, J. J., Venkatraman, S. & Wilson, J. R. (1988), 'Least–squares estimation of distribution functions in Johnson's translation system', *Journal of Statistical Computation and Simulation* **29**, 271–297.

Terry, D. B., Theimer, M. M., Petersen, K., Demers, A. J., Spreitzer, M. J. & Hauser, C. H. (1995), Managing update conflicts in bayou, a weakly connected replicated storage system, *in* 'Proceedings of the Fifteenth Symposium on Operating Systems Principles'.

AUTHOR BIOGRAPHIES

PETER P. WARE is a PhD student in the Computer and Information Science Department at The Ohio State University. He is interested in distributed operating systems and replicated file systems. See `http://www.cis.ohio-state.edu/~ware` for more information about him and this paper.

THOMAS W. PAGE, JR. is an Assistant Professor in the Computer and Information Science Department at The Ohio State University. He received his Ph.D. in Computer Science and Engineering from UCLA in 1989. His areas of interest include distributed operating systems, distributed and replicated file systems, concurrency control for replicated data, and distributed shared memory.

BARRY L. NELSON is an associate professor in the Department of Industrial Engineering and Management Sciences at Northwestern University. He is interested in the design and analysis of computer simulation experiments, particularly statistical efficiency, multivariate output analysis and input modeling. He is the simulation area editor for *Operations Research* and will be Program Chair for the 1997 Winter Simulation Conference.

Proceedings of the 1996 Winter Simulation Conference
ed. J. M. Charnes, D. J. Morrice, D. T. Brunner, and J. J. Swain

A HYBRID SIMULATION-QUEUEING MODULE FOR MODELING UNIX I/O IN PERFORMANCE ANALYSIS

Barry L. Nelson

Department of Industrial Engineering
and Management Sciences
Northwestern University
Evanston, Illinois 60208, U.S.A.

William S. Keezer
Thomas F. Schuppe

LEXIS-NEXIS
Dayton, Ohio 45401, U.S.A.

ABSTRACT

LEXIS®-NEXIS® frequently develops simulation models to estimate the computer system capacity required to deliver on-line information services. The load imposed by Unix Input/Output (I/O) processes is a key factor in many of these models. However, the I/O processes themselves are seldom of interest, and explicitly modeling them results in simulations executing *much slower than real time*. This paper presents a hybrid simulation-queueing module that can be inserted into any simulation to accurately model I/O resource consumption and queueing delays without explicitly modeling each individual I/O process. This module is in use at LEXIS-NEXIS today (LEXIS and NEXIS are registered trademarks of Reed-Elsevier Properties, Inc., and are used under license).

1 INTRODUCTION

LEXIS-NEXIS uses a diverse collection of computer systems to deliver on-line information services worldwide. An important part of the development of new information services is estimating the system capacity required to support the product. Even for existing products, it is important to estimate capacity requirements for services experiencing growth.

At LEXIS-NEXIS, capacity estimates are typically based on a simulation of the processes involved in preparing and delivering a new service. Usually simulation models describe processes in great detail and include such tasks as computation of specific subprocesses, input/output (I/O) of data, remote procedure calls (RPC's), etc. The data used to describe time delays and resource consumption (CPU, memory, etc.) are based on historical data for familiar, well-understood processes. For an unfamiliar or completely new process, it is often necessary to estimate time delays and resource consumption based on expert opinion, since no data are available.

For I/O operations in Unix environments, no systematic collection of data had been done prior to 1994, so I/O times and resource consumption data were based on expert estimates. Most estimates were constants or simple functions of I/O size. These methods were used because they produced reasonable results and no better information was available. In 1994, however, extensive controlled testing was done on both HP and Sun servers to gather precise data on I/O times and resource consumption. At about the same time the modeling effort described in this paper was initiated to develop a better method for simulating resource consumption and time delays caused by I/O. The concept was to produce a portable module that could be inserted into any LEXIS-NEXIS performance simulation that required modeling local (as opposed to network) Unix I/O. The module needed to be customizable to the actual physical configuration of the hardware, and to simulate resource consumption and delays with high fidelity but minimal additional computational burden on the simulation. Ultimately, the HP and Sun laboratory tests were used to parameterize and validate this module. The remainder of the paper describes the processes used to develop, validate and apply this improved modeling process.

2 SYSTEMS MODELED

Both the HP and Sun systems of interest to us had similar overall architectures: Multiple CPUs are connected to memory modules via a high-speed backplane (about 240–960 megabytes per second). All I/O data passes through a converter that contains buffering to match the speed and the protocol of the peripheral bus to that of the backplane. SCSI controllers are connected to a peripheral bus with speeds on the order of 30–50 megabytes per second. The actual I/O devices are then attached to the SCSI devices. The DASD (disk) devices for both systems had very similar architectures with on-board CPUs and buffers, as well as look-ahead capability. The details of each of the main components differs considerably between the two systems despite the similarity of the overall architectures.

The details of the architecture of the HP9000 Model T500 may be found in the article by Alexander, et al. (1994). The main concerns for our work were not the internals of the system, but rather the rates at which various modules process data and various buses and paths transmit data. The HP9000 backplane operates at a rate of 240–960 MB/s (megabytes per second), and our system had the maximum capability. Because of the high speed and the use of a packet-switching protocol on the bus to transfer data, we assumed that there would never be any queueing on this device, only transfer delays. Hewlett-Packard (HP) uses a device called a bus converter to provide an interface for the peripherals to the backplane. This device has two 64 MB/s Precision Buses, each of which can have 1–7 SCSI adapters attached to it. The actual data transfer rate after protocol overhead is about 42 MB/s. In our experience, this data bandwidth is sufficient to present no detectable queueing delays.

The DASD devices are connected to fast, wide SCSI adapters that have a nominal transfer rate of 20 MB/s. These adapters can accept up to 12 devices daisy-chained along a single data path. The data path from the device to the adapter has the capability of handling up to 20 MB/s for data buffered at the device. The details of the HP DASD devices may be found in Ruemmler and Wilkes (1993). Their paper gives the necessary details for calculating the device level delays and accounting for the effects of the on-board cache and the look-ahead algorithm of the device. More importantly, it provides an algorithm for calculating seek time that is much better than anything used previously.

At the operating system level, the HP9000 uses HP's version of Unix called HP-UX. This is a version of Berkeley Unix. The most important consideration here is that I/O is handled by a set of cache buffers that have their own management system. This became a major performance concern in trying to validate our module against the benchmark data. Based upon our results, the space devoted to I/O cache is divided into blocks of predetermined size, in our case 8172 bytes, and evenly distributed among buffer queues. The number of queues appears to be 128 plus the number of application files. To find out if a block of data is in cache, the operating system hashes to the correct queue and then sequentially searches the queue for the desired block. The implications are that the larger the cache is relative to the number of application files, the more time will be spent in sequentially searching the queues.

The Sun Microsystems SPARCcentera 2000 (SPARC 2000) is fully described in a white paper produced by Sun (1992). The main difference of importance in our model is that the backplane transfer rate is 640 MB/s. This still appears to be sufficiently fast that it creates no measurable queueing delay in the system. The internal architecture of the conversion module is considerably different from the HP9000, but only the timing is of importance. The Sun equivalent of the HP Precision Bus is the SBus. This bus has a transfer rate of 50 MB/s and can accept up to four SCSI adapters.

The SCSI adapters are labeled as differential, wide devices and have a transfer rate of 22 MB/s. Each adapter can accept up to 6 devices on a data path that transfers buffered data at 22 MB/s. From our viewpoint, there are no differences in the disk devices between the HP9000 and the SPARC 2000.

The operating system is Sun's proprietary version of Unix, SUNOS. It is different from HP-UX in its handling of I/O cache. In the SPARC 2000, cache is simply a part of memory and is handled by the memory manager. The overhead to access any block of cached data is a constant and the same as the overhead to access any other block of memory. This means that performance does not depend on the amount of cache versus the number of application files.

3 BENCHMARK DATA

The benchmark data used to validate the module and provide the values for some of the parameters was originally obtained during a stress test of the two different systems that were being considered for I/O-intensive applications. Nominally, the benchmark created a series of totally random READS of random length (1–64 kilobytes with an average of 32 kilobytes) at varying rates up to the maximum sustainable rate. A number of different configurations of SCSI adapters and DASD devices were tested. I/Os per second, CPU utilization, and response times were measured. In the case of the HP9000, several relatively complete throughput-response curves were obtained for the various configurations, and eventually a set of parameters was developed for our module that gave an error of 5% or less compared to the benchmark data. In the case of the SPARC 2000 tests, the system was greatly over-stressed, the goal being to see what the maximum throughput was. As a consequence, not all parameters could be as carefully defined as for the HP9000. However, useful data were obtained.

4 THE UNIX I/O MODULE

This section describes the Unix I/O module. The module is a collection of Fortran subroutines that can be used in isolation, or can (more typically) be added to simulation models that need to account for local I/O resource consumption and queueing delays. A conceptual presentation of the module is shown in Figure 1; each box in the figure corresponds to a queue in the module. The philosophy of the module is to randomly sample processing times (service times, in queueing terminology) as a function of I/O size, but compute expected queueing delays based on system load as characterized by recent CPU utilization, read/write rates and the current probability of a memory cache hit. In this way the module achieves a high degree of fidelity without incurring the prohibitive computational cost of explicitly simulating each I/O.

Each subsection below describes one component of Figure 1, including the inputs that the component requires. Each component returns a (possibly randomly sampled) processing time and an expected queueing delay. Throughout the section, notation such as **iolam** indicates the name of a Fortran variable, and "ms" indicates milliseconds.

4.1 Cache

The CPU cache model has the following inputs:

$\lambda_{I/O} \equiv$ overall rate of reads and writes in MB/ms; this parameter may be dynamic (**iolam**).

$\lambda_{ru} \equiv$ rate at which data is reused in MB/ms; can also view $1/\lambda_{ru}$ as the expected time between requests for the same data; this parameter is not dynamic (**rulam**).

$c \equiv$ cache size in units of the average I/O size, $a_{I/O}$ (**avsize**); in Fortran cache size is **c**.

$p_{hit} \equiv \Pr\{\text{memory cache hit}\}$; this parameter can be specified or computed dynamically (**phit**).

Whether or not there is a local cache hit is modeled as a Bernoulli (0 or 1) random variable, where the probability of a hit, p_{hit}, on each trial is independent of the previous trial. This random variable is sampled.

The modeler can specify a value for p_{hit}, or it will be computed by a simple Markov-process model. The

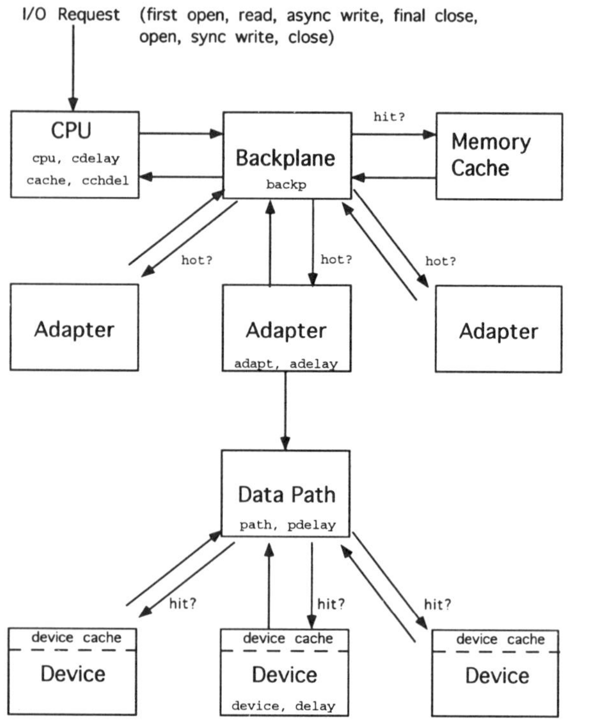

Figure 1: Conceptual View of the Unix I/O Module

Markov-process model assumes that data requests arrive to cache according to a Poisson process at a rate of $\lambda_{I/O}$, and data are reused at a rate λ_{ru}, with the time between reuse being exponentially distributed. A data quantum is pushed out of cache when it is the oldest quantum in cache and something new is pushed in. These assumptions allow a steady-state value of p_{hit} to be calculated analytically,

$$p_{hit} = 1 - \left(\frac{\lambda_{I/O}}{\lambda_{ru} + \lambda_{I/O}}\right)^{c}$$

4.2 CPU

The CPU model has the following inputs:

$\rho_{cpu} \equiv$ utilization per CPU for tasks with which I/O has to contend (i.e., other system calls); this parameter is dynamic (**crho**).

$k \equiv$ number of CPUs (**k**).

$\mu_{cpu} \equiv$ service rate of CPU in MIPS (**cmu**).

$\lambda_{cpu} = k\mu_{cpu}\rho_{cpu} \equiv$ the effective arrival rate to the CPU in the same units as μ_{cpu} (**clam**).

For each task performed, the constants γ_0 and γ_1 such that the total CPU processing time consumed by that task is given by $t(\text{I/O size}) = \gamma_0 +$

(I/O size)/γ_1, where there are different values of γ_0 and γ_1 depending on task and whether or not there was a cache hit, if appropriate (in Fortran **gamma0(task,hit)** and **gamma1(task,hit)**).

$1/\mu_{\text{cache}}$ $\equiv$ the expected time to read the cache buffer (**cacherd**), which is a function of the number of pages in a string (**cachstrg**) and the time to read one page (**pagerd**).

$\lambda_{\text{cache}} = \lambda_{\text{I/O}}/a_{I/O}$ $\equiv$ arrival rate to the cache-buffer queues in units of I/O's per time (**cchlam**).

The model returns a total CPU processing time and an expected total delay waiting for the CPU; the same quantities are returned for contention and use of the cache-buffer queues.

The collection of CPUs is modeled as an M/G/k processor-sharing queue. Processing times at the CPU are calculated as a deterministic function of overhead plus a linear function of I/O size, depending on I/O task and whether or not there is a cache hit. Expected delays waiting for the CPU are proportional to the total processing time for a processor-sharing model, and depend only on the service rate and not the distribution (Cooper, 1981).

Contention for the cache-buffer queue is modeled as an M/G/1 queue. The actual processing time depends on whether or not there is a cache hit or miss; when there is a hit then the time to search out the appropriate page is sampled.

4.3 SCSI Host Adapter

The adapter model has the following inputs:

n $\equiv$ number of adapters (**n**).

n^* $\equiv$ number of "hot" adapters (**nhot**).

f_{hot} $\equiv$ is as measure of imbalance between "hot" and not "hot" adapters, given as a fraction of all I/O requests routed to the "hot" adapters (**fhot**).

λ_{adpt} $\equiv$ arrival rate to the adapter that is used in the current calculation in MB/ms (**alam**). This value is computed as an appropriate fraction of $\lambda_{\text{I/O}}$.

μ_{adpt} $\equiv$ service rate of adapter in MB/ms (**amu**).

For each task performed, the constants α_0 and α_1 such that the total adapter processing time consumed is given by $t(\text{I/O size}) = \alpha_0 + (\text{I/O size})/\alpha_1$ (in Fortran **alpha0(task)** and **alpha1(task)**).

Whether or not the current I/O requires a hot adapter is a Bernoulli random variable with f_{hot} as the probability of requiring a hot adapter. The concept of a "hot" adapter is included to account for nonuniform load across devices. If a hot adapter is required, then

$$\lambda_{\text{adpt}} = f_{\text{hot}}\lambda_{\text{I/O}}/n^*$$

otherwise

$$\lambda_{\text{adpt}} = (1 - f_{\text{hot}})\lambda_{\text{I/O}}/(n - n^*)$$

The model returns a total adapter processing time and an expected total delay waiting for the adapter. Each adapter is modeled as an M/D/1 queue. Processing times at the adapter are calculated as a deterministic function of overhead plus a linear function of I/O size, depending on I/O task. Since two delays are experienced at the adapter (outbound to the data path and inbound from the data path), the total delay is

$$2 \times \frac{\lambda_{\text{adpt}}/\mu_{\text{adpt}}}{2(\mu_{\text{adpt}} - \lambda_{\text{adpt}})} = \frac{\lambda_{\text{adpt}}/\mu_{\text{adpt}}}{\mu_{\text{adpt}} - \lambda_{\text{adpt}}}$$

4.4 Data Path

The data-path model has the following inputs:

λ_{adpt} $\equiv$ arrival rate to the adapter that is used in the current calculation in MB/ms (**alam**). This value is computed as an appropriate fraction of $\lambda_{\text{I/O}}$.

For each task performed, the constants ω_0 and ω_1 such that the total data-path processing time consumed is given by $t(\text{I/O size}) = \omega_0 + (\text{I/O size})/\omega_1$ (in Fortran **omega0(task)** and **omega1(task)**).

μ_{path} $\equiv$ service rate of the data path in MB/ms (**pmu**).

The model returns a total data-path processing time and an expected total delay waiting for the data path. The data path is modeled as an M/D/1 queue. Processing times are calculated as a deterministic function of overhead plus a linear function of I/O size, depending on I/O task. Since two delays are experienced (outbound to the device and inbound from the device), the total delay is

$$2 \times \frac{\lambda_{\text{adpt}}/\mu_{\text{path}}}{2(\mu_{\text{path}} - \lambda_{\text{adpt}})} = \frac{\lambda_{\text{adpt}}/\mu_{\text{path}}}{\mu_{\text{path}} - \lambda_{\text{adpt}}}$$

4.5 Backplane

For each task performed, the backplane model requires constants β_0 and β_1 such that the total backplane processing time consumed is given by $t(\text{I/O size}) = \beta_0 + (\text{I/O size})/\beta_1$ (in Fortran **beta0(task)** and **beta1(task)**).

The backplane is modeled as pure processing time (overhead plus a linear function of I/O size) with no contention (queueing). Backplane time accounts for all transmission time down to the adapter level and back up from the adapter level.

4.6 Device

The disk storage device model has the following inputs:

$d \equiv$ number of devices per adapter (**d**).

$p_{\text{dhit}} \equiv$ probability of a device cache hit (**pdhit**).

$r \equiv$ total rotation time, in ms (**rotate**).

$p_{\text{short}} \equiv$ probability of a short head movement (**pshort**).

$t_{\text{short}} \equiv$ time to make a short head movement and settle, in ms (**short**).

$t_{\text{long}} \equiv$ time to make a non-short head movement and settle, in ms (**long**).

$a_{\text{I/O}} \equiv$ size of the average I/O (**avsize**).

$b \equiv$ size of a block in the same units as $a_{\text{I/O}}$.

$\delta_0 \equiv$ seek time + rotational latency; this value is sampled.

$\delta_1 \equiv$ time to transmit data in MB/ms on a physical read/write (**delta1**).

$\eta_0 \equiv$ set-up time for I/O when a device cache hit occurs, in ms (**eta0**).

$\eta_1 \equiv$ marginal transmission time when a device cache hit occurs, in MB/ms (**eta1**).

The device is modeled as an M/G/1 queue for the purpose of calculating delay. The service time at the device is composed of seek time, rotational latency and transmission time when there is a cache miss; it is composed of a set-up time and marginal transmission time when there is a cache hit.

The seek time is modeled as a two-point distribution on t_{short} and t_{long}. The expected value and variance of this random variable are used in the queueing model, but a sampled value is returned.

The rotational latency is modeled as a random variable that is uniformly distributed on the time to complete one rotation of the disk. The expected value and variance of this random variable are used in the queueing model, but a sampled value is returned.

The transmission time for a physical I/O is modeled as a linear function of the I/O size. The size of a block is used in the queueing model, but the actual transmission time is returned. Specifically, $t(\text{I/O size}) = \delta_0 + (\text{I/O size})/\delta_1$, where δ_0 = seek + rotational latency.

When a cache hit occurs, the device processing time is $t(\text{I/O size}) = \eta_0 + (\text{I/O size})/\eta_1$.

4.7 Task Models

The Unix I/O module can simulate the seven I/O tasks described below.

1. Open an unopened file without accounting for prefetch:
   (a) CPU processing and delay; always a memory cache miss
   (b) adapter processing and delay
   (c) data path processing and delay
   (d) device processing and delay; always a device cache miss
2. Read
   (a) determine if memory cache hit or miss
   (b) if memory cache hit, CPU processing and delay
   (c) if cache miss...
      i. adapter processing and delay
      ii. data path processing and delay
      iii. device cache hit or miss
      iv. device processing and delay
3. Write; write-back version with no housecleaning. That is, only a write to cache. Includes CPU processing and delay.
4. Close; concluding close.
   (a) CPU processing and delay; always a memory cache miss
   (b) adapter processing and delay
   (c) data path processing and delay
   (d) device processing and delay; always a device cache miss

5. Open an already open file. Treated as always a CPU cache hit. Includes only CPU processing and delay.
6. Write; synchronous version, so always a CPU cache miss.
   (a) CPU processing and delay
   (b) adapter processing and delay
   (c) data path processing and delay
   (d) device delay and processing time to write to cache
7. Close, but not concluding close. Treated as always a cache hit. Includes only CPU processing and delay.

Values are obtained by making calls to `subroutine io` as follows:

```
 subroutine io(cp,task,size,crho,iolam,p,
$   cache,cchdel,cpu,cdelay,backp,adapt,
$   adelay,device,ddelay,path,pdelay,hit,hot)
```

The inputs are the configuration platform `cp`, the `task` type (as listed above), the `size` of the I/O in MB, the current input/output rate `iolam` in MB/ms, and the current CPU utilization $0 < \texttt{crho} < 1$. The dynamic parameters `crho` and `iolam` should not be based on instantaneous snapshots, however, since the delay values returned by `subroutine io` are based on steady-state queueing models. Rather, average values for a recent time period should be passed. In this way, the I/O module can reflect current load on the system. Some experimentation is required to determine an appropriate time window for averaging.

5 PARAMETRIZATION

In its final form, the Unix I/O module requires two different types of parameters: (1) those that change with each simulation scenario or group of scenarios and deal with the configuration of the system and the data layout on the devices, and (2) those that deal with the internal operating parameters of the platform. In a multi-system test, a different configuration parameter file is required for each computer system in the model, and a different internals file for each type of system. For example, if a system of two HP9000s and three SPARC 2000s was being simulated, five different configuration files would be required and two different internals files (one for HP and one for Sun).

The system configuration values that might change from scenario to scenario of the model are the amount of cache defined (in the case of the SPARC 2000 it is 0); the reuse rate of data in cache (used to calculate cache-hit probabilities if that value is not supplied); the effective number of CPUs (since these are multi-CPU systems); the total number of application files (required for calculating buffer queues); the number of adapters and the fraction defined as "hot" (this allows the model to account for heavily-used devices); the probability of a device cache hit (very important in response-time modeling); the transfer block size; and the total number of devices. All of these parameters may be defined by the Systems Administrator in a real system or are part of the what-if testing when trying to design a system. The only one that requires calculation is the probability of a device cache hit. In our applications, the size of the I/O and the totally random access patterns created a probability of a device cache hit of near zero, and we often used zero for this value. In cases where there are a high proportion of records shorter than a block and a high degree of sequential access, the probability of a device cache hit becomes quite high. Ruemmler and Wilkes (1992) discuss this issue in detail.

System internals are values that change only with a change of platform. They are implemented as permanent reference files that are called as needed when the overall computing environment changes its configuration. Specifically, they are matrices of values, based on various divisions of the model: CPU, backplane, cache manager, SCSI adapter, data path from device to adapter, and the device. For each component of the Unix I/O module there is a set of seven parameters, one for each of the I/O tasks. Each set of values includes a base service time and a data-size dependent rate for cache miss situations and the same pair for cache hit situations. For those elements that do not have dependence on the task or for those elemental parameters that are not task dependent, a single set of parameters was created.

The internal parameters for consumption of CPU, transfer on the backplane, adapter processing and transfer on the data path from device to the SCSI adapter each require a parameter for the fixed delay (in some cases zero) and another for the transfer rate for calculating data-size dependent delays. In the case of CPU processing, an additional set of parameters is created to account for the differences between memory cache hit and cache miss. There are a pair of these values for each of the seven I/O operations modeled. The CPU values are for setting up the I/O and not for doing the cache search, which was calculated separately then added to the calculated CPU time. These parameters were obtained primarily from the manufacturer's literature on the system. In the case of the CPU, estimates were made of the fixed delays for operations other than READ based on the amount of work required to accomplish the operation. The amount of CPU for the READ operation was determined from the benchmark data. The module is not critically sensitive to any of these values.

Caching of I/O data is an important constraint on the performance of the systems in reality, and therefore requires considerable care to be represented properly in the module. To maintain integrity of the data, only one processor may access the cache data at a time. This means that when the CPU consumption of cache access and management is equivalent to one fully utilized processor, no higher throughput is possible.

In the case of the HP9000, the cache processing was particularly important for obtaining a good fit of the module to the benchmark data. Early plots of throughput response curves from the benchmark data indicated different curves at the device level for different configurations. We found that the curves were essentially the same for the same number of files. Coupling this information with the cache buffer management scheme, we were able to understand that this effect was due to the decrease in the sequential search time on a buffer queue. After considerable study of the benchmark data it was found that a good approximation for the amount of CPU time due to cache buffer processing could be obtained via a linear function of the number of application files. The y-intercept of this equation is the time to process a buffer when only one application file is present, and the slope is negative indicating a decrease in processing per buffer with an increase in the number of files. The slope is multiplied by the number of files and subtracted from the intercept to determine the delay per buffer. This value is then multiplied by the number of buffers in a cache buffer queue to determine the overall delay to search a single queue. This value is corrected for the probability of a cache hit and the fact that on average half of the queue is required to be searched to find the correct buffer.

In the case of the SPARC 2000, data were found in the benchmarks that indicated that the limit had been reached on cache processing. This information was used to determine the cost per I/O for cache handling. This is fixed for each I/O, regardless of configuration, because cache in Sun systems is handled as part of general memory. Therefore, the parameters for calculating cache buffer management were set to zero, and only a fixed time was used to determine the service time and queueing for this part of the module.

The individual disk devices presented the most important and sensitive portions of the module. Disk access is usually described by three parameters, seek time, rotational delay, and transfer time. However, as Ruemmler and Wilkes (1993) point out, the seek time is not linear, and is itself composed of several parts, acceleration, linear travel, deceleration, and settling. They provide an algorithm based on the distance the arm has to travel for calculating the seek time. This algorithm depends on the distance traveled, using one equation for less than one third of the full width of the disk, and a second for traveling the full width of the disk. We chose to approximate this as a two-point probability distribution using the values for a short seek, a long seek, and the probability of a short seek. The short seek time was set equal to that required to move one track. The long seek time was set equal to the time to move the distance from the middle of the file nearest to the spindle to the middle of the file farthest from the spindle in the benchmark test. The probability of a short seek (which implies the probability for a long seek) was calculated using the average response time for a very lightly loaded disk (essentially no queueing), the average rotation and transfer times, and the short and long seek times. The module is extremely sensitive to this value, because it directly affects the variance of the service time and therefore the queueing at heavy loads. Despite the sensitivity, once this parameter is established and when the I/O is distributed relatively evenly across the disk surface, the module will work for simulations other than the one used to create it. The values we calculated work quite well if the short and long seeks are correctly estimated. Thus, if the device changes or the data layout changes greatly, this parameter should be recalculated. Rotational delay and transfer rates are obtained from the manufacturer's literature. Other than the differences in rotational delay and transfer rates, both HP9000 and SPARC 2000 devices used the same calculations for disk device performance.

6 VALIDATION

Validation of the parameters was coupled with validation of the module in an iterative process. The data for validation came from the benchmarks, and the comparisons were made to throughput-response curves plotted from that data. The goal was to be able to reproduce all the curves from a benchmark run with one set of parameters. The process was essentially trial-and-error. Multiple runs would be made for a set parameters with varied throughput and an empirical best fit obtained. When the behavior of the module did not correspond to the benchmark system, further research into the details of Unix and the hardware was done, and the module modified to account for the new findings. The process of fitting parameters was then repeated. The module required two modifications from the first edition, one to add queueing behavior to the connection between the device and the SCSI adapter, and the second and more important one to add the cache manager behavior. Our final validation gave results that could be superimposed on the HP9000 benchmark data with no more than a 5% error at any data point.

Because of the difference in the quality of the SPARC 2000 benchmark data when used for our purposes, a fit such as we obtained for the HP data was not possible. However, as mentioned above, the cache processing delay could be obtained as well as the CPU cost per READ I/O. The manufacturer's literature and experience were used to complete the set of parameters for this platform.

7 AN APPLICATION

After the basic modeling concept was developed and validated against the HP and Sun benchmark data, the module was implemented in an existing simulation model. At LEXIS-NEXIS the primary simulation modeling tool used is SLAMSYSTEM. The basic simulation package has been supplemented with about two dozen FORTRAN subroutines known as the Data Driven Modeling System (DDMS). DDMS was created at LEXIS-NEXIS to allow very large models to be represented by data files and to allow easy modification of these models. A detailed description of the process used at LEXIS-NEXIS is provided by Robinson (1994).

DDMS operates by using a basic network diagram that allows numerous options (await a resource, delay, free a resource, return to a network diagram, etc.) for each step in a process. The option selected is controlled by attribute values read in from a data file. This data file also provides all parameters required to accomplish a step or call additional subroutines to compute other required values.

The I/O module is a collection of three Fortran subroutines, and calling `subroutine io` one time provides estimates of the queueing delays and resource consumption for one I/O process. All information passes between subroutines via calling arguments and common blocks. To make the I/O module work in the SLAMSYSTEM/DDMS structure, it was necessary to incorporate these three subroutines into this structure by specifying five new attributes to carry additional information to the I/O subroutines.

The simulation model chosen to test the Unix I/O module was one that had been developed to provide capacity estimates for a new product. This new product requires that a large number of documents be processed and stored and that sets of relevant documents be prepared and shipped to customers each day. When the number of I/O's per document is scaled up to the number of documents and customers processed per day, we obtain a very large number of I/O's, on the order of 1,000,000 per day. These I/Os are accomplished on two servers, an HP hosting a Sybase database and a Sun acting as a file server.

Four different scenarios representing different levels of business volume were run for both the old I/O method and the new I/O module. While each run of the simulation using the new I/O module took longer than the corresponding run using the old I/O method, the time delays and resources consumed as generated by the new Unix I/O module were much more realistic. For example, the time required to process a batch of documents was now sensitive to the overall loading (utilization of the CPUs) of the system. Formerly, I/O processing times were assumed to be independent of current CPU utilization rates.

8 DISCUSSION

The potential for the Unix I/O module is substantial, since the simulation of every I/O to the detail accounted for in the module requires prohibitive amounts of simulation execution time. Because the module is parameter-driven, considerable attention must be paid to establishing the parameters. Many of the parameters may be obtained from manufacturers' literature. However, for critical parameters, thorough testing and understanding of the workings of the operating system and the hardware is required. The most critical areas are cache management and the disk device. In our experience, a combination of manufacturers' literature, a detailed description of the operating system (Leffler, et al. 1990), and careful analysis of benchmark data to understand cache management was required. Ruemmler's work (1992, 1993) provided the needed information on modeling the physical devices.

To fully and adequately parametrize the module, good benchmark data is essential. We were fortunate to obtain the HP9000 benchmarks, since they were designed for a different purpose. In the case of the Sun data, we obtained usable data but not as complete as the data for HP. We strongly recommend that specially designed benchmarks be commissioned for proper parameter determination. The characteristics of a good benchmark include systematically varied loads and configurations, with careful measurements of CPU utilization, response times, I/O rates, service versus user time, interrupt counts and memory utilization.

When the module was tested by including it in the DDMS model, the run times for a single simulation of each scenario increased by 33–70%. However, the I/O module replaced two lines of SLAM code with a large subroutine. Furthermore, the results obtained with the Unix I/O module could not have been obtained in any reasonable computation time with a fully simulated I/O model. There are also a number of improvements that could be made to increase the efficiency of the module in the DDMS environment, most of them focusing on the interface between DDMS and the module. We estimate that about 50% of the current increase in run time could be eliminated, resulting in a 15–35% increase in run time over deterministic methods.

9 SUMMARY

A hybrid queueing-simulation module was developed for modeling Unix I/O. This module is parameter-driven and allows great flexibility in modeling both varying system hardware configurations and varying workloads. We found that cache management and the physical disk devices were the most important factors influencing the accuracy of the module. We also found that manufacturers' literature could provide many, but not all, of the necessary parameters. The remaining parameters had to be obtained by iterative validation and modification of the module against benchmarks. As might be expected, benchmarks are a vital link, and should be custom designed and run for this purpose. The hybrid module runs longer than a deterministic model of I/O, but provides results that could only be obtained from prohibitively long simulation runs.

REFERENCES

Alexander, T. B., K. G. Robertson, D. T. Lindsay, D. L. Rogers, J. R. Obermeyer, J. R. Keller, K. Y. Oka, and M. M. Jones, II. 1994. Corporate business servers: An alternative to mainframes for business computing. *Hewlett-Packard Journal* 45: 8–30.

Cooper, R. B. 1981. *Introduction to queueing theory*, 2d ed. New York: North Holland.

Leffler, S. J., M. K. McKusick, M. J. Karels, and J. S. Quarterman. 1990. *The design and implementation of the 4.3SD Unix operating system*. New York: Addison-Wesley.

Robinson, J. N. 1994. Capacity and performance analysis of computer systems. In *Proceedings of the 1994 Winter Simulation Conference*, ed. J. D. Tew, S. Manivannan, D. A. Sadowski, and A. F. Selia, 34–41. Institute of Electrical and Electronics Engineers, Piscataway, New Jersey.

Ruemmler, C., and J. Wilkes. 1992. Unix disk access patterns. In *USENIX Winter 1993 Technical Conference Proceedings*, San Diego, California.

Ruemmler, C., and J. Wilkes. 1993. Modeling disks. HP Laboratories Technical Report HPL-93-68, revision 1.

SPARCcentera 2000. 1992. Technical White Paper, Sun Microsystems, Inc.

AUTHOR BIOGRAPHIES

WILLIAM S. KEEZER has been with LEXIS-NEXIS for ten years and is currently a Staff Systems Engineer focusing on mainframe system storage management and performance issues. He has also done considerable work at LEXIS-NEXIS on mainframe and Unix platform application and communication performance. Before coming to LEXIS-NEXIS, he was an in-house consultant on OLTP system performance problems for the Data Pathing Division of NCR. He holds B.S. and Ph.D. degrees from the University of Oklahoma, and is a member of ACM and CMG.

BARRY L. NELSON is an associate professor in the Department of Industrial Engineering and Management Sciences at Northwestern University. He is interested in the design and analysis of computer simulation experiments, particularly statistical efficiency, multivariate output analysis and input modeling. He is the simulation area editor for *Operations Research* and will be Program Chair for the 1997 Winter Simulation Conference.

THOMAS F. SCHUPPE is the Manager of System Integration at LEXIS-NEXIS. He received a B.S. in Mechanical Engineering from the University of Wisconsin, an M.S. in Systems Engineering from the Air Force Institute of Technology, and a Ph.D. in Operations Research from The Ohio State University. His primary research interest is in simulation modeling of complex man-machine systems. He is currently a member of INFORMS and SCS.

Proceedings of the 1996 Winter Simulation Conference
ed. J. M. Charnes, D. J. Morrice, D. T. Brunner, and J. J. Swain

MULTITASKING AND RESEQUENCING IN A TWO-STAGE MULTIPROCESSING SYSTEM

Helen D. Karatza

Department of Informatics
Aristotle University of Thessaloniki
54006 Thessaloniki, GREECE

ABSTRACT

The effects of multitasking and resequencing on system and program performance of a two-stage multiprocessing system are studied using simulation techniques. A closed queueing network model is considered. It consists of the CPU and the I/O channel. The CPU consists of a FCFS single-server sequential processing center and a multiprocessing center with two FCFS independent processors each serving its own queue. Resequencing of jobs after CPU service ensures that jobs leave the CPU on a first-in-first-out basis. We examine how the synchronisation of tasks in the multiprocessing center in conjunction with the subsequent resequencing of jobs affects the overall performance for various coefficients of variation of the CPU service times (exponential and Branching Erlang distributions) and for different degrees of multiprogramming.

1 INTRODUCTION

In this paper we study the effects of multitasking and resequencing on system and program performance of a two-stage multiprocessing system. The results are obtained using simulation techniques.

A closed queueing network model is considered which consists of the CPU and the I/O unit. The CPU consists of a FCFS single-server sequential processing center and a multiprocessing center with two FCFS independent processors each serving its own queue. Therefore the CPU system consists of two stages, Stage 1 and Stage 2, called the "job stage" and the "task stage", respectively.

In center Stage 2, each service time of a job (program) consists of two tasks. In half of the jobs the tasks are independent and can be processed in parallel. Synchronisation between tasks is required. In the other half of jobs both tasks of the same job must be executed sequentially on the same processor. These jobs join the shorter queue.

An alternative approach we use to avoid the synchronisation of tasks is to run sequentially both tasks of all jobs. These two approaches are studied and compared.

Resequencing of jobs after CPU service ensures that jobs leave the CPU on a first-in first-out basis.

The resequencing problem can be found in several system contexts such as distributing computing systems and computer communication systems. In those systems customers who arrive at the system should depart from the system in the same order as their arrivals. Therefore, after the completion of their services, customers who go out of order are forced to wait in a special buffer—the so-called resequencing buffer—in order to rearrange their sequences, i.e. until the overtaken job of the same class completes its service. The delay due to resequencing is called resequencing delay.

The resequencing delay in multiserver queues has been studied by many authors as Iliadis, and Lien (1988); Lien (1986); Sasase, and Mori (1991); Takine, and Hasegawa (1990); Varma (1991). Load balancing in a system of two queues with resequencing has been studied by Jean-Marie (1988). All these works study open queueing network models and exponential CPU service times.

Multitasking and resequencing in a homogeneous distributed system is studied in Karatza (1995a). This work studies closed queueing network models.

Two-stage parallel processing systems with parallel servers at the second CPU center are studied in Persone, and Iazeolla (1994) and in Karatza (1995b).

In this work we study a two-stage multiprocessing system where there are two independent processors each serving its own queue at the second CPU center. Our intention is to examine if the synchronisation of tasks in the multitasking case, in conjunction with the subsequent resequencing of jobs is rewarded with increased parallelism for various coefficients of variation of the CPU service times (exponential and Branching Erlang distributions) and for different degrees of multiprogramming.

2 MODEL DESCRIPTION

We consider a closed queueing network model which consists of the CPU and the I/O channel. The CPU consists of a FCFS single-server sequential processing center and a multiprocessing center with two FCFS independent processors each serving its own queue. Therefore the CPU system consists of two stages, Stage 1 and Stage 2, called the "job stage" and the "task stage" respectively.

In center Stage 2, each service time of a job consists of two tasks. In half of the jobs the tasks are independent and can be processed in parallel. In the other half of jobs both tasks of the same job must be executed sequentially on the same processor.

In the case of jobs with parallel tasks, upon arrival at the center Stage 2 a job forks into two tasks. Task i, $i = 1, 2$ is assigned to the ith queue. Tasks corresponding to same job are joined before departing the CPU system, i.e. the task that completes its service first waits for its sibling to finish execution. Therefore in our model synchronisation between tasks is required. The price that one pays for the increased parallelism is the synchronisation delay that occurs when tasks wait for their siblings to finish execution.

In the case of jobs with sequential tasks they join the shorter queue. If both queues have an equal number of jobs (including empty), the arriving job at center Stage 2 is dispatched randomly to one of the two queues with equal probability.

Therefore in this paper we consider multiprocessing systems in which programs are composed of two stages which must be processed in sequence. For each program there is some preparatory sequential computation, processed by center Stage 1, followed by a second computation which is parallel for half of the jobs and is processed by processors in center Stage 2.

The important requirement of the model is that jobs must leave the CPU system in the order of their arrival. For this, each job enters the resequencing buffer after completion of its service, where it eventually waits until all jobs that entered the system before it have been served. We furthermore assume that the time taken by its resequencing process is negligible: if this job has no one to wait for, it leaves instantaneously the system. On the other hand, it leaves as soon as the last customer it has to wait for enters the resequencing buffer.

Both tasks of a given job are to be completed at center Stage 2 before the job enters the resequencing buffer.

An alternative approach we use in this paper to avoid the synchronisation of tasks is to run sequentially both tasks of all jobs. In this case the only delay that incurs is due to resequencing of jobs.

In both cases, the FCFS queueing discipline is assumed.

The model is closed as the degree of multiprogramming N is constant during the simulation experiment. Neither arrivals nor departures are permitted while the system is under observation. The configuration of the model is shown in Figure 1. A fixed number of jobs N is circulating within the system alternatively between the CPU and the I/O channel.

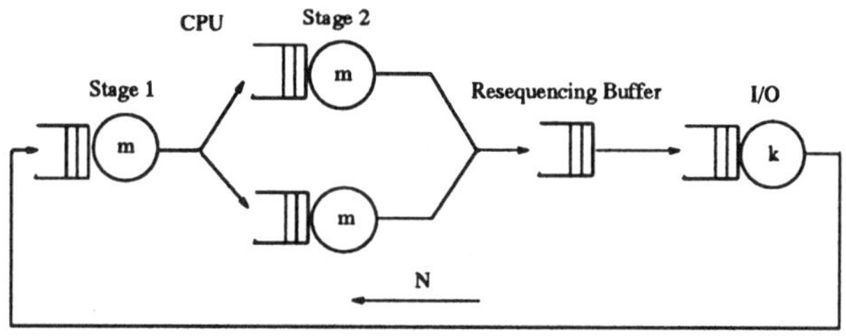

Figure 1: The Queueing Network Model

We examine two cases of CPU service time distributions:

(1) CPU service times are independent and identically distributed (IID) exponential random variables with mean m at each processor.

(2) CPU service times have a Branching Erlang distribution with two Stages (Sauer and Chandy 1981) and are IID. The coefficient of variation is C, where $C > 1$ and the mean is m at each processor.

A job after leaving the CPU requests service from the I/O unit. The I/O queueing discipline is FCFS. The I/O service times are exponentially distributed with mean k and are IID.

3 PERFORMANCE PARAMETERS

We define:

Stage i response time of a random job as the interval of time measured from an arrival of this job at center Stage i, i=1,2 to service completion of this job at center Stage i.

Cycle time of a random job as the time between two successive CPU service requests of this job.

We also denote the following:

$RT1$: mean Stage 1 response time
$RT2$: mean Stage 2 response time
D : mean resequencing delay
$RT2D$: mean Stage 2 response time plus resequencing delay
K : mean cycle time
N : degree of multiprogramming
R : system throughput rate

The performance of our model is indicated by the mean cycle time (program performance) and the system throughput rate (system performance).

When the model works first with tasks of all jobs running sequentially, and then with multitasking at center Stage 2, we define the relative performance parameters calculated on a percentage basis as follows:

D_{RT2} : relative decrease in $RT2$
D_{RT2D} : relative decrease in $RT2D$
D_K : relative decrease in K
D_R : relative increase in R

4 SIMULATION RESULTS

We considered a balanced system with $m = 1.0$ and $k = 1.0$. The system was examined in cases of exponential CPU service times (C=1) and Branching Erlang for $C = 2, 4$. We have not examined cases with $C > 4$ because in real systems jobs service times tend to have low variations. Degree of multiprogramming N was taken as 2, 4, 6, 8, 10.

We simulated the queueing network model with discrete event simulation models using the method of independent replications (Law, and Kelton, 1991). For every mean value a 95% confidence interval was evaluated. All confidence intervals were less than 5% of the mean values.

In Tables 1–6 performance parameters of all cases are presented. Tables 7–9 represent relative performance parameters for C = 1, 2, 4.

In Figures 2 and 3 relative performance parameters are plotted versus N in all cases examined.

Table 1: Sequential Tasks Case, C=1

N	$RT1$	$RT2$	D	$RT2D$	K	R
2	1.259	2.000	0.240	2.240	4.854	0.412
4	1.792	2.473	0.334	2.807	6.583	0.608
6	2.356	2.993	0.399	3.392	8.364	0.717
8	2.944	3.518	0.432	3.950	10.221	0.783
10	3.588	4.105	0.427	4.532	12.091	0.827

Table 2: Multitasking Case, C=1

N	$RT1$	$RT2$	D	$RT2D$	K	R
2	1.255	2.121	0.042	2.162	4.707	0.425
4	1.807	2.645	0.193	2.838	6.561	0.610
6	2.349	3.256	0.315	3.572	8.493	0.706
8	2.934	3.877	0.399	4.276	10.414	0.768
10	3.492	4.458	0.529	4.987	12.380	0.808

Table 3: Sequential Tasks Case, C=2

N	$RT1$	$RT2$	D	$RT2D$	K	R
2	1.282	2.004	0.609	2.613	5.280	0.379
4	1.868	2.567	1.061	3.628	7.653	0.523
6	2.447	3.094	1.298	4.392	9.851	0.609
8	2.960	3.574	1.533	5.108	11.956	0.669
10	3.414	4.127	1.616	5.743	13.933	0.718

Table 4: Multitasking Case, C=2

N	$RT1$	$RT2$	D	$RT2D$	K	R
2	1.280	2.302	0.187	2.489	5.096	0.392
4	1.875	3.013	0.528	3.541	7.470	0.535
6	2.455	3.658	0.792	4.450	9.711	0.618
8	3.006	4.286	0.992	5.278	11.872	0.674
10	3.384	5.089	1.135	6.224	13.982	0.715

Table 5: Sequential Tasks Case, C=4

N	$RT1$	$RT2$	D	$RT2D$	K	R
2	1.313	1.934	0.857	2.791	5.497	0.364
4	1.999	2.489	1.872	4.361	8.560	0.467
6	2.658	3.101	2.440	5.541	11.327	0.530
8	3.264	3.443	3.234	6.677	14.137	0.566
10	3.841	3.738	3.983	7.721	16.804	0.595

Table 6: Multitasking Case, C=4

N	$RT1$	$RT2$	D	$RT2D$	K	R
2	1.317	2.390	0.321	2.712	5.381	0.372
4	1.998	3.253	0.990	4.243	8.408	0.476
6	2.668	4.056	1.551	5.607	11.339	0.529
8	3.256	4.938	1.867	6.805	14.069	0.569
10	3.756	5.657	2.299	7.956	16.744	0.597

Table 7: Relative Performance, C=1

N	D_{RT2}	D_{RT2D}	D_K	D_R
2	-6.04	3.46	3.02	3.11
4	-6.95	-1.10	0.33	0.33
6	-8.81	-5.31	-1.54	-1.52
8	-10.21	-8.27	-1.89	-1.86
10	-8.61	-10.04	-2.39	-2.34

Table 8: Relative Performance, C=2

N	D_{RT2}	D_{RT2D}	D_K	D_R
2	-14.87	4.74	3.48	3.60
4	-17.38	2.42	2.39	2.45
6	-18.24	-1.32	1.41	1.44
8	-19.92	-3.33	0.70	0.70
10	-23.31	-8.37	-0.35	-0.35

From the results we observe the following.

In all systems, both cases of multitasking and of sequential tasks perform almost the same as D_K and D_R are less than 4%.

In all cases $RT2$ is lower with the sequential case. This is due to the following.

(a) In the multitasking case, half of the jobs after arriving at the center Stage 2 split into two tasks where each one is assigned to a different processor queue independently of the system state. However, in the sequential tasks case there is an optimal job allocation to two queues as all jobs join the shortest queue.

(b) In the multitasking case there is the advantage of parallel processing of tasks of the same job at center Stage 2, but this incurs in synchronisation delay of tasks.

For all N, the relative difference in $RT2$ increases with increasing C. This is due to the increase in the variability of task sizes with increasing C which results in unbalanced processor queues at Stage 2 and

Table 9: Relative Performance, C=4

N	D_{RT2}	D_{RT2}	D_K	D_R
2	-23.62	2.84	2.12	2.16
4	-30.71	2.70	1.77	1.81
6	-30.80	-1.20	-0.10	-0.10
8	-43.43	-1.92	0.48	0.48
10	-51.36	-3.05	0.35	0.35

therefore in better exploitation of the "join the shortest queue" policy. However, because of the subsequent resequencing delay of jobs, for all C, the relative difference in $RT2D$ is much smaller than the relative difference in $RT2$.

For $C = 1$ and for all N the mean resequencing delay is relatively small compared to $RT2$. However, for $C > 1$ and for all N, D increases with increasing C and it becomes a considerable part of K especially in the case of jobs with sequential tasks. For example for $C = 4$ and $N = 10$, D is close to $K/4$ and it is bigger than $RT2$ (Table 5).

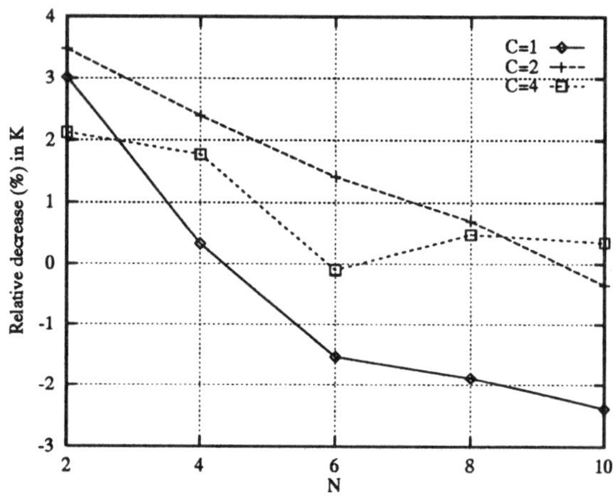

Figure 2: D_K versus N

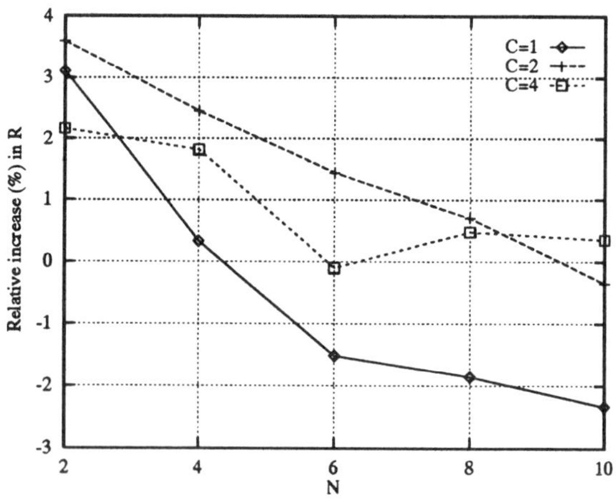

Figure 3: D_R versus N

This is due to the fact that the higher C is, the more the CPU service times vary from the mean service time m. Furthermore, more service times are produced that are much shorter than m and fewer ones that are much longer than m. So, when a processor at Stage 2 is busy for a long time processing a task with a long service time, the other processor can serve both tasks of other subsequent jobs which, however, must wait in the resequencing buffer until the big task is finished.

The resequencing delay is more serious for high degrees of multiprogramming because it is more probable at high N more subsequent jobs to go out of order.

In all systems for all N the performance deteriorates with increasing C (K increases and R decreases with increasing C).

5 CONCLUSIONS AND FURTHER RESEARCH

The effects of multitasking and resequencing on system and program performance of a two-stage multiprocessing system were studied using simulation techniques. The simulation results reveal the following.

- In the multitasking case the synchronisation of tasks in the multiprocessing center results in higher response times in this center compared with the case of sequential tasks. However, because the resequencing delay is higher with the case of sequential tasks, the two cases do not differ significantly.
- The resequencing delay of jobs increases with increasing C and also with increasing N.

In this research one class jobs have been considered. It should be extended to the following direction.

- Jobs of different classes to be considered depending on whether they consist of parallel or sequential tasks, and preemption.

ACKNOWLEDGMENTS

This work was carried out while the author was on sabbatical in the Department of Computer Science, at York University, Ontario, Canada.

REFERENCES

Iliadis, I., and Y. C. Lien. 1988. Resequencing Delay Distribution for a Queueing System with Two Heterogeneous Servers under Threshold Scheduling. In *Data Communication Systems and Their Performance*, ed. L. F. M. de Moraes, E. de Souza e Silva, and L. F. G. Soares, 359–373. Elsevier Science Publishers B.V., North-Holland, IFIP.

Jean-Marie, A. 1988. Load Balancing in a System of Two Queues with Resequencing. In *Performance '87*, ed. P.-J. Courtois, and G. Latouch, 75–88. Elsevier Science Publishers B.V., North-Holland.

Karatza, H. D. 1995a. Simulation Study of Multitasking and Resequencing in a Homogeneous Distributed System. In *Proceedings of the Eurosim Congress '95*, ed. F. Breitenecker, I. Husinsky, 541–546. Elsevier Science Publishers B.V., North-Holland.

Karatza, H. D. 1995b. Simulation Study of Two-Stage Parallel Processing Systems with Resequencing. In *Proceedings of the 7th European Simulation Symposium, ESS 95*, ed. M. DalCin, U. Herzog, G. Bolch, A. Riza Kaylan, 464–468. Society for Computer Simulation.

Law, A., and D. Kelton 1991. *Simulation Modeling and Analysis*. 2d ed. New York: McGraw-Hill.

Lien, Y. C. 1986. Evaluation of the Resequencing Delay in a Poisson Queueing System with Two Heterogeneous Servers. In *Computer Networking and Performance Evaluation*, ed. T. Hasegawa, H. Takagi, and Y. Takahashi, 189–197. Elsevier Science Publishers B.V., North-Holland, IFIP.

Persone, V. de Nitto, and G. Iazeolla. 1994. Performance Analysis of the Parallel Cyclic Two-Stage Queueing Model. *Performance Evaluation* 19: 167–193.

Sasase, I., and S. Mori. 1991. Resequencing Delay for a Queueing System with Multiple Servers under Threshold-Type Scheduling. In *Proceedings of the Tenth Annual Joint Conference of the IEEE Computer and Communications Societies*, 391-399, Vol.1, Institute of Electrical and Electronics Engineers.

Sauer, C. H., and K. M. Chandy. 1981. *Computer Systems Performance Modelling*, Englewood Cliffs, New Jersey: Prentice-Hall.

Takine, F., and T. Hasegawa. 1990. Resequencing Delay in Preemptive Priority M/M/2 Queues. In *Performance '90*, ed. P. J. B. King, I. Mitrani, and R. J. Pooley, 109–121. Elsevier Science Publishers B.V., North-Holland.

Varma, S. 1991. Optimal Allocation of Customers in a Two Server Queue with Resequencing, *IEEE Transactions on Automatic Control* 36 (11) :1288–1293.

AUTHOR BIOGRAPHY

HELEN D. KARATZA is an Assistant Professor in the Department of Informatics at Aristotle University of Thessaloniki, Greece. She has a B.S. degree in Mathematics and a Ph.D. degree in Computer Science from Aristotle University. Her research interests mainly include Performance Evaluation of Parallel and Distributed Systems and Simulation.

Proceedings of the 1996 Winter Simulation Conference
ed. J. M. Charnes, D. J. Morrice, D. T. Brunner, and J. J. Swain

SCHEDULING WITH SIMULATION IN THE FOOD & DRINKS INDUSTRY

Mike Novels

The CIMulation Centre Ltd
Cornbrash Park, Bumpers Way
Chippenham, Wiltshire
UNITED KINGDOM
SN14 6RA

Stan Jonik

SFJ Systems Ltd
Fernleigh House, Church
Street Trowbridge, Wiltshire
UNITED KINGDOM
BA14 8DY

ABSTRACT

The ever increasing pressures from the market place has caused companies to re-assess how the production planning function is carried out and to look for tools that will support the decision making process and generate achievable schedules in a timely manner. This paper describes the use of capacity planning and scheduling decision support tools that combine a simulation model generated with ARENA, with PREACTOR, a PC based finite scheduler. The use of this combination is illustrated by their use in food processing plant in the UK and then discusses how these may influence the way that variable costs are measured.

KEY WORDS

Manufacturing Applications
Scheduling
Food & Drinks Sector

INTRODUCTION

This paper describes the increasing use of decision support tools such as simulation to help in understanding problems in the production environment and to generate achievable schedules on a day-to-day basis. The text firstly covers the term 'finite capacity planning' and explains how this compares to traditional planning systems. This is followed by a case study of the installation of a simulation based process line scheduler installed in a food processing plant in the UK . Finally the benefits of using finite capacity scheduling and how this is translated into savings in variable costs is discussed.

CURRENT STATUS

Production planning and scheduling have always been an important element in maximizing the efficiency of plant. Inefficient planning and unclear objectives can lead to significant increases in variable costs. However the emphasis in terms of the way that schedules are produced, is changing. Currently, production planners see their goal as making best use of resources since standard methods of costing invariably makes 'maximum utilization of plant' the key cost driver. The pressure in the market place is now shifting to providing 100% service levels at shorter and shorter notice. This has shifted the production planner's emphasis from 'maximum resource utilization' to the elimination of stock-outs by making the right products at the right time. This may require a re-evaluation of traditional techniques such as batch consolidation, sequencing for minimum set-up times etc.. Now producers will be forced to examine the impact of running smaller and smaller production batches in order to react to variable demand. Attention is thus turning to decision support tools that can help production planners make decisions on a daily, if not hourly, basis.

The food and drinks industry, particularly the small to medium business, has undergone significant changes over the last 10 years. Many of these were controlled by the original owners of the business who had an intimate knowledge of the products and processes. As they retired or their businesses were acquired by larger organizations, they were replaced by managers with a more financial background, and in trying to deal with the day-to-day control of production, they tried the tools with which they were most familiar, invariably spreadsheets.

These managers then realized that a spreadsheet could not deal with the complexities of their business and deal adequately with such issues as capacity planning driven by forecast distributions that are used for setting stock control levels, shelf-life expiry of products in storage, and then provide a total vision of current stock at SKU (stock keeping unit) and total pallet level, detailed production work-to-lists, and dispatch waiting times.

Other systems on the market have proved to be expensive, require extensive consulting activity to install and maintain, and often do not provide an interactive interface that the planner can use to make fast changes in the event of unexpected changes in demand or plant performance.

FINITE CAPACITY PLANNING/SCHEDULING?

Finite capacity planning and scheduling is a process whereby a production plan consisting of a sequence of operations to fulfill orders is generated based on the *real* capacity of resources. Resources can be machines, tanks, piping, labor, storage areas, fork-lift trucks, delivery lorries, tooling or anything that could constrain production processes.

Most planning systems assume sufficient resources are available to produce batches when required, that is, resources have infinite capacity. Traditional planning systems take the orders, break them down into component parts or ingredients and then calculate when to launch the batch based on the individual lead times (perhaps adding adjustments for waiting time etc.) and when raw materials are required taking into account current stocks. No account is taken of the real capacity of resources to produce batches in a timely manner so that it does not matter if resources have a high utilization or not, the same lead time is used to calculate the launch time for any batch.

At the same time that the launch time for batches are calculated, the materials required are also ordered to arrive in time for work to start. If there is a delay in production upstream of a particular operation then the material will be ordered too early. With no concept of bottlenecks available to the planning system, resources become overloaded, queues of work get longer and work in progress increases. Because now, batches must join queues for resources at each process step, orders take longer to make progress through the process route, expected lead times are too optimistic and deliveries are late.

Using finite capacity scheduling, operations are only planned when resources are available. Materials are ordered only when they are needed for the operation to be carried out. Case studies have shown that inventory levels fall dramatically, key resources are better utilized and flow of work more controlled. Work in progress remains at a relatively constant level, lead times are more predictable and delivery dates more reliable. In this way production management spends less time progress chasing and fire fighting and can concentrate on the every day job of balancing, often variable demand, with the capacity available.

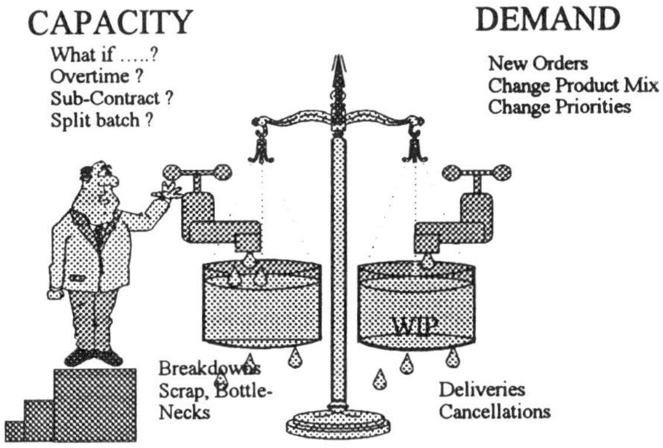

Figure 1 A question of balance

This is the art of the production controller. He/she must constantly try to maintain the balance between demand and capacity. On the demand side the 'flow' of work can vary due to new orders arriving, changes in priority, and stock orders launched against forward forecasts of demand. 'Leakages' take place as deliveries are made, forecasts are translated into real demand and orders canceled. The options available to the scheduler to increase capacity might include overtime, subcontracting, splitting batches, and avoiding bottlenecks, while leakages in capacity will occur due to machine breakdowns, operator absence, scrap, re-work and excessive changeover times. Any decision support tool must be able to support this balancing task.

CASE STUDY: PROCESS LINE SCHEDULER

This is an example of how Preactor, in combination with a discrete event simulation model, is used in a food processing plant in the UK. St. Ivel is a major food processing company. Management had recognized a problem with the scheduling of work at a plant making low fat spreads, particularly in the packaging and storage facilities. After processing, the product is passed to one of a number of filling lines. The packaged product is then passed to a quality assurance then held in a chill store. The product is removed from the chill store as orders are received.

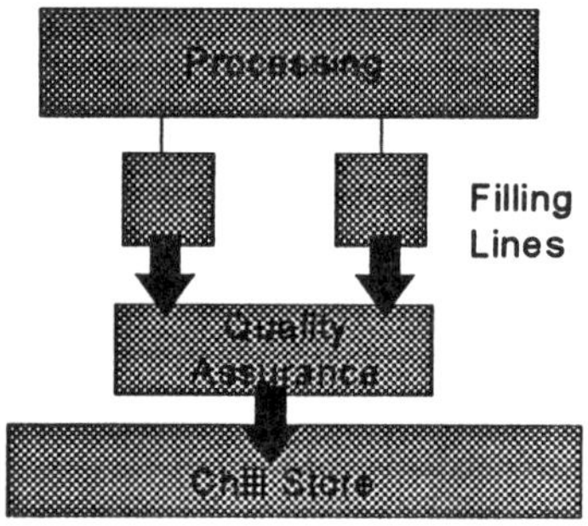

Figure 2. Process Schematic

The planner's primary task was to ensure that the utilization of the machines was high over a 66 hour cycle since this was the main constraint to production. Unfortunately the decisions taken took little account of current stock levels of the various product lines nor the capacity of the chill store. Sometimes the wrong products were being made at the wrong time. This resulted in unacceptable levels of stock out and/or overloading of storage space. The use of spreadsheets had been attempted but this could not take into account the complexities of the system. The requirements of the scheduler was to:-

- plan pallet storage
- control stock at case or part pallet level
- control stock replacement either by dynamic or fixed demand
- schedule changeover times between products
- plan manpower requirements

The solution was a combination of Preactor linked to an Arena** simulation model. The Preactor database holds information on the forecast orders, real orders and opening stock for each product. This information was held on another database and downloaded to Preactor over a PC network. An example of the initial stock view/edit screen is shown in Figure 3. This shows the number of SKUs at each stage of the Quality Control process and in the cold store.

The database also holds information on how to control the stock for each product. In this system three options were available:-

- fixed quantity replacement (FQR)
- dynamic replacement using minimum economic batch quantity (MEQ)
- Just-in-time replacement (JIT)

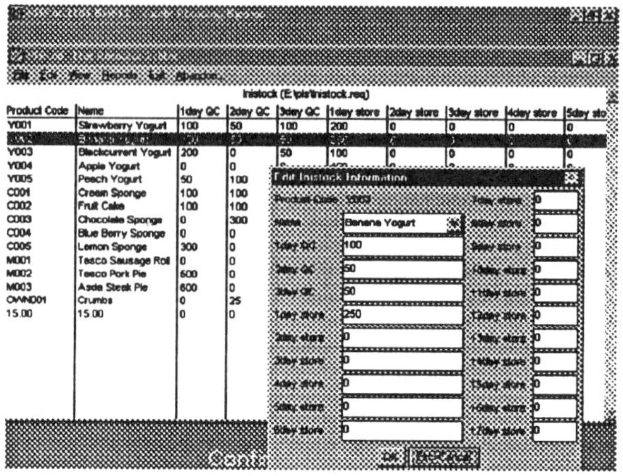

Figure 3. Stock Initialization View/Edit Screen

The FQR control rule is a conventional stock replacement rule where a threshold trigger level is defined and a fixed replacement quantity, MEQ, set. When stock is below this threshold, a demand equal to the replacement quantity is added to the manufacturing requirement.

The dynamic MEQ control again uses a threshold trigger and replacement MEQ value, but here it is calculated as multiples of MEQ (nearest to the order requirements) and in addition will use a forward forecast period, set by the user, to estimate the total manufacturing requirement over the period and generate demand accordingly.

The JIT control is used for controlling the processing of product through a Kanban system. Typically the Kanban level would be set at much lower thresholds than the MEQ used in the other two methods of control. As orders arrive which use up the products then this will trigger the Kanban quantity replacement in the manufacturing requirement that is generated.

Having set the current stocks, orders and forecast demand in the database, as well as the period to schedule in advance, the user runs the ARENA model to decide which batches should be scheduled using the rules described. These requirements are then used by Preactor to load them onto the production lines. The appearance of the scheduling screen is shown in figure 4.

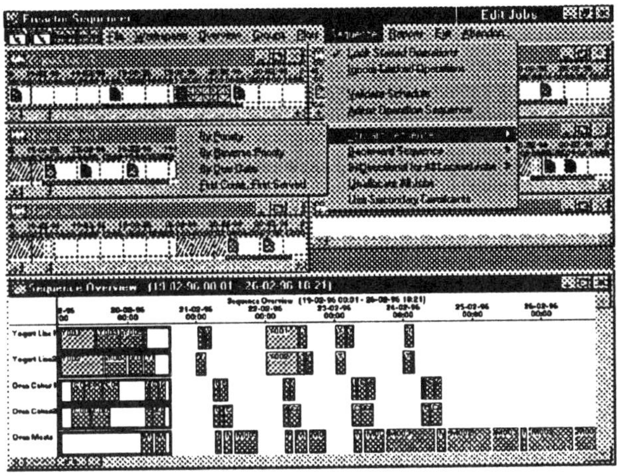

Figure 4. Scheduling Screen

The user has a range of batch sequencing options e.g. by priority, by due date etc. and it is here that various 'what if ...?' scenarios can be tried out, alternatives tried, alterations made by picking up batches with the mouse and moving them, before the decision is made.

Reports are available to the user on required manning, any problems of shelf-life, material requirements to meet the schedule, late dispatch and so on. A product demand report is shown in figure 5. This shows the total volume required by product base and the time span over which it is required.

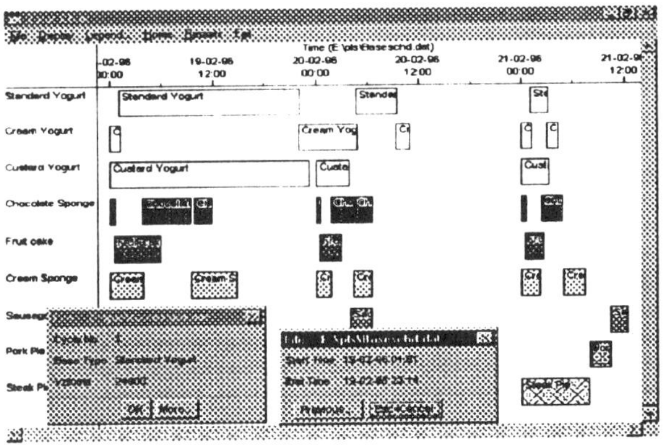

Figure 5. Product Demand Report

The user could also use the scheduler on a long term forecast basis typically a year ahead to assess the implications on production and to take decisions early to correct an imbalance between demand and capacity and in particular how they could maintain a high service level to customers.

COST v VALUE

The case studies are good examples of how management is having to re-assess traditional ways of doing things and introduce new methods that may be quite radical for the production planner but increasingly required in the ever more competitive world. No longer is the goal one of maximum utilization of machines. Now the key is giving customers a 100% service level which requires making the right products at the right time even if utilization of packaging machinery is lower than in the past. Using simulation and finite capacity scheduling has provided a vast improvement in decision making and enabled service levels to be improved dramatically.

Control of variable costs is obviously important in meeting the need to maintain good profit margins in any company. The cost of holding inventory is an important contributory element to variable costs and case studies have shown a major reduction in inventory when production is planned using finite resource constraints. This is an obvious tangible benefit.

However how can we measure the advantage of on-time deliveries ?

We have seen that pressures in the market place are changing and how some companies are reacting to this. The challenge now, perhaps, is how do you put a value on service. The old cliché that accountants '*know the cost of everything but the value of nothing*' springs to mind. What is the value of service ? Lost business, perhaps, or the ability to differentiate in the market place, not on price, but on ability to react to your client's changing demand. Some advantages of changing the approach to planning and scheduling of production may well be tangible while others, of equal importance, may be intangible. There will be pressure on managers to put a value on these intangible benefits as well as quantifying the cost of providing an increased quality of delivery.

AUTHOR BIOGRAPHIES

MICHAEL D. NOVELS is Chairman and Managing Director of The CIMulation Centre Ltd., a company based in the UK that provides software, support, training and consultancy in simulation and scheduling applications throughout Europe in a wide range of industrial and commercial sectors. He has an Honours degree in Engineering from the University of Bath, and has led a number of European and National initiatives in the use of simulation for finite scheduling applications in manufacturing and process industries.

STAN JONIK is an independent consultant who specializes in decision support tools for the food industry such as simulation, stock control and finite scheduling. Stan started his career in the engineering industry including a period with Hawker Siddeley where he took part in design studies for the implementation of manufacturing control systems such as MRP and Kanban. His experiences required extensive use of discrete event simulation for system design and extended into bespoke applications of simulation models used as a day-to-day operational tool. This led to the development of the PLS (Process Line Scheduler), a stock control and scheduling tool for food and drinks processing company sector.

Proceedings of the 1996 Winter Simulation Conference
ed. J. M. Charnes, D. J. Morrice, D. T. Brunner, and J. J. Swain

ACE: A DECISION TOOL FOR RESTAURANT MANAGERS

Todd C. Whyte
Darrell W. Starks

I•SYS Engineering
14419 Broad Oak Place
Louisville, Kentucky 40245-5136, U.S.A.

ABSTRACT

A simulation based decision support tool called ACE has been developed for Pepsico Restaurants International (PRI) allowing restaurant managers to quickly and accurately analyze a given restaurant. ACE gives the restaurant manager a tool to evaluate the customer point of contact within a given restaurant. This simulation based tool uses Arena for the model and Lotus 123 for the interface. PRI has installed this tool in two of the six international regions and will by early 1997 have it in all of the international regions.

1 INTRODUCTION

PRI needed a decision tool which could be used by the KFC restaurant managers. This tool had to be user friendly and at the same time had to produce results that could be used by a restaurant manager to determine the impact of changes in the customer arrivals upon the service time for the customers. The challenge was to produce a model that was accurate for a variety of restaurants and an interface that would allow a non operations research individual to parameterize the model and use the model results to analyze a given restaurant.

2 OBJECTIVE

The primary objective of this project was to develop a decision tool for the KFC restaurant manager which would be accurate for a given restaurant and would be useable by the restaurant manager to make specific decisions for that restaurant. In particular, a manager would be able to interface with the model in a user friendly environment, parameterize the model, run the model, and see the results in a timely manner. The primary decision to be made would be the number of restaurant personnel required to achieve a given service level for the customers, in particular the percent of customers that achieved this service level.

3 SYSTEM DESCRIPTION

The system to be modeled included the arrival of customers to the restaurant either at the inside front counter or at the drive through window. The front counter and drive through personnel performed the following tasks: take the order, pack the order, exchange money and hand-off the order.

Three different restaurant setups were considered at the front counter and at the drive through. At the front counter there could be only cashiers, cashiers with packers, or a pick system. In the case of the cashiers only the cashiers performed all the above tasks. In the case of the cashiers with packers the cashiers performed the take order and money exchange tasks while the packers performed the packing and hand-off tasks. In this case the ratio of cashiers to packer could be one to one or two to one respectively. The pick system case was the same as the cashiers only except instead of packing, the items were prepacked, requiring less time than the packing operation in the first case. For the drive through window the first case had only one cashier who performed all the tasks. In the second case a cashier performed all the tasks except the packing which was performed by a packer. Finally in the third case called split cashiering, a cashier took the order, a packer did the packing of the order, and a third person did the money exchange and hand-off. For each individual restaurant the front counter and drive through window can be configured in any of the three ways.

4 ARENA MODEL DESCRIPTION

The models were developed in the Arena modeling framework. There are six basic models: a cashiers only model with a straight counter configuration and an angled counter configuration, a cashiers and packers model with a straight counter configuration and an angled counter configuration, and a pick model with a straight counter configuration and an angled counter configuration. Figures 1 and 2 illustrate the two different counter configurations.

The model looked only at the front counter and drive through operations with the customers. The model assumed that product is always available. Each customer transaction consisted of one to five orders based upon historical probabilities. Because packing times depend upon the type of order, eight meal type categories were developed:
small meal, large meal, small snack, large snack, sandwich, strips/wings, pot pie, and other. Depending upon the region all may or may not be used. The probability for each category was dependent upon the menu mix which is input by the user.

Each model consists of two primary modules: the front counter module and the drive through module. Two other modules read the model parameters from a file created by the interface and write out specific output values to a file to be read by the interface to create the user output.

The front counter module models the arrival of customers to the front counter, the placing of the order, the money exchange, the packing(picking) of the order, and the hand-off of the order. The times for each of the above tasks were developed from time studies performed at various restaurants from various regions. Some of the task times have been found to be regionally dependent. The user inputs the following data to parameterize the front counter: length of peak period, number of arrivals per ten minute interval, type of model, number of registers, and capacity of queuing area.

The drive through module models the arrival of cars to the menuboard where the customer places the order, the movement of the

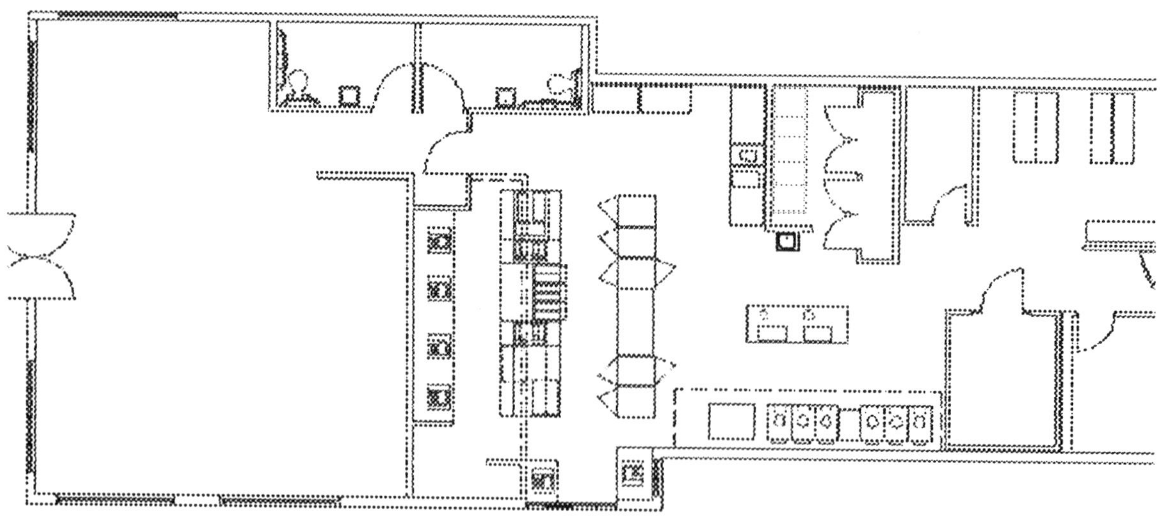

Figure 1 - Straight Front Counter

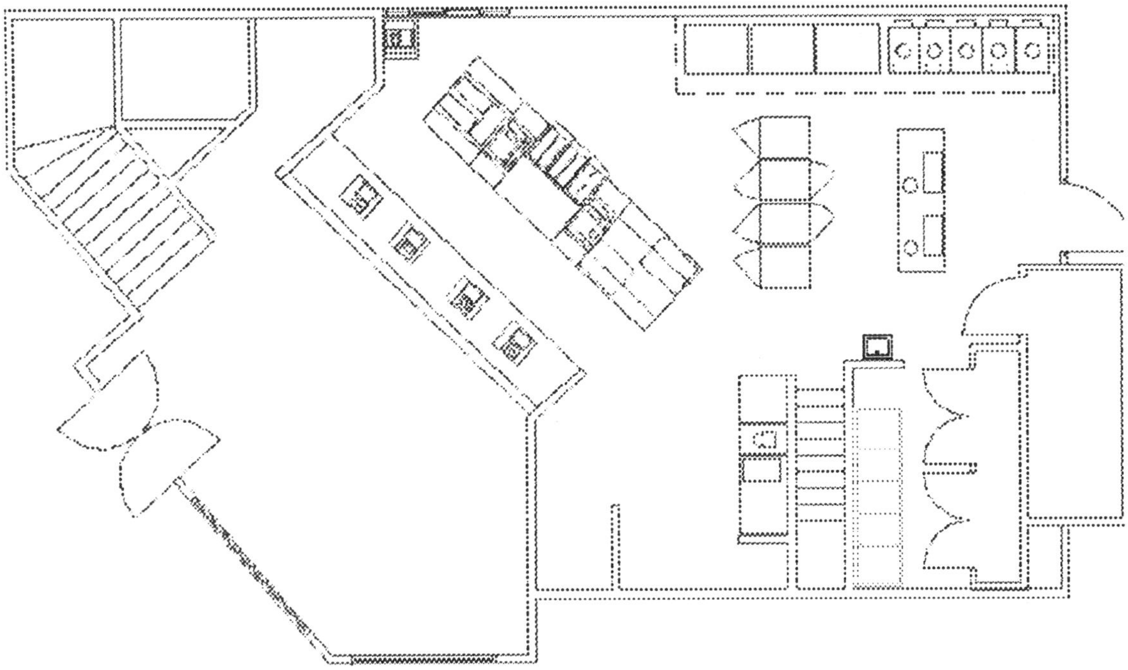

Figure 2 - Angled Front Counter

cars to the drive through window, the packing of the order, the money exchange, the hand-off of the order, and the movement of the car out of the system. As in the front counter logic the times for the above tasks were developed from time studies and some have been found to be regionally dependent. The user inputs the following data to parameterize the drive through window: length of peak period, number of arrivals per ten minute interval, type of model, capacity of queue before the menuboard, and the capacity of the queue between the menuboard and the window.

The model animation was easily achieved by importing a cad drawing as the background and building the model around this drawing. The user has the choice of either animating or not. The animation is also illustrated in Figures 1 and 2.

5 USER INTERFACE

As stated previously the intended user of the simulation will be restaurant managers and other field personnel with little or no knowledge of simulation modeling. Thus the people who could benefit from simulation should be able to perform a simulation experiment when and where it is needed. In lieu of developing an expensive training program to instruct personnel on how to load the parameters, execute and evaluate simulations, our philosophy was to develop an interface to reduce the skill level required to use the models.

All of the intended users were assumed to have a working knowledge with various Windows applications, through normal job routines (e.g., corporate reports, communications, etc.) The software that every user can access is Lotus SmartSuite, in particular the Lotus 123 package. Thus, if an interface could be developed with Windows constructs (pop-up windows, dialog boxes, list boxes, etc.) then the conjecture was that implementation and acceptance in the field should be greatly enhanced. The language that was selected for the initial development of the interface was the Lotus macro programming language.

A general purpose programming language was initially proposed for the development of the interface; however, the spreadsheet macro was ultimately selected by the design team. The advantages of the macro language compared to a general purpose language were rapid development, lack of additional software costs, and increase user acceptance (i.e., the application will be perceived as "just another spreadsheet"). The development of the interface using macro approach proved to be more difficult than originally thought due in part to several limitations in the macro language itself. However, despite the few difficulties encountered in the design phase, an effective interface was developed and implemented.

The interface is driven by the user selecting one of the various buttons that call a particular routine. The input screens (a total of two exist) are composed of approximately 16 buttons and has been illustrated in Figure 3. To begin an input session the user simply clicks on the following buttons:

- **Peak Duration** - activates a list box that allows the user to specify the duration of the simulation (from 10 minutes to three hours in 10 minute increments).
- **Menu Mix** - reveals a worksheet that allows the user to input actual menu items sold
- **Ticket Average** - average transaction amount used to calculate potential income based on the number of balks.
- **Registers** - activates a list box that allows a user to select the number of registers on the front counter.
- **Drive-through** - activates a dialog box to specify the presence of a drive-through window.
- **Worker Assignment** - activates a dialog box that is used to specify customer service worker assignments (a separate button is used for both front counter and drive-through).
- **Self - Serve Drinks** - activate a dialog box to specify yes/no
- **Queue Lengths** - activates text boxes that allow the user to specify queue capacities for front counter, drive-through menuboard and drive-through window.
- **Arrival Patterns** - reveals a worksheet that is

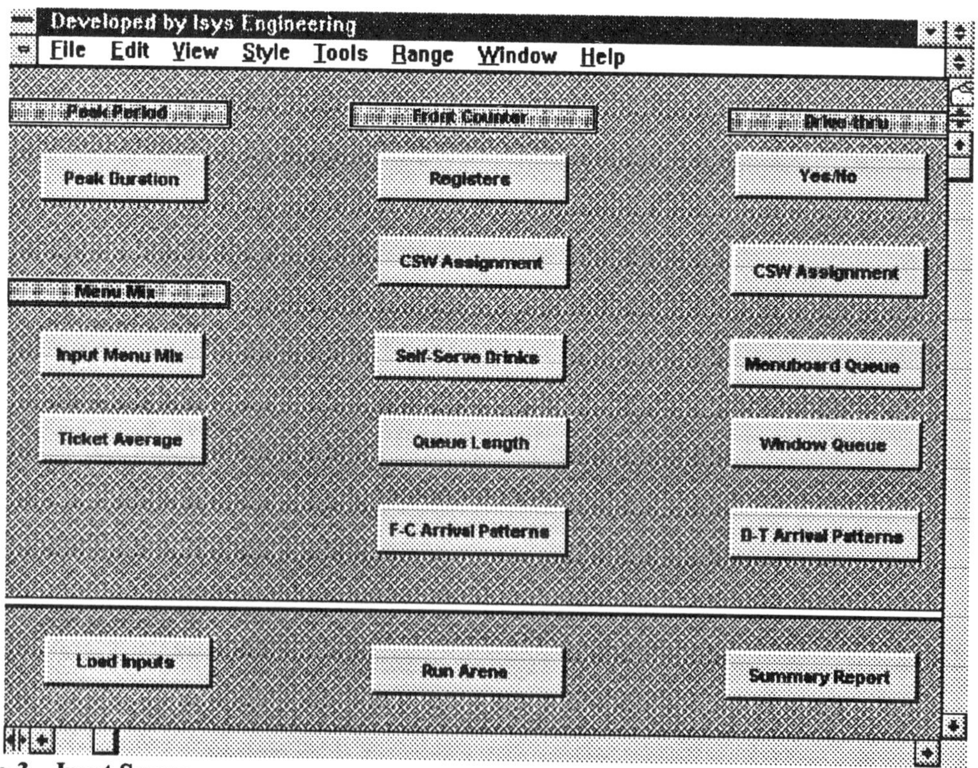

Figure 3 - Input Screen

used to input the number of customers that arrived in 10 minute increments, for the duration of the simulation.

- **Load Inputs** - takes the inputs that were input and writes to an ASCII data file.
- **Run Arena** - based upon the inputs, the button selects the proper model and launches Arena .
- **Summary Report** - used to review the statistics after Arena has been executed

Once Arena has been successfully launched with the run Arena button, the user simply clicks on the appropriate Arena control to execute the model. After the simulation has run the specified number of runs, the simulation session terminates and the user can exit Arena. The interface resumes control and the input screen (Figure 2) reappears, permitting the user to select the Summary Report button to expose the customer statistics that were collected, as illustrated by Figure 4.

In addition to the customer statistics, server statistics are collected (by position description) as illustrated in Figure 5. An additional section was created on the output report to attempt to capture lost revenue. The total number of balks are multiplied by the ticket average and then annualized. This monetary measure serves as an additional performance measure for performing a what-if style of analysis. To facilitate this iterative style of analysis, a button has been created to input a new session (e.g., alter menu mix, customer service assignments, number of registers, etc.). This iterative analysis gives the restaurant manager a tool to determine staffing levels and worker assignments to better serve the customer.

6 Conclusion

The initial implementations have been successful, with minimal amount of training time (usually much less than an eight hour day). The user

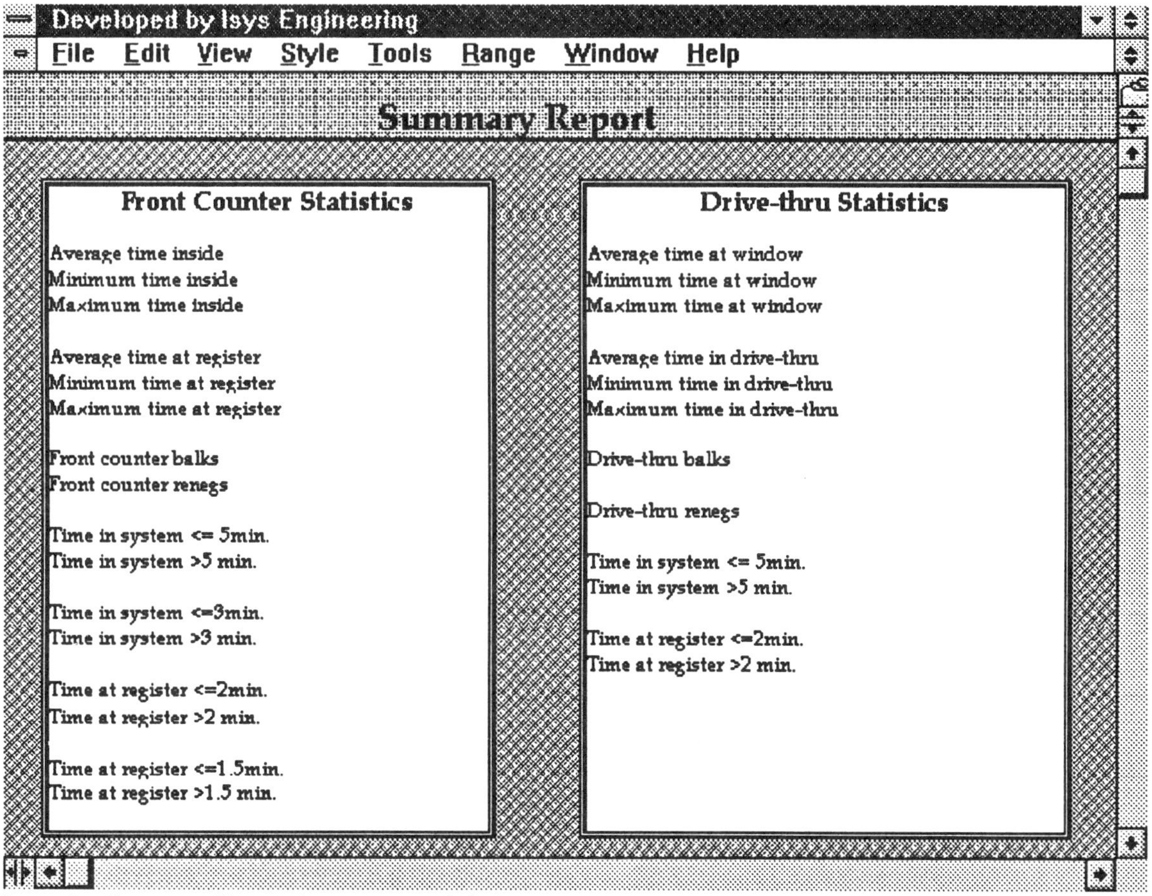

Figure 4 - Summary Report (customer statistics)

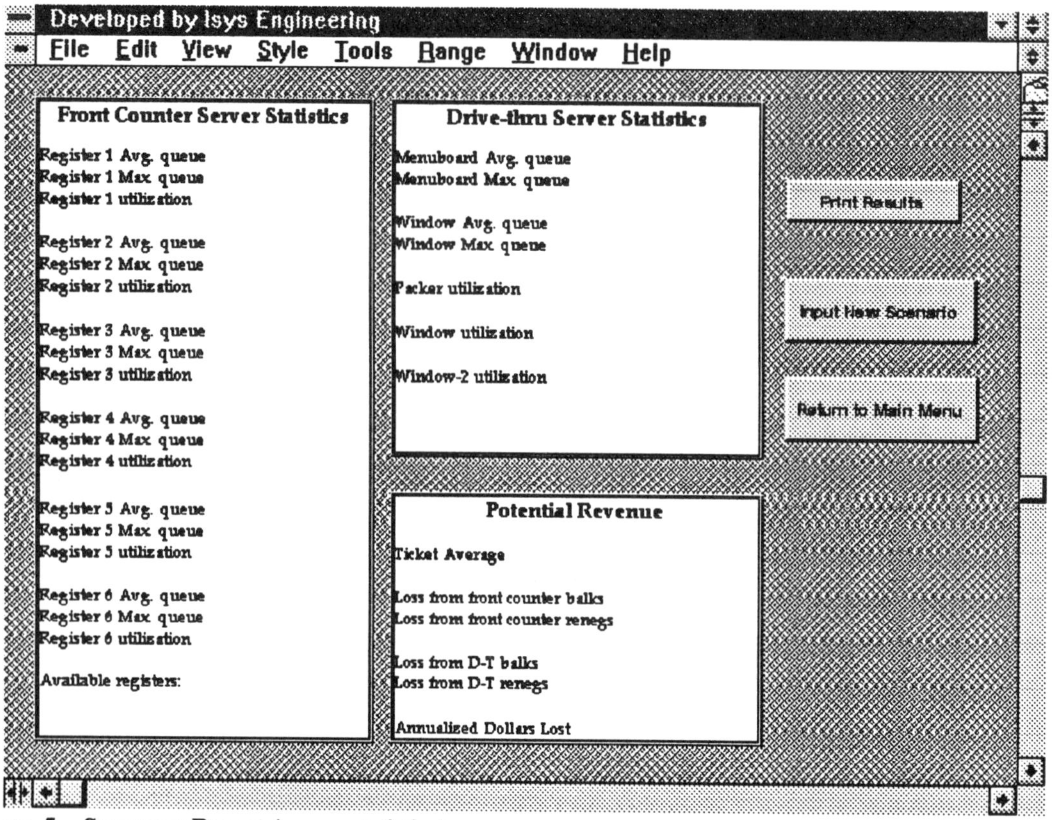

Figure 5 - Summary Report (server statistics)

responses have been extremely favorable, and on occasion have helped shaped the final appearance of the interface. The design team also was privileged enough to experience complete acceptance (and in some instances enthusiasm) throughout the senior management of PRI.

Although this initial effort was considered a success, there is room for improvement. The first improvement which is currently underway is a complete restaurant simulation for an entire day of operations. This added detail will encompass aspects to normal restaurant operations; that is, product shortages, congestion in the kitchen, blocking of critical equipment, etc.

The second area for continuing improvement is the interface. Given the modifications that will be necessary to interface with the new model, a general purpose language was selected for the interface development. The language that was ultimately selected was VisualBasic for Windows 95. Despite changing the language, the interface should be similar to the current version so any retraining will be minimal.

The final area for further development is the statistical analysis of the output data. Currently the summary report consists of means and does not yet compute confidence intervals, or measures of dispersion. In addition to enhanced summary statistics, a procedure is being developed to compare the statistical significance between alternative scenarios.

ACKNOWLEDGEMENTS

The success of this work would not have been possible without the funding and guidance provided by Mr. Ray Lantz and Mr. Ron McGinley, both of PRI.

AUTHOR BIOGRAPHIES

Todd C. Whyte is a senior project engineer with I•SYS Engineering of Louisville and is currently a doctoral candidate at the University of Louisville, Department of Industrial Engineering. His research interests include simulation, and facilities layout

Darrell W. Starks is a senior consultant with I•SYS Engineering of Louisville and is currently a doctoral candidate at the University of Louisville. Mr. Starks has worked extensively in the application of simulation technology to real world applications while at McDonald-Douglas Corporation, Pritsker Corporation, Biles & Associates, and I•SYS Corporation. Application areas include manufacturing, transportation, logistics, hospitals, and restaurants. Mr. Starks research interests are in the area of interfacing simulation to optimization techniques for real world application.

Proceedings of the 1996 Winter Simulation Conference
ed. J. M. Charnes, D. J. Morrice, D. T. Brunner, and J. J. Swain

SIMULATION AND ANIMATION OF THE OPERATION OF A FAST FOOD RESTAURANT

Kambiz Farahmand

Texas A&M University-Kingsville
Department of Mechanical & Industrial Engineering
Campus Box 191
Kingsville, Texas 78363

Alejandro Francisco Garza Martinez

Componentes Mecanicos
De Matamoros, S.A. De C.V.
P.O. Box 4447
Brownsville, Texas 78523

ABSTRACT

A model was developed for the Simulation and Animation of the Drive-Thru (DT) and the lobby sections of a fast food restaurant. A discrete, non-continuous, parallel simulation of the operation using animation in the Witness environment has made it possible to utilize real time data or forecasted data to optimize scheduling and maximize operation efficiency. Performance measures such as the number of customers balking from the system, the average time each customer spends in the system, the average waiting time for the customers, the average queue length before and after placing an order, the average time to fill an order and even the average utilization of each employee could be determined.

The results could serve as basis to provide recommendations on how to improve efficiency and throughput for the fastest growing industry and the largest employer in the US. The modeling methodology and the tracking mechanism could also prove useful for the manufacturing or other service industries employing similar queuing techniques.

1 INTRODUCTION

In its broadest sense, computer simulation is the process of designing a mathematical-logical model of a real system and experimenting with this model on a computer. Thus simulation encompasses a model building process as well as the design and implementation of an appropriate experiment involving that model. These experiments, or simulations, permit inferences to be drawn about systems and their operation without actual construction of the system. Decisions regarding the capacity requirements are difficult, since it is often impossible to accurately predict when units will arrive to seek service and/or how much time will be required for the service provided. Allocating more resources than necessary would involve excessive costs. On the other hand not providing enough resources would increase queue size. Therefore our goal is to achieve an economic balance between the cost of service and the cost associated with waiting for that service.

In the DT operation, customers (cars) arrive at the station, pull up to the DT sign, place an order, form a queue before and after placing an order if there are any customers seeking service, pay and receive the order and leave the system. In the lobby section, customers go to the lobby, walk to the cashiers area, form a queue to order if there are any other customers seeking service, place an order, pay, wait to receive the order, go to tables area, eat and leave the system. This paper presents the modeling of the above scenario in a realistic but simplified animation network. Operation recommendations have been made to optimize operation, labor and resource utilization, improve quality and customer service, and increase efficiency.

2 MODEL BUILDING

WITNESS was selected as the simulation software to be used for the simulation and animation of the operation of a fast food restaurant due to its applications strengths such as manufacturing, service and process environments, hardware requirements such as a personal computer running the Microsoft Windows and relatively low cost. WITNESS provides facilities for building models of great complexity, but the principles of building a WITNESS model are simple. Using three commands from the Model menu, you can start building useful models.

After defining, displaying and detailing the elements of the model (see Figure 1), it can run immediately and modified by adding, changing or deleting elements.

This ability to build a model incrementally, testing each section as you go, is a powerful aid to productivity, and generates confidence in the validity of the model. The model could be run in a variety of modes, from step-by-step with full screen display to a "batched" time in the future, with no screen display.

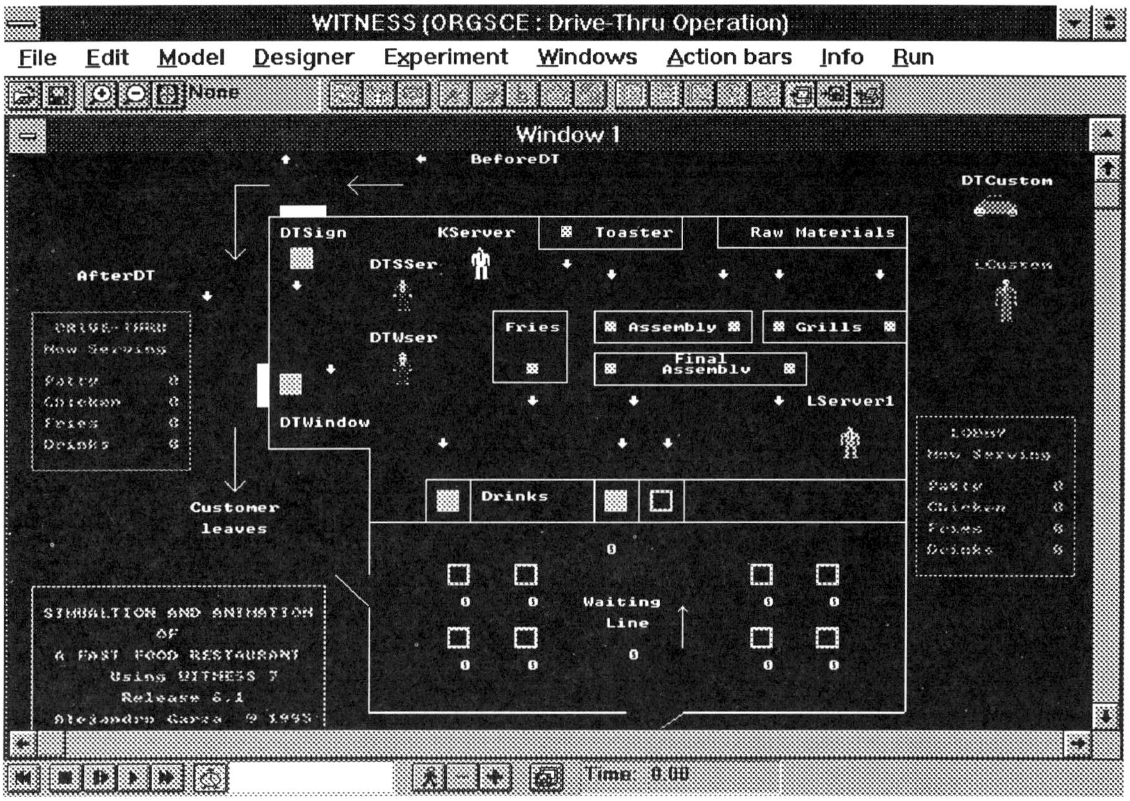

Figure 1: Main Screen

2.1 Model Structure and Details

2.1.1 Parts

DTCustom (Drive-Thru customers)
Part defined to identify Drive-Thru customers.
LCustom (Lobby customers)
Part defined to identify Lobby customers.
Patty (Patty burger)
Part defined to identify Patty burgers.
Chicken (Chicken burger)
Part defined to identify Chicken burgers.
Drinks (Soft drinks)
Part defined to identify soft drinks.
Fries (French fries)
Part defined to identify French fries.
Buns (Buns)
Part defined to identify Buns

2.1.2 Machines or Stations

PMac (Patty machine)
Machine or station where the Patties are cooked or grilled.
CMac (Chicken machine)
Machine or station where the Chicken is cooked or grilled.
AMac (Assembly machine)
Machine or station where initial assembly is made of bun sets and patties
AMac2 (Assembly machine 2)
Machine or station where initial assembly is made of bun sets and Chicken
FAMac (Final assembly machine)
Machine or station where the final assembly of Patty burgers is made.
FAMac2 (Final assembly machine 2)
Machine or station where the final assembly of Chicken burgers is made.
FMac (Fries machine)
Machine or station where French fries are cooked or fried.

TMac (Toaster machine)
Machine or station where the buns are cooked preparing them for burgers assembly.
DTSign (Drive-Thru sign)
Machine or station where Drive-Thru customers place their orders.
DTWindow (Drive-Thru window)
Machine or station where Drive-Thru customers pay and pick up orders.
LCash (Lobby cashier)
Machine or station where Lobby customers place and pay their orders.
LDel (Lobby delivery)
Machine or station where Lobby customers receive the orders placed after paying.
Lobby (Lobby)
Machine or station where Lobby customers eat after paying and receiving their orders.
DMac (Drinks machine)
Machine or station where soft drinks are served or dispensed.

2.1.3 Buffers or Queues

BeforeDT (Before Drive-Thru)
Waiting line or buffer before placing an order at the Drive-Thru sign.
AfterDT (After Drive-Thru)
Waiting line or buffer after placing an order at the Drive-Thru sign.
LCBuf (Lobby)
Waiting line or buffer formed by the Lobby customers before placing an order at the Lobby cashier.

2.1.4 Labor or Employees

LServer1 (Lobby server 1)
Server or employee dedicated to operate or attend the Lobby cashier, Fries machine, Drinks machine, Final Assembly of Patty and Chicken burgers and delivery of orders to Lobby customers.
KServer (Kitchen server)
Server or employee dedicated to operate or attend the Patty machine or grill, Chicken machine or grill, Toaster machine, initial assembly Patty machine and initial assembly Chicken machine.
DTSSer (Drive-Thru sign server)
Server or employee dedicated to operate or attend the Drive-Thru sign, Fries machine and Toaster machine.
DTWser (Drive-Thru window server)
Server or employee dedicated to attend the Drive-Thru window, Fries machine, Toaster machine and final assembly of Patty burgers.

2.2 Measuring Performance

2.2.1 Drive-Thru

- Drive-Thru customers leaving because of full queues
- Average time Drive-Thru customer spent in the system
- Average waiting time for the Drive-Thru customer before ordering
- Average waiting time for the Drive-Thru customer after ordering
- Average queue length before placing an order at the Drive-Thru
- Average queue length after placing an order at the Drive-Thru
- Average time to fill an order at the Drive-Thru

2.2.2 Lobby

- Lobby customers leaving because of full queues
- Average time Lobby customers spent in the system
- Average waiting time for the Lobby customer before ordering
- Average queue length before placing an order at the Lobby
- Average time to fill an order at the Lobby

2.2.3 Labor

- Average utilization of Lobby employee
- Average utilization of Kitchen employee
- Average utilization of Drive-Thru window employee
- Average utilization of Drive-Thru sign employee

3 SENSITIVITY ANALYSIS

This sensitivity analysis contains the original scenario and seven alternate scenarios where modifications to the original assumptions were made. The file name is the name under which each scenario was saved. The conditions presented are only the outstanding conditions that have been changed from the original scenario. Table 1, sensitivity analysis summary, does not specify units for the input data. The model is valid as long as all units are presented consistently. The following fields in this summary contain the evaluated factors and their outputs under each scenario provided by the corresponding simulation reports.

- Drive-Thru customers leaving because of full queues
- Lobby customers leaving because of full queues

- Average time Drive-Thru customer spent in the system
- Average time Lobby customers spent in the system
- Average waiting time for the Drive-Thru customer before ordering
- Average waiting time for the Drive-Thru customer after ordering
- Average waiting time for the Lobby customer before ordering
- Average queue length before placing an order at the Drive-Thru
- Average queue length after placing an order at the Drive-Thru
- Average queue length before placing an order at the Lobby
- Average time to fill an order at the Drive-Thru
- Average time to fill an order at the Lobby
- Average utilization of Lobby employee
- Average utilization of Kitchen employee
- Average utilization of Drive-Thru window employee
- Average utilization of Drive-Thru sign employee

Table 1: Sensitivity Analysis Summary

Simulated Scenario	Original	Case 1	Case 2	Case 3	Case 4	Case 5	Case 6	Case 7
File Name	ORGSCE	CASE1	CASE2	CASE3	CASE4	CASE5	CASE6	CASE7
Conditions		No Balking	No Balking 2 DT Window Servers 2 DT Windows	Balking 2 Lobby Servers 2 DT Window Servers	Balking LBuf=14 BeforeDT = 14	Balking 2 DT Window Servers 2 DT Windows	Under capacity	Over capacity
Drive-Thru customers leaving because of full queues	166	309	269	134	161	155	449	0
Lobby customers leaving because of full queues	233	84	79	218	220	221	2412	0
Average time Drive-Thru customer spent in the system	53.41	165.24	106.64	46.27	78.85	51.06	72.84	10.10
Average time Lobby customers spent in the system	63	56.01	54.8	54.94	86.4	65.26	62.98	59.11
Average waiting time for the Drive-Thru customer before ordering	112.69	114.24	71.05	98.06	178	84.71	114.00	0
Average waiting time for the Drive-Thru customer after orderin	33.06	33.62	19.5	28.34	31.99	23.89	33.06	0.21
Average waiting time for the Lobby customer before ordering	39	28.02	25.78	31.79	66.17	41.55	47.44	0

Average queue length before placing an order at the Drive-Thru sign	7.78	7.77	7.67	7.75	13.53	7.71	7.87	0
Average queue length after placing an order at the Drive-Thru sign	1.98	1.98	1.95	1.98	1.98	1.96	1.98	0.01
Average queue length before placing an order at the Lobby	7.02	4.88	4.69	6.93	12.44	7.06	7.78	0
Average time to fill an order at the Drive-Thr	42.18	42.77	28.78	37.62	41.13	32.94	156.21	9.37
verage time to fill an order at the Lobby	13.08	13.34	13.2	13.42	13.47	13.4	60.69	13.42
Average utilization of Lobby employee	100	100	99.98	59.35	100	100	99.93	75.77
Average utilization of Kitchen employee	14.94	14.61	20.47	19.07	14.72	19.2	15.59	11.14
Average utilization of Drive-Thru window employee	100	99.98	88.69	62.87	99.95	74.03	100	67.26
Average utilization of Drive-Thru sign employee	38.34	38.12	50.71	40.6	37.12	44.24	36.66	26.47

4 ORIGINAL SENARIO

Scenario Description: This project consists of the Simulation and Animation of the Drive-Thru (DT) and the lobby sections of a fast food restaurant. In the DT operation, customers (cars) arrive at the station, pull up to the DT sign, place an order, form a queue before and after placing an order if there are any customers seeking service, pay and receive the order and leave the system. In the lobby section, customers go to the lobby, walk to the cashiers area, form a queue to order if there are any other customers seeking service, place an order, pay, wait to receive the order, go to tables area, eat and leave the system.

This operation is served by four employees or servers which are assigned to the following main areas of responsibility: Drive-Thru sign, Drive-Thru window, Kitchen and Lobby. However, there are shared responsibilities for each station or machine in the restaurant based on pre-assigned priorities. There are also other stations with a dedicated server.

Results: Simulation results show that the amount of Lobby customers rejected is larger than Drive-Thru customers. Average time in the system is larger for Lobby customers than Drive-Thru customers. All queues lengths are almost always at full capacity. The Lobby is by far the station which is idle the larger amount of percentage of time. The kitchen and Drive-Thru sign employees are idle the largest amount of percentage of time. The Lobby and Drive-Thru window employees are always busy.

4.1 Case 1

Scenario Description: This scenario presents the case where the Drive-Thru customers do not balk to the Lobby if the queue in the Lobby is full. This condition is to be maintained even if the queue is not full to the capacity.

Results: Simulation results show that the amount of Drive-Thru customers rejected is larger than Lobby customers. The average time in the system is larger for the Drive-Thru customers than Lobby customers. The queue lengths for the Drive-Thru section are almost always full but the queue at the Lobby is not full all the time. The Lobby is by far the station which is idle most of the time. The kitchen and Drive-Thru sign employees are idle the largest amount of percentage of time. The Lobby and Drive-Thru window employees are always busy.

4.2 Case 2

Scenario Description: This case presents a scenario where Drive-Thru customers are not allowed to go to the Lobby if the queue for ordering is full. There are two Drive-Thru windows and two employees to serve them. These conditions are to be maintained even if the queue at the Lobby can accept more customers.

Results: Simulation results show that the amount of Drive-Thru customers rejected is larger than Lobby customers. The average time in the system is larger for the Drive-Thru customers than Lobby customers. The queue lengths for the Drive-Thru section are almost always full but the Lobby queue is not full all the time. The Lobby is the station with the largest percentage of idle time followed by the Lobby delivery station and Drive-Thru windows. The kitchen employee utilization is very low followed by the Drive-Thru sign employee. The Lobby server is busy almost all the time.

4.3 Case 3

Scenario Description: This case presents a scenario similar to the original problem but there are two employees at the Lobby and two employees at the Drive-Thru window. Balking from the Drive-Thru sections to the Lobby is allowed again.

Results: Simulation results show that the amount of Lobby customers rejected is larger than Drive-Thru customers. The average time in the system is slightly larger for Lobby customers than Drive-Thru customers. All queue lengths are almost full all the time. The Lobby station is idle about half the time. All employees utilization are very poor.

4.4 Case 4

Scenario Description: This case presents a scenario where the queues before ordering at the Drive-Thru and the Lobby have capacity for 14 customers. Balking from the Drive-Thru sections to the Lobby is allowed again.

Results: Simulation results show that the amount of Lobby customers rejected is larger than Drive-Thru customers. The average time in the system is slightly larger for Lobby customers than Drive-Thru customers. All queue lengths are almost full all the time. The Lobby station is idle about half the time. The Lobby and Drive-Thru window employees are always busy. The Kitchen and Drive-Thru sign employees are very poorly utilized. See Appendix E for more detailed simulation results.

4.5 Case 5

Scenario Description: This case presents a scenario where Drive-Thru customers are allowed to go to the Lobby if the ordering queue is full. There are two windows at the Drive-Thru and two employees to serve them. The remaining details of the original scenario are maintained.

Results: Simulation results show that the amount of Lobby customers rejected is larger than Drive-Thru customers. The average time in the system is slightly larger for Lobby customers than Drive-Thru customers. All queue lengths are almost full all the time. The Lobby and Drive-Thru window stations are idle a very significant amount of time. The Lobby employee is always busy. The Kitchen and Drive-Thru sign employees are very poorly utilized.

4.6 Case 6

Scenario Description: This case presents a scenario where Drive-Thru customers and Lobby customers arrive in such high frequency rate that the operation becomes insufficient due to under capacity in relation to the demand. The remaining details of the original scenario are maintained.

Results: Simulation results show that the amount of Lobby customers rejected is larger than Drive-Thru customers. However average time in the system is larger for Drive-Thru customers than Lobby customers. All queue lengths are almost at maximum capacity. The Lobby station has the highest percentage of idle time from all stations. The Kitchen and Drive-Thru sign employees are idle most of the time and Lobby and Drive-Thru window employees are almost always busy..

4.7 Case 7

Scenario Description: This case presents a scenario where Drive-Thru customers and Lobby customers arrive in such low frequency rate that the operation becomes oversized due to over capacity in relation to the demand. The remaining details of the original scenario are maintained.

Results: Simulation results show that there are not customers rejected because of full queues. The average time in the system is larger for Lobby customers than Drive-Thru customers. All queues are normally empty. The Drive-Thru sign, Lobby cashier and Lobby station have the highest percentage of idle time. The Kitchen and Drive-Thru sign employees have the highest percentage of idle time.

5 CONCLUSIONS

5.1 Lobby Section

The number of Lobby customers leaving the system because of full queues or waiting lines is larger when the Drive-Thru customers are allowed to balk to the Lobby. The original scenario, Case 3, Case 4 , Case 5 and Case 6 represent this condition. The scenario introduced in Case 7 is an exception of this behavior due to the over capacity.

The average time Lobby customers spent in the system is not significantly affected unless the queues or waiting lines are made larger such as in Case 4. Scenarios represented in Case 2 and Case 3 generate the smallest times in the system.

The average waiting time for the Lobby customers before ordering is smaller if Drive-Thru customers are not allowed to balk to the Lobby. Scenarios represented in Case 1 and Case 2 generate the smallest times in the system.

The average time to fill an order at the Lobby does not change significantly under all simulated scenarios at the given conditions. Other factors need to be modified and analyzed in order to reduce the average time.

5.2 Drive-Thru Section

The number of Drive-Thru customers leaving the system because of full queues or waiting lines is larger when balking to the Lobby is not allowed such as in Case 1 and Case 2 and when the under capacity scenario is presented such as in Case 6. This means that the Drive-Thru section of the restaurant is not able to satisfy the demand for the Drive-Thru arrival rate.

The average time Drive-Thru customers spent in the system is larger when balking is not allowed to the Lobby. The average time in the system is reduced when balking is allowed because of the smaller times spent in the system by the Drive-Thru customers becoming Lobby customers. The average time is minimized when there are 2 Lobby servers and 2 Drive-Thru window servers such as in Case 3. Scenarios represented by Case 1 and Case 2 show these worst case scenarios.

The average waiting time for the Drive-Thru customer before ordering is not dependent on balking or not balking to the Lobby. This observation can be explained by the spread presented by cases with and without balking. If 2 Drive-Thru windows and 2 Drive-Thru window servers are available this time is reduced such as in Case 2. This time is directly proportional to the queue length. If the queue length is increased then this time also increases.

The average waiting time for the Drive-Thru customers after ordering is not dependent on balking or not balking to the Lobby. If 2 Drive-Thru windows and 2 Drive-Thru window servers are available this time is reduced. Scenario represented in Case 2 generates the smallest time in the system. The average queue lengths before and after placing an order at the Drive-Thru are full in all simulated scenarios at the given conditions. Other factors need to be modified and analyzed in order to reduce the queue lengths. The over capacity scenario presented by Case 7 is an exception to this condition.

The average queue length before placing an order at the Lobby is reduced if Drive-Thru customers are not allowed to balk to the Lobby. Otherwise, this queue length is almost full in all simulated scenarios. Other factors need to be modified and analyzed in order to reduce the queue length. The over capacity scenario presented by Case 7 is an exception to this condition.

The average time to fill an order at the Drive-Thru is independent of balking or not balking. The average time is reduced if 2 Drive-Thru windows and 2 Drive-Thru window servers are available. Scenario presented in Case 5 represents this condition.

5.3 Labor

The average utilization of the Lobby server is normally 100 %. If 2 Lobby servers are present, individual utilization is about 60 %. This means that 2 Lobby servers are needed to maintain the operation more effectively, unless work load is reassigned to somebody else.

The average utilization of the Kitchen server runs about 20 % in all simulated scenarios. The Kitchen server can carry additional workload easily. If additional workload is transfer from the Lobby server, overall utilization is higher and only one Lobby server may be needed.

The average utilization of the Drive-Thru window server is independent of balking or not balking since queues are normally full. Average utilization is about 100 % under the original scenario, Case 1and Case 4. Scenario represented in Case 3 generates the lowest server utilization.

The average utilization of the Drive-Thru sign server is independent of balking or not balking. Average utilization is under 50 % under all simulated scenarios. The Drive-Thru server assigned to the sign can carry additional workload. Scenario represented in Case 2 generates the highest server utilization.

REFERENCES

AT&T ISTEL. 1995. Introduction to Witness.

AT&T ISTEL. 1995. Witness Training Manual.

COLLINS, M.J. 1995. Benchmarking with simulation: how it can help your production operations. Production Journal, 50-52.

LENZ, J. and NEITZEL, R. 1995. Cost modeling: An effective means to compare alternatives. Journal of Industrial Engineering, 18-19.

PRISTKER B. ALAN. 1986. Introduction to Simulation and SLAM II, John Wiley.

SMITH, D. J. 1994. Computer simulation applications in service operations. The Service Industries Journal, 395-408.

AUTHOR BIOGRAPHIES

KAMBIZ FARAHMAND is currently an Associate Professor of Industrial Engineering at Texas A&M University-Kingsville. His primary teaching and research activities are in the areas of design, implementation, and control of manufacturing systems, production and inventory control, and simulation. He earned a B.S. in Petroleum Engineering from University of Oklahoma. He completed his M.S. and Ph.D. in Industrial Engineering at University of Texas at Arlington. He is currently the president of the Coastal Bend Chapter of IIE, member of ORSA, HFES, ASEE, and APICS.

ALEJANDRO FRANCISCO GARZA MARTINEZ completed his M.S. in Industrial Engineering from Texas A&M University-Kingsville. He attended Institute Tecnologico de Matamoros where he obtained a Bachelor of Science degree in Electromechanical Engineering with minor in Production in 1983. He joined Delphi Interiors and Lighting Systems, former Inland Fisher Guide, a division of General Motors Corporation in 1985. He has worked in several Manufacturing Operations and Engineering Management positions.

Proceedings of the 1996 Winter Simulation Conference
ed. J. M. Charnes, D. J. Morrice, D. T. Brunner, and J. J. Swain

DISCRETE EVENT SIMULATION AS A TOOL TO DETERMINE NECESSARY NUCLEAR POWER PLANT OPERATING CREW SIZE

Ron Laughery
Beth M. Plott
Thomas H. Engh
Shelly Scott-Nash

Micro Analysis and Design, Inc.
4900 Pearl East Circle, Suite 201E
Boulder, Colorado 80301, U.S.A.

ABSTRACT

There are not always sufficient resources or time available to identify human factors issues early enough for development of detailed technical bases using empirical experimentation with human subjects. Consequently, analytical approaches are needed to augment the experimental approach for human factors regulatory decision making at the US Nuclear Regulatory Commission. One analytical approach, computer modeling of human performance, is being investigated by the NRC Office of Nuclear Regulatory Research. As an example of the types of supporting research required, we discuss two specific studies pertaining to the use of Micro Saint, a discrete event simulation package, as a means of evaluating the effects of crew size on safety in a nuclear power plant setting. Both studies provided data that permit an evaluation of the practicality and validity of using models built in Micro Saint for the specific purpose of studying staffing issues, as well as the value of modeling of human performance in general.

1 INTRODUCTION

Current shift staff requirements in nuclear power plants in the United States have been based primarily upon experience. While computer models have been used to predict thermal hydraulic and hardware performance of the plant, no such tools were available for considering the human element. There are many reasons to consider human performance when making predictions of any system's performance. For the NRC, the need to make sound licensing decisions is paramount and a key determinant of plant safety is human performance. The number of operators in a control room, the tasks those operators perform, and how well operators perform tasks are all significant factors in plant safety. Experience, expert judgment, experiments, and modeling are all tools used to make predictions of human performance in a variety of non-nuclear environments. When these tools are used together, the resulting "consensus" prediction carries more weight than a prediction derived from only one of these tools. To strengthen both the analytical power and believability of human factors analysis, it is our position that if models of human performance can be shown to be valid predictors , these tools should be one of the methods used for predicting human performance in nuclear power plants.

A specific area of NRC human factors analysis that is ripe for the use of human performance modeling is the area of determining shift staffing requirements. In 1993, the Nuclear Regulatory Commission (NRC) began several research projects seeking to establish a technical basis for criteria for minimum shift staffing levels for licensed and non-licensed operators at nuclear power plants. Specifically, the NRC sought to either confirm the adequacy of requirements of 10 CFR 50.54 (m) or establish a basis for modification. These requirements deal with the minimum number of licensed operators required to operate a nuclear power plant. The required number of licensed operators varies based upon the plant design. As we will discuss, there are computer modeling tools available that permit the analysis of staff requirements as a function of task and equipment design. Because this was an area of specific interest to the NRC, laboratory research was planned to address the issue empirically. This combination of 1) a need for answers on the staffing issue, 2) the availability of models that were directly relevant, and 3) the potential for empirical research that could be used to validate the models provided an ideal opportunity to seriously explore the use of human/system modeling at the NRC.

The research program that the NRC and Micro Analysis and Design have been pursuing involves leveraging the shift staffing empirical research with the goals of 1) the development of models designed to predict human/system performance as a function of staff makeup and then 2) comparing the predictions from these models to experimental data. The comparison of model predictions to experimental data is the essence of any model validation exercise. If the models predict reasonably accurately, then the models can be used to significantly augment empirical research and, when empirical research is impossible, to serve as a means for performing sensitivity studies.

The remainder of this paper provides an overview of this model validation research program. First, we will discuss the human/system modeling technology that is being used and evaluated - Micro Saint. Second, we will provide overviews of two of the research projects that are underway to evaluate the validity of models built in Micro Saint to predict human/system performance in the nuclear power plant environment.

The goal of this paper is to describe a framework for the investigation into the use of modeling as a tool to support regulatory decision making and to discuss some experiences we have had in attempting to fill out this framework.

2 A BRIEF DESCRIPTION OF THE APPROACH TO HUMAN/SYSTEM PERFORMANCE MODELING - TASK NETWORK MODELING AND MICRO SAINT

One technology that has evolved over the past 20 years for predicting human/system performance is *task network modeling.* In a task network model, human performance of an individual performing a function (e.g., performing a procedure) is decomposed into a series of subfunctions which are then decomposed into tasks. This is, in human factors engineering terms, the task analysis. The sequence of tasks is defined by constructing a *task network.* This concept is illustrated in Figure 1 which presents a simple task network for dialing a telephone.

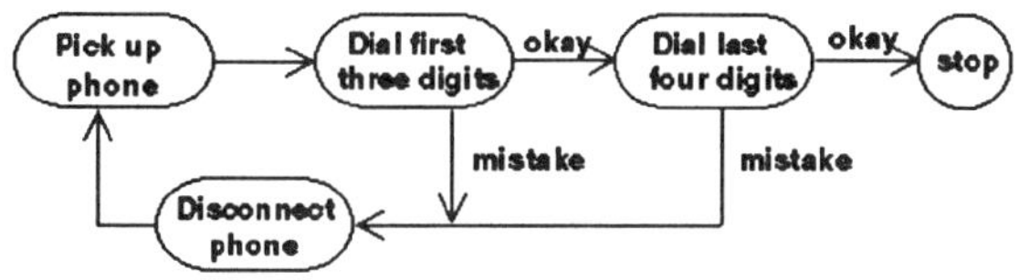

Figure 1: Example of a Task Network for Dialing a Phone

Task network modeling is an approach to modeling human performance in complex systems that has evolved for several reasons. First, it is a reasonable means for extending the human factors staple - task analysis. Task analyses organized by task sequence are the basis for the task network model. Second, as we will show in the example below, task network models can include sophisticated submodels of the plant hardware and software to create a closed-loop representation of the human/machine system. Third, task network modeling is relatively easy to use and understand. With Micro Saint, a human factors analyst can build computer models of task networks that can involve many operators interacting with a complex system in complex ways. For more details of task network models and how they perform, please see McMillan et al. (1989) and Micro Analysis and Design (1993).

2.1 An Example of a Task Network Model of a Process Control Operator

This simple hypothetical example illustrates how many of the basic concepts of task network modeling can be applied to studying human performance in a process control (e.g., nuclear power plant operation) environment. The simple human task for our model is of an operator responding to an annunciator. The procedure requires that the operator compare two meter readings. Based on the relative values of these readings, the operator must either open or close a valve until the two meter values are nearly the same. The operator activities for this model are represented by the task network in Figure 2. To allow the study of the effects of different plant dynamics (e.g., control lags), a simple one node model of the line in which the valve is being opened is included in Figure 3.

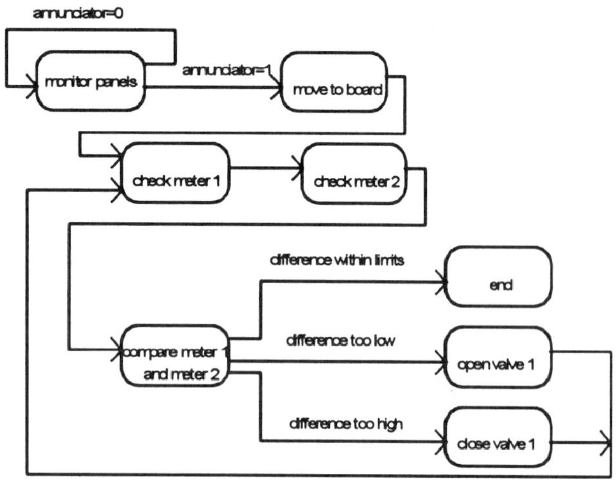

Figure 2: Sample Task Network Model of a Process Control Operator Responding to an Annunciator

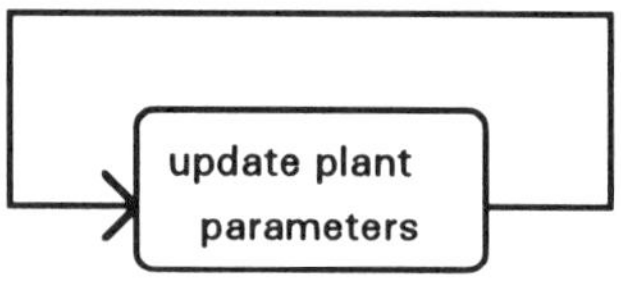

Figure 3: One Node Model of the Plant

The operator portion of the model will run the "monitor panels" task until the values of the variables "meter 1" and "meter 2" are different. The simulation could begin with these values being equal and then precipitate a change in values based on what is referred to as a scenario event (e.g., an event representing the effects of a line-break on plant state). This event could be as simple as:

meter 1 = meter 1 + 2.0;

or as complex as an expression defining the change in the meter as a function of line break size, flow rates, etc.

When the transient occurs and the values of "meter 1" and "meter 2" start to diverge, the annunciator flag will be set in the model to "on." Then, the operator will move to the appropriate board. The operator model will continue through a loop where the operator checks the values for "meter 1" and "meter 2" and either opens "valve 1," closes "valve 1," or makes no change. The determination of whether to make a control input is determined by the difference in values between the two meters. If the value is less than the acceptable threshold, then the operator would open the valve further. If the value is greater than the threshold, then the operator would close the valve. This opening and closing of the valve would be represented by changes in the value of a model parameter. In this simple model, operators do not consider rates of change in values for the meter they are viewing and, therefore, would get into an operator induced oscillation if there was any response lag. A more sophisticated operator model could use rates of change meter values when deciding whether to open or close valves.

Again, this is a very small model reflecting simple operator activity on one control via a review of two displays. However, it illustrates how large models of operator teams looking at numerous controls and manipulating many displays could be built via the same building blocks used in this model. The central concepts of a task network and shared variables reflecting human/system dynamics remains the same.

2.2 Addressing Staffing Issues with Task Network Modeling and Micro Saint

The technology described above is the basis for evaluating many aspects of human/system design. The specific nuclear power plant issue in which we were interested was the evaluation of staffing requirements both within and outside of the nuclear power plant control room.

The idea of using Micro Saint to address staffing and manpower issues in not new. Drews et al. (1985) reported the use of Micro Saint to evaluate the number of crew members required in a helicopter being designed by the Army. The Army had hoped that the helicopter could be safely and effectively operated with fewer crewmembers with extensive use of automation in the cockpit. However, the Micro Saint analysis of crew workload helped to ascertain that this staffing reduction would not be possible. The methods that were used during this research were later embedded into several Micro Saint-based tools for evaluating staffing requirements for Army systems and missions (e.g., Fontenelle and Laughery, 1988). These tools are now being used routinely to evaluate staffing requirements in the Army (e.g., Plott, 1995) and, in fact, have successfully undergone the Army's rigorous process for model verification, validation, and accreditation (Allender et al., 1995).

The details of the specific techniques for studying staffing issues with Micro Saint are too detailed to delve into in this paper, but are well documented in the references above. Briefly, we have developed a method for estimating individual and crew workload by building workload estimation metrics into a Micro Saint model. The underlying theory of workload that we employ is the Wickens' Multiple Resource Theory (Wickens, Sandry, and Vidulich, 1983). By embedding a representation of this theory within the task network, we can determine points at which each human's workload is unreasonably high and, therefore, when task performance can be expected to change and potentially suffer (e.g., task delays, increased error rates). Through the model, we can study the possibility of reallocating tasks to other available, qualified operators when workload appears to be too high in an attempt to "level out" workload across the crew. However, if this leveling out still cannot solve problems of excessive workload demands or undesired task delays because of high workload, then the model tells us that we are designing a system that will probably not perform as the designers intended because they are expecting more of the human operators than the operators have the capacity to provide.

This approach, which is identical to that developed and accredited by the US Army, is the approach that we were testing in the series of studies described briefly below. Because these studies are still ongoing, we can not state definitive success. However, the trends that we are seeing in the data are consistent with what we have found in the other applications discussed above.

3 VALIDATION STUDY 1 - THE USE AND EVALUATION OF MICRO SAINT MODELS IN ASSESSING OUTSIDE THE CONTROL ROOM STAFFING REQUIREMENTS

Background - In August 1993, the Nuclear Regulatory Commission (NRC) contracted with Brookhaven National Laboratories (BNL) to establish a technical basis for criteria for minimum shift staffing levels of licensed and non-licensed operators at nuclear power plants. As part of this effort, a detailed, on-site data collection effort was performed. The goal of the data collection was to assess the adequacy of current nuclear power plant (NPP) staffing practices for performing the activities necessary for responding to and mitigating emergency events. A total of seven NPPs were visited and information was collected regarding current staffing practices for both licensed and non-licensed shift personnel. The plants visited represented different plant types, vendor types, NRC regions, number of units, and plant ages. The two person research teams sent to each site consisted of both an operational and a human factors specialist. Four methods were used to collect data: plant documentation review, table-top analyses, walk-throughs of specific outside of the control room tasks, and interviews with individuals from different groups and organizational levels at the site. At the conclusion of the data collection exercise, the data was synthesized into a timeline or GANTT chart for each plant and scenario. Two different scenario events were examined at each of the seven NPPs.

The next phase of the effort was to develop Micro Saint models of these events. The models were built and modified to reflect system, personnel, timing, and resource changes. This effort was divided into two phases. For the first phase, five base models covering different scenario events were developed from the task analysis and timing data collected during the site visits. Often, this involved a more detailed analysis of the process followed by the operators. In the second phase, the baseline models were modified to reflect different plant parameters and staffing configurations. Only subject matter expertise and plant specific procedures were used to modify the models -- the data collected on-site was not used in the model modification process. The modified models were then executed and compared to the collected data to conduct model validation. Figure 4 illustrates the process used. This evaluation provided an excellent test of the models' validity for evaluating staffing.

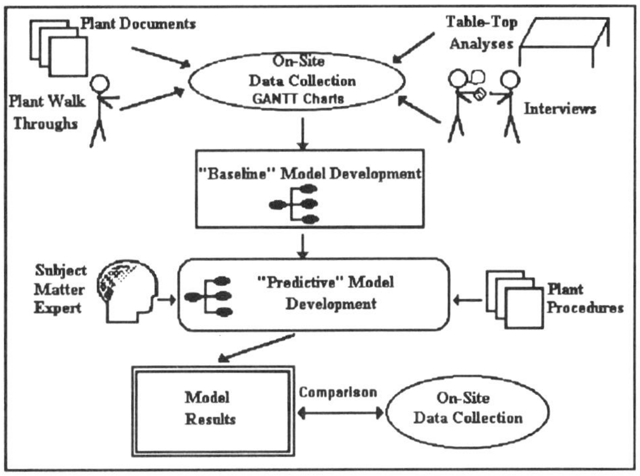

Figure 4: Model Development Process

Results -

<u>Baseline Model Development</u>

As an indication of possible staffing shortages, the models were constructed to predict and record when tasks were delayed in starting due to operator unavailability. Since the collected data was used as input to construct the baseline models it was expected that the model output would be closely aligned with the on-site data. This was largely true. When the model predicted task delays due to operator unavailability for more tasks than the data collection was able to capture, we conducted an item-by-item analysis to determine why. We found that the variability between the data collected and the model output was due to:

- the GANTT chart analysis represented a synthesis of table top data and walk-through data from two separate observers who sometimes disagreed

- the model development task analysis of the scenarios was conducted at a more detailed level than was used for the GANTT chart analysis

- when building the model, the source data from the data collection sometimes had to be redefined into a task framework that more clearly defined operator tasks, task sequences, and task interdependencies

- task data that were not obtained in the data collection were estimated by a subject matter expert (SME)

In sum, the construction of the baseline models required a more detailed analysis of the process than the table-top analyses and walk-throughs. Therefore, the opinion of the SME was that, in all likelihood, the model predictions were more realistic than the "data."

Predictive Model Development

After the baseline models were built, we altered them to reflect different plant parameters and staffing conditions using only subject-matter expertise and plant specific procedures for the other plants and scenarios. The actual data collected on-site was not used to construct the models. The new predictive models were then executed and the results compared to the data collected at the site visits.

The feasibility of using the modeling approach was evaluated by comparing the simulated response of crews at different plants with the data collected at those plants. Plant-specific task network models were developed for the new plants by modifying the existing baseline models for a similar type of plant (PWR, BWR) and scenario. The task networks were modified on the basis of expert judgment to account only for differences in plant-specific equipment or procedural requirements between the baseline plant and the new plant. The completion times for the baseline plant were used for corresponding tasks on the modeled plant. Completion times were only modified when there were plant-specific differences between the baseline plant and the modeled plant that would clearly impact the performance of the task. In all cases, the task networks were modified without referring to the on-site data collected from the new plant for operator tasks. After the task network was constructed, the on-shift personnel resources were selected for the modeled plant.

After the task network was constructed and the shift staffing level was established, the Micro Saint model was run for 100 iterations using normalized statistics for task completion times. The results were tabulated and then compared to the on-site data that had been collected by the Brookhaven team at the modeled plant. This analysis compared the observed start times of the various tasks from the on-site data collection effort to the predicted mean start times and standard deviations from the Micro Saint model.

Again, we found that virtually all observed differences were attributable to the higher resolution of the model's analysis than the analysis that was performed at the site. Both phases of the model development process provided insight into the staffing practices and procedures used at the plants that were not readily apparent from the collected data. In general, three types of insights were gained that were not apparent from either the on-site table-top analysis or the high level GANTT chart analysis. These areas are:

- *Deviation from the procedures* - The extended "predictive" models were built assuming that the crews would follow the operating procedures of the plant. Instances were identified where operators deviated from station procedures when the table-top and walk-through analyses were conducted.

- *Delay of tasks due to lack of staff availability* - During the data collection, the Shift Supervisor (SS) occasionally lost track of the personnel resources available to respond to the event. In some cases, the SS thought that there were available Auxiliary Operators (AOs) and directed them to perform tasks when all AOs were engaged in completing previously assigned tasks. In other instances, the SS didn't realize that an operator was available. Some of these instances were caused by the SS under-estimating the time that it would actually take to complete a task in the plant (as compared to the plant walk-through data). In addition, the models simulated several administrative tasks that were not included in the table-top simulation and on-site data collection. Although these tasks have been observed to occur during actual events they were beyond the capability of the table-top methodology, but easily addressed by the model.

- *Close consideration of the link between inside and outside control room activities* - The activities in a nuclear power plant during an event are closely tied to the status of the plant. The nature of the table-top procedure does not allow for the close link between plant status, control room activities, and outside control room activities to be considered in detail whereas the model did.

Summary - This study of the validity of task network modeling to study staffing issues demonstrated how models can be used in conjunction with empirical data to improve the basis for decision making. The models were able to use the empirical data as a starting point and then *immediately improve the value and accuracy of the empirical data.* Additionally, when subjected to the predictive validation test with new data, the model predictions were, in some ways, better representations of expected human/system performance than the empirical data.

4 STUDY 2 - THE USE AND EVALUATION OF MICRO SAINT MODELS IN ASSESSING INSIDE THE CONTROL ROOM STAFFING REQUIREMENTS

The work reported below is ongoing as of this writing, so the discussion below will serve as more of a status report than a final description of the results. For more information on the final results of this study, please contact any of the authors.

Background - In 1994, the NRC entered into a bilateral research agreement with Halden Reactor Project in Halden, Norway. The purpose of this agreement was to conduct research into the staffing requirements of advanced control rooms. In 1995 and continuing into 1996, Halden staff are conducting empirical studies with operating crews at the Lovissa Nuclear Power Plant in Finland. The first phase of these studies, which were completed in 1995, involved the comparison of 3 vs. 4 person operating crews in a conventional control room under five different emergency scenarios. Measures of human/system performance for these crews were collected and are being analyzed.

Method - As with the other study, we are using one set of the data as a baseline set for building and calibrating the models. This set of data is the 4-person crew configuration for two of the four scenarios. The scenarios that were calibrated involved a steam generator tube rupture (SGTR) and a loss of off-site power (LOOP). We have completed this phase. Then, using analyst and subject-matter expert input, we will change these models to reflect what we *expect* to occur when the staff is reduced to a 3-person crew configuration. The subject matter expert inputs to the model will reflect expected changes in task time and allocation to operators. The model will then be used to predict overall human/system performance as a function of these expected changes when reducing the crew from four to three people. Finally, we will compare our model predictions to the actual data that was obtained in the 3-person phase of the study. This second portion of the phase is expected to be completed in mid 1996.

The measures of crew performance that we are predicting with the models and that were collected in the Halden study include 1) the time required to perform critical groups of tasks in the scenario and 2) subjective workload. These measures were selected for modeling due to their relevance in assessment of the changes in crew size. The time a critical task is delayed due to a lack of resources can impact the probability of a serious plant problem occurring. The subjective workload is a relevant measure because of links shown between workload and the error rate of humans. By modeling the control room operators' workload, it is hoped to determine portions of the scenario where the operator is likely to make errors.

Results - Since only the first phase of this study - model development and calibration have been completed, we report only those results here.

The initial model was developed with purely SME input. In other words, we performed a table top analysis using Lovissa operators and developed the task networks, task times and error rates, and the task interdependencies. Then, we used the four person crew empirical data from Lovissa (i.e., the baseline data set) to calibrate the model. Therefore, one measure of the model's validity was the amount of calibration that was required to make the model match the baseline data. In this study less than 5% of the tasks in the models required any modification.

With respect to getting the model time lines to match the data time lines, very little calibration had to be done. Calibration of the models primarily consisted of synchronizing plant events from the experimental data, from the Lovissa project, to events that occurred in the models. The task times predicted by the SMEs proved to be highly accurate. The majority of calibration involved removing tasks that were in the model as a result of the task analysis and were not actually performed in the experiment. Adjustments were also made with regard to the number of times repeating tasks were performed so the number of repetitions more closely reflect the data obtained experimentally.

The portions of the model that predicted crew performance, measured by task completion time, were statistically the same as the data we had obtained from the SMEs when building the models. After calibration, the correlation coefficient for the SGTR scenario times was 0.97 in comparing the model times to the times collected for the 4 person crew. For the LOOP scenario, a correlation coefficient of 0.99 between the model and experimental data was obtained.

With respect to the workload predictions, the model workload scales were different from the data collected in the study. The workload values in the model were predicted by SMEs using McCracken and Aldrich's (1984) visual, auditory, cognitive and psychomotor (VACP) scales. The experimental workload values were obtained by administration of the NASA Task Load Index (TLX) scale. The TLX asks subjects to rate their workload based upon mental demand, physical demand, temporal demand, performance, effort and frustration. Therefore, the calibration of the model has involved using regression analysis in an attempt to build a map defining the relationships between model workload predictions (VACP) and the scales used by Halden (TLX). Preliminary results show that the regression models provide a statistically significant estimate of the individual TLX scales. The VACP scores only predict a percentage of the subjective TLX workload scales. The multiple correlation coefficients are on the order of 0.4 to 0.6 for the individual TLX scales.

The next step in this study is to make predictions about crew performance when the number of operators in the control room is reduced from four to three. Modifications in the models related to which operators perform specific tasks, and problem detection probabilities will be made in conjunction with the reduction in the number of operators.

Summary - Preliminary results of this study are quite encouraging with respect to being able to predict task timelines without any empirical data as input. It appears that in this case SME input alone was a valid basis for model design and parameter estimation. The validity of the models to predict workload is yet to be determined. Early results indicate that the VACP and TLX scales are related, but not highly.

5 SUMMARY

In all, we find these results quite encouraging with respect to the potential of modeling human/system behavior as a means of developing technical bases for regulatory decisions. We have shown how we can improve and supplement the empirical data collection process and provide sound results. However, significantly more research and evaluation will be required to assess whether this modeling technology is valid for assessing all staffing issues, much less the many other human factors regulatory issues that are faced by the NRC. The answer, undoubtedly will be "for some things it works, and, for others, probably not."

REFERENCES

Allender, L., T. Kelley, L. Salvi, D. B. Headley, D. Promisel, D. Mitchell, C. Richer, and T. Feng. 1995. Verification, Validation, and Accreditation of a Soldier-System Modeling Tool. In the *Proceedings of the 39th Human Factors and Ergonomics Society Meeting, October 9-13, San Diego, CA* 1219-1223. Human Factors and Ergonomics Society, Santa Monica, CA.

Drews, C., K. R. Laughery, K. Kramme, and R. Archer. 1985. LHX Cockpits: Micro SAINT Simulation Study and Results. Report prepared for Texas Instruments - Equipment Group, Dallas, Texas.

Fontenelle, G.A. and K.R. Laughery. 1988. A Workload Assessment Aid for Human Engineering Design." In *Proceedings of the 32nd Annual Human Factors Society Meeting, Anaheim, CA.* 1122-1125. Human Factors and Ergonomics Society, Santa Monica, CA.

McCracken, J. H. and T. B. Aldrich. 1984. Analysis of selected LHX mission functions: Implications for operator workload and system automation goals. Technical Note ASI479-024-84. Anacapa Sciences, Inc., Fort Rucker, AL.

McMillan, G. R., D. Beevis, E. Salas, M. H. Strub, R. Sutton, and L. Van Breda. 1989. *Applications of Human Performance Models to System Design*. New York: Plenum Press.

Micro Analysis and Design. 1993. *Micro Saint for Windows User's Guide*. Boulder, Colorado: Micro Analysis and Design.

Plott, B. 1995. Software User's Manual for WinCrew, the Windows-Based Workload and Task Analysis Tool. Aberdeen Proving Ground, MD: U.S. Army Research Laboratory,

Wickens, C. D., D. L. Sandry, and M. Vidulich. 1983. Compatibility and Resource Competition Between Modalities of Input, Central Processing, and Output. *Human Factors* 25:227-248.

AUTHOR BIOGRAPHIES

Ron Laughery is the President of Micro Analysis and Design in Boulder, Colorado. He received B.S., MS., and Ph.D. degrees from the State University of New York at Buffalo. For over twelve years, he has led the development of the discrete-event modeling product Micro Saint. He has also applied Micro Saint to modeling human performance, health care systems, manufacturing systems, and process reenigneering.

Beth M. Plott is a Principal Engineer at Micro Analysis and Design. Ms. Plott holds a Bachelor's degree in Chemical Engineering from the University of Colorado and is currently pursuing a Master's degree in Environmental Engineering from Colorado State University. She has eight years experience in software development and computer modeling. Ms. Plott has developed several simulation models in diverse areas such as manufacturing, health care, retail systems, and nuclear power plants.

Thomas H. Engh is a Human Factors Engineer with Micro Analysis and Design in Boulder, Colorado. He received B.S. and M.S. degrees in industrial engineering from Purdue University in 1991 and 1993 respectively. His interests include the modeling of human performance and user interface design.

Shelly Scott-Nash is a software analyst with Micro Analysis and Design in Boulder, Colorado. She received a B.A. in Math and Computer Science from the University of Colorado in 1987. Her interests include simulation software design and application.

Proceedings of the 1996 Winter Simulation Conference
ed. J. M. Charnes, D. J. Morrice, D. T. Brunner, and J. J. Swain

VALUATION OF POTASH RESERVES AT THE WASTE INSTALLATION PILOT PLANT

Peter C. Anselmo

Department of Management
New Mexico Institute of Mining and Technology
Socorro, New Mexico 87801, U.S.A.

ABSTRACT

Monte Carlo simulation was used in conjunction with several random walk price and cost models to value potash reserves at the Waste Installation Pilot Plant (WIPP) site near Carlsbad, New Mexico. Selection of market price and product processing cost models is the critical issue discussed in this paper. Key issues and inputs to the thirty five year simulation are briefly discussed. Results for one type of potash found in the WIPP region are presented. Present value distributions for revenues from potash operations and estimated cash flows to a single firm conducting potash extraction were generated using several price/cost model combinations. A simple Wiener process was selected for annual potash market prices, and processing costs were generated using random draws from a uniform distribution. The rationale for selection of this model combination is discussed.

1 OVERVIEW

The Waste Installation Pilot Plant (WIPP) is located in Eddy County, Southeastern New Mexico near the town of Carlsbad. Projected as an underground burial area for low level radioactive waste, the WIPP site has been in various stages of completion for the last twenty years. As part of the process of activating the site, personnel from the New Mexico Bureau of Mines and Mineral Resources (NMBMMR) evaluated oil, gas and two types of potash reserves in the area (NMBMMR, 1995). The primary end product use for Eddy County potash is agricultural fertilizer.

More specifically, resources within the sixteen square mile WIPP site, a one mile buffer zone (known as the "additional area") around the site and the two combined areas were assessed. NMBMMR provided mineral reserve estimates that were used in the value estimation process. The concern in this paper is description of the process, results and sensitivity analysis associated with evaluation of combined assessed potassium chloride (KCl) reserves within the WIPP and additional area. If WIPP indeed opens and begins storing waste, all mineral resources within the combined area are assumed to be inaccessible. Thus the need for valuation of the mineral resources at the site.

The policy implications of both this study and the overall NMBMMR effort are potentially very significant. Although mining as a whole is declining in New Mexico, the industry still provides some of the highest wage jobs in the state and is the employment sector with the highest average wages. Southeastern New Mexico, with (among other minerals) potash and oil and gas resources, is an area of the state that is very sensitive to changes in the mining industry and its economic climate. Potash operators have been in the area since the 1920s. Oil and gas producers are enjoying something of a resurgence in the area, and the issue of mining access to public lands has been the cause of some conflict between regional oil and potash producers.

Potash mining in the Eddy County area is conducted underground. Burrowing machines extract the raw ore, which is processed into potash product on site. Several operators are currently mining in the vicinity of the WIPP site. For this study, it was assumed that potash mining within the combined area would begin one year after a decision to *not* open WIPP. No new development costs are directly associated with this scenario. All costs associated with mining WIPP area potash are included in the per-ton rate obtained by NMBMMR consultants from area operators.

Data presented here represent results from 1560 simulation runs each. An individual simulation run consists of a 35 year sequence of data resulting from randomly generated annual prices and costs in conjunction with NMBMMR-provided mineral extraction data. These simulations were conducted using Microsoft Excel on a personal computer with a 133Hz Pentium processor and 16 megabytes (mb) of RAM. The main simulation models each occupy about 10 mb of memory.

A discount rate of 15% was used as the base case for evaluation of potash resources, which was and is a point of some controversy. This issue is discussed below.

Operation cash flow net present values (NPV) were calculated from the perspective of a single firm. They represent the average of the present values of all the cash flows associated with each simulation run.

Total expected revenue present values (PV) were calculated as an indication of the overall worth of KCl deposits. Revenue values are also expected present values, as they are the average of the expected present values generated by each simulation run. Bureau of Land Management (BLM) specifications were used to determine mining variables which in turn impacted estimates of the tonnages of various grades of KCl present and the raw material tonnage required for a given level of market demand. The full potash valuation analysis is available in Anselmo (1996).

2 ISSUES AND ASSUMPTIONS

KCl reserve estimates for the WIPP area were provided by consultants and specialists at the New Mexico Bureau of Mines and Mineral Resources, and the simulation model was constructed. Key model inputs, in addition to reserve data, included the discount rate, annual potash production level, market price of the resource, the unit cost of extraction, and tax and royalty rates. Development of a method to anticipate future market prices and operating costs for potash was a key issue. Time units were years, and the time frame simulated was 1995-2030.

Forecasting in cases like this is as much an inexact art as a science, particularly when the forecasting horizon is thirty five years. Thus, although historical prices for Eddy County potash were modeled using time series methods, a simulation approach was used to value these resources. Annual market prices and costs were simulated using several random walk methodologies (Karlin and Taylor, 1975), which are discussed below. Depletion was calculated using the standard percentage methodology, which may be found in Stermole and Stermole (1993).

Key assumptions and features associated with the potash cash flows used in the simulation include:

All calculations are performed from 1 January 1995. KCl extraction activity is treated as a capital project that was evaluated from 1 January 1995 and undertaken beginning 1 January 1996. The decision to go forward with any potash mining and development venture was considered from the perspective of 1 January 1995.

Revenues were treated as if realized monthly, and taxes and royalties were treated as if they are paid on a quarterly basis.

The Monte Carlo simulations generated numbers for each year for the present market value of the reserves and the present value of the total cash flow for each simulation run. As is standard practice in financial analysis (Levy and Sarnat, 1994), cash flows attributable to the decision to mine potash are the sum of income after taxes, depreciation and depletion less annual investment (in this case, the annual investment is zero). Data were also generated for expected present values for severance taxes, state and corporate taxes and royalties. These may be found in Anselmo (1996). We briefly discuss tax and royalty issues below because of their effect on expected cash flows.

2.1 Discount Rate

Results presented here are associated with a 15% nominal discount rate. 15% was also used in the last major analysis of WIPP mineral reserves (Weisner, Lemons and Coppa, 1980). Estimation of discount rates for risky investment projects (the perspective taken in this study was one of viewing potash exploration activity as a risky investment project) is generally a difficult and (potentially) inexact process. Though there is detailed knowledge regarding the location and grade of potash deposits in the area, there is some uncertainty regarding the future market price of potash products. Along with the long time horizon, price uncertainty may be considered a major source of potash mining operation risk.

The 15% rate used in the results presented above is reasonable in light of the short and long-term market uncertainties faced by potash producers (Searls, 1992). Although the development risk associated with potash deposits in the WIPP area is low, there is market price risk that is faced by operators. Extensive KCl (also known as sylvite) deposits in various stages of development exist around the world. Further, any investment situation as long-term in duration as a potash mining operation faces considerable interest rate risk - the risk that more favorable return opportunities will arise over the life of a project.

A precise discount rate for different firms operating in the WIPP area is difficult to estimate (particularly in the absence of debt/equity ratio and other data for said firms). So, although pinpointing a discount rate for a 35 year project is quite risky in and of itself, current levels of activity in the region and market factors point to a 15% nominal discount rate as a reasonable estimate for Carlsbad area potash operations.

2.2 Annual Production Levels

Because potash is mined in a linear fashion over many

years, present value-based analyses of the viability of potash operations will be controversial because large future cash flows are severely discounted. Market demand is a major factor impacting annual production levels. In the original study, 450,000 tons of KCl was obtained as a reasonable average annual demand figure from area operators. Product output levels of 600,000 and 1,000,000 tons (annual demand) were also used to assess the sensitivity of the original analysis to annual market demand. At the time this sensitivity analysis was conducted, we were interested only in the sensitivity of the base case results to increases in annual production levels.

Increasing annual demand levels results, as one might expect, in higher expected values for the relevant variables. The key is whether assuming a particular level of market demand is appropriate. Demand for KCl as a major component of fertilizer is far from certain at any level, a point which is discussed further below.

2.3 Market Prices

Annual prices and mining costs per ton for sylvite (KCl) were generated using random walk methods. Two types of random walk processes and a Geometric Brownian Motion model were used. Historical, confidential Eddy County sylvite prices were analyzed using time series techniques to show that these historical prices may be modeled as a random process. However, the amount of available Eddy County data was limited, which precluded the use of a time series model to actually estimate thirty five years of future market prices. A national KCl price index covering 1965-1993 was provided by the NMBMMR. Historical Eddy County potash prices tracked the index. However, a limited number of years of Eddy County prices were available and the prices and index figures were not the same. An estimate for the end of 1994 market price was obtained from area operators (and later verified as reasonable), and was used as the 1995 market price and the point of departure for the price simulations.

As with other commodities, the weak-form market efficiency assumption is appropriate here, which partially justifies the use of Markov processes. The independence of year-to-year price changes assumption is not as evidently congruous, but in the longer term (over a 35 year period, for instance) it probably holds. The third key assumption is that annual price changes are normally distributed. Based on the analysis of historical annual price data conducted for this study, we cannot refute this assumption. The trouble with this assumption is that very large price (and cost) swings are possible. In fact, negative prices and costs are theoretically possible. However, since all data reported are the result of 1560 simulation runs, potential negative price (and cost) impacts in individual simulation runs are negated.

Individual price paths have a tendency to wander and vary considerably. However, the number of runs conducted resulted in a average KCl price and cost figures for all price/cost models considered that were nearly identical to the time zero price and cost. Price and cost standard deviations depended on the model used. Annual changes in potash prices (as measured using Eddy County KCl data and the 1965-1993 national KCl index) exhibit a 4% average nominal return and a 0% average real return. These averages were essentially the same regardless of whether a geometric or arithmetic mean was computed.

2.4 Operating Costs

After much discussion (in the absence of historical data), the uncertainty associated with annual operating costs was expressed as plus or minus 10% of the previous year's per ton operating cost. Thus, a uniform distribution was initially used for annual per-ton operating cost, with the mean for any one year the cost generated for the previous year. All non capital annual costs were aggregated into the operating cost figures provided by consultants and area operators.

As noted above, the average, inflation adjusted change in the analyzed historical potash prices was zero. Although mine productivity (in terms of tons produced per potash mine worker) has been essentially flat since 1965 (*Economic Report of the President*, 1994), area operators have expressed confidence in their abilities to offset the effects of inflation on operating costs in both the near and distant future.

2.5 Taxes and Royalties

The state of New Mexico assesses severance taxes on revenues attributable to minerals extracted within the state. Rather than attempt to predict factors which might contribute to alterations in the severance tax rate, a rate of 1.25% (based on current tax law) of potash revenues was used for the study period. A royalty rate of 2.5% of revenues, provided as a current estimate by experts in this area, was used in the simulations for similar reasons.

Capital investment and other tax incentives that mining companies may periodically receive from political entities were likewise ignored in this work. In addition to presenting a major limited data prediction problem, consideration of tax incentives would involve acquisition of additional proprietary data from area producer firms. An average corporate tax rate of 34% was therefore used.

All taxable income is assumed to be New Mexico income for state tax purposes. New Mexico tax rates are 4.8% of taxable income under \$500,000, \$24,000 plus 6.4% of the excess over \$500,000 for amounts between \$500,000 and \$1,000,000 and \$56,000 plus 7.6% of the excess over \$1,000,000 for taxable income over \$1,000,000.

Of course, the key here is that the expected tax values generated in this study represent expected opportunity losses to the various governmental entities under the specified mining and development circumstances if WIPP comes on line. These numbers were also essential in computation of expected annual cash flows.

3 MODELS AND OUTPUT

3.1 Models

A Monte Carlo sampling method was used to generate random walk price and mining cost data for the time period 1995-2030, and three price/cost models were used in this evaluation. Model 1 is a version of a simple Wiener price process combined with a Monte Carlo method for operating costs that is based on information provided by area potash firms. This was used as the base case for several reasons. First, historical operating cost data of any kind are proprietary and were unavailable. Second, limited and inconsistent data made price variance estimation difficult. Third, firms did indicate to NMBMMR consultants that annual per-ton operating costs tended to be between 90% and 110% of the previous year's per-ton cost.

We define P_t and C_t as per ton price and cost in year t, respectively. z_t is a normally distributed random variable with mean 0 and standard deviation 1 that is generated using random numbers in the simulation runs (this value is estimated as the sum of twelve uniform [0,1] random variables minus 6) and σ denotes standard deviation. Random variable y_t is uniformly distributed on the [0,1] interval, and is generated using the RAND() comment in Excel. Price/cost model 1 is written as follows:

$$P_t = P_{t-1} + z_t\sqrt{t}$$

$$C_t = .9C_{t-1} + .2y_t C_{t-1} \tag{1}$$

In model 2, where P_0 (\$75/ton) and C_0 (\$18/ton) are the time zero market price and per-ton processing cost, both prices and costs were analyzed using Wiener processes, as shown below:

$$P_t = P_0 + z_t\sqrt{t}$$

$$C_t = C_0 + z_t\sqrt{t} \tag{2}$$

Geometric Brownian Motion is often used to assess commodity prices (Dixit and Pindyck, 1994). In order to compute parameters for the price component of model 3, which were also used in the cost component, real returns on Eddy County (and the national index) KCl prices were calculated. The real return over the time period for which we had data was 0 with a standard deviation of about .167. Model 3 therefore had no mean coefficient, as follows:

$$P_t = P_{t-1} + z_t\sigma_P P_{t-1}$$

$$C_t = C_{t-1} + z_t\sigma_C C_{t-1} \tag{3}$$

Although estimates for σ_P were obtained from the limited data, estimates of σ_C were unavailable. In the absence of historical cost data, $\sigma_C = \sigma_P$ was assumed. Two values for price and cost standard deviation were used in valuation simulations, .167 and .085. As noted above, the first value represents the standard deviation of the limited Eddy County KCl price data that were available.

3.2 Output

Key variables were the expected present value of KCl potash revenues and the expected net present value of cash flows to a single (imaginary) firm from potash operations in the WIPP area. As expected, these values did not vary according to the model used. Distribution standard deviations did vary considerably with the model used. Expected present values E(PV) and expected net present values E(NPV) and their standard deviations (noted as SD) are summarized below in Table 1 for the base production case of 450,000 tons per year. For model 3a, the standard deviation used was .167; it was .085 for 3b. Note that although expected values are all in the \$193 million range, estimate standard deviations vary considerably. In addition to the potential profitability of potash operations, a major concern throughout this process was the variability of all estimates.

Table 1: Base Case PV and NPV Results

Model	E(PV)	SD(PV)	E(NPV)	SD(NPV)
1	$193m	10m	52m	14m
2	193m	1.4m	52m	4.6m
3a	197m	72m	50m	68m
3b	192m	32m	51m	32m

Additional data for the Table 1 expected values are provided below in Tables 2 and 3. In Table 2, in which revenue data are provided (and in subsequent tables), Max and Min refer to the maximum and minimum values obtained, Med refers to the data median and Skew refers to skewness. Numbers in all Tables are rounded off.

Table 2: Additional Base Case Revenue Data

Model	Max	Min	Med	Skew
1	225m	161m	193m	-.045
2	197m	188m	193m	-.034
3a	651m	54m	187m	1.23
3b	307m	111m	190m	.41

Table 3 contains distribution data for the expected present value of cash flows to a single mining operation from combined area potash operations.

Table 3: Additional Base Case Cash Flow Data

Model	Max	Min	Med	Skew
1	90m	-14m	66m	-.338
2	67m	36m	52m	.006
3a	362m	-270m	49m	.02
3b	152m	-46m	51m	.015

Sensitivity analysis was conducted with respect to annual production levels. Tables 4 and 5 contain E(PV), E(NVP) and standard deviations for production levels of 600,000 and 1,000,000 tons per year respectively.

Table 4: 600,000 Tons Annual Production

Model	E(PV)	SD(PV)	E(NPV)	SD(NPV)
1	$257m	13m	65m	19m
2	257m	1.9m	66m	6.1m
3a	257m	91m	63m	90m
3b	257m	45m	66m	42m

Table 5: 1,000,000 Tons Annual Production

Model	E(PV)	SD(PV)	E(NPV)	SD(NPV)
1	$424m	22m	94m	33m
2	424m	3m	94m	11m
3a	428m	146m	90m	148m
3b	427m	80m	92m	80m

The models provided essentially consistent estimates for expected present values for total revenues and cash flows over the thirty five year life of the hypothetical mining project. Model selection for the final estimate is a critical issue, though, as examination of the divergent standard deviations for each estimate suggests.

Although expected present and net present values were the "bottom line" concern for most interested parties, standard deviations vary considerably across models. Model 1 was the base case. There is no evidence that any interested parties (it was necessary throughout this process to receive and consider input from any and all interested individuals and groups) are interested in either minimum or maximum variance price/cost models. This, combined with data availability and accuracy, supports the use of model 1 as the model to use for WIPP site KCl valuation analysis.

Only model 2 produced results indicating no chance of the mining operation having a negative net present value. However, the low variance associated with model 2 estimates also preclude statements about the upside potential of potash operations. The high variances associated with model 3 estimates could cause a risk averse firm to avoid WIPP area potash mining altogether. Furthermore, the uncertainty associated with model 3 variance estimates makes use of that model difficult to justify.

4 DISCUSSION

From a KCl valuation perspective, a major decision impacting the analysis once input variables are specified was price/cost model selection. The concern here has been examination of the effects of alternate models on the variance of some of the relevant simulation output data. This is a critical issue because the expectations derived from the three models under different production scenarios are close in value. The notion that model 1 is the appropriate base case model is supported by the data and circumstances surrounding this project; a major objective of this study was assessment of the validity of model 1 as the base model. As may be seen in Tables 1, 4 and 5, expected present values for the assessed variables are comparable across models. Distribution variability depends on the model selected, which makes the variance of each estimate a key variable - along with the data availability issue discussed above - in model selection.

Choice of a discount rate is also made a subjective decision because Eddy County area potash operators are either subsidiaries of large corporations or privately held entities. In either instance, firm information that would aid in the use of a conventional process (Copeland, Koller and Murrin, 1990) to determine an appropriate discount rate was unavailable. However, there was (and is) no shortage of macro-level data to aid in the discount rate decision.

There is little doubt that global demand for food, fueled by both the large and expanding global population and the rapid industrialization occurring in areas where much of the world's population lives will be increasing for the foreseeable future. Demand for food will not decline in the near or distant future unless a global catastrophe occurs. However, as Searls (1992) notes, near term demand for potash products as food-producing fertilizer is far from certain.

The random walk methods used in this study do not allow for any drastic year to year planned upward (or downward) non-random movements in these commodities prices and operating costs. This is a conservative approach that is based on several factors which impact the national and global demand for fertilizer. Farm productivity in the United States has been increasing slowly but steadily since 1965 along with the use of agricultural inputs. However, the use of agricultural inputs relative to all other inputs to the overall in the United States agricultural process (such as labor and capital equipment such as farm machinery) has remained steady since 1985 (*Economic Report of the President to Congress*, 1994). Stable domestic demand for potash based fertilizer products is a likely feature of the potash market for the (at least) near future. Searls (1992) notes that future demand increases are likely to come from outside the U.S.

Although the world population is rapidly increasing, passage of international and regional treaties such as the General Agreement on Tariffs and Trade (GATT) and the North American Free Trade Agreement (NAFTA) may have a negative impact on agriculture subsidies around the world. Stabilization or rolling back of agricultural subsidies in countries such as Japan may have a stabilizing (or, in the short term, negative) impact on global demand for potash based fertilizer products. Certainly, the near future for potash prices (despite the historical upward trend in per ton product prices) is uncertain; the more distant future is uncertain by definition.

ACKNOWLEDGMENTS

The support of Mr. James Barker and Dr. Charles Chapin of the New Mexico Bureau of Mines is gratefully acknowledged. George Griswold provided the mining data; and was also a valuable interface with the potash operators. David Sivils and Jennie Wong provided data collection assistance.

REFERENCES

Anselmo, P.C. 1996. "Valuation of Langbeinite and Sylvite Reserves at the WIPP Site, Additional Area and Combined Area", New Mexico Bureau of Mines.

Copeland, T., T. Koller and J. Murrin. 1990. *Valuation: Measuring and Managing the Value of Companies*. New York: Wiley.

Dixit, A. K., and R. S. Pindyck. 1994. *Investment Under Uncertainty*. Princeton, N.J.: Princeton Press.

Economic Report of the President to Congress. February, 1994.

Karlin, S. and H. M. Taylor. 1975. *A First Course in Stochastic Processes*. New York: Academic Press.

Levy, H. and M. Sarnat. 1994. *Capital Investment and Financial Decisions*. Englewood Cliffs, N.J.: Prentice-Hall.

New Mexico Bureau of Mines and Mineral Resources, 31 March 1995. "Evaluation of Mineral Resources at the Waste Installation Pilot Plant (WIPP) Site".

Searls, J., 1992. "Potash", United States Department of the Interior, Bureau of Mines.

Stermole, F. J. and J. M. Stermole. 1993. *Economic Evaluation and Investment Decision Methods*. Golden,Colo.: Investment Evaluations Corporation.

Weisner, R. C., J. F. Lemons and L. V. Coppa. 1980. Valuation of Potash Occurrences Within the Nuclear Waste Isolation Pilot Plant Site in Southeastern New Mexico", United States Department of the Interior, Bureau of Mines.

AUTHOR BIOGRAPHY

Peter C. Anselmo is an assistant professor in the department of management at the New Mexico Institute of Mining and Technology. His primary research interests are in decision analysis and the connection between decision analysis and finance. He has been involved with the WIPP project for two years. He is a member of INFORMS.

Proceedings of the 1996 Winter Simulation Conference
ed. J. M. Charnes, D. J. Morrice, D. T. Brunner, and J. J. Swain

SIMULATION IN POLICY ANALYSIS: THE TRANSFER OF FACILITIES WITHIN THE DEPARTMENT OF ENERGY'S ENVIRONMENTAL MANAGEMENT PROGRAM

Samuel M. Hart

Project Performance Corporation
46030 Manekin Plaza, Suite 180
Sterling, Virginia 20166, U.S.A.

ABSTRACT

This paper explains how process simulation can be applied as a policy analysis tool. While many simulations are used to evaluate and improve specific processes, few are currently used as a policy analysis or decision-making tool. For the purpose of choosing from several potential policy choices, a simulation model can describe the probable outcomes under each alternative in a concise and easily understood manner. Such a model can combine the work of several more traditional policy analysis tools into a single analysis that, once designed, can be performed in a fraction of the time. The specific case presented in this paper is the transfer of deactivated facilities from one organization (the Office of Nuclear Material and Facility Stabilization) to another (the Office of Environmental Restoration) within the Department of Energy's Environmental Management program. The model was developed using VENSIM, an off-the-shelf modeling software package. With the completed model, several potential policy choices regarding the rate of facility transfer were assessed to identify funding and scheduling problems with each policy.

1 INTRODUCTION

Computer simulation has become an established tool for analyzing and improving processes in the manufacturing industry as well as other production-oriented industries such as the chemical industry. Directors at DuPont and Dow Chemical believe that simulation modeling is essential to the progress of their companies (Krieger 1995). More recently, simulation models have been entering the arena of business management. James Krieger (1995) indicated:

> *"Companies in the [chemical] industry are at the beginning of a major transformation in how they operate: how they focus their research and engineering design activities, how they run their plants, how they decide which products to market, and how they make tactical and strategic business decisions."*

Simulating how to best manufacture a product, as well as simulating its performance, are important to the success of that product (Dvorak 1994). However, what about considering the importance of modeling in areas other than production? Businesses have been making the transition from using models strictly as a production tool to employing them as a decision tool as well. Models are now used as business tools focusing primarily on cost estimating, scheduling, and logistics.

Evaluating the costs of new products and market acceptance of those products is important to business for obvious reasons. The ability to predict the success or failure of a product helps companies capitalize on a success or avoid costly failures. A strict cost evaluation; however, could be completed without simulation. The value of a simulation in cost estimating stems from the fact that non-economic factors can play an important role in the way costs behave in reality. As Max Weber stated (Breiner 1995):

> *"The explanation of everything by economic causes alone is never exhaustive in any sense whatsoever, in any sphere of cultural phenomena, not even in the economic sphere itself."*

Using simulations for scheduling and logistics enables businesses and manufacturing plants to operate more efficiently. As James Trainham, director of engineering research and development at DuPont, declared (Krieger 1995):

"Outstanding supply-chain management cannot make a second-rate manufacturing process world-class, but second-rate supply-chain management can make a best-in-the-industry manufacturing plant perform no better than a mediocre one."

If the private sector can use simulation tools for such a wide array of applications, why shouldn't the federal government use simulation tools in a similar manner? Some organizations within the government, such as the military, are already using simulation models for a wide array of uses. As Major General Joseph J. Redden (1995) explained:

"Simulation has been extensively and successfully applied to a wide range of military problems including

- *operations of weapons*
- *establishing requirements*
- *acquisition*
- *logistics*
- *communications*
- *war-gaming"*

One such model used by the Department of Defense is the Cost Analysis Strategy Assessment model. The model serves as a Life Cycle Cost model to be used as a decision tool (Manary 1995). Some government organizations are now beginning to use simulation models on a large scale. Within the Department of Energy's Office of Environmental Management (EM), a simulation tool is used to provide a life-cycle cost and scheduling estimate for the entire program. However, few models are used on a smaller scale to evaluate specific problems in EM. In 1995, Idaho National Engineering Laboratory designed a simulation to analyze different scenarios of storing high-level nuclear waste (Kristofferson 1995). The model evaluated five scenarios varying the availability of a waste repository facility. It provided a useful scheduling policy analysis tool for the transfer of high-level waste, although the different scenarios were assumed outcomes with varying funding levels, and cost data was not directly integrated into the model.

The objective of the model described in this paper is to develop a policy analysis model that integrates budgetary and scheduling concerns. The process chosen was the transfer of facilities from the Office of Nuclear Material and Facility Stabilization to the Office of Environmental Restoration. This simulation is also unique to other policy analysis models in that the transfer that is modeled is a "paper" transfer. In other words, it is not a physical transfer of materials, but rather an administrative transfer.

An additional goal of this simulation was to use an off-the-shelf software program to design the model. The software chosen was VENSIM, a discrete event, system simulation program.

2 THE MODEL

The model shown in Figure 1 simulates the process that a facility scheduled to be decommissioned follows through the Office of Nuclear Material and Facility Stabilization (EM-60) and the Office of Environmental Restoration (EM-40). The majority of facilities will go through five phases in EM-60 and two phases in EM-40. The EM-60 phases include Pre-Stabilization Surveillance and Maintenance, Stabilization, Post-Stabilization Surveillance and Maintenance, Deactivation, and Post-Deactivation Surveillance and Maintenance. The EM-40 phases are Decontamination and Decommissioning. Two additional phases in the EM-40 program that are not included above are: 1) a general Surveillance and Maintenance phase, shown as Transfer Holding in the model flowchart, and 2) Long Term Surveillance and Maintenance. The Transfer Holding is identical to EM-60's Post-Deactivation Surveillance and Maintenance with the exception that facilities in this phase are under EM-40 management instead of EM-60. The Long Term Surveillance and Maintenance phase includes facilities that have been decommissioned but are still owned by EM-40 and must be maintained. Surveillance and Maintenance includes basic operations and costs necessary to maintain the facility such as security, power, etc.

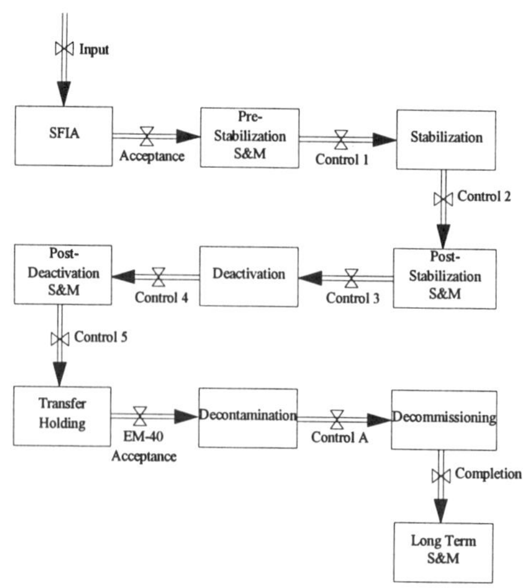

Figure 1: Facility Transfer Model Flowchart

It is possible for additional facilities to enter the process at a later date. Most of these facilities are logged in the Surplus Facilities Inventory Assessment (SFIA) database. It is also possible; however, that some facilities not currently in the SFIA database will be added. Therefore, the model was designed to include both SFIA additions as well as any other additions to the current SFIA database.

A facility spends a given amount of time in each phase and then moves on to the next phase in the process; however, the rate at which facilities proceed is also somewhat determined by the costs of processing those facilities and the budget allocated to the program. Therefore, cost information as well as program budget information were included (for each phase) in the model.

The facilities follow a linear flow from one phase to the next until completion. The facilities included range from a large production reactor to storage tanks to retention basins. Obviously, the costs and time to complete these different types of facilities may vary. As organized by the Office of Environmental Management, there are 22 possible facility types. They are based on the type of building, the size of the building, and waste type. Therefore, the model was subscripted to include different cost and time duration information of each phase and each facility type. Since the program is already several years old, those facilities already into the process were included in the phases where they are currently located.

The model was also designed to remain as flexible as possible. It is likely that the amount of time required to complete a phase may vary over the course of the several decades required to complete all facilities. Therefore, the model includes a time dilation factor. This variable will increase the time required to complete each stage by a certain percentage each year. While this variable is not required, it is helpful for evaluating different scenarios. Similarly, a cost inflation rate is factored into the model. Finally, it must also be remembered that budgets will fluctuate from year to year. The VENSIM software allows a step function to be used that increases or decreases the budget figures at specified years. It is even possible to set the model to automatically pause during the simulation run to adjust the budget.

3 MODEL ASSUMPTIONS

Three basic assumptions had to be made in the model. First, those facilities already in the process could not start in the middle of a given phase. Therefore, facilities were placed at the beginning of their current phase if less than half way to completion of that phase, or at the beginning of the next phase if more than half way through the phase. This rule does not apply to facilities in SFIA or Transfer Holding since there is no time constraint on those phases.

The second assumption applies to determining the method in which facilities are accepted from SFIA into EM-60 and from Transfer Holding to the Decontamination phase of EM-40. There is not a standard program-wide policy on the method of selecting facilities into the process. Therefore, a system was devised that would calculate a proportionate number of each facility type to be accepted into the next phase (Pre-Stabilization Surveillance and Maintenance or Decontamination) based on the number of facilities of that type remaining and the cost of accepting each facility. This is explicitly stated by the following equations:

$$F_tP = F_t/F$$

$$B_t = B * F_tP$$

$$F_{t0} = B_t/C_t$$

where F_t is the number of facilities of type t remaining in SFIA or Transfer Holding, F is the total number of facilities, F_tP is the percentage of facilities of type t remaining, B is the total budget for accepting new facilities, B_t is the budget for accepting facilities of type t, C_t is the cost of facilities of type t in the next phase, and F_{t0} is the number of facilities of type t accepted into the next phase. While this is not the method that EM-40 and EM-60 necessarily follow in accepting facilities into the process, it is the best standard method available at this time.

Finally, the model assumes that EM-60 will receive a minimum funding level equal to that which is necessary to process the facilities that have already been accepted into the program.

4 OPERATING THE MODEL

While the model is designed to simulate the entire program, it can also be used on a site specific basis. Here the model was run using test data for a single site in the Environmental Management program. The site includes eight facility types out of the possible 22. Three different policy scenarios were examined: increased funding for EM-60 to complete facilities in the minimum amount of time with constrained EM-40 funding at 50% of EM-60 funding, equal funding for both EM-60 and EM-40 with 50 percent more facilities being added from SFIA, and equal funding for both EM-60 and EM-40 as the control scenario. The model runs from fiscal year 1995 through fiscal year 2070.

5 RESULTS

In the first scenario, EM-60 completes all the facilities in the minimum time allowed; however, a bottleneck forms in the transfer holding phase since EM-40 does not have the funding to process all the facilities when they are received. As seen in Figure 2, the low EM-40 funding causes a delay and the last facility is not accepted into the Decontamination phase until 2045.

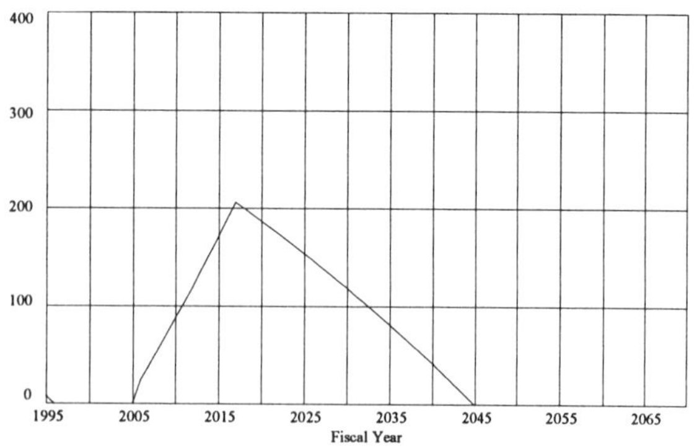

Figure 2: Total Facilities in Transfer Holding Phase

In the second scenario, EM-60 has only enough funding to process the facilities already in the program. Therefore, as seen in Figure 3, the additional facilities from SFIA do not enter the Pre-Stabilization Surveillance and Maintenance phase until 2007. This

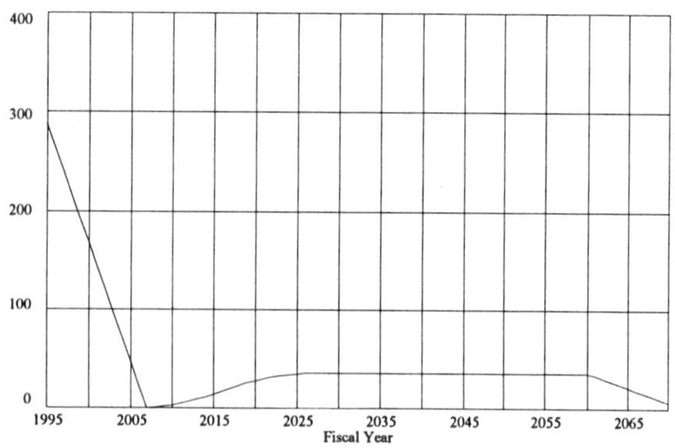

Figure 3: Total Facilities in Pre-Stabilization Surveillance and Maintenance Phase

influx of facilities at such a late date makes it impossible to complete the facilities by 2070. In fact, as shown in Figure 4, some facilities are still in the Transfer Holding phase in 2070.

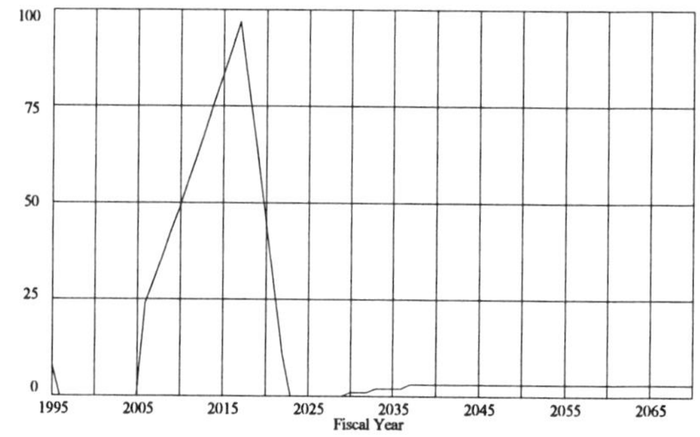

Figure 4: Total Facilities in Transfer Holding Phase

In the control scenario, EM-40 is able to accept facilities almost as quickly as EM-60 completes them as illustrated in Figure 5. The backlog in the first scenario is avoided, and instead of a peak of over 200 facilities, the Transfer Holding phase never reaches 100. Furthermore, all facilities have been absorbed into EM-40 by 2025 instead of 2045 in the first scenario or later than 2070 as in the second scenario.

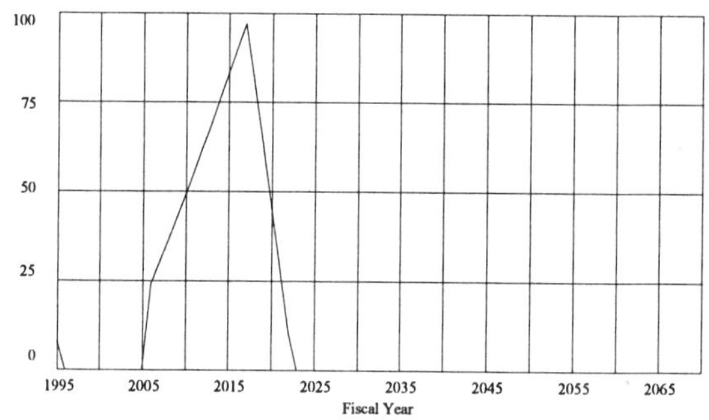

Figure 5: Total Facilities in Transfer Holding Phase

6 FUTURE REVISIONS

In future revisions of the model, additional variables will be added to aid in analysis of the best methods by which to accept facilities into the process. Such methods will include relative risk and difficulty level. Relative risk refers to the risk posed by a facility to the workers, the public, and the environment. Risk levels will be assigned to each facility type. This variable will enable the policy of completing high risk facilities to be evaluated. For the difficulty level, assigning such a score could show whether it would be cost effective to process the most difficult facilities first or last.

7 CONCLUDING REMARKS

This model has shown that simulation models can be successfully used as policy analysis tools for specific processes. Once the model was designed, the results for several different scenarios were obtained in a fraction of the time it would take to obtain them from more traditional means. The model was able to combine cost and schedule analyses in evaluating several policy alternatives. Furthermore, the model shows that non-physical transfer processes can be simulated as well as physical processes. Finally, the model successfully used an off-the-shelf software package for a unique government process.

ACKNOWLEDGMENTS

The author would like to thank Dr. Rick Shangraw for his guidance and support.

REFERENCES

Manary, Joel M. 1995. A Useful and Popular DoD Life Cycle Cost Model. *Estimator*. Fall: 23-45.

Krieger, James H. 1995. Process Simulation Seen As Pivotal in Corporate Information Flow. *Chemical and Engineering News*. 3/27: 50-61.

Breiner, Peter. 1995. The Political Logic of Economics and the Economic Logic of Modernity in Max Weber. *Political Theory*. 2: 25-47.

Dvorak, Paul. 1994. Yielding to Analysis. *Industry Week*. 7/18: 325-363.

Redden, Joeseph J. 1995. Military Keynote Address and Panel Discussion: Military Simulation and Modeling - Today - Tomorrow. *Proceedings of the 1995 Winter Simulation Conference*, ed. C. Alexopoulos, K. Kang, W.R. Lilegdon, and D. Goldsman, 1133.

Kristofferson, K., T.P. Oholleran, R.H. Powell, and E.C. Thiel. 1995. *National High-Level Waste Systems Analysis Plan.* Idaho National Engineering Laboratory.

AUTHOR BIOGRAPHY

SAMUEL M. HART is currently a consultant in the field of environmental management with Project Performance Corporation. He received his B.A. in Physics and Political Science from Ripon College in 1994 and his M.P.A. in Technology and Information Policy from Syracuse University in 1995. His areas of interest include simulation, process improvement, database design, and technology development.

Proceedings of the 1996 Winter Simulation Conference
ed. J. M. Charnes, D. J. Morrice, D. T. Brunner, and J. J. Swain

INTEGRATING PROCESS MAPPING AND SIMULATION

Charles R. Harrell
Kevin C. Field

Department of Manufacturing Engineering and Engineering Technology
Brigham Young University
Provo, Utah 84602, U.S.A.

ABSTRACT

This paper presents two different yet successful approaches that have been taken for integrating process mapping technology and simulation technology. The discussion is based on two actual implementations in which two different process mapping tools were separately integrated with ProModel simulation software. The process mapping tools used in the integration were Design/IDEF from Meta Software Corporation and ABC FlowCharter from Micrografx, Incorporated. Both of these tools have been widely adopted in government and private industries for process reengineering. Also, both tools are based on different paradigms which present unique challenges for use with simulation. Design/IDEF is based on the IDEF-0 modeling paradigm which is highly structured with limited adaptability. ABC FlowCharter is an unstructured process diagramming tool which is adaptable for doing all types of diagramming. In effect, they both plug into ProModel as a front-end interface. The implementation of the bridge is different in both instances, however. One approach uses a modified version of the process mapping tool while the other utilizes OLE automation without having to modify either the process mapping software or the simulation software.

This presentation deals with the issues and challenges encountered in each integration effort. We will also discuss how these challenges were addressed and how a successful solution was implemented. Finally, we will assess the suitability of integrating process mapping and simulation and compare the relative outcome of each. The successful integration of process mapping and simulation is a major step towards a complete integration of process reengineering technologies.

1 INTRODUCTION

With all the improvements in simulation software to simplify the modeling process, simulation is still largely perceived as a complex and sophisticated technology by most managers and engineers. Process reengineering tools such as process mapping software, on the other hand, have proliferated in recent years due largely to their ease of use and comprehensibility.

One of the reasons for the lag in application of simulation technology is that it has not been effectively integrated with more general purpose process mapping tools. Consequently, even though much of the process definition used in a simulation model is contained in a process map, a decision maker must start over from scratch in building a simulation model. The successful marriage of process mapping and simulation would extend the usefulness of process mapping and make simulation more acceptable among those doing process reengineering since it would be based on a familiar interface and a more simple paradigm.

Advances in software technology to support product interoperability has opened the door for integrating technologies such as process mapping and simulation that previously functioned only as stand-alone applications. Of particular importance is the development and acceptance of Microsoft's OLE technology which is opening the way for massive software integration. Many business planning tools are beginning to take advantage of these integrating technologies. This new capability holds the promise of being able to seamlessly integrate process mapping and simulation software without having to modify either product.

2 INITIAL EFFORTS TO COMBINE PROCESS MAPPING AND SIMULATION

The concept of combining process mapping with simulation technology is not new. Several attempts have been made to integrate process mapping with commercially available simulation products. Unfortunately, none of them has met with the degree of success that was anticipated. This is due largely to incompatibilities in both purpose and paradigm. Since insufficient data is provided in process map for running a simulation, additional information had to be manually added on the simulation side. This inevitably required some knowledge of the simulation product as well as the process mapping product.

In addition to a partial integration of process mapping and simulation technology, some process mapping products have recently been introduced with extremely limited simulation capability. Unfortunately, trying to model real world situations with such limited capability is almost futile. It is likely that simulation capability will continue to be enhanced over time. Meanwhile, the opportunity for tapping into the power of existing simulation products would be a real benefit to those doing process reengineering.

3 PROCESS MAPPING

One of the most widespread practices employed among leading businesses today is process reengineering. Process reengineering is the activity of radically rethinking how business processes are performed and even whether they should be performed at all. All aspects of the process are questioned and carefully designed. Many tools are becoming available for supporting business process reengineering activities including benchmarking software, CASE tools and process mapping tools.

Process mapping is perhaps the most widely practiced reengineering method, simply because it so quickly and easily captures and communicates a process flow. Process mapping essentially provides a graphic depiction of a process. A process is defined by process reengineering pioneers, Hammer and Champy (1993), as a collection of activities that create an output based on one or more inputs. There may be many uses of process mapping including documentation, visualization, analysis and communication.

Process maps are *static* models in that they are unable to capture the dynamics of the system. They are also *qualitative* in that their intended use is for conceptualization and documentation, but provide little or no quantitative analysis. Many commercial tools are available for use in process mapping. Inexpensive and easy-to-use flow charting software has made process mapping very popular.

Process mapping tools can generally be classified as either structured or unstructured. Structured process mapping tools impose a specific methodology for representing a process while unstructured tools leave it up to the modeler as to how to represent a process.

4 STRUCTURED PROCESS MAPPING

Several methodologies have been developed to structure process mapping. One established method of flow diagramming, defined by the American Society of Mechanical Engineers (ASME), utilizes the symbols shown in Table 1.

Table 1: Standard Process Flow Symbols

Symbol	*Description*
○	Operation
⇨	Transportation
▽	Storage
D	Delay
□	Inspection
▣	Combination Operation and Inspection

Another structured technique, which is the one used in this study, is the ICAM DEFINITION language (IDEF) developed by the US Air Force through its ICAM program. This language is essentially a schematic modeling tool used to provide a universal modeling language. The techniques developed by the Air Force pertain to the concepts of planning (IDEF-0), information system design (IDEF-l), and stochastic system evaluation (IDEF-2). The IDEF-0 methodology is the one used for process mapping and is based on a paradigm of activities, inputs, outputs, controls and mechanisms or resources (see Figure 1).

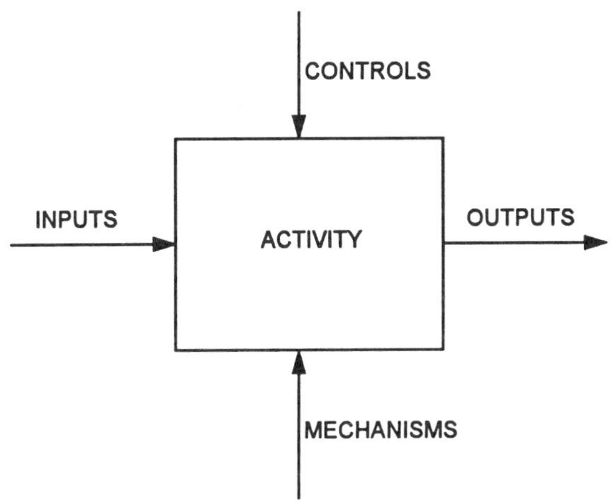

Figure 1: IDEF-0 Modeling Paradigm

As illustrated in Figure 1, inputs, outputs, controls and mechanisms must always enter or leave on the sides indicated. According to the paradigm, all processes can be defined in terms of these components. Figure 2 shows an example of a simple IDEF-0 diagram.

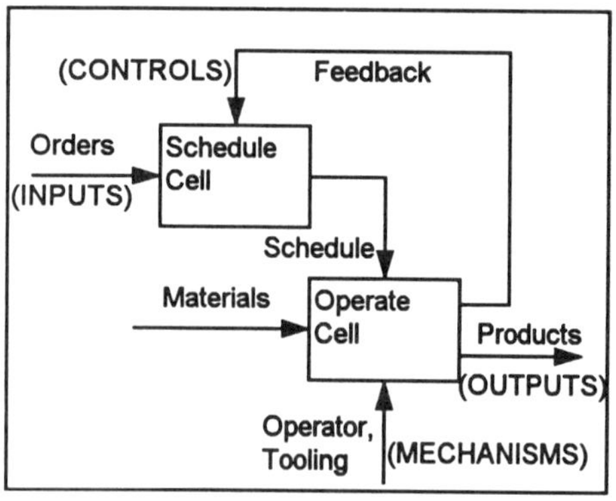

Figure 2: Example of a Flow Diagram

Diagrams can be constructed hierarchically so that a process can be decomposed into subprocesses.

There are certainly advantages to having a formalized method for diagramming a process. For one thing, it imposes structure on the process and provides categories for all of the elements of the process. It also standardizes process mapping so that diagrams are easily communicated and are consistent. The problem with structured modeling methods such as IDEF from a simulation point of view is that they may provide more or less information than what is needed for running a simulation. In order to use a structured process mapping tool directly with simulation, some modifications to the methodology may be necessary.

On the positive side, there are many close parallels between the IDEF-0 paradigm and common simulation paradigms. They may be summarized as follows:

- Both utilize inputs and outputs
- Both have activities
- Both utilize mechanisms or resources in performing activities
- Both use controls to determine where, when and under what conditions the activity is performed.

While many of the same components are present in both paradigms, they are not always thought of in the same way due to the differences in use of the models being created.

5 UNSTRUCTURED PROCESS MAPPING

Many unstructured process mapping or flow charting tools are available which provide a simple and versatile interface for diagramming business processes. Most of these flow charting tools support free diagramming which follows no strict rules as long as the process relationships are communicated clearly. The most simple diagrams consist only of boxes representing the activities performed, diamonds representing decisions, and arrows depicting the activity sequence (see Figure 3). These diagrams can become extremely complex and can be enhanced by built-in or user-defined data fields to include additional information such as costs, activity times, etc.

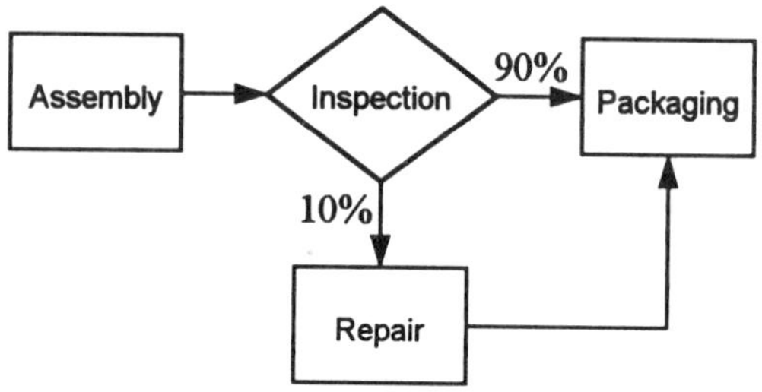

Figure 3: Example of a Flow Diagram

Obviously, an unstructured diagramming tool is more suitable for integration with simulation since it can be more easily adapted to a simulation paradigm.

6 PROCESS SIMULATION VS. SYSTEM SIMULATION

Regardless of whether a structured or unstructured process mapping tool is used, there is still a fundamental difference in perspective when shifting between process mapping and simulation (Harrell and Tumay 1994). This difference lies in the fact that process mapping by nature is *process* oriented while simulation tends to be *system* oriented. A process orientation focuses on the *logical flow* of entities or work items through a series of *activities*. A system orientation is based on the *physical flow* of entities through a series of *work stations*. This difference is not arbitrary but rather is a direct reflection of the nature and purpose of the activity. Process mapping is generally a high level depiction of a process defining only *what* happens to entities. It doesn't have to deal with *how*, *where* and *when*. Simulation, on the other hand, must have information on the mechanics of the process in order to imitate the actual drivers of the process. The most natural way to define the physical operation of a process is at a system level which is the detailed implementation of the process level (see Figure 4).

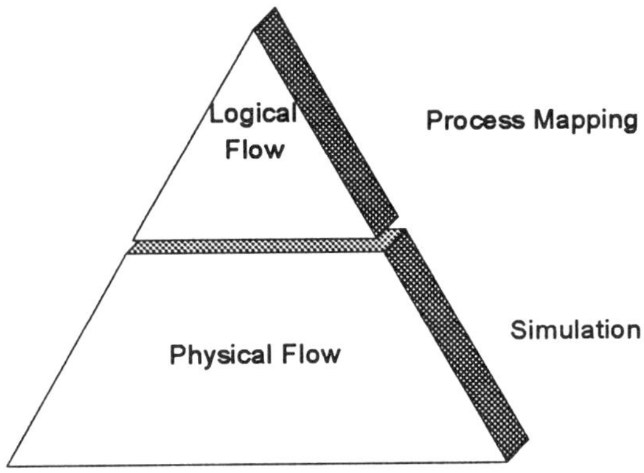

Figure 4: Relationship between Process Mapping and Simulation

7 INTEGRATING DESIGN/IDEF AND PROMODEL

Since Design/IDEF is based on a structured modeling methodology, the challenge was to extend the definition capability without deviating from the basic IDEF methodology. Some of the extensions made are as follows:

- The addition of entity attributes.
- The addition of input buffers to every activity.
- A way to model different entity types.
- The addition of data fields to capture dynamic information.

The implementation of the integration was to have the entire model built using Design/IDEF and then create a simulation model file that could be read in and executed by ProModel. The downside to this approach is that the Design/IDEF code had to be modified to allow input of simulation data within Design/IDEF. On the positive side, the user doesn't need to get into ProModel and hence deals with only a single user interface for both process mapping and simulation.

8 INTEGRATING ABC FLOWCHARTER AND PROMODEL

The ABC FlowCharter integration with ProModel ended up providing a more intuitive interface than the Design/IDEF integration because ABC FlowCharter is not based on a structured methodology. A paradigm for process mapping was developed to be able to support dynamic modeling.

The integration was performed using OLE automation technology which provided the advantage of not having to modify either application. A separate application was written that communicated with ABC FlowCharter using OLE automation to display simulation-related property sheets that are displayed as shapes and connections are created in ABC FlowCharter. Once the model is defined, it is translated by this separate application to a ProModel model and simulated.

The methodology used for defining a simulation model is very similar to that used to define a process flow diagram. A simulation model consists of shapes and connections where shapes define the objects in the model (i.e. activities, resources and entities) and connections represent relationships between the objects (i.e. arrivals, routings and resource usages). The OLE automation enabled intelligent connections to be made based on the object types connected. The graphic properties of connections could be modified automatically to provide feedback to the user as to the type of connection made.

9 CONCLUSIONS

The problems encountered in integrating process mapping and simulation were due primarily with attempting to maintain a process mapping paradigm. This was necessary in order to build utilize process mapping as the basis for doing process simulation. Since Design/IDEF enforces a particular paradigm, it didn't provide as much latitude to work with.

While the integration works well for modeling even complex process logic, process mapping tools are not well suited for capturing the dimensional and many other physical aspects at the system level.

In conclusion, the integration of process mapping with simulation technology at the process level was successful. Both implementations, while based on different methodologies and implemented differently, provide seamless, single-interface modeling. Both implementations also provide considerable modeling flexibility including activity-based costing, parallel processing and hierarchical modeling. The growing use of these two applications attests to their suitability for use as an effectively integrated process mapping and simulation tool.

REFERENCES

Hammer, M., and J. Champy. 1993. *Reengineering the Corporation*. New York: HarperCollins.

Harrell, C., and K. Tumay. 1994. *Simulation Made Easy.* Norcross, Georgia: IIE Press.

AUTHOR BIOGRAPHIES

CHARLES R. HARRELL is an Associate Professor in the Department of Manufacturing Engineering and Engineering Technology at Brigham Young University. He received his Ph.D. degree in Manufacturing Engineering from the Technical University of Denmark in 1988. He has worked for Ford Motor Company and Eaton Kenway in simulation and systems design and has performed many simulation consulting projects. He is founder and chairman of PROMODEL Corporation. Dr. Harrell is a senior member of IIE and SME.

KEVIN C. FIELD is an M.S. student in Manufacturing Engineering at Brigham Young University. He received his B.S. degree in Mechanical Engineering from Brigham Young University in 1994. He is currently working as a simulation support engineer for PROMODEL Corporation.

Proceedings of the 1996 Winter Simulation Conference
ed. J. M. Charnes, D. J. Morrice, D. T. Brunner, and J. J. Swain

SIMULATION FOR INTRA- AND INTER-ORGANISATIONAL BUSINESS PROCESS MODELLING

George M. Giaglis
Ray J. Paul

Department of Computer Science and Information Systems
Brunel University, West London
Uxbridge UB8 3PH, UNITED KINGDOM

Georgios I. Doukidis

Department of Informatics
Athens University of Economics and Business
Patission 76, Athens 104 34, GREECE

ABSTRACT

Business process modelling (BPM) is an increasingly emerging field of simulation application. Although it has been practically demonstrated that simulation can be an effective tool for business redesign, there does not exist a comprehensive framework to explain the characteristics of business processes and identify specific requirements for their modelling. Furthermore, hardly any attention has been paid to the modelling of inter-organisational business systems. In this paper, we examine the nature of business processes in the light of modern change management approaches and propose a set of requirements for their modelling. We then concentrate on inter-organisational processes and argue that modelling problems can be much more difficult to overcome when more than one business is involved, mainly due to the multiplicity of decision making levels involved and the subsequent need for multi-level output analysis. Based on an empirical study, we illustrate the practical problems of modelling inter-organisational business systems and suggest desirable characteristics of simulation packages for that purpose.

1 INTRODUCTION

A multitude of change management concepts have emerged during recent years to help modern enterprises in their efforts to adapt in a constantly changing business, social, and technological environment. The most popular of these approaches include:

- Business Process Re-engineering (BPR) (Hammer 1990, Davenport and Short 1990)
- Continuous Process Improvement (CPI) (Harrington 1991)
- Total Quality Management (TQM) (Oakland 1993)
- Organisational Transformation (OT) (Adams 1984)

Each approach differs significantly in the scope and range of the anticipated changes, the management tools utilised to achieve change, and the business contexts in which they can be used. However, they all have in common that they require businesses to model the ways in which they currently operate, to identify opportunities for change, and to design and implement alternative ways of carrying out business processes.

In view of the above, Business Process Modelling (BPM) has recently received widespread attention and has been acknowledged as an integral part of any change management project. Different tools and techniques have been proposed for BPM, and simulation has been identified as an extremely useful tool for this purpose (see for example Tumay 1995, Swami 1995, Bhaskar et al 1994). Despite the availability of these tools, it has been reported that companies are usually facing significant practical problems when trying to model in detail the way they operate (Hansen 1994) or to implement changes in existing environments (Galliers 1994).

Many reasons can be attributed to this difficulty, the primary ones being:

- the complexity of most real-world business processes
- their stochastic and usually unpredictable behaviour
- the interdependencies between individual tasks
- the informal nature of many tasks which makes their analysis and documentation profoundly problematic
- the different perceptions of users regarding the way in which work is being done.

Such modelling problems can become very significant in large, complex organisational settings, especially in cases where more than one business is involved (inter-organisational systems). For example, Business Process Re-engineering projects, where inter-organisational processes almost always play an important role (see for example Riggins and Mukhopadhyay 1994), have been reported to have a large proportion of failures in practice. In this paper, we first examine the nature of the problems of business modelling in general and identify a set of requirements that should be met if a process modelling exercise is to be successful. We then concentrate on the modelling of Inter-Organisational Business Processes and examine their distinct characteristics and requirements for modelling.

Simulation modelling is assessed as a potential tool for BPM (at intra- and inter-organisational levels). Finally, an example of inter-organisational business modelling is presented to help clarify some practical modelling problems that might arise.

2 BUSINESS PROCESSES: CHARACTERISTICS AND MODELLING REQUIREMENTS

2.1 Process-Based Organisational Approach

Most of the change management concepts identified above (especially BPR) adopt a new look to organisations based on the processes they perform rather than the functional units, divisions or departments they are divided into. Processes are defined as 'structured, measured sets of activities designed to produce a specified output' (Davenport 1993).

Typical examples of major business processes include the purchasing of raw materials from suppliers, the use of these materials to produce goods and/or services, the delivery of these goods/services to customers, the acquisition of new customers, the development of new products according to customer needs, etc. It is obvious that each of these processes requires the co-operation and synchronisation of different functional units in order to be successfully performed. It is also typical for a business process to cross organisational boundaries and extend to third parties (customers, suppliers, etc.). Of course, processes can be described at different levels of detail depending on the abstraction put into analysing the organisation. Typical business processes like those identified above, can be divided into sub-processes which can be further split until the level of individual business tasks.

2.2 Requirements for Business Process Modelling

It has been argued (Willcocks and Smith 1995, Galliers 1993) that businesses and business processes are sufficiently complex systems and therefore carefully developed models are necessary to understand their behaviour in order to be able to design new systems or improve the operation of existing ones. As businesses are essentially 'socio-technical' systems, we can distinguish the basic requirements of the decision makers regarding the modelling process in two separate areas: 'Technical' requirements which refer to those needs that call for the application of engineering principles in process analysis and design, and 'Political' requirements which refer to the needs that emerge from the social nature of business systems. These requirements include:

Technical Requirements

Formal Modelling. Formal engineering principles should be adopted during the modelling process in order to enable the development of models that can be readily understood and agreed upon by all parties, thus providing a common basis for decision making.

Quantitative Modelling. Managers need to have quantitative information that will allow for informed decision making (e.g. cost-benefit analysis) and for direct comparison between alternative system designs.

Stochastic Modelling. Modelling should take into account the stochastic nature of business processes, especially the way in which they are triggered by external factors and should allow for representation of and experimentation with situations where a great degree of uncertainty exists. Sensitivity analysis of business models becomes a significant issue in this case.

Model Documentation. Models should be easy to document for exchanging information between modellers, analysts, and decision makers. Model documentation can also be used as a reference in subsequent modelling exercises and/or if the model development teams change.

Model Adaptability/Reusability. Models should be easily updatable to follow changes in actual processes. Thus, they can be continuously used for future modelling exercises. Reusable models could assist in reducing the cost of model building and can provide an additional means of justifying the initial investment.

Objective-driven Modelling. BPM is usually performed having in mind specific business goals to be achieved through the process redesign exercise. The evaluation of alternative configurations is therefore highly dependent on the objectives of the particular study. Business models should reflect this requirement of decision makers and allow for output analysis that can be configured according to objectives so as to provide alternative views of measuring business performance.

Political/Social Requirements

Feasibility of alternative designs. Modelling and decision making in business contexts should take into account such factors as legislation restrictions, user acceptance of changes, etc. It is not sufficient to simply derive a particular system configuration that seems to optimise business performance, if the changes required in business processes cannot be practically implemented for this configuration.

Communication of Models. Business models are often used in brainstorming sessions of business management in order to assist in deciding changes. The models should therefore allow for easy communication of results between different parties. Moreover, generating alternatives and modifying the model during the decision making process is another very important aspect of business modelling, as managers clearly need to be able to interact with the models as new information or ideas emerge during brainstorming sessions.

User Friendliness. Modelling tools should be easy to use to allow users of the processes to be personally involved in the modelling exercise. The personal involvement of users should increase the confidence of the whole enterprise in the change initiative, thus enabling a greater degree of acceptability of the derived results.

Business Process Modelling approaches should combine the requirements identified above if they are to be proven useful tools for business decision making. In the next section we will assess the potential of simulation modelling as a suitable BPM technique.

2.3 Simulation as a Tool for Business Process Modelling

Simulation can be an invaluable tool for BPM, as it incorporates characteristics and capabilities that can accommodate all the requirements identified above.

Formal Modelling: Simulation is a formal and robust technique. It does not rely heavily on mathematical abstraction therefore it is suitable for modelling even complicated management systems (Pidd 1992).

Quantitative Modelling: Simulation is basically a numerical technique, therefore it can be used to generate quantitative output data on various parameters that influence a business system performance. Output Data Analysis, Experimental Design, and other techniques can be employed to ensure a significant degree of mathematical robustness at every stage of a simulation project.

Stochastic Modelling: Statistical representation of real-world uncertainty is an integral part of simulation models. Indeed, simulation is perhaps the most suitable modelling technique regarding its ability to capture the dynamic and stochastic behaviour of real systems. Sensitivity Analysis for example can be employed to assess a simulation model's validity with respect to variations in the values of (unknown) system parameters.

Model Documentation: The development of a simulation model requires certain assumptions about the real system which can be documented as the model is being developed. Therefore, documentation of simulation models can be a relatively easy task. However, users in practice do not always pay enough attention to the documentation process. Simulation packages which allow for automatic documentation of models can prove particularly useful for this purpose, although they cannot entirely replace the modeller's role.

Model Adaptability/Reusability: Simulation models can easily be updated to reflect changes in real world processes. With respect to BPM, a great opportunity exists to integrate workflow capabilities in simulation environments to support not only the modelling and redesigning exercise, but also the actual carrying out of business tasks, thus resulting in highly flexible and continuously reusable models.

Objective-driven Modelling: Due to their flexibility, simulation models can be customised to serve multiple purposes of management. Modellers can specify alternative output measures and apply different output data analysis techniques to simulation models, thus allowing for multiple uses of a single business model according to management requirements.

Feasibility of alternative designs: Simulation as a process is meant to help with identifying appropriate solutions to complex decision problems. The feasibility of alternative designs in a business context is essentially a step that has to be built into the assumptions made during model development. If certain managerial, legislative, or other restrictions occur, they should be taken into account during the experimentation phase by adhering to these assumptions. In this way, 'political' requirements can be easily accommodated by simulation models.

Communication of Models: Simulation models, especially when combined with graphical, animation, and/or interactive characteristics are probably the best means of communicating the essence of a model to managers and decision makers.

User Friendliness: Simulation models for business process analysis can be as friendly as their developers choose them to be. In general, simulation allows for a great degree of user friendliness (e.g. through graphical user interfaces) which is supported by the majority of existing simulation packages.

In the next section we will concentrate on inter-organisational business settings and try to understand the characteristics of inter-organisational business processes, to identify additional modelling requirements, and to assess the potential of simulation in this context.

3 INTER-ORGANISATIONAL BUSINESS PROCESSES: CHARACTERISTICS AND ADDITIONAL MODELLING REQUIREMENTS

3.1 Characteristics of Inter-organisational Processes

Problems of BPM become even more difficult to tackle when considering processes which extend beyond the boundaries of a single organisation and include multiple businesses in the value chain. Examples of such processes include purchasing (supplier involvement), shipping (when third parties are employed to transport goods), sales (customer involvement), etc. Inter-organisational involvement might exist even in processes which seem at first to be internal, but require (explicitly or implicitly) the co-operation of third parties.

A single organisation cannot control the behaviour of external parties in the way it can with its own resources (people, equipment, etc.). Therefore, the degree of uncertainty is substantially increased with possible implications for the validity of the derived models. Modelling of inter-organisational processes must be exercised with great care to avoid such pitfalls and sensitivity analysis becomes an extremely important issue in this case.

Furthermore, modelling requires extensive data collection and organisation in order to understand and structure the behaviour of the system under study. In the case of external players, such data might not be available, so businesses usually have to rely on additional assumptions that might further jeopardise the validity of the business model.

Decision making becomes extremely difficult in situations where multiple players are involved, since the decisions made by managers in one firm are affected by the (uncontrollable) behaviour of outside parties. For example, a company might decide to adopt a Just-In-Time method of production based on results derived by a, perhaps simulation, modelling exercise showing that such a strategy could streamline the company's operation, cut down on operating costs, and increase timeliness and quality of service to customers. However, the company's suppliers might not be able or willing to deliver raw material at short notice and in small quantities (an essential prerequisite of JIT implementation). If managers decide to adopt such a JIT approach and redesign the respective business processes without a prior assessment of the possible outcomes according to alternative supplier reactions, results might be unexpected. Interactions between players should always be taken into account in inter-organisational modelling and decision making.

There are cases where the decision to model inter-organisational business processes comes from all the parties involved. For example, a company might agree with its suppliers to develop a common model of their trading communication in order to identify opportunities for change. Such a case is presented in the following section. In such a setting, most of the aforementioned problems become less important, since the behaviour of all parties is generally more controlled and input data can be more easily available. The new problem that arises in this case is the multiplicity of decision making levels. When for example, two companies (Company A and Company B) develop a single model to represent their trading interface, decision making can be performed either at a single-site level (e.g. Company A experiments with the model to identify opportunities for change within its own operations) or at a multi-site level (i.e. joint brainstorming sessions of the two companies).

In the first case, the model should allow for experimentation only with those parts (sub-models) which represent processes that are performed by Company A (since A cannot control or change the processes of B), and should also allow for multiple output analyses, both at an intra-organisational level (to assess impacts of changes within the company) and at an inter-organisational level (to understand the consequences of individual decisions to the whole system, as these might influence the reactions and future behaviour of Company B).

In the second case, multi-level output analysis is again of paramount importance, since companies need to assess the impact of changes both on their individual performance and in the efficiency of the value chain as a whole.

We can conclude that inter-organisational business processes differ significantly from processes limited to one organisation. Although the requirements reported in previous sections for BPM are still valid, Inter-organisational Business Process Modelling (IBPM) requires additional considerations which will be outlined below, after presenting a small-scale case study of Inter-organisational modelling.

3.2 A Case Study Of Inter-Organisational Simulation Modelling

The work presented here was part of a wider BPR project aiming at redesigning trading communication between a major pharmaceutical company in Greece (Company A) and its main distributors. The sub-project that is described here involved the medical division of Company A which sells small equipment and medical consumables to hospitals and healthcare institutions, via a number of distributors throughout Greece. Each distributor covers a specific geographic area and is responsible for delivering products to customers from its own inventories which are replenished by the main inventory of Company A at regular intervals.

One of the problems identified by the management of Company A (due to space constraints we will be limited only to one of the areas that was actually examined in the project) was the long time it took for a hospital to receive goods from the time it had placed an order. The long lead times were causing customer dissatisfaction and their reduction was considered as a top priority by the management of Company A.

The ordering process was analysed and found to be unnecessarily complicated: Customers placed their orders *either* with Company A *or* with Company A's salesmen who visited the customer sites *or* directly to the distributors. However, all orders had to be authorised by Company A before the distributors proceeded with order execution. This policy resulted in unnecessary delays as the distributors and the salesmen had to forward the orders to Company A, which gave the authorisation (usually with no amendments) and forwarded the orders again to the distributors. Figure 1 illustrates the process.

A simulation model to represent the aforementioned ordering procedure will necessarily include activities performed and controlled by Company A (including Order Receipt, Order Processing, and Order Forward to Distributors), but also activities performed by the distributors (e.g. Forward Order to Company A), by the salesmen, and by the customers. Furthermore, certain activities require a degree of co-operation between parties, thus reducing the flexibility of individual firms to initiate changes in the respective processes.

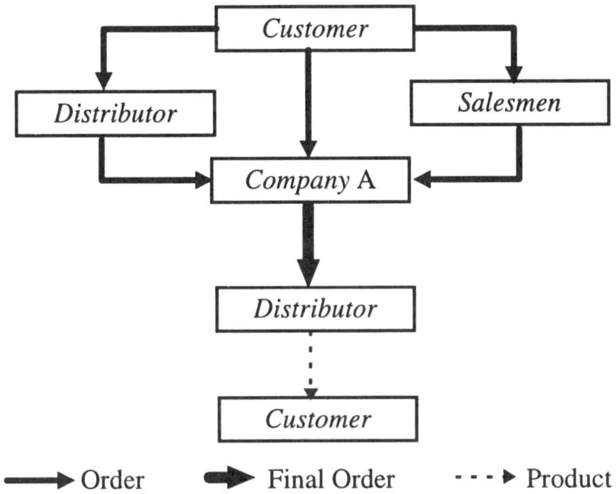

Figure 1: Ordering Process

For example, looking at the model from Company A's perspective, one might want to assess the possibility of changing the ordering process in such a way that salesmen have greater independence and can authorise orders and forward them directly to the distributors, thus relieving Company A from the additional administrative burden. If this scenario is simulated and seems to improve performance in Company A, it might still not be applicable in practice because the distributors might not accept having too many communication channels open to receive orders, as this could influence their own throughput. In this case, the burden of order authorisation in Company A is simply transferred down the value chain to individual distributors. The requirement is for a simulation model that will enable Company A not only to identify the impact of changes in their own performance, but also to measure the influence that the same changes might have on other parties (such as the distributors) in order to be able to more safely 'predict' their reaction to changes.

Another solution might be for distributors to proceed with the orders they receive from customers without waiting for an authorisation from Company A. Such a policy would surely eliminate the delay of orders delivered to distributors and waiting idle until they are authorised, thus resulting probably in reducing the overall order lead time (which is the objective of the whole exercise). However, this is again a policy that has to be implemented by a party outside of the decision sphere of Company A. Distributors might not be willing to adopt such a policy as it would mean an additional level of responsibility for them. Therefore, they have to be persuaded that such a policy would be in their interest too. How can this happen if the model used does not provide output data to assess the performance of the individual players? And what would happen if distributors wanted to experiment only with the parts of the whole model that describe their own behaviour to improve their understanding about the impact of proposed changes on them? If the simulation model is not modular and easy to decompose, they would probably have to build a new model from scratch just to represent the same workings which are already included in the initial model.

A third option for Company A and its distributors might be to co-operate in a joint effort to improve the quality of service they offer to their customers. They might, for example, want to investigate the possibility of adopting EDI to electronically exchange orders, thus reducing unnecessary delays during communications. In this case, the requirement would be for a simulation model that will allow them (a) to fully understand the interactions implied by the new way of communication before committing to changes, (b) to assess their individual performances under the new scheme, (c) to be able to measure the overall improvement in the value chain in terms of total order lead time reduction, and (d) possibly to communicate the results with their customers to persuade them to adopt EDI as a means of order submission as well.

Surely, there are a lot of alternative ways to implement the ordering process in a way that can fit Company A's management objectives. These ways need to be modelled keeping an eye on the influence changes might have on the performance of individual players as well as the whole system. Such a level of modelling and analysis cannot easily be implemented with existing simulation packages, as it requires modular model design and multiple levels of output analysis.

A way to overcome the modelling problems would be to use a general purpose programming language to implement a modular model. An approach is to implement each player (customers, distributors, salesmen, and Company A) as an independent sub-model. Each sub-model communicates with others when necessary (for example when an order is forwarded) via a message passing mechanism which initiates the creation of an entity in the TO model, while it places the respective entity in an idle state in the FROM model. For example, when distributors forward orders to Company A for authorisation, orders remain idle in the distributor submodel, while a 'new' order is created in the Company A submodel. When the order is authorised, it is 'sunk' in the Company A submodel and the respective entity in the distributor submodel becomes busy again (forwarded to be processed).

Although this message passing mechanism facilitates 'independent' modelling of the various levels of decision making of the system and allows for output analysis of submodels, the whole process is clearly not user-friendly and cannot provide the necessary degree of adaptability and reusability required for business models. A far better solution might be to use a user-friendly simulation package for model development, provided that a package to satisfy the aforementioned requirements actually exists on the market.

3.3 Additional Requirements for Inter-Organisational Business Modelling

Based on the analysis of the case study presented above, we can derive the following additional requirements for Inter-organisational Business Modelling:

Modular Model Design. A holistic approach to business modelling is necessary to identify implicit interdependencies among processes, even when they are performed by different organisations. On the other hand, different parties should be able to use suitable sub-models to assess their own performance (but, at the same time, keeping an eye on the influence of 'local' changes on 'global' performance). There is clearly a need for modelling conventions that will allow for modular model implementation and for experimentation with selected sub-models.

Modular Model Analysis. Models that represent the workings of more than one business unit should also allow for a multi-level output data analysis to accommodate the decision making needs of the individual parties involved as well as any of their combinations. Business process configurations that are derived from these models should optimise the performance of individual firms and, at the same time, streamline the efficiency of the whole system. Improvements achieved by individual players should not be lost downstream or upstream in the value chain.

Model Decomposition and Integration. Implementation of modular models should be achievable even if this is not the initial target of the modelling exercise. For example, two firms might develop models independently of each other and at a later stage wish to link these models into a concrete inter-organisational model. To enhance model reusability, individual models should be easy to link, without extensive modifications. In the same way, a single model might need to be decomposed to sub-models, when for example departments of an organisation need to assess their individual performance. Perhaps the only way to achieve problem-free model decomposition and integration, is by defining standard interfaces between models.

3.4 Simulation as a Tool for Inter-Organisational Business Modelling

Simulation models have the potential to become powerful tools for modelling inter-organisational business processes. However, in contrast to intra-organisational simulation, existing commercial products do not generally provide the necessary functionality to meet the requirements identified above.

Modular Model Design: Although simulation as a technique can theoretically be used for modular model development and use, existing simulation packages do not generally include such characteristics. At the time being, the requirement for modular model design can only be met if a general-purpose programming language or a simulation programming language is used to develop the model. We are aware of no existing simulation packages (i.e. software that allows model development and use without or with very limited programming) that include advanced functionalities to assist the user in modular model design. This does not mean that modular model development cannot be done with simulation packages. Rather it means that the user is left alone in performing this task, i.e. the package does not guide him/her towards defining appropriate sub-models that will *clearly* indicate the 'decision territories' of the firms involved in modelling.

Modular Model Analysis: Things become even more difficult when considering the issue of analysing inter-organisational models for the purpose of decision making. Again the problem seems to be that existing simulation platforms do not provide multiple levels of output analysis. Even worse, in the majority of cases output analysis is not provided at aggregate levels at all, and is usually limited to performance indicators within single functional units of the model (resources, activities, or queues). This however, cannot satisfy the requirement for assessing whole, inter-function extended, business processes which is usually needed in business change projects.

Model Decomposition and Integration: At the current status of non-existence of industrial standards to define the interconnectivity issues between simulation model components, this requirement cannot be easily satisfied. Only if individual models are developed on the same platform, can they be connected in a relatively painless way. If different simulation environments are used, then model integration is usually simply not possible, and a new model should be developed from scratch. Regarding model decomposition, things are slightly better since a single model can always be truncated to include only a subset of its initial range. Even in this case however, the user will probably need to carry out sub-model definition, to define new output analysis measurements, to 'fill' gaps generated by model truncation, etc.

4 DISCUSSION: THE ROAD AHEAD

Business process modelling carries several distinct characteristics that differentiate it from other types of modelling problems that simulation has been traditionally used for. This highlights the need for a focused approach to the development of software packages for business process simulation.

The requirements identified in this paper are basically meant as guidelines for prospective users or developers of business simulation models. Of course, each individual requirement carries a different weight depending on the objectives and characteristics of a particular change management project.

Two avenues for further research can emerge from the aforementioned analysis. The first refers to the development of a formal framework to explain the characteristics of business modelling and assess simulation under a variety of practical business problems. The second refers to a more holistic concept that might be necessary for an integrated approach to business modelling. This approach should combine simulation modelling with other computer-supported modelling, experimentation, and decision making tools in an effort to integrate business process modelling (as part of a redesign exercise) with business performance monitoring (as part of the actual carrying out of business tasks). Computer Aided Business Modelling and Monitoring (CABMM) can be the next target of applications to integrate Simulation, Expert Systems, Workflow Software, and internal business applications in an effort to institutionalise process modelling and measurement in modern businesses.

ACKNOWLEDGEMENTS

The case study work described in this paper was partially funded by Johnson & Johnson Hellas S.A. as part of a wider Business Process Re-engineering project undertaken by the Electronic Commerce/EDI research group of the Athens University of Economics and Business.

REFERENCES

Adams, J.D. (1984) *Transforming Work.* Miles River Press, Alexandria, VA, USA.

Bhaskar, R., Lee, H.S., Levas, A., Petrakian, R., Tsai, F. and Tulskie, B. (1994) Analysing and Re-engineering Business Processes Using Simulation. In the *Proceedings of the 1994 Winter Simulation Conference*, Lake Buena Vista, FL, USA, December 1994, pp. 1206-1213.

Davenport, T.H. and Short, J.E. (1990) The new industrial engineering: information technology and business process redesign, *Sloan Management Review*, Summer, pp.11-27.

Davenport, T.H. (1993) *Process Innovation: Reengineering Work Through Information Technology*, Harvard Business School Press, Boston, Massachusetts.

Galliers, R.D. (1993) Towards a Flexible Information Architecture: Integrating Business Strategies, Information Systems Strategies and Business Process Redesign, *Journal of Information Systems*, vol.3, no.3, pp.199-213.

Galliers, R.D. (1994) Information Systems, Operational Research and Business Reengineering, *International Transactions of Operational Research*, vol.1, no.2, pp.1-9.

Hammer, M. (1990) Re-engineering work: don't automate - obliterate, *Harvard Business Review*, July/August, pp.104-112.

Hansen, G.A. (1994) *Automating Business Process Reengineering: Breaking the TQM Barrier*, Prentice-Hall, Englewood Cliffs, New Jersey.

Harrington, H.J. (1991) *Business Process Improvement: The Breakthrough Strategy for Total Quality, Productivity and Effectiveness*, McGraw-Hill, New York.

Oakland, J.S. (1993) *Total Quality Management: The route to improving performance*, 2nd ed., Nichols Publishing, New Jersey.

Pidd, M. (1992) *Computer simulation in management science*, 3rd ed., John Wiley, Chichester.

Riggins, F.J. and Mukhopadhyay, T. (1994) Interdependent Benefits from Interorganisational Systems: Opportunities for Business Partner Reengineering, *Journal of Management Information Systems*, vol.11, no.2, pp.37-57.

Swami, A. (1995) Building the Business Using Process Simulation. In the *Proceedings of the 1995 Winter Simulation Conference*, Arlington, VA, USA, December 1995, pp. 1081-1086.

Tumay, K. (1995) Business Process Simulation. In the *Proceedings of the 1995 Winter Simulation Conference*, Arlington, VA, USA, December 1995, pp. 55-60.

Willcocks, L. and Smith, G. (1995) IT-enabled Business Process Reengineering: From Theory to Practice, In *The Proceedings of the 3rd European Conference on Information Systems*, Athens, Greece, pp.471-485.

AUTHOR BIOGRAPHIES

GEORGE M. GIAGLIS is a research student in the Department of Computer Science and Information Systems, at Brunel University, UK. He received his B.Sc. in Informatics from the Athens University of Economics and Business. He has published in the areas of simulation modelling for business systems, Business Process Re-engineering, Electronic Data Interchange, and Information Systems Investments Evaluation. His current research interests include Dynamic Modelling of Business Systems, Inter-Organisational Information Systems (IOS), BPR, and EDI. He is project manager in the Electronic Commerce/EDI research group in the Athens University of Economics and Business, he has participated in various EC funded research projects, and has also acted as a consultant for a number of companies in Greece.

RAY J. PAUL holds the first UK Chair in Simulation Modelling in Brunel University. He previously taught Information Systems and Operational Research at the London School of Economics. He received a B.Sc. in Mathematics, and a M.Sc. and a Ph.D. in Operational Research from Hull University. He has published widely (two books, over 100 papers in journals, books, and conference proceedings), mainly in the areas of the simulation modelling process and in software environments for simulation modelling. He has also acted as a consultant for UK Government departments, software companies, and private organisations. He is the Director of the Centre for Applied Simulation Modelling, and Head of the Department of Computer Science and Information Systems in Brunel University.

GEORGIOS I. DOUKIDIS is a Professor in Information Systems at the Athens University of Economics and Business. He received a B.Sc. in Mathematics from the University of Thessaloniki, a M.Sc. in Operational Research, and a Ph.D. in Simulation Modelling and Artificial Intelligence from the London School of Economics, where he also worked as a lecturer in the IS department. He has published 8 books and over 60 papers in the areas of Simulation, Expert Systems, and Information Systems Management and Strategy. His current research interests also include Electronic Commerce, EDI, and DSS. He is the Director of the Laboratory of AI & DSS as well as the Head of the Electronic Commerce/EDI research group of the University.

Proceedings of the 1996 Winter Simulation Conference
ed. J. M. Charnes, D. J. Morrice, D. T. Brunner, and J. J. Swain

A METHODOLOGY FOR INTEGRATING BUSINESS PROCESS AND INFORMATION INFRASTRUCTURE MODELS

Michael K. Painter
Ronald Fernandes
Natarajan Padmanaban
Richard J. Mayer

Knowledge Based Systems, Inc.
1500 University Drive East
College Station, TX 77840, USA

ABSTRACT

Many organizations today undertake Business Process Reengineering (BPR) and information infrastructure (i.e., network hardware, communications, and applications infrastructure) modernization efforts to drastically reduce costs and improve performance. While these efforts would appear to be mutually supportive and complementary in nature, they are rarely conducted jointly. Although it is known that changes to one of these two facets of the organization can produce significant impacts on the other, analysts and decision-makers have had only limited support for predicting the impacts of change to one or both of these two facets of the enterprise. BPR efforts are generally conducted using business process simulation tools. Information infrastructure modernization efforts are generally conducted using network simulation tools. What is lacking is an integrated method and the automated environment for studying the impacts of change to one of these two aspects of the enterprise on the other. This paper presents a simulation-based methodology enabling simultaneous consideration of changes to core business processes and the infrastructure mechanisms that support those processes.

1. INTRODUCTION

Many organizations today undertake Business Process Reengineering (BPR) and information infrastructure (II) modernization efforts to drastically reduce costs and improve performance. While these efforts would appear to be complementary, they are rarely conducted jointly. That is, although it might make sense to conduct BPR and II modernization efforts in a highly coordinated fashion, there has been little success to date in making the attempt. The importance of coordinating such efforts is obvious. Making changes to the logic and structure of a business process may introduce new requirements on the supporting infrastructure. Likewise, making changes to the network hardware, communications, and application infrastructure can have dramatic impacts on business process performance.

Although it is known that changing one of these two facets of the organization can produce significant impacts on the other, analysts and decision-makers have limited support for predicting the impacts of change. There are stand-alone methods and tools to assist with reengineering business processes. There are also stand-alone tools to help communication engineers analyze alternative hardware, software, and network configurations. What is missing is the ability to bridge these two analysis domains to determine new infrastructure requirements arising through changes to the business process. Similarly, there is little support for translating the impacts of infrastructure change on the processes the infrastructure is intended to support.

Without effective change impact analysis methods and tools there is an enormous risk that proposed infrastructure designs will have little benefit, or possibly even severely detrimental cost and/or performance impacts. Likewise, proposed business process changes, considered outside the context of their potential impact on the supporting II, may be equally fraught with risk. Lacking the means to integrate business process and II models, decision-makers are forced to rely on intuition and personal experience while committing the organization to massive process infrastructure change.

In this paper, we present a simulation-based methodology for change impact assessment enabling simultaneous consideration of changes to core business processes and the infrastructure mechanisms that support those processes. We call this methodology the BPR-II methodology. We also present preliminary findings from exploring the use of commercial off-the-shelf (COTS) business process modeling, process simulation, and network simulation tools to provide automated support for the methodology. Together, the BPR-II methodology and an accompanying automated support environment provide the ability to seamlessly link models of the business process models with models and supporting infrastructure. These models may then be used to conduct rapid, realistic cost/benefit studies of alternative process and infrastructure design concepts before they are implemented.

2. CHANGE IMPACT ASSESSMENT

The BPR-II methodology provide a holistic approach for BPR and II modernization analysis. Three model types (business process, application, and network models) are used to represent objects and relations that correspond to real world elements (e.g., network topology models representing actual computer and communications hardware). In addition to the three model types, there are physical and conceptual relationships and constraints between objects of different model types. For example, business process **X** requires application **Y**, application **Y** is only available on machines **M1** and **M2**, etc. These relationships and constraints are also represented in the BPR-II methodology.

Changes made within a model type can be analyzed within that model type to determine cost, cycle time, manpower implications, and so forth. However, because of interdependencies between the model types, they cannot be treated as though they are independent. It thus becomes necessary to somehow integrate the models to support more reliable analyses (See Figure 1).

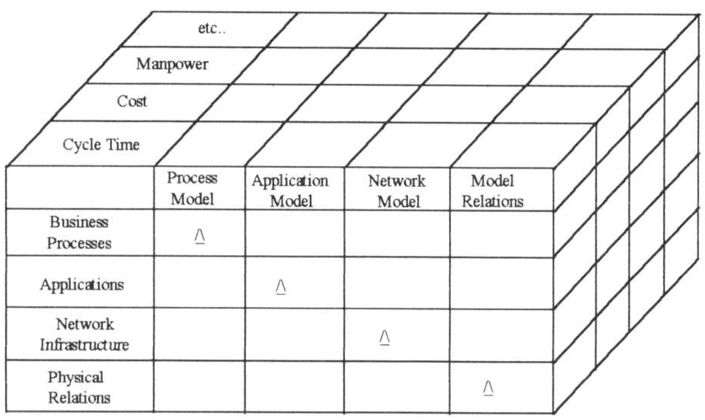

Figure 1. Dimensions of the Physical World Represented through Models

Table 1 shows three dimensions of the change impact analysis support by the BPR-II methodology. Delta symbols (Δ) in the table indicate areas of change. Question marks (?) represent the dependent variables of the analysis; i.e., the areas an analyst would be responsible for investigating in a change situation. For each situation and model type, the only independent variable that is modified is the model itself. For example, in Situation 1, we may wish to propose changes to the structure of one or more business processes. The analyst then determines the resulting impacts (model, performance, cost, and manpower) on the other three model types, as well as business process performance, cost, and manpower impacts.

Table 1. Change Types and their Unknown Impacts

	Situation		
Infrastructure Dimension	**1**	**2**	**3**
Business process	Δ	?	?
Applications	?	Δ	?
Network H/W and Comm	?	?	Δ

Thus, the methodology supports impact assessments of changes to business process structure and logic, application process change, network hardware and communications infrastructure change (e.g., topology changes, parameter set changes), changes in relationships or constraints (referred to above as inter-model mappings), and so forth. Listing and organizing proposed changes by the type of change involved helps the modeler/analyst isolate the changes that need to be made to the models, and devise an appropriate analysis strategy. For example, if the network hardware and communications infrastructure is the only aspect to be changed, and if an ambient traffic file representing the typical traffic generated by the existing business processes is available, there is no need to develop any process models. The analyst can simply elect to "run" the ambient traffic file on the proposed topology. By understanding the possible set of changes that may be considered and their implications, the methodology can be tailored to meet the more narrowly scoped analysis requirements. The following list briefly describes some of the change types that may be considered.

I. Changes in the business process model (e.g., eliminating artifacts of the process).
II. Changes in the application model (e.g., replace application).
III. Changes in the network computing strategies (e.g., centralized to distributed).
IV. Changes in the network hardware and communications infrastructure (e.g., routing, bandwidth)
V. Changes in physical or conceptual relationships among system components (e.g., machine used for an activity, applications residing on a machine).

Developing a methodological framework involved defining the procedures, techniques, heuristics, and languages of expression needed to analyze, predict, and quantify the impacts change for different situations. The use of simulation tools as part of the methodology was motivated by the need for mechanisms to help determine the impacts of shared resources and of probabilistic situations.

3. BPR-II METHODOLOGY

The IDEF3 Process Description Capture [Mayer 92] method serves as the key mechanism for process knowledge capture and organization in the BPR-II methodology. IDEF3 process descriptions are used to capture a definition of the process at the business, application, and network processing levels and to directly generate the structure and logic of simulation models reflecting these levels.

3.1 IDEF3 Method Overview

IDEF3 is a scenario-driven process flow modeling method designed to capture descriptions of the precedence and causality relations between situations and events in a form that is natural to domain experts. IDEF3 provides a structured method for expressing knowledge about the "behavioral" aspects of an existing or proposed system. IDEF3 helps structure user descriptions of enterprise processes in a way that facilitates the construction of analysis and design models.

Two modeling modes exist within IDEF3: process flow descriptions and object state transition descriptions. Process flow descriptions are intended to capture knowledge of "how things work" in an organization; object state transition descriptions summarize the transitions an object may undergo during a particular process. Both descriptions contain units of information that form the basic units of an IDEF3 description.

IDEF3 uses the "scenario" as the basic organizing structure for establishing the focus and boundaries for the process description. The basic syntactic unit of IDEF3 graphical descriptions is the Unit of Behavior (UOB) box, which represents a real world process. Each UOB is described in terms of other UOBs (called a decomposition) and in terms of a set of participating objects and their relations (called a UOB elaboration). UOBs are connected to one other via junctions and links. Junctions provide the semantic facilities for expressing synchronous and asynchronous behavior among a network of UOBs. There are three types of IDEF3 links: 1) temporal precedence, 2) object flow, and 3) relational. Relational links are provided to permit constraint capture not accommodated by the default semantics of the precedence and object flow links. Figure 2 presents an example IDEF3 diagram.

IDEF3 based descriptions are used to automatically generate WITNESS™ simulation code in the target language using PROSIM™, an intelligent simulation modeling tool developed at KBSI (Benjamin et al. 93, Padmanaban et al. 95). This description-driven approach significantly reduces the time and effort involved in simulation model development.

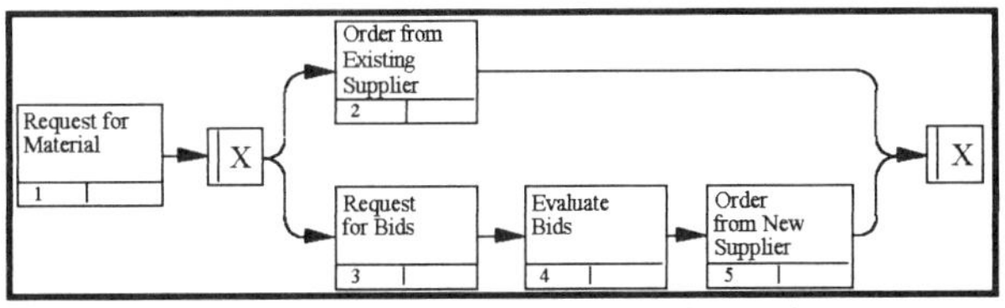

Figure 2. Example of a Process Flow Diagram

The IDEF3 method provides the foundation for acquiring and documenting process knowledge at the business, application, and network processing levels. This task is one of several included in the BPR-II methodological framework. The following section expands on how IDEF3 is used and describes the additional tasks involved in the methodology.

3.2 BPR-II Change Impact Analysis Steps

Developing a methodological framework involves defining the procedures, techniques, heuristics, and languages of expression needed to analyze, predict, and quantify the impacts change. In this section we describe the methodology developed to analyze the impacts of change depicted in Table 1. In general, the following steps are applied.

I. Establish the boundaries of the domain.

II. Catalog domain assets, their relevant performance characteristics, and their fixed and variable costs. Six general categories of assets are listed as part of the domain: computer and communications equipment (e.g., computers, communication links, routers), applications, data sets, employees, sites, and facilities.

III. Capture information about network layout, topology, and performance characteristics.

IV. Catalog and define AS-IS domain business processes using IDEF3.

V. For each application that interfaces directly with the user and that is required to satisfy a business process, model (as a decomposition) the steps that the application performs. This constitutes a process description of the application.

VI. For each step of the application process that requires the use of a remote application or database (i.e., involving distributed processing and/or delay), model the communication process required. Model remote application processes as a decomposition.

VII. Classify and label all UOBs in the resulting IDEF3 descriptions. For convenience, the lower right-hand corner of the IDEF3 UOB box is used for a UOB classification label. Classification categories are as follows:

A. *Scenario (SCN):* UOBs with decompositions, where all child UOBs are classified as involving manual activity (MAN), man-machine interaction (MMI), or local application processing (L_APP).

B. *Manual (MAN):* UOBs involving manual activity and whose decompositions, if they exist, are comprised of UOBs involving manual activity only.

C. *Man-Machine Interaction (MMI):* UOBs involving man-machine interaction and whose decompositions are comprised of both manually accomplished UOBs and local application processes (L_APP UOBs).

D. *Process Data (P_DATA):* UOBs involving Central Processing Unit (CPU) activity.

E. *Transport Message (T_MSG):* UOBs involving message transport (data transmission) across communication lines.

F. *Local Application (L_APP):* UOBs involving local application processing and whose decompositions are comprised of UOBs involving Central Processing Unit (CPU) delay (P_DATA), possible message transport (T_MSG), and possible remote application processing (R_APP).

G. *Remote Application (R_APP):* UOBs that involve remote application processing and whose decompositions are comprised of UOBs involving CPU delay (P_DATA), possible message transport (T_MSG), and potentially additional remote application processing (R_APP).

IX. For UOBs involving message transport (T_MSG), specify source and destination nodes and a unique message identifier distinguishing messages originating from that UOB. For UOBs involving CPU delay (P_DATA), local application processing (L_APP), and remote application processing (R_APP), specify the node where processing occurs. For the leftmost UOB in the decomposition of a parent UOB labeled as involving local or remote application processing (L_APP or R_APP), specify the event type (i.e., interarrival time or message received) that triggers an activation. For activations triggered by messages, specify the message identifier that initiates the activation.

X. Use data developed in steps II through VIII to develop specifications for network and business process simulation models.

XI. Develop or collect baseline performance data for the AS-IS. This may be accomplished by developing and validating simulation models of the AS-IS or through instrumentation of the AS-IS process (e.g., statistics from workflow manager, application program monitors, network managers and monitoring devices, etc.).

XII. Identify and classify the set of proposed changes to the AS-IS. Collect cost data associated with the proposed changes.

XIII. Design the set of simulation experiments to analyze the impact of proposed changes. Using the experiment design, perform the following steps:

A. Develop and run TO-BE simulation models of the network. Collect simulated utilization rates and total processing time statistics for activations spanning the start of the leftmost UOB through the completion of the rightmost UOB in the decomposition of each L_APP UOB. Plug the resulting cycle-time distribution values into the simulation data specification for each L_APP UOB.

B. Develop and run TO-BE simulation models representative of the business process level using the values obtained from the network level simulation. Collect simulated utilization rates and processing times.

XIV. Analyze simulation results.

XV. Calculate costs from actual and simulated process performance data.

XVI. Contrast the AS-IS and each candidate TO-BE configuration in terms of their respective cycle-time, manpower, and cost implications. Prepare high-level business case.

4. INTEGRATING BUSINESS, APPLICATION, AND NETWORK LEVEL SIMULATION MODELS

The most significant element of automated support required of a BPR-II environment is that which integrates process models at the business, application, and network levels. This integration is accomplished by recognizing that network simulation models are designed to model traffic flows. That traffic is generated by the business processes that operate applications, which in turn operate on the network. Network simulation models do not include any graphical mechanisms to explicitly represent applications or their corresponding traffic-generating business processes. Nevertheless, network simulation modelers begin by understanding and later implicitly representing those processes in the network simulation model.

Since business and application processes are not graphically represented in network simulation models, those who are not intimately familiar with network simulation modeling are largely unable to visualize how network processes operate or how they relate to business-level views of the process. This graphical limitation of network simulation models also prevents all but well-trained network analysts from developing and using network simulation models.

The graphical limitation of network simulation was overcome in two ways. First, by using IDEF3 to develop an explicit graphical representation of business, application, and network processes, it is easy for people to see what is happening at any level of the process. Second, the methodology creates an explicit link between the business process level and the network processing level by defining the application process in the context of the business process. The result is a single, integrated model. This strategy not only provides high visibility to the uninitiated but enables one to program the network simulation model directly from the process description. More specifically, in COMNET III, UOBs involved in the decomposition of an L-APP UOB define the command set and sequence for an application source at the node where the application resides. Additional application sources are similarly defined for each R_APP UOB in the hierarchy. Since the entire hierarchy below an L_APP UOB represents a linear process, the L_APP application source run length represents the end-to-end delay that will be experienced by the end user. This value, once obtained, can be used in the L_APP UOB as part of the specification for developing a business process-level simulation model.

To support the BPR-II methodology, we developed a prototype BPR-II environment that integrates tools supporting IDEF3 with simulation tools. The BPR-II environment integrates KBSI's ProSim™ IDEF3 process modeling tool with two simulation tools: WITNESS® by AT&T ISTEL™ and COMNET III™, by CACI Products Company. WITNESS and COMNET III are both discrete event simulation tools. WITNESS has a user interface tailored primarily for those involved in modeling and simulating manufacturing processes. WITNESS also provides an experiment management utility enabling users to design and manage simulation experiments. COMNET III has a user interface that is tailored for network simulation. COMNET III provides an extensive library of objects with default characteristics (or parameter sets) representative of typical network hardware and communications equipment. The BPR-II environment was developed using Microsoft® application development and database tools.

The BPR-II prototype consists of four main modules: the business case notebook (BCN), the situation modeling environment, an experiment manager, and a business case analyzer. The role of the business case notebook is twofold: 1) to capture a description of the domain including a characterization of the assets included in the domain, and 2) to maintain static and variable cost information with which to perform business case analyses. The situation modeling environment guides users through the series of modeling and simulation steps required for analysis. The experiment supports the user in designing a set of simulation experiments to be applied within the environment. By applying this module, users can help ensure the reliability of simulation-generated data and its interpretation. The business case analyzer is used to generate graphical and tabular displays contrasting the cost and cycle-time implications of alternative configurations.

5. EXAMPLE

Consider a simple example of a network that includes three nodes connected by three communication links. That is, assume that the network topology is drawn in the shape of a triangle with nodes resting at each corner and communication links defining the edges of the triangle.

One modeling situation of interest involves altering the supporting network hardware and communications infrastructure while leaving the business process unchanged. The resulting models may reflect contingency situations or business decisions aimed at achieving savings (e.g., wartime loss of dedicated communication lines or satellite uplinks). A second situation involves leaving the supporting network hardware and communications infrastructure unchanged while introducing a reengineered business process. Each analysis situation provides cost and cycle-time information that can be displayed through the Business Case Analyzer for side-by-side comparison.

Since we have already developed a description of the network topology through the BCN, we can now turn our attention toward defining the business process(es) supported by the II. The top portion of Figure 3 illustrates one such process. The process depicted involves data retrieval from a remote site, such as an engineer using a local application to assemble product data distributed across multiple nodes in the network. It should be obvious, even in this simple process description, that changes in infrastructure performance directly impact the user. Likewise, it is clear that the structure of the process and its activation can directly impact the infrastructure.

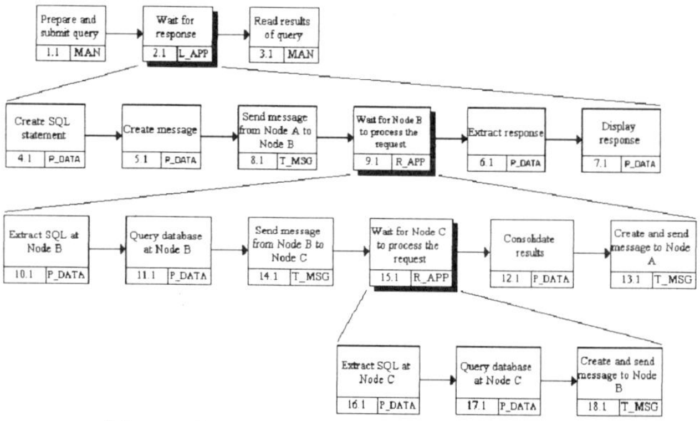

Figure 3. Example Business Process

Further definition of this process is developed through process decomposition or detailing. Using the multiple viewpoint feature of IDEF3, each decomposition level assumes a different viewpoint corresponding to the key agent of the process at that level. For example, at the highest level, the viewpoint assumed is that of the user. From the user's viewpoint, a manual activity is accomplished after which he must wait for a local application to respond with the requested data. The local application process is detailed as a decomposition of the L_APP UOB. At this level, the viewpoint assumed is not that of the user. Rather, the viewpoint assumed is that of the local application. Thus, the local application performs a series of activities until it must wait for a remote application to respond. The remote application process is detailed, in turn, as a decomposition of the R_APP UOB. At this level, the viewpoint assumed is that of the remote application. If each of the decompositions were expanded in place, a serial network process would be revealed representing a query originating with the user at Node A passing to Node B. All the data required is not available at Node B prompting a query to Node C. Node C responds with data to Node B. The data at Node B and that coming from Node C is then assembled at Node B and shipped to Node A and presented to the user.

The decomposition of each L_APP and R_APP UOB represents the sequence of commands executed at corresponding nodes of the network. This process model may be used to program a model of the network topology with the set of commands that are executed by individual nodes, including the sequence in which they are executed. In fact, this is exactly what the BPR-II prototype does. By doing so, the network model is not only configured to run simulations, but is explicitly tied to the current or proposed definition of the business process. Unlike typical network models, those generated by the BPR-II system provide a convenient graphical representation of the application and network process integrated with the business process.

Figure 4 is a picture of the COMNET III model generated from the process description above coupled with the domain assets information provided to the BPR-II repository.[1] As can be seen, three computers were specified with the names Node A, Node B, and Node C. Three communication links (Link 1, Link 2, and Link 3) are connected to the nodes. Traffic is generated on the network model topology using application sources. Three traffic (application) sources are depicted. Each application source is prefaced by the label L_APP or R_APP and the name of the corresponding UOB in the IDEF3 description whose decomposition defines the segment of the process accomplished at the node to which the application source is connected.

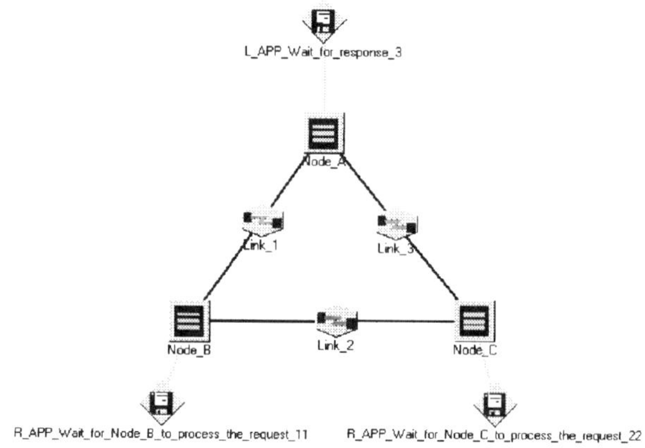

Figure 4. COMNET III Model of AS-IS network

Defining the AS-IS situation provides a baseline against which to compare alternative TO-BE configurations. Rather than defining the entire TO-BE situation from scratch, the BPR-II environment allows the user to describe the set of changes involved in realizing the proposed situation. A number of possible changes may be involved in transitioning from the AS-IS to a given TO-BE situation.

For demonstration purposes, we chose to examine the impacts of removing Link 1. In the physical world, removal of a link might correspond to a business decision aimed at reducing costs or a contingency situation in which the line is destroyed. By removing this link, it should be noted that the previously defined process must query Node B using Links 3 and 2 in place of Link 1. The response from Node B will also require traversing Links 3 and 2 in place of Link 1. Only changes to the network topology affects the network model. By regen-

[1] COMNET III model icons are placed on the screen using an algorithm permitting rapid manual adjustment. The model shown was organized into the shape of a triangle after initial placement of the icons on the screen.

erating a COMNET III model from the information in the BPR-II repository for the TO-BE situation, we generated the model depicted in Figure 5.

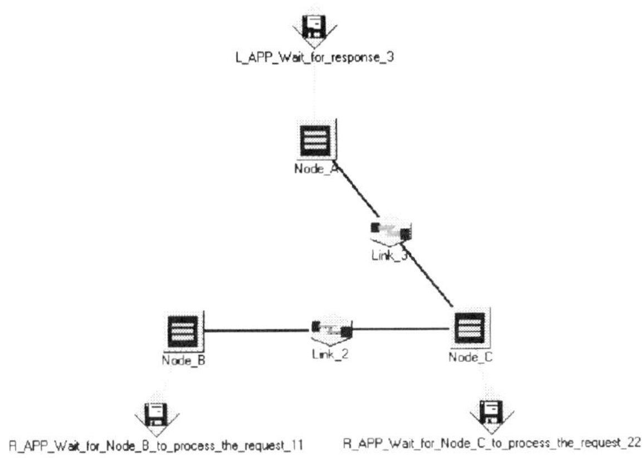

Figure 5. COMNET III Model without Link 1

Similarly, there may be structural changes in the network model that arise through changes in the process. Suppose, for example, that the data set residing at Node C is moved to Node B's database. The resulting process would not require Node B to issue a query to Node C. The bottom two levels of the IDEF3 process description would be changed by eliminating the T_MSG UOB, the R_APP UOB, and the R_APP UOBs' decomposition (Figure 3). Furthermore, the application source attached to Node C would not be generated in the corresponding COMNET III model.

Each of the resulting COMNET III models are used, in turn, to populate the corresponding L_APP UOBs in the IDEF3 process description with the appropriate simulation results. The IDEF3 process descriptions corresponding to each situation are then used to generate a WITNESS simulation model. WITNESS simulation produces cycle time and utilization rate statistics for each situation. (Figure 6.)

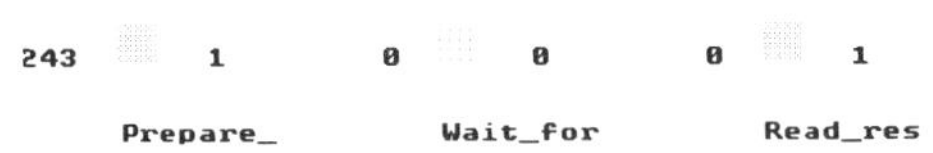

Figure 6. WITNESS Model

Cost and simulated cycle-time performance data is collected and stored in the BPR-II repository. This data may then be used to conduct comparative analyses of alternative situations. The business case analyzer component of BPR-II is intended for this purpose.

The costs associated with each of the three simulated situations described above may be displayed to the analyst using the data in the repository. Figure 7 illustrates how the BPR-II prototype presents this information in a graphical fashion that shows contributing costs by type (e.g., employee move cost, asset change cost) and total costs for each situation.

Tabular and graphical representations of the cycle times associated with process activations for each situation are similarly depicted (Figure 8).

The BPR-II prototype demonstrates the feasibility of using current technology to rapidly analyze the impacts of II change on the business processes it supports, and vice versa. A robust BPR-II system would be capable of integrating business process and II models to perform simple and/or sophisticated "What-If" analyses. AS-IS models required as a baseline for analysis could be maintained semi-automatically by integrating them with sources of information about the assets comprising the domain. Using a BPR-II system, analysts, managers, and decision-makers would have significantly improved visibility of the constraints governing how business is done and how it is supported by the underlying network hardware and communications infrastructure. Finally, with a BPR-II capability, users could more rapidly and reliably generate the data needed for high-level business case development.

Situation	Employee Move Cost	Asset Change Cost	Total Cost
AsIs	$0.00	$0.00	
To Be Alternative 1	$5,000.00	$300.00	$5,300.00
To Be Alternative 2	$0.00	$450.00	$450.00

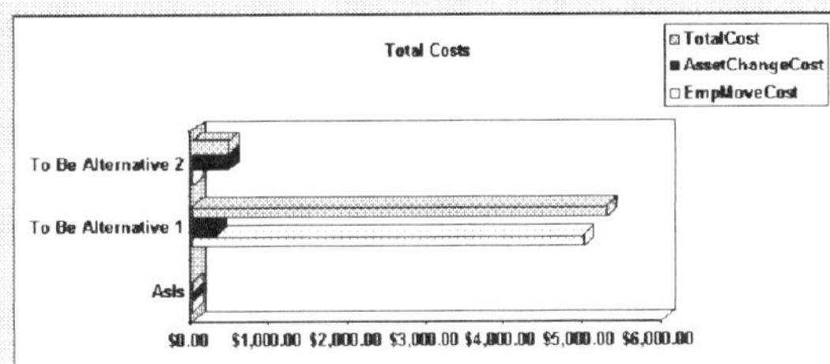

Figure 7. Cost Comparison among Situations

Situation	Domain	Time in System (sec)
AsIs	Domain 1	89.567
To Be Alternative 1	Domain 1	98.813
To Be Alternative 2	Domain 1	87.964

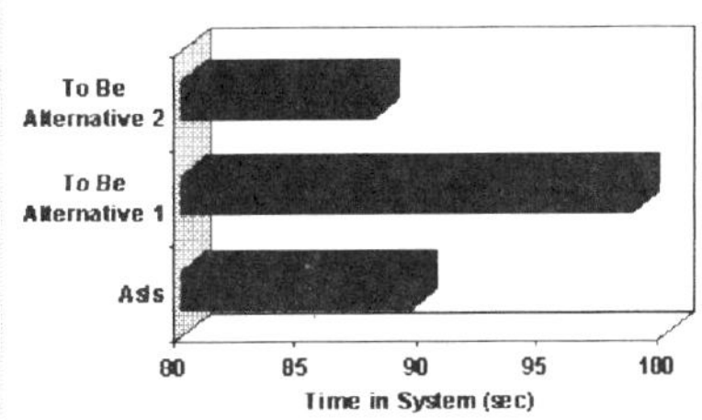

Figure 8. Time Comparison among Situations

6. SUMMARY

There is clearly a strong need for a robust BPR-II capability. Both the commercial and defense sectors stand to gain tremendous benefits from such technology. As illustrated through the effort described in this paper, it is clear that business process models can be integrated explicitly with models of the supporting network hardware and communications infrastructure. Furthermore, doing so provides an effective means for rapidly acquiring a realistic, quantitative assessment of the impacts of proposed change. These assessments can be used to support the development of a business case or to facilitate contingency analyses.

Central to this capability is an effective methodology that provides increased management visibility and "What-If" analysis capability. A key element of this methodology is the provision for effective automated support which appears to be available with current COTS tools.

ACKNOWLEDGMENTS

This work was conducted under subcontract to Lockheed-Martin Corporation in support of their contract to the Defense Information Systems Agency (DISA). The sponsor and government lead for this effort was Lt. Col. Paul Condit of DISA. The Lockheed-Martin program manager was Mr. Ed Johnson.

REFERENCES

[Benjamin et al., 93] Benjamin, P. C., F. Fillion, R. J. Mayer, and T. M. Blinn. 1993. Intelligent support for simulation modeling: A description-driven approach. In *Proceedings of the 1993 Summer Simulation Conference*: 273-277.

[Mayer 92] Mayer, R.J., Cullinane, T.P., deWitte, P.S., Knappenberger, W.B., Perakath, B., and Wells, M.S. *IDEF3 Process Description Capture Method Report, AL-TR-1992-0057.* Information Integration for Concurrent Engineering Project, United States Air Force AL/HRGA, Wright-Patterson Air Force Base, OH, May, 1992.

[Padmanaban et al., 95] Padmanaban, N., Benjamin, P.C., Mayer, R.J., "Integrating Multiple Descriptions in Simulation Model Design: A Knowledge Based Approach", in *Proceedings of the 1995 Winter Simulation Conference*: 714-719.

AUTHOR BIOGRAPHIES

MICHAEL K. PAINTER is a Program Manager and Senior Systems Analyst at Knowledge Based Systems, Inc. (KBSI). He manages business process re-engineering and systems development efforts which focus on re-engineering aircraft maintenance, material acquisition, and reverse engineering processes. He received a B.S. in Mechanical Engineering from Utah State University in 1985. He is a former Air Force officer and program manager on efforts to develop and demonstrate technologies for large-scale process and information systems integration.

RONALD FERNANDES is a research scientist at KBSI He received a Ph.D. in Computer Science from Texas A&M University (TAMU) in 1994. His research interests include network design and simulation, network management, communication protocols, and parallel & distributed processing.

NATARAJAN PADMANABAN is a research scientist at KBSI. He received his M.S. in Electrical Engineering from the University of Rhode Island in 1991, and is currently pursuing his Ph.D. in Industrial Engineering (IE) at TAMU. His research interests include knowledge based simulation modeling, computer aided software engineering, information systems integration, information modeling methods, data warehousing, on-line analytical processing, and intelligent manufacturing systems.

RICHARD J. MAYER received a Master of Science in IE from Purdue University in 1977. In 1988, he received a Ph.D. in IE from TAMU. From 1984 to 1989, he was Project Manager and Principal Investigator on thirty-nine funded research efforts in the Knowledge Based Systems Laboratory. He founded KBSI in 1988 and has received funding for applications in engineering design assistance, systems analysis and concurrent engineering methods and tools. Currently, he is a Professor of IE at TAMU.

Proceedings of the 1996 Winter Simulation Conference
ed. J. M. Charnes, D. J. Morrice, D. T. Brunner, and J. J. Swain

BUILDING A SPECIAL PURPOSES SIMULATION TOOL FOR EARTH MOVING OPERATIONS

Dany Hajjar
Simaan AbouRizk

Department of Civil Engineering
220 Civil Electrical Building
University of Alberta
Edmonton CANADA T6G 2G7

ABSTRACT

This paper discusses the process of building a special purpose simulation environment. By making the environment specific for a given industry many advantages are gained including wider acceptance and use in practical settings. Special purpose simulation tools help bring simulation to the desks of construction engineers who have little or no experience with simulation theory. By isolating the user from the low level constructs and presenting a modeling interface that more closely represents the actual system, the overall environment becomes more effective. Also, through specialization, the simulation reports can be created in the standard format that engineers in the intended domain are accustomed to. Various reports can be made to include varying levels of information summarized in a manner that can be used directly for estimating and bidding purposes. This paper shares some of the design and implementation issues used when building special purpose simulation tools.

1 INTRODUCTION

A simulation tool can be conveniently described as composed of three components: a pre-processor, a simulation algorithm (we will refer to that as "simulation engine") and a post-processor as depicted in Figure 1. The pre-processor allows for the modeling of a scenario using user level constructs involving some form of a user interface. The other function of the pre-processor is to convert the model into a format that can be used by the simulation engine. The simulation engine then performs the required analysis by performing the required number of runs, generating raw results and statistics to output files. The post-processor's main function is to read the raw information and generate the required statistics which are grouped and displayed in organized user level reports.

Traditional construction simulation tools strive to be general in order to be flexible and comprehensive.

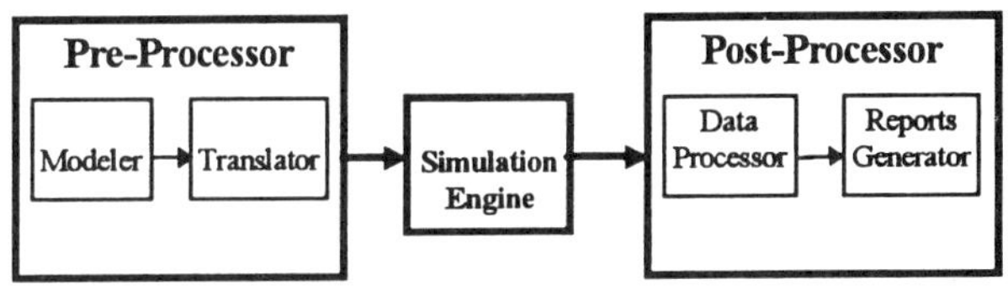

Figure 1: Simulation Components

This allows for the modeling and analysis of a wide range of application areas. While these tools are very powerful, they require the user to enter and manipulate information in a counter-intuitive manner. Construction engineers who have little experience with simulation theory must use abstract simulation level elements during the modeling phase. The correspondence between real system and the computer model is rarely similar. Also, the simulation results from these tools usually require some added level of analysis to get the necessary information. This translates into lost time and productivity. Special purpose tools remedy these deficiencies by presenting a highly visual preprocessor with a graphical user interface (GUI). This interface is tailored to the specific needs of the application domain and the end users.

Researchers in construction have used object oriented programming (OOP) to enhance the process interaction modeling construct while preserving the graphical modeling strategy of process interaction systems (e.g. COOPS (Liu and Ioannou, 1992); and CIPROS (Odeh et. al. 1992)). Although one can argue that the level of effort associated with model development greatly depends on the modeler's background (i.e. whether more effort is associated with object oriented simulation (OOS) versus traditional graphical process interaction modeling) advantages of OOS are accentuated when masses of information must be handled (e.g. modeling an entire construction project or a complex operation). The one-to-one correspondence between the model and the physical entity available through OOS may be appealing and prove very competitive as long as the modeled process is within a reasonable level of detail. Modeling more complex projects may require significant model abstraction, however, which may necessitate the use of some form of process interaction modeling elements.

The advantages of OOS include 1) reduced development time due to re-usability of model components as outlined by Odeh et.al, (1992), and 2) modularity and hierarchical structuring which would facilitate focus on high level models while maintaining the accurate representation at the process level.

Ziegler (1987) described the use of OOP as basis for modularity in simulation applications. Luna (1991, and 1992) presented concepts pertaining to hierarchical modeling for simulation. Model definition is achieved through a successive combination of two types of building blocks, the atomic model component and the coupled model component. The hierarchical structuring of the model is achieved through successive combinations of atomic and coupled models. The essence of the modeling strategy is based on input/output links between the various component. Luna uses the Smalltalk language (Goldberg and Robson 1983) to implemented the derived concepts.

2 DESIGN METHODOLOGIES

Building a special purpose simulation tool involves the design and implementation of the components described in Section 1.

The works of Ziegler and Luna formed the basis of the work described in this paper. The main focus of the described developments were on the modeling side since simulation algorithms are abundant in the literature.

In general, information about a given system is represented through processes and their interactions within the overall system. In this approach, a project is broken down into independent processes each with a distinct function. These processes interact together though shared and well defined interaction points. By modularizing these processes and defining how they can interact, a library of processes can be built, manipulated and extended to suit the needs of construction engineers. Once linked to iconic representations we achieve our objective of a flexible specialty tool for a given industry

3 BUIDLING PROCESS

This section describes the various design stages of a special purpose simulation tool.

3.1 Preliminary Conceptual Design

The design process begins with identifying the general processes that make up the entire project. The initial problem statement, the description of the high level processes and their relationships define the of the simulation tool. The next step is to list the various active and passive elements of each process. Active elements are those that cause a transition in the state of the system; they request and free up resources and drive the simulation experiment. Passive elements, on the other hand, do not actively participate in the experiment.

The sequence of events for active elements can be defined through high level flow diagrams. The flow diagrams will form the initial conceptual simulation model. they determine the required events, properties and methods for each process.

Any particulars, assumptions, and constraints are then stated as part of each process. Constraints determine how and when the simulation events will take place, for example. The assumptions and constrains will further limit the scope of the application domain.

3.1.1 Interaction Points

After the definition of each process, the nature of the interaction and relationships between the processes are conceptualized. The final simulation model will consists of a set of processes interacting in some manner. There are generally two types of interactions: production based and entity based.

Production based interactions are those that involve a product quantity being manipulated. Some processes will increase the amount of this product while others will decrease it. The dependency and, therefore, the interaction comes from the fact that the active entities within each process will wait if their actions could result in a production level that is below the minimum level or above maximum level. If the two processes are not "production compatible" (i.e. they use different measures or require different forms of production), a transfer function is defined. For example, when one process interacting with another while one is producing a product with a cubic measure while the other producing a product with a surface measure require the application of a cubic to square conversion function.

In entity based interactions, active elements pass from one process to another. Again, conversion function on the attributes of the entities might need to be specified in order to make the two processes "entity compatible".

The conceptual design stage is completed when all other elements falling outside of all the individual processes are defined. These include any global resources that are shared between processes or external interactions that can affect more than one process at a time (examples are weather processes and breakdown processes).

3.2 Simulation Level Design

At this stage, the models needed to drive the simulation of each process are formally described using entity flow diagrams. These diagrams will serve as the basis for implementing the models using discrete event simulation libraries. They are drawn from the simple flow diagram defined for each active element in the previous steps. These diagrams must be specific enough to be easily translated.

3.3 Data Structure Design

Data structure design involves the definition of the data objects or classes that will represent the elements and processes of the tool. A preliminary class hierarchy is first defined and the functionality of each one class is described. These structures will represent:

- Project level information such as global resource and container structures.
- Process level information.
- Linking information between processes.
- Active and passive element representation

This structure is then revised to implement specialized behavior requirements. For simulation support, basic extended functionality includes integrity checking, simulation initialization, processing and clean-up. Graphical user interaction behavior is then added to support the creation of a GUI.

3.4 Preprocessor Design

The graphical interface simplifies the task of constructing and running a simulation model. However, users are still working with low level simulation constructs such as resources, interaction points, and processes. As stated earlier, the objective of a special purpose simulation tools is to allow end users such as project managers or even on site engineers to use the tools effectively. This means that there is a need to add another layer of abstraction that brings the modeling environment to an intuitive and direct level. A preprocessor must be designed to take advantage of the assumptions and fundamentals of the model and reduce the required amount of input data.

3.4 Post Processor Design

The post processor objective is to analyze and present the raw simulation results in a familiar and immediately useful format. This is best done with in simple database environment where data can be quickly analyzed using queries. The final results can then be formatted and displayed in customizable database reports.

4 EXAMPLE APPLICATION: EARTH MOVING SIMULATION

Simulation studies are well suited for the analysis of earth moving for many reasons including repetition of given operations, dynamics of resource interactions, external factors that need to be included in the analysis such as weather and the general randomness associated with such systems. The ultimate objective is to build a highly visual and flexible modeling environment where a wide range of scenarios can be analyzed with a great deal of ease. The purpose of this example is not to detail the lowest level of programming or design specifications; rather it will focus on design methodologies discussed earlier and how it is applied.

4.1 Preliminary Conceptual Design

Earth moving operations involve a number of hauling trucks loading earth material from a preparation area and transporting it to a dump location. Other supporting activities that also take place are earth preparation at the source and spreading at the placement location. The entire operation can be represented as a number of interacting processes: Preparation, loading, hauling, dumping and spreading.

4.1.1 Preparation

Preparation is the process of breaking up and collecting soil so that it can be easily loaded into trucks. It involves a number of dozers cycling between the collection area and the loading pile. With each cycle, the amount of prepared earth in the loading pile increases. Preparation is not necessary if the earth is adequate for immediate loading .

4.1.2 Loading

During the loading process, several trucks wait in line to be loaded with earth from the loading pile. Several loaders could be present at one loading location. Loading durations are usually obtained through on site studies or from equipment handbooks. Waiting will result when loaders' production is less than that of the trucks or when loading production is greater than the preparation production.

4.1.3 Hauling

The hauling process involves a number of trucks hauling earth from the preparation area or "source" to the destination or "placement". Roads are normally modeled as road segments with properties such as grade, length and maximum travel speed. The haul route could contain various intersections and junctions where trucks must stop and give way to other traffic which might have higher priority. The trucks follow a static path from the source to the placement known as the "travel loaded" path and return on the "travel empty" path. The source and placement in both paths are the same therefore trucks always follow a closed loop. Different trucks could follow different paths since earth moving operations could involve several sources and placements as shown Figure 2.

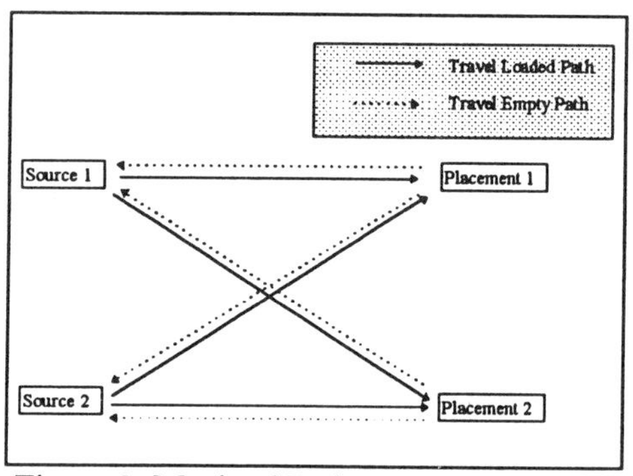

Figure 2: Maximal Path Arrangement Between Two Sources and Two Placements

The travel times of trucks can be obtained from empirical travel time curves. These curves give an approximate travel duration given the length and total effective grade of the road segment for both loaded and empty trucks. Total effective grade is composed of grade resistance and rolling resistance. Grade resistance is a measure of the force that must be overcome to move trucks over uphill slopes. Rolling resistance is a measure of the force that must be overcome to roll or pull a wheel off the ground.

Within earth moving projects, intersections serve to regulate the traffic between arriving entities. Trucks must stop at intersections and check before proceeding. Intersection are not simple resources since the maximum number of trucks or other types of traffic allowed at any one time is variable and dependent on the "crossing path" of entities. As illustrated in Figure 3, Case 1 presents no conflict between the crossing behavior of incoming traffic. The intersection serves as a relay station and no waiting is involved. On the other hand, In Case 2 traffic could queue at A or B since the desired paths cross.

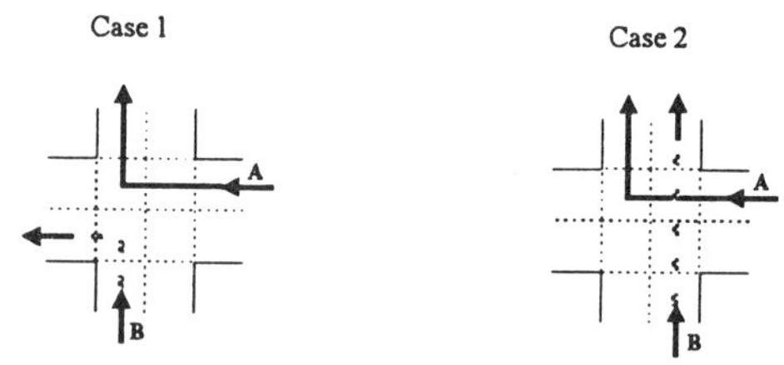

Figure 3: Two Crossing Scenarios

4.1.4 Dumping

Dumping is a simple process whereby trucks wait for a position at the dump location, then dump their payload into the spreading pile.

4.1.5 Spreading

The spreading process is the reverse of the preparation process. One or more dozers spread the hauled earth from the dump pile across a given area.

4.1.6 Interaction Points

The processes described so far are well contained and independent. To construct a useful scenario, they must be linked to form a representative model of the whole project. Production based interactions are present between preparation and loading as well as dumping and spreading. Entity based interaction points are present between loading and hauling, and between hauling and dumping. The entities that are being transferred between the two processes represent trucks.

4.1.6.1 Production Based Interactions

The preparation and dumping processes interact through the loading pile. On one side, dozers in the preparation area continually increase the level in the pile. While, on the other side, trucks are being loaded using that same pile. Queuing is possible only from the loading side as trucks will wait if not enough earth is present. A "swell" ratio is a conversion factor used in order to make the preparation and the preparation processes production compatible. This is because during preparation, quantities are tracked in *bank* cubic meters; while in loading, they are tacked as a process to a another. The choice of destinations is specified by the source process when a request for a transfer is made.

4.1.7 Sample Scenario

An example project is shown in Figure 4. It consists of several possible preparation processes, loading, hauling, dumping, and spreading processes.

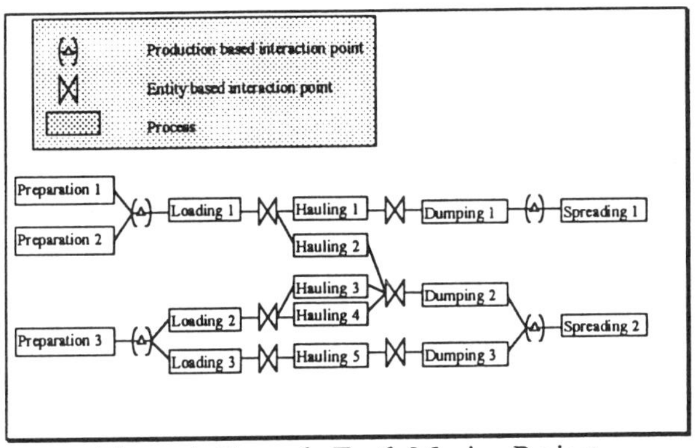

Figure 4: Example Earth Moving Project

4.2 Simulation level design

The event flow diagrams of each process can now be designed. These diagrams define the event sequence of each active element and can be directly translated into discrete-event simulation models. The symbols used are explained in Table 1. The following subsections present sample diagrams for the loading and hauling processes.

4.2.1 Loading

The loading simulation module is illustrated in Figure 5. Duration A and B represent loader holdup time and loading time consecutively. Interaction point 1 is the loading pile can be manipulated by other loading or preparation processes (see Figure 4). Interaction points 2 and 3 represent entity transfer points from and to the hauling process.

Table 1: Model Design Symbols

Symbol	Description
Event Name	Represents a simulation event or transition in simulation.
Cond A → Dur A	Represents an activity with a defined duration and a crietieria in the case of conditional branching
Resource	Allocated or releases a certain resource.
	Accumulate a production level for any desired purpose. Used to track the cumulative production level.
+ (Int. 1	Represents a production interaction points.
→⋈ Int. 1	Represents an entity interaction point. The arrow indicates whether entities are arriving to the process or exiting from it.
* →	Indicates the starting event of a process.

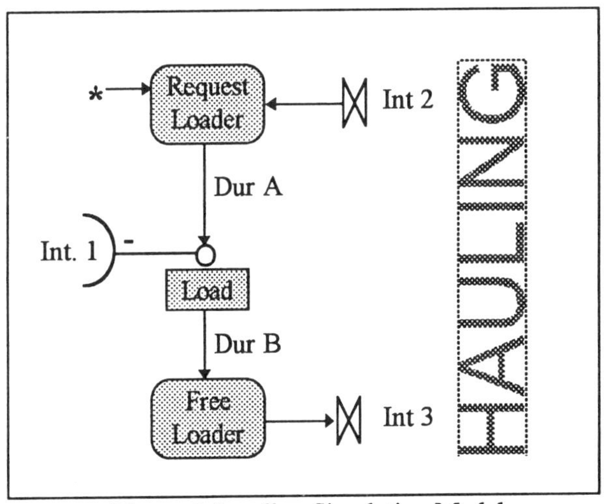

Figure 5: Loading Simulation Model

4.2.2 Hauling

The hauling simulation module is illustrated in Figure 8. Duration descriptions are shown in Table 2, Conditions in Table 3 and interaction points in Table 4.

Table 2: Durations For Hauling Model

Duration	Description
A	Segment Travel Time
B	Intersection Wait Time
C	Crossing time
D	Junction Wait Time

Table 3: Conditional Branching For Hauling

Cond.	Description
1	Travel Empty path complete
2	Next object in path is a junction
3	Next object in path is an intersection
4	Travel Loaded path complete
5	Actual hauled amount $<$ Desired hauling amount
6	Actual hauled amount $\geq$ Desired hauling amount

Table 4: Interaction Information for Hauling Model

Int.	Interaction Point	Interaction Command
1	Truck departure to loading area	transfer(destination loading process)
2	Truck arrival from loading area	
3	Truck departure to dumping area	transfer(destination dumping process)
4	Truck arrival from dumping area	

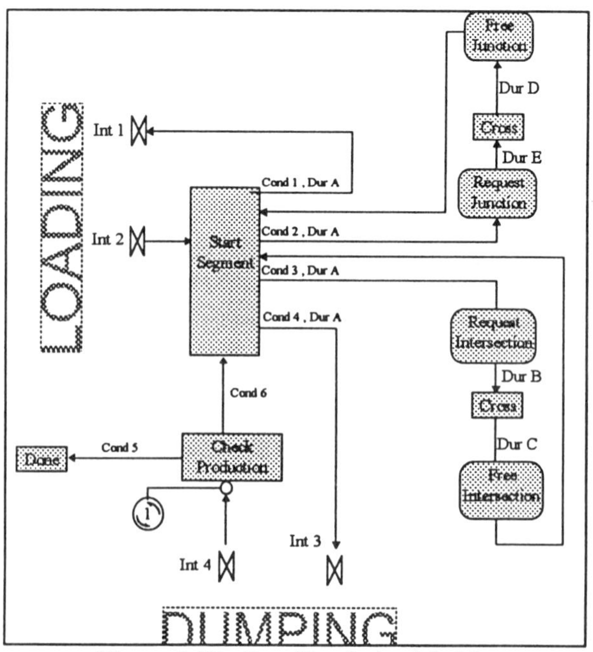

Figure 6: Hauling Simulation Model

4.2.2.1 Intersection Modeling

As explained before, intersection modeling is not a simple task. The traffic behavior at an intersections depends on several factors including priority, "crossing path" and crossing duration of incoming traffic. One way of modeling this behavior is to use a state grid. The grid determines whether a certain traffic block is currently busy or free. Traffic entities arriving at an intersection cross only when all the blocks constituting the crossing path are free; otherwise, they will queue up. When an entity crosses the intersection and releases the blocks, waiting entities are allowed to proceed based on priority. If priorities are equal, they are served on first come first serve basis. The result is a sort of four way intersection.

4.3 Data Structure Design

The class hierarchy is presented in Figure 9. It is developed through continuous refinement until all the required behaviors and objects have been represented. Following is a brief overview of some of the base classes:

4.3.1 CGraphicalBase

Implements basic graphical state and position representation and manipulation functionality.

4.3.2 CInteractivePositioning

"click and drag" interactive positioning functionality is supported through this class Classes which can take advantage of this behavior such CProcess and CCStructuralBase use multiple inheritance. This added behavior simplifies data entry for position information of all classes. As well, connection information for structural classes can be implicitly defined by their positions relative to the site and other objects.

4.3.3 CAPEBase

Provides basic object information, serialization, container membership, unique sequence number assignment, and attribute access interface.

4.3.4 CSimBase

This class will define basic virtual functions for simulation purposes. These functions provide various "user hooks" that allow the specialized classes to perform custom manipulation at the start and end of simulation and before and after each run. Virtual functions can be "overridden" by child classes in order to specialize.

4.3.5 CStructuralBase

This class adds connection management functionality. Classes like CRoadSegment and CIntersections, which constitute the structural elements of the model, derive from this class.

4.3.6 CProcess

A CProcess class defines basic functionality of a high level process such as ability to connect to interaction points and the base structures needed to define a simulation model. Child classes implement the specific simulation models.

4.3.7 CProject

Encapsulates all project level information and behavior. This includes the list of all processes, interaction point definitions, as well as any global resources that are available to all the processes.

4.4 Design Remarks

The classes provided above are used to construct the initial structure of the application. However, as the difficult task of coding a design progresses, a number of classes might still be added resulting in a slightly different class hierarchy.

5 CONCLUSION

Throughout the design process, every effort was made to enforce tight encapsulation and integration at all levels. The simulation design of each process is consistent and structured. Therefore, constructing a lower level representation such as CYCLONE (Halpin, 1990) is possible. At the project level, inter-process interaction through well defined points allows the tool to be easily extended. For example, if a compaction or a paving module is desired, another process is defined and linked without any changes being required of the remainder of the processes.

The design complexity in this paper is not very deep. The reason is that any more time spent on design will prove unproductive since any further design decisions are highly dependent on the implementation platform. The level of design discussed is general enough to be implemented using any object oriented framework with graphical user interface support. The design stage serves to examine all the various requirements and how they translate into modules and objects that will interact together. A good design will allow the application to gradually grow into a complex but well structured system.

The approach described in this paper was used in developing two special purpose simulation tools AP2_Earth (see Figure 10) and CRUISER. The first is a specialty tool for design and analysis of earth moving operations while the second is a specialty tool for analysis of aggregate production. Both tools are available via on ftp cem.civil.ualberta.ca/ public. They were described in the proceeding of the 1995 WSC (see AbouRizk et.al., 1995).

REFERENCES

AbouRizk, S. and Dozzi P. (1993). "Application of Computer Simulation in Resolving Construction Disputes". Journal of Construction Engineering and Management, ASCE. In press.

AbouRizk S. Shi, J., McCabe B. and Hajjar D.(1995) *"Automating the process of Building Simulation Models"*. Winter Simulation Conference, December 1995 Washington D.C.

Goldberg, A. and Robson, D. (1983). SmallTalk-80: The language and Its Implementation. Addison-Wesley Publishing Company, Reading, MA.

Halpin, D. W. (1977) "CYCLONE: Method for Modeling of Job Site Processes." *Journal of the Construction Division*, ASCE, 103(3), 489-499.

Halpin, D., Woodhead, R., *(1976). Design of Construction Process Operations,* John Wiley and Sons, Inc., New York, NY, 1976.

Liu L.Y., and Ioannou P.G., (1992). "Graphical Object-Oriented Simulation System for Construction Oriented Discrete-Event Simulation System." *Proceedings of the 1992 Winter Simulation Conference* 1285-1291.

Luna, J. (1991). "Object Framework for Application of Multiple Analysis Paradigms." Object-Oriented Simulation 1991, Simulation Series Volume 23, Number 3, 81-86.

Luna, J.L. (1992) "Hierarchical, modular concepts applied to an object-oriented simulation model development environment." Proceedings of the 1992 Winter Simulation Conference 694-699.

Miller A., Glenda G. and Bird C. (1991). "Simulation and Analysis of Computer Systems With Multiple Supercomputers", SCS Multi-Conference, New Orleans, Apr 1-5, 1991.

Odeh, A. Tommelein I, and Carr R. (1992). "Knowledge-Based Simulation of Construction of Construction Plans". Proceedings of the Eighth

Conference on Computing in Civil Engineering, ASCE, Dallas, Texas. 1042-1049

Zeigler, B.P. (1987). Hierarchical, Modular Discrete-Event Modeling in an Object-Oriented Environment, Simulation, November 1987, 219-230.

ACKNOWLEDGEMENT

This work was supported by the Natural Sciences and Engineering Research Council of Canada under Grant CRD 166 483.

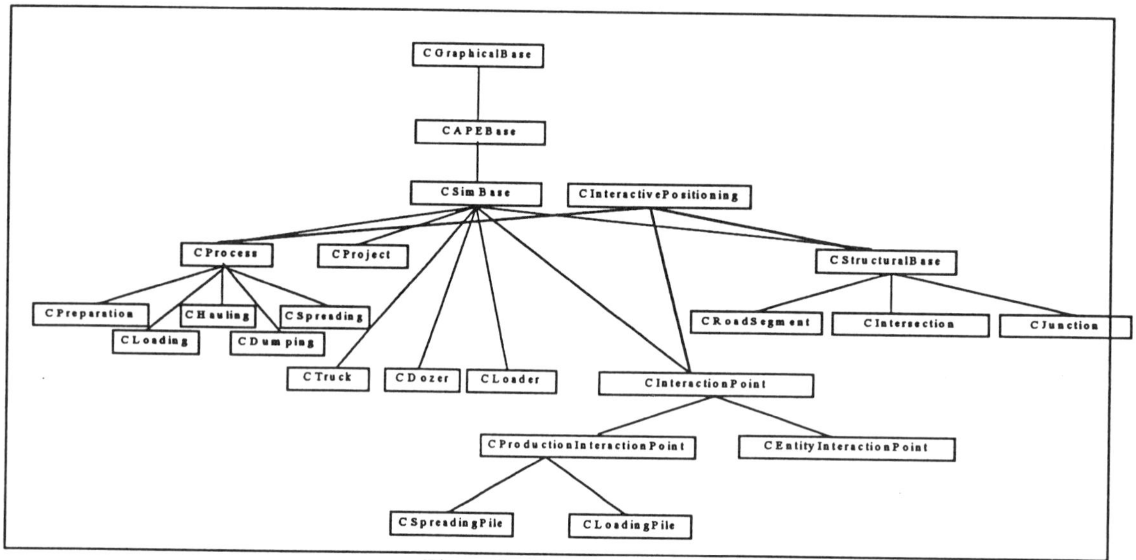

Figure 7: Class Hierarchy

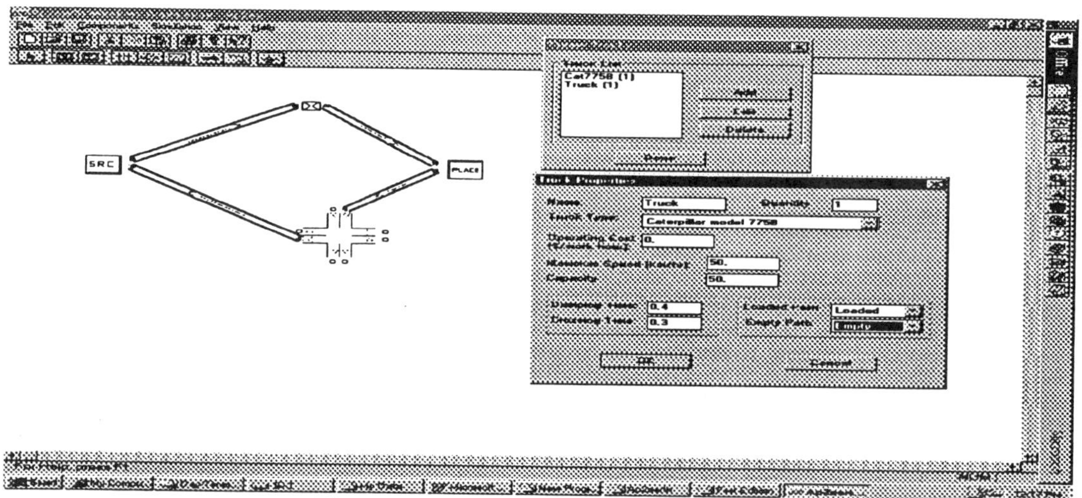

Figure 10: Sample screen from AP2_Earth

AUTHOR BIOGRAPHIES

DANY HAJJAR is a provisional Ph.D. candidate in Civil Engineering at the University of Alberta. He received his Bachelor of Science with specialization in Computing Science from the University of Alberta in 1995. His research interest are focused on applying automation and object oriented methodologies to the management and control of construction projects..

SIMAAN M. ABOURIZK is a Professor in the Department of Civil Engineering at the University of Alberta. He received his BSCE, and MSCE in Civil Engineering from Georgia Institute of Technology in 1984 and 1985 respectively. He received his Ph.D. degree from Purdue University in 1990. His research interests are mainly focused on the application of computer methods and simulation techniques to the management of construction projects.

Proceedings of the 1996 Winter Simulation Conference
ed. J. M. Charnes, D. J. Morrice, D. T. Brunner, and J. J. Swain

SIMULATION OF COMPLEX CONSTRUCTION PROCESSES

Photios G. Ioannou
Julio C. Martinez

Civil & Environmental Engineering Department
University of Michigan
Ann Arbor, Michigan 48109-2125, U.S.A.

ABSTRACT

Eight successively refined simulation models for the earthmoving operations involved in the construction of a dam provide the foundation for illustrating the ease and effectiveness of modeling complex construction processes by using STROBOSCOPE, a modern simulation system based on a programming language that follows the activity-scanning paradigm. This approach and the use of characterized resources in a stochastic environment make the typical engineering calculations for heavy equipment performance relatively easy to implement and significantly more realistic and accurate.

1 INTRODUCTION

Traditionally there have been three major impediments to the wide-spread use of simulation to model complex construction processes. The first was that available simulation languages employed the process interaction paradigm and were developed to suit the needs of manufacturing and not construction. This difficulty was partially addressed by the advent of construction-oriented simulators, such as CYCLONE, RESQUE, COOPS, etc. All these systems have been based on activity-scanning, a paradigm suitable for modeling construction processes because of its use of the notion of activities and thus the similarity to CPM. However, these systems are not languages and thus lack the power to model complex construction processes (Martinez and Ioannou, 1995).

Today, the first two problems have been largely resolved with the advent of STROBOSCOPE (STate and ResOurce Based Simulation of COnstruction ProcEsses), a system developed by the authors that combines activity-scanning with a powerful programming language. However, the third difficulty still remains. Construction is a conservative industry with very high stakes that has not yet been sold on the ease and effectiveness of simulation modeling or the accuracy of its results. Thus, model development, validation and verification are extremely important, especially when the real decision makers do not have any prior experience or confidence in the capabilities of simulation modeling.

This paper describes how a construction simulation model can be designed and implemented at eight levels of increasing complexity and accuracy and then animated to verify that it is indeed a faithful representation of reality. The example used is that of earthmoving for the construction of a dam. This example goes through a process of successive model refinement that serves both as a modeling case study and as an illustration of the fact that it is relatively easy to model the performance of heavy construction equipment by performing complex engineering calculations under stochastic conditions that exceed the capabilities of traditional methods. The final model is then verified through animation.

2 EARTHMOVING FOR DAM CONSTRUCTION

An earthmoving contractor needs to move 200,000 m^3 of soil for the construction of a dam. Soil will be loaded at the borrow area using two types of heavy loaders. The soil is loaded into two types of heavy trucks that travel to the dam embankment, dump, and return empty. The density of the loose soil moved is 1690 kg/m^3. A total of 3 loaders and 11 trucks will be used on this job. Tables 1 and 2 show the number and characteristics of the loaders and trucks by type and model.

Table 1: Loader Data

Model	**CAT 936E**	**CAT 950E**
Machines available	2	1
Bucket Size (m^3)	2	2.4
Soil loaded in loader's bucket, S (m^3)	Bucket Size * N[1,0.05]	Bucket Size * N[1,0.04]
Time to load a scoop of soil in truck (min)	0.40+0.05S	0.43+0.035S
Cost ($/hr)	45	62

Table 2: Truck Data

Model	CAT 769C	CAT 773B
Machines available	8	3
Capacity (m^3)	23.6	34.1
Mass when empty (kg)	31,178	39,396
Flywheel Power (kW)	336	485
Transmission efficiency	0.82	0.84
Retarding power (kW)	434	526
Max. speed (km/hr)	75	62
Actual speed (km/hr)	Calc. speed *N[1,0.03]	Calc. speed *N[1,0.02]
Dumping time (min)	1	1.12
Maneuvering time at loading area (min)	0.6	0.72
Cost ($/hr)	117	175

Trucks at the borrow area queue up and are loaded in FIFO order. Each truck is loaded by one loader. Before loading can begin, a truck must leave the waiting queue and maneuver into position. The maneuvering time depends on the truck as shown in Table 2. After this time passes, the loader starts putting scoops of material into the truck.

The amount of soil in each scoop, *S*, is random and depends on the size of the loader's bucket as shown in Table 1. The time for a loader to place one scoop of material into a truck has a fixed and a variable component. The latter is proportional to the amount of soil actually loaded, *S*, as shown in Table 1.

Soil continues to be loaded for as long as the amount of empty space in the truck exceeds one half of the loader's bucket size. Thus, the last scoop brings the loaded material to at least within half a loader's bucket from the truck's capacity. This means that the total material loaded may occasionally exceed the nominal capacity of the truck.

When a truck is loaded it starts hauling the soil to the dam embankment. The haul/return route from the borrow area to the dam crosses a small river. The road from the borrow to the river is 2.3 km long. It descends at a 4% grade and has a rolling resistance of 7%. The road from the river to the dam is 1.7 km long. It ascends at a 6% grade and has a rolling resistance of 4%.

The speed of a truck on any given route is determined on the basis of engineering calculations using the data in Table 2 (instead of using rimpull or retarding curves) and the slope and rolling resistance characteristics of the road. Trucks are assumed to travel at constant speed (acceleration and deceleration times are disregarded). The calculated speed is adjusted for random variability as shown in Table 2.

Trucks cannot dump material at the dam without the help of a spotter. There is only one spotter and she can only pay attention to one truck at a time.

3 BRIDGE DECISION PROBLEM

The existing bridge cannot support heavy trucks and the contractor must provide one or more new temporary low-cost bridge(s) for his trucks to cross the river.

There are two types of temporary bridges to choose from. A Type I bridge costs $95,000 and can hold only one truck at a time. Moreover, trucks must pass very slowly (it takes 1.75 minutes to cross over). A Type II bridge costs $165,000 and can hold up to six trucks at the same time. Because the bridge is narrow, however, these trucks must be traveling in the same direction, separated at least 0.30 minutes from each other. The time to cross a Type II bridge is also 1.75 minutes.

The earthmoving contractor must determine the type and number of bridges that gives the lowest project cost and the corresponding total cost and time to complete the project.

4 BRIDGE CROSSING POLICIES

Based on the contractor's past experience with one-way bridges, it appears that the best crossing policy is to let all trucks going in one direction cross the bridge without reversing the flow of traffic for as long as there is a queue. When no more trucks need to cross the bridge in the current direction and the bridge is clear, the trucks in the other direction are allowed to cross in the same manner. If the bridge is empty and no trucks are waiting to cross the bridge (on either side), then the next direction of travel is established by the first truck to arrive at the bridge.

It must be pointed out that under very heavy traffic it is customary to impose a limit on the time available for crossing the bridge in any one direction to allow traffic to cross in the other direction. When this time limit is reached (e.g., three minutes), no more trucks going in that direction are allowed to enter the bridge. When all the trucks finish crossing, then the trucks at the other end are allowed to pass in the same manner. It is easy to show that this situation does not apply to the problem at hand because the truck cycle-time is long (relative to the time to cross the bridge) and the number of trucks is limited.

5 MODELING METHODOLOGY

Successively refined simulation models for this problem are presented below. These models were developed using the STROBOSCOPE system.

STROBOSCOPE is a general-purpose discrete-event simulation system based on the activity-scanning paradigm that is particularly suited for modeling complex construction processes. STROBOSCOPE models are networks of nodes and links (similar to activity cycle diagrams). The heart of the system is a simulation programming language that provides full access to dynamic variables, the dynamic properties of resources, and the state of the simulation. It allows modeling stochastic resource production, utilization, and consumption; smart resource allocations; characterization of resources created at runtime by combining other resources; and dynamic decisions regarding the sequence of activities. STROBOSCOPE is a 32-bit program for Windows NT and Windows 95.

A complete description of STROBOSCOPE appears in (Martinez 1996). Example applications can be found in (Martinez, Ioannou, and Carr 1994; Martinez and Ioannou 1994, 1995; Ioannou and Martinez 1995, 1996a, 1996b). The STROBOSCOPE program, its documentation, and several examples are available via anonymous ftp from "grader.engin.umich.edu".

The networks for the simulation models shown in this paper were developed using STROBOSCOPE's graphical user interface (GUI). This GUI is built on top of a commercial drawing program (Visio 3 or 4) that has been enhanced by the authors with specially programmed "drag-and-drop" graphics elements (nodes and links) that serve as primitives for constructing STROBOSCOPE networks interactively. A custom dynamic-link-library (DLL) provides a plethora of tabbed dialog-boxes for data input, extensive error-checking, and the ability to compile and communicate to STROBOSCOPE all the information entered. This is accomplished either "live" via OLE automation or by saving ready-to-use simulation input files to disk.

6 EARTHMOVING SIMULATION MODELS

A series of eight simulation models for the earthmoving problem are presented below. At first, these models will ignore the bridge decision problem and concentrate on how to model the earthmoving operation as accurately as possible using generic resources that have no properties or attributes (as in CYCLONE). We later introduce characterized loader and truck resources to allow for engineering calculations and finally we extent that model to solve the bridge decision problem.

6.1 Model 1—Simple Earthmoving

The network for the first model is shown in Figure 1. It shows that the *combi* (conditional) activity *Load* cannot start unless there are resources in the queues *LoadersWait* and *TrucksWait* that precede it. When *Load* starts, it removes one resource from each of these two queues (a loader and a truck) via the links LD1 and TR1. At the end of *Load*, the loader is returned to the queue *LoadersWait* through link LD2 and the truck passes through link TR2 to the *Haul* activity which is then allowed to start. At the end of the *Haul* activity, the truck is passed to the *Dump* activity and then to the *Return* activity. Finally, it returns to the queue *TrucksWait*.

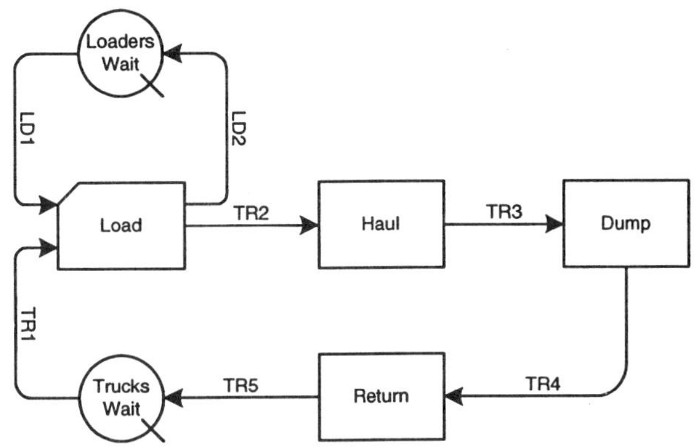

Figure 1: STROBOSCOPE Network for Dam Construction Simulation Models 1 and 2

To have a complete model, we need (in addition to the network) to define the problem parameters, resource types, and provide simulation control statements.

The following define the problem parameters:

```
VARIABLE NumLoaders 3;
VARIABLE NumTrucks 11;
VARIABLE LdrHrCst  50; /($/hr)
VARIABLE HlrHrCst 135; /($/hr)
```

We also need to define *Loader* and *Truck* as generic resource types and initialize the queues *LoadersWait* and *TrucksWait*:

```
GENTYPE Loader;
GENTYPE Truck;
INIT LoadersWait NumLoaders;
INIT TrucksWait NumTrucks;
```

In our first model, the duration of the activity *Load* is the total time it takes the average loader to load an average truck completely. (In reality, this time depends on the specific properties of the loader and truck actually used.) We shall assume that the duration of *Load* follows a Normal distribution with a mean of 7 and a standard deviation of 1.2 minutes.

```
DURATION Load Normal[7,1.2];
```

These parameters for the Normal distribution were chosen so that the results from the first model will be close to those of the more detailed models that follow.

The Haul activity represents the total time it takes to haul soil to the dump location and should thus have a duration that depends on the specific properties of the truck, the road, and the material hauled. Here we will again assume a Normal distribution with appropriately chosen parameters for the average truck. The duration of the *Return* activity is specified in a similar manner:

```
DURATION Haul    Normal[11.66,0.3];
DURATION Return Normal[7.68,0.33];
```

Based on the problem data, the average truck has a capacity of 26.67 m^3 and moving 200,000 m^3 requires a total of 7,500 dumps. Thus, simulation should run until the total number of instances of *Dump* reach this limit.

```
SIMULATEUNTIL Dump.TotInst>=7500;
```

The total cost of the project depends on the time to move the soil and the hourly equipment costs:

```
VARIABLE TotalFleetCost
          '(NumLoaders*LdrHrCst+
          NumTrucks*HlrHrCst)*SimTime/60;
DISPLAY "Total Working Time  : "
          SimTime/60 " Hours";
DISPLAY "Total Equipment Cost: "
          TotalFleetCost " ($)";
```

6.2 Model 2—Hard-wired Resource Drawing Loop

As a first refinement to the above model, the duration of activity *Load* will be redefined as the sum of twelve IID random variables, each representing the time to load an average scoop of material (i.e., 2.22 m^3) into the truck. Thus, each of these distributions depends only on the properties of the loader (i.e., its "size") and not on the truck being loaded. The capacity of the average truck (i.e., 26.67 m^3) is reflected in the number of random variables that must be added up (i.e., twelve).

```
DURATION Load
         'Normal[1,0.44]+Normal[1,0.44]+
          Normal[1,0.44]+Normal[1,0.44]+
          Normal[1,0.44]+Normal[1,0.44]+
          Normal[1,0.44]+Normal[1,0.44]+
          Normal[1,0.44]+Normal[1,0.44]+
          Normal[1,0.44]+Normal[1,0.44]';
```

As expected, the results of this model are by design equivalent to those of the first model.

6.3 Model 3—Resource Drawing Amount

We will now model soil as another resource and allow it to flow through the network via links SL1-SL4, as shown

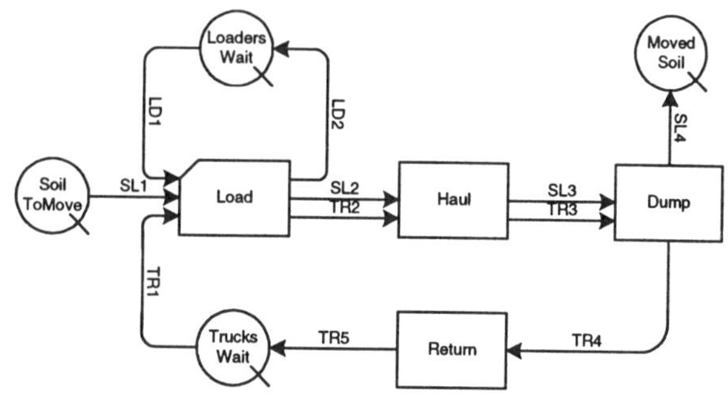

Figure 2: STROBOSCOPE Network for Dam Construction Simulation Models 3, 4, and 5

in Figure 2. To do so we define *Soil* as a generic resource and initialize the *SoilToMove* queue.

```
VARIABLE AmountOfSoil 200000; / m3
GENTYPE  Soil;
INIT     SoilToMove AmountOfSoil;
```

Each instance of the *Load* activity should not simply remove (draw) one unit of *Soil* from the *SoilToMove* queue (one is the default amount). Instead, the amount of soil drawn must equal the size of the average truck and must be specified explicitly by defining the *draw amount* attribute of link SL1:

```
DRAWAMT SL1 26.67;/ Truck capacity in m3
```

Simulation should now continue and stop only when the *SoilToMove* queue becomes empty.

```
SIMULATE;
```

The *actual* amount of soil drawn by the combi *Load* travels through the network via links SL1-SL4. Thus, every time *Dump* finishes, it *automatically* releases a truckload of soil (26.67 m^3) into the queue *MovedSoil.*

6.4 Model 4—Resource Drawing Loop

The network in Figure 2 can be refined by taking advantage of the resource drawing attributes of link SL1. These attributes allow us to model the duration of activity *Load,* as well as the drawing of resources out of the *SoilToMove* queue, the way it really happens, i.e., one scoop at a time. This is best shown by example:

Drawing through link SL1 should continue until the number of "draws" (i.e., scoops) reaches 12:

```
DRAWUNTIL SL1 SL1.nDraws>=12;
```

On average, each of these scoops removes 2.22 m^3 of soil from *SoilToMove* and transfers it to *Load:*

```
DRAWAMT SL1 2.22;/ m3 per scoop
```

The distribution of the time it takes to load a scoop is defined as follows:

```
DRAWDUR SL1 Normal[0.58,0.33];
```

We can now define the duration of activity *Load* to be equal to the sum of all (i.e., twelve) sampled draw durations through link SL1:

```
DURATION Load SL1.SumDrawDur;
```

The elegance of this model should now be obvious. The number of random variables that must be added to give the duration of a single instance of activity *Load* is no longer hard-wired in a long formula that adds up twelve random variables. Instead, the drawing process is treated like a loop that in this case is repeated twelve times (i.e., it is controlled by the number 12 in the DRAWUNTIL statement).

The draw duration that is sampled in each loop does not make simulated time go by. Only activities can do that. Instead, it provides a mechanism by which we can sample the duration associated with each draw and collect statistics over the entire set of samples (sum, average, standard deviation, minimum, and maximum). The number of samples that are collected in this manner (in this case, twelve) depends on the number of draws needed to complete an operation (in this case, to load an entire truck scoop by scoop). These statistics (e.g., the sum of draw durations) can then be used to establish the duration of a real activity (such as *Load*) that makes simulated time go by.

6.5 Model 5—Soil Loaded and Dumped is Random

Now that soil is drawn from the *SoilToMove* queue scoop by scoop, it is possible to refine the network in Figure 2 further so that the amount of soil drawn in each scoop is random. This will also make the total amount of soil hauled and dumped by each truck different, and presents us with the opportunity to make the duration of the *Haul* activity sensitive to the truck's total weight.

This statement defines the amount of soil drawn through SL1 link as a random variable:

```
DRAWAMT SL1 2.22*Normal[1,0.05]; / m3
```

The duration of activity *Haul* can now be adjusted based on the amount of soil actually loaded in the truck relative to the average truck capacity (i.e., 26.67 m^3):

```
DURATION Haul 'Normal[11.66,0.3]*
                  Haul.Soil.Count/(26.67)';
```

This statement makes a dynamic calculation by using a *predefined resource variable* to access the state of the simulation at run time. The predefined variable is *Haul.Soil.Count*. It returns the amount of *Soil* that is currently held by the instance of activity *Haul* whose duration we are about to compute.

6.6 Model 6—Characterized Resources

A more realistic model requires that loaders and trucks be modeled as characterized resources. Characterized resources must belong to a resource type (e.g., *Loader*) that defines the *properties* (i.e., static attributes) that all resources of that type must have (e.g., *Power*, *BucketSize*, etc.). Within a type, such as *Loader*, we must define *subtypes*, such as *S936E* and *S950E*, for which we can specify the specific values of these properties. Thus, the characterized types and subtypes for the loaders and trucks in this problem can be defined as follows:

```
CHARTYPE Loader BucketSize
            LoadVariability FixedLdTm
            VarLdTmFact CostPerHr;
SUBTYPE Loader S936E 2 0.05 0.40 0.05 45;
SUBTYPE Loader S950E 2.4 0.04 0.43 0.035
                 62;
CHARTYPE Truck Capacity EmptyWgt Power
               Eff RetPower MaxSpeed ...;
SUBTYPE Truck S769C 23.6 31178 336000
                   0.82 434000 75/3.6...;
SUBTYPE Truck S773B 34.1 39396 485000
                   0.84 526000 62/3.6...;
```

In addition to the above CharType (static) properties, characterized resources can have *SaveProps*, i.e., properties whose values can be assigned and changed during simulation run-time. Thus, individual resources that belong to the same subtype have exactly the same static properties but have *SaveProps* that have different values from one resource to the next. For the problem at hand it is convenient to define truck *SaveProps* to hold the amount of soil loaded into a truck (*AmtLoaded*), and the rolling resistance (*RR*) and grade of the road (*GR*):

```
SAVEPROPS Truck AmtLoaded RR GR;
```

The definition of characterized resource types and subtypes allows modeling the loading process with complete accuracy. As shown by the superscripts on the links LD1 and TR1 in Figure 3, activity *Load* first draws the next truck (via TR1), then the next loader (via LD1) and then the soil (via SL1). Thus, the *Load* instance being created already holds a specific loader and truck when soil is about to be drawn. This means that drawing through SL1 should continue until the amount of soil deposited into *Load* is within 1/2 of *the* loader's bucket from *the* truck's capacity ("*the*" means the *loader* and *truck* held by the *Load* instance *being created*):

```
DRAWUNTIL SL1 'Load.Soil.Count >=
               Load.Truck.Capacity-
               Load.Loader.BucketSize/2';
```

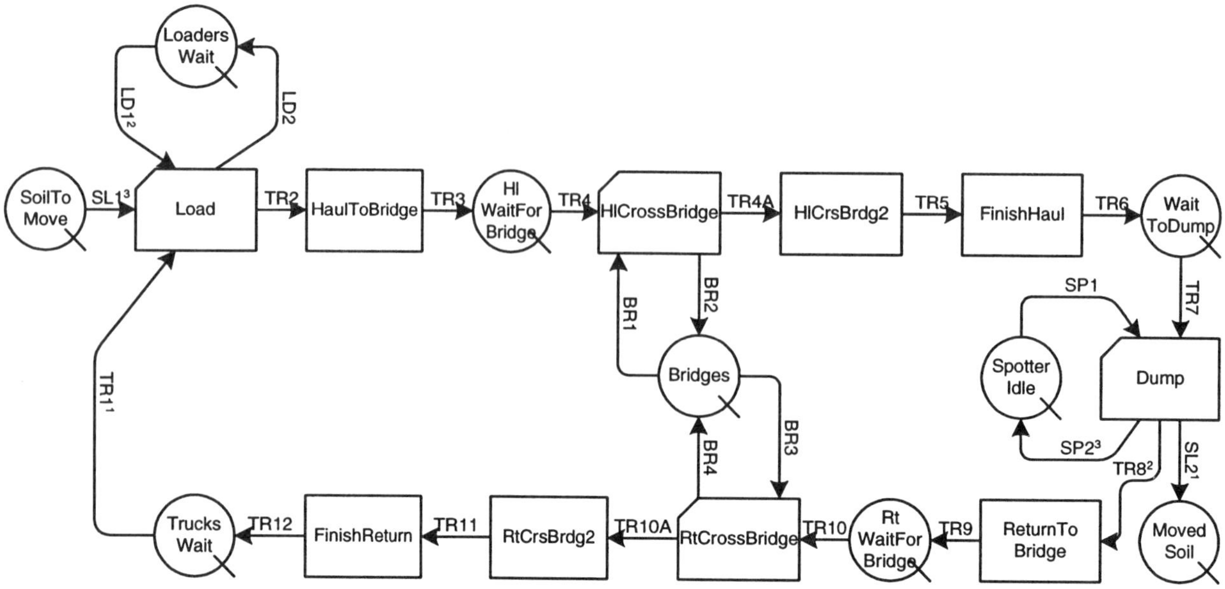

Figure 3: STROBOSCOPE Network for Dam Construction Simulation Model

The amount of soil drawn by each scoop should equal the size of *the* loader's bucket modified by its random load variability:

```
DRAWAMT SL1 'Load.Loader.BucketSize*
  Normal[1,Load.Loader.LoadVariability]';
```

The draw duration consists of a fixed and a variable component that depends on *the* amount of soil drawn:

```
DRAWDUR SL1 'Load.Loader.FixedLdTm+
                SL1.LastAmtDrawn*
                Load.Loader.VarLdTmFact';
```

The duration of *Load* equals the time for *the* truck to maneuver into loading position plus *the* sum of the sampled draw durations:

```
DURATION Load 'Load.Truck.ManeuverTm
              +SL1.SumDrawDur';
```

The amount of soil loaded in the truck is stored (for later use) in the truck's SaveProp *AmtLoaded*:

```
ONRELEASE TR2 ASSIGN AmtLoaded
                        Load.Soil.Count;
```

The duration of *Haul* depends on *AmtLoaded* relative to the truck's capacity:

```
DURATION Haul 'Normal[11.66,0.3]*
               Haul.Truck.AmtLoaded/
               Haul.Truck.Capacity'
```

Given that each truck has *AmtLoaded* as a *SaveProp*, there is no reason for soil to flow through the network. Instead, the amount of soil dumped into the *MovedSoil* queue is set equal to *the* truck's *AmtLoaded*:

```
RELEASEAMT SL2 'Dump.Truck.AmtLoaded'
```

6.7 Model 7—Engineering Calculations

The effective grade *EG* is the sum of the road properties stored in RR and GR:

```
VARPROP Truck EG RR+GR;
```

A truck's effective power depends on whether the effective grade is positive or negative:

```
VARPROP Truck EffectivePow 'EG>=0 ?
              Power*Eff : RetPower'; /Watts
```

The force that the engine has to overcome to move up/down the grade (in Newtons) is given by:

```
VARPROP Truck Force '(EmptyWgt
     +AmtLoaded*SoilDensity)*Abs[EG]*9.8';
```

The truck's speed (m/sec) equals its effective power over force (but must not exceed its maximum speed):

```
VARPROP Truck Speed
        Min[EffectivePow/Force,MaxSpeed];
```

Road properties are assigned to truck properties while it flows through the links (similarly for TR5, TR8, TR11):

```
ONRELEASE TR2 ASSIGN RR 0.07;
ONRELEASE TR2 ASSIGN GR -0.04;
ONRELEASE TR2 ASSIGN AmtLoaded
                        Load.Soil.Count;
```

The duration of *HaulToBridge* equals distance over speed (converted from seconds to minutes)

```
DURATION HaulToBridge 'DistSourceToBridge
          /HaulToBridge.Truck.Speed/60';
```

The duration of *Dump* depends on the type of truck dumping (only one truck dumps at a time):

```
DURATION Dump Dump.Truck.DumpTm;
```

6.8 Model 8—Final Model with Type I or II Bridges

The final network is shown in Figure 3. At first we need to define a few problem parameters:

```
VARIABLE NumBridges 1; / No. of bridges
VARIABLE TimeToCrossBridge 1.75;/minutes
/LeadTime: TypeI:1.75min, TypeII:0.30 min
VARIABLE LeadTime 0.30;
VARIABLE CostOfABridge 165000;/or $95,000
```

At the end of *HlCrossBridge* the bridge is made available to trucks going in the same direction. The duration of *HlCrossBridge* is (I) 1.75, or (II) 0.30 minutes for proper spacing between Trucks.

```
DURATION HlCrossBridge LeadTime;
DURATION RtCrossBridge LeadTime;
```

The duration of *HlCrsBrdg2* is the remaining time:

```
DURATION HlCrsBrdg2 'TimeToCrossBridge-
                              LeadTime';
DURATION RtCrsBrdg2 'TimeToCrossBridge-
                     LeadTime';
```

The direction of traffic on a Type II bridge cannot reverse (i.e., *HlCrossBridge* cannot start), even if the bridge is available, if there are trucks finishing a crossing in the other direction (i.e., if *RtCrsBrdg2* has any current instances). This assumes that only 1 Type II Bridge is available:

```
SEMAPHORE HlCrossBridge
                     !RtCrsBrdg2.CurInst;
SEMAPHORE  RtCrossBridge
                     !HlCrsBrdg2.CurInst;
```

7 CONCLUSION

The results for models 1-7 are shown in Table 3 and the final results in Table 4. As expected, the results for models 1-7 are very close and much less than the final results (they do not include the cost of any bridges). Table 4 shows that using two Type I bridges is the best alternative with a an expected total cost of $710,000 and a duration of 322 hours. This time is very close to that for models 1-7 so no further improvement is possible.

Table 3: Results of Preliminary Simulation Models

Model	Cost	Time (hr)
1	$521,954	319
2	$522,236	319
3	$522,339	319
4	$520,400	318
5	$521,156	319
6	$517,858	321
7	$511,180	317

Table 4: Simulation Results for Bridge Selection

No of	Type I Bridges		Type II Bridges	
Bridges	Total Cost	Time	Total Cost	Time
1	$802,467	439 hr	$715,433	341 hr
2	$709,787	322 hr	$869,101	334 hr
3	$798,318	318 hr		

To verify that the final model was correct and to illustrate its results to others it was animated using PROOF Animation. A snapshot of the animation is shown in Figure 4.

ACKNOWLEDGMENTS

The development of STROBOSCOPE has been supported by the University of Michigan Horace H. Rackham School of Graduate Studies and the National Science Foundation (Grant No CMS-9415105).

REFERENCES

Ioannou, P. G., and J.C. Martinez. 1995. Evaluation of alternative construction processes using simulation. In *Proceedings of the 1995 Construction Congress*, San Diego, CA, October 22-26, ASCE.

Ioannou, P. G., and J.C. Martinez. 1996a. Animation of complex construction simulation models. In *Proceedings, Third Congress on Computing in Civil Engineering*, Anaheim, CA, June 17-19, ASCE.

Ioannou, P. G., and J.C. Martinez. 1996b. Comparison of construction alternatives using matched simulation experiments. *Journal of Construction Engineering and Management*, ASCE, (122) 3.

Martinez, J. C., Ioannou, P. G., and R. I. Carr. 1994. State and resource based construction process simulation. In *Proceedings of the First Congress on Computing in Civil Engin.*, ASCE, Washington, DC.

Martinez, J.C., and P.G. Ioannou. 1994. General purpose simulation with STROBOSCOPE, In *Proceedings,*

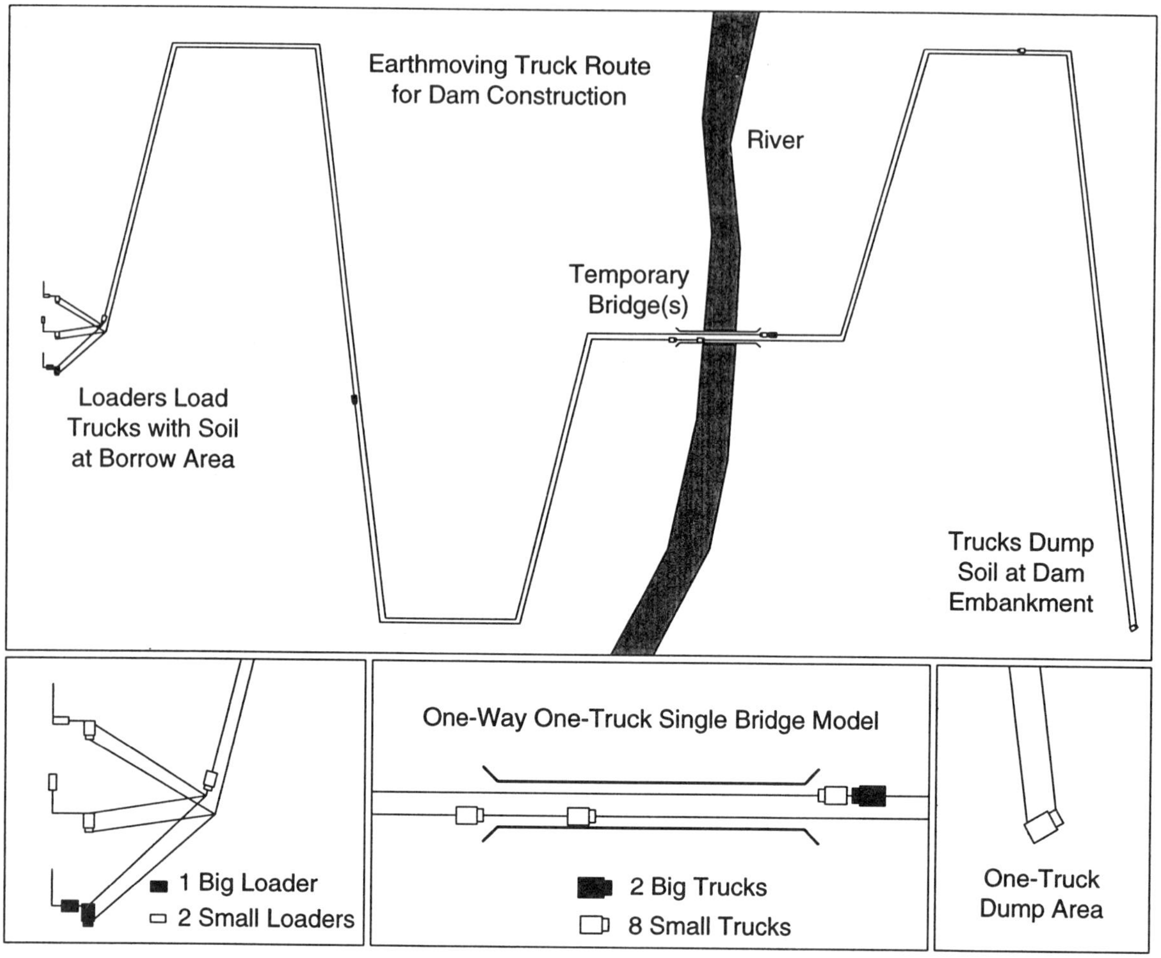

Figure 4: Earthmoving Layout for Dam Construction (PROOF Animation)

1994 Winter Simulation Conference, Orlando FL, December 11-14, IEEE, Piscataway, New Jersey.

Martinez, J.C., and P.G. Ioannou. 1995. Advantages of the activity scanning approach in modeling complex construction processes. In *Proceedings, 1995 Winter Simulation Conference*, Washington, DC, December 3-6, IEEE, Piscataway, New Jersey.

Martinez J.C. 1996. STROBOSCOPE—State- and resource-based simulation of construction processes, Ph.D. thesis, Civil and Environmental Engineering Dept., Univ. of Michigan, Ann Arbor, Michigan.

AUTHOR BIOGRAPHIES

PHOTIOS G. IOANNOU is Associate Professor in the Dept. of Civil and Environ. Engin. at the Univ. of Michigan. He received a Diploma in Civil Engin. from the National Technical Univ., Athens, Greece, in 1979; and a SMCE and Ph.D. from MIT in 1981 and 1984. He has just completed a six-year term as Chairman of the Computing in Construction Technical Committee of the ASCE. He co-developed the UM-CYCLONE construction process simulation system with R.I. Carr, supervised the design and development of COOPS by L.Y. Liu, and served as chairman of J.C. Martinez's Ph.D. dissertation committee. His research interests are primarily focused on the areas of construction decision support systems and construction process modeling.

JULIO C. MARTINEZ is Post Doctoral Research Fellow in Civil Engineering at the University of Michigan. He designed and implemented the STROBOSCOPE simulation language as part of his Ph.D. dissertation research. He received a Civil Engineer's degree from Universidad Catolica Madre y Maestra (Dominican Republic) in 1986, an MS in Civil Engin. from the University of Nebraska in 1987, and an MSE and a Ph.D. in Construction Engineering and Management from the University of Michigan in 1993 and 1996. His research interests are in computer applications to construction engineering and management.

Proceedings of the 1996 Winter Simulation Conference
ed. J. M. Charnes, D. J. Morrice, D. T. Brunner, and J. J. Swain

SCALABLE SIMULATION MODELS FOR CONSTRUCTION OPERATIONS

Photios G. Ioannou
Julio C. Martinez

Civil & Environmental Engineering Department
University of Michigan
Ann Arbor, Michigan 48109-2125, U.S.A.

ABSTRACT

Construction operations are often repetitive not only in terms of time (the same tasks are performed over and over) but also in terms of space (the same tasks are repeated at several places, such as different floors in a high-rise building). Thus, construction simulation models in general must be cyclic to represent temporal repetitions, but also scaleable to represent spatial repetitions. This paper presents the mechanisms for preprocessor replacement and automatic code generation that have been designed and implemented to facilitate the development of scaleable simulation models in STROBOSCOPE, a general-purpose discrete-event simulation system developed by the authors. A relatively complex simulation model for the vertical transportation of people serves as an example to illustrate how to develop a completely scaleable model for the operation of an elevator in a building with any number of floors.

1 INTRODUCTION

Construction operations are typically repetitive in nature and have traditionally been modeled using simulation systems based on cyclic activity networks. Thus, repetitive construction processes are most often represented as cycles through the same set of nodes of a simulation network.

There is an entire class of problems, however, for which this approach does not work. A typical example is the vertical transportation of people (e.g., construction labor) in a high-rise building (e.g., with more than 50 floors) using multiple elevators that may or may not span the entire height of the building. In such cases, the queuing behavior at each floor, although similar to that of other floors, needs to be modeled separately. For example, to collect statistics about the number of people waiting for elevator service at each floor and for each direction of travel, the model must include two queues per floor, one for the people waiting to go up and another for those waiting to go down. Thus, a building with 50 floors would require the arduous task of defining 98 queues (the 1st and 50th floors only need one queue) and at least 196 links (one link for entering and another for leaving each queue). Lumping all these queues into two global queues (one for all the people waiting to go up and one for all the people waiting to go down) trades this problem for another. While it eliminates the need to define multiple queues and links, it also prevents the automatic collection of any per-floor statistics. Thus, the problem now requires the definition of two statistics collectors per floor, while ensuring that these collectors receive the correct data every time a person enters or leaves the global waiting queues at a particular floor.

The same type of problem occurs when modeling a horizontal transportation system, such as the inter-terminal bus service at an airport. As the number of repetitive units increase (such as the floors in a building or the number of terminals at an airport), simulation models must be augmented with new modeling elements to represent the new floors or terminals.

A convenient solution to this type of problems is to create scaleable simulation models at the "meta" level that can create detailed simulation models at the simulation language level. Thus, a scaleable simulation model must have the capability to generate simulation code that defines any number of simulation elements (e.g., activities, queues, links, etc.) as necessary to match their real counterparts. This paper describes the facilities for statement preprocessing and automatic code generation provided within STROBOSCOPE (STate and ResOurce Based Simulation of COnstruction ProcEsses), a simulation language developed by the authors based on activity-scanning. The elevator problem is used as an example to illustrate these capabilities.

2 MODELING ELEVATOR OPERATIONS

The facilities for creating scaleable simulation models will be described by using as an example a building served by a single elevator. This fairly complex problem has been adapted from (Law and Kelton 1991) and is used to illustrate the effectiveness of preprocessing and automatic code generation to create a scaleable model that once verified (and even animated) can provide an accurate representation of the elevator's complicated control logic for any number of floors.

The problem is of interest in high-rise building construction (50+ stories) as subcontractors add money to their bid to account for lost time in hoisting personnel. If at the bidding stage a general contractor can show the subcontractors a video (or an animation) and supporting documentation that the proposed hoisting system has been optimized and that waiting times will be as short as possible, there may be cost benefits to all parties (including the owner). In order to do this, project managers have expressed the need for models depicting the movement of labor that can provide queue-related statistics for different hoisting policies. In general, such models should be able to analyze and optimize the operation of multiple elevators and hoists (some of which may or may not span the entire building height) in very tall buildings.

In addition to being interesting, this problem was also selected because the "classical" elevator problem (even with simplified control logic) is one of the most difficult problems for simulation languages to model and verify. It has been estimated, for example, that the elevator problem (with much simpler control logic than the example described below) requires an expected modeling effort of 20-30 hours of work (Chisman 1996).

3 EXAMPLE PROBLEM STATEMENT

A five-story building is served by a six-person elevator. People arrive to the ground floor (floor 1) with independent exponential interarrival times having a mean of 1 minute. A person will go to each of the four upper floors with probability 0.25. It takes the elevator 15 seconds to travel up or down one floor. For simplicity it will be assumed that the elevator loading and unloading time at any floor is zero. The length of stay of a person at a particular floor is distributed uniformly between 15 and 120 minutes. When a person leaves floor i = 2, 3, 4, 5, he or she will go to floor 1 with probability 0.7, and will go to each of the other three floors with probability 0.1. A person coming down to floor 1 departs from the building immediately.

When the elevator is going up, it will continue in that direction if a current passenger wants to go to a higher floor or if a person on a higher floor wants to get on the elevator. When the elevator is going down, it will continue in that direction if it has at least one passenger or if there is a waiting passenger at a lower floor. When the elevator stops at floor i = 2, 3, 4 while going up (down), it picks up only those people at that floor that want to go up (down). If the arriving elevator does not have enough room to get all the people waiting at a particular floor, the excess remain in queue.

The elevator decides at each floor what floor it will go to next. It does not change directions between floors. At the start of the simulation the elevator is at rest at its base floor. This is the same floor the elevator returns to whenever it is idle. The best choice for the base floor is made by minimizing the average of individual delays over all floors and all people. An integral part of the modeling requirements for this problem is the collection of statistics about the number of people waiting and the corresponding waiting times at each floor and for each direction of travel.

4 SIMULATION METHODOLOGY

A scaleable simulation model for this problem has been developed using STROBOSCOPE, a general-purpose simulation programming language based on activity-scanning.

A complete description of STROBOSCOPE appears in (Martinez 1996). Example applications can be found in (Martinez, Ioannou, and Carr 1994; Martinez and Ioannou 1994, 1995; Ioannou and Martinez 1995, 1996a, 1996b). The STROBOSCOPE program, its documentation, and several examples are available via anonymous ftp from "grader.engin.umich.edu."

5 SIMULATION MODEL NETWORK

The STROBOSCOPE cyclic activity network for the elevator problem is shown in Figure 1. From a tactical point of view, this network can be divided into two parts.

The eight queues enclosed by the dashed lines (*Q1U*, *Q2U*, *Q3U*, *Q4U*, *Q2D*, *Q3D*, *Q4D*, and *Q5D*) as well as the 16 links that enter and depart these queues represent the scaleable portion of the network. The four queues ending in *U* represent the people waiting at the corresponding floor for the elevator to go up. Similarly, the four queues ending in *D* represent the people waiting for the elevator to go down. Thus, each of these queues is specific to a particular floor and a direction of movement (up or down). To increase the number of floors in the building it is necessary to define additional network elements: two queues and four links per floor.

The remainder of the simulation network represents the cycle of loading and moving the elevator from one floor to the next. The main network elements in this cycle are the activities *EnterElev* and *MoveElev*. Notice that this cycle is not specific to any particular floor, nor does it depend on the number of floors.

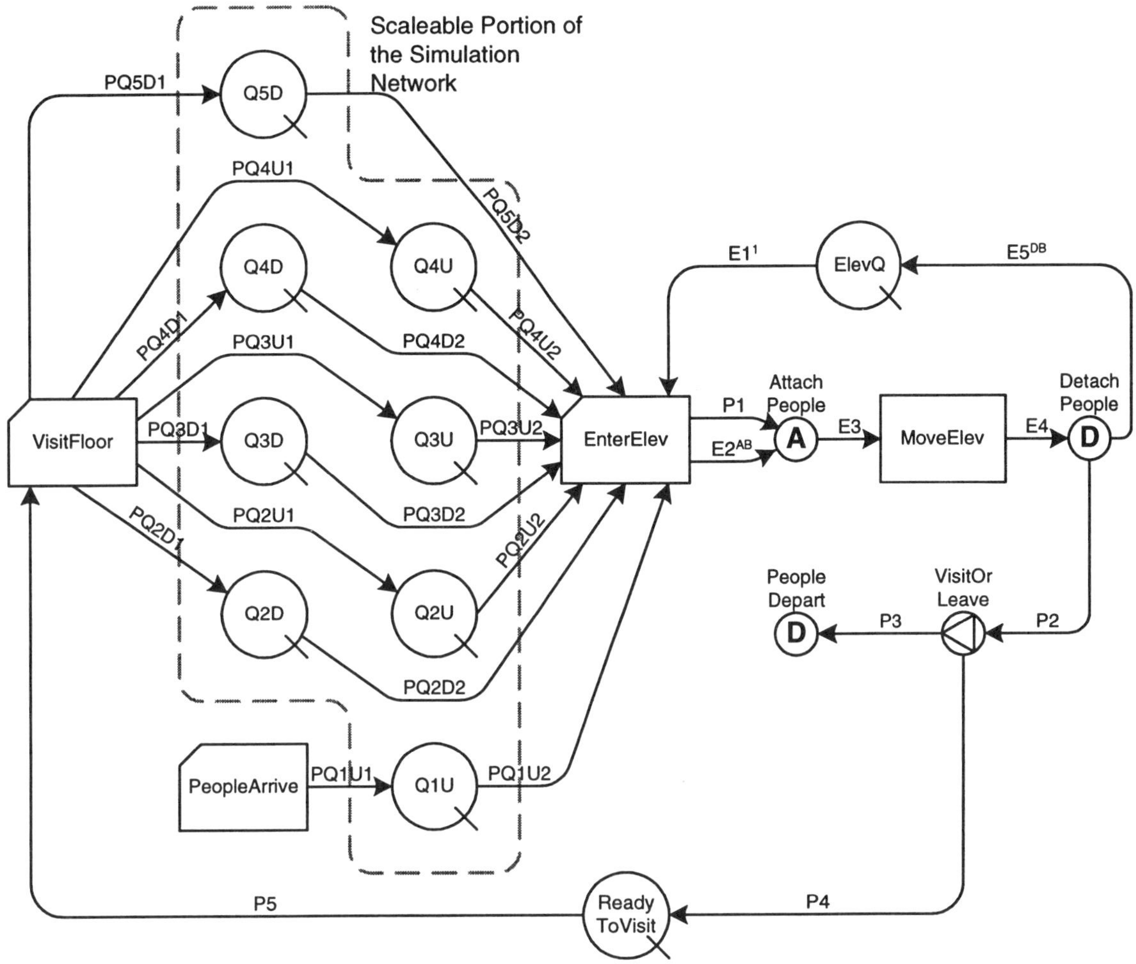

Figure 1: STROBOSCOPE Network for the Elevator Model

There are two types of resources in this model, the *elevator* and *people*. For convenience both are defined as compound characterized resources. Although people could have been modeled just as well as simple characterized resources with no static properties, compound characterized resources are easier to define in this case because there is no need to define subtypes. Furthermore, compound resources provide a base to which other resources (generic, simple characterized, or compound characterized) can be attached and detached. Thus, defining the elevator as a compound resource makes it extremely simple to model a loaded elevator as an elevator with people attached.

The attachment of people (a compound resource) to the elevator (another compound resource that serves as the assembly base) is performed by the assembly node *AttachPeople*. At the end of activity *MoveElev*, the people that have reached their destination and need to unload are detached from the elevator by the disassembly node *DetachPeople*. The rest of the people inside the elevator are in transit and remain attached to the elevator which is then routed to the queue *ElevQ* ready to start another cycle.

The people detached from the elevator are routed to either the disassembly node *PeopleDepart* (i.e., to floor 1 and out of the building), or to the queue *ReadyToVisit*. From there each person that exited the elevator starts a separate instance of the combi (conditional) activity *VisitFloor* where it spends time visiting the associated floor.

The most crucial element of this model is to specify the precise conditions under which the combi activity *EnterElev* can start. This combi is preceded by nine queues (eight *people* queues and one *elevator* queue). Obviously, the only hard requirement for *EnterElev* to be able to start is the availability of the elevator (i.e., that the *ElevQ* queue be non-empty). This requirement, however, does not hold true for the eight people queues. Because the elevator should be able to move to another floor even when all these queues are empty (this occurs when the elevator is idle and returns to its base floor). To allow this, the *Enough* attributes of the links joining the eight people queues to the *EnterElev* combi must be set so that they are always *true*. Thus, from a resource availability point of view, only the *Enough* attribute of link E1 matters (by default this is determined by the contents of queue *ElevQ*).

However, the availability of the elevator should not be sufficient for *EnterElev* to start. In addition, at least one of three conditions must also be satisfied: either the elevator contains people in transit, or the total contents of the eight people queues are nonzero (there are somewhere people waiting for the elevator), or the elevator is not at its base floor (and should return there). These three conditions constitute the *semaphore* for *EnterElev*. The *semaphore* for a combi activity is the first logical condition that is evaluated every time the combi activity attempts to start. If the *semaphore* is *false* then the attempt fails. If the *semaphore* is true, then STROBOSCOPE continues checking and evaluates the *Enough* attributes for the links joining the combi to its preceding queues. A combi is able to start only if its *semaphore* and the *Enough* expressions for all these links evaluate to *true*.

Once the decision to start *EnterElev* is made (because its *semaphore* is *true* and the elevator is waiting in *ElevQ*), the elevator is certain to move to another floor. The issue now is to determine the direction of travel: *up* or *down*? This decision is made on the *beforedraws* event of the *EnterElev* combi (i.e., before the combi *EnterElev* draws the elevator and possibly any people as new passengers). On this event, the model sets the *EDirection* property of the elevator to one of three values: "1" to indicate "going up", "-1" to indicate "going down" and "0" to indicate "idle at base floor".

The expression that gets evaluated to determine *EDirection* at this event is a complex conditional statement that implements the following logic: The elevator keeps its current direction (up or down) if it contains people in transit, or if there are people waiting at higher/lower floors and the current direction is up/down. If none of these conditions are true, then if there are any people waiting to board, the elevator switches direction; otherwise, it moves in the direction of its base floor.

Once the decision has been made to move the elevator in a specific direction (i.e., *EDirection* is set), the next issue is to decide which one (if any) of the eight people queues will be allowed to have the people waiting there enter the elevator. This is accomplished by evaluating the *drawwhere* attribute of the links from each of the eight people queues to the *EnterElev* activity. The *drawwhere* attribute of a link is a filter that is evaluated for each of the characterized resources residing in the preceding queue to determine those that can be drawn through the link. Only those characterized resources that pass the filter can pass though the link. Clearly, the statements specifying the *drawwhere* attributes for these links are repetitive and should be generated automatically as described in the next section.

Two *SaveProps* (i.e., assignable properties) for the *people* compound resource are *CurFloor* and *Direction* (i.e., the floor on which the person is now and the direction he/she wants to move to). The *drawwhere* attribute of the links simply states that only those people that are on the same floor as the elevator and want to move in the elevator's current direction should be allowed to board. The number of people allowed to board is regulated by the link's *drawuntil* attribute. The expression for this attribute is the same for all links and allows loading people until the number of passengers equals the capacity of the elevator. Again, the *drawuntil* statements are repetitive and their specification in a generic manner is presented in the next section.

After the elevator loads the people currently waiting at the same floor and wishing to go in its own direction (up to its capacity), the combi activity *EnterElev* can start. Upon start-up, the elevator property *EPeopleCount* is updated to reflect the current number of passengers inside the elevator.

The *EnterElev* combi has zero duration and terminates immediately after it starts. Its termination leads to the assembly node *AttachPeople* where the newly loaded passengers are attached to the elevator compound resource (and join the rest of the passengers already attached). The loaded elevator is then routed as one compound resource to the *MoveElev* activity where it spends an amount of time equal to the duration of a one-floor move.

Just before the *MoveElev* activity ends and releases the elevator, the elevator property *ECurrentFloor* is updated to reflect its new current floor position. The elevator is then routed to the disassembly node *DetachPeople*. Here all passengers are for a moment detached from the elevator compound resource to allow those people that have reached their destination to depart and flow through link P2. The *releasewhere* attribute of link P2 compares each person's *NextFloor* property to the elevator's *ECurrentFloor* property. Those that match flow through P2 and leave the elevator. The remaining passengers are in transit. They are reattached to the elevator compound object and are released to the *ElevQ* queue via link E5 to start another cycle.

If the elevator is on floor 1 then the departing passengers are routed by the fork *VisitOrLeave* to the disassembly node *PeopleDepart* where they are destroyed (there are no exit links). Otherwise, they are routed to the *ReadyToVisit* queue where each person starts a separate instance of the *VisitFloor* combi. On release through link P2, each person's *CurFloor* property is updated to reflect the floor he/she is currently on. On flow through link P4, each person's next floor destination (at the conclusion of the visit to the current floor) is determined through simple Monte Carlo sampling and is assigned to its *NextFloor* property. At the same time, a person's *Direction* property is determined based on the values of *NextFloor* and *CurFloor*.

When an instance of the *VisitFloor* combi activity finishes, it releases the person it holds to one of the seven people queues that follow. The recipient queue is determined by the *releasewhere* attributes of the seven links out of *VisitFloor*. The *releasewhere* attribute for each of these eight links allows a person to flow through only if its current floor (*CurFloor*) and direction of movement (*Direction*) match those of the succeeding queue. Thus, exactly one of these links will allow the person to flow through. It should be obvious that the specification of the *releasewhere* attributes of the links out of *VisitFloor* are repetitive and can be generated automatically as described below.

6 STATEMENT PREPROCESSING AND AUTOMATIC CODE GENERATION

For the elevator model to be scaleable, there is clearly a need to generate automatically the STROBOSCOPE statements that define all the *up* and *down* queues (two queues per floor), as well as all the links in out of these queues (two links per queue). The number of floors in the building should be the only variable that controls the size of the model. However, in addition to these statements (that have an obvious effect on the *size* of the simulation network), we must also be able to generate statements that define the necessary queue or link *attributes* (*drawwhere*, *drawuntil*, etc.) as well as those statements necessary for the collection of statistics and the communication of the results.

In STROBOSCOPE this is accomplished by using the preprocessor operator $<*Arg*>$, where *Arg* stands for the preprocessor replacement expression. A simulation model file is a text file that is read, pre-processed, and executed one statement at a time. As each statement is read, STROBOSCOPE searches for the two-character sequence "$<" (i.e., a *dollar sign* immediately followed by a *less than sign*). If the sequence "$<" is found, then what follows is treated as the preprocessor replacement expression. This expression extends to but does not include the next occurrence of ">$" (i.e., a *greater than sign* followed by a *dollar sign*). The entire preprocessor replacement expression is treated as a mathematical formula, which is then evaluated, truncated to an integer number, and used to replace the entire sequence $<*Arg*>$ as a string. After this substitution, the resulting statement is parsed and executed like any other ordinary STROBOSCOPE statement.

For example, the following statements:

```
QUEUE Q$<Log[10]>$U People;
LINK PQ$<Sin[3.14/2]>$U2
            Q$<Cos[0]>$U EnterElev;
```

are equivalent to:

```
QUEUE Q1U People;
LINK PQ1U2 Q1U EnterElev;
```

Statement preprocessing when coupled with control statements that implement *While-Wend* loops provide STROBOSCOPE with a powerful mechanism for automatic code generation. This mechanism is best illustrated with an example. The following short snippet of STROBOSCOPE code uses a *While-Wend* loop (controlled only by the number of floors in a building) to define the scaleable elements of the simulation network. These are the 4 "going up" queues (Q1U to Q4U), the 4 "going down" queues (Q2D to Q5D), the 8 incoming links (PQ1U1 to PQ4U1, and PQ2D1 to PQ5D1), and the 8 outgoing links (PQ1U2 to PQ4U2, and PQ2D2 to PQ5D2). Thus, for a building with *N+1* floors this short set of statements defines *2N* queues and *4N* links:

```
VARIABLE nFloors 5;
SAVEVALUE I 1;
WHILE I<=nFloors-1;
   QUEUE Q$<I>$U People;/ Ith Up Queue
   IF I==1; /1st floor links (exception)
      LINK PQ1U1 PeopleArrive Q1U;
      LINK PQ1U2 Q1U EnterElev;
   ELSE; /Intermediate floor links (rule)
      LINK PQ$<I>$U1 VisitFloor Q$<I>$U;
      LINK PQ$<I>$U2 Q$<I>$U EnterElev;
   ENDIF;
   QUEUE Q$<I+1>$D People;/ (I+1) down Q
   LINK PQ$<I+1>$D1 VisitFloor Q$<I+1>$D;
   LINK PQ$<I+1>$D2 Q$<I+1>$D EnterElev;
ASSIGN I I+1;
WEND;
```

Similarly, the following *While-Wend* loop defines several STROBOSCOPE statements that specify the behavior of the above nodes and links. In particular, these statements set the *Enough* attributes of all the links entering activity *EnterElev* to *true*; define *drawwhere* filters that allow only the correct people to board the elevator; define *drawuntil* attributes to allow people to board for as long as there is room in the elevator; define statistics collectors for each queue and supply data for the number of people that could not board the elevator because it was full; collect statistics about each person's overall waiting time; and define filters that route each person that wants to leave a floor to the correct waiting queue at the elevator. The following statements include detailed comments and should be self-explanatory:

```
ASSIGN I 1;
WHILE I<=nFloors-1;
 /set the ENOUGH for links PQxU2 to TRUE
 ENOUGH PQ$<I>$U2 !FALSE;
 /Define statistics collectors NoFitxU
 COLLECTOR NoFit$<I>$U;

 /Before EnterElev draws any resources
 / collect in NoFitxU the number of people
 /  at QxU that do not fit to board.
 /   (provided people do wait at QxU
 /     and the elevator is going up
 /      and the elevator is at floor x)
 BEFOREDRAWS EnterElev COLLECT NoFit$<I>$U
  PRECOND 'Q$<I>$U.CurCount &
         EDirection==1 & ECurFloor==$<I>$'
  Max[Q$<I>$U.CurCount-SpaceLeftInElev,0];

 /Draw via link PQxU2 only those people
 / whose floor and direction match those
 /  of the elevator:
 DRAWWHERE PQ$<I>$U2
             'EDirection==Direction &
              ECurFloor==CurFloor';

 /When drawing a person through link PQxU2
 / collect in OverallWait the waiting time
 /  for that person in queue QxU
 ONDRAW PQ$<I>$U2 COLLECT OverallWait
              SimTime-TimeIn;

 /Keep drawing people through link PQxU2
 / until there is no space in the elevator
 DRAWUNTIL PQ$<I>$U2 !SpaceLeftInElev;

 / Release through link PQxU1 (x<>1)
 /  (to queue QxU) only those
 /  people that are currently
 /  at floor x and want to go up
 IF I!=1;
  RELEASEWHERE PQ$<I>$U1  'CurFloor==$<I>$
              & Direction== 1';
 ENDIF;

 /****From here down "x" means "I+1"****

 /set the ENOUGH for links PQxD2 to TRUE
 ENOUGH PQ$<I+1>$D2 !FALSE;

 /Define statistics collectors NoFitxD
 COLLECTOR NoFit$<I+1>$D;

 /Before EnterElev draws any resources
 / collect in NoFitxD the number of people
 /  at QxD that do not fit to board.
 /   (provided people do wait at QxD
 /     and the elevator is going down
 /      and the elevator is at floor x)
 BEFOREDRAWS EnterElev
   COLLECT NoFit$<I+1>$D
      PRECOND 'Q$<I+1>$D.CurCount &
      EDirection==-1 & ECurFloor==$<I+1>$'
    'Max[Q$<I+1>$D.CurCount-
         SpaceLeftInElev,0];

 /Draw via link PQxD2 only those people
 / whose floor and direction match those
 /  of the elevator:
 DRAWWHERE PQ$<I+1>$D2
              EDirection==Direction &
              ECurFloor==CurFloor;

 /When drawing a person through link PQxD2
 / collect in OverallWait the waiting time
 /  for that person in queue QxD
 ONDRAW PQ$<I+1>$D2 COLLECT OverallWait
              SimTime-TimeIn;

 /Keep drawing people through link PQxD2
 / until there is no space in the elevator
 DRAWUNTIL PQ$<I+1>$D2 !SpaceLeftInElev;

 / Release through link PQxD1
 /  (to queue QxD) only those
 /  people that are currently
 /  at floor x and want to go up
 RELEASEWHERE PQ$<I+1>$D1
      'CurFloor==$<I+1>$ & Direction==-1';

 / Increment the counter I and loop
 ASSIGN I I+1;
WEND;
```

The above examples are not the only places where the elevator model requires the use of preprocessing and automatic code generation. This capability is also needed to produce the required output at appropriate points during simulation run-time and after the simulation is completed. For example, automatic code generation is used to generate STROBOSCOPE statements that produce animation instructions for PROOF Animation (by writing to a text file at appropriate points during simulation run-time). Similarly, at the end of simulation,

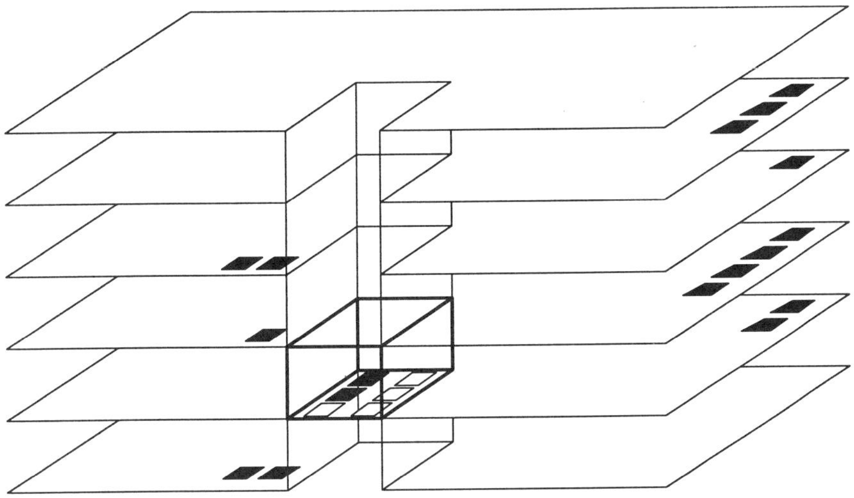

Figure 2 - PROOF Animation of Elevator Model

automatic code generation provides a convenient way to create statements that print the desired statistical results on a *per-floor*, or *per-queue* basis. In both cases, the necessary STROBOSCOPE statements are repetitive in nature (i.e., they repeat over the queues, their links, or both). Thus, generating these statements with a *While-Wend* loop is a natural solution to the problem of making the model scaleable.

7 ANIMATION

The STROBOSCOPE simulation model for the elevator problem described above was designed to be scaleable with respect to animation as well. By simply changing a single parameter that represents the number of floors in a building, it is possible to produce the correct animation trace file (ATF) that could then be processed in playback mode by Proof Animation to show the operation of the elevator for the selected number of floors.

Figure 2 shows a snapshot of the animation for a building with 5 floors. The building in this figure is part of the layout background, whereas the elevator (shown as a box with six slots) and the people (shown as colored squares) are instances of classes defined in the animation layout file. Every time a new person enters the building a new square is created and placed on a "line path" (a predefined trajectory) where it queues waiting for the elevator. When the elevator arrives, a person going in the same direction is moved from its "path queue" to one of the empty six slots inside elevator. A person exiting the elevator is placed on another "path queue" on the right-hand side of the corresponding floor where it stays for the duration of the person's visit to that floor. At the end of this visit, a person is removed from the "path queue" on the right-hand side of the floor and is placed on another "path queue" on the left-hand side of the floor where it waits for the elevator to pick it up. Figure 2 shows two people inside the elevator, two people waiting for the elevator at floors 1 and 4 and one person waiting at floor 3. In addition, there are two, four, one, and three people still visiting floors 2, 3, 4, and 5. Even a short playback of the animation makes it obvious that the elevator policy has been implemented correctly.

The only and rather obvious difficulty in scaling the entire animation is that the background drawing representing the building itself must also be scaled to have the correct number of floors. Fortunately, PROOF Animation stores the layout file as a CAD drawing in text format and this makes it possible to produce a layout file with the correct number of floors in the building as part of the output of a completely scaleable simulation run.

8 CONCLUSION

After a parametric model's accuracy is verified through statistics collection and animation, scaling it to represent, for example, a 50+ story building becomes just a matter of changing a single number to define the required number of floors.

The model presented above is not the only way to make the elevator problem scaleable. As mentioned earlier, it is possible to lump the separate "up" and "down" queues for each floor into two *global* "up" and "down" queues for the entire building (or even just one queue with unspecified direction). In this case, automated code generation would be necessary to define many statistics collectors that provide the same statistical information on a per floor basis as provided automatically by the deleted queues.

Alternatively, it is also possible to eliminate entirely the use of *collectors* (i.e., STROBOSCOPE objects) for gathering per floor statistics. For the elevator problem, for example, it is possible to store statistical information on a per floor basis in one- or two-dimensional arrays. In essence, this approach uses arrays to perform manually the functions that collectors do automatically. No matter what approach is used, however, it is very difficult to create completely scaleable models that do not have any need for statement preprocessing and automated code generation.

ACKNOWLEDGMENTS

The development of STROBOSCOPE has been supported by the University of Michigan Horace H. Rackham School of Graduate Studies and the National Science Foundation (Grant No CMS-9415105).

REFERENCES

Chisman J.A. 1996. *Industrial cases in simulation modeling*, Duxbury Press, ITP, Inc.

Ioannou, P. G., and J.C. Martinez. 1995. Evaluation of alternative construction processes using simulation. In *Proceedings of the 1995 Construction Congress*, San Diego, CA, October 22-26, ASCE.

Ioannou, P. G., and J.C. Martinez. 1996a. Animation of complex construction simulation models. In *Proceedings, Third Congress on Computing in Civil Engineering*, Anaheim, CA, June 17-19, ASCE.

Ioannou, P. G., and J.C. Martinez. 1996b. Comparison of Construction Alternatives Using Matched Simulation Experiments. *Journal of Construction Engineering and Management*, ASCE, (122) 3.

Law A.M., and W.D. Kelton. 1991. *Simulation Modeling and Analysis*, 2nd ed. McGraw Hill, New York, NY.

Martinez, J. C., Ioannou, P. G., and R. I. Carr. 1994. State and resource based construction process simulation. In *Proceedings of the First Congress on Computing in Civil Engin.*, ASCE, Washington, DC.

Martinez, J.C., and P.G. Ioannou. 1994. General purpose simulation with STROBOSCOPE, In *Proceedings, 1994 Winter Simulation Conference*, Orlando FL, December 11-14, IEEE, Piscataway, New Jersey.

Martinez, J.C., and P.G. Ioannou. 1995. Advantages of the activity scanning approach in modeling complex construction processes. In *Proceedings, 1995 Winter Simulation Conference*, Washington, DC, December 3-6, IEEE, Piscataway, New Jersey.

Martinez J.C. 1996. STROBOSCOPE—State- and resource-based simulation of construction processes, Ph.D. thesis, Civil and Environmental Engineering Dept., Univ. of Michigan, Ann Arbor, Michigan.

AUTHOR BIOGRAPHIES

PHOTIOS G. IOANNOU is Associate Professor in the Dept. of Civil and Environ. Engin. at the Univ. of Michigan. He received a Diploma in Civil Engin. from the National Technical Univ., Athens, Greece, in 1979; and a SMCE and Ph.D. from MIT in 1981 and 1984. He has just completed a six-year term as Chairman of the Computing in Construction Technical Committee of the ASCE. He co-developed the UM-CYCLONE construction process simulation system with R.I. Carr, supervised the design and development of COOPS by L.Y. Liu, and served as chairman of J.C. Martinez's Ph.D. dissertation committee. His research interests are primarily focused on the areas of construction decision support systems and construction process modeling.

JULIO C. MARTINEZ is Post Doctoral Research Fellow in Civil Engineering at the University of Michigan. He designed and implemented the STROBOSCOPE simulation language as part of his Ph.D. dissertation research. He received a Civil Engineer's degree from Universidad Catolica Madre y Maestra (Dominican Republic) in 1986, an MS in Civil Engin. from the University of Nebraska in 1987, and an MSE and a Ph.D. in Construction Engineering and Management from the University of Michigan in 1993 and 1996. His research interests are in computer applications to construction engineering and management.

Proceedings of the 1996 Winter Simulation Conference
ed. J. M. Charnes, D. J. Morrice, D. T. Brunner, and J. J. Swain

SIMULATION WORKS: A PANEL DISCUSSION

CHAIR

William R. Lilegdon
Pritsker Corporation
8910 Purdue Road, Suite 600
Indianapolis, Indiana 46268, U.S.A.

PANELISTS

Stuart Gittlitz
Kraft Foods
250 North Street
White Plains, N.Y. 10625,USA

Randy D. Moore
SONY Display Tube Company
16450 West Bernardo Drive
San Diego, California 92127

Patrick Reardon
Los Alamos National Laboratories
Los Alamos, NM 87545

ABSTRACT

Simulation works. This panel will discuss the issues relating to successful simulation within their respective organizations. The panel will present successful simulation projects and the key issues that led to the success. Sample questions to be addressed by the panel include:

- How to introduce simulation to an organization?
- How to complete a project with a diverse project team? How to select or scope projects?
- How much technology is enough to solve a problem?

The diverse background of the participants will provide a broad view of what makes simulation work and how to repeat the process. Example projects will be discussed in addition to other issues.

1 INTRODUCTION

The use of simulation to support the decision making process within commercial organizations has been growing consistently for several years. New software technologies and more simulation education has led to an environment where simulation is easier, less expensive, and performed more often. The focus of many simulation conferences and discussions is often the technology applied to a given area. However, in practice, the technology often takes a back seat to other implementation issues. These issues include the introduction of the technology to the organization, the project approach, the dynamics of the project team, and many others.

These are issues that new and existing users need to examine. This panel will begin the discussion of these issues relying on the varied experience of the three active simulation practitioners.

2 SELECTING A SIMULATION STRATEGY

An organization must select an implementation strategy prior to implementation. This may or may not be a conscience selection. A continuum of options exists. At one end of the spectrum is the purchase of a simulation tool and the use of the tool by current facility staff. At the other end of the spectrum is the simulation consultant that is hired for point solutions. Both approaches have advantages and disadvantages. In between is a mix of the two approaches. Many larger organizations employ the consultative approach using in-house or corporate expertise. The panel has representatives from the entire spectrum.

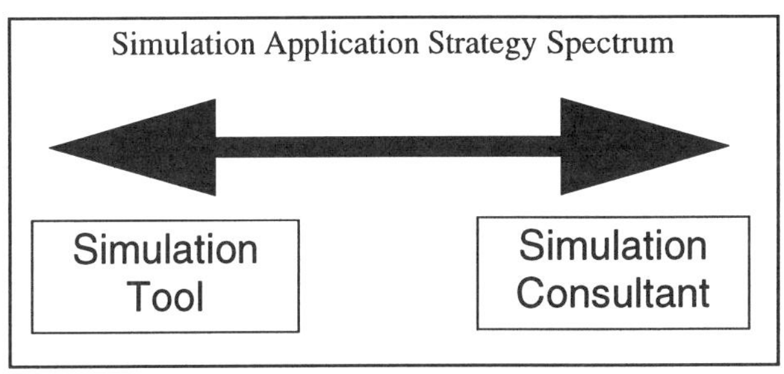

Figure 1. Simulation Implementation Strategy Spectrum

The simulation tool approach has the highest up-front costs, including the purchase of the software, hardware, training, and labor. But is also offers a high degree of flexibility and opportunity for application. Depending on the volume of simulation projects, in the longer term analysis this approach could be more cost effective.

The consultive approach provides a quick solution. Getting the proper design for a new facility to be started in the next four weeks requires that this approach be used by an organization. Many simulation consultants can support multiple tools providing a broader range of skills that can be acquired in a short time by company staff. The premium expense for these consultants is often offset by the increased efficiency. From a strategic point of view, an organization must anticipate the volume of simulation work to be required before they chose a strategy. And most importantly they need to understand the continuum of services and the ability to move from a consultive approach towards more staff capabilities.

3 A SAMPLE PROJECT

Successful simulation, in this panel's viewpoint, means a positive result for the sponsoring organization. This result is usually measured in financial terms, although the successful simulations may also result in "soft" benefits. The "soft" benefits generated from a simulation project include better group communication and understanding of the processes being modeled. These benefits are also commonly associated with business re-engineering projects. Reduce operating expenses are the easiest to measure and communicate to organization management. However, one of simulations key benefits is often cost avoidance. In these projects a proposed design is evaluated. The design is refined to meet production targets with a lower investment. The avoidance of the expense is a cost savings but requires the publication of the less than favorable first design. Often these issues can not be communicated as widely within an organization and reduce the perceived benefit of simulation.

Other measures that impact the success of a simulation project include timeliness, cost, and team support. Project methodology can dramatically impact these measures. The following brief project description provides an overview of the typical simulation project.

Kraft utilizes simulation to design new facilities and improve existing ones. In this example, a Kraft plant was considering adding process capacity to a facility that processed and packaged a grain product. The existing operation steps included the processing of the grain, the in-process storage of the grain in one of 12 large bins, and the packaging of the finished product.

An engineering team was scoping the project to determine cost, timing, layout, etc. The simulation group was asked to help determine the right number of in-process storage bins to support the added production capacity. Additionally, less expensive methods to achieve the required production capacity, other than a major equipment installation, were to be considered.

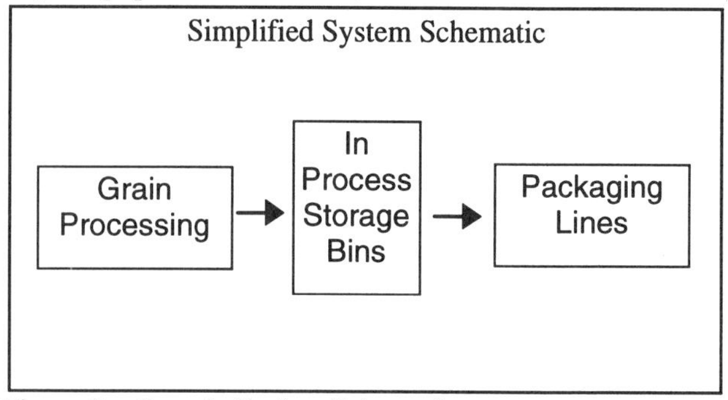

Figure 2. - Sample Project Schematic

To begin the modeling process the plant personnel were interviewed. From this information a simulation was developed showing the current process and packaging production for an entire week. The simulation showed that the existing grain process over-produced the packaging lines capacity. In fact, by mid-week of the simulation, the grain process was forced to shut down because all the in-process storage bins were full. When plant personnel reviewed the simulation results and animation, they said, "That's just what happens to us. When we run well, by mid-week, we have to shut down the process and let packaging catch up."

By checking the flow rates through the entire operation, it was discovered that a conveyor between the in-process storage bins and the packaging area limited the production capacity. While there were several packaging lines, the conveyor flow limited the amount of grain, so that only one or two packaging lines could be operated concurrently. When the amount of in-process grain was at high levels, additional packaging lines could not be supplied to help reduce the levels in storage. Eventually the grain production process had to be stopped. The original scope of the project was the addition of new process equipment to add capacity above what was already available.

After simulating several options for capacity increases, the final project added a larger transfer conveyor between in-process storage and the packaging lines. Added packaging line flexibility and higher transfer rates out of the in-process bins were proposed. The capital cost of this alternative was about 20 percent of the originally planned equipment addition.

Simulation worked to reduce the cost of a proposed design in this example. This project also describes the dynamic nature of the scope of some simulation projects. The original scope to determine the equipment needs in the grain production area was changed to evaluate the identified bottleneck occurring between the in-process bin storage and the packaging lines. These re-directions of simulation projects are not uncommon and are very difficult to manage.

4 MAINTAINING SIMULATION SUCCESSES

In most organizations the use of simulation is sporadic. Budget and market cycles require that facilities be re-designed or developed on a less than continuous basis. To continue to provide an organization with the benefits of the simulation technology several issues need to be addressed.

First, the primary focus on every project should be the benefit to the user. Many projects allow focus to turn the advanced nature of the technological solution. These projects may be successful, but it is due more to luck than actions. The simulation process begins with the identification of the problem to be addressed. The benefits or key questions to be answered must be foremost in every phase of the simulation process.

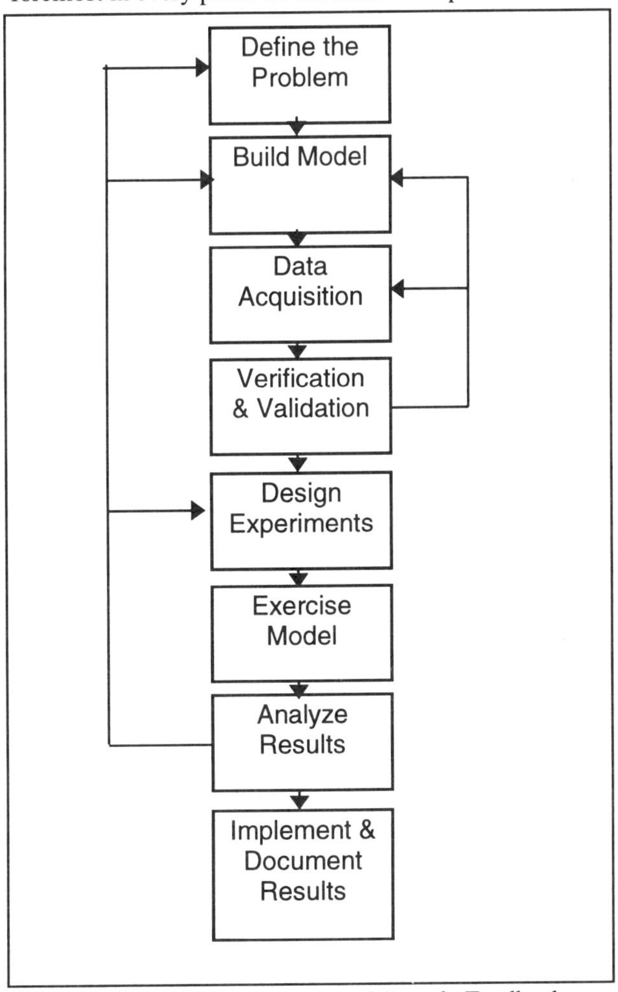

Figure 3. - Simulation Process and Sample Feedback

Focusing on this goal provides a measurable result with each project. These results are necessary to justify the investment in the next project.

Second, the accurate and timely documentation of results from previous projects is of the utmost importance. Many organizations have performed numerous successful simulation projects, but can point to few concrete advantages of the project twelve months after completion. This is caused not by the lack of contribution of the simulation technology but in large part by the dynamics of most organizations. In this twelve month period staffing and job assignment changes have erased or lost any corporate memory relative to the projects. To build an active simulation presence in an organization the demonstrated benefit of the technology must be documented and communicated.

Finally, simulation technology is changing rapidly. To contribute to an organization in the most effective manner requires an investment of time to stay abreast of the latest technology. A corollary to this idea is that as technology evolves and revolutionizes the application of simulation, the tools most familiar to the practitioner may become outdated. As the tools are developed for specific simulation applications it may also become necessary for practitioners to extend their "toolkit" to include other tools.

5 CONCLUSION

Simulation can work for many problems. The panelists will discuss how they make simulation work for their organizations. There will also be the opportunity for audience members to contribute and challenge the panelist's ideas.

REFERENCES

Mussleman, Kenneth J. (1994), Guidelines for Simulation Project Success. *Proceedings 1994 Winter Simulation Conference*, ed. J. Tew, S. Mannivannan, D. Sadowski, A. Seila., 88-95, Lake Buena Vista, Florida.

Pritsker Corporation, (1993), *Simulation: A Decision Support Tool.* West Lafayette, Indiana: Pritsker Corporation.

AUTHOR BIOGRAPHY

WILLIAM R. LILEGDON is a Vice President for Pritsker Corporation. He is responsible for the promotion, development direction, and delivery of Pritsker's engineering software and services. He received a Bachelors Degree in Industrial Engineering from Purdue University. Since joining Pritsker Corporation he has worked in the consulting, development, support and training areas. William led the development of SLAM II/PC, SLAMSYSTEM, and FACTOR/AIM software products. He has published numerous papers and articles on simulation products and their application. William is currently the Vice Chair of Sim.TECH an association of simulation software vendors. He was General Chair of the 1995 Winter Simulation Conference.

STUART GITTLITZ is a Staff Industrial Engineer with Kraft Foods. He is responsible for the simulation of process and packaging lines throughout Kraft's manufacturing facilities. He teaches in-house simulation courses, hosts user meetings, and provides simulation consulting for new and existing production lines. Mr. Gittlitz earned a B.S. in Industrial Engineering from the Polytechnic Institute of New York. He is a member of the Institute of Industrial Engineers and the Society for Computer Simulation.

RANDY D. MOORE is Senior Industrial Engineer at SONY Display Tube Company. At SONY he is responsible for the design and coordination of major capital equipment projects for Sony's Trinitron picture tube manufacturing plant. Randy has twenty two years of manufacturing experience. He received a bachelor of science degree in business and management from the University of Redlands. Randy is currently enrolled full time in graduate studies in Strategic Management at United States International University.

Proceedings of the 1996 Winter Simulation Conference
ed. J. M. Charnes, D. J. Morrice, D. T. Brunner, and J. J. Swain

AN AIRCRAFT OPERATIONS SIMULATION MODEL FOR THE UNITED PARCEL SERVICE LOUISVILLE AIR PARK

Kay K. Cornett
W. Swain Ottman

United Parcel Service
1400 North Hurstbourne Parkway
Louisville, Kentucky 40223, U.S.A.

Scott A. Miller

Systems Modeling Corporation
504 Beaver Street
Sewickley, Pennsylvania 15143, U.S.A.

ABSTRACT

This paper presents the development of a flexible data-driven model of the aircraft operations at the United Parcel Service Louisville Air Park. The model describes the unloading, refueling, and loading of aircraft by dedicated crews and equipment during the daily sortation operations. Inputs to the model include aircraft schedules, crew schedules and equipment allocation. Customized outputs include aircraft departure information, crew utilization, and equipment requirements. The model allows planners to develop crew schedules and allocation strategies to employ resources efficiently while maintaining rapid aircraft turnaround time. Preliminary analysis has identified aircraft arrival patterns and the specific crew availability as parameters to which the system is extremely sensitive.

1 INTRODUCTION

Each day over 100 aircraft arrive at the UPS Louisville Air Park carrying over 700,000 packages for Next Day Air and Second Day Air delivery. Ensuring that all of the packages are unloaded, sorted, and loaded within tight time constraints is a formidable task. With customers demanding earlier deliveries, competition increasing and the market expanding rapidly, the effects of poor planning or decision making can be disastrous. Just a single aircraft departing late can result in thousands of dissatisfied customers and a potential loss of millions of dollars. A natural reaction in this type of environment might be to add more and more personnel and equipment with less regard for cost and efficiency.

At the Louisville Air Park, the equipment and personnel used to service the aircraft have become difficult to manage using conventional methods. Studies of aircraft operations and package handling tasks involve analysis of interdependent resources that cannot be accurately studied with static models. Since 1987, the IE Department has used simulation to study and manage the modern resources required to provide on-time delivery to the customer. Earlier simulations focused on single aspects of the aircraft operation, like refueling or maintenance, without considering the interaction between the resources. The Total Air Park Simulation (TAPS) was designed to represent the overall operation accurately by including all aircraft-related tasks in order to capture the interaction of the crews and equipment involved. Experiments to examine crew allocation, equipment utilization and system capacity may be performed without disturbing the actual operation.

2 PROBLEM DEFINITION

2.1 Objectives

United Parcel Service initially envisioned the TAPS project as a tool to allow planners to experiment with resource requirements for the aircraft servicing operations. The effects of factors such as aircraft arrival patterns, aircraft departure patterns, incoming package volume, and crew allocation strategies on aircraft turnaround time may be investigated. Additional goals for the simulation included developing new aircraft servicing strategies, justifying new equipment purchases, and investigating the effects of increased aircraft and package volume on the facility. The model enables analysts to schedule personnel and equipment more efficiently, resulting in cost savings without adversely affecting customer service.

A longer-range goal of the project is to evolve the use of the simulation from a planning tool to an operational tool. Ultimately, the model will be run on a daily basis with near real time data for inputs. Execution of the model immediately prior to the start of the actual operation will highlight potential trouble areas, which will allow adjustments to be made.

2.2 Model Scope

The TAPS model simulates the aircraft servicing operations at the UPS Louisville Air Park. The TAPS model is designed to evaluate the effect of aircraft arrival and departure schedules, inbound volume, and crew allocation strategies on the aircraft servicing operations. The model scope is limited to operations directly related to aircraft servicing.

The daily operations at Louisville include both a Next Day Air and a Second Day Air sort. The two sort operations are separate and are not linked in the simulation model. Rather, the model is designed to run either a Next Day scenario or a Second Day scenario.

The model begins with aircraft arrivals generated from an input schedule. The aircraft taxis down the ramp to a parking position where a marshaling crew directs the plane into location. After the aircraft powers down, unload, maintenance, and fuel crews are requested. Inbound packages arrive either packed into containers on the main decks of the aircraft or loose in one of the belly sections. Upon unloading, the inbound cargo is transferred to one of several locations based on its cargo type. Most of the inbound containers are transferred to the hub where they are emptied and the individual packages are sorted. Sorted packages are loaded into outbound containers based on destination. When a container is filled, it is weighed and then transported to its outbound aircraft. As soon as an aircraft is ready for the loading operation, a load crew is requested. After all the containers have been loaded, the aircraft leaves its parking position with the aid of a marshaling crew, taxis down the ramp, and exits the system.

3 MODEL DEVELOPMENT AND PROJECT ORGANIZATION

The TAPS simulation model was developed using Arena® from Systems Modeling Corporation. In order to facilitate model development and validation, the model was divided into three segments:

- Inbound Aircraft Operations
- Hub Sortation
- Outbound Aircraft Operations

Each segment of the project was designed to capture the overall sensitivity of the system without burdening the model with needless detail. Because the primary focus was the aircraft operations, both the inbound and the outbound segments were modeled in detail. While the hub sortation segment was not an integral part of the overall project objective, it was critical that it accurately tie together the inbound and outbound operations. However, because of the incredible volume of packages moving through the hub, creation of a detailed model of the sortation would have greatly increased the model size and run length. To balance these two considerations, the hub was modeled with a series of historical distributions that break the inbound volume into outbound destinations. The sortation model includes a depletion rate for containers and packages arriving at the hub and a travel time to the outbound destination. The actual conveyor sortation process was not modeled explicitly.

In addition, each segment of the model includes two possible scenarios, Next Day Air and Second Day Air operations. While the general processes are the same for each, the two scenarios are different in scope. Because of the small window for on-time delivery in the Next Day Air operation, the primary goal is to get aircraft in and out of Louisville on time. The resources needed in terms of equipment and crews are secondary. Second Day has more options for delivery, and as such, is more concerned with balancing cost with service. Taking these and other fundamental differences into account, operational rules and strategies also differ for the two services. These differences are incorporated in the model by separate Next Day and Second Day logic in each segment.

Due to the size and complexity of the operation, many different departments within UPS participated in the definition of the model. Extensive interviews were held with the operations personnel responsible for the tasks to be modeled. Many of the algorithms used in the model for crew and equipment allocation come directly from actual operation guidelines. However, in a few situations, an algorithm was developed in order to approximate the human decision making in the operation.

4 MODEL DESCRIPTION

4.1 Model Inputs

The model is designed to be as flexible as possible so the user can change system parameters in external data files without having to alter the actual model. The TAPS model is driven by 15 external data files:

- Arrival file: Contains each inbound aircraft, as well as its arrival time, origin, cargo, fuel requirements, and outbound destination.
- Volume file: Defines inbound cargo mix and additional outbound cargo for each aircraft.
- Schedule file: Defines schedule for 17 crew types. Each crew type is dedicated to a specific task and may contain up to 50 individual crews. Each crew can have from one to eight crew

members. Crew schedules are based on half-hour increments.

- Process Time file: Defines each aircraft operation time as a function of crew size and aircraft type.
- Equipment file: Defines the number and initial staging location for all equipment required for the load and unload operations.
- Parameter file: Defines overall system parameters like start time, hub pull time, and model scenario.
- Parking file: Defines ramp parking positions for each aircraft type.
- Sortation files: These files define up to five inbound-to-outbound mixes, belt assignments, sortation times, outbound flight options, and "hot" flights.

4.2 Model Features

The TAPS model that was developed details the allocation, movement, and operation of crews and equipment from aircraft to aircraft as well as to and from the hub. Accurate descriptions of crew and equipment movements between all ramp locations required a detailed distance map that included every possible point-to-point combination. These large matrices were also necessary to display the crew movements for animation.

An arriving aircraft requires a series of tasks to be completed before it can depart. Some of these tasks may be performed in parallel while others may only occur after a preliminary task is completed (see Figure 1). Each task requires a specific type of crew. When a crew is allocated, it first procures the correct equipment for the aircraft type that it was assigned and then moves to the aircraft. Each crew operation consists of a setup time, an operation time, and a wrap-up time. After completion of its task, the crew is then available for assignment to another aircraft.

Incoming packages are either packed into containers located on the main decks of the aircraft or stored loosely in the belly sections. Packages in the belly sections are initially unloaded into empty containers. Upon unloading, most containers are moved to the hub where the individual packages are removed and sorted to outbound destinations.

Outbound containers generated from the hub are transported to an appropriate outbound aircraft where they are either loaded onto the main decks or broken into loose packages and loaded into the belly sections. Because loading individual packages into a belly section is extremely time consuming, load crews first will attempt to fill the main deck of the aircraft. However, if the expected volume for an outbound flight is large, the analyst has the option of pre-loading the belly sections to reduce the chance of a late departure. Both the belly

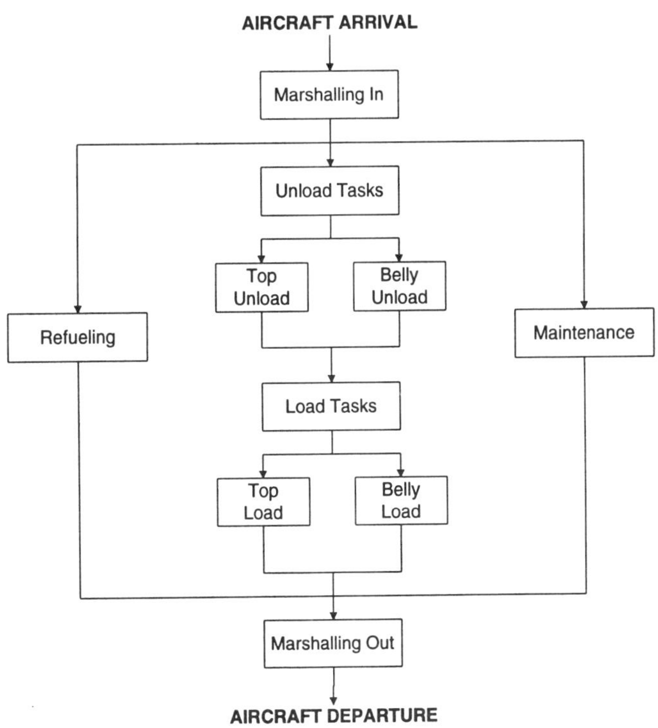

Figure 1: Aircraft Operational Flow Chart

unloading and belly loading tasks may be accomplished by the regular load and unload crew assigned to the aircraft or by special belly-only crews that work in parallel with the regular crews.

Crews are allocated to an aircraft based on allocation algorithms developed to approximate the human decision involved in the dispatching process. The simplest allocation algorithm involves taking the closest available crew for a given task. More advanced algorithms can be used to allocate unload and load crews to specific aircraft.

In order to describe the decision process to assign multiple unload crews to an aircraft, an allocation algorithm was developed utilizing a "Look Ahead" window. The decision to assign additional unload crews is then based on current crew availability and expected arrivals within the window. If enough crews are available to service all of the aircraft expected in the "Look Ahead" window, any additional crews may be assigned to the current aircraft.

4.3 Model Outputs

Extensive performance parameters are collected and reported in multiple output reports. General overall statistics are directed to a customized output file designed for less advanced users. Statistics reported in the generic output file include:

- Equipment Utilization (by ramp area)
- Overall Aircraft Arrivals and Departures (by ramp area)
- Average Aircraft Task Duration (by ramp area)
- Container and Dolly Statistics
- Overall Crew Utilization (by crew type)

More detailed statistics are included in the Arena output report. In addition to those previously mentioned, these statistics are also included:

- Extensive counters for tracking container and package movement through each segment of the system.
- Individual crew statistics. These statistics can track up to 30 different tasks performed by over 500 crews.

Collection of these statistics were facilitated through the use of the Arena Frequencies and Stateset features.

Additional output files are also generated for specific operations:

- Belly Load and Unload
- Holdovers
- Aircraft Departures

5 MODEL VALIDATION

The validation of TAPS was begun concurrently with the model development and is still ongoing. There are two main methods being used in the validation process: the examination of output reports and model traces, and visual validation through observation of the model's animation (see Figure 2). Validation has followed the same basic order as the model development:

- Input Data
- Inbound Aircraft Operations
- Hub Sortation
- Outbound Aircraft Operations

Authentication of the input data for the external data files was the first step in the validation process. Because the input data came from a number of different sources, it was thoroughly examined for consistency and accuracy. Initially, some of the information provided was out of date and was subsequently updated.

Validation of the TAPS unload process included examination of the following processes:

- the arrival of an aircraft.
- the requesting of marshaling crews, maintenance crews, and fuel trucks upon arrival.
- gathering and movement of various crews and equipment.
- removal of containers from the decks of an aircraft as well as loose belly volume.

Validation of the unload operation began with the confirmation of aircraft arrivals and the inbound package volume. The volume was broken down into two main categories: location of the volume in the aircraft, (topside or bellies); and the type of volume, such as domestic, export, import, cargo and empty containers. Utilization statistics for the various crews and processes were validated by comparing output reports to standard work measurement and documented process times.

The allocation of crews, the gathering and movement of various equipment, and the actual unloading and loading of containers at the aircraft were certified through observation of the animation.

The simple counter statistics for the unload operation were extremely helpful in validating the unload process. The model is very sensitive to changes in the scheduling of crews as well as the process times required for each task. Slight variations in the scheduling of crews resulted in the incorrect number of arrivals and/or departures and the incorrect amount of package volume being unloaded. These deviations also directly affected whether the aircraft would complete the unload process. In turn, this had an effect on the total package volume sent to the hub.

Validation of the hub sortation operation involved examining the total number of containers transferred to the hub and the total number of containers generated from the hub. Simple counter statistics are used to track containers into and out of the hub. Container and package volume to and from sortation was then validated by comparison to the expected volumes based on the model inputs and archived data. The outbound destination for a package sorted at the hub is determined by one of five statistical distributions. Each distribution represents a different regional mix of inbound packages to the Louisville hub. Additional validation was performed by comparing the mix of the containers generated for each destination to historical data.

Validation of the load operation included the examination of the following processes:

- the shifting and staging of containers to various yards,
- dispatching of load crews and equipment,
- loading of containers and loose belly volume,
- departure of aircraft.

Validation began by confirmation of the shifting and staging of outbound containers to outbound aircraft. The allocation of load crews and equipment, as well as the actual loading of containers and loose packages, were validated through observation of the animation and numerous statistical counters. Aircraft departure information was checked against historical records. Utilization statistics for the load crews and processes were validated by comparing output reports to standard work measurement and documented process times.

The validation of the load process again indicated that slight changes had a noticeable effect on the results. These changes included the scheduling of selected crews, the process times for various tasks, aircraft arrivals and departures, or changes to the constraints within the hub. Changes in the crew schedules and process times affected the time required to load the aircraft, which ultimately had an effect on whether the aircraft was able to depart on time.

Overall, the segments of the model are extremely interdependent. Modifications to the aircraft arrival or departure schedule have a significant effect on the number of outbound containers that can be generated by the hub. One minor change can have a domino effect on the rest of the operation—affecting the unloading and loading of other aircraft, the number of empty containers loaded onto the aircraft, and utilization of various crews. Alterations to the arrival times of aircraft also had an enormous effect on the number of packages that the hub processed by the end of the sort.

Various users, including UPS industrial engineers and planners, as well as operational personnel have been closely involved in validating the model. This collaboration has been critical to the success of the validation

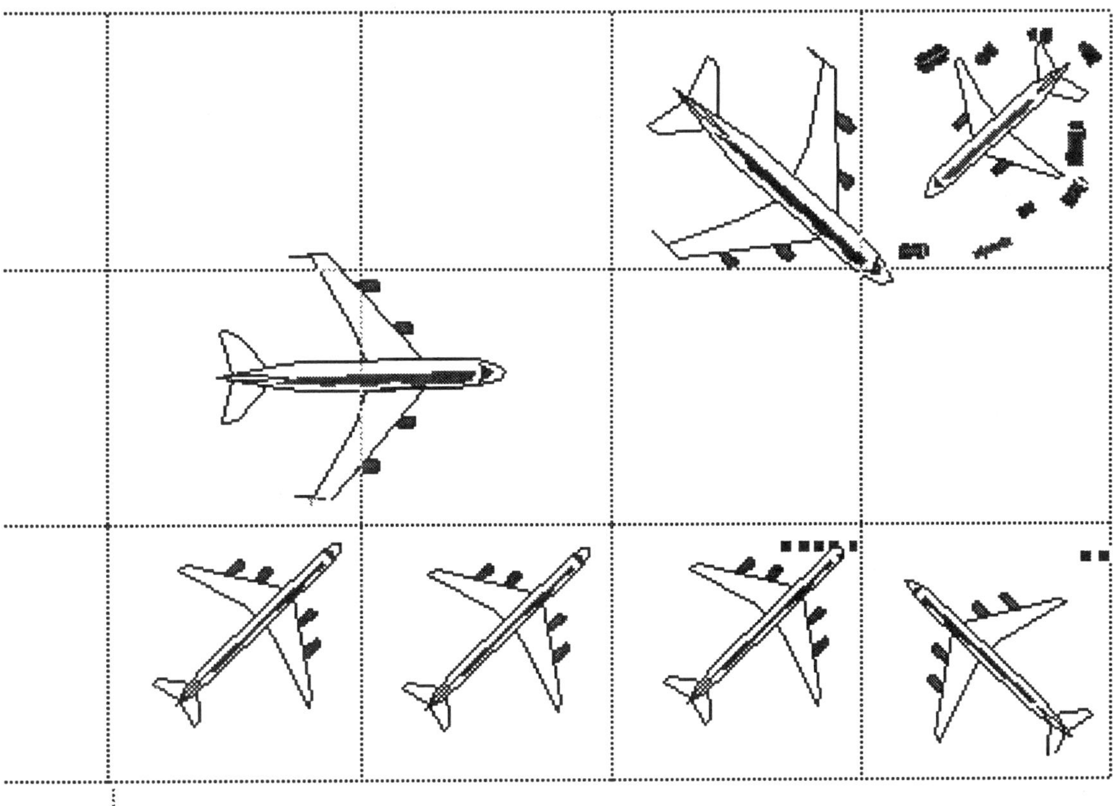

Figure 2: TAPS Model Animation

process. The model animation allows users to visualize aircraft arriving, being unloaded and loaded, and finally departing. Observing the model animation also increases awareness of the various operations (marshaling, maintenance, fueling, unloading, loading, etc.) and raises questions that are extremely helpful during validation. In addition, this involvement creates excitement and instills confidence in the model.

6 SUMMARY

The Total Air Park Simulation was developed as a flexible, data driven model to investigate the aircraft operations at the UPS Louisville Air Park. The model includes all aircraft servicing tasks and was designed to capture the interaction between multiple resources working in parallel on the aircraft. At the time of the writing, preliminary analysis has begun with experimentation on different aircraft schedules and examination of the sensitivity of model inputs. The initial success and acceptance of the TAPS model for scheduling and allocating crews and equipment has laid the groundwork for its transition into an operational tool. In the future, execution of the model with near real-time data prior to the start of the actual operation will highlight potential problem areas and facilitate adjustments to the crew schedules and allocation strategies.

KEYWORDS

General applications, transportation systems, logistics, resource allocation, resource scheduling

BIOGRAPHIES

KAY K. CORNETT is an Operations Research Specialist in the Industrial Engineering department at United Parcel Service. She received a B.S. in Mathematical Science in 1990 from the University of Kentucky, an M.A. in Mathematics in 1992 from the University of Louisville, and an M.A. in Computer Resource and Information Management in 1995 from Webster University. She has been with UPS for three years, and joined the Operations Research department in 1995. She is a member of Pi Mu Epsilon.

SCOTT A. MILLER is a Consultant in the Simulation Services Division of Systems Modeling Corporation. He has provided model development, analysis, and training support in the transportation, steel, communications, electronics, manufacturing, and service industries. He received a B.S. in 1991 and an M.S. in 1993 in Industrial Engineering from The Pennsylvania State University. He is a member of IIE.

W. SWAIN OTTMAN is a Senior Programmer Analyst in the Operations Research department of United Parcel Service Airlines. Besides working on TAPS, he has developed simulation models of an aircraft parts distribution warehouse, an aircraft wheel and brake maintenance shop, the aircraft fueling operation and a UPS air hub. He received a Bachelor of Engineering Science in 1989 and a Master of Engineering in 1991 in Engineering Mathematics and Computer Science from the University of Louisville.

Proceedings of the 1996 Winter Simulation Conference
ed. J. M. Charnes, D. J. Morrice, D. T. Brunner, and J. J. Swain

UTILIZING AN EXPERT SYSTEM FOR CONFLICT RESOLUTION IN A "TRAIN SIDING SELECTION" SIMULATION MODEL

Gregory J. Allen
Khaled M. Mabrouk

The Model Builders, Inc.
343 South Main Street
Suite 217
Ann Arbor, MI 48104

Michael L. Weigel

Union Pacific Railroad
116 Dodge Street
Room 406
Omaha, NE 68179

ABSTRACT

Simulation is an excellent tool for evaluating alternative configurations for system flow and/or logic. Occasionally, the complexity of a system presents problems with respect to simulating and improving the performance of that system. When confronted with these complexities, the logic requirements can easily become more difficult to develop than the effort to understand the actual system logic.

Expert systems, on the other hand, are better suited for reflecting complex decision making. Many times this means solving complex problems which can only be stated in written words, therefore extending computing power beyond the mathematical and statistical constructs of simulation and conventional programming. As a result, expert systems can be significantly easier to modify, for experimentation purposes, than a set of complex "IF-THEN" statements.

Through the development of an expert system and its interface to a simulation, the problem of modeling complex systems can be encountered with improved efficiency. The end result is an increased level of simulation performance and customer satisfaction.

INTRODUCTION

The railroad industry is typical of the complexities which often prevent the utilization of computer simulation as an effective decision support tool. As such, it has been selected as the focus industry for testing the feasibility of an interfaced expert and simulated system.

The success of computer simulation as a decision support tool in other industries, coupled with the advances of expert systems in the railroad industry has indicated that it would be advantageous to interface the two respective tools. Previous attempts at combining these two technologies have been performed with relative success. However, none of the previous attempts dynamically connected the expert system and the simulation while the simulation program was operating. Rather, the expert systems were only called before or after the simulations had executed. Specifically, The Model Builders' expert/simulation interface performs real time decision support and analysis for a simulated railway system.

SCOPE AND OBJECTIVES

Conflict resolution is an integral part of most railroad industry simulation efforts. The issue becomes even more important when developing a simulation model for a train corridor where decisions need to be made continuously for each train on whether it should take a siding or stay on the main line. The simulation model that was developed, consisted of a decision support tool capable of reflecting some very basic routing procedures used in a railway system. Furthermore, a small scale expert system was developed and interfaced to provide the simulation's more complex, real time decision logic. It was the interaction between these two systems that was monitored to determine whether the potential value of sharing these applications does indeed exist.

METHODS AND MATERIALS

This project was initiated with the research of information pertaining to simulation, expert systems, software interfaces, and the decision logic behind a simple train conflict resolution module. The collection methodologies consisted of a material search at the University of Michigan libraries, GMI library, and the

Ann Arbor Public Library, as well as conversations with people working in the railroad and expert system industries. Based upon the information gathered through these searches, as well as other economic considerations, CLIPS™ was selected as the type of expert system software to reflect the decision logic needed for the railroad industry. The ProModel™ software package was selected as the tool for simulating the rail system. Visual Basic™ was chosen to provide the interface between CLIPS™ and ProModel™. Upon completing these preliminary stages, the simulated railroad model was constructed first, followed by the conflict resolution logic, and finally the Visual Basic interface for connecting ProModel™ with CLIPS™ was developed.

DESIGN OF THE TRAIN SIDING SYSTEM

A mainline rail and siding system is used to process train meets that occur when two trains traveling on the same main line are moving towards one another. The simulated system is structured so that one train can divert from the main track to a siding and let the other train pass on the main line. A train exits the main by moving into an available siding.

A satisfactory simulation model was built based on the design structures indicated through literature searches and interviews with railroad industry experts. These inquiries helped determine the appropriate level of detail to adequately test the simulation and expert system interface. Ultimately, a physical design was suggested and implemented consisting of the following elements:

Main Line: The main line is the portion of the track upon which trains primarily travel. The trains only leave the main line in the event of a traffic conflict, e.g. another train. The main line selected for the model was 26 miles in length.

Sidings: The length of the main rail allowed enough room to include four sidings. Each siding spanned a distance of 1.5 miles. These sidings were located at equal intervals but on alternating positions along the main line. Siding number one was placed 2.5 miles from the western endpoint on the main rail. The west end of each of the remaining three sidings were placed at 5 mile intervals from the east end of the previous siding, with the east end of the fourth siding located 2.5 miles from the east-most endpoint along the main rail. These intervals between sidings were intentionally given enough capacity for the entire length of a train entity. The multiple sidings modeled created the complexity needed to adequately test the capabilities of the decision logic being utilized.

Switches: At each of the eight points connecting the sidings to the main rail were the switching devices. These switches were where trains could merge with and exit from the main rail and track sidings.

Blocks: Tracking a train's simulated movement through the main rail and sidings, was accomplished by assigning one of 16 possible block numbers that identified a train's current location. The blocks were used to help communicate the status and location of a train at any given time during the model's execution. Essentially the blocks served as points of reference, that were checked by the expert system and implemented by the simulation. All of the information pertaining to trains traveling within a particular block was stored in an array call Block_Array. It was the Block_Array which contained the information communicated between the simulation and expert system.

Intermediate Signals: Intermediate Signals were added to the model at the middle of each 5 mile main rail sections between each siding, creating two block occupancies. Although these signals had no functional capabilities they were necessary to provide a more precise tracking of a train's location within the model.

Measurable Units: Time units were simulated in terms of minutes and distance units were measured in terms of feet.

Entities: In this particular model, the entities being processed were the trains. There were two types of trains for this model: loaded and empty. Every train arriving into the rail system was either assigned an attribute called "empty", or assigned an attribute called "loaded". These attributes were the identifiers used to evaluate the train conflict resolution logic and return a decision based on the current and conflicting trains attributes. The loaded trains were assigned the highest priority and always traveled in a easterly direction. The empty trains were given lower priority and always traveled in a westerly direction. Both the empty trains and loaded trains were given a length of 6,500 feet.

Train Performance: For the purposes of simplification, the model did not account for numerous train performance measures. Other models exist which can duplicate train performance, including velocity and acceleration for a given set of operating conditions such as horse power per trailing ton, track grade, track curvature, track conditions, and locomotive characteristics. However, this level of detail was not necessary to complete the proof of concept for an interfaced expert and simulated system. As a result, the velocities for both empty and loaded trains were maintained at constant speeds of 45 miles per hour on the main line and 10 miles per hour through switches and sidings (until the last car cleared the block).

Arrivals: Any time a new entity was introduced into the system, it was called an arrival. The number of train entities for each arrival was one. The total number of arrivals was 50 per day (25 for loaded trains and 25 for empty trains). The frequency of the arrivals was exponentially distributed over a 24 hour period, e.g., $E^{(24/25)}$. The loaded trains arrived at the western point on the main line and the empty trains arrived at the eastern most point on the main line. The first arrival occurred at time zero of the model execution.

Figure 1 contains a graphical representation of the entire track and siding system. Each of the numbers (1 through 16) represent a block. The track blocks numbered 3, 7, 11, and 15 were the sidings used in train meets and passes. Figure 2 contains the dimension details for a specific section of that simulated rail system. Finally, in Figure 3 the labels for the primary elements of the track and sidings are shown.

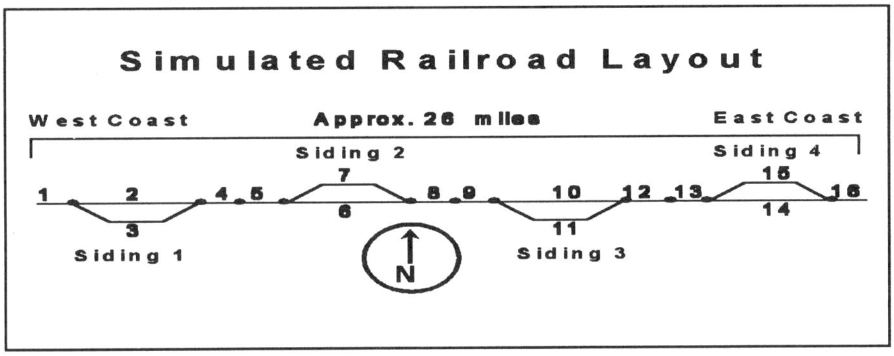

Figure 1: Graphical Representation of the Track and Siding System

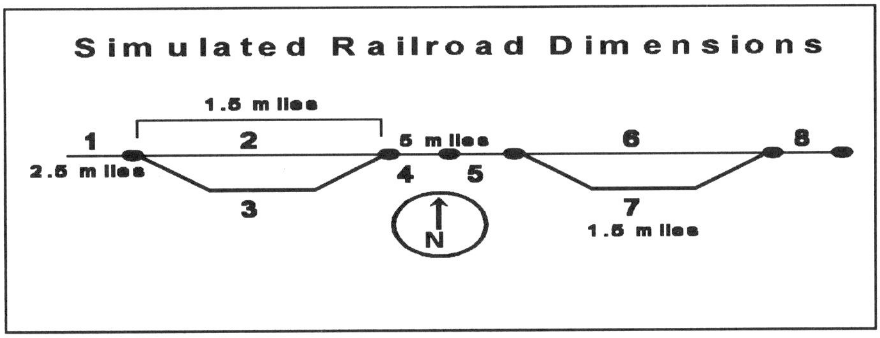

Figure 2: Dimensions of the Track and Siding System

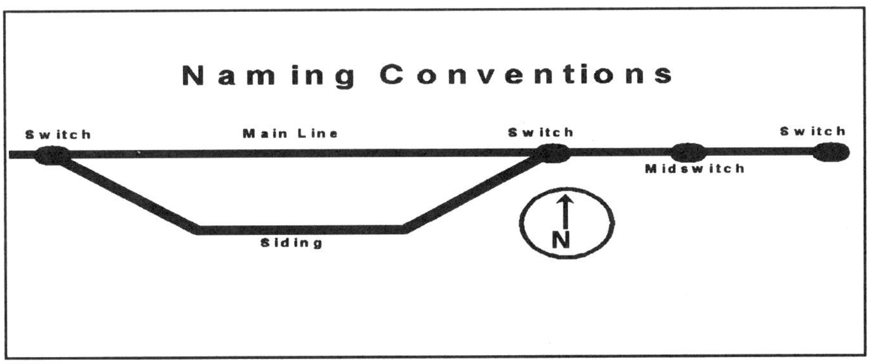

Figure 3: Naming Conventions of the Primary Elements

DESIGN OF THE DECISION LOGIC

The rule sets were developed based on three, isolated facts. Those facts were train type, train location, and train status. Train type indicated whether the train was loaded or empty. Train location indicated the block in which the train was traveling. Train status indicated whether or not the train needed a decision returned from the expert system.

The rule sets were divided into three modules. Each module was referenced at various points throughout the simulation and represented a different problem/solution set. Referencing a particular solution set was based on the facts triggered by the "current" facts for a train's type, location, and status. As a result, the simulation would only refer to a rule set if a train's attributes indicated a rule set was appropriate under the current circumstances. Thus, limiting the event of unnecessarily processing decision attributes.

Module One: Module one pertained to any trains on the main that were approaching a switch exiting a siding and mainrail connection. Its purpose was to keep trains, which were traveling in the same direction, from sharing the same block at any given time during the simulation.

Module Two: Module two pertains to each of the five mile long rail sections located on the main rail between one siding's entrance and the next siding's exit. Each of these sections were divided into two blocks using an intermediate signal. The division was made to create smaller track sections, making the pinpointing of a train's location more precise. Because these sections of rail lacked any siding outlets, trains were instructed to travel the length of the two rail sections without stopping.

Module Three: Module three contained the most complex decision logic to be evaluated. Its purpose was to issue instructions to trains traveling east and west on the main line that were approaching a switch at the entrance of a siding. In these instances the logic would look ahead two sidings to check the approaching blocks for their contents. Anything outside of the two siding cushion was not close enough to cause an eventual conflict, since there was still enough unoccupied main rail remaining given the level of traffic volume modeled. Loaded trains received a higher priority than empty trains, and in the case of a meet, the siding that minimized the delay to the higher priority train was chosen.

HOW THE APPLICATION WORKS

The simulation model defined some of the decision and execution logic for the train entities whenever they would enter a block's location. Essentially, the logic can be separated into operation logic and routing logic. Operation logic specified anything that needed to happen to an entity while in a location. Routing logic defined the outputs for each of the processes defined. Both the operation logic and routing logic were differentiated by location. Those locations were arrival/exit locations, switching locations, siding locations, and intermediate signal locations.

Within the arrival/exit locations, trains would enter and exit the model. The operation logic at these locations specified the appropriate routing logic based on the current train's attributes. If these attributes indicated that the train needed to exit the system, the appropriate Block_Array positions were updated and the train departed from the system immediately. However, if the train was a new arrival, it required that some decision logic to be carried out in the routing logic. The routing logic for new arrivals would wait until the Block_Array indicated that the arrival block was available before moving the new arrival forward into the system. Once the arrival block became available, the arriving train could move forward and update the Block_Array with information specific to its nature.

The switching locations were where the most complex processing logic was carried out. As a result, these locations used the expert system for the more sophisticated decision rules. Depending on the direction of the train traveling through a switch, the operation logic could assume two forms. One was for trains traveling through a switch to a main rail and siding entrance, and the other was for trains traveling through a switch exiting a main rail and siding.

Because the trains traveling through a switching exit were only looking one block ahead, they did not require the assistance of the expert system. These trains merely had to wait until the Block_Array indicated that the next block was available for routing. Once there was a vacancy, the simulation deleted the train's information from the current Block_Array position and updated the new position in the Block_Array with information.

Any trains traveling through a switching entrance had to look ahead several blocks before determining if they could stay on the main rail or if they had to take a siding and allow a on coming train to pass. Because they needed to process several blocks at a single time, their logic would have been difficult to develop with the simulation's typical operation constructs. Rather, the simulation's current status, maintained in the Block_Array, was written to a general text file every time a train arrived at a switching entrance. This text file would later become the information evaluated by the expert system. This process of passing information between systems required using ProModel's™ most powerful feature, the X-Sub command, to call a DLL file allow for communication with the expert system. A DLL is a library of functions that Windows™ reads and executes as a .EXE program requires them.

Once CLIPS™ had finished reporting its routing decisions to a general text file, ProModel™ reset the previous input file, and read-in the new information for execution. Actually, these routing decisions were integer values which were assigned to variable ROUTE statements within ProModel's™ operation logic. If the integer instructed the train to take ROUTE 1, allowing the train to stay on the main rail, ProModel™ waited until the designated route was available before moving forward. For example, if a loaded train was traveling east in Block 2 then it would have to wait until Block 4 AND Block 5 were both empty OR Block 4 was empty and Block 5 contained an east bound train. If the integer instructed the train to take ROUTE 3, telling the train to take a siding, ProModel™ merely updated the Block_Array and moved the train forward.

The procedures and logic mentioned above were used repeatedly at the switching locations throughout the model.

Those trains traveling within a siding location were merely instructed by the operation logic to route to a switch based on the trains' attributes. The resultant routing rule waited until the two blocks prior to the next switch were both empty before moving forward.

The final piece of processing logic dealt with the trains traveling through an intermediate signal. Remember, the intermediate signal's only purpose was to divide a large section of main rail into smaller, more manageable, distances. The signal's decision logic was relatively straight forward and was also performed by the ProModel™ simulation. There was no waiting time incurred by trains passing through an intermediate signal. The simulation simply updated the Block_Array to show the train leaving one block and entering another.

DOES THE APPLICATION MAKE SENSE?

The design and development of a real-time simulation and expert system interface have proven to be both a feasible and practical approach to modeling complex systems. A virtually seamless interface was developed between the two technologies. Embellishing the programming constructs of a simulation package with the functionality and flexibility of an exert system tool will significantly improve the ability to develop decision logic that reflects the random behavior and dynamic interactions of complex systems.

The remainder of the conclusions formulated from this project are based on a specific set of system performance measures. These measures were established prior the design and development phases of this thesis and were used to determine the future of the resultant application. Specifically, the performance criteria applied were:

1. Whether or not the interface could operate with real time capabilities, and
2. The relative effort required to use an interfaced application.

The following section contains each conclusion based on the aforementioned evaluation criteria.

Real Time Capabilities: Real time capabilities refer to the ability of the simulation application to communicate seamlessly, and not distort the simulator's clock, with the expert system during the execution of the model. Additionally, the communication link must occur with relatively little delay in clock time. With this particular simulation interface, a direct line of communication was initiated by the simulation and received by the expert system at any time during the model's execution. However, the interface was not seamless. In other words, there was about a second of delay in the simulator's clock experienced by the entity making a call to the expert system. This delay consisted of the simulated entity calling the expert system, the expert system evaluating the current situation, the entity waiting for a response, the entity receiving back a decision rule from the expert system, and the entity executing the resultant routing rule. Although the interfaced application didn't operate in a purely real time fashion, its delay was insignificant. Especially given the fact that the model was based on a train track and siding system, where a second of delay to the simulator's clock had little affect on the accuracy of such a large system. One conclusion could be that the interface "virtually" performed in real time. This virtual real time capability allowed the expert system to enhance the ability of the simulation to process complicated decision logic with very little affect on the model's performance. Refer to Appendices One and Two for graphical representation of the current and future real time capabilities.

Effort: The majority of the effort expended during this project related to the development of the interface between the simulation and expert system. This effort was the result of having to establish an interfaced line of communication using three different software products: ProModel™, CLIPS™, and Visual Basic™. The major challenge was finding a communication medium that all three packages could use jointly. However, the eventual interface was generic in nature, virtually eliminating the interface development and coding phase for any future projects where this type of technology would be applicable. That fact alone justified the amount of effort required to make the initial communication interfere work.

AUTHOR BIOGRAPHIES

GREGORY J. ALLEN is a systems engineer with The Model Builders, Inc. He is a graduate from the GMI Engineering & Management Institute, where he received his Bachelor of Science in Management with a concentration in information systems. Mr. Allen's current responsibilities include simulation modeling and analysis, the development of graphical user interfaces, and the development of expert systems.

KHALED M. MABROUK is a Simulation Strategist for The Model Builders. He has spent the last ten years assisting many companies in the U.S. & Canada to successfully implement simulation as a productivity improvement tool. His contributions include assisting these companies in choosing the appropriate software tool, training their employees on how to properly implement this tool, and managing simulation projects for them.

MICHAEL L. WEIGEL is a manager in the Network Planning Group at Union Pacific Railroad. he has developed simulation models in the areas of Rail Transportation, Aerospace Logistics and Manufacturing. He is currently involved in the development of Simulation Models used in the analysis of Railroad Terminal Operations. He received a B.S. degree in Industrial Engineering from the University of Nebraska and is a member of IIE.

Proceedings of the 1996 Winter Simulation Conference
ed. J. M. Charnes, D. J. Morrice, D. T. Brunner, and J. J. Swain

MODELING ISSUES IN A SHIPPING SYSTEM

Edward J. Williams

206-2 Engineering Computer Center
Mail Drop 3
Ford Motor Company
Dearborn, Michigan 48121-2053 U.S.A.

Susan Khoubyari

Production Modeling Corporation
Three Parklane Boulevard
Suite 910 West
Dearborn, Michigan 48126 U.S.A.

ABSTRACT

We describe the application of simulation and statistical analyses to the improvement of a shipping and distribution system supporting a component-fabrication plant and two automotive assembly plants interconnected by truck lines and railroads. These analyses enabled managers to determine the number and location of loading docks required, optimize the inventory level and distribution of racks (containers) circulating throughout this shipping system, determine the sizes of turnaround areas for truck and rail cargo at each plant, and predict annual truck and rail shipping volumes and costs.

1 INTRODUCTION

A simulation model is an alternative *realization* of a real-world system, such as a manufacturing, distribution, computer/communications, or service system, which permits timely, cost-effective experimentation directed toward improvement of the real-world system being modeled (Seila 1995). Such models provide strong advantages to the analyst responsible for design and implementation of a transportation, shipping, or distribution system, such as the abilities to evaluate flow of vehicles (in this context, trucks and trains) and trace entities (in this context, racks either empty or loaded with components) at minimal economic risk (Fishburn, Golkar, and Taaffe 1995). Simulation provides these advantages in situations where problem complexity and intractability frustrate use of queueing theory, linear and integer programming formulation and optimization (Subramanian et. al. 1994), or "closed-form" mathematical equations. Furthermore, simulation analyses provide these advantages irrespective of whether the shipping system is wholly contained in the production facility (Angers, Gagnon, and Villeneuve 1995) or, as in this study, physically external to it (Koh et al. 1994).

2 OVERVIEW OF SHIPPING SYSTEM

This shipping system supports three plants, a component-fabrication plant (Plant #1) in the United States, an automotive assembly plant (Plant #2) in the United States, and another automotive assembly plant (Plant #3) in Mexico. This system, comprising "northern" and "southern" truck routes and two railroad yards, one in the United States and one in Mexico, holds responsibility for transport of components from Plant #1 (an "upstream" plant) to Plant #2 and Plant #3 ("downstream" plants) in proper quantities and at proper times to meet production schedules set by management at Plant #2 and Plant #3.

The primary objective of this simulation study was to determine the minimum number of racks (these racks are expensive to purchase, transport, store, and maintain) required to meet these production schedules. Such racks are often a "workhorse" within a storage and/or distribution system, due to their abilities to both thrift floor space ("footprint") and provide precise access to either unit loads or large items (Kulwiec 1994). Related objectives were the confirmation of any increases in capacity needed at plant shipping and receiving docks or at cargo storage/transfer points, and increased accuracy and lead time relative to the prediction of annual shipment volumes and costs. Additionally, transportation and supply managers were keenly interested in the identification of methods of negotiating those costs (particularly via railroad) downward by decreasing variability of shipment volumes and dates within operational constraints of the system. Hence, migration from judgmental forecasting methods to analytical, quantitative ones (Sanders and Manrodt 1994) was already a management goal awaiting support from this simulation study.

3 INPUT DATA AND ASSUMPTIONS

Input data collection represented a significant portion (correctly estimated by the simulation project team, during preliminary project feasibility evaluation, estimation, and scheduling at 30% of the total project effort and time) (Williams 1996) due to the high complexity and broad scope of collecting data from multiple sources (three plants, two railroad yards, and trucking operations). Also, after initial data collection, these data required integration and cross-checking against both study objectives and intended simplifying modeling assumptions (Robinson and Bhatia 1995). To manage this complexity conveniently, the modeling team subdivided the data into three categories: general system input, plant-specific input, and railyard-specific input.

3.1 Input Data by Category

General system input answered these questions:

- initially, how many racks, truck trailers, and truck cabs were available?
- how many components does one rack carry?
- how many racks does one truck trailer (of each of two sizes) carry?
- how much time is required to attach a trailer to a cab, or detach it?
- how much time is required to load or unload a trailer?
- what are truck and train transit times, allowing for issues such as passing through customs, traveling across three time zones, and accommodating different holidays in the United States and Mexico?
- what are rates of damage and needed repair times to racks and to trailers, both of which must be returned to the United States for repair?

Plant-specific input answered these questions:

- what are the hours of operation at each plant?
- what is the production rate and schedule at each plant?
- what is the material-rejection rate at each plant?
- how many racks can the inbound and outbound storage areas accommodate at each plant?
- how many truck trailers can the staging areas at each plant accommodate?
- how many truck docks does each plant have, and what is their schedule of availability?
- at what times must trucks leave each plant to "catch" the next train departure at the local railroad yard?

Railroad-yard specific input answered these questions:

- what are the hours of operation at each railyard?
- what are the inbound and outbound storage capacities at each railyard?
- how much time does loading or unloading a train require?
- what are the train schedules and cutoff times for train departures from each railyard?

3.2 Modeling Assumptions

With the customer's concurrence, the modeling team chose to assume that no plant shutdowns or major production blockages would occur, and that customs-clearance procedures and times would remain identical with historical observations. Additionally, the complete absence of cab, trailer, or railroad breakdowns, and of border-crossing delays, was assumed. This early agreement on documented assumptions, and establishment of a contingency plan for sensitivity analyses concerning them, removed the otherwise ominous danger of eventual users' misinterpretation or unfounded extrapolation of modeling results (Harrell and Tumay 1995). As trenchantly stated by (Musselman 1994), "It is easier to correct an expectation *now* than to change a belief *later*" (italics added).

4 BUILDING, VERIFYING, AND VALIDATING THE MODEL

4.1 Choice of Modeling Environment

Consistent with prior discussions between the simulation users and the model-building team, the simulation model was built using the SIMAN/ARENA analysis and animation tool. SIMAN is a general-purpose simulation language of high power, often used to model manufacturing, service, computer, and transportation systems (Profozich and Sturrock 1995). ARENA, an object-oriented simulation system, simplifies the building of complex SIMAN models and also provides animation capability (Hammann and Markovitch 1995).

4.2 Model Run Length, Time Unit, and Initialization

Since the variations in production schedules and the holiday breaks in both countries (United States and Mexico) both had yearly periodicities, the model run length was chosen to be one year. Accordingly, and in keeping with the consideration that the real system runs "around the clock, seven days a week," the simulation time increment was chosen to be one hour. Significant system activity durations, such as trip times, were then convenient multiples of one hour. The system being modeled is steady-state, not transient; therefore, removing the initial transient from model runs was necessary. In view of the long, computationally intensive run times implied by the year of simulated time, the modeling team chose to initialize the simulation model to typical, actually observed, conditions (Banks, Carson, and Nelson 1996), rather than begin model runs "empty and idle" and discard data collected during a warm-up period.

4.3 Choices of Probability Densities

As noted in Section 3.2, damage to trailers was assumed rare enough to ignore.

Extensive empirical data available from the modeling clients were used to choose an appropriate probability density for the number of damaged racks per interval of time. Those data were analyzed via heuristic examination of histograms, the chi-square test, the Kolmogorov-Smirnov test, and the Anderson-Darling test (Law and Kelton 1991). The software tool UniFit [now ExpertFit] (Vincent and Law 1995) proved convenient for running these tests and assessing their results. The first three of these tests indicated normality of these data; the Anderson-Darling test demurred. Since the Anderson-Darling test emphasizes behavior in the tails, and since the users and modelers agreed that extreme values for the number of damaged racks were rare and hence unimportant, the normal distribution was selected to model the number of damaged racks.

Likewise, extensive empirical data were used to choose probability densities for the durations pertinent to loading and unloading of trailers, attaching truck cabs to and detaching them from trailers, loading and unloading of trains, and transit times of trucks and trains. The four density-fitting analyses mentioned above concurred on use of the uniform density to model these input data.

In accordance with user requests based on both contractual provisions related to the outsourcing of repair service for damaged racks and observed invariance of repair-time duration, the number of days required for rack repair was represented as a constant.

All parameters of these probability densities (including the constant value for rack repair time) were changeable directly by the users via the convenient SIMAN "Experiment Frame" interface.

4.4 Model Verification and Validation

To assist in verifying and validating the model, the modeling team began by building a model of the existing system, with the intention of incorporating proposed modifications subsequently. This "base case" model was verified by use of walkthroughs, use of animation as a verification aid, and extensive checking of entity traces (Pegden, Shannon, and Sadowski 1995). Validation and establishment of model credibility, during which the modeling team worked closely with the model users, employed further use of animation while presenting the model to the users, removing all stochastic variation and arithmetically checking the results, varying the input data and confirming the model responded correctly to the change, and the use of Turing tests (Harrell et al. 1995). For example, model results pertinent to the use of two different sizes of truck trailers were compared to those obtained by analytical methods in (Bausch, Brown, and Ronen 1995), with close agreement achieved. These Turing tests, in which model predictions and actual system data agreed within 3½%, thus becoming well-high indistinguishable, established high credibility of and users' confidence in the model as a vehicle to explore alternative system configurations.

5 STUDY RESULTS AND CONCLUSIONS

Four scenarios, a base case and three competing methods of shipping deployment, were analyzed in detail, beginning, as indicated above, with model validation via the base case. In all four scenarios, certain parameters were fixed, namely: each rack held eight components, each 40-foot truck trailer carried fourteen racks, and each 53-foot truck trailer carried eighteen racks. Also, all trains were "unit trains." Traditionally, shipments had always been made in lot sizes filling at least one railroad freight car; such shipments avoid the time and cost overhead of processing within railroad freight terminals (Van Metre 1939). Unit trains naturally extend these economies of time and cost by their complete dedication to moving the goods of *one* shipper from *one* loading point to *one* unloading point (Alexander 1967).

To provide for the construction of confidence intervals, five replications of each scenario were run. Intuition (the appeal of testing "competing proposals under equal loads") suggested the use of common random numbers (CRN) as a variance reduction technique (VRT) among the four scenarios. Re-examination of test runs verified the monotonicity of system performance metrics as functions of uniform random variates. These monotonicities confirmed the modelers' expectation that CRN would indeed reduce variance, as formally proved in (Bratley, Fox, and Schrage 1983). Specific results were:

Table 1: Starvation Results of Four Scenarios

Scenario	# racks in system	# docks at Plant #1	# instances starvation/year
Base	1699	1	21
1	1299	2	0
2	1055	2	0
3	1030	2	10

Highlights of these results presented to management were that dedication of an additional dock at Plant #1 to these shipments was urgently needed, and, given that dedication, the initial estimate of 1699 racks required could be safely thrifted to 1055 racks, a savings of three-eighths the initial estimate and translatable to significant reductions in capital investment, storage, and maintenance costs. Using Scenario 2, the model further predicted:

- average utilization of 40-foot trailers is 20; of 53-foot trailers, 5
- plant #1 requires yard space to hold 13 trailers loaded with empty racks and 19 trailers loaded with full racks
- plant #2 requires yard space to hold 2 trailers loaded with empty racks and 5 trailers loaded with full racks
- plant #3 requires yard space to hold 2 trailers loaded with empty racks and 17 trailers loaded with full racks
- the average number of trailers per train is 4; the maximum, 6.

Management implemented Scenario 2 on the bases of no instances of starvation coupled with a highly thrifted number of racks in the system. After eighteen months of actual operation under Scenario 2, all predictions held to within 4%.

6 DIRECTIONS FOR FURTHER WORK

The successful predictions described in the previous section have spawned significant customer interest in further work. Sensitivity analyses will be undertaken to determine:

- the number of truck trailers required to maintain continuous production at each plant during a year
- the impact of damaged trailers on those trailer requirements
- the number of truck cabs required to maintain production in the total system
- the impact on container requirements of various percentages (e.g., 5%, 10%, 15%) of racks damaged per year.

Interest in these system parameters is especially high due to changes in the trucking industry spawned by federal deregulation, ongoing North American Free Trade Agreement [NAFTA] negotiations on border crossings, and the increasing tendency of buyers to ask freight carriers to provide value-added services incremental to freight movement (Minahan 1996a). Similarly, ongoing development and use of the model will help assess economic and timing opportunities available through increased use of intermodal transport service providers (Minahan 1996b). Further, management is eager to extend improvements conceived and implemented through this study to shrinking the time lapse between vehicle-order placement and delivery within the retail distribution system (Keller 1996).

Additionally, the users have requested modeling of scenarios involving presumably rare but surely high-impact events, such as plant shutdowns or strikes, train wrecks blocking rail traffic, or temporary but dramatic increases in border inspection and crossing times. Analysis of such rare yet highly significant events will be undertaken using the "external unknown unknowns" methods explained in (Okashah and Goldwater 1994), in view of the mathematical intractability of the "rare event problem" acknowledged there and in, for example, (van Moorsel, Kant, and Sanders 1996).

ACKNOWLEDGMENTS

John M. Dennis, Simulation Analyst, Ford Motor Company, provided criticisms valuable to the presentation and clarity of this paper.

APPENDIX: TRADEMARKS

SIMAN/CINEMA and ARENA are trademarks of Systems Modeling Corporation.

REFERENCES

Alexander, Robert E. 1967. The Shipper Views the Unit Train. In *Unit Train Operations*, 81-85. Chicago, Illinois: Railway Systems and Management Association.

Angers, Pierre-E., Gaétan Gagnon, and Laurent Villeneuve. 1995. An Analogical Approach to Optimize a Large Scale Job Shop Transportation System. In *Proceedings of the 1995 Summer Computer Simulation Conference*, eds. Tuncer I. Ören and Louis G. Birta, 396-399.

Banks, Jerry, John S. Carson, II, and Barry Nelson. 1996. *Discrete-Event System Simulation.* Englewood Cliffs, New Jersey: Prentice-Hall, Incorporated.

Bausch, Dan O., Gerald G. Brown, and David Ronen. 1995. Consolidating and Dispatching Truck Shipments of Mobil Heavy Petroleum Products. *Interfaces* 25(2):1-17.

Bratley, Paul, Bennett L. Fox, and Linus E. Schrage. 1983. *A Guide to Simulation.* New York, New York: Springer-Verlag.

Fishburn, Paul T., Javad Golkar, and Kevin M. Taaffe. 1995. Simulation of Transportation Systems. In *Proceedings of the 1995 Winter Simulation Conference*, eds. Christos Alexopoulos, Keebom

Kang, William R. Lilegdon, and David Goldsman, 51-54.

Hammann, John E., and Nancy A. Markovitch. 1995. Introduction to ARENA. In *Proceedings of the 1995 Winter Simulation Conference*, eds. Christos Alexopoulos, Keebom Kang, William R. Lilegdon, and David Goldsman, 519-523.

Harrell, Charles R., Robert E. Bateman, Thomas J. Gogg, and Jack R. A. Mott. 1995. *System Improvement Using Simulation*, 3d ed. Palos Verdes, California: JMI Consulting Group.

Harrell, Charles R., and Kerim Tumay. 1995. *Simulation Made Easy a Manager's Guide*. Norcross, Georgia: Industrial Engineering and Management Press.

Keller, Maryann. 1996. Distribution: the Next Battle Ground. *Automotive Industries* 176(2):17.

Koh, Peng-Hong, Hak-Soon Ng, Hwei-Chiat Ng, and Jimmy L. K. Goh. 1994. Using Simulation to Preview Plans of a Container Port Operations. In *Proceedings of the 1994 Winter Simulation Conference*, eds. Jeffrey D. Tew, Mani S. Manivannan, Deborah A. Sadowski, and Andrew F. Seila, 1109-1115.

Kulwiec, Ray. 1994. Storage Rack Selection Guide. *Modern Materials Handling* 49(5):49-50.

Law, Averill M., and W. David Kelton. 1991. *Simulation Modeling and Analysis*, 2d ed. New York, New York: McGraw-Hill Incorporated.

Minahan, Tim. 1996a. The New Face of Trucking. *Purchasing* 120(4):38-43.

Minahan, Tim. 1996b. The Little Engine That Will. *Purchasing* 120(4):46-49.

Musselman, Kenneth J. 1994. Guidelines for Simulation Project Success. In *Proceedings of the 1994 Winter Simulation Conference*, eds. Jeffrey D. Tew, Mani S. Manivannan, Deborah A. Sadowski, and Andrew F. Seila, 88-95.

Okashah, Lobna A., and Paul M. Goldwater. 1994. Unknown Unknowns: Modeling Unanticipated Events. In *Proceedings of the 1994 Winter Simulation Conference*, eds. Jeffrey D. Tew, Mani S. Manivannan, Deborah A. Sadowski, and Andrew F. Seila, 689-694.

Pegden, C. Dennis, Robert E. Shannon, and Randall P. Sadowski. 1995. *Introduction to Simulation Using SIMAN*, 2d ed. New York, New York: McGraw-Hill, Incorporated.

Profozich, David M., and David T. Sturrock. 1995. Introduction to SIMAN/CINEMA. In *Proceedings of the 1995 Winter Simulation Conference*, eds. Christos Alexopoulos, Keebom Kang, William R. Lilegdon, and David Goldsman, 515-518.

Robinson, Stewart, and Vinod Bhatia. 1995. Secrets of Successful Simulation Projects. In *Proceedings of the 1995 Winter Simulation Conference*, eds. Christos Alexopoulos, Keebom Kang, William R. Lilegdon, and David Goldsman, 61-67.

Sanders, Nada R., and Karl B. Manrodt. 1994. Forecasting Practices in US Corporations: Survey Results. *Interfaces* 24(2):92-100.

Seila, Andrew F. 1995. Introduction to Simulation. In *Proceedings of the 1995 Winter Simulation Conference*, eds. Christos Alexopoulos, Keebom Kang, William R. Lilegdon, and David Goldsman, 7-15.

Subramanian, Radhika, Richard P. Scheff, Jr., John D. Quillinan, D. Steve Wiper, and Roy E. Marsten. 1994. Coldstart: Fleet Assignment at Delta Air Lines. *Interfaces* 24(1):104-120.

Van Metre, T. W. 1939. *Trains, Tracks, and Travel*, 5th ed. New York, New York: Simmons-Boardman Publishing Corporation.

van Moorsel, Aad P. A., Latha A. Kant, and William H. Sanders. 1996. Computation of the Asymptotic Bias and Variance for Simulation of Markov Reward Models. In *Proceedings of the 29th Annual Simulation Symposium*, 173-182.

Vincent, Stephen, and Averill M. Law. 1995. ExpertFit: Total Support for Simulation Input Modeling. In *Proceedings of the 1995 Winter Simulation Conference*, eds. Christos Alexopoulos, Keebom Kang, William R. Lilegdon, and David Goldsman, 395-400.

Williams, Edward J. 1996. Making Simulation a Corporate Norm. In *Proceedings of the 1996 Summer Computer Simulation Conference*, eds. V. Wayne Ingalls, Joseph Cynamon, and Annie Saylor, 627-632.

AUTHOR BIOGRAPHIES

EDWARD J. WILLIAMS holds bachelor's and master's degrees in mathematics (Michigan State University, 1967; University of Wisconsin, 1968). From 1969 to 1971, he did statistical programming and analysis of biomedical data at Walter Reed Army Hospital, Washington, D.C. He joined Ford in 1972, where he works as a computer software analyst supporting statistical and simulation software. Since 1980, he has taught evening classes at the University of Michigan, including both undergraduate and graduate simulation classes using GPSS/H, SLAM II, or SIMAN. He is a member of the Association for Computing Machinery [ACM] and its Special Interest Group in Simulation [SIGSIM], the Institute of Electrical and Electronics Engineers [IEEE], the Society for Computer Simulation [SCS], and the Society of Manufacturing

Engineers [SME]. He serves on the editorial board of the *International Journal of Industrial Engineering – Applications and Practice*.

SUSAN KHOUBYARI is a Systems Consultant at Production Modeling Corporation, Dearborn, Michigan. She holds B.S. and M.S. degrees in Industrial Engineering from the State University of New York at Buffalo. She has been working at PMC for five years, where she is responsible for managing discrete event simulation projects and finite-capacity scheduling systems, as well as training in project management concepts and software packages. She is a senior member of the Society of Manufacturing Engineers [SME].

Proceedings of the 1996 Winter Simulation Conference
ed. J. M. Charnes, D. J. Morrice, D. T. Brunner, and J. J. Swain

MODELING AND ANALYSIS OF AIRCRAFT OFFLOADING OPERATIONS

S. Manivannan

Department of Intelligent Systems Engineering
Consolidated Freightways, Inc
Portland, OR 97209, U.S.A

M. Zeimer

Engineering Department
Schneider Logistics, Inc.
Green Bay, WI 54306, U.S.A

ABSTRACT

This paper describes a successful application of discrete-event simulation techniques to evaluate and improve the plane offloading operations in a central aircargo Hub. The simulation model was built after a thorough analysis of plane offloading procedures, freight flow, and work rules involving the usage of equipment and personnel. Experimentation and extensive statistical analysis were performed to determine the (i) best operating rules, (ii) strategic changes to tug and dolly operations, and (iii) ramp configuration. The final recommendations based on a rigorous analysis to improve the ramp performance and use of models in planning and control are discussed.

1 INTRODUCTION

Application of discrete-event simulation to improve the ramp operations of aircargo companies has been gaining enormous impetus during this decade. A typical aircargo offloading ramp includes (a) runways, (b) gates or park locations, (c) K-loaders, (d) tug and dollies, (e) forklifts, (f) sortline conveyors, (g) belt/roller conveyors, and (h) movement lanes for tugs, planes, etc. The aircargo ramp discussed in this paper is divided into north, south, east and west sides with a capacity to handle 77 planes at any time. During each night, a scheduled number of aircraft arrive at the Hub between 11:00 p.m and 3:00 a.m.

A tailsheet provides the data on the number and types of aircraft, origin, destination, and scheduled arrival and departure times. The gate (park location) for each plane is assigned by the aircraft control tower and it stays at that gate until departure. Containerized freight is transported by 6 different models of aircraft and are offloaded by 14 K-loaders onto a set of dollies linked to a tug. Nearly 40 tugs and 120 dollies are used to offload and move the containers to one of 62 radio-controlled sortline conveyor locations at north and south transfer areas. At each transfer area, 16 forklifts offload the containers from the dollies and transport to breakdown sortline conveyors. The ramp encompasses a centralized park location for all tug and dolly units and two outside conveyors to handle belly freight (see Figure 1).

Essentially, three types of containers are offloaded from the aircraft. These are categorized as (i) *top side huts* (ULDs), (ii) *top side pallets* (PNs), and (iii) *belly freight* transferred to a container for efficient handling. In the existing system, two tugs each connected with 2, 3 or 4 dollies, are assigned to an aircraft to transport the containers. The tugs, once assigned to an aircraft, tend to stay as a team until all the aircraft on the ramp are offloaded. As the tugs approach the Hub building, they are assigned to one of the two transfer areas. Once an aircraft is fully offloaded, the tugs and the crew move as a group to the next nearest plane. Several work rules are followed by tug operators in offloading aircraft. These rules apply to both *top side* and *belly* containers.

The containers from aircraft are transported to one of two locations on the ramp before they are opened, sorted, consolidated, and loaded back into aircraft for delivering to destinations all over the world; (a) huts and PNs are offloaded to sortline conveyors at transfer areas, and (b) loose freight from the aircraft belly is offloaded first to a container and then moved to outside conveyors.

Forklifts at transfer areas take different pickup or dropoff time depending on the number of dollies hooked to a tug. Hence, the total offloading time as well as the congestion levels at the transfer areas vary depending upon the number of dollies hooked to a tug. The more number of dollies per tug tends to increase the dropoff time at the transfer area whereas, a fewer number of dollies per tug tends to cut down the pickup time for the K-loader process at the aircraft.

Huts and PNs are offloaded by forklifts from tug and dollies to any one of 62 sortline conveyors and are moved from the transfer areas to sortation area. The sortline conveyor is an indexing conveyor and has the capacity to accumulate five containers. The conveyor for offloading is decided by a dispatcher based on (i) availability, (ii) proximity, and (iii) the least number waiting on the slat conveyor. A traffic light mounted at the top of each sortline conveyor is used to indicate its availability to forklift operators. The general rule is to load each conveyor with at least one container to ensure uniform loading and minimum delays on the conveyor. The huts and PNs are indexed to the end of the sortline where the freight inside the container is offloaded by forklifts, sortation devices and human operators. Once a container becomes empty, it is removed from the sortline and transported for reuse at one of 16 buildup modules where the sorted freight is consolidated and reloaded to aircraft.

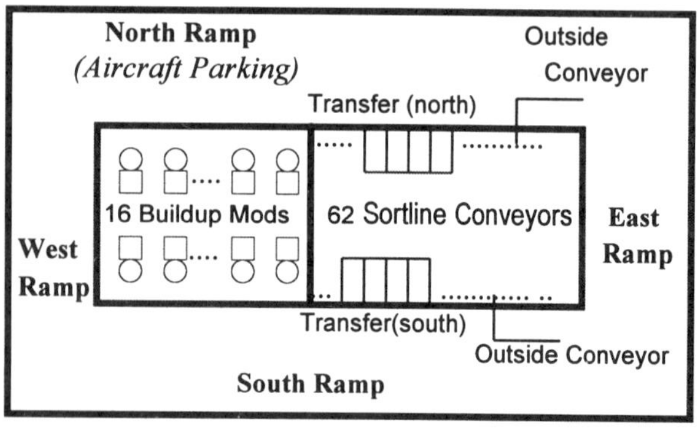

Figure 1: Block Layout of the Central Hub

Two outside conveyors at the north and south end of the Hub building are used to offload the belly containers. In general, the freight in each belly fills up a container. Forklifts are used to move the loaded belly containers from the aircraft. Once these containers are brought to the outside conveyor areas, human operators transfer the belly freight on to the conveyor belt. The freight is conveyed to sortation areas and then to buildup modules.

In this paper, we focus our attention primarily to plane offloading operations at the aircargo Hub in order to accomplish the following objectives;

(a) Build a simulation model to visualize the problems, trouble spots, and changes in behavior of the aircraft offloading operations to the ramp crew as the Hub undergoes changes in the usage of K-loaders, tugs and dollies, and other auxiliary equipment.

(b) Study the impact of varying the number of planes and gates and accordingly find the *best* number of K-loaders and tug and dollies. Identify changes in work rules and communication system to maximize throughput and number of aircraft offloaded and minimize the average time to offload an aircraft.

(c) Study the impact of changing the number of dollies per tug, number of tugs per plane on the average time required to offload a plane. Study the impact of these operation parameters on varying number of planes that arrive at the Hub during a night.

(d) Study the impact of speeding up the K-loader process and changing the number of tugs per plane to five and dollies per tug to three.

(e) Study the impact of assigning a variable number of tugs (instead of two tugs) to aircraft and dispatching tugs by radio on the completion time and average offloading time per aircraft.

2 MODEL DATA

Two types of input data were collected and used in the simulation model. The first type involved the technical specifications associated with the freight, aircraft, gates, tug and dollies, forklifts, sortline conveyors, and outside conveyors along with the assignment rules, work rules, freight flow behavior, control rules for tugs and forklifts, etc. More specifically, the following data were obtained from the Hub; (a) number of gates, (b) ramp maps, (c) technical specifications of conveyors, tugs, dollies, and forklifts, (d) flight profile, (e) aircraft models or types, (f) chronological plane arrival/departure information, (g) number of *topside* containers by plane type, (h) number of *bellies* by plane type, (i) number of containers per belly by plane type, (j) number of dollies per tug, (k) team makeup for offloading containers from a plane, (l) total number of tugs and dollies, (m) number of transfer forklifts, (n) outside conveyor locations/their physical characteristics, (o) aircraft parking/offloading rules, (p) runway characteristics, and so forth.

The second type of information involved a set of input distributions associated with the night-to-night (or hour-to-hour) variations found in the plane offloading operations. These included; (a) plane arrival process at different locations of the ramp, (b) pickup times and dropoff times for tugs and forklifts based on freight type and number of containers, (c) operation times at the slat conveyor, and (d) the plane contents.

3 EXISTING PLANE OFFLOADING MODEL

The model logic was implemented using *Automod II* on an IBM-PC. Ten processes, *P1* to *P10* were developed to represent accurately the behavior of plane offloading operations *(base model)* on the ramp. These include;

P1: Aircraft Arrival at the Hub

P2: Aircraft Parking at the Park Location

- Assign park locations using a rulebase
- Send aircraft directly to next-available gate

P3: Aircraft Arrive at the Gates and Request for Tugs

- Enter the park location for offloading
- Turn off the engine and open cargo doors
- Request tugs/K-loaders/conveyors/crews

P4: Plane Location Specification to Tugs

- Assign two tugs to a plane
- Assign rampid and park location id to the tug

P5: Tug Scheduling, Movement and Control

- Specify the plane park location to begin offloading

- Move the tug to the park location
- Determine the number of huts/PNs picked up
- Transport to container destination
  - North transfer area
  - South transfer area
  - Outside conveyors
- Dropoff the container at its destination
- Move the tug back to the aircraft being offloaded
- Repeat until all containers are offloaded

P6: K-loader Transfer, Tug Pickup and Dropoff
- Specify transfer time based on the K-loader process
- Specify tug pickup time at the aircraft
- Specify tug dropoff time at the transfer area

P7: Transfer Area Selection for Tugs

P8: Pickup and Dropoff by Forklifts at Transfer Areas
- Specify the pickup time based on number of dollies
- Specify the dropoff time per container

P9: Slat Conveyor Selection
- Select the slat conveyor to offload containers
- Selection rules
  - Nearest neighbor rule
  - Next available segment rule
  - Waiting rules if no segment available

P10: Container Movement on a Sortline Conveyor
- Entry station rules
- Exit station rules
- Travelling on segment rules
- Waiting rules on the segment

All ten processes were built with intelligent rules and algorithms, implemented in *Automod II*, to capture the exact behavior of plane offloading operations.

The simulation model was linked with input data files through an efficient interface to perform *"what-if"* scenarios. The aircraft arrival and parking process was implemented with a rulebase to mimic the tower rules used to park the aircraft. The rulebase consists of twenty rules and utilizes the flight profile to obtain all pertinent information on the arriving planes during a night. The rulebase was validated using several flight profiles and arrival times to ensure that all the planes are parked at north, south, east, and west ends of the ramp.

The simulation model was built with a ramp map to incorporate tug and dolly movement lanes, parking locations for planes and tugs, waiting areas for tugs, interaction zones for transfer forklifts to pickup huts and PNs from the tug and dollies, capacity limitations at the intersections to avoid collision between tugs, and control points to keep sufficient space between tugs.

With respect to tug and dolly system, the model was built to depict explicit and accurate representation of its behavior. A smart algorithm was implemented to incorporate the plane assignment rules to tug and dollies (belonging to a team), tugs per plane rules, tug waiting rules, offloading crew rules, collision avoidance rules, right-of-way rules, parking rules, and passing rules. The algorithm uses the tug specification such as acceleration, deceleration, loaded/unloaded speeds, forward, reverse and curve speeds, and ramp map to compute the precise travel times between aircraft and transfer areas.

The processes at the north and south transfer areas were modelled very precisely as well. Eight forklifts on each side of the transfer areas move the containers from the dollies to sortline conveyors. The sortline conveyor selection process was modelled in detail. Another smart algorithm was implemented to ensure uniform container offloading at sortline conveyors such that no single conveyor would be overloaded during a night. Also, the algorithm was built to reroute the tugs to south (or north) transfer areas in case of overloading. At the Hub, this was carried out by a human dispatcher who observed the container movement along the sortlines on a computer screen. Depending upon the overload conditions, the dispatcher communicated with the tug operators using a radio and rerouted the containers.

4 FACTORS AND RESPONSES

Controllable factors and responses were identified to evaluate alternative system configurations based on a list of issues that the Hub management was interested in. For the plane offloading model, the major factors were (a) number of planes (flight profile and plane arrival time distributions), (b) total number of tugs, (c) number of tugs per plane, (d) number of dollies per tug, and (e) number of K-loaders. Likewise, the responses that were of interest include; (a) plane offload capacity in terms of total number of containers offloaded and total number of planes offloaded (b) average time to offload a plane, and (c) completion time for offloading all planes. A report generator and an experimental setup were developed using *AutoStat* and interfaced with the *base model* to conduct *"what-if"* scenarios and statistical analysis.

5 WHAT-IF SCENARIOS

The *base model* was modified to study the impact of alternative ways of (a) assigning tug and dollies to the aircraft, (b) communication, and (c) using equipment, people and facilities. These are described below;

Alternative System I encompasses a change in the existing plane offloading model in which the number of tugs per plane is set to be five, number of dollies per tug is set to be three, and the K-loader process is speeded up to reduce the transfer time from the aircraft to dollies.

Alternative System II encompasses a major change in the existing plane offloading process. The work rules associated with the tug and dolly system are modified to provide more flexibility and independence to tugs. This system is referred as *radio-dispatched tug operations* (RDTO). In RDTO system, once an aircraft arrives, the dispatcher is informed via radio about its park location and number of containers in the plane. The dispatcher determines the number of tugs required (each tug is fitted with three dollies) and then identifies at least one tug that is closest to the plane that can be dispatched and send to that aircraft (similar to dispatching taxi cabs). This means that this tug is currently idle and waiting to be dispatched. The maximum number of tugs dispatched at any time is equal or less than the exact number of tugs required to offload the entire plane. This depends upon the availability of tugs for offloading as well as the fact that there are no other plane is currently requesting for a tug. If there are several planes waiting for tugs at the same time, then the dispatcher sends one or more tugs per plane based on the proximity rules and number of tugs currently idle. Several parking areas are setup on the ramp where the tug and dollies wait for work when they become idle with no planes to offload.

In RDTO system, it is essential that the dispatcher is knowledgeable about two facts; (i) plane information once it is ready for offloading, and (ii) tug-id and its location when a tug completes offloading a container at transfer areas. Based on these facts, the dispatcher can identify the aircraft where there is immediate need for a tug. The next available tug closest to that aircraft is dispatched.

The RDTO system breaks down all the barriers of 2-tug setup found in the existing operation. It offers the flexibility to move K-loaders to the next plane once the last topside container is transferred from the aircraft to the dolly. Further, the belly crews are separated from the top side crews which cuts down the 'waiting for each other to finish offloading' syndrome.

6 EXPERIMENTATION

A simulation experiment was conducted using the *base model* encompassing the current operating procedures and existing levels of aircraft and equipment (59 planes, 40 total tugs, 2 tugs per plane, and 2, 3, or 4 dollies per tug). The purpose was to bench-mark the existing ramp operations in order to provide a basis of comparison for the alternative scenarios investigated.

We made ten replications and computed the mean and standard deviation for the average offloading time for an aircraft using the simulation outputs. The mean and standard deviation were: $y = 31.802$ minutes and $s = 0.8694$ minutes respectively. Hence, any configuration with an average offloading time significantly less than 31.802 minutes would be considered as an improvement over the existing configuration.

The following sets of experiments were conducted to determine the best configuration(s);

(a) Using the *base model*, 37 different configurations were tested for three sets of aircraft (59,68,77) arriving nightly, five levels of total number of tugs operating on the ramp (40,60,80,100,120), four levels of tugs assigned to each plane (2,3,4,5), and three levels of dollies hooked to a tug (2,3,4).

(b) Using the revised *base model* to depict *Alternative System I*, 11 configurations were tested for three sets of aircraft (59,68,77) arriving nightly and five levels of total number of tugs (40,60,80,100,120).

(c) Using the revised *base model* to depict *Alternative System II*, 33 configurations were tested for three sets of aircraft arriving nightly and various levels of tugs (40,60,80,100,120) and K-loaders (8,14,20,26,32).

The first set of experiments (i) establishes a relationship between the overall ramp performance and the resource capacity (tugs, dollies, K-loaders), and (ii) estimates the best configuration even if the total number of aircraft offloaded vary widely (between 59 and 77). The second set of experiments pertains to a configuration in which 5 tugs are assigned to each plane, 3 dollies are assigned to each tug, and modified K-loader process. This series of experiments (i) compares the overall performance with the *base model*, and in particular, to the best equipment configuration without major changes to existing ramp operations; and (ii) determines the total number of tugs required to achieve this best performance. The third set of experiments pertains to the analysis of RDTO system.

In all experiments, the overall system performance was measured in terms of average plane offloading time. Plane offloading time is defined as the amount of time between the arrival of an aircraft at the hub and the completion of the offloading process for that aircraft.

6.1 Existing Plane Offloading (Base Model)

The *base model* to depict the existing plane offloading operation was utilized to accommodate varying number of tugs, tugs per plane, and dollies per tug. The purpose of this experiment was to estimate the best number of tugs to achieve the least offloading time per aircraft. Consider the set of factors shown in Table 1 and their impact on the average plane offloading time (y).

Table 1: Factors for Base Model Experiments

Factor	Description
x_1	Number of aircraft arriving nightly
x_2	Number of tugs
x_3	Number of tugs per plane
x_4	Number of dollies per tug

To assess the impact of varying levels of these factors on the average plane offloading time, we conducted a set of experiments with the *base model* and used the results to fit a linear statistical model of the general form:

$$y = b_o + \Sigma_i b_i x_i + \Sigma_i\Sigma_j b_{ij}x_ix_j + \Sigma_i\Sigma_j\Sigma_k b_{ijk}x_ix_jx_k + \Sigma_i b_{ii}x_i^2 + \Sigma_i b_{iii} x_i^3 + \Sigma_i b_{iiii}x_i^4 + \varepsilon \quad (1)$$

$i = \{1, 2,.., 4\}, j = \{1, 2,.., 4\}$
$k = \{1, 2,.., 4\}, l = \{1, 2, ., 4\}$

The ranges of experimental factors studied in these experiments are given in Table 2. Coded factor levels were used to estimate expression (1) to minimize the variance of prediction (i.e., in order to obtain the best possible predictions with the least amount of data).

Table 2: Factor Levels for Base Model

Factor	Factor Levels	Coded Factor Levels(ξ_i)
x_1	59, 68, 77	-1.0, 0.0, 1.0
x_2	40, 60, 80, 100, 120	-1.0, - 0.5, 0.0, 0.5,1.0
x_3	2, 3, 4, 5,6	-2.0, -1.0, 0.0, 1.0, 2.0
x_4	2, 3, 4	-1.0, 0.0, 1.0

In order to obtain the best possible fit and to stabilize the error variance, y was transformed by taking the natural log of y. Thus, the form of the fitted equation is:

$$y = \exp(b_o + \Sigma_i b_i\xi_i + \Sigma_i\Sigma_j b_{ij}\xi_i\xi_j + \Sigma_i\Sigma_j\Sigma_k b_{ijk}\xi_i\xi_j\xi_k + \Sigma_i b_{ii}\xi_i^2 + \Sigma_i b_{iii} \xi_i^3 + \Sigma_i b_{iiii}\xi_i^4 + \varepsilon) \quad (2)$$

$i = \{1, 2, ..., 4\}, j = \{1, 2, ..., 4\},$
$k = \{1, 2, ..., 4\}, l = \{1, 2, ..., 4\}.$
$\xi_1 = (x_1 - 68) / 9, \xi_2 = (x_2 - 80) / 40,$
$\xi_3 = (x_3 - 4) / 1, \xi_4 = (x_4 - 3) / 1$

Multiple regression procedure in *Statistica* software was used to estimate model (2).

Table 3 provides a list of statistically significant coefficients and factors for model (2). The p-value for each model component is the probability that the effect estimated by the analysis procedure is due to chance and not to a systemic behavior. The lower the p-value the more significant (statistically) the component is. All components with p-values greater than 0.1 were selected to be insignificant and excluded from Table 3. The signs of the model coefficients (b's) indicate whether or not the component has a positive or negative effect on the response. The percentage of total variation in the data that is accounted for by the estimated linear model is given by R^2.

Model (2) accounts for 94.42% of all of the variation in the data. High R^2 value generally indicates a comprehensive statistical model as well as a good fit. Subsequent to the estimation, model (2) was used to locate the (near) *optimal* equipment configurations for varying numbers of aircraft arriving nightly at the Hub.

Table 3: Estimated Linear Regression Model Model (2) (R^2 = .9442)

Component	b	p-value
Intercept	3.485979	0.000000
x_1	0.175055	0.000000
x_2	-0.046504	0.088576
x_3	0.166626	0.000000
x_4	-0.153704	0.000000
x_1x_2	-0.027974	0.000150
x_1x_3	0.030806	0.000002
x_1x_4	-0.054193	0.000000
x_2x_3	-0.069812	0.000000
x_2x_4	-0.038229	0.000001
$x_1x_2x_3$	0.031979	0.000019
$x_2x_3x_4$	-0.054050	0.000000
$x_1x_2x_3x_4$	-0.056447	0.000000
x_2^2	0.163786	0.000000
x_4^2	0.115800	0.000000
x_2^3	-0.088960	0.002268
x_3^3	-0.052137	0.000000

6.2 Alternative System I

In the *base model*, the number of tugs assigned to each plane was increased to five and the K-loader process was modified to speed up the operations at the plane to set up the *Alternative System I* scenario. Further, the dollies were retrofitted with ball tables so that the operator can simply push the container off the K-loader on to a dolly thereby eliminating intermediate operation stages. The purpose of this experiment was (I) to estimate the best number of tugs and K-loaders to be used on the ramp to achieve the least offloading time per plane and (ii) reduction in the overall ramp offloading time for all the planes. Consider the set of factors given in Table 4 used to examine their impact on the average time to complete the plane offloading process.

Table 4 - Factors for Alternative System I Experiments

Factor	Description
x_1	Number of aircraft arriving nightly
x_2	Total number of tugs

To assess the impact of varying levels of tugs on the average offloading time for a given configuration, let us consider the following linear statistical model:

$$y = b_0 + \Sigma_i b_i x_i + \Sigma_j\Sigma b_{ij}x_ix_j + \Sigma b_{ii}x_i^2 + \Sigma b_{iii}x_i^3 + \Sigma b_{iiii}x_i^4 + \varepsilon \quad (3)$$

$i = \{1, 2\}, j = \{1,2\}$

The ranges and levels of controllable factors investigated in this experiment are provided in Table 5.

Table 5: Factor Levels for Alternative System I

Factor	Factor Levels	Coded Factor Levels(ξ_i)
x_1	59, 68, 77	-1.0, 0.0, 1.0
x_2	40, 60, 80, 100, 120	-1.0, -0.5, 0.0, 0.5, 1.0

In order to obtain the best possible fit y was transformed by taking the natural log of y. Thus, the form of the fitted equation is:

$$y = \exp(b_o + \Sigma_i b_i\xi_i + \Sigma_i\Sigma_j b_{ij}\xi_i\xi_j + \Sigma_i b_{ii}\xi_i^2 + \Sigma_i b_{iii}\xi_i^3 + \Sigma_i b_{iiii}\xi_i^4 + \varepsilon) \quad (4)$$
$$i = \{1, 2, ..., 4\}, j = \{1, 2, ..., 4\}.$$
$$\xi_1 = (x_1 - 68) / 9, \xi_2 = (x_2 - 80) / 40$$

Multiple regression procedure in *Statistica* was used to estimate model (4). Table 6 shows a list of statistically significant coefficients and factors for model (4).

Table 6: Estimated Linear Regression Model
Model (4) ($R^2 = 0.9085$)

Component	b	p-value
Intercept	3.432598	0.000000
x_1	0.188586	0.000000
x_1x_2	-0.088700	0.000002
x_2^2	0.217485	0.000000
x_2^3	-0.227007	0.000000

Subsequent to estimation, model (4) was used to locate optimal (or near-optimal) number of tugs for varying number of planes.

6.3 Alternative System II

The RDTO system was modelled to accomplish a greater flexibility. In particular, RDTO dictates that tugs be dispatched individually to arriving aircraft as opposed to twin tug dispatching. The tug operators have more independence and are always directed to move towards the aircraft where the most work load is. The K-loaders are dispatched independently to the next nearest plane (or the critical one) as well.

In RDTO, the existing work rules in assigning tug and dollies were revised to accomplish a major reduction in offloading time per aircraft. The purpose of these experiments was to estimate the best number of tugs and K-loaders to achieve the least offloading time per plane and reduction in overall ramp offloading time for all the planes. Table 7 shows the set of factors selected for this experiment.

Table 7: Factors for RDTO Experiments

Factor	Description
x_1	Number of aircraft arriving nightly
x_2	Total number of tugs
x_3	Number of K-loaders

In RDTO system, the number of dollies per tug is three. The number of tugs per aircraft depends upon the plane type, number of containers to be offloaded, and the availability of tugs at that time. In order to assess the impact of varying numbers of tugs and K-loaders on the ability of the RDTO system to complete offloading a varying number of aircraft (59 to 77) arriving nightly, let us consider the following linear regression model:

$$y = b_0 + \Sigma_i b_i x_i + \Sigma_j\Sigma_i b_{ij}x_i x_j + \Sigma_j\Sigma_i\Sigma_k b_{ij}x_i x_j x_k\ \Sigma_i b_{ii}x_i^2 + \Sigma_i b_{iii}x_i^3 + \Sigma_i b_{iiii}x_i^4 + \varepsilon \quad (5)$$
$$i = \{1, 2, 3\}, j = \{1,2, 3\}, k = \{1,2,3\}$$

The ranges of the controllable factors studied in these experiments are given in Table 8. Coded factor levels were used to estimate model (5) in order to minimize the variance of prediction (i.e., to obtain the best possible predictions with the least amount of data).

Table 8: Factor Levels for RDTO Experiments

Factor	Factor Levels	Coded Factor Levels(ξ_i)
x_1	59, 68, 77	-1.0, 0.0, 1.0
x_2	40, 60, 80, 100, 120	-1.0, -0.5, 0.0, 0.5, 1.0
x_3	8, 20, 32	-1.0, 0.0, 1.0

In order to obtain the best possible fit y was transformed by taking the natural log of y. Thus, the fitted equation is expressed by:

$$y = \exp(b_o + \Sigma_i b_i\xi_i + \Sigma_i\Sigma_j b_{ij}\xi_i\xi_j + \Sigma_i b_{ii}\xi_i^2 + \Sigma_i b_{iii}\xi_i^3 + \Sigma_i b_{iiii}\xi_i^4 + \varepsilon) \quad (6)$$
$$i = \{1, 2, ..., 4\}, j = \{1, 2, ..., 4\}, \xi_1 = (x_1 - 68) / 9,$$
$$\xi_2 = (x_2 - 80) / 40, \xi_3 = (x_3 - 20) / 12$$

Multiple Regression Procedure in *Statistica* was used to estimate model (6). Table 9 shows a list of statistically significant coefficients and factors for model (6).

Table 9 - Estimated Linear Regression Model
Model (6) ($R^2 = .8000$)

Component	b	p-value
Intercept	2.455574	0.000000
x_1	0.116782	0.000000
x_1x_3	-0.143195	0.000000
x_2^2	0.050264	0.005523
x_3^2	0.196400	0.000000
x_3^3	-0.187527	0.000000

Subsequent to estimation, model (6) was used to locate optimal (or near optimal) equipment combinations for varying numbers of planes.

6.4 Selection of Best Configurations

Linear regression models in Sections 6.1 - 6.3 were used to demonstrate the relationships between the factors and accordingly identify the equipment combination that has yielded least (average) offloading time/aircraft. In this section, we discuss the results of the statistical analysis. Table 10 shows the short-listed configuration set chosen from a large number of feasible configurations for fifty nine planes arriving at the Hub per night. These results were obtained using the response surfaces and contour plots generated to show the interrelationships between the number of planes, total number of tugs, and number of dollies per tug. The surface plots were created using the regression equations (models 2, 4 and 6).

Table 10. Short-listed Configurations for Number of Planes/night = 59 (All time units are in minutes)

C	S	N_t	n	d	K	N_k	C_t	X_t
C1	B	40	2	2,3,4	O	-	300	31.8
C2	B	70	3	3	O	-	288	24.9
C3	B	40	3	3	O	-	295	28.0
C4	A1	80	5	3	N	-	299	26.9
C5	A2	40	-	20	N	20	261	11.6
C6	A2	80	-	3	N	17	292	10.2
C7	A1	74	5	3	N	-	307	25.4
C8	B	76	3	4	O	-	269	25.5

In Table 10, S refers to the scenarios evaluated, (where B = *base model*, A1 = *Alternative System I*, A2 = *Alternative System II*), N_t corresponds to the number of tugs on the ramp, n refers to the number of tugs assigned to a plane, d denotes the number of dollies hooked to a tug, K denotes the K-loader process (O refers to existing and N refers to the modified process), N_k corresponds to the number of K-loaders, C_t denotes the completion time (or time window for plane offloading) in minutes, and X_t refers to the mean of average offloading time in minutes per aircraft.

Eight short-listed configurations (as denoted by C) are provided in Table 10. C1 configuration corresponds to the current plane offloading process at the Hub. C2 and C3 refer to *"what-if"* scenarios performed with the current system except (i) tugs were fitted with 3 and 4 dollies respectively, (ii) 3 tugs were assigned to a plane instead of 2, and (iii) total number of tugs on the ramp were 70 and 40 respectively.

C4 and C7 refer to *Alternative System I* in which (i) all tugs were fitted with 3 dollies, (ii) 5 tugs were assigned to a plane, (iii) number of tugs was 80 and 74 respectively, and (iv) K-loader process was speeded up.

C5 and C6 refer to *Alternative System II* with an exception that the number of tugs and K-loaders varied. C8 refers to the scenario in which the total number of tugs was 76 with the number of tugs per plane was 3 and the number of dollies per tug was 4.

Analogous to the analysis conducted for 59 planes per night, best-feasible configurations were identified for 68 and 77 planes arriving each night at the Hub. The short-listed configurations were examined carefully and a list of final recommendations were identified.

7 RECOMMENDATIONS

Three recommendations were made to the management of the Hub based on our interactions with simulation/ animation models and extensive statistical analysis. Each recommendation improved the performance of the ramp and any one, if implemented, would bring in reduction in X_t and C_t. The recommendations in Table 11 were ranked based on the tradeoffs between the anticipated performance improvement, changes needed in operation procedures, and capital investment.

Table 11. Final Recommendations

Rank	Recommended Operation Setup	C_t	X_t
I	Radio Dispatched Tug Operation (40 Tugs, 20 K-loaders)	261	11.6
II	Existing System (76 tugs, 3 Tugs/Plane, 4 Dollies/Tug)	269	25.5
III	Existing System (70 Tugs, 3 Tugs/Plane, 3 Dollies/Tug)	288	24.9

Recommendation I was highly ranked as it reduced immensely the average offloading time per aircraft when compared to existing plane offloading operations. This behavior was found to be consistent even when the total number of planes arriving each night at the Hub was increased. In addition, this recommendation required major changes in work rules for assigning tugs to the aircraft, flexibility and independence to tugs, as well as dispatching by the radio.

Recommendation II was the second choice as it involved no changes to work rules and K-loader process while offloading containers from the aircraft to dollies; however, the number of dollies hooked to each tug was increased to 4.

Recommendation III involved no changes to K-loader process, work rules, and tug setup. Overall, these recommendations showed a significant improvement in the ramp performance at the least cost.

8 CONCLUSIONS

In summary, the benefits of the ramp simulation study in the aircargo company are two-fold;

(a) Pinpoint both strategic and operation improvements on the current plane offloading processes, equipment usage, and facility layouts due to changes in the behavior of freight handled at the Hub. This helps the decision maker to continuously improve productivity and throughput.

(b) Estimate the appropriate levels of equipment and suitable facility layouts *a priori,* due to an increase in the number of planes offloaded. This helps the decision maker to better utilize the top dollars before it is being spent on new equipment purchases.

Further, the ramp model along with experimental setups are useful in conducting operations planning and control on a nightly basis at the Hub. For a given plane arrival process during a night, the Hub management can estimate (a) completion time to offload all the planes, (b) peak hours and lean hours and accordingly plan for the operation crew, equipment, and service requirements, (c) number of gates required at the ramp, (d) number of tugs and forklifts needed to finish offloading all the planes within a specified time window, and (e) number of containers expected to be offloaded by the hour and the anticipated work loads.

AUTHOR BIOGRAPHIES

S. MANIVANNAN earned his Ph.D from Pennsylvania State University in 1988. He specializes in simulation of transportation, logistics and manufacturing systems, artificial intelligence, and database design. Since 1993, he is a member of senior research staff in the department of Intelligent Systems Engineering at the Consolidated Freightways, Inc. He is the Project Leader for building a series of simulation models to improve and redesign Emery Worldwide Dayton Hub and a pioneer of object-oriented transportation simulation model, SIMTRAN 2000. Prior to coming to CF. Inc., Dr. Manivannan worked as an Assistant Professor for five years in the School of Industrial and Systems Engineering at Georgia Tech. He received over one million dollars in research grants from industry and federal government and solved a variety of complex logistics problems in Ford, Torrington, Kodak, IBM and General Motors. He established the *Intelligent Manufacturing Lab* at Georgia Tech with grants from NSF and Torrington Company. He published numerous articles in reputed journals and conference proceedings. He is the recipient of the 1991 *Outstanding Young Manufacturing Engineer of the Year* Award from SME. He received the 1996 *Ray O'Brien Award of Excellence* from the CF Board of Directors. He serves as a *Panel Member* at DDM, NSF and a *Program Chair* for WSC'98 Conference. He is an active member of IEEE, SCS, ACM, ASEE and SME.

MICHAEL E. ZEIMER is a Logistics Engineer with Schneider Logistics, Inc. Prior to this position he was a member of the Intelligent Systems Engineering Group at Consolidated Freightways and a Product Analyst for Manugistics, Inc. He received his M.S. in Industrial and Systems Engineering from Virginia Polytechnic Institute and State University in 1992 where he conducted research on the design and analysis of simulation experiments. This research included his Thesis, entitled *The Selection of Pure Error Generators for Simulation Experiments*, and projects from Rome Labs and the Naval Air Warfare Center for metamodeling techniques and applications in air combat simulation. His current research interests include simulation of transportation and logistics systems, metamodeling, and simulation optimization.

Proceedings of the 1996 Winter Simulation Conference
ed. J. M. Charnes, D. J. Morrice, D. T. Brunner, and J. J. Swain

DISTRIBUTION LOGISTICS IN THE PROCESS INDUSTRIES: ESTABLISHING RAILCAR REQUIREMENTS

Charles H. White, Ph.D.

DuPont Process Engineering
DuPont Company
Wilmington, Delaware 19898, USA

ABSTRACT

The Manufacturing World is often classified into Discrete Manufacturing and Process Operations.Process Industry Operations generally are classified as 'Batch' or 'Continuous'; we can think of these as producing 'stuff' as opposed to the discrete individual 'things' produced in Discrete Manufacturing Operations. These Manufacturing World's clearly have much in common, but there are areas in which they can be quite different. They certainly have overlap, but at the extremes they can be fundamentally different and present unique challenges. Continuity of Operations is one of these areas; some Continuous Chemical plants may operate 'around-the-clock' for several years between 'shutdowns'. This means that they literally continue to operate even when there are maintenance difficulties. They may slow down for grade changes or unusual operating conditions, but they do not stop altogether as this could force the entire production train down for a major overhaul/restart which can be very expensive and/or time consuming. For some chemical plants a total shutdown can poison the reaction catalyst. For some polymer plants a total shutdown can mean the molten polymer 'freezes' in the lines. Each of these will result in a long period of total outage while the plant is refurbished and restarted. When a Process is 'Continuous' in this sense, it certainly can create some unusual challenges from a "Logistics Viewpoint". Aspects of these challenges are JIT/WIP Inventory, Material Handling, and Product Distribution. This paper will (1) discuss these challenges with emphasis on Final Product Distribution and Railcar Requirements, and (2) briefly describe two cases where careful Analysis and Modeling helped lead to substantial improvements in Operations Understanding and to significant Financial Savings (both Investment and Cost).

1. INTRODUCTION

In the Process Industries there are often many steps along the production chain as we move from original raw materials (such as unrefined oil) through to finished products (such as refined petroleum products, plastics, pharmaceuticals, synthetic fibers, and food additives). At each step in the production chain, in addition to the main product we are trying to produce, we can also have both co-products and by-products which need to be further processed (for sale) or treated (for disposal). Many Process Industry plants are large complex production systems which carry out a series of integrated chemical processing steps. These plants often generate very large production flows which can present significant handling, storage, and distribution challenges. As was once observed by one of my colleagues "a flow of a few tons per hour of chlorine can really get your attention!"

Large integrated chemical companies often operate large-scale plants serving both to supply external customers and to be sources of intermediate feedstocks for internal down-stream plants which then further process these intermediates into one or several finished products. From a Manufacturing Logistics viewpoint the objective is to operate the intermediates plant as a reliable source of high quality intermediates to both external and internal customers at the lowest possible cost. In this increasingly competitive world marketplace 'reliable source', 'high quality', and 'lowest possible cost' are all key business drivers and important to remaining in business.

A relatively simple Manufacturing Chain is illustrated in Figure I where we see two intermediates plants supplying four internal finished product plants and several external customers. For simplicity all co-product and by-product streams have been omitted (at both the intermediates producing plants as well as for the finished products plants) and the production chain has been limited to two steps (many real world systems are indeed much longer).

From a Manufacturing Logistics viewpoint some of the key goals are to (1) operate each of the internal finished product plants so as to meet the allocated production tasks, (2) operate each of the intermediates plants so as to supply the demand requirements of all of the external and internal customers, and (3) plan and

DISTRIBUTION LOGISTICS NETWORK:
Two Producer Plants & Four Consuming Sites

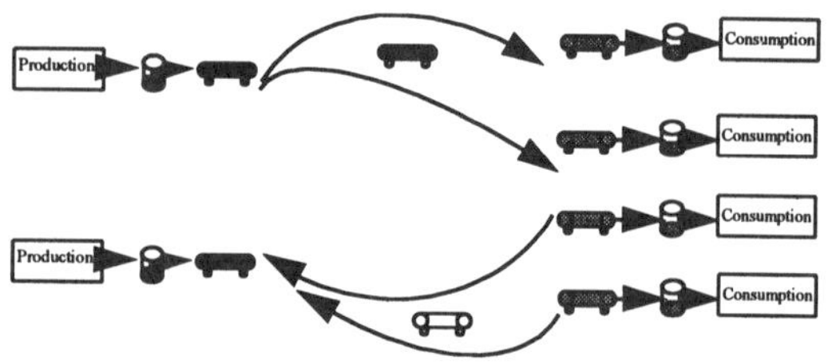

Figure I

manage the logistics system so that all materials are stored and distributed in a timely and cost effective manner. It is not an easy task to plan and operate such Manufacturing Logistics Chains for a number of reasons, including: (a) all of these plants may be subject to periodic production upsets and/or maintenance slowdowns, (b) the production tasks (such as weekly or monthly production quantities) are not always smooth and constant, (c) the railcar transit times on the distribution routes may vary significantly, and (d) the railcars used for making many of these moves are themselves subject to both unplanned and planned maintenance downtimes. It is in this context we approach the problem of determining the Railcar Requirements for effective low cost logistics operation of Process Industry Production Systems.

At this point you may ask "Why the big focus on Railcar Requirements in Process Industry Operations?". Many of the railcars used to move chemicals, refined petroleum products, and even foodstuffs are solely dedicated to a single product (sometimes even to a single grade of product) in order to strictly avoid any possible cross-contamination problems. These railcars often represent both a large long-term investment as well as a significant annual operating cost. Many of these fleets are large (sometimes many hundreds of cars in a single fleet supporting a multi-plant system), complex (serving several plants with disparate needs, conditions, and scheduling policies), old (some of these fleets have grown over many years as operations expanded and new customers were added), and they often operate in highly dynamic and variable conditions. This certainly can result in both a challenge and an opportunity. On a positive note our experience has shown that a program of careful analysis, simulation modeling, and coordinated fleet improvements can both reduce costs and improve fleet operations service levels.

2. ANALYSIS AND MODELING

Simulation Modeling has proven to be a valuable tool for helping to understand and improve Railcar Fleets. However we need to start by emphasizing that preceding the development of any simulation model there needs to be a careful Operations Analysis that first determines the key questions to be addressed, and then guides information gathering about how the system operates. This information gathering should focus on the system components: how they individually behave, how they interact, the rules for operating the both the individual components and the total system, and all of the sources of uncertainty and variability impacting the performance of the total system. The major operating and scheduling assumptions need to be agreed upon and documented along with all of the relevant data and parameters. The variable parameters need to be studied, the available data needs to be thoroughly analyzed, and additional data often must be gathered and analyzed. It should however be noted that the whole effort doesn't need to wait until all of the data has been gathered and analyzed; some of the model structuring and early development tasks can go along in parallel once the key data gathering tasks have been planned. Approximate data can be used during early model development once the task of gathering and analyzing better data has been agreed upon and people assigned to the task.

When planning for modeling a stochastic (dynamic with variability) system and planning the information gathering tasks, we must strongly focus on the questions the model will be used to address. The issue of level of detail to be in the model is a critical one; this can be the difference between a successful and mediocre modeling activity. Too little detail can lead to an incomplete or even inaccurate model that could lead to erroneous conclusions (or confusion altogether); whereas too much detail can lead to the modeling taking too long, costing too much or even failing altogether by getting bogged down in the detail. Correct balance is important; this can be the difference between success and failure. Certainly this is the case as seen from the eyes of the decision makers we are trying to help. This is an area that is sometimes not carefully planned; all novice modelers are cautioned to discuss this with their colleagues before getting 'in too deep'. It is generally better to build a reasonable model fast and to then incorporate additional details as needed to gain realism and get implementation acceptance. However, it is also important that to the best extent possible these details should be considered and weighed early. This means the information can at least be roughly estimated and the model can be built in a fashion that will allow these details to be added without undue difficulty. This is certainly one area where consulting with others early is time well spent. Cost effective models are desirable, but avoiding inaccurate models is even more important. There are a number of excellent references available on how to go about planning and carrying out a good valid simulation modeling study. In addition to papers and sessions in the *Winter Simulation Conference Proceedings,* the reader is directed to (1) seek out help from your fellow WSC attendees and (2) to start your literature search with the books Askin, Banks, Harrell, Law, Schriber and Shannon. (See References)

3. GENERAL OBJECTIVES

In modeling these railcar fleets what are the key questions we are attempting to address? Generally we want to know the best fleet size under a range of different conditions (different market demand rates, different plant configurations and rates, different travel transit time and variabilities, ...). Do we need to buy/lease more railcars as the business grows? Can we reduce the fleet and still provide continuing good service levels? If we improve maintenance operations can we reduce the fleet? What if we can get better travel transit time? What is the impact of the times cars wait to be unloaded and turned around at the consuming sites - will improvements allow us to reduce fleet sizes ? What are the major sources of variability and what is their impact of fleet size requirements? What is the impact of buffer tanks - what is the impact if we reduced inventory levels at either or both of the ends of the pipeline?

Many of these questions and our overall objectives can be discussed in the framework of Figure II.

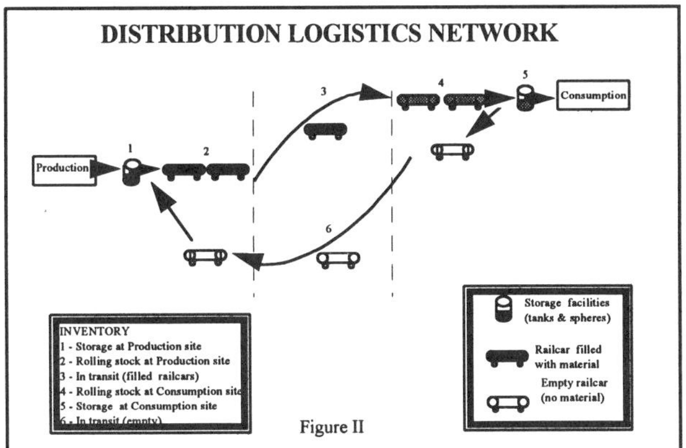

Figure II

Here we see a simple two plant-site system with just one producing and one consuming plant with buffer tanks at each site; a simple railcar fleet services the material movements between them, and a single maintenance branch-site is present to handle all maintenance on the railcars. Real-world systems, such as in our two case studies discussed later, are certainly bigger and more complex, but in principal they operate essentially the same way. Material is produced, stored in a tank, loaded into a railcar, moved to a consuming site (internal plant or an external customer), unloaded, and used while the railcar is sent back to the/a producing site (in some cases the material is consumed directly from the railcar and the car then can be sent back only after it is emptied and sealed off). In multi-plant systems the issue of where the empty car is sent can be an important one; it may be that the car should go back to where it was originally loaded, or it may be better to send it to another producing site that needs it more urgently. Use of a fleet operating policy that determines where best to send a car each time one is unloaded and available for routing back to a producing site can be powerful. Another important area is often the one of "when" to dispatch a loaded car from a producing site to a consuming site; the specific decision rules and criteria that are used can have a significant effect on service levels for customers throughout the whole system. We could end up with loaded cars in locations where we don't want them if we are not careful about our scheduling and dispatch policies, and as a result, not have the cars available to send to other locations where we do urgently need them.

4. TWO CASES STUDIES

Case studies will be used to illustrate two rather different situations where simulation modeling proved to be a valuable tool for helping to achieve significant investment and operating cost savings. Before getting into the specifics of these two cases, I want to strongly note that these successes were very much the result of good team efforts with many varied and valuable contributions. Simulation modeling was the guiding process and a tool for quickly and objectively comparing alternatives, but the system knowledge as well as the ideas to be evaluated came from the manufacturing and logistics members of the team. I am pleased to report that through this good work each of these illustrative cases had savings over $1 million.

4.1 Tool Set

Our Operations Research Group has been doing simulation modeling for many years. We have used a wide range of simulation modeling tools across the full range of computers. We started on the early engineering and technical computers, moved to mainframes, then to VAX's, and now work primarily on high-end PC's. There are many good simulation modeling tools available these days and we use several of them. However, in undertaking any important modeling job, it is important to first focus on a good understanding of the problem needs, and, to then be sure to select a tool with enough power to deal with all aspects of the problem. By all aspects, I include both technical capability and user interface features.

We have found that Pro-Model is a good tool for our Manufacturing Logistics modeling work. It has many "user friendly" features and, even more importantly, it has sufficient modeling power (both modeling constructs and language capabilities) to allow us to model very complex systems. In addition to these modeling features three additional important capabilities for us are: the ability to read external files, to generate graphic presentations, (via its own features and by exporting to spreadsheet packages), and very importantly the ability to present concurrent dynamic animations. These dynamic animations have three advantages that we find important: (1) they clearly help us with initial management buy-in and help the team "sell the job", (2) they help the whole team work together by serving as excellent communication aids and debugging tools and then (3) they help the team efficiently evaluate alternatives and gain good acceptance for the conclusions. Certainly I must note animation is not a substitute for the careful review and analysis needed to thoroughly debug a model and reach statistically valid conclusions from a stochastic simulation. This solid analysis definitely must be carried out by modelers, but the whole team will often not be a part of this. Dynamic animations can greatly help other team members understand the complexity and variability of the system interactions, this can result in their feeling much more comfortable with the statistical based conclusions. The active support, based on this visual evidence, by the non-modeling members of the team can be an enormous strength during the implementation phase. It is often their strong recommendations that really 'wins the day' with the key decision makers.

4.2 CASE 1: Evaluating Expansion Requirements

In this case the situation involved two large production sites (with a total annual capacity of over 200 million pounds) supplying five internal consuming sites plus a significant number of external customers. The business was doing well and a planned expansion was scheduled to come on-stream with a 20% annual capacity increase. There were people who believed this would require also adding 20% more railcars to handle the distribution logistics; others believed the needs would be smaller, and some who felt that because of the complexity the railcar expansion needs would be even higher.

Based on previous successes involving modeling work we were asked to join the logistics team to help them determine the railcar needs under a number of different scenarios. This was important to the business as a 20% increase would add close to 35 more cars to the fleet at a cost that could be $ 4.4 MM. In addition to the investment required for more railcars, a fleet increase would also increase the annual operating costs (more cars to be maintained, etc...), and there would be a significant dollar value tied up in the material that would be in these cars. Clearly there was incentive to find the best fleet size under different possible operating scenarios.

The final version of the model is quite large and detailed, but the essence is still quite well captured in Figure I (which shows a multi-plant system as discussed above). A very noteworthy point is that for this specific model we now have tied it to a separate Production Planning/Scheduling Model (a Mathematical Programming based Optimization Model) that produces a time phased Master Schedule which is used to drive the Logistics Simulation Model. Thus for each time period (week, month,...) we have both a planned production task for each of the two producing sites and planned demand quantities for each of the consuming sites, internal plants and external customers.

The simulation then is a more detailed dynamic model that is driven by the Master Schedule, but it also takes into account additional details such as equipment outages, transit time delays, plant car loading limitations, care maintenance activities/delays, and other sources of variability too detailed to explore further in this forum. Suffice it to say that each of the internal producing and consuming plants is modeled in some detail and each of the external customers is modeled focusing on the demand and the 'time spent at the location for car movements and unloading'. As you can well imagine it required a real a team effort to gather this information, and there was significant variability in the data so uncertainty was modeled carefully.

One of the areas that we found requiring a thorough team exploration is the one of dispatching policies. Specifically, the consuming plants have monthly production goals, finite storage capacity, periodic production slow-downs (for maintenance type activities), and they are supplied by railcars that have rather long and variable transit times. Specifically, for instance, we had good transit time data showing times ranging from 5 to 29 days along one of the frequently traveled routes. In general these routes pass through several rail yards and most of the time delays are at these "pass off" points; the cars can be significantly delayed when passing from one rail carrier to another. Now the dispatch policy question is simply "what set of conditions(rules) should cause us to send a car from a producing site to the consuming site?"

A set of overlapping and sometimes contradictory rules had evolved since this was a large and complex system (there were two large plants supplying several consuming sites). All of these were well intentioned and shared the goal of keeping the system running smoothly. However, taken altogether they sometimes did not produce the best results. Sometimes cars were sent to consuming sites that were already well supplied (with adequate material on hand and cars on the way) while other plants were running low. Some people felt more cars were necessary to insure smooth operations and timely distribution; others felt it was possible to meet service needs without buying additional cars.

In the model we explicitly treated: (1) periodic plant slow-downs and shut-downs (i.e. planned and unplanned), (2) plant car load/unload limitations (i.e. only x cars could be handled per day) , (3) car maintenance delays (i.e. cars were taken out of service for both planned and unplanned maintenance), (4) random car transit times (i.e. these were often not standard well-behaved distributions), (5) plant storage capabilities (these were varied via cases) and (6) car dispatch scheduling rules.

After running many scenarios the team concluded that (1) we did not need to add more railcars to the 200+ car fleet in order to meet distribution needs as the new capacity came on-stream, (2) that specific improvements in transit times and customer stay times were possible, and (3) that further fleet improvements were possible by using a better set of coordinated dispatch policies. These, plus other fleet utilization ideas, have improved asset utilization with a bottom line impact of \$4.4 MM of avoided new investment, \$225,000 of reduced operating cost, and the added benefit of avoiding the additional material that would have been sitting in those new railcars, plus the team gained a better understanding of the distribution system to the point that new opportunities on the "spot market" could be pursued (considered a significant benefit by the business).

4.3 CASE II: Fleet Consolidation

The second case involves a simpler system with one producing plant feeding four consuming plants (three internal and one external). These plants were for historical reasons fed with cars "owned" by the consuming plants. The key business driver for digging into this situation was that a significant number of the cars were reaching the replacement point and had to be replaced at a potential cost of several million dollars. The business leaders were open to a thorough analysis and recommendation on the most economic way to maintain service levels. Maintaining service was critical, but reducing the needed new investment was also important if possible.

Each of the five plants was individually modeled with focus on outages, rate variations, and car movement and load/unload capabilities. Much of the focus was on upset conditions at the plants and on the rail dispatch and transit time aspects. The cars were not all identical; due to differences in size (some cars were too big for some plants), filling attachments (some cars had unusual placement of the filling nozzles), some cars could not be used at some of the plants.

The model was fairly detailed so that cars were only sent on 'legal routes' where the car both met the size restrictions and had appropriate load/unload attachments. Car maintenance was incorporated and was actually based on car age and condition. In addition to the size/attachment, the model dealt with the policy issue of which cars were available to be used for each of the plants. This way cases could be run with different fleet consolidation rules. For instance, the cars servicing the external customer were kept in a separate pool in almost all of the cases analyzed; whereas, the cars originally dedicated to the other three plants were consolidated in different ways as the team evaluated different alternatives. It was possible when running cases to allow plant A's cars to be used by plant B, but not vice-versa.

In this model there was a strong focus on individual car characteristics and differences so each car was actually modeled as a separate entity. This way there was no ambiguity about which cars could service the different plants or pass over specific rail routes. In addition, this allowed for maintenance activity to focus on the individual cars which differed in age and total miles traveled.

Good work by the entire team on analyzing the system, determining car and plant differences, establishing innovative ways to deal with upset conditions, and modeling the entire system lead to great success. By consolidating part of the fleet, finding creative ways of handling unusual plant upsets, and improving fleet operations, the team was able to recommend a 25% fleet reduction and thus avoid a \$1 MM investment in new railcars.

5.0 CONCLUSION

Process Industry Operations are frequently large scale plants with continuity characteristics quite different from much of the discrete parts manufacturing and assembly plants. Further, the materials consumed and produced often need to be moved in dedicated railcars across long distances to service plants which do not have large storage tank buffers. This often results in large railcar fleets that are difficult to plan and manage due to many sources of variation and interaction between the system components.

Experience has shown that the first step toward success is to assemble a good team with logistics, plant operations, and modeling capabilities. This team should study the situation at hand, determine the specific needs and establish what are the opportunities and where are the system boundaries. They should carefully work to describe the entire system: what are the components, how do they operate, how do they interact, where are the sources of variability, what are the concerns of the people running the day-to-day operations? If we can adequately describe these logistics systems, then we know that we can indeed model them. The key is to build the minimum essential model: one that contains enough detail and realism to adequately address the questions at hand but not so detailed that it takes too long or costs too much. Finding that correct balance is often the key to successfully modeling these often complex systems. It is clear also that benefits flow from the increased operations understanding that the team develops even before the final model results are generated; these can be as significant as any specific results from the model. In the two cases we have briefly visited the modeling activity was the main guiding process, but the team learning's were the real key to success. In both cases it was the logistics people whose advice was the 'final convincer' in changing how the businesses were going to operate the systems. We have seen that the results were significant and certainly it can be said that "Modeling Does Pay".

REFERENCES

Askin, R. G., and Standridge, C. R. 1993. *Modeling and Analysis of Manufacturing Systems*. New York; John Wiley and Sons

Banks, J., and Carson, J. S. II. 1984. *Discrete-Event Simulation*. Englewood Cliffs, NJ; Prentice Hall

Harrell, C., and Tumay, K. 1996. *Simulation Made Easy: A Manager's Guide*. Norcross, GA; Industrial Engineering and Management Press

Law, A. M., and Kelton, D. W. 1991. *Simulation Modeling and Analysis*. New York; McGraw Hill

Schriber, T. J. 1990. *An Introduction to Simulation Using GPSS/H*. New York; John Wiley & Sons

Shannon, R. E. 1975. *System Simulation: The Art and Science*. Englewood Cliffs, NJ; Prentice Hall

AUTHOR BIOGRAPHY

Charles H. White is a Principal Consultant in Operations Research with the Process Engineering Section of the DuPont Company. He has 30+ years experience in Consulting and Modeling with a strong focus on Simulation and Mathematical Programming as applied to Business Planning, Plant Design, Research Guidance, Productivity/Asset Improvement, and Distribution Planning. Dr. White has degrees in Physics, Engineering, Mathematics, and Operations Research all from the University of Michigan. During his 30 years with DuPont he has also taught courses at the University of Michigan, the University of Delaware, DuPont, General Electric, and several short courses through the AIChE. He is a member of IIE, Informs, AIChE, for several years was the DuPont representative to the Materials Handling Research Center at Georgia Tech and was Business Chair for WSC in 1993.

Proceedings of the 1996 Winter Simulation Conference
ed. J. M. Charnes, D. J. Morrice, D. T. Brunner, and J. J. Swain

SIMULATION-BASED SUPPLY-CHAIN OPTIMIZATION FOR CONSUMER PRODUCTS

David R. Kalasky

Process Analysis & Improvement
1540 Stoddard Avenue
Wheaton, IL 60187

ABSTRACT

This paper presents an application of discrete-event simulation in modeling a supply-chain for consumer products. Optimization of the simulation model provides an efficient way to produce operating recommendations from the model. The use of ProModel and SimRunner in this application describes a modeling and optimization software suite which can be used for supply-chain planning and operation.

1 SUPPLY-CHAIN MODELING

The concept of supply-chain management has become a complex and important facet of manufacturing and delivery systems. Companies desire supply-chains that are lean on inventory, responsive to customer demands with global markets and suppliers. These three seemingly contradictory goals need to be monitored and input factors managed to anticipate and plan for any uncertainty in supply, demand and costs. Through observational data and simulation, companies have identified "bull whip" effects in supply-chains where small movements in any key link in the chain cause dramatic negative consequences rippling throughout the chain.

In this application, a linear program had been employed to provide cost models for supply-chain management which attempt to satisfy the three objectives of low cost, high service and inventory control. The company sited that the LP models did not account for variability and must be maintained and analyzed as input data changed. While simulation is a proven tool to account for variability and data fluctuations, the iterative nature of simulation experimentation has been a shortcoming which has limited the use of simulation for supply-chain modeling due to the number of possible combinations and permutations.

2 THE SUPPLY-CHAIN DESCRIPTION & GOALS

An international consumer product company was agreeable to developing the application and testing the suitability of simulating and optimizing their supply-chain. The model consisted of two factories, two distribution centers, four products and four shipment routes. Each factory was simplified into capacity constrained workcenters with two workcenters for products A & B and three workcenters for products C & D. Each workcenter had production rates (cases/shift) and schedules (shifts/day and days/month) which provided the plant-level simulation input data.

The company provided forecasted monthly production requirements for each distribution center. Due to market competition, the commodity nature of the products and the management policies of the company, the supply-chain had to provide 100% of the forecasted requirements. Service levels below 100% were not acceptable.

The company's goals in this effort were to:

- Build a model which accurately represents the data
- To run the model through a forecast year
- To derive production costs by factory
- To derive delivered costs by product/factory/distribution center and route
- To identify the least-cost sourcing solution (capacity constrained)

At this point in the project, sufficient information was available to build the ProModel simulation model of the plant operations. This first model provided an easy way to verify and validate the modeled plant operations. The model was used to evaluate the aggregate production requirements (for both distribution centers) before adding further detail to insure sufficient plant capacity. Another finding from the initial model was that both plants were required since neither plant had capacity to provide a single source of supply for any of the four products.

Confident that the model was performing correctly with sufficient supply capability, the two distribution centers, shipping delays and costing information were added to the model. This step-wise approach to incremental model building is recommended to provide rapid modeling, accurate results and customer-approved simulation models. Adding two distribution centers increased the complexity of the model and defined the number of variables to be used in the optimization. Production demands (N) ranged from 48K to 60K cases per month for each product. With four products (i), two plants (j) and two distribution centers (k) the number of ways to schedule this system was:

$$Nijk = 50K(4)(2)(2) = 800{,}000 \qquad (1)$$

The capacity of the distribution centers was considered as infinite and therefore did not enter into the capacity constraints. Also, there were no plant-to-distribution center rules other than varying shipping delays based on which plant was supplying which distribution center.

Costs were entered to produce each product at each manufacturing plant (dollars per case) for both fixed and variable costs. Freight costs including applicable taxes were also entered into the model. The animation layout is shown in Figure 1.

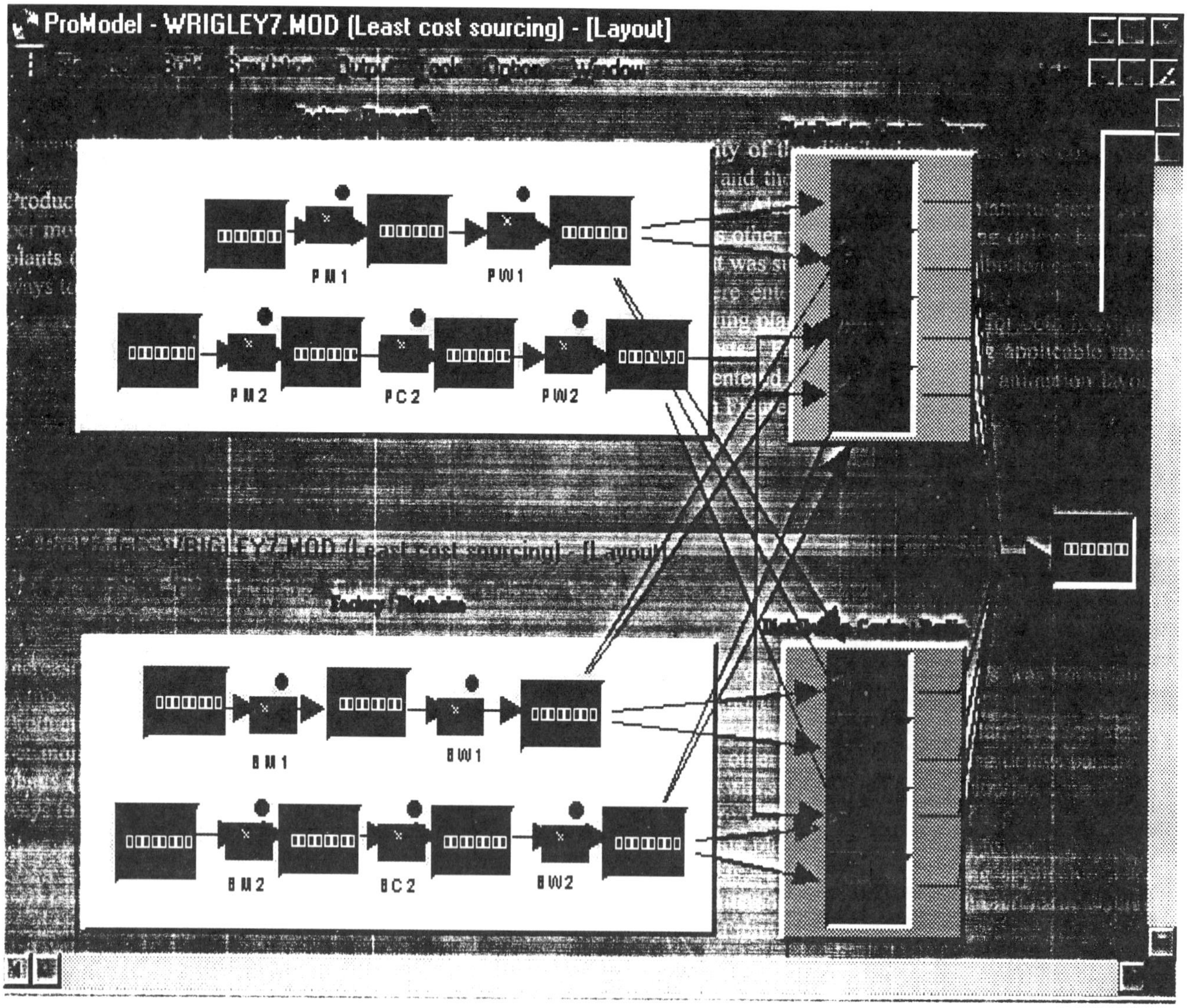

Figure 1: Layout of Supply-Chain Model

3 MODELING CONSIDERATIONS

The ProModel for Windows package provides many convenient features for modeling constrained supply chains. Also, the optimization capabilities and seamless linkage with SimRunner makes for easy, yet complex analysis to be performed. The model was designed to generate product demand with an attribute to designate the final distribution center location. Macros were established in the model to vary the amount of each product produced at each factory. By using macros, SimRunner was then able to run experiments on the supply-chain model, varying and eventually optimizing the four key variables.

Capacity was modeled using finite queues for each production line. Each of these queues had overflow capacity at the other factory. During the optimization, SimRunner could generate a schedule which exceeded the capacity of a particular production line. However, the alternate routing capability in ProModel allowed the model to maintain capacity constraints. Following production, the product was routed to the appropriate distribution center based on the predetermined attribute. Therefore, plant capacity and distribution center requirements were maintained.

As each case of product was produced, costs based on factory and product were assigned to the product category and the factory. Based on the routing chosen, transportation costs and taxes were calculated and summed as product was delivered to the distribution centers. Upon completion of each simulation experiment, a variable (TOTAL_COST) was calculated in the termination logic of the model. This was a sum of all the four (4) product category costs and represented the objective function for SimRunner to optimize (minimize).

4 OPTIMIZATION OF THE SUPPLY-CHAIN MODEL

Figure 2 shows the input for the SimRunner optimizer. The four (4) factors listed are ProModel macros which are varied during SimRunner's DOE and optimization. In this application, the objective function includes only one (1) term, TOTAL_COST as described above. It is possible to construct a multi-variable, weighted objective function.

In the scenario presented in Figure 2, one month's production was analyzed. SimRunner ran 246 experiments in optimizing the schedule. The results of the SimRunner optimization are shown in Figure 3.

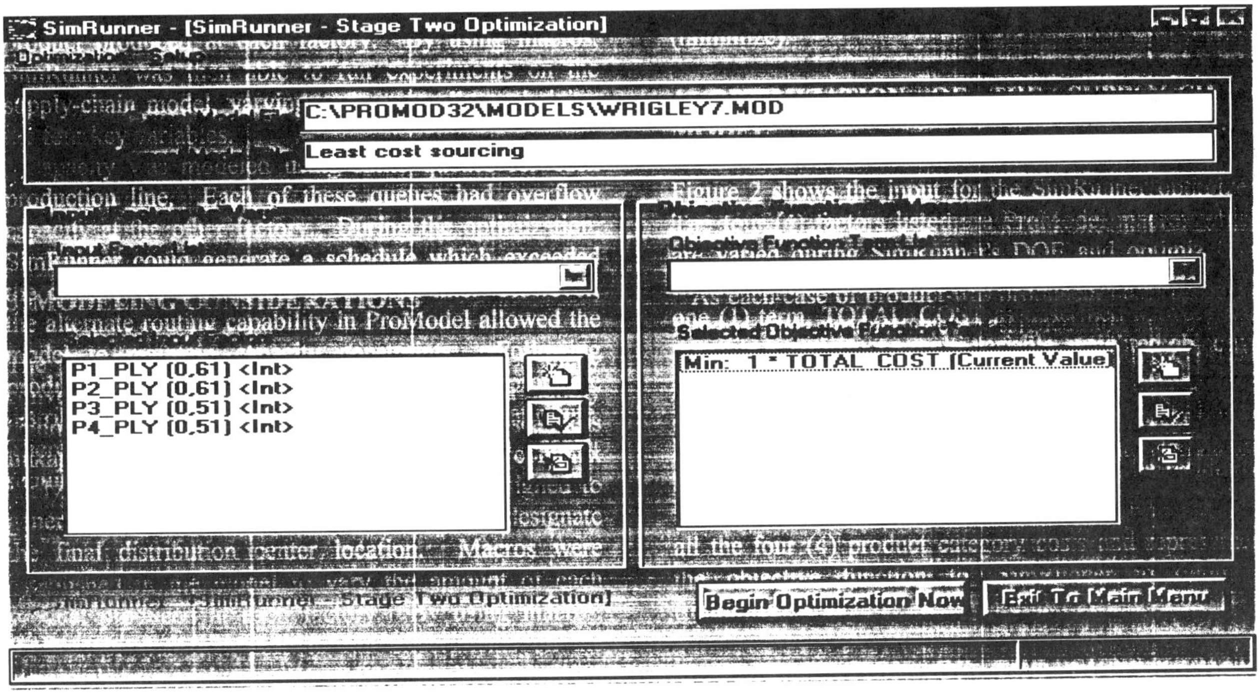

Figure 2: Data Entry Screen for the SimRunner Optimizer

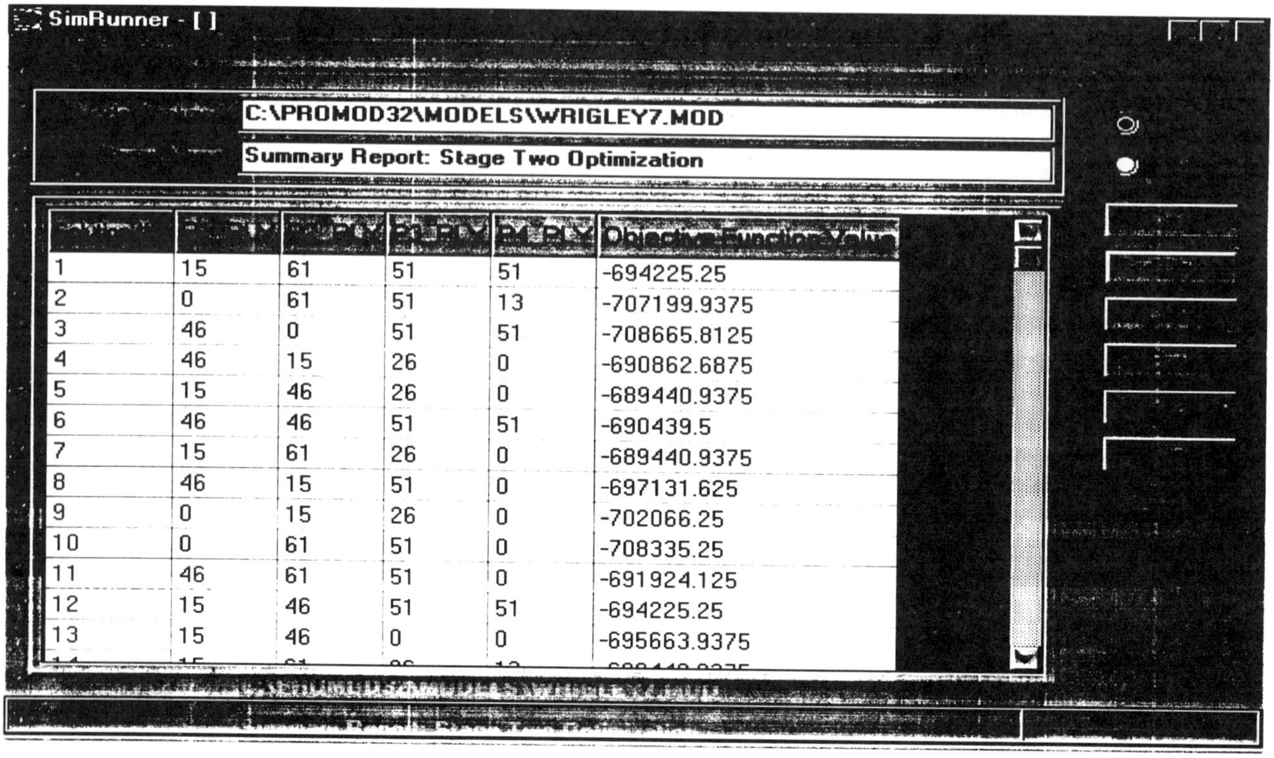
SimRunner - [ ]

C:\PROMOD32\MODELS\WRIGLEY7.MOD

Summary Report: Stage Two Optimization

			P3_PLY	P4_PLY	Objective Function Value
1	15	61	51	51	-694225.25
2	0	61	51	13	-707199.9375
3	46	0	51	51	-708665.8125
4	46	15	26	0	-690862.6875
5	15	46	26	0	-689440.9375
6	46	46	51	51	-690439.5
7	15	61	26	0	-689440.9375
8	46	15	51	0	-697131.625
9	0	15	26	0	-702066.25
10	0	61	51	0	-708335.25
11	46	61	51	0	-691924.125
12	15	46	51	51	-694225.25
13	15	46	0	0	-695663.9375

Model Optimized: C:\PROMOD32\MODELS\WRIGLEY7.MOD

Input Factors that were optimized:
P1_PLY (0 to 61)
P2_PLY (0 to 61)
P3_PLY (0 to 51)
P4_PLY (0 to 51)

Objective Function Terms that were measured:
1 * TOTAL_COST (Current Value)

246 experiments were run.

The best solution found:
P1_PLY = 27
P2_PLY = 52
P3_PLY = 28
P4_PLY = 42

The best objective function value is -678544.5

Figure 3: Results of the SimRunner Optimizer

The recommended (optimal) schedule for the first factory is as shown below:

- 27 units of product A
- 52 units of product B
- 28 units of product C
- 42 units of product D

Products A & B are produced on the same line in each factory. Products C & D are produced on the same line in each factory. Both of these combined production requirements exceeds the capacity of the lines at factory 1. Therefore, we must look to the ProModel output report to see how alternate capacity at factory 2 was allocated. In section 3 we explained this anticipated capacity issue and used the modeling and optimization features of the products to minimize modeling effort and optimization run-time.

5 VALIDATION AND OUTPUT ANALYSIS

The optimized model recommended that all of the capacity from factory 1, the low cost provider, be used to supply distribution center 1 for all four products. Additional supply from factory 1, up to its capacity, was used to supply products 2 and 4 to the second distribution center. The capacity from the second factory was used to supply the second distribution center for all products where the cost differences were minimal between the two sources. This recommendation matched the expected results, when calculated manually and noting that the factory capacity and distribution center requirements were met.

6 CONCLUSIONS AND OBSERVATIONS

This application of discrete event simulation and optimization proved that the combined technologies are a viable and useful tool for planning and operation of supply-chains. This approach allows for an accurate model with variability to be constructed and then optimized in an efficient algorithm. SimRunner establishes a designed experiment (DOE) to begin convergence on the optimal answer from the ProModel simulation model. The algorithm is adaptive in that the results from the DOE direct the final local optimization search.

While the example problem was small, it was a real problem with a fair degree of complexity (800,000 possible solutions) and was executed in a matter of minutes on a personal computer. The effort did require a competency in the use of the simulation package as well as familiarity with the optimizer to insure valid results. Akbay (1996) has a good description of other uses of optimizing discrete-event simulations in the areas of manpower planning, line balancing and scheduling. As computing power becomes more available and affordable, further uses of these technologies will be developed and existing static and deterministic methods will be displaced.

ACKNOWLEDGMENTS

Messrs. Dodman, Heinz and Moraski of the Wm. J. Wrigley Corporation provided the example problem and data. Mr. Harald Mutzke of Siemens and Mr. Peter Maslowski of EBERLE SYSTEME provided the baseline ProModel simulation model. Mr. Don Hicks of Decision Science provided assistance in the model development and SimRunner support. Microsoft Windows is a trademark of Microsoft Corporation. ProModel for Windows is a trademark of PROMODEL Corporation. SimRunner is a trademark of Decision Science, Inc.

REFERENCES

Akbay, Kunter S. 1996. Using SimulationOptimization to Find the Best Solution. *Industrial Engineering Solutions* 28: 24-29.

Decision Science, Inc. 1996. SimRunner Version 1.0 User's Guide. Ypsilanti, MI: Decision Science, Inc.

Harmonosky, Catherine M. 1995. Simulation-Based Real-Time Scheduling: Review of Recent Developments. In *Proceedings of the 1995 Winter Simulation Conference*, ed. C. Alexopoulos, K. Kang, W.R. Lilegdon, and D. Goldsman, 220-225. Institute of Electrical and Electronics Engineers, Piscataway, NJ.

Harrell, Charles R. and Jeffrey J. Leavy. 1993. ProModel Tutorial. In *Proceedings of the 1993 Winter Simulation Conference*, ed. Gerald W. Evans, Mansooreh Mollaghasemi, Edward C. Russell, and William E. Biles, 184-189. Institute of Electrical and Electronics Engineers, Piscataway, NJ.

Harrell, Charles R. and Kerim Tumay. 1995. *Simulation Made Easy a Manager's Guide*. Norcross, GA: Industrial Engineering and Management Press.

Kalasky, David R. 1995. Manufacturing Systems: Modeling and Simulation. In *Systems Modeling and Computer Simulation*, 2d ed. ed. Naim Kheir, 279-310. .New York: Marcel Dekker, Inc.

PROMODEL Corporation. 1995. ProModel Version 3.0 Manufacturing Simulation Software Reference Guide. Orem, UT: PROMODEL Corporation.

PROMODEL Corporation. 1995. ProModel Version 3.0 Manufacturing Simulation Software User's Guide. Orem, UT: PROMODEL Corporation.

Sargent, Robert G. 1994. Verification and Validation of Simulation Models. In *Proceedings of the 1994 Winter Simulation Conference*, eds. Jeffery D. Tew,

Mani S. Manivannan, Deborah A. Sadowski, and Andrew F. Seila, 77-87. Institute of Electrical and Electronics Engineers, Piscataway, NJ.

AUTHOR BIOGRAPHY

DAVID R. KALASKY is the regional manager for the PROMODEL simulation products based out of Chicago, IL. He received his BS degree in industrial engineering from the University of Kentucky and his MS degree in operations research from the Pennsylvania State University. David has been involved with simulation modeling and operation research all of his professional career. He has held engineering and management positions with Conoco, Westinghouse Electric and Systems Modeling Corporation prior to founding Process Analysis & Improvement, Inc. He is a registered Professional Engineer, an active member of ASQC and IIE, CPIM and instructor within APICS, invited author and speaker for SME, and was the business chairman for the 1990 Winter Simulation Conference. His publications include over 20 books, journal articles and proceedings papers in the areas of simulation, quality and process improvement.

Proceedings of the 1996 Winter Simulation Conference
ed. J. M. Charnes, D. J. Morrice, D. T. Brunner, and J. J. Swain

TEACHING INTRODUCTORY SIMULATION IN 1996: FROM THE FIRST ASSIGNMENT TO THE FINAL PRESENTATION

Peter Lorenz

Faculty of Informatics
Institute for Simulation and Graphics
Otto von Guericke University of Magdeburg
D-39106 Magdeburg, Germany

Thomas J. Schriber

School of Business
Computer and Information Systems
The University of Michigan
Ann Arbor, Michigan 48109-1234, U.S.A.

ABSTRACT

Rapid developments in computing and simulation are providing continuing challenges for the design and delivery of courses in discrete-event simulation. Such courses not only need to keep up with relevant developments, but also have the potential for taking advantage of some of the developments in the structuring of the courses. For example, animation can be the topic initially taken up in a course that covers both simulation and animation. We report here on a teaching approach in which a simulation course starts with the topic of animation, then introduces the programming-language subset of a modeling language to support animation at a fundamental level, then returns to animation to introduce more sophisticated aspects of that material, then goes back to the modeling language to investigate the modeling of queuing networks, and so on. Introducing animation first has the effect of capturing intense student interest and dedication at the outset of the course and has the additional benefit of helping students visualize and deal with parallel processes of the type they will be studying and modeling later in the course. Recent technological developments are further reflected by including a World Wide Web-based support system in the course.

1 FACTORS INFLUENCING THE TEACHING OF SIMULATION COURSES

The teaching of simulation is influenced by factors both external to and internal to the subject itself. Some of these factors are indicated below.

1.1 External Factors

The *audience* consists of people with differing backgrounds in computing, statistics, stochastic processes and practical work experience. Members of the audience often have a wide range of interests, from the general to the highly specialized. The audience can consist of undergraduate or graduate students majoring in such areas as engineering, computer science, business, health care, and public policy studies, and can include on-the-job professionals.

The available *time* for teaching a simulation course varies from one or several hours for a self-contained tutorial to one or two full-semester courses for students specializing in simulation.

The available *software* ranges from general programming tools to highly developed simulation and animation software with graphical user interfaces.

The available *hardware* ranges from standalone PCs to workstations with access to a global network of computers.

The skill and experience of the *instructor* is another factor that bears importantly on the teaching process.

1.2 Internal Factors

Internal factors involve topics intrinsic to the subject of simulation. Some of these internal factors are indicated below, in no particular order of importance.

Random Numbers are basic components of most discrete-event simulation models. The simulation specialist should know how random numbers are generated, tested, and transformed.

Event Handling with suitable algorithms and data structures is central to discrete-event simulation. It is easy to understand how a "future events list" is managed. It is not as easy to understand such things as the way time-tied events are resolved during the course of a simulation.

Input Modeling requires some skill in statistics and is a pre-requisite for building valid models.

Model Construction is as much an art as a science. The process of model construction is strongly supported by some simulation systems, less so by others, and the instructor has to decide what degree of support to make available to the students.

Verification and Validation of simulation models must be understood to support interpretation of modeling results in the context of real systems.

Experimental Design is an important skill for problems in which even a moderate number of experimental parameters is involved.

Output Analysis involves mathematical statistics and is an essential component of a simulation study.

Result Presentation is an area of increasing importance. Presentations typically include graphs and animations.

Simulation Software is available for supporting various phases of a simulation study. An understanding of the types of simulation software available and the development of operational capabilities with representative software is central to the study of simulation.

Simulator Construction is a subject appropriate for computer science students studying simulation.

The preceding topics, although not all inclusive, do reflect the content of representative simulation textbooks. Individual textbooks each seem to have their own bias, however, covering some of the topics in depth and providing only brief or no treatment of other topics.

Apart from textbooks, the teaching of simulation has also been the subject of discussion in some recent Winter Simulation Conferences, e.g. Romeu 1995; Withers et al. 1994.

2 TECHNOLOGICAL INFLUENCES OF THE LAST DECADE

The last decade has witnessed major new developments and trends in simulation methodology, in computing and communication, and in economic and business processes. Which technological trends impact simulation and should be reflected, included or exploited in a simulation course? Which trends are only of transient interest? We comment on these matters in the following subsections, while remaining open to constructive discussion and debate in this regard.

2.1 Presentation: Graphics and Animation

Graphical presentation of results has long been an essential aspect of simulation. Pseudo graphic presentation of simulation results dates back to simulation systems of the 1960s. In 1976 the PC appeared on the scene, but even by 1986 most PCs still provided only an alphanumeric display. Now in 1996 it is perhaps impossible to buy a PC without graphical display capabilities. And it is almost impossible to conduct a simulation study that does not include graphical presentation of results.

Animation is intimately involved with simulation. Animation depicts simultaneously moving objects and simultaneously running processes, and so animation algorithms are closely related to the basic underlying concepts of simulation. In fact, animation can be regarded as a sister or a child of simulation. Animation is a favorite candidate for inclusion in a simulation course.

2.2 Visual Interactive Modeling

The increased graphics performance of computers has been the basis for establishing Visual Interactive Modeling, which is "the use of icons on the screen to represent the system in a way that closely resembles the physical system" (Seila 1995), and involves an approach to modeling in which icons are linked together to represent network flows.

If a modeler has a VIM simulator with the predefined classes of objects needed in a particular model and with simple connections between the instances of the classes, then it can be relatively easy to build a model and still have time remaining to work on other aspects of a simulation project. VIM simulators can work well for specialized applications and can promote fast modeling of standard situations. They seem however not to be well suited for a general simulation course: they are not really *general* simulation systems and they hide essential parts of model behavior by hiding the supporting internal processes. A general simulation course should (we believe) promote understanding of these internal processes.

2.3 Multi-Purpose Models

Computer performance is growing by a factor of 10 every 7 years. That is a factor of about 1000 in the last 20 years! However, the increasing sophistication and complexity of simulation software might mean that in some cases simulations take more time to execute than in the past, despite the dramatic increases in computer performance. Nevertheless, fast simulators do exist and it is appropriate to ask: "How we can take advantage of fast, efficient simulators to enhance the quality of simulation modeling?"

One potential answer to this question lies in the possibility of building multipurpose models with a scalable degree of detail. Rather than following the old paradigm of simply "building models as detailed as necessary for a special purpose," we now have enough computing and modeling power to build models which answer not only predefined questions but also other (future) questions which cannot be foreseen by the model developer.

There is a general tendency in the technology toward more multipurpose hardware and software. But most simulation models are still only single-purpose. They could alternatively be multipurpose and be equipped with features that might be needed for a range of special-purpose uses. For example, a given multipurpose model could be useful for gaining insight into characteristics of the modeled system, for presenting aspects of system behavior to a variety of interested parties, and for training people who work within the system.

2.4 Business Processes and Global Economy

The last decade and probably to a much greater extent the next decade was/will be characterized by global operating organizations, global products and global markets. That fact should influence the content of simulation models and studies.

Material flow in a single factory is typical of the type of problem on which simulation has focused in the past. In contrast, the global flow of material and information might be a central subject of current and future simulation models. Should simulation courses include study of large scale and global models?

2.5 Communication Networks

Communication networks are growing dramatically. The Internet has already started to influence our entire society. How should the simulation community react to this? Where are there new ideas forthcoming for simulation software and simulation applications involving the World Wide Web? How should the Web be used for teaching support and how should it be included in a simulation course?

2.6 The Challenge

How should developments such as those listed above be reflected in simulation courses? It is clear that there are more questions at this time than there are answers. Over time it will be necessary for many simulation educators and practitioners to help contribute answers to questions impinging on future simulation instruction.

3 AN EXAMPLE OF ONGOING COURSE EXPERIMENTATION

This section describe procedures and experiences resulting from a teaching approach that is being developed and used to seek constructive responses to some of the preceding questions and issues. The approach is one that has been followed in three settings:

1. A one-term course offered in 1995 to computer science undergraduates at the University of Wisconsin at Stevens Point;
2. A one-term course offered in 1995 and 1996 to computer science graduate students at the University of Magdeburg; and
3. Tutorials on "Simulation and Animation" offered at EUROSIM '95 in Vienna; at a 1996 conference at the University of Magdeburg (Lorenz and Schriber 1996); for industrial people in Wisconsin in 1995; and for upper high school ("Gymnasium") instructors in Nordrhein-Westfalen (Germany) in 1996.

The content and the concept of the courses indicated under 1 and 2 above and their underlying teaching approach are called the *Reference Course* here. The following subsections are closely related to this Reference Course, which is based on a course duration of about 14 or 15 weeks.

3.1 Starting with Animation

A simulation course might reasonably start by explaining goals, terms, history and applications, and by giving introductory examples. But what comes next?

Historically, many simulation courses have then continued with the topic of *random number generators*. But this topic is now often deferred to a later course position (and moved to later chapters or even to appendices in textbooks) and might only be given very brief treatment or no treatment at all in a first course.

An alternative approach is to proceed with *basic algorithms and data structures* for simulation, explaining alternative control approaches, e.g., polling; related waiting; and so on.

A third possible approach is to proceed with *animation*. Moving the shapes of various objects around on a computer screen is a new experience for most students. Arranging for the smooth movement of objects along paths involves using a polling algorithm with small time steps and this goes hand in hand with basic aspects of simulation.

Proof Animation software (Henriksen et al. 1995) is well suited for this phase of a simulation course. Working with Proof Animation, the following basic concepts of animation can be introduced and explained:

1. Drawing or importing and using a scene or background or layout;
2. Defining classes and objects (which in turn provides a possible starting point for object-oriented programming and simulation);
3. Using the animation clock;
4. Creating objects, placing them on the layout, and moving them; and
5. Creating, writing and reading animation trace files and layout files.

No more than one lecture is needed to explain these basic concepts and bring students to the point of being given a first assignment. Here are examples of potential Assignment #1 exercises:

1. Objects are to be placed on the screen at various starting positions and then moved smoothly to final resting positions. Plan and implement the sequential creation and simultaneous movement of the objects.
2. Display a triangle using three different types of dots for the three sides.
3. Trace out a sine curve in a coordinate system.
4. Place 100 objects taken from five different object classes into a square.
5. Trace out a circle with small objects from a variety of object classes.
6. Build a chessboard, using red and white squares.
7. Make up your own assignment involving the movement of objects!

Some students come up with really good ideas when

given a chance to compose their own Assignment #1. They often devise better exercises than those suggested by the instructor. For example, Figure 1 shows a snapshot taken from the animation of the assembly of a jigsaw puzzle developed for Assignment #1 after the first animation lecture. (Figures shown here look much better in their original color than in the black and white of the Winter Simulation Conference Proceedings.)

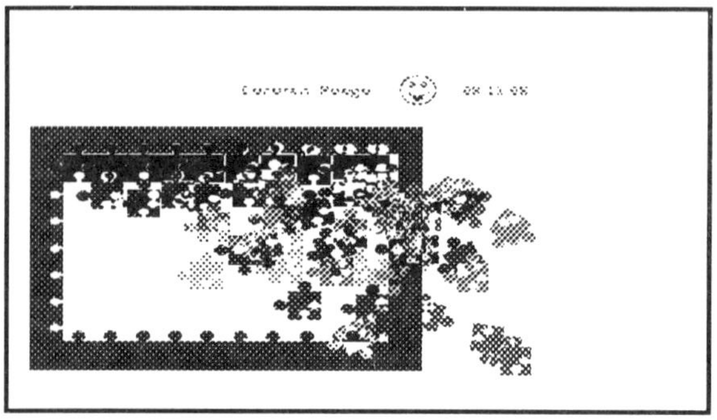

Figure 1: An Example from Assignment #1 - "Jigsaw-Puzzle Assembly Animation" (Carsten Poege, Magdeburg, Germany, 1995)

Starting with animation not only has methodological and educational advantages. It also has the advantage of exciting a high level of student interest and involvement with the material early in the course. Most students react very favorably to lectures on animation and look forward with anticipation to doing their animation homework.

In the first assignment the students learn to use software to create and write animation trace files. It is useful in lecture to show examples for trace file generation with a variety of tools: with a text editor, for example, and with Basic and/or Pascal and/or C++. *But it is recommended that students also use whatever software the instructor has chosen for building simulation models later in the course.* A simulation system with elementary algorithmic and file handling capabilities is needed to support Assignment #1.

Note that no simulation-specific features, functions or capabilities are needed to carry out Assignment #1

3.2 Intermediate Animation and Basic Simulation Tools

The next segment of the Reference Course deals with intermediate animation and basic simulation tools. Intermediate animation involves the dynamic display of messages, bars and plots. On the simulation side, the items of choice for providing information to be animated can include random number generators and procedures for the transformation of uniform random numbers into random numbers sampled from theoretical or empirical distributions.

Assignment #2 can then be an exercise in using simulation and animation tools of the indicated kind. First, all students are given the preliminary assignment to write their name and the current date on the screen and display a running analog clock (with hour and minute hands). Individual problems of the following type are then assigned, one for each student:

1. Generate exponentially distributed random numbers over time and display the relative frequencies of realized values in a dynamically changing histogram as well as in a table. Ask the user for a random number seed, a sample size, and the end points of the frequency class intervals.
2. Repeat 1, but with alternative theoretical or empirical distributions.
3. Display the plot and the corresponding tabled values for a trigonometric function of choice. Ask the user for the range of angles to be covered and the angle increments to be used.
4. Display the plot and the corresponding tabled values for a specified polynomial. Ask the user to provide the domain of the function to be displayed and the size of argument increments to be used.

Figure 2 shows a snapshot from a student solution to a typical Assignment #2 problem.

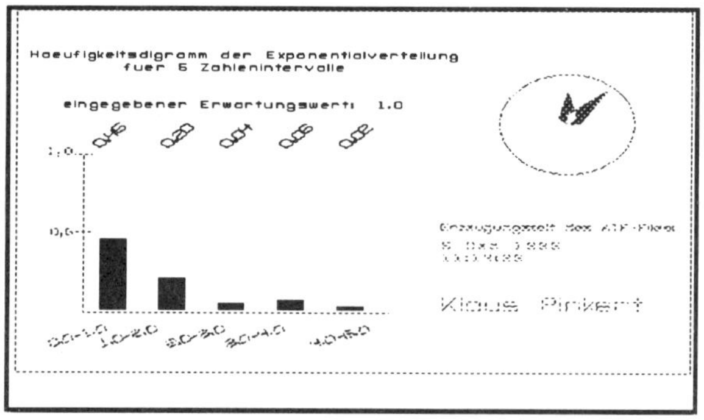

Figure 2: An Example from Assignment #2 - "Frequency Histogram, Exponential Distribution" (Klaus Pinkert, Magdeburg, Germany, 1995)

The objective in Assignment #2 is to gain practice in displaying plots and histograms while using the algorithmic capabilities of the simulation software to produce the information being animated. After completion of the assignment, students are familiar with operational aspects of handling the animation and simulation software. They are then ready to jump into a more traditional area of discrete event simulation: the simulation of queuing systems and networks.

3.3 HTML-Based Lecture Notes (WWW Support)

Customized materials developed and written by an instructor to support a course are traditionally produced on paper and distributed to students in that form. A relatively new alternative to this approach is to use HTML (the HyperText Markup Language) to make such materials available to students on the World Wide Web. Some of the considerations involved in taking this approach have been articulated by Antchev et al. 1996.

HTML-based lecture notes and course materials are used in the Reference Course. The course support system can be found at:

http://isgnw.cs.uni-magdeburg.de/~usergrps/sim/simos2/docs/welcome.htm

German and English versions of these HTML-based lecture notes are under continuing development. The support system is intended to escort and guide students throughout the entire course. Above and beyond completely replacing the former paper-based course materials, the HTML-based course support system has these characteristics:

1. It is easily accessed and handled by students, providing them with needed course materials and updates at low (or no) cost.
2. It is a hypertext system with links between related subjects.
3. It facilitates easy and timely modification, extension, and correction of materials by the instructor.
4. It is a tool for providing a suite of graduated simulation models that reflect increasing degrees of modeling complexity. The models are provided for more than viewing purposes. These models can be freely edited *in situ* on their WWW page and simulations can be performed with the models without leaving the Web browser.
5. The materials can be easily and inexpensively enhanced with colored pictures, experimental frames, and so on.

Among the Web support materials, for example, is the full text of all assignments that exist to date for the Reference Course. These assignments are in German at:

http://isgnw.cs.uni-magdeburg.de/~usergrps/sim/simos2/docs/assign.htm

and can be found in English at:

http://isgnw.cs.uni-magdeburg.de/~usergrps/sim/simos2/edocs/assignmt/ass.htm

Currently the simulation models provided as part of the support materials are based on GPSS/H (Banks et al. 1995; Schriber 1991) as a general purpose simulation system. But other text-based simulation systems could be used as alternatives. Good choices for using the Web are those simulation modeling systems that have a complete alphanumeric interface.

Figure 3 shows a snapshot of the introductory simulation model included in the support materials for the Reference Course. This is a model for the one-line, one-server system known as Joe's Barbershop. Anyone and everyone can see, modify, and run this model at:

http://isgnw.cs.uni-magdeburg.de/~usergrps/sim/simos2/docs/sa4/sa41/welcome.htm#model

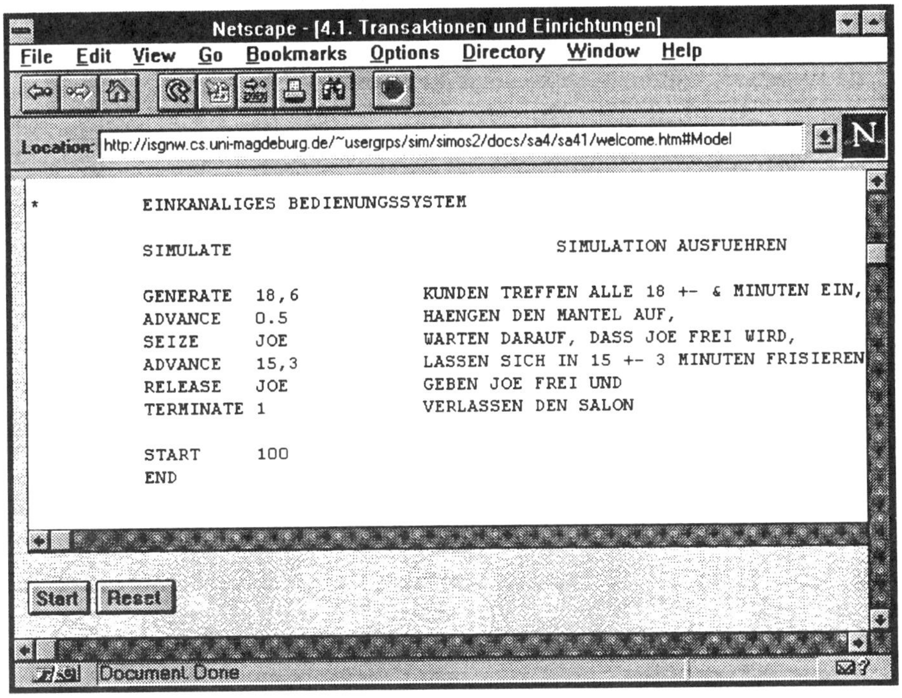

Figure 3: A Snapshot of a Web Page
Showing the Introductory Editable and Executable Simulation Model

Joe's Barbershop is the first in a series of about 20 simulation models currently included in the Web materials for the Reference Course. Entree to these models is provided at:

http://isgnw.cs.uni-magdeburg.de/~usergrps/sim/simos2/docs/appendix/progexpl.htm

The Reference Course does not have full WWW-based support for *animation* as of this writing (July, 1996). Ongoing developments in Web technology lead to good expectations for the near future in this regard, however (Dorwarth et al. 1996).

3.4 Modeling Queuing Networks

As suggested above, simulation of queuing networks is introduced in the Reference Course with Web-based accompaniment of live exercises. The students by now have experience in using the algorithmic statements of the simulation software. This means they only have to learn a few simulation-specific object classes and statement types at this point to move into queuing-network simulation. And, thanks to their earlier work with animation, when students leave their first lecture on queuing system simulation they already have specific operational insights into techniques for animating such systems.

The WWW-based course support does currently contain an example of a single-server animation to reinforce and extend student insights into animation. The example is available at:

http://isgnw.cs.uni-magdeburg.de/~usergrps/sim/simos2/docs/sa4/sa41/singlpr.htm

Taking the single-server model as a starting point, students build models for more interesting queuing networks in Reference Course Assignment #3. Here are typical examples of Assignment #3 problems:

1. Modify the model of Joe's Barbershop by including one of the following conditions:
   a. One barber (for men only) and two hairdressers (for women only) work in the shop. Women do not wait for a particular hairdresser.
   b. In a shop with one barber and two hairdressers, the men can be served by the hairdressers too.
2. A fast-food drive-in has three windows. The first window is for placing orders. The second is to pay for the food and the third is to pick up the food. Between consecutive windows there is enough room for five cars. Cars arrive every 4 +/- 3 minutes. It takes 2 +/- 1 minutes to order, 1.0 +/- 0.5 minutes to pay for and 3 +/- 2 minutes to pick up the food. How many cars arrive in a day? What can you say about the queues in front of each window?
3. A cable car moves between two points to take people up and down a mountain. The car holds 20 people. It starts moving only if it contains at least 15 people. The trip takes 15 minutes each way. How often does the cable car move in a day if people arrive every 4 +/- 2 minutes at one point and every 6 +/- 3 minutes at the other point?

Figure 4 shows an animation snapshot taken from an animation produced by a student for Assignment #3.

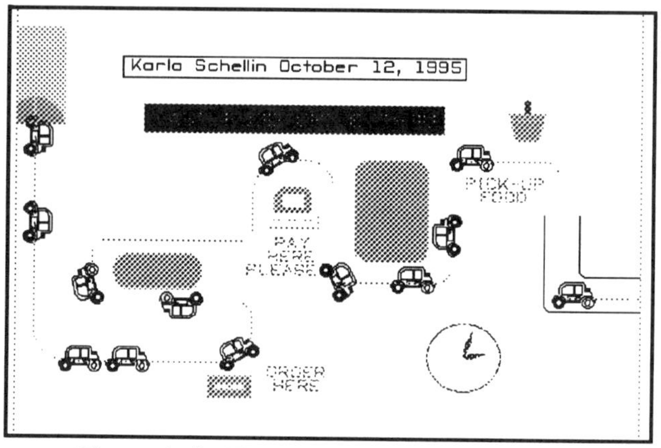

Figure 4: An Example from Assignment #3 - "Fast-Food Drive-Thru" (Karla Schellin, Stevens Point WI, 1995)

3.5 Final Presentation

After finishing Assignments #3, the students are prepared to begin work on their Final Presentation (final assignment) in the Reference Course. The Final Presentation is a simulation project. The students are entering the last segment of the Reference Course at this point and are poised to learn how to model more complicated queuing situations and use animation to demonstrate system operation. There is strong student motivation now to absorb the content of the concluding lectures to support work on the Final Presentation.

The lectures in this last segment of the course provide an introduction to modeling complex control strategies and the corresponding control mechanisms and underlying logic provided by the simulation software. This insight into "how simulation software works" (Schriber and Brunner 1995) is important for correctly capturing the logical complexities of the real or envisaged system in the associated simulation model and animation.

Work on the Final Presentation is demanding. The students have to be innovative and imaginative and "fill in the gaps" in terms of material that has not been given explicit treatment in the Reference Course. The following text is a lightly edited extract from the Web-based specifications for the Final Presentation. The full text can be found at:

http://isgnw.cs.uni-magdeburg.de/~usergrps/sim/ simos2/edocs/syllabus.htm

The following collection contains exercises. You can select one of these exercises as the subject of the Final Presentation. (See "Grading Policy" in the Syllabus.) You must build a simulation model with an animation for the Final Presentation.

The Final Presentation will be evaluated as follows:

1. *Simulation and animation models (60%)*
2. *Written documentation (20%)*
3. *Verbal presentation (10 minutes; 20%)*

The main criteria for evaluating the models are the correctness of mapping the given system into the model, readily comprehensible code, and the visual quality of the animation.

Every Final Presentation has to include the following:

1. *A Proof Animation presentation file that describes the system being modeled, the input data, the running processes and the results.*
2. *A display of current simulated time on an animated clock.*
3. *A title page with your name, a title for the assignment, and the date of the presentation.*

If work on the Final Presentation is limited to the points mentioned above, the work will be evaluated with a maximum grade of "C." At least two additional components taken from the following are needed for the maximum grade to be a "B":

1. *The source code must be well documented by building easily-read comments into it.*
2. *Multiple experimental runs must be specified in an experimental frame that provides values for model parameters and random number seeds.*
3. *A report file presenting the results in an easily readable format must be included.*

Additional considerations used to evaluate the animation are these:

1. *Have plots and bars been used for the presentation of results?*
2. *Does the presentation file show a variety of animation views?*
3. *Does the presentation includes images (PCX files)?*

The written documentation must contain these sections:

1. *Assignment*
   This section contains the description of the assignment. This section should also indicate the requirements for animating the model.
2. *Analysis of the input data*
   The given input data must be analyzed and statistically fitted. How did you transform the input data into computer-readable form? In case no input data is given, you have to create your own data by e.g. observing the system.
3. *Model conception*
   This section provides the description of the model. Describe the mapping of the real objects to object classes in GPSS/H.
4. *Model description*
   This section contains a listing of the GPSS/H model and explains the structure of the model.
5. *Animation description*
   This section describes the mapping of model elements into animation system object classes and objects.
6. *Description of the experiment*
   In this section the content and organization of experiments are to be described.
7. *Evaluation and discussion*
   The results of the experiments are to be discussed and evaluated here.

A snapshot taken from the animation for a Final Presentation is shown in Figure 5.

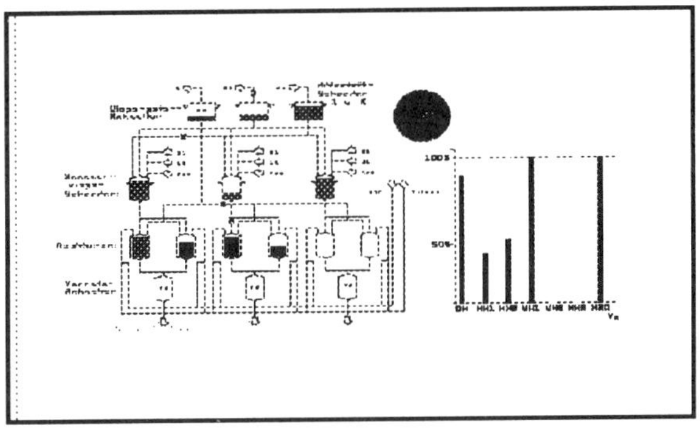

Figure 5: An Example from a Final Presentation - "Polymerization Reactors in a Chemical Plant" (Marcel Goetze, Magdeburg, 1996)

4 BENEFITS OF STARTING A SIMULATION COURSE WITH ANIMATION

Some of the benefits realized from starting an introductory simulation course with animation are summarized here.

The benefits for the instructor include these:

1. Intense student interest and dedication is captured from the very beginning of the course.
2. Animation provides a straightforward and interesting way for the instructor to gain a preliminary impression of a student's work.
3. A high level of student interest develops in the possibility of taking an advanced simulation course.

Among the benefits for the student are these:

1. The slope of the simulation/animation learning curve in the course is flattened substantially.
2. Knowledge and insights about the *visualization* of concurrent processes are gained.
3. Valuable experience is obtained in the *presentation* of systems, models, and processes.

5 ADVANCED COURSES

The graduate of the Reference Course has worked intensely on model construction and animation and is capable of building simulation and animation models at an intermediate level. But s/he lacks knowledge about the theoretical foundations of simulation. Many approaches and techniques have been explained and learned strictly by example, without benefit of a general or theoretical foundation. A student coming out of the Reference Course is poised to take an advanced simulation course. The student has been primed for such a course, having encountered many questions that could not be answered precisely (for lack of time) during the Reference Course.

Included among potential advanced-course topics are:

1. How do *Random Number Generators* work? What are the measures of generator quality? Which quality attributes might influence the validity of a model?
2. How can Real System Data can be transformed into *Input Model Data?* Which statistical methods and which programming tools are available for modeling input distributions?
3. What approaches and what body of experience can be brought to bear on *the construction of large, multipurpose models* of complicated systems?
4. What methods are available for *Verification and Validation* of simulation models?
5. How can *experiments best be designed and conducted* to investigate system behavior with models characterized by large numbers of parameters?
6. What methods and tools are available for the *Analysis of Output*?

6. CONCLUSION

The pattern experimented with in structuring an introductory discrete-event simulation course that starts with animation and then alternates between simulation and animation topics has been described. Experience gained in teaching such a course has been reported, and glimpses of several examples of student work produced in such a course have been presented. Inclusion of a World Wide Web-based support system for such a course has been described. The success of this teaching approach has been so encouraging that continued offerings of introductory courses along this line are planned.

REFERENCES

Antchev, K., N. Luhtalahti, J. Multisilta, and S. Pohjolainen. 1996. A WWW microworld for mathematics. In: *Proceedings of ED-TELECOM 96* (World Conference on Educational Telecommunications), 5-10. Charlottesville, Virginia: Association for the Advancement of Computing in Education.

Banks, J., J. S. Carson, and J. N. Sy. 1995. *Getting Started with GPSS/H*, Second Edition. Annandale, Virginia: Wolverine Software Corporation.

Dorwarth, H., H. Graessner, and P. Lorenz. 1996. Simulation and animation in the WWW. In: *Fachtagung Simulation und Animation Magdeburg 1996, ASIM-Mitteilungen Heft Nr. 54*, 173-184, Vienna, Austria: Arbeitsgemeinschaft Simulation.

Henriksen, J., N. Earle, J. Lopacki, and E. Tucker. 1995. *Using Proof Animation*, Second Edition. Annandale, Virginia: Wolverine Software Corporation.

Lorenz, P., and T. J. Schriber. 1996. Simulation und animation tutorium. In: *Fachtagung Simulation und Animation Magdeburg 1996, ASIM-Mitteilungen Heft Nr. 54*, 419-420, Vienna, Austria: Arbeitsgemeinschaft Simulation.

Romeu, J. L. 1995. Simulation and statistical education. In: *Proceedings of the 1995 Winter Simulation Conference*, 1371-1375. LaJolla, California: Society for Computer Simulation.

Schriber, T. J. 1991. *An Introduction to Simulation Using GPSS/H.*. New York, New York: John Wiley & Sons.

Schriber, T. J., and D. T. Brunner. 1995. Inside simulation software: how it works and why it matters. In: *Proceedings of the 1995 Winter Simulation Conference*, 110-117. LaJolla, California: Society for Computer Simulation.

Seila, A. 1995. Introduction to simulation. In: *Proceedings of the 1995 Winter Simulation Conference*, 7-15. LaJolla, California: Society for Computer Simulation.

Withers, D. H., E. Yucesan, and D. W. Kelton (panelists). 1994. Teaching simulation. In: *Proceedings of the 1994 Winter Simulation Conference*, 1378-1381. LaJolla, California: Society for Computer Simulation.

AUTHOR BIOGRAPHIES

PETER LORENZ is a Professor in the Institute for Simulation and Graphics within the Faculty of Informatics at the Otto-von-Guericke University of Magdeburg in Magdeburg, Anhalt-Sachsen, Germany. He teaches discrete simulation, computer animation and graphics. His research interests include layout-based simulation model generation, advanced Web supported teaching concepts and applications of simulation and animation in manufacturing, logistics and traffic.

THOMAS J. SCHRIBER is a Professor of Computer and Information Systems at The University of Michigan. Long active in the Winter Simulation Conferences, he teaches a range of subjects while doing research and consulting in discrete-event simulation. He is a member of ASIM (the German-language simulation society), DSI, IIE, and INFORMS.

Proceedings of the 1996 Winter Simulation Conference
ed. J. M. Charnes, D. J. Morrice, D. T. Brunner, and J. J. Swain

TEACHING THE FUNDAMENTALS OF SIMULATION IN A VERY SHORT TIME

Ingolf Ståhl

Stockholm School of Economics
Box 6501
S-113 83 Stockholm, Sweden

ABSTRACT

This paper informs about the development of the micro-GPSS system on the basis of feed-back from some 5000 students during a period of two decades, providing a stream-lined simulation system which within ten class room hours of teaching make students prepared to write fairly advanced simulation programs of practical interest.

1. INTRODUCTION

Simulation has proved to be a very powerful tool not only in engineering but also in business administration. Against this background it is surprising that most business schools and quite a few engineering schools do not give their students any substantial amount of teaching in simulation. The main reason for this it that student time is an increasingly scarce resource in a very crowded curriculum. Often only a handful of hours can be spent on simulation. This, in turn, often leads to the teaching of simulation being limited to a very broad overview, without allowing for any "hands on" experience. Most teachers with this approach believe that the learning of a simulation language or package would take too much time. If any computer is used, it is limited to the input of data into an already existing model.

This approach has several draw backs compared to allowing the students to actually work with a simple simulation package on concrete problems. In this latter way the student can actively learn the whole process of doing simulation for a problem, from delimiting the actual problem, formulating the question to be answered by simulation, gathering data, outlining the simulation program graphically, coding the program, verifying, validating and documenting the program, running the program a sufficient number of times, doing a statistical analysis for drawing significant conclusions, and presenting the results in a form suitable for a potential user, with a focus on the implementation aspects. In particular, it is important to allow the students to be involved in some kind of modelling, since the modelling aspect, in my view, is of fundamental importance for understanding simulation.

To allow for such an approach to be feasible in a short time, a very simple simulation system is required. Let us already at this stage exemplify the time span that we are discussing, namely ten class room hours and all together at most a week of student time including reading and project work. (We are hence not discussing a full semester course, involving perhaps all together a month of student time.) The question is then what can be accomplished in such a very short time.

I shall in this paper discuss what we have done at two universities in Sweden: namely one business school, the Stockholm School of Economics (SSE), and one technical university, the Chalmer's Institute of Technology (CIT).

At SSE all 300 students will in their third year study simulation, having 10 class room hours of instruction in the micro-GPSS system to be discussed below, spending roughly 10 hours on reading and preparing shorter exercises and spending about 20 hours on a larger GPSS program. This course of ten hours, called *Integrative Computer Applications,* has the particular aim of integrating elements of Operations Research, Corporate Finance, Accounting and Statistics. The focus is on a student project, roughly 100 blocks in size, regarding some kind of cash flow forecast and involving decisions on the purchasing, production, inventory and borrowing policy of a smaller corporation. An example of such a project is presented in detail in Ståhl (1996a).

At the department of Transport Technology at CIT around one hundred students annually do roughly the same thing, also using micro-GPSS, with the main difference that the project work is more directly focused on transportation technology.

2. WHY GPSS?

The question then naturally arises why we in the teaching discussed above use a version of such an "old-fashioned" language as GPSS. I shall divide this discussion into two parts: a. Why we use a GPSS-based language, to be discussed in this section and b. why we use our own special version of GPSS, micro-GPSS, to be discussed in the next section. The discussion in this section hence applies to all versions of GPSS, be it GPSS/H, GPSS/PC, GPSS/R, HGPSS, GPSS/V, GPSS/VI, SIMPC, GRAMOS-GPSS, micro-GPSS, etc.

As I see it, the following factors are favorable for GPSS when it comes to choosing a simulation system for an educational task like that one specified above.

1. **Ease of learning**. The very short time available requires a special simulation language. We can hence directly rule out the alternative of doing simulation in a General Programming Language, like C++ or Pascal, or a simulation language that is built on a GPL, like Simula, SLX or YANSL. This point is further strengthened by the poor background in programming, in particular as regards the business students

2. **The GPSS World View**, involving temporary components, like customers, which move through the system to be served by various permanent servers provides a very problem oriented approach which has proved to have great benefits both from the modelling point of view and of making GPSS fun to learn.

3. The provision of a great amount of **automatic output**, like block statistics, server statistics, waiting line statistics, tables, etc. This automatic provision is very important since the novice will generally not know what kind of output is relevant. The study of this standard output is an important part of the learning process. Furthermore one avoids having to spend much of the initial time on print commands.

4. The block diagrams make it easy for students to study, discuss and document the logic of a program. The teaching can also be more visually oriented. One can start by looking at the main structure of model before looking at the details of the program syntax. The availability of standardized block symbols is a great advantage. Compared to other simulation systems with block symbols, GPSS has the advantage of having many blocks coming in pairs, with one block being the mirror picture of the block doing the "opposite" thing, e.g. SEIZE and RELEASE. The GPSS symbols are easy to read and generally clearly distinct from each other.

5. GPSS is a **"fun" language**. Simple things can be done in a simple manner. (As discussed below micro-GPSS focuses even more on this.) One important factor in this respect is the unity of program and experiment conditions which is an advantage for the novice, but perhaps a disadvantage for the professional user. Among other special features that are regarded by some as disadvantages, but are advantageous when it comes allowing the student to do interesting things at a very early stage, one can mention the concept of a facility, providing service to only one customer at a time and hence not needing a capacity definition.

6. GPSS allows for very **compact programs**. With only a few blocks one can produce quite interesting programs. The advantage of GPSS in this respect is self-evident when comparing with languages like SIMSCRIPT, MODSIM and SIMULA, but it appears that also compared with languages like SIMAN and SLAM, GPSS will generally lead to shorter programs. The shorter code has the advantage among other things of making the programs easier to "overview".

7. GPSS has, due to its long history, the great advantage that there are **many program examples and much literature** available, in fact over a dozen text books, (Ståhl 1993b and Schriber 1995), including what many teachers regard as the best text book ever on simulation, Schriber's famous Red Book from 1974. The availability of well-documented programs is important for the novice, when doing a project work, since he is then more likely to find a text book example to build upon.

8. There are reasons to believe that GPSS with its more than 35 years' tradition is **better debugged** than other simulation languages. The constructors of one GPSS version can use other GPSS versions for checking correctness. Furthermore, there has been a much higher degree of scientific scrutiny of GPSS than of other simulation languages. The internal mechanism of GPSS is better documented and therefore better understood. The user is therefore less likely to run into strange and unwanted effects in certain situations. (Schriber and Brunner 1995 and Ståhl 1996b) .

9. Moderately limited versions of GPSS (allowing for programs of up to 150 blocks) are available at very **low cost** ($20 - $40) in several GPSS versions (e.g. GPSS/H, and micro-GPSS). The student therefore has the additional incentive of working in a simulation system that he knows he can run on his own PC and continue to use also after the simulation course is over.

10. GPSS is a **good starting point** for later moving to many other simulation systems The existence of GPSS similar packages in object oriented languages like Simula (e.g. GPSSS and DEMOS) makes a switch to some object oriented languages simpler. Since SIMAN contains many basic structures similar to that of GPSS, several of our GPSS students at CIT later learnt SIMAN much faster than their fellow students not having taken the GPSS course.

11. Finally it should be stressed that GPSS is a **General Purpose** Simulation System and that for simulation problems, not only focused on manufacturing, but e.g. combining physical and financial streams, and where one wants to investigate the full implications of uncertainty by many runs, GPSS is often more suitable than systems that are very good for animation of production systems, but otherwise fairly weak.

3. WHY MICRO-GPSS?

The development of the micro-GPSS system has been a gradual process, starting already in 1978. The first versions of micro-GPSS were virtually proper subsets of GPSS V, with the main difference being the internal mechanism (Ståhl 1996b). Having taught 200 - 300 students a year for almost 20 years, in courses with a decreasing amount of time allotted to them, but with the ambition of still providing the students with the same amount of simulation knowledge, we have, however, step by step changed the original system into the present one, simplifying it to make it easier to learn and use.

This development of micro-GPSS can be seen as a reaction against the development of increased complexity of the GPSS language over time. One measure of this increased complexity is the ever higher number of block types. GPSS I (1961) had some 20, GPSS III (1965) around 40, GPSS/360 (1967) 44, GPSS V (1970) 48, GPSS/H (version 1, 1977) 58 and GPSS/H (version 2, 1989) 62 block types. This development towards more and more block types has in turn lead to increased redundancy in the way that several blocks carry out the same job. This complexity has also made GPSS increasingly more difficult to learn and has lead to students learning a steadily decreasing share of the full language and therefore making more logical errors. Although many of the developers themselves (Gordon 1979 and Henriksen 1983 and 1985) have raised critique against this development, it has yet continued due to the developers' interest in compatibility with older versions of GPSS. Such compatibility is of interest to old users who have already done a lot of programming in earlier versions of the language, but it is of no interest to the novice who desires to have as easy a learning process as possible with regard to the goal of being able to do a certain kind of simulation.

During the process of development, we have spent much time reflecting on the design. Questions have arisen not only with regard to how to cut out a pure subset of GPSS/360, GPSS V or GPSS/H, but also whether or not we should change this or that feature of GPSS. I have in the process tried to formulate the general principles that have guided us in this work (Ståhl 1993a). The main ones are the following, divided into three groups, dealing with A. ease of learning, B. ease of use and C. safe programming.

A. Ease of learning

A1. The main aim is that the students after a very short time shall be able to do simulation projects of practical value. First, as mentioned above, we have only a very limited time of teaching at our disposal.. This has lead to our aim that one should be able to learn most of micro-GPSS in less than ten hours so that one after that can do some student projects of practical interest. Some time should also be left for the issues of collection and evaluation of input data, the main principles of experimental design, statistical analysis of the output, aspects of implementation, etc., i.e. issues that are left out in many courses where all time is spent on the mechanics of a difficult-to-learn simulation language.

A2. As we want the students to focus on modelling (and experimentation), and not on syntax detail, the language should be such that one in the course does not have to learn a new block type every time that a new and different thing shall be done. It is from a pedagogical point of view often preferable that the new aspects can be handled using already known block types, even if the program thereby becomes a few blocks longer.

A3. When students frequently make the same mistake, one must always consider the alternative of changing the syntax instead of forcing them to learn strange features that, for example, might depend on hardware limitations of computers in the 60's or on pure mistakes made by developers in the early years of GPSS. Since we, in contrast to GPSS/H, are not bound to compatibility with earlier GPSS versions, we have in this way over the 20 years of teaching allowed our 5000 students to change the language syntax. Some examples are given in section 5 below.

A4. We have in micro-GPSS been very keen to keep and increase the “fun aspect” of GPSS. If the simulation system is fun to learn from the beginning, this will in itself provide strong incentives for further learning. It is here important that the language provides a possibility for the students to do interesting things after only a very short period of learning. Preferably the students should already after one or two class room hours be able to write some non-trivial simulation programs, i.e. students should be able to do simple things in a very simple fashion. One should not sacrifice the ease of introduction for the sake of having sophisticated features for the advanced user. One should in particular restrict what appears as unnecessary details, e.g. avoid commas that are not absolutely essential. (See section 5 below). It is important to "encourage users to forge ahead and experiment rather than present barriers that lead to discouragement" (Banks 1995). It should be mentioned that this positive aspect of learning has given the micro-GPSS courses increasingly favorable student ratings, much higher than for comparable computer courses.

A5. We have also developed micro-GPSS with the aim of facilitating the teaching in computer labs as well as facilitating self-studies in front of the computer, instead of learning by lectures in ordinary class rooms or text book studies. In this connection it is important that micro-GPSS allows several programs to be run in a stream, with both program listing and output presented one screen at a time, without the student having to leave the GPSS system.

A6. micro-GPSS was developed with the aim of being so simple that it could be completely covered in a pedagogical manner, with many examples in a book of reasonable size, say a maximum of 400 pp, and hence with a moderate price. This was accomplished with Ståhl (1990). The student should not have any need for a difficult-to-read manual in order to find features not covered in the text book or in class.

A7. For micro-GPSS it has been very important not to presuppose any pre-knowledge of programming. This is one difference between how micro-GPSS and GPSS/H are positioned. As mentioned in Henriksen (1985), the focus of GPSS/H is on persons with considerable programming experience. Most of our business students do not have any programming knowledge except possibly some very rudimentary knowledge of (Visual) BASIC. Even for my technical students at CIT, who some year earlier had programming in Pascal, this presupposition has appeared to be advantageous.

B. Ease of use

B1. It should be very easy to do the coding of the programs. Since there are great variations between students as regards their preferences for input systems, one should provide a choice between conventional coding using an editor and coding by choices in a menu, either GUI based or text based. For easy coding by an editor, micro-GPSS has in contrast to other GPSS versions a **completely** free format so that students do not have to worry about the fact that certain words have to start in a certain column; every line can start in column 1. In contrast to GPSS/H, no distinction is made between upper-case and lower-case letters. As regards coding by use of a menu in an interactive dialogue mode, there are two systems for micro-GPSS, our own GPSSMENU and GPSSGUI for DOS and GPSSEDIT for Windows (Nywall 1992). Finally, these input systems aim at making syntax rules self-evident or easily available, so that the student does not have to refer to a text book..

B2. It must be easy to read the program listing and the output. micro-GPSS does not provide a lot of advanced output that the novice does not know how to read and would find confusing. In contrast to GPSS/H the output always fits into an 80 column screen. It should also be easy to read the extended program listing provided by the system. micro-GPSS provides a neat and easy-to-read program listing, with an automatic line-up of operators and operands, independent of the appearance of the original code

B3. Since it in teaching is important to focus on block diagrams, it is first of all essential that every block type has a corresponding block symbol. This is the case of micro-GPSS, but not of GPSS/H. Furthermore it is in this context important to have facilities for the automatic generation of block diagrams, in micro-GPSS by GPSSDIA, as well as systems for coding by clicking in a menu of block symbols (see B1).

B4. Since we also want to focus on experimentation, it is essential that it is very easy to make replications of the runs and also to make a statistical analysis of these repeated runs. Running a program 20 times with different random numbers is done by SIMULATE 20.

C. Safe programming

C1. Closely related to ease of learning, but also to ease of use, is the principle of **safe programming.** We want to minimize the risk of logical errors, i.e. that that program produces unwanted and erroneous output. In this way a great amount of student time spent on debugging can be saved. If the simulation language is made as safe as possible with regard to logical errors, this also reduces the need for an extensive debugging system which, in turn, requires a lot of student time to learn. To secure safe programming we want to stress the "Lead us not into temptation" principle, implying that the simulation language should not be excessively permissive, allowing constructions that with a significant probability lead to logical errors. It is better that an unsuitable construction leads to a syntax error message and execution stops right away than have it lead to a difficult-to-find logical error. micro-GPSS is therefore much less permissive than ordinary GPSS; e.g. servers must have symbolic names, parameter usage is restricted, etc.

C2. It is also important that the simulation language has an extensive error trapping system with as clear error codes as possible. This error system is very developed in micro-GPSS. All of my 5000 students have been asked to report on errors with no or an unclear error message. This has lead to constant improvements.

C3. Closely connected with the idea of C1 is the aim that students should not run into surprises and unexpected logical errors due to not having learnt the full language. It is of special importance that the language does not have any reserved words, in particular reserved words that lead to strange logical errors. We must avoid problems such as, e.g. that SEIZE XID1 and SEIZE XJD1 in GPSS/H lead to completely different results. Unless the student knows that XID1 is a reserved word with special meaning in GPSS/H, he can make a serious logical error.

4. SELECTION OF MICRO-GPSS BLOCKS

On the basis of the principles discussed above we have developed a GPSS version with much simpler syntax. For example, instead of the 62 block types of GPSS/H we have arrived at only 22. These 22 block types can be seen in figure 1 below. The question is then how we arrived at these 22 block types. This is discussed in detail in Ståhl (1992b), but the main ideas are as follows:

With few block types there is less to learn. It is also in line with the "modelling" principle of point A2 above. Speaking, to the contrary, for the inclusion of more block types is the idea that short and compact programs in general are preferable to longer programs as regards solving a specific task. With a new block type, one can sometimes with one block do what would require several blocks to do with the earlier existing block types. This is one factor that has lead to the continued increase in the size of the GPSS language mentioned above.

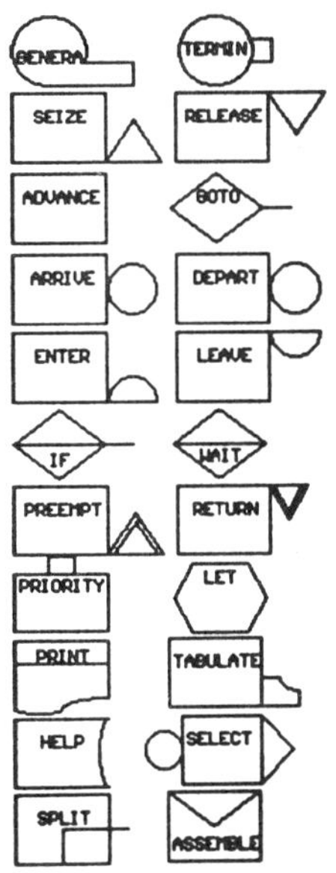

Figure 1: Menu of micro-GPSS blocks

A critical issue is then when to adopt a new block type, if this leads to shorter programs. Our general principle was that we started with a quite small "compulsory" core set of block types. We would extend this set with another block type only if this adoption would lead to significantly shorter programs as regards applications that appear to have a frequent usage. When establishing what kind of applications have a fairly frequent usage, the only alternative available to us was to look at the examples in existing GPSS text books. If applications were not dealt with in these text books they were not regarded to be in frequent usage. Furthermore, we did not adopt a new block type, like LOOP of "old" GPSS, if the use of one block of such a type could easily be replaced with two blocks of already adopted types.

Another consideration speaking for limiting the number of block types was our interest in easy-to-use programs for coding via a dialogue menu, with, for example, block symbols on which to click. With 62 block types like in GPSS/H, the menu itself would hardly fit readably into one computer screen. The 22 block symbols of micro-GPSS fit nicely into the left fourth of the screen as seen in figure 1, produced by GPSSGUI. A final consideration favoring the limitation of the number of blocks is the possibility for the micro-GPSS user to create his own new block types from one or several existing block types (Ståhl 1993c).

The set of 22 block types has proved to be quite powerful and micro-GPSS programs are in general not longer than corresponding programs in other GPSS dialects. The reason for this is that the increase in the number of blocks due to fewer block types is compensated by some new micro-GPSS features. The most important one is that for collecting statistics in a queue, e.g. in front of a facility, we can just add ',Q'. Thus the micro-GPSS block SEIZE SAL,Q does the same thing as the following blocks in ordinary GPSS: QUEUE SALQ, SEIZE SAL, DEPART SALQ.

We have hence been able to rewrite 99 percent of the programs in leading GPSS text books with almost the same amount of code (at most 20 percent difference). For example, for the 29 programs in Schriber's famous "red book" from 1974 the average number of blocks used is virtually the same (18.6 in Standard GPSS and 18.8 in micro-GPSS).

5. SOME OTHER MICRO-GPSS FEATURES

It should first be mentioned that of the 22 block types, five do not exist in "old GPSS", namely ARRIVE, GOTO, IF, LET and WAITIF. ARRIVE, the natural counterpart of DEPART, replaces QUEUE and has (almost) the same syntax; GOTO replaces TRANSFER, but with a simpler syntax. Instead of TRANSFER ,BYE (with BYE as B operand) and TRANSFER .4,,BYE (with BYE as C operand), micro-GPSS uses GOTO BYE and GOTO BYE,.4, with BYE always as A operand.

LET replaces ASSIGN and SAVEVALUE (and BLET in GPSS/H). IF replaces TEST and GATE blocks **with** an address, while WAITIF replaces TEST and GATE blocks **without** an address. In contrast to TEST, IF works with a "straight logic", namely that we proceed to the address in the C operand, if the stated condition is **true** (not false as in "old" GPSS). In micro-GPSS we write IF P1<0,BYE instead of TEST GE P1,0,BYE as in "old" GPSS versions, when we go to BYE if P1<0.

Among the many improvements as regards the block syntax , one can here mention that the PRINT statement is much more versatile in micro-GPSS, allowing us e.g. to write PRINT 'PROFITS ',X$PROF. Matrix handling as well as reading and writing the contents of the matrix to or from a file is also handled in a very compact and simple manner in micro-GPSS. Furthermore, the production of an output graph can be handled by a single block. Standard Numerical Attributes can be given easy-to-remember names. The SELECT block has a more understandable syntax.

Also as regards control statements, micro-GPSS requires a lot less learning. Instead of the 34 control statements of GPSS/H, micro-GPSS has only 13, where CAPACITY replaces STORAGE for the definition of capacity and SEEDS replaces RMULT, since it changes the seed, not the multiplier. SIMULATE has, as mentioned, an A operand determining the number of runs. Micro-GPSS also has facilities for automatic statistical analysis (e.g. of Student's *t*-distribution) and for optimization. It has several built-in statistical functions. Micro-GPSS also allows for animation by a very simple interface to Proof Animation (Ståhl 1992a). and for simple tracing and step by step animation through the block diagram.

The whole of micro-GPSS syntax is presented in Ståhl (1990). A shorter introduction to micro-GPSS is provided in Ståhl (1995) for which there is a tutorial diskette with restricted versions of the micro-GPSS system and GPSSDIA as well as 49 program examples. This contains roughly the material covered in the ten hour course presented above.

Due to these and several other simplifications, micro-GPSS is **much** easier to learn than traditional GPSS. We have thus with micro-GPSS been able to cover the same material in 10 hours that required 22 hours when using "old" GPSS. Micro-GPSS has also in tests compared favorably as regards learning time with other systems such as SIMAN and WITNESS.

6. TWO PROGRAM EXAMPLES

To give some idea of the simplicity of micro-GPSS we shall present two program examples. The first program, which will produce block statistics, facility statistics, queue statistics and a table on waiting times so that we, for example, can say what percentage of the 30 patients had to wait more than two hours for the doctor. I have not found any other simulation language which with so little code can do so much. The program is run 5 times with different random number streams. This is a program that students can write already after two hours of teaching.

```
simulate    5
qtable      doc,0,10,20
generate    18,6
seize       doc,q
advance     25,5
release     doc
terminate   1
start       30
end
```

Program 1

The next program will illustrate a problem that appears to be much more difficult to program in many "modern" simulation systems.

"At a small store customers arrive at a rate of every 4 to 10 minutes (7 $\pm$ 3 minutes; assume a rectangular distribution for all time data.) In the store there are two people working, Boris and Naina. Customers first go to Boris and choose the goods and find out how much they have to pay. This takes between 3 and 7 minutes. Next they go to Naina to pay for the goods and obtain a receipt. This also takes between 3 and 7 minutes. Finally, they return to Boris to pick up their goods after presenting the receipt, which is then stamped. This takes between 1 and 3 minutes. There is one waiting line in front of Boris and one in front of Naina. Customers returning to Boris to pick up the goods have to start at the end of this line again.

The program should be written so that times spent by customers in the store can be easily measured. How many customers will spend more than 15, 20, 25 minutes, etc. in the store? Assume that the store is closed after eight hours and that the mentioned statistics refer to customers having left the store at this closing time. Calculate also by repeated runs whether there is any significant risk (e.g. happening in one out of ten cases) that a customer has to spend more than an hour in the store."

```
simulate  20
qtable    store,0,5,20
generate  7,3
arrive    store
seize     boris
advance   5,2
release   boris
seize     naina
advance   5,2
release   naina
seize     boris
advance   2,1
release   boris
depart    store
terminate
generate  480
terminate 1
start     1
end
```

Program 2

This problem has been solved in micro-GPSS in ten minutes by students who have studied micro-GPSS for three hours. I have at earlier WSCs asked sales representatives of various modern animation based simulation packages to solve this problem and it has taken them over half-an-hour to do so in their system. The difference is that in GPSS you only need a simple sequence of three SEIZE-ADVANCE-RELEASE, while these other systems require Boris to be located at **one** spot and the program must then for each customer keep track of whether she comes to Boris for the first or the second time, something which is difficult, obviously not only for the novice.

7. HOW TO PROCEED FROM MICRO-GPSS

As discussed above, students can proceed fairly far in simulation after ten class room hours of micro-GPSS. The most clear proof that our business students after these studies can do fairly interesting simulation is given by the project work mentioned in section 1 above. The students here write a GPSS program involving on average one hundred blocks regarding the operations of a smaller corporation. An example of such a project with the micro-GPSS program is presented in Ståhl (1996a).

The question is then what long term benefits the students can have of this. Even if the students never do any more simulation there are clearly several valuable lessons to be learnt. One clear benefit is that simulation illuminates the connection between the actual physical processes and its counterparts in the accounting system. This has a pedagogical merit and has appreciably increased our business students' understanding of the relationship between accounting procedures and the physical and financial flows in a company. Another benefit is that the students in the course get a clearer understanding of concepts such as pseudo-random numbers, exponential, normal and Erlang distributions, experimental design, etc. than they could get by just reading about it. They also become better informed buyers of simulation services, knowing that simulation in certain situations can be a powerful, but yet fairly inexpensive, tool.

Many of our students have, however, proceeded in simulation. We have at SSE an optional longer course in simulation, focused on a larger project done in a Swedish corporation. In this course, the remaining parts of micro-GPSS are covered as well as Proof Animation with the micro-GPSS interface to this. There is also a focus on verification, validation, documentation and implementation. Quite a few of these student projects have been continued on a consultant basis or as a Master's thesis. It should be mentioned that all students have stayed with micro-GPSS. There has obviously not been any need to migrate to GPSS/H, in spite of the fact that we supply a program GPHM which can translate almost any micro-GPSS program into GPSS/H code.

Most of these projects have been fairly small in size, 200 - 300 blocks. The longest micro-GPSS program is 500 blocks. This gives an idea of how we see micro-GPSS positioned. Since micro-GPSS, as well as GPSS/H, in contrast to e.g. HGPSS (Claeys *et al.*, 1995), does not allow any hierarchical structure, we see micro-GPSS positioned as not only a pure teaching device, but also as suitable for rapid prototyping in simulation, for smaller models in line with Woolsey's "Quick and dirty" approach. When it comes to larger models than the 300 - 500 blocks discussed here, object oriented simulation languages like SIMULA or MODSIM III are clearly preferable. However, before committing oneself to a major effort in such a language a rapid prototype in GPSS can be a suitable starting point, as in cases like that of the Swedish ice-breaking operations (Jennergren *et al.* 1995).

8. FURTHER INFORMATION ON MICRO-GPSS

Micro-GPSS is available on the PC and Macintosh as well as on SUN and VAX workstations.

A demonstration diskette with scaled down versions of the micro-GPSS interpreter and GPSSDIA and with 50 program examples is available free of charge, together with a short tutorial, from the author. Professor R. Born (1995) has produced a very pedagogical computer slide show to accompany this tutorial.

More information on micro-GPSS and its use in the discussed ten hour course at the SSE is also available on the web at WWW.HHS.SE/5e/7784.htm.

REFERENCES

Banks, J. 1995. Semantics of Simulation Software. In *OR/MS Today*, December 1995, pp. 38 - 40.

Born, Richard. 1995. *Computer Slide Presentation to Accompany Ingolf Ståhl's Simulation Made Simple with micro-GPSS.* Northern Illinois University. DeKalb.

Claeys, F. , H. Vangheluwe and G.C. Vansteenkiste. 1995. HGPSS: A Hierarchical Extension to GPSS. In *Proceedings of the 1995 European Simulation Multiconference*, ed. M. Snorek, M. Sujansky and A. Verbraek. Prague.

Gordon. G. 1979. The Design of the GPSS Language in Adams, R.N. and Dagramici, A. (eds) *Current Issues in Simulation*. Wiley, N.Y.

Henriksen, J.O. 1983. *State-of-the-Art GPSS.* Paper presented at the 1983 Summer Computer Simulation Conference, 1983, Vancouver, B.C, Canada.

Henriksen, J.O. 1985. *The Development of GPSS/85.* Paper presented at the 18th Annual Simulation Symposium, Tampa, Florida.

Jennergren, L.P., L. Lundh, U. Törnqvist and S. Wandel. 1995. Icebreaking Operations in the Northern Baltic. In Miser, H.J. (ed) *Handbook of Systems Analysis: Cases.* Wiley, N.Y.

Nywall, J. 1992. *GPSSEDIT Manual.* Mimeo. Karlstad College, Karlstad, Sweden.

Schriber, T. J. 1974. *Simulation Using GPSS*, Wiley, N.Y.

Schriber, T. J. 1995. Perspectives on Using GPSS. In *Proceedings of the 1995 Winter Simulation Conference*, ed. C. Alexopoulos, K.Kang, W. Lilegdon and D. Goldsman, 110-117. Piscataway, New Jersey: Institute of Electrical and Electronics Engineers.

Schriber, T. J. and D. T. Brunner. 1995. Inside Simulation Software: How It works and Why It Matters. In *Proceedings of the 1995 Winter Simulation Conference*, ed. C. Alexopoulos, K.Kang, W. Lilegdon and D. Goldsman, 451-456. Piscataway, New Jersey: Institute of Electrical and Electronics Engineers.

Ståhl, I. 1990. *Introduction to Simulation with GPSS: On the PC, Macintosh and VAX,* Prentice Hall International, Hemel Hempstead, U.K., 1990.

Ståhl, I. 1992a, Animation with micro-GPSS and Proof In *Visualisierung und Präsentation von Modellen und Resultaten der Simulation,* ASIM, Heft. Nr. 31.

Ståhl, I. 1992b, *Principles behind the Design of an Easy-to-Learn Simulation Language: Implications for the Selection of Block Types.* EFI Research Paper 6485.

Ståhl, I. 1993a, Principles Behind the Design of an Easy-to-Learn Simulation Language. In Roberts, R.S. and S. Monroe (eds) *Simulation Applications in Business Management and MIS.* SCS, San Diego.

Ståhl, I. 1993b. GPSS Will Prevail - Some Reasons for the Resilience of the GPSS Simulation Ideas. In *GPSS-Users' Group Europe - Gruendungveranstaltung,* ASIM Heft nr. 36, Magdeburg.

Ståhl, I. 1993c. Recent Developments of micro-GPSS. In *GPSS-Users' Group Europe - Gruendundgveranstaltung*, ASIM Heft nr. 36, Magdeburg.

Ståhl, I. 1995. *Simulation Made Simple with micro-GPSS: A Short Tutorial with Seven Lessons,* Stockholm School of Economics, Stockholm.

Ståhl, I. 1996, Simulation of the Business Operations of a Small Furniture Company. In Javor, A., A. Lehman and I. Molnar (eds) *Modelling and Simulation,. ESM '96.* Budapest

Ståhl, I. 1996b. *Steps Towards a Better Internal GPSS Mechanism.* In this volume.

AUTHOR BIOGRAPHY

INGOLF STÅHL is Professor at the Stockholm School of Economics, Stockholm, and has a chair in Computer Based Applications of Economic Theory. He was visiting Professor, Hofstra University, N.Y., 1983-1985 and leader of research project on inter-active simulation at the International Institute for Applied Systems Analysis, Vienna, 1979-1982. He has taught GPSS for twenty years at universities and colleges in Sweden and the USA. He has on the basis of this experience led the development of the micro-GPSS system. He is also consultant in simulation to Swedish banks and industry.

Proceedings of the 1996 Winter Simulation Conference
ed. J. M. Charnes, D. J. Morrice, D. T. Brunner, and J. J. Swain

SIMULATION FOR COMPUTER SCIENCE MAJORS: A PRELIMINARY REPORT

Ruth Silverman

Department of Computer and Information Science
University of the District of Columbia
Washington DC 20008
and
Center for Automation Research
University of Maryland
College Park, MD 20742-3275

ABSTRACT

With the support of the NSF Grant listed below, the author is revising and restructuring an existing simulation course designed primarily for senior computer science majors by: 1) developing an integrated set of laboratory exercises based on computer science topics using commercially available software (GPSS/H); 2)incorporating these materials into a formal laboratory manual along with related computer science reference materials and instructions in the use of the software; 3) implementing a pilot course using this manual together with a single text in the theory of simulation; 4) preparing a syllabus and a detailed annotated course outline for the instructor, keyed to the manual and the text. The materials developed will be flexible and highly modular allowing their adoption or adaptation at other institutions.

1 INTRODUCTION

With the support of the National Science Foundation under Grant #DUE- 9554863, the author is revising and restructuring an existing simulation course designed primarily for senior computer science majors by: 1) developing an integrated set of laboratory exercises based on computer science topics using commercially available software (GPSS/H); 2)incorporating these materials into a formal laboratory manual along with related computer science reference materials and instructions in the use of the software; 3) implementing a pilot course using this manual together with a single text in the theory of simulation; 4) preparing a syllabus and a detailed annotated course outline for the instructor, keyed to the manual and the text. The materials developed will be flexible and highly modular allowing their adoption or adaptation at other institutions.

This project is being undertaken at the University of the District of Columbia (UDC).

The instructional materials are designed to be especially helpful to computer science students by providing them with marketable skills in computer simulation which can be expected to enhance their ability to obtain professional positions with a bachelor's degree.

2 THE PROBLEM AND COMPUTER SCIENCE CURRICULUM NEEDS

In COMPUTER CURRICULA 91, a major report on the problems and needs of computer science curricula, Tucker (Curricula 91 1991) notes that existing computer science curricula often lack an established infrastructure for the senior year. He identifies a number of special needs in the computer science curricula, including: the need for a capstone course; the need for more hands-on laboratory experience; the need to integrate abstraction, theory and practice at an upper division level in the curriculum for computer science majors; and the need for collaborative learning, small teams, and learning by doing. The report also defines knowledge units and subject areas of the field of computing.

Tucker recommends establishing an advanced undergraduate elective course in Modeling and Simulation, but does not establish syllabus and content for the course.

The proposed senior course in simulation for computer science majors has been specifically designed and focused on meeting the needs of a capstone course, as well as the other requirements discussed above.

Most computer science majors have very limited exposure to simulation. When they elect a course in simulation, the course is often largely theoretical. When hands-on experience in use of simulation software is provided, it is often limited to applications to business and manufacturing. Computer Science graduates are thus not adequately prepared for employment opportunities involving the use of simulation as a tool in solving problems that require a computer science background, e.g., problems in networks, operating systems, computer architecture, performance analysis and scheduling.

Optimized instructional materials are needed for an updated course in simulation for senior computer science majors. Suitable instructional materials also need to be made available nationally, so instructors will not be required to develop their own materials.

The author has examined available instructional materials in the areas of simulation and simulation software for use with the course she teaches at the University of the District of Columbia. She has been unable to locate a single text that covers both theoretical and experimental aspects in a comprehensive fashion. Schriber (Schriber 1990), for example, which is aimed at graduate students in business, has coverage of GPSS/H (General Purpose Simulation Software), but does not cover the theoretical material. In contrast, Ross (Ross 1990) covers the theory but has no software coverage.

The usefulness of the available texts on the theory of simulation for courses for computer science seniors would be improved if they were used with a brand-new substantial laboratory component based on the use of simulation software on personal computers, with an emphasis on applications of the computer science "knowledge units" recommended in Computing Curricula 91 which will meet the special needs of computer science majors.

The most pressing need is for a laboratory manual containing carefully selected experiments which would furnish hands-on computer science applications of computer simulation theory. It should be usable either with a text or as an independent instructional aid that would stress practical applications of simulation to areas of computer science rather than to business and manufacturing. The manual needs to include sufficient summarized reference material on operating systems, computer architecture, and other computer science areas to enable students to develop the simulation models and run them on simulation software.

The manual needs to have the experiments based on a simulation package such as GPSS/H. Also, the laboratory manual needs to have step-by-step guidance in use of the software, at least to the level of modeling queues.

Also needed are instructor's materials for this enhanced course, including a syllabus and a detailed outline, keyed to the theoretical text and the laboratory manual. In addition, a workshop for instructors is needed, to furnish aid in teaching both the software and the technical material.

3 GOALS

The author proposes to meet the needs outlined above by developing a new course in computer simulation with GPSS/H for computer science seniors. As discussed in more detail below, the major goals of this pilot project are:

- to develop a laboratory manual
- to establish a syllabus and detailed outline for the course

The overall goal of this project is to provide instructional materials which will be key components of a capstone course in computer simulation with GPSS/H for senior computer science majors. These materials will integrate the student's previous experience in computer science, mathematics, physics and engineering.

The first major goal is to develop a new laboratory manual containing computer simulation exercises of special interest and value to majors in computer science. The manual will be designed for use either as a supplement to existing simulation texts or as a standalone manual for students and professionals who already have a background in simulation.

The manual will contain laboratory exercises which encourage computer science students to apply their knowledge to specific, constrained problems and to produce solutions. Section 5.2 provides further details concerning the proposed exercises.

The manual will incorporate enough introductory and background material on the use of GPSS/H to obviate the need for a textbook for teaching use of the software. The manual will also incorporate a summary of background information and reference materials on the knowledge units defined in Computer Curricula 91 in areas such as operating systems, computer architecture and computer networks, to the extent that they are relevant to the laboratory exercises and helpful in solving the problem presented.

The second major goal of the project is to prepare an annotated course syllabus and outline showing appropriate integration of the accompanying laboratory exercises with the material in a theoretical text. The laboratory exercises must be synchronized both with the theory of simulation and the simulation software experience.

Emphasis will be placed on the following innovative aspects of this proposal:

- development of a new laboratory manual which is usable with existing textbooks, so that the novel aspects of the manual can be easily and quickly incorporated into existing courses

- incorporation of theory and practice in computer science, into the laboratory exercises linked to "knowledge units" in major subject areas
- orientation of the laboratory exercises towards applying the techniques of simulation to topics of interest and value to computer science majors
- incorporation of qualities desirable in a capstone course which will summarize and reinforce the student's undergraduate experience.

4 POTENTIAL IMPACT AND SIGNIFICANCE

The optimized computer simulation course is expected to be a strong pedagogical tool because the processes of theory, abstraction and design, as used in simulation, require that students integrate their knowledge of mathematics, probability, computer science and physics. In addition students will gain experience in hypothesis formation and testing, evaluating alternatives and trade-offs, costs and performance, written presentation of findings and technical writing, and statistical analysis of results in computer science problems, while learning simulation and simulation software.

The proposed manual will provide students with hands- on experience in applying the theory of simulation and will reinforce computer science knowledge units from throughout the curriculum. Student problem-solving ability, analytic skills and professional judgment will also increase as a result of laboratory exercises based on the computer science curriculum.

Because this manual will be based on a low-cost, widely used and easily available commercial software, it will be of particular value to students and educators in underfunded colleges. The additional proposed course materials will support teachers in integrating software into current simulation courses. This will avoid the labor-intensive need for teachers individually to develop their own laboratory materials which would otherwise be necessary because of the lack of a comprehensive computer science oriented laboratory manual.

This project will also have a major impact on the overall computer science curriculum. Student laboratory experience in the upper division computer science curriculum will extend and consolidate students' knowledge and improve their ability to apply the information gained during their four years of study. The author believes that an enhanced simulation course using the instructional materials produced in this project will be a model for the type of computer simulation course that might be required for all computer science majors.

Ultimately, the improvement of courses in simulation and computer science curricula is expected to contribute to the placement of Bachelor's degree candidates in better professional positions.

See 5.3 for a further discussion of the impact of this course on students, including underrepresented groups.

5 PROCEDURES AND METHODS

5.1 The Plan and Specific Products

The overall plan is to create optimized instructional materials for an expanded and revised course in simulation with GPSS/H for computer science seniors. The starting point for this project isthe current course in computer simulation.

The course will be thoroughly revised as the laboratory manual is developed and integrated with the course work during this project.

The two major products to be prepared during this project are a laboratory manual and other instructional materials.

The laboratory manual will contain: 1) introductory material about GPSS/H software, including new implementations of the software; 2)laboratory exercises using GPSS/H; 3) reference material in relevant computer science knowledge units. The manual will be designed for use with current published textbooks in the theory of simulation.

The other instructional materials will include a course guide, a detailed course outline, and keys linking the theory of simulation, the level of use of GPSS/H and the appropriate knowledge units.

A preliminary draft of the table of contents for the proposed laboratory manual in Appendix A identifies the major knowledge units to be included in the manual.

In each exercise in the laboratory manual the student will: start with a problem; clearly define and analyze the problem, developing a "lumped model" or abstraction; solve the problem; implement the solution by writing the GPSS/H program file to run the simulation; use the GPSS/H software to run the program file; and answer questions about the simulation. Appendix B is a preliminary draft of a sample laboratory exercise.

The exercises will be designed for use by groups of two or three students so as to promote collaborative learning and experience in working as a team.

5.2 The Current Status of the Course at UDC

The current version of the simulation course with GPSS/H for computer science majors was developed by the author, and has been taught at UDC for the last several years.

The prerequisites for the current course are: the calculus, physics and upper division standing in computer science. To avoid additional prerequisites, the author includes sufficient material on probability and statistics before presenting the core material on queueing theory. Our curriculum includes probability and statistics in the second semester of the senior year. By my including this material in the course, I am able to allow juniors to take the course as well as seniors. This gives UDC more flexibility in scheduling the course.

For the laboratory part of the course, the author selected GPSS/H, (Wolverine 1995), because the manufacturer of this widely used simulation software has been very cooperative with academic institutions. The textbook used with the GPSS/H software is Schriber, which includes diskettes for IBM-compatible machines with the book. No site license need be purchased by the college. However, this book is aimed at candidates for a master's in business administration, rather than undergraduates in computer science, and it contains little material on the theory of simulation.

Aburdene (Aburdene 1988), an undergraduate text for science majors, was previously used for the more theoretical aspects of simulation. However, since it is currently out of print, beginning Fall 1993, the author has substituted Ross.

The pair of books selected encompassed the material for the course. However, the textual and laboratory emphasis on computer science material needs to be significantly expanded to fulfill Computing Curricula 91 guidelines.

The current version of the simulation course is integrated into the curriculum at UDC as an upper division elective for computer science majors.

5.3 Impact of the Course On Students

This course is expected to be especially beneficial to those students who have a protracted undergraduate experience, helping them to improve their skills and integrate their knowledge. However, this integration of knowledge is an important reason why every computer science major should take the course.

Most of our graduates go directly to industrial jobs or to teaching, since they lack the resources for full-time graduate study. This course will enhance their marketability for employment. The low cost exposure to important commercial simulation software is expected to be of major benefit to students in underfunded colleges, giving them a marketable skill.

6 ANTICIPATED RESULTS

The goal is to produce instructional materials which improve instruction in computer simulation at UDC and will furnish instructors in computer science departments throughout the United States with a defined method for teaching computer science seniors simulation with GPSS/H.

The specific instructional materials to be produced are:

1. A student laboratory manual consisting of laboratory exercises for a semester of computer simulation instruction. The laboratory manual will include; background information from courses in operating systems, architecture and algorithms; and an introduction to GPSS/H; sufficient for students to do the exercises. A study of Proof Animation (Wolverine1992) may be included in the final course. The background material will summarize the reference information from other computer science courses so that students with limited exposure in a particular area will be able to perform the exercises. This is intended to make the material more flexible for use in departments with various objectives and curricula.

2. A course guide for instructors including: an annotated course syllabus, an outline correlating the laboratory exercises to the specific knowledge units proposed for computer science majors in Computing Curricula 91; and suggested teaching time frames. The background material and course guide will be organized in a fashion that will facilitate their use with standard texts in the theory of simulation.

3. Preliminary and final results will be presented at meetings such as ACM SIGSCE and ACM SIGSIM.

7 DISSEMINATION OF PROJECT RESULTS

During the implementation of the project, the initial results will be disseminated informally via world wide web and internet news groups journals such as comp.simulation, along with input from participants. As the project progresses, preliminary and final results will be presented at professional meetings.

The laboratory manual and teaching materials will be disseminated to schools nationally, both through bookstores (publisher) and as above.

The author plans to apply for funding under the NSF UFE (Undergraduate Faculty Enhancement) Program to organize a faculty workshop which will be designed to enhance the skills of faculty who are teaching or who plan to teach computer simulation. This workshop could be scheduled for the summer of 1998, when the written materials for the current project are complete.

Arrangements will be made for publication of the laboratory manual by established publishing companies.

ACKNOWLEDGMENTS

Acknowledgment is gratefully made of the support of the National Science Foundation through Grants #s DUE-9554863, CCR-9310707 and CDA-9401151, and of the University of the District of Columbia.

APPENDIX A

Preliminary Table of Contents
for "A Laboratory manual for Simulation with GPSS/H for Computer Science Majors: an Integrated Approach"

Section 1. Getting started with GPSS/H simulation software

Introduction to GPSS/H. Software installation; transactions, generate and terminate blocks.

Continuation of GPSS/H. Advance, Seize and Release blocks; practice in setting up and running program files. Modelling a server: queues.

Section 2. Laboratory exercises using GPSS/H
A preliminary draft of a sample exercise is shown in Appendix 2.
Section 3. Computer Science Reference Material: subject areas and knowledge units

Operating Systems; Tasking and Processes
Scheduling and Dispatch
Device Management
Communications and Networking
Distributed and Real-time Systems

Computer Architecture; Interfacing and Communication
Alternative Architectures

Algorithms and Data Structures; Complexity Classes
Recursive Algorithms

Computer Networks
Performance Prediction and Analysis

APPENDIX B

Preliminary Draft of a Sample Laboratory Exercise in Simulation with GPSS/H

Exercise

Topics: Networks, Scheduling, Queues

Problem. An office has a local area network. There are 20 workers on the staff, each with a PC on the LAN. However, there is only one printer. Each worker will want to print 1 file. Assume the files are of various lengths. Make any reasonable assumption about printer speed. Refer to your materials on queues and modeling of single servers, both in your text and in handouts, to answer the questions in Part I. Only then, proceed to Part II.

Part I. Setting up model and theory.

1. Draw a spatial representation of the system, showing arrivals, waiting time, server and departures.

2. Choose a suitable distribution for your arrival times from: deterministic, random. Explain your choice.

3. When you model the server, should you use FCFS (first come, first served) or should you establish priority levels? Explain your choice. Make any reasonable assumptions about the office use of the printer. You might want to solve the problem both ways.

Part II. Build and run a GPSS/H model file for this system.

1. Consider the GPSS/H operations: Generate, Advance, Seize, Release, Transfer. What is the sequence of steps for your model file? What are suitable values for your Operands?

2. What is the default distribution in GPSS/H? Should you use that distribution, or is there a better one for this problem? What is the server and what are the transactions for your system?

3. Set up your GPSS/H file.

4. Simulate your model (run GPSS/H) until all 20 files have left the system. Run your system three times, and furnish statistical data about the runs: show file arrival times at the printer, start of printer time, end of printer time, and average waiting time for each file.

Hand in your results.

REFERENCES

Aburdene, M. 1988. *Computer Simulation of Dynamic Systems*. Brown, Out of print.

Banks, J., J. S. Carson, and J. N. Sy. 1995. *Getting Started with GPSS/H*, 2nd ed, Wolverine Press.

Curricula 91. 1991. Tucker, A., editor, *Communications of the ACM*, June 1991, v. 34 #6, pp. 69-84. A summary of the ACM/IEEE-CS joint curriculum task force report Computing Curricula.

Ross, Sheldon M. 1990. *A Course in Simulation*. Macmillan,

Schriber, T. J. 1990. *An Introduction to Simulation using GPSS/H*. Wiley.

Schruben, L. 1991. *Sigma*. Scientific Press.

Wolverine. 1992. *Proof Animation*. Wolverine Software Corp, 4115 Annandale Rd, Annandale, Va. 22003-2500.

Wolverine. 1995. *GPSS/H Student Version*, Release 3.0. Wolverine Software Corp, 4115 Annandale Rd, Annandale, Va. 22003-2500.

AUTHOR BIOGRAPHY

RUTH SILVERMAN is a full professor in the Department of Computer Science of the University of the District of Columbia, as well as a Visiting Professor at the Center for Automation Research, University of Maryland-College Park. She received a B.S. in physics from Syracuse University, an M.S. in mathematics from Cornell University, and a Ph.D. in mathematics from the University of Washington (Seattle). Her research interests include computational geometry and algorithm development.

Proceedings of the 1996 Winter Simulation Conference
ed. J. M. Charnes, D. J. Morrice, D. T. Brunner, and J. J. Swain

MODELING AND SIMULATION EDUCATION: IS THERE A NEED FOR GRADUATE DEGREES IN MODELING AND SIMULATION?

Ralph V. Rogers

University of Central Florida
P.O. Box 25000
Orlando, FL USA 32816

Mike Zyda

Naval Postgraduate School
Monterey, CA USA 93943-5118

Edward Mykytka

Department of Operational Sciences.
Air Force Institute of Technology
Wright-Patterson, AFB
Dayton, Ohio USA

Lt Col. David Bartlett

Defense Modeling and Simulation Office
1901 N. Beauregard St., Suite 504
Alexandria, VA USA 22311

Michael Bathe

Operations Research Unit
School of Defence Management
Cranfield University
RMCS Shrivenham
Shrivenham, Swindon, UK
SN68LA

ABSTRACT

The authors represent educational institution that currently offer simulation programs and non-educational organizations with vested interests in simulation education. The programs represented by the panel members provide a varied perspective on the answer to the questions "What skills do simulation professionals need?" Are separate simulation programs needed?"; "Is there a sufficient body knowledge to warrant programs in simulation?"; "Do existing programs in computer science, operations research, and engineering provide sufficient simulation education to meet simulation demand?". Panel discussions will center on these and other issues related to the growth and need of degree programs in simulation.

INTRODUCTION

The increasing success and promise of simulation methodologies and technologies to contribute to the solution of the complex of challenges confronting industry, government, military, and education has created an unprecedented demand for skilled simulation professionals at multiple levels. Simulation professionals are needed to:

- Discover, design and develop basic simulation principles and methodologies.
- Design, develop, and manufacture simulation product.
- Apply and use simulation products to meet specified goals and objectives of an enterprise.
- Manage and integrate simulation into projects and enterprise wide development plans.
- Integrate simulation into the decision processes of managers and leaders..

However, the questions confronting the demand sources are, "Where do simulation professional come from?" and "Exactly what skills do simulation professionals need?"

A number of academic institutions have tried to answers these questions through programs which provide tracks, options, and even degrees in simulation. These academic institution's answers have, in turn, raised more questions, questions such as: "Are separate simulation programs needed?"; "Is there a sufficient body knowledge to warrant programs in simulation?"; "Do existing programs in computer science, operations research, and engineering provide sufficient simulation education to meet simulation demand?"; "What is a simulation professional?"

BACKGROUND

Simulation is ubiquitous in today's leading-edge concepts and applications in entertainment, training, design, planning, and decision making. Current realities include virtual environments where soldiers conduct exercises with teams around the world without leaving their home base. No new auto assembly line or microchip fabrication facility would be built today without extensive simulation of every aspect of the proposed operation. In the development of new aircraft, parts are tested for form, fit, and function in a simulation before they are turned over to manufacturing to build. Simply, simulation reduces risk in endeavors where failure cannot be afforded even once. For better or worse, in today's competitive and dynamic environment, the number of failures an enterprise can or will accept is dropping and, in the process, creating an increasing demand for simulation products and services.

The U.S. Military alone has committed tens of billions of dollar to simulation and simulation based technologies and processes over the last decade. Further, simulation is one of few areas of DoD's budget which has been increasing over the last few years. The May 1996 issue of *Industrial Engineering Solutions* provided a buyer's guide for simulation software targeted to manufacturing and service organizations that identified 30 companies selling simulation software. The combined sales of these companies are in the hundreds of millions of dollars. The Winter Simulation Conference's growth from a few hundred attendees a decade ago to close to a 1000 attendees today reflects the increasing interest and importance of simulation to a variety of segments of the world economy. While these anecdotal stories do not tell the full story, the simulation industry is clearly an integral and critical part the world's military and industrial base. Further, the simulation industry represent a significant economic force in itself through its direct employment of tens of thousands if not hundreds of thousands of people.

Given the importance of simulation both as a fundamental tools for enterprise operations and as a significant economic segment of the economy, it is surprising simulation exposure in most formal education programs has been through adjunctive or elective courses and projects. Few formal programs exists which purport to emphasize simulation or integrate it thoroughly in the philosophy of the curriculum. Those that do claim a simulation emphasis have only begun to do so in recent years.

Today's simulation professionals are typically a product of the equivalent of a high-tech apprenticeship and on-the-job training. Simulationist have typically evolved under the mentoring/apprenticeship of others such as a university researcher or organizational leader or simply by successfully responding to pressures of competition and opportunity. While the simulationist's education has typically been grounded in traditional science and engineering degrees, their breadth and depth of simulation knowledge has come primarily through direct trial and error empiricism or through a corporate oral tradition of lessons learned. Not surprisingly, this hap-hazard approach has led to what Dr. Anita Jones, Director of Defense Research and Engineering an the U.S. Department of Defense, refers to as "...a lack of a gene pool [in simulation]".

Whether there is a true lack of professionals in simulation or merely a lack correct classification of professionals can surely be debated. However, simulation is an important and growing industry and industry and government perceive a shortage of qualified and educated simulation personnel. What is not clear is what the market expects from academia in simulation education.

The educational institutions represented in this panel are pursuing educational strategies to meet the perceived needs and markets for advanced education in modeling and simulation. The following provides an indication of the academic discipline base, the philosophy and directions of their efforts. The panel hopes this sessions is the begining of sustain dialogue on need, merits, and directions of simulation education.

AIR FORCE INSTITUTE OF TECHNOLOGY

The Department of Operational Sciences at the Air Force Institute of Technology (AFIT) offers M.S. degrees in both Operations Research and Operational Analysis and the Ph.D. degree in Operations Research. The program offers a specific concentration in modeling and simulation (M&S). The purpose of these programs is to educate Air Force officers and both Department of Defense (DoD) and non-DoD civilians in the theory and practice of operations research, with special emphasis on their application to defense-related problem solving and decision making.

The M.S. program in Operational Analysis is designed to prepare military officers with operational backgrounds to conduct analyses of military forces. It provides its students with a broad exposure to operations research techniques along with coursework that relates specifically to operational Air Force (AF) issues. Graduates of this program among the premier analysts in the AF and are capable of mathematically modeling and simulating various operational situations, including war or peace-time scenarios. Research conducted by these students has examined force-structure tradeoffs, force employment, campaign planning, and engagement modeling.

The graduate programs in Operations Research are designed to provide mission-ready, advanced-degreed officers and civilians with more specific exposure to operations research techniques that support both AF Force and DoD M&S initiatives. Graduates of this program are typically assigned to analysis groups at the Pentagon, major command headquarters, AF product centers, and other DoD agencies. Research conducted by these students has supported resource allocation, cost and operational effectiveness analyses, test and evaluation, and logistics. Many students in this program also tailor their programs to provide a concentration in modeling and simulation.

The department also houses AFIT's Center for Modeling, Simulation, and Analysis which provides the resources necessary for research and consulting activities in all areas of M&S. The Center is AFIT's focal point for modeling, simulation, and analysis and, like the department, actively supports the Air Force's new vector in modeling and simulation. It provides a broad range of faculty expertise that crosses traditional department boundaries and fosters research and consulting efforts in M&S that enable the Air Force to more fully develop and exploit its joint warfighting capability.

NAVAL POSTGRADUATE SCHOOL

The proposed Modeling, Virtual Environments and Simulation (MOVES) Curriculum of the Naval Postgraduate School provides the MS and Ph.D. students both fundamental and specialized courses in applied computer simulation technology and the application of quantitative analyses to human-computer interaction in simulation technology. The MS program is a two year, eight quarter program whose core covers the fundamentals of computer science and human-computer interaction. These topics include object-oriented programming, artificial intelligence, software methodology, computer communications and networks, computer graphics, virtual worlds and simulation systems, probability, statistics, stochastic modeling, data analysis, and human performance evaluation.

Specialization by the MS student is accomplished by choosing a track and completing a sequence of courses providing depth in the selected area. There are two tracks that support the curriculum's research efforts, the Visual Simulation Track and the Human-Computer Interaction Track.

Once the MOVES Curriculum core courses have been taken and while the specialization courses are underway, the final step in the MS degree is the completion of a written thesis. This thesis is usually conducted on a research problem specified by a thesis advisor attached to a MOVES associated laboratory. Current laboratories working with the MOVES Curriculum are the NPSNET Research Group, a leading developer of networked, large-scale virtual environments, and the Information Infrastructure Research Group (IIRG), whose focus is on advanced network issues such as asynchronous transfer mode (ATM), multicast backbone (MBONE) and internetworking regional research institutions.

The Laboratory for Human Interaction in the Virtual Environment (HIVE Lab), a part of the Naval Postgraduate School's Computer Science Department, is the premier site for the study of insertion of humans into virtual environments. The HIVE lab is an off-shoot of the NPSNET Research Group - a leading developer of real-time, interactive 3D networked virtual worlds for the combat simulation arena. In recent years, this experience has been utilized to create human-centered training systems in the NPSNET virtual environment (VE); systems to provide shipboard Officer-of-the-Deck (OOD) training, shipboard walkthrough and damage control training, submarine operations training, and naval Anti-Submarine Warfare (ASW) helicopter flight training.

This work is now being expanded for placing humans in the VE in order to deliver more advanced interactive training and analysis systems. The design of human based simulation systems is an interactive process. Work initially focuses on the detailed analysis of the user requirements for these systems. It is then decided what can be accomplished within the present limits of simulation technology. Prototyping of the desired features is done using a variety of tools in order to facilitate analysis of the system components and to allow for user testing. This process is repeated in order to allow for modification of the deployed systems based on previous learning.

Areas of Human Study: development and promotion of standards, user-centered design issues, development and evaluation of tools, CBone and vrtp integration, development of training systems, definition of hardware requirements, development of software architectures, development of network architecture, advancement of DIS and HLA standards, and analysis of existing software.

CRANFIELD UNIVERSITY AT THE ROYAL MILITARY COLLEGE OF SCIENCE

The Cranfield University at Royal Military College of Science's (RMCS) MSc program in Defence Modeling and Simulation aims to explain the main principles of hardware and software, including the underlying models, used in creating a synthetic battlefield and to consider the issues associated with procuring, using and managing such facilities for the training, testing, and assessment of military forces and equipment. The program provides opportunities for the modeling and simulation of specific systems and practical experience with relevant computer software; lectures, seminars, worked examples and visiting speakers are used as necessary. Modules in the program include: Foundations of Modeling and Simulation, War Gaming and Combat Modeling, Reusable Software, Computer Graphics, Expert Systems, Neural Networks, Parallel Processing, Systems Dynamics, and Modeling and Simulation in Engineering Systems. The RMCS has seen a major investment in simulation by the Royal Ministry of Defence in the setting up of the Simulation and Synthetic Environment Laboratory and the sponsorship of the one week Introduction to Defence Simulation course.

The MSc program in Defence Modeling and Simulation is run by the Department of Applied Mathematics and Operational Research (AMOR) within the School of Defence Management, with substantial contributions coming from other departments within the College. AMOR already runs a set of modular MSc programs in Mathematical Modeling, Military Operational Research, Numerical Methods and Software Systems, and Scientific Applications Software.

UNIVERSITY OF CENTRAL FLORIDA

The simulation focus at the University of Central Florida (UCF) is on the Department of Industrial Engineering and Management Systems' internationally recognized M.S. program in Simulation Systems. Developed in 1984 to support the Center of Excellence in Simulation and Training concept for Central Florida, this program is designed to prepare individuals with undergraduate degrees in engineering, mathematics, or science for careers in the modeling and simulation field.

The Department of Industrial Engineering and Management Systems (IEMS) is a world leader in preparing engineers, scientists and managers to design integrated simulation systems. Its M.S. in Simulation Systems provides the most comprehensive graduate degree program available in simulation. The Masters of Science in Simulation Systems provides a choice of two program tracks: Simulators and Training Systems Track and Simulation and Analysis Track.

The Simulators and Training Systems Track responds to the needs of those who wish to pursue or are currently pursuing careers in the training simulation/simulator industries. The underlying theme and philosophy of this track is Systems Engineering and Project Management. Graduates of the Simulators and Training System Track possess the basic tools to create system designs for simulators and simulator based training systems and provide project management for the development of such systems. Typical required courses include: Simulator Engineering, Simulation of Real-Time Processes, and Project Engineering. The IEMS Department's close relationship with UCF's Institute for Simulation and Training (IST) and the simulation industry presence in Central Florida provides the opportunity to experience the training simulation/simulator first-hand.

The Simulation Modeling and Analysis Track caters to those who desire to gain expertise in simulation as an analysis and design tool for effective planning and decision making. The emphasis of this track is on problem definition, model formulation, and model based analysis. A graduate will be prepared to work with developers, planning offices and decision makers as they model and evaluate the impacts of proposed policies and systems designs. Typical courses include: Discrete System Simulation, Experimental Design, Object-Oriented Simulation, and Intelligent Simulation. Opportunities are available to work with researchers and faculty on simulation projects.

The University of Central Florida is at the center of the world's largest modeling and simulation community, with a current work force of over 8,000 people. This work force includes over 100 private companies, as well as, the Department of Defense's major commands associated with simulation and training including the Army's Simulation, Training and Instrumentation Command (STRICOM), the Navy's Naval Airwarfar Center-Training Systems Division (NAWC-TSD), and the Air Force's Agency for Modeling and Simulation (AFAMS). A vitial component is the Central Florida Research Park. Adjacent to our main campus, the park was established by the Florida Legislature in 1978 and is the home the STRICOM, NAWC-TSD, and AFAMS as well as many of the nation's principle training and simulation/simulator companies.

DEFENSE MODELING AND SIMULATION OFFICE

On June 21, 1991 the Undersecretary of Defense for Acquisition, Mr. Don Yockey, established the Defense Modeling and Simulation Office (DMSO) to serve as the executive secretariat for the Executive Council on Modeling and Simulation (EXCIMS) and to provide a full-time focal point for information concerning DoD modeling and simulation (M&S) activities. Currently the DMSO promulgates M&S policy, initiatives, and guidance to promote cooperation among DoD components to maximize efficiency and effectiveness. DMSO is a staff activity reporting to the Director, Defense Research and Engineering, office of the Undersecretary of Defense for Acquisition and Technology (USD(A&T)).

The DMSO's functions include, but are not limited to:

- Promulgating policies, at the direction of the USD(A&T), that facilitate the application of M&S among joint education and training, research and development, test and evaluation, and operations and cost analysis disciplines.
- Distributing USD(A&T) approved guidelines to assist in military Service Component development of consistent M&S plans in the areas of: configuration management, verification, validation, accreditation, and releasability.
- Developing USD(A&T) approved liaison processes to coordinate and assist in the development, acquisition, and sharing of M&S technology, standards, verification, validation, and accreditation processes among DoD Components and the Defense industry.
- Developing USD(A&T) approved mechanism to foster cooperation among DoD components to maximize M&S interoperability while eliminating redundant development of advanced M&S technologies.
- Advising USD(A&T) on matters relating to improving the use of models and simulations that support input to the Joint Requirements Oversight Council, Defense Planning and Resource Board, and the Defense Acquisition Board.
- Initiating and sustaining a research and development program aimed at improving M&S tools, data bases, associated network concepts, standards, interoperability, interfaces, and accreditation technologies that are applicable to the operations analysis, RDT&E, and training elements of Defense business.
- Documenting requirements for, and developing programs to establish, a clearinghouse network to facilitate M&S information exchange among appropriate users.
- Determining requirements, delineating procedures, and identifying research and development efforts to improve effectiveness and efficiency among existing systems and facilities where appropriate.
- Identifying high priority M&S technology objectives for future investment.
- Documenting resources required and available to execute candidate M&S improvement projects and providing the EXCIMS a basis to approve joint investment plans for M&S projects.
- Conducting evaluations of funded M&S projects to verify progress and validate candidacy for continued resourcing or termination as appropriate.

AUTHOR BIOGRAPHIES

RALPH V. ROGERS is Program Coordinator for the Modeling and Simulation Academic Initiative at the University of Central Florida and is an Associate Professor in the Industrial Engineering and Management Systems Department. He is also a member of the Intelligent Simulation Training Systems Laboratory at Central Florida and has performed research for NASA and the FAA on modeling autonomous objects in air traffic control simulation. Prior to joining the faculty of Central Florida in 1986, Dr. Rogers was on faculty of the Industrial and Systems Engineering Department of Ohio University (1984-1989). He has also worked as a senior project engineer for U. S. Navy's Naval Electronics Engineering Center and for Booze-Allen Applied Research in Bestheda, Maryland. Dr. Rogers received his B.S. degree in Electrical Engineering and his M.S. in Industrial and Systems Engineering from Ohio University, Athens, Ohio in 1971 and 1983, respectively, and received his Ph.D. in Systems Engineering from the University of Virginia, Charlottesville, Virginia in 1987.

MICHAEL ZYDA is a Professor in the Department of Computer Science at the Naval Postgraduate School, Monterey, California. Professor Zyda is also the Academic Associate and Associate Chair for Academic Affairs in that department. He has been at NPS since February of 1984. Professor Zyda's main focus in research is in the area of computer graphics, specifically the development of large-scale, networked 3D virtual environments and visual simulation systems. Professor Zyda is a member of the National Research Council's Committee on Virtual Reality Research and Development. Professor Zyda is also the Senior Editor for Virtual Environments for the MIT Press quarterly PRESENCE, the journal of teleoperation and virtual environments. For that journal, Professor Zyda has co-edited special issues on "Pacific Rim Virtual Reality and Telepresence", on "The Application of Virtual Environments to Architecture, Building and Large Structure Design", and on "Networked Virtual Environments and Teleoperation". Professor Zyda has been active with the Symposium on Interactive 3D Graphics and was the chair of the 1990 conference, held at Snowbird, Utah and the chair of the 1995 Symposium, held in Monterey, California. Professor Zyda began his career in Computer Graphics in 1973 as part of an undergraduate research group, the Senses Bureau, at the University of California, San Diego. Professor Zyda received a BA in Bioengineering from the University of California, San Diego in La Jolla in 1976, an MS in Computer Science/Neurocybernetics from the University of Massachusetts, Amherst in 1978 and a DSc in Computer Science from Washington University, St. Louis, Missouri in 1984.

EDWARD F. MYKYTKA is an Associate Professor of Operations Research and serves as the Acting Head of the Department of Operational Sciences in the Graduate School of Engineering at the Air Force Institute of Technology. He received a B.S. in Mathematics from the University of Dayton in 1976; an M.S. in Industrial Engineering from the University of Iowa in 1978; and a Ph.D. in Systems Engineering from the University of Arizona in 1983. His research interests include discrete-event system simulation (particularly random variate modeling and generation), quality improvement and statistical process control, stochastic modeling, and applied statistics.

LtCol BARTLETT is the Chief of Operations for the Defense Modeling and Simulation Office. A graduate of Texas A. & M. University, LtCol Bartlett entered the Marine Corps in March of 1972 through the officer Commissioning program and was commissioned in July of 1974. He was designated a naval aviator in May of 1976. During 1986 he was assigned as Kilo Company Commander, OCS, MCDEC prior to reporting for Command and Staff College at Quantico, Va. Upon completion of Command and Staff College he continued his professional education with the completion of the Defense Systems Management College, Fort Belvoir, VA. LtCol Bartlett was then assigned to Naval Air Systems Command as the Marine Liaison and as Program Manager for Training Systems and Simulation. During this period he was responsible for acquisition of all training systems and simulators for the Marine Corps. In 1991 he was assigned as the Logistic Officer for MAG-36, 1st MAW futenma, Okinawa and upon completion of the overseas tour was assigned to the Marine Corps Presentation Team. In January 1993, he was assigned to his current position with the Defense Modeling & Simulation Office

MICHAEL BATHE is Head of Management and Systems Group in the School of Defence Management in Cranfield University at the Royal Military College of Science in Shrivenham, England, UK. His research interest include logistics modeling, decision analysis and combat modeling.

Proceedings of the 1996 Winter Simulation Conference
ed. J. M. Charnes, D. J. Morrice, D. T. Brunner, and J. J. Swain

DYNAMIC SIMULATION FOR TIME SERIES MODELING

Harriet Black Nembhard

Department of Industrial and Systems Engineering
Auburn University
Auburn, Alabama 36849-5346, U.S.A.

David A. Nembhard

Department of Management
Auburn University
Auburn, Alabama 36849-5241, U.S.A.

ABSTRACT

There has been considerable interest in investigating time series in the fields of economics, business, and engineering. This paper illustrates that dynamic simulation can be used as an instructional tool to introduce time series concepts. We design two simulation models using SIMULINK to represent dynamic models for autoregressive moving average (ARMA) time series data given by a state-space representation. SIMULINK is a program for simulating dynamic systems. It offers a graphical user interface that allows easy development of block diagrams and hierarchical models. This paper presents an overview of how SIMULINK can be used for time series modeling.

1 INTRODUCTION

Time series modeling is widely used in economics, business, and engineering. The topic is often covered in undergraduate courses in statistics or operations research. Many graduate programs in these areas offer an entire course on time series modeling and time series analysis.

Dynamic simulation is a powerful tool for introducing time series concepts. In this paper we will investigate time series modeling using SIMULINK (Simulink 1994). SIMULINK is a program for simulating dynamic systems. A part of its appeal as a learning tool is that it offers a visual (or graphical) user interface that allows the development of block diagrams and hierarchical models. The result of the simulation model can be viewed in real-time as the model progresses.

We consider practical areas of time series modeling application (Section 2), design a simulation model of the time series process (Section 3) and demonstrate it on two illustrative examples (Section 4). The model can be obtained until December 1997 from http://www.eng.auburn.edu/~nembhard.

2 PRACTICAL PROBLEMS

A *time series* is commonly defined as a sequence of observations taken sequentially in time on a variable of interest (*e.g.*, Box, Jenkins, and Reinsel 1994). The variable is usually discrete, observed at equally spaced intervals, and adjacent observations are typically dependent. *Time series modeling* is concerned with representing the behavior of the process by a mathematical model that can be extended into the future. *Time series analysis* is concerned with techniques for analyzing the variable dependence.

Examples of time series in economics and business may include a daily series of corporate stock prices and a monthly series of the level of inventory in a factory. Examples of time series in engineering include an hourly series of uncontrolled chemical concentration and a minutely series of random disturbances. Box, Jenkins, and Reinsel (1994) identify and illustrate the use of time series models in four important practical problems: forecasting time series; estimation of transfer functions; analysis of effects of unusual intervention events to a system; and discrete control systems.

With this broad applicability, there is considerable interest in time series. Recently, there has been more interest in representing time series in the state space form because it lends itself well to simulation analysis as well as to other modeling involving numerical computation. The state space representation provides a convenient way to represent complex mathematical systems.

In Appendix A, we show how the ARMA time series can be represented in the state-space form as a set of equations that describe a dynamic process. We will use this representation in examples of characteristic time series shown in Section 4 of this paper. Several other characteristic patterns are shown in Montgomery, Johnson, and Gardiner (1988) (Chapter 1) and Box, Jenkins, and Reinsel (1994) (Chapter 4).

3 CONSTRUCTING SIMULINK MODELS

SIMULINK is a simulation program designed to model and analyze dynamic systems and can be executed on the Sun, Macintosh, or PC platform. A typical session starts by invoking SIMULINK and defining a model or loading a previous model. To facilitate model definition, SIMULINK has a graphical user interface containing ***block diagram*** windows that provide libraries of functional blocks. The parameters of each block are evaluated in MATLAB which is an interactive environment for numeric computation. (SIMULINK and MATLAB are written in C by The MathWorks, Inc.)

The standard block library is shown in Figure 1. The SOURCES, SINKS, and CONNECTIONS libraries allow input and output representation. The DISCRETE, LINEAR, and NONLINEAR libraries allow the representation of the system or process. The EXTRAS library provides software demos and other help.

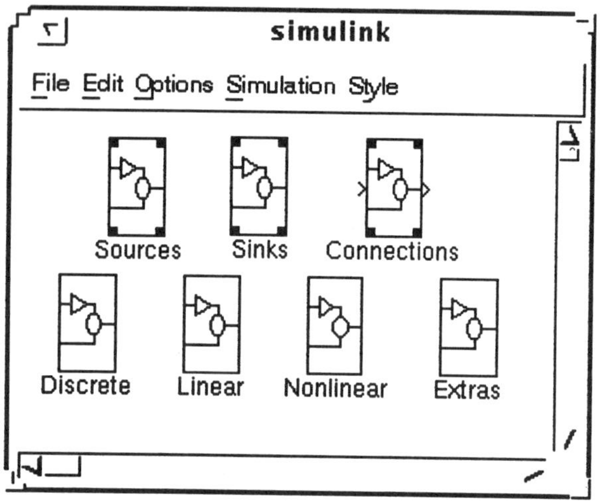

Figure 1: Standard SIMULINK Block Library

Opening any of the libraries shows a subsystem of blocks. For example, the SOURCES library contains the RANDOM NUMBER block (see Figure 2). To define a model, blocks are simply copied from the libraries to draw a block diagram of the system. Such a model is shown for the ARMA Process in Figure 3. Most blocks have an input environment as shown in Figure 4.

Next, parameters of the simulation run may be specified. These parameters include when to start and stop the simulation run, limits on the step size taken during numerical integration, and method of integration (i.e., algorithm) desired.

Once the model has been designed and the parameters specified, ***start*** is chosen from the ***simulation*** pull-down menu to make the model run. The progress of a simulation run can be viewed in real time and the final results made available in the MATLAB workspace. Analysis of the results can be accomplished by selecting built-in options from the SIMULINK menu or by entering commands in the MATLAB command window. In the next section, we construct a SIMULINK simulation model for the ARMA time series.

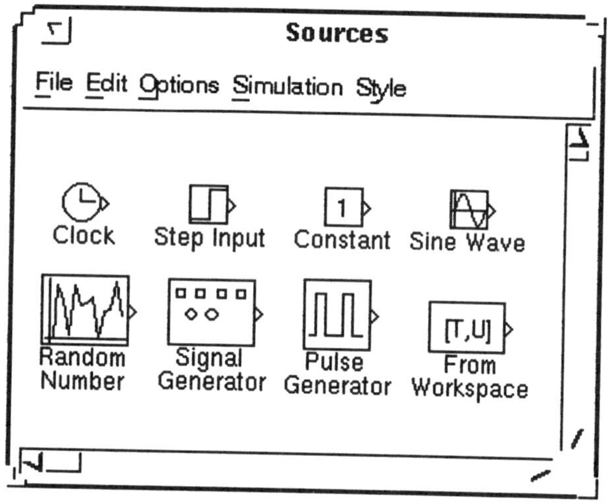

Figure 2: SOURCES Block Library

4 SIMULINK SIMULATION OF THE TIME SERIES

In this section, we present the SIMULINK simulation model that represents the state-space form of the ARMA process. As stated above, Appendix A gives the state-space form of the ARMA time series. Simulation is a powerful tool for solving these equations. Simulation can also approximate solutions to system equations that are intractable by analytic methods.

Figure 3 shows the block diagram of a SIMULINK model that simulates the ARMA process. The STANDARD NORMAL RANDOM NUMBER GENERATOR block provides random shocks from a $N(0,1)$ distribution. At the STANDARD DEVIATION block the output of the random shock is multiplied by the standard deviation. The LINEAR FILTER block provides the discrete state-space form of the linear filter process. The LEVEL block provides the process level (mean). The SUM block combines the linear filter with the process level; the output of the sum block is an ARMA(p,q) process. The ARMA process is plotted during the simulation run via the GRAPH block.

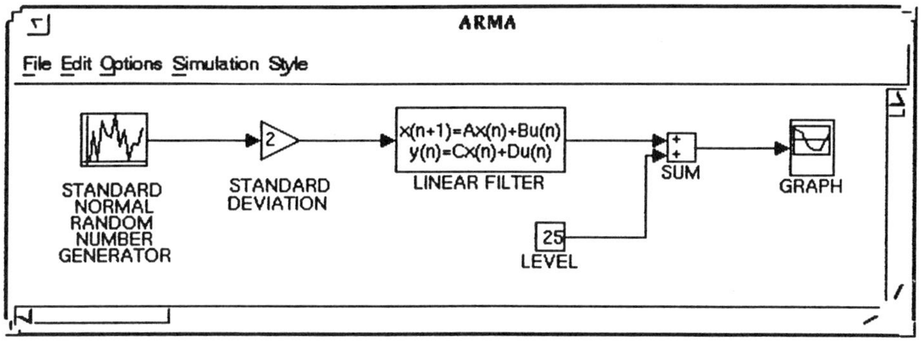

Figure 3: Block Diagram of an ARMA Process Simulation Model

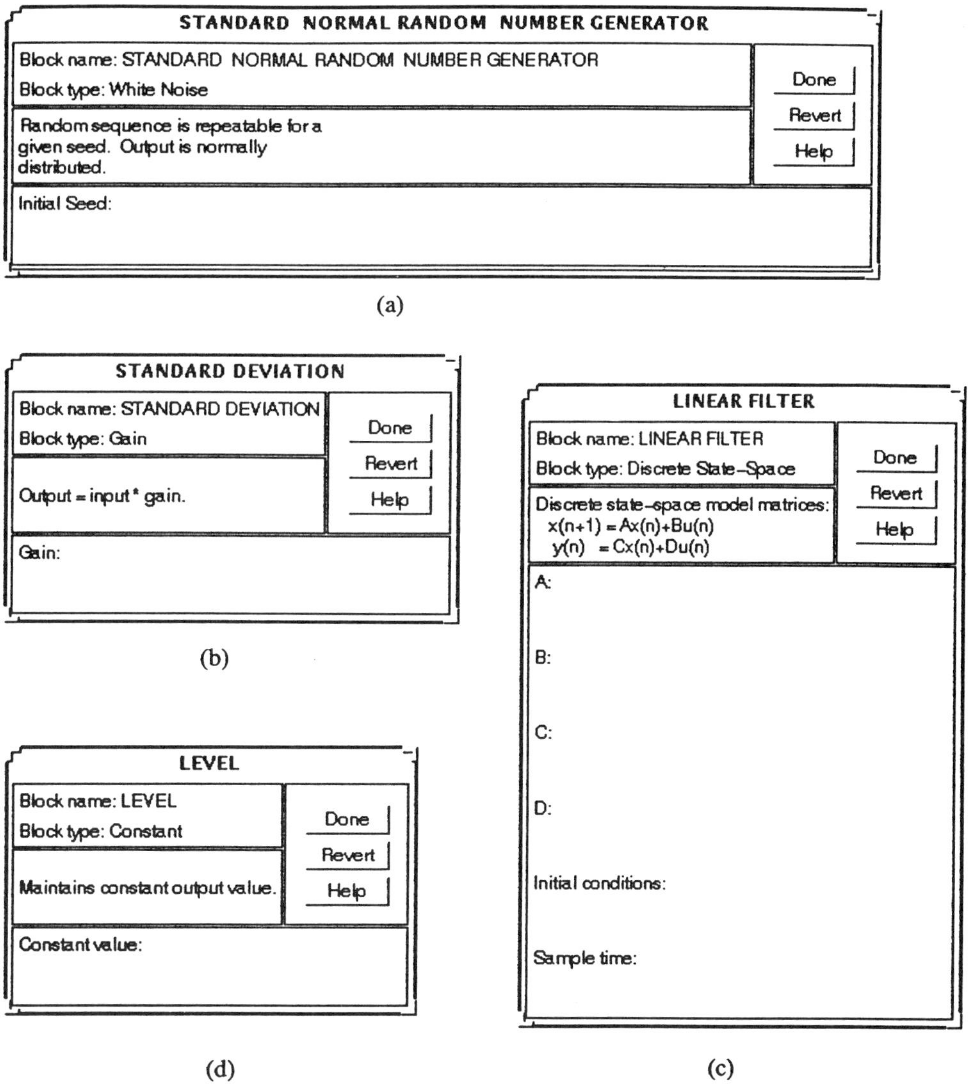

Figure 4: Input Environments for Selected Blocks

Selecting any of these blocks reveals an input environment for the model parameters. For example, Figure 4 shows the input environment for the first four blocks of the model: the STANDARD NORMAL RANDOM NUMBER GENERATOR (Figure 4(a)); the STANDARD DEVIATION (Figure 4(b)); the LINEAR FILTER (Figure 4(c)) and the LEVEL (Figure 4(d)).

4.1 Example 1: ARMA(1, 1) Model

The ARMA (1, 1) process is a special case of the ARMA (p, q) process of Equation (A-1) that has been widely used in practical economic and engineering applications. For comparison with previous results, we will consider the following ARMA (1, 1) process used by Montgomery, Johnson, and Gardiner (1988):

$$y_n = 10 + 0.6y_{n-1} + u_n + 0.9u_{n-1} \tag{1}$$

with $\sigma^2 = 4$. By Equation (A-2) the mean (level) of the process is 25. By Equations (A-3) and (A-4), the matrices for the discrete state-space representation of the disturbance are

$$A = \begin{bmatrix} 0.6 & 0.9 \\ 0 & 0 \end{bmatrix} \qquad B = \begin{bmatrix} 1 \\ 1 \end{bmatrix}$$

$$C = [1\ 0] \qquad D = [0] \ .$$

The standard deviation is specified in the Figure 4(b) window, the values of the matrices in the Figure 4(c) window and the mean in the Figure 4(d) window.

To keep this example brief, we made only two simulation runs of 50 time periods. The state-space representation of the process was used for each of the runs yielding two different realizations of the ARMA(1, 1) process. Specifying random streams 1, and 2 (in the Figure 4(a) window) result in Figures 5(a) and 5(b), respectively. Of course, to make better probabilistic statements about the process we must make additional replications. To do so, we could change the random number stream manually for each run or use a small MATLAB program to implement them in a batch.

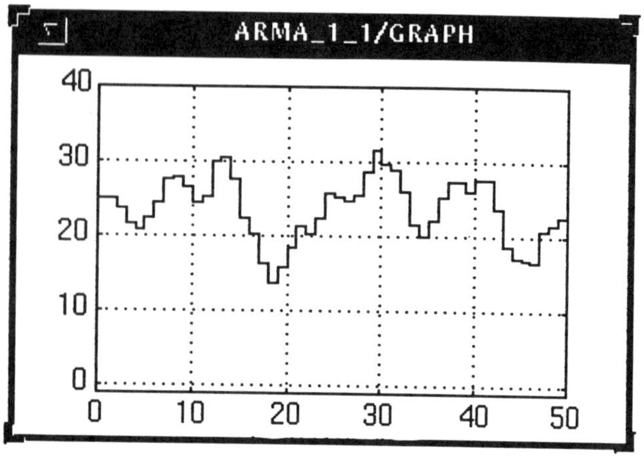

(a)

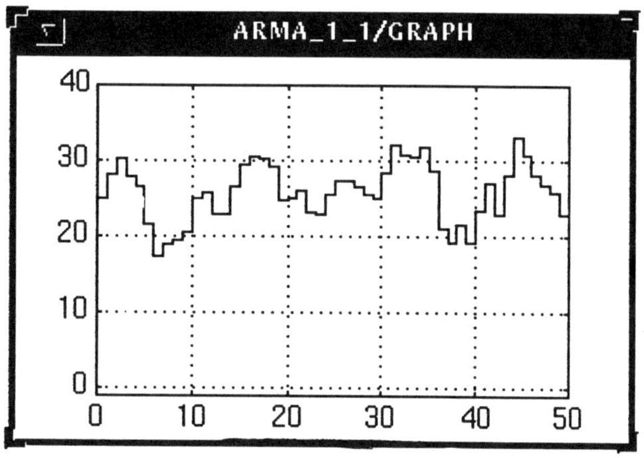

(b)

Figure 5: Two realizations of the ARMA(1, 1) process $y_n = 10 + 0.6y_{n-1} + u_n + 0.9u_{n-1}$

4.2 Example 2: ARMA(2, 2) Model

Consider two continuous stirred-tank reactors in series used to transform a comonomer (which is a chemical building block used to make film for products like plastic bags and cups). The disturbance in the system can be represented by the ARMA (2, 2) process

$$\begin{aligned} y_n &= 5 + 0.9y_{n-1} - 0.8y_{n-2} \\ &\quad + u_n + 0.6u_{n-1} + 0.2u_{n-2} \end{aligned} \tag{2}$$

with $\sigma^2 = 9$. By Equation (A-2) the mean (level) of the process is 5.56. By Equations (A-3) and (A-4),

$$A = \begin{bmatrix} 0.9 & -0.8 & 0.2 & 0 \\ 1 & 0 & 0 & 0 \\ 0 & 0 & 0 & 0 \\ 0 & 0 & 0 & 1 \end{bmatrix} \qquad B = \begin{bmatrix} 1 \\ 0 \end{bmatrix}$$

$$C = [1\ 0\ 0\ 0] \qquad D = [0] \ .$$

Using the same methods as given in Example 1, two simulation runs were made of this process. The graphical results from these runs are shown in Figure 6.

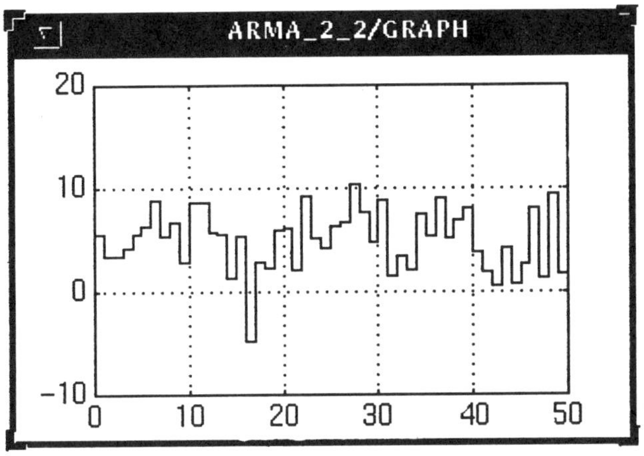

(a)

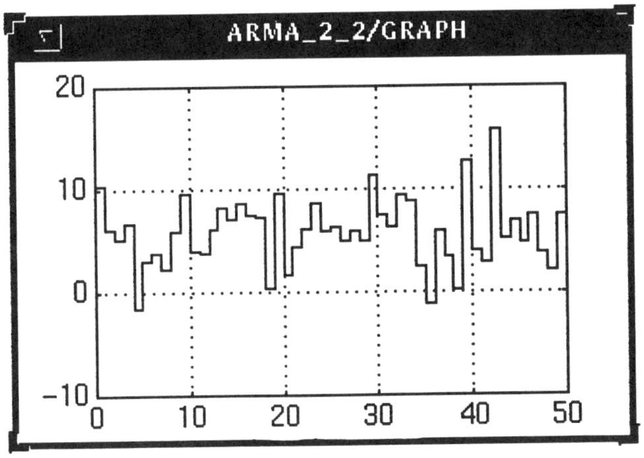

(b)

Figure 6: Two realizations of the ARMA(2, 2) process $y_n = 5 + 0.9y_{n-1} - 0.8y_{n-2} + u_n + 0.6u_{n-1} + 0.2u_{n-2}$

5 SUMMARY

Dynamic simulation using SIMULINK is a very powerful tool and has been frequently used in the chemical engineering, electrical engineering, and applied physics disciplines. Part of the motivation for this paper is to increase the exposure of dynamic simulation software as a teaching tool within the industrial engineering and operations management disciplines.

This paper illustrates that dynamic simulation can be used as a learning tool for time series concepts. We developed a SIMULINK simulation model for the state-space representation of a general ARMA time series process. We illustrated the use of the model on examples of an ARMA (1, 1) process and an ARMA (2, 2) process. (They can be down-loaded from the internet address given in Section 1.) The model can, of course, be used to represent other types of systems such as those with cyclic, ramp, and impulse characteristics. It can also be used to illustrate other concepts in the statistics area such as forecasting, prediction, model identification, and maximum likelihood estimation of parameters.

Dynamic simulation has also been used as a research tool to investigate control and monitoring policies for noisy dynamic systems (Nembhard 1996 and Nembhard and Mastrangelo 1996). Future plans include exploring its application in organizational learning and in the economic design of Kalman filtering models.

APPENDIX A: STATE-SPACE REPRESENTATION OF THE TIME SERIES

A state-space representation lends itself well to simulation analysis. It provides a convenient way to represent complex mathematical systems that include multiple inputs or outputs, nonlinearity, or time-varying components. A state-space representation is used in the simulation model to represent a stationary stochastic auto-regressive-moving average (ARMA) model. The ARMA (p, q) process can be represented as a time series (Montgomery, Johnson, and Gardiner 1988) by

$$y_n = \xi + \sum_{i=1}^{p} \phi_i y_{n-i} - u_n - \sum_{j=1}^{q} \theta_j u_{n-j} \qquad \text{(A-1)}$$

where y_n is the current observation and is regressed on previous realizations y_{n-1}, y_{n-2}, ..., y_{n-p}; ϕ_1, ϕ_2, ..., ϕ_p are the unknown process parameters; u_n, u_{n-1}, ..., u_{n-q} are the independent random variables; and θ_1, θ_2, ..., θ_q are a finite set of weights.

The mean of the ARMA (p, q) is

$$\mu = \frac{\xi}{1 - \sum_{i=1}^{p} \phi_i} . \qquad \text{(A-2)}$$

The expanded form of Equation (A-1)

$$y_n = \xi + \phi_1 y_{n-1} + \phi_2 y_{n-2} + \cdots + \phi_p y_{n-p} - u_n \\ - \theta_1 u_{n-1} - \theta_2 u_{n-2} - \cdots - \theta_q u_{n-q}$$

allows the state vector to be written as

$$[y_{n-1} \; y_{n-2} \; \cdots \; y_{n-p} \; u_{n-2} \; u_{n-3} \; \cdots \; u_{n-q}]^T .$$

We note that the set of state variables is not unique for a given system (see Ogata 1987 for further discussion of this point) so another set could be defined with u_n, u_{n-1}, and u_{n-q-2} if preferred. Hence the discrete state-space representation

$$\begin{aligned} x_{n+1} &= Ax_{n+1} + Bu_n \\ y_n &= Cx_n + Du_n \end{aligned}$$

for the ARMA(p, q) process is given by

$$x_n = \begin{bmatrix} \phi_1 & \cdots & \phi_p & -\theta_2 & \cdots & -\theta_{q-1} & -\theta_q \\ 1 & \cdots & 0 & 0 & \cdots & 0 & 0 \\ 0 & \cdots & 0 & 0 & \cdots & 0 & 0 \\ \cdots \\ \cdots \\ 0 & \cdots & 0 & 0 & \cdots & 0 & 0 \\ 0 & \cdots & 0 & 1 & \cdots & 0 & 0 \\ 0 & \cdots & 0 & 0 & \cdots & 0 & 0 \\ \cdots \\ \cdots \\ 0 & \cdots & 0 & 0 & \cdots & 1 & 0 \end{bmatrix} * x_{n-1} + \begin{bmatrix} 0 \\ 0 \\ 0 \\ \cdots \\ \cdots \\ -\theta_1 \\ 0 \\ 0 \\ \cdots \\ \cdots \\ 0 \end{bmatrix} u_{n-1} \qquad \text{(A-3)}$$

$$y_n = \begin{bmatrix} 1 & 0 & \cdots & 0 & 0 & 0 & \cdots & 0 & 0 \end{bmatrix} x_n - [\theta_0] u_n . \qquad \text{(A-4)}$$

There would be little reason to use p or q are greater than 2 in a simulation model because time series model identification methods often show little differentiation between second-order and third-order processes. Further, many industry processes can be represented by first or second-order models.

REFERENCES

Box, G. E. P., Jenkins, G. M., and Reinsel, G. C. 1994. *Time series analysis: forecasting and control.* Third Edition. Prentice Hall, Inc., Englewood Cliffs, NJ.

Montgomery, D. C., Johnson, L. A., and Gardiner, J. S. 1988. *Forecasting and time series analysis.* Second Edition. McGraw-Hill, Inc., New York, NY.

Nembhard, H. B. 1996. Simulation using the state-space representation of noisy dynamic systems to determine effective integrated process control designs. In review with *IIE Transactions.*

Nembhard, H. B. and Mastrangelo, C. 1996. Integrated process control for startup operations. ASQC 1996 Fall Technical Conference. Scottsdale, AZ, October.

Ogata, K. 1987. *Discrete-time control systems.* Prentice-Hall, Inc., Englewood Cliffs, NJ.

SIMULINK™. A Program for Simulating Dynamic Systems. Computer Software. 1994. The MathWorks, Inc.

AUTHOR BIOGRAPHIES

HARRIET BLACK NEMBHARD is an Assistant Professor of Industrial and Systems Engineering at Auburn University. She has previously held positions with Pepsi-Cola, General Mills, and Dow Chemical. Her B.A. is in Management from Claremont McKenna College and her B.S.E. is in Industrial and Management Systems Engineering from Arizona State University. Her Ph.D. and M.S.E. degrees are in Industrial and Operations Engineering from the University of Michigan. Professor Nembhard's current research interests involve formulating solving models for manufacturing systems to address control, production, quality, and economic concerns that arise while a process is in a transient phase. She is a member of ASQC, IIE, and INFORMS.

DAVID A. NEMBHARD is an Assistant Professor of Operations Management at Auburn University. He has previously held positions with Dow Chemical. He holds a B.S.E. and an M.S. in Systems and Control Engineering from Case Western Reserve University and an M.S. and a Ph.D. in Industrial and Operations Engineering from the University of Michigan. Professor Nembhard's current research interest include Bayesian Estimation, organizational learning, and the transportation of hazardous materials. He is a member of ASQC and INFORMS.

Proceedings of the 1996 Winter Simulation Conference
ed. J. M. Charnes, D. J. Morrice, D. T. Brunner, and J. J. Swain

SIMULATION-BASED LEARNING TOOL FOR FACILITIES DESIGN

Bala Ram

Department of Industrial Engineering
North Carolina A&T State University
Greensboro, NC 27411

Rajiv Girdhar

General Electric Corporation
PO Box 414
Milwaukee, WI 53201

ABSTRACT

This paper describes a tool, LAYSIM, to emphasize to the students the important role of material handling cost in choosing the type of layout; i.e., Product, Jobshop or Group Technology-based layout for a certain set of products. It addresses how through experimentation with LAYSIM students understand and appreciate the manufacturing costs associated with each type of layout. The paper describes the underlying learning model, development of the LAYSIM, the features of LAYSIM and the proposed dissemination technique.

1 INTRODUCTION

Allowing students to experience the practical and realistic situations of their profession has to be the thrust of the curriculum in engineering. The design process is iterative, the analyzer has to go back and forth between the different steps of defining, developing, analyzing, testing, and evaluation until a satisfactory design is obtained. Design is also a creative process of devising a product or a system to meet the ultimate goal of satisfying the customer needs (Steudel 1993). To aid the process of designing, the student has to be provided with enough tools in the form of fundamental concepts, which can be embedded into their minds by using the experiential learning approach. Experiential learning is a pedagogical approach that offers students an opportunity to link classroom theory with practical applications. People learn by many different means, but prefer to gain knowledge and skills through experiential opportunities (Richardson 1994).

Learning tools that are interactive are a cost-effective way of giving several different experiences to a student. Simulation is the art and science of creating a representation of a process or system for the purpose of experimentation and evaluation (Gogg and Mott 1993). It is a powerful analytical tool that can significantly facilitate the problem solving process. The capability of the simulation environment to allow the user to experiment by changing the different parameters is unique, and in the process helps the user discover what additional information is needed to solve the problem. With the increased power of the personal computers and software, simulation - the interactive teaching tool - has a much wider scope in the engineering curriculum.

This paper describes a tool, LAYSIM, to impart to the students the impact of choosing a type of layout; i.e., Product, Jobshop or Group Technology-based layout for a certain set of products on manufacturing costs including material handling cost. The simulation tool developed is meant to be used in introducing the topic of Facilities Design in an engineering or management curriculum. The module is intended to give the students a first-hand experience in understanding the relationship between the type of layout, number of components being manufactured in the factory and the resulting manufacturing costs.

2 LEARNING MODEL

LAYSIM has been developed as a means of providing the students an environment for experimentation with different plant layout options. It presents the students with three different types of plant layouts: 1) Product Layout, 2) Group Technology-based Layout,

Table 1 Layouts used in LAYSIM

Jobshop Layout	Group Technology-Based Layout	Product Layout
2 components		2 components
6 components	6 components	6 components
12 components	12 components	12 components

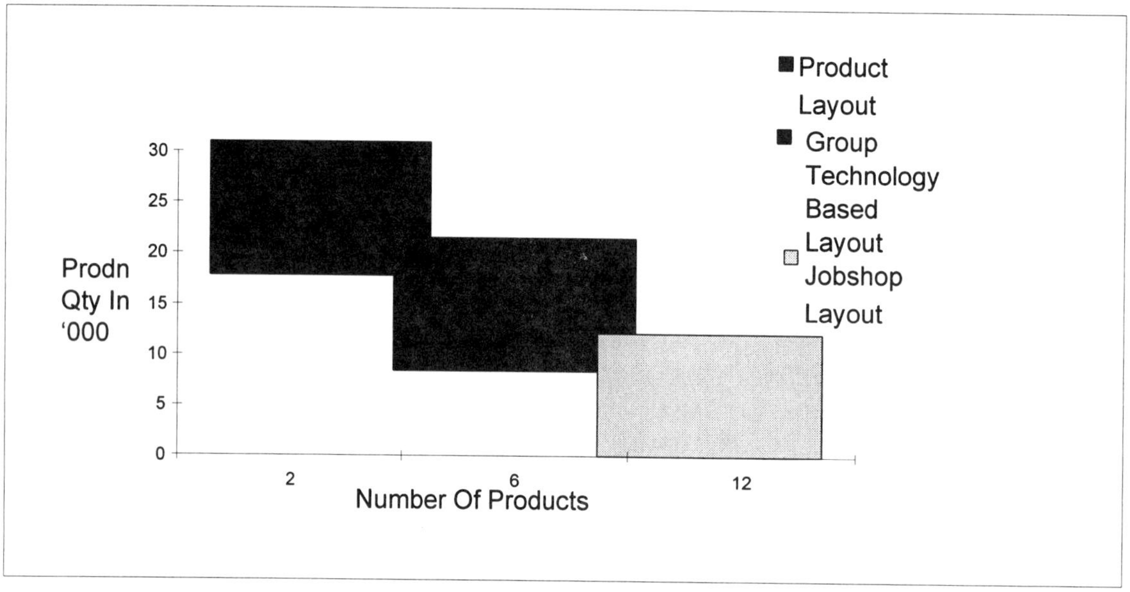

Figure 1 P - Q Chart

3) Jobshop Layout and a common set of products. LAYSIM incorporates eight layouts as shown in the Table 1. The students have an opportunity to compute the material handling costs with each type of layout for different levels of production volume and different number of products. (This is explained in more detail in the next section.)

LAYSIM will enable students to perform the following experiential exercise. They would be able to change certain parameters and perform a number of simulation runs to collect enough data based on which they would be able to construct the P - Q Chart, where P represents the number of products being manufactured in the factory and Q represents the volume of production. For example, one simulation run will be performed for each of the layout types for 2 components with "large" production volumes. The simulation runs would provide the manufacturing costs in each case; the layout type with the least manufacturing costs will be the "preferred layout". A typical P - Q Chart is shown in Figure 1.

The above exercise will enable the students to understand and appreciate the manufacturing costs associated with each type of layout. The layouts and the number of components chosen has been designed in such a way that after going through the exercise the student is able to come up with results which support some of the theory behind the P - Q Chart (Tompkins et al. 1996).

3 DEVELOPMENT OF THE LEARNING TOOL

3.1 Development of Software

ProModel 3.0 was chosen as the simulation software for LAYSIM. ProModel 3.0 is a PC-based software which runs under the Windows environment. This software was chosen because of the following capabilities:

- a run-time model can be made for distributing the module for feedback and dissemination;

- animation of the material handling equipment, material and personnel is possible;
- appropriate icons for production environment such as machines, material handling equipment; etc. are built into the software;
- importing of AutoCAD drawings into the layout background is possible;
- shifts can be defined and assigned, and down-time for machines can be specified;
- global variables can be defined: this is used to compute the manufacturing costs during simulation;
- subroutines can be built: this has been used to make the models interactive; and
- a number of views of different parts of the plant can be defined for the factory layout and can be presented to the user in a defined sequence during the simulation run, or can be used by the user in a random manner;

Multimedia ToolBook 3.0 was chosen as a driver software for LAYSIM. It is possible to run any Windows-based software, such as ProModel, from within Multimedia ToolBook 3.0 and ToolBook can help in building aesthetic screens with buttons for navigating through the module. A sample of the driver screens used in the module are shown in Figures 2 and 3 as seen by the user of the module.

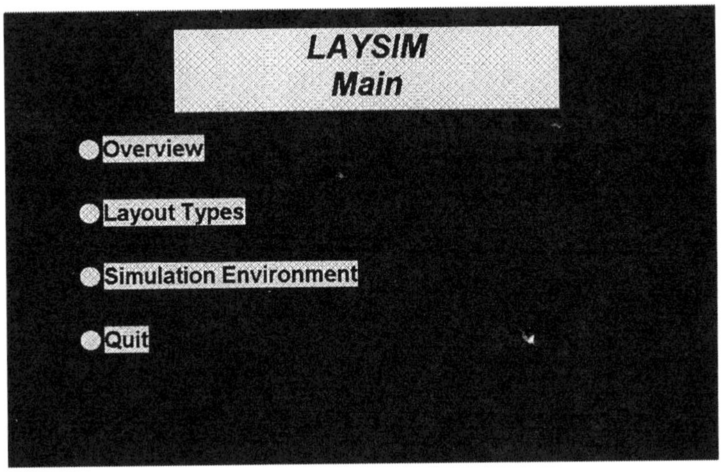

Figure 2 LAYSIM Main Screen

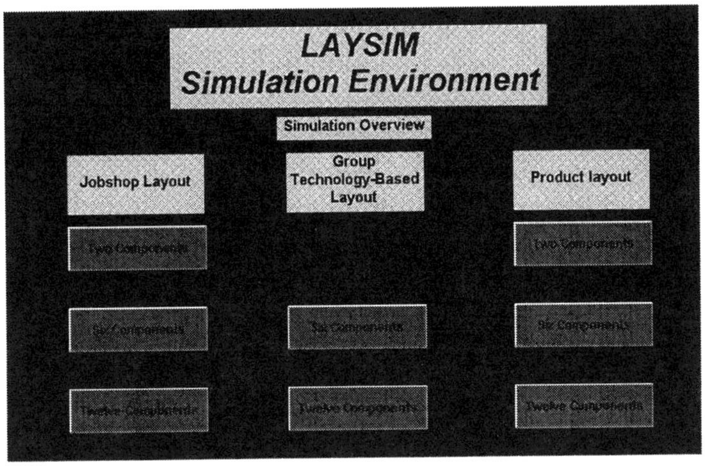

Figure 3 LAYSIM Simulation Environment

The floorplan of the factories used for the simulation model have been built using FactoryCAD which runs on a AutoCAD software platform and provides additional features for developing floorplans. The floorplan includes the boundary walls for the factory, office, break rooms and also the workcenter/machine boundaries as laid out on the floor of the factory. These boundaries are then used as a reference to place the icons of the various machines in ProModel. The layout drawing developed using FactoryCAD is saved as a Windows metafile and then imported into ProModel. A sample of the layout imported into ProModel is shown in Figure 4, and in Figures 5a and 5b is shown the interrelationship of all the software which has been used in developing the learning tool.

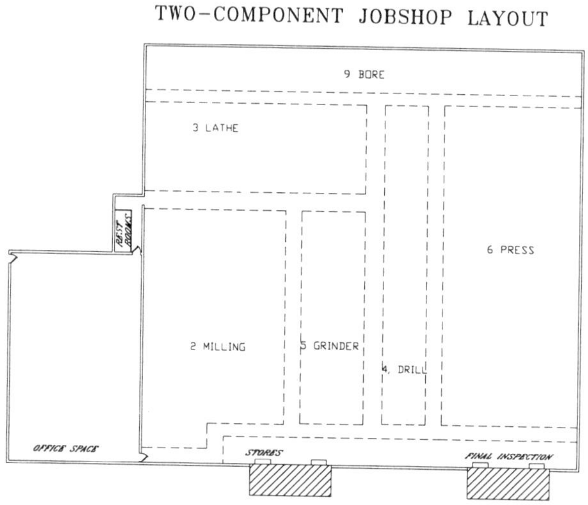

Figure 4 An Example Floorplan in FactoryCAD

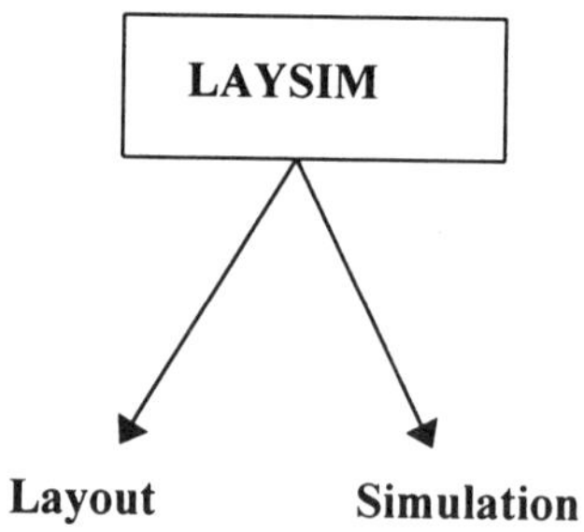

Figure 5a Components of LAYSIM

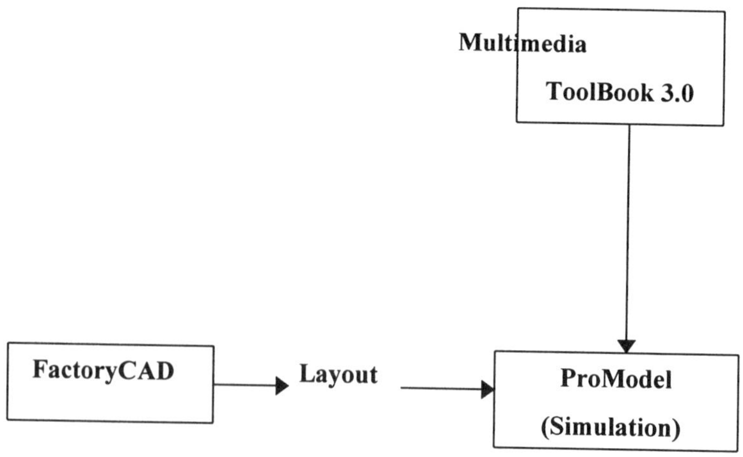

Figure 5b Relationship between FactoryCAD, Multimedia ToolBook and ProModel

Table 2 - Components to be Manufactured

Part No.	Part Name	Wt/Pc in lbs.	Production Routing(process times in minutes)
1	Crank Case	1.1000	1- 2 - 3 - 6 - 4 - 10 (5) (3) (2) (2)
2	Cylinder	0.5000	1 - 3 - 4 - 7 - 10 (2) (1) (0.5)
3	Cylinder Head	0.3000	1 - 3 - 4 - 10 (3) (2)
4	Crankshaft	0.1500	1 - 8 - 3 - 2 - 4 - 10 (0.8) (3) (3) (1)
5	Connecting Rod	0.0750	1 - 5 - 4 - 9 - 4 - 2 - 6 - 4 - 6 - 10 (2) (1) (3) (1) (1) (1) (0.5) (0.5)
6	Outside Bearing	0.0125	1 - 3 - 10 (0.5)
7	Inside Bearing	0.0150	1 - 3 - 10 (0.5)
8	Breather	0.0500	1 - 2 - 4 - 10 (1) (0.5)
9	Cover plate	0.0125	1 - 6 - 10 (0.1)
10	Suction Fitting	0.0125	1 - 3 - 10 (0.3)
11	Discharge Fitting	0.0125	1 - 3 - 2 - 10 (0.3) (0.5)
12	Breather Plate	0.0125	1 - 6 - 10 (0.1)

3.2 Example

A set of compressor components have been chosen for manufacturing in the example factory. These components are listed in Table 2. The table has details on the Weight of the components in pounds and the processing sequence along with the process times. The processing sequence is shown by numbers which represent the machines, the reference for the number representing the machine is found in Table 3.

The machines required for manufacturing the components and the areas per machine are shown in Table 3. Three different choices are possible for moving material. Table 4 provides data for selection of appropriate material handling method.

Table 5 presents the calculation for the calculation for the number of machines and floor space for one of the eight layouts used in LAYSIM. Table 5 is for the 2 component Job Shop case.

Table 3 - Machines and their area requirements

Machine Number	Machine Name	Reqd. Area(Sq. Ft.)
1	Mill	100
2	Lathe	90
3	Drill	50
4	Grinder	100
5	Press	200
6	Hone	45
7	Saw	50
8	Bore	100

Table 4 Material Handling Alternatives

Wt. Of Unit Load (Lbs.)	Method Of Handling	Cost/Move /Ft. Traveled
0 To 25	Manual Handling	$0.010
26 To 500	Walkie Pallet Lift	$0.015
501 And Up	Fork Lift Truck	$0.025

The first step in developing the simulation models was to import the FactoryCAD layout into the ProModel model; the procedure for doing this has already been described earlier. The layout forms a static background for the simulation. The next step is to introduce machine icons in the space allocated in the layout, this is specifying of the locations, beyond that entities, resources, path networks, and processing sequence are specified for running the model. Salient features of the module are described below.

It is possible to define regular shift hours and shift breaks for the full week and save it as a separate file. A number of such shift times can be defined and stored in separate files. These files can then be assigned to the various locations and resources in the simulation model. For the models developed for LAYSIM three different shift files have been created and assigned to different locations and resources.

By defining the distance traveled by the material handling resources and the material handling cost as "global variables" and incrementing their value the material handling cost is dynamically presented on the simulation screen. A few parameters such as the batch size of the components being processed are variable and the user will have an opportunity to specify the value before the simulation run.

The run time model has been made for distributing the tool to get feedback about its effectiveness. The run time model can be used on any personal computer without having the need to buy the simulation software. ProModel 3 has a provision to create a model package which can then be distributed to the users'; the model package consists of the models and ProModel which has to be installed at the users' site before the packaged model can be run. The user, however, cannot modify the model or develop other models.

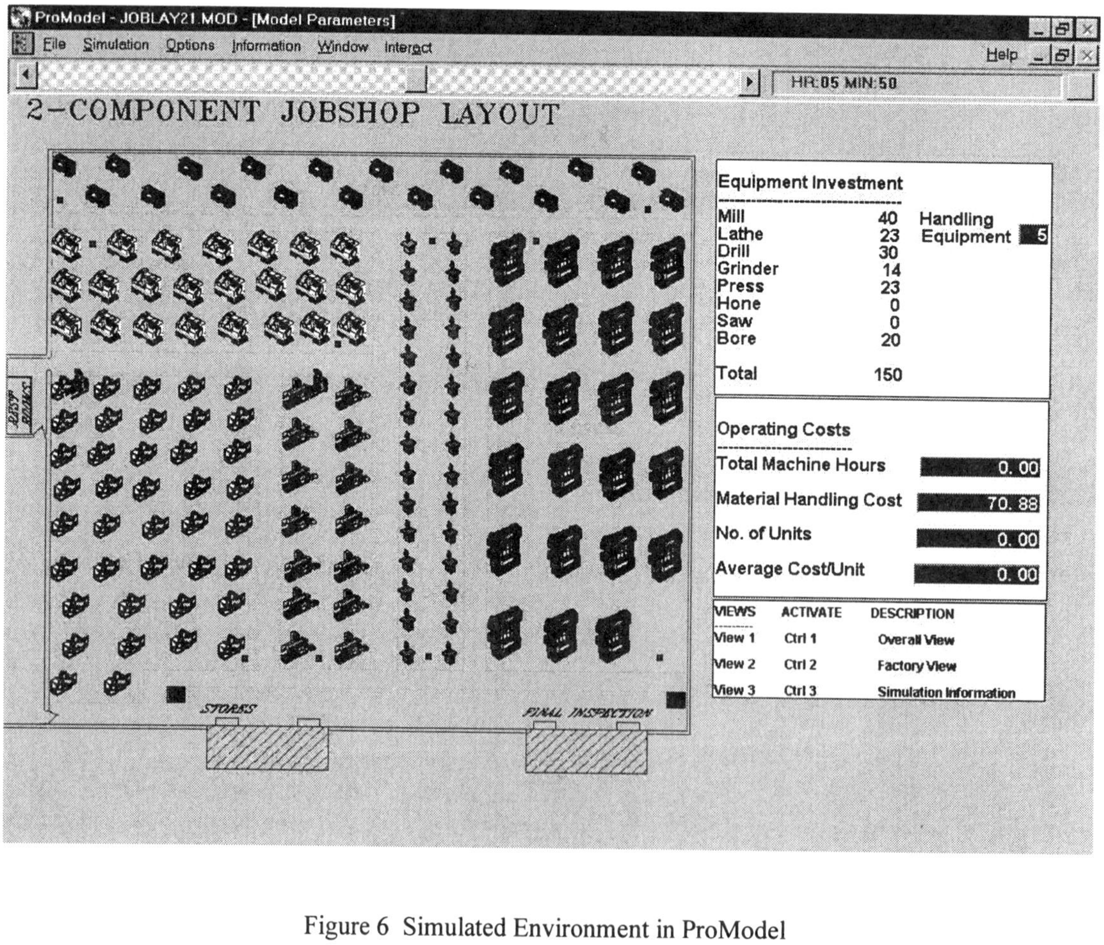

Figure 6 Simulated Environment in ProModel

Table 5 - Calculations For Machine Determination And Space Allocation For Two Component Jobshop Layout

Part # & Name	No. of Comp. per shift	Rough Stores	Milling	Lathe	Drill	Grinder	Press	Hone	Saw	Bore	Final Insp.	Total Area
		1	2	3	4	5	6	7	8	9	10	
1 Crank Case	2500		12500	7500	5000		5000					
2 Connecting Rod	2500		2500	1250	6250	5000	3750			7500		
Total Minutes Needed		0	15000	8750	11250	5000	8750	0	0	7500	0	
Total No. of Components	5000											
Machines Needed (Total mins./480 mins.)		0	31.25	18.23	23.44	10.41	18.23	0	0	15.63	0	
Efficiency Rate 80%*			39.06	22.79	29.3	13.02	22.79	0	0	19.53	0	
Machines Needed		0	40	23	30	14	23	0	0	20	0	
Area (Sq. Ft.)*		800	5000	2588	1875	1750	5750	0	0	2500	500	20763

* Assumptions:

1. An additional 25% was added to sq. footage required for machines for:
   - Maneuverability (i.e. work-in-process, aisleways, etc.)
   - Storage of parts, tools, misc.

2. 80% Efficiency Rate assumed for all machines

The user is able to see the movement of the material through the various stages of production in the simulated factory. The material is moved from the stores to the first stage of production and from there to the various departments/machines before it is finally routed to the final inspection and then dispatched. A number of different icons are used to identify the components in various stages of production, material handling icons are used to simulate the movement of material between the departments. There is also an indicator for machines in use. All this gives the students a better understanding of the material handling occurring with each type of layout. A sample screen for a model in ProModel shown in Figure 6.

On the simulation screen several different views have been defined. A few of these get activated from within the model during the simulation run to highlight the various regions of the factory and present a better view to the user, a list of these views also appears on the simulation screen with information on accessing these via the keyboard during the simulation.

4 DISSEMINATION

The learning tool discussed here has been developed for the Product Realization Consortium. This consortium is funded by the Technology Reinvestment Program and involves Cornell University, Massachusetts Institute of Technology, Tuskegee University, Worcester Polytechnic Institute, Society of Manufacturing Engineers, and North Carolina A&T State University as partners. The objective of the consortium is to revitalize manufacturing engineering education. One significant task identified by the consortium is the development of appropriate courseware. The courseware developed will be tried at multiple campuses to establish its usefulness.

ACKNOWLEDGMENTS

The work reported here was partially supported by the National Science Foundation Cooperative Agreement DMI-9413089 under the interagency Technology Reinvestment Project (TRP).

REFERENCES

Steudel, H. J., "Integrating Simulation into Team-Based Class Projects", Proceedings of the 1993 Winter Simulation Conference, eds. G. W. Evans, M. Mollaghasemi, E. C. Russell, W. E. Biles.

Richardson, J. G., "Learning Best Through Experience", Journal of Extension, vol.32, No. 2, Aug.1994.

Gogg, T. J., Mott, J. R. A., "Introduction to Simulation", Proceedings of the 1993 Winter Simulation Conference, eds. G. W. Evans, M. Mollaghasemi, E. C. Russell, W. E. Biles.

Tompkins, J. A., White, J. A., Bozer, Y. A., Frazelle, E. H., Tanchoco, J. M. A., Trevino, J., "Facilities Planning", 2nd Edition, John Wiley, NY, pp. 54-55, 1996.

AUTHOR BIOGRAPHIES

BALA RAM is a Professor in the Department of Industrial Engineering at North Carolina A&T State University in Greensboro. He received his BS in Mechanical Engineering and MS in Industrial Engineering from Indian Institute of Technology, Madras in 1975 and 1977 respectively, and he received his PhD in Industrial Engineering from the State University of New York at Buffalo in 1983. His research interests are in Material Handling and Simulation. He is a member of ASEE, IIE and SME.

RAJIV GIRDHAR is a Manufacturing Engineer with General Electric Corp. in Milwaukee, WI. He received his BS in Mechanical Engineering from the University of Bombay in 1992. He received his MS in Industrial Engineering from North Carolina A&T State University in Greensboro in July 1996. He is a member of SME.

Proceedings of the 1996 Winter Simulation Conference
ed. J. M. Charnes, D. J. Morrice, D. T. Brunner, and J. J. Swain

INJECTING SIMULATION AND MODELING UPSTREAM IN THE SYSTEM ENGINEERING PROCESS

Shahar Yarden

Whiting School of Engineering, The Johns Hopkins University
Baltimore, MD 21218, U.S.A

ABSTRACT

This paper describes the application of simulation and modeling to enhance the understanding of a communication networking solution for the Trucking Locator Control System (TLCS) case study at the Johns Hopkins University Masters Program in Technical Management. The four major steps in this process were: developing a model of the TLCS system; generating assumptions about the TLCS system; selecting a simulation tool using a trade-study; and implementing the selected simulation tool. The selected simulation package, COMNET III, was used to establish common understanding about the system under consideration, to provide insight into its performances, and to generate data for further analysis. The same process can be utilized in other System Engineering applications, not only for educational purposes, but also to provide tools for conceptual system design and to enhance the design and integration phase in the System Engineering life-cycle.

1 INTRODUCTION

The traditional roles of simulation and modeling were to conceptualize thoughts, to define alternatives, and to evaluate and predict the performances of complex systems. Until recently, most of the simulation tools, that were used to achieve these goals, were developed in high level languages (e.g., FORTRAN) and required high level of user expertise. Therefore, these simulation tools could not be used, in a timely manner, as helpful educational tools.

The introduction of new powerful and user-friendly simulation tools, such as COMNET III and ARENA, provided the means to expand the use of simulation and modeling beyond its traditional roles. For example, it has been suggested to use discrete event simulation tools to teach and motivate statistics students, and to help to present interdisciplinary case studies more effectively (Romeu 1995). This paper describes the application of communication networking simulation to enhance a System Engineering scenario that is used at the Johns Hopkins University Technical Management Masters Program in the Whiting School of Engineering.

The technical management masters program instructional methodology employs a mixture of lectures on theory and practice and realistic scenarios. One of these scenarios is the Trucking Locator Control System (TLCS) scenario, in which students assume the role of system engineers in a company that was awarded a contract to develop an information system for a trucking company.

2 THE TLCS SCENARIO

2.1 The TLCS Case Study

The TLCS scenario is part of a *System Design and Integration* course. In this course, competing teams of students are required to develop a TLCS design, based on a the following high-level requirements:

1. Present the truck location to the driver on an on-board display system
2. Send location and status updates from the trucks to a central control center
3. Present the trucks location at the control center
4. Send messages from the central control center to the trucks
5. Support a fleet of 10,000 trucks that operates all over the continuos 48 states and parts of Canada and Mexico

2.2 The TLCS Architecture

The TLCS consists of Mobile Truck Systems (MTS) that will be mounted on the trucks, a Central Control Center (CCC), and a communication service. The MTS unit consists of a navigation sub-system, a communication sub-system, an onboard computer that provides user interface to the truck driver, and an application software that include a digital map. At the CCC, a LAN of workstations loaded with a digital map and application software will be operated by a team of dispatchers. The MTSs will send location and status updates from the CCC. Using this information, the dispatchers will track loads, reroute trucks, maintain a database of trucks and drivers performances, and disseminate weather, traffic advisory, and other relevant information to truck drivers as required.

2.3 Networking Solutions for the TLCS

According to the TLCS scenario, the communication between the CCC and the MTSs will be provided by a commercial communication service. The requirement to provide full coverage of the continuos 48 states implies that this service must be provided by a mobile satellite communication system. Several mobile Satellite communication systems were considered: The American Mobile Satellite Corporation (AMSC) circuit-switched mobile satellite SKYCELL telephony service, the QUALCOMM packet-switched mobile satellite messaging service, and the ORBCOMM packet-switched mobile satellite messaging service. The simulation model that was developed to enhance the TLCS scenario is based on the SKYCELL mobile satellite service. This model can be modified in the future to represent other mobile satellite services.

2.4 The Need for a Communication Network Simulation

One of the key TLCS design issues is the selection of a communication network for the system. Due to the technical complexity and cost that are associated with communication system of that scale, it has been decided to select a commercial communication service rather than to develop an independent communication sub-system. The selection of an optimal communication networking service was identified as an area where the application of a discrete-event simulation may significantly improve the understanding of the system engineering aspects of the problem. The following requirements for a simulation and modeling tool were identified:

1. Model the TLCS users profile
2. Model candidate networking solutions
3. Develop Quality of Service (QOS) parameters
4. Evaluate the performances of candidate networking solutions
5. Assess changes in the users profile
6. Provide data for further analysis

3 SIMULATION TOOLS

3.1 Selecting a Simulation Tool

Four categories of simulation tools were considered: special-purpose Commercial On The Shelf (COTS) communication simulation packages, general-purpose COTS simulation packages, general-purpose simulation languages and libraries, and developing a simulation tool using a standard programming language.

The products in the first category , special-purpose communication simulation tools have built-in modules that represent communication network building blocks, such as nodes and links. Two COTS products were considered in this category: COMNET III by CACI and OPNET by MIL 3.

The products in the second category, general-purpose simulation tools, provide means for quick model building using graphical user interface. Some of the COTS products in this category are: ARENA by Simulation Modeling, EXTEND by Imagine That, and Workbench by SES.

The third category, Simulation Languages, consists of a large variety of languages and libraries that vary in their capabilities and level of complexity. Two of the products in this category are Mathematica by Wolfram Research and SIMSCRIPT by CACI.

The fourth category, standard programming language, was excluded from being a viable solution due to the typically long development time and high risk that are typically associated with it (Canitz 1994).

The Measures of Effectiveness (MOEs) that were used to compare between the candidate simulation tools and their relative weight are presented in table 1. These MOEs cover both the level of performance (effectiveness) and the required amount of effort (cost).

Table 1: MOEs for the Trade-Study

MOEs	Weight	Alt. 1	Alt. 2	Alt. 3
Built-in functions	1			
User friendly	3			
Easy to modify	2			
Easy to interpret	3			
required training	2			
Hardware req.	1			
Software cost	2			
Total score				

3.2 COMNET III

After completing the trade-study and after prototyping the TLCS simulation tool with both COMNET III, ARENA, and EXTEND, it has been decided to use the COMNET III simulation package. COMNET III is a discrete-event communication networks simulation environment that contains pre-defined hardware and protocol objects and built-in graphics and reporting. This enables fast and simple prototyping. COMNET III runs on both PC and UNIX workstations (Law and McComas 1995, Jones 1995).

4 ASSUMPTIONS

The following assumptions about the communication networking service were made prior to the development of the simulation model:

1. The mobile satellite service users are represented using "user groups" based on the system topology. For example, the SKYCELL system "user groups" represent users that have the same incoming calls profile that are located in geographical coverage zones that correspond to the four time zones.
2. Situations where users cross geographical coverage borders shall be ignored.
3. The of out-of-band communications can be excluded from the simulation - The out-of-band communications are the command and signaling transmissions that are used to manage the communication service, and they should never utilize more than 5 percent of the communication system resources.
4. Blockage by building and vegetation, rain attenuation, and other RF disturbances shall be represented by a simple MTBF/MTTR model
5. The candidate networking systems have pre-existing communication load that represent other users. For example, the SKYCELL system already provide telephony services to subscribers in rural areas.
6. Call inter-arrival time has a negative exponential distribution with mean λ
7. Call duration has a negative exponential distribution with mean μ

5 THE SIMULATION MODEL

5.1 The Communication Network Model

The TLCS communication network model consists of a model of the communication network and a model of the TLCS system user profile. The TLCS communication network model is shown in figure 1. The initial model of the TLCS users profile is based on the following assumptions: only 40% of the trucks are on the road at any given time, on the average each MTS sends 10 messages per hour to the CCC, and the CCC send two messages per hour to each MTS. This initial TLCS users profile is given in table 2.

Table 2: The initial TLCS users profile

	From trucks	From CCC
λ [seconds]	0.1	0.5
μ [minutes]	0.01	0.01

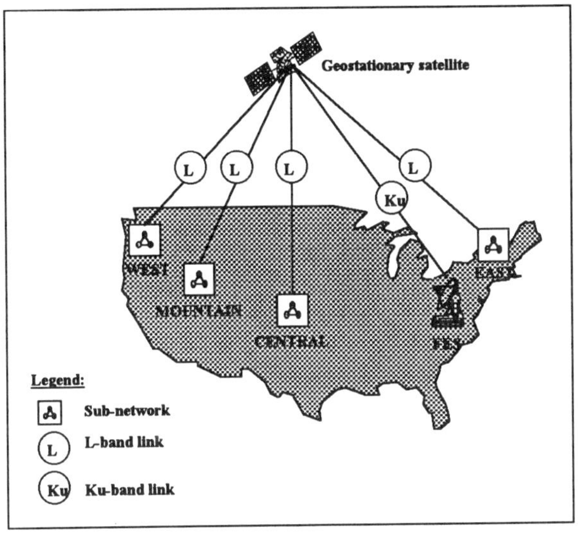

Figure 1: The TLCS communication network

5.2 Quality of Service (QOS)

A set of QOS parameters is needed to evaluate and interpret the simulation results. From the user perspective, the QOS parameters should represent the probability that a placed call go through and that it is completed successfully. From the provider perspective, the QOS parameters should represent the use of network resources. The following QOS parameters represent these requirements:

- The number of carried calls
- The usage of scared network resources
- The number of disconnected calls
- The number of preempted calls
- The blockage probability

6 A SUMMARY OF THE SIMULATION RUNS

The simulation model was run in class to evaluate the performances of a communication networking solution for the TLCS. The students actively participated in the simulation runs, by suggesting "what if" scenarios, by modifying the user profile, by trying to predict the simulation results, and by analyzing the meaning of the simulation runs output. Table 3 describes the input for a set of simulation runs: First, the simulation model was run without the TLCS users, and then it was run five more times with different TLCS users profiles. Figure 2 and Figure 3 show the resulting simulation output.

It should be noted that the number of carried calls in case 2, case 3, case 4, and case 5 is by orders of magnitude larger than the number of carried calls in case 1, and yet the blockage probability in these four cases in significantly smaller than the blockage probability in case 1. These results are different from the intuitive expected results. The reason for this behavior is that the additional load in cases 2 through 5 is made of short calls, compared with the other users of the mobile satellite service (μ = 0.01 minutes .vs. μ = 3 minutes). Case 6, on the other hand, demonstrates a situation where the incoming calls overflow the communication network.

Table 3: Simulation runs summary

Run	User profile description
1	Without the TLCS users
2	With the initial TLCS users profile
3	Longer calls (0.1 minute) from trucks
4	With 60 calls/truck/hour
5	Case 3 + Case 4
6	Longer (1 minute) calls from CCC

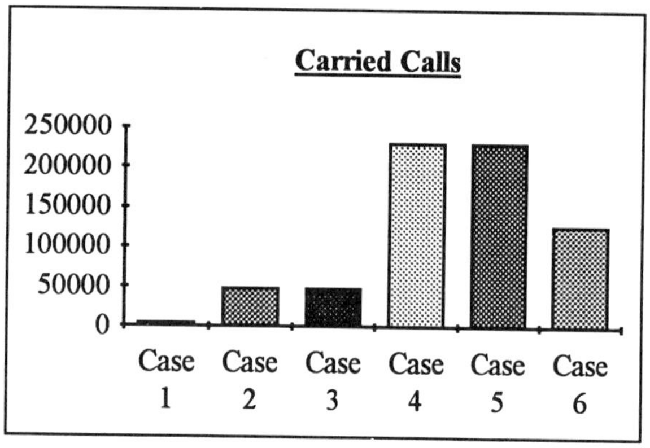

Figure 2: Number of carried calls

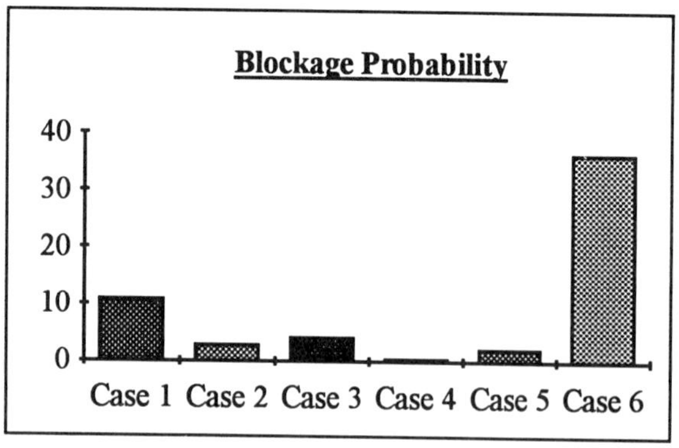

Figure 3: Blockage probability

7 CONCLUSIONS

This paper described the application of simulation and modeling to evaluate a communication network solution for the TLCS system, as part of a System Engineering case study at the Johns Hopkins University Masters Program in Technical Management.

The use of a powerful and user-friendly simulation tool (COMNET III in this case) provided an opportunity for student with no prior experience, to focus on the system under consideration, rather than on the simulation development environment. In fact, it took these students less than an hour to understand the TLCS simulation model. Although no quantitative data was collected about the use of Simulation and Modeling for this purpose, the overall response from the students who participated in this exercise was that using simulation and modeling improved their understanding of the TLCS scenario; facilitated better communication about the TLCS communication needs and the candidate solutions; and helped them to identify the TLCS communication sub-system constraints and un-used resources.

The introduction of powerful and easy to use simulation tools such as ARENA and COMNET III enables the use of Simulation and Modeling in the early phases of the System Engineering process. The use of Simulation and modeling in the early phase of the System Engineering process enhances the understanding of complex System Engineering problems, provide means to mitigate risks without having to build costly prototypes, and can also be used to generate data for further analysis (for example, cost analysis).

The System Engineering process of evaluating the performances of a communication networking solution for the TLCS consisted of four major steps: developing a model of the TLCS system; generating assumptions about the TLCS system; selecting a simulation tool using a trade-study; and implementing the selected simulation tool. The same process can be used in other applications, not only for educational purposes, but also to provide tools for conceptual system design and to enhance the design and integration phase in the System Engineering life-cycle.

ACKNOWLEDGMENTS

The author wishes to thank Mr. David Buscher, Johns Hopkins University Applied Physics Laboratory, who advised him through this project, and Mr. Dean Lamiano, American Mobile Satellite Corporation, for many helpful discussions.

REFERENCES

H. Canitz. July 1994. ARENA Dynamic Animation Simulation System. APICS - the performance advantage.

A. M. Law and M. G. McComas. 1995. Simulation of Communication Networks. *Proceedings of the 1995 Winter Simulation Conference*, ed. C. Alexopoulos, K. Kang, W. R. Lilegdon, D. Goldsman, 46 - 50.

J. Jones. 1995. COMNET III: Object-Oriented Network Performance Prediction. *Proceedings of the 1995 Winter Simulation Conference*, ed. C. Alexopoulos, K. Kang, W. R. Lilegdon, D. Goldsman, 545 - 547.

J. L. Romeu. 1995. Simulation and Statistical Education. *Proceedings of the 1995 Winter Simulation Conference*, ed. C. Alexopoulos, K. Kang, W. R. Lilegdon, D. Goldsman, 1371 - 1375.

AUTHOR BIOGRAPHY

Shahar Yarden has a BS. degree in Aeronautical Engineering from the Technion Institute of Technology in Haifa, Israel in 1983. He received an MS degree in Technical Management from Johns Hopkins University in 1995, specializing in System Engineering and Project Management. His current position is a Test Manager, working for Computer Sciences Corporation (CSC) on a software development and maintenance program for the Department of Defense. Prior to that he worked for CSC as project integrator on NASA's Hubble Space Telescope control center re-engineering program. Mr. Yarden is also a guest lecturer at Johns Hopkins University Whiting school of Engineering.

Proceedings of the 1996 Winter Simulation Conference
ed. J. M. Charnes, D. J. Morrice, D. T. Brunner, and J. J. Swain

SENSITIVITY ANALYSIS OF A LARGE SCALE TRANSPORTATION SIMULATION USING DESIGN OF EXPERIMENTS AND FACTOR ANALYSIS

Glenn G. Rousseau

Studies and Analysis Flight
Headquarters Air Mobility Command
402 Scott Drive, Unit 3L3
Scott Air Force Base, Illinois 62225, USA

Kenneth W. Bauer, Jr.

Department of Operational Sciences
Graduate School of Engineering
Air Force Institute of Technology, 2950 P Street
Wright-Patterson Air Force Base, Ohio 45433-7765, USA

ABSTRACT

This paper describes how design of experiments and factor analysis were used to conduct sensitivity analysis on multivariate output from a large scale transportation simulation model. Specifically, this research focused on the sensitivity of airlift system performance to changes or errors in a list of transportation requirements. The general approach included perturbing a time-phased list of transportation requirements according to an experimental design and using a simulation model to estimate the airlift system performance response. We used factor analysis to reduce the dimensionality of the multivariate output data and to generate sensitivity plots, which proved to be valuable graphical tools for sensitivity analysis. Additionally, we identified how factor analysis can be used as a verification and validation tool for large stochastic simulation models.

1 INTRODUCTION

The Airlift Flow Module (AFM) is a large-scale transportation simulation model used primarily for estimating strategic military airlift system performance. Some of the basic airlift system features modeled in AFM include: aircraft, aircraft crews, air bases, routes, aerial refueling, ground refueling, materiel handling equipment, cargo, and passengers. The performance of an airlift system, as a whole, is the result of a complex interaction of all these features. This paper describes how AFM performance sensitivities provide insight for improving airlift system performance.

This paper is organized as follows. In Section 2, we provide background information on the AFM simulation, the modeled scenarios, and airlift system performance measures. In Section 3, we discuss how we perturbed four characteristics of time-phased force deployment data (TPFDD) according to a 2^{4-1} fractional factorial experimental design. In Section 4, we present a quick review of factor analysis and then discuss specific results from an airlift scenario example. In Section 5, we highlight implications pertinent to verification and validation. Finally, in Section 6, we recapitulate our main findings. This paper is based on Rousseau (1996).

2 BACKGROUND

2.1 The AFM Simulation

The AFM simulation contains over 50,000 lines of FORTRAN code and models to the individual aircraft and pallet level of detail. AFM was selected for this research because of its flexibility in representing airlift system activities and its ability to estimate many detailed aspects of airlift system performance. Currently, this flexibility and detail come at the costs of long setup and run times.

Being demand driven, AFM needs an airlift requirement for cargo or personnel in order to plan a mission. The AFM mission planning algorithm is deterministic, starting at the top of the TPFDD and reading sequentially to plan a mission for the first available and compatible load with the next available aircraft. During AFM mission execution, the aircraft's simulated flight times and ground times are drawn from random distributions defined by the user.

2.2 The Modeled Scenarios

Two different airlift scenarios were modeled and they will be generically referred to as the "small" and the "large" scenarios, respectively. Only pertinent results from the small scenario are presented in this paper. The small scenario models a minor conflict in the Caribbean. It requires 75,854 tons of cargo and 139,480 passengers to be delivered over a fairly simple air base network using 138 total aircraft. The large scenario models a major conflict in Southwest Asia. It requires 402,796 tons of cargo and 498,369 passengers to be delivered over a complex air base network using 448 aircraft.

2.3 Multivariate Output Data

Airlift system performance was broken down into three major categories: aircraft performance, throughput, and timeliness. For each run, the AFM simulation was stopped at day 30 to record the following performance measures for later analysis:

Aircraft Performance Measures (by aircraft type)
- utilization rate
- ground time per cycle
- average payload
- average million-ton-miles per day

Throughput Measures
- outsize tons delivered
- total cargo tons delivered
- total passengers delivered

Timeliness Measures
- percent of shipments delivered ontime
- average number of days late

Some of these variables are further explained for those not familiar with airlift. The aircraft performance measures represent per aircraft averages across all aircraft of a homogeneous type (e.g., C-141, C-5, B-747, etc.) modeled in the simulation. Utilization rate is a gross estimate of aircraft productivity and is a number between 0.0 and 24.0, representing the average number of hours per day an aircraft of that type flew (Kowalsky 1977). Conversely, ground time per cycle is a gross estimate of non-productivity, representing the average number of hours each aircraft of that type was on the ground during an average mission cycle. A mission cycle starts when an aircraft takes off to start a mission and ends when the aircraft lands at the recovery base after mission completion (Kowalsky 1977). The average million-ton-miles per day measure is used as gross estimator of the productive workload accomplished by each aircraft. Most of the large tanks, missile batteries, and other firepower that need to be airlifted are classified as outsize cargo. Therefore, total outsize tons delivered is an indirect measure of airlift effectiveness with respect to the capability to conduct combat operations.

3 EXPERIMENTAL DESIGN

Four general TPFDD characteristics were selected and perturbed according to a 2^{4-1} fractional factorial design to identify their effects on airlift system performance. The four characteristics were: location, timeline, total demand, and cargo flavor. A short definition of these characteristics and brief descriptions of the perturbation strategies are presented in the following paragraphs.

The location characteristic refers to the onload and offload locations, or in airlift terms, the aerial ports of embarkation (APOE) and debarkation (APOD). To perturb this location characteristic, new TPFDD files were generated that were identical to the original TPFDD except for the specified APOEs and APODs. For each line of the original TPFDD, the specified onload and offload locations were replaced by randomly selected locations, then output to the new TPFDD file. The random selection process was constrained to keep the new APOE-APOD distance within 10 percent of the old distance. The intent of this constraint was to keep the airlift workload, estimated in ton-miles, fairly constant across all generated TPFDD files.

The timeline characteristic refers to available-to-load (ALD) and required delivery dates (RDD) for each requirement line in the TPFDD. To perturb this timeline characteristic, new TPFDD files were generated that were identical to the original TPFDD except for the specified ALDs and RDDs. For each line of the original TPFDD, the specified ALD and RDD were replaced by randomly selected dates, then output to the new TPFDD file. The random selection was accomplished by choosing -1, 0, or 1, with equal probabilities, and adding it to both the ALD and RDD. Since the AFM mission planning algorithm reads a TPFDD sequentially, from the beginning, each time it plans a mission, there is an implicit assumption of priority for each line of the TPFDD based on its order in the file during execution. In reality, changing the RDD usually implies a change in airlift priority. To effect the same conceptual change in priority during AFM execution, the lines in the newly generated TPFDD were sorted appropriately.

The total demand characteristic refers to the total amount of cargo and passengers needing airlift. The cargo portion is broken down further into outsize, oversize, and bulk in AFM to determine compatibility with the different aircraft types. To perturb this total demand characteristic, new TPFDD files were generated that were identical to the original TPFDD except for the specified cargo and passenger requirement amounts. For each line of the original TPFDD, the specified demand requirement values were increased by 10 percent, then output to the new TPFDD file.

Cargo flavor is not really an airlift term, but seemed to be an appropriate simple label for this characteristic. For this paper, the cargo flavor characteristic refers to the relative amounts of cargo within each requirement. To perturb this cargo flavor characteristic, new TPFDD files were generated that were identical to the original TPFDD except for the outsize and oversize cargo requirement amounts. For each line of the original TPFDD, the outsize cargo requirement was doubled and an equal tonnage was subtracted from the oversize cargo requirement, then output to the new TPFDD file.

An orthogonal 2^{4-1} fractional factorial design formed the basis for the experimental design shown in Table 1, allowing for unbiased estimation of the effects of the four perturbation schemes and all possible 2-way interactions (Box and Draper 1987). Two more design points were added to directly observe the effects of the two random perturbation schemes. Ten replications of each design point were then fed into the AFM simulation and the desired output variables from all 100 runs were recorded in a flat file for later analysis.

Table 1: Experimental Design for Evaluating Four TPFDD Perturbation Effects

Design Point	*Location*	*Timeline*	*Total Demand*	*Cargo Flavor*
0	0	0	0	0
1	0	0	1	1
2	0	1	0	1
3	0	1	1	0
4	1	0	0	1
5	1	0	1	0
6	1	1	0	0
7	1	1	1	1
8	0	1	0	0
9	1	0	0	0

4 FACTOR ANALYSIS

Factor analysis was used to interpret the effects of the four TPFDD perturbations on the selected AFM output. Factor loadings matrices provided pertinent information on the relationships between the variables as they responded to the TPFDD perturbations. In addition, visual pictures of output sensitivity to the perturbations were found by plotting two factors against each other. These "sensitivity plots" proved to be useful tools for sensitivity analysis.

Factor analysis is a multivariate data reduction technique that identifies common factors underlying a set of observed variables. The formal factor analysis model describes each original observable variable in terms of a linear sum of unobservable common factors and a single latent unique factor, as shown by Equation (1) (Dillon and Goldstein 1984).

$$\mathbf{X} = \Lambda \mathbf{f} + \mathbf{e} \qquad (1)$$

where

$\mathbf{X} \equiv$ p-dimensional vector of observed responses
$\mathbf{f} \equiv$ q-dimensional vector of unobservable common factors ($q \ll p$)
$\Lambda \equiv p \times q$ matrix of weights or factor loadings
$\mathbf{e} \equiv$ p-dimensional vector of unobservable unique factors

Assuming the common and unique factors are uncorrelated with themselves and each other, the covariance matrix of the response vector **X**, denoted by Σ_{xx}, can be expressed as shown in Equation (2) (Dillon and Goldstein 1984). In simplified terms, the object of factor analysis is to replace many variables, **X**, with much fewer variables, or factors, **f**, preserving as much information as possible during the transformation to the new variables (factors). The preserved information is represented in the common variance, $\Lambda\Lambda'$.

$$\Sigma_{xx} = \Lambda\Lambda' + \Psi \qquad (2)$$

where

$\Sigma_{xx} \equiv p \times p$ variance-covariance matrix
$\Lambda \equiv p \times q$ matrix of weights or factor loadings
$\Lambda\Lambda' \equiv$ common part of variance-covariance matrix
$\Psi \equiv p \times p$ diagonal matrix containing p unique variances

Once a solution to Equation (2) is found, Λ represents a matrix of possible correlation coefficients between the new factors and the original variables. The patterns of loadings contained in Λ identify which variables each factor represents and are subject to interpretation by the analyst. This interpretation can usually be enhanced by a rotational transformation of the loadings matrix using matrix multiplication. The rotated matrix of factor loadings represents an alternative interpretation of the data and allows the analyst a degree of flexibility (Dillon and Goldstein 1984). The number of such rotational transformations, orthogonal or oblique, is infinite, leading to criticism that any solution chosen is, mathematically speaking, arbitrary (Basilevsky 1994). However, selection of certain criterion functions can define unique rotations and confront this criticism with more generally accepted statistical practice (Basilevsky 1994).

A commonly used rotation criterion results in a unique rotation called the varimax rotation. The varimax rotation seeks to maximize the variation of the squared factor loadings within each factor, thereby forcing the loading coefficients to either really high or low values. With not many of the loadings falling into the middle "gray area" it is usually easier to find an interpretation from the loadings matrix structure.

Table 2 shows the results of a varimax rotation and highlights the significant factor loadings from the factor analysis of the small scenario's data. (All loadings were not shown to simplify this discussion.) These loadings were deemed to be significant if they were the largest (in magnitude) value in any particular row of the table. Selecting the significant loadings in this manner identified which factors correlated most closely with the original variables. Notice that one variable, C-17 MTMs/acft/day, has two entries in this table because the loadings were fairly equal. All this means is the information contained in C-17 MTMs/acft/day was split between factors 1 and 4, which are orthogonal dimensions in the new factor space.

Table 2: Small Scenario Factor Loadings Matrix

Factor	1	2	3	4
C-141 MTMs/acft/day	-.91			
C-141 use rate	-.89			
C-5 MTMs/acft/day	.89			
C-141 average payload	-.88			
C-5 use rate	.85			
B-747P use rate	.84			
KC-10 use rate	.83			
KC-10 MTMs/acft/day	.77			
C-17 ground cycle	.76			
C-141 ground cycle	.76			
Total Passengers	.75			
C-5 average payload	.73			
C-17 MTMs/acft/day	.63			.59
KC-10 average payload	-.59			
Total outsize tons		.93		
C-17 use rate		.84		
C-17 average payload		-.80		
% Shipments ontime			-.81	
Average days late			.78	
Total cargo tons				.94
Eigenvalues	9.46	2.98	2.39	1.71
% of Total Variance	47%	15%	12%	9%
Cumulative %	47%	62%	74%	83%

The dimensionality of the original data was reduced from 20 to 4, reflecting the simplicity of the small scenario. The 4 new dimensions account for 83 percent of the information. For comparison, though not listed in this article, the dimensionality of the more complex large scenario data was reduced from 29 to 8 factors, accounting for 80 percent of the variance/covariance information. For a mild increase in the number of original variables, the number of factors doubled, indicating the impact of increasing the complexity of the model greatly increases the dimensionality and complexity of the output.

Table 2 further indicates, in general, a positive correlation between factor 1 and positive widebody aircraft (B-747, C-5, C-17 and KC-10) performance. Passenger throughput is also positively correlated with factor 1, however, C-141 performance is negatively correlated. One possible valid interpretation of the factor 1 loadings indicate that the changes made to the TPFDD caused an increase in widebody aircraft performance at the expense of C-141 performance. The dimension of throughput most highly affected by this change in aircraft performance was passenger throughput. Factor 1 could be labeled as a widebody and passenger **index**, which is the label chosen for the rest of this discussion. It could also be labeled as a widebody and C-141 passenger performance **contrast**. In either case, positive values for this factor indicate positive widebody and passenger throughput performance, and negative values indicate positive C-141 performance.

A similar interpretation of factor 2 loadings identifies a positive relationship between total outsize tons throughput and C-17 usage. In other words, variance in the C-17 use rate accounted for the majority of the variance in total outsize tons delivered. The negative correlation with C-17 payload amounts reflects the less dense nature of outsize cargo (in tons/unit volume). Factor 2 is most definitely an index reflecting total outsize tons throughput. Similarly, factor 3 is some sort of lateness index, and factor 4 is a total cargo tons throughput index.

Most statistical packages that can solve Equation (2) can also solve Equation (1) for **f**, the factor scores. In terms of the small scenario, for each of the 100 runs conducted for the experimental design, the 20 output variable values were transformed into 4 factor variable scores. These factor scores were used to graphically represent the AFM simulation's output for sensitivity analysis.

The sensitivity plot shown in Figure 1 is a result of plotting the factor 2 scores against the factor 1 scores for all 100 runs from the small scenario. The plotted points were coded to identify the experimental design points listed in Table 1. The axes were labeled as indices to remind the reader of the general information represented by each factor. The cluster of points representing design point 0 provides a focus for interpreting this plot because the TPFDD was not perturbed for these 10 runs. The only difference among these 10 runs was the random seed selected at the start of the simulation. The size (area) of the reference cluster in Figure 1 approximately represents the variance, or random noise, expected from the stochastic nature of the AFM simulation. Design point clusters larger (covering more area) than the reference cluster show an effect on variance that is attributable to the design point. Clusters for design points 4, 5, 6, 7, and 9 were larger, indicating that perturbing the location characteristic greatly increased the variance of some airlift system performance measures.

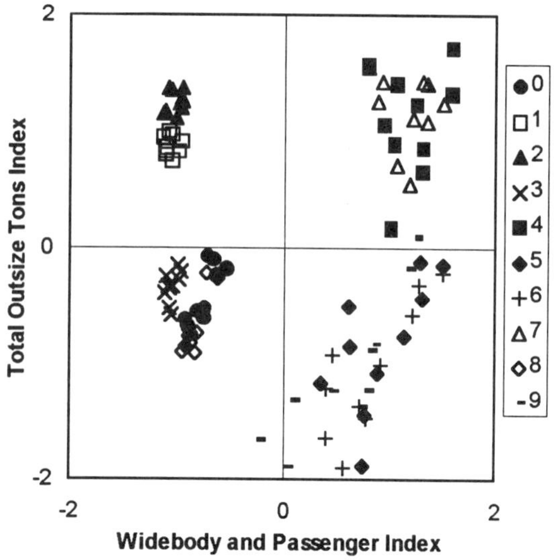

Figure 1: Sensitivity Plot of All Runs

Another observation from this sensitivity plot (Figure 1) has to do with the shape of the design point clusters. The cluster for design point 0 seems to be broken into at least two distinct groups. Furthermore, the clusters for design points 6, 8, and 9 also seem to exhibit the same phenomenon. The investigation into this phenomenon led to a discovery about the instability of one of the algorithms in the AFM simulation. The overall effect of this instability was not considered significant by the users of the AFM simulation, but the discovery did identify one potential model improvement area.

The plot shown in Figure 2 is a result of plotting the centroid of each design point cluster. Displacements between the plotted centroids indicate a change in airlift system performance, as measured by the AFM simulation. Once again, the reference is design point 0. Noticeable displacements to total outsize tons throughput are identified by design points 1, 2, 4, and 7. These 4 design points reflect the increase in outsize throughput resulting from perturbing the cargo flavor characteristic, where the demand for outsize cargo was increased. Similarly, the centroids for design points 4, 5, 6, 7, and 9 are displaced from the reference point, reflecting an increase in widebody aircraft performance and an increase in passenger throughput performance.

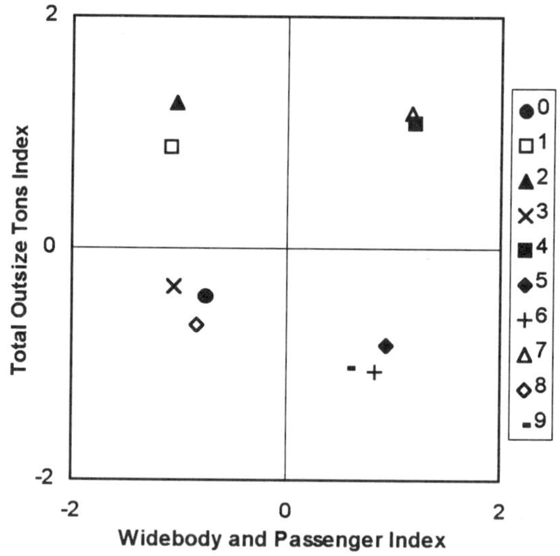

Figure 2: Sensitivity Plot of Design Point Centroids

5 VERIFICATION AND VALIDATION

The following definitions of verification and validation from Pritsker (1986) are presented to support the following discussion. Verification is the process of establishing that the computer program executes as intended. Validation is the process of establishing that a desired accuracy or correspondence exists between the simulation model and the real system.

A slightly different look at the factor analysis model in Equation (1) leads to implications supporting stochastic model verification and possibly validation. The view assumed so far sees a factor analysis model which attempts to account for variation in multivariate data by identifying a small number of common factors which encompass as much of the original variation as possible. A slightly different view is analogous to putting on factor analysis glasses to look inside the multivariate data and see what factors, or processes, are causing the data to change. If the latter view is accepted, factor analysis can be seen as a tool to identify the source of variation in multivariate data.

By changing the random seed from run to run in a complex, stochastic simulation model, the simulation's internal stochastic processes are the known source of variation in output data. If we then put on factor analysis glasses to look at the simulation's multivariate output data, we should be able to see the simulation's stochastic processes as the source of output data variation. If factor analysis results adequately reflect the simulation's stochastic processes then the program is working as intended and we have a method of verification. The choice of output variables selected determines the level of verification, allowing the analyst to tailor this methodology to focus on whatever process or combination of processes needs verification.

Using the small scenario example, Table 3 shows the loadings matrix for the output from 50 runs, where the only difference between the runs was the random seed. All the relationships between variables that loaded on the same factor were explainable through knowledge of AFM's internal processes and knowledge of how those process are supposed to work. With the aggregate variables selected to measure airlift system performance, an aggregate, surface-level verification of AFM was the result. From the information contained in Table 3, the AFM simulation's processes seem to work as designed.

Table 3: Factor Loadings Matrix for 50 Seed Runs

Factor	1	2	3	4	5
Total cargo delivered	0.97				
C-17 MTMs/acft/day	0.95				
Passengers delivered	0.91				
C-141 use rate	0.87				
C-141 MTMs/acft/day	0.86				
C-17 use rate	0.85				
C-17 ground cycle	-0.74				
C-17 avg payload	0.71				
KC-10 use rate		0.92			
KC-10 MTMs/acft/day		0.85			
KC-10 avg payload		-0.61			
C-5 MTMs/acft/day			0.88		
C-5 use rate			0.82		
C-141 avg payload			-0.68		
Average days late				0.83	
% Shipments ontime				-0.72	
B-747P use rate				-0.56	
C-5 avg payload					0.75
Outsize delivered					0.56
C-141 ground cycle					0.41
Eigenvalues	6.24	2.40	2.34	1.96	1.44
% of Total Variance	31%	12%	12%	10%	7%
Cumulative %	31%	43%	55%	65%	72%

Furthermore, the small scenario was designed to favor the smaller C-17 and C-141 aircraft, which is reflected in the factor 1 loadings. As can be seen in Table 3, passenger and cargo throughput are positively correlated with the C-17 and C-141 aircraft performance variables. Since the AFM formulation of the small scenario appears to execute similar to expected performance of the real world airlift system, there appears to be some potential for validation with this method, as well.

Another tool to use during a validation effort would be a plot of the factor scores similar to what was done to create the sensitivity plots in the previous section. In this example, Figure 3 is the plot of the factor 2 scores against the factor 1 scores for all 50 random seed runs. The distinct grouping of the output into at least two groups led to a discovery of a weakness in one of the model's mission planning algorithms. In this case, some of the cargo was ignored by the AFM simulation if it was too small, or trivial, to justify dispatching an aircraft. The size of a trivial cargo load is aircraft dependent and therefore the smaller aircraft, with the smaller trivial load limits will pick up more cargo throughout the duration of the simulation scenario. In this case, some of the runs were running out of cargo to deliver before the simulation ended because cargo was being ignored. The source of this effect was a combined function of the size of each requirement line in the TPFDD and the trivial load limit algorithm. Graphical plots similar to that shown in Figure 3 should be considered an essential part of a verification effort using this factor analysis technique, along with a factor loadings matrix similar to that shown in Table 3.

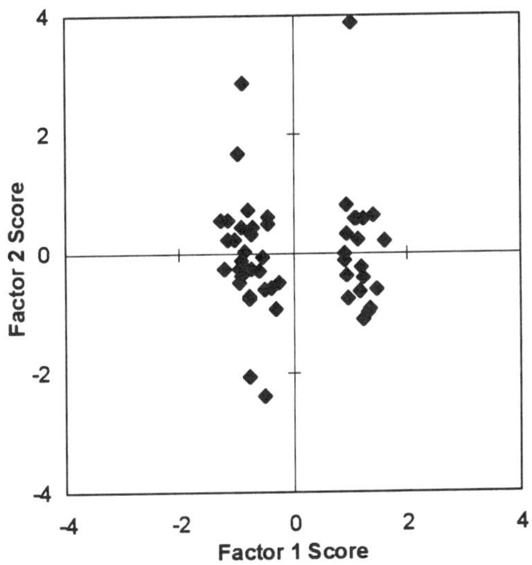

Figure 3: Plot of Factor Scores for 50 Seed Runs

6 SUMMARY AND CONCLUSIONS

Design of experiments and factor analysis techniques provide a powerful combination for analyzing large, stochastic simulations. This paper demonstrated an approach to conduct sensitivity analysis on the output of a large, stochastic airlift simulation model. We demonstrated how the factor loadings matrix can be used as a tool to interpret what happened in the simulation. In our example, the sensitivity plots created from the factor scores graphically depicted the relative magnitude and variance of the simulation's response to each experimental design point. We also successfully highlighted a new application area for factor analysis in stochastic simulation model verification and validation. The potential to tailor this verification and validation approach to any desired level of detail seems promising and worthy of further research.

REFERENCES

Basilevsky, A. 1994. *Statistical factor analysis and related methods: Theory and applications.* New York: Wiley.

Box, G. E. P., and N. R. Draper. 1987. *Empirical model-building and response surfaces.* New York: Wiley.

Dillon, W. R., and M. Goldstein. 1984. *Multivariate analysis.* New York: Wiley.

Kowalsky, T. 1977. Utilization rates and productivity. Study report, Studies and Analysis Flight, Air Mobility Command, Scott Air Force Base, Illinois.

Pritsker, A. A. B. 1986. *Introduction to simulation and SLAM II.* 3d ed. New York: Halstead.

Rousseau, G. G. 1996. Airlift system sensitivity to perturbed time-phase force deployment data. Master's thesis, Department of Operational Sciences, Air Force Institute of Technology, Wright-Patterson Air Force Base, Ohio.

AUTHOR BIOGRAPHIES

GLENN G. ROUSSEAU, Captain, USAF, is an airlift pilot currently serving as a mobility analyst for Air Mobility Command at Scott Air Force Base, Illinois. He received a B.S. degree in engineering physics from the United States Air Force Academy in 1985, and he received an M.S. degree in operations analysis from the Air Force Institute of Technology in 1996. His research interests are in statistics and simulation modeling, primarily with application to transportation problems. He is a member of Tau Beta Pi and INFORMS.

KENNETH W. BAUER, JR., Lt Col, USAF, is an Associate Professor in the Department of Operational Sciences, School of Engineering, at the Air Force Institute of Technology (AFIT). He received a B.S. in Mathematics from Miami University (Ohio) in 1976, a M.E.A. from the University of Utah in 1980, a M.S. from AFIT in 1981, and a Ph.D. from Purdue University in 1987. His research interests are in the statistical aspects of neural networks and simulation modeling. His mailing address is: AFIT/ENS, WPAFB, OH, 45433-7765.

Proceedings of the 1996 Winter Simulation Conference
ed. J. M. Charnes, D. J. Morrice, D. T. Brunner, and J. J. Swain

REGRESSION METAMODELS AND DESIGN OF EXPERIMENTS

Willem J.H. van Groenendaal
Jack P.C. Kleijnen

Department of Information Systems and Auditing (BIKA)/Center for Economic Research (CentER)
School of Management and Economics (FEW)
Tilburg University (KUB)
5000 LE Tilburg, NETHERLANDS

ABSTRACT

Simulation models are often used to support decision making for problems with uncertain inputs and parameters. Three types of models are used: deterministic, risk, and uncertainty models. Risk models are popular with researchers, but can be used only when the joint probability distribution of the inputs and parameters is known. In many real-life situations, however, this is not the case. Uncertainty models are too restrictive for real-life situations. Therefore deterministic models are then used. The sensitivity of the results is often analyzed by changing one factor at a time or by simulating a few scenarios. This paper, however, shows that in case of uncertainty it might be better to apply design of experiments (DOE) in combination with regression metamodels.

1 INTRODUCTION

Several methods may be used to analyze the sensitivity of model output with respect to changes in input variables, parameters, and model structure. (In the sequel we shall refer to all these changes as factors.) The models we have in mind are simulation models that support decision making when no analytical models are available to describe the problem and to find an optimal solution. In economic theory three types of models are distinguished: certainty or deterministic, risk, and uncertainty models. Which type of model is used, depends on what information is available about the real-life problem. In case of certainty, each set of factor values invariably leads to a specific outcome. Economists speak of risk if the factors are stochastic but their joint probability distribution is known. In case of uncertainty the joint probability distribution is unknown. We restrict uncertainty here to uncertainty of nature; that is, we exclude situations where the managers are competing against an intelligent opponent (game theory).

The fact that a model is classified as one of these three types does not mean that the real-life problem falls in the same category. For large econometric and environmental models it is quite common to ignore the stochastic nature of the actual problem, once (part of) the model is estimated. But for relatively simple models, such as net present value (NPV) models, users may also stick to the deterministic analysis. There seems to be a gap between the research community and the other model users, such as businesses and governmental organizations. The former are more interested in risk and uncertainty models, whereas the latter usually stick to deterministic models. We are interested in real-life problems with uncertainty, so we can perform either a risk analysis or a deterministic analysis of the situation.

As case study we use an investment analysis for a gas transmission project on the island of Java in Indonesia (Van Groenendaal, 1996). For the evaluation of this project the World Bank's guidelines are followed (Ward and Deren, 1991). The NPV is the criterion used to judge the investment. This paper aims at methods that can be used within the restricted time available in applied work.

It turns out that in the base-case scenario the project is feasible (NPV = 1,751 billion Rupiahs or 815 million US$ in 1991 prices). Ten factors play a role in the NPV calculation (also see Table 1).

In § 2 we review deterministic, risk, and uncertainty models in relation to robust decision making and their applicability in applied work. In § 3 we use DOE and regression metamodels to obtain the information required by decision makers. In § 4 we present conclusions.

2 ROBUST DECISIONS

When modeling is used to support decision making, the goal is often to support the choice of a policy from a set of alternatives. Particularly in the context of long term decision making (such as large scale investment problems), the uncertainty of future circumstances is a major concern. In practice, often a particular set of values or circumstances is chosen and the problem is analyzed for this set only. The set is either a base-case (most likely), or a worst-case scenario. Based on this analysis, a policy is recommended. What decision makers want to know, is how robust their decision is. "Robust" means that commitments are made as late as possible, and plans can be changed as much as possible when the future is not as anticipated.

Broens (1995, p. 41) uses the term flexible planning as opposed to robust planning. He claims that in robust planning the plan is fixed, but has to cover as many alternative scenarios as possible. This implies that the intent is *not* to change the plan, and to stick to the original commitment. Broens's view is not generally accepted; we use the term robust decision to indicate that a decision has to facilitate as many scenarios as possible, and if possible this decision should include opportunities to alter the plan if circumstances change during execution.

2.1 Deterministic models

When using deterministic models, analysts often rely on some form of what-if analysis to determine the robustness of a decision. Often they change one variable at a time, and they analyze (main) effects. The one-factor-at-a-time approach is popular among economists (Ward and Deren, 1991; Van Groenendaal and Vingerhoets, 1995). This approach, however, is misleading when used to support a decision, because it does not take into account interactions. It is well known among simulationists that it can easily give the decision makers the wrong impression about the problem; in our example no single factor leads to an NPV smaller than zero (§ 3).

In policy related studies, analysts often construct scenarios, that is, a particular choice of factor values. The analysts examine the differences among scenarios. Normally, they compare results with respect to the base case scenario. At best they use a few scenarios; see for example Grübler and McDonald (1996). Managers, however, are often interested in the worst case scenario only. It is difficult to formulate scenarios on which all decision makers agree, especially when the evaluation period is long and the problem complex. The result of the analysis is then weakened by the discussions about the likelihood or even the feasibility of scenarios; alternative scenarios will be proposed. But even if there are no discussions on the scenarios, this approach results only in descriptions of a limited number of futures. In applications often two or three scenarios are used; hardly ever, more than five scenarios are analyzed.

To avoid discussions on scenarios or neglect interactions between factors, we propose to use DOE in combination with regression metamodels. The advantage of this approach is that it gives information on the whole experimental area of interest, to be used by decision makers.

2.2 Risk models

Instead of using a deterministic model to analyze a problem with uncertainty, analysts may introduce a joint probability distribution for the factors. Especially with the growth of software (such as @RISK), this option is becoming more and more popular. Often marginal distributions are formulated based on minimum information per factor: minimum, maximum, and most likely values. The marginal distributions are then combined into a joint distribution, assuming independent factors (for dependent factors see the references in Kleijnen 1996). However, such an approach will not necessarily improve results compared to deterministic models: garbage in garbage out. It does not produce more useful information for the decision makers. As was already shown by Iman and Conover (1982), if information on the joint probability distribution of the factors is wrong, the results obtained from rather arbitrary assumptions on correlations between factors are useless. Furthermore, in many cases the tail of the output distribution is most important for decision support, but this is also the part about which the reliability of the information is lowest; see Kunreuther et al. (1983) for an example.

Instead of using "objective" data, the analysts may use subjective probabilities (expert judgements). Expert judgements, however, will be useful only in case of a limited number of variables and decisions that resemble past decisions. And even then, mistakes are easily made; see Granger and Herion (1990). Again the correlation between factors will be a major problem.

Another approach to getting a robust overall result is to combine the results for a set of scenarios by asking the decision makers to assign subjective probabilities to the individual scenarios (Draper, 1995). These probabilities can then be used to calculate the expected scenario.

Note that in case the information on the joint probability distribution is available but unreliable, the analysts also need to perform a robustness analysis for the distribution, for example, by using the Score Function method; see Rubinstein and Shapiro (1993).

2.3 Uncertainty models

A third type of models is decision models under uncertainty, such as stochastic dominance models (Copeland and Weston, 1992, pp 92-6). This type relies on the axioms of utility theory, which unfortunately are not met in practice (Tversky and Kahneman, 1988).

Most methods using probability or utility are rejected by practitioners. They prefer simple methods, such as one-factor-at-a-time or "switching" values (factor value that causes the NPV to become negative; see Gittinger, 1982, pp 371-3), in combination with the worst case scenario.

Table 1: A Plackett-Burman design for ten factors in the NPV simulation model

combination factor	1	2	3	4	5	6	7	8	9	10	11	12
1 investment costs	+	+	-	+	+	+	-	-	-	+	-	-
2 construction time	-	+	+	-	+	+	+	-	-	-	+	-
3 reserves West Java	+	-	+	+	-	+	+	+	-	-	-	-
4 real GVA	-	+	-	+	+	-	+	+	+	-	-	-
5 energy prices	-	-	+	-	+	+	-	+	+	+	-	-
6 relative gas/oil price	-	-	-	+	-	+	+	-	+	+	+	-
7 purchase prices	+	-	-	-	+	-	+	+	-	+	+	-
8 coal prices	+	+	-	-	-	+	-	+	+	-	+	-
9 other costs	+	+	+	-	-	-	+	-	+	+	-	-
10 discount rate	-	+	+	+	-	-	-	+	-	+	+	-

For a robust decision the worst case approach is probably the best of the naive methods, because then the actual result will always be better. A problem is how to find the worst case scenario. Moreover, the worst case approach seems rather pessimistic. As Broens (1995) points out, too much attention is given to the worst case; all other information is left unused. Furthermore, the worst case is usually chosen from the likely scenarios; unlikely scenarios are excluded, despite the fact that information on these scenarios might be useful.

A final problem is that risk and uncertainty models often require the problem to fit into a certain form, for example, dynamic programming. Many real-life problems do not meet this requirement.

In summary, many methods for sensitivity analysis do not meet the needs of practitioners. These models are either too rigorous and do not result in the information required by the decision makers, or they require too much information, which is not available. When uncertainty about factors is a real problem and no information on the joint probability distribution can be obtained, DOE combined with regression metamodelling can be a solution, as we shall show next.

3 DESIGN OF EXPERIMENTS

Whatever experimental approach is used, the first problem is to find the appropriate experimental area. Our approach is to start with a base case scenario, that is, a scenario based on the most likely values of the factors. In our investment case study the values concern the amount of gas reserves available in West Java, the growth paths for the Indonesian economy, energy prices, etc.; see Table 1. The base case scenario we obtained through interviews with experts in the various Indonesian ministries and industries involved in the project. For this scenario the project is feasible; that is, it has a positive NPV. Once we had this information, we discussed with our counterparts to what extend the factors could deviate from their base case values. Our analysis focusses on those conditions that will jeopardize the positive advise for the investment project (which follows from the positive value for the NPV in the base case). The reason for this focus is that decision makers are more interested in what can cause the project to become infeasible than in windfall profits. Hence all results in our sensitivity analysis are expected to be worse than the base case result. Identifying the base case values with the 10-dimensional unit vector $\boldsymbol{e}_{10}$ = (+1,+1, ... ,+1), the situation where all factor values deviate from their base case value is (-1,-1, ... ,-1). Opposite to popular believe, (-1,-1, ... ,-1) is not necessarily the worst case scenario, because of interactions; see Tables 1 and 2, combinations 2 and 12 .

Sensitivity analysis should result in information on both main effects (as the one-factor-at-a-time approach does) and interactions. Moreover, we want to execute a minimum number of simulation runs. Therefore we apply a *Plackett-Burman* design (Kleijnen and Van Groenendaal, 1992, pp. 175-7). These designs require a number of runs equal to a multiple of four. Hence, for ten factors a design with twelve runs is used. The design we use is given in Table 1, where + is interpreted as +1 and - as -1; we use standardized factors (see Kleijnen and Van Groenendaal, 1992, pp 177-9). Every column in that table represents the input scenario of a simulation run. It is easily checked that the columns of this design matrix are orthogonal: $(D^T D)^{-1} = 12^{-1} I$ where D is the 12x10 design matrix, the superscript T denotes the transposed matrix, and I denotes the 10x10 identity matrix. Furthermore, the design satisfies one linear constraint: the sum of the first eleven rows of D equals minus row twelve. The augmented matrix $X = (\boldsymbol{e}_{12} : D)$, has the same properties as D has.

When we execute the twelve simulation runs of this design, we get twelve values for the NPV. This is sufficient information to obtain OLS estimates of the main effects only; that is, we fit the first-order metamodel:

$$\underline{y}_i = \sum_{h=0}^{10} \beta_h x_{ih} + \underline{\varepsilon}_i \,, \tag{1}$$

where $\underline{y}_i$ denotes the response for combination i, x_{ih} are elements of the matrix X, $\underline{\varepsilon}_i$ is an independent and identical distributed error term; and the underscore denotes random variables. However, if interactions are important, then these estimates are biased. In the sequel we assume that there are only two-factor interactions.

Let $\beta_M = (\beta_0, \beta_1, \dots, \beta_{10})^T$ be the vector of coefficients of model (1), and let $\beta_A = (\beta_{1,2}, \beta_{1,3}, \dots, \beta_{9,10})^T$ be the vector of two-factor interactions. Let the matrix of independent variables associated with β_A be denoted by V (with $V = (V_1, \dots, V_9) \in R^{12\times 45}$ and $V_i = (x_i x_{i+1}, \dots, x_i x_{10})$ where, e.g., $x_1 x_2$ denotes the 12-dimensional vector with elements $x_{j1} x_{j2}$, j = 1, 2, ... , 12). Then the expected value of $\underline{\hat{\beta}}_M$ is

$$E\{\underline{\hat{\beta}}_M\} = \beta_M + 12^{-1} X^T V \beta_A \,, \tag{2}$$

where $12^{-1}X^T V$ is called the *alias* or bias matrix, which shows how the main effects are confounded with the interactions; see Raktoe, Hedayat, and Federer (1981). Hence the estimator for the main effects would be unbiased if either $X^T V = 0$ or $\beta_A = 0$.

Unfortunately, $X^T V = 0$ does not hold for the Plackett-Burman design. Fortunately this equality can be achieved by applying the *foldover* theorem (see Kleijnen, (1987, p. 303): add $-D$ to the original design matrix D. Hence, twenty-four instead of twelve simulation runs are executed. The resulting design is a resolution IV design; that is, no main effect is confounded with any other main effect or any two-factor interaction; the two-factor interactions, however, are confounded with each other (Kleijnen, 1987, Chapter 25). (Obviously, since there are 1 + 10 + 45 effects and only 24 runs, unbiased estimators of all main effects and two-factor interactions are impossible.) Adding two-factor interactions to (1) and applying the foldover technique leads to

$$\underline{y} = \left(e_{24}, \begin{pmatrix} D \\ -D \end{pmatrix}, \begin{pmatrix} V \\ V \end{pmatrix} \right) \begin{pmatrix} \beta_0 \\ \beta_M \\ \beta_A \end{pmatrix} + \underline{\varepsilon} = Z\beta + \underline{\varepsilon} \,. \tag{3}$$

This gives the OLS estimator $\hat{\beta} = (Z^T Z)^{-1} Z^T y$ where

$$Z^T Z = \begin{pmatrix} 24 & 0 & 0 \\ 0 & 24I & 0 \\ 0 & 0 & 2V^T V \end{pmatrix} . \tag{4}$$

Note that the rank of V is eleven, which implies that up to eleven two-factor interactions can be estimated. Next eleven independent columns from V have to be selected, to form the matrix (say) V_I that corresponds with the interactions. The remaining columns of V are combined in the matrix (say) V_A. The resulting alias matrix $(V_I^T V_I)^{-1} V_I^T V_A$ for the eleven interactions can be formed in the same way as we did for $\underline{\hat{\beta}}_M$ in (2).

Note that if unbiased estimators of more than eleven interactions are needed, the design must be further augmented (beyond the foldover); that is, more simulation runs are required.

Table 2: NPV results (in billions of Indonesian Rupiahs) for the Plackett-Burman design and its foldover

comb.	NPV	comb.	NPV
1	-1,252.8	13	-1,268.0
2	-3,033.8	14	-67.9
3	-1,132.5	15	-1,006.0
4	-1,210.0	16	-1,084.0
5	-1,301.7	17	-1,339.6
6	359.0	18	-2,639.2
7	-997.8	19	-1,362.2
8	454.2	20	-1,353.7
9	-602.6	21	-2,045.6
10	-175.6	22	-1,324.4
11	-1,341.7	23	-278.1
12	-2,985.3	24	1,750.9

The results of the twenty-four simulation runs are shown in Table 2. In this table, combination 13 is the "foldover" of combination 1, ..., combination 24 is the foldover of combination 12 (hence combination 24 is the base case scenario). Because we analyze only factors that cause a decrease in the NPV, there are many negative entries in Table 2. Since each minus value of a single factor lowers the NPV, these negative effects are mitigated only if *interactions* have positive influences. In case the interactions strengthen the main effects, the results of the simulation will be even more negative. The base case gives the NPV result 1,750.9.

The analysis of the data of Table 2 starts with the OLS estimation of the first-order approximation in (1) using the design with 24 runs. The estimates β_0, β_1, β_3, β_5, β_6, β_7, and β_8 are significant at the level $\alpha = 0.05$. The adjusted coefficient of determination R^2_{adj} is 0.88. Because $D^T D = 24I$ (see eq. 4) is a diagonal matrix, the estimates of the main effects do not change when we delete or add main effects.

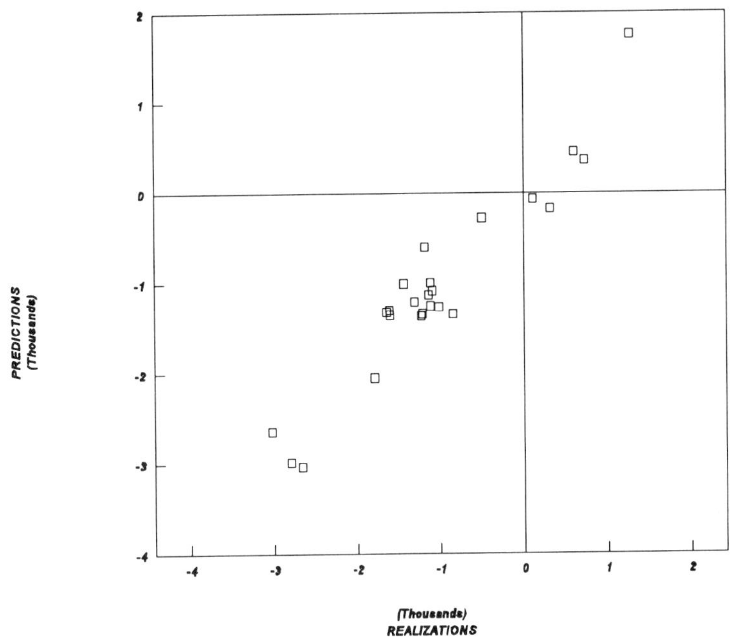

Figure 1: Scatter plot of NPV regression predictions and simulation realizations

Once the significant main effects are known, the significant two-factor interactions may be determined. There are many possible ways to augment (1) with individual interactions. So determining which individual interactions are significant, is problematic, without making further assumptions or executing more simulation runs. A popular assumption is that there are only interactions between factors with *significant* main effects. Estimating this model and testing it against the model without interactions shows that there are indeed some significant interactions: the F-statistic is 3.14, which, however, is barely significant ($F^{6}_{13;0.05} = 2.92$). (Note, that not rejecting H_0 does not prove that there are no interactions.)

Taking into account those interactions that are related to significant main effects is indeed a reasonable approach, when no other information is available (the simulation model is then treated as a black box). However, in our case-study we derive clues from the simulation model itself and from the intermediate simulation results that lead to the outputs in Table 2. We know that in the simulation model, economic growth plays an important role. Nevertheless, Indonesia's economic growth (factor 4) has no significant main effect, β_4. This seems odd, and is also contrary to what economic theory tells us. Studying the detailed simulation results of the twenty-four simulation runs shows that economic growth does strengthen the effect of a change in the West Java reserves (factor 3). Moreover, it strengthens the effects of some of the changes in prices (factors 5, 6, 7, and 8). Therefore we restrict the search for interactions to these six variables (factors 3-8). After testing several alternative specifications of the metamodel, the following model showed the best test results:

$$\begin{aligned}\hat{y} = & -1051.6 + 142.5x_1 + 461.2x_3 + 659.2x_5 + \\ & + 447.2x_6 + 242.7x_7 + 236.4x_8 + \\ & + 226.0x_3x_4 + 112.1x_3x_5 - 107.7x_3x_6 + \\ & + 128.6x_3x_8 .\end{aligned} \tag{5}$$

Its R^2_{adj} is 0.98. The hypothesis $H_0: \beta_{34} = \beta_{35} = \beta_{36} = \beta_{38} = 0$ yields $F^{4}_{13_4} = 10.81$, which is significant even at the 0.5% level ($F^{4}_{13;0.005} = 6.23$).

The validity of the metamodel (5), relative to the underlying simulation NPV model, can be tested through *cross-validation*: eliminate combinations one by one, re-estimate the regression model, and use that model's prediction $\hat{y}_{-i}$ for the simulation result for the i-th combination eliminated; see Kleijnen and Van Groenendaal (1992, pp. 156-7). To indicate the quality of these predictions $\hat{y}$ we use a scatter plot; see Figure 1. If (5) were perfect, the scatter plot would be a straight line through the origin with an angle of 45°. Actually, the correlation coefficient between $\hat{y}_{-i}$ and y_i is 0.996, which we find satisfactory.

4 CONCLUSIONS

Practitioners often stick to simple methods, such as changing one factor at a time, studying a few scenarios or the worst case scenario only. We think that design of experiments (DOE) in combination with regression metamodelling can be a fruitful method for problems with uncertainties that do not fit into the standard uncertainty models and that have no reliable information on the joint probability distribution. In many areas these circumstances are the rule rather than the exception. Then deterministic models in combination with DOE and regression analysis will provide the decision makers with the information they require. They want to know explicitly what the most important factors are, and how these factors are related. These factors deserve special attention during project execution. Decision makers will use this information, when monitoring the progress of the project, and when designing adjustments whenever progress is not as expected.

We illustrated DOE and regression analysis by a large practical investment problem. It might be argued that our interpretation of interactions is not supported by a complete statistical analysis, since not *all* two-factor interactions were systematically checked. However, analysts who understand their problem will in many cases be able to qualitatively derive which interactions are the most important ones, as we did. The resulting metamodel may be statistically sound and supported by knowledge about the problem at hand. After all, mathematical statistics is only an auxiliary science.

The results obtained for our case-study, using DOE, allow the estimation of the main effects for the ten factors that influence the NPV plus the important interactions between factors. The results have given the Indonesian Government clear insight into the robustness of the project.

REFERENCES

Broens D-F. (1995) *Investment evaluation with respect to commercial uncertainty*, Labyrint Publication, Capelle a/d IJssel, The Netherlands.

Copeland T.E. and J.F. Weston (1992) *Financial theory and corporate policy* (third edition), Addison-Wesley Publishing Company, Reading, Massachusetts, U.S.A.

Draper D. (1995) Assessment and propagation of model uncertainty, *Journal of the Royal Statistical Society*, 57(1):45-97.

Gittinger J.P. (1982) *Economic analysis of agricultural projects*, The John Hopkins University Press, Baltimore and London.

Granger M. M. and M. Henrion (1990) *Uncertainty: a guide to dealing with uncertainty in quantitative risk and policy analysis*, Cambridge University Press, Cambridge.

Grübler A. and A. McDonald (1996) The drive to cleaner energy, *Options*, IIASA, Laxenburg, Austria.

Iman R.L. and W.J. Conover (1982) A distribution-free approach to inducing rank correlation among input variables, *Communications in Statistics - Simulation and Computation*, 11(3):311-34.

Kleijnen J.P.C. (1987) *Statistical tools for simulation practitioners*, Marcel Dekker, New York, U.S.A.

Kleijnen J.P.C. and W.J.H. Van Groenendaal (1992) *Simulation: a statistical perspective*, John Wiley & Sons, Chichester, England.

Kleijnen J.P.C. (1996) Five-stage Procedure for the Evaluation of Simulation Models through Statistical Techniques, Proceedings of the 1996 Winter Simulation Conference (edited by J.M. Charnes, D.J. Morrice, D.T. Brunner, and J.J. Swain)..

Kunreuther H.C., J. Linnerrooth, J. Lathrop, H. Atz, S. Macgill, C. Mandl, M. Schwartz, and M. Thompson (1983) *Risk analysis and decision processing: the sitting of liquefied energy gas facilities in four countries*, Springer-Verlag, Berlin, Germany.

Raktoe B.L., A Hedayat, and W.T. Federer (1981) *Factorial designs*, John Wiley & Sons, New York, U.S.A.

Rubinstein R.Y. and A. Shapiro (1993) *Discrete event simulation: sensitivity analysis and stochastic optimization by the score function method*, John Wiley & Sons, Chichester.

Tversky A. and D. Kahneman (1988) Rational choice and the framing of decisions. In DE Bell, H. Raiffa, and A. Tversky (ed.) *Decision making: descriptive, normative, and prescriptive interactions*, Cambridge University Press, Cambridge, Great Britain.

Van Groenendaal W.J.H. and J. Vingerhoets (1995) Can international commodity agreements work? *Journal of Policy Modelling*, 17(3):257-278.

Van Groenendaal W.J.H. (1996) *Investing in energy infrastructures: methodological aspects*, Oxford University Press, Oxford, England. (Forthcoming.)

Ward W.A. and B.J. Deren (1991) *The economics of project analysis: a practitioner's guide*, The World Bank, Washington, D.C.

AUTHOR BIOGRAPHIES

WILLEM J.H. VAN GROENENDAAL is Assistant Professor in the Department of Information Systems and Auditing, and Fellow at the Center for Economic Research (CentER). Both the Department and the Center are within the School of Management and Economics of Tilburg University in Tilburg, The Netherlands. He is also a consultant for several international development agencies. He has a Ph.D. from Tiburg University. His research interests are in simulation, decision support, and investment analysis.

JACK P.C. KLEIJNEN is Professor of Simulation and Information Systems in the Department of Information Systems and Auditing; he is also associated with the Center for Economic Research (CentER). Both the Department and the Center are within the School of Management and Economics of Tilburg University (Katholieke Universiteit Brabant) in Tilburg, Netherlands. He received his Ph.D. in Management Science at Tilburg University. His research interests are in simulation, mathematical statistics, information systems, and logistics. He published six books and more than 130 articles; lectured at numerous conferences throughout Europe, the USA, Turkey, and Israel; he was a consultant for various organizations; and is a member of several editorial boards. He spent some years in the USA, at different universities and companies. He was awarded a number of fellowships, both nationally and internationally.

Proceedings of the 1996 Winter Simulation Conference
ed. J. M. Charnes, D. J. Morrice, D. T. Brunner, and J. J. Swain

SENSITIVITY AND SCENARIO ANALYSIS FOR SIMULATION METAMODELS

Susan M. Sanchez
L. Douglas Smith
Edward C. Lawrence

School of Business Administration
University of Missouri–St. Louis
8001 Natural Bridge Road
St. Louis, Missouri 63121–4499, U.S.A.

ABSTRACT

We use simple orthogonal and non-orthogonal designs to analyze a multi-tiered model for forecasting performance of a large-scale home mortgage portfolio. The experiments are used to assess the sensitivity of performance to projected changes in economic conditions, as well as the sensitivity of the model to coefficients estimated from historical data. Our results attribute the variation in loan performance to variation in individual factors or factor combinations, indicating which are crucial to monitor or forecast accurately. The results are at times counter-intuitive, indicating the benefits of a systematic approach to sensitivity assessment and scenario generation.

1 INTRODUCTION

Managing credit risk in financial institutions requires the ability to forecast aggregate losses on existing loans, to predict the length of time that loans will be on the books before prepayment or default, to analyze the expected performance of particular segments in the existing portfolio, and to project payment patterns of new loans. To accomplish these tasks, Smith et al. (1996) developed a comprehensive forecasting model in SAS for a major California financial institution. The comprehensive model consists of three components, as summarized in Table 1. The first component is a model of transition probabilities among five financial states. Current loans are those with outstanding balances and payments on schedule, delinquent loans (30 to 89 days or 90+ days) have outstanding balances and payments overdue, paid-off loans have had the outstanding balance paid in full at maturity or earlier, and defaulted loans are those for which partial or complete charge-off has occurred and/or the title has been acquired through foreclosure.

The transition component of the comprehensive model is structured as a Markovian recursion to predict the likelihoods that a loan will be in each alternative state at annual intervals. In aggregate, it predicts the expected number of accounts in the portfolio that will fall in each financial state. At the beginning of the forecast period, the state of an active account will either be current, delinquent 30 to 89 days, or delinquent 90+ days. Each time a payment is due, the debtor may make a payment to remain current, delay payment, or prepay the loan to maturity. Over the scheduled term of the loan, it is possible that the loan can be in any of the five financial states. The loan will either be in the paid-off state or defaulted state after maturity. Let

$$\begin{aligned} P_j(t) &= \Pr\{\text{loan is in state } j \text{ at time } t\} \\ P_{k,j}(t) &= \Pr\{\text{loan is in state } j \text{ at time } t+1 \\ &\qquad \text{given it is in state } k \text{ at time } t\} \end{aligned}$$

Allowing for non-stationary transitional probabilities, the likelihood that a loan is in state j at time $t+1$ is expressed as a Markov chain as follows:

$$P_j(t+1) = \sum_{k=1}^{5} P_k(t) \times P_{k,j}(t).$$

Transitions can be made from the three active states (current, delinquent 30–89 days, or delinquent 90+ days) to each of the other five states. However, the defaulted and paid-off states are absorbing (terminal) states from which no transitions can be made.

The second component models the likelihood of a loss occurring after transition to the defaulted state. The magnitude of the expected loss depends on the amount of equity in the property and the position of the lien holder. The net equity is affected by changes in market price of the property and by the extent to which the principal has been paid off. Because there is no loss on a substantial portion of loans which terminate with default, and because loss patterns may

Table 1: Structure of the Comprehensive Forecasting Model

	Component		
	Transition	Loss	Severity
Description	Yearly transition probabilities from the three active states to the five potential states	Probabilities of defaulted loan transitions into four potential states	Proportion of outstanding balance charged off
Model States	1. Current 2. Delinquent 30-89 days 3. Delinquent 90+ days 4. Paid-off 5. Defaulted	ORE/Loss incurred ORE/No loss incurred Non-ORE/Loss incurred Non-ORE/No loss incurred	Total loss Partial loss
Methodology	Multinomial logistic	Logistic	Regression

depend on whether or not the bank acquires title, we analyze the severity of loss in stages. We first predict the probability that the title of the loan is acquired through foreclosure, i.e., the loan is transferred to *owned real estate* (ORE). We then predict the probability that a loss is incurred on the loan. This categorization was found to provide better predictions of loss, and it is consistent with the financial institution's internal reporting and monitoring procedures.

The final component models the severity of the loss. We first predict the probability that the remaining balance is totally lost. For loans which incur partial losses, we then model the proportion of the outstanding balance lost.

The output of the forecasting model describes many aspects of model behavior, both on a yearly basis, a five year interval, and over the remaining life of the portfolio (30 years). We will focus on four performance measures in this paper which reflect the lifetime performance of the portfolio:

- Adjusted Severity (ADJSEV): average loss incurred on defaulted loans with non-zero losses;
- Percent Default (PCTDFLT): expected percent of loans defaulted;
- Percent Loss (PCTLOSS): expected percent of current net receivables lost;
- Time on books (ONBOOKS): expected length of time the loans in the portfolio remain active.

The models are refit on a quarterly basis to allow for seasonal patterns in delinquency rates, as well as changes in portfolio composition as active loans go off the books, new loans are initiated, and blocks of loans are purchased from or sold to other lenders.

Markovian models of mortgage performance have been proposed as alternatives to discrete event simulation in evaluating mortgage-backed securities, although the modeling tradeoff between simplicity and realism is acknowledged (Zipkin 1993). This tradeoff is important for our application since the modeling and analysis is not a one-time project, but is conducted quarterly basis after refitting the component models. A better understanding of the trade-off will either support the use of expected values or demonstrate the need to estimate performance variability.

In this paper, we investigate three causes of performance variation: variation attributable to uncertainty about future economic conditions and loan characteristics, variation due to the use of estimated coefficients in the component models, and the inherent variation which results from individual loans' discrete state transitions rather than expected value performances. In Section 2, we show how elliptical experimental designs can be used in a tolerance analysis framework to model performance variability for new loans in the portfolio. In Section 3, we present results from a highly fractionated factorial experiment to assess the model's sensitivity to the estimated coefficients of the component submodels. We run the model in a discrete event simulation mode in Section 4 to provide bounds on the mean performance measures computed from the base model. Our conclusions are provided in Section 5.

2 TOLERANCE EXPERIMENT: SCENARIO ANALYSIS

Throughout this paper, we concentrate on one small segment (6508 loans) of the portfolio: fixed-rate uninsured loans in southern California with an initial loan balance of $150,000 or less. A single run of the fore-

Table 2: Coded Factor Levels for Scenario Analysis

DELUNEM	MKTAPPR	CONVRATE	LTV	AMOUNT	INTRATE	TERM
−0.692	−0.692	−1	+1.344	+1.344	+1	−1
+1.233	−1.233	−1	−1.344	−1.344	+1	+1
−1.233	+1.233	−1	−0.439	−0.439	−1	+1
+0.692	+0.692	−1	+0.439	+0.439	−1	−1
−0.692	−0.692	+1	+0.439	+0.439	−1	+1
+1.233	−1.233	+1	−0.439	−0.439	−1	−1
−1.233	+1.233	+1	−1.344	−1.344	+1	−1
+0.692	+0.692	+1	+1.344	+1.344	+1	+1

casting model for this segment takes just under four minutes of CPU time on a Sparc 20 with four co-processors at the University of Missouri–St. Louis, and substantially longer on the financial institution's older, IBM mainframe. Previously, scenario analysis efforts were performed in an *ad hoc* manner because of the time involved. Full portfolios may contain over 90,000 loans.

2.1 Initial Experiment

Monte Carlo sampling for scenario analysis is a time-consuming activity. It is complicated by the facts that the economic variables are correlated, both with each other and over time, and the coefficients in the component models are related to each other. Instead, we adapt the so-called tolerance analysis of Taguchi (1986, 1987) by using response surface metamodeling methods to examine the robustness of the comprehensive forecasting model.

Response surface metamodels are efficient scenario analysis tools. Once constructed, they permit an analyst to examine many alternatives without re-running the simulation. In the tolerance analysis context, response surface metamodels can be used to assess the system's overall performance, to attribute the overall performance variation to variation in the factors, and to determine whether or not changes in the variability of the factors would increase or decrease the performance variability (Myers et al. 1992; Sanchez et al. 1994a, 1996).

We consider seven factors in our experiment. The first three represent sources of variation in the general economic environment, and are not controllable by the financial institution, and so can be referred to as external noise factors. The latter four are internal sources of noise in that they represent characteristics which vary from loan to loan across the portfolio. These factors can be influenced by lending policies for new loans. For example, new loans with low down-payments (high LTVs) may be charged higher interest rates. Regardless, the level to which the institution segments the portfolio for projection purposes will influence the noise factors' variability. The factors are:

- DELUNEM: A constant yearly percent change in unemployment rate for 1–4 years in the future (for 5 years and out, this was held at zero);
- MKTAPPR: An index value representing a constant yearly change in market appreciation rates for 1–4 years in the future (for 5 years and out, this was held at its average value);
- CONVRATE: The conventional interest rate on new 30 year fixed-rate mortgages;
- LTV: The loan-to-value ratio at origination;
- AMOUNT: The loan amount;
- INTRATE: The interest rate on the loan;
- TERM: The term of the loan.

Two sets of jointly correlated factors exist. The first set is DELUNEM and MKTAPPR. These environmental factors vary by year and by principal metropolitan statistical area, but had a correlation of -.521 over the prior five year history when DELUNEM was lagged by one year. In other words, increases in local unemployment rates led to subsequent decreases in market appreciation. LTV and AMOUNT constitute the second set of correlated factors. Since we are interested in predicting the performance for new loans, and inflation makes comparisons of dollar amounts questionable over long periods of time, we calculated the correlation for only the newest loans (with a vintage of one year or less.) These had a strong positive correlation of .807. Higher loan-to-value ratios tended to be associated with larger loans, perhaps indicating more difficulty in raising money for the down payment.

Table 3: Transmitted Variances as Percentages of Total Performance Measure Variances

Factor Groupings	Performance Measure							
	ADJSEV		PCTDFLT		PCTLOSS		ONBOOKS	
DELUNEM	0.00%		2.56%		6.67%		2.23%	
MKTAPPR	0.07%		3.25%		5.47%		1.20%	
Cross term	-0.01%		-3.03%		-6.55%		-1.72%	
Net Effect		0.06%		2.78%		5.59%		1.72%
CONVRATE		0.00%		0.01%		.02%		.11%
LTV	0.18%		44.52%		26.20%		87.26%	
AMOUNT	92.77%		5.35%		11.08%		2.58%	
Cross term	6.65%		24.91%		27.50%		-24.23%	
Net Effect		99.60%		74.78%		64.78%		65.61%
INTRATE		.09%		21.16%		27.40%		12.28%
TERM		.25%		1.27%		2.21%		20.28%
Total		100.00%		100.00%		100.00%		100.00%

With only seven factors, we can construct a saturated two-level factorial experiment. The design matrix (in natural units) is shown in Table 2. The levels for the non-correlated factors (CONVRATE, INTRATE, and TERM) were set at approximately one standard deviation above and below the mean. (The levels for TERM were rounded slightly since most mortgages have terms in multiples of 12 months.) These levels are recommended for discrete sampling of a normal distribution, since they result in a two-point distribution with the same mean and standard deviation as the underlying distribution.

For the correlated factor pairs, the factor levels were calculated using the experimental designs of Sanchez (1994b). Design points are located at the major and minor axes of an ellipse, rather than at corners or faces of the cubes typically used to graphically represent factorial designs. Let μ_i and σ_i denote the mean and standard deviation of factor i ($i = 1, 2$). The ellipse's shape is a function of the correlation ρ between the two factors:

Major axes: $(\mu_1 + \sigma_1\sqrt{1+\rho},\ \mu_2 + \sigma_2\sqrt{1+\rho})$
$(\mu_1 - \sigma_1\sqrt{1+\rho},\ \mu_2 - \sigma_2\sqrt{1+\rho})$

Minor axes: $(\mu_1 - \sigma_1\sqrt{1-\rho},\ \mu_2 + \sigma_2\sqrt{1-\rho})$
$(\mu_1 + \sigma_1\sqrt{1-\rho},\ \mu_2 - \sigma_2\sqrt{1-\rho})$

Means and standard deviations can be estimated from historical data or expert opinion.

After running the experiment, we can construct models for performance measure PM_j as functions of the noise factors W_h ($h = 1, \ldots, w$):

$$PM_j \approx \hat{\beta}_{0,j} + \sum_{h=1}^{w} \hat{\beta}_h, jW_h \qquad (1)$$

Treating the regression coefficients in equation (1) as constants, we obtain

$$\sigma_j^2 \approx \sum_{h=1}^{w} \hat{\beta}_{h,j}^2 \sigma_h^2 + 2\sum_{h=1}^{w}\sum_{k=h+1}^{w} \rho_{h,k}\hat{\beta}_{h,j}\hat{\beta}_{k,j}\sigma_h\sigma_k. \qquad (2)$$

Equation (2) estimates the overall variation in performance measure j. For a noise factor h which is uncorrelated with all other noise factors, its contribution to this overall variance is called the *transmitted variance* and is equal to $\hat{\beta}_{h,j}^2\sigma_h^2$. For correlated factor pairs, the transmitted variance cannot be separated completely. For example, if factors 1 and 2 are correlated then their joint transmitted variance to performance measure j is equal to

$$\hat{\beta}_{1,j}^2\sigma_1^2 + \hat{\beta}_{2,j}^2\sigma_2^2 + 2\rho_{1,2}\hat{\beta}_{1,j}\hat{\beta}_{2,j}\sigma_1\sigma_2. \qquad (3)$$

Noise factor assessment refers to the attribution of variance in a performance measure to known sources of variation. The noise factor assessments for all four performance measures are provided in Table 3. (Although the transmitted variances are approximate, percentages are shown to two decimal places to facilitate reassessment as in Section 3.2.) For this experiment, the net transmitted variance from the correlated loan characteristics (AMOUNT and LTV) dominates for all four performance measures. This is not surprising for ADJSEV: one would expect the magnitude of the loss to depend on the size of the loan. LTV appears to be the most influential of the pair for predicting PCTDFLT and ONBOOKS. INTRATE impacts the overall loss, but through the transition and loss behavior rather than through the severity component. External market conditions had surprisingly little impact on the overall performance, and loan term was only a factor for predicting the expected time on the books.

Note that a traditional factorial design and analysis would be inappropriate for two reasons. First, certain combinations of the factors are not likely to occur, thus the goal of obtaining the same variability in the sampling distribution as in the underlying distribution would not be realized by setting all factor levels at $\mu \pm \sigma$. Second, the original forecasting model (metamodel) was not constructed using similar combinations. The results would be highly suspect for such dramatic extrapolations. The analysis differs as well. For example, without considering the high positive correlation between LTV and AMOUNT, one would underestimate the overall variance in PCTDFLT and PCTLOSS by over 24%. Conversely, the variance of ONBOOKS would be overestimated if the correlation of LTV and AMOUNT were ignored.

2.2 Model use and refinement

Metamodels can often be simplified by removing statistically insignificant terms. Normal probability plots are useful for this activity if few or no degrees of freedom are available for error estimation, as in this experiment. However, we advise caution in oversimplification when two or more noise factors are correlated. The sampling scheme is not orthogonal, so if $|\rho|$ is near one, then the standard errors of the correlated regression coefficients may be inflated due to multicollinearity. The effect of correlation on the overall variance is computed through the metamodel coefficients, so excluding a marginally insignificant main effect from the model may have noticeable impacts on system evaluation and noise factor assessment due to the exclusion of interaction terms. Thus, we elect to retain both terms of a correlated factor pair in the metamodel even if only one is significant. We used a cutoff of $\alpha = .20$ because of low degrees of freedom and our desire to achieve models with accurate fits. For the 8-run tolerance experiment, reduced metamodels with relatively high R^2 values were constructed for each of the four performance measures:

- ADJSEV: $R^2 = .999$ with LTV, AMOUNT and TERM.
- PCTDFLT: $R^2 = .941$ with LTV, AMOUNT and INTRATE.
- PCTLOSS: $R^2 = .889$ with LTV, AMOUNT and INTRATE.
- ONBOOKS: $R^2 = .982$ with LTV, AMOUNT, CONVRATE, INTRATE and TERM.

If the reduced metamodels are used for noise factor assessment, the transmitted variance percentages are zero for factors excluded from the metamodels, and not materially different from those reported in Table 3 for the factors retained.

Note that statistical significance in the metamodels does not necessarily imply practical importance. Since the transmitted variances depend on the component variances (and covariances), a term may be statistically significant in the metamodel yet correspond to a factor with transmitted variance near zero. For example, TERM is statistically significant in the model of ADJSEV, but accounts for only 0.25% of the variability of this performance measure.

One benefit of tolerance analysis is the ease of modifying the results to predict the effects of changes in the underlying sources of noise. For example, the financial institution might be interested in forecasting the future loss percentage for a smaller portfolio segment: new loans with half the standard deviations of loan-to-value and amount. The transmitted variance of this factor grouping would be reduced by a factor of four. After re-standardizing, the net effect of DELUNEM and MKTAPPR would account for 10.87% of the variance in PCTLOSS; the net effect of LTV and AMOUNT would account for 31.50%, the INTRATE would account for 53.30% and the TERM would account for 4.29%. (CONVRATE's contribution remains negligible at 0.04%.) Similarly, if the volatility in market conditions increased, one could re-estimate the transmitted variances to determine the impacts (if any) on performance. Such transmitted variance reassessments require no additional computer runs and may help guide scenario analysis efforts. For example, from Table 3 we see that rerunning the base model with different conventional rates will have no appreciable effect on any of the four performance measures if the conventional rate ranges from 7.45% to 8.45%. Even doubling the range would increase performance variability by at most 0.33%, according to our metamodels. However, it is important to remember that the metamodels are only approximations of the true response surface. If large changes in noise factor variance are anticipated, a confirmation run using new factor settings should be made. If the confirmation run's performance measures are not close to those predicted, the new information can be used to develop into more appropriate metamodels of performance.

Table 4: Transmitted Variance Percentages: Coefficient Sensitivity Experiment

Coefficient Grouping	# Coeffs. Perturbed	Performance Measure: ADJSEV	PCTDFLT	PCTLOSS	ONBOOKS
Transition from Current	32	57%	86%	55%	66%
Transition from Del. 30-89	8	2%	0%	0%	0%
Transition from Del. 90+	8	15%	14%	5%	32%
Loss	11	27%	0%	40%	1%
Severity	2	0%	0%	0%	0%
Total %:		100%	100%	100%	100%

2.3 Validation Experiment

A common concern when using first-order metamodels is that they may not adequately approximate the response surface. To validate the results of Section 3.1 and 3.2 we performed a 64-run half-fraction experiment, which allows us to estimate all main effects and two-way interactions. The larger number of data points provides ample degrees of freedom for assessing statistical significance, although the p-values for positively correlated factor pairs will still be inflated. Both main-effect models and second-order models were examined. For ADJSEV, the reduced model contained the same three terms as the 8-run metamodel: no interaction effects were present. MKTAPPR was significant for the other three performance characteristics, and CONVRATE was significant for PCTLOSS and ONBOOKS. The effects due to variation in economic conditions were thus better identified in the 64-run experiment than in the 8-run experiment. Few interactions appeared to influence performance. Only three are significant using $\alpha = .05$ (LTV by MKTAPPR, LTV by AMOUNT, and LTV by INTRATE, all in the metamodel for PCTDFLT). One more was present at $\alpha = .10$ (LTV by AMOUNT in the metamodel for ONBOOKS).

The transmitted variances are small for the factors not found to be statistically significant in the 8-run experiment, so leaving them in the model has little impact on the noise factor assessment for the present portfolio. However, statistical significance is difficult to determine with so few degrees of freedom available for error estimation, and the full model provides more accurate results if the economic forecasts change or if the portfolio is segmented differently. The latter is of interest for evaluating changes to current lending policies. In summary, it appears that the 8-run metamodels suffice for noise factor assessment of our mortgage portfolio example if all factors are retained in the metamodels and transmitted variances are estimated using equation (3).

3 TOLERANCE EXPERIMENT: MODEL SENSITIVITY

Recall from Table 1 that the comprehensive model makes use of fitted values from multinomial logistic, logistic, and regression analysis. For the 6508 loan segment we analyze, a total of 61 coefficients are used in the various model components. (Another 85 coefficients are needed to run the forecasting model for the full portfolio.) These coefficients are themselves estimated from previous data, so the sensitivity of the responses to errors in their estimation (or changes in the relationships over time) is of interest. However, since the model coefficients are updated quarterly to reflect more recent loan histories and environmental information, we prefer a method which is easily automated, even at the expense of some accuracy.

To this end, we modified the base forecasting program so that the coefficients could be perturbed using a nearly saturated two-level fractional factorial design with 64 runs. Each of the model coefficients is coded as $c_i + f_i s_i$ where c_i and s_i are, respectively, the mean and standard deviation of the ith coefficient and f_i is the coded factor level for the run. (For the base model calculations, all f_i can be set to zero.) A shell script loops through all 64 design points and iteratively calls the SAS forecasting program. Note that this design treats the coefficients as independent factors although they are correlated. Because of the nature of the multinomial logistic structure, we expect the results from the tolerance experiment to be conservative (i.e., show wider variation) relative to results that would be achieved by a method which made use of the multivariate correlation structures.

The market conditions (unemployment, market appreciation, and conventional interest rate) were fixed at the nominal (center point) levels of the 8-run tolerance design experiment. This left four factors: original loan-to-value, loan amount, interest rate, and term. For efficiency purposes, we chose to run the experiment using the four factor level combinations of the scenario tolerance experiment augmented by a center point, rather than using the full portfolio.

The results are summarized in Table 4. The coefficients for the 'transitions from current' state appear to be the major source of variation for all four performance measures, with transmitted variance percentages ranging from 55% to 86% of the totals. This represents the majority of the transitions which occur over the life of loans in the portfolio. Coefficients from the 'transitions from 90+ days delinquent' component account for 15% of the transmitted variance for ADJSEV, 14% for PCTDFLT, and 5% for PCTLOSS. Since the majority of loans entering default first pass through this late delinquency stage, the impact on the overall model is not surprising. The severity coefficients appear to impact both ADJSEV and PCTLOSS, but have little effect on the PCTDFLT or ONBOOKS. The transmitted variance of the severity coefficients is negligible relative to the other model components.

The transmitted variance percentages in Table 4 are based on full regression models. For comparison purposes, stepwise regression was used to obtain simpler models. The numbers of terms in the reduced metamodels dropped from 61 to 38, 30, 27 and 36 for ADJSEV, PCTDFLT, PCTLOSS, and ONBOOKS, respectively, but the transmitted variance percentages did not materially change.

4 INHERENT VARIANCE ESTIMATION

Recall that the base program relies on expected values rather than discrete event simulation to reduce the computational effort. We wished to estimate the variability of the means to determine whether or not reliance on expected behavior was reasonable. To accomplish this task, we modified the forecasting program to run as a discrete event simulation. For example, in the base program we calculate probabilities for each of the five states one year out, two years out, and so forth up to 30 years out. For the discrete-event version, we calculate the state probabilities on an individual loan basis to create multinomial distributions for the the loan state after one year ha: elapsed. SAS's internal random numbers routines are then used to determine which multinomial outcome is realized. Similarly, loss transitions are randomly determined with probabilities calculated from the comprehensive forecasting model. If the loan does not enter a terminal state at the beginning of year t, then transition probabilities for the year $t+1$ are recalculated after accounting for the amortization.

We provide estimates of the performance measures' variability obtained using discrete event simulation in Table 5. Three different portfolios are used as the basis for simulation: (1) 1,000 replications of the center point loan from Section 3. (2) 200 replications for each of the five loans used in the sensitivity analysis experiment of Section 3, and (3) 16 replications of the 6508 loan portfolio. In all cases, standard deviations are normalized to represent the standard error of the mean for 6508 loans.

Table 5: Standard Error Estimates for 6508 Loan Portfolio

Performance	Simulated Portfolio		
Measure	Center Pt. Loan	5 Design Pt. Loans	6508 Loans
ADJSEV	60.60	141.06	518.35
PCTDFLT	0.234	0.669	0.327
PCTLOSS	0.007	0.065	0.058
ONBOOKS	0.045	0.101	0.056

We anticipated that the standard deviations computed by discrete event simulation of very small portfolios (the center point loan, or the five design point loans) would be conservative estimates of the standard deviations resulting from the entire 6508 loan simulation, in part because the larger portfolio has loans closer to maturity and thus fewer potential paths through their remaining life. If so, and if the resulting bounds were sufficiently narrow, then future estimation of the inherent variability could be accomplished efficiently using a reduced portfolio. Unfortunately, the standard deviation of ADJSEV was not approximated closely by either the center point simulation or the five design point simulation. The standard deviations estimated from the five design points are conservative for the other three performance measures. Discrete-event simulation of the center point loan underestimated the standard deviations for all four performance measures, so for our example it would not provide adequate bounds on the expected performance variability.

5 CONCLUDING REMARKS

We have examined three causes of variation for four performance measures in the base forecasting model: variation due to uncertainties about future economic conditions and new loan characteristics, variation due to the use of estimated coefficients for component models, and inherent variation due to the stochastic nature of loan behavior within the portfolio. The central limit theorem means the inherent variation is less a concern for large portfolios. Similarly, if the coefficients in the component models are estimated from a large number of loans, their standard errors will tend to be smaller and their effect will be dampened. However, the noise factor assessments from tolerance analysis do not directly depend on the size of the portfolio for which scenario analysis is conducted.

Taken together, these three causes of variation can be used to provide bounds around the expected performance measures computed from the base forecasting model. When this is done for the five design point loans, the overall performance variance computed during tolerance analysis was the major source of uncertainty for ADJSEV, the inherent variance was the major source of difference for PCTDFLT and PCTLOSS, and the uncertainty concerning the coefficients had the major impact on ONBOOKS. This indicates that all types of variation should be considered in order to obtain bounds on the expected portfolio performance.

We have investigated methods for efficient response variability assessment. Small experimental designs were be used to augment or replace extensive discrete event simulation to predict the performance of home mortgage portfolios. We found that strong correlations between factors must be considered when designing and analyzing these experiments in order to obtain accurate noise factor assessments. In addition to one-time insights into a particular portfolio segment's behavior, these methods hold promise for adaptation to an ongoing process of model refitting and revision.

REFERENCES

Myers, R. H., A. I. Khuri and G. Vining. 1992. Response surface alternatives to the Taguchi robust parameter design approach. *The American Statistician* 46(2): 131–139.

Sanchez, S. M. 1994a. Experiment designs for system assessment and improvement when noise factors are correlated. In *Proceedings of the 1994 Winter Simulation Conference*, ed. J. D. Tew, M. S. Manivannan, D. A. Sadowski and A. F. Seila, 290–296. Institute of Electrical and Electronic Engineers, Orlando, Florida.

Sanchez, S. M. 1994b. A robust design tutorial. In *Proceedings of the 1994 Winter Simulation Conference*, ed. J. D. Tew, M. S. Manivannan, D. A. Sadowski and A. F. Seila, 106–113. Institute of Electrical and Electronic Engineers, Orlando, Florida.

Sanchez, S. M., P. J. Sanchez and J. S. Ramberg. 1996. A simulation framework for robust system design. In *Concurrent design of products, manufacturing processes and systems*, ed. B. Wang. New York: Gordon and Breach, forthcoming.

Smith, L. D., S. M. Sanchez and E. C. Lawrence. 1996. A comprehensive model for managing credit risk and forecasting losses on home mortgage portfolios. *Decision Sciences*, forthcoming.

Taguchi, G. 1986. *Introduction to quality engineering*, White Plains, New York: UNIPUB/Krauss International.

Taguchi, G. 1987. *System of Experimental Design, Vols. 1 and 2.* White Plains, New York: UNIPUB/Krauss International.

Zipkin, P. 1993. Mortgages and Markov chains: a simplified valuation model. *Management Science* 39: 2–16.

AUTHOR BIOGRAPHIES

SUSAN M. SANCHEZ is Associate Professor of Management Science and Information Systems at the University of Missouri–St. Louis. Her research interests include the design and analysis of simulation experiments, selection, and data-intensive statistical methods in business and manufacturing. She is an associate editor for *Operations Research* and *Naval Research Logistics.*

L. DOUGLAS SMITH is Professor of Management Science and Information Systems at the University of Missouri–St. Louis. He holds a Ph.D. in Management Sciences from the University of Minnesota and an M.B.A. and B.Sc. (Physics) from McMaster University. Smith is a former Associate Editor of *Decision Sciences.* His research has emphasized the development of models and computer-based systems for the solution of managerial problems.

EDWARD C. LAWRENCE is Professor of Finance in the School of Business Administration at the University of Missouri–St. Louis. He has written extensively in the areas of banking and corporate finance. He received his Ph.D. in Finance from the Pennsylvania State University.

Proceedings of the 1996 Winter Simulation Conference
ed. J. M. Charnes, D. J. Morrice, D. T. Brunner, and J. J. Swain

RECENT DEVELOPMENTS IN INPUT MODELING WITH BÉZIER DISTRIBUTIONS

Mary Ann Flanigan Wagner

Boeing Information Services
7990 Boeing Court MS CV-82
Vienna, VA 22183-7000, U.S.A.

James R. Wilson

Department of Industrial Engineering
North Carolina State University
Raleigh, NC 27695-7906, U.S.A.

ABSTRACT

New methods are presented for estimating univariate and bivariate Bézier distributions. A likelihood ratio test is used to estimate the number of control points for a univariate Bézier distribution fitted to sample data. To estimate the control points of a bivariate Bézier distribution with fixed marginals based on either sample data or subjective information about the joint dependency structure, a linear-programming approach is formulated. These methods are implemented in the Windows-based software system called PRIME—PRobabilistic Input Modeling Environment. Several examples illustrate the application of these estimation procedures.

1 INTRODUCTION

One of the central problems in the design and construction of stochastic simulation experiments is the selection of valid input models—that is, probability distributions that accurately mimic the behavior of the random input processes driving the system. In many applications, it is critical not only to capture the shape of the marginal distribution of each major input random variable but also to accurately represent the stochastic dependencies between those variates.

Although many practitioners appreciate the need for valid models of multivariate simulation inputs, they lack effective and widely available tools for building such input models. Stanfield et al. (1996) developed a technique for fitting a multivariate distribution when the correlation matrix and the first four moments for each marginal distribution have been specified or estimated by the user. Because the fitted joint distribution is built from univariate marginals belonging to the Johnson translation system (Johnson 1949a; Swain, Venkatraman, and Wilson 1988), the multivariate input-modeling technique of Stanfield et al. has substantial flexibility. Unfortunately, the fitted joint distribution does not belong to the multivariate Johnson translation system (Johnson 1949b, Johnson 1987); moreover, the corresponding conditional distributions do not belong to the Johnson system—and this lack of "closure" makes it impossible to obtain convenient closed-form expressions for the conditional distributions that naturally arise in some applications.

Other approaches to multivariate input modeling can be based on TES (Transform-Expand-Sample) processes (Jagerman and Melamed 1992a, 1992b; Melamed, Hill, and Goldsman 1992) and ARTA (AutoRegressive To Anything) processes (Cario and Nelson 1996). Both methodologies enable the user to specify the autocorrelation function out to an arbitrary lag for a univariate stochastic process with a user-specified marginal distribution, but ARTA processes seem to be substantially easier to use. Unfortunately the conditional distributions associated with TES and ARTA processes do not appear to possess any advantages in analytical or numerical tractability when compared to multivariate processes based on the Johnson translation system.

In this paper we extend the input-modeling methodology of Wagner and Wilson (1993, 1994, 1995, 1996) for representing continuous univariate and bivariate populations using Bézier distributions. The remainder of this paper is organized as follows. In Section 2 we summarize the main properties of univariate and bivariate Bézier distributions, and we establish some basic notation that is used throughout the paper. In Section 3 we develop a likelihood ratio test to estimate the number of control points (that is, the number of parameters) for a univariate Bézier distribution fitted to sample data. To estimate a configuration of control points for a bivariate Bézier distribution with fixed marginals that will yield a desired covariance structure based either on sample data or subjective information, in Section 4 we present a fitting procedure that is formulated as a linear-programming problem. These methods are implemented in the Windows-based software system called PRIME—PRobabilistic Input Modeling Environment. A public-domain version of PRIME is available upon request; and in Section 5 an application of PRIME illustrates the effectiveness of these distribution-fitting techniques. Finally in Section 6 we summarize the main contributions of this work, and we make recommendations for future research.

2 OVERVIEW OF BÉZIER DISTRIBUTIONS

2.1 Definition of Bézier Curves

A Bézier curve of degree n with *control points* $\{\boldsymbol{p}_i \equiv (x_i, z_i)^{\mathrm{T}} : i = 0, 1, \ldots, n\}$ is given parametrically by

$$\boldsymbol{P}(t) = \boldsymbol{P}(t; n, \boldsymbol{x}, \boldsymbol{z}) = \sum_{i=0}^{n} B_{n,i}(t)\, \boldsymbol{p}_i \tag{1}$$

for $t \in [0, 1]$, where $\boldsymbol{x} \equiv (x_0, x_1, \ldots, x_n)^{\mathrm{T}}$ and $\boldsymbol{z} \equiv (z_0, z_1, \ldots, z_n)^{\mathrm{T}}$ respectively denote the vectors of x- and z-coordinates of the control points $\{\boldsymbol{p}_i\}$, and where the *blending function* $B_{n,i}(t)$ is the Bernstein polynomial

$$B_{n,i}(t) \equiv \frac{n!}{i!\,(n-i)!}\, t^i (1-t)^{n-i} \tag{2}$$

for $t \in [0, 1]$ and $i = 0, 1, \ldots, n$. (Throughout this paper, all vectors will be column vectors unless otherwise stated; and the roman superscript ^{T} will denote the transpose of a vector or matrix so that each control point is understood to be a column vector. Moreover, we will use the simpler the notation $\boldsymbol{P}(t)$ in (1) when no confusion can arise from this usage.)

In the definition (1) of the Bézier curve, we note that the control points act like "magnets"; and the "magnetic attraction" exerted on the Bézier curve $\{\boldsymbol{P}(t) : t \in [0, 1]\}$ by the ith control point $\boldsymbol{p}_i$ is strongest at the value $t = i/n$ for the parameter t so that the corresponding point $\boldsymbol{P}(t)$ on the curve is "in the vicinity" of $\boldsymbol{p}_i$. If the weight (magnetic attraction) of a control point is 1, then the Bézier curve is forced to pass through that control point exactly (Farin 1990).

2.2 Definition of Bézier Surfaces

If selected control points are represented by the column vectors $\{\boldsymbol{q}_{i,j} \equiv (x_{i,j},\ y_{i,j},\ z_{i,j})^{\mathrm{T}} : i = 0, 1, \ldots, n_x;\ j = 0, 1, \ldots, n_y\}$, then the corresponding two-dimensional Bézier surface in three-dimensional Euclidean space is given parametrically as

$$\begin{aligned} \boldsymbol{Q}(t_x, t_y) &= \boldsymbol{Q}(t_x, t_y; n_x, n_y, \boldsymbol{x}, \boldsymbol{y}, \boldsymbol{z}) \\ &= \sum_{i=0}^{n_x} \sum_{j=0}^{n_y} B_{n_x,i}(t_x) B_{n_y,j}(t_y)\, \boldsymbol{q}_{i,j} \end{aligned} \tag{3}$$

for all $t_x, t_y \in [0, 1]$, where $\boldsymbol{x}$, $\boldsymbol{y}$, and $\boldsymbol{z}$ are the $(n_x + 1) \times (n_y + 1)$ matrices of the x-, y-, and z-coordinates of the given control points $\{\boldsymbol{q}_{i,j}\}$. As in (1) and elsewhere in this paper, we will suppress the dependence of Bézier functions like $\boldsymbol{Q}(t_x, t_y)$ on the parameters n_x, n_y, $\boldsymbol{x}$, $\boldsymbol{y}$, and $\boldsymbol{z}$ when no confusion can arise from this simplification.

2.3 Univariate Bézier Distributions

In this subsection we summarize briefly some key properties of univariate Bézier distributions. For a detailed development of these properties, see Wagner and Wilson (1993, 1996). Given a continuous random variable X with bounded support $[x_*, x^*]$ and unknown cumulative distribution function (c.d.f.) $F_X(\cdot)$, we can approximate $F_X(\cdot)$ arbitrarily closely by a Bézier curve of the form (1) with sufficiently high degree n, where

$$\left.\begin{aligned} x(t) &= \sum_{i=0}^{n} B_{n,i}(t)\, x_i \\ F_X[x(t)] &= \sum_{i=0}^{n} B_{n,i}(t)\, z_i \end{aligned}\right\}, \tag{4}$$

for all $t \in [0, 1]$. If $F_X(\cdot)$ is given parametrically by (4), then the corresponding probability density function (p.d.f.) $f_X(\cdot)$ is given parametrically by $x(t)$ in (4) together with

$$f_X[x(t)] = \frac{\sum_{i=0}^{n-1} B_{n-1,i}(t)\Delta z_i}{\sum_{i=0}^{n-1} B_{n-1,i}(t)\Delta x_i} \tag{5}$$

for all $t \in [0, 1]$, where

$$\left.\begin{aligned} \Delta x_i &\equiv x_{i+1} - x_i \\ \Delta z_i &\equiv z_{i+1} - z_i \end{aligned}\right\} \text{ for } i = 0, 1, \ldots, n-1. \tag{6}$$

In Wagner and Wilson (1993, 1996), we present several applications of the Bézier family of univariate distributions for modeling simulation inputs.

2.4 Bivariate Bézier Distributions

In this subsection we summarize briefly some key properties of bivariate Bézier distributions. For a detailed development of these properties, see Wagner and Wilson (1994, 1995). If $(X, Y)^{\mathrm{T}}$ is a continuous random vector with bounded support $[x_*, x^*] \times [y_*, y^*]$, unknown c.d.f. $F_{XY}(\cdot, \cdot)$, and unknown p.d.f. $f_{XY}(\cdot, \cdot)$, then we can approximate $F_{XY}(\cdot, \cdot)$ arbitrarily closely with an appropriate Bézier surface of the form (3) that has sufficiently large values of n_x and n_y (Farin 1990), where

$$x(t_x) = \sum_{i=0}^{n_x} B_{n_x,i}(t_x)\, x_i \quad \text{for all } t_x \in [0,\, 1], \tag{7}$$

$$y(t_y) = \sum_{j=0}^{n_y} B_{n_y,j}(t_y)\, y_j \quad \text{for all } t_y \in [0,\, 1], \tag{8}$$

and

$$F_{XY}[x(t_x),\, y(t_y)] = \sum_{i=0}^{n_x}\sum_{j=0}^{n_y} B_{n_x,i}(t_x) B_{n_y,j}(t_y)\, z_{i,j} \tag{9}$$

for all $t_x, t_y \in [0,\, 1]$. In Wagner and Wilson (1994, 1995) we describe special properties of the $\boldsymbol{x}$, $\boldsymbol{y}$, and $\boldsymbol{z}$ matrices that are required to ensure the validity of the parametric representation (7)–(9) of a bivariate probability distribution.

To represent the control points for each marginal distribution of $F_{XY}(\cdot, \cdot)$, we observe that the $\{x_i\}$ in (7) and the $\{y_j\}$ in (8) respectively serve as the x- and y-coordinates of the control points for the marginal c.d.f.'s $F_X(\cdot)$ and $F_Y(\cdot)$. We reserve the symbols $\{z_i^{(X)} : i = 0, 1, \ldots, n_x\}$ to denote the z-coordinates of the control points of $F_X(\cdot)$; and we reserve the symbols $\{z_j^{(Y)} : j = 0, 1, \ldots, n_y\}$ to denote the z-coordinates of the control points of $F_Y(\cdot)$.

If $F_{XY}(\cdot, \cdot)$ is given parametrically by Equations (7)–(9), then the corresponding bivariate density function $f_{XY}(\cdot, \cdot)$ is given parametrically by $x(t_x)$ in (7) and $y(t_y)$ in (8) together with

$$f_{XY}[x(t_x),\, y(t_y)] = \frac{\displaystyle\sum_{i=0}^{n_x-1}\sum_{j=0}^{n_y-1} B_{n_x-1,i}(t_x) B_{n_y-1,j}(t_y)\, \Delta_i \Delta_j z_{i,j}}{\left[\displaystyle\sum_{i=0}^{n_x-1} B_{n_x-1,i}(t_x)\, \Delta x_i\right]\left[\displaystyle\sum_{j=0}^{n_y-1} B_{n_y-1,j}(t_y)\, \Delta y_j\right]} \tag{10}$$

for all $t_x, t_y \in [0,\, 1]$, where

$$\left.\begin{aligned} \Delta_j z_{i,j} &= z_{i,j+1} - z_{i,j} \\ \Delta_i \Delta_j z_{i,j} &= z_{i+1,j+1} - z_{i,j+1} - z_{i+1,j} + z_{i,j} \end{aligned}\right\} \tag{11}$$

for $i = 0, 1, \ldots, n_x - 1$ and $j = 0, 1, \ldots, n_y - 1$.

2.5 Estimating Bézier Distributions Using PRIME

PRIME is a graphical Windows-based software system that is used to construct both univariate and bivariate Bézier distributions. PRIME is designed to be easy and intuitive to use. The construction of a Bézier distribution is performed through the actions of the mouse, and several options are conveniently available through menu selections. To manipulate a c.d.f., the user may move any of the control points by clicking on a chosen control point and then dragging that control point to the desired location by moving the mouse. Control points are represented as small black squares, and each control point is given a label corresponding to its index i in Equation (4). The user may also add or delete control points via the mouse and the keyboard. Any movement, addition or deletion of a control point causes the displayed distribution to be updated (nearly) instantaneously so that the user gets immediate feedback on the effects of editing that distribution. See Wagner and Wilson (1993, 1994, 1995, 1996) for more information on PRIME.

3 ESTIMATING THE NUMBER OF CONTROL POINTS

A likelihood ratio test is used to estimate the number of control points of a univariate Bézier distribution. If the random sample $\boldsymbol{X} \equiv \{X_j : j = 1, \ldots, m\}$ has been taken from a univariate Bézier p.d.f. $f_X(\cdot; n, \boldsymbol{x}, \boldsymbol{z})$ that is given parametrically by (5) with a known value of n, then the maximum likelihood estimates $\widehat{\boldsymbol{x}}_n$, $\widehat{\boldsymbol{z}}_n$ of the x- and z-coordinates of the control points $\{\boldsymbol{p}_i : i = 0, \ldots, n\}$ in (1) are obtained by solving the following nonlinear optimization problem:

$$\left.\begin{array}{ll} \underset{\boldsymbol{x},\, \boldsymbol{z}}{\min.} & \mathcal{L}_n(\boldsymbol{x}, \boldsymbol{z}|\boldsymbol{X}) \equiv \displaystyle\prod_{j=1}^{m} f_X(X_j; n, \boldsymbol{x}, \boldsymbol{z}) \\ \text{s. t.} & f_X[x(t); n, \boldsymbol{x}, \boldsymbol{z}] > 0 \text{ for } t \in (0,\, 1) \\ & z_0 = 0 \\ & z_n = 1 \\ & x_0 < X_{(1)} \\ & x_n > X_{(m)} \end{array}\right\}, \tag{12}$$

where

$$X_{(1)} \le X_{(2)} \le \cdots \le X_{(m)}$$

are the order statistics for the sample $\{X_j\}$. Thus the optimal value of the likelihood function $\mathcal{L}_n(\boldsymbol{x}, \boldsymbol{z}|\boldsymbol{X})$ for the given sample $\boldsymbol{X}$ is $\mathcal{L}_n\left(\widehat{\boldsymbol{x}}_n, \widehat{\boldsymbol{z}}_n \middle| \boldsymbol{X}\right)$.

Next we consider the hypothesis that the sample $\boldsymbol{X}$ was actually taken from a Bézier distribution with one additional control point—that is, with a total of $n+2$ control points rather than the $n+1$ control points postulated in (12). For a hypothesized Bézier p.d.f. $f_X(\cdot; n+1, \boldsymbol{x}, \boldsymbol{z})$ involving one additional control point, let $\widehat{\boldsymbol{x}}_{n+1}$ and $\widehat{\boldsymbol{z}}_{n+1}$ respectively denote the maximum likelihood estimators of the x- and z-coordinates of the corresponding control points; and let $\mathcal{L}_{n+1}\left(\widehat{\boldsymbol{x}}_{n+1}, \widehat{\boldsymbol{z}}_{n+1} \middle| \boldsymbol{X}\right)$ denote the corresponding maximum value of the likelihood function. Using the matrix representation of Bézier curves of degree n and $n+1$, we can show that any nth degree Bézier curve can be represented as an $(n+1)$st degree Bézier curve whose control-point coordinates are subject to two linear constraints (see §4.6 of Farin 1990). Thus under the null hypothesis that n is the true degree of the underlying Bézier distribution from which the sample $\boldsymbol{X}$ was taken, Theorem 4.4.4 of Serfling (1980) ensures that the likelihood ratio test statistic asymptotically has a chi-square distribution with two degrees of freedom as the sample size m grows large:

$$2\left[\mathcal{L}_{n+1}\left(\widehat{\boldsymbol{x}}_{n+1}, \widehat{\boldsymbol{z}}_{n+1} \middle| \boldsymbol{X}\right) - \mathcal{L}_{n}\left(\widehat{\boldsymbol{x}}_{n}, \widehat{\boldsymbol{z}}_{n} \middle| \boldsymbol{X}\right)\right] \underset{m\to\infty}{\xrightarrow{\mathcal{D}}} \chi^2(2). \tag{13}$$

We exploit (13) to assess the importance of successive increments of the likelihood function as the number of control points is repeatedly incremented by one. The final estimate of the number of control points for the Bézier distribution fitted to the sample $\boldsymbol{X}$ is determined to be the smallest value of n for which the difference on the left-hand side of (13) is not significant at a prespecified level of significance. The corresponding vectors $\widehat{\boldsymbol{x}}_n$ and $\widehat{\boldsymbol{z}}_n$ respectively provide the final estimates of the x- and z-coordinates for the fitted Bézier c.d.f.

4 ESTIMATING BIVARIATE BÉZIER DISTRIBUTIONS

Wagner and Wilson (1994, 1995) present a methodology for estimation of bivariate Bézier distributions. This technique requires fitting each marginal distribution separately; then the user must employ trial and error in manipulating the control points either for the joint p.d.f. or for selected conditional distributions in order to match the joint dependency structure between the two components of the target random vector $(X, Y)^{\mathrm{T}}$. This trial-and-error approach, however, is extremely difficult to perform in practice; and this difficulty motivated the development of the bivariate fitting procedure given in this section.

A detailed derivation of the covariance between two Bézier variates X and Y is given in Flanigan (1993) and Wagner and Wilson (1995). For completeness, we summarize the final result. The covariance between X and Y, $\mathrm{Cov}(X, Y)$, is

$$\mathrm{Cov}(X, Y) = \sum_{i=0}^{n_x-1} \sum_{j=0}^{n_y-1} \vartheta_i^{(X)} \vartheta_j^{(Y)} \Delta_i \Delta_j z_{i,j}\,, \tag{14}$$

where

$$\vartheta_i^{(X)} = \left[\frac{1}{2} \sum_{l=0}^{n_x} \frac{\binom{n_x}{l}\binom{n_x-1}{i}}{\binom{2n_x-1}{i+l}} x_l\right] - \mathrm{E}[X] \tag{15}$$

for $i = 0, 1, \ldots, n_x$ and

$$\vartheta_j^{(Y)} = \left[\frac{1}{2} \sum_{k=0}^{n_y} \frac{\binom{n_y}{k}\binom{n_y-1}{j}}{\binom{2n_y-1}{j+k}} y_k\right] - \mathrm{E}[Y] \tag{16}$$

for $j = 0, 1, \ldots, n_y$. Notice that the covariance defined by (14)–(16) depends on the coordinates of the control points $\{\boldsymbol{q}_{i,j} : i = 0, 1, \ldots, n_x; = 0, 1, \ldots, n_y\}$.

If we are given a target covariance $\widehat{C}$ that has been elicited from experts or has been estimated from sample data and if we have previously estimated the marginals so that all of the following quantities have fixed values

$$\left\{x_i,\, z_i^{(X)},\, \vartheta_i^{(X)} : i = 0, 1, \ldots, n_x\right\}$$

and

$$\left\{y_j,\, z_j^{(Y)},\, \vartheta_j^{(Y)} : j = 0, 1, \ldots, n_y\right\},$$

then we can match the value $\widehat{C}$ as closely as possible by minimizing the absolute deviation

$$\left|\left\{\sum_{i=0}^{n_x-1} \sum_{j=0}^{n_y-1} \vartheta_i^{(X)} \vartheta_j^{(Y)} \Delta_i \Delta_j z_{i,j}\right\} - \widehat{C}\right| \tag{17}$$

as a linear function of the $\{z_{i,j} : i = 1, \ldots, n_x - 1; j = 1, \ldots, n_y - 1\}$ subject to linear constraints that ensure a valid joint distribution. In particular the following linear programming problem must be solved:

$$\underset{\substack{W_1, W_2;\ \text{all } z_{i,j}:\\ 1 \le i \le n_x - 1\\ 1 \le j \le n_y - 1}}{\text{minimize}} \quad W_1 + W_2 \tag{18}$$

subject to

$$\left.\begin{aligned}
&W_1 - W_2 \\
&\quad -\sum_{i=1}^{n_x-1}\sum_{j=1}^{n_y-1}\left[\Delta_i\Delta_j\vartheta_{i-1}^{(X)}\vartheta_{j-1}^{(Y)}\right] z_{i,j} \\
&\quad = \widehat{C} - \vartheta_0^{(X)}\vartheta_{n_y-1}^{(Y)} z_1^{(X)} - \vartheta_{n_x-1}^{(X)}\vartheta_0^{(Y)} z_1^{(Y)} \\
&\quad -\vartheta_{n_x-1}^{(X)}\vartheta_{n_y-1}^{(Y)}\left[1 - z_{n_x-1}^{(X)} - z_{n_y-1}^{(Y)}\right] \\
&\quad - \sum_{j=1}^{n_y-2} \vartheta_{n_x-1}^{(X)}\vartheta_j^{(Y)}\Delta_j z_j^{(Y)} \\
&\quad - \sum_{i=1}^{n_x-2} \vartheta_i^{(X)}\vartheta_{n_y-1}^{(Y)}\Delta_i z_i^{(X)}
\end{aligned}\right\} \quad (19)$$

as well as

$$\left.\begin{aligned}
&z_{1,j+1} - z_{1,j} \geq 0 \\
&z_{1,n_y-1} \leq z_1^{(X)} \\
&z_{i+1,1} - z_{i,1} \geq 0 \\
&z_{n_x-1,1} \leq z_1^{(Y)} \\
&z_{i+1,j+1} - z_{i,j+1} - z_{i+1,j} + z_{i,j} \geq 0 \\
&z_{n_x-1,j+1} - z_{n_x-1,j} \leq z_{j+1}^{(Y)} - z_j^{(Y)} \\
&z_{i+1,n_y-1} - z_{i,n_y-1} \leq z_{i+1}^{(X)} - z_i^{(X)} \\
&z_{n_x-1,n_y-1} \geq z_{n_x-1}^{(X)} + z_{n_y-1}^{(Y)} - 1
\end{aligned}\right\} \quad (20)$$

for $i = 1, \ldots, n_x - 2$ and $j = 1, \ldots, n_y - 2$; and finally

$$W_1, W_2, z_{i,j} \geq 0 \; . \quad (21)$$

for $i = 1, \ldots, n_x - 1$ and $j = 1, \ldots, n_y - 1$. In this formulation W_1 and W_2 respectively represent the positive and negative parts of the estimation error as defined in (19); the constraints in (20) ensure that the joint density $f_{XY}(\cdot,\cdot)$ is positive; and finally (21) represents the standard nonnegativity constraints. Notice that in (19), the first and second differences $\Delta_i z_i^{(X)}$, $\Delta_j z_j^{(Y)}$, and $\Delta_i\Delta_j\vartheta_{i-1}^{(X)}\vartheta_{j-1}^{(Y)}$ are defined in the same way as the analogous quantities $\Delta_i x_i$, $\Delta_j y_j$, and $\Delta_i\Delta_j z_{i,j}$ in (6) and (11).

The linear programming problem (18)–(21) must be solved to complete the estimation of a bivariate Bézier distribution. This problem consists of $(n_x - 2)(n_y - 2) + 2(n_x - 2) + 2(n_y - 2) + 4 = n_x n_y$ structural constraints involving $(n_x - 1)(n_y - 1) + 2 = n_x n_y - n_x - n_y + 3$ decision variables. Thus it follows that the target covariance $\widehat{C}$ may not be exactly achieved by the final fitted bivariate Bézier distribution, but the target covariance $\widehat{C}$ will be matched as closely as is mathematically possible.

5 EXAMPLES

To illustrate the application of the likelihood ratio test (13) and the covariance-matching procedure based on (18)–(21), we consider a sample data set consisting of $m = 672$ bivariate observations that were generated in a simulation-based forestry study. Figures 1 and 2 depict the empirical and fitted marginal distributions for X and Y, respectively. Table 1 displays the sample statistics for each marginal distribution together with the corresponding population characteristics for the marginal Bézier distributions that were fitted using PRIME.

The histogram in Figure 1 clearly reveals that the distribution of the first coordinate X has two modes; and in our experience some manual intervention beyond routine application of the likelihood ratio test (13) is often required to obtain adequate fits to multimodal data sets. The fitted c.d.f. in Figure 1 was obtained by using PRIME to position 11 control points interactively so that $n_x = 10$. An alternative approach is to automatically obtain a bimodal Bézier p.d.f. via the likelihood ratio procedure (13). Using this approach, we had to start the test procedure with 11 control points; and with a significance level of 20% for each iteration, the likelihood ratio test procedure yielded a final estimate of 13 control points whose associated c.d.f. and p.d.f. closely resemble their counterparts in Figure 1. On a 66 Mhz 80486-based microcomputer running Windows 3.1, four iterations of the likelihood ratio test procedure (13) starting with 11 control points ($n_x = 10$) and stopping with 13 control points ($n_x = 12$) required 4.6 minutes of execution time for this univariate data set. Note that the execution time reported here is for a nonoptimized, debugger-enabled version of PRIME rather than a production version of the software; and substantially better execution times are expected for the final production version of the software.

Because the histogram in Figure 2 strongly suggests a unimodal distribution for the second coordinate Y, we applied the likelihood ratio test (13) starting with three control points (so that $n_y = 2$ initially) and using a 20% significance level for each iteration of the test procedure. Figure 2 also displays the final fitted c.d.f. and p.d.f. with 6 control points (so that the final estimate is $n_y = 5$). On a 66 Mhz 80486-based microcomputer running Windows 3.1, five iterations of the likelihood ratio test procedure (13) starting with 3 control points ($n_y = 2$) and stopping with 6

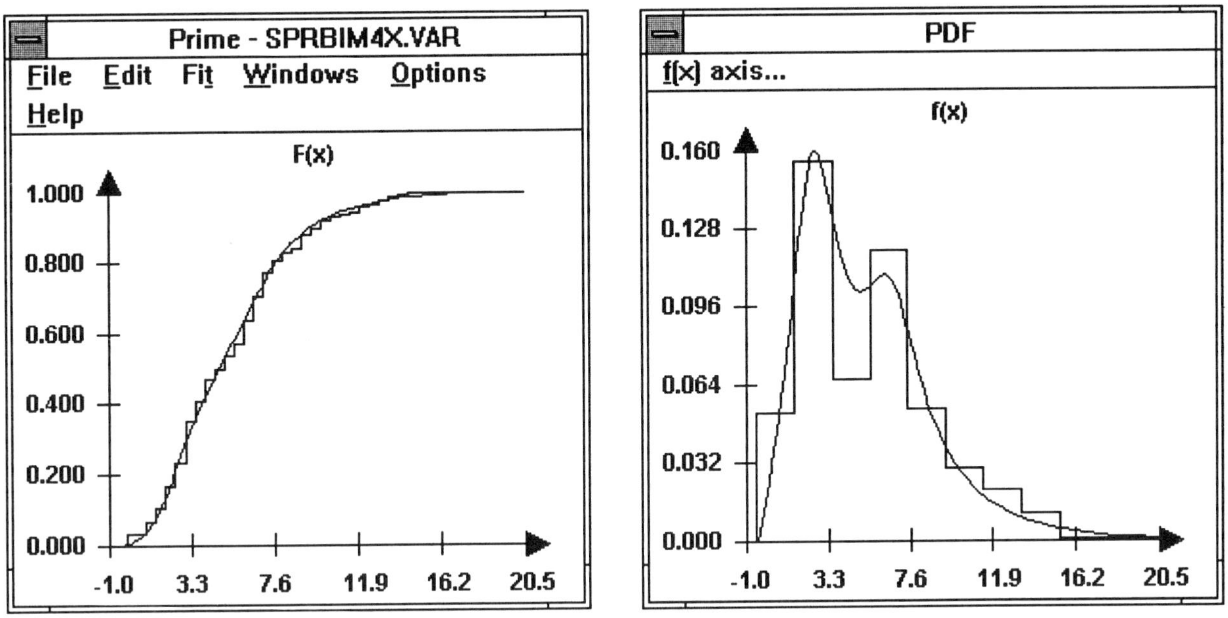

Figure 1: Fitted and Empirical Marginal Distributions of X for Sample Data Set

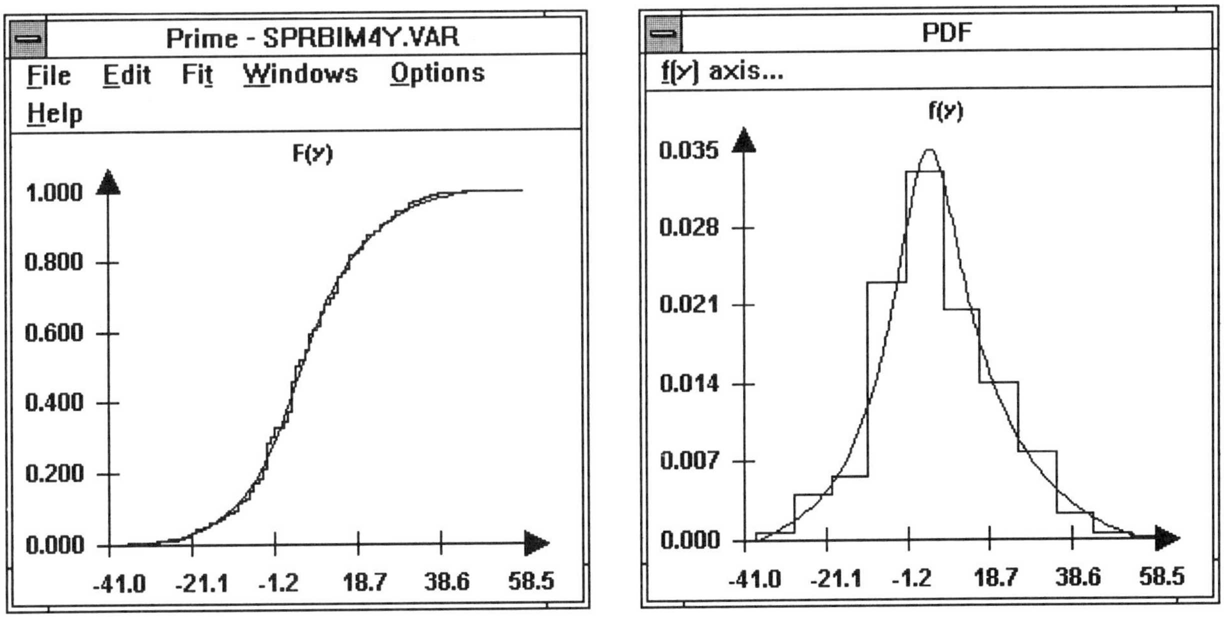

Figure 2: Fitted and Empirical Marginal Distributions of Y for Sample Data Set

Table 1: Fitted vs. Empirical Marginal Distributions of Sample Data Set

Characteristic	X Fitted	X Empirical	Y Fitted	Y Empirical
Mean	5.230	5.310	6.117	6.097
Standard Deviation	3.227	3.314	14.210	13.781
Skewness	0.997	0.845	0.252	0.155
Kurtosis	4.162	3.545	3.259	3.276
Minimum	−0.323	0.000	−37.678	−36.000
Maximum	19.824	19.500	55.000	54.000

control points ($n_y = 5$) required about 49 seconds of execution time for this univariate data set.

Starting from the fitted marginal distributions described above with $n_x = 10$ and $n_y = 5$, we applied the covariance-matching procedure (18)–(21) to the entire bivariate data set. The sample covariance and correlation for this data set were

$$\widehat{\text{Cov}}(X,Y) = \widehat{C} = 12.535 \quad \text{and} \quad \widehat{\text{Corr}}(X,Y) = 0.274,$$

respectively; and the corresponding values for the fitted bivariate Bézier distribution were

$$\text{Cov}(X,Y) = 12.535 \quad \text{and} \quad \text{Corr}(X,Y) = 0.273.$$

The resulting bivariate density is displayed in Figure 3. Clearly the fitted bivariate p.d.f. is also bimodal. On a 66 Mhz 80486-based microcomputer running Windows 3.1, solution of the linear programming problem (18)–(21) required about 3 seconds of execution time.

6 CONCLUSIONS AND RECOMMENDATIONS

The likelihood ratio test (13) provides a means for automatic determination of an appropriate number of control points in fitting a univariate Bézier distribution to sample data. However in our computational experience, the log-likelihood function $\mathcal{L}_n(\boldsymbol{x}, \boldsymbol{z}|\boldsymbol{X})$ corresponding to a univariate Bézier c.d.f. $F_X(\cdot; n, \boldsymbol{x}, \boldsymbol{z})$ is often extremely flat in the neighborhood of the exact maximum likelihood estimates. From the standpoint of visual closeness between the empirical c.d.f. $\widehat{F}_X(\cdot)$ and the fitted c.d.f. $\widehat{F}_X(\cdot; n, \widehat{\boldsymbol{x}}_n, \widehat{\boldsymbol{z}}_n)$, we have consistently obtained better fits with PRIME by using the methods of least squares or minimum L_1 norm estimation rather than the method of maximum likelihood to calculate the parameter estimates $\widehat{\boldsymbol{x}}_n$, $\widehat{\boldsymbol{z}}_n$ used in each stage of the test procedure (13). With this modification, the likelihood ratio test (13) yields excellent fits to many types of sample data with 4 control points ($n = 3$) so that effectively the fitted univariate Bézier distribution frequently has 6 parameters.

The covariance-matching procedure based on (18)–(21) often yields excellent fits to the joint dependency structure exhibited by bivariate sample data sets, but there is no guarantee that the target covariance or correlation between the two components of the random vector $(X, Y)^{\mathrm{T}}$ will be achieved even approximately. It appears that a relaxation of the system (20) of constraints may sometimes be required to match a given target covariance; but in such cases it is unclear how to ensure that the fitted bivariate Bézier c.d.f. has a legitimate p.d.f. Although further investigation of this issue is required, the general approach outlined in Section 4 of this paper promises to yield highly effective methods for fitting bivariate Bézier distributions to sample data or subjective information.

REFERENCES

Cario, M. C., and B. L. Nelson. 1996. Autoregressive to anything: Time series input processes for simulation. Working paper, Department of Industrial, Welding and Systems Engineering, The Ohio State University, Columbus, Ohio.

Farin, G. 1990. *Curves and surfaces for computer aided geometric design: A practical guide.* 2d ed. New York: Academic Press.

Flanigan, M. A. 1993. A flexible, interactive, graphical approach to modeling stochastic input processes. Ph.D. diss., School of Industrial Engineering, Purdue University, West Lafayette, Indiana.

Jagerman, D. L., and B. Melamed. 1992a. The transition and autocorrelation structure of TES processes, Part I: General theory. *Communication in Statistics–Stochastic Models* 8 (2): 193–219.

Jagerman, D. L., and B. Melamed. 1992b. The transition and autocorrelation structure of TES processes, Part II: Special cases. *Communication in*

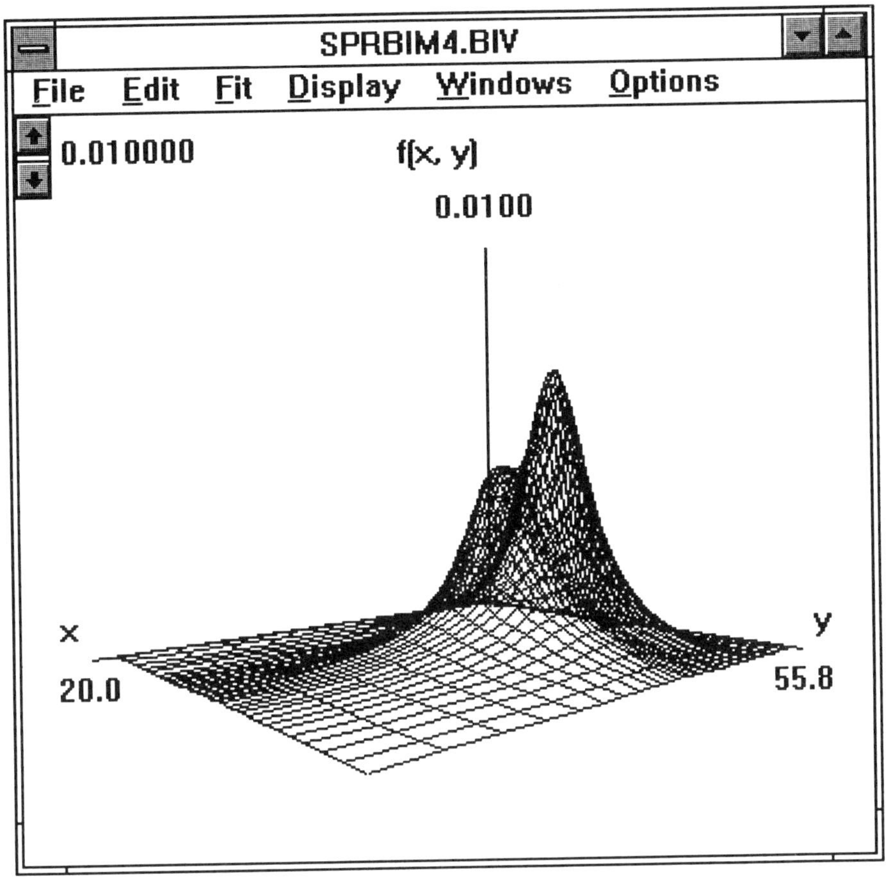

Figure 3: The Joint Density Fitted to the Sample Data Set

Statistics–Stochastic Models 8 (3): 499–527.

Johnson, M. E. 1987. *Multivariate statistical simulation.* New York: John Wiley & Sons.

Johnson, N. L. 1949a. Systems of frequency curves generated by methods of translation. *Biometrika* 36:149–176.

Johnson, N. L. 1949b. Bivariate distributions based on simple translation systems. *Biometrika* 36:297–304.

Melamed, B., J. R. Hill, and D. Goldsman. 1992. The TES methodology: Modeling empirical stationary time series. In *Proceedings of the 1992 Winter Simulation Conference*, ed. J. J. Swain, D. Goldsman, R. C. Crain, and J. R. Wilson, 135–144. Piscataway, New Jersey: Institute of Electrical and Electronics Engineers.

Serfling, R. J. 1980. *Approximation theorems of mathematical statistics.* New York: John Wiley & Sons.

Stanfield, P. M., J. R. Wilson, G. A. Mirka, N. F. Glasscock, J. P. Psihogios, and J. R. Davis. 1996. Multivariate input modeling with Johnson distributions. In *Proceedings of the 1996 Winter Simulation Conference*, ed. J. M. Charnes, D. J. Morrice, D. T. Brunner, and J. J. Swain, to appear. Piscataway, New Jersey: Institute of Electrical and Electronics Engineers.

Swain, J. J., S. Venkatraman, and J. R. Wilson. 1988. Least-squares estimation of distribution functions in Johnson's translation system. *Journal of Statistical Computation and Simulation* 29:271–297.

Wagner, M. A. F., and J. R. Wilson. 1993. Using univariate Bézier distributions to model simulation input processes. In *Proceedings of the 1993 Winter Simulation Conference*, ed. G. W. Evans, M. Mollaghasemi, E. C. Russell, and W. E. Biles, 365–373. Piscataway, New Jersey: Institute of Electrical and Electronics Engineers.

Wagner, M. A. F., and J. R. Wilson. 1994. Using bivariate Bézier distributions to model simulation input processes. In *Proceedings of the 1994 Winter Simulation Conference*, ed. J. D. Tew, S. Manivannan, D. A. Sadowski, and A. F. Seila, 324–331. Piscataway, New Jersey: Institute of Electrical and Electronics Engineers.

Wagner, M. A. F., and J. R. Wilson. 1995. Graphi-

cal interactive simulation input modeling with bivariate Bézier distributions. *ACM Transactions on Modeling and Computer Simulation* 5 (3): 163–189.
Wagner, M. A. F., and J. R. Wilson. 1996. Using univariate Bézier distributions to model simulation input processes. *IIE Transactions* to appear.

AUTHOR BIOGRAPHIES

MARY ANN FLANIGAN WAGNER is currently working at Boeing Information Services, located in Vienna, Virginia. Her responsibilities include the simulation and analysis of large-scale telecommunications models. Her principal interests are in simulation development, modeling and analysis. Her undergraduate and graduate degrees are in the field of Industrial Engineering, and in May 1993, she received her Ph.D. from Purdue University. From 1987 to 1989 she held a research position at the Regenstrief Institute, where she was responsible for the development and analysis of simulation models. Dr. Wagner is a member of: Alpha Pi Mu, ACM, IIE, INFORMS, Omega Rho, SIGGRAPH, and SIGSIM.

JAMES R. WILSON is Professor and Director of Graduate Programs in the Department of Industrial Engineering at North Carolina State University. He received a B.A. degree in mathematics from Rice University, and he received M.S. and Ph.D. degrees in industrial engineering from Purdue University. His current research interests are focused on the design and analysis of simulation experiments. He also has an active interest in applications of operations research techniques to all areas of industrial engineering. From 1988 to 1992, he served as Departmental Editor of *Management Science* for Simulation. He was *Proceedings* Editor for WSC '86, Associate Program Chair for WSC '91, and Program Chair for WSC '92. He has also held various offices in TIMS (now INFORMS) College on Simulation. He is a member of ASA, ACM/SIGSIM, IIE, and INFORMS.

Proceedings of the 1996 Winter Simulation Conference
ed. J. M. Charnes, D. J. Morrice, D. T. Brunner, and J. J. Swain

MULTIVARIATE INPUT MODELING WITH JOHNSON DISTRIBUTIONS

Paul M. Stanfield

ABCO Automation, Inc.
6202 Technology Drive
Brown Summit, NC 27214, U.S.A.

James R. Wilson

Department of Industrial Engineering
North Carolina State University
Raleigh, NC 27695-7906, U.S.A.

Gary A. Mirka
Naomi F. Glasscock
Jennie P. Psihogios
Joseph R. Davis

Ergonomics Laboratory
Department of Industrial Engineering
North Carolina State University
Raleigh, NC 27695-7906, U.S.A.

ABSTRACT

This paper introduces a new method for multivariate simulation input modeling based on the Johnson translation system of probability distributions. This technique matches the first four marginal moments and the correlation structure of a given set of sample data, allowing computationally efficient parameter estimation and random-vector generation. Applications of the technique in ergonomics and production scheduling are discussed. The proposed method is compared to traditional multivariate input-modeling techniques based on the Johnson translation system.

1 INTRODUCTION

In statistical and simulation applications, one is often faced with the task of representing empirical data with a parameterized distribution. Many such distributions exist for modeling univariate data. Few of these distributions are easily extended to model multivariate populations. The most commonly used multivariate distribution, the multivariate normal, has an inflexible shape and is often inappropriate for data modeling.

In this paper we present a more flexible method for multivariate simulation input modeling. The proposed method is based on the Johnson translation system of univariate probability distributions, and it exploits an appropriate affine transformation of independent standardized Johnson variates to generate a multivariate random vector with the desired first four marginal moments and correlation structure. Furthermore, parameter estimation and variate generation are computationally efficient with this procedure.

There are numerous potential applications for this multivariate input-modeling technique. In this paper two applications are discussed in some detail.

Ergonomics Application. We construct a biomechanical model of the forces exerted by the trunk musculature during lifting. Ten trunk muscles are used to describe the trunk's moment generating mechanism. Each lifting task is characterized by the angular kinematics of the torso as well as the total moment generated during the lift. The forces exerted by the ten-muscle system can be modeled as a ten-dimensional random vector. The muscular force is estimated through the use of electromyography (EMG) and is expressed as a percentage of the muscle's capacity. The degree of muscle coactivation (correlation) can be high, so it is inappropriate to use independent random variables to model the joint behavior of these EMG measurements. Additionally, the shape of the marginal distributions can be significantly skewed so it is inappropriate to use a normal multivariate distribution.

Production-Scheduling Application. A simulation model is used to support real-time scheduling of a repair shop. Each job entering the shop for repair must pass through the same series of operations. The operation times are stochastic and nonnormal. In addition, the operation times corresponding to a particular job are stochastically interdependent. Once again, independent univariate and normal multivariate distributions are inappropriate for modeling such a situation. Moreover, component variates are realized sequentially in this system. As a result, it is desired to determine the distribution of remaining operation times conditioned on the completed repair times.

This paper is organized as follows. Section 2 gives a concise introduction to the Johnson translation system of univariate distributions. Section 3 contains a brief survey of multivariate distributions including a previously developed multivariate Johnson distribution. Section 4 outlines the proposed multivariate model, discusses its capabilities and limitations, and gives a method for fitting multivariate distributions to sample data. Section 5 discusses the ergonomics application of the proposed input-modeling procedure. Finally, conclusions and recommendations for future work are given in Section 6.

2 JOHNSON TRANSLATION SYSTEM OF UNIVARIATE DISTRIBUTIONS

Starting from a continuous random variable X whose distribution is unknown and is to be approximated and subsequently sampled, Johnson (1949a) proposed a set of four normalizing translations. These translations have the general form

$$Z = \gamma + \delta \cdot g\left(\frac{X-\xi}{\lambda}\right), \tag{1}$$

where Z is a standard normal random variate, γ and δ are shape parameters, λ is a scale parameter, ξ is a location parameter, and $g(\cdot)$ is a function whose form defines the four distribution families in the Johnson translation system,

$$g(y) = \tag{2}$$
$$\begin{cases} \ln(y), & \text{for } S_L \text{ (lognormal) family,} \\ \ln\left[y+\sqrt{y^2+1}\right], & \text{for } S_U \text{ (unbounded) family,} \\ \ln[y/(1-y)], & \text{for } S_B \text{ (bounded) family,} \\ y, & \text{for } S_N \text{ normal family.} \end{cases}$$

The translation (1) should approximately transform the continuous random variate X into a standard normal variate. The process of fitting a Johnson distribution to sample data involves first selecting a fitting method and the desired translation function $g(\cdot)$ and then obtaining estimates of the four parameters γ, δ, λ, and ξ. The fitting method utilized in this paper is moment matching. The Johnson translation system of distributions has the flexibility to match any feasible set of sample values for the mean, variance, skewness, and kurtosis. Additionally, the skewness and kurtosis uniquely identify the appropriate translation function $g(\cdot)$. As a result, fitting a data set using moment matching is reduced to the problem of finding the values of γ, δ, λ, and ξ which approximately transform X into a standardized normal variate. Although there are no closed-form expressions for the parameter estimates based on the method of moments, these parameter estimates can be accurately approximated using an iterative procedure of Hill, Hill, and Holder (1976). Moreover, other methods may used to fit each marginal distribution—for example, any of the estimation procedures implemented in the FITTR1 software package (Swain, Venkatraman, and Wilson 1988).

After the data set has been fitted with a Johnson distribution, variate generation is straightforward. First, a standardized normal variate Z should be generated. The corresponding realization of the Johnson variate X is found by applying to Z the inverse translation

$$X = \xi + \lambda \cdot g^{-1}\left(\frac{Z-\gamma}{\delta}\right),$$

where

$$g^{-1}(z) = \tag{3}$$
$$\begin{cases} e^z, & \text{for } S_L \text{ (lognormal) family,} \\ \left(e^z - e^{-z}\right)/2, & \text{for } S_U \text{ (unbounded) family,} \\ 1/\left(1+e^{-z}\right), & \text{for } S_B \text{ (bounded) family,} \\ z, & \text{for } S_N \text{ (normal) family.} \end{cases}$$

If normal variates are generated by an approximation to the method of inversion, then the generated Johnson variates can enhance the effectiveness of correlation-induction techniques such as common random numbers or antithetic variates; and this can be reflected in improved efficiency of simulation-based performance measures for the target system.

3 MULTIVARIATE DISTRIBUTIONS

3.1 Overview

Many univariate distributions have been generalized to form bivariate distributions. These include numerous discrete (binomial, hypergeometric, Poisson) and continuous (uniform, normal, exponential, beta, and gamma) distributions (Mardia 1970). However, very few of these can be practically extended to higher dimensions. The multivariate normal distribution (Johnson and Kotz 1972) is the most easily manipulated and most frequently used multivariate distribution. However, the component normal distributions have a fixed shape and are often inappropriate for data fitting. In addition to the inflexibility of the multivariate normal, Johnson (1987) cites the following limitations of existing multivariate distributions:

- Some distributions (e.g., the Bessel function distributions) present significant computational problems.
- The support of some distributions (e.g., the beta-Stacy) is too limited to be of interest.
- Some multivariate distributions (e.g., Morgenstern's distribution) are able to represent only weak correlation structures.
- Computational methods for distribution fitting and variate generation have not been developed for some multivariate distributions.

Although the bivariate Bézier distribution family (Wagner and Wilson 1995) seems to have the potential for accurately representing many commonly occurring forms of bivariate dependence, the extension of this family to three or more dimensions appears to be cumbersome and computationally prohibitive. Other approaches to multivariate input modeling can be based on TES (Transform-Expand-Sample) processes (Jagerman and Melamed 1992a, 1992b; Melamed, Hill, and Goldsman 1992) and ARTA (AutoRegressive To Anything) processes (Cario and Nelson 1996). Both methodologies enable the user to specify the autocorrelation function out to an arbitrary lag for a univariate stochastic process with a user-specified marginal distribution, but ARTA processes seem to be substantially easier to use. In the following subsections we consider flexible multivariate distributions that are based on the Johnson family of univariate distributions.

3.2 Multivariate Johnson Distributions

Johnson (1949b) proposed a bivariate distribution based on the univariate Johnson distributions. The parameterized model matches the first four moments for each marginal distribution and then attempts to approximate the correlation between component variates. As detailed below, the technique is easily extended to higher dimensions. Consider a continuous multivariate random vector $\mathbf{X}$ with ν components,

$$\mathbf{X} = (X_1, \ldots, X_\nu)^{\mathrm{T}},$$

which is to be modeled with some parameterized distribution. The Johnson multivariate modeling method determines a normalizing translation such that

$$\mathbf{Z} = \boldsymbol{\gamma} + \boldsymbol{\delta}\mathbf{g}[\boldsymbol{\lambda}^{-1}(\mathbf{X} - \boldsymbol{\xi})] \sim N_\nu(\mathbf{0}_\nu, \boldsymbol{\Sigma}) . \quad (4)$$

This is accomplished as follows:

1. Identify the transformation
$$\mathbf{g}\left[(y_1, \ldots, y_\nu)^{\mathrm{T}}\right] \equiv [g_1(y_1), \ldots, g_\nu(y_\nu)]^{\mathrm{T}}$$
such that the marginal distribution of X_i is approximated by an appropriate univariate Johnson distribution, where $i = 1, \ldots, \nu$ and $g_i(\cdot)$ is one of the translation functions in (2)
2. Estimate the matrices of shape parameters,
$$\boldsymbol{\gamma} \equiv (\gamma_1, \ldots, \gamma_\nu)^{\mathrm{T}}, \quad \boldsymbol{\delta} \equiv \mathrm{diag}(\delta_1, \ldots, \delta_\nu),$$
and the matrices of the respective location and scale parameters,
$$\boldsymbol{\xi} \equiv (\xi_1, \ldots, \xi_\nu)^{\mathrm{T}}, \quad \boldsymbol{\lambda} \equiv \mathrm{diag}(\lambda_1, \ldots, \lambda_\nu),$$
using the method of moments on each marginal distribution separately.
3. Estimate correlation matrix $\boldsymbol{\Sigma}$ by (a) inserting each sample value $\{\mathbf{X}_j : j = 1, \ldots, n\}$ into the estimated normalizing translation (4) to obtain the corresponding sample $\{\mathbf{Z}_j : j = 1, \ldots, n\}$ of estimated standard normal random vectors; and (b) computing the sample correlation matrix of the $\{\mathbf{Z}_j\}$ as the approximate moment-matching estimator of $\boldsymbol{\Sigma}$.

Random vector generation consists of generating $\mathbf{Z}$ from a ν-dimensional multivariate normal distribution $N_\nu(\mathbf{0}_\nu, \boldsymbol{\Sigma})$ and then applying the inverse translation,

$$\mathbf{X} = \boldsymbol{\xi} + \boldsymbol{\lambda}\mathbf{g}^{-1}[\boldsymbol{\delta}^{-1}(\mathbf{Z} - \boldsymbol{\gamma})],$$

using the previously determined parameter vectors and the vector-valued inverse translation function

$$\mathbf{g}^{-1}\left[(z_1, \ldots, z_\nu)^{\mathrm{T}}\right] \equiv \left[g^{-1}(z_1), \ldots, g^{-1}(z_\nu)\right]^{\mathrm{T}},$$

where $g_i^{-1}(\cdot)$ is defined by (3) for $i = 1, \ldots, \nu$. This method will generate random vectors with exactly the same marginal moments as the original sample data (at least to the limits of machine accuracy); and if each of the empirical marginal distributions of the original sample data is nearly symmetric about its mean, then the intercomponent correlations of the fitted multivariate Johnson distribution will nearly match the sample correlations of the original sample data. However, if some of the empirical marginal distributions of the original sample data (or the corresponding underlying theoretical marginals) possess marked skewness, then the correlation matrix of the fitted multivariate Johnson distribution will not match the sample correlation matrix of the original data set.

These multivariate Johnson distributions have been shown in several applications to achieve an improved fit when compared to previously used multivariate distributions. Schreuder and Hafley (1977) used the bivariate bounded Johnson distribution (S_{BB}) to fit tree height and diameter data. Due to its flexibility, this distribution produces a consistently better fit than the beta, gamma, Weibull, lognormal, and normal distributions. The implied relationships between the component variates can be interpreted based on the results of a well-known height-diameter regression model. Subsequently, Schreuder, Bhattacharyya, and McClure (1982) successfully used a trivariate Johnson Bounded distribution (S_{BBB}) to fit tree diameter, height, and volume data.

4 A NEW MULTIVARIATE EXTENSION OF THE JOHNSON SYSTEM

An improved method to model multivariate distributions is now presented. Suppose the continuous random vector $\mathbf{X}$ has mean $\boldsymbol{\mu}_{\mathbf{X}}$ and correlation matrix $\mathbf{C}_{\mathbf{X}}$. Define the lower triangular matrix

$$\boldsymbol{\Theta}_{\mathbf{X}} \equiv [\theta_{ij}] = \mathbf{C}_{\mathbf{X}}^{1/2}$$

based on the Cholesky decomposition $\mathbf{C}_{\mathbf{X}} = \boldsymbol{\Theta}_{\mathbf{X}}\boldsymbol{\Theta}_{\mathbf{X}}^{\mathrm{T}}$ together with the matrix of standard deviations

$$\boldsymbol{\sigma}_{\mathbf{X}} \equiv \mathrm{diag}\left[\mathrm{Var}^{1/2}(X_1), \ldots, \mathrm{Var}^{1/2}(X_\nu)\right] .$$

If $\mathbf{Y} = (Y_1, \ldots, Y_\nu)^{\mathrm{T}}$ consists of *independent* standardized Johnson variates so that each component Y_i has mean zero and variance one for $i = 1, \ldots, \nu$, then

$$\mathbf{W} \equiv \boldsymbol{\mu}_{\mathbf{X}} + \boldsymbol{\sigma}_{\mathbf{X}}\boldsymbol{\Theta}_{\mathbf{X}}\mathbf{Y} \tag{5}$$

has the same mean vector and covariance matrix as $\mathbf{X}$. Notice that $\mathbf{W}$ does not have a multivariate Johnson distribution as defined in Subsection 3.2. The parameters $\gamma_i, \delta_i, \lambda_i$, and ξ_i of Y_i are set so that the ith component W_i of the random vector $\mathbf{W}$ has the same skewness and kurtosis as X_i for $i = 1, \ldots, \nu$.

Let $\mathbf{a}_{\mathbf{X}}$ and $\mathbf{b}_{\mathbf{X}}$ be $\nu \times 1$ vectors whose ith elements are the skewness and kurtosis of X_i, respectively. Similarly, let $\mathbf{a}_{\mathbf{Y}}$ and $\mathbf{b}_{\mathbf{Y}}$ denote the skewness and kurtosis vectors for $\mathbf{Y}$. Finally, define the k-fold Hadamard product of $\boldsymbol{\Theta}_{\mathbf{X}}$ as

$$\boldsymbol{\Theta}_{\mathbf{X}}^{(k)} \equiv \left[\theta_{ij}^{k}\right] \quad \text{for} \quad k = 3, 4 \tag{6}$$

together with the auxiliary vector

$$\boldsymbol{\Psi}_{\mathbf{X}} \equiv (\psi_1, \ldots, \psi_\nu)^{\mathrm{T}} , \tag{7}$$

where

$$\psi_i = 6 \sum_{j=1}^{\nu} \sum_{\ell=j+1}^{\nu} \theta_{ij}^2 \theta_{i\ell}^2 \quad \text{for} \quad i = 1, \ldots, \nu. \tag{8}$$

Now if the random vector $\mathbf{W}$ is generated according to the affine transformation (5), then it is easily shown that the skewness and kurtosis vectors $\mathbf{a}_{\mathbf{W}}$ and $\mathbf{b}_{\mathbf{W}}$ describing the components of $\mathbf{W}$ have the following relationship to the skewness and kurtosis vectors $\mathbf{a}_{\mathbf{Y}}$ and $\mathbf{b}_{\mathbf{Y}}$ describing the components of $\mathbf{Y}$,

$$\left.\begin{array}{rcl} \mathbf{a}_{\mathbf{W}} & = & \boldsymbol{\Theta}_{\mathbf{X}}^{(3)}\mathbf{a}_{\mathbf{Y}} \\ \mathbf{b}_{\mathbf{W}} & = & \boldsymbol{\Theta}_{\mathbf{X}}^{(4)}\mathbf{b}_{\mathbf{Y}} + \boldsymbol{\Psi}_{\mathbf{X}} \end{array}\right\}. \tag{9}$$

Thus, the Johnson parameter matrices $\boldsymbol{\gamma}$, $\boldsymbol{\delta}$, $\boldsymbol{\lambda}$, and $\boldsymbol{\xi}$ that determine the distribution of $\mathbf{Y}$ are selected to satisfy the moment-matching conditions

$$\left.\begin{array}{l} \mathbf{a}_{\mathbf{Y}} = \left[\boldsymbol{\Theta}_{\mathbf{X}}^{(3)}\right]^{-1} \mathbf{a}_{\mathbf{X}} \\ \mathbf{b}_{\mathbf{Y}} = \left[\boldsymbol{\Theta}_{\mathbf{X}}^{(4)}\right]^{-1} (\mathbf{b}_{\mathbf{X}} - \boldsymbol{\Psi}_{\mathbf{X}}) \end{array}\right\}. \tag{10}$$

It follows from (5), (9), and (10) that the transformed random vector $\mathbf{W}$ has the same second-order moment structure as $\mathbf{X}$; moreover the skewness and kurtosis of W_i match the skewness and kurtosis of X_i for $i = 1, \ldots, \nu$.

This method requires relatively little computational effort. Data fitting requires determining $4 \times \nu$ marginal moments and $\nu(\nu-1)/2$ correlation values, finding the fourth power of $\nu(\nu+1)/2$ numbers, inverting two $\nu \times \nu$ matrices, finding ν ψ_i's and fitting ν univariate Johnson distributions. Random vector generation requires generating ν independent Normal variates, applying ν Johnson inverse translations of the form (3), and two $\nu \times \nu$ matrix multiplication operations.

4.1 Limitations of the Procedure

The ability to match the skewness of $\mathbf{W}$ to that of $\mathbf{X}$ depends only on finding a random vector $\mathbf{Y}$ of standardized Johnson variates with the vector $\mathbf{a}_{\mathbf{Y}}$ of marginal skewness values specified by (10). A standardized univariate Johnson distribution can always be found whose skewness matches any target value. As a result, the proposed multivariate input-modeling procedure always matches the first three marginal moments and correlation structure of the modeled data.

The ability to match the kurtosis of $\mathbf{W}$ to that of $\mathbf{X}$ depends only on finding a vector $\mathbf{Y}$ of independent standardized Johnson variates with the vector $\mathbf{a_Y}$ of marginal skewness values and the vector $\mathbf{b_Y}$ of marginal kurtosis values specified by (10). Now for every choice of marginal distributions for $\mathbf{Y}$, we must have

$$\mathbf{b_Y} \geq \mathbf{a_Y^{(2)}} + \mathbf{1}_\nu; \tag{11}$$

see, for example, page 216 of Stuart and Ord (1994). Unfortunately, the requirement (11) is not guaranteed to hold for every solution of (10); and in some applications of the proposed method for multivariate simulation input modeling, we have observed that (11) fails to hold.

To handle the (relatively rare) anomalous situations in which (11) fails to hold for the solution to (10), we match each marginal skewness and "adjust" the corresponding marginal kurtosis values to feasible levels that satisfy (11). Therefore, if $\mathbf{a_Y}$ and $\mathbf{b_Y}$ completely consist of feasible skewness/kurtosis pairs, then the transformed multivariate $\mathbf{W}$ will have the same first four marginal moments and correlation structure as the original variate $\mathbf{X}$. If some infeasible pairs exist, then the corresponding kurtosis is increased to a feasible value. The resulting random vector $\mathbf{W}$ will have the first three marginal moments and correlation structure of $\mathbf{X}$ with some marginal kurtosis values slightly above those of $\mathbf{X}$. It is easy to apply the transformation to the corrected kurtosis values to determine the marginal kurtosis of $\mathbf{W}$ and the deviation from the marginal kurtosis of $\mathbf{X}$. Notice that the order in which the multivariate components are determined has some influence on the adjustment required for infeasible kurtosis pairs.

4.2 Conditional Marginal Distributions

Consider the specific case where the multivariate distribution is used to model operation times for successive jobs passing through a repair shop. Each job requires the same operations to be performed. Additionally, the operation times associated with a single job are correlated. Information about the job repair-time distributions is used in a real-time scheduling system. As a result, it is desired to utilize information from partially complete jobs to improve the estimate of remaining work. For example, we seek to determine the conditional probability distribution of the current job's service time for operation k given the service times of completed operations 1 through $k-1$. Now the density function for the ith independent standardized Johnson variate Y_i is

$$f_{Y_i}(y_i) = \frac{\delta_i}{\lambda_i\sqrt{2\pi}} g_i'\left(\frac{y_i - \xi_i}{\lambda_i}\right) \times \exp\left\{-\frac{1}{2}\left[\gamma_i + \delta_i \cdot g_i\left(\frac{y_i-\xi_i}{\lambda_i}\right)\right]^2\right\}, \tag{12}$$

where $g_i(\cdot)$ is the Johnson translation associated with i and $g_i'(\cdot)$ is the corresponding first derivative. Let

$$\mathbf{Y}(k) \equiv (Y_1, \ldots, Y_k)^{\mathrm{T}} \quad \text{for} \quad k = 1, \ldots, \nu,$$

denote the subvector consisting of the first k components of $\mathbf{Y}$. Since the components of $\mathbf{Y}$ are independent, the joint probability density function (p.d.f.) for $\mathbf{Y}(k)$ is

$$f_{\mathbf{Y}(k)}[\mathbf{y}(k)] = \prod_{i=1}^{k} f_{Y_i}(y_i)\,. \tag{13}$$

Using transformation (5) and the change-of-variables formula, we see that $\mathbf{X}(k)$, the subvector of $\mathbf{X}$ consisting of the first k operation times (components) of $\mathbf{X}$, has joint p.d.f.

$$f_{\mathbf{X}(k)}(\mathbf{x}(k)) = \frac{f_{\mathbf{Y}(k)}\Big([\boldsymbol{\sigma}_{\mathbf{X}(k)}\boldsymbol{\Theta}_{\mathbf{X}(k)}]^{-1}[\mathbf{x}(k) - \boldsymbol{\mu}_{\mathbf{X}(k)}]\Big)}{\det[\boldsymbol{\sigma}_{\mathbf{X}(k)}\boldsymbol{\Theta}_{\mathbf{X}(k)}]}, \tag{14}$$

where $\boldsymbol{\mu}_{\mathbf{X}(k)}$ is a $k \times 1$ subvector consisting of the first k components of $\boldsymbol{\mu}_{\mathbf{X}}$; and $\boldsymbol{\Theta}_{\mathbf{X}(k)}$ and $\boldsymbol{\sigma}_{\mathbf{X}(k)}$ are the $k \times k$ analogues of $\boldsymbol{\Theta}_{\mathbf{X}}$ and $\boldsymbol{\sigma}_{\mathbf{X}}$, respectively. Thus, the conditional p.d.f. of the kth operation time X_k given the vector $\mathbf{X}(k-1) = \mathbf{x}(k-1)$ of the preceding $k-1$ operation times may be computed as follows,

$$f_{X_k|\mathbf{X}(k-1)}[x_k \mid \mathbf{x}(k-1)] = \frac{f_{\mathbf{X}(k)}[\mathbf{x}(k-1), x_k]}{\int_{-\infty}^{\infty} f_{\mathbf{X}(k)}[\mathbf{x}(k-1), w_k]\, dw_k},$$

where the joint p.d.f. $f_{\mathbf{X}(k)}(\cdot)$ is evaluated by combining (12), (13), and (14). This expression may be numerically determined and manipulated. Through similar analysis, nearly any measure of the job service time distribution can be accurately estimated using this procedure.

5 ERGONOMICS APPLICATION

Consider now the ergonomic model of trunk muscle coactivation. The muscle exertion of a random subject on a particular type of lifting task is modeled as follows:

- The sample data set of size $n = 230$ yielded the marginal moments shown in Table 1. The corresponding matrix $\mathbf{C_X}$ of sample correlations given in Table 2.
- Extracting the "square root" $\mathbf{\Theta_X}$ of the correlation matrix yields the results shown in Table 3.
- The skewness vector $\mathbf{a_Y}$ and the kurtosis vector $\mathbf{b_Y}$ are computed from (6)–(10) as shown in Table 4.
- Finally, we fit each Y_i with the appropriate Johnson distributions as shown in Table 5.
- This model can now be used to generate random vectors that drive a biomechanical simulation model.

6 CONCLUSIONS AND RECOMMENDATIONS

In this paper we have developed a method for multivariate simulation input modeling based on the Johnson translation system of probability distributions. The method matches the first four moments and correlation structure of a given set of sample data. The distribution-fitting procedure allows computationally efficient parameter estimation and random-vector generation. An application to the field of ergonomics illustrates the procedure.

Future work should include a comparison of the proposed method for multivariate simulation input modeling versus the approaches based on TES and ARTA processes. Moreover, the approach outlined in this paper could be adapted to Bézier distributions; and the results achieved with such an adaptation should be compared to other techniques for fitting multivariate distributions with Bézier marginals.

In the ergonomics application briefly described in this paper, interest ultimately centers on how the ten-dimensional distribution of trunk-muscle coactivation varies over time. In this context the computational complexity of the parameter-estimation and variate-generation procedures appears to become a dominant consideration.

ACKNOWLEDGMENTS

This research was sponsored in part by Office of Naval Research Contract N00014-90-J-1009 and by Grant No. KO1 OH00135-03 from the National Institute for Occupational Safety and Health.

REFERENCES

Cario, M. C., and B. L. Nelson. 1996. Autoregressive to anything: Time series input processes for simulation. Working paper, Department of Industrial, Welding and Systems Engineering, The Ohio State University, Columbus, Ohio.

Hill, I. D, R. Hill, and R. L. Holder. 1976. Fitting Johnson curves by moments. *Applied Statistics* 25 (2): 180–189.

Jagerman, D. L., and B. Melamed. 1992a. The transition and autocorrelation structure of TES processes, Part I: General theory. *Communication in Statistics–Stochastic Models* 8 (2): 193–219.

Jagerman, D. L., and B. Melamed. 1992b. The transition and autocorrelation structure of TES processes, Part II: Special cases. *Communication in Statistics–Stochastic Models* 8 (3): 499–527.

Johnson, M. E. 1987. *Multivariate statistical simulation.* New York: John Wiley & Sons.

Johnson, N. L. 1949a. Systems of frequency curves generated by methods of translation. *Biometrika* 36:149–176.

Johnson, N. L. 1949b. Bivariate distributions based on simple translation systems. *Biometrika* 36:297–304.

Johnson, N. L., and S. Kotz. 1972. *Distributions in statistics: Continuous multivariate distributions.* New York: John Wiley & Sons.

Law, A. M., and W. D. Kelton. 1991. *Simulation modeling and analysis.* 2d ed. New York: McGraw Hill.

Mardia, K. V. 1970. *Families of bivariate distributions.* London: Griffin.

Melamed, B., J. R. Hill, and D. Goldsman. 1992. The TES methodology: Modeling empirical stationary time series. In *Proceedings of the 1992 Winter Simulation Conference*, ed. J. J. Swain, D. Goldsman, R. C. Crain, and J. R. Wilson, 135–144. Piscataway, New Jersey: Institute of Electrical and Electronics Engineers.

Schreuder, H. T., and W. L. Hafley. 1977. A useful bivariate distribution for describing stand structure of tree heights and diameters. *Biometrics* 33:471–478.

Schreuder, H. T., H. T. Bhattacharyya, and J. P. McClure. 1982. Towards a unified distribution theory for stand variables using the S_{BBB} distribution. *Biometrics* 38:137–142.

Stuart, A., and J. K. Ord. 1994. *Kendall's advanced theory of statistics.* Vol. 1, *Distribution theory.* 6th ed. London: Edward Arnold.

Swain, J. J., S. Venkatraman, and J. R. Wilson. 1988. Least-squares estimation of distribution functions

Table 1: Marginal Moments of Trunk Muscle Coactivation Data

Component	Mean	Standard Deviation	Skewness	Kurtosis
1	0.00645	0.00431	0.50456	2.62549
2	0.00947	0.00713	0.61072	3.04158
3	0.00831	0.00822	0.45553	2.80806
4	0.01141	0.01151	0.67521	3.01325
5	0.02709	0.02096	0.73188	3.03418
6	0.02938	0.01781	0.49942	2.76829
7	0.07834	0.02973	0.47293	2.71599
8	0.09164	0.03254	0.20482	2.33007
9	0.17535	0.05683	0.42247	2.83503
10	0.20070	0.06729	0.54304	3.20566

Table 2: Sample Correlation Matrix $\mathbf{C_X}$ for Trunk Muscle Coactivation Data

$$\begin{bmatrix} 1.0000 & 0.6157 & 0.3277 & 0.4067 & 0.3611 & 0.5264 & 0.3811 & 0.3122 & 0.2987 & 0.2310 \\ 0.6156 & 1.0000 & 0.2612 & 0.3960 & 0.2449 & 0.4421 & 0.3138 & 0.2956 & 0.3179 & 0.2179 \\ 0.3276 & 0.2612 & 1.0000 & 0.5391 & 0.1671 & 0.2779 & 0.2171 & 0.2914 & 0.1923 & 0.1994 \\ 0.4067 & 0.3960 & 0.5391 & 1.0000 & 0.2152 & 0.3677 & 0.2965 & 0.2682 & 0.2303 & 0.1459 \\ 0.3611 & 0.2449 & 0.1671 & 0.2152 & 1.0000 & 0.6195 & 0.4547 & 0.4016 & 0.3293 & 0.3736 \\ 0.5264 & 0.4421 & 0.2779 & 0.3677 & 0.6195 & 1.0000 & 0.4718 & 0.4971 & 0.4833 & 0.4514 \\ 0.3811 & 0.3138 & 0.2171 & 0.2965 & 0.4547 & 0.4718 & 1.0000 & 0.6490 & 0.7334 & 0.7031 \\ 0.3122 & 0.2956 & 0.2914 & 0.2682 & 0.4016 & 0.4971 & 0.6490 & 1.0000 & 0.6928 & 0.7582 \\ 0.2987 & 0.3179 & 0.1923 & 0.2303 & 0.3293 & 0.4833 & 0.7335 & 0.6928 & 1.0000 & 0.7499 \\ 0.2310 & 0.2179 & 0.1994 & 0.1459 & 0.3736 & 0.4514 & 0.7031 & 0.7582 & 0.7499 & 1.0000 \end{bmatrix}$$

Table 3: Sample Estimate of $\mathbf{\Theta_X}$ for Trunk Muscle Coactivation Data

$$\begin{bmatrix} 1.0000 & 0.0000 & 0.0000 & 0.0000 & 0.0000 & 0.0000 & 0.0000 & 0.0000 & 0.0000 & 0.0000 \\ 0.6156 & 0.7880 & 0.0000 & 0.0000 & 0.0000 & 0.0000 & 0.0000 & 0.0000 & 0.0000 & 0.0000 \\ 0.3276 & 0.0754 & 0.9417 & 0.0000 & 0.0000 & 0.0000 & 0.0000 & 0.0000 & 0.0000 & 0.0000 \\ 0.4067 & 0.1848 & 0.4161 & 0.7919 & 0.0000 & 0.0000 & 0.0000 & 0.0000 & 0.0000 & 0.0000 \\ 0.3611 & 0.0287 & 0.0495 & 0.0535 & 0.9292 & 0.0000 & 0.0000 & 0.0000 & 0.0000 & 0.0000 \\ 0.5264 & 0.1497 & 0.1000 & 0.1065 & 0.4460 & 0.6929 & 0.0000 & 0.0000 & 0.0000 & 0.0000 \\ 0.3811 & 0.1005 & 0.0899 & 0.1080 & 0.3271 & 0.1295 & 0.8372 & 0.0000 & 0.0000 & 0.0000 \\ 0.3122 & 0.1312 & 0.1902 & 0.0477 & 0.2940 & 0.2278 & 0.4406 & 0.7171 & 0.0000 & 0.0000 \\ 0.2987 & 0.1700 & 0.0867 & 0.0521 & 0.2254 & 0.2681 & 0.5741 & 0.2480 & 0.5980 & 0.0000 \\ 0.2310 & 0.0960 & 0.1237 & 0.0218 & 0.3039 & 0.2449 & 0.5559 & 0.3637 & 0.1862 & 0.5414 \end{bmatrix}$$

in Johnson's translation system. *Journal of Statistical Computation and Simulation* 29:271–297.

Wagner, M. A. F., and J. R. Wilson. 1995. Graphical interactive simulation input modeling with bivariate Bézier distributions. *ACM Transactions on Modeling and Computer Simulation* 5 (3): 163–189.

Table 4: Skewness and Kurtosis Vectors $\mathbf{a}_Y$ and $\mathbf{b}_Y$ for Trunk Muscle Coactivation Data

Component	Skewness	Kurtosis
1	0.004244	2.087854
2	1.097259	4.179445
3	-0.287684	1.702515
4	-0.204774	7.474299
5	0.658706	3.054686
6	0.033781	6.183929
7	1.503758	3.289309
8	1.032861	2.103015
9	-0.787371	3.155087
10	-0.499475	3.865285

Table 5: Parameters of Fitted Marginal Johnson Distributions for Trunk Muscle Coactivation Data

Muscle i	Type	γ_i	δ_i	ξ_i	λ_i
1	S_B	0.808	1.139	−2.004	5.692
2	S_B	1.019	0.727	−1.220	4.685
3	S_B	0.966	1.283	−2.155	6.350
4	S_B	0.818	0.340	−0.794	3.290
5	S_B	0.994	0.819	−1.375	4.914
6	S_B	0.633	0.159	−0.707	2.606
7	S_B	0.670	0.654	−1.305	3.992
8	S_B	0.161	0.205	−1.065	2.421
9	S_B	0.677	0.152	−0.678	2.639
10	S_B	1.189	0.161	−0.453	3.568

AUTHOR BIOGRAPHIES

PAUL M. STANFIELD is currently General Manager of ABCO Automation, Greensboro, North Carolina, and Adjunct Assistant Professor of Industrial Engineering at North Carolina Agricultural and Technical State University. He received the following degrees from North Carolina State University: a B.S. degree in electrical engineering, an M.S. degree in industrial engineering and operations research, and a Ph.D. degree in industrial engineering. He also received an M.B.A. degree from the University of North Carolina at Greensboro. He is a registered professional engineer in North Carolina. He is a member of Alpha Pi Mu, IIE, INFORMS, and Omega Rho.

JAMES R. WILSON is Professor and Director of Graduate Programs in the Department of Industrial Engineering at North Carolina State University. He received a B.A. degree in mathematics from Rice University, and he received M.S. and Ph.D. degrees in industrial engineering from Purdue University. His current research interests are focused on the design and analysis of simulation experiments. He also has an active interest in applications of operations research techniques to all areas of industrial engineering. From 1988 to 1992, he served as Departmental Editor of *Management Science* for Simulation. He was *Proceedings* Editor for WSC '86, Associate Program Chair for WSC '91, and Program Chair for WSC '92. He has also held various offices in TIMS (now INFORMS) College on Simulation. He is a member of ASA, ACM/SIGSIM, IIE, and INFORMS.

GARY A. MIRKA is an Assistant Professor in the Department of Industrial Engineering at North Carolina State University. He received B.S.I.E., M.S.I.E., and Ph.D. degrees from The Ohio State University. He is a member of the Human Factors and Ergonomics Society.

NAOMI F. GLASSCOCK is a Ph.D. student in the Department of Industrial Engineering at North Carolina State University. She received B.S. and M.S. degrees from North Carolina State University. She is a member of the Human Factors and Ergonomics Society.

JENNIE P. PSIHOGIOS is a Ph.D. student in the Department of Industrial Engineering at North Carolina State University. She is a member of the Human Factors and Ergonomics Society.

JOSEPH R. DAVIS is a Ph.D. student in the Department of Industrial Engineering at North Carolina State University. He received B.S. and M.S. degrees from North Carolina State University. He is a member of the Human Factors and Ergonomics Society.

Proceedings of the 1996 Winter Simulation Conference
ed. J. M. Charnes, D. J. Morrice, D. T. Brunner, and J. J. Swain

VISUALIZING GENERALIZED SEMI-MARKOV PROCESSES

Lee W. Schruben
Eric L. Savage

School of Operations Research and Industrial Engineering
Cornell University
Ithaca, New York 14853, U.S.A.

ABSTRACT

Generalized Semi-Markov Processes (GSMPs) are usually described by sets of variables, events and clock distributions. This kind of representation often lacks intuitive appeal. In this paper we propose a mapping from GSMPs to Event Graph Models. This mapping allows us to use an event graph to visualize a GSMP model as an intermediate step to implementation. By examining the event graph model, we can perform logic checking and verification more easily than if we try to interpret the GSMP description.

1 THE TWO MODELING PARADIGMS

Before presenting the mapping algorithm, we first introduce the two modeling schemes. S will denote a countable set of "physical" states and A equal a finite set of integers enumerating the events. Generic states are *s* and *s'*; a generic event is α.

1.1 Event Graph Models

Event graph models (EGMs) were first described by Schruben (1983) and later enriched by others, including Som and Sargent (1989). Pictorially the vertices of an EGM represent the various events in the simulation. The edges of the graph represent relationships between events. Basically, the edges define the conditions under which and the time delay after which one event will schedule another event to occur. Suppose the following edge is part of a simulation graph,

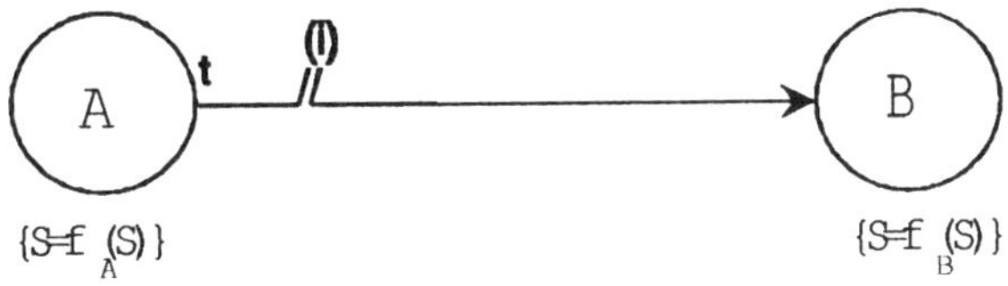

This edge is read as follows:

"whenever event A occurs, the system state, S, changes to $f_A(S)$. Then, if condition (i) is true, event B will be scheduled to occur after a delay of t."

Appropriate labels are omitted if the inter-event edge delay is zero or if the scheduling is unconditional.

One of the simplest examples of an EGM is a single server queue. Here the single state variable, Q, is the number of customers in the system (waiting in line or in service). The random time between customer arrivals is denoted as t_a and the random time of customer service is t_s. The EGM for a generic queue is as follows:

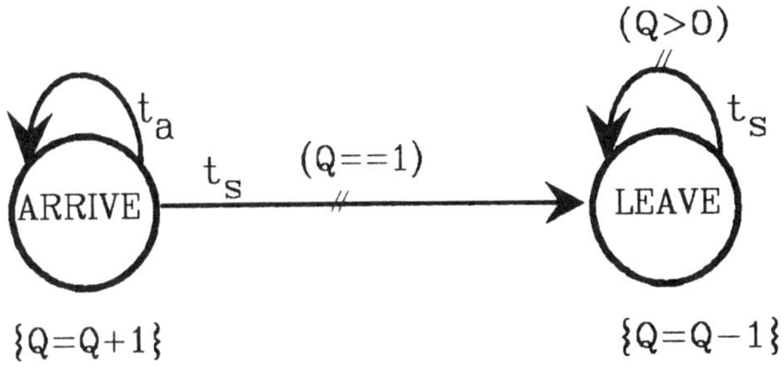

The ARRIVE event simply increments the queue and the LEAVE event decrements the queue.

We formally define an EGM using a directed graph G = {E,V} with edge set E and vertex set V and an associated state space, S. Generic vertices are denoted by v (perhaps with a subscript). Generic edges are denoted as $e = (v_o, v_d)$, which specifies the origin and destination of a directed edge. We label the graph with the following sets:

$F = \{f_v: \forall v \in V\}$ are the state changes associated with each event.

$C = \{c_e: S \rightarrow \Re, \forall e = (v_o, v_d) \in E\}$ when $c_e = 0$, the edge condition is false (as in the C programming language).

$T = \{t_e: \forall e = (v_o, v_d) \in E\}$ are the inter-event delay times.

$P = \{p_e: \forall e = (v_o, v_d) \in E\}$ are execution priority expressions used to break time ties.

The conditions in C specify whether or not an edge's destination event will be scheduled after the edge's origin event occurs. At any given time in the execution of the simulation, those edges where $c(s) \neq 0$ (i.e. the edge conditions are true) are referred to as *active* edges. Edges where $c(s) = 0$ can be thought of as being temporarily missing from the graph.

The basic notion of an event graph model, M = (V,E,S,F,C,T,P), is to represent the *indices* for above sets with the edges and vertices of a directed graph. It is this graph of indices that organizes the above sets into a simulation model.

1.2 Generalized Semi-Markov Processes

GSMPs are a useful probability model for discrete event simulation: Comprehensive treatments appear in (Glynn and Iglehart, 1988) and in (Glynn, 1989). We define GSMPs following the development in (Glasserman, 1991). Define the following: $E(s)$ = non-empty set of possible events in state $s \in S$; $p(s';s,\alpha)$ = probability of jumping from state s to s' when event, α, occurs; $F_\alpha(.)$ = distribution of interval until event α occurs. The dynamics of a particular GSMP model may require the two doubly-indexed independent sequences of random variables: (i) $\{X(\alpha,k);\ \alpha \in A,\ k=1,2,...\}$ distributed according to $F_\alpha(.)$ (these are the successive inter-event times for the GSMP sample path) and (ii) $\{U(\alpha,k);\ \alpha \in A,\ k=1,2,...\}$ uniformly distributed on the semi-open interval (0,1] (these are used for random state changes when events occur). Glasserman [1991] also defines the mappings ϕ for each $\alpha \in A$, $s' = \phi_\alpha(s,U(\alpha,k))$ so that $\text{Prob}(s' = \phi_\alpha(s,U(\alpha,k)) = p(s';s,\alpha)$. In an EGM of the same system both these input processes might be mappings of the random number sequence, U. The algorithm or scheme by which the dynamics of a GSMP evolve is typically specified by mimicking the execution of a typical event-scheduling discrete event simulation code.

2. A PROPOSED GSMP→EGM MAPPING

We start with a directed graph G = (V,E) and use the same physical state space, S, with a unique vertex, $v \in V$ in the event graph corresponding to each event, α, in the GSMP. The vertices are labeled with the state changes associated with each event $f_v = \phi_\alpha, \alpha = v$. For each pair of vertices, $e = (v_o, v_d)$, define:

$C_e = \{s \to s' \in S : v_o \in E(s) \text{ and } v_d \in \{E(s') - (E(s) - \{v_o\})\}\}$ where $s \to s'$ indicates that the present state of the system is s' and the state immediately preceding s' was s.

In other words if v_o can cause a transition out of state s and v_d is a new event when the state changes to s' then it must be possible for v_o to schedule v_d (i.e. if the systems enters state s' from state s).

If s' and v_o uniquely determine s (i.e f_v is invertible) the we can write a *static* definition:

$C_e = \{s' \in S : v_o \in E(s) \text{ and } v_d \in \{E(s') - (E(s) - \{v_o\})\}\ \forall s \in S \ni p(s';s,v_o) > 0\}$.

Here, if it is possible to get from state s to s' when v_o is executed and v_d is a new event in state s' then v_o must schedule v_d if the system is in state s'.

The edges of G are $E = \{e = (v_o, v_d): C_e \neq \emptyset\}$. Each edge is labeled with:

$t_e = \{X(v_d,k), k=1,2,...\}$, its delay time

$p_e: S \to R$, its execution priority,

and $c_e: S \to \{\text{true,false}\}$, its edge condition, where $c_e(s') = \text{true}$ iff $s' \in C_e$; that is we label each edge with a membership rule, C_e. Note that E can be countably infinite if S is, typically $|E| < |S|$.

3 EXAMPLE 1: A SINGLE-SERVER QUEUE

GSMP:

$S = \{0,1,2,...\} = Q$ = number of customers in the system

$A = (1, 2)$ = (Arrive, Leave)

$F_1(.)$ is the distribution of the inter-arrival time,t_a,

$F_2(.)$ is the distribution of the service time,t_s.

$E(s) = \{1,2\}$ if $s>0$, and $E(0) = \{1\}$, finally $p(s+1,s,1) = 1$, $p(s-1,s,2) = 1$

EGM:

V = {Arrive, Leave}

note that the state changes f_1 and f_2 are invertible so we can use the static definition of C_e. $f_1^{-1}(s)=s-1$ for $s > 0$, $f_2^{-1}(s)=s+1$

$C_{(1,1)} = \{s' \in Q : 1 \in E(s'-1) \text{ and } 1 \in \{E(s') - (E(s'-1) - \{1\})\}, s>0\} = \{Q>0\}$

$C_{(1,2)} = \{s' \in Q : 1 \in E(s'-1) \text{ and } 2 \in \{E(s') - (E(s'-1) - \{1\})\}, s>0\} = \{Q=1\}$

$C_{(2,1)} = \{s' \in Q : 2 \in E(s'+1) \text{ and } 1 \in \{E(s') - (E(s'+1) - \{2\})\}\} = \{\emptyset\}$

$C_{(2,2)} = \{s' \in Q : 2 \in E(s'+1) \text{ and } 2 \in \{E(s') - (E(s'+1) - \{2\})\}\} = \{Q>0\}$

The resulting event graph is identical to the single server queue model pictured in section 1.1 with the addition of the redundant condition (Q > 0) on the edge generating the arrivals.

4 EXAMPLE 2: A COLLISION-FREE BUS NETWORK

An interesting example of a larger GSMP is the collision-free bus network modeled by Iglehart & Shedler (1983). In this model there are N ports connected by a passive bilateral bus on which message packets are transmitted and received. In addition to the bus, there is a one-way logic control wire that links the ports. This wire transmits send flipflops which are essentially requests by a port to use the bus for transmission. The signal tapped at the control wire input to a port is the inclusive OR of the send flipflops from upstream ports, i.e., the port sees whether some upstream port is waiting to transmit.

When a message packet arrives at a port for transmission, the port sets the send flipflop (requests the bus), waits for a time interval to make sure that its flipflop is received and that it has been notified of any occurring transmission. After that delay, the port waits until the bus is observed to be idle and there are no bus requests from upstream. When those conditions are satisfied, it begins transmission and resets its send flipflop to 0.

Propagation times between ports i and j are denoted as T(i,j) for the actual propagation time along the bus and R(i,j) for the propagation time along the control wire. S(j) represents the send flipflop at port j and P(j) is the inclusive OR signal tapped at port j. For simplicity the model assumes that there is at most one packet in queue at each port; A(j), a random variable, represents the time between completion of transmission and arrival of the next message packet at port j. L(j) is a random variable representing the transmission time for port j. Since event interval times are constant, we omit their description by sampling function.

The variables for this GSMP are:
$W(t) = (W_1(t), ..., W_N(t))$ where $W_j(t)$ is the state of port j at time t. $W_j(t)=1$ if port j has set the flipflop but is not ready to transmit (because it is still waiting), $W_j(t)=2$ if port j has completed the waiting time but has not begun transmitting, $W_j(t)=3$ if port j is transmitting, $W_j(t)=4$ if transmission is complete and port j is waiting for the next message packet.
$U(t) = (U_1(t), ..., U_N(t))$ where $U_j(t)=1$ if port j observes the bus to be busy at time t, and equals 0 otherwise.
$V(t) = (V_{2,1}(t), V_{3,1}(t), V_{3,2}(t), V_{4,1}(t), ..., V_{N,N-1}(t))$ where $V_{j,k}(t)=1$ if port j has observed that port k has set its flipflop and equals zero otherwise.

The following are the events of the GSMP: "arrival of new packet at port j", "end of waiting period for port j", "end of transmission by port j", "observation by port j of the setting (to 1) of the flipflop by port k upstream (k<j)", "observation by port j of the resetting (to 0) of the flipflop by port k upstream (k<j)", "observation by port j of the start of transmission", "observation by port j of the end of transmission." Table 1 displays the events (with abbreviated names) along with their state changes, interval times and the subset of S for which the event is a member of E(s).

Table 1: Attributes of the Events of the Collision-Free Bus Network GSMP

Event	State Changes	time delay	s: event ∈ E(s)
New Packet(j)	W(j) = 1	A(j)	W(j) = 4
Clear(j)	W(j) = 2 or 3*	R(j,N)+T(1,N)	W(j) = 1
End Transmission(j)	W(j) = 4	L(j)	W(j) = 3
Observe Set(j,k)	V(k,j) = 1	R(j,k)	∃ k<j ∋ W(k) = 1, V(k,j) = 0
Observe Reset(j,k)	V(k,j) = 0, W(j) = 2 or 3*	R(j,k)	∃ k<j ∋ W(k) = 3, V(k,j) = 1
Observe Start(j)	U(j) = 1	T(k,j)	∃ k ∋ W(k) = 3, U(j) = 0
Observe End(j)	U(j) = 0, W(j) = 2 or 3*	T(k,j)	∃ k ∋ W(k) = 4, U(j) = 1

* W(j) = 3 if P(j) = 0 and U(j) = 0, W(j) = 2 otherwise

4.1 The Event Graph Translation

In this GSMP all of the state changes are invertible so we can use the static definition of C_e, $C_e = \{s' \in S : v_o \in E(s)$ and $v_d \in \{E(s') - (E(s) - \{v_o\})\}\ \forall s \in S \ni p(s';s,v_o) > 0\}$. For example consider v_o=Clear(j), v_d=Observe Start(k). Now Clear(j) $\in$ E(s) if and only if W(j)=1 and Observe Start (k) $\in$ E(s) if and only if W(j)=3 for some j and U(k)=0. So the only possible combinations of s and s' are those where s includes W(j)=1 and s' includes W(j)=3 and U(k)=0 and $p(s';s,v_o) > 0$. C_e therefore contains {s: W(j)=3 and U(k)=0}. C_e is not empty so we would draw an edge from Clear(j) to Observe Start(k); its edge condition would be (W(j)=3 and U(k)=0) and its time delay, from the GSMP, would be T(j,k). Table 2 lists all of the edges for which C_e is non-empty with the edge conditions and time delays. The state changes are the same as given in Table 1. Figure 1 shows the event graph.

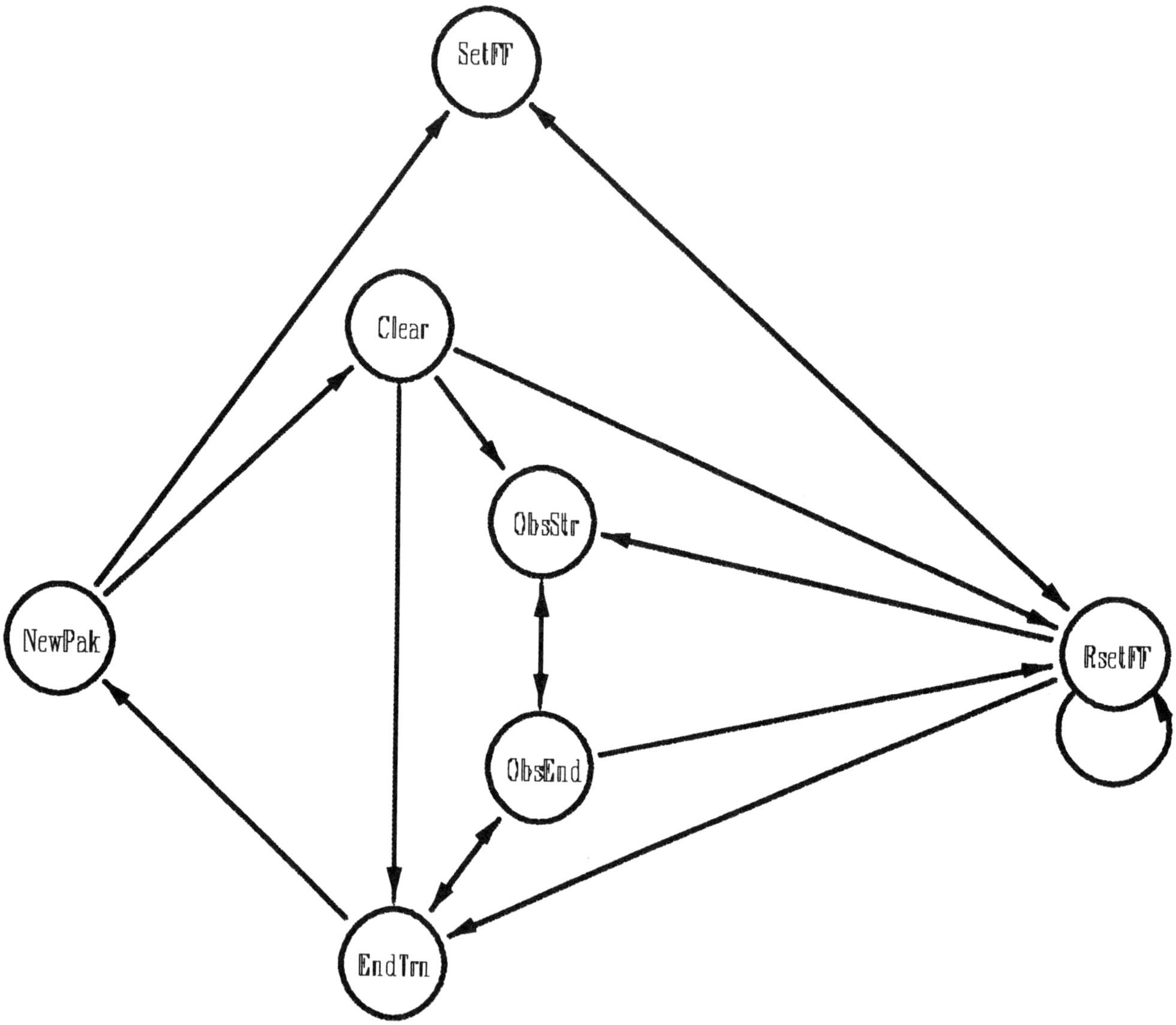

Figure 1: Event Graph of the Collision-Free Bus Network

Table 2: Edges of the Event Graph Translation of the Collision-Free Bus Network GSMP

Edge	Ce	time delay
New Packet(j) → Clear(j)	W(j) = 1	R(j,N)+T(1,N)
New Packet(j) → Observe Set(j,k), k>j	W(j) = 1, V(k,j) = 0	R(j,k)
Clear(j) → End Transmission(j)	W(j) = 3	L(j)
Clear(j) → Observe Reset(j,k), k>j	W(j) = 3, V(k,j) = 1	R(j,k)
Clear(j) → Observe Start(k), ∀k≠j	W(j) = 3, U(k) = 0	T(j,k)
End Transmission(j) → New Packet(j)	W(j) = 4	A(j)
End Transmission(j) → Observe End(k), ∀k≠j	W(j) = 4, U(k) = 1	T(j,k)
Observe Set(j,k) → Observe Reset(l,k), l<k	W(l) = 3, V(k,l) = 1	R(l,k)
Observe Reset(j,k) → End Transmission(j)	W(j) = 3	L(j)
Observe Reset(j,k) → Observe Set(l,k), l<k	W(l) = 1, V(k,l) = 0	R(l,k)
Observe Reset(j,k) → Observe Reset(k,l), l>k	W(k) = 3, V(l,k) = 1	R(k,l)
Observe Reset(j,k) → Observe Start(l), ∀l≠k	W(k) = 3, U(l) = 0	T(k,l)
Observe Start(j) → Observe End(j)	∃ k ∍ W(k) = 4, U(j) = 1	T(j,k)
Observe End(j) → End Transmission(j)	W(j) = 3	L(j)
Observe End(j) → Observe Reset(j,k), k>j	W(j) = 3, V(k,j) = 1	R(j,k)
Observe End(j) → Observe Start(j)	∃ k ∍ W(k) = 3, U(j) = 0	T(j,k)
Observe End(j) → Observe Start(k), ∀k≠j	W(j) = 3, U(k) = 0	T(j,k)

4.2 Checking the Logic

Now that we have an event graph model, we can code it up using available software. In this case we used SIGMA (Schruben 1995). In debugging this model a number of implementation issues arise.

The first concern was that some of the events would become inactive without being executed. For example, when port j begins transmission, an Observe Start event is scheduled for the other ports (because W(j) = 3). If the transmission is finished before port k observes the start then Observe Start(k) disappears from the active event set (W(j) ≠ 3 for any j) although it was never executed.

This means that events are being canceled in the original GSMP. While some GSMP definitions explicitly allow for canceling, others do not. If this were a problem for the modeler then certain assumptions would have to be made to prevent this occurrence

Iglehart and Shedler explicitly state that R(i,j)>T(i,j) ∀i,j. Additional assumptions are required however to prevent cancellation in the GSMP. L(j)>R(j,N) will allow the reset flipflop to finish propagating before the port finishes transmission and $A(j)>\max_i T(j,i)$ will allow the end of transmission to propagate before a new packet arrives at the port.

Another possibility is to redefine the event sets in the original GSMP. In fact without the above assumptions or the introduction of new state variables it is possible that the model will fail to be a GSMP at all; the system can reach a state where the active event set cannot be fully determined.

For example, suppose port j ends transmission and very soon afterwards a new packet arrives at port j. There may be a port k that has not observed that end of transmission yet; W(j) = 1 but Observe Start(k) should still be an active event. We also have the usual case however, where if W(j) ≠ 4 for any j then there are no Observe Start(k) events active. Therefore if W(j) = 1 we can't tell whether Observe Start(k) should be active or not.

Another situation is the existence of simultaneous events. In running the event graph of this GSMP it became clear that this situation exists in the model despite efforts and assumptions aimed at preventing simultaneous events. Further inspection revealed that the problem was caused by the fact that the end of transmission and start of transmission propagate at the same speed. If a start of transmission by port j is enabled by the observation that port k has ended transmission then the end of k's transmission and start of j's transmission now travel together down the bus. If they are executed in the wrong order (Observe Start then Observe End) then an error results because the other ports will think the bus is free when it is not.

In running the event graph version of the model, it also became clear that there were other problem in the definitions of the event sets for certain states. Specifically Observe End(j) is supposed to be an active event whenever W(k)=4 for some k and U(j)=1. What happens is that whenever port j observes a start of transmission, U(j) will become 1 and if any other port is waiting for a new packet (W(k)=4) then an Observe End(j) will be scheduled. If port j observes a start of transmission while the transmission is still happening an Observe End may be scheduled because of some other port which is waiting for a new packet, clearly this should not happen.

5 IMPLICATIONS FOR SIMULATION MODELING

Mapping from a GSMP to an EGM may uncover logical errors that are not apparent in the GSMP description. In addition, implementing and running the event graph model may uncover other concerns that need to be addressed in the original GSMP.

We can see in this example that EGMs offer an effective and efficient method for model development and structural analysis and permits development and enrichment of GSMP results and models. GSMPs offer a path to proofs of validity of simulation methodologies and potential algorithms for implementation. A GSMP→EGM mapping bridges a basic application-theory gap in that EGMs are very popular for model building and GSMPs are very popular for developing DEDS theory.

ACKNOWLEDGMENT

The authors are grateful to the Division of Design, Manufacture and Industrial Innovation of the National Science Foundation for their generous support through a research grant to Cornell University.

REFERENCES

Glasserman, P., (1991). "Structural Conditions for Perturbation Analysis Derivative Estimation: Finite Time Performance Indices." *Operations Research* **39**, pp. 724-738

Glynn, P. W., and D. L. Iglehart, (1988) "Simulation Methods for Queues: an Overview", *Queueing Systems*, Vol. 3, pp. 221-255, 1988.

Glynn, P., (1989). "A GSMP Formalism for Discrete Event Systems." *Proceedings of the IEEE*, Vol. 77, No. 1, pp. 14-23.

Iglehart, D. L. and G. S. Shedler, (1983). "Simulation of Non-Markovian Systems." *IBM Journal of Research and Development*, Vol. 27, No.5, pp. 472-479.

Schruben, L. W., (1983). "Simulation Modeling with Event Graphs." *Communications of the ACM*, Vol. 26, No. 11, pp. 957-963.

Schruben, L. W., (1995). *Graphical Simulation Modeling and Analysis: Using SIGMA for Windows*. Boyd & Fraser. Danvers, MA.

Som, T. K. and R. G. Sargent, (1989). "A Formal Development of Event Graphs as an Aid to Structured and Efficient Simulation Programs." *ORSA Journal on Computing*, Vol. 1, No. 2, pp. 107-125.

AUTHOR BIOGRAPHIES

LEE W. SCHRUBEN is a Professor in the School of Operations Research at Cornell University. He received his undergraduate degree in engineering from Cornell and his Ph.D. from Yale. His research interests are in statistical design and analysis of simulation experiments and in graphical simulation modeling methods.

ERIC L. SAVAGE is a Ph.D. candidate in the School of Operations Research at Cornell University. He received his undergraduate degree in mathematics from Rensselaer Polytechnic Institute and an M.S. degree in Operations Research from Cornell. His research interests include modeling logic and methodology in graphical representations.

Proceedings of the 1996 Winter Simulation Conference
ed. J. M. Charnes, D. J. Morrice, D. T. Brunner, and J. J. Swain

ON THE DEVELOPMENT OF A COMPUTER SYSTEM TO SIMULATE PORT OPERATIONS CONSIDERING PRIORITIES

José Holguín-Veras

Department of Civil Engineering
Universidad Autónoma de Santo Domingo
3573-D Lake Austin Blvd.,
Austin, Texas 78703, U.S.A.

C. Michael Walton

Department of Civil Engineering
The University of Texas at Austin
Austin, Texas 78712-1076, U.S.A.

ABSTRACT

This paper describes the simulation system developed at The University of Texas at Austin for the performance analysis of port operations considering container's priority. The paper discusses the current trends toward service differentiation and provides a brief description of port operations and the historical evolution of performance analysis in this area. The conceptual and computational characteristics of the simulation system are described, as well as the calibration process. Finally, conclusions and recommendations for further research are discussed.

1 INTRODUCTION

Increasingly over the last twenty years, developments in electronics and computer control are allowing production of goods with higher added value, smaller unit size and relatively low volume. Concurrently, the growing popularity of Just-in-Time (JIT) production systems have increased the importance of the cargoes' logistic value. On the other hand, the advent of intermodalism has provided container carriers with the opportunity to target non-traditional markets. As part of these efforts, container carriers are trying to attract low-valued cargoes as a way to reduce the number of empty movements, e.g., cotton movements from Texas to the West Coast. If these attempts to attract low-valued cargoes succeed, container carriers and intermodal terminals may be handling a potentially high number of containers carrying low-valued cargoes.

The combined effect of the aforementioned trends is to increase the relative importance of both ends of the cargo value distribution. In this context, an operational policy that does not distinguish containers according to cargo value is likely to penalize the segments of users located at the ends of the cargo value distribution, i.e., the low-valued cargoes may be charged for a service that they do not need, and the high-valued cargoes may receive a quality of service below their needs. Container carriers have responded to this new challenge by implementing "hot hatch programs," where the high priority containers are located on the ship hatches that will be unloaded first.

In this context, the implementation of priority systems will provide a level of service consistent with the container's priority. The objective of this research is to analyze the implementation of priority systems at container ports where network effects are not considered. The complexity of this problem required the research to be divided into three major areas: information systems, performance analysis and pricing policy.

This paper focuses on describing the simulation system developed for the performance analysis of priority systems. The system is able to simulate a wide range of operational policies ranging from the current "hot hatch" programs to more complex systems in which service differentiation is done at all the stages (i.e., movement to storage yard, storing yard operations, gate processing in/out of the storage yard and container retrieval).

The paper is comprised of five major sections in addition to the introduction. Section 2 presents a brief overview of port operations in general. In Section 3, the different approaches used for the performance analysis of port operations are discussed, and the needs for a simulation system considering priorities are analyzed. Section 4 is dedicated to describing the simulation system. Section 5 focuses on describing the calibration process. Finally, conclusions and recommendations for further research are presented.

2 OVERVIEW OF PORT OPERATIONS

The operation of transportation terminals is one of the most challenging tasks in the transportation industry. The amount of investment involved, the conflicting objectives of the agents involved, the ever pressing needs for increased efficiency, and the multitude of constraints that affect the operations are some of the elements that make terminal operations such a difficult task. The operation of marine container terminals is no exception.

The investment in a medium size container terminal frequently surpasses one hundred million dollars. Interacting at these expensive facilities are a number of different agents: (a) shipping companies, (b) container terminal operator, (c) railroads and motor carriers, (d) brokers, (e) shippers; (f) forwarders, and (g) regulatory agencies. Balancing the frequently conflicting objectives of these agents, and their different operational criteria is, to say the least, a difficult task.

The term "port operations" encompasses the set of service processes that takes place at marine ports. The characteristics of these processes depend upon the type of flow (e.g., import containers, export containers); the type of terminal (e.g., dedicated container terminal, multipurpose terminals), and the handling technology being used, that is frequently determined by managerial styles, financial capability and labour agreements.

Although there is not a unique way of conducting "port operations," the alternative systems share some common characteristics. In a very simplistic way, four major systems are identified: (a) marine side interface, (b) transfer system, (c) container storage system, and (e) land side interface. A "typical" U.S. terminal has the following characteristics:

a. marine side interface: two or three gantry cranes move containers to/from the ship. Since the efficiency of the gantry cranes is the single most important factor in determining ship turnaround time, maintaining a continuous operation of the gantry cranes is a primary objective.

b. transfer system: yard trucks and straddle carriers are the two most popular types of equipment. In the former case, the gantry cranes put the containers onto trailers towed by yard trucks to/from the container yard. Usually seven or eight yard trucks serve a gantry crane. When straddle carriers are used, no trailers are needed because the straddle carrier works as a mobile crane that lifts the containers from the apron, moves it to the container yard, and places it at the final destination.

c. container storage system: there are two basic schemes, i.e., stacking containers on the ground and storing the containers on chassis. In the former case, special handling equipment, e.g., yard cranes or straddle carriers are needed. In the latter, no special handling equipment is needed, though more land is required. Additionally, yard operations involve a number of supporting activities like updating container location and reorganizing the container yard.

d. land side interface: seven to ten lanes operated by clerks process related paperwork and clearances of incoming and exiting trucks. An increasing number of terminals are implementing Automatic Equipment Identification (AEI) devices and other information technologies (e.g., electronic tags, bar coded and magnetic strip cards) to facilitate the identification of vehicles and drivers, though the market penetration of these technologies is still relatively low. For a full discussion on information technology, see Holguín-Veras and Walton (1995d and 1995e).

3 PERFORMANCE ANALYSIS OF PORT OPERATIONS

Three different approaches have been used for the performance analysis of port operations: ship distribution at ports (SDP), queueing theory (QT) and simulation (Frankel, 1987; Holguín-Veras and Walton, 1995a). Historically, the first approach used was SDP; followed by QT and simulation.

SDP relies on the assumption that the berth occupancy analysis can be performed using the observed ship distribution at ports and, consequently, the number of ships at port is an independent random variable. The weakest point of this approach is that the number of ships present at the port, at any given day, is not independent of the number of ships that were present the day before and, consequently the independence assumption does not hold. In addition, since it is assumed that the SDP remains the same, irrespective of the number of berths and the demand, the influence of the service characteristics is not properly considered. Because of these flaws, the use of SDP has been abandoned.

The use of QT was suggested in the 20's and 30's for the capacity analysis of ports, but it was not until the early 60's when it became widely used (Agerschou and Korsgaard, 1969). The first application is commonly attributed to Mettam (1967) but an earlier application by Gould (1963) was found. Mettam did demonstrate the practicality of QT by stating the basic principles of this type of application, and highlighted the potential benefits that could be obtained from it. As it was to become typical of this approach, Mettam considered only the ship-berth interface. His paper was influential in attracting other analysts to QT. In general, the majority of QT applications consider only the ship-berth interface. In these applications basic QT, i.e., birth-death processes in equilibrium, has been used to provide performance estimates. Other classes of QT models, e.g., queueing network and cyclic queues, have been only sporadically applied. For instance, see Daskin and Walton (1983).

The number of papers on simulation applications is enormous; however, the number of innovative applications is much less. Most papers do not describe the details of the models. At most, a simplified description is provided in the context of actual applications. This situation arises because most of the simulation models are developed by private companies, being for most cases, proprietary materials. Among them it is worth mentioning the simulation systems developed by August Design, Inc., Liftech Consultants, Inc., and Vickerman, Zachary and Miller. Some of the traditional models are described by Frankel (1987).

After analyzing the different methodological options, the research team achieved the following conclusions:

a. QT formulations would be extremely complicated because priorities, at the user level, translate to a subdivision of the user population that significantly increases the number of states of the system. So far, most QT models that consider priorities deal with relatively simple problems. In addition, the big differences in the service times for the different ships make it difficult to justify the assumption of a homogeneous population of users that is required by QT.

b. simulation offers numerous advantages. It allows the explicit consideration of the geometry of the system and their interactions with the operational policy. Secondly, since it is not likely that the priority systems under consideration have been implemented in practice, there is no data with which to construct the empirical distributions needed in QT models. This is not a problem for simulation because the service time can be estimated by simulating the micro-movements of the equipment.

Thus, it was decided to use simulation as the performance analysis tool to model the different operational policies. Since the available simulation systems were not able to deal with different priority levels, it was decided to develop a new simulation system, PRIOR, which is described in the next section.

4 DESCRIPTION OF THE SYSTEM

The simulation system is comprised of two programs, PRIOR, which perfoms the simulation, and ECON, which calculates economic indicators of performance.

4.1 Description of PRIOR

The terminal is modelled using arrays to represent the storage location (on ship and in the yard) and networks to represent travel times for the different servers. Figure 1 shows a 3-D representation of the networks. The truck network is represented using a directed network. The yard crane and gantry crane networks are represented using non-directed networks. The program has the capability of simulating stochastic travel times, though this option was not used in the simulations. PRIOR uses the following operating principles:

String of ships: At the beginning of the simulation run, the file containing ship description, arrival time, and the fractions of high and low-priority containers is read, and the corresponding ship arrivals are scheduled. When all gantry cranes are idle and the previous ship has already departed, the queue list is scanned to begin service for the next ship.

Beginning of service (ships): When this event is processed the containers are created and the corresponding retrieval events are scheduled.

Creation of containers: The simulation system allows the user to specify the characteristics of the containers to be created (i.e., the fractions of containers of each kind and dwelling times for high and low-priority containers and the location of high-priority containers on the ship). After the containers are created, the control of the program is transferred to the subroutine in charge of simulating the unloading process.

Lot assignment: After the containers on ship have been created, the lot assignment process takes place. By virtue of this process the ship hatches are assigned to specific destination lots on the storage yard.

Gantry crane operations: The subroutines in charge of simulating gantry crane operations estimate the corresponding service times according to different operational rules. The first subroutine simulates the base case, in which the containers are evenly unloaded from top to bottom, regardless of priority. The second one simulates the operational rules for cases in which the gantry crane operator unloads high-priority containers first. Low-priority containers are unloaded only after all high-priority containers have been unloaded.

Yard truck operations: It is assumed that the yard trucks serve all gantry cranes. Whenever a gantry

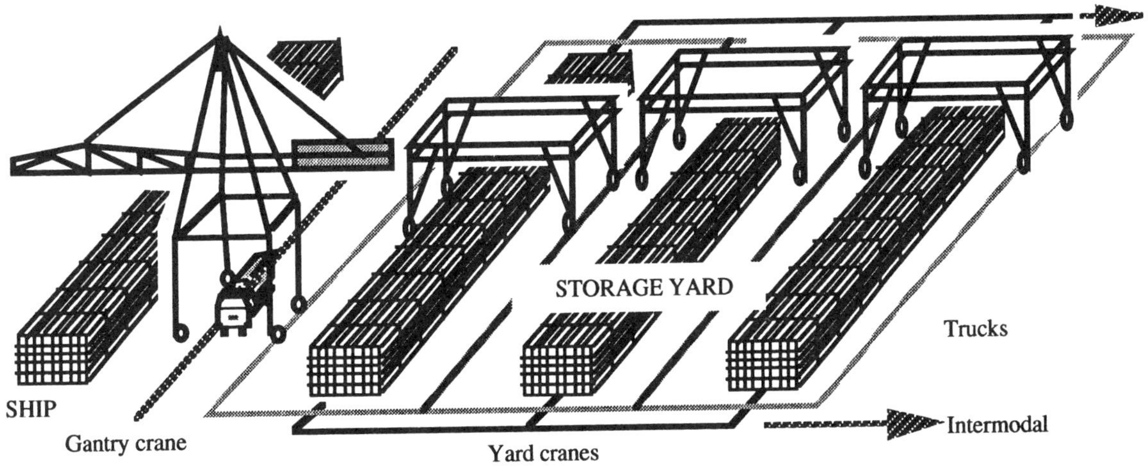

Figure 1: 3-D Representation of the Networks

crane needs a yard truck, the first yard truck of the pool of trucks moves over. When the container is loaded onto the truck, the end of service event is scheduled for the gantry crane and a beginning of service is scheduled for the truck. When processing a beginning of service event:

1. the simulation system determines the destination lot for the container.

2. a suitable slot is found in the corresponding yard lot. A suitable slot meets the following requirements: it is close to the current position of the corresponding yard crane (if present in the lot) and it is empty and not reserved for another container. When a suitable slot is found, it is reserved.

3. the links connecting the origin and destination nodes are updated to represent both the current truck location and the future container location. Then, the shortest paths are calculated from the origin node (current truck position) to the destination node (slot assigned to the container) and from the destination node to the node representing the pool of trucks. The travel times from origin to destination node will be used to schedule the arrivals at the storage yard, while the travel times from the destination node to the pool of trucks will be used to schedule the end of the truck's reposition.

Yard operations: Yard operations are quite complex because yard cranes interact with several processes. First, yard cranes are the last link of the unloading process. Secondly, they are a key component of the container retrieval process. Thirdly, they are in charge of reorganizing the storage yard. The current version of the simulation system includes only the first two roles of yard cranes.

The operational rules are the following:

1. unloading has a higher priority than container retrieval. External trucks arriving to retrieve containers are served only after all yard trucks (loaded with containers unloaded from the ship) are served.

2. service is non-preemptive. If a yard truck needs to be unloaded while the yard crane is serving an external truck, the yard truck waits until the yard crane finishes serving the external truck.

3. matching is performed to guarantee that both truck (external or yard truck) and yard crane are assigned to the same container.

Yard crane allocation rules: Two yard crane allocation rules are considered, static and dynamic. Static allocation refers to the case in which the list of lots served by a yard crane does not change over the simulation. In this case, a crane having an extremely long queue will not be helped by idle yard cranes. In the dynamic allocation scheme, idle yard cranes collaborate in tackling the longest queue. Helping yard cranes are assigned to help a needy yard crane (the one with the longest queue) provided that the queue of the needy yard crane exceeds a given threshold, and the helping yard crane is idle. In this scheme, the allocation is re-assessed at a fixed time interval specified by the user. Since dynamic allocation produces more realistic results, it is used in all runs of the simulation system.

Gate operations (in): When an external truck arrives to retrieve a specific container, the external truck is assigned to a gate. If no gate is available, the truck is placed in a queue list. After processing the truck at the gate, it is necessary to determine if the container is already in its corresponding slot at the storage yard. If the container is in its slot, a beginning of service event is scheduled for the external truck and a request is sent to the corresponding yard crane to retrieve the container. Otherwise, the truck is placed in a queue list to wait for the corresponding yard crane.

First movement of retrieval (gate to yard): When the beginning of service event is processed, the computer system checks the container's availability. If available, the shortest paths from the gate to the storage yard and from the storage yard to the gate are calculated. The arrival at the yard is scheduled using the travel time from the gate to the storage yard. If the container is not available, the truck is placed in a queue list.

Match: Since the yard cranes are assigned to specific lots, trucks must be matched to the corresponding yard crane. Matching requires determining the identification number of the target container, which yard crane is assigned to the container lot and which truck is delivering or retrieving the target container. After both truck and yard crane are identified, it is necessary to determine their status. If one of the servers is busy or repositioning, the other server waits. By matching servers, the system is able to provide a realistic representation of port operations and it also provides useful cross-statistics, e.g., mean waiting times for yard cranes waiting for trucks and vice versa.

Container retrieval at the storage yard: Once the yard crane has been assigned to retrieve a particular container, the system simulates all necessary processes. If the yard crane is not located in the corresponding lot, a reposition event is scheduled. If the target container is obstructed by other containers, the system simulates the clearing process. Finally, the target container is loaded on the truck (after proper matching).

Second movement of retrieval (yard to gate): After the yard crane is released, the beginning of service for the movement to the gate is scheduled. The travel time (previously determined using shortest path) is used to schedule the arrival at the gate.

Gate operations (out): Upon arrival at the gate, the trucks are assigned to the different gates. If no gate is available, the truck is placed in a queue list. Otherwise the service time is estimated and departure is scheduled.

PRIOR has a hierarchy structure. The system was written in FORTRAN and is comprised of more than 16,000 instructions and more than 150 different subroutines. PRIOR is based on the next event approach. Figures 2 and 3 show some examples of subroutines at different levels of the hierarchy. The majority of the second level subroutines are in charge of processing the different events, transferring control to the corresponding third level subroutine. The subroutines called by the subroutine in charge of processing beginning of service, for instance, estimate service times for the different service stages.

The simulation system reads the variables and parameters controlling the run from an ASCII control file. The control file is comprised of five different sections. The first section provides the program with the parameters that specify the operational policy. The second section contains the global control parameters. The third and fourth sections focus on the specification of input and output files. The fifth section contains the parameters that specify the printing interval. In its current version, fifteen different input files provide the program with the detailed information the program needs. Since most of the file's contents are self-explanatory only a brief description will be provided here.

The first file contains the ship characteristics (e.g., maximum number of containers, maximum number of containers per hatch and number of containers on deck, and the geometric information about the gantry cranes). The second file contains the characteristics of the incoming ships. The third file specifies the geometric characteristics of the storage yard. The information provided by this file is used to calculate link lengths and link travel times associated with truck and yard crane networks. The fourth file contains the list of high priority lots. The fifth and sixth file provide the morphology of truck and yard crane network, respectively. The seventh file contains the structure of the simulation network (i.e., the server's identification number, the stage to which the server belongs and the type of operation). The eigth and ninth files contain the characteristics of gantry and yard cranes respectively (e.g., job assignment, initial location of cranes). The tenth file provides the initial configuration of the storage yard, i.e., number of containers stacked at each spot. The eleventh file specifies the memory locations in which the simulation statistics and the queue list will be collected

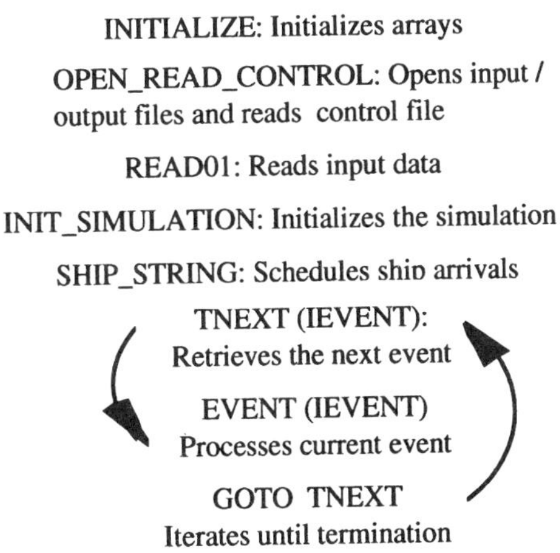

Figure 2: First Level of Subroutines (called by main program)

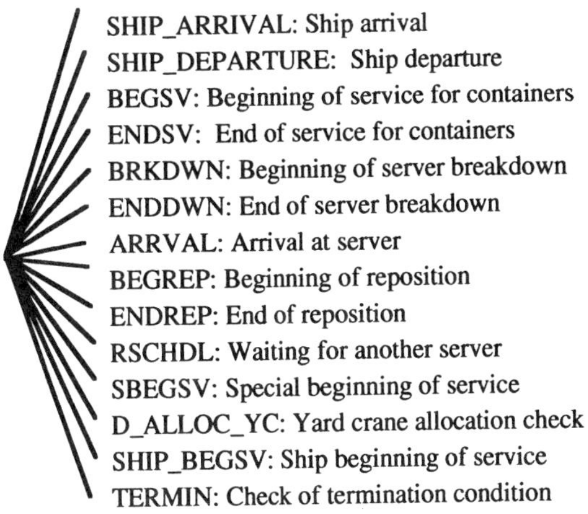

Figure 3: Examples of Second Level Subroutines

and maintained. The last three files provide the titles and headers that will be used in the printout.

The specification of the simulation output requires two sets of variables, file names and print codes. Output file names must be specified according to user's needs to avoid overwriting old outputs. Print codes specify whether or not the output file will be generated and its format. Additional parameters help to keep the output size to a manageable size, e.g., lower and upper bounds to the simulation clock.

4.2 Description of ECON

ECON is in charge of post-processing the PRIOR's output to produce economic indicators of performance. This section focuses on providing a brief description of the operational principles ECON uses.

It is important to highlight that ECON does not take into consideration the influence of labour agreements on operational costs. Instead, ECON considers exclusively the direct cost of equipment and labour. The reasons to make this decision are two-fold. First, although considering labour agreements may lead to potentially more realistic results, considering them may masquerade the relative advantage of different operational systems. For instance, it is common practice to pay the stevedors for the full shift, even if the work is finished in half an hour after the beginning. From this real-life perspective, there is no difference between an operational system that requires the gang to work for the full shift and another more efficient system in which the same job only takes half an hour. Thus, the costs estimated by ECON must be interpreted as estimates of economic costs. Second, since labour agreements vary significantly from terminal to terminal, considering labour agreements on cost calculations would impede the generalization of the conclusions of this research.

The following paragraphs provide brief descriptions of the elements considered in the cost calculations. Costs are calculated for the different types of equipment, namely, ship, gantry cranes, yard cranes, gates (in and out), and external trucks (in and out). Additionally, costs are broken down by server status, namely, idle, busy, repositioning, waiting for another server.

Ship costs: The ship costs are comprised of the fixed cost of the ship, plus the direct cost associated with the service. The latter component was calculated by assuming that two longshoremen and two lashers are required for each gantry crane, plus one lash leader. The service and the waiting time for the ship, in conjunction with the unit ship costs and unit service costs, are used to calculate the total unit cost per container.

Gantry cranes, yard trucks, yard cranes, external trucks and gates: Using the output of the simulation system, the time the servers spend in each status is determined. The corresponding total costs are calculated by using the unit costs of operating the server (equipment + labour). Then, the unit costs for each status are calculated by dividing total costs by the output, measured in containers.

The computational structure of ECON is relatively simple. The main program initializes the arrays and reads the file containing the information about unit costs. The unit costs are stored and the input file produced by PRIOR is read, beginning with high priority containers. The data is transferred to the subroutine that calculates the operating costs. The process is repeated for low priority containers. Then the output file containing operating costs for both priorities is printed. The process is repeated for all observations. ECON uses two input files. The first one is the output of the simulation provided by PRIOR and the second is the control file containing economic information about equipment and labour costs. The unit costs for labour and equipment were taken from "Assessment of Cargo Handling Technology" (PRC Inc., 1993). The output of ECON consist of two files containing detailed costs and a summary of results, respectively.

5 CALIBRATION OF THE SIMULATION SYSTEM

The calibration of the simulation system required the use of two different approaches (i.e., combined models and empirical distributions). "Combined models" refer to models that have a systematic and a random component. The systematic component expresses service time as a function of task's characteristics, while the random component represents random noise, i.e., the non-explained component. This approach was used to model gantry crane operations, yard crane operations and yard crane movements. On the other hand, traditional empirical distributions were used in the cases in which the characteristics of the service process were not suitable for analytical modelling, e.g., yard gate operation.

The parameters of the systematic component of the service time were estimated using multiple regression. After choosing the final models, the residuals were analyzed to determine which statistical distribution can be used to describe them. In the simulation system, both components (i.e., systematic and random) are used to estimate the service time. The task's characteristics (e.g., distance travelled, type of container) are inputs to the systematic component. The statistical distributions representing the random component of the service times are used to generate random numbers that are added to the systematic component to obtain the service time. The service time models described in this section focus on the following processes: (a) yard crane service times; (b) yard crane movements; and (c) gantry crane service time. The data set was obtained from video tapes taken at the container terminals in the Port of Houston.

5.1 Yard crane operations

This model provides estimates of the service times of yard cranes. The final model is:

$$Time = \underset{(5.652)}{4.42 Dhyp} + \underset{(3.777)}{27.38 Picking} - \underset{(-4.278)}{28.87 Empty} \quad (1)$$

Adjusted $R^2 =$ 0.322

(t-statistics in parentheses),
Where:
Time: time to move the spreader (in seconds);
D_{hyp}: hypotenuse of the triangle formed by the vertical and horizontal distances travelled by the spreader (in ms);
Empty: equal to 1 if spreader is empty, otherwise 0
Picking: equal to 1 if picking up container, otherwise 0

5.2 Yard crane movements

This model estimates the time required by the yard crane to move from one container lot to another. The variables used were the following:
Time: travel time (in seconds);
Distance: total distance travelled (in ms);
Stop: equal to 1 if the crane stops, otherwise 0;
Turning: equal to 1 if turning, otherwise 0
The final model is:

$$Time = \underset{(9.929)}{0.61 Distance} + \underset{(4.490)}{19.97 Stop} + \underset{(15.779)}{133.22 Turning} \quad (2)$$

Adjusted $R^2 =$ 0.930

5.3 Gantry crane operations

The gantry crane operation model estimates the gantry crane's service time. The variables used are:

Time: time required to move the spreader (in seconds);
Dx: horizontal distance of the movement (in ms)
Dy: vertical distance of the movement (in ms);
Empty: equal to 1 if spreader is empty, otherwise 0;
On deck: equal to 1 if container is on deck, otherwise 0
The final models is:

$$\frac{Time}{Dx} = 0.23 + 1.72\frac{Dy}{Dx} - 10.76\frac{Empty}{Dx} + 7.41\frac{OnDeck}{Dx} \quad (3)$$

(25.734) (7.165) (3.488)

Adjusted $R^2 = 0.856$

For a comparison between the observed average service time per layer and the average service time per layer calculated by the model the reader is referred to Holguín-Veras and Walton (1995d).

6 APPLICATION

PRIOR was used to assess the performance of a set of priority systems (PS) that differ in the degree in which service differentiation was implemented. Service differentiation can be implemented in different ways: (a) locating high priority containers (HPC) on special hatches, (b) at the storage yard i.e., storing HPC on chassis, anc (c) at the gates. Table 1 shows the different systems.

Table 1: Description of Operational Policies

	Location of HPCs	Yard crane operations	Yard gate operations
Base case	Random	No priority service	No priority service
PS-I	Hot hatches	No priority service	No priority service
PS-II	Random	HPC are wheeled	No priority service
PS-III	Random	No priority service	Priority service
PS- IV (I + II)	Hot hatches	HPC are wheeled	No priority service
PS-V (I+II+III)	Hot hatches	HPC are wheeled	Priority service

PRIOR and ECON were used to assess the performance of each of these systems under various combinations of demand and percentage of high priority containers. The output of the system was translated to a multi-criteria decision matrix including: total service time at unloading and at retrieval, probability of non-compliance and operating costs (for both priorities).

Using this decision matrix in conjunction with a multicriteria decision model, the range of applicability of the aforementioned systems was determined. The reader is referred to Holguín-Veras and Walton (1996b) for more details. Table 2 shows the range of values of the multi-criteria model for which each of the systems is the preferred option. In Table 2 only the decision criteria associated to high priority containers are shown.

Table 2: Range of Applicability

Operational system	Time at unloading	Time at retrieval	Non-compliance	Total cost
Base case	0	0	0/+	0
PS-III	0	0	0	0/+
PS-I	0	0	+	0
PS-II	0	0/+++	0/++	0/+
PS-IV	0/++++	0/++	0/++++	0/+
PS-V	0/++++	0/++++	0/++++	0/++++

Where:
0: No importance
+: Weak importance
++: Essential importance
+++: Very strong importance
+++++: Absolute importance

As can be seen, for the level of demand examined in this research, the base case would be the preferred option if the level ot service associated with priority systems is of no importance to the decision maker. Similarly, PS-III would be the preferred option if operating costs for high priority containers are of weak importance. Systems that articulate service differentiation at different stages (PS-IV and PS-V) are the preferred options only if all the performance indicators associated with priority systems are of considerable importance to the decision maker.

7 CONCLUSIONS AND FURTHER RESEARCH

The simulation system proved to be an invaluable decision aid that provided the information needed to assess the range of applicability of priority systems. Although in its current version PRIOR is a research tool, its potential as a decision aid could be enhanced if other feaures are added. Among them, the developers would like to incorporate animation capabilities that would facilitate the interpretation of the output.

ACKNOWLEDGMENT

This research was conducted at The University of Texas at Austin with support provided by a grant from the U.S. Department of Transportation, University Transportation Centers Program to the Southwest Region University Center.

The authors want to express their gratitude to Professor Douglas Morrice, from The University of Texas at Austin, for his invaluable support and encouragement.

REFERENCES

Agerschou, H. and J. Korsgaard. 1969. Systems analysis for port planning. *Dock Harbor Authority*, 49: 411-415.

Daskin M. and C.M. Walton. 1983. An approximate analytic model for supertanker lightering operations. *Transportation Research B* 17:201-219.

Frankel, E. 1987. *Port Planning and Development.* John Wiley and Sons.

Gould, P.R. 1963. Dar-es-Salaam: a geographical and queueing analysis of congestion. *East African Economic Review.*

Holguín-Veras, J. and C.M. Walton. 1995a. *A Categorized and Annotated Bibliography on the Performance Analysis of Port Operations.* Research Report SWUTC/95/721912-1, Center for Transportation Research, The University of Texas at Austin.

Holguín-Veras, J. and C. M. Walton. 1995b.*The Calibration of PRIOR, A Computer System for the Simulation of Port Operations Considering Priorities.* Research Report SWUTC/96/721928-1, Center for Transportation Research, The University of Texas at Austin.

Holguín-Veras, J. and C. M. Walton. 1995c. *PRIOR, A Computer System for the Simulation of Port Operations Considering Priorities.* Research Report SWUTC/96/ 721928-2, Center for Transportation Research, The University of Texas at Austin.

Holguín-Veras, J. and C. M. Walton. 1995d. *The Role of Information Technology on the Implementation of Priority Systems and The State of the Practice of Information System and Information Technology on Marine Container Ports.* Research Report SWUTC/96/721928-3, Center for Transportation Research, The University of Texas at Austin.

Holguín-Veras, J. and C. M. Walton. 1996a. The State of the Practice of Information System and Information Technology on Marine Container Ports. Published by the *Transportation Research Board*, paper no. 960261; Washington, D.C. Forthcoming in Transportation Research Record.

Holguín-Veras, J. and C. M. Walton. 1996b. *Range of Applicability of Priority Systems.* Research Report (forthcoming) Center for Transportation Research, The University of Texas at Austin.

Muller, G. 1995. *Intermodal Freight Transportation* 3rd. edition, ENO Foundation for Transportation.

PRC, Inc. 1993. *Assessment of Cargo Handling Technology. Vol. I and II.* Report no. MA-RD-840-93004; written for the Maritime Administration.

AUTHOR BIOGRAPHIES

JOSE HOLGUIN-VERAS has degrees from the Universidad Autónoma de Santo Domingo (Civil Engineering, Magna Cum Laude, 1981), the Universidad Central de Venezuela (Magister Scientiarum, 1984) and The University of Texas at Austin (Ph.D., 1996). He has worked as international consultant for private companies and international organizations. He is Adjunct Professor at the Department of Civil Engineering, Universidad Autónoma de Santo Domingo, and has taught courses in Transportation Planning and Port Engineering.

C. MICHAEL WALTON, P.E., received his B.S. in Civil Engineering from Virginia Military Institute in 1963. He received his M.S. (1969) and Ph.D. (1971) from North Carolina State University. He joined the faculty at The University of Texas at Austin in 1971 and currently researches and teaches as the Ernest H. Cockrell Centennial Chair in Engineering. He is Chairman of the Department of Civil Engineering and holds a joint academic appointment in the Lyndon B. Johnson School of Public Affairs. He is actively involved in developing and defining transportation policy through his research, publications and national service to government and industry.

Proceedings of the 1996 Winter Simulation Conference
ed. J. M. Charnes, D. J. Morrice, D. T. Brunner, and J. J. Swain

DERIVATION OF THE INITIAL STOCK OF BAGS – A MONTE-CARLO APPROACH

Zubin Dowlaty

United Parcel Service
55 Glenlakes Parkway NE
Atlanta, GA 30328

Chong Loo

United Parcel Service
55 Glenlakes Parkway NE
Atlanta, GA 30328

ABSTRACT

The objective of this paper is to calculate the number of bags needed to successfully supply a large package delivery operation. These bags are used to consolidate many small packages into an easier to transport large container -- thus decreasing the handling cost associated with them. Presently, disposable plastic bags are being used. The hypothesis is that permanent bags may make economical sense if the net present value of this modification is positive. The major determinant in the financial success of this proposition is the number of bags needed to feed the operation. Through monte-carlo simulation techniques, we were able to accurately model the inherent dynamic uncertainty in this calculation. Thereby enabling the financial calculation to be performed.

Keywords: Monte-Carlo Simulation, Transportation Systems, Selection Procedures

1 INTRODUCTION

Plastic disposable bags are presently used at United Parcel Service (UPS) to consolidate small packages into an easier to transport large container. Plastic bags have a very limited life and contribute to a large disposal expense. A durable bag having more permanent characteristics than the disposable bag may make economical sense. To calculate the net present value of this capital budgeting project the initial outlay must be estimated. The overwhelming determinant of the initial outlay is the amount of these durable bags that are needed to feed the operation. If the operation needs an excessive amount of durable bags, the investment will not be profitable. Given the magnitudes of dollars involved and the importance of the estimated initial stock of bags on the capital budgeting decision, the calculation of bags needed must approach reality as accurately as possible. The complexity of the UPS delivery system, that is capable of delivering a package to any address in the US, is enormous; building a mathematical model to estimate the stocks and flows of every hub would be an extremely time consuming task. Compounding the problem is the everyday potential of random shocks to the system, weather, breakdowns, etc.. Another scenario would be to examine the UPS system at the Macro level, abstracting away from examining every hub's interrelationship and viewing the system as one large stock. This results in loss of accuracy, however the gain is in simplicity and significantly less time in model construction. The proposed solution is in using a dynamic multiple run monte-carlo simulation of the initial stock of bags viewing the UPS system at the Macro level. This simulation approach has the benefit of having the capability of being an empirically based model, further the inherent random shocks to the system are naturally modeled. The combination of using empirical data in the simulation while controlling for the uncertainty in everyday operations makes this approach extremely appealing. This paper is organized into the following sections: Section 2 describes the variables that influence the flow of bags. Section 3 develops the simulation method. Section 4 discusses the implementation of the simulation. Section 5 concludes the paper with a discussion of the method used and its appeal.

2 STOCK-FLOW VARIABLES

To determine the number of bags needed to support the package operation, we must first derive the equations that will enable us to intuitively understand the variables that influence the calculation. This problem can be understood in a stock-flow framework, see figure one.

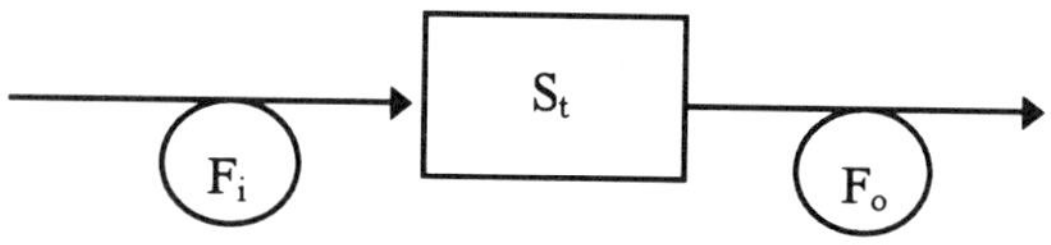

Figure 1: Bag Inflow, Initial Stock of Bags, Bag Outflow. Stock-Flow framework

$BagOutflow: Fo = \frac{Vto}{C}$; where V_{to} = Outflow of Packages, C = Bag Package Capacity

$BagInflow: Fi = \frac{Vti}{C}$; where V_{ti} = Inflow of Packages

Let $Z = T_l + T_r$; Lag Time = Transit Time Lag + Restocking Lag

Transit Time lag is defined as the time the bags are in transit. Restocking lag is the time to actually restock the bags for use after arrival at the hub.

Lets assume that the inflow of bags equals the outflow of bags with a lag of Z. This assumption is valid for the operation as a whole and yields:

$Fi_{(t)} = Fo_{(t-z)}$.

Therefore:

$$S_T = S_{T-1} + (Fi)_T - (Fo)_T$$

Substitute for F_i,

$$S_T = S_{T-1} + (Fo)_{T-Z} - (Fo)_T$$

Discrete aggregation of stock over time yields:

$$\sum S_T - S_{T-1} = \sum (Fo)_{T-Z} - \sum (Fo)_T$$

Evaluate at T = 1 to Z

$$= \sum_{T=1}^{Z} (Fo)_{(T-Z)} - \sum_{T=1}^{Z} (Fo)_T$$

$$= 0 - \sum_{T=1}^{Z} (Fo)_T \quad (1)$$

Evaluate at T = Z + 1 to Infinity:

$$= \sum_{T=Z+1}^{\infty} (Fo)_{T-Z} - \sum_{T=Z+1}^{\infty} (Fo)_T \quad (2)$$

Result (1) states that the total stock of bags needed on or before time period Z is equal to the bag outflow summed up over time until time period Z. In other words the number of bags needed before the inflow of bags is received is simply the sum of outflows.

Result (2) presents three different interpretations given the basic trend of bag outflow:

1. If the slope of the trend of bag outflow over time is zero or bag outflow is constant then result (2) reduces to zero.
2. If the slope of the trend of bag outflow over time is negative then result (2) implies a surplus of bags will accumulate, therefore reducing the number of bags needed from result (1).
3. If the slope of the trend of bag outflow over time is positive then result (2) implies a deficit of bags will accumulate, therefore more bags are needed than is stated in result (1).

3 SIMULATION METHOD

United Parcel Service(UPS) volume can be assumed to grow in the future therefore it follows from result (2) that a deficit of bags would occur. More bags are needed than simply the sum of outflows until time Z. Because the objective is to calculate the total number of bags that will be required over a span of time a projection of future exogenous values are required. This introduces additional uncertainty into the analysis. The identified variables that drive the initial bag calculation and impart uncertainty into the analysis are:

1. Package Volume : Number of Packages
2. Physical Package Size
3. Bag Shrinkage: Damages, loss, etc..
4. Transit Time Lag
5. Restocking Lag

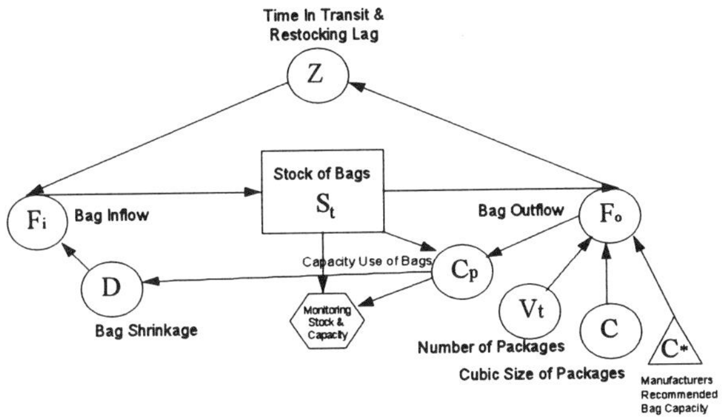

Diagram 1: Simulation Structure

The simulation design chosen is a monte-carlo simulation. Monte-carlo is used because we can place precise probability distributions around the uncertain variables, therefore accounting for all possible scenarios. The structure of the simulation is displayed graphically in Diagram 1.

The variables in Diagram 1 not described in Section 2 are as follows:

C_p = : *Capacity Use of Bags*
- $S_t >= F_o$, $C_p = 100\%$
- $S_t < F_o$, $C_p = F_o / S_t$

D = : *Bag Shrinkage, reduction in bag inflow*
- $C_p = 100\%$, D = Random Variable
- $C_p > 100\%$, D = Random Variable with an increase in the mean.

4 SIMULATION IMPLEMENTATION

Using the model described in the Simulation Methodology section, the simulation of the initial stock of bags can be performed. The simulation model is designed to analyze only one initial stock at a time, set the initial stock and monitor performance throughout the simulation. The time period of the simulation is 254 working days (approximately 1 year). Each working day is simulated. The initial stock of bags is fixed and the simulation is run for thousands of iterations during that year; the model stops when the standard error of the mean deficient days reaches a preset limit (The software used to build this model is Excel 5.0 with Visual Basic Implementation, Crystal Ball 4.0 which is a monte-carlo simulation package, and Eviews [econometric views] for the forecasting of the exogenous variables).

The exogenous variables have been calibrated for the most empirical accuracy as possible. The following are the exogenous variables and explanation of how the values for the simulation were determined:

1. Package Volume = Projected UPS package volume using econometric models. A Normal Probability Distribution with mean zero and standard deviation of the residuals derived from the econometric model are used and added to the forecast per simulation day.

2. Cube Size of Packages = Projected UPS, system average package cube size derived by econometric models. A Normal Probability Distribution with mean zero and standard deviation of the residuals derived from the econometric model are used and added to the forecast per simulation day.

3. Bag Shrinkage = Manufacturers recommendation of bag failure, with an added Probability distribution accounting for unforeseen damages. This Probability distribution is designed to be dynamic through the simulation – the mean increases when the Capacity Utilization of the bags exceeds 100%.

4. Time in Transit and Restocking Lag = Projected UPS average time in transit derived by econometric models. A Discrete Normal Probability Distribution with mean zero and standard deviation of the residuals derived from the econometric model are used and added to the forecast per simulation day. Estimated mean restocking lag with a Discrete Normal Probability Distribution with standard deviation equal to the standard error of the estimated restocking lag. A seasonality adjustment is integrated into the restocking lag estimate.

The variables throughout the simulation being monitored are the mean number of days the bags were unavailable during the year and the mean number of deficient bags on those days. A *successful* scenario would be when the mean number of days and bags during those deficient days are within acceptable bounds. For example: A simulation of an initial stock of 10,000 bags yielded a mean of 2 days out of 254 where there was not enough bags available, with a mean of 500 bags deficient. The acceptable bound could be set allowing for zero days of deficiency, or X days deficient. This model provides the ability to quantify the jump from less then 100% success to 100% success, and assess the magnitude of the number of bags needed to realize this increased coverage.

The first step in identifying the minimum number of bags required was to find the minimum value of the initial stock that meets the requirements of acceptability. This was accomplished by running multiple iterations of the monte-carlo model, capturing mean deficient days and deficient bags given the initial stock, then incrementing the initial stock and rerunning the model. The advantage of this is to thoroughly test each initial stock for performance. Given the nature of monte-carlo simulation all realistic scenarios are probed. The start value of the initial stock was derived from result (1). The data is then plotted (see Chart 1, for a typical plot using fictitious data). This plot has the initial stock on the X axis and the number of deficient days on the Y. One can also plot initial stock on the X axis and mean number of bags deficient on the Y axis. The final minimum number of bags would be determined by selecting the minimum initial stock that achieves an acceptable range.

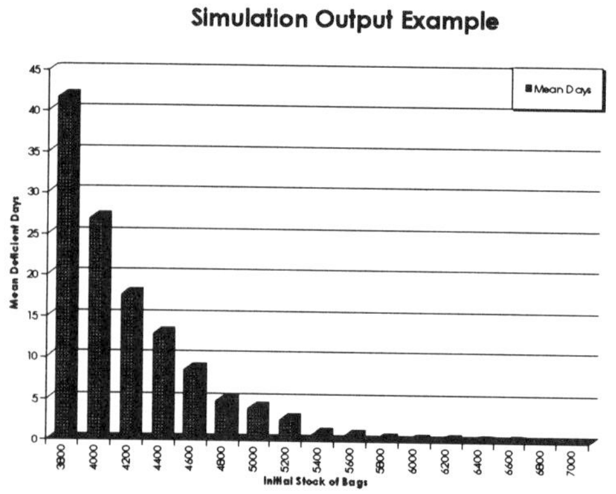

Chart 1: Plot of Mean Deficient Days for multiple initial stocks.

5 CONCLUSION

A multi-run monte-carlo model was successfully developed to estimate the initial stock of bags to support an entire package delivery operation. Through simulation a realistic representation of the operation was derived. The model incorporates empirical exogenous variables and adds the uncertainty that accompanies future projections. Using dynamic probability parameters (e.g. in the case of mean bag shrinkage changing when bag capacity exceeds 100%), real life dynamics are modeled. Furthermore, it provides users with modeling flexibility, ranging from assumptions regarding uncertainty, the time span of the simulation, and quantifying the level of confidence in determining the initial stock of bags.

ACKNOWLEDGEMENTS

The authors would like to thank United Parcel Service for supporting this project.

REFERENCES

Crystal Ball Users Manual. 1996. Decisioneering Inc.

Hilderbrand and Ott. 1987. *Statistical Thinking For Managers*. Duxbery Press.

Law, Averell. 1991. *Simulation Modeling and Analysis*. McGraw Hill.

Pindyck & Rubinfeld. 1981. *Econometric Models and Economic Forecasts*. McGraw Hill.

Richmond, Barry. 1987. *An Academic Users Guide to Stella*. High Performance Systems.

Roberts, Nancy. 1994. *Introduction to Computer Simulation*. Productivity Press.

AUTHOR BIOGRAPHIES

ZUBIN DOWLATY is a quantitative methods supervisor at the UPS corporate office. He has a B.S. degree in Finance from the University of Florida and a M.A. in Economics with an emphasis on Econometrics from the University of South Carolina. Research interests are in simulation of dynamic systems, cybernetic principles in simulation, and multivariate statistics. He is a member of the ASA.

CHONG LOO is a financial controller at the UPS corporate office. He has a B.S. in Electrical Engineering and a M.S. in Industrial Engineering from Louisiana Tech. His research interests are in cost models and applications of activity based costing in transportation systems. He has previously published papers in the field of Ergonomics. He is a member of I.I.E.

Proceedings of the 1996 Winter Simulation Conference
ed. J. M. Charnes, D. J. Morrice, D. T. Brunner, and J. J. Swain

SIMULATION OF THE ILLINOIS WATERWAY LOCKS SYSTEM

D. Brent Bandy

Department of Information Systems and Operations Management
College of Business Administration
University of Wisconsin Oshkosh
Oshkosh, WI 54901, U.S.A.

ABSTRACT

A computer simulation model of the locks system on the Illinois Waterway is described. The model was developed using ProModel for Windows and includes boats and barge tows traveling through the locks system and operation of the locks. Entities with attributes are used to simulate the boats and barge tows that enter the system, travel either upstream or downstream through the locks, and then leave the system. Locations are used for the locks and for entry and exiting of the boats and barge tows. The operation of the locks is simulated through the use of variables, resources, and operation logic in the processing module. The handling of both boats and barge tows in the locks greatly complicates the simulation model, since their differences in size is significant. For example, a lockage can easily handle ten or more boats, while a barge tow with fifteen barges can require a double lockage (typically nine barges for the first lockage and six barges and the tow boat for the second lockage). At times both barge tows and boats are included in a single lockage.

1 INTRODUCTION

The Illinois Waterway is part of the waterway system that links the Great Lakes with the Gulf of Mexico. It is the largest and most important water route in Illinois, extending from Lake Michigan at the Chicago and Calumet Harbors to the mouth of the Illinois River where it empties into the Mississippi River at Grafton, Illinois, about 50 miles north of St. Louis. The Illinois Waterway is composed of the Illinois River; the Chicago Sanitary and Ship Canal; the Calumet Sag Channel; and parts of the Calumet, Little Calumet, Chicago, and Des Plaines Rivers. There are nine locks on the Illinois Waterway. The total drop in water elevation from Lake Michigan to Grafton is about 160 feet.

ProModel is a discrete event simulation software package developed primarily for manufacturing and service systems (ProModel Corporation 1992). The Windows version is very easy to use, and its capabilities allow for building simulation models quickly and efficiently. One of the primary features of ProModel for Windows is its animation capabilities. It uses the graphic capabilities of Windows to build both the model and its animation representation interactively. Thus, animation is not an add-on feature. It is an integral part of the model building process. ProModel has been used extensively in various application areas; however, using it to simulate a river locks system is a challenging, unique application.

2 THE ILLINOIS WATERWAY LOCKS SYSTEM

Eight of the locks on the Illinois Waterway are U.S. Government locks operated by the Army Corps of Engineers (U.S. Army Corps of Engineers 1974). They are at the following locations (relative to the mouth of the Illinois River at Grafton): La Grange, Mile 80.2; Peoria, Mile 157.7; Starved Rock, Mile 231.0; Marseilles, Mile 244.6; Dresden Island, Mile 271.5; Brandon Road, Mile 286.0; Lockport, Mile 291.1; and Thomas J. O'Brien, Mile 326.5. The other lock is in Chicago Harbor and is operated by the Metropolitan Sanitary District of Greater Chicago. The Thomas J. O'Brien Lock is the largest of the locks on the Illinois Waterway, with usable dimensions of 110 feet in width by 1000 feet in length. The other seven U.S. Government locks all have usable dimensions of 110 feet in width by 600 feet in length. The lock operated by the Metropolitan Sanitary District of Greater Chicago has usable dimensions of 80 feet in width by 600 feet in length.

The northern part of the Illinois Waterway starts at Lake Michigan, which has an elevation of about 580 feet, and has two branches. The southern branch starts at the mouth of the Calumet River at Calumet Harbor in Lake Michigan on the south side of Chicago, east of 91st Street. From there it consists of the Calumet River for eight miles to where it becomes the Little Calumet River near 141st Street and Torrence Avenue; the Little Calumet River for six miles to where it joins the Calumet Sag Channel at Blue Island; and the Calumet Sag Channel for 16 miles to where it joins the Chicago Sanitary and Ship Canal near Lemont.

The northern branch starts at the mouth of the Chicago River at Lake Michigan in downtown Chicago, essentially at the lock in Chicago Harbor. From there it consists of the Chicago River for two miles to where it branches near Lake Street and Franklin Street in downtown Chicago; the South Branch of the Chicago River for four miles to where it is connected to the Chicago Sanitary and Ship Canal near 31st Street and Ashland Avenue; and the Chicago Sanitary and Ship Canal for 18 miles to where the Calumet Sag Channel joins it near Lemont.

The Chicago Sanitary and Ship Canal project was completed in 1900 and the canal carries Chicago's treated sewage southwest, away from Lake Michigan. Before its completion, Chicago dumped its sewage in Lake Michigan. The lock in Chicago Harbor was built as part of the project, which included reversing the flow of the Chicago River from east to west. It was the first river in the world to flow away from its mouth. The Chicago Sanitary and Ship Canal is 30 miles long, 202 feet wide, and 24 feet deep. Its rate of flow is controlled by sluice gates at Chicago Harbor and at the Thomas J. O'Brien Lock and Controlling Works.

There are two locks on the northern part of the Illinois Waterway, the lock in Chicago Harbor at the beginning of the northern branch and the Thomas J. O'Brien Lock and Controlling Works on the southern branch. The O'Brien Lock is at about the midpoint of the 14-mile stretch of the Calumet and Little Calumet Rivers that connects Calumet Harbor to the Calumet Sag Channel. It is on the Calumet River seven miles from Calumet Harbor. The entrance to the lock by car is on 130th Street, just west of Torrence Avenue on the south side of Chicago. The change in the water level at the Thomas J. O'Brien Lock varies somewhat, as it does at all of the locks, but is approximately four feet.

The location near Lemont where the Calumet Sag Channel joins the Chicago Sanitary and Ship Canal is 303.5 miles by water from the mouth of the Illinois River at Grafton. Of course, the actual distance is much less, since there are many bends in the Illinois and Des Plaines Rivers. From its junction with the Calumet Sag Channel, the Chicago Sanitary and Ship Canal extends 13.5 miles further, joining the Des Plaines River between Joliet and Lockport. There is one lock on this part of the Chicago Sanitary and Ship Canal, the Lockport Lock and Dam, which is one mile east of the junction of the Chicago Sanitary and Ship Canal with the Des Plaines River. The change in elevation at the Lockport Lock and Dam is about 40 feet.

The Des Plaines River is next to the Chicago Sanitary and Ship Canal for 24 miles, from their junction between Joliet and Lockport northeast to Summit. From its junction with the Chicago Sanitary and Ship Canal, the Des Plaines River flows southwest for 17 more miles until it joins with the Kankakee River near Dresden Heights in Will County to form the Illinois River. The Brandon Road Lock and Dam is on this part of the Des Plaines River, four miles southwest of its junction with the Chicago Sanitary and Ship Canal. The entrance to this lock by car is at the intersection of Brandon Road and U.S. Highway 6 near Rockdale, which is just west of Joliet. The change in elevation at the Brandon Road Lock and Dam is about 35 feet.

The Illinois River is 273 miles long and makes up the longest stretch of the Illinois Waterway. There are five locks on the Illinois River. Going downstream, the first one is Dresden Island Lock and Dam, which is near Kankakee Bluffs 1.5 miles from the confluence of the Des Plaines and Kankakee Rivers. The change in elevation at Dresden Island is about 25 feet.

The Marseilles Lock is 27 miles further downstream, while the Marseilles Dam is only 25 miles downstream near downtown Marseilles. The Marseilles Canal is two miles long next to the Illinois River, extending from the dam to the lock. The change in elevation at Marseilles is about 20 feet.

Proceeding downstream the Starved Rock Lock and Dam is 13.5 miles from the Marseilles Lock, eight miles west of Ottawa and five miles east of LaSalle. It has a change in elevation of about 20 feet, and is the easiest of the Illinois River locks to view, since Starved Rock State Park, with its high bluffs, is adjacent to the dam on the south shore. The lock is next to the north shore and has a visitor center.

The Peoria Lock and Dam is 73 miles downstream of the Starved Rock Lock and Dam. It has a change in elevation of about ten feet, but varies greatly. It can be seen from Interstate 494, just south of Peoria, where it crosses the Illinois River.

The La Grange Lock and Dam is 68 miles downstream of the Peoria Lock and Dam and 80 miles upstream of the mouth of the Illinois River at Grafton, about half way between Beardstown and Meredosia. By car it is on some back country roads, pretty much "in the middle of nowhere." The change in elevation is about five feet, but varies greatly.

The elevation of the Mississippi River at Grafton is approximately 420 feet. However, in floods the elevation increases significantly. On the other hand, during droughts the elevation is lower. In recent floods both the Peoria and the La Grange Lock and Dams have been completely under water, with the water level reaching to the second floor of the buildings next to the locks.

3 USE OF THE LOCKS

River locks provide a means for bypassing rapids, dams, and other locations on the river where there are significant changes in elevation. Sometimes canals are associated with locks system, connecting the locks to the river either upstream or downstream of the elevation change, as is the case with the Marseilles Canal. Canals can also provide connections between rivers and other waterways, including both the Calumet Sag Channel and the Chicago Sanitary and Ship Canal. Of course, there can be locks and dams directly on canals, such as the Lockport Lock and Dam on the Chicago Sanitary and Ship Canal.

Whether vessels are traveling upstream or downstream, the procedure for a lockage is essentially the same. Before vessels enter the lock, the elevation of the water inside the lock has to be the same as that where the vessels are located. The water elevation inside the locks is controlled by underwater valves at both the upstream and downstream gates. When the lock gates are opened, vessels can enter the lock. Of course, the lock gates and underwater valves must be closed at the other end of the lock to maintain the proper water level.

The number of vessels that will fit in the lock depends on the usable dimensions of the lock and the size and shape of the vessels. Safety considerations also play a role. As many as twenty or more small boats can easily fit into a large lock. On the other hand, barge tows are often too big to fit into a lock and must be moved through in two parts, which is referred to as a double lockage.

Once the vessels are in the lock, the underwater valves and the lock gates are closed. Then the underwater valves at the other end of the lock are opened, which allows the water elevation in the lock to eventually equalize with that on the other side of the exit gates. The length of time required for this is determined by the size of the valve openings and the total change of elevation required for the water inside the lock. Often the valves are not used wide-open, due to safety considerations with respect to water turbulence, both inside the lock and on the other side of the lock gates when other vessels are waiting there to use the lock. Once the water elevation inside the lock is sufficiently close to that on the other side of the exit gates, the gates are opened. The vessels then exit the lock and proceed on their trip.

Vessels traveling through locks systems are obviously self-propelled. However, often barge tows must be split into two parts to get them through a lock. When this happens, the part with the tow boat is still self-propelled; however, the part that has only barges must be moved by some other external force. This force is provided by electric wenches that are located on shore both upstream and downstream of the lock.

4 BOAT AND BARGE TRAFFIC

Traffic on the Illinois Waterway consists of both boats and barges, which must be moved by tow boats. A barge tow consists of a tow boat and from one to sixteen or more barges, depending on the size of the barges. The barges carry a variety of products, including grain, chemicals, coal, sand and gravel. Barge tows move up and down the waterway at approximately five miles per hour. For lockages there are essentially three types associated with barge tows: 1) a single lockage without having to reconfigure the barge tow; 2) a single lockage with some reconfiguration of the barge tow being required; and 3) a double lockage.

For the first type of lockage, the barge tows have overall dimensions that are less than the usable dimensions of the lock. It is possible that these could consist of nine or more barges and the tow boat, especially at the Thomas J. O'Brien Lock, which is the largest lock in the system. Typically they consist of one or two barges (or three small barges) in single file, or of six barges, three abreast, followed by the tow boat.

For lockages of the second type, the barge tows have an overall length that exceeds the usable length of the lock. They could have as few as three barges, but often have four barges, in single file, followed by the tow boat. Of course, it would be possible to configure the barge tow to fit into the lock (and thus into the first type of lockage); however, they are configured in single file for navigational purposes. These barge tows move through the lock in the following manner: 1) The barge tow is moved into the lock and the front barges are tied to the side of the lock. 2) The barge tow is untied in such a way that the two parts will fit into the lock lengthwise. 3) The back part of the barge tow backs away from the front part, moves to the side and enters the lock next to the front part. 4) The lockage is completed. 5) The front part of the barge tow is pulled out of the lock using one of the electric wenches on shore. 6) The back part of the barge tow moves into position behind the front part. 7) The barge tow is tied back together again. 8) The barge tow continues the trip.

Double lockages are required for barge tows that are so big that they cannot fit into the lock for a single lockage. Usually they consist of nine or more barges, three abreast, followed by the tow boat. These barge tows require two lockages and they move through the lock in the following manner: 1) The barge tow is moved into the lock and the front barges are tied to the side of the lock. 2) The barge tow is untied in such a way that each of the two parts will fit into the lock. 3) The back part of the barge tow backs out of the lock. 4) The lockage for the front part is completed. 5) The front part of the barge tow is pulled out of the lock using an electric wench. 6) The water level in the lock is returned to the initial level (essentially an empty lockage, which does not count). 7) The back part of the barge tow moves into position into the lock. 8) The lockage is completed for the back part of the barge tow. 9) The back part of the barge tow moves into position behind the front part. 10) The barge tow is tied back together again. 11) The barge tow continues the trip.

Commercial boats are typically much smaller than barge tows. In addition, the Illinois Waterway is used extensively for recreation, which results in numerous small boats using the system. The recreational boating season is primarily during the summer months, especially on weekends and holidays. Several small boats can be moved through a lock on a single lockage. In addition, a few small boats will often fit into the lock with a barge tow, or part of a barge tow. This is often done, at the discretion of the lockmaster, subject to safety considerations. In cases where a double lockage is required for the barge tow, boats can be included in the second lockage.

5 DESCRIPTION OF THE MODEL

The model, which was developed using ProModel for Windows, simulates the operation of the Illinois Waterway in terms of the boats and barge tows traveling through the locks system and operation of the locks. Entities and locations were the starting points for developing the model. Entities in the model represent the boats and barge tows that enter the system, travel either upstream or downstream through the locks, and then leave the system. Locations within the model include the locks, locations where barge tows and boats enter and/or leave the system, and queues for barge tows and boats where they wait to use the locks. Operation of the locks is simulated through the use of variables, resources, and operation logic in the processing module. The northern branch of the northern part of the Illinois Waterway, which extends from the mouth of the Chicago River at Lake Michigan to the junction of the Chicago Sanitary and Ship Canal and the Calumet Sag Channel near Lemont, is not included in the model. In addition, there are a great many barge tows and boats that use the Illinois Waterway without going through a lock. These barge tows and boats are not included in the simulation, since its primary focus is on the locks.

There are only two entities in the model, one for boats and one for barge tows. They have five attributes as follows: 1) direction in which the boat or barge tow is traveling (1 = downstream, 2 = upstream); 2) speed of the boat or barge tow in miles per hour; 3) size of the boat or barge tow for use in determining how many will fit into the lock for a lockage; 4) additional duration required for the lockage due to the size of the boat or barge tow; and 5) final destination for the boat or barge tow (from 1 for the La Grange Lock at the southern part to 8 for the Thomas J. O'Brien Lock in the northern part).

There are 44 locations in the model, as shown in Figure 1. Of these 44 locations, 24 are directly associated with the locks, eight locations for the eight locks and 16 locations for queues for the boats and barge tows that are waiting for the locks. Two queues, which look like ladders in Figure 1, are needed for each lock to handle the vessels going upstream and the vessels going downstream. Six locations are used to identify the rivers and canals for the animation aspects of the models: Illinois River, Des Plaines River, Little Calumet

Figure 1. Locations in the ProModel Simulation Model of the Illinois Waterway Locks System

River, Calumet River, Chicago Sanitary and Ship Canal, and Calumet Sag Channel. The other 14 locations are also used for animation to identify locations that are part of or near the Illinois Waterway: Calumet Harbor, Blue Island, Lemont, Joliet, Rockdale, Dresden Heights, Kankakee Bluffs, Marseilles, Ottawa, LaSalle, Peoria, Beardstown, Meredosia, and Grafton. Squares are shown in Figure 1 for all locations that are not queues. When the simulation model is run, the squares and "ladders" disappear, and the animation depicts the boats and barge tows moving up and down the Illinios Waterway, waiting for the locks, moving through the locks, and proceeding to their final destination. It should be noted that Figure 1 is not to scale. For example, the distance from Grafton to the Peoria Lock is 158 miles, while the distance from the Starved Rock Lock to the Lockport Lock is only 60 miles.

Of the twenty locations in the model that are used to identify locations for animation, nine are also used to handle arrivals of the entities that represent boats and barge tows. These include Calumet Harbor and Grafton, the two end points of the Illinois Waterway, as well as one location between each pair of adjacent locks. There are 18 sets of arrivals in the model, for each of the two entities arriving at each of the nine arrival locations. For each set of arrivals several aspects are included: the time of the first arrival, the total number of arrivals in the model run, the interarrival frequency, and logic for calculating values for all of the attributes for the entity. The exponential probability distribution is used interarrival times, thus assuming a Poisson process for arrivals.

Operation of the locks and routing of the barge tows and boats from location to location is one of the most involved aspects of the model. This part of the model is carried out through the use of resources, global variables, priorities, and operation logic in the processing section of ProModel. For each lock four resources are required, for boats and barge tows traveling upstream and traveling downstream. These resources are used to move the boats and barge tows from the corresponding location queue to the location for the lock. The operation logic for the boats and barge tows at the lock locations also uses these resources, in order to prevent them from moving any boats or barge tows into the lock location during a lockage that is already in progress. Seven global variables are used in this logic for each lock for the following aspects: 1) the total amount of time required for the lockage; 2) the number of boats involved in the lockage; 3) the direction the barge tow and/or boats are traveling for the lockage; 4) an indicator variable for whether or not a barge tow is involved in the lockage; 5) an indicator variable for whether or not the lock is in use or is reserved for a lockage; 6) a counter to keep track of the boats going upstream as they enter and leave the lock; and 7) a counter to keep track of the boats going downstream as they enter and leave the lock. Barge tows have priority over boats for lockages, so their resources are freed first when a lockage ends. Thus, if any barge tows are waiting in the queues, one will enter the lock before any boats are allowed to enter. If there is room for boats in the lockage along with the barge tow, they enter the lock following the barge tow. If no barge tow is involved in the lockage, all of the boats that are waiting and are going in the direction of the lockage will be included, up to the capacity of the lock, which is assumed to be 33 boats for the Thomas J. O'Brien Lock and 20 boats for the other locks. Resources are also freed in a manner that results in alternating between lockages going upstream and lockages going downstream, when boats going in both directions are present. However, barge tows still take priority over boats, even if consecutive barge tows are going in the same direction.

REFERENCES

ProModel Corporation. 1992. *ProModel PC User Manual - Production Systems Planning and Analysis*. ProModel Corporation. Orem, Utah.

U.S. Army Corps of Engineers. 1974. *Charts of the Illinois Waterway - from Mississippi River at Grafton, Illinois to Lake Michigan at Chicago and Calumet Harbors*. U.S. Army Engineer District, Chicago, Illinois.

AUTHOR BIOGRAPHY

D. BRENT BANDY is an assistant professor in the Information Systems and Operations Management Department at the University of Wisconsin Oshkosh. Prior to coming to UW Oshkosh in 1984, he worked for sixteen years for Amoco as a systems analyst and operations research consultant. He received his B.S. in Chemical Engineering from the University of Illinois, his M.S. and Ph. D. in Chemical Engineering from Northwestern University, and his M.B.A. from the University of Chicago. He is a member of SCS, INFORMS, and APICS.

Index

Author Directory

Simon M. AbouRizk
University of Alberta, Edmonton
Construction Engr. & Management
304D Civil/Electrical Building
Edmonton, Alberta T6G 2G7
CANADA
PH: 403-492-8096
FAX: 403-492-0249
E-M: abourizk@civil.ualberta.ca

Moyeen Ahmad
Kansas State University
Dept. of Industrial & Manufacturing Sys. Engr.
Manhattan, KS 66506 USA

Christos Alexopoulos
Georgia Institute of Technology
School of Industrial & Systems Engr.
Atlanta, GA 30332-0205 USA
PH: 404-894-2361
FAX: 404-894-2301
E-M: christos@isye.gatech.edu

Greg J. Allen
The Model Builders, Inc.
Systems Engr.
343 South Main, Suite 217
Ann Arbor, MI 48104 USA
PH: 313-998-5560
FAX: 313-998-5561
E-M: mbldr@izzy.net

Mahmoud H. Alrefaei
University of Wisconsin - Madison
Dept. of Industrial Engr.
1513 University Ave.
Madison, WI 53706 USA
PH: 770-319-8251
FAX: 404-894-2301
E-M: alrefaei@isye.gatech.edu

Eugenio M. Alvarez
Nine West Group, Inc.
9 West Broad St.
Stamford, CT 06902 USA
PH: 203-978-5938
E-M: gene_alvarez@ninewest.com

Sigrún Andródottir
Georgia Institute of Technology
School of ISyE
Atlanta, GA 30332-0205 USA
PH: 404-894-3933
FAX: 404-894-2301
E-M: sa@isye.gatech.edu

Michael Angel
CACI Products Company
3333 North Torrey Pines Court
La Jolla, CA 92037 USA
PH: 619-455-6300 x161
FAX: 619-458-0031
E-M: mike@cacisl.com

Peter C. Anselmo
New Mexico Institute of Mining & Technology
Dept. of Management
Socorro, NM 87801 USA
PH: 505-835-5438
FAX: 505-835-5887
E-M: anselmo@mailhost.nmt.edu

F. Bradley Armstrong
ABB Power T & D Company, Inc.
Transmission Technology Institute
1021 Main Campus Dr.
Raleigh, NC 27606 USA
PH: 919-856-3896
FAX: 919-856-2459
E-M: Brad.F.Armstrong@ustra101.mail.abb.com

James D. Arthur
The Dept. of Computer Science
Virginia Tech.
Blacksburg, VA 24061-0106 USA
PH: 540-231-7538
FAX: 540-231-6075
E-M: arthur@vt.edu

Tajudeen A. Atolagbe
Brunel University
Dept. Computer Science & Info. Systems
Uxbridge, Middlesex UB8 3PH
UNITED KINGDOM
PH: + 44-1895-274-000
FAX: +44-1895-251-686
E-M: tayudeen.atolagve@brunel.ac.uk

Farhad Azadivar
Kansas State University
Dept. Industrial & Manufacturing Systems Engr.
Manhattan, KS 66506 USA
PH: 913-532-6329
FAX: 913-532-7072
E-M: fazad@ksu.ksu.edu

Rajive L. Bagrodia
University of California, Los Angeles
Computer Science Dept.
3532 Boelter Hall
Box 951596
Los Angeles, CA 90095 USA
PH: 310-825-0956
FAX: 310-825-7578
E-M: rajive@cs.ucla.edu

Mary L. Bailey
Rincon Research Corp.
101 N. Wilmot, Suite 310
Tucson, AZ 85711 USA
PH: 520-745-0400
FAX: 520-745-1436
E-M: mlb@ rincon.com

Jay Bakst
Automation Associates, Inc.
13333 Bel-Red Road
Bellevue, WA 98005 USA
PH: 206-228-5586
FAX: 206-228-4340
E-M: jbakst@aaisim.cts.com

D. Brent Bandy
University of Wisconsin Oshkosh
Dept. of Info. Systems & Operations Management
Oshkosh, WI 54901 USA
PH: 414-722-7869
FAX: 414-424-7413
E-M: bandy@vaxa.cis.uwosh.edu

Jerry Banks
Georgia Institute of Technology
School of ISyE
Atlanta, GA 30332-0205 USA
PH: 404-894-2312
FAX: 404-894-2301
E-M: jerry.banks@isye.gatech.edu

Martin Barnes
Deneb Robotics, Inc.
3285 Lapeer Road West
P. O. Box 214687
Auburn Hills, MI 48321 USA
PH: 810-377-6900
FAX: 810-377-8125
E-M: barnes@deneb.com

David Bartlett
Defense Modeling & Simulation Office
1901 North Beauregard St.
Suite 504
Alexandria, VA 22311 USA

Joseph C. Barto III
Joint Warfighting Center
United States Army
Fort Monroe, VA 23651 USA
PH: 804-726-6645
FAX: 804-726-6433
E-M: Bartoj@jwfc.army.mil

Michael Bathe
Cranfield University
Operations Research Unit
School of Defence Management
RMCS Shrivenham
Shrivenham, Swindon SN6 8LA
UNITED KINGDOM

Kenneth W. Bauer, Jr.
Air Force Institute of Technology
Dept. of Operational Sciences
Graduate School of Engr.
2950 P St.
Wright Patterson AFB, OH 45433-7765
USA
PH: 513-255-6565 x4326
FAX: 513-476-4943
E-M: kbauer@afit.af.mil

Terrence G. Beaumariage
Oregon State University
Dept. Industrial & Manufacturing Engr.
Covell Hall 118
Corvallis, OR 97331-2407 USA
PH: 503-737-3342
E-M: beaumart@ccmail.orst.edu

Perakath C. Benjamin
Knowledge Based Systems, Inc
One KBSI Place
1500 University Dr. East
College Station, TX 77840-2335 USA
PH: 409-260-5274
FAX: 409-260-1965
E-M: pbenjamin@kbsi.com

Deborah Benson
PROMODEL Corp.
1875 South State St.
Suite 3400
Orem, UT 84097 USA
PH: 801-223-4600
FAX: 801-226-6046
E-M: deborah@promodel.com

Michael R. Berthold
University of Karlsruhe
Institute for Computer Design & Fault Tolerance
P. O. Box 6980
76128 Karlsruhe
GERMANY
PH: +49-721-608-4219
FAX: +49-721-370455
E-M: berthold@ira.uka.de

Mohamed Bettaz
University of Constantine
Computing Science Institute
Constantine 25000
ALGERIA
PH: +213-469-38-61
FAX: +213-469-38-61
E-M: bettaz@ist.cerist.dz

Bert Bettonvil
BIKA/Center
Postbox 90153
5000 LE Tilburg
NETHERLANDS
E-M: bert@kub.nl

Maya Binun
CACI Products Company
3333 North Torrey Pines Court
La Jolla, CA 92037 USA
PH: 619-455-6300 x112
FAX: 619-458-0031
E-M: maya@caciasl.com

Christopher G. Blood
Naval Health Research Center
Operations Research Division
P. O. Box 85122
San Diego, CA 92186-5122 USA
PH: 619-553-8386
FAX: 619-553-8607
E-M: BLOOD@nhrc.navy.mil

C. J.M. Booth
DRA Malvern
St. Andrews Road
Room E210
Great Malvern, Worcs. WR14 3PS
ENGLAND
PH: +44-(0)1684-896400
FAX: +44-(0)1684-894389
E-M: cjmb@signal.dra.hmg.gb

John E. Brennan
ALPHATECH, Inc.
50 Mall Road
Burlington, MA 01803-4562 USA
PH: 617-273-3388 x260
FAX: 617-273-9345
E-M: john.brennan@alphatech.com

Andrew L. Brook
University of Illinois at Urbana-Champaign
Dept. of General Engr.
104 S. Matthews Ave.
Urbana, IL 61801 USA
PH: 217-333-0531
FAX: 217-244-5705
E-M: w-davis@uiuc.edu

Douglas R. Brouse
Science Applications International Corp.
4001 North Fairfax Dr. Suite 300
Arlington, VA 22203 USA
PH: 703-558-2776
FAX: 703-524-1643
E-M: douglas.r.brouse@cpmx.saic.com

D. I. Bruce
DRA Malvern
St. Andrews Road
Room E210
Great Malvern, Worcs. WR14 3PS
ENGLAND
PH: +44-(0)1684-895112
FAX: +44-(0)1684-894389
E-M: dib@dra.hmg.gb

Daniel T. Brunner
Systemflow Simulations, Inc.
6366 Guilford Ave. Suite 310
Indianapolis, IN 46220-1750 USA
PH: 317-466-0664
FAX: 317-466-0821
E-M: brunner@systemflow.com

Gary N. Bundy
The MITRE Corp.
1820 Dolley Madison Blvd.
McLean, VA 22102-3481 USA
PH: 703-883-5560
FAX: 703-883-6435
E-M: gbundy@mitre.org

Carl D. Burke
The MITRE Corp.
1820 Dolly Madison Blvd.
McLean, VA 22102-3481 USA
PH: 703-883-6278
FAX: 703-883-6435
E-M: cburke@mitre.org

Arnold H. Buss
Naval Postgraduate School
Operations Research Dept.
Code OR/BU
Monterey, CA 93943 USA
PH: 408-656-3259
FAX: 408-656-2595
E-M: bussa@or.nps.navy.mil

Walter Callahan
New York Power Authority
Indian Point Nuclear Power Plant Unit 3
Buchanon, NY 10511 USA
PH: 914-736-8104
E-M: icallaw@ip3gate.usa.com

William A. Cann
Science Applications International Corp.
4001 North Fairfax Dr. Suite 300
Arlington, VA 22203 USA
E-M: William.A.Cann@cpmx.saic.com

Thomas Caraco
State University of New York at Albany
Dept. of Biological Sciences
Albany, NY 12222 USA
PH: 518-442-4343
FAX: 518-442-4767

Bruce Caraway
Knowledge Based Systems, Inc.
1500 University Dr. East
College Station, TX 77840 USA
PH: 409-260-5274
FAX: 409-260-1965
E-M: bcaraway@kbsi.com

Scott A. Carey
Logicon RDA
317 South 2nd St.
Leavenworth, KS 66048 USA
E-M: scarey@logicon.com

Marne C. Cario
Delphi Packard Electric Systems
Station 12C
408 Dana St.
P. O. Box 431
Warren, OH 44486 USA
PH: 330-373-3464
FAX: 330-373-5649

Diana F. Carroll
PROMODEL Corp.
Healthcare Systems Division
13333 Blanco Road, Suite 302
San Antonio, TX 78216 USA
PH: 210-493-1509
FAX: 210-408-1529
E-M: medmodel@aol.com

John S. Carson II
Auto Simulations, Inc.
1827 Powers Ferry Road
Building 17, Suite 100
Atlanta, GA 30339 USA
PH: 770-955-1501
FAX: 770-955-1592
E-M: john@autosim.com

Don Caughlin
Mission Research Corp.
102 South Tejon, Suite 200
Colorado Springs, CO 80903-2200 USA
PH: 719-633-2344 x282
FAX: 719-633-2543
E-M: donc@mrcos.com

David B. Cavitt
Old Dominion University
Dept. Computer Science
Norfolk, VA 23529-0162 USA
PH: 757-683-3915
FAX: 757-683-4900
E-M: cavitt@cs.odu.edu

Martha A. Centeno
Florida International University
Industrial Engr.
Miami, FL 33199 USA
PH: 305-348-3531
FAX: 305-348-3721
E-M: centeno@eng.fiu.edu

Yanick Champoux
Université de Montréal
C. P. 6128
Succursale Centre-Ville
Montréal, Québec H3C 3J7
CANADA
E-M: champoux@iro.umontreal.ca

Frank Chance
Chance Industrial Solutions
6900G Lake Dr.
Dublin, CA 94568 USA
PH: 510-803-7759
FAX: 510-803-7759
E-M: frank@chances.com

John M. Charnes
U. of Kansas School of Business
345 Summerfield Hall
Lawrence, KS 66045 USA
PH: 913-864-7591
FAX: 913-864-5328
E-M: john.charnes@ukans.edu

Chun-Hung Chen
University of Pennsylvania
Dept. of Systems Engr.
Philadelphia, PA 19104-6315 USA
PH: 215-898-3967
FAX: 215-573-2065
E-M: chchen@eniac.seas.upenn.edu

Hsiao-Chang Chen
University of Pennsylvania
Dept. of Systems Engr.
Philadelphia, PA 19104-6315 USA
PH: 215-898-5951
E-M: hcc@eniac.seas.upenn.edu

Russell C. H. Cheng
University of Kent at Canterbury
Institute of Mathematics & Statistics
Canterbury, Kent CT2 7NF
ENGLAND
PH: +44-1227-764-000 x3665
FAX: +44-1227-827-932
E-M: R.C.H.Cheng@ukc.ac.uk

Ching-Chuan Chiang
University of California, Los Angeles
Computer Science Dept.
Box 951596
Los Angeles, CA 90095 USA
PH: 310-825-1888
FAX: 310-825-2273
E-M: ccchiang@cs.ucla.edu

Eunmi Choi
Michigan State University
Dept. of Computer Science
A714 Wells Hall
East Lansing, MI 48824-1027 USA
PH: 517-355-0844
E-M: choie@cps.msu.edu

Leon Bing-Shiun Chuang
University of Washington
Industrial Engr. 352650
Seattle, WA 98195 USA
E-M: leonc@u.washington.edu

Gordon M. Clark
System Solutions, Inc.
400 Longfellow Avenue
Worthington, OH 43085-3074 USA
PH: 614-847-1394
FAX: 614 847-1454
E-M: clark17@osu.edu

Patrick D. Clark
Joint Warfighting Center
United States Navy
Fort Monroe, VA 23651 USA
E-M: clarkp@jwfc.army.mil

John Cleary
University of Waikato
Dept. of Computer Science
Te Whare Wananga o Waikato
Private Bag 3105
Hamilton, NEW ZEALAND
PH: 64-7-838-4378
FAX: 64-7-838-4155
E-M: jcleary@cs.waikato.ac.nz

Jeffery K. Cochran
Arizona State University
Industrial & Management Systems Engr.
Systems Simulation Laboratory
Tempe, AZ 85287-5906 USA
PH: 602-965-2758
FAX: 602-965-2910
E-M: atjkc@asuvm.inre.asu.edu

Kay K. Cornett
United Parcel Service Airlines
IE AGB-4A
1400 North Hurstbourne Parkway
Louisville, KY 40223 USA
PH: 502-329-6603
FAX: 502-329-3199
E-M: air1kkc@airaxp.air.ups.com

Robert C. Crain
Wolverine Software Corp.
7617 Little River Turnpike
Suite 900
Annandale, VA 22003-2603 USA
PH: 703-750-3910
FAX: 703-642-9634
E-M: wolverine@intr.net

Robert Crosslin
Technology Economics International
11212 Stephalee Lane
Rockville, MD 20852 USA
PH: 301-984-1334
FAX: 301-984-0816
E-M: rcrosslin@aol.com

James Curry
Texas A&M University
Dept. of Industrial Engr.
College Station, TX 77843 USA

Lawrence J. Dacunto
Logicon RDA
3620 "A" 96th St SW
Tacoma, WA 98499-4430 USA
PH: 206-964-1611
FAX: 206-964-3576
E-M: ldacunto@logicon.com

Liyi Dai
Washington University
Dept. of Systems Science & Mathematics
St. Louis, MO 63130 USA
PH: 314-935-7240
FAX: 314-935-6121
E-M: liyi@zach.wustl.edu

O. Patrick Daily
UNOS
1100 Boulders Parkway, Suite 500
Richmond, VA 23225 USA
PH: 804-330-8520
FAX: 804-330-8593
E-M: dailyop@gwwpm1.unos.org

Halim Damerdji
North Carolina State University
Dept. of Industrial Engr.
Box 7906
Raleigh, NC 27695-7906 USA
PH: 919-515-3598
FAX: 919-515-5281
E-M: damerdji@eos.ncsu.edu

Samir R. Das
Unveristy of Texas at San Antonio
Division of Computer Science
6900 North Loop, 1604 West
San Antonio, TX 78213-0667 USA
PH: 210-458-5537
FAX: 210-458-4437
E-M: samir@ringer.cs.utsa.edu

Wayne J. Davis
University of Illinois at Urbana-Champaign
Dept. of General Engr.
104 South Mathews Ave.
Urbana, IL 61801-2996 USA
PH: 217-333-0531
FAX: 217-244-5705
E-M: w-davis@uiuc.edu

Joseph R. Davis
North Carolina State University
College of Engr.
Industrial Ext. Serv.
Box 7513
Raleigh, NC 27695-7513 USA
PH: 919-515-8584
FAX: 919-515-8585
E-M: Joe_Davis@ncsu.edu

Ewa Deelman
Rensselaer Polytechnic Institute
Computer Science Dept.
110 - 8th St.
Troy, NY 12180-3590 USA
PH: 518-276-8377
FAX: 518-276-4033
E-M: deelmane@cs.rpi.edu

Dursun Delen
Oklahoma State University
School of Industrial Engr. & Management
322 Engr. North
Stillwater, OK 74078 USA
PH: 405-744-6055
FAX: 405-744-6187
E-M: delen@ceatlabs.okstate.edu

Yogesh L. Deshpande
University of Western Sydney, Macarthur
Faculty of Business & Technology
P. O. Box 555
Campbelltown, NSW 2560
AUSTRALIA
PH: +61-046-203-233
FAX: +61-046-284-289
E-M: Y.Deshpande@uws.EDU.AU

Luc Devroye
McGill University
School of Computer Science
Montreal, H3A 2A7
CANADA
PH: 514-398-3738
FAX: 514-398-3883
E-M: luc@cs.mcgill.ca

Karim Djemame
University of Glasgow
Computing Science Dept.
Glasgow, Scotland G12 8QQ
UNITED KINGDOM
PH: +44-141-339-8855 x8009
FAX: +44-141-330-4913
E-M: karim@dcs.gla.ac.uk

Georgios I. Doukidis
Athens University of Economics & Business
Dept. of Informatics
Patission 76 Athens 104 34
GREECE
E-M: simon.taylor@brunel.ac.uk

Zubin Dowlaty
United Parcel Service
Product Cost Dept.
55 Glenlake Parkway, NE
Atlanta, GA 30328 USA
PH: 404-828-6388
FAX: 404-828-6005
E-M: nat2zxd@is.ups.com

Glenn R. Drake
Systems Modeling Corp.
504 Beaver St.
Sewickly, PA 15143 USA
PH: 412-741-3727
FAX: 412-741-5635
E-M: gdrake@mail.sm.com

Catherine E. Drury
Micro Analysis & Design, Inc.
4900 Pearl East Circle
Suite 201 E
Boulder, CO 80301 USA
PH: 303-442-6947
FAX: 303-442-8274
E-M: cdrury@madboulder.com

Julie N. Ehrlich
Pritsker Corp.
1305 Cumberland Ave.
P. O. Box 2413
West Lafayette, IN 47906-0413 USA
PH: 317-420-6063
FAX: 317-420-6170
E-M: ehrlich@pritsker.com

Thomas H. Engh
Micro Analysis & Design, Inc.
4900 Pearl East Circle
Suite 201E
Boulder, CO 80301 USA
PH: 303-442-6947
FAX: 303-442-8274
E-M: engh@madboulder.com

Katherine B Ensor
Rice University
Dept. of Statistics
Houston, TX 77251-1892 USA
PH: 713-527-4687
FAX: 713-285-5476
E-M: ensor@rice.edu

Gerald W. Evans
University of Louisville
Dept. Industrial Engr.
Louisville, KY 40292 USA
PH: 502-852-6342
FAX: 502-852-5633
E-M: gwevan01@ulkyvm.louisville.edu

Gregory Ewing
University of Canterbury
Dept. of Computer Science
Christchurch
NEW ZEALAND
PH: +64-3-3642366
FAX: +64-3-3642569
E-M: greg@cosc.canaterbury.ac.nz

Kambiz Farahmand
Texas A&M University – Kingsville
Dept. Mechanical & Industrial Engr.
Campus Box 191
Kingsville, TX 78363 USA
PH: 512-595-2288
FAX: 512-595-2371
E-M: k_farahmand@tamuk.edu

Lawrence S. Farey
193 SW Seminole Dr.
Aloha, OR 97006 USA
PH: 503-629-9690
FAX: 503-693-2025
E-M: lfarey@fcpa.fujitsu.com

Steven D. Farr
Rome Laboratory / IRAE
32 Hangar Rd.
Rome, NY 13441-4114 USA
PH: 315-330-3037
FAX: 315-330-2728
E-M: farrs@rl.af.mil

Phillip A. Farrington
University of Alabama at Huntsville
Center for Automation & Robotics
Research Institute A-11
Huntsville, AL 35899 USA
PH: 205-890-6100
FAX: 205-890-6970

Ronald Fernandes
Knowledge Based Systems Inc.
1500 University Dr. East
College Station, TX 77840-2335 USA
PH: 409-260-5274
FAX: 409-260-1965
E-M: rfernandes@kbsi.com

Alois Ferscha
Institute for Applied Computer Science & Info. Sys.
Universität Wien
Lenaugasse 2/8
A-1080 Vienna
AUSTRIA
PH: +43-1-408-63-66-19
FAX: +43-1-408-04-50
E-M: ferscha@ani.univie.ac.at

Kevin C. Field
Brigham Young University
Dept. of Manufacturing Engr. & Engr. Technology
Orem, UT 84058 USA
PH: 801-223-4663
FAX: 801-226-6040
E-M: kevin@promodel.com

Pamela K. Fink
Medical Science Systems, Inc.
100 N.E. Loop 410, Suite 1350
San Antonio, TX 78216 USA
PH: 210-340-1496
FAX: 210-349-6900
E-M: mss@worldnet.att.net

Paul A. Fishwick
University of Florida
Dept. of CISE
Room 301, CSE Bldg.
Gainesville, FL 32611 USA
PH: 352-392-1414
FAX: 352-392-1414
E-M: fishwick@cise.ufl.edu

Bennett Foster
Dupont
1007 Market St.
Wilmington, DE 19898 USA
PH: 302-774-2242

John W. Fowler
Arizona State University
Dept. of Industrial & Management Sys. Engr.
Box 875906
Tempe, AZ 85287-5906 USA
PH: 602-965-3727
FAX: 602-965-8692
E-M: john.fowler@asu.edu

Michael C. Fu
University of Maryland
College of Business & Management
Van Munching Hall
College Park, MD 20742 USA
PH: 301-405-2241
FAX: 301-314-8787
E-M: mfu@umd5.umd.edu

Richard M. Fujimoto
Georgia Institute of Technology
College of Computing
Atlanta, GA 30332-0280 USA
PH: 404-894-5615
FAX: 404-894-9846
E-M: fujimoto@cc.gatech.edu

Deborah Fullford
MäK Technologies
185 Alewife Brook Parkway
Cambridge, MA 01238 USA
PH: 617-876-8085 x1103
FAX: 617-876-9208
E-M: deb@mak.com

Carlos Gandarillas
Florida International University
Industrial Engr.
Miami, FL 33199 USA
PH: 305-821-4630
FAX: 305-348-3721
E-M: cganda01@fiu.edu

Marelys L. Garcia
Florida International University
Industrial Engr.
Miami, FL 33199 USA
PH: 305-348-3452
FAX: 305-348-3721

Mario Gerla
University of California, Los Angeles
Computer Science Dept.
Box 951596
Los Angeles, CA 90095 USA
PH: 310-825-4367
FAX: 310-825-2273
E-M: gerla@cs.ucla.edu

George M. Giaglis
Brunel University
Dept. of Computer Science & Info. Systems
Uxbridge, UB8 3PH
UNITED KINGDOM
PH: +44-1895-274000
FAX: +44-1895-251686
E-M: ray.paul@brunel.ac.uk

Randall Gibson
Automation Associates, Inc.
512 Via De La Valle, Suite 209
Solana Beach, CA 92675 USA
PH: 619-481-6050
FAX: 619-481-7427
E-M: rgibson@aaisim.cts.com

Dennis C. Gilles
University of Glasgow
Computing Science Dept.
Glasgow, Scotland G12 8QQ
UNITED KINGDOM
PH: +44-141-339-8855
FAX: +44-141-330-4913
E-M: dcg@dcs.gla.ac.uk

Rajiv Girdhar
North Carolina A&T State University
Dept. of Industrial Engr.
1601 E. Market St.
Greensboro, NC 27411 USA
PH: 910-334-7780
FAX: 910-334-7729
E-M: girdhar@ncat.edu

Stuart Gittlitz
Kraft Foods
250 North St.
White Plains, NY 10625 USA

Naomi F. Glasscock
North Carolina State University
Ergonomics Laboratory
Dept. of Industrial Engr.
Raleigh, NC 27695-7906 USA
PH: 919-828-0693
FAX: 919-515-5281
E-M: nfglassc@eos.ncsu.edu

Paul Glasserman
Columbia University
403 Uris Hall
3022 Broadway
New York, NY 10027 USA
PH: 212-854-4102
FAX: 212-316-9180
E-M: pglassser@research.gsb.columbia.edu

Fred Glover
University of Colorado
School of Business, CB 419
Boulder, CO 80309 USA
PH: 303-492-8589
FAX: 303-492-5962
E-M: Fred.Glover@colorado.edu

Peter W. Glynn
Stanford University
Dept. of EES/OR
Stanford, CA 94305-4022 USA
PH: 415-725-0554
FAX: 415-723-4107
E-M: glynn@leland.stanford.edu

Carmelita Görg
Aachen University of Technology
Communication Networks
Kopernikusstr. 16
D-52074 Aachen
GERMANY
PH: +49-241-8890322
FAX: +49-241-8888242
E-M: cg@comnets.rwth-aachen.de

David Goldsman
Georgia Institute of Technology
School of Industrial & Systems Engr.
Atlanta, GA 30332-0205 USA
PH: 404-894-2365
FAX: 404-894-2301
E-M: sman@isye.gatech.edu

Fabian Gomes
University of Calgary
Dept. of Computer Science
2500 University Dr. N. W.
Calgary, Alberta T2N 1N4
CANADA
PH: 403-220-6320
FAX: 403-284-4707
E-M: gomes@cpsc.ucalgary.ca

Lissette R. Gonzalez
Florida International University
9060 N. W. 8th St. #106
Miami, FL 33172 USA
PH: 305-221-9272
FAX: 305-348-3721
E-M: lissettg@eng.fiu.edu

Fernando G. Gonzalez
University Illinois at Urbana-Champaign
Dept. of Electrical & Computer Engr.
1406 W. Green St.
Urbana, IL 61801 USA
PH: 217-351-2069
FAX: 217-244-5705
E-M: fernando@ww.ge.uiuc.edu

Tesham B. Gor
University of Louisville
Dept. of Industrial Engr.
Louisville, KY 40292 USA
PH: 502-852-6342
FAX: 502-852-5633
E-M: tbgor001@starbase.spd.louisville.edu

Eric S. Grajo
Production Modeling Corp.
Three Parklane Blvd. Suite 910 W.
Dearborn, MI 48126 USA
PH: 313-441-4460 x127
FAX: 313-441-6098
E-M: ericg@pmcorp.com

Ali K. Gunal
Production Modeling Corp.
Three Parklane Blvd. Suite 910 W.
Dearborn, MI 48126 USA
PH: 313-441-4460 x109
FAX: 313-441-6098
E-M: agunal@pmcorp.com

Gül Gürkan
University of Wisconsin - Madison
Dept. of Industrial Engr.
1513 University Ave.
Madison, WI 53706-1539 USA
E-M: ggurkan@cs.wisc.edu

Jorge Haddock
Rensselaer Polytechnic Institute
Dept. of Decision Sciences & Engr. Systems
Troy, NY 12180-3590 USA
PH: 518-276-8099
FAX: 518-276-8099
E-M: Haddock@rpi.edu

Dany Hajjar
University of Alberta, Edmonton
Construction Engr. & Management
220 Civil/Electrical Building
Edmonton, Alberta T6G 2G7
CANADA
PH: 403-492-8096
FAX: 403-492-2722
E-M: hajjar@cem.civil.ualberta.ca

James F. Hallas
Texas Instruments
East Building
Mail Station 385
P. O. Box 655012
Dallas, TX 75265 USA

Fang Hao
Georgia Institute of Technology
College of Computing
Atlanta, GA 30332-0280 USA
PH: 404-894-1155
FAX: 404-894-0272
E-M: fang@cc.gatech.edu

Charles Harrell
Brigham Young University
Manufacturing Engr.
435 CTB
Provo, UT 84602 USA
PH: 801-223-4630
FAX: 801226-6046
E-M: charley@promodel.com

Samuel M. Hart III
Project Performance Corp.
46030 Manekin Plaza, Suite 180
Sterling, VA 20166-6518 USA
PH: 703-404-1514
FAX: 703-406-8972
E-M: shart@ppc.com

Philip Heidelberger
IBM T. J. Watson Research Center
P. O. Box 704
Yorktown Heights, NY 10598 USA
PH: 914-784-7002
FAX: 914-784-6031
E-M: berger@watson.ibm.com

Joseph A. Heim
University of Washington
Industrial Engr.
P. O. Box 352650
Seattle, WA 98195-2650 USA
PH: 206-543-2308
FAX: 206-685-3072
E-M: jheim@u.washington.edu

Jon C. Helton
Sandia National Laboratories
Dept. 6848
P. O. Box 5800
Albuquerque, NM 87185-1328 USA
PH: 505-848-0701
FAX: 505-848-0705

James O. Henriksen
Wolverine Software Corp.
7617 Little River Turnpike
Suite 900
Annandale, VA 22003 USA
PH: 703-750-3910
FAX: 703-742-9634
E-M: wolverine@intr.net

Henry Herper
Otto-von-Guericke-Universitat of Magdeburg
Dept. of Computer Simulation & Graphics
P. O. Box 4120 / ISG
D-39106 Magdeburg
GERMANY
PH: +49-391-67-12866
FAX: +49-391-67-11164
E-M: nenry@isg.cs.uni-magdeburg.de

L. Tandy Herren
Medical Science Systems, Inc.
100 N. E. Loop 410, Suite 1350
San Antonio, TX 78216 USA
PH: 210-340-1435
FAX: 210-349-6900
E-M: ltherren@aol.com

John A. Hixson
Logicon RDA
P. O. Box 607
Leavenworth, KS 66048 USA
PH: 913-684-8778
FAX: 913-684-8758
E-M: Jack@leav-156-100.army.mil

Vlatka Hlupic
Brunel University
Dept. of Computer Science & Info. Systems
Uxbridge, Middlesex UB8 3PH
UNITED KINGDOM
PH: +44-1895-274-000
FAX: +44-1895-251-686
E-M: vlatka.hlupic@burnel.ac.uk

P. R. Hoare
DRA Malvern
St. Andrew's Road
Malvern, Worcestershire WR14 3PS
UNITED KINGDOM

Peter Hoefer
Pace University
Lubin School of Business
One Pace Plaza
New York, NY 10038 USA
PH: 212-346-1990
FAX: 212-346-1872
E-M: hoefer@pacevm.dac.pace.edu

Joel Hoffner
E & E Corp.
225 West Second St.
Bethlehem, PA 18015 USA
PH: 610-866-6008
FAX: 610-974-8049

José Holguín-Veras
California Polytechnic Institute
Civil & Environmental Engr. Dept.
San Lius Obispo, CA 93407 USA
PH: 805-756-2947
FAX: 805-756-6503
E-M: Jose_Holquin-Veras.CEMAIL@cemailgate.ce.utexas.edu

Wayne Holland
University of Kent at Canterbury
Institute of Mathematics & Statistics
Canterbury, Kent CT2 7NF
ENGLAND
PH: +44-1227-764-000 x4051
FAX: +44-1227-827-932
E-M: W.Holland@ukc.ac.uk

Mehdi Hoseyni-Nasab
Georgia Institute of Technology
School of Industrial & Systems Engr.
Atlanta, GA 30332 USA
PH: 404-303-1447
FAX: 404-894-2301
E-M: hoseyni@isye.gatech.edu

Jian-Qiang Hu
Boston University
Dept. of Manufacturing Engr.
Boston, MA 02215 USA
PH: 617-353-9696
FAX: 617-353-5548
E-M: hu@enga.bu.edu

Klaus-Peter Huber
University of Karlsruhe
Institute for Computer Design & Fault Tolerance
P. O. Box 6980
76128 Karlsruhe
GERMANY
PH: +49-721-608-4257
FAX: +49-721-370455
E-M: kphuber@ira.uka.de

Neil A. Hughes
University of Kent at Canterbury
Institute of Mathematics & Statistics
Canterbury, Kent CT2 7NF
ENGLAND
PH: +44-1227-764-000 x3135
FAX: +44-1227-827-932
E-M: N.A.Hughes@ukc.ac.uk

Ricki G. Ingalls
Compaq Computer Corp.
MS040801
20555 SR 249
Houston, TX 77070 USA
PH: 713-518-4766
FAX: 713-518-3216
E-M: RIngalls@bangate.compaq.com

Carolyn Internicola
SEMATECH
2706 Montopolis Dr.
Austin, TX 78741-6499 USA

Photios G. Ioannou
University of Michigan
Civil & Environmental Engr. Dept.
Ann Arbor, MI 48109-2125 USA
PH: 313-764-3369
FAX: 313-764-4292
E-M: photios@umich.edu

Michael Iuliano
NIST
Building 220 Room A127
Gaithersburg, MD 20899 USA
PH: 301-975-3891
FAX: 301-258-9749
E-M: michael.iuliano@nist.gov

Theodora Ivanova
Lucent Technologies
9333 S. John Young Parkway
Orlando, FL 32819 USA
PH: 407-345-6587
FAX: 407-345-6082
E-M: 0rthi@micro.lucent.com

Sheldon H. Jacobson
Virginia Polytechnic Institute & State University
Dept. of Industrial & Systems Engr.
274 Whittemodre Hall
Blacksburg, VA 24061-0118 USA
PH: 540-231-7099
FAX: 540-231-33232
E-M: jacobson@tioli.ise.vt.edu

Kris Jacoby
Random House, Inc.
400 Hahn Rd.
Westminister, MD 21154 USA
PH: 410-386-7725
FAX: 410-386-7727

Linda Jankauskas
Palisade Corp.
3740 N. W. Hayes Ave.
Corvallis, OR 97330 USA
PH: 541-757-6352
FAX: 541-727-6634
E-M: lindaj@proaxis.com

Thomas Jefferson
Intel Corp.
CH3-84
5000 W. Chandler Blvd.
Chandler, AZ 85226 USA
PH: 602-554-7162
FAX: 602-554-5323
E-M: Thomas_Jefferson@ccm.ch.intel.com

Roger Jenkins
University of Western Sydney, Macarthur
Faculty of Business & Technology
P. O. Box 555
Campbelltown, NSW 2560
AUSTRALIA
PH: 61-46-203-240
FAX: 61-46-266-683
E-M: r.jenkins@uws.edu.au

Paul A. Jensen
The University of Texas at Austin
Mechnical Engr.
Mailcode C2200
Austin, TX 78712 USA
PH: 512-471-6495
FAX: 512-471-8727
E-M: pjensen@mail.utexas.edu

Vikas Jha
University of California, Los Angeles
Computer Science Dept.
Box 951596
Los Angeles, CA 90095 USA
PH: 310-825-4885
FAX: 310-825-2273
E-M: jha@cs.ucla.edu

James Johnson
Insitute for Applied Computer Science & Info. Sys
Universität Wien
Lenaugasse 2/8
A-1080 Vienna
AUSTRIA
PH: +43-1408-6366-19
FAX: +43-1408-0450
E-M: johnson@ani.univie.ac.at

Jeffery A. Joines
North Carolina State University
Dept. of Industrial Engr.
Campus Box 7906
Ralaeigh, NC 27695-7906 USA
PH: 919-515-5188
FAX: 919-515-1543
E-M: jjoine@eos.ncsu.edu

Albert Jones
NIST
Building 220 Room A127
Gaithersberg, MD 20899 USA
PH: 301-975-3554
FAX: 301-258-9749
E-M: albert.jones@nist.gov

Stan Jonik
SFJ Systems, Ltd.
Office 2 Fernleigh House
Church St.
Trombridge, Wilts BA14 8DY
UNITED KINGDOM
PH: 01225-775103

Thomas Kagerl
Engr. Office for Project Management
Work Structuring & Plant Organisation
Schleinufer 14
D-39104 Magdeburg
GERMANY
PH: ++49-391-5419209
FAX: ++49-391-5419210
E-M: kagerl@isg.cs.uni-magdeburg.de

David R. Kalasky
Process Analysis & Improvement
1540 Stoddard Ave.
Wheaton, IL 60187 USA
PH: 630-690-1699
FAX: 630-690-1614
E-M: pai4sim@aol.com

Manjunath Kamath
Oklahoma State University
School of Industrial Engr. & Management
322 Engr. North
Stillwater, OK 74078 USA
PH: 405-744-6055
FAX: 405-744-6187
E-M: mkamath@okway.okstate.edu

Sungho Kang
Yonsei University
Dept. Electrical Engr.
134 Shinchon-Dong
Seodaemun-ku
Seoul, KOREA
PH: +82-2-361-2775
FAX: +82-2-312-7735
E-M: shkang@bubble.yonsei.ac.kr

S. Cem Karacal
Southern Illinois University at Edwardsville
Dept. Mechanical & Industrial Engr.
Science Building Rm 0319
Edwardsville, IL 62026 USA
PH: 618-692-2435
FAX: 618-692-2555
E-M: karacal@red.soe.siue.edu

Zeynep A. Karacal
Electronic Data Systems
Communications Industry 1180
13736 Riverport Dr.
St. Louis, MO 63146 USA
PH: 314-344-5984
FAX: 314-344-8399
E-M: gzn2d@fmspa7.stl.gtd.edu.com

Helen D. Karatza
Aristotle University of Thessaloniki
Dept. of Informatics
54006 Thessaloniki
GREECE
PH: 031-997974
FAX: 031-429784
E-M: karatza@olymp.ccf.auth.gr

Ganesh Kateel
Oklahoma State University
School of Industrial Engr. & Management
322 engineering North
Stillwater, OK 74078 USA
PH: 405-744-6055
FAX: 405-744-6187
E-M: kganesh@okstate.edu

Randy H. Katz
University of California at Berkeley
Computer Science Division
637 Soda Hall
Berkeley, CA 94720-1776 USA
E-M: randy@cs.berkeley.edu

Richard Keeble
Brunel University
Dept. of Computer Science & Info. Systems
Uxbridge, UB8 3PH
ENGLAND
PH: +44-1895-274000
FAX: +44-1895-251686
E-M: ray.paul@brunel.ac.uk

William S. Keezer
LEXIS-NEXIS
P. O. Box 933
Dayton, OH 45401 USA
PH: 513-865-1015
FAX: 513-865-6808
E-M: wsk@lexis-nexis.com

Thomas Kelley
Florida A&M University - FSU
Dept. of Industrial Engr.
2525 Pottsdamer St.
Tallahassee, FL 32310 USA
E-M: kelley@eng.fsu.edu

Christian Kelling
Technische Universität Berlin
Institute für Tehcnishe Informatik
Franklinstrasse 28/29
10587 Berlin
GERMANY
PH: +49-30-314-73-116
FAX: +49-30-314-21-116
E-M: ck@cs.tu-berlin.de

James F. Kelly
Florida A&M University - FSU
Dept. of Industrial Engr.
2525 Pottsdamer St.
Tallahassee, FL 32310 USA
E-M: kelly@eng.fsu.edu

James P. Kelly
University of Colorado
Graduate School of Business
Boulder, Colorado 80309-0419 USA
E-M: James.Kelly@Colorado.Edu

W. David Kelton
University of Cincinnati
Dept. of Quant. Analysis & Operations Management
College of Busines Administration
Cincinnati, OH 45221-0130 USA
PH: 513-556-0404
FAX: 513-556-4999
E-M: david.kelton@uc.edu

Karl Kempf
Intel Corp.
5000 W. Chandler Blvd.
CH3-68
Chandler, Arizona 85226 USA
PH: 602-554-3325
FAX: 602-554-7311
E-M: kkempf@fa.intel.com

Charles Kerns
Simons Technologies
P. O. Box 1286
Atlanta, GA 30301 USA
PH: 404-370-7761

Susan Khoubyari
Production Modeling Corp.
Three Parklane Boulevard
Suite 910 West
Dearborn, MI 48126 USA
PH: 313-441-4460 x118
FAX: 313-441-6098
E-M: susan@pmcorp.com

Jane D. Kim
SEMATECH
2706 Montopolis Dr.
Austin, TX 78741-6499 USA

Ben C. King
The MITRE Corp.
1820 Dolly Madison Blvd.
McLean, VA 22102 USA
PH: 703-883-6688
FAX: 703-883-6435
E-M: bking@mitre.org

Cliff B. King
F & H Simulations
P. O. Box 658
Orem, UT 84059-0658 USA
PH: 801-224-6914
FAX: 801-224-6984
E-M: cliff@taylorii.com

M. J. Kirton
DRA Malvern
St. Andrews Road
Room E204
Great Malvern, Worcs. WR14 3PS
ENGLAND
PH: +44-(0)1684-894419
FAX: +44-(0)1684-894389
E-M: kirton@signal.dra.hmg.gb

Jack P. C. Kleijnen
Tilburg University
Dept. of Info. Systems & Auditing
Center for Economic Research
School of Management & Economics
5000 Le Tilburg,
NETHERLANDS
PH: +31-13-4662029
FAX: +31-13-4663377
E-M: kleijnen@kub.nl

Martin S. Kleiner
Logicon RDA
133 Kings Way, Suite 102
Hampton, VA 23669 USA
PH: 804-723-6878
FAX: 804-723-6985
E-M: mkleiner@logicon.com

J. Frederick Klingener
Brock Engr. PC
18 Old Roxbury Rd.
P. O. Box 302
Roxbury, CT 06783 USA
PH: 860-354-7069
FAX: 860-354-7069
E-M: klingener@aol.com

Felipe Knop
Purdue University
Dept. of Computer Sciences
West Lafayette, IN 47907-1398 USA
PH: 317-494-0361
FAX: 317-494-0739
E-M: knop@cs.purdue.edu

Toshio Komatsu
Nippon Telegraph & Telephone Corp.
High-Speed Computer Networking Laboratory
NTT Info. & Communication Systems Lab.
1-2356 Take
Yokosuka-shi Kanagawa 238-03
JAPAN
PH: +81-468-59-2616
FAX: +81-468-59-3125
E-M: komatsu@balss21.isl.ntt.jp

David Krahl
Imagine That, Inc.
6830 Via Del Oro
Suite 230
San Jose, CA 95119 USA
PH: 408-365-0305
FAX: 408-629-1251
E-M: davek@imaginethatinc.com

Jasna Kuljis
Roehampton Institute London
Dept. of Applied Technology & Computing
West Hill
London, SW15 3SN
UNITED KINGDOM
PH: +44-181-392-3672
FAX: +44-181-392-3271
E-M: jkuljis@roehampton.ac.uk

Anand S. Kunnathur
University of Toledo
Dept. Info. Systems & Operations Management
Toledo, OH 43606 USA
PH: 419-530-2420
FAX: 419-530-7744
E-M: fac0066@uoft01.utoledo.edu

Indira Kuraganti
University of Virginia
Dept. of Systems Engr.
Charlottesville, VA 22903 USA
PH: 804-924-5393
FAX: 804-982-2972
E-M: ik2k@virginia.edu

Michael J. Kwinn Jr.
United States Military Academy
Dept. of Systems Engr.
West Point, NY 10996 USA
PH: 914-938-2700
FAX: 914-938-5919
E-M: fm9536@se.usma.edu

Pierre L'Ecuyer
Université de Montréal
Département IRO
C. P. 6128
Succursale Centre-Ville
Montréal, Québec H3C 3J7
CANADA
PH: 514-343-2143
FAX: 514-343-5834
E-M: lecuyer@iro.umontreal.ca

Lynn S. Lachenmaier
Arizona State University
Industrial & Management Engr. Systems
Tempe, AZ 85287-5906 USA
PH: 317-883-7453
FAX: 317-883-8520
E-M: lachenma@asu.edu

Manuel Laguna
University of Colorado
Graduate School of Business
Boulder, CO 80309-0419 USA
E-M: Manuel.Laguna@Colorado.Edu

Cynthia LaJimodiere
Texas A&M University
Dept. of Industrial Engr.
College Station, TX 77843 USA

Ron Laughery
Micro Analysis & Design, Inc
4900 Pearl East Circle, Suite 201E
Boulder, CO 80301 USA
PH: 303-442-6947
FAX: 303-442-8274
E-M: ron@madboulder.com

K. R. Laughery
Micro Analysis & Design, Inc
4900 Pearl East Circle, Suite 201 E
Boulder, CO 80301 USA

Averill M. Law
Averill M. Law & Associates
P. O. Box 40996
Tucson, AZ 85717 USA
PH: 502-795-6265
FAX: 520-795-6302
E-M: averill@ix.netcom.com

Edward C. Lawrence
University of Missouri - St. Louis
School of Business Administration
1102 Tower
8001 Natural Bridge Road
St. Louis, MO 63121-4499 USA
PH: 314-516-6148
FAX: 314-516-6420
E-M: c1975@umslvma.umsl.edu

Don Lazzari
Systems Modeling Corp.
504 Beaver St.
Sewickley, PA 15143 USA
PH: 412-741-3727
FAX: 412-741-5635
E-M: don@mail.sm.com

Jae-wook Lee
Yonsei University
Dept. of Electrical Engr.
Seoul, 120-749
KOREA
PH: 02-361-2775
FAX: 02-312-7735
E-M: jwlee@sleepy.yonsei.ac.kr

Kangsun Lee
University of Florida
1212 SW 5th Ave. #7
Gainsville, FL 32601 USA
PH: 352-392-1435
E-M: kslee@cis.ufl.edu

Michael S. Lee
University of Illinois at Urbana-Champaign
Dept. of General Engr.
104 S. Matthews Ave.
Urbana, IL 61801 USA
PH: 217-333-0531
FAX: 217-244-5705
E-M: w-davis@uiuc.edu

Lawrence M. Leemis
The College of William & Mary
Dept. of Mathematics
Williamsburg, VA 23187-8795 USA
PH: 757-221-2034
FAX: 757-221-2988
E-M: leemis@math.wm.edu

Brian Lehaney
University of Luton
Faculty of Business
Park Square, Luton
Beds, LU1 3JU
ENGLAND
E-M: simon.taylor@brunel.ac.uk

Gerald A. Levasseur
AESOP Corp.
233 South Wacker Dr.
Sears Tower, Suite 9604
Chicago, IL 60606 USA
PH: 312-559-9880
FAX: 312-559-9910
E-M: jerry@aesop-us.com

William R. Lilegdon
Pritsker Corp.
8910 Purdue Road, Suite 600
Indianapolis, IN 46268-1170 USA
PH: 317-471-6530
FAX: 317-471-6525
E-M: lilegdon@pritsker.com

Donald H. Liles
University of Texas at Arlington
Automation & Robotics Research Institute
7300 Jack Newell Blvd. S.
Fort Worth, TX 76118 USA
PH: 817-272-5902
FAX: 817-272-5952
E-M: dliles@arri.uta.edu

Mahhavi Lingineni
Knowledge Based Systems, Inc.
1500 University Dr. East
College Station, TX 77840 USA
PH: 409-260-5274
FAX: 409-260-1965
E-M: mlingineni@kbsi.com

Winston W. Liu
University of California, Los Angeles
Computer Science Dept.
Box 951596
Los Angeles, CA 90095 USA
PH: 310-825-1888
FAX: 310-825-2273
E-M: lwinston@cs.ucla.edu

Chong Loo
United Parcel Service
55 Glenlakes Parkway NE
Atlanta, GA 30328 USA
PH: 404-828-7621
FAX: 404-828-6005
E-M: nat1chl@is.ups.com

Peter Lorenz
Otto-von-Guericke University of Magdeburg
Institute for Simulation & Graphics
Post Box 4120
D-39106 Magdeburg
GERMANY
PH: 49-391-67-18534
FAX: 49-391-67-11164
E-M: pelo@isg.cs.uni-magdeburg.de

Julie C. Lowery
Center for Practice Mngt.& Outcomes Research
VA Medical Center
P. O. Box 130170
Ann Arbor, MI 48113-0170 USA
PH: 313-930-5111
FAX: 313-930-5159
E-M: jlowery@umich.edu

Kahaled M. Mabrouk
The Model Builders, Inc.
343 South Main, Suite 217
Ann Arbor, MI 48104 USA
PH: 313-998-5560
FAX: 313-998-5561
E-M: mbldr@izzy.net

Lewis M. Mackenzie
University of Glasgow
Computing Science Dept.
Glasgow, Scotland G12 8QQ
UNITED KINGDOM
PH: +44-141-330-4262
FAX: +44-141-330-4913
E-M: lewis@dcs.gla.ac.uk

Robert Macredie
Brunel University
Dept. of Computer Science & Info. Systems
Uxbridge, UB8 3PH
ENGLAND
PH: +44-1895-27400 x2126
FAX: +44-1895-251686
E-M: robert.macredie@burnel.ac.uk

Joseph G. Macro
University of Illinois at Urbana-Champaign
Dept. of General Engr.
104 S. Matthews Ave.
Urbana, IL 61801 USA
PH: 217-333-0531
FAX: 217-244-5705
E-M: w-davis@uiuc.edu

Brian A. Malloy
Clemson University
Dept. of Computer Science
416 Edwards Hall
Clemson, SC 29634-1906 USA
PH: 864-656-0808
FAX: 864-656-0145
E-M: malloy@cs.clemson.edu

Linda Malone
University of Central Florida
Dept. of Industrial Engr. & Management Sys.
Orlando, FL 32816 USA
PH: 407-823-2833
FAX: 407-823-3413

Kurt J. Maly
Old Dominion University
Dept. of Computer Science
Norfolk, VA 23529-0162 USA
PH: 757-683-3915
FAX: 757-683-4900
E-M: maly@cs.odu.edu

Hans Manansang
University of Washington
Industrial Engr. Program
P. O. Box 352650
Seattle, WA 98195-2650 USA
E-M: hansm@u.washington.edu

S. Manivannan
Consolidated Freightways, Inc
Dept. of Intelligent Systems Engr.
1717 NW 21ST Ave.
Portland, OR 97209 USA
PH: 503-499-3948
FAX: 503-499-3947
E-M: mani@eniac.cnf.com

Nancy Markovitch
Systems Modeling Corp.
504 Beaver St.
Sewickley, PA 15143 USA

Jeffery S. Marks
Naval Health Researach Center
P. O. Box 85122
San Diego, CA 92186-5122 USA
PH: 619-553-8393
FAX: 619-553-8607
E-M: marks@nhrc.navy.mil

Alejandro Francisco Garza Martinez
Componentes Mecanicos
De Matamoros, S.A. De C.V.
P. O. Box 4447
Brownsville, TX 78523 USA

Julio C. Martinez
University of Michigan
Civil & Environmental Engr. Dept.
2354 G. G. Brown
Ann Arbor, MI 48109-2125 USA
PH: 313-764-3369
FAX: 313-764-4292
E-M: martjuli@engin.umich.edu

Edward Mascarenhas
Purdue University
Dept. of Computer Sciences
West Lafayette, IN 47907-1398 USA
PH: 317-494-0361
FAX: 317-494-0739
E-M: edm@cs.purdue.edu

Scott J. Mason
Advanced Micro Devices
5204 East Ben White Blvd.
M/S 612
Austin, TX 78741 USA
PH: 512-602-2816
FAX: 512-602-7430
E-M: scottj.mason@amd.com

Serge A. Matheny
Medical ISSO
Research Dept.
Naval Health Research Center
P. O. Box 85122
San Diego, CA 92186-5122 USA
PH: 619-553-8399
FAX: 619-553-8607
E-M: matheny@nhrc.navy.mil

Richard J. Mayer
Texas A&M University
Dept. of Industrial Engr.
College Station, TX 77843 USA
PH: 409-845-9363
E-M: ric@kbssun1.tamu.edu

Michael G. McComas
Averill Law & Associates
P. O. Box 40966
Tucson, AZ 85717 USA
PH: 520-795-6265
FAX: 520-796-6302
E-M: averill@ix.netcom.com

Michael L. McGinnis
United States Military Academy
Operations Research Center
Dept. of Systems Engr.
West Point, NY 10996 USA
PH: 914-938-5529
FAX: 914-938-5665
E-M: fm0768@se.usma.edu

Sam McLafferty
Palisade Corp.
31 Decker Road
Newfield, NY 14867 USA
PH: 607-277-8000
FAX: 607-277-8001
E-M: s_mcl@palisade.com

Donald C McNickle
University of Canterbury
Dept. of Management
Christchurch
NEW ZEALAND
PH: 64-3-364-2666
FAX: 64-3-364-2020
E-M: dcm@mang.canterbury.ac.nz

Scott A. Miller
Systems Modeling Corp.
504 Beaver St.
Sewickley, PA 15143 USA
PH: 412-741-3727
FAX: 412-741-5635
E-M: Scott@mail.sm.com

J. O. Miller
AFIT/ENS
Dept. of Operational Sciences
WPAFB, OH 45433-7765 USA
PH: 513-236-7339
FAX: 513-236-7497
E-M: jmiller@afit.af.mil

John A. Miller
University of Georgia
Dept. of Computer Science
415, Boyd GSRC
Athens, GA 30602-7404 USA
PH: 706-542-3440
FAX: 706-542-2966
E-M: jam@cs.uga.edu

K. R. Milner
DRA Malvern
St. Andrews Road
Room E103
Great Malvern, Worcs. WR14 3PS
ENGLAND
PH: +44-(0)1684-894383
FAX: +44-(0)1684-894389
E-M: milner@signal.dra.hmg.gb

Dugki Min
KonKuk University
Dept. of Computer Engr.
MoJin Dong KwangJin Gu 93-1
Seoul, KOREA
PH: 011-82-2-450-3490
FAX: 011-82-2-458-1395
E-M: dkmin@pluto.konkuk.ac.kr

Gary A. Mirka
North Carolina State University
Dept. of Industrial Engr.
Raleigh, NC 27695-7906 USA
PH: 919-515-6399
FAX: 919-515-5281
E-M: mirka@eos.ncsu.edu

Paul Mitchell
New York Power Authority
123 Main St.
White Plains, NY 10601 USA
PH: 914-287-3249
E-M: wmitchp@ip3gate.usa.com

Louis A. Molina
Florida International University
Industrial Engr.
Miami, FL 33199 USA
PH: 305-348-3452
FAX: 305-348-3721
E-M: molinal@eng.fiu.edu

Mansooreh Mollaghasemi
Unviersity of Central Florida
Dept. Industrial Engr & Management Systems
Orlando, FL 32816 USA
PH: 407-823-0216
FAX: 407-823-3413
E-M: mollag@iems.engr.ucf.edu

Randy D. Moore
SONY Display Tube Company
16450 West Bernardo Dr.
San Diego, CA 92127 USA

Kendra E. Moore
ALPHATECH, Inc.
50 Mall Road
Burlington, MA 01803-4562 USA
PH: 617-273-3388
FAX: 617-273-9345
E-M: kendra.moore@alphatech.com

Douglas J. Morrice
University of Texas at Austin
MSIS Dept.
CBA 5.202
Austin, TX 78712-1175 USA
PH: 512-471-7857
FAX: 512-471-0587
E-M: morrice@mail.utexas.edu

Charles T. Mosier
Clarkson University
School of Business
Box 5790
Potsdam, NY 13699-5790 USA

Alasdar Mullarney
CACI Products Company
3333 North Torrey Pines Court
La Jolla, CA 92037 USA
PH: 619-457-9681
FAX: 619-458-0031
E-M: alasdar@caciasl..com

Daniel J. Muller
AutoSimulations, Inc.
1917 Stirling Dr.
Lansdale, PA 19446 USA
PH: 610-222-0601
FAX: 610-222-0602
E-M: danm@autosim.com

Edward Mykytka
Air Force Institute of Technology
Dept. of Operational Sciences
Wright-Patterson, AFB
Dayton, OH USA

Rajesh S. Nair
University of Georgia
Dept. of Computer Science
223 A, Boyd GSRC
Athens, GA 30602-7404 USA
PH: 706-542-3445
FAX: 706-542-2966
E-M: rajesh@cs.uga.edu

Yukihiro Nakamura
Nippon Telegraph & Telephone Corp.
High-Speed Computer Networking Laboratory
NTT Info. & Communication Systems Lab.
1-2356 Take
Yokosuka-shi Kanagawa, 238-03
JAPAN
PH: +81-468-59-2600
FAX: +81-468-59-3726
E-M: nakamura@isl.ntt.jp

Marvin K. Nakayama
New Jersey Institute of Technology
Dept. of Computer & Info. Science
University Heights
Newark, NJ 07102 USA
PH: 201-596-3398
FAX: 201-596-5777
E-M: marvin@cis.njit.edu

Richard E. Nance
Virginia Polytechnic Institute & State University
Systems Research Center
562 McBryde Hall
Blacksburg, VA 24061-0251 USA
PH: 540-231-6144
FAX: 540-231-4330
E-M: nance@vt.edu

Barry L. Nelson
Northwestern University
Dept. of Industrial Engr. & Management Sciences
2225 N. Campus Dr.
Evanston, IL 60208-3119 USA
PH: 847-491-3747
FAX: 847-491-8005
E-M: nelsonb@random.iems.nwu.edu

David A. Nembhard
Auburn University
Dept. of Management
Auburn, AL 36849-5241 USA

Harriet Black Nembhard
Auburn University
Dept. Industrial & Systems Engr.
310 Dunstan Hall
Auburn, AL 36849-5346 USA
PH: 334-844-1426
FAX: 334-844-1381
E-M: nembhard@eng.auburn.edu

Giao T. Nguyen
University of California at Berkeley
Computer Science Division
473 Soda Hall
Berkeley, CA 94720-1776 USA
PH: 510-642-8919
FAX: 510-642-5775
E-M: gnguyen@cs.berkeley.edu

David Nicol
Dartmouth College
Dept. Computer Science
6211 Sudikoff Laboratory
Hanover, NH 03755-3510 USA
FAX: 603-646-1672
E-M: nicol@cs.dartmouth.edu

Syoiti Ninomiya
IBM Research
Tokyo Research Laboratory
1623-14
Shimotsuruma
Yamato-shi Kanagawa-ken, 242
JAPAN
E-M: ninomiya@trl.ibm.co.jp

Brian Noble
Carnegie Mellon University
School of Computer Science
Pittsburgh, PA 15213-3891 USA
E-M: bnoble@cs.cmu.edu

William B. Nordgren
F & H Simulations
P. O. Box 658
Orem, UT 84059-0658 USA
PH: 801-224-6914
FAX: 801-224-6984
E-M: bill@taylorii.com

Junro Nose
NTT Sortware Corp.
Yokohama-Daiichi Bldg.
4-40 Hon-cho Naka-ku
Yokohama-shi Kanagawa, 231
JAPAN
PH: +81-45-212-7547
FAX: +81-45-212-8030
E-M: nose@sofia.ntts.cae.ntt.jp

Michael Novels
CIMULATION CENTRE LTD
Cornbrash Park, Bumpers Way
Chippenham, Wiltshire SN14 6RA
UNITED KINGDOM
PH: +44-1249-650316
FAX: +44-1249-443413

Ed O'Donnell
Naval Health Research Center
Medical Info. Systems & Operations Research Dept.
P. O. Box 85122
San Diego, CA 92186-5122 USA

Jean J. O'Reilly
Pritsker Corp.
1305 Cumberland Ave.
P. O. Box 2413
West Lafayette, IN 47906 USA
PH: 317-420-6080
FAX: 317-420-6170
E-M: oreilly@pritsker.com

Michael S. Odowick
Naval Health Research Center
P. O. Box 85122
San Diego, CA 92186-5122 USA
PH: 619-553-8386
FAX: 619-553-8607
E-M: blood@nhrc.navy.mil

W. Swain Ottman
United Parcel Service Airlines
IE AGB-4A
1400 North Hurstbourne Parkway
Louisville, KY 40223 USA
PH: 502-329-3253
FAX: 502-329-3199
E-M: air1wso@airaxp.air.ups.com

C. Michael Overstreet
Old Dominion University
Dept. of Computer Science
Norfolk, VA 23529-0162 USA
PH: 757-683-4545
FAX: 757-683-4900
E-M: cmo@cs.odu.edu

A. Yonca Özge
University of Wisconsin - Madison
Dept. of Industrial Engr.
1513 University Ave.
Madison, WI 53706-1539 USA
E-M: ozge@cs.wisc.edu

Nataranjan Padmanaban
Knowledge Based Systems Inc
1701 Directors Blvd. Suite 220
Austin, TX 78744 USA
PH: 512-416-1161
FAX: 512-416-1165
E-M: nat@kbsi-aus.com

Ernest H. Page
The MITRE Corp.
1820 Dolly Madison Blvd.
McLean, VA 22102-3481 USA
PH: 703-883-5728
FAX: 703-883-1279
E-M: epage@mitre.org

Thomas W. Page
Ohio State University
Dept. Computer Science & Info. Science
2015 Neil Ave. Dreese Labs 774
Columbus, OH 43210 USA
PH: 614-292-5835
FAX: 614-292-2911
E-M: page@cis.ohio-state.edu

Michael K. Painter
Knowledge Based Systems, Inc.
1500 University Dr. East
College Station, TX 77840-2335 USA
PH: 409-260-5274
FAX: 409-260-1965
E-M: mpainter@kbsi.com

Ray J. Paul
Brunel University
Dept. Computer Science & Info. Systems
Kingston Lane, West London
Uxbridge, Middlesex UB8 3PH
UNITED KINGDOM
PH: +44-1895-203-374
FAX: +44-1895-203-391
E-M: Ray.Paul@brunel.ac.uk

Krzysztof Pawlikowski
University of Canterbury
Dept. of Computer Science
Christchurch
NEW ZEALAND
PH: +64-3-3642-351
FAX: +64-3-3642-569
E-M: krys@cosc.canterbury.ac.nz

Brett A. Peters
Texas A&M University
Dept. of Industrial Engr.
College Station, TX 77843 USA

Robert G. Phelan Jr.
United States Military Academy
Operations Research Center
West Point, NY 10996 USA
PH: 914-938-5941
FAX: 914-938-5665
E-M: fr0161@se.usma.edu

Michael Pidd
Lancaster University
Dept. Management Science
The Management School
Lancaster, LA1 4YX
UNITED KINGDOM
PH: +44-1524-593870
FAX: +44-1524-592417
E-M: M.Pidd@lancaster.ac.uk

Beth M. Plott
Micro Analysis & Design, Inc.
4900 Pearl East Circle, Suite 201E
Boulder, CO 80301 USA
PH: 303-442-6947
FAX: 303-442-8274
E-M: bplott@madboulder.com

David B. Pratt
Oklahoma State University
Center for Integrated Manufacturing
Dept. of Industrial Engr. & Management
322 Engr. North
Stillwater, OK 74078-0533 USA
PH: 405-744-6055
FAX: 405-744-6187
E-M: dpratt@okway.okstate.edu

Adrien R. Presley
University of Texas at Arlington
Automation & Robotics Research Institute
7300 Jack Newell Blvd. S.
Fort Worth, TX 76118 USA
PH: 817-272-5907
FAX: 817-272-5952
E-M: apresley@arri.uta.edu

Kenneth D. Pritsker
Pritsker Planning Group
2304 Eldridge St.
Pittsburgh, PA 15217 USA
PH: 412-521-2122
FAX: 412-422-3455

A. Alan B. Pritsker
Pritsker Corp.
8910 Purdue Rd. Suite 600
Indianapolis, IN 46268 USA
PH: 317-471-6501
FAX: 317-471-6525
E-M: pritsker@pritsker.com

David M. Profozich
Systems Modeling Corp.
504 Beaver St.
Sewickley, PA 15143 USA
PH: 412-741-3727
FAX: 412-741-5635
E-M: dmp@mail.sm.com

David J. Prybyla
Logicon, Inc.
3620 "A" 96th St. SW
Tacoma, WA 98499-4430 USA
PH: 206-964-1611
FAX: 206-964-3576
E-M: dprybyla@logicon.com

Jennie P. Psihogios
North Carolina State University
Dept. of Industrial Engr.
Box 7906
Raleigh, NC 27695-7906 USA
PH: 919-515-7210
FAX: 919-515-5281
E-M: jppsihog@eos.ncsu.edu

Bala Ram
North Carolina A&T State University
Dept. of Industrial Engr.
1601 E. Market St.
Greensboro, NC 27411 USA
PH: 910-334-7780
FAX: 910-334-7729
E-M: ram@ncat.edu

Madhav Rangaswami
Intel Corp.
CH3-84
5000 W. Chandler Blvd.
Chandler, AZ 85226 USA
PH: 602-554-3056
FAX: 602-554-5323
E-M: madhav.rangaswami@ccm.ch.intel.com

Patrick Reardon
Los Alamos National Laboratories
Los Alamos, NM 87545 USA

Vernon Rego
Purdue University
Dept. of Computer Sciences
West Lafayette, IN 47907-1398 USA
PH: 317-494-7835
FAX: 317-494-0739
E-M: rego@cs.purdue.edu

Charles H. Reilly
University of Central Florida
Dept. of Industrial Engr. & Management Systems
P. O. Box 162450
Orlando, FL 32816-2450 USA
PH: 407-823-5306
FAX: 407-823-3413
E-M: reilly@iems.engr.ucf.edu

I. J. Relf
DRA Malvern
St. Andrews Road
Room E211
Great Malvern, Worcs. WR14 3PS
ENGLAND
PH: +44-(0)1684-896401
FAX: +44-(0)1684-894389
E-M: relf@signal.dra.hmg.gb

Wilkey Richardson
Hughes Missile Systems Co.
MS M20/P8
P. O. Box 11337
1151 E. Hermans Rd.
Tucson, AZ 85734-1337 USA
PH: 520-794-6432
FAX: 520-794-0872
E-M: gwrichardson@ccgate.hac.com

Stephen D. Roberts
North Carolina State University
Dept. Industrial Engr.
Campus Box 7906
Raleigh, NC 27695-7906
USA
PH: 919-515-2362
FAX: 919-515-5281
E-M: roberts@eos.ncsu.edu

Stephen M. Robinson
University of Wisconsin-Madison
Dept. of Industrial Engr.
1513 University Ave.
Madison, WI 53706-1539 USA
PH: 608-263-6862
FAX: 608-262-8454
E-M: smr@cs.wisc.edu

Jennifer Robinson
University of Massachusetts
581 VFW Parkway
Chestnut Hill, MA 02167 USA
PH: 617-323-2417
FAX: 617-323-2417
E-M: jkr@world.std.com

Ralph V. Rogers
University of Central Florida
Dept. of Industrial Engr. & Management Systems
P. O. Box 25000
Orlando, FL 32708 USA
PH: 407-823-5912
FAX: 407-823-3413
E-M: rogers@iems.engr.ucf.edu

Matthew W. Rohrer
Autosimulations, Inc
655 East Medical Dr.
Bountiful, UT 84010 USA
PH: 801-298-1398 x335
FAX: 801-298-8186
E-M: mattr@autosim.com

Robert Rönngren
Royal Institute of Technology (KTH)
Dept. of Teleinformatics
Electrum 204
Kista, S-164 40
SWEDEN
PH: +46-8-752-1380
FAX: +46-8-751-1793
E-M: robertr@it.kth.se

Dan Rotblatt
Naval Health Research Center
Medical Info. Systems & Operations Research Dept.
P. O. Box 85122
San Diego, CA 92186-5122 USA
E-M: ROTBLATT@VAX309.NHRC.NAVY.MIL

Glenn G. Rousseau
Headquarters Air Mobility Command
Studies & Analysis Flight
402 Scott Dr. Unit 3L3
Scott Air Force Base, IL 62225 USA
PH: 618-256-3450
FAX: 618-256-2502
E-M: rousseg@hqamc.safb.af.mil

Shigeru Sadakane
Production Modeling Corp.
Three Parklane Boulevard
Suite 910 West
Dearborn, MI 48126 USA
PH: 313-441-4460 x112
FAX: 313-441-6098
E-M: sadakane@pmcorp.com

Stacy Marie Sammons
Arizona State University
Dept. of Industrial & Management Systems Engr.
Tempe, AZ 85287-5906 USA
PH: 602-585-4905
FAX: 602-965-2910
E-M: sammonsm@t-bird.edu

Sriram Sampath
Cap Gemini America
25800 Science Park Dr. Suite 180
Cleveland, OH 44122 USA

Susan M. Sanchez
Univeristy of Missouri - St. Louis
School of Business Administration
232 CCB
8001 Natural Bridge Rd.
St. Louis, MO 63121-4499 USA
PH: 314-516-6169
FAX: 314-516-6827
E-M: susan@whimsy.umsl.edu

Raka Sandell
Motorola, Inc.
MOS12 Die Manufacturing
Semiconductor Products Sector
1300, North Alma School Rd.
Chandler, Arizona 85224 USA

William H. Sanders
University of Illinois at Urbana-Champaign
Coordinated Science Laboratory
1308 W. Main St. MC-228
Urbana, IL 61801 USA
PH: 217-333-0345
FAX: 217-244-3359
E-M: whs@crhc.uiuc.edu

Robert G. Sargent
Syracuse University
Simulation Research Group
College of Engr. & Computer Science
439 Link Hall
Syracuse, NY 13244 USA
PH: 315-443-4348
FAX: 315-443-4936
E-M: rsargent@syr.edu

Hessam S. Sarjoughian
University of Arizona
AI & Simulation Research Group
Electrical & Computer Engr. Dept.
1230 E. Speedway Blvd.
Tucson, AZ 85721-0104 USA
PH: 520-621-6184
FAX: 520-621-8076
E-M: hessam@ece.arizona.edu

Mahadev Satyanarayanan
Carnegie Mellon University
School of Computer Science
Pittsburgh, PA 15213-3891 USA
E-M: satya@cs.cmu.edu

Eric L. Savage
Cornell University
School of Operations Research & Industrial Engr.
272 Rhodes Hall
Ithaca, NY 14853-1801 USA
PH: 607-255-9139
FAX: 607-255-9129
E-M: savage@orie.cornell.edu

Bruce W. Schmeiser
Purdue University
School of Industrial Engr.
West Lafayette, IN 47907-1287 USA
PH: 317-494-5422
FAX: 317-494-1299
E-M: schmeise@ecn.purdue.edu

Friedrich Schreiber
Aachen University of Technology
Communication Networks
Kopernikusstr. 16
D-52074 Aachen
GERMANY
PH: +49-241-807910
FAX: +49-241-8888-242
E-M: fschr@comnets.rwth-aachen.de

Thomas J. Schriber
University of Michigan
School of Business
701 Tappan St.
Ann Arbor, MI 48109-1234 USA
PH: 313-764-1398
FAX: 313-936-8716
E-M: schriber@umich.edu

Bernard J. Schroer
University of Alabama in Huntsville
Center for Automation & Robotics
Research Institute A-11
Huntsville, AL 35899 USA
PH: 205-890-6100
FAX: 205-890-6970
E-M: schroerb@email.uah.edu

Lee W. Schruben
Cornell University
School of Operations Research & Industrial Engr.
219 Rhodes Hall
Ithaca, NY 14853-1801 USA
PH: 607-255-9133
FAX: 607-255-9129
E-M: lee@orie.cornell.edu

Thomas F. Schuppe
LEXIS-NEXIS
P. O. Box 933
Dayton, OH 45401 USA
PH: 513-865-6518
FAX: 513-865-6808
E-M: tschuppe@lexis-nexis.com

Herb Schwetman
Mesquite Software
3925 West Braker Lane
Austin, TX 78759-5321 USA
PH: 512-305-0080
FAX: 512-305-0009
E-M: hds@mesquite.com

Marian Scott
University of Glasgow
Dept. of Statistics
Glasgow, G12 8QW
UNITED KINGDOM
E-M: marian@stats.gla.ac.uk

Shelly Scott-Nash
Micro Analysis & Design, Inc.
4900 Pearl East Circle, Suite 201E
Boulder, CO 80301 USA
PH: 303-442-6947
FAX: 303-442-8274
E-M: ssn@madboulder.com

David W. Seidel
The MITRE Corp.
1820 Dolly Madison Blvd.
McLean, VA 22102 USA
PH: 703-883-6828
FAX: 703-883-6535
E-M: dseidel@mitre.org

Andrew F. Seila
University of Georgia
Terry College of Business
Athens, GA 30602-6255 USA
PH: 706-542-3586
FAX: 706-542-4290
E-M: aseila@cba.uga.edu

Perwez Shahabuddin
Columbia University
Dept. of Industrial Engr. & Operations Research
331 S.W. Mudd Building
New York, NY 10027 USA
PH: 212-854-1477
FAX: 212-854-8103
E-M: perwez@ieor.columbia.edu

Alexander Shapiro
Georgia Institute of Technology
School of Industrial & Systems Engr.
Atlanta, GA 30332-0205 USA
PH: 404-894-6544
FAX: 404-894-2301
E-M: ashapiro@isye.gatech.edu

Chien-Chung Shen
Bellcore
331 Newman Strings Rd.
NVC3B-110
Red Bank, NJ 07701 USA
PH: 908-758-2524
FAX: 908-758-4371
E-M: cshen@nyquist.bellcore.com

Leyuan Shi
University of Wisconsin - Madison
Dept. of Industrial Engr.
1513 University Ave.
Madison, WI 53706 USA
PH: 608-265-5969
FAX: 608-262-8454
E-M: leyuan@ie.engr.wisc.edu

Junhong Shu
Kansas State University
Dept. of Industrial & Manufacturing Systems Engr.
Manhattan, KS 66506 USA

Julika Siemer
London School of Economics & Political Science
Dept. of Info. Systems
Houghton St.
London, WC2A 2AE
ENGLAND
PH: +44-171-955-6029
FAX: +44-171-955-7385
E-M: j.siemer@ise.ac.uk

Ruth Silverman
University of the District of Columbia
Dept. of Computer and Info. Science
Washington, DC 20008 USA
PH: 301-384-3135
FAX: 301-384-9115
E-M: ruth@cfar.umd.edu

Alex F. Sisti
Rome Laboratory / IRAE
32 Hangar Rd.
Rome, NY 13441-4114 USA
PH: 315-330-4518
FAX: 315-330-2728
E-M: sistia@rl.af.mil

Sven Sköld
Royal Institute of Technology (KTH)
Dept. of Teleinformatics
Electrum 204
Kista, S-16S-164 40
SWEDEN
PH: +46-8-752-1372
FAX: +46-8-751-1793
E-M: sven@it.kth.se

David P. Sly
Cimtechnologies Corp.
2501 North Loop Dr. Suite 700
Iowa State Research Park
Ames, IA 50010 USA
PH: 515-296-9914
FAX: 515-296-9909
E-M: daves@cimtech.com

Jeffrey S. Smith
Texas A & M University
Industrial Engr. Dept.
Zachry Engr. Building
College Station, TX 77843-3131 USA
PH: 409-845-4335
FAX: 409-847-9005
E-M: jsmith@tamu.edu

L. Douglas Smith
University of Missouri - St. Louis
School of Business Adminstration
220 CCB
8001 Natural Bridge Rd.
St. Louis, MO 63121-4499 USA
PH: 314-516-6108
FAX: 314-516-6827
E-M: ldsmith@umslvma.umsl.edu

Robert L. Smith
University of Michigan
Dept. Industrial & Operations Engr.
Ann Arbor, MI 48109 USA
PH: 313-763-2060
FAX: 313-764-3451
E-M: rlsmith@umich.edu

Roger D. Smith
Mystech Associates
20131 Maxim Parkway
Orlando, FL 32833 USA
PH: 407-249-5124
FAX: 407-823-7529
E-M: smithr@mystech.com

Wheyming Tina Song
National Tsing Hua University
Dept. of Industrial Engr.
Hsinchu, Taiwan 300
REPUBLIC OF CHINA
PH: +866-35-722685 x3932
FAX: +866-35-722685
E-M: wheyming@wsong.ie.nthu.edu.tw

Sudhir Srinivasan
Mystech Associates, Inc.
5205 Leesburg Pike, Suite 1200
Falls Church, VA 22041 USA
PH: 703-671-8680
FAX: 703-671-8932
E-M: sudhirs@mystech.com

Krishna Srinivasan
Motorola Manufacturing Systems
Worldwide Enterprise Modeling Group
1299 East Algonquin Road
Schaumberg, IL 60196 USA
PH: 847-538-9449
E-M: krishna_srinivasan_AKS010@email.mot.com

Ingolf Ståhl
Stockholm School of Economics
Box 6501
S-113 83 Stockholm
SWEDEN
PH: +46-8-736-9425
FAX: +46-8-30-4762
E-M: iis@hhs.se

Charles R. Standridge
Florida A&M University - FSU
Dept. of Industrial Engr.
College of Engr.
2525 Pottsdamer St.
Tallahassee, FL 32310 USA
PH: 904-487-6160
FAX: 904-487-6342
E-M: stand@eng.fsu.edu

Paul M. Stanfield
ABCO Automation, Inc.
6202 Technology Dr.
Brown Summit, NC 27214 USA
PH: 910-375-6400
FAX: 910-375-0090
E-M: stanfield@garfield.ncat.edu

Darrell W. Starks
University of Louisville
Dept. of Industrial Engr.
2553 Glenmary Ave.
Louisville, KY 40204 USA
PH: 502-473-1194
FAX: 502-852-5633

George F. Stone III
US Army STRICOM
389 Blue Spruce Ct.
Winter Springs, FL 32708 USA
PH: 407-384-3621
FAX: 407-366-8961
E-M: stoneg@stricom.army.mil

Gregory Stoner
Intel Corp.
CH3-84
5000 W. Chandler Blvd.
Chandler, AZ 85226 USA
PH: 602-554-2097
FAX: 602-554-5323
E-M: gregory-w_stoner@ccm.ch.intel.com

Kirk A. Stork
Operations Research Dept.
Naval Postgraduate School
Monterey, CA 93943 USA
E-M: kstork@or.nps.navy.mil

Stephen G. Strickland
University of Virginia
Dept. of Systems Engr.
Thornton Hall
Charlottesville, VA 22903 USA
PH: 804-982-2742
FAX: 804-982-2972
E-M: sgs8r@virginia.edu

David Sturrock
Systems Modeling Corp.
504 Beaver St.
Sewickley, PA 15143 USA
PH: 412-741-3727
FAX: 412-741-5635
E-M: dsturrock@mail.sm.com

Jeffrey E. Sumner
TASC, Inc.
2555 University Blvd.
Fairborn, OH 45324 USA
PH: 513-426-1040
FAX: 513-426-8888
E-M: jesumner@tasc.com

P. S. Sundararaghavan
University of Toledo
ISOM Dept.
Toledo, OH 43606-3390 USA
E-M: fac0066@uoft01.utoledo.edu

Scott C. Sundstrom
Naval Health Research Center
P. O. Box 85122
San Diego, CA 92186-5122 USA
PH: 619-553-8389
FAX: 619-553-8607
E-M: sundstrom@nhrc.navy.mil

James J. Swain
University of Alabama at Huntsville
Lakeside Dr. & South Loop Road
Dept. of ISE
EB 236
Huntsville, AL 35899 USA
PH: 205-890-6749
FAX: 205-890-6608
E-M: jswain@ebs330.eb.uah.edu

David A. Sykes
ObjectKnowlogy, Inc.
11 Oakview Dr.
Greenville, SC 29605-1816 USA
E-M: dsykes@cs.clemson.edu

Helena Szczerbicka
University of Bremen
Dept. of Computer Science
P. O. Box 330440
28334 Bremen
GERMANY
PH: +49-421-218-7389
FAX: +49-421-218-7385
E-M: helena@informatik.uni-bremen.de

Boleslaw K. Szymanski
Rensselaer Polytechnic Institute
Dept. of Computer Science
110 - 8th St.
Troy, NY 12180-3590 USA
PH: 518-276-2714
FAX: 518-276-4033
E-M: szymansk@cs.rpi.edu

Akira Tajima
IBM Research
Tokyo Research Laboratory
1623-14 Shimotsuruma
Yamoto-shi Kanagawa-ken, 242
JAPAN
PH: +81-462-73-4886
FAX: +81-462-73-7413
E-M: tajima@trl.ibm.co.jp

Soemon Takakuwa
Nagoya University
School of Economics
Furo-cho, Chikusa-ku
Nagoya, Aichi, 464-01
JAPAN
PH: +81-52-789-2389
FAX: +81-52-789-4924
E-M: Takakuwa@nsec.soec.nagoya-u.ac.jp

Simon J. E. Taylor
Brunel University
Dept. of Computer Science & Info. Systems
Uxbridge, UB8 3PH
ENGLAND
PH: +44-1895-251686 x2126
FAX: +44-1895-274000
E-M: simon.taylor@brunel.ac.uk

Raymond Telerole
New York Power Authority
123 Main St.
White Plains, NY 10601 USA
PH: 914-681-6544
E-M: wtellarr@ip3gate.usa.com

Shu Tezuka
IBM Research
Tokyo Research Laboratory
1623-14
Shimotsuruma
Yamato-shi, Kanagawa-ken, 242
JAPAN
E-M: tezuka@trl.ibm.co.jp

Arne Thesen
University of Wisconsin - Madison
Dept. of Industrial Engr.
1513 University Ave.
Madison, WI 53706 USA
PH: 608-262-8456
FAX: 608-262-8454
E-M: thesen@ie.engr.wisc.edu

W. Bruce Thompson
AT&T Istel, Inc
60 Mall Road
Burlington, MA 01803 USA
PH: 617-272-7333
FAX: 617-272-4218
E-M: WBruceT@aol.com

John A. Tufarolo
The MITRE Corp.
1820 Dolley Madison Blvd.
McLean, VA 22102-3481 USA
PH: 703-883-7950
FAX: 703-883-1279
E-M: tufarolo@mitre.org

Kerim Tumay
CACI Products Company
3333 North Torrey Pines Court
La Jolla, CA 92037 USA
PH: 619-455-6300 ex 150
FAX: 619-458-0031
E-M: kerim@cacisl.com

Brian Unger
University of Calgary
Dept. of Computer Science
2500 University Dr. N. W.
Calgary, Alberta T2N 1N4
CANADA
PH: 403-220-8259
FAX: 403-284-4707
E-M: unger@cpsc.ucalgary.ca

Edward Unger
Baptist Hospital East
4000 Kresge Way
Louisville, KY USA
PH: 502-897-8164

Dawn R. Utley
University of Alabama at Huntsville
Center for Automation & Robotics
Research Institute A-11
Huntsville, AL 35899 USA
PH: 205-890-6100
FAX: 205-890-6970

Willem J.H. van Groenendaal
Tilburg University
Dept. of Info. Systems & Auditing
Center for Economic Research
5000 Le Tilburg
NETHERLANDS
PH: +31-13-4662423
FAX: +31-13-4663377
E-M: w.j.h.vgroenendaal@kub.nl

Pieter A. Voss
Rensselaer Polytechnic Institute
Dept. Decision Science & Engr. Systems
Troy, NY 12180 USA
PH: 518-276-2803
E-M: vossp@rpi.edu

Mary Ann Flanagan Wagner
Boeing Info. Services
7990 Boeing Court
MS CV-82
Vienna, VA 22183-7000 USA
PH: 703-827-2853
FAX: 703-821-6342
E-M: maflanig@gmu.edu

Jack Walther
Florida A&M University - FSU
Dept. of Industrial Engr.
2525 Pottsdamer St.
Tallahassee, FL 32310 USA
E-M: walther@eng.fsu.edu

C. Michael Walton
University of Texas at Austin
Dept. of Civil Engr.
Ernest H. Cockrell Hall, Suite 4.20
Austin, TX 78712-1076 USA
PH: 512-471-1414
FAX: 512-471-4995
E-M: michael_walton.cemail@
cemailgate.ce.utexas.edu

Peter P. Ware
Ohio State University
Dept. of Computer & Info. Science
2015 Neil Ave. Dreese Labs 774
Columbus, OH 43210 USA
PH: 614-292-8501
FAX: 614-292-2911
E-M: ware@cis.ohio-state.edu

Michael L. Weigel
Union Pacific Railroad
Manager Terminal Planning
1416 Dodge St.
Omaha, NE 68179-0700 USA
PH: 402-271-3066
FAX: 402-271-2420

Andrew B. Whinston
University of Texas at Austin
MSIS Dept.
CBA 5.202
Austin, TX 78712-1175 USA
PH: 512-471-3322
FAX: 512-471-0587
E-M: abw@uts.cc.utexas.edu

Charles H. White
Dupont Company
EN99-N6506
1007 Market St.
Wilmington, DE 19898 USA
PH: 302-774-2342
FAX: 302-774-1347
E-M: white@pccvax.dnet.dupont.com

Todd C. Whyte
University of Louisville
1325 Lydia St.
Louisville, KY 40217 USA
PH: 502-253-0027
FAX: 502-253-0028

Ralf Wieting
Informatik-Werkzeuge & -Systeme
OFFIS
Escherweg 2
D-26121 Oldenburg
GERMANY
PH: +49-441-9722-171
FAX: +49-441-9722-242
E-M: wieting@offis.uni-oldenburg.de

A. Martin Wildberger
Electronic Power Research Institute
Strategic Research & Development
3412 Hillview Ave.
Mountain View, CA 94304-1395 USA
PH: 415-855-1043
FAX: 415-855-2287
E-M: mwildber@eprinet.epri.com

Thomas R. Willemain
Rensselaer Polytechnic Institute
Dept. of Decision Sciences & Engr. Systems
Troy, NY 12180-3590 USA
E-M: willet@rpi.edu

Edward J. Williams
Ford Motor Company
206-2 Engr. Computer Center
Mail Drop 3
P. O. Box 2053
Dearborn, MI 48121-2053 USA
PH: 313-594-6776
FAX: 313-390-4565
E-M: willilame@umdsun2.umd.umich.edu

Karen Wilson
Georgia Institute of Technology
College of Computing
Atlanta, GA 30332-0280 USA
FAX: 404-894-0272
E-M: kcw@cc.gatech.edu

James R. Wilson
North Carolina State University
Dept. of Industrial Engr.
Box 7906
Raleigh, NC 27695-7906 USA
PH: 919-515-6415
FAX: 919-515-5281
E-M: jwilson@eos.ncsu.edu

Hsiao-Kuang Wu
University of California, Los Angeles
Computer Science Dept.
Box 951596
Los Angeles, CA 90095 USA
PH: 310-825-1888
FAX: 310-825-2273
E-M: hsiao@cs.ucla.edu

Shahar Yarden
Computer Sciences Corp.
M/C 376
6565 Arlington Blvd.
Falls Church, VA 22042 USA
PH: 301-320-8579
FAX: 703-607-9666
E-M: syarden@explorer.csc.com

Xiaoning Yu
Brunel University
Dept. of Computer Science & Info. Systems
Uxbridge, UB8 3PH
ENGLAND
PH: +44-1895-274000
FAX: +44-1895-251686
E-M: ray.paul@brunel.ac.uk

Eric A. Zahn
TASC, Inc.
2555 University Blvd.
Fairborn, OH 45324 USA
PH: 513-426-1040 x414
FAX: 513-426-8888
E-M: eazahn@tasc.com

Tim Zajic
Columbia University
IEOR Dept.
302 Mudd Building
New York, NY 10027-6699 USA
PH: 212-854-4345
FAX: 212-854-8103
E-M: zajic@ieor.columbia.edu

Ellen W. Zegura
Georgia Institute of Technology
College of Computing
Atlanta, GA 30332-0280 USA
PH: 404-894-1403
FAX: 404-894-0272
E-M: ewz@cc.gatech.edu

Bernard P. Zeigler
University of Arizona
AI & Simulation Research Group
Electrical & Computer Engr. Dept.
1230 E. Speedway Blvd.
Tucson, AZ 85721-0104 USA
PH: 520-626-4846
FAX: 520-621-8076
E-M: zeigler@ece.arizona.edu

Michael E. Zeimer
Schneider Logistics, Inc.
Dept. of Engr.
3101 S. Packerland Dr.
P. O. Box 2666
Green Bay, WI 54306-2666 USA
PH: 414-592-2000 Ext. 2533
FAX: 414-592-2338

Zhiwei Zhang
University of Georgia
Dept. of Computer Science
219D, Boyd GSRC
Athens, GA 30602-7404 USA
PH: 706-542-3445
FAX: 706-542-2966
E-M: zhizhang@cs.uga.edu

Guoyan S. Zhou
University of Illinois at Urbana-Champaign
Dept. of General Engr.
104 S. Matthews Ave.
Urbana, IL 61801 USA
PH: 217-333-0531
FAX: 217-244-5705
E-M: w-davis@uiuc.edu

Dietrich Ziems
Otto-von-Guericke-University of Magedeburg
Dept. of Logistics & Mechanical Handling Engr.
P. O. Box 4120 / IFSL
D-39016 Magdeburg
GERMANY
PH: +49-391-671804
FAX: +49-391-6712646
E-M: dietrich.ziems@mb.uni-magdeburg.de

Mike Zyda
Naval Postgraduate School
Monterey, CA 93943-5118 USA